Fundamental Physical Constants

Constant	Symbol	Value
Speed of light	c	2.99792458×10^8 m/s
Permittivity of vacuum	ϵ_0	$8.8541878 \times 10^{-12}$ C²/J m
Permeability of vacuum	$\mu_0 = 1/\epsilon_0 c^2$	$1.25663706014 \times 10^{-6}$ H/m
Elementary charge	e	$1.6021892 \times 10^{-19}$ C or 4.803242 esu
Planck's constant	h	6.626176×10^{-34} J s or J/Hz
	$\hbar = h/2\pi$	$1.0545887 \times 10^{-34}$ J s
	hc	1.986478×10^{-25} J m
Avogadro's constant	N_A	6.022045×10^{23} mol⁻¹
Atomic mass unit	amu	$1.6605655 \times 10^{-27}$ kg
Electron rest mass	m_e	9.109534×10^{-31} kg
Proton rest mass	m_p	$1.6726485 \times 10^{-27}$ kg
Ratio of proton mass to electron mass	m_p/m_e	1836.15152
Electron charge to mass ratio	e/m_e	1.7588047×10^{11} C/kg
Bohr magneton	$\mu_B = e\hbar/2m_e$	9.274078×10^{-24} J/T or 5.788378×10^{-9} eV/G
Nuclear magneton	$\mu_N = e\hbar/2m_p$	$5.0508248 \times 10^{-27}$ J/T or 3.152515×10^{-12} eV/G
Free-electron g factor	g_e	2×1.0011596567
Rydberg constant	R_∞	1.0973731×10^7 m⁻¹
Bohr radius	a_0	$0.52917706 \times 10^{-10}$ m
Boltzmann constant	k_B	1.380662×10^{-23} J/K
Faraday constant	$\mathcal{F} = N_A e$	96484.56 C/mol
Gas constant	R	8.31441 J/mol K
Standard volume of a perfect gas (1 atm, 273.15 K)	V_m	22.41383×10^{-3} m³/mol
Loschmidt constant	$L = N_A/V_m$	2.686754×10^{25} m⁻³
Gravitational constant	G	6.6720×10^{-11} N m²/kg²

A=ampere, C=coulomb, G=gauss, H=henry, Hz=hertz or cycles per second, J=joule, K=kelvin or degrees kelvin, N=newton, T=tesla, amu=atomic mass unit, eV=electron volt.

Fundamental Physical Constants [From E. R. Cohen and B. N. Taylor, J. Phys. Chem. Ref. Data 2, 663 (1973)].

PHYSICAL CHEMISTRY

PHYSICAL CHEMISTRY

R. Stephen Berry
The University of Chicago

Stuart A. Rice
The University of Chicago

John Ross
Massachusetts Institute of Technology

With the assistance of

George P. Flynn
Massachusetts Institute of Technology

Joseph N. Kushick
Amherst College

John Wiley & Sons
New York · Chichester · Brisbane · Toronto

To our families

This book was printed and bound by Halliday. It was set in Times Roman by Universities Press. It was designed by Blaise Zito Associates. The Drawings were designed and executed by John Balbalis with the assistance of the Wiley Illustration Department. Deborah Herbert was the copyeditor. Claire Egielski supervised production.

Library of Congress Cataloging in Publication Data:

Berry, R. Stephen, 1931-
    Physical chemistry.
    Includes index.
    1. Chemistry, Physical and theoretical. I. Rice,
Stuart Alan, 1932- joint author. II. Ross, John,
1926- joint author. III. Title.
QD453.2.B48 541 79-790
ISBN 0–471–04829–1

Printed in the United States of America

10 9 8 7 6 5 4 3

Preface

Soul of the world, inspired by thee,
The jarring seeds of matter did agree,
Thou didst the scatter'd atoms bind,
Which, by the laws of true proportion joined
Made up of various parts one perfect harmony

Text by Nicholas Brady for the *Ode on
St. Cecilia's Day, 1692* by Henry Purcell

We started thinking about this book almost twenty-five years ago. As graduate students we began to realize that undergraduate physical chemistry as it had been presented to us, and physical chemistry as a field of scholarly scientific endeavor, differed greatly in content, emphasis, style, expectation of achievement, and even in the nature of what was considered explanation. We began then to outline an approach that would bring the undergraduate into the subject in a manner consistent with the conceptual structure and the values that characterize physical chemistry as a contemporary discipline. Over the years that followed we talked, planned, and then started to write. There were fruitful periods of gestation, and exciting periods when new research showed that entire sections had to be revised or written from viewpoints altogether different from our original conceptions. Even as we complete this book we see new results appearing in the scientific journals that would have been incorporated in the text had they appeared in time.

Physical chemistry is an empirical science. A science is a set of constructs, called theories, that link fragments of experience into a consistent description of natural phenomena. The adjective "empirical" refers to the common experiences from which the theories grow, that is, to experiments. Simple working hypotheses are guessed by imaginative insight or intuition or luck, usually from a study of experiments. This repetitive interplay in time leads to the formulation of theories that correlate the accumulated experimental information, and that can predict new phenomena with accuracy. As scientists we have, throughout our careers, endeavored to combine both experimental and theoretical work. In this book we try to knit these two inseparable parts of the science into a coherent structure that represents accurately the way they interact in physical chemistry today.

Our goal is the presentation of the three major areas of physical chemistry: molecular structure, the equilibrium properties of systems, and the kinetics of transformations of systems. The theoretical foundations of these subjects are, respectively, quantum mechanics, thermodynamics and equilibrium statistical mechanics, and chemical kinetics and kinetic theory. These theories, firmly based on experimental findings, constitute the structure required for the understanding of past accomplishments and a basis for recognition and development of significant new areas in physical chemistry.

The presentation of the theories of physical chemistry requires careful discussions at several levels of exposition. Our approach aims toward depth of understanding of fundamentals more than toward breadth of recognition of the multitude of activities that go on under the name of physical chemistry. The organization of the book, with its three principal sections, should make this clear. The mathematical level begins with elementary calculus, and rises to the use of simple properties of partial differential equations and the special functions that enter into their solutions. Our intention is to keep the reader's mind on the science rather than on the mathematics, especially at the beginning. This procedure also corresponds to the pattern, followed by many students, of taking physical chemistry and advanced calculus concurrently. Appendices develop the details of the mathematical tools as they are needed.

The text discussion contains more material than can be covered in the traditional one-year physical chemistry sequence; it is designed to fulfill the dual purpose of providing a clear and incisive treatment of fundamental principles at a level accessible to all students while broadening the perspectives and challenging the minds of the best students. Individual instructors will wish to make their own selections of material for inclusion and exclusion, respectively. We have provided guidance on this matter by having the more advanced sections of the book printed in smaller type on shaded paper. These sections can be omitted without breaking the flow of argument from chapter to chapter. It is also easy to use the material discussed in a different order than in the text, and to omit or downplay classes of topics deemed unsuitable to a particular group of students. For example, we have taught the junior-level physical chemistry course both as organized in this text and with the material of Part Two preceding that of Part One. To invert the order of Parts Two and One, it is only necessary to ask the student to accept the existence of quantized energy levels and a few specific examples of energy level spectra. This has proved quite easy for students who have at least heard of these matters in current freshman chemistry courses. Students with strong backgrounds will have seen the material in the first chapter and parts of the second chapter. Chapter 10, on intermolecular forces, and Chapter 11, on the structure of solids, contain large blocks of material that could be passed over in a traditionally oriented one-year course. The thermodynamic description of matter can be emphasized and the statistical molecular description de-emphasized, or vice versa, by selection of the relevant sections of Chapters 21 to 26. Chapter 20, dealing with hydrodynamic phenomena and negative temperature, can easily be omitted if so wished by the instructor. In Part Three, the elements of physical kinetics or transport theory are contained in Section 27.1 to 28.8 and 29.1; the elements of chemical kinetics are contained in Chapter 30. The specialized topics in the remaining sections may be used for more extensive treatments of these subjects, either in a three-semester physical chemistry sequence or in a senior-graduate course on these topics.

At the other end of the scale there is sufficient material in this book for a first-year graduate course for students whose undergraduate preparation in physical chemistry did not emphasize modern aspects of the subject.

At the end of each chapter we have suggested extra reading for the interested student. The book contains about 700 problems. A few of these problems are designed to acquaint the reader with dimensions, units, and simple manipulations. More are intended to develop intellectual skills, to enable students to master the material of the text discussions by the difficult process of thinking through the kinds of questions encountered in the laboratory. Some of the problems are designed to extend the theoretical analysis of the text to special but interesting situations.

Part One opens with a review of the elementary quantities of the atomic world and how they are measured. Many readers will find some or all of the material in Chapter 1 familiar; they may choose to read quickly through this chapter and start their more intensive study with Chapter 2, where we develop the experimental evidence for the quantum structure of matter at the molecular level. In the process, we examine the harmonic oscillator, the primitive model for many kinds of behavior examined later, and the concept of action. Chapter 3 is more theoretical and mathematical than its predecessors, but begins at a level the well-prepared student will find quite elementary. Waves and wave equations are introduced and are used to describe only very simple situations: the states of particles in boxes of various sorts, and of rigid rotators.

Chapter 4 brings us back to the more realistic problems of the quantum-mechanical oscillator and the simplest atom, hydrogen. In Chapter 5, we further develop the concepts such as orbitals and transitions between quantum states that are introduced for hydrogen in Chapter 4. Chapter 5 treats atoms with more than two electrons, in particular their electronic states and the interactions among the electrons.

Molecules are first introduced in Chapter 6, which is almost analogous to Chapter 4, in that Chapter 6 goes in depth into the description and behavior of the simplest molecules, H_2^+ and H_2, just as Chapter 4 examined the H atom. The concepts developed in Chapter 6 are then extended to more complex diatomic molecules in Chapter 7. Chapter 6 deals almost exclusively with electronic states; Chapter 7 introduces molecular vibration and rotation, and discusses concepts used to correlate and unify our observations regarding diatomic molecules.

Chapter 8 begins with the primitive triatomic species H_3^+ and H_3 and then goes on to ideas that begin to be important with three or more nuclei: hybrid orbitals and delocalized molecular orbitals at the level of electronic states, and normal modes of vibration. Larger molecules are discussed in Chapter 9; here we introduce the concepts of chirality and optical activity, and explore some aspects of ligand field theory and the magnetic properties of molecules.

Chapter 10 is the first in which we go beyond the properties of individual molecules. The discussion of intermolecular forces describes the interactions between charge distributions and how one molecule behaves when it collides with another. The material in this chapter is based wholly on the framework of molecular structure, but becomes especially useful to us in Parts Two and Three, where we study the behavior of matter in the aggregate. Part One concludes with another structural aspect of aggregated matter, the structure of solids. Here, we extend the various concepts of bonding previously developed to include the concept of metallic bonding, to describe the structure and states of periodic condensed phases.

Part Two is concerned with the equilibrium properties of bulk matter. Our presentation simultaneously develops the statistical molecular theory and the classical thermodynamic theory in a mutually reinforcing fashion. Despite use of this "mixing" of microscopic and macroscopic points of view, no compromise is made with respect to the rigor of classical thermodynamics, and if desired the two points of view can be separated.

In Chapter 12, we begin with a discussion of the zeroth law of thermodynamics and the concept of temperature. By examining the phenomenological bases for the equation of state and the definition of temperature, together with the elements of the kinetic theory of perfect gases, we establish a first connection between macroscopic and microscopic descriptions. The building of bridges between the two classes of description is a prinicipal theme of succeeding chapters.

Chapters 13 and 14 treat the first law of thermodynamics and some of its many applications. Particular care is devoted to the precise definition of work and heat, and to how the nature of these quantities exemplifies the differences between the thermodynamic and mechanical descriptions of matter.

Chapter 15 introduces the concept of entropy by way of the microscopic structure of matter. Given only that every sample of matter has an energy-level spectrum, it is shown that there exists a function of the density of states, the entropy, that behaves like a property only of macroscopic variables of the system, despite its definition in terms of the microscopic energy-level spectrum.

Chapter 16 develops the second law of thermodynamics via the classical Clausius and Kelvin principles, and Chapter 17 is devoted to examples of the use of the second law to solve problems of chemical interest. These chapters contain a careful discussion of the nature of irreversibility and its interpretation in terms of thermodynamic and statistical molecular theories.

Chapter 18 introduces, discusses, and gives applications of the third law of thermodynamics.

In Chapter 19 we examine the central problem of describing equilibrium as a function of the external constraints on the system. The thermodynamic theory of open systems is developed and is used to derive the several criteria of equilibrium that are suitable to different external constraints. With this background, the notion of ensemble is introduced, and the classical thermodynamics of equilibrium is related to the development of the grand canonical, canonical, and microcanonical partition functions of statistical mechanics. The theory is illustrated by analyzing the velocity distribution in a perfect gas.

Chapter 20 introduces a new point of view into the analysis—it deals with the description of systems whose properties vary slowly in time, and with the extension of thermodynamics to systems with negative temperature.

Chapters 12 through 19 establish the principles and develop the tools needed to study the bulk properties of matter. This study is carried out systematically in Chapters 21 through 26. Chapter 21 deals with gases, Chapter 22 with solids, Chapter 23 with liquids, Chapter 24 with phase transformations, Chapter 25 with solutions of nonelectrolytes, and Chapter 26 with solutions of electrolytes. In each chapter the thermodynamic theory is developed first, then the statistical molecular theory. Extensive use is made of the details concerning molecular behavior that are provided by computer simulation studies. In addition, the principles developed are illustrated with data from experimental situations wherever that is appropriate.

The simultaneous development of classical and statistical thermodynamics, employed through Part Two, is designed to overcome the difficulties associated with the very abstract nature of purely thermodynamic reasoning, and also to illustrate the richness of the phenomena that can arise from molecular interactions. Nevertheless, for any given problem, the generality of the thermodynamic approach is made evident, as are the wealth of detail and dependence on assumed models of the statistical molecular approach.

Part Three is concerned with time-dependent processes, especially the approach to equilibrium. The topic of physical kinetics (transport processes) is introduced in Chapter 27 with a discussion of the mechanics of molecular collisions, mostly binary collisions. We present as simply as possible the elements of kinematics and dynamics, including the concept of scattering, which is illustrated with the hard-sphere model. In Chapter 28 we consider the kinetic theory of gases, beginning with how velocity distribution functions change with time because of collisions. We present an elementary discussion of the time-dependent transport equations and show how they govern a gas's approach to equilibrium. With these results we can discuss fluxes of mass, momentum, and energy, and study the process of effusion and the simple transport properties (diffusion, viscosity, and thermal conduction) in dilute gases. We conclude with a brief treatment of energy exchange processes, and sound propagation and absorption in gases.

The transport properties of dense phases are taken up in Chapter 29. Transport in liquids is approached with a discussion of Brownian motion, leading to the relation of transport coefficients to autocorrelation functions. A brief discussion of transport in solids concludes the chapter.

Chemical kinetics is treated in a manner parallel to physical kinetics, with an elementary development followed by selected advanced applications. We begin in Chapter 30 with a presentation of the mechanics of reactive collisions, including both kinematics and dynamics. The emphasis is again on the simple hard-sphere model. The collision-theory approach is compared with the activated-complex theory, and both theories are used for an analysis of kinetics in gases and solutions. After a brief survey of experimental methods, we discuss complex reactions and provide an elementary discussion of chemical reaction mechanisms. Chapter 31 is devoted to various advanced topics in chemical kinetics, including the RRKM theory of unimolecular reactions, symmetry rules in chemical reactions, chain reactions, oscillatory reactions, photochemistry, and homogeneous and heterogeneous catalysis.

Our educational and professional associations have obviously influenced the way we wrote this book, as have a few remarkable volumes. In particular, the books by H. Reiss (*Methods of Thermodynamics*, Blaisdell, N.Y., 1965) and F. Reif (*Statistical Physics*, McGraw-Hill, N.Y., 1965) helped to clarify our independently developed presentations. Dr. G. P. Flynn (M.I.T.) and Professor J. N. Kushick (Amherst College) painstakingly read, corrected, and made innumerable improvements to the methods and details of presentation. John Hansen, Donald Jordan and David J. Zvijac assisted Professor Kushick in working and checking many of the problems following the chapters. We are grateful to Professor E. Heller (U.C.L.A.), to Professor R. Jarnigan (University of North Carolina), Professor Rodney J. Sime (California State University, Sacramento), Professor Edward I. Solomon (M.I.T.), Professor Jeffrey I. Steinfeld (M.I.T.), Professor Mark S. Wrighton (M.I.T.) and to many students who used this text during its development, for helpful comments.

To the readers, we say that we hope that you will have as much delight in using and creating physical chemistry as we have.

R. Stephen Berry
Stuart A. Rice
John Ross

Suggested Topic Selections for Various Courses

The following table illustrates possible choices of subject matter for a variety of physical chemistry courses.

Chapter	One-Year Course (initial emphasis on macroscopic approach)	One-Year Course (initial emphasis on microscopic approach)	Three-Semester Course	Senior or Graduate Course
1	Start with Chapters 12 through 26, with emphasis placed on the thermodynamic analysis. Keep only such elements of the statistical molecular analysis as fits with the instructional goal. Then return to Chapter 2 and follow the outline for the one-year course with initial emphasis on microscopics, going rapidly to Section 3.3.	Start with Chapter 1, setting pace as fast as students can absorb Chapters 1 and 2, and Sections 3.1 and 3.2. Touch lightly or omit Section 3.10 and Appendix 3A	All, as introduction	Omit
2			All	Read quickly
3			All	Read Sections 3.1 and 3.2 quickly
4		All	All	All
5		All, but Section 5.5 may be omitted for very well-prepared students	All	Touch lightly on Section 5.5
6		All	All	All
7		Section 7.8 may be dropped	All	All
8		All	All	All
9		Omit Sections 9.6 and 9.7	All	All
10		Omit	All	All
11		Sections 11.1 through 11.5, 11.8, and 11.10	All	All
12		All	All	All
13		All	All	All
14		Omit Section 14.6. Only touch lightly on Sections 14.9, 14.10	All	All
15		Omit optional Section 15.1	Omit optional Section 15.1	All
16		Omit optional Sections and Appendix 16A	Omit Appendix 16A	All
17		All	All	All
18		All	All	All
19		Omit Section 19.3	All	All
20		Omit	Touch lightly on Sections 20.1–20.4, keep 20.5	All
21		Omit Appendices 21A, 21B, 21C	All	All
22		Omit Section 22.5 and Appendix 22A	All	All
23		Omit Section 23.4	All	All
24		Omit Section 24.3. Touch lightly on Section 24.4	All	All
25		Omit latter half of Section 25.8. Touch lightly on Section 25.9	All	All
26		Omit Section 26.6	All	All
27		27.1–4	All	All
28		28.1, 4, 6, 7, 8	28.9, 10 optional	All
29		29.1	29.2–6	All
30		Omit 30.5, 6, 9	All	All
31		As time permits Assign as special topics	31.3, 6–11 and as time permits	All

Contents

Contents

xiv

PART ONE

The Structure
of Matter

1

The Microscopic World: Atoms and Molecules

We begin our study of the microscopic properties of matter by considering the properties of atoms and molecules. But first, what *are* atoms and molecules? For the moment, let us simply use a working definition: For any of the hundred-odd substances (the chemical elements) listed on the inside cover of this book, an *atom* is the smallest bit of matter that can be identified as that substance; for any other chemical substance, a *molecule* (made up of atoms) is the smallest recognizable bit of the substance. Admittedly, definitions like these are somewhat vague, so we shall begin by reviewing some of the evidence that led to the formulation of these concepts.

Granting the existence of atoms and molecules, what do we want to know about them? In this chapter we deal with some of the simplest properties that can be described in gross terms. When one investigates a macroscopic object, it is natural to ask first about its size and its weight. By the same token, how big and how heavy is an atom or a molecule? On the atomic scale, the electrical charges carried by the various bits of matter are central to their behavior, so we must also consider the charges of the atom's components. We shall describe a number of methods for determining these various atomic magnitudes. Although these are not the only possible methods, nor in general the most precise,[1] they are the most straightforward conceptually, in that the quantities measured are related in very simple ways to the properties desired. Finally, we review some of the basic facts of macroscopic chemistry, the atomic weight scale and the periodic table, which must be related to the properties of atoms.

Chemistry began as a qualitative study of the properties and transformations of the substances found in nature. It evolved into an exact science when quantitative regularities (laws) were found to underlie these properties and transformations. Its goal became the construction of theories to correlate and predict the regularities. The most important of these regularities was the *law of conservation of mass*, which states that the total mass (quantity of matter) present is the same before and after a chemical reaction. Once this law was recognized, one could keep track of the amounts of various kinds of matter involved in reactions and thus determine the composition of substances. A small number of substances—elements—could not be broken down into anything simpler, whereas other substances—compounds—could be separated into two or

1.1 Development of the Atomic Theory: Relative Atomic Weights

---

[1] In practice it often turns out that the most direct way of doing something is not in fact the most accurate. For example, the review that established the fundamental constants listed in Table 1.1 actually used none of the methods cited here, although some were examined in great detail before it was decided to use less direct but more accurate methods.

more elements. But it was not a trivial matter to distinguish elements from compounds in the early days of chemistry.

Every pure[2] sample of a given compound, regardless of its source, always shows the same relative amounts by weight (in modern terms, by mass) of each of its elements. Thus the composition of each substance is uniquely specified in terms of the relative amounts of its component elements; there is no continuous change from pure element A to pure element B through a succession of intermediate compositions, but rather one or more compounds characterized by fixed ratios of the weights of A and B. This is the *law of definite composition*, and it implies that there may be basic units of each element having definite masses. Furthermore, if the elements X, Y, and Z combine in pairs to form three binary compounds, then the weight ratio of Y to Z in the Y–Z compound is related by some simple fraction to the X–Y and X–Z weight ratios; that is, we can write

$$\frac{\text{wt. of Y in Y–Z cpd.}}{\text{wt. of Z in Y–Z cpd.}} = \frac{m(\text{wt. of Y in X–Y cpd.})/(\text{wt. of X in X–Y cpd.})}{n(\text{wt. of Z in X–Z cpd.})/(\text{wt. of X in X–Z cpd.})},$$

(1.1)

where m and n are small integers. This implies that the units involved in chemical combination are the same for various compounds, in other words that these units are characteristic of the element rather than the compound.

The "units" just mentioned, of course, are what we now know as *atoms*. John Dalton's atomic theory (1808) was based on the hypothesis that matter is made of such atoms, those of each element having a fixed mass, which combine in relatively simple proportions to form *molecules*, the simplest particles of compounds. The strongest argument then available for this theory was based on the empirical *law of multiple proportions*. Suppose that elements A and B form several compounds with each other, say compounds 1, 2, 3, and so on. If we form the ratios of the weights of A and B in the various compounds, such as $w_A(1)/w_B(1)$, in which we read $w_A(1)$ as "weight of A in compound 1," then the ratios of these ratios, such as

$$R_{12} = \frac{w_A(1)/w_B(1)}{w_A(2)/w_B(2)},$$

(1.2)

are all simple fractions. For example, oxygen and sulfur form two binary compounds, with w_O/w_S equal to 1.00 in one and 1.50 in the other, giving a ratio $R_{12} = \frac{2}{3}$ (the compounds are in fact SO_2 and SO_3, respectively).

How does the law of multiple proportions provide such strong evidence for the atomic theory, in spite of its dealing with ratios of weights rather than ratios of numbers of particles? If we knew only the *relative* masses of the atoms of A and B, that is, the value of the ratio m_A/m_B, where m is an atomic mass, then we could determine the relative numbers (n) of A and B atoms in a given compound,

$$\frac{n_A}{n_B} = \frac{w_A/m_A}{w_B/m_B} = \frac{w_A}{w_B}\frac{m_B}{m_A},$$

(1.3)

---

[2] By a pure substance we mean a quantity of matter that has the same composition throughout and that cannot be separated into simpler substances by purely physical means, such as evaporating a solution. The statement of constant composition is not precisely accurate because there are variations in isotopic composition of materials with the same formula but different cosmic origin. Such differences offer a powerful tool for the study of cosmochemistry.

since w_A/m_A and w_B/m_B must be the total numbers of A and B atoms, respectively, in our sample of the compound. In Dalton's time the ratios of atomic masses (or *atomic weights*, to use what is still the conventional term) were not yet known. However, if we form the ratio of ratios for compounds 1 and 2, we have

$$R_{12} = \frac{w_A(1)/w_B(1)}{w_A(2)/w_B(2)} = \frac{[w_A(1)/w_B(1)](m_B/m_A)}{[w_A(2)/w_B(2)](m_B/m_A)} = \frac{n_A(1)/n_B(1)}{n_A(2)/n_B(2)}. \quad (1.4)$$

That is, the relative-atomic-weight factors cancel, and the quantity R_{12} also gives the ratio of the ratios of atoms in the two compounds. If there were no fundamental atomic structure to matter (in which case m_A and m_B would simply be arbitrary mass units), one might expect any value for the ratio R_{12}. That R_{12} is a ratio of small integers is consistent with A and B's being composed of atoms which combine in simple proportions, so that the n_A/n_B ratios (and *their* ratio R_{12}) are simple fractions.

This argument showed that the existence of atoms was a reasonable hypothesis, but alternative models were still possible. More convincing evidence came much later, when it became possible to study directly the atomic properties with which we shall deal in this chapter. As late as 1900, a few prominent scientists still rejected the idea that atoms could really exist, accepting them only as a useful computational device.

Another thing the law of multiple proportions does not tell us is how one can deduce the relative weights of different atoms. Consider, for example, the two compounds of hydrogen and oxygen, water and hydrogen peroxide. Analysis of their composition gives the weight ratios $w_H/w_O = \frac{1}{8}$ for water and $w_H/w_O = \frac{1}{16}$ for hydrogen peroxide. It is clear that, if matter is composed of atoms, the hydrogen peroxide molecule must contain twice as many oxygen atoms per hydrogen atom as does the water molecule. But this does not tell us whether water should be written with the formula HO (one oxygen atom per hydrogen) and hydrogen peroxide as HO_2, or as H_2O and HO, respectively, and so on. We *can* tell, from the fact that the weight ratios are (approximately) $\frac{1}{8}$ and $\frac{1}{16}$, that it is reasonable to assume that an oxygen atom weighs 4, or 8, or 16, or even 32 times as much as a hydrogen atom (the corresponding formulas for water being HO_2, HO, H_2O, and H_4O). In fact, water was long assumed (on grounds of simplicity) to be HO. By comparing the relative combining weights of elements in many compounds, chemists such as Berzelius were able to develop self-consistent sets of relative atomic weights for many elements, long before anyone conceived of the more precise microscopic methods available today. But all these results were uncertain by various integral factors, since they were based on guesswork about the formulas of simple compounds like water.

This difficulty was not overcome until it became possible to obtain relative *molecular weights*. The key came from the theory of gases. It was observed that when gases reacted, the ratios of their volumes were also simple fractions. In 1811 Avogadro proposed that equal volumes of gases at the same temperature and pressure must contain the same numbers of molecules. We shall show in Chapter 12 how this result (*Avogadro's principle*) can be derived from microscopic theory. Avogadro's conclusions were long resisted, since it was felt that the "molecules" of the elements must be single atoms, whereas measurements indicated that the common gases had to be composed of diatomic molecules (hydrogen H_2, oxygen O_2, etc.) for Avogadro's principle to be valid. Finally, in 1860, Cannizzaro showed how one could unify the existing atomic weight data into a consistent system by accepting Avogadro's principle. For example, since two volumes of hydrogen combined with one volume of oxygen to give two volumes of water (as steam), the reaction was correctly described by the equation

$$2H_2 + O_2 \rightarrow 2H_2O,$$

in which the numbers of atoms balance and the numbers of molecules are proportional to the volumes involved.

Other methods of determining molecular weights (especially for substances not available in gaseous form) were developed; we shall mention some of these in later chapters. When these results were combined with the measurements of the combining weights of elements, it became possible to develop a complete set of relative atomic weights. Such a table is given on the inside cover of this book; we shall discuss in Section 1.6 the arbitrary scale to which the numbers are referred.

1.2
Atomic Magnitudes

We conclude, then, that matter is made up of atoms, those of each element having a characteristic mass, and that from macroscopic data alone one can obtain the relative values of these masses. What other things do we want to know about the atom? What are the atomic properties that are ultimately the basis of chemistry, and what experiments or theories will tell us their magnitudes? These magnitudes of course include the absolute values of the atom's mass, its linear dimensions, and the electrical charges of its components. Going beyond this, we wish to determine the internal structure of the atom, the way in which mass and charge are distributed within its volume.

To discuss these matters conveniently, it is best to anticipate some of the results to be developed in this and subsequent chapters. We take for granted the following assumptions: The atom is composed of a tiny, positively charged nucleus around which orbit negatively charged electrons. The nucleus contains over 99.9% of the atom's mass, but occupies less than 10^{-12} (one-trillionth) of its volume. The nucleus in turn is an aggregation of positively charged protons and uncharged neutrons. The simplest atom, hydrogen, contains one proton and one electron. The atom as a whole is normally uncharged, but can become ionized by the loss or gain of negative charge in the form of electrons. Experiment shows that the charge on an atom or molecule is always a multiple of a small fixed quantity: Charge as well as mass comes in discrete units. This quantity is the charge of the electron or the proton, which are equal in magnitude but opposite in sign.

Now that we have indicated what we are talking about, let us look at some numerical values for the quantities in which we are interested. These numbers cannot be deduced from some *a priori* theory; they can only be obtained, directly or indirectly, from experiment. Some are given in Table 1.1, and a more extensive list can be found on the inside cover of this book. Constants such as these, which are applicable to all forms of matter and turn up in many fields of science, are referred to as the fundamental constants of nature.

Numbers with such extreme exponential factors as those in Table 1.1 tend to be meaningless to us. Let us try to clarify their magnitudes by comparison with other, more familiar quantities. For example, the mass of the electron stands in about the same relation to a $2\frac{1}{3}$-g weight as the weight does to the mass of the earth (6×10^{24} kg). From the table, the proton (or the neutron, or the hydrogen atom) weighs about 1800 times as much as the electron, so that on our gram–earth scale the proton corresponds to roughly 4 kg. Atomic nuclei can contain as many as 250

TABLE 1.1
SOME ATOMIC CONSTANTS[a]

Electron rest mass (m_e)	9.109534×10^{-31} kg
Proton rest mass (m_p)	$1.6726485 \times 10^{-27}$ kg
Neutron rest mass (m_n)	$1.6749543 \times 10^{-27}$ kg
Elementary electronic charge (e)	$1.6021892 \times 10^{-19}$ C
Bohr radius (a_0)	$5.2917706 \times 10^{-11}$ m

From E. R. Cohen and B. N. Taylor, *J. Phys. Chem. Ref. Data* **2**, 663 (1973).

protons and neutrons, so that absolute atomic masses run up to about 4×10^{-25} kg (about a ton on our scale).

Next, how big is the charge of the electron? If the earth had an excess of about 4×10^{15} electrons (weighing less than 4×10^{-15} kg, remember), or for that matter 4×10^{15} protons, then it would have a potential of about 1 V, or two-thirds the voltage of a flashlight battery. By this standard the electron's charge may seem rather large, but this is a result of the ease with which a sphere can be charged. Perhaps a more vivid comparison is this: 1 microampere (μA), a small but readily detectable current, corresponds to a flow of 6×10^{12} (6 trillion) electrons per second past a measuring device. Finally, to visualize the linear dimensions of the atom, consider the Bohr radius. We shall say a good deal about this quantity in the next chapter, but for now it can be considered loosely as the radius of the hydrogen atom. It is convenient at this stage to introduce a special unit for atomic dimensions, the *angstrom* (abbreviated Å), which is 10^{-10} m (0.1 nm). The Bohr radius is about 0.5 Å, which is to 1 cm roughly as 1 cm is to 2000 km. The largest known atoms have radii of about four times the Bohr radius. As for the atomic nucleus, its radius is less than 10^{-14} m (10^{-4} Å), or less than one ten-thousandth of the atom's radius.

Let us now begin to consider the experimental determination of the quantities we have been looking at. The mass of the electron can be determined almost directly from the optical spectra of hydrogen and deuterium. (See Problem 8 at the end of Chapter 2.) However, most atomic or molecular masses are not measured directly. Instead, one converts the species of interest into *ions* (atoms or groups of atoms carrying a charge) by adding or removing electrons. Then, by measuring the deflection of the ions in an electric or magnetic field, one can find the charge-to-mass ratio. The charge is most often identical to the electronic charge e, and so far as is known is always an integral multiple of e. This technique was introduced by J. J. Thomson, whose determination of e/m for the electron will be discussed in the next section; in Section 1.5 we shall examine the general application of the technique (mass spectrometry).

Before one can determine masses from a charge-to-mass ratio, one must know the value of e. Historically, the ratio e/m_e was measured well before either e or m_e separately. The charge of the electron was first determined by R. A. Millikan, from observation of the motion of charged droplets in an electric field; we discuss this method in Section 1.4. Today a more straightforward method is possible, in which individual electrons are counted. This method simply involves collecting a beam of electrons with two detectors, which gather the same fraction of the electron current but measure it in different ways. One detector measures the number of individual electrons striking it per unit time, and the other measures the electric current (charge per unit time) that it receives; the ratio of the charge per unit time to the number of electrons per unit time is the charge per electron.

There are at least two straightforward ways to measure atomic dimensions. The more accurate is to scatter x rays of known wavelength from a crystal, and from the pattern produced by the scattered x rays to deduce the interatomic distances in the crystal. The other is to measure the density of a substance in the bulk, to find independently the mass of a single atom or molecule, and from the two to compute the volume available to one atom or molecule. We shall postpone further discussion of these methods until our treatment of crystal structure in Chapter 11. One can also determine the distances between atoms in a molecule by various spectroscopic techniques. The "size" of an individual atom is a somewhat vague concept, since atoms have no sharp boundaries; the best one can do is to find the spatial distribution of electrons, as we shall see when we take up the theory of atomic structure.

The charge-to-mass[3] ratio of the electron, e/m, was the first atomic parameter to be determined, by J. J. Thomson in 1897. This is also in many ways the most straightforward of the experiments we shall discuss, and our description follows Thomson's method. His experiments were performed on what were then known as cathode rays—rays that we now know consist of streams of electrons. The purpose of these experiments, in fact, was to show that the cathode rays *were* streams of particles. The apparatus used is shown schematically in Fig. 1.1.

In the experiment the tube is evacuated and the cathode A is charged negatively with respect to the anode B, causing electrons to stream from A toward B. The anode is constructed with a narrow slit in its center, allowing some of the electrons to pass through; downstream is a second slit S. The two slits act to produce a narrow beam of cathode rays (electrons). These travel from A through B and S to reach the luminescent screen Z, where they produce a small spot, just an enlarged image of the hole in S. If S is replaced by a small object that only partially blocks the beam, then one sees on the screen a shadow image of this object. Such a shadow was taken as evidence that the cathode rays did indeed behave like particles. (But those who thought otherwise could point to the fact that the rays passed through thin metal foil.) More significant for us now, however, are the electrical properties of these rays. If plate X is charged positively and plate Y is charged negatively, then the beam moving from S strikes Z at some point closer to X than to Y. In other words, the positive charge on X has attracted the beam of cathode rays, and we conclude that the cathode rays consist of something with a negative charge.

We describe the force exerted on the particles traveling between plates X and Y as the force of an *electric field*. This field of force is a mental construct, if you like, in which one can imagine literal lines of force, like strings, pulling on a charged particle. The lines of force simply describe the trajectories that particles (initially at rest) would follow under the influence of the field. The total force $\mathbf{F}$ on the particle is directly proportional to its own charge q. The proportionality constant, the electrical force per unit charge, is called the electric field strength $\mathbf{E}(=\mathbf{F}/q)$. The electric field between X and Y consists essentially of

1.3
The Charge-to-Mass Ratio of the Electron: Thomson's Method

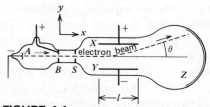

FIGURE 1.1
Schematic diagram of Thomson's apparatus for determining e/m. (See text for explanation of symbols.)

---

[3] We shall designate the mass of the electron simply by m rather than m_e whenever no ambiguity is involved.

parallel lines of force pointing from X toward Y; that is, a positively charged particle placed between X and Y would be forced from X toward Y. If we neglect the regions near the ends of plates X and Y, where the lines of force curve outward, the field strength between the plates is everywhere the same. We can say that the space between the plates is a region of uniform field, and thus of constant force.

To clarify the notion of a field a bit more, let us consider another familiar force, namely, that exerted between one charged point particle and another in free space. Suppose that the two particles have charges q_1 and q_2. Then the force between them in a vacuum is directed along the line joining them, and by Coulomb's law has the magnitude

$$F = \frac{q_1 q_2}{4\pi\epsilon_0 r^2}, \tag{1.5}$$

where r is the distance between the two charges and ϵ_0 is a constant (the *permittivity of free space*); in SI units[4] (with q in coulombs, C; r in meters, m, and $\mathbf{F}$ in newtons, N), we have

$$\epsilon_0 = 8.85418 \times 10^{-12} \text{ C}^2/\text{Nm}^2.$$

The electric field strength at particle 2 (defined as $\mathbf{F}/q_2$) thus has a magnitude $q_1/4\pi\epsilon_0 r^2$, which varies with position. Similarly, the electric field in any region of space can be mapped by measuring at various points the force it exerts on some test object of known charge, such as an electron or a charged pith (or plastic) ball. In general, a field is simply a property, extending through a volume of space, whose magnitude is a function of position.

The force $\mathbf{F}$ on a charged particle is in fact a *vector* (identified by boldface type), a quantity with both magnitude and direction. It is expressible as a set of three quantities F_x, F_y, F_z, corresponding to the components of force parallel to the x, y, z axes of a Cartesian coordinate system. The magnitude of $\mathbf{F}$, denoted by F, is given by the relation

$$|\mathbf{F}| = F = (F_x^2 + F_y^2 + F_z^2)^{1/2}. \tag{1.6}$$

Similarly, any vector $\mathbf{A}$ is really the set of three quantities A_x, A_y, A_z, and has the magnitude

$$A = (A_x^2 + A_y^2 + A_z^2)^{1/2}. \tag{1.7}$$

We may think of $\mathbf{A}$ as the generalization in three dimensions of the hypotenuse of a right triangle. In fact, if one of the components A_x, A_y, A_z is zero, the analogy to a right triangle holds exactly. We shall develop other properties of vectors as we need them.

The electric field strength $\mathbf{E}$ is of course also a vector, since it is defined as the force per unit charge. That is, a test object with charge q will be subjected to a force

$$\mathbf{F} = q\mathbf{E} \tag{1.8}$$

at a point where the field strength is $\mathbf{E}$. What this vector equation means is that $F_x = qE_x$, $F_y = qE_y$, $F_z = qE_z$; thus the magnitudes of the vectors

---

[4] We should also mention the electrostatic and Gaussian systems of units, in which the constant in Coulomb's law is suppressed—$|\mathbf{F}| = q_1 q_2/r^2$—otherwise cgs (centimeter-gram-second) units are used. The *electrostatic unit* of charge (esu) thus has a value such that there is a force of 1 dyne between two 1-esu charges 1 cm apart: 1 esu $= 3.33564 \times 10^{-10}$ C. The esu is also known as the *statcoulomb* when Coulomb's law is taken as its definition, 1 esu $\equiv 1$ dyn$^{1/2}$ cm, or franklin (Fr) (when taken as an independent unit.)

are also proportional, $F = qE$. The electric field strength can be measured in newtons per coulomb, or in the equivalent units of volts per meter.

If a charged particle moves under the influence of a force, then the force has done work on the particle, and thus has changed its energy. When the force is constant and the motion is in a straight line, the work done is the magnitude of the force multiplied by the displacement. Let us now return to the motion of an electron between charged plates, as shown in Fig. 1.2. Since the electron's charge is negative ($q_e = -e$), by Eq. 1.8 the vectors **F** and **E** must have opposite signs, that is, opposite directions; the electron is drawn toward the positive plate, so the electric field vector must point toward the negative plate. If an electron moves from y_1 to y_2 under the influence of a constant field **E**, then the work done is

$$W = F(y_2 - y_1) = eE(y_2 - y_1). \qquad (1.9)$$

(For force in newtons and distance in meters, W must be expressed in joules.) The electron has moved "downhill" in the electric field, that is, to a point of lower potential energy; the decrease in its potential energy is the same as the work done, $eE(y_2 - y_1)$. Just as we picture the force on a charged particle as the product of charge and electric field strength, so we can describe the particle's potential energy as the product of charge and a quantity we call the *potential*. The potential is the potential energy per unit charge, and is thus measured in joules per coulomb, better known as volts.[5] If the distance between our two plates is h, then the potential energy difference between them is eEh, and the potential (voltage) difference is Eh.

Now we have the vocabulary to give a quantitative description of Thomson's measurement of e/m for the electron. Consider again the diagram of the apparatus in Fig. 1.1. We define a coordinate system with the x axis in the electrons' initial direction of motion and the y axis normal to plates X and Y. Between these plates exists an electric field, which we assume to have the constant value **E** everywhere in a region of length l. (That is, we assume that all the lines of force are vertical, neglecting the curvature at the ends.) Let us say that all the electrons enter the field from the left with initial velocity **v** in the x direction. At any time when an electron is between the plates, it feels a force in the positive y direction of magnitude eE. This force, by Newton's second law, must be the product of the electron's mass and its acceleration **a**:

$$\mathbf{F} = -e\mathbf{E} = m\mathbf{a} \qquad (F_y = -eE = ma_y). \qquad (1.10)$$

Note that the force and acceleration are exerted in the y direction only, so the x component of the electron's velocity, v_x, is unchanged. Therefore the time the electron spends in the field is

$$t_f = \frac{l}{v_x}. \qquad (1.11)$$

During this time the y component of velocity, v_y, changes from zero to $a_y t_f$; the force and thus the acceleration being constant, the velocity is merely the acceleration multiplied by the time it is applied. Consequently, the y component of the velocity reaches the value

$$v_y = a_y t_f = \frac{eE}{m} \frac{l}{v_x}. \qquad (1.12)$$

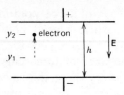

FIGURE 1.2
Motion of an electron, initially at rest, in a uniform electric field **E**. The electron's position y is measured from the negative toward the positive plate.

---

[5] The Gaussian unit of potential is the erg per esu, or *statvolt*; 1 statvolt = 299.793 V.

After the electron leaves the region between the plates, it is no longer accelerated, so v_x and v_y are both constant. Over a time t, the electron then moves a distance $v_x t$ in the x direction, and $v_y t$ in the y direction. Then θ, the angle of deflection[6] from the x axis, is given by

$$\tan \theta = \frac{v_y}{v_x}. \qquad (1.13)$$

Substituting Eq. 1.12, we obtain

$$\tan \theta = \frac{eE}{m} \frac{l}{v_x^{\,2}}. \qquad (1.14)$$

Thus if we could measure θ, $\mathbf{E}$, l, and v_x, we could immediately find e/m.

The electric field $\mathbf{E}$ is measurable with a voltmeter and a ruler. It is just the voltage difference between the charged plates divided by their separation. The length l is again a measurement we make when we build the apparatus. The angle θ we find by doing the experiment, for example with a fluorescent screen. The initial speed v_x is a bit more difficult to determine. In principle one can measure it by placing in the path of the beam two shutters that can be opened for brief periods. A small burst of electrons is allowed to pass through the first shutter; one then determines when the second shutter must be opened to permit the electrons to reach a collector. If the separation between the shutters is d, and the time interval that just allows the burst from shutter 1 to pass through shutter 2 is τ, then the speed is simply d/τ. For shutters one could use two wheels, each with a single slot, mounted on a common shaft with variable speed of rotation; this is a method commonly used to determine the velocities of atomic beams,[7] but it is inconvenient for electron beams because they move too rapidly. We shall see how fast electrons actually move when we reach the point of evaluating the electron's mass and charge. Fortunately, there are better ways to determine the electron's velocity.

An especially convenient way to deal with v_x is by eliminating it from Eq. 1.14. This method, the one actually used by Thomson, also forms the basis for much of modern mass spectrometry (see Section 1.5). It is based on a second way of deflecting moving charged particles, namely, the use of a *magnetic field*. One could introduce the idea of a magnetic field in a way analogous to Eq. 1.5, by writing an expression for the force between two (hypothetical) magnetic poles. However, the "magnetic pole" is a clumsy and only approximately meaningful concept. It is more logical, as well as more pertinent to our present needs, to define a magnetic field in terms of its interaction with a moving charge.

The magnetic force on a moving charged particle is known as the *Lorentz force*. If a particle with charge q moves with velocity $\mathbf{v}$ through a magnetic field $\mathbf{B}$ perpendicular to $\mathbf{v}$, the Lorentz force $\mathbf{F}$ has the magnitude

$$F = qvB. \qquad (1.15)$$

In SI units, $\mathbf{F}$ is measured in newtons, q in coulombs, $\mathbf{v}$ in meters per second, and $\mathbf{B}$ in volt seconds per meter squared; the latter unit is called the *tesla* (T). This equation could be taken as a definition of the *magnetic*

---

[6] The angle θ is measured from the center of the region between the plates. Given the uniform field we have defined, the electron will leave the field moving directly away from this point.

[7] Cf. Problem 10 at the end of this chapter.

induction[8] (or *magnetic flux density*) **B**, except that we must also establish the direction of the vector **B**. If a coordinate system is defined such that **v** points in the positive x direction and the force on a positively charged particle in the positive z direction, then the magnetic field vector is defined to point in the positive y direction.[9] These directional relationships are shown schematically in Fig. 1.3; they are sometimes called the *right-hand rule*, for reasons indicated in the figure.

All the information given in the previous paragraph is summarized compactly in the vector equation[10]

$$\mathbf{F} = q(\mathbf{v} \times \mathbf{B}). \qquad (1.16)$$

The expression **v**×**B** is what is known as a *cross product* (or *vector product*). It is itself a vector, with three components related to the components of **v** and **B** by the equations

$$(\mathbf{v} \times \mathbf{B})_x = v_y B_z - v_z B_y, \qquad (1.17a)$$

$$(\mathbf{v} \times \mathbf{B})_y = v_z B_x - v_x B_z, \qquad (1.17b)$$

$$(\mathbf{v} \times \mathbf{B})_z = v_x B_y - v_y B_x. \qquad (1.17c)$$

These equations can readily be extended to the general case. The cross product of any two vectors $\boldsymbol{\alpha}$ and $\boldsymbol{\beta}$ has the magnitude $\alpha\beta |\sin \phi|$, where ϕ is the angle between them. Note that $\boldsymbol{\alpha} \times \boldsymbol{\beta}$ and $\boldsymbol{\beta} \times \boldsymbol{\alpha}$ have the same magnitude but opposite directions; that is, their corresponding components have opposite signs. In setting up Fig. 1.3 we simply chose the axes and the orientation of **v** and **B** so as to make the x and y components of force, Eqs. 1.17a and 1.17b, identically zero. In practice, it is often advantageous to build one's apparatus so that the equipment itself can fulfill simplifying conditions like these.

Now we can derive an expression for the deflection of an electron by a magnetic field, just as we obtained Eq. 1.14 for its deflection by an electric field. We wish to use a set of axes consistent with those in Fig. 1.1, in which the positive z axis can be taken as coming forward out of the paper (right-handed coordinates). We again allow the electron beam to move in the x direction; however, since we have already put an electric field along the (negative) y axis, let us direct the magnetic field along the positive z axis. Then the only nonvanishing term on the right-hand sides of Eqs. 1.17 is the $-v_x B_z$ in Eq. 1.17b, so that **v**×**B** and thus the Lorentz force must be directed along the y axis. Since for the electron q in Eq. 1.16 is negative, the force will be in the positive y direction, with magnitude

$$F_y = (-e)(-v_x B_z) = e v_x B_z. \qquad (1.18)$$

The above is true only initially: Once the force has been applied, v_y is no longer zero, and a component $F_x = -e v_y B_z$ appears, changing v_x in turn.

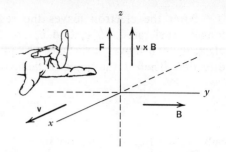

FIGURE 1.3
Lorentz force on a positively charged particle moving in a uniform magnetic field. The small drawing illustrates the right-hand rule: thumb, index finger, and middle finger point respectively in the directions of **v**, **B**, and **F** (or of the x, y, and z axes, in a right-handed coordinate system).

---

[8] It is common to refer to **B** as simply the "magnetic field." Strictly, however, the *magnetic field strength* **H** is a different vector. In a vacuum the two are related by $\mathbf{B} = \mu_0 \mathbf{H}$, where $\mu_0 \equiv 4\pi \times 10^{-7}$ N/A^2 or henry/m (the *permeability of free space*); **H** is measured in amperes per meter.

In the Gaussian system μ_0 is suppressed, like the $4\pi\epsilon_0$ in Coulomb's law. The Lorentz force equation then takes the form $F = qvB/c$, where c is the speed of light. With **F** in dynes, q in esu, **v** and c in centimeters per second, the unit of **B** is the gauss (G): $1\,\text{G} = 10^{-4}$ T. The Gaussian unit of **H** is the *oersted* (Oe), which has the same dimensions as the gauss. Since in these units **B** and **H** have the same numerical value (in a vacuum), most chemists speak indiscriminately of "magnetic field **H** in gauss."

[9] The direction of **B** is that in which a positive (north) magnetic pole would move under the influence of the field.

[10] In the Gaussian system, $\mathbf{F} = (q/c)(\mathbf{v} \times \mathbf{B})$.

We shall see later how this problem is handled, but for now let us simply assume that the deflection region is short, so that the change in v_x is negligible. Following steps just like those leading from Eq. 1.10 to Eq. 1.14, we then obtain from Eq. 1.18 the result

$$\tan \theta' = \frac{eB_z}{m} \frac{l}{v_x}, \qquad (1.19)$$

where θ' is the angle of deflection caused by the magnetic field. We can design the apparatus so the length l is essentially the same for the electric and magnetic deflection regions. For small deflection angles one can approximate $\tan \theta$ by θ and $\tan \theta'$ by θ'; in this approximation we can see that both deflection angles are proportional to field strength but that the electric field deflection θ varies with v_x^{-2} whereas the magnetic field deflection θ' varies with v_x^{-1}.

We were stopped after Eq. 1.14 because we had two unknowns: the ratio e/m, which we want to know, and the electron speed v_x, which we don't particularly care about at this point. But now we have both Eq. 1.14 and Eq. 1.19, two simultaneous equations in these two unknowns, based on two independent experiments. Combining the two equations to eliminate v_x, and then solving for e/m, we obtain

$$\frac{e}{m} = \frac{\tan^2 \theta'}{\tan \theta} \frac{E}{B_z^2 l}. \qquad (1.20)$$

All the quantities on the right side of this equation can be determined experimentally, so e/m can now be obtained directly.

Given an apparatus in which the electric and magnetic fields can be applied either simultaneously or separately, there are various ways in which the experiment can be performed. We could, of course, simply pick arbitrary values of E and B_z and measure θ and θ' separately. Or we could choose E and B_z so that the angles θ and θ' are equal (so that $\tan^2 \theta'/\tan \theta$ becomes simply $\tan \theta$); that is, we can deflect the electrons with an electric field and find the spot to which the deflected beam moves, then apply a magnetic field just large enough to bring the beam to exactly the same spot. Finally, we can measure θ with the electric field alone, then turn on the magnetic field simultaneously and adjust it so that the beam is again undeflected; in this case we have $\theta + \theta' = 0$, or $\theta' = -\theta$.

Thomson obtained only an approximate value of e/m. Modern measurements, using more accurate techniques, give the currently accepted value

$$e/m = 1.7588047 \times 10^{11} \text{ C/kg}.$$

This number, the charge-to-mass ratio of the electron, is by itself not very useful to us. We cannot guess from it alone how many electrons there are per gram of carbon, for example, or even what the charge of a single electron is. It becomes useful only if we obtain an independent measure of the charge of the electron itself, a measurement to which the next section is devoted.

1.4
The Charge of the Electron: Millikan's Method

We now turn to the determination of the electronic charge e, by the method R. A. Millikan used in 1909. We should point out once again that, although this method is one of the simplest conceptually and experimentally, and the most important historically, it is no longer among the most precise ways of evaluating e. (It is, nonetheless, one of the ways people have searched for *quarks*, predicted particles with charges of $\frac{1}{3}e$ or $\frac{2}{3}e$.) Later, when we have developed some of the quantum properties of matter, we shall point out other measurements that can yield an accurate value of the charge on the electron. The essence of Millikan's method lies in examining the behavior of small but macroscopic charged particles under the combined influence of gravity, electrical fields, and friction. From a knowledge of the various force laws one can deduce the size of the charge on such a particle. Determinations of many such charges, on particles of different kinds, show that all the particle charges can be expressed as integral multiples of a constant value, identified with the elementary charge e.

In practice, the particles one observes are either droplets of oil or (today) small plastic balls (Millikan tried water first, but water droplets evaporate too fast). These droplets or balls manage to pick up negative charges (electrons) spontaneously as they drift in air, and one can hasten the process by bringing a bit of radioactive material nearby, so that there are quite a lot of charges in the neighborhood. It is convenient to watch the charged particles with a microscope having calibration lines, so that one can see how long it takes for them to drift downward through a known distance under the influence of gravity. If the region of observation is between two condenser plates to which a charge can be applied, a particle carrying a charge can be forced upward against gravity by an electric field. If one is lucky one can watch the same particle move up and down many times, changing its charge by bringing the radioactive source near. It is convenient to shine light on the particles at right angles to the direction of observation so that the particles shine as bright spots, much like dust motes in a sunbeam. The apparatus is sketched in Fig. 1.4.

We begin by considering the motion of the particle in the absence of the electric field, falling under the influence of gravity alone. Remember that the particle is moving in air, not in vacuum. Since it is very small, its fall is significantly retarded by friction with the surrounding atmosphere. Under this condition, the motion is best described by *Stokes's law*, which we simply state without proof:[11]

$$F = 6\pi\eta r v, \qquad (1.21)$$

where v is the final speed attained by a spherical body of radius r, drawn slowly by a force of magnitude F through a medium of viscosity[12] η. In

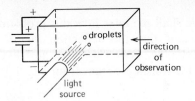

FIGURE 1.4
Schematic diagram of the Millikan oil-drop experiment. The top and bottom of the chamber are condenser plates charged by the battery at left. The negatively charged droplets will tend to move upward in the electric field, but will fall when the battery is disconnected.

---

[11] This equation may appear at first sight to contradict Newton's second law, which states that force is proportional to *acceleration*; according to Eq. 1.21, in this situation the force is proportional to *velocity*. The contradiction is only apparent, of course, because Stokes's law is derivable from the laws of mechanics.

Stokes's law is actually a statement of the *retarding force* that a medium exerts on an object moving through it. Similar laws determine the maximum velocity of a space vehicle reentering the atmosphere, or of a skier in a long schuss. The frictional force increases with velocity, because the moving object collides with more molecules of the impeding medium as its velocity increases, and thereby experiences more drag. A falling object reaches a terminal velocity when the gravitational force mg upon it is balanced by the frictional force. If the frictional force is proportional to velocity, $F_{fric} = bv$, then the terminal speed must be $v_{term} = b/mg$. Since the *net* force is zero, there is no further acceleration, and Newton's second law is in fact satisfied.

An object falling at its terminal velocity has a constant kinetic energy and a potential energy that decreases at a constant rate. Where does this energy go? It is transformed into random kinetic and internal energy of the molecules of the medium and the object—into *heat*. If the velocity is high enough, this heat may be sufficient for the object to become incandescent, to melt, or even to react with the medium. This phenomenon is responsible for the visibility of meteors. It also required the development of the ablative nose cone for reentry vehicles: Ablation is the process in which surface material on the space vehicle is heated by friction up to its softening or melting point, and then bit by bit blows away from the vehicle, carrying much of the frictional heat with it.

[12] For a discussion of viscosity, see Chapter 28. The viscosity of atmospheric air (as in Millikan's experiment) is about 2×10^{-5} N s/m^2.

the Millikan experiment, in the absence of an electric field, F is the net gravitational force *while the sphere is in air*—in other words, the gravitational force in vacuum minus the buoyant force of the air. If ρ_s is the density of the sphere and ρ_{air} is the density of air, then we can write

$$F_{grav} = (m_s - m_{air})g$$
$$= \left(\frac{4\pi}{3}r^3\rho_s - \frac{4\pi}{3}r^3\rho_{air}\right)g, \tag{1.22}$$

where m_s is the actual mass of the sphere and m_{air} is the mass of an equal volume of air. Substituting in Eq. 1.21 for the gravitational force given by Eq. 1.22, which must equal the retarding frictional force, we obtain

$$(F_{fric})_1 = 6\pi\eta rv_1 = F_{grav} = \frac{4\pi r^3}{3}g(\rho_s - \rho_{air}), \tag{1.23}$$

where v_1 is the limiting (downward) speed under these conditions.

Now suppose that we have an electric field **E** exerting an upward force on the particle strong enough to produce a net *upward* speed v_2. In this way we can bring a specific particle back above some calibration mark. Under these conditions the frictional force must equal the net upward force applied to the particle (electrical force minus gravity). If the charge on the particle is ne and $|\mathbf{E}| = E$, we have

$$(F_{fric})_2 = 6\pi\eta rv_2 = F_{elec} - F_{grav} = neE - \frac{4\pi r^3}{3}g(\rho_s - \rho_{air}). \tag{1.24}$$

This expression contains the unknown elementary charge e multiplied by the also unknown n, the number of such charges that happen to be on the particle observed. Combining Eqs. 1.23 and 1.24 and solving for ne, we obtain

$$ne = \frac{4\pi r^3}{3}(\rho_s - \rho_{air})\frac{g}{E}\left(\frac{v_2}{v_1} + 1\right); \tag{1.25}$$

v_1 and v_2 must, of course, be measured with the same particle. If we had a way of measuring r independently, then we could use Eq. 1.25 exactly as it is. The density of the sphere, whether it be plastic or oil, can be determined from the bulk material; the density of air is known accurately; the electric field, as in Thomson's experiment, is the measured voltage divided by the distance between the two metallic plates; and the speeds v_1 and v_2 are just the quantities we measure in the experiment. In practice, however, it is not always possible to measure the radius r; this was the case, for example, in Millikan's original experiments. Fortunately, we can use Eq. 1.23 to determine r from the measured downward speed under gravity alone; rearrangement gives

$$r^2 = \frac{9}{2}\frac{v_1\eta}{g(\rho_s - \rho_{air})}. \tag{1.26}$$

Substituting this result in Eq. 1.25, we obtain the final expression

$$ne = \frac{4\pi}{3}\left[\frac{9}{2}\frac{v_1\eta}{g(\rho_s - \rho_{air})}\right]^{3/2}(\rho_s - \rho_{air})\frac{g}{E}\left(\frac{v_2 + v_1}{v_1}\right)$$
$$= \frac{36\pi}{E}\left[\frac{\eta^3}{8g(\rho_s - \rho_{air})}\right]^{1/2}(v_2 + v_1)v_1^{1/2}. \tag{1.27}$$

Now all the quantities on the right are measurable; that is, they were all measurable for Millikan. Some of Millikan's own original data are reproduced in Table 1.2. It can be seen that the measured charges were all

TABLE 1.2
SAMPLE DATA FROM MILLIKAN'S REPORT OF THE OIL-DROP
EXPERIMENT[a,b]

q (10^{-10} esu)	19.66	24.60	29.62	34.47	39.38	44.42	49.41
n	4	5	6	7	8	9	10
e (10^{-10} esu)	4.915	4.920	4.937	4.923	4.931	4.936	4.941

q (10^{-10} esu)	53.92	59.12	63.68	68.65	78.34	83.22
n	11	12	13	14	16	17
e (10^{-10} esu)	4.902	4.927	4.900	4.904	4.897	4.894

[a] From R. A. Millikan, *Phys. Rev.* **32**, 349 (1911).
[b] These data were all obtained with a single oil droplet; they are not in chronological order, and most values were obtained several times. The first line of the table lists the values obtained for the droplet's charge q at various times, calculated with Eq. 1.27. The integer n is the number of elementary charges assumed to make up q, selected so as to give the best fit for the value of $e(=q/n)$. The average for this droplet was $e = 4.917 \times 10^{-10}$ esu; after hundreds of measurements with many droplets, Millikan obtained an overall average of 4.891×10^{-10} esu (1.631×10^{-19} C).

almost exactly integral multiples of about 4.9×10^{-10} esu (1.63×10^{-19} C), which Millikan took to be the value of e.

With plastic spheres of uniform diameter available, one can now do the experiment with Eq. 1.25 as the basis of the calculations, rather than the more cumbersome Eq. 1.27 involving the viscosity of air. Actually, Millikan's original value of e differed from the present value primarily because of an error in the then-accepted value of the viscosity of air. The best value now available for the electronic charge is

$$e = 1.6021892 \times 10^{-19} \text{ C.}$$

At this point we can combine the value of e with the charge-to-mass ratio e/m of the last section to derive a value for the mass of the electron (strictly speaking, the rest mass, but in this text we shall not consider the effects of relativistic velocities). The value thus found is

$$m_e = 9.109534 \times 10^{-31} \text{ kg.}$$

1.5 Mass Spectrometry

Given the possibility of determining the charge-to-mass ratio of the electron, it is clear that by similar techniques one can just as well (or at least almost as well) determine the charge-to-mass ratio for other charged particles, such as ions—atoms or molecules with charges. Since the charges must always be integral multiples of e, in this way one can measure almost directly the mass of any ion. The mass of one or a few electrons is so small that these masses are essentially identical to the masses of the uncharged atoms or molecules. Such measurements are in fact among the most precise ways used to determine actual atomic weights (as opposed to the relative atomic weights discussed in Section 1.1). One need only be sure that the ions are singly charged, or else that the number of charges can be identified, a relatively simple task in either case. For example, the hydrogen atom yields only one positively charged ion, presumably with a single positive charge (H^+), and with $q/m = 9.57876 \times 10^7$ C/kg; assuming that $q = +e$, we obtain 1.67265×10^{-27} kg for the mass of the proton. The apparatus used for such measurements is usually not precisely of the design used by Thomson; almost invariably, however, the principles and fundamental equations for the process are the same as those described in Section 1.3.

The most common device for measuring precise masses of individual atoms or molecules is called the *mass spectrometer*. Several kinds of instruments go under this name; they have in common their capability for selecting particles with specific values of the charge-to-mass ratio, and thus (for singly charged ions) of the mass. The word "spectrometer" is used because one can select or measure which values actually occur, out of the entire spectrum of possible values of the physical quantity of interest, in this case mass.

Virtually all mass spectrometers have the following features in common:

1. A source region, where neutral atoms or molecules are introduced, converted to ions, and passed on to

2. An analyzing region, where particles of different masses are actually separated from each other in space, and

3. A detecting region, where the ions of a particular mass reach a detector.

The source usually generates ions by either ultraviolet light or electron bombardment. The detector may be an electrical device for counting ions (as particles or as electric current), designed to select only ions of a single mass, or it may be a photographic plate capable of collecting particles of many masses simultaneously at different places. If the information is recorded, either photographically or on a chart, in any form equivalent to the number of ions striking the detector as a function of the mass of the ions, then the instrument is called a mass spectro*graph*. As for the analyzing region, there are several means of separating or *dispersing* the ions according to their masses. We shall now examine two of them, to see how the operations of the instruments are related to the equations of motion and to measurable physical quantities.

One class of instruments, the *magnetic mass spectrometer*, includes Thomson's method and its variations: Devices in this category use a combination of electric and magnetic fields to put moving ions into curved trajectories; the curvature of any ion's trajectory is a function of its charge-to-mass ratio. Commonly, an electric field draws the ions out of the source region and into a region where the trajectories are bent by a magnetic field. Only ions on one specific trajectory can reach the detector; others strike the walls of the chamber. By varying the strength of the magnetic field or the initial electric "drawout" field, one can vary all the trajectories and allow ions of any desired mass to strike the detector. A diagram of such a mass spectrometer is shown in Fig. 1.5. This is the general type of instrument used for accurate measurement of absolute masses; admittedly one now could use a rather fancy version.

The spectrometer of Fig. 1.5 and others of the same general type depend on the Lorentz force given by Eq. 1.16. In such a machine, the acceleration of an ion of mass m and charge q in the magnetic field $\mathbf{B}$ is

$$\mathbf{a} = \frac{\mathbf{F}}{m} = \frac{q}{m}(\mathbf{v} \times \mathbf{B}). \tag{1.28}$$

However, we can no longer make the simplifying assumption that the deflecting region is small and v_x essentially constant; rather, we typically have a situation in which the deflection is 60 or 180°, with nearly all the trajectory within the deflecting region. We must thus solve the complete equation for the trajectory. Let us use the coordinate system of Fig. 1.3, with the magnetic field in the y direction and the ions initially moving in

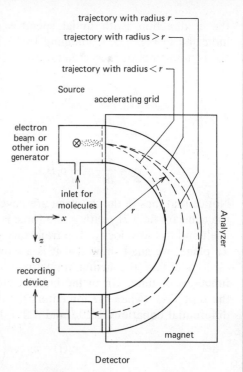

FIGURE 1.5
Schematic diagram of a magnetic mass spectrometer (180° variety). The trajectory leading to the detector is a semicircle of radius *r*. The magnetic field is directed forward out of the page; this corresponds to the positive *y* direction in the coordinate system of Fig. 1.3.

the x direction (with initial speed v_0); these coordinates have been indicated in Fig. 1.5. Combining Eq. 1.28 with Eqs. 1.17, we obtain for the components of the acceleration

$$a_x = \frac{d^2x}{dt^2} = -\frac{q}{m} v_z B_y, \tag{1.29a}$$

$$a_y = \frac{d^2y}{dt^2} = 0 \qquad \text{(so that } v_y \text{ is always zero),} \tag{1.29b}$$

$$a_z = \frac{d^2z}{dt^2} = \frac{q}{m} v_x B_y. \tag{1.29c}$$

Notice that ions in the xz plane are never accelerated out of this plane by the constant field B_y; rather, the force is always perpendicular to the field itself and to the velocity. The trajectory generated thus lies entirely in the xz plane; we shall show that it is a circular arc.

To obtain the actual trajectory, we need to know v_x and v_z as functions of time. Given the initial conditions that $v_x = v_0$ and $v_z = 0$ at the time $t = 0$ (when the ion enters the deflecting field), the simultaneous differential equations 1.29a and 1.29c have the solution

$$
\begin{aligned}
v_x(t) &= v_0 \cos\left(\frac{qB_yt}{m}\right), \\
v_z(t) &= v_0 \sin\left(\frac{qB_yt}{m}\right),
\end{aligned}
\tag{1.30}
$$

as can be confirmed by substitution ($a_x = dv_x/dt$, $a_z = dv_z/dt$). These are the equations for the velocity of a body moving in a circle at a constant angular velocity (radians per unit time) of qB_y/m. The speed of such a body is the product of the circle's radius r and the angular velocity ω; since the magnitude of the velocity here is always v_0, we can obtain the radius as

$$r = \frac{v_0}{\omega} = \frac{mv_0}{qB_y}. \tag{1.31}$$

The initial speed v_0 is a function of the kinetic energy given to the ion, which must equal the work done by the electric field in the source region. Usually this field is constant; in the coordinates of Fig. 1.5, it is in the x direction. If the magnitude of the electric field is E_x, and the distance through which it accelerates the ions is d, then the kinetic energy of an ion will be

$$\frac{mv_0^2}{2} = qE_xd, \tag{1.32}$$

so that

$$v_0 = \left(\frac{2qE_xd}{m}\right)^{1/2}. \tag{1.33}$$

In conjunction with Eq. 1.31, this tells us that an ion with mass m and charge q, accelerated through a distance d by an electric field E_x and then moving through a magnetic field B_y, will follow a circular trajectory whose radius r is expressible entirely in terms of measurable quantities and the ion's charge-to-mass ratio:

$$r = \frac{m}{qB_y}\left(\frac{2qE_xd}{m}\right)^{1/2} = \frac{(2E_xd)^{1/2}}{B_y}\left(\frac{m}{q}\right)^{1/2}. \tag{1.34}$$

Thus one can fix r, and adjust either B_y or E_x to make ions of any desired

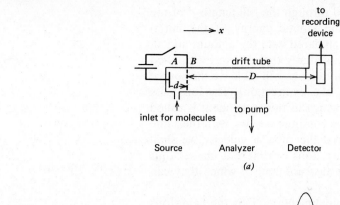

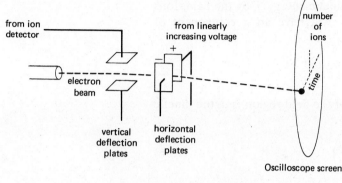

FIGURE 1.6

(a) Schematic diagram of a time-of-flight mass spectrometer. Molecules are ionized near the plate A; the ions are accelerated through a distance d until they pass through the grid B; they then travel at constant velocity a distance D to the detector. (b). Schematic diagram of the interior of an oscilloscope tube being used to display the mass spectrum from a time-of-flight mass spectrometer. (The tube is normally evacuated.)

q/m ratio come around to a fixed detector;[13] or one can fix E_x and B_y, and move the detector to observe ions moving in circles of different radii. Given q/m, one need only estimate the value of q to obtain the ionic mass; in practice, singly charged ions ($q = e$) are by far the most common, and atomic weights are known well enough for most other cases to be easily distinguished.

The above analysis gives the essence of how a magnetic mass spectrometer operates. In practice, certain fixed deflection angles such as 60 or 180° are ordinarily chosen. This is because, with these specific deflection angles, the magnetic field focuses ions that enter the field non-normally (i.e., slightly off the center of the beam) at the same place in space as those (with the same q/m value) that enter along the axis of the beam. Instruments of this type are used for absolute mass determinations because one can determine the quantities r, d, B_y, and E_x with great accuracy, while minimizing other sources of uncertainty such as off-axis trajectories. Such machines are also useful for relative mass measurements, for steady monitoring of the intensities of single-mass beams, and even as sources of beams of ions with a single known mass.

The second type of instrument is called the *time-of-flight mass spectrometer*. Conceptually it is simpler than the magnetic variety. It can scan mass spectra rapidly, but does not provide as precise separation or resolution of masses as do some types of magnetic mass spectrometers. In the time-of-flight instrument, shown schematically in Fig. 1.6a, a sudden and short pulse of force accelerates the ions that happen to be in the source chamber when the pulse is applied. This is done by making plate A positive and grid B negative, so that the positively charged ions are accelerated through the grid. The pulse duration is long enough for all the

---

[13] Alternatively, given a beam of identical ions with different initial velocities, one can use an instrument of this type as a velocity selector.

ions in the resulting beam to have passed through the full length d of the electric field, and thus to gain the same kinetic energy. If the initial kinetic energy of the ions is very small compared with the amount added, then their total kinetic energy must be essentially that due to the applied field. Furthermore, if the field is uniform, the ions will all leave the source region going in the same direction (say, the x direction). The ions pass through the grid and into the analyzer, which in this instrument is called the drift tube. Once in the drift tube, they continue moving with the same constant velocity with which they entered the tube, a velocity we shall now show to be dependent on their mass. Ions of different masses thus reach the detector at different times, by measurement of which they can be distinguished.

Suppose that all the ions are given kinetic energy T by the field. Ions of type j, with mass m_j and charge q_j, will have an x component of velocity determined by the condition

$$T = \tfrac{1}{2} m_j v_{xj}^2 \quad \text{or} \quad v_{xj} = \left(\frac{2T}{m_j} \right)^{1/2}. \tag{1.35}$$

If the applied field is E_x and the length of the field region is d, the kinetic energy must be $q_j E_x d$. Hence we have

$$v_{xj} = \left(\frac{2 q_j E_x d}{m_j} \right)^{1/2}, \tag{1.36}$$

of the same form as Eq. 1.33. In the time-of-flight instrument, no further fields are required to separate the ions of different masses. Rather, the detector has the capability of distinguishing the pulses of ions reaching it at different times. The time required for an ion of type j to travel the distance D from the source to the detector, after the accelerating pulse has been applied at the source, is then

$$t_j = \frac{D}{v_{xj}} = D \left(\frac{m_j}{2 q_j E_x d} \right)^{1/2}. \tag{1.37}$$

As in the magnetic mass spectrometer, the equations of motion give us a relation from which the ratio q_j/m_j can be obtained, since all other quantities in Eq. 1.37 can be determined experimentally. In practice, one rarely measures the quantities D, d, and E_x precisely for each new set of experiments. Rather, the apparatus is calibrated by measuring t_j for one or more ions of known mass; the times of flight for two singly charged ions of masses m_i and m_j are then in the ratio

$$\frac{t_i}{t_j} = \left(\frac{m_i}{m_j} \right)^{1/2} \tag{1.38}$$

if the other quantities are held constant.

The time of arrival of an ion is measured, typically, as the time interval from the instant when the accelerating pulse is delivered. At that moment, a voltage begins to build at a constant rate. This increasing voltage can be used to displace the spot horizontally on the screen of an oscilloscope, generating a time axis. The arrival of ions of a single mass at the detector produces a pulse of voltage to the oscilloscope that displaces the spot vertically for a moment. Thus one obtains a graph on the oscilloscope screen that may be photographed: The horizontal axis represents time, and the vertical axis, the appearance of ions. If the voltage of the pulse from the ion detector is proportional to the number of ions of a single mass, then the vertical axis becomes an indicator of the number of ions of that mass. Figure 1.6b shows the display system schematically.

There are still other types of mass spectrometers; some operate with time-varying fields, others use pure electrostatic separation. The two varieties we have described, however, are among the most widely used and are relatively simple to analyze.

In Section 1.1 we discussed how a scale of relative atomic weights can be obtained from macroscopic chemistry, and in subsequent sections we have shown how absolute atomic masses can be measured. It remains for us to obtain the conversion factor between the relative scale and absolute units. We could now abandon the use of the relative scale, of course, but this has not been done—partly because the relative atomic weights were used so long and became so firmly established before this became possible, and partly because they roughly correspond to the numbers of nucleons in the atoms. To explain the latter statement, let us now briefly introduce a few more facts about atomic nuclei.

The nucleus of any atom contains integral numbers of protons and neutrons (the latter usually somewhat in excess, except for light atoms). The electron weighs so little that the nuclear and atomic masses of a given atom do not differ significantly. Since the masses of the proton, the neutron, and the hydrogen atom are almost identical, one might expect all atomic masses to be nearly integral multiples of the mass of the hydrogen atom. This is not quite true, because of the mass defect associated with the nuclear binding energy, which amounts to nearly 1% of the total mass for most nuclei. But it is close enough to true that a scale can be set up on which the masses of all stable atoms are integral within 0.1 unit. This scale corresponds to our relative atomic weight scale.

It is apparent from the table on the inside cover, however, that many of the relative atomic weights obtained are far from integral. This is because a given element can have atoms of several different masses, known as *isotopes*. The number of protons in an atom's nucleus equals the number of electrons in the neutral atom and thus determines the atom's chemical behavior, so that this number (the *atomic number, Z*) defines the element. Isotopes of a given element are thus atoms with the same number of protons but different numbers of neutrons. Since isotopes have (nearly) the same chemistry, they are thoroughly mixed in nature,[14] and the observed atomic weight of an element is the average of the isotopic atomic masses weighted by their relative abundances. Isotopic ions have different q/m values, and can thus be separated in a mass spectrometer.

None of these facts about nuclei, of course, were known when the atomic weight scale was first established. Its basis then was simply the convenience of setting the lightest atomic mass approximately equal to unity, yielding near-integral values for most of the more common elements (H, C, O, N, Na, S, etc.). Since nearly all the elements form simple compounds with oxygen, this element was chosen as the standard, the atomic weight of oxygen being defined as exactly 16.0000...; this gave the *chemical scale* of atomic weights, which was used until 1961. Since natural oxygen consists of three isotopes and can vary slightly in composition, a scale based on a single isotope is inherently more precise (and was needed anyway for mass spectrometric measurements). Physicists therefore introduced the *physical scale*, on which the mass of the

1.6 The Atomic Mass Scale and the Mole

---

[14] There are small natural variations in isotopic abundance, however, so that the mean atomic mass may differ slightly between samples from different sources. This is especially true, of course, for radioactive elements and their decay products, accounting for the poor precision of the atomic weight of lead.

isotope[15] of oxygen ^{16}O was defined as exactly 16.0000.... The use of two scales often led to confusion, and in 1961 both were replaced by the *carbon-12 scale*, based on the standard $^{12}C = 12.0000$.... This scale retains the advantage of being based on a single isotope, but it is much closer to the old chemical scale than the physical scale had been.[16] The atomic weights listed on the inside cover are referred to the ^{12}C scale.

The unit in which atomic masses are expressed, the *atomic mass unit* (amu, or sometimes u), is thus one-twelfth of the mass of a ^{12}C atom. By making an absolute measurement of the ^{12}C mass one can obtain the conversion factor $1.6605655 \times 10^{-27}$ kg/amu; its reciprocal is known as *Avogadro's number*, and has the value 6.022045×10^{26} amu/kg (6.022045×10^{23} amu/g). This is a fundamental quantitative relationship between microscopic and macroscopic chemistry.

Atoms and electrons are such incredibly tiny objects that it would be terribly inconvenient to do chemistry with macroscopic objects while counting particles one at a time. It is much easier to count them many at a time, and to use a general convention defining a basis for the counting system. We thus define a physical quantity called *amount of substance*, the SI unit of which is the *mole* (abbreviated mol). What is a mole? A mole is a number of units of a given substance, exactly the number of elementary units as there are carbon atoms in[17] 12.0000···g of ^{12}C. A mole of atomic mass units has a mass of 1 g, and Avogadro's number N_A, the conversion factor between amu and g, gives the number of entities per mole. The terms "gram-atom" and "gram-molecule" are sometimes used for a mole of atoms or molecules, respectively, but the general term "mole" is sufficient as long as the elementary entities are specified. They need not be particles; for example, in "a mole of sodium chloride," the elementary entity is the Na^+–Cl^- ion pair. However, it is important to avoid ambiguity by specifying the species we are counting 1 mol at a time.

In dealing with physical and chemical properties, we shall find it convenient to use both the macroscopic scale, in which we count particles a mole at a time and energies likewise on a molar basis, and the microscopic scale, in which we count particles one at a time and energies on a per-particle basis. For example, when dealing with the absorption and emission of light by atoms or with collisions between atoms, we shall find it convenient to speak on the scale of single atoms and to define energy units and distances appropriate to this scale. On the other hand, when we deal with bulk chemical phenomena, such as equilibria between phases or the heat evolved in a chemical reaction, we shall consider particles a mole at a time and speak of such quantities as molar volume or energy evolved per mol of reaction. The two methods of counting are entirely equivalent; we need only be careful to realize which we are using.

The values of Avogadro's number and the electronic charge can be related to each other through Faraday's laws of electrolysis. The experiment to establish this relationship is still one of the fundamental methods of determining the physical constants. What Faraday found was that a

---

[15] By ^{16}O we mean oxygen with *mass number* 16, the nucleus containing 16 nucleons (8 protons and 8 neutrons); it is more completely described by the notation $^{16}_{8}O$, with the atomic number at the lower left.

[16] The numerical values assigned to a given mass M on the three scales are related by the equations

$$M_{chem} = 1.000049 M_{C\text{-}12}, \qquad M_{phys} = 1.000321 M_{C\text{-}12}.$$

[17] In the SI it would be more consistent to relate this unit to kilograms rather than grams, but the mole (like the gram itself) was too firmly established for the terminology to be changed.

constant quantity of electricity was required to carry out one equivalent[18] of electrochemical reaction. By the definition of the equivalent, this quantity—now known as the faraday—must equal the total charge on 1 mol of electrons. Precise measurements of the faraday usually involve the electroplating of silver from solution. Chemical and physical evidence shows that silver exists in aqueous solution as Ag^+ ions, so that one electron is required for each atom of silver deposited ($Ag^+ + e^- \rightarrow Ag$) and the equivalent and the mole of silver are the same. From extremely precise mass spectrometric measurements, the atomic weight of silver is known[19] to be 107.868. The faraday is thus the quantity of electricity required to deposit 107.868 g of metallic silver; from precise measurements of weight and charge (current times time), its value is found to be 96484.56 C. *Faraday's constant* ($\mathscr{F}$) is the conversion factor 96484.56 C/mol. Since $\mathscr{F} = N_A e$, dividing this value by Avogadro's number gives the familiar 1.60219×10^{-19} C for the charge of the electron.

1.7
The Periodic Table

So far in this chapter we have sought to develop a feeling for atomic magnitudes, and to introduce the means for dealing with these magnitudes on both microscopic and macroscopic scales. We have emphasized the common properties of atoms, introducing just enough information about the differences between atoms to explain the mass scale. But as we said at the outset, it is these differences that constitute the essential subject matter of chemistry. In subsequent chapters we shall examine the structure of atoms and molecules in detail, eventually relating these properties to macroscopic chemical behavior. At this point let us briefly review the facts we must account for, as summarized in the periodic table of the elements.

At the present time 106 chemical elements are known (although the number may have changed by the time this book is published); about 90 of these exist in appreciable quantities in nature. One obvious way of ordering the elements is in the sequence of increasing atomic weights, beginning with hydrogen and extending to the artificial transuranium elements. It became increasingly apparent in the nineteenth century that many properties of the elements vary periodically over this sequence. This is graphically illustrated by the curve of atomic volumes (shown in Fig. 1.7), which Lothar Meyer used in 1870 to establish the principle of periodicity. Dmitri Mendeleev in 1869 had reached the same conclusion on the basis of periodicities in chemical properties, such as valence and metallic character. Mendeleev imposed a further order on the sequence of elements, changing it from a linear array into a sort of spiral on which elements with similar properties were grouped together. This required the assumption of some gaps in the sequence, and Mendeleev correctly predicted the properties of the elements later found to occupy those positions. The resulting arrangement became the *periodic table;* a modern version (somewhat refined beyond Mendeleev's) is given as Table 1.3.

As the reader is aware, in the periodic table the groupings of similar elements generally occupy vertical columns. The most clearly defined of

---

[18] If an elementary electrochemical process involves the transfer of n electrons, then there are n equivalents per mole of reaction. Thus in the reaction $Cu^{2+} + 2e^- \rightarrow Cu$ we have $n = 2$, so that in this case an equivalent of Cu is half a mole.

[19] What follows is a circular argument, of course, since we need the value of e to get the atomic weight. In practice, one combines the results of a wide variety of experiments involving the different physical constants, selecting the set of values that best fits all the data simultaneously.

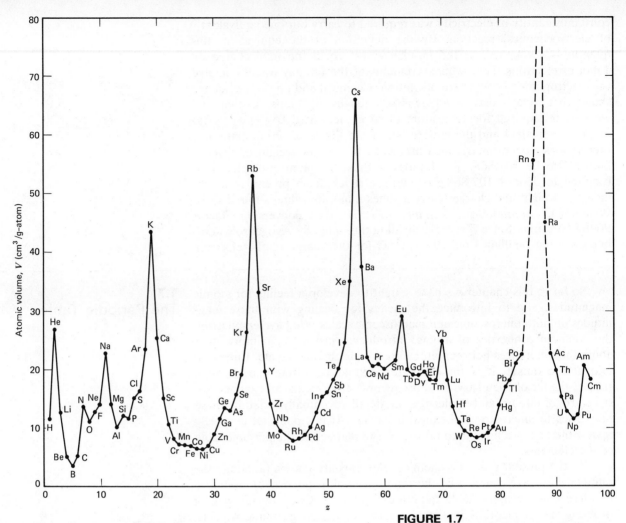

FIGURE 1.7
Atomic volumes of the elements, from solid-state density data extrapolated to 0 K (room-temperature data used for a few of the less common elements). Note the periodic behavior, especially the alkali metal peaks.

such groupings (although unknown to Mendeleev) is that of the "inert" gases, which occupy the last column: The elements helium, neon, argon, krypton, xenon, and radon are all gases at room temperature, and so nonreactive that until a few years ago they were believed to form no chemical compounds. These elements are immediately preceded in the sequence by the halogens (fluorine, chlorine, bromine, iodine, astatine) and immediately followed by the alkali metals (lithium, sodium, potassium, rubidium, cesium, francium), both well defined by their chemical similarities. Other groups, though less clear-cut, similarly fall into place vertically. Several other natural groupings are horizontal or diagonal; one derives from the close resemblance among the elements in sequence from lanthanum to lutetium (the lanthanides) and another is the diagonal relationship among aluminum, germanium, and antimony. The table as finally assembled is an empirical arrangement giving such correlations for all the elements.

To obtain this ordering, it was necessary to take some liberties with the sequence of atomic weights, even after all the vacancies had been filled. For example, tellurium by its properties had to precede the lighter iodine, and similarly potassium and argon had to be interchanged. Elements not found in nature can be assigned the atomic weights of their most long-lived or most stable isotopes. The order actually used is not that of the atomic weights, but rather that of the *atomic numbers*. For

TABLE 1.3
PERIODIC TABLE OF THE ELEMENTS[a]

1 H 1.0079																	2 He 4.0026
3 Li 6.941	4 Be 9.0122											5 B 10.81	6 C 12.011	7 N 14.0067	8 O 15.9994	9 F 18.9984	10 Ne 20.179
11 Na 22.9898	12 Mg 24.305											13 Al 26.9815	14 Si 28.086	15 P 30.9738	16 S 32.06	17 Cl 35.453	18 Ar 39.948
19 K 39.098	20 Ca 40.08	21 Sc 44.9559	22 Ti 47.90	23 V 50.9414	24 Cr 51.996	25 Mn 54.9380	26 Fe 55.847	27 Co 58.9332	28 Ni 58.71	29 Cu 63.546	30 Zn 65.38	31 Ga 69.72	32 Ge 72.59	33 As 74.9216	34 Se 78.96	35 Br 79.904	36 Kr 83.80
37 Rb 85.4678	38 Sr 87.62	39 Y 88.9059	40 Zr 91.22	41 Nb 92.9064	42 Mo 95.94	43 Tc (97)	44 Ru 101.07	45 Rh 102.9055	46 Pd 106.4	47 Ag 107.868	48 Cd 112.40	49 In 114.82	50 Sn 118.69	51 Sb 121.75	52 Te 127.60	53 I 126.9045	54 Xe 131.30
55 Cs 132.9054	56 Ba 137.34	57–71 *	72 Hf 178.49	73 Ta 180.9479	74 W 183.85	75 Re 186.2	76 Os 190.2	77 Ir 192.22	78 Pt 195.09	79 Au 196.9665	80 Hg 200.59	81 Tl 204.37	82 Pb 207.2	83 Bi 208.9804	84 Po (209)	85 At (210)	86 Rn (222)
87 Fr (223)	88 Ra (226)	89–103 **	104 (257)	105 (262)	106 (263)												

*

57 La 138.9055	58 Ce 140.12	59 Pr 140.9077	60 Nd 144.24	61 Pm (145)	62 Sm 150.4	63 Eu 151.96	64 Gd 157.25	65 Tb 158.9254	66 Dy 162.50	67 Ho 164.9304	68 Er 167.26	69 Tm 168.9342	70 Yb 173.04	71 Lu 174.97

**

89 Ac (227)	90 Th 232.0281	91 Pa (231)	92 U 238.029	93 Np (237)	94 Pu (244)	95 Am (243)	96 Cm (247)	97 Bk (247)	98 Cf (251)	99 Es (254)	100 Fm (257)	101 Md (258)	102 No (255)	103 Lr (256)

[a] Atomic weights are given on the ^{12}C scale. For artificial radioactive elements, the longest-lived known isotope is given in parentheses.

the moment, we can take the atomic number Z as giving nothing more than the empirical position of an element in the sequence. In fact, however, as mentioned earlier, Z is the number of positive charges in the nucleus and thus the number of electrons in the atom; in Chapter 2 we shall give the evidence for this statement. In subsequent chapters we shall develop the theory of atomic structure, in the light of which we can add more to our interpretation of the patterns of chemistry associated with the periodic table.

Further Reading

A. D'Abro, *The Rise of the New Physics* (Dover Publications, Inc., New York, 1951).

G. K. T. Conn and H. D. Turner, *The Evolution of the Nuclear Atom*, (American Elsevier Publishing Company, Inc., New York, 1965).

V. Kondratyev, *The Structure of Atoms and Molecules*, (P. Noordhoff N. V., Groningen, The Netherlands), Chapter 1.

O. Oldenberg and W. G. Holladay, *Introduction to Atomic and Nuclear Physics*, (McGraw-Hill Book Co., New York, 1967).

F. Richtmyer and E. H. Kennard, *Introduction to Modern Physics* (McGraw-Hill Book Co., New York, 1947), Chapters 1 and 2.

Problems

1. Suppose that one were trying to apply the law of multiple proportions when weights could not be measured with modern accuracy and separations were often not quantitative. In particular, suppose that one were trying to use this law to infer molecular formulas of two salts of Mg, S, and O, and the atomic weight of magnesium by comparing the percent Mg in $MgSO_3$ and $MgSO_4$, but that the apparent weight of sulfite is approximately 20% lower than the true value, whereas the apparent weight of sulfate is within 2% of the correct value. What would one take for the formulas of magnesium sulfite and sulfate, and for the atomic weight of Mg with this error in the data? Assume atomic weights of 16 and 32 for oxygen and sulfur, respectively.

2. Estimate each of the following: the volume of 1 mol of pinheads; the time it takes 1 mol of electrons to pass a monitoring point if the current is 0.1 A; the radius of a hydrogen atom if the nucleus has a radius of 0.1 mm; the mass of all the electrons striking a target in 1 s, from a current of 0.1 A.

3. Design an experiment, by giving a diagram and suitable dimensions, voltages, and magnetic field strengths, to determine the ratio e/m by Thomson's method for a muon. Assume you know that the charge is approximately that of the electron and that the mass falls between 200 and 500 times that of the electron. Assume the muons can be injected into the apparatus with energies of 100 ± 0.1 eV. Finally, assume that the detector has an area of 1 mm^2.

4. Design an experiment to search for quarks, with charge $+\frac{1}{3}e$, by Millikan's method but with uniform plastic spheres whose density is 0.6 g/cm^3 or $6 \times 10^5 \text{ g/m}^3$. Would you choose a radius $r = 0.1$ mm or 0.01 mm for the spheres? The quark might have a mass as large as that of the proton. Would this affect the experiment?

5. Compute the following: the force on an electron in a uniform electric field of 10^5 V/m in the z direction and the corresponding acceleration; the force on

an electron moving 10^5 m/s in the x, y plane at exactly 45° to a magnetic field whose magnetic field strength **H** lies along the y axis and has a magnitude of 10^4 A/m, and the acceleration of that electron; the force on an electron moving at 10^5 m/s along the x axis simultaneously in an electric field of 10^4 V/m along the y axis and a magnetic field in the z direction whose magnetic flux density is 0.1 T, and the acceleration of that electron. Be sure to indicate the direction of each force.

6. Suppose that ions with masses between 1 and 200 amu are to be analyzed by a magnetic mass spectrometer of the type described in Section 1.5. The radius r has been selected to be 0.015 m. It is impractical to use electric fields E_x below about 2×10^5 V/m because, with lower fields, ions emerging from the source in which they are generated leave so slowly that they inhibit the formation and extraction of more ions. This condition is called a "space-charge-limited" condition. It is also impractical to use electric fields much above 3×10^6 V/m because of the problem of electric discharges; avoiding these discharges becomes expensive. Assume that the mass spectrometer will operate with a fixed bending angle of 60° and that the paths of the ions will be determined by the electric accelerating field and the magnetic deflecting field. Select suitable ranges for the electric and magnetic fields, and specify what fields would be used to focus masses of 1, 50, and 200 amu on the detector.

7. Sometimes unstable ions are formed in the source of a mass spectrometer; these break into fragments before they pass through the entire mass-analyzing region. Suppose that an ion of mass 150 is produced in the source of a magnetic mass spectrometer, passes through the accelerating field, and then decomposes into fragments with masses 135 and 15. The light fragments are nearly all neutral and furthermore are lost, but the heavy fragments that carry the charges continue to move with approximately their initial velocity as they travel into the region of magnetic deflection. At what apparent mass will these fragments appear at the detector?

8. In an experiment designed by Classen, electrons are emitted from the hot filament cathode labeled C in Fig. 1.8. The electrons are then accelerated toward an anode A by a known potential difference V. A known magnetic field **H** acting at right angles to the plane of the drawing deflects the electrons along a semicircular path EL_1 onto photographic plate P_1. On reversal of the direction of **H** the electrons follow EL_2 and strike plate P_2. Show that e/m can be calculated if V, **H**, and the radii of curvature of the orbits EL_1 and EL_2 are measured.

9. Consider the apparatus sketched in Fig. 1.9. The electric field E_y is along the y axis only. A stream of Li^+ ions enters the slit with velocity in the x direction of $v_x = 5 \times 10^7$ cm/s. What must E_y be if, on the photographic plate, the lines corresponding to ^6Li and ^7Li are to be separated by 1 cm? Assume the slit is 10 cm from the photographic plate.

10. A velocity selector for molecules is made of two slotted disks on a common rotatable shaft. The speed of the shaft is variable. The disks are 0.1 m in diameter and have rim slots 2 mm wide and 0.01 m (10 mm) high. They are 0.1 m apart. It is hoped that the selector can be used to choose velocities in the range 5×10^2 to 5×10^3 m/s. How fast must the selector rotate to achieve such selection? Plot the rotation speed (revolutions/s) against the transmitted velocity, for angles ϕ of 10, 45, and 90° between the slots, when they are viewed along the rotational axis of the shaft. If the angle is 45° and the velocity of rotation is set to pass particles traveling at 10^3 m/s, what are the upper and lower limits of the velocities that also pass, due to the finite widths of the slots, if we assume that velocities are all parallel to the axis of the shaft?

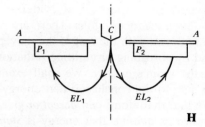

FIGURE 1.8

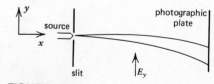

FIGURE 1.9

Origins of the Quantum Theory of Matter

In Chapter 1 we dealt with the gross properties of atoms; now we begin to study their internal structure. The broad purpose of this chapter is to develop the experimental evidence leading to what is known as the *quantum theory of matter.* There are two principal ways by which an atom can interact with the rest of the universe: by colliding with other particles, and by absorbing or emitting radiation; either process results in a change in the atom's energy. We shall examine experiments of both kinds, all leading to the conclusion that energy associated with atoms is *quantized.* We find that atoms can *accept or give up* energy only in discrete amounts known as *quanta,* that energy is *stored* in quantized units, and that the energy *in light* is also found only in quantized units. We shall examine the experimental basis for each of these aspects of the quantum theory.

Although quantization is the main theme of the chapter, several other important topics are introduced. These include the evidence for the nuclear structure of the atom; the mechanical concept of action (in terms of which quantization can most easily be formulated); the remarkably useful model of the harmonic oscillator; and phenomena such as spectra and heat capacities, which will later be examined in detail. To close the chapter, we use much of this material in formulating the first and simplest model of atomic structure based on quantization, the model of Niels Bohr.

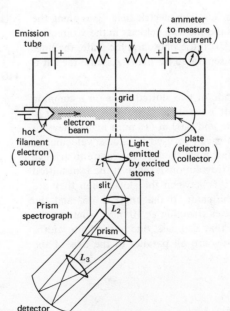

FIGURE 2.1
Schematic diagram of Franck–Hertz apparatus. The emission tube is filled with a gas at low pressure. The batteries and resistances at the top are adjusted to give the desired electrostatic potentials (cf. Fig. 2.2). The light emitted by excited atoms near the grid is collected by the lens L_1, the focus of which is at the spectrograph slit. The lens L_2 collimates (makes parallel) the light rays, the prism separates the light of different wavelengths, and the lens L_3 focuses the light of each wavelength at a different point on the detector, usually a photographic film.

Some of the clearest evidence for our interpretation of the structure of matter comes from the classic experiment performed in 1914 by James Franck and Gustav Hertz. Although historically this came after much of the other evidence we shall discuss, it is a particularly straightforward example, both conceptually and practically, of a quantum process. In brief, one bombards a chosen gas with electrons of known energy; the electrons collide with the atoms of the gas, losing energy to them; the excited atoms emit light upon returning to their original state. One studies both the light emitted and the energy lost by the electrons.

A schematic diagram of the Franck–Hertz apparatus is shown in Fig. 2.1. Electrons leave the hot filament both because of its high temperature and because the grid is positively charged relative to the filament. We define the grid-filament and grid-plate potential differences as V_1 and V_2, respectively, as illustrated in Fig. 2.2. The electric field between filament and grid can accelerate the electrons to a maximum kinetic energy[1] of

$$T_0 = \tfrac{1}{2} m_e v^2 = eV_1; \qquad (2.1)$$

V_1 and thus T_0 can be measured fairly accurately. (It will be recalled from Section 1.3 that electrostatic potential is just the potential energy per unit charge.) If the plate were more positive than the grid ($V_2 < 0$), all the electrons passing through the grid would reach the plate and be counted as part of the plate current. However, making the plate more negative than the grid ($V_2 > 0$) allows the apparatus to act as an analyzer of electron energies. The grid-to-plate region is then one of steadily increasing potential energy, analogous to a hill that the electrons must climb. If the plate is more positive than the filament (as in Fig. 2.2), the electrons will have acquired enough kinetic energy in their "downhill" slide to the grid to travel back up the lower hill to the plate—unless they have lost energy along the way. It is, of course, precisely the energy lost along the way in which we are interested. For given values of the potential differences, only electrons with an energy loss less than ΔT will reach the plate, where $\Delta T = e(V_1 - V_2)$. By measuring the plate current as a function of ΔT, one can determine what fraction of the electrons have lost any given amount of energy.

In practice, one usually holds V_2 constant and varies V_1. Figure 2.3 illustrates the kind of results that one obtains. Whenever the maximum electron energy eV_1 is less than some threshold value E_{thr}, the plate current rises continuously with increasing V_1, as more electrons are drawn from the filament. After the threshold energy is reached, the current drops sharply, and thereafter shows successive peaks corresponding to energy losses in fixed amounts. We interpret these current measurements as follows: As long as the electrons cannot excite the internal structure of the atom, the kinetic energy must be conserved; that is, electron–atom collisions are virtually elastic, with the electron losing little energy, and the current rises as V_1 rises. But when an electron has enough energy to excite an internal state or states of the atom, it can lose a fixed large amount of energy equal to the atomic excitation energy. At the threshold this excitation requires virtually all of the electron's energy, so that electrons undergoing collisions of this type cannot reach the plate,

---

[1] In measurements of this kind it is most convenient to measure the energy in *electron volts* (eV), where 1 eV is the energy of one electron accelerated through a potential of 1 V. We thus have

$$1\,\text{eV} = (e)(1\,\text{V}) = (1.60219 \times 10^{-19}\,\text{C})(1\,\text{V})\left(\frac{\text{J}}{\text{V C}}\right) = 1.60219 \times 10^{-19}\,\text{J}.$$

2.1
The Franck–Hertz Experiment

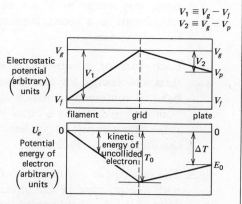

FIGURE 2.2
Electrostatic potential (V) and potential energy of an electron (U_e) in the Franck–Hertz emission tube. If U_e is set equal to zero at the filament, the two are related by $U_e = -e(V - V_f)$, and the kinetic energy of an electron is $-U_e$ if no energy has been lost by collision. Near the grid the electron that has not collided should thus have kinetic energy $T_0 = eV_1$. The total energy of the electron is initially zero on this scale, but may be lowered by collisions with gas atoms. It must be at least E_0 for the electron to reach the plate, where

$$E_0 = -\Delta T = -e(V_p - V_f) = -e(V_1 - V_2);$$

ΔT is thus the maximum allowable energy loss.

and the plate current drops. The potential at which such a drop begins is known as a *resonance potential*. As V_1 continues to increase, the electrons have enough kinetic energy left over to reach the plate once more, and the current resumes its rise. The thresholds after the first correspond to successive excitations by the same electrons, or to the appearance of different kinds of excitation.

In addition to measuring the energy lost by the electrons, one also looks at the light emitted by the atoms with which they collide. The simplest observable properties of light are its wavelength and its intensity. The light of different wavelengths can be separated with a *spectrometer* or *spectrograph*, much as the mass spectrometers of Section 1.5 spatially separate ions of different mass. In an optical spectrometer the light is usually dispersed with a prism (as in Fig. 2.1) or a diffraction grating of closely spaced parallel grooves. Auxiliary lenses or mirrors then focus the light at the detector, which may be a photographic film, a photoelectric cell, a thermocouple (for infrared light), or any other device sensitive to incident radiation. The intensity of the light can be determined from the strength of the signal observed (darkening of film, photoelectric current, etc.). In the Franck–Hertz experiment, one studies how the wavelength and intensity of the emitted light vary with the grid potential; the most important conclusions come from a comparison of these data with the electron energy measurements.

The essential results of the Franck–Hertz experiment are these:

1. Every gas introduced into the tube emits light with its own characteristic spectrum of wavelengths; such an emission spectrum may occur as a set of lines of very sharply defined wavelength, as a broad, continuous spectrum, or as a combination of the two.

2. Each part of a substance's emission spectrum has its own threshold energy for excitation. That is, each spectral line or band of continuous emission appears only if the energy of the bombarding electrons is equal to or greater than some minimum value characteristic of that line. More than one line may have the same energy threshold.

3. The occurrence and intensity of each spectral line are correlated with one or more of the electron-current resonance potentials. That is, the threshold energy for the appearance of the line is the same as the threshold energy for one of the sharp drops in current (or the difference between two such energies). The intensity of the line is proportional to the number of electrons that have lost just this amount of energy as indicated by the plate current.

4. The shorter the wavelength of a spectral line, the higher is the threshold electron energy E_{thr} required to excite the line.

5. In a large number of cases, the wavelength λ_{min} of the line of shortest wavelength associated with a given E_{thr} is inversely proportional to the threshold energy; that is, frequently we have

$$\lambda_{min}(E_{thr}) \propto (E_{thr})^{-1}. \qquad (2.2a)$$

Since the wavelength λ and the frequency ν of light are inversely proportional to each other (cf. Appendix 2A), a minimum wavelength λ_{min} corresponds to a maximum frequency ν_{max}, and

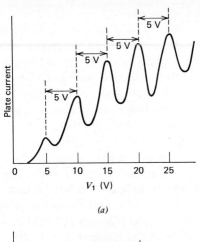

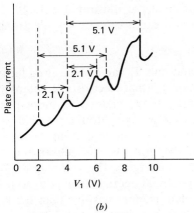

FIGURE 2.3
Observed plate currents as functions of grid voltage in the Franck–Hertz experiment: (*a*) Hg vapor, resonance potential 4.9±0.1 V; (*b*) Na vapor, first resonance potential 2.1 V.

we can rewrite Eq. 2.2a as the direct proportionality

$$\nu_{max}(E_{thr}) \propto E_{thr}. \qquad (2.2b)$$

The foregoing information tells us that atoms are capable of picking up energy from collisions with electrons and of returning this energy (or at least part of it) as light. The light may be returned as spectral lines or bands characteristic of the substance, and not necessarily as a simple continuous spectrum. Result 3 suggests that each spectral line associated with a single threshold energy is also associated with the acquisition by the atom of the same well-defined amount of energy from the electrons; it is the amount of energy absorbed, rather than the total energy of the electrons, that determines if a particular line appears. In short, the results suggest that any given atom can accept energy only in well-defined (quantized) amounts.

We can use an analogy to interpret this result and emphasize how much it differs from expectations based on classical physical laws. If the electrons in atoms could move about as freely as classical laws would allow, we would expect an atom to be capable of having any energy at all. Instead, we observe that atoms can possess only discrete amounts of energy. A classical analogue of this model is a rock on a hill. The gravitational potential energy of the rock is analogous to the total internal energy of the atom. With no knowledge of the shape of the hill, we expect the rock to roll continuously down the hill. However, a rock behaving like an atom would be found only on one of a set of well-defined ledges. Figure 2.4 illustrates the contrast between the two views. Even as late as 1925, we had no basis except the observation of discrete frequencies of light on which to postulate "ledges" for electrons. Interpretation of the Franck–Hertz experiment, and the photoelectric effect and the discrete lines of atomic spectra as well, required an audacious assumption that defied classical physical ideas.

Let us extend our analogy further. If the atom is initially in its "ground state," the threshold energy of the Franck–Hertz experiment corresponds to the potential energy needed to lift a rock to a higher ledge. The emission of light then corresponds to the rock's falling to a lower ledge or the ground. Gravitational potential energy, of course, is proportional to height ($V_{grav} = mgz$). Result 5 above suggests that the frequency ν_{max} corresponds to the height of the first ledge above the ground, whereas the lower frequencies associated with the same E_{thr} give its height above various intermediate ledges. In other words, the frequency of any emission line should be proportional to the energy difference between the excited state (reached by adding E_{thr}) and the atom's state after emission. Thus, if we number the levels as in Fig. 2.4, we would expect an atom excited to level 2, by an amount of energy we can call $E_{thr}(2)$, to emit two spectral lines—one with a frequency (ν_{max}) proportional to $E_{thr}(2)$, the other with a frequency proportional to $E_{thr}(2) - E_{thr}(1)$.

The test of this hypothesis is the proportionality between frequency and energy difference. If we introduce into Eq. 2.2b a proportionality constant h, so that for the ith threshold of a given substance we have

$$E_{thr}(i) = h\nu_{max}(i), \qquad (2.3a)$$

then we would expect the other observed frequencies to be given by some combination

$$E_{thr}(i) - E_{thr}(j) = h\nu_{ij}, \qquad (2.3b)$$

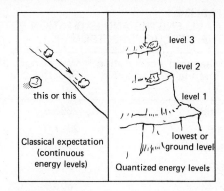

FIGURE 2.4
Analogue of the quantized energy levels suggested by the Franck–Hertz experiment.

TABLE 2.1
WAVELENGTHS AND EXCITATION POTENTIALS FOR SELECTED LINES OF THE MERCURY SPECTRUM[a]

Measured Accelerating Potential at Threshold (V)	Wavelength of Spectral Line (Å)	Theoretical Potential[b] (V)
4.68	2656.5	4.67
4.9	2537.0	4.89
5.47	2270.6	5.46
6.73	1849.6	6.70
7.73	1603.9	7.73
9.37	2656.5	$9.33 = 2 \times 4.67$ (excitation of two atoms)

[a] From J. Franck and E. Einsporn, *Z. Physik* **2**, 18 (1920).
[b] Setting the energies in Eqs. 2.1 and 2.3a equal to each other, we have

$$V_1 = \frac{h\nu}{e} = \frac{hc}{e\lambda}$$

for the theoretical threshold potential corresponding to wavelength λ. (The values in the third column are calculated with our current values of the physical constants; Franck and Einsporn themselves used the values $e = 1.592 \times 10^{-19}$ C, $h = 6.545 \times 10^{-34}$ J s, and thus obtained results about 0.6% lower.)

with the same constant h. This is indeed what one finds. Not all $i-j$ combinations give an observed spectral line (some lines are naturally weak, for reasons we shall discuss later), but the lines that are observed satisfy Eq. 2.3b. In fact, not only does this equation hold for a given substance, but the same proportionality constant h is found for *all* substances.

We shall have much to say later about the constant h (known as *Planck's constant*). It is another fundamental constant like those in Table 1.1, with the value

$$h = 6.626176 \times 10^{-34} \text{ J s.}$$

This value can be used to calculate the threshold energy corresponding to a given spectral line; Table 2.1 gives some early data illustrating this point.

What, then, can we conclude from the Franck–Hertz experiment? It seems that the internal structure of atoms allows for the absorption and emission of energy only in well-defined amounts. For the emission of light by gaseous atoms, at least, the energy given off is related to the frequency of the light by $\Delta E = h\nu$, where h is a universal constant. This relationship will turn up again and again, eventually leading us to a general quantum theory of matter. Quantized behavior of this sort occurs not only with visible light, but in all parts of the electromagnetic spectrum; to illustrate this, in the next two sections we consider the properties of ultraviolet light and x rays.

2.2 The Photoelectric Effect

One of the first major phenomena accounted for by the hypothesis of quantization was the *photoelectric effect*. This consists of the emission of electrons when light shines onto a material. The interpretation of this process generated the idea that light contains energy in discrete quantized units.

Photoelectrons are produced when light of frequency higher than some threshold value strikes any substance. (Historically, the effect was first studied for solids.) The threshold frequency is a characteristic property of the substance, and sometimes of the condition of its surface. The alkali metals yield photoelectrons with visible light, but for most substances the threshold lies in the ultraviolet.

The rate of electron emission can be determined by measuring the current through a collector plate, as in the Franck–Hertz experiment; the electrons' kinetic energy is measured by the voltage required to prevent them from reaching the collector, as in Eq. 2.1.

How do the number of electrons and their energies vary with the frequency and intensity of the incident light? Below the threshold frequency ν_0, no electrons at all are emitted. Above ν_0, electrons with a variety of energies are obtained; the energy distribution is independent of the light intensity, but varies with frequency. In particular, the *maximum* electron energy for a frequency ν obeys the equation

$$E_{max} = h\nu - h\nu_0, \qquad (2.4)$$

where h is again Planck's constant; $h\nu_0$ is the so-called *work function*. For a given frequency, the rate of electron emission is proportional to the light intensity.

These results were puzzling in terms of the wave theory of light. Since the energy delivered by a wave increases with the wave's amplitude, one would expect the energy of the liberated electrons to increase with the light intensity. Furthermore, if the energy were distributed uniformly over the wavefront, it would take a rather long time for any one electron to obtain the observed energy, but no such lag is observed. The answer to the problem was to assume that light can behave as a stream of particles as well as a wave—that is, that light is an entity with both particle and wave aspects. This explanation, like that of the heat capacity of solids, was first given in 1905 by Einstein, one of the very few scientists who never really accepted the quantum theory.

Einstein's model, then, amounted to the quantization of light. He assumed that the light striking a photoelectric surface (and, by extension, all light) consists of discrete particles or quanta, each with a definite energy proportional to the frequency of the light:

$$E(\text{per quantum}) = h\nu. \qquad (2.5)$$

Each electron emitted from the surface then corresponds to the absorption of a single quantum of light[2]. It must take a certain minimum energy, the work function $h\nu_0$, to remove an electron from a solid surface;[3] quanta with less energy (lower frequency) have no effect. If the quantum absorbed has more than the requisite amount of energy, then the extra energy is retained by the newly freed electron as kinetic energy. The maximum energy given by Eq. 2.4 is that of electrons that needed just $h\nu_0$ for liberation; the photoelectrons with less energy are those that experienced one or more collisions in which they lost some part of $h\nu - h\nu_0$ on their way to free flight. The more quanta of a given frequency strike the surface, the more electrons of the corresponding energy distribution are emitted, giving the observed dependence on light intensity. Einstein's model thus explains all the experimental facts about the photoelectric effect in a clear, simple, and vivid way.

---

[2] According to modern quantum mechanics, this is not absolutely true. But the probability of a single quantum's removing two electrons, or of two quanta together exciting a single electron, is so small that it can be disregarded here.

[3] The same energy is found to be required for *thermionic* emission from a hot filament.

Even more conclusive evidence of the particle nature of light was provided later (1923) by the *Compton effect*. When x rays strike a target, some of the scattered radiation is of lower frequency (and thus energy) than the incident beam; at the same time electrons are emitted by the target. What happens is that the incident *photon* (light particle) gives up some but not all its energy to an electron with which it collides. This interpretation is confirmed by the observation that both energy and momentum[4] are conserved in the process.

The idea that light consists of quanta of energy $h\nu$ also explains many other phenomena. Both the Franck–Hertz experiment and x-ray emission, described in Section 2.3, can be regarded as examples of the *inverse* photoelectric effect: Electrons strike a target, giving up energy which is released as quanta of radiation. In the ordinary absorption and emission of light, Section 2.4, electrons *within* atoms gain or lose energy by means of quanta. In all these cases the energy transferred is related to the frequency of the radiation by $E = h\nu$; the particular frequencies involved are determined by the energy levels of the atoms.

The idea that light behaves like a stream of particles seemed at first like a resurrection of the old quarrel between the rival seventeenth- and eighteenth-century schools that supported the corpuscular and wave theories of light. The wave theory had been thought to be proved by such phenomena as interference and diffraction, but now new evidence supported the particle interpretation. The split between these two camps now seems to be only a historical accident. As we shall see in more detail in the next chapter, the particlelike and wavelike properties are merely complementary aspects of the same phenomenon.

2.3
x Rays and Matter

x Rays are electromagnetic waves, similar to light and radio waves but with much shorter wavelengths, of the order of 10^{-10} m. (See Appendix 2A for a general survey of the electromagnetic spectrum, its range of magnitudes, and some characteristics of its various regions.) x Rays are produced when electrons, accelerated to high velocities from a cathode, bombard a solid target. The process is similar to that in the Franck–Hertz experiment, except that the electrons have higher energies so that higher-frequency radiation is emitted, and the target is a solid rather than a gas. Figure 2.5 shows a schematic diagram of such an x-ray source.

x Rays are useful for studying atomic and molecular properties in several ways. In this section we shall examine the emission spectra produced by an x-ray source (which are precisely analogous to the emission spectra produced in the Franck–Hertz experiment) and the ways in which a beam of x rays can be attenuated by passing through matter. These experiments give us further indications of the quantization of atomic energies, and suggest the physical interpretation of the atomic number Z. In later chapters we shall see how x rays can be used to study the structure of solids and liquids.

The most precise and useful relationship between the intrinsic structural properties of atoms and their interaction with x rays involves the x-ray emission spectra of the elements. These spectra are the distributions of intensities, over wavelengths, of the x rays emitted when a target made of a particular element is bombarded with electrons. For x rays the dispersion by wavelength is done by using a crystal in the role of a diffraction grating, as we shall see in Chapter 11. In general, the x-ray emission spectra of the elements resemble that shown in Fig. 2.6, consist-

[4] See Section 3.1 for the definition of a photon's momentum.

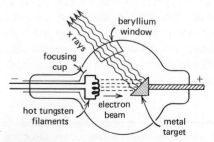

FIGURE 2.5
Schematic diagram of x-ray tube.

ing of a continuous emission on which are superimposed a number of characteristic lines. The shape of the continuous spectrum is independent of target, depending only on the voltage applied to the x-ray tube, that is, on the kinetic energy of the electrons that hit the target. The line emissions, on the other hand, are different for each target, and thus presumably arise from excitation of the target atoms, just as in the Franck–Hertz experiment.

H. G. J. Moseley, between 1912 and 1914, made a very systematic and dramatic series of observations of the characteristic x-ray emission lines produced by various elements. For each element the lines were known to fall into several closely bunched groups, which are called the $K, L, M, \ldots$ series (proceeding toward higher wavelengths); the K and L series were found for all the elements studied, the later series only for the heavier elements. In each series Moseley found a rather good correlation between the frequency of the x rays and the atomic weight of the element. However, the correlation with atomic *number* was so remarkable as to prove the real physical significance of Z. Figure 2.7 displays this correlation for the K series.

In every series the relationship between the frequency ν and the atomic number Z was found to be of the form

$$\nu = A(Z - S)^2, \tag{2.6}$$

where A and S are (very nearly) constants. For the *screening constant S*, Moseley found a value close to unity for the K series so that $\nu^{1/2}$ is nearly proportional to Z. Later measurements gave S of about 3 for the K series and 10 to 20 for the L series. As for the constant A, its value for the various series is approximately given by

$$A = cR_\mathrm{H}\left(\frac{1}{n_1{}^2} - \frac{1}{n_2{}^2}\right), \tag{2.7}$$

where c is the speed of light, R_H is the Rydberg constant (see the next section), and n_1 and n_2 are integers. For the K series n_1 is 1, whereas n_2 is 2 for the K_α (lowest-frequency) lines, 3 for K_β, and so on; in the L series, n_1 is 2 and $n_2 = 3, 4, \ldots$. We shall see in the next section that the form of Eq. 2.7 is common to many types of spectra, clearly illustrating the quantized nature of atomic energy levels.

What did the appearance of Z in Eq. 2.6 prove? Remember that the atomic number originally gave only the element's place in the empirical periodic table. As we shall see, several lines of evidence already suggested that Z was equal to the number of electrons in the atom, and Bohr's theory of the atom (1912) was based on this assumption. Not until Moseley's work, however, was an exact relationship involving Z shown to apply to an extensive sequence of elements. A firm basis was thus given to the periodic table, eliminating the anomalies that appeared when the sequence was based on atomic weights.

We have mentioned the similarity between x-ray emission and the Franck–Hertz experiment. This similarity extends to the appearance of threshold energies. To produce any given x-ray emission line from a particular target material, the bombarding electrons must be accelerated to a minimum energy characteristic of the particular line. As the electron energy increases, lines of higher and higher frequency are produced, with the threshold energies obeying Eqs. 2.3. Note also that for a given electron energy E continuous emission occurs only with wavelengths greater than some wavelength λ_min, which is such that $E = Ve = hc/\lambda_\mathrm{min} = h\nu_\mathrm{max}$; this is the point at which *all* of the electron's energy is transferred to the emitted x ray. All this evidence reinforces the hypothesis that

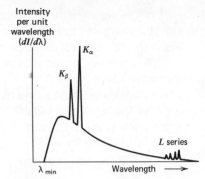

FIGURE 2.6
A typical x-ray emission spectrum.

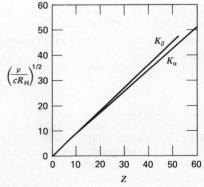

FIGURE 2.7
Dependence of x-ray emission frequencies on atomic number for the K series. Note that the square root of frequency, $\nu^{1/2}$, is very nearly linear in Z. Each of the curves shown is actually mde up of subsets (e.g., K_{α_1} and K_{α_2}) too close together to be distinguished on this scale.

radiation of frequency ν is associated with an energy change $h\nu$ in the material absorbing or emitting the radiation.

We shall see that threshold energies also exist for the absorption of x rays. First, however, we must discuss the general laws governing the attenuation of x rays by matter.

If one passes a beam of x rays through a sample of some material, the intensity I of the beam will fall off with the thickness of the sample. This loss of intensity is usually found to obey the equation

$$\frac{dI}{I} = -\sigma n \, dx, \qquad (2.8)$$

where dI is the change in intensity over a small distance dx, n is the *number density* (in units such as atoms per cubic meter), and σ is a proportionality constant characteristic of the substance. Note that the dimensions of σ are those of area; in fact, σ is called the *attenuation cross section*; the product σn is the *attenuation coefficient*. Integrating over the total thickness x of the sample, we find the total loss of intensity to be given by

$$\ln I - \ln I_0 = -\sigma n x \quad \text{or} \quad I = I_0 e^{-\sigma n x}, \qquad (2.9)$$

where I_0 is the intensity of the originally impinging beam. In other words, the intensity of the beam decreases exponentially with distance as the radiation penetrates the material. This result, usually referred to as *Beer's law*,[5] is a very powerful and general equation, which applies to many kinds of radiation and even to beams of particles.

What causes a beam of radiation to be attenuated? We are concerned here primarily with the process of *absorption*, a term implying that the energy of the beam goes into producing some kind of internal excitation in the atoms (or molecules) of the material. We saw one effect of absorption with the photoelectric effect. However, some of the beam can also be deflected or *scattered* in different directions and thus fail to reach the detector.[6] We shall have a great deal to say later about the scattering of particles. It is not always easy to distinguish between scattering and absorption; in some cases (*inelastic scattering*), the incident radiation is scattered but some of its energy is absorbed. Usually, however, one can analyze the various effects and obtain separate absorption and scattering coefficients, or the corresponding cross sections. For x rays the absorption effect is much greater at the energies with which we are concerned, and attenuation measurements effectively give the absorption cross section.[7]

---

[5] Or the *Beer–Lambert law*; Lambert discovered the distance effect and Beer (1852) the concentration (density) dependence for the absorption of light by liquid solutions.

[6] Beer's law is strictly valid only in the limit of a very thin sample, and any system for which it does hold is thus called *optically thin*. One reason is the possibility of *double* scattering (out of the beam and back in), which increases with thickness. Reemission in the direction of the detector can also distort the results, as can the heating of the sample caused by energy absorption.

[7] x Ray scattering also had some relevance to the development of our theory. The scattering cross section can be seen to be proportional to the scattering effect of a single atom. According to J. J. Thomson's model of the atom (see Section 2.5), the scattering was done by individual electrons moving nearly freely within the atom, so that σ should be proportional to the number of such electrons per atom. The experiments of Barkla (1909) gave a number equal to about half the atomic weight—and thus approximately equal to the atomic number. This result was largely fortuitous (Thomson's model was wrong, and Barkla's results would have been quite different with x rays of longer wavelength), but nevertheless it was the first indication that Z gives the number of electrons in the atom.

If one measures the absorption of x rays by a given material as a function of wavelength, the results resemble those shown in Fig. 2.8. In general, the absorption coefficient rises steeply with increasing wavelength. At a series of points known as *absorption edges*, however, it drops precipitously to a lower value before rising again to a new peak. In the direction of higher frequencies (and thus higher energies), each of these points represents an absorption increase. In short, each of the absorption edges corresponds to the threshold energy for a new type of absorption. We thus have threshold energies, indicating well-defined energy levels within the atom, for both emission and absorption. The absorption edges, like the emission lines, can be assigned to various series (again labeled $K, L, \ldots$). It turns out that the absorption edge frequencies for different elements vary with atomic number in just the same way as do the emission frequencies, that is, in accordance with Eq. 2.6. Moreover, when one follows a series of x ray absorption lines obeying Eq. 2.7 with n_1 fixed and n_2 $(>n_1)$ increasing, one finds that when $n_2 \to \infty$, electrons are emitted. In short, the photoelectric effect sets in.

All that we have said about x rays and matter thus reinforces two main conclusions: that the energies of atoms are quantized in some way (involving the constant h), and that the atomic number Z is of fundamental significance to atomic structure. But these conclusions were well accepted before either Moseley or Franck and Hertz, even if the evidence was perhaps more circumstantial. In the next section we examine what was historically the first evidence of quantization, the nature of atomic spectra.

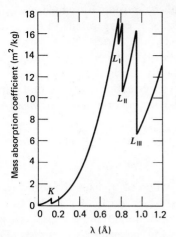

FIGURE 2.8
x Ray absorption of lead as a function of wavelength. The "mass absorption coefficient" is the absorption coefficient divided by the density. Note the sharp breaks (absorption edges) labeled K, L_I, L_{II}, and L_{III}; additional sets of absorption edges (M, N, O, ... series) appear at longer wavelengths.

2.4
The Emission Spectra of Atoms

The atoms of any element will emit light when sufficiently excited, whether by bombardments as in the Franck-Hertz experiment or simply by heating to a high temperature. As with x rays, the light emitted consists of a continuous spectrum (black-body radiation, which we shall discuss in Section 2.6) and a spectrum of lines characteristic of the element. We are concerned here with the line spectra; these are actually of the same nature as the x-ray emission line spectra, but at much longer wavelengths—in the ultraviolet, visible, and infrared regions. They can thus be excited by relatively low energies, for example, by an electric discharge in a gas or vapor at low pressure. We shall discuss mainly the spectrum of atomic hydrogen, which is by far the simplest.

The hydrogen spectrum consists of several series of lines, with those in each series apparently closely related to one another. The series first discovered (by J. J. Balmer in 1885) lies primarily in the visible region, a similar series is in the far ultraviolet, and several more are in the infrared. Balmer found that the nine lines he observed in the visible series (now known as the *Balmer series*) have wavelengths given very precisely by the formula

$$\lambda = b\frac{n^2}{n^2 - 4} \qquad (n = 3, 4, 5, \ldots), \qquad (2.10a)$$

Balmer's value for the constant b being 3645.6 Å. This equation is now usually written in the inverted form

$$\frac{1}{\lambda} = R_H\left(\frac{1}{4} - \frac{1}{n^2}\right) \qquad (n = 3, 4, 5, \ldots); \qquad (2.10b)$$

$1/\lambda$ is known as the *wave number* (nearly always expressed in cm^{-1}). The constant R_H, which we mentioned in the last section, is known as the

Rydberg constant (for hydrogen), and has a value of 109678 cm^{-1}. Each value of n corresponds to one spectral line. The ultraviolet series, called the *Lyman series*, fits a formula resembling Eq. 2.10b, but with the fraction $\frac{1}{4}$ replaced by the integer 1, and the values of n beginning with 2. Similarly, each successive series[8] in the infrared region can be fitted to such an equation with the first constant in brackets replaced by $\frac{1}{9}$, $\frac{1}{16}$, and so on. In general, one can write

$$\frac{1}{\lambda} = R_{\mathrm{H}}\left(\frac{1}{n_1^2} - \frac{1}{n_2^2}\right) \qquad (n_1 = 1, 2, 3, \ldots),$$
$$(n_2 = n_1 + 1, n_1 + 2, \ldots). \qquad (2.11)$$

The value of n_1 fixes the series to which the line belongs, the value of n_2 fixes which particular line it is. Note that each series converges to a low-wavelength limit R_{H}/n_1^2 as $n_2 \to \infty$; beyond this limit a continuous emission spectrum is observed. Equation 2.11 is a remarkably accurate and succinct expression for all the lines in the spectrum of atomic hydrogen.

What can be deduced from the form of Eq. 2.11? The very existence of line spectra, of course, is suggestive of quantized energy levels. If the energy of light is proportional to its frequency, $E = h\nu = hc/\lambda$ (as we concluded from the Franck-Hertz experiment), then Eq. 2.11 can be satisfied if the hydrogen atom has quantized states with energies given by $E_n = -hcR_{\mathrm{H}}/n^2$ (n_2 = initial state, n_1 = final state). The problem, however, was how to formulate a model of atomic structure that would have such energy levels. As we shall see in Section 2.11, this was the achievement of Niels Bohr.

Expressions similar to Eq. 2.11 also describe the emission spectra of vapors of the alkali metals (Li, Na, K, Rb, Cs). The alkali spectra are more complex than that of hydrogen, however, in that they contain several series of rather weak lines in addition to the strong and well-defined series analogous to hydrogen. Furthermore, the integers n_1 and n_2 of Eq. 2.11 are replaced by numbers n_j^*, which are usually nearly but not quite integers:

$$n_j^* = n_j - \delta \qquad (j = 1, 2). \qquad (2.12)$$

For higher members of any one series in an element's spectrum, the *quantum defect* δ is essentially a constant. Also, the constant R_H of Eq. 2.11 is replaced in each of the corresponding alkali formulas by another constant differing from it only very slightly (by less than 0.1%, in fact). From the similarity of their spectra, we might assume that hydrogen and the alkalis share some very striking common property. We therefore expect that, given a theoretical explanation of the spectrum and structure of the hydrogen atom, it should be relatively easy to extend this explanation to the alkali atoms.

When we go on to the spectra of other elements, the situation immediately becomes far more complex, but not completely intractable. Some regularities were recognized long before the spectra could be fully interpreted. These were largely in the form of repeating sets of lines having the same frequency intervals. These repeating sets, for example, triplets or quartets of lines, can sometimes be recognized even in very complex patterns like those in Fig. 2.9. It is not surprising that each element's spectrum is unique and easily distinguishable from any other

FIGURE 2.9
A portion of the emission spectrum of iron, in the region between approximately 2600 and 3300 Å, or 260 and 330 nm.

---

[8] The series with $n_1 = 3, 4, 5$ are known as the *Paschen, Brackett,* and *Pfund series,* respectively.

element's, so that one can use these spectra to recognize the presence of specific elements, even in the atmospheres of stars.

One can generate emission spectra in several ways. Electric discharges in gases at low pressures are one way, as we already mentioned. Electric arcs also excite a number of lines of most elements, and it is easy to introduce solid or liquid samples into such arcs (struck, for example, between two pieces of high-purity carbon). Many more lines are excited by a spark discharge, in which the electric current is subjected to a much higher accelerating voltage than in the arc. In the spark, because of the higher electric field, the accelerated ions and electrons reach much higher energies than in an arc, and one can recognize lines characteristic not only of neutral atoms, but of the positive ions of many elements, even of those with several electrons removed. Given still higher excitation energies, one can obtain emission lines at shorter and shorter wavelengths, eventually reaching the x ray spectra of the last section; it should be clear from the similarity of Eqs. 2.7 and 2.11 that the processes involved are related.[9]

So far we have spoken only of emission, but of course each element also has a characteristic absorption spectrum. The two spectra are in fact the same: Any substance that absorbs light of a given wavelength can be stimulated to emit light of that wavelength. We conclude that absorption is simply the inverse process to emission, with atoms going from lower to higher energy levels. The absorption spectrum can be observed by passing continuous radiation through a sample and determining which wavelengths are removed; the amount of absorption usually obeys Beer's law, with the absorption coefficient a function of wavelength. The dark (Fraunhofer) lines in the spectra of the sun and other stars result from absorption by gases in the outer layers of their atmospheres.

2.5
The Nuclear Atom

We have now introduced most of the experimental facts needed to formulate a theory of atomic structure. One key piece of information remains: the evidence for the nuclear structure of the atom. Let us review what we already know about the nature of atoms, corresponding to what was known before 1910. We know approximately the sizes and weights of individual atoms. We know that they can be ionized into particles with positive and negative charges, and that, because atoms are electrically neutral, the total amounts of positive and negative charge must be the same in an intact atom. We know from measurements of charge-to-mass ratios that the negatively charged electrons weigh far less than the rest of the atom. Various measurements we have not discussed suggest that the total number of electrons (and thus of positive charges) in an atom is of the order of its atomic number Z. The final link in this chain is the evidence, which we consider in this section, that the atom has a nuclear structure—that the positive charge and its associated large mass are concentrated in a very small volume at the atom's center, the *nucleus*, whereas the light electrons occupy the remainder of the atom's volume. In the simplest formulation, we say that the atom is like a miniature solar system, with the electrons orbiting the nucleus as the planets orbit the sun.

What alternatives are there to the nuclear model of atomic structure? For example, could it be that the positive charges are in orbit around the

---

[9] In fact, the first line of the Lyman series ($n_1 = 1$, $n_2 = 2$) is actually the hydrogen K_α line, shifted into the ultraviolet by the low value of Z (the screening constant is zero for hydrogen).

lighter negative charges? Probably not; such a model is physically unreasonable. If the atom does consist of particles moving about one another, some very light and some very heavy, then obviously the lighter particles must move about the heavier ones. Indeed, if the positive charge is concentrated in a small volume, any volume smaller than that of the atom as a whole, then the model with the positive particles at the nucleus must surely be the correct one. The other principal alternative is one in which the positive charge and the mass associated with it are spread in some moderately uniform way throughout the entire volume occupied by the atom. This model, sometimes called the "plum pudding model," was proposed by J. J. Thomson, who favored it until experiment gave incontrovertible proof that the nuclear model was the correct one.

Just how does one establish that one model is right and another wrong? Each model can be used to predict the results of various experiments. One finds an experiment for which the two predictions are different, then carries out the experiment. One hopes that the results will be clearly inconsistent with one of the hypotheses, which can then be rejected. This does not *prove* the alternative hypothesis, but merely demonstrates that it can be accepted for want of other equally valid models. (This is the ideal case, of course; sometimes no experiment can be performed to distinguish between two models—and sometimes *all* the models turn out to be wrong!) It is almost always in this way that new scientific theories replace old ones. The crucial experiment that established the nuclear theory of atomic structure involved the scattering of particles by atoms; it was first performed in 1909 by H. Geiger and E. Marsden, and the results were then explained by Ernest Rutherford.

Specifically, the experiment consisted of the scattering of a beam of α particles (He^{2+} ions) by a thin metal foil, for example, gold foil about 2 μm thick. Figure 2.10 illustrates the type of apparatus used in measurements of this sort. (The detectors shown in the figure are more sophisticated than Geiger and Marsden's. They measured the actual scattered intensity with a ZnS screen that showed a flash of light when an α particle struck it. The flashes were counted by observing the screen through a microscope.) The basic quantity measured is the number of particles scattered in various directions, that is, as a function of the *scattering angle* θ; from this measurement can be obtained what we shall call the *differential cross section.*

First, however, let us consider the *total cross section*, which corresponds to the quantity σ introduced in Eq. 2.8. Particle scattering and x-ray attenuation are so different that it is worthwhile to look at the general expression again, with a slightly different terminology. We refer to Fig. 2.10. Suppose that the beam of projectile particles emerging from B has a *flux density* of f_0 particles per unit area per unit time. Suppose also that in the scattering region C there are n target particles per unit volume, so that a thin layer of thickness dx contains $n\,dx$ particles per unit area. Let us assume that each of these targets has an effective cross-sectional area σ for scattering; that is, whenever a projectile passes through the area σ around a target particle, it will be scattered away from the beam and thus not reach the detector D. Then the fraction of the beam flux scattered out of the beam will be just the fraction of the target area effectively covered by target particles. The fractional change in the beam's intensity in the layer of thickness dx will thus be

$$\frac{df}{f} = -(\text{area per particle})(\text{particles per unit area}) = -\sigma n\,dx. \quad (2.13)$$

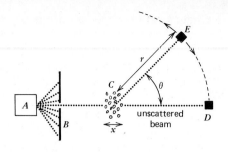

FIGURE 2.10

Schematic diagram of a scattering apparatus. The source A, which may be a hot oven, radioactive material, and so on, emits particles (black dots) in random directions. The collimating aperture B forms a reasonably well-defined beam of these projectile particles. The projectiles strike target particles (open circles) in the scattering region C, with thickness x; this may be a confined chamber, a solid film or foil, or just the intersection of two beams (projectiles and targets). The unscattered projectiles continue on to the detector D, which measures attenuation and total cross section. Other projectiles are scattered in all directions, one trajectory being chosen for illustration. The movable detector E is used to determine the number of particles scattered in a given direction; usually the scattering is symmetric around D, and only the scattering angle θ need be varied. To determine the differential cross section as a function of θ, one must move the detector E over the full range of angles into which a significant amount of scattering takes place.

This is the same expression that we stated empirically as Eq. 2.8, but now we can see why σ is called a cross section. As before, we can integrate to obtain

$$f_D = f_0 e^{-\sigma n x} \tag{2.14}$$

for the flux density of the beam emerging from the scattering region and reaching the detector D. In the x-ray case the beam attenuation was due primarily to absorption; here it is due entirely to scattering, and we refer to σ as the *total scattering cross section*.

The detector D measures the loss of intensity of the direct beam, the number of particles scattered *out* of the incident direction; one can then use the detector E to measure the number of particles scattered *into* a particular scattering angle. Experimentally, θ is the angle between E and the original beam, measured from the center of the scattering region. Of course, the angle θ defines not a single trajectory but a whole cone of trajectories, symmetric around the unscattered beam; ordinarily, however, the same number of particles will be scattered into all these directions, so that the orientation of E and D does not matter.[10]

Since the detector and the scattering region both have nonzero areas, the particles entering the detector at any given position have actually been scattered through a range of angles (Fig. 2.11); if the two are far enough apart, however, the average scattering angle will still be θ. We can thus make our calculations as if all the particles radiated from the center of the scattering region, relative to which the detector E subtends the solid angle

$$\Omega_E = \frac{A_E}{r^2}, \tag{2.15}$$

where A_E is the detector's area. (See Appendix 2B for the definition of solid angle.) Given the flux of particles reaching the detector and the solid angle subtended by the detector, one can obtain[11] $df(\theta)/d\Omega$, the flux density scattered into θ per unit solid angle—that is, the number of particles per unit area of the incident beam that are scattered per unit time into unit solid angle in the vicinity of the angle θ. The *fraction* of the incident beam flux scattered into θ per unit solid angle by the layer dx is then given by an expression analogous to Eq. 2.13,

$$\frac{1}{f} d\left[\frac{df(\theta)}{d\Omega}\right] = \sigma(\theta)n\, dx; \tag{2.16}$$

as before, f is the flux density of the beam incident on the layer dx, and n is the number density of targets. This equation defines the quantity $\sigma(\theta)$, the *differential scattering cross section* per unit solid angle, which is

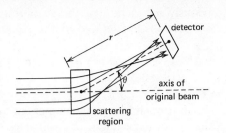

FIGURE 2.11
Uncertainty in the scattering angle. The angle θ is defined by the centers of the scattering region and the detector, but the trajectories in the figure show that some particles scattered through angles greater or less than θ will also be detected. The greater the distance r, the smaller will be the range of angles.

---

[10] To obtain scattering that is not symmetric around the original beam, one must use target particles with nonrandom orientation. An example of this is the scattering of x rays by crystals (Chapter 11).

[11] If the detector records n_E particles per unit time, we have

$$\frac{df(\theta)}{d\Omega} = \frac{n_E}{A_0 \Omega_E},$$

where A_0 is the cross-sectional area of the original beam. Integration over all solid angles should give

$$\int \frac{df(\theta)}{d\Omega}\, d\Omega = f_0 - f_D$$

for the total scattered flux density.

obviously a function of the scattering angle. The differential and total cross sections are related by the equation[12]

$$\sigma = \int \sigma(\theta) \, d\Omega = 2\pi \int_0^\pi \sigma(\theta) \sin \theta \, d\theta, \qquad (2.17)$$

so that $\sigma(\theta) = (d\sigma/d\Omega)_\theta$.

Both σ and $\sigma(\theta)$ are numerical quantities that must be determined either from experiment or from microscopic information about how the projectiles and targets interact. They are parameters of the system, and cannot be derived from any line of deductive reasoning or purely mathematical construction; some real physical information is needed. Equations 2.13 and 2.16 are precise mathematical statements of some very general relationships, but they tell us nothing about any specific system. For example, Eq. 2.16 says that the fraction of the incident particles scattered into our detector is proportional to the number of scatterers (targets) available, and that there exists a function $\sigma(\theta)$ that gives this proportionality. We can identify $\sigma(\theta)$ with some fraction of an effective cross section σ for purposes of guiding our ideas, but to say that either is a specific and recognizable physical area may often be misleading. Equation 2.16 is in fact valid under a wide variety of circumstances in which $\sigma(\theta)$ can be treated as no more than a proportionality factor. The graphic interpretations suggested by the names "cross section" and "differential cross section" are strictly valid only for a few classical systems. Equations like 2.13 and 2.16 are what we call *phenomenological* relationships. They express physical phenomena in very general and precise mathematical terms, but at the same time leave the characteristic behavior of individual and specific systems contained in a small number of parameters such as the cross section. These parameters must be derived from some entirely different theory or from experiment. We shall frequently make use of phenomenological relationships throughout this text. We shall also encounter and make use of the contrasting approach, often called a microscopic theory—the kind of approach in which one derives properties such as the value of a cross section, or its dependence on the scattering angle, from the structures of the colliding particles.

All this discussion of cross sections has told us nothing about atomic structure, but it has given us a vocabulary with which we can examine Geiger and Marsden's results for α-particle scattering. Remember that there were two fundamentally different microscopic pictures of atomic structure, namely, the plum pudding model and the nuclear model. For each of these models one can calculate the dynamics of atomic collisions and obtain the dependence of the differential cross section $\sigma(\theta)$ on the scattering angle θ. If the density and thickness of the scattering foil are known, one thus obtains from Eq. 2.16 a prediction of the fraction of a beam that will be scattered into an angle θ, the quantity measured directly in the experiment. According to *both* hypothetical models, the heavy α particles must be deflected primarily by the heavy positively charged part of the atom; the light electrons, rather than deflecting the α particles, must themselves be deflected from their positions or trajectories as the α particles pass near them. The plum pudding model leads to the conclusion that the scattered intensity should fall off with the angle θ according to a Gaussian distribution, that is,

$$\sigma(\theta) \propto e^{-\theta^2/\theta_m^2}, \qquad (2.18)$$

---

[12] Experimentally, the integration of Eq. 2.17 must begin, not at $\theta = 0$, but at some small angle just outside the unscattered beam.

where θ_m is the mean deflection of the particles passing through the scattering region. The mean angle θ_m is proportional to the mean deviation due to a single atom but includes the effects of scattering by more than one target atom. The form of expression 2.18 is characteristic of the accumulation of many small effects—here, many small deviations in trajectory, with an average of θ_m for each—and is often called a random distribution. A ball cascading through a pegboard is an example of how a single α particle would behave if this model were correct. With many balls cascading from a single slot into many receptacles, one gets a distribution like 2.18 (Fig. 2.12). If one can determine θ_m experimentally, then one can predict quantitatively the intensity at any other angle θ relative to that at θ_m. The other model, the nuclear model, assumes the nuclei to be so small that more than one encounter is very unlikely, so that double or multiple scattering events can be disregarded. On the other hand, any individual encounter in the nuclear model is likely to be a much more violent experience for a projectile α particle than would be the total effect of passage through several plum puddings. Specifically, according to the nuclear model the differential cross section should obey the relation

$$\sigma(\theta) \propto \frac{1}{\sin^4(\theta/2)} ; \qquad (2.19)$$

note that this quantity is divergent, becoming infinite as $\theta \to 0$. The cross section for the nuclear model is also predicted to be proportional to Z^2, the square of the atomic number, and inversely proportional to the square of the initial kinetic energy of the α particles. Rutherford's derivation of expression 2.19 is given in Appendix 2C.

Expression 2.19 predicts that the differential cross section in the nuclear model should drop sharply as θ increases from zero ($\theta = 0$ being the forward direction of the beam) and tend to level off as θ approaches its maximum of π. The infinite cross section for the forward direction reflects the very long range of the Coulomb potential, the longest-range potential we know; in crude terms, this means that every particle in the beam feels some effect from at least one single scatterer. The flatness of $\sigma(\theta)$ at large θ reflects the very strong interaction associated with head-on collisions. These are the collisions that scatter particles backward—into the region for which $\theta > \pi/2$—and they can exist only if there is a concentrated center of force responsible for the scattering. The plum pudding model, without concentrated scatterers, predicts much weaker backward scattering. For gold foil 2 μm thick, θ_m is inferred from experiment to be about 1°. According to expression 2.18, for the plum pudding model a vanishingly small number of α particles could be deflected to all angles very much larger than 1°; one certainly would not expect to see particles deflected through 45° or more. The nuclear model and expression 2.19, however, open the possibility of seeing large-angle deflections.

Which model is correct? The original experiments of Geiger and Marsden showed that about one α particle in 8000 was deflected through an angle greater than 90°. This in itself is strongly suggestive,[13] but it is not conclusive evidence that the nuclear atom is the correct model. We must have something firmer and more quantitative if we are going to be convinced. The convincing proof came from further measurements by Geiger and Marsden. The intensity scattered into unit area, or better still

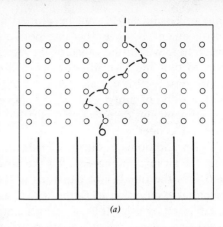

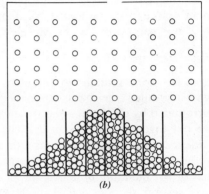

FIGURE 2.12
A ball-and-pegboard demonstration of the effect of multiple scattering and how it gives rise to a Gaussian distribution: (a) A single ball and its trajectory, (b) The result of dropping many balls through the board.

---

[13] Rutherford, who had thought the plum pudding model to be correct, later said that it was "as if you fired a 15-inch shell at a piece of tissue paper and it came back and hit you." It was in fact only *after* this experiment that he began to devise the nuclear model.

into unit solid angle, can be plotted against $1/\sin^4(\theta/2)$. This is a little inconvenient, however, because both quantities vary over many orders of magnitude within the range of the experimental data. It is more convenient to plot the logarithm of one against the logarithm of the other. If expression 2.19 is correct, then this log–log plot must give a straight line with a slope of unity. Figure 2.13 shows clearly that this is the case. The experiment thus demonstrated clearly that the nuclear model must be the preferable one.

An additional result of Geiger and Marsden's work was to confirm that the atomic number Z gives the positive charge on the atomic nucleus and thus the number of electrons in the neutral atom, since this was an essential assumption in Rutherford's derivation (Appendix 2C). This was established by using foils of various materials as targets: As predicted, the scattering cross section was proportional to Z^2, where Z was the element's place in the periodic table. Further confirmation of the significance of Z was soon provided by Moseley's x-ray measurements.

One stone remains to be set in the foundation of our theory. We have identified the basic components of the atom—a heavy nucleus surrounded by electrons—and we must still explain the way in which they are put together. We began this chapter by describing several types of evidence suggesting that atoms can gain or lose energy only in certain well-defined amounts. The key concept explaining these phenomena is that of the *quantum of action*, an idea foreign to classical physics. In the next several sections we shall develop its meaning and show how it constrains the energy and angular momentum in microscopic systems.

The quantum concept dominated the development of atomic and molecular physics in the early years of this century. A dramatic change occurred when three very different phenomena, all inexplicable within the framework of classical physics, yielded to rather thorough interpretation within less than 12 years; all three interpretations were intimately grounded in the concept of the quantum of action. One is the photoelectric effect, discussed in Section 2.2; another, the heat capacity of solids, will be considered in Section 2.9. We now turn to the problem of what is best known as *black-body radiation*.

Imagine a solid object with a hollow interior. Now suppose that the exterior of this object is completely insulated from the outside world, except for some heating wires and some telephone wires to bring us information such as the object's temperature. If the object is heated to, say, 700 or 800°C, it should glow bright orange. This is a property we know well from common experience, from looking at "red hot" solids. But the object of our new and peculiar experiment must glow only in its own interior; after all, we have insulated its exterior, so it can lose no energy to the outside world. Although we cannot see the glowing surface, we know that the inside walls are emitting bright orange light (and also infrared radiation) into the empty volume or *cavity* in the object's interior. If our object is really *very* well insulated, it will soon reach a state of thermal equilibrium, when it will require essentially no additional energy from the heating wires to maintain a constant temperature. But those inside walls are still radiating. Will the solid object lose all its energy to the cavity simply by emission of radiation? Clearly it will not. Some of that radiation must be reabsorbed by the walls. As soon as our perfectly insulated object draws no more energy from the heating wires, that is, as soon as it has reached a steady temperature, then no *net* energy enters the cavity. Every bit of energy that appears as radiation must be

2.6
The Problem of Black Body Radiation

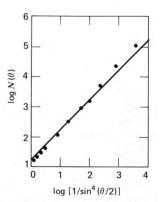

FIGURE 2.13
Geiger and Marsden's data for the scattering of α particles by silver foil [replotted from intensities reported in *Phil. Mag.* **25**, 604 (1913)]. $N(\theta)$ is the number of particles scattered per minute into unit area of the detector screen at a scattering angle θ. The data can be seen to give a reasonably good fit to the 45° straight line predicted by expression 2.19.

paid for by an equal amount of radiation energy absorbed by the walls. The object is called a *black body*, and is characterized by its radiation, which is in equilibrium with the container at a precise temperature.

The classical problem, the very first problem to involve the concept of the quantum, was this: How much energy is present *as radiation* in a given volume of "empty" space within such a simple hot object held at any definite temperature T? A subsidiary question is this: What is the distribution of the radiation energy in the cavity, as a function of the wavelength? For example, is the radiation all of the same frequency? This is a conceivable answer, but a simple experiment tells us that it is not correct. All one has to do is steal just a bit of radiation from the cavity (through a small hole) and examine it with a spectrometer. As shown in Fig. 2.14, the radiation intensity gives a broad, smooth distribution that varies with the temperature of the black body. This is quite similar to the radiation given off by any hot object; the importance of the ideal (perfectly insulated) black body, whose radiation is independent of the material from which the walls are made, is that it can be treated by relatively simple theories. The curves in the figure are primarily in the infrared, but with increasing temperature the peak shifts to lower wavelengths; the light given off by the sun (ca. 6000°C) or an incandescent lamp has its peak in the visible region.

What one wants to know, naturally, is why the energy distribution of black-body radiation has this particular form. The answer to this question lies primarily in the domain of thermodynamics, the subject that comprises Part Two of this book. Despite the division that is sometimes drawn between thermodynamics and structure, the beginnings of all our modern theories of atomic and molecular structure lay in Max Planck's exposition of the thermodynamic problem of black-body radiation in 1901. Because its analysis would take us prematurely into thermodynamics, we shall not go into Planck's theory in detail. We merely indicate why there was a problem and what key assumptions Planck made to solve it.

The work of James Clerk Maxwell and Heinrich Hertz during the latter part of the nineteenth century made it clear that light and many other electromagnetic phenomena behave like waves. What this means in detail we shall examine in Chapter 3. At the moment we can rely simply on our intuitive concepts of waves (in strings or on oceans) to obtain a graphic feeling for the processes involved. Rayleigh and Jeans made a great stride toward the solution of the black-body radiation problem when they suggested that one think of the radiation within the black body as standing waves, forced to remain in the cavity by the condition that there be no wave displacement in any direction at the boundary walls. These waves are analogous to the vibrations in a jump rope or a violin string whose ends are held tight so that they cannot be displaced. This picture creates a model on which the physicist or chemist can proceed. Rayleigh and Jeans computed how many modes of wave motion[14] can be contained in a given cavity and satisfy the stated boundary condition; they found that the number of modes should increase steadily with frequency, becoming infinite in the short-wavelength limit. According to the classical theory applied by Rayleigh and Jeans, the radiation energy should be distributed evenly over all these modes, and thus an infinite amount of

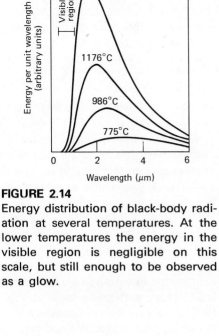

FIGURE 2.14
Energy distribution of black-body radiation at several temperatures. At the lower temperatures the energy in the visible region is negligible on this scale, but still enough to be observed as a glow.

[14] That is, how many possible standing waves with different frequencies and/or directions of propagation. This number is limited, since the waves must have an integral number of wavelengths in the distance between opposite walls.

energy should be stored in the short-wavelength modes. By contrast, the experimental data show that the energy distribution drops sharply at short wavelengths. The theory did give a satisfactory description of the long-wavelength end of the spectrum, but was patently wrong in the short-wavelength region. This discrepancy was known as the *ultraviolet catastrophe*, and could not be explained by classical means.

Planck resolved the problem by restricting the way in which the radiation energy could be distributed over the modes of wave motion in the cavity. Classically, one assumed that this distribution could be carried out by a method like that used to define a derivative or an integral in elementary calculus: One begins by chopping a system into small but discrete bits, then imagines that the bits get smaller and smaller as their number gets greater and greater, all the way to the limit of infinitely small bits. Planck found that to explain the observed black-body spectrum, he had to assume that the radiation field could not be subdivided into infinitely small bits—that it was made up of some kind of indivisible elements. He called the modes of wave motion *oscillators* (for reasons that we shall see later) and said that their properties were *quantized*. By making this assumption, Planck was able to calculate a theoretical energy distribution in excellent agreement with the experimental data for a black body, giving curves like those in Fig. 2.14. To see why Planck's assumption made so great a difference, we must begin by describing the properties of his oscillators. We can carry out this description on two levels.

In the simpler formulation, Planck assumed that the possible energies of the oscillators must be separated by finite amounts, or *quanta*. The energy associated with any oscillatory or wave motion is proportional to the frequency; Planck proposed that each mode of the radiation field could have only specific and precise energies, and not any energy between these characteristic values. In the classical model the increment between any two allowed energies could be arbitrarily small, but Planck fixed the increment at a constant small value, proportional to the oscillator frequency ν. If we call the proportionality constant h (*Planck's constant*), then the increment, or quantum of energy, between allowed energies of a given oscillator is given by

$$\Delta E = h\nu. \tag{2.20}$$

The constant h is the same as that we introduced in Eqs. 2.3, with the value of 6.626176×10^{-34} J s. On the basis of then-existing data, Planck estimated a value of 6.55×10^{-34} J s. This was the first formulation of the idea of quantization of energy.

Now we can understand the black-body energy distribution. The number of possible oscillators (modes of wave motion) still becomes infinite as $\nu \to \infty \, (\lambda \to 0)$, but so does the minimum nonzero energy that each oscillator can have. Classically, even if each of the high-frequency oscillators had only a tiny amount of energy, their great number meant that most of the energy would be at that end of the spectrum. In Planck's model, however, an oscillator of frequency ν that has any energy at all must have at least the amount $h\nu$. As ν increases, $h\nu$ becomes a greater and greater fraction of the available energy, and it becomes more and more likely for the same energy to be divided among many oscillators of lower ν. This is why the energy density falls off sharply at high ν (low λ). As the temperature increases, more total energy is available and the fall-off occurs at higher frequencies.

We said that the quantization of energy was one of two ways to interpret Planck's quantization process. The other and perhaps more illuminating interpretation is in terms of the quantity called *action*, the quantum of action being *h* itself. To understand what this means, we must discuss the concept of action at some length.

Action is a quantity with the same dimensions as Planck's constant. These dimensions can be expressed as energy×time or as momentum×distance, and are thus the same as those of angular momentum. Action is a standard variable of classical mechanics; it is related to momentum in the same way that work is related to force. To express these relationships in their most general forms, we need to introduce more information about vectors.

2.7
The Concept of Action

A vector, it will be recalled from Section 1.3, can be expressed as a set of three quantities, the *x*, *y*, and *z*-components. In Eqs. 1.17 we introduced the cross product of two vectors, $\boldsymbol{\alpha} \times \boldsymbol{\beta}$, which is itself a vector. A second way of multiplying vectors gives a different quantity, the *dot product* (or *scalar product*) $\boldsymbol{\alpha} \cdot \boldsymbol{\beta}$. This is not a vector, but a scalar (pure number) given in terms of the components of the vectors $\boldsymbol{\alpha}$ and $\boldsymbol{\beta}$ by

$$\boldsymbol{\alpha} \cdot \boldsymbol{\beta} = \alpha_x \beta_x + \alpha_y \beta_y + \alpha_z \beta_z. \tag{2.21}$$

Alternatively, the dot product can be expressed as

$$\boldsymbol{\alpha} \cdot \boldsymbol{\beta} = \alpha \beta \cos \phi, \tag{2.22}$$

where α, β are the magnitudes of the vectors $\boldsymbol{\alpha}, \boldsymbol{\beta}$, and ϕ is the angle between them; it is thus the magnitude of $\boldsymbol{\alpha}$ times the component of $\boldsymbol{\beta}$ in the direction of $\boldsymbol{\alpha}$ (or vice versa, since $\boldsymbol{\alpha} \cdot \boldsymbol{\beta} = \boldsymbol{\beta} \cdot \boldsymbol{\alpha}$).

Now how does this apply to work and action? Suppose that an object moves an infinitesimal distance; since this displacement has both magnitude and direction, it is expressed as a vector *d*s with components *dx*, *dy*, *dz*. If this motion is produced by a force **F**, the infinitesimal increment of work performed by the force[15] is defined as

$$dW = \mathbf{F} \cdot d\mathbf{s}, \tag{2.23}$$

or the displacement times the component of force in the direction of the displacement. (The elementary formulation "work = force×distance," as in Eq. 1.9, applies only when force and displacement are parallel.) The total work done when an object moves from an initial point 1 to a final point 2 is thus the sum of such infinitesimal increments,

$$W_{12} = \int_1^2 dW = \int_1^2 \mathbf{F} \cdot d\mathbf{s}. \tag{2.24}$$

An expression of the form $\int_1^2 \mathbf{F} \cdot d\mathbf{s}$ is known as a *line integral*, since the variable of integration is the displacement along some path. The value of the integral in general depends on the route chosen between points 1 and 2, and this path must be specified before one can calculate it.

We can now define the action, in which the momentum **p** plays the role that the force **F** plays in defining work. Specifically, the infinitesimal increment of action is

$$dA = \mathbf{p} \cdot d\mathbf{s}, \tag{2.25}$$

---

[15] Here *W* is the work performed *on* the object *by* the force, and is thus positive when force and displacement are in the same direction. This is not important here, but later we shall have to be careful in defining the sign of work.

and the total action associated with motion from point 1 to point 2 is the line integral

$$A_{12} = \int_1^2 dA = \int_1^2 \mathbf{p} \cdot d\mathbf{s}. \tag{2.26}$$

It is all very well to define action by its formal analogy to work, but we need a better understanding of its physical meaning. Let us examine a specific case where the action becomes identical with a more familiar quantity. The momentum of a particle, of course, is the product of its mass and velocity,

$$\mathbf{p} = m\mathbf{v}. \tag{2.27}$$

Suppose that we compute the action given by Eq. 2.26 for a particle moving at constant speed on a circle of radius r. The displacement along the circle is given by

$$ds = r\,d\theta, \tag{2.28}$$

where θ is the angular displacement in radians. Since the displacement and the momentum at any given time are both in the same direction, the $\cos\phi$ of Eq. 2.22 is always unity and we have simply

$$dA = \mathbf{p} \cdot d\mathbf{s} = mvr\,d\theta. \tag{2.29}$$

To obtain the total action for one complete orbit around the circle, we integrate to obtain

$$A_{\text{orbit}} = \int_0^{2\pi} mvr\,d\theta = mvr \int_0^{2\pi} d\theta = 2\pi mvr. \tag{2.30}$$

The action is therefore simply 2π times the magnitude of the angular momentum,

$$L = mvr. \tag{2.31}$$

This illustrates the point made earlier that the dimensions of action are the same as those of angular momentum.

2.8
The Harmonic Oscillator

We shall calculate the action explicitly for one other classical example, that of the *harmonic oscillator*. First, however, we must examine at some length the properties of the ubiquitous harmonic oscillator, which is far and away the most powerful and most widely used single model in all of physical science. It underlies our theories of solids and, to some degree, of liquids; it is the basis of much of our understanding of the behavior of molecules and of electromagnetic radiation. The reason for this generality is that a great variety of physical systems are described by equations that are mathematically equivalent to (or can be transformed into) those of the harmonic oscillator. In fact, the oscillator model is an obvious starting point for describing any system that remains near but not at some position or state of equilibrium.

The entire physical description of the harmonic oscillator is contained in its force law, which is simpler than those we have encountered earlier. To understand the nature of this law, consider a spring rigidly fixed at one end and constrained to move along a straight line (Fig. 2.15a). The spring has some equilibrium length l_0: if stretched, it will spontaneously contract; if compressed, it will expand again. The spring's length l will then oscillate back and forth around its equilibrium value (until friction stops the process). For small deformations in either direction, the restoring force is proportional to the deformation (Hooke's law):

$F \propto (l - l_0)$. The harmonic oscillator is an idealization of this model in which a similar force law always holds.

The ideal one-dimensional harmonic oscillator, then, is a mass point moving frictionlessly along a straight line; it is subject to a force that always acts to return it toward the same specific position; and the force is proportional to the particle's displacement from this equilibrium position. If we define the equilibrium position as $x = 0$, the displacement is x (corresponding to $l - l_0$ in the spring model), and the oscillator's force law can be written as

$$F = -kx, \qquad (2.32)$$

where F is the force in the positive x direction (cf. Fig. 2.15b). When x is positive, the force is negative, meaning that it tends to push the particle in the direction of $-x$, that is, back toward the origin. The proportionality constant k is called the *force constant* or spring constant.

We wish to find an equation for the oscillator's displacement x as a function of time. Using Newton's second law, we can rewrite Eq. 2.32 as

$$F = ma = m\frac{dv}{dt} = m\frac{d^2x}{dt^2} = -kx, \qquad (2.33)$$

where m is the mass of the particle. Equating the last two terms, we have a differential equation in x and t to solve. To obtain such a solution, we must add some more information: For example, we can give the particle's position at the arbitrary time $t = 0$ and state its maximum displacement. (Pieces of information of this sort are called *boundary conditions*, about which we shall have much to say in the next chapter; we mentioned another example in our discussion of the black-body problem.) For convenience, we set $t = 0$ when the particle is at its equilibrium position, $x = 0$; and we denote by x_1 the particle's maximum displacement from equilibrium (the *amplitude* of the oscillation). The equation of motion and its boundary conditions can then be written compactly together as

$$\frac{d^2x(t)}{dt^2} = -\frac{k}{m}x(t), \qquad \text{subject to} \begin{cases} x(0) = 0, \\ |x|_{\max} = x_1. \end{cases} \qquad (2.34)$$

There are formal methods for solving such an equation (two integrations are required, with the two boundary conditions giving the integration constants); however, we proceed in a more direct and intuitive way. We need to know what function $x(t)$ has a second derivative equal to itself times a negative constant. (Since k and m are both positive by definition, $-k/m$ must be negative.) Pulling a rabbit out of the hat, we recognize that the sine or cosine of t, or of any multiple of t, is such a function. Moreover, $\sin t$ would be equal to zero at $t = 0$. The maximum displacement is x_1, so $x_1 \sin t$ would satisfy the equation completely except for the constant k/m. If the argument of the sine is not t itself, but some multiple[16] αt, then each differentiation of x with respect to t introduces a multiplicative factor α. Our equation contains two differentiations, so we must have $\alpha^2 = k/m$. We conclude that the displacement as a function of time is given by

$$x(t) = x_1 \sin\left(\frac{k}{m}\right)^{1/2} t, \qquad (2.35)$$

with x oscillating continually between positive and negative values (Fig.

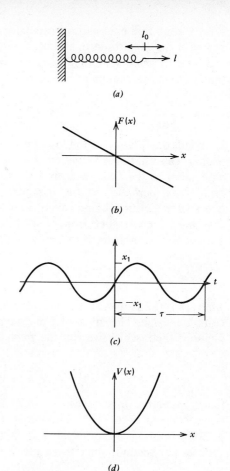

FIGURE 2.15
The one-dimensional harmonic oscillator. (*a*) The spring model: The spring's length l oscillates around its equilibrium value l_0, (*b*) The force law, Eq. 2.32, where F is the force in the positive x direction, (*c*) The displacement as a function of time, Eq. 2.35, assuming that $t = 0$ at $x = 0$, (*d*) The potential energy, Eq. 2.48.

---

[16] Since the sine and cosine functions can have only dimensionless arguments, t *must* have some multiplier to appear in such an argument; the multiplier must have the dimensions of a frequency, with units such as s^{-1}.

2.15*c*). (In a real macroscopic oscillator, of course, friction would steadily decrease the amplitude of the oscillation.)

The displacement of the oscillator is given by Eq. 2.35 as a sine function, whose argument is naturally an angle. This angle,

$$\theta = \left(\frac{k}{m}\right)^{1/2} t, \tag{2.36}$$

is called the *phase* of the oscillator. Since $(k/m)^{1/2}t$ is dimensionless, proper cancellation of the units should give θ in radians; θ is thus equivalent to the time measured in units of $(m/k)^{1/2}$. The motion of the oscillator is *periodic*—that is, it repeats the same motions over and over, one such repetition being called a *cycle*. In the cycle beginning at $t = 0$, the sine function goes from 0 to 1 to 0 to -1 and back to 0, as its argument θ goes from 0 to 2π. We thus have

$$2\pi = \left(\frac{k}{m}\right)^{1/2} \tau \quad \text{or} \quad \tau = 2\pi \left(\frac{m}{k}\right)^{1/2}, \tag{2.37}$$

where τ is the *period* of the oscillator, the time it takes to go through one complete cycle. The inverse of the time per cycle is the number of cycles per unit time, which we call the *frequency*,

$$\nu = \frac{1}{\tau} = \frac{1}{2\pi} \left(\frac{k}{m}\right)^{1/2}; \tag{2.38}$$

if t is measured in seconds, the units of ν are s^{-1}, also written as cycles per second or hertz (Hz). Sometimes it is more convenient to describe a periodic motion in terms of the *angular frequency* ω, the number of radians per second; since there are 2π radians to a cycle, we have

$$\omega = 2\pi\nu = \left(\frac{k}{m}\right)^{1/2}. \tag{2.39}$$

The terminology introduced here is applicable to any periodic phenomenon. However, the value of ω or ν is a property of the particular system; for the oscillator it depends on the values of k and m, and cannot be derived from any of the physical laws we have so far written down. We shall see later that frequencies can be estimated for particular physical models, but to do this one must always have additional information, including the masses of the particles and something about the strength of the forces binding them.

We can now draw a parallel between the harmonic oscillator and the last section's example of a particle moving in a circular orbit. Suppose that the center of the circle is the origin of a Cartesian coordinate system. Equation 2.35 then gives exactly the variation of x with time for a circular orbit of radius x_1 and constant angular velocity $(k/m)^{1/2}$; $y(t)$ is the corresponding cosine function. All the quantities introduced in the last paragraph then have the same values for the circular orbit, for which the "angular" terms can be taken in their literal sense. However, the two systems are not equivalent in all respects, as we can show by calculating the action for the harmonic oscillator.

As with the particle moving in a circle, we wish to know the total action for a complete cycle. Combining Eqs. 2.35 and 2.39, we have

$$x(t) = x_1 \sin \omega t; \tag{2.40}$$

the velocity of the particle at any given time is thus

$$v(t) = \frac{dx(t)}{dt} = \omega x_1 \cos \omega t = v_0 \cos \omega t \qquad (v_0 \equiv \omega x_1), \tag{2.41}$$

where v_0 is the maximum velocity. Note that when the displacement is zero, that is, when $\sin \omega t = 0$, then $\cos \omega t$ is a maximum in one direction or the other; in other words, the particle reaches its maximum speed when it passes its equilibrium position. Similarly, since the two quantities are 90° out of phase with each other, the velocity is zero when the displacement is at its maximum. By Eq. 2.40, the increment of the particle's action is

$$dA = mv \, dx, \tag{2.42}$$

since the momentum $m\mathbf{v}$ is always in the same direction as the displacement. By the "chain rule" we can write $dx = (dx/dt) \, dt = v \, dt$, and substitution of Eq. 2.41 then gives

$$dA = mv^2 \, dt = mv_0^2 \cos^2 \omega t \, dt. \tag{2.43}$$

Note that the increment of action is twice the kinetic energy times the time increment dt—again, energy × time. Integrating over the time from 0 to τ, we obtain[17] the total action associated with one cycle of the oscillator,

$$
\begin{aligned}
A_{\text{cycle}} &= \int_0^\tau mv_0^2 \cos^2 \omega t \, dt \\
&= \frac{mv_0^2}{\omega} \int_0^{2\pi} \cos^2 \theta \, d\theta \qquad (\theta \equiv \omega t) \\
&= \frac{mv_0^2 \pi}{\omega}. \tag{2.44}
\end{aligned}
$$

Since $v_0 = \omega x_1$, we have $A_{\text{cycle}} = \pi m v_0 x_1$, which is only half as much as Eq. 2.30 gives for the circular orbit of radius x_1.

We can rewrite our expression for the action in several equivalent ways:

$$A_{\text{cycle}} = \frac{mv_0^2 \pi}{\omega} = \frac{mv_0^2}{2} \frac{1}{\nu} = \frac{mv_0^2 \tau}{2}, \tag{2.45a}$$

$$A_{\text{cycle}} = \frac{kx_1^2 \pi}{\omega} = \frac{kx_1^2}{2} \frac{1}{\nu} = \frac{kx_1^2 \tau}{2}. \tag{2.45b}$$

We can interpret these expressions by recognizing the physical meaning of certain factors that appear in them. The harmonic oscillator, like any mechanical system with no frictional losses, must conserve its energy. Its kinetic energy is instantaneously $\frac{1}{2}mv^2$, with a maximum value of $\frac{1}{2}mv_0^2$, which we find in Eq. 2.45a. At the position of maximum displacement the kinetic energy is zero, and all the oscillator's energy is stored as potential energy. To calculate this we need a definition of potential energy.

The potential energy of a particle is a function of its position and the forces acting on it. It is strictly defined only for *conservative* systems, those with no frictional or other dissipative forces. If in such a system a particle moves from point 1 to point 2 under the influence of a total force $\mathbf{F}$, the change in the particle's potential energy is the negative of the work done on the particle,

$$V_2 - V_1 = -W_{12} = -\int_1^2 \mathbf{F} \cdot d\mathbf{s}. \tag{2.46}$$

[17] The last line of Eq. 2.44 makes use of the definite integral

$$\int_0^{2\pi} \cos^2 \theta \, d\theta = \pi.$$

(Note that in a conservative system the work done is independent of path.) The zero of potential energy can be chosen arbitrarily, since only *differences* in potential energy are physically meaningful.

Let us apply this definition to the harmonic oscillator. Since the force on the particle and its displacement both lie along the x axis, $\mathbf{F} \cdot d\mathbf{s}$ reduces to $F_x \, dx$, where F_x is simply the F of Eq. 2.32. The change in potential energy as the particle goes from $x = a$ to $x = b$ is thus

$$V(b) - V(a) = -\int_a^b F(x) \, dx = \int_a^b kx \, dx = \tfrac{1}{2}kb^2 - \tfrac{1}{2}ka^2. \qquad (2.47)$$

It is most convenient to set the potential energy equal to zero at the equilibrium point $x = 0$; if $V(0) = 0$, the potential energy of the oscillator at any other point x is simply

$$V(x) = \tfrac{1}{2}kx^2. \qquad (2.48)$$

This is the equation of a parabola, as illustrated in Fig. 2.15d. The maximum potential energy is $\tfrac{1}{2}kx_1^2$, the value at the point of maximum displacement, where the kinetic energy vanishes. The total energy E of the oscillator is the sum of the kinetic and the potential energy;[18] since each of these vanishes when the other has its maximum, each of the two maxima must be equal to the total energy:

$$E = \tfrac{1}{2}mv^2 + \tfrac{1}{2}kx^2 = \tfrac{1}{2}mv_0^2 = \tfrac{1}{2}kx_1^2. \qquad (2.49)$$

The information in Eqs. 2.45 can therefore be summarized in the single equation

$$A_{\text{cycle}} = \frac{E}{\nu} = E\tau, \qquad (2.50)$$

that is, the action per cycle is the energy per unit frequency or the energy times the period of oscillation. The total energy E is a constant of the oscillator's motion.

There is one further way in which we can visualize the action, with a geometric interpretation that contrasts with the more mechanical picture considered thus far. The geometric concept must be drawn by analogy. Consider a particle moving in a plane, in which we define a Cartesian coordinate system. If the only constraint on its motion is that it remain within the region $0 < x < x_1$, $0 < y < y_1$ (with x and y independent), its position can be anywhere in the rectangle of area $x_1 y_1$ shown in Fig. 2.16a. If there were an additional constraint that the sum $(x/x_1)^2 + (y/y_1)^2$ always be less than unity, then the particle would be restricted to the quadrant of the ellipse shown in Fig. 2.16b. One can imagine innumerable other restrictions. Basically, however, if there are two variables describing the behavior of a particle, perhaps with some constraints, then one can describe an available region in a plane within which the particle can move.

Now we can construct our analogy for the harmonic oscillator. We choose as one of the two variables the displacement x itself. Then, instead of displacement in another spatial direction (since none exists for our model), we choose as our second variable the momentum p. For any object restricted to finite velocities and to a finite region of space, that is, for anything we can conceivably observe for an arbitrarily long time, each component of both position and momentum is a quantity bounded above and below—it has a maximum and a minimum. Each component of the motion (and our one-dimensional oscillator has only one such compo-

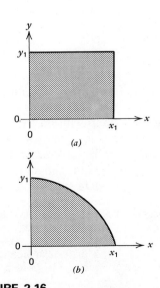

FIGURE 2.16
Areas available to particles moving in a real plane, constrained to stay within limits $0 < x < x_1$, $0 < y < y_1$. (a) x, y independent, (b) Added constraint $(x/x_1)^2 + (y/y_1)^2 < 1$.

[18] Note that E, like V, has an arbitrary zero.

nent) can thus be thought of as defining an area in a plane. Such a plane—a two-dimensional space—is the simplest example of what is called a *phase space*, a space whose dimensions are spatial coordinates and the corresponding momentum components. The "area" in this plane has the dimensions of momentum × distance, which it will be recalled are the same as those of action. In other words, we can associate with every bounded one-dimensional system an arealike quantity with the dimensions of action which describes the ranges of momentum and displacement available to the system.

Now that we have defined the "action plane," let us use it to describe the motion of the harmonic oscillator. We need an equation for the motion of the harmonic oscillator. We need an equation for the motion in terms of x and p $(= mv)$. We have just such an equation available in Eq. 2.49, which we can rearrange to read

$$\frac{p^2}{(mv_0)^2} + \frac{x^2}{x_1^2} = 1 \tag{2.51}$$

(where we have divided through by $mv_0^2/2 = kx_1^2/2$). But from analytic geometry we recognize immediately that this is the equation of an ellipse in the px plane, with its center at the origin and semiaxes of mv_0 and x_1. This ellipse is illustrated in Fig. 2.17. The displacement and momentum of the oscillator are represented by a point moving around the ellipse in the clockwise direction: One trip around the ellipse represents one cycle of the oscillator. The motion of the point is along the curve itself rather than within it (as in Fig. 2.16*b*); nevertheless, the area within the curve is significant. If we combine the definition of the action as $\int p\,dx$ with the familiar interpretation of the definite integral as an area, we realize immediately that the area within the ellipse must equal the action over a cycle. (There is nothing remarkable about this: Any integral over a cycle equals the area inside the closed curve of the integrand.) The area of an ellipse is πab, where a and b are the two semiaxes; the action over a cycle of the harmonic oscillator is thus

$$A_{\text{cycle}} = \pi(mv_0)x_1 = \frac{\pi mv_0^2}{\omega}, \tag{2.52}$$

in agreement with Eq. 2.44. That is, the geometric interpretation of the action gives the same value as the direct calculation.

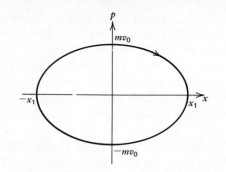

FIGURE 2.17
Ellipse in the action plane describing the motion of the one-dimensional harmonic oscillator. The two semiaxes are related to each other and the energy by

$$mv_0 = (km)^{1/2}x_1 = (2mE)^{1/2}.$$

The area within the ellipse is the action per cycle.

2.9
Action Quantized: The Heat Capacity of Solids

After our long digression to clarify the nature of action, we can now return to our main line of argument, broken off in Section 2.6. We spoke then of three phenomena that were explained in terms of the quantum of action; now we can complete our treatment of black-body radiation and proceed to the heat capacities of solids.

Remember that Planck's model for the black-body radiation field was an array of "oscillators," by which he meant harmonic oscillators such as we have just been describing. To the extent that this model is valid, we should be able to extract some useful insight from what we have learned about oscillators. Sure enough, we have Eq. 2.50, which says that an oscillator with energy E and frequency ν has an action per cycle of E/ν. But for Planck's oscillators the allowed values of E are separated by increments of $h\nu$; this means that their action per cycle can only have values separated by increments of $\Delta E/\nu = h$. In terms of Fig. 2.17, the allowable ellipses are only those with areas of exactly $h, 2h, 3h, \ldots$. One can now see why h is called the "quantum of action."

Our second example of the quantization of action involves the classic problem of the heat capacity of a crystal lattice. Again, because a detailed analysis would take us too far afield, we shall limit ourselves here to a simple qualitative description. (See Chapters 11, 14, and 22.) A crystal is composed of an orderly array (lattice) of atoms or ions. The equilibrium position of each atom is called a *lattice site*. Because of the repulsion between adjacent atoms, the structure with each atom occupying its lattice site is one of minimum potential energy. If the atoms have any kinetic energy (as they do at any temperature above absolute zero), then they do not rest at their lattice sites but move about them, striking each other and rebounding. To a good approximation one can describe this vibration by considering the atoms as harmonic oscillators. (Here they are again!) To be precise, since each atom has three independent directions of motion, one represents it by three one-dimensional oscillators, all oscillating around the lattice site. One can calculate the energy of such an array of oscillators as a function of temperature, and this is where the trouble arose.

The problem is best described in terms of the physical quantity called the *heat capacity* (see Chapter 14). Simply defined, the heat capacity of a given quantity of matter is the amount of energy (in the form of heat) required to increase its temperature by 1°C; that is, it is the temperature derivative of the substance's energy. In microscopic terms, the heat capacity must thus be directly related to the capacity of the atom-oscillators for storing energy. According to classical mechanics the heat capacity of a crystal should be independent of temperature, with a constant value of approximately 25 J/mol °C. This result agrees with the empirical *law of Dulong and Petit* (1819), which holds fairly well for many solids (especially metals) at room temperature, and can thus be used to estimate molecular weights from specific heats. However, a number of solid substances, particularly nonmetals, have heat capacities much lower than 25 J/mol °C at room temperature; furthermore, all heat capacities appear to fall off toward zero at sufficiently low temperatures. The classical theory thus appears to be correct only as a limiting case. The key to the correct theory was found by Albert Einstein in 1907, and of course involved the quantization of action.

The problem of the vibrational energy of solids is cleared up by essentially the same assumption as that used for the problem of black-body radiation. Indeed, the same sweeping hypothesis can be applied to any bounded system undergoing periodic motion. It consists of one simple but very severe constraint: The action integral for for a single cycle of the system's motion must equal an integral multiple of Planck's constant h. That is, we must have

$$A_{\text{cycle}} = \int_{\text{cycle}} \mathbf{p} \cdot d\mathbf{s} = nh, \tag{2.53}$$

where n is an integer, both for the system as a whole and for each individual oscillator. In the geometric interpretation of Fig. 2.17, this equation again means that the ellipse in the action associated with an individual oscillator must have an area that is an integral multiple of h. Classically, any value was permissible for the ellipse's area. Referring back to Eq. 2.50, we recognize immediately that the energy of a single oscillator of frequency ν can only be

$$E_{\text{osc}} = \nu A_{\text{cycle}} = nh\nu. \tag{2.54}$$

The energy of an oscillator is thus restricted to integral multiples of the product $h\nu$, and the oscillator can only increase its energy in discrete steps of magnitude $h\nu$. Using this assumption, as we shall see in Chapter 22, Einstein was able to derive a qualitatively correct temperature dependence for the heat capacity of crystals.

We can give here an outline of how quantization of the oscillator energies affects the macroscopic properties of crystals. Energy as heat is transferred to a solid by means that need not be specified. That energy is distributed among the oscillators of the solid. Classical mechanics assumed that the oscillators could exchange energy in any amount. Now, however, Eq. 2.54 tells us that the oscillators are allowed to increase their energy only in jumps of size $h\nu$. If an oscillator's $h\nu$ is larger than the average energy per oscillator, then that oscillator must be rather unaccommodating as a storehouse for energy: It hardly ever gets enough energy all at once to provide an entire quantum. When the temperature is low, the average energy per oscillator is low, so most of the oscillators cannot contribute to the capacity of the solid to store energy. Hence the heat capacity is low at low temperatures. The amount of available energy of course increases with temperature, and at a high enough temperature the heat capacity c approaches the classical (Dulong–Petit) value of 25 J/mol °C. The temperature required for this to occur is proportional to the value of ν, which can thus be estimated from the temperature dependence of c. The highest values of ν are found in hard substances like diamond, for which temperatures of about 2000°C are required for c to approach the classical limit. For most solids, however, the value of ν is low enough to give classical behavior at room temperature, resulting in the Dulong–Petit law. At still lower temperatures the available thermal energy per atom becomes smaller and smaller relative to $h\nu$, eventually becoming so small that the heat capacity of all solids decreases to zero.

Our equations for the energy of light quanta and for the energy of an oscillator, Eqs. 2.20 and 2.54, look very similar. Their content is also similar, but not quite identical. Each quantum has a specific and well-defined energy $h\nu$; the oscillator has a definite frequency ν, but can absorb any number of quanta, changing its state with each energy transfer.[19] What about Planck's oscillators in the black-body radiation field? The space within the black body is filled with a vast number of light quanta with frequencies all over the spectrum. The number of quanta of any given frequency ν must always be some integer n, and the total energy of these quanta is $nh\nu$, just as in Eq. 2.54. The "oscillator" of frequency ν is thus just a convenient way of summing up the quanta of frequency ν; the "oscillator" gains or loses energy when one quantum at a time leaves or enters the walls, yielding Eq. 2.20.

We have now introduced the fundamental concepts of the quantum theory—quantum levels of energy, the quantization of action, and the particulate or quantum character of light. We are ready to go on to study the structure of the atom. First, however, as we did in Section 1.2, let us examine some numerical magnitudes, in order to put some of the esoteric quantities we have been discussing into context. We begin by recalling in familiar units the magnitudes usually associated with energy and with action. The mechanical unit of energy, the joule ($1\,\text{J} = 1\,\text{kg}\,\text{m}^2/\text{s}^2$), is not a very large quantity: 1 J is just about the energy required to lift a 100-g weight 1 m; it takes about 4.2 J (1 cal) to heat a gram of water by 1°C. As for action (the unit of which has no special name), for that 100-g weight

2.10
Some Orders of Magnitude

---

[19] The heating of a solid involves the absorption of quanta of vibrational energy (*phonons*) rather than quanta of electromagnetic radiation (photons), but the principle is the same.

to drop 1 m would involve an action of about 0.3 J s; to take a cyclic phenomenon, at 30 mph an automobile tire might have an action of 200 J s per revolution.

What about microscopic quantities? To begin with, what do the phenomena just mentioned amount to on a microscopic scale? Some things we can calculate easily enough: A mole of water weighs 18 g and contains Avogadro's number of H_2O molecules, so it takes about

$$\frac{4.2 \text{ J}}{\text{g}} \times \frac{18 \text{ g}}{\text{mol}} \times \frac{\text{mol}}{6 \times 10^{23} \text{ molecules}} \approx \frac{1.3 \times 10^{-22} \text{ J}}{\text{molecule}}$$

to heat water 1°C. It is a little more complicated to find the action per atom in the automobile tire. Say that the tire weighs 10 kg. What about the atomic weight? Most of the tire is made of rubber (or some synthetic equivalent), which is essentially a hydrocarbon polymer; the rayon or cotton cords are basically cellulose, $(C_6H_{10}O_5)_x$. Hydrogen's atomic weight is about 1, carbon's 12, and oxygen's 16; a little playing with formulas shows that the average atomic weight in the tire should not be far from 6, which will make the arithmetic easy. Some parts of the tire are moving faster than others, but the average action per atom for the whole tire will suffice for our purposes. This is easily calculated as

$$\frac{200 \text{ J s}}{10 \text{ kg}} \times \frac{\text{kg}}{1000 \text{ g}} \times \frac{\text{g}}{6 \times 10^{23} \text{ amu}} \times \frac{6 \text{ amu}}{\text{atom}} \approx \frac{2 \times 10^{-25} \text{ J s}}{\text{atom}}$$

per turn of the tire.

Something should be said about these calculations, which are typical of the way one often approaches a scientific problem. That is, we don't care what the exact action per atom really is; we just want to know roughly how much it is. To do this we can make rough guesses of the sizes of the numbers. It would be an incredible waste of time to get a tire, measure its exact weight, carry out a chemical analysis, make a detailed calculation of its moment of inertia,[20] and so on. All we want is a simple *order-of-magnitude* estimate. One cannot overemphasize the importance of order-of-magnitude calculations in science. They are our sounding lines, the guides that tell us whether it is reasonable or nonsensical to continue in a given direction. They are the guideposts we use to give us a feeling for unfamiliar quantities, in just the way we are using them now. They are the quick checks that tell us if we have made a gross error in calculation, that give us an on-the-job diagnosis of whether or not an experiment is likely to be working properly. With a little practice, and a few natural constants and conversion factors in one's head, one develops a facility to make these estimates quickly and painlessly at the slightest provocation.

All right, how do the quantities we have calculated compare with the quanta of energy and action? Planck's constant h, the quantum of action, is only about 6.6×10^{-34} J s. This means that one revolution of our automobile tire involves about 300 million quanta of action *per atom*! Obviously h is a small quantity indeed, and it is clear why quantum effects are not observed in macroscopic phenomena. What about the energy of a light quantum? Yellow light, with a wavelength of about 5000 Å, has a frequency of

$$\nu = \frac{c}{\lambda} \approx \frac{(3 \times 10^8 \text{ m/s})}{5 \times 10^{-7} \text{ m}} = 6 \times 10^{14} \text{ s}^{-1};$$

[20] To determine the total action. We got our figure by guessing an average radius and applying Eq. 2.30, another order-of-magnitude calculation.

multiplying this frequency by Planck's constant, we obtain about 4×10^{-19} J for the energy of a single quantum of yellow light. This is not so small: It is 3000 times our figure for heating one water molecule 1°C, and a mole of these quanta[21] would amount to 240 kJ, enough energy to heat 57 liters of water 1°C. But it is well to keep in mind the vast range of the electromagnetic spectrum when one thinks of the energy associated with radiation. x-Ray quanta are usually more than a thousand times as energetic as visible-light quanta, and gamma rays are more energetic still. On the other hand, if we proceed to lower frequencies, in the radio region we commonly find frequencies in megahertz, that is, mere millions of cycles per second; this corresponds to energies of the order of 10^{-27} J per quantum, a very small number even by microscopic standards.

2.11
Bohr's Model of the Atom

This chapter concludes with a discussion of the simplest model of the atom containing the essence of quantization, the model developed by Niels Bohr in 1913. It is not really correct in all its details, or even in some of its most fundamental ideas; however, its utility as a simple device for computing atomic magnitudes is unsurpassed. There is no other computational tool known to which we can turn so quickly to estimate whether a new effect will be observable or beyond our detectable range. We shall make no attempt to explore the model in detail, but even by examining its simplest form we can derive all the atomic properties and magnitudes we have been introducing throughout this chapter. In particular, we wish to establish that atoms are stable entities (i.e., that electrons do not fall into nuclei), that they exhibit sharp spectral lines associated with definite excitation energies, and that the intervals between these spectral lines are remarkably regular.

Our derivation of Bohr's model will not be carried out in quite the way Bohr himself did it, but by a somewhat shorter and, for our purposes, more efficient route. (You can usually improve a derivation *after* you know how it comes out.) We begin with a few simple assumptions about the energy of the system, its mechanical stability, and, above all, the quantization of its action. From this simple model we can derive the characteristic lengths, velocities, frequencies, and times associated with the motion of electrons in atoms. We shall also see, in a surprisingly quantitative way, how the general character of atomic and molecular spectra has its origins in the structure of the atom, and we shall be able to infer from the model many of the general characteristics of more complex atoms and molecules.

Bohr's model was derived for the hydrogen atom, or for any other system in which a single electron orbits a positively charged nucleus. It thus took for granted (and immediately followed) Rutherford's model of the nuclear atom. In most of our derivation we shall make the simplifying assumption that the nucleus is so heavy that it can be considered stationary, with only the electron moving; strictly speaking, both particles orbit about the center of mass.[22] We follow Bohr's original model in assuming that the electron moves in a circular orbit (elliptical orbits were also introduced to account for the fine structure of spectra). We consider

---

[21] Incidentally, photochemists call a mole of light quanta an *einstein*, for obvious reasons.

[22] The model can in fact easily be generalized to apply to any two oppositely charged particles. Of particular interest is the case where the two particles have equal mass, and orbit about each other like a binary star. There are real physical examples of such species, for example, *positronium*, composed of a positron (positively charged electron) and an ordinary negative electron.

the general case of a nucleus with charge $+Ze$; the hydrogen atom corresponds to $Z=1$.

We begin by defining the energy of the system. The total energy is the sum of the electron's kinetic energy and its potential energy in the field of the nucleus. Let the electron's mass be m, its velocity v, and its distance from the nucleus r. The kinetic energy is simply $\frac{1}{2}mv^2$, while the potential energy is some function of r. We can evaluate Eq. 2.46 for the potential energy between any two charges, using Eq. 1.5 for the force. Since the force is directed radially, we have $\mathbf{F}\cdot d\mathbf{s}=F(r)\,dr$ for a radial displacement, and the potential energy is given by[23]

$$V(r)-V(\infty)=-\int_r^\infty F(r)\,dr=\frac{q_1 q_2}{4\pi\epsilon_0}\int_r^\infty \frac{dr}{r^2}=\frac{q_1 q_2}{4\pi\epsilon_0 r}. \qquad (2.55)$$

The zero of potential energy can as usual be set arbitrarily, and it is clearly most convenient to set $V(\infty)=0$. For our atomic system we have $q_1=+Ze$, $q_2=-e$, $V(r)=-Ze^2/4\pi\epsilon_0 r$, and the total energy is

$$E=\frac{mv^2}{2}-\frac{Ze^2}{4\pi\epsilon_0 r}. \qquad (2.56)$$

Our choice of the zero of energy corresponds to setting $E=0$ for an electron at rest ($v=0$) at an infinite distance from the nucleus ($r=\infty$). The advantage of this choice is that all negative values of E then correspond to bound states of the electron, states in which it cannot escape from the nucleus. This is so because the kinetic energy $mv^2/2$ is necessarily a positive quantity; for E to be negative we must have $Ze^2/4\pi\epsilon_0 r>mv^2/2$, which sets an upper bound on the value of r. By contrast, a positive value of E corresponds to a free electron, which can have as large a value of r as one wishes; at sufficiently great distances, the potential energy becomes negligible.

The next step in the derivation comes from the assumption that the electron's orbit is stable. By Coulomb's law, the inward attractive force on the electron is $Ze^2/4\pi\epsilon_0 r^2$. The centrifugal force on a particle moving in a circular orbit is mv^2/r. For the electron to continue moving in a stable circular orbit, these two forces must be equal. Consequently, we have

$$\frac{mv^2}{r}=\frac{Ze^2}{4\pi\epsilon_0 r^2}. \qquad (2.57)$$

The assumption of stability may appear trivial, but this was one of the most revolutionary aspects of Bohr's model. According to classical electromagnetic theory, an orbiting charge would continuously emit radiation, thereby losing energy and spiraling inward to the nucleus. By contrast, the quantum theory assumes not only that the orbit is stable, but that the electron can gain or lose energy only in quantized amounts.

The introduction of this quantization constitutes our final major assumption. As in previous sections, it is the action that must be quantized: We require that the action integral over each cycle be an integral multiple of h. Introducing our earlier result for the action in a circular orbit, Eq. 2.30, we have

$$A_{\text{orbit}}=2\pi mvr=nh. \qquad (2.58)$$

The product mvr is the electron's angular momentum, which must be an integral multiple of $h/2\pi$. The latter quantity appears so often in subse-

---

[23] This and all the equations that follow assume the use of SI units. Many texts give the corresponding expressions in electrostatic or Gaussian units; these expressions can be generated by substituting $(4\pi)^{-1}$ for ϵ_0.

quent expressions that it is convenient to introduce a special symbol for it,

$$\hbar \equiv \frac{h}{2\pi} = 1.0545887 \times 10^{-34} \, \text{J s.} \tag{2.59}$$

Such shorthand ways of combining symbols, especially those for natural constants, make our equations less cumbersome and thus tend to prevent mistakes.

We are now ready to combine our assumptions. Multiplying both sides of Eq. 2.57 by r/mv, we obtain an expression for the velocity of the electron,

$$v = \frac{Ze^2}{4\pi\epsilon_0 mvr}. \tag{2.60}$$

But the mvr in the denominator of this equation must, by Eq. 2.58, be an integral multiple of $\hbar$. We immediately have an explicit expression for the allowed values of the velocity,

$$v = \frac{Ze^2}{4\pi\epsilon_0 n\hbar} = \frac{Ze^2}{2\epsilon_0 nh}. \tag{2.61}$$

This equation tells us that the velocity of the electron is directly proportional to the nuclear charge Ze and inversely proportional to the integer n, which we call the *principal quantum number*. The essence of Bohr's theory is that only those orbits corresponding to integral values of n are physically allowed. Later we shall obtain numerical values for v and other atomic quantities, but not until we have completed the formal structure of the theory.

Another quantity whose value we can easily derive is the radius of the electron's orbit. From Eq. 2.58 we immediately have

$$r = \frac{n\hbar}{mv}, \tag{2.62}$$

and substitution of the value of v from Eq. 2.61 then gives

$$r = \frac{4\pi\epsilon_0 n^2\hbar^2}{Ze^2 m} = \frac{\epsilon_0 n^2 h^2}{\pi Ze^2 m} \tag{2.63}$$

in terms of fundamental constants. The radius depends inversely on the velocity and thus on the nuclear charge; a nucleus with a larger positive charge will hold the electron in a tighter orbit, but with a greater orbital velocity. Note also that the radius depends explicitly on the mass m; a charged particle heavier than the electron would orbit closer to the nucleus.[24]

Next we compute the period of the orbit, the time it takes the electron to go around the nucleus once. This time is simply the distance (the circumference) divided by the velocity, or from Eqs. 2.61 and 2.63,

$$\tau = \frac{2\pi r}{v} = \frac{(4\pi\epsilon_0)^2 2\pi n^3 \hbar^3}{Z^2 e^4 m} = \frac{4\epsilon_0^2 n^3 h^3}{Z^2 e^4 m}. \tag{2.64}$$

The period is inversely proportional to the mass and to the square of the nuclear charge.

---

[24] This property can be used to probe the structure of nuclei. Short-lived "atoms" can be made in which negatively charged mesons move in orbits about nuclei. For a meson whose mass is 200 times that of the electron, the stable radii are about 1/200 of those for electron orbits. Such an orbit is close enough to the nucleus to be affected by its detailed internal structure.

It can now be seen that our model of the atom exhibits a series of discrete states, each corresponding to a different integral value of the quantum number n. The values of n begin with 1 and have no upper limit. (A value $n = 0$ would correspond to a particle moving with infinite velocity at zero radius; negative values would merely correspond to reversing the direction of rotation.) Each successive larger value of n defines a new stable orbit, with lower velocity, a larger radius, and a much longer period than the preceding orbit. The orbit for which $n = 1$ must be the closest to the nucleus and, as we shall now see, corresponds to the most tightly bound state.

By "most tightly bound," of course, we mean the state with the lowest energy, which thus requires the greatest amount of work to remove the electron from the atom. Let us then consider the energy of our system. Equation 2.57 can be rewritten as

$$\frac{mv^2}{2} = \frac{1}{2}\left(\frac{Ze^2}{4\pi\epsilon_0 r}\right); \tag{2.65}$$

what this says is that, for any stable circular orbit, the kinetic energy is half the negative of the potential energy (relative to the energy zero previously defined). Substituting this relation into Eq. 2.56, we immediately obtain a simple expression for the total energy,

$$E = \frac{Ze^2}{4\pi\epsilon_0 r}\left(\frac{1}{2} - 1\right) = -\frac{Ze^2}{8\pi\epsilon_0 r}. \tag{2.66}$$

Note that this is still a classical expression. We obtain the quantized energy levels by substituting the value of r from Eq. 2.63:

$$E = -\frac{Z^2 e^4 m}{2(4\pi\epsilon_0)^2 n^2 \hbar^2} = -\frac{Z^2 e^4 m}{8\epsilon_0^2 n^2 h^2}. \tag{2.67}$$

The energy of any stable orbit is of course negative, approaching the limit $E = 0$ as n (and thus r) becomes infinite. The absolute value of the energy is proportional to the mass and to the square of the nuclear charge.

The above expression for the energy is rather cumbersome; it is time to introduce another simplification. Designating the energy of the nth quantum state as E_n, we can write Eq. 2.67 in the form

$$E_n = -hcR_\infty \frac{Z^2}{n^2}, \tag{2.68}$$

where hcR_∞ is the energy defined by

$$hcR_\infty \equiv \frac{e^4 m}{8\epsilon_0^2 h^2} = 2.179907 \times 10^{-18}\,\text{J} = 13.605804\,\text{eV}. \tag{2.69}$$

Since the joule is an inconveniently large quantity for atomic energies, we also give the value in electron volts (cf. footnote 1 on page 29). Spectroscopists often express energies in terms of the corresponding wave numbers ($1/\lambda$); since for a transition between two energy levels (see below) we have $|\Delta E| = h\nu = hc/\lambda$, the constant corresponding to hcR_∞ is

$$R_\infty \equiv \frac{e^4 m}{8\epsilon_0^2 h^3 c} = 109737.3177\,\text{cm}^{-1}. \tag{2.70}$$

Note that R_∞ itself is known more accurately than the other physical constants in the equation. Like the R_H of Eq. 2.10, R_∞ is known as the *Rydberg constant* (the reason for the subscript ∞ will be explained later); the energy hcR_∞ is sometimes called one *rydberg*.

We can now determine the magnitudes of the energies involved in actual interactions. Quanta of visible light, with wavelengths between 4000 and 8000 Å, have energies in the range 2–4 eV. In the hydrogen atom ($Z = 1$), the lowest energy level is $E_1 = -hcR_\infty$, so that about 13.6 eV is required to remove the electron from the atom and leave it at rest at a great distance; this energy corresponds to a wavelength of 911 Å, far in the ultraviolet. If the electron is initially in the state with $n = 2$, only one-fourth as much energy is required to remove it, corresponding to light of 3645 Å, just over the edge into the ultraviolet; and so forth for the higher energy levels. (The energy levels of the hydrogen atom are shown in Fig. 2.18). Suppose that an electron in the nth state absorbs a quantum with energy greater than $-E_n$: Then the electron is not only freed from the atom but given a kinetic energy equal to the difference between the actual energy of the quantum and $-E_n$. Note that at positive values of E the energy is no longer quantized, so this kinetic energy can have any value. It should be obvious that the process we have been describing is simply the photoelectric effect for free atoms.

Of course, an electron can also absorb a quantum of light without being set completely free. For example, an electron in a state with $n = 1$ can go to a state with $n = 2$ if it absorbs a quantum with just the right amount of energy. A transition between any other two states can be brought about by a quantum whose energy equals the energy difference between them. An atomic electron can thus absorb a quantum with any of a set of very specific energies, but none with energies different from these. If continuous radiation passes through a sample of matter, the specific wavelengths corresponding to these energies (characteristic of the sample) will be selectively absorbed, giving a spectrum of sharply defined absorption lines. Quanta with enough energy to liberate an electron completely give a continuous absorption spectrum beyond the limit of the line spectrum.

Electrons can also fall back from higher to lower energy levels, emitting quanta in the process. The energy of the quantum emitted again just equals the energy difference between the two states. This is why any element's emission spectrum exactly corresponds to its absorption spectrum. Since the ground state is most stable, the emission spectrum will be observed only when an atom has first been excited to some higher state. This excitation may be due to earlier absorption of a quantum, to electron bombardment (as in the Franck–Hertz experiment), or simply to high temperatures (which increase the rate of collisions between atoms).

Note that in all the processes just described we mention only the initial and final states (the bound states corresponding to quantized energies are called *stationary states*). We say that the electron goes from one to the other by absorbing or emitting a quantum, but do not discuss the mechanism by which this occurs. In fact, Bohr simply assumed an instantaneous "jump," and it took a more advanced theory to explain how the process could occur (cf. Section 4.5). But it is only the energy differences between the various states that concern us now.

We can now interpret the spectrum of atomic hydrogen in terms of quantum states of the hydrogen atom. Let us apply Eq. 2.68 (setting $Z = 1$) to the case of a transition between two states. If the electron in a hydrogen atom goes from an initial state n_2 to a final state n_1, its change in energy is

$$\Delta E = E_{\text{final}} - E_{\text{initial}} = E_{n_1} - E_{n_2} = -hcR_\infty \left(\frac{1}{n_1^2} - \frac{1}{n_2^2} \right). \qquad (2.71)$$

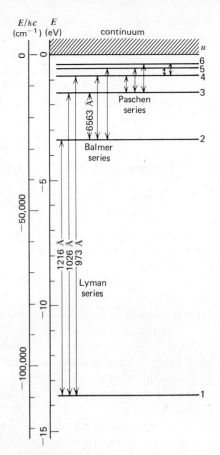

FIGURE 2.18
Energy levels of the hydrogen atom, with some of the principal spectral lines. The levels are almost exactly given by

$$E_n = \frac{-hcR_\text{H}}{n^2},$$

where $R_\text{H} = 109678$ cm^{-1} and $hcR_\text{H} = 13.60$ eV. The wavelengths of the spectral lines are related to the energy levels by

$$\lambda = \frac{hc}{\Delta E}.$$

If the atom loses energy in this process ($\Delta E < 0$, $n_2 > n_1$), then this energy must reappear somewhere; if it is not removed by a collision of some kind, it must appear as emitted radiation, that is, as light. The quantum of this radiation then has an energy exactly equal to the energy lost by the atom. This fact enables us to relate the energy change to the wavelength of the emitted light,

$$\Delta E_{\text{atom}} = -E_{\text{emitted light}} = -h\nu = -\frac{hc}{\lambda}. \tag{2.72}$$

Combining Eqs. 2.70 and 2.71, then, we have for emission

$$\frac{1}{\lambda} = -\frac{\Delta E}{hc} = R_\infty \left(\frac{1}{n_1^2} - \frac{1}{n_2^2} \right). \tag{2.73}$$

Comparing this result with the experimental Eq. 2.11, we see that the two are identical except for the tiny difference (about 0.05%) between R_H and R_∞, which we account for below. The absorption spectrum of hydrogen is described by the same equations with the roles of n_1 and n_2 reversed ($\Delta E > 0$, $n_1 > n_2$). Figure 2.18 shows the major series of hydrogen spectral lines in relation to the energy levels of the atom.

We must now account for the discrepancy between the two Rydberg constants: the experimental $R_H = 109678 \text{ cm}^{-1}$ and the theoretical $R_\infty = 109737 \text{ cm}^{-1}$. This discrepancy is not real, but due merely to a simplification introduced into our derivation for the sake of conciseness. This simplification was the assumption that the nucleus is stationary and only the electron moves. In fact, both particles must revolve about their center of mass. Now, according to classical mechanics, the motion of two bodies around their center of mass can be expressed in terms of the motion of one body around a force center, the one body having a mass equal to the *reduced mass* of the two bodies. For individual masses m_1 and m_2, the reduced mass is defined by

$$\mu \equiv \frac{m_1 m_2}{m_1 + m_2} \quad \text{or} \quad \frac{1}{\mu} = \frac{1}{m_1} + \frac{1}{m_2} \tag{2.74}$$

(by the latter expression, μ is the harmonic mean of m_1 and m_2). Suppose now that the two particles are the electron and the proton; we have for

TABLE 2.2
CHARACTERISTIC PROPERTIES OF CIRCULAR ORBITS*

Property	Expression	Value
Energy	$E_n = \dfrac{e^4 m_e}{8\epsilon_0^2 h^2 n^2} \dfrac{Z^2}{n^2}$	$\left. \begin{array}{l} 2.180 \times 10^{-18} \text{ J} \\ 13.61 \text{ eV} \end{array} \right\} \times \dfrac{Z^2}{n^2}$
Radius	$r_n = \dfrac{\epsilon_0 h^2}{\pi e^2 m_e} \dfrac{Z^2}{n}$	$\left. \begin{array}{l} 5.292 \times 10^{-11} \text{ m} \\ 0.5292 \text{ Å} \end{array} \right\} \times \dfrac{Z^2}{n}$
Velocity	$v_n = \dfrac{e^2}{2\epsilon_0 h} \dfrac{Z}{n}$	$2.188 \times 10^6 \dfrac{\text{m}}{\text{s}} \times \dfrac{Z}{n}$
Period	$\tau_n = \dfrac{4\epsilon_0^2 h^3}{e^4 m_e} \dfrac{n^3}{Z^2}$	$1.520 \times 10^{-16} \text{ s} \times \dfrac{n^3}{Z^2}$

 a The expressions given are for nuclei of infinite mass. For the hydrogen atom, m_e should be replaced by $\mu_H = 0.999456 m_e$.

the hydrogen atom

$$\mu_{\mathrm{H}} = \frac{m_e m_p}{m_e + m_p} = \frac{m_e(1836.15 m_e)}{m_e(1 + 1836.15)} = 0.9994557 m_e. \qquad (2.75)$$

If we carry through our derivation of the Bohr theory with μ_{H} replacing the electron mass m (i.e., m_e), we find that the constant R_∞ in Eq. 2.73 is replaced by

$$\frac{e^2 \mu_{\mathrm{H}}}{8\epsilon_0^2 h^3 c} = 0.9994557 R_\infty = R_{\mathrm{H}}; \qquad (2.76)$$

the agreement with the experimental value of R_{H} is exact. For an atom in which an electron revolves about a heavier nucleus, the value of μ would be still closer to m_e and the corresponding Rydberg constant closer to R_∞. In the limit of a nucleus of infinite mass, the constant would be exactly R_∞, which is therefore more properly called the *Rydberg constant for infinite mass*.

The Bohr theory is thus able to reproduce exactly the spectrum of hydrogen as we described it in Section 2.4.[25] From an extremely simple (although *ad hoc*) model we have easily and directly derived the remarkable Rydberg equation, which elegantly describes hundreds of individual spectral lines. (In principle the number is infinite, since n can be as large as one likes.) One stands in awe at the beauty, clarity and power of Bohr's simple model.

Of course, this magnificent model *is* only a model. It is quite wrong in representing electrons as ordinary physical objects orbiting around nuclei, even if this assumption is sufficient to explain the spectra. More seriously, it ceases to be accurate when we try to apply it to atoms with more than one electron. This breakdown is due to the fact that each electron moves in the field of all the other electrons as well as the nucleus. In the alkali metals one electron is much farther from the nucleus than all the others, and thus sees the rest of the atom as a net charge of $+1$ spread over the atom's interior; the Bohr model then applies approximately, with n replaced by the effective quantum number n^* of Eq. 2.12. In general, however, the simplicity of the theory was lost in attempts to explain complex atoms; as a result a still more revolutionary theory had to be devised, as we shall see in the next few chapters.

Nevertheless, the Bohr model is still useful. It gives us expressions from which we can estimate the magnitudes of most atomic quantities, and exhibits in simple form the dependence of these quantities on nuclear charge and on the principal quantum number. Some of the most important of these quantities are given in Table 2.2 for our simple model of one-electron atoms with circular orbits. The energy we have already discussed. We can immediately see that atoms must be of the order of 10^{-10} m (1 Å) in radius. The radius of the first orbit for $Z=1$, 0.52918 Å, is sometimes called the *Bohr radius*, or 1 *bohr*; it was this value that we listed in Table 1.1. The nuclear charge Z, of course, increases with atomic mass, but so does the value of n for the outermost electron; these effects largely cancel, so that all atoms are of much the same size. (To the very rough extent that this is true, the densities of solid materials should be proportional to their atomic weights.) We also see

[25] Greater resolution reveals a "fine structure," with each spectral line actually consisting of two or more very closely spaced lines. Much of this can be accounted for within the Bohr theory by assuming elliptical orbits, but modern quantum mechanics is needed for a full explanation.

that a bound electron moves along its orbit at a speed of the order of 1000 km/s; by comparison, a rifle bullet moves at less than 1 km/s, a satellite orbiting the earth at up to 8 km/s. Finally, the period for a single revolution of the electron around the nucleus is of the order of 10^{-16} s; events occurring much faster than this can be considered crudely as if the electron were not moving, whereas much slower events will occur as if the electron were a charge distribution smeared over the entire orbit.

At this point we have completed our overview of the origins of the microscopic theory of matter. We have examined a range of experimental evidence, from which one can infer the basic structure of the atom and the quantized nature of its energy, and we have rough estimates for the characteristic magnitudes involved. To go further we must modify our ideas of the fundamental nature of matter itself, a nature that must be regarded as having wavelike properties. In the next chapter we continue to develop the quantum theory by introducing the concept of matter waves.

Appendix 2A

The Electromagnetic Spectrum

An electromagnetic wave is an oscillating force field that acts on charged particles. In the simplest sort of electromagnetic wave, the force at a point in space oscillates in time with a well-defined frequency ν and period τ $(=1/\nu)$; that is, at time t the force is proportional to

$$\sin(2\pi\nu t + \text{const.}).$$

The force field has a certain direction of propagation in space, which for simplicity we may call the x direction. At any instant of time, the spatial dependence of the force is also described by a sine wave with a well-defined wavelength λ; that is, the force is proportional to

$$\sin\left(\frac{x}{\lambda} + \text{const.}\right).$$

The electromagnetic force field is made up of coupled electric and magnetic fields in oscillation together; at any point in space, the vectors $\mathbf{E}$ and $\mathbf{B}$ are perpendicular to each other and to the direction of propagation (cf. Fig. 3.4).

The nature and properties of waves are discussed in greater detail in Section 3.2. Here we wish only to give some physical idea of the phenomena associated with the various kinds of electromagnetic waves, which can be distinguished by their wavelengths or frequencies. For all electromagnetic waves in a vacuum, the wave crests move with the universally constant speed

$$c = 2.99792458 \times 10^8 \text{ m/s}$$

$(3 \times 10^8$ m/s for most purposes); the frequency ν and wavelength λ are thus related by the equation

$$\nu\lambda = c. \tag{2A.1}$$

The frequency is ordinarily expressed in s^{-1}, also known as cycles per second or hertz (Hz); any convenient unit can be used for the wavelength, but the wave number $1/\lambda$ is almost always given in cm^{-1}. Figure 2A.1 gives the values of these quantities for the various kinds of waves, together with the energy per quantum ($E = h\nu$) in both joules and electron volts.

We shall briefly describe each type of radiation shown in Fig. 2A.1. The regions shown are not sharply defined (except that for visible light); where they overlap, the distinction is in terms of the methods of production and application. Note that the high- and low-wavelength ends of the spectrum are most commonly described in terms of frequency and energy, respectively.

The longest electromagnetic waves commonly encountered are in the region of audible frequencies (between 15 and 20,000 Hz). We are more directly familiar with sound waves in this frequency range, since we happen to have sensory organs to detect them. However, the loudspeakers of our audio systems are driven by electromagnetic waves of the same frequency, and we also recognize the humming sound associated with ordinary 60-Hz alternating current.

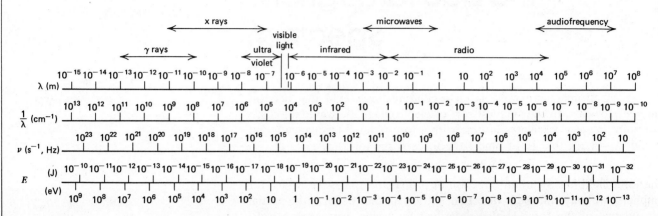

FIGURE 2A.1
The electromagnetic spectrum.

Proceeding to the left in Fig. 2A.1, we come next to radio waves. These are used for broadcasting since they travel rather freely through the atmosphere; furthermore, the shorter wavelengths bounce off the ionosphere and can thus travel great distances. The standard AM and FM radio bands are in the vicinity of 1 and 100 MHz, respectively, whereas television uses several bands between 50 and 1000 MHz. The microwave region largely coincides with the short-wavelength end of the radio spectrum; microwaves are used for radar, for some information and energy transmission, and (relatively recently) for cooking.

The infrared region includes the radiations that our bodies feel as heat. All objects emit radiation, and at temperatures below about 4000°C most of this radiation is in the infrared (cf. Section 2.6). Infrared and microwave spectroscopy are important in determining the structure of molecules, since the spacings of vibrational and rotational energy levels correspond to radiation in these regions. The area where the infrared and microwave regions overlap ("millimeter waves") has only recently been made accessible by the development of suitable sources and detectors.

Between 4000 Å (violet) and 7000 Å (red) lies the narrow band of radiation to which our eyes are sensitive. We evolved in this way, we suppose, because the sun's radiation output has its peak in this region (ca. 4900 Å), and because the atmosphere is transparent to it. This transparency extends only a limited distance into the ultraviolet: The sun's ultraviolet radiation is largely absorbed by ozone in the upper atmosphere, and beyond about 2000 Å even O_2 and N_2 become so opaque that one must work with evacuated apparatus (the "vacuum ultraviolet"). The visible and ultraviolet regions are associated primarily with electronic transitions in atoms and molecules.

The term "x rays" is usually applied to the radiation produced by electron bombardment of targets, as described in Section 2.3; it results from electronic transitions between the inner electron shells of atoms. Gamma rays are the similar photons emitted by radioactive nuclei. Radiation more energetic than about 10 MeV is produced only by cosmic rays or in high-energy particle accelerators; in this region the predominant interaction with matter is the creation or annihilation of electron–positron pairs.

Appendix 2B

Spherical Coordinates

Many physical problems involve the motion of objects around some center of force—the planets around the sun, electrons around an atomic nucleus, colliding molecules around one another. Such problems are most conveniently described in terms of *spherical coordinates*, which give the distance from the origin and the angular orientation around it. Spherical coordinates are used in many places in this text; this appendix summarizes their definition and a number of the most useful relationships involving them.

Spherical coordinates are illustrated in Fig. 2B.1, relative to a Cartesian coordinate system with the same origin. For an arbitrary point P, the radial variable r gives the distance from P to the origin O; θ is the angle between the line OP and the z axis; and ϕ is the angle between the x axis and the projection of OP on the xy plane. (In geographical terms, ϕ is the longitude and θ the colatitude on a sphere of radius r.) The spherical coordinates are related to the Cartesian coordinates of the same point by the conversion equations

$$x = r \sin \theta \cos \phi,$$
$$y = r \sin \theta \sin \phi, \qquad (2B.1)$$
$$z = r \cos \theta,$$

which can easily be deduced from the figure.

Suppose that we wish to integrate some function over a region of space. By the definition of integration, what this amounts to is a summation over infinitesimal volume elements. In Cartesian coordinates, the volume element is a simple parallelepiped with volume $dx\,dy\,dz$. What is the corresponding element in spherical coordinates? It cannot be simply $dr\,d\theta\,d\phi$, which does not even have the dimensions of a volume. Although there is a formal method for obtaining volume elements in any coordinate system, we can obtain our answer more easily by considering Fig. 2B.2. We define our volume element as the space between r and $r + dr$, θ and $\theta + d\theta$, ϕ and $\phi + d\phi$. The edges in the radial direction are of length dr; the edge along which θ varies is a circular arc of radius r and angle $d\theta$, and thus has the length $r\,d\theta$; and the edge along which ϕ varies equals its projection in the xy plane, which is a circular arc of radius $r \sin \theta$ and angle $d\phi$, and thus of length $r \sin \theta\,d\phi$. In the limit of small dimensions this irregular volume element can be approximated by a parallelepiped; the volume is then the product of the three edge lengths, or

$$dV = r^2 \sin \theta\,dr\,d\theta\,d\phi. \qquad (2B.2)$$

Since a surface of constant r is a sphere, the inner face of the volume element is an element of area on a sphere of radius r. The volume dV can thus be written as $dr\,dA$, where the area element is

$$dA = r^2 \sin \theta\,d\theta\,d\phi. \qquad (2B.3)$$

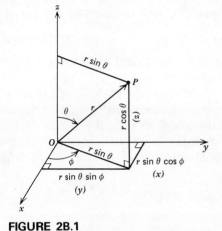

FIGURE 2B.1
Spherical coordinates and their relation to Cartesian coordinates. The point P has the spherical coordinates r, θ, ϕ and the Cartesian coordinates x, y, z; the two sets of coordinates are related by Eqs. 2B.1.

Integration over the angles gives the familiar formula for the total area of a sphere,

$$A_{\text{sphere}} = r^2 \int_0^\pi \sin\theta\, d\theta \int_0^{2\pi} d\phi = r^2(2)(2\pi) = 4\pi r^2; \qquad (2\text{B}.4)$$

the volume of a spherical shell of thickness dr is thus $dV_{\text{shell}} = 4\pi r^2\, dr$. The fraction of a sphere covered by an area element dA is

$$\frac{dA}{A} = \frac{r^2 \sin\theta\, d\theta\, d\phi}{4\pi r^2} = \frac{\sin\theta\, d\theta\, d\phi}{4\pi}, \qquad (2\text{B}.5)$$

which is independent of r. The *solid angle* Ω subtended at the origin by any object is defined to equal the area of the object's projection on a sphere of unit radius. The element of solid angle is thus

$$d\Omega = \sin\theta\, d\theta\, d\phi, \qquad (2\text{B}.6)$$

and the total solid angle around a point is 4π, in units called *steradians* (sr).

Many problems have cylindrical symmetry, in which rotation around some symmetry axis makes no physical difference. In such cases it is most convenient to take the z axis of Fig. 2B.1 as the symmetry axis and immediately integrate over ϕ. The element of solid angle is then the total solid angle between θ and $\theta + d\theta$, which is

$$d\Omega(\theta) = \sin\theta\, d\theta \int_0^{2\pi} d\phi = 2\pi \sin\theta\, d\theta. \qquad (2\text{B}.7)$$

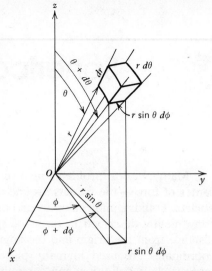

FIGURE 2B.2
Volume element in spherical coordinates. The volume of the element is the product of the three edge lengths,

$$dV = (dr)(r\, d\theta)(r \sin\theta\, d\phi)$$
$$= r^2 \sin\theta\, dr\, d\theta\, d\phi.$$

The element of solid angle subtended by dV is defined by the lines radiating from the origin, and has the value

$$d\Omega = \sin\theta\, d\theta\, d\phi.$$

Appendix 2C

Rutherford Scattering

The names *Rutherford scattering* and *Coulomb scattering* are used to describe the classical scattering of one particle by another when the force between the particles varies as the inverse square of the distance between them. Hence both electrostatic and gravitational interactions lead to Rutherford scattering, although we consider here only the electrostatic case. This appendix is a brief derivation of the expression for the flux of scattered particles as a function of the scattering angle. (For the terminology of scattering see Section 2.5, or the more extensive discussion in Chapter 27.)

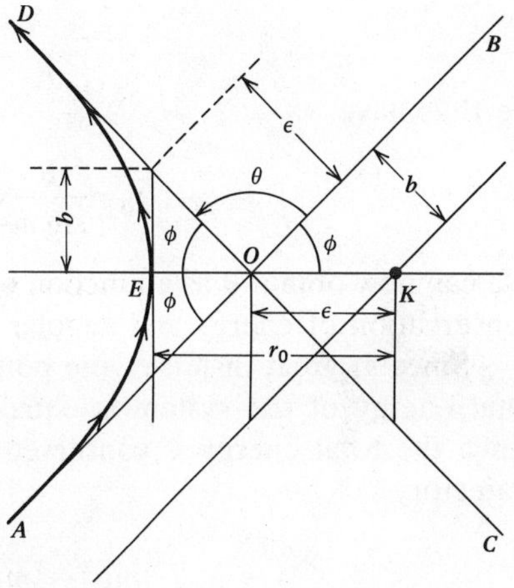

For simplicity, let us assume that one particle is sufficiently heavy, compared to its collision partner, to be considered at rest; this is the particle shown at the point K in Fig. 2C.1. We call this heavy particle the *scatterer*, and assume that it has a charge Ze. The other particle, the *projectile*, has a charge Q, an initial speed v, and a mass[26] m. The distance between projectile and scatterer at any given time is r; the force of electrostatic interaction between them is then

$$F(r) = \frac{ZeQ}{4\pi\epsilon_0 r^2}, \tag{2C.1}$$

and the corresponding potential energy is

$$V(r) = \frac{ZeQ}{4\pi\epsilon_0 r} \tag{2C.2}$$

by Eq. 2.55.

If there were no force of interaction, the projectile would pass K in a straight line; the distance of its closest approach to the point K along this hypothetical "unscattered" trajectory is b, called the *impact parameter*. The actual trajectory, however, is deflected through the scattering angle θ, as shown in Fig. 2C.1. For a given trajectory, θ is uniquely determined by the impact parameter and the initial speed of the projectile. One can fairly easily obtain a beam of projectiles with essentially the same speed, but any beam of nonzero width must have a range of values of b. We must therefore determine what fraction of the entire beam will be scattered into a given angle θ; this fraction will be expressed in terms of the differential scattering cross section $\sigma(\theta)$, defined in Eq. 2.16.

We need not go through the derivation of the actual trajectory (which can be found in most texts on classical mechanics). It is sufficient to know that the trajectory is a hyperbola with the scatterer at one focus. It is drawn in Fig. 2C.1 on the assumption that the scattering is repulsive, which is the case when both charges are of the same sign (as in Rutherford's measurements). As indicated above, θ is the scattering angle; we

FIGURE 2C.1
Geometry of Rutherford scattering. The heavy scatterer is at point K; the projectile's trajectory is the curve with arrows. This trajectory is initially along the line AOB, which it would follow if there were no interaction; after scattering, the trajectory becomes asymptotic to the line COD. The scattering angle θ is the angle between these two lines, and the impact parameter b is the distance from K to the line AOB. See the text for definitions of the other quantities shown in the figure. The trajectory is a hyperbola with center at O, focus at K, and vertex at E.

[26] Strictly, we should use the system's reduced mass μ and take the center of mass as point K; cf. Eq. 2.74 and the accompanying discussion. The angle θ that appears in our equations is then the scattering angle in relative (center-of-mass) coordinates, for which see Section 27.1.

also define the angle

$$\phi \equiv \tfrac{1}{2}(\pi - \theta). \tag{2C.3}$$

We need to know the actual distance of closest approach, called r_0, which is the distance from the hyperbola's vertex E to its focus K. From the properties of a hyperbola, this is

$$r_0 = \epsilon + \epsilon \cos \phi, \tag{2C.4}$$

where ϵ, the distance between focus and center, is given by

$$\epsilon = \frac{b}{\sin \phi}; \tag{2C.5}$$

we thus have

$$r_0 = \frac{b}{\sin \phi}(1 + \cos \phi). \tag{2C.6}$$

We can now obtain θ as a function of b and v, using only the principles of conservation of energy and angular momentum.

Since at great distances the potential energy becomes negligible, the total energy of the system is equal to the initial kinetic energy, $\tfrac{1}{2}mv^2$. Since the total energy is conserved, we must have at any point on the trajectory

$$\tfrac{1}{2}mv^2 = \tfrac{1}{2}mu^2 + \frac{ZeQ}{4\pi\epsilon_0 r}, \tag{2C.7}$$

where the instantaneous speed u is a function of r. In particular, at the point of closest approach ($r = r_0$), the velocity has an extremum value u_0 (maximum for attraction, minimum for repulsion); we can then write

$$\tfrac{1}{2}mv^2 = \tfrac{1}{2}mu_0^2 + \frac{ZeQ}{4\pi\epsilon_0 r_0}. \tag{2C.8}$$

This can be rearranged, with the help of Eq. 2C.6, to give

$$\frac{u_0^2}{v^2} = 1 - \frac{ZeQ}{2\pi\epsilon_0 mv^2 r_0} = 1 - \frac{ZeQ \sin \phi}{2\pi\epsilon_0 mv^2 b(1 + \cos \phi)}. \tag{2C.9}$$

Next we consider the angular momentum around point K, which at any given instant is the linear momentum multiplied by the "lever arm."[27] Conservation of angular momentum requires that its initial value equal the value at the point of closest approach:

$$mvb = mu_0 r_0. \tag{2C.10}$$

We therefore have

$$\frac{u_0}{v} = \frac{b}{r_0} = \frac{\sin \phi}{1 + \cos \phi}, \tag{2C.11}$$

and squaring this result gives

$$\frac{u_0^2}{v^2} = \frac{\sin^2 \phi}{(1 + \cos \phi)^2} = \frac{1 - \cos^2 \phi}{(1 + \cos \phi)^2} = \frac{1 - \cos \phi}{1 + \cos \phi}. \tag{2C.12}$$

[27] That is, by the shortest distance between point K and a line drawn tangent to the trajectory at that instant. More formally, the angular momentum is defined as

$$\mathbf{L} \equiv \mathbf{r} \times m\mathbf{u},$$

where $\mathbf{r}$ is the radius vector (here drawn from point K) and $\mathbf{u}$ is the instantaneous velocity.

Combining Eqs. 2C.9 and 2C.12 to eliminate u_0^2/v^2, and multiplying through by $(1+\cos\phi)$, we have

$$1-\cos\phi = 1+\cos\phi - \frac{ZeQ}{2\pi\epsilon_0 mv^2 b}\sin\phi, \qquad (2C.13)$$

which can be further rearranged to

$$\frac{4\pi\epsilon_0 mv^2 b}{ZeQ} = \frac{2\sin\phi}{2\cos\phi} = \tan\phi. \qquad (2C.14)$$

We want our result in terms of the scattering angle θ, of course. From Eq. 2C.3 we have

$$\tan\phi = \tan\left(\frac{\pi}{2} - \frac{\theta}{2}\right). \qquad (2C.15)$$

Using the standard trigonometric relation

$$\tan(x-y) = \frac{\sin x \cos y - \cos x \sin y}{\sin x \sin y + \cos x \cos y}, \qquad (2C.16)$$

with $x=\pi/2$, $y=\theta/2$, we obtain

$$\tan\phi = \frac{\cos(\theta/2)}{\sin(\theta/2)} = \cot\left(\frac{\theta}{2}\right). \qquad (2C.17)$$

Substituting this result in Eq. 2C.14, we have direct relationship between b and θ for a given trajectory,

$$b = \frac{ZeQ}{4\pi\epsilon_0 mv^2}\cot\frac{\theta}{2}. \qquad (2C.18)$$

We also need the differential of this equation,

$$db = \frac{ZeQ}{4\pi\epsilon_0 mv^2}d\left(\cot\frac{\theta}{2}\right) = -\frac{ZeQ}{8\pi\epsilon_0 mv^2}\frac{d\theta}{\sin^2(\theta/2)}. \qquad (2C.19)$$

To obtain the scattering cross section, we must know what fraction of particles will be scattered into a given range of scattering angles. Consider a plane normal to the incident beam, at a point where no appreciable scattering has yet occurred. If the flux density of incident particles across this plane is everywhere uniform, the flux[28] across a given area must be proportional to that area. If f is the incident flux density, and $dn(b)$ the flux of particles with impact parameters between b and $b+db$, then the ratio $dn(b)/f$ must equal the area of the ring shown in Fig. 2C.2. This differential area is simply the radius increment multiplied by the circumference, so that we have

$$\frac{dn(b)}{f} = 2\pi b\, db. \qquad (2C.20)$$

Since b and θ are directly related, the range of impact parameters from b to $b+db$ corresponds to a range of scattering angles from θ to $\theta+d\theta$ (cf. figure). Substituting Eqs. 2C.18 and 2C.19, we find

$$\frac{dn(\theta)}{f} = \frac{dn(b)}{f} = 2\pi\left(\frac{ZeQ}{4\pi\epsilon_0 mv^2}\right)\cot\frac{\theta}{2}\left(\frac{ZeQ}{8\pi\epsilon_0 mv^2}\right)\frac{d\theta}{\sin^2(\theta/2)}$$

$$= \pi\left(\frac{ZeQ}{4\pi\epsilon_0 mv^2}\right)^2\frac{\cos(\theta/2)}{\sin^3(\theta/2)}\,d\theta, \qquad (2C.21)$$

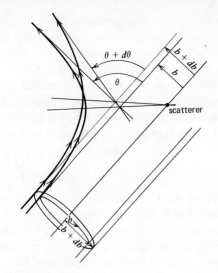

FIGURE 2C.2
Trajectories of particles with impact parameters between b and $b+db$ and scattering angles between θ and $d\theta$. Note that θ decreases as b increases, so that db and $d\theta$ must have opposite signs (i.e., $db/d\theta$ is negative). The ring shown at lower left is in a plane normal to the incident beam, and has an area $2\pi b\, db$.

---

[28] In our terminology, *flux* means the number of particles per unit time, whereas *flux density* is the flux per unit area; however, many authors use "flux" with the latter meaning.

where $dn(\theta)$ is the number of particles scattered per unit time into angles between θ and $\theta + d\theta$; we omit the minus sign of Eq. 2C.19, since we only want the absolute value of $dn(\theta)$.

We can now introduce the cross section, by analogy with Eq. 2.16. In that equation $n\, dx$ was the number of scatterers per unit area (of a plane normal to the incident beam) in thickness dx; here we are considering only one scatterer, so we multiply both sides by unit area and eliminate the differential in x. We can therefore write

$$\frac{1}{f}\frac{dn(\theta)}{d\Omega} = \sigma(\theta), \qquad (2C.22)$$

where $dn(\theta)/d\Omega$ is the number of particles scattered per unit time into unit solid angle in the vicinity of θ. The differential scattering cross section $\sigma(\theta)$ in this equation is the same as that in Eq. 2.16; we see that $\sigma(\theta)$ is the number of particles per scatterer scattered per unit time into unit solid angle near θ, per unit flux density of the incident beam. The $dn(\theta)$ of Eq. 2C.21 is the flux into the range of solid angles between θ and $\theta + d\theta$, which we know from Eq. 2B.7 to be $d\Omega(\theta) = 2\pi \sin\theta\, d\theta$. We can thus substitute into Eq. 2C.21 to obtain

$$\sigma(\theta) = \frac{dn(\theta)/f}{2\pi \sin\theta\, d\theta} = \pi\left(\frac{ZeQ}{4\pi\epsilon_0 mv^2}\right)^2 \frac{\cos(\theta/2)}{\sin^3(\theta/2)}\frac{1}{2\pi \sin\theta}$$

$$= \left(\frac{ZeQ}{8\pi\epsilon_0 mv^2}\right)^2 \frac{1}{\sin^4(\theta/2)}, \qquad (2C.23)$$

where we have used the fact that

$$\sin\theta = 2\sin\frac{\theta}{2}\cos\frac{\theta}{2}. \qquad (2C.24)$$

Equation 2C.23 is the complete Rutherford scattering formula that we summarized in Eq. 2.19. In combination with the experimental data of Fig. 2.13, it confirms the nuclear model of atomic structure. Our heavy scatterer is the concentrated mass of the atomic nucleus, with a charge of $+Ze$ (where Z is the atomic number). The projectiles in Geiger and Marsden's experiments were α particles, with charge $Q = +2e$; since all had virtually the same energy, the initial speed v could be taken as constant.

Further Reading

A. D'Abro, *The Rise of the New Physics*, Vol. 2 (Dover Publications, Inc., New York, 1951), chaps. 16–18.

O. Oldenberg and W. G. Holladay, *Introduction to Atomic and Nuclear Physics*, (McGraw-Hill Book Co., Inc., New York, 1967).

F. K. Richtmyer and E. H. Kennard, *Introduction to Modern Physics*, 6th Ed. (McGraw-Hill Book Co., Inc., New York, 1969).

Problems

1. Explain why, in the Franck–Hertz experiment, one never ordinarily observes spectral lines with frequencies greater than E_{thr}/h. Find a rationalization for

the fact that sometimes the maximum frequency of the emission lines associated with a given E_{thr} is less than E_{thr}/h.

2. When a Franck–Hertz experiment is conducted with atomic sodium vapor and the energy of the electrons is increased from zero, the first visible spectral emission occurs when the energy of the electrons is 2.103 eV. What is the minimum wavelength at which one might observe atomic spectral lines in this experiment? Sodium vapor lamps in fact emit radiation in lines whose wavelengths are 589.18 and 588.99 nm. The first spectral lines of atomic lithium to appear in such an experiment have wavelengths of 670.791 and 670.761 nm. What is the minimum voltage of the electrons at which these lines will appear?

3. The lowest electron voltage at which neon exhibits visible emission in a Franck–Hertz experiment is 18.72 V. The emission occurs at 585.24 nm. On this basis, predict the wavelengths of *two* other spectral lines that one might observe for neon.

4. The Franck–Hertz experiment is based on the collision of electrons with atoms and molecules. In some ranges of collision energies, the quantities that best characterize the probability of excitation are the duration of the interaction or, alternatively, the velocity of the electron as it passes a target atom. Presumably the transient electric field of the passing electron is the source of the coupling that passes energy from this electron to others in the target. If so, then passing protons having the same *velocity* as the passing electrons should cause similar excitation, and they often do. Compute the voltage through which a proton must be accelerated to have the same velocity as a 300-V electron. Give a general formula for the voltage V' required to give the proton the same velocity as that of a proton accelerated through a voltage V.

5. Compute the energy emitted by one atom emitting a spectral line with frequency $10^{15}\,s^{-1}$; what is the energy of the light emitted by 1 *mole* of such atoms? Compare this with the kinetic energy of an average person walking (60 kg, 6 km/h), and with the energy of a 60-W light bulb burning 1 h. (1 W is 1 J/s.)

6. If σ of Eq. 2.8 is approximately the same as the cross section one derives from the packing of spherical atoms in a crystal, and the number density n is that of the same spheres in contact, estimate the distance required to attenuate the intensity I of an x-ray beam by 10%. Do this evaluation both by supposing $dI \approx \Delta I$, $dx \approx \Delta x$, and by carrying out the integration and using the integrated form of Eq. 2.9.

7. In an experiment of the type of Geiger and Marsden's, with α particles scattering from gold foil 2 μm thick, what fraction of the α particles would be expected to be found at angles greater than $\pi/2$, if θ_m is 1° and if Eq. 2.18 described the scattering?

8. The correction for the fact that the electron and nucleus revolve around their center of mass is made very simply; one carries through all the equations 2.47–2.65 with the mass m taken as the reduced mass. The *reduced mass* is defined, for two particles of mass m_1 and m_2, by Eq. 2.74.
Calculate the wavelengths of the spectral lines corresponding to the $n = 3 \rightarrow n = 2$ and $n = 4 \rightarrow n = 2$ transitions in the atoms of hydrogen (1.007825 amu) and of deuterium (2.01410 amu), which both have nuclear charges of $+e$. Note that these masses include the electron mass. (See Table 1.1.) Show that a comparison of the positions of atomic spectral lines can give a value for the electron mass.

9. Mercury atoms may be excited by absorbing light of wavelength 2537 Å. When excited Hg atoms are allowed to collide with N_2 molecules, light is emitted with a wavelength of 4047 Å, and no other wavelengths appear in the spectrum. Assume that the kinetic energies of excited Hg and of N_2 may be

neglected prior to a collision. What is the total kinetic energy of the Hg and N_2 after a collision in which 4047 Å light is emitted?

10. Assume that the one electron ion C^{5+} behaves like a hydrogen atom with a nuclear charge of +6. Compute the ionization energy and the wavelength of the first lines in the Lyman and Balmer series for this atom. In what spectral regions do they lie? What is the quantum number n of the lowest state of a transition in which visible light is emitted in a process $n_2 \rightarrow n_1$?

11. The two outer electrons of the magnesium atom are in the same shell. The first ionization potential of Mg is 7.644 eV and the second is 15.031 eV. Using these two values, estimate an approximate average distance of the two outermost electrons from the nucleus in the ground state of this atom.

12. Transitions to the ground state from several excited states of the sodium atom produce spectral lines that form a regular series with wavelengths 589.18, 330.26, 285.28, 268.04, and 259.39 nm. If one replaces the integral quantum number n with $n - \delta$, where δ lies between 0 and 1, this series can be described by a formula analogous to the Balmer formula. Evaluate δ for the lines given; how much variation in δ is necessary to fit these spectral lines within the accuracy quoted? (Note that excited atoms become more hydrogen-like as n increases; the Bohr model may not fit the long-wavelength lines.)

13. Suppose that a detector such as E in Fig. 2.10 subtends a solid angle of 10^{-3} steradians and is used in a scattering experiment in which a beam of fast light particles scatters from a stationary collection of heavy gaseous molecules. The system has cylindrical symmetry as in Fig. 2.10. Suppose it is believed that the scattering cross section σ for this experiment behaves as $A \cos(\theta/2)$, and at $\theta = 5°$ or 0.87 radians, the differential cross section is measured to be 2.584×10^{-20} cm^2/10^{-3} sr or 2.584×10^{-17} cm^2/sr. Compute A and the total cross section from this measured value.

14. Estimate the action per cycle of
    (a) A 33-rpm phonograph record;
    (b) A piston in an automobile engine oscillating (approximately) harmonically at 1000 rpm;
    (c) A 1-g ball bouncing perfectly elastically under the action of gravity, reaching a maximum height of 1 m on every rebound.
    What is the action associated with a free hydrogen atom moving through free space for 1 s at 10^6 cm/s?

15. A laser emits 1 W of infrared radiation whose wavelength is 10.7 μm $(10.7 \times 10^{-6}$ m). How many quanta per second are being emitted in this beam? How long would it take to raise the temperature of 200 g of water from room temperature to its normal boiling point if all these quanta were absorbed?

16. Suppose that in another universe, $h = 10$ J s, $m_e = 1$ g, and $e = 0.1$ C. What would be the approximate diameter of a hydrogen atom in this universe? (Assume that all other masses scale as does the electron mass.)

17. Positronium is a short-lived species composed of an electron and its positive counterpart, the positron, whose mass is equal to that of the electron, and whose charge is equal in magnitude but opposite in sign. Use the Bohr model to compute the frequency of the first Lyman line for positronium. Compute the ionization energy of positronium also.

18. Ionization potentials of several atoms and ions are

    Li: 5.363 eV   Na: 5.12 eV
    Be^+: 18.12 eV Mg^+: 14.96 eV
    B^{2+}: 37.75 eV Al^{2+}: 28.31 eV
    C^{3+}: 64.22 eV Si^{3+}: 44.93 eV
    N^{4+}: 97.4 eV P^{4+}: 64.70 eV

Plot the square roots of these energies against the corresponding nuclear charges, and explain the relationship as well as you can in terms of the Bohr model.

19. What must be the force constant of a harmonic oscillator if its mass is 10 amu and its frequency is $10^{14}\,s^{-1}$? What is the natural frequency of an oscillator whose mass is $10^{-6}\,kg\,(1\,mg)$ and whose force constant is $0.1\,N/m\,(100\,dyn/cm)$?

20. A spherically symmetric harmonic oscillator is defined by the equivalent conditions

$$F_x = -kx, \qquad F_y = -ky, \qquad F_z = -kz$$

and/or

$$V(x, y, z) = \frac{k}{2}(x^2 + y^2 + z^2).$$

Calculate the action for this oscillator. Assuming that the action is quantized, what are the possible energy levels of the oscillator?

21. Use the Bohr–Sommerfeld–Wilson quantization rule, that action occurs only in integral multiples of h, to show that a particle moving along a line of length L (repulsive barriers at the ends of this segment) has only a discrete set of energy levels. Show that the possible energies are proportional to n^2 where n is an integer.

22. Use the Bohr model to find the spectrum of a single electron bound to a nucleus of neon. Calculate the wavelengths of the lowest five transitions connecting the ground state with excited states of this species.

23. When the velocity of a particle approaches the speed of light, classical mechanics no longer describes the particle accurately and one must use relativistic mechanics. As the nuclear charge Z increases, the velocity of an electron in a Bohr atom increases. At what value of Z is the velocity of an electron in a Bohr atom 1% of the speed of light? At what Z is the first excitation energy equal to 1% of the rest mass, $m_e c^2$, of the electron?

24. Suppose that one were to use the Bohr model to describe the motion of the earth around the sun. What would be the principal quantum number of the earth if the sun's mass is assumed infinite?

3

Matter Waves
in
Simple Systems

We have now examined most of the evidence that led to the modern quantum theory of matter. In this chapter we begin to develop that theory itself, the theory known as *quantum mechanics* or (in the form presented here) *wave mechanics*.

The fundamental hypothesis of this theory is that matter, as well as light, has both wavelike and particlelike properties. The whole of classical mechanics dealt with matter as made of particles, whereas nineteenth-century physics generally assumed light to be simply a wave phenomenon. In the last chapter we introduced the evidence that light has particlelike (quantized) properties; now we must describe the wavelike properties of matter. We begin by developing the nature and properties of waves in general, then formulate the equation that describes "matter waves"—the famous Schrödinger equation. As we shall see, this equation cannot be derived, but must be postulated; it is justified only by agreement with experiment.

In principle (but not in practice), it should be possible to derive all the properties of matter from the Schrödinger equation. Even for the simplest atoms, however, the calculations involved are formidable. In the present chapter we consider only some idealized systems simple enough to treat in detail: the free particle in space, and particles constrained by "boxes" of various shapes. These solutions introduce many of the principles of quantum mechanics and illustrate the kinds of calculations required to describe matter waves. In particular, we show how the quantization of energy is related directly to the boundary conditions on a system.

3.1
The de Broglie Hypothesis

We have already seen some strong reasons for assuming that light has particlelike as well as wavelike properties—in particular, the photoelectric effect and the concept of the quantum of radiation. In 1923, in an elegant and daring extrapolation, Louis de Broglie proposed that matter should exhibit a parallel to this dual character of light. Just as light can show both particlelike and wavelike properties, matter, he suggested, might well exhibit wavelike properties. de Broglie proposed this hypothesis despite the fact that only the particlelike properties of matter were actually known at the time. Indeed, the theory of quantum mechanics was well developed before experimental proof of "matter waves" was obtained.

This proof was found in 1927 by Davisson and Germer, who showed that electrons impinging on a crystal exhibit a property that had always been associated with waves: *diffraction*. Diffraction appears as a pattern of light and dark areas or areas of high and low intensity, often regular, produced when light waves or other waves encounter some obstacle as they travel toward a screen or detector. (An easy way to observe diffraction is this: With one eye closed, look at a lamp some distance

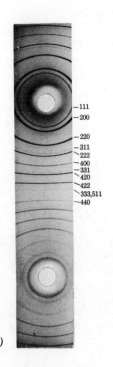

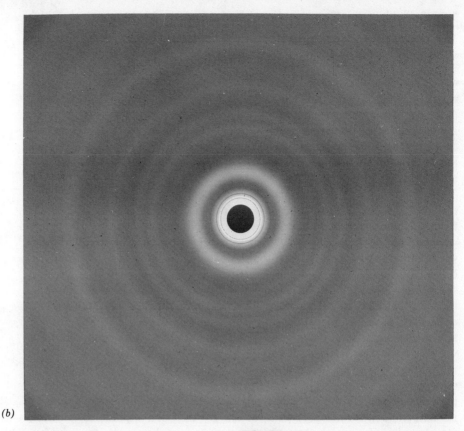

(a)   (b)

FIGURE 3.1
Diffraction patterns for (a) x rays, (b) electrons. The x ray photograph is a print of a strip of film from a cylindrical film holder. The x ray beam enters through the hole at the bottom and leaves through the hole at the top. The sample is sodium chloride. The electron diffraction photograph of chlorobenzene is a print from a plate large enough to show the full circles of the diffraction pattern; most of these circles are cut off in the x ray picture. (x ray picture courtesy of Prof. S. Bailey; electron diffraction picture courtesy of Prof. Denis Kohl.)

away. Then, holding your hands just beyond your nose, bring the tips of your forefingers *almost* together in front of your open eye. When the blurred edges of your out-of-focus fingertips seem to merge, you will see rather sharply defined lines of light and dark, roughly parallel to the silhouetted edges of the fingertips. This light and dark pattern is a diffraction pattern.) As can be seen in Fig. 3.1, electrons are diffracted by crystals in much the same way as are x rays; the dimensions of the patterns are related to crystal structure, as we shall explain in Chapter 11. Similar effects can be obtained with much heavier particles, and neutron diffraction is particularly useful in structural analysis.

But let us return to the actual formulation of de Broglie's hypothesis. If there are waves associated with particles of matter, what are their wavelengths? De Broglie drew an analogy to the particles of radiation (photons). By Einstein's relation $E = mc^2$, the mass associated with a photon of frequency ν must be $E/c^2 = h\nu/c^2$; the momentum of the photon is thus mass $\times$ velocity, or $h\nu/c = h\lambda$. What de Broglie proposed was that the same equation applied to material particles. The *de Broglie wavelength* of a particle is thus simply Planck's constant divided by the particle's momentum p:

$$\lambda = \frac{h}{p}. \qquad (3.1)$$

Note that it is the wavelength, not the frequency of the wave, that we relate explicitly to the momentum; we must return to this after discussing waves in general, for it is a very important point, which forms part of the basis of our physical picture of atoms.

What sort of wavelengths do material particles have? In gaseous hydrogen at room temperature an average molecule moves at about

1.77 km/s; since the H_2 molecule weighs 3.35×10^{-27} kg, its wavelength is

$$\lambda = \frac{h}{mv} = \frac{6.63 \times 10^{-34} \text{ J s}}{(3.35 \times 10^{-27} \text{ kg})(1.77 \times 10^3 \text{ m/s})} \times 1 \frac{\text{kg m/s}^2}{\text{J}} = 1.12 \times 10^{-10} \text{ m},$$

or about 1.1 Å. High-speed electrons also have wavelengths in the x-ray range, accounting for the similarity of x-ray and electron diffraction patterns. The slower or lighter a particle, the longer is its wavelength; the heavier or faster the particle, the shorter is its wavelength. Macroscopic objects thus have extremely small wavelengths; for example, a thrown baseball would have $\lambda \approx 2 \times 10^{-26}$ m, compared with which even an atomic nucleus is huge.

By analogy with the familiar properties of light, we should expect to observe wavelike phenomena, such as diffraction, when matter waves are forced to interact with something having dimensions comparable to their wavelength. A beam of light will be strongly diffracted by a slit if the width of the slit is roughly equal to the wavelength of the light. Naturally, diffraction does occur even with much wider slits, or even with a single sharp edge. However, if a slit is significantly wider than the wavelength, then the light streams through in effectively straight rays with negligible deflection. In the same way, as long as all the dimensions of an apparatus are much greater than the wavelength of the particles under examination, we can treat the particles as though they obeyed the laws of classical mechanics. The estimates above tell us that material "particles" of atomic size will show wavelike properties only when they interact with other particles of roughly the same size, whereas macroscopic objects will have no detectible wavelike properties at all. By contrast, visible light, with wavelengths of several thousand angstroms, can be made to exhibit wavelike properties with such simple pieces of apparatus as a slit made from a pair of razor blades (or even two fingertips!).

Before we can explore how de Broglie's hypothesis of matter waves leads us to the structure of the atom, we must turn aside to study the properties of waves in general—what they are, how they are represented, and how they interact.

3.2
The Nature of Waves

When most people think of waves, they think first of water waves. Unfortunately, despite our familiarity with them, water waves are in some ways rather complicated. Nevertheless, one of their characteristics is common to most types of waves: The wave is a moving disturbance but not a gross motion of the medium. A small bit of water at the surface of a pond does not travel across the pond as a wave passes by; rather, it moves up, down, and a bit backward and forward. It is easy to recognize this when one thinks of the motion of a small stick floating on the surface of the pond. The stick bobs with the waves, but it does not travel any distance.

Now let us speak in more general terms. In the broadest definition, a wave is a disturbance that varies regularly with time. We can immediately think of a classification into two types of waves, *traveling waves* and *standing waves*. The ripples that spread on a pond when a stone is thrown in are traveling waves: The waves themselves propagate outward in concentric circles, regardless of the motion of the floating stick. A simple standing wave almost as familiar as water waves is the wave in a moving jump rope held at both ends and swung in a loop, as shown in Fig. 3.2a. As the rope moves around, any given point on the rope describes a circle about the axis of rotation, always maintaining a fixed distance from the axis. Ordinarily a jump rope has only its end points on the axis of rotation. However, most people have noticed that it is possible to set the rope in motion so that one, two, or more points on the rope are fixed on the axis. In other words, one can have the rope rotating in two parts, as in Fig. 3.2*b*, or even in three or more parts. The point or points that remain fixed on the axis, that is, that have no displacement from the axis, are called *nodes*, and the points of maximum displacement from the axis, *antinodes*. In a standing wave the nodes remain fixed in space, and the antinodes at most oscillate (in this case circularly) around fixed positions.

The standing-wave motion just described happens to be circular motion: As a given point on the rope moves about the axis, its behavior is completely described by a statement giving its angle about the axis as a function of time. However, motion need not be circular to be standing-wave motion. For example, an elastic cord with its ends fixed could be stretched into one of the shapes of Fig. 3.2; on being released, the cord would oscillate up and down, but only within the plane of the figure. The characteristic that makes this a standing-wave motion is that the nodes always remain nodes, the antinodes always remain antinodes, and all the sections between them always retain their own positions of displacement relative to neighboring sections. The vibrations of the strings in a violin or other stringed instrument are of this type.

One can also easily set up a traveling wave in a rope. If one end is free and the other is given a shake or twirl, a wave will move down the length of the rope. If one shakes the end repeatedly in a regular way, a steady flow of waves down the rope can be established. A traveling wave also has its nodes and antinodes, but they move steadily in some direction rather than remaining fixed.

Let us now consider one other variety of waves in material media, best exemplified by sound waves. These arise from the motion of the atoms or molecules of the medium, and give rise to the sensations we call sounds when they strike our ears. In a sound wave the molecules are displaced primarily back and forth along the direction in which the wave moves forward (propagates). Such a wave is called a *longitudinal* wave. The waves in a jump rope are examples of *transverse* waves, in which the individual particles are displaced in a direction perpendicular to the axis of propagation. Water waves are partially transverse and partially longitudinal.

Thus far all our examples have been waves in a medium: a rope, water, or air. But there are other waves that we can describe without reference to a medium. The example that comes to mind first is, of course, a light wave. What we speak of as a wave of light or other electromagnetic radiation is in fact a field of force that propagates as a wave. One can detect such an electromagnetic wave by observing the oscillations of suitable test charges in the space where the wave is supposed to be. For example, we use the electrons in a metal aerial to probe for the electromagnetic waves that we call radio waves. Both radio and light waves propagate perfectly well through the best vacuum known, namely, outer space, where there is no question of a material medium. One of the great scientific challenges at the turn of the century was to discover if there was in fact an all-permeating medium, the so-called luminiferous ether, which was supposed to act as the carrier of light waves. However, experiments[1] showed that there was no detectable medium associated with electromagnetic waves, so the idea of the luminiferous ether has gone the way of phlogiston and the *élan vital*.

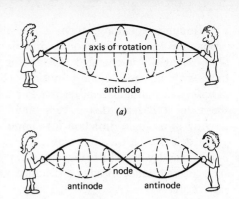

FIGURE 3.2
Standing waves in a jump rope (*a*) with no nodes except at the ends; (*b*) with one node in the middle.

---

[1] The key experiment was that of Michelson and Morley (1887), showing that the earth's motion relative to the hypothetical ether did not affect the observed velocity of light. This result was an important foundation for Einstein's theory of relativity.

Roughly speaking, then, a traveling wave is a disturbance that propagates in some direction or directions, leaving the medium (if there is one) essentially unchanged. A standing wave is one that repeats itself in time with no net forward motion. How do we describe these more precisely in mathematical terms? For a traveling wave that maintains its shape as it moves, this is very simple indeed. Let us choose a wave moving in a single direction for simplicity. Any function of the variable $x - vt$ describes a traveling wave, where x is the distance, t the time, and v the speed with which the wave moves forward. Figure 3.3 illustrates such a disturbance moving a distance $x_2 - x_1$ in the time interval $t_2 - t_1 = (x_2 - x_1)/v$.

We have expressed the wave in terms of the variable $x - vt$, which has the dimension of length. However, it is sometimes more convenient to multiply this by a constant with the dimension of $(length)^{-1}$ to obtain a dimensionless variable with two characteristic scale factors, one for distance and one for time. The representation commonly used has the form

$$\frac{1}{\lambda}(x - vt) = \frac{x}{\lambda} - \nu t \qquad \left(\nu \equiv \frac{v}{\lambda}\right), \tag{3.2}$$

in which the scale of length is defined by the *wavelength* λ and the scale of time by the *frequency* ν. If λ is expressed in meters and v in meters per second, then ν has the units s^{-1} (hertz). The reciprocal of the wavelength is the *wave number* $\tilde{\nu} \equiv 1/\lambda$, usually expressed in cm^{-1}. It is often convenient (cf. our jump-rope example) to describe a wave in terms of some equivalent circular motion. For this purpose we can represent the wave as a function of the variable

$$k(x - vt) = kx - \omega t \begin{cases} \left(k \equiv \dfrac{2\pi}{\lambda} = 2\pi\tilde{\nu}\right), \\[2mm] (\omega \equiv 2\pi\nu = kv) \end{cases} \tag{3.3}$$

in which k is the *circular wave number* and ω is the *angular frequency*, with SI units of radians per meter and radians per second, respectively.

The words "wavelength" and "frequency" naturally suggest a *periodic* wave, one that repeats itself in space or time. Until now, however, we have made no assumption of periodicity for traveling waves. We have considered only a disturbance that maintains its shape as it moves forward, and that may or may not be periodic. If it is *not* periodic, of course, the quantities λ, ν, and so on, are simply arbitrary scale factors with no particular physical meaning (except that one must always have $v = \lambda\nu$). Nevertheless, any such disturbance can be described mathematically in terms of a single variable, which can be $x - vt$ or the forms of Eqs. 3.2 and 3.3. In the circular representation the single variable is most conveniently expressed as

$$\varphi \equiv kx - \omega t, \tag{3.4}$$

called the *phase* of the wave. Any disturbance described by a function $f(x, t)$ that is also a unique function of some phase,

$$f(x, t) = g(\varphi), \tag{3.5}$$

is a wave, no matter what the form of $f(x, t)$.

What about a wave that *is* periodic—as are all those with which we are concerned? All the symbols in Eqs. 3.2 and 3.3 are then physically meaningful. The wavelength is the distance over which the wave repeats itself, and the frequency is the number of such replicas that pass a given point per unit time. The time required for a single wavelength to pass—that is, the time between two successive wave crests—is the inverse

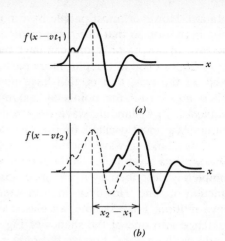

FIGURE 3.3
A wave whose amplitude is described by the function $f(x - vt)$, shown at two times: (*a*) at time t_1; (*b*) at time t_2, when each point of the wave has moved a distance $x_2 - x_1 = v(t_2 - t_1)$, as illustrated for the highest peak.

of the frequency and is called the *period*, $\tau \equiv 1/\nu$. In the circular representation, the period of course corresponds to a single trip around the circle, at the angular velocity ω; the phase φ is an angle and thus varies through 2π in the course of a period.

All this can be recognized as the same language that we used to describe the harmonic oscillator in Section 2.8. The oscillator's motion is given by a sine function, Eq. 2.40. One can show that *any* periodic motion—including waves—can be analyzed into a sum of such sine (or cosine) functions; this is why the harmonic oscillator is such a useful model. A wave of the simple form

$$f(x, t) = A \sin(kx - \omega t + \delta),\qquad(3.6)$$

where A and δ are constants, is thus called a *harmonic* wave (or sine wave). Electromagnetic waves are of this form, with the electric and magnetic fields in phase but at right angles to each other, as shown in Fig. 3.4.

Standing waves can also be periodic. In our example of a rope fixed at the ends, the waves at any time have the shape of a sine curve. The wavelength can thus be defined as twice the distance between nodes. Since there are nodes at the ends, the rope must contain an integral number of half-wavelengths; this gives a hint of how quantization will enter our theory. We shall see in Section 3.5 that a standing wave is simply the resultant of two identical traveling waves going in opposite directions.

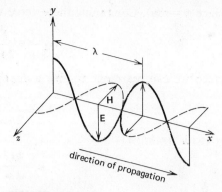

FIGURE 3.4
An electromagnetic wave. The wave shown is plane-polarized, with the electric-field vector **E** restricted to the *xy* plane and the magnetic-field vector **H** restricted to the *xz* plane. (In general, both vectors will rotate about the *x* axis, but remain at right angles and in phase.) Both **E** and **H** are harmonic (sine) waves of wavelength λ.

3.3 Dispersion Relations and Wave Equations: The Free Particle

The functions $f(x, t)$ that satisfy Eq. 3.5 are characterized primarily by their values of the physical parameters k and ω—that is, by their wavelengths and frequencies. The wavelength and frequency are not independent, of course, but are connected by the relation

$$\nu\lambda = \frac{\omega}{k} = v,\qquad(3.7)$$

where v is the speed at which the wave propagates. For light waves in a vacuum this speed is the universal constant $c = 2.99792458 \times 10^8$ m/s, and the relation between λ and ν is straightforward. But what is the velocity of propagation of a matter wave?[2] Fortunately, we do not need to know this. Whenever a given phenomenon can be described in terms of waves, one can always obtain a relation between the wavelengths and the frequencies that does not explicitly involve the wave velocity. Such a relation is called a *dispersion relation*.

What, then, is the dispersion relation for matter waves? Let us first examine a free particle in motion. According to de Broglie's hypothesis, the wavelength of the matter waves must be $\lambda = h/p$, where p is the momentum of the particle. Suppose we also assume that the frequency of these matter waves is related to the energy of the particle by the same law that applies to a photon, the Planck-Einstein relation $E = h\nu$. For a free particle of mass m and velocity v, the energy E is simply the kinetic energy,

$$E = \frac{mv^2}{2} = \frac{p^2}{2m}\qquad(3.8)$$

[2] Note that the velocity of the waves that constitute a piece of matter (*phase velocity*) is in general not the same as the velocity at which the matter as a whole moves (*group velocity*); v stands for phase velocity in Eq. 3.7, and for group velocity in $\lambda = h/mv$.

(since $p = mv$). Combining these three conditions, we obtain

$$h\nu = \frac{p^2}{2m} = \frac{1}{2m}\left(\frac{h}{\lambda}\right)^2, \qquad (3.9)$$

which we can rearrange to give a dispersion relation in the form

$$\nu = \frac{h}{2m\lambda^2} \qquad (3.10)$$

or (with $\hbar \equiv h/2\pi$)

$$\omega = \frac{\hbar k^2}{2m}. \qquad (3.11)$$

In the more general case, if a particle is subject to external forces that can be expressed in terms of a potential energy $V(x, y, z)$—for example, an electron in the field of a nucleus—then we can generalize Eq. 3.8 to

$$E = \frac{p^2}{2m} + V(x, y, z), \qquad (3.12)$$

from which we obtain the slightly more complicated dispersion relation

$$\omega = \frac{\hbar k^2}{2m} + \frac{V(x, y, z)}{\hbar}. \qquad (3.13)$$

Do not get the idea that we have "derived" these equations. Let us review just what we have done. Experiment shows that matter has wavelike properties; we interpret this as meaning that *something* associated with matter can be described as a wave. We do not know what that "something" is, and we are only guessing at the properties of the wave. The guesses are plausible—that matter waves obey the equations $\lambda = h/p$ and $E = h\nu$ that apply to light waves—but still unproven. In fact, they are *postulates*, which can be justified only by the conclusions drawn from them. This is a point that we shall emphasize again and again.

Of what use is a formal dispersion relation when we want to obtain physical information about a system? That is, how do we make use of a relation like Eq. 3.13 to find out how an electron behaves in an atom? We must discover the answer to this question in stages. To begin with, since we have decided to describe matter as a wave phenomenon, we must be interested in knowing the amplitude of a matter wave at an arbitrary point in space and time. Therefore we must find some way to obtain an expression for this wave amplitude, or *wave function*, which is usually designated by the Greek letter Ψ (psi). The amplitude must be a function of position and time; we need an equation whose solution is this function. Again, we do not derive this wave equation; we postulate it.

Why do we not work directly with the algebraic equations, Eq. 3.11 or Eq. 3.13, rather than use the more complex differential equation? The answer is clearly that these equations make no reference to the amplitude of the matter wave, the quantity we are trying to find. Nevertheless, we can use them to determine the form of a differential equation[3] whose solution will give the wave amplitude we seek. Specifically, given any dispersion relation, one can find an equation involving derivatives of the wave amplitude Ψ from which the dispersion relation can be derived. In general, there may be more than one wave equation consistent with the

---

[3] One can also formulate quantum mechanics in terms of integral equations, but the differential equation approach is simpler and was developed first. Ever since Newton, most physical laws had been expressed as differential equations and thought of in terms of changes in variables, and a great deal was known about how to solve such equations.

dispersion relation we want to satisfy; if this happens, we try to pick the simplest one.

Let us see what kind of wave equation we can obtain for the simple case of a free particle in motion. For simplicity, we restrict the particle to motion in only one dimension, along the x axis. We then need an equation in terms of a wave amplitude $\Psi(x, t)$, describing a wave that obeys the dispersion relation of Eq. 3.11. We must try to find the simplest equation that relates a space derivative of Ψ, hopefully only the first or second derivative, to a corresponding time derivative of Ψ. The equation should have a solution that is physically meaningful and consistent with Eq. 3.11.

This assertion, that the wave equation should involve only simple derivatives, may seem arbitrary. Remember, however, the fundamental principle that the validity of our theories rests only upon our obtaining results in agreement with experiment. Although we cannot derive this restriction, we can show why it is plausible. One obvious reason why it is desirable to restrict ourselves to first and second derivatives of Ψ is that it makes the problem reasonably simple mathematically. More significantly, from classical mechanics we are used to the idea that one needs to specify only two conditions, for example, the initial position and velocity, to determine the entire future history of a particle. This is because Newton's second law, $F = ma$, is an equation involving a *second* derivative, so that two integration constants must be supplied to fix the solution. We would like our quantum mechanics to be economical and as nearly parallel to classical mechanics as possible, because we know that classical mechanics does apply to a vast part of our experience. Invoking third or higher derivatives would be unpleasant on both counts. We shall quickly see that it is possible to restrict ourselves to first and second derivatives.

How, then, must the derivatives enter into a differential equation? Let us first consider the forms of the derivatives themselves. Let us assume that the wave amplitude can be written as an explicit function of the phase,

$$\Psi(x, t) = \Psi(\varphi) = \Psi(kx - \omega t); \tag{3.14}$$

as mentioned before, such an equation describes any traveling wave that maintains its shape as it propagates. (Note that the sign of k determines the direction in which the wave moves.) We can thus apply the simple "chain rule" of differentiation to take partial derivatives[4] of Ψ with respect to x or t. If we want to differentiate Ψ with respect to x at constant t, we take the derivative with respect to the phase φ and multiply this by the partial derivative of φ with respect to x, that is,

$$\left(\frac{\partial \Psi}{\partial x}\right)_t = \frac{d\Psi}{d\varphi}\left(\frac{\partial \varphi}{\partial x}\right)_t. \tag{3.15}$$

Similarly, the derivative with respect to t at constant x can be written as

$$\left(\frac{\partial \Psi}{\partial t}\right)_x = \frac{d\Psi}{d\varphi}\left(\frac{\partial \varphi}{\partial t}\right)_x. \tag{3.16}$$

But the partial derivatives of φ are very simple; differentiation of Eq. 3.4 gives

$$\left(\frac{\partial \varphi}{\partial x}\right)_t = k \quad \text{and} \quad \left(\frac{\partial \varphi}{\partial t}\right)_x = -\omega. \tag{3.17}$$

---

[4] If you are not familiar with partial derivatives, see Appendix II at the back of the book.

Equations 3.15 and 3.16 therefore become

$$\left(\frac{\partial \Psi}{\partial x}\right)_t = k\frac{d\Psi}{d\varphi} \quad \text{and} \quad \left(\frac{\partial \Psi}{\partial t}\right)_x = -\omega\frac{d\Psi}{d\varphi}. \qquad (3.18)$$

Notice that, since Ψ is a function only of φ, we can use ordinary derivative notation (d rather than ∂) for the derivative of Ψ with respect to φ.

How do we combine these derivatives in a wave equation? We know from the dispersion relation, Eq. 3.11, that ω must be proportional to k^2. Our differential equation must therefore involve the first time derivative of Ψ, to introduce a factor of ω, and the second space derivative of Ψ, to introduce a factor of k^2. The simplest relationship that will give the right proportionality is

$$\left(\frac{\partial \Psi}{\partial t}\right)_x \propto \left(\frac{\partial^2 \Psi}{\partial x^2}\right)_t, \qquad (3.19)$$

which by Eq. 3.18 is equivalent to

$$-\omega\frac{d\Psi}{d\varphi} \propto k^2\frac{d^2\Psi}{d\varphi^2}. \qquad (3.20)$$

This is not quite a wave equation, but is very close to it.

We still need to know how Ψ depends on φ. Since no other variable appears in Eq. 3.11, $d\Psi/d\varphi$ and $d^2\Psi/d\varphi^2$ must cancel out (except perhaps for a constant factor). What kind of function $\Psi(\varphi)$ has essentially the same form when it is differentiated once as when it is differentiated twice? The only such function is the exponential: The first and second derivatives of e^φ with respect to φ are both just e^φ itself. Could Ψ then be just e^φ or $e^{-\varphi}$? Formally this is possible, but physically it would be nonsense. The reason is that the function e^φ becomes infinite as x goes to $+\infty$, and $e^{-\varphi}$ as x goes to $-\infty$, regardless of the value of t. This would mean that the wave amplitude Ψ would be larger at infinity than at any point in accessible space. We anticipate by saying that the physical significance of Ψ must somehow be related to the chance of finding the particle at a particular point in space. If Ψ were larger at infinity than at any accessible x, then the particle would be found more probably at infinity than at any other place; that is, one could not find it at all. To avoid this problem, we require for Ψ a functional form that remains *bounded*—a functional form that prohibits $|\Psi|$ from exceeding some maximum value.

Fortunately, there does exist a simple function that is bounded yet still exponential. Remember that the complex exponential $e^{i\varphi}$ is equivalent to $\cos\varphi + i\sin\varphi$, in which both terms are clearly bounded. The function $\Psi(\varphi)$ can thus have the form $e^{i\varphi}$ or $e^{-i\varphi}$ and satisfy our mathematical conditions. But what is the physical meaning of a wave with a complex amplitude? On this point let us wait and see what kind of results we can obtain. We therefore write

$$\Psi \propto e^{\pm i\varphi}. \qquad (3.21)$$

In the equations that follow, the upper sign (in $\pm$ or $\mp$) corresponds to the choice of $e^{+i\varphi}$ in Eq. 3.21. Given this form for $\Psi(\varphi)$, we obtain from Eqs. 3.18 the partial derivatives

$$\left(\frac{\partial \Psi}{\partial t}\right)_x = -\omega(\pm i\Psi) = \mp i\omega\Psi \qquad (3.22)$$

and

$$\left(\frac{\partial^2 \Psi}{\partial x^2}\right)_t = k^2\frac{d^2\Psi}{d\varphi^2} = -k^2\Psi. \qquad (3.23)$$

Substituting for ω in terms of k from Eq. 3.11, we obtain the wave equation[5]

$$\left(\frac{\partial \Psi}{\partial t}\right)_x = \mp i \Psi \frac{\hbar k^2}{2m} = \pm \frac{i\hbar}{2m}\left(\frac{\partial^2 \Psi}{\partial x^2}\right)_t. \tag{3.24}$$

Changing the sign of $(\partial \Psi / \partial t)_x$ corresponds to reversing the direction of propagation of the wave. Since we can describe this by simply allowing k to have both positive and negative values, the lower signs in the above equations are redundant and can be omitted. Thus our final wave equation for a free particle moving in one dimension is

$$\left(\frac{\partial \Psi(x,t)}{\partial t}\right)_x = \frac{i\hbar}{2m}\left(\frac{\partial^2 \Psi(x,t)}{\partial x^2}\right)_t. \tag{3.25}$$

Remember that we still have no proof that this wave equation is correct. We have made one assumption after another. First we postulated that matter waves obeyed the equations $\lambda = h/p$ and $E = h\nu$, on the basis of which we obtained the dispersion relation of Eq. 3.11. Then we assumed that the waves were described by a simple differential equation. The simplest form that *could* give Eq. 3.11 was that of Eq. 3.19, and $e^{\pm i\varphi}$ is its simplest solution that *does* give Eq. 3.11. All this is plausible, but all it proves is the *possibility* that our result is correct. The real proof of its correctness is that it predicts no results inconsistent with experiment.

Let us now consider the wave function itself. It has the form

$$\Psi(x,t) = Ae^{i\varphi} = Ae^{i(kx-\omega t)}, \tag{3.26}$$

direct substitution of which in Eq. 3.25 gives us back Eq. 3.11 directly. The constant A is the scale factor or normalization factor for the function Ψ, and its value is of no concern to us as yet; the functional form of Ψ *is* of great interest to us. It is very important to recognize that it takes a *complex* function, a function with an imaginary as well as a real part, to yield the dispersion relation that we inferred from the Planck-Einstein and de Broglie relationships. We can see this by writing Ψ in the form

$$\Psi(\varphi) = A(\cos\varphi + i\sin\varphi). \tag{3.27}$$

Here $A\cos\varphi$ is the real part and $A\sin\varphi$ is the imaginary part, the part multiplied by i; each of these terms is a harmonic or sine wave like Eq. 3.6. But Ψ itself is not just a simple sine wave. It is composed of two pieces, a cosine term and a sine term, which have the same shape but are 90° out of phase with each other. The cosine term is real and the sine term imaginary. (Strictly speaking, the constant A can also be complex—that is, it can have both real and imaginary parts. But we are at liberty to choose A as real, and we do so.) Thus, at a particular time t_1, the real part of Ψ (Re Ψ) has a spatial variation proportional to $\cos(kx - \omega t_1)$, whereas the imaginary part of Ψ (Im Ψ) varies as the sine of the same argument, $\sin(kx - \omega t_1)$. Conversely, at a fixed point in space x_1, Re Ψ varies with time as $\cos(kx_1 - \omega t)$, whereas Im Ψ varies with time as $\sin(kx_1 - \omega t)$.

---

[5] Since $(\partial^2 \Psi / \partial t^2)_x = -\omega^2 \Psi$, we can also write

$$\left(\frac{\partial^2 \Psi}{\partial x^2}\right)_t = \frac{k^2}{\omega^2}\left(\frac{\partial^2 \Psi}{\partial t^2}\right)_x = \frac{1}{v^2}\left(\frac{\partial^2 \Psi}{\partial t^2}\right)_x,$$

which is the general equation obeyed by *any* kind of wave Ψ moving along the x axis at a speed $v = \omega/k$. This equation is the basis of our statement that any unique function of $\varphi (\equiv kx - \omega t)$ is a wave.

To summarize, we have constructed a wave equation, Eq. 3.25, describing the wave associated with a free particle moving in the x direction. We have found a solution of this equation, $\Psi(x, t)$, given by Eq. 3.26. Equation 3.25 is a form of what is usually called the *time-dependent Schrödinger equation*, after Erwin Schrödinger, who in 1926 first developed it from the stimulus provided by de Broglie's relation. For a particle moving freely in a space of three dimensions, the Schrödinger equation would have the form

$$\left(\frac{\partial \Psi}{\partial t}\right)_{x,y,z} = \frac{i\hbar}{2m}\left[\left(\frac{\partial^2 \Psi}{\partial x^2}\right)_{y,z,t} + \left(\frac{\partial^2 \Psi}{\partial y^2}\right)_{x,z,t} + \left(\frac{\partial^2 \Psi}{\partial z^2}\right)_{x,y,t}\right], \qquad (3.28)$$

in which the partial derivatives of Ψ with respect to x, y, and z correspond to the kinetic energy associated with the three components of momentum. The wave function Ψ must then be a function of x, y, z, and t. We shall examine the three-dimensional problem in more detail in connection with confined particles.

3.4 Operators

Let us now approach the wave equation from a somewhat different direction. We want to develop a complete theory of matter, including particlelike as well as wavelike properties; indeed, the theory must yield the laws of classical mechanics in the macroscopic limit. We must therefore relate the wave equation to the mechanical properties of particles. For simplicity we continue to consider the free particle moving in one direction, but the principles we shall introduce have quite general validity.

We obtained the wave equation (Eq. 3.25) by requiring that it satisfy the dispersion relation 3.11, which was written in terms of the wavelike properties described by k and ω (wavelength and frequency). However, we could as well have emphasized the particlelike properties of energy and momentum, since our fundamental assumption is that a "particle" has both sets of properties. From the Planck-Einstein and de Broglie relations, we have for the energy and momentum

$$E = h\nu = \hbar\omega \quad \text{and} \quad p = \frac{h}{\lambda} = \hbar k. \qquad (3.29)$$

Substituting for ω and k in Eqs. 3.22 (upper sign only) and 3.23, we obtain

$$i\hbar\left(\frac{\partial \Psi}{\partial t}\right)_x = E\Psi \qquad (3.30)$$

and

$$-\hbar^2\left(\frac{\partial^2 \Psi}{\partial x^2}\right)_t = p^2\Psi. \qquad (3.31)$$

Now we go one step beyond the dispersion relation. Let us *identify* the operations on the left side of Eq. 3.30, time differentiation of Ψ followed by multiplication by $i\hbar$, with the operation on the right side, multiplication of Ψ by the energy E. To generalize, let the symbol E represent the set of operations on Ψ that are equivalent to multiplying Ψ by the energy. Thus we define

$$\mathsf{E} \equiv i\hbar\frac{\partial}{\partial t}, \qquad (3.32)$$

with the partial derivative understood to be with all spatial variables held constant. Here E is an example of a symbolic representation of a set of instructions to be carried out by operation on some understood thing, in this case the function Ψ representing the wave amplitude. As such, it is an example of an *operator*.[6] The concept of operators is fundamental to quantum mechanics. First we identify operators with various classical quantities like E, then we can rewrite the equations of classical mechanics in terms of the corresponding operators. We shall use sans serif type to denote operators in general; thus E is the operator counterpart of the total energy E, and V will be the operator counterpart of the potential energy V.

First let us say a little about the algebra of operators. When more than one operator is applied in succession, their symbols can be written together as in multiplication: the expression ABf means that first the operator B is applied to a function f, then the operator A to the result. One must be cautious in such manipulations, since in general the order of operations cannot be interchanged without affecting the result. For example, if A means "multiply by 2" and B means "add 2," then AB$f = 2f + 4$, whereas BA$f = 2f + 2$. If AB *is* equivalent to BA,

$$ABf = BAf \quad \text{or} \quad (AB - BA)f = 0, \tag{3.33}$$

we say that A and B *commute*. It is generally true, however, that

$$Af + Bf = Bf + Af. \tag{3.34}$$

We call A a *linear* operator if

$$A(f + g) = Af + Ag \tag{3.35}$$

for any functions f and g; we shall deal only with linear operators. Any sequence of operations can be combined into a single operator, as we combined differentiation and multiplication to define E; similarly, the second-derivative operator d^2/dx^2 is equivalent to one differentiation followed by another.

The basic variables of classical mechanics are positions, momenta, and time; what are the corresponding operators? Since we express Ψ as a function of x and t, we do not want to tamper with those variables; thus the operator corresponding to x (or any Cartesian coordinate) or t is simply multiplication by that variable itself. The same applies to any quantity (such as potential energy) that is a function of only these variables. Momentum is another matter. Just as we identified the energy with $i\hbar(\partial/\partial t)$ on the basis of Eq. 3.30, we can obtain an operation equivalent to multiplication by the momentum p from Eq. 3.31. If we define the operator

$$\mathsf{p} \equiv -i\hbar \frac{\partial}{\partial x}, \tag{3.36}$$

we find that

$$\mathsf{p}^2 = -\hbar^2 \frac{\partial^2}{\partial x^2}, \tag{3.37}$$

consistent with Eq. 3.31. We know that the *quantities* E and p are related by $E = p^2/2m$, where E is the kinetic energy of the free particle. Multipli-

---

[6] More familiar examples of operators are "+" or "÷," denoting the operations of addition and division. Addition and division happen to be defined as operators acting on real or complex numbers; addition is also defined as an operation on vectors, but division is not. Both addition and division are also defined for a function such as Ψ, whose value must be a real or complex number. For multiplication we have the operators "·" and "×," which have the same meaning for numbers but different meanings for vectors: cf. Eqs. 1.17 and 2.21.

cation of both sides by the wave function Ψ gives

$$E\Psi = \frac{p^2\Psi}{2m}, \tag{3.38}$$

which is a simple algebraic equation. But we have implied that operators can be identified with their corresponding quantities, which suggests that E can be replaced by E and p by p. If we make this replacement in Eq. 3.38, using Eqs. 3.32 and 3.37 to express E and p^2, we obtain

$$i\hbar\frac{\partial\Psi(x,t)}{\partial t} = -\frac{\hbar^2}{2m}\frac{\partial^2\Psi(x,t)}{\partial x^2}, \tag{3.39}$$

which is, of course, the same as Eq. 3.25.

In classical mechanics the energy expressed as a function of coordinates and momenta—in this case, as $p^2/2m$—is known as the *Hamiltonian* (after Sir William Rowan Hamilton, the nineteenth-century Irish mathematician). The corresponding quantum mechanical operator, with p replacing p, is called the *Hamiltonian operator* H, which for the free particle in one dimension is given by

$$\mathsf{H} = \frac{\mathsf{p}^2}{2m} = -\frac{\hbar^2}{2m}\frac{\partial^2}{\partial x^2}. \tag{3.40a}$$

The corresponding Schrödinger equation can therefore be written in operator notation as

$$\mathsf{H}\Psi = \mathsf{E}\Psi = -i\hbar\frac{\partial\Psi}{\partial t}. \tag{3.41}$$

To anticipate, when a classical particle in one dimension is subject to a potential $V(x)$, then H becomes

$$\mathsf{H} = \frac{\mathsf{p}^2}{2m} + \mathsf{V}(x) = -\frac{\hbar^2}{2m}\frac{\partial^2}{\partial x^2} + \mathsf{V}(x) \tag{3.40b}$$

and this new Hamiltonian must be used in Eq. 3.41. Equation 3.41 is the fundamental equation of quantum mechanics. We now postulate that it applies not just to the free particle in one dimension, but (with the appropriate Hamiltonian) to *any* physical system. In other words, for any system described by a wave function Ψ, the Hamiltonian operator H acts on Ψ to give $-i\hbar(\partial\Psi/\partial t)$. Since H (in terms of p) corresponds formally to the energy (in terms of p), it is H and not E that is commonly called the "energy operator."

Our derivation of Eq. 3.39 may have seemed trivial, since we defined the operators E and p in just such a way as to give the Schrödinger equation; furthermore, our "identification" of E, p with E, p was quite arbitrary. This is true: Rather than "deriving" Eq. 3.39, we have merely showed its plausibility in a new way. The real point is the formal expression 3.41, which is still a postulate—but now a general postulate, of which Eq. 3.39 is one special case. We make a very far-reaching claim for this equation: If we write the energy of any classical system in Hamiltonian form and obtain the corresponding operator H, then the solution of Eq. 3.41 with this H is the wave function Ψ that describes the same system in quantum mechanics. If this statement is valid, we can transcribe the whole of classical mechanics into quantum mechanical form. *Is* it valid? Apparently so. How do we know? Once again, because the results obtained in this way agree with experiment.

This question—what it means to say that a theory is valid—deserves some further comment. We should always try to distinguish our assumptions about the nature of matter from mere mathematical manipulations or deductive reasoning. The real substance of any physical theory lies in the choice of variables and in the relations between them that we *assume* to be valid. In classical Newtonian mechanics, the choice of variables—force, acceleration, and momentum—leads us to think in terms of particles and rigid bodies, and to express our assumptions in the form of differential equations. These assumptions are the three laws of motion: the law of inertia, stating that a free body moves with a constant velocity; the law relating externally applied forces to velocity changes, stated as $F = ma$; and the law of reaction. In quantum mechanics, the basic assumptions can be taken as the relations $E = h\nu$ and $\lambda = h/p$, and the validity of classical Newtonian physics in the limit of macroscopic phenomena. These assumptions, together with a series of shrewd guesses, led us to the Schrödinger equation, but in no way "proved" it. In fact, there are other wave equations that could be used to describe matter waves, in that they satisfy the same assumptions. The Schrödinger equation was the first one tried because it is the simplest and therefore the most economical; so far as it works, there is no need to go any further. We have assumed, then, that economy or simplicity is a valid criterion for scientific choice, in the absence of any contradictory evidence. Whether or not this assumption is justified is a question of epistemology, metaphysics, or possibly aesthetics.

So we have a method for converting classical equations into quantum mechanical operator equations. Once we have such an equation, what do we do with it? The answer, of course, is to make predictions about the behavior of physical systems. The rules for making these predictions involve still more postulates; to understand what these are, we must again consider the nature of the wave function.

We are still talking about the free particle moving in one dimension. The wave function we have obtained, Eq. 3.26, is the amplitude of a traveling wave. Can we construct a standing wave capable of describing a free particle? The answer to this question, strangely enough, is no, at least so long as we require the particle to have a precise momentum. The function with which we have been dealing is associated with a definite momentum p, with a value given by Eq. 3.29. This momentum has a definite sign, that is, a definite direction: A positive value of p corresponds to motion in the positive x direction. Suppose that we now give up the requirement that the particle have definite momentum, and relax our conditions to finding any function that remains bounded and satisfies Eq. 3.25. Can we find a standing wave to meet these conditions?

To obtain such a standing wave, we must somehow separate the space-dependent and time-dependent parts of the wave function. Our previous mathematical arguments are still valid: The function $Ae^{i(kx-\omega t)}$ is a solution of Eq. 3.25 for any value of k and ω; each such value gives a different solution. But since differentiation is a linear operation, any sum or difference of these solutions must also satisfy Eq. 3.25. This *principle of superposition* applies to all linear wave phenomena. Let us then consider an arbitrary wave function for which k has the value k' (which we can assume without lack of generality to be positive); we can identify this wave function as $\Psi_{k'}$. There is, of course, another wave function, $\Psi_{-k'}$, for which k has the value $-k'$; we assume the same value of ω for both functions so they correspond to identical traveling waves moving in opposite directions. We can now construct two new wave functions, still

3.5
Eigenfunctions and Eigenvalues

satisfying Eq. 3.25, by adding and subtracting $\Psi_{k'}$ and $\Psi_{-k'}$:

$$\Psi' = \frac{\Psi_{k'} + \Psi_{-k'}}{2} = \frac{A}{2}(e^{ik'x} + e^{-ik'x})e^{-i\omega t} = A\cos k'x\, e^{-i\omega t},$$

$$\Psi'' = \frac{\Psi_{k'} - \Psi_{-k'}}{2i} = \frac{A}{2i}(e^{ik'x} - e^{-ik'x})e^{-i\omega t} = A\sin k'x\, e^{-i\omega t}. \tag{3.42}$$

The factors 2 and $2i$ are included to give the cosine and sine functions, by the trigonometric equalities

$$\cos\theta = \frac{e^{i\theta} + e^{-i\theta}}{2} \quad \text{and} \quad \sin\theta = \frac{e^{i\theta} - e^{-i\theta}}{2i}. \tag{3.43}$$

We have thus achieved our goal. By combining the amplitudes of two traveling waves moving in opposite directions—that is, by superposing the two waves—we have obtained a pair of standing waves, each of which oscillates in time with the frequency ω. More precisely, Ψ' and Ψ'' both have real and imaginary parts, each of which is a standing wave varying as $\cos\omega t$ and $\sin\omega t$, respectively. Direct differentiation of Ψ' and Ψ'' shows that both satisfy Eq. 3.25. Nevertheless, neither Ψ' nor Ψ'' is associated with a specific value of the momentum. To see this, let us compare how the momentum operator $\mathbf{p}$, defined in Eq. 3.26, acts on our original Ψ and on Ψ'. With Ψ we obtain

$$\mathbf{p}\Psi = -i\hbar\frac{\partial\Psi}{\partial x} = \hbar k A e^{i(kx - \omega t)} = \hbar k\Psi = p\Psi, \tag{3.44}$$

Ψ itself multiplied by the momentum $\hbar k$; but with Ψ' we obtain

$$\mathbf{p}\Psi' = -i\hbar\frac{\partial\Psi'}{\partial x} = \frac{\hbar}{i}(-Ak'\sin k'x\, e^{-i\omega t}) = i\hbar k'\Psi''. \tag{3.45}$$

In other words, applying $\mathbf{p}$ to Ψ' not only does not give us Ψ' multiplied by the value of the momentum, it gives us an altogether different function, proportional to Ψ''. A system described by Ψ can be assigned a value ($\hbar k$) equivalent to the classical momentum, but this cannot be done for a system described by Ψ'. Let us now investigate the difference between the two kinds of wave functions.

Actually, for an operation to change one function into some other is the usual situation, not the exception. A case like Eq. 3.44, in which an operation leaves a function unchanged except perhaps for some multiplicative factor, is the unusual situation in mathematics, physics, and chemistry. However unusual it may be, this situation is an exceedingly important one. We have already seen other examples of such behavior, for example, in Eqs. 3.30 and 3.31. This special property, the ability of an operator to change a function only by a numerical factor, is peculiar to specific combinations of operations and functions. The simplest case is perhaps the differentiation of an exponential: $d(e^{ax})/dx = ae^{ax}$. The operator d/dx leaves the function e^{ax} unchanged except to multiply it by the constant a. By contrast, differentiation changes a sine or a cosine into a different function. A function left unchanged except for multiplication by a constant when it is acted on by a particular operator is said to be an *eigenfunction* (*characteristic function*) of that operator. Thus by the equations of Section 3.4 we know that $\Psi = A e^{i(kx - \omega t)}$ is an eigenfunction of the operators of momentum and energy. We have shown that the functions Ψ' and Ψ'' of Eqs. 3.42 are not eigenfunctions of momentum; however, they are eigenfunctions of the energy ($H\Psi' = E'\Psi'$, with $E' = \hbar^2 k'^2/2m$), since a sine or cosine does become a multiple of itself when it is differentiated twice.

What physical interpretation can we give to the concepts of eigenfunctions and their corresponding operators? Suppose, for example, that a particular wave function is an eigenfunction of the energy operator. This means that if we allow the energy operator (**H** or **E**) to operate on the wave function, then the result will be the wave function multiplied by a constant. This result contains one very specific piece of physical information, the value of the characteristic constant. This value is the energy corresponding to that particular eigenfunction, and is thus known as an *eigenvalue* (*characteristic value*) of the energy. If instead of **H** we used the momentum operator **p**, and the wave function were an eigenfunction of momentum, then we would obtain the wave function multiplied by the corresponding momentum eigenvalue. Whenever we wish to describe a system in a state having a definite value of some variable, we must use a wave function that is an eigenfunction of the corresponding operator,[7] that particular eigenfunction which corresponds to the given value; only such a function can correctly describe the system as we have assumed it to exist. Thus if we wish to specify that a free particle has an energy E_0, its wave function must be that eigenfunction of the energy operator which has the eigenvalue E_0, that is, the function $Ae^{i(kx-\omega t)}$ for which k and ω have the values given by

$$E = \hbar\omega = \frac{\hbar^2 k^2}{2m} = E_0. \qquad (3.46)$$

On the other hand, any wave function that is not an eigenfunction of a given operator (as with Ψ' and the momentum operator) cannot be assigned a definite value of the variable corresponding to that operator.

By this time the reader may be confused. Are we saying that the wave function of a system depends on how we choose to describe it? In a sense this is true, but "how we describe it" has a real physical meaning. To say that a system has a definite energy, one must carry out some measuring process that gives a particular value of the energy; similarly, any statement specifying the properties of a system implies some kind of measurement. But one of the fundamental principles of quantum mechanics is that the process of measurement itself affects the state of a system—basically by imposing new boundary conditions. This is reflected in what is known as the *uncertainty principle*, which we shall consider in detail in Section 3.7. All that needs to be understood at this point is that specifying (i.e., measuring) different properties does in general cause the system to have different wave functions. It is useless to ask what the state of the system would be in the absence of any measurement, since there is no way for us to know this.

We have perhaps implied that the specification of a single quantity, such as the energy or the momentum, is enough to define the wave function of a system. In general this is not true. There are ordinarily several characteristic quantities with precisely defined values, usually the same quantities that are conserved in classical mechanics: energy, linear momentum, and angular momentum. These properties are constants of the motion, remaining unchanged in the absence of external interference, and are associated with some general symmetry of the physical system itself. For example, the angular momentum about a particular axis is associated with cylindrical symmetry about that axis. This does not mean that the system itself need be cylindrical, but that its orientation about the axis makes no physically significant difference in its description. Thus, if a

---

[7] Note that the eigenvalues of the Cartesian coordinates are continuous: The operator corresponding to x is itself x, so we have $\mathbf{x}\Psi = x\Psi$ for any value of x.

particle moves in an elliptical orbit about an attractive center of force, the properties of the orbit (its shape, the angular momentum, the total energy) would be the same for any orientation of the ellipse; if there is no external torque or other dissipative mechanism, the angular momentum and energy will remain constant. The symmetry associated with the conservation of linear momentum involves the fact that we can locate the origin of our coordinate system anywhere in space without affecting the value of the momentum. The three Cartesian components of momentum are conserved separately, so that an external force applied in one direction will change the momentum in that direction but leave the other components unaffected. As for the energy, note that it is the total energy that is conserved; the kinetic energy and the potential energy separately can vary, but their sum remains constant. Although the conservation theorems are ordinarily derived from Newton's laws, they can also be shown to follow from the principles of quantum mechanics.

Let us suppose that we can enumerate the quantities that we are sure will be conserved. For the free particle in one dimension, these are just the energy and the momentum. Does specifying these give enough information for a complete determination of the particle's wave function? The answer is almost, but not quite enough. We know that the wave function $\Psi = Ae^{i(kx-\omega t)}$ is a satisfactory solution to the wave equation 3.25, inasmuch as it has well-defined values of the energy and momentum, satisfies the dispersion relation 3.11, and has the aesthetic simplicity of being a very straightforward and uncomplicated kind of wave. But this wave function contains one other very important piece of information to which we have not yet turned our attention. We refer to the constant A, which gives the maximum amplitude of the wave. The real part of Ψ is restricted to values between $+A$ and $-A$, and the imaginary part between $+iA$ and $-iA$, no matter how large or small x or t may become. It will be recalled that we deliberately chose the form of Ψ to be such that the amplitude would be everywhere finite. We can now expand on the reasons for this restriction, in connection with a physical interpretation of Ψ itself.

We have said that the value of Ψ must be somehow related to the probability of finding the particle at a given position. By the "probability" of a given value we mean the fraction of all measurements that should yield that value. (See Section 15.2 for a more extensive discussion.) This probability is analogous to the intensity of a light wave (which is proportional to the density of light quanta). But the intensity of a light wave is proportional not to the amplitude of the wave itself, but to its square; the same is true of the energy carried by a water wave. We can thus interpret the intensity of our matter wave, the probability of finding a particle, as proportional to the square of Ψ. This is another postulate, of course. But the value of Ψ is a complex number, whereas the probability (like any other measurable quantity) must be a real number; to obtain a real result, we need simply take the square of the absolute value of Ψ. This is why it does not matter that Ψ is complex: We never observe Ψ directly, and it is only for mathematical convenience that Ψ rather than $|\Psi|^2$ is the "wave function."

The intensity of the matter wave is thus given by

$$\mathscr{I}(x, t) = |\Psi(x, t)|^2 = \Psi^*(x, t)\Psi(x, t), \tag{3.47}$$

where Ψ^* is the complex conjugate of Ψ, the function identical to Ψ except that every i ($\equiv\sqrt{-1}$) is replaced by $-i$:

$$\Psi^*(x, t) = A^*e^{-i(kx-\omega t)}, \tag{3.48}$$

in which we write A^* to allow for the possibility that A is also complex. The last part of Eq. 3.47 is justified by the fact that $|z|^2 = z^* z$ for any complex number.[8] But when we multiply Ψ^* by Ψ the exponential factors cancel, and we obtain

$$\mathcal{I}(x, t) = A^* A = |A|^2; \tag{3.49}$$

that is, the intensity of the wave is simply given by the square of the absolute value of A.

The physical interpretation of the matter wave is thus clear. For the free particle moving in one dimension, the probability of finding the particle at a given place and time is proportional to the intensity of the wave at that place and time. Consider an infinitesimal region of length dx around the point x; the probability of finding the particle in this region at time t is given by

$$\mathcal{P}(x, t)\, dx \propto \mathcal{I}(x, t)\, dx = \Psi^* \Psi\, dx = |A|^2\, dx, \tag{3.50}$$

where $\mathcal{P}(x, t)$ is what we call a *probability density*. The most striking fact about this result is that the probability density is a constant, and not a function of x or t at all. For the particular case of the free particle, all points are equally probable; like Kipling's "cat that walked by himself," all places are alike for it. This result should not be surprising: If all we know about a particle is that it is moving along a line with a given energy and momentum, we can say nothing about *where* on the line it is; this can be determined by a measurement, of course, but the performance of such a measurement will change the wave function.

We wrote Eq. 3.50 as a proportionality, but we might as well define A so as to make it an equality. Given this assumption, what is the absolute value of A? To answer this question we must carry out what is called a *normalization*. By the definition of probability (see Section 15.2), the total probability over the entire available range must be unity. Since the range of a completely free particle is unlimited, we have

$$\int_{-\infty}^{\infty} \mathcal{P}(x, t)\, dx = \int_{-\infty}^{\infty} \Psi^* \Psi\, dx = \int_{-\infty}^{\infty} |A|^2\, dx = 1, \tag{3.51}$$

a result that can be true only if $|A| = 0$. The physical meaning of this is also quite simple: If the particle can be *anywhere*, the probability of its being at any particular position is vanishingly small. By contrast, if a measurement of the particle's position localized it within some small range Δx, integration over that range would give a finite value $|A|^2 = 1/\Delta x$. The normalization process tells us nothing new here, but we shall see that it is quite useful for bounded systems. Equation 3.51 does illustrate more clearly why Ψ must remain bounded at infinity, the reason for which we rejected forms like $e^{\pm\varphi}$.

Before ending our consideration of the free particle, let us briefly take up the case of the free particle in three dimensions. We have already given the wave equation that must be satisfied, Eq. 3.28. In three dimensions the constant k must be replaced by three constants k_x, k_y, k_z, corresponding to the three components of the momentum. The wave function thus has the form

$$\Psi(x, y, z, t) = A e^{i(k_x x + k_y y + k_z z - \omega t)} = A e^{i(\mathbf{k} \cdot \mathbf{r} - \omega t)}. \tag{3.52}$$

---

[8] Any complex number has the form $z = a + ib$; its complex conjugate is thus $z^* = a - ib$, and we have

$$z^* z = (a - ib)(a + ib) = a^2 - i^2 b^2 = a^2 + b^2 = |z|^2,$$

since $|z|$ is $(a^2 + b^2)^{1/2}$ by definition.

This still has the form $Ae^{i\varphi}$, since $\mathbf{k} \cdot \mathbf{r} - \omega t$ is in the form of a phase. The function 3.52 has real and imaginary oscillations in all three directions, x, y, and z. It is still an eigenfunction of the energy operator, so $H\Psi = E\Psi = E\Psi$, and the corresponding energy eigenvalue can easily be shown to be

$$E = \hbar\omega = \frac{\hbar^2}{2m}\,(k_x{}^2 + k_y{}^2 + k_z{}^2) = \frac{\hbar^2 k^2}{2m}, \qquad (3.53)$$

where $k^2 \equiv k_x{}^2 + k_y{}^2 + k_z{}^2$. The momentum $\mathbf{p}$, a vector with components $\hbar k_x$, $\hbar k_y$, $\hbar k_z$, is an eigenvalue of the three-dimensional momentum operator[9]

$$\mathbf{p} = -i\hbar\left(\mathbf{i}\,\frac{\partial}{\partial x} + \mathbf{j}\,\frac{\partial}{\partial y} + \mathbf{k}\,\frac{\partial}{\partial z}\right), \qquad (3.54)$$

in which $\mathbf{i}$, $\mathbf{j}$, $\mathbf{k}$ are unit vectors in the x, y, z directions, respectively. The arguments of the last paragraphs still apply, so that the particle is equally likely to be found anywhere in space. We must add some physical restriction to get a system for which $|\Psi|^2$ exhibits some spatial dependence; in the next section we begin to consider such restrictions.

The simplest way to generate spatial dependence for $|\Psi|^2$, to produce a varying probability of finding a particle at different points in space, is to leave the particle as free as possible, but within the confines of a box from which it cannot escape. Physically, this means that within the box the particle's potential energy is independent of its position, but that no matter how much kinetic energy it has it cannot get beyond the walls.

We again deal first with a one-dimensional system for simplicity. As before, we consider a particle of mass m moving along the x axis. Let us suppose that this particle is constrained to be between the points $x = 0$ and $x = a$. For convenience we assume that the potential energy between these two limits is identically zero, so that the total energy inside the box is pure kinetic energy. The potential energy outside the box must be so high that the particle cannot escape, no matter how large the kinetic energy may be; in short, it must be infinite. A sketch of such a box is given in Fig. 3.5.

The physical problem before us is clearly to determine the allowable values that the energy of the confined particle may have, and to find the wave function (and thus the spatial probability distribution) associated with each of these energy eigenvalues. The corresponding mathematical problem requires that we solve the Schrödinger equation subject to the specific *boundary conditions* appropriate to this particular problem. These are that the probability of finding the particle, and therefore the value of Ψ, must be identically zero for all values of x beyond the boundaries of the box. This means we require that

$$\Psi(x, t) = 0 \quad \text{for all } x < 0 \text{ and } x > a. \qquad (3.55)$$

We must also assume that the wave function is *continuous* through the walls (and everywhere else in space). Like the assumptions that Ψ be finite and single-valued, this is necessary if the results are to make physical sense; a function that satisfies these three conditions is called

3.6
The Particle in a One-Dimensional Box

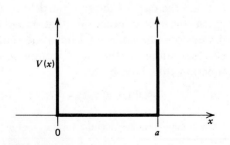

FIGURE 3.5
The potential energy of the one-dimensional box.

---

[9] For any scalar function f, the vector whose components are $\partial f/\partial x$, $\partial f/\partial y$, and $\partial f/\partial z$ is called the *gradient* of f, written **grad** f or ∇f (the operator "∇" is read "del"). Thus we can write $\mathbf{p} \equiv -i\hbar\nabla$.

"well-behaved." In general, the first derivatives of Ψ must also be continuous, but this rule does not apply at an infinite potential jump such as we have here. Given the assumption of continuity, we must have

$$\Psi(0, t) = 0 \quad \text{and} \quad \Psi(a, t) = 0, \tag{3.56}$$

whether we approach the wall from the inside or the outside. This is our first example of a boundary condition.

Our treatment in previous sections assumed that the potential energy was zero everywhere, with the dispersion relation 3.11 yielding the wave equation 3.39. For the general case of nonzero potential energy, we start instead with Eq. 3.13, from which we easily obtain

$$i\hbar\left[\frac{\partial\Psi(x, t)}{\partial t}\right]_x = -\frac{\hbar^2}{2m}\left[\frac{\partial^2\Psi(x, t)}{\partial x^2}\right]_t + \mathsf{V}(x)\Psi(x, t) \tag{3.57}$$

for the one-dimensional time-dependent Schrödinger equation. However, since we have assumed the potential energy $\mathsf{V}(x)$ to be zero inside the box, we can still use Eq. 3.39 to describe the behavior of the wave function there. It is up to us to find the solution of this equation that satisfies the boundary conditions of Eq. 3.56.

The best way to begin is by separating the spatial dependence of the wave function from its time dependence. We wish our wave function to be an eigenfunction of the energy operator, which means that it must satisfy the equation $\mathsf{E}\Psi = E\Psi$, or

$$i\hbar\left(\frac{\partial\Psi}{\partial t}\right)_x = \hbar\omega\Psi, \tag{3.58}$$

with the usual substitution $E = \hbar\omega$. But this equation can hold only if Ψ has the form

$$\Psi(x, t) = \psi(x)e^{-i\omega t}, \tag{3.59}$$

in which $\psi(x)$ is a function of only the spatial coordinate x. Our solution for the free particle was also of this form, with $\psi(x) = Ae^{ikx}$. In fact, the argument is valid for *any* wave function that is an eigenfunction of the energy, and these are what we usually want to obtain. Substituting Eqs. 3.58 and 3.59 into the Schrödinger equation 3.57, we obtain

$$\hbar\omega\psi(x)e^{-i\omega t} = -\frac{\hbar^2}{2m}\frac{d^2\psi(x)}{dx^2}e^{-i\omega t} + \mathsf{V}(x)\psi(x)e^{-i\omega t}. \tag{3.60}$$

Cancelling $e^{-i\omega t}$ from both sides, again replacing $\hbar\omega$ by E, and rearranging, we have

$$-\frac{\hbar^2}{2m}\frac{d^2\psi(x)}{dx^2} = [E - \mathsf{V}(x)]\psi(x), \tag{3.61}$$

the *time-independent Schrödinger equation*, involving only the spatial variable x. This equation is usually written in the form

$$\mathsf{H}\psi = E\psi, \tag{3.62}$$

where the Hamiltonian operator H, representing the energy in terms of coordinates and momenta, is given by

$$\mathsf{H} = \frac{\mathsf{p}^2}{2m} + \mathsf{V}(x) = -\frac{\hbar^2}{2m}\frac{\partial^2}{\partial x^2} + \mathsf{V}(x). \tag{3.63}$$

Equation 3.62, like Eq. 3.41, is quite general: With the appropriate Hamiltonian, it applies to the spatial part of any energy eigenfunction. Since $\mathsf{H}\psi$ must be a function of spatial variables only, the Hamiltonian cannot be an explicit function of time: If $V(x)$ also varies with time, there is no $\psi(x)$ (and thus no E) that can satisfy Eq. 3.61.

We included $V(x)$ in Eq. 3.61 for generality, but it can immediately be dropped again for the region inside the box. We must thus solve the equation

$$\frac{d^2\psi(x)}{dx^2} = -\frac{2mE}{\hbar^2}\psi(x) \tag{3.64}$$

to obtain the spatial part of the wave function. As we have done several times before, we can simplify the equation by removing the dimensional dependence of the variable. We thus make the transformation

$$\xi \equiv \left(\frac{2mE}{\hbar^2}\right)^{1/2} x, \tag{3.65}$$

which allows us to rewrite Eq. 3.64 as

$$\frac{d^2\psi(\xi)}{d\xi^2} = -\psi(\xi). \tag{3.66}$$

The only simple functions that are the negatives of their own second derivatives are the complex exponential, which is never zero, and the sine and cosine functions. Both sine and cosine do of course pass through zero, at $n\pi$ and $(n+\frac{1}{2})\pi$, respectively, and can thus satisfy the boundary conditions. A general form for $\psi(\xi)$ should then be

$$\psi(\xi) = A \sin \xi + B \cos \xi. \tag{3.67}$$

Although this wave function is an eigenfunction of the energy, it clearly cannot be an eigenfunction of momentum: Since ψ is real, the equation $-i\hbar(d\psi/dx) = p\psi$ can be satisfied for no real value of p. One can see a "reason" for this in simple terms. The presence of a wall means that the particle must in effect bounce, thereby changing its momentum; hence the momentum cannot be a constant of the motion. The real reason is complex but very important, and will be discussed in the next section.

The general procedure for solving differential equations such as Eq. 3.66 consists of writing down the formal solution, as we have done in Eq. 3.67, and then using the boundary conditions, in our case those of Eq. 3.56, to determine the integration constants, here A and B. The formal solution 3.67 is by no means *the* solution to the problem. A differential equation by itself does not contain all the information required to specify a solution. Each time we carry out an integration, we necessarily introduce an integration constant; fixing the value of these constants is just as much a part of finding a solution as is finding a general form like Eq. 3.67. It is the boundary conditions that fix the integration constants, and changing them sometimes completely changes the entire character of the problem, either in the form of the solution or in the difficulty of obtaining it. We can see this explicitly by comparing the solution for the particle in a box with that obtained for the free particle. After all, in these two cases the Hamiltonians and therefore the wave equations are exactly the same; the problems differ *only* in their boundary conditions.

Now let us turn to obtaining this solution. The boundary conditions tell us that $\Psi = 0$ at $\xi = 0$ $(x = 0)$ and $\xi = (2mE/\hbar^2)^{1/2}a$ $(x = a)$. Where $\Psi = 0$ for all values of t, we must also have $\psi = 0$. We can require that $\psi(0)$ be zero only by setting the constant B in Eq. 3.67 equal to zero. Therefore the solution has the form

$$\psi(\xi) = A \sin \xi \tag{3.68}$$

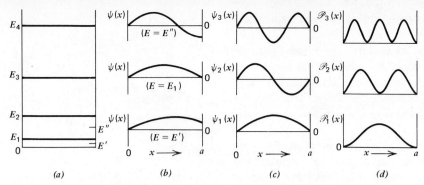

FIGURE 3.6
Energy levels and wave functions for a particle in a one-dimensional box. (a) The first four energy eigenvalues; the energies E' and E'' that are not eigenvalues correspond to the top and bottom wave functions shown in part (b). (b) Illustration of how only the eigenvalues give wave functions satisfying the boundary condition $\psi(a) = 0$. (c) Eigenfunctions corresponding to the first three eigenvalues. (d) The corresponding probability densities, $\mathcal{P}_n(x) = |\psi_n(x)|^2$, for $n = 1, 2,$ and 3.

or

$$\psi(x) = A \sin\left(\frac{2mE}{\hbar^2}\right)^{1/2} x. \tag{3.69}$$

This brings us to the crux of the problem of quantization. How can we satisfy the boundary condition at the point $x = a$? With the function 3.69 there is only one way to accomplish this goal: We must require the argument of the sine to be an integral multiple of π; that is, the condition

$$\left(\frac{2mE}{\hbar^2}\right)^{1/2} a = n\pi \qquad (n = 1, 2, \ldots) \tag{3.70}$$

must be satisfied. (The boundary condition could also be met by setting $n = 0$—and thus $E = 0$—or even by letting both A and B be zero, but in either case the wave function would be identically zero everywhere in the box; this would be of no help to us, because it would correspond to an empty box. Since $-n$ and $+n$ correspond to the same value of E, negative values of n can also be discarded.) Of the quantities in Eq. 3.70, only E and n are not fixed beforehand. The equation thus relates the allowed values of the energy E to the integral values of n. To each n there corresponds an energy E_n, which we find by rearrangement to be

$$E_n = \frac{n^2\pi^2\hbar^2}{2ma^2} \qquad (n = 1, 2, \ldots). \tag{3.71}$$

The boundary conditions at the two ends of the box—and note that it takes two boundaries, not one—have forced us to accept only a discrete set of possible values for the energy E. These energies vary inversely with the square of the box length, and directly with the square of the quantum number n. The separation between consecutive energy levels, $E_{n+1} - E_n$, thus also increases directly with n, as illustrated in Fig. 3.6a.

How do the energy levels vary with the size of the box? If we squeeze the length of the box to $a/2$, we raise the energy of the nth level to four times its original value. If we stretch the box, we correspondingly lower the energy levels. Squeezing the box by a factor of two can be interpreted in another very straightforward way, as equivalent to placing a boundary at the point $x = a/2$; we can then apply the same kind of boundary conditions at $x = a/2$ that we originally applied at $x = a$. The method of solution is exactly the same as before, so the only acceptable wave functions are those that manage to vanish at $x = 0$ and $x = a/2$. In other words, we have in the new box the condition

$$\left(\frac{2mE}{\hbar^2}\right)^{1/2} \frac{a}{2} = n\pi \qquad (n = 1, 2, \ldots), \tag{3.72}$$

which is exactly equivalent to taking every other energy level of the set found for the original box, those levels corresponding to even values of n in the box of length a.

Now let us examine just how the boundary conditions lead to quantization of the energy levels. Equation 3.66 tells us that the curvature of the wave function ψ is equal to the negative of the function itself. (Recall that the first derivative of a function is its slope, whereas the second derivative is the rate of change of the slope, which we may loosely refer to as the function's "curvature." Strictly, the curvature is $(d^2\psi/dx^2)[1+(d\psi/dx)^2]^{-3/2}$.) If ψ is positive, then its curvature must be negative, so that the function must be curving back downward toward the x axis. Correspondingly, wherever ψ is negative, the curvature must be positive and the function curving upward toward the x axis. These are conditions that produce a sine- or cosinelike wave. The function 3.69 has zero displacement at the origin. According to Eq. 3.64, the curvature of ψ is proportional to the energy of the state with which it is associated. If the energy E has a value E' near zero, the curvature of ψ is so slight that the function (which must be positive near the origin) never returns to the x axis in the interval between 0 and a. If the value of E is somewhat higher, say E'', then ψ increases with increasing x, passes through a maximum, and curves downward, crossing the x axis before the point $x = a$ is reached. After it crosses the axis, ψ must start curving upward again, but let us assume that E'' is not high enough for it to reach the axis a second time before $x = a$. For exactly one value of E between the two arbitrary values E' and E'', the function ψ will just reach the axis at the point $x = a$. This value E_1 is the first permissible value that the energy can assume, according to the boundary condition we have imposed. The behavior just described is illustrated in Fig. 3.6b. As E increases beyond E'', eventually an energy will be reached at which ψ just reaches the axis a second time at $x = a$; this is the second allowed energy value, E_2, corresponding to $n = 2$ in Eq. 3.71. Similarly, there are specific values of E for all the higher values of n. Each value of E that satisfies the boundary conditions is an eigenvalue of the energy operator, and the wave function

$$\psi_n(x) = A \sin\left(\frac{2mE_n}{\hbar^2}\right)^{1/2} x \qquad (3.73)$$

is the spatial eigenfunction corresponding to E_n. The first three eigenfunctions are plotted in Fig. 3.6c.

Using Eq. 3.71, we can rewrite the wave function in the form

$$\psi_n(x) = A \sin\frac{n\pi x}{a}. \qquad (3.74)$$

This expression is identical to that for the possible standing waves in a string fastened at both ends (as in Fig. 3.2). This identity should not be surprising, since the two systems are mathematically equivalent: Both wave amplitudes must satisfy a wave equation, with the boundary condition that the amplitude must vanish at the ends. Only the *meanings* of the wave amplitudes are different, and this does not affect their mathematical form.

We still have to evaluate the constant A, which fixes the maximum amplitude of the wave function. This can be done quite simply, as soon as we recognize that there is one more physical condition to be satisfied. We refer to the normalization process introduced in the preceding section; it did us no good there, but here it can give us the value of A. The probability of finding the particle in a region of length dx around x is

$$\mathcal{P}(x, t)\, dx = \Psi^*\Psi\, dx = \psi^*(x)e^{i\omega t}\psi(x)e^{-i\omega t}\, dx = |\psi(x)|^2\, dx. \quad (3.75)$$

Note that the probability density is a function of x only, as is true whenever Eq. 3.59 holds. For our eigenfunctions we have

$$\mathcal{P}_n(x) = |\psi_n(x)|^2 = |A|^2 \sin^2 \frac{n\pi x}{a}; \qquad (3.76)$$

this probability distribution is plotted in Fig. 3.6d. The total probability of the particle's being anywhere in its range is again unity. Since we have specified that there is zero probability of finding the particle outside the box, we need integrate only over the length of the box, to obtain

$$\int_0^a \mathcal{P}_n(x)\, dx = |A|^2 \int_0^a \sin^2 \frac{n\pi x}{a}\, dx = 1. \qquad (3.77)$$

Again making the substitution

$$\xi = \left(\frac{2mE}{\hbar^2}\right)^{1/2} x = \frac{n\pi x}{a}, \qquad (3.78)$$

we can convert Eq. 3.77 to[10]

$$|A|^2 \left(\frac{a}{n\pi}\right) \int_0^{n\pi} \sin^2 \xi\, d\xi = |A|^2 \left(\frac{a}{n\pi}\right)\left(\frac{n\pi}{2}\right) = |A|^2 \frac{a}{2} = 1, \qquad (3.79)$$

which immediately tells us that

$$|A| = \left(\frac{2}{a}\right)^{1/2}. \qquad (3.80)$$

There is no reason not to assume A real, so we set $A = |A|$ and finally obtain

$$\psi_n(x) = \left(\frac{2}{a}\right)^{1/2} \sin \frac{n\pi x}{a} \qquad (3.81)$$

for the spatial part of the nth eigenfunction.

Note that the nth eigenfunction, $\psi_n(x)$, has $n-1$ nodes, not counting the walls. The energy of the system thus increases with the number of nodes in its wave function. We shall see that this relationship between energy and nodes is generally true. As $n \to \infty$, the nodes become closer and closer together; eventually the oscillations of the wave function blur together, and the probability density over any measurable distance averages out[11] to the constant value $\langle \mathcal{P}(x) \rangle = 1/a$. This is the *classical limit*: A classical particle would be equally likely to be anywhere in the box. In the same limit E_n becomes so large that the fractional difference between

[10] The integral

$$\int_0^\pi \sin^2 \theta\, d\theta = \frac{\pi}{2}$$

can be found in any table of definite integrals. Here the range is from 0 to $n\pi$, but the integral over each range of π is of course the same, so that

$$\int_0^{n\pi} \sin^2 \theta\, d\theta = \frac{n\pi}{2}.$$

[11] The average of a function $f(x)$ over a range Δx is

$$\langle f(x) \rangle = \frac{1}{\Delta x} \int_{\Delta x} f(x)\, dx.$$

Since $\mathcal{P}_n(x)$ is a periodic function with period a/n, its average is simply

$$\langle \mathcal{P}_n(x) \rangle = \frac{n}{a} \int_0^{a/n} \mathcal{P}_n(x)\, dx = \left(\frac{n}{a}\right)\left(\frac{2}{a}\right)\left(\frac{a}{n\pi}\right) \int_0^\pi \sin^2 \xi\, d\xi = \frac{1}{a}.$$

successive energy levels is negligible; the energy is thus effectively continuous. This behavior illustrates the general rule that quantum mechanical systems behave classically in the limit $n \to \infty$ (the *correspondence principle*).

The form of Eq. 3.81 also tells us that the wave $\psi_n(x)$ has n half-waves in a length a, or that $\psi_n(x)$ has a wavelength λ of $2a/n$. The de Broglie condition $p\lambda = h$ implies that the momentum p of the particle in a box is h/λ or $nh/2a$ or $\pi n\hbar/a$. (There is a subtle point being glossed over here: The momentum may be positive or negative for a free particle; by putting up a box, we allow only the absolute value of the momentum to be meaningful.) The kinetic energy, $p^2/2m$, is therefore $n^2\pi^2\hbar^2/2ma^2$, in agreement with Eq. 3.71. In short, our development of the particle in a box is indeed consistent with the de Broglie condition, as it should be. The wavelengths go down with n, the (absolute values of) momenta go up with n, and the energies go up with n^2.

What is the separation between energy levels for the particle in a box? We wish eventually to develop a theory of atomic structure, and we can make the present model a very crude approximation to an atom if we consider an electron ($m = m_e$) in a box of atomic dimensions. From the results in Section 2.11 we can guess in what range the energy levels will lie, but let us make our estimate anew. Let us say that $\pi^2 \approx 10$, $h \approx 10^{-34}$ J s, $m_e \approx 10^{-30}$ kg, and $a \approx 10^{-10}$ m (1 Å). Substituting these numbers in Eq. 3.71, we have

$$E_n \approx \frac{n^2 \times 10 \times (10^{-34}\,\text{J s})^2}{2 \times 10^{-30}\,\text{kg} \times (10^{-10}\,\text{m})^2} \times \frac{1\,\text{kg m/s}^2}{\text{J}} = 5n^2 \times 10^{-18}\,\text{J},$$

equivalent to about $30n^2$ eV (since $1\,\text{eV} \approx 1.6 \times 10^{-19}$ J). As we expected, for small values of n these numbers are indeed of the same order of magnitude as those given in Table 2.2 for the Bohr atom. Another example, for which the particle-in-a-box model is more realistic, is that of a gas molecule in a real macroscopic container. (Extending the problem to three dimensions does not change the orders of magnitude involved.) If we set $m \approx 3.3 \times 10^{-27}$ kg (a molecule of H_2) and $a \approx 1$ cm, we obtain $E_n \approx 3n^2 \times 10^{-37}$. This is such a small spacing, even by atomic dimensions, that the energies can be regarded as continuous for most purposes; this is why nearly all the macroscopic properties of gases can be explained in terms of classical mechanics (see Part Two).

3.7
The Uncertainty Principle

We have already mentioned the uncertainty principle in Section 3.5, where we pointed out that the wave function of a system depends on which of its properties one measures. Now we can understand the reason for this: Any measuring process imposes a new boundary condition on the system, and thus affects the wave function. The time has now come to examine this principle in detail. The uncertainty principle, derived by Werner Heisenberg in 1927, is perhaps the most general and fundamental point on which the conclusions of quantum mechanics diverge from those of classical mechanics. In classical mechanics it is assumed that one can simultaneously know as many properties of a system as one wishes, to any desired degree of accuracy. It is already apparent that this is not true in quantum mechanics; the most obvious example is that the wave function gives only the probability of a particle's various possible locations, rather than specifying a single location. The uncertainty principle gives the limitations on the accuracy with which any quantity can be known.

The nature of these limitations can best be understood by considering an example. Suppose that we wish to know simultaneously the position and momentum of a particle in motion. We can specify the momentum initially by launching the particle with a known kinetic energy. Its wave function is then an eigenfunction of energy and momentum, which we have shown to have the form $Ae^{i(kx-\omega t)}$; however, this function gives the same probability for any value of x, and thus tells us nothing about the particle's position. We can in principle measure the position by dividing the trajectory into short segments and applying some kind of test to each to determine which contains the particle. (Don't worry about the practicality of this; the difficulty is more fundamental.) But such a test is tantamount to looking for the particle in a highly localized state, that is, confining it to a very narrow box. We must thus apply boundary conditions rather like those of the last section. But confining a wave to a box forces the wave to oscillate through at least a half-cycle within the box. The narrower the box, the more rapid is the oscillation of ψ, the higher the average value of $|\partial\psi/\partial x|$, and thus (since $\partial\psi/\partial x$ is proportional to $\mathbf{p}\psi$) the higher the average momentum of the particle. Another way to express the same idea is in terms of Eq. 3.71: Decreasing the box length increases the energy and thus the momentum. In other words, the more accurately we try to specify the particle's position, the more we increase (and thus make uncertain) its momentum. The best we can do is to balance off the uncertainties in position and momentum at some level.

We can make a rough estimate of this level. Suppose that a particle is at rest, and we wish to find its exact position by looking through a microscope. To see the particle we must of course bounce one or more photons off it. If the light we use has a wavelength λ, we cannot expect to determine the position within a distance much shorter than λ because of diffraction; so let us say that λ is the uncertainty in the position measurement. But a photon of wavelength λ has a momentum h/λ, and in the collision it may transfer roughly this much momentum to the particle. The particle is then no longer at rest, and its momentum is uncertain to within h/λ. The product of the position and momentum uncertainties is thus of the order of $\lambda(h/\lambda)=h$, with Planck's constant again making a significant appearance. We shall state this principle more accurately, but first we must define our terms.

Position and momentum are what is called a *conjugate* pair of variables. This is a concept from classical mechanics that we shall not consider in detail. Other examples of conjugate pairs include energy and time, and angular momentum and angle. In each case the pair consists of two variables whose product has the dimensions of action (cf. Section 2.7), and that are associated with the same coordinate or degree of freedom. What this means is that position on the x axis and momentum in the x direction (x and p_x) are conjugate, as are y and p_y, z and p_z, but that x and p_y are not conjugate. When one member of the pair is used as a coordinate, the other is called the *generalized momentum* conjugate to that coordinate even though it may not have the dimensions of momentum. It can be shown by reasoning like that above that the two members of any conjugate pair of variables cannot be simultaneously measured accurately.[12] The uncertainty principle is a quantitative statement of this conclusion.

We must also define just what we mean by the "uncertainty" in a variable. The closest quantum mechanical approximation to a classical particle localized at a point is a sharply peaked probability distribution—a wave function something like that for a particle in a narrow box, but trailing off gradually at the sides (since there are no sharp boundary conditions). Such a distribution, or *wave packet*, is shown in Fig. 3.7. There are various ways in which Δx, the uncertainty in x, can be defined; we have chosen the root-mean-square (rms) deviation, defined by

$$(\Delta x)^2 \equiv \langle (x - x_0)^2 \rangle = \int_{-\infty}^{\infty} (x - x_0)^2 \mathscr{P}(x)\, dx, \tag{3.82}$$

where x_0 is the average value of x for the wave packet. The probability distribution for any other variable, when plotted against that variable, will resemble that in Fig. 3.7. We can thus define the momentum uncertainty Δp by

$$(\Delta p)^2 \equiv \langle (p - p_0)^2 \rangle = \int_{-\infty}^{\infty} (p - p_0)^2 \mathscr{P}(p)\, dp, \tag{3.83}$$

and other uncertainties in the same way.

We can now formulate the uncertainty principle. (It can be derived from what we already know about quantum mechanics, but we shall simply give it without proof.) Suppose that p and q are a conjugate pair of variables, with uncertainties Δp and Δq as defined above. The uncertainty principle states that

$$(\Delta p)(\Delta q) \geq \frac{\hbar}{2}. \tag{3.84}$$

In other words, our simultaneous knowledge of any two conjugate variables is limited by the condition that the product of their uncertainties must be greater than or at best equal to $\hbar/2$. The better we know p, that is, the smaller we make Δp, the larger must Δq be, and vice versa.

Let us apply the uncertainty principle to the particle in a box. The average position is, of course, the center of the box, and the uncertainty in position must be something less than half the box's width. In fact, a detailed calculation with Eq. 3.82 shows that

$$\Delta x = \left[\int_{-\infty}^{\infty} (x - x_0)^2\, \mathscr{P}(x)\, dx \right]^{1/2}$$

$$= \left[\int_{0}^{a} \left(x - \frac{a}{2} \right)^2 \left(\frac{2}{a} \right) \sin^2 \left(\frac{n\pi x}{a} \right) dx \right]^{1/2} = \frac{a}{2\sqrt{3}} \tag{3.82a}$$

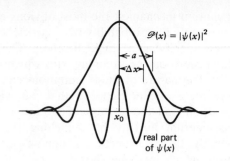

FIGURE 3.7
Wave packet representing a particle localized in space. The figure is drawn for a wave function

$$Ae^{-(x-x_0)^2/2a^2}\, e^{ik(x-x_0)}$$

giving the Gaussian probability distribution

$$\mathscr{P}(x) = |\psi(x)|^2 = A^2 e^{-(x-x_0)^2/a^2},$$

where x_0 is the average value of x. The scale factor a, defined by $\mathscr{P}(x_0 \pm a) = \mathscr{P}(x_0)/e$, is equal to $\sqrt{2}$ times the root-mean-square deviation Δx.

---

[12] An equivalent (and more general) formulation is that two variables cannot simultaneously be determined exactly if their corresponding operators do not commute with each other. For example, we have

$$\mathsf{x}\mathsf{p}_x\psi = -i\hbar x \frac{\partial \psi}{\partial x} = -i\hbar x \frac{\partial \psi}{\partial x}$$

but

$$\mathsf{p}_x \mathsf{x}\psi = -i\hbar \frac{\partial}{\partial x}(x\psi) = -i\hbar \left(x \frac{\partial \psi}{\partial x} + \psi \right),$$

so that $\mathsf{x}\mathsf{p}_x \neq \mathsf{p}_x\mathsf{x}$, and the operators x and p_x do not commute. On the other hand, x does commute with p_y. For any conjugate position and momentum one obtains

$$(\mathsf{q}_i\mathsf{p}_i - \mathsf{p}_i\mathsf{q}_i)\psi = i\hbar\psi;$$

$\mathsf{q}_i\mathsf{p}_i - \mathsf{p}_i\mathsf{q}_i$ is called the *commutator* of q_i and p_i.

for all values of n. (That $x_0 = a/2$ is proved in the next section.) We can obtain the momentum uncertainty much more simply. We know that the particle's kinetic energy in the nth eigenstate is $E_n = n^2\pi^2\hbar^2/2ma^2$; since $E = p^2/2m$, the absolute value of the momentum must be $|p_x| = n\pi\hbar/a$. Since we have no information on which way the particle is going, the actual value of the momentum may be either $+|p_x|$ or $-|p_x|$; the two values are equally likely, so the average momentum p_0 is zero (as we shall shortly prove analytically). The uncertainty in p_x is the rms deviation from the average, which, from Eq. 3.83, becomes

$$\Delta p_x = \left[\int_{-\infty}^{\infty} (p-0)^2 \mathscr{P}(p)\, dp \right]^{1/2}$$

$$= [(+p_x - 0)^2 \times \tfrac{1}{2} + (-p_x - 0)^2 \times \tfrac{1}{2}]^{1/2}$$

$$= |p_x| = \frac{n\pi\hbar}{a}. \tag{3.85}$$

The smallest possible momentum uncertainty is therefore in the lowest state, for which $n = 1$ and $\Delta p_x = \pi\hbar/a$. The product of the uncertainties is thus

$$\Delta p_x\, \Delta x = \left(\frac{\pi\hbar}{a}\right)\left(\frac{a}{2\sqrt{3}}\right) = 0.907\hbar, \tag{3.85a}$$

which, as Eq. 3.84 predicts, is greater than $\hbar/2$.

Suppose that we did not know the wave function for the particle in the box. We could still say that Δx must be less than $a/2$, so by the uncertainty principle Δp_x must be greater than $\hbar/a$. In an energy eigenstate $|p_x|$ has a definite value, which must be at least as much as Δp_x. Even the lowest energy level must thus satisfy the inequality

$$E = \frac{|p_x|^2}{2m} \geq \frac{(\Delta p_x)^2}{2m} > \frac{\hbar^2}{2ma^2}. \tag{3.86}$$

In short, assuming only that the system is restricted to a region of length a, we conclude that the kinetic energy cannot be less than some minimum value. This illustrates an important general principle, that all confined systems have a positive *zero-point energy* in their lowest energy level; we shall discuss other examples in Chapter 4. The existence of zero-point energy is a direct consequence of the uncertainty principle.

Since energy and time are conjugate variables, we can write the uncertainty principle in the form

$$\Delta E\, \Delta t \geq \frac{\hbar}{2}, \tag{3.87}$$

where Δt is of the order of the minimum time required to measure the energy with an accuracy of ΔE. This means that to determine the energy exactly would require an infinite time of observation ($\Delta t = \infty$). How does this affect our derivation of wave functions corresponding to exact energy eigenvalues? A system in an eigenstate with energy E must obey Eq. 3.62, $\mathsf{H}\psi = E\psi$, with the Hamiltonian H including all the forces acting on the system. But since ψ is a function of spatial coordinates only, this can be true only if H does not explicitly include the time—that is, if all the forces on the system, including those we use to observe it, are independent of time. Equation 3.87 thus expresses the fact that any measurement taking a nonzero time requires the exertion of some outside force on the system. The time-independent results are limiting values that cannot be exactly obtained in any actual measurement.

Although one cannot determine a particle's position and momentum exactly, it is possible to determine their average values in a given system. By "average value" we mean the value that we would expect to find as the average of a large number of measurements. We thus call this average the *expectation value*. To find the expectation value of a given variable, one must sum or integrate over all possible values of the variable, weighting each value by the probability of its being observed. The expectation value of Q is thus defined as

$$\langle Q \rangle \equiv \sum_i Q_i \, P(Q_i), \tag{3.88}$$

where $P(Q_i)$ is the probability of obtaining the ith value. For example, in the throwing of two dice, Q_i may be 2, 3, 4, 5, 6, 7, 8, 9, 10, 11, or 12. The total number of possible throws is $6 \times 6 = 36$. The probabilities are $P(2) = P(12) = \frac{1}{36}$; $P(3) = P(11) = \frac{2}{36}$; $P(4) = P(10) = \frac{3}{36}$; $P(5) = P(9) = \frac{4}{36}$; $P(6) = P(8) = \frac{5}{36}$, and $P(7) = \frac{6}{36}$. A few moments with a pencil or a calculator will show that $\langle Q \rangle = 7$. If the possible values of Q are continuous, we write

$$\langle Q \rangle \equiv \int Q \, \mathcal{P}(Q) \, dQ, \tag{3.89}$$

where $\mathcal{P}(Q)$ is the probability density within the range dQ. (All this will be discussed in greater detail in Sections 15.1 and 15.2.) Equations 3.88 and 3.89 are true whether the system is classical or quantum mechanical. Where quantum mechanics enters, of course, is in determining the probabilities.

For simplicity, we still consider only a one-dimensional system. Suppose that we wish to know the expectation value of x. If $\mathcal{P}(x) \, dx$ is the probability of finding a particle within the region dx around x, then the expectation value is

$$\langle x \rangle \equiv \int x \, \mathcal{P}(x) \, dx, \tag{3.90}$$

with the integral taken over the entire possible range of x. But for a quantum system we know that the probability density in the coordinates is given by the absolute square of the normalized wave function,

$$\mathcal{P}(x, t) = \Psi^*(x, t)\Psi(x, t) = |\Psi(x, t)|^2. \tag{3.91}$$

The quantum mechanical expectation value of x at time t is thus

$$\langle x(t) \rangle = \int x \, |\Psi(x, t)|^2 \, dx. \tag{3.92}$$

We are interested primarily in time-independent systems; if the Hamiltonian is time-independent, then $\Psi(x, t)$ is an energy eigenfunction of the form $\psi(x)e^{-i\omega t}$, and $\mathcal{P}(x)$ and $\langle x \rangle$ are also time-independent:

$$\langle x \rangle = \int x \, |\psi(x)|^2 \, dx. \tag{3.93}$$

The expectation value of x is obtained directly from our interpretation of $|\psi|^2$; for other variables we must introduce a new postulate. It turns out that the expectation value of any other variable Q is given by an equation similar to Eq. 3.93, an integral weighted by the square of the wave function. However, one must integrate not the variable Q itself, but the quantum mechanical operator Q that represents it (for x these are the same thing). A complication arises when Q is not a simple multiplicative operator, since the order of the factors in the integral then makes a difference. The correct way to write the integral must be taken as part of the postulate: The expectation value of Q is

$$\langle Q \rangle = \int \Psi^* Q \Psi \, dx. \tag{3.94}$$

Why this way? Again, because it has to be done this way to make the answers physically reasonable. We are still assuming that Ψ is normalized; if it were not normalized, the above integral would have to be divided by $\int \Psi^* \Psi \, dx$. If the system has more than one dimension, the integral is taken over all the coordinates.

Just what is the meaning of a quantum mechanical expectation value? All the physically admissible information about a system is contained in its wave function Ψ. If Ψ is an eigenfunction of the operator Q, then the system has a definite value Q' of the variable Q such that $Q\Psi = Q'\Psi$, and a measurement of Q can give only the value Q'; Eq. 3.94 then reduces to $\langle Q \rangle = Q' \int \Psi^* \Psi \, dx = Q'$. If Ψ is not an eigenfunction of Q, however, all we can say about Q is that it has a certain probability distribution. Any single measurement must give one of the eigenvalues of Q, but a series of measurements on identical systems (systems with the same Ψ) will give results distributed among these values.[13] As the number of such measurements increases, the average value of Q should approach the limiting value given by Eq. 3.94.

Let us apply the averaging process to the particle in a box. Using the normalized eigenfunctions of Eq. 3.81, we have[14]

$$\langle x \rangle = \int_0^a x \, |\psi(x)|^2 \, dx = \left(\frac{2}{a}\right) \int_0^a x \sin^2 \frac{n\pi x}{a} \, dx = \left(\frac{2}{a}\right)\left(\frac{a^2}{4}\right) = \frac{a}{2}; \tag{3.95}$$

just as we expected, the average position of the particle is in the middle of the box. For the momentum we must use Eq. 3.94, which yields[15]

$$\langle p_x \rangle = \int_0^a \psi^*(x) \, p_x \, \psi(x) \, dx = \left(\frac{2}{a}\right) \int_0^a \sin \frac{n\pi x}{a} \left(-i\hbar \frac{d}{dx}\right) \sin \frac{n\pi x}{a} \, dx$$

$$= -i\hbar \left(\frac{2}{a}\right)\left(\frac{n\pi}{a}\right) \int_0^a \sin \frac{n\pi x}{a} \cos \frac{n\pi x}{a} \, dx = 0. \tag{3.96}$$

This is what we had deduced from the fact that the particle is equally likely to be traveling in either direction. On the other hand, $\langle p_x^2 \rangle$ is simply $2mE_n$, so that $\langle p_x^2 \rangle \neq \langle p_x \rangle^2$; this illustrates that one must be careful in manipulating averages. We shall show in Section 15.2 that one always has

$$\langle (\Delta Q)^2 \rangle \equiv \langle (Q - \langle Q \rangle)^2 \rangle = \langle Q^2 \rangle - \langle Q \rangle^2, \tag{3.97}$$

where $\langle (\Delta Q)^2 \rangle$ corresponds to the mean-square deviation introduced in Eq. 3.82. For the particle in a box we immediately obtain $(\Delta p_x)^2 = 2mE_n$, in agreement with Eq. 3.85.

---

[13] The probability of obtaining a value Q' (that is, the fraction of measurements that should give Q') can be shown to be

$$P(Q') = \left| \int \psi_{Q'}{}^* \psi \, dx \right|^2,$$

where $\psi_{Q'}$ is the eigenfunction corresponding to Q', and ψ is the system's actual wave function; when $\psi = \psi_{Q'}$, of course, we have $P(Q') = 1$.

[14] From a table of integrals we can obtain

$$\int x \sin^2 \alpha x \, dx = \frac{x^2}{4} - \frac{x \sin 2\alpha x}{4\alpha} - \frac{\cos 2\alpha x}{8\alpha^2}.$$

We have $\alpha = n\pi/a$ and x varying from 0 to a; thus $2\alpha x$ runs from 0 to $2n\pi$, and the sine and cosine terms drop out.

[15] Since $\int_0^\pi \sin \theta \cos \theta \, d\theta = 0$.

We have now introduced all the major principles of quantum mechanics that we shall be using. However, the introduction has been piecemeal, and most of the principles have been formulated for the one-dimensional case only. This is a good point to summarize our results with a formal statement of our postulates in general form. The set of postulates we give here is not complete, but covers everything needed in this text.

POSTULATE I. Every physical system is completely described by a wave function $\Psi(q_1, \ldots, q_N, t)$, where the q_i are the coordinates[16] that define the system. The function Ψ and its first derivatives must be everywhere finite, continuous, and single-valued (except that the derivative may be discontinuous at an infinitely high potential barrier); Ψ may be real or complex.

POSTULATE II. The probability that the coordinates of the system are in the range dq_1 around $q_1, \ldots, dq_N$ around q_N is given by

$$\mathscr{P}(q_1, \ldots, q_N, t)\, dq_1 \cdots dq_N = \Psi^*(q_1, \ldots, q_N, t)\Psi(q_1, \ldots, q_N, t)\, dq_1 \cdots dq_N. \tag{3.98}$$

For a single particle moving in three dimensions, for example, this probability becomes $\Psi^*\Psi\, dx\, dy\, dz$. This definition assumes a normalized wave function, one such that $\int \Psi^*\Psi\, dq_1 \cdots dq_N$ over all possible values of the coordinates is unity; the integral must in any case be finite.

POSTULATE III. To every variable Q of classical mechanics there corresponds a linear operator Q. These operators are constructed by the following rules: (1) If Q is one of the coordinates q_i or the time, the operator is simply multiplication by Q. (2) If Q is the momentum p_i conjugate to the coordinate q_i, the operator is

$$\mathsf{p}_i \equiv -i\hbar \frac{\partial}{\partial q_i}. \tag{3.99}$$

(3) If Q is some function of coordinates and momenta, the operator is obtained by substituting Eq. 3.99 for all the momenta.[17]

We must call special attention to the Hamiltonian operator H. As mentioned earlier, the classical Hamiltonian is the total energy expressed as an explicit function of only coordinates, momenta, and (in general) time. Since kinetic energy is given by $p^2/2m$, a single particle moving in a potential field $V(x, y, z)$ has the Hamiltonian operator[18]

---

[16] One can formulate quantum mechanics with other quantities, for example momenta, as the independent variables. However, we shall restrict our treatment to the formulation in terms of coordinates.

[17] There may be some ambiguity in this process; for example, xp_x and $p_x x$ are algebraically the same, but for a general function f the operations $-i\hbar x(\partial f/\partial x)$ and $-i\hbar[\partial(xf)/\partial x]$ will not give the same result. We thus need an additional rule: (4) The order of factors must be such that the resulting operator is *Hermitian*. A Hermitian operator is one for which

$$\int \cdots \int \Phi^*(\mathsf{Q}\Psi)\, dq_1 \cdots dq_N = \int \cdots \int \Psi(\mathsf{Q}^*\Phi^*)\, dq_1 \cdots dq_N$$

for any two wave function Φ and Ψ. (It turns out that $\mathsf{x}\mathsf{p}_x$ is Hermitian and $\mathsf{p}_x\mathsf{x}$ is not, as one can show by setting $\Phi = \Psi = e^{ix}$.) This condition will create a problem for us only when we deal with non-Cartesian coordinates.

[18] The operator

$$\frac{\partial^2}{\partial x^2} + \frac{\partial^2}{\partial y^2} + \frac{\partial^2}{\partial z^2}$$

is called the *Laplacian*, and is usually abbreviated as ∇^2 ("del squared" or "nabla squared").

$$H = \frac{\mathbf{p} \cdot \mathbf{p}}{2m} + V(x, y, z) = \frac{\mathbf{p}_x{}^2 + \mathbf{p}_y{}^2 + \mathbf{p}_z{}^2}{2m} + V(x, y, z)$$

$$= -\frac{\hbar^2}{2m}\left(\frac{\partial^2}{\partial x^2} + \frac{\partial^2}{\partial y^2} + \frac{\partial^2}{\partial z^2}\right) + V(x, y, z), \qquad (3.100)$$

where $\mathbf{p}$ is the vector whose components are the three momentum operators. This Hamiltonian is written in terms of Cartesian coordinates, but other coordinate systems are often more convenient; we shall discuss later how to express the Hamiltonian in those coordinates.

POSTULATE IV. Any possible measurement of the variable Q for a single atom or molecule can only yield one of the eigenvalues of the corresponding operator Q, that is, a value Q_n such that

$$Q\Psi_n = Q_n\Psi_n, \qquad (3.101)$$

where Ψ_n is the eigenfunction of Q corresponding to Q_n. To put it the other way around, any system whose wave function is an eigenfunction of Q has a definite value of Q.

POSTULATE V. All wave functions must satisfy the time-dependent Schrödinger equation,

$$H\Psi(q_1, \ldots, q_N, t) = i\hbar \frac{\partial \Psi(q_1, \ldots, q_N, t)}{\partial t}. \qquad (3.102)$$

If Ψ is an eigenfunction of the energy, and thus of $i\hbar(\partial/\partial t)$, by Eq. 3.101 it must have the form

$$\Psi(q_1, \ldots, q_N, t) = \psi(q_1, \ldots, q_N)e^{-iEt/\hbar}, \qquad (3.103)$$

where E is the energy; the spatial part of Ψ then satisfies the time-independent Schrödinger equation,

$$H\psi(q_1, \ldots, q_N) = E\psi(q_1, \ldots, q_N). \qquad (3.104)$$

Remember that this is possible only when H does not explicitly contain the time.

POSTULATE VI. The expectation value of the variable Q (in a system described by the normalized wave function Ψ) is

$$\langle Q \rangle = \int \cdots \int \Psi^* Q\Psi \, dq_1 \cdots dq_N, \qquad (3.105)$$

with the integral taken over all possible values of the coordinates. From this rule and Eq. 3.97 one can derive the uncertainty principle.

We have devoted a great deal of effort to showing the plausibility of these postulates, for purely pedagogical reasons. Logically, we could have just set them down and then gone on from there; advanced texts do just that. The whole elaborate structure of quantum mechanics is justified only by its agreement with experiment; the postulates are simply the most concise statement of that structure's building blocks.

In the remainder of this chapter we deal with particles in boxes of two or three dimensions. We shall have something to say about how one sets up and solves the Schrödinger equation in various systems, but our main concern will be with how the boundary conditions—the shapes of the boxes—affect the solutions. We begin in this section with the relatively simple case of a rectangular two- or three-dimensional box; this is a

3.9
Particles in Two- and Three-Dimensional Boxes

straightforward extension of our one-dimensional-box theory, but it illustrates some new concepts.

As before, we assume a potential energy that is zero inside the box and infinitely large outside. It is most convenient to treat a box with boundaries perpendicular to the Cartesian axes; we therefore assume a rectangular box extending from $x = 0$ to $x = a$, from $y = 0$ to $y = b$, and (for the three-dimensional case) from $z = 0$ to $z = c$. We shall write most of the following equations in terms of three dimensions, with the understanding that the term or factor involving z (or c) drops out in the two-dimensional case. As before, the wave function must be identically zero outside the box, giving boundary conditions analogous to Eq. 3.56:

$$\psi(0, y, z) = \psi(a, y, z) = 0 \qquad \text{for all } y, z;$$
$$\psi(x, 0, z) = \psi(x, b, z) = 0 \qquad \text{for all } x, z; \qquad (3.106)$$
$$\psi(x, y, 0) = \psi(x, y, c) = 0 \qquad \text{for all } x, y.$$

The Hamiltonian operator is given by Eq. 3.100. Inside the box, where $V(x, y, z) = 0$, one can see that H is a sum of three terms, each involving only one of the spatial variables; we can thus write

$$\mathsf{H} = \mathsf{H}_x + \mathsf{H}_y + \mathsf{H}_z. \qquad (3.107)$$

Any operator that can be divided into single-variable operators in this way is called *separable*. An operator may be separable in only one coordinate system, or in several; some are not separable at all. Clearly, if the Hamiltonian included a potential energy such as $V(x, y, z) = xyz$, it would not be separable in Cartesian coordinates. The importance of a separable Hamiltonian is that it allows us to separate the wave equation into equations in the individual variables, which are of course much easier to solve. When the Hamiltonian is separable, the energy eigenvalue of Eq. 3.104 can obviously also be written as a sum,

$$E = E_x + E_y + E_z \qquad (3.108)$$

(where $\mathsf{H}_x\psi = E_x\psi$, etc.). In such a case the spatial part of the wave function can always be expressed as a *product* of single-variable functions,

$$\psi(x, y, z) = f(x)g(y)h(z), \qquad (3.109)$$

as we shall now demonstrate.

Given Eqs. 3.107 and 3.109, the time-independent Schrödinger equation can be written as

$$\mathsf{H}_x f(x)g(y)h(z) + \mathsf{H}_y f(x)g(y)h(z) + \mathsf{H}_z f(x)g(y)h(z)$$
$$= Ef(x)g(y)h(z). \quad (3.110)$$

Since the operator H_x acts only on $f(x)$, the functions g and h can be brought outside it as multiplicative factors; treating the H_y and H_z terms similarly and rearranging, we have

$$g(y)h(z)\mathsf{H}_x f(x)$$
$$= Ef(x)g(y)h(z) - f(x)h(z)\mathsf{H}_y g(y) - f(x)g(y)\mathsf{H}_z h(z). \quad (3.111)$$

We now divide both sides of the equation by $f(x)g(y)h(z)$:

$$\frac{1}{f(x)}\mathsf{H}_x f(x) = E - \frac{1}{g(y)}\mathsf{H}_y g(y) - \frac{1}{h(z)}\mathsf{H}_z h(z). \qquad (3.112)$$

The left side of this equation is a function of x only, whereas the right side is a function of y and z only. Here are two expressions that are functions of entirely different variables, yet are always equal, regardless of the values of the variables. Such a condition can be satisfied only if both are equal to one and the same constant; since this constant is uniquely determined by H_x, $f(x)$, and the boundary conditions on x, it is logical to call it E_x:

$$\frac{1}{f(x)} H_x f(x) = E_x. \tag{3.113}$$

This is in fact the same E_x as was defined in Eq. 3.108, since $(H_x f)/f = (H_x \psi)/\psi$. We see now, however, that $f(x)$ is an eigenfunction of H_x and E_x is the corresponding eigenvalue:

$$H_x f(x) = E_x f(x). \tag{3.114}$$

The corresponding derivations for y and z are straightforward, yielding

$$H_y g(y) = E_y g(y) \quad \text{and} \quad H_z h(z) = E_z h(z); \tag{3.115}$$

E_y and E_z are also the same as in Eq. 3.108, and we have thus proved the validity of Eq. 3.109.

It should be apparent that the above technique is applicable to any eigenvalue equation in which the operator is separable, and we shall have many further occasions to apply it. In each case, an equation in all the variables is replaced by separate equations in the individual variables. Physically, we can say that the behavior of each variable is not affected by any other variable when there are no interactions (i.e., terms in the Hamiltonian) involving both variables. Separability is a property that is particularly useful when we wish to consider systems of several particles, such as many-electron atoms and molecules.

We can now obtain the wave functions and energy levels for a particle in our box. Consider first Eq. 3.114. Since $H_x \equiv (-\hbar^2/2m)(\partial^2/\partial x^2)$, the form of this equation is identical to that of Eq. 3.64. The boundary conditions in x are also the same, and we can immediately write down the solution

$$f_{n_1}(x) = \left(\frac{2}{a}\right)^{1/2} \sin\frac{n_1 \pi x}{a} \qquad (n_1 = 1, 2, \ldots), \tag{3.116}$$

corresponding to Eq. 3.81; the corresponding eigenvalues are

$$(E_x)_{n_1} = \frac{n_1^2 \pi^2 \hbar^2}{2ma^2} \qquad (n_1 = 1, 2, \ldots). \tag{3.117}$$

The solutions for y and z are similar, and substitution in Eq. 3.109 gives

$$\psi_{n_1 n_2 n_3}(x, y, z) = \left(\frac{8}{abc}\right)^{1/2} \sin\frac{n_1 \pi x}{a} \sin\frac{n_2 \pi y}{b} \sin\frac{n_3 \pi z}{c}$$

$$(n_1, n_2, n_3 = 1, 2, \ldots) \tag{3.118}$$

for the complete spatial wave function, with a total energy of

$$E_{n_1 n_2 n_3} = \frac{\pi^2 \hbar^2}{2m}\left(\frac{n_1^2}{a^2} + \frac{n_2^2}{b^2} + \frac{n_3^2}{c^2}\right) \qquad (n_1, n_2, n_3 = 1, 2, \ldots). \tag{3.119}$$

The energy-level spectrum of a particle in a two- or three-dimensional box is naturally more complex than that for a one-dimensional box, because each of the quantum numbers n_1, n_2, n_3 can assume any positive integral value. In a box of arbitrary dimensions, the lengths a, b, c are usually all different; more specifically, they are usually incommensurate with one another, that is, no one length can be expressed as a rational fraction multiplying either of the others. So long as this is the case, every energy level is uniquely defined by a specific set of quantum numbers n_1, n_2, n_3. Change or rearrange any of the numbers, and you are referring to a different energy level. As with the particle in a one-dimensional box, the momentum and its components are not constants of the motion because of the boundaries on the box. But as previously, p_x^2, p_y^2, and p_z^2 and the absolute values of p_x, p_y, and p_z are preserved. Again, we can use the de Broglie condition to relate the wavelength λ of the matter wave, $\lambda = h/p$, to the energy $p^2/2m$, but here, separability of x, y, and z motion allows us to apply our earlier argument for one dimension to each of the components independently.

A very important special situation arises if two or three of the lengths a, b, c become equal. Let us suppose that $a = b$. Then, with a little rearrangement, Eq. 3.119 becomes

$$E - \frac{\pi^2\hbar^2 n_3^2}{2mc^2} = \frac{\pi^2\hbar^2}{2ma^2}(n_1^2 + n_2^2). \tag{3.120}$$

In this situation, for a given value of n_3 there is clearly more than one way of obtaining certain of the energy levels, namely, those for which n_1 is not equal to n_2. Consider the state for which $n_1 = N$ and $n_2 = N'$; the state with $n_1 = N'$ and $n_2 = N$ must have the same energy, since such an interchange leaves the sum $n_1^2 + n_2^2$ in Eq. 3.120 unchanged. If we assumed $a = b = c$, then any states with the same total $n_1^2 + n_2^2 + n_3^2$ would have the same energy; if n_1, n_2, n_3 are all different, there are at least six such states. A situation in which two or more states lie at exactly the same energy is called a *degeneracy*. The n states at the same energy are said to be *degenerate*, and the corresponding energy level to exhibit an n-fold degeneracy. Degeneracy always arises from some natural symmetry of the system. In the first example, we said that the x and y dimensions of the box were indistinguishable. We do not change the physics of the problem if we interchange the names on the x and y axes and thereby interchange n_1 and n_2. A state containing an amount of kinetic energy proportional to n_1^2 in the x direction, and an amount proportional to n_2^2 in the y direction, is physically indistinguishable from the state in which the two kinetic energy terms are interchanged. The energy levels of the particle in a two-dimensional box, for the cases $a \neq b$ and $a = b$, are illustrated in Fig. 3.8.

The eigenfunctions corresponding to degenerate states have the property that we can add or subtract them as we choose and never produce a new combination having an energy different from the original value.[19] This is exactly parallel to what we were able to do for the free particle in one dimension, where a twofold degeneracy is associated with the two directions the momentum can have: We saw that the traveling waves associated with states of definite momentum in the positive and negative x directions could be combined to give a pair of standing waves, Eqs. 3.42, which had the same energy as the traveling waves but were not eigenfunctions of momentum. In the present case, we could combine the two solutions for one of the degenerate energy levels to form a new pair of standing-wave functions.

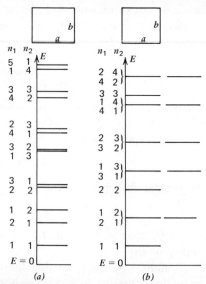

FIGURE 3.8
Energy levels for a particle in a two-dimensional box. (a) Unequal sides (levels drawn for $a = 1.25b$), so that all states are uniquely associated with their own energies, (b) Equal sides ($a = b$), so that all energy levels with $n_1 \neq n_2$ are degenerate. The energy scales of parts (a) and (b) are the same if the two boxes have the same area. [The reader may be amused to realize that the example with $a = b$ is richer and more subtle than has been indicated thus far. Additional degeneracies occur with this system because of the existence of "doubly magic" numbers, numbers that can be written in two essentially different ways as the sum of two squares. The states with $n_1 = 7$, $n_2 = 1$, with $n_1 = 1$, $n_2 = 7$, and with $n_1 = n_2 = 5$ are all degenerate, for example. See G. B. Shaw, *J. Phys. A* **7**, 1537 (1974) for a discussion of this problem.]

A rectangular box of particles is a reasonable model for a macroscopic gas, but hardly for the structure of an atom or molecule. We frequently find in such systems that a particular point forms the natural center of the system, so that distance from the center—say the distance of an electron

3.10
Particles in Circular Boxes

---

[19] If $H\psi_1 = E_1\psi_1$ and $H\psi_2 = E_1\psi_2$, then for any function ψ_3 given by $\psi_3 = \alpha\psi_1 \pm \beta\psi_2$ we have $H\psi_3 = E_1\psi_3$.

from a nucleus, or of one atom from another in a molecule—is important but orientation is irrelevant: The electron's energy does not depend on the side of the nucleus where it is found. Any such problem has *circular* or *spherical symmetry*, depending on whether two or three dimensions are involved. We shall want to treat systems like this, and a good way to approach them is by considering particles in round boxes. This treatment will introduce the coordinate systems used and give some physical intuition for the symmetries and constants of motion. We therefore consider the two-dimensional circular box in this section, the spherical box in the next.

For the two-dimensional system, we wish to examine the effects of circular symmetry. We therefore assume that the potential energy of our particle is a function only of the distance r from some origin, $V = V(r)$, independent of any angular variables giving the orientation about the origin. We must have a circular boundary, of course, which we set at $r = R$. This is to be a box similar to those of the preceding sections, so we define the potential energy as zero for $r < R$, infinite for $r \geq R$. The next step would be to state the Hamiltonian in circular coordinates, but before doing this let us first consider the corresponding classical problem, which can give us some useful insights.

The coordinate system we use is ordinary circular polar coordinates, the radius r from the origin and the angle ϕ measured from the x axis. These coordinates are illustrated in Fig. 3.9, along with their relationships to Cartesian coordinates. The total energy of a particle of mass μ (we shall need m for a quantum number) is simply

$$E = \frac{1}{2\mu}(p_x{}^2 + p_y{}^2) + V(x, y), \tag{3.121}$$

which we wish to convert to circular coordinates. We have already assumed that $V(x, y)$ becomes simply $V(r)$, but how do we express the momenta? We can obtain these by first putting the transformation equations in differential form. We have

$$dx = d(r \cos \phi) = \cos \phi \, dr - r \sin \phi \, d\phi,$$
$$dy = d(r \sin \phi) = \sin \phi \, dr + r \cos \phi \, d\phi. \tag{3.122}$$

Substituting these into the expressions for the momenta, we obtain

$$p_x = \mu \frac{dx}{dt} = -\mu r \sin \phi \frac{d\phi}{dt} + \mu \cos \phi \frac{dr}{dt} \tag{3.123}$$

and

$$p_y = \mu \frac{dy}{dt} = \mu r \cos \phi \frac{d\phi}{dt} + \mu \sin \phi \frac{dr}{dt}. \tag{3.124}$$

When we write $p_x{}^2 + p_y{}^2$, the two cross terms $[\pm \mu^2 r \sin \phi \cos \phi (d\phi/dt) \times (dr/dt)]$ cancel each other, and the energy becomes

$$E = \frac{\mu}{2}\left[r^2(\sin^2 \phi + \cos^2 \phi)\left(\frac{d\phi}{dt}\right)^2 + (\cos^2 \phi + \sin^2 \phi)\left(\frac{dr}{dt}\right)^2\right] + V(r)$$

$$= \frac{\mu r^2}{2}\left(\frac{d\phi}{dt}\right)^2 + \frac{\mu}{2}\left(\frac{dr}{dt}\right)^2 + V(r), \tag{3.125}$$

since $\sin^2 \phi + \cos^2 \phi = 1$. The first term in Eq. 3.125 can be written in several ways:

$$\frac{\mu r^2}{2}\left(\frac{d\phi}{dt}\right)^2 = \frac{\mu}{2}\left(r\frac{d\phi}{dt}\right)^2 = \frac{\mu}{2}v_\phi{}^2 = \frac{(\mu v_\phi r)^2}{2\mu r^2} = \frac{p_\phi{}^2}{2I}. \tag{3.126}$$

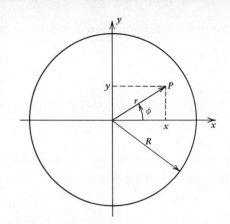

FIGURE 3.9

Circular polar coordinates and the circular box. The point P can be described by the circular coordinates r, ϕ or the Cartesian coordinates x, y. The two coordinate systems are connected by the conversion equations

$$x = r \cos \phi, \qquad y = r \sin \phi$$

or

$$r = (x^2 + y^2)^{1/2}, \qquad \phi = \arctan\left(\frac{y}{x}\right).$$

The heavy line is the boundary of a circular box of radius R.

We have introduced three definitions in these equations:

$$v_\phi \equiv r\frac{d\phi}{dt} \tag{3.127}$$

is the instantaneous velocity perpendicular to the radial direction;

$$p_\phi \equiv \mu v_\phi r \tag{3.128}$$

is our old friend the angular momentum, written with the subscript ϕ to indicate the angle that changes during the rotational motion; and

$$I \equiv \mu r^2 \tag{3.129}$$

is the *moment of inertia*, of which we shall say more later. In the second term of Eq. 3.125 we simply define

$$p_r \equiv \mu\frac{dr}{dt}, \tag{3.130}$$

the momentum in the radial direction. Substituting these definitions in Eq. 3.125, we obtain

$$E = \frac{p_\phi{}^2}{2I} + \frac{p_r{}^2}{2\mu} + V(r). \tag{3.131}$$

The first term in Eq. 3.131 appears to concern only the kinetic energy of angular motion, the second term is the kinetic energy of radial motion, and the third term is the potential energy. So long as V is a function only of r, and not of angle, there is no force acting to affect the angular momentum p_ϕ. Hence, if $V = V(r)$, then p_ϕ must be a constant of the motion. This is a specific example of a completely general and very powerful tool in physics and chemistry, whether in the classical or the quantum mechanical description: *If the energy of a system is independent of some coordinate q_i, then p_i is a constant of the motion of the system.* This principle can be restated in other ways. For example, if the energy is completely independent of some coordinate q, then the uncertainty Δq in this variable is arbitrarily large, and therefore the uncertainty Δp in the conjugate momentum p can be made as small as we choose. For the free particle we can choose the origin $x = 0$ anywhere we wish without affecting the energy, and the momentum is thus a constant. In the present case, energy is independent of the axis from which ϕ is measured. These examples indicate further how the constants of the motion reflect some type of symmetry inherent in the system, as pointed out earlier.

Let us now proceed to the quantum mechanical problem, for which we first need the Hamiltonian operator. One might think that we could simply substitute $-i\hbar(\partial/\partial\phi)$ for p_ϕ and $-i\hbar(\partial/\partial r)$ for p_r in Eq. 3.131, as we have previously done with momenta in Cartesian coordinates. Unfortunately, with non-Cartesian coordinates this method often does not give the correct operators, for mathematical reasons that we need not discuss here.[20] However, we already know the Hamiltonian operator in Cartesian coordinates, in this case

$$\mathsf{H} = -\frac{\hbar^2}{2\mu}\left(\frac{\partial^2}{\partial x^2} + \frac{\partial^2}{\partial y^2}\right) + \mathsf{V}(x, y), \tag{3.132}$$

[20] The basic problem is that the resulting operators may not be Hermitian (see footnote 17 on page 106). A Hermitian operator, Eq. 3A.7, is obtained if one replaces p_r by $r^{-1}\mathsf{p}_r r$ and *then* substitutes $-i\hbar(\partial/\partial r)$ for p_r.

and this can be transformed directly to the equivalent expression in circular coordinates. The details of this transformation are given in Appendix 3A; here we need only the result,

$$\mathbf{H} = -\frac{\hbar^2}{2\mu}\left(\frac{\partial^2}{\partial r^2} + \frac{1}{r}\frac{\partial}{\partial r} + \frac{1}{r^2}\frac{\partial^2}{\partial \phi^2}\right) + \mathbf{V}(r). \qquad (3.133)$$

Inside the circular box we have $\mathbf{V}(r) = 0$, and the time-independent Schrödinger equation becomes

$$-\frac{\hbar^2}{2\mu}\frac{\partial^2\psi(r,\phi)}{\partial r^2} - \frac{\hbar^2}{2\mu r}\frac{\partial\psi(r,\phi)}{\partial r} - \frac{\hbar^2}{2\mu r^2}\frac{\partial^2\psi(r,\phi)}{\partial \phi^2} = E\psi(r,\phi), \quad (3.134)$$

where $\psi(r,\phi)$ is the spatial part of the wave function.

So far, of course, we have done nothing but transform coordinates; the shape of the box does not enter the problem until we introduce the boundary conditions. As in our previous models, the wave function must vanish at the wall of the box where $r = R$, giving the condition

$$\psi(R,\phi) = 0 \qquad \text{for all } \phi. \qquad (3.135)$$

We also need a boundary condition in the angular variable. The wave function must be a single-valued function of position in space, but real space corresponds only to values of ϕ between 0 and 2π. Formally, ϕ can go beyond these limits, but this merely represents going around the same circle again. Two values of ϕ whose difference is exactly 2π (or any integral multiple of 2π) describe the identical physical location. The wave function must therefore be periodic in ϕ, with a period of 2π; otherwise a particular point in real space would correspond to more than one possible value of ψ. Mathematically, this means we must require that

$$\psi(r,\phi) = \psi(r,\phi \pm 2\pi) \qquad \text{for all } r, \phi. \qquad (3.136)$$

The Hamiltonian of Eq. 3.133 is not directly separable; however, multiplying through by r^2 gives an operator that *is* separable. We can therefore write the wave function as a product of functions of r and ϕ separately,

$$\psi(r,\phi) = f(r)g(\phi). \qquad (3.137)$$

We could substitute this in the Schrödinger equation and proceed as we did following Eq. 3.109; however, a simpler method is available to evaluate the angular function. We said above that, since ϕ does not appear explicitly in the energy, the angular momentum p_ϕ must be a constant of the motion. In quantum mechanical terms, this means that p_ϕ is an eigenvalue of the angular momentum operator $\mathbf{p}_\phi$, which is given by

$$\mathbf{p}_\phi \equiv -i\hbar\frac{\partial}{\partial \phi} \qquad (3.138)$$

(see Appendix 3A for proof). It seems logical to assume that $g(\phi)$, the angular part of the wave function, is an eigenfunction of $\mathbf{p}_\phi$. We can thus write

$$-i\hbar\frac{\partial g(\phi)}{\partial \phi} = p_\phi g(\phi), \qquad (3.139)$$

where p_ϕ is a constant. The solution to this equation is simply

$$g(\phi) = Ae^{ip_\phi\phi/\hbar}, \qquad (3.140)$$

as can be verified by direct differentiation.

Now we can apply the angular boundary condition, which in terms of $g(\phi)$ becomes

$$g(\phi) = g(\phi \pm 2\pi). \qquad (3.141)$$

Since the function e^{ix} has a period of 2π in x, we can satisfy this condition with Eq. 3.140 only if $p_\phi \phi/\hbar$ changes through an integral multiple of 2π while ϕ is changing through 2π. In other words, p_ϕ must be an integral multiple of $\hbar$,

$$p_\phi = m\hbar \qquad (m = 0, \pm 1, \pm 2, \ldots), \qquad (3.142)$$

and the angular part of the wave function has the form

$$g(\phi) = Ae^{im\phi} \qquad (m = 0, \pm 1, \pm 2, \ldots). \qquad (3.143)$$

(Keep in mind that m is a quantum number, whereas the mass is μ.) The real part of this function behaves as $A \cos m\phi$, and the imaginary part as $A \sin m\phi$; the two have identical shapes, but differ by 90° in phase. Note that the angular part of the probability density, $g^*(\phi)g(\phi)$, is a constant, giving the expected result that all values of ϕ are equally likely.

Equation 3.142 states that the angular momentum is quantized.[21] The allowable values (except for $m = 0$) occur in pairs, corresponding to clockwise and counterclockwise motion with the same absolute value of the angular momentum. Since the angular momentum appears only as p_ϕ^2 in the expression for the energy, $+m$ and $-m$ must correspond to the same energy level; all the levels except that with $m = 0$ are thus doubly degenerate.

We have not yet considered the radial part of the wave function, $f(r)$. The technique used in evaluating $f(r)$ is straightforward but tedious. One substitutes Eq. 3.137 in the Schrödinger equation, reduces to obtain an equation involving r alone, and solves that equation for $f(r)$. Neither the details of this derivation nor the exact form of the solution need be considered here, though some additional information is given in Appendix 3A. The following points about the solution are significant: The function $f(r)$, of course, describes a standing wave in the radial direction. The form of this function depends explicitly on $|m|$, the absolute value of the angular momentum quantum number; and for each value of $|m|$ there is an infinite set of energy eigenvalues that we can enumerate by a second quantum number n, which takes on the values of all the positive integers. The complete spatial wave function can thus be written as[22]

$$\psi_{n,m}(r, \phi) = Af_{n,|m|}(r)e^{im\phi}. \qquad (3.144)$$

As we explain in Appendix 3A, there is no way to express the radial function $f(r)$ directly in terms of simple (algebraic, trigonometric, or exponential) functions, as we have been able to do with other wave functions. We can nevertheless say something about this solution. It turns out that the forms of the functions $\psi(r, \phi)$ described by Eq. 3.118 are exactly the same as the possible forms of a standing wave on a circular drumhead. This parallels our earlier analogy between the wave functions in a one-dimensional box and the vibrations of a string with both ends

[21] It will be recalled from Section 2.7 that in circular motion the action per cycle is 2π times the angular momentum. Here this would give an action of mh, in agreement with our earlier assumptions on the quantization of action.

[22] The constant A can be evaluated by normalization over the area of the box. The area element in circular coordinates is $r \, dr \, d\phi$, so for one particle in the box we must have

$$\int_{\phi=0}^{2\pi} \int_{r=0}^{R} |\psi(r, \phi)|^2 \, r \, dr \, d\phi = 1.$$

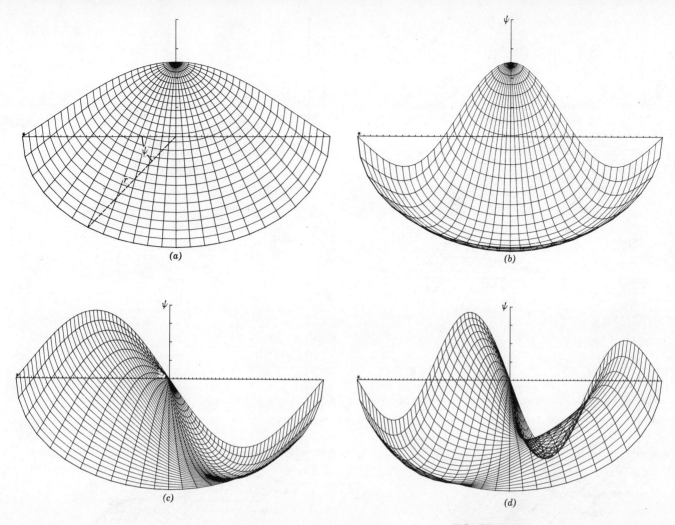

(a)

(b)

(c)

(d)

FIGURE 3.10
Standing waves in a circular box (perspective views). These are the wave functions of Eq. 3.144, with the quantum numbers m and n of: a) 0, 1; b) 0, 2; c) 1, 1; d) 1, 2. Radial nodes are circles, angular nodes are radii in the plane corresponding to $\Psi = 0$. Only half the circle is shown; the semicircle "behind" the plane of the page is omitted.

fixed: In each case, the boundary conditions are mathematically equivalent. Figure 3.10 illustrates some of the simpler standing waves given by Eq. 3.144; for comparison, Fig. 3.11 shows some standing waves for a rectangular box (or drumhead), that is, the functions of Eq. 3.118 in two dimensions. Physical analogies such as those introduced here are common in physics and chemistry: In innumerable cases one finds systems that are very different physically, yet are described by the same mathematical equations. The recognition of these analogies frequently enables us to economize our efforts in solving new problems; in addition, we often find quick and penetrating insights into new systems as we discover exact or nearly exact analogies for them in systems already known to us.

The energy eigenvalues cannot be expressed in simple form either, but some of them are plotted in Fig. 3.12. Each value of $|m|$ has its own set of energy levels, and all those except $|m| = 0$ are doubly degenerate. The quantum number n is an index number that is necessary but not sufficient to specify the state and its energy; n is not directly related to the energy by any simple equation like Eq. 3.71. In general, any two levels with different angular momenta (different $|m|$) also have different energies. The quantity actually plotted in Fig. 3.12 is the dimensionless $2\mu R^2 E/\hbar^2$, which is a function of only the quantum numbers; the energy of a given state is thus inversely proportional to the square of the box's radius R.

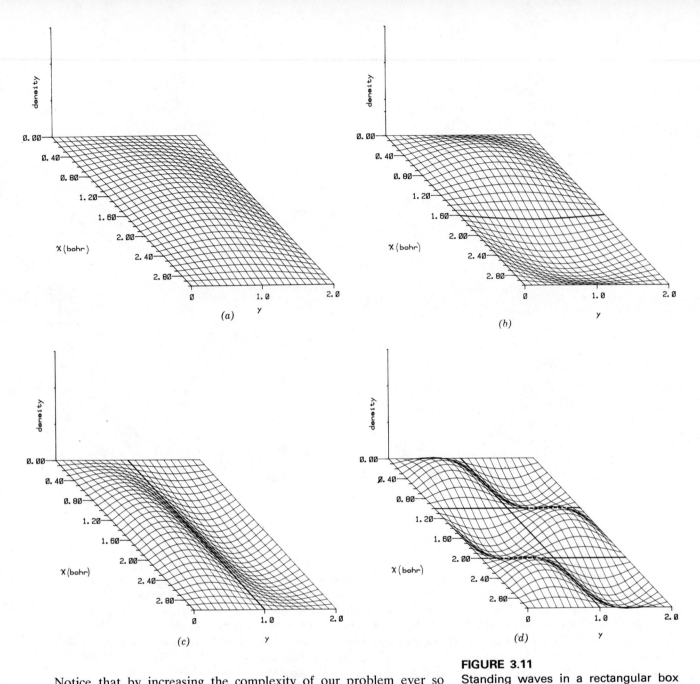

(a)

(b)

(c)

(d)

FIGURE 3.11
Standing waves in a rectangular box (perspective views). The quantum numbers n_1 and n_2 are those used in Eq. 3.118 with a) $n_1 = n_2 = 1$; b) $n_1 = 2$, $n_2 = 1$; c) $n_1 = 1$, $n_2 = 2$; d) $n_1 = 3$, $n_2 = 2$; see Fig. 3.8a for the energy levels corresponding to these states. The lines drawn across the rectangle indicate the nodes of the wave function.

Notice that by increasing the complexity of our problem ever so slightly, simply by going from a square box to a round box, we have entered a realm where the mathematics begins to look unfamiliar and much more complex. Yet the basic physics of the problem remains very similar. In the square and circular cases, and in any other two-dimensional box with a simple boundary, we have every expectation that the wave function of the lowest energy state will have no nodes in the plane, and that the next higher state will have one node. There may be more than one way of placing this node, and the different ways of placing the node may correspond to somewhat different energy values. The next functions will have two nodes, and so on.[23] The number of levels will be

---

[23] These sets will in general overlap in energy: For example, the state $|m| = 2$, $n = 1$, with two nodes, has a lower energy than $|m| = 0$, $n = 2$, with one node. But for each value of $|m|$ the number of nodes increases directly with energy.

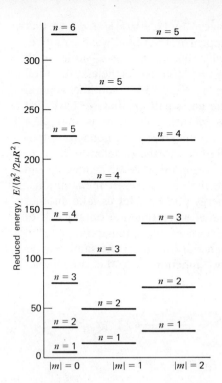

countably infinite, and in general their spacing will grow as the energy gets higher and higher. This similarity is not surprising, since the various boxes form part of a continuous set. Suppose we replace the drumhead of our analogy by a rubber sheet stretched on a flexible rim. As we deform the rim from one shape to another, the standing-wave patterns will vary continuously. In the same way, deformation of our box boundary causes the wave functions and their corresponding energies to vary continuously, but without changing the overall pattern. Degeneracy, of course, occurs only when the box has some kind of symmetry, but this may be subtle; see Fig. 3.8. caption.

Relating the nodal character of a wave function in a circular box to the de Broglie relation is a bit more complicated than it was for a rectangular box. The angular nodes in the real and imaginary parts of $e^{im\phi}$ correspond to the radii, the lines of constant ϕ, marking well-defined wavelengths for angular motion. These wavelengths correspond precisely to the constant angular momentum of the particle in a circular box. However the radial "wavelength" is not a constant; the momentum operator does not give back a multiple of $f_{n,|m|}(r)$ when it operates on this function, and the energy is not simply a sum of independent radial and angular parts. Nevertheless, we can recognize that the more radial nodes $f_{n,|m|}(r)$ has, the more will be the average radial kinetic energy of the particle. More nodes and more waves in a fixed interval along a given direction always mean more curvature, more mean-square average momentum, and more average kinetic energy along that direction.

So far we have considered only problems in which the potential energy is zero (or infinite). Let us briefly examine what happens when this limitation is removed. We again consider a particle in a circular box, but now suppose that the classical potential energy $V(r)$ or its quantum counterpart $V(r)$ within the box is not constant. We retain the assumption of circular symmetry, so that V depends only on the radial distance r and not on the angle ϕ. We can then still make the separation of Eq. 3.137 and obtain the same angular wave function as before, Eq. 3.143. However, the right side of Eq. 3.134 is replaced by $[E - V(r)]\psi(r, \phi)$, so that an additional term involving $V(r)$ will appear in the radial wave equation. The resulting equation will be somewhat harder to solve—how much harder depends on the form of the function $V(r)$. Nevertheless, the total wave function will still be describable by Eq. 3.144, and the general behavior of the solutions will follow the basic pattern that we have just described. In this case, the analogous vibration problem is that of a vibrating circular drumhead of nonuniform thickness. The nonzero potential energy thus makes no essential difference to the problem (except to complicate the mathematics); here and in general, it is the symmetry and the boundary conditions that determine the basic form of the solutions.

Returning to the zero-potential-energy problem, let us take another look at the form of Eq. 3.144. The spatial wave function obtained is of course complex (except for $m = 0$), because of the angular factor $e^{im\phi}$. However, we can generate real wave functions in much the same way as we obtained Eqs. 3.42, by combining two functions with the same values of $|m|$ and n:

$$\psi'_{n,|m|}(r, \phi) = \frac{\psi_{n,m}(r, \phi) + \psi_{n,-m}(r, \phi)}{2}$$

$$= \frac{A}{2} f_{n,|m|}(r)(e^{im\phi} + e^{-im\phi})$$

$$= A f_{n,|m|}(r) \cos m\phi;$$

$$\psi''_{n,|m|}(r, \phi) = \frac{\psi_{n,m}(r, \phi) - \psi_{n,-m}(r, \phi)}{2i}$$

$$= \frac{A}{2i} f_{n,|m|}(r)(e^{im\phi} - e^{-im\phi})$$

$$= A f_{n,|m|}(r) \sin m\phi.$$

(3.145)

The functions in 3.145 are not eigenfunctions of angular momentum, just as those in Eqs. 3.42 were not eigenfunctions of linear momentum. Nevertheless, they are still just as good eigenfunctions of *energy* as the original function, 3.144. At the moment, these functions may seem like nothing but mathematical curiosities. However, we shall later face problems in which the angular momentum no longer has a good quantum number, that is, is not conserved for a particular particle. In such a case functions like Eq. 3.145 are more useful than those like Eq. 3.144.

3.11
Particles in Spherical Boxes

The next step to take beyond the circular box is obviously a consideration of its three-dimensional analog, the spherical box. We are interested primarily in what new features are introduced by the added dimension. As just pointed out, the symmetry of a problem governs its solution in a more fundamental way than does the detailed form of the potential energy (although, of course, the potential and the boundary conditions fix the symmetry). The spherical box is therefore of particular interest, since we expect an atom to have spherical symmetry around its nucleus. We shall not give the calculations in as great detail as in the previous section, since they are for the most part similar in form but more complicated.

The system we consider is a particle of mass μ in a spherical box of radius R. The potential energy is $V(r)$, a function only of the distance from the center of the sphere; we assume $V(r)$ to equal zero for $r < R$ and to be infinite for $r \geq R$. We shall use the spherical coordinates r, θ, ϕ

defined in Appendix 2B, which should be consulted for a diagram. As with the circular box, we begin by outlining the relevant classical relationships.

In three dimensions the kinetic energy of a particle of mass μ is

$$T = \frac{1}{2\mu}(p_x{}^2 + p_y{}^2 + p_z{}^2) = \frac{1}{2\mu}\mathbf{p} \cdot \mathbf{p}, \qquad (3.146)$$

where $p_x = \mu(dx/dt)$, and so on. Using the conversion relations from Appendix 2B, we can transform each of the momenta to spherical coordinates by a method like that of Eqs. 3.96ff. The kinetic energy then becomes

$$T = \frac{1}{2\mu}\left(p_r{}^2 + \frac{p_\theta{}^2}{r^2} + \frac{p_\phi{}^2}{r^2 \sin^2 \theta}\right), \qquad (3.147)$$

where

$$p_r \equiv \mu \frac{dr}{dt} \qquad (3.148)$$

is the *classical* radial momentum,

$$p_\theta \equiv \mu r^2 \frac{d\theta}{dt} \qquad (3.149)$$

is the angular momentum corresponding to rotation in the plane formed by the radius vector $\mathbf{r}$ and the z axis, and

$$p_\phi \equiv \mu(r \sin \theta)^2 \frac{d\phi}{dt} \qquad (3.150)$$

is the angular momentum around the z axis (i.e., for rotation in the xy plane, in which $r \sin \theta$ is the projection of $\mathbf{r}$). Equation 3.147 corresponds to Eq. 3.131 for the two-dimensional case; the resemblance is closer if we write μr^2 as I, the moment of inertia.

The total angular momentum around the origin is defined as the vector

$$\mathbf{L} \equiv \mathbf{r} \times \mathbf{p}, \qquad (3.151)$$

in terms of the cross product introduced in Eqs. 1.17. If the motion of the particle is in a plane (as it must be in the absence of external forces), the vector $\mathbf{L}$ is directed perpendicular to that plane. In terms of the spherical coordinates, the square of its magnitude can be shown to be given by

$$L^2 = \mathbf{L} \cdot \mathbf{L} = p_\theta{}^2 + \frac{p_\phi{}^2}{\sin^2 \theta}. \qquad (3.152)$$

The kinetic energy can thus also be written as

$$T = \frac{p_r{}^2}{2\mu} + \frac{L^2}{2I} \qquad (I \equiv \mu r^2), \qquad (3.153)$$

i.e. the sum of a part associated with radial momentum and a part associated with the total angular momentum.

Besides the total angular momentum, it is convenient to single out one particular axis and consider the angular momentum around that axis. In our spherical coordinate system the mathematics is simplest if we choose this axis to be the z axis, the angular momentum around which (L_z) is simply p_ϕ. Since the kinetic energy is independent of the angle ϕ, its conjugate momentum p_ϕ must be a constant of the motion. Another such constant is of course the total angular momentum, which is conserved as long as no external torque is applied to the system.

Let us now consider the quantum mechanical problem. The Hamiltonian operator in spherical coordinates turns out to be[24]

$$H = -\frac{\hbar^2}{2\mu}\left[\frac{1}{r^2}\frac{\partial}{\partial r}\left(r^2\frac{\partial}{\partial r}\right) + \frac{1}{r^2\sin\theta}\frac{\partial}{\partial\theta}\left(\sin\theta\frac{\partial}{\partial\theta}\right)\right.$$
$$\left. + \frac{1}{r^2\sin^2\theta}\frac{\partial^2}{\partial\phi^2}\right] + V(r), \quad (3.154)$$

which can be compared with Eq. 3.133 for the circular case. We give this merely for purposes of illustration, since we have no intention of describing how the wave equation is solved. However, it is worthwhile to consider certain general characteristics of the solution that depend on the spherical symmetry. Whenever the potential energy is a function of r only, the Schrödinger equation can be split into equations separately involving the radial coordinate r, the angle θ, and the angle ϕ. The wave function is then a product of functions involving the three variables separately,

$$\psi(r, \theta, \phi) = f(r)\Theta(\theta)\Phi(\phi). \quad (3.155)$$

The radial function $f(r)$ will in general depend strongly on the form of the potential energy; even for $V(r) = 0$ it will be a more complicated function than in the circular box, subject to the same boundary condition $f(R) = 0$. We shall say no more about it. The angular functions, however, have several interesting aspects.

Consider first the function involving the angle ϕ. Since ϕ plays exactly the same role here as in the circular coordinate system, we can expect the function $\Phi(\phi)$ to behave in the same way as did our $g(\phi)$ in the circular box. It must thus be an eigenfunction of the operator corresponding to p_ϕ, which is still $-i\hbar(\partial/\partial\phi)$, and the corresponding eigenvalues will be the quantized values of $p_\phi = L_z$, the angular momentum around the z axis. The boundary condition is again a periodicity of 2π in ϕ, and we obtain

$$p_\phi = L_z = m\hbar \qquad (m = 0, \pm 1, \pm 2, \ldots) \quad (3.156)$$

just as before. The term in the kinetic energy involving p_ϕ then becomes

$$\frac{p_\phi{}^2}{2\mu r^2\sin^2\theta} = \frac{m^2\hbar^2}{2\mu r^2\sin^2\theta}. \quad (3.157)$$

We can consider this term as an effective potential energy (corresponding to centrifugal force around the z axis), with a minimum in the xy plane and an infinite maximum all along the z axis; only for $m = 0$ can a particle with finite total energy reach the z axis. This is intuitively reasonable: In Cartesian coordinates we have $p_\phi = xp_y - yp_x$ (cf. Appendix 3A), so that the line $x = y = 0$ can be reached only by achieving infinite values of p_x or p_y (unless p_ϕ, and thus m, is zero).

We also expect the wave function or, more precisely, its angular parts, to be an eigenfunction of the operator corresponding to the total angular momentum, since this quantity is conserved. We might expect the eigenvalues to be defined by an equation similar to Eq. 3.156, for example,

$$L^2 \overset{?}{=} l^2\hbar^2 \qquad (l = 0, 1, 2, \ldots). \quad (3.158)$$

(We are interested only in the magnitude of the angular momentum,

---

[24] This can be derived from the Hamiltonian in Cartesian coordinates by a method like that used in Appendix 3A.

which is why we write an equation for L^2 rather than **L**.) We shall see that this is not quite correct, but it is sufficiently accurate for illustrative purposes. If we substitute our quantized equations for L^2 and p_ϕ into Eq. 3.152, we find that the remaining component of angular momentum is given approximately by

$$p_\theta = \left(L^2 - \frac{p_\phi^2}{\sin^2\theta}\right)^{1/2} \approx \left(l^2\hbar^2 - \frac{m^2\hbar^2}{\sin^2\theta}\right)^{1/2}. \qquad (3.159)$$

Since p_θ is a measurable quantity, it must have a real value; this means that we require

$$l^2 \sin^2\theta - m^2 \geq 0 \qquad (3.160)$$

for all θ, which in turn yields

$$-l \leq m \leq l. \qquad (3.161)$$

In short, the length $l\hbar$ of the vector **L** must be at least as great as the length $|m|\hbar$ of its projection on the z axis—a quite logical result.

As we said, Eq. 3.158 is not actually the correct result. Just as we can obtain the eigenvalues of p_ϕ from the wave equation in ϕ, we can expect to obtain the eigenvalues of L^2 from the wave equation in θ. This is a fairly complicated equation, but its solutions can be written in closed form (i.e., in terms of a finite number of simple functions). The boundary condition that must be satisfied is governed by the fact that θ goes only from 0 to π, so that, for example, $\Theta(\pi + \theta)$ must equal $\Theta(\pi - \theta)$; as we shall see, the solutions are all sums of sines and cosines, which automatically satisfy this condition. As with the radial function of Eq. 3.144, the solution depends on the value of $|m|$, each $|m|$ corresponding to an infinite series of solutions numbered with the new quantum number l. It turns out that the eigenvalues of total angular momentum are given by

$$L^2 = l(l+1)\hbar^2 \qquad (l = 0, 1, 2, \ldots) \qquad (3.162)$$

rather than Eq. 3.158. However, this still yields Eq. 3.161 for the relationship between m and l; for each value of l there are $2l+1$ possible values of m.

The replacement of l^2 by $l(l+1)$ is essentially due to the uncertainty principle. Suppose that Eq. 3.158 were correct. Then for the case $|m| = l$, and thus $m^2 = l^2$, Eq. 3.160 would allow only $\sin^2\theta = 1$, $\theta = \pi/2$, which in turn yields $p_\theta = 0$. But by the uncertainty principle we cannot simultaneously have exact knowledge of both θ and p_θ, which are a conjugate pair of variables. This problem is removed when we replace l^2 by $l(l+1)$, which is always greater than m^2 (unless $l = m = 0$), allowing a range of possible values of θ rather than the single value $\pi/2$. The physical meaning of this result is that the component of angular momentum along a particular axis can never be quite as large as the total (nonzero) angular momentum. (In the case $l = m = 0$, we have total information about all the components of the angular momentum, but the wave function is spherically symmetric, so we have absolutely no information about θ or ϕ.)

Let us consider the term $L^2/2\mu r^2$ in Eq. 3.153, which gives the part of the kinetic energy associated with rotation, the *rotational* or *centrifugal* energy T_{rot}. If L is conserved, as it must be in the absence of external torque, the centrifugal energy increases as the particle comes closer to the origin, which is the center of force in a potential field $V(r)$. For example, a satellite in an elliptical orbit moves fastest when it is closest to the earth. From Eq. 3.162 we can immediately write the quantum expression for the centrifugal energy,

$$T_{\text{rot}} = \frac{l(l+1)\hbar^2}{2\mu r^2} \ . \tag{3.163}$$

This would be the total energy for a particle in a circular orbit (r constant). Note that it depends on l but not on m, so that the $2l + 1$ values of m allowable for a given l correspond to a $(2l+1)$-fold degenerate energy level. The same result is found in the general case, with the energy eigenvalues depending on l and a radial index n, but not on m. Note that the energy levels grow farther and farther apart as l increases, resembling the behavior we have found for other quantum systems.

Let us now take a closer look at the properties of the quantized angular momentum. Remember that the angular momentum is a vector, directed along the axis around which rotation is occurring. Its magnitude is given by

$$L = |\mathbf{r} \times \mathbf{p}| = rp \sin \chi, \tag{3.164}$$

where χ is the angle between the vectors $\mathbf{r}$ and $\mathbf{p}$. The vector $\mathbf{L}$ forms the positive z axis in a right-handed coordinate system (cf. Fig. 1.3) in which $\mathbf{r}$ and $\mathbf{p}$ define the x and y axes, respectively. If the motion is in a circle, then $\mathbf{r}$ and $\mathbf{p}$ are always perpendicular ($\sin \chi = 1$), p is constant, and L has the familiar value $\mu v r$ (or $\mu r^2 \omega$, where $\omega = v/r$ is the angular velocity). In an arbitrary coordinate system, $\mathbf{L}$ will in general have a nonvanishing component (i.e., projection) along any axis chosen at random. Classically, one could specify the components along any such axes, for example, the three Cartesian axes; in quantum mechanics this is no longer true.

We have outlined how the angular parts of the Schrödinger equation give two quantized values of angular momentum. The solution for $\Theta(\theta)$ gives us the magnitude of the total angular momentum, whereas the solution for $\Phi(\phi)$ gives us its component along the z axis. If we were to change the axis about which we measure ϕ, that is, if we were to redefine the z axis, we could obtain the component of angular momentum along the new axis instead. However, no matter how we place the axes, we can determine only one component at a time; the other components must remain unknowable. We cannot simultaneously have precise values for three, or even two, components of the angular momentum unless all are zero, i.e., unless there is no rotation. This can be demonstrated by obtaining the operators (designated by $\mathbf{L}$ rather than L) corresponding to the total angular momentum and its components: $\mathbf{L}^2$ commutes with $\mathbf{L}_x$, $\mathbf{L}_y$, or $\mathbf{L}_z$, but no two of the components commute with each other. According to the uncertainty principle, then, we can simultaneously measure L and either L_x, L_y, or L_z, but measurement of any one component disturbs the values of the other components.

Let us expand on the consequences of this limitation on our knowledge. In a system with spherical symmetry, all axes are equivalent; thus we can arbitrarily choose any axis as our z axis. We *assume* that this is the axis with respect to which the angular momentum is quantized, that is, we assume that the wave function is an eigenfunction of p_ϕ, with Eq. 3.156 giving the eigenvalues of p_ϕ. We can also obtain the magnitude of the total angular momentum, which is quantized according to Eq. 3.162. It has already been pointed out that a nonzero $|\mathbf{L}|$ must be larger than L_z, so L_x and L_y (which furnish the rest of the length of $\mathbf{L}$) cannot both be zero. In fact, we can evaluate

$$L_x^2 + L_y^2 = L^2 - L_z^2 = [l(l+1) - m^2]\hbar^2, \tag{3.165}$$

which is a constant for any given values of m and l. But the uncertainty principle tells us that we cannot know L_x or L_y separately. In fact, just as all values of ϕ are equally likely, so all directions are equally likely for the projection of **L** in the xy plane, the magnitude of which is $(L_x{}^2 + L_y{}^2)^{1/2}$. The best we can do is to obtain a probability distribution for the values of L_x and L_y.

We can clarify the quantization of angular momentum by a geometric interpretation, as illustrated in Fig. 3.13. Each allowed value of L^2, corresponding to a value of the quantum number l, fixes the length of the vector **L** and thus defines a sphere about the origin; only one such sphere is shown in the figure. Each allowed value of L_z, corresponding to a value of m, fixes the z component of **L** and defines a plane parallel to the xy plane. When we specify particular values of both l and m, we choose both the sphere and the plane corresponding to $|\mathbf{L}|$ and L_z. The intersection of the sphere and the plane is a circle, which defines the locus of all vectors compatible with the quantized values of $|\mathbf{L}|$ and L_z. If we think of the vector **L** as a movable arrow of fixed length attached by a swivel to the origin, this circle is the base of the cone along whose side **L** may lie.[25] The altitude of the cone is L_z, and the radius of its base is $(L_x{}^2 + L_y{}^2)^{1/2}$. All vectors lying on this cone are equally likely representations of **L**. We thus see that not only the magnitude but also the direction of **L** is quantized, with the allowed directions restricted to the cones of Fig. 3.13. For each value of l there are $2l+1$ such cones allowed, corresponding to the possible values of m from $-l$ to l. This phenomenon is called *space quantization*.

So far we have not looked in any detail at the form of the spherical-box wave functions. The radial functions we can disregard, but the angular functions have a significance much wider than this particular problem. Since the only angular boundary conditions are those imposed by the coordinate system, we must obtain the *same* angular wave functions for any system with spherical symmetry—and all atoms have spherical symmetry.

Until now we have written the angular function as a product of separate functions of θ and ϕ. These are often combined into a single function, the *spherical harmonic*

$$Y_{l,m}(\theta, \phi) = \Theta_{l,|m|}(\theta)\Phi_m(\phi), \qquad (3.166)$$

where l and m are the quantum numbers previously introduced. We already know that $\Phi_m(\phi)$ is $e^{im\phi}$, but the derivation of $\Theta_{l,|m|}(\theta)$ is too complicated to justify our including it here. However, since these functions are of such general applicability, it is desirable to have some familiarity with their properties. We therefore include in Table 3.1 a list of all the spherical harmonics with values of l up to 3. The quantum number l specifies the total angular momentum, and a finite number $(2l+1)$ of m values correspond to each possible value of l; it is thus natural to classify the solutions primarily by the value of l.

We have pointed out earlier that increased momentum is associated with increasing oscillation of the wave function; the same is true of angular momentum and oscillation with angle. We might thus expect that the higher the values of l and m, the more "wiggly" the spherical harmonic will be as a function of θ and ϕ. The nodes of the wave functions for a particle in a spherical box are of three kinds: radial nodes, analogous to the radial nodes in the circular box; nodes in $\Theta(\theta)$, and

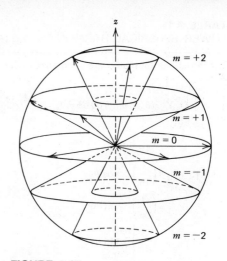

FIGURE 3.13
A graphic representation of quantized angular momentum. Each cone is the locus of the total angular momentum vector **L** for a particular value of m consistent with $l = 2$. The magnitude of **L** here is $[l(l+1)]^{1/2}\hbar = \sqrt{6}\hbar$; its z component L_z is $m\hbar$.

---

[25] If in addition to $|\mathbf{L}|$ and L_z we also knew L_x and L_y, the circle would be reduced to a single point. But this would be more information than the uncertainty principle allows us to have.

TABLE 3.1
LOWER SPHERICAL HARMONICS, $Y_{l,m}(\theta, \phi)$

The functions tabulated are normalized over the surface of a sphere (cf. Appendix 2B):

$$\int_{\phi=0}^{2\pi} \int_{\theta=0}^{\pi} Y_{l,m}(\theta, \phi) \sin \theta \, d\theta \, d\phi = 1.$$

$l = 0$: $Y_{0,0} = \left(\dfrac{1}{4\pi}\right)^{1/2}$

$l = 1$: $Y_{1,0} = \left(\dfrac{3}{4\pi}\right)^{1/2} \cos \theta$

$\qquad Y_{1,1} = \left(\dfrac{3}{8\pi}\right)^{1/2} \sin \theta \, e^{i\phi}$

$\qquad Y_{1,-1} = \left(\dfrac{3}{8\pi}\right)^{1/2} \sin \theta \, e^{-i\phi}$

$l = 2$: $Y_{2,0} = \left(\dfrac{5}{16\pi}\right)^{1/2} (3 \cos^2 \theta - 1)$

$\qquad Y_{2,1} = \left(\dfrac{15}{8\pi}\right)^{1/2} \cos \theta \sin \theta \, e^{i\phi}$

$\qquad Y_{2,2} = \left(\dfrac{15}{32\pi}\right)^{1/2} \sin^2 \theta \, e^{2i\phi}$

$\qquad Y_{2,-1} = \left(\dfrac{15}{8\pi}\right)^{1/2} \cos \theta \sin \theta \, e^{-i\phi}$

$\qquad Y_{2,-2} = \left(\dfrac{15}{32\pi}\right)^{1/2} \sin^2 \theta \, e^{-2i\phi}$

$l = 3$: $Y_{3,0} = \left(\dfrac{63}{16\pi}\right)^{1/2} (\tfrac{5}{3} \cos^3 \theta - \cos \theta)$

$\qquad Y_{3,1} = \left(\dfrac{21}{64\pi}\right)^{1/2} (5 \cos^2 \theta - 1) \sin \theta \, e^{i\phi}$

$\qquad Y_{3,2} = \left(\dfrac{105}{32\pi}\right)^{1/2} \sin^2 \theta \cos \theta \, e^{2i\phi}$

$\qquad Y_{3,3} = \left(\dfrac{35}{64\pi}\right)^{1/2} \sin^3 \theta \, e^{3i\phi}$

$\qquad Y_{3,-1} = \left(\dfrac{21}{64\pi}\right)^{1/2} (5 \cos^2 \theta - 1) \sin \theta \, e^{-i\phi}$

$\qquad Y_{3,-2} = \left(\dfrac{105}{32\pi}\right)^{1/2} \sin^2 \theta \cos \theta \, e^{-2i\phi}$

$\qquad Y_{3,-3} = \left(\dfrac{35}{64\pi}\right)^{1/2} \sin^3 \theta \, e^{-3i\phi}$

nodes in the real and imaginary parts of $\Phi(\phi)$. The radial nodes are spherical surfaces; more radial nodes means higher average kinetic energy of motion through the origin, even though the radial momentum or even its square is not constant. The nodes in $\Theta(\theta)$ are cones, surfaces of constant θ. The greater is l for a given m, the greater is the average angular momentum over the poles. The nodes in the real and imaginary parts of $\Phi(\phi)$ are planes of constant longitude. The greater is m for a given l, the greater is the angular momentum around the polar axis.

In more detail, the real part of $e^{im\phi}$ has m nodes between 0 and π (the m nodes between π and 2π are on the same nodal planes extended through the origin, and are not counted separately). The function $\Theta(\theta)$, a polynomial of order l in $\sin \theta$ and $\cos \theta$, has $l - m$ nodes ($\theta = 0$ and $\theta = \pi$

do not count, since the nodal planes in ϕ all pass through the z axis). Let us look at specific cases. For the simplest case, $l = m = 0$, the angular momentum is zero; this means that the system has no angular preferences, and the angular part of the wave function is simply a constant. The functions with $l = 1$ have a single angular node: for $\cos \theta$ the node is on the equatorial plane $\theta = \pi/2$; for $\sin \theta \, e^{\pm i\phi}$ there is a node on the plane defined by $\phi = 0$ and $\phi = \pi$. Similarly, the functions with $l = 2$ have two nodes, and so forth. In general, for a given value of l there are $2l + 1$ different spherical harmonics, each of which has l angular nodes.

We shall see that the angular behavior of the spherical harmonics—especially the *location* of the nodes—is fundamental to the distribution of electrons in atoms, and thus to the shapes of molecules. The electrons in each atom, and thus the bonds they form, are most likely to be at those angles where the wave function has its maxima. We return to this subject in the next chapter, where we consider the wave functions of the hydrogen atom; diagrams of the spherical harmonics will be given at that point.

3.12 The Rigid Rotator

The particle of the previous section was in a spherically symmetric environment, but was free to move both radially and rotationally. One can also envision a simpler system, known as a *rigid rotator*, whose only form of motion is rotation. An example would be a point mass rigidly fixed at a constant distance from a center about which it is free to swing. There are no pure rigid rotators in nature, but the rotation of molecules can be described quite well by this model.

We consider first the elementary system just described, a mass μ at a constant distance R from the origin, with no external forces. Since the mass can have no radial motion, the kinetic energy of Eq. 3.153 reduces to $T = L^2/2\mu R^2$. The wave function is a function of θ and ϕ only, and must have exactly the same form as the angular solutions in the spherical box—that is, the spherical harmonics of Table 3.1. Since the total energy consists of only the energy of rotation, its Hamiltonian is a simplified form of Eq. 3.154,

$$H = \frac{\hbar^2}{2\mu R^2}\left(\frac{1}{\sin \theta}\frac{\partial}{\partial \theta}\sin \theta \frac{\partial}{\partial \theta} + \frac{1}{\sin^2 \theta}\frac{\partial^2}{\partial \phi^2}\right), \tag{3.167}$$

with no radial contribution to the kinetic energy. The Schrödinger equation for the rigid rotator is therefore

$$H\psi(\theta, \phi) = \frac{\hbar^2}{2\mu R^2}\left(\frac{1}{\sin \theta}\frac{\partial}{\partial \theta}\sin \theta \frac{\partial}{\partial \theta} + \frac{1}{\sin^2 \theta}\frac{\partial^2}{\partial \phi^2}\right)\psi(\theta, \phi) = E\psi(\theta, \phi), \tag{3.168}$$

which simplifies drastically when we use Eqs. 3.153, 3.163, and 3.166:

$$H\psi(\theta, \phi) = \frac{L^2}{2\mu R^2} Y_{l,m}(\theta, \phi)$$

$$= \frac{\hbar^2}{2\mu R^2} l(l+1) Y_{l,m}(\theta, \phi). \tag{3.169}$$

Consequently, the eigenvalues of energy for the rigid rotator are those given in Eq. 3.163:

$$E = \frac{l(l+1)\hbar^2}{2\mu R^2} \qquad (l = 0, 1, 2, \ldots), \tag{3.170}$$

with each energy level $(2l+1)$-fold degenerate.

We know that the spacing between energy levels increases with increasing l, but how does the total number of quantum states vary with energy? By a "state" we mean a solution with a particular set of quantum numbers, so that the $2l+1$ values of m corresponding to a given l define distinct states. Let us determine the number of such states per unit energy, the *density of states*, in a given energy range. For a given value of l this is $2l+1$ times the number of energy levels (values of l) per unit energy. In most molecules the spacing between rotational energy levels is very small, and at ordinary temperatures the average value of l is so large that l can be taken as a continuous variable. In this limit ($l \gg 1$) we can obtain the density of states by differentiating Eq. 3.170:

$$\text{density of states} = (2l+1)\frac{dl}{dE} = (2l+1)\left(\frac{dE}{dl}\right)^{-1}$$

$$= (2l+1)\left[\frac{\hbar^2}{2\mu R^2}(2l+1)\right]^{-1} = \frac{2\mu R^2}{\hbar^2}. \quad (3.171)$$

That is, for the rigid rotator the density of states at high l is a constant, independent of energy or quantum number and depending only on the moment of inertia μR^2. By contrast, for a particle in a three-dimensional box[26] the density of states is proportional to $E^{1/2}$. We shall see in Chapter 15 that the density of states is a very important characteristic of any system, affecting the distribution of particles over the available energy levels and thus the system's thermodynamic properties.

The result of Eq. 3.170 is valid not only for the simple mass-point rotator, but for any system that can be reduced to it. By "reducing" we mean the sort of thing we spoke of in Section 2.11: The motion of the two particles in a hydrogen atom is equivalent to the motion of a single particle around a center of force. Similar reductions can be carried out for more than two particles if their arrangement is sufficiently symmetric. The systems of particles with which we are concerned here are of course molecules, the particles being their constituent atoms.

Let us see what "sufficiently symmetric" means in this context. In general, the rotational energy of a molecule is a very complicated function of the positions and momenta of the individual atoms. However, one can to a fairly good approximation treat a rotating molecule as a rigid body (neglecting vibrations and electronic motion). For any rigid body one can always find a Cartesian coordinate system in terms of which the rotational energy has the simple form

$$E_{\text{rot}} = \frac{L_x^2}{2I_x} + \frac{L_y^2}{2I_y} + \frac{L_z^2}{2I_z}, \quad (3.172)$$

where the L's are the components of the total angular momentum $\mathbf{L} \equiv \sum_i \mathbf{r}_i \times \mathbf{p}_i$, summed over all particles, and the I's are constants. The axes of this coordinate system are called the *principal axes*, and have their origin at the center of mass. The *principal moments of inertia* I_x, I_y, I_z

---

[26] For example, in a cubical box the energy is $n^2\pi^2\hbar^2/2ma^2$, where $n^2 \equiv n_1^2 + n_2^2 + n_3^2$. As n becomes large, the number of states between n and $n+dn$ approaches the limit $4\pi n^2\, dn$. The density of states thus becomes

$$4\pi n^2\left(\frac{dn}{dE}\right) = 4\pi n^2\left(\frac{dE}{dn}\right)^{-1} = 4\pi n^2\left(\frac{n\pi^2\hbar^2}{ma^2}\right)^{-1} = \left(\frac{4ma^2}{\pi\hbar^2}\right)n,$$

proportional to n and thus to $E^{1/2}$.

then have the form

$$I_x \equiv \sum_i m_i(y_i^2 + z_i^2),$$

$$I_y \equiv \sum_i m_i(x_i^2 + z_i^2), \qquad (3.173)$$

$$I_z \equiv \sum_i m_i(x_i^2 + y_i^2)$$

($m_i =$ mass), with each sum taken over all the atoms in the molecule.

Suppose now that all three moments of inertia are equal, $I \equiv I_x = I_y = I_z$; such a molecule is called a *spherical top*. Symmetry of this type is found for all molecules with tetrahedral, octahedral or cubic symmetry, such as CH_4, SF_6 or cubane, C_8H_8, respectively. For a spherical top the rotational energy reduces to simply $L^2/2I$, just as for the single-particle rotator. The rotational energy levels obtained are thus those of Eq. 3.170, with μR^2 replaced by I.

The next case in order of complexity is the *symmetric top*, a molecule in which only two of the moments of inertia are equal: $I_0 \equiv I_x = I_y \neq I_z$. Molecules of this type include the pyramidal NH_3 (ammonia) and CH_3Cl (methyl chloride), the triangular BCl_3, and the hexagonal C_6H_6 (benzene); each has an axis of at least threefold symmetry, which is defined as the z axis. Note that a spherical top can be converted to a symmetric top by a simple distortion that reduces its symmetry. For example, the six bonds in SF_6 are identical; but if we stretch the two bonds along one axis, spinning the molecule about that axis will make it look something like a football rather than a sphere. Alternatively, changing one of the H atoms in methane to Cl lowers the symmetry by shifting the center of mass (the bond length changes, but the added mass has a much greater effect). Let us see what effect such a reduction in symmetry has on the energy levels.

The rotational energy of the symmetric top is of course

$$E_{\text{rot}} = \frac{1}{2I_0}(L_x^2 + L_y^2) + \frac{L_z^2}{2I_z}. \qquad (3.174)$$

If we add $L_z^2/2I_0$ to the first term and subtract it from the second, we have the equivalent equation

$$E_{\text{rot}} = \frac{L^2}{2I_0} + \frac{1}{2}\left(\frac{1}{I_z} - \frac{1}{I_0}\right)L_z^2 \qquad (3.175)$$

Replacing L^2 and L_z^2 by their equivalent operators, we have for the angular part of the Schrödinger equation

$$\mathsf{H}_{\text{rot}} Y(\theta, \phi) = \frac{1}{2I_0}\mathsf{L}^2 Y(\theta, \phi) + \frac{1}{2}\left(\frac{1}{I_z} - \frac{1}{I_0}\right)\mathsf{L}_z^2 Y(\theta, \phi)$$

$$= E_{\text{rot}} Y(\theta, \phi), \qquad (3.176)$$

where $Y(\theta, \phi) = \Theta(\theta)\Phi(\phi)$ is the angular part of the wave function. We would like this function to be of the type already discussed for the case of spherical symmetry, that is, a spherical harmonic. It turns out that this is indeed the case, since the spherical harmonics are specifically defined to be eigenfunctions of both L^2 and L_z (and thus L_z^2). We can thus obtain the eigenvalues of L^2 and L_z separately, using Eqs. 3.162 and 3.156, and substitute to obtain the total rotational energy:

$$E_{\text{rot}} = \frac{\hbar^2}{2}\left[\frac{l(l+1)}{I_0} + m^2\left(\frac{1}{I_z} - \frac{1}{I_0}\right)\right], \quad \left(\begin{array}{l} l = 0, 1, 2, \ldots; \\ m = -l, \ldots, l \end{array}\right). \qquad (3.177)$$

Compare this result with Eq. 3.170. The most obvious change is that the energy has now become a function of m, that is, of the angular momentum about the z axis. This axis is no longer arbitrary, of course, but the main symmetry axis of the molecule. The energy levels can now be no more than doubly degenerate (for $+m$ and $-m$, corresponding to clockwise and counterclockwise rotation about the z axis), rather than the $(2l+1)$-fold degeneracy of the earlier case. It is generally true that a reduction in symmetry tends to result in a removal of degeneracy, a splitting of energy levels (in this case those of a given l). The extent of this splitting depends on how asymmetric the molecule is: If I_z and I_0 are nearly the same, the coefficient of m^2 and thus the splitting are small, but if one moment of inertia is much larger than the other, the energy spectrum will be substantially altered.

The *linear molecule*, with all its atoms aligned on a single axis, is an important special case of the symmetric rotor. This class includes all diatomic molecules, as well as some others, such as CO_2 and C_2H_2 (acetylene). In such a molecule the only rotation about the z axis is that of each atom about its own nucleus. As long as we are considering the atoms as mass points, we have $L_z = 0$; that is, there is no z term in the rotational energy,[27] which reduces to $L^2/2I_0$. This is again of the same form as for the single-particle rotator, giving $l(l+1)\hbar^2/2I_0$ for the eigenvalues of rotational energy. For a diatomic molecule, if R is the bond length and μ the reduced mass $m_1 m_2/(m_1 + m_2)$, the moment of inertia is μR^2, as in Eq. 3.170; this demonstrates the equivalence of the one-body and two-body problems, which we stated earlier without proof.

Finally, we have the *asymmetric top*, a molecule in which all three moments of inertia are different. This class includes molecules as simple as H_2O and SO_2. In the asymmetric top there is no axis around which angular momentum is conserved, so the angular wave function can no longer be a spherical harmonic. The solution is very complicated, and the energy levels cannot be expressed by a simple equation; there are still $2l+1$ levels for each value of the total angular momentum, but none of the levels is degenerate.

[27] Of course, in a real molecule the motions of the electrons in general produce a nonzero L_z; but this goes beyond what we can treat in terms of a rigid-rotator model. (See Section 7.2.)

Appendix 3A

More on Circular Coordinates and the Circular Box

Circular polar coordinates are introduced in Section 3.10. In this appendix we give supplementary information on the following topics: (1) the expression of the Hamiltonian and angular momentum operators in circular coordinates; (2) the solution of the radial wave equation for the circular box. This material is given to illustrate the types of manipulations required when one uses non-Cartesian coordinates.

As mentioned in the main text, the most straightforward way to obtain differential operators in non-Cartesian coordinates is to obtain the operators in Cartesian form and then transform the coordinates. For a particle moving in a plane, we know that the Hamiltonian operator in Cartesian coordinates is given by Eq. 3.132. Given circular symmetry, we can replace $V(x, y)$ by $V(r)$; this leaves the problem of transforming the partial derivatives.

In Appendix I at the end of the text, we discuss the transformation of coordinates in general, with the Cartesian-circular transformation used as an example. There we obtain Eqs. I.16 for the operators $(\partial/\partial x)_y$ and $(\partial/\partial y)_x$ in terms of circular coordinates. The second derivatives that appear in the Hamiltonian can be obtained by applying these equations twice (we omit subscripts to avoid cluttering the equation):

$$\frac{\partial^2}{\partial x^2} = \cos^2\phi\,\frac{\partial^2}{\partial r^2} - \cos\phi\sin\phi\,\frac{\partial}{\partial r}\left(\frac{1}{r}\frac{\partial}{\partial\phi}\right) - \frac{\sin\phi}{r}\frac{\partial}{\partial\phi}\left(\cos\phi\,\frac{\partial}{\partial r}\right) + \frac{\sin^2\phi}{r^2}\frac{\partial^2}{\partial\phi^2},$$

$$(3A.1)$$

$$\frac{\partial^2}{\partial y^2} = \sin^2\phi\,\frac{\partial^2}{\partial r^2} + \sin\phi\cos\phi\,\frac{\partial}{\partial r}\left(\frac{1}{r}\frac{\partial}{\partial\phi}\right) + \frac{\cos\phi}{r}\frac{\partial}{\partial\phi}\left(\sin\phi\,\frac{\partial}{\partial r}\right) + \frac{\cos^2\phi}{r^2}\frac{\partial^2}{\partial\phi^2}.$$

When we add these, the second terms cancel, the first and last terms are simplified by $\sin^2\phi + \cos^2\phi = 1$, and in the third terms we have

$$-\frac{\sin\phi}{r}\left(\cos\phi\,\frac{\partial^2}{\partial\phi\,\partial r} - \sin\phi\,\frac{\partial}{\partial r}\right) + \frac{\cos\phi}{r}\left(\sin\phi\,\frac{\partial^2}{\partial\phi\,\partial r} + \cos\phi\,\frac{\partial}{\partial r}\right)$$

$$= \frac{1}{r}(\sin^2\phi + \cos^2\phi)\frac{\partial}{\partial r} = \frac{1}{r}\frac{\partial}{\partial r}. \quad (3A.2)$$

Combining these results gives

$$\frac{\partial^2}{\partial x^2} + \frac{\partial^2}{\partial y^2} = \frac{\partial^2}{\partial r^2} + \frac{1}{r}\frac{\partial}{\partial r} + \frac{1}{r^2}\frac{\partial^2}{\partial\phi^2}, \quad (3A.3)$$

yielding Eq. 3.133 for the Hamiltonian operator. Similar procedures can be followed to obtain H in other coordinate systems.

For the angular momentum operator we first need the classical form in terms of Cartesian coordinates. From Eqs. 3.127 and 3.128 and the transformation relations (Fig. 3.9), this is

$$p_\phi = \mu v_\phi r = \mu r^2 \frac{d\phi}{dt} = \mu(x^2 + y^2)\frac{x(dy/dt) - y(dx/dt)}{x^2 + y^2} = xp_y - yp_x.$$

$$(3A.4)$$

The quantum equivalent of this equation is

$$\mathsf{p}_\phi = -i\hbar\left(x\frac{\partial}{\partial y} - y\frac{\partial}{\partial x}\right). \qquad (3A.5)$$

Using Eqs. I.16 again, we transform the partial derivatives to obtain

$$\mathsf{p}_\phi = -i\hbar\left[r\cos\phi\left(\sin\phi\frac{\partial}{\partial r} + \frac{\cos\phi}{r}\frac{\partial}{\partial\phi}\right) - r\sin\phi\left(\cos\phi\frac{\partial}{\partial r} - \frac{\sin\phi}{r}\frac{\partial}{\partial\phi}\right)\right]$$

$$= -i\hbar\frac{\partial}{\partial\phi}, \qquad (3A.6)$$

as might be expected. However, a similar calculation starting with Eq. 3.130 gives for the radial momentum operator not $-i\hbar(\partial/\partial r)$ but

$$\mathsf{p}_r = -i\hbar\left(\frac{\partial}{\partial r} + \frac{1}{r}\right), \qquad (3A.7)$$

illustrating the point that the form of an operator depends very much on the choice of coordinates. Substituting the last two equations in Eq. 3.131 is another way to obtain the Hamiltonian operator.

Finally, we consider the solution of the radial wave equation for a particle in a circular box. If we substitute the factored form of the wave function, Eq. 3.137, into Eq. 3.134, we have

$$-\frac{\hbar^2}{2\mu}\left[g(\phi)\frac{d^2f(r)}{dr^2} + \frac{g(\phi)}{r}\frac{df(r)}{dr} + \frac{f(r)}{r^2}\frac{d^2g(\phi)}{d\phi^2}\right] = Ef(r)g(\phi). \qquad (3A.8)$$

Dividing through by $f(r)g(\phi)$, multiplying by r^2, and rearranging to separate the terms in r and ϕ gives

$$\frac{r^2}{f(r)}\frac{d^2f(r)}{dr^2} + \frac{r}{f(r)}\frac{df(r)}{dr} + \epsilon r^2 = -\frac{1}{g(\phi)}\frac{d^2g(\phi)}{d\phi^2}, \qquad \left(\epsilon \equiv \frac{2\mu E}{\hbar^2}\right). \qquad (3A.9)$$

As in Eq. 3.112, each side of this equation must equal the same constant; given the form of $g(\phi)$ from Eq. 3.143, we can use the right-hand side to identify the constant as m^2, where m is the angular momentum quantum number. Further rearrangement then gives

$$\frac{d^2f(r)}{dr^2} + \frac{1}{r}\frac{df(r)}{dr} + \left(\epsilon - \frac{m^2}{r^2}\right)f(r) = 0 \qquad (3A.10)$$

for the radial wave equation.

Equation 3A.10 is a form of *Bessel's equation*, one of the most thoroughly studied equations in all of mathematical physics. It is beyond the scope of this text to develop the methods for its solution, and in any case the solutions cannot be expressed in terms of simple functions (except as infinite series). However, these solutions are of such widespread application that they are defined as functions in their own right, called *Bessel functions*; like the trigonometric or other simple functions, their values can be found in tables. There are several different kinds of Bessel functions: Some look rather like sines or cosines, others like exponentials, still others like hyperbolic sines or cosines. Those appropriate to the circular box are the Bessel functions of the first kind, which resemble sine curves with slowly decreasing amplitude (Fig. 3A.1); they are designated as $J_{|m|}(x)$, where the index number m is the same as in Eq. 3A.10.

The boundary condition we must satisfy is given by Eq. 3.135, and means that $f(r)$ must go to zero at $r = R$. The radial wave function is then given by

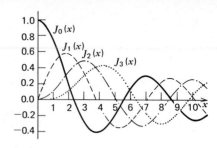

FIGURE 3A.1
Some Bessel functions of the first kind: $J_n(x)$ for $n = 0$, 1, 2, and 3. Note that these functions are drawn with a common scale for x, not with a scale to make their first zeros appear at the same point on the abscissa, as one would in order to show the radial functions with $n = 1$ for particles in a circular box.

$$f(r) = J_{|m|}(r\epsilon^{1/2}), \tag{3A.11}$$

where ϵ (i.e., the energy) must be such that

$$J_{|m|}(R\epsilon^{1/2}) = 0. \tag{3A.12}$$

For any particular $|m|$, Eq. 3A.12 is satisfied only by a specific discrete (but infinite) set of values of ϵ, which we number by the additional index n. These correspond to the values of x ($= R\epsilon^{1/2}$) for which $J_{|m|}(x) = 0$, given by the points at which the curves of Fig. 3A.1 cross the x axis. The radial wave function $f_{n,|m|}(r)$ thus has the shape of $J_{|m|}(x)$ out to its nth zero, which corresponds to the box wall; for a given n, the two states with $m = \pm|m|$ have different angular functions $g(\phi)$, but the same $f(r)$. Since R is a constant, going to higher values of ϵ (higher n) simply squeezes more waves together in the radial direction; this squeezing increases the average value of $|df(r)/dr|$, and therefore increases the absolute value of the radial momentum and kinetic energy.

This problem parallels that of the one-dimensional box, for which we also have an equation in one variable that is satisfied only by a specific set of energy eigenvalues. Mathematically, the difference between the two problems is the result of the two terms in Eq. 3A.10 that were not present in Eq. 3.64. The resulting solution, Eq. 3A.11, is certainly less familiar to the reader than is the sine function of Eq. 3.69. But the difference is not just one of familiarity. From Eq. 3.69 we were able to derive an *explicit* expression, Eq. 3.71, giving the values of E for which the wave function satisfies the boundary conditions—that is, the energy eigenvalues. We could write down this expression because we know where the sine function has its zeros, at integral multiples of π. By contrast, the zeros of the Bessel function follow no such simple analytic pattern; in general, they must be derived by numerical estimation (or looked up in a table or graph like Fig. 3A.1). Some of the circular-box energy eigenvalues have been shown in Fig. 3.12, in which the quantity actually plotted is $R^2\epsilon$.

Further Reading

J. Avery, *Quantum Theory of Atoms, Molecules and Photons* (McGraw-Hill Book Co., Inc., New York, 1972), Chapters 2–4.

M. Karplus and R. N. Porter, *Atoms and Molecules* (W. A. Benjamin, Inc., Menlo Park, Calif., 1970), Chapter 2.

W. Kauzmann, *Quantum Chemistry* (Academic Press, Inc., New York, 1957), Chapters 1–5, 6.I., 7, 8.

H. A. Kramers, *Quantum Mechanics* (North-Holland Publishing Company, Amsterdam, 1958; Dover Publications, Inc., New York, 1964).

E. Merzbacher, *Quantum Mechanics*, 2nd Ed. (John Wiley and Sons, Inc., New York, 1970), Chapters 1–6, 8.

A. Messiah, *Quantum Mechanics*, Vol. 1, translated by G. M. Temmer (John Wiley and Sons, Inc., New York, 1958).

M. A. Morrison, T. L. Estle and N. F. Lane, *Quantum States of Atoms, Molecules and Solids* (Prentice-Hall, Inc., Englewood Cliffs, N.J., 1976), Chapters 1 and 2.

L. I. Schiff, *Quantum Mechanics*, 3rd Ed. (McGraw-Hill Book Co., Inc., New York, 1968), Chapters 2–4.

Problems

1. Compute the de Broglie wavelengths of
   (a) A hydrogen atom with a velocity of 10^3 m/s,
   (b) An electron with an energy of 0.05 eV,
   (c) An electron with an energy of 5×10^6 eV,
   (d) A xenon atom with an energy of 0.05 eV,
   (e) A proton with an energy of 200 GeV (2×10^{11} eV),

2. An expression for the amplitude of a transverse wave might be

   $$\mathbf{f}(x, y, z) = [A \cos(kz - \omega t), 0, 0].$$

   What is the direction of the displacement of this wave? What is its direction of propagation? Is this a traveling wave or a standing wave? What is its wavelength? Write the expression for a longitudinal wave propagating in the same direction as the transverse wave given above.

3. The internuclear spacings of most diatomic molecules are between 1 Å and 3 Å.
   (a) Give a crude estimate of the wavelengths of the electrons that bind atoms into molecules.
   (b) Using the answer to (a), give an estimate of the kinetic energies of these binding electrons.
   (c) Using the answer to (b), and the information that the lowest ionization potentials are in the range 5 eV to 10 eV, give an estimate of the potential energy of the binding electrons.

4. Suppose that the relation connecting the wavelength of a particle with its momentum depended on c, the speed of light, and on m, the mass of the particle, so that

   $$p = \left(\frac{hcm}{\lambda}\right)^{1/2}.$$

   What dispersion relation would follow from this? What implications would such a dispersion relation hold for quantum theory?

5. Show that if $\Psi(x, t)$ is a wave function as in Eq. 3.14, then

   $$\left(\frac{\partial^n \Psi}{\partial x^n}\right)_t = k^n \frac{d^n \Psi}{d\varphi^n},$$

   and

   $$\left(\frac{\partial^n \Psi}{\partial t^n}\right)_x = (-\omega)^n \frac{d^n \Psi}{d\varphi^n}.$$

6. Show that the entries in column (a) are eigenfunctions of the operators in column (b). In each case determine the eigenvalue.

(a)	(b)
(i) $a \cos(bt + c)$	d^2/dt^2
(ii) e^{ibt}	d^2/dt^2
(iii) $ze^{-z^2/2}$	$-\dfrac{d^2}{dz^2} + z^2$
(iv) ce^{-ax}	d^n/dx^n

7. The function e^{-ax^2} is an eigenfunction of the operator $(d^2/dx^2) - bx^2$ only under certain conditions. What are these conditions, and what is the eigenvalue when they are satisfied?

8. State, in general terms, under what conditions one may expect the diffraction of matter waves to be important in determining the dynamics of a particle, and under what conditions diffraction is unimportant.

9. Using your knowledge of the boundary conditions the wavefunction must satisfy, explain what happens to the energy levels of a particle in a one-dimensional box if the box length is changed from L to L/n ($n = 2, 3, \ldots$).

10. What is the spacing of the energy levels of a particle in a one-dimensional box if
    (a) The mass is m_p, the mass of the proton, and the box side is 5 Å
    (b) The mass is that of the proton, m_p, and the box is made much larger, to correspond to the volumes in which gases are customarily confined, for example, 0.1 m on a side
    (c) The mass is m_p and the box is made much smaller, to a length of 10^{-13} cm, roughly the size of atomic nuclei

    The second of these will show very quickly how, in large containers, the quantization of energy levels is not observable because the levels are so closely spaced. The third problem, the nuclear problem, will quickly give some feeling for why nuclear physics experiments require high-energy machines for the study of nuclear reactions, although one can use ordinary light to carry out chemical excitation processes.

11. The molecules H_2C—$(CH{=}CH)_n$—CH_2, $n = 2, 3, \ldots$, can be considered successively longer and longer one-dimensional boxes for the electrons. Suppose that an electron can move freely the whole length of the molecule. If each bond has length $b_0 = 1.5$ Å, and the end CH bonds are neglected, what are the wavelengths of absorption of the lowest transitions when $n = 2, 3,$ and 4?

12. Set up the functions corresponding to the sum and difference combinations based on the solutions of the form of Eq. 3.118 for the case in which $a = b$, but $n_1 \neq n_2$.

13. What degeneracies arise in the three-dimensional box if $a = b = c$ in Eq. 3.119? Plot the lowest five of these levels on an energy scale and give their degeneracies.

14. Consider a particle in a cylindrical box with radius R and length L. The mass of the particle is μ. Derive an expression analogous to Eq. 3.121 for the classical energy of the particle, beginning with the general expression for the energy of a particle in three dimensions,

$$E = (2\mu)^{-1}(p_x{}^2 + p_y{}^2 + p_z{}^2) + V(x, y, z).$$

Show that the problem is separable in the cylindrical coordinate system $r = (x^2 + y^2)^{1/2}$, $\phi = \arctan(y/x)$, and z, when z is chosen along the cylinder axis.

15. Carry out the quantum mechanical counterpart of Problem 14. Find the first several energy eigenvalues and show how they depend on the relationship between R and L. Show that the problem (i.e., the Hamiltonian) is separable into axial and circular parts and that the energy is a sum of independent axial and circular contributions. What relation between R and L makes the energy of the first excited state of circular motion ($n = 1$, $|m| = 1$ in Fig. 3.12) equal to the energy of the first excited state of motion along the cylinder axis? What relation between R and L makes the state $n = 2$, $|m| = 0$ degenerate with (have energy equal to that of) the first excited state of motion along the cylinder axis?

16. Show that if, in Eq. 3.140,

$$\int_0^{2\pi} |g(\phi)|^2 \, d\phi = 1,$$

then

$$|A| = (2\pi)^{-1/2}$$

17. On radial graph paper, plot the real part of g (of Eq. 3.140), for $p = \hbar, 2\hbar,$ and $4\hbar$, or $m = 1, 2,$ and 4.

18. Prove that

$$\mathbf{p} \cdot \mathbf{p} = \frac{1}{r^2} (\mathbf{p} \cdot \mathbf{r})^2 + \frac{1}{r^2} (\mathbf{r} \times \mathbf{p}) \cdot (\mathbf{r} \times \mathbf{p})$$

in classical mechanics but not in quantum mechanics.

19. Using tables of Bessel functions and estimates based on the vertical scale of Fig. 3.12, plot on circular graph paper the nodes of the real and imaginary parts of the functions 3A.11 for the lowest five energy levels shown in Fig. 3.12. Note that for m different from zero the functions all have nodal points at the origin. This is a result of the fact that their angular momenta are greater than zero, so that at $r = 0$ the velocity of the particle must become infinite. But if the velocity were infinite, then the slope of the wave function would be infinite, which would make the wave function infinite, which contradicts our requirement for the wave function. The answer to the dilemma is that the particle can never appear at the origin if it has any angular momentum. Therefore the amplitude of its wave function is zero at the origin for those states with $m \neq 0$.

20. For an electron in a circular Bohr orbit with $n = 2$, what is the centrifugal kinetic energy in joules, electron volts, and cm^{-1}?

21. Suppose that the methane molecule is a rigid tetrahedron with the carbon atom located at its center. What is the moment of inertia about any one axis? Hint: Pick an axis along one of the CH bonds. Show that the moment of inertia about either of the other perpendicular axes is equal to that about the first, and therefore that the moment of inertia is completely independent of the choice of axis in this molecule. Such a molecule is called a spherical top, because its rotations are just as simple as those of a sphere. Calculate the energies of the first three rotational levels of this molecule. What are the separations between them? In what part of the electromagnetic spectrum do these differences lie? Is this a region in which one uses optical methods, microwave or radar equipment, transmitters and detectors, or is it in still another region of the spectrum?

22. Calculate the components of moment of inertia and show graphically the energy level spectrum for each of these two examples: (a) sulfur hexafluoride, SF_6, in which four of the six bonds are 1.58 Å and the other two, chosen opposite each other to lie along a single axis, are 1.60 Å; (b) methyl chloride, CH_3Cl, a tetrahedron in which the CH bonds are 1.1 Å and the C—Cl bond is 1.77 Å. Assume that the chlorine atoms are all isotopes of atomic mass number 35. Our examples have been chosen to show the magnitudes associated with molecular rotation and, perhaps more important, the way in which successively larger deviations from an original simple model lead to successively larger deviations in the "results," that is, in the observables we calculate from these models. In this case, where the rigid rotator is the simple model with a simple spectrum and the symmetric rotator represents the deviation, we are fortunate in that Eq. 3.174 is still both relatively simple and essentially correct for many, many molecules.

4

Particles in Varying Potential Fields; Transitions

Thus far we have dealt only with particles moving in constant potential fields: Both for the free particle and within our various boxes, the potential energy was constant wherever the particles could go. In this chapter we begin to consider situations in which the potential energy varies from point to point. After examining the general behavior of the wave function when the potential energy changes, we treat two basic models of particles bound within "potential wells."

The first of these models is the ubiquitous harmonic oscillator, which we introduced in Chapter 2 to illustrate quantization. We shall now see how the assumptions of Planck and Einstein ($\Delta E = h\nu$) can be derived from the Schrödinger equation. The results obtained here will turn up again when we treat the vibrations of molecules and solids.

The second model is that of the hydrogenlike atom: an electron in the Coulombic field of a positively charged nucleus. Although the physical picture of the atom is very different from that of Bohr's theory, the energy levels obtained are the same. The principles introduced here will later be extended to many-electron atoms, to show that the forms of the atomic wave functions largely govern the shapes of molecules.

We close the chapter with a look into how transitions occur between atomic energy levels, introducing the fundamental problem of how a quantum mechanical system undergoes change.

4.1 Finite Potential Barriers

In this section we discuss some general characteristics of the wave function's dependence on potential energy. For simplicity, we consider only the one-dimensional case, for which the general time-independent Schrödinger equation is Eq. 3.61. We can rewrite this in the form

$$\frac{d^2\psi(x)}{dx^2} = \frac{2m}{\hbar^2}[V(x) - E]\psi(x). \tag{4.1}$$

On the left-hand side of this equation is the second derivative of $\psi(x)$, which for brevity we shall call the "curvature" of $\psi(x)$. On the right-hand side, a positive constant multiplies the energy difference $V(x) - E$ and the function $\psi(x)$ itself. Now let us examine the behavior of $\psi(x)$ in different regions of space. In classical mechanics the total energy E is the sum of kinetic and potential energy; since the classical kinetic energy ($p^2/2m$) is necessarily a positive quantity, E is always greater than the potential energy alone. As we shall see, this restriction no longer applies in quantum mechanics.

If E is larger than V, then the multiplier of $\psi(x)$ on the right-hand side of Eq. 4.1 is negative. When $\psi(x)$ is positive, its curvature must be negative; when $\psi(x)$ is negative, its curvature is positive. Any function $f(x)$ whose sign is opposite to the sign of $d^2f(x)/dx^2$ is *concave* toward the x axis. The type of function with this property that arises in real physical problems is normally one whose slope changes sign, bringing the function back across the axis and therefore changing the direction of its concavity. In short, it must be oscillatory or sinusoidal. Figure 4.1*a* illustrates such a function. The sine and cosine are the simplest examples of functions that are everywhere concave toward the axis. The wave function must be of this type wherever $E > V$.

The other situation, in which a function has the same sign as its curvature, is illustrated in Fig. 4.1*b*; here there are two possible cases. Let us assume, without losing generality, that the function is positive at some point x_0. If the function has a positive slope at x_0, then for $x > x_0$ the slope must be greater than at x_0, so that the function itself must grow steadily larger as x increases. Such a wave function is physically permissible for a region from $-\infty$ to some fixed value of x, but clearly will not do for arbitrarily large values of x. If we allowed $\psi(x)$ to grow infinitely large in this way, the probability density $|\psi(x)|^2$ would accumulate at $x = +\infty$, leaving essentially zero probability of a particle's being found at any real location. The other case is one in which a positive function has negative slope but positive curvature. In this case, the slope grows smaller and smaller as $x \to +\infty$. Such a wave function can be used for x greater than some fixed value, but diverges as $x \to -\infty$, with the same catastrophic result as in the first case. It was to eliminate such possibilities that we postulated that the integral $\int \cdots \int \Psi^* \Psi \, dq_1 \cdots dq_N$ over all allowed values of the coordinates—in this case, $\int_{-\infty}^{\infty} |\psi(x)|^2 \, dx$—must be finite. The simplest examples of functions with everywhere the same signs as their curvatures are e^x and e^{-x}, which correspond to our first and second cases, respectively. Wave functions of this type will appear wherever $E < V$.

Until now, however, we have not encountered the possibility that E could be less than V. The free particle has $V = 0$ and $E > 0$ everywhere; the particle in a box is the same where V is finite, and the wave simply does not exist where V is infinite. As we saw in Chapter 3, the wave functions in these systems are indeed sines and cosines. But our various boxes had impenetrable walls, whereas real matter waves (just as light waves) can actually penetrate a little way into a wall, even when the wall acts as an *almost* perfect reflector. For a high (but not infinitely high) potential barrier, a particle with finite energy can have $E < V$ in the region *within* the wall itself. We must therefore examine the consequences of having $E < V$. In this case the multiplier of $\psi(x)$ in Eq. 4.1 is positive, and $d^2\psi(x)/dx^2$ has the same sign as $\psi(x)$ itself. By the arguments of the preceding paragraph, we cannot deal with a wave function whose eigenvalue E is *everywhere* less than $V(x)$; such a function would become infinite in one of the directions $x \to +\infty$ or $x \to -\infty$. On the other hand, there is no objection whatever to having a wave function whose eigenvalue E is greater than $V(x)$ for some range of x and less than $V(x)$ for another range of x.

To make this discussion concrete, we have illustrated in Fig. 4.2 a potential having the form of a step. We wish to determine the eigenfunction for a state with energy E less than the barrier height V_0, so that $E > V(x)$ for negative x and $E < V(x)$ for positive x. For $x < 0$ the Schrödinger equation is simply Eq. 3.64, the most general solution of which is

$$\psi(x) = Ae^{i\alpha x} + Be^{-i\alpha x}, \qquad \alpha \equiv \left(\frac{2mE}{\hbar^2}\right)^{1/2} \qquad (x < 0). \qquad (4.2)$$

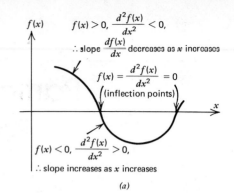

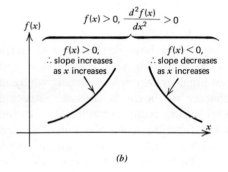

FIGURE 4.1

Examples of the relationships between functions and their second derivatives: (*a*) The function $f(x)$ and $d^2f(x)/dx^2$ have opposite signs. (*b*) The function $f(x)$ and $d^2f(x)/dx^2$ have the same sign.

For $x > 0$ we must solve Eq. 4.1 with $V(x) = V_0$; it is not difficult to show that a solution is

$$\psi(x) = Ce^{\beta x} + De^{-\beta x}, \qquad \beta \equiv \left[\frac{2m(V_0 - E)}{\hbar^2}\right]^{1/2} \qquad (x > 0). \quad (4.3)$$

By the reasoning already described, we must have $C = 0$ if $\psi(x)$ is not to become infinite as $x \to \infty$. (The dashed line in Fig. 4.2 indicates how the wave function might behave without this restriction.) The boundary condition to be satisfied is that both $\psi(x)$ and its first derivative must be continuous across the boundary. That is, one sets $x = 0$ in Eqs. 4.2 and 4.3 and equates the results, and does the same with the two equations obtained by differentiation. The details of this calculation constitute Problem 2 at the end of the chapter. The wave function is in general complex, but qualitatively it must resemble the function plotted in Fig. 4.2: sinusoidal where $E > V$, decaying exponentially as one moves into the potential barrier.

The one-dimensional box of Chapter 3 is simply a region with infinitely high potential barriers on both sides. How does the wave function for a finite-step potential differ from that of the particle in a box? Here the wave function has a nonzero value within the barrier, vanishing only in the limit as $x \to \infty$. For the infinitely high barrier, however, we have $V_0 \to \infty$, $\beta \to \infty$, and $De^{-\beta x} \to 0$ for all $x > 0$; this agrees with our assumption that $\psi(x) = 0$ beyond the box walls. In both cases the wave function itself must be continuous at the boundary; for the finite step we have the additional boundary condition that $d\psi(x)/dx$ be continuous.[1]

The region where $V > E$ is called the classically forbidden region. By Eq. 4.1, the larger the factor $V - E$, the larger is the magnitude of $d^2\psi(x)/dx^2$, and thus the more rapidly $\psi(x)$ decays with distance into the barrier. In a classically allowed region, by contrast, the more E exceeds V, the more rapidly the wave function must oscillate. Whenever E is very close to V, whether above or below, the wave function must be relatively flat. In Chapter 3 we interpreted behavior like this in terms of the kinetic energy $(E - V)$: The higher the kinetic energy, the more rapidly the wave function changed. This is still true if we substitute the *magnitude* of the kinetic energy, because in the classically forbidden region we have the remarkable phenomenon of *negative* kinetic energy.

The wave function is nonzero in the classically forbidden region. This means that, whenever a real particle approaches a finite potential barrier, the particle (i.e., the wave that describes it) can penetrate some distance into the classically forbidden region. The larger the mass of the particle (and thus the value of β), the shorter the distance it is able to penetrate; macroscopic objects have masses too great for this effect to be observed. Suppose now that the potential barrier has only a limited extent in space, as shown in Fig. 4.3. In this case, the particle can actually penetrate *through* the barrier, even though it has insufficient energy to go over it classically. Mathematically, this is a consequence of the continuity conditions on the wave function and its derivative. Even though $\psi(x)$ may be a dying exponential coming from the left side of the potential barrier

---

[1] If $d\psi/dx$ were discontinuous at some boundary, then $d^2\psi/dx^2$ would be either infinite or undefined. By Eq. 4.1, $d^2\psi/dx^2$ can become infinite only if either ψ or V becomes infinite at the boundary. The latter condition is met at an infinitely high potential barrier; for all other types of potentials, ψ itself would have to become infinite. We cannot allow this, so we require the derivative to be continuous everywhere except at infinite potential jumps. (This was included in Postulate I of Section 3.8.)

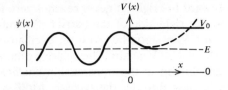

FIGURE 4.2
The step potential:

$$V(x) = 0 \qquad (x < 0),$$
$$V(x) = V_0 \qquad (x \geqslant 0).$$

Superimposed on the potential energy curve we have plotted a possible wave function for energy E, using the line corresponding to energy E as the x axis for $\psi(x)$. The dashed curve $(---)$ is an example of the forbidden class of functions that diverge as $x \to \infty$.

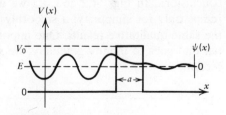

FIGURE 4.3
Penetration of a potential barrier by a particle wave. As in Fig. 4.2, the wave function is superimposed on the potential energy curve.

toward the right, it never reaches zero for any finite value of x. Therefore, at the right side of the barrier, this small but still nonzero wave function must join smoothly with an oscillatory function for the region beyond where $E > V$ again. The relative amplitudes of the two sinusoidal waves are related to the transmission coefficient, the probability that a particle will pass through the barrier; for $E \ll V_0$ this is approximately proportional to $e^{-2\beta a}$, where a is the barrier thickness and β is defined by Eq. 4.3.

What we have described here, of course, is a time-dependent process. How, then, is it described by the time-independent wave function $\psi(x)$? We can think of $\psi(x)$ as describing the steady-state behavior of a continuous stream of particles striking the barrier. At any given time, some particles will be moving toward the barrier from the left and some will be rebounding toward the left; these correspond to the $Ae^{i\alpha x}$ and $Be^{-i\alpha x}$ terms, respectively, in Eq. 4.2. For a barrier of finite thickness, other particles will be moving away from the barrier toward the right; in this region $\psi(x)$ must be of the form $Ge^{i\alpha x}$. The relative values of $|A|^2$, $|B|^2$, and $|G|^2$ give the steady-state fractions of incident, reflected, and transmitted particles. A single particle is described by the same wave function if we know nothing about its position at a given time: $|B|^2/|A|^2$ and $|G|^2/|A|^2$ then represent the probabilities of reflection and transmission. If we wanted to follow the process as a function of time, we would have to solve the time-dependent Schrödinger equation for a wave packet like that in Fig. 3.7, but there is no need to do this when we are interested only in the initial and final states of the system. The same simplification can be made for any collision or scattering process.

The interpretation may be more straightforward if we think in terms of matter waves rather than particles. The fact that the barrier is finite means that it is only an imperfect reflector: Part of the wave is transmitted through the region in which $E < V$, the rest of it is reflected. A simple analogy is the behavior of light striking a thin sheet of metal. Most of the light is reflected, but some is transmitted; if the metal is thin enough, one can even see through it. But the fraction of the light transmitted decreases rapidly with thickness; for most sheets of metal the amplitude of the transmitted wave is indetectably small. The classically forbidden transmission of a particle through a barrier is known as *tunneling*. The situations where this process is important include the escape of α particles from atomic nuclei in radioactive decay, the "umbrella inversion" of pyramidal molecules like NH_3, and the so-called tunnel diode (in which the barrier is the interface between two semiconductors). Tunneling also affects the rates of chemical reactions, as we shall point out in Chapter 30.

So far we have considered two kinds of potential barriers: a potential that rises discontinuously from zero to infinity, and a barrier of finite height. The finite barrier can be of arbitrary shape without affecting our conclusions. In Figs. 4.2 and 4.3 we used potentials with discontinuous jumps only for simplicity; a smoothly rising potential would have given the same qualitative results. One important possibility remains—a potential that goes to infinity gradually; we shall consider this case in the next section.

In Section 2.8 we introduced the harmonic oscillator, a point mass bound to an equilibrium position by a springlike restoring force whose strength is directly proportional to the particle's displacement. As before, we consider a one-dimensional oscillator. If the equilibrium point is taken

4.2
The Quantum Mechanical Harmonic Oscillator

as the origin, the force and the classical equation of motion are given by Eq. 2.33 as

$$F(x) = -kx = m\frac{d^2x}{dt^2}. \qquad (4.4)$$

If we take our arbitrary zero point of energy to be the energy of the particle at rest at the origin, then by Eq. 2.48 the potential energy is

$$V(x) = \tfrac{1}{2}kx^2. \qquad (4.5)$$

To derive the quantum mechanics of the harmonic oscillator, we must solve the Schrödinger equation with this value of $V(x)$.

The one-dimensional potential before us differs in two significant ways from that of our earlier one-dimensional box (Fig. 3.5). First, the oscillator potential is smooth: At no point does it have an abrupt change of value or an infinite derivative, as the box potential does at the walls. Second, it remains finite for all finite values of x. But the oscillator and the one-dimensional box are alike in one very important way: As $x \to \pm\infty$, both potentials approach a positive infinite value, so that no matter how large the energy of a particle, it is always confined within a "potential well."

We recall that Planck assumed the energy levels of an oscillator to be equally spaced integral multiples of $h\nu$, Planck's constant multiplied by the oscillator's fundamental frequency. We certainly expect to find in the wave picture that the energy levels of the oscillator are quantized. Moreover, from what we have already seen, we can make an intuitive statement about how these levels should be spaced. For a particle in a box, the larger the box, the more closely spaced are the levels; if we enlarge the box, each level moves accordingly to a lower energy. The oscillator is a particle in a box that grows wider as the energy increases. We therefore expect that with increasing energy the spacing of the harmonic oscillator energy levels should increase less rapidly than the spacing in a simple square box. In fact, it turns out that the spacing does not increase at all, in agreement with Planck's assumption. To find the energy levels explicitly, we must solve the one-dimensional Schrödinger equation, which for the harmonic oscillator becomes

$$\mathsf{H}\psi(x) = -\frac{\hbar^2}{2m}\frac{d^2\psi(x)}{dx^2} + \frac{kx^2\psi(x)}{2} = E\psi(x). \qquad (4.6)$$

This equation, like those for the particles in circular and spherical boxes, is an extensively studied equation whose eigenvalues and eigenfunctions are quite well known.

Without solving Eq. 4.6 explicitly, what can we deduce about its solutions from the general principles of Section 4.1? Here we have a potential for which $V(x)$ becomes infinite as x approaches $\pm\infty$. Therefore, no matter what a given state's energy eigenvalue E may be, E becomes less than V for sufficiently large $|x|$. In the regions where $E < V$, every wave function must fall off more or less exponentially, as shown in Fig. 4.2, approaching zero as $x \to \pm\infty$. As the wave function for a given state comes in toward the origin, the sign of its second derivative changes at the point where E and V become equal. In the region around the origin, $d^2\psi(x)/dx^2$ must be negative and $\psi(x)$ oscillatory. By analogy with the particle in a box, we expect the lowest state to be one with no nodes, the next state to have one node, and so on as the energy increases. Thus, from purely qualitative considerations, we can infer the general form of all the wave functions of the harmonic oscillator—or indeed of any particle bound in a smooth and infinitely high potential well.

Planck was correct in supposing that the energy spectrum has the remarkable property of equally spaced levels. The levels are in fact given by

$$E_n = (n + \tfrac{1}{2})h\nu \qquad (n = 0, 1, 2, \ldots), \tag{4.7}$$

which differs from Eq. 2.54 only in the extra $\tfrac{1}{2}h\nu$. Since the energy retains this value even in the lowest state (with $n = 0$), $\tfrac{1}{2}h\nu$ is called the *zero-point energy*.

As we pointed out in Section 3.7, the uncertainty principle requires that all confined systems have a zero-point energy. We can rationalize this result for the harmonic oscillator in a different way. If an oscillator had exactly zero energy, it would of course have $E < V$ everywhere except at the origin. This means that the wave function would resemble the curves of Fig. 4.1b: If we required $\psi(x)$ to go to zero at both $+\infty$ and $-\infty$, it would become steeper as $x \to 0$ from both sides. There would thus be a discontinuous slope at the origin, a point where $V(x)$ is perfectly smooth. This is simply not an acceptable wave function, because it violates the continuity condition. This argument shows that we just cannot localize a particle at a point in space, even if we try to do it with a smooth potential.

We can generalize from the harmonic oscillator to see how energy-level spacings occur in other potential wells. If a potential is flatter than that of the harmonic oscillator, then the energy levels must come closer together as the energy increases; if a potential is steeper than that of the harmonic oscillator, then the spacing increases as the energy increases. (The particle in a box is an example of the latter case.) The simplest test of "steepness" is the second derivative of the potential energy, which for the harmonic oscillator is a constant. We shall return to this conclusion when we examine the vibrations of molecules.

We shall not carry out a systematic derivation of the solutions of Eq. 4.6, but we can pursue the foregoing qualitative comments a bit further. The relation of Eq. 4.6, stating that the second derivative of $\psi(x)$ is proportional to $x^2\psi(x)$, implies that the argument x appears in an exponent—in fact, that it is x^2, not x, that is in the exponent. The Gaussian, with the form e^{-cx^2}, has just this form. It has the required nodeless behavior of the lowest eigenfunction, it is concave toward the axis from its maximum out to its two points of inflection and it falls off at larger $|x|$, convex toward the x axis. By substituting this form into Eq. 4.6, one can readily show that the Gaussian has the correct form:

$$\frac{d}{dx} e^{-cx^2} = -2cx e^{-cx^2}, \tag{4.6.a}$$

and

$$\frac{d^2}{dx^2}(e^{-cx^2}) = -2c e^{-cx^2} + 4c^2 x^2 e^{-cx^2}. \tag{4.6.b}$$

Hence we can identify the constant c with $mE/\hbar^2$ of Eq. 4.6 and $4c^2x^2$ with $mkx^2/\hbar^2$ for the lowest state of the harmonic oscillator. This is adequate to fix the energy of the state in terms of the parameters of the system—its force constant k and mass m—as

$$E = \frac{\hbar}{2}\left(\frac{k}{m}\right)^{1/2}. \tag{4.7.a}$$

Recall that $(k/m)^{1/2}$ is the angular frequency ω or $2\pi \times$ the circular frequency ν, so the energy of the lowest state of the harmonic oscillator is $\hbar\omega/2$ or $h\nu/2$, in agreement with Eq. 4.7.

TABLE 4.1
HARMONIC OSCILLATOR WAVE FUNCTIONS

n	$H_n(z)$	$\psi_n(z)$
0	1	$(m\omega/\hbar\pi)^{1/4}e^{-z^2/2}$
1	$2z$	$(m\omega/\hbar\pi)^{1/4}(2 \cdot 1)^{-1/2}e^{-z^2/2}2z$
2	$4z^2-2$	$(m\omega/\hbar\pi)^{1/4}(4 \cdot 2)^{-1/2}e^{-z^2/2}(4z^2-2)$
3	$8z^3-12z$	$(m\omega/\hbar\pi)^{1/4}(8 \cdot 6)^{-1/2}e^{-z^2/2}(8z^3-12z)$
4	$16z^4-48z^2+12$	$(m\omega/\hbar\pi)^{1/4}(16 \cdot 24)^{-1/2}e^{-z^2/2}(16z^4-48z^2+12)$

The lowest state of the quantum oscillator is quite different from what one expects for a classical oscillator: The highest amplitude is at the point $x=0$, where the potential is deepest and therefore where the kinetic energy is greatest. In classical systems, particles spend most of their time in the places where their velocities are low. Quantum mechanical systems with several quanta behave in the same way, but the states that deviate furthest from classical mechanical behavior, the states of lowest energy, do not conform to our classical expectations for particles.

The states above the ground state have wave functions whose forms are products of polynomials and Gaussians. The one-quantum state has a wave function with one node at $x=0$, and is simply of the form xe^{-cx^2}; the next state has a wave function that is a product of a quadratic and a Gaussian, and thus has two nodes. In general,

$$\psi_n(z) = \left(\frac{m\omega}{\hbar\pi}\right)^{1/4}\frac{e^{-z^2/2}}{(2^n n!)^{1/2}}H_n(z), \qquad (4.8)$$

where[2] $z \equiv (m\omega/\hbar)^{1/2}x$, $\omega = 2\pi\nu$, and the *Hermite polynomial* H_n is defined by

$$H_n(z) \equiv (-1)^n e^{z^2}\frac{d^n}{dz^n}(e^{-z^2}). \qquad (4.9)$$

The first five harmonic oscillator wave functions are listed in Table 4.1 and plotted in Fig. 4.4a, along with the energy levels. The correctness of these solutions can be checked by substituting them in Eq. 4.6.

These wave functions have a number of interesting properties. The first thing to note is that the energy levels are nondegenerate: Only one wave function corresponds to each level. The Hermite polynomials are alternately functions of even and odd powers of z (and thus of x); this property makes the functions themselves alternately even and odd with respect to $x=0$. That is, for $n=0, 2, 4, \ldots$, the function for $-x$ is exactly the same as that for $+x$, but for $n=1, 3, 5, \ldots$, the two functions have opposite signs:

$$\begin{aligned}\psi_n(x) &= \psi_n(-x) &&(n \text{ even}),\\\psi_n(x) &= -\psi_n(-x) &&(n \text{ odd}).\end{aligned} \qquad (4.10)$$

We say that the even-n and odd-n wave functions have even and odd *parity*, respectively. Finally, as usual, the number of nodes increases with energy, with $\psi_n(x)$ having n nodes.

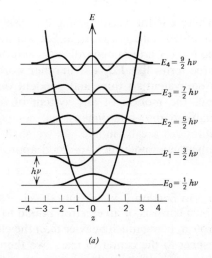

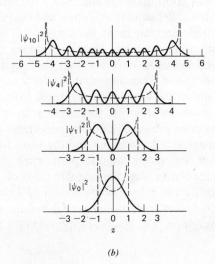

(b)

FIGURE 4.4
The quantum mechanical harmonic oscillator. (a) Potential energy curve, energy levels E_n, and superimposed wave functions $\psi_n(z)$, where $z \equiv (m\omega/\hbar)^{1/2}x$. (b) Probability densities for $n = 0, 1, 4, 10$, with the dashed lines (– – –) giving the probabilities for classical oscillators of the same energies.

---

[2] Note that $(\hbar/m\omega)^{1/2}$ is the classical amplitude x_0 for an oscillator with $E = \frac{1}{2}h\nu = \frac{1}{2}\hbar\omega$ (the zero-point energy); thus we have $z = x/x_0$.

Let us now look at the oscillator probability distributions, some of which are plotted in Fig. 4.4b. Here we can see how much the Gaussian distribution of the lowest state of the quantum oscillator differs from that of the classical oscillator with the same energy. And we can see too how the higher states approach their classical counterparts, whose probability distributions are proportional to $(x_0^2 - x^2)^{-1/2}$. As n increases, however, the probability maxima move outward and $|\psi_n|^2$ is more and more closely approximated by the classical probability, as can be seen in the figure. In the classical limit $(n \to \infty)$, the oscillations in $|\psi|^2$ become too close together to distinguish, and the probability over any finite range of x averages out to the classical value.

So far we have considered only the one-dimensional harmonic oscillator. The spherically symmetric three-dimensional case, for which

$$\mathbf{F} = -k\mathbf{r} \quad \text{and} \quad V(r) = \tfrac{1}{2}kr^2 = \tfrac{1}{2}k(x^2 + y^2 + z^2) \tag{4.11}$$

(where $\mathbf{r}$ is the vector giving the displacement from the origin), presents no added difficulties. As in Section 3.9, the Hamiltonian is separable in the three Cartesian coordinates, and the total energy is simply the sum of three terms like Eq. 4.7. In other words, the three-dimensional oscillator is equivalent to three independent one-dimensional oscillators. In general, the motion of any system of oscillating masses can be similarly analyzed into one-dimensional components, provided only that all the vibrations are harmonic. As we shall see in later chapters, crystals and polyatomic molecules are good approximations to such systems.

4.3
The Hydrogen Atom

In Section 2.11 we introduced Bohr's model of the hydrogen atom, which consists of an electron bound to a proton by a Coulomb force. The proton, being much heavier than the electron, is essentially immobile with respect to the center of mass; we therefore take the proton as the origin of our coordinate system.[3] Since the potential energy depends only on the distance between the electron and the proton, that is, on the radial distance of the electron from the origin, we are confronted with a problem of spherical symmetry. Bohr's condition that the action per cycle be quantized led us to conclude that the electron's angular momentum is quantized. Now, knowing what we do about waves in spherical potentials, we come to the same conclusion just from the symmetry of the potential. As soon as we describe the electron by a wave, we find ourselves applying the same periodic boundary conditions that we applied to the particle in a spherical box: In order that the wave function be a single-valued function of θ and ϕ, its angular parts must close on themselves. We conclude immediately that the angular parts of the hydrogen wave functions are exactly the spherical harmonics of Table 3.1.

Let us spell out in more detail the mathematical statement of the last paragraph. The classical expression for the energy of a hydrogen atom is

$$E_{\text{classical}} = T + V(r), \tag{4.12}$$

where the potential energy is

$$V(r) = -\frac{Ze^2}{4\pi\varepsilon_0 r} \tag{4.13}$$

---

[3] Strictly, one takes the center of mass as the origin and replaces the two particles by a single particle of reduced mass μ, as in Eqs. 2.74ff.; for the H atom, $\mu = 0.9994557 m_e$. Although we shall not go into the details of this refinement, we use μ for mass anyway, to avoid confusion with the quantum number m.

and the kinetic energy is already familiar to us:

$$T = \frac{1}{2\mu} \mathbf{p} \cdot \mathbf{p} \qquad (3.146)$$

$$= \frac{1}{2\mu} p_r^2 + \frac{L^2}{2I} . \qquad (3.153)$$

The quantum mechanical analog of expression 4.12 is the Hamiltonian operator whose eigenvalues are the allowed energies of this atom (Postulate III):

$$\mathsf{H} = \frac{-\hbar^2}{2\mu} \left[\frac{1}{r^2} \frac{\partial}{\partial r} \left(r^2 \frac{\partial}{\partial r} \right) + \frac{1}{r^2 \sin \theta} \frac{\partial}{\partial \theta} \left(\sin \theta \frac{\partial}{\partial \theta} \right) + \frac{1}{r^2 \sin^2 \theta} \frac{\partial^2}{\partial \phi^2} \right] - \frac{Ze^2}{4\pi\varepsilon_0 r} .$$

$$(3.154)$$

The wave functions $\psi(\mathbf{r})$ for the hydrogen atom, whose existence is supposed in our first postulate, satisfy a time-independent Schrödinger equation if they correspond to eigenstates of energy. According to Postulate V:

$$\mathsf{H}\psi(\mathbf{r}) = E\psi(\mathbf{r}) \qquad (\mathbf{r} \equiv (r, \theta, \varphi)). \qquad (4.14)$$

The form of Eq. 3.153 tells us that the classical Hamiltonian is the sum of a radial part, $[p_r^2/2\mu + V(r)]$, and an angular part, $L^2/2I$. This in turn assures us that the quantum mechanical operator, Eq. 3.154, is separable into corresponding radial and angular parts, with nothing but the square of the angular momentum operator, L^2 (and a constant scale factor) in the angular part. The wave function $\psi(\mathbf{r})$ can be written as a product $\psi(\mathbf{r}) = R(r)\Theta(\theta)\Phi(\phi)$, and the Schrödinger equation for the hydrogen atom is separable into a radial equation for $R(r)$ (to which we shall return) and an angular equation for $\Theta(\theta)\Phi(\phi)$, involving only the rotational kinetic energy. This latter is identical to the equation discussed in Section 3.11 but never written explicitly there:

$$\mathsf{L}^2\Theta(\theta)\Phi(\phi) = \left(p_\theta^2 + \frac{p_\phi^2}{\sin^2 \theta} \right) \Theta(\theta)\Phi(\phi) = L^2\Theta(\theta)\Phi(\phi). \qquad (4.15)$$

We asserted previously that the eigenvalues of L^2, the constants that may appear in Eq. 4.15, have the form

$$\langle \mathsf{L}^2 \rangle = l(l+1)\hbar^2, \qquad (3.162)$$

and the eigenfunctions of Eq. 4.15 are the spherical harmonics $Y_{l,m}(\theta, \phi)$, such as those in Table 3.1.

We thus know the angular behavior of the wave function, and therefore the allowable values of angular momentum, for an electron in a hydrogen atom. But this is still a far cry from knowing either the energy-level spectrum or the spatial distribution of charge in the atom. As we have done before, let us make some general inferences about the relationships between the eigenvalues of angular momentum and energy, and about the general shapes of the wave functions themselves. Then, without going through a derivation, we can examine some of the actual solutions to the hydrogen wave equation, in particular the energy-level spectrum.

The total energy of a hydrogenlike atom is still given by Eq. 2.56, with the zero of energy chosen to correspond to particles at rest infinitely far apart. (For generality, we shall continue to let the nuclear charge be $+Ze$, rather than just $+e$ as in the hydrogen atom itself.) In contrast with the harmonic oscillator or the particle in a box, the potential energy of the Coulomb field given by Eq. 4.13 has no lower bound: It becomes negatively infinite at the origin. Again in contrast with the earlier examples, the Coulomb potential does have an upper bound: It approaches zero as r becomes infinite. This is why our choice for the zero of the energy scale is the most convenient and natural one to deal with the Coulomb potential.

Because the potential has a range $-\infty < V < 0$, any state of a hydrogenlike atom with a negative energy value ($E < 0$) must have $E > V$ near the origin and $E < V$ for large r. We say that all states with negative energy must be *bound states*. Following the reasoning of Section 4.1, we immediately recognize that the amplitude of the wave function for any such state must approach zero asymptotically as $r \rightarrow \infty$; the particle is effectively confined, and will be found mainly in the region where $E > V$ and the wave function is oscillatory.

By contrast, all states with $E > 0$ have $E > V$ everywhere in space, and are thus *free states*. Their wave functions are oscillatory for all values of r and are therefore analogous to the very first matter waves we examined, those describing a free particle. The positive-energy wave functions of the Coulomb potential are much like those of the free particle, in that they exhibit no discrete quantization and can be represented by either traveling or standing waves. In other ways, however, the two cases differ. We know that the second space derivative of a wave function measures the kinetic energy of the particle to which the wave function corresponds. The free particle feels a constant potential (arbitrarily set equal to zero), and in a state of fixed energy has exactly the same kinetic energy at every point in space. As a result, $\partial^2 \psi / \partial x^2$ is constant and the free-particle wave is a simple sinusoidal function. In a Coulomb field, because the potential energy varies smoothly with r, the kinetic energy must also vary with r for a particle with a given total energy. As one follows a particular wave function from large r inward toward the origin, the potential energy drops off toward $-\infty$, the kinetic energy must increase correspondingly, and the magnitude of $\partial^2 \psi / \partial r^2$ must increase. (When the variables are separated, V appears only in the radial wave equation; thus only the radial derivative is affected.) Instead of behaving like simple sinusoidal functions, the waves describing an unbound particle in a Coulomb field thus oscillate more rapidly at small r than at large r. The same property is of course exhibited by the bound states: They too show more rapid oscillations at small r.

We have already seen, from the harmonic oscillator and the particles in various boxes, that the bound states of simple systems show a very regular relationship between their energies and the number of nodes in the wave function. The lowest bound state has no nodes, the next-to-lowest state one node, and so on. For a system with spherical symmetry (like the hydrogen atom), this order holds rigorously for states of the same total angular momentum; however, for an arbitrary spherically symmetric potential, the energy corresponding to a given number of nodes varies with angular momentum (as in Fig. 3.12). If the ground state of the hydrogen atom is to be nodeless, it must be a state with $l = 0$, that is, with no angular momentum (see Table 3.1). The next energy level should correspond to a wave function with one node, which may be either angular or radial. In the former case we have $l = 1$, and the single nodal surface is a plane; in the latter case $l = 0$, and the node is a spherical surface in the radial part of the wave function. Higher energy levels have more nodes; for a state with a given value of l, there will be l angular nodes, the remainder if any being radial.

What can one say about the relative energies of the two single-node states—or of any states with the same number of nodes? Let us define a quantum number n as the number of nodes plus one. In general, as we said above, for a spherically symmetric potential, the energy levels for a given n vary with the angular momentum quantum number l. However, it is a unique property of the pure Coulomb potential that all bound states with the same n have exactly the same energy. Since the energy depends only on n and not on the angular momentum or its orientation, we call n the *principal quantum number*. Remarkably, this turns out to be the same as the quantum number n that appeared in Eq. 2.67 for the energy of the nth Bohr orbit. In fact, solution of the Schrödinger equation gives exactly the same expression for the bound-state energy eigenvalues of a hydrogen-like atom:

$$E_n = -\frac{Z^2 e^4 \mu}{8 \epsilon_0^2 n^2 h^2} \qquad (n = 1, 2, \ldots). \qquad (4.16)$$

(We shall not attempt to derive this result.)

One can now see why the Bohr model is so useful. Unfortunately, things are not really this simple: A number of effects that we have neglected perturb the energy-level structure of Eq. 4.16 and split each level into a group of very closely spaced levels. One speaks of the *fine structure* of the energy spectrum. We shall consider these effects in the next chapter; for our purposes here, we can go on assuming the validity of Eq. 4.16, which is very nearly correct.[4]

Assuming that all states of the same n have the same energy, how many such states are there for a given eigenvalue E_n? The wave function corresponding to E_n has a total of $n-1$ nodes. Since a state with a given value of l has l angular nodes (and thus $n-l-1$ radial nodes), it is clear that l can never exceed $n-1$; however, all values of l from 0 to $n-1$ are permitted. Now recall our discussion of space quantization in Section 3.11: For any value of l (which gives the total angular momentum), there are $2l+1$ possible orientations of the angular momentum vector (cf. Fig. 3.13), that is, $2l+1$ possible values of the quantum number m. Incidentally, l and m are usually called the *azimuthal quantum number* and the *magnetic quantum number* (because the degenerate states with different m will split in a magnetic field), respectively. The total number of states with energy E_n is obtained by summing over all the states described above:

$$\begin{aligned} N(E_n) &= \sum_{l=0}^{n-1} (2l+1) = 2\left(\sum_{l=0}^{n-1} l\right) + n \\ &= 2[\tfrac{1}{2}n(n-1)] + n \\ &= n^2. \end{aligned} \qquad (4.17)$$

If we neglect the fine structure, then, the nth energy level of a hydrogen-like atom is n^2-fold degenerate. The state with $n=1$ is nondegenerate, the $n=2$ state is quadruply degenerate, and so on.

For purely historical reasons, based on the appearance of the spectral lines with which they are associated, the states of specific angular momentum values have come to be known by a set of letters. Illogical as they now seem, these designations are in such common use that we must introduce them here. One refers to states with $l=0$ as s states, those with

---

[4] If we write Eq. 4.16 for the hydrogen atom as $E_n/hc = -R_H/n^2$, where $R_H = 109677.58 \text{ cm}^{-1}$, the maximum error is 1.18 cm^{-1} in the ground state; the greatest spread of fine structure levels in hydrogen is 0.36 cm^{-1} in the $n=2$ state.

$l = 1$ as p states, those with $l = 2$ as d states, and those with $l = 3, 4, 5, 6, \ldots$ as $f, g, h, j, \ldots$ states (note that i is not used). A state with $n = 1, l = 0$ is called a $1s$ state, one with $n = 2, l = 1$ a $2p$ state, and so on. The letters s, p, d, f originally stood for "sharp," "principal," "diffuse," and "fundamental." A wave function describing a bound state of a single electron is frequently called an *orbital*, a term coined to suggest the connection between the wave function and the classical orbit or trajectory. Small letters, as above, are generally used to denote the quantum numbers of individual orbitals, and capital letters for the total quantum numbers indicate the state of an atom. For example, an atom with zero total electronic angular momentum has $L = 0$, and is said to be in an S state. We shall see in the next chapter just what these total quantum numbers are. But in a hydrogenlike atom, with only one electron, the small-letter notation is sufficient.

The energy spectrum of the hydrogenlike atom is clearly very different indeed from that of the particle in a box or the harmonic oscillator, in that its bound states are more and more closely spaced as the energy approaches zero from below. This can be clearly seen in Fig. 2.18. There are, in fact, an infinite number of negative energy levels in a Coulomb potential well. Moreover, since the degeneracy increases with n, the number of bound states at a given energy also increases as $E \to 0$. For $E > 0$ the energy is no longer quantized; that is, the eigenvalues are continuous. This is called the *continuum* region, and corresponds to a free electron passing through the field of a nucleus.

Having discussed the energy levels, let us now see what we can say about the wave functions of the hydrogenlike atom and the distributions implied by Postulate II. If we separate the Schrödinger equation 4.14 according to the variables r, θ, and ϕ, we can immediately write down the radial wave equation as

$$\left[\frac{\mathbf{p}_r^2}{2\mu} + \frac{\hbar^2 l(l+1)}{2\mu r^2} - \frac{Ze^2}{4\pi\epsilon_0 r} \right] R_{nl}(r) = E R_{nl}(r). \tag{4.18}$$

where $\mathbf{p}_r \equiv -i\hbar[(\partial/\partial r) + r^{-1}]$ (see Eq. 3A.7); note that the radial part of the wave function depends on the quantum numbers n and l.

The negative infinity of the Coulomb potential at the origin—or the *singularity* there—places special constraints on the wave function, which must itself remain finite. There are two possible cases, depending on the value of l. When $l > 0$, the kinetic energy includes a contribution from the angular momentum, the centrifugal energy $\hbar^2 l(l+1)/2\mu r^2$. This term, which is a function of the coordinate r only, looks like a contribution to the potential energy, and can be referred to as the *centrifugal potential*. Its derivative is the so-called centrifugal force. The electron thus moves radially as if it were in an effective potential field,

$$V'_{\text{eff}} = \frac{L^2}{2\mu r^2} + V(r) = \frac{\hbar^2 l(l+1)}{2\mu r^2} - \frac{Ze^2}{4\pi\epsilon_0 r}. \tag{4.19}$$

Because the centrifugal term is always positive and increases as r^{-2} when $r \to 0$, it always dominates the effective potential for sufficiently small r. The effective potential thus always becomes *positively* infinite and acts as a repulsive potential in the vicinity of the origin. This means that for $l > 0$ the wave function $R_{nl}(r)$ must go to zero at the origin, representing the fact that an electron with nonzero angular momentum cannot reach the origin.

We can expand upon this conclusion by another route. Without going through the details of the calculation, we can reasonably assume that only the first two terms of Eq. 4.18 are significant at very small r. If this is so, then d^2R_{nl}/dr^2 (part of $\mathbf{p}_r^2 R_{nl}$) will be proportional to $l(l+1)R_{nl}(r)/r^2$, so that $R_{nl}(r)$ itself must vary as either r^{l+1} or r^{-l}. The function r^{-l} becomes infinite when r goes to zero, so this mathematical possibility is physically inadmissable. We conclude, then, that for $l>0$ the radial wave function must vary as r^{l+1} at small r. This in turn means that, the higher the angular momentum, the lower is the initial slope $[dR(r)/dr]_{r=0}$. In simple terms, centrifugal force pushes the electron away from the nucleus.

We said that there were two possible cases for the behavior of the wave function at the origin, and the second case is that of $l=0$, that is, of s states. There is then no centrifugal potential and the electron can in principle penetrate to the nucleus. This case is not as simple to analyze without going through the mathematics, but it turns out that $R_{n0}(r)$ rises to a *cusp*, varying as $e^{-r/n}$ as $r \to 0$. The wave function is continuous and finite at the origin, but its slope changes discontinuously along any line through the nucleus; as with the particle in a box, this discontinuity is a result of the potential becoming infinite. Since the wave function is nonzero at the origin, what keeps the electron from collapsing into the nucleus? Here we can appeal to the uncertainty principle: If the electron collapsed to a radius less than some value r, then by Eq. 3.84 the uncertainty in p_r would be greater than about $\hbar/2r$, and the average kinetic energy greater than $\hbar^2/4\mu r^2$. Like the centrifugal term of the earlier argument, this kinetic energy goes to $+\infty$ faster than V goes to $-\infty$, so that no electron with finite total energy can collapse all the way to $r=0$. In fact, the product $rR_{n0}(r)$, which determines the electron's probability distribution, does vanish as $r \to 0$.

The behavior that we have deduced can be confirmed by looking at the actual radial wave functions, listed in Table 4.2 and plotted in Fig. 4.5a. These are the solutions of Eq. 4.18. Each $R_{nl}(r)$ consists of an exponential $e^{-\rho/n}$ multiplied by a polynomial of degree n in ρ, where $\rho \equiv Zr/a$ (where a is approximately[5] the Bohr radius a_0, or $4\pi\epsilon_0\hbar^2/e^2m_e$). In agreement with our earlier conclusions, there are $n-l-1$ radial nodes (not counting the origin, through which pass the angular nodes for $l>0$).

The next question to be answered is where the electron is most likely to be in any given state. The point of highest probability is, of course, where $|\psi|^2$ has its maximum value. However, we are really more interested in knowing the *radius* at which the electron is most likely to be found. This radius is greater than that of maximum $|\psi|^2$, since a thin spherical shell at large r has a greater volume than an equally thin shell at small r. The probability of finding the electron in a shell of thickness dr is equal to the integral of $|\psi|^2$ over the volume of the shell,

$$\mathscr{P}_{nlm}(r)\,dr = \int_{\text{shell}} |\psi_{nlm}|^2\,dV = \int_{\phi=0}^{2\pi} \int_{\theta=0}^{\pi} |\psi_{nlm}|^2\, r^2\,dr\sin\theta\,d\theta\,d\phi,$$

(4.20)

where dV is given by Eq. 2B.2. We separate the variables as usual,

$$\psi_{nlm}(r, \theta, \phi) = R_{nl}(r)\Theta_{lm}(\theta)\Phi_m(\phi) = R_{nl}(r)Y_{l,m}(\theta, \phi),$$
(4.21)

and take for the $Y_{l,m}(\theta, \phi)$ the normalized spherical harmonics of Table

---

[5] Strictly, the radius appearing in the solution to Eq. 4.16 is inversely proportional to the reduced mass μ, equal to $m_e m_{\text{nucleus}}/(m_e + m_{\text{nucleus}})$, thus:

$$a \equiv \frac{4\pi\epsilon_0\hbar^2}{e^2\mu},$$

whereas the Bohr radius a_0 is based on a nucleus with infinite mass, so the electron mass m_e appears in a_0, instead of the reduced mass μ.

TABLE 4.2
NORMALIZED RADIAL WAVE FUNCTIONS, $R_{nl}(r)$, FOR HYGROGENLIKE ATOMS[a]

			$\displaystyle\int_0^\infty |R_{nl}(r)|^2\, r^2\, dr = 1 \qquad \rho \equiv \dfrac{Zr}{a_0}$
Orbital	n	l	
$1s$	1	0	$2\left(\dfrac{Z}{a_0}\right)^{3/2} e^{-\rho}$
$2s$	2	0	$\dfrac{1}{2\cdot 2^{1/2}}\left(\dfrac{Z}{a_0}\right)^{3/2}(2-\rho)e^{-\rho/2}$
$2p$	2	1	$\dfrac{1}{2\cdot 6^{1/2}}\left(\dfrac{Z}{a_0}\right)^{3/2}\rho e^{-\rho/2}$
$3s$	3	0	$\dfrac{2}{81\cdot 3^{1/2}}\left(\dfrac{Z}{a_0}\right)^{3/2}(27-18\rho+2\rho^2)e^{-\rho/3}$
$3p$	3	1	$\dfrac{4}{81\cdot 6^{1/2}}\left(\dfrac{Z}{a_0}\right)^{3/2}(6\rho-\rho^2)e^{-\rho/3}$
$3d$	3	2	$\dfrac{4}{81\cdot 30^{1/2}}\left(\dfrac{Z}{a_0}\right)^{3/2}\rho^2 e^{-\rho/3}$

[a] As written, with the Bohr radius a_0 instead of a based on a general reduced mass μ, these expressions describe an atom with an infinitely massive nucleus.

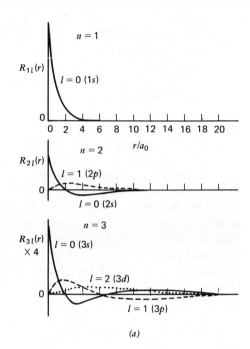

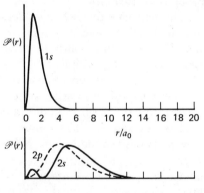

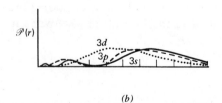

(a)

(b)

FIGURE 4.5
Radial wave functions and probabilities for the first few states of the hydrogen atom. (a) The radial wave function, $R_{nl}(r)$. (b) The radial distribution function, $\mathcal{P}_{nl}(r) = r^2|R_{nl}(r)|^2$. All the graphs are plotted on the same radial scale, r/a_0, where a_0 is the Bohr radius (0.529 Å). Since the wave functions are normalized, the total area under each of the $\mathcal{P}(r)$ curves is the same, corresponding to one electron.

3.1. Since the integral of $|Y_{l,m}|^2$ over θ and ϕ is unity, Eq. 4.20 reduces to

$$\mathcal{P}_{nl}(r)\, dr = r^2|R_{nl}(r)|^2\, dr, \qquad (4.22)$$

with the probability proportional to the square of $rR(r)$ rather than the square of the radial wave function itself. The factor r^2 can be interpreted as being due to the fact that the volume of the spherical shell is $4\pi r^2\, dr$ (remember that $|\psi|^2$ is the probability per unit volume).

Note that in the above equations we have kept explicit the subscripts n, l, m. In fact, the radial wave functions (and thus the probability densities) do depend on both n and l, and the spherical harmonics depend on l and m, even though the energy is a function of n only. In Fig. 4.5b we plot the radial probability densities for the first few states of the hydrogen atom. When the radial probability density has more than one *lobe* (i.e., has one or more nodes), it is always the outermost lobe that has the largest area and the highest amplitude. The $\mathcal{P}_{nl}(r)$ functions with the same n and different l have their outermost maxima in roughly the same region. In particular, the function with $l = n-1$ has only a single maximum,

$$r_{\max}[\mathcal{P}_{n,n-1}(r)] = \frac{4\pi\epsilon_0 n^2\hbar^2}{Ze^2\mu}, \qquad (4.23)$$

which is the same as the radius of the nth orbit in the Bohr theory, Eq. 2.63. The most probable radius for the hydrogen $1s$ electron ($Z = 1$, $n = 1$, $l = 0$) is thus the Bohr radius a_0, as can be seen in Fig. 4.5b.

Thus, in a rough way, the conclusions based on the wave picture of the hydrogen atom are in agreement with the results of the simple Bohr model. The energy levels are the same (if we neglect the fine structure), and the electron is most likely to be at a radius near the value predicted by Bohr. The physical picture is entirely different, however. In Bohr's model the electron remained in a fixed orbit restricted to a single plane.

In the present theory the electron can be anywhere in space except at the nodes of ψ; some places are just more probable than others. We have discussed the radial probability density; in the next section we shall take another look at the angular distribution.

4.4
The Shapes of Orbitals

The potential energy is spherically symmetric for any isolated atom, not just the hydrogen atom. The angular part of all atomic wave functions must thus be given by the spherical harmonics of Table 3.1. As we pointed out earlier, the angular distribution of electrons in atoms governs the shapes of the molecules formed when these atoms join together. We must therefore become acquainted with the actual shapes of the spherical harmonics. Let us consider several ways of illustrating those shapes.

First, of course, we could simply plot $Y(\theta, \phi)$ as a function of the angles. However, we have the problem that all the $Y_{l,m}$ for $|m| > 0$ are complex, containing the factor $e^{im\phi}$. We can obtain real functions by combining the eigenfunctions for $+m$ and $-m$:

$$Y_{l,\cos m\phi} = \frac{Y_{l,m} + Y_{l,-m}}{\sqrt{2}} = \Theta_{l,m}\left(\frac{e^{im\phi} + e^{-im\phi}}{\sqrt{2}}\right)$$
$$= \sqrt{2}\Theta_{l,m}\cos m\phi,$$

$$Y_{l,\sin m\phi} = \frac{Y_{l,m} - Y_{l,-m}}{\sqrt{2}i} = \Theta_{l,m}\left(\frac{e^{im\phi} - e^{-im\phi}}{\sqrt{2}i}\right) \tag{4.24}$$
$$= \sqrt{2}\Theta_{l,m}\sin m\phi.$$

This is the same method that we used to obtain standing-wave functions for the free particle in Eq. 3.42. (The factor $\sqrt{2}$ here maintains the normalization.) The new functions satisfy the Schrödinger equation as well as the old ones, and thus are valid wave functions of the system. If we look at the spherical-coordinate Hamiltonian, Eq. 3.154, we can see that both must give the same energy eigenvalues for a given m: In both cases $\partial^2 Y/\partial \phi^2 = -m^2 Y$. The functions 4.24 are thus eigenfunctions of the energy and the total angular momentum; unlike the original spherical harmonics, however, they are not eigenfunctions of L_z (cf. Section 3.11).

Using the real forms, we can now plot the angular wave functions in the form of polar graphs, Fig. 4.6. In such a graph, the magnitude of the function at a particular value of the angles is given by the radial distance from the origin. The graphs in the figure are cross sections of the complete three-dimensional graphs, and thus show the behavior of each $Y(\theta, \phi)$ in only a single plane, as a function of either θ or ϕ alone. However, the planes selected for the figure are those that most clearly display the shapes of the individual wave functions. (See Fig. 2B.1 if you need to review the coordinate relationships.)

Let us examine the shapes of these angular wave functions. As expected, the function with $l = 0$ (s orbital) has no dependence on the angles and is thus spherically symmetric. There are three p ($l = 1$) orbitals, each symmetric around one of the Cartesian axes and thus labeled p_x, p_y, p_z. In a three-dimensional polar graph, each p orbital consists of two spherical lobes tangent at the origin. The function $Y_{2,0}$ is symmetric around the z axis, and thus consists of two elongated axial lobes with a toroidal "collar" around the middle; each of the other four d ($l = 2$) orbitals consists of four lobes whose axes are at right angles in a single plane. The latter four orbitals vary with θ and ϕ in the same way as do the functions xy, xz, yz, and $x^2 - y^2$ over the surface of a unit sphere, and are thus labeled d_{xy}, and so on (See Fig. 4.6); $Y_{2,0}$ varies as $3z^2 - r^2$,

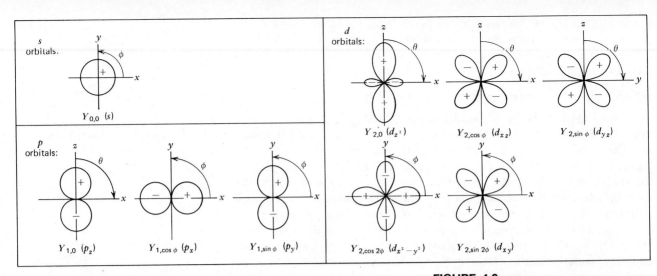

FIGURE 4.6
Polar graphs of spherical harmonics in real form; see Table 3.1 and Eqs. 4.24 for definitions. The s, p, d designations are discussed in the text. The radial distance gives the magnitude of $Y(\theta, \phi)$; the sign of each lobe of the function is indicated by a + or − sign.

and is labeled d_{z^2}. The $f, g, \ldots$ orbitals have even more complicated shapes, but one seldom needs to consider these.

The distribution of electrons is governed not by the wave function itself but by its square. We therefore plot in Fig. 4.7 the squares of the various kinds of real[6] spherical harmonics. This time we give perspective views of the complete three-dimensional polar graphs. It is clear that the squares of the p and d orbitals are even more strongly directional than the functions themselves since squaring weights more heavily those angles for which Y was already large.

Finally, to obtain the electron distribution in an atom, we must multiply the $|Y(\theta, \phi)|^2$ of Fig. 4.7 by the square of the radial wave function $R(r)$. For the hydrogen atom the radial wave functions are the $R_{nl}(r)$ of Fig. 4.5a. Combining the two sets of functions, we obtain the hydrogen probability densities[7] shown in Fig. 4.8. The probability densities $|\psi|^2$ here are represented by perspective views; the radial coordinate is now the actual radial distance from the nucleus in the atom. As in Fig. 4.6, Figs. 4.8a–f show density as functions of r and θ, and can be interpreted as cross sections of the complete three-dimensional density functions.

We now have a comprehensive idea of the electron distribution in the various states of the hydrogen atom. In any other one-electron species (He^+, Li^{2+}, etc.) the distribution will be the same except for the radial scale (with a_0/Z replacing a_0). As we shall see in the next chapter, the individual electrons in more complex atoms can still be considered to have wave functions like those in hydrogen, though distorted by interaction with one another. More significantly, the angular functions $Y(\theta, \phi)$ remain exactly the same, and the basic shapes of $s, p, d, \ldots$ orbitals are thus universal.

[6] Note that only the real functions of Eq. 4.24 yield the complete angular variation. If we used the original spherical harmonics containing $e^{im\phi}$, we would have

$$|Y_{l,m}|^2 = Y_{l,m}^* Y_{l,m} = \Theta_{l,m} e^{-im\phi} \Theta_{l,m} e^{im\phi} = \Theta_{l,m}^2,$$

with no ϕ dependence.

[7] The radial distribution function $\mathcal{P}(r)$ of Fig. 4.5b is related to the probability density $|\psi|^2$ by Eq. 4.20.

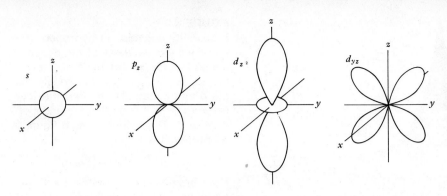

FIGURE 4.7
Angular probability distributions: perspective views of squared spherical harmonics, $|Y(\theta, \phi)|^2$. Except for orientation, the p_x and p_y orbitals have the same shape as the p_z; similarly, the d_{xy}, d_{xz}, and $d_{x^2-y^2}$ orbitals have the same shape as the d_{yz}.

4.5
Transitions between Energy Levels

We now examine how energy is absorbed or emitted by a hydrogen atom. Any such process must involve a change in the atom's wave function from one standing wave (eigenstate) to another. This change of state must always be associated with emission or absorption of a quantum of energy ΔE; if the energy is in the form of radiation, its frequency must satisfy the Einstein condition $\Delta E = h\nu$. Since the energy levels of the hydrogen atom are correctly given by the Bohr theory, we know that our wave description of the hydrogen atom is consistent with the spectroscopic data on transitions between energy levels, as summarized in Eq. 2.73. Thus we can calculate the exact energy difference or frequency in a transition between any two levels. Up to this point, however, the transition process itself has been just a black box. We have not examined, in either classical or quantum mechanical terms, the mechanism that produces the transition. In this section we shall consider the general nature of the transition process and look at some orders of magnitude for the quantities involved. Here we discuss only the hydrogen atom in any detail, and that on a quite elementary level, but one can use the same ideas in examining more complicated transitions—in molecules, for example.

Let us begin with the crudest sort of model, a classical electron in a circular Bohr orbit in the xy plane. Suppose that a beam of light is incident in the z direction, with its electric field[8] polarized along the x axis. This electric field, which exerts a force on the electron, has the form of a wave of frequency ν (cf. Appendix 2A). Imagine for the moment that ν is very much lower than the frequency of the electron's revolution about the nucleus, so that the electron goes round the nucleus many times before the force on it changes appreciably. In this case, the force exerted by the electric field has the effect of polarizing the atom, pulling the nucleus in one direction and shifting the *average* position of the electron slightly in the opposite direction, just as a constant electric field would do. The oscillations of the field simply change the direction of polarization slowly from one side to the other. The electron distribution is distorted, and a strong enough field may even pull the electron away from the nucleus altogether (ionize the atom). The net effect of a slowly varying field is to shift the average energy of the atom just so long as the field is applied; no energy is permanently transferred from the field to the atom.

At the other extreme of the frequency scale, suppose that the electric field oscillates much faster than our Bohr electron moves about the nucleus. In this case the atom cannot respond to the fast vibrations of the

[8] We neglect the magnetic field, which also exerts a force on the moving electron.

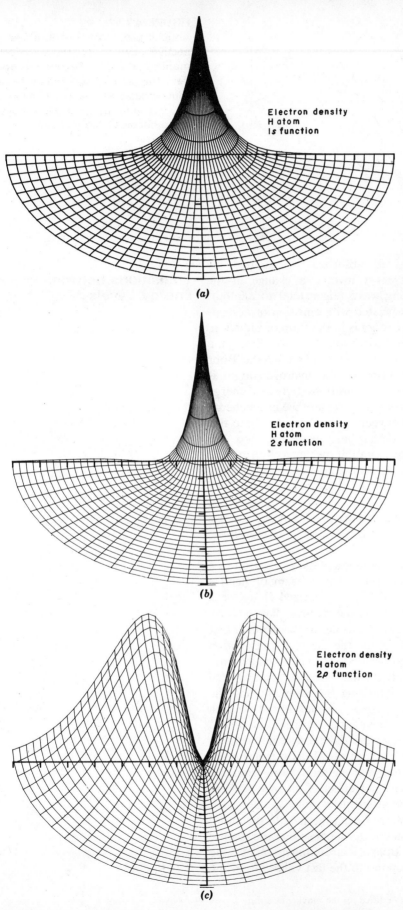

Electron density
H atom
1s function

(a)

Electron density
H atom
2s function

(b)

Electron density
H atom
2p function

(c)

FIGURE 4.8
Probability densities in hydrogen orbitals. Perspective views of the densities are shown for the (a) 1s, (b) 2s, (c) 2p, (d) 3s, (e) 3p, and (f) 3d orbitals. For the p and d orbitals, the densities correspond to $m_l = 0$, so they are independent of ϕ. The height in each figure is proportional to

$$|\psi_{nlm}(r)|^2 = |R_{nl}(r)|^2 |Y_{10}(\theta)|^2.$$

Note how all the s orbitals have densities peaking at the origin (the nucleus), the p orbital functions have densities rising rapidly from zero at the nucleus, and the 3d orbital density rises slowly from zero at the nucleus. Functions with principal quantum number n exhibit $n-1$ surfaces of zero density. These can be picked out easily except for the 3s function, whose outer node is almost indiscernable because it occurs at a value of r so large that the density is low everywhere in its vicinity. (These figures were supplied through the courtesy of Mary E. Dolan.)

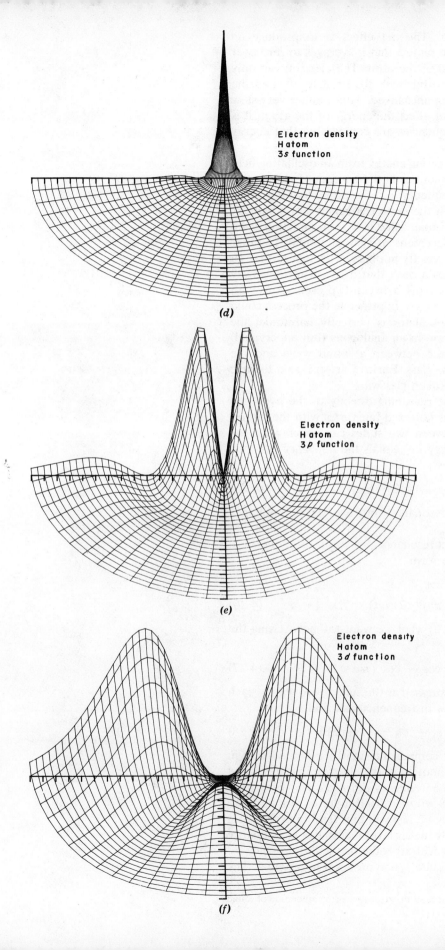

Electron density
H atom
3s function

(d)

Electron density
H atom
3p function

(e)

Electron density
H atom
3d function

(f)

force field; its inertia is too great. The net effect is to produce an instantaneous polarization varying so rapidly that it averages to zero over a time much shorter than the period of the orbit. The electron can only quiver a bit as it goes round in its orbit, with the orbit itself—and the electron's average energy—remaining unchanged. Thus neither very slow nor very fast electric field oscillations affect the energy of the atom; it is only when the orbital and field frequencies are comparable that energy transfer can occur.

We seek a mechanism for transferring energy from an oscillating field to an atom. The key to this mechanism lies in the concept of *resonance*.[9] Consider a harmonic oscillator of frequency ν, and suppose that it is acted upon by a force field oscillating with the same frequency. For simplicity we assume that the force has its maximum value in a given direction when the oscillator has its greatest displacement in that direction—in other words, that the two oscillations are exactly in phase. This means that on every oscillation the particle receives a push that drives it farther away from the origin. Such a transfer of energy between two oscillations with the same frequency is known as resonance. In principle the process could go on without limit, but no real oscillator is perfectly harmonic; the model must break down in some way. In an analogous (but apocryphal) situation, by establishing a resonance between a sound wave and the natural vibration frequency of a wine glass, Enrico Caruso is said to have covered a good many tables with broken glassware.

Now we cannot apply the same reasoning directly to the hydrogen electron's orbital frequency, which is not fixed but varies with the radius. But we know that a transition between two states can occur when we apply light whose frequency ν (energy $h\nu$) equals the *difference* between the two orbital frequencies (energies):

$$E = h\nu = h(\nu_2 - \nu_1) = E_2 - E_1. \qquad (4.25)$$

In what sense can the light be in resonance with two states simultaneously?

As in Eq. 4.25, we assume exact resonance, $\nu = \nu_2 - \nu_1$ (with $\nu_2 > \nu_1$). The oscillating electric field has the form

$$\mathbf{E} = \mathbf{E}_0 \sin 2\pi\nu t$$
$$= \mathbf{E}_0 \sin(2\pi\nu_2 t - 2\pi\nu_1 t). \qquad (4.26)$$

The second form of Eq. 4.26 leads to what we want to know. Using the trigonometric identity

$$\sin(x - y) = \sin x \cos y - \cos x \sin y, \qquad (4.27)$$

we find that the total field can be expressed as the sum of two waves, each constructed from two components with frequencies ν_1 and ν_2:

$$\mathbf{E} = \mathbf{E}_0(\sin 2\pi\nu_2 t \cos 2\pi\nu_1 t - \cos 2\pi\nu_2 t \sin 2\pi\nu_1 t). \qquad (4.28)$$

Thus the field of frequency ν has components that are individually in exact resonance with the oscillatory motion of the electron in states 1 and 2, which can be any two states whose frequency *difference* is ν. Either the first or the second term of Eq. 4.28 will find itself more or less in phase with electrons revolving at frequency ν_1, and will thus tend to excite them to a higher energy level. But the only higher level that is also in resonance with the field (and thus can interact with it) is the one with frequency ν_2. The field pushes the electron out of state 1, and pushes it into state 2. The

---

[9] The term "resonance" is, of course, derived from the resonance or reverberation of sound waves.

process can also occur in reverse, with the electron giving up energy to the field and falling from state 2 to state 1. This is why, as we recall from Chapter 2, any system capable of absorbing radiant energy at a particular frequency must be just as capable of emitting radiation of the same frequency.

But we still have not answered the question of what happens to the electron *between* the two states. This is where a classical approach breaks down, and we must appeal to wave mechanical concepts for some understanding. Rather than being a body that undergoes an oscillation, the electron itself *is* an oscillation, that is, a standing wave. What happens is that an oscillation dies out at one place as a new oscillation appears at another place. One can observe something similar with, say, two tuning forks of the same resonant frequency: Set one in motion, and the sound waves from it will start the other vibrating. But the tuning fork continues to exist when it is not oscillating, and the electron does not; its presence is always in the form of *some* wave. Since mass and charge must be conserved, the new oscillation is still an electron, not some other beast, but strictly speaking it is meaningless to call it "the same" electron. The actual excitation process, the transfer of energy from one oscillation to the other, requires many cycles of the radiation field. To get a feeling for why this is so, we must examine the magnitude of the interaction energy between the electric field and the atom.

For a first estimate, let us fall back upon the Bohr model, in which the electron and the proton are considered as a pair of point charges. The hydrogen atom thus acts like a rotating *electric dipole*, a pair of equal but opposite charges some distance apart. For charges $+q$ and $-q$, the *dipole moment* is defined as

$$\boldsymbol{\mu} \equiv q\mathbf{r}, \tag{4.29}$$

where $\mathbf{r}$ is the radius vector from the negative to the positive charge. For a Bohr hydrogen atom in its ground state, the instantaneous dipole moment thus has a magnitude

$$\mu = ea_0 = (1.602 \times 10^{-19}\,\text{C})(5.292 \times 10^{-11}\,\text{m}) = 8.478 \times 10^{-30}\,\text{C m}.$$

Dipole moments are most often reported in *debyes*: 1 debye is the moment of two charges of 10^{-10} esu (about $0.2e$) 1 Å apart, so that

$$1\ \text{debye (D)} \equiv 10^{-18}\,\text{esu cm} = 3.33564 \times 10^{-30}\,\text{C m}$$

and $ea_0 = 2.542\,\text{D}$.

The potential energy of interaction between a microscopic dipole and an external field is given by

$$V_{\text{dipole-field}} = -\mathbf{E} \cdot \boldsymbol{\mu} = -E\mu \cos\theta, \tag{4.30}$$

where θ is the angle between the dipole and field directions; the interaction energy has its greatest magnitude when the dipole is lined up with the field. (We shall derive this equation in Section 10.1.) The electric fields of light waves vary over many orders of magnitude. In a brightly lit room, the electric field strength of the light may be of the order of 25 V/m. On the other hand, a fairly intense monochromatic source like a laser—not by any means the most intense available, but a rather powerful one—can easily produce a beam with a field of 50 million V/m. Thus the instantaneous interaction energy between the field and the atomic dipole should be at most

$$|E\mu| = Eea_0 \approx (25\ \text{to}\ 5 \times 10^7)\text{V/m} \times e \times 5.3 \times 10^{-11}\,\text{m}$$
$$\approx (1.3 \times 10^{-9}\ \text{to}\ 2.6 \times 10^{-3})\,\text{eV}$$

or about 2×10^{-28} to 4×10^{-22} J. The actual energy must be somewhat less, since **E** and **μ** will not always be perfectly aligned. However, the total energy that must be absorbed in a single hydrogen atom transition (cf. Fig. 2.18) is of the order of 10 eV, thousands or even billions of times as great. It is hard to conceive in classical terms how such a transfer could occur, unless the field is applied for, say, a million cycles, transferring a millionth of the necessary total energy on each cycle. This reasoning supports our earlier statement that many cycles of the radiation field are required to bring about a transition. This is still not a long time, since for visible light a cycle is only about 2×10^{-15} s.

The calculation just carried out was entirely classical, in so far as it considered the electron as a point charge with a definite position at any given time. How can we reconcile this with the wave model of the electron? We have already interpreted a transition as a replacement of one oscillation by another, analogous to the resonant oscillation of two tuning forks. A different analogy may give some feeling for the mechanism of the process. Imagine that you hold one end of a rope, with the other end attached firmly to a fixed swivel. You can swing the rope so that, although your hand remains practically at rest, the rope takes the shape of a rotating standing wave with no nodes—in other words, the conventional way to swing a jump rope (Fig. 3.2a). Now suppose that you want to make the rope oscillate in a standing-wave mode with one node (Fig. 3.2b). With a very slight extra wiggle of your hand, you can in a few cycles transform the original motion into the desired new form. What you have done is to apply a weak force in resonance with the rope's motion, thereby changing its mode of oscillation. This is essentially how the rotating electric field transfers energy from one oscillation to another, except that many more cycles are involved.

This section has dealt almost exclusively with analogies and simple models, since the actual quantum mechanical approach to transitions is too complicated to describe at this level. Still, we can outline the basic concepts involved. Consider first an isolated hydrogen atom in its initial state, before any electromagnetic field is applied. This state is time-independent and thus an eigenstate of the energy, which has the value $E_0 = h\nu_0$. Now we apply the oscillating field. Since the system is varying with time, the uncertainty principle (in the form $\Delta E \, \Delta t \geq \hbar/2$) tells us that the energy can no longer be definitely fixed. There is in fact a nonzero probability of the system's being in *any* of its energy eigenstates, and the total time-dependent wave function can be written as a sum over the corresponding eigenfunctions,

$$\Psi(\mathbf{r}, t) = \sum_n a_n(t) \Psi_n(\mathbf{r}, t) = \sum_n a_n(t) \psi_n(\mathbf{r}) e^{-2\pi i \nu_n t} \qquad (4.31)$$

where $\nu_n \equiv E_n/h$. This general, time-dependent wave function must, according to Postulate V, be a solution to the general time-dependent Schrödinger equation, whose Hamiltonian operator H includes the kinetic energy T, the time-independent potential interactions V, and the time-dependent fields $H'(t)$ such as that of a light wave:

$$H\Psi(\mathbf{r}, t) = [T + V(\mathbf{r}) + H'(\mathbf{r}, t)]\Psi(\mathbf{r}, t) = i\hbar \frac{\partial \Psi(\mathbf{r}, t)}{\partial t}. \qquad (4.32)$$

If $\Psi(\mathbf{r}, t)$ is expressed as in Eq. 4.31, then Eq. 4.32 can be recast so that the solutions are the expressions for the time-dependent amplitudes $a_n(t)$. We shall not attempt to carry out this calculation. The important result, however, is the same as in the corresponding classical wave problem: $a_n(t)$ is negligibly small except when the field frequency ν satisfies the condition $\nu \approx \pm(\nu_n - \nu_0)$, that is, when the field is in resonance with both the initial state and the nth state. Given the $a_n(t)$, one can go on to calculate the probability and rate of transitions between any two states.

The probability of finding the system in the nth eigenstate at time t is $|a_n(t)|^2$, that is, $|a_n(t)|^2$ is the probability that we find the system with energy E_n and density distribution $|\psi_n(\mathbf{r})|^2$ at time t. The $a_n(t)$'s have the property that the sum of their absolute squares is unity. This is a consequence of the conditions that the eigenstates of energy are mutually exclusive—a system observed to have energy E_n cannot simultaneously be in a state with energy E_m—and that every possible eigenstate of energy for the system is included in the set $\psi_n(\mathbf{r})$. For the hydrogen atom, the transition rate is of the same order of magnitude as that we estimated from the rotating-dipole model.

Of course, not all transitions are brought about by an external electromagnetic field. Consider what happens when two atoms or molecules collide with each other. Each molecule is a collection of charged particles in motion and thus generates its own highly localized field, which can induce a transition in the other molecule when the two are close enough. But since the total energy is conserved, this can happen only if both molecules undergo transitions simultaneously. However, one or both of these transitions can be merely between two eigenstates of the molecule's overall motion—in other words, just a change in the molecule's kinetic energy and momentum.[10] Since the energy spectrum for free motion is continuous (or nearly so), a collision between two molecules can induce a wide variety of transitions in their internal states—or even the shuffling of atoms between molecules that we call a chemical reaction.

The relative probability of transitions depends rather sensitively on the system and the two states involved. Some transitions, logically named *forbidden transitions*, almost never occur unless an atom or molecule is so violently disturbed that it virtually loses its identity. These are transitions for which the $a_n(t)$ of Eq. 4.26 are very small even when $\nu = |\nu_n - \nu_0|$, usually because of the symmetry relationships between the initial and final states. For a given atom or molecule, the probability of a transition between states n and n' by light absorption is proportional to $|\boldsymbol{\mu}_{n'n}|^2$, where[11]

$$\boldsymbol{\mu}_{n'n} \equiv \int \cdots \int \psi_{n'}{}^* \boldsymbol{\mu} \psi_n \, dq_1 \ldots dq_N, \qquad (4.33)$$

$\boldsymbol{\mu}$ being the system's instantaneous dipole moment ($e\mathbf{r}$ for the hydrogen atom); the integral is, as usual, taken over all the coordinates of the system.

For many simple systems the integral of Eq. 4.33 is nonzero only for certain changes in the quantum numbers. For the hydrogen atom, for example, the integral over the angular coordinates vanishes unless $\Delta l = \pm 1$ and $\Delta m = 0$ or ± 1. These are examples of what are called *selection rules*. For the harmonic oscillator, with $E_n = (n + \frac{1}{2})h\nu$, the selection rule is $\Delta n = \pm 1$; thus ΔE must equal $\pm h\nu$, and only radiation of the oscillator frequency ν should induce a transition. Similarly, the rigid rotator of Eq. 3.168 also must have $\Delta l = \pm 1$. You may wonder how forbidden transi-

---

[10] A collision in which the total kinetic energy of relative motion is unchanged and both molecules change only their individual kinetic energies and momenta is called *elastic*. We shall treat elastic collisions at length in Part Three.

[11] Compare the expectation value of $\boldsymbol{\mu}$ in a single stationary state,

$$\langle \boldsymbol{\mu} \rangle_n = \int \cdots \int \psi_n{}^* \boldsymbol{\mu} \psi_n \, dq_1 \cdots dq_N.$$

tions manage to occur at all. Actually, the transition probability is given by a series, ordinarily dominated by the $|\mathbf{\mu}_{n'n}|^2$ term; the later terms, involving such quantities as magnetic dipoles (cf. Section 5.1) and electric quadrupoles, are still present (though very small) even when $\mathbf{\mu}_{n'n}$ vanishes.

In addition to transitions induced by external fields or collisions, it is possible for an atom or molecule to undergo a *spontaneous transition*. That is, rather than changing its energy by $h\nu$ when stimulated by a field of frequency ν, the system itself emits radiation of frequency ν. Of course, such a transition can go only to a state of lower energy. Whereas the rate of induced transitions depends directly on the intensity of the applied field, the rate of spontaneous emission depends only on the initial and final states. The fraction of emissions that occur spontaneously increases rapidly with the energy difference. For spontaneous radiation of visible light, the rate can be as much as 10^8 or $10^9 \, \text{s}^{-1}$ (per atom); the inverse of this rate, about 10^{-8}–$10^{-9}\,\text{s}$, is the average lifetime of the excited state. Forbidden transitions are usually considered to be those that occur spontaneously more slowly than $10^6 \, \text{s}^{-1}$

If induced transitions are triggered by resonance with the applied field, just what makes a spontaneous transition occur? Of course, for an excited system to give up its energy without inducement is inherently plausible, and in agreement with classical mechanics. (Remember that Bohr had to explain why the orbiting electron didn't radiate *all* its energy.) But the actual wave mechanical mechanism is very complicated. Perhaps the simplest way to describe it is to say that the transition is induced by the zero-point oscillations of the radiation field: If the field is made up of Planck's harmonic oscillators, it must have a zero-point energy, even when no external field is applied. This mechanism accounts for transitions in isolated atoms, but of course no atom is truly isolated from the fields of other atoms or stray radiation.

Consider an excited state from which spontaneous emission is relatively slow. One can often by various means "pump" a large number of atoms into such a state. If the system is then subjected to radiation of the emission frequency ν, a massive amount of induced emission will occur in a short time. The emitted radiation is all of the same frequency ν and in phase with the inducing field; such radiation is called *coherent*. This is the phenomenon of *laser* (or *maser*) action, the name standing for *l*ight (or *m*icrowave) *a*mplification by *s*timulated *e*mission of *r*adiation. The typical laser has mirrors at the ends, so that most of the emitted light travels back and forth through the system and induces more and more emission.

To sum all this up, consider a large collection of molecules with two states separated by an energy $E_2 - E_1 = h\nu$, in the presence of radiation of frequency ν. Transitions of all the kinds we have discussed go on simultaneously: transitions induced by the field, transitions induced by collisions, and spontaneous transitions. The first two can go to either the higher or the lower state (the probability per molecule being the same in both directions), whereas spontaneous transitions can go only to the lower state. A dynamic equilibrium is established among all these processes when the total rate of $1 \rightarrow 2$ transitions equals the total rate of $2 \rightarrow 1$ transitions. The ratio of atoms in states 1 and 2 when this occurs is a function of the kinetic energy per molecule, which is proportional to the parameter we call the temperature. As we shall show in Part Two, the equilibrium distribution is given by

$$\frac{N_2}{N_1} = \exp\left[\frac{-(E_2 - E_1)}{k_B T}\right] = \exp\left(\frac{-h\nu}{k_B T}\right), \qquad (4.34)$$

where T is the absolute temperature and k_B is a constant. The rates of upward and downward field-induced transitions vary with the intensity of radiation of frequency ν, which for black-body radiation increases with the temperature of the radiation source (cf. Fig. 2.14). Collisions become both more frequent and more likely to produce transitions as the temperature increases (see Part Three), but spontaneous transitions are independent of temperature.

Let us make this discussion concrete by considering a gas of sodium atoms, sufficiently dilute that we can ignore collisions. The sodium atom can undergo a transition between two states by absorbing or emitting a quantum of yellow light ($\lambda = 5890$ Å, $\nu = 5.09 \times 10^{14}\,\text{s}^{-1}$); it is this transition that produces the illumination in sodium-vapor street lights. If the gas is placed in a box whose temperature is about 3000°C, then the rate of induced transitions per atom is about $5000\,\text{s}^{-1}$ in each direction. However, the rate of spontaneous emission from the higher state is about $10^7\,\text{s}^{-1}$, so that virtually all the excited atoms will return to the lower state spontaneously rather than by induced emission. Since the downward rate is so much greater than the upward rate, at any time only about 0.05% of the atoms are in the excited state. Now suppose that we increase the intensity of black-body radiation by raising the box temperature from 3000 to 35,000°C. The rate of induced transitions in each direction will then become essentially equal to the rate of spontaneous emission, so that the total emission will be half spontaneous and half induced. With the total downward rate just twice the upward rate, we can expect to find twice as many atoms in the lower state as in the upper state at any time; in other words, at 35,000°C, about one-third of the sodium atoms should be in the excited state.

Further Reading

J. Avery, *Quantum Theory of Atoms, Molecules and Photons*, (McGraw-Hill Book Co., Inc., New York, 1972) Chapters 3, 4.

A. S. Davydov, *Quantum Mechanics* (Pergamon Press, Oxford, England and Addison-Wesley Publishing Co., Reading, Mass., 1965), Chapters I, II, III, V and VI.

M. Karplus and R. N. Porter, *Atoms and Molecules* (W. A. Benjamin, Inc., Menlo Park, Calif., 1970), Chapter 3.

W. Kauzmann, *Quantum Chemistry* (Academic Press, Inc., New York, 1957), Chapters 6.11, 9 and 15.

E. Merzbacher, *Quantum Mechanics*, 2nd Ed. (John Wiley and Sons, Inc., New York, 1970), Chapters 6, 8, 9.

A. Messiah, *Quantum Mechanics*, Vol. 1, translated by G. M. Temmer (John Wiley and Sons, Inc., New York, 1958).

M. A. Morrison, T. L. Estle and N. F. Lau, *Quantum States of Atoms, Molecules and Solids* (Prentice-Hall, Inc., Englewood Cliffs, N. J., 1976), Chapter 3.

L. I. Schiff, *Quantum Mechanics*, 3d Ed. (McGraw-Hill Book Co., Inc., New York, 1968), Chapter 4.

Problems

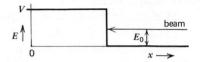

1. Complete the details of the argument indicated in Section 4.1, that there can be no physically admissible state for which the total energy E is everywhere less than the potential energy V.

2. The penetration of a wave into a steplike barrier as outlined in Eqs. 4.2 and 4.3 is much like the penetration of an electromagnetic wave into a metal. The depth at which the amplitude of the wave is e^{-1} of its amplitude at the surface is called the *skin depth*. Find the general expression for the skin depth of a one-dimensional matter wave for a particle of mass m and energy E_0 striking a barrier of height V, with $0 < E_0 < V$, as shown in the accompanying diagram. Suppose that $m = 10^{-28}$ g, $E_0 = 1$ eV, and $V = 2$ eV. What is the skin depth? Below what energy is the skin depth greater than 1 Å?

3. Explain why both the wave function and its first derivative must be continuous at the boundaries of a square potential well of finite depth.

4. In Chapter 3, we saw that the energies of the states of a particle in a one-dimensional infinite square well increase as n^2, the square of the quantum number. For the harmonic oscillator, the energies go as n; for the hydrogen atom, as n^{-2}. From these observations, what qualitative statement can one make about the relation between the slope of the walls of a box and the level spacing? What form would you expect for the spacings of the levels if V varied as x^4 instead of x^2? as $x^{3/2}$ instead of x^2?

5. A particle of mass m is bound in a one-dimensional potential $V(x)$ that satisfies

$$V = V_0, \qquad x < -a;$$
$$V = 0, \qquad -a \leq x \leq a;$$
$$V = V_0, \qquad x > a;$$

and $V_0 = 49h^2/128ma^2$.
   (a) How many bound states does such a particle have?
   (b) What are the energies of all the bound states?
   (c) Construct a curve showing each of the bound-state eigenfunctions.

6. A particle of mass m is confined in a two-dimensional box with sides of lengths l_1 and l_2. The potential is zero everywhere inside the box and infinite elsewhere. Calculate the number of energy eigenstates per unit energy interval for such a particle. A particle of mass m is confined inside a three-dimensional box with sides l_1, l_2, and l_3. The potential, like that of the problem above, is zero inside the box and infinite outside. Calculate the number of energy eigenstates per unit energy interval for this particle.

7. Show by substitution that the first three functions of Table 4.1 satisfy the wave equation 4.6, that they are normalized to unity, and that they are orthogonal, that is, that $\int_{-\infty}^{\infty} \psi_i{}^*(z)\psi_j(z)\,dz$ is zero if $i \neq j$.

8. A particle with mass 12 amu is bound in a harmonic potential well with a force constant k of 5×10^5 dyn/cm. What are the zero-point energy and the interval between adjacent energy levels for this system? What happens to these numbers if the force constant is doubled?

9. Graph the five functions of Table 4.1 for a particle with mass of 12 amu and a force constant k of 5×10^5 dyn/cm (500 kg/s^2).

10. In the discussion of the states of the hydrogen atom, with $-\infty < V < 0$, it was pointed out that states with energy $E < 0$ are bound states. What is the form of the wave function for a physically reasonable state whose wavefunction has an amplitude bounded everywhere, and with $E > 0$?

11. Calculate the radii for which the probability $\mathscr{P}$ is a maximum for $2p$, $4f$, and $6h$ levels of the hydrogen atom. How much energy is required to remove a $5p$

electron from a hydrogen atom and leave it infinitely far away, with no kinetic energy? Li^{2+} has a hydrogenic structure with nuclear charge $Z = 3$. What are the energies of the $1s$ and $2p$ states of Li^{2+}? What are the radii of maximum probability for these two states?

12. Compute the average value of r, the most probable value of r, and the root-mean-square value of r for the $1s$, $2s$, and $2p$ levels of the hydrogen atom. Compare the three kinds of values and explain the origin of their differences.

13. A hydrogenlike atom can be formed from a proton and a negative muon whose mass is approximately 206 times that of the electron. What are the energies and most probable radii for the $1s$ and $2p$ levels of this atom?

14. Find the general expression for the distance r_0 at which the effective potential V_{eff} of the Coulomb Hamiltonian, Eq. 4.19, becomes zero. How do this distance and the slope of V_{eff} at r_0 depend on the quantum number l? What is the distance at which V_{eff} is a minimum?

15. Show by substitution that the first three functions of Table 4.2 satisfy Eq. 4.18.

16. At what value of r do the hydrogenic $3s$, $3p$, and $3d$ orbitals have their outermost maximum amplitudes?

17. The radial equation for the hydrogen atom, Eq. 4.18, can be analyzed in part by finding the form of its solutions near $r = 0$ and as $r \to \infty$.

(a) Transform the equation to eliminate the first derivative. This is done by making explicit the condition that $|\psi(\mathbf{r})|^2$ for any wave function describing a particle in a central potential in three dimensions must fall off as r^{-2}, or that $\psi(\mathbf{r})$ must have the form $r^{-1}f(r)Y(\theta, \phi)$.

(b) Transform Eq. 4.18 by substituting $r^{-1}f(r)$ for $R_{nl}(r)$. Then let $r \to 0$ in the Hamiltonian and drop all but the largest term, to show that $f(r)$ behaves as r^l or r^{-l-1} near the origin. Only one of these two possibilities is physically admissable; which one, and for what reason?

(c) Now let $r \to \infty$ and drop all but the largest terms of the differential equation for $f(r)$. Show that $f(r)$ drops off exponentially with r and derive the relationship between the energy of the state and the coefficient α multiplying r in the exponential $e^{-\alpha r}$.

(d) Conclude by writing

$$f(r) = r^l e^{-\alpha r} g(r)$$

and find the differential equation for $g(r)$. Write the function $\psi(\mathbf{r})$ in terms of all the factors you have now derived, with the oscillatory function $g(r)$ left undetermined.

18. Compute the approximate number of cycles of the electric field and the time interval required for an atom to absorb energy equivalent to one quantum, $h\nu$, of light from a laser exerting a field of 5×10^7 V/m, whose light has wavelength 589 nm (5890 Å), if the atom absorbs the maximum interaction energy on every cycle of the light field.

19. Compute all the *transition moments* of the electric dipole operator $e\mathbf{r}$ (really the three components of this vector quantity, corresponding to ex, ey, and ez), where $\psi_0(\mathbf{r})$ and $\psi_1(\mathbf{r})$ are the wave functions for the ground and (threefold degenerate) first excited state of a three-dimensional isotropic harmonic oscillator whose mass is 1 amu and whose force constant is 5×10^5 dyn/cm (500 kg/s^2).

20. A proton passes an atom at a velocity of 5×10^6 cm/s, with a distance of closest approach of 1 nm (10 Å). What is the proton's electric field at the atom when the two are at their closest point? At what distance is the field e^{-1} times as strong as at its maximum? What is the time interval during which the field rises from its first "e^{-1}" point and then falls to its second "e^{-1}" point? To approximately what frequency of oscillation would this correspond, and what wavelength would correspond to a spectral line at this frequency?

21. Refer to the wave functions of the harmonic oscillator of Table 4.1 and Eq. 4.8. Prove the footnote to Eq. 4.8 (footnote 2) and extend it to the state $n = 1$. That is, find the classical turning points for the states with $n = 0$ and 1, the values of z at which the potential energy is exactly equal to the kinetic energy for these states. Show that the curvature of the wave function is zero at a classical turning point for *any* system satisfying the Schrödinger equation.

22. Using Eqs. 4.8 and 4.9, derive the expressions for $\psi_0(z)$, $\psi_1(z)$, and $\psi_2(z)$ of Table 4.1.

23. The peak electric field of a beam of light applied to a hypothetical model system is 25,000 V/m. The system consists of an electron harmonically bound to an infinitely heavy nucleus; the force constant is 10^6 dyn/cm or 10^3 kg/s^2. Calculate the maximum distance of displacement due to this applied electric field.

5

The Structure
of
Atoms

The last chapter gave us a comprehensive picture of the nature of the hydrogen atom, or of any atom containing a single electron. However, most of our world (though not most of the universe) is made up of atoms containing more than one electron. It is the purpose of this chapter to describe these more complex atoms.

How might we construct a many-electron atom? We naturally begin with a nucleus with the desired positive charge $+Ze$. We know from Chapter 4 what is obtained when we add a single electron to the nucleus: a species whose radial wave functions and energy levels differ from those of hydrogen only by scale factors. The natural step is to add a second electron, then a third, and so on until Z electrons have been added to give a neutral atom. But what rules govern the addition of these electrons? Each interacts not only with the nucleus, but with all the other electrons. These interactions involve not only the electrostatic forces but the magnetic properties of electrons as well. An accurate description of a many-electron atom must thus be quite complex. Fortunately, in many cases one can construct a useful approximation by assigning each electron its own set of quantum numbers and its own wave function, or orbital, analogous to the wave functions of the hydrogen atom. We must determine the sequence in which these orbitals are "occupied." This sequence is governed by the Pauli exclusion principle, which says that each orbital has only a limited capacity to contain electrons. In fact, no two electrons can have the same values of the one-electron quantum numbers n, l, m, and m_s, where m_s is a quantum number for a property we shall introduce in the next section, called the electron spin.

Given the basic concepts of atomic structure, we go on to discuss several effects that influence the details of electron configurations and energies in individual atoms. Finally, we begin to explain how chemical behavior can be interpreted in terms of the microscopic structure of atoms.

Until now we have treated electrons as though they were simple charged mass points. This is an oversimplification. In reality, electrons (and other particles) behave in some ways as though they are spinning about their own axes. We shall examine some of the evidence for this conclusion, but first let us explain what it means by comparing spin and orbital motion.

We designate the spin angular momentum by the vector **S**, just as we use **L** to represent the electron's orbital angular momentum about the atomic nucleus. The spin angular momentum behaves in most ways like other angular momenta. Its magnitude $|\mathbf{S}|$ for a single electron is given by

$$|\mathbf{S}|^2 = s(s+1)\hbar^2, \tag{5.1}$$

5.1
Electron Spin; Magnetic Phenomena

where s is a quantum number; this is completely analogous to the relation $|\mathbf{L}|^2 = l(l+1)\hbar^2$ for the orbital angular momentum (Section 3.11). Thus $|\mathbf{S}|^2$ is an eigenvalue of an operator $\mathbf{S}^2$, which we shall not discuss further. As with the other quantum numbers, we use the lowercase s for a single electron and the capital letter S for the total spin of an atom. Equations 5.1 and 3.162 still apply to the total atom, with S, L replacing s, l. (Note that we must now write the magnitudes of the angular momenta as $|\mathbf{S}|, |\mathbf{L}|$, because S, L are reserved for the quantum numbers.) One major difference between the spin and orbital angular momenta is in the values that the quantum numbers may have. For a single electron, s always has the value $\frac{1}{2}$, whereas you will recall that l may be any non-negative integer. When a system contains two or more electrons, the total spin (or orbital) angular momentum is the *vector* sum of the individual electronic angular momenta; we shall return later to this point.

The components of the spin angular momentum are also analogous to those of the orbital angular momentum. Only one component is a constant of the motion, characterized by a quantum number; as usual, we define this as the z component. Its value is given by

$$S_z = m_s \hbar, \tag{5.2}$$

analogous to Eq. 3.156, with the quantum number m_s (or M_S). Usually m_s is called the *spin quantum number* and s simply the *spin* ("the electron has spin $\frac{1}{2}$"). The magnetic quantum number that we called m in the hydrogen atom can be designated as m_l (or M_L) for uniformity. Just as m_l can range from $+l$ to $-l$ in integral steps, the values allowed for m_s are $s, s-1, \ldots, -s+1, -s$. Hence, for a single electron, the only possible values are $m_s = +\frac{1}{2}$ and $m_s = -\frac{1}{2}$; the corresponding values of S_z are $+\hbar/2$ and $-\hbar/2$.

The total angular momentum is designated as $\mathbf{J} \equiv \mathbf{L} + \mathbf{S}$, with quantum numbers j, m_j (or J, M_J). We postpone a full discussion of $\mathbf{J}$ until Section 5.7.

Being charged, any spinning or orbiting electron constitutes an electric current moving in a loop; like any such loop current (as in a solenoid), it acts as a bar magnet or magnetic dipole. From this magnetic behavior, the existence of spin or orbital angular momentum can be deduced. In the presence of an external magnetic field $\mathbf{B}$, the potential energy of a microscopic magnetic dipole is given by

$$V_{\text{dipole-field}} = -\boldsymbol{\mu}_m \cdot \mathbf{B} = -\mu_m B \cos\theta, \tag{5.3}$$

where $\boldsymbol{\mu}_m$ is the *magnetic dipole moment*, analogous to the electric dipole moment $\boldsymbol{\mu}_e$ of Eq. 4.30. The potential energy thus has its minimum when $\boldsymbol{\mu}_m$ and $\mathbf{B}$ are parallel; since $\boldsymbol{\mu}_m$ by definition is directed from the "south" to the "north" (north-seeking) pole of the magnet, this occurs when the north pole points in the direction of the external field.

Both the electron's orbital motion and its spin about its own axis produce magnetic dipole moments. The orbital moment $\boldsymbol{\mu}_l$ in the Bohr model is given by

$$\boldsymbol{\mu}_l = -\frac{e}{2m_e} \mathbf{L}, \tag{5.4}$$

from the classical equation for a current moving in a circular loop. Let us take our z axis in the direction of $\mathbf{B}$. The component of $\boldsymbol{\mu}_l$ in the direction of the magnetic field is then

$$(\boldsymbol{\mu}_l)_z = \mu_l \cos\theta = -\frac{e}{2m_e} L_z = -\frac{e\hbar}{2m_e} m_l. \tag{5.5}$$

The quantized unit of $(\boldsymbol{\mu}_l)_z$ is thus

$$\mu_B \equiv \frac{e\hbar}{2m_e} = 9.274078 \times 10^{-24} \text{ J/T} \qquad (5.6)$$

(T = tesla), known as the *Bohr magneton*.[1] Equation 5.5 is not exactly correct, because m_e should be replaced by the reduced mass and there is a slight interaction with the spin moment.

The spin magnetic dipole moment $\boldsymbol{\mu}_s$ does not obey the classical Eq. 5.4, but rather, an equation with an additional factor:

$$\boldsymbol{\mu}_s = g_s \left(\frac{-e}{2m_e} \right) \mathbf{S}, \qquad (5.7)$$

where g_s (the *Landé g factor*) is approximately 2. The actual value of g_s for the electron is 2.002319, differing from exactly 2 because of relativistic and radiative effects. By reasoning like that leading to Eq. 5.5, the component of $\boldsymbol{\mu}_s$ in the direction of the external field must equal $-g_s m_s$ Bohr magnetons; since $m_s = \pm\frac{1}{2}$, we have $(\boldsymbol{\mu}_s)_z \approx \pm \mu_B$. If the electron were literally a charged particle spinning about its axis, the spin magnetic dipole moment and spin angular momentum would be related by the equivalent of Eq. 5.4; that is, g_s would have to be 1. The fact that this is not so shows that the electron spin is an essentially quantum mechanical phenomenon. Strictly, one can say only that the electron somehow has an intrinsic angular momentum $\mathbf{S}$; the idea of "spin" is meaningful only as a convenient visualization, but the language has stuck.

We can now see why m_l is called the "magnetic" quantum number. In the absence of a magnetic field, all states with the same n and l have the same energy (except for fine-structure splitting). Once we apply an external field $\mathbf{B}$, however, the energy of each state changes by the amount given by Eq. 5.3. The total energy of an atom in a given state is thus shifted by an amount

$$E - E_0 = -(\boldsymbol{\mu}_L + \boldsymbol{\mu}_S) \cdot \mathbf{B} = -(\boldsymbol{\mu}_L + \boldsymbol{\mu}_S)_z B = -\mu_B (M_L + g_s M_S) B; \qquad (5.8)$$

the splitting of spectral lines due to this process is called the *Zeeman effect*.[2] In the "normal Zeeman effect," for atoms with $S = 0$ (and thus $M_S = 0$), the adjacent energy levels are separated by an amount $\mu_B B$. In a moderately large field of 1 T (10^4 G), this splitting is about 9.3×10^{-24} J (or 5.8×10^{-5} eV), equivalent to less than 0.5 cm^{-1}. For nonzero spin the spectrum is more complicated ("anomalous Zeeman effect"), and spin–orbit interactions produce splitting even with no external field. It was to account for this behavior that Uhlenbeck and Goudsmit (1925) proposed the hypothesis of electron spin.

The electron is not the only particle with spin. In particular, both the proton and the neutron also have spin $\frac{1}{2}$. The total nuclear spin I is the vector sum of the individual nucleon spins; I is seldom large, and often is zero. Nuclear magnetic moments are far smaller than those of electrons, being measured in terms of the *nuclear magneton*,

$$\mu_N \equiv \frac{e\hbar}{2m_p} = \mu_B \left(\frac{m_e}{m_p} \right) = \frac{\mu_B}{1836.15} = 5.050824 \times 10^{-27} \text{ J/T.} \qquad (5.9)$$

---

[1] In Gaussian units one writes $\boldsymbol{\mu}_l = -(e/2m_e c)\mathbf{L}$, and the Bohr magneton is $e\hbar/2m_e c = 9.27410 \times 10^{-21}$ erg/G (G = gauss). The SI unit can be written as either joules per tesla or ampere meters^2.

[2] An external electric field produces a similar splitting known as the *Stark effect*.

The proton and neutron have moments $\mu_p = 2.793\mu_N$ and $\mu_n = 1.913\mu_N$, respectively. The energy-level splitting of these particles in a magnetic field is correspondingly small, about 10^{-3} of that for an electron; heavier nuclei have even smaller splittings.

In spite of the minuscule size of these energy shifts, one can measure them quite accurately by the method of *nuclear magnetic resonance* (*nmr*). One places the system in a strong magnetic field (≈ 1 T), passes a radiofrequency (rf) signal through it ($\nu \approx 60$ MHz), and varies **B** slowly until ΔE for some magnetic transition is equal to $h\nu$; the energy difference is then in *resonance* with the applied radiation, and absorption occurs. By sweeping the field strength $|\mathbf{B}|$ and recording the absorption of rf power, one obtains a *magnetic resonance* spectrum. Nuclear magnetic moments observed this way can act as probes to tell us a good deal about the structure of atoms and molecules. This is possible because a nucleus is not merely acted upon by an external field, but is affected in a delicate and sensitive way by the fields of all the other particles (nuclei and electrons) in its vicinity. For example, a proton in an —OH group is in a slightly different field than a proton in a —CH$_3$ group, and protons in these two environments have slightly different spacings of their energy levels. Hence they exhibit resonances at slightly different magnetic field strengths in an oscillating field of fixed frequency. The *chemical shifts* of lines in the magnetic resonance spectrum are very small indeed, of the order of 10^{-6} of the total energy-level splitting, but they can still be measured. The detailed shape of the absorption spectrum can be interpreted to reveal much information about the environment of the nucleus studied, whether it is a proton or a heavier, more complex nucleus.

One can, of course, observe similar effects in the energy-level splittings due to electron spin. These are studied by the method of *electron spin resonance* (*esr*), which gives information mainly on the electron distribution in molecules. Because of the larger magnetic moment of the electron, the esr transitions occur at higher frequencies, mainly in the microwave region. However, atoms or molecules with net electron spins different from zero are relatively uncommon, whereas nonzero nuclear magnetic moments are ubiquitous.

The astute reader will have recognized by now that although we have introduced several kinds of magnetic moments into our model of the atom, associated with electron spin, electron orbital motion, and nuclear spin, we have said little about the obvious problem that such magnetic moments must interact with one another. Indeed they do, but their interactions are weak and need not concern us in detail at this point. The interaction of electron spin and orbital moments, called spin–orbit interaction, is important for the energy levels of heavy atoms, and is measurable but relatively unimportant for the chemical properties of light atoms. Spin–orbit interaction is discussed in Section 5.7. The interactions among nuclear moments and between nuclear and electron moments are crucial for the use of nuclear and electron spin magnetic resonance as analytic and diagnostic tools; the identification of lines in nuclear resonance spectra generally depends on knowing how one nuclear magnetic moment interacts with those of its neighbors.

Something should be said here about the origins of *macroscopic* magnetism. All the individual electrons in a material are tiny magnets, but ordinarily they are aligned at random and, in large numbers, cancel one another on average. An external magnetic field, however, tends to align the microscopic magnets; the net effect is to make **B** slightly higher within the material than in a vacuum.[3] This phenomenon is called *paramagnetism*. Both orbital and spin magnetic moments can contribute to paramagnetism. However, because the orbital moments are largely fixed in space by the molecular structure, it is mainly the spin moments that change their alignment in the presence of a field. Paramagnetism is thus rarely found in substances whose total molecular spin is zero, that is, for which the spin magnets cancel out within the molecule; such substances are called *diamagnetic*. In diamagnetic materials **B** is actually a little less than in a vacuum because the field interferes with the orbital motions, but the effect is much smaller than it is in paramagnetism. Paramagnetic substances are weakly attracted, diamagnetic substances weakly repelled, by a magnet. As for the very large permanent magnetism that can be induced in metals like iron (*ferromagnetism*), this results from a cooperative interaction among spins on different atoms: For certain substances the electron spins are thus aligned in parallel over large regions or *domains*, appreciably reducing the total energy of the crystal.

We have introduced electron spin primarily because it is an important property in its own right: The total state of an electron is specified only when its spin is added to all the other properties characterized by quantum numbers. Moreover, it must be introduced to provide a full basis for the exclusion principle, which, in turn, underlies atomic structure. Spin is one of the properties that could not have been predicted from classical mechanics. It is illustrated by one of the most famous and fundamental experiments on which quantum theory rests, the Stern–Gerlach experiment. Appendix 5A describes this experiment.

5.2 The Pauli Exclusion Principle; the Aufbau Principle

The set of quantum numbers n, l, m_l, and m_s, and the physical attributes to which they correspond, give us the basis for our first approach to the structure of complex atoms. In this section we shall see how electrons fit together to give each atom a sort of shell-like structure; this will lead us to a systematic interpretation of the periodic table. The model we discuss here is only approximate. In subsequent sections we shall consider the validity of our approximations and the refinements necessary for a detailed explanation of atomic structure.

Consider first an ion made up of a nucleus with charge $+Ze$ $(Z > 1)$ and a single electron bound to that nucleus. This is a hydrogenlike ion of the type discussed in Chapter 4, where we characterized the electron by the quantum numbers n, l, m (or m_l). We must now add the spin quantum number m_s to give a complete description of the state of the electron. Let us denote these four quantum numbers of the first electron by n_1, l_1, m_{l1}, m_{s1}, because we shall add another $Z - 1$ electrons, one by one, to neutralize the total nuclear charge.

Suppose that we add a second electron to the ion, and that there are states in which both electrons are bound to the nucleus. To a first approximation, known as the *central field approximation*, the second electron can be treated as though it moves in the field of the nucleus and the *average* field produced by the first electron. In other words, we

---

[3] In a vacuum we have $\mathbf{B} = \mu_0\mathbf{H}$, where **H** is the magnetic field strength (see footnote 8, p. 12, Chapter 1); **H** is determined only by the external field source and any permanent magnetic movements of the matter in the field. Within a material we write $\mathbf{B} = \mu_0(1 + \chi)\mathbf{H}$, defining the *magnetic susceptibility* χ; χ is positive in paramagnetic substances, negative in diamagnetic substances. One usually tabulates the molar susceptibility $\chi_M \equiv \chi M/\rho$ (where M is molecular weight and ρ is density); a typical value for a paramagnetic metal would be about $10^{-10} \text{m}^3/\text{mol}$. We shall discuss magnetic susceptibility in greater detail in Section 9.6.

pretend that the second electron sees only a smeared-out charge cloud equivalent to the average spatial distribution of the first electron. Carrying the approximation a bit further, we also assume that the first electron responds to the *average* field of the second, rather than to its field at each instant.

The central field approximation permits us to assign to the second electron a one-electron wave function with a set of quantum numbers n_2, l_2, m_{l2}, m_{s2}. The energy levels and the shapes of the spatial wave functions or orbitals for the atom or ion with two electrons are somewhat different from those of the one-electron ion, for reasons we shall see very shortly. However, the quantum numbers retain the same basic meaning in both situations. The principal quantum number n is one more than the number of radial nodes in the one-electron wave function, and is the first guide to the energy; l specifies the electron's orbital angular momentum, and gives the number of angular nodes; m_l specifies the orientation of the electron's orbital angular momentum $\mathbf{L}$; and m_s specifies the orientation of its spin angular momentum $\mathbf{S}$.

Suppose that the two-electron species, for example, the helium atom with $Z = 2$, is in its state of lowest energy. It would be natural to assume that both electrons then have the same quantum numbers, presumably those of the lowest state in the one-electron atom. This is correct for the quantum numbers n, l, and m_l, which for both electrons have the values $n = 1$, $l = 0$, and $m_l = 0$. (Remember that $n = 1, 2, \ldots$, whereas $0 \leq l \leq n - 1$ and $|m_l| \leq l$.) However, both electrons cannot have the same value of m_s: One has $m_s = +\frac{1}{2}$, the other $m_s = -\frac{1}{2}$, and the total electron spin is zero. Here is a point where our classical intuition fails us, and we must introduce a quantum mechanical postulate to describe the way in which electrons behave. It is found that any two electrons in a single atom with the same values of n, l, and m_l always have different values of m_s. More generally, in any atom or ion, *no two electrons ever share the same four quantum numbers*. This rule, proposed empirically by Wolfgang Pauli in 1925, is known as the *Pauli exclusion principle*. For the time being we shall take the exclusion principle as a postulate in its own right; in Section 6.7 we shall give a more general rule from which it can be derived.

The exclusion principle imposes strict conditions on the assignment of quantum numbers. We have assigned $n = 1$, $l = 0$, $m_l = 0$ to the first two electrons, which must have $m_s = +\frac{1}{2}$ and $m_s = -\frac{1}{2}$, respectively. Now suppose that $Z \geq 3$, and we add a third electron, as, for example, in Li with $Z = 3$. There is no way to assign this electron to a state with $n = 1$, because all the possible combinations of n, l, m_l, and m_s correspond to states already occupied. We must thus assign the third electron a quantum number n_3 greater than 1. If we set $n_3 = 2$, then we may have $l_3 = 0$ or $l_3 = 1$. If $l_3 = 0$, then m_{l3} is necessarily 0; if $l_3 = 1$, then m_{l3} may be 0, $+1$, or -1; whatever values we assign l_3 and m_{l3}, the quantum number m_{s3} may be either $+\frac{1}{2}$ or $-\frac{1}{2}$. Which of all these states will the electron occupy? It *may* occupy any of them, but the most stable state is that of lowest energy, the ground state. To begin with, the energy nearly always increases with the value of n; an electron will thus ordinarily take the lowest available value of n, in this case 2. Among states with the same value of n, the states with $l = 0$ normally have energies lower than those with $l = 1$, and so forth; the energies increase with increasing l. In the absence of external fields, and with internal electron–electron interactions neglected, states with the same n and l but different m_l or m_s differ only in orientation, and therefore must have the same energy.

In the hydrogen atom, where the electron moves in the Coulomb field of a single proton, the energy depends only on the principal quantum number n; it makes no difference energetically what value we assign to l. This is a special property of two particles interacting through a Coulomb field. In more complex atoms, the field felt by each electron is a combination of the field of the nucleus and the fields of the other electrons, which have probability distributions extending over considerable regions of space. The effect of the other electrons is really twofold. First, each of them exerts a Coulomb force of the classical type; second, the mere presence of an electron at a point $\mathbf{r}$ reduces the probability that another electron of the same spin be in the vicinity of $\mathbf{r}$. The latter is a purely quantum mechanical phenomenon, with no classical equivalent; it is a direct result of the exclusion principle. The net effect is as if there were an additional repulsive force (sometimes called *exchange force*) added to the Coulomb repulsion between electrons of the same spin. This is not a real force, represented by a term in the Hamiltonian; rather, we introduce the effect as a constraint on the form of the wave function, which has the effect of changing the charge distribution from that of the unconstrained wave function. As a result of these interactions the field on each electron is significantly different from that of a simple, central Coulomb field. Because of the more complex field in a many-electron atom, the electron's energy depends on its angular momentum; thus the energies of states with the same n but different l are separated (split). In general, as we said previously, for a given n the energy increases with l. The reason should be clear from Fig. 4.5: The higher the value of l, the less likely an electron is to be found near the nucleus.

At this point we must say something about the word "orbital." We introduced this term in Section 4.3 to refer to a single-electron eigenfunction, but it is also loosely used for the corresponding state: One says that an electron is *in* a given orbital. By referring to one-electron states as orbitals, one can maintain a clear distinction between the true state of the entire atom, a concept that need imply no approximations, and the state of an individual electron, a concept associated with a particular approximate description of nature and therefore limited in its precision and applicability.

The complete wave function of an electron must include the spin as well as the spatial coordinates; we might write it as $\psi(r, \theta, \phi; m_s)$. To a very good approximation, however, the spin–orbit interaction can be neglected, and the Hamiltonian is separable into spatial and spin terms. We can then as usual factor the wave function, say, as $\psi(r, \theta, \phi)\alpha(m_s)$. It is the spatial part of the eigenfunction that is ordinarily called an "orbital." Two electrons with the same values of n, l, m_l have the same spatial wave function, and are thus said to occupy the same orbital.[4] Every orbital can thus contain two electrons, which by the exclusion principle must have different values of m_s. An orbital is characterized by its values of n, l, m_l, but usual notation ignores m_l (which, to the extent that the simple orbital picture is valid, does not affect the energy) and uses the $s, p, d, \ldots$ notation to identify l. The first two electrons in an atom are thus said to be in a $1s$ $(n = 1, l = 0)$ orbital, and the third should go into a $2s$ $(n = 2, l = 0)$ orbital, which has a lower energy than a $2p$ $(n = 2, l = 1)$ orbital.

The conditions imposed by the exclusion principle, together with our knowledge about energy levels, provide us with a rather accurate guide for describing the structure of atoms of any complexity in their ground states. This guide is a way of assigning quantum numbers, and implicity

---

[4] The wave function including spin is called a *spin orbital*, and this *is* restricted to a single electron. In this text we do not consider the nature of the function $\alpha(m_s)$ or the corresponding operator, for which a relativistic theory must be used.

wave functions, to each electron in the atom. In effect, we build up the atom by assigning one electron at a time to the available orbital of lowest energy. This is called the *Aufbau* (building-up) *principle*. The method is rather like adding marbles, one at a time, to a conical cup. The first marble goes to the bottom (the state of lowest energy in the earth's gravitational field), and successive marbles pile up as they can, each in the lowest place where there is room.

We have already considered the ground states of the first few atoms. Now we shall extend this analysis to the entire periodic table.

For most purposes one can classify any state of an atom by giving the principal quantum number n and the angular momentum quantum number l for each of the electrons in the atom. It is these quantum numbers that largely determine the atom's energy and chemical behavior; varying the value of m_l or m_s makes only a very slight difference, which in first approximation we neglect altogether. A given assignment of n and l to all the electrons in an atom is called a *configuration*. Although it is not strictly correct to say that any atomic state is identical with a particular configuration, it is accurate to say that almost all known atomic states can be characterized by, and exhibit properties dominated by single configurations.

For an example of a configuration, let us consider the ground state of the nitrogen atom, which can be built up by adding seven electrons, one at a time, to a nucleus of charge $+7e$. As before, the first two electrons go into a $1s$ orbital. The third electron goes into a $2s$ orbital, with $n = 2$, $l = 0$, $m_l = 0$, and $m_s = \pm\frac{1}{2}$. The fourth can go into the same orbital, with the same values of n, l, m_l and whichever value of m_s is left over. The next higher energy level is that with $n = 2$, $l = 1$, the $2p$ level, which can hold six electrons in three orbitals (with $m_l = +1, 0, -1$, respectively). The fifth, sixth, and seventh nitrogen electrons can thus all go into $2p$ orbitals. Later we shall have to consider *which* $2p$ orbitals they enter, but we do not need to know this here. Having said this much, we have specified the ground-state configuration of the nitrogen atom: There are two electrons in a $1s$ orbital, two in a $2s$ orbital, and three in $2p$ orbitals. The shorthand representation for this configuration is $1s^2 2s^2 2p^3$. In this notation each large number specifies the value of n, and the letter following it the value of l for a particular electronic state; the superscript gives the number[5] of electrons with those values of n and l.

If one neglects electron–electron interactions and fine-structure effects, all the orbitals with the same values of n and l but different m_l have the same energy (in the absence of external fields). They differ only in the orientation of the electron's angular momentum with respect to an arbitrarily chosen axis. As in the hydrogen atom, their probability densities differ in orientation but not in radial distribution. The energies of the $np, nd, \ldots$ orbitals are moderately close to that of the ns orbital, but not equal to it except in the hydrogen atom. The radial distributions of the $np, nd, \ldots$ orbitals are also somewhat similar to that of the corresponding ns orbital at fairly large radii (cf. Fig. 4.5). Because of this similarity in both size and energy, one frequently refers to all the orbitals with the same n as a *shell*, and to those with the same n and l as a *subshell*; this is the "shell model" of the atom. The shells with $n = 1, 2, 3, \ldots$ are also referred to as the $K, L, M, \ldots$ shells. This terminology

5.3
Electronic Configurations of Atoms

[5] The superscript 1 is usually omitted.

derives from x-ray spectroscopy (Section 2.3): the K series of emission lines are produced by electrons dropping into vacancies in the K shell, and so forth.

When all the quantum numbers available for a given shell or subshell have been assigned to electrons, it is said to be *filled* or *closed*. In the ground-state configuration of the nitrogen atom, the $n = 1$ or K shell (consisting of only the $1s$ subshell) is filled, the $2s$ subshell is filled, but the $2p$ subshell is only half-filled. There are still three vacancies in the $2p$ subshell, no matter how we assign the quantum numbers m_l and m_s, provided that we satisfy the exclusion principle. These vacancies are filled as we add more electrons, until in the neon atom ($Z = 10$) the shells with $n = 1$ and $n = 2$ are both completely filled.

To proceed any further we must know the sequence of orbital energy levels. The next section describes some of the ways these are derived; for the present, we can examine the levels as they turn out. One might assume that the energy increases in order from one shell to the next, and in the order $s, p, d, \ldots$ within each shell. This would give the sequence $1s, 2s, 2p, 3s, 3p, 3d, 4s, 4p, 4d, 4f, 5s, \ldots$. This scheme is valid through the $3p$ subshell, which is filled in the ground state of the argon atom ($Z = 18$). In the potassium atom ($Z = 19$), however, the nineteenth electron goes into a $4s$ rather than a $3d$ orbital. What happens is this: As the number of electrons increases, the energy difference between orbitals in the same shell becomes greater than the average energy difference between shells—in other words, the energy ranges spanned by the shells begin to overlap. One thus ought to know just how the energy levels vary as a function of Z, and we shall discuss this in the next section.

Fortunately, this detailed knowledge is not necessary to obtain a good approximation to the electronic configuration. The sequence in which orbitals are filled in the ground states of the elements is given with remarkable accuracy by the simple mnemonic device shown in Fig. 5.1. One need only remember that any subshell with angular momentum quantum number l can hold up to $2(2l+1)$ electrons ($2l+1$ possible values of m_l, each with two possible values of m_s): two in s subshells, six in p subshells, 10 in d subshells, and so forth. Given this information, and filling orbitals in the sequence indicated in Fig. 5.1, one can predict the ground-state configuration for any number of electrons.

The actual ground-state electronic configurations of the elements, so far as they are now known, are given in Table 5.1. The configurations are obtained by analyses of electronic spectra, and in some cases (where two or more subshells are very close in energy) they are uncertain. Table 5.1 shows that the sequence predicted by Fig. 5.1 is indeed quite reliable. The exceptions (marked by $\star$) are those cases in which detailed interactions override the gross features of the shell model; we shall examine these effects later. But already we have essentially explained the form of the periodic table (Table 1.3). The first two columns contain those elements in which s subshells are being filled; the last six columns correspond to the filling of p subshells; the 10 short columns in the middle (transition metals) to d subshells; and the lanthanide and actinide series to f subshells.

Thus far we have considered only ground states. Although highly excited states sometimes require more elaborate descriptions, the concept of electronic configuration still gives a satisfactory representation of many excited states. For example, one can obtain a very good representation of the first excited state of helium by assigning one electron to the $1s$ orbital and the other to the $2s$, the next orbital on the energy ladder. The configuration is then He*$(1s2s)$, where the asterisk indicates an excited

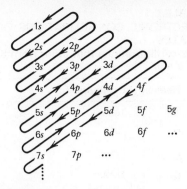

FIGURE 5.1
Approximate filling order of orbitals for the ground states of the elements.

TABLE 5.1

ELECTRONIC CONFIGURATIONS OF THE ELEMENTS IN THEIR GROUND STATES

The configurations that differ from those predicted by Fig. 5.1 are indicated by a star (*).

Z	Element	Configuration	Z	Element	Configuration	Z	Element	Configuration
1	H	$1s$	36	Kr	$[Ar]3d^{10}4s^24p^6$	71	Lu	$[Xe]4f^{14}5d6s^2$
2	He	$1s^2$				72	Hf	$[Xe]4f^{14}5d^26s^2$
3	Li	$1s^22s$	37	Rb	$[Kr]5s$	73	Ta	$[Xe]4f^{14}5d^36s^2$
4	Be	$1s^22s^2$	38	Sr	$[Kr]5s^2$	74	W	$[Xe]4f^{14}5d^46s^2$
5	B	$1s^22s^22p$	39	Y	$[Kr]4d5s^2$	75	Re	$[Xe]4f^{14}5d^56s^2$
6	C	$1s^22s^22p^2$	40	Zr	$[Kr]4d^25s^2$	76	Os	$[Xe]4f^{14}5d^66s^2$
7	N	$1s^22s^22p^3$	41	Nb	$[Kr]4d^45s$ *	77	Ir	$[Xe]4f^{14}5d^76s^2$
8	O	$1s^22s^22p^4$	42	Mo	$[Kr]4d^55s$ *	78	Pt	$[Xe]4f^{14}5d^96s$ *
9	F	$1s^22s^22p^5$	43	Tc	$[Kr]4d^55s^2$	79	Au	$[Xe]4f^{14}5d^{10}6s$ *
10	Ne	$1s^22s^22p^6$	44	Ru	$[Kr]4d^75s$ *	80	Hg	$[Xe]4f^{14}5d^{10}6s^2$
11	Na	$[Ne]3s$	45	Rh	$[Kr]4d^85s$ *	81	Tl	$[Xe]4f^{14}5d^{10}6s^26p$
12	Mg	$[Ne]3s^2$	46	Pd	$[Kr]4d^{10}$ *	82	Pb	$[Xe]4f^{14}5d^{10}6s^26p^2$
13	Al	$[Ne]3s^23p$	47	Ag	$[Kr]4d^{10}5s$ *	83	Bi	$[Xe]4f^{14}5d^{10}6s^26p^3$
14	Si	$[Ne]3s^23p^2$	48	Cd	$[Kr]4d^{10}5s^2$	84	Po	$[Xe]4f^{14}5d^{10}6s^26p^4$
15	P	$[Ne]3s^23p^3$	49	In	$[Kr]4d^{10}5s^25p$	85	At	$[Xe]4f^{14}5d^{10}6s^26p^5$
16	S	$[Ne]3s^23p^4$	50	Sn	$[Kr]4d^{10}5s^25p^2$	86	Rn	$[Xe]4f^{14}5d^{10}6s^26p^6$
17	Cl	$[Ne]3s^23p^5$	51	Sb	$[Kr]4d^{10}5s^25p^3$			
18	Ar	$[Ne]3s^23p^6$	52	Te	$[Kr]4d^{10}5s^25p^4$	87	Fr	$[Rn]7s$
			53	I	$[Kr]4d^{10}5s^25p^5$	88	Ra	$[Rn]7s^2$
19	K	$[Ar]4s$	54	Xe	$[Kr]4d^{10}5s^25p^6$	89	Ac	$[Rn]6d7s^2$ *
20	Ca	$[Ar]4s^2$				90	Th	$[Rn]6d^27s^2$ *
21	Sc	$[Ar]3d4s^2$	55	Cs	$[Xe]6s$	91	Pa	$[Rn]5f^26d7s^2$ *
22	Ti	$[Ar]3d^24s^2$	56	Ba	$[Xe]6s^2$	92	U	$[Rn]5f^36d7s^2$ *
23	V	$[Ar]3d^34s^2$	57	La	$[Xe]5d6s^2$ *	93	Np	$[Rn]5f^46d7s^2$ *
24	Cr	$[Ar]3d^54s$ *	58	Ce	$[Xe]4f5d6s^2$ * (or $4f^26s^2$)	94	Pu	$[Rn]5f^67s^2$
25	Mn	$[Ar]3d^54s^2$	59	Pr	$[Xe]4f^36s^2$	95	Am	$[Rn]5f^77s^2$
26	Fe	$[Ar]3d^64s^2$	60	Nd	$[Xe]4f^46s^2$	96	Cm	$[Rn]5f^76d7s^2$ *
27	Co	$[Ar]3d^74s^2$	61	Pm	$[Xe]4f^56s^2$	97	Bk	$[Rn]5f^86d7s^2$ * (or $5f^97s^2$)
28	Ni	$[Ar]3d^84s^2$ (or $3d^94s$ *)	62	Sm	$[Xe]4f^66s^2$	98	Cf	$[Rn]5f^96d7s^2$ * (or $5f^{10}7s^2$)
29	Cu	$[Ar]3d^{10}4s$ *	63	Eu	$[Xe]4f^76s^2$	99	Es	$[Rn]5f^{10}6d7s^2$ * (or $5f^{11}7s^2$)
30	Zn	$[Ar]3d^{10}4s^2$	64	Gd	$[Xe]4f^75d6s^2$ *	100	Fm	$[Rn]5f^{11}6d7s^2$ * (or $5f^{12}7s^2$)
31	Ga	$[Ar]3d^{10}4s^24p$	65	Tb	$[Xe]4f^85d6s^2$ * (or $4f^96s^2$)	101	Md	$[Rn]5f^{12}6d7s^2$ * (or $5f^{13}7s^2$)
32	Ge	$[Ar]3d^{10}4s^24p^2$	66	Dy	$[Xe]4f^{10}6s^2$	102	No	$[Rn]5f^{13}6d7s^2$ * (or $5f^{14}7s^2$)
33	As	$[Ar]3d^{10}4s^24p^3$	67	Ho	$[Xe]4f^{11}6s^2$	103	Lw	$[Rn]5f^{14}6d7s^2$
34	Se	$[Ar]3d^{10}4s^24p^4$	68	Er	$[Xe]4f^{12}6s^2$	104		$[Rn]5f^{14}6d^27s^2$
35	Cl	$[Ar]3d^{10}4s^24p^5$	69	Tm	$[Xe]4f^{13}6s^2$	105		$[Rn]5f^{14}6d^37s^2$
			70	Yb	$[Xe]4f^{14}6s^2$	106		$[Rn]5f^{14}6d^47s^2$

state. At a slightly higher energy one finds the configuration He*$(1s2p)$. One can proceed all the way up the series of configurations in which one electron is excited to the normally empty orbitals with any values of n and l. Finally, at the limit of the series, sufficient energy removes the electron from the atom altogether; one then has $He^+(1s) + e^-$, a helium ion in its ground state plus a free electron.[6]

To obtain an excited configuration, of course, one need not excite just a single electron. For example, if we excite a helium atom by giving it enough energy to put one electron in the $2s$ orbital and the other electron in the $2p$ orbital, we have the doubly excited configuration $He^*(2s2p)$; or both could be excited to the $2s$ orbital (with opposite spins), giving $He^*(2s^2)$. Such states are actually known experimentally, but their energies are very high indeed compared with states having only one excited electron. We might expect this, particularly for a small atom like helium, because we know that in the hydrogen atom the states with $n = 2$ are three-quarters of the way up from the ground state to the ionization limit. This should be roughly true for helium also, and in fact the first excited state of He is about 80% of the way from the ground state to the energy of $He^+ + e^-$. Therefore the energy required to excite two electrons from the K shell to the L shell would be over one and one-half times the energy required to remove a single electron from the atom, leaving the other in its lowest state. This is indeed the case: A doubly excited helium atom has more than enough total energy to become a helium positive ion and a free electron. However, much of the time, the energy is divided between the two electrons and is not readily available to just one for purposes of ionization. As a result, when doubly excited helium atoms are produced, it takes some time (corresponding to many Bohr periods of revolution in the $n = 2$ orbit) for an average doubly excited atom to convert itself into a singly charged helium ion and a free electron.

Transitions between one electronic configuration and another are responsible for most of the lines in the emission and absorption spectra of atoms. Generally speaking, transitions involving the outer (valence) electrons lie in the visible and adjacent regions (the highest atomic ionization energy for removal of one outer electron—or *first ionization potential*—is that of helium, corresponding to $\lambda = 504 \text{ Å}$). The more tightly bound inner-shell electrons give rise to the x-ray spectrum. Tens of thousands of assignments of atomic spectral lines have been made, meaning that their initial and final states have been determined. Much of modern science and technology could hardly exist without these assignments. (Some examples: The design of lasers requires a knowledge of the specific states of individual atoms and the selection rules governing transitions among them. The industrial analysis of steel uses atomic spectroscopy for the quantitative determination of trace materials.)

5.4 Calculation of Atomic Structures

Now we shall briefly survey the methods used to determine energy levels, quantum states, and wave functions for complex atoms. Unlike those for the hydrogen atom, however, the results obtained are only approximate (though of high accuracy for light atoms).

To describe the state of any quantum mechanical system, one need "simply" write down the Hamiltonian and solve the Schrödinger equation for the wave function. But let us look at such a Hamiltonian. For an atom with N electrons we have the operator

$$H = \sum_{i=1}^{N} \left(\frac{p_i^2}{2m} - \frac{Ze^2}{4\pi\epsilon_0 r_i} \right) + \frac{1}{2} \sum_{i=1}^{N} \sum_{j \neq i}^{N} \frac{e^2}{4\pi\epsilon_0 r_{ij}}$$
$$+ \text{magnetic moment interactions}, \quad (5.10)$$

---

[6] The energy levels of the helium atom are plotted in Fig. 5.10.

where r_i is the distance of the ith electron from the nucleus and r_{ij} is the distance between the ith and jth electrons. The factor $\frac{1}{2}$ is to prevent counting each interaction twice. Even in classical mechanics one cannot solve exactly the equations of motion for the three-body problem; here we have an $(N+1)$-body problem, and an exact solution of the Schrödinger equation is clearly impossible. One must therefore resort to approximate methods.

One of the most important ways of obtaining approximate solutions is the *variation method*. Suppose that one guesses a trial wave function ψ, which can be any well-behaved and normalized function of the coordinates. One can show that, for any such ψ,

$$E' \equiv \int \cdots \int \psi^* H \psi \, dq_1 \cdots dq_N \geq E_0, \qquad (5.11)$$

where H is the Hamiltonian operator of the system and E_0 is its lowest eigenvalue. The equality, of course, holds only when ψ is the true ground-state wave function of the system. The more closely ψ resembles this true wave function, the closer the integral E' will come to the value E_0. If ψ contains one or more adjustable parameters, it is a straightforward calculation to find the values of these parameters that minimize E'. By either trial-and-error or systematic procedures, one can obtain functions that give lower and lower values of E'; when no further improvement can be made, ψ is presumably the best possible approximation to the true wave function. There are ways to extend the method to excited states.

To see how the variation method works, let us consider the ground state of the helium atom. If the two electrons did not interact with each other at all, each would independently be in a hydrogenlike 1s orbital; since the 1s wave function is proportional to e^{-Zr/a_0}, the total wave function for the helium atom would then be of the form

$$\psi^{(1)} = A e^{-Zr_1/a_0} e^{-Zr_2/a_0}, \qquad (5.12)$$

with $Z = 2$. (When the Hamiltonian is separable, the wave function is a product of functions of the individual coordinates.) If $\psi^{(1)}$ is taken as a trial wave function and the operator of Eq. 5.10 as the Hamiltonian (with the magnetic terms omitted), then the integral Eq. 5.11 gives $E' = -74.83$ eV. The actual ground-state energy (E_0) is known from spectroscopic data to be -78.99 eV, so the wave function needs considerable improvement. Since each helium electron is partially screened from the nucleus by the other electron, it is logical to try replacing Z by an adjustable parameter Z':

$$\psi^{(2)} = A e^{-Z'r_1/a_0} e^{-Z'r_2/a_0}. \qquad (5.13)$$

Varying the value of Z', one finds the integral Eq. 5.11 to be minimized when $Z' = \frac{27}{16}$, yielding the value $E' = -77.49$ eV; the error in energy is only one-third as much as before.

The next step carries us beyond the simple configurational wave function, by introducing a term that takes explicit account of the repulsion between the electrons. One way is to use a wave function with terms containing the interelectronic distance r_{12}, for example

$$\psi^{(3)} = A(1 + cr_{12}) e^{-Z'r_1/a_0} e^{-Z'r_2/a_0} \qquad (5.14)$$

with $c > 0$, which grows larger as the electrons move farther apart; varying both Z' and c, one obtains $E' = -78.66$ eV, only 0.33 eV higher than the experimental value. The function of Eq. 5.14 cannot be labeled by a single configuration as 5.12 or 5.13 can be. Obviously one can carry this process as far as one wants, by using more and more complicated polynomials in the r's. With enough adjustable parameters—and enough computer time—one can get arbitrarily close to E_0. Calculations have in fact been made with ψ's containing over 1000 adjustable parameters, giving an E' that not only agrees with E_0 within experimental accuracy, but is indeed far more precise (and presumably accurate) than experiment. With 26 parameters, one can determine E' to within about 1 part in 10^7. Such results are, for all practical purposes, equivalent to an exact solution. Later in this section we shall give a physical interpretation to the complex wave functions needed to obtain such accuracy.

Because we can compute very accurate wave functions for two-electron atoms, we can construct very reliable pictures of the electron distributions in these atoms. Figure 5.3 (below) shows several such distributions. Note that the atoms and ions in their ground states have very high concentrations of charge density near their nuclei. The total charge at distances far from a nucleus is significant, but in the figures showing the charge density itself (without a weighting factor of r^2), the charge density away from the nucleus is quite small. The total charge at about the distance r is spread out over the entire spherical shell with volume $4\pi r^2\, dr$, so the amount at any point may be quite small although the amount in the entire shell is not.

Unfortunately, such elaborate and accurate computations are not feasible for most atoms. When one goes beyond three or four electrons, even the largest computers do not have the capacity to do the job with methods now known. But although one cannot achieve spectroscopic accuracy for the total wave function and energy, it is possible to get useful representations in terms of the central field approximation and one-electron orbitals.

This method was introduced by D. R. Hartree. The basic assumption is that the wave function can be well represented as a product of one-electron orbitals,

$$\psi = \varphi_1 \varphi_2 \cdots \varphi_N, \tag{5.15}$$

where each of the φ_i is a function of the coordinates of the ith electron only. Suppose that this wave function is substituted in Eq. 5.11, using the Hamiltonian 5.10 without the magnetic terms; one can show that the lowest possible value of E' is then obtained when each of the φ_i is a solution of the equation

$$\mathsf{H}_i \varphi_i = \left(\frac{\mathsf{p}_i^2}{2m} - \frac{Ze^2}{4\pi\epsilon_0 r_i} + \sum_{j \neq i} \int |\varphi_j|^2 \frac{e^2}{4\pi\epsilon_0 r_{ij}}\, dV_j \right) \varphi_i = \epsilon_i \varphi_i, \tag{5.16}$$

where $dV_j \equiv r_j^2 \sin\theta_j\, d\theta_j\, d\phi_j$ and ϵ_i is the orbital energy eigenvalue. Hartree's approximation is thus equivalent to the central field approximation. The instantaneous electron–electron repulsions appearing in Eq. 5.10 are replaced by integrals giving the *average* repulsion over all possible positions of the other electrons (since $|\varphi_j|^2\, dV_j$ is the probability of finding the jth electron in the volume element dV_j).

It is still necessary to solve simultaneously N equations of the form of Eq. 5.16; this is done by successive approximation. One starts with a trial set of orbitals $\varphi_i^{(1)}$. These could be hydrogenlike orbitals, for example, although other functions are known that lead to more rapid convergence; the angular part of each orbital, of course, is always taken to be a spherical harmonic. In each of the N Eqs. 5.16 one substitutes the $\varphi_j^{(1)}$ (for $j \neq i$) and solves to obtain a new function $\varphi_i^{(2)}$ (which in general can be expressed only in numerical form). These are again substituted in Eqs. 5.16 and the equations solved to obtain a set of $\varphi_i^{(3)}$. The process is repeated until no further change occurs, that is, until the $\varphi_i^{(n)}$ differ negligibly form the $\varphi_i^{(n-1)}$. In this limit the orbitals consistently reproduce the average field in which each electron moves, and are thus called *self-consistent field* (*SCF*) orbitals.

The initial set of trial orbitals must correspond to the desired electronic configuration of the atom. For example, in the nitrogen atom, two of the φ_i must be 1s-type functions, two 2s-type, and three 2p-type. (What is a "1s-type" orbital? As in any other spherically symmetric problem, the wave function with quantum numbers n, l must have l angular nodes and $n-l-1$ radial nodes.) If one carries out such SCF calculations for a variety of possible configurations, the configuration of the atomic ground state should be the one that gives the lowest total energy. Although the calculations are in fact too approximate to reproduce all the observed sequences of electronic states, they do give the gross features of the orbital-filling sequence.

The Hartree approximation still does not take full account of the interaction between electrons. It includes the average Coulombic repulsion in the e^2/r_{ij} integrals of Eq. 5.16, but ignores the *exchange interaction* associated with the exclusion principle. As we shall see in the next chapter, to obtain a wave function that satisfies the exclusion principle, one must replace the simple product of Eq. 5.15 with a determinant, containing all the possible permutations of the electrons among the orbitals. We shall wait until Chapter 6 to go into the details of this approach. One still applies the iterative procedures of the self-consistent field method, leading to what are called *Hartree–Fock* SCF orbitals.

Hartree–Fock SCF calculations have been carried out for all the elements, leading to the results summarized in Fig. 5.2. The quantity

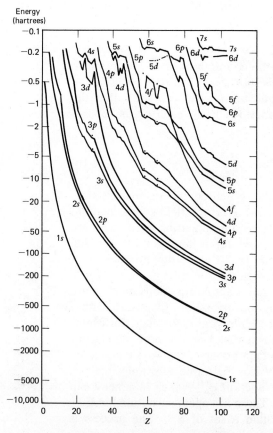

FIGURE 5.2

Hartree–Fock orbital energies of the elements. The energies plotted are those calculated by F. Herman and R. Skillman, *Atomic Structure Calculations* (Prentice–Hall, Englewood Cliffs, N.J., 1963), for the configurations listed in Table 5.1. Spin–orbit splitting (Section 5.7) is neglected, but relativistic corrections are included. The energy units (hartrees) are defined at the end of Section 5.4.

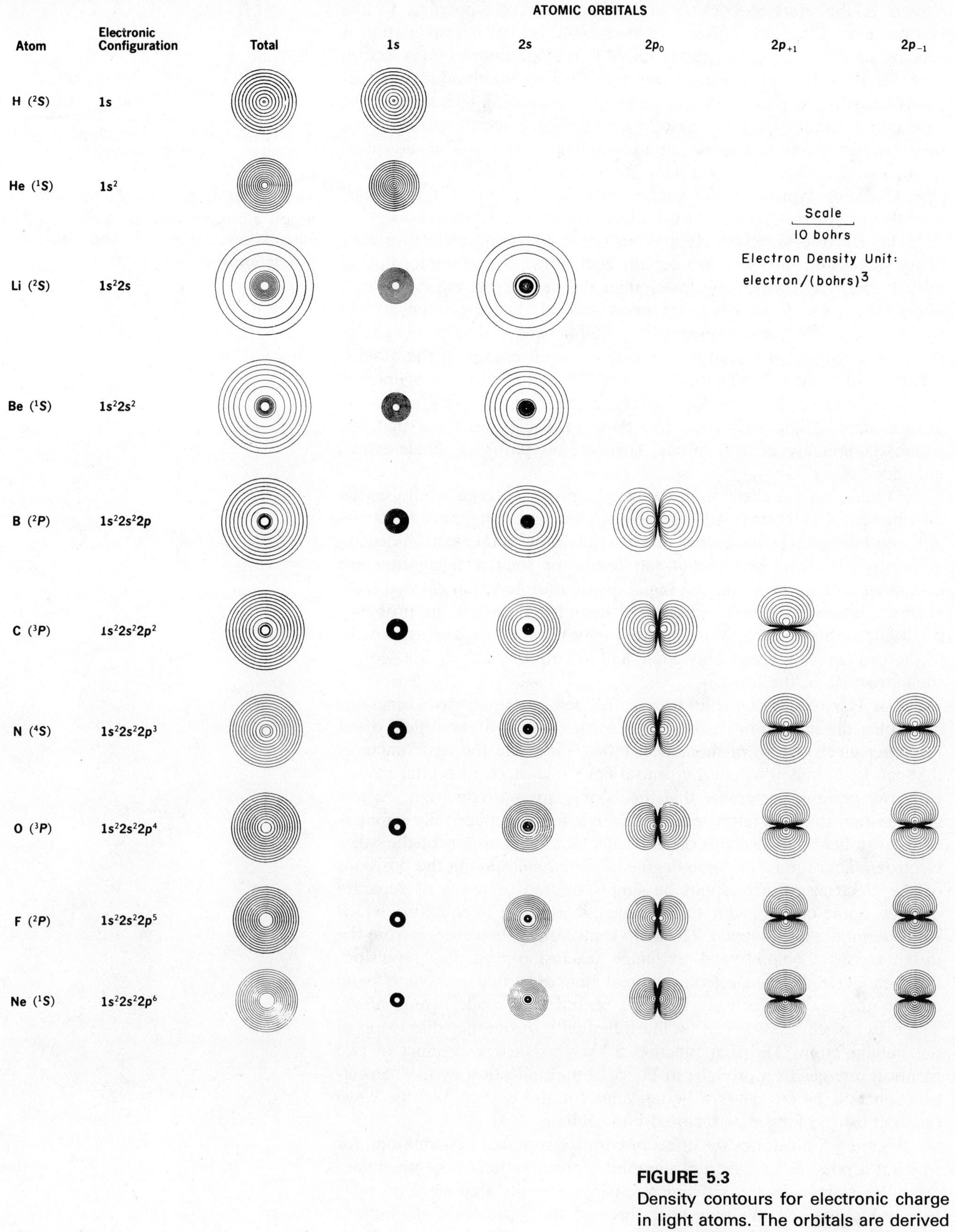

ATOMIC ORBITALS

Atom	Electronic Configuration	Total	1s	2s	$2p_0$	$2p_{+1}$	$2p_{-1}$
H (^2S)	1s						
He (^1S)	$1s^2$						
Li (^2S)	$1s^22s$						
Be (^1S)	$1s^22s^2$						
B (^2P)	$1s^22s^22p$						
C (^3P)	$1s^22s^22p^2$						
N (^4S)	$1s^22s^22p^3$						
O (^3P)	$1s^22s^22p^4$						
F (^2P)	$1s^22s^22p^5$						
Ne (^1S)	$1s^22s^22p^6$						

Scale
───────
10 bohrs

Electron Density Unit:
electron/(bohrs)^3

FIGURE 5.3
Density contours for electronic charge in light atoms. The orbitals are derived from Hartree–Fock orbitals. Plots were made by A. C. Wahl, and published in *Scientific American*, April 1970, p. 55.

plotted is the Hartree–Fock orbital energy ϵ_i, corresponding to the Hartree ϵ_i of Eq. 5.16. To the extent that the orbital approximation is valid, ϵ_i should equal the energy required to remove (ionize) one electron from the ith orbital (*Koopmans' theorem*). Such orbital binding energies can be measured by spectroscopic or electron-scattering techniques. (See especially Section 7.10 for a description of photoelectron spectroscopy, the most direct way to measure orbital binding energies.) The calculated ϵ_i's of Fig. 5.2 generally agree with experimental values to within a few percent. Some typical atomic charge distributions, both for individual orbitals and for total charge distributions, are shown in Fig. 5.3.

The Hartree–Fock calculations also give the approximate orbital-filling sequence, but there are certain anomalies. For example, the $3d$ orbital energy is consistently lower than the $4s$ energy, indicating correctly that a $4s$ electron is more easily ionized. The predicted ground states of Sc and Sc$^+$ are, respectively, $\cdots 3d4s^2$ and $\cdots 3d4s$. In spite of this, the $4s$ orbital is filled first, because the *total* energy of the atom is lower in the $3d4s^2$ configuration. For Sc again, the configurations $\cdots 3d^24s$ and $\cdots 3d^3$ are, respectively, 2.11 eV and 4.19 eV above the ground state. These anomalies, like those in Table 5.1, show that one cannot completely explain atomic structure in terms of single-orbital energies.

Another way to check the SCF calculations is in terms of the spatial distribution of electrons. Adding together the squared orbital functions $|\varphi_i|^2$ and integrating over angles, one can obtain the total electron density as a function of r. In Fig. 5.4 the results of such a calculation are compared with an experimental radial distribution function derived from electron-diffraction measurements. The quantity plotted is the probability of finding an electron in the shell between r and $r + dr$. The agreement of the two curves is reasonably good, and in both one can clearly see the shell structure of the atom.

The Hartree–Fock method gives the best possible wave functions describing the motion of each electron in the *mean* potential field of all the other electrons. In mathematical terms, these are the best functions that can be obtained within the central field, one-electron orbital model. But now we must remember that this is only an approximation. At any given instant the field felt by each electron is *not* the spherically symmetrical mean field, but depends on the instantaneous positions of the other electrons. There must be some degree of *correlation* among the positions of the electrons which cannot be simply treated in terms of separate orbitals. Some of this is due to the magnetic moment interactions, which we take up in a later section. The main contribution, however, is from the instantaneous Coulomb and exchange (exclusion-principle) repulsions between electrons; the electrons in a real atom must thus be able to avoid one another more effectively than the central field model would allow. The effect of correlation can be illustrated with our earlier discussion of the helium atom. The trial function 5.13 is a Hartree product of two identical one-electron orbitals; in Eq. 5.14 multiplication by the correlation function $1 + cr_{12}$ gives a better value for the energy, but the wave function can no longer be factored into orbitals.

Figure 5.5 illustrates the effect of correlation in the helium atom, for which it is possible to construct a graphic representation of the phenomenon. We start with the six-dimensional probability density $\mathscr{P}(\mathbf{r}_1, \mathbf{r}_2) = |\psi(\mathbf{r}_1, \mathbf{r}_2)|^2$; one can integrate over three of the variables and obtain a three-dimensional probability density $\rho(r_1, r_2, \theta_{12})$ that depends only on the distances r_1 and r_2 of electrons 1 and 2 from the nucleus, and on θ_{12}, the angle between the vectors $\mathbf{r}_1$ and $\mathbf{r}_2$. From this function of three variables, one can construct the *conditional* probability density for finding

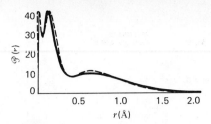

FIGURE 5.4
Radial distribution of electrons in the argon atom, according to Hartree SCF calculations (– – – –) and electron-diffraction measurements (———).

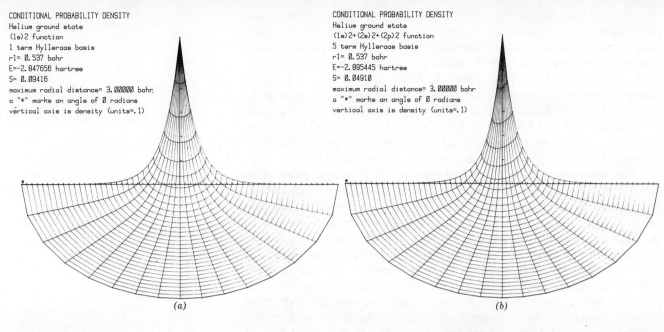

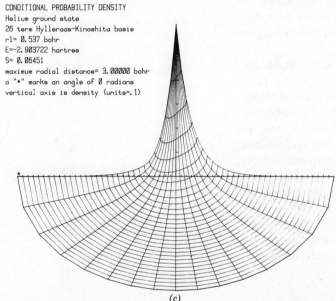

CONDITIONAL PROBABILITY DENSITY
Helium ground state
26 term Hylleraas-Kinoshita basis
r1= 0.537 bohr
E=-2.903722 hartree
S= 0.06451
maximum radial distance= 3.00000 bohr
a "*" marks an angle of 0 radians
vertical axis is density (units=.1)

(c)

FIGURE 5.5

Conditional probability densities for the ground state of helium, for three wave functions. (a) Probability density for finding electron 2 at any point in space when electron 1 is 0.537 bohr to the left of the nucleus, based on a simple, single-configuration $1s^2$ wave function. (b) The corresponding conditional probability density based on a best superposition of $1s^2$, $2s^2$, and $2p^2$ configurations. (c) The corresponding conditional probability density based on an accurate (26-term) wave function. The nucleus is located where the distribution peaks; the distance 0.537 bohr is the most probable distance of electron 1 from the nucleus. Note from the tilt of the contours of constant radius that there is no angular correlation in (a), some in (b), and more in (c). Graphs supplied by Paul Rehmus.

electron 2 at r_2 and θ_{12} with r_1 at a chosen value r_1'. This conditional probability is the function

$$d(r_2, \theta_{12} \mid r_1') = \frac{\rho(r_1', r_2, \theta_{12})}{\rho(r_1')}, \qquad (5.17)$$

where $\rho(r_1')$ is the probability density for finding r_1 at r_1' whatever r_2 and θ_{12} may be. By constructing graphs of $d(r_2, \theta_{12} \mid r_1')$ for various values of r_1', and—more important in our present context—for wave functions of different levels of refinement, we can learn how the distribution of probability for one electron is affected by the position of another, and how different wave functions represent the effects of spatial correlation. Figure 5.5a shows the conditional probability $d(r_2, \theta_{12} \mid r_1')$ for the ground state of helium according to a wave function based on a $1s^2$ configuration; Fig. 5.5c is the conditional probability for the same atom and for the same value of r_1' but based on a very accurate wave function.

Another way to improve wave functions beyond the level of single-configuration representations is to write them as sums of terms, in which each term corresponds to a different configuration. This method, called configuration interaction, will be discussed in detail in Chapter 6, in connection with the H_2 molecule. For the present, we note only that a wave function composed mostly of a $1s^2$ term, but with a bit of $2s^2$ configuration, is a better representation of the ground state of He than the pure $1s^2$ function, and that addition of some $2p^2$ configuration makes it better still; Fig. 5.5b shows the conditional probability for such a function, which is still not as accurate as the 26-term function used for Fig. 5.5c.

The difference between the total Hartree–Fock energy of an atom and the true energy in a given state is defined as the *correlation energy*; the Hartree–Fock method always gives an energy higher than the true value, in accordance with Eq. 5.11. The correlation energy is usually of the order of 1 eV per electron pair; this is of about the same magnitude as chemical bond energies, and thus cannot be neglected if one wishes to proceed accurately from atomic to molecular properties. Various methods exist for estimating correlation energies. To a good approximation, the total correlation energy can be taken as a sum of two-electron correlation energies, which can be calculated by methods like those used in the helium atom; most of the effect seems to be due to the interaction of electron pairs in the same orbital. In Chapter 14 we shall see how this approximation can be used to estimate *molecular* correlation energies from thermochemical data.

An additional error in the Hartree–Fock method arises from the effects of relativity. In atoms with fairly large Z, the inner electrons have such high kinetic energies that their velocities approach the speed of light; their masses thus become significantly larger than the mass of a stationary electron. This results in a shrinkage of the electrons' orbits and an increase in their binding energy. Although Hartree–Fock calculations can include this effect in only an approximate way,[7] the relativistic energy correction is known to be larger than the nonrelativistic correlation energy for most atoms. For light atoms this is of little significance. For heavy atoms, the relativistic shrinkage of s and p shells makes them especially effective screens of the nuclear charge for d and f shells, which thus enlarge. Closed sub shell atoms such as Hg become somewhat inert.[8]

In Fig. 5.3 we expressed the energies in *hartrees*. This unit, named for D. R. Hartree, is defined as

$$1 \text{ hartree} \equiv 2 \text{ rydbergs} \equiv 2hcR_\infty \equiv \frac{e^4 m_e}{4\epsilon_0^2 h^2}$$

$$= 4.359814 \times 10^{-18} \text{ J} = 27.21161 \text{ eV}.$$

The ground-state energy of the hydrogen atom is thus -0.5 hartree. Published calculations on atoms and molecules most commonly give energy in hartrees and distance in Bohr radii ($a_0 = 0.5292$ Å), a combination known as *atomic units*.[9] The utility of such units is obvious: The ubiquitous $e^2/4\pi\epsilon_0 r$ in hartrees is numerically equal to a_0/r. Similarly, m_e is the atomic unit of mass, e of charge, and $\hbar$ of angular momentum. The

---

[7] A completely relativistic quantum mechanics requires the revision of the Schrödinger equation itself. In this theory (originated by Dirac), the electron spin appears automatically in the solution of the wave equation; the spin is thus inherently not only a quantum mechanical, but also a relativistic, phenomenon.

[8] For relativistic effects, see P. Pyykko and J. P. Desclaux, *Acc. Chem. Res.* **12,** 276 (1979) and K. Pitzer *ibid.*, 282. For a more detailed discussion of atomic orbitals see R. S. Berry, *J. Chem. Educ.* **43,** 283 (1966).

[9] Often written as "a.u." with no further explanation, or not specified at all. This can be confusing, especially since some people use rydbergs instead of hartrees.

constants are usually suppressed, leading to equations like $V = -1/r$; this is certainly convenient for making calculations, but is not recommended unless you know just what is missing. In this text we write all such equations out in full.

The physical and chemical properties of the elements vary systematically with atomic number, in a way best summarized by the periodic table. We may now relate this behavior to atomic structure. To do this, at least for light elements, we do not need refinements such as term splitting (Sec. 5.6). The gross features of the periodic table can be explained quite simply in terms of electron configurations and shell structure. In this section we shall look briefly at the chemical behavior of the various families of elements, and at a number of physical properties associated with individual atoms.

The most strikingly similar group of elements is probably the so-called inert gases (rare gases, noble gases): helium, neon, argon, krypton, xenon, and radon. Until 1962 these elements were thought to form no compounds whatsoever; now krypton, xenon, and radon are known to form a limited number of compounds with fluorine and oxygen. Still, they remain by far the most unreactive family of elements. Even the forces between their monatomic molecules are very weak, so that they have very low melting and boiling temperatures. (Radon, the heaviest, boils at $-62°C$.) Usually one glibly attributes the inertness of these elements to their "closed-shell structure." In fact they have only closed subshells; all but helium have outer shells with the configuration $\cdots ns^2 np^6$, and all their occupied inner subshells are filled. But other atoms have closed-subshell structures also, yet are hardly in the same category chemically. Palladium, for example, has the configuration $\cdots 4s^2 4p^6 4d^{10}$, and is clearly not inert; even more reactive are the alkaline earth metals, with $\cdots ns^2$ outside an inert-gas configuration, and the family of zinc, cadmium, and mercury. What, then, is really responsible for the inertness?

For a closed-shell atom to form a chemical bond, the outer shell must be "cracked open" in some way. One indication of the difficulty of doing this is the energy of the lowest excited state, in which one electron has been promoted to the lowest empty orbital. In the alkaline earths and the zinc group this excitation is simply $ns \to np$, between two orbitals that lie quite close together in energy. But for the inert gases the lowest excitation is $np \to (n+1)s$, all the way to the next higher shell, and the energy required is much greater (cf. Fig. 5.10). To illustrate this point, here are the first excitation energies (in electron volts):

He 19.8	Ne 16.6	Ar 11.5	Kr 9.9	Xe 8.3	Rn 6.8
Be 2.7	Mg 2.7	Ca 1.9	Sr 1.8	Ba 1.1	Ra 1.6
		Zn 4.0	Cd 3.7	Hg 4.7	

(As for palladium, the excitation is $4d \to 5s$ and takes only $0.8\,eV$.) The energy gained by forming a chemical bond is usually in the range 2–6 eV, which clearly is enough to "crack the shell" in all but the inert gases.

Another way for an atom to undergo chemical reaction is ionization, in which one electron is removed from the atom altogether. The energy required to remove one electron from a neutral atom is usually called the first *ionization potential*[10] (I or I_1); the second ionization potential (I_2) is

---

[10] Strictly, "potential" implies electric potential or voltage. But for a single electron the ionization potential in volts is numerically the same as the ionization energy in electron volts, and the two terms are often used interchangeably.

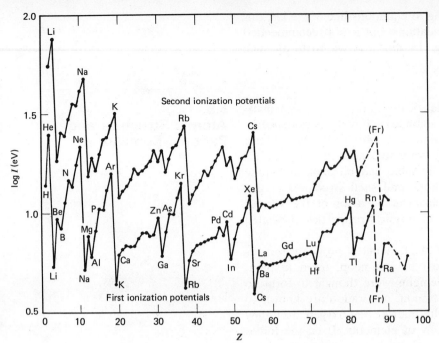

FIGURE 5.6
First and second ionization potentials
of the elements. A number of the more
significant peaks in the lower curve
have been identified. The broken lines
(– – – –) give the probable behavior of
the curves where no data are available.

the energy of removing a second electron, and so forth. The ionization potential is among the properties that show the most clearly periodic behavior, as can be seen in Fig. 5.6. Note that the second ionization potential is always higher than the first (since there is a greater net positive charge holding the second electron) and has its peaks shifted one unit to higher Z.

Let us examine the structure of the first-ionization-potential curve. The inert gases, as we might expect, have the highest first ionization potentials: it is very hard to remove an electron from a closed p subshell. (There are small peaks for such other closed-shell atoms as the zinc family and palladium.) The lowest values are those for the alkali metals (Li, Na, K, Rb, Cs, Fr), with one loosely bound s electron outside a closed shell; similarly, the next most prominent minima in the curve are for those atoms with a single outer p electron (B, Al, Ga, In, Tl). The small peaks for the lighter $\cdots np^3$ atoms (N, P, As) illustrate the stability of a half-filled subshell. All these details are superimposed on two grosser features, the overall downward trend and the upward trend within each period; these effects are closely associated with the variation in atomic size, which we shall discuss shortly.

Figure 5.6 can help us to understand a good deal about chemical behavior. After the inert gases, the most clearly defined family of elements is that of the alkali metals. Their ionization potentials are all about 4–5 eV, comparable to the energies normally available in chemical reactions. The result is that these atoms readily lose one electron to form the M^+ ions, and their chemistry is primarily the chemistry of these ions. However, the alkali metals have the *highest* second ionization potentials (ca. 25–75 eV), so that one virtually never sees alkali M^{2+} ions in chemical systems. This is again a matter of closed-shell structure: The alkali M^+ ions are isoelectronic[11] with the inert gas atoms, with the outer electrons even more tightly bound.

[11] Two atoms or ions are *isoelectronic* if they have the same number of electrons. Isoelectronic systems usually have the same configuration and certain similarities in physical properties. But their chemical behavior is quite different—compare the reactivity of Ar and K^+—because of the dominant effect of the ionic charge and its Coulomb field.

The alkaline earth metals (Be, Mg, Ca, Sr, Ba, Ra) have closed-subshell configurations, but we have already shown that this does not lead to inertness. On the contrary, the outer s electrons are relatively weakly bound (the alkaline earths have the lowest second ionization potentials), and the atom readily gives both up to form the M^{2+} ion. Beryllium is an exception: Having the highest ionization potentials of the family, it forms mainly covalent compounds. The elements of the zinc family, with the configurations $\cdots (n-1)d^{10}ns^2$, also generally form M^{2+} ions; however, the ionization potentials are relatively high, and the metals are much less reactive than the alkaline earths.

Since it is easier to remove one electron than two, why do we never see compounds containing ions like Ca^+? The principal reason is the intense Coulomb field of the M^{2+} ion, which is of course twice as strong as that of an M^+ ion at the same distance. This field makes the lattice (bonding) energy of, say, solid $CaCl_2$ much greater than that of $CaCl$ would be—enough greater to outweigh a second ionization potential of over 10 eV. Similar considerations apply to the hydration energy in solution, where we again find only M^{2+} alkaline-earth ions. We shall consider the energy balance of processes such as these in Chapter 14. One *can* obtain the M^+ ions by applying energy to isolated atoms, as in electrical discharges or very hot gases.

Now let us look at the other side of the periodic table, where most of the nonmetallic elements are found. Here the atoms have vacancies in nearly filled p subshells, and can attain more stable configurations by *adding* electrons to form negative ions. The *electron affinity* (A) of a given atom is the energy gained when one electron is added. The electron affinity of X is thus the same as the first ionization potential of X^-, and a plot of electron affinities should resemble the lower curve in Fig. 5.6 but shifted one unit *lower* in Z. Such a plot is given in Fig. 5.7. Note that even the highest atomic electron affinities are less than the lowest atomic ionization potentials. The periodic pattern is clear.

As expected, the highest electron affinities are those of the halogens (F, Cl, Br, I, At), which readily add electrons to form the familiar X^- (halide) ions. Most of the elements near the right side of the periodic table have relatively large electron affinities, and can generally be regarded as electron-accepting species. These elements complement the electron-donating tendency of the alkali and alkaline earth metals, with which they form ionic compounds. The second-highest peaks, surprisingly, are the coinage metals (Cu, Ag, Au), but these form negative ions only in hot gases.[12] At the other extreme, many elements actually have *negative* electron affinities—that is, force would be required to make an electron stick to the neutral atom. The dips in the curve are mainly those atoms with exceptionally stable configurations, having either closed subshells (inert gases, alkaline earths, Zn family) or *half*-filled outer subshells ($\cdots np^3$: N, P, etc.; $\cdots nd^5$: Mn, Re).

Unlike the case of the ionization potentials, no atoms have positive second electron affinities. The Coulomb repulsion between an X^- ion and an additional electron is so great that no stable X^{2-} ion is known to exist in isolation. Nevertheless, species such as O^{2-} and S^{2-} do appear to exist in ionic crystals, where they are stabilized by the high lattice energy; all these compounds hydrolyze readily in aqueous solution, forming more stable covalently bonded species (OH^-, HS^-, etc.).

Whereas the alkali and alkaline earth metals form almost exclusively ionic compounds, this is not true of the elements at the other side of the

[12] However, the bonding in CsAu (and perhaps RbAu) is largely ionic.

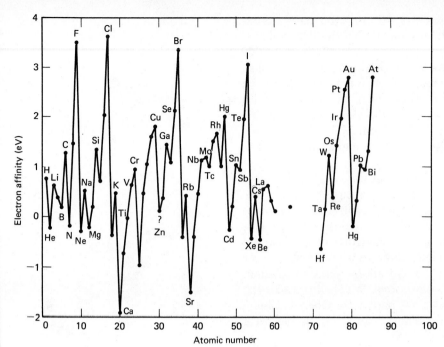

FIGURE 5.7
Electron affinities of the elements. The values are those selected in the review by H. Hotop and W. C. Lineberger, *J. Phys. Chem. Ref. Data* **4,** 539 (1975); others are those of R. J. Zollweg, *J. Chem. Phys.* **50,** 4251 (1969).

periodic table. Besides adding electrons to form negative ions, they can also share their outer electrons to form covalent bonds. In most simple compounds this results in the equivalent of an inert-gas configuration around each atom, as illustrated by the familiar Lewis formulas in which dots represent the outer-shell electrons (e.g., $2\mathrm{H}\cdot + \cdot\ddot{\mathrm{O}}\cdot = \mathrm{H}\!:\!\ddot{\mathrm{O}}\!:\!\mathrm{H}$). We shall see in the next few chapters that such electron-sharing can be described in terms of orbitals extending over more than one atom. Covalent bonding occurs primarily between atoms not too different in *electronegativity*. The latter is a rather vague concept, defined by Pauling as "the power of an atom in a molecule to attract electrons to itself." Electronegativity is roughly equivalent to "nonmetallic character," and generally increases toward the upper right of the periodic table. Since atomic ionization potentials and electron affinities both tend to increase in the same direction, one method of defining numerical electronegativities takes the average of these two quantities. We shall look at this and other methods in Section 7.7, where electronegativities will be found in Table 7.7.

We can now account for the most obvious division of the periodic table, that between metals and nonmetals. Basically, a metallic element is one whose atom has one or more loosely bound valence (outer-shell) electrons; this description fits all the elements whose outermost subshell is ns ($n > 1$), as well as many with partially filled p subshells. The valence electrons are easily removed to form positive ions, as we have seen in the alkalis and alkaline earths. Metals ordinarily exist as solids in which these electrons are relatively free to move throughout the crystal; the mobility of the electrons accounts for the high electrical and thermal conductivity characteristic of metals. (We shall have more to say about the structure of metals in Chapter 11.) By contrast, the typical nonmetals have nearly full p subshells, and can form closed-shell configurations by sharing only a few electrons in small, covalently bonded molecules (Cl_2, O_2, P_4, etc.). The bonding forces within these molecules are strong, but (as in the inert gases) the forces between them are quite weak, so that many of the nonmetals are gases at ordinary temperatures. Along the borderline between metals and nonmetals, the elements often exist in solid forms containing infinite networks of covalent bonds; the best-known examples are the diamond and graphite forms of carbon. The great majority of the elements are metals, since all the transition elements (in which d or f subshells are being filled) have ns^2 or ns outer shells.

The trends in physical and chemical properties across the periodic table can best be understood by considering atomic sizes. The size of an atom cannot have a precise meaning, since at large distances the electron density merely trails off exponentially. However, fairly self-consistent sets of "atomic radii" can be defined for various purposes. In principle, the radii should be such that $r_A + r_B$ gives the actual internuclear distance in a bond between atoms A and B. Unfortunately, there exists no set of radii that can reproduce all bond lengths with experimental accuracy. In Table 5.2 we list *covalent radii*, which should add to give the lengths of single (two-electron) bonds between atoms not too different in electronegativity. These values are most easily obtained by halving the lengths of homonuclear single bonds (e.g., r_{Cl} is half the bond length in Cl_2): in many cases they reproduce other bond lengths within a few percent. A graph of covalent radii would closely resemble the atomic volume graph of Fig. 1.7, except that those elements made up of distinct molecules (most of the nonmetals) have relatively large volumes.

However, there are also many bond lengths quite different from what Table 5.2 predicts; let us digress to consider some of the reasons. A drastic shortening occurs when more than two electrons are shared, that is, in multiple bonds (C—C, 1.54 Å; C=C, 1.34 Å; C≡C, 1.20 Å); we shall see that this results from an increase in bonding energy. There is also a shortening associated with the ionic character (electronegativity difference) of the bond; for example, the C—O single bond is 1.43 Å rather than 1.51 Å $(r_C + r_O)$. For crystals essentially composed of ions rather than neutral atoms—the alkali halides, for example—one can derive an altogether different set of *ionic radii*, as we shall see in Chapter 11; the ionic radii depend strongly on charge, positive ions being smaller and negative ions larger than the neutral atoms. Finally, the effective radius of an atom increases with its *coordination number*, the number of atoms to which it is bonded. Since an atom in a solid metal usually has 6 to 12 nearest neighbors, the interatomic distances in metals are typically about 10% more than twice the covalent radii. With corrections for all these effects, one can reproduce most bond lengths with reasonable accuracy.[13]

Regardless of all these reservations, Table 5.2 is quite adequate for studying the relative sizes of atoms. To a rough approximation, the atomic radius corresponds to the outermost peak in the electron radial distribution (Fig. 5.4), that is, to the most probable radius of the valence electrons. Since this radius increases with n, the atomic size increases with each period in most families of the periodic table.

In an alkali metal atom we have a single outer-shell electron; if it were completely outside the closed-shell core, it would simply feel the Coulomb field of a net +1 charge, and its most probable radius would be that of a hydrogen ns orbital, $r_n = n^2 a_0$. The actual increase is of course much slower: The $6s$ electron in cesium has its outermost peak at about 2.7 Å, compared with 26 Å for a hydrogen $6s$ electron. The reason is that

---

[13] For detailed discussions of the various kinds of atomic radii, see L. Pauling, *The Nature of the Chemical Bond*, 3rd ed. (Cornell University Press, Ithaca, N.Y., 1960), and J. C. Slater, *Quantum Theory of Molecules and Solids*, Vol. 2 (McGraw-Hill, New York, 1965).

TABLE 5.2
SINGLE-BOND COVALENT RADII OF THE ELEMENTS (Å)

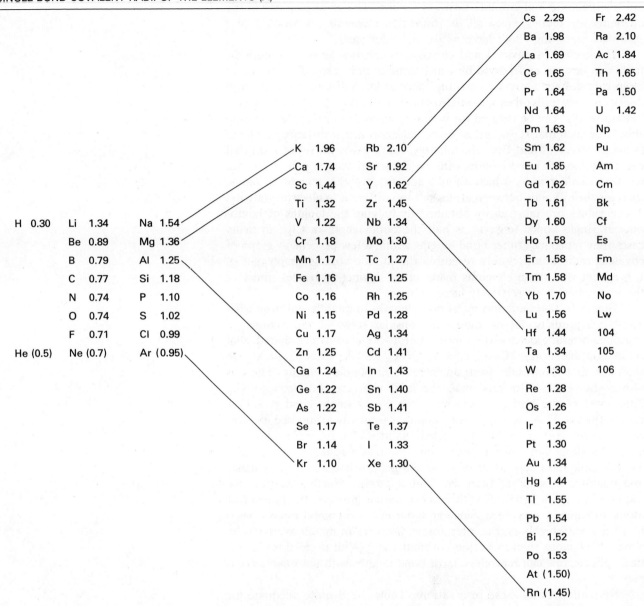

						Cs 2.29	Fr 2.42
						Ba 1.98	Ra 2.10
						La 1.69	Ac 1.84
						Ce 1.65	Th 1.65
						Pr 1.64	Pa 1.50
						Nd 1.64	U 1.42
						Pm 1.63	Np
			K 1.96	Rb 2.10		Sm 1.62	Pu
			Ca 1.74	Sr 1.92		Eu 1.85	Am
			Sc 1.44	Y 1.62		Gd 1.62	Cm
			Ti 1.32	Zr 1.45		Tb 1.61	Bk
H 0.30	Li 1.34	Na 1.54	V 1.22	Nb 1.34		Dy 1.60	Cf
	Be 0.89	Mg 1.36	Cr 1.18	Mo 1.30		Ho 1.58	Es
	B 0.79	Al 1.25	Mn 1.17	Tc 1.27		Er 1.58	Fm
	C 0.77	Si 1.18	Fe 1.16	Ru 1.25		Tm 1.58	Md
	N 0.74	P 1.10	Co 1.16	Rh 1.25		Yb 1.70	No
	O 0.74	S 1.02	Ni 1.15	Pd 1.28		Lu 1.56	Lw
	F 0.71	Cl 0.99	Cu 1.17	Ag 1.34		Hf 1.44	104
He (0.5)	Ne (0.7)	Ar (0.95)	Zn 1.25	Cd 1.41		Ta 1.34	105
			Ga 1.24	In 1.43		W 1.30	106
			Ge 1.22	Sn 1.40		Re 1.28	
			As 1.22	Sb 1.41		Os 1.26	
			Se 1.17	Te 1.37		Ir 1.26	
			Br 1.14	I 1.33		Pt 1.30	
			Kr 1.10	Xe 1.30		Au 1.34	
						Hg 1.44	
						Tl 1.55	
						Pb 1.54	
						Bi 1.52	
						Po 1.53	
						At (1.50)	
						Rn (1.45)	

even the outermost electron has much of its charge density within the electron core (cf. Fig. 4.5), and is thus not perfectly "shielded" from the nucleus. The binding force on each electron thus increases steadily with the nuclear charge Z, as we already know from Fig. 5.2. The $6s$ electron in cesium ($Z = 55$) is bound almost as tightly as the $5s$ electron in rubidium ($Z = 37$), and the atomic size is only slightly larger.

We can now understand the trend across each period. One might expect that adding a second ns electron would increase the atomic size, because of the repulsion between the two valence electrons. In fact, the increase of Z by 1 has a much greater effect, and the alkaline earth atoms are uniformly smaller than their alkali neighbors. Electrons in an open shell generally shield one another from the nucleus quite poorly, since they tend to be in different regions of space (for two electrons, on opposite sides of the nucleus). Only the completion of a closed shell, with its spherically symmetric charge density, gives a significant improvement in shielding. Thus the atomic sizes generally decrease with Z all the way across each period (but slower than Z, since the electron–electron repulsion does increase as the outer shell fills up), and increase again only when a new electron shell is added outside a closed shell.

There are some exceptions to this trend. In the transition metals the $(n-1)d$ subshell is being filled, whereas the outer shell remains virtually unchanged at ns^2; since inner-shell electrons shield one another poorly, these atoms are relatively small, and a slight size increase accompanies the resumption of outer-shell filling. The shrinkage due to poor shielding is even more marked for the lanthanides, in which the $4f$ subshell is filled[14] inside a $6s^2$ outer shell. This *lanthanide contraction* is so great that hafnium has virtually the same size as its relative zirconium, and the two elements are extremely difficult to separate by ordinary chemical techniques. A similar contraction appears to occur in the actinide series, in which the $5f$ subshell is filled.

The trends in atomic size are directly correlated with those in ionization potential and electron affinity. Both the latter increase across each period, as the atomic size shrinks and the outer-shell electrons become more tightly bound. They decrease sharply at the beginning of a new period, as a larger, more weakly bound outer shell is begun. And each period has values lower than the preceding, since the binding energy decreases with distance from the nucleus. This effect too is less than it would be with perfect shielding: Cesium has $I_1 = 3.89 \text{ eV}$, compared with the $0.38 \text{ eV} (= 13.6 \text{ eV}/n^2)$ required to ionize a $6s$ hydrogen electron; the effective nuclear charge seen by the valence electron in its ground state is thus $Z^* = (3.89/0.38)^{1/2} = +3.2$, rather than the $+1$ of perfect shielding.[15]

We have not yet discussed the chemical behavior of the transition metals. Although these elements have only two outer-shell ns electrons, the partially filled $(n-1)d$ subshell is quite near the "surface" of the atom and relatively loosely bound. As a result, transition metals ordinarily form compounds in which both the s and the d electrons are involved in bonding, with a wide variety of possible oxidation states. For example, manganese $(\cdots 3d^5 4s^2)$ and oxygen form the species MnO, Mn_2O_3, MnO_2, MnO_4^{2-}, and MnO_4^{-}. Even the unoccupied orbitals close to the outer shell can accept electrons from other species (*ligands*), giving rise to the complexes typical of the transition elements, such as $[Fe(H_2O)_6]^{2+}$. These complexes usually have low-lying excited states resulting from the splitting of d-orbital levels; since the energy differences often correspond to visible radiation, most such complexes are colored. We shall have much to say about these species in Chapter 9. Unlike the $(n-1)d$ subshells, the lanthanide $4f$ subshell is fairly deep within the atom, and the M^{3+} ions that dominate these elements' chemistry involve the loss of only one $4f$ electron. The actinide $5f$ electrons are not as deeply buried, and the lower actinides (especially uranium) commonly have high oxidation states. Since the binding energies of both d and f electrons increase with Z more rapidly than those of outer-shell electrons (cf. Fig. 5.2), the lower oxidation states again predominate toward the high-Z ends of the transition and actinide series.

As was mentioned in Section 5.1, paramagnetism occurs primarily in substances with unpaired electron spins. In most compounds of the

---

[14] Note the anomalously large sizes of $Eu(\cdots 4f^7 6s^2)$ and $Yb(\cdots 4f^{14} 6s^2)$, with their exceptionally stable configurations.

[15] For the alkali metal atoms we have $Z^* = n/n^*$, in terms of the effective quantum numbers of Eq. 2.12.

nontransition elements, either ionization or the formation of covalent bonds leaves all the electrons paired. This is not true of the transition metals, in which some or all of the d electrons often remain unpaired; thus many transition metal compounds are paramagnetic. We shall discuss this behavior (with special reference to complex ions) in Section 9.6. The most intense paramagnetism is found among the lanthanides, with their many unpaired f electrons.[16]

We could look at other properties of the elements, nearly all showing some periodic behavior, but our point has been made. A vast amount of "chemistry"—that is, the observed differences between chemical substances—can be well understood in terms of a simple description of atomic structure. We have only scratched the surface of this subject, and further details can be found in advanced texts on inorganic chemistry. We turn in the next two sections to more detailed interactions that determine the states of atoms, in order to complete the groundwork we need to see how atoms are put together to form molecules. Then we shall devote the next four chapters to this latter task.

We have gone far by assuming that the electronic configuration of an atom dominates its energy and other properties. We have examined some aspects of correlation, for which the configuration model is inadequate, but the configuration model is a very good representation of atoms with closed shells, and is a good approximation for many other real atoms as well. The set of closed-shell species includes the ground states of helium ($1s^2$), the other inert gases ($\cdots ns^2 np^6$), and the alkaline earths ($\cdots ns^2$); these configurations are distinguished by having all the occupied subshells filled. Another group that is well represented by single configurations includes hydrogen and the alkalis in their ground states ($\cdots ns$), in which the atom has a single s electron outside a closed-shell structure, and the halogens ($\cdots np^5$), with only a single vacancy in an otherwise closed shell. In all these cases the electron configuration corresponds to only a single state or energy level of the atom.

In most atoms, however, a single configuration may describe several different states. These are real, physically distinguishable states, which can differ from one another by as much as several electron volts in energy. One speaks of the splitting of the configuration into *terms*;[17] in this context a "term" is equivalent to a state or group of states. The basic difference between terms of the same configuration lies in the angular momentum of the atom. The configuration defines the magnitudes of the vectors **L** and **S** for each electron, but not their orientations. There is thus more than one possible way to couple (add) these vectors to give the total angular momentum. These different ways of coupling correspond to different assignments of the quantum numbers m_l and m_s, hence to different spatial distributions of the electrons. Since the spatial relationships of the electrons are different, the terms must differ in energy.

5.6
Term Splitting and the Vector Model

---

[16] Some typical room-temperature molar susceptibilities (χ_M) in $10^{-12}\,\mathrm{m^3/mol}$: nontransition metals, Na 16, Zn -11; transition metals, Ti 153, Mn 529; lanthanides, Pr 5010, Er 44300; nonmetals, S -15, Xe -44. These values are for the elements in their normal states; higher values are often found in compounds (e.g., $MnCl_2$ 14350).

[17] This language derives from spectroscopy: For a transition between states 1 and 2, the frequency of the absorbed or emitted radiation is given by

$$\nu = \left| \frac{E_2}{h} - \frac{E_1}{h} \right|,$$

the difference of two *terms* characteristic of the two states.

We can describe this term splitting most simply with the *vector model of the atom*. This is a method for organizing the possible states of an atom into terms, by selecting only the most important of the many types of interaction. It also serves as a bookkeeping device for counting and classifying the terms corresponding to a given configuration. The essence of the vector model lies in the ways of adding angular momentum vectors, the topic we postponed discussing in Section 5.1.

In contrast to vector addition in classical physics, the addition of quantum mechanical angular momenta cannot be carried out by determining the *x*, *y*, and *z* components of the individual vectors and adding these separately. This is because the uncertainty principle forbids us to know more about any angular momentum vector than its magnitude and *one* of its components, usually defined as the *z* component. Knowing only the lengths and the *z* components, however, is sufficient to define the quantum states we need. For any quantized angular momentum vector, say **J**, the two are related by the usual space-quantization rules: the magnitude of **J** is given by an equation of the form of Eq. 5.1,

$$|\mathbf{J}| = [J(J+1)]^{1/2}\hbar, \tag{5.18}$$

and the *z* component by an equation of the form of Eq. 5.2,

$$J_z = M_J \hbar \qquad (M_J = J, J-1, \ldots, -J+1, -J), \tag{5.19}$$

with appropriate quantum numbers J and M_J.

We can approach the classification of atomic states by a series of approximations. The first, of course, is the central field approximation, in which the energy of the atom depends only on the configuration—that is, on the values of n and l (and thus $|\mathbf{L}|$) for each of the electrons, but not on the values of m_l and m_s. This would be valid only if each electron moved in a spherically symmetric field; in the real atom, the electron–electron interactions split up each configuration into many states of somewhat different energies. The next approximation rests on the fact that some kinds of interactions are much stronger than others. For the lighter atoms, all states with the same values of the quantum numbers L and S (for the entire atom) have nearly the same energy, whereas states with different values of L or S are appreciably different in energy. One can thus group together all states with the same L, S as a single term; when this is a good approximation, one speaks of *LS coupling*.[18]

Why is the energy so strongly dependent on the values of L and S? The reasons are those we introduced when discussing the exclusion principle: the Coulomb and exchange interactions between electrons. The quantum number L defines the total orbital angular momentum of the atom; as in the hydrogen atom, the value of L governs the symmetry of the atomic wave function. States with different L obviously have different charge distributions, and thus different electron–electron interaction energies. The effect of the spin quantum number S is more indirect. Each value of S corresponds to a particular combination of the electronic m_s's, which by the exclusion principle can be combined only with particular values of m_l (e.g., if two electrons in a given subshell have the same m_s, their values of m_l must differ). Thus a change in S also implies a change in the charge distribution and energy.

The total angular momentum vectors for an atom can be obtained by adding together (*coupling*) the vectors for the individual electrons:

$$\mathbf{L} = \sum_i \mathbf{L}_i \quad \text{and} \quad \mathbf{S} = \sum_i \mathbf{S}_i. \tag{5.20}$$

---

[18] Also known as *Russell-Saunders coupling*.

The z components of $\mathbf{L}$ and $\mathbf{S}$ are then given by

$$L_z = M_L \hbar \left(M_L = \sum_i m_{li} \right) \quad \text{and} \quad S_z = M_S \hbar \left(M_S = \sum_i m_{si} \right). \quad (5.21)$$

One can readily show that both $\mathbf{L}$ and $\mathbf{S}$ sum to zero over any closed subshell;[19] thus only those electrons outside closed subshells ("open-shell electrons") need to be taken into account. For example, the inert gases and alkaline earths have $\mathbf{L} = \mathbf{S} = 0$, and can thus have only a single term with $L = S = 0$. The alkalis have one s electron outside a closed shell and therefore have $L = 0$, $S = \frac{1}{2}$. Group III elements boron, aluminum, etc., with one p electron outside a closed s^2 subshell, can have only $L = 1$, $S = \frac{1}{2}$, and also show no term splitting. The same is true for the halogens, with only one vacancy in a subshell: Again we have $L = 1$, $S = \frac{1}{2}$, because m_l and m_s cancel out for all but one of the electrons. It is only in configurations with more than one unpaired electron[20] that things get more complicated and term splitting appears.

The simplest atom for which these complications appear in the ground state is that of carbon, with the configuration $1s^2 2s^2 2p^2$. We need consider only the two open-shell $2p$ electrons. A single $2p$ electron has $n = 2$, $l = 1$, three possible choices of m_l, and two possible choices of m_s, a total of six possible sets of quantum numbers; because of the exclusion principle, only five of these are available to a second electron. There are thus $(6 \times 5)/2 = 15$ possible assignments of the two electrons. (Division by 2 is required because the electrons are indistinguishable, and we cannot tell which came first.) Not all of these 15 assignments are physically distinguishable; to see how many separate terms they give rise to, we must determine what values of L and S can correspond to each assignment.

In Fig. 5.8 we show the possible ways in which the angular momentum vectors of two p electrons can be added to give different values of L and S. The value of L for an atom, like that of l for an electron, must be a positive integer or zero ($L = 0, 1, 2, \ldots$); the value of S is integral ($S = 0, 1, 2, \ldots$) for an even number of electrons, half-integral ($S = \frac{1}{2}, \frac{3}{2}, \ldots$) for an odd number. In the LS-coupling approximation, each physically distinct state (term) of the atom corresponds to a specific combination of L and S. These terms can be found in several ways, but the most straightforward method is sufficient for our purposes.

For an electron with given quantum numbers l_i, m_{li}, m_{si} we know both the magnitudes and the z components of the vectors $\mathbf{L}_i$, $\mathbf{S}_i$. Although these vectors can add to give several different atomic values of L or S, there is no ambiguity in the addition of the z components:

$$M_L = \sum_i m_{li}, \qquad M_S = \sum_i m_{si}. \quad (5.22)$$

One can thus make a preliminary classification of atomic states according to the values of M_L and M_S. This is done by tabulating all the possible assignments of m_l and m_s for the open-shell electrons, then obtaining the sums of Eq. 5.22 for each such assignment. In Table 5.3 we have illustrated how this is done for the carbon atom (or any other $\cdots np^2$ configuration). In order to satisfy the exclusion principle, any state in

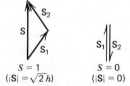

FIGURE 5.8
Possible ways of coupling angular momentum vectors for two p electrons: $\mathbf{L} = \mathbf{L}_1 + \mathbf{L}_2$, $\mathbf{S} = \mathbf{S}_1 + \mathbf{S}_2$. Each electron has $l = 1$, $s = \frac{1}{2}$, so that

$$|\mathbf{L}_1| = |\mathbf{L}_2| = [l(l+1)]^{1/2}\hbar = \sqrt{2}\hbar,$$

$$|\mathbf{S}_1| = |\mathbf{S}_2| = [s(s+1)]^{1/2}\hbar = \frac{\sqrt{3}}{2}\hbar.$$

The diagrams show the only ways of adding these vectors consistent with

$$|\mathbf{L}| = [L(L+1)]^{1/2}\hbar, \qquad |\mathbf{S}| = [S(S+1)]^{1/2}\hbar.$$

The vectors are drawn in their own plane; the z axis is in general not in this plane, but must be at an angle such that $L_{zi} = m_{li}\hbar$, $L_z = M_L\hbar$, $S_{zi} = m_{si}\hbar$, $S_z = M_S\hbar$. There are six possible combinations of L and S, but for equivalent electrons (same n, l) the exclusion principle allows only $L = 1$, $S = 1$; $L = 2$, $S = 0$; and $L = 0$, $S = 0$.

---

[19] From Table 3.1 one can see that $\sum_{m_l} |Y_{l,m_l}(\theta, \phi)|^2$ for any l is independent of θ and ϕ. That is, the total electron distribution for a closed subshell is spherically symmetric, and thus has no orbital angular momentum. It is also clear that the z components of $\mathbf{L}$ and $\mathbf{S}$ must vanish, since positive and negative values of m_l and m_s cancel.

[20] That is, in which at least one subshell contains more than one electron *and* more than one vacancy.

TABLE 5.3
ADDITION OF z COMPONENTS OF ANGULAR MOMENTA FOR TWO EQUIVALENT p ELECTRONS (AS IN THE CARBON ATOM, $1s^2 2s^2 2p^2$): $l_1 = l_2 = 1$

m_{l1}	m_{l2}	$M_L = \sum_i m_{li}$	m_{s1}	m_{s2}	$M_S = \sum_i m_{si}$	Term
1	1	2	$\frac{1}{2}$	$-\frac{1}{2}$	0	^1D
1	0	1	$\frac{1}{2}$	$\frac{1}{2}$	1	^3P
			$\frac{1}{2}$	$-\frac{1}{2}$	0 }	$^1D, {}^3P$
			$-\frac{1}{2}$	$\frac{1}{2}$	0 }	
			$-\frac{1}{2}$	$-\frac{1}{2}$	-1	^3P
1	-1	0	$\frac{1}{2}$	$\frac{1}{2}$	1	^3P
			$-\frac{1}{2}$	$-\frac{1}{2}$	-1	^3P
			$\frac{1}{2}$	$-\frac{1}{2}$	0	
			$-\frac{1}{2}$	$\frac{1}{2}$	0 }	$^1D, {}^3P, {}^1S$
0	0	0	$\frac{1}{2}$	$-\frac{1}{2}$	0 }	
0	-1	-1	$\frac{1}{2}$	$\frac{1}{2}$	1	^3P
			$\frac{1}{2}$	$-\frac{1}{2}$	0 }	$^1D, {}^3P$
			$-\frac{1}{2}$	$\frac{1}{2}$	0 }	
			$-\frac{1}{2}$	$-\frac{1}{2}$	-1	^3P
-1	-1	-2	$\frac{1}{2}$	$-\frac{1}{2}$	0	^1D

Number of Assignments

M_L \ M_S	1	0	-1
2		1	
1	1	2	1
0	1	3	1
-1	1	2	1
-2		1	

which both electrons have the same m_l must have $m_{s1} \neq m_{s2}$. Since the two electrons are equivalent, two assignments that differ only by interchange of the subscripts "1" and "2" correspond to the same state, which we list only once.[21] We thus obtain the predicted 15 assignments.

Next we decide which values of L and S go with each of these assignments. Since for any state we must have $M_L \leq L$, the existence of states with $M_L = \pm 2$ implies that there must be a term with $L = 2$, that is, a D term. This term must also include states with the other possible orientations of $\mathbf{L}$ relative to the z axis—that is, it includes one each of the states with $M_L = 1, 0, -1$ in Table 5.3. We do not need to specify which states these are, and within the scope of this text one cannot make such a specification. Finally, since the states with $M_L = \pm 2$ have only $M_S = 0$, we conclude that S must be zero for the D term. To sum up, five of the states

---

[21] Thus $m_{l1} = 1$, $m_{l2} = 1$, $m_{s1} = \frac{1}{2}$, $m_{s2} = -\frac{1}{2}$ is the same state as $m_{l1} = 1$, $m_{l2} = 1$, $m_{s1} = -\frac{1}{2}$, $m_{s2} = \frac{1}{2}$. But $m_{l1} = 1$, $m_{l2} = 0$, $m_{s1} = \frac{1}{2}$, $m_{s2} = -\frac{1}{2}$ and $m_{l1} = 1$, $m_{l2} = 0$, $m_{s1} = -\frac{1}{2}$, $m_{s2} = \frac{1}{2}$ are different states, since only the m_s's are interchanged. In Chapter 6 we shall designate the spin states of individual electrons by the *Pauli spin functions*: If electron 1 has $m_s = +\frac{1}{2}$, we write $\alpha(1)$ as its spin function; if electron j has $m_s = -\frac{1}{2}$, we write its spin function as $\beta(j)$.

in Table 5.3 belong to a term with $L = 2$, $S = 0$. We designate any atomic term by a capital letter corresponding to the value of L, with a left superscript equal to $2S + 1$ (called the *multiplicity*); in the present case the term is ^1D, read "singlet D."

That is one term accounted for. There are no other states with $M_L = 2$, and thus no other terms with $L = 2$. Three states each remain unassigned with $M_L = 1, -1$; these must belong to a term (or terms) with $L = 1$, a P term. Each of these groups of three (triplets) is made up of states with $M_S = 1, 0, -1$, which correspond to the possible orientations of **S** for $S = 1$. The term containing these states thus has $L = 1$, $S = 1$, with the designation[22] ^3P ("triplet P"). To complete the possible orientations of **L**, the ^3P term must also include a set of states with $M_L = 0$ and $M_S = 1, 0, -1$; these account for three of the as-yet-unassigned states with $M_L = 0$. The ^3P term thus comprises a total of nine states, and only one of the original 15 states remains unassigned, one of the states with $M_L = 0$, $M_S = 0$. This state can only be the single component of a term with $L = 0$, $S = 0$, a ^1S ("singlet S") term.

We have thus shown that the ground-state configuration of the carbon atom contains three terms: ^1D, ^3P, and ^1S. (Note that three of the geometrically possible L, S combinations in Fig. 5.8 do not exist; they are ruled out by the exclusion principle.) Among them these terms account for all 15 possible assignments of quantum numbers: nine states in the ^3P term, five in the ^1D, one in the ^1S. For seven of these 15 states, however, we cannot tell which assignment goes with which term (see Table 5.3); in these states we cannot assign a complete set of quantum numbers to the individual electrons. For example, there are two states with $M_L = 1$, $M_S = 0$, one in the ^1D term and one in the ^3P term; but there exists no way of specifying in which of these the electron with $m_l = 1$ has $m_s = \frac{1}{2}$. In cases such as this, the orientations of $\mathbf{L}_i$ and $\mathbf{S}_i$ for each electron are not fixed, and the z components of $\mathbf{L}_i$ and $\mathbf{S}_i$ ($m_l \hbar$ and $m_s \hbar$) are not constants of the motion. We say that m_l and m_s are no longer "good quantum numbers." One can interpret this by saying that the electrons interact in such a way as to exert torques on one another, changing the orientation of each electron's angular momentum. This picture explains why we speak of "coupled" angular momenta. What the LS-coupling model assumes is that the total angular momenta **L** and **S** *are* conserved, so that L, M_L, S, and M_S are good quantum numbers.

If the electrons do not have definite quantum numbers, how does one set up the wave function for an atom in a given term? In the orbital model the atomic wave function, Eq. 5.15, is a product (or a determinant) of the orbital wave functions φ_i, each of which involves the quantum numbers of a particular electron. What one must do is to obtain such a product, call it $\psi^{(i)}$, for each of the possible assignments in the term; the atomic wave function is then a linear combination of these products, $\sum_i c_i \psi^{(i)}$. The problem, which we shall not go into, is to obtain the c_i such that the resulting wave function gives the correct values of L and S, i.e., is an eigenfunction of the operators $\mathbf{L}^2$ and $\mathbf{S}^2$.

The three terms in the carbon atom's ground configuration are physically distinct, each with its own energy, electron distribution, and magnetic properties. The terms ^3P and ^1D, however, are *multiplets*, consisting of nine and five states, respectively. In the LS-coupling approximation the states within a given term are degenerate, all having the

---

[22] The multiplicity $2S + 1$ gives the number of possible quantized orientations of the vector **S**. Terms with $2S + 1 = 1, 2, 3, 4, \ldots$ are called *singlets, doublets, triplets, quartets*, etc.

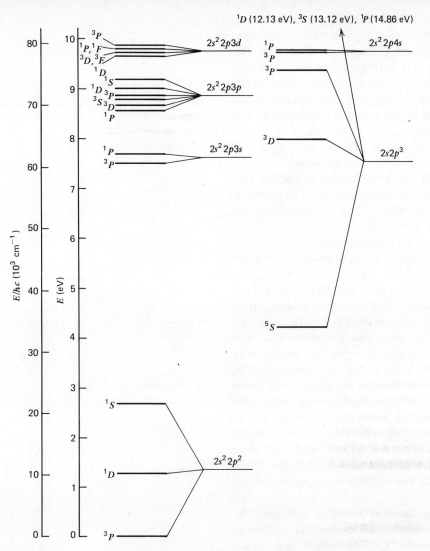

^1D (12.13 eV), ^3S (13.12 eV), ^1P (14.86 eV)

FIGURE 5.9
Low-lying energy levels of the carbon atom, with term designations. The terms belonging to each configuration have been bracketed together. The energy is measured above the ground state.

same energy. In reality there is a further fine-structure splitting, for reasons we shall discuss in the next section.

One can obtain from spectroscopic data the energy levels corresponding to the various terms. In the carbon atom the ^3P term is lowest in energy, the ^1D next, and the ^1S highest (cf. Fig. 5.9). For most ground-state configurations the order of term energies is conveniently given by *Hund's rules*:

1. Within a given configuration, the term energies increase as S decreases.

2. Among terms with the same S, the term energies increase as L decreases.

This ordering implies that the differences in electron–electron interactions associated with different values of S affect the energy more strongly than the corresponding differences associated with L; this is indeed true when the LS-coupling approximation applies. Since the energy is lowest when S has its maximum value, the ground states of open-shell atoms usually have as many electrons as possible with unpaired spins.

It was long believed that the physical basis for Hund's first rule is a simple effect: that the Pauli exclusion principle keeps two electrons of the same spin far enough away from each other to make their repulsive Coulomb interaction smaller than it is in a corresponding singlet, where the exclusion principle does not affect the spatial correlation. When careful studies were done,[23] it became clear that the interelectronic energy of Coulomb repulsion is *larger* in triplet states of simple atoms and molecules than it is in the corresponding singlets based on the same configurations. The reason the energies of the triplets are lower and Hund's first rule holds is that the larger electron–electron repulsion of the triplet is accompanied by an attractive electron–nuclear attraction that more than compensates for the electron–electron repulsion. The electrons in the triplets have larger average values of r_{12}^{-1}, by coming closer, on average, to their nuclei and thus to one another.

The method we have developed for the carbon atom can be applied to other atoms as well. Reviewing the single-term configurations, we see that the closed-shell atoms, with $L = 0, S = 0$, have only ^1S terms; the alkalis $(\cdots ns, L = 0, S = \frac{1}{2})$ only ^2S terms; and the boron–aluminum group $(\cdots np, L = 1, S = \frac{1}{2})$ only ^2P terms. If a subshell is more than half-filled, it is simpler to work with the vacancies ("holes") in the subshell as if they were electrons; one can show that the values of L and S are the same when calculated either way. For example, the ground state of the oxygen atom, $1s^2 2s^2 2p^4$, with two holes in the $2p$ subshell, can be treated as a case of $2p^2$. The oxygen atom thus has the same three terms as the carbon atom, as do the other $\cdots np^2$ and $\cdots np^4$ elements. Similarly, the halogens $(\cdots np^5)$ have only ^2P terms like the boron group. The only nontransition elements remaining are the nitrogen group $(\cdots np^3)$, for which the ground state must have $S = \frac{3}{2}$ by Hund's first rule; the rest of the term analysis is left as a problem. (Note that the filling sequence $ns^2, ns^2 np, \ldots, ns^2 np^6$ has the ground-state multiplicities 1, 2, 3, 4, 3, 2, 1.)

As for the transition elements, one can now see what is implied by most of the anomalies in those electron configurations marked by stars in Table 5.1. Consider the chromium atom $(Z = 24)$, which one would expect to be $\cdots 3d^4 4s^2$. By Hund's first rule, the ground state would have $S = 2$, with four parallel spins. But the actual ground-state configuration is $\cdots 3d^5 4s$, which by Hund's rule should have six parallel spins $(S = 3)$. Thus the effect of the increase in S on the energy must be greater than the very small difference between the $3d$ and $4s$ orbital energies. In the copper atom $(Z = 29)$, the shift from $\cdots 3d^9 4s^2$ to $\cdots 3d^{10} 4s$ does not affect S; however, the exchange interaction is particularly large for a filled (or half-filled) subshell, in which the electrons of each spin have a spherically symmetric charge density, and the resulting term splitting again outweighs the orbital energy difference. Nearly all the anomalous configurations contain (or at least tend toward) filled or half-filled d or f subshells, which are stabilized by these effects.

So far we have discussed only configurations with *equivalent* open-shell electrons, electrons in the same subshell. If the electrons are nonequivalent, there is a slight difference in the bookkeeping. Let us consider the $1s^2 2s^2 2p 3p$ excited state of carbon, in which the $2p$ and $3p$ electrons are nonequivalent. The table of quantum number assignments, Table 5.4, is now somewhat longer than in the equivalent-electron case.

---

[23] See, for example, E. R. Davidson, *J. Chem. Phys.* **42,** 4199 (1965); J. P. Colpa and R. E. Brown, *Mol. Phys.* **26,** 1453 (1973); E. A. Colbourn, *J. Phys. B* **8,** 1926 (1975). J. Katriel and R. Pauncz, *Adv-Quantum Chem.* **10,** 145 (1977)

TABLE 5.4

ADDITION OF z COMPONENTS OF ANGULAR MOMENTA FOR TWO NONEQUIVALENT p ELECTRONS (AS IN $1s^2 2s^2 2p 3p$, THE EXCITED STATE OF CARBON): $l_1 = l_2 = 1$

m_{l1}	m_{l2}	$M_L = \sum_i m_{li}$	m_{s1}	m_{s2}	$M_S = \sum_i m_{si}$
1	1	2	$\frac{1}{2}, -\frac{1}{2}$	$\frac{1}{2}, -\frac{1}{2}$	$1, 0, 0, -1$
1	0	1	$\frac{1}{2}, -\frac{1}{2}$	$\frac{1}{2}, -\frac{1}{2}$	$1, 0, 0, -1$
1	-1	0	$\frac{1}{2}, -\frac{1}{2}$	$\frac{1}{2}, -\frac{1}{2}$	$1, 0, 0, -1$
0	1	1	$\frac{1}{2}, -\frac{1}{2}$	$\frac{1}{2}, -\frac{1}{2}$	$1, 0, 0, -1$
0	0	0	$\frac{1}{2}, -\frac{1}{2}$	$\frac{1}{2}, -\frac{1}{2}$	$1, 0, 0, -1$
0	-1	-1	$\frac{1}{2}, -\frac{1}{2}$	$\frac{1}{2}, -\frac{1}{2}$	$1, 0, 0, -1$
-1	1	0	$\frac{1}{2}, -\frac{1}{2}$	$\frac{1}{2}, -\frac{1}{2}$	$1, 0, 0, -1$
-1	0	-1	$\frac{1}{2}, -\frac{1}{2}$	$\frac{1}{2}, -\frac{1}{2}$	$1, 0, 0, -1$
-1	-1	-2	$\frac{1}{2}, -\frac{1}{2}$	$\frac{1}{2}, -\frac{1}{2}$	$1, 0, 0, -1$

Number of Assignments

M_L \ M_S	1	0	-1
2	1	2	1
1	2	4	2
0	3	6	3
-1	2	4	2
-2	1	2	1

Note: The tabulation of the m_s's is condensed. For each combination of m_{l1} and m_{l2}, there are four possible assignments:

m_{s1}	$\frac{1}{2}$	$\frac{1}{2}$	$-\frac{1}{2}$	$-\frac{1}{2}$
m_{s2}	$\frac{1}{2}$	$-\frac{1}{2}$	$\frac{1}{2}$	$-\frac{1}{2}$
M_S	1	0	0	-1

Since the two p electrons have different values of n, the exclusion principle no longer limits the possible values of m_s, and all 36 of the possible combinations are distinct assignments. The smaller table shows how these assignments are distributed over the possible values of M_L and M_S. By reasoning like that we used before, it is not difficult to find the terms: These are ^1D (5 states), ^3D (15 states), ^1P (3 states), ^3P (9 states), ^1S (1 state), and ^3S (3 states), the number of states again adding up to 36. This time we have used all possible combinations of L and S (Fig. 5.8).

Other excited configurations can be similarly analyzed into their terms; Fig. 5.9 shows all the energy levels of the carbon atom up to 10 eV above the ground state, each with its configuration and term designation. In general, both the energy differences between configurations and those between terms of the same configuration tend to grow smaller with increasing excitation energy. This is apparent here in the sequence $2p^2$, $2p3s$, $2p3p$, $2p3d$, $2p4s$ (as usual, the $3d$ and $4s$ energies are practi-

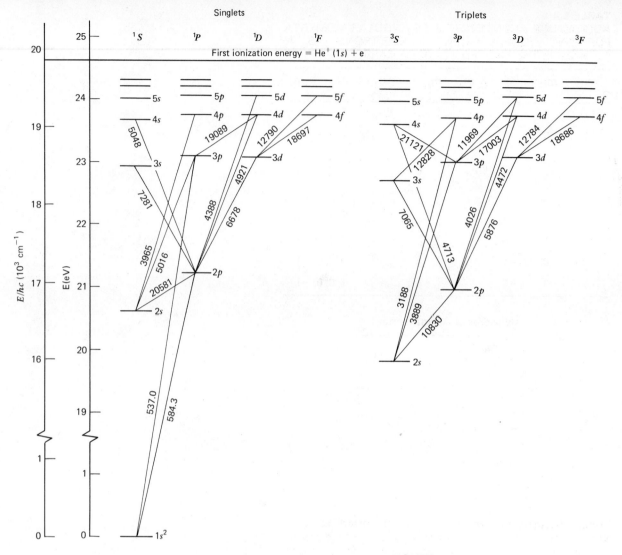

FIGURE 5.10
Energy levels of the helium atom, arranged by term designations. Except for the ground state, each level is identified by the configuration of only the excited electron. Some of the principal (most intense) spectral lines are shown, each with its wavelength in angstroms.

cally the same): The more a given electron is excited, the greater is its average distance from the nucleus, the less it interacts with the inner electrons, and thus the smaller is the term splitting. In highly excited states, however, the configurations tend to overlap in energy, indicating that the differences between orbital energies in the central field decrease even more rapidly than the term splitting due to deviations from the central field.

For most atoms term assignments have been made for hundreds or even thousands of states, where energy differences are obtained from spectroscopic data. Yet only a fraction of the conceivable transitions between states are observed, many transitions being "forbidden" for reasons outlined in Section 4.5. In the *LS*-coupling approximation the principal selection rules for allowed transitions can be shown to be

$$\Delta l = \pm 1 \quad \text{(for an electron)},$$
$$\Delta L = 0, \pm 1, \quad \Delta S = 0 \quad \text{(for the atom)};$$

(5.23)

these rules hold quite well for the lighter atoms. They are illustrated for the helium atom by Fig. 5.10, which groups the energy levels by their term designations and shows the principal spectral lines. The most striking feature is that transitions between singlet and triplet states are forbidden, in accordance with the rule $\Delta S = 0$. (A line corresponding to $1s^2\,{}^1S$—$1s2p\,{}^3P$ does appear, but very weakly.) Both $1s2s$ excited states are metastable: Since they can decay to the ground state only by forbidden transitions, they have relatively long lifetimes. Note also how all the energy levels for a given n approach one another as $n \to \infty$; each series of terms converges to the same limit, the ground state of the He^+ ion.

5.7
Fine Structure and Spin–Orbit Interaction

We still have not come to the end of our analysis of atomic energy levels. Not only do configurations split into terms, but the terms themselves often split into closely spaced levels. This is the fine structure of the energy spectrum, which we have already mentioned on several occasions. Some examples are illustrated in Fig. 5.11.

What is the cause of the fine-structure splitting? The term splitting is due to the nonspherical part of the electrostatic interactions between electrons. Reviewing the Hamiltonian of Eq. 5.10, we see that we still have not considered the magnetic moment interactions. Each electron has both an orbital magnetic moment and a spin magnetic moment, which interact with each other and with the moments of all the other electrons. Since it is precisely in the orientation of $\mathbf{L}$ and $\mathbf{S}$ that the states within a given term differ from one another, the electrons in each of these states see a slightly different magnetic field. The states of a multiplet term are thus not really degenerate after all.

The interaction between the spin and orbital magnetic moments is ordinarily far greater than the spin–spin and orbital–orbital interactions. We can thus neglect the latter two and speak only of the *spin–orbit interaction*. The logical extension of the *LS*-coupling model is to assume that this interaction depends only on the total L and S for an atom. To a fairly high level of approximation, the interaction energy is given by

$$E_{so} = \zeta \mathbf{L} \cdot \mathbf{S} = \langle \zeta \mathsf{L} \cdot \mathsf{S} \rangle. \tag{5.24}$$

where the multiplier ζ is a function of the specific atomic state. The value of ζ is notably difficult to obtain from theory, and is often treated as an experimental parameter. Equation 5.24 then represents the fact that the interaction of two magnetic dipoles must involve their scalar product, in this case $\boldsymbol{\mu}_l \cdot \boldsymbol{\mu}_s$, which is proportional to $\mathbf{L} \cdot \mathbf{S}$. Leaving ζ as a parameter is often useful when one wishes to make spectral assignments or predict magnetic properties.

The spin–orbit energy can be put in more convenient form in terms of the total angular momentum $\mathbf{J}$, which is the vector sum of $\mathbf{L}$ and $\mathbf{S}$:

$$\mathbf{J} \equiv \mathbf{L} + \mathbf{S}. \tag{5.25}$$

Since

$$\mathbf{J}^2 \equiv \mathbf{J} \cdot \mathbf{J} = \mathbf{L}^2 + \mathbf{S}^2 + 2\mathbf{L} \cdot \mathbf{S}, \tag{5.26}$$

substitution in Eq. 5.24 gives

$$E_{so} = \frac{\zeta}{2} (\mathbf{J}^2 - \mathbf{L}^2 - \mathbf{S}^2). \tag{5.27}$$

All three of these squares of angular momenta are capable of being measured simultaneously. Hence it is no surprise that $\mathbf{J}^2$ for an atom is a constant of the motion, with eigenvalues given by

$$\mathbf{J}^2 = J(J+1)\hbar^2; \tag{5.28}$$

J is sometimes called the *inner quantum number*. Like S, J is integral for an even number of electrons, half-integral for an odd number. The z component of $\mathbf{J}$ is

$$J_z = M_J \hbar \qquad (M_J = J, J-1, \ldots, -J), \qquad (5.29)$$

so that for each J there are $2J+1$ possible space-quantized orientations of J, giving Zeeman-effect splitting in an external field.

Just as in a given configuration the $\mathbf{L}_i$ and $\mathbf{S}_i$ can add to give various values of L and S (cf. Fig. 5.8), so in a given term $\mathbf{L}$ and $\mathbf{S}$ can add to give several values of J. The maximum value of the z component for given L, S is clearly

$$(J_z)_{\max} = (L_z + S_z)_{\max} = (M_L + M_S)_{\max}\hbar = (L+S)\hbar, \qquad (5.30)$$

or $(M_J)_{\max} = L+S$; since M_J runs up to J, the maximum value of J is also $L+S$. One can similarly show that the minimum possible value of J is $|L-S|$, so that J can have the values $L+S, L+S-1, \ldots, |L-S|$. For example, in a ^3D term ($L=2, S=1$) we have $J=3, 2, 1$. The term thus splits into states of slightly different energy, designated by a right subscript giving the value of J: $^3D_3, ^3D_2, ^3D_1$. Note that no such splitting occurs if either L or S is zero, that is, in all S or singlet terms; this is why the inert gases (^1S), alkalis (^2S), and alkaline earths (^1S) have no fine structure in their ground states. In the LS-coupling approximation the selection rules (Eq. 5.23) are supplemented by

$$\Delta J = 0, \pm 1 \qquad (5.31)$$

(with $J=0 \to J=0$ forbidden); the various values of J in a given pair of initial and final terms give rise to what is called a spectral multiplet, several closely spaced lines.

For light atoms the spin–orbit interactions lead to relatively small differences between the energies of states with the same L, S but different J. In the first row of the periodic table (Li to Ne), these energy differences are at most comparable to mean thermal energies at room temperature (ca. 0.04 eV/molecule). However, the interaction energy increases rapidly with nuclear charge, roughly as Z^4/n^3. The halogen atoms offer a clear illustration of this effect. The ground configuration of a halogen atom is $\cdots np^5$, giving only a ^2P term ($L=1, S=\frac{1}{2}$). This term may have either of two values of J, $\frac{1}{2}$ or $\frac{3}{2}$. The energy separations, $E(^2P_{1/2}) - E(^2P_{3/2})$, for fluorine, chlorine, bromine, and iodine are 0.05, 0.11, 0.46, and 0.94 eV, respectively. This means that both states are effectively part of the ground state in the fluorine atom at ordinary temperatures, but that the higher-energy $^2P_{1/2}$ states must really be considered as excited states for the other halogens. The upper state of iodine was used to store the energy in the first chemical laser, with the radiation due to the transition $^2P_{1/2} \to$ $^2P_{3/2}$. Some other typical spin–orbit splittings are shown in Fig. 5.11;[24] note that the effect occurs even in the hydrogen atom, breaking up the degeneracy of states with the same n.

The LS-coupling approximation assumes that all states of the same term have virtually the same energy, regardless of J, with the spin–orbit

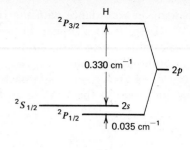

(a)

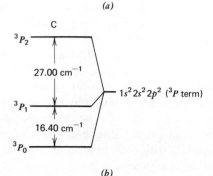

(b)

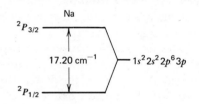

(c)

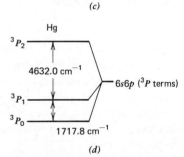

(d)

FIGURE 5.11

Some fine-structure splittings (not drawn to the same scale). Energy differences are expressed as $\Delta E/hc$ in cm^{-1}: 1 eV/hc = 8066 cm^{-1}. (a) The $n=$ 2 level of hydrogen. (b) The ground-state term of carbon (cf. Fig. 5.9). (c) The first excited state of sodium: Sodium-vapor lamps emit a doublet line at 5890 and 5896 Å, corresponding to the transitions $2p\ ^2P_{3/2} \to 2s\ ^2S_{1/2}$ and $2p\ ^2P_{1/2} \to 2s\ ^2S_{1/2}$. (d) The first excited state of mercury, showing the effect of large spin–orbit interaction in heavy atoms.

---

[24] With sufficient resolution one can often observe even finer splittings, known as *hyperfine structure*; the difference between hyperfine levels is usually of the order of a few tenths of a cm^{-1} ($\Delta E/hc$). There are two principal causes: (1) each isotope of an element has a different nuclear mass. Thus the reduced masses of the electrons differ slightly among the isotopes, as do their energy levels; cf. Eq. 4.16 and footnote 5 of Chap. 4 for hydrogenlike atoms. (2) The spin magnetic moment of the nucleus interacts with the electron magnetic moments, with an effect similar to that of spin–orbit splitting. In the hydrogen atom both spins are $\frac{1}{2}$, giving rise to two states; the transition between them produces the famous 21 cm line used by radio astronomers to detect hydrogen.

interaction only a minor perturbation. This clearly breaks down in the heavier atoms, where an approximation known as *jj coupling* is more useful. In this scheme the $\mathbf{L}_i$ and $\mathbf{S}_i$ of each electron are first coupled to form a $\mathbf{J}_i$ (with quantum numbers j_i, m_{ji}), and the $\mathbf{J}_i$ are then coupled to form $\mathbf{J}$ for the atom. Each set of j_i then gives a "*jj* term," a group of closely spaced levels with the Coulomb-exchange interaction producing the fine structure. The selection rules for L and S break down and are replaced by $\Delta j = 0, \pm 1$ (for an electron). In the previous section we pointed out how the electronic interactions spoil the quantum numbers m_l and m_s, since each electron's angular momenta no longer have constant orientations. What the spin–orbit interaction does is to make L and S no longer good atomic quantum numbers. Once $\mathbf{L}$ and $\mathbf{S}$ interact with each other, neither is separately conserved.

Strictly speaking, things are even worse than this. Once the electrons interact at all, the energy and total angular momentum of each electron cannot be conserved, which means that n and l are not really good quantum numbers either. The assignment of atoms to specific configurations (not to mention terms) is thus only an approximation, although for most purposes a very good one. In the most accurate calculations including electron correlation one must take the actual atomic wave function to be a linear combination ("mixture") of the wave functions corresponding to various configurations; this is an extension of the method we described earlier for the wave functions of individual terms. In such a rigorous approach one cannot meaningfully assign any quantum numbers or energies to individual electrons.

Why do we emphasize these limitations on our assumptions when one can get quite good results with the electron configuration model? Unfortunately, there are important quantities for which these results are not good enough, in particular those quantities that depend on small differences between very large numbers. The total electronic binding energy in a large atom may be some thousands of electron volts, but the energy *change* when two such atoms join to form a molecule—the bond energy—is typically of the order of 5 eV. The best Hartree–Fock calculations for atoms are no better than 0.1–0.2% in the energy. In molecules one cannot do even that well, since there is not spherical symmetry and interatomic forces spoil the configuration model still more. It is thus clear that one must use different techniques to obtain bond energies and other quantities of chemical interest for molecules. In the next chapter we shall begin to see how this is done.

Appendix 5A

The Stern-Gerlach Experiment

Among the experiments that illustrate the nature of quantization, perhaps the most vivid is the one first performed by O. Stern and W. Gerlach in 1921. This experiment clearly demonstrated that angular momenta (and the associated magnetic moments) are quantized, and that space quantization makes possible the physical separation of atoms or molecules in different quantum states. It also furnished the evidence from which the existence of electron spin was later deduced. The experiment could be done with atoms having either spin or orbital angular momentum; in fact, Stern and Gerlach worked with silver atoms, which have a spin of $\frac{1}{2}$ (the same as the electron) and zero orbital angular momentum.

In essence, the Stern–Gerlach experiment consists of passing a beam of atoms down a collimating axis and into an *inhomogeneous* magnetic field, whose field strength is greater on one side of the beam than on the other. Such a system is shown in Fig. 5A.1. A uniform field would merely tend to align the atomic magnets; the inhomogeneous field also exerts a net translational force on the atoms. The field as shown in the figure is stronger toward the north pole of the magnet. Any particle entering the region of the field with its own north pole toward the north pole of the field is thus repelled more by the north–north repulsion than by the south-south repulsion, and is driven toward the south pole of the field. Similarly, any particle entering with its own south pole toward the field's north pole is driven toward the north pole of the field. The extent of this deflection depends on the angle between the magnetic moment and the field, and on the field gradient.

If the atomic magnets were classical particles, their moments could initially be directed at any angle to the field. One would then observe a continuous and symmetrical distribution of deflected atoms leaving the region of the inhomogeneous field; some would be deflected toward the field's north pole, some toward the south pole, and some would be undeflected. But what is actually observed is one group of atoms all deflected the same amount toward the north pole, and an equal number deflected the same amount toward the south pole. The distribution is discontinuous. The reason, as we now know, is that the spin moments are

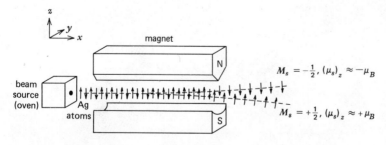

$$M_s = -\tfrac{1}{2}, \ (\mu_s)_z \approx -\mu_B$$

$$M_s = +\tfrac{1}{2}, \ (\mu_s)_z \approx +\mu_B$$

FIGURE 5A.1
Schematic diagram of the Stern–Gerlach apparatus. The magnetic field is stronger near the pointed north pole than near the smoothly curved south pole. The "atomic magnets" emerging from the oven are driven downward or upward according to whether the z components of the magnetic moments point up or down (as indicated by arrows). The two emerging beams are in distinct quantum states.

not randomly oriented. If the applied field vector defines the z axis, then for $S = \frac{1}{2}$ the z component of $\boldsymbol{\mu}_s$ can only have the values $+\mu_B$ and $-\mu_B$. Atoms in each state are deflected by a characteristic fixed amount, and two distinct beams leave the region of the field. Each of these beams contains only one of the spin states.

Note that the experiment as described assumes no special preparation of the entering beam. One can prepare the beam in ways that affect the relative intensities of the two emerging beams, but one cannot increase the number of emerging beams[25] or produce a continuous distribution of emerging particles. One way of preparing the beam, however, is to select for the entering beam *one* of the beams emerging from an identical apparatus having the same spatial orientation. This choice amounts to selecting only those atoms in a particular quantum state, all with the same value of M_S, and passing only those into the second magnet. Since these atoms all have magnetic moments with the same orientation to the field (which again is directed along the z axis), only a single beam will emerge from the second apparatus. Figure 5A.2a illustrates this phenomenon.

The ability to separate particles in different quantum states has made the Stern–Gerlach experiment particularly important in providing a conceptual basis for quantum mechanics. With some kind of trap to remove the other beam (or beams), the apparatus is like a filter that passes only those particles in one precisely known quantum state. Once selected, the particles remain in that state so long as they are undisturbed. A second filter that passes the same quantum state leaves the beam unaffected. On the other hand, suppose that the second filter is set to pass a different quantum state—for example, that the first filter passes only the state with $M_S = +\frac{1}{2}$ and the second only the state with $M_S = -\frac{1}{2}$ (both defined with respect to the same z axis). If the filters are so adjusted, then *none* of the beam atoms will get through the second filter. This is shown schematically in Fig. 5A.2b. The two possible quantum states, $M_S = +\frac{1}{2}$ and $M_S = -\frac{1}{2}$, are mutually exclusive; an atom in one state cannot be in the other. All this illustrates the distinct identity of quantum mechanical eigenstates, in this case eigenstates of the z component of angular momentum.

The Stern–Gerlach experiment yields a second kind of insight into the quantum nature of matter when one uses a second selector in another way. In our discussion so far, we have spoken about testing for the quantum state transmitted by the first filter, or for another of the quantum states distinguished by the apparatus. As we have described the experiment, these are states with different quantized values of the z component of spin angular momentum, $S_z = M_S \hbar$. The act of measuring S_z by applying an inhomogeneous magnetic field in the z direction forces the atoms to be in one or another of the eigenstates of S_z. But suppose that one uses a second apparatus to test for quantization relative to the x or y axis, perpendicular to the z axis defined by the first apparatus. Recall from our discussion of the quantum properties of angular momentum that one *cannot* simultaneously know the components of angular momentum along two axes. To do so would violate the uncertainty principle. The knowledge that a particular beam emerging from the first apparatus has $M_S = +\frac{1}{2}$ (and thus $S_z = \frac{1}{2}\hbar$) with respect to the z axis is essentially complete knowledge. One might know less, but one

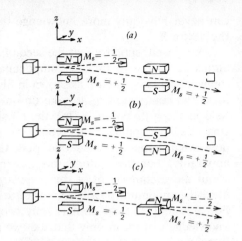

FIGURE 5A.2

Effects of successive Stern–Gerlach experiments. (*a*) Both apparatuses set to pass only atoms with $M_S = +\frac{1}{2}$: the first "filter" removes half of the original beam, the second "filter" passes all of the residual beam. (*b*) The first apparatus passes only $M_S = +\frac{1}{2}$, the second (with magnetic field reversed) passes only $M_S = -\frac{1}{2}$; none of the beam passes through the second "filter." (*c*) First apparatus selects for z component of spin, passing the beam with $S_z = +\frac{1}{2}\hbar$ (and S_y indeterminate); second apparatus selects for y component of spin, splitting that beam into two beams with $S_y = \pm\frac{1}{2}\hbar$ (and S_z indeterminate).

---

[25] For a given kind of atoms, that is. In general there will be one beam for each possible value of $M_J = M_L + M_S$; for the silver atom in its ground state we have $L = 0$ (S state), $M_L = 0$, $M_J = M_S = \pm\frac{1}{2}$.

can never have any more knowledge than this, about the orientation of the vector **S**.

Very well, suppose that the second apparatus is arranged to separate atoms according to their y components of angular momentum (in the same coordinate system), as shown in Fig. 5A.2c. This measurement must force the component S_y to take on one of its eigenvalues; relative to a field in the y direction, S_y has the possible values $M'_S\hbar$, where $M'_S = \pm\frac{1}{2}$, since Eq. 5.2 applies to the component of S in *any one* direction. But now what happens when one puts the beam with $S_z = \frac{1}{2}\hbar$ into this apparatus? For these atoms one has the maximum possible knowledge about the z component of spin angular momentum; one must therefore have the minimum possible knowledge about the x and y components. For a system that can assume only two states, a condition of minimum knowledge can mean only that the two states are equally probable. This in turn means that the beam with $S_z = \hbar/2$ ($M_S = \frac{1}{2}$) must split into two beams of equal intensity, corresponding to $S_y = \hbar/2$ ($M'_S = \frac{1}{2}$) and $S_y = -\hbar/2$ ($M'_S = -\frac{1}{2}$). But what is the value of S_z for these new beams? We cannot know this. The beam entering the second apparatus has a definite value of S_z, and thus is indeterminate in S_y; the outgoing beams have definite values of S_y and are indeterminate in S_z. We have traded information about one kind of quantization for information about another kind of quantization, and the two kinds are mutually exclusive.

The uncertainty principle is particularly well illustrated by the experiment just described. A "complete" description of the system, within the constraints of this principle, consists of the specification of eigenvalues for as many variables as nature allows to be measured simultaneously. The classical notion, that one can simultaneously determine each and every property with unlimited accuracy, is simply invalid. Similar results should be obtained in any other experiment, real or conceptual, in which particles in different quantum states are physically separated.

Further Reading

H. A. Bethe and E. E. Salpeter, *Quantum Mechanics of One- and Two-Electron Atoms* (Springer-Verlag, Berlin and Academic Press, New York, 1957).

E. U. Condon and G. H. Shortley, *The Theory of Atomic Spectra* (Cambridge University Press, Cambridge, England, 1953).

A. S. Davydov, *Quantum Mechanics* (Pergamon Press, Oxford, England and Addison-Wesley Publishing Co., Reading, Mass.), Chapter X.

D. R. Hartree, *The Calculation of Atomic Structures* (John Wiley and Sons, Inc., New York, 1957).

M. Karplus and R. N. Porter, *Atoms and Molecules* (W. A. Benjamin, Inc., Menlo Park, Calif., 1970), Chapter 4.

W. Kauzmann, *Quantum Chemistry* (Academic Press, Inc., New York, 1957), Chapters 9 and 10.

V. Kondratyev, *The Structure of Atoms and Molecules* (P. Noordhoof N. V., Groningen, The Netherlands, 1964), Chapters 5–8.

N. H. March, *Self-Consistent Fields in Atoms* (Pergamon Press, Oxford and New York, 1975).

M. A. Morrison, T. L. Estle and N. F. Lane, *Quantum States of Atoms, Molecules and Solids* (Prentice-Hall, Inc., Englewood Cliffs, N. J., 1976), Chapters 5–10.

L. I. Schiff, *Quantum Mechanics*, 3d Ed., (McGraw-Hill Book Co., Inc., New York, 1968), Chapter 12.

Problems

1. A magnetic dipole with strength 1.2 Bohr magnetons is in a uniform magnetic field whose strength is 500 oersted. What is the difference in energy between the orientations of the dipole parallel (north-to-north) and antiparallel (north-to-south) to the applied magnetic field? Suppose the dipole has the low-energy orientation; the application of an oscillatory magnetic field at the appropriate frequency can induce absorption of energy and the eventual transition to the high-energy orientation. Sketch the system, indicating the dipole, the external field and the oscillatory field, at three specified times during a single cycle. Be careful to show the *directions* of all vectors. What is the resonant frequency of the dipole in the given field?

2. Using a classical expression analogous to Eq. 5.4, compute the magnetic moment of a 500-g sphere of metal on a 1-m string, whirling at a rate of 1 revolution per second, and carrying an excess negative charge of 10^{-12} $\mathscr{F}$. (One faraday, $\mathscr{F}$, is 1 mol of electrons, or 96,485 C.) What is the energy of interaction of this magnetic dipole with the earth's magnetic dipole if the two are parallel?

3. Compute the frequency of the electromagnetic radiation associated with a transition of an electron between the state with $m_s = +\frac{1}{2}$ and the state with $m_s = -\frac{1}{2}$, if the electron is in a field of 0.5 T (5000 gauss). Compute the corresponding frequency if the particle is a proton instead of an electron.

4. A Stern–Gerlach experiment is conducted with silver atoms passing through a magnetic field whose gradient is 3×10^2 T/m. The length of the region of the field gradient is 0.05 m. The particles travel 0.5 m to their target surface after leaving the field with initial speed 10^5 m/s. How far apart are the points where the two emerging beams strike the target? (Recall the analysis of Thomson's experiment to determine e/m.)

5. Sodium atoms are put through a Stern–Gerlach apparatus and separated into two beams of equal intensity as shown in Fig. 5A.2a. The beam corresponding to $M_S = +\frac{1}{2}$ is then put through a second pair of magnets whose orientation is rotated about the beam axis through an angle θ with respect to the first pair of magnets. Figures 5A.2b and 5A.2c correspond to special cases with $\theta = 180°$ and $\theta = 90°$, respectively. Calculate the relative intensities of the beams with $M'_S = +\frac{1}{2}$ and $M'_S = -\frac{1}{2}$ for arbitrary θ. Ignore the small nonzero angle at which the beam emerges from the first separating device.

6. Explain, in terms of the symmetry of the forces exerted on an electron, why l and m are good quantum numbers for each electron in the central field model, but not in real atoms.

7. The most probable value of each electron–nuclear distance in H^- is 1.178 bohr; the average value of the electron–nuclear distance is 2.707 bohr.

What qualitative inferences can one make about the shape of the electron probability density from a comparison of these two numbers?

8. Within a series of isoelectronic ions such as Li, Be^+, B^{2+}, C^{3+}, and so on, the relative contribution of the correlation energy to the total energy of the system diminishes with increasing nuclear charge. Explain.

9. Prove that $\frac{27}{16}$ is the value of the "effective nuclear charge" Z' of Eq. 5.13 that minimizes the expectation value E' of energy (Eq. 5.11) when $\psi^{(2)}(r_1, r_2)$ of Eq. 5.13 is used to evaluate E'.

10. Put the entries in the following sets of configurations in order of increasing energy. Then check your intuition by comparing the results with the experimental results from atomic spectroscopy, as given, for example, by C. E. Moore, *Atomic Energy Levels* (National Bureau of Standards Circular 467).
    (a) Li, $1s^2 2s$
       $1s 2s^2$
       $1s 2s 2p$
       $1s^2 2p$
    (b) C, $1s^2 2s^2 2p^2$
       $1s^2 2s 2p^3$
       $1s 2s^2 2p^3$
       $1s^2 2s^2 2p 3s$

11. Using the tables given by C. E. Moore in *Atomic Energy Levels* (National Bureau of Standards Circular 467), compute the energies of the following atomic configurations by taking the appropriately weighted averages of the atomic state energies in the tables. (The configuration energies can be given relative to the ground state of the atom.)
    (a) C, $1s^2 2s^2 2p^2$;
    (b) N, $1s^2 2s^2 2p^3$;
    (c) N, $1s^2 2s^2 2p^2 3s$.

12. Explain why the second ionization potential of lithium, for the process $Li^+ \to Li^{2+} + e^-$, is less than that predicted for a hydrogenlike atom with $Z = 3$.

13. Consider the effect of the nuclear charge on the importance of relativistic effects in atoms, first by estimating from a Bohr model the velocity of a $1s$ electron in an atom of lead. The effective mass of a particle with rest mass m and speed v is $m' \simeq m/\sqrt{1-(v^2/c^2)}$. What is v/c for a $1s$ electron of Pb? What is its effective mass, based on the simple expression above for m'? What velocity v' does one obtain if one carries out the Bohr calculation with the above expression for m', instead of a velocity-independent m?

14. The outer electron of an alkali atom may be treated in an approximate way, as if it were in a hydrogenic orbital. Suppose that one takes the quantum number for the outer electron to be 2, 3, 4, 5, and 6, respectively, for Li, Na, K, Rb, and Cs. What values must Z be given to account for the observed first ionization potentials of these atoms? Explain why they differ from unity.

15. Use the Pauli exclusion principle and Hund's rules to find the number of unpaired electrons and the term of lowest energy for the following atoms:
    (a) P
    (b) S
    (c) Ca
    (d) Br
    (e) Fe

16. Hydrogen atoms are placed in a strong magnetic field and excited; their emission spectrum is recorded under conditions of moderately high resolution (0.1 cm^{-1} or better). What is the appearance of the portion of the spectrum due to $H(3p) \to H(2s)$ transitions, and how does it differ from the spectrum of the same transition in the absence of a magnetic field?

17. Show that the lowest configuration of the nitrogen atom, $1s^2 2s^2 2p^3$, gives rise to the terms ^2D, ^2P, and ^4S.

18. Derive the terms of the configuration $1s^2 2s^2 2p^6 3s^2 3p\, 3d$, of the silicon atom.

19. Explain this observation: The energy difference between $1s^2 2s$, $^2S_{1/2}$ and $1s^2 2p$, $^2P_{1/2}$ states of Li is 14904 cm^{-1}, whereas that between the $2s$, $^2S_{1/2}$ and $2p$, $^2P_{1/2}$ states of Li^{2+} is only 2.4 cm^{-1}.

20. The first excited state (and higher excited states as well) of He exhibits a positive electron affinity, although the electron affinity of He in its ground state is negative.
    (a) What are the configuration and term designations of the first two excited states of the helium atom?
    (b) What is the configuration and term designation of the lowest-energy bound state of the negative ion He$^-$? Note that this cannot be the same as any state built by adding an electron to He in its ground state.
    (c) Interpret why atoms in excited configurations generally have positive electron affinities, whereas atoms in closed-shell ground-state configurations seem to have only negative electron affinities.

21. What are the lowest terms for the ions Ti^{2+}, Mn^{2+}, and Fe^{3+}, if the first two electrons to be lost by the corresponding neutrals are the two $4s$ electrons? What are the possible values of the quantum number J for each of these ions?

22. What J values are possible for the ^3P, ^1D, and ^1S states of the silicon atom? Look up the observed splittings of these states in C. E. Moore, *Atomic Energy Levels* (National Bureau of Standards Circular 467) and estimate the spin–orbit splitting parameter ζ for as many of these states as the data permit.

23. Given the following data:

	Li	Na	K	Rb	Cs	Fr
Z	3	11	19	37	55	87
Atomic weight	6.94	23.00	39.10	85.47	132.91	?
Density (g/cm^3)	0.53	0.97	0.86	1.53	1.90	?
Atomic volume (cm^3)	13.0	23.7	45.5	55.9	70.0	?
Melting point (°C)	186	97.5	62.3	38.5	26	?
Boiling point (°C)	1200	880	760	700	670	?

Predict the numerical values for the atomic weight, density, atomic volume, melting and boiling points of francium, element number 87. What will be its characteristic chemical properties?

24. Show from Eqs. 5.26 and 5.27 that the energy associated with spin–orbit interaction is

$$E_{so} = \tfrac{1}{2}\zeta\hbar^2[J(J+1)-L(L+1)-S(S+1)].$$

25. (Difficult) Substitute $\psi^{(1)}$ of Eq. 5.12 and $\psi^{(2)}$ of Eq. 5.13 into the expression for the expectation value of the energy, Eq. 5.11, and show that the corresponding values given in the text, -74.83 eV and -77.49 eV, respectively, are correct.

6

The Chemical Bond in the Simplest Molecules: H_2^+ and H_2

In this chapter we begin to address the problem of the chemical bond. What makes atoms stick together to form a molecule? Why are some bonds strong and others weak? Can we predict the properties of a molecule—the energy levels, the spectra resulting from transitions among them, the spatial distribution of nuclei and electrons? And how are these properties related to those of the separated atoms? These are typical of the questions we must try to answer.

The simplest molecules, of course, are those with only two atoms, in which we can examine a single chemical bond in isolation. We begin with a general discussion of the forces between two atoms and how they are responsible for bonding. It will become clear that the key to bonding is how the electron distribution changes when the atoms are brought together. In the rest of the chapter we deal with ways to obtain this electron distribution, in terms of various approximations to the wave function.

The present chapter is concerned only with the two simplest molecules. Despite their simplicity, they are sufficient to illustrate most of the main principles of bonding. The H_2^+ ion, with only one electron, is the simplest possible molecule; we use it to introduce the most widely employed approximation of bonding theory, the concept of the molecular orbital. From H_2^+ we go on to H_2, in which the presence of two electrons gives us the first complete picture of a "normal" bond. We consider these molecules in both ground and excited states, and in terms of several approximate descriptions. In Chapter 7 we shall extend our theory to diatomic molecules in general.

6.1
Bonding Forces between Atoms

What forces hold a molecule together as an entity? Within any molecule there are attractive forces between electrons and nuclei, and repulsive forces between pairs of nuclei and between pairs of electrons. We must examine how these forces reach a delicate balance that enables molecules to have stable and well-defined structures, structures that neither fall apart at the slightest touch nor collapse into a united atom. We must ask how the attractive forces can just balance the repulsive forces for some particular molecular geometry and make that geometry a stable one. If the molecule is stretched, squeezed, or twisted from its most stable shape, then the forces within the molecule will tend to restore it to

206

that shape. Each molecular structure is thus in true stable equilibrium, corresponding to the bottom of a reasonably deep potential energy well.[1]

But in speaking of molecular structures we are already making an important assumption. The term "structure" implies a rather rigid arrangement of the atomic nuclei, close to the equilibrium positions for which the molecule's energy is minimized. The molecule cannot be totally rigid; it would violate the uncertainty principle for any particle to be at rest at a fixed point. However, the assumption of near-rigid structure is very good indeed. Since nuclei have thousands of times the mass of electrons, Δp for a nucleus can be a quite small fraction of its total momentum and still large enough to give a negligible Δq. To put it another way, the light electrons move so much faster than the heavy nuclei that they have a smooth average distribution over a time during which the nuclei hardly move at all. We can thus obtain a good description of the molecule's electronic structure even if we assume the nuclei to be at rest.

This assumption is part of what is known as the *Born–Oppenheimer approximation* (after Max Born and J. Robert Oppenheimer). What is this approximation? Let us approach it by stages. Consider a diatomic molecule composed of nuclei A and B, with charges of $+Z_A e$ and $+Z_B e$, respectively, and some electrons. We define the distance between the two nuclei as R (if there were more atoms we would have to specify R_{AB}), and the distances of the ith electron from the two nuclei as r_{Ai} and r_{Bi}. These coordinates are illustrated in Fig. 6.1. The Hamiltonian operator (neglecting all magnetic interactions) is then

$$H = \left[\frac{\mathbf{p}_A^2}{2m_A} + \frac{\mathbf{p}_B^2}{2m_B}\right] + \left[\sum_i \frac{\mathbf{p}_i^2}{2m_e} + \frac{e^2}{4\pi\epsilon_0}\left(\frac{Z_A Z_B}{R} - \sum_i \frac{Z_A}{r_{Ai}} - \sum_i \frac{Z_B}{r_{Bi}} + \frac{1}{2}\sum_i \sum_{j\neq i}\frac{1}{r_{ij}}\right)\right]$$

$$= H_{nucl} + H_{elec}, \tag{6.1}$$

where H_{nucl} and H_{elec} are defined by the two sets of square brackets, r_{ij} is as usual the distance between electrons i and j, and the sums are taken over all the electrons; H_{nucl} is the kinetic energy operator for the nuclei, and H_{elec} corresponds to the energy with the nuclei fixed. The molecular wave function ψ is the solution of the equation $H\psi = E\psi$. Now we make the Born–Oppenheimer approximation, which is simply that the nuclear and electronic motions are separable (cf. Section 3.9). Thus, having collected the kinetic energy of the heavy—and presumably slowly moving—nuclei in the first set of brackets in Eq. 6.1, and all the terms governing the motion of the light, fast electrons in the second set of brackets, we can write

$$\psi(r, R) = \psi_{nucl}(R)\psi_{elec}(r, R). \tag{6.2}$$

Here r stands for all the r_i and r_{ij}, and $\psi_{elec}(r, R)$ satisfies the equation

$$H_{elec}\psi_{elec}(r, R) = E(R)\psi_{elec}(r, R). \tag{6.3}$$

That is, $\psi_{elec}(r, R)$ is defined as the eigenfunction of the operator H_{elec}, whose eigenvalue $E(R)$ gives the energy of the molecule for a given value of the parameter R. Although ψ_{nucl} must be a function of *all* the nuclear coordinates, only R affects the energy and appears in ψ_{elec}. In other

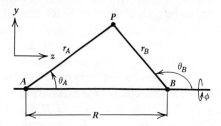

FIGURE 6.1
Coordinates in a diatomic molecule. The nuclei A and B (charges $+Z_A e$ and $+Z_B e$) are a distance R apart. Any point P can be described in terms of r_A and r_B, the distances from the two nuclei; θ_A and θ_B, the angles relative to the A–B axis; and an angle ϕ describing rotation about that axis.

---

[1] One could imagine a different kind of molecule, in which the energy of the molecule would be quite insensitive to the location of the constituent atoms. Such a molecule would be more like a liquid drop than the conventional ball-and-stick structure. Although this is not a good description of most molecules, many properties of atomic nuclei, especially heavy nuclei, are nicely described by a liquid drop model.

words, $|\psi_{elec}(r, R)|^2$ and $E(R)$ describe the electron distribution (as a function of r) and energy that the molecule would have if the nuclei were fixed a distance R apart. These are the quantities we want to know.

In this chapter we shall consider only the solution of Eq. 6.3, but let us briefly survey the remainder of the problem. To the extent that the Born–Oppenheimer approximation is valid, one can solve Eq. 6.3 to obtain $E(R)$ for any value of R. Combining these values gives a function $E(R)$ from $R = 0$ to $R = \infty$, which for the ground state of a stable diatomic molecule has a form like that in Fig. 6.2. Given such a function $E(R)$, one can obtain the part of the wave function describing nuclear motion by solving[2]

$$H_{nucl}\psi_{nucl}(R) = [E - E(R)]\psi_{nucl}(R); \qquad (6.4)$$

in this equation E is the total energy, $E(R)$ is the effective potential energy in which the nuclei move, and $E - E(R)$ is the kinetic energy of the nuclei. Since a system in a potential well always has a zero-point energy, the lowest eigenvalue of E must be some distance above the bottom of the well. We shall come back to this problem in Section 7.1.

When atoms A and B are very far apart, there is essentially no interaction between them; we define the energy of infinitely separated $A + B$ without kinetic energy to be our zero. At the other extreme, when the nuclei are close to each other, the repulsive energy $Z_A Z_B e^2/4\pi\epsilon_0 R$ becomes much larger than all the other terms in Eq. 6.1, and $E(R)$ appears to become infinite as R goes to zero.[3] This repulsion is what keeps molecules from collapsing into single atoms. We are interested in what happens in the region between these limits. A stable molecule can exist only if $E(R)$ has a minimum at some value of R, as in Fig. 6.2. Otherwise the nuclear repulsion would make the atoms fly apart without limit. The depth of this minimum, relative to the energy at infinite separation, is called the *dissociation energy* (D_e) of the molecule, and the value of R at which it occurs is the *equilibrium distance* (R_e). To see how such a minimum can occur, let us examine the balance of forces in a molecule.

---

[2] Substituting Eqs. 6.1 and 6.2 into $H\psi = E\psi$, we have

$$(H_{nucl} + H_{elec})\psi_{nucl}(R)\psi_{elec}(r, R) = E\psi_{nucl}(R)\psi_{elec}(r, R)$$

and

$$\psi_{elec}(r, R)H_{nucl}\psi_{nucl}(R) + \psi_{nucl}(R)H_{nucl}\psi_{elec}(r, R) + \psi_{nucl}(R)H_{elec}\psi_{elec}(r, R)$$
$$= E\psi_{nucl}(R)\psi_{elec}(r, R),$$

since H_{nucl} and H_{elec} act only on the nuclear and electronic coordinates, respectively. Substituting Eq. 6.3 in the third term and rearranging gives us

$$\psi_{elec}(r, R)H_{nucl}\psi_{nucl}(R) + \psi_{nucl}(R)H_{nucl}\psi_{elec}(r, R) = \psi_{elec}(r, R)[E - E(R)]\psi_{nucl}(R).$$

In the second term we find $H_{nucl}\psi_{elec}(r, R)$, which contains the usually small interactions between electronic and nuclear motions; the Born–Oppenheimer approximation says that this term can be neglected. If we do this, dividing through by $\psi_{elec}(r, R)$ yields Eq. 6.4.

[3] At *very* small distances—about 10^{-15} m—the strong nuclear forces become significant. Since these forces are attractive, $E(R)$ goes not to infinity, but to a very high peak bounding a potential well. If we were to push the two nuclei together hard enough, they would cross this barrier and combine into a single nucleus of charge $(Z_A + Z_B)e$; this would be a nuclear *fusion* process, such as occurs in stars and thermonuclear bombs. But since the barrier height is in the MeV range, there is little danger of producing fusion with ordinary chemical energies. As we mentioned in Section 4.1, the emission of α particles by radioactive nuclei corresponds to the reverse process, tunneling through the barrier from inside.

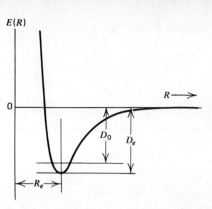

FIGURE 6.2
Internal energy $E(R)$ of a stable diatomic molecule as a function of the internuclear distance R. The zero of energy is the two separated atoms at rest in the limit $R \to \infty$. The dissociation energy can be defined as D_e, relative to the bottom of the well, at the equilibrium distance R_e, or as D_0, relative to the ground state; $D_e - D_0$ is the zero-point energy of the molecule. In a typical molecule D_e is of the order of 5 eV, and R_e is about 1–2 Å.

In our diatomic molecule with nuclei A and B a distance R apart, the repulsive force between the nuclei is, of course, of magnitude

$$F_{AB} = \frac{Z_A Z_B e^2}{4\pi\epsilon_0 R^2}, \tag{6.5}$$

acting directly along the A–B axis. Suppose now that there is a bit of electronic charge, say an amount $-q$, at the arbitrary point P (Fig. 6.1). This charge exerts an attractive force $\mathbf{F}_{AP}$ on nucleus A, in the direction AP, and an attractive force $\mathbf{F}_{BP}$ on nucleus B, in the direction BP. The magnitudes of these forces are

$$F_{AP} = \frac{Z_A eq}{4\pi\epsilon_0 r_A^2} \quad \text{and} \quad F_{BP} = \frac{Z_B eq}{4\pi\epsilon_0 r_B^2}. \tag{6.6}$$

We can define a *bonding* force as one that tends to draw the nuclei together, and an *antibonding* force as one that tends to push them apart. The repulsive force F_{AB} is clearly antibonding; what about the forces exerted by the charge at point P?

Only the force exerted along the A–B axis has any effect on the bonding. We thus break the forces $\mathbf{F}_{AP}$ and $\mathbf{F}_{BP}$ into their components parallel to the A–B axis (our z axis) and perpendicular to it—that is, into the interesting z components and the irrelevant y components. In terms of the coordinates shown in Fig. 6.1, the z components are

$$(\mathbf{F}_{AP})_z = \frac{Z_A eq}{4\pi\epsilon_0 r_A^2} \cos\theta_A \quad \text{and} \quad (\mathbf{F}_{BP})_z = \frac{Z_B eq}{4\pi\epsilon_0 r_B^2} \cos\theta_B. \tag{6.7}$$

The bonding force on nucleus A is the force that tends to push it toward B, that is, the component of force in the positive z direction. The bonding force on nucleus B, however, is the force in the *negative* z direction. The net classical bonding force on the nuclei due to the charge $-q$ at point P is therefore

$$
\begin{aligned}
F_{\text{bonding},P} &= (\mathbf{F}_{AP})_z - (\mathbf{F}_{BP})_z \\
&= \frac{eq}{4\pi\epsilon_0}\left(\frac{Z_A \cos\theta_A}{r_A^2} - \frac{Z_B \cos\theta_B}{r_B^2}\right).
\end{aligned}
\tag{6.8}
$$

The quantum mechanical bonding force operator $\mathbf{F}_{\text{bonding},P}$ looks exactly the same.

To obtain the total bonding force exerted by the electrons, we would have to sum Eq. 6.8 over the entire electron distribution, which we can obtain only by solving the Schrödinger equation. Assume that we have solved Eq. 6.3 to obtain the wave function $\psi_{\text{elec}}(r, R)$ for a given R. The expectation value of the total bonding force is then

$$\langle \mathsf{F}_{bonding}(R) \rangle = \langle \mathsf{F}_{bonding,P} \rangle - F_{AB}$$

$$= \int \psi^*_{elec} \mathsf{F}_{bonding,P} \psi_{elec}\, d\tau - \frac{Z_A Z_B}{4\pi\epsilon_0 R}, \qquad (6.9)$$

with the integral taken over all possible positions of all the electrons. (The symbol "$d\tau$" is shorthand for what we earlier called "$dq_1 \cdots dq_N$," and is meant to indicate three coordinates for each electron.)

One does not have to evaluate the integral in Eq. 6.9 explicitly to obtain the total bonding force. The operator corresponding to $\mathsf{F}_{bonding}(R)$ is $\partial\mathsf{V}(R, r)/\partial R$, where $\mathsf{V}(R, r)$ is the potential energy part of H_{elec}. We thus have

$$\langle \mathsf{F}_{bonding}(R) \rangle = \left\langle \frac{\partial \mathsf{V}(R, r)}{\partial R} \right\rangle = \int \psi^*_{elec} \frac{\partial \mathsf{V}(R, r)}{\partial R} \psi_{elec}\, d\tau$$

$$+ \int \frac{\partial \psi_{elec}}{\partial R} E(R) \psi^*_{elec}\, d\tau \qquad (6.10)$$

One can show that

$$\int \psi^*_{elec} \frac{\partial \mathsf{H}_{elec}}{\partial R} \psi_{elec}\, d\tau = \frac{\partial}{\partial R} \int \psi^*_{elec} \mathsf{H}_{elec} \psi_{elec}\, d\tau = \frac{dE(R)}{dR}, \qquad (6.11)$$

giving the final result

$$\langle \mathsf{F}_{bonding}(R) \rangle = \frac{dE(R)}{dR}, \qquad (6.12)$$

as follows. We have

$$\frac{\partial}{\partial R} \int \psi^*_{elec} \mathsf{H}_{elec} \psi_{elec}\, d\tau$$

$$= \int \frac{\partial \psi^*_{elec}}{\partial R} \mathsf{H}_{elec} \psi_{elec}\, d\tau + \int \psi^*_{elec} \mathsf{H}_{elec} \frac{\partial \psi_{elec}}{\partial R}\, d\tau + \int \psi^*_{elec} \frac{\partial \mathsf{H}_{elec}}{\partial R} \psi_{elec}\, d\tau.$$

Since H_{elec} is a real Hermitian operator (cf. Appendix 6B), we can write

$$\int \psi^*_{elec} \mathsf{H}_{elec} \frac{\partial \psi_{elec}}{\partial R}\, d\tau = \int \frac{\partial \psi_{elec}}{\partial R} \mathsf{H}_{elec} \psi^*_{elec}\, d\tau.$$

Substitution of Eq. 6.3 then removes the first two terms on the right-hand side above:

$$\int \frac{\partial \psi^*_{elec}}{\partial R} E(R) \psi_{elec}\, d\tau + \int \frac{\partial \psi_{elec}}{R} E(R) \psi^*_{elec}\, d\tau$$

$$= E(R) \frac{\partial}{\partial R} \int \psi^*_{elec} \psi_{elec}\, d\tau = 0,$$

since $\int \psi^*_{elec} \psi_{elec}\, d\tau$ has the constant value unity. The same substitution gives the last term of Eq. 6.11.

Equation 6.12 is known as the *Hellmann–Feynman theorem*. This confirms our earlier statement that $E(R)$ is the effective potential energy for nuclear motion.

Without solving the Schrödinger equation for $\langle \mathsf{F}_{bonding}(R) \rangle$, we can get a good deal of insight into the nature of the bonding forces simply by looking at the form of Eq. 6.8. Following a method introduced by T. Berlin and K. Fajans, let us see how $F_{bonding,P}$ varies with the location of the point P. First, whenever P lies between A and B ($z_A < z_P < z_B$), we have $\theta_A < \pi/2$ and $\theta_B > \pi/2$; thus $\cos\theta_A$ is positive, $\cos\theta_B$ is negative, and

$F_{bonding,P}$ must be positive. This is a very fundamental (if obvious) point, which we must recognize from the outset: Any electronic charge lying between two nuclei is necessarily bonding because it pulls the two nuclei toward each other. In order for charge to be antibonding, it must lie "beyond" the nuclei.

Now suppose we place our test point P to the right of nucleus B ($z_P > z_B$), so that $\cos \theta_A$ and $\cos \theta_B$ are both positive. A negative charge at P then pulls both nuclei to the right, pulling A toward B but B away from A. Which force wins? For simplicity, let us suppose that point P lies on the A–B axis ($\cos \theta_A = \cos \theta_B = 0$). Suppose that the charges Z_A and Z_B are equal. A charge to the right of B must be closer to B than to A ($r_B < r_A$), so the right-hand term in Eq. 6.8 is larger than the left-hand term, and the net force is antibonding. For equal nuclear charges, then, any electronic charge on the axis to the right of B or the left of A is antibonding. This is not true in every case, however. Suppose, for example, that $Z_A > Z_B$, with point P still on the A–B axis to the right of B. For P close to B, the small value of r_B will outweigh the charge difference ($Z_B/r_B^2 > Z_A/r_A^2$) and $F_{bonding,P}$ will be negative. As we move P to the right, the ratio r_A/r_B comes closer and closer to unity. Eventually a point is reached at which the larger charge on nucleus A makes the first (bonding) term in Eq. 6.8 exceed the second (antibonding) term, even though the second term has a smaller denominator; for all points on the axis beyond this the charge is bonding. Similar calculations can be made for points not on the A–B axis.

We thus recognize that regions of space in the vicinity of nuclei are inherently bonding or antibonding; whether or not a bond is formed depends on how the electrons are distributed among these regions. To find the positions and shapes of the bonding and antibonding regions, one need merely locate the surfaces that divide them. These boundary surfaces are obtained by solving Eq. 6.8 for those values of r_A, r_B, θ_A, θ_B at which $F_{bonding,P} = 0$. Although we shall not go through the algebra, it is not difficult to simplify this equation to

$$\alpha(1 + \rho^2 - 2\rho\alpha)^{3/2} = \frac{Z_B}{Z_A}(\rho - \alpha), \qquad (6.13)$$

where $\rho \equiv R/r_A$ and $\alpha \equiv \cos \theta_A$; Eq. 6.13 contains only one physical parameter (Z_B/Z_A) and two independent variables (ρ and α). Illustrations of the boundary surfaces for several values of Z_A/Z_B are given in Fig. 6.3. Note that the larger the ratio Z_A/Z_B, the smaller is the antibonding region near the low-Z nucleus B. In a molecule such as HCl, this region includes only that charge very close behind the H nucleus; because of the large ratio Z_{Cl}/Z_H, a charge almost anywhere except beyond the Cl nucleus is in the bonding region.

How much charge is needed to balance the repulsive force between the two nuclei? Suppose that a negative charge $-q$ is at the midpoint of the bond between A and B, so that $r_A = r_B = R/2$, $\cos \theta_A = 1$, $\cos \theta_B = -1$. How large must this bit of charge be for the binding force it exerts to equal the repulsive force F_{AB}? Combining Eqs. 6.5 and 6.8, we require that

$$\frac{Z_A e q}{(R/2)^2} + \frac{Z_B e q}{(R/2)^2} = \frac{Z_A Z_B e^2}{R^2}, \qquad (6.14)$$

or, with a bit of rearrangement,

$$q = \frac{(Z_A + Z_B)e}{4 Z_A Z_B}. \qquad (6.15)$$

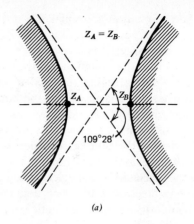

(a)

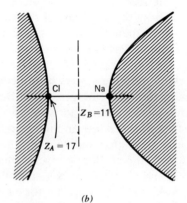

(b)

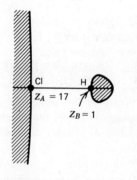

(c)

FIGURE 6.3
Bonding (unshaded) and antibonding (shaded) regions of space for diatomic molecules with various values of Z_A/Z_B. (a) Homonuclear molecule (for example, H_2, Cl_2), $Z_A = Z_B$. (b) NaCl molecule, $Z_A/Z_B = 17/11 = 1.545$. (c) HCl molecule, $Z_A/Z_B = 17$. From Hirschfelder, Curtiss, and Bird, *Molecular Theory of Gases and Liquids* (Wiley, New York, 1954), p. 936.

If the two charges are equal, $Z_A = Z_B = Z$, then q need be only $-Ze/8$ to balance the nuclear repulsion. If Z_A is larger than Z_B, then q must be somewhat larger, but even in the limit $Z_A/Z_B \to \infty$ it need not exceed $Z_Be/4$. From this rough model we see that the amount of negative charge in the bonding region required to stabilize a molecule is very small, at least when the charge lies on the bond axis. The bonding effectiveness of a given charge of course decreases as it moves away from the axis.

Thus we have a rationalization of how electrons draw nuclei together and counter the mutual repulsion of the nuclei of a diatomic molecule. But closer examination shows that we only have part of the picture. Thus far, the discussion has neglected the electronic kinetic energy, and surely, when an electron with a well-defined total energy passes into a region of low potential energy, its kinetic energy must increase. This implies that if the potential energy of the electron in the region between the nuclei is lowered by the approach of the two nuclei, then the electron must spend less time there. And if that happens, how can we expect charge to accumulate between the nuclei?

This question was examined in detail by Ruedenberg and co-workers, and the results have been discussed by Mulliken and Ermler.[4] The picture that emerges is this: Because the potential between the nuclei shows a maximum, and is relatively flat in that region, the force on an electron there is low and the computed local electronic kinetic energy T_e is low near the midpoint of the internuclear axis. Near the nuclei, however, the electronic kinetic energy is increased above that of the free atom, and this increase is greater than its decrease between the nuclei, so the average, $\langle T_e \rangle$, is greater for electrons in the diatomic molecule than in the separated atoms. The total energy of a bound system with only Coulombic forces is related to the average kinetic energy and to the average potential energy: $E = -\langle T \rangle = \frac{1}{2}\langle V \rangle$ (the virial theorem as applied to a Coulombic system), and if the diatomic molecule is at its equilibrium internuclear distance, $\langle T \rangle = \langle T_e \rangle$. Therefore, because $\langle T_e \rangle_{molecule} > \langle T_e \rangle_{free\ atoms}$, $E(R_e) < E(\infty)$. In short, the simple model based on the statics of electrostatic forces hides the delicate balance between potential and kinetic energy. Without the electrostatic forces of attraction between the nuclei and the electronic charge there would be no chemical bond, but the *effective* potential, the internal energy $E_{elec}(R)$, takes its form—slow drop with decreasing R to a minimum at R_e and a sharp rise for $R < R_e$—from the increasing kinetic energy of the electrons, as much as from their decreased potential.

We are now in a position to ask the following questions: First, under what conditions does negative charge distribute itself so as to create a stable bond? Second, is there a meaningful way to subdivide the electron density so that we can speak of bonding and antibonding electrons (or orbitals), rather than bonding and antibonding regions of space? And third, can one calculate the bonding forces with sufficient detail and accuracy to make reliable predictions of bond energies and molecular geometries?

The first of these three questions is really a rather large one. To know how the charge density in a molecule is distributed, we must presumably know the electronic wave function of the molecule. This means that we must solve the Schrödinger equation for the molecule at

[4] K. Ruedenberg, in O. Chalvet, ed., *Localization and Relocalization in Quantum Chemistry*, Vol. 1, (Reidel, Dordrecht, 1975), pp. 223–245; K. Ruedenberg, *Rev. Mod. Phys.* **34,** 326 (1962); M. J. Feinberg, K. Ruedenberg, and E. L. Mehler, *Adv. Quantum Chem.* **5,** 27 (1970); R. S. Mulliken and W. C. Ermler, *Diatomic Molecules, Results of ab Initio Calculations* (Academic, New York, 1977), pp. 38–43.

many internuclear distances, a task appreciably more formidable than that for an atom. Nevertheless, in strongly bound molecules the problem can be solved with reasonable accuracy by an approach similar to the one used for atoms, the method of one-electron orbitals. This method, in so far as it is valid, suffices to answer our second question in the affirmative. If each electron has its own wave function, we can indeed classify electrons and speak meaningfully of individual electrons as being responsible for particular bonds. The orbital method has its limitations, even with regard to electronic energies and charge distributions. In particular, it is not sufficiently accurate to describe weakly bound molecules in which the bonding forces are due in large part to correlation between the electrons. To the third question, on the possibility of predicting molecular properties, we can only answer, "Sometimes." Given a wave function, one can make such calculations; their accuracy depends on how good the wave function is. For very simple molecules, those with no more than five or six atoms, the theoretical calculations are of useful accuracy. For the vast majority of molecules, however, the prediction from theory of such things as bond angles and lengths still lies beyond our technical ability. We have seen how small a charge is needed to produce a net bonding force; it should not be surprising that the delicately balanced attractive and repulsive forces must be known with considerable accuracy before one can say how the balance actually works out in any particular case. With this overview of the problem, we are now prepared to examine its solution in the simplest possible molecule.

6.2
The Simplest Molecule: The Hydrogen Molecule-Ion, H_2^+

Naturally, the simplest molecule is the one that contains the smallest number of particles. There must be two nuclei if we are to have a molecule at all, the simplest nuclei are protons, and there must be at least one electron to provide a bonding force. The species made from these constituents is called the *hydrogen molecule-ion*, H_2^+. It is a real species, known to exist in electric discharges; it is readily detected in a mass spectrometer. In contrast to most of the molecules we shall discuss, however, no spectrum of H_2^+ has been observed, and only a little is known of its chemistry. We thus have very little experimental information about it. Nevertheless, a good deal is known about H_2^+ because it is simple enough to be analyzed by reliable theoretical methods. We shall examine the hydrogen molecule-ion in several ways, successively more precise and quantitative, each giving a slightly different viewpoint on how such a molecule can exist. In the present section we make a preliminary survey, before getting down to the business of finding the wave function.

In studying a molecule, a good starting point is to consider its properties when the atoms are very far apart, the limit at which we define the zero of energy. In the case of H_2^+, then, we begin with a hydrogen atom in its $1s$ ground state and a bare proton far away from it. Let us consider what happens when the distance R between the two protons is decreased.[5] When the two are very far apart, the energy $E(R)$ must be essentially independent of R. As the free proton approaches the atom,

---

[5] This may sound as if we are talking about a dynamic (time-dependent) process. Actually, although we are considering a sequence of values of R, in each case we use the Born–Oppenheimer approximation and treat the nuclei as motionless. What we want is the solution to the time-independent Eq. 6.3 for each value of R.

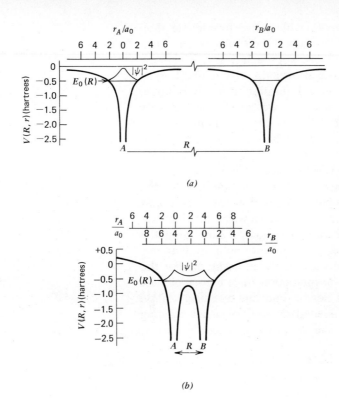

FIGURE 6.4
Potential energy of H_2^+,

$$V(R, r) = \frac{e^2}{4\pi\epsilon_0}\left(\frac{1}{R} - \frac{1}{r_A} - \frac{1}{r_B}\right),$$

along the internuclear axis. (a) $H + H^+$, nuclei far apart, electron localized around one nucleus. (b) H_2^+, $R = 4a_0$, electron equally likely to be around either nucleus. In each case the ground-state energy $E_0(R)$ and corresponding $|\psi|^2$ (arbitrary units) are shown; in part (a) these differ only slightly from those of a 1s H atom. Note that $V(R, r)$ and $E(R)$ include the internuclear repulsion, which in 6.4(b) equals +0.25 hartree; this is the limit of $V(R, r)$ as r_A, $r_B \to \infty$.

however, the repulsive force between the two protons and the attractive force between the electron and the free proton must begin to make themselves apparent. The electron will be more likely to be found on the side of the atom toward the free proton—in quantum mechanical terms, its wave function will be distorted from the symmetric 1s function to one with a greater amplitude in this direction. The centers of positive and negative charge no longer coincide, and we say that the atom is *polarized* by the free proton. The attractive force is greater than the repulsive force (since the electron is on the average closer to the free proton), and a net bonding force appears. According to Eq. 6.12, $dE(R)/dR$ is then positive, and $E(R)$ must be negative, since we have defined $E(\infty) = 0$. One can show (see Section 10.1) that $E(R) \propto -1/R^4$ in this fairly long-range region.

In the long-range limit it is still legitimate to speak of the H atom and the H^+ ion as two distinct entities. Consider Fig. 6.4a, which shows the potential energy along a line through the two nuclei: There are two potential energy wells, separated by a high, very wide barrier. Although the probability of tunneling from one well to the other is not quite zero, at large R (say, over 20 Å) it is low enough to be neglected for practical purposes. We can thus treat a state with the electron localized near one nucleus as a stationary (time-independent) state of the system. The electron's potential energy and wave function differ only slightly from those in an isolated hydrogen atom.

Now suppose that the nuclei are fairly close together. As shown in Fig. 6.4b, there is no longer an effective barrier between the two potential wells. The electron is thus free to travel anywhere in the combined well, and is as likely to be found near one nucleus as the other in any given observation. In such a situation the two protons are completely equivalent with regard to their interaction with the electron. More than that, they are *indistinguishable*. This is a very important concept. Two macroscopic charged spheres might also exert identical forces, but we could distinguish them from one another in various ways—say, by painting them different colors. But there is no way to do this with protons. Two protons—or two electrons, or two of any fundamental particle—are absolutely indistinguishable by any kind of measurement. In our H_2^+ system, then, neither we nor the electron can distinguish proton A from proton B. Even if we prepare the H_2^+ molecule by bringing together atom A and free proton B, there is no experiment that can distinguish A from B once they are together. If we then draw the two apart again, there is no way to tell if the electron has ended up around the same nucleus as at the start.[6] Like the simultaneous position and momentum of a particle, these things are simply unknowable.

What does this indistinguishability imply about the wave function of H_2^+? We may "label" the nuclei A and B for convenience, but any physically measurable quantity must not be affected when the labels are interchanged. Since $|\psi|^2$, the probability density of finding an electron at a given point, is a measurable quantity, the value of $|\psi|^2$ at $r_A = a$, $r_B = b$ must be the same as at the corresponding point $r_A = b$, $r_B = a$. This can be summarized by writing

$$|\psi(A, B)|^2 = |\psi(B, A)|^2. \tag{6.16}$$

For Eq. 6.16 to be valid, $|\psi|^2$ must be symmetric with respect to a plane equidistant from the nuclei, the plane normal to the bond at its midpoint.

We can say a bit more about the form of ψ. Given the Born–Oppenheimer approximation, the electronic wave function must satisfy Eq. 6.3, which for H_2^+ becomes

$$H\psi = \left[\frac{p^2}{2m} + \frac{e^2}{4\pi\epsilon_0}\left(\frac{1}{R} - \frac{1}{r_A} - \frac{1}{r_B}\right)\right]\psi = E(R)\psi. \tag{6.17}$$

(From here on H and ψ will stand for H_{elec} and ψ_{elec}, unless we indicate otherwise.) The potential energy goes to $-\infty$ at two points, the two nuclei. Near each nucleus it diverges in exactly the same way as does $V(r)$ in the hydrogen atom. We can thus carry over the reasoning of Section 4.3 to find the behavior of ψ in the vicinity of the nuclei. In particular, the ground-state wave function at each nucleus must decay exponentially from a cusp, like the corresponding hydrogen $1s$ function. The ground-state $|\psi|^2$ thus has two equal peaks, one centered on each nucleus, as shown in Fig. 6.4b. However, since $\int |\psi|^2 \, d\tau$ over the molecule must equal unity, each of these peaks must be only about half the size of the peak in a hydrogen atom. Note that most of the electron density in Fig. 6.4b appears to lie in the bonding region between the nuclei. We shall see that this is indeed the case, and that the net bonding force is sufficient for H_2^+ to be a stable molecule in its ground state.

Now let us look again at the long-range case. Since the arguments of the previous paragraph still apply, should we not expect to find two equal

---

[6] Unless the atoms passed each other so fast that their de Broglie wavelengths were much less than the shortest distance between them (though even then there is a small chance of what is called *exchange scattering*). Here we see the fundamental difference between classical and quantum mechanics. If classical mechanics were valid, one could distinguish the protons by following the trajectory of each particle with as much precision as required. But the uncertainty principle makes this impossible.

peaks even at large R? The answer is yes, as long as the two nuclei are still indistinguishable. If all we know about the system is that an electron is somewhere in the field of two protons, the probability density $|\psi|^2$ must be symmetric. But ordinarily we do know more than this. In Fig. 6.4a the presumption is that we have prepared the system in such a way that the electron is known to be around nucleus A—so that the system is H(A)+H$^+$(B), rather than just H$_2^+$. This amounts to our imposing an additional condition that $r_A \ll r_B$. One can obtain an approximate solution of Eq. 6.17 with such a condition, giving a wave function ψ' like that in Fig. 6.4a. This solution is only as good as the approximation on which it is based, and obviously cannot be valid over the whole of space. Still, for large R it is at least as good as many of our other assumptions. One cannot expect the approximate ψ' to be valid at small R, however; it becomes a poor description when the potential barrier is so low that the electron is not likely to stay around one nucleus for the duration of a measurement.

Another way to look at the problem, then, is in time-dependent terms. Suppose that we have the Hamiltonian of Eq. 6.17, with fixed nuclei, but that we no longer try to assume a stationary state. We must solve the time-dependent Schrödinger equation, H$\psi = -i\hbar(\partial\psi/\partial t)$. One possible solution is a wave function $\psi(r, t)$ that at $t = 0$ is almost entirely located around nucleus A. The function ψ' of the previous paragraph will be a good approximation to $\psi(r, t)$ in the region where its magnitude is significant. Both functions must have cusps at nucleus B, but the amplitudes of these cusps are far too small to be observed. Now suppose that we leave the system alone for a long time. Since tunneling from A to B has a nonzero (if very low) probability, the likelihood of the electron's being found at B will steadily increase. Eventually the system will approach a stationary state in which the electron is equally likely to be found on A or B, and the wave function will have two equal peaks. For large R, however, "eventually" may easily mean many times the age of the universe. On the other hand, when R is a few angstroms, it may take only 10^{-16} s (about a Bohr period) for the wave function to equalize itself, and the time-dependent process can be ignored. The distinction between our two cases is thus simply one of time scale: A stationary state is one that does not change measurably over the time in which we are interested.

All this raises another interesting question. Our three particles continue to interact, even if they are a meter or a mile apart. But what about *other* particles? Must the wave function of each electron somehow contain information on the positions of all the other particles in the universe? Stated in these grandiose terms, the question may be considered a philosophical one, but it symbolizes a real problem. As far as we know, all particles do interact with one another, no matter how far apart they are. If the principles of quantum mechanics are universally valid—and there is no reason to think otherwise—then strictly speaking the wave function of the universe (or as much of it as we care to consider) is not separable. It is only an approximation to speak of a wave function for any particular piece of matter; how good an approximation this is depends on how weak the interactions are. For a molecule in a gas the approximation is excellent; for an atom in a molecule it is rather poor, unless the atoms are very far apart; for an individual electron in an atom it is moderately good (the orbital approach, which we shall see is also useful in molecules). In general, one makes successive approximations of this sort until the interactions left out are too small to worry about.

We can now proceed to find the actual wave function of the H_2^+ molecule. This wave function is what we call a *molecular orbital* (abbreviated MO): Just as an atomic orbital describes a single electron in an atom, a molecular orbital describes a single electron in a molecule. In H_2^+ the molecular orbital is identical with the true molecular wave function, but in many-electron molecules the orbital approach is again only an approximation.

As we have already seen, in describing molecules we have an important added complication that we did not face in atoms, the dependence of the wave function on the internuclear distance. We again start with the long-range case, where the problem is relatively simple; however, we wish to obtain a wave function that will join smoothly with the short-range solution. This means that we want the stationary-state solution to the time-independent Eq. 6.17 with indistinguishable nuclei. Even our long-range solution, then, must satisfy Eq. 6.16 and have equal amplitudes around the two nuclei: It must describe a stationary state of H_2^+, not of $H + H^+$. This is not a state that one would ordinarily observe, but its properties can be easily described.

Suppose, then, that we look at the system in the limit of infinitely large R. In the vicinity of nucleus A, the Schrödinger equation 6.17 then reduces to

$$\mathsf{H}\psi = \mathsf{H}_A\psi = \left(\frac{\mathbf{p}^2}{2m} - \frac{e^2}{4\pi\epsilon_0 r_A}\right)\psi = E(\infty)\psi. \qquad (6.18)$$

Since this is of the same form as the hydrogen atom wave equation, it will be satisfied by any function of the form $c_A\varphi_A$, where c_A is a constant and φ_A is a normalized hydrogen wave function centered at A. In other words, as $R \to \infty$, the wave function near A becomes identical to that in a hydrogen atom, except for a constant multiplier. By the same reasoning, the wave function near nucleus B must have the form $c_B\varphi_B$, with φ_B a normalized hydrogen wave function centered at B. If A and B are infinitely far apart, the regions "near A" and "near B" are the only places where ψ will differ significantly from zero; anywhere that φ_A has a reasonable amplitude, φ_B will be negligibly small, and vice versa, since each decreases exponentially with distance from its nucleus. We can thus describe the wave function everywhere in space by the equation

$$\psi = c_A\varphi_A + c_B\varphi_B. \qquad (6.19)$$

What this says is that the matter wave described by ψ is constructed by superposition (adding together the amplitudes) of two waves corresponding to states of atomic hydrogen. We say that the molecular orbital ψ is formed by *linear combination of atomic orbitals*, abbreviated LCAO. The atomic orbitals are what are called *basis functions*.

What can we say about the constants c_A and c_B? In accordance with Eq. 6.16, $|\psi|^2$ must be unaffected by interchange of the nuclei. This means that φ_A and φ_B must be the same function (e.g., both $1s$ functions), and that we must have $|c_A|^2 = |c_B|^2 \equiv |c|^2$. If we restrict the constants to real values, there are only two possible combinations, $c_A = c_B$ and $c_A = -c_B$, and ψ must have the form $c(\varphi_A \pm \varphi_B)$. The constant c is readily evaluated by normalizing the wave function, that is, setting the integral of $|\psi|^2$ over all space equal to unity:

$$\int |\psi|^2 \, d\tau = 1 = c^2 \int |\varphi_A \pm \varphi_B|^2 \, d\tau$$

$$= c^2\left(\int |\varphi_A|^2 \, d\tau \pm \int \varphi_A{}^*\varphi_B \, d\tau \pm \int \varphi_A\varphi_B{}^* \, d\tau + \int |\varphi_B|^2 \, d\tau\right). \qquad (6.20)$$

In the limit $R \to \infty$, either φ_A or φ_B is negligibly small at every point in space; we can thus write

$$\int \varphi_A{}^* \varphi_B \, d\tau = \int \varphi_A \varphi_B{}^* \, d\tau = 0, \qquad (6.21)$$

since the integrands are effectively zero. We defined φ_A and φ_B as normalized wave functions, so we have

$$\int |\varphi_A|^2 \, d\tau = \int |\varphi_B|^2 \, d\tau = 1. \qquad (6.22)$$

Substituting these results into Eq. 6.20, we find that

$$1 = c^2(1+0+0+1) = 2c^2 \quad \text{or} \quad |c| = \frac{1}{\sqrt{2}}. \qquad (6.23)$$

Thus $|\psi|^2$ around each nucleus is half as much as in a lone hydrogen atom, in agreement with our premise that the electron is equally likely to be around either nucleus.

What we have said thus far applies to any state of the H_2^+ molecule. The only restriction we have placed on φ_A and φ_B is that they must describe the same state of the hydrogen atom. The ground state of the molecule will of course be one in which φ_A and φ_B are $1s$ functions. The ground-state wave function at $R = \infty$ must thus be either

$$\psi_0(R = \infty) = \frac{\varphi_A(1s) + \varphi_B(1s)}{\sqrt{2}} \quad \text{or} \quad \psi_1(R = \infty) = \frac{\varphi_A(1s) - \varphi_B(1s)}{\sqrt{2}}. \qquad (6.24)$$

These two functions are in fact degenerate, with the energy

$$
\begin{aligned}
E_0(\infty) &= \int_{\text{all space}} \psi^* \mathsf{H} \psi \, d\tau \\
&= \int_{\text{near A}} \frac{\varphi_A{}^*(1s)}{\sqrt{2}} \mathsf{H}_A \frac{\varphi_A(1s)}{\sqrt{2}} \, d\tau + \int_{\text{near B}} \frac{\varphi_B{}^*(1s)}{\sqrt{2}} \mathsf{H}_B \frac{\varphi_B(1s)}{\sqrt{2}} \, d\tau \\
&= \tfrac{1}{2} E_H(1s) + \tfrac{1}{2} E_H(1s) = E_H(1s), \qquad (6.25)
\end{aligned}
$$

where H_A, H_B are defined as in Eq. 6.18 and $E_H(1s)$ is the energy of a $1s$ hydrogen atom; just as in Eq. 6.20, the cross terms vanish. The ground-state energy of H_2^+ at $R = \infty$ is thus the same as that of $H(1s) + H^+$ in the same limit, the energy we defined to be zero for the molecule. Similarly, each higher-energy atomic wave function $\varphi(nl)$ gives rise to a degenerate pair of molecular wave functions with the energy $E_H(nl)$.

Thus we have a complete solution—both wave functions and energies—for the H_2^+ molecule at infinite R. This is all very well, but what does it tell us about the real molecule at finite R? The answer is that it gives us a starting point for an approximate solution. For large but finite R, the wave function near each atom is very much like a hydrogen orbital, so we write the molecular wave function in the form of Eq. 6.19, which is exactly correct only in the limit $R \to \infty$. At finite R, however, it seems plausible that the wave function is still *something* like a sum of atomic orbitals. Suppose, then, that we retain the LCAO approach and assume the molecular wave function to be still given by Eq. 6.19, but with constants that vary with R. Since $|\psi|^2$ must be symmetric, we continue to have $|c_A|^2 = |c_B|^2$. The two functions formed from $1s$ orbitals are thus

$$\psi_0 = c_0[\varphi_A(1s) + \varphi_B(1s)] \qquad (6.26)$$

and

$$\psi_1 = c_1[\varphi_A(1s) - \varphi_B(1s)], \qquad (6.27)$$

corresponding to Eqs. 6.24. Unlike Eqs. 6.24, however, these are only *approximate* wave functions; later we shall see how good the approximation is.

The LCAO approximation is by far the most widely used method of constructing molecular orbitals. It has the advantage of combining maximum simplicity with reasonable accuracy. The use of atomic orbitals as a basis not only has a natural and intuitive appeal, but also greatly simplifies the calculations that must be made. This is because the atomic orbitals have already been computed with considerable accuracy, even for quite complex atoms; they merely have to be looked up and substituted into the appropriate equations. We shall see that the LCAO method also yields an important insight into how chemical bonds are (or are not) formed, emphasizing the way in which constructive or destructive interference develops between the atomic wave functions. But remember that it is only an approximation. If one calculates molecular orbitals as accurately as possible, one finds that, even near nuclei, the functions are not really quite like atomic orbitals. In complex molecules, particularly in regions midway between nuclei, the molecular orbitals responsible for chemical bonds simply do not look like superpositions such as Eq. 6.19. Nevertheless, the LCAO method is nearly always adequate for an approximate understanding of chemical bonding. After all, the origin of the chemical bond lies in the fact that an electron can be in a region where it simultaneously attracts two nuclei closer together. Crudely, it is like an atomic electron held simultaneously in the field of each nucleus, and this is just what the LCAO approximation describes.

Analysis of the LCAO wave functions can give us a simple way of looking at how chemical bonds are formed. Consider the form of Eq. 6.26. Since s functions have no nodes, both $\varphi_A(1s)$ and $\varphi_B(1s)$ are everywhere positive. The sum of two positive numbers is a larger positive number. This means that the function ψ_0 corresponds to *constructive interference* of the two atomic wave functions. In any molecular orbital of this type the electron density is proportional to

$$|\varphi_A + \varphi_B|^2 = |\varphi_A|^2 + |\varphi_B|^2 + \varphi_A{}^*\varphi_B + \varphi_B{}^*\varphi_A$$
$$= |\varphi_A|^2 + |\varphi_B|^2 + 2\,\mathrm{Re}(\varphi_A{}^*\varphi_B) \qquad (6.28)$$

("Re" ≡ real part of). For $1s$ functions, $\varphi_A{}^*\varphi_B$ is a positive real number, and Eq. 6.28 is everywhere greater than $|\varphi_A|^2 + |\varphi_B|^2$ (which would give the electron density if there were no interaction between the atoms). The increase is greatest where φ_A and φ_B both have appreciable magnitude; this occurs mainly in the region between the nuclei. Thus a higher fraction of the total electronic charge is found between the nuclei than would be there if the atoms did not interact. Since the molecular wave function is normalized ($\int |\psi|^2\,d\tau = 1$), this effect must be balanced by a relatively lower electron density in the region beyond the nuclei; in that region the increase in $|\varphi_A + \varphi_B|^2$ is outweighed by the decrease in the normalization constant c (which we shall evaluate later).

In contrast, let us now look at Eq. 6.27. Here the wave function ψ_1 is the difference of two positive numbers, and corresponds to *destructive interference* of the atomic orbitals. The electron density is proportional to

$$|\varphi_A - \varphi_B|^2 = |\varphi_A|^2 + |\varphi_B|^2 - 2\,\mathrm{Re}(\varphi_A{}^*\varphi_B), \qquad (6.29)$$

which is everywhere less than $|\varphi_A|^2 + |\varphi_B|^2$; again the effect is greatest between the nuclei. Thus the electron density is decreased between the nuclei and increased beyond the nuclei, relative to the density in noninteracting atoms. In fact, since φ_A and φ_B are identical, they exactly cancel each other at all points where $r_A = r_B$; this means that the function ψ_1 has a nodal plane midway between the nuclei, a surface on which the value of ψ_1 is zero.

What is the significance of all this? Recall the bonding and antibonding regions of Fig. 6.3. We have shown that the wave function ψ_0 has a relatively high electron density in the bonding region between the nuclei, whereas ψ_1 has excess electron density in the antibonding region beyond the nuclei. It is thus clear that ψ_0 describes a "more bonding" state than ψ_1, that on the average the bonding force is greater in the state ψ_0. The Hellmann–Feynman theorem, Eq. 6.12, thus requires that $dE(R)/dR$ be greater for ψ_0 than for ψ_1. The two states have the same $E(\infty)$, Eq. 6.25, so ψ_0 must have a lower energy than ψ_1 for all finite values of R: $E_0(R) < E_1(R)$. We can reach the same conclusion by considering the potential energy curve of Fig. 6.4b. Since the electron in state ψ_1 is more likely to be found in the high-V regions beyond the nuclei, its total energy is higher in this state. We said earlier that one of the states ψ_0 and ψ_1 (which were degenerate at $R = \infty$) must be the ground state of the H_2^+ molecule. Now we have established that ψ_0 is the true ground state (or rather an approximation to it), whereas ψ_1 is an excited state of higher energy.

Note that we still have not proved that H_2^+ is a stable molecule. Although ψ_0 describes the state in which the bonding force is greatest, we do not yet know if that bonding force is sufficient to overcome the internuclear repulsion—if there is *enough* excess charge in the bonding region. To put it another way, we know that ψ_0 corresponds to the lowest $E(R)$ curve; but does that curve have the minimum necessary for a stable molecule (cf. Fig. 6.2)? To answer these questions, one must actually solve the Schrödinger equation for $E(R)$.

First we must evaluate the normalization constants in our wave functions. It is convenient to define the integral

6.4
H_2^+: Obtaining the Energy Curve

$$S_{AB} \equiv \int \varphi_A{}^* \varphi_B \, d\tau, \qquad (6.30)$$

called the *overlap integral* of the orbitals φ_A and φ_B. We pointed out earlier that this integral vanishes when A and B are infinitely far apart. The principal contribution to the integral comes only from the regions where φ_A and φ_B are both appreciable, that is, the regions that dominate the bonding. Note that $S_{BA} = S_{AB}{}^*$ for any φ_A, φ_B.

We can normalize the ground-state wave function ψ_0, defined by Eq. 6.26. Integrating over all space, we have

$$\int |\psi_0|^2 \, d\tau = 1 = c_0{}^2 \left(\int |\varphi_A|^2 \, d\tau + \int |\varphi_B|^2 \, d\tau + \int \varphi_A{}^* \varphi_B \, d\tau + \int \varphi_A \varphi_B{}^* \, d\tau \right)$$

$$= c_0{}^2 (1 + 1 + S_{AB} + S_{AB}{}^*)$$

$$= 2c_0{}^2 (1 + S_{AB}). \qquad (6.31)$$

The first two terms equal unity because we choose φ_A, φ_B to be normalized. We have $S_{AB} = S_{AB}{}^*$ in this case because $1s$ orbitals are real functions, so that S_{AB} is also real. Solving for the constant c_0 gives

$$c_0 = [2(1 + S_{AB})]^{-1/2}, \qquad (6.32)$$

and our ground-state molecular orbital is

$$\psi_0 = \frac{\varphi_A(1s) + \varphi_B(1s)}{[2(1 + S_{AB})]^{1/2}}. \tag{6.33}$$

If we carry out a similar calculation with Eq. 6.27, we obtain

$$\psi_1 = \frac{\varphi_A(1s) - \varphi_B(1s)}{[2(1 - S_{AB})]^{1/2}} \tag{6.34}$$

for the first-excited-state orbital.

Let us compare these results with our earlier discussion of bonding. In a hypothetical "nonbonded" molecule with no interaction between the atomic orbitals, the electron density would be simply $\frac{1}{2}(|\varphi_A|^2 + |\varphi_B|^2)$, as in the long-range case. By contrast, in the ground state, Eqs. 6.28 and 6.33 give

$$|\psi_0|^2 = \frac{|\varphi_A|^2 + |\varphi_B|^2 + 2\varphi_A\varphi_B}{2(1 + S_{AB})}, \tag{6.35}$$

since $\mathrm{Re}(\varphi_A{}^*\varphi_B) = \varphi_A\varphi_B$ for $1s$ orbitals. In this case S_{AB} is also necessarily positive, and the effect of division by $1 + S_{AB}$ is to lower the electron density everywhere. Counteracting this, however, is the positive term $2\varphi_A\varphi_B$ in the numerator, which tends to increase the electron density. The latter effect is dominant between the nuclei, where $|\psi_0|^2$ is higher than the nonbonded value. This reinforces our earlier conclusion that ψ_0 corresponds to excess electron density in the bonding region. We need not repeat the argument for ψ_1. These conclusions are illustrated by Fig. 6.7, which compares the actual electron densities.

Given the normalized LCAO wave functions, one can determine the corresponding energies $E(R)$. If ψ_0 were a true wave function, its energy (for a given R) would be given exactly by $\int \psi_0{}^*\mathsf{H}\psi_0\,d\tau$, where H is the electronic Hamiltonian of Eq. 6.17. But in fact our ψ_0 and ψ_1 are only approximations to the true wave functions; we can still carry out the integration, but we must realize that the result will be an approximate energy. The variation principle, Eq. 5.11, tells us that this energy must be higher than the true value; thus if we obtain a minimum in the $E(R)$ curve, we can be sure that there is really an even deeper minimum. Bearing this in mind, let us proceed.

It is useful to define another kind of integral,

$$H_{ij} \equiv \int \varphi_i{}^*\mathsf{H}\varphi_j\,d\tau, \tag{6.36}$$

where H is the electronic Hamiltonian and the i, j stand for nuclei; when H is real, we have $H_{ji} = H_{ij}{}^*$. For a given internuclear distance, the energy corresponding to ψ_0 is

$$E_0(R) = \int \psi_0{}^*\mathsf{H}\psi_0\,d\tau$$

$$= \frac{\int \varphi_A{}^*\mathsf{H}\varphi_A\,d\tau + \int \varphi_A{}^*\mathsf{H}\varphi_B\,d\tau + \int \varphi_B{}^*\mathsf{H}\varphi_A\,d\tau + \int \varphi_B{}^*\mathsf{H}\varphi_B\,d\tau}{2(1 + S_{AB})}$$

$$= \frac{H_{AA} + H_{AB} + H_{BA} + H_{BB}}{2(1 + S_{AB})} = \frac{H_{AA} + H_{AB}}{1 + S_{AB}}. \tag{6.37}$$

Since H is symmetric in A and B, we must have $H_{AA} + H_{BB}$ and

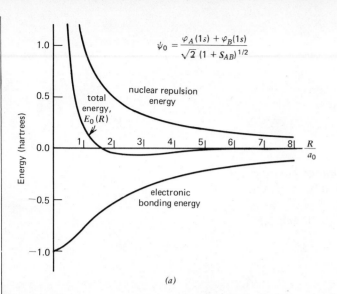

$$\psi_0 = \frac{\varphi_A(1s) + \varphi_B(1s)}{\sqrt{2}\,(1 + S_{AB})^{1/2}}$$

(a)

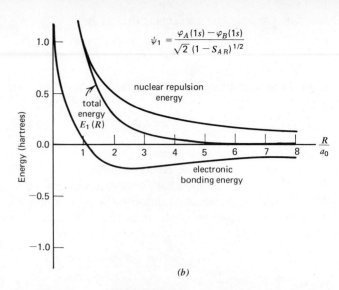

$$\psi_1 = \frac{\varphi_A(1s) - \varphi_B(1s)}{\sqrt{2}\,(1 - S_{AB})^{1/2}}$$

(b)

FIGURE 6.5
Energy of H_2^+ in the LCAO approximation. (a) The ground state. (b) The first
excited state. In each graph we show
the nuclear repulsion energy, $e^2/4\pi\epsilon_0 R$;
the electronic bonding energy, the last
term in Eqs. 6.41 and 6.42; and the total
energy $E(R)$, the sum of the previous
two. The energy zero is the energy
of $H(1s) + H^+$ at infinite separation.

$H_{AB} = H_{BA}$. Evaluating the H_{ij}, with H from Eq. 6.17, we have[7]

$$H_{AA} = \int \varphi_A{}^* \left(\frac{\mathbf{p}^2}{2m} - \frac{e^2}{4\pi\epsilon_0 r_A} \right) \varphi_A\, d\tau + \frac{e^2}{4\pi\epsilon_0 R} \int \varphi_A{}^* \varphi_A\, d\tau$$

$$- \int \frac{e^2}{4\pi\epsilon_0 r_B}\, \varphi_A{}^* \varphi_A\, d\tau$$

$$= E_H(1s) + \frac{e^2}{4\pi\epsilon_0 R} + J \qquad (6.38)$$

and

$$H_{AB} = \int \varphi_A{}^* \left(\frac{\mathbf{p}^2}{2m} - \frac{e^2}{4\pi\epsilon_0 r_B} \right) \varphi_B\, d\tau + \frac{e^2}{4\pi\epsilon_0 R} \int \varphi_A{}^* \varphi_B\, d\tau$$

$$- \int \frac{e^2}{4\pi\epsilon_0 r_A}\, \varphi_A{}^* \varphi_B\, d\tau$$

$$= \int \varphi_A{}^* E_H(1s) \varphi_B\, d\tau + \frac{e^2}{4\pi\epsilon_0 R}\, S_{AB} + K$$

$$= S_{AB} \left[E_H(1s) + \frac{e^2}{4\pi\epsilon_0 R} \right] + K, \qquad (6.39)$$

where

$$J \equiv - \int \frac{e^2}{4\pi\epsilon_0 r_B}\, \varphi_A{}^* \varphi_A\, d\tau \quad \text{and} \quad K \equiv - \int \frac{e^2}{4\pi\epsilon_0 r_A}\, \varphi_A{}^* \varphi_B\, d\tau; \quad (6.40)$$

J is called the *Coulomb integral* (it gives the Coulomb interaction between
the orbital around one atom and the nucleus of the other atom) and K
the *exchange integral*. Substituting in Eq. 6.37, we obtain

$$E_0(R) = \frac{E_H(1s) + e^2/4\pi\epsilon_0 R + J + S_{AB}[E_H(1s) + e^2/4\pi\epsilon_0 R] + K}{1 + S_{AB}}$$

$$= E_H(1s) + \frac{e^2}{4\pi\epsilon_0 R} + \frac{J + K}{1 + S_{AB}}. \qquad (6.41)$$

---

[7] The first terms in Eqs. 6.38 and 6.39 are, respectively, $\int \varphi_A{}^* H_A \varphi_A\, d\tau$ and $\int \varphi_A{}^* H_B \varphi_B\, d\tau$,
where H_A and H_B are hydrogen-atom Hamiltonians. We can thus substitute $H_A \varphi_A = E_H \varphi_A$
and $H_B \varphi_B = E_H \varphi_B$ and take the constant E_H outside the integrals.

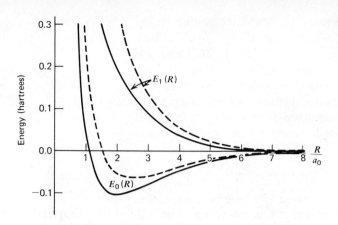

FIGURE 6.6
Energy curves for the two lowest states of H_2^+: ———, exact solution; – – – –, LCAO approximation.

The first term is the ground-state energy of a hydrogen atom, which we have taken as the energy zero for the H_2^+ molecule, the second term is simply the internuclear repulsion at distance R, and the third term gives the electronic bonding energy. A similar calculation for the first excited state gives

$$E_1(R) = \frac{H_{AA} - H_{AB}}{1 - S_{AB}} = E_H(1s) + \frac{e^2}{4\pi\epsilon_0 R} + \frac{J-K}{1-S_{AB}}. \qquad (6.42)$$

In order to obtain the $E(R)$ curves, one must evaluate the integrals S_{AB}, J, K for each value of R and substitute the results in the above equations. This is a straightforward calculation, and we shall not bother with the details.[8] The results for $E_0(R)$ and $E_1(R)$ are plotted in Fig. 6.5, with the terms in Eqs. 6.41 and 6.42 shown separately. In the ground state the electron has a net bonding effect at all values of R, and at moderate distances this is enough to overcome the nuclear repulsion. Thus there is a shallow potential well, and H_2^+ can indeed exist as a stable molecule.[9] As R grows smaller, the electronic bonding energy increases, but only to a finite value; at the same time the nuclear repulsion increases without limit, so that the total energy curve must become repulsive. As for the excited state ψ_1, it is interesting that the electron still has a net bonding effect over most of the range, even though there is less charge in the bonding region. Here, however, the nuclear repulsion is greater for all values of R; thus the $E_1(R)$ curve is everywhere repulsive, and the excited state cannot be stable.

How good is the LCAO approximation? It can be improved by variational methods; for example, one can vary Z in the atomic orbitals, as in Eq. 5.13 for the helium atom. However, since there is only one electron to worry about, the H_2^+ molecule is one of the few quantum mechanical systems for which the Schrödinger equation can be solved exactly (in the Born–Oppenheimer approximation). In Fig. 6.6 we compare the LCAO and exact energies for the first two states; Fig. 6.7 gives the corresponding electron densities along the internuclear axis. In accordance with the variation principle, the exact energy is everywhere lower. The exact solution gives an equilibrium distance $R_e = 2.00a_0$ (1.06 Å) and

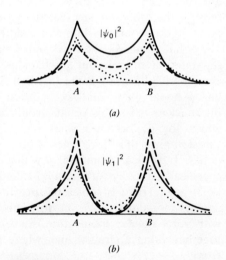

(a)

(b)

FIGURE 6.7
Electron density $|\psi|^2$ along the internuclear axis for the two lowest states of H_2^+. (a) The bonding ground state, ψ_0. (b) The antibonding excited state, ψ_1. The wave functions are calculated for the true equilibrium distance, $R_e = 2.00a_0$.

· · · · · ·, noninteracting 1s atomic orbitals, $|\psi|^2 = \frac{1}{2}(|\varphi_A|^2 + |\varphi_B|^2)$;
– – – –, LCAO approximation, Eqs. 6.33 and 6.34;
———, exact solution.

[8] If $\rho \equiv R/a_0$, one obtains

$$S_{AB} = e^{-\rho}\left(1 + \rho - \frac{\rho^2}{3}\right), \qquad J = -\frac{e^2}{4\pi a_0}\left[\frac{1}{\rho} - e^{-2\rho}\left(1 + \frac{1}{\rho}\right)\right], \qquad K = -\frac{e^2}{4\pi a_0}e^{-\rho}(1+\rho).$$

[9] Stability is relative, of course. The reason one cannot keep H_2^+ around is that it reacts very readily to form some *more* stable species. For example, if one brings H_2^+ into collision with H_2, the system immediately yields $H_3^+ + H$.

a dissociation energy $D_e = -0.103$ hartree (-2.79 eV), compared to the LCAO values of $2.50a_0$ (1.32 Å) and -0.065 hartree (-1.78 eV). Since the nuclear repulsion is known exactly, the error of the LCAO method is entirely in the electronic binding energy (cf. Fig. 6.5): At $R = 2a_0$ the latter is off by only 8%, but the dissociation energy by 48%. This illustrates the basic problem in calculating dissociation energies, that they are small differences of large quantities. For other molecules the problem is even more difficult, because interactions between the electrons also come into play.

Thus far we have discussed only the lowest two energy levels of the H$_2$⁺ molecule, but there are many higher levels. Our interest is mainly in how these states, and indeed those of any diatomic molecule, can be classified. We can best approach this problem by looking at what happens in the long-range and short-range limits.

We have already discussed the long-range or separated-atom limit. As $R \to \infty$ the molecular wave function goes smoothly into a sum or difference of atomic orbitals; for the two H$_2$⁺ states we have studied, these are hydrogen $1s$ orbitals. At the other extreme, as $R \to 0$, the two nuclei must eventually collapse into one; this we call the *united-atom limit*. The united atom obtained from the H$_2$⁺ molecule would have a nucleus with charge $+2e$ and a single electron, and would thus be a He⁺ ion. What happens to the wave function in the united-atom limit? In the ground state, the two peaks in the function ψ_0 must coalesce as $R \to 0$. In the limit we have a nodeless wave function with a single cusp at the united nucleus; this clearly describes the $1s$ state of the hydrogenlike He⁺ ion. Since there is a continuous transition of the molecular state from one limit to the other, we say that the ground state *correlates* smoothly with H($1s$) in the separated-atom limit and He⁺($1s$) in the united-atom limit.

What about ψ_1, the first excited state? This also goes into the H($1s$) separated-atom limit, but to what state of He⁺ does its united-atom limit correspond? The key is the fact that for all values of R the function ψ_1 has a nodal plane midway between the nuclei, perpendicular to the internuclear axis. This results from the symmetry condition (Eq. 6.16), which for ψ_1 becomes $\psi_1(A, B) = -\psi_1(B, A)$. The united-atom wave function must thus have a nodal plane through the nucleus, but no other nodes. It is clear from Fig. 4.8 that this must be a $2p$ orbital—more specifically, that $2p$ orbital whose long axis corresponds to the original A–B axis. There is something puzzling here: a p orbital has $l = 1$ and angular momentum $\sqrt{2}\hbar$, yet this orbital was constructed from $1s$ orbitals with zero angular momentum. Since we have only pushed the nuclei together along a straight line, where did the angular momentum come from?

The answer to this question can give us a good deal of insight into the symmetry properties of a diatomic molecule. Recall the principle we stated in Section 3.10: If the energy is independent of some coordinate q_i, the conjugate momentum p_i is a constant of the motion. In an isolated atom the potential energy is spherically symmetric, and thus independent of all the angular coordinates, whose conjugate momenta are components of the angular momentum. In a classical atom, then, the angular momentum and all its components would be constants of the motion. As we showed in Section 3.11, the uncertainty principle makes it impossible for all these components to be known simultaneously, so they cannot all be constants of the motion in a real quantum mechanical system. However, the magnitude of the total angular momentum and any one of its

**6.5
H$_2$⁺: Correlation of
Orbitals; Excited States**

components do remain constants of the motion in a system with spherical symmetry. Corresponding to these we have the two good quantum numbers l, m_l (for an electron) or L, M_L (for an atom).

In a diatomic system, which has cylindrical rather than spherical symmetry, the above argument no longer applies. The energy of an electron depends very much on angular coordinates like θ_A and θ_B (Fig. 6.1). Thus the components of angular momentum corresponding to these angles cannot be constants of the motion, and neither can the magnitude of the orbital angular momentum: l (or L) is no longer a good quantum number. There is one exception to this breakdown. Since the AB or z axis is an axis of cylindrical symmetry, the energy is independent of the angle ϕ that describes rotation about this axis, and the component of angular momentum along this axis (p_ϕ or L_z) is thus a constant of the motion. The operator corresponding to L_z has the same form as for one component of angular momentum in an atom, $-i\hbar(\partial/\partial\phi)$, and the eigenvalues of L_z are given by

$$L_z = \pm\lambda\hbar \qquad (\lambda = 0, 1, 2, \ldots), \qquad (6.43)$$

where λ is a quantum number corresponding to the absolute value of our earlier m_l. As in the atom, we use lowercase letters for a single electron and capital letters for a many-electron system. The axial angular momentum quantum number is designated by λ for an electron, Λ for a molecule, for which $L_z = \pm\Lambda\hbar$.

In spite of this formal similarity, there is a fundamental difference between L_z in an atom and L_z in a diatomic system. In an atom our choice of a z axis is entirely arbitrary, and L_z is quantized along whatever axis we choose. In the diatomic system this freedom of choice is gone. The z axis must be the system's unique symmetry axis, the axis joining the nuclei. Another difference involves the degeneracy of states. In a spherically symmetric potential the energy depends only on the quantum number l (or L), with $2l+1$ degenerate states for each value of l. In the cylindrically symmetric potential we no longer have a quantum number l, and only λ (or Λ) is available to classify the energy levels. These levels are either nondegenerate, for $\lambda = 0$ ($\Lambda = 0$), or doubly degenerate, for any other value of λ (or Λ).

Now we can answer our original question, "Where did the angular momentum come from?" In the separated-atom H($1s$) orbitals the angular momentum component is zero along any one axis, including the one that becomes the molecular z axis. Thus the ψ_1 state of H$_2^+$ has $L_z = 0$ in the separated-atom limit. Since L_z is a constant of the motion, it must also be zero for finite R, and even in the united-atom limit He$^+$($2p$). This is the case because a $2p$ orbital has zero angular momentum along its long axis, here taken as the z axis; that is, the orbital is $2p_z$, with $m_l = 0$, $L_z = 0$ (cf. Fig. 4.6). The $\sqrt{2}\hbar$ we mentioned above is the *total* angular momentum of the electron of He$^+$($2p_z$), which is no longer a constant when the spherical symmetry is removed.

It is customary to use the value of the axial angular momentum quantum number to classify the states of diatomic molecules. Just as atomic orbitals with $l = 0, 1, 2, 3, \ldots$ are respectively called $s, p, d, f, \ldots$, so in diatomic molecules orbitals with $\lambda = 0, 1, 2, 3, \ldots$ are designated by the corresponding Greek letters $\sigma, \pi, \delta, \varphi, \ldots$. Capital letters are again used for the molecule as a whole: Molecular states with $\Lambda = 0, 1, 2, 3, \ldots$ are called $\Sigma, \Pi, \Delta, \Phi, \ldots$ states. The spin multiplicity ($2S + 1$) of these states is indicated by a left superscript as in atoms. The two states of H$_2^+$ we have described both have a single electron in an orbital with $\lambda = 0$, a σ orbital; thus each of these states has $\Lambda = 0$, $S = \frac{1}{2}$, and can be called a $^2\Sigma$

state. When, as here, the molecule has a center of symmetry, a further distinction must be made. If the wave function is unchanged by reflection through the center of symmetry, that is, if $\psi(x, y, z) = \psi(-x, -y, -z)$, it is said to have *even parity*; if the reflection changes only the sign, $\psi(x, y, z) = -\psi(-x, -y, -z)$, the wave function has *odd parity*. States with even and odd parity are designated by the subscripts *g* and *u* (German *gerade* and *ungerade*), respectively. It is clear from Eqs. 6.26 and 6.27 that ψ_0 is even and describes a σ_g orbital, whereas ψ_1 is odd and describes a σ_u orbital; the corresponding molecular states are $^2\Sigma_g$ and $^2\Sigma_u$, respectively.

A molecular orbital may be labeled by either the united-atom orbital or the separated-atom orbital with which it correlates. The united-atom label precedes the species designation ($\psi_0 \rightarrow 1s\sigma_g$, $\psi_1 \rightarrow 2p\sigma_u$, etc.), whereas the separated-atom label follows it ($\psi_0 \rightarrow \sigma_g 1s$, $\psi_1 \rightarrow \sigma_u 1s$, etc.). This nomenclature is used mainly when one is emphasing the two limits. For example, the "$\sigma_g 1s$" notation is appropriate for an LCAO wave function. When discussing the molecule itself, however, one can simply number the orbitals of each species (σ_g, σ_u, π_g, etc.) in order of increasing energy. Thus ψ_0 and ψ_1 correspond to the $1\sigma_g$ and $1\sigma_u$ orbitals, respectively. Figure 6.8 gives further examples of all these notations.

Now we are ready to examine the higher-energy orbitals of H$_2^+$. For variety, we shall start with the united-atom limit instead of the separated-atom limit. This makes good physical sense since the orbitals in question are very large. They have appreciable magnitudes in regions far beyond the two nuclei, regions in which the electron density must be quite similar to that in a He$^+$ ion. For this reason the LCAO approximation, based on the separated atoms, does not give an accurate representation of the excited states.[10] We start with He$^+$ in a given state, pull the nucleus apart into two singly charged nuclei, and see what kind of wave function we should obtain. We expect the higher states to be unstable, that is, to have repulsive $E(R)$ curves (curves with no minima), since they have even less electron density in the binding region of space than the state ψ_1. (Some states in fact have shallow $E(R)$ minima at large R, but for most practical purposes these are also unstable.)

We know that the $1s$ orbital of He$^+$ transforms into the $1\sigma_g$ molecular orbital (ψ_0), which at long range transforms into the sum of two $1s$ orbitals of atomic hydrogen. Into what states do the other united-atom orbitals transform? In speaking of such a "transformation" at all, we are invoking a very general principle of quantum mechanics, the *adiabatic principle*. Let us assume that the energy levels of a system vary smoothly with some parameter R, which in our case is the internuclear distance. This gives us a series of $E(R)$ curves, each of which we can label as a particular state of the system. The adiabatic principle states that, if a system is initially in the *i*th energy level at a particular value of R, then for *sufficiently slow* changes in R it will remain in the same level unless otherwise disturbed.[11] For the particular case of the internuclear distance, this reduces to the Born–Oppenheimer approximation: The $E(R)$ curves

---

[10] For an illustration of how the LCAO approximation breaks down in the united-atom limit, consider the ground state. Since $\varphi(H, 1s)$ varies as e^{-r/a_0}, the LCAO wave function (Eq. 6.33) also approaches e^{-r/a_0} when $R \rightarrow 0$ (and $r_1, r_2 \rightarrow r$). But in this limit the true wave function must approach that of He$^+$, which varies as e^{-2r/a_0} (since $Z = 2$). Similarly, at $R = 0$ the LCAO electronic bonding energy is 1.0 hartree (Fig. 6.5), whereas the true value must be

$$E_1(\text{H}) - E_1(\text{He}^+) = -0.5 \text{ hartree} - (-2.0 \text{ hartrees}) = 1.5 \text{ hartrees}.$$

[11] "Adiabatic" comes from the Greek word for "uncrossable"; i.e., under adiabatic conditions the system will not "cross" from one $E(R)$ curve to another.

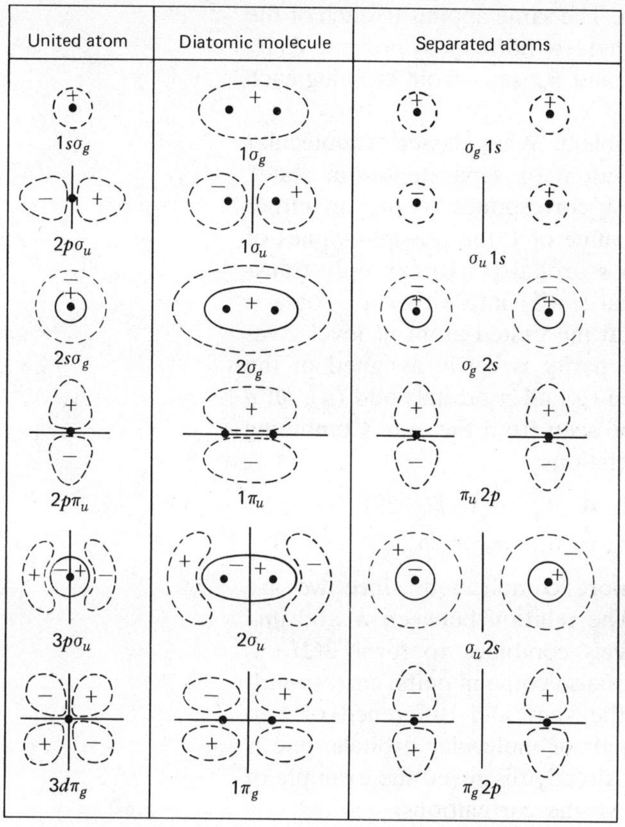

United atom	Diatomic molecule	Separated atoms
$1s\sigma_g$	$1\sigma_g$	$\sigma_g 1s$
$2p\sigma_u$	$1\sigma_u$	$\sigma_u 1s$
$2s\sigma_g$	$2\sigma_g$	$\sigma_g 2s$
$2p\pi_u$	$1\pi_u$	$\pi_u 2p$
$3p\sigma_u$	$2\sigma_u$	$\sigma_u 2s$
$3d\pi_g$	$1\pi_g$	$\pi_g 2p$

FIGURE 6.8

Correlation and shapes of orbitals in homonuclear diatomic molecules (for orbitals whose ϕ-dependent factors are real). The diagrams are purely schematic, with dashed lines bounding roughly the regions where the electron density is significant. The solid curves and lines are nodal surfaces; the sign of the wave function is given in each of the regions they separate. Each orbital is labeled by the three systems described in the text. Note that for H_2^+ the separated-atom limit of $2\sigma_g$ could be either $\sigma_g 2s$ or $\sigma_g 2p$, and that of $2\sigma_u$ either $\sigma_u 2s$ or $\sigma_u 2p$.

calculated for fixed nuclei give the energy of the actual molecule in the limit of sufficiently slow nuclear motion. For rapid nuclear motion, as in a high-speed collision, both the adiabatic approximation and the very concept of an energy curve cease to apply, and transitions between electronic states become likely.

What we wish to know, then, is the behavior of the orbitals when the nuclei are pulled apart very slowly (adiabatically). There should be an $E(R)$ curve connecting each united-atom state with a corresponding separated-atom state; the two limiting states are said to *correlate* with each other. Correlating states, as well as the sequence of molecular states joining them, must be identical in those properties that are constants of the motion; these include the value of λ (or Λ) and the parity of the wave function. Thus a σ_g state correlates only with a σ_g state, π_u only with π_u, and so for each class of states. But each limit has many states of each symmetry class; how do we know which ones correlate with each other? The best answer is given by the *noncrossing rule*: Two $E(R)$ curves with the same invariant properties (λ, parity, spin, etc.) never cross each other. This is only another approximation, valid to the same extent as the Born–Oppenheimer approximation (from which it can be derived), but it is sufficiently accurate to be our main tool for analyzing the correlation of states.

Given these assumptions, the analysis of H_2^+ is easy. The lowest σ_g orbital derived from the united atom must correlate with the lowest σ_g orbital formed from the separated atoms, since it can cross no other σ_g curve; this is the ground state, for which we had already tacitly assumed the noncrossing rule. Similarly, the second lowest united-atom σ_g orbital correlates with the second lowest separated-atom σ_g orbital, the third lowest with the third lowest, and so forth. The same applies to each of the other symmetry classes (but note that there is nothing to prevent $E(R)$ curves of *different* symmetry classes—σ_g and π_u, say—from crossing each other).

But now we immediately have a problem: What classes of molecular states do we obtain from each united-atom or separated-atom state? Remember that the quantum number λ corresponds to $|m_l|$ in either limit. For an atomic state with a given value of l, the possible values of $|m_l|$ are $0, 1, \ldots, l$. Thus a united-atom s orbital ($l = 0$) can only transform into a σ orbital ($\lambda = 0$), a p orbital ($l = 1$) into either a σ or a π orbital ($\lambda = 0, 1$), and so forth; in general the united-atom nl level gives rise to l distinct molecular states. The parity is easily assigned in the united-atom limit: All s orbitals are even (g), all p orbitals odd (u), all d orbitals even, and so on, as can easily be seen from Fig. 4.6. Combining these rules, we have the following correlations:

United-atom level: $\quad s \qquad p \qquad\quad d \qquad\qquad\quad f$

Molecular orbitals: $\quad \sigma_g \quad \sigma_u, \pi_u \quad \sigma_g, \pi_g, \delta_g \quad \sigma_u, \pi_u, \delta_u, \varphi_u \quad \cdots$

The separated-atom limit is a little more complicated, since we are dealing with *pairs* of atomic orbitals. The relation between λ and $|m_l|$ remains, but now each pair of nl levels combines to form $2(2l+1)$ molecular orbitals. The molecular orbitals also come in pairs, corresponding in the LCAO approximation to the sum and difference of two identical atomic orbitals. In each such pair of molecular orbitals, one is even (g) and the other odd (u); we have already discussed the example of the $\sigma_g 1s$ and $\sigma_u 1s$ orbitals. We thus have the correlations:

Separated-atom level: $\quad s \qquad\qquad p \qquad\qquad\qquad d$

Molecular orbitals: $\quad \sigma_g, \sigma_u \quad \sigma_g, \sigma_u, \pi_g, \pi_u \quad \sigma_g, \sigma_u, \pi_g, \pi_u, \delta_g, \delta_u \quad \cdots$

Later we shall discuss the nature of the molecular orbitals in greater detail, but now we have enough information for a complete description of the orbital correlation.

The simplest way to display such a correlation is with a *correlation diagram*; Fig. 6.9a gives such a diagram for H₂⁺. On the left and right sides of the diagram are the orbitals of the united atom and the separated atoms, respectively, each arranged in order of increasing energy. The energy scale is purely schematic; only the sequence of orbitals is significant. Lines are drawn connecting each united-atom orbital to the separated-atom orbital with which it correlates, following the correlation rules described above. Each of these lines can then be labeled with its appropriate molecular orbital designation; given the noncrossing rule, such designations as $1\sigma_g, 2\sigma_g, \ldots$ for the lowest, second lowest, $\ldots \sigma_g$ states can be given unambiguously. Since the two limits correspond to $R = 0$ and $R = \infty$, the vertical sequence of orbitals across the diagram should approximately represent the sequence of orbital energies as a function of R. For H₂⁺ it is possible to test this, since the exact $E(R)$ curves have been calculated for many of the molecular states; these are plotted in Fig. 6.9b. The picture is similar to that in the schematic diagram, but there are a number of differences in detail. In particular, the noncrossing rule is apparently violated by a number of the higher energy levels at large R, but this does not affect the correlation, since the states that thus cross are degenerate in one or the other limit.

Degeneracy leads to a special problem in the cases of H₂⁺ and H₂. The $2\sigma_g$ orbital, derived from He⁺($2s$), should correlate with the second lowest σ_g orbital of the separated atoms. But σ_g orbitals can be formed from both the $2s$ and the $2p$ levels of H + H⁺, and in the hydrogen atom

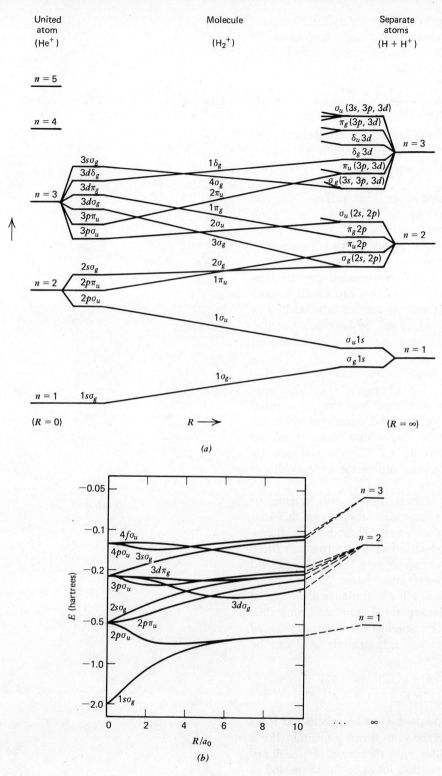

United atom (He^+)   **Molecule (H_2^+)**   **Separate atoms (H + H^+)**

(a)

(b)

FIGURE 6.9
Correlation of energy levels for H_2^+. (*a*) Schematic correlation diagram, with degenerate energy levels spread apart for clarity. Correlations are shown for all united-atom levels through $n = 3$. See the text for a discussion of separated-atom levels such as "$\sigma_g(2s, 2p)$." (*b*) Actual $E(R)$ curves for some of the energy levels, as calculated by Bates, Ledsham, and Stewart. The energy is measured relative to the free-electron limit, $H^+ + H^+ + e^-$. Since the calculations have not been carried beyond $R = 10a_0$, the curves are labeled in united-atom notation only.

these levels are degenerate.[12] We have thus used the notation "$\sigma_g(2s, 2p)$" to label the separated-atom limits of both the $2\sigma_g$ and the $3\sigma_g$ molecular orbitals; similar notation is used for other such degenerate orbitals of the same species. What this means is that, as far as satisfying the correlation rules goes, the separated-atom limit of the $2\sigma_g$ orbital

[12] In heavier atoms $2s$ has lower energy than $2p$, so the $2\sigma_g$ orbital correlates with $\sigma_g 2s$; cf. Fig. 7.14.

could be any linear combination of the form $a(\sigma_g 2s) + b(\sigma_g 2p)$. The exact calculations have not been carried far enough to reveal the true nature of this limit. But the simplest approximation to the $2\sigma_g$ orbital is an LCAO representation using only the $2s$ atomic orbitals:

$$\psi(\sigma_g 2s) = \frac{\varphi_A(2s) + \varphi_B(2s)}{[2 + 2S_{AB}(2s)]^{1/2}}, \tag{6.44}$$

analogous to Eq. 6.33. This approximation gives a fair description of the actual $2\sigma_g$ orbital at moderate R; if it is correct, then the $He^+(2s)$ radial node eventually splits into two $H(2s)$ radial nodes, but only at very large R (cf. Fig. 6.8).

Let us now look at the nature of some other orbitals. There are two degenerate σ_g orbitals in the $n = 3$ level of He^+, but here there is no ambiguity. The calculations clearly show that the $3d\sigma_g$ has the lower energy and thus becomes the $3\sigma_g$ molecular orbital. It of course correlates with the *other* separated-atom orbital formed from the $\sigma_g(2s, 2p)$ degeneracy, which in the simplest LCAO approximation can be represented by $\psi(\sigma_g 2p)$. But the LCAO representations are hardly important for these higher-energy states, which near R_e are better described by the united-atom orbitals. As for the $3s\sigma_g$ level, it becomes the $4\sigma_g$ molecular orbital and correlates with the separated-atom $\sigma_g(3s, 3p, 3d)$ degenerate set of levels.

The first few σ_u orbitals derive from the p orbitals of He^+—specifically, those p orbitals with $m_l = L_z = 0$ (where the z axis becomes the internuclear axis). We have already described how the $1\sigma_u$ orbital (ψ_1) derives from the $He^+(2p, m_l = 0)$ orbital and correlates with the difference of two $H(1s)$ orbitals. The similar $2\sigma_u$ orbital derives from the $He^+(3p, m_l = 0)$ orbital and correlates with the $\sigma_u(2s, 2p)$ degeneracy. Its simplest LCAO representation is $\sigma_u 2s$, the difference of two $H(2s)$ orbitals.

Thus far we have described only σ orbitals, with $L_z = 0$. In order to construct H_2^+ orbitals with nonzero angular momentum about the axis, we must begin with He^+ orbitals having such angular momentum themselves. For $\lambda = 1$ ($L_z = \pm\hbar$), the simplest such orbitals are obviously the p orbitals with $m_l = \pm 1$ relative to the internuclear axis. Since the real and imaginary parts of each of these orbitals must have a nodal plane including the axis, symmetry requires that such nodal planes also exist in the molecule and the separated atoms; the separated atom limit must thus also consist of p orbitals with $m_l = \pm 1$. The simplest molecular orbitals of this kind are formed from the $He^+(2p, m_l = \pm 1)$ orbitals, and can be given the LCAO representation

$$\psi(\pi_u 2p, \lambda = 1) = \frac{\varphi_A(2p, m_l = \pm 1) + \varphi_B(2p, m_l = \pm 1)}{[2 + 2S_{AB}(2p, m_l = \pm 1)]^{1/2}}. \tag{6.45}$$

Since the value of λ is 1, this is a π orbital—$1\pi_u$ to be specific. It is an odd (u) orbital because it is formed from the sum of two p orbitals. If the orbitals are chosen real, or if we consider only their real or imaginary parts, the two atomic orbitals have their positive lobes on the same side of the axis, so reflection through the center of symmetry changes the sign of ψ. The difference of two $2p(m_l = \pm 1)$ atomic orbitals would give a π_g orbital, which turns out to correlate with a $3d$ orbital of the united atom. This may be made clearer by Fig. 6.8, which shows schematically the shapes of these and other orbitals.

Note that Eq. 6.45 defines two orbitals, one with $L_z = +\hbar$ and one with $L_z = -\hbar$. Such a pair of wave functions exists for any molecular state in which $\lambda \neq 0$. Since $\lambda\hbar$ is the magnitude of the angular momentum about the internuclear axis, the states with $L_z = \pm\lambda\hbar$ differ (in classical terms) only in the direction of the electron's rotation about the axis. If we changed our coordinate system from right-handed to left-handed, simply by reversing the direction of the z-axis, we would interchange the functions $\psi(L_z = +\hbar)$ and $\psi(L_z = -\hbar)$. But this would in no way change the physics of the situation, and the two states must thus be physically indistinguishable.[13] Being indistinguishable, they must have the same energy. In short, for every nonzero value of λ in a diatomic molecule there are two degenerate states, with identical energies and with wave functions differing only in orientation. This degeneracy, like that of the hydrogen atom's three $2p$ states, derives directly from the symmetry of space.

As usual when we have degenerate wave functions, we can take linear combinations of them to define new orbitals. For example, we can take the sum and difference of $\psi(L_z = +\hbar)$ and $\psi(L_z = -\hbar)$ to give functions analogous to Eqs. 3.145. These vary respectively as $\cos\lambda\phi$ and $\sin\lambda\phi$, where ϕ is the angle of rotation about the z axis, while the original functions varied as $e^{\pm i\lambda\phi}$. The new orbitals are no longer eigenfunctions of even one component of angular momentum, but they still have the same degenerate energy as the original orbitals. For axially symmetric molecules we are at liberty to use either description. As usual, the real functions are more convenient for graphic purposes, and have been shown in Fig. 6.8. Only in molecules that have no axis of cylindrical symmetry does the difference become physically significant. As for nomenclature, in common usage any linear combination of, say, π_u orbitals is still called a π_u orbital.

This completes our study of the H_2^+ molecule-ion. What have we learned that can be applied to other molecules? Like the electron in a hydrogen atom, that in H_2^+ has a series of distinct quantum states or orbitals, with wave functions and energies that vary with the internuclear distance. Only in the ground state is the electron density in the bonding region high enough to produce a very stable molecule. We have introduced the remarkably simple and useful LCAO approximation, which represents the molecular orbitals as sums or differences of atomic orbitals, and shown how it can be used to obtain the $E(R)$ curve. But we have also seen that the LCAO method has its limitations for even this simplest of molecules. Finally, we have outlined how molecular orbitals can be classified according to their symmetry and angular momentum, and how they can be organized in a correlation diagram. Next we must find out what complications arise when a molecule has more than one electron.

6.6 The H₂ Molecule: Simple MO Description

The H_2^+ molecule-ion served as a simple example to help us grasp the physical concepts associated with chemical bonding. However, it is a rather exotic molecule compared with most of those we meet in everyday chemistry. (This is often the case in science: The most useful or familiar examples of a phenomenon are often terribly complex—water waves, for example—whereas those easiest to interpret tend to be rather abstruse.) The hydrogen molecule, H_2, is quite another matter. This is a perfectly stable species that exists as isolated gaseous molecules over a wide range

---

[13] This would not be true if we had a way of measuring the sign of L_z, that is, the direction of the electron's rotation. One can do this, for example, with a magnetic field, which can sense the direction of an electric current in a loop. Such a field would remove the degeneracy of the two states. However, we take it for granted in our discussion that external fields are absent. (There is a weak field from the various magnetic moments of the atoms, and this as usual produces a fine-structure splitting of energy levels.)

of temperatures. Indeed, it is the most abundant molecule in the universe. Its physical and chemical properties are thoroughly known from experiment, and it is second only to H_2^+ in simplicity, having only two electrons. Thus it should be relatively easy both to apply a theory of molecular structure to H_2 and to test the results.

The H_2 molecule is similar in several ways to H_2^+. The symmetry is the same, the potential energy due to the nuclei is the same for a given R, and the molecule still resembles two H atoms in the limit $R \to \infty$. The fundamental difference is that there are two electrons rather than one. It is natural to make the same initial assumption as in a many-electron atom: that each electron in a molecule sees only the average field of the other electron(s), and that one can describe such an electron by a one-electron wave function (molecular orbital). This corresponds to the central-field approximation in an atom. Here, however, the field is no longer central, and we must do without the advantages of spherical symmetry. As before, the best wave functions in this model are obtained by self-consistent-field methods; to improve on these results, one must apply corrections for the correlation between electrons.

Assume that we can get one-electron wave functions; how do we combine them to get the total molecular wave function? The simplest assumption is that one can assign to each electron a unique orbital, the aggregate of such assignments defining a unique electronic configuration of the molecule. For example, just as the boron atom has the configuration $1s^2 2s^2 2p$, we shall see (in Chapter 7) that the B_2 molecule has the ground-state configuration $(1\sigma_g)^2 (1\sigma_u)^2 (2\sigma_g)^2 (2\sigma_u)^2 (1\pi_u)^2$. Even in an atom this is only an approximation, though quite a good one, as we pointed out at the end of Section 5.7. Strictly, the interactions between electrons prevent the energy and angular momentum of a single electron from being conserved. One would not expect the single-configuration model to work any better for molecules. However, we can certainly use this model as a first approximation; later we shall introduce an approximation involving more than one configuration.

To the extent that electronic configurations are meaningful in molecules, the Pauli exclusion principle holds for individual electrons: If the electron spin is taken into account, only one electron can be assigned to a given quantum state. Each distinct spatial molecular orbital can then contain only two electrons, which must have opposite spins. One can thus apply the Aufbau principle, filling each orbital in turn in the order of increasing energy. Since H_2 has only two electrons, these rules would not seem to concern us here; yet the law behind them does entail a serious restriction on the possible wave functions of H_2, as we shall see in the next section.

The simplest kind of molecular wave function made up of one-electron orbitals is a Hartree product like Eq. 5.15. For a given state of H_2 such a wave function would have the form

$$\psi_{MO}(1, 2) = \varphi_i(1)\varphi_j(2), \tag{6.46}$$

where i, j designate orbitals and 1, 2, electrons; each φ_i must include both spatial and spin coordinates. Spectroscopy tells us[14] that the ground state of H_2 is $^1\Sigma_g$. We thus say that the lowest-energy orbital is $1\sigma_g$ (as in H_2^+)

[14] It is a $^1\Sigma$ state because it exhibits no Zeeman splitting at all in a magnetic field, and thus must have both zero orbital angular momentum (so $\Lambda = 0$) and zero spin angular momentum (so $S = 0$, $2S + 1 = 1$). The g character is actually inferred from the intensities of spectral lines associated with specific rotational states, but we shall not pursue this point. For details, see G. Herzberg, *Molecular Spectra and Molecular Structure, Volume I, Spectra of Diatomic Molecules*, 2nd ed. (Van Nostrand, Reinhold, New York, 1950).

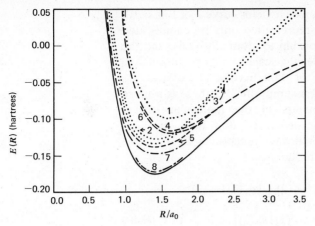

FIGURE 6.10
Potential energy curves for the ground state of H_2, calculated by various approximations; the energy zero corresponds to $H(1s) + H(1s)$. The heavy solid line is the experimental $E(R)$ curve. Molecular orbital calculations ($\cdots\cdots$):
1. Simple LCAO, Eq. 6.48, with $1s = (\pi a_0^3)^{-1/2} e^{-r/a_0}$,
2. Same with scaled atomic orbitals: $1s = (\alpha^3/\pi)^{1/2} e^{-\alpha r}$, with α varied to minimize $E(R)$,
3. Self-consistent-field solution.

Valence bond calculations ($----$):
4. Heitler–London, Eq. 6.72,
5. Same with scaled atomic orbitals.

Other calculations ($-\cdot-\cdot-\cdot-$):
6. Mixed MO and VB, Eq. 6.76,
7. Same with scaled atomic orbitals,
8. Eleven-term variation function, including terms in r_{12}.

and the ground-state configuration is $(1\sigma_g)^2$, with the two electrons assigned opposite spins. If we use the orbital designation to stand for the spatial part of the wave function, Eq. 6.46 for the ground state becomes

$$\psi_{MO}[(1\sigma_g)^2] = 1\sigma_g(1)\alpha(1)1\sigma_g(2)\beta(2), \tag{6.47}$$

where α and β are the two different Pauli spin wave functions introduced briefly in Chapter 5: α for $m_s = +\frac{1}{2}$ and β for $m_s = -\frac{1}{2}$. Because we are neglecting electron correlation, each of the spatial orbitals must be identical to the corresponding H_2^+ orbital, and in the LCAO approximation will be given by Eq. 6.33. Combining these assumptions, we have

$$\psi_{MO}[(\sigma_g 1s)^2] = \sigma_g 1s(1)\sigma_g 1s(2)\alpha(1)\beta(2)$$
$$= \frac{[1s_A(1) + 1s_B(1)][1s_A(2) + 1s_B(2)]\alpha(1)\beta(2)}{2(1 + S_{AB})}. \tag{6.48}$$

Here $1s_A$, $1s_B$ are the same as our earlier $\varphi_A(1s)$, $\varphi_B(1s)$, normalized hydrogen atom eigenfunctions: $1s_A(1) = (\pi a_0^3)^{-1/2} e^{-r_{A1}/a_0}$, where r_{A1} is the distance from nucleus A to electron 1.

Equation 6.48 is the simplest molecular orbital wave function for the ground state of the H_2 molecule. It can be used to obtain an approximate $E(R)$ curve by the same method as in Section 6.4. In this case one evaluates $\iint \psi^* H\psi \, d\tau_1 \, d\tau_2$ for each value of R, with the electronic Hamiltonian of Eq. 6.1 (including the $1/r_{12}$ term); the integration is lengthy but feasible. The calculated $E(R)$ is shown as curve 1 in Fig. 6.10. One obtains $R_e = 1.57a_0$ (0.84 Å) and $D_e = 0.0974$ hartree (2.65 eV), the latter relative to two $1s$ H atoms. The experimental values are $1.40a_0$ (0.74 Å) and 0.1744 hartree (4.75 eV). The electron density looks quite similar to that in the ground state of H_2^+ (Fig. 6.7), multiplied by 2. In the energy at least, this solution gives appreciably worse agreement than does the LCAO solution for H_2^+. This is not surprising, since here the flaws of the LCAO approach are combined with those of the one-electron-orbital and single-configuration assumptions.

One can improve upon this solution and still remain within the framework of the molecular orbital model. The greatest single improvement can be made by the technique we illustrated in Eq. 5.13, inserting a scale factor in the atomic orbitals. In this case one still uses Eq. 6.48 but sets $1s_A(1) = (\alpha^3/\pi)^{1/2} e^{-\alpha r_{A1}}$, and so on; α is then varied to minimize the energy for each value of R. This solution is shown as curve 2 in Fig. 6.10, and is obviously a considerable improvement over the simple LCAO solution. One obtains $R_e = 1.38a_0$, $D_e = 0.1282$ hartree, with $\alpha(R_e) = 1.197/a_0$. The best possible wave function of the form of Eq. 6.46 is, of course, the self-consistent-field solution, which yields curve 3 of Fig. 6.10; but the SCF molecular orbitals can be represented only by complicated functions with no simple analytic relationship to atomic orbitals, and the $E(R)$ curve is only slightly improved over the scaled LCAO solution. We are still left with a correlation energy of 0.04 hartree (1.1 eV) at R_e, illustrating the limitations of the orbital assumption. But there is a more fundamental flaw that sometimes appears in the molecular orbital method, as we can see by looking at the separated-atom limit for H_2.

What happens to the molecular orbital wave function as $R \to \infty$? The overlap integrals vanish, and Eq. 6.48 reduces to

$$\lim_{R \to \infty} \psi_{MO} = \frac{\alpha(1)\beta(2)}{2} [1s_A(1)1s_A(2) + 1s_A(1)1s_B(2)$$
$$+ 1s_B(1)1s_A(2) + 1s_B(1)1s_B(2)]. \qquad (6.49)$$

In this limit, of course, there is negligible probability of finding an electron anywhere but near one of the two nuclei. If we square Eq. 6.49, each of the cross terms contains a factor like $1s_A(1)1s_B(1) \propto e^{-r_{A1}/a_0}e^{-r_{B1}/a_0}$, which vanishes since everywhere either r_{A1} or r_{B1} must be very large. Thus all the cross terms drop out, and we have

$$\lim_{R \to \infty} (\psi_{MO})^2 \propto [1s_A(1)]^2[1s_A(2)]^2 + [1s_A(1)]^2[1s_B(2)]^2$$
$$+ [1s_B(1)]^2[1s_A(2)]^2 + [1s_B(1)]^2[1s_B(2)]^2. \qquad (6.50)$$

We have assumed that the location of each electron is independent of that of the other. The probability that two independent events occur simultaneously is the product of the probabilities of the two events. Thus in the integral of Eq. 6.50, $\iint \psi^2 \, d\tau_1 \, d\tau_2$, the first term gives the probability that both electrons are in atom A; the next two terms give the probability that one electron is in each atom; and the fourth gives the probability that both electrons are in atom B. Since all the $1s$ functions have the same form, each of the four terms contributes the same amount to the integral, and each of the four electron arrangements has the same probability, 25%. This means that the wave function describes a state in which one is equally likely to observe two neutral atoms, $H(1s) + H(1s)$, or a negative ion and a bare proton, $H^-(1s^2) + H^+$.

But for two hydrogen atoms infinitely far apart, the state of lowest energy is clearly one in which each atom is in *its* ground state, $H(1s) + H(1s)$. It takes 13.60 eV or 0.5 hartree, the ionization potential, to remove an electron from a hydrogen atom, and only 0.75 eV, the electron affinity, is gained by adding that electron to another neutral hydrogen atom. Thus the state $H^-(1s^2) + H^+$ has an energy 12.85 eV higher than $H(1s) + H(1s)$, and the mixed state described by Eq. 6.49 should have an energy halfway between the two, +6.42 eV. If one extends the $E(R)$ curve obtained from the LCAO function (Eq. 6.48) to $R = \infty$, one in fact obtains an energy 8.50 eV (0.312 hartree) higher than the ground state; the 2.08-eV discrepancy is half the correlation energy of H^-, for which the Hartree function $1s(1)1s(2)$ is only an approximation. As one can guess from Fig. 6.10, the SCF solution is not much better. The molecular orbital approach is thus clearly inadequate at large R. A wave function of this type may be reasonably accurate when R is near its equilibrium value,[15] but it certainly does not describe the ground state—or any stationary state—of two separate H atoms.

[15] The wave function near R_e does have a significant amount of ionic character (contribution of terms corresponding to H^-H^+), because Coulomb attraction drastically lowers the energy of the ionic state. Cf. Section 6.9.

What, then, is the wave function really like in the separated-atom limit? Since the electron densities on the two atoms are independent, we might expect to find a simple product of two atomic orbitals like

$$\lim_{R \to \infty} \psi[(\sigma_g 1s)^2] \propto 1s_A(1)1s_B(2). \qquad (6.51)$$

This wave function will obviously give the correct energy for two separate $1s$ H atoms. It describes a state in which there is a very high degree of correlation between the two electrons: The probability density $|\psi|^2$ has a significant magnitude only when electron 1 is near nucleus A and electron 2 is simultaneously near nucleus B. This is in contrast to a molecular orbital wave function like Eq. 6.48, in which there is by definition no correlation between the electrons.

Here we have two mathematical descriptions representing sharply contrasting physical situations, yet each is a reasonably good description of the H_2 molecule in one range of R. The MO description (Eq. 6.48) is never exactly correct, but for small R it does give a qualitatively correct $E(R)$ curve; as $R \to \infty$ it becomes completely wrong. Equation 6.51 is an exact solution at $R = \infty$, but is inadequate to describe bonding at finite R. The real ground-state wave function must somehow make a transition between these two extremes, and we wish to find a function that behaves in the same way. Before we begin this search, however, we must study the consequences of an apparently trivial fact—that all electrons are identical.

6.7
Symmetry Properties of Identical Particles

We have several times discussed the importance of symmetry in quantum mechanics. The constants of motion of a system are directly determined by the symmetry of the Hamiltonian; thus in diatomic molecules the equivalence of all orientations about the z axis requires that L_z be conserved. In H_2, as in H_2^+, the indistinguishability of the two nuclei means that $|\psi|^2$ must be symmetric about a plane midway between the nuclei. It should not be surprising that the indistinguishability of the electrons also has important consequences. We suggested some of these consequences in our discussion of atoms (which also contain identical electrons); now we shall look at the fundamental principles involved.

Here is the situation that confronts us. All electrons are absolutely indistinguishable. The same is true for all protons, and for all particles of any particular kind. The value of any real (measurable) physical quantity must thus be independent of the way we name or number such identical particles. Now, the physical description of a system depends only on the absolute square of the wave function. This means that, if $\psi(1, 2)$ is a wave function involving the identical particles 1 and 2, we must have

$$|\psi(1, 2)|^2 = |\psi(2, 1)|^2, \qquad (6.52)$$

where $\psi(2, 1)$ is obtained from $\psi(1, 2)$ by interchanging particles 1 and 2—that is, rewriting the wave function with the names of particles 1 and 2 exchanged. Equation 6.16 says the same thing for the special case of identical nuclei.

What limitations does Eq. 6.52 place on the wave function? It tells us that $\psi(2, 1)$ must differ from $\psi(1, 2)$ by no more than a phase factor of the form $e^{i\delta}$, where δ is a constant. This is the only kind of factor that has no effect on $\psi^*\psi$, in that $(e^{i\delta})^* e^{i\delta} = e^{-i\delta}e^{i\delta} = e^0 = 1$. We can be more specific about the value of $e^{i\delta}$. If we define an operator P_{12} to mean "permute (exchange) particles 1 and 2," then by what we have just said we must have

$$\psi(2, 1) \equiv P_{12}\psi(1, 2) = e^{i\delta}\psi(1, 2). \qquad (6.53)$$

If we carry out the permutation of 1 and 2 a second time, we must be back where we started. We can thus write

$$\psi(1, 2) = P_{12}\psi(2, 1) = P_{12}[e^{i\delta}\psi(1, 2)] = e^{i\delta}P_{12}\psi(1, 2)$$
$$= e^{2i\delta}\psi(1, 2). \tag{6.54}$$

This means that

$$e^{2i\delta} = 1 \quad \text{or} \quad e^{i\delta} = \pm 1, \tag{6.55}$$

or, in terms of the wave functions,

$$\psi(2, 1) = \pm\psi(1, 2). \tag{6.56}$$

This argument is completely general: It holds for systems with not just two, but any number of identical particles, because we can always interchange them two at a time.

According to Eq. 6.56, the wave function for any system containing identical particles must either remain unchanged or change sign if we interchange two of the identical particles. There are two physically distinct cases here. If the wave function is unchanged, $\psi(2, 1) = \psi(1, 2)$, we say that it is *symmetric* with respect to particles of the type in question; if the wave function changes sign, $\psi(2, 1) = -\psi(1, 2)$, we say it is *antisymmetric* with respect to those particles. The particles found in nature are sharply divided into two classes, those with symmetric and those with antisymmetric wave functions. A given particle never changes from one class to the other. The names of these classes are based on the statistical behavior of the particles, which we shall discuss in Part Two. Those with symmetric wave functions obey what is called Bose–Einstein statistics, and are thus named *bosons*. These include photons, deuterons, α particles, and all other particles with integral spin (photons have spin 1). The particles with antisymmetric wave functions obey Fermi–Dirac statistics, and are called *fermions*. These include electrons, protons, neutrons, and all other particles with half-integral spin.

We are concerned principally with the wave functions describing electrons. Since the electron is a fermion, any wave function $\psi(1, 2, \ldots)$ involving two or more electrons must change sign if we interchange the numbers on any two of the electrons. This is clearly not true of our molecular orbital wave functions for the H_2 molecule. Equations 6.47 through 6.49 all contain the factor $\alpha(1)\beta(2)$, specifically describing a state in which electron 1 has the spin wave function α and electron 2 has the spin wave function β. If we interchange the electrons we get a new wave function, containing $\alpha(2)\beta(1)$, rather than the same wave function with reversed sign. Since α and β here correspond to different spins, $\alpha(1)\beta(2)$ and $\alpha(2)\beta(1)$ describe what appear to be physically distinct states. Thus these wave functions are not really physically admissible descriptions of a two-electron system.

Very well, then, let us write a wave function that *is* antisymmetric with respect to interchange of electrons. We assume that the spatial and spin coordinates are separable, so that $\psi = \psi_{\text{spatial}}\psi_{\text{spin}}$. (For a two-electron system, this can always be done.) For the product to be antisymmetric, if one of the two factors is symmetric the other must be antisymmetric. We already have symmetric spatial functions, such as $1\sigma_g(1)1\sigma_g(2)$ in Eq. 6.47; thus we must look for an antisymmetric spin function. The trouble with our original $\alpha(1)\beta(2)$ was that interchange of electrons transformed it to the quite different function $\alpha(2)\beta(1)$. But suppose that we try a linear combination of these two functions, with opposite signs to give the antisymmetry:

$$\psi_{\text{spin,anti}} = \alpha(1)\beta(2) - \alpha(2)\beta(1). \tag{6.57}$$

236

This function does indeed change sign when we exchange 1 and 2. We can thus combine it with any symmetric spatial function to give an antisymmetric total wave function. The molecular orbital wave function for the ground state of H_2 then becomes

$$\psi_{MO,anti}[(1\sigma_g)^2] = \frac{1\sigma_g(1)1\sigma_g(2)}{\sqrt{2}}[\alpha(1)\beta(2) - \alpha(2)\beta(1)], \qquad (6.58)$$

in which the $\sqrt{2}$ maintains the normalization. This function is antisymmetric with respect to interchange of the electrons and is thus physically admissable. Unfortunately, since the spatial part is the same as before, it is just as poor an approximation as Eq. 6.47. One can also obtain an antisymmetric wave function by combining an antisymmetric $\psi_{spatial}$ with a symmetric ψ_{spin}; we shall see that such a function describes the first excited state of H_2.

What physical interpretation can we give to the wave function of Eq. 6.58? Both $\alpha(1)\beta(2)$ and $\alpha(2)\beta(1)$ are legitimate solutions of the wave equation, each describing a conceivable eigenstate of the system. The principle of superposition assures that a linear combination of such solutions is also a solution, with the squared coefficients of the individual terms giving the relative probabilities of observing the corresponding eigenstates.[16] Here the coefficients of the two terms are equal in magnitude, so the two states should be equally likely. In other words, if we make measurements on an H_2 molecule in its ground state, half the time we should find electron 1 with $m_s = +\frac{1}{2}$ and electron 2 with $m_s = -\frac{1}{2}$; the other half of the time we should find electron 1 with $m_s = -\frac{1}{2}$ and electron 2 with $m_s = +\frac{1}{2}$. Since we cannot tell electrons 1 and 2 apart anyway, this result is just what we should expect.

Given the antisymmetry of electrons one can derive the Pauli exclusion principle, which states that no two electrons can occupy the same quantum state. To see why this is so, let us try writing a wave function for two electrons in exactly the same state. This means that we assign to both electrons the same orbital φ_i, including both spatial and spin coordinates; for example, $\varphi_i(1)$ might be $1\sigma_g(1)\alpha(1)$. The antisymmetrized orbital-product wave function corresponding to Eq. 6.58 is then

$$\psi_{anti}(1, 2) = \frac{1}{\sqrt{2}}[\varphi_i(1)\varphi_i(2) - \varphi_i(2)\varphi_i(1)] = 0. \qquad (6.59)$$

A wave function of zero magnitude describes a state of zero probability; in other words, though the state (Eq. 6.59) satisfies the antisymmetry requirement, it can never be observed. This proves the exclusion principle, which applies not just to electrons, but to all fermions. (It is possible to infer antisymmetry from the exclusion principle as well, but the argument is beyond the scope of this book.) By contrast, there is no such restriction on the quantum states of bosons. Since the wave function for two bosons must be symmetric, suitable spin functions to accompany a symmetric spatial function include

$$\psi_{spin,symm} = \begin{cases} \alpha(1)\alpha(2), \\ \beta(1)\beta(2), \\ \alpha(1)\beta(2) + \beta(1)\alpha(2). \end{cases} \qquad (6.60)$$

---

[16] That is, if for some state of a system $\psi = \sum_i c_i \varphi_i$, where the φ_i are eigenfunctions of an operator Q, then in that state the probability of observing the ith eigenvalue Q_i is

$$P(Q_i) = |c_i|^2 = c_i{}^* c_i.$$

The symmetric wave function for two bosons in exactly the same state can thus be written, for example, as

$$\psi_{\text{symm}}(1, 2) = \varphi_i(1)\varphi_i(2)\alpha(1)\alpha(2). \qquad (6.61)$$

Since this function is in general nonzero, there is nothing to prevent two bosons, or indeed any number of bosons, from occupying the same quantum state.

Note that the exclusion principle as we have stated it is meaningful only when electrons can be assigned to individual atomic or molecular orbitals, with the wave function written as a linear combination of orbital products. This is logical, since only in the orbital approximation does each electron have its own set of quantum numbers. But this form of the exclusion principle is only a special case of the general rule, which we can add to the postulates of Section 3.8:

POSTULATE VII: Any eigenfunction of a many-particle system must be antisymmetric with respect to interchange of any two identical fermions, and symmetric with respect to interchange of any two identical bosons.

Which particles are fermions and which bosons must at present be determined by experiment, but may ultimately be derivable from a theory of elementary particles.

How does one write a wave function for more than two electrons? We again use the orbital approximation, with φ_i including both spatial and spin coordinates. The antisymmetric wave function for two electrons in the orbitals φ_i and φ_j, respectively, must be of the form

$$\psi_{\text{anti}}(1, 2) = \frac{1}{\sqrt{2}}[\varphi_i(1)\varphi_j(2) - \varphi_i(2)\varphi_j(1)] = \frac{1}{\sqrt{2}}\begin{vmatrix} \varphi_i(1) & \varphi_i(2) \\ \varphi_j(1) & \varphi_j(2) \end{vmatrix}, \qquad (6.62)$$

equivalent to a determinant of the orbital functions, in which the orbitals fix the row indices and the electrons the column indices. This can be generalized to any number of electrons: The appropriate wave function for N electrons in N orbitals is

$$\psi_{\text{anti}}(1, 2, \ldots, N) = (N!)^{-1/2}\begin{vmatrix} \varphi_1(1) & \varphi_1(2) \cdots \varphi_1(N) \\ \varphi_2(1) & \varphi_2(2) \cdots \varphi_2(N) \\ \cdots\cdots\cdots\cdots\cdots \\ \varphi_N(1) & \varphi_N(2) \cdots \varphi_N(N) \end{vmatrix}, \qquad (6.63)$$

a determinant in which each row corresponds to a given orbital and each column to a given electron. The $(N!)^{-1/2}$ is needed as a normalization constant, since the expansion of an $N \times N$ determinant has $N!$ terms. This function clearly satisfies the antisymmetry requirement, since any determinant changes sign when two columns are interchanged.

Now we can understand certain cryptic statements in Chapter 5. The trouble with a Hartree wave function like 5.15 is that it is not antisymmetric; instead one writes the determinantal function of expression 6.63. If the orbitals are chosen to give the lowest possible energy, within the approximation that each electron moves in the *average* potential of all the others, the result is called the *Hartree–Fock* wave function for N electrons.[17] (Further discussion of the Hartree–Fock method is postponed until Chapter 8.) We can also see why the phenomena associated with the exclusion principle are often called "exchange" effects; the apparent repulsion ("exchange force") between electrons of the same spin merely

---

[17] In an open-shell atom or molecule, with N electrons in *more than* N orbitals, one must replace Eq. 6.63 by a linear combination of determinants, one for each possible distribution of the electrons among the orbitals.

expresses the fact that they must have different spatial orbitals, or else the total wave function would vanish by Eq. 6.59. We could have explained all this in Chapter 5, but we preferred to use the two-electron case of H_2 to illustrate the concepts, which are equally applicable to atoms and molecules.

Now we can return to the problem of the H_2 molecule. Making the molecular orbital wave function antisymmetric does not keep it from going wrong as $R \to \infty$, so we still need another approach. Suppose we try to find an approximate wave function that joins smoothly with the separated-atom limit. What is the wave function in this limit? We suggested earlier that it should resemble Eq. 6.51 (which does give the correct energy), but that function is obviously not antisymmetric. Nevertheless, it does illustrate correctly that the wave function at $R = \infty$ must be made up of atomic orbitals. The simplest method of doing this was devised by W. Heitler and F. London in 1927, in what was historically the first quantum mechanical treatment of the chemical bond.

We begin by considering two completely independent hydrogen atoms, with one electron in each atom. Since the electrons are indistinguishable, we may have either electron 1 around nucleus A and electron 2 around nucleus B, or vice versa. Neglecting for the moment the symmetry problem, we have two possible spatial wave functions, $1s_A(1)1s_B(2)$ and $1s_A(2)1s_B(1)$. As for the spin, when the atoms are completely separated, we may assign either spin ($m_s = \pm\frac{1}{2}$) to each electron, independent of the other. This gives us four possible two-electron spin wave functions: $\alpha(1)\alpha(2)$, $\beta(1)\beta(2)$, $\alpha(1)\beta(2)$, and $\beta(1)\alpha(2)$. But in fact none of these spatial or spin functions is antisymmetric. How can we obtain an antisymmetric total wave function?

Let us write the total wave function as a product of spatial and spin functions. As in our molecular orbital functions, this form is chosen for mathematical and physical simplicity;[18] it certainly is not a correct form for the true wave function. For the product of spatial and spin functions to be antisymmetric, one must be symmetric and the other antisymmetric. As before, let us make symmetric and antisymmetric linear combinations of our simple functions. For the spatial functions we have

$$\psi_{\text{spat,symm}} = 1s_A(1)1s_B(2) + 1s_A(2)1s_B(1) \qquad (6.64)$$

and

$$\psi_{\text{spat,anti}} = 1s_A(1)1s_B(2) - 1s_A(2)1s_B(1). \qquad (6.65)$$

There are three simple symmetric spin functions,

$$\psi_{\text{spin,symm}} = \begin{cases} \alpha(1)\alpha(2), & (6.66a) \\ \beta(1)\beta(2), & (6.66b) \\ \alpha(1)\beta(2) + \alpha(2)\beta(1), & (6.66c) \end{cases}$$

and one antisymmetric spin function,

$$\psi_{\text{spin,anti}} = \alpha(1)\beta(2) - \alpha(2)\beta(1). \qquad (6.67)$$

Combining these to give an antisymmetric product, we have, in this

6.8
H₂: The Valence Bond Representation

[18] In fact, this is possible only for the two-electron system. For three or more electrons there is no way to write an antisymmetric total wave function that can be factored into spin and spatial parts. There are systematic methods for dealing with such cases, but we need not consider them here.

approximation, four possible wave functions generated from hydrogen $1s$ orbitals:

$$\psi_1 = C_1[1s_A(1)1s_B(2) + 1s_A(2)1s_B(1)][\alpha(1)\beta(2) - \alpha(2)\beta(1)], \quad (6.68)$$

$$\psi_2 = C_2[1s_A(1)1s_B(2) - 1s_A(2)1s_B(1)]\alpha(1)\alpha(2), \quad (6.69)$$

$$\psi_3 = C_3[1s_A(1)1s_B(2) - 1s_B(2)1s_B(1)]\beta(1)\beta(2), \quad (6.70)$$

$$\psi_4 = C_4[1s_A(1)1s_B(2) - 1s_B(2)1s_B(1)][\alpha(1)\beta(2) + \alpha(2)\beta(1)]. \quad (6.71)$$

The C_i's are normalization constants; for $R = \infty$ one readily evaluates them as $C_1 = C_4 = 1/2$, $C_2 = C_3 = 1/\sqrt{2}$.

What is the significance of the four functions 6.68–6.71? They are all satisfactory wave functions for two $1s$ hydrogen atoms infinitely far apart—satisfactory in that they both give the correct energy and are antisymmetric. There are other such functions, but these are the only simple ones that are constructed from H($1s$) orbitals and can be factored into spatial and spin parts. Now we make the *approximation* that these exact solutions for $R = \infty$ can be applied to the H₂ molecule at finite R, with appropriate normalization constants. This is exactly analogous to the reasoning by which we wrote the MO wave functions of Eqs. 6.26 and 6.27 for H₂'. Obviously the present wave functions cannot be factored into one-electron molecular orbitals. They are basically two-electron functions, and were specifically designed by Heitler and London to represent the electron-pair bond. Their purpose was to find a quantum mechanical basis for G. N. Lewis' concept that chemical bonds consist of electron pairs shared between two atoms. Similar approximations can be made for more complicated molecules if one assumes that the molecular wave function is a product of two-electron functions, each representing a conventional chemical bond. Wave functions of this type are called *valence bond* (VB) functions, since only the valence electrons and their bonds are taken into account.

The functions of Eqs. 6.68 through 6.71 all correspond to the same energy at $R = \infty$, but this can no longer be true at finite R. The spatial functions of Eqs. 6.64 and 6.65 obviously correspond to different electron densities at any given point in space, and thus to different energy levels. We can neglect the effect of the electron spins on the energy. We conclude that, of our four wave functions, ψ_1 corresponds to one state of the H₂ molecule, whereas ψ_2 ψ_3, and ψ_4 all correspond to a single other state, or rather to three degenerate states. Two $1s$ hydrogen atoms thus give rise to both a singlet and a triplet term of the H₂ molecule.

We can analyze these terms in much the same way that we treated atomic terms in Section 5.6. Let α correspond to $m_s = +\frac{1}{2}$, so that in the state corresponding to ψ_2 the molecule has $M_s = \sum m_s = 1$. Similarly, ψ_3 describes a state in which each electron has $m_s = -\frac{1}{2}$ and the molecule $M_S = -1$. These two wave functions differ only in the direction of the spin vector, which is not physically significant in the absence of an external field. Thus they are degenerate, corresponding to the same energy at any value of R. Since a two-electron system can never have $|M_S| > 1$, the states described by ψ_2 and ψ_3 must be components of a term with $S = 1$. This is, of course, a triplet term, with $2S + 1 = 3$, and must thus have a third component.

The three components of a triplet have $M_S = 1, 0, -1$ ($S_z = \hbar, 0, -\hbar$). The third component must therefore be a state with $M_S = 0$, corresponding to one electron with $m_s = +\frac{1}{2}$ and one with $m_s = -\frac{1}{2}$. Since both the electrons and the atoms are indistinguishable, the wave function for this state should logically contain an equal mixture of $\alpha(1)\beta(2)$ and $\alpha(2)\beta(1)$. This condition is satisfied by the function ψ_4, which is also degenerate

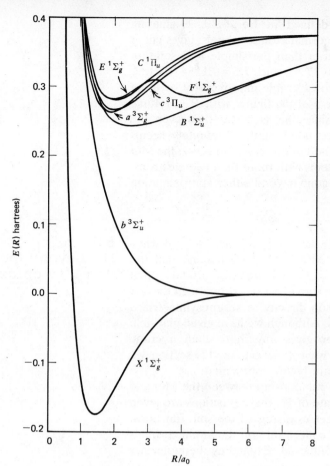

FIGURE 6.11
Potential energy curves for the ground state and the first few excited states of H_2; the energy zero corresponds to two separated H(1s) atoms. Each state is labeled with its spectroscopic term symbol, the initial letter of which is an arbitrary identifier (X always designates the ground state, excited singlets are usually assigned A, B, . . . , in order of increasing energy, and excited triplets a, b, . . . , similarly).

with ψ_2 and ψ_3 since it has the same spatial wave function. According to our approximate wave functions, then, the components of this triplet are exactly degenerate if one neglects spin–orbit interactions. To complete our term analysis, we are left with the function ψ_1, which also has $M_S = 0$; this can obviously describe the single component of a singlet term ($S = 0$).

Which of these two terms, the triplet or the singlet, has the lower energy? One could, of course, find out by solving for the $E(R)$ curves, but we can deduce the answer without doing this. Note that the spatial part of ψ_1 is the sum of two positive terms, which add constructively at every point in space. In contrast, the spatial part of the triplet wave functions has a nodal plane midway between the nuclei. Thus the singlet state has excess electron density in the bonding region, and can be assumed to have a minimum in the $E(R)$ curve, whereas the triplet state is certainly antibonding. This conclusion is confirmed by both calculation and experiment. The ground state is indeed a singlet state, with a fairly deep energy minimum. The first excited state is a triplet state in which the two atoms repel each other very strongly. Because both these states correlate with 1s atomic states, they are both Σ states ($\Lambda = 0$); the singlet is $^1\Sigma_g$ and the triplet is $^3\Sigma_u$. The exact $E(R)$ curves for these states are shown in Fig. 6.11; we shall discuss the excited states of H_2 in Section 6.10.

As for the ground state, one can obtain the valence bond $E(R)$ curve in the usual way. Normalization of Eq. 6.68 gives for the wave function

$$\psi_{VB} = \frac{[1s_A(1)1s_B(2) + 1s_A(2)1s_B(1)][\alpha(1)\beta(2) - \alpha(2)\beta(1)]}{2(1 + S_{AB}^2)}, \quad (6.72)$$

where S_{AB} is the overlap integral defined in Eq. 6.30. Evaluating $\iint \psi^* H\psi \, d\tau_1 \, d\tau_2$ for this function, with normal atomic orbitals, gives curve 4 of Fig. 6.10. This is appreciably better than the simple MO solution (curve 1), and yields $R_e = 1.64a_0$ (0.87 Å) and $D_e = 0.1154$ hartree (3.14 eV). As before, one can obtain considerable improvement by using scaled atomic orbitals. This gives curve 5 of the figure, with $R_e = 1.40a_0$, $D_e = 0.1383$ hartree, $\alpha(R_e) = 1.166$. Since this is better than even the self-consistent-field result, the valence bond method definitely seems preferable to the molecular orbital method in this case. However, the MO method is more easily extended to systems with more than two electrons. But the state of the art has long since gone beyond either approximation in its simple form, as we shall now see.

Thus far we have what seem to be two quite distinct kinds of wave function for the H_2 molecule. In the MO method we assume that the molecular wave function is a product of one-electron orbitals, then approximate each of these orbitals by a sum of atomic orbitals. In the VB method we form a two-electron function directly as an antisymmetrized combination of atomic orbital products. Although we have given plausible arguments for both methods, neither gives any more than a crude approximation to the true molecular wave function. In this section we shall look at some of the ways to obtain better approximations.

To begin with, let us examine the relationship between the MO and VB wave functions. For the ground state of H_2 these functions are given by Eqs. 6.58 and 6.72, respectively. Let us disregard the spin functions, which are the same in both cases, and concentrate on the atomic orbitals from which the spatial functions are composed. Expanding the molecular orbitals according to Eq. 6.33, we have

6.9
H_2: Beyond the Simple MO and VB Approximations

$$\psi_{MO,spat} = \frac{[1s_A(1) + 1s_B(1)][1s_A(2) + 1s_B(2)]}{2(1 + S_{AB})}$$

$$= \frac{1s_A(1)1s_A(2) + 1s_B(1)1s_B(2) + [1s_A(1)1s_B(2) + 1s_B(1)1s_A(2)]}{2(1 + S_{AB})},$$

(6.73)

whereas the corresponding VB function is

$$\psi_{VB,spat} = \frac{1s_A(1)1s_B(2) + 1s_B(1)1s_A(2)}{\sqrt{2}(1 + S_{AB}^2)}.$$

(6.74)

We can ignore the denominators of these equations, which are merely normalization constants. If we compare the two functions, we see at once that the numerator of Eq. 6.74 is identical to the two terms in brackets in the numerator of Eq. 6.73. Thus the MO function contains the VB function.

How can we interpret this relationship? Each of the first two terms in Eq. 6.73 describes a state in which both electrons are on the same atom, $H^- + H^+$, whereas the two terms in brackets describe states with one electron on each atom. It is customary to call these "ionic" and "covalent" terms, respectively. Since all the terms in each function have equal coefficients, one says that the MO function is an equal mixture of covalent and ionic terms, whereas the VB function contains only the covalent terms. We have gone through this argument before, in discussing Eq. 6.50: At $R = \infty$ the MO function literally describes a state that is 50% covalent and 50% ionic. This is *not* true at finite R, since the ionic–covalent cross terms in $(\psi_{MO})^2$ do not vanish. Thus in this approximation one cannot divide the bond (i.e., the electron density) in any unique way into "covalent" and "ionic" parts. If we say that the wave function is 50% ionic, this is merely a convenient shorthand to describe the mathematical form of our approximate function.

To compare the two approximations further, let us consider how they treat the correlation between the positions of the two electrons. The valence bond function (Eq. 6.74) is highly correlated. For $R \gg R_e$, ψ_{VB} is negligibly small unless the two electrons are near different nuclei; as we pointed out in connection with Eq. 6.51, this indeed describes the correct long-range behavior. But even at values of R near R_e, ψ_{VB} is quite small whenever the two electrons are close to each other.[19] In contrast to this, the ionic terms in Eq. 6.73 ensure that ψ_{MO} has appreciable magnitude even when both electrons are near the same nucleus. In fact, writing ψ as an orbital product explicitly assumes that there is *no* Coulombic correlation between the electrons. (On the other hand, both wave functions do contain what we have called exchange correlations between electrons of the same spin, as a direct result of the wave functions' being anti-symmetric.) So one approximation to the singlet state contains no Coulombic correlation, whereas the other assumes strong correlation that becomes absolute at long range. A naive but sensible guess would say that the true extent of correlation is somewhere between these extremes.

Assuming that this is so, can we devise a wave function intermediate between the two extreme representations? This is quite easy to do. If we define

$$\psi_{\text{covalent}} \equiv 1s_A(1)1s_B(2) + 1s_B(1)1s_A(2),$$
$$\psi_{\text{ionic}} \equiv 1s_A(1)1s_A(2) + 1s_B(1)1s_B(2), \tag{6.75}$$

then it is immediately clear that both ψ_{MO} and ψ_{VB} are of the form

$$\psi_{\text{mixed}} = a(R)\psi_{\text{covalent}} + b(R)\psi_{\text{ionic}}, \tag{6.76}$$

where $a(R)$ and $b(R)$ are constants for any given value of R. In the MO approximation we impose the condition $a(R) = b(R)$; in the VB approximation we set $b(R) = 0$. But in accord with the variation principle, we can obviously obtain a better wave function by varying the ratio of $a(R)$ to $b(R)$ until $E(R)$ is minimized. As usual, the more variable parameters one introduces into the wave function, the more closely one can approximate the true solution of the Schrödinger equation. The results of this calculation are shown in Fig. 6.10 as curve 6 (using normal atomic orbitals) and curve 7 (using scaled atomic orbitals). The latter gives $R_e = 1.415a_0$ (0.749 Å), $D_e = 0.1470$ hartree (4.00 eV), our closest approach yet to the true values. The ratio $b(R)/a(R)$ for curve 7 is 0.256 at R_e, and of course approaches zero as $R \to \infty$.

We are still a long way from the true wave function, but already we have gone beyond the simple MO and VB approaches. However, one can extend the language of either method to cover mixed functions like Eq. 6.76. In VB language, one describes the wave function as a linear combination of functions corresponding to hypothetical "structures," a structure being a particular arrangement of electrons and bonds. Thus ψ_{covalent} describes the covalent structure written as H:H or H—H, ψ_{ionic} is a combination of the ionic structures $H_A^- H_B^+$ and $H_A^+ H_B^-$ (which must be equally likely), and ψ_{mixed} combines all three structures.

---

[19] Each term in the VB function has the form $e^{-\alpha r_{1A}}e^{-\alpha r_{2B}}$, or its equivalent with 1 and 2 exchanged. If the nuclei are not close together, then at any point in space either r_A or r_B (or both) must be large, and the corresponding $e^{-\alpha r}$ small. So if both electrons are near the same point, at least one of the factors in $e^{-\alpha r_{1A}}e^{-\alpha r_{2B}}$ is small.

We can also reach Eq. 6.76 by an extension of the MO method. In the LCAO approximation, the ground state is one in which both electrons are in the orbital $1\sigma_g$, defined by Eq. 6.33 as a sum of $1s$ orbitals. But the difference of the same atomic orbitals gives rise to the excited (and antibonding) orbital $1\sigma_u$, defined by Eq. 6.34. Consider a state in which both electrons are in the orbital $1\sigma_u$. The LCAO molecular wave function, corresponding to Eq. 6.48, is then

$$\psi_{\mathrm{MO}}[(\sigma_u 1s)^2] = \frac{[1s_A(1) - 1s_B(1)][1s_A(2) - 1s_B(2)]}{2(1 + S_{AB})}$$
$$\times \frac{[\alpha(1)\beta(2) - \alpha(2)\beta(1)]}{\sqrt{2}}. \qquad (6.77)$$

As in the ground state, the spatial function is symmetric in the electrons, so the spin function must be antisymmetric. Expanding the spatial function as in Eq. 6.73, we have

$$\psi_{\mathrm{MO,spat}}[(\sigma_u 1s)^2]$$
$$= \frac{[1s_A(1)1s_A(2) + 1s_B(1)1s_B(2) - 1s_A(1)1s_B(2) - 1s_B(1)1s_A(2)]}{2(1 + S_{AB})}$$
$$= \frac{\psi_{\mathrm{ionic}} - \psi_{\mathrm{covalent}}}{2(1 + S_{AB})}, \qquad (6.78)$$

which is again of the form of Eq. 6.76. Like the ground state, the state described by Eq. 6.77 is $^1\Sigma_g$, the product of two odd (u) functions giving an even (g) function. Since functions 6.58 and 6.77 have the same electronic symmetry, either *could* describe the ground state—and so, of course, could any linear combination of the two,

$$\psi_{\mathrm{MO,mixed}} = x(R)\psi_{\mathrm{MO}}[(\sigma_g 1s)^2] + y(R)\psi_{\mathrm{MO}}[(\sigma_u 1s)^2]. \qquad (6.79)$$

We can again improve the wave function by varying the ratio $x(R)/y(R)$ to minimize $E(R)$. But substitution of Eqs. 6.73 and 6.78 in Eq. 6.79 gives

$$\psi_{\mathrm{MO,mixed}} = \frac{\alpha(1)\beta(2) - \alpha(2)\beta(1)}{2\sqrt{2}(1 + S_{AB})}[x(R)(\psi_{\mathrm{ionic}} + \psi_{\mathrm{covalent}})$$
$$+ y(R)(\psi_{\mathrm{ionic}} - \psi_{\mathrm{covalent}})]$$
$$= \frac{\alpha(1)\beta(2) - \alpha(2)\beta(1)}{2\sqrt{2}(1 + S_{AB})}\{[x(R) + y(R)]\psi_{\mathrm{ionic}}$$
$$+ [x(R) - y(R)]\psi_{\mathrm{covalent}}\}$$
$$= a'(R)\psi_{\mathrm{covalent}} + b'(R)\psi_{\mathrm{ionic}}, \qquad (6.80)$$

an equation of the same form as Eq. 6.76. So since ψ_{covalent} and ψ_{ionic} are exactly the same functions in both cases, the variation to minimize $E(R)$ must give identical results in both cases: $a(R) = a'(R)$, $b(R) = b'(R)$. Thus the mixed wave function obtained is the same whether it is derived according to Eq. 6.76 or Eq. 6.79. It is purely a matter of taste and convenience which form one chooses for computation.

In MO language, a wave function like Eq. 6.79 is said to involve *configuration interaction* (CI). A "configuration" here has the same meaning as in our theory of atomic structure,[20] a particular distribution of

---

[20] One can also carry out CI calculations on atoms, writing the total wave function as a sum of determinantal functions. This is the usual way of improving on the ordinary SCF calculations. As we saw in Chapter 5, however, the single-configuration model works rather well for atoms—at least for those with closed shells or a single electron outside closed shells.

electrons among orbitals. In this case our ground-state wave function is a mixture of the two configurations $(\sigma_g 1s)^2$ and $(\sigma_u 1s)^2$. One can improve the wave function still further by adding additional configurations, as long as they have the correct symmetry properties. Given enough configurations, one should be able to approach arbitrarily close to the true wave function, although this convergence is usually rather slow. One is not restricted to LCAO orbitals, of course, and it is possible to define other orbitals that give more rapid convergence.

With orbitals chosen to give the most rapid convergence of a CI series, the so-called *natural* orbitals, the wave function for H_2 accounts for over 90% of the energy associated with electron correlation.[21] By using wave functions containing the interelectronic distance explicitly, it is possible to predict energies of dissociation and ionization for H_2 that lie within about 10^{-4} eV or less of the experimental values.[22]

There were, for a period, extended controversies over whether MO or VB wave functions were better for any particular molecule. In retrospect these arguments lose their force, because the two methods converge with the next approximation. This was recognized very early, of course, but for molecules larger than H_2 people thought, until the mid-1950s, that the next level would be too difficult to carry out. Since about 1960, however, computers have made quite extensive CI calculations feasible. Yet there are still many instances—in large molecules, or when only an approximate result is needed—in which the simple MO and VB approximations are used. Each has its advantages, and we shall use both approaches in the next few chapters. The simple VB method gives more accurate results in cases where correlation is very important; the MO calculations are much easier, and offer convenient ways to interpret spectra and bonding pictorially. Excited states are quite awkward to handle by VB methods. The "structures" of the VB method are easily visualizable chemical species, so that one can often use "chemical intuition" to guess the properties of the ground states of molecules that involve several such structures. On the other hand, electronic spectra are more easily interpreted in terms of single-electron transitions between molecular orbitals. The choice of method thus depends on what one wants to know, how accurate a result one wants, and how much work one is prepared to do.

Before leaving the ground state of H_2, we should say something about the most accurate calculations that have been performed. The best available CI calculations account for only about 97% of the correlation energy. One can do better by writing a function that explicitly contains electron correlation in the form of r_{12} terms, like Eq. 5.14 for the He atom. Given enough adjustable parameters to vary, one can get quite close to the true $E(R)$ curve, and thus presumably to the true wave function. Curve 8 of Fig. 6.10 was obtained by varying an 11-parameter function. Calculations have been made with as many as 50 parameters, giving an $E(R)$ curve that is correct within experimental error. Just as in atoms, unfortunately, it is impractical to extend calculations of this type to more than a few electrons.

[21] S. Hagstrom and H. Shull, *Rev. Mod. Phys.* **35**, 624 (1963); E. R. Davidson and L. L. Jones, *J. Chem. Phys.* **37**, 2966 (1962); W. D. Lyons and J. O. Hirschfelder, *J. Chem. Phys.* **46**, 1788 (1967).

[22] W. Kolos and L. Wolniewicz, *J. Chem. Phys.* **41**, 3663, (1964); **43**, 2429 (1965); **49**, 404 (1968); **51**, 1417 (1969); **45**, 509 (1966); **48**, 3672 (1968); *Chem. Phys. Lett.* **24**, 457 (1974).

Thus far we have examined in detail only the ground state of the H_2 molecule. We have mentioned two excited states: the $^3\Sigma_u$ state whose VB approximation is given by Eqs. 6.69–6.71, and the $^1\Sigma_g$ state whose MO approximation is given by Eq. 6.77. In this section we shall consider the excited states of H_2 more systematically; $E(R)$ curves for the ground state $(X\,^1\Sigma_g^+)$ and several excited states are given in Fig. 6.11.[22]

To begin with, let us find the MO representation of the first excited state. This is obviously a state in which one electron is in the lowest-energy orbital, $\sigma_g 1s$, whereas the other electron is in the lowest excited orbital, $\sigma_u 1s$. Previously we have considered the configurations $(\sigma_g 1s)^2$ (the ground state) and $(\sigma_u 1s)^2$, both of which can give only singlet states. The exclusion principle tells us that if both electrons are in the same orbital they must have opposite spins, giving $S = 0$. However, for the configuration $(\sigma_g 1s)(\sigma_u 1s)$ both singlet and triplet states are possible. The symmetries of these states follow the same rules as we developed in Section 6.8: The singlet state has a symmetric spatial function and antisymmetric spin function, the triplet states the reverse. The singlet state is thus represented by

$$\psi_1[(\sigma_g 1s)(\sigma_u 1s)]$$
$$= C_1[\sigma_g 1s(1)\sigma_u 1s(2) + \sigma_g 1s(2)\sigma_u 1s(1)][\alpha(1)\beta(2) - \alpha(2)\beta(1)], \quad (6.81)$$

and the triplet state by

$$\psi_2[(\sigma_g 1s)(\sigma_u 1s)]$$
$$= C_2[\sigma_g 1s(1)\sigma_u 1s(2) - \sigma_g 1s(2)\sigma_u 1s(1)]\alpha(1)\alpha(2),$$

$$\psi_3[(\sigma_g 1s)(\sigma_u 1s)]$$
$$= C_3[\sigma_g 1s(1)\sigma_u 1s(2) - \sigma_g 1s(2)\sigma_u 1s(1)]\beta(1)\beta(2), \quad (6.82)$$

$$\psi_4[(\sigma_g 1s)(\sigma_u 1s)]$$
$$= C_4[\sigma_g 1s(1)\sigma_u 1s(2) - \sigma_g 1s(2)\sigma_u 1s(1)][\alpha(1)\beta(2) + \alpha(2)\beta(1)].$$

These functions are analogous to Eqs. 6.68–6.71, but with the atomic orbitals of the latter replaced by molecular orbitals. In fact, if we expand the spatial part of Eqs. 6.82 in atomic orbitals, we obtain

$$\sigma_g 1s(1)\sigma_u 1s(2) - \sigma_g 1s(2)\sigma_u 1s(1)$$
$$= \frac{[1s_A(1) + 1s_B(1)][1s_A(2) - 1s_B(2)] - [1s_A(2) + 1s_B(2)][1s_A(1) - 1s_B(1)]}{2(1 + S_{AB})}$$
$$= \frac{1s_A(2)1s_B(1) - 1s_A(1)1s_B(2)}{1 + S_{AB}}, \quad (6.83)$$

which differs only by a constant multiplier from the spatial part of Eqs. 6.69–6.71; all the "ionic" terms like $1s_A(1)1s_A(2)$ cancel out. In short, for the first excited state of H_2 the simplest MO and VB wave functions turn out to be identical.

As we pointed out earlier, this first excited state has the symmetry $^3\Sigma_u$. Its $E(R)$ curve, labeled $b\,^3\Sigma_u^+$ in Fig. 6.11, is everywhere repulsive. Like the ground state it gives $H(1s) + H(1s)$ in the limit $R \to \infty$. The repulsive nature of this state is vividly exhibited by a hydrogen discharge lamp, the ultraviolet light from which consists largely of continuous radiation. Any state with a minimum in the potential energy curve is a

6.10
H_2: Excited Electronic States

---

[22] The superscript $^+$ in the Σ term symbols means that the wave function is unchanged by reflection in a plane containing the internuclear axis. We shall say more about this symmetry property in the next chapter.

bound state, and thus has quantized energy levels (in this case vibrational-rotational energy levels, which we shall discuss in the next chapter); but a state with no minimum is free and unquantized. Electrons dropping from higher orbitals into the $\sigma_u 1s$ orbital to give this triplet state thus produce continuous emission.

What about the singlet state represented by $(\sigma_g 1s)(\sigma_u 1s)$? If we expand the spatial part of Eq. 6.81 in atomic orbitals, we obtain

$$\sigma_g 1s(1)\sigma_u 1s(2) + \sigma_g 1s(2)\sigma_u 1s(1) = \frac{1s_A(1)1s_A(2) - 1s_B(1)1s_B(2)}{1 + S_{AB}}; \quad (6.84)$$

here the "covalent" terms cancel out, and we have the odd-parity equivalent of the ψ_{ionic} of Eq. 6.75. This state has the symmetry $^1\Sigma_u$, and is labeled $B\,^1\Sigma_u^+$ in Fig. 6.11. Its energy is relatively high: Eq. 6.84 predicts that it should give $H^-(1s^2) + H^+$ in the long-range limit. Actually, it gives $H(1s) + H(2s)$, the energy of which is lower than that of two ions, but still quite high. Nevertheless, there is a minimum in the $E(R)$ curve, so that this configuration can exist as a stable excited state.

As for the $^1\Sigma_g$ state whose simplest MO representation is the doubly excited $(\sigma_u 1s)^2$, its energy is obviously still higher. The simple MO expansion (Eq. 6.78) again gives the wrong energy as $R \to \infty$, in fact the same wrong energy as we obtained for the ground state. The lowest excited $^1\Sigma_g$ state actually has *two* minima in the $E(R)$ curve, which appear spectroscopically to be distinct states (since each has its own set of vibrational-rotational levels); these are labeled $E\,^1\Sigma_g^+$ and $F\,^1\Sigma_g^+$ in Fig. 6.11. A single MO configuration cannot give a complicated curve like this; a much better representation is obtained by taking a mixture of the configurations $(\sigma_g 1s)(\sigma_g 2s)$ and $(\sigma_u 1s)^2$. If we wrote such a wave function in a form like Eq. 6.79, we would find that the coefficients $x(R)$ and $y(R)$ are rather rapidly varying functions of R. States of this type are fairly rare and can be considered as pathological cases. In most such cases, including this one, the outer minimum is described by a wave function with a large amount of ionic character—in this case the configuration $(\sigma_u 1s)^2$, which like $(\sigma_g 1s)^2$ is half "ionic" at long range.

We have now described the ground state and three excited states (one of them a triplet), and these are only the states that can be constructed from $1s$ atomic orbitals. This by no means exhausts the excited states of the H_2 molecule; there are, in fact, an infinite number of such states. Obviously we can also construct molecular states from H atoms one or both of which are themselves in excited states. Each combination of atomic orbitals can give two or more distinct molecular wave functions, corresponding to all the possible symmetric and antisymmetric combinations of electrons and nuclei.[23] Thus there are many more possible

[23] For example, consider the states formed from $H(1s) + H(2s)$. There are four possible spatial wave functions, which in the valence bond approximation can be written:

$$1s_A(1)2s_B(2) + 1s_A(2)2s_B(1) + 1s_B(1)2s_A(2) + 1s_B(2)2s_A(1)$$
$$\text{(symmetric in both electrons and nuclei, } ^1\Sigma_g),$$

$$1s_A(1)2s_B(2) - 1s_A(2)2s_B(1) + 1s_B(1)2s_A(2) - 1s_B(2)2s_A(1)$$
$$\text{(symmetric in nuclei, antisymmetric in electrons, } ^3\Sigma_g),$$

$$1s_A(1)2s_B(2) + 1s_A(2)2s_B(1) - 1s_B(1)2s_A(2) - 1s_B(2)2s_A(1)$$
$$\text{(symmetric in electrons, antisymmetric in nuclei, } ^1\Sigma_u),$$

$$1s_A(1)2s_B(2) - 1s_A(2)2s_B(1) - 1s_B(1)2s_A(2) + 1s_B(2)2s_A(2)$$
$$\text{(antisymmetric in both electrons and nuclei, } ^3\Sigma_u).$$

Each of these describes a physically distinct state of the molecule (three degenerate states for the triplets).

electronic states of the molecule than of the H atom. Of course, only a limited number of the excited states have actually been observed, and not all of those have been fully analyzed.

To the extent that the molecular orbital approximation is valid, we can speak of singly and doubly excited states of H_2, the singly excited states being those in which one electron remains in the lowest $(1\sigma_g)$ orbital. We have seen that the first excitation requires a large amount of energy, even for the lowest stable excited states. Just as in the helium atom (Section 5.3), a doubly excited state would have an energy well above the ionization limit, which in this case is the ground-state energy of $H_2^+ + e^-$. On the scale of Fig. 6.11, the minimum in the H_2^+ $E(R)$ curve lies at $+0.397$ hartree. Since all the stable singly excited states have energies not much below this,[24] one would expect them to bear some resemblance to the H_2^+ molecule-ion. In fact the $1\sigma_g$ orbital is very like the H_2^+ wave function, extending over a quite small region of space around the nuclei, whereas the excited electron occupies a very large orbital, which must resemble an excited orbital of the helium atom.[25] The higher the excitation, the less the orbitals overlap and the better the orbital approximation works. In effect, one has an H_2^+ ion with an electron orbiting it at long range. A highly excited electron must see the H_2^+ core as something close to a point charge. The orbital energies are then given reasonably well by a "Rydberg formula" like Eq. 2.68, and we speak of these highly excited states as molecular *Rydberg states*. There must be an infinite number of Rydberg states for any molecule having a stable molecule-ion like H_2^+. The excited electron is far from the nuclei, so it has very little effect on the bonding force between them. Thus the $E(R)$ curves for excited states of H_2 are often quite similar to that for H_2^+, as can be seen by comparing Figs. 6.6 and 6.11.

What about the doubly excited states? Not just for H_2 but for most molecules, doubly excited states generally have energies higher than the energy needed to remove a single electron. In the case of H_2, this means that the excited state has an energy higher than at least some states of H_2^+. Such an excited state can thus lose energy by getting rid of (ionizing) one of its electrons, the excess energy becoming kinetic energy of the ionized electron. This process is called *autoionization* or *preionization*,[26] and normally takes place with no emission of radiation. Autoionization thus competes with the various kinds of radiative transitions (Section 4.5) as a way for some excited states to decay. Although the relative efficiency of these decay processes varies widely with the states involved, it is clear that the possibility of autoionization shortens the lifetime of an excited state. In practice, it is mostly doubly excited states that can undergo autoionization, and many such states have very short lifetimes ($\approx 10^{-11}$–10^{-12} s) compared with typical singly excited states (10^{-7}–10^{-8} s). When it can occur, ionization is usually the mode selected by an excited molecule or atom to relieve itself of excess energy.

This completes our discussion of the electronic states of the H_2 molecule. But each of these states encompasses many vibrational and rotational energy levels. In the next chapter we investigate these levels and the spectra to which they give rise, and we extend our analysis to diatomic molecules more complicated than H_2.

[24] Those shown in Fig. 6.11 have $E(\infty) = +0.375$ hartree, corresponding to $H(1s) + H(2s$ or $2p)$, whereas H_2^+ has $E(\infty) = +0.500$ hartree, corresponding to $H(1s) + H^+ + e^-$.

[25] As we pointed out for H_2^+, highly excited states are best described in terms of the united-atom limit. The united-atom descriptions of the states in Fig. 6.11 are as follows (cf. Fig. 6.8):

$X\,^1\Sigma_g^+$	$E\,^1\Sigma_g^+, a\,^3\Sigma_g^+$	$B\,^1\Sigma_u^+, b\,^3\Sigma_u^+$	$c\,^1\Pi_u, c\,^3\Pi_u$
$(1s\sigma_g)^2$	$(1s\sigma_g)(2s\sigma_g)$	$(1s\sigma_g)(2p\sigma_u)$	$(1s\sigma_g)(2p\pi_u)$

[26] The corresponding process in excited atoms is called the *Auger effect*.

Appendix 6A

Orthogonality

The concept of orthogonality is a generalization of the geometric concept of perpendicularity. Two vectors **a** and **b** with real components are perpendicular, or orthogonal, if their scalar product is zero:

$$\mathbf{a} \cdot \mathbf{b} = a_x b_x + a_y b_y + a_z b_z$$
$$= |\mathbf{a}| \, |\mathbf{b}| \cos \theta_{ab}$$
$$= 0 \quad \text{if and only if (iff) } \mathbf{a} \text{ and } \mathbf{b} \text{ are orthogonal.} \quad (6A.1)$$

The first of these equations suggests (but does not prove) that the two vectors have no basis vectors in common. The value of a scalar product is independent of the choice of orientation of the coordinate system. If **a** and **b** are orthogonal, one can always choose an orientation for the coordinate system so that **a** and **b** have no common basis vectors; for example, **a** and **b** can be chosen to lie along the x and y axes.

The generalization of orthogonality from ordinary vectors in three-dimensional space to vectors in an n-dimensional abstract space is straightforward: If $\mathbf{a} = \{a_1, a_2, \ldots, a_n\}$ and $\mathbf{b} = \{b_1, b_2, \ldots, b_n\}$, then $\mathbf{a} \cdot \mathbf{b} = a_1 b_1 + a_2 b_2 + \cdots + a_n b_n$; **a** and **b** are orthogonal iff $\mathbf{a} \cdot \mathbf{b} = 0$. If the components of the vectors are complex numbers instead of real numbers, then one writes $\mathbf{a} \cdot \mathbf{b} = a_1{}^* b_1 + \cdots + a_n{}^* b_n$ for the scalar product. Again, if **a** and **b** are orthogonal, they are built from different, mutually exclusive sets of basis vectors of the space.

To generalize the idea of orthogonality from vectors to functions we replace the summation $a_1 b_1 + \cdots + a_n b_n$ with an integration over the continuous variable or variables of the functions. Thus, if $\phi_A(x)$ and $\phi_B(x)$ are real and exist for $-\infty < x < \infty$, then the equivalent of the scalar product of ϕ_A and ϕ_B is the integral of their product over the range of their argument. If x ranges from $-\infty$ to ∞,

$$S_{AB} = \int_{-\infty}^{\infty} \phi_A(x) \phi_B(x) \, dx. \quad (6A.2)$$

If $\phi_A(x)$ and $\phi_B(x)$ are complex and $-\infty < x < \infty$, then we write, for the equivalent of their scalar product, Eq. 6.30:

$$S_{AB} = \int_{-\infty}^{\infty} \phi_A{}^*(x) \phi_B(x) \, dx. \quad (6.30)$$

The functions $\phi_A(x)$ and $\phi_B(x)$ are orthogonal if and only if $S_{AB} = 0$.

Sometimes the range of the argument is finite. For example, consider $\phi_A = \sin mx$ and $\phi_B = \sin nx$, for $0 \le x \le 2\pi$:

$$S_{AB} = \int_0^{2\pi} \sin mx \sin nx \, dx = 0 \quad \text{if } m \ne n,$$
$$= \tfrac{1}{2} \quad \text{if } m = n, \quad (6A.3)$$

so $\sin mx$ and $\sin nx$ are orthogonal on the interval $0 \le x \le 2\pi$.

The second equality of Eq. 6A.1 shows that $\mathbf{a} \cdot \mathbf{b}$ is the projection of the vector **a** on the vector **b**, which, as the first equality shows, is the same as the projection of **b** on **a**. We can likewise think of Eq. 6A.1 or Eq. 6.30 as the projection of function ϕ_A on ϕ_B. The projection of one vector or function on another is the same, geometrically, as the overlap of one vector or function with the other. But the overlap of functions has a physical interpretation that goes beyond the simple geometric model. In physical terms, if two functions representing states of a system are orthogonal, then the states are mutually exclusive: a $1s$ state and a $2s$ state of the hydrogen atom represent two mutually exclusive states of that atom. The wave functions ψ_0 and ψ_1 of Eqs. 6.26 and 6.27 or 6.33 and 6.34 are orthogonal, and correspond to mutually exclusive states of an electron in the H_2^+ molecule.

Appendix 6B

Hermitian Operators

In Section 3.8, Postulate III supposes that every variable of classical mechanics can be represented by a linear mathematical operator. The statement of that postulate now needs to be made a bit stricter: To every variable of classical mechanics there corresponds a linear *Hermitian* operator (after the French mathematician of the nineteenth century, Charles Hermite). The reason: Hermitian operators have two properties that we would like observables to exhibit. For Hermitian operators, the eigenvalues, which correspond to the values the property can assume (Postulate IV), are *real*, and the eigenfunctions corresponding to different eigenvalues are *orthogonal*. Thus the observables are real rather than complex numbers, as we want them to be, and, from the interpretation of orthogonality in Appendix 6A, the eigenstates with different eigenvalues must be mutually exclusive states of the system.

An operator R is Hermitian iff (if and only if)

$$\int \phi_A{}^* R\phi_B \, d\tau = \int (R\phi_A)^* \phi_B \, d\tau. \qquad (6B.1)$$

To give this a physical interpretation, note that $R\phi_B$ is some new function ϕ_C, corresponding to the state of the system initially described by ϕ_B but then acted upon by the process represented by the operator R. Hence $\int \phi_A{}^* R\phi_B \, d\tau$ is the projection of the state of the system in state C, represented by ϕ_C, onto the state A, represented by ϕ_A. If C and A are mutually exclusive, $\int \phi_A{}^* \phi_C \, d\tau \equiv \int \phi_A{}^* R\phi_B \, d\tau = 0$; otherwise A and C share some aspects and have nonzero overlap.

The state represented by $R\phi_A$ corresponds to a system initially in A and then acted upon by the operation corresponding to the operator R. Call this state D, and represent it with ϕ_D. Then $\int (R\phi_A)^* \phi_B \, d\tau$ is the same as $\int \phi_D{}^* \phi_B \, d\tau$, the projection or overlap of state B onto state D. The operator R is Hermitian if the projection of C on A, $\int \phi_A{}^* \phi_C \, d\tau$, is the same as the projection of B on D, $\int \phi_D{}^* \phi_B \, d\tau$. In other words, an operator is Hermitian if and only if the same result occurs (but with proper allowance for complex conjugation, where $\sqrt{-1}$ appears) whether the operation R is performed on B and the result is projected onto A, or B is projected onto the result of performing R on A.

Let us show that the eigenvalues of a Hermitian operator are real. If Eq. 6B.1 holds and if $R\phi_A = R_A\phi_A$, so ϕ_A is an eigenfunction of R, then

$$\int \phi_A{}^* R\phi_A \, d\tau = R_A \int \phi_A{}^* \phi_A \, d\tau$$

$$= \int (R\phi_A)^* \phi_A \, d\tau = R_A{}^* \int \phi_A{}^* \phi_A \, d\tau, \qquad (6B.2)$$

so $R_A = R_A{}^*$ and R_A must be real. Similarly, if $R\phi_B = R_B\phi_B$, so that both ϕ_A and ϕ_B are eigenfunctions of R, and if $R_A \neq R_B$, then $\int \phi_A{}^* \phi_B \, d\tau = 0$, since

$$\int \phi_A{}^* R\phi_B \, d\tau = R_B \int \phi_A{}^* \phi_B \, d\tau$$

$$= \int (R\phi_A)^* \phi_B \, d\tau = R_A \int \phi_A{}^* \phi_B \, d\tau. \qquad (6B.3)$$

The equalities of 6B.3 can hold only if the overlap integral of ϕ_A and ϕ_B vanishes. Therefore the Hermitian operators do indeed have eigenvalues that are real and can correspond to observable properties, and the physical states corresponding to different observable values, that is, to different eigenvalues, are mutually exclusive.

Further Reading

Herzberg, G., *Molecular Spectra and Molecular Structure, Volume I, Spectra of Diatomic Molecules*, 2nd ed. (Van Nostrand Reinhold, New York, 1950).

Hurley, A. C., *Introduction to the Electron Theory of Small Molecules* (Academic, London, 1976).

Karplus, M., and Porter, R. N., *Atoms and Molecules* (Benjamin, Menlo Park, Calif., 1970), esp. chaps. 5, 6 and 7.

Kauzmann, W., *Quantum Chemistry* (Academic, New York, 1957), chap. 11A, B.

Kondratyev, V., *The Structure of Atoms and Molecules* (Noordhoof, Groningen, The Netherlands, 1964), chap. 7.

Mulliken, R. S., and Ermler, W. C., *Diatomic Molecules, Results of ab initio Calculations* (Academic, New York, 1977), esp. chaps. I–III.

Slater, J. C., *Quantum Theory of Molecules and Solids*, Volume I (McGraw-Hill, New York, 1963), esp. chaps. 1–4.

Problems

1. In making the Born–Oppenheimer approximation, one supposes that the molecular wave function is separable (Eq. 6.2), and thus that the electronic wave function $\psi_{elec}(r, R)$ is an eigenfunction of an electronic Hamiltonian H_{elec} for each fixed R. If one supposes that R is a quantum mechanical variable, then H_{nuc} can act on $\psi_{elec}(r, R)$ as well as on $\psi_{nuc}(R)$. When $\psi(r, R)$ is separable but H_{nuc} is allowed to act on ψ_{elec}, what terms occur in $\int \psi^*(r, R) H \psi(r, R)\, d\tau$ that do not appear if H_{nuc} is only allowed to act on $\psi_{nuc}(R)$?

2. The binding energy of the deuteron, ^{2}H (D), is 2.2 MeV. The binding energy of the α particle is approximately 28.5 MeV. The first ionization potential of He is 24.5 eV; the second is 54.14 eV. Using a logarithmic scale for the energy, draw a curve of the total energy, including the nuclear binding energy, for the D_2 molecule going from the equilibrium internuclear distance down to $R = 0$. Follow the nuclear repulsion up to $R \sim 10^{-6} R_e$ and sketch the rest of the curve schematically. You may wish to put R on a nonlinear scale.

3. Show that if, in the diatomic molecule AB, the nuclear charges Z_A and Z_B are unequal, with $Z_A > Z_B$, then the region far enough from the molecule and out beyond B is a bonding region. At what distance beyond B is any electronic charge on the internuclear axis exactly nonbonding?

4. Construct the surfaces on which the bonding force is zero for the diatomic molecule LiF, that is, for $Z_A = 3$, $Z_B = 9$.

5. Figure 6.3 shows contours on which the bonding force of Eq. 6.8 is zero. For the system of two protons, calculate enough points to sketch contours for curves on which the bonding force is not zero; develop the curves for two negative (antibonding) values of $F_{bonding}$ and two positive values of $F_{bonding}$. Express your values in terms of $F_{bonding} \times (4\pi\epsilon_0/eq)$; give the value of $(4\pi\epsilon_0/eq)$, so a reader could translate your plot into SI units of force. (Eliminate one distance, r_A or r_B, and one cosine, $\cos\theta_A$ or $\cos\theta_B$, from Eq. 6.8. You may want to use a programmable calculator or computer to map $F_{bonding}$, and then construct contours from the map.)

6. Compute the charge density in electrons per cubic angstrom and the expectation value of the electron's kinetic energy in the neighborhood of a nucleus and at the midpoint of the H–H axis in H_2^+, for the two normalized molecular orbitals Eqs. 6.33 and 6.34. Estimate the local kinetic energy from the slopes of the functions at their cusps and at $R/2$.

7. A proton B approaches a hydrogen atom A. At what internuclear distance R is the most probable electron–nuclear distance r_A of an undisturbed hydrogen atom equal to R? Where is the distance R equal to the mean value of r_A? (You may wish to look back to Chapter 4.)

8. The graphs of Fig. 6.7 show the charge density of the orbitals of Eqs. 6.33 and 6.34 along the internuclear axis. Construct similar curves for a line parallel to the internuclear axis but a distance $0.1a_0$ away from it.

9. The overlap integral of Eq. 6.30 can be evaluated easily by making the coordinate transformation $\lambda = (r_A + r_B)/R$, $\mu = (r_A - r_B)/R$.
   (a) Show that surfaces of constant λ are ellipses and surfaces of constant μ are hyperbolas.
   (b) Show that Eq. 6.33 and 6.34 take the forms

$$\psi_0 = \text{constant} \times e^{-\lambda R/2}(e^{\mu R/2} + e^{-\mu R/2})$$

and

$$\psi_1 = \text{constant} \times e^{-\lambda R/2}(e^{\mu R/2} - e^{-\mu R/2}),$$

or

$$\psi_0 = \text{constant} \times e^{-\lambda R/2} \cosh\left(\frac{\mu R}{2}\right)$$

and

$$\psi_1 = \text{constant} \times e^{-\lambda R/2} \sinh\left(\frac{\mu R}{2}\right).$$

   (c) Show that the variable λ ranges from 1 to ∞, and that μ ranges from -1 to $+1$.

10. Show that the overlap integral

$$S_{AB} = \frac{1}{\pi} \int_{\substack{\text{all} \\ \text{space}}} e^{-\alpha r_A} e^{-\alpha r_B} \, d\tau$$

$$= e^{-\alpha R}(1 + \alpha R + \tfrac{1}{3}\alpha^2 R^2).$$

Use the fact that

$$\int_0^\infty x^n e^{-ax} \, dx = \frac{n!}{a^{n+1}};$$

to evaluate the term involving $e^{-\alpha(r_A + r_B)}$, use the transformation of Problem 9 and the fact that the volume element $dx\,dy\,dz$, when transformed to the coordinates λ, μ, ϕ, is $(R^3/8)(\lambda^2 - \mu^2)\,d\lambda\,d\mu\,d\phi$.

11. Show that atomic orbitals follow the rule that they are even (g) with respect to their nuclei if l is zero or even, and odd (u) if l is odd. In an H_2^+ molecule we can construct both g and u molecular orbitals from sums and differences of $p\sigma$ orbitals on the separated atoms. Resolve this apparent paradox.

12. Draw a correlation diagram connecting the orbital energies of the separated-atom limit with those of the united-atom limit for the one-electron molecule HeH^{2+}.

13. Write the analogs of Eqs. 6.58, 6.62, 6.68, and 6.69 if particles 1 and 2 are bosons.

14. What are the charge densities of the molecular orbital wave function, Eq. 6.73, and the valence bond wave function, Eq. 6.74, at the midpoint between A and B and at the nucleus A, when R has the value $1.415a_0$? Which function would you say delocalizes the charge more?

15. The molecule-ion HeH^+ has been studied both experimentally and theoretically. Its equilibrium internuclear distance is $1.4632a_0$. Construct a potential curve for the electrons along the He–H axis analogous to Fig. 6.4a and indicate the energies of the ground-state ($1s$) levels of the separated atoms. Sketch the free-atom $1s$ orbitals of H and of He^+, each centered at its appropriate nucleus. The lowest molecular orbital of HeH^+ is *not* well expressed by the form constant $\times [\varphi_H(1s) + \varphi_{He^+}(1s)]$. Explain in physical terms why the lowest state of this species contains much more of one $1s$ orbital than of the other. Which predominates? What is the energy of the orbitals of He^+ with $n = 2$? What atomic orbitals of hydrogen and of the helium ion are likely to be important in the first excited state of HeH^+?

16. Draw a curve of the interference term of Eq. 6.28, $2\,\mathrm{Re}(\varphi_A^* \varphi_B)$, for the midpoint between the two nuclei in H_2^+, as a function of the internuclear distance R.

17. From Eqs. 6.41 and 6.42, and footnote 8 on page 223, compute the internuclear distance at which $E_0(R) = 0$, and the distances at which $E_0(R) = +1$ eV and $E_1(R) = +1$ eV. The latter correspond to the classical distances of closest approach for a proton colliding with a hydrogen atom along the $1\sigma_g$ and $1\sigma_u$ potential curves, respectively.

18. One of the dissociation limits of H_2 is the ion pair $H^+ + H^-$. Locate this dissociation limit on an energy scale, relative to the neutral states of the form $H(1s) + H(nl)$ and $H(2s) + H(nl)$. Where does the ionic curve of $-e^2/4\pi\epsilon_0 R$ fall on the diagram of Fig. 6.11? Can you associate any of the potential curves or portions of the curves of Fig. 6.11 with an electrostatic attraction $E(R)$ in Eq. 6.11?

19. If two nuclei move at a relative velocity of order 1% or more of the average speed of the electrons attached to them, one can be concerned about the validity of the adiabatic or Born–Oppenheimer approximation. What energy must a proton have, in its motion relative to a hydrogen atom, to achieve roughly 1% of the average speed of the electron in the transient H_2^+ molecule? Use the energy of the H_2^+ molecule at its equilibrium distance to estimate the electron's speed. (For a particle bound by Coulomb forces, $\langle \mathrm{K.E.}\rangle = -\frac{1}{2}\langle \mathrm{P.E.}\rangle$.)

20. Recall from Section 5.5 that the hydrogen atom forms a stable negative ion, H^-, and that the electron affinity of the hydrogen atom is approximately 0.75 eV. The H_2 molecule in its ground electronic and vibrational state does not form a stable negative ion. Give a physical interpretation of why no stable H_2^- can be made by attaching a slow electron to H_2, and explain what occurs

when a hydride ion (H^-) and a hydrogen atom undergo a close, slow collision. Sketch a potential curve for $H^- + H$ together with the curve for the ground state of $H + H + e^-$ as given in Fig. 6.11.

21. By using the curves of Fig. 6.11, assuming that electronic excitation of a molecule of H_2 in its ground state occurs from the lowest point of the ground-state potential and that neither the position nor the momentum of the nuclei can change during the excitation process, estimate the wavelengths at which one might expect to observe transitions to the $b^3\Sigma_u^+$ and the $B^1\Sigma_u^+$ states.

22. From your knowledge of the spatial dimensions of the $1\sigma_g$ orbital of H_2 and of the atomic orbitals of the hydrogen atom, estimate at approximately what principal quantum number n the excited-state orbitals of H_2 could be well represented by atomic orbitals. How will the accuracy of this representation depend on the angular momentum $l\hbar$?

7

More About Diatomic Molecules

Our study of H_2^+ and H_2 has given us some idea of the basic principles of chemical bonding and the electronic properties associated with the simple chemical bond. There is much more to learn about molecular structure, even for the relatively simple case of diatomic molecules. In this chapter we consider the general properties of diatomic molecules.

In Chapter 6 we were concerned only with the distribution and energy of the electrons; we completely neglected the motion of the atomic nuclei. But nuclei are not fixed in space. They can oscillate around their equilibrium positions as if the bond were a spring; they can revolve about the molecule's center of mass; and the molecule as a whole can move in space. There are energy levels associated with all these motions, and we must find out what they are. Fortunately, we have done most of the work already, in terms of simple models. As we develop a description of these energy levels, we shall be able to analyze the full richness of molecular spectroscopy.

We are not through with electrons either. Although H_2 does illustrate the nature of the chemical bond, there are many phenomena that appear only in more complicated molecules. We shall discuss the general properties of diatomic molecules, mainly in terms of the molecular orbital approach, and go on to study several interesting classes of molecules in some detail. These include homonuclear molecules such as N_2, which have the same symmetry properties as H_2; first-row hydrides such as OH, the simplest heteronuclear molecules; and several series of related molecules in which trends can be discerned.

7.1 Vibrations of Diatomic Molecules

Let us look a little more carefully at a molecule of H_2 in its ground electronic state. We know the $E(R)$ curve, but just what does this curve mean? According to our interpretation in Chapter 6, $E(R)$ is the electronic energy of the molecule at internuclear distance R, calculated *with the nuclei fixed*. But the uncertainty principle ensures that no real molecule can have fixed nuclei. The more precisely we know the positions of the nuclei, the more uncertain become their momenta. Thus the best we can say (in classical language) is that the nuclei oscillate or vibrate within a certain range about their equilibrium positions. This language still assumes the molecule as a whole (i.e., its center of mass) to be fixed in space, but this cannot be generally true either: The molecule almost certainly has some translational momentum. Finally, the orientation of the molecule cannot be fixed in space, or the uncertainty in its angular momentum would be infinite. Thus we must also consider the molecule's rotation about its center of mass. We wish to describe all these types of motion.

Our starting point is again the Born–Oppenheimer approximation, which says that the nuclear and electronic motions are separable. This is justified by the fact that electrons move much faster than nuclei. Just as we could consider the nuclei at rest when treating the electronic motion, we can consider the electrons as a smoothly averaged distribution that adjusts at once to the nuclear motion. The Born–Oppenheimer approximation is expressed by Eqs. 6.2 and 6.3; for the nuclear motion it yields Eq. 6.4, which we can rewrite as

$$\left(\frac{\mathbf{p_A}^2}{2m_A} + \frac{\mathbf{p_B}^2}{2m_B}\right)\psi_{\text{nucl}}(R_n) = [E - E(R)]\psi_{\text{nucl}}(R_n). \tag{7.1}$$

Here R_n is shorthand for all the nuclear coordinates, E is the molecule's total energy, and $E(R)$ is its electronic energy—which is also the *effective* potential energy field in which the nuclei move.

Now we make another assumption, that the translational, vibrational, and rotational motions of the nuclei are also separable from one another. (This is not quite accurate, and later we shall mention some corrections to this approximation.) We thus have

$$\mathbf{H}_{\text{nucl}} \equiv \frac{\mathbf{p_A}^2}{2m_A} + \frac{\mathbf{p_B}^2}{2m_B} = \mathbf{H}_{\text{trans}} + \mathbf{H}_{\text{vib}} + \mathbf{H}_{\text{rot}}, \tag{7.2}$$

$$E - E(R) = E_{\text{trans}} + E_{\text{vib}} + E_{\text{rot}}, \tag{7.3}$$

and

$$\psi_{\text{nucl}}(R_n) = \psi_{\text{trans}}\psi_{\text{vib}}\psi_{\text{rot}}; \tag{7.4}$$

$\mathbf{H}_{\text{trans}}$ and ψ_{trans} depend only on the center-of-mass coordinates, $\mathbf{H}_{\text{vib}}$ and ψ_{vib} only on the internuclear distance R, and $\mathbf{H}_{\text{rot}}$ and ψ_{rot} only on the angles giving the molecule's orientation in space. For each type of motion we obtain a wave equation in the form $\mathbf{H}_i\psi_i = E_i\psi_i$. In a gas, the translational wave equation is simply that for a free particle in three dimensions, the solution of which we already know. The rotational motion will be considered in the next section; let us now analyze the vibrational motion.

Classically, the total energy of the nuclei can be written as

$$E - E(R) = \left[\frac{p_X^2 + p_Y^2 + p_Z^2}{2M}\right] + \left[\frac{p_R^2}{2\mu} + V(R)\right]$$
$$+ \left[\frac{1}{2\mu R^2}\left(p_\theta^2 + \frac{p_\phi^2}{R^2\sin^2\theta}\right)\right], \tag{7.5}$$

where X, Y, Z are the coordinates of the center of mass, M is the total molecular mass, μ is the reduced mass $m_A m_B/(m_A + m_B)$, and p_R, p_θ, p_ϕ have the same meaning as in Eq. 3.147. When the p's are converted to the corresponding operators, the three expressions in square brackets become $\mathbf{H}_{\text{trans}}$, $\mathbf{H}_{\text{vib}}$, and $\mathbf{H}_{\text{rot}}$, respectively. According to the Hellmann–Feynman theorem, the potential energy $V(R)$ for vibration (motion along the internuclear axis) must be the same as our $E(R)$. Separating the variables in Eq. 7.1 in the usual way, we find that the wave equation for vibrational motion is simply

$$\mathbf{H}_{\text{vib}}\psi_{\text{vib}}(R) = \left[\frac{\mathbf{p_R}^2}{2\mu} + E(R)\right]\psi_{\text{vib}}(R) = E_{\text{vib}}\psi_{\text{vib}}(R), \tag{7.6}$$

where $\mathbf{p_R} \equiv -i\hbar(\partial/\partial R)$, the operator for the relative momentum of the nuclei. Given the $E(R)$ curve for a molecule, one can solve Eq. 7.6 to obtain the vibrational eigenfunctions $\psi_{\text{vib}}(R)$ and eigenvalues E_{vib}. But the function $E(R)$ is in general quite complicated and known only in numerical form. Although Eq. 7.6 can be solved by numerical methods, it is useful to obtain a simpler function that will give a good approximate solution.

As we have indicated before, one of the scientist's most powerful tools for understanding a system is to find a parallel to it in a system already understood. By "parallel" we mean "similar in mathematical structure." If the parallelism is exact, then one has solved the new problem. Even if the parallelism is only approximate, one can generally learn a great deal about the new system by asking just how it deviates from the old one. The point of all this, of course, is that such a parallel model exists for molecular vibrations; not surprisingly, the model is our old friend the harmonic oscillator. Let us see why this is a good model.

The most significant characteristic of a molecule like H_2 is the existence of a stable minimum in the potential energy curve. Remember that the bonding force is $dE(R)/dR$; thus at the minimum R_e, where the slope of the $E(R)$ curve is zero, there is no net bonding force on the nuclei. If we squeeze or stretch the molecule, there will be a force tending to restore the internuclear distance to R_e. This restoring force is what gives rise to molecular vibrations, with the distance oscillating back and forth around the value R_e. This motion is mathematically equivalent to that of a single particle of mass μ in a one-dimensional potential well defined by $E(R)$. We have described the properties of such potential wells in Chapter 4. Bound states (those with $E < 0$, in terms of Fig. 6.2) have quantized energies; their wave functions are oscillatory where $E > V$, but decay rapidly toward zero in the classically forbidden region where $E < V$. The actual number of bound states and the energies at which they fall depend very specifically on the form of the potential energy curve. A molecule in a given electronic state may exist in any one of the vibrational states associated with the potential curve for that state. For the H_2 molecule in its ground electronic state, for example, spectroscopy reveals the existence of 14 bound vibrational states,[1] shown in Fig. 7.1.

We still have not introduced our model. We know that there is a restoring force tending to bring R to its equilibrium value. The simplest assumption one can make about the restoring force is that it is proportional to the displacement from equilibrium, $F(R) \propto |R - R_e|$. This is just what we have defined as a harmonic oscillator, for which Eqs. 4.4 and 4.5 give us

$$F(R) = -k(R - R_e) \quad \text{and} \quad V(R) = \tfrac{1}{2}k(R - R_e)^2, \qquad (7.7)$$

with $R - R_e$ replacing the displacement x; here $F(R)$ is the magnitude of the force tending to increase R, and $V(R)$ is zero at the bottom of the potential well. To the extent that the harmonic oscillator model is valid, we can apply all the results of Section 4.2 to molecular vibrations. Obviously it is not really valid for any molecule, because $E(R)$ always levels off at large R, where the atoms no longer interact. However, it is approximately valid for all stable molecular states near equilibrium. In other words, any $E(R)$ curve with a stable minimum can be approximated by a parabola in the vicinity of the minimum; this is illustrated for the H_2 molecule in Fig. 7.1.

How good this approximation is depends on the spacing of the vibrational energy levels. For bound states at energies for which the true and parabolic curves nearly coincide, the harmonic oscillator model should give an excellent description. But how many states fall in this

[1] In the case of H_2, the complete (non-Born–Oppenheimer) wave equation can be solved with sufficient accuracy to confirm this experimental result.

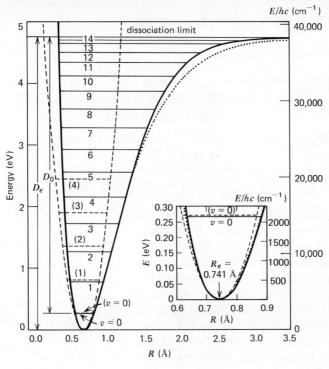

FIGURE 7.1
Vibrational energy levels of the H_2 molecule. Solid lines (———) represent the experimental $E(R)$ curve and energy levels; dashed lines (– – – –) give the parabolic approximation described in the text and the first few of the corresponding harmonic oscillator levels; and the dotted line (· · · · ·) is a Morse potential fit. The inset shows an enlargement of the region near the minimum. The two "dissociation energies" are indicated: $D_e = 4.748$ eV, $D_0 = 4.477$ eV.

range of energies, and how likely is the molecule to be in these states? To find out, let us make an order-of-magnitude calculation.

For most stable diatomic molecules in their ground states, the potential well has a depth of about 5 eV, or 8×10^{-19} J. To keep the numbers simple, let us say that the best parabola fitted to the minimum is 1 Å wide ($R - R_e = 0.5$ Å) when $V(R) = 8 \times 10^{-19}$ J; this is approximately correct for H_2. Solving Eq. 7.7 for the force constant k, we obtain

$$k = \frac{2V(R)}{(R - R_e)^2} = \frac{2 \times 8 \times 10^{-19}\,\text{J}}{(0.5 \times 10^{-10}\,\text{m})^2} = 640\,\text{N/m}.$$

The force constant of a harmonic oscillator is related to the frequency by Eq. 2.38; here, for the mass m we must use the molecule's reduced mass μ. For the H_2 molecule μ is just half the mass of a single hydrogen atom, or about 8×10^{-28} kg. The oscillator frequency in our approximate model should thus be

$$\nu(H_2) = \frac{1}{2\pi}\left(\frac{k}{\mu}\right)^{1/2} \approx \frac{1}{2\pi}\left(\frac{640\,\text{N/m}}{8 \times 10^{-28}\,\text{kg}}\right)^{1/2}$$

$$= \frac{(80 \times 10^{28}\,\text{s}^{-2})^{1/2}}{2\pi} \approx 1.5 \times 10^{14}\,\text{s}^{-1},$$

equivalent to a wavenumber $\tilde{\nu} \approx 5000$ cm^{-1}. Since the spacing of harmonic oscillator energy levels is $h\nu$, this corresponds to

$$\Delta E(H_2) = h\nu(H_2) \approx (6.6 \times 10^{-34}\,\text{J s})(1.5 \times 10^{14}\,\text{s}^{-1})$$

$$\approx 10^{-19}\,\text{J},$$

which is one-eighth of our assumed well depth. The lowest energy level is $\frac{1}{2}h\nu$ above the bottom of the well (the zero-point energy), the next level is $h\nu$ higher, and so on. Looking at Fig. 7.1, we can see that the harmonic oscillator model should be reasonably good for at least the first level or

TABLE 7.1
TEST OF HARMONIC OSCILLATOR MODEL WITH VIBRATIONAL CONSTANTS OF H_2 ISOTOPES

(atomic masses: ^{1}H, 1.00782 amu; D, 2.01410 amu)

Molecule	Reduced Mass, μ(amu)	Vibrational Frequency, $\nu_e \times 10^{-14}$(s^{-1})	$\dfrac{\mu^{1/2}\nu_e}{10^{14}\text{amu}^{1/2}\,s^{-1}}$
$^1H_2$	0.50391	1.3192	0.9365
^1HD	0.67171	1.1429	0.9367
D_2	1.00705	0.9345	0.9378

two. For many processes at room temperature it is only these levels that will be significantly occupied, as we shall see in Part Two.[2] All other molecules have greater reduced masses than H_2, and none have significantly larger force constants. Thus the harmonic oscillator model gives a fairly good description of the energy-level spacing in nearly all stable diatomic molecules. (There are a few exceptions—molecules with very shallow or highly nonparabolic potential wells, such as HgH or Ar_2.)

To check the accuracy of our calculations, let us now look at some data. The vibrational energy levels of a diatomic molecule can be fitted (see below) to a power series in the quantum number v, beginning with a harmonic oscillator term $(v+\frac{1}{2})h\nu_e$. For $^1H_2$ the frequency ν_e is found to be $1.3192 \times 10^{14}\,s^{-1}$, which is quite close to our crude estimate.[3] The various isotopic forms of the H_2 molecule presumably have the same electronic structure and thus the same potential curve.[4] Then the best parabolic fits to $E(R)$ should all have the same k, and Eq. 2.38 shows the harmonic oscillator frequency should be proportional to $\mu^{-1/2}$. We can thus test the validity of the harmonic oscillator model with the measured values of the vibrational constant ν_e. If the model were valid, ν_e would be the oscillator frequency and all the isotopes would have the same value of $\mu^{1/2}\nu_e$. The data are given in Table 7.1. The deviations from exact correspondence to $\nu_e \propto \mu^{-1/2}$ are largely due to the deviation of the real curve from its parabolic approximation.

Since the harmonic oscillator model does work rather well for most diatomic molecules, we can apply the results of Section 4.2. The vibrational energy levels should thus be given by

$$E_{\text{vib}}(v) = (v + \tfrac{1}{2})h\nu_0 \qquad (v = 0, 1, 2, \ldots), \tag{7.8}$$

[2] The relative population of two energy levels in thermal equilibrium is given by

$$N_2/N_1 = e^{-(E_2-E_1)/k_BT},$$

where $k_B = 1.38 \times 10^{-23}$ J/K and T is the absolute temperature. At room temperature (300 K) we have, for two levels 10^{-19} J apart,

$$\frac{N_2}{N_1} = e^{-10^{-19}\text{J}/(1.38\times10^{-23}\text{J/K})(300\text{K})} \approx e^{-24} \approx 4 \times 10^{-11};$$

in other words, each level will contain only 4×10^{-11} as many molecules as the one below it. Since there are about 3×10^{22} gas molecules/liter under ordinary conditions, only about 50 molecules/liter will be in the third vibrational level.
[3] Working backward from this frequency gives a force constant

$$k = 4\pi^2\mu\nu_e^2 = 4\pi^2(8.3676 \times 10^{-28}\text{ kg})(1.3192 \times 10^{14}\,s^{-1})^2$$
$$= 574.9\text{ N/m},$$

the value used to draw the parabolic curve in Fig. 7.1.
[4] This is not quite true. As in atoms (see footnote 24 on page 198), there is a very small isotopic shift in the electronic energy, but this effect can be ignored here.

where ν_0 is the oscillator frequency and v is the usual symbol for the vibrational quantum number. The vibrational wave functions in this model are $\psi_v(z)$, as listed in Table 4.1, in terms of the dimensionless displacement variable $z \equiv (4\pi^2\mu\nu_0/h)^{1/2}(R - R_e)$. For example, the ground vibrational state is described by a wave function with a single maximum at R_e; the first excited state, by a wave function with two maxima and a node at R_e, and so on. The vibration is not strictly harmonic, so these are only approximations to the true wave functions (for which the ground state's maximum and the next state's node will not lie exactly at R_e); but the qualitative behavior is given correctly. As v increases, the harmonic model becomes less accurate. But as v increases, the molecule behaves more and more like a classical oscillator, in that the difference between adjacent energy levels becomes a small fraction of the total vibrational energy.

Note in Fig. 7.1 the distinction between the two "dissociation energies," D_e and D_0. The well depth D_e is fundamentally more interesting to the theoretician. However, a measurement of the minimum energy required to dissociate a molecule in its ground state yields D_0, since no molecule can have less than the zero-point vibrational energy. For the same reason, when one calculates the energy change of a chemical reaction in which a given bond is formed or broken, it is D_0 that must be taken into account. For example, the energy required to bring about the reaction

$$H_2(g) + Cl_2(g) \rightarrow 2HCl(g),$$

with reactants and product in their ground states, is

$$\Delta E_{\text{reac}} = D_0(H_2) + D_0(Cl_2) - 2D_0(HCl).$$

Thus only D_0 is properly called the dissociation energy, and, to be strict, we should but will not always refer to it by that name. When a distinction must be made, D_e can be called the well depth. In the harmonic oscillator model, the two quantities are related by the equation $D_e = D_0 + \frac{1}{2}h\nu_0$.

Suppose that we want to describe molecular vibrations with more accuracy than the harmonic oscillator model can yield, but without going all the way to a numerical solution. How can we do this? The harmonic model approximates the potential curve by a parabola. The actual curve, however, rises more steeply at small R and falls off more slowly at large R. We can thus improve the fit by adding to $V(R)$ a correction that is positive for $R < R_e$ and negative for $R > R_e$. An obvious choice is a term proportional to $(R - R_e)^3$, which changes sign at $R = R_e$. We can thus replace Eqs. 7.7 by

$$F(R) = -kx + k'x^2 \quad \text{and} \quad V(R) = \tfrac{1}{2}kx^2 - \tfrac{1}{3}k'x^3, \tag{7.9}$$

with $x \equiv R - R_e$. This is an *anharmonic oscillator* model, with the added term called the *anharmonicity*. Equations 7.9, of course, fit the experimental data appreciably better near R_e than does the harmonic model. For most vibrational levels observed at room temperature the anharmonicity effect simply shifts E_{vib} by a small amount, generally less than 1%. However, this model is clearly unsuitable for describing the vibrational levels high in the potential well (which might be occupied in a very hot gas), since the predicted $V(R)$ diverges at large R instead of leveling off. One can obtain a still better fit by adding a term in x^4, a term in x^5, and as many terms as the data can justify; given enough terms, one can approximate the potential curve to any desired accuracy. In general, the energy eigenvalues can be written in the form

$$E_{\text{vib}}(v) = h\nu_e[(v+\tfrac{1}{2}) - x_e(v+\tfrac{1}{2})^2 + y_e(v+\tfrac{1}{2})^3 + \cdots] \qquad (v = 0, 1, 2, \ldots),$$

$$\text{(7.10)}$$

where $\nu_e, x_e, y_e, \ldots$ are constants that can be fitted to the spectroscopic data;[5] this equation defines the ν_e we introduced in Table 7.1.

The experimental $E(R)$ curve can be obtained by numerical calculation from the observed $E(v)$ values, but this is difficult. However, there are a number of empirical functions in closed form that can be used to give quite good approximate potential energy curves. The best known of these is the *Morse potential*,[6]

$$V(R) = D_e[1 - e^{-a(R-R_e)}]^2, \qquad \text{(7.11)}$$

where D_e is the well depth and a is a constant; to a good approximation one has $a = \pi\nu_e(2\mu/D_e)^{1/2}$. Note that this function correctly rises steeply at small R and levels off at large R, giving $V(0) = \infty$ and $V(\infty) = D_e$. In fact the Schrödinger equation can be solved with $V(R)$ given by Eq. 7.11. The energy eigenvalues of the Morse potential are given by the first two terms of Eq. 7.10 with $x_e = h\nu_e/4D_e$, and can be used to approximate the true energy levels all the way to the dissociation limit. A Morse potential curve for H_2 is included in Fig. 7.1; note that the agreement is rather poor at large R.

One can learn much about the vibrational energy levels of molecules by observing transitions between these levels. Such transitions can be induced by radiation, giving rise to discrete spectral lines in the usual manner (Section 4.5). Molecular vibrations are most commonly studied by observing the vibrational absorption spectra, which for diatomic molecules are found primarily in the near-infrared region. Most observed values of ν_e lie in the range $(0.6-12) \times 10^{13}\,\text{s}^{-1}$, corresponding to wavenumbers of $200-4000\,\text{cm}^{-1}$ or wavelengths of $2.5-50\,\mu\text{m}$ (cf. Table 7.2). The molecules observed can be in gas, pure liquid, solution, or solid form; the effects of the surrounding medium on ν_e are usually small. But not all molecules display such spectra. In fact, isolated homonuclear molecules such as H_2 or O_2 have no pure vibrational spectra at all. To see the reasons for this and other features of vibrational spectra, let us examine the excitation process in greater detail.

We consider excitation in terms of a classical model, essentially the same as that we applied to the hydrogen atom in Section 4.5. Any heteronuclear diatomic molecule has a dipole moment, since no two elements have exactly the same electronegativity. Thus such a molecule can be considered as an oscillating dipole, as illustrated in Fig. 7.2a (where $+q$ and $-q$ indicate the positive and negative ends of the dipole, respectively). Suppose that the molecule is bathed in infrared radiation, the wavelength of which is far greater than any dimension of the molecule. We can thus assume the electric field of the radiation to be uniform in space near the molecule, but oscillatory in time. At a given time, the component of this field parallel to the bond axis[7] exerts an instantaneous force on the dipole that tends to either stretch or compress the bond (Fig. 7.2a).

---

[5] Since most spectroscopists express their data in wavenumbers, one usually finds tabulated the values of $\bar{\nu}_e \equiv c\nu_e, \bar{\nu}_e x_e, \bar{\nu}_e y_e, \ldots$. The wavenumber corresponding to $E_{\text{vib}}(v)$ is called $G(v) \equiv E_{\text{vib}}(v)/hc$.

[6] P. M. Morse, *Phys. Rev.* **34,** 57 (1929).

[7] Since the force exerted by an electric field acts in the same direction as the field itself ($\mathbf{F} = q\mathbf{E}$), the field components perpendicular to the bond axis do not affect the bond length and can be neglected here.

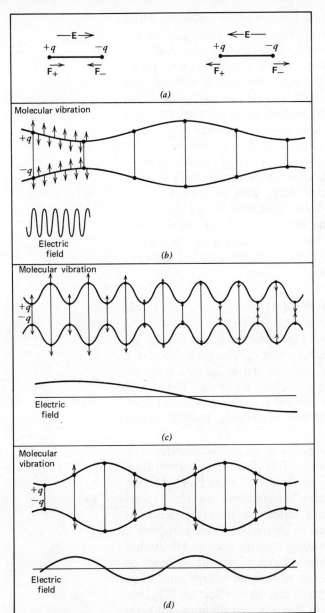

FIGURE 7.2

Interaction between an electric field and an oscillating (molecular) dipole. (a) The forces exerted on the dipole by the field, which alternately tend to compress or stretch the dipole as the field direction changes. (b) Field varies much faster than the molecular vibration ($\nu \gg \nu_e$); here and in the subsequent diagrams, the instantaneous forces on the dipole are indicated by arrows. (c) Field varies much slower than the molecular vibration ($\nu \ll \nu_e$). (d) Field and dipole oscillate at the same frequency ($\nu = \nu_e$), but 90° out of phase. As shown here, the dipole absorbs energy from the field; for a field phase 180° different, the forces are reversed and the dipole gives up energy to the field.

The effect of this oscillating force on the molecular vibration is the same as in our earlier analysis. Let the field frequency be ν and the molecule's natural vibration frequency be ν_e. If $\nu \gg \nu_e$ (Fig. 7.2b), the field reverses itself many times during a single vibration period, and can only impose a slight quivering onto the normal vibration. If $\nu \ll \nu_e$ (Fig. 7.2c), the field varies slowly compared to the vibration rate, and can only slightly increase or decrease the average value of R for some number of cycles. In either of these extremes, the field cannot easily transfer energy to the molecule. But when $\nu = \nu_e$ and the phase relationship is right (Fig. 7.2d), the field stretches the bond when it is expanding and compresses it when it is contracting, or vice versa. These are just the conditions that maximize energy exchange between the field and the dipole, which absorbs energy in the first case and gives it up in the second. Such an exchange of energy is an *electric dipole* transition.

Thus far we have been considering a heteronuclear molecule, which is necessarily an oscillating dipole. But what if the two ends of the molecule are identical, as in H_2 or any other homonuclear diatomic? Can such a molecule, with no dipole moment, absorb electromagnetic radiation? Our model says no, provided that the electric field is uniform over the length of the molecule; for then the forces on the two ends of the molecule are equal at all times, and the oscillations of the field can have no effect on the bond length. In our simple model, this is the explanation for the absence of vibrational spectra of homonuclear molecules.

But this classical model is inadequate for a full explanation of vibrational spectra. For one thing, a classical dipole can absorb energy in any amount from an oscillating electric field, with its own vibration amplitude varying continuously; we know that this is not true of a quantized oscillator. To go beyond the simple model, we refer to Eq. 4.33, which gives the probability of a transition between any two quantum states. Here we are interested only in the vibrational part of the wave function, for which Eq. 4.33 reduces to

$$\boldsymbol{\mu}_{v'v} = \int_0^\infty \psi_{v'}{}^*(R)\boldsymbol{\mu}(R)\psi_v(R)\, dR, \qquad (7.12)$$

where $\boldsymbol{\mu}$ is the molecule's instantaneous dipole moment and ψ_v and $\psi_{v'}$ are the vibrational wave functions in states v and v'. The probability of a transition between these two states is proportional to $|\boldsymbol{\mu}_{v'v}|^2$. It is immediately clear what happens in a homonuclear molecule: The dipole moment is zero for all values of R, the integral $\boldsymbol{\mu}_{v'v}$ vanishes for all v, v', and there is thus zero probability of any vibrational electric dipole transition.

We have thus shown (on two levels) why homonuclear diatomic molecules have no pure vibrational spectra. But if that is the case, then how does one obtain vibrational energy levels like those in Fig. 7.1? One way is to look at the vibrational "structure" of *electronic* spectra. Transitions between different bound electronic states can be associated with specific initial and final vibrational states. In particular, transitions from a single vibrational level in an upper electronic state to a set of vibrational levels in a lower electronic state involve a series of quanta (a *band* of spectral lines) whose energies differ by the vibrational energy interval in the lower state. We shall consider spectra of this type in Section 7.3. One can also observe vibrational transitions in homonuclear diatomics directly by bombarding the molecules with electrons of precisely known energy and measuring the characteristic energy losses of the scattered electrons; this is essentially a variation of the Franck–Hertz experiment. In such a collision, of course, the electric field due to the electron does vary over molecular dimensions, and our previous arguments against vibrational transitions no longer apply.

Now let us return to Eq. 7.12 and see what transitions are allowed in heteronuclear molecules. For the integral to be nonzero it is not enough that $\boldsymbol{\mu} \neq 0$; the dipole moment of the molecule must vary[8] with R. But this is no problem, because the dipole moments of all polar molecules do vary with the internuclear distance R. An especially simple result is obtained with the harmonic oscillator model, the vibrational wave functions of which are defined by Eq. 4.8. Setting $|\boldsymbol{\mu}| = qR$ (with q constant), one can show without much difficulty that the integral of Eq. 7.12 is nonzero only when v and v' differ by 1. In other words, for the harmonic

---

[8] If $\boldsymbol{\mu}$ is constant, it can be taken out of the integral, which then becomes $\int_0^\infty \psi_{v'}{}^*\psi_v\, dR = 0$, vanishing because the vibrational eigenfunctions are orthogonal (see Appendix 6A).

oscillator we have the *selection rule* for electric dipole transitions

$$\Delta v = \pm 1; \tag{7.13}$$

all other transitions are forbidden. This means that the quantum mechanical harmonic oscillator can absorb or give up only one quantum at a time.

The mathematical basis for the selection rule 7.13 can be seen easily if we consider the situation of small-amplitude oscillations, which are the only ones for which the harmonic approximation and the $\Delta v = \pm 1$ rule apply. If we so restrict our model, then we can write our wave functions $\psi_v(R)$ as functions of the displacement of R from the equilibrium distance R_e (μ is the reduced mass, not to be confused with the dipole moment μ):

$$z = (\mu\omega/\hbar)^{\frac{1}{2}}(R - R_e), \tag{7.14}$$

and the vibrational wave function is given by Eq. 4.8. Moreover, $\mu(R)$ can also be expanded about R_e, in a Taylor series:

$$\mu(R) = \mu(R_e) + \frac{d\mu(R)}{dR}\bigg|_{R_e} z + \frac{1}{2}\frac{d^2\mu(R)}{dR^2}\bigg|_{R_e} z^2 + \cdots. \tag{7.15}$$

Note that $\mu(R_e)$ and all the derivatives of $\mu(R)$ evaluated at $R - R_e$ are constants. Hence the *transition dipole*, Eq. 7.12, becomes

$$\mu_{v'v} = \mu(R_e)\int \psi_{v'}(z)\psi_v(z)\,dz$$
$$+ \frac{d\mu(R)}{dr}\bigg|_{R_e}\int \psi_{v'}(z)z\psi_v(z)\,dz$$
$$+ \frac{1}{2}\frac{d^2\mu(R)}{dR^2}\bigg|_{R_e}\int \psi_{v'}(z)z^2\psi_v\,dz + \cdots. \tag{7.16}$$

The limits of integration can be taken as $\pm\infty$ so long as we make the harmonic approximation. Because $\psi_{v'}(z)$ and $\psi_v(z)$ are eigenfunctions of the same Hermitian operator and have different eigenvalues (if we are considering a transition), they are orthogonal. Hence $\int \psi_{v'}(z)\psi_v(z)\,dz$ vanishes. The next term does not vanish. First, note that $\int \psi_v^*\psi_v\,dz$ or $\int e^{-z^2}H_v^2(z)\,dz$ does not vanish. Next, from the definition of the Hermite polynomials, we find by direct substitution that

$$2zH_v(z) = 2vH_{v-1}(z) + H_{v+1}(z). \tag{7.17}$$

In words, multiplication of $H_v(z)$ by z transforms $H_v(z)$ into a sum of one higher and one lower $H_{v'}(z)$, on the scale of eigenvalues. This in turn means that the integral multiplying $d\mu/dR$ in Eq. 7.16 does not vanish provided that either $v' = v+1$ or $v' = v-1$. Therefore the harmonic oscillator satisfies Eq. 7.13. We shall not show here why the harmonic oscillator does not exhibit transitions with $|\Delta v| > 1$. Real molecules, which are never strictly harmonic, do exhibit transitions with $|\Delta v| > 1$, but these are generally much less probable than those with $\Delta v = \pm 1$ provided that the molecule has a nonvanishing dipole moment. The higher terms of Eq. 7.16 for general anharmonic oscillators are not zero but are small.

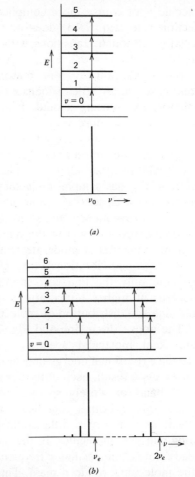

FIGURE 7.3
Schematic representation of molecular vibrational absorption spectra (neglecting rotational effects). (a) Harmonic oscillator model: $\Delta v = \pm 1$ only, all transitions at the same frequency ν_0. (b) Real diatomic molecule: The energy levels are not evenly spaced, so one can observe a band of lines beginning near ν_e; since the selection rule is not rigorous, there is a weaker band ($\Delta v = \pm 2$) near $2\nu_e$, and still weaker bands at $3\nu_e$, $4\nu_e$, ... Within each band, the intensity of a given line is proportional to the population of the initial state; the intensities as drawn here correspond to a gas with $h\nu_e/k_B T \approx 1.8$ (cf. Chapter 21).

The form of the vibrational spectrum is governed by the spacing of the energy levels. Since the energy levels of the harmonic oscillator are equally spaced, the selection rule 7.13 limits its vibrational spectrum to a single line, with $\Delta E = \pm h\nu_0$; this is illustrated in Fig. 7.3a. The spectrum of a real molecule is of course more complicated. First, as mentioned above, the selection rule (Eq. 7.13) does not hold rigorously for an anharmonic oscillator, although transitions with $\Delta v = \pm 1$ do remain far more likely than those with $\Delta v = \pm 2, \pm 3, \ldots$. Also, the spacing of the energy levels is no longer constant. By our reasoning in Section 4.2, since the real potential is "flatter" (i.e., widens more rapidly) than that of the harmonic oscillator, the spacing should decrease with increasing energy; this is indeed what one observes (cf. Fig. 7.1). Thus the transition $(v = 0) \rightarrow (v = 1)$ requires more energy than the transition $(v = 1) \rightarrow (v = 2)$, and so on upward. The vibrational spectrum therefore consists of many lines rather than one, as shown in Fig. 7.3b. If one takes an absorption spectrum of a gas sample, molecules in the ground state absorb light at one frequency (close to ν_e), molecules in the first excited state absorb at a slightly lower frequency, and so forth. The intensity of an absorption line is of course proportional to the number of molecules in the initial state. For light molecules at moderate temperatures virtually all the molecules are in the ground vibrational state (cf. Chapter 21 and footnote 2 on page 260), and only the line corresponding to $(v = 0) \rightarrow (v = 1)$ is normally observed; at higher temperatures one can use the intensity ratios to determine the distribution of molecules among the vibrational states.

The above discussion and Fig. 7.3 refer to a pure vibrational spectrum. Real molecular spectra are complicated by the fact that a molecule's rotational energy may also change when the vibrational energy changes. As a result, each of the lines in the vibrational spectrum resolves into a band of closely spaced rotational–vibrational lines. We shall consider this and other complications in Section 7.3.

Let us now say a little about how the vibrational constants vary among diatomic molecules. In the harmonic oscillator model, as we have already noted, the oscillator frequency is proportional to $\mu^{-1/2}$, where μ is the molecule's reduced mass. This is a fairly good approximation for real molecules. For homonuclear molecules μ is one-fourth the molecular weight, so heavy molecules have relatively low vibrational frequencies. However, the reduced mass of a heteronuclear molecule cannot exceed the mass of the lighter atom, which does most of the actual vibrating relative to the center of mass; all diatomic hydrides thus have $\mu < 1$ amu, and high vibrational frequencies. The variation of force constants (k) among molecules is not nearly so great as that of μ. In spite of these variations, the actual magnitudes of molecular vibrational frequencies (or energies) fall within a rather short and well-defined range. This can be seen in Table 7.2, in which $\tilde{\nu}_e (\equiv \nu_e/c)$, x_e, y_e are the constants in Eq. 7.10: The wavenumber $\tilde{\nu}_e$ ranges from over $4000 \, \text{cm}^{-1}$ for $^1\text{H}_2$ to below $50 \, \text{cm}^{-1}$ in some heavy molecules, but this covers only two decades of the whole electromagnetic spectrum (see Appendix 2A).

Finally, as can also be seen in Table 7.2, there is at least a qualitative correlation between the length of a chemical bond (R_e) and its fundamental vibration frequency ν_e. This is not difficult to rationalize. Chemical bonds fall within a relatively narrow range of energies and lengths, so most ground-state $E(R)$ curves are similar in shape. This is why they can almost all be approximated by a given function like a Morse potential. The bond length is the distance at which the short-range repulsive force balances the more slowly varying attractive force. The smaller the value of R_e, the more steeply both forces vary with R, and the narrower is the potential well near R_e. But a narrow well is one with a high force constant and thus a high vibration frequency. There we have our qualitative

TABLE 7.2
VIBRATIONAL AND ROTATIONAL CONSTANTS OF SOME
DIATOMIC MOLECULES

Molecule	μ (amu)	R_e (Å)	$\tilde{\nu}_e$ (cm^{-1}) ($\tilde{\nu}_e \equiv \nu_e/c$)	$\tilde{\nu}_e x_e$ (cm^{-1})	$\tilde{\nu}_e y_e$ (cm^{-1})	D_0 (eV)	B_e (cm^{-1})	α_e (cm^{-1})
$^1H_2$	0.50391	0.7412	4400.39	120.815	0.7242	4.4773	60.864	3.0764
HD ($^1H^2H$)	0.67171	0.7412	3812.29	90.908	0.504	4.5128	45.663	2.0034
D_2 ($^2H_2$)	1.00705	0.7412	3117.0	61.82	0.562	4.5553	30.457	1.0786
First-row homonuclear molecules								
$^7Li_2$	3.50800	2.6725	351.44	2.592	−0.0058	1.12	0.6727	0.00704
$^{11}B_2$	5.50465	1.590	1051.3	9.4		2.9	1.212	0.014
$^{12}C_2$	6.00000	1.2425	1854.71	13.340	−1.172	6.24	1.8198	0.01765
$^{14}N_2$	7.00154	1.094	2358.07	14.188	−0.0124	9.7598	1.9987	0.01781
$^{16}O_2$	7.99745	1.2075	1580.19	11.98	0.0475	5.1156	1.4456	0.01593
$^{19}F_2$	9.49910	1.409	919.0	13.6		1.604	0.8901	0.0146
Other homonuclear molecules								
$^{23}Na_2$	11.4949	3.0786	159.23	0.726	−0.0027	0.75	0.1547	0.00079
$^{39}K_2$	19.48185	3.923	92.64	0.354		0.51	0.0562	0.00022
$^{85}Rb_2$	42.4558	4.20	57.28	0.96	−0.0008	0.47	0.0127	0.0000264
$^{133}Cs_2$	66.9525	4.58	41.99	0.080	−0.0002	0.45		
$^{35}Cl_2$	17.48222	1.9878	559.71	2.70		2.484	0.2441	0.00153
$^{79}Br^{81}Br$	39.9524	2.2809	323.33	1.081		1.9708	0.0811	0.00032
$^{127}I_2$	63.4502	2.6666	214.52	0.607	−0.0013	1.5437	0.0374	0.00012
Hydrides								
$^7Li^1H$	0.88123	1.5954	1405.65	23.200	0.1633	2.429	7.5131	0.2132
$^{12}C^1H$	0.92974	1.124	2859.1	63.3		3.47	14.448	0.530
$^{16}O^1H$	0.94808	0.9706	3735.21	82.81		4.392	18.871	0.714
$^1H^{19}F$	0.95705	0.9168	4139.04	90.05	0.932	5.86	20.9560	0.7958
$^1H^{35}Cl$	0.97959	1.2746	2991.09	52.82	0.2244	4.4361	10.5936	0.3072
$^1H^{81}Br$	0.99511	1.4145	2649.21	45.22	−0.0029	3.755	8.4651	0.2333
$^1H^{127}I$	0.99988	1.6090	2308.09	38.981	−0.1980	3.053	6.5108	0.1686
Other heteronuclear molecules								
$^7Li^{19}F$	5.12381	1.5638	910.34	7.929		5.94	1.3454	0.02030
$^9Be^{16}O$	5.76432	1.3310	1487.32	11.830	0.0224	4.60	1.6510	0.0190
$^{11}B^{14}N$	6.16351	1.281	1514.6	12.3		3.99	1.666	0.025
$^{12}C^{14}N$	6.46219	1.1720	2068.70	13.144		7.567	1.8991	0.01735
$^{12}C^{16}O$	6.85621	1.1283	2169.82	13.294	0.0115	11.09	1.9313	0.01751
$^{14}N^{16}O$	7.46676	1.1508	1904.03	13.97	−0.0012	6.50	1.7046	0.0178
$^{23}Na^{35}Cl$	13.8707	2.3606	366	2.05		4.25	0.2181	0.00161
$^{39}K^{79}Br$	26.0850	2.8207	213	0.80	0.0011	3.925	0.0812	0.00040

correlation.[9] Potential wells at small R tend to be not only steeper, but also deeper, because almost all the bonding forces fall off with distance. Thus we also have a correlation between bond energy and vibration frequency. To sum up: Long bonds are weak and correspond to slow vibrations; short bonds are strong, with deep potential wells, and correspond to high vibration frequencies.

7.2 Rotations of Diatomic Molecules

Next we consider the rotational behavior of diatomic molecules. We continue to assume that the translational, vibrational, and rotational motions of the nuclei are separable, Eqs. 7.2–7.4. Thus we begin by considering rotation in the absence of vibration, that is, with fixed internuclear distance. Just as vibration is well described by the harmonic oscillator model, the rotational motion of the molecule is well described by another model familiar to us, that of the rigid rotator (see Section 3.12).

More specifically, to a good approximation we can treat any diatomic molecule as a rigid symmetric top. This, it will be recalled, is a rigid body with two equal moments of inertia, $I_0 \equiv I_x = I_y \neq I_z$; the z axis is as usual the bond axis. We have already analyzed this case in detail. The rotational Hamiltonian of the rigid symmetric top is

$$\mathsf{H}_{\text{rot}} = \frac{1}{2I_0}\mathsf{L}^2 + \frac{1}{2}\left(\frac{1}{I_z} - \frac{1}{I_0}\right)\mathsf{L}_z^{\,2}, \qquad (7.18)$$

where L^2 and $\mathsf{L}_z^{\,2}$ are angular momentum operators, and its eigenvalues are given by Eq. 3.177. Now, in a diatomic molecule the only contributions to $I_z[\equiv \sum_i m_i(x_i^2 + y_i^2)]$ are those due to the electrons and the nonzero radii of the nuclei, both very small; I_0, on the other hand, is essentially μR^2, where μ is the molecule's reduced mass and R is the internuclear distance. Thus we have $I_z \ll I_0$, and the coefficient of $\mathsf{L}_z^{\,2}$ in Eq. 7.18 is *much* larger than that of L^2. The energy term associated with $\mathsf{L}_z^{\,2}$ is either very large (when $L_z \neq 0$ for the electrons, i.e., in all except $^1\Sigma$ states) or negligibly small (in $^1\Sigma$ states). In either case it is a constant in any given electronic state, and can conveniently be included in the electronic energy.

Thus we can disregard the second term in Eq. 3.177 and write the rotational energy of the molecule as simply

$$E_{\text{rot}} = \frac{J(J+1)\hbar^2}{2I_0} = \frac{J(J+1)\hbar^2}{2\mu R_e^{\,2}} \qquad (J = 0, 1, 2, \ldots), \qquad (7.19)$$

[9] There are a number of empirical formulas correlating vibrational constants with bond lengths. These are accurate enough to be used in estimating bond lengths in new compounds from their infrared spectra. One of the most widely used is *Badger's rule*,

$$k = a(R_e - d_{ij})^{-3},$$

where k is the force constant, R_e is the usual equilibrium internuclear distance, $a \approx$ 186 (N/m)Å^3 (a universal "constant"), and d_{ij}, which is not universal, depends on the rows of the periodic table in which the atoms i and j fall. The following table gives d_{ij} in angstroms for various rows of the periodic table:

	j			
i	H	Row 1	Row 2	Row 3
H	0.025	0.335	0.585	0.650
Row 1	0.335	0.680	0.900	
Row 2	0.585	0.900	1.180	

where J (replacing the earlier l) is the conventional symbol for the rotational quantum number of a molecule; we have evaluated I_0 at the equilibrium internuclear distance, R_e. This equation is identical in form to Eq. 3.170, which gives the energy levels of the pure rigid rotator. Each energy level is thus $(2J+1)$-fold degenerate: The angular momentum about a given axis through the center of gravity is $M_J\hbar$, where for each J the quantum number M_J can assume the values $J, J-1, \ldots, -J+1, -J$. And the rotational wave functions are simply the spherical harmonics $Y_{J,M_J}(\theta, \phi)$.

This is all quite straightforward, but how valid is the model? The assumption of fixed internuclear distance seems questionable, since the molecule is certainly vibrating (at least with the zero-point energy) at the same time as it rotates. But let us compare the rates of the two motions. If the molecule is rotating with angular velocity ω, its angular momentum for fixed R is $\mu R_e^2 \omega$; combining this with the eigenvalue equation $L^2 = J(J+1)\hbar^2$, we obtain

$$\omega = \frac{[J(J+1)]^{1/2}\hbar}{\mu R_e^2} = \left(\frac{2E_{\rm rot}}{\mu R_e^2}\right)^{1/2} \tag{7.20}$$

for the angular velocity. We shall see in Chapter 21 that the average rotational energy of gaseous diatomic molecules is $k_B T$, where $k_B = 1.381 \times 10^{-23}$ J/K and T is the absolute temperature. For $^1H_2$ molecules at room temperature, the average period of a rotation is thus

$$\tau_{\rm rot} = \frac{2\pi}{\omega} \approx 2\pi\left(\frac{\mu R_e^2}{2k_B T}\right)^{1/2} = 2\pi\left[\frac{(8.368 \times 10^{-28}\text{ kg})(0.7412 \times 10^{-10}\text{ m})^2}{2(1.381 \times 10^{-23}\text{ J/K})(300\text{ K})}\right]^{1/2}$$
$$= 2\pi(5.55 \times 10^{-28}\text{ s}^2)^{1/2} = 1.48 \times 10^{-13}\text{ s}.$$

On the other hand, the period of a harmonic oscillator is simply ν^{-1}, so for $^1H_2$ we have (cf. Table 7.1)

$$\tau_{\rm vib} \approx \nu_e^{-1} = (1.3192 \times 10^{14}\text{ s}^{-1})^{-1} = 7.58 \times 10^{-15}\text{ s}.$$

Thus the $^1H_2$ molecule at room temperature goes through about 20 vibrational periods in the course of a single rotation; for most molecules the ratio $\tau_{\rm rot}/\tau_{\rm vib}$ is even greater.[10] As a first approximation, then, it seems reasonable to assume that the vibrational effects average out, justifying our calculation of rotational energy with fixed R. This is exactly the same reasoning as that leading to the Born–Oppenheimer assumption.[11]

Although the simple rigid-rotator model is adequate for qualitative purposes, more accurate expressions are needed to meet the demands of quantitative spectroscopy. One must allow for the effect of vibration on the rotational motion. Strictly, the R_e^{-2} in Eq. 7.19 should be replaced by the average value of R^{-2} over a vibration; because of the asymmetry of the $E(R)$ curve, the average value of R must increase with v, so that we have $\langle R \rangle_v > R_e$, $\langle R^{-2} \rangle_v < R_e^{-2}$. The actual rotational energies are thus

[10] For example,

	$^1H^{35}Cl$	$^{35}Cl_2$	$^{127}I_2$
$\tau_{\rm rot}(300\text{ K})/s$	4.05×10^{-13}	2.34×10^{-12}	5.98×10^{-12}
$\tau_{\rm vib}/s$	1.12×10^{-14}	5.90×10^{-14}	1.55×10^{-13}

[11] Note that both $\tau_{\rm rot}$ and $\tau_{\rm vib}$ are appreciably longer than the Bohr period of about 10^{-16} s (Table 2.2), which is typical of the time scale of electronic motions.

somewhat less than those given by Eq. 7.19, and can be approximated by[12]

$$E_{rot} = J(J+1)hcB_v = J(J+1)hc[B_e - \alpha_e(v+\tfrac{1}{2}) + \cdots], \qquad \left(B_e \equiv \frac{\hbar}{4\pi c\mu R_e^2}\right),$$

(7.21)

where α_e is a constant, B_v is the *rotational constant* for vibrational state v, and B_e is the rigid-rotator value. The factor hc is inserted to give B_e and α_e the units of wavenumbers. For $^1H_2$ we readily obtain

$$B_e = \frac{1.0546 \times 10^{-34}\,J\,s}{4\pi(2.9979 \times 10^{-8}\,m/s)(8.3676 \times 10^{-28}\,kg)(0.7412 \times 10^{-10}\,m)^2}$$
$$= 60.9\,cm^{-1},$$

in agreement with experiment, whereas α_e is found empirically to be about $3.0\,cm^{-1}$. (Actually, as we shall see, the value of R_e is obtained from the measured B_e.) Other molecules have larger moments of inertia ($I = \mu R_e^2$) and thus smaller B_e's; values of B_e and α_e for a number of molecules are listed in Table 7.2.

In writing Eq. 7.19 for a diatomic molecule, we have effectively assumed the atoms to be point masses, so that we can neglect rotation about the internuclear axis. This is valid enough for a molecule in a $^1\Sigma$ state, where the electrons have zero angular momentum relative to the nuclei, but suppose that this is not the case. Let us go back to the symmetric-top model. We can rewrite the eigenvalue equation (Eq. 3.177) as

$$E_{rot} = J(J+1)hcB_v + M_J^2 hc(A - B_v), \qquad \left(\begin{array}{l} J = 0, 1, 2, \ldots; \\ M_J = J, J-1, \ldots, -J+1, -J \end{array}\right),$$

(7.22)

where $A \equiv \hbar/4\pi cI_z$ and B_v is the same as in Eq. 7.21. The quantum numbers J and M_J are defined by the equations

$$|\mathbf{J}|^2 = J(J+1)\hbar^2 \quad \text{and} \quad J_z = M_J\hbar,$$

(7.23)

which respectively give the square and the z component of the molecule's total angular momentum $\mathbf{J}$. Although both electronic and nuclear motions contribute to $\mathbf{J}$, its z component consists almost entirely of the angular momentum of the electrons about the bond axis. There are a number of ways in which the orbital and spin electronic angular momenta (which we call $\mathbf{L}$ and $\mathbf{S}$, respectively) can couple with each other and with the nuclear rotation; we shall look at only the simplest case. In most electronic states the orbital electronic angular momentum is strongly coupled to the bond axis, and only its z component is quantized; thus we have $|L_z| = \Lambda\hbar$, as we assumed in the last chapter. In singlet states ($\mathbf{S} = 0$), L_z is the only contribution to J_z, so we also have $|J_z| = \Lambda\hbar$, $|M_J| = \Lambda$. Thus for most singlet states Eq. 7.22 can be written in the form

$$E_{rot} = J(J+1)hcB_v + \Lambda^2 hc(A - B_v), \qquad \left(\begin{array}{l} \Lambda = 0, 1, 2, \ldots; \\ J = \Lambda, \Lambda+1, \Lambda+2, \ldots \end{array}\right).$$

(7.24)

Similar equations can be derived for other cases. Since the second term can be included in the electronic energy, what is the point of obtaining these equations? The answer is that they reveal restrictions on the

---

[12] There is also a smaller correction for centrifugal distortion, i.e., the stretching of the bond by centrifugal force. This appears in E_{rot} as a term $D_v hcJ^2(J+1)^2$, where $D_v \lesssim 10^{-4}B_v$.

allowable energy levels. In an electronic state described by Eq. 7.24, for example, we can only have $J \geqslant \Lambda$, since the total angular momentum, $[J(J+1)]^{1/2}\hbar$, cannot be less than its z component $\Lambda\hbar$. Since $A \gg B_v$, electronic states with different Λ's can have very different total rotational energies, often differing by several electron volts. Within a given electronic (and vibrational) state, however, adjacent rotational levels may differ by as little as 10^{-5} eV.

This brings us to rotational spectra. In the next section we shall consider the complications that arise when a molecule simultaneously undergoes rotational and vibrational (or electronic) transitions; here we are concerned only with pure rotational spectra, which are found in the far infrared and microwave regions. As in the case of vibration (pages 262–265), rotational electric dipole spectra appear only for heteronuclear diatomic molecules, since a homonuclear molecule has $\boldsymbol{\mu} = 0$ in all orientations. For heteronuclear molecules $\boldsymbol{\mu}$ varies in direction, though not magnitude, in the course of a rotation. The selection rule for electric dipole transitions of a heteronuclear rigid rotator is

$$\Delta J = \pm 1, \tag{7.25}$$

which also usually holds for a real molecule in a given electronic state. Substituting in Eq. 7.21, we find that the only allowed pure rotational transitions are those with

$$\Delta E = 2J'hcB_v \quad \text{or} \quad \tilde{\nu} = 2J'B_v, \tag{7.26}$$

where J' is the quantum number of the *upper* state. We thus obtain a series of equally spaced spectral lines, beginning at $\tilde{\nu} = 2B_v$ and proceeding to higher wavenumbers at intervals of $2B_v$. This is illustrated in Fig. 7.4. Since $2B_e$ ranges from 120 cm^{-1} (^{1}H$_2$) down to less than 0.1 cm^{-1}, corresponding to 0.08 mm–10 cm in wavelength, at least the beginning of this series may be in the microwave region. However, one frequently observes transitions in which the quantum number J' is quite large, giving rise to higher-frequency lines well into the infrared.[13]

The spacing of rotational lines is one of the best methods for determining internuclear distances (bond lengths). It is worth describing here how one carries out such a determination. The pure rotational spectrum, as we have said, lies in the far infrared or microwave region. To obtain such a spectrum, one can pass radiation through a rather rarefied gaseous sample of the molecule in question, vary the frequency of the applied radiation, and measure the fraction of energy transmitted from the power source to a detector.[14] The source may be either electrons oscillating in a vacuum (a klystron tube) or electrons accelerated in a nonmetallic solid by a sufficiently high voltage; the detector is ordinarily a crystal-diode rectifier. One can carry out such measurements with exceedingly high accuracy—so high, in fact, as to outstrip the ability of existing

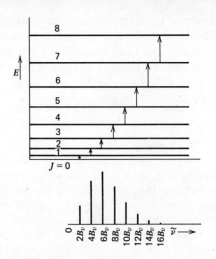

FIGURE 7.4
Pure rotational absorption spectrum of a diatomic molecule. Given the selection rule $\Delta J = \pm 1$, there is a series of lines with the constant wavenumber spacing $2B_v$. The intensities of the lines are proportional to the populations of the initial states, and as drawn here correspond to a gas with $B_v hc/k_B T \approx 0.1$ (cf. Chapter 21).

---

[13] As we mentioned earlier, in a gas the average value of E_{rot} is $k_B T$, which at 300 K is 4.14×10^{-21} J (0.0258 eV), corresponding to a wavenumber of 208 cm^{-1}. Thus at room temperature the average value of $J(J+1)$ for ^{1}H$_2$ is about 3.5 ($J \approx$ 1–2), for I$_2$ about 5500 ($J \approx 75$).

[14] In practice, rotational spectroscopy usually involves a technique somewhat more complicated than direct absorption. In a laboratory system, the amount of energy absorbed directly is always an extremely small fraction of the total incident radiation, so that the direct spectral signal would be very weak. Instead, one alternately applies and turns off an external electric or magnetic field, which splits and shifts the spectral lines (by the Stark or Zeeman effect), moving them on and off the frequency of the applied radiation. The observed signal is then the difference between the amounts of microwave power detected with the external field on and off. The effect of taking such a difference signal is a strong enhancement of the signal strength relative to the background "noise" reaching the detector.

theories to interpret all the details, which include interactions with vibration, electronic motion, and nuclear spins. In first approximation, however, the lines in a given band are equally spaced, and B_v is simply half the wavenumber spacing. Given such experimental values of B_v for several vibrational states, one simply uses Eq. 7.21 to calculate B_e and thus R_e.

We mention in passing the other principal method of determining bond lengths of gaseous molecules, which utilizes electron diffraction (see Sections 3.1 and 11.6). This method is less precise and often less accurate than microwave spectroscopy, but it does give a direct measure of R_e. A beam of electrons is allowed to impinge on a gas of diatomic molecules; because the electrons are deliberately given quite high energies, they are scattered primarily by the massive atomic nuclei. Since the molecules move randomly, one obtains a quite complex pattern of scattering intensity averaged over all possible molecular orientations. Nevertheless, this pattern contains information on the internuclear spacing. Each nucleus in a molecule produces a set of scattered wavelets, and the wavelets from each pair of nuclei interfere with each other. A single diatomic molecule thus scatters electrons much as two pinholes in a screen produce a diffraction pattern with light from a point source. Despite the random orientation, the net effect of all the molecules is to produce a diffraction pattern which depends on the scattering angle in a very specific way. One can compare this pattern with that corresponding to an assumed R_e, and adjust the latter until the two match. Similar calculations can be made for even very complex molecules, but for diatomic molecules the computation is simple and limited only by the accuracy with which one can measure the shapes of the diffraction peaks.

Most of our knowledge of molecular structures derives from spectroscopic measurements of one kind or another. Suppose that one subjects a gas of diatomic molecules to electromagnetic radiation, varying the frequency ν across the spectrum. Whenever $h\nu$ equals the difference between two molecular energy levels, a transition may be induced. What kinds of transitions are these? In the microwave and far infrared regions, as just described, one observes pure rotational transitions, in which only the quantum number J changes. In the near and middle infrared one sees vibrational transitions. In Section 7.1 we described the pure vibrational spectrum, in which only v changes; unfortunately, things are not really that simple, since rotational transitions of various kinds usually occur at the same time. In the visible and ultraviolet regions there occur transitions between different electronic states, again complicated by simultaneous vibrational and rotational transitions. At still higher energies, x rays produce transitions among the energy levels of the atomic cores (see Chapter 2). Figure 7.5, which illustrates some of the energy levels in a diatomic molecule, should give an idea of the energy differences involved in various kinds of transitions.

7.3
Spectra of Diatomic Molecules

These transitions are observed by the standard techniques of experimental spectroscopy, which we shall review here briefly. In *emission spectroscopy*, the sample is subjected to thermal, electric, or other excitation, and the excited atoms or molecules emit energy as they drop to lower energy levels. The emitted radiation is collected at a detector and its intensity is measured as a function of

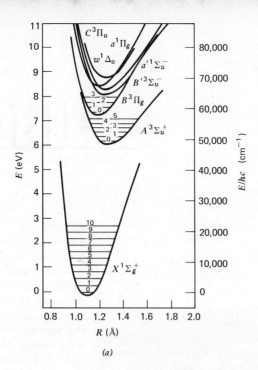

(a)

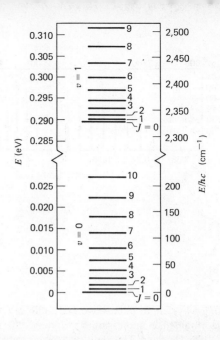

(b)

FIGURE 7.5
Energy levels of the N_2 molecule. (*a*) Electronic states, with vibrational levels ($v = 0, 1, 2, \ldots$) shown in the lowest three. [Not shown is the $W^3\Delta_u$ state; this state has about the same minimum energy as the $B^3\Pi_g$ (see Table 7.6), but its R_e is not known.] The energy zero is the ground electronic-vibrational state ($X^1\Sigma_g^+$, $v = 0$). After W. Benesch, J. T. Vanderslice, S. G. Tilford, and P. G. Wilkinson, *Astrophys. J.* **142**, 1227 (1965). (*b*) Rotational structure of the two lowest vibrational levels of the ground state; the energy zero is the same as in (*a*).

wavelength. An emission spectrum of atomic iron was shown in Fig. 2.9. In *absorption spectroscopy*, a sample of the material under study is subjected to continuous radiation in an appropriate part of the spectrum, and the molecules absorb energy in undergoing transitions to higher energy levels. This energy transfer depletes the incident radiation at specific wavelengths (one for each transition), giving rise to a set of spectral lines. The radiation passing through the sample is collected at a detector, where its intensity is measured. The signal of interest is the amount of radiation absorbed, that is, the difference between incident and transmitted intensities. What we usually call a "spectrum" is a representation of the amount of absorption as a function of wavelength or frequency; two regions of the infrared spectrum of HCl are shown in Fig. 7.6.

All experimental spectrometers have the same basic components, but their details can vary greatly, especially with the region of the spectrum under study. These components include the following.

1. *Sample.* When the sample is gaseous or liquid, it must be placed in a holder of some kind; such a sample holder must be reasonably transparent to the radiation one is using. In the ultraviolet below 2000 Å even air becomes opaque, and the whole apparatus must be placed in a vacuum chamber.

2. *Radiation source.* For absorption spectra in the visible and ultraviolet regions the source can be any of the standard types of light sources: tungsten-filament lamps, electric discharges and arcs, and so on. The usual infrared source is a heated rod of some refractory material such as SiC. Klystron tubes are used in the microwave region, and standard radiofrequency generators at longer wavelengths. In all regions of the spectrum, lasers are increasingly used as high-intensity radiation sources. In emission spectroscopy the radiation source is the sample itself; it may be excited by simple heating, by electric discharge, by chemical reaction (as in a flame), or by undergoing laser action.

3. *Optical system.* The optical system consists of whatever devices are used to disperse radiation of different wavelengths and collect it at appropriate detecting locations. In most spectral regions of interest, the separation can be performed by either prisms or diffraction gratings; the separated radiation can be focused with mirrors and lenses. Prisms and lenses can be used, of course, only where they are transparent. Such "optical" methods are not practical in the

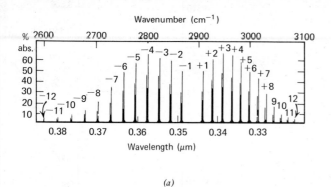

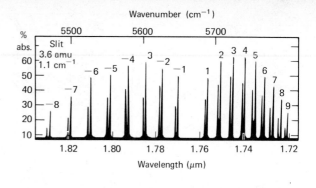

(a)

(b)

FIGURE 7.6
The infrared spectrum of HCl vapor associated with transitions (a) $v = 0 \rightarrow v = 1$ and (b) $v = 0 \rightarrow v = 2$. The individual rotation-vibration lines are all doublets because of the presence of the isotopic species $H^{35}Cl$ and $H^{37}Cl$; the former is the more intense. Numbers above spectral lines specify ΔJ. From C. F. Meyer and A. A. Levin, *Phys. Rev.* **34**, 44 (1929).

long-wavelength microwave and radio regions, where different wavelengths must be obtained in succession by "tuning" the radiation source.

4. *Detector.* The simplest type of radiation detector is a photographic emulsion; the intensity is determined by measuring the extent of darkening. For more accurate intensity measurements one can use photoelectric cells (visible and ultraviolet), thermocouples and bolometers (infrared), and crystal diodes (microwave). The apparatus is usually so designed as to present an intensity-versus-wavelength or similar curve in graphical form.

We saw that any transition between bound states of a molecule has a definite energy and frequency ($\Delta E = h\nu$), and should thus give rise to a sharply defined spectral line. Yet in practice one always observes "lines" of nonzero width (in wavelength or energy), at best resembling narrow spike-like curves. Sometimes this is an apparatus effect, due to the limited resolution of the optical system, but with sufficient resolving power one obtains line widths characteristic of the sample itself. This broadening of spectral lines has several causes, principally the following. (1) The *natural line width* is the consequence of the uncertainty principle, as stated in Eq. 3.87. Any measurement carried out in a finite time Δt has an energy uncertainty of the order of $\hbar/\Delta t$. But the natural line width is usually extremely small, about 10^{-4} Å for spontaneous emission of visible light. (2) *Doppler broadening* results from the fact that the molecules of the sample are in motion in various directions, so that the frequency of the radiation absorbed or emitted is shifted up or down by the well-known Doppler effect. (3) *Pressure broadening* results from the perturbation of energy levels by intermolecular forces, especially during collisions. In a gas, pressure broadening increases rapidly with density, and is the dominant contribution to line widths at high pressures. Normally, visible and ultraviolet lines from gaseous samples have linewidths dominated by Doppler broadening. Infrared and microwave lines have widths due primarily to pressure broadening at all but the lowest pressures.

In the remainder of this section we shall discuss some of the major aspects of spectroscopy that we have not yet described: (1) simultaneous vibration–rotation spectra; (2) the Raman effect, in which light is scattered inelastically rather than absorbed; (3) electronic spectra.

Since a real molecule may vibrate and rotate simultaneously, it can undergo transitions in which both quantum numbers change. In fact, in the ground electronic states of most diatomic molecules a pure vibrational transition is forbidden. In any such simultaneous transition ΔE_{vib} is likely to be far greater than ΔE_{rot} (because of the energy-level spacings), so vibration–rotation spectra are found in the same part of the spectrum as pure vibrational spectra. For Σ electronic states, which include most stable ground states, the selection rules are those we have already introduced:

$$\Delta v = \pm 1 \quad \text{and} \quad \Delta J = \pm 1, \tag{7.27}$$

with $\Delta v = \pm 1$ holding only approximately for anharmonic oscillators.

Figures 7.6 and 7.7 illustrate the kind of spectrum obtained when molecules undergo such transitions. Each line of the pure vibrational spectrum is replaced by a whole band of vibration–rotation lines. These lines are nearly equally spaced, as in the pure rotational spectrum; however, there is a gap in the center of the band. This gap corresponds to the pure vibrational transition with $\Delta J = 0$, which is forbidden for Σ states by the selection rule of Eq. 7.27. The position of the missing line is referred to as the *null line* or *band origin*, with wavenumber $\tilde{\nu}_0$. The parts of the band on each side of the origin are called the *P branch* and the *R branch*, with $\tilde{\nu} < \tilde{\nu}_0$ and $\tilde{\nu} > \tilde{\nu}_0$, respectively. It is conventional to designate the upper state by a prime (v', J', etc.) and the lower state by a double prime (v'', J'', etc.). As is shown in Fig. 7.7, the *P* branch corresponds to transitions with $J' = J'' - 1$ ($\Delta E < hc\tilde{\nu}_0$), and the *R* branch to transitions with $J' = J'' + 1$ ($\Delta E > hc\tilde{\nu}_0$). With the rotational energy in each state given by Eq. 7.21, the wavenumbers of the lines in the band are given by

$$\tilde{\nu} = \frac{E' - E''}{hc} = \tilde{\nu}_0 + B_v' J'(J'+1) - B_v'' J''(J''+1), \tag{7.28}$$

where $\tilde{\nu}_0 \equiv (E_{vib}' - E_{vib}'')/hc$. The relative intensities of the lines are as usual proportional to the populations of the initial rotational states; the intensity distribution in each branch (Fig. 7.6b) resembles that in a pure rotational band (Fig. 7.4).

Equation 7.28 can be rewritten in the form

$$\tilde{\nu} = \tilde{\nu}_0 + (B_v' + B_v'')m + (B_v' - B_v'')m^2, \tag{7.29}$$

where the *running number m* equals $-J''$ in the *P* branch and $J'' + 1$ in the *R* branch. For a rigid rotator the spectral lines are equally spaced, but because of the quadratic term in Eq. 7.29 this cannot be true in a vibrating rotator. From Eq. 7.21 we know that B_v decreases with increasing v within a given electronic state, so B_v' must be a little smaller than B_v'' in any infrared transition.[15] Thus the quadratic term is negative, and the spacing between successive lines must decrease with increasing m; this effect is illustrated in Fig. 7.7 in slightly exaggerated form (since B_v' and B_v'' are nearly equal). Eventually a point should be reached at which the spacing decreases to zero ($d\tilde{\nu}/dm = 0$), after which $\tilde{\nu}$ decreases with increasing m. There should thus be a sharp upper limit to the spectrum at some wavenumber $\tilde{\nu}_b$, which we call the *band head*. Band heads in vibration–rotation spectra always occur in the *R* branch. However, B_v' and B_v'' are usually so close together that these heads appear at very high values of m (or J), and thus are usually too weak to observe in spectra taken at ordinary temperatures. We shall see that the situation is different for electronic spectra.

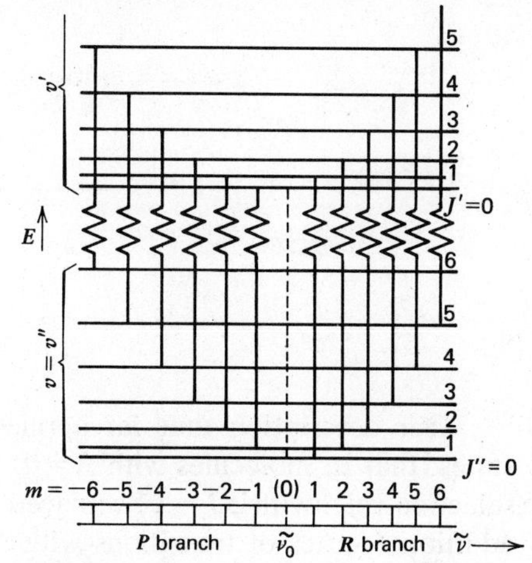

FIGURE 7.7
Schematic representation of a vibration-rotation band in the infrared spectrum of a diatomic molecule in a Σ electronic state. Each line in the band corresponds to a transition between a lower level v'', J'' and an upper level v', J', with $J' = J'' - 1$ in the *P* branch and $J' = J'' + 1$ in the *R* branch. The lines are labeled with the running number m, defined in Eq. 7.29. The band origin $\tilde{\nu}_0$ corresponds to the forbidden transition v'; $J' = 0 \leftrightarrow v''$; $J'' = 0$. The spectrum is drawn for $B_{v'} < B_{v''}$, so that the band converges to a head (not shown) in the *R* branch.

---

[15] Another way to say this is that the upper state has a larger moment of inertia, since $\langle I \rangle = \mu \langle R^2 \rangle$ increases with v.

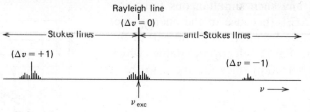

FIGURE 7.8
Schematic Raman spectrum of a diatomic gas. Three vibrational bands are shown, each with its rotational fine structure. The intense line at the exciting frequency ν_{exc} is due mainly to Rayleigh scattering; in the other bands the center line is the unresolved Q branch.

Our description thus far applies only to molecules in Σ electronic states, that is, molecules with $\Lambda = 0$. When $\Lambda \neq 0$ there is a change in the selection rule, with $\Delta J = \pm 1$ replaced by $\Delta J = 0, \pm 1$. One thus observes an additional series of transitions with $\Delta J = 0$, known as the Q *branch*. The best-known example is in the ground state of NO, which is $^2\Pi$. For the Q branch, Eq. 7.29 is replaced by

$$\tilde{\nu}_Q = \tilde{\nu}_0 + (B'_v - B''_v)J + (B'_v - B''_v)J^2. \tag{7.30}$$

Since the spacing increases steadily in both directions, there is no band head. The coefficient $B'_v - B''_v$ is very small, so all the observed lines of the infrared Q branch are very close to $\tilde{\nu}_0$, and under low or moderate resolution appear as a single intense line.

Up to this point we have been talking only about absorption and emission spectra, but one can also observe the spectrum of radiation *scattered* by molecules. By scattered radiation (cf. Section 2.3) we simply mean radiation that leaves the sample in a direction different from its incident direction. This involves a two-step process on the molecular level: An incident photon strikes a molecule and excites it to a highly unstable condition (in classical terms, sets up a forced, nonresonant oscillation); the molecule then attains a new stationary state by emitting a second photon, which may depart in any direction. But the two steps are virtually simultaneous, and no stationary excited state exists in the interval between them. Scattering should thus not be confused with fluorescence or phosphorescence, in which the molecule absorbs one photon, forms an excited state that lasts long enough to be characterized as a stationary state (anything from picoseconds to hours), then decays by emitting a second photon. In either kind of process the two photons need not have the same energy; if they do not, of course, the final state of the molecule is different from the initial state, and a spectrum is observed. If the incident and scattered photons do have the same energy, one speaks of *Rayleigh scattering*, corresponding to the *elastic* scattering of particles. The effect is the same as if a single photon bounced off the molecule. If the energies are different, the process is called *Raman*[16] *scattering*, corresponding to *inelastic* scattering of particles. It is the Raman effect that gives rise to a spectrum.

A schematic Raman spectrum is illustrated in Fig. 7.8. To keep the spectrum simple, one ordinarily uses a monochromatic beam of incident light, that is, light of a single wavelength, usually in the visible or ultraviolet. The scattered light is customarily observed at right angles to the incident beam. Since Rayleigh scattering is by far the more likely effect,[17] the scattered light is dominated by a single intense line, the

[16] The phenomenon was predicted in 1923 by A. Smekal and first demonstrated by C. V. Raman and K. S. Krishnan in 1928.

[17] Incidentally, the probability of Rayleigh scattering is proportional to λ^{-4}, so that from a beam of white light far more blue than red will be scattered. This is why the sky is blue.

Rayleigh line, with the same wavelength as the incident light. All the other lines in the spectrum are the result of Raman scattering. If ν_{exc} is the exciting frequency, a line of frequency ν corresponds to a molecular transition of energy

$$\Delta E = h(\nu_{exc} - \nu). \tag{7.31}$$

Note that the frequency shift $\nu - \nu_{exc}$ for a given transition is independent of ν_{exc} itself, depending only on ΔE. The shift may be to either lower or higher frequency, corresponding respectively to net absorption and net emission of energy by the molecule. The lines with $\nu < \nu_{exc}$ are called *Stokes lines*, and those with $\nu > \nu_{exc}$ are called *anti-Stokes lines*. As in the infrared spectrum, each vibrational transition gives rise to a closely spaced band of lines corresponding to different rotational transitions. Since at ordinary temperatures most molecules are in the vibrational ground state, the Stokes band for a given $|\Delta v|$ is much more intense than the corresponding anti-Stokes band.

What kinds of transitions do we see in the Raman spectrum? Let us look again at Eq. 4.33, from which we find the transition probability. This equation involves the molecular dipole moment $\boldsymbol{\mu}$, which we have thus far taken to be the moment in the absence of external fields; let us call the latter $\boldsymbol{\mu}_0$. But the incident light beam has an electric field oscillating at frequency ν_{exc}, and this must in general induce an additional dipole moment in the molecule. (The electrons follow the field more easily than the heavy nuclei, so the centers of positive and negative charge oscillate relative to one another.) The total molecular dipole moment is thus $\boldsymbol{\mu}_0 + \boldsymbol{\mu}_{ind}$, with the induced moment given by

$$\boldsymbol{\mu}_{ind} = \alpha \mathbf{E} = \alpha \mathbf{E}_0 \cos 2\pi\nu_{exc} t, \tag{7.32}$$

where $\mathbf{E}$ is the oscillating field and α is the molecular *polarizability* (cf. Section 10.1). We can thus rewrite Eq. 4.33 as

$$\boldsymbol{\mu}_{n'n} = \int \cdots \int \psi_{n'}{}^* \boldsymbol{\mu}_0 \psi_n \, d\tau + \mathbf{E} \int \cdots \int \psi_{n'}{}^* \alpha \psi_n \, d\tau. \tag{7.33}$$

(Remember that the transition probability between states n and n' is proportional to $|\boldsymbol{\mu}_{n'n}|^2$). Consider a vibrational–rotational transition for which we define $\nu_{nn'} \equiv |E_{n'} - E_n|/h$. This transition can occur in either of two ways. If the first term in Eq. 7.33 is nonzero, there can be ordinary absorption or emission of radiation with frequency $\nu_{nn'}$ in the infrared or microwave regions; but this can occur only when there is a permanent dipole moment $\boldsymbol{\mu}_0$, and thus is not possible for homonuclear molecules. Here we are interested in the second possibility. If the second term in Eq. 7.33 is nonzero, a Raman scattering process can occur in which the scattered light has the frequency $\nu_{exc} \pm \nu_{nn'}$. When the incident light is in the visible or ultraviolet, we have $\nu_{exc} \gg \nu_{nn'}$, and the scattered light is in the same region. Just as the first integral vanishes unless $\boldsymbol{\mu}_0$ changes in the course of the vibration or rotation, the second integral vanishes unless α changes. But all diatomic molecules have a nonzero polarizability, which varies with both bond length and orientation to the field: Thus both homonuclear and heteronuclear molecules display Raman spectra.

Since the mechanisms are so different, we need not be surprised if different selection rules apply for Raman and ordinary infrared transitions. The selection rule for vibration is in fact the same: $\Delta v = \pm 1$ for the harmonic oscillator, with weaker bands corresponding to $\Delta v = \pm 2, \pm 3, \ldots$ for anharmonic molecules; we also have a band with $\Delta v = 0$, corresponding to the pure rotational spectrum. But the rotational selection rule for

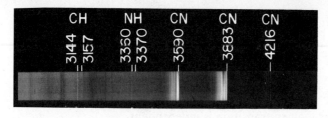

FIGURE 7.9
A typical electronic absorption spectrum of a gas undergoing a rapid reaction. The spectrum shows the presence of three diatomic molecules, CH, NH, and CN, none of which is stable under ordinary conditions. The numbers are wavelengths, in angstroms, of the band heads. This spectrum was taken with what is called a "medium-dispersion" prism spectrograph, exhibiting the spectral range from about 200 to 700 nm (2000 to 7000 Å) on a 10-in. photographic plate. The photograph is printed in negative so absorption lines appear white. From D.W. Cornell, R.S. Berry, and W. Lwowski, *J. Am. Chem. Soc.* **88**, 544 (1966).

diatomic molecules in Σ states is quite different from the rule for simple absorption or emission; for Raman transitions,

$$\Delta J = 0, \pm 2. \tag{7.34}$$

Thus each band contains an *S branch* ($\Delta J = +2$), an *O branch* ($\Delta J = -2$), and a *Q* branch ($\Delta J = 0$); the *Q*-branch lines are again given by Eq. 7.30 and are hard to resolve. For molecules with $\Lambda \neq 0$ the selection rule becomes $\Delta J = 0, \pm 1, \pm 2$, and *P* and *R* branches are also observed.

Raman spectroscopy offers a useful complement to infrared spectroscopy for studying molecular vibrations and rotations. Raman spectra are observed for transitions that do not appear at all in the infrared, including the entire vibration–rotation spectrum of homonuclear diatomic molecules. This technique has one significant drawback, in that Raman intensities are relatively weak. This is to be expected, since a Raman transition involves two fairly unlikely processes rather than one. Note also that the measurement of Raman spectra is not limited to the infrared region. Any convenient exciting line can be used, since the frequency shift is independent of the exciting frequency. In fact, ν_{exc} is usually chosen to be in the visible or ultraviolet regions, where $\nu_{exc} \gg \nu_{nn'}$ and the whole Raman spectrum can be found close to the exciting line.

Finally we come to electronic spectra. Since the electronic states of a molecule typically differ in energy by several electron volts, transitions between them are generally observed in the visible and ultraviolet regions. To a given electronic transition there corresponds a series of bands, one for each accessible vibrational transition; within a band there is a line for each possible rotational transition. A typical electronic spectrum is shown in Fig. 7.9.

The most prominent features of an electronic band spectrum are the band heads, the sharply defined edges on one side of the bands. We have already explained the origin of band heads in the context of vibration–rotation spectra. The most important selection rule for electronic transitions is that for the rotational quantum number *J*, namely

$$\Delta J = 0, \pm 1, \tag{7.35}$$

with the following exceptions: $\Delta J = 0$ (the *Q* branch) does not occur if $\Lambda = 0$ in both initial and final states, that is, if both are Σ states; and the transition $(J' = 0) \leftrightarrow (J'' = 0)$ is always forbidden. Thus there are always *P* and *R* branches, and in most cases also a *Q* branch. The wavenumbers in these branches are again given by Eqs. 7.29 and 7.30, but the physical situation is quite different. In infrared transitions B'_v and B''_v are nearly equal, so heads appear only at high *m* and the *Q* branch is very narrow. In electronic spectra this is no longer the case. Since two electronic states are involved, with completely different $E(R)$ curves and values of R_e, there is no reason why B'_v and B''_v should be close to each other. Thus sharp band heads can appear at low values of *m*, and the *Q* branch can be as broad as the *P* and *R* branches; an example is analyzed in Fig. 7.10.

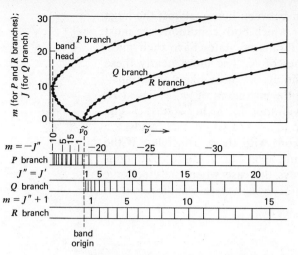

FIGURE 7.10

Structure of an electronic band spectrum. The upper part of the figure is what is known as a *Fortrat diagram*, showing the parabolic curves describing the three branches according to Eqs. 7.29 and 7.30. Below are shown the corresponding spectral lines of the three branches (which would be superimposed in the actual spectrum). The spectrum illustrated has a head in the P branch, at about $m = -9$.

Another difference from the infrared spectrum is that either B'_v or B''_v may be the larger, depending on which state has the higher moment of inertia, so a head may appear in either the P or the R branch. Because the band heads are so well defined, they are commonly used for identification and analysis of spectra.

There is no vibrational selection rule in electronic transitions; its place is taken by the *Franck–Condon principle*. This assumes that, during an electronic transition, the nuclei tend to retain their initial positions and momenta. The concept is based on the rapidity with which the very light electrons make their transition, relative to the time and impulse required for the nuclei to change their positions and momenta. Retention of position means that the transition in a single molecule can be represented by a vertical line between the potential curves of the upper and lower electronic states, $E'(R)$ and $E''(R)$, respectively. Retention of momentum means that the lower terminus of this "transition line" should be as far above $E''(R)$ as the upper terminus is above $E'(R)$. The vertical lines in Fig. 7.11a illustrate such transitions. Strictly, position and momentum

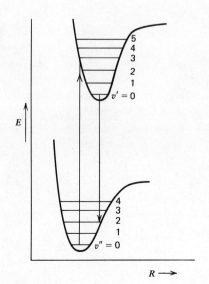

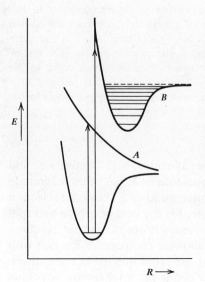

(a)

(b)

FIGURE 7.11

Illustration of the Franck–Condon principle. (*a*) Stable excited state: Any v'–v'' transition is possible, but the most likely are between states having maximum values of $|\psi_{vib}|^2$ (cf. Fig. 4.4) at nearly the same R. Here, for example, an excitation from $v'' = 0$ is most likely to yield $v' = 3$, and the state $v' = 0$ is most likely to decay to $v'' = 2$, as indicated by the vertical arrows. (*b*) Two ways to obtain a continuous spectrum: excitation to an unstable state (*A*), or to a stable state (*B*) at a level above its dissociation limit.

cannot both be preserved in a transition between bound states; the most probable transitions should be those for which both conditions are most nearly satisfied. In addition, different vibrational states have their maximum values of $|\psi_{vib}|^2$ at different values of R. The more nearly these values coincide for a given v'—v'' transition, the more likely that transition is to be observed. (Remember that the maximum of $|\psi_{vib}|^2$ lies at R_e for $v = 0$, but near the extremes of the vibration for higher states; cf. Fig. 4.4.) Each electronic transition gives rise to a set of vibrational bands, the relative intensities of which vary in accord with these principles. If the final state lies above the dissociation limit, as in Fig. 7.11b, a continuous spectrum will be observed. This is always the case when the electronic state in question is unstable, like the H_2 $b^3\Sigma_u^+$ state we discussed in Section 6.10.

To make the concept of the Franck–Condon principle more precise, we write out the electronic counterpart of Eq. 7.12 for a diatomic molecule. The wave function for the initial state is $\psi_{nucl}(R)\psi_{elec}(r, R)$, and for the final state, $\psi'_{nucl}(R)\psi'_{elec}(r, R)$. We neglect the rotational wave functions, which are not relevant at this point, and use r to represent all electronic coordinates, as we did in Eqs. 6.2 and 6.3. The transition is electronic; for convenience, suppose that it is induced by a uniform electromagnetic field, so that the transition operator is the electronic dipole moment operator $\boldsymbol{\mu}(r)$. Then the transition amplitude is

$$A = \int dR \int \psi^*_{nucl}(R)\psi^*_{elec}(r, R)\boldsymbol{\mu}(r)\psi'_{nucl}(R)\psi'_{elec}(r, R)\,dr. \tag{7.36}$$

We might assume that the electronic factor—the electronic transition dipole—is given by its value at $R = R_e$, or

$$\langle\boldsymbol{\mu}(r)\rangle_{R_e} = \int \psi^*_{elec}(r, R_e)\boldsymbol{\mu}(r)\psi'_{elec}(r, R_e)\,dR. \tag{7.37}$$

This is called the Condon approximation. Alternatively, we can call upon the mean value theorem and our knowledge that the expectation of $\boldsymbol{\mu}(r)$ is bounded to tell us that there is some mean value for which

$$A = \langle\boldsymbol{\mu}(r)\rangle \int \psi^*_{nucl}(R)\psi'_{nucl}(r)\,dR. \tag{7.38}$$

In words, the transition amplitude can be written to be proportional to the overlap of the vibrational wave functions of the initial and final states (see Appendix 6A). If $\psi_{nucl}(r)$ and $\psi'_{nucl}(R)$ have amplitudes distributed in the same range of R *and* if they have similar wavelengths, that overlap is large. The first of these conditions corresponds to the preservation of location; the second, to the preservation of momentum. If either is not well met, the integral A of Eq. 7.37 is small and the transition probability, proportional to $|A|^2$, is correspondingly low.

Not all electronic transitions are allowed; here again we find selection rules. The total angular momentum $\mathbf{J}$ has the magnitude $[J(J+1)]^{1/2}\hbar$, where the rotational quantum number J obeys the selection rule of Eq. 7.35. For a diatomic molecule, $\mathbf{J}$ is the sum of orbital and spin electronic components and the angular momentum of nuclear rotation. As in the atom (Sections 5.6 and 5.7), all these components interact with one another, but some kinds of coupling are more important than others. We shall assume that one can separately define the total electronic orbital angular momentum $\mathbf{L}$, with z component $\Lambda\hbar$, and the total electronic spin angular momentum **S**, of magnitude $[S(S+1)]^{1/2}\hbar$. When spin–orbit coupling can be neglected,[18] one finds the selection rules

$$\Delta\Lambda = 0, \pm 1 \tag{7.39}$$

($\Sigma \leftrightarrow \Sigma$ or $\Sigma \leftrightarrow \Pi$ allowed, but $\Sigma \leftrightarrow \Delta$ forbidden) and

$$\Delta S = 0 \tag{7.40}$$

($^1\Sigma \leftrightarrow {}^1\Sigma$ allowed, $^1\Sigma \leftrightarrow {}^3\Sigma$ forbidden).

There are additional selection rules for various coupling conditions, but we shall ignore these. For homonuclear molecules one must also consider the symmetry of the wave function, since the parity of the wave function always changes in an electronic transition; we write this rule as

$$g \leftrightarrow u, \qquad g \nleftrightarrow g, \qquad u \nleftrightarrow u, \tag{7.41}$$

where "$\nleftrightarrow$" designates a forbidden transition. (In the orbital approximation, the same rule holds for the orbitals if only one electron takes part in the transition.) On the other hand, the wave function's symmetry under interchange of electrons or nuclei is fundamentally associated with the indistinguishability of identical particles, and must always be conserved:

$$s \leftrightarrow s, \qquad a \leftrightarrow a, \qquad s \nleftrightarrow a \tag{7.42}$$

(s = symmetric, a = antisymmetric). This consequence of particle identity is probably the strictest of all molecular selection rules.

We have been able to give only the barest outline of molecular spectroscopy. To finish our treatment of the subject, let us mention one additional complication: the *isotope effect*. The spacing of both vibrational and rotational energy levels depends on the molecule's reduced mass μ, which of course varies from one isotopic species to another. The effect is quite large for hydrogen, as we illustrated in Table 7.1. It is still present in even the heaviest molecules. For example, a given vibration–rotation band of BrCl appears in the spectrum at four different (but overlapping) positions, corresponding to $^{79}\mathrm{Br}^{35}\mathrm{Cl}$, $^{81}\mathrm{Br}^{35}\mathrm{Cl}$, $^{79}\mathrm{Br}^{37}\mathrm{Cl}$, and $^{81}\mathrm{Br}^{37}\mathrm{Cl}$; in the pure rotational spectrum there are four closely spaced lines for each transition. This is one more complexity that the spectroscopist must take into account when analyzing a spectrum. But this very complexity makes it possible to excite selectively a spectral line or band of a single isotopic species. Such selective excitation can be used as a first step for starting a photo-induced chemical reaction or ionization that allows one to separate one isotope from another, a process that is otherwise difficult and costly.

7.4 The Ionic Bond

We can now return to the problem of chemical bonding. In Chapter 6 we introduced the basic principles of the covalent bond, for the simple cases of one- and two-electron molecules; later in this chapter we shall extend these principles to the general problem of the electronic structure of diatomic molecules. First, however, we must examine a class of molecules constituting such an extreme case that they are best described in terms of a different (and simpler) model. These are the molecules, best typified by the diatomic alkali halides, in which we can say that there is an *ionic bond*.

---

[18] Spin–orbit coupling does occur, of course, and as in atoms gives the spectrum a fine structure. In addition, when $\Lambda > 0$ electronic–rotational interaction splits the otherwise degenerate energy levels with $M_J = \pm|M_J|$ (*Λ-type doubling*). We shall say more about these effects in Section 7.9.

The characteristic feature of such a molecule is that its properties—especially electron distribution and bonding energy—closely resemble those of two oppositely charged ions in close proximity. The NaCl molecule, for example, is quite like a Na^+Cl^- ion pair. Of course, some redistribution of electronic charge occurs in the formation of any diatomic molecule from the atoms. In a covalent bond this redistribution is rather subtle: A quite small amount of charge is shifted from antibonding to bonding regions, increasing the net bonding force enough to hold the molecule together. But in a homonuclear molecule like H_2 the total amount of charge in each half of the molecule must be the same as in the separated atoms. In the alkali halides we have a charge redistribution of quite a different order. Charge roughly equivalent to one electron shifts from one atom to the other, i.e., from the region surrounding one nucleus to that surrounding the other. This is clearly illustrated by Fig. 7.12, which shows that the electron distribution in the LiF molecule is much more like the ion pair Li^++F^- than the neutral atoms $Li+F$.

Why does a charge shift of this magnitude occur? As always in bonding problems, because the resulting charge distribution gives the lowest energy for the molecule as a whole. Thus ionic bonding is most likely to occur when an electron can be easily removed from one atom and easily added to the other—"easily" in both cases refers to the energy involved. In the terminology we introduced in Section 5.5, one atom must have a low ionization potential and the other a high electron affinity. It is obvious from Figs. 5.6 and 5.7 that these conditions are best satisfied for the alkali halides. This reasoning is confirmed by detailed calculations of the molecular wave functions, such as those leading to Fig. 7.12. But calculations of this type are essentially the same as those used in covalently bonded molecules. We brought up ionic bonding at this point mainly because there is a *simpler* way to get useful results. Let us see now what it is.

The model we use could not be simpler. We assume that the molecule *is* an ion pair, and calculate the energy of interaction between two ions at close range. Why is this easier than calculating the interaction between two neutral atoms? The answer is clear when we consider the two cases at long range. There is virtually no interaction between two neutral atoms at long range;[19] the dominant terms in the bonding energy result from the reorganization of charge that occurs when the atoms come close together, and must thus be calculated in terms of the forces between individual electrons and nuclei. But two oppositely charged ions do interact strongly at quite long range because of our old friend, the attractive Coulomb force. In first approximation the ions are spherically symmetric charge distributions, which interact as if all their charge were concentrated at the nuclei. Thus the long-range force between the ions is the same as that between two point charges: The force is given by Eq. 1.5 and the corresponding potential energy by Eq. 2.55. Since this interaction involves the whole charge of the ions, even at R_e it is much larger than the corrections resulting from charge reorganization. One can readily confirm this assertion, since the Coulomb energy ($e^2/4\pi\epsilon_0R$) is remarkably easy to calculate. It will be seen in Table 7.3 that the Coulomb energy at R_e by itself gives a fairly good approximation to the observed bond energy (last column).

But at R_e there must be more to the bond than the attractive Coulomb force alone. As in all molecules, the equilibrium bond length R_e

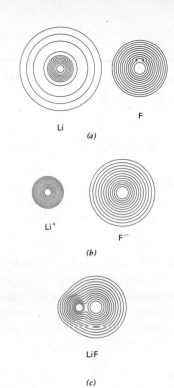

FIGURE 7.12
Electron distribution in the lithium–fluorine system. (*a*) Separated atoms, $Li+F$. (*b*) Separated ions, Li^++F^-. (*c*) The LiF molecule at the equilibrium internuclear separation, $R_e = 1.564$ Å. Diagrams taken from A. C. Wahl, *Sci. Am.* **322,** No. 4, 54 (April, 1970).

---

[19] The van der Waals attraction that exists between all atoms or molecules at long range (see Chapter 10, especially Section 10.1) is negligible in comparison with the energies involved in bonding.

is the distance at which there is a balance between attraction and repulsion. The repulsion here is essentially the same as that which occurs when any two atoms or ions "come in contact," that is, when there is significant overlap between the atomic (ionic) wave functions. There is no simple expression that would enable us to calculate this repulsive force as easily as the attractive Coulomb force. A detailed calculation can of course be made in terms of individual electrons and nuclei, but that is just what we are trying to avoid. Fortunately, there is an empirical expression that gives a good description of the short-range repulsion: The potential energy is assumed to fall off exponentially with distance,

$$V_{\text{rep}}(R) = Be^{-R/\rho}, \tag{7.43}$$

where B and ρ are constants. This is the repulsive part of the *Born–Mayer potential*, originally introduced to describe the similar interaction in ionic crystals. One can at least get a qualitative idea of why the repulsive potential has this steep form. For very small R the potential is dominated by the direct internuclear repulsion, which has the relatively slowly decreasing Coulombic form ($V \propto R^{-1}$); yet for $R \gtrsim R_e$ the repulsive part of the potential becomes extremely small. The repulsive potential must thus decrease appreciably faster than R^{-1} in the intermediate region, and the exponential is one of the simplest functions that has this behavior.[20] The reason for this steep decrease is the increased screening of the nuclei from one another as the amount of electronic charge between them increases with increasing R. This process is essentially complete at a distance slightly greater than R_e, as the overlap of the atomic wave functions becomes negligible.

In the ionic bond model, then, the total potential energy of a diatomic molecule must include a Coulombic ion-attraction term and a short-range repulsion term like Eq. 7.43. Additional corrections can be added; the most important of these is an attractive term due to the *polarization* of each ion by the charge on the other. That is, the field of each ion tends to push the other ion's nucleus in one direction and its electrons in the opposite direction, creating an induced dipole on which the first ion exerts an additional attractive force. This process is analyzed in Section 10.1. The magnitude of the polarization energy for each ion is given by Eq. 10.14. Combining this with the other terms, we find a total ionic bond interaction energy of

$$E(R) = -\frac{q_1 q_2}{4\pi\epsilon_0 R} + Be^{-R/\rho} - \frac{\alpha_1 q_2{}^2 + \alpha_2 q_1{}^2}{32\pi^2 \epsilon_0{}^2 R^4}, \tag{7.44}$$

where q_i is the charge and α_i the polarizability of the ith ion; for the alkali halides we of course have $q(M^+) = -q(X^-) = e$.

The validity of the ionic bond model can be tested by evaluating Eq. 7.44 at $R = R_e$ and comparing the result with the experimental dissociation energy. But here we must be careful. Equation 7.44 gives the energy of the ionic molecule MX relative to the separated ions $M^+ + X^-$; this corresponds to the "ionic dissociation energy"

$$D_e(MX \rightarrow M^+ + X^-) = E(M^+ + X^-, R = \infty) - E(MX, R = R_e). \tag{7.45}$$

---

[20] An even more suitable (but less tractable) repulsive potential would be $V_{\text{rep}}(r) = BR^{-1}e^{-R/\rho}$, which gives the correct Coulomb behavior as $R \rightarrow 0$. This "screened Coulomb potential" is also known as the *Debye potential* in the context of ionic solutions, as we shall see in Chapter 26.

TABLE 7.3
TEST OF THE IONIC BOND MODEL FOR SOME ALKALI
HALIDE MOLECULES*

Molecule	R_e (Å)	$e^2/4\pi\epsilon_0 R_e$ (eV)	Ionic Dissociation Energy, D_e(MX → ions) (eV)	
			Calculated	Observed
LiF	1.564	9.21	7.9996	7.983
LiCl	2.021	7.14	6.513	6.648
NaCl	2.361	6.10	5.616	5.750
KF	2.171	6.63	5.993	6.036
KI	3.048	4.72	4.458	4.601
RbCl	2.7869	5.17	4.835	4.917
CsCl	2.906	4.96	4.692	4.870

*Data taken from P. Brumer and M. Karplus, *J. Chem. Phys.* **58**, 3903 (1973) and references therein.

However, the ground state of the system at large R corresponds to the neutral atoms in their ground states, not to the ions. This is true for any pair of atoms: Since the lowest ionization potentials (I) are greater than the highest electron affinities (A), it always requires energy to make a pair of separated ions from the ground-state neutral atoms. This energy of ion-pair formation is

$$E(M^+ + X^-, R = \infty) - E(M + X, R = \infty) = I(M) - A(X). \quad (7.46)$$

The true dissociation energy[21] of the MX molecule is thus that into the separated neutral atoms,

$$D_e(MX \rightarrow M + X) = D_e(MX \rightarrow M^+ + X^-) - I(M) + A(X). \quad (7.47)$$

The predictions of the ionic bond model for a number of alkali halides are compared with experiment in Table 7.3. The "ionic dissociation energies" are those defined by Eq. 7.45. The calculated values are essentially the values of $-E(R_e)$ predicted by Eq. 7.44, with some minor corrections (van der Waals interaction, dipole–dipole interaction, etc.). The values of B and ρ are fitted to the repulsive part of the $E(R)$ curve, and the values of the α's to the observed dipole moment (see below). It turns out that the repulsive and polarization terms are both so small near R_e that the ionic Coulomb potential alone ($e^2/4\pi\epsilon_0 R_e$) gives a fairly good approximation to D_e, as we noted earlier. With the other terms included, the agreement with experiment is within a few percent. We conclude that the ionic bond model gives a satisfactory description of the alkali halide molecules. This is not surprising, since Eq. 7.44 is basically the same as the equation used to calculate the lattice energy of an MX crystal, which is well known to be a lattice of M^+ and X^- ions (see Chapter 11).

For an additional test of the ionic bond model, let us consider the dipole moments of ionic molecules. If such a molecule did consist simply of two spherical nonoverlapping ions with charges $\pm q$ a distance R_e apart, the magnitude of the dipole moment would be given by Eq. 4.29 as $\mu = qR_e$, or eR_e for the alkali halides. As can be seen from Table 7.4, the

[21] That is, the well depth, measured from the $E(R)$ minimum, not from the ground vibrational state.

TABLE 7.4
DIPOLE MOMENTS OF SOME IONIC MOLECULES

Molecule	Ionic Charge, q	R_e (Å)	qR_e (D)	Measured Dipole Moment, μ (D)
LiH	$\pm e$	1.595	7.66	5.882
LiF	$\pm e$	1.564	7.51	6.325
LiCl	$\pm e$	2.018	9.69	7.126
LiBr	$\pm e$	2.170	10.42	7.265
LiI	$\pm e$	2.392	11.49	7.428
NaCl	$\pm e$	2.361	11.34	9.001
KCl	$\pm e$	2.667	12.81	10.269
RbCl	$\pm e$	2.787	13.39	10.510
CsCl	$\pm e$	2.906	13.96	10.387
HCl	$\pm e$	1.275	6.12	1.109
TlCl	$\pm e$	2.485	11.94	4.543
SrO	$\pm 2e$	1.920	18.44	8.900
BaO	$\pm 2e$	1.940	18.64	7.954

actual dipole moments[22] are appreciably less. This discrepancy is due to the polarization effect already mentioned. By attracting the electrons of a negative ion away from the nucleus and toward itself, a positive ion has the effect of reducing the dipole moment; the negative ion attracts the positive ion's nucleus and repels its electrons. The more polarizable the ions are, the more the molecular dipole moment is reduced. In fact, this effect can be used to evaluate the α_i's of Eq. 7.44. The polarizability of an ion is roughly proportional to its size. In most of the alkali halides the X^- ions are significantly larger and thus more polarizable than the M^+ ions, so the principal effect is the distortion of the halide ion; the alkali ion is effectively very stiff.

The extent of polarization is of course also affected by the charge doing the polarizing. The larger the value of q, the more the dipole moment is reduced. This is made strikingly clear by the example of SrO and BaO in Table 7.4. Naive chemical notions (based on the stability of closed-shell structures) suggest that these molecules might have the structure $M^{2+}O^{2-}$, but examination of the second ionization potentials (Fig. 5.6) shows the unlikelihood of this idea. In fact, the alkaline earth oxides have dipole moments close to what one would expect for pairs of singly charged ions. One can still describe these compounds with an ionic bond model, but it takes a more elaborate version than we have used for the alkali halides. The reason is that the ions involved, Ba^+ and O^-, for example, have open shells and are thus far more easily distorted by polarization than closed-shell ions.

[22] Dipole moments can be measured in several ways. If an electric field is applied to a gas of polar molecules, the molecules will tend to align themselves with the field. The extent of this alignment can be related to the gas's dielectric constant, and thus to the (measured) capacitance of the system. The electric field also splits the degenerate rotational energy levels with the same J and different M_J (Stark effect); this splitting can be observed either directly in the microwave spectrum or by measuring the deflection of molecular beams by the field.

It might seem that the ionic bond model is of limited importance, since it is directly applicable to only a small class of diatomic molecules. But the utility of the concept actually extends over a far wider range. For one thing, it can be used with little change to describe the bonding in a vast number of ionic crystals. (In fact, it works better for crystals than for single molecules, since the presence of neighbors on all sides of each ion reduces the polarization effects.) The model displays very clearly the distinction between the long-range attractive and short-range repulsive forces, whose balance is crucial for the existence of stable chemical bonds. Moreover, the ionic bond provides us with the extreme case of a chemical bond having the maximum asymmetry of charge distribution. The homonuclear covalent bond, with its perfectly symmetric charge distribution, lies at the other end of this scale, and the bonds in other heteronuclear diatomic molecules fall between these extremes. One can usefully interpret most bonds as some kind of mixture or superposition of ionic and symmetric covalent contributions.

Finally, we can use the ionic model to gain an insight into the nature of bonding that applies even to covalent molecules. Consider any two atoms that form a bond. In the long-range limit, each atom consists of one or more valence electrons outside a closed-shell core—"outside" in the sense that the core electrons have very low charge density at the radii where the valence electrons have their maxima. As the atoms move closer together, bonding effects occur as the valence electrons begin to overlap. But even near R_e there is no significant overlap between each atom and the other atom's core. An atomic core is essentially the same as a spherically symmetric ion. If an ionic bond model is valid for two ions in "contact" with each other, it should also work reasonably well for the core part of the interaction in an ordinary bond. Given the distance between the cores, only the Coulombic part of the interaction should be important; in other words, the cores act much like point charges. This means that to a fairly good approximation one can neglect the atomic cores and consider only the valence electrons[23] in describing chemical bonding. This is of course precisely the approximation made in elementary treatments of chemistry, and now we can see why it works as well as it does.

To study the electronic structure of covalent diatomic molecules, we must further develop the theory introduced in Chapter 6. We shall base our discussion on the molecular orbital approach, which is the simplest and clearest for our present purposes. Thus we assume that each electron sees only the average field of the others and can be described by a one-electron wave function. Since each orbital can "contain" only two electrons (with opposite spin), one must in general use several orbitals to describe a molecule.

Suppose that one finds a suitable set of molecular orbitals; what does one do with them? The most important thing to know is how the energies and shapes of the orbitals vary as some characteristic parameter of the molecule is changed. For a diatomic molecule the key parameter is the internuclear distance R. Given the sequence of orbital energies, one can develop an Aufbau principle like that used for atoms and derive the equilibrium electronic configuration of the molecule. The orbital energies as functions of R are most simply displayed in a correlation diagram, like

7.5
Homonuclear Diatomic Molecules: Molecular Orbitals and Orbital Correlation

[23] In a transition metal atom the "valence electrons" must include the partly filled inner d or f subshells, which are quite close to the atom's "surface."

the one we drew for H_2^+ in Section 6.5. We shall see that these and other properties of orbitals vary systematically in series of molecules.

The first tool we need for the description of molecular orbitals is a systematic way of classifying them. Just as atomic orbitals are given the names of the analogous states of the one-electron H atom, molecular orbitals are named after the states of the one-electron H_2^+ molecule. We have already introduced most of the terminology needed, but we shall review it here. Both the orbitals and the states of the molecule as a whole are classified in terms of the wave function's symmetry properties.

We have introduced two kinds of symmetry thus far, the *permutational* symmetry associated with the indistinguishability of elementary particles, and the *spatial* symmetry associated with the indistinguishability of coordinate systems. It is spatial symmetry that concerns us here; we classify molecules according to the invariance of their physical description under specific kinds of coordinate changes—rotation, reflection, and inversion. We must keep in mind that a change in a molecule's coordinate frame is equivalent to an equal but opposite change in the molecule itself; translating the origin of coordinates a distance $+x_0$ along the x axis with the molecule fixed is equivalent to translating the molecule a distance $-x_0$ in a fixed coordinate system. For convenience we describe spatial transformations in terms of moving the molecule rather than the coordinates, but the two are equivalent. Each of us can think of symmetry and invariance in whichever way seems clearer. Now let us list the kinds of symmetry properties that apply to diatomic molecules.

1. The most important symmetry property, at least for the molecular energy, describes the behavior of the electronic wave function under rotation about the internuclear axis. This is given by the quantum number λ (Λ for the many-electron wave function), which specifies the angular momentum about the internuclear axis: $L_z = \pm \lambda \hbar$. The corresponding part of the wave function is $e^{\pm i\lambda\phi}$, where ϕ is the angle of rotation about the z axis (Fig. 6.1). Any rotation through an angle ω simply changes the wave function to $e^{\pm i\lambda(\phi+\omega)}$, introducing a constant phase factor $e^{\pm i\lambda\omega}$ but leaving the angular momentum unchanged. This corresponds to the fact that all values of ϕ are physically indistinguishable. States with $L_z = +\Lambda\hbar$ and $L_z = -\Lambda\hbar$ (for $\Lambda > 0$) are doubly degenerate. We identify states with $\lambda, \Lambda = 0, 1, 2, 3, \ldots$ by the already familiar designations $\sigma, \pi, \delta, \varphi, \ldots$ (orbitals), $\Sigma, \Pi, \Delta, \Phi, \ldots$ (molecules).

2. The molecule is also physically unchanged by reflection in a plane containing the z axis, the axis of the bond. The electronic wave function must either remain unchanged or reverse its sign; the two cases are designated by the superscripts $+$ and $-$, respectively. For $\Lambda > 0$, one of the degenerate molecular states with $L_z = \pm\Lambda\hbar$ is $+$, the other $-$, and the superscript is usually omitted; Σ states must be either Σ^+ or Σ^-, but all those we discussed in Chapter 6 are Σ^+.

Symmetry properties (1) and (2) apply to all diatomic (or other linear) molecules. In a homonuclear molecule the two ends are indistinguishable, imposing an additional symmetry condition:

3. The electronic wave function must either remain unchanged (even parity) or reverse its sign (odd parity) when the molecule is inverted through its center of symmetry, $\psi(-x, -y, -z) = \pm\psi(x, y, z)$. As described in Section 6.5, we designate even and odd states by the subscripts g and u, respectively.

We have spoken here only of the electronic wave function (assuming the Born–Oppenheimer approximation), but there are also symmetry conditions on the total molecular wave function. The most important of these is its symmetry or antisymmetry (denoted by s or a) under interchange of the nuclei, which divides the rotational levels into two groups. We shall postpone analysis of this subject to Chapter 21, which treats the relation between symmetry and the distribution of molecules over energy levels.

We wish to know how the orbitals vary as a function of internuclear distance. The limiting cases are simple enough: as $R \to \infty$ the molecule behaves like two separated atoms; as $R \to 0$ it collapses into a simple united atom. We generally know the states, orbitals, and orbital energies of both the separated-atom and united-atom limits. Given the Born–Oppenheimer approximation, we know that the total electronic energy of a molecule varies smoothly with R. If the orbital approximation holds for all R, we can say the same about the orbital energies $\epsilon_i(R)$. Thus we can join the united-atom and separated-atom orbital energies by a set of smooth $\epsilon(R)$ curves, making up what we have already defined as a correlation diagram. In the one-electron diagram for H_2^+ (Fig. 6.9) many of the energy levels are degenerate. A more typical example is the N_2 molecule (Fig. 7.13), whose united-atom limit is the silicon atom. It is often unnecessary in such a diagram to give the energies quantitatively for intermediate values of R. Frequently the ordering of connnecting lines as functions of R, with the limiting energies, is sufficient for analysis of physical phenomena.

How, then, does one know where to draw the connecting lines between the separated-atom and united-atom limits? The rules are quite straightforward, all taking the form of conservation laws:

1. Angular momentum about the internuclear axis (defined by the quantum number λ) is conserved, so that a π orbital, for example, remains a π orbital for all values of R. Since $\lambda\hbar$ gives the absolute value of L_z, in the two limits λ goes over into $|m_l|$, the absolute value of the atomic quantum number m_l. In the united-atom limit, for each subshell with quantum numbers n, l there are $2l+1$ degenerate atomic orbitals, whose angular momentum components along a given axis have the values $L_z = m_l\hbar$ ($m_l = -l, -l+1, \ldots, l-1, l$); we of course take the z axis to be the same as the internuclear axis for $R > 0$. As soon as R does have a value greater than zero, these degenerate orbitals become physically distinguishable and have distinct energies, except that each pair with $m_l = \pm|m_l|$ remains degenerate. Thus the $2l+1$ molecular orbitals have only l distinct energy levels, with values of λ $(=|m_l|)$ ranging from 0 to l. For example, the $3d$ subshell of the united atom has five degenerate orbitals ($l=2, m_l=-2, \ldots, 2$) which in the molecule become a single $3d\sigma$ orbital ($\lambda=0$), a degenerate pair of $3d\pi$ orbitals ($\lambda=1$), and another degenerate pair of $3d\delta$ orbitals ($\lambda=2$).

2. In homonuclear molecules, the parity of each orbital is also conserved. Consider the separated-atom limit. Each atomic subshell yields the same orbitals as in the united atom, and each of these orbitals has its exact counterpart on the other atom. By exactly the same reasoning as in our analysis of H_2^+ (Section 6.3), the symmetry of a homonuclear molecule requires that at $R = \infty$ all its orbitals have the form of a sum or difference of two atomic orbitals. Thus for each atomic orbital φ_i there are two molecular orbitals, $(\varphi_i)_A \pm (\varphi_i)_B$, which at $R = \infty$ have the same energy; the sum is obviously of even parity (g) and the difference of odd parity (u). As R decreases the LCAO approximation no longer holds exactly and the two orbitals become nondegenerate, but the parity remains the same. Corresponding to our example above, the $3d$ subshells of identical separated atoms yield 10 molecular orbitals: $\sigma_g 3d$, $\sigma_u 3d$, and a degenerate pair each of $\pi_g 3d$, $\pi_u 3d$, $\delta_g 3d$, $\delta_u 3d$. In the united-atom limit the parity of the orbitals is easily derived from the form of the

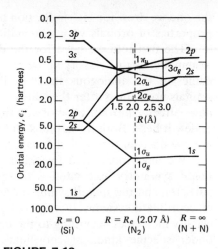

FIGURE 7.13
Correlation diagram for orbitals of N_2. The orbital energy curves in the inset rectangle are those calculated by P. E. Cade, K. D. Sales, and A. C. Wahl, *J. Chem. Phys.* **44**, 1973 (1966); elsewhere the correlation curves are straight lines connecting the atomic and molecular orbital energies.

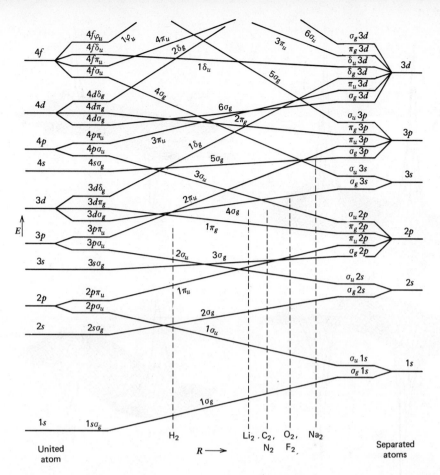

FIGURE 7.14
Orbital correlation diagram for homonuclear diatomic molecules. Each orbital is labeled by the three systems described in Section 6.5. The energy scale is schematic, but adjusted so that the sequence of orbital energies in several molecules can be shown by vertical dashed lines (corresponding roughly to $R = R_e$). The orbitals whose lines slant upward from left to right are bonding; those whose lines slant downward are antibonding.

spherical harmonics (Table 3.1): All orbitals with even l are g, since they are polynomials of even degree in $\sin \theta$ and $\cos \theta$; similarly, all orbitals with odd l are u. Thus our five united-atom $3d$ orbitals, for example, are all g (having $l = 2$).

Now we can define all the molecular orbitals near the united-atom and separated-atom limits, with notations of the form $1s\sigma_g$ and $\sigma_g 1s$, respectively, as illustrated at the left and right sides of Fig. 7.14. We still need to know how to correlate these limits. Given that λ and parity are conserved, we must correlate only orbitals of the same symmetry type—σ_g with σ_g, π_u with π_u, and so on. But there are many orbitals (an infinite number, in fact) of each type. To decide which correlates with which, we need a third and final rule:

3. Orbitals are connected in order of increasing energy. In the two limits, the order of orbital energies is that of the isolated atoms (cf. Fig. 5.2). Two connecting lines may cross if they refer to orbitals that differ in λ or parity, but must *not* cross if they refer to orbitals of the same type. This is the *noncrossing rule*, which we introduced in Section 6.5. It has its origin in the assumption that there are no *accidental* degeneracies in nature: There is always some perturbation that keeps two states of a system from having exactly the same energy, unless some fundamental symmetry underlies the degeneracy. (The spherical symmetry of an atom is reponsible for the $(2l + 1)$-fold degeneracy of states with quantum number l, and the axial symmetry of a diatomic molecule for the twofold degeneracy of states with $\lambda > 0$; in the fine structure, even these de-

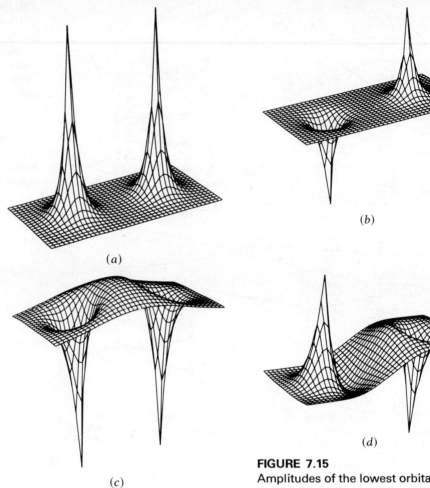

(a)

(b)

(c)

(d)

FIGURE 7.15

Amplitudes of the lowest orbitals of the Li$_2$ molecule. The graphs are all drawn for an Li–Li distance of 2.67 Å or 5.047 bohrs, the equilibrium distance. The representation gives a projection of a three-dimensional plot with the vertical scale linear in (electron density)$^{1/2}$. The first three orbitals, shown as (a), (b), and (c), are the 1σ_g (1s_A + 1s_B, nearly), the 1σ_u (1s_A − 1s_B, nearly), and the 2σ_g (roughly 2s_A + 2s_B) and are normally occupied. The 2σ_u orbital, (d), is normally empty. Calculations were carried out by Mary Dolan and plotted by Mary Dolan and David Campbell.

generacies are removed.) The noncrossing rule is itself only approximate, since the Born–Oppenheimer approximation is required for its proof.

Given the noncrossing rule, the construction of the correlation diagram is simple. We connect the lowest σ_g orbital on one side with the lowest σ_g orbital on the other, giving the 1σ_g molecular orbital, which is always the lowest-energy orbital. Similarly, we connect the second-lowest σ_g orbitals to give the 2σ_g molecular orbital, and so forth. The same process is carried out for each of the other symmetry types. We thus obtain the diagram of Fig. 7.14. The energy scale of Fig. 7.14 is schematic only, and not intended to represent the true spacing of energy levels.

The quantum number λ, equivalent to $|m_l|$, is a good quantum number within the orbital approximation for all values of R, and corresponds to the one conserved component of angular momentum. By contrast, the quantum numbers n and l for a given orbital are often different in the united-atom and separated-atom limits. The angular momentum quantum number l simply loses its meaning in the molecule because the total angular momentum of an individual electron is not conserved when the electron moves in a nonspherical potential. We could say that the quantum number n retains some validity, as an index to the number of nodes ($n-1$) in the orbital wave function at a given R; but this number varies with R, and is not ordinarily used to identify orbitals. An orbital for which n is higher in the united atom than in the separated atoms is said to be *promoted*. The energies of promoted orbitals are generally also higher in the united-atom limit, because their promotion more than compensates for the larger nuclear charge of the united atom. Correspondingly, the energies of nonpromoted orbitals are always lower in the united-atom limit. The reason for calling attention to the distinction is this. Electrons in nonpromoted orbitals tend to stabilize, and those in promoted orbitals to destabilize the formation of a bond. In other words, promoted orbitals orbitals are usually antibonding, and nonpromoted orbitals, bonding.

Finally, we should say something about the shapes of the molecular orbitals. These are basically the same as the orbitals in H_2^+, several of which were shown in Fig. 6.8. Figures 7.15 and 7.16 show the amplitudes of several orbitals for Li_2 and for N_2 and CO. The nodal surfaces of orbitals are of particular interest because they are closely related to the symmetry of the orbital. For example, an orbital with $\lambda \neq 0$ has λ nodal planes containing the internuclear axis, which remain nodal planes for all values of R. All $\sigma_u, \pi_g, \delta_u, \varphi_g, \ldots$ orbitals have an additional nodal plane through the center of symmetry, perpendicular to the bond axis. The shapes of other nodal surfaces, such as those radial nodes that are spheres in free atoms, vary with R, usually in a complicated manner. Two nodal surfaces of the separated atoms are likely to merge into one at small R, as in the $2\sigma_g$ orbital (Fig. 6.8).

To the extent that the molecular orbital model is valid, we can describe the electronic structure of a diatomic molecule in terms of an Aufbau principle. Just as in atoms, we assign the electrons to orbitals in order of increasing energy until all the electrons are accounted for. But it is clear from Fig. 7.14 that the order of orbital energies varies with R. The sequence for a given molecule must be determined empirically, using spectroscopic and chemical evidence. We have drawn Fig. 7.14 so that the orbital sequences for a number of molecules (at R_e) can be indicated by vertical lines: One simply reads up each dashed line to find the orbitals in order of increasing energy. The representation is still only schematic and tells us nothing about the relative spacing of the energy levels, but note the agreement with the more quantitative Fig. 7.13.

The Pauli exclusion principle still applies: Each orbital can hold only two electrons, which must be assigned opposite spins. Remember that the lines in Fig. 7.14 represent not individual orbitals but orbital energy levels. Each σ level represents a single orbital, holding two electrons, but each $\pi, \delta, \ldots$ level represents two degenerate orbitals (with $L_z = \pm\lambda\hbar$) and can thus hold four electrons. Applying this rule, we obtain the ground-state configurations listed in Table 7.5 for the homonuclear diatomic molecules through the first row of the periodic table.

The fact that we can write a configuration for a molecule does not mean that such a molecule really exists. For example, the Aufbau principle implies that the ground state of He_2 should clearly be $(1\sigma_g)^2(1\sigma_u)^2$; yet there is no stable He_2 molecule corresponding to this configuration.[24] We must thus look into the conditions governing the stability of a molecule.

The key point here is that some orbitals are bonding, others antibonding. The principles developed in the last chapter indicate that if $R > R_e$, the bonding orbitals normally have a higher electron density between the nuclei than in the separated atoms, whereas the antibonding orbitals have a lower electron density between the nuclei than in the

7.6
Homonuclear Diatomic Molecules: Aufbau Principle and Structure of First-Row Molecules

---

[24] But there are many stable excited states, the lowest of which is $(1\sigma_g)^2(1\sigma_u)(2\sigma_g)$.

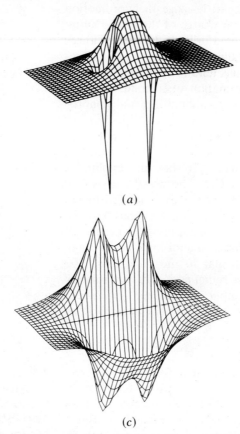

(a)

(c)

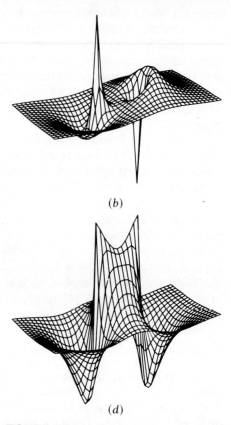

(b)

(d)

separated atoms. It is clear that all the orbitals with nodal planes midway between the nuclei (cf. Fig. 6.8) must be antibonding. Thus all $\sigma_u, \pi_g, \delta_u, \ldots$ orbitals are antibonding. These are molecular orbitals in which there is destructive interference between the atomic orbitals, and may be referred to as "minus" orbitals. The "plus" orbitals with constructive interference ($\sigma_g, \pi_u, \delta_g, \ldots$) are ordinarily bonding. Note that the

FIGURE 7.16
Amplitudes of the normally occupied orbitals of N_2 and CO that correlate with 2s and 2p orbitals of the separated atoms. The N–N bond distance is 1.10 Å or 2.08 bohrs; the C–O bond distance is 1.13 Å or 2.14 bohrs. All amplitudes are plotted with a vertical scale linear in ψ itself. (a) N_2, $2\sigma_g$ (largely $2s_A + 2s_B$). (b) N_2, $2\sigma_u$ (largely $2s_A - 2s_B$). (c) N_2, $1\pi_u$ (largely $2p\pi_A + 2p\pi_B$). (d) N_2, $3\sigma_g$ (roughly $2p\sigma_A - 2p\sigma_B$ if both atoms are drawn with right-handed coordinate systems). (e), (f), (g), (h) are the corresponding orbitals for CO. Calculations were carried out by Mary Dolan and plotted by Mary Dolan and David Campbell.

TABLE 7.5
GROUND STATES OF THE HOMONUCLEAR DIATOMIC MOLECULES FROM H_2 TO Ne_2*

Molecule	Electron Configuration[a]	Term Symbol	Bond Order	D_e (eV)	R_e (Å)
H_2	$(1\sigma_g)^2$	$^1\Sigma_g^+$	1	4.7478	0.7412
He_2	$(1\sigma_g)^2(1\sigma_u)^2$	$^1\Sigma_g^+$	0	—	—
Li_2	$KK(2\sigma_g)^2$	$^1\Sigma_g^+$	1	1.14±0.3	2.6725
Be_2	$KK(2\sigma_g)^2(2\sigma_u)^2$	$^1\Sigma_g^+$	0	—	—
B_2	$KK(2\sigma_g)^2(2\sigma_u)^2(1\pi_u)^2$	$^3\Sigma_g^-$	1	3.0±0.2	1.590
C_2	$KK(2\sigma_g)^2(2\sigma_u)^2(1\pi_u)^4$	$^1\Sigma_g^+$	2	6.24±0.22	1.2425
N_2	$KK(2\sigma_g)^2(2\sigma_u)^2(1\pi_u)^4(3\sigma_g)^2$	$^1\Sigma_g^+$	3	9.7599±0.0017	1.094
O_2	$KK(2\sigma_g)^2(2\sigma_u)^2(3\sigma_g)^2(1\pi_u)^4(1\pi_g)^2$	$^3\Sigma_g^-$	2	5.116±0.004	1.2075
F_2	$KK(2\sigma_g)^2(2\sigma_u)^2(3\sigma_g)^2(1\pi_u)^4(1\pi_g)^4$	$^1\Sigma_g^+$	1	1.604±0.1	1.409
Ne_2	$KK(2\sigma_g)^2(2\sigma_u)^2(3\sigma_g)^2(1\pi_u)^4(1\pi_g)^4(3\sigma_u)^2$	$^1\Sigma_g^+$	0	—	—

* Taken from B. de B. Darwent, *Bond Dissociation Energies in Simple Molecules*, NSRDS-NBS 31, U.S. Dept. of Commerce (U.S. Govt. Printing Office, Washington, D.C., 1970).
[a] The designation KK refers to the filled inner-shell configuration $(1\sigma_g)^2(1\sigma_u)^2$, corresponding to filled K ($n=1$) shells in the separated atoms.

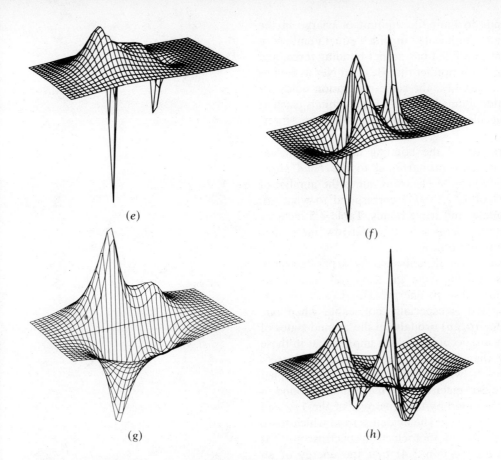

(e)

(f)

(g)

(h)

"plus" and "minus" orbitals can be classified into pairs, with each orbital of the separated atoms producing one of each type, $\varphi_A \pm \varphi_B$ in the simple LCAO approximation. This pairing is precise and unambiguous in homonuclear molecules, but loses some of its clarity in molecules composed of two very different atoms. The electronic energies of bonding and antibonding orbitals generally increase and decrease, respectively, with increasing R; the energy levels in Fig. 7.14 have been drawn to reflect this.

Let us apply this information to the ground states of the simplest few molecules. The H_2^+ molecule-ion has one electron in the bonding $1\sigma_g$ orbital, and a dissociation energy (D_e) of 2.79 eV. The H_2 molecule has the configuration $(1\sigma_g)^2$, with both electrons in the bonding orbital; the dissociation energy, $D_e = 4.75$ eV, is nearly twice that for H_2^+. But the excited state $(1\sigma_g)(1\sigma_u)$ (i.e., $b^3\Sigma_u^+$, cf. Fig. 6.11), with one bonding and one antibonding electron, is unstable. There exists a He_2^+ molecule-ion, which is stable in the sense that its $E(R)$ curve has a true minimum, at $R_e = 1.08$ Å. The dissociation energy of He_2^+, the difference between $E(R_e)$ and the energy of $He(1s^2) + He^+(1s)$, is about 2.5 eV, close to that of H_2^+. The configuration should be $(1\sigma_g)^2(1\sigma_u)$, which, like H_2^+, has one more bonding than antibonding electron. And He_2, $(1\sigma_g)^2(1\sigma_u)^2$, with two bonding and two antibonding electrons, does not exist as a stable molecule.

We can draw a fairly obvious generalization from these data: The bonding power of an electron in a bonding ("plus") orbital is roughly equal to the antibonding power of an electron in the complementary antibonding ("minus") orbital. In fact, the antibonding effect is usually slightly greater. There are a number of plausible ways to account for this imbalance. Perhaps the most physically explicit explanation is that electron–electron repulsion tends to limit the amount of charge in the strongly bonding region. Thus any "molecule" in which equal numbers of bonding and antibonding orbitals are filled has no net bonding force, and does not exhibit a chemical bond. We predict that Be_2 and Ne_2 as well as He_2 should not exist as stable molecules; the same conclusion holds for any species in which the separated atoms have only filled orbitals (such as Ar_2, Kr_2, Xe_2). This is the molecular orbital explanation for the "inertness" of closed-shell atoms.

We conclude that the strength of the bonding in a homonuclear diatomic molecule should be roughly proportional to the *bond order*, defined as the number of bonding pairs of electrons minus the number of antibonding pairs. Bond orders of 1, 2, and 3 correspond to what are conventionally called single, double, and triple bonds. Table 7.5 includes the bond orders and dissociation energies of the first-row molecules, which can be seen to support our hypothesis.

In general, when a molecule is best described in the separated-atom scheme, the molecular orbitals deriving from the same subshell of the separated atoms tend to be very close in energy. This is the case for $R \geq R_e$ in first-row molecules, and is especially noticeable when one compares the $1\pi_u$ ($\pi_u 2p$) and $3\sigma_g$ ($\sigma_g 2p$) orbitals. In the ground states of B_2 and C_2 only the $1\pi_u$ orbitals are occupied, which implies that in these species $1\pi_u$ has lower energy than $3\sigma_g$. The experimental evidence[25] indicates that this is also true for N_2 where both levels are first occupied in the ground state, but that the $3\sigma_g$ electrons are more tightly bound in O_2 and F_2. This behavior is represented by the crossover of the $1\pi_u$ and $3\sigma_g$ lines between N_2 and O_2 in Fig. 7.14. The evidence from which these conclusions are drawn is primarily that of photoelectron spectroscopy.[26] It follows from Koopmans' theorem (Section 5.4) that the energy of an orbital should equal the energy required to ionize an electron from it.

Let us survey how the physical and chemical properties of the molecules in Table 7.5 reflect their configurations. Hydrogen has been sufficiently covered in the last chapter. As already noted, He_2, Be_2, and Ne_2 do not exist as stable molecules in their ground states; at ordinary temperatures helium and neon are monatomic gases, whereas beryllium is a metal that vaporizes (b.p. 2970°C) to a monatomic gas.[27] But we shall now consider each of the other stable species individually.

Lithium is also a metal at ordinary temperatures. Its vapor is primarily monatomic, containing only a few percent of Li_2 molecules. This can be attributed to the relative weakness of the Li–Li bond. Similar behavior is found for the other alkali metals; the bonds become longer and weaker in successively heavier molecules in the series. The bonds are weak for the same reason that the alkali atoms are large (Section 5.5). The valence (ns) electrons are loosely bound and diffuse in the atoms, and even in the molecules the electron density between the nuclei is low. In contrast to H_2^+ and H_2, the one-electron bonds in the Li_2^+, Na_2^+, ... ions are stronger than the bonds in the corresponding neutral molecules. Apparently the electron–electron repulsion (including that between valence and core electrons) outweighs the bonding effect of the second valence electron.

[25] The calculation shown in Fig. 7.13 predicts that $3\sigma_g$ should be lower than $1\pi_u$ in N_2, but by a very small amount; the experimental energies are in the reverse order.

[26] In this technique a sample is ionized by irradiation with photons of known energy, usually in the vacuum ultraviolet or x ray regions, and the energies of the emitted electrons are measured; subtraction gives the ionization energy for each electron. See Section 7.10.

[27] However, there are stable Mg_2, Ca_2, ... molecules, indicating that the simple MO theory is inadequate for heavier molecules.

Not as much is known about the vapor of boron, which remains a black semiconducting solid to well over 2000°C. However, spectroscopy has shown that the B_2 molecule is a reasonably stable species at high temperatures. The bond is of a "normal" length, and correspondingly much stronger than the bond in Li_2.

As for carbon, C_2 molecules are well known in flames, shock waves, arcs, or indeed any hot system containing an excess of carbon. But the study of carbon vapor is complicated by the presence of larger polymers. Up to around 5000°C the vapor is mainly C_3, with significant amounts of $C_4, C_5, \ldots$; it is not easy to get a large concentration of C_2. This situation reflects the ease with which carbon forms chains or networks of bonds, as in the solid forms of diamond and graphite and the enormous variety of organic compounds. One reason for this behavior is revealed by the energy spectrum of C_2 (Table 7.6): the energies of the $1\pi_u$ and $3\sigma_g$ orbitals are so close together that C_2 always contains a significant number of molecules in the lowest excited configuration, $KK(2\sigma_g)^2(2\sigma_u)^2(1\pi_u)^3(3\sigma_g)$.[28] The unpaired electrons can readily form bonds with other C atoms, gaining more than enough energy to compensate for the excitation.

This illustrates the general principle that low-lying excited states facilitate bond formation. We introduced this idea (in a negative sense) in Section 5.5 to account for the "inertness" of the rare gas atoms, but it also applies to molecules. The molecules Li_2, B_2, and C_2 all have fairly low excited states (Table 7.6), and can thus exist only in the vapor phase; the solid forms of these elements (and of most other elements) have structures in which each atom is bonded to many nearest neighbors. In contrast, the first excited state of N_2 is very high (cf. Fig. 7.5), and nitrogen exists as discrete diatomic molecules even in the solid phase, which is thus called a *molecular crystal*. The ground state of N_2 is effectively a closed-shell structure, with all the occupied orbitals filled, and the physical properties of nitrogen are in many ways similar to those of an inert gas. In particular, N_2 exists as a gas to well below room temperature.

The properties of oxygen are unusual in several respects. The ground state of O_2 has the configuration $\ldots (1\pi_g)^2$. Since there are two degenerate $1\pi_g$ orbitals, the $1\pi_g$ electrons can go in either the same or different orbitals, leading to singlet and triplet states, respectively. In analogy to Hund's rule for atoms, and for the same reasons, the triplet state is of lower energy. Thus the ground state of O_2 has $S = 1$ and is paramagnetic; the explanation of this phenomenon was one of the early triumphs of molecular orbital theory.[29] Additional confirmation is provided by the ions derived from O_2, in which the bond length (and thus strength) varies in accordance with the MO bond order:

Species	Ground-State Configuration	Bond Order	R_e (Å)	D_0 (eV)
O_2^+	$\ldots (3\sigma_g)^2(1\pi_u)^4(1\pi_g)$	$2\frac{1}{2}$	1.117	6.662
O_2	$\ldots (3\sigma_g)^2(1\pi_u)^4(1\pi_g)^2$	2	1.208	5.116
O_2^-	$\ldots (3\sigma_g)^2(1\pi_u)^4(1\pi_g)^3$	$1\frac{1}{2}$	1.33	4.07
O_2^{2-}	$\ldots (3\sigma_g)^2(1\pi_u)^4(1\pi_g)^4$	1	1.49 (in solid Na_2O_2, etc.)	

[28] It is much easier to observe transitions to or from this excited state than transitions involving the closed-shell ground state. As a result, for many years the $\ldots (1\pi_u)^3(3\sigma_g)$ configuration was thought to be the true ground state. The correct analysis of the spectrum and the ordering of states was only made in 1963 by Ballik and Ramsay.

[29] The simple valence bond theory would predict one of the structures :Ö=Ö:, which is not paramagnetic, or :Ö—Ö:, which is too weakly bonded. The MO interpretation straightforwardly accounts for both the paramagnetism and the bond order.

TABLE 7.6

SOME LOW-LYING EXCITED STATES OF FIRST-ROW HOMONUCLEAR DIATOMIC MOLECULES

Molecule	State	Configuration	Energy Above Ground State (eV)
Li_2	$A\,^1\Sigma_u^+$	$KK(2\sigma_g)(2\sigma_u)$	1.744
	$B\,^1\Pi_u$	$KK(2\sigma_g)(1\pi_u)$	2.534
	$C\,^1\Pi_u$	$KK(2\sigma_g)(2\pi_u)$ or $KK(1\pi_u)^2$	3.788
	$D\,^1\Pi_u$	?	≤4.233
B_2	$A\,^3\Sigma_u^-$	$KK(2\sigma_g)^2(2\sigma_u)^2(3\sigma_g)(3\sigma_u)$	3.791
C_2	$a\,^3\Pi_u$	$KK(2\sigma_g)^2(2\sigma_u)^2(1\pi_u)^3(3\sigma_g)$	0.089
	$b\,^3\Sigma_g^-$	$KK(2\sigma_g)^2(2\sigma_u)^2(1\pi_u)^2(3\sigma_g)^2$	0.798
	$A\,^1\Pi_u$	$KK(2\sigma_g)^2(2\sigma_u)^2(1\pi_u)^3(3\sigma_g)$	1.040
	$c\,^3\Sigma_u^+$	$KK(2\sigma_g)^2(2\sigma_u)(1\pi_u)^4(3\sigma_g)$	1.651
	$d\,^3\Pi_g$	$KK(2\sigma_g)^2(2\sigma_u)(1\pi_u)^3(3\sigma_g)^2$	2.482
	$C\,^1\Pi_g$	$KK(2\sigma_g)^2(2\sigma_u)(1\pi_u)^3(3\sigma_g)^2$	4.248
	$C'\,^1\Pi_g$	$KK(2\sigma_g)^2(2\sigma_u)^2(1\pi_u)^2(3\sigma_g)(1\pi_g)$	4.643
	$e\,^3\Pi_g$	$KK(2\sigma_g)^2(2\sigma_u)^2(1\pi_u)^2(3\sigma_g)(1\pi_g)$	5.058
N_2 (cf. Fig. 7.5)	$A\,^3\Sigma_u^+$	$KK(2\sigma_g)^2(2\sigma_u)^2(1\pi_u)^3(3\sigma_g)^2(1\pi_g)$	6.169
	$B\,^3\Pi_g$	$KK(2\sigma_g)^2(2\sigma_u)^2(1\pi_u)^4(3\sigma_g)(1\pi_g)$	7.353
	$W\,^3\Delta_u$	$KK(2\sigma_g)^2(2\sigma_u)^2(1\pi_u)^3(3\sigma_g)^2(1\pi_g)$	7.356
	$B'\,^3\Sigma_u^-$	$KK(2\sigma_g)^2(2\sigma_u)^2(1\pi_u)^3(3\sigma_g)^2(1\pi_g)$	8.165
O_2	$a\,^1\Delta_g$	$KK(2\sigma_g)^2(2\sigma_u)^2(3\sigma_g)^2(1\pi_u)^4(1\pi_g)^2$	0.977
	$b\,^1\Sigma_g^+$	$KK(2\sigma_g)^2(2\sigma_u)^2(3\sigma_g)^2(1\pi_u)^4(1\pi_g)^2$	1.627
	$c\,^1\Sigma_u^-$	$KK(2\sigma_g)^2(2\sigma_u)^2(3\sigma_g)^2(1\pi_u)^4(1\pi_g)^2$	4.050
	$C\,^3\Delta_u$	$KK(2\sigma_g)^2(2\sigma_u)^2(3\sigma_g)^2(1\pi_u)^4(1\pi_g)^2$	4.255
	$A\,^3\Sigma_u^+$	$KK(2\sigma_g)^2(2\sigma_u)^2(3\sigma_g)^2(1\pi_u)^3(1\pi_g)^3$	4.340
	$B\,^3\Sigma_u^-$	$KK(2\sigma_g)^2(2\sigma_u)^2(3\sigma_g)^2(1\pi_u)^3(1\pi_g)^3$	6.120
F_2	$^3\Pi_u$	$KK(2\sigma_g)^2(2\sigma_u)^2(3\sigma_g)^2(1\pi_u)^4(1\pi_g)^3(3\sigma_u)$	~1–1.5 (not observed)
	$A\,^1\Pi_u$	$KK(2\sigma_g)^2(2\sigma_u)^2(3\sigma_g)^2(1\pi_u)^4(1\pi_g)^3(3\sigma_u)$	repulsive
	$B\,^1\Pi_g$	$KK(2\sigma_g)^2(2\sigma_u)^2(3\sigma_g)^2(1\pi_u)^3(1\pi_g)^4(3\sigma_u)$	?

But in O_2 the ground state itself has unpaired electrons. How is it that additional bonds do not form[30] as in carbon? The answer may be that the electrons in question are in the antibonding $1\pi_g$ orbitals; the next available bonding orbital is the much higher $4\sigma_g$. Yet S_2, with a similar orbital structure, readily polymerizes to ring molecules such as S_8 and S_6. Whatever the reason, the O_2 molecule is the stable form of oxygen at all temperatures below those at which the molecules dissociate. Like nitrogen, oxygen is a room-temperature gas and a low-temperature molecular crystal.

Finally, the F_2 molecule is similar to N_2 in that it has a closed-shell ground state and relatively inaccessible excited states. Fluorine shows the same "inert" behavior as nitrogen and oxygen—inert with regard to

---

[30] Oxygen does form the O_3 (ozone) molecule, but no further polymerization has been observed. Indeed, O_3 is quite unstable relative to O_2; the reaction $2O_3 \rightarrow 3O_2$ occurs readily if a suitable catalyst is present.

physical properties only, since F_2, like O_2, is highly reactive chemically. The energy of the first excited state is probably quite low, but transitions from the ground state are strongly forbidden.

Thus far we have discussed only electron configurations and the qualitative inferences one can draw from them. Of course, the configurations are only the starting point for molecular orbital calculations. We need not go into the details of these calculations here; the methods are those we outlined in Chapter 6. One sets up a trial wave function in some way and computes the energy as a function of R by Eq. 5.11. The trial function ordinarily contains adjustable parameters which are varied to minimize the energy. Within the orbital approximation, the molecular wave function has the form of the antisymmetrized Hartree–Fock product (Eq. 6.63), in which the φ_i's are molecular orbitals. The simplest trial MOs are LCAO functions of the type already discussed, for example

$$\varphi(2\sigma_g) = \varphi(\sigma_g 2s) = C[\varphi_A(2s) + \varphi_B(2s)], \qquad (7.48)$$

but such simplistic forms give unsuitably inaccurate results for anything more complicated than H_2. The next step is to go beyond the correlation diagram and use mixtures of atomic orbitals of appropriate symmetry, as in

$$\varphi(2\sigma_g) = C_1\varphi(\sigma_g 2s) + C_2\varphi(\sigma_g 1s) + C_3\varphi(\sigma_g 2p) + \cdots. \qquad (7.49)$$

This process is called *hybridization* and will be discussed in greater detail in the next chapter. However the trial MOs are set up, the energy is then minimized by the self-consistent-field method. As in atomic calculations, the speed of convergence depends largely on what functions are chosen to represent the atomic orbitals. The best SCF MO calculations now give essentially accurate Hartree–Fock results for first-row molecules, and are being extended to larger molecules as larger computers are introduced. The Hartree–Fock solution, we recall, is the best possible solution within the one-electron-orbital approximation; it is adequate for many purposes. The correlation energy, the difference between the true energy and the Hartree–Fock energy, can be calculated by configuration interaction or similar techniques (Section 6.9). For simple molecules, the best calculations lead to predictions of spectral line positions within about 10–20 cm^{-1} of the observed lines. This is accurate enough to be very useful for identification, but considerably less accurate than the measurement of spectral lines.

The energy is not the only molecular property that can be computed. We can gain insights from examining the electron distributions in homonuclear diatomic molecules. In Figs 6.7, 7.15, and 7.16, we looked at orbital amplitudes. Now we turn to total densities. Figure 7.17 shows typical "three-dimensional" pictures of total densities for three molecules. Figure 7.18 shows electron-density contour maps for a larger set. All were obtained from SCF calculations. Electron densities are given in Fig. 7.18 for both individual orbitals and the molecule as a whole. In the H_2 molecule all but the innermost few contours are nearly circular, that is, the outer portions of the electron distribution resemble that in the spherical He atom; even between the nuclei the electron density is only slightly below that at the two nuclear peaks. In contrast, the weakly bound Li_2 molecule is quite like two separate atoms, with a very deep valley between the two peaks. For the other first-row molecules, the stronger the bond, the more nearly the total electron density resembles that in a single atom. The "saddle" between the nuclei is still quite low in B_2, reaches about half the peak height in N_2, and has fallen again by the

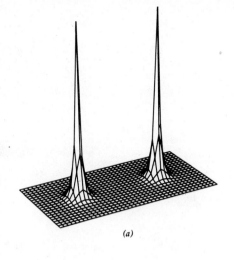

(a)

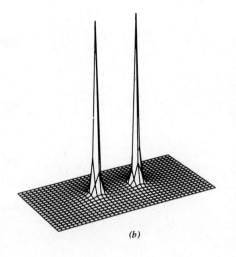

(b)

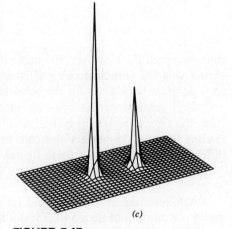

(c)

FIGURE 7.17
Total electron densities for (a) Li_2; (b) N_2; (c) CO. Each molecule is taken at its equilibrium internuclear distance. Calculations were carried out by Mary Dolan and plotted by Mary Dolan and David Campbell.

MOLECULAR ORBITAL

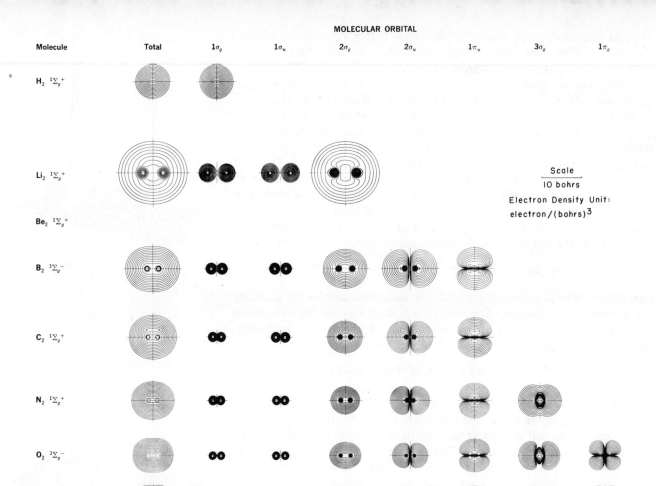

FIGURE 7.18
Contour maps of electron densities of orbitals and of total electron densities for the seven lightest stable homonuclear molecules. From A. C. Wahl, *Sci. Am.* **322**, No. 4, 54 (April 1970).

time we reach F_2. However, the outer reaches of the electron distribution shrink steadily with increasing Z, the same trend one finds for atomic sizes across a period. As for the orbitals, note that all the inner-shell ($1\sigma_g$ and $1\sigma_u$) orbitals of the first-row molecules do not differ significantly from the core densities in two separate atoms. Although the calculations leading to Fig. 7.18 neglect electron correlation, one could hardly tell the difference on the scale of these figures.

7.7 Introduction to Heteronuclear Diatomic Molecules: Electronegativity

A heteronuclear molecule has no end-to-end symmetry. Thus symmetry condition 3 of Section 7.5 is not applicable: neither the orbitals nor the molecular wave function can be described as g or u. Otherwise our previous nomenclature can be retained. In the orbital approximation, the molecular orbitals tend to localize predominantly around one nucleus or the other; the ionic molecules of Section 7.4 are an extreme case of this. Only when the molecule is very close to the united-atom limit do the normally occupied orbitals have comparable amplitudes around both

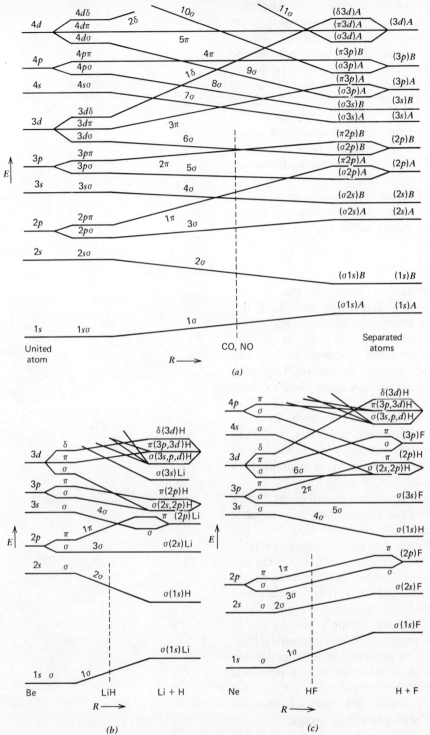

FIGURE 7.19
Correlation diagrams for heteronuclear diatomic molecules: (a) for atoms of nearly equal size. (b) for LiH; (c) for HF. (In the latter two diagrams the degenerate H levels are grouped as in Fig. 6.9.)

nuclei. This asymmetric charge distribution carries over to the molecule as a whole, and heteronuclear molecules are therefore polar, that is, have nonzero dipole moments.

As in homonuclear molecules, we can use a correlation diagram to represent the sequence of orbital energies. Figure 7.19a is a general correlation diagram for heteronuclear diatomic molecules. Since the two atoms are different, they have different energy levels in the separated-atom limit. Our assumption is that in this limit each molecular orbital correlates only with an atomic orbital of one or the other atom (the atom around which its amplitude is greatest). As we shall see, this is an oversimplification. The resulting correlation diagram is somewhat simpler than that for homonuclear molecules, primarily because there are no orbital parities that must be matched. However, it contains an additional element of arbitrariness, in that the relationship between A and B separated-atom levels varies with the nature of these atoms: The higher an atom's nuclear charge Z, the lower are that atom's energy levels of a given n and l. The effect is especially pronounced when one of the atoms is hydrogen. In Figs. 7.19b and 7.19c we illustrate this with correlation diagrams for LiH and HF; the other first-row diatomic hydrides are intermediate between these two.

What is the nature of the orbitals in heteronuclear molecules? In the LCAO approximation they consist of sums or differences of atomic orbitals in unequal amounts. Thus one might write a given molecular orbital φ_n as

$$\varphi_n = \alpha(R)(\varphi_i)_A + \beta(R)(\varphi_j)_B, \tag{7.50}$$

where $(\varphi_i)_A$ and $(\varphi_j)_B$ are atomic orbitals of atoms A and B. The coefficients $\alpha(R)$ and $\beta(R)$, like those in Eqs. 6.76 and 6.79, are constants for any given value of R. The correlation diagrams imply that either $\alpha(R)$ or $\beta(R)$ vanishes in the separated-atom limit; at R_e one coefficient is usually much greater than the other.[31] Results such as this are obtained by the usual SCF techniques, and can be improved by using hybridized atomic orbitals, as in Eq. 7.49. One can again obtain orbital and whole-molecule electron densities: Compare the "charge densities"— strictly, total probability densities—shown in Fig. 7.20 for the first-row diatomic hydrides with Fig. 7.18 for homonuclear molecules. Although one cannot yet determine the electron distribution experimentally with comparable precision, these results are generally in agreement with chemical intuition and the measured properties of molecules.

In Section 5.5 we said that "electronegativity" describes an atom's power to attract electrons. In Fig. 7.20 we see a clear illustration of what this means. In a heteronuclear molecule the electrons that form the chemical bond—that is, those in the highest occupied bonding orbital— are found primarily near the more electronegative atom.[32] Among the first-row hydrides, calculations like those leading to Fig. 7.20 show clearly that hydrogen is intermediate in electronegativity between metals and nonmetals. In HF the 3σ bonding orbital is mainly centered around the F atom, whereas in LiH the 2σ orbital is almost entirely around the H atom. In fact, the alkali metals are so much more electropositive than hydrogen that all the alkali hydrides can be regarded as ionic molecules (M^+H^-), just like the alkali halides; in the solid state these compounds consist of separate M^+ and $H^-(1s^2)$ ions. Similar behavior is found for CaH_2 and the heavier alkaline earth hydrides. We shall see that the empirical electronegativity scales agree with these conclusions.

The simplest electronegativity scale is that proposed by R. S. Mulliken. It is based on the idea that an atom's ability to hold its outermost

---

[31] In particular, inner-shell MOs are almost pure atomic orbitals: For example, in all the first-row hydrides (HA) the 1σ orbital is found to be over 99% $(1s)_A$. This further confirms our conclusion in Section 7.4 that atomic cores take little part in bonding.

[32] The lowest normally empty orbital—such as 3σ in LiH or 4σ in HF—is commonly centered around the electropositive atom. Such an orbital is normally antibonding: Since it is usually a promoted orbital, the separated-atom electrons must be forced "uphill" to occupy it. If overlap is neglected and the bonding orbital is approximated by $\alpha\varphi_A + \beta\varphi_B$, orthogonality requires that the corresponding antibonding orbital be $\beta\varphi_A - \alpha\varphi_B$.

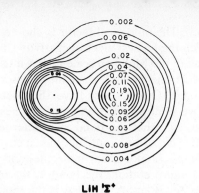

LiH $^1\Sigma^+$

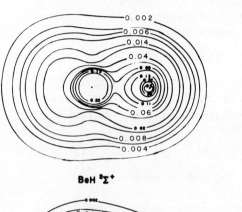

BeH $^2\Sigma^+$

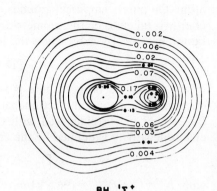

BH $^1\Sigma^+$

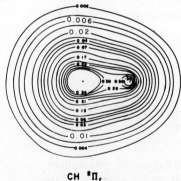

CH $^2\Pi_r$

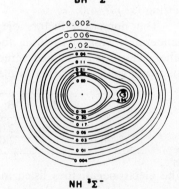

NH $^3\Sigma^-$

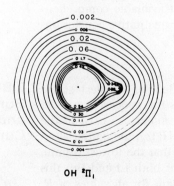

OH $^2\Pi_i$

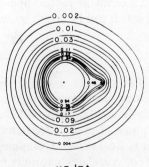

HF $^1\Sigma^+$

FIGURE 7.20

Electron-density contours for total probability densities of the first-row diatomic hydride molecules. Energies are given in hartrees. From R. F. W. Bader, I. Keaveny, and P. E. Cade, *J. Chem. Phys.* **47**, 3381 (1967).

electrons is proportional to its ionization potential (I), whereas its ability to attract additional electrons is proportional to its electron affinity (A). The Mulliken electronegativity is thus defined by the average of the two quantities,

$$x^M = K^M\left(\frac{I+A}{2}\right),$$ (7.51)

where $K^M = (3.15 \text{ eV})^{-1}$. The numerical factor is chosen to give a convenient scale, one in which the highest electronegativity (that of F) is about 4.0. The chief drawback of the Mulliken scale is that it uses isolated-atom properties, whereas "electronegativity" is meant to describe the behavior of an atom *in a molecule*. Other scales have thus been devised using molecular properties; we shall describe only one of these.

Linus Pauling suggested that the bond in a heteronuclear molecule could be regarded as the sum of two contributions: a covalent part (which is the only contribution in a homonuclear molecule) and an ionic part. In Pauling's valence bond interpretation, the bond in the molecule AB is a hybrid of the structures A—B and A^+B^-, if A is the more electropositive atom. On purely empirical grounds, the covalent contribution to the bond energy is taken to be the geometric mean of the covalent A—A and B—B bond energies. By subtraction the ionic contribution is

$$\Delta = D_e(\text{A—B}) - [D_e(\text{A—A}) D_e(\text{B—B})]^{1/2}, \tag{7.52}$$

where the D_e to be used here is the measured or inferred single-bond energy.[33] The apparent electronegativity difference between atoms A and B is found to correlate well with the square root of Δ, so the Pauling electronegativity is defined by the equation

$$(x^P)_B - (x^P)_A = K^P \Delta^{1/2}, \tag{7.53}$$

with $K^P = (1 \text{ eV})^{-1/2} = (96.5 \text{ kJ/mol})^{-1/2}$; the scale is anchored by again setting $(x^P)_F \approx 4.0$.

The numerical factors of the Pauling and Mulliken scales were deliberately chosen to bring the two scales into conformity with each other. The two do indeed agree quite closely, as can be seen from the following values for first-row elements:

	Li	Be	B	C	N	O	F
x^M	0.94	1.46	2.01	2.63	2.33	3.17	3.91
x^P	0.98	1.57	2.04	2.55	3.04	3.44	3.98

The electronegativities listed in Table 7.7 are obtained by yet another method involving the Coulomb force actually felt by an electron at the atom's covalent radius, and are probably the best available comprehensive set. It can be seen that the values confirm our qualitative conclusions above.

7.8
Bonding in LiH; Crossing and Noncrossing Potential Curves

Let us take a closer look at lithium hydride, the simplest stable heteronuclear molecule. At ordinary temperatures this species is a white crystalline solid (m.p. 680°C), but individual LiH molecules exist in the gas phase. The separated-atom limit is $\text{Li}(1s^2 2s) + \text{H}(1s)$; the united-atom limit is $\text{Be}(1s^2 2s^2)$. Examining Fig. 7.19b, we see that the lowest orbital, the 1σ, correlates naturally with the $1s$ orbitals of both Li and Be; this orbital is thus closely centered around the Li nucleus for all values of R. The next orbital, the 2σ, is much higher on the energy scale (about 60 eV at R_e). In the separated-atom limit, the $\text{H}(1s)$ orbital has a binding energy (ionization potential) of 13.6 eV, whereas the $\text{Li}(2s)$ orbital has a binding energy of only 5.36 eV (cf. Fig. 7.21). Because of this large energy difference, the 2σ orbital even near R_e is primarily like $\text{H}(1s)$;

[33] This is not necessarily the dissociation energy of the diatomic molecule. For example, N_2 has $D_e = 9.8$ eV but contains a triple bond; the N—N single-bond energy of 1.65 eV is obtained from measurements on molecules such as $H_2N—NH_2$. See Chapter 14 for a discussion of bond-energy calculations.

TABLE 7.7
ELECTRONEGATIVITIES OF THE ELEMENTS[a]

1	2	3	4	5	6	7	8	9	10	11	12	13	14	15	16	17	18
H 2.20																	He
Li 0.97	Be 1.47											B 2.01	C 2.50	N 3.07	O 3.50	F 4.10	Ne
Na 1.01	Mg 1.23											Al 1.47	Si 1.74	P 2.06	S 2.44	Cl 2.83	Ar
K 0.91	Ca 1.04	Sc 1.20	Ti 1.32	V 1.45	Cr 1.56	Mn 1.60	Fe 1.64	Co 1.70	Ni 1.75	Cu 1.75	Zn 1.66	Ga 1.82	Ge 2.02	As 2.20	Se 2.48	Br 2.74	Kr
Rb 0.89	Sr 0.99	Y 1.11	Zr 1.22	Nb 1.23	Mo 1.30	Tc 1.36	Ru 1.42	Rh 1.45	Pd 1.35	Ag 1.42	Cd 1.46	In 1.49	Sn 1.72	Sb 1.82	Te 2.01	I 2.21	Xe
Cs 0.86	Ba 0.97	57–71 *	Hf 1.23	Ta 1.33	W 1.40	Re 1.46	Os 1.52	Ir 1.55	Pt 1.44	Au 1.42	Hg 1.44	Tl 1.44	Pb 1.55	Bi 1.67	Po 1.76	At 1.96	Rn
Fr 0.86	Ra 0.97	89–103 **	104	105	106												

*

La 1.08	Ce 1.06	Pr 1.07	Nd 1.07	Pm 1.07	Sm 1.07	Eu 1.01	Gd 1.11	Tb 1.10	Dy 1.10	Ho 1.10	Er 1.11	Tm 1.11	Yb 1.06	Lu 1.14

**

Ac 1.00	Th 1.11	Pa 1.14	U 1.22	Np 1.22	Pu 1.22	Am	Cm	Bk	Cf	Es 1.2 (est.)	Fm	Md	No	Lr

[a] From A. L. Allred and E. G. Rochow, *J. Inorg. Nucl. Chem.* **5**, 264 (1958).

that is, by far the largest coefficient[34] in Eq. 7.50 is that of $(\varphi_{1s})_H$. The 3σ orbital, on the other hand, is mainly like Li($2s$). The LiH molecule has four electrons; applying the Aufbau principle, we conclude that the ground-state configuration near R_e should be $1\sigma^2 2\sigma^2$, corresponding to a $^1\Sigma$ state. Since the valence (2σ) electrons are centered near the H nucleus, the molecule should be largely ionic, with the effective structure Li$^+$H$^-$. Spectroscopic and dipole-moment measurements confirm these conclusions.

As usual, however, the simple MO method breaks down at large R. Suppose that we take the molecule at R_e and slowly move the nuclei away from each other. If the electrons remained in the same orbitals (as represented by the correlation diagram), the separated-atom limit would be the ionic Li$^+$($1s^2$)$+$H$^-$($1s^2$). But we know that any two neutral atoms A and B have a lower total energy than the corresponding ions A$^+$ and B$^-$. Eq. 7.46 gives the energy of ion-pair formation as

$$Q(A^+B^-) \equiv E(A^+ + B^-) - E(A + B) = I(A) - A(B), \qquad (7.54)$$

which is always positive. For LiH we have $I(\text{Li}) = 5.36$ eV, $A(\text{H}) = 0.75$ eV, $Q(\text{Li}^+\text{H}^-) = 4.61$ eV. The problem here is rather akin to that we found for H$_2$ in Section 6.6. There (and for any other homonuclear molecule) a single-configuration MO method predicts a 50%-ionic limit; here (and for any other heteronuclear molecule with an even number of electrons) it predicts a 100%-ionic limit. Within the MO framework, the paradox can again be resolved by the use of configuration interaction. Such a description of the LiH molecule would be given by the wave function

$$X(R)\psi(1\sigma^2 2\sigma^2) + Y(R)\psi(1\sigma^2 2\sigma 3\sigma),$$

in which we must have $X(R) \gg Y(R)$ for $R \lesssim R_e$, $Y(R) \gg X(R)$ as $R \to \infty$, so that at long range we have the configuration $1\sigma^2 2\sigma 3\sigma \to$ Li($1s^2 2s$)$+$H($1s$).

Although accurate multiconfiguration calculations have been carried out for LiH, similar results have not yet been achieved for most other heteronuclear molecules. It is thus useful to know the limits of the single-configuration model. Fortunately, we can get a good idea of these limits from a very simple and crude calculation of the potential energy. Consider the two long-range states, Li$+$H and Li$^+$+H$^-$; what happens in each case if we decrease R slowly, neglecting configuration interaction? In the ionic case the model of Section 7.4 should be applicable, with the interaction energy given by Eq. 7.44. This energy is dominated by the Coulomb term, $-e^2/4\pi\epsilon_0 R$. Even at R_e this term gives a fairly good approximation to D_e (cf. Table 7.3), and at a somewhat larger distance, large enough to neglect overlap of the electron clouds, it should be the

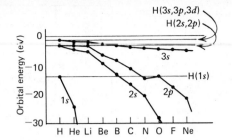

FIGURE 7.21
Orbital energies of first-row atoms. The energy plotted is that required to remove an electron from the orbital in question, as obtained from spectroscopic data.

---

[34]Here are the results of a SCF-MO calculation on LiH, using hybridized atomic orbitals on the Li atom. The numbers tabulated are the coefficients c_{ni} in the expression $\varphi_n = \sum_i c_{ni}\varphi_i$, where φ_n is a molecular orbital and the φ_i are atomic orbitals.

		φ_i		
φ_n	Li($1s$)	Li($2s$)	Li($2p\sigma$)	H($1s$)
1σ	0.997	0.016	-0.005	0.006
2σ	0.131	-0.323	-0.231	-0.685
3σ	0.134	-0.805	0.599	0.148

Better calculations have been performed, using many more atomic orbitals, but the basic pattern remains the same.

only significant term. Let us then assume that $E_{\text{ionic}}(R) = -e^2/4\pi\epsilon_0 R$ for large R. To the same order of approximation, the two neutral atoms should have no significant interaction energy at large R. In Fig. 7.22 we have plotted these approximate potential-energy curves, which can be seen to cross at about $2R_e$. A single-configuration MO calculation gives essentially the same result, demonstrating the validity of this extremely crude model.

What is the significance of these results? Both single-configuration states must have the symmetry $^1\Sigma^+$. (There is also a $^3\Sigma^+$ state derived from the neutral atoms, but as in H_2 it should be entirely repulsive.) But the noncrossing rule is applicable to molecular states as well as individual orbitals: Two nondegenerate states of the same symmetry cannot have exactly the same energy, so their potential-energy curves cannot cross. Thus a single-configuration calculation must break down at the point where it predicts such a crossing, though it may be reasonably accurate elsewhere. One can obtain reasonably good results by joining the curve segments away from the crossing, as shown in the inset to Fig. 7.22. Curves obtained in such a way would be qualitatively similar to the actual curves for the two lowest states of LiH, shown by the solid lines in Fig. 7.22.[35] Similar effects occur in any of the "ionic molecules" of Section 7.4: Although the ground state is predominantly ionic near R_e, it goes over to a pair of neutral atoms at long range.

When we introduced the noncrossing rule, we mentioned that it is only approximately true; let us see why. To determine a potential-energy curve we must assume the nuclei to be at rest for each value of R (the Born–Oppenheimer approximation). The energy of a real system is thus given exactly by the $E(R)$ curve only in the limit of infinitely slow nuclear motion, that is, in an *adiabatic* process (to use the term we introduced in Section 6.5). In an adiabatic process the noncrossing rule can be shown to hold rigorously. But no real molecule is an adiabatic system. Since vibration is always present, R is always changing at a nonzero rate. How does this affect the noncrossing rule? In most cases where the rule applies, the two potential curves involved approach very close to each other (cf. the $A\,^1\Sigma^+ - B\,^1\Sigma^+$ "intersection" in Fig. 7.22). Let us say that they are within ΔE of each other over a range ΔR. Now the uncertainty principle enters the picture: If the uncertainty in a molecule's energy is greater than the separation between two such curves, there is a good chance of the molecule's crossing from one curve to the other, that is, entering the transition region ΔR in one electronic state and leaving it in another. According to Eq. 3.87, the noncrossing rule should apply only when $\Delta t \gtrsim \hbar/\Delta E$, where Δt is the time the molecule spends within the transition region; this is called the *Massey adiabatic criterion*. If the relative speed of the two atoms is $v = |dR/dt|$, we have $\Delta t = \Delta R/v$, and the criterion becomes $v \lesssim \Delta E\,\Delta R/\hbar$. The speed v, of course, increases with increasing vibrational excitation. For a given crossing, the adiabatic criterion is usually satisfied for low vibrational states but not for higher states. In terms of the Born–Oppenheimer approximation, the separability of nuclear (vibrational) and electronic motions breaks down when the vibrational energy is sufficiently great.

Transitions of the type just described can also occur between states of different symmetry classes, whose potential curves *can* cross one another. The usual conservation laws and selection rules of course apply in such transitions. Processes of either kind give rise to a number of

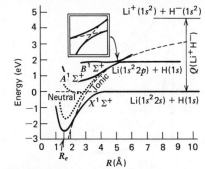

FIGURE 7.22
Potential energy curves for LiH: $- - - -$ simple model described in text,

$$E_{\text{ionic}}(R) = E(\text{Li}^+ + \text{H}^-) - e^2/4\pi\epsilon_0 R,$$

$$E_{\text{neutral}}(R) = E(\text{Li} + \text{H});$$

single-configuration MO calculation for the configurations $1\sigma^2 2\sigma^2$ (ionic) and $1\sigma^2 2\sigma 3\sigma$ (neutral); ——— experimental curves for the two lowest states ($X\,^1\Sigma^+$ and $A\,^1\Sigma^+$), and the approximate location of the next state ($B\,^1\Sigma^+$). The inset illustrates how one can join segments of single-configuration curves (----) to obtain curves that do not violate the noncrossing rule (———).

---

[35] Similar crossings (only one of which is shown) occur every time the "ionic" curve intersects the energy of an excited $^1\Sigma^+$ state of Li+H.

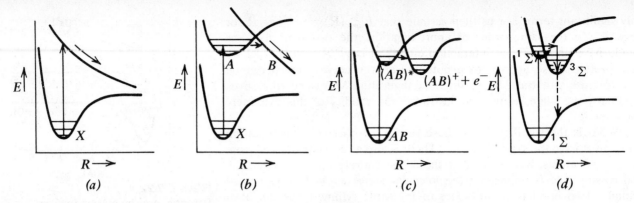

interesting phenomena, some of which are illustrated in Fig. 7.23. In ordinary dissociation (Fig. 7.23*a*) a molecule is excited into a repulsive state and the atoms immediately fly apart; the process takes about as long as a single vibrational period. *Predissociation* (Fig. 7.23*b*) is a slower process, in which the molecule is excited to a bound state *A*, remains there at least long enough to execute a few vibrations, then undergoes a transition to a repulsive state *B* which crosses[36] the bound state, and dissociates. *Autoionization* or *preionization* (Fig. 7.23*c*), which we discussed at the end of Section 6.10, is a similar transition from a bound neutral state to a state of molecule-ion plus free electron. Finally, in *phosphorescence* (Fig. 7.23*d*) the transition is to a lower-energy stable excited state from which decay to the ground state is forbidden (usually because it is a triplet–singlet transition, violating the rule $\Delta S = 0$). Since "forbidden" is not an absolute term, the decay does occur with emission of light, but over a long period after the initial excitation.[37]

FIGURE 7.23

Excited-state transitions. (*a*) shows ordinary dissociation, whereas the other figures show processes involving crossing potential curves: (*b*) predissociation; (*c*) autoionization; (*d*) phosphorescence. See discussion in the text.

7.9
Other First-Row Diatomic Hydrides

Except for LiH and HF, all the first-row diatomic hydrides are highly reactive species observed mainly in high-temperature systems. They are common reaction intermediates: CH can be detected in nearly all hydrocarbon flames, and OH (hydroxyl) in all flames containing oxygen and hydrogen in any form. They can also be detected by radio astronomy in interstellar space, where molecules are usually too far apart to react. The first molecule of any kind so detected was OH, in 1963. Since most of space is very cold, virtually all the molecules there must be in their ground electronic states. The states thus observed are indeed the ground states predicted by *a priori* calculations. The member of this series most commonly encountered in the laboratory is of course HF, the only one that exists as stable diatomic molecules at room temperature: HF is a colorless liquid boiling at 19.4°C.

By combining the information in Figs. 7.19 and 7.21, it is not difficult to deduce the ground-state MO configurations of the other first-row hydrides as we have done for LiH. These configurations are given in Table 7.8, along with other data on the molecules. Over this series the energy sequence of the occupied orbitals does not change, but the atomic orbitals with which they correlate do. For example, in LiH and BeH the 2σ orbital correlates with and thus largely resembles the H(1*s*) orbital. Since H(1*s*) and B(2*s*) have nearly the same energy, the 2σ orbital in BH

---

[36] The figure has been drawn for two states that actually cross. If they have the same symmetry, then the lower state instead has a maximum at the "intersection," and dissociation occurs by ordinary tunneling through this potential barrier.

[37] Phosphorescence should not be confused with *fluorescence*, which is the emission of light by nonforbidden transitions, most commonly singlet–singlet, and thus occurs mainly within nanoseconds (or at most microseconds) after excitation.

TABLE 7.8
GROUND STATES OF FIRST-ROW DIATOMIC HYDRIDES

Molecule	Electron Configuration	Term Symbol	D_e (eV)	R_e (Å)	$\tilde{\nu}_e$ (cm^{-1})	$|\mu|$ (D)
LiH	$1\sigma^2 2\sigma^2$	$^1\Sigma^+$	2.515	1.5954	1406	5.88
BeH	$1\sigma^2 2\sigma^2 3\sigma$	$^2\Sigma^+$	2.4±0.3	1.297	2058	(0.3 calc.)
BH	$1\sigma^2 2\sigma^2 3\sigma^2$	$^1\Sigma^+$	3.54	1.236	2367	1.27
CH	$1\sigma^2 2\sigma^2 3\sigma^2 1\pi$	$^2\Pi$	3.65	1.124	2859	1.46
NH	$1\sigma^2 2\sigma^2 3\sigma^2 1\pi^2$	$^3\Sigma^-$	3.40	1.045	3126	(1.0–1.9 calc.)
OH	$1\sigma^2 2\sigma^2 3\sigma^2 1\pi^3$	$^1\Pi$	4.621	0.9706	3735	1.66
HF	$1\sigma^2 2\sigma^2 3\sigma^2 1\pi^4$	$^1\Sigma^+$	6.11	0.9168	4139	1.82

is a roughly equal mixture of the two, and in CH and beyond the 2σ orbital is predominantly the heavy-atom $2s$, which lies below the H($1s$). Similar analyses can be made for the other orbitals. Although the 3σ and 1π orbitals (from CH on) are degenerate in both limits, the 3σ in each case is found to have a lower energy at R_e and thus is occupied first.

What is the bonding nature of these orbitals? Remember that a bonding orbital is expected to have a higher electron density in the bonding region than one finds in the separated atoms. The 1σ orbital in all these molecules is virtually identical to the heavy-atom $1s$, and like most inner-core orbitals is essentially nonbonding. The 1π orbital is also nonbonding, since it consists of only the heavy-atom $2p$ with no significant contribution from hydrogen. (The lowest p orbital of hydrogen is at a much higher energy.) The 2σ orbital is clearly bonding in LiH, but takes less and less part in the bonding as it becomes more like the heavy-atom $2s$; in HF it can also be considered nonbonding. But the slack is taken up by the 3σ orbital, which from CH on is the main constituent of the bond. In simple chemical terms, one can think of the combination $2\sigma^2 3\sigma^2$ as adding up to a single bond and a nonbonding lone pair, as in the Lewis formulas for :B:H or H:F̈:, but the bond is really made up of contributions from both orbitals.

The strength of the bonding, as indicated by the value of D_e, increases fairly steadily from LiH to HF. Even more clear-cut is the upward trend in the vibrational frequency $\tilde{\nu}_e$. This cannot be due to the relation $\tilde{\nu}_e \propto \mu^{-1/2}$, since the reduced mass μ varies very little over this series ($\mu_{\text{LiH}} = 0.88\, m_{\text{H}}$; $\mu_{\text{HF}} = 0.96\, m_{\text{H}}$). But $\tilde{\nu}_e$ does correlate well with the decrease in the bond length R_e, which is of course due to the decreasing size of the heavy atoms (Section 5.5). An increase in $\tilde{\nu}_e$ corresponds to a more sharply curved potential minimum, and we conclude that both the depth and the curvature of the potential wells tend to increase as R_e decreases.[38] The long-range attractive forces have much the same form between any two atoms, depending mainly on the electronegativity difference, but the short-range repulsive forces depend strongly on the atomic "sizes." Thus the position of the potential minimum in the hydrides must be sensitive mainly to the repulsive forces. This is in fact generally true, and is what makes the concept of a "covalent radius" meaningful.

Now let us look at the dipole moments. It can be seen that the magnitude of μ decreases sharply from LiH, then increases slightly toward HF. Presumably this is due to a change in polarity, as the heavy

[38] This agrees with the empirical Badger's rule (see footnote 9 on page 268).

atom changes from being more electropositive to more electronegative. However, there are no direct measurements of the direction of $\boldsymbol{\mu}$ in these molecules. Apart from LiH, how can we tell whether a given molecule is primarily H^+X^- or X^+H^-? A good idea of this can be obtained from the long-range limit. If we set the energy of two separated atoms $A+B$ as zero, then the energy of the ion pair $A^+ + B^-$ is simply the $Q(A^+B^-)$ of Eq. 7.54, that is, $I(A) - A(B)$. If we calculate $Q(H^+X^-)$ and $Q(X^+H^-)$, the smaller of these quantities must correspond to the lowest ionic state at $R = \infty$, which is probably the principal ionic contribution to the ground state at $R = R_e$. For example, in CH we have $Q(C^+H^-) = I(C) - A(H) = 11.26\,\text{eV} - 0.75\,\text{eV} = 10.51\,\text{eV}$, $Q(H^+C^-) = I(H) - A(C) = 13.60\,\text{eV} - 1.25\,\text{eV} = 12.35\,\text{eV}$; thus $C^+ + H^-$ has an energy 1.84 eV lower than $H^+ + C^-$, and the CH molecule should be primarily C^+H^-. For the series of first-row hydrides such calculations give:

X	Li	Be	B	C	N	O	F
$Q(X^+H^-)$ (eV)	4.61	8.57	7.55	10.51	13.78	12.86	16.67
$Q(H^+X^-)$ (eV)	12.98	13.22	13.42	12.35	13.8	12.13	10.15

Thus the heavy atom should be the positive end of the molecule in LiH, BeH, BH, and CH, the negative end in OH and HF, whereas the NH molecule should be very nearly nonpolar. These results are plausible, and are supported by the electron-distribution calculations of Fig. 7.20.

So far we have spoken only of the ground states of the diatomic hydrides. Although we shall make no detailed study of the excited states, it is worthwhile to survey the relationship between molecular states and those of the separated atoms. This can be done systematically by extending the vector model of Section 5.6 from atoms to molecules. The method is applicable to both heteronuclear and homonuclear molecules, though complicated in the latter case by degeneracies.

As an example, let us consider the manifold of states of the CH molecule derived from the ground states of the separated atoms. As we showed in Section 5.6, these ground states are $C(^3P)$ $(L = 1, S = 1)$ and $H(^2S)$ $(L = 0, S = \frac{1}{2})$. The spin and orbital angular momenta of the atoms add as vectors to give those of the molecule. The reasoning is the same as we applied to obtain the values of J in connection with Eq. 5.30. Since S is quantized in integral steps, its possible values for a diatomic molecule are

$$S = S_A + S_B, S_A + S_B - 1, \ldots, |S_A - S_B|, \tag{7.55}$$

where S_A and S_B are the atomic values. In this case we can have $S = \frac{3}{2}$ or $S = \frac{1}{2}$, that is, quartet or doublet states. As for the orbital angular momentum, we are concerned only with its component along the internuclear axis. The quantum number Λ is given by

$$\Lambda = |(M_L)_A + (M_L)_B|, \tag{7.56}$$

where each M_L can have any of its possible values $L, L - 1, \ldots, -L$. In CH we have $(M_L)_C = 1, 0, -1$ and $(M_L)_H = 0$, giving the possible values $\Lambda = 0$ (Σ states) or $\Lambda = 1$ (degenerate Π states). Since the spin and orbital angular momenta add independently, we should have all told from $C(^3P) + H(^2S)$ the states $^2\Pi$, $^2\Sigma$, $^4\Pi$, and $^4\Sigma$, in order of increasing energy. The order is that predicted by Hund's rules, with Λ replacing L. The $^2\Pi$ state is in fact the ground state of CH, and $^2\Sigma$ is the lowest observed excited state; the other two states are as yet unobserved. Similar analyses can be made for any other pair of separated-atom states.

It is generally true, as it is for CH, that some of the states generated from a given separated-atom limit remain unobserved. The reason for this is that many of the states have no stable minima in their $E(R)$ curves. Such states have no characteristic band spectra by which they can be identified, but only continuum radiation. Indeed, their potential curves are sometimes so strongly repulsive that molecules with R near the ground-state R_e are not encountered at all. To see how such repulsive states come about, let us consider the $^4\Pi$ state of CH. This state has $S = \frac{3}{2}$, and thus must have at least three singly occupied orbitals; to have $\Lambda = 1$, there must be an odd number of electrons in π orbitals. The lowest-energy configuration meeting these specifications is $1\sigma^2 2\sigma^2 3\sigma 1\pi 4\sigma$, where one electron is promoted from the 3σ to the 4σ orbital, which is so strongly antibonding that it overcomes the remaining bonding forces. (In the $^2\Sigma$ state, with configuration $1\sigma^2 2\sigma^2 3\sigma 1\pi^2$, the promotion of one electron to the *non*bonding 1π orbital is enough to reduce the binding energy to only 0.4 eV.) Despite the repulsive nature of the $^4\Pi$ state, it must become the lowest-energy state at some small value of R, since it correlates with the united-atom ground state, $N(^4S)$, which also has three singly occupied orbitals. As $R \to 0$ we have $1\sigma \to 1s$, $2\sigma \to 2s$, $3\sigma \to 2p\sigma$, $1\pi \to 2p\pi$.

We mentioned previously that the diatomic hydrides have been observed by radio astronomy. But just what kind of transitions can these molecules have in the radiofrequency region of the spectrum? For OH, for example, even the pure rotational spectrum is well into the infrared region, with $B_e = 18.9 \text{ cm}^{-1}$. First of all, the energy levels of molecules show fine-structure splitting, which, as in atoms (Section 5.7), is due primarily to spin–orbit interaction. When the spin and orbital angular momenta are strongly coupled, analogous to LS coupling in atoms, the component of total angular momentum along the internuclear axis is $\Omega\hbar$, where

$$\Omega = |\Lambda + \Sigma| \qquad (\Sigma = S, S-1, \ldots, -S). \tag{7.57}$$

The quantum numbers Λ, Σ, Ω correspond to the atomic M_L, M_S, M_J, respectively. By "strongly coupled" we mean that the orbiting electrons produce a magnetic field that tends to align the spin magnetic moment with the axis. There are other types of coupling, but we need consider only this one. The ground-state term of the OH molecule is $^2\Pi$ ($S = \frac{1}{2}$, $\Lambda = 1$), with the possible values $\Sigma = \frac{1}{2}, -\frac{1}{2}$ and $\Omega = \frac{3}{2}, \frac{1}{2}$. Thus there are two states, designated as $^2\Pi_{3/2}$ (the ground state) and $^2\Pi_{1/2}$. But the energy difference between these states is 0.017 eV (corresponding to 140 cm^{-1}), which is even greater than the rotational spacing.[39] One thus observes two distinct though overlapping rotational bands in laboratory spectra, so we must seek further for a radiofrequency transition. Besides the coupling between spin and orbital electronic angular momenta, there is a much weaker interaction between electronic and *rotational* angular momenta. This interaction destroys the degeneracy between the two states differing only in the orientation of **L** (with $L_z = \pm\Lambda\hbar$; cf. Section 6.5), and is thus known as *Λ-type doubling*.[40] The splitting is proportional to $J(J+1)$, where J is the rotational quantum number. For the $J = 1$ state of OH ($^2\Pi_{3/2}$) it equals 7.8×10^{-6} eV or 0.063 cm^{-1}, which is indeed in the radiofrequency region. Thus we see how successively weaker interactions can give us finer and finer probes of molecular structure.

[39] The average temperature of interstellar space is probably about 3 K, corresponding to an average molecular energy of only 4×10^{-4} eV (3 cm^{-1}); thus virtually all OH molecules in space should be in the $^2\Pi_{3/2}$ state.

[40] When $\Lambda = 0$, that is, in Σ states, there can still be an interaction between spin and rotational angular moments. But $^1\Sigma$ states (like atomic ^1S states) have no fine structure except that due to nuclear spin.

The final topic we shall discuss in this section takes us out of the realm of pure intramolecular forces; this is the *hydrogen bond*, a phenomenon of which HF provides the simplest illustration. The subject is introduced here, rather than in Chapter 10, because it is so characteristic of HF. A number of compounds in which hydrogen is bonded to very electronegative elements (mainly F, O, and N) show strong attractive forces between molecules. For example, although one expects boiling points in a family of compounds to increase with molecular weight as in the inert gases, HF, H_2O, and NH_3 go counter to this trend:

NH_3	−33.4	H_2O	0.0	HF	19.5	Ne	−245.9
PH_3	−133	H_2S	−60.7	HCl	−84.9	Ar	−185.7
AsH_3	−55	H_2Se	−41.5	HBr	−67.0	Kr	−152.3
SbH_3	−17.1	H_2Te	−2.2	HI	−35.4	Xe	−107.1

(boiling points in °C). The heats of vaporization vary in the same way; those for HF, H_2O, and NH_3 are higher than expected by 20–40 kJ/mol (0.2–0.4 eV/molecule). This is about one order of magnitude weaker than a normal covalent bond, but still much stronger than ordinary intermolecular attractions (see Chapter 10); for example, the potential well for two Ar atoms is about 0.01 eV deep. Data from spectroscopy, neutron diffraction, and other sources clearly show that in these substances the hydrogen atoms are normally located between two electronegative atoms, but closer to one than the other. This is usually indicated by a formula such as H—F$\cdots$H—F, with the "long bond" shown by dots.

How does a hydrogen atom between two electronegative atoms have a bonding effect? We know that in an HF molecule the bond is strongly polar, so that there is an excess of negative charge on the F atom and an excess of positive charge on the H atom. As in our model for ionic bonds, one can to a good approximation treat each atom as a point charge but with a charge less than e. Just as negative charge in the region between two positive nuclei tends to draw them together (see Section 6.1), so a positively charged H atom between two negatively charged F atoms exerts a bonding force—but a much weaker one, because the charges are only partial and the distances greater. In an MO treatment, one might consider the ordinarily nonbonding 2σ orbital of one HF molecule to include an H($1s$) component from the other molecule. The hydrogen bond tends to pull the H atom away from its nearest-neighbor fluorine, with the net effect of weakening the restoring force in the H—F oscillator (in which the H does nearly all the moving). This effect lowers $\tilde{\nu}_e$ from 4138 cm^{-1} in free HF to about 3400 cm^{-1} in (HF)$_2$, showing that the environment of the H atom is rather drastically changed by hydrogen bonding.

The simplest example of a "molecule" showing hydrogen bonding is the dimer (HF)$_2$, which exists in the gas at relatively low temperatures. Two possible structures come to mind, neither of which corresponds to the true structure. One might expect either the linear H—F$\cdots$H—F, or the ring

with two hydrogen bonds. In reality, the HF dimer appears to have the

bent structure

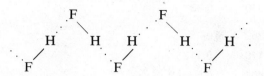

$$
\begin{array}{c}
H \\
| \\
F \cdots H - F
\end{array}
$$

The location of the central H may be on or off the F—F axis—nobody knows yet. More highly polymerized forms also exist, and liquid and solid HF are basically made up of long zigzag chains:

In the next chapter we shall see why the chains are bent rather than linear. If equimolar amounts of HF and KF are crystallized together, one obtains a well-defined crystalline species KHF_2, made of K^+ and FHF^- ions; in the FHF^- ion the hydrogen atom is found to be exactly midway between the two F atoms. This carries the hydrogen bond to its ultimate form, with the H atom associated with no particular molecule.[41]

7.10
Isoelectronic and Other Series

To interpret the differences among molecules, we must find or invent concepts that characterize the important changes from one molecule to another. Which concepts we choose will depend on the molecules under consideration. Thus far the key factor in our analysis of diatomic molecules has been the number of electrons, added one by one to a relatively stable set of orbitals. In heteronuclear molecules we needed the additional concept of orbital polarity, leading to the distinction between ionic and covalent bonding. To study the polarity effect in relative isolation, let us now consider some sets of molecules with the same total number of electrons—what we call *isoelectronic series* of molecules.

Among the simplest and most informative of such series is that isoelectronic with C_2, including BN, BeO, and LiF, in order of increasing polarity. All are known as diatomic molecules in the vapor phase, with properties varying from C_2, which is homonuclear and thus covalently bound, to LiF, a very ionic molecule which we described in Section 7.4. It is worth noting that a similar transition from covalent to ionic bonding is found in the solid forms of these species, from the covalent structures of carbon (diamond and graphite) to the ionic lattice of Li^+F^-.

One would expect all these molecules to have ground-state electron configurations equivalent to that of C_2, that is, $1\sigma^2 2\sigma^2 3\sigma^2 4\sigma^2 1\pi^4$. The ground states of C_2, BeO, and LiF are all $^1\Sigma$, corresponding to this configuration. Although the lowest known state of BN is $^3\Pi$, corresponding to what was long thought to be the ground state of C_2 (see footnote 28 on page 295), the low-lying $^1\Sigma$ state has not yet been observed, and the ground state may yet prove to be this $^1\Sigma$ state. Within the series, the orbitals change character regularly as the difference between the nuclear charges grows. The 1σ and 2σ orbitals, which we approximate by $1s_A \pm 1s_B$ in C_2, become the $1s$ orbitals of the high-Z and low-Z atom, respectively: They do not mix significantly with the $n = 2$ orbitals, since

---

[41] The valence bond theory treats FHF^- as a hybrid of the structures

$$(F - H \cdots F)^- \quad \text{and} \quad (F \cdots H - F)^-;$$

in MO theory one must use "three-center orbitals," with contributions from all three atoms.

TABLE 7.9
PROPERTIES OF SOME SERIES OF DIATOMIC MOLECULES

	Ground State	D_0 (eV)	R_e (Å)	$\tilde{\nu}_e$ (cm^{-1})	$|\mu|$(D)
Isoelectronic Series					
C_2	$^1\Sigma_g^+$	6.24	1.2425	1855	0
BN	$^3\Pi$ (?)	3.99	1.281	1515	(1.4 calc.)
BeO	$^1\Sigma^+$	4.60	1.331	1487	(7.3 calc.)
LiF	$^1\Sigma^+$	5.94	1.564	910	6.33
N_2	$^1\Sigma_g^+$	9.760	1.094	2358	0
CO	$^1\Sigma^+$	11.09	1.128	2170	0.112
BF	$^1\Sigma^+$	7.85	1.262	1401	0.5±0.2
Families of the Periodic Table					
HF	$^1\Sigma^+$	5.86	0.917	4139	1.826
HCl	$^1\Sigma^+$	4.446	1.275	2991	1.109
HBr	$^1\Sigma^+$	3.755	1.414	2649	0.828
HI	$^1\Sigma^+$	3.053	1.609	2308	0.448
Li_2	$^1\Sigma_g^+$	1.12	2.672	351.4	0
Na_2	$^1\Sigma_g^+$	0.75	3.079	159.2	0
K_2	$^1\Sigma_g^+$	0.51	3.923	92.6	0
Rb_2	$^1\Sigma_g^+$	0.47	4.20	57.3	0
Cs_2	$^1\Sigma_g^+$	0.45	4.58	42.0	0
F_2	$^1\Sigma_g^+$	1.604	1.409	919.0	0
Cl_2	$^1\Sigma_g^+$	2.484	1.988	559.7	0
Br_2	$^1\Sigma_g^+$	1.971	2.281	323.3	0
I_2	$^1\Sigma_g^+$	1.544	2.667	214.5	0

even in LiF the Li(1s) orbital has appreciably lower energy than the F(2s). The 3σ orbital changes from $2s_A + 2s_B$ in C_2 to the high-Z (N, O, or F) $2s$, and the 1π from $2p_A + 2p_B$ to the $\pi 2p$ orbital on the high-Z atom. But the 4σ orbital, which correlates with the atomic $2s$ in C_2, becomes more and more like the $\sigma 2p$ of the high-Z atom as one goes to LiF. (Cf. the atomic orbital energies in Fig. 7.21.)

Some of the properties of molecules in this and other series are listed in Table 7.9. From C_2 to LiF the bond lengths increase and the vibrational frequencies decrease, but the dissociation energy varies in a more complicated manner. This is the result of a balance between two opposing trends: The covalent bonding power weakens as the bond grows longer and the orbitals become more nonbonding, but the ionic contribution to the bond increases as the charge distribution becomes more polarized. The dipole moment of course increases with the electronegativity difference between the two atoms.

Another interesting set of isoelectronic molecules consists of N_2, CO, and BF. Both N_2 and CO exist as stable diatomic gases at room temperature, and even consist of diatomic molecules in the solid state; BF, however, is unstable.[42] The properties listed in Table 7.9 show the same

[42] Even CO has a tendency to disproportionate, $2CO \rightarrow C + CO_2$, but the reaction is *very* slow at room temperature.

conflicting trends as in the previous series. All three compounds have $^1\Sigma$ ground states, corresponding to the configuration $1\sigma^2 2\sigma^2 3\sigma^2 4\sigma^2 1\pi^4 5\sigma^2$. In CO, calculations show that the 1σ and 2σ orbitals are essentially atomic $1s$ orbitals, the 3σ and 5σ orbitals are largely concentrated on O and C, respectively, whereas the 4σ and doubly degenerate 1π orbitals furnish the bulk of the bonding. This corresponds approximately to the Lewis formula :C≡O:. That the dissociation energy of CO is higher than that of N_2 can be attributed to a small amount of ionic character added to what is still essentially a triple bond. In BF, ionic bonding is presumably important, but not enough to make ionic B^+F^- crystals stable.

There are a number of striking similarities in the physical properties of N_2 and CO. (Their chemical properties are of course rather different, but even there the two exhibit comparable inertness.) The gas densities, boiling points, viscosities, and thermal conductivities of the two species are almost the same. This similarity is due in part to their near-identical molecular weights, but the intermolecular forces also reflect the similar internal structure of the two molecules. The latter is particularly apparent if one looks at the orbital binding energies, which have been measured by photoelectron spectroscopy (values in eV):

	1σ	2σ	3σ	4σ	1π	5σ
N_2	409.9	409.9	37.3	18.6	16.8	15.5
CO	542.1	295.9	38.3	20.1	17.2	14.5

Except for the atomic-core 1σ and 2σ orbitals, the two sets of energies are almost identical; the energies obtained by molecular orbital calculations are somewhat different, but show the same pattern. Results of this sort clearly justify our treating the members of an isoelectronic species as closely related.

We must say something more here about photoelectron spectroscopy, a conceptually simple and powerful method for determining molecular energy levels. The nature of the photoelectric effect has been described in Section 2.2; here we apply it to free molecules. Specifically, suppose that radiation of frequency ν (energy $h\nu$) strikes a molecule and releases an electron with binding energy ϵ. If we neglect the small recoil effects, the binding energy should be given by

$$\epsilon = h\nu - T, \tag{7.58}$$

where T is the kinetic energy of the released electron. By measuring or selecting ν and then measuring T as in the Franck–Hertz experiment, for example, one can determine ϵ, which by Koopmans' theorem should equal the electron's orbital energy in the molecule. The energy $h\nu$ must of course be greater than ϵ. One most commonly uses radiation in the vacuum ultraviolet to release the valence electrons, and x rays for the core electrons. Typically one irradiates a sample, gaseous or solid, with monochromatic radiation such as that of the He $2p \rightarrow 1s$ transition at 584 Å, and measures the kinetic energies of the photoelectrons. Ultraviolet radiation is associated with states whose natural lifetimes are of order 10^{-9} s and so, according to the uncertainty principle, can provide energy resolution at best with $\Delta E = \hbar/\Delta t$ or about 4×10^{-6} eV (0.03 cm^{-1}), considerably better than the energy of the electrons can be determined. With x rays, the energy resolution is typically 1–5 eV, due in large part to the short lifetimes of the excited states from which they are emitted. Hence x-ray photoelectron spectroscopy can locate the approximate energy of shells, but one must use ultraviolet radiation to probe the separations of valence orbitals. With ultraviolet photoelectron spectroscopy it is quite straightforward to distinguish different vibrational levels, especially of the final ion, and even rotational levels of very light molecules have been resolved.

Thus far we have not carried our analysis beyond the first row of the periodic table. According to the principles outlined in Section 5.5 we expect the elements below the first row to exhibit bonding behavior similar to that of the first element in each family (alkali metals, halogens, etc.). And in fact, each family does bond in generally similar ways. But what systematic changes may we expect to find as we go down a family of diatomic molecules (varying one or both atoms)? The properties of several such series are given in Table 7.9. One can readily name others (the alkali halides, interhalogen compounds such as ClF, interalkali compounds such as NaK, the analogs of N_2 and O_2, etc.), but essentially the same trends are found in all cases.

These trends largely reflect the effects of atomic size, which of course increases slowly as one goes down each column of the periodic table. In a similarly bonded series of molecules, the bonds must become longer and thus weaken with increasing atomic size. An additional effect is found in heteronuclear molecules: Since the nuclei become better shielded as more electrons are added, the ionization potentials, electron affinities, and thus electronegativities also decrease with atomic size. Depending on the nature of the series, these changes will either increase or decrease the ionic contribution to the bonding. There is one other effect of atomic size: The core electrons are by no means completely shielded or nonbonding, and do take some part in bond formation; as we noted earlier, in many transition metals the inner-shell d electrons are almost as important as the valence electrons in bonding.

The trends in Table 7.9 are for the most part clear and consistent, with the exception of the anomalously low dissociation energy of F_2. This recalls the fact that the electron affinity of F is less than that of Cl. The low dissociation energy of F_2 has puzzled scientists for many years. The explanation, as with the electron affinity, seems to be that the inner core electrons ($1s$ and, to some degree, $2s$) form a relatively more important fraction of the total electron cloud in fluorine than in larger halogen atoms, so that the repulsive contribution to the F—F bond is more important than in other halogen–halogen bonds.

It may be noted that the macroscopic properties of molecular families often also show clear-cut trends—for example, the boiling points tabulated in the last section. However, these depend on intermolecular forces, which have more to do with the overall size of a molecule than with the nature of its bonding except where hydrogen bonding or other electrostatic effects are significant. We shall consider intermolecular forces in Chapter 10, and macroscopic behavior in Part Two. For now, though, let us proceed to molecules with more than two atoms.

Further Reading

Gaydon, A. G., *Dissociation Energies and Spectra of Diatomic Molecules*, 3rd Ed. (Chapman and Hall, London, 1968).

Herzberg, G., *Molecular Spectra and Molecular Structure, Volume I. Spectra of Diatomic Molecules*, 2nd Ed. (Van Nostrand-Reinhold, Princeton, N.J., 1950).

Hurley, A. C., *Introduction to the Electron Theory of Small Molecules* (Academic Press, London, 1976).

Karplus, M., and R. N. Porter, *Atoms and Molecules* (W. A. Benjamin, Inc., Menlo Park, Calif., 1970), Chapters 5, 6, and 7.

Kauzmann, W., *Quantum Chemistry* (Academic Press, Inc., New York, 1957) Chapters 11C–G and 12.

Kondratyev, V., *The Structure of Atoms and Molecules* (P. Noordhoff N. V., Groningen, The Netherlands, 1964), Chapters 7–10.

Morrison, M. A., T. L. Estle, and N. F. Lane, *Quantum States of Atoms, Molecules and Solids* (Prentice-Hall, Inc., Englewood Cliffs, N.J., 1976), Chapters 12, 14–17.

Mulliken, R. S., and W. C. Ermler, *Diatomic Molecules, Results of* ab initio *Calculations* (Academic Press, New York, 1977), esp. Chapters IV, V, and VI.

Slater, J. C., *Quantum Theory of Molecules and Solids*, Volume I (McGraw-Hill Book Co., Inc., New York, 1963) esp. Chapters 5, 6, and 7.

Streitweiser, A., and P. H. Owens, *Orbital and Electron Density Diagrams. An Application of Computer Graphics* (The Macmillan Company, New York, 1973).

Problems

1. Because the vibrational spacings of diatomic molecules generally diminish with increasing energy, it is possible to extrapolate the vibrational spacing, as a function of vibrational energy, to zero, and thereby obtain a moderately accurate estimate of the dissociation energy of the molecule. Such graphs are known as Birge–Sponer plots, after R. Birge and H. Sponer. Using values of $\tilde{\nu}_e$, $\tilde{\nu}_e x_e$, and $\tilde{\nu}_e y_e$ from Table 7.2, construct such plots for Na_2, CH, and HCl, then evaluate D_e and D_0, the dissociation energies from the bottom of the potential and from the ground vibrational state. Compare these values of D_0 with those given in Table 7.2.

2. Find the outer classical turning points and thus the classical zero-point amplitudes of vibration for H_2, LiH, and HCl, from the data in Table 7.2. (Refer to Problems 21 and 22 in Chapter 4 if you need help finding the classical turning point.)

3. Using Table 7.2, determine the vibrational quantum numbers at which the actual vibrational energy level spacings of F_2 and Cl_2 deviate 1% and 10% from the harmonic spacings based on $\tilde{\nu}_e$ alone.

4. The dipole moment of HCl is 1.109 D (Table 7.4), its equilibrium bond length is 1.27 Å, and its vibration frequency is approximately 2991 cm^{-1} (Table 7.2). Suppose that a spatially uniform laser beam of 1 W, with precisely this frequency and a cross section of 0.01 cm^2, is incident on a sample of HCl vapor. What is the maximum instantaneous stretching force exerted by this field on an HCl molecule with $R = R_e$? Based on the force

constant of Table 7.2 and the assumption that the dipole moment μ increases directly with R very near R_e, compute the effective restoring force of the chemical bond on the nuclei at their classical outer turning point in the ground vibrational state. Compare the restoring force of the bond with the stretching force of the electric field.

5. Compute the average moment of inertia for the ground vibrational-electronic state of O_2, assuming that this molecule is a harmonic oscillator, so that the lowest vibrational function of Table 4.2 can be used for the probability amplitude. Compare this moment of inertia with the value implied by the B_e of 1.4456 cm^{-1} in Table 7.2.

6. Most common diatomic molecules have lowest vibrational spacings that are much larger than their lowest rotational spacings, yet vibrational spacings diminish and rotational spacings increase with increasing quantum number. At what vibrational and rotational quantum numbers do these spacings become roughly equal for H_2, N_2, and HBr? What are the total energies of the molecules at these levels? Compare these total energies with the corresponding dissociation energies.

7. Calculate the frequencies of the vibration–rotation transitions of OH from data in Table 7.2, for the transitions $v = 0 \rightarrow 1$ and $v = 1 \rightarrow 2$, for $\Delta J = -1, 0$, and $+1$ for J from 0 to 10. Plot the transitions on an energy scale that displays the entire set of lines and still allows one to resolve them. What spectral resolution would be required to distinguish the lines, if one requires resolution of at least half the spacing of the most closely spaced lines?

8. As the rotational energy of a molecule increases, the centrifugal force on the nuclei adds a repulsive term to the effective potential as given in Eq. 7.19. Find the value of J for which the centrifugal potential brings the lowest point of the rotationless curve up to D_0, that is, find the first rotational state for which $E_{rot} + V(R) > 0$ everywhere, for H_2, Na_2, and I_2.

9. Compute a rotational analog of Problem 4. That is, assume that an assembly of HCl molecules is in a laser beam whose power density is 1 W/cm^2 and whose frequency is precisely resonant with the first rotational transition of HCl. Given that its dipole moment is 1.109D, compute the total force on the nuclei and the torque (force times lever arm from the center of mass) when the force is a maximum.

10. From the rotational line frequencies given below, compute the bond length of the diatomic molecule NaCl in its ground and first excited vibrational state. [Data are from A. Honig, M. Mandel, M. L. Stitch, and C. H. Townes, *Phys. Rev.* **96**, 629 (1954).] All transitions are $J = 1$ to $J = 2$.

	$v = 0$ (MHz)	$v = 1$ (MHz)
$Na^{35}Cl$	26051.1 ± 0.75	25857.6 ± 0.75
$Na^{37}Cl$	25493.9 ± 0.75	25307.5 ± 0.75

How much effect do the uncertainties in the spectral line frequencies have on the inferred bond lengths?

11. The rotational spectrum of a polar diatomic molecule is observed in the microwave region of the electromagnetic spectrum. In the case of RbBr, in the vibrational state $v = 0$, the $J = 8 \rightarrow 9$ transition is observed at:

Molecule	(MHz)
$^{85}Rb^{79}Br$	25 596.03
$^{87}Rb^{79}Br$	25 312.99
$^{85}Rb^{81}Br$	25 268.84

Assume that RbBr behaves as a rigid rotator and calculate the internuclear separations in the various isotopic molecules. The atomic masses are $^{79}Br = 78.94365$ amu, $^{81}Br = 80.93232$ amu, $^{85}Rb = 84.93920$ amu, $^{87}Rb = 86.93709$ amu. Are you surprised by your results? How do you interpret them?

12. The bond lengths of Na_2 are 3.078 Å and 3.63 Å in the ground and first excited electronic states, respectively. Based on this difference, compute the separation in cm^{-1} between the band origin ($J' = 0$ to $J'' = 0$) and the band head (where the rotational lines turn around and start returning on themselves) for a transition between these states.

13. One speaks of rotational bands in electronic spectra as being "degraded toward the red" or "degraded toward the blue" depending on whether the rotational line spacings eventually increase toward longer or shorter wavelengths. Such "degradation" is usually immediately apparent to the eye. Show that one can tell immediately whether R_e is greater in the ground or excited state depending on the direction of the degradation or shading.

14. Doppler broadening is very important in many regions of the spectrum and in a variety of situations, including spectroscopy of stars and the interstellar medium. The Doppler shift is the result of wave crests and peaks reaching the observer faster or slower than they would if the source of radiation and the observer were at rest relative to each other. Show that the shift in wavelength $\Delta\lambda$ of radiation sent by a source moving with velocity v relative to the observer is given approximately by

$$\Delta\lambda = \frac{\lambda v}{c}.$$

Calculate the Doppler shift for radiation of 1000 MHz and 1000 cm^{-1} if $v = 10^5$ cm/s.

15. Compare the bonding in Cl_2 and Cl_2^+. How does it differ? Would you expect Cl_2^- to exist as a stable species? Why?

16. Show the forms of the nodal surfaces for the σ, π, and δ orbitals constructed as sums and differences of $3d$ orbitals in the homonuclear diatomic molecule Si_2.

17. The dissociation energy of F_2 is 1.60 eV, whereas those of Cl_2 and Br_2 are 2.48 eV and 1.97 eV, respectively. Give at least one interpretation of this apparent anomaly.

18. The molecules N_2 and C_2H_2 are isoelectronic. From considerations based on the electronic structure of N_2, predict the geometry (i.e., bent or linear) of C_2H_2, and discuss its electronic structure in terms of the types of orbitals occupied, nature of the orbitals, and so on.

19. Write the electron configurations for the ground states of Mg_2, Fe_2, MgH, and HCl. Which of these would you expect to be stable on the basis of the electron configuration alone?

20. Predict the dissociation energy, equilibrium internuclear distance, and vibration frequency $\bar{\nu}_e$ by extrapolation from the data in Table 7.9, for each of the following:

   (a) At_2
   (b) DAt
   (c) Fr_2

21. Rationalize the empirical facts that the ionization potential of H_2 is greater than that of atomic H, whereas that of O_2 is less than that of atomic O.

22. Rationalize the exponential form of the repulsive contribution to the energy of an ionic molecule, Eq. 7.43. Recall that the united-atom limit has a finite energy, achieved when the two nuclei coalesce. Hint: Consider the effect of

the Pauli principle. Explain why a form $Be^{-R/\rho}/R$ might be even more plausible.

23. Consider the reaction

$$K + HI = KI + H.$$

Suppose that the initial velocity of the K atom is 5×10^4 cm/s, and that of the HI molecule is negligibly small. Suppose that the rotational energy of HI is negligibly small. Further, suppose that the K would have approached to within 5 Å of the center of mass of the HI if no reaction occurred. Finally, assume that the velocity of the H atom leaving after the reaction is negligibly small. The energy of the KI bond is 3.34 eV and that of the HI bond is 3.06 eV. The internuclear separation in KI is 3.048 Å, and the vibrational frequency is 173 cm^{-1}. What do the conditions for conservation of energy, linear momentum, and angular momentum imply about the vibrational and rotational state of the product molecule KI? Estimate the rotational quantum number J and the vibrational quantum number v that characterize the product KI. (Hint: Look back at Rutherford scattering to see how the angular momentum of a particle moving toward another molecule is related to its velocity and the impact parameter; cf. Appendix 2C.)

24. The hydrogen halides have continuous electronic absorption spectra, the onsets being about 2500 Å for HCl, 2650 Å for HBr, and 3270 Å for HI.
    (a) How do you interpret these observations? Draw likely potential energy curves for the states in question in HI.
    (b) If the difference in energy between the $^2P_{1/2}$ and $^2P_{3/2}$ states of the I atom is 1 eV, must the onsets of production of these two dissociated iodine atoms from HI also be separated by 1 eV?
    (b) What is the likely consequence of irradiating HI with light of wavelength 2537 Å? 1849 Å?

25. The occurrence of predissociation is often inferred from spectra by the sudden appearance of wide, diffuse lines where at lower energies the lines are narrow and sharp. Explain why this is an indication of predissociation or autoionization. How would you distinguish predissociation from auto-ionization experimentally?

26. Compare the dissociation energies, bond lengths, and vibrational frequencies of a number of diatomic molecules, for example, from Table 7.2. Examine these for correlations among bond length, dissociation energy, and vibration frequency. What correlations do you find? (Other sources of data should also be consulted.)

Triatomic
Molecules

Starting with the fundamental quantum laws, we have examined the properties of first atoms, then diatomic molecules. Now it is time to take another step in building up our picture of the microstructure of matter. In this chapter we extend our analysis to triatomic molecules. These molecules are interesting in themselves, including as they do some of the species most important in chemistry and biology, especially H_2O and CO_2. But in addition they display most of the complexity involved in considering larger molecules.

The principal feature of this added complexity is, of course, *molecular geometry*. Diatomic molecules have only one internal coordinate, the internuclear distance; in triatomic and larger molecules we must consider the distances and directions between many pairs of atoms—in simple language, we wish to know the bond lengths and bond angles. In the triatomic case there are three independent internal coordinates: We can expect the values of all these variables to depend on the molecule's electronic structure, and in the first part of the chapter we consider this relationship. The problem of the relation between molecular geometry and electronic structure is a partially solved puzzle that continues to tantalize theorists, but is well enough understood to permit making many powerful generalizations.

In the second part of the chapter we shall be concerned with the vibrations and rotations of triatomic molecules. In the triatomic molecule we have three or four vibrational *degrees of freedom*—that is, three or four independent types of vibration going on at the same time. This results in a complicated energy spectrum, which plays a crucial role in all fields concerned with the capacity of molecules to absorb, emit, or store energy.

We shall begin to depart from fundamental concepts in this chapter. Our discussion has reached a level where the complexity of the problems forces us to develop semiempirical guidelines, particularly for the interpretation of structure. This is even more necessary in larger molecules, but the triatomic molecules are still simple enough for us to see the connections between these semiempirical rules and the underlying physical laws responsible for their validity.

We begin our study of triatomic molecules with the simplest examples, which (as in diatomic molecules) are those made up of hydrogen atoms. The one-electron H_3^{2+} ion is quite unstable and need not be considered. However, the two-electron H_3^+ ion is a stable species, produced whenever an H_2 molecule collides with an H_2^+ molecule-ion, as occurs in electric discharges in gaseous hydrogen. The neutral molecule H_3 is unstable with respect to decomposition into $H_2 + H$, but since the exchange reaction

$$H + H_2 \rightarrow H_2 + H$$

8.1
Electronic Structure and Geometry in the Simplest Cases: H_3 and H_3^+

probably involves transient configurations similar to H_3, its nature is of considerable interest to students of chemical reaction mechanisms. Let us therefore consider the properties of the H_3^+ and H_3 molecules, in particular their potential-energy surfaces and the resulting equilibrium shapes.

The H_3^+ and H_3 molecules—and any other triatomic molecule—may be either linear or triangular. To know the "structure"—by which we mean the equilibrium geometry—of a triatomic or larger molecule, one must determine not only the bond lengths but also the angles between the bonds. Chemical intuition leads us to describe the geometry of the H_3 molecule in terms of two H—H bonds and a single angle between them. Our interest here is primarily in the equilibrium value of this angle. Thus we need to examine how the molecule's energy varies as a function of the bond angle.

For a straightforward approximate interpretation we again employ the molecular orbital approach, based on the use of one-electron wave functions and energies. Furthermore, we again employ the LCAO approximation, which treats the molecular orbitals as sums and differences of atomic orbitals. Even though much more elaborate calculations are necessary to describe molecular properties quantitatively, the crude LCAO model is easy to use and gives results that are vivid and, for the most part, qualitatively correct. In particular, we shall see that it gives the correct answer for the shape of H_3^+.

We are still using the Born–Oppenheimer approximation, so we wish to know the electronic energy for each possible position of the nuclei. Let us begin with the most symmetric linear structure, that with the nuclei equally spaced, so that we have $R_{AB} = R_{BC}$, $\angle ABC = 180°$, where A, B, C designate the three nuclei (cf. Fig. 8.1). In the LCAO approximation the lowest-energy molecular orbital should be one formed entirely from hydrogen $1s$ orbitals, and with as few nodes as possible. The simplest orbital that satisfies these requirements is

$$\psi_1 = a_1(1s_A) + b_1(1s_B) + c_1(1s_C), \qquad (8.1a)$$

where a_1, b_1, c_1 are constants. We can immediately simplify this further: Given the equal nuclear spacing, the nuclei A and C are exactly equivalent physically, so their atomic orbitals must contribute equally to any molecular orbital. Thus we have $a_1 = c_1$ and

$$\psi_1 = a_1(1s_A + 1s_C) + b_1(1s_B). \qquad (8.1b)$$

As in the diatomic molecule we could form two molecular orbitals from a given pair of atomic orbitals, here we can form three MOs from a given set of three AOs—in this case $1s_A$, $1s_B$, $1s_C$. Since the lowest orbital, ψ_1, has no nodes, we expect the other two to have respectively one and two nodes. Since in our model nuclei A and C are equivalent, the one-node orbital must have its node exactly midway between them, that is, on a plane bisecting the A–C axis. But this nodal plane passes directly through the B nucleus, so the coefficient of $1s_B$ in this orbital must be zero. Thus we can write the one-node orbital as

$$\psi_2 = a_2(1s_A - 1s_C). \qquad (8.2)$$

A suitable and general way to write the two-node orbital built from $1s$ atomic orbitals is

$$\psi_3 = a_3(1s_A + 1s_C) - b_3(1s_B), \qquad (8.3)$$

with the A and C nuclei again equivalent. The values of a_1, b_1, a_3, and b_3 are not independent. We normally require ψ_1 and ψ_3 to be orthogonal,

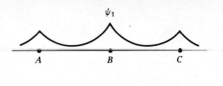

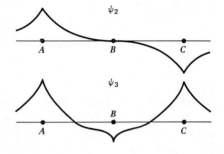

FIGURE 8.1
Orbitals in linear H_3^+ or H_3 (equally spaced, $R_{AB} = R_{BC}$). The graphs show schematically the variation of each orbital along the internuclear axis.

which allows us to reduce the number of coefficients to three; normalization allows us to remove two more. Hence ψ_1 and ψ_3 have only one independent coefficient once the atomic orbitals are chosen. Although we have made no systematic attempt to select the best LCAO orbitals, Eqs. 8.1–8.3 are convenient functions with the correct symmetry and nodal properties. They can be optimized by varying the constants a_i, b_i to minimize the energy. The orbitals ψ_1, ψ_2, ψ_3 are shown schematically in Fig. 8.1.

As in all our previous discussions of orbital models, we obtain the ground state of the molecule as a whole by an Aufbau-principle approach. That is, the electrons are assigned in succession to the one-electron states (orbitals) in accordance with the exclusion principle, up to two electrons, of opposite spin, in each orbital, starting with the lowest energy level and proceeding upward in energy until all the electrons are assigned. Thus the ground-state configuration of the linear H_3^+ ion is $(\psi_1)^2$, with both electrons in the lowest orbital; and the ground state of the linear neutral H_3 molecule is $(\psi_1)^2\psi_2$. The actual molecular wave functions must of course be properly antisymmetrized, with electron spin included, as we explained in Section 6.7. The energy of each of these species can then be calculated in the usual way, by evaluation of the integral $\int \psi^* H \psi \, d\tau$.

Now we can ask how the molecular energy varies with the bond angle ($\theta \equiv \angle ABC$; cf. Fig. 8.2a). Let us assume for simplicity that we continue to have $R_{AB} = R_{BC}$, so that the molecule passes from a linear structure ($\theta = 180°$) into an isosceles triangle and eventually an equilateral triangle ($\theta = 60°$, $R_{AB} = R_{AC}$). We can continue to use the orbitals Eqs. 8.1–8.3: Only the coefficients a_i, b_i and the orbital energies vary as θ changes. What predictions can we make about the energy changes?

Consider first the ground state of H_3^+, in which only the orbital ψ_1 is occupied. Recall our naive concept of bonding force as arising from the constructive interference of electron waves in regions between nuclei. The orbital ψ_1, with all its atomic orbital coefficients positive, exhibits such constructive interference between all three pairs of nuclei. If we bend the molecule, the overlap between the electrons on atoms A and C increases, creating an incipient bond between these two atoms. There is thus a net bonding force that lowers the orbital (and molecular) energy as R_{AC} decreases. For sufficiently small R_{AC}, as in any other bond, the repulsive forces between the nuclei become dominant. Hence an equilibrium structure occurs for the value of θ and R_{AC} at which the energy has its minimum value. Given the symmetry of the problem, it should not surprise us if the equilibrium geometry of H_3^+ is an equilateral triangle, and detailed calculations show that this is indeed the case. Figure 8.2a shows how the potential energy of H_3^+ varies with the bond angle when the bond lengths are fixed. Diagrams such as those in Fig. 8.2, displaying orbital or molecular energies as functions of bond angles, are often called *Walsh diagrams*, after A. D. Walsh, an early exploiter of such diagrams for systematically interpreting molecular structures.

Although the equilateral triangle is the equilibrium structure of H_3^+, the linear arrangement is also one of special symmetry. The reason is that any displacement made by varying θ from linearity has a physically equivalent counterpart on the "other side." That is, bending H_3^+ (or H_3) to reduce θ from 180° to 170° is exactly the same physical operation as increasing θ from 180° to 190°; similarly, any angle $180° - x$ is physically equivalent to $180° + x$. Because of this symmetry, any such molecule's orbital and molecular energies as functions of θ for fixed internuclear distance must be symmetric around 180°, which must thus be an *extremal*

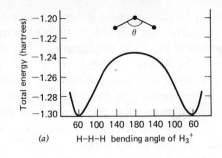

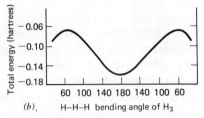

FIGURE 8.2

Potential energy of bending in H_3^+ and H_3. (*a*) Definition of the bond angle θ and $V(\theta)$ in H_3^+. Energy values from R. E. Christofferson, *J. Chem. Phys.* **41**, 960 (1964). H—H bond distance is 1.625 bohrs. (*b*) $V(\theta)$ in H_3. Figures from graphs given by C. W. Eaker and C. A. Parr, *J. Chem. Phys.* **65**, 5155 (1977). H—H bond distance is 1.72 bohrs. Both potential energy curves calculated for fixed bond lengths, with $R_{AB} = R_{BC}$.

point—a maximum or minimum. In the case of H_3^+ we have argued that the extremum should be a maximum, and the calculations used to generate Fig. 8.2*a* confirm this.

The linear arrangement of H_3^+ is thus what would be called a position of *unstable equilibrium* in classical mechanics. At $\theta = 180°$ there is no force tending to bend the molecule, but once θ is even infinitesimally different from 180° the "force"[1] $-\partial V/\partial\theta$ tends to increase the bending. The equilateral triangle, in contrast, is a position of *stable equilibrium*, in that the potential energy has a minimum, any displacement from which creates a restoring force. That the equilibrium position is an *equilateral* triangle (rather than some other isosceles triangle) is not something that we can infer from considerations of symmetry alone. The equilateral arrangement does have a higher symmetry, but symmetry is not the only factor governing molecular geometries. The equilibrium shape is the outcome of the detailed balance among the various interactions (electron–electron, electron–nucleus, and nucleus–nucleus). There are other triatomic molecules in which the three atoms are identical but the equilibrium geometry is only an isosceles triangle; an example is ozone (O_3), for which the equilibrium value of θ is 117°.

Although we were able to deduce easily that H_3^+ should be bent, the neutral H_3 molecule is more problematical. By the same reasoning as before, the linear arrangement must correspond to an extremum in the energy—but an extremum of which kind? The ground-state electronic configuration is $(\psi_1)^2\psi_2$, with a molecular energy crudely given by $2\epsilon_1 + \epsilon_2$. (Equation 8.26 will give a much more accurate expression.) As in H_3^+, the energy of ψ_1 decreases with bending, because ψ_1 is bonding between atoms A and C. But ψ_2 is antibonding between A and C, having a node midway between these atoms; hence the energy of ψ_2 goes *up* when the A–C distance is reduced. Whether the energy of the whole molecule goes up or down upon bending depends on whether the increase in ϵ_2 is greater or less than twice the decrease in ϵ_1 ("twice," because of the two electrons in ψ_1). But we cannot deduce this by simple arguments; one must actually carry out the energy calculations. Detailed calculations in fact show that the total energy increases as θ deviates from 180°, so that the linear structure is the equilibrium geometry for H_3. The potential energy of H_3 as a function of θ is shown in Fig. 8.2*b*.

However, the important point here is not the result itself, but what it shows us about the limitations of various approaches. In the case of H_3^+ we could infer the structure correctly from qualitative considerations. In H_3, such a simple approach does not suffice, and either measurements or elaborate calculations are necessary to learn the true structure. (Indeed, the molecular orbital model itself does not suffice. One must take electron correlation into account to obtain a conclusive result.) One who studies molecular structures must become skilled in judging which of these kinds of situations applies in a given case. We shall see that simple MO considerations, combined with some generalizations based on experiment, often do provide a powerful base for inferring molecular structures—but that one must be careful to avoid applying the method beyond its range of validity.

---

[1] Although $-\partial V/\partial\theta$ does not have the dimensions of a true force, it can be considered the *generalized force* associated with θ. If a mechanical system is completely described by a set of variables q_i (distances, angles, etc.), then the work performed in an infinitesimal displacement can always be written in the form $\sum_i Q_i\,dq_i$, where the Q_i are generalized forces. For systems whose energy is conserved (those for which a potential energy can be defined) this means that $Q_i = -\partial V/\partial q_i$, and there is an effective "force" tending to lower the value of V.

Thus far we have considered only the case $R_{AB} = R_{BC}$, but a complete determination of the molecular geometry must include the dependence of the energy on bond lengths as well as on bond angle. Such calculations have also been made. The results are similar to those we found for diatomic molecules. For each bond length the energy becomes infinite as $R \to 0$, levels off as $R \to \infty$, and may have a minimum somewhere between. For a given bond angle, the energy as a function of R_{AB} and R_{BC} can best be represented by a "contour map"; Fig. 8.3 is such a map for linear neutral H_3. (In Part Three we shall see how such diagrams are used in the study of chemical kinetics.) Note that $R_{AB} = R_{BC}$ is not any kind of equilibrium position, but rather a *saddle point* (named after the shape of the potential energy surface): $V(R_{AB}, R_{BC})$ decreases as one goes down into either "valley," that is, as the H_3 molecule splits into $H_2 + H$. This is consistent with the observation that H_3 does not exist as a stable species.

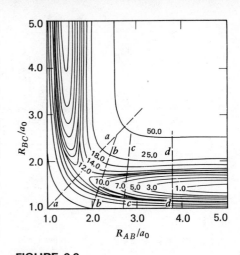

FIGURE 8.3
Contour map for the potential energy surface of linear H_3. The energy contours are labeled in kilocalories per mole (1 eV = 23.06 kcal/mol), relative to $H_2 + H$ at infinite separation. The calculated saddle point is at $R_{AB} = R_{BC} = 1.765a_0$ (0.934 Å), $V(R_{AB}, R_{BC}) = 11.35$ kcal/mol.

We have used the term "symmetry" a number of times in this section. Symmetry is extremely useful in analyzing the properties of polyatomic molecules, and we must say something more about the subject. The symmetry properties of a system are constraints, additional knowledge about a system beyond what we know about the "ordinary" unconstrained system. Consequences of symmetry—the theorems of *group theory*—often provide us with the tools to make complicated problems much easier, making the difference between a problem that can be solved with pencil and paper and a problem requiring a large computer. We shall not develop any of these theorems here; we shall only provide enough illustrations of spatial symmetries to allow us to draw some simple, qualitative (but mathematically correct) inferences from the structures of some symmetrical molecules.

Of the two kinds of symmetry, permutational and spatial, that were discussed in Section 7.5, it is clearly spatial symmetry with which we are concerned here. By the symmetry properties of a given molecular structure we mean the spatial transformations one can perform on the molecule (or, equivalently, on the coordinate system to which it is referred) that leave the molecule in a condition physically indistinguishable from its initial state. Let us see what these properties are for an H_3 molecule.

Suppose that we restrict ourselves to the case that $R_{AB} = R_{BC}$. There are three possible H_3 structures with at least two equivalent hydrogen atoms: isosceles triangle, equilateral triangle, and linear. We have listed these in what we say is the order of increasing symmetry, on the following basis. The isosceles triangle, as shown in Fig. 8.4a, could be transformed in three ways without our being able to detect that the transformation had been made: rotation by 180° about an axis through the central H atom, reflection in a plane passing through the same atom (and equidistant between the other two), and reflection in the plane of the molecule. In the equilateral triangle, Fig. 8.4b, all three atoms are equivalent, so we have a 180° rotation axis and a transverse reflection plane through *each* of them. The plane of the molecule remains a symmetry plane, as it must be in all triatomic molecules; in addition we have a threefold symmetry axis perpendicular to this plane, passing through the center of the triangle. Thus the equilateral triangle has a higher symmetry (more symmetry operations) than the isosceles triangle. But the linear structure, Fig. 8.4c, has even greater symmetry, indeed, an infinite number of symmetry operations: *Any* plane including the bond axis is a symmetry plane; *any* axis through the central atom perpendicular to the bond axis is a twofold symmetry axis; and the bond axis itself is an ∞-fold symmetry axis, in that rotation through *any* angle about it leaves the molecule unchanged. There is also a single symmetry plane perpendicular to the bond axis, and the central atom is a center of symmetry through which the coordinates of all the particles in the molecule can be reflected. In fact, the linear H_3 molecule has all the symmetry properties of the diatomic molecule.

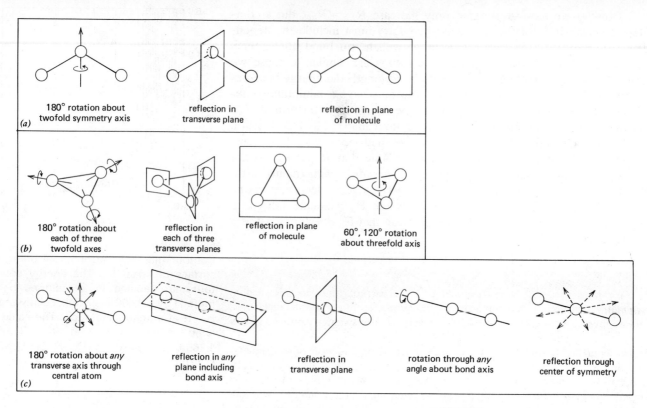

180° rotation about
twofold symmetry axis
(a)

reflection in
transverse plane

reflection in plane
of molecule

180° rotation about
each of three
(b)   twofold axes

reflection in
each of three
transverse planes

reflection in plane
of molecule

60°, 120° rotation
about threefold axis

180° rotation about *any*
transverse axis through
central atom
(c)

reflection in *any*
plane including
bond axis

reflection in
transverse plane

rotation through *any*
angle about bond axis

reflection through
center of symmetry

FIGURE 8.4
Symmetry operations of the H_3 molecule for various structures. (a) isosceles triangle; (b) equilateral triangle; (c) linear.

How does symmetry allow us to apply constraints to the solution of the quantum mechanical problem? These constraints often sharply reduce the amount of calculation required, and in some cases—those of continuous symmetry, such as rotation about the axis of a linear molecule—allow us to infer constants of the motion, quantum numbers, and even energy-level schemes without having to solve complicated equations explicitly. As we have mentioned several times, the conservation of linear and angular momentum can be interpreted as consequences of the translational and rotational symmetry of physical systems. Because the physically measurable properties of an isolated system are unchanged by any translation or rotation of external coordinates, the momenta conjugate to those coordinates must be conserved. We have seen a similar effect of continuous symmetry in the internal coordinates, with cylindrical symmetry (in diatomic and other linear molecules) leading to conservation of one component of angular momentum. *Point* symmetry, such as that of the equilateral triangle, is not powerful enough to force the constancy of any dynamical quantity, but it does place a limitation on the possible wave functions, which must have forms that keep the physically equivalent sites indistinguishable. For example, we were able to make the inference that the linear configuration of a triatomic molecule A—B—C must correspond to either a maximum or a minimum energy with respect to changes in θ. This deduction is based on the observation that all directions in which the molecular A—B—C angle can bend are equivalent, which is a way of saying that the linear triatomic molecule has symmetry higher than that of the bent triatomic. If our triatomic molecule is homonuclear, the equilateral triangle is also a geometry of high symmetry and therefore its energy must be a maximum or minimum or *saddle point*, a point that is a minimum along one direction and a maximum along a perpendicular direction, like the center of a saddle or the top of a mountain pass. More general inferences concerning the forms of orbitals, of vibrational motions, or of magnetic interactions among atomic nuclei, for example, can be drawn from knowledge (or supposition) of the symmetries of molecules or solids.

Another sort of inference concerns what sorts of wave functions can be coupled by a physical process represented by an operator. If ψ_A and ψ_B are the wave functions of states A and B and R is the operator, then the process corresponding to R can couple states A and B if the corresponding integral $I = \int \psi_A {}^* R \psi_B \, d\tau$ does not vanish. But this, in turn, puts a condition on the integrand, $\psi_A {}^* R \psi_B$: For I to differ from zero, it must not have equivalent positive and negative parts that cancel one another. But in systems of relatively high symmetry, such as an atom or a small molecule, for which atomic or molecular orbitals are natural functions, the integrands $\psi_A {}^* R \psi_B$ themselves often do have cancelling parts, such as the equivalent positive and negative regions of the $1\sigma_u$, $1\pi_u$, $2\sigma_u$, and $1\pi_g$ functions of Fig. 6.8. These integrands give vanishing integrals I. When an integral between ψ_A and ψ_B is zero for general reasons of symmetry, one says that a *selection rule* prohibits the coupling of states A and B. By contrast, integrands with forms such as that of the $1\sigma_g$ or $2\sigma_g$ orbitals of Fig. 6.8 give rise to nonzero integrals, in general, because they do not have equivalent cancelling parts. Such integrands are said to be totally symmetric or invariant, meaning that they are unchanged, even in sign, if the coordinate system or the orientation of the molecule is transformed by any of the symmetry operations (rotation, reflection, etc.) that leave the molecule itself indistinguishable from what it was prior to the transformation. If ψ_A and ψ_B give rise to a nonzero integral with the operator R, then we say that the R-type coupling of states A and B is allowed. For example, the electric dipole coupling of vibrational states of a heteronuclear diatomic molecule is allowed for states separated by one vibrational quantum, but is forbidden for the corresponding states of a homonuclear diatomic molecule. The coupling of vibrational states separated by two quanta in a heteronuclear diatomic molecule is also forbidden in the approximation that the molecule is harmonic, but this rule does not apply for real anharmonic molecules, as Fig. 7.6b shows. In Section 8.8 we shall make use of the symmetry properties of the H_2O molecule to determine the form of its molecular orbitals.

8.2
Dihydrides: Introduction to the Water Molecule

Now we consider the next major step in molecular complexity beyond H_3, the triatomic molecules of the form AH_2. These dihydrides are all symmetrical, with the A atom in the center; some are linear and some are bent. Many of the dihydrides are known only as transient, unstable, or at least very reactive species. Among these are BH_2, CH_2, and NH_2, which have been detected only by spectroscopy, mass spectrometry, or the appearance of specific products testifying to their presence during a chemical reaction. The alkaline earth metals form ionic hydrides (CaH_2, etc.) much like those of the alkalis, and some other metals have dihydrides in the solid states, but we are not concerned with these here. However, the elements of one family of the periodic table do form stable, covalently bound dihydrides. These are, of course, the well-known series H_2O, H_2S, H_2Se, H_2Te (and H_2Po, which has been prepared in only minute quantities). Our discussion here will concentrate on the water molecule, H_2O, which is both the simplest member of the series and by far the most important to chemistry.

Water, as we all know, is a liquid between 0 and 100°C at atmospheric pressure. The other dihydrides in the series are all evil-smelling, toxic gases under the same conditions. At least one important reason for the differences in the volatility of these species is the tendency of H_2O molecules to form hydrogen bonds with one another. This property is also found in HF (cf. Section 7.9) and in NH_3, but not to the extent observed in water. In solid water (ice) the strength of the hydrogen bonds is sufficient to create a completely ordered crystal structure: Each oxygen atom is surrounded by a tetrahedron of four other oxygen atoms; one hydrogen atom lies between each pair of oxygens, joined by a normal covalent bond to one and a hydrogen bond to the other (O—H···O). Much of the order of this structure is retained in liquid water. We shall discuss the structure of water in more detail in Chapters 23 and 26.

How is it that H_2S, H_2Se, and H_2Te are so much less likely to form hydrogen bonds? As usual in trends within a family of the periodic table, the effect is basically one of atomic size, both directly and as reflected in electronegativity: As the size of the central atom (A) increases, its nucleus becomes better shielded by the core electrons and the atom becomes less electronegative. Thus from H_2O to H_2Te the H—A bond becomes less ionic, the electron density around the H atom greater, and the H nucleus itself better shielded. Being less like a bare proton than in H_2O, the H atom is less likely to bond to another atom. In addition, with increasing size of A both H—A and H···A bonds necessarily become longer and thus weaker.

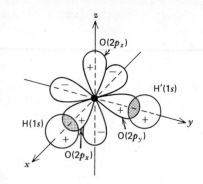

FIGURE 8.5
Localized-bond picture of H_2O. The two bonds are formed by overlap (indicated by shading) of the oxygen $2p_x$ and $2p_y$ orbitals with one hydrogen 1s orbital each. The oxygen $2p_z$ orbital is nonbonding, as are the oxygen 1s and 2s orbitals (not shown).

Let us now turn to the electronic structure of the water molecule. The molecule has 10 electrons, one from each of the hydrogen atoms and eight from the oxygen atom, which has the ground-state electronic configuration $1s^2 2s^2 2p^4$. It is thus isoelectronic with the neon atom ($1s^2 2s^2 2p^6$) and the hydrogen fluoride molecule. The HF molecule (Section 7.9) is much like neon. Its configuration is $1\sigma^2 2\sigma^2 3\sigma^2 1\pi^4$, but most of the orbitals are essentially the same as in the fluorine atom. Only the 3σ bonding orbital is primarily a mixture of hydrogen and fluorine orbitals, so that the configuration of HF can be approximated as[2]

$$(1s_F)^2 (2s_F)^2 (2p\pi_F)^2 (2p\sigma_F + 1s_H)^2.$$

One might expect the H_2O molecule to have a closely related structure. Specifically, it is reasonable to suppose that two equivalent O—H σ bonds are formed when different 2p orbitals of oxygen, say, the $2p_x$ and $2p_y$, combine with the 1s orbitals of the two hydrogen atoms (Fig. 8.5). The configuration of H_2O in terms of *localized* bonding orbitals would then be

$$(1s_O)^2 (2s_O)^2 (2p_{z,O})^2 (2p_{x,O} + 1s_H)^2 (2p_{y,O} + 1s_{H'})^2$$

(by "localized" we mean that each bond is described by an orbital of its own).

The assumption we make in writing this description of H_2O is that the 1s, 2s, and $2p_z$ orbitals of oxygen are nonbonding. How reasonable is this? The 1s electrons are certainly deep in the core and thus out of the picture. The same is largely true of the 2s orbital. The binding energy of an O(2s) electron is 28.5 eV, compared with 13.6 eV for O(2p), whereas the O—H bond energy in H_2O is only about 5 eV; thus the 2s orbital is too deep to take much part in the bonding. The $2p_z$ orbital, on the other hand, is essentially nonbonding since it has a nodal plane in the plane of the molecule (Fig. 8.5), and thus provides very little charge density in the O—H bonding regions.

The angle between the O—H bonds in this localized-bond model would be expected to be 90°, corresponding to the right angle between any two of the oxygen p orbitals. But spectroscopic measurements that

---

[2] The notation $2p\sigma_F + 1s_H$ implies only some linear combination of the two orbitals, in general not an equal mixture. A more exact notation for the orbital would be

$$\varphi = c_1(2p\sigma_F) + c_2(1s_H),$$

with the constants c_1, c_2 varied as usual to minimize the energy.

determine the moments of inertia show that the bond angle in H_2O is actually 104.52°. Although this is not far from 90°, it is surely enough to make us ask about the cause of the deviation. One can rationalize the larger angle by considering the interaction between the two bonds. The electrons in one bond repel those in the other, and the two hydrogen nuclei also repel each other; this repulsion should cause the bond angle to be somewhat larger than in the pure localized-bond model. The attraction between the electrons in one bond and the proton of the other bond acts in the opposite direction, but the proximity of the electron clouds makes the repulsion dominate, and the net effect should tend to increase the angle. This model is intuitively pleasing, and can even be quantified with sufficiently elaborate calculations. It also lends itself to the kind of argument that rationalizes experimental observations such as the value of the dipole moment, for example. But the method requires skill and judgment for effective application, and it is not very useful for obtaining more information than one puts in. We shall see that there are better ways to go beyond the localized-bond model.

Although the assumption of simple localized $p\sigma$ bonds is thus not quite adequate for the water molecule, it does a better job of accounting for the structures of the other molecules in the series. As the size of the central atom increases, the bond angle becomes closer to the predicted 90° value. This can again be interpreted as a size effect. The bonds are longer and thus farther apart, so that the interbond repulsion just described becomes less important. The following table illustrates these trends:

Molecule (AH_2)	H_2O	H_2S	H_2Se	H_2Te
A–H bond length (Å)	0.9572	1.33	1.46	1.69
Bond angle (deg)	104.52	92.3	91.0	89.5

8.3 Hybrid Orbitals

To improve the simple localized-bond model, one commonly uses what are called *hybrid orbitals*—atomic orbitals that are mixtures (linear combinations) of the orbitals used in describing isolated atoms. The hybrid atomic orbitals can then be combined in the usual way to construct molecular orbitals, usually still of the localized variety corresponding to individual bonds. We mentioned this method previously in connection with Eq. 7.49; now we shall see how it works.

In a systematic, rigorous theory, it is difficult to find a place for hybridized bond orbitals. The more straightforward course is to construct delocalized molecular orbitals extending over the entire molecule; this is what we did for H_3, and in the next section we shall outline a similar method for H_2O. But in such calculations one loses sight of the intuitively appealing concept of the isolated chemical bond. Like an individual orbital, an individual bond in a polyatomic molecule has no "real" existence; yet the concept of the individual bond, with moderately well-defined characteristics of its own, is too useful to chemistry to be abandoned. In particular, experience shows that many molecular properties can be well represented by adding contributions associated with individual bonds, bond energies, for example; cf. Section 14.8. Thus there is still a useful role for a theory involving localized bond orbitals. More advanced methods of this type can be developed as extensions of either the MO or the VB method, with the localized orbitals chosen subject to the constraint that individual orbitals have as little exchange interaction with one another as possible. The construction of hybrid orbitals is in effect a way of approximating and short-cutting these approaches, by making educated guesses as to the atomic orbital contributions to the localized molecular orbitals. In practice this is rather effective, provided that the molecular geometry is known.

For an example of the simplest type of hybrid orbitals, let us consider a linear molecule of the form AH_2, where A is a first-row atom. (There may be no stable example of this species: It used to be thought that the ground state of CH_2 was linear, but CH_2 is now known to be bent. However, we shall see that the orbitals introduced here are useful in more complicated molecules.) We wish to generate two collinear A—H bonding orbitals, each made up of a H(1s) orbital and some orbital of the A atom. If we use one of the three A(2p) orbitals, the other two have their maximum densities at right angles to the internuclear axis and cannot take part in the bonding; thus to make a pair of bonds we must also call upon the A(2s) orbital. We could say that the 2s orbital forms one bond and the 2p orbital the other, but this would be unrealistic. Such bonds would have different properties, and experiment shows that the bonds in any AH_2 molecule are identical. Thus we take the next step and say that each bond contains contributions from both A(2s) and A(2p), that is, that the A orbitals taking part in the bonds are hybridized. This is not unreasonable if the 2s and 2p orbitals have fairly similar binding energies.

How do we construct such hybrid orbitals? Although any linear combination of 2s and 2p might do, let us take for simplicity the case of an equal mixture. The simplest such orbitals that satisfy the requirements of equivalence and collinearity are

$$\varphi_+ = \frac{1}{\sqrt{2}}(2s + 2p) \quad \text{and} \quad \varphi_- = \frac{1}{\sqrt{2}}(2s - 2p), \quad (8.4)$$

where "2s" and "2p" stand for the ordinary atomic orbitals. The two hybridized orbitals are identical except in orientation, as can be seen in Fig. 8.6. They are strongly localized on opposite sides of the atom, with their greatest extensions lying exactly 180° apart. Thus each can combine with a H(1s) orbital, to produce two collinear A—H bonds.[3] Since the orbitals φ_+ and φ_- contain equal contributions from s and p orbitals, they are called sp hybrids.

We see here the characteristic directionality that motivates us to use hybrid orbitals. This property makes them especially useful in interpreting stereochemistry. Frequently a particular atom in a molecule forms several bonds symmetrically distributed in space. In such cases one can always define a set of hybrid orbitals projecting in the bond directions; however accurate they may be otherwise, they at least give the correct geometry. To construct such orbitals, one must in general start with as many conventional atomic orbitals as the number of hybrids desired.

Consider next the case of three equivalent orbitals. The BH_3 molecule is known as a reactive transient species (detected by mass spectrometry); although its structure has never been determined experimentally, theoretical calculations point to its having a planar, equilateral-triangle geometry. We thus wish to construct three equivalent boron orbitals whose axes point to the vertices of an equilateral triangle. We can use the 2s orbital and two different 2p orbitals, say, $2p_x$ and $2p_y$ (defining the xy plane as the plane of the molecule). The resulting hybrid orbitals are called trigonal or sp^2.

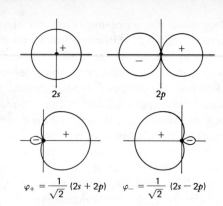

FIGURE 8.6
Construction of sp hybrid orbitals. The diagrams represent the angular parts of the wave functions, shown as in Fig. 4.6.

---

[3] Or with suitable orbitals of other atoms; $MgCl_2$ is one example of a stable linear AX_2 molecule for which sp hybridization should give a good description.

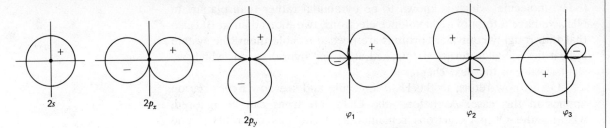

It is instructive to go through the construction of trigonal orbitals. To begin with, if the orbitals are equivalent, all three must have the same proportions of s- and p-orbital characteristics. Thus each should have an s–p ratio of 1:2 in the electron density, so that the ratio of s to p amplitudes must be $1:\sqrt{2}$ in the wave function itself. Similarly, since each orbital contains one-third of the total s-orbital contribution, the $2s$ coefficient in each must have an absolute value of $1/\sqrt{3}$ for normalization. Combining this information, we can immediately write one of the hybrid orbitals as

$$\varphi_1 = \frac{1}{\sqrt{3}}(2s) + \frac{\sqrt{2}}{\sqrt{3}}(2p_x). \tag{8.5}$$

We are left with the $2p_y$ orbital and the remaining third of the $2p_x$ orbital to make up the p contribution in the still-undefined hybrids φ_2 and φ_3. If each is divided evenly between the two hybrids, the coefficients of $2p_x$ and $2p_y$ should have absolute values of $1/\sqrt{6}$ and $1/\sqrt{2}$, respectively. It remains to determine the signs of these coefficients. If φ_1 points along the positive x axis, then by symmetry φ_2 and φ_3 must both extend toward negative x, one above and the other below the x axis. Thus both $2p_x$ coefficients should be negative, whereas the $2p_y$ coefficients should have opposite signs. We can thus write the remaining hybrids as

$$\varphi_2 = \frac{1}{\sqrt{3}}(2s) - \frac{1}{\sqrt{6}}(2p_x) + \frac{1}{\sqrt{2}}(2p_y) \tag{8.6}$$

and

$$\varphi_3 = \frac{1}{\sqrt{3}}(2s) - \frac{1}{\sqrt{6}}(2p_x) - \frac{1}{\sqrt{2}}(2p_y). \tag{8.7}$$

The trigonal orbitals of Eqs. 8.5–8.7 are illustrated in Fig. 8.7. It can be seen that they are indeed equivalent and symmetrical, with their long axes 120° apart.

The localized-bond representation of the BH_3 molecule is obtained by combining each of the trigonal boron orbitals with a H($1s$) orbital, to obtain a bond orbital of the form

$$\psi_i = \alpha(\varphi_{i,B}) + \beta(1s_H). \tag{8.8}$$

The constants α and β are unequal, simply because the electronegativities of boron and hydrogen are unequal. Just as in heteronuclear diatomic molecules, these localized bonds are somehat polar. Although BH_3 itself is unstable with respect to B_2H_6 or the elements, the bonding is similar in the boron trihalides and their analogs (BF_3, $AlCl_3$, etc.) and some other compounds: In each case the observed plane-triangular structure can be described in terms of sp^2-hybridized orbitals on the central atom. On the other hand, one must not use plane trigonal orbitals for the ammonia

(NH$_3$) molecule, which is known to be pyramidal rather than planar. In NH$_3$ we have a total of eight valence electrons, two more than in BH$_3$, so that four roughly equivalent orbitals are needed to hold them; the system is best described in terms of tetrahedral or sp^3 hybridization, which we shall discuss in the next chapter.

Let us now return to the H$_2$O molecule and see how hybridization applies in this case. As before, the O($1s$) electrons can be ignored, whereas the O($2p_z$) electrons remain out of the molecular plane and nonbonding. This leaves us with six electrons, four from the oxygen atom and one from each of the hydrogens, which we wish to assign to three orbitals. Suppose that we assume trigonal hybridization on the oxygen atom, combining the $2s$, $2p_x$, and $2p_y$ orbitals to obtain three equivalent sp^2 orbitals. Then two of these orbitals can combine with H($1s$) orbitals to form the O—H bonds, whereas the third contains a nonbonding "lone pair" of oxygen electrons. There is one major difficulty with this interpretation: it again predicts the wrong bond angle. Bonds formed with sp^2 hybrids would be 120° apart, whereas the pure p bonds discussed in the previous section would be 90° apart. Since the bond angle is actually 104.5°, the bonding is apparently best described by something midway between the sp^2 and pure-p models.

There is a way to handle this problem within the hybrid orbital model, since we are not restricted to equivalent hybrids. The bond angle obviously increases with the amount of s-orbital character (pure p 90°, sp^2 120°, sp 180°). Suppose, then, that the two bond-forming hybrids are intermediate between pure p and sp^2, say, about 20% s and 80% p. The nonbonding hybrid would then have to be 60% s and 40% p to make the sums come out right; this is reasonable, since we expect the more strongly bound $2s$ electrons to take a smaller part in the bonding than the $2p$ electrons. By varying the s–p ratio, one can adjust the bond angle to exactly match the observed value; the percentages just stated happen to be about right. But we have apparently extracted more information from the model only by feeding more information in; the next section will show us a better way to analyze the H$_2$O molecule.

We have not yet exhausted the varieties of hybridization. Besides the hybrids of s and p orbitals that we have discussed, one can also construct hybrids with contributions from d orbitals. These are useful in analyzing the bonding of the larger atoms, especially the transition metals. With suitable orbital combinations one can reproduce a wide range of molecular geometries. We shall have more to say on this subject in the next chapter.

8.4
Delocalized Orbitals in H$_2$O; the General MO Method

An alternative method of analyzing the electron distribution in polyatomic molecules involves the use of *delocalized* molecular orbitals extending over the entire molecule (or some large section thereof) rather than orbitals localized in individual bonds. In this model it is assumed that each electron responds to the field of the entire set of nuclei and other electrons; the one-electron orbitals are the solutions of the one-electron Schrödinger equation corresponding to this field. This approach has become increasingly common with the development of large electronic computers. In particular, it has been used in almost all extensive *ab initio* calculations on the electronic structure of H$_2$O. The first such calculations that included all the occupied orbitals were those of F. O. Ellison and H. Shull, completed in 1955; we shall refer to their results in the following illustrative discussion.

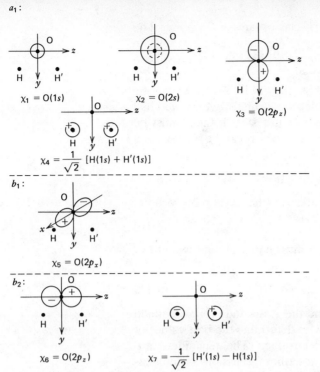

$\chi_1 = O(1s)$

$\chi_2 = O(2s)$

$\chi_3 = O(2p_z)$

$\chi_4 = \dfrac{1}{\sqrt{2}}[H(1s) + H'(1s)]$

$\chi_5 = O(2p_x)$

$\chi_6 = O(2p_z)$

$\chi_7 = \dfrac{1}{\sqrt{2}}[H'(1s) - H(1s)]$

FIGURE 8.8
Basis set of orbitals for the H_2O molecule, classified by symmetry types (a_1, b_1, b_2). All orbitals except the $O(2p_x)$ are shown in cross section in the plane of the molecule; the $O(2p_x)$ orbital extends perpendicular to the molecular plane.

As usual we construct the molecular orbitals by taking linear combinations of a *basis set* of orbitals. For a basis set in the H_2O molecule we could use the fundamental s and p atomic orbitals, the hybrid orbitals described in the last section, or some other combination of localized bond orbitals. The net result would be the same by any of these ways, provided that one carried out full calculations to obtain the best delocalized orbitals possible. Accepting this, we may just as well use the basic atomic orbitals themselves: the $1s$, $2s$, and three $2p$ oxygen orbitals directly, and two orbitals (χ_4 and χ_7) representing the sum and difference of the two $H(1s)$ orbitals. These basis orbitals are shown in Fig. 8.8.

We want both the basis set and the eventual molecular orbitals to be *symmetry orbitals*, orbitals that share some of the symmetry properties of the molecule. The basis orbitals in Fig. 8.8 are thus classified into three symmetry classes, called a_1, b_1, and b_2. The labels indicate properties thus: An a orbital remains unchanged (i.e., looks exactly the same) when the molecule is rotated 180° about the y axis, whereas a b orbital changes sign under this rotation. Similarly, and orbital with subscript 1 remains unchanged when the molecule is reflected in the xy plane, whereas one with subscript 2 changes sign in this operation. Since the molecule (and thus $|\psi|^2$) remains unchanged under these operations, the wave functions must either remain unchanged or reverse their signs. As can be seen in the figure, the yz plane is defined as the plane of the molecule, and the xy plane is the symmetry plane midway between the two hydrogen atoms; the origin is at the oxygen nucleus, with the hydrogen atoms at $y > 0$.

The significance of this classification is that symmetry orbitals of a given class can mix only with others of the same class to generate *molecular symmetry orbitals* of that class. In Section 8.1, we outlined the geometric basis of this statement, and we shall not attempt to prove it in any more detail. We earlier assumed the validity of this theorem for diatomic molecules, where the symmetry classes are σ_g, π_u, etc. The noncrossing rule is one of its consequences. We shall now see how, in polyatomic molecules, the use of symmetry orbitals makes possible a considerable simplification in our calculations.

The seven independent basis orbitals shown in Fig. 8.8 can be combined to give seven independent molecular orbitals, which we shall designate as φ_i. According to the mixing rule just stated, there must be four orbitals of the symmetry class a_1, each of which can be written as a linear combination of the a_1 basis orbitals,

$$\varphi_i = c_{i1}\chi_1 + c_{i2}\chi_2 + c_{i3}\chi_3 + c_{i4}\chi_4 \qquad (i = 1, 2, 3, 4); \qquad (8.9)$$

one is of the class b_1, which must be identical to the oxygen $2p_x$ orbital,

$$\varphi_5 = c_{55}\chi_5 \qquad (8.10)$$

(although we include c_{55} for consistency, it must equal 1); and two are of the class b_2,

$$\varphi_i = c_{i6}\chi_6 + c_{i7}\chi_7 \qquad (i = 6, 7). \qquad (8.11)$$

In these equations the c_{ij} are constants and the χ_j are the orbitals defined in Fig. 8.8. The computational task is now to determine the best values of the c_{ij} and the corresponding energy eigenvalues. The techniques are the standard ones we have described in our earlier discussions of molecular wave functions, but it may be worthwhile to review the process.

First we must define the molecular (many-electron) wave function. We have seven molecular orbitals and only 10 electrons to assign to them, so we must specify a configuration. From spectroscopy one can deduce that in the ground state of H_2O three a_1 orbitals, one b_1, and one b_2 are occupied by two electrons each. Let us define these occupied orbitals to be the φ_1, φ_2, φ_3, φ_5, and φ_6. To each of these orbitals we assign one electron with spin function α ($m_s = +\frac{1}{2}$) and one with spin function β ($m_s = -\frac{1}{2}$). We designate a spin orbital (product of spatial and spin functions) with α spin by φ_i alone, the corresponding spin orbital with β spin by $\bar{\varphi}_i$ (with a superscript bar). The total wave function corresponding to a given assignment of electrons to orbitals could then be a Hartree product like Eq. 5.15, for example,

$$\psi = \varphi_1(1)\bar{\varphi}_1(2)\varphi_2(3)\bar{\varphi}_2(4)\varphi_3(5)\bar{\varphi}_3(6)\varphi_5(7)\bar{\varphi}_5(8)\varphi_6(9)\bar{\varphi}_6(10) \quad (8.12)$$

where the numbers in parentheses designate electrons: $\bar{\varphi}_2(4)$ is the wave function of electron 4 in spatial orbital 2 with β spin.

But Eq. 8.12 is not suitable for a molecular wave function. As we showed in Section 6.7, the indistinguishability of electrons requires that the total wave function be antisymmetric under interchange of any pair of electrons. We must thus include in the wave function all the possible products like Eq. 8.12. There are 10! ($= 3,628,000$) such products,[4] each of which can be constructed from Eq. 8.12 by a series of permutations (exchanges) of pairs of electrons. To make the total wave function antisymmetric, we must put a *minus* sign before each product obtained by an *odd* number of permutations, a *plus* sign before each product obtained by an *even* number of permutations, and add all the results together. Fortunately for the simplicity of our notation, this unwieldy sum of products is identical to the expansion of the 10×10 determinant $\|\varphi_i(k)\|$, in which the rows (i) are indexed according to the spin orbital and the

---

[4] Any of the 10 electrons can be assigned to φ_1; for each of these choices any of the remaining nine can be assigned to $\bar{\varphi}_1$; and so forth, giving a total number of assignments $10! \equiv 10 \cdot 9 \cdot 8 \cdot \cdots \cdot 2 \cdot 1$.

columns (k) according to the label on the electron. The total molecular wave function thus becomes

$$\psi = (10!)^{-1/2}\mathbf{A}\varphi_1(1)\cdots\bar{\varphi}_6(10) = (10!)^{-1/2}\det\varphi_1(1)\cdots\bar{\varphi}_6(10)$$

$$= (10!)^{-1/2}\begin{vmatrix} \varphi_1(1)\varphi_1(2)\cdots\varphi_1(10) \\ \bar{\varphi}_1(1)\bar{\varphi}_1(2)\cdots\bar{\varphi}_1(10) \\ \cdots\cdots\cdots\cdots\cdots \\ \cdots\cdots\cdots\cdots\cdots \\ \bar{\varphi}_6(1)\bar{\varphi}_6(2)\cdots\bar{\varphi}_6(10) \end{vmatrix}, \qquad (8.13)$$

where $\mathbf{A}$ is the antisymmetrization operator and the factor $(10!)^{-1/2}$ normalizes the wave function.[5] This is of the same form as the general many-electron wave function we wrote as Eq. 6.63, and is called a *Slater determinant*.

Although we thus have a formal expression for the total wave function in terms of the $\varphi_i(k)$, we still do not know the orbitals themselves. As usual, this problem is solved by applying the variation principle. In accord with Eq. 5.11, we wish to find the orbitals that minimize the energy $\int\psi^*\mathbf{H}\psi\,d\tau$, where ψ is given by Eq. 8.13. The integration is carried out over all 10 sets of electron coordinates, with the electronic Hamiltonian

$$\mathbf{H} = \sum_{k=1}^{10}\frac{\mathbf{p}_k^2}{2m_e} - \sum_{K=1}^{3}\sum_{k=1}^{10}\frac{Z_K e^2}{4\pi\epsilon_0 r_{Kk}} + \frac{1}{2}\sum_{k=1}^{10}\sum_{l\neq k}\frac{e^2}{4\pi\epsilon_0 r_{kl}}, \qquad (8.14)$$

where the index K is used for nuclei and the indices k, l for electrons. (We assume the Born–Oppenheimer approximation, with the nuclei fixed in some chosen geometry.)

When the wave function is of the form of Eq. 8.13, one can show that the orbitals that minimize the total energy are the *Hartree–Fock orbitals*. These are defined as the solutions of the equations

$$\mathbf{H}_{HF}(1)\varphi_i(1) = \epsilon_i\varphi_i(1) \qquad (i = 1, 2, \ldots, 10), \qquad (8.15)$$

where $\mathbf{H}_{HF}$ is the Hartree–Fock effective Hamiltonian defined below, φ_i is a one-electron orbital, ϵ_i is the corresponding energy eigenvalue ("orbital energy of φ_i"), and the argument "1" of $\mathbf{H}_{HF}$ and φ_i indicates the coordinates of a single electron. The operator $\mathbf{H}_{HF}(1)$ is defined by

$$\mathbf{H}_{HF}(1)\varphi_i(1) = \left\{\frac{\mathbf{p}_1^2}{2m_e} + \sum_{K=1}^{3}\frac{Z_K e^2}{4\pi\epsilon_0 r_{K1}} + \frac{1}{2}\sum_j\left[\int\int\varphi_j^*(2)\frac{e^2}{4\pi\epsilon_0 r_{12}}\varphi_j(2)\,d\tau_2\right]\right\}\varphi_i(1)$$

$$-\frac{1}{2}\sum_j\left[\int\int\varphi_j^*(2)\frac{e^2}{4\pi\epsilon_0 r_{12}}\varphi_i(2)\,d\tau_2\right]\varphi_j(1); \qquad (8.16)$$

the sums in j are taken over all the occupied orbitals, unbarred (spin α) and barred (spin β), and the argument "2" designates the coordinates of a second electron over which the two kinds of integrals in square brackets are taken. The first two terms on the right-hand side of Eq. 8.16 are simply the kinetic energy of electron 1 and its Coulomb interaction with the nuclei. The remaining two sums give, respectively, the Coulomb and exchange interactions of electron 1 with all the other electrons. Each of these sums contains a term, with $j = i$, representing the electron's interaction with itself, but these two terms cancel each other; they are retained to assure that the operator $\mathbf{H}_{HF}$ is identical for all the orbitals, which are thus all part of the same set of solutions to Eq. 8.15. Note that the jth

---

[5] If all the $\varphi_i(j)$ are orthonormal (as we assume), each product like Eq. 8.12 would contribute unity to $\int|\psi|^2\,d\tau$; since there are 10! such products in our total wave function, we must divide $|\psi|^2$ by 10!, ψ by $(10!)^{1/2}$, to obtain $\int|\psi|^2\,d\tau = 1$.

exchange term differs from the jth Coulomb term precisely in the exchange of places and arguments between φ_j and φ_i. Also note that if φ_i and ψ_j correspond to different spin functions, their *exchange* interaction vanishes[6] because $\alpha(1)\beta(1) = 0$.

As we explained in Section 5.4, the Hartree–Fock equations treat the motion of each electron as if it were governed by the average field of all the other electrons and the nuclei, with no correlation between the motions of different electrons. This neglect of correlation results from the replacement of the exact wave function by a product, albeit anti-symmetrized, of one-electron orbitals. The correlation energy can be allowed for by configuration interaction, writing the total wave function to include electron configurations (i.e., Slater determinants) in addition to the initial Hartree–Fock configuration: cf. Section 6.9. In the present section, however, we shall restrict ourselves to the Hartree–Fock model.

In principle, then, all one has to do is solve Eq. 8.15 for the orbitals. Although this is manageable with high-speed computers, it remains a substantial task. Since the equation for each φ_i contains all the other orbitals (in the Hartree–Fock operator), it can only be done by successive approximation; an exact solution is impossible for systems with more than one electron. What one always does is to expand the orbitals in a set of basis functions; when the basis functions are atomic orbitals, this is the LCAO approximation. The more basis functions one uses, the better is the approximation. For a given set of basis functions, the best molecular orbitals are obtained by the self-consistent-field method we outlined in Section 5.4.

In general, we assume that our molecular orbitals are of the form

$$\varphi_i(1) = \sum_j c_{ij}\chi_j(1), \tag{8.17}$$

where the sum is taken over all the basis functions χ_j. We have already specified these orbitals for H_2O, Eqs. 8.9–8.11, but let us develop the general case. In the first stage of the calculation, one must guess the values of the c_{ij}, and give a trial set of orbitals $\varphi_i^{(0)}$. Suppose now that we define the quantity

$$\epsilon_i' \equiv \frac{\int \varphi_i^*(1)H_{HF}(1)\varphi_i(1)\,d\tau_1}{\int \varphi_i^*(1)\varphi_i(1)\,d\tau_1}. \tag{8.18}$$

If the φ_i were the solutions of the Hartree–Fock equations (Eq. 8.15), the ϵ_i' would be the eigenvalue ϵ_i; but when the φ_i are only trial functions, the variation principle tells us that $\epsilon_i' \geqslant \epsilon_i$. If one uses the trial orbitals $\varphi_i^{(0)}$ to define $H_{HF}(1)$ by Eq. 8.16, the next approximation $\varphi_i^{(1)}$ can be obtained by varying the φ_i to minimize the ϵ_i'. Then a new $H_{HF}(1)$ is defined in terms of the $\varphi_i^{(1)}$, and the ϵ_i' are minimized again to obtain the $\varphi_i^{(2)}$. The process is continued until no further change in the φ_i is observed.

The key step in the above process is the variation of the φ_i—that is, of the c_{ij}—to minimize the energy. How is this done? Let us expand the orbitals in Eq. 8.18 in terms of the basis functions, giving

$$\epsilon_i' = \frac{\int \left[\sum_j c_{ij}\chi_j^*(1)\right]H_{HF}(1)\left[\sum_k c_{ik}\chi_k(1)\right]d\tau_1}{\int \left[\sum_j c_{ij}\chi_j^*(1)\right]\left[\sum_k c_{ik}\chi_k(1)\right]d\tau_1} = \frac{\sum_j\sum_k c_{ij}c_{ik}H_{jk}}{\sum_j\sum_k c_{ij}c_{ik}S_{jk}}, \tag{8.19}$$

[6] This means that the exchange interaction vanishes for the ground states of two-electron systems such as He, H_2, and H_3^+. This, in turn, implies that the total energies of such states can be written as the sum of two orbital energies, something that does not hold for systems of three or more electrons.

where

$$H_{jk} \equiv \int \chi_j^*(1) \mathsf{H}_{HF}(1) \chi_k(1)\, d\tau_1 \quad \text{and} \quad S_{jk} \equiv \int \chi_j^*(1) \chi_k(1)\, d\tau_1.$$

$$(8.20)$$

We assume that all the c_{ij} are real, $c_{ij}^* = c_{ij}$, and that all the χ_j are expressed as functions of electron 1; the indices j and k are used only to distinguish the double summations over a common set of functions. We can rearrange Eq. 8.19 as

$$\sum_j \sum_k c_{ij} c_{ik} (H_{jk} - \epsilon_i' S_{jk}) = 0. \qquad (8.21)$$

We now vary the c_{ij} to minimize ϵ_i'. Let us differentiate Eq. 8.21 with respect to *one* of the c_{ij}, say c_{il}, holding all the others constant. We obtain

$$\sum_j c_{ij} H_{jl} + \sum_j c_{ij} H_{lj} - \epsilon_i' \left(\sum_j c_{ij} S_{jl} + \sum_j c_{ij} S_{lj} \right) - \frac{\partial \epsilon_i'}{\partial c_{il}} \sum_j \sum_k c_{ij} c_{ik} S_{jk}$$

$$= 2 \sum_j c_{ij} H_{jl} - 2\epsilon_i' \sum_j c_{ij} S_{jl} - \frac{\partial \epsilon_i'}{\partial c_{il}} \sum_j \sum_k c_{ij} c_{ik} S_{jk} = 0, \quad (8.22)$$

where we have used the relationships $H_{jl} = H_{lj}$ (because $\mathsf{H}_{HF}(1)$ is a Hermitian operator; cf. Appendix 6B) and $S_{jl} = S_{lj}$. If ϵ_i' is a minimum with respect to variation of c_{il}, it must satisfy $\partial \epsilon_i' / \partial c_{il} = 0$. Eq. 8.22 can then be valid only if

$$\sum_j c_{ij} (H_{jl} - \epsilon_i' S_{jl}) = 0. \qquad (8.23)$$

We obtain an equation like 8.23 for each value of l, that is, for each of the basis functions that make up the orbital φ_i. If there are n basis functions, we have n equations in n unknowns, the c_{ij} ($j = 1, 2, \ldots, n$). It can be shown (Cramer's rule) that such a set of linear homogeneous equations has a nontrivial solution if and only if the determinant of the coefficients vanishes; thus we must have

$$\| H_{jl} - \epsilon' S_{jl} \| = 0. \qquad (8.24)$$

A determinant of this form is called a *secular determinant*. Note that we have dropped the subscript on ϵ': Since all the orbitals are of the same form (8.17), varying the c_{ij} to minimize the energy must lead to the same determinant in every case. Expansion of Eq. 8.24 gives an nth-degree equation in ϵ', which of course has n roots. These roots are the orbital energies corresponding to our n molecular orbitals φ_i—or rather the approximations to the orbital energies at this stage in the SCF process. Once these ϵ_i' are obtained, one substitutes each of them in Eq. 8.23 to get n sets of c_{ij}, then substitutes the c_{ij} in Eq. 8.17 to get the new set of φ_i; given the new φ_i, one recalculates $\mathsf{H}_{HF}(1)$ and the H_{jl}, solves Eq. 8.24 over again, and so forth.

In the preceding paragraphs our language has been completely general, applying to any molecular orbitals of the form of Eq. 8.17. Now let us make a number of simplifications. First, it is convenient to choose the basis functions to be normalized and orthogonal (cf. Appendix 6A). This means that $S_{jk} = \delta_{jk}$, where δ_{jk} is a Kronecker delta. The diagonal elements in the secular determinant become $H_{jj} - \epsilon'$, and the off-diagonal elements are simply H_{jl}. Next we assume that the basis functions (and the molecular orbitals constructed from them) are symmetry orbitals. It can be shown by direct integration or group theory that the integrals H_{jk}, defined in Eq. 8.20, must vanish whenever χ_j and χ_k are of different symmetry types. As a result, the only nonzero terms in the secular determinant appear in blocks, one for each symmetry type. For the H_2O

molecule, with the basis functions defined in Fig. 8.8, these simplifications reduce Eq. 8.24 to[7]

$$
\begin{Vmatrix}
H_{11}-\epsilon' & H_{12} & H_{13} & H_{14} & 0 & 0 & 0 \\
H_{21} & H_{22}-\epsilon' & H_{23} & H_{24} & 0 & 0 & 0 \\
H_{31} & H_{32} & H_{33}-\epsilon' & H_{34} & 0 & 0 & 0 \\
H_{41} & H_{42} & H_{43} & H_{44}-\epsilon' & 0 & 0 & 0 \\
0 & 0 & 0 & 0 & H_{55}-\epsilon' & 0 & 0 \\
0 & 0 & 0 & 0 & 0 & H_{66}-\epsilon' & H_{67} \\
0 & 0 & 0 & 0 & 0 & H_{76} & H_{77}-\epsilon'
\end{Vmatrix}
$$

$$
= \begin{Vmatrix}
H_{11}-\epsilon' & H_{12} & H_{13} & H_{14} \\
H_{21} & H_{22}-\epsilon' & H_{23} & H_{24} \\
H_{31} & H_{32} & H_{33}-\epsilon' & H_{34} \\
H_{41} & H_{42} & H_{43} & H_{44}-\epsilon'
\end{Vmatrix}
(H_{55}-\epsilon')
\begin{Vmatrix}
H_{66}-\epsilon' & H_{67} \\
H_{76} & H_{77}-\epsilon'
\end{Vmatrix} = 0.
$$

$$(8.25)$$

Expansion of the original determinant in minors thus enables us to factor it into three smaller determinants, each of which can separately be set equal to zero. Thus, instead of having to solve a seventh-degree equation in ϵ', one need solve only one linear equation, one quadratic, and one quartic. This greatly reduces the work of calculation, and illustrates the advantage of using symmetry orbitals. (If we do *not* assume the χ_i orthogonal, then the nonzero terms in Eq. 8.25 are all of the form $H_{jl}-\epsilon' S_{jl}$, but the determinant still factors.)

What we have outlined here is the molecular orbital method used in the great majority of molecular calculations. One sets up the MOs, evaluates the integrals, constructs the secular determinant, solves the latter for the orbital energies, and repeats the process as many times as needed to yield self-consistency. In practice, not only the c_{jk} but constants within the χ_i themselves may be varied. (It should now be obvious why large computers are needed for calculations on many-electron systems!) Once the SCF orbitals and energies are obtained, one applies the Aufbau principle, putting the available electrons into the orbitals in order of increasing energy; in the H_2O molecule the 10 electrons fill the lowest five of our seven spatial orbitals. The molecular wave function is then given by Eq. 8.13 or its equivalent, using only the occupied spin orbitals. The final step is to determine the total molecular energy. This cannot be taken as simply the sum of the orbital energies, since we would then be counting each of the electron–electron interactions twice. When the orbitals are orthogonal, the total energy is actually

$$
E = 2\sum_i \epsilon_i + \sum_i \sum_{j>i} (2J_{ij} - K_{ij}), \tag{8.26}
$$

---

[7] There may be some confusion as to how we got from the 10×10 determinant (Eq. 8.13) to the 7×7 determinant (Eq. 8.25). The orders of the two determinants are actually quite independent. If we choose to form our molecular orbitals from n basis functions, we can have n independent linear combinations of the form of Eq. 8.17, leading to an $n\times n$ secular determinant (Eq. 8.24); n can be as large as one wishes (and can handle), as long as it is greater than the number of actually occupied spatial orbitals. Here we have $n=7$, giving seven MOs, of which only five are occupied in the ground state of H_2O. The molecular wave function for a given configuration, Eq. 8.13, is formed from only the occupied orbitals—in this case the 10 electrons give us five doubly occupied spatial orbitals and thus 10 spin orbitals, leading to a 10×10 determinant.

where ϵ_i is the orbital energy,

$$J_{ij} \equiv \iint \varphi_i^*(1)\varphi_j^*(2) \frac{e^2}{4\pi\epsilon_0 r_{12}} \varphi_i(1)\varphi_j(2)\, d\tau_1\, d\tau_2, \qquad (8.27)$$

and

$$K_{ij} \equiv \iint \varphi_i^*(1)\varphi_j^*(2) \frac{e^2}{4\pi\epsilon_0 r_{12}} \varphi_i(2)\varphi_j(1)\, d\tau_1\, d\tau_2; \qquad (8.28)$$

J_{ij} and K_{ij} are referred to as the two-electron Coulomb and exchange integrals, respectively. The sums in Eq. 8.26 are taken over all the occupied *spatial* orbitals, with the factors of 2 appearing because of the double occupancy of these orbitals; K_{ij} remains undoubled, because the exchange interaction takes place only between electrons of the same spin.

The results of Ellison and Shull's SCF LCAO calculation on the H₂O molecule are shown in Table 8.1, for a bond angle of 105° (close to the experimental value). The basis orbitals are labeled as in Fig. 8.8; the molecular orbitals are given their spectroscopic designations, with the notation of Eqs. 8.9–8.11 in parentheses. The orbital energies are in the order $1a_1 < 2a_1 < 1b_2 < 3a_1 < 1b_1 < 4a_1 < 2b_2$, with the first five occupied by two electrons each; this agrees with the spectroscopic results we mentioned earlier. Of the occupied orbitals, the $1a_1$ is essentially the O(1s); the $2a_1$ is predominantly O(2s), with a bit of O($2p_y$) and H(1s)+H'(1s); the $3a_1$, which is strongly bonding, is mainly O($2p_y$) with large admixtures of O(2s) and H(1s)+H'(1s); the $1b_2$, also strongly bonding, is a mixture of O($2p_z$) and H'(1s)−H(1s) with the latter predominating; and the $1b_1$ is the nonbonding O($2p_x$) orbital.

Ellison and Shull's calculations were not accurate enough to give the equilibrium bond angle in H₂O, since the total energy varies very slightly with angle; they made calculations at several angles from 90° to 180°, and found the energy minimum to be at somewhat over 120°. More recent calculations have improved considerably on this result. Figure 8.9 gives a Walsh diagram based on one such calculation, which used a total of 22 basis functions; this calculation gave 108.4° for the bond angle, in reasonably good agreement with experiment.

The variation of orbital energies with bond angle can be predicted approximately by a simple qualitative argument due to A. D. Walsh. The energies of the $1a_1$, $2a_1$, and $1b_1$ orbitals (like the orbitals themselves) are rather insensitive to the bond angle. The $3a_1$ and $1b_2$ energies, on the

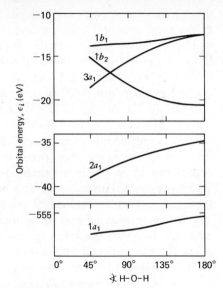

FIGURE 8.9
Energies of the occupied orbitals of H₂O as functions of bond angle.

TABLE 8.1
MOLECULAR ORBITALS FOR THE GROUND STATE OF THE H₂O MOLECULE WITH R(O—H) = 0.9581 Å, ∡H—O—H = 105°

Molecular Orbital	Coefficients in $\varphi_i = \sum_i c_{ij}\chi_i$							Orbital Energy
	σ_1(1s)	σ_2(2s)	σ_3($2p_y$)	σ_4(H+H')	σ_5($2p_x$)	σ_6($2p_z$)	σ_7(H'−H)	ϵ_i (eV)
$1a_1(\varphi_1)$	1.0002	0.0163	0.0024	−0.0033	0	0	0	−557.27
$2a_1(\varphi_2)$	−0.0286	0.8450	0.1328	0.1781	0	0	0	−36.19
$3a_1(\varphi_3)$	−0.0258	−0.4601	0.8277	0.3341	0	0	0	−13.20
$4a_1(\varphi_4)$	−0.086	−0.833	−0.642	1.061	0	0	0	13.7
$1b_1(\varphi_5)$	0	0	0	0	1	0	0	−11.79
$1b_2(\varphi_6)$	0	0	0	0	0	−0.5428	0.7759	−18.55
$2b_2(\varphi_7)$	0	0	0	0	0	−1.013	1.230	15.9

[a] From F. O. Ellison and H. Shull, *J. Chem. Phys.* **23**, 2348 (1955).

other hand, are strongly angle-dependent and chiefly govern the variation of the total energy. If the molecule were linear, the $H+H'$ basis orbital would overlap as much with the negative as with the positive lobe of the $O(2p_z)$. Thus these orbitals could not mix at all in forming the $3a_1$ MO, the energy of which would therefore be quite high at 180°. As the bond angle decreases from 180°, we expect the $3a_1$ energy to decrease monotonically as the mixing of $H+H'$ and $O(2p_z)$ increases. Conversely, we expect $H'-H$ and $O(2p_y)$ to exhibit a maximum of mixing at 180°, so that the $1b_2$ energy should increase with decreasing bond angle. It can be seen that Fig. 8.9 bears out these predictions approximately, although the $2a_1$ orbital energy in particular seems to vary with angle much more than Walsh expected. However, Walsh's method seems to work better for the total energy than for individual orbitals.

Similar reasoning can be applied to other dihydrides. In any AH_2 molecule in which the bonding involves only s and p orbitals of the A atom, the valence-shell MOs should resemble the $2a_1$ and higher orbitals of H_2O. Thus a dihydride with only four valence electrons, such as BeH_2, has the $3a_1$ orbital empty; the $1b_2$ orbital dominates the angular dependence of the energy, and the total energy has its minimum at 180°—in other words, BeH_2 should be linear. On the other hand, in all dihydrides with five or more valence electrons (BH_2, CH_2, NH_2, H_2S, etc.) the $3a_1$ orbital has a sufficiently great effect to shift the minimum energy to some lesser angle—these molecules should all be bent. These conclusions, known as *Walsh's rules*, are at best semiempirical, but they seem to work in virtually all cases; they can also be extended to predict the shapes of molecules in excited states.

One can also consider the variation of orbital energies with bond length, the only parameter we had to worry about in diatomic molecules. For H_2O or any other AX_2 molecule, one natural way to do this is to assume a fixed bond angle and hold the two bond lengths equal while varying their magnitude; keeping the lengths equal ensures that the orbitals retain their symmetry properties. One can readily obtain correlation diagrams similar to those we drew for diatomics. In dihydrides the heavy-atom core orbitals—the $1a_1$ and $2a_1$ in H_2O, corresponding to the oxygen $1s$ and $2s$ in the separated-atom limit—of course remain at the bottom of the energy scale for all values of R; as in diatomics, their energies decrease as they approach the united-atom limit (because of the higher Z). The $3a_1$, $1b_1$, and $1b_2$ orbitals have energies fairly close together at all R, eventually converging to become the triply degenerate $2p$ level of the united atom, but the energies of the bonding $3a_1$ and $1b_2$ orbitals drop more rapidly with decreasing R than does that of the nonbonding $1b_1$, showing the effect of increased bond formation on the energy. The first excited orbital, the $4a_1$, is promoted to the relatively high-energy united-atom $3s$ level and is thus antibonding. All this is illustrated in the correlation diagram of Fig. 8.10, which is generally applicable to dihydrides.

To close this section, let us see how the orbital energies calculated for H_2O compare with those determined experimentally. Photoelectron spectroscopy (Section 7.10), with either ultraviolet or x-ray radiation, is the most straightforward technique for making such measurements. In Table 8.2 the measured binding energies of the occupied orbitals of H_2O are compared with two sets of theoretical values, those of Ellison and Shull and a more recent calculation by Neumann and Moskowitz,[8] both for approximately the experimental geometry.

[8] D. Neumann and J. W. Moskowitz, *J. Chem. Phys.* **49,** 2056 (1968).

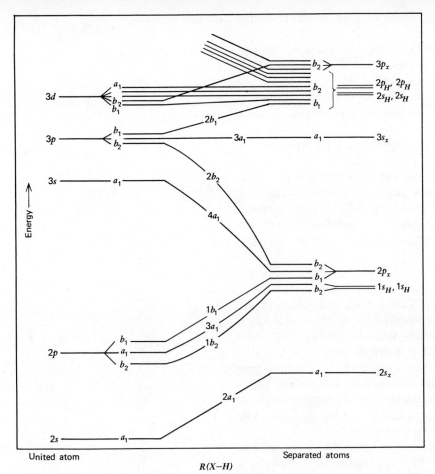

FIGURE 8.10

Correlation diagram for bent XH_2 molecules with equal X—H bond lengths. The $1a_1$ orbital (united-atom $1s$, separated-atom $1s_x$) has a very low energy and is omitted. It is assumed that the $X(2p)$ electron has a smaller ionization energy than the $H(1s)$ electron, and so on. From G. Herzberg, *Electronic Spectra and Electronic Structure of Polyatomic Molecules*, (van Nostrand Reinhold, *New York*, 1966)

TABLE 8.2
OBSERVED AND CALCULATED BINDING ENERGIES OF H_2O ORBITALS

Orbital	Measured Energy, ϵ_i (eV)		Calculated Energy, ϵ_i (eV)	
	X ray	Ultraviolet	Ellison and Shull	Neumann and Moskowitz
$1a_1$	−539.7	insufficient energy	−557.27	−559.40
$2a_1$	−32.2	insufficient energy	−36.19	−36.79
$1b_2$	−18.4	−18.55	−18.55	−19.56
$3a_1$	−14.7	−14.73	−13.20	−15.84
$1b_1$	−12.6	−12.61	−11.79	−13.80
Total molecular energy	−2080.6		−2064.3	−2069.54

The electronic structures of more complex triatomic molecules can be analyzed in ways analogous to those we have described for H_2O. The calculations are more demanding and extensive, but the principles remain the same. In this section we shall merely give a rapid survey of such molecules, with comments on the nature of the bonding. The most

8.5
Bonding in More Complex Triatomic Molecules

interesting question to be considered here is why some triatomic molecules are linear and others bent.

Next in simplicity after the dihydrides are the triatomic monohydrides. Those that are known, like most dihydrides, are largely unstable or highly reactive transient species such as HNO or HCO (which are found in flames).[9] One very important exception to this generalization is the molecule HCN, hydrogen cyanide. This substance is a stable, volatile liquid with an almondlike odor; its moderately high toxicity is well known. As in H_2O, HF, and so on, the anomalously high boiling point (25.6°C) is due to hydrogen bonding. In aqueous solution HCN behaves as a weak acid, again like HF. In fact, the tightly bound CN (cyanide) group acts in many ways like a halogen atom; its electron affinity of 3.8 eV is higher than those of the halogens (Fig. 5.7), so CN⁻ ions are readily formed. Cyanide salts such as NaCN are quite similar to the corresponding halides, and cyanogen halides such as ClCN resemble interhalogen compounds.

The isolated HCN molecule bears a close relationship to the isoelectronic N_2 molecule. The molecule is linear, corresponding to *sp* hybridization on the carbon atom (Section 8.3). That is, one can interpret the C—H bond as a localized σ orbital composed of the H(1s) orbital and a mixture of the carbon 2s and $2p\sigma$ orbitals, with the latter orbitals also combining with $N(2p\sigma)$ to form the C—N σ bond; the remaining C and N 2p orbitals form two π bonds as in N_2. Delocalized MO calculations give an electron distribution not far from this description. The excited states are also similar to those of N_2; the absorption spectra for excitation in both molecules lie in the vacuum ultraviolet, in similar patterns. The known excited states of HCN all have bent structures.

There are many stable triatomic species containing no hydrogen. The most common (and most important to life) is surely carbon dioxide, a relatively unreactive substance made up of linear CO_2 molecules. Other well-known stable triatomic molecules include those of ozone (O_3), sulfur dioxide (SO_2), carbonyl sulfide (OCS), carbon disulfide (CS_2), nitrous oxide (N_2O), and nitrogen dioxide (NO_2). All of these are gases under ordinary conditions, as is usual for small molecules without hydrogen bonding. Their structures follow patterns consistent with the similarity of elements in the same family of the periodic table.[10] Thus O_3, SO_2, and S_2O all have bent structures, with bond angles of 116.8°, 119.5°, and 118.0°, respectively; and CO_2, OCS, CS_2, OCSe, and so on, are all linear. The N_2O molecule, isoelectronic with CO_2, is also linear, but NO_2 has a bond angle of 134.1°.

To interpret the relationship between geometry and electronic structure in these molecules, let us consider only the three examples CO_2, NO_2, and O_3. We can neglect the 1s and 2s atomic electrons, which in each case give rise to six molecular orbitals with relatively little net bonding power (four σ_g and two σ_u in the linear geometry, four a_1 and two b_2 in the bent geometry). The orbitals formed from the atomic 2p orbitals offer a richer and structurally more important picture. In Fig. 8.11 we show one set of symmetry orbitals that can be formed from atomic s and p orbitals in bent triatomic molecules; the best MOs, as indicated, would be mixtures of these.

---

[9] The hypohalous acids (HOCl, HOBr, HOI) are well known in aqueous solution, but are so unstable that they cannot be prepared in the pure state.

[10] The geometry (and presumably the bonding) is also little changed when halogen atoms replace hydrogen. One finds the bond angles OF_2 103.2°, OCl_2 110.8°, SCl_2 101°, all similar to H_2O; whereas ClCN, etc., are linear like HCN.

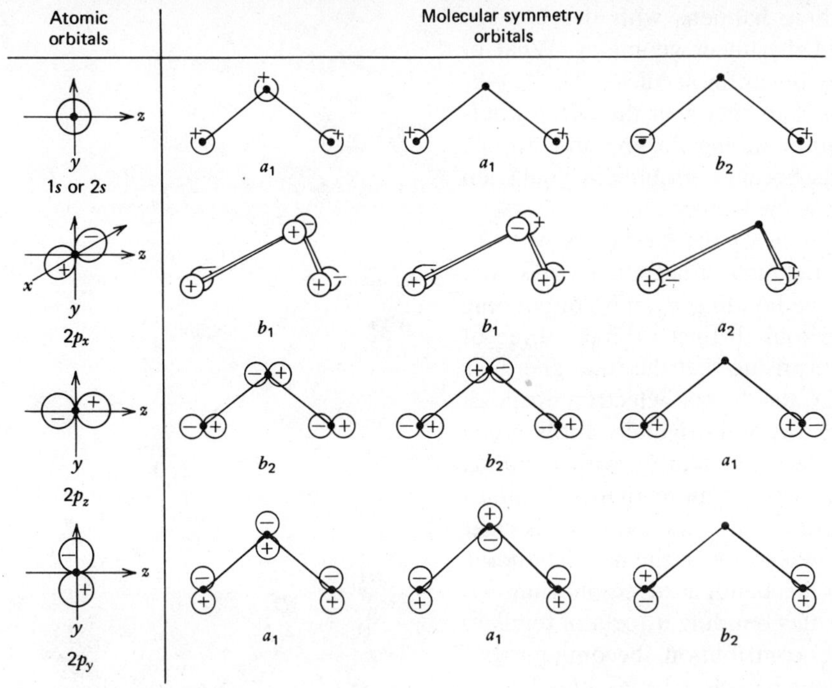

Atomic orbitals	Molecular symmetry orbitals

FIGURE 8.11

Schematic representation of a possible set of molecular symmetry orbitals for first-row bent triatomic molecules. Those shown here would be suitable as basis orbitals; the most realistic MOs of each symmetry type would be a mixture of all those shown here of that type. The coordinate system is the same as that used in Fig. 8.8. In the linear geometry (with all three atoms along the z axis), the symmetries of the orbitals corresponding to those shown are as follows: $1s$, $2s-\sigma_g$, σ_g, σ_u; $2p_x-\pi_u$, π_u, π_g; $2p_y-\pi_u$, π_u, π_g; $2p_z-\sigma_u$, σ_g, σ_g.

Consider first the $2p$-derived orbitals in the linear geometry. Those formed from $2p_z$ ($2p\sigma$) orbitals become σ MOs, including: a bonding σ_u orbital with large overlap along both bond aces; a σ_g orbital that is normally nonbonding because its maxima occur near the end atoms; and an antibonding σ_u orbital with nodes crossing each bond. The atomic $2p_x$ and $2p_y$ orbitals similarly each give rise to a strongly bonding π_u orbital, a nonbonding π_g orbital, and an antibonding π_u orbital; in the linear case the $2p_x$ and $2p_y$ sets are degenerate with each other. Now consider the molecules CO_2, NO_2, and O_3 in turn in linear form. In CO_2 we have four $2p$ electrons from each of the two oxygens and two from the carbon, a total of 10. Let us apply the Aufbau principle. Six of the 10 electrons occupy the bonding σ_g and two π_u orbitals, leaving four to go into the nonbonding σ_g and two π_g orbitals. In the ground state the last four electrons in fact occupy the π_g orbitals, consistent with the idea that these are predominantly atomic electrons localized on the oxygen atoms (Lewis formula $:\ddot{O}=C=\ddot{O}:$). In a linear NO_2 molecule we would presumably have the same configuration with one more electron, which would go into the nonbonding σ_g orbital; and in linear O_3 this σ_g orbital would be filled by two electrons.

Let us compare these descriptions with the corresponding electronic structures for molecules with bent geometries. As Fig. 8.11 indicates, the orbitals that form degenerate π pairs in the linear molecule split upon bending. For example, the first orbitals shown as formed from $2p_x$ and $2p_y$ atomic orbitals are physically indistinguishable, and thus degenerate π_u orbitals in the linear molecule. In the bent geometry they are both physically inequivalent and of different symmetry classes, and thus have different energies. The $2p_x-b_1$ orbital continues to show π-like bonding (with a node on the bond axis, and its maximum amplitude perpendicular to the bond axis), and its energy is only weakly affected by bending until the outer atoms begin to overlap (lowering the orbital energy). In contrast, the $2p_y-a_1$ orbital starts to lose bonding character as soon as bending begins, so that its energy increases sharply as the bond angle decreases from 180°. Much the same thing happens with the $2p_z - b_2$ orbital which is the bonding σ_g orbital in the linear geometry. What of the orbitals that are nonbonding in the linear molecule? The $2p_z - a_1$ orbital becomes more bonding and decreases in energy as the outer atoms begin to overlap; so do the originally antibonding $2p_x - b_1$ and $2p_y - a_1$ orbitals; the $2p_x - a_2$ and $2p_y - b_2$ orbitals become antibonding between the outer pair of atoms and thus increase in energy.

This analysis is enough to let us account for the structures of CO_2, NO_2, and O_3. In the linear CO_2 molecule the highest occupied levels (and the only ones very sensitive to angle) are the bonding π_u and nonbonding π_g orbitals. Upon bending these become four distinct orbitals, three of which increase in energy. Thus it is not surprising that the total energy is lowest in the linear geometry. In bent NO_2 the "extra" electron occupies a mixture of the originally nonbonding $2p_z - a_1$ and originally antibonding $2p_y - a_1$ orbitals, both of whose energies decrease sharply with bending. This lowering apparently affects the total energy more than the simultaneous energy increase in the π-like orbitals, since the evidence is clear that NO_2 is bent. In bent O_3 there are *two* electrons in the newly bonding a_1 orbital, so the energy lowering—and the bending at equilibrium—is even more pronounced. (The structure of this bonding a_1 orbital tends to be dominated by the central atom's $2p_y$ contribution, becoming more localized on this atom as its atomic number increases.)

Reasoning such as that we have described here was introduced by R. S. Mulliken and fully developed by A. D. Walsh.[11] It is summarized by a set of "Walsh's rules" like those we discussed for the dihydrides: Molecules of the types AB_2 and BAC should be linear if they have up to 16 valence electrons (CO_2 and N_2O have 16); those with 17 to 20 valence electrons should be bent, with progressively smaller bond angles (17: NO_2, 134°; 18: O_3, 117°; 19: ClO_2, 117°; 20: OF_2, 103°); and those with 22 valence electrons should again be linear (I_3^-). Similarly, for HAB molecules Walsh obtained the rules: 10 valence electrons, linear (HCN); 11–14, bent (HCO, HNO, HO_2, HOCl); 16, linear (FHF^-).

An alternative way of rationalizing molecular geometries is based on the localized-orbital model. Let us look again at CO_2, NO_2, and O_3. In CO_2 we can say that the $2p$ electrons of the central carbon atom, sp-hybridized with the $2s$ electrons, are used in σ bonding orbitals, with the oxygen atoms supplying electrons for the π orbitals. In NO_2 and O_3 the a_1 orbital holding the extra electron(s) is interpreted as an atomic $2p$ orbital on the central atom; we have just noted that this is not far from the truth. Now we apply an empirical rule of thumb: A nonbonding atomic p electron tends to remain localized and repel electrons in adjacent chemical bonds. The effect is even more pronounced with two such electrons, which occupy a localized hybrid orbital and repel adjacent bonds even more strongly than the bonding pairs repel one another. Thus in NO_2 one electron is repulsive enough to reduce the bond angle to 134°; in O_3 the "lone-pair" reduces the angle to less than 120°, the angle one would expect for equivalent sp^2 orbitals. Like the ordinary hybridization model, this concept of lone-pair repulsion is useful for qualitative interpretation but not for exact calculations.

In Table 8.3 we summarize the geometric properties of a number of triatomic molecules, including examples of all the classes we have discussed. Also tabulated are the fundamental vibration frequencies, which we shall take up in the remainder of this chapter.

[11] Walsh's rules for triatomic and larger molecules were derived in a series of articles in *J. Chem. Soc.* (1953), p. 2260ff.

TABLE 8.3
CONSTANTS OF SOME TRIATOMIC MOLECULES[a]

Molecule, A—B—C	R_{AB} (Å)	R_{BC} (Å)	Bond Angle	Fundamental Vibration Frequencies		
				$\tilde{\nu}_1$ (cm⁻¹)	$\tilde{\nu}_2$ (cm⁻¹)	$\tilde{\nu}_3$ (cm⁻¹)
Dihydrides						
H—O—H	0.9572	0.9572	104°31′	3657	1595	3776
H—O—D	0.9572	0.9572	104°31′	2724	1403	3708
D—O—D	0.9572	0.9572	104°31′	2666	1179	2787
H—S—H	1.334	1.334	92°16′	2611	1183	2626
H—Se—H	1.46	1.46	91°0′	2260	1074	2350
Monohydrides						
H—C—O	1.08	1.20	119°30′	2700	1083	1820
H—N—O	1.063	1.212	108°36′	3596	1562	1110
H—O—Cl	0.975	1.689	102°30′	3626	1242	739
H—C—N	1.064	1.156	180°	2096	712	3312
Other Linear Molecules						
O—C—O	1.62	1.162	180°	1388	667	2349
O—C—S	1.164	1.558	180°	859	522	2050
S—C—S	1.554	1.554	180°	655	397	1510
Cl—C—N	1.629	1.163	180°	714	396	2213
N—N—O	1.126	1.191	180°	1285	589	2224
Other Bent Molecules						
O—N—O	1.197	1.197	134°15′	1306	755	1621
O—O—O	1.278	1.278	116°49′	1110	705	1043
O—S—O	1.433	1.433	119°33′	1151	518	1362

[a] The geometric constants are for the ground vibrational state rather than the bottom of the potential well (R_0, not R_e).

In the remainder of this chapter we shall consider the dynamical behavior of triatomic molecules—that is, the phenomena associated with the motions of the atomic nuclei, especially the molecular vibrations. This behavior is more complex than the simple vibrations of diatomic molecules, but can be thought of in a similar way. The concepts we shall introduce here are generally applicable to larger molecules as well as to triatomics.

We begin as usual by splitting up the overall problem of molecular structure into tractable pieces. Thus we assume that the electronic energy of the molecule can be separated from the energy of nuclear motion (the Born–Oppenheimer approximation), and that the electronic structure and energy can be calculated to any desired degree of accuracy for any assignment of nuclear coordinates along lines such as those we described in the preceding sections. We further assume that the translational, vibrational, and rotational motions of the nuclei are separable. Without these assumptions, the problem can be solved approximately but quite accurately. However the calculations are messy.

8.6
Normal Coordinates and Modes of Vibration

In general, an N-atom molecule has $3N$ nuclear coordinates, but we know that not all of these affect the molecular energy. In the diatomic molecule we needed to consider only one coordinate, the internuclear distance R. The reason is that, in the absence of external fields, the internal energy of a free molecule is independent of the location of the molecule's center of mass, and of the molecule's orientation in space. It takes three coordinates to locate the center of mass, whereas the orientation is defined by two coordinates for a linear molecule and three coordinates for a nonlinear molecule. This is illustrated in Fig. 8.12 with possible sets of such coordinates. Thus the internal energy of the N-atom molecule is a function of $3N-5$ (linear) or $3N-6$ (nonlinear) coordinates. It is these coordinates, which can be chosen in many ways, that we call the vibrational coordinates of the molecule; we say that the molecule has $3N-5$ or $3N-6$ vibrational *degrees of freedom*.

If we were dealing with a very large molecule like a protein (or with a crystal), then N would be so large that the difference between $3N-6$ and $3N$ could be neglected. On the other hand, in a diatomic molecule (which is of course linear) we have $N=2$, $3N-5=1$, the single vibrational coordinate R. Our present concern is with the case $N=3$, and we find that we have three vibrational coordinates for a nonlinear molecule, four vibrational coordinates for a linear molecule. This is clearly not as simple as the diatomic case. However, we shall see that the vibrations of a triatomic molecule present a quite tractable problem, containing all the essentials for understanding more complex species.

Let us digress to dispose of the nonvibrational degrees of freedom. The three coordinates of the center of mass of course describe the translational motion of the molecule as a whole, which is again that of a free particle. A rigid linear triatomic molecule is a highly prolate symmetric top with the same rotational properties as a diatomic molecule, and the results of Section 7.2 can be applied with little change; only the definition of I_0 ($\equiv \sum_i m_i z_i^2$, where the z axis is the symmetry axis) is different. We again have the selection rule $\Delta J = \pm 1$, and observe a pure rotational spectrum only when the molecule has a permanent dipole moment, as do HCN or OCS, but not CO_2. As for nonlinear triatomic molecules, they are all asymmetric tops (Section 3.12); although many are close to being symmetric tops, the others have very complex sets of rotational energy levels, about which we shall say no more.

In a triatomic molecule, then, the internal energy is a function of three (or four) independent coordinates R_j. By "internal energy" we mean the equivalent of the $E(R)$ of the diatomic molecule, the total electronic energy plus the potential energy of the nuclei. Recall that this energy can be interpreted as the effective potential energy of nuclear motion. Instead of a potential energy curve we now have a *surface* in a space of four (or five) dimensions, with peaks and valleys and saddles displaying the behavior of E as a function of the R_j. We cannot show such a surface in full, but we can give cross sections showing how E varies with one or two coordinates when the others are held constant. Examples of these for CO_2 are shown in Figs. 8.13 and 8.14.

But we are getting ahead of ourselves. The first question we must ask is, what is the best way to select the vibrational coordinates R_j? Out of the infinite number of coordinates we could define, is there one set preferable to all the others? The answer is yes. We want a set of coordinates that are linearly independent so none are redundant, and in terms of which the vibrational motion can be described in the simplest possible way. Such a set exists, and can be found in a relatively straight-forward way, provided that we make one assumption about the molecular

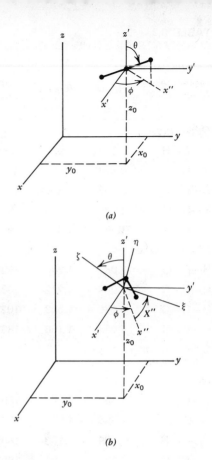

FIGURE 8.12
Possible sets of coordinates for triatomic molecules. The Cartesian coordinates x_0, y_0, z_0 give the location of the center of mass, and the angles θ, ϕ, χ give the molecule's orientation relative to the Cartesian axes for (a) linear molecules, (b) nonlinear molecules. In (a) the angles θ, ϕ are the standard spherical coordinates giving the direction of the molecule's symmetry axis. In (b) the *Eulerian angles* θ, ϕ, χ are obtained as follows: Rotate the coordinate system through an angle ϕ about the z' axis, with the x' axis becoming the x'' axis; then through an angle θ about the x'' axis (the *line of nodes*), with the z' axis becoming the ζ axis; and finally through an angle χ about the ζ axis. The new center-of-mass-fixed coordinates ξ, η, ζ thus defined are chosen to fit the symmetry of the molecule, in this case assumed to lie in the $\xi\eta$ plane.

structure: We must assume that the molecule has a well-defined equilibrium geometry and that the nuclei perform only small oscillations about their equilibrium positions, in a coordinate system fixed with respect to that equilibrium geometry, a so-called molecular frame.

In the space defined by the vibrational coordinates, the complete potential energy surface for a polyatomic molecule has a many-dimensional well; the bottom of this well is the equilibrium point corresponding to the geometry in which the energy is a minimum. If the potential were one-dimensional (as in a diatomic molecule), we could approximate the energy levels and wave functions by assuming the well to be parabolic. This is the harmonic oscillator model we described in Section 4.2. We extend this model by assuming that an n-dimensional well can be approximated by a set of n noninteracting harmonic oscillators—in other words, that the cross section along each of the coordinates in our "best set" can be approximated by a parabola. The coordinates we call the *normal coordinates* of the molecule are the set for which this approximation is most nearly true. The directions of the normal coordinates in the n-dimensional space are thus the directions along which the bottom of the well is best represented by parabolic curves. The basic problem of describing molecular vibrations is to find that combination of stretches, bends, torsions, and vibrations that defines these directions.

We can state the properties of normal coordinates more precisely in mathematical language. The potential energy V can be expanded as a power series in any complete set of linearly independent coordinates. Generally, for an N-atom molecule with $3N-6$ vibrational coordinates X_i, the potential energy can be written as the Taylor series

$$V(X_1, X_2, \ldots, X_{3N-6}) = V_0 + \sum_i \left(\frac{\partial V}{\partial X_i}\right)_{X_{io}} (X_i - X_{i0})$$
$$+ \frac{1}{2} \sum_i \sum_j \left(\frac{\partial^2 V}{\partial X_i \, \partial X_j}\right)_{X_{io}, X_{jo}} (X_i - X_{i0})(X_j - X_{j0})$$
$$+ \cdots, \tag{8.29}$$

where the X_{i0} are the coordinates for the equilibrium geometry, and $V_0 = V(X_{i0})$. The linear terms drop out at once, since the equilibrium point—the bottom of the well—is by definition the point where all first derivatives vanish, $(\partial V/\partial X_i)_{X_{io}} = 0$. The expression becomes even simpler if we choose our coordinates to have their origins at the equilibrium values, $X_{i0} \equiv 0$ (the corresponding procedure for a diatomic molecule would be to define a coordinate $X \equiv R - R_0$), and take V_0 to be the energy zero; Eq. 8.29 then becomes

$$V(X_1, X_2, \ldots, X_{3N-6}) = \frac{1}{2} \sum_i \sum_j \left(\frac{\partial^2 V}{\partial X_i \, \partial X_j}\right)_0 X_i X_j + \cdots. \tag{8.30}$$

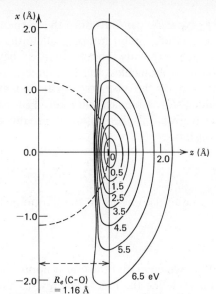

FIGURE 8.13
Potential energy contours for CO_2, drawn for the special case in which one C—O distance is fixed at its equilibrium value of 1.16 Å. The oxygen on the right is allowed to move relative to the clamped carbon atom at the origin and left-hand oxygen atom. Motion of the oxygen atom along the dashed semicircle would correspond to pure bending of the O—C—O angle, with no change in bond length. Note that the natural motion of the oxygen along the bottom of the potential trough would not follow the dashed semicircle; it would involve a small but significant amount of bond stretching as the O—C—O angle decreases from 180°. From G. Herzberg, *Electronic Spectra and Electronic Structure of Polyatomic Molecules* (van Nostrand Reinhold, New York, 1966).

FIGURE 8.14
Potential energy curves showing how the energy of the CO_2 molecule varies with (*a*) stretching of one C—O bond (variation along z in Fig. 8.13); (*b*) displacement of one O atom perpendicular to the symmetry axis (variation of x in Fig. 8.13); (*c*) change in the O—C—O bond angle for fixed bond lengths (minimum at 180°).

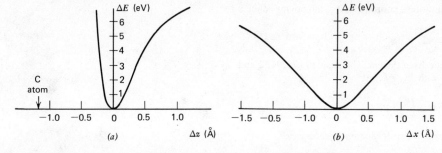

Now we can introduce the harmonic oscillator approximation, which consists of simply dropping all terms in the Taylor series beyond those written explicitly above, that is, beyond the quadratic terms. As in diatomic molecules, the harmonic approximation is useful only near the bottom of the potential well, and anharmonicity corrections must be added for an exact solution. For convenience we define the second derivatives of the potential energy at the equilibrium point as force constants,

$$k'_{ij} \equiv \left(\frac{\partial^2 V}{\partial X_i \, \partial X_j} \right)_{X_i = X_j = 0} ; \qquad (8.31)$$

the potential energy in the harmonic approximation then has the form

$$V(X_1, X_2, \ldots, X_{3N-6}) = \frac{1}{2} \sum_i \sum_j k'_{ij} X_i X_j. \qquad (8.32)$$

What we have said thus far is applicable to any set of nuclear coordinates, but to define the kinetic energy we must be more specific. Let us say that the coordinates X_j are linear combinations of the Cartesian coordinates of the nuclei in the molecule-fixed system. This is rigorously valid for coordinates that are internuclear distances, and also valid in the limit of small displacements for any other choice of coordinates. It is not hard to show that the vibrational kinetic energy must then have the form

$$T = \frac{1}{2} \sum_i \sum_j m'_{ij} \left(\frac{dX_i}{dt} \right) \left(\frac{dX_j}{dt} \right), \qquad (8.33)$$

where the m'_{ij} are constants corresponding to the reduced mass in the case of the diatomic molecule.[12]

One could obtain the Hamiltonian operator by combining Eqs. 8.32 and 8.33 and converting all the X_i to the corresponding operators. But the resulting Schrödinger equation would be very awkward to solve, because of all the cross terms involving two variables. As always, we prefer a situation in which the variables are separable. If there is a set of variables Q_α in terms of which the Hamiltonian can be written as a sum of noninteracting parts,

$$\mathsf{H}(Q_1, Q_2, \ldots, Q_{3N-6}) = \sum_\alpha \mathsf{H}_\alpha(Q_\alpha), \qquad (8.34)$$

then the wave function can be written as a product,

$$\psi(Q_1, Q_2, \ldots, Q_{3N-6}) = \prod_\alpha \psi_\alpha(Q_\alpha), \qquad (8.35)$$

and the energy as a sum,

$$E \equiv \sum_\alpha \epsilon_\alpha,$$

where $\qquad\qquad \mathsf{H}_\alpha(Q_\alpha)\psi_\alpha(Q_\alpha) = \epsilon_\alpha \psi_\alpha(Q_\alpha). \qquad (8.36)$

[12] In the diatomic molecule we have only one vibrational coordinate, X_1; the obvious choice for this coordinate is

$$X_1 \equiv R - R_0 = |z_A - z_B| - R_0,$$

where the A–B axis is the z axis of the molecule-fixed system. The sums in Eqs. 8.32 and 8.33 then reduce to one term each ($j = k = 1$), with $k'_{11} = k$, $m'_{11} = \mu$:

$$V = \tfrac{1}{2}k X_1^2, \qquad T = \tfrac{1}{2}\mu \left(\frac{dX_1}{dt} \right)^2.$$

Since X_1 is already a normal coordinate, no further transformation is necessary.

All this is very well, but can it be done? The answer is yes. Given any coordinate system in which Eqs. 8.32 and 8.33 hold, one can always transform to a new coordinate system in which the cross terms vanish and the Hamiltonian is separable. (We shall not attempt to prove this general theorem,[13] but in the next section we show how the transformation is performed in a specific case.)

The new coordinates Q_α thus defined are the normal coordinates of the molecule. When the kinetic and potential energies of vibration are expressed in terms of them, they take the diagonal forms

$$T = \frac{1}{2} \sum_\alpha m_\alpha \left(\frac{dQ_\alpha}{dt} \right)^2 = \sum_\alpha \frac{p_\alpha^2}{2m_\alpha}, \qquad \left(p_\alpha \equiv m_\alpha \frac{dQ_\alpha}{dt} \right) \tag{8.37}$$

and

$$V = \frac{1}{2} \sum_\alpha k_\alpha Q_\alpha^2, \tag{8.38}$$

where the m_α and k_α are new reduced masses and force constants, respectively.[14] The Hamiltonian can then clearly be separated according to Eq. 8.34, with

$$\mathsf{H}_\alpha = \frac{\mathsf{p}_\alpha^2}{2m_\alpha} + \tfrac{1}{2} k_\alpha Q_\alpha^2 \qquad \left(\mathsf{p}_\alpha \equiv i\hbar \frac{\partial}{\partial Q_\alpha} \right). \tag{8.39}$$

But this is identical in form to the Hamiltonian of the one-dimensional harmonic oscillator, Eq. 4.6. Thus the wave functions $\psi_\alpha(Q_\alpha)$ are all simple harmonic oscillator functions of the sort with which we are familiar, and the energy eigenvalues are given by

$$\epsilon_\alpha = (v_\alpha + \tfrac{1}{2}) h \nu_\alpha = (v_\alpha + \tfrac{1}{2}) \hbar \omega_\alpha, \qquad (v_\alpha = 0, 1, 2, \ldots), \tag{8.40}$$

where

$$\omega_\alpha = 2\pi \nu_\alpha = \left(\frac{k_\alpha}{m_\alpha} \right)^{1/2}. \tag{8.41}$$

For sufficiently small displacements, then, the vibrations of a polyatomic molecule or any other system of coupled oscillators reduce to those of a set of independent harmonic oscillators, one for each vibrational degree of freedom. Each of these independent oscillations is referred to as a *normal mode of vibration*.

But what do normal coordinates look like? In general, each of the Q_α is a linear combination of the positions of all the nuclei in terms of, say, their Cartesian coordinates. They are thus most effectively represented by diagrams showing the relative motions of the nuclei. In Fig. 8.15 we illustrate the three normal modes of vibration of the SO_2 molecule. In each normal mode the nuclei oscillate in the directions indicated by the arrows, with relative displacements from their equilibrium positions indicated by the lengths of the arrows, and all with the same frequency ν_α. Any actual vibration of the molecule is a superposition (linear combination) of the normal modes.

In general, it is a routine calculation to determine the normal modes of vibration if one knows the potential energy surface. But potential energy surfaces cannot be observed directly. What one measures spectros-

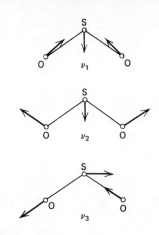

FIGURE 8.15
Normal modes of vibration of the SO_2 molecule.

---

[13] The transformation involved here is identical to the principal-axis transformation of rigid-body mechanics. Texts on mechanics may be consulted for details.

[14] Note that Eq. 8.38 corresponds to our earlier definition of the normal coordinates as those in which the potential energy well is most nearly parabolic. (Remember that this result holds exactly only in the limit of small displacement.)

copically are the frequencies corresponding to energy-level spacings, which can be analyzed to yield the normal vibration frequencies ν_α. The usual problem is thus to obtain as much of the surface as possible from the normal modes, rather than vice versa. Furthermore, the normal coordinates themselves are not always the most interesting physically; we are frequently concerned more with how the energy varies with particular bond lengths and angles. This means that in first approximation we want to know the "valence" force constants k'_{ij} of Eq. 8.32, which give the curvature of the potential surface for such variations.

But here a major difficulty arises. For a nonlinear molecule there are $3N-6$ independent vibrational frequencies, but $(3N-6)^2$ parameters k'_{ij} (in a given set of coordinates). Not all the latter are independent, since by Eq. 8.31 we must have $k'_{ij} = k'_{ji}$, but we are still left with a total of[15] $\frac{1}{2}(3N-6)(3N-5)$ independent force-constant parameters. Thus in a nonlinear triatomic molecule one must try to determine six parameters from three fundamental vibration frequencies, in a nonlinear four-atom molecule, 21 parameters from six frequencies, and so forth. How, then, can one obtain enough information about the force constants to construct the potential energy surface from experimental data?

The answer to this seemingly insoluble problem lies in the experimenter's control over the *masses* of the nuclei, through the use of isotopes. One assumes that each potential energy surface depends only on the nuclear charges, and not on the masses. We know that this is a good assumption in the H_2 molecule (Table 7.1), and it works even better for molecules containing heavier atoms. Replacing one isotope with another should not change any of the force constants. But such isotopic substitution does change the reduced mass m_α for any vibrational mode involving motion of the substituted nucleus, and Eq. 8.41 implies that the vibration frequency ν_α must also be changed. For example, replacing one ^{16}O atom in CO_2 by ^{18}O produces a molecule whose three fundamental vibration frequencies are different from those of $^{12}C^{16}O_2$. Each such substitution provides a new set of frequencies,[16] and it is not difficult to obtain enough data to determine all the unknown force constants. Indeed, for small molecules one can often overdetermine the force constants, and thus check the validity of assuming them mass-independent.

The CO_2 molecule is one case in which the forms of the normal modes of vibration can be easily deduced from the molecular structure. The treatment here is simplified, but it can be done rigorously. We

8.7
A Solvable Example: The Vibrational Modes of CO_2

[15] One k'_{ii} for each coordinate and one k'_{ij} for each pair of different coordinates; for $3N-6$ coordinates, this adds up to

$$3N-6+\frac{(3N-6)[(3N-6)-1]}{2} = \frac{(3N-6)(3N-5)}{2}.$$

[16] To illustrate isotopic effects, here are the wavenumbers corresponding to the fundamental vibration frequencies for some of the isotopic forms of CO_2:

	$\tilde{\nu}_1$ (cm^{-1})	$\tilde{\nu}_2$ (cm^{-1})	$\tilde{\nu}_3$ (cm^{-1})
$^{12}C^{16}O_2$	1388	667	2349
$^{13}C^{16}O_2$	1370	648	2283
$^{16}O^{12}C^{18}O$	1259	663	2332
$^{16}O^{13}C^{18}O$	1342	643	2266
$^{12}C^{18}O_2$	1230	657	2314

(See also the isotopic forms of H_2O in Table 8.3.)

assume that at equilibrium the molecule is linear and symmetric, with both C—O distances equal to R_0 (1.162 Å). We define the coordinate system shown in Fig. 8.16, with the equilibrium molecular axis taken as the x axis. The *local displacement coordinates* x_i, y_i, z_i for each nucleus are then its Cartesian coordinates measured from its equilibrium position.

We have a total of nine local displacement coordinates, but only four ($= 3N - 5$) of these can vary independently. Thus there must exist five relationships among the local displacement coordinates, which we can derive from the momentum conservation laws. With our coordinate system fixed in the molecule, the total linear momentum of the nuclei is not only conserved but equal to zero. Thus we have for the three Cartesian components

$$\sum_{i=1}^{3} p_{xi} = \sum_{i=1}^{3} p_{yi} = \sum_{i=1}^{3} p_{zi} = 0, \qquad (8.42)$$

with the sums taken over the three atoms. Similarly, since the coordinate system rotates with the molecule, conservation of angular momentum about the center of mass gives us

$$\sum_{i=1}^{3} L_{xi} = \sum_{i=1}^{3} L_{yi} = \sum_{i=1}^{3} L_{zi} = 0. \qquad (8.43)$$

The first of Eqs. 8.43 is of no use to us: Since the nuclei remain near the x axis at all times, the individual L_{xi} are negligibly small. But the other five of the above equations can give us useful interrelationships.

Note that the momentum components are defined by $p_{xi} \equiv m_i(dx_i/dt)$, and so on, and that for a harmonic oscillation the velocity dx_i/dt is proportional to the displacement x_i a quarter-cycle later. Thus we can replace Eqs. 8.42 by $\sum_i m_i x_i = 0$, and so on, which for the CO_2 molecule become

$$m_O(x_1 + x_3) + m_C x_2 = 0,$$
$$m_O(y_1 + y_3) + m_C y_2 = 0, \qquad (8.44)$$
$$m_O(z_1 + z_3) + m_C z_2 = 0.$$

Similarly, the only significant angular momentum components are those that can be approximated for small displacements as $L_{yi} = \pm m_i R_0 (dz_i/dt)$ and $L_{zi} = \pm m_i R_0 (dy_i/dt)$ (for the end atoms only). Making the same substitution as above, we convert the second and third of Eqs. 8.43 to

$$m_O R_0(z_1 - z_3) = 0,$$
$$m_O R_0(y_3 - y_1) = 0. \qquad (8.45)$$

Equations 8.44 and 8.45 are the five relationships among the local displacement coordinates that we sought. Combining them to solve for the displacements, we obtain unique solutions for the y and z coordinates,

$$y_1 = y_3, \qquad y_2 = -\frac{2m_O}{m_C} y_1, \qquad (8.46)$$

and

$$z_1 = z_3, \qquad z_2 = -\frac{2m_O}{m_C} z_1. \qquad (8.47)$$

For the x coordinates we have $x_2 = -(m_O/m_C)(x_1 + x_3)$, which yields two solutions consistent with the symmetry of the molecule (since atoms 1 and 3 are indistinguishable):

$$x_1 = -x_3, \qquad x_2 = 0, \qquad (8.48)$$

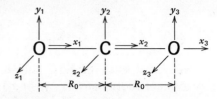

FIGURE 8.16
Coordinate system for the CO_2 molecule. The local coordinates x_i, y_i, z_i for each nucleus are measured relative to its equilibrium position. The molecular center of mass is at the equilibrium position of the C nucleus.

or

$$x_1 = x_3, \qquad x_2 = -\frac{2m_O}{m_C} x_1. \qquad (8.49)$$

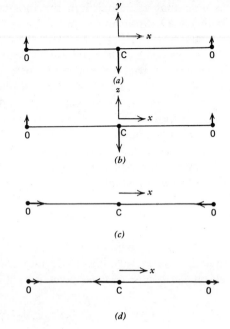

FIGURE 8.17
Normal modes of vibration of the CO_2 molecule: (a), (b) bending, Eqs. 8.46 and 8.47, respectively; (c) symmetric stretching, Eq. 8.48; (d) asymmetric stretching, Eq. 8.49.

Note that each of Eqs. 8.46–8.49 involves only a single set of displacement coordinates (x_i, y_i, or z_i). This implies that the x's, y's, and z's can vary independently of one another, and that each of these equations alone can describe a possible vibration of the molecule for small displacements from equilibrium. We can thus identify Eqs. 8.46–8.49 with the four normal modes of vibration of the CO_2 molecule; a detailed calculation would show that they indeed have all the properties of normal vibrations outlined in the previous section. The normal modes of CO_2 are illustrated in Fig. 8.17. Two of these modes, those in which only the x coordinates vary, are called *stretching* modes, one symmetric and the other asymmetric. The other two modes are *bending* modes in the y and z directions; since the two bending modes differ from each other only in their direction, they are physically indistinguishable and thus degenerate.

In each mode the relative motions of the nuclei are given by the appropriate one of Eqs. 8.46–8.49. Thus in the y-bending mode, if the oxygen atom in position 1 vibrates away from equilibrium by an amount described by the vector

$$(x_1, y_1, z_1) = (0, y_1, 0),$$

then Eq. 8.46 tells us that the other oxygen atom must simultaneously be at

$$(x_3, y_3, z_3) = (0, y_1, 0)$$

and the carbon atom at

$$(x_2, y_2, z_2) = \left(0, -\frac{2m_O}{m_C} y_1, 0\right) \approx (0, -\tfrac{8}{3} y_1, 0).$$

The normal coordinate for this mode is thus a combination of y_1, y_2, and y_3 in the ratio $1 : (-\tfrac{8}{3}) : 1$. Similar analyses are easily made for the other modes. The symmetric stretching mode involves motion of only the oxygen atoms, but in all three of the other modes the light carbon atom moves nearly three times as far as the heavier oxygens. In each mode the three nuclei remain in phase with one another, oscillating with the same frequency ν_α. Remember, however, that all this is true only for the normal modes themselves: an ordinary molecular vibration is in general a linear combination of all the normal modes, with the nuclei not in phase.

To calculate the normal frequencies ν_α themselves, one must know the potential energy surface, or at least the force constants, which give the curvature of the surface along the normal coordinates. However, our previous knowledge of the behavior of oscillators at least permits us to order the modes by frequency. The two bending modes must have identical frequencies, since they are physically indistinguishable. For a given amount of atomic displacement, the stretching modes involve greater changes in nearest-neighbor distances, and thus in bond energies, than do the bending modes. Hence the stretching modes must have higher frequencies than the bending modes. Finally, since lighter particles can move more rapidly, the asymmetric stretching mode, mainly a motion of the C atom, has a higher frequency than the symmetric stretching mode, motion of O atoms only. Thus we expect the order of the frequencies to be

$$\nu(\text{asymm. stretch}) > \nu(\text{symm. stretch}) > \nu(y\text{-bending}) = \nu(z\text{-bending}).$$

Experiment bears this out. The characteristic frequencies of $^{12}C^{16}O_2$ obtained from spectroscopy are assigned as follows[17] ($\tilde{\nu} = \nu/c$):

$$\tilde{\nu}_3 \text{ (asymmetric stretch)} = 2349 \text{ cm}^{-1},$$
$$\tilde{\nu}_1 \text{ (symmetric stretch)} = 1388 \text{ cm}^{-1},$$
$$\tilde{\nu}_2 \text{ (bending)} = 667 \text{ cm}^{-1}.$$

This kind of ordering is generally unambiguous. Occasionally one cannot predict whether the stretching motion of a heavy atom will have a higher or lower frequency than the bending motion of one or more hydrogen atoms. But the rules of thumb that

$$\nu(\text{stretch}) > \nu(\text{bend}) \quad \text{and} \quad \nu(\text{light}) > \nu(\text{heavy}) \tag{8.50}$$

are reliable generalizations for most molecules.

In the harmonic oscillator approximation, the individual normal modes of oscillation act independently. Each mode may contain any number of vibrational quanta from zero upward. Thus the vibrational energy level of the CO_2 molecule is defined by specifying all four quantum numbers, the v_α of Eq. 8.40. There are clearly many ways to distribute a given amount of energy among several modes, so the polyatomic molecule has a much higher energy-level density or number of levels per unit energy than does the diatomic molecule. This property is illustrated in Problem 5 at the end of the chapter; its significance will become clear in Chapter 21, where we discuss heat capacities in terms of the distribution of energy over molecular energy levels.[18] On simple probability grounds, in most of the possible states the quanta are approximately evenly distributed among the normal modes, just as if we threw a large number of dice the great majority of possible results would have the dice evenly distributed among their six "modes." This is the so-called equipartition of energy. In the next section we shall look into the rules governing transitions among these numerous vibrational energy levels.

8.8
Transitions and Spectra of Polyatomic Molecules

We have already had a good deal to say about the mechanisms governing transitions from one energy level to another. The general principles were outlined in Section 4.5, and in Section 7.1 we applied these principles to the vibrations of diatomic molecules. Let us now see how they can be extended to molecules with three or more atoms.

Recall that a pure vibrational transition can occur in a diatomic molecule only when the dipole moment varies in the course of a vibration. A similar rule holds for polyatomic molecules, except that one must speak of individual modes of vibration. Pure vibrational transitions can occur only in those modes for which the molecular dipole moment varies with the normal coordinate. If there is no such change, then an oscillating electromagnetic field exerts a net force of zero on the molecule at all times, just as in a homonuclear diatomic molecule, and there can be no interaction between the vibrational mode and the field. But if the dipole moment does vary with a given normal coordinate, then the molecule can exchange energy with a field whose frequency is the same as that of the normal mode in question, by a mechanism like that illustrated in Fig. 7.2.

[17] The numbering of the fundamental modes (ν_1, ν_2, ν_3) is conventional, corresponding to that shown for SO_2 in Fig. 8.15, and tabulated for other molecules in Table 8.3: ν_1 and ν_3 are primarily bond-stretching modes, with the end atoms moving in opposite directions in ν_1, the same direction in ν_3; whereas ν_2 is primarily a bending mode. Symmetric vibrations are always listed first, in order of decreasing frequency.

[18] Another application will be in our treatment of the kinetics of unimolecular reactions (Sections 30.14 and 31.1).

Modes in which the dipole moment varies are called *infrared-active*; those that leave the dipole moment unchanged are *infrared-inactive*.

To illustrate the foregoing argument, let us again consider the CO_2 molecule. Of the normal modes shown in Fig. 8.17, the only one in which the dipole moment does not change with the normal coordinate is the symmetric stretching mode, for which μ is at all times identically zero. This mode thus cannot absorb electromagnetic radiation, at least not within the approximations we have been using. The asymmetric stretching mode, on the other hand, clearly has a time-varying molecular dipole moment. The average value of the moment is, of course, zero, corresponding to the equilibrium configuration about which the oscillation takes place. If the molecule is placed in an electromagnetic field oscillating at $\tilde{\nu} = 2349 \text{ cm}^{-1}$ ($\nu = 7.04 \times 10^{13} \text{ s}^{-1}$), then the field reverses direction at the same frequency as the transient dipole moment, and energy can be transferred between the molecule and the field. This energy transfer must, of course, involve a gain or loss of vibrational quanta in the mode in question. The two bending modes also involve variation of the instantaneous molecular dipole moment, and can thus exchange energy with a field oscillating at $\tilde{\nu} = 667 \text{ cm}^{-1}$.

The selection rules for such vibrational transitions are again derived from Eq. 7.12, and are similar to those for diatomic molecules. That is, to the extent that the harmonic oscillator approximation applies, we have for each normal mode the rule

$$\Delta v_\alpha = \pm 1, \tag{8.51}$$

corresponding to Eq. 7.13 for the diatomic case. Also as in the diatomic case, each allowed vibrational transition gives rise to a band of vibration-rotation lines. Since the vibrations are not truly harmonic, one also observes *overtone bands* with $\Delta v_\alpha = \pm 2$, ± 3, and so on, as well as something not possible for diatomic molecules, *combination bands* in which two or more normal modes undergo transitions simultaneously,[19] but these are in general much weaker than the fundamental bands.

In Section 7.3 we described the nature of Raman scattering. Analogous to the rule for ordinary vibrational transitions, a given normal mode of a polyatomic molecule is Raman-active only if the polarizability varies with the normal coordinate. For any molecule with a center of symmetry, infrared-active modes are Raman-inactive, whereas Raman-active modes are infrared-inactive (the *rule of mutual exclusion*). Thus in CO_2 and other linear AB_2 molecules the symmetric stretching mode (ν_1) is Raman-active only, and the other modes (ν_2 and ν_3) are infrared-active only. This rule does not apply if the molecule has no center of symmetry. The selection rule of Eq. 8.51 also applies to Raman vibrational transitions.

As for rotational transitions, the same rule holds as for diatomic molecules: A pure rotational spectrum is observed only for molecules with a permanent dipole moment. We have already noted (Section 8.6) that linear polyatomic molecules have the same rotational properties as diatomic molecules; the selection rules are again $\Delta J = \pm 1$ for the infrared spectrum, $\Delta J = 0, \pm 2$ for the Raman spectrum. Other types of molecules (Section 3.12) have more complicated rotational spectra; spherical tops (such as CH_4) and nonlinear symmetric tops (such as NH_3) are of course not found among triatomic molecules. Since a molecule's rotational

---

[19] A mode that is inactive by itself can give a band in combination with another mode. For example, CO_2 has a band at 2076 cm^{-1} corresponding to transitions with $\Delta v = +1$ in both the bending mode and the (normally inactive) symmetric stretching mode.

constants are inversely proportional to its moments of inertia, one can determine bond lengths by comparing the rotational spectra of isotopically-substituted molecules, just as one uses the vibrational spectra of such molecules to determine the force constants.

Finally, of course, we come to transitions between electronic states of a molecule. We know that each such state can be represented by a potential energy surface in multidimensional space. Let us say more about the properties of such surfaces.

The classical vibration of a diatomic molecule corresponds to the motion of a particle of mass μ back and forth along the molecule's $V(R)$ curve. A horizontal line represents the molecule's fixed total energy; the height of this line above the $V(R)$ curve gives the instantaneous kinetic energy as a function of R. All this can be represented in two dimensions because the potential energy of a diatomic molecule is a function of a single variable, the distance R.

Polyatomic molecules require us to be more sophisticated in visualizing the vibrational energy levels. If a molecule has n degrees of freedom, its potential energy "surface" requires a space of $n+1$ dimensions: n for the normal coordinates or equivalent variables, one for the energy. For an example that we can visualize, consider a hypothetical system with only two degrees of freedom, a CO_2 molecule somehow constrained to prevent bending. Such a molecule would have only the two stretching modes, and its energy could be represented in three dimensions. If we choose the x axis to represent displacement in the symmetric stretching mode and the y axis to represent the asymmetric stretch displacement, leaving the z axis for the energy, we obtain a surface like that shown in Fig. 8.18. Within the harmonic approximation the surface is an elliptic paraboloid, extending upward from a single minimum at the molecule's equilibrium position. Each horizontal cross section of the paraboloid is an ellipse bounding the classically allowed region for a given total energy. The principal axes of all these ellipses correspond to the directions of maximum and minimum vertical curvature, and thus to the normal coordinates of the molecule. Quantum mechanically, only certain energy levels are allowed, and in each level the vibrational wave function is a product of harmonic oscillator functions for each coordinate. All these results are easily extended mathematically (though not pictorially) to the case of three or more degrees of freedom.

What we have been describing, of course, is the potential surface for only the ground state of CO_2, and indeed only an approximation to that. Each electronic state of a molecule has its own potential surface, just as each state of a diatomic molecule has its own potential curve. As in the diatomic case, we know that the vibrations must become anharmonic for large displacements; the true potential surfaces are thus not parabolic, but either "level off" or go toward infinity at extreme values of the normal coordinates. The various surfaces need not have even approximately the same shapes. For example, some of the excited states of CO_2 are bent rather than linear in their equilibrium geometries.

All this should make it clear that the spectra associated with electronic transitions in polyatomic molecules can be extremely complicated. Whether a given transition is allowed depends, at the lowest level of interpretation, on the relationship between the symmetry classes of the two electronic states. This is a subject we cannot go into here, except to say that the rules are generalizations of those for diatomic molecules. As in diatomic molecules, the spectrum of a given electronic transition consists of a band with a complex vibrational and rotational fine structure. Electronic transitions can often be interpreted well in molecular orbital

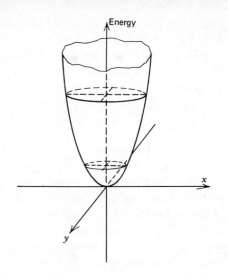

FIGURE 8.18
Potential energy surface for a system with two vibrational degrees of freedom.

terms, on the assumption that one electron at a time changes orbitals, with the others left unchanged.

In our discussion of diatomic molecules we mentioned a number of phenomena that can occur when two potential curves cross (Section 7.8). The same can occur in polyatomic molecules when two surfaces intersect: The molecule can move from one surface to the other, that is, make a radiationless transition from one electronic state to the other. Such crossings are in fact far more important in polyatomic molecules than in diatomics. One reason for this, which we have already mentioned, is the high density of vibrational energy levels in a polyatomic molecule. For a transition between two bound states, the smaller the interval between energy levels, the more likely it is that levels in two electronic states will interact. In addition, if the new electronic state has a lower potential surface than the initial state, it also has many more vibrational levels into which the molecule can drop (by emission or collision) after the transition. Such subsidiary processes tend to make the initial transition irreversible. Processes like predissociation and autoionization are also more likely for polyatomic molecules, since a given molecule can usually split into many different sets of fragments.

The mechanism just described is one of the most important ways in which energy is degraded, that is, becomes less available. Suppose that an excited state is initially produced by a single quantum of high-energy radiation; a radiationless transition to a lower bound state (*internal conversion*) then transforms much of the electronic energy into vibrational energy, and collisions further degrade the vibrational energy into rotational and finally translational energy, that is, into heat. The original energy increment is split into smaller and smaller quanta, the reassembly of which becomes progressively less likely. Furthermore, the conversions become easier as the spacing between the energy levels involved becomes smaller. Vibrational energy can be converted into thermal motion much more readily than could the original electronic excitation.

We have touched here upon collisions, which are one of the two major mechanisms for inducing transitions between energy levels (the other, of course, being absorption or emission of electromagnetic radiation). Indeed, collisions are the dominant mechanism for vibrational and rotational energy exchange in solids, liquids, and even gases at atmospheric pressure. Radiative processes, which we have stressed because of their importance in studying molecular energy levels, become dominant only in extremely rarefied gases (as in the upper atmosphere and outer space). We shall discuss the mechanisms of energy transfer by collision at some length in Part Three of this book; here let us note only some general principles. Any kind of molecular transition can be induced by collision; the greater the energy difference involved, the more impulsive a collision is required, that is, the more energy, either kinetic or internal, the molecules must have and the closer they must approach each other. Transfer of energy can also occur between one type of motion (translational, rotational, vibrational, electronic) in one molecule and the same or another type in the other molecule. As we have already indicated, the probability of such energy transfer is greatest when the energy levels involved are closest together; thus in general the probabilities fall in the order (T = translation, R = rotation, V = vibration)

$$T\text{–}T > T\text{–}R > R\text{–}R > T\text{–}V \approx R\text{–}V > V\text{–}V.$$

One noteworthy exception to this ranking occurs when two transition processes are *resonant*, that is, have ΔE's that match very closely. Thus vibrational energy transfer between identical molecules is often very fast.

Similarly, there may be "accidental" resonances, such as that between the vibrational frequency of N_2 ($\tilde{\nu} = 2360$ cm^{-1}) and the asymmetric stretching mode of CO_2 ($\tilde{\nu} = 2349$ cm^{-1}), and energy transfer between such modes is easy.

Further Reading

Avery, J., *Quantum Theory of Atoms, Molecules and Photons* (McGraw-Hill Book Co., Inc., New York, 1964), Chapter 7.

Herzberg, G., *Electronic Spectra and Electronic Structure of Polyatomic Molecules* (D. van Nostrand and Co., Princeton, N.J., 1966).

Herzberg, G., *Infrared and Raman Spectra* (D. van Nostrand and Co., Princeton, N.J., 1945).

Kondratyev, V., *The Structure of Atoms and Molecules* (P. Noordhoff N. V., Groningen, The Netherlands, 1964), Chapter 8.

Molecular Spectroscopy, A Specialist Periodical Report, Vols. 1 ff. (The Chemical Society, London, 1973 et seq.).

Slater, J. C., *Quantum Theory of Molecules and Solids, Volume 1. Electronic Structure of Molecules* (McGraw-Hill Book Co., Inc., New York, 1963), Chapters 5–7.

Wilson, E. B., J. C. Decius and P. C. Cross, *Molecular Vibrations* (McGraw-Hill Book Co., Inc., New York, 1955), Chapters 2, 4, 6, and 8.

Problems

1. What are the symmetry operations of an equilateral H_3 molecule? Which correspond to the symmetry operations of an isosceles H_3 molecule? In what sense is the symmetry of the equilateral triangle higher than that of the isosceles?

2. Show how the orbitals φ_{2s}, φ_{2px}, φ_{2py}, φ_{2pz}, and $\varphi_{2px} \pm i\varphi_{2py}$ transform if we take

    a) $x \rightarrow -x, y \rightarrow -y, z \rightarrow -z$ (inversion);
    b) $x \rightarrow x, y \rightarrow y, z \rightarrow -z$ (reflection in the x, y plane);
    c) $x \rightarrow -x, y \rightarrow y, z \rightarrow z$ (reflection in the y, z plane).

3. Construct three normalized hybrid orbitals of the form $a\varphi_{2s} + b\varphi_{2px} + c\varphi_{2p\pi}$ with two of the orbitals equivalent and the third different. Give a, b, and c explicitly in terms of the angle between the two equivalent hybrids.

4. Figure 8.8 shows orbitals of types a_1, b_1, and b_2, but none of type a_2. Construct or sketch an orbital that would be of symmetry type a_2 for the H_2O molecule.

5. Construct molecular orbitals with symmetry appropriate to a linear geometry for the molecule MgF_2, using the atomic orbitals with $n = 3$ for Mg and with $n = 2$ for F as the basis set. First suppose that only the $3s$ and $3p$ orbitals of Mg need to be included; then suppose that the $3d$ orbitals should also be included. Which orbitals have the same symmetry properties? If the orbitals are required to be normalized and orthogonal, how many mixing parameters still remain to be computed?

6. Construct molecular symmetry orbitals for a nonlinear PF_2 radical (FPF). Use only the atomic $2p$ orbitals for fluorine and the orbitals with $n = 3$ for

phosphorus. First assume only the $3p$ orbitals are important; then include the $3d$ orbitals as well.

7. Write out the full meaning—that is, the explicit sum—represented by the abbreviated notation for the determinantal representation of the electronic ground state of the H_3 molecule

$$\Psi(\mathbf{r}_1, \mathbf{r}_2, \mathbf{r}) = \text{const.} \cdot \|\varphi_A(1)\bar{\varphi}_B(2)\varphi_C(3)\|,$$

where φ_A, φ_B, and φ_C are $1s$ atomic orbitals on hydrogen atoms A, B, and C, respectively.

8. What is the geometric meaning of the statement

$$\frac{\partial^2 V(X_1, X_2)}{\partial X_1 \, \partial X_2} = \frac{\partial^2 V(X_1, X_2)}{\partial X_2 \, \partial X_1}$$

for all X_1, X_2 in some region? How is this connected to the smoothness of V?

9. Show from the properties of determinants or by explicit calculation that the determinantal form of an n-electron wave function automatically satisfies the Pauli exclusion principle, in that no determinantal wave function exists in which two or more electrons occupy the same orbital and have the same spin.

10. Write out the explicit Hartree–Fock equations for the normally occupied molecular orbitals of a triangular H_3 molecule. How many independent equations are there? Be sure to include all the terms of the potential, with each orbital, spin, and variable of integration specified. (You may use vector notation for the coordinate variables.)

11. Obtain the exact solutions to the equation

$$\begin{Vmatrix} H_{11} - E & H_{12} \\ H_{21} & H_{22} - E \end{Vmatrix} = 0.$$

Now suppose $H_{12} = H_{21}$. What is the form of E if $H_{11} = H_{22}$? Suppose H_{12} can be controlled by the observer; how does the difference between the two values of E depend on H_{12}? Now suppose $|H_{12}| \ll H_{11} - H_{22}$. How does the difference between the two values of E depend on $|H_{12}|$ in this case? How do the two kinds of behavior connect? (Use your general solution.)

12. A subject of some controversy at one time was the pair of related questions regarding CH_2: is the ground state a singlet state or a triplet state, and is the ground state linear or bent? Explain, by using an argument based on the molecular orbitals of this molecule, why both pairs of alternatives are plausible and why one might expect the answer to the first question to imply the answer to the second.

13. Consider the 4×4 factor of Eq. 8.25. Make the approximating assumption that one root can be found by supposing that only the diagonal elements and the first row and column of the array are important. By expanding the determinant so obtained, find the approximate value of E. Explain the physical assumption inherent in this approximation by discussing the meaning of the off-diagonal quantities H_{ij}, with $i \neq j$. The formula you obtain is called the second-order perturbation expression for E.

14. Using Walsh's rules and the molecular orbitals of NH_2, predict the spin and structure—linear or bent—of this molecule. Would you expect any change in the bond angle in the series NH_2, PH_2, AsH_2, SbH_2?

15. When it was discovered that the water molecule has a dipole moment, electrostatic models were proposed to account for a low enough symmetry in this molecule to permit a nonzero moment. Show that a "suitable" model is one in which the protons are simple positive point charges at fixed equal distances from the oxygen, which is a doubly-negative polarizable ion, and that the interaction between the two induced dipoles together with the

Coulomb attractions and repulsion give rise to a finite H—O—H angle α, as shown.

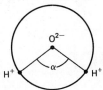

16. Construct the occupied molecular orbitals of the bent ozone molecule and show qualitatively how the energy of each orbital changes as the molecule is a) bent toward a liner configuration and b) bent further away from a linear configuration.

17. Make a diagram showing all the vibrational levels of the CO_2 molecule for which the total vibrational energy is less than $5000\ cm^{-1}$. What is the total number of levels in the neighborhood of $5000\ cm^{-1}$ of excitation? (Each combination, such as 1 quantum in the y-bend, 2 quanta in the z-bend or 1 quantum in the symmetric stretch, counts as a *single energy level*.) Plot the mean density of levels, choosing a wide enough interval of energy for averaging to make a reasonably smooth curve. Compare this with the mean density of levels of a single one-dimensional harmonic oscillator with frequency $667\ cm^{-1}$.

18. Describe the potential surface for a CO_2 molecule constrained to bend in the xy and xz planes. Show that, with two quanta appropriately assigned, the motion can be construed as a sort of rotation. What degeneracy is associated with this rotation? What is the classical picture of the way this rotation occurs? How would the potential surface change if the molecule were permitted to undergo symmetric stretching also?

19. Describe the potential surface for a molecule such as NO_2 whose equilibrium geometry is that of an isosceles triangle and which is constrained to undergo only bending (nondegenerate for a nonlinear molecule) and symmetric stretching.

20. Show, by using a classical argument, that, within the approximation of a spatially uniform field, a diatomic molecule whose dipole moment is independent of internuclear distances does not absorb electromagnetic radiation to undergo a vibrational transition.

21. Which modes of vibrational motion of the NO_2 molecule can absorb energy from an oscillating, spatially-uniform electromagnetic field?

22. Show that CO_2 may undergo rotational transitions and exchange energy with an oscillatory electromagnetic field if the field has significant variation in strength over a range comparable with the size of the molecule.

23. The pure rotational absorption spectrum of the linear molecule OCS shows the following transitions:

		ν(MHz)	B_0 (MHz)
$^{16}O^{12}C^{32}S$	$J = 1 \rightarrow 2$	24325.92	6081.480
	$3 \rightarrow 4$	48651.7	
	$4 \rightarrow 5$	60814.1	
$^{16}O^{12}C^{34}S$	$J = 1 \rightarrow 2$	23731.33	5932.840
	$3 \rightarrow 4$	47462.3	

Use these data to deduce the O—C and C—S bond lengths in OCS. In the case of this linear triatomic molecule,

$$I = \frac{1}{m_O + m_C + m_S}[m_O m_C d_{CO}^2 + m_C m_S d_{CS}^2 + m_O m_S (d_{CO} + d_{CS})^2],$$

where d_{CO} and d_{CS} are the CO and CS bond lengths.

24. Sketch the potential for the stretching modes of a linear triatomic molecule ABA, whose equilibrium geometry has two different A—B distances.

9

Larger
Polyatomic
Molecules

In this chapter we shall survey the properties of molecules with four or more atoms. We shall emphasize the structural characteristics of these polyatomic molecules, especially the ways in which bonding affects the spatial arrangement of atoms or groups. One of the key concepts in the study of polyatomic molecules is *separability*. Such molecules can often be treated as being composed of subunits (radicals, functional groups, ligands) bonded together in such a way as to have a chemical identity almost independent of their surroundings. We shall discuss how this separability comes about, and why it sometimes does not.

We begin by considering small polyatomic modecules, such as the tetrahedral molecule CH_4 (methane). Species like this can be understood easily if we extend the ideas we developed in Chapter 8. Larger molecules have extended structures that can be classified into two main groups, although some species have characteristics of both. These groups are the *catenated* and the *polyhedral* molecules.

Catenated ("chainlike") molecules are those in which a chain of atoms linked together gives the molecule its essential structure. The chain may be straight, branched, or bent into a ring, and usually other atoms or groups are attached to it. Typical examples are the alkanes, C_nH_{2n+2}, and the simple alcohols, $C_nH_{2n+1}OH$, and a large fraction of the other compounds of carbon, the subject matter of organic chemistry. Among inorganic substances, catenated species include the extended structures formed by boron, and by silicon or aluminum with oxygen. Some of the discussion of very extended structures will be reserved for Chapter 11, on solids, because some kinds of solids are nothing more than indefinitely large catenated molecules. A polyhedral compound is a species in which a central atom (or ion) is bound to several *ligands*, which may be atoms, ions, molecules, or molecular fragments. Usually the central atom can be thought of as being at the center of a polyhedron, the vertices of which are defined by the atoms to which the central atom is bound. Many simple molecules and ions like CH_4 or SO_4^{2-} are of this polyhedral type; the term "coordination compound" is sometimes used for weakly bound polyhedral species—in fact, those in which the central atom and the ligands are separable in solution reactions near room temperature. The best known of these are the complexes composed of a metal cation surrounded by a nearest-neighbor shell of neutral or charged ligands, for example, the ion $Pt(NH_3)_4Cl_2^+$. We shall have much to say about the bonding and other properties of these complexes. Proceeding to still weaker bonds, we shall look at such loosely defined species as solvated ions, and finally (in the next chapter) at the interactions between distinct molecules.

The first group of polyatomic molecules we shall examine are analogous to those discussed in Chapter 8, where, in fact, we gave some consideration to the bonding in species like BH_3. These are what we have been calling polyhedral molecules, with a central atom bound to other atoms at the vertices of a polyhedron. Some examples are the plane triangle BH_3, the tetrahedron CH_4, the triangular pyramid NH_3, and the trigonal bipyramid PF_5, each with a characteristic set of bond angles. As in Chapter 8, we wish to be able to interpret and predict molecular shapes in order to see what factors determine the bond angles.

The most familiar and in many ways the simplest interpretation of bond angles is that based on hybridization. The bonding electrons around the central atom are assumed to occupy hybrid orbitals, linear combinations of the atomic orbitals so defined as to have a strong directional character. In Section 8.3 we outlined this method and described the construction of sp (linear) and sp^2 (trigonal) orbitals; the latter gives a good description of the bonds and structure of the BH_3 molecule. Extending the method to a combination of the $2s$ and all three $2p$ orbitals (sp^3 hybridization), one can show that the four orbitals

$$\varphi_1 = \tfrac{1}{2}(2s + 2p_x + 2p_y + 2p_z),$$
$$\varphi_2 = \tfrac{1}{2}(2s + 2p_x - 2p_y - 2p_z),$$
$$\varphi_3 = \tfrac{1}{2}(2s - 2p_x + 2p_y - 2p_z),$$
$$\varphi_4 = \tfrac{1}{2}(2s - 2p_x - 2p_y + 2p_z),$$

$$(9.1)$$

are completely equivalent and symmetric, and directed toward the corners of a regular tetrahedron. Figure 9.1 indicates their distribution. Thus sp^3 orbitals are appropriate to describe the bonding in the tetrahedral CH_4 molecule. In the localized-bond approximation each of these orbitals can be combined with a $H(1s)$ orbital to form two σ-type pair orbitals, one bonding and normally occupied, the other antibonding and normally empty. Experimental determination of the structure confirms that the four C—H bonds are equivalent and arranged tetrahedrally, with a bond angle of $109°28'$. Nearly all other compounds[1] in which a carbon atom forms four single bonds have bond angles within a few degrees of this value, and can be described in terms of sp^3 hybridization.

One can similarly devise hybridization schemes to correspond to all the other bond geometries commonly observed. For larger central atoms this often requires the use of d orbitals; this is clearly the case when the atom forms more than four bonds because more than four atomic orbitals must participate. For the hybridization model to be applicable, all the atomic orbitals contributing to the hybrid must be fairly close in energy. Figure 9.2 illustrates some of the most common molecular geometries and the corresponding types of hybridization.

9.1
Small Molecules

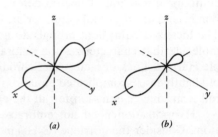

(a) *(b)*

FIGURE 9.1

(a) The spatial coordinates for the orbitals of Eq. 9.1 and the spatial distribution for the $2p_x$ orbital, to exemplify the basic p orbital. (b) The spatial distribution of the hybrid φ_2 of Eq. 9.1.

FIGURE 9.2

Some common types of small symmetric molecules. In each diagram the solid lines represent the bonds formed by the central atom (●), whereas the dashed lines indicate the polyhedron (or plane figure) formed by the atoms (○) to which the central atom is bound. Bond angles are indicated; in each case except the trigonal and pentagonal bipyramids, all the angles are the same and the bonds equivalent. Each symmetry type is labeled with its name, a hybridization scheme that gives the appropriate bond directions, and an example of a molecule with that symmetry.

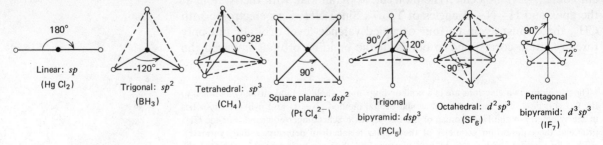

Linear: sp
($HgCl_2$)

Trigonal: sp^2
(BH_3)

Tetrahedral: sp^3
(CH_4)

Square planar: dsp^2
($PtCl_4^{2-}$)

Trigonal bipyramid: dsp^3
(PCl_5)

Octahedral: d^2sp^3
(SF_6)

Pentagonal bipyramid: d^3sp^3
(IF_7)

[1] The exceptions are discussed in Section 9.4.

Combinations of hybrid orbitals are ordinarily used to portray localized bond orbitals. This is basically a valence bond approach, and it was in the early development of the VB theory that hybrid orbitals were first introduced. In the usual molecular orbital treatment, as we outlined it in Section 8.4, one obtains delocalized orbitals extending over the whole molecule. A molecular orbital calculation for tetrahedral CH_4 shows that the occupied orbitals of the valence shell consist of one s-like orbital ψ_1, of the form $\alpha\varphi_1 + \beta(h_1 + h_2 + h_3 + h_4)$ where h_j is the $1s$ orbital on hydrogen j, and, at another energy, three degenerate p-like orbitals ψ_2, ψ_3, ψ_4 each comprised of the orbitals φ_2, φ_3, and φ_4, with the hydrogen $1s$ orbitals.[2] What has happened to the tetrahedral symmetry? We can regenerate it by taking four linear combinations of the MOs, of exactly the same form as the combinations of atomic orbitals in Eq. 9.1, but with ψ_1 replacing $2s$ and ψ_2, ψ_3, and ψ_4 in place of the atomic $2p$ orbitals. The resulting *equivalent orbitals* are functions that have their contours of maximum electron density pointed along the lines of the four bonds, and are thus more appropriate for calculations of bond properties. The localized equivalent orbitals do not have the symmetry of the entire molecule that characterizes the natural stationary-state solutions of simple Hartree–Fock one-electron Schrödinger equations. They are solutions of slightly more complicated equations, which are still consistent with the full Schrödinger equation for all the electrons.

Hybridization need not embrace all the bonds formed by a given atom. Consider the ethylene (C_2H_4) molecule, which is found to have the planar skeletal structure

$$\underset{H}{\overset{H}{\diagdown}} C - C \underset{H}{\overset{H}{\diagup}},$$

with three 120° bond angles around each carbon. This structure is interpreted as the result of sp^2 hybridization like that in BH_3 (Section 8.3), using the $2s$ and two of the $2p$ orbitals on each carbon to form three trigonal σ bonds. The remaining $C(2p)$ orbitals, which have their axes perpendicular to the plane of the molecule, combine to form a bonding π orbital like that in the isoelectronic O_2 molecule. Thus we have one σ bond and one π bond between the two carbon atoms, corresponding to the double bond of the conventional formula $H_2C{=}CH_2$. Similarly, the linear acetylene molecule ($H{-}C{\equiv}C{-}H$) is interpreted as having sp (linear) hybridization on each carbon; the leftover $2p$ orbitals form two bonding π orbitals at right angles to each other, as in the isoelectronic N_2, and the single $C{-}C$ σ bond and two π bonds constitute the conventional triple bond. The orbitals in ethylene and acetylene are shown in Fig. 9.3.

Thus far we have spoken only of equivalent hybrids, but these are applicable only in the most highly symmetric molecules. Consider the ammonia (NH_3) molecule. The molecule is pyramidal, with the N atom at the apex and H–N–H angles of 106.7°. Since NH_3 is isoelectronic with CH_4, there must again be four occupied valence-level orbitals, but only three of these correspond to bonds. If the bonding orbitals involved the

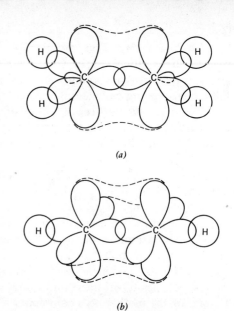

(a)

(b)

FIGURE 9.3

Schematic representation of orbitals in (a) ethylene and (b) acetylene, with the C—H and C—C σ bonds interpreted in terms of hybrid orbitals and the π-bonding orbitals described by molecular Hartree–Fock-like orbitals.

---

[2] The remaining two electrons are in a nonbonding "inner-shell" orbital virtually identical to the $C(1s)$ orbital. The reason the molecular orbital method gives s- and p-like orbitals lies in the symmetry of the Hamiltonian of the one-electron Schrödinger equation. For the CH_4 problem, the equilibrium geometry of the regular tetrahedron determines that symmetry. With only s and p orbitals of the central atom and the four hydrogenic $1s$ orbitals, the stationary-state functions must have the same angular dependence as the free atom.

N(2p) electrons only, without the participation of the N(2s) orbital, the bond angle would be only 90°. On the other hand, an equivalent mixture as in CH_4 (each orbital $\frac{1}{4}s$ and $\frac{3}{4}p$) would give the tetrahedral angle[3] 109.5°. Thus the hybrid orbitals should be intermediate between these two extremes. A mixture of about 22% s and 78% p for the bonding orbitals gives the observed angle. We used a similar argument for the H_2O molecule in Section 8.3. In a quantitative treatment, one should properly vary the amount of mixing to optimize the energy, then see how well this predicts the bond angle. In practice, one must frequently go to extensive MO calculations—approaching the Hartree–Fock model—to make quantitatively accurate predictions of bond angles (cf. Fig. 8.9 for H_2O).

The hybridization model is generally useful in describing simple first-row compounds. When boron is the central atom, one usually picks the same sorts of hybrids as for carbon. In compounds of nitrogen, oxygen, and fluorine, the contribution of the 2s orbital becomes successively less important. The reason is that, as the nuclear charge increases, the 2s electrons become more tightly bound and farther from the 2p in energy (cf. Fig. 7.21). In HF, isoelectronic with CH_4, NH_3, and H_2O, one can to a good approximation ignore hybridization and speak of a simple $\sigma 2p$ bond.

Hybridization without optimization does not really explain anything. It gives us a convenient way to interpret molecular shapes, but we must know the geometry in order to determine the hybridization. One of the simplest techniques for actually predicting shapes (but not necessarily values of angles) was developed by Mulliken and Walsh: One estimates how the energies of the molecular orbitals vary with bond angle, then judges on semiempirical grounds how these variations balance out to minimize the total energy. In Sections 8.4 and 8.5 we outlined the reasoning leading to "Walsh's rules" in triatomic molecules; here we shall merely state some of the results[4] (in terms of the number of valence electrons):

> AH_3: ⩽6 valence electrons, planar (BH_3); 7–8 valence electrons, pyramidal (CH_3, NH_3, H_3O^+); 10 valence electrons, planar (none known).
> HAAH: 10, linear (C_2H_2); 12, bent planar (N_2H_2); 14, bent nonplanar (H_2O_2).
> AB_3: ⩽24, planar (BF_3, CO_3^{2-}, NO_3^-, SO_3); 25–26, pyramidal (ClO_3, IO_3^-, PF_3); 28, planar (ClF_3).
> H_2AB: ⩽12, planar (H_2CO); 14, nonplanar (H_2NF).

The concepts involved here are at best somewhat diffuse,[5] but Walsh's rules are a useful guide, and do lead us to the correct answer far more often than not.

---

[3] In fact, the NH_4^+ ion, with the same number of electrons but one more nucleus, is tetrahedral.

[4] See A. D. Walsh, *J. Chem. Soc.* (1953), p. 2260ff.

[5] The problem is, as we pointed out in connection with Fig. 8.9, that the "energy" of Walsh's arguments often does not correspond closely to the actual orbital energy. Moreover, the total energy is not simply a sum of orbital energies: cf. Eq. 8.25. Recently, Coulson and Deb have argued that the quantity most appropriate to be represented by the vertical axis in a Walsh diagram is the work done when, say, a bond angle is varied. This approach has been used to obtain a set of molecular-shape rules including but more extensive than Walsh's; see B. M. Deb et al., *J. Am. Chem. Soc.* **96**, 2030, 2044 (1974). The highest occupied MO seems to have a dominant effect on the molecular geometry.

An alternative approach, based on the localized-electron-pair viewpoint, is also useful in predicting molecular shapes qualitatively, particularly for larger molecules and molecules with very low symmetry. This method (which we also mentioned at the end of Section 8.5), was developed primarily by Gillespie and Nyholm,[6] and is sometimes called the valence-shell electron-pair repulsion (VSEPR) theory. Its basic premises are (1) that electrons in an atom's valence shell can be considered as localized pairs, whether bonding or nonbonding (lone pairs); (2) that all these electron pairs repel one another, and are distributed in space so as to minimize the repulsion; (3) that lone pairs repel more strongly than bonding pairs, so the repulsive interactions decrease in the order lone pair–lone pair > lone pair–bonding pair > bonding pair–bonding pair.

Premise (2) predicts that an AX_n molecule in which all the electron pairs around A are bonding will have its bonds as far apart as possible. One can easily show that the molecule must then have one of the following symmetric structures:

Number of Electron Pairs	Structure	Example
2	Linear	$HgCl_2$
3	Triangular	BCl_3
4	Tetrahedral	CH_4
5	Trigonal bipyramid	PCl_5
6	Octahedral	SF_6
7	Pentagonal bipyramid	IF_7

Most of these are illustrated in Fig. 9.2. Molecules with lone pairs are assumed to have essentially the same structures, but with the lone pairs occupying some of the vertices; the shapes of most such types of simple molecules are shown in Fig. 9.4. The structures are of course distorted by the greater repulsion of the lone pairs; this effect is in accord with the decreasing bond angles in the sequences

$$CH_4 \ 109.5°, \ NH_3 \ 106.7°, \ H_2O \ 104.5°;$$

$$SiH_4 \ 109.5°, \ PH_3 \ 93.3°, \ H_2S \ 92.2°;$$

$$GeH_4 \ 109.5°, \ AsH_3 \ 91.8°, \ H_2Se \ 91.0°;$$

$$SnH_4 \ 109.5°, \ SbH_3 \ 91.3°, \ H_2Te \ 89.5°.$$

The decrease in angle as one goes from NH_3 to SbH_3, or from H_2O to H_2Te, is associated with the increasing size (or decreasing electronegativity) of the central atom: The farther out the bonding pairs are, the smaller the angle at which their mutual repulsion reaches equilibrium.

Useful as the VSEPR theory is, it cannot predict molecular structures in all cases without additional information. The model naturally tends to fail when two or more alternative structures lie rather close in energy. A very good example of this is the molecule ClF_3. The valence shell of the central Cl atom, according to a localized-pair picture, consists of three Cl—F bonding pairs and two lone (nonbonding) pairs. Applying premise (2) above, we would predict that the five pairs are arranged approximately along the axes of a trigonal bipyramid, as in the PCl_5 molecule (Fig. 9.2). But not all the positions in a trigonal bipyramid are equivalent: In PCl_5, for example, the three *equatorial* P—Cl bonds are 2.04 Å long, whereas the two *axial* bonds are 2.19 Å long. Taking this difference into account, we can formulate three possible structures for ClF_3, as shown in Fig. 9.5; these structures should differ in energy and other properties. The angles between lone pairs in structures I, II, and III are about 180°, 120°, and 90°, respectively. If the lone pair–lone pair repulsion were the only significant effect, we would expect I to be the true structure. But the lone pair–bonding pair and bonding pair–bonding

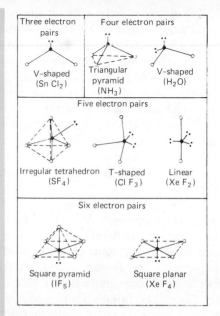

FIGURE 9.4
Shapes of small molecules with lone pairs of electrons. The diagrams are drawn as in Fig. 9.2, with the lone pairs indicated by —: symbols. The cases are classified by the number of electron pairs around the central atom, and for each shape a molecular example is given. The shapes shown here are those actually observed, and do not include all the conceivable configurations; see the text for discussion of this point.

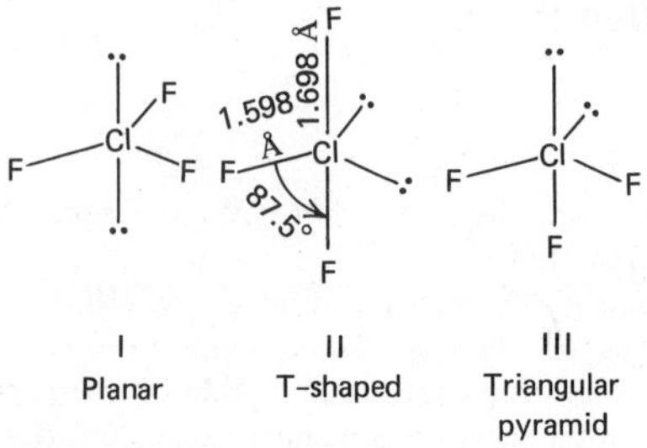

FIGURE 9.5
Possible structures of the ClF_3 molecule. The experimentally observed structure is closest to II, with the dimensions shown on the figure.

---

[6] See R. J. Gillespie and R. S. Nyholm, *Quart. Rev.* **11**, 339 (1957); R. J. Gillespie, *J. Chem. Educ.* **40**, 295 (1963).

pair repulsions are less in structure II, and there are more of these; there is no way to tell *a priori* from a qualitative model which of these effects is dominant. In fact, experiment shows that the true structure is II, with all the atoms in a plane but one F atom distinctly different from the other two. Measurements on other five-electron-pair molecules show that we can interpret the structure of ClF_3 as though the lone pairs always occupy the equatorial positions in the trigonal bipyramid, as we have indicated in Fig. 9.4. Apparently the lone pair–lone pair repulsion does not vary significantly at angles beyond 120°. In six-electron-pair molecules, by contrast, the six positions are all equivalent, and two lone pairs always occupy *trans* positions as shown for XeF_4.

The localized-pair model has yet to be quantified in any comprehensive way. We do know that both Coulomb and exchange forces contribute to the repulsion between electron pairs. It is also known from calculations in specific cases that the exchange contribution is often very small when a molecule is near its equilibrium geometry. This suggests that a pure Coulomb repulsion model might be adequate for predicting molecular structures. The simplicity of such a model makes it attractive for computations. One simply treats the ligands as point charges sliding over the surface of a sphere, and determines the geometry of lowest energy. This method usually does give a correct picture, in agreement with experiment or more elaborate calculations.

Either Walsh's rules or the VSEPR method is adequate if one merely wishes to know the approximate equilibrium structure of a molecule. But more accurate calculations are required for many purposes, especially in studying the dynamics of rearrangements, as in molecular vibrations or reaction mechanisms. In such cases one most commonly uses delocalized molecular orbitals, which are convenient for computations. The variations of the MO method should be familiar by now. In order of increasing accuracy, one can use a semiempirical approach with variable parameters, *ab initio* self-consistent-field calculations, or a model including configuration interaction to account for electron correlation. Yet in some ways, the more one refines the calculation for quantitative accuracy, the more one loses the interpretive simplicity of elementary models. For instance, the complete electron density (that is, $|\psi|^2$) of a molecule cannot be partitioned in any obvious way into bonds and lone pairs, nor even into one-electron orbitals. Much research in recent years has been devoted to finding methods of extracting simple interpretive concepts and generalizations from the exact wave function.

In the future one can expect to see the structures of larger and larger molecules interpreted in terms of the microscopic forces between electrons and nuclei. The calculation of accurate bond lengths requires primarily a knowledge of the very strong interactions between nearest-neighbor atoms. Bond angles are governed largely by the interactions between next-nearest neighbors, and structural factors like the "staggered" ends of the H_2O_2 and C_2H_6 molecules require consideration of still more distant interactions. As the distance increases the strength of the interaction decreases, and the computations needed to account for it become more lengthy as small effects add up. Meanwhile, chemists and spectroscopists can do much to interpret these structural factors with simple models like the stretching and bending motions described in the last chapter. But we must remember that bonds are the result of electron distributions around nuclei in space, not the springs holding rigid balls together that molecular-vibration theory implies. The simplifications of such models are powerful tools, but they merely extract the most important factors; they do not provide an accurate representation of reality.

Most of the molecules we have examined thus far are simple polyhedra. However, there are several classes of compounds whose molecules have large, extended structures, made possible by the capacity of atoms to link to one another in long chains of consecutive bonds. The compounds to which we refer have definite structures and compositions; they consist of well-defined molecules, and are not to be confused with solids, some of which are infinitely extended molecules. We begin with the compounds of carbon, for which the problems of bonding and electronic structure have been studied extensively. In the next section we shall look at a variety of inorganic extended structures, the study of which is less well developed.

It is a commonplace that more compounds of carbon are known than of any other element except hydrogen, itself contained in most carbon compounds. This myriad of carbon compounds exists primarily because carbon atoms can bond to one another to produce chains of any length, and because a variety of other atoms or groups can be attached at many positions along such a chain. Such compounds are referred to as *catenated* or *chainlike*. Only sulfur exhibits an ability for "self-catenation"—bonding to like atoms to form chains—comparable to carbon's. Liquid sulfur consists mainly of a mixture of S_8 rings and very long chains of ($\sim 10^5$) S atoms which also comprise "plastic sulfur." But since such sulfur chains have no free valences to attach other groups, their chemistry is comparatively impoverished. One might expect silicon to form structures like those of carbon, but silicon so readily binds oxygen that Si—O—Si rather than Si—Si is the basis for its extended structures.[7] Carbon is unique in its chemical versatility—and fortunately so, for we literally live on its chemistry.

The bonding in carbon compounds is most commonly described in terms of localized bonding orbitals. For useful approximate descriptions of the contribution of each carbon atom to these bond orbitals, we turn to the various hybrid atomic orbitals already discussed: sp^3 for 4-coordinated carbon, with four bonds arranged tetrahedrally as in CH_4; sp^2 for three of the bonds around a 3-coordinated (planar) carbon, with the remaining electrons forming a $2p\pi$ bond as in C_2H_4; and sp for two bonds on a 2-coordinated (linear) carbon, with two $2p\pi$ bonds as in C_2H_2. The hybrid orbitals on adjacent carbon atoms overlap constructively to produce the localized bond orbitals. The structures of most organic compounds can be understood without going beyond this level of approximation.

We know that electrons are not really localized in bonds, and that a correct model of a molecule's electronic structure should allow the electrons to wander anywhere in the molecule. If we wanted to use a delocalized MO description, a different set of orbitals would be required for every compound. Yet the localized description is extremely powerful. For example, the C—C and C—H bonds in ethane (C_2H_6) are quite similar to those in pentane (C_5H_{12}), decane ($C_{10}H_{22}$), or any other hydrocarbon with no double or triple bonds. As long as one compares groups of atoms that have the same kinds of bonds, the properties of a group in one compound can be used to predict the properties of the same group in another compound, almost without regard to the rest of either molecule. Properties of this sort are said to exhibit *transferability*.

9.2
Catenated Carbon Compounds; Transferability

---

[7] Some long-chain silicon compounds (up to at least $Si_{14}F_{30}$) are known, but these react with oxygen so readily that they are unstable in the presence of air or water.

TABLE 9.1
CHARACTERISTIC PARAMETERS OF COMMON STRUCTURAL UNITS

Group	Molecule	Bond Length (Å) or Angle (°)	Vibration Frequency (cm^{-1})
C—C	C_2H_6	1.54	993
	C_2H_5Cl	1.54	972
	CH_2Cl—CH_2Cl	1.55	1052
C—H	C_2H_6	1.10	2975
	C_2H_5Cl	1.1	2890–2983
	CH_2Cl—CH_2Cl	(assumed 1.09)	2950–3005
	CH_3OH	1.096	2834–2980
	H_2C=O	1.09	2843
C—O	CH_3OH	1.43	1030
C=O	H_2C=O	1.22	1746
O—H	CH_3OH	0.96	3328
C—C—Cl	C_2H_5Cl	110°	336 (bend)
	CH_2Cl—CH_2Cl	112°	300 (bend)

Some examples of transferability are given in Table 9.1. The bond lengths and angles are established by the usual methods, based on either rotational spectra or x-ray and electron diffraction. The vibration frequencies come directly from the infrared spectrum, with their assignments to particular groups based partly on normal coordinate analyses, but mainly on correlations among the spectra of related compounds. Indeed, the latter is one of the clearest and most important applications of transferability. The vibration frequencies associated with many functional groups are so characteristic, and often so easy to measure, that one can readily use them for purposes of identification. For example, carbonyl ($>$C=O) groups always have a vibration-rotation band at about 1800 cm^{-1}, as indicated in Table 9.1; other clear-cut examples include the hydroxyl stretching frequency at about 3600 cm^{-1}, the nitrile (—C≡N) stretching band in the region 2100–2250 cm^{-1}, and the acetylenic (—C≡C—) band near 2100 cm^{-1} but varying from about 2020 to 2300 cm^{-1}. In fact, all the groups common in organic chemistry have their own characteristic patterns of infrared bands. Infrared spectroscopy has thus become an indispensable physical tool for the rapid characterization of materials. Besides the properties in Table 9.1, others that show transferability include bond energies (see Section 14.8) and even chemical reactivities. Many macroscopic properties such as heat capacity, critical constants and surface tension have also been analyzed empirically as sums of group contributions.[8]

How does transferability come about? How can it be that electrons behave as though they were localized when we know that ultimately they are not? The answer lies largely in the ease with which delocalization occurs—more specifically, in the smallness of the energy differences

---

[8] Many such empirical correlations can be found in Robert C. Reid and Thomas K. Sherwood, *The Properties of Gases and Liquids*, 2nd ed., (McGraw-Hill, New York, 1966).

between the delocalized orbitals. For example, the valence-electron MOs in CH_4 have an energy range not much greater than that between the carbon $2s$ and $2p$ orbitals. If certain of the occupied delocalized MOs are not only built from the same types of atomic orbitals but are very similar in energy as well, then to a reasonably good approximation one can consider the delocalized orbitals as degenerate. The significance of this lies in an important property of any degenerate set of eigenfunctions of an operator: One can always construct from those functions linear combinations that are also eigenfunctions—but that may have quite different spatial properties from the original functions. This is just the method used to construct the localized "equivalent orbitals" of CH_4, which we mentioned in the last section. If the valence MOs were actually degenerate, the localized and delocalized descriptions would be equally valid. As it is, the energies are close enough for the localized description to be a useful approximation. In all saturated hydrocarbons and many other organic compounds, the normally occupied MOs based primarily on carbon $2s$ and $2p$ orbitals have energies similarly close together, and the properties of (for example) the C—C and C—H bonds are largely transferable.

We can express this reasoning in more formal terms, in a way that will be useful when we study the structure of solids. Suppose that we begin our description of a molecule with a set of atomic orbitals (probably hybridized), from which we construct a set of localized bonding orbitals φ_j. These localized orbitals are clearly different from the delocalized molecular orbitals ψ_k, which are the eigenfunctions of a one-electron Hartree–Fock operator $H(1)$ like that defined in Eq. 8.16: $H(1)\psi_k = \epsilon_k \psi_k$. The localized φ_j's, on the other hand, are not eigenfunctions of $H(1)$; thus the operation of $H(1)$ on φ_j does not multiply φ_j by an eigenvalue, but creates a new function $H(1)\varphi_j$. The degree to which $H(1)$ "spoils" the functional form of φ_j gives the "amount" by which φ_j is not a true eigenfunction. This "amount" is measured by the values of integrals of the form $\int \varphi_i^* H(1)\varphi_j \, d\tau$, which would vanish[9] (for $i \neq j$) if φ_i and φ_j were eigenfunctions of $H(1)$. One can then determine the true (stationary-state) eigenfunctions ψ_k by finding linear combinations of the φ_j's for which the integrals $\int \psi_k^* H(1)\psi_j \, d\tau$ vanish. This is just the standard MO calculation with a different kind of basis function. The process described in the previous paragraph is simply the same calculation carried out in reverse, the construction of localized φ_j's as linear combinations of the delocalized ψ_k's.

The interesting thing about the localized bond orbitals φ_j is that those between two atoms of any given elements typically have nearly the same expectation value of the energy. This is particularly so if the hybridization or coordination number is specified. For example, the integral $\int \varphi_j^* H(1)\varphi_j \, d\tau$ has nearly the same value for all single-bond orbitals φ_j between pairs of carbon atoms, especially when the carbons have the same coordination number. The corresponding integrals for φ_j's representing C—H bonds have a different value, but are again all similar to one another. In short, the localized orbitals fall into groups corresponding to different kinds of bonds, with the functions in each such group nearly degenerate with one another. This corresponds to the empirical observation of characteristic bond energies.

Thus we can form localized orbitals φ_j that fall into nearly degenerate sets. If the delocalized MOs are written as linear superpositions of

---

[9] This is the property of orthogonality of eigenfunctions, for which see Appendices 6A and 9A.

these φ_j's,

$$\psi_k = \sum_j c_{jk}\varphi_j, \qquad (9.2)$$

then the expectation values of the MO energies are given by

$$\epsilon_k = \frac{\int \psi_k{}^*\mathsf{H}(1)\psi_k \, d\tau}{\int \psi_k{}^*\psi_k \, d\tau} = \frac{\sum_i \sum_j c_{ik}c_{jk}\int \varphi_i{}^*\mathsf{H}(1)\varphi_j \, d\tau}{\sum_i \sum_j c_{ik}c_{jk}\int \varphi_i{}^*\varphi_j \, d\tau} = \frac{\sum_i \sum_j c_{ik}c_{jk}H_{ij}}{\sum_i \sum_j c_{ik}c_{jk}S_{ij}}, \qquad (9.3)$$

with H_{ij} and S_{ij} defined as in Eq. 8.20. The magnitude of the "off-diagonal" H_{ij}'s $(i \neq j)$ corresponds roughly to the rate at which electrons exchange between bonds. The smaller these integrals are, on the other hand, the more nearly the localized orbitals approach stationary states. But if the φ_j's for a given set of bonds are nearly degenerate and the off-diagonal H_{ij}'s can be neglected, then Eq. 9.3 implies that the ψ_k's built mainly from those bonds must also be nearly degenerate. In other words, the better the approximation that the electrons behave like localized pairs, the smaller are the energy separations between the delocalized MO states.

What happens as the size of the molecule increases? The more atoms there are in a molecule, the more φ_j's and ψ_k's we have. Now the exchange integrals $H_{ij}(i \neq j)$ are usually of appreciable magnitude only for adjacent (nearest-neighbor) bonds; and the number of adjacent-bond pairs increases much more slowly than the total number of possible pairings of bonds. For example, in a straight-chain hydrocarbon C_nH_{2n+2}, there are $n-1$ C—C bonds, $n-2$ pairs of adjacent C—C bonds, but a total of $(n-1)(n-2)$ pairs of C—C bonds, adjacent or not. Since the energy spacing between the delocalized MOs (of a given type) depends mainly on the H_{ij}, the average spacing between the ϵ_k decreases as the size of the molecule increases. When the energy levels in a given set become very closely spaced (compared, say, with the mean thermal energy), they can be treated as a continuum—or rather as a band of states, a continuum with upper and lower energy limits. This is just what happens in a solid, and we shall see in Chapter 11 that this is the key to the interpretation of the electronic structure of solids.

Localized bond orbitals, then, are rather good substitutes for the delocalized stationary-state molecular orbitals ψ_k. If one can construct localized orbitals representing not individual bonds but entire fragments of molecules, these "group orbitals" should be still better approximations to the ψ_k. In fact, it is possible to obtain good wave functions for the polyatomic building blocks of large molecules. These functions not only describe the group properties well, but as a set are quite good for representing all but the most global properties of the entire molecule. One is not even restricted to one-electron functions; it is quite feasible to use many-electron functions that include electron correlation. This type of calculation is called "molecules in molecules," and is an extension of the similar "atoms in molecules" method in which accurate wave functions for free atoms are used as basis functions. Both methods lend themselves well to semiempirical parameterization when accurate wave functions for the building blocks are not available. The formulation of wave functions of large molecules in terms of the wave functions of well-defined constituents is the mathematical expression of the assumption of transferability of group properties.

TABLE 9.2

BINDING ENERGIES OF ELECTRONS IN PARTICULAR GROUPS AS DETER-
MINED BY PHOTOELECTRON SPECTROSCOPY[a]

Molecule	Group	Binding Energy (eV)
Ethane C_2H_6	C—C bond	11.5
	C—H bond	14.7
n-Butane C_4H_{10}	C—C bond	10.6 (probably several C—C bonds)
	C—H bond	14.4
n-Hexane C_6H_{14}	C—C bond	10.3
Ethylene C_2H_4	C=C π electron	10.5
	C—C σ electron	12.4
	C—H	14.3
1-Butene C_4H_8	C=C π electron	9.6
	C—C (probably both true single bonds and σ bond of double bond)	11.3
	C—H	14.5
cis-2-Butene C_4H_8	C=C π electron	9.1
	C—C	11.2
	C—C	12.4
Benzene C_6H_6	C=C (aromatic π electrons)	9.24, 11.5
Methanol CH_3OH	O—H	10.8
Ethanol C_2H_5OH	O—H	10.5
Phenol C_6H_5OH	C=C (aromatic π electrons) O—H	8.5, 9.4, 11.3
Acetone $(CH_3)_2C=O$	C=O	9.7
2-Butanone $CH_3COCH_2CH_3$	C=O	9.5

[a] From M. J. S. Dewar and S. D. Worley, *J. Chem. Phys.* **50**, 694 (1969).

For a further illustration of the transferability of properties of
chemical subunits, let us consider the electronic binding energies, as
measured by photoelectron spectroscopy. Table 9.2 gives some examples.
As with vibration frequencies, the assignments to particular groups are
based on correlation of measurements for many compounds, supplemented by observing the effects of specific chemical changes, and
reinforced by some theoretical interpretation of why the spectra appear
where they do. According to Koopmans' theorem the binding energies
should approximately equal the MO energies, and thus should be quite

close to the energies of group orbitals. In fact, the energies for particular groups remain clustered about a common average energy throughout a series, shifting only by small, systematic amounts.

In certain classes of molecules the concept of localized properties breaks down badly. Some of these breakdowns are due to what are called steric strain effects, consequences of the physical interference or blockage of one group by another, which will be discussed in Section 9.4. In other instances a significant part of the electron density belongs to one-electron states that spread over many atoms and offer high mobility for electrons to move from one region of the molecule to another. This phenomenon is generally a property of molecules that have both multiple and single bonds in a formal valence bond picture; one might say that some of the electron density "leaks" out of the multiple bonds.

Delocalization is most striking in the so-called *aromatic* carbon compounds, and occurs to some degree throughout the broader class of unsaturated compounds of carbon. The classic example is, of course, the simplest aromatic hydrocarbon, benzene (C_6H_6). The cyclic benzene molecule contains six equivalent carbons, one hydrogen atom attached to each carbon, and *six equivalent carbon–carbon bonds*. The simplest valence picture of benzene is the classic pair of Kekulé structures, written with alternating double and single bonds:

Yet experiment shows clearly that all the carbon–carbon bonds have identical properties; for example, all are 1.397 Å long, intermediate between the "normal" lengths of C—C (1.537 Å) and C=C (1.335 Å) bonds. The simplest way to represent this equivalence theoretically is to make the molecular wave function an equal mixture (superposition) of the wave functions corresponding to the two Kekulé structures. The equivalence of the bonds is taken to mean that the π electrons constituting the formal double bonds in these structures are so mobile that, on any conceivable experimental time scale for an experiment, the π electrons are best represented as a superposition of wave functions corresponding to the two Kekulé structures.

Our most significant clue to the properties of benzene comes from the molecule's energy. The total energy of all the bonds (the energy required to separate the molecule into its constituent atoms) is measured to be about 1.6 eV greater than the total energy of the bonds in a single Kekulé structure, calculated from "standard" C—C and C=C bond energies derived from molecules such as propylene (CH_3—CH=CH_2). In Section 14.8 we describe how these standard bond energies are obtained. The difference between the real and hypothetical energies of all the bonds is called the *resonance energy*. The historical basis is that, in a valence bond representation, the most important contribution to the difference of real and hypothetical energies is the delocalization achieved by superposing two physically equivalent—and therefore resonant—Kekulé structures. It is generally accepted that delocalization is the most important factor giving rise to resonance energies. As in the H_2 molecule, however, the superposition of only two valence bond structures, although a good first approximation, is quite inadequate for a quantitative description of benzene.

A second clue to the special character of benzene and other aromatic hydrocarbons is their relative chemical inertness compared with hydrocarbons containing "normal" double bonds, such as ethylene ($CH_2{=}CH_2$) or butadiene ($CH_2{=}CH{-}CH{=}CH_2$). Molecules such as ethylene, the unsaturated or olefinic hydrocarbons, are quite reactive toward the addition of halogen molecules,

$$CH_2{=}CH_2 + Br_2 \rightarrow CH_2Br{-}CH_2Br,$$

or of hydrogen halides,

$$CH_2{=}CH{-}CH_3 + HBr \rightarrow CH_3{-}CHBr{-}CH_3.$$

Although olefinic molecules undergo such reactions readily, aromatic molecules such as benzene, naphthalene, and anthracene add halogens, hydrogen halides, and other species only under rather severe conditions. In other words, the π electrons of the aromatic systems appear on chemical grounds to be more strongly held than those of olefins.

A third property of aromatic systems that suggests a special character for their π electrons comes from the behavior of these molecules in magnetic fields. As we mentioned in Section 5.1, one can use the protons of hydrogen atoms in molecules as probes of the local magnetic field in their vicinity (proton magnetic resonance or nuclear magnetic resonance spectroscopy). Such measurements show that the protons of benzene experience a local magnetic field significantly smaller than that observed at protons in typical nonaromatic hydrocarbons. The interpretation that best describes the origin of this lower field is that the external magnetic field induces a ring current, in which the delocalized π electrons circulate rapidly around the ring. This ring current in turn induces a local magnetic field that acts to *oppose* the external field outside the current loop, where the hydrogen atoms of benzene lie. Aromatic compounds are virtually unique in exhibiting this property, which is so characteristic of delocalized electrons.

Before we go on to interpret the electronic structure of aromatic molecules, we should point out two quantities that can be used only with great caution as indications of aromatic behavior. One is the ionization potential; the other is bond length. Ionization potentials of aromatic and olefinic molecules are in fact rather similar. By careful choice one can arrive at suitable comparisons. For example, the first ionization potential of a hypothetical cyclohexatriene, corresponding to a single Kekulé structure, would surely be less than the first ionization potentials of cyclohexene, cyclopentene, or cyclopentadiene (8.9, 9.1, and 8.5 eV, respectively), yet the first ionization potential of benzene is 9.25 eV. This further illustrates the higher binding energy (of the π electrons) as a result of delocalization. Yet care must be used in such comparisons; in particular, one must compare compounds whose sizes are comparable. The reason is that the ionization potentials of small molecules tend to be higher than those of their higher homologs. For example, ethylene has a first ionization potential of 10.5 eV, propylene 9.7 eV.

As for bond lengths, we have noted that the 1.397 Å carbon–carbon bond of benzene is shorter than the 1.537 Å C—C bond of ethane but longer than the 1.338 Å C—C bond of ethylene. It is thus frequently said that the amount of double-bond character (or π-bond order) in benzene is intermediate between that in ethane (single bond, or π-bond order of 0) and that in ethylene (double bond, corresponding to a π-bond order

TABLE 9.3
CARBON–CARBON BOND LENGTHS FOR VARIOUS COMPOUNDS[a]

Molecule	Bond Type	C—C Bond Length (Å)
Ethane, CH_3—CH_3	sp^3-sp^3	1.537
Propylene, CH_2=CH—CH_3	sp^3-sp^2	1.51
Methyl acetylene, CH≡C—CH_3	sp^3-sp	1.459
Butadiene, CH_2=CH—CH=CH_2	sp^2-sp^2	1.476
Vinyl cyanide, CH_2=CH—C≡N	sp^2-sp	1.426
Cyanoacetylene, CH≡C—C≡N	sp-sp	1.376
Benzene, C_6H_6 (any bond)	sp^2-sp^2	1.397
Graphite (any bond)	sp^2-sp^2	1.421
Ethylene, CH_2=CH_2	sp^2-sp^2	1.338
Allene, CH_2=C=CH_2	sp^2-sp	1.309
Butatriene, CH=C=C=CH	sp-sp	1.285
Acetylene, CH≡CH	sp-sp	1.205

[a] From H. J. Bernstein, *Trans. Faraday Soc.* **57**, 1649 (1961).

of 1). The π-bond order is the bond order p minus the σ-bond order, normally 1. There is a very good correlation between bond length R_e and MO bond order p (cf. Section 7.6) proposed by C. A. Coulson,

$$p = \frac{1.02 - 0.53 R_e}{0.235 R_e - 0.16},$$
(9.4)

where R_e is expressed in angstroms. In benzene the MO calculation gives a bond order of $1\frac{2}{3}$, in agreement with Eq. 9.4. The difficulty in the present context is in the level of subtlety required to isolate a connection between bond length or bond order and aromatic character. To some extent the gross behavior of carbon–carbon bond lengths can be interpreted in terms of the hybridization of the two carbon atoms, as Table 9.3 indicates. Several "single" C—C bonds, such as those of vinyl cyanide, CH_2=CH—C≡N, and cyanoacetylene, HC≡C—C≡N, are comparable to the "aromatic" bond lengths of benzene and graphite. Presumably this is because the π electrons nominally in the adjacent multiple bonds are actually somewhat delocalized. Delocalization is especially likely to occur when "single" and "multiple" bonds alternate (*conjugated* multiple bonds), as in butadiene, vinyl cyanide, and cyanocetylene—and in the nominal structures of benzene and other aromatic ring compounds.

The simplest theory for aromatic hydrocarbons, the *Hückel molecular orbital* (HMO) *theory*, is virtually an archetype for simple, phenomenological models. One treats explicitly only the π electrons, which are assumed to be responsible for such chemical and physical properties as the optical spectrum, the ease of attack of particular sites by other reagents, the local charge densities on atoms and in bonds, and the bond orders. The rest of the electrons and the nuclei, which define the geometry of the usually planar molecular skeleton but not necessarily the precise values of bond lengths, appear only as part of an effective potential field for the π electrons. The π electrons are described by molecular orbitals ψ_j which are represented as linear combinations of the carbon $2p\pi$ orbitals φ_k, the

$2p$ orbitals extending perpendicular to the molecular plane. For benzene we have

$$\psi_j = \sum_{k=1}^{6} c_{jk}\varphi_k ; \tag{9.5}$$

note that each of these π MOs extends over the entire ring. The coefficients c_{jk} must be determined by the variation method, as outlined in Section 8.4. To simplify the process, the HMO theory makes further assumptions. The only interactions large enough to require inclusion in the Hamiltonian are those (1) between an electron on a given carbon atom and the other electrons and nucleus of that atom, and (2) between an electron on a given carbon atom and the electrons and nuclei of adjacent atoms to which that atom is bonded. In benzene each of these classes of interactions is the same for all the carbon atoms, so we can define

$$\alpha \equiv \int \varphi_j{}^* \mathsf{H}_{\mathrm{eff}}\varphi_j \, d\tau \qquad \text{(for all } j), \tag{9.6}$$

$$\beta \equiv \int \varphi_j{}^* \mathsf{H}_{\mathrm{eff}}\varphi_k \, d\tau \qquad \text{(atoms } j \text{ and } k \text{ bonded)}, \tag{9.7}$$

where $\mathsf{H}_{\mathrm{eff}}$ is the effective one-electron Hamiltonian for a π electron. In terms of Eq. 8.20, we set $H_{jj} = \alpha$ for all j, $H_{jk} = \beta$ for bonded atoms, $H_{jk} = 0$ otherwise. The values of α and β are empirical parameters to be determined from experiment. The overlap integrals S_{jk} of Eq. 8.20 are taken as unity on a given atom, zero between different atoms, or $S_{jk} = \delta_{jk}$. Substituting these approximations into the secular equation 8.24, we have for benzene

$$\begin{Vmatrix} \alpha-\epsilon & \beta & 0 & 0 & 0 & \beta \\ \beta & \alpha-\epsilon & \beta & 0 & 0 & 0 \\ 0 & \beta & \alpha-\epsilon & \beta & 0 & 0 \\ 0 & 0 & \beta & \alpha-\epsilon & \beta & 0 \\ 0 & 0 & 0 & \beta & \alpha-\epsilon & \beta \\ \beta & 0 & 0 & 0 & \beta & \alpha-\epsilon \end{Vmatrix} = 0, \tag{9.8}$$

the lowest three roots of which are $\epsilon_1 = \alpha + 2\beta$, $\epsilon_2 = \epsilon_3 = \alpha + \beta$, because $\beta < 0$. Putting six electrons in these three orbitals gives a total π-electron energy of $6\alpha + 8\beta$; in the same approximation a localized π bond has the energy[10] $\alpha + \beta$, so one of the Kekulé structures has the π-electron energy $6\alpha + 6\beta$. The *delocalization energy* of benzene is thus -2β, which should equal the empirical resonance energy; thus we estimate $\beta \approx -0.8\,\mathrm{eV}$. Similar calculations for a number of aromatic compounds give values of β that cluster closely around $-0.69\,\mathrm{eV}$ ($-67\,\mathrm{kJ/mol}$). This indicates clearly that in these compounds delocalization is too important to be neglected.

Benzene is of course not the only type of aromatic compound. Many such compounds consist of fused benzene rings, with the π electrons effectively delocalized over the entire ring system; some examples are shown in Fig. 9.6a. Note that the fraction of hydrogen atoms decreases as the size of the system increases. In the limit we have pure carbon in the

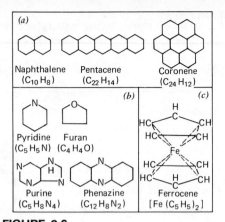

FIGURE 9.6
Some aromatic compounds. (*a*) Fused benzene rings. (*b*) Heterocycles. (*c*) Ferrocene.

---

[10] The localized-bond secular equation is

$$\begin{Vmatrix} \alpha-\epsilon & \beta \\ \beta & \alpha-\epsilon \end{Vmatrix} = 0,$$

giving $\epsilon = \alpha \pm \beta$, with $\epsilon = \alpha + \beta$ lower in energy.

form of graphite, which consists of parallel infinite sheets of fused six-carbon rings (see Chapter 11). The C—C distance in graphite is 1.421 Å, close to the 1.395 Å typical of aromatic compounds. The parallel sheets, on the other hand, are 3.35 Å apart and bound together very weakly by overlap of the π-electron clouds. Hence the layers slide over or off one another quite easily, the property that makes graphite so suitable for pencil leads. There are also aromatic systems with rings containing one or more atoms other than carbon (heterocyclic compounds); Fig. 9.6b shows several, including pyridine (isoelectronic with benzene), furan (also with six π electrons), and some fused systems. Also isoelectronic with benzene is the $C_5H_5^-$ (cyclopentadienyl) ion, which forms the remarkable series of "sandwich" compounds typified by *ferrocene* (dicylopentadienyl-iron), Fig. 9.6c; the bonding in the rings is aromatic, and the π electrons bond to iron by overlapping with empty $3d$-orbitals on the iron atom. Benzene forms similar "sandwich" compounds, such as $Cr(C_6H_6)_2$.

We have outlined the principles that underlie the great number and variety of carbon compounds. A particularly rich assortment of these compounds are found in living systems, each evolved to fit a specific function. In addition to the variations in the carbon skeleton, the many kinds of active functional groups that can be attached to that skeleton result in a wide range of chemical behavior. Another significant factor is the stability of so many carbon compounds under what we call "ordinary conditions." Chemistry would surely be very different in an atmosphere containing much hydrogen rather than oxygen, or at an ambient temperature of 500°C. But as things are, only carbon appears to form such a myriad of compounds with such a varied chemistry.

9.3
Other Extended Structures

Carbon may be unique in the variety of its extended structures, but it is not the only element that forms them. Let us now look at some examples from among the inorganic substances.

We have already mentioned the Si—O—Si linkage so important to silicon chemistry. Many silicon–oxygen compounds have structures built from interlinked SiO_4 tetrahedra. Such compounds typically have a framework of polymeric silicate anions, with cations simply inserted in gaps in the structure. These anions may form infinite chains, flat or pleated sheets, or three-dimensional networks. Typical structures for the first two cases are shown in parts a and b of Fig. 9.7. These structures are reflected in the physical properties of the materials. Asbestos, a chain silicate, readily splits into thin fibers, which is apparently the reason it is such a health hazard, whereas mica, a sheet silicate at the molecular level, cleaves into thin sheets. A three-dimensional silicon–oxygen network in which every oxygen atom is bound to two silicon atoms is simply silica, SiO_2.

If some of the silicon atoms are replaced by aluminum atoms, the resulting structures are polymeric anions. The charge-balancing cations in these structures are rather loosely attached to the oxide sites, whereas the anionic framework is both rigid and rather open. As a consequence, these aluminosilicates make good cation exchangers. Silicate and aluminosilicate minerals make up most of the earth's crust.

Other extended structures are formed by organosilicon compounds, in which carbon atoms are directly bonded to silicon. These are called *siloxanes* when the framework is still built up from the basic Si—O—Si linkage, with organic groups replacing the singly bound oxygen atoms of the polysilicates. In Fig. 9.7c we show such a linear polysiloxane, the structure of which is quite analogous to the linear silicate of Fig. 9.7a; similarly, there are sheet polysiloxanes with structures like Fig. 9.7b. These polymers are the materials known commercially as "silicones."

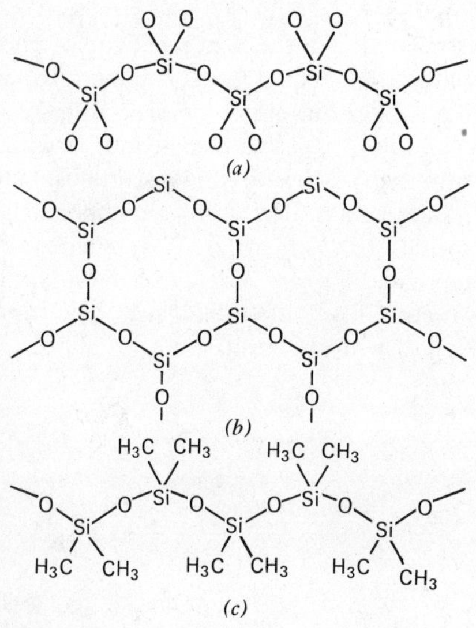

FIGURE 9.7
Structures of some silicon compounds. (a) A chain silicate, $(SiO_3^{2-})_n$. (b) A sheet silicate, $(Si_2O_5^{2-})_n$. Each Si atom is bound to four O atoms, the fourth being alternately above or below the sheet. (c) A linear polysiloxane, $[(CH_3)_2SiO]_n$.

Let us mention briefly some extended structures formed by other nonmetallic elements, starting with sulfur. We have already referred to the polymeric forms of the element. Sulfur vapor contains all the species from S_2 to S_{10}, most of which are probably rings. Liquid sulfur at temperatures near the normal melting point consists primarily of S_8 rings which, with increasing temperature, polymerize into very long chains (—S—S—S—S—). In aqueous solution sulfur forms polysulfide ions, S_n^{2-}, and polythionate ions, $(O_3S—S_n—SO_3)^{2-}$, both of which have extensive chemistry. "Silicon disulfide," $(SiS_2)_n$, has a structure rather like the linear polysilicates, consisting of SiS_4 tetrahedra joined by pairs of sulfur atoms:

$$\cdots \overset{\displaystyle S}{\underset{\displaystyle S}{Si}} \overset{\displaystyle S}{\underset{\displaystyle S}{Si}} \overset{\displaystyle S}{\underset{\displaystyle S}{Si}} \cdots$$

Although neither ozone nor SO_2 polymerizes, the analogous SeO_2 forms infinite chains,

$$\cdots \overset{\displaystyle O}{\underset{\displaystyle O}{Se}} \overset{\displaystyle O}{\underset{\displaystyle O}{Se}} \overset{\displaystyle O}{\underset{\displaystyle O}{Se}} \cdots ,$$

and solid SO_3 has a number of polymeric forms including the silicate-like

$$-O-\overset{\displaystyle O}{\underset{\displaystyle O}{S}}-O-\overset{\displaystyle O}{\underset{\displaystyle O}{S}}-O-\overset{\displaystyle O}{\underset{\displaystyle O}{S}}-O-,$$

which, as one might expect, resembles asbestos in its properties.

Phosphorus forms a number of interesting structures. The vapor, the liquid, and the "white" form of the solid all consist of tetrahedral P_4 molecules, Fig. 9.8a; the other solid forms are polymers of various kinds. Since the P—P—P angle in P_4 is only 60°, the structure is highly strained (see Section 9.4); this is associated with the high reactivity of phosphorus. The tetrahedral arrangement of P atoms is retained in the common oxides, shown in parts b and c of Fig. 9.8, and in a similar but more complex series of phosphorus sulfides. The orthophosphate anion, PO_4^{3-}, is also tetrahedral; as with the silicates, these tetrahedra can join to form many types of condensed phosphates with P—O—P linkages, some forming infinite chains. There are a number of phosphonitrilic compounds of the general formula $(PNX_2)_n$, where X is a halogen atom, an organic group, and so on; for large n these are linear polymers with alternating P and N atoms, having a structure like the polysiloxanes. The trimers such as $(PNCl_2)_3$ are cyclic and nearly planar, with the structure

$$
\begin{array}{c}
Cl \qquad\quad N \qquad\quad Cl \\
\diagdown \; \diagup \quad\; \diagdown \; \diagup \\
P \qquad\qquad\quad P \\
\diagup \; \diagdown \qquad\quad \diagdown \\
Cl \qquad N \qquad\qquad N \\
\diagdown \qquad\quad \diagup \\
P \\
\diagup \quad \diagdown \\
Cl \qquad\quad Cl
\end{array}
$$

in which the ring is quasi-aromatic, with delocalized π bonding rather like that in benzene.

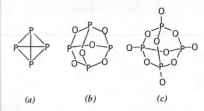

FIGURE 9.8
Phosphorus and its oxides. (a) The tetrahedral P_4 molecule. (b) "Phosphorus trioxide," P_4O_6. (c) "Phosphorus pentoxide," P_4O_{10}. The "trioxide" and "pentoxide" names derive from the empirical formulas, P_2O_3 and P_2O_5, respectively.

Another quasi-aromatic compound is borazine, $B_3N_3H_6$, with the structure

$$
\begin{array}{c}
\text{H} \\
| \\
\text{N} \\
\diagup \;\;\;\; \diagdown \\
\text{H—B} \qquad \text{B—H} \\
| \qquad\qquad | \\
\text{H—N} \qquad \text{N—H} \\
\diagdown \;\;\;\; \diagup \\
\text{B} \\
| \\
\text{H}
\end{array}
$$

The compound is isoelectronic with benzene and has somewhat similar properties, but is more reactive because of the polar B—N bonds. There are many other compounds in which carbon is replaced by the boron–nitrogen combination. The most notable of these is boron nitride, BN, which in the solid state has two structures exactly paralleling those of diamond and graphite; the diamondlike material ("borazon") is even harder than diamond itself.

Although all the compounds discussed thus far are entirely nonmetallic, many metallic elements can also form extended structures. For example, tungsten forms a series of elaborate anions with oxygen; some examples are $W_6O_{21}^{6-}$ and $W_{12}O_{41}^{10-}$. These are called *isopolytungstates*, and compounds of this type are generally referred to as anions of *isopoly acids*. (The "iso" means that only a single element is present besides oxygen and hydrogen.) Similar *heteropoly* anions also exist in which two or more metals (or nonmetals such as Si, As, and P) are present, for example, $Co_2W_{12}O_{42}^{8-}$ or $PMo_{12}O_{40}^{3-}$; however, a single metal always predominates, most commonly tungsten or molybdenum. These anions are typically built up from 6-coordinated metal atoms at the centers of octahedra, with the WO_6 or MoO_6 octahedra joined in much the same way as the SiO_4 and PO_4 tetrahedra described earlier: shared vertices correspond to metal atoms sharing single oxygen atoms (W—O—W), whereas shared edges correspond to the sharing

of pairs of oxygen ligands (W $\begin{smallmatrix}\text{O}\\ \diagup\;\diagdown\\ \diagdown\;\diagup\\ \text{O}\end{smallmatrix}$ W).

Remarkably, a simple electrostatic model largely accounts for the relative sizes and stabilities of these aggregates of octahedra, some of which are shown in Fig. 9.9, and even for the nonexistence of certain species. The electrostatic calculations similar to those we shall describe for ionic crystals in Sections 11.8 and 11.9 indicate that structures in which octahedra share edges increase in stability as more octahedra join on, themselves arranged into polyhedra. The increasing amount of attractive energy from oppositely charged nearest-neighbor atoms then outweighs the repulsive contributions from next-nearest neighbors— up to the point at which a closed "superpolyhedron" is formed, with maximum symmetry and the largest possible number of shared edges, but with much less stability than its less complex congeners. These predictions are in excellent accord with what is observed in nature. "Open" clusters of polyhedra are common, whereas closed, simple polyhedra are very rare.

Although the species we have discussed in this section have a variety of strange structures, the bonding in all of them is relatively conventional. That is, in each case one can construct bonding orbitals corresponding to the bonds in ordinary chemical formulas, and there are enough valence electrons available to fill all these bonding orbitals. Sometimes the best orbitals are delocalized, but this does not affect their capacity. For example, in benzene the three delocalized π orbitals hold the same six electrons as the three π bonds of the Kekulé formula. Now we shall look at some species for which this is no longer true, the so-called electron-deficient molecules, best typified by the boron hydrides.

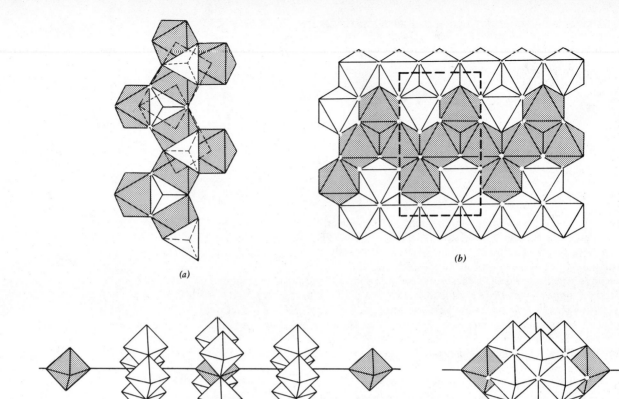

(a)

(b)

(c)

(d)

FIGURE 9.9
Typical structures that are well represented by octahedra with shared edges; (a) and (b) represent olivine, the basaltic material comprising approximately 65% of the earth's upper mantle, whose composition is $(Mg, Fe)_2SiO_4$—that is, a variable ratio of Mg/Fe, but always two metal ions per silicate; (c) and (d) represent an exploded view and a condensed picture of the synthetic mineral $Mg_3V_{12}O_{40}$, in which the shaded octahedra represent the magnesium ions with their oxygen ligands and the unshaded octahedra represent the vanadium ions. In (a), the repeat units of olivine are shown; in (b), one sees the extended structure. Figure provided by P. B. Moore, with permission.

The boron hydrides (or *boranes*) range from the unstable BH_3 through B_2H_6 (diborane), B_4H_{10}, B_5H_9, B_5H_{11}, B_6H_{10}, B_6H_{12}, B_8H_{12}, B_8H_{18}, B_9H_{15}, $B_{10}H_{14}$ (decaborane, the most stable member of the series), $B_{10}H_{16}$, and $B_{18}H_{22}$ (two isomers) to $B_{20}H_{16}$. At room temperature B_2H_6 and B_4H_{10} are gases, the other species below B_{10} are liquids, and the rest are solids. Although diborane is thermally stable at room temperature, most others below B_{10} are not, and all the species below B_8 react readily—sometimes violently—with air or water, largely because of boron's strong affinity for oxygen. Boron and hydrogen also form such anions as BH_4^- (the tetrahedral borohydride ion), $B_3H_8^-$, $B_{10}H_{10}^{2-}$, $B_{12}H_{12}^{2-}$, and $B_{20}H_{18}^{2-}$.

The structures of some of the boron hydrides are illustrated in Fig. 9.10. These structures have been interpreted as based on an icosahedral frame of 12 boron atoms, most clearly seen in the regular icosahedron of the $B_{12}H_{12}^{2-}$ ion. The smaller species are largely incomplete icosahedrons, whereas the larger ones consist of these fragments fused in various ways. We may also note here the *carboranes*, compounds in which the framework is made of both boron and carbon atoms. The best known is *ortho*-carborane, $B_{10}C_2H_{12}$, which is isoelectronic with $B_{12}H_{12}^{2-}$ and has essentially the same structure.

Now what do we mean by calling the boron hydrides "electron-deficient molecules"? Consider the B_2H_6 molecule: Since there are eight atoms, at least seven conventional two-electron bonds—14 electrons—would be needed to join the molecule together; but there are only 12 valence electrons available. Thus some of the bonding orbitals must either be unfilled or extend over more than two atoms; we shall see that the latter is the case in B_2H_6. Such species naturally tend to be less tightly bound than molecules with conventional bonds, that is,

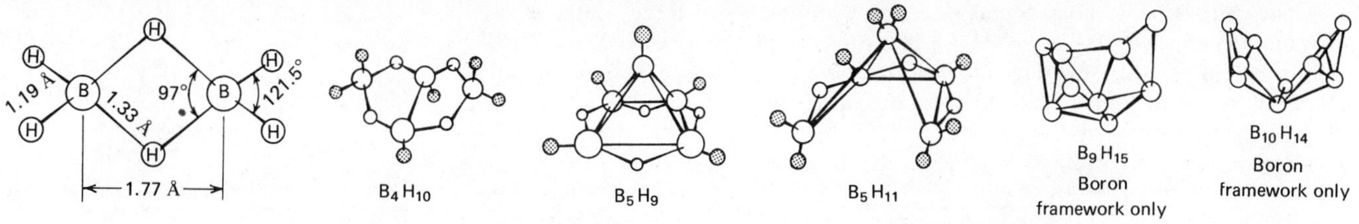

○ Boron
◉ Hydrogen, terminal
○ Hydrogen, bridge

FIGURE 9.10
Structures of representative boron hydrides. [Cf. F. A. Cotton and G. Wilkinson, *Advanced Inorganic Chemistry*, 2nd ed., (Wiley-Interscience, New York, 1966), p. 276; and W. N. Lipscomb, in H. J. Emeléus and A. G. Sharpe (Eds.), *Advances in Inorganic Chemistry and Radiochemistry*, Vol. 2 (Academic, New York, 1960), p. 279.]

molecules in which all the bonding orbitals are normally occupied. To interpret how bonding is accomplished in the boron hydrides, W. N. Lipscomb found it useful to distinguish three kinds of bonding orbitals. One of these is the conventional two-electron bond between adjacent atoms, thus called a *two-center bond*. These bonds are formed in the same way as any of the other two-center bonds we have discussed. The terminal B—H bonds in B_2H_6 are of this type.

The second kind of bond is the three-center, two-electron bond involving two borons and a bridging hydrogen atom. Bonding of this type is necessary to account for the B—H—B bridges in B_2H_6. Normally, two electrons occupy the lowest-energy molecular orbital that can be constructed from the participating atomic orbitals of the three atoms. We have already looked at one such three-center orbital in detail, the lowest orbital of the H_3 molecule (Section 8.1), which is simply the sum of $1s$ orbitals on the three H atoms. The lowest-energy three-center MO has only as many nodes as the constituent atomic orbitals (in H_3, no nodes at all); that is, the atomic orbitals combine to give only *constructive* interference.[11] The B—H—B bonding orbital is best considered as the sum of two tetrahedral sp^3 orbitals, one from each boron atom, and the $1s$ orbital of the bridging hydrogen, as shown in Fig. 9.11a; the B_2H_6 molecule has two such orbitals, each occupied by two electrons. The same atomic orbitals can combine to give two higher-energy orbitals with interatomic nodes, but these are normally empty. The average energy of the B—H—B bond is only about 0.75 eV (72 kJ/mol), so that, doubly occupied, it contributes the 1.5 eV needed to hold two BH_3's as a B_2H_6. Note how weak such bonds are relative to the typical two-center bond (~3–5 eV).

[11] Cf. Fig. 8.11 for examples of possible MOs in a homonuclear three-center system.

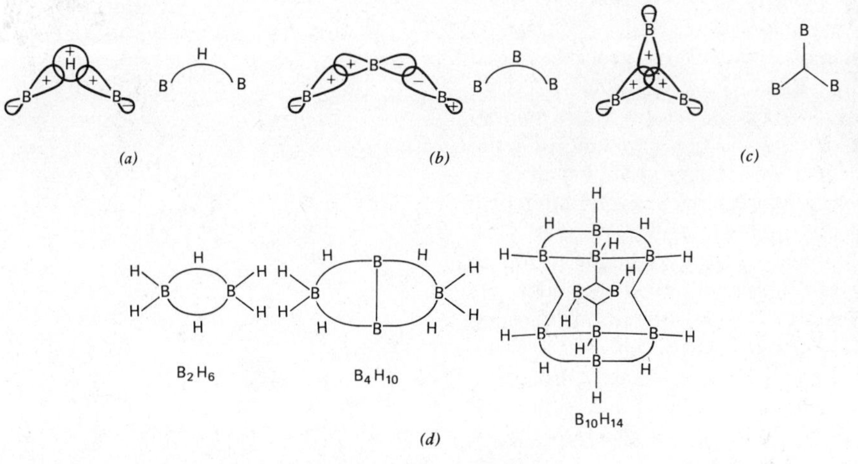

FIGURE 9.11
Bridge bonding in boron hydrides. (*a*) B—H—B bonding orbital, formed from B sp^3 orbitals and an H $1s$ orbital, as in B_2H_6. (*b*) B—B—B bonding orbital with properties of the central atom different from those of the outer two, as in B_4H_{10}. (*c*) B—B—B bonding orbital between three nearly equivalent atoms, as in $B_{10}H_{14}$. (*d*) Bonding diagrams of some boron hydrides, using the notation indicated in parts (*a*)–(*c*).

377

The third type of bond found in boron hydrides is the three-center, two-electron homonuclear (B—B—B) bond. Two forms of this bond are distinguished. In one form, Fig. 9.11*b*, the three atoms form an isosceles triangle, and the central atom has properties quite different from the other two. In the other form, Fig. 9.11*c*, the three atoms are nearly equivalent. In Fig. 9.11*d* we illustrate how the bonding in some of the boron hydrides is accounted for in terms of the various types of bonds we have described.

Three-center bridge bonds are by no means restricted to the boron hydrides. The ordinary hydrogen bond (Section 7.9) can be interpreted in terms of a three-center orbital to which one of the end atoms makes a much greater contribution than the other—except in the symmetric FHF⁻ ion, which has a bonding energy (~1.2 eV) comparable to that of the B—H—B bond. In the free H_2F_2 molecule, apparently the two hydrogen atoms act alternately as bridges, exchanging roles rapidly; this contrasts with the boron hydrides, where the hydrogen atoms are rather rigidly fixed. Three-center bonds also occur in a variety of dimeric and polymeric species in which halogen atoms act as bridges. The dimers Al_2Cl_6 and Be_2Cl_4 are examples of this, with two Cl atoms in each forming bridges and the others forming normal metal-halogen bonds; and solid beryllium chloride has the polymeric structure

made up of joined tetrahedra like the (conventionally bound) SiS_2.

9.4 Some Steric Effects

In this section we consider a number of effects that influence molecular structure on a scale larger than that of localized bonding. Small molecules such as H_2O or CH_4 have nearly rigid structures, in that the distances between all pairs of atoms, not just those directly bonded to each other, remain constant, except for small variations due to vibration. In larger molecules this is frequently not the case, and nonbonded atoms can assume various positions relative to one another. Such molecules can assume different geometric *conformations* of the atoms in space that can be reached from one another without breaking bonds. Sometimes the molecules are quite flexible, and can assume many different conformations with nearly equal ease. Other large molecules prefer quite specific conformations as the result of many individually weak interactions, such as hydrogen bonds or the simple fact that atoms get in each other's way. When one conformation is appreciably more stable than others, it is useful to think of the molecule as having a *secondary structure*, as opposed to the *primary structure* defined by the chemical bonds. Let us see what factors govern the capacity of molecules to take on different conformations.

Certain general patterns are immediately obvious: In general, closed structures—rings, three-dimensional polyhedra, etc.—are particularly rigid, whereas open chainlike structures are flexible. What is the nature of the internal motions associated with this flexibility? It certainly is not a matter of bond stretching, since the force constants for stretching are the largest among all types of molecular vibrations; because of this "stiffness," the energy levels for stretching modes are typically quite widely spread, of the order of 0.1–0.3 eV apart. The bending of bond angles is easier, but not enough so to account for the loose nature of open-chain molecules. Conversely, bending modes must be moderately stiff to account for the rigidity of closed structures. The answer to our question lies in a third kind of motion, which we have not considered up to this point, *internal rotation*, the rotation of two groups of atoms relative to each other about the bond joining them. If the electron density in the bond has cylindrical symmetry we would expect this rotation to occur freely; this is generally true of

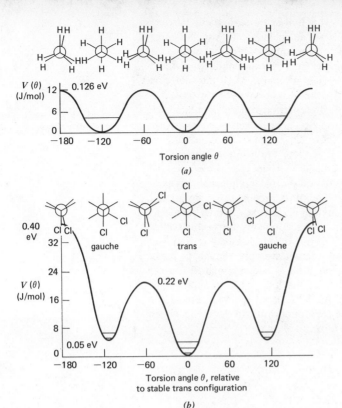

FIGURE 9.12

(a) Hindered rotation in ethane. The graph shows the internal effective potential energy $V(\theta)$ as a function of the torsion angle θ. The diagrams above the graph show the conformation at intervals of 60°, as viewed down the C—C axis. Only the "near" carbon is shown; the other is hidden behind it. The horizontal lines indicate the lowest torsional vibrational levels, but do not show the splittings associated with the multiple wells. The torsional vibration frequency $\bar{\nu}$ is 289 cm^{-1} or 0.0358 eV. (b) Hindered rotation in 1,2-dichloroethane, CH_2Cl—CH_2Cl. Two of the potential wells in this molecule are equivalent; the third is 0.05 eV deeper, enough to support three vibrational levels lying below the lowest point of the other two wells.

single (σ-type) bonds,[12] the electron density in which is concentrated symmetrically along the bond axis.

However, internal rotation is never completely free except in linear molecules. Even if the bond itself is cylindrically symmetric, the interactions between more distant atoms cannot be. These interactions are always small, but not always negligible. Consider the ethane (H_3C—CH_3) molecule, in which the relevant interactions are the repulsions between the hydrogen atoms of the two methyl groups. These repulsions create a potential barrier to rotation about the C—C bond, as shown in Fig. 9.12a. The equilibrium conformation is one in which one methyl group is "staggered" with respect to the other, so that we could see all six hydrogen atoms if we looked down the C—C axis. The "eclipsed" configuration, in which the hydrogen atoms are lined up, has an energy about 0.127 eV higher (1024 cm^{-1} in E/hc). Thus there is a potential energy barrier to rotation every 120°. This does not mean that rotation cannot occur. As with any barrier of

[12] But not of multiple bonds: Rotation cannot occur about a π bond without destroying the overlap of p orbitals that creates the bond, i.e., without breaking the π bond. This is why

$$
\begin{array}{ccc}
H & \quad & H \\
\quad\diagdown\!C\!=\!C\diagup\quad & \text{and} & \quad\diagdown\!C\!=\!C\diagup\quad \\
Cl & \quad Cl & Cl & \quad H
\end{array}
$$

are chemically distinct species with different properties (geometrical isomers), whereas at ordinary temperatures we distinguish only one 1,2-dichlorethane,

$$
\begin{array}{c}
H \quad\quad H \\
H\!-\!C\!-\!C\!-\!H. \\
Cl \quad\quad Cl
\end{array}
$$

finite height, the system can cross over or even tunnel through these barriers (Section 4.1). In each of the potential wells we have harmonic oscillator-like energy levels corresponding to the energy of torsional oscillation about the minimum. The true stationary states of the system are combinations of these localized states. In ethane, the barrier is so low that we can think of this molecule as having rather free rotation about the C—C bond—except at very low temperatures, where only the ground state for torsional oscillation will be populated (cf. Chapter 21).

What we have said about the ethane molecule is generally applicable to internal rotation about single bonds. However if there are at least two kinds of substituents in each end of the bond, as in 1,2-dichloroethane, the three potential minima analogous to those of Fig. 9.12a are not equivalent; the internal energy of 1,2-dichloroethane is shown in Fig. 9.12b. The three *rotamers*, or rotational isomers, are different, even though they are interconvertible at ordinary temperatures. The first has an energy 0.5 eV lower than the second and third, which differ only in being mirror images of one another. (The mirror image aspect is discussed further in Section 9.6.) The substituents may interact strongly enough to make one rotamer truly stable, in the sense that its potential is deep enough to support a vibrational quantum state below the minima of the less stable rotamers. This situation occurs with 1,2-dichloroethane, as the vibrational levels in Fig. 9.12b indicate. A more extreme example is the ring compound *trans*-cyclooctene, which has a stable *cis* form and an unstable *trans* form:

cis   trans

Rotation about the double bond is highly hindered in this system, as it is in ethylene, but here we also see the additional feature of an asymmetric situation in which the ground state of the stable form has an energy well below that of the unstable form.

Even if the rotameric forms of a molecule have the same energy, it may be that we cannot observe the conversion of one form into another. If the groups are very large, the quantum levels for torsional oscillation (libration) must be closely spaced, so that many levels lie well down in the potential well. Recall from our discussion in Chapter 4 that the quantum mechanical wave function penetrates the classically forbidden region where the total energy is less than the potential energy, but that the wave amplitude must diminish as the wave penetrates. If the potential is relatively steep, the wave amplitude diminishes to a small fraction, say e^{-1} of its amplitude at entry, within a range of about half a wavelength or less. Therefore, for a quantum state with a low quantum number, and thus a large spatial extent, the wave function can penetrate relatively far into the classically forbidden region in a potential such as that of Fig. 9.12b. However, the wave function of a quantum state with a high quantum number but the same total energy, corresponding to a larger mass and momentum than in the first case, must penetrate only a short distance before its amplitude becomes very small. This in turn implies that massive substituents are much less effective than very light substituents at penetrating barriers far enough to tunnel through them. The only way to observe conversion of one rotamer into another in such a case is to excite the species to an energy higher than the potential barrier, where it can rotate essentially as a free rotor. Hence, if the barrier is high, rotational motion about one bond may well be a rare event.

A long-chain molecule may nevertheless change its shape if it can incur small rotations about many bonds. (Molecular ball-and-and-stick models give persua-

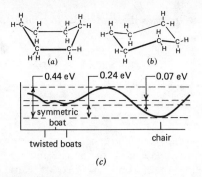

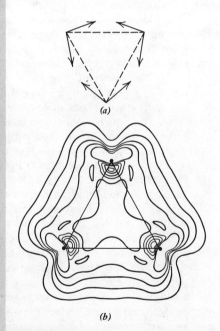

FIGURE 9.13
The two conformations of cyclohexane. (a) Boat. (b) Chair. (c) The effective potential for boat–chair interconversion, showing the slight relative minimum for twisting in the boat conformation and the large absolute minimum for the chair, with a high barrier between the minima. The coordinate of the abscissa is a complicated mixture of bond bending and bond stretching, corresponding to the lowest-energy path from one minimum to the next.

sive demonstrations of this behavior.) In general, unless there are specific stiffening forces between formally nonbonded atoms, such as intramolecular hydrogen bonds, long-chain molecules tend to be quite flexible in the liquid or gaseous state.

As the example of cyclooctene suggests, even ring compounds need not have rigid structures. In such molecules as cyclopentane or cyclohexane, rotation about the C—C bonds is sufficiently free to allow the molecules to take on a range of conformations. Bond angles and bond lengths are well preserved throughout these motions. But although open-chain compounds often prefer no definite conformation until they are crystallized, ring compounds usually have certain well-defined shapes that are more stable than others. The best-known examples are the "boat" and "chair" forms of cyclohexane, shown in Fig. 9.13. These two forms have almost equal energies, with the "chair" form about 0.24 eV (23 kJ/mol) more stable. The "chair" and "boat" forms can also interconvert quite readily; the barrier between them is only 0.44 eV (42 kJ/mol) high. This barrier results largely from the fact that interconversion cannot occur without *some* change in bond angles. The difference between the energies of "boat" and "chair" forms is thought to be due partly to the greater repulsion between nonbonded hydrogen atoms in the former. Repulsive interactions of this type (one speaks of "steric forces") play an increasingly important role in determining the preferred conformations as molecules become larger and less rigid.

The "natural" bond angle for the carbon atom is the tetrahedral 109°28′ (or 120° for trigonal carbon). If rotation about single bonds can occur easily, then rings of almost any size can be constructed with all the C—C—C angles close to the natural value. However, there are a few exceptions. For example, the carbon skeleton of the C_3H_6 (cyclopropane) ring is necessarily planar; in a ball-and-stick picture cyclopropane thus has internal bond angles of only 60°. As a result, the molecule is severely strained, that is, it has considerably more internal energy per CH_2 group than does a larger ring such as cyclohexane. This strain energy is equivalent to the work required to bend the bond angle from its natural value of 109°28′ to 60°. One may speculate that the bending energy is stored in the system of bonding electrons in two ways: partly as an increase in electron–electron repulsion due to the proximity of the skeletal C—C bonds, and partly as a spatial redistribution of electron density away from the bond axis. In unstrained molecules we assume that the electron density in a bond is largely in an ellipsoid whose axis is the internuclear axis of the bonded atoms. But apparently in the C_3H_6 ring the "ridge" of the C—C bonding electron density contours follows a curve bent outward from the ring, as shown in Fig. 9.14. This displacement reduces both electron–electron repulsions and electron–nuclear attractions. Strained bonds of this type are sometimes called "banana bonds" or "bent bonds" because of the shape of their electron density contours.[13]

The cyclobutane (C_4H_8) ring is also quite strained. In rings with more than four atoms, however, the strain is generally unimportant, because they can attain strain-free conformations by bending out of the plane. Planar cyclobutane would be square, with 90° bond angles; the equilibrium conformation is actually highly puckered, with about the same carbon strain—bond angles of 88°—but the hydrogen atoms are farther apart in the bent structure than in the planar. The benzene ring is planar but quite strain-free, because the C—C—C angle is the natural bond angle of 120° for trigonal carbons. The planarity of the ring is apparently necessary for its aromatic behavior, because this creates the maximum overlap of the $2p\pi$ orbitals and therefore the maximum delocalization of the

FIGURE 9.14
The bonds of cyclopropane, showing how the regions of maximum electron density fall outside the ring of C—C internuclear axes. (a) The directions of the optimized hybrid C—C bonding orbitals. (b) Contour maps of the corresponding electron density, with the C—C internuclear axes drawn in to emphasize the "bent" nature of the bonds, especially near the carbon atoms.

---

[13] The use of "banana bonds" is not limited to strained molecules. For example, the triple bond in N_2 can be interpreted as three equivalent "banana bonds" rather than the conventional one σ and two π bonds. For a symmetric case such as N_2, in the orbital approximation the two descriptions are mathematically equivalent: It is simply a matter of how one finds it convenient to partition the total electron density. There is, however, a real physical meaning to the "bent bonds" of cyclopropane.

electrons in these orbitals. In contrast, the cyclooctatetraene ring,

$$HC=CH$$
$$HC \quad\quad CH$$
$$HC \quad\quad CH$$
$$HC=CH$$
,

which one might expect to resemble benzene, is nonplanar (since a planar ring would have 135° bond angles and much strain) and is not like aromatic molecules in its properties. This is both because of its nonplanarity and because it has $4n$ electrons, rather than $4n+2$, which affects the orbital energy-level structure.

We close this section with a brief discussion of the structures that arise when molecules are so large that there are important interactions between segments far from each other along any sequence of bonds. The simplest effect to discuss—though one of the most difficult to describe quantitatively—is that of *excluded volume*. Suppose that we grow a long-chain polymer in solution. As the molecule grows, so long as it has the flexibility associated with internal rotation about single bonds, it will form what is called a *random coil*. That is, it will have no particular preferred shape, and can be compared to a piece of soft spaghetti in water. The only constraint on its shape and growth is then the prohibition against two pieces of the chain occupying the same space—the excluded-volume effect. This is the simple result of the short-range repulsion always present between nonbonded atoms. With random-coil molecules, one is forced to use statistical calculations to determine the volume occupied per unit chain length and other properties related to the physical form of the molecules. However, there are other large molecules with quite rigid shapes, and these we consider next.

Three kinds of interactions are particularly important in giving a large molecule a well-defined rigid shape. One is the presence of a network of strong bonds, as in condensed aromatic ring compounds; we have already discussed interactions of this type in some detail. If the network is essentially unlimited in extent, then the "molecule" is in fact a covalently bound solid; examples we have mentioned include graphite and the polysilicates. A second type is weak bonding between one part of a large molecule and another that is near in space but far away along the sequence of (strong) bonds; the most common interactions of this type involve hydrogen bonds. And third is the effect of interactions between the large molecule and the solvent medium that contains it, especially when different parts of the solute molecule interact in different ways.

The possibilities of forming hydrogen bonds between otherwise unbonded parts of large molecules are rich indeed, whenever the molecules contain both acidic (proton-donating) and basic (proton-accepting) functional groups. Of particular interest are biological macromolecules, in which the proton acceptors are most commonly carbonyl (—C=O) groups, whereas the hydrogen atoms are most commonly furnished by hydroxyl (—OH), amino (—NH$_2$), or imino (—NH) groups. Perhaps the most important of such interactions for us are those in deoxyribonucleic acid (DNA), the material making up the genes that determine our heredity. The DNA molecule consists of two long chains, each made of alternating ribose (a five-carbon sugar) and phosphate groups, with one of the purines or pyrimidines adenine, thymine, guanine, or cytosine attached to each ribose ring. The two chains are held together only by hydrogen bonds joining the amino acids—adenine to thymine, and guanine to cytosine—in such a way that the entire molecule forms the famous double helix (Fig. 9.15). The hydrogen bonding is weak enough to allow a living cell's reproduction system to separate the strands gently and thus replicate each chain one link at a time, yet strong enough to keep the DNA in the form of a rigid rod of coiled strands when it is not being replicated. (At high temperatures the hydrogen bonds are broken, and the rigid helix degenerates into a random coil.) Hydrogen bonds perform similar functions in many other biological systems, notably the helical proteins. Their function, it seems, is to stabilize well-defined structures of very large molecules, both for storage purposes and to keep specific active sites in suitable positions for performing specific tasks such as catalyzing reactions.

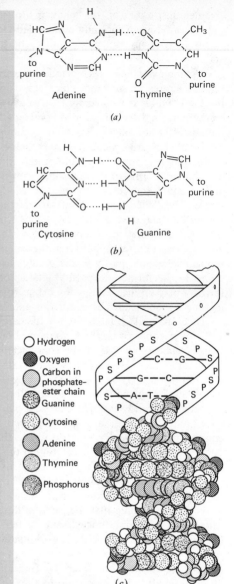

FIGURE 9.15
Hydrogen bonding between (a) adenine and thymine, (b) cytosine and guanine, resulting in (c) the double-helix structure of deoxyribonucleic acid.

As for interactions between solvent molecules and large solute molecules, these are often similar in their effects to intramolecular hydrogen bonding, but far less specific. A familiar old saw in chemistry is *similia similibus solvuntur*—"like dissolves like," where "like" usually refers to the classes of polar and nonpolar molecules. This is the result of the attractive interactions between polar groups (often involving hydrogen bonds) and of the much weaker attractions or even repulsions between polar and nonpolar groups. Consider a molecule containing both polar and nonpolar regions—for example, a long hydrocarbon chain with a polar functional group at one end, such as $CH_3(CH_2)_{16}COOH$—in a polar solvent such as water. There will be a strong tendency for the molecule to arrange itself so as to maximize the contact between its polar segments and the solvent, and simultaneously to minimize the contact between its nonpolar segments and the solvent. This may result in the solute molecules' aggregating into droplets or films with their polar segments forming the interface with the solvent. Such interactions can have strong effects on the conformation of the solute molecules. A hydrogen-bonded solvent such as water has a structure of its own (Chapter 10 and Chapter 26), and this will also be affected, in general, by the *hydrophilic* (water-attracting) or *hydrophobic* (water-repelling) nature of the solute molecules. These systems are complex enough that molecular mechanics is insufficient to describe their behavior; to go further one needs the statistical methods that we shall consider in Part Two.

9.5
Complex Ions and Other Coordination Compounds: Simple Polyhedra

Now we shall begin our consideration of the second major class of polyatomic molecules, the so-called *coordination compounds* or *complexes*. Structurally, these species are characterized by a central atom or ion and a set of surrounding *ligands*, which may be atoms, ions, or neutral molecules. Most complexes have well-defined geometric structures, and even those that can rearrange internally have strongly preferred geometries. These structures are ordinarily polyhedral, with each of the ligand atoms occupying a vertex and bonded to the central atom. In coordination compounds the bonds between ligands and the central atom are generally weaker than the corresponding bonds in such polyhedral species as CH_4 or SO_4^{2-}; the ligands thus more readily undergo substitution reactions. In these respects the coordination compounds are intermediate between ordinary covalent molecules and the loose ion-molecule clusters we shall discuss in Section 10.3.

Virtually all positive ions exist as coordination compounds of some kind in aqueous solution (and in a number of other solvents); many of these species also exist as complex ions in the solid phase. At one extreme we have the weakly bound, labile structures of the hydrated alkali metal ions. The larger ones, of Rb^+ and Cs^+, are particularly labile. At the other extreme are the ordinary covalent molecules. Many complexes differ from covalent molecules in that the central ion and the ligands are capable of independent stable existence, usually in solution. The most studied complexes are those formed by the transition metal cations, which combine with a host of ligands, including all the common polar solvents. Transition metal complexes offer especially rich challenges, in that they have several properties—most notably magnetism and electronic absorption spectra in the visible and near-ultraviolet regions—that can be used to probe the electronic levels involved in their bonding. In subsequent sections we shall use these properties to analyze the electronic structures of the complex ions, and then see how this information can help us to interpret their geometric structures.

First, however, we must survey the types of structures that these compounds exhibit. The most important factor in classifying structures is the *coordination number*, the number of ligands bonded directly to the central atom (or the number of nearest neighbors, if they are all equivalent); these ligands constitute the *coordination shell*. The most common coordination numbers in the solid or liquid phase are 4 and 6, although 5 is reasonably common and values ranging from 2 to 9 are known. We shall for the most part limit ourselves to the species with values of 4, 5, and 6.

The geometric structures of complexes were first deduced by counting the number of *isomers*—species with the same formula, but different chemical or physical properties—of a given complex. This analysis was carried out primarily by Alfred Werner, who showed that (1) certain complexes formed series from which the number of ligands could be inferred, and (2) the number of isomers of each complex could be used to determine the geometry, if one assumed that all the possible isomers were known.

Werner's reasoning is best illustrated by the classic case of the complexes of Pt(IV)[14] with Cl^- and NH_3. These were formed by adding ammonia to aqueous solutions of $PtCl_4$. Five such complexes were known to Werner; they were isolated and studied by direct chemical analysis, by conductance measurements to find the number of ions present in the solution, and by titration with Ag^+ to find the number of free chloride ions present. The five complexes crystallized into salts with the compositions $PtCl_4 \cdot 2NH_3$, $PtCl_4 \cdot 3NH_3$, $PtCl_4 \cdot 4NH_3$, $PtCl_4 \cdot 5NH_3$, and $PtCl_4 \cdot 6NH_3$. Their solutions respectively had conductivities corresponding to 0, 2, 3, 4, and 5 ions per platinum atom, whereas the amounts of AgCl precipitated immediately corresponded to 0, 1, 2, 3, and 4 "free" Cl^- ions per platinum atom. All the Cl^- will be precipitated as AgCl if one waits long enough, but Werner was wise enough to recognize that chlorine was present in two states of binding, one that reacted slowly with silver ions and one that reacted immediately. The conclusions he drew were that there are six ligands coordinated with each platinum and that the five complexes can be represented as

$$[Pt(NH_3)_2Cl_4],$$

$$[Pt(NH_3)_3Cl_3]^+ \cdot Cl^-,$$

$$[Pt(NH_3)_4Cl_2]^{2+} \cdot 2Cl^-,$$

$$[Pt(NH_3)_5Cl]^{3+} \cdot 3Cl^-,$$

$$[Pt(NH_3)_6]^{4+} \cdot 4Cl^-.$$

The species in square brackets presumably maintain their identities in solution, with the Cl^- made available for precipitation only by slow exchange reactions. These are referred to as complex ions. The complex ions move about freely as entities in the solution. The "free" chloride ions, shown outside the brackets, are also separate entities in solution. Werner had thus shown that in these complexes Pt(IV) has a coordination number of 6. By his reasoning complex ions with the formulas $[Pt(NH_3)Cl_5]^-$ and $[PtCl_6]^{2-}$ should also exist, and these are indeed now

[14] Here we use the notation now common in inorganic chemistry for the *formal* positive charge on the ion; thus Pt(IV) indicates a formal charge or oxidation state of +4 on platinum. The electric fields around a real Pt^{4+} ion are far too strong for such a highly charged ion to exist as such near other atoms or molecules. Although highly charged ions are found in sparks and plasmas, they are not found in condensed matter. Note, however, that we continue to write the charge in the usual way for ions such as $[Pt(NH_3)_6]^{4+}$ because the complex ion is an identifiable species even with its charge of 4+.

known. Werner extended his analysis to show correctly that Cr(III) and Co(III) also have coordination numbers of 6, whereas Pt(II) and Pd(II) have coordination numbers of 4.

The geometries of the complexes were then inferred from the numbers of isomers. For example, two distinct species were known to share the formula $[Co(NH_3)_4Cl_2]^+$; one is green, the other lavender. Werner supposed correctly that the difference lies in the geometric arrangement of the ligands. If the structure were highly irregular, there ought to be more than two isomers; indeed, if all six sites were geometrically different, the number of possible isomers would be $\frac{1}{2}(6 \times 5) = 15$. Only if all the sites are essentially equivalent can the number of isomers be as low as two. A polyhedron with six equivalent vertices is an octahedron, and an octahedral arrangement of the ligands does give rise to only two isomers, the *cis* and *trans* structures of Fig. 9.16a (*cis* and *trans* refer to the relative positions of the Cl$^-$ ligands). Similar pairs of isomers are found for most other 6-coordinate complexes with ligands in 4:2 ratio. If Co(III) complexes are octahedral, then $[Co(NH_3)_3Cl_3]$ should also have two isomers, with the structures shown in Figs. 9.16b; this is indeed the case. Similarly, the 4-coordinate complex $[Pt(NH_3)_2Cl_2]$ has two isomers and is thus assumed to be square-planar, Fig. 9.16c. Evidence of this sort has been used to determine the structures of a great number of complexes.

Thus far we have discussed only *monodentate* ("one-toothed") ligands, those that can occupy only one site in a complex's coordination shell. An example of a *bidentate* ligand is ethylenediamine, $H_2NCH_2CH_2NH_2$, which behaves almost like two ammonia molecules linked by a short, flexible chain. Both the —NH$_2$ groups are able to act as ligands, so that the ethylenediamine molecule occupies two sites in the coordination shell, not necessarily around the same central atom. Other common bidentate ligands include propylenediamine, $H_2N(CH_2)_3NH_2$, and acetylacetonate, $(CH_3COCHCOCH_3)^-$. Still other species exist that can act as ligands of three, four, or more sites. When a polydentate ligand can occupy two (or more) sites around the *same* central atom, thus forming a ring, it is called a *chelate* (from the Greek *chele*, "claw"), and the central atom is said to be chelated. Chelation is a good method of stabilizing ions in solution, since the bulky chelate rings make it difficult for any reactive species to get at the central atom.

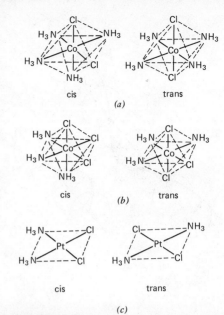

FIGURE 9.16
Geometric isomerism in complexes: *cis* and *trans* isomers of (a) the octahedral complex $[Co(NH_3)_4Cl_2]^+$; (b) the octahedral complex $[Co(NH_3)_3Cl_3]$; (c) the square-planar complex $[Pt(NH_3)_2Cl_2]$. Note that in each case all the vertices are equivalent.

The existence of bidentate ligands makes a different kind of isomerism possible. Consider the complex $[CoCl_2en_2]^+$ ("en" is the standard abbreviation for ethylenediamine), which is found to have three isomers. One of the three is chemically, and in most ways physically, different from the other two. The second and third isomers are identical in almost every way—spectroscopically, chemically, and with regard to most physical measurements. They differ only in properties that distinguish the left- or right-handedness, the *chirality* of the molecule. Before seeing what this means for the structure, we must explain what chirality is.

Left- and right-handed gloves of a pair are a commonplace example of two objects that differ only in their chirality. In other words, they are mirror images of each other. To detect such a difference on the microscopic level, one must use a physical phenomenon that is itself inherently chiral. The most straightforward such phenomenon is interaction with circularly polarized light. "Circular polarization" means that the electric field vector **E** at a point in space rotates in the plane perpendicular to the direction of propagation of the light. This is in contrast to plane polariza-

9.6 Chirality and Optical Rotation

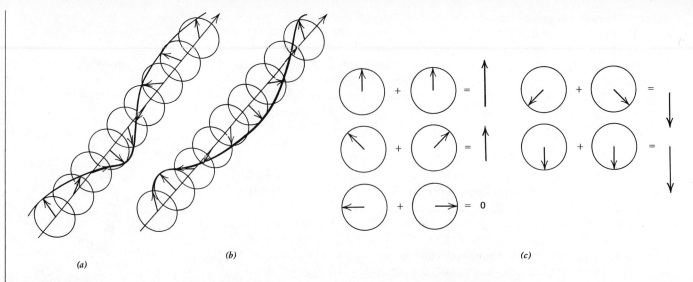

(a)   (b)   (c)

FIGURE 9.17
Right- (a) and left- (b) circularly polarized light: The field vectors **E** at points along the light ray are indicated by the small arrows, the points of which are joined to form a helix. Part (c) of the figure shows how left- and right-circularly polarized light in phase can add to give plane-polarized light; the **E** vectors are again shown from the viewpoint of one looking down the direction of propagation.

tion (Fig. 3.4), in which **E** at a point in space merely oscillates back and forth along a single axis. At a given instant, the field vectors at successive points along a ray of circularly polarized light describe a helix, either right-handed or left-handed. This is illustrated in Fig. 9.17.

How does one use circularly polarized light to study chirality? The physically significant fact is that the combination of a left-handed thing together with a right-handed thing is physically different from a pair of things of the same handedness—where "physically different" means a difference in more than just handedness. An example may make this clearer: A left-handed glove glued palm-first to a right-handed glove makes a quite different-shaped object than the result of the same operation on two left-handed gloves. Similarly, a right-handed molecule interacts with right-polarized light in just the same way that a left-handed molecule interacts with left-polarized light; neither of these interactions is physically equivalent to the way a right-handed molecule interacts with left-polarized light, but the latter is the same as the interaction of a left-handed molecule with right-polarized light.

The optical properties most commonly used to study chirality are the differences, either in absorption or refraction, between left- and right-circularly polarized light interacting with molecules of a specific chirality. The difference between the optical absorption (extinction) coefficients for left- and right-polarized light is called the *circular dichroism* (CD); it is a function of the frequency of the light, and may have either sign. The other property, the difference between the refractive indices for left-circular and right-circular polarized light has a much longer history of use in physical chemistry. It is still sometimes called *optical activity*, but a more meaningful and precise name is *optical rotatory dispersion* (ORD).[15] In the most common form of this method, plane-polarized light is passed through a chiral sample. Plane-polarized light can be described as the resultant of two circularly polarized beams with equal amplitudes and opposite directions of rotation. If the two components are exactly in phase, they will add constructively when both **E** vectors point in the same direction, but exactly cancel when both are at right angles to that direction (Fig. 9.17c). But if the refractive index of the sample differs for the two components, then one will have a higher velocity of propagation

[15] ORD and CD together are referred to as the *Cotton effect.*

than the other, and the phase relationship between the components will constantly change with distance along the propagation axis.[16] The resultant light remains linearly polarized, since there is always some axis along which the two rotating **E** vectors are parallel; but this axis of polarization rotates as the light passes through the sample. The angle of rotation α is given by

$$\alpha = \alpha_n lc, \qquad (9.9)$$

where α_n is a constant (the *optical rotatory power*) characteristic of the sample, l is the path length through the sample, and c is the molar concentration of chiral molecules in the sample; α_n depends on wavelength, and is most commonly determined with light of the sodium D lines, at 589.0 and 589.6 nm.

9.7
Chiral and Other Complex Ions

Now that we know what chirality is, let us return to the complex $[CoCl_2en_2]^+$. The three isomers must be those shown in Fig. 9.18. Note that the *trans* species has no chiral character, whereas the two *cis* species are identical except for their chirality: Like the left and right gloves, they are mirror images of each other. The *trans* species is a *geometric isomer* (or *diastereoisomer*) of either *cis* species; the two *cis* species are *optical isomers* of each other (such pairs are called *enantiomorphic*—"opposite-shaped"—isomers or *enantiomers*). Any molecule that has a plane of symmetry is identical with its mirror image, and thus can have no optical isomers. Conversely, any molecule that cannot be superimposed on its mirror image must have an optical isomer of opposite chirality. Optical isomers are also commonly found among organic molecules; they are usually, but not always, associated with an "asymmetric carbon" (or chiral center), a tetrahedral carbon atom bound to four different substituents, as in

$$\begin{array}{ccc} & F & CH_3 \\ & | & | \\ H\!-\!C\!-\!Cl & or & CH_3CH_2\!-\!C\!-\!COOH, \\ & | & | \\ & Br & H \end{array}$$

since a tetrahedron with four differently labeled corners is not identical to its mirror image.

The existence of the set of isomers shown in Fig. 9.18 is one of the clearest demonstrations of the octahedral structure of this complex and,

---

[16] One can also use the difference in refractive index to separate the two components of polarization. This is just how circularly polarized light is obtained. Both plane and circular polarization are ordinarily produced by passing light through suitably cut prisms of an anisotropic crystal such as quartz (SiO_2), in which the crystal structure rather than the individual molecule is chiral.

by extension, of other Co(III) complexes. A great number of complexes with chelate ligands are optically active, and this is nearly always useful in determining their structures. Optical isomers can often be separated or *resolved*, usually by precipitating the ion in combination with another optically active species; again, the combination of two chiral properties results in a nonchiral physical difference.

Thus far we have discussed isomerism almost exclusively in octahedral complexes, but similar analyses can be applied to determine the structure of complexes with other geometries. We mentioned earlier that 4-, 5-, and 6-coordinate complexes are the most common; Fig. 9.19 illustrates the usual structures found for each of these. (As one might expect, the symmetries of these structures are often the same as those observed in small covalent molecules: cf. Fig. 9.2.) We shall briefly survey here some types of complexes with each of these structures.

The principal structures known for 4-coordination are the tetrahedral and the square-planar. Many of the tetrahedral complexes are tetrahalides, such as $[ZnCl_4]^{2-}$, $[CdBr_4]^{2-}$, $[FeCl_4]^-$, $[CoI_4]^{2-}$, and similar species such as $[Hg(CN)_4]^{2-}$. Square-planar complexes are more common for noble metals, as in $[PtCl_4]^{2-}$ and $[Pd(NH_3)_4]^{2+}$, as well as for Ni(II) and Cu(II), as in $[Ni(CN)_4]^{2-}$ and $[CuCl_4]^{2-}$. (Cu(II) is also found as a distorted tetrahedron.) The square-planar geometry is sometimes interpreted as an extreme case of 6-coordination in which two *trans* vertices of a tetragonally distorted octahedron have extremely weak bonding capacity.

Five-coordinate complexes are relatively uncommon. They usually have trigonal-bipyramidal form, like PCl_5; the simplest instances are pentacarbonyls such as $[Fe(CO)_5]$ or $[Mn(CO)_5]^-$. Rare examples, such as $[Ni(CN)_5]^{3-}$, are square-pyramidal.

Six-coordinate complexes, probably the most common of all (at least the most studied), are almost always either octahedral or distorted-octahedral. We have already given many examples of the normal octahedron, in which all six vertices are equivalent. The most common distortions are either *tetragonal*, with a lengthened (or shortened) axis perpendicular to the square equatorial plane of the octahedron; or *trigonal*, with the octahedron stretched (or compressed) along an axis perpendicular to two opposite triangular faces. A noteworthy example of tetragonal distortion is the $[Cu(H_2O)_6]^{2+}$ complex (which produces the bright blue color of Cu(II) solutions): two of the Cu—O bonds are longer than the other four. Ferrous salts, such as ferrous ammonium sulfate and $FeSiF_6$, display trigonal distortions, particularly in the solid state.

Now let us summarize the more exotic coordination numbers. The coordination number 2 occurs in a few linear complexes of singly charged cations, such as $[CuCl_2]^-$ and $[Ag(NH_3)_2]^+$. Most species once thought to have coordination number 3 have turned out to be polymeric;[17] one real example appears to be the triangular $[HgI_3]^-$. The species with apparent coordination numbers of 2 and 3 may actually have higher values, with very weakly bound solvent molecules occupying additional sites in their coordination shells. Coordination numbers greater than 6 are fairly rare

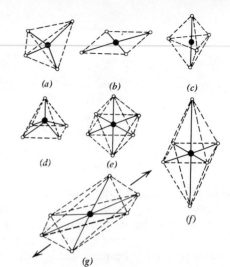

FIGURE 9.19
Common geometries for complexes: 4-coordinate (*a*) tetrahedron, (*b*) square-planar; 5-coordinate (*c*) trigonal bipyramid, (*d*) square pyramid; 6-coordinate (*e*) octahedron, (*f*) tetragonally distorted octahedron, (*g*) trigonally distorted octahedron; (axis of elongation indicated by arrows).

---

[17] For example "$CuCl_3$" is really

$$-Cl-\underset{\underset{Cl}{|}}{\overset{\overset{Cl}{|}}{Cu}}-Cl-\underset{\underset{Cl}{|}}{\overset{\overset{Cl}{|}}{Cu}}-Cl-\underset{\underset{Cl}{|}}{\overset{\overset{Cl}{|}}{Cu}}-Cl-,$$

with coordination number 4.

and limited almost entirely to large cations; some examples include $[UF_7]^{3-}$ (pentagonal bipyramid), $[Mo(CN)_8]^{4-}$ (dodecahedron), $[TaF_8]^{3-}$ (square antiprism—a cube with one face rotated $45°$ relative to the opposite face), and $[ReH_9]^{2-}$.

We note in passing that there also exist *polynuclear* complexes, which have two or more central atoms connected by bridges. These bridges may be ordinary monodentate ligands shared by two coordination shells, or bidentate ligands with one end in each coordination shell. Some examples of the first kind are

and

the platinum compound is planar and has *cis* and *trans* isomers. There are also polynuclear complexes with direct metal-to-metal bonds, some with quite elaborate structures; for example, in $[Mo_6Cl_8]^{4+}$ and $[Rh_6(CO)_{16}]$ the metal atoms form an octahedron!

This completes our preliminary survey of complexes. In the next two sections we shall see how their properties are related to their electronic structure.

9.8 Magnetic Properties of Complexes

Among the most telling diagnostic properties of complexes are their magnetic and optical characteristics, from which one can infer much about their electronic structures. It is true that we can learn little in this way about tightly bound closed-shell molecules such as PF_5, or about the complexes of closed-shell ions such as Zn^{2+}, which are colorless and essentially nonmagnetic. However, the complexes of ions with partially empty d shells, such as Cr^{3+} or Fe^{3+}, are often strongly colored and paramagnetic; this class includes most complex ions of the transition metals. One of the attractions of this field has always been the range of hues and intensities that the transition metal complexes exhibit. For example, copper forms the light blue $[Cu(H_2O)_6]^{2+}$ complex in water, and the more strongly bound, intensely deep blue $[Cu(NH_3)_6]^{2+}$ complex in liquid ammonia, plus all the intermediate species; and we have already mentioned the lavender and green isomers of $[Co(NH_3)_4Cl_2]^+$. The colors of these species are generally associated with low-lying energy levels to which electrons can be excited by visible light; in the next section we shall look into the nature of these levels. Let us turn first, however, to the magnetic properties.

The magnetic properties of complexes have been studied by two rather different methods. One of these measures the bulk property called magnetic susceptibility; the other probes the magnetic moments of individual molecules by the spectroscopic technique generally called electron spin resonance. We mentioned both of these in our study of atomic magnetism (Section 5.1), but we shall go into more detail here. In both cases the quantity one seeks to determine is $\boldsymbol{\mu}_m$, the magnetic dipole moment of the complex.

Magnetic susceptibility is a quantity that measures the capacity of a substance to develop magnetization when subjected to a magnetic field. In a vacuum the magnetic induction (magnetic flux density) $\mathbf{B}$ is related to the magnetic field strength $\mathbf{H}$ by the equation $\mathbf{B} = \mu_0\mathbf{H}$, where μ_0 is the permeability of free space ($\mu_0 \equiv 4\pi \times 10^{-7}$ H/m, where H means henrys). Within a piece of matter $\mathbf{B}$ is in general not equal to $\mu_0\mathbf{H}$ (where $\mathbf{H}$ is the external field), but may be greater or less. We thus write the equation

$$\mathbf{B} = \mu_0(1+\chi)\mathbf{H}, \tag{9.10}$$

which defines the magnetic susceptibility χ. The *magnetization* is $\mathbf{M} = \chi\mathbf{H}$. If χ is negative ($\mathbf{B} < \mu_0\mathbf{H}$), the substance is said to be *diamagnetic*; if χ is positive ($\mathbf{B} > \mu_0\mathbf{H}$), it is *paramagnetic*.

All materials have the capacity for diamagnetism, that is, for responding to an external magnetic field by setting up a weak field that tends to cancel the applied field. Diamagnetism is due in large part to the precession of the electronic orbital angular momentum vectors induced by the applied field, and in lesser part to actual distortion of the orbital wave functions. Diamagnetism is generally a small effect, and is by far outweighed by paramagnetism when the latter is present. In paramagnetic species the diamagnetic effects can be more than adequately taken into account by approximation methods in reducing experimental data.

Paramagnetism, on the other hand, results from the existence of permanent magnetic moments within the molecular units of a substance. An external magnetic field tends to align these microscopic magnets parallel to itself, and thus produces a net positive magnetization on the macroscopic level. We have outlined the theory of the magnetic moment in Section 5.1. When a molecule has both orbital angular momentum $\mathbf{L}$ and spin angular momentum $\mathbf{S}$ (and the two are separable), its magnetic dipole moment is given by

$$\boldsymbol{\mu}_m = \boldsymbol{\mu}_L + \boldsymbol{\mu}_S = -\frac{e}{2m_e}(\mathbf{L} + g_s\mathbf{S}). \tag{9.11}$$

The Landé g-factor g_s is found experimentally to have the value 2.002319 for the free electron; the value of g_s varies slightly with environment, but for many purposes can be approximated by 2. The magnitude of $\boldsymbol{\mu}_m$ depends on the quantum state of the molecule. Some special cases can be evaluated immediately. For a state with only orbital angular momentum ($S = 0$) we have

$$|\boldsymbol{\mu}_m|_{\text{orb}} = |\boldsymbol{\mu}_L| = \frac{e\hbar}{2m_e}[L(L+1)]^{1/2} = \mu_B[L(L+1)]^{1/2}, \quad \mu_B \equiv e\hbar/2m_e, \tag{9.12}$$

and for a spin-only state ($L = 0$) we have

$$|\boldsymbol{\mu}_m|_{\text{spin}} = |\boldsymbol{\mu}_S| = g_s\mu_B[S(S+1)]^{1/2}, \tag{9.13}$$

where L and S are the orbital and spin quantum numbers and μ_B is the Bohr magneton. The situation in states with both orbital and spin angular momentum is more complicated, depending on both the coupling conditions and the distribution of molecules over the term levels (which in a complex, as we shall see in the next section, are split by the presence of the ligands). For *LS* coupling, with spin-orbit interaction negligible relative to the ligand field splitting, one can show that the magnetic moment should be

$$|\boldsymbol{\mu}_m| = \mu_B[4S(S+1) + L(L+1)]^{1/2}, \qquad (9.14)$$

assuming that $g_s = 2$. This equation is in principle applicable to most transition metal complex ions, but we shall see that there are complications. In general, the effective g-factor g_{eff}, defined by

$$|\boldsymbol{\mu}_m| = g_{\text{eff}}\mu_B, \qquad (9.15)$$

may or may not be given by Eq. 9.14.

Before seeing how $|\boldsymbol{\mu}_m|$ can be related to the macroscopic susceptibility, let us digress to consider *electron spin resonance* (esr)—more properly called *electron paramagnetic resonance* (epr) for complex ions, because more than electron spin is usually involved. This method utilizes the Zeeman-effect splitting of energy levels in a magnetic field, described by Eq. 5.8 for a free atom. Complications can again occur; to encompass these, one can write the Zeeman shift as

$$E - E_0 = -\boldsymbol{\mu}_m \cdot \mathbf{B} = -g_{\text{eff}}M_J\mu_B B \qquad (M_J = M_L + M_S), \qquad (9.16)$$

where M_J is the quantum number for the component of total angular momentum in the direction of the field $\mathbf{B}$. The separation of adjacent Zeeman levels is then

$$\Delta E = g_{\text{eff}}\mu_B B. \qquad (9.17)$$

To measure this quantity, one normally applies a magnetic field of variable strength and an oscillatory radiation field of precisely known frequency ν in the microwave region. The variable field is adjusted until the Zeeman splitting $g_{\text{eff}}B$ coincides with the energy $h\nu$ of the microwave quantum, at which point resonant absorption of energy is observed. This sort of spectroscopy, in which one tunes the energy-level spacing rather than the frequency of the radiation, is usually more convenient in the radiofrequency and microwave regions. The reason is that it is easier to maintain a very precisely known, fixed frequency and then measure the magnetic field accurately than to fix the magnetic field and then make accurate measurements of a variable frequency.

We can now return to the magnetic susceptibility χ, which is defined by Eq. 9.10. This quantity is usually measured with a *magnetic balance:* A sample is suspended half in and half out of a strong magnetic field, and the force exerted on the sample is measured directly. This force is given by

$$F = \tfrac{1}{2}\chi A\mathbf{B} \cdot \mathbf{B}, \qquad (9.18)$$

where A is the cross-sectional area of the sample and $\mathbf{B}$ is the magnetic field. A paramagnetic material $(\chi > 0)$ is pulled into the field; a diamagnetic material $(\chi < 0)$ is pushed out. One usually expresses results in terms of the chemically more significant *molar susceptibility,*

$$\chi_M \equiv \frac{M\chi}{\rho}, \qquad (9.19)$$

where M is the molecular weight and ρ is the density. Although χ_M for a paramagnetic sample contains both paramagnetic and diamagnetic contributions, the diamagnetic part is very small and can be estimated with sufficient accuracy; subtracting this out, one obtains the "corrected" molar susceptibility χ_M^{corr}. This quantity is usually found to be inversely proportional to the absolute temperature T,

$$\chi_M^{corr} = \frac{C}{T},$$ (9.20)

where C is a constant: Eq. 9.20 is known as the *Curie law*.

One can derive the Curie law directly from the microscopic theory. The magnetization $\mathbf{M}$ ($= \chi \mathbf{H}$) of a bulk sample is simply the average magnetic moment per unit volume. For paramagnetic materials it is zero in the absence of an external field, since the microscopic magnets are then randomly oriented. The external field tends to align the individual magnetic moments with itself, but because of molecular motion this alignment is only partial; what results is a distribution over a range of angles. If a given molecule's moment $\boldsymbol{\mu}_i$ is directed at an angle θ_i away from the direction of $\mathbf{H}$, its contribution to the net moment is $|\boldsymbol{\mu}_i| \cos \theta_i$. One obtains the average moment of the bulk material (and thus $\mathbf{M}$ and χ) by averaging over all possible angles; Part Two deals with the way in which one obtains such averages. As a result of this calculation, one finds that at thermal equilibrium the corrected molar susceptibility should be

$$\chi_M^{corr} = \frac{N_A |\boldsymbol{\mu}_m|^2}{3 k_B T},$$ (9.21)

where N_A is Avogadro's number, k_B is Boltzmann's constant, and $\boldsymbol{\mu}_m$ is as usual the magnetic moment of an individual molecule. If the sample contains more than one kind of molecule or ion, $|\boldsymbol{\mu}_m|^2$ is replaced by an average over the various species present. Equation 9.21 is of the same form as the Curie law, with $C = N_A |\boldsymbol{\mu}_m|^2 / 3 k_B$; this equation is used to evaluate $|\boldsymbol{\mu}_m|$ from a measurement of χ_M^{corr}.

Whether one measures the spin resonance or the susceptibility, one obtains a value of $|\boldsymbol{\mu}_m|$. This observed moment is expressible, as in Eq. 9.15, in terms of the Bohr magneton multiplied by an effective g-factor g_{eff}. For gaseous atoms g_{eff} has the values predicted by Eq. 9.14. But for complex molecules and ions, and even for "free" atoms and monatomic ions in liquid and solid media, the magnetic moments are generally not as large as Eq. 9.14 would indicate. Very often the observed moments are much closer to the values predicted for the electron spins alone by Eq. 9.13.

Let us first consider some species that have only spin angular momentum. The ions Mn^{2+} and Fe^{3+} in their ground-state configurations have five $3d$ electrons, corresponding to a half-filled $3d$ subshell; all other occupied shells in these ions are filled. According to Hund's first rule, the lowest-energy state of this configuration should be that with the maximum value of S; here this means that all the $3d$ spins should be parallel (have the same value of m_s), giving $S = \frac{5}{2}$ for the ion. But if all five $3d$ electrons have the same m_s, they must all be assigned different values of m_l (from $+2$ to -2). This means that all five orbitals of the $3d$ subshell are equally populated with one electron each: Thus the $3d$ charge distribution, like that of the filled shells, is spherically symmetric, and the orbital angular momentum of the ion is zero. With $L = 0$ and $S = \frac{5}{2}$, the ground states of these ions must be ^6S. Since they have no orbital angular momentum, the magnetic moments of the Mn^{2+} and Fe^{3+} should be given by Eq. 9.13:

$$|\boldsymbol{\mu}_m| = 2[\tfrac{5}{2}(\tfrac{5}{2}+1)]^{1/2} \mu_B = 5.92 \mu_B.$$

The observed values are, in fact, approximately 5.9 Bohr magnetons for these species.

Most ions with permanent magnetic moments have both spin and orbital angular momenta. Even in most of those cases the observed moments are strikingly close to the spin-only values. For example, the Cu^{2+} ion has the ground-state configuration $3d^9$, with a single unpaired d electron, and thus is ^2D ($L = 2$, $S = \frac{1}{2}$). The spin-plus-orbital magnetic moment predicted by Eq. 9.14 is $3.00\mu_B$, whereas the spin-only value predicted by Eq. 9.13 is $1.73\mu_B$. The experimental values for Cu(II) in various complexes with ligands that are not themselves paramagnetic range between 1.7 and 2.2 Bohr magnetons. Similarly, for Ni^{2+} ($3d^8$) the ground state of the free ion is ^3F ($L = 3$, $S = \frac{3}{2}$), the spin-plus-orbital and spin-only calculations give $4.47\mu_B$ and $2.83\mu_B$, respectively, and the observed moments for various Ni(II) complexes in solution span the range $2.8–4.0\mu_B$.

In cases in which the observed magnetic moment is less than the full amount expected from the combination of spin and orbital angular momenta, we say that the orbital angular momentum is quenched. Quenching occurs because the electrons are in a potential energy field that is not spherically symmetric; this is the polyhedral field of the ligands in a complex, or that of the surrounding ions in a crystal. As we pointed out in Chapter 5, every departure from symmetry removes some degeneracy and causes the "spoiling" of some conservation law. In complexes, the deviations from spherical symmetry ("quenching potential") stimulate interchange of orbital angular momentum between an ion and its surroundings. Classically, one might say that torques are exerted on the orbiting electrons. Thus **L** is no longer a well-conserved quantity. In particular, its component L_z in the direction of an external field is not conserved, and it is this component that interacts with the field to produce magnetization: The orbital part of the interaction energy is $-\boldsymbol{\mu}_L \cdot \mathbf{B} = -(e/2m_e)L_z B$. In some cases the interchange of angular momentum is so rapid that L_z averages to zero, and quenching is complete. This corresponds to a quenching potential much greater than the Zeeman splitting due to the external field. In the next section we shall inquire into the nature of the quenching potential in complexes.

Thus far we have tacitly assumed that the magnetic behavior of a complex can be interpreted in terms of the magnetic properties of the central atom or ion in isolation. This is indeed true in many instances and provides us with yet another example of the separability of properties. For example, our analysis of the Fe^{3+} ion, with its ^6S ground state, corresponds well to the condition of Fe(III) in the colorless $[FeF_6]^-$ complex, in the brown $[Fe(H_2O)_6]^{3+}$ complex, in alums,[18] in beryl, and as an impurity in rutile (TiO_2). But the red $[Fe(CN)_6]^{3-}$ (ferricyanide) complex has a magnetic moment of only about $2.3\mu_B$, corresponding to a single unpaired electron with some unquenched orbital angular momentum (since for $S = \frac{1}{2}$ the spin-only moment is $1.73\mu_B$). The isoelectronic Mn(II) complexes exhibit the same pattern, with $[Mn(CN)_6]^{4-}$ an anomalous "low-spin complex" like $[Fe(CN)_6]^{3-}$. For another example, consider the Co^{3+} ion, which has six $3d$ electrons and a ^5D ($L = 2$, $S = 2$) ground state. The blue $[CoF_6]^{3-}$ complex has a magnetic moment of $5.2\mu_B$, corresponding to the full spin-plus-orbital value of Eq. 9.14, but the yellow $[Co(NH_3)_6]^{3+}$ complex is diamagnetic, corresponding to $L = 0$, $S = 0$. Clearly such large deviations from the isolated-ion model cannot be accounted for merely by quenching, especially when spin as well as orbital angular momentum is involved. We are dealing here with a

---

[18] A typical alum is $KFe(SO_4)_2 \cdot 12H_2O$, which contains $[K(H_2O)_6]^+$ and $[Fe(H_2O)_6]^{3+}$ ions in its crystal lattice.

Some complexes have electronic structures and chemical bonds that are well described by a classical electrostatic model, that is, a central ion to which the ionic or dipolar ligands are attracted by Coulomb forces only. For example, the hydrated ions of the heavier alkali metals approach this limit rather closely. The classical electrostatic model, similar to that we introduced for ionic molecules in Section 7.4, works best with closed-shell central ions and simple ligands. At the other extreme are covalent polyhedral molecules (Section 9.1), which are best described in terms of delocalized molecular orbitals. Most of the species ordinarily considered as complexes are intermediate between these limits. To describe their properties, especially those arising from partially filled electron shells of the central atoms, we must use an approach involving at least some of the quantum mechanical apparatus.

One of the simplest methods is basically electrostatic, in that the ligands are treated as point charges or dipoles located at the vertices of a polyhedron; one then considers the effect of this electrostatic field on the orbitals of the central atom. This simple but often effective approach was first used to interpret the spectra of ions in crystals, and is thus called *crystal field theory* (CFT); most of this section is written on the level of crystal field theory. For more exact results one must introduce at least some aspects of the molecular orbital approach, with orbitals that may encompass both the central atom and the ligands. The term *ligand field theory* (LFT) is generally applied to the range of models intermediate between the CFT and the full molecular orbital treatment.

Because of the high symmetries of polyhedral complex ions, the analysis of their properties becomes particularly elegant when it is done with the techniques of group theory, that part of mathematics concerned specifically with symmetry. Space is too limited here to permit us to develop the elements of group theory; several of the references at the end of this chapter present and apply its methods. Here, we concentrate on the physical properties of complexes and their origins in electronic structure, without drawing extensively on the symmetries of the ions.

What is the role of the ligands in determining the properties of a complex? First, they shield the central atom or ion from the environment; more important, they create a potential field within which the valence electrons of that central atom move. If this field is weak, the central atom behaves much as it would in isolation. But if the field is strong, the energy levels and orbitals of the central atom are no longer like their counterparts in the free atom or ion. Fortunately, one can analyze the properties of complexes, treating both weak-field and strong-field limits, without performing elaborate calculations. The problem lends itself to an empirical representation in terms of simple parameters. Our discussion here will be qualitative, but indicates how the quantitative treatment proceeds.

One usually assumes that the equilibrium geometries of the complexes of interest are known. It is possible but costly to find the geometry from direct computations of the potential energy surface; in practice, it is frequently more convenient to infer the structure from experiments or, for the more elaborate complexes, to assume the geometry from chemical analogs. Fortunately, the experimental knowledge of the geometry of complexes is usually sound. Structural information can be obtained from chemical methods such as Werner's, from x-ray diffraction (Chapter 11),

9.9
Electronic Structure of Complexes

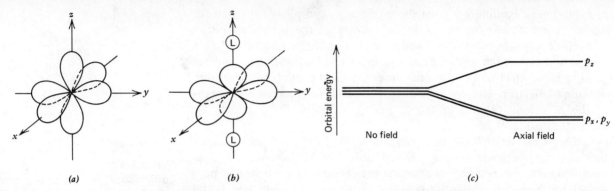

(a)   *(b)*   *(c)*

FIGURE 9.20
Behavior of p orbitals in an axially symmetric field. (*a*) The three degenerate orbitals with no external field. (*b*) The effect of ligands (L) along the z axis, splitting the orbitals into a twofold-degenerate pair (p_x, p_y) and a nondegenerate (p_z) orbital. (*c*) Schematic illustration of the splitting of orbital energies (the variation of which is presumably a continuous function of the strength of the axial field).

from spectroscopy, or from magnetic measurements. We shall see why the spectroscopic and magnetic properties depend sensitively on the shape of the complex.

Given the geometry, then, what can we say about the electronic structure? Consider the outer electrons of the central atom or ion. In the free atom these electrons would be subject to a potential field that is spherically symmetric, except for the interactions between the electrons themselves, because of the isotropy of free space. There is no physical distinction between different orientations of the electrons and those of the atom itself. Remember that a symmetric Hamiltonian leads to degeneracy. In this case, for example, an atomic state[19] with a given value of J corresponds to $2J+1$ degenerate wave functions with the same energy. But in a complex the situation is quite different. The polyhedron of ligands produces a potential field that is less than spherically symmetric, and the usual result is some removal of degeneracy. For a specific polyhedral geometry, it may become possible to observe a physical distinction between different orientations.

The latter statement does not apply to an orbital or state with zero angular momentum, an s orbital or ^1S state: Since such states are nondegenerate and have spherically symmetric wave functions, they have no capacity to distinguish orientations. The presence of a set of ligands may shift the energy of such a state and distort its wave function from spherical symmetry, but the directional character will still be the most symmetric allowed by the ligand field. Thus an s orbital in a pure octahedral field may distort, but only into octahedral symmetry; it will still have the same amplitude for all the equivalent directions of the octahedral environment.

Orbitals and states with nonzero angular momentum do not respond to the presence of ligands in so simple a manner. For the most elementary example of the directionality induced by ligands, let us consider what happens to the central atom's p orbitals in a linear "complex" such as XeF$_2$ (coordination number 2). We call the internuclear axis the z axis, with the origin of coordinates at the central atom (Fig. 9.20). In the free atom one could define the usual p_x, p_y, p_z orbitals extending along the three coordinate axes, and all three of these orbitals would be equivalent in every physically distinguishable way. But if we put a ligand on each side of the central atom along the z axis, the fields acting on the different p orbitals are no longer equivalent. The p_z orbital becomes physically different from the p_x and p_y orbitals, which remain equivalent to each other. By "physically different" we mean that the shape and energy of the

[19] By "state" here we mean, say, a particular component of a multiplet term, such as $^2P_{1/2}$: cf. Section 5.7.

p_z orbital must be different from those of the p_x and p_y. We say that the ligand field has split the threefold degeneracy of the p orbitals into a twofold-degenerate pair (p_x, p_y) and a single nondegenerate orbital (p_z). One would expect the central atom's p_z orbital to be somewhat distorted so as to move electron density away from the filled orbitals of the ligands because of electron–electron repulsion. The result is an increase in energy of the p_z orbital relative to the p_x and p_y orbitals. These effects are illustrated in Fig. 9.20.

If we considered XeF_2 as a complex made up of Xe^{2+} and F^- ions, the four $5p$ electrons of Xe^{2+} would probably be best described as occupying the lower-energy degenerate orbitals. However, this is a poor model for XeF_2, in which the bonding is strongly covalent and better treated in terms of three-center MOs. We have gone through this analysis with p orbitals only for illustrative purposes: As we shall see, CFT and LFT are ordinarily most powerful when they are applied to d and f orbitals, which partake less in covalent bond formation than do p orbitals.

Here is another way to look at the splitting. In a field with only axial symmetry (with the z axis taken as the symmetry axis), the space-quantized orbitals with $m_l = \pm 1$ (or their real combinations, p_x and p_y) differ physically from the orbital with $m_l = 0$ (p_z). Clearly a similar splitting must occur for d, f, ... orbitals in an axially symmetric field. The splitting always takes the same form: From a set of orbitals that are degenerate in the isolated atom, the one with $m_l = 0$ becomes nondegenerate, whereas those with any given value of $|m_l|$ greater than zero form a degenerate pair with its own energy. The twofold degeneracy that remains is the result of the equivalence of the x and y directions.

In an octahedral ligand field, on the other hand, the x, y, and z directions are all physically equivalent just as in the isolated atom. To see that this is so, we need merely choose the coordinate axes to coincide with the bonds to the ligands, that is, along the lines connecting opposite vertices of the octahedron, as shown in Fig. 9.21a. Because of this equivalence the three components of a p state remain equivalent and unsplit in an octahedral environment. It may be slightly less obvious that a tetrahedral ligand field is also incapable of splitting p levels. To demonstrate this, we inscribe the tetrahedron in a cube and choose the axes as shown in Fig. 9.21b: One can see that the x, y, and z axes are again equivalent. Note that in these cases, as with s orbitals, the energy level may be shifted and the wave function distorted, but the degeneracy remains.

In order for a degenerate set of atomic orbitals to display splitting in an octahedral or tetrahedral field, the orbitals must be numerous enough, and have enough angular dependence, to distinguish physically different directions in space. Although the p orbitals do not have this capacity, we shall now see that the five components of a set of d orbitals do satisfy this requirement. The same is true a fortiori of f, g, ... orbitals. A great many complexes have octahedral or tetrahedral symmetry or can be treated in terms of distortions of these structures, and we expect splitting to occur in such complexes when there is a partially filled d or f subshell on the central atom—that is, in complexes of transition metal or rare earth ions. The d and f electrons of these ions are largely responsible for the bonding, optical, and magnetic properties of the complexes. It is to complexes of this type that CFT and LFT are mainly applied.

The most useful way to define five d orbitals is to pick the real functions we described in Section 4.4, five independent functions that vary respectively as xy, xz, yz, $x^2 - y^2$, and $z^2 - \frac{1}{2}(x^2 + y^2)$. (The last is called $d_z{}^2$ for convenience.) These are suitable analogs of the real p

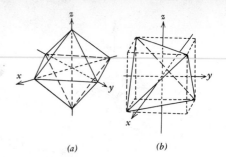

FIGURE 9.21
Equivalence of x, y, and z axes in (a) octahedral and (b) tetrahedral ligand fields. In (b) the coordinate axes pass through the centers of opposite faces of the circumscribed cube, and thus through the midpoints of opposite edges of the tetrahedron.

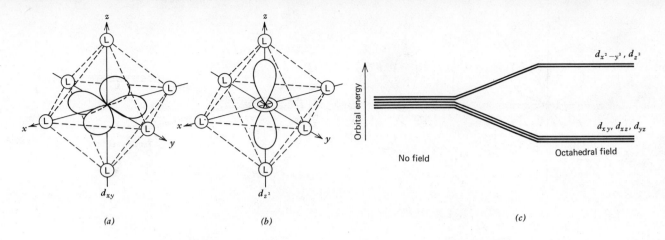

(a) (b) (c)

FIGURE 9.22
Splitting of d orbitals in an octahedral field. (a) The d_{xy} orbital (d_{xz} and d_{yz} have the same shape and also extend between the coordinate axes). (b) The d_{z^2} orbital ($d_{x^2-y^2}$ resembles d_{xy}, but extends along the x and y axes). (c) Splitting of orbital energies; in no field, all five orbitals are degenerate. In an octahedral field, these split into a set of three degenerate levels and a degenerate pair of levels.

functions that vary as x, y, z. We are concerned here only with their shapes, that is, their angular dependence,[20] which has been illustrated in Fig. 4.7.

In an octahedral ligand field the d_{xy}, d_{xz}, and d_{yz} orbitals are physically equivalent. Each has two equivalent axes of maximum amplitude that bisect the angles between pairs of coordinate axes. In the coordinate system of Fig. 9.21a, they are directed toward the midpoints of the octahedron's edges, that is, midway between pairs of ligand sites. The $d_{x^2-y^2}$ and d_{z^2} orbitals, in contrast, have their maxima along the coordinate axes—toward the vertices of the octahedron, and thus toward the ligand sites. If the valence orbitals of the ligands are filled, as is usually the case, then Coulomb and exclusion repulsion will tend to drive electrons of the central atom away from the regions near the ligands. This is especially true when the ligand is negatively charged, or is a dipole with its negative end inward, both common cases. This effect is strongest in the $d_{x^2-y^2}$ and d_{z^2} orbitals, in which the electrons are on the average closer to the ligands to begin with. As a result, the $d_{x^2-y^2}$ and d_{z^2} orbitals generally have higher energies than the d_{xy}, d_{xz}, and d_{yz} orbitals in an octahedral field. The five d orbitals that are degenerate in the free atom split into three degenerate low-energy orbitals and two degenerate high-energy orbitals.[21] The splitting is illustrated in Fig. 9.22.

We have thus far dealt only with one-electron orbitals. However, the same splitting rules apply to the many-electron atomic wave functions. Thus in an octahedral field an atomic D term (a set of states with $L = 2$) splits into a triply degenerate low-energy set and a doubly degenerate high-energy set, whereas a P term ($L = 1$) remains unsplit.

[20] The angular parts of these functions are given by the following table:

d_{z^2}	d_{xz}	d_{yz}	$d_{x^2-y^2}$	d_{xy}
$Y_{2,0}$	$Y_{2,\cos\phi}$	$Y_{2,\sin\phi}$	$Y_{2,\cos2\phi}$	$Y_{2,\sin2\phi}$

where the Y's are defined by Eq. 4.24 and Table 3.1. The magnitude of the angular function $Y_{2,\sin2\phi}$ varies as xy/r^2 for d_{xy}, etc. (division by r^2 removes the radial dependence).
[21] It may not be obvious why the $d_{x^2-y^2}$ and d_{z^2} orbitals remain degenerate, since their shapes are quite different (Fig. 4.7). However, since

$$z^2 - \tfrac{1}{2}(x^2 + y^2) = \tfrac{1}{2}[(z^2 - x^2) + (z^2 - y^2)],$$

the d_{z^2} orbital is formally equivalent to a sum of $d_{z^2-x^2}$ and $d_{z^2-y^2}$ orbitals, each with the same shape as the $d_{x^2-y^2}$ and thus degenerate with it when the x, y, and z directions are equivalent (as is the case in both octahedral and tetrahedral fields).

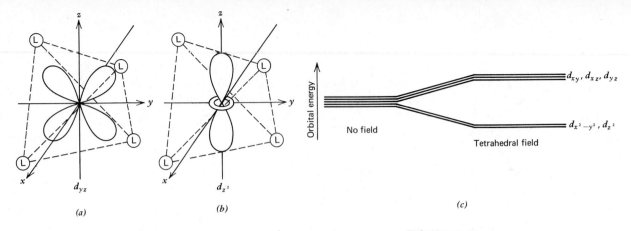

d_{yz}

(a)

d_{z^2}

(b)

(c)

No field

Tetrahedral field

d_{xy}, d_{xz}, d_{yz}

$d_{x^2-y^2}, d_{z^2}$

Orbital energy

FIGURE 9.23
Splitting of d orbitals in a tetrahedral field. (a) The d_{yz} orbital (the lobes extend close to but not directly toward the ligand sites, none of which are in the yz plane). (b) the d_{z^2} orbital (the lobes point midway between ligand sites). (c) Splitting of orbital energies. The five degenerate d orbitals again split into a triply degenerate set and a degenerate pair, but, in contrast to the octahedral case, the pair usually has lower energy.

A tetrahedral ligand field also splits a set of d orbitals, but not in quite the same way. In the tetrahedral field, as shown in Fig. 9.23, it is the d_{xy}, d_{xz}, and d_{yz} orbitals that have their maxima relatively close to the directions of the ligand sites, whereas the $d_{x^2-y^2}$ and d_{z^2} orbitals point between the ligand sites. Thus the d orbitals again split into sets of two and three, but the energy relationships are reversed from those in an octahedral field. Most commonly the interaction between d electrons and ligands is predominantly repulsive, so the doubly degenerate pair of $d_{x^2-y^2}$ and d_{z^2} orbitals are the low-energy set and the other three are the high-energy set. This splitting is shown in Fig. 9.23; for similar ion–ligand interactions, the energy splitting of the d orbitals in the tetrahedral complexes is only about half as great as in the octahedral cases.

Distortions, if they occur, lower the symmetry of a tetrahedral or octahedral complex. We mentioned earlier the tetragonal and trigonal distortions of the octahedron. The result of such distortions is that the energy levels split even further, as the orbitals that are equivalent in octahedral or tetrahedral symmetry become nonequivalent. In a tetragonal distortion along the z axis, for example, the energy of the orbitals that extend in the z direction (d_{xz}, d_{yz}, and especially d_{z^2}) are shifted relative to those in the xy plane (d_{xy} and $d_{x^2-y^2}$). The 4-coordinate square-planar complex, as we have pointed out, is an extreme case of this distortion. In tetragonal complexes the d_{z^2} orbital has a binding energy considerably greater than the d_{xy} and $d_{x^2-y^2}$ pair; sometimes the d_{z^2} orbital lies lower than the d_{xz} and d_{yz} levels, which in this case remain degenerate. Other distortions can be similarly analyzed (cf. Problem 20 at the end of the chapter). Distortions are relatively common, though often small. One can show that any complex in which an orbital degeneracy occurs can lower its total energy by undergoing a geometric distortion that splits the degeneracy (the *Jahn–Teller effect*).

Our analysis has now gone far enough to allow us to give some interpretation of the relationships between the structures of complexes and their spectroscopic and magnetic properties.

As we mentioned earlier, the characteristic colors of transition metal complexes must be due to the absorption of light in the visible region of the spectrum. What is the nature of the transitions involved? Most transitions involving the valence electrons of free atoms—and especially of positive ions—are associated with absorption or emission of ultraviolet radiation. Relatively few such transitions have energies low enough to give rise to spectral lines in the visible region. But in complexes the ligand field is relatively weak compared to the strong central field of the atom. The bonding between the central atom and the ligands is relatively weak for the same reason. Thus the ligand field splitting of d levels is very often small enough to put the absorption bands due to d–d transitions into the visible region, and thus make the complexes colored. The spectra of complexes can usually be explained in considerable detail with the full apparatus of ligand field theory. The spectra may be obtained under many different conditions, and their appearances change accordingly. In solutions at room temperature broad bands rather than discrete spectral lines are usually observed, primarily because of the vibrational motion of the ligands. In crystals at very low temperatures very sharp spectra often appear, in which vibrational levels can be assigned with confidence.

As for the magnetic properties, it is easy to see how they can be interpreted in terms of energy-level diagrams like those in Figs. 9.22 and 9.23. Let us examine the magnetic and optical properties for several cases of increasing complexity.

The simplest transition metal ion in which d orbitals are occupied is Ti^{3+}, with a single $3d$ electron outside an argon closed shell. The magnetic moments of Ti(III) complexes average $1.7\mu_B$, corresponding to the spin-only value with quenching of orbital angular momentum. We can now see that the "quenching potential" we spoke of in the last section is the ligand field, which has a much greater effect on the electron's motion than does the external magnetic field; detailed calculations of this effect are possible, but will not be discussed here. For an example of the spectroscopic properties, consider the hexaquo ion $[Ti(H_2O)_6]^{3+}$. It has a broad absorption band whose peak lies at about 4930 Å, giving the complex a violet color. (The transmitted light on both sides of the band is mainly red and blue, combining to produce violet.) The complex is presumably octahedral, with an energy-level diagram like that shown in Fig. 9.22c. We assume that the single d electron occupies the lower, triply degenerate level in the ground state, and that the visible absorption corresponds to a transition to the upper, doubly degenerate level.

The V^{4+} ion is isoelectronic with Ti^{3+}, and forms some octahedral complexes with similar properties. It also forms some distorted tetrahedral compounds, of which VCl_4 is the best known. As we noted above, the ligand field splittings in tetrahedral complexes are typically about half as great as those in octahedral complexes for ligands of the same sort. Thus VCl_4 is brown because of an absorption band centered near 9000 Å in the near infrared, presumably due to a transition from the lower, doubly degenerate to the upper, triply degenerate level in Fig. 9.23c. Both types of complexes have magnetic moments of 1.7–$1.8\mu_B$, again the spin-only value for one electron.

Ions with more d electrons offer richer examples for interpretation. The electron–electron interactions play a role that, in effect, competes with the ligand field effects. A good and rather simple example is the Fe^{3+} ion, which has the configuration $[Ar]3d^5$, with a half-filled d shell. As we indicated in the last section, the ground state of the isolated ion is ^6S, with maximum spin and zero orbital angular momentum. In a weak ligand field—one whose effects are small compared with the electron–electron interactions responsible for Hund's rules—the designation ^6S remains characteristic of the ground state of an Fe(III) complex. That is, the total spin is $\frac{5}{2}$, the magnetic moment is $5.9\mu_B$, and we can assume that each of the five $3d$ orbitals still contains one electron despite the splitting of the orbital levels by the ligand field. We have mentioned a number of complexes, such as $[Fe(H_2O)_6]^{3+}$, in which this behavior is observed. The electron configuration in these complexes is presumably that shown in

Fig. 9.24*a*. The best-known *d–d* absorption bands[22] in these systems involve either transfer of an electron from the doubly degenerate level to the formally lower-energy triply degenerate level, at the expense of increased electron–electron repulsion, of course, or transitions between different terms of the ground-state configuration. In either case the energy difference is relatively small and the bands weak.

In a strong ligand field, on the other hand, the splitting between the two sets of *d* levels can become so large that the difference in orbital energies becomes greater than the separation of the ^6S and higher terms arising from the electron–electron repulsion. In these circumstances all five 3*d* electrons of the Fe^{3+} ion can be assigned to the lower-energy triply degenerate level, as shown in Fig. 9.24*b*. In this configuration there should be only one unpaired electron—and, as we mentioned earlier, the $[Fe(CN)_6]^{3-}$ complex, in which the CN^- ligand exerts a very strong field, does have a magnetic moment characteristic of only one unpaired spin, instead of five.

Many other ions exhibit a similar distinction between weak-field (also called *high-spin*) and strong-field (also called *low-spin*) complexes. In general these have the maximum and minimum possible numbers of unpaired spins, respectively; it is exceedingly rare to find an intermediate case. Of the examples we cited at the end of Section 9.8, the Mn(II) complexes have the same configuration as those of Fe(III), whereas for Co(III) $(\ldots 3d^6)$ the low-spin complexes have all six *d* electrons in the triply degenerate lower level, leaving no electrons unpaired and making the complex diamagnetic. One can easily see from Fig. 9.22 that high-spin and low-spin configurations should be possible for all octahedral complexes with four to seven *d* electrons in the outermost shell. Similarly, for tetrahedral complexes (Fig. 9.23) the phenomenon appears with three to six *d* electrons. But which complexes are high-spin and which low-spin? This depends on the strength of the ligand field, and thus on the nature of the ligands. For a number of common ligands the ability to split *d* orbitals is found to fall in the order:

$$CN^- > NO_2^- > en > NH_3 > H_2O > OH^- > F^- > Cl^- > Br^- > I^-,$$

(the *spectrochemical series*).

Thus far we have not faced the question of how bonding occurs between the central atom and the ligands. In the original crystal field theory the bonding was interpreted in terms of classical electrostatics. As useful as the simple ionic CFT is in giving a qualitative interpretation of magnetic and spectral properties, it is inadequate to account for bonding. When it became possible to interpret spectra in terms of microscopic interactions, it was clear that the CFT predictions were not in satisfactory quantitative agreement with the observations. The reason, of course, is that the metal–ligand bonds have a significant covalent contribution. To some extent this can be accounted for by inserting empirical parameters into the theory, but this procedure is not altogether satisfactory either.

It was a major step forward when the concept of molecular orbitals was extended to the bonding in complexes. The MOs used, of course, must be symmetry orbitals with the symmetry of the complex as a whole. In general, all the valence-shell orbitals of the central atom (*s* and *p* as well as *d*) are used in constructing the MOs, together with at least one orbital from each ligand. Central-atom orbitals that extend toward the

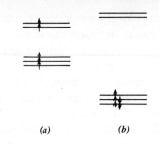

FIGURE 9.24
Configurations of the five 3*d* electrons in the ground states of Fe(III) complexes: (*a*) weak field (high spin); (*b*) strong field (low spin).

---

[22] Often there are much stronger charge-transfer bands, involving the transfer of an electron from a metal orbital to a ligand orbital or vice versa. These bands have their peaks in the ultraviolet, but are often strong enough in the visible region to have a dominant effect on the color and other optical properties.

ligands (Figs. 9.22 and 9.23) usually form σ bonds, whereas those that extend between the ligands can form π bonds. This approach has made it possible to bring the experiments and the theory together, at least as well as contemporary measuring and computing techniques can justify. We already know that complexes span a range from highly "electrostatic" and weakly bound to highly covalent, with strong short-range attractive forces. Experiment shows, and the MO calculations confirm, that the more covalent complexes tend to be low-spin (strong-field); this is to be expected, since the ligands whose orbitals overlap the central-atom orbitals to the greatest extent are naturally those that produce the strongest ligand fields. Thus the pure d character of the central atom's orbitals is maintained best in the high-spin complexes.

Before we close this section, a few words are in order about complexes of the rare earth (lanthanide) elements. The partially filled $4f$ shell of these elements is buried deeper within the atom than are the outer d shells of the ordinary transition metals. Hence the ligand field splittings of the f levels are generally quite small so there are no low-spin states, and the bands associated with transitions between these split levels often lie in the infrared. The f–f bands are also relatively narrow, because of the shielding effect of the outer electrons. Since spin–orbit interaction is strong in these atoms, the magnetic moments should be given by $g_J \mu_B [J(J+1)]^{1/2}$, rather than Eq. 9.13 or 9.14. Quenching is not significant, again because of shielding, and the experimental values agree well with these equations. In the actinide elements, by contrast, the $5f$ shell is relatively poorly shielded, so the complexes and other compounds of these elements have properties intermediate between those of the lanthanides and the transition metals.

Appendix 9A
Schmidt Orthogonalization

The most convenient way to generate a set of n orthogonal functions $\psi_i (i = 1, \ldots, n)$ from a set of n arbitrary but independent functions $\varphi_j (j = 1, \ldots, n)$ is called *Schmidt orthogonalization*. Suppose that the functions are listed serially. In essence, one picks the first function φ_1 arbitrarily. Then one makes the second function φ_2 orthogonal to the first by removing any component of φ_1 from it; next, one makes the third function φ_3 orthogonal to the first two by removing any component of φ_1 and the now-orthogonalized version of φ_2. One continues this way through the list until every function has been made orthogonal to all the preceding functions in the list. Finally, it is convenient to renormalize the functions so they all give unity for the integrals of their (absolute) squares. The set that one obtains in the end obviously depends on the sequence in which the initial functions φ_j were chosen.

Formally, the procedure is this. Assume that the set φ_j is normalized; otherwise normalize each function to unity. Let

$$\psi_1 = \varphi_1 \tag{9A.1}$$

and construct

$$\psi_2' = \varphi_2 - \psi_1 \int \psi_1^* \varphi_2 \, d\tau, \tag{9A.2}$$

which is φ_2 without its ψ_1 component. Hence ψ_2' is orthogonal to ψ_1, thus

$$\int \psi_1^* \psi_2' \, d\tau = \int \psi_1^* \varphi_2 \, d\tau - \left(\int \psi_1^* \psi_1 \, d\tau \right) \left(\int \psi_1^* \varphi_2 \, d\tau \right)$$

$$= \int \psi_1^* \varphi_2 \, d\tau - 1 \cdot \int \psi_1^* \varphi_2 \, d\tau$$

$$= 0. \tag{9A.3}$$

Now normalize ψ_2',

$$\psi_2 = \frac{\psi_2'}{\left(\int \psi_2'^* \psi_2' \, d\tau \right)^{1/2}}, \tag{9A.4}$$

so that

$$\int \psi_2^* \psi_2 \, d\tau = 1. \tag{9A.5}$$

Next, construct

$$\psi_3' = \varphi_3 - \psi_1 \int \psi_1^* \varphi_3 \, d\tau - \psi_2 \int \psi_2^* \varphi_3 \, d\tau, \tag{9A.6}$$

which is orthogonal to ψ_1 and ψ_2, and normalize to get

$$\psi_3 = \frac{\psi_3'}{\left(\int \psi_3'^* \psi_3' \, d\tau \right)^{1/2}}. \tag{9A.7}$$

Continue the procedure through ψ_n.

Further Reading

Ballhausen, C. J., *Introduction to Ligand Field Theory* (McGraw-Hill Book Co., Inc., New York, 1962).

Cotton, F. A. and G. Wilkinson, *Advanced Inorganic Chemistry*, Second Ed. (Interscience Publishers, New York, 1966).

Dewar, M. J. S., *The Molecular Orbital Theory of Organic Chemistry* (McGraw-Hill Book Co., Inc., New York, 1969).

Doggett, G., *The Electronic Structure of Molecules: Theory and Application to Inorganic Molecules* (Pergamon Press, Oxford and New York, 1972).

Dunn, T. M., D. S. McClure, and R. G. Pearson, *Some Aspects of Crystal Field Theory* (Harper and Row, New York, 1965).

Herzberg, G., *Molecular Spectra and Molecular Structure. III. Electronic Spectra and Electronic Structure of Polyatomic Molecules* (D. van Nostrand Co., Inc., New York, 1966).

Jørgensen, C. K., *Absorption Spectra and Chemical Bonding in Complexes* (Pergamon Press, Oxford, England, 1962).

Orville-Thomas, W. J., editor, *Internal Rotation in Molecules* (John Wiley & Sons, Inc., New York, 1974).

Shubnikov, A. V. and V. A. Koptsik, *Symmetry in Science and Art*, translated by G. D. Archart, edited by D. Harker (Plenum Press, New York, 1974), esp. Chapter 3.

Streitweiser, A., *Molecular Orbital Theory for Organic Chemists* (John Wiley & Sons, Inc., New York, 1961).

Tables of Interatomic Distances, Special Publications Nos. 11 and 18, The Chemical Society, London, 1958 and 1965. (This is a standard reference for data on molecular structures.)

Wells, A. F., *Structural Inorganic Chemistry*, 4th Ed. (Oxford University Press, Oxford, 1975).

Problems

1. The claim was made that the orbitals of Eq. 9.1 are equivalent, but with different spatial orientations. Show that they are equivalent, and determine the angles θ, ϕ along which they have maximum amplitude.

2. What structure do you predict for UF_5 monomer?

3. The ammonia molecule has a bond angle (HNH) of 106.7°. Suppose the N—H bond orbitals are symmetric about their respective N—H axes and are expressible as sums of hydrogenic $1s$ orbitals and nitrogen hybrids of the form $\alpha\varphi(2s) + \beta\varphi(2p)$. Find α and β from the value of the bond angle. If the lone pair orbital contains the rest of the $2s$ and $2p$ orbitals, what is its form?

4. Use single-lobed Slater-type orbitals with $Z = 3.18$ to compute the charge density in a radial direction along the direction of maximum charge for a) a $2p$ orbital; b) an sp hybrid; c) an sp^2 hybrid; d) an sp^3 hybrid. These are traditional approximations for the atomic orbitals of carbon. Their explicit

forms are

$$\varphi(2s) = (Z^3/32\pi)^{1/2}(2Zr)\exp(-Zr/2)$$

$$\varphi(2p_z) = (Z^3/96\pi)^{1/2}Zr\exp(-Zr/2)\cos\theta$$

$$\varphi(2p_x) = (Z^3/96\pi)^{1/2}Zr\exp(-Zr/2)\sin\theta\cos\varphi$$

$$\varphi(2p_y) = (Z^3/96\pi)^{1/2}Zr\exp(-Zr/2)\sin\theta\sin\varphi.$$

Suppose the length of a carbon-carbon bond can be determined approximately by superposing the points of maximum charge density for the appropriate hybrid orbitals from the two separate atoms. Use this supposition to make a table of C—C bond lengths for the various hybrid orbital combinations. Compare your results with the values given in Table 9.3.

5. Hydrogen peroxide, HO—OH, has a structure in which the hydrogens lie off the O—O axis, making a dihedral angle of roughly 90°, as shown. In ethane, H_3C—CH_3, the hydrogens at one end of the molecule lie between those at the other end, when the molecule is viewed along the C—C axis.

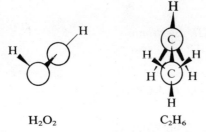

$$H_2O_2 \qquad\qquad C_2H_6$$

What physical effects could be responsible for these structures?

6. When two identical groups such as those in Table 9.1 are near each other, their oscillations—strictly, their oscillating charge distributions—induce a coupling between the two oscillators. The result is that the pair of vibrating groups oscillate synchronously in their stationary states. Show that the first two excited states would be degenerate in the absence of any coupling, and that with the nonzero coupling the stationary states correspond to motion exactly in phase and exactly out of phase. If two carbonyl ($>C=O$) groups are coupled by an interaction of $100\ \mathrm{cm}^{-1}$, what is the resulting splitting in the infrared spectrum? (The form of the calculation should be suggested by footnote 10; justify the use of such a form.)

7. Of the binding energies derived from photoelectron spectroscopy that are given in Table 9.2, which could be determined by photoionization with a lamp in which the light must pass through a lithium fluoride window passing light with $\lambda > 120\ \mathrm{nm}$? a sapphire window passing $\lambda > 150\ \mathrm{nm}$? Which could be studied by excitation with the light from the first allowed transition $(2p \rightarrow 1s)$ of helium? from the first allowed transition $(3s \rightarrow 2p)$ of neon? from the first allowed transition $(4s \rightarrow 3p)$ of argon?

8. Predict the ultraviolet (58.5 nm) photoelectron spectra of
   a) allyl alcohol, $H_2C=CH$—CH_2OH;
   b) 1,5-hexadiene, $H_2C=CH$—CH_2—CH_2—$CH=CH_2$;
   c) benzaldehyde, C_6H_5CHO.
   Compare your predictions with the data given in the reference of Table 9.2.

9. Set up and solve the secular equations for the Hückel molecular orbitals of
   a) the allyl radical, $H_2\dot{C}$—CH—$\dot{C}H_2$;
   b) the cyclopropenyl radical,
   c) cyclobutadiene.
   (Hint: use the molecular symmetry to factor the 3×3 and 4×4 matrices so you have nothing larger than a quadratic to solve.)

10. Find an *approximate* value for the lowest value of the unknown ϵ in Eq. 9.8 by expanding the determinant in minors. Terminate the expansion successively after one, two and three steps.

11. Consider a ring of $2n$ carbon atoms, in a molecule $C_{2n}H_{2n}$. Show that the Huckel molecular orbital model implies that the energies ϵ_j of the molecule's π-orbitals have the form

$$\epsilon_j = \alpha + 2\beta \cos(j\theta), \quad \text{where} \quad \theta = \pi/n \quad \text{and} \quad j = 0, \pm 1, \ldots, \pm(n-1), n.$$

Be careful to note how the range of j is defined.

Show this implies that if n is odd, so that the molecule contains $4m+2$ carbons (with m an integer), then all the orbitals occupied in the ground state are fully occupied, but that if n is even, so the number of carbons is divisible by four, the highest normally occupied orbitals are degenerate and only two electrons occupy these two degenerate orbitals.

12. Construct a set of molecular orbitals for B_2H_6 from the $1s$ atomic orbitals of the hydrogens and the $2s$ and $2p$ orbitals of the boron atoms. Assume that the outer B—H bonds are localized but that the bonds in the central boron-hydrogen ring are delocalized over the four atoms. What is the number of electrons in each orbital in the molecular ground electronic state? Indicate the bonding and antibonding character of each orbital; be specific with respect to which atoms the orbital is bonding or antibonding.

13. Estimate the torsional oscillation frequency for ethane on the basis of the $0.126 \, eV$ barrier shown in Fig. 9.12. Use a quadratic approximation for the minimum of the sinusoidal well and neglect any tunneling. Approximately how many vibrational levels lie below the maximum of the barrier to internal rotation?

14. The barrier to internal rotation in nitromethane, CH_3NO_2, is roughly a hundredfold smaller than the barrier in ethane. Give a physical explanation in terms of the structures of these molecules why this is so. The NO_2 group defines a plane that contains the C—N axis.

15. How many isomers are there for each of the following complexes? Assume the six-coordinate species are regular octahedra and the five-coordinate species are trigonal bipyramids. (The abbreviation "en" is used for ethylene diamine, $H_2NCH_2CH_2NH_2$.) $[Co(NH_3)_2(H_2O)_2Cl_2]^+$; $[Co(NH_3)_3(H_2O)Cl_2]^+$; PF_2Cl_3; PF_2ClH; $[Co\,en_2ClF]^+$; $Co\,enCl_3$.

16. The argument leading to Eq. 9.9 implies that an optically active material rotates the direction of polarization for plane-polarized light by having different velocities of propagation for the left- and right-circularly polarized components of the light. Show that this phenomenon can be explained by supposing that the material has different absorption cross sections for left- and right-circularly polarized light.

17. If every atom in an ideal gas had a magnetic moment of 0.1 Bohr magneton, what would be the magnetic susceptibility of the gas at room temperature, when the density is 10^{19} atoms/cm^3?

18. Prove equation 9.14.

19. Potassium chromium alum, containing Cr^{3+}, has a magnetization of 1.00 Bohr magneton when H/T is 3.5×10^3 oersted/K or $3.5 \times 10^6/4\pi$ A/mk. What is its effective magnetic moment and how does this value compare with the pure spin-only value for Cr^{3+}?

20. What distortions of an octahedron leave it with threefold and twofold symmetry but no fourfold symmetry? On the basis of physical equivalence or inequivalence, indicate what further splittings one could expect for the $3d$ orbitals beyond those shown in Fig. 9.22, if a threefold (trigonal) distortion took place in an octahedral complex? Indicate which functions—d_{z^2}, d_{xy}, d_{xz}, d_{yz}, $d_{x^2-y^2}$—remain degenerate. (Be careful in your choice of axes.)

21. Assume a charge $-q$ on each ligand L of the octahedron in Fig. 9.22. Compute the electrostatic energy of a charge $-\epsilon$ interacting with the six ligand charges when a) the negative charge is midway between the center of

the octahedron and one of the ligands (i.e. one of the vertices) and b) when the negative charge is midway between the center of the octahedron and the center of one of its edges. This is a very rough model for estimating the difference in energy between the upper two and lower three orbitals of Fig. 9.22c.

22. Carry out an electrostatic energy calculation analogous to that of Problem 21 for the tetrahedron of Fig. 9.21. Compare the splitting with that of the octahedral complex.

23. Consider a hypothetical octahedral complex of a transition metal ion and six hydrogen atoms. From the six hydrogen $1s$ orbitals, concept six molecular orbitals that have symmetries corresponding to s, p and d-like orbitals in an octahedral structure. Show which (if any) of these will mix with each of the $3d$-orbitals of the central ion.

24. Estimate the strength of an external magnetic field large enough to decouple the electrons in an Fe^{3+} complex from a low-spin configuration (Fig. 9.24b) and align them as in the high-spin configuration of Fig. 9.24a, if the ion absorbs light of 900 nm in the absence of any field. Assume the absorption is due to the d–d transition.

10

Intermolecular Forces

In the last four chapters we have examined the bonding within isolated molecules. Now we shall consider the interaction between one molecule and another. By a "molecule" we mean a group of atoms linked by chemical bonds strong enough for the entity to retain its identity over a significant period of time at ordinary temperatures. Except in very dilute gases, this means that the energy required to break the bonds must be large compared with the average energy available in a molecular collision. For example, we have shown that the system of two hydrogen atoms has a ground state with a binding energy of 4.5 eV (432 kJ/mol) relative to the isolated atoms. Since an average gas molecule at 25°C has a kinetic energy of only 0.04 eV, the H_2 molecule is not likely to be dissociated by collisions. In contrast, although the system of two hydrogen molecules also has bound states, the binding energy is only about 0.004 eV, and the system cannot hold together at ordinary temperatures. Interactions of the latter type are too weak to produce stable molecules, but they nevertheless play important roles in the physics and chemistry of materials. In Parts Two and Three we shall have much to say about the effects of these interactions; in this chapter we consider their origin.

That there are forces between molecules can be deduced from two simple macroscopic observations: (1) All substances form condensed phases at sufficiently low temperatures; this indicates the existence of intermolecular attractions that are strong enough to hold together molecules with low kinetic energy. (2) All condensed phases strongly resist further compression; this indicates the existence of short-range repulsive forces that keep the molecules from coalescing once they are in "contact." The ranges of the repulsive forces should roughly equal the average intermolecular distance in a liquid. Since attraction and repulsion are dominant at long and short ranges respectively, the intermolecular potential energy must look something like the interatomic energy in a diatomic molecule (cf. Fig. 6.2), but the scale and exact shape of the curve are different, and these are what we now investigate.

Ultimately all the forces we need to consider between atoms are either Coulombic or exchange forces—that is, due either to the direct interaction of charged particles or to the quantum effects that give rise to the Pauli exclusion principle. At long range the Coulomb forces are dominant, and we examine these first, looking at the hierarchy of interactions from the electrostatic force between two ions to the "dispersion forces" that exist between any two atoms. These long-range forces are for the most part attractive, but the overlap of electron clouds causes molecules to repel one another (and thus keep their identity) at short distances. The repulsive interaction is the more difficult to calculate, and to a large extent must be treated in terms of empirical models, a number of which we shall describe.

The distinction between intramolecular and intermolecular forces is somewhat arbitrary, since the entire range of possible interaction energies can be found in one system or another. We have already discussed one intermediate case, the hydrogen bond. This and other types of "weakly associated" species are surveyed in the final section of the chapter.

Any atom, molecule, or ion can be considered as a spatial distribution of electric charge. At long range—that is, distances great enough that the overlap of electron distributions can be neglected—the interaction of two such species can be treated largely in terms of classical electrostatics, although one major component, the "dispersion force," requires quantum mechanics for its description. All atoms and molecules exert attractive forces on one another at sufficiently great distances; generally the attractive region extends down to the point at which the atoms "touch," that is, at which the electron clouds overlap significantly. The most important exception is the repulsive Coulomb force between two ions of like sign; even there, in the presence of other ions of opposite sign to shield the Coulomb force, a weak attraction becomes dominant at *very* long distances. It is the purpose of this section to examine the origins and strengths of these attractive forces.

First, let us introduce the notion of the *moments* of a charge distribution. The electric dipole moment is one example that we have already considered; now we shall generalize the concept. Consider a set of point charges Q_i at various points in space. The total charge,

$$Q = \sum_i Q_i, \tag{10.1}$$

is called the *zeroth moment* or *monopole moment* of the charge distribution. If we have a continuous charge distribution, with a total charge density (charge per unit volume) of $\rho(x, y, z)$ at each point in space, Eq. 10.1 should be replaced by

$$Q = \iiint \rho(x, y, z) \, dx \, dy \, dz, \tag{10.2}$$

with the integration taken over all space. In the vicinity of an atom or molecule it is usually appropriate to consider the nuclei as point charges, and the electrons as continuous charge distributions. (The electronic ρ could be written as $\sum_i \rho_i$, where ρ_i is the charge density in the ith orbital; but remember that this is only an approximation.)

The (electric) *dipole moment* or *first moment* has been defined in Section 4.5 for a system of two point charges. The general definition for a point-charge distribution is

$$\boldsymbol{\mu} = \sum_i Q_i \mathbf{r}_i, \tag{10.3}$$

where $\mathbf{r}_i$ is the radius vector of the ith particle from some origin, in a molecule usually taken to be the center of mass. For a continuous distribution the dipole moment becomes

$$\boldsymbol{\mu} = \iiint \mathbf{r}(x, y, z) \rho(x, y, z) \, dx \, dy \, dz; \tag{10.4}$$

$\mathbf{r}(x, y, z)$ is the vector whose Cartesian components are x, y, z. Thus the electric dipole moment is the set of averages of x, y, and z over the charge distribution $\rho(x, y, z)$. Note that the dipole moment of a distribution with zero total charge is simply the absolute charge of either the positive or the

10.1
Long-Range Forces: Interactions Between Charge Distributions

negative part, multiplied by the vector difference between the centers of positive and negative charge:

$$\boldsymbol{\mu} = Q^+(\mathbf{r}^+ - \mathbf{r}^-) = |Q^-|(\mathbf{r}^+ - \mathbf{r}^-). \tag{10.5}$$

In this case the dipole moment is independent of the origin of coordinates. Thus a permanent dipole moment exists in all molecules in which the centers of positive and negative charge do not coincide. We have already shown that such *polar molecules* include all heteronuclear diatomics (HCl, CO) and many more complicated species (H_2O, NH_3); their dipole moments are usually of the order of 1 debye (1 D = 3.33564×10^{-30} C m, which is $0.2e \times 1$ Å).

The higher moments are called the second, third, ..., nth moments of the charge distribution, or more commonly the *electric quadrupole*, *octupole*, ..., 2^n-pole moments. They are not quite as easily described as the dipole moment, but fortunately we shall not need to evaluate them. The quadrupole moment is essentially the set of averages of all pairs of the form x^2, xy, xz, y^2, yz, and z^2 over the distribution ρ. The quadrupole moment is what is called a second-rank tensor, which has six distinct components, five of which are in general independent. For nonpolar molecules with cylindrical symmetry, such as in H_2 or CO_2, the quadrupole moment consists of only one independent quantity, which can then be called "the" quadrupole moment. The quantity usually reported is this moment divided by the electronic charge,

$$|\mathbf{q}| \equiv \frac{\sum_i Q_i(3z_i{}^2 - r_i{}^2)}{e}, \tag{10.6}$$

with the dimensions of an area. (The z axis is taken as the symmetry axis.) For homonuclear diatomic molecules $|\mathbf{q}|$ is typically about 3×10^{-21} m^2 (0.3 Å^2).

It is the structure of a molecule that determines whether it has a nonzero dipole, quadrupole, or higher moment when it stands in isolation—that is, whether it has permanent moments. A spherically symmetric neutral charge distribution, such as that of any isolated atom, has no permanent multipole moments at all.[1] Any heteronuclear diatomic molecule can be expected to have an electric dipole moment: A difference in nuclear charge implies differences in electronegativity, which in turn places the center of electronic charge toward one or the other nucleus, on average, relative to the center of nuclear charge. A homonuclear diatomic molecule has no permanent dipole moment, but does have a permanent quadrupole moment. And in a tetrahedral molecule such as CH_4 both dipole and quadrupole moments are zero, but the permanent octupole moment is nonzero. All nonspherical species must exhibit some permanent electric moments. The first nonzero moment is independent of the choice of origin, but the values of any subsequent moments do depend on where the origin lies.

In any electric field an atom or molecule suffers some distortion of its charge cloud. As we noted in Section 6.2, the effect is to separate the centers of positive and negative charge, thus creating an induced dipole moment. In a uniform field $\mathbf{E}$ the induced moment can be written as

$$\boldsymbol{\mu} = \alpha\mathbf{E}, \tag{10.7}$$

[1] In saying that an atom's charge distribution is spherically symmetric, we neglect the internal structure of the nucleus. Nuclei do not have dipole moments, but many do have nonzero quadrupole moments, with $|\mathbf{q}|$ of the order of 10^{-30} m^2. The quadrupole energy levels split in an electric field, and the splittings can be measured by the technique called *nuclear quadrupole resonance*.

an equation that defines the *polarizability* α. In principle α is a function of the field strength, but in practice it can usually be treated as a constant. Only in very strong fields, such as those of focused laser beams, of the order of 10^8 V/m, can one observe significant variation in α. Fields of even greater magnitude may be involved when two molecules collide, but α can safely be taken as constant[2] when considering long-range interactions. Polarizability is a measure of the "softness" of a charge cloud, that is, the ease with which it can distort. In classical electrostatics, a perfectly conducting sphere of volume V has a polarizability of exactly $4\pi\epsilon_0 V$; similarly, for molecules the quantity $\alpha/4\pi\epsilon_0$ is usually of the order of the molecular volume, a few cubic angstroms. The polarizabilities of the elements in fact correlate well with their atomic sizes (Section 5.5). The large alkali atoms are highly polarizable, whereas the relatively small inert gas atoms have low polarizabilities; a graph of α versus atomic number thus looks much like Figure 1.7. Typical polarizabilities are given in Table 10.1.

Positive ions generally have polarizabilities much smaller than the corresponding neutral atoms—both because of their smaller sizes and because the decreased shielding of the nuclei makes it harder to overcome the electron–nuclear attractions. Negative ions are much more polarizable than neutral atoms for the converse reasons: They are large, and the outer electrons are not strongly bound to their nuclei.

A nonuniform electric field induces not only a dipole moment, but higher moments as well. Just such a nonuniform field is exerted on one charge distribution by another nearby—that is, on one molecule by another—unless at least one charge distribution is spherically symmetric and neutral. Hence we can expect all atoms and molecules to exhibit induced moments when they are either in an external field or interacting with other nonspherical molecules. The induced moments are added to whatever permanent moments the species may have.

When molecules are free to orient themselves in the presence of an electric field, they of course tend to occupy the lowest-energy orientations. In a uniform field, for example, a dipolar molecule will tend to align itself so that $\boldsymbol{\mu}$ and $\mathbf{E}$ are parallel. The same is true when the field is due to another molecule, and at moderately large distances (beyond the region where the repulsive forces are significant) the lowest-energy orientations are those in which the forces acting on the electric moments are attractive.

What now concerns us is finding the dependence of these forces on the distance between the molecules. In general, the strength of the interaction between two moments is directly proportional to the product of their magnitudes. The moments are all defined so that an nth moment (2^n-pole moment) has the dimensions of charge $\times$ (length)^n, whereas the interaction energy must have the dimensions of $\epsilon_0^{-1} \times$ (charge)$^2 \times$ (length)$^{-1}$ as in Coulomb's law. Combining these relationships, we surmise that the long-range interaction energy between two moments M_1 and M_2, of orders n_1 and n_2, respectively, should have the form

$$V(R) \propto \epsilon_0^{-1} |M_1| |M_2| R^{-n_1 - n_2 - 1}, \tag{10.8}$$

where R is the distance between molecules, the length we expect to be

TABLE 10.1a
POLARIZABILITIES OF NEUTRAL ATOMS IN THEIR ELECTRONIC GROUND STATES[a]

Atom	α (Å³)
H	0.666793
He	0.204956
Li	24.3
Be	5.6
B	3.03
C	1.76
N	1.10
O	0.802
F	0.557
Ne	0.3946
Na	23.6
Cl	2.18
Ar	1.64
K	43.4
Kr	2.48
Rb	47.3
Sr	27.6
I	3.9
Xe	4.04
Cs	59.6
Hg	5.1

[a] From T. M. Miller and B. Bederson, *Adv. At. Mol. Phys.* **13**, 1 (1977).

---

[2] For a nonspherical molecule α is in general a tensor with nine components,

$$\mu_i = \sum_{j=1}^{3} \alpha_{ij} E_j \qquad (i, j = x, y, z),$$

so that $\boldsymbol{\mu}$ and $\mathbf{E}$ need not point in the same direction.

TABLE 10.1*b*
POLARIZABLES AND ANISOTROPIES OF
POLARIZABILITIES OF NEUTRAL MOLECULES
IN THEIR GROUND STATES[a,b]

Molecule	α (Å^3)	$\alpha_{\parallel} - \alpha_{\perp}$ (Å^3)
H_2	0.819	0.314
D_2	0.809	0.299
N_2	1.77	0.70
CO	1.98	0.53
O_2	1.60	1.10
HCl	2.60	0.311
CO_2	2.63	2.10
NH_3	2.22	0.288
Cyclopropane	5.64	−0.81
Benzene	10.4	−5.6
CH_3Cl	4.53	1.55
$CHCl_3$	8.50	−2.68

[a] From N. J. Bridge and A. D. Buckingham, *Proc. Roy. Soc.* (*London*) **A295**, 334 (1966).
[b] Note that $\alpha = \frac{1}{3}(\alpha_{\parallel} + 2\alpha_{\perp})$, the average of parallel and perpendicular polarizability components, for a linear molecule or a molecule with at least a threefold rotational symmetry axis.

most significant for long-range interactions. More detailed calculations confirm this law, as we shall shortly illustrate.

The longest-range interactions (by which we mean those falling off most slowly with distance) are those between two monopoles ($n_1 = n_2 = 0$), for which Eq. 10.8 predicts an energy proportional to R^{-1}. For point charges this is simply a restatement of Coulomb's law in the form of Eq. 2.55, and we have shown in Section 7.4 that two ions interact in the same way at large R. Monopole–monopole interactions of course exist only when both partners are charged. If one molecule is charged and the other is neutral but dipolar, the monopole–dipole interaction is longest-range, varying as R^{-2}. Similarly, dipole–dipole interactions fall off as R^{-3}, dipole–quadrupole as R^{-4}, quadrupole–quadrupole as R^{-5}, and so forth. For most systems the longest-range interaction is also the strongest at all distances greater than that at which the repulsive forces become dominant.[3]

Let us now derive some of the exact laws corresponding to Eq. 10.8, beginning with the direct interaction between an ion and a molecule with a permanent dipole moment. At long range the field of the ion will be essentially that of a point charge, whereas the field of the polar molecule will be essentially that of an ideal dipole: two point charges $+q$ and $-q$ a distance δ apart, such that

$$\boldsymbol{\mu} = q\boldsymbol{\delta}; \qquad (10.9)$$

the vectors $\boldsymbol{\mu}$ and $\boldsymbol{\delta}$ both point from $-q$ toward $+q$. Now suppose that the center of the dipole lies a distance R from the ion of charge Q, in a direction that forms an angle θ with $\boldsymbol{\mu}$ (Fig. 10.1). If $R \gg \delta$, we can take

---

[3] There are exceptions such as HD, which has a tiny dipole moment (about 10^{-3} D) but an appreciable quadrupole moment.

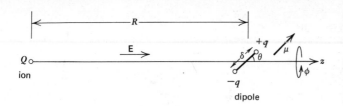

FIGURE 10.1
Ion–dipole interaction. The ion has a radial electric field **E**, which near the dipole is parallel to the z axis; since **E** is the force per unit positive charge, it points away from the ion (as shown) for $Q > 0$, toward the ion for $Q < 0$.

the distances from Q to $\pm q$ to be simply $R \mp \frac{1}{2}\delta \cos \theta$, and the Coulomb force on the dipole then has the z component[4]

$$
\begin{aligned}
F_z = F_+ + F_- &= \frac{qQ}{4\pi\epsilon_0 (R + \frac{1}{2}\delta \cos \theta)^2} - \frac{qQ}{4\pi\epsilon_0 (R - \frac{1}{2}\delta \cos \theta)^2} \\
&= \frac{qQ}{4\pi\epsilon_0 R^2}\left[\left(1 - \frac{\delta \cos \theta}{R} + \cdots\right) - \left(1 + \frac{\delta \cos \theta}{R} + \cdots\right)\right] \\
&\approx -\frac{qQ\delta \cos \theta}{2\pi\epsilon_0 R^3} = \boldsymbol{\mu} \cdot \frac{\partial \mathbf{E}}{\partial R},
\end{aligned}
\tag{10.10}
$$

where **E** is the electric field of the ion, of magnitude

$$
|\mathbf{E}| = \frac{Q}{4\pi\epsilon_0 R^2}.
\tag{10.11}
$$

For a molecule with permanent dipole $\boldsymbol{\mu}$, this force corresponds to an interaction energy

$$
V(R, \theta) = -\boldsymbol{\mu} \cdot \mathbf{E} = -\frac{\mu Q \cos \theta}{4\pi\epsilon_0 R^2}
\tag{10.12}
$$

since $F_z = -\partial V / \partial R$, in agreement with Eqs. 4.30 and 10.8.

Now consider the interaction between an ion and a nonpolar molecule. The ion's field will induce a dipole moment in the molecule as given by Eq. 10.7, in which we assume α constant; in general, α must depend on the molecule's orientation to the field, but one can take an average value (which is what is usually measured anyway). The result of Eq. 10.10, giving the instantaneous force on the dipole, is valid whether the dipole is permanent or induced. We thus have, for $R \gg \delta$,

$$
F_z = \boldsymbol{\mu} \cdot \frac{\partial \mathbf{E}}{\partial R} = \alpha \mathbf{E} \cdot \frac{\partial \mathbf{E}}{\partial R} = \frac{\alpha}{2}\frac{\partial (\mathbf{E} \cdot \mathbf{E})}{\partial R} = \frac{\alpha}{2}\frac{\partial |\mathbf{E}|^2}{\partial R} = -\frac{\alpha Q^2}{8\pi^2\epsilon_0^2 R^5}.
\tag{10.13}
$$

In this case $\boldsymbol{\mu}$ varies with R, so to find the interaction energy at a given R we must integrate over the distance from infinity (where V is zero by definition) to R. Applying Eq. 2.46, we have

$$
V_{\text{ind}}(R) = -\int_\infty^R \mathbf{F} \cdot d\mathbf{s} = -\int_\infty^R F_z \, dR = \frac{\alpha Q^2}{8\pi^2\epsilon_0^2}\int_\infty^R \frac{dR}{R^5} = -\frac{\alpha Q^2}{32\pi^2\epsilon_0^2 R^4}.
\tag{10.14}
$$

Note that no angle θ appears, since the induced dipole is taken to be parallel to the field. Here $V(R)$ corresponds to the Born–Oppenheimer $E(R)$ of the ion-molecule system; this result confirms our statement in Section 6.2 that $E(R) \propto -1/R^4$ for $H + H^+$ at long range. Note that an induction effect of this kind is also present when the molecule has a permanent dipole

---

[4] The second line uses the binomial expansion

$$
(1 \pm x)^{-2} = 1 \mp 2x + 3x^2 \mp \cdots.
$$

moment, but at long range is much smaller than the direct interaction given by Eq. 10.12.

One can carry out similar calculations for the direct and induced interactions between dipoles, quadrupoles, and so on—calculations necessarily more complicated, since the orientations of both molecules must be taken into account. We shall mention only the result for two dipolar molecules. For two permanent dipoles A and B, each oriented at angles θ, ϕ relative to the A–B axis (cf. Fig. 10.1), the long-range interaction energy is

$$V(R, \theta_A, \theta_B, \phi_A, \phi_B)$$

$$= \frac{\mu_A \mu_B}{4\pi\epsilon_0 R^3} [-2 \cos\theta_A \cos\theta_B + \sin\theta_A \sin\theta_B \cos(\phi_B - \phi_A)], \quad (10.15)$$

in agreement with Eq. 10.8. A dipolar molecule, as well as an ion, can induce a dipole moment in another molecule, whether or not the latter has one to begin with. The magnitude of this interaction should again be proportional to the product of the two dipoles divided by R^3, but the induced dipole is as usual $\alpha\mathbf{E}$, where the field is that of the first dipole, varying as μ/R^3. Thus the induced dipole has a magnitude proportional to $\alpha\mu/R^3$, and the interaction energy between the permanent and induced dipoles must vary as $\alpha\mu^2/R^6$. A detailed calculation gives

$$V_{ind}(R, \theta_A) = -\frac{\alpha_B \mu_A^2 (3 \cos^2\theta_A + 1)}{2(4\pi\epsilon_0)^2 R^6}. \quad (10.16)$$

In the interaction of two permanent dipoles there are two such induction terms to be added to Eq. 10.15.

Let us estimate the magnitudes of the interactions we have been talking about, for typical values of molecular quantities and a reasonable distance R. Thus if we take $Q = \pm e$ (a singly charged ion), $\mu = 1.5$ D (about that of NH_3), $\alpha/4\pi\epsilon_0 = 3$ Å^3 (typical for small molecules), and set all the θ's equal to zero (dipoles aligned along the A–B axis), we obtain the following interaction energies:

	$R = 5$ Å	$R = 10$ Å
ion–ion (Coulomb's law), Eq. 2.55	2.88 eV	1.44 eV
ion–dipole, Eq. 10.12	0.180 eV	0.045 eV
ion–induced dipole, Eq. 10.12	0.0346 eV	0.0022 eV
dipole–dipole, Eq. 10.15	0.0225 eV	0.0028 eV
dipole–induced dipole, Eq. 10.16	5.4×10^{-4} eV	8.4×10^{-6} eV

The dipole–dipole term is the largest that can appear in the interaction of two neutral molecules, and even at $R = 3$ Å (a typical "touching" distance for small molecules) this term only reaches about 0.10 eV. The interaction energy of neutral molecules is thus at best comparable to ordinary thermal energies ($\frac{3}{2}k_B T$, or 0.039 eV for a gas molecule at 300 K), and much smaller than the energies of chemical bonds.

Note that Eqs. 10.12, 10.15, and 10.16 give the instantaneous interaction energy at a particular orientation. The average interaction energy among a large number of such molecules (as in a gas) will fall off with R more rapidly, since the tendency for dipoles to align with the field also decreases with distance. Alignment also decreases with increasing temperature, that is, with increasing random motion of the molecules.

These average energies are important, since many of the methods for investigating intermolecular forces involve measurements on bulk samples. For example, it is possible to show, by methods such as those to be introduced in Part Two, that the average direct interaction energy between two dipoles at distance R is

$$\langle V(R) \rangle = -\frac{2}{3 k_B T} \left(\frac{\mu_A \mu_B}{4\pi\epsilon_0} \right)^2 \frac{1}{R^6}, \tag{10.17}$$

where k_B is Boltzmann's constant and T is the absolute temperature. This is equivalent to Eq. 10.15 averaged over all possible values of the angles θ_A, θ_B, ϕ_A, ϕ_B. If we again take $\mu = 1.5$ D and set $T = 300$ K, Eq. 10.17 gives us average direct dipole–dipole energies of 3.3×10^{-3} eV at 5 Å and 5.1×10^{-5} eV at 10 Å, considerably smaller than the values for perfectly aligned dipoles. The corresponding average for the induction energy of two dipoles, from Eq. 10.16, is

$$\langle V_{ind}(R) \rangle = -\frac{\alpha_A \mu_B{}^2 + \alpha_B \mu_A{}^2}{(4\pi\epsilon_0)^2 R^6}; \tag{10.18}$$

here the average is just half the value for $\theta = 0$, and for weakly polar molecules such as CO or HI may be larger than the average direct interaction. Note that no temperature factor appears here, since the induced moment is always parallel to the field. It is interesting that both average dipole–dipole energies vary as R^{-6}, since there is an additional interaction that varies as R^{-6} for *all* molecules; we shall now look at this interaction.

The final kind of long-range attractive interaction is that arising from the so-called *dispersion forces* (or *London forces*). These forces are always present between any two molecules whether or not there are permanent multipoles of any order. The dispersion forces are often said to be quantum mechanical in origin, and they are indeed, in the sense that they result from the correlated motion of the electrons in the two molecules, or, more specifically, from correlation between the amplitudes of the electronic wave functions. This correlation can be thought of crudely in a semiclassical way: Although the positions of the electrons in a nonpolar molecule average out over time, at any instant their positions give rise to a transient dipole moment. This moment generates a transient field that tends to polarize the electrons in the other molecule, producing (or rather modifying) a transient dipole moment there. The fluctuations of the charge distributions in the two molecules tend to be correlated, since the field of each transient dipole must influence the other. This argument is a convenient way of visualizing the interaction, but it should not be taken too literally; the actual effect is not time-dependent in the sense of classical fluctuations occurring.

The accurate calculation of the dispersion forces is extremely complicated and will not be discussed here. One relatively simple model gives an approximate result which can furnish some insight into how the forces arise:

$$V(R) = -\frac{3}{2} \left(\frac{h\nu_A \nu_B}{\nu_A + \nu_B} \right) \frac{\alpha_A \alpha_B}{(4\pi\epsilon_0)^4} \frac{1}{R^6}. \tag{10.19}$$

That is, the interaction energy for the dispersion force between two molecules is proportional to the product of their polarizabilities and varies as the inverse sixth power of their separation[5] independent of

---

[5] At very long distances—a few hundred angstroms or more—this dependence changes from R^{-6} to R^{-7} because of the finite velocity of light: It takes a significant time for the transient fields to propagate from one molecule to the other and back again.

orientation. The frequencies ν_A and ν_B are parameters characteristic of the two molecules, and are approximately equal to the frequencies of their first allowed electronic transitions. Thus Eq. 10.19 suggests that the distortion of each molecule in the fluctuating field of the other is expressed as a tendency to undergo an optical transition to the first excited state. A real transition of this sort is impossible, because energy would not be conserved; but one can say that the effect of the mutual polarization is to mix a little of the excited state into each molecule's wave function. More detailed calculations also give an R^{-6} dependence, with additional terms varying as $R^{-8}, R^{-10}, \ldots$; if one writes the leading term as $-C/R^6$, a typical value of C would be 5×10^{-78} J m^6 (for molecules such as Ar, N_2, O_2). For comparison with our earlier calculations, this value gives energies of 2×10^{-3} eV at 5 Å and 3×10^{-5} eV at 10 Å. Thus the dispersion energy is comparable to the dipole–dipole energy for polar molecules, and is the dominant contribution to the long-range interaction energy for nonpolar molecules.

The term *van der Waals* forces is often used to include all the attractive forces between neutral molecules—the dispersion forces in all cases, and the dipole–dipole and other multipole interactions whenever these are present. The name refers to J. D. van der Waals, who was concerned with intermolecular forces in his studies of the equation of state (Chapter 21).

10.2 Empirical Intermolecular Potentials

The discussion in the previous section applies only to the interactions of molecules at long range, that is, at distances appreciably greater than the dimensions of the molecules themselves. At shorter distances things become much more complicated, as the electron distributions of the molecules begin to overlap. Under these conditions one can no longer use simple electrostatic models for the molecules, just as within a single molecule quantum mechanics must be used to calculate the interactions. In an exact calculation, indeed, one cannot meaningfully separate such a system into two molecules, but must treat all the electrons and nuclei as a single system. In this sense, for example, "the interaction energy of two (monatomic) Ar molecules" and "the energy of the Ar$_2$ molecule" are but two ways to describe the same quantity, the ground-state energy of the system of two argon nuclei and 36 electrons as a function of internuclear distance. Although theoretical calculations can be and have been made for many two-atom systems such as He—He, Ne—Ne, Ar—Ar, and H_2—H_2, it will always be important to determine intermolecular potentials from experiment. Such determinations were being made long before *ab initio* theoretical calculations were possible, so it is not surprising that many empirical formulas have been devised to correlate and summarize the inferences from experimental observation.

Without attempting to examine the calculations, what can we say about the nature of the potential energy curve? We take the interaction energy to be zero when the particles are infinitely far apart. At long range the potential must be attractive ($E < 0$) because of the effects described in the previous section. At short range, however, two stable molecules ordinarily repel each other strongly. (The exceptions are those pairs of molecules that readily undergo chemical reaction, and even there a "hard core" eventually takes control.) This is easy to understand in terms of the crude MO theory that we outlined for atom–atom interactions in Section 7.6. Like the inert gas atoms, most individual stable molecules have only filled orbitals. Thus the "supermolecule" made of two such molecules would have equal numbers of bonding and antibonding orbitals filled, giving a net antibonding effect. In short, the H_2—H_2 and N_2—N_2 systems have repulsive potential energy curves for much the same reasons as the He—He and Ne—Ne systems. In all these cases, however, the repulsive energy, which falls off roughly exponentially with distance (cf. Section 7.4), is eventually overtaken at large distances by the long-range attraction. And since we are interested primarily in the region where attractive and repulsive forces are comparable, let us look at the experimental evidence.

For simplicity let us consider only the interaction of two monatomic molecules, so that the internuclear distance R is the only parameter of significance. One then finds that the interaction energy $V(R)$ is of a form familiar to us: steep repulsion at small R, attraction trailing off gradually at large R, with a minimum between—in short, quite similar to the $E(R)$ curve for the ground state of a strongly bound diatomic molecule. The difference is principally one of scale; we illustrate this with Fig. 10.2, which compares the energy curves for the H—H and Ne—Ne systems, with well depths of about 5 eV and 0.003 eV, respectively. That we do not ordinarily consider Ne_2 as a molecule is because the average room-temperature kinetic energy of two neon atoms is far more than enough to dissociate such a molecule. Thermal energies of molecules are usually of the order of $k_B T$ ($k_B = 1.3807 \times 10^{-23}$ J/K $= 8.617 \times 10^{-5}$ eV/K). It is thus customary to describe intermolecular energies for systems having no "chemical bond" in terms of the parameter ϵ/k_B, where ϵ is the well depth, which has the units of temperature. All the inert gas diatomic systems have extremely shallow potential wells: ϵ is 9.5×10^{-4} eV for He—He, 0.0027 eV for Ne—Ne, and 0.012 eV for Ar—Ar; the corresponding ϵ/k_B values are 11 K, 31 K, and 144 K, and at room temperature these systems are almost totally monatomic. For comparison, the rather weakly bound alkali metal diatomic molecules have ϵ's of 0.5–1 eV, yet alkali vapors are only about 1% diatomic, largely because of entropy effects; see Chapter 21.

How does one obtain intermolecular potential energy curves? In some cases spectroscopy can be used, as with strongly bound molecules. The He—He well is so shallow that it contains no bound vibrational states at all, but the Ne—Ne well contains three vibrational levels and the Ar—Ar well six; the shapes of the curves can be obtained by the usual methods from vibration-rotation or electronic spectra. But since so few two-molecule systems are likely to occupy bound states, other methods are more common. Scattering of molecular beams off one another gives a direct measurement of the interaction energy, especially its repulsive part (see Chapter 27). Such bulk properties of gases as the second virial coefficient and the viscosity are related to the intermolecular potential, as we shall discuss in Chapters 21 and 28, respectively; these measurements used to be the most important source of intermolecular energy data. When individual molecules retain their identity in the solid state (molecular crystals), the cohesive energy and compressibility of the crystal give much information on the shape of the curve near its minimum. And x-ray scattering can give the average distances between molecules, and thus tell something about their interactions, in both solid and liquid states. In a few cases, as we have already noted, theoretical calculations are also available. It is a sort of triumph that all these approaches usually give consistent results, in the sense that they agree within their expected errors, which are small relative to ϵ in cases where several data are available.

The relationships between the intermolecular potential and the various measurable properties are quite complex, and the properties are

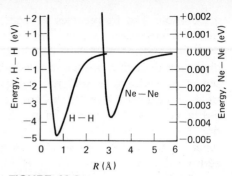

FIGURE 10.2
Potential energy curves for the H—H and Ne—Ne systems; note the difference in energy scales. The first curve is usually called "the binding energy of the H_2 molecule," the second "the interaction energy of two Ne molecules," yet they are essentially the same except for their scale.

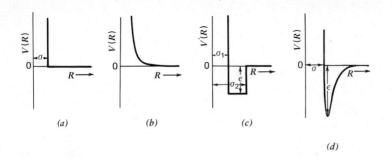

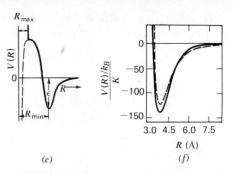

FIGURE 10.3

Some empirical intermolecular potential functions: (*a*) hard spheres; (*b*) point centers of repulsion ($\delta = 4$ shown); (*c*) square-well potential; (*d*) Lennard–Jones potential; (*e*) exp-6 potential (schematic); (*f*) comparison of the Dymond–Alder numerical potential (———) with the Lennard–Jones potential (– – – –) for the Ar—Ar interaction.

usually quite insensitive to the exact form of the potential. That is, a large variation in the shape of the potential well may make a difference too small to measure in the value of a given macroscopic property. For this reason it is customary to approximate the potential by some simple and mathematically tractable function with only a few adjustable parameters, which are then fitted as well as one can manage to the experimental data. There are many such empirical potential functions in use, some of which are important enough to deserve discussion here; these are illustrated in Fig. 10.3.

(a) *Hard spheres.* This is the simplest possible model, representing molecules as rigid spheres of diameter σ: Two such molecules will bounce off each other elastically if their centers approach to a distance σ, but otherwise do not interact at all. The potential energy can thus be written in the form

$$V(R) = \infty \quad (R < \sigma),$$
$$V(R) = 0 \quad (R \geq \sigma), \qquad (10.20)$$

as shown in Fig. 10.3*a*. This model takes account of the short-range repulsion only, and even that in idealized form, since the real repulsive force cannot be infinitely large. But it has the great advantage of mathematical simplicity, so that calculations with this model can be carried out with relative ease. In many cases such calculations give a good qualitative picture of the effects of molecular collisions. This is especially true at relatively high temperatures, where molecules have enough kinetic energy that they encounter mainly the upper portion of the repulsive curve. Most of the principles of kinetic theory can be developed with only this model, which we thus use almost exclusively in Part Three of this book.

(b) *Point centers of repulsion.* A somewhat more realistic representation of the repulsive energy is given by the potential

$$V(R) = dR^{-\delta}, \qquad (10.21)$$

Fig. 10.3*b*, where d is a constant and the index of repulsion δ must be larger than 3 (values between 9 and 15 are usually found).[6] This model recognizes the fact that more energetic molecules can approach each other more closely before being repelled, and thus can represent some of the temperature dependence of properties. But it again neglects the attractive part of the molecular interaction, and has mainly mathematical convenience to recommend it.

(c) *Square well.* The simplest potential model that includes both attractive and repulsive contributions is that shown in Fig. 10.3*c*, a

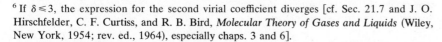

[6] If $\delta \leq 3$, the expression for the second virial coefficient diverges [cf. Sec. 21.7 and J. O. Hirschfelder, C. F. Curtiss, and R. B. Bird, *Molecular Theory of Gases and Liquids* (Wiley, New York, 1954; rev. ed., 1964), especially chaps. 3 and 6].

hard-sphere core surrounded by an attractive well of constant depth. The square-well potential is defined by the equations

$$V(R) = \infty \qquad (R \leqslant \sigma_1),$$
$$V(R) = -\epsilon \qquad (\sigma_1 < R < \sigma_2), \qquad\qquad (10.22)$$
$$V(R) = 0 \qquad (R \geqslant \sigma_2),$$

with the three adjustable parameters σ_1, σ_2, and ϵ. This model is a good compromise between mathematical simplicity and realism.

(d) *Lennard-Jones potential.* The most widely used empirical potential is that named for J. E. Lennard-Jones, the first to use it extensively in analyzing the properties of gases. It is the simplest of the models considered here that gives a potential energy curve qualitatively similar to the correct one. The Lennard-Jones (12–6) potential, shown in Fig. 10.3d, is

$$V(R) = 4\epsilon \left[\left(\frac{\sigma}{R} \right)^{12} - \left(\frac{\sigma}{R} \right)^6 \right], \qquad\qquad (10.23)$$

where ϵ is the well depth and σ the distance at which $V(R) = 0$. One can easily show that the minimum energy, $V(R) = -\epsilon$, occurs at the distance $R = 2^{1/6}\sigma$. The attractive term is proportional to R^{-6}, thus giving the correct long-range dependence of the dispersion energy.[7] The repulsive term is more arbitrary, in that a wide range of exponents would fit the available data about as well; 12 is chosen for convenience (since R^{-12} is the square of R^{-6}). The Lennard-Jones parameters ϵ and σ have been determined for a wide range of molecules; some of the most accurate of these values are listed in Table 10.2. Calculations of many macroscopic properties have been made on the basis of the Lennard-Jones model, partly because of its mathematical convenience.

(e) *Exp-6 potentials.* Since the repulsive part of the potential energy is best described by an exponential term (cf. Section 7.4), a number of potential functions have been devised to make use of this. One problem is that any simple function of the form $Ae^{-\alpha R} - BR^{-6}$ has a maximum at small R and plunges to $-\infty$ as $R \to 0$; this is usually overcome by introducing a hard-sphere core. One of the most commonly used potentials of this type is

$$V(R) = \frac{\epsilon}{1 - 6/\alpha} \left\{ \frac{6}{\alpha} \exp \left[\alpha \left(1 - \frac{R}{R_{min}} \right) \right] - \left(\frac{R_{min}}{R} \right)^6 \right\} \quad (R > R_{max}),$$
$$V(R) = \infty \hspace{8cm} (R \leqslant R_{max}), \qquad (10.24)$$

where R_{min} is the usual potential minimum, R_{max} is the distance at which the first function has its maximum, ϵ is the well depth, and α is an adjustable parameter (usually in the range 12–15). This exponential-six (exp-6) potential is shown schematically in Fig. 10.3e, with the rejected function for $R \leqslant R_{max}$ indicated by a dashed line. The maximum would actually be a much higher energy than shown here, a typical value being $V_{max} = 3 \times 10^4 \epsilon$ at $R_{max} = 0.2 R_{min}$ (for $\alpha \approx 14$). This potential has three adjustable parameters, ϵ, R_{min}, and α.

Many other potential models exist besides those we have described here. The double-exponential Morse potential, Eq. 7.11, gives a good fit near the minimum but falls off too rapidly at large R, where the

---

[7] However, when the dispersion energy $-CR^{-6}$ is calculated directly from molecular polarizability—by expressions such as Eq. 10.19—the value of C obtained is usually only half the Lennard-Jones value ($C = 4\epsilon\sigma^6$) for the same molecule. This demonstrates that the Lennard-Jones potential is not quite adequate to represent the true shape of the inter-molecular potential.

TABLE 10.2
LENNARD-JONES POTENTIAL PARAMETERS FOR
SELECTED PAIRS OF ATOMS

Atom Pair	ϵ (10^{-21} J)	ϵ/k_B (K)	σ (Å)
LiAr	0.92	67	3.7
LiKr	1.5	110	3.0
LiXe	2.4	175	3.8
NaAr	1.0	72	3.8
NaKr	1.4	100	4.5
NaXe	2.1	150	4.4
Ar_2	2.0	145	3.8
Kr_2	2.7	201	3.6
Xe_2	3.9	288	3.9

[a] From W. R. Hindmarsh and J. M. Farr, *Progress in Quantum Electronics* **2**, 141 (1972); J. A. Barker, R. O. Watts, J. K. Lee, T. P. Schafer, and Y. T. Lee, *J. Chem. Phys.* **61**, 3081 (1974); J. M. Parson, P. E. Siska, and Y. T. Lee, *J. Chem. Phys.* **56**, 1511 (1972).

Lennard-Jones potential falls off too slowly. Nonspherical molecules are sometimes represented by hard ellipsoids or cylinders rather than hard spheres; however, the Lennard-Jones and other spherically symmetric models work reasonably well for small molecules such as N_2 or CO_2. For polar molecules the dipole–dipole interaction must be taken into account; the most commonly used model is the *Stockmayer potential*, which is simply Eq. 10.15 added to the Lennard-Jones potential.

However, it is unlikely that any simple function with a limited number of adjustable parameters can represent the true intermolecular potential accurately over all distances. In recent years attempts have been made to compute the true potential directly from a wide range of experimental data. One such potential for the Ar—Ar interaction, obtained in numerical form by Dymond and Alder,[8] is shown in Fig. 10.3*f* in comparison with the Lennard-Jones potential; the difference in the shape of the attractive potential is quite apparent. And even the Dymond–Alder potential, which was based on all the data available for the thermodynamic and transport properties of gaseous argon, gives a poor representation of the Ar_2 vibrational spectrum. Only the most precise experiments, such as high-resolution spectroscopy or measurement of the angular distribution of particles from the scattering of colliding beams of fast atoms, coupled with very flexible mathematical representations of potential curves, are accurate enough to reproduce the properties of species such as Ar_2.

Most determinations of intermolecular potentials derive from measurements on one-component systems (which are relatively easy to analyze), and thus give information only on the interaction between two molecules of the same kind. The entry under Ar_2 in Table 10.2, for example, gives the parameters for the Ar—Ar interaction. Some potential parameters have been determined for unlike molecules, but in only a limited number of cases. In the absence of such information, one often estimates the Lennard-Jones parameters by the empirical combining laws

$$\sigma_{AB} = \tfrac{1}{2}(\sigma_{AA} + \sigma_{BB}), \qquad (10.25)$$

[8] J. H. Dymond and B. J. Alder, *J. Chem. Phys.* **51**, 309 (1969).

which would be exactly correct for hard-sphere molecules, and

$$\epsilon_{AB} = (\epsilon_{AA}\epsilon_{BB})^{1/2}. \qquad (10.26)$$

The latter equation would follow from Eq. 10.19 for the dispersion energy if the parameters ν_A and ν_B had the same value; for many pairs of molecules this is approximately the case. But these combining rules and similar ones for other potentials are only approximate, and it is unlikely that any universal relationship of this type actually exists.

10.3
Weakly Associated Molecules

In this section we survey a number of species in which atoms or groups of atoms are attracted to one another by bonding forces intermediate in strength between the strong interactions we call "chemical bonds" and the weak "intermolecular" interactions just discussed. The "molecules" thus formed tend to be loose, fluctuating aggregations more like liquids than the rigid geometric structures characteristic of strongly bound molecules. We usually refer to such loosely bound species as "associated."

The most weakly bound "compounds" we know are those formed by two inert gas atoms. The only contribution to the bonding is that from dispersion forces, without even any permanent multipoles to help hold the atoms together. Yet as we noted in the last section, a molecule such as Ar_2 is stable enough for its vibrational spectrum to be measured. The fact that the inert gases do condense to form liquids implies that they must also be capable of forming triatomic and larger associations. But so far as we know, these species exist in measurable concentrations over an exceedingly narrow range of conditions, and thus are difficult to observe. These polyatomic associations are called "clusters" in the theory of gas–liquid condensation; in Chapter 24 we shall examine why the transition from a gas of isolated molecules to the indefinitely large cluster we call a liquid is so abrupt.

The next species we consider are those in which both dispersion and multipole forces contribute to the attraction, giving molecules somewhat more tightly bound than the inert gas dimers. The study of these species has a very short history because they can exist for extended periods of time only under conditions of low temperature and low density—low temperature, to keep the kinetic and vibrational energy low enough to prevent molecules from breaking up with every collision, and low density so that collisions are infrequent. It is difficult to find these conditions on or near the earth, but they do occur in interstellar space. Among the examples of these weakly bound molecules that have been observed and characterized are $XeHCl$, HeI_2, He_2I_2, NeI_2, ArI_2, Ar_2I_2, $(C_6H_6)_2$, $(HF)_2$ (cf. Section 7.9), $(HCl)_2$, and $(HCN)_2$. The last three of these are best described in terms of hydrogen bonds, of which we shall have more to say. It is quite certain that vastly more species of this type will be studied in the coming years.

In some cases of "cluster formation" the participating molecules may have one or more unpaired electrons, which are available to form weak chemical bonds. (Cf. our discussion of electron-deficient bonds in Section 9.3.) A classic example is the dimer of nitrogen dioxide, which appears to have the structure

the weakness of the N—N bond is indicated by its length of 1.78 Å, compared with 1.47 Å for the normal single bond in hydrazine (H_2N—NH_2). The NO_2 monomer is a brown gas; the color is associated with the presence of an odd or unpaired electron, and thus of a partially occupied orbital. The N_2O_4 dimer, by contrast, has no unpaired electrons and is a colorless gas.[9] The color difference makes it easy to measure the relative amounts of NO_2 and N_2O_4 in a mixture of the two: The solid and liquid are almost pure N_2O_4, but the gas consists mostly of NO_2 molecules.

An even weaker association of odd-electron molecules occurs in the solid form of nitric oxide, made up of dimers with the structure

$$
\begin{array}{c}
N \cdots O \\
| \quad\quad | \\
O \cdots N
\end{array} \; ;
$$

the short and long N—O bonds have lengths of 1.10 Å and 2.38 Å, respectively. Yet unpaired electrons do not always lead to association: The oxygen molecule, with two unpaired electrons in its ground state, shows little tendency to form O_4 dimers.

Hydrogen-bonded molecules are a special case of association due to multipole interactions. The poor shielding of the hydrogen nucleus allows relatively strong interactions to take place, in some cases leading to quite well-defined structures. One such example is the dimer of acetic acid, with the structure

Among other examples is the dimer of HF (Section 7.9), with its F—H$\cdots$F bond; the F—F distance is fixed at about 2.79 Å, but the hydrogen atoms oscillate and wobble easily. And in Section 9.4 we have described how internal hydrogen bonding stabilizes the structures of DNA and other biological macromolecules.

Thus far we have discussed only association between neutral molecules, but a stronger electrostatic interaction is obviously possible when one component is an ion. In some cases such an "association" actually leads to a quite strongly bonded molecule. For example, the H_3O^+ (hydronium) ion might be thought of as an association between the H^+ ion (proton) and a water molecule, but the energy required to split H_3O^+ into $H_2O + H^+$ in the gas phase is 7.34 eV, so one can hardly think of this as a case of weak bonding. This is why bare protons are virtually nonexistent in aqueous solutions of acids. The properties these solutions share in common are due to the presence of H_3O^+ or even larger ions. The NH_4^+ (ammonium) ion is also strongly bonded, with about 9.0 eV needed to split it into $NH_3 + H^+$. The NH_4^+ ion plays a role in liquid-ammonia solutions like that of H_3O^+ in water.[10] Both H_3O^+ and NH_4^+ ions are stabilized by delocalization, since all the hydrogen atoms in each are equivalent.

But whereas the H^+ ion has so intense a field that its "associations" are strongly bonded, other ions do form weakly associated species in combination with neutral molecules. These are usually called *ion-molecule clusters* in the gas phase, *solvated ions* in solution. They are normally composed of a single ion with one or more neutral molecules in a shell around it. The relationship to the complex ions of the last chapter is obvious; the difference is simply the strength of the bonding. An ion-molecule cluster or solvated ion can be thought of as a coordination complex in which the ligands are loosely held—"loosely" being

[9] A large number of odd-electron compounds (including many of those called "free radicals") are colored. The usual rationalization is that the unpaired electron is somewhat weakly bound, and thus undergoes electronic transitions in the visible rather than the ultraviolet. But the phenomenon is not universal; one obvious exception is NO, which is colorless in the gas phase.

[10] There are many solid hydronium salts ("acid hydrates") completely analogous to the corresponding ammonium salts, except that most of the hydronium salts melt at much lower temperatures. For example hydronium chloride ($H_3O^+Cl^-$, but usually written $HCl\cdot H_2O$) melts at $-15.4°C$, whereas NH_4Cl remains solid up to 340°C.

421

arbitrarily defined by wherever one wants to draw the line between the two classes.

Definitely on the weakly bound side of such a line are the species formed by the alkali metal ions in association with water molecules. These hydrated ions are held together mainly by the attraction between the positive charge of the central alkali ion and the permanent dipoles of the water molecules, augmented by the additional dipole moments that the central ion induces in the water molecules. The bonding in these clusters may be so weak that, unlike the more tightly bound complexes, they do not maintain definite shapes for long intervals of time. Even the number of associated water molecules may fluctuate. The fluctuations result from the balance between the attractive force of the ion and the random motion of the solvent molecules. The association is nevertheless real on the average. Rather than evidence for a definite structure, one finds the weaker evidence that at certain distances from the central ion there is a higher than random concentration of water molecules. That is, there are strong peaks in the radial distribution function, which we discuss in Chapters 23 and 26.

The attachment of water molecules does not stop with a coordination shell of nearest neighbors; a second shell, and perhaps even more, can be bound somewhat less strongly. Some of the properties of the hydrated alkali ions are given in Table 10.3. Because Li^+ is the smallest alkali ion, it can exert the strongest binding forces on surrounding water molecules which can approach more closely than with the larger ions; the result is that Li^+ has the largest effective radius as a hydrated ion, the greatest hydration energy (energy released by the formation of the hydrated ion), and the highest average number of associated water molecules. In the Li^+ hydrated ion the inner shell of water molecules appears to be bound tightly enough to form a well-defined tetrahedron; Na^+ and K^+ also have four molecules in the first hydration shell, but Rb^+ and Cs^+ may have six. Similar solvated ions are formed in polar solvents other than water; in liquid ammonia, for example, the $[Na(NH_3)_4]^+$ ion is a well-defined tetrahedral complex.

Wherever ions and neutral molecules are both abundant, it is usual to find associations between the two. This is true not only in liquid solutions but in ionized gases, where the resulting ion-molecule clusters can be readily detected by mass spectrometry. The binding energies of these clusters are usually strongly dependent on the charge and size of the ion, but not on the detailed structure of the neutral molecule(s). Both $K^+(CO_2)$ and $NO^+(CO_2)$ have binding energies of about 0.5 eV, but the corresponding values for $NO^+(N_2)$ and $NO^+(O_2)$ are only about 0.2 eV. The binding energy of Ar_2^+ is considerably greater, about 1.5 eV; like such other homonuclear ions as H_2^+, He_2^+, Li_2^+ (cf. Chapter 7), it should be considered a strongly bound molecule-ion rather than a cluster. For most clusters of one ion and one nonpolar molecule, the binding energies are consistent with the values predicted by Eq. 10.14 for the equilibrium separation. Clusters with more than two entities also occur: The Ar_3^+ ion has been observed, as have such aggregates of nitrogen molecules as N_4^+, N_6^+, and N_8^+.

TABLE 10.3
PROPERTIES OF HYDRATED ALKALI IONS

Property	Li^+	Na^+	K^+	Rb^+	Cs^+
Radii of central ions (in crystals) (Å)	0.68	0.98	1.33	1.48	1.67
Radii of hydrated ions (Å)	3.40	2.76	2.32	2.28	2.28
Hydration energy[a] (eV)	5.3	4.2	3.3	3.0	2.7
Average number of associated water molecules[b]	25.3	16.6	10.5	—	9.9

[a] These energies are slightly lower than the *heats of* hydration in Table 26.4.
[b] These numbers differ from those derived by computer simulation and quoted in Chapter 26. The difference is a measure of the ambiguity inherent in the concept of hydration.

Especially large clusters can be formed with molecules capable of forming hydrogen bonds. Additional water molecules readily attach themselves to the H_3O^+ ion to form $H_5O_2^+$, $H_7O_3^+$, and $H_9O_4^+$; even higher numbers of water molecules can be bound in the gas phase, apparently going into an outer shell of the cluster. The $H_9O_4^+$ cluster is believed to have the polyhedral structure shown in Fig. 10.4; the smaller clusters are assumed to be fragments of this polyhedron, with additional H_2O molecules going on its exterior. However, the binding energies for successive addition of one, two, three, four, and five H_2O molecules to H_3O^+ are approximately 1.5, 1.0, 0.75, 0.65, and 0.55 eV, respectively, with no sharp break at the completion of the assumed first shell. The limiting value is simply the energy of vaporization of water, which is about 0.46 eV per molecule. By extending the hydration energy measurements in the gas phase to larger and larger clusters, it is possible to estimate the total energy of hydration of the proton in aqueous solution. The value obtained in this way is 12.0 ± 0.3 eV, in rather good agreement with the accepted value of 11.35 ± 0.15 eV, based on a variety of other measurements.

The $H_9O_4^+$ ion in Fig. 10.4 has been drawn as made up of one H_3O^+ ion (at the top of the figure) and three H_2O molecules. But the protons involved in hydrogen bonds are extremely mobile, so that either of the bonding protons in the H_3O^+ ion could move readily to the oxygen atom in the adjacent H_2O molecule. A rotation of the entire $H_9O_4^+$ complex would give a result indistinguishable from this permutation, but would be far slower. The exterior protons can also exchange places with bridging protons, but again more slowly than the latter can "jump across" the hydrogen bonds. As a result of this mobility, the lifetime of an individual H_3O^+ ion in aqueous solution is estimated to be only about 10^{-13} s.

Ammonia also forms clusters with the ammonium ion, and species of the type $NH_4^+(NH_3)_n$ with n as large as 20 have been observed. The notion of a shell structure is somewhat better supported by experiment here than in the case of the $H_3O^+(H_2O)_n$ clusters. The energies for attachment of NH_3 molecules to NH_4^+ drop smoothly from about 1 eV for the first NH_3 to about 0.75 eV for the fourth, and then abruptly to about 0.4 eV for the fifth. The limit for $n \rightarrow \infty$ is in this case the energy of vaporization of NH_3, which is roughly 0.2 eV. There is no evidence for a well-defined second shell.

Negative-ion clusters can also be formed. For example, the hydroxyl (OH^-) ion can be hydrated to form $H_3O_2^-$, $H_5O_3^-$, and $H_7O_4^-$; the energies associated with these steps are very similar to those for hydration of the H_3O^+ ion. Although neutral $(O_2)_2$ associations seem not to exist, the O_4^- cluster is observed; the energy required to dissociate it into $O_2 + O_2^-$ is 0.54 eV. The hydrate $O_2^-(H_2O)$ is slightly more tightly bound, with a dissociation energy of about 0.8 eV; the energies for adding successive water molecules are 0.6–0.8 eV each through the pentahydrate.

The tendency for ions to form large clusters—especially those involving H_2O molecules—is quite in line with the accepted notion that ions are good nuclei for initiating condensation. The formation of such clusters, growing into aerosol droplets, is naturally important as a major mechanism in the production of clouds and rain. (But it is not the only mechanism: Solid particles—which are often themselves electrostatically charged—can also act as condensation nuclei.) The ions that initiate the process are formed in the upper atmosphere by the action of cosmic rays, natural radioactivity, and ultraviolet light. Ions may be formed from any of the neutral species present in the atmosphere, but a series of exchange reactions leads primarily to the formation of $H_3O^+(H_2O)_n$ clusters.

All the ion-molecule clusters we have mentioned thus far are singly charged. Solvated ions with charges greater than 1 are well known in solution, where solvent molecules can bind to and thus stabilize the ion. In gases, however, no such clusters are known. There is invariably some mechanism that allows the splitting of the charge between separate fragments, which, by escaping, lower the repulsive energy of the system. For example, a hypothetical $[Fe(H_2O)_2]^{2+}$ ion would presumably decompose at once to $FeOH^+$ and H_3O^+.

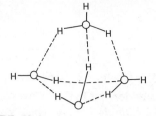

FIGURE 10.4
Presumed structure of the $H_9O_4^+$ ion in the gas phase.

Further Reading

Blaney, B. L., and G. E. Ewing, "Van der Waals Molecules," *Ann. Rev. Phys. Chem.* **27,** 553 (1976).

Hirschfelder, J. O., C. F. Curtiss, and R. B. Bird, *Molecular Theory of Gases and Liquids* (Wiley, New York, 1954; rev. ed., 1964).

Kauzmann, W., *Quantum Chemistry* (Academic, New York, 1957), chap. 13.

Kondratyev, V., *The Structure of Atoms and Molecules* (Noordhoff, Groningen, The Netherlands, 1964), chap. 9.

Margenau, H., and N. R. Kestner, *Intermolecular Forces* (Pergamon Press, Oxford, 1969).

Amdur, I. and J. E. Jordan, Chapter 2, "Elastic Scattering of High-Energy Beams: Repulsive Forces," in *Molecular Beams*, J. Ross, ed., Vol. X of *Advances in Chemical Physics* (Wiley, New York, 1966).

Buck, U., "Elastic Scattering," in *Molecular Scattering: Physical and Chemical Applications*, K. P. Lawley, ed., Vol. XXX of *Advances in Chemical Physics* (Wiley, New York, 1975).

Dalgarno, A. and W. D. Davison, "The Calculation of van der Waals Interactions," *Advances in Atomic and Molecular Physics* **2,** 1 (1966).

Pauly, H. and J. P. Toennies, "The Study of Intermolecular Potentials with Molecular Beams at Thermal Energies," *Advances in Atomic and Molecular Physics* **1,** 195 (1965).

Maitland, G. C. and E. B. Smith, "The Forces Between Simple Molecules," *Chem. Soc. Rev.* **2,** 181 (1973).

Problems

1. Prove that the value of the dipole moment $\boldsymbol{\mu}$ of a charge distribution is independent of the choice of origin of coordinates if the total charge is zero, but depends on the choice of origin if the net charge is nonzero.

2. Estimate the equilibrium distance R_e and the polarizability of the neutral partner, assuming that atomic polarizabilities are approximately $4\pi\epsilon_0 V$ and that molecular polarizabilities are approximately the sums of the polarizabilities of the atoms. Then evaluate the dissociation energy for Ar_2^+, K^+CO_2 and N_4^+.

3. Carry the approximation of Eq. 10.13 one step further by deriving the term involving the induction of a dipole in the ion by the field of the induced dipole of the neutral. Show by examples based on data of Chapter 5 that this effect is far more important in negative ion clusters than in positive ion clusters.

4. Polarizabilities, the derivatives of induced moments with respect to increases in an applied electric field, are usually treated as constants. In reality, polarizabilities are not constant; for example in sufficiently strong uniform fields, polarizabilities change linearly with the field strength, implying that the electric dipole moment is changing quadratically with the field strength. Which way does the polarizability change with increasing field and why?

5. (a) Evaluate the quadrupole moment of a charge distribution in which the charges are as follows: $+2e$ at $(0, 0, 1)$ and $(0, 0, -1)$, and $-1e$ at $(1, 0, 0)$, $(-1, 0, 0)$, $(0, 1, 0)$ and $(0, -1, 0)$, with all distances in Å.
   (b) Evaluate the quadrupole moment of a distribution in which the positive

charges of $+2e$ lie at $(0, 0, 1)$ and $(0, 0, -1)$ and the negative charge of $-4e$ is distributed uniformly throughout a sphere of radius 4 Å.

6. Estimate the magnitudes of electric dipole–dipole, dipole–quadrupole and quadrupole–quadrupole interaction energies for separation distances of 4 Å and 8 Å, basing your values on magnitudes of the molecular moments. Estimate the magnitude of the dispersion energy at the same distance on a similar basis.

7. Compute the electric dipole–dipole interaction energy for two HCl molecules 1 nm (10 Å) apart when the two molecules are aligned as shown: (see Table 7.4).

a) $\begin{array}{cc} \text{Cl} & \text{Cl} \\ | & | \\ \text{H} & \text{H} \end{array}$,

b) $\begin{array}{cc} \text{H} & \text{Cl} \\ | & | \\ \text{Cl} & \text{H} \end{array}$,

c) $\text{H—Cl} \quad \begin{array}{c} \text{Cl} \\ | \\ \text{H} \end{array}$,

d) $\text{H—Cl} \quad \text{H—Cl}$.

8. Compute the interaction energy between a polar diatomic molecule with electric dipole $\boldsymbol{\mu}$ and an ion with charge Q when the molecule is rotating in a state with $J = 1$ quantized along the ion-molecule axis with (a) $M_J = 0$; (b) $M_J = \pm 1$. [The wave functions for rotational motion can be taken here to be the normalized spherical harmonics $Y_L^M(\theta, \phi)$.]

9. When two alkali halide molecules are far from each other, their interaction potential is that of two permanent dipoles. When the diatomic molecules are close together, their interaction is the sum of the separate ion–ion potentials that eventually give rise to a stable compound, for example for 2NaCl, as shown:

From the ionic radii, estimate the geometry of Na_2Cl_2. Then sketch the form of the effective potential as the dimer is separated into two diatomic molecules. Indicate the analytic form of the potential for the short-range and long-range regions, and show how the former transforms to the latter as the intermolecular distance grows.

10. Evaluate and graph the dipole moment induced in an atom of argon by the field of a proton approaching from 20 Å to 5 Å. Assume the polarizability of Ar is 1.64 Å^3, as given in Table 10.1.

11. The dipole moment of every diatomic molecule in its ground state is zero when the internuclear distance $R = 0$ and when $R \to \infty$. Sketch the behavior of $|\boldsymbol{\mu}(R)|$ and explain why it goes to zero at both limits. Be sure to indicate an approximate scale of distance.

12. Compute the force constants, vibrational frequencies and equilibrium internuclear distances for NaAr, NaXe and Ar$_2$, based on the parameters of their Lennard–Jones potentials.

13. What are the coefficients of the first nonvanishing terms beyond the quadratic if one expands the Lennard–Jones and exp-6 potentials around their minima at $R = R_{min}$? Express these in terms of the parameters of the Lennard–Jones and exp-6 potentials, respectively. Evaluate the former for Ar$_2$ and compare it with the observed value.

14. The polarizability of CO_2 has two independent components, one parallel to the O—C—O axis and the other perpendicular. These have values of 4.03

and 1.93 Å^3, respectively. The first allowed optical transition of CO_2 falls at 46,000 cm^{-1} or 5.70 eV. Estimate the potential energy of interaction of the two CO_2 molecules in a linear configuration, a T-shape and side by side, when the C—C distances are 5 Å and 10 Å, if the dispersion forces are the dominant component of the internuclear force.

15. Compare the magnitudes of the three kinds of attractions behaving as R^{-6} for the following pairs:

a) Ar—Ar;
b) Ar—HCl;
c) He—HCl;
d) HCl—HCl.

16. The quadrupole moment of CO_2 is 4.32×10^{-26} esu cm^2 or 4.32×10^{-30} esu m^2. For the same configurations and distances considered in Problem 7, compute the interaction energy of two CO_2 molecules, assuming the interactions are pure quadrupole–quadrupole. Show that the T-shaped geometry has the lowest energy. Compare the energies based on quadrupole–quadrupole interactions with those based on dispersion forces as derived in Problem 7.

17. Construct a square-well potential for Ar_2 that has the same depth (0.012 eV) as the actual potential well of Ar_2 and contains the same number (6) of bound vibrational states with no rotation. Are there any remaining free parameters in this square-well potential?

18. Rationalize the observation that He_2 has no bound states but He_3 does.

19. Compare the bonding of Ar_2^+ with that of Cl_2 and Ar_2, in terms of the occupied molecular orbitals and their bonding or antibonding character. On this basis, interpret the differences between the dissociation energies of these three molecules.

20. The hydrogen fluoride molecule forms hydrogen-bonded dimers. Referring to Table 7.2, we find that the equilibrium H—F distance in the monomer is 0.9168 Å and the vibrational frequency $\bar{\nu}_e$ is 4139 cm^{-1}. The $(HF)_2$ dimer has a linear F—H $\cdots$ F arrangement with the F atoms 2.79 Å apart at equilibrium. The other H is off at an angle of about 108° from the F—H—F axis. Suppose the bonding H atom can be attached to either F, so that the configurations F—H $\cdots$ F and F $\cdots$ H—F are equivalent. Construct the potentials for these two structures assuming that only the H moves, and find the energy of their intersection, thereby setting an approximate upper limit for the potential barrier between the two structures.

21. Estimate the total binding energy of a complex $[Na(H_2O)_4]^+$, taking into account the interaction of the charge of the Na^+ ion with the dipoles of the water molecules and the interaction of the dipoles with each other. What distances and geometry do you choose? Explain your choices. Compare your estimates with the values given in Table 10.3.

22. Rationalize the difference between the "hydration energy" of the proton, 7.34 eV, that is, its binding energy to a single water molecule, and the total hydration energy of a lithium ion of 5.3 eV, as given in Table 10.3.

23. Alkali halide molecules, particularly those of the lighter alkalis and halogens, have a strong tendency to form dimers and higher polymers in the gas phase. What geometries do you expect to correspond to the configurations of minimum energy for Li_2F_2 and Li_3F_3? Estimate the energy required to dissociate these into separated molecules. For Li_2F_2, compare this with the energy to dissociate into $Li_2F^+ + F^-$. Use the data from Tables 7.3 and 7.4.

24. Construct molecular orbital representations of O_4 and O_4^-, assuming that they are plane, rectangular molecules.

11

The Structure
of
Solids

Solids and liquids are the states of matter called "condensed phases." On the microscopic scale they are characterized by the proximity of each atom or molecule to its neighbors. In a solid or a liquid every particle is always moving in the force fields due to several neighbors, usually near the minimum of the net potential energy surface produced by all these interactions. The equilibrium intermolecular distance is in the region in which both attractive and repulsive forces are significant. Gases, by contrast, are so rarefied that each molecule spends most of its time outside the range of strong interaction with any neighbor—when molecular interactions do occur in gases under ordinary conditions, they occur almost invariably only through two-body collisions dominated by long-range attractive forces.

In this chapter we shall examine some bulk properties of solids from a phenomenological viewpoint, study the crystalline and electronic structure of solids, and begin to interpret the dramatic differences between the properties of solids and gases. This will take us into the most important method for determining structures, into the connections among bonding, structure, and stability, and finally into the electronic structure of periodic solids.

In a solid the simultaneous attractions of each particle for its several neighbors are the cause of *cohesion*. The *cohesive energy* of a solid is the energy that would be required to separate it into its constituent atoms, ions, or molecules at distances so large that they effectively no longer interact with one another. The cohesive energy is the analog for solids of the dissociation energy of a diatomic molecule, and is often of similar magnitude. Indeed, in many cases a solid can be thought of as an infinitely large molecule.

11.1
Some General Properties of Solids

To begin with, what is the difference between a solid and a liquid? Both are condensed phases with comparable values of intermolecular distances (although for a given substance the liquid is usually slightly less dense), and the cohesive energy of a liquid is not much less than that of a solid. The great difference is that a solid has a fixed structure: Two molecules in a given piece of solid can maintain their relative positions, except for vibrations, over a long period of time, and the solid as a whole thus maintains a fixed shape. In a liquid the molecular positions are not fixed, and the liquid readily changes its shape under even a weak external force such as gravity; we say that the liquid assumes the shape of its container.

Solids fall into two classes according to the degree of order in their structure. Those in which the constituent atoms or molecules are arranged in a *repeating* pattern that may be continued arbitrarily far are called *crystalline*. Those in which no regular pattern of repetition can be found are called *amorphous*. The distinction between amorphous solids and very viscous liquids is not a sharp one, naturally. A crystalline material may occur in a single crystal or as a collection of many small crystals; in the latter case the material is called *polycrystalline* or, if the crystals are very small, *microcrystalline*.

When we speak of a repeating pattern of atoms or molecules in a crystal, we of course mean that their *equilibrium positions* generate a repeating pattern. Just as the nuclei of a molecule vibrate about their equilibrium positions, the constituents of a crystal vibrate about their equilibrium positions. And also as in a molecule, analysis of these vibrations can give us the shape of the potential energy surface—in simplest terms, the variation of the cohesive energy with interparticle distance. Information about the interparticle forces and potentials can be obtained from such bulk properties as the isothermal compressibility and the coefficient of thermal expansion. But these topics are better treated as part of the thermodynamics of solids, and have thus been postponed to Chapter 22. In the remainder of this chapter we shall treat solids as if the nuclei were at rest at their equilibrium positions.

Let us return to the distinction between crystalline and amorphous solids. A substance is crystalline if its microscopic structure consists of units that are replicated along three independent directions in space (although the term "two-dimensional crystal" is sometimes used); the replication must be capable of extending arbitrarily far. In the next section we shall examine the nature of the replicated units. An amorphous material contains no replicating unit. Liquids are amorphous; so are glasses and some other solids. It is not sufficient for a material to have local regions of ordered structure to be crystalline. Liquid water has some degree of local order at any instant, but because the structure continually changes, there is no replicating pattern. Amorphous solids often have regions of local order but no such order over macroscopic distances. We shall devote most of this chapter to crystalline solids; the structure of liquids and other amorphous materials will be considered in Chapter 23.

Four types of crystals can be distinguished readily. One consists of molecules bound together by forces that are so weak, compared with the intramolecular forces, that the molecules retain their identity. In such a crystal, called a *molecular crystal*, the constituents are recognizable entities, and, at least in their lowest electronic states, exhibit properties very similar to the properties of the free gaseous molecules. The infrared spectra, bond lengths, and magnetic properties of the constituents all change only slightly when the molecules condense into molecular crystals. The forces responsible for the existence of molecular crystals are the weak attractions we examined in Chapter 10: dipole–dipole and other multipole interactions and van der Waals forces, especially the latter. Some examples of molecular crystals are those of argon, benzene, and naphthalene, typically weakly bound and relatively volatile.

A second type is an *ionic crystal*, such as that of sodium chloride. The best simple model of such a crystal supposes that it is composed of a three-dimensional array of alternating positive and negative ions. The cohesive energy comes almost entirely from the strong Coulomb attraction of the oppositely charged ions, so that ionic crystals are tightly bound. Because the binding forces are strong, ionic crystals are hard and involatile. The ions are packed together as tightly as possible, so as to minimize the total energy, which is composed of three kinds of interactions: the attractions of oppositely-charged ions, the repulsions of ions of

TABLE 11.1

COMPARISON OF PROPERTIES OF TYPICAL CRYSTALS OF VARIOUS SORTS

Material	Type of Crystal	Cohesive Energy (eV)	Electrical Conductivity (MS/m)[a]	Near-Neighbor Distance or Molecular Dimension (Å)
Argon	Molecular	0.08	(insulator)	3.83
Methane (CH_4)	Molecular	0.10	(insulator)	4.49 (nearest neighbor)
Naphthalene ($C_{10}H_8$)	Molecular	0.39	(insulator)	5.10 (nearest neighbor)
Rock salt (NaCl)	Ionic	7.9	(insulator)	5.628 (Na–Na)
Cesium fluoride	Ionic	7.48	(insulator)	6.008 (Cs–Cs)
Diamond (C)	Covalent	7.32	(insulator)	1.5445
Graphite (C)	Covalent	7.34	7.27×10^{-2}	1.421
Quartz (SiO_2)	Covalent	6.17	(insulator)	1.50 (Si–O)
Silicon	Valence-metallic	4.34	10	2.352
Sodium	Metallic	1.1	23.8	3.716
Copper	Metallic	1.0	59.8	2.556

[a] MS = megasiemens; 1 siemens is 1 ohm^{-1}.

the same charge, and the hard-core repulsions exhibited by all pairs of nearest neighbors.

Covalent crystals such as diamond constitute the third category. These crystals are also hard, tightly bound, and fairly close-packed. The forces that bind them are virtually identical with the forces that bind large molecules together (cf. Chapter 9).

Metallic crystals constitute the last of our categories. They are often soft, ductile, and malleable, but nevertheless have large cohesive energies. Metals as we usually find them are composed of very small crystals pressed together into polycrystalline bulk materials. However, large single crystals of metals can be grown and studied. In metals we normally cannot distinguish molecules or whole neutral atoms, but the cores—nuclei and inner shells of electrons—retain their identities.

Examples of some of the properties that distinguish the types of crystals are collected in Table 11.1. In addition to the cohesive energy, we have cited the electrical conductivity and the nearest-neighbor distance or some other typical molecular dimension. Molecular solids and covalent crystals are usually insulators for both electricity and heat. (The two conduction processes are related, as we shall see.) Ionic and metallic crystals are normally electrical conductors, but of different kinds. Conduction in ionic crystals occurs by motion of the ions themselves under the influence of an applied electric field, as in electrolytic solutions. In metallic conduction only the electrons move under the influence of the electric field, and the conductivity is thus much greater. We shall return to these properties when we discuss particular kinds of solids in more detail.

We have said that the structure of a crystal is made up of replicated (duplicated and repeated) units. Let us explain just what this means. Suppose that we wish to construct a model of a crystal, and have a basic physical unit with which to build; the crystal consists of copies of this unit. We can create a three-dimensional structure as follows. Through the basic unit we lay out three axes, not coplanar and not necessarily orthogonal. Then we copy the basic unit along one of the axes, say the x axis, at regular intervals, maintaining its orientation. This gives us a row of physical units extending from $-\infty$ to $+\infty$. Then we lay out an infinite set of identical x axes at regular intervals along the y axis, and replicate the row of physical units at each new x axis; this gives us a two-dimensional structure, an infinite planar array of identical physical units. Finally we repeat this plane at regular intervals along the z axis and replicate the physical units throughout to create a three-dimensional array, our crystal. We define three vectors, one along each axis, that define a parallelepiped enclosing one physical unit. Then we let these vectors have the shortest lengths consistent with the condition that periodic repetition of their parallelepiped at each physical unit covers the whole crystal. These vectors, $\mathbf{a}$, $\mathbf{b}$, and $\mathbf{c}$, are called the *primitive vectors* of the crystal and their lengths, a, b, and c, are the *primitive distances* of the crystal. The primitive lengths may be equal but they need not.

The parallelepiped defined by the primitive vectors $\mathbf{a}$, $\mathbf{b}$, and $\mathbf{c}$ is the *primitive cell*. Note that the primitive cell is a parallelepiped in space, not the physical unit that the cell encloses. If we repeat the primitive vectors throughout space, we define an infinite abstract periodic structure, called a *lattice*, that has the same translational symmetry as the physical crystal. The directions and distances of translation that superimpose the abstract lattice on itself are the same as the directions and distances of translation that leave the physical crystal looking unchanged. The special lattice that has the same periodicity as the physical crystal is the *Bravais lattice* corresponding to that crystal. We may think of the Bravais lattice either as the space-filling set of all the nonoverlapping parallelepipeds or as the sets of vertices that define these parallelepipeds.

A primitive elementary parallelepiped is the simplest set of vertices of the Bravais lattice incorporating all the information necessary to construct the entire Bravais lattice by translations. The primitive elementary parallelepiped can contain no points of the Bravais lattice in its body or on its faces; its only Bravais lattice points are its eight vertices.

In three dimensions, there are 14 Bravais lattices, in the sense that there are exactly 14 sets of relations of equality or inequality among the primitive distances and the angles between the primitive vectors. These are illustrated in Fig. 11.1, as the parallelepipeds defined by the primitive vectors. The figures shown are not in general the primitive elementary parallelepipeds, because several of the units shown in Fig. 11.1 have vertex points of the Bravais lattice at the centers of their bodies or faces. In this presentation it is easier to see the spatial relations among the lattice vertices than it is if only the eight lattice points of the primitive elementary parallelepiped are shown. The hexagonal right prism of Fig. 11.1, for example, is composed of three primitive elementary parallelepipeds, one of which is indicated in the drawing.

The Bravais lattices fall into seven types, called the seven *crystal systems*. These are classified as follows in terms of relations among the primitive distances a, b, and c, and the angles α (between $\mathbf{b}$ and $\mathbf{c}$), β

11.2
Space Lattices and Crystal Symmetry

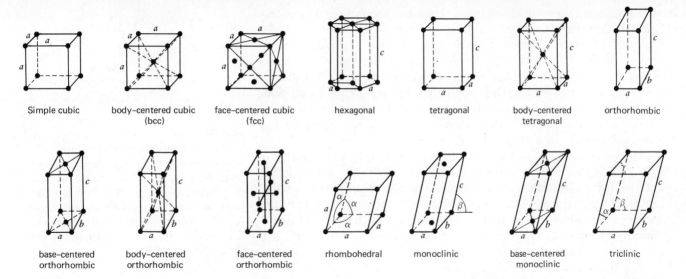

Simple cubic   body-centered cubic (bcc)   face-centered cubic (fcc)   hexagonal   tetragonal   body-centered tetragonal   orthorhombic

base-centered orthorhombic   body-centered orthorhombic   face-centered orthorhombic   rhombohedral   monoclinic   base-centered monoclinic   triclinic

FIGURE 11.1
The 14 Bravais lattices. The first three belong to the cubic system; the next, to the hexagonal; the next two, to the tetragonal. The orthorhombic system has three possible lattices; the rhombohedral, one; the monoclinic, two; and the triclinic, only one.

(between **a** and **c**) and γ (between **a** and **b**):

Cubic: $a = b = c$; $\alpha = \beta = \gamma = 90°$.

Hexagonal: $a = b \neq c$; $\gamma = \beta = 90°$, $\gamma = 120°$ (Note that this corresponds to three equal coplanar axes in the ab plane intersecting at $60°$, and a fourth, the c axis perpendicular to the ab plane.)

Tetragonal: $a = b \neq c$; $\alpha = \beta = \gamma = 90°$.

Orthorhombic: $a \neq b \neq c$; $\alpha = \beta = \gamma = 90°$.

Rhombohedral: $a = b = c$; $\alpha = \beta = \gamma \neq 90°$.

Monoclinic: $a \neq b \neq c$; $\alpha = \beta = 90°$; $\gamma \neq \alpha$.

Triclinic: $a \neq b \neq c$; $\alpha \neq \beta \neq \gamma$.

Crystals themselves have *structures*. Any ideal crystal structure can be associated with—and therefore considered as a realization of—a Bravais lattice belonging to one or another crystal system. In some crystals each atom or molecule lies on a lattice point. The alkali metals, for example, crystallize with a structure that lies on a body-centered cubic lattice. In other crystals we must consider pairs or larger groups of atoms, ions, or molecules as the basic units that define the Bravais lattice. The smallest arrangement of basic constituents that both represents the chemical composition of the material and can generate the Bravais lattice by replication at integral multiples of the displacement vectors **a**, **b**, **c** is called the *unit cell* of the physical crystal. The unit cell is not necessarily the same as the primitive elementary parallelepiped; a unit cell may contain one or more primitive cells.

Let us consider some examples of real crystals. All of the alkali metals have body-centered cubic (bcc) lattices; one way of constructing the unit cell of sodium locates one sodium atom in the center and eight in the corners. Since the corner atoms are shared among eight adjoining cells, the unit cell consists of $1 + 8(\frac{1}{8}) = 2$ sodium atoms. If we neglect the differences between different kinds of atoms, cesium chloride would also be bcc, with a Cs^+ ion in the center and Cl^- ions at the corners, or vice versa (see Section 11.8), and would have one Cs^+Cl^- unit per unit cell. Because Cs^+ and Cl^- are different, we must consider the unit cell as something containing at least one atom of each. This implies that the lattice of CsCl is simple cubic. However, the *crystal structure* of CsCl is more important for many purposes than its Bravais lattice, and gives its name to a general class of like structures. Other crystals are more complicated, sometimes involving two interpenetrating lattices. For example, if the difference between Na^+ and Cl^- were negligible, the NaCl crystal would be face-centered cubic (fcc) (see Section 11.8); its unit cell contains four Na^+Cl^- units. The diamond crystal (Section 11.10) has a structure built on the tetrahedral bonds of carbon (Fig. 11.2), but its unit cell contains only two atoms. The naphthalene crystal is monoclinic, with two *molecules* per unit cell, and so forth. Once the structure is known, whatever it is, the unit cell dimensions can be easily related to the macroscopic density of a single crystal.

Crystals exhibit symmetry. The simplest symmetry of a crystal is always translational: If the crystal (supposed infinite in extent in all directions) is translated through an integral multiple of any lattice vector, the displaced crystal is physically indistinguishable from the initial crystal. Thus we can translate the crystal by the vector $n_a\mathbf{a} + n_b\mathbf{b} + n_c\mathbf{c}$ and find it apparently unchanged. Most crystals also have rotational symmetry. If a cubic crystal is rotated through 90° about any of the three mutually perpendicular axes $\mathbf{a}$, $\mathbf{b}$, $\mathbf{c}$, it is left unchanged. Similarly, a cubic crystal is invariant under a rotation of 120° about any of the body-diagonal axes of the cube. Because a fourfold repetition of rotation through 90° returns the crystal to its original orientation, the symmetry of the cubic structures with respect to rotation about the $\mathbf{a}$, $\mathbf{b}$ or $\mathbf{c}$ axes is said to be fourfold; similarly, rotation about the body diagonals of the cube exhibits threefold symmetry.

The tetragonal lattice retains fourfold symmetry with respect to rotation about the $\mathbf{c}$ axis, but has only twofold symmetry with respect to rotation about the $\mathbf{a}$ and $\mathbf{b}$ axes, and has no threefold symmetry. The rhombohedral lattice has the same threefold symmetry as the cubic lattice, but exhibits only twofold symmetry with respect to rotation about the $\mathbf{a}$, $\mathbf{b}$, $\mathbf{c}$ axes. The orthorhombic crystal has twofold symmetry about the $\mathbf{a}$, $\mathbf{b}$, $\mathbf{c}$ axes but no threefold axes. Monoclinic crystals have only twofold symmetry with respect to rotation about the $\mathbf{b}$ axis (the axis perpendicular to the plane containing the angle not equal to 90°), and the triclinic lattice has no symmetry other than translation.

What we have just described is the symmetry of the Bravais lattice, or of any of the solid models having the forms of the drawings of Fig. 11.1. Only when there is a single molecule per unit cell is the symmetry of the crystal simply the symmetry of the lattice. Otherwise the unit cell has its own symmetry. (The molecules or atoms comprising such a crystal may have symmetries of their own, but normally such internal symmetries do not correspond to transformations that leave the crystal in an unchanged condition, and thus are not part of the symmetry of the crystal.) In a crystal containing two or more identical constituents per unit cell, however, the symmetry of the crystal is greater than that of the lattice. By this we mean that there are operations that cannot be generated by combinations of the symmetry operations of the lattice, but that nevertheless leave the crystal unchanged. These additional operations are the symmetry operations of the unit cell. Often the unit cell has some symmetries that do correspond to operations already included in the lattice symmetry; if we find that all the symmetries of the unit cell are also lattice symmetries, then we must go back and redefine the unit cell. It is the presence of extra symmetry operations that are not among those of the lattice that gives rise to the concept of unit cell. The extra kinds of operations may include reflections, specific combinations of translations, specific translations plus rotations, or other combinations of these operations. The lattice symmetry and the unit cell symmetry together define the symmetry of the entire crystal.

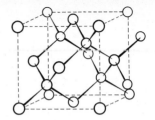

FIGURE 11.2
The crystal structure of diamond. The figure shows how the structure consists of a tetrahedron inside a face-centered cubic frame. The unit cell consists of only two carbon atoms. From A. F. Wells, *Structural Inorganic Chemistry*, 2nd ed. (Oxford University Press, Oxford, 1950).

The most powerful tool for determining the structures of crystals (and of the molecules that constitute them) is the analysis of the patterns we record when their regular structures diffract waves. Waves are scattered by any particles they pass. If the wavelength λ is comparable to the distances between the scattering particles, then the scattering can be explained in terms of Huygens' theory of diffraction: Each scatterer produces a spherical wave going outward, and all these spherical waves add together to produce a scattered wavefront. Each kind of atom has its own characteristic capacity to scatter waves of any particular kind, defined by a quantity called a *scattering factor,* to which the scattered amplitude is proportional. The kinds of radiation of interest to us are x rays, electrons, and neutrons, because their wavelengths are comparable to atomic dimensions. In general, x rays are by far the most widely used for structural work, and neutrons are the most recently exploited for problems of chemical interest.

One carries out diffraction experiments by measuring the intensity of scattered radiation as a function of the scattering angle θ, the angle of deviation from the propagation axis of the incident beam (cf. Section 2.5). If the scattering material is amorphous, such as a glass, a liquid, or a gas, then θ is the only significant angle. In such a case the intensity of scattered radiation $I(\theta)$ is symmetrical about the propagation axis and tends to fall off as θ increases. From the behavior of $I(\theta)$ one can calculate the probability of finding a particle a given distance R away from any other selected particle: if $\mathscr{P}(R)\,dR$ is the probability of finding a particle between R and $R + dR$, given that one particle is at the origin, then one can show that (for an amorphous medium)

$$I(\theta) \propto \int_0^\infty \mathscr{P}(R)\frac{\sin kR}{kR}\,dR, \quad \text{where} \quad k \equiv \frac{4\pi}{\lambda}\sin\left(\frac{\theta}{2}\right). \quad (11.1)$$

Since $\mathscr{P}(R)$ would be proportional to $4\pi R^2$ in a completely uniform medium, it is convenient to introduce a function called the *pair correlation function* $g_2(R)$ defined by

$$\mathscr{P}(R) = 4\pi R^2 n g_2(R), \quad (11.2)$$

where n is the material's number density. The more nearly random the distribution of particles, the less $g_2(R)$ varies with R; if the medium were continuous and homogeneous, $g_2(R)$ would be constant. For liquids $g_2(R)$ exhibits oscillations such as those in Fig. 11.3a: There is a most probable nearest-neighbor distance corresponding to the first peak, but all other distances in this vicinity also occur; the next maximum in $g_2(R)$ corresponds to the most probable second-nearest-neighbor distance, and so forth. As the distance R grows larger, it is increasingly difficult to identify maxima; as the temperature increases, fewer maxima can be found. We shall have more to say about the pair correlation functions of liquids in Chapter 23.

Pair correlation functions can be drawn for crystals, too, as in Fig. 11.3b. If there were no vibrational motion in the crystal, its pair correlation function would be a series of lines, one at each of the regular interparticle distances. The height of each bar would be proportional to the number of occupied lattice sites at the corresponding distance from a selected reference site. But real atoms in crystals do vibrate, so crystals must also have pair correlation functions that are smooth curves. Equation 11.1 applies only to an amorphous material, so how does one

11.3
x-Ray Diffraction from Crystals: The Bragg Model

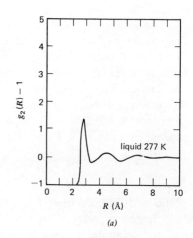

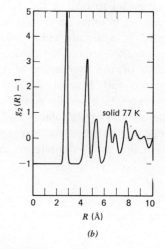

FIGURE 11.3
Pair correlation functions $g_2(R)$ for (a) liquid water, (b) solid ice. From A. H. Narten, C. G. Venkatesh, and S. A. Rice, *J. Chem. Phys.* **64**, 1106 (1976).

obtain the pair correlation function for a crystal? One way is to use a *powder* sample, made up of many small and randomly oriented crystals: Averaging over the random orientations again gives Eq. 11.1. When a beam of x rays (or electrons or neutrons) is scattered from such a powder sample, the scattered x rays come out in cones as shown in Fig. 11.4. The intensity $I(\theta)$ is independent of the angle ϕ about the propagation axis of the x-ray beam but varies with the angle θ of the cone from the propagation axis. In the apparatus most commonly used for such measurements, $I(\theta)$ is recorded on a strip of photographic film or by a counter sweeping across a line or arc over a range of θ. The $I(\theta)$ so obtained has a series of narrow peaks, which appear as circular arcs on a photographic film. These *powder patterns* are characteristic of the crystal structure, and can thus be used to identify a powdered substance by comparison with the powder patterns of known substances. An example is shown in Fig. 11.5.

Diffraction and scattering from crystals yield much more information than just the pair correlation function $g_2(R)$. This is suggested by the fact that a crystal has an ordered structure with symmetry axes of its own. We should expect the intensity of scattered radiation to depend on the angles of orientation of the crystal to the axis of the incident beam, as well as on the angle *about* the latter axis. There are two ways to see how this occurs; before we can introduce them, however, we must develop some nomenclature.

The structure of a space lattice can be described with reference to the various planes of atoms that can be constructed in the lattice. Each plane is defined by a set of indices, customarily designated as h, k, l, which in turn are defined with respect to the primitive distances a, b, c. If a plane cuts the principal axes of the crystal at the intercepts pa, qb, rc (measured from a corner of the unit cell), then the *Miller indices* of that plane are the smallest integers h, k, l such that

$$h:k:l = \frac{1}{p} : \frac{1}{q} : \frac{1}{r}. \tag{11.3}$$

The plane with indices h, k, l is conventionally designated as an (hkl) plane. A negative index is indicated by a bar over the corresponding number; for example, a plane with indices 1, -2, 3 is a $(1\bar{2}3)$ plane. A plane parallel to one of the principal axes is thought of as intercepting that axis at infinity, so the corresponding Miller index is $1/\infty$ or 0. For example, in a simple cubic lattice the (100), (010), and (001) planes correspond to faces of the cubic primitive cell. A set of planes equivalent (because of the symmetry of the crystal) to a given plane (hkl) is designated with braces: $\{hkl\}$. Thus we say that all the planes of the square faces in a simple cubic lattice are $\{100\}$ planes. In Fig. 11.6 we illustrate some crystal planes and their corresponding Miller indices. The distance d_{hkl} between successive planes of a set $\{hkl\}$ can be expressed in terms of the lattice parameters and the Miller indices. In a cubic lattice,

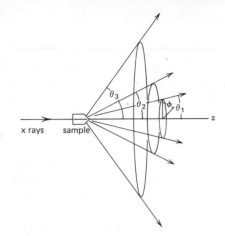

FIGURE 11.4

Diffraction of x rays from a powder sample. The x-ray beam propagates from the left along the z axis. Three cones of diffracted x rays are shown, corresponding to angles θ_1, θ_2, and θ_3; the intensity is independent of the angle ϕ about the z axis for a powder or any other sample that is, on average, isotropic, meaning a sample having no distinguishable internal directions.

FIGURE 11.5

A typical x-ray powder pattern: olivine. (Supplied by Prof. Joseph Smith, The University of Chicago.)

Olivine

for example, we have

$$d_{hkl} = a(h^2 + k^2 + l^2)^{-1/2}, \qquad (11.4)$$

where a, as usual, is the lattice parameter.

We can now resume our discussion of diffraction in crystals. The simplest picture of x-ray diffraction, conceptually, is that introduced in 1913 by W. H. and W. L. Bragg. The physical picture underlying the usual derivation of this model (which we shall use) is rather crude, and might leave one feeling that the conclusions rest on very insecure ground. In fact, the results are equivalent to those obtained by more painstaking methods. The Braggs' concept supposes that x rays are reflected by the planes of atoms on which they impinge. This reflection is supposed to follow *Snell's law* for the optics of specular (mirrorlike) reflection: The propagation vectors of the incident and reflected beams and the normal to the surface of the crystal all are coplanar, and the angles of incidence and reflection are equal. This behavior is illustrated in Fig. 11.7. In general, the beams reflected from successive planes parallel to the crystal surface interfere with each other so that no scattered radiation emerges for most angles of incidence. At certain angles, however, constructive interference of all the reflected beams produces net outgoing rays with intensities well above zero. Let us analyze how this phenomenon comes about.

For constructive interference to occur, the emerging beams A', B', C', D',... of Fig. 11.7 must all come out of the crystal with a common phase; the emerging beams then regenerate a plane wavefront corresponding to the incident wavefront $ABCD$.... But the ray BB' that strikes the second plane travels farther than the ray AA', and the ray CC' goes farther still. If the reflected rays are to emerge in phase, then the extra distances must all be integral multiples of λ, the wavelength of the radiation. Because we are dealing with a crystal, where planes of any given sort are precisely equally spaced, we know that if the first and second planes generate rays in phase with each other, so will all the others; the third plane thus has exactly the same relation to the second as the second has to the first, and so forth. So we need consider only the first two planes. The extra distance traveled by BB' relative to AA' is $2d \sin \theta$, where d is the interplanar distance and θ is the angle of incidence. Hence we require that

$$2d \sin \theta = n\lambda, \qquad (11.5)$$

where n is an integer.

What Eq. 11.5 tells us is that x rays reflected from a given series of crystal planes with separation d will give a diffraction pattern with maxima at the angles

$$\theta = \sin^{-1}\left(\frac{n\lambda}{2d}\right), \qquad (n = 1, 2, \ldots). \qquad (11.6)$$

The integer n tells us the number of "extra" wavelengths traveled by BB', or the number of cycles that BB' is retarded relative to AA'. We call n the *order* of the spectrum. Because of Eq. 11.4, we can write for a cubic lattice

$$\frac{2a \sin \theta}{\lambda} = n(h^2 + k^2 + l^2)^{1/2}$$
$$= [(nh)^2 + (nk)^2 + (nl)^2]^{1/2}, \qquad (11.7)$$

where θ is the angle at which we obtain nth-order reflection from the (hkl) plane. This expression is the form in which we shall derive the diffraction conditions from another viewpoint.

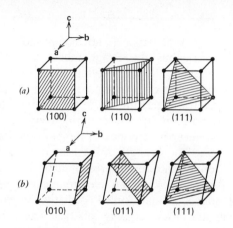

FIGURE 11.6
Miller indices of some crystal planes. (*a*) Cubic lattice. (*b*) Monoclinic lattice.

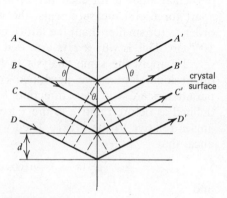

FIGURE 11.7
Bragg reflection of x rays from parallel planes of atoms. The incident beam *A* is reflected as the beam *A'*, *B* as *B'*, and so on; in each case the angles of incidence and reflection have the same value θ.

A real lattice is not composed of reflecting planes, but of discrete atoms; thus the Bragg model is useful and conceptually simple but fundamentally unsatisfying. Max von Laue (1912) used the more realistic model that crystals are composed of small scattering centers that act on wavefronts passing by. In the tradition of the Huygens construction, Laue supposed that each scattering center generates a spherical outgoing wave whenever it responds to the periodic force of an applied electromagnetic wave. The strength or intensity of the scattered wave is a property of each atomic scatterer, represented by a quantity called the *atomic scattering factor* f_s. When an electromagnetic wave passes through a crystal it induces a small forced oscillation in each atom, which then becomes the center of an outgoing spherical wave, the scattered wave. The amplitudes of the waves coming from all the atoms add together algebraically. In almost every direction coming out of the crystal these individual amplitudes add to give exceedingly small total amplitudes, because the positive contributions at any instant are almost always nearly equal to the negative contributions. However, Laue showed that in a few special directions the amplitudes of the individual scattered waves add constructively, producing narrow pencils of scattered radiation. More important still, these directions depend intimately on the structure of the crystal.

To understand the scattering process in a quantitative way, we can proceed stepwise from a one-dimensional array to two- and then three-dimensional periodic structures. A little trigonometry will suffice to describe the process. Suppose that, as shown in Fig. 11.8a, a wavefront is incident on a line of atoms spaced regularly a distance a apart; let **a** be a vector of length a directed along the line of atoms. Let α_0 be the angle of incidence, and α one of the infinitely many possible angles of refraction; $\mathbf{s}_0$ and $\mathbf{s}$ are unit vectors in the respective directions of propagation. Then the refracted waves from two adjacent atoms X and Y will interfere constructively only for those values of α such that the path lengths $\mathbf{a} \cdot \mathbf{s}_0 = a \cos \alpha_0$ and $\mathbf{a} \cdot \mathbf{s} = a \cos \alpha$ differ by an integral number of wavelengths:

$$a(\cos \alpha - \cos \alpha_0) = \mathbf{a} \cdot (\mathbf{s} - \mathbf{s}_0) = n\lambda. \tag{11.8}$$

For a fixed wavelength λ, Eq. 11.8 defines a cone of outgoing rays at a particular angle α for each integer n. If these cones are to be far enough apart for useful measurements, the wavelength λ must be of the same order as (or smaller than) the lattice parameter a, which is of the order of 10^{-8} cm. This is why x rays are used.

To apply the same analysis to both x and y directions of a two-dimensional lattice, we need two conditions. Assume that the lattice parameters are a in the x direction, b in the y direction. Let α_0 and α be defined as before relative to the x axis, with β_0 and β the corresponding angles relative to the y axis. Then we require, for constructive interference, that

$$\mathbf{a} \cdot (\mathbf{s} - \mathbf{s}_0) = a(\cos \alpha - \cos \alpha_0) = n_1\lambda \tag{11.9a}$$

and

$$\mathbf{b} \cdot (\mathbf{s} - \mathbf{s}_0) = b(\cos \beta - \cos \beta_0) = n_2\lambda \tag{11.9b}$$

be fulfilled *simultaneously*. For each n_1, n_2 these conditions define the intersection of two cones, one for the solution of each equation, as in Fig. 11.8b. But the intersection of two cones with the same vertex is a pair of lines. Thus for the two-dimensional lattice we obtain constructive interference along a set of rays corresponding to all the pairs of integers n_1, n_2

11.4
The Laue Model

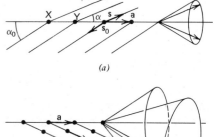

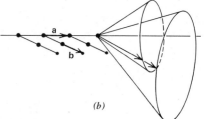

FIGURE 11.8

Diffraction from linear and planar lattices. (*a*) For a line of atoms, reinforcement occurs only for the cones of radiation defined by $\mathbf{a} \cdot (\mathbf{s} - \mathbf{s}_0) = n\lambda$. (*b*) Reinforcement occurs for a two-dimensional array only along the lines of intersection of the two sets of cones from the simultaneous one-dimensional diffraction. The double-headed arrows indicate these directions.

for any incident direction of radiation $\mathbf{s}_0$. Incidentally, this two-dimensional portrayal of diffraction is the one used to represent diffraction of low-energy, long-wavelength *electrons* by the surface layers of solids, the process known as *LEED* (*low-energy electron diffraction*).

If the lattice is three-dimensional, we must have a third condition for constructive interference, associated with periodicity in the z direction. Let the lattice parameter along the z axis be c, with γ_0 and γ the angles of incident and refracted beams relative to the z axis. Then constructive interference requires that

$$\mathbf{c} \cdot (\mathbf{s} - \mathbf{s}_0) = c(\cos \gamma - \cos \gamma_0) = n_3 \lambda \tag{11.9c}$$

be satisfied simultaneously with Eqs. 11.9a and 11.9b. With three conditions to be satisfied simultaneously, the diffracted radiation can no longer exhibit constructive interference for almost all directions of incident radiation. Rather, $\mathbf{s}_0$ itself must be in one of a set of special directions if Eqs. 11.9a, 11.9b, 11.9c are all to apply. To see how this follows, we square and add the three equations, obtaining

$$a^2(\cos^2 \alpha + \cos^2 \alpha_0 - 2 \cos \alpha \cos \alpha_0) + b^2(\cos^2 \beta + \cos^2 \beta_0 - 2 \cos \beta \cos \beta_0)$$
$$+ c^2(\cos^2 \gamma + \cos^2 \gamma_0 - 2 \cos \gamma \cos \gamma_0) = (n_1{}^2 + n_2{}^2 + n_3{}^2)\lambda^2. \tag{11.10}$$

For simplicity let us restrict our discussion to cubic crystals, for which $a^2 = b^2 = c^2$; applying the law of cosines to both incident and refracted angles,

$$\cos^2 \alpha + \cos^2 \beta + \cos^2 \gamma = 1,$$
$$\cos^2 \alpha_0 + \cos^2 \beta_0 + \cos^2 \gamma_0 = 1, \tag{11.11}$$

we obtain

$$2a^2(1 - \cos \alpha \cos \alpha_0 - \cos \beta \cos \beta_0 - \cos \gamma \cos \gamma_0) = (n_1{}^2 + n_2{}^2 + n_3{}^2)\lambda^2. \tag{11.12}$$

We now define the angle between the unit vectors $\mathbf{s}$ and $\mathbf{s}_0$ as ϕ,

$$\mathbf{s} \cdot \mathbf{s}_0 = \cos \phi, \tag{11.13}$$

so that

$$\cos \phi = \cos \alpha \cos \alpha_0 + \cos \beta \cos \beta_0 + \cos \gamma \cos \gamma_0$$
$$= 1 - 2 \sin^2\left(\frac{\phi}{2}\right). \tag{11.14}$$

Hence Eq. 11.12 becomes

$$2a^2(1 - \cos \phi) = 4a^2 \sin^2\left(\frac{\phi}{2}\right)$$
$$= (n_1{}^2 + n_2{}^2 + n_3{}^2)\lambda^2 \tag{11.15}$$

or

$$2a \sin\left(\frac{\phi}{2}\right) = (n_1{}^2 + n_2{}^2 + n_3{}^2)^{1/2}\lambda. \tag{11.16}$$

Since $\phi/2$ is half the angle between $\mathbf{s}$ and $\mathbf{s}_0$, it can be used to define the direction with respect to which $\mathbf{s}$ and $\mathbf{s}_0$ are symmetrical. This is the direction relative to which $\mathbf{s}_0$ and $\mathbf{s}$ correspond to *incidence* and *specular reflection*, respectively, with a reflecting plane perpendicular to the direction of $\mathbf{s} - \mathbf{s}_0$. Hence $\phi/2$ is equivalent to the θ of Eq. 11.5. By identifying n_1, n_2, n_3, respectively, with the nh, nk, nl of Eq. 11.7, we establish the equivalence of Eqs. 11.7 and 11.16. This means that n, the order of the reflection, is the largest common factor of n_1, n_2, and n_3.

Now that we have derived the basic formula for x-ray diffraction in crystals by two different routes, let us see how this can be used to obtain crystal structures from diffraction patterns.

Two structural characteristics enter into the formation of diffracted beams: the phase differences among the waves scattered in a given direction by the various atoms in the unit cell, and the amplitude of the wave scattered by each atom. We shall now see how these characteristics determine the directions (scattering angles) and intensities of the beams one detects, and how these directions and intensities in turn enable us to infer the crystal structure. Our discussion is necessarily rudimentary, but will illustrate the general character of the diffraction process.

The preceding discussions of Bragg and Laue diffraction were phrased in terms of waves that emerge from a crystal exactly in phase, in the sense that all such waves have either the same phase or phases differing by an integral multiple of 2π. Yet the general case of scattering in arbitrary directions involves a phase difference that need not be an integral multiple of 2π. The phase difference φ between two rays is related to l, the difference in path length, by the equation

$$l = \frac{\varphi\lambda}{2\pi}. \tag{11.17}$$

This path difference is $2d \sin \theta$ in the Bragg picture (Fig. 11.7), where d is the spacing between Bragg planes. Hence we have

$$\varphi = \frac{4\pi d}{\lambda} \sin \theta. \tag{11.18}$$

In the Laue picture we must express l in terms of spacings between *points*, rather than spacings between planes of atoms. We therefore write the position vector of the jth atom of the unit cell relative to the origin (at an atom of the cell) in terms of the lattice vectors $\mathbf{a}$, $\mathbf{b}$, $\mathbf{c}$ and the coordinate values u_j, v_j, and w_j of the nucleus of atom j:

$$\mathbf{r}_j = u_j\mathbf{a} + v_j\mathbf{b} + w_j\mathbf{c}. \tag{11.19}$$

We suppose that the atoms defining the coordinate origins of unit cells already satisfy the Laue or Bragg conditions. Then the phase difference between waves diffracted from the origin and atom j of the same cell is

$$\varphi_j = \frac{2\pi}{\lambda}\mathbf{r}_j \cdot (\mathbf{s} - \mathbf{s}_0), \tag{11.20a}$$

where $\mathbf{r}_j \cdot (\mathbf{s} - \mathbf{s}_0)$ is the difference in path lengths, or

$$\varphi_j = 2\pi(u_j n_1 + v_j n_2 + w_j n_3), \tag{11.20b}$$

because we require that Eq. 11.9 be satisfied.

Consider once more a cubic lattice, with atoms only at the corners of the unit cell. Let us choose $\mathbf{s}$ and $\mathbf{s}_0$ such that $n_1 = 1$, $n_2 = n_3 = 0$, that is, such that these corner atoms give a first-order diffracted beam in the direction corresponding to reflection from the (100) plane; the phase of this beam is by definition zero. Suppose now that additional atoms are placed in the unit cell, in particular at the positions $(a/2, 0, 0)$, where $u_j = \frac{1}{2}$, $v_j = w_j = 0$. For the same $\mathbf{s}$ and $\mathbf{s}_0$, these atoms produce waves with the phase $\varphi_j = 2\pi(\frac{1}{2} \cdot 1) = \pi$, which are exactly out of phase with, and thus cause destructive interference with, the waves scattered by the corner atoms. If the two sets of atoms scatter with the same strength, then the first-order beam in the given direction simply disappears. The first non-vanishing beam from the (100) plane is the one for which $n_1 = 2$, rather than $n_1 = 1$, for then $\varphi_j = 2\pi$, and the two sets of waves add constructively. Hence the angle at which we see the lowest-order beam from the (100) plane is

$$\theta = \sin^{-1}\left(\frac{2\lambda}{2a}\right) = \sin^{-1}\left(\frac{\lambda}{a}\right), \tag{11.21}$$

from Eq. 11.7 or 11.16 (with $\theta = \phi/2$). This agrees with our previous conclusion, in connection with Eq. 11.8, that to observe diffraction we must have $\lambda \lesssim a$.

The intensity of scattered radiation from an individual atom with Z electrons is most conveniently expressed in terms of the ratio of the *amplitude* of a scattered wave to the amplitude that would be produced by one free electron at the position of the nucleus. This ratio is roughly proportional to the total electronic charge. More specifically, it is the sum of contributions of all parts of the electronic charge cloud, each with its own appropriate phase φ relative to the nucleus. At a point P defined by the radius vector $\mathbf{r}$, the charge density $\rho(\mathbf{r})$ is, as usual, proportional to the square of the wave function $\psi(\mathbf{r})$, whereas the phase is given by Eq. 11.18 as

$$\varphi(\mathbf{r}) = \frac{4\pi d(\mathbf{r})}{\lambda} \sin\theta, \tag{11.22}$$

where now $d(\mathbf{r})$ refers to the distance between the zero-phase reflecting plane and the point P; here θ is again half the angle between the incident and diffracted beams. Figure 11.9 shows these relations and defines the angle ζ; we have

$$d(\mathbf{r}) = r \cos\zeta, \tag{11.23}$$

and thus

$$\varphi(r) = \frac{4\pi}{\lambda} r \cos\zeta \sin\theta = kr \cos\zeta. \tag{11.24}$$

We have here introduced the quantity

$$k \equiv \frac{4\pi}{\lambda} \sin\theta, \tag{11.25}$$

which has the dimensions of a wavenumber. It is not hard to show that $\hbar k$ is the magnitude of the momentum transferred to the scattered photon.[1]

The amplitude of the waves scattered from the point P is proportional to the electron density, and thus to $|\psi(\mathbf{r})|^2$, weighted by the phase factor $e^{i\varphi(\mathbf{r})}$. The atomic scattering factor f_s is thus defined as the integral of $e^{i\varphi(\mathbf{r})} |\psi(\mathbf{r})|^2$ over the whole volume of the atom:

$$f_s = \int_{\phi=0}^{2\pi} \int_{\zeta=0}^{\pi} \int_{r=0}^{\infty} e^{i\varphi(\mathbf{r})} |\psi(\mathbf{r})|^2 r^2 \sin\zeta \, dr \, d\zeta \, d\phi. \tag{11.26}$$

We now substitute the value of $\varphi(\mathbf{r})$ from Eq. 11.24 and assume that the

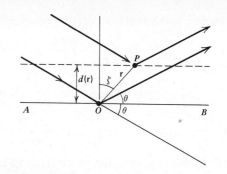

FIGURE 11.9
Scattering within an atom, from a point P defined by the radius vector $\mathbf{r}$ from the nucleus O. Here AOB is the base plane for Bragg reflection, ζ is the angle between $\mathbf{r}$ and the normal to AOB, $d(\mathbf{r}) = r \cos\zeta$ is the distance from P to AOB, and θ is the scattering angle.

---

[1] A photon of wavelength λ has momentum $h\nu/c = 2\pi\hbar/\lambda$; since the wavelength is not changed by scattering, the momentum change is that required to rotate the momentum vector through an angle 2θ without changing its magnitude.

atomic wave function $\psi(\mathbf{r})$ is spherically symmetric, so that

$$f_s = 2\pi \int_{r=0}^{\infty} \left[\int_{\zeta=\pi}^{0} e^{ikr\cos\zeta}\, d(\cos\zeta) \right] |\psi(\mathbf{r})|^2\, r^2\, dr$$

$$= 2\pi \int_{0}^{\infty} \frac{e^{-ikr} - e^{ikr}}{-ikr} |\psi(r)|^2\, r^2\, dr$$

$$= 4\pi \int_{0}^{\infty} |\psi(r)|^2 \frac{\sin kr}{kr}\, r^2\, dr. \tag{11.27}$$

Note that a similar derivation leads to Eq. 11.1, in which θ corresponds to the angle we now call ϕ. The scattering factor f_s is a function of k, and thus of both θ and λ; it becomes equal to the total number of electrons in the atom when $(\sin kr)/kr \approx 1$, that is, when $\sin\theta \approx 0$. It is a decreasing function of k, approaching zero as $k \to \infty$ or $\lambda \to 0$; that is, all atoms become weak scatterers when the wavelength of the incident radiation becomes small enough. Typical x-ray scattering factors for several atoms are given in Table 11.2.

Atomic wave functions give rather good representations of the scattering factors f_s. Strictly, if one is dealing with scattering from atoms in a molecule, the appropriate wave function $\psi(\mathbf{r})$ for use in Eq. 11.27 is the part of the *molecular* wave function around the atom in question. Then, when the total scattering of the crystal is calculated, one should integrate over all the charge density in the unit cell. In practice, however, one frequently uses atomic wave functions to find f_s and sums over atomic scattering factors to obtain the total *structure factor F*:

$$F = \sum_{j} f_{sj} e^{i\varphi_j}$$

$$= \sum_{j} f_{sj} e^{2\pi i(u_j n_1 + v_j n_2 + w_j n_3)}, \tag{11.28}$$

TABLE 11.2
TYPICAL ATOMIC SCATTERING FACTORS FOR X RAYS[a]

Element	$(\sin\theta)/\lambda$			
	0	0.2	0.5	1.0
H	1	0.481	0.071	0.007
Li	3	1.741	1.032	0.320
C	6	3.581	1.685	1.114
N	7	4.600	1.944	1.263
O	8	5.634	2.338	1.374
Na	11	8.34	4.29	1.78
Na^+	10	8.390	4.328	1.785
Cl	17	12.00	7.29	4.00
Cl^-	18	12.20	7.28	4.00
K	19	13.73	7.87	4.84
K^+	18	13.76	7.86	4.84
Fe	26	20.09	11.47	6.51

[a] The units of the atomic scattering factor f_s are electrons per atom; f_s is a function of the wavelength λ of the x ray and the scattering angle θ, through the variable $(\sin\theta)/\lambda$. When $(\sin\theta)/\lambda$ is zero, f_s is the number of electrons in the atom. (From *International Tables for Crystallography, Vol. III, Physical and Chemical Tables*, published for the International Union of Crystallography by the Kynoch Press, Birmingham, England, 1962).

where f_{sj} and φ_j are the scattering factor and phase (at the nucleus) of the jth atom. It has been possible in a few cases to distinguish between the scattering predicted by Eq. 11.28 and the more accurate value one would obtain if one were to use

$$F = \int_{\phi=0}^{2\pi} \int_{\zeta=0}^{\pi} \int_{r=0}^{\infty} e^{i\varphi(\mathbf{r})} |\psi_{\text{crystal}}(\mathbf{r})|^2 \, r^2 \sin \zeta \, dr \, d\zeta \, d\phi. \quad (11.29)$$

However, one usually does not know the wave function ψ_{crystal}, and the simple atomic $\psi(\mathbf{r})$ serves quite well for almost all purposes. Moreover, the atomic $\psi(\mathbf{r})$'s have the computational advantage of being spherically symmetric.

Finally, one can calculate the total relative scattered intensity, defined in terms of the structure factor F. The relative scattered intensity (the absolute magnitude is of no concern to us) is directly proportional to the quantity

$$|F|^2 = \left| \sum_j f_{sj} e^{i\varphi_i} \right|^2$$
$$= \left(\sum_j f_{sj} \cos \varphi_j \right)^2 + \left(\sum_j f_{sj} \sin \varphi_j \right)^2. \quad (11.30)$$

Here $|F|^2$ is the absolute square of the sum of the amplitudes of the waves scattered by individual atoms, and thus displays the interferences among these waves.

To see how the structure factor depends on the indices n_1, n_2, n_3 and how structures can be inferred from diffraction patterns, we consider some examples. First, in a simple cubic structure (of which there are no known examples in nature) there is only one atom per unit cell, so we have

$$|F|^2 = |f_s|^2, \quad (11.31)$$

and all rays appear for any integers n_1, n_2, n_3.

Next we consider a body-centered-cubic structure. The unit cell (Fig. 11.1) has atoms at $(0, 0, 0)$ and at $(a/2, a/2, a/2)$; thus in Eq. 11.28 we have $u_1 = v_1 = w_1 = 0$ and $u_2 = v_2 = w_2 = \frac{1}{2}$. Suppose that the atoms at $(0, 0, 0)$ and equivalent points have scattering factors f_1 and the other atoms have scattering factors f_2. Then we obtain

$$F = f_1 + f_2 e^{\pi i (n_1 + n_2 + n_3)} \quad (11.32)$$

and

$$|F|^2 = f_1^2 + 2f_1 f_2 \cos[\pi(n_1 + n_2 + n_3)]$$
$$+ f_2^2 \{\cos^2[\pi(n_1 + n_2 + n_3)] + \sin^2[\pi(n_1 + n_2 + n_3)]\}$$
$$= f_1^2 + f_2^2 + 2f_1 f_2 \cos[\pi(n_1 + n_2 + n_3)]. \quad (11.33)$$

If f_1 and f_2 are equal, then $|F|^2$ vanishes whenever $\cos[\pi(n_1 + n_2 + n_3)]$ is -1, or whenever $n_1 + n_2 + n_3$ is odd. Hence the "odd" reflections vanish, but the even reflections appear. If f_1 and f_2 are not equal, then the odd reflections appear. If f_1 and f_2 are similar in magnitude but not identical, then the odd reflections are weak, but not absent.

Face-centered-cubic crystals have atoms of type 1 at the eight corners, and atoms of type 2 at the centers of the six sides. We leave as an exercise (see Problem 11) the determination of the diffracted beam intensities in terms of f_{s1} and f_{s2}, the scattering factors for these two sorts of atoms. When $f_{s1} = f_{s2}$, two kinds of interference are possible: If n_1, n_2, n_3 are all even or all odd, constructive reinforcement occurs. Otherwise (two even, one odd, or vice versa) the reflections vanish.

We have spoken of two types of x-ray diffraction by crystals. One, considered only briefly, is powder diffraction by monochromatic x rays; the other, which we have examined in some detail, is diffraction of monochromatic x rays by single crystals. We implicitly assumed in discussing the single-crystal method that the crystal could be oriented at any desired angle relative to the direction of propagation of the x rays, so as to meet the diffraction conditions. Both powder and single-crystal techniques are used, the first for classifying and identifying solids, the second for determining structures of molecules by finding the locations of atoms within unit cells. A third method, also used for classification, uses polychromatic ("white") x rays for determining crystal symmetry.

In addition to x rays, both electrons and neutrons are used for diffraction experiments. *Electron diffraction* is also used for studying gaseous molecules. Scattering of electrons by atoms differs from scattering of x rays in three ways. First, nuclei contribute significantly to electron scattering; often they furnish the dominant part of the atomic scattering factor, especially at large angles. Second, electrons are scattered far more effectively than x rays of the same energy. Third, the wavelengths are different. According to the de Broglie relationship, Eq. 3.1, the wavelength of an electron is

$$\lambda = \frac{h}{p} = \frac{h}{(2mE)^{1/2}}, \qquad (11.34)$$

where p and E are the electron's momentum and kinetic energy, respectively. Substituting the values of h and m_e, and manipulating units, we find that

$$\lambda \ (\text{Å}) \approx [150/E(\text{eV})]^{1/2},$$

so that a 15-keV electron has a wavelength of about 0.1 Å. This is well below the wavelengths one can use for x-ray diffraction, so in principle one can obtain much better resolution with electrons than with x rays. In practice, however, the ease with which electrons are scattered restricted this technique for many years to gaseous (or easily volatilized) samples. In crystals it is difficult for electrons to penetrate more than a few tens of atomic layers, and electron diffraction is used primarily to study surface phenomena.

The atomic scattering factor for electrons contains contributions from both nuclei and bound electrons because electrons, unlike x rays, can transfer momentum moderately effectively to nuclei as well as to other electrons. The presence of the two contributions with opposite signs causes the electron scattering factor to have a minimum in the forward direction. The electronic contribution falls off with angle faster than the nuclear part, so that the high-angle scattering is due primarily to the nuclei.

Neutron diffraction is the third diffraction technique used for structure determination. Neutrons are scattered in two ways; through close encounters with nuclei, or through interaction with *magnetic* scatterers. Because neutrons are essentially as massive as protons, their de Broglie wavelengths are 1800 times smaller than those of electrons with the same kinetic energy. Their scattering cannot be described by a simple systematic function of the Z of the scatterer because the process depends in detail on the nuclear structure. However, even isolated protons are capable of scattering neutrons effectively; hence neutron diffraction offers one way to locate hydrogen atoms. Also, because neutrons scatter from nuclei, they are particularly suited to the determination of the amplitudes of vibrational motions. X rays can also be used to locate hydrogen atoms, and to determine, or at least to estimate, vibration amplitudes, but only with considerable refinement of the data beyond what is necessary to find molecular structures.

Neutrons are unique in their capability to detect magnetic order superposed on the chemical order of the crystal. If, for example, a crystal is antiferromagnetic, then the spins of the component atoms or molecules are paired to generate an order whose unit cell is typically twice the size of the "chemical" unit cell. This is because the unit cell in the magnetically paired crystal must have equal numbers of subunits with spin α and spin β. Doubling the size of the unit cell increases the number of terms entering into F, and hence the number of reflections. One can actually detect the onset of antiferromagnetic order by watching new diffracted beams appear when a sample is cooled below the temperature of its normal-to-antiferromagnetic transition. An example is MnO, whose transition temperature is 122 K.

Thus far we have surveyed the types of symmetry found in crystals and described methods for the experimental determination of crystal structure. In the remainder of this chapter we shall study the types of bonding and electronic structure that exist in crystals. We begin with those solids in which the bonding is most like what we have described in earlier chapters.

Some solids have properties that can be described easily in terms of their separate component molecules. Each molecule retains its identity, and is held to the aggregate structure by the weak dispersion forces we described in Chapter 10. The simplest examples of these *van der Waals crystals* or *molecular crystals* are solid phases of the inert gas elements, in which the molecules are single atoms. The dispersion forces, it will be recalled, are associated with the fluctuating, transient moments of individual atoms, moving in an only partially correlated way. They are thus weak and only instantaneously directional. Because of this lack of directionality, the atoms crystallize in *close-packed* structures.

There are two close-packed structures, the face-centered cubic (fcc) and the hexagonal close-packed (hcp). Both can be constructed by making three-dimensional arrays of hard spheres: Around one central atom, we can make a planar hexagon of nearest neighbors. In the interstices below this plane, we can place three more hard spheres of the same radius, by using every other one of the six interstices next to the central atom. Then, we can place three more atoms above the original plane, again over alternating interstices. There are two options at this stage: We can put the last three spheres above the three covered interstices, or above the open interstices. If we choose the first option, we get the hexagonal close-packed structure; the second gives us the face-centered cubic structure. These are illustrated in Figs. 11.10a and 11.10b.

The actual structures of the solids of neon, argon, krypton, and xenon are face-centered cubic. Understanding why these are close-packed structures is simple: Hard spheres pack this way, and the inert gas atoms behave almost like hard spheres when they condense. The reasons for the choice of the fcc structure are more subtle, involving the balance between the effective interactions at the level of next-nearest and third-nearest neighbors. Model calculations are just about accurate enough to indicate that the preference for the fcc structure is consistent with all one can include in the best theoretical descriptions. However, predicting the

11.7
Molecular Crystals

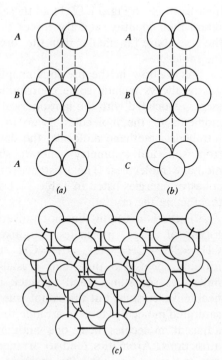

FIGURE 11.10
Close-packed crystal structures: (a) hexagonal close-packed (hcp); (b) face-centered cubic (fcc); (c) fcc drawn to show the hexagonal structure in a "cubic picture."

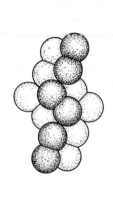

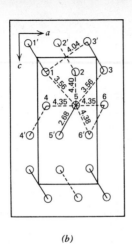

(a)   (b)   (c)

FIGURE 11.11
Crystal structures of typical linear molecules. (a) The array of molecules in the halogens. (b) The projection of the iodine structure on the ac plane of the I_2 crystal, showing the shortest intermolecular distances. (c) Packing of CO_2 molecules in the carbon dioxide crystal. From A. I. Kitiagorodskii, *Organic Chemical Crystallography*, translated by Consultants Bureau, New York, 1961.

structures of these crystals is actually a far more difficult problem than predicting the structures of most other solids, simply because the inert gas interactions are weak, and the two alternative structures differ only very slightly in energy. It is interesting to note that no species of any kind has been found that exhibits both close-packed structures. Each atomic or molecular system that forms a close-packed crystal chooses one or the other.

Molecular crystals of diatomic or larger molecules are rarely found to be close-packed. The structures are varied, and simple molecules do not necessarily form crystals with simple structures. Structures of some of the crystals of simple linear molecules are shown in Fig. 11.11, and of some hydrocarbons in Fig. 11.12. In all these examples the individual molecules retain their identity, and the forces that hold the molecules together in the crystal are far smaller than the forces that hold the atoms together in the molecules.

As we saw in the previous chapter, the forces between molecules nearly always include some contributions from permanent electric moments associated with the nonspherical shapes and permanent local bond moments of the molecules. Even in a species as symmetrical as the tetrahedral methane molecule, the deviations of the charge distribution from spherical symmetry generate short-range forces on neighboring methane molecules. The existence of these forces is reflected in the cohesive energies listed in Table 11.1 and in the way the molecules pack together in the crystal.

There is a useful (but not universal) rule of thumb for the crystal forms of large molecules. Rodlike molecules such as polyenes, $CH_2=CH—(CH=CH)_n—CH=CH_2$, tend to lie in close, parallel rows, forming sheets, so that their crystals are frequently flat plates. Flat molecules sometimes tend to stack into columns, which give rise to needlelike crystals, but this is not true for aromatic hydrocarbons. Large saturated molecules tend to arrange themselves so as to put large areas of adjacent molecules near one another, maximizing the near-neighbor attractions. Aromatics tend to arrange themselves so the hydrogens of one molecule are near the π-electron cloud of its neighbors. These generalizations rationalize the primary structure of many crystals; we find rodlike arrays of saturated flat molecules or flat arrays of rodlike molecules, and structures like those of Fig. 11.12 for aromatic molecules. These arrays then combine into secondary structures that reflect the primary crystal structure.

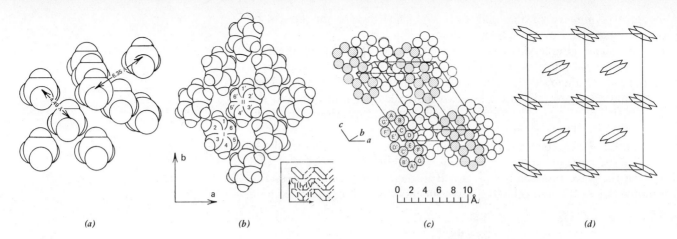

(a)   (b)   (c)   (d)

FIGURE 11.12
Crystal structures of hydrocarbons. (a) Methane. (b) Benzene. (c) Anthracene, projected along [010]. (d) Naphthalene, projected along [001]. Figures are from D. Fox, M. M. Labes and A. Weissberger, *Physics and Chemistry of the Organic Solid State* (Interscience, New York, 1963); (a), (b) from Chap. 11, by D. A. Dows; (c), (d) from Chap. 6, by P. Hartman.

The dynamical properties of molecular crystals reflect those of their component molecules. The vibrational modes of these crystals separate into two quite distinct groups, particularly for large, stiff molecules such as the aromatic hydrocarbons. One set consists of the intramolecular vibrations, which are very much like the vibrations of the atoms in the free molecules. The other set consists essentially of the vibrations of the molecular centers of mass about their equilibrium positions; these include the *librations*, the rocking motions that would be rotations if the intermolecular forces produced no barrier to molecular rotation in the crystal. The intramolecular vibrations, as one might expect, have the high frequencies characteristic of chemical bonds and atomic masses, with $\tilde{\nu}$ of the order of $1000\,cm^{-1}$. The vibrations of the molecules as a whole, however, have lower frequencies characteristic of the weaker intermolecular forces and the larger molecular masses. If we think of the entire crystal as a single weakly bound molecule, these vibrations are its normal modes. Since they are characteristic of the crystal lattice as a whole, they are called *lattice modes*. The highest frequencies of lattice modes typically correspond to wavenumbers well below $100\,cm^{-1}$. The frequencies of the very lowest modes, in principle, extend almost to zero, because they consist of one-half of the crystal oscillating against the other half. Lattice modes occur even in crystals having single-atom unit cells. Lattice modes are the modes of oscillation responsible for the ordinary temperature dependence of the heat capacity of solids. The discussion in Section 2.9 used an oversimplified representation, in which all the lattice modes were taken to be degenerate, but they were nonetheless lattice modes. In Chapter 22 we shall give a full explanation of this relationship, with a more complete discussion of the vibrational spectra of solids.

Electronic excitation in molecular crystals is, in some respects, like that in the free molecules, at least in a dense gas. The electronic spectral band systems of many molecular crystals can be identified with similar transitions in the gaseous molecules. However, the bands in spectra of crystals are usually broader than those of the free molecules (at least at room temperature), partly because of the varying amounts of lattice vibrational energy that may be superposed on the molecular electronic and vibrational energy, and partly because of electronic perturbations of one molecule by another. These perturbations can be systematized, and even estimated rather accurately, because their form is strongly dominated by the symmetry of the crystal. The general phenomenon involves two kinds of symmetries: the translational symmetry of the entire crystal, and the local symmetry of the unit cell. We shall discuss the effects of translational symmetry in a later section.

Molecular interactions cause each electronic transition of the free molecule to split into closely spaced multiplets, with as many components as there are distinguishable molecules in the unit cell of the crystal. Thus in the anthracene crystal there are two molecules per unit cell, and the spectral bands responsible for molecular electronic transitions are split into doublets. The reason for this, first elucidated by Davydov, is as follows. Let Φ_A be the wave function for the ground state of molecule A, and Φ_B the corresponding wave function for molecule B. Let $\Phi_A{}^\star$ and $\Phi_B{}^\star$ be the wave functions for A and B in their first excited states. The ground state of the unit cell has a wave function that we can approximate well as

$$\Psi_0 = \Phi_A \Phi_B, \tag{11.35}$$

because A and B retain most of their molecular identity. In lowest approximation we could write two degenerate excited-state wave functions, $\Phi_A{}^\star\Phi_B$ and $\Phi_A\Phi_B{}^\star$. We can reestablish this equivalence in the usual way by writing the functions

$$\Psi_1 = \frac{1}{\sqrt{2}} (\Phi_A{}^\star\Phi_B + \Phi_A\Phi_B{}^\star) \tag{11.36}$$

and

$$\Psi_2 = \frac{1}{\sqrt{2}} (\Phi_A{}^\star\Phi_B - \Phi_A\Phi_B{}^\star), \tag{11.37}$$

in each of which A and B are interchangeable. The functions Ψ_1 and Ψ_2 are wave functions for the *excited unit cell* that reflect the equivalence of A and B. But now the functions Ψ_1 and Ψ_2 themselves *are* physically distinguishable because of the difference in sign; hence Ψ_1 and Ψ_2 describe states with different energies. The splitting between these two excited states is just twice the "interaction energy" between $\Phi_A\Phi_B{}^\star$ and $\Phi_A{}^\star\Phi_B$, that is, the Hamiltonian term $\int(\Phi_A\Phi_B{}^\star)H(\Phi_A{}^\star\Phi_B)\,d\tau$. It is known that such crystals retain their molecular character, even in their excited states, when this splitting is small relative to the energy intervals between excited electronic states of the free molecules. To describe the entire crystal of N cells, we multiply Ψ_1 and Ψ_2 by the ground-state wave function for all the other unit cells of the crystal, and then add to this all the other $N-1$ functions in which the excitation is localized in another unit cell. Then, to normalize, we multiply the sum by $N^{-1/2}$. This wave function represents the state that is usually the first electronically excited state of the crystal.

There are certain exceptions to this simple picture of electronic excitation in molecular crystals. Sometimes an excited molecule can react with a neighboring molecule to form a dimeric molecule. These excited dimers may persist as chemical entities called *photodimers*. Alternatively the excited dimers can radiate their energy and return the system of two molecules to the ground state. Such transient species are called *excimers*. When excimers form, they give a little vibrational energy to the lattice around them in order to gain stability. When they radiate, they generally emit light of considerably longer wavelength than that of the exciting light. This shift occurs because the downward, emissive transition usually takes the dimer to a very repulsive region high on the potential curve of the ground state of the pair. Hence excimer formation and emission shift the light to the red, and transfer large amounts of the remaining excitation energy to the lattice when the ground-state pair flies apart. This is shown schematically in Fig. 11.13 for the diatomic Xe_2 system. This particular pair occurs as an excimer in liquid or high-pressure gaseous xenon when excited by the light of the first allowed electronic transitions of the xenon atom, at 147.0 and 129.5 nm in the vacuum ultraviolet region. The emission occurs throughout the region between about 160.0 and 185.0 nm, also in the vacuum ultraviolet region. Other species that are known to form excimers include small molecules such as argon and krypton (which give rise to bound excited states of Ar_2 and Kr_2, even when their ground states exhibit only van der Waals interactions), as well as larger molecules such as benzene, anthracene, and pyrene. There are also examples of mixed excimers, in which the two partners need not be of the same species.

Photodimers are formed by species such as anthracene and many of its derivatives. The tendency for neighboring molecules to combine when one is photoexcited can be exploited very nicely, particularly when the crystal structure holds the molecules in an orientation especially suited to the formation of one desired product.

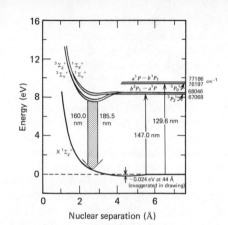

FIGURE 11.13
Potential curves and absorption and emission frequencies for formation and decay of the excimer Xe_2 in condensed xenon. From R. S. Mulliken, *J. Chem. Phys.* **52**, 5170 (1970).

11.8
Structures of Ionic Crystals

After molecular crystals, the easiest to understand are the *ionic crystals*. The alkali halides are the simplest examples, but many other materials form crystals that are essentially ionic. These include such typical species as CaCN, CaF_2, Na_2O, and most other substances composed of something strongly electropositive combined with something strongly electronegative. In contrast to molecular crystals, it is quite easy to calculate the cohesive energy of ionic crystals because the forces can be approximated well by a simple classical model. For the simplest crystals, even *ab initio* quantum mechanical calculations are feasible and accurate. We begin by describing some of the common structural types found in ionic crystals.

The basic structures of all ionic crystals ensure that each ion is surrounded by a shell of ions of opposite charge. The *rock salt* structure, Fig. 11.14a, puts each positive ion in the center of a regular octahedron of negative ions and each negative ion in the center of an octahedron of positive ions. This is *not* the same as a simple cubic lattice; the ions of each kind do not define a simple cubic lattice. (What kind of lattice do they define?) The rock salt structure occurs in all the lithium, sodium, potassium, and rubidium halides, in CsF, in the silver halides (except AgI), and in the alkaline earth oxides and sulfides (except BeO and BeS). This is the most common structure for ionic crystals with the empirical formula MX.

The other cesium halides exhibit a structure called simply the *cesium chloride* or *CsCl* structure, Fig. 11.14b; the thallium salts TlCl, TlBr, and TlI also have this structure. The CsCl structure is based on interpenetrating simple cubic lattices of positive and negative ions, with each ion surrounded by a cubic shell of oppositely charged nearest neighbors. We refer to the number of nearest neighbors as the *coordination number*, as in complex ions (see Chapter 9). The rock salt lattice has a coordination number of 6, and the cesium chloride lattice has a coordination number of 8.

Two common ionic structures combine the empirical formula MX and the coordination number 4 associated with tetrahedral coordination. These are the *zincblende* and *wurtzite* structures shown in Figs. 11.14c and 11.14d, respectively, and named after the two crystalline forms of ZnS. A number of similar compounds, such as CdS and ZnTe, also exhibit both forms. Some other II–VI compounds, for example, BeO and MgTe, exhibit only the wurtzite structure; still others, such as BeS, BeSe, BeTe, only the zincblende structure; and some, as already noted, crystallize in the rock salt structure.

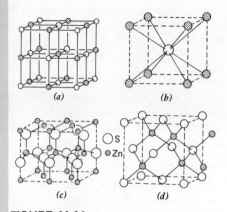

FIGURE 11.14
Structures of MX-type ionic crystals. (*a*) Rock salt. (*b*) Cesium chloride. (*c*) Wurtzite. (*d*) Zincblende.

The rock salt, cesium chloride, zincblende, and wurtzite structures are by no means the only structures exhibited by MX compounds. In the tenorite structure of CuO, for example, the oxygens are tetrahedrally 4-coordinated, but the coppers are square planar. There are still other structures, but those illustrated in Fig. 11.14 are the most important.

Two other examples will illustrate the kinds of structures encountered when the number of ions of one kind is unequal to the number of the other. We have chosen two of the most common: Figure 11.15a shows the *rutile* structure, named for a form of TiO_2. Figure 11.15b shows the *fluorite* structure, named for the crystalline form of CaF_2. In rutile the more highly charged ions are octahedrally 6-coordinated, whereas the ions with lesser charge are at the centers of triangles, and hence 3-coordinated. In fluorite, on the other hand, the ions of lesser charge are tetrahedrally 4-coordinated, whereas the ions of greater charge are 8-coordinated at the centers of cubes. The names "fluorite" and "rutile" are reserved for compounds of the type MX_2, in which the positive ions are more highly charged; the corresponding structures for M_2X compounds are called *antifluorite* and *antirutile*.

A great variety of structures are exhibited by compounds with more complex formulas, but we have given enough examples for illustrative purposes. The next question we must ask ourselves is, why does a given ionic compound crystallize in one structure rather than another? As with the structures of molecules, the complete answer to this question requires a comparison of the binding energies of the various possible structures. But in ionic solids the simplicity of the forces between the ions makes it possible to guess the correct structure by an elementary but usually accurate method.

In our considerations of ionic diatomic molecules (Section 7.4) we saw that the interaction between two ions is represented well by a point-charge Coulomb force combined with a short-range repulsion. The same model is useful in describing ionic crystals. Since the short-range repulsion falls off quite steeply, to a good approximation we can represent each ion by a charged sphere of definite radius. In other words, just as one can assign approximate "sizes" to neutral atoms (Section 5.5), it is similarly possible to define ionic sizes. One determines the ionic radii in much the same way as covalent radii, by measuring the nearest-neighbor distances in a large number of crystals and selecting a set of radii that most nearly reproduce these distances. Table 11.3 gives such sets of ionic radii calculated by Zachariasen[2] and by Pauling and Sherman,[3] with some atomic radii for comparison.

Given this model, how can we determine the structure that maximizes the binding energy in an ionic crystal? Clearly we want ions to be as close as their radii allow to ions of opposite charge (the radii are of course defined to satisfy this condition), and be as far as possible from ions of the same charge. If we pack ions as though they were charged hard spheres, each ion must necessarily be in contact with its nearest neighbors (of opposite charge). But to minimize the repulsive energy it is best that these nearest neighbors (which have a common charge) not be in contact with one another. The situation in which an ion is in contact both with nearest neighbors (of opposite charge) and with its next-nearest neighbors (of the same charge) is a limit of stability.[4] We illustrate this in Fig. 11.16, with which we introduce the concept of the *radius-ratio effect*.

The stability principle (sometimes called *Pauling's first stability criterion*) that we are now proposing is this: Consider a "central" ion in an ionic crystal; the most stable structure is that which maximizes the number of its nearest-neighbor ions of opposite charge, subject to the condition that the central ion be "in

[2] W. H. Zachariasen, *Z. Krist.* **80,** 137 (1931).

[3] L. Pauling, *Proc. Roy. Soc.* (London) **A114,** 181 (1927); L. Pauling and J. Sherman, *Z. Krist.* **81,** 1 (1932).

[4] This condition implies that the truly close-packed structures of Fig. 11.8—hcp and fcc—*cannot exist* in ionic crystals, because the next-nearest neighbors touch and the repulsive Coulomb forces are thus too strong for stability.

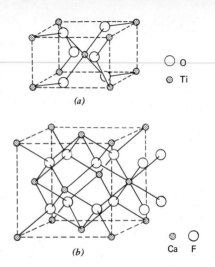

FIGURE 11.15
Two common structures for MX_2 crystals. (a) Rutile. (b) Fluorite.

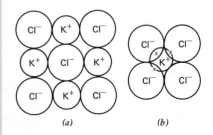

FIGURE 11.16
The radius-ratio problem. (a) Packing of KCl in one layer of a rock salt structure, stable for this salt; $R(K^+) = 1.33$ Å, $R(Cl^-) = 1.81$ Å, so the Cl^- ions around each K^+ ion can all touch the K^+ ion without touching one another. (b) Packing of KCl in a CsCl structure, unstable for this species. Here, even with the Cl^- ions touching one another, only three of the four required by the structure can simultaneously be in contact (at the points marked x) with a given K^+ ion in the next layer; thus Cl^-–Cl^- repulsion would be high, K^+–Cl^- attraction low, and the structure is unstable.

TABLE 11.3
IONIC AND ATOMIC RADII OF SELECTED ELEMENTS (Å)

	Pauling–Sherman	Zachariasen	Corresponding Neutral Atom[a]
Li^+	0.60	0.68	1.45
Na^+	0.95	0.98	1.80
K^+	1.33	1.33	2.20
Rb^+	1.48	1.48	2.35
Cs^+	1.69	1.67	2.60
Be^{2+}	0.31	0.39	1.05
Mg^{2+}	0.65	0.71	1.50
Ca^{2+}	0.99	0.98	1.80
Sr^{2+}	1.13	1.15	2.00
Ba^{2+}	1.35	1.31	2.15
F^-	1.36	1.33	0.50
Cl^-	1.81	1.81	1.00
Br^-	1.95	1.96	1.15
I^-	2.16	2.19	1.40
O^{2-}	1.40	1.40	0.60
S^{2-}	1.84	1.85	1.00
Se^{2-}	1.98	1.96	1.15
Te^{2-}	2.21	2.18	1.40

[a] From J. C. Slater, *J. Chem. Phys.* **41**, 3199 (1964); these are slightly different from those of Table 5.2, by Pauling, and are not as widely used, even though they are more recent. The differences reflect the arbitrariness and ambiguities associated with defining atomic and ionic radii.

contact" with all these nearest neighbors. Here "in contact" means that the average distance between the nucleus of any nearest neighbor and the nucleus of the central ion is equal to the sum of their ionic radii. Thus, simple geometry shows that a square of A ions surrounding a smaller B ion satisfies the limiting condition

$$\sqrt{2}R_A \leqslant R_A + R_B, \quad \text{or} \quad \frac{R_B}{R_A} \geqslant \sqrt{2} - 1 = 0.414.$$

The equality applies to the case shown in Fig. 11.17a, in which all the possible contacts between ions are made. By convention we write the radius ratio with the smaller radius, here R_B, in the numerator. The square geometry of this condition is that of a rock salt crystal, so that salts of equally charged ions with radius ratios greater than 0.414 can be expected to have rock salt structures—unless they can also satisfy the criterion for 8-coordination, as indicated in Fig. 11.17b for the CsCl structure. The condition for stability of this structure is

$$\sqrt{3}R_A \leqslant R_A + R_B, \quad \text{or} \quad \frac{R_B}{R_A} \geqslant \sqrt{3} - 1 = 0.732.$$

Where both criteria are satisfied, the CsCl structure is more stable than the rock salt structure, since each ion has eight nearest neighbors rather than six. Thus, for radius ratios between 0.732 and 1, we expect CsCl structures; for radius ratios

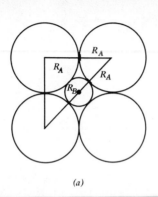
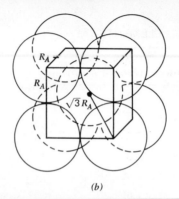

(a)   (b)

FIGURE 11.17 ·
Radius-ratio limits for (a) square planar and (b) cubical nearest-neighbor geometries.

between 0.414 and 0.732, we expect rock salt structures. If the radius ratio is below 0.414, we expect coordination numbers of 4 (as in the ZnS structures) or 3.

The minimum radius ratios for several structures are given in Table 11.4, where "coordination polyhedron" refers to the arrangement of nearest neighbors about a given central ion. Note that the radius-ratio limit for the most stable 5-coordinate arrangement, a trigonal bipyramid, is that of the equilateral triangle, 0.155. The tetrahedron has a higher limit of 0.225 (see Problem 12), so that any truly ionic species that might be expected to 5-coordinate would really exist as a 4-coordinate tetrahedron (unless it had a ratio as large as 0.414, in which case it would occur as an octahedron).

Similar limiting conditions can be derived for crystals of other types. For example in $(B^+)_2A^{2-}$ crystals the fluorite structure is the more stable when $R_B/R_A > 0.73$, but the rutile structure is predicted to be more stable for lower ratios.

The radius-ratio rule must be used with great caution, and generally more in the spirit of a first guess than as a predictive tool. The radius-ratio conditions do not even apply to all the A^+B^- crystals that we might consider ionic. The lithium halides all have rock salt structures, but only the fluoride has a radius ratio $R(Li^+)/R(A^-) > 0.414$. This is true for both the Pauling-Sherman and Zachariasen values of ionic radii. Similarly, KF, KCl, RbF, RbCl, RbBr, and CsF all have rock salt structures but radius ratios $R_B/R_A > 0.732$. Attempts have been made to define other sets of ionic radii that satisfy radius-ratio rules, but these radii tend to improve agreement near one ratio boundary at the expense of agreement at the other boundary. We shall find other examples later (Section 11.10) where the radius-ratio rules are not met by presumably ionic crystals.

11.9
Binding Energy of Ionic Crystals

The binding energy of an ionic crystal, which is commonly called the *cohesive energy* or *lattice energy*, is a readily calculable quantity. One can write it, to an excellent approximation, in terms of a sum of pair interactions; and if one is clever about it, this sum can quite easily be evaluated accurately.

TABLE 11.4
MINIMUM RADIUS RATIOS FOR VARIOUS COORDINATION POLYHEDRA[a]

Polyhedron	Coordination Number	Minimum Radius Ratio
Cubo-octahedron	12	1.000
Cube	8	0.732
Square antiprism	8	0.645
Octahedron	6	0.414
Tetrahedron	4	0.225
Triangle	3	0.155

[a] From A. F. Wells, *Structural Inorganic Chemistry*, 2nd ed. (Oxford University Press, Oxford, 1950).

Let us suppose, as in Section 7.4, that the pair interaction between ions consists of a point-charge Coulomb interaction and a short-range repulsion. For simplicity we shall neglect polarization (which, because of the symmetric environment around each ion, is much less in crystals than in ionic molecules), van der Waals forces, and other weak interactions, but these could readily be included in the model. The repulsive part of the potential energy can be approximated as an exponential function.[5] The potential energy between two ions (pair potential) is then given by the *Born-Mayer* potential

$$V(R_{ij}) = \frac{Z_i Z_j e^2}{4\pi\epsilon_0 R_{ij}} + \lambda_{ij} e^{-R_{ij}/\rho}, \qquad (11.38)$$

where R_{ij} is the distance between ions i and j, Z_i and Z_j are the ionic charges (in units of e), λ_{ij} is in general different for each kind of ionic pair, and ρ is the same for all pairs.

The total potential energy of the crystal is obtained by summing the pair potential $V(R_{ij})$ over all i, j and then dividing by 2 (to avoid counting each pair twice):

$$U = \frac{1}{2}\sum_i\sum_j V(R_{ij}) = \frac{1}{2}\sum_i\sum_j \left(\frac{Z_i Z_j e^2}{4\pi\epsilon_0 R_{ij}} + \lambda_{ij} e^{-R_{ij}/\rho}\right). \qquad (11.39)$$

The evaluation of this sum is then easily carried out for any particular lattice, provided that one knows the values of the λ_{ij} and ρ. To take a concrete example, let us consider the rock salt lattice, Fig. 11.14a. We can scale the interaction in terms of the equilibrium nearest-neighbor distance R_0 (which in this case is half the unit cell length). If we start with a cation as origin, there are six anions at a distance R_0, twelve cations at $\sqrt{2}R_0$, eight anions at $\sqrt{3}R_0$, and so forth; with an anion as origin, we obtain the same distances with cations and anions exchanged.

The Coulomb potential sum in Eq. 11.39 is easily evaluated, since R_0 can be factored out to leave only a numerical sum. Consider a crystal with the rock salt lattice and containing $2N$ ions (or N ion pairs). Since each ion (whether cation or anion) has exactly the same Coulomb interaction energies with its nearest neighbors, next-nearest neighbors, and so on, we can replace the double sum over i, j with $2N$ identical sums over j:

$$U_{\text{Coul}} = \frac{1}{2}\sum_i\sum_j \frac{Z_i Z_j e^2}{4\pi\epsilon_0 R_{ij}} = \frac{1}{2}\left(2N\sum_j \frac{Z_i Z_j e^2}{4\pi\epsilon_0 R_{ij}}\right) = \frac{Ne^2}{4\pi\epsilon_0}\sum_j \frac{Z_i Z_j}{R_{ij}}. \quad (11.40)$$

And since Z_+ and Z_- have the same magnitude Z in the rock salt lattice, we can write

$$U_{\text{Coul}} = \frac{NZ^2 e^2}{4\pi\epsilon_0}\left(-\frac{6}{R_0} + \frac{12}{\sqrt{2}R_0} - \frac{8}{\sqrt{3}R_0} + \cdots\right)$$

$$= -\frac{NZ^2 e^2}{4\pi\epsilon_0 R_0}\left(6 - \frac{12}{\sqrt{2}} + \frac{8}{\sqrt{3}} - \cdots\right) \equiv -\frac{NMZ^2 e^2}{4\pi\epsilon_0 R_0}, \qquad (11.41)$$

where the constant M, equal to the numerical sum in parentheses, is known as the *Madelung constant*. This constant has a value of 1.747558 for the rock salt lattice. For any other lattice one obtains an expression of the same form as Eq. 11.41, but with a different sum and value of M. The evaluation of Madelung constants is tedious, since these sums converge

[5] As we noted in Section 7.4, the screened Coulomb potential

$$V_{\text{rep}}(R_{12}) = \frac{Be^{-R_{12}/\rho}}{R_{12}}$$

is more accurate.

TABLE 11.5
MADELUNG CONSTANTS FOR SEVERAL LATTICE TYPES[a]

Lattice	Example	M	M/ν	Average Coordination No.
Cesium chloride	CsCl	1.7627	0.8813	8
Rock salt	NaCl	1.7476	0.8738	6
Fluorite	CaF_2	2.5194	0.8398	$5\frac{1}{3}$
Wurtzite	ZnS	1.641	0.820	4
Zincblende	ZnS	1.6381	0.8190	4
Rutile	TiO_2	2.408	0.803	4
Cuprite	Cu_2O	2.0578	0.6859	$2\frac{2}{3}$

[a] The Madelung constant is the numerical factor M obtained when one evaluates the Coulomb energy in the form

$$U_{Coul} = -\frac{NMZ_+Z_-e^2}{4\pi\epsilon_0 R_0},$$

where R_0 is the nearest-neighbor distance. In the fourth column, ν is the number of ions per "molecule" (formula unit).

very slowly, but the calculation only has to be done once for a given lattice. Values of the Madelung constant[6] for a number of common lattices are given in Table 11.5. (The last two columns of Table 11.5 will be considered later.)

The calculation of the repulsive energy is somewhat more complicated. Since there are only two kinds of ions in the rock salt lattice, we have three λ_{ij}'s: λ_{+-}, λ_{++}, and λ_{--}; we cannot in general make the assumption of Eq. 11.40, that cations and anions have the same total energy. There are N cations and N anions in our crystal, so we write the repulsive part of Eq. 11.39 as

$$U_{rep} = \frac{1}{2}\sum_i\sum_j \lambda_{ij}e^{-R_{ij}/\rho} = \frac{N}{2}\left[\sum_j(\lambda_{ij}e^{-R_{ij}/\rho})_{i=cation} + \sum_j(\lambda_{ij}e^{-R_{ij}/\rho})_{i=anion}\right]$$

$$= \frac{N}{2}[6\lambda_{+-}e^{-R_0/\rho} + 12\lambda_{++}e^{-\sqrt{2}R_0/\rho} + 8\lambda_{+-}e^{-\sqrt{3}R_0/\rho} + \cdots$$

$$+ 6\lambda_{+-}e^{-R_0/\rho} + 12\lambda_{--}e^{-\sqrt{2}R_0/\rho} + 8\lambda_{+-}e^{-\sqrt{3}R_0/\rho} + \cdots]$$

$$= 6N\left[\lambda_{+-}e^{-R_0/\rho} + (\lambda_{++}+\lambda_{--})e^{-\sqrt{2}R_0/\rho} + \frac{4}{3}\lambda_{+-}e^{-\sqrt{3}R_0/\rho} + \cdots\right].$$

$$(11.42)$$

Bear in mind that the repulsive energy falls off very rapidly with distance: In a typical ionic crystal ρ is only about $0.1R_0$. If λ_{+-}, λ_{++}, and λ_{--} are all approximately the same, the ratio between second-nearest-neighbor and nearest-neighbor terms in Eq. 11.42 should be

$$\frac{\lambda_{++}+\lambda_{--}}{\lambda_{+-}}e^{-(R_0/\rho)(\sqrt{2}-1)} \approx 2e^{-10(\sqrt{2}-1)} \approx 0.032;$$

for the next term the ratio is only 0.0009, and the series converges rapidly. To a good approximation, then, we can neglect all but nearest-neighbor repulsions in calculating the lattice energy. If z is the number of

[6] The actual value of M depends on how it is defined: some authors subsume Z_+Z_- into M, and others use a different unit length than R_0 (the unit cell length, for example).

nearest neighbors (6 for the rock salt lattice), we can write

$$U_{\text{rep}} \approx zN\lambda_{+-}e^{-R_0/\rho}. \tag{11.43}$$

We shall use this approximation to simplify certain later results, but as with the Madelung constant the complete sum can be used in exact calculations.

Combining Eqs. 11.41 and 11.43, we have, for the total potential energy of an ionic crystal,

$$U = -\frac{NMZ_+Z_-e^2}{4\pi\epsilon_0 R_0} + zN\lambda_{+-}e^{-R_0/\rho}, \tag{11.44}$$

where Z_+ and Z_- are the magnitudes of the positive and negative charges. Since R_0 is the equilibrium value of the nearest-neighbor distance, the energy U must be a minimum with respect to R; we therefore have

$$\left(\frac{\partial U}{\partial R}\right)_{R=R_0} = \frac{NMZ_+Z_-e^2}{4\pi\epsilon_0 R_0{}^2} - \frac{zN\lambda_{+-}e^{-R_0/\rho}}{\rho} = 0. \tag{11.45}$$

Combining these two equations to eliminate the exponential, we obtain

$$U = -\frac{NMZ_+Z_-e^2}{4\pi\epsilon_0 R_0}\left(1 - \frac{\rho}{R_0}\right), \tag{11.46}$$

Note that this result assumes only that the total repulsive energy is of the form 11.43, varying exponentially with R_0; the actual value of z drops out of the calculation. In order to calculate U we still need to know the value of ρ, and for this we require thermodynamics; such a derivation will be found in Section 14.9, including numerical results for the NaCl lattice. (The latter section also contains a description of how lattice energies are obtained experimentally.) The values obtained for ρ in various crystals cluster around 0.33 Å.

Let us now take another look at the Madelung constant M, the calculation of which we passed over rather perfunctorily above. The problem of finding M is merely one of adding the positive and negative contributions to the Coulomb energy due to interactions with more and more distant neighbors, until the sum has converged to the desired accuracy. The most obvious, naive way to do this sum would be first to add the contributions from the nearest neighbors, then the contributions of the next-nearest neighbors, and so on. Unfortunately this is a very inefficient method, because the terms of the series alternate in sign, and although the distances increase, so do the numbers of contributions to each successive term. The most straightforward way of getting round the convergence problem was presented by H. M. Evjen in 1932. Rather than taking the discrete charges of successive shells of neighbors as the terms of the sum, one looks at groups of ions (or parts of ions) with zero total charge. To evaluate the Madelung constant for the rock salt crystal, one makes a first approximation by combining all the charges within the unit cell cube of Fig. 11.14a: The six nearest-neighbor ions (on the cube faces) are half within the cube, the 12 next-nearest-neighbors (cube edges) one-fourth within, and the eight ions in the next shell (cube corners) one-eighth within. Thus the first estimate of the Coulomb energy for the rock salt lattice is, by analogy with Eq. 11.41,

$$(U_{\text{Coul}})_0 = \frac{NZ^2e^2}{4\pi\epsilon_0}\left[-\frac{6(\frac{1}{2})}{R_0} + \frac{12(\frac{1}{4})}{\sqrt{2}R_0} + \frac{8(\frac{1}{8})}{\sqrt{3}R_0}\right] = -\frac{NZ^2e^2}{4\pi\epsilon_0 R_0}\left(3 - \frac{3}{\sqrt{2}} + \frac{1}{\sqrt{3}}\right)$$

$$= -\frac{NZ^2e^2}{4\pi\epsilon_0 R_0}(1.4560), \tag{11.47}$$

TABLE 11.6

COMPARISON OF THEORETICAL AND EXPERIMENTAL VALUES FOR COHESIVE ENERGIES AND COMPRESSIBILITIES OF ALKALI HALIDES[a]

	NaCl		KCl	
	Theoretical	Experimental	Theoretical	Experimental
Lattice constant (Å)	5.50	5.58	6.17	6.23
Cohesive energy (eV)	7.89	7.87	7.19	7.08
Compressibility (10^{-10} Pa^{-1})	4.6	3.3	6.0	4.8

[a] Based on the calculations of P.-O. Löwdin, *Ark. Mat. Astron. Fysik* **35A**, No. 9, 30 (1947), and cited by C. Kittel, *Introduction to Solid State Physics*, (Wiley, New York, 1953).

corresponding to a Madelung constant of 1.4560. If one carries out the same process for a cube with twice the edge length, one obtains a Madelung constant of 1.750; successive calculations of this type converge on the limiting value of 1.747558. There are other methods that converge even more rapidly, and these have been used in calculating the values of M in Table 11.5. Note that M/ν, the Madelung constant per ion, increases with the number of nearest neighbors; this is consistent with our earlier discussion of Pauling's first stability criterion.

The cohesive energies calculated by this simple model from measurements of lattice constants, structures, and compressibilities are remarkably close to the experimental values for the alkali halides, the silver halides, and the chlorides, bromides, and iodides of thallium and copper. The largest discrepancies among these compounds are about 4–6% in the copper halides, for which one might question the model because of covalent contributions.

Calculations of the cohesive energy of alkali halides from the Hartree–Fock self-consistent-field model have also been quite successful; the results for NaCl and KCl are given in Table 11.6. These are *ab initio* calculations, not based on fits to empirical parameters. The results indicate that the Hartree–Fock independent-particle model is a good one for alkali halide crystals, in which one might expect the electrons to be rather well localized around individual atoms. This conclusion justifies the use of independent-particle models for covalent and metallic crystals, where localization and therefore correlation should be even less important than in ionic crystals.

We now turn to the solids in which bonding between the atoms is essentially the same as bonding in covalent molecules. At one extreme are the homopolar solids such as diamond and germanium, in which the atoms are all of the same kind. From these there is a regular gradation of properties in crystals made up of increasingly dissimilar atoms. An example is the progression from Ge to the binary compound indium antimonide (In Sb), to zinc sulfide (ZnS), to copper chloride (CuCl). Qualitatively, this takes us through the range from near one extreme of covalency to near the other extreme of ionic character. Naturally, this parallels the progression we saw in Section 7.10 for series such as C_2, BN, BeO, and LiF, which first illustrated the regular transition from covalent to ionic bonding.

11.10
Covalent Solids

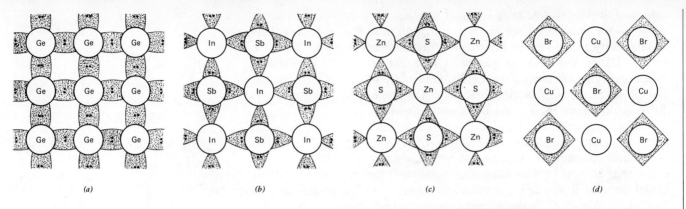

(a)   (b)   (c)   (d)

FIGURE 11.18

Projections of the (schematic) electron distributions in (a) germanium, (b) indium antimonide (which has the same structure as gallium arsenide), (c) zinc sulfide, and (d) copper(I) bromide. All four are tetrahedral structures. Think of the two nearest neighbors above and below the central atom as projecting up out of the page, and the two neighbors at the sides as projecting down below the page. From J. M. Ziman, *Theory of Solids* (Cambridge University Press, London, 1964).

Let us examine the diamond crystal, Fig. 11.2. Each carbon atom in the lattice occupies the center of a regular tetrahedron, with another carbon atom at each vertex. The crystal is therefore a three-dimensional molecule, but with no limits on its extension; in other words, diamond is a three-dimensional polymer of carbon atoms. The molecule neopentane, $C(CH_3)_4$, might be thought of as a precursor.

The bonds in diamond are very similar to the carbon–carbon bonds of saturated hydrocarbons (cf. Section 9.2). We can think of them as sp^3 hybrids rather localized along the internuclear axes. However, just as the electrons in extended molecules are actually free to travel throughout the molecule, the electrons in diamond are free to travel throughout the entire crystal. Again, as with molecules, we can think of the electrons as occupying extended molecular orbitals. With the possible exception of a (relatively) minute number of localized sites at the surface, all the bonding orbitals of diamond are fully occupied; this statement must be true whether we conceive of the crystal in terms of localized bonds or delocalized bonding orbitals. Hence, although an electron is free to move from one localized orbital φ_a to another φ_b, an electron must simultaneously move out of φ_b and into φ_a for the Pauli exclusion principle to be obeyed. The same statement can be put in terms of delocalized orbitals: The filled condition of the bonding orbitals assures that for any electron moving in the direction **r** there is a complementary electron moving in the direction $-\mathbf{r}$, with the same average rate $\langle |d\mathbf{r}/dt| \rangle$. (We neglect transient fluctuations, which average to give no net contribution to the motion.)

The implication of the Pauli principle is stringent: It says that electrons in covalent solids move only in a correlated way, so that there is no way for current to flow when a small voltage is applied to a crystal such as diamond. In other words, covalent crystals are *insulators* because their bonding orbitals are all fully occupied. Electrons can be free to move from one site to another without a compensating "back-flow" only if they are excited up to unoccupied orbitals. Hence covalent crystals might be expected to be photoconductors—solids that can carry current when they absorb radiation of the appropriate wavelength to excite electrons.

Now compare diamond and germanium (Fig. 11.18a) with their "III–V" counterparts, the structurally similar forms of BN and GaAs. Both borazon and gallium arsenide (Fig. 11.18b) are tetrahedral like diamond: Each boron has four nitrogen nearest neighbors; each gallium has four arsenic nearest neighbors. The chemical difference between BN or GaAs and diamond, silicon, or germanium is that we have replaced a uniform group IV solid with group III and group V elements from the same row of the periodic table in equal numbers. This produces what we call a III–V compound. They are true compounds, in the sense that their stoichiometry is 1:1 and their structures are perfectly regular; there are no III—III or V–V nearest neighbors in a perfect crystal of a III–V compound. The difference between diamond and borazon, or between Ge and GaAs, in terms of electronic structure, is precisely analogous to the difference between the C_2 and BN diatomic molecules.

When we come to zinc sulfide, we seem to have reached a paradox. Here we are calling ZnS an example of a covalent solid with the tetrahedral structure of the diamond lattice, since it is isoelectronic with Ge. But in Section 11.8 we cited ZnS as meeting the criteria of the ionic model because it assumes a tetrahedral structure with a radius ratio of 0.40 (see Table 11.4). Is one of these interpretations more nearly correct than the other? Probably not; ZnS does have a closed-shell electronic structure, and a zincblende or wurtzite lattice, in either model. If we were to use an ionic model to solve the Schrödinger equation for the ZnS crystal, the model would take as its starting point the wave functions for Zn^{2+} and S^{2-} ions, and then allow these functions to be distorted by the presence of the other ions. This corresponds to allowing the charge on the negative ions to be polarized in the directions of the Zn^{2+} ions—in other words, to be localized along the tetrahedral bond directions of the diamond structure. If we were to solve the Schrödinger equation for ZnS as a covalent crystal, we would start by generating some form of localized bond orbital between Zn and S—which would lie primarily in the neighborhood of the sulfur atom, with its maximum density along the Zn–S axis. As with other approximations we have described, the two pictures become quite equivalent as one refines them. In truth, ZnS is almost an ideal example of the fortunate kind of intermediate case: In this instance the models for both extremes describe the intermediate case correctly.

Even copper(I) bromide, CuBr, the next member of the sequence from Ge, GaAs, and ZnS, has the zincblende structure, as shown in Fig. 11.18d. Here, however, the covalent model and the ionic model do not give the same results. The radius ratio $R(Cu^+)/R(Br^-)$ is approximately 0.5, which implies that CuBr should have the octahedral arrangement of the rock salt crystal. This case shows the limitations, interestingly, of the *ionic* model, rather than of the covalent model—the opposite to what one might expect for an A^+B^- compound. One might think of rationalizing this disparity by invoking the ideas of the ligand field model (Sections 9.5–9.9) to account for stabilization of a tetrahedral rather than an octahedral geometry around Cu^+. Unfortunately, this approach offers no help: Since Cu^+ has a filled $3d$ shell, neither a tetrahedral nor an octahedral field causes a net energy change of the $3d$ shell, at least so long as one neglects mixing with higher, unoccupied orbitals. As much energy is lost by the electrons in orbitals destabilized by the ligands as is gained by the orbitals whose energy is lowered. If one admits higher orbitals and electron correlation into the ionic description, one naturally moves this model closer to reality, but at the expense of simplicity. Most self-consistent models become more and more accurate if they are refined indefinitely, but at some point they lose the utility of the simple physical concept inherent in the simplest form of the model.

Our discussion of covalent and ionic solids has had molecular orbitals, localized or delocalized, at its conceptual basis. Such a picture is useful for qualitative considerations, and for quantitative calculations as well. However, some of the properties of solids, especially the relationship between insulators and metallic solids and the origins of the electronic properties of metals, are clearer in terms of another set of models, the free-electron and band models. We now turn to this subject; once we have developed it for metals, we shall return to add a little reinterpretation to our picture of molecular, ionic, and covalent solids.

The most significant characteristics of metals are the properties that we associate with mobile electrons: high electrical conductivity, high reflectance, and strong bonding coupled with rather easy deformability. The conductivity comes directly from the ease with which electrons in metals can move under the influence of a static or slowly oscillating electric field. The reflectance involves the capability of nearly free electrons to be accelerated by a high-frequency electromagnetic field and then reradiate the energy absorbed; since the electrons very near the surface of a metal absorb and reradiate virtually all the energy of an incident light beam, the metal is opaque and (because of relations governing the directions of reradiation, which we shall not derive) gives rise to *specular* or mirror-like reflection, with an angle of reflection equal to the angle of incidence. The strong bonding coupled with deformability is the consequence of the relatively nondirectional character of bonds in metals, by comparison with the bonds in covalent solids. That is, when one changes a "bond angle" (or on a macroscopic scale, when one introduces a shear deformation), the resulting change in crystal energy is much smaller than the corresponding effect in a covalent solid.

Ease of motion and lack of localized character are two properties suggesting that electrons in metals might be considered as particles free to move everywhere inside the metal. This picture is the starting point for the *free-electron model*, which we shall quickly refine into the more general band model of electrons in solids.

The starting point for the free-electron model is the notion that each valence electron moves independently of all the others, and that the positive atomic cores—composed of nuclei and tightly bound core electrons—can be treated approximately as a continuous background of massive positive charge density. Thus each electron is treated as a particle in a three-dimensional box the size of the metal crystal. From Eq. 3.119 we know that the energy levels of a particle of mass m in a cubical box of side L (within which the potential energy is constant) are given by

$$E = \frac{\hbar^2}{2m}\left(\frac{\pi}{L}\right)^2 (n_x{}^2 + n_y{}^2 + n_z{}^2)$$

$$= \frac{\hbar^2 \pi^2 n^2}{2mL^2} = \frac{\hbar^2 k^2}{2m}, \tag{11.48}$$

where n_x, n_y, n_z ($= 1, 2, \ldots$) are the quantum numbers for motion in the x, y, z directions, with

$$n^2 \equiv n_x{}^2 + n_y{}^2 + n_z{}^2, \tag{11.49}$$

and k^2 is defined by

$$k^2 \equiv \frac{\pi^2 n^2}{L^2}. \tag{11.50}$$

We can think of the quantity k^2 as the squared magnitude of a vector $\mathbf{k}$, with components k_x, k_y, k_z ($k_x \equiv \pi n_x/L$, etc.); the magnitude of $\mathbf{k}$ has the dimensions of a wavenumber, whereas $\hbar\mathbf{k}$ is the momentum of the free particle. The possible values of $\mathbf{k}$ define points in what we call a k *space*, which is equivalent to a momentum space. Each point in such a "space" represents a particle's momentum rather than its position.

11.11
The Free-Electron Theory of Metals

Now let us determine the *density of states* of the free-electron gas, that is, the number of states per unit volume (in real space) with momentum magnitude between $\hbar k$ and $\hbar k + \hbar \, dk$, or with wavenumber between k and $k + dk$. This is the number of states lying in k space in the spherical shell with radius k and thickness dk. We know from Appendix 2B that the k-space volume of this shell is

$$dV_k = 4\pi k^2 \, dk. \tag{11.51}$$

However, since k_x, k_y, k_z must all be positive,[7] only one-eighth of this volume actually contains possible states of the system. Equations 11.48–11.50 show that the energy eigenstates correspond to points evenly spaced π/L apart (in a simple cubic lattice) in k space, with a k-space volume of $(\pi/L)^3$ per eigenstate. For each of these points, however, there are actually two possible states of the electron, corresponding to the two values of the spin. We take the real-space volume of our crystal to be L^3. Combining all these results, we find that the number of one-electron states with momentum magnitude between $\hbar k$ and $\hbar k + \hbar \, dk$ per unit volume of real space is

$$G(k) \, dk = \frac{2}{L^3} \cdot \frac{1}{8} \cdot \frac{4\pi k^2 \, dk}{(\pi/L)^3} = \frac{k^2}{\pi^2} \, dk. \tag{11.52}$$

The function $G(k)$ is known as the density-of-states function for free electrons; strictly, it is a density of momentum states. We can similarly define a density-of-states function $g(\epsilon)$, where $g(\epsilon) \, d\epsilon$ is the number of one-electron states with energy between ϵ and $\epsilon + d\epsilon$. We differentiate Eq. 11.48,

$$d\epsilon = \frac{\hbar^2}{m} k \, dk, \tag{11.53}$$

and substitute to obtain

$$g(\epsilon) = G(k) \frac{dk}{d\epsilon}$$

$$= \frac{k^2}{\pi^2} \cdot \frac{m}{\hbar^2 k} = \frac{mk}{\pi^2 \hbar^2}$$

$$= \frac{m^{3/2} \epsilon^{1/2}}{\sqrt{2} \pi^2 \hbar^3}. \tag{11.54}$$

If we let $C \equiv m^{3/2}/\sqrt{2}\pi^2\hbar^3$, we can write

$$g(\epsilon) = C\epsilon^{1/2}. \tag{11.55}$$

Thus the density of momentum states $G(k)$ of a free-electron gas rises as the square of the wavenumber, or the density of energy states $g(\epsilon)$ as the square root of the energy. This behavior is shown in Fig. 11.19, in parts (a) and (b) of which the area under the curve represents the number of states.

Now consider the metal at 0 K, where the electrons should have no excitation at all. The Pauli exclusion principle requires that no more than two electrons have the same **k**, and the temperature condition requires that each electron be in as low an energy level as possible. Together these imply that all the levels are doubly occupied, from the lowest on upward, until all the electrons have found a state. This means that there is a maximum energy of the occupied states; all states with less energy than this maximum are occupied, whereas states with more energy are empty.

[7] Recall that we are dealing with standing waves only, and thus with only positive n_x, n_y, n_z.

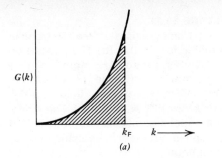
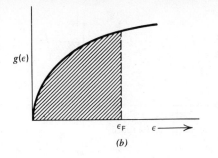
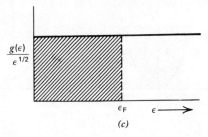

This maximum energy of the occupied levels in a very cold metal is called the *Fermi energy*, ϵ_F, and the set of levels below ϵ_F is often called the "Fermi sea."

In Part Two we shall show how knowledge of the density of states and the fundamental nature of the particles—classical, Fermi–Dirac, or Bose–Einstein—is all we need to know, in principle, to determine the average occupation of every quantum state for a system in thermal equilibrium at a given temperature. At this point we shall merely quote the result, which we can express in the form

$$d\!\left(\frac{N}{V}\right) = f(\epsilon, T)g(\epsilon)\,d\epsilon, \tag{11.56}$$

where $d(N/V)$ is the number of *occupied* states in the energy range between ϵ and $\epsilon + d\epsilon$, $g(\epsilon)$ is still the density of states (whether or not occupied), and $f(\epsilon, T)$ is called the *distribution function*. In general $f(\epsilon, T)$ is a function of both the energy and the absolute temperature T; it is defined in such a way as to approach unity as $T \to 0$. For the class of particles that includes electrons (called Fermi–Dirac particles or fermions), we shall show in Chapter 21 that

$$f(\epsilon, T) = \frac{1}{e^{(\epsilon - \mu)/k_B T} + 1}, \tag{11.57}$$

where μ is a constant at any given temperature. The value of μ can be evaluated by integration of Eq. 11.56 over all energies, which must give the total number of effectively free electrons per unit volume (for example, one electron per atom for alkali metals). Since $f(\epsilon, T) = \frac{1}{2}$ when $\epsilon = \mu$, we conclude that μ equals the energy at which $f(\epsilon, T)$ has half its maximum value. At absolute zero we have

$$\begin{aligned}
f(\epsilon, 0) &= 1 \quad (\epsilon < \epsilon_F), \\
f(\epsilon, 0) &= 0 \quad (\epsilon > \epsilon_F),
\end{aligned} \tag{11.58}$$

which is consistent with Eq. 11.57 only if $\lim_{T\to 0} \mu = \epsilon_F$; at other temperatures it is a good approximation to set $\mu \approx \epsilon_F$, as long as $k_B T \ll \epsilon_F$ (for most metals ϵ_F is of the order of $5\,\mathrm{eV}$, equivalent to a temperature of about $6 \times 10^4\,\mathrm{K}$). Making this approximation, we can write the distribution of electrons over energy levels as

$$d\!\left(\frac{N}{V}\right) = \frac{C\epsilon^{1/2}\,d\epsilon}{e^{(\epsilon - \epsilon_F)/k_B T} + 1}, \tag{11.59}$$

using the value of $g(\epsilon)$ from Eq. 11.55.

At low temperatures $(\epsilon - \epsilon_F)/k_B T$ is large and negative for $\epsilon < \epsilon_F$ and large and positive for $\epsilon > \epsilon_F$. Hence $f(\epsilon, T)$ is slightly less than 1 for levels below ϵ_F and goes rapidly to zero as the energy goes above ϵ_F. In Fig. 11.20 we show how the distribution of electrons over energy levels varies

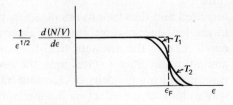

FIGURE 11.20
The electron distribution in a metal at temperatures above $0\,\mathrm{K}$; $0 < T_1 < T_2$. The dashed line indicates the distribution at $0\,\mathrm{K}$.

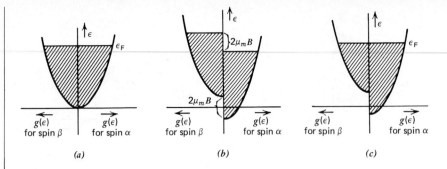

FIGURE 11.21
The paramagnetism of free electrons in a metal, for $T \approx 0$ K. (a) The population distribution for spins α (right side of y axis) and β (left side of y axis) in the absence of a field: $g(\epsilon) = C(\hbar^2 k^2/2m)^{1/2}$, $N_\alpha = N_\beta$. (b) The transient population after a magnetic field **B** is applied but before the spins relax to equilibrium,

$$g(\epsilon) = C\left(\frac{\hbar^2 k^2}{2m} \pm \mu_m B\right)^{1/2},$$

$$N_\alpha = N_\beta.$$

(c) The population of the spin system in the presence of **B** after equilibrium: $g(\epsilon)$ as in (b), $N_\alpha > N_\beta$.

with temperature. In a normal metal at room temperature, the distribution differs only slightly from that at 0 K; only a very small fraction of the electrons are normally out of their states in the "Fermi sea," occupying levels above ϵ_F. But it is just these electrons that have left the Fermi sea, and the "holes" they leave behind, that are responsible for ordinary metallic electrical conductivity and the electronic contribution to the heat capacity of metals.

Because electrons excited out of the Fermi sea are sparse relative to the number of levels available for their occupancy, the constraints of the Pauli exclusion principle hardly affect them at all. They can move rather freely under the influence of an applied voltage, and thereby give rise to a current. So also can the holes left in the Fermi sea; they can move freely, just as their electron counterparts can. Electrons and holes both contribute to conductivity.

When we introduced the heat capacity of solids in Section 2.9, we described the empirical law of Dulong and Petit, which states that many solids have a temperature-independent heat capacity of about 25 J/mol K. We shall see in Chapter 22 how this result can be derived from models (those of Einstein and Debye) in which the heat capacity arises from the vibrational motions of the nuclei in the lattice. In these models the contribution of the electrons to the heat capacity is entirely neglected, and we can now see why. Only those electrons that can absorb energy contribute to the heat capacity, but at ordinary temperatures (as we see in Fig. 11.20) only a small fraction of the electrons, far less than one per atom, have their energies significantly affected by a temperature change. Actually, the Einstein and Debye models predict a constant heat capacity of $3R$ per mole, where $R \equiv N_A k_B = 8.314$ J/mol K. Metallic heat capacities average slightly higher—Dulong and Petit's own value was equivalent to $3.1R$—because the electronic contribution is not completely negligible.

The model of the nearly filled Fermi sea with a few excited electrons also accounts for the slight paramagnetism of most metals. The overwhelming majority of electrons in the Fermi sea are in states with their spins paired. However, some of the electrons with energies near ϵ_F are unpaired and thus free to orient, so that their spins can orient themselves in the low-energy direction if an external magnetic field is applied. The application of the magnetic field thus causes a redistribution of electrons among α and β spin states until the system has achieved a new equilibrium in which a majority of electrons have spins parallel to the field. One way to look at the effect on the electrons of a uniform external magnetic field **B** is as though it shifts the energy of the entire set of electrons of one spin upward by an amount $\mu_m B$ (where μ_m is the effective electronic magnetic moment), and the energy of the electrons with the opposite spin a like amount downward. The system then comes to a new equilibrium, with both sets of spin states filled *up to the same energy level*. Figure 11.21

TABLE 11.7
TYPICAL MOLAR MAGNETIC SUSCEPTIBILITIES OF SOME ELEMENTS AND COMPOUNDS[a]

Element	Susceptibility
Aluminum	$+16.5\,(\times 10^{-6}/\text{mol})$
Argon	-19.6
Bismuth	-280.1
Carbon (diamond)	-5.9
Cerium (α phase, 80 K)	$+5,160.0$
Chromium	$+180.0$
Chromium(II) chloride, $CrCl_2$	$+7,230.0$
Copper	-5.46
Copper(II) chloride, $CuCl_2$	$+1,030.0$
Gadolinium	$+755,000.$
Helium	-1.88
Hydrogen (g)	-3.98
Nitrogen	-12.0
Nitric oxide, NO (147 K)	$+2,324.0$
Oxygen (g)	$+3,449.0$
Potassium	$+20.8$
Sodium	$+16.0$
Water (273 K)	-12.65
(373 K)	-13.09

[a] Positive values correspond to paramagnetic substances and negative values, to diamagnetic substances. In SI units, the magnetic susceptibility is in mol^{-1}; in cgs units, the values are the same but χ_M has units of cm^3/mol.

shows this process as if it occurred in the two steps just described: first a separation due to the Zeeman effect, and then a redistribution to achieve a common highest filled level. Since there is an excess of one spin over the other, the metal has a net magnetization as a result of this process. As with the heat capacity, however, it is only the small fraction of the electrons with energies near ϵ_F that actually change spin; for $k_B T \ll \epsilon_F$ this fraction can be shown to be approximately equal to $k_B T/\epsilon_F$. Thus the molar magnetic susceptibility of a metal in the free-electron model is found to be

$$\chi_M = \frac{N_A \mu_m^{\,2}}{k_B T_F} \qquad (T_F \equiv \epsilon_F/k_B), \qquad (11.60)$$

which unlike Eq. 9.21 is temperature-independent; most metals do follow Eq. 11.60 approximately. (The quantitative evaluation of the magnetic susceptibilities of metals is considerably more involved than the evaluation of the spin contribution; the orbital motion must also be included, and is often quite complicated.) Some magnetic susceptibilities are given in Table 11.7, to show the range of magnitudes they span.

We shall not treat the problem of the electrical conductivity of metals here, except in outline. The free-electron model would imply that the conductivity is infinite for a perfectly free electron gas because free electrons could attain arbitrarily high velocities under the influence of a constant field. To obtain a finite conductivity the model must be extended to include a mechanism for electrons to lose energy more and more effectively as they are accelerated. In effect, the existence of electrical resistance demands that there be some kind of friction to retard the electrons. This friction can be put into the free-electron theory if one supposes that electrons scatter occasionally from vibrating atoms of the lattice, and that on the average the electrons have no velocity component parallel to the electric field after collision. With this addition, the theory can give a good representation of the electrical conductivity of a metal.

The free-electron theory has several major limitations. First, the electrons in a real metal see not the perfectly uniform potential of a particle in a box, but rather the potential of a regular assembly of ion cores and other electrons. Second, the free-electron theory gives no clue about the relationship between metals and other solids. Third, the theory gives a totally inadequate picture of all the properties of metals except those that depend on the excited electrons and corresponding holes that exist near the surface of the Fermi sea at temperatures above 0 K. The stimulus of these limitations led to the development of the *band theory* of solids. We shall not try to pursue the band theory of metals to the point of actually deriving properties, but we shall develop the theory enough to make the model consistent with our previous ideas about the structure of solids, and to show how metals and insulators are related.

11.12
The Band Theory of Solids

The mathematical basis of the band model is the periodicity of the crystal lattice. Periodicity is important not because it is responsible for the properties of solids, but because it furnishes an easy way to solve the mathematical problem of the electronic structure. In fact, the physical processes in an amorphous solid do not differ very much from those in a chemically similar material with a periodic structure. Even properties of liquid metals are much like those of solid metals, such as the reflectivities and electrical and thermal conductivities. Nevertheless, the practical and conceptual value of the periodic model is extremely high, and we certainly should not slight the model merely because it is not general enough to describe all condensed materials.

The band theory is an extension of the Hartree–Fock model, and assumes that one can write one-electron wave functions for the electrons in a solid. In essence, we first suppose that we can find *local* solutions to the Hartree–Fock equations—solutions that describe the electrons well when they are in the vicinity of atoms—and then we build linear combinations of these functions that have the periodicity of the lattice. The process is similar to that we introduced in Chapters 6 and 7 for diatomic molecules, but now the number of atomic wave functions combined is essentially infinite.

The most striking consequence of the periodic potential is the appearance of *energy bands*. When we combine two identical atoms into a diatomic molecule, each pair of identical (and therefore degenerate) energy levels of the separated atoms splits into two distinct levels, as illustrated in Figs. 6.9 or 7.14. Much the same thing happens when we combine a large number of identical atoms into a periodic solid; however, since the number of atoms is effectively infinite, each level of the separated atoms now becomes a continuous band of energy levels over some range of energy. More precisely, as the number of states becomes infinite the spacing of the levels becomes indefinitely small, that is, the spectrum becomes more and more nearly continuous. But these bands of energy levels never cover the entire energy spectrum, and there are always intervals of energy (*gaps*) within which no physically allowable states exist. The formation of energy bands from separated-atom energy levels is illustrated schematically in Fig. 11.22. We shall see how the distribution of allowed and forbidden energy intervals determines the nature of a solid.

Some features of the free-electron theory are retained in the band model, in particular the meaning of the wavenumber k. The energy spectrum of the band model differs from that of the free-electron model by being discontinuous. Nevertheless, in both models, all values of k from zero up to the inverse of the lattice parameter remain physically allowed. Thus it is clear that the electron energy can no longer be such a simple function of k as that given for free electrons by Eq. 11.48, and in what follows we shall direct our attention to the derivation of the function $\epsilon(k)$ for a periodic potential. But these remarks are premature; let us proceed to the actual development of the band model in a relatively simple form.

We treat the crystal as made up of mobile but not completely free electrons and a periodic lattice of fixed atomic cores. Let $V(\mathbf{r})$ be the total potential acting on an electron at the point $\mathbf{r}$. Then the periodicity condition requires that the potential must be invariant under translation through a lattice vector $\mathbf{a}$, so that

$$V(\mathbf{r}) = V(\mathbf{r}+\mathbf{a}). \tag{11.61}$$

Consider now the simple case of a one-dimensional circular lattice, a very large ring with N members. Thus $2\pi/N$ is the angle subtended by the arc between neighbors. The periodicity condition requires that

$$V(\theta) = V\left(\theta + \frac{2\pi}{N}\right), \tag{11.62}$$

where the angle θ defines the location in the ring. The Schrödinger equation for an electron in this ring is

$$\left[\frac{\mathbf{p}^2}{2m} + V(\theta)\right]\psi(\theta) = \epsilon\psi(\theta). \tag{11.63}$$

Since two positions $2\pi/N$ apart are physically equivalent, we must have

$$|\psi(\theta)|^2 = \left|\psi\left(\theta + \frac{2\pi}{N}\right)\right|^2; \tag{11.64}$$

in other words, rotation of the ring through an angle $2\pi/N$ can at most change the wave function by a phase factor:

$$\psi\left(\theta + \frac{2\pi}{N}\right) = C\psi(\theta), \tag{11.65}$$

where $C^*C = 1$. There is an additional condition that $C^N = 1$, since N-fold repetition of the rotation through $2\pi/N$ must restore the original state. These conditions are satisfied if C has the form $e^{2\pi i/N}$.

Let us write the wave function in terms of distance; also, let j be an indexing integer that we shall relate to the location of the atoms, and then to a sort of momentum. Call the distance along the ring x and the spacing between neighbors a. The function $\psi(x)$ could therefore have a form as simple as

$$\psi_j(x) = e^{2\pi i j x/Na} \qquad (j = 0, 1, \ldots, N-1), \tag{11.66}$$

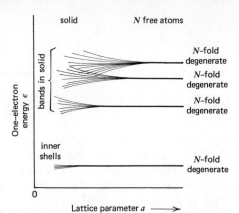

FIGURE 11.22
Schematic representation of the formation of bands of energy levels as the component atoms approach one another.

or

$$\psi_k(x) - e^{ikx}, \qquad (11.67)$$

where we have anticipated the following discussion by defining

$$k \equiv \frac{2\pi j}{Na}. \qquad (11.68)$$

But these are merely the wave functions for an electron on a uniform ring. We can make our model much more general, taking into account local structure, because we can multiply $\psi_k(x)$ by any other function $u_k(x)$ that has the periodicity of the system, that is, for which $u_k(x+a) = u_k(x)$. This gives us our general solution

$$\psi_k(x) = e^{ikx} u_k(x). \qquad (11.69)$$

If the ring consisted of a set of N atoms, the natural first approximation to a solution would take for u_k a sum of atomic functions $\varphi_j(x)$, one from each atom j in our ring model:

$$u_k(x) = \sum_{j=1}^{N} \varphi_j(x). \qquad (11.70)$$

This would make $\psi_k(x)$ a linear combination of atomic orbitals, modulated by the sinusoidal wave e^{ikx}. For $k = 0$ there is no modulation. For the function with $j = N/2$, k becomes π/a so that e^{ikx} changes sign, and the wavelength of the modulation is $2a$.

Now let us move on from this simple periodic ring structure to see how energy bands arise. We carry out a calculation for a particle in a special sort of one-dimensional box. We suppose that the box is the limit of the model shown in Fig. 11.23; we obtain that limit by letting the higher value V_0 of the potential grow arbitrarily higher, and the width of the regions under the teeth diminish, so that the area under each tooth remains constant. This potential, the *Kronig–Penney potential*, is given by

$$\lim_{\substack{b \to 0 \\ V_0 b = \text{const.}}} \begin{cases} V(x) = 0, & n(a+b) \leqslant x \leqslant n(a+b)+a, \\ V(x) = V_0, & n(a+b)-b < x < n(a+b). \end{cases} \qquad (11.71)$$

This form makes it particularly easy to develop the banded character of the energy spectrum, because in the limit $b \to 0$ the wave functions can be required to have no change in wavelength at the boundaries of the teeth of $V(x)$. For convenience let us call each periodic unit a cell.

We set up the Schrödinger equation for the general case $b > 0$ and take the limit in which the teeth are so narrow and high that we can require the wave function $\psi(x)$ that solves the equation to have the same wavelength everywhere. We assume, that is, that the width of each barrier is negligible. The Schrödinger equation has its usual form:

$$\frac{d^2\psi(x)}{dx^2} + \frac{2m}{\hbar^2}(\epsilon - V)\psi(x) = 0. \qquad (11.72)$$

On the basis of the insight we just obtained concerning periodicity, we substitute

$$\psi(x) = e^{ikx} u_k(x) \qquad (11.73)$$

to get equations for each local atomic function $u_k(x)$:

$$\frac{d^2 u(x)}{dx^2} + 2ik \frac{du(x)}{dx} + \frac{2m}{\hbar^2}\left(\epsilon - V - \frac{\hbar^2 k^2}{2m}\right) u(x) = 0. \qquad (11.74)$$

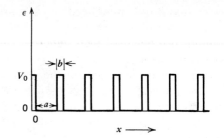

FIGURE 11.23
The Kronig–Penney model potential for a one-dimensional solid, prior to the limiting process $b \to 0$ subject to $V_0 b = \text{const.}$

The quantity k is a kind of momentum for motion from cell to cell, which appears explicitly in the effective potential $(\epsilon - V - \hbar^2 k^2 / 2m)$ because we have made the separation of Eq. 11.73.

Let us define

$$\alpha \equiv \left(\frac{2m\epsilon}{\hbar^2}\right)^{1/2}, \tag{11.75}$$

the momentum in the region where $V = 0$ for a particle with energy ϵ, and

$$\beta \equiv \left(\frac{2m(V_0 - \epsilon)}{\hbar^2}\right)^{1/2}, \tag{11.76}$$

the momentumlike quantity in the region where $V(x) = V_0$. Then in the region $0 < x < a$, where $V(x)$ is zero, the solution to Eq. 11.74 is

$$u(x) = Ae^{i(\alpha - k)x} + Be^{-i(\alpha + k)x}, \tag{11.77}$$

and in the region $a < x < a + b$, where $V(x) = V_0$, the solution is

$$u(x) = Ce^{(\beta - ik)x} + De^{-(\beta + ik)x}. \tag{11.78}$$

(Note that these solutions are exactly those of Eqs. 4.2 and 4.3.) We set the conditions that $u(x)$ and its derivative be continuous at a and b. When we let $b \to 0$, the condition of periodicity on the function is

$$u(0) = A + B = u(a) = Ae^{i(\alpha - k)a} + Be^{-i(\alpha + k)a}, \tag{11.79}$$

and the corresponding condition on the derivative is

$$\left(\frac{du}{dx}\right)_0 = \left(\frac{du}{dx}\right)_{a+b}. \tag{11.80}$$

We may now approximate the derivative at $a + b$ by the leading terms of its Taylor expansion at a:

$$\left(\frac{du}{dx}\right)_{a+b} = \left(\frac{du}{dx}\right)_a + \left(\frac{d^2u}{dx^2}\right)_a b$$

$$= \left(\frac{du}{dx}\right)_a + b\beta^2 u(0), \tag{11.81}$$

because $u(0) = u(a)$. Now let us state the limiting process more precisely. We define the area under a tooth but above the particle energy to be

$$P = \frac{ab\beta^2}{2}, \tag{11.82}$$

and require that

$$\lim_{\substack{V_0 \to \infty \\ b \to 0}} \frac{ab\beta^2}{2} = \frac{2mab}{\hbar^2}(V_0 - \epsilon) = P. \tag{11.83}$$

If we now take the limit of $b \to 0$, the derivative equation can be transformed into an expression for the A and B of Eq. 11.78, the relative amplitudes of waves moving to the right and left:

$$\left[i(\alpha - k) - \left(\frac{2P}{a}\right)\right]A - \left[i(\alpha + k) + \left(\frac{2P}{a}\right)\right]B$$

$$= i(\alpha - k)Ae^{i(\alpha - k)a} - i(\alpha + k)Be^{-i(\alpha + k)a}. \tag{11.84}$$

We can solve the continuity equations 11.78 and 11.82 simultaneously for A and B by requiring the determinant of the coefficients of A and B to

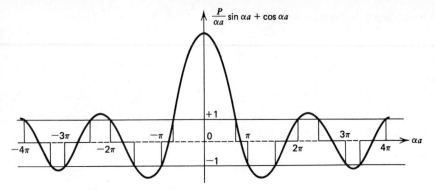

FIGURE 11.24
The function $Q = P$ (sin αa)/αa + cos αa, as a function of αa, and for a value of P that generates forbidden ranges of this function. Energy values are allowed when αa takes on values for which $-1 \leq Q(\alpha a) \leq 1$. From C. Kittel, *Introduction to Solid State Physics* (Wiley, New York, 1953).

be zero. That is, when we write these equations as

$$a_1 A + b_1 B = 0, \qquad (11.85a)$$

$$a_2 A + b_2 B = 0, \qquad (11.85b)$$

the solution is given by the condition

$$\begin{Vmatrix} a_1 & b_1 \\ a_2 & b_2 \end{Vmatrix} = 0, \qquad (11.86)$$

where

$$a_1 = 1 - e^{i(\alpha-k)a}, \qquad (11.87a)$$

$$b_1 = 1 - e^{-i(\alpha+k)a}, \qquad (11.87b)$$

$$a_2 = i(\alpha - k)(1 - e^{i(\alpha-k)a}) - \frac{2P}{a}, \qquad (11.87c)$$

$$b_2 = -i(\alpha + k)(1 - e^{-i(\alpha+k)a}) - \frac{2P}{a}. \qquad (11.87d)$$

Equation 11.84 leads to the condition on P and the momentum

$$P\frac{\sin \alpha a}{\alpha a} + \cos \alpha a = \cos ka. \qquad (11.88)$$

This equation has solutions only when the left-hand side has an absolute value less than or equal to unity, because $|\cos ka| \leq 1$. There are regions in which this is true, but in general, for large enough P (which is always positive), there are also values of αa for which the absolute value of the left-hand side is greater than 1. Figure 11.24 illustrates this point; here we show P (sin αa)/αa + cos αa versus αa.

When $abm/\hbar^2$ is small or ϵ approaches V_0, then P is small and there may be no forbidden regions. However, in general there are, and the energy as a function of k takes on the form shown in Fig. 11.25 for the Kronig–Penney model. There are regions for which the energy is a continuous function of the wave vector **k**; these are the allowed *energy bands*. There are also values of $k \to$ **k** at which the energy is a discontinuous function of **k**, leaving ranges of ϵ for which there is no physical solution. These ranges of ϵ are called *band gaps*.

We saw how sets of degenerate levels split when isolated atoms or groups are brought into proximity and interaction. We can think of each band of the Kronig–Penney model as originating with one discrete level of a square potential well, repeated many times. The special character of

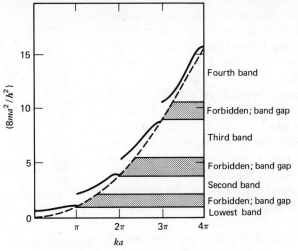

FIGURE 11.25
The energy as a function of the wave number k, for the Kronig–Penney potential with $P = 3\pi/2$. Adapted from C. Kittel, *Introduction to Solid State Physics* (Wiley, New York, 1953).

infinite systems or infinitely long chains, which makes them differ from ordinary molecules, is that their level structure becomes continuous. Like the free particle, the electron in a Kronig–Penney potential can go on out to infinity. Its wave function is not constrained to go to zero at large distances from the origin; hence its spectrum is continuous. However, it is constrained by the boundary conditions imposed by the potential, which continue to appear at all distances. This constraint is responsible for closing off some values of energy, but the continuous bands remain.

The band spectrum is not unique to the Kronig–Penney potential or to one-dimensional potentials. The phenomenon occurs in two-dimensional and three-dimensional situations as well. In one form or another, we find it in all solids, whether they be insulators or conductors, ionic, molecular, covalent, or metallic. The differences among these are (1) the widths of the bands relative to the excitation energies of the separated atoms or molecules that comprise the crystal, and (2) the way the electrons fill the bands.

We stated that covalent solids are insulators because their energy levels consist of filled bands of bonding orbitals, which we call *valence bands*. Metals, by contrast, are *conductors* because they have only partly filled bands called *conduction bands*. The highest occupied band of a metal is in almost every way like the highest occupied set of orbitals of a large molecule. In the metal there are not enough electrons to fill that band. The difference between the bands of metals and the orbital levels of finite molecules is the continuous range of allowed energy of the band, in contrast to the sharp discrete energies of orbital levels. In any solid at 0 K, electrons fill the bands up to the Fermi level ϵ_F. In an insulator, the electrons are like electrons in molecules, and cannot move freely. In a metal, the occupied states at ϵ_F are only infinitesimally below empty levels just above ϵ_F. Hence any extra energy, however small, can promote an electron from a state locked by the exclusion principle into immobility in the Fermi sea, into a state above ϵ_F where there are essentially no constraints on the motion of the electron. The availability of states at ϵ_F (or infinitesimally above it) makes it possible for electrons of a metal to move under the influence of any external electric field, as weak as we wish.

11.13
Conductors, Insulators, and Semiconductors

Each band of a metallic crystal can be associated with one particular level of the individual atoms. Thus, the conduction band of sodium is the $3s$ band; each atom contributes its $3s$ orbital toward the formation of the band, and one electron to occupy a state in the band. The result, therefore, is a $3s$ band exactly half-filled.

This argument suggests that magnesium would have a filled $3s$ band and an empty $3p$ band. In fact, the low-energy limit of the $3p$ band lies below the upper limit of energy of the $3s$ band, as indicated schematically in Fig. 11.26. Hence the electron population is as shown in Fig. 11.26a, not as in Fig. 11.26b, so that the empty parts of both the $3s$ and $3p$ bands contribute to the conductivity of magnesium. It is not at all rare for bands to overlap in energy in this way; the $4s$ band is very broad, and overlaps about half the $3d$ band in most of the transition metals, as indicated in Fig. 11.27.

In some cases the levels of a band all crowd near a common energy E_0; in others, they spread over a broad range of energy. In the former case, the density of states of the band is high near E_0, but necessarily low at other energies. In the latter case the density of states is a rather smooth function of energy or momentum.

We have now seen qualitatively how a partially filled band is responsible for metallic conduction. We have also seen how the bands have their origins in the interactions between equivalent states on the separate atoms of the crystal. Previously, we recognized how the filled orbitals of covalent crystals give rise to the insulating properties of these crystals. Now it only remains to identify the sets of delocalized orbitals of our earlier picture with the bands of the section just preceding. The band picture puts emphasis on the set of levels, rather than on the individual atomic parentage of each, but the concept of individual atomic components lurks behind all band models except the pure free-electron picture. Recall how we wrote the one-electron function in Eqs. 11.69 and 11.70 as a modulated periodic function, described locally at each site by an atomic orbital $u_k(\mathbf{r})$ there. The picture of solids based on covalent bonds emphasizes the microscopic, local structure of the orbitals that go to make up the band, rather more than the band structure itself. The split degeneracy of the local pairs or small sets of orbitals occurs because of the delocalization and interaction between sites. We saw a similar situation previously with the splitting of degenerate orbital energy levels in the formation of bonds in molecules.

The most successful modern treatments of energy levels and electronic structures of metals and covalent solids have converged, from both covalent and band model directions, to a form in which the electronic wave functions are superpositions of free-electron waves and localized atomic orbitals. The result is a consistent and rather accurate description of the electronic properties—conductivity, magnetic susceptibility, electronic spectra—of both conductors and insulators.

The particular electrical character of insulators is the immobility of their electrons, because their highest occupied bands are normally filled. This is much like the closed-shell structure of inert gases. For an electron to be free, in the band picture of an insulator, it must gain an energy of excitation large enough to carry it to the next band. The spacing between bands of an insulator is large compared with mean thermal energies. Typically, the radiation required to promote an electron from a filled valence band to an empty conduction band of an insulator lies in the ultraviolet region. Hence the population of electrons in the conduction band is normally almost nil.

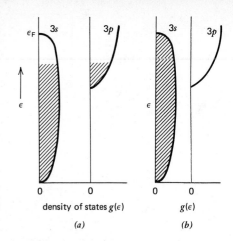

density of states $g(\epsilon)$ $g(\epsilon)$

(a)   (b)

FIGURE 11.26
Bands of magnesium metal (schematic): (a) as actually occupied; (b) as they could be occupied if the $3s$ levels were filled and the $3p$ band were empty.

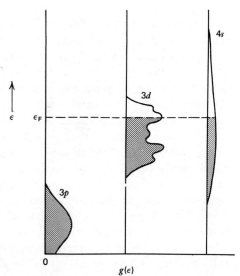

FIGURE 11.27
Schematic representation of the density of states $g(\epsilon)$ for a transition metal such as cobalt. The $3d$ band has multiple peaks and covers a range of energies much narrower than the $4s$ band. Only a portion of the $3p$ band is shown, and lower bands are omitted.

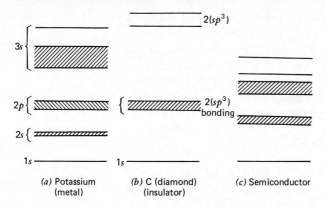

FIGURE 11.28
Schematic diagram of levels of (*a*) a metal; (*b*) an insulator; (*c*) a semiconductor. Hatched areas indicate filled levels.

However, there are insulators whose filled valence bands are close to the next empty conduction bands, close enough that a few electrons occupy conduction levels at room temperature, or perhaps a few hundred degrees higher. Pure elements such as silicon and germanium, and oxides such as ZrO_2, have this character. These materials are called *semiconductors*. They differ from conductors in that their highest occupied level coincides with the top of a band at 0 K; they differ from insulators in that their conduction bands can be populated thermally. (Insulators may melt or vaporize at temperatures lower than those required to populate their conduction bands.) The three types of level structures are shown schematically in Fig. 11.28.

There are also impurity semiconductors—in which foreign atoms in low concentration provide occupied levels that contribute electrons to conduction bands (*n*-type semiconductors), or empty levels that trap electrons from the valence band (highest filled insulator band) leaving mobile holes (*p*-type semiconductors). These are the basis for transistors and other solid-state electronic devices.

The conductivity of semiconductors increases with temperature because the number of conduction electrons increases rapidly with temperature, roughly as $\exp(-\Delta E_{gap}/k_B T)$. (The Boltzmann constant k_B is not to be confused with the k of Eqs. 11.68 and 11.73.) The number of conducting electrons increases with temperature in a metal also, but not so rapidly. Moreover, the scattering of electrons by vibrating nuclei, which is responsible for positive electrical resistance in all conductors, also increases with temperature. In a semiconductor the increase in resistance is less important to the conduction than the increase in the number of mobile electrons, and the resistance drops with increasing temperature. In a metal, where electrons can be set free with any amount of energy, the increased scattering dominates the temperature-dependent aspects of conductance and the resistance increases with increasing temperature.

We close this chapter with a few comments about the band structure of a *molecular* crystal, and use this to introduce the concepts of *excitons* and *effective mass*. Here the basic building blocks are molecules, and the interactions between these building blocks are weak. Hence we expect, and find, that (1) the electronic and vibrational structures of molecular crystals are characteristic of the separated molecules, only slightly perturbed, and (2) the *bands* of electronic states are very narrow, because of the weak intermolecular forces. Often the widths of bands in molecular crystals are much less than the vibration frequencies of the free molecules. When this is the situation, the bands can be characterized not only by their parent electronic states, but by parent vibrational states as well. In cases of this sort we envision a crystal in which a quantum of vibrational-plus-electronic excitation is delocalized, traveling through the crystal with a local wave function that we call $\varphi_j{}^\star(\mathbf{r})$, an excited-state function (hence the star) analogous to a component of Eq. 11.70. The state function for this case is best constructed as $\Psi(\mathbf{r}_1, \ldots, \mathbf{r}_N)$, an N-electron function in which excitation may appear on any molecule:

$$\Psi_k{}^\star(\mathbf{r}_1, \ldots, \mathbf{r}_N) = \text{const.} \sum_i A_j \Phi_j{}^\star(\mathbf{r}_1, \ldots, \mathbf{r}_N). \qquad (11.89)$$

From our examination of wave functions in periodic structures, we now expect a phase term $\exp(i\mathbf{k} \cdot \mathbf{r}_j)$ in the constant A_j. (We avoid explicit sorting out of the antisymmetrization, for simplicity.) Each term of $\Psi_k{}^\star$ has one localized excitation and $N-1$ unexcited sites; thus one such term is

$$\Phi_j{}^\star = \|\varphi_1(\mathbf{r}_1)\varphi(\mathbf{r}_2) \ldots \varphi_j{}^\star(\mathbf{r}_j) \ldots \varphi_N(\mathbf{r}_N)\|, \qquad (11.90)$$

so that, altogether, Ψ_k has in it one component with the excitation on each atom, and the excitation propagates with the modulation $\exp(i\mathbf{k} \cdot \mathbf{r})$. As in the case of one-electron functions, $\mathbf{k}$ acts like a momentum. The quantum of excitation behaves much like a particle, but the part associated with $\mathbf{k}$ travels from electron to electron. The propagating excitation is called a *quasi-particle*; in particular, a quasi-particle of electronic excitation in a crystal is called an *exciton*. This leads us to the ingenious concept of *effective mass* for a quasi-particle.

To associate excitons and electrons in bands with velocities and effective mass, we must know the relation between the energy ϵ and the wave vector $\mathbf{k}$, the so-called *dispersion relations* of the bands. This is a generalization of the dispersion relation discussed in Chapter 2. We turn to classical mechanics to generate our definitions. For a free particle, $\epsilon = h\nu = 2\pi\hbar\nu$ and $p = \hbar k$, so,

$$\epsilon = \frac{\hbar^2 k^2}{2m}. \qquad (11.91)$$

Hence the energy and momentum are related:

$$\frac{1}{\hbar}\frac{d\epsilon}{dk} = \frac{\hbar k}{m}, \qquad (11.92)$$

which is just the velocity, and

$$\left(\frac{1}{\hbar^2}\frac{d^2\epsilon}{dk^2}\right)^{-1} = m, \qquad (11.93)$$

the mass.

Now we apply Eqs. 11.92 and 11.93 to any system for which we know $\epsilon(k)$. We can define an effective velocity

$$v_{\text{eff}}(k) = \frac{1}{\hbar}\frac{d\epsilon(k)}{dk}, \qquad (11.94)$$

and an effective mass

$$m_{\text{eff}}(k) = \frac{1}{\hbar^2}\frac{d^2\epsilon(k)}{dk^2}. \qquad (11.95)$$

The behavior of $v_{\text{eff}}(k)$ and $m_{\text{eff}}(k)$ are displayed in Fig. 11.29 for a typical $\epsilon(k)$ such as one obtains from the first band of the Kronig–Penney model. The velocity is greatest in the middle of the band and is zero at

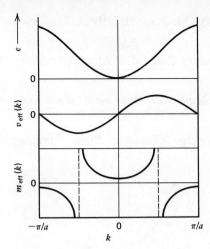

FIGURE 11.29
Energy, velocity, and effective mass for a particle in the first band of a one-dimensional solid. (a) Energy ϵ. (b) Effective velocity $v_{eff}(k)$. (c) Effective mass $m_{eff}(k)$.

the edges. The effective mass, on the other hand, goes off to $\pm\infty$ at these points. Moreover m_{eff} is positive for $0 \leq |k| < \pi/2a$, and negative for $\pi/2a < |k| \leq \pi/a$, where π/a is the upper limit of the band. Hence the direction of acceleration of an electron is *toward* a negative electrode when $|k| > \pi/2a$, because the sign is reversed from the usual one in the expression $\mathbf{F} = m\mathbf{a}$. This means that an electron in the high-k half of the band acts like a positive particle, not a negative one.

At this point, we have just completed Part One by abstracting from a complicated many-body problem the simplifying generalizations of quasi-particles, effective velocities, and effective masses. In Part Two we shall treat the properties of complex systems at equilibrium by creating far more general and powerful variables with the science of thermodynamics and, by relating these variables to the microscopic structural information of Part One, with statistical mechanics.

Further Reading

Blakemore, J. S., *Solid State Physics*, 2nd Ed. (W. B. Saunders Company, Philadelphia, 1974).

Bunn, C. W., *Chemical Crystallography* (Oxford University Press, London, 1961).

Hannay, B. N., *Treatise on Solid State Chemistry*, especially Vol. 1 Plenum Press, New York, 1975).

Kittell, C., *Introduction to Solid State Physics*, John Wiley & Sons, Inc., New York, 1953).

Kittell, C., *Quantum Theory of Solids* (John Wiley & Sons, Inc., New York, 1967).

Koerber, G. G., *Properties of Solids* (Prentice-Hall, Inc., Englewood Cliffs, N.J., 1962).

Morrison, M. A., T. L. Estle and N. F. Lane, *Quantum States of Atoms, Molecules, and Solids* (Prentice-Hall Inc., Englewood Cliffs, N.J., 1976), Chapters 18–24.

Peierls, R. E., *Quantum Theory of Solids* (Oxford University Press, London, 1965).

Seitz, F., *The Modern Theory of Solids* (McGraw-Hill Book Co., Inc., New York, 1940).

Wells, A. F., *Structural Inorganic Chemistry*, 2nd Ed. (Oxford University Press, London, 1950).

Wheatley, P. J., *The Determination of Molecular Structure* (Oxford University Press, London, 1968).

Ziman, J. M., *Theory of Solids* (Cambridge University Press, London, 1964).

Problems

1. Show that the base-centered tetragonal, face-centered hexagonal, and body-centered monoclinic structures do *not* define additional space lattices beyond the 14 of Fig. 11.1.

2. Show that the diamond crystal defines a Bravais lattice in which there are two carbon atoms per unit cell.

3. Sketch the following planes of a cubic lattice and indicate which of them are equivalent:
   (a) (102);
   (b) ($\bar{1}$11);
   (c) ($\bar{1}\bar{1}$1);
   (d) (021);
   (e) (003).

4. How do the translational symmetries of the bcc and fcc structures differ from that of the simple cubic structure?

5. A plane parallel to the x axis of a simple cubic lattice, with lattice vector a, intersects the y and z axes at $3a$ and $4a$. What are its Miller indices? Sketch the plane (222) of a simple cubic lattice. Do the same for a base-centered monoclinic lattice.

6. Show that for a cubic lattice the spacing between adjacent (hkl) planes is

$$d = \frac{a}{(h^2 + k^2 + l^2)^{1/2}}.$$

7. Calculate the x-ray scattering pattern (i.e., the angles of the diffracted beams) for a powder of body-centered cubic crystals of plutonium. The lattice parameter is 3.638 Å; assume that the x-ray line is the Cu K_α line, whose wavelength is 1.54 Å. Identify each maximum in the powder pattern by the Miller indices of the reflecting plane, up to at least the eighth reflection. Compare this pattern to the pattern that would be generated by a simple cubic lattice with the same lattice parameter.

8. The derivation for x-ray diffraction based on Fig. 11.5 and the Bragg model as given in Eqs. 11.5 and 11.6 implies that constructive interference occurs at discrete angles θ, so that the diffracted x rays come out in sheets or planes. Extend the derivation to show that a three-dimensional crystal produces pencils or rays of diffracted x rays. (Note that the derivation from the Laue model, Section 11.4, gives this result.)

9. Calculate the scattering pattern from a single body-centered cubic crystal of Ag, with the lattice parameter 4.086 Å, and with Cu K_α x rays of wavelength 1.54 Å. Suppose that the crystal is oriented to give Bragg reflection from the (100) plane; from the (111) plane. Give the value of a suitable orientation angle. Identify reflections by their order and, if more than one plane can give reflections at this angle, identify the plane associated with each reflection.

10. Iron has three modifications with body-centered cubic structures, called α-Fe, β-Fe, and δ-Fe. Their lattice parameters are, respectively, 2.8665 Å, 2.91 Å, and 2.94 Å. What are the quantitative differences of the diffraction patterns of single crystals of these materials, when the Mo K_α radiation, 0.75 Å, is used to study the diffraction?

11. Show that a face-centered cubic crystal composed of only one kind of atom has reflections for n_1, n_2, n_3 all even or all odd, but not if one integer differs in oddness from the other two.

12. Show that salts of equally charged ions should exhibit zincblende or wurtzite structures when the ratio of radii lies between 0.225 and 0.414. Devise a structure that could exist for a ratio below 0.225 and find its limits. Check the radius ratios of zinc sulfide and other species known to exhibit 4-coordination against the limits 0.225 and 0.414.

13. The Born-Mayer model of the ionic crystal, Section 11.9, gives the energy $U(R_0)$. The compressibility of an isotropic crystal can be obtained from this expression, because the compressibility κ_T is the fractional decrease in volume per unit increment of pressure, or of force unit area:

$$-\kappa_T = \frac{1}{V}\left(\frac{\partial V}{\partial p}\right)_T = \left(\frac{\partial \ln V}{\partial p}\right)_T$$

(Strictly, as indicated, the derivative is taken at constant temperature.) Noting that $\partial U(R_0)/\partial R = -F(R_0)$, an effective force, derive an expression for κ_T from the expression 11.44 for $U(R_0)$. Evaluate κ_T for a rock salt crystal.

14. The volume of a crystal always has the form $V = cNR^3$, where R is a characteristic interparticle distance. If R is taken to be the metal–metal distance, what is the value of c for the rock salt structure? For the CsCl structure? For the zincblende structure?

15. Find the equilibrium nearest-neighbor distance for a linear chain or particles, alternately positive and negative, with pairwise interactions of the form $V_{ij} = \pm e^2/4\pi\epsilon_0 R_{ij} + Ae^{-\beta R_{ij}}$.

16. Evaluate the Fermi energy ϵ_F for a free-electron metal with the lattice spacing of sodium. Estimate the fraction of valence electrons with energies above ϵ_F when the temperature is 300 K.

17. Consider a ring with the potential of Eq. 11.71, with $a = b$ rather than with $b \to 0$. Suppose the ring has length $2Na$. Write the wavefunction in the nth valley (with $l = a + b$) as

$$\psi(x) = A_n e^{ik(x-nl)} + B_n e^{-ik(x-nl)}.$$

Show that the coefficients A_{n+1} and B_{n+1} of the exponential in the next valley can be expressed as

$$A_{n+1} = (\alpha_1 - i\beta_1)e^{ikl}A_n - i\beta_2 e^{ikl}B_n,$$
$$B_{n+1} = i\beta_2 e^{-ikl}A_n + (\alpha_1 + i\beta_1)e^{-ikl}B_n.$$

18. Compute the effective mass as a function of k for an electron in the first band of a line of "atoms" with a Kronig-Penney potential and a lattice spacing of 1.5 Å. Compute the density of states for the first band of the same system.

19. Compare the eigenfunctions and eigenvalues for an electron on a one-dimensional ring of length L with those of an electron in a one-dimensional box of the same length. In what ways do these two examples differ and why?

20. Show that Eq. 11.88 follows from Eq. 11.84 and find ϵ at the lower and upper limits of the first three bands. At what values of ϵ(eV) do these *band edges* occur for an electron in a lattice with $a = b = 1.5$ Å and $V_0 = 3$ eV?

PART TWO

Matter in Equilibrium: Statistical Mechanics and Thermodynamics

In Part One we treated the detailed structure of microscopic systems: individual atoms and molecules and their interactions. Now we shall consider the macroscopic systems of our common experience.

We have seen that an exact description of matter requires the use of quantum mechanics. The solution of the Schrödinger equation (with appropriate boundary conditions) yields as complete a description of a system as it is possible to obtain, consisting of the wave function and the energy levels. In principle all properties of a system, whatever its size, can be calculated with this information. In practice, however, as we saw throughout Part One, this method is tractable only for systems with a few particles or degrees of freedom. The hydrogen atom, with only two particles, can be treated exactly; the hydrogen molecule presents considerably more difficulty, with even the most extensive calculations giving at best only a nearly exact solution; for larger atoms or molecules one must increasingly resort to approximate methods. The situation is still worse when more than one molecule is involved: Even two interacting argon atoms form a "complex system" that can be treated only with relatively crude approximations. What, then, can one do with a system containing a *mole* of argon atoms?

Consider such a system. A mole of argon is not a large quantity by macroscopic standards: It occupies about 24 liters at room temperature and atmospheric pressure, and weighs a mere 40 g. Yet this paltry amount of gas contains approximately 6×10^{23} argon atoms, each with a nucleus and 18 electrons. Even if we neglect the electrons and consider each atom as a mass point, we have 1.8×10^{24} translational degrees of freedom. There is no conceivable way in which we could solve the Schrödinger equation (or the classical equations of motion) for that many particles. And even if we could somehow obtain such a solution, we would have no way of accumulating and using the enormous quantity of information involved.[1] Thus we must content ourselves with a less than complete description of a macroscopic system; fortunately, such an incomplete description is satisfactory for most purposes.

---

[1] For systems of moderate size, containing up to a few thousand molecules, it is possible to solve the classical equations of motion with high-speed computers. Such calculations do yield some information useful in understanding much larger systems.

How much information can we really use? A complete description of our gas in terms of classical mechanics (which is sufficiently accurate to describe the translational motion) would require knowledge of the positions and velocities of all the particles. One reduced level of analysis is that using only the *distributions* of position and velocity, that is, the relative numbers of particles with positions and velocities falling within given ranges; one can also define such distributions for other microscopic properties of a system, such as energy, dipole moment, and so on. The subject of *statistical mechanics* (also known as *kinetic theory*, from its concern with the motions of molecules) is the proper formulation of such distributions, the calculation of various average quantities, and the connection of these averages with the properties of macroscopic systems. For example, the average kinetic energy of the molecules in a gas is found to be identifiable with the macroscopic quantity we call temperature. Statistical mechanics is so called because its models are those of mechanics (whether classical or quantum); it is in obtaining the averages that additional assumptions must be made, primarily with regard to the relative likelihood of distributions of various kinds.

A still greater reduction in the level of description is found in a theory based *only* on macroscopic properties. The properties referred to are of various kinds: some, such as temperature and pressure, are familiar to all (and thus all the more needful of careful analysis); others, such as internal energy and entropy, must be newly defined. The subject of *thermodynamics* is the proper formulation of the relationships among these quantities, relationships that can be derived from a small set of empirical principles called the *laws of thermodynamics*. Since these laws are statements of observation, they are independent of any assumptions about the structure of matter, which could be either atomic or continuous as far as the principles of thermodynamics are concerned. Thermodynamics was in fact largely developed before the atomic nature of matter was established, and can be expounded axiomatically in complete isolation from any microscopic considerations; but since we do have our microscopic theory, it would be foolish not to use it. As already noted, one can in fact identify the averages of statistical mechanics with the macroscopic variables of thermodynamics, and such juxtapositions can often serve to illuminate our understanding of both theories. We shall thus undertake a parallel development of the principles of both statistical mechanics and thermodynamics.

The macroscopic properties of matter can be classified into two broad categories, equilibrium and nonequilibrium. In an isolated system at equilibrium the macroscopic properties (thermodynamic variables or statistical averages) are time-independent. In a nonequilibrium system one observes spontaneous and irreversible natural processes, such as flows of matter or heat. Part Two of this text is devoted to matter in equilibrium, with the nonequilibrium behavior reserved for Part Three. The first several chapters of Part Two cover the general principles of our two theories and some of their simpler applications; in the remaining chapters we study the specific properties of the various classes of macroscopic systems (gases, liquids, solids, solutions, etc.).

12

The Perfect Gas at Equilibrium and the Concept of Temperature

We begin the study of the properties of bulk matter with the simplest type of matter: the dilute gas. It is simple because its molecules are far apart and hardly interact with one another, so that the macroscopic properties are directly related to those of the individual molecules. As indicated in the introduction to Part Two, we approach the same properties from the viewpoints of statistical mechanics and thermodynamics.

In both cases we deal in this chapter not with a real gas but with an abstraction known as a "perfect gas". In the microscopic theory the perfect gas is a model with *no* interaction between the molecules. We consider the properties of a collection of such molecules, in particular the pressure they exert on the walls of a container, and develop an equation of state connecting the gas's pressure, volume, and energy.

In thermodynamics the perfect gas is an abstraction from the limiting behavior of real gases at low pressure. Here the equation of state is experimental, as embodied in the familiar ideal gas law, and involves the nonmechanical concept of temperature. We shall devote an extensive discussion to the meaning of temperature, which is much less simple than one might imagine.

Finally, comparing the two equations of state, we shall build our first bridge between microscopic and macroscopic properties. The two equations have the same form if temperature is proportional to average molecular kinetic energy. By this correspondence we can say that the "perfect gases" of statistical mechanics and thermodynamics are the same, so that the results of each theory can be used to throw light upon the other.

According to the *kinetic hypothesis*, the individual molecules of which matter is composed are continually in motion, regardless of whether or not the piece of matter as a whole is moving. These individual motions take place in all directions and with a variety of speeds, so that, as far as gross motion is concerned, the contributions of individual molecules tend to cancel. From the macroscopic viewpoint, however, the molecular motion has two major consequences: (1) The kinetic energy of the molecules contributes to the internal energy of the material; and (2) the impacts of the moving molecules on the container wall contribute to the pressure exerted by the material on its surroundings. Although the molecular motion is also presumed to determine all other properties of the medium, for our present purposes it is sufficient to consider only the two cited, the internal energy and the pressure, for the case of a dilute gas.

12.1
The Perfect Gas: Definition and Elementary Model

We must first define what we mean by a dilute gas. We assume for clarity that the molecules of the gas are monatomic (such as He, Ne, or Ar), to avoid having to consider the relative motions of the atoms in a polyatomic molecule. We also assume that the molecules are independent and exert no forces on one another, except during occasional binary collisions—collisions in which only two molecules come sufficiently close together to exert a significant force on each other. In an *elastic collision* the translational momentum and kinetic energy of the pair of molecules are conserved. For monatomic molecules we can safely neglect any internal energy changes, e.g., those due to electronic excitation, and regard any collision as elastic. A gas that satisfies all these conditions will be called a *perfect gas*. The model is a fairly good approximation for most real gases at moderate pressures (as in the atmosphere), and such gases can be called "dilute."

We make the additional assumption that the gas is macroscopically at *equilibrium*. Equilibrium, as we use the term here, means that the properties of the system do not change measurably with time. This is a purely macroscopic concept, since the positions and velocities of individual molecules are continually changing at all times; yet experiment shows that macroscopic properties such as pressure and temperature can and do become time-independent in a real gas. Suppose that we have a quantity of dilute gas in a container of fixed volume, which we can isolate in such a way that no interaction of any kind occurs between the gas and the rest of the universe.[1] No matter what the initial state of the gas—even if all the gas is initially in one corner of the container, with some of it hotter than the rest—we find that in time a state is reached in which the properties of the gas are uniform throughout the container and thereafter do not vary measurably. We shall see in this chapter that such quantities as pressure and temperature can be related to averages over the molecular velocities. Although these averages also change whenever a molecular collision occurs, the number of molecules is so enormous that these changes can be—and at equilibrium must be—negligibly small fluctuations in the total quantities. Thus we shall assume that at equilibrium the velocity distribution—the relative numbers of molecules in different velocity ranges—remains invariant with time.[2]

Additional properties of the equilibrium state can be deduced from the observation that the gas becomes uniform throughout its container. Measurements of such properties as density and temperature then give the same results if performed at any point within the gas. The gas is also found to be at rest relative to the container, with no macroscopic flows or currents: Any flows that may have been initially present die away with time. What are the consequences of these conditions for the equilibrium distribution of molecular velocities and positions? Clearly the density of gas molecules must be the same in every volume element large enough to perform a measurement on, that is, in every volume element containing very many molecules. The kinetic energy per unit volume (which we shall see is proportional to temperature) must also be the same in every such volume element; for practical purposes, this is possible only if the

---

[1] We cannot actually isolate a gas from gravity, and the Earth's gravitational field produces a density gradient with height, but this lack of uniformity is ordinarily too small to measure over laboratory distances.

[2] Because the binary collisions are infrequent, a molecule can move a long distance along a linear trajectory without interruption. For the purpose of calculating the properties of the gas at equilibrium, the collisions can be neglected. However, the existence of collisions permits us to describe qualitatively how a gas arrives at an equilibrium state.

distribution of velocities among the molecules is the same everywhere. If the gas as a whole is at rest, the average velocity of all the molecules must be zero,

$$\langle \mathbf{v} \rangle \equiv \frac{\sum_{i=1}^{N} \mathbf{v}_i}{N} = 0 \qquad \left(\text{and thus} \sum_{i=1}^{N} \mathbf{v}_i = 0 \right). \qquad (12.1)$$

(Remember that velocity is a vector, so that velocities in opposite directions cancel one another; obviously the average molecular speed is not zero.) Furthermore, at any point in the gas there must be equal numbers of molecules traveling in every possible direction, since any imbalance would constitute a net flow of gas; this must hold not just overall but within each range of speeds, if the velocity distribution is to remain the same from point to point. All of the preceding can be summarized as follows: In a perfect gas at equilibrium the distributions of molecular velocities and positions must be *homogeneous* (uniform in space) and *isotropic* (uniform in all directions).

A word here about the types of "densities" we shall use. The ordinary density ρ, or *mass density*, is simply the mass per unit volume. The *number density* n is the number of molecules per unit volume. If there are N molecules, each of mass m, in the volume V, we have

$$n = \frac{N}{V} \quad \text{and} \quad \rho = nm. \qquad (12.2)$$

One can also define a kinetic energy density (kinetic energy per unit volume) and other quantities of the same type.

Having defined our terms, let us consider first a very elementary description of the relationship between the pressure exerted by a gas and the kinetic energy of its molecules. The argument is essentially one of dimensional analysis, but yields a result of the right order of magnitude. Suppose that a perfect gas is contained in a cubical box with edge length l, and define a Cartesian coordinate system with axes normal to the walls. Imagine that the walls of the box have the property of reflecting molecules elastically; that is, when a molecule strikes a face of the box, its motion perpendicular to that face is reversed and its motion parallel to that face is unaltered (see Fig. 12.1). Thus, if a molecule initially has a velocity $\mathbf{v}$ with components v_x, v_y, v_z, its velocity at any later time must always be some combination of the components $\pm v_x, \pm v_y, \pm v_z$. Let S be a face of the box normal to the z axis, with area l^2. Between two successive collisions with S, a molecule must move to the opposite face (which is a distance l away from S) and back. Since the molecule's velocity component perpendicular to S is v_z, the time interval between successive collisions with S is $2l/v_z$, and the frequency of these collisions is $v_z/2l$. At each collision the molecule undergoes a change of momentum in the z direction of amount $2mv_z$, from $+mv_z$ to $-mv_z$; an equal and opposite momentum is transferred to the wall. We conclude that the total change per unit time in the z component of a single molecule's momentum due to collisions with S is

$$\left[\frac{d(mv_z)}{dt} \right]_S = 2mv_z \cdot \frac{v_z}{2l} = \frac{mv_z^2}{l}. \qquad (12.3)$$

We now assume the simplest possible velocity distribution. We imagine that all the molecules have the same velocity components $\pm v_x$, $\pm v_y, \pm v_z$ along the x, y, z axes. Then, if there are N molecules in the volume V $(= l^3)$, the total change in momentum per unit time arising from

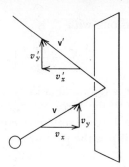

FIGURE 12.1

Elastic reflection of a particle from the wall of a box. The initial velocity is $\mathbf{v}$, with components v_x perpendicular to the wall and v_y parallel to the wall. After reflection the velocity is $\mathbf{v}'$, with components $v_x' = -v_x$ and $v_y' = v_y$.

all collisions with S is

$$\left[\sum_{i=1}^{N} \frac{d(mv_z)_i}{dt}\right]_S = \frac{Nmv_z^2}{l} = \frac{nVmv_z^2}{l} = nml^2v_z^2, \quad (12.4)$$

where n is the number density. Now, by definition, the pressure on a wall is equal to the force exerted per unit area. In turn, by Newton's second law, the force exerted is equal to the rate of change of momentum. Hence, if p is the pressure exerted by the gas, pl^2 (the total force on S) is the rate at which momentum is exchanged by collisions with S, and we have

$$p = nmv_z^2. \quad (12.5)$$

But we know that the pressure in a gas is the same in all directions (Pascal's law), so we must have

$$v_x^2 = v_y^2 = v_z^2$$
$$= \tfrac{1}{3}(v_x^2 + v_y^2 + v_z^2) = \tfrac{1}{3}v^2, \quad (12.6)$$

where v is the common speed of all the molecules. We finally obtain

$$p = \tfrac{1}{3}nmv^2. \quad (12.7)$$

Since $\tfrac{1}{2}mv^2$ is the translational kinetic energy of a single molecule and n is the number per unit volume, we conclude that the pressure exerted by the gas equals two-thirds of the kinetic energy density.

12.2
General Kinetic Theory of the Perfect Gas

The arguments just given are too restrictive. In general, molecules are not reflected elastically from a vessel wall (which is, after all, also composed of molecules); the molecular speeds are not all the same, but are distributed over a wide range; the directions of motion are random; and the vessel need not be cubical. We now proceed to give a more general derivation of the relationships between the distribution of molecular velocities in a gas and the pressure and energy of the gas. We shall see that for many purposes we need not know the detailed form of the velocity distribution, but only some of its properties.

Consider a perfect gas at equilibrium, with N molecules in a volume V. We expect that the molecules will move with different velocities. Let the number of molecules per unit volume with velocities between $\mathbf{v}$ and $\mathbf{v}+d\mathbf{v}$ (that is, with x components between v_x and $v_x + dv_x$, etc.) be denoted as $f(\mathbf{v})\,d\mathbf{v}$. The function $f(\mathbf{v})$ is known as the *velocity distribution function*, and it plays an important role in the kinetic theory of gases. Because the total kinetic energy of all N molecules is finite, there must be some maximum speed of molecular motion that cannot be exceeded. Thus we expect that $f(\mathbf{v}) \to 0$ as $|\mathbf{v}| \to \infty$. Furthermore, since $f(\mathbf{v})\,d\mathbf{v}$ is the number of molecules per unit volume with velocities in a given range $d\mathbf{v}$, if we sum over all possible velocity ranges, we must obtain the *total* number of molecules per unit volume, which is just the number density n. This last condition can be formalized by writing the integral

$$\int_{\mathbf{v}} f(\mathbf{v})\,d\mathbf{v} = \int_{-\infty}^{\infty} \int_{-\infty}^{\infty} \int_{-\infty}^{\infty} f(\mathbf{v})\,dv_x\,dv_y\,dv_z = n. \quad (12.8)$$

Now, we have already specified that the molecular motion at equilibrium is isotropic—that equal numbers of molecules in any given speed interval must travel in every direction. Thus $f(\mathbf{v})$ must be independent of the direction of $\mathbf{v}$ and can be a function only of its magnitude, the speed v. It is therefore useful to introduce a new distribution function: We define the number of molecules per unit volume with speeds between v and $v + dv$ as $f(v)\,dv$. Suppose that we represent the molecular velocities by points in a "velocity space," a Cartesian space whose coordinates are v_x, v_y, v_z; in this space $d\mathbf{v}\ (=dv_x\,dv_y\,dv_z)$ is a volume element, and v is a radial coordinate. The molecules with speeds between v and $v + dv$ then occupy a spherical shell of radius v and thickness dv; by the formulas of Appendix 2B, the "volume" of this shell is $4\pi v^2\,dv$. Thus the relation between the two distribution functions must be

$$f(v) = 4\pi v^2 f(\mathbf{v}). \tag{12.9}$$

(See Section 28.1 for a more systematic proof.) Although we shall not at this time derive the form of $f(v)$, the results of such a derivation (cf. Chapter 19) are illustrated in Fig. 12.2. Note that there is some speed, at the peak of the curve shown, that is characteristic of more molecules than is any other speed; this most probable speed is called the *mode* of the distribution. Recall also that at equilibrium the gas is homogeneous, each volume element being the same as every other volume element, so $f(v)$ must be independent of position within the gas.

We can now formally calculate the internal energy of the gas arising from the molecular motion. A molecule with mass m and speed v has kinetic energy $\frac{1}{2}mv^2$. To obtain the total kinetic energy of the gas, we multiply the kinetic energy of a single molecule with speed v by the number of molecules per unit volume with speeds between v and $v + dv$, then sum contributions from all possible speeds and from all possible volume elements in the total volume. For the perfect monatomic gas (neglecting the internal structure of the atoms) the kinetic energy is the same as the total internal energy, which we shall call U. This internal energy is thus

$$U = \int_V \int_{v=0}^{\infty} \frac{mv^2}{2} f(v)\,dv\,dV. \tag{12.10}$$

But since $f(v)$ must be independent of position, Eq. 12.10 can be immediately integrated over volume to give

$$U = \frac{mV}{2} \int_0^{\infty} v^2 f(v)\,dv \tag{12.11}$$

for the internal energy of the perfect monatomic gas. Other types of gases (which we consider in Chapter 21) may contain polyatomic molecules, or may have significant interactions between molecules. In these cases there are also contributions to the internal energy U from molecular vibration and rotation, or from the intermolecular potential energy.

We wish now to obtain the relationship between internal energy and pressure, corresponding to Eq. 12.7 in the simple model. As we mentioned before, the pressure on the container wall is the force exerted by the gas per unit area of wall or, equivalently, the rate at which momentum is transferred to a unit area of wall. But we need not restrict our definition to the walls: By the assumptions of homogeneity and isotropy, the pressure at equilibrium must be the same anywhere within the gas (this again is equivalent to Pascal's law). Picture a plane surface S fixed at an arbitrary position within the gas; we define a coordinate system with the z axis normal to S (Fig. 12.3). We assume that gas molecules collide elastically with S, in the same way as shown in Fig. 12.1. Thus whenever a molecule whose z component of momentum is mv_z strikes S, an amount of momentum $2mv_z$ is transferred to S. The pressure on S is the total momentum so transmitted per unit time per unit area of S, and at equilibrium must be the same on both sides of S.

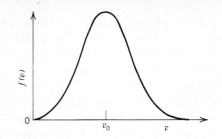

FIGURE 12.2
The distribution of molecular speeds, $f(v)$. The speed v_0, which is the most probable, is the *mode* of the distribution.

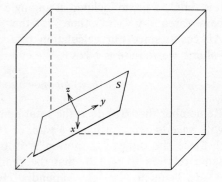

FIGURE 12.3
The plane surface S is fixed at an arbitrary position in a container of gas. The xyz coordinate system is defined so that the z axis is normal to S.

Now suppose that the real surface S is replaced by an imaginary plane in the same position. The molecules that would rebound from a real surface are just those that in this model cross S. But a molecule that would transfer z-momentum $2mv_z$ to the real S carries only mv_z (half as much) across the imaginary S. We *define* the pressure to be the same in both cases, thus obtaining a definition independent of the presence of a wall: The pressure normal to an imaginary plane S is *twice* the momentum transported across S per unit time per unit area of S, by molecules coming from one side of S (since the two sides of a real surface are different areas).

How do we evaluate this rate of momentum transport? The molecules we shall consider are those crossing S in the positive z direction. Let θ be the angle between a given molecule's trajectory and the z axis (Fig. 12.4). We define $F(\theta, v)\, d\theta\, dv$ as the number of molecules with speeds between v and $v + dv$ that cross S per unit area per unit time at angles between θ and $\theta + d\theta$. Each such molecule has a momentum of magnitude mv, with z component $mv \cos \theta$. Only the z component of momentum contributes to the pressure. Because of the gas's isotropy, the other components add to give a null result; that is, for a given range of θ there are equal numbers of molecules with x components $+mv_x$ and $-mv_x$, or with y components $+mv_y$ and $-mv_y$, so that the net transport of these components across S is zero. Thus the contribution to the pressure from the molecules in the ranges dv, $d\theta$ is the flux density (number per unit time per unit area) of such molecules multiplied by twice the z component of momentum transported per molecule, or

$$dp = 2mv \cos \theta\, F(\theta, v)\, d\theta\, dv. \qquad (12.12)$$

Integration over all possible values of θ and v—that is, over all molecules crossing S in the positive z direction—gives a total pressure of

$$p = \int_{v=0}^{\infty} \int_{\theta=0}^{\pi/2} 2mv \cos \theta\, F(\theta, v)\, d\theta\, dv \qquad (12.13)$$

(note the limits of integration for θ).

To use Eq. 12.13 we must evaluate $F(\theta, v)$. The expression $F(\theta, v)\, d\theta\, dv$ may be thought of as the product of (1) the number of molecules per unit volume with speeds between v and $v + dv$ moving at angles between θ and $\theta + d\theta$, which we call $f(\theta, v)\, d\theta\, dv$, and (2) the volume occupied by all such molecules capable of crossing unit area of S in unit time. The latter volume is easily evaluated: As shown in Fig. 12.4, it is simply a prism inclined at an angle θ from the normal to its base. For base area A and time t, the volume of such a prism is $Avt \cos \theta$; since our calculation refers to unit area and unit time, the volume we require is $v \cos \theta$. We can thus write

$$F(\theta, v)\, d\theta\, dv = v \cos \theta\, f(\theta, v)\, d\theta\, dv. \qquad (12.14)$$

Remember that $f(v)\, dv$ is the total number of molecules per unit volume with speeds between v and $v + dv$, or $f(\theta, v)$ integrated over all values of θ. Since the molecular motion is isotropic, $f(\theta, v)\, d\theta\, dv$ must bear the same relation to $f(v)\, dv$ that the range of solid angles between θ and $\theta + d\theta$ bears to 4π (cf. Appendix 2B). By Eq. 2B.7, the solid angle between θ and $\theta + d\theta$ is $2\pi \sin \theta\, d\theta$, so we have

$$\frac{f(\theta, v)\, d\theta\, dv}{f(v)\, dv} = \frac{2\pi \sin \theta\, d\theta}{4\pi} = \frac{1}{2} \sin \theta\, d\theta. \qquad (12.15)$$

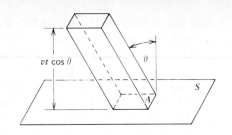

FIGURE 12.4

All the molecules within the inclined prism that have velocity **v** in the direction θ will strike the area A within time t. The volume of the prism is $Avt \cos \theta$.

Substituting Eqs. 12.14 and 12.15 into Eq. 12.13, we find the pressure to be

$$p = \int_{v=0}^{\infty} \int_{\theta=0}^{\pi/2} (2mv \cos \theta)(v \cos \theta)(\tfrac{1}{2} \sin \theta \, d\theta) f(v) \, dv$$

$$= m \int_{0}^{\infty} v^2 f(v) \, dv \int_{0}^{\pi/2} \cos^2 \theta \sin \theta \, d\theta \qquad (12.16)$$

$$= \frac{m}{3} \int_{0}^{\infty} v^2 f(v) \, dv.$$

But, comparing this result with Eq. 12.11, we obtain for a perfect monatomic gas

$$p = \frac{2}{3} \frac{U}{V}, \qquad (12.17)$$

where U/V is the internal energy density.

Thus the hypotheses we have introduced lead to the conclusion that the pressure in a perfect monatomic gas is just equal to two-thirds of the internal energy density. This result is the same as that obtained in Eq. 12.7, from analysis of our original, very crude model. Note that we have been able to obtain this interesting relationship without specifically calculating $f(v)$. Subsequent calculations, especially in Part Three of this book, will require knowledge of the form of $f(v)$, which we shall derive in Chapter 19. Later in the present chapter (Section 12.6) we shall examine critically the analysis just given, but first we begin our study of the macroscopic theory.

12.3 General Principles of Thermodynamics

As we explained in the introduction to Part Two, thermodynamics deals with relationships among the macroscopic properties of matter—more specifically, among a particular class of macroscopic variables that we must define. These relationships can be derived from the *laws of thermodynamics*, a small set of abstract principles based on observations of the behavior of bulk matter. The laws of thermodynamics are analogous to the postulates of quantum mechanics (Chapter 3). Both are sets of general principles inferred (guessed, if you like) from a wide variety of observations, and ultimately justified by the agreement of the conclusions drawn from them with experiment; and both are extraordinarily fruitful in the range of such conclusions that one can draw. Thermodynamics in fact provides an accurate and useful description of the equilibrium properties of any piece of matter containing a very large number of molecules.

The methods ordinarily used to formulate the principles of thermodynamics make no assumptions about the microscopic structure of matter. An internally consistent and logically complete theory can be constructed without reference to the existence of atoms and molecules. Naturally, the specific properties of real matter must appear in the theory, in the form of material constants such as heat capacities or compressibilities, but thermodynamics concerns itself only with the relations among these quantities, not with why they have the values they do; to answer the latter question one must resort to microscopic theories of the properties and interactions of molecules. By considering the same quantities from the viewpoints of both microscopic and macroscopic theories, one can gain significant insights into the macroscopic consequences of molecular behavior—and the microscopic implications of macroscopic measurements.

As a concrete example of how the theories can be interrelated on various levels, the pressure of a real (imperfect) gas can be expressed empirically as a power series in the density n; the coefficients of the powers of n are determined from experimental data. Thermodynamics gives the relationships between pressure and other macroscopic quantities (internal energy, entropy, etc.), and thus enables one to calculate these quantities in terms of the coefficients in the density expansion of the pressure. All this is on the macroscopic level; what of the microscopic approach? Here we begin with the concept of a potential energy of interaction between molecules, the $V(R)$ described in Chapter 10. Using the methods of statistical mechanics, one can average $V(R)$ over all the molecular pairs in a gas, over all the triples of molecules, and so on, obtaining average quantities that can be identified with the coefficients in the density expansion of the pressure. Combining the two theories gives one a complete chain of connections between the molecular interactions and the thermodynamic properties. In principle one could obtain $V(R)$, and by extension the macroscopic properties, by solving the Schrödinger equation for the molecular interaction. In practice, as we know, this is extraordinarily difficult, and one ordinarily works in the opposite direction, deducing $V(R)$ from macroscopic measurements of the pressure or other properties, or more directly from molecular scattering.

But we have a long way to go before we can draw such connections as these (which will be developed in Chapter 21). Let us now get down to actually developing the principles of thermodynamics, on what must initially be a much more abstract level. We begin by examining some basic notions, starting with definitions of what it is that thermodynamics studies.

A *system* is that part of the physical world which is under consideration. It may have fixed or movable boundaries of arbitrary shape and may contain matter, radiation, or both. All the rest of the universe is defined to be the *surroundings*. A system is said to be *open* if matter can be exchanged between the system and the surroundings. A *closed system*, which does not exchange matter, may still be able to exchange energy with its surroundings. An *isolated system* has no interactions of any kind with its surroundings.

To describe the properties of a system, one needs to be able to measure certain of its attributes. In thermodynamics the system is described by a set of macroscopic variables or "coordinates," the determination of which requires only measurements over regions containing very many molecules, carried out over periods long enough for very many molecular interactions, and involving energies very large compared with individual quanta. The thermodynamic properties of the system are assumed to be entirely described by the set of macroscopic coordinates. To anticipate a bit, the coordinates to which we refer are most often the temperature, pressure, and volume. When a macroscopic coordinate is fixed in value by the boundary conditions defining the system, there is said to be a *constraint* on the system. There can be other forms of constraints, such as one allowing no deformation of the system. Many of these, and the roles they play in thermodynamics, will be discussed later, in Chapter 19.

The classical theory of thermodynamics deals with the properties of systems at *equilibrium*. This has essentially the meaning already outlined in our treatment of the perfect gas: A thermodynamic system is in equilibrium when none of its thermodynamic properties change with time at a measurable rate. (This should not be taken to include a *steady state*, in which matter or energy flows steadily through the system but the properties measured at any given point are time-independent. An equilibrium state is a state of rest.) The equilibrium state of a system depends on the constraints placed on it, which thus must be specified. In general, there is only one true equilibrium state for a given set of constraints. However, many systems have states that show no change with time, and thus can be regarded for practical purposes as equilibrium states, even though they are not the true equilibrium state. An example is a room-temperature mixture of H_2 and O_2, for which the reaction $H_2 + \frac{1}{2}O_2 \rightarrow H_2O$ is so slow (in the absence of a catalyst) that the mixture can be considered nonreactive for most purposes. Such a condition is called a *metastable equilibrium*. Thermodynamics can also be used to describe some of the properties of systems not at equilibrium; this aspect of the theory is still under active development.

The thermodynamic state of a system is assumed to be uniquely determined when a complete set of independent macroscopic coordinates is given. But what constitutes a "complete" set of thermodynamic coordinates? They are just those variables that must be independently specified for one to reproduce a given thermodynamic state, *as determined by experiment*. It should be noted that the number of thermodynamic coordinates necessary to specify the state of a system is very small (frequently only two), whereas the total number of microscopic coordinates defining the same system is of the order of the number of atoms present.

The number of variables required for a complete set differs from system to system. Let us concentrate on the properties of a *fluid*, a form of matter that assumes the shape of its container (i.e., a gas or liquid). In general, the shape is found to be of no importance in determining the bulk properties of the fluid. Experiment shows that, for a given mass of pure fluid (fluid containing only a single chemical species) in the absence of external fields, the specification of the pressure and volume uniquely defines the thermodynamic state; for a fluid mixture one must also specify the composition.[3] For example, given the pressure and volume of a certain mass of oxygen, one finds that all equal masses of oxygen whose pressures and volumes are adjusted to be the same are in identical thermodynamic states. That is, for the given mass of oxygen there is a unique relationship among pressure, volume, and any other thermodynamic variable; this relationship can be expressed by an empirical *equation of state*,

$$f(p, V, X) = 0, \qquad (12.18)$$

where X is any third variable. In the next section we shall consider the properties of one such variable, the quantity we call temperature.

We wish to describe the properties of matter, beginning with the dilute gas, from a thermodynamic point of view. One concept important in this description is the thermodynamic coordinate known as *temperature*, the nature of which we must now clarify. Everyone is familiar with the qualitative notion that temperature is associated with the "hotness" of a

12.4
Temperature and the Zeroth Law of Thermodynamics

---

[3] Of course, should one be concerned with, say, the electric or magnetic properties of a system, other variables must also be used. There are also certain anomalous cases, the best known being that of liquid water just above its freezing point: The density increases from 0 to 4°C, then decreases again, so that, say, $p = 1$ atm and $v = 1.000060$ cm^3/g can correspond to either 2°C or 6°C. Thus in this region p, v do not constitute a complete set of variables (but p, T do).

body. But we can easily fool ourselves in this regard: For example, try placing both your hands in a bowl of tap water after one has been in ice water and the other in very hot water. This experiment shows that we cannot trust our simple intuitive or physiological reactions. A quantitative definition suitable for scientific work must be made with more attention to the logical foundations of the concept, and this we shall now proceed to do.

Consider again a mass of pure fluid, the thermodynamic state of which depends on the pressure and volume. We can define a temperature θ by writing an equation of state in the form

$$f(p, \rho, \theta) = 0, \tag{12.19}$$

where we have replaced the volume by the density ρ to make the equation independent of the particular mass of fluid.[4] We emphasize that Eq. 12.19 is to be interpreted as a *definition* of θ: We choose some function of pressure and density and call it the temperature. How do we decide what function to choose? Well, certainly θ should be well behaved (finite, continuous, and single-valued); it ought to bear some relationship to the intuitive notion of temperature (increasing θ should make a system "hotter"); and it should be mathematically convenient to use. But there are still infinitely many functions that satisfy these conditions. For example, we shall see that for an ideal gas a convenient temperature function is $\theta = p/\rho$; but $(p/\rho)^2$, $\log(p/\rho)$, and so on (or any of these times a constant multiplier) would do equally well. We can express this generality by "solving" Eq. 12.19 for temperature in the form

$$g(\theta) = \varphi(p, \rho), \tag{12.20}$$

where $g(\theta)$ is any of the alternative possible temperature functions and $\varphi(p, \rho)$ is the function of pressure and density that defines it. Which of the functions we choose as *the* temperature is strictly a matter of convention, as we shall see in the next section. But first we must introduce some concepts that apply to all temperature scales.

We begin by defining the properties of two idealized walls. Consider two systems 1 and 2, initially isolated and separately at equilibrium. Let the two systems initially contain the same pure fluid at different temperatures—that is, have values of the pressure and density such that, for a given choice of temperature function (Eq. 12.20), we have

$$\varphi(p_1, \rho_1) \neq \varphi(p_2, \rho_2). \tag{12.21}$$

Suppose now that the isolation of the two systems from each other is reduced to their separation by a rigid barrier B, both remaining isolated from the rest of the universe (Fig. 12.5). If the thermodynamic states of the two systems subsequently remain unchanged from the initial states, then the barrier B is called an *adiabatic wall*. In less precise language, an adiabatic wall is a pure insulator across which heat cannot flow ("adiabatic" means uncrossable), so that the temperatures of the two systems remain different. We are all familiar with real approximations to adiabatic walls—for example, the silver-coated, evacuated, double glass walls of a Thermos bottle, or a thick, rigid sheet of asbestos. Clearly, the concept of a perfect adiabatic wall is an abstraction, but one that is linked to the observed behavior of a number of materials. Obviously, an isolated system must be surrounded by adiabatic walls.

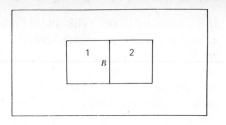

FIGURE 12.5
The two systems labeled 1 and 2 are separated by a rigid wall B. Both systems are isolated from the remainder of the universe.

---

[4] This can be done only for those properties, such as temperature, that are in fact independent of the total mass. We shall discuss this distinction in Section 13.4, on intensive and extensive variables.

There is another—and more likely—outcome of the experiment. Suppose again that the two systems, initially isolated and at different temperatures, are separated from one another only by a rigid barrier B. The usual result is that the thermodynamic states of the two systems change with time, eventually approaching a limit in which they are time-independent. Such a barrier is called a *diathermal wall*, and in the final time-independent state the two systems are said to be in *thermal equilibrium* with each other. In ordinary language, a diathermal wall (*dia*, "through," + *therme*, "heat") allows heat to pass through until the two systems are at the same temperature. All real material walls are diathermal, although in some cases the rate at which heat can cross is very small, so that the approach to thermal equilibrium can be very slow.

The definitions of adiabatic and diathermal walls are not limited to fluid systems, nor to systems of the same chemical composition. In general, the thermodynamic states of the systems may be defined by more variables than p and ρ, but the principle is the same: If the states of two otherwise isolated systems change when they are separated only by a rigid barrier B, then B is diathermal; if both states remain unchanged, then B is adiabatic (except for the trivial case that the systems are in thermal equilibrium to begin with). The requirement that the barrier be rigid is necessary to prevent the performance of work, which we shall discuss in the next chapter.

We said that the original isolated systems were separately in equilibrium, the equilibrium defined by the constraint of isolation. When the constraint of isolation is replaced by the constraint of contact through a diathermal wall, the new constraint defines a new equilibrium state. The overall system is initially out of equilibrium with respect to the new constraint, but eventually approaches the equilibrium state. The way in which this state is reached is immaterial at present, only the final equilibrium itself being of concern to us. Experiment shows that, once two systems have reached thermal equilibrium with each other, one cannot change the state of system 1 without simultaneously changing the state of system 2. Thus the overall system at thermal equilibrium is not described by separate equations of state like Eq. 12.18, but by some single relationship involving the independent coordinates of both systems. For simplicity we shall restrict ourselves again to fluid systems, for which this single relationship has the form

$$F_{12}(p_1, \rho_1, p_2, \rho_2) = 0. \tag{12.22}$$

We need not know the precise form of the function F_{12}, which will depend on the nature of the two fluids and their individual equations of state.

At this point it is appropriate to introduce the first postulate of thermodynamic theory, which is known as the *zeroth law of thermodynamics*. It is called "zeroth" because, although it was formulated after the first, second, and third laws, it is logically prior to them in a rigorous development of the subject. The zeroth law is stated in the form:

Two systems, each separately in thermal equilibrium with a third system, are in thermal equilibrium with each other.

Like the other laws of thermodynamics, this statement is an abstraction from experiment. It clearly embodies everyday experience: For example, if each of two systems is separately found to "have the same temperature" with respect to, say, a mercury-in-glass thermometer, we do not expect to observe any change when the two systems are brought into contact with each other. But why do we even need to state what seems so obvious a fact? Thus far we have only examined how to define a temperature for a particular system, by some such means as Eq. 12.19. But what we wish to obtain is a universal definition of temperature applicable to *all* systems. For such a universal temperature to exist, there must be some property that all systems "at the same temperature" (i.e., in thermal equilibrium with one another) have in common. Since we cannot actually investigate all systems, for logical completeness we must *assume* that this is true; the zeroth law has been chosen as the simplest way of formulating this assumption. We shall now see just how the zeroth law leads to a unique definition of temperature.

Consider three fluids, 1, 2, and 3, for each of which p and ρ are the only independent variables. When fluids 1 and 2 are in mutual thermal equilibrium, their thermodynamic states are related by an equation of the form 12.22, whereas for equilibrium between fluids 1 and 3 we have the equation

$$F_{13}(p_1, \rho_1, p_3, \rho_3) = 0. \tag{12.23}$$

Equations 12.22 and 12.23 both contain the variables p_1 and ρ_1. In principle, each of these equations can be solved for, say, the pressure of fluid 1. As a result of solving these two equations, p_1 can be expressed as some function w_{12} of ρ_1, p_2, ρ_2 or as some function w_{13} of ρ_1, p_3, ρ_3:

$$p_1 = w_{12}(\rho_1, p_2, \rho_2) \quad \text{or} \quad p_1 = w_{13}(\rho_1, p_3, \rho_3). \tag{12.24}$$

But since p_1 is the same physical quantity in both of these equations, we can immediately write

$$w_{12}(\rho_1, p_2, \rho_2) = w_{13}(\rho_1, p_3, \rho_3). \tag{12.25}$$

Equation 12.25 is an alternative representation of the conditions placed on the states of systems 1, 2, and 3 by the requirements that 1 and 2 be in mutual equilibrium and that 1 and 3 be in mutual equilibrium.

We now introduce the new element required by the zeroth law. According to this law, if fluids 2 and 3 are separately in thermal equilibrium with fluid 1, they are also in thermal equilibrium with each other. Thus the relationship between the states of fluids 2 and 3 must be completely described by an equation of the form

$$F_{23}(p_2, \rho_2, p_3, \rho_3) = 0. \tag{12.26}$$

But Eq. 12.25 contains the density ρ_1, whereas Eq. 12.26 does not. If the two equations are to hold simultaneously, as the zeroth law requires, the variable ρ_1 must drop out of Eq. 12.25. That is, the functions w_{12} and w_{13} must be of such a form that we can write a new equation

$$w_2(p_2, \rho_2) = w_3(p_3, \rho_3), \tag{12.27}$$

in which each of the functions w_2 and w_3 depends on the state of a single system only. By an obvious extension of the argument we obtain the more general equation

$$w_1(p_1, \rho_1) = w_2(p_2, \rho_2) = w_3(p_3, \rho_3), \tag{12.28}$$

which can be further extended to any number of fluids in mutual thermal equilibrium. Note that the $w_i(p_i, \rho_i)$ may be quite different functions of pressure and density for different fluids; it is only their numerical values that must be the same at equilibrium.

Example: These arguments are all very abstract, and can perhaps be clarified by a concrete example. Consider the ideal gas, a hypothetical system that obeys the equation of state $pV = nRT$; we shall explain this equation in the next section, but for now take it as a given. If we put ideal gases 1 and 2 in thermal equilibrium and determined the relationship between their states, the form of Eq. 12.22 that would obtain is

$$F_{12}(p_1, \rho_1, p_2, \rho_2) \equiv \frac{p_1 M_1}{\rho_1} - \frac{p_2 M_2}{\rho_2} = 0, \qquad (12.22')$$

where M is the molecular weight. Similarly, we have for ideal gases 1 and 3,

$$F_{13}(p_1, \rho_1, p_3, \rho_3) \equiv \frac{p_1 M_1}{\rho_1} - \frac{p_3 M_3}{\rho_3} = 0. \qquad (12.23')$$

Solving both equations for p_1, we obtain

$$p_1 = w_{12}(\rho_1, p_2, \rho_2) \equiv \frac{\rho_1 p_2 M_2}{\rho_2 M_1}, \qquad p_1 = w_{13}(\rho_1, p_3, \rho_3) \equiv \frac{\rho_1 p_3 M_3}{\rho_3 M_1}, \qquad (12.24')$$

and thus

$$\frac{\rho_1 p_2 M_2}{\rho_2 M_1} = \frac{\rho_1 p_3 M_3}{\rho_3 M_1}. \qquad (12.25')$$

The zeroth law tells us that we must also have

$$F_{23}(p_2, \rho_2, p_3, \rho_3) \equiv \frac{p_2 M_2}{\rho_2} - \frac{p_3 M_3}{\rho_3} = 0, \qquad (12.26')$$

which does not contain ρ_1. By the reasoning in the text, this means that ρ_1 must drop out of Eq. 12.25', as it obviously does, giving

$$w_2(p_2, \rho_2) \equiv \frac{p_2 M_2}{\rho_2} = w_3(p_3, \rho_3) \equiv \frac{p_3 M_3}{\rho_3}. \qquad (12.27')$$

By extension the function pM/ρ has the same value for all ideal gases in thermal equilibrium; in the next section we shall see the significance of this fact in defining a temperature scale.

We have shown that, given the zeroth law, there must exist a set of functions w_i, each depending only on the state of the ith fluid, but all having the same value for any number of fluids in thermal equilibrium. For this to be true, the w_i must all depend on some single property that all fluids in thermal equilibrium have in common, regardless of their natures. This single property is defined to be their common temperature. Furthermore, the argument is not restricted to fluids: If an arbitrary system is described by the set of independent variables X_i, Y_i, $Z_i, \ldots$, then the F_{12} of Eq. 12.22 becomes $F_{12}(X_1, Y_1, Z_1, \ldots, X_2, Y_2, Z_2, \ldots)$ and the rest of the derivation proceeds along exactly the same lines. Thus it is possible to define a common temperature for all systems of whatever kind in thermal equilibrium.

The functions $w_i(p_i, \rho_i)$ are clearly equivalent to the $\varphi(p, \rho)$ of Eq. 12.20, which says that for any fluid one can define a temperature function $g(\theta)$ on the basis of that particular fluid's equation of state. What we are now saying is that each such $g(\theta)$ is a function of the property that all systems in thermal equilibrium have in common. This tells us how to obtain a universal temperature scale: We choose a particular system as a standard, select one of that system's possible temperature functions $g(\theta)$, and define the numerical value of this $g(\theta)$ as the temperature θ; the same numerical value of θ is assigned to any other system in thermal equilibrium with our standard system. Such a standard system is an example of what we call a *thermometer*, and the function used to define a temperature is called a *thermometric property*; in the next section we shall see how such a system and property are chosen.

The arguments of the preceding section seem a long way from our intuitive understanding of "temperature." What is the relationship between these abstract considerations and a usable temperature scale?

Any system property that varies with temperature in a well-behaved way can be used to define a temperature scale. The choice among the various possible thermometric properties is arbitrary and based essentially on convenience. One important requirement is that significant changes in the thermometric property be accurately measurable in a small system (a thermometer should be small enough to come into thermal equilibrium with a system without changing the system's temperature significantly). The best-known thermometric property is the volume of a liquid, most commonly mercury; the volume changes are magnified by being measured in a narrow tube attached to a larger reservoir. The first thermometers invented were of this *liquid-expansion* type, with temperature scales defined as linear functions of the liquid column's length. Among the other thermometric properties in common use are the *electrical resistance* of a fine metal wire (usually platinum); the electromotive force generated by a *thermocouple* (the junction between two wires of different metals); and the intensity of the radiation emitted by a hot body, assumed to obey the black-body radiation law (Section 2.6), as measured with a *pyrometer*. But none of these properties is actually used as the basis of our temperature scale; rather, instruments of all these types are calibrated with respect to the scale we shall now describe.

A particularly elegant and useful temperature is that based on the low-pressure properties of a very dilute gas, to which we now finally return. Since the easily measured properties of a gas are its volume (or density) and pressure, we seek a convenient way of relating temperature to these properties, as in Eq. 12.20. The most straightforward procedure is to define the temperature as proportional to one of these properties, with the other held constant. The two alternatives lead to constant-volume and constant-pressure gas thermometers, and instruments of both types in fact exist; but as we shall see, both temperatures converge to the same low-pressure limit.

Suppose that we measure the pressure of a fixed volume of gas in thermal contact with an equilibrium mixture of ice, liquid water, and water vapor (i.e., water at its *triple point*) and call this pressure p_3.[5] We use the pressure p of the *same* volume of gas under other conditions as a thermometric property, defining a constant-volume temperature scale by the equation

$$\theta(p) = \frac{p}{p_3} \theta_3, \quad (V = \text{constant}), \qquad (12.29)$$

where θ_3 is an arbitrary constant, the temperature of the triple point of water. By convention (for reasons that will become apparent later) we set the value of θ_3 equal to 273.16 K, the "K" standing for the temperature

12.5
Empirical Temperature:
The Perfect Gas
Temperature Scale

[5] We shall show, in Chapter 24, that in a one-component system, solid, liquid, and vapor are simultaneously in equilibrium with each other at only one temperature and pressure. This unique temperature and pressure is known as the triple point of the substance.

unit, the *kelvin* (formerly *degree Kelvin*, °K). Alternatively, we can measure the volume V_3 of a given mass of gas at the triple-point temperature and gas pressure p, then use the volume at the same pressure to define the constant-pressure temperature scale

$$\theta(V) = \frac{V}{V_3} \theta_3, \qquad (p = \text{constant}). \tag{12.30}$$

The empirical temperature scales defined in Eqs. 12.29 and 12.30 are functions of both the mass of gas in the thermometer and the nature of the gas. Experimentally, however, as one lowers the quantity of gas (and thus the values of both p_3 and p) in a constant-volume thermometer or the constant pressure p in a constant-pressure thermometer, both temperatures are found to approach the same limiting value,

$$\lim_{p,p_3 \to 0} \theta(p) = \lim_{p \to 0} \theta(V) = \theta^*, \tag{12.31}$$

which is, furthermore, the same for all gases. This limiting process is illustrated in Fig. 12.6. The limit θ^* is called the *perfect gas temperature*.

Although the elaborate limiting procedure described here is not itself convenient for ordinary temperature measurements, it can be used as a primary standard to calibrate practical thermometers of the various types we have mentioned. The *International Practical Temperature Scale of 1968* (IPTS-68) is a set of reproducible fixed points[6] and interpolation formulas that have been reliably measured in terms of the perfect gas temperature scale. The IPTS-68 is defined in terms of a platinum-resistance thermometer from 13.81 K to 903.89 K; of a platinum/platinum–rhodium thermocouple from 903.89 K to 1337.58 K; and of Planck's black-body radiation law above 1337.58 K. These can in turn be used to calibrate thermometers for ordinary measurements, so that a perfect gas temperature can be determined, even if indirectly, for any system of interest.

Equation 12.31 is a direct consequence of the experimental observations embodied in Boyle's law and Charles's law, which we can now formulate as follows: If the temperature θ^* of a mass of gas is held fixed and its pressure p is lowered, then the product pV/n, where V/n is the volume per mole (cf. Section 1.6), approaches for all gases the same limiting value,

$$\lim_{p \to 0} pV/n = \beta(\theta^*) \tag{12.32}$$

where the common limit $\beta(\theta^*)$ is a function of the temperature θ^* only (Boyle's law). Furthermore, the limits obtained at two different tempera-

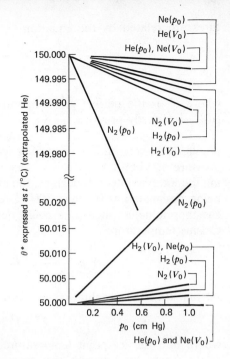

FIGURE 12.6

As the pressure is lowered in a constant-volume thermometer [$\theta(p)$], or a constant-pressure thermometer [$\theta(V)$], both empirical temperatures approach a common value, namely, the perfect gas temperature θ^*. In the figure p_0 and V_0 are, respectively, the pressure and volume of the gas in the thermometer at the ice point. From F. H. Crawford, *Heat, Thermodynamics and Statistical Physics*, Harcourt Brace Jovanovich, New York, 1963.

[6] The fixed points of the IPTS-68 are as follows ("boiling point" and "freezing point" imply a pressure of 1 atm):

13.81 K (−259.34°C) triple point of equilibrium hydrogen (see Chapter 21);

17.042 K (−256.108°C) liquid–vapor equilibrium point of equilibrium hydrogen at 33330.6 N/m² pressure;

20.28 K (−252.87°C) boiling point of equilibrium hydrogen;

27.102 K (−246.048°C) boiling point of neon;

54.361 K (−218.789°C) triple point of oxygen;

90.188 K (−182.962°C) boiling point of oxygen;

273.16 K (0.01°C) triple point of water;

373.15 K (100.00°C) boiling point of water;

692.73 K (419.58°C) freezing point of zinc;

1235.08 K (961.93°C) freezing point of silver;

1337.58 K (1064.43°C) freezing point of gold.

tures are related by the equation

$$\frac{\beta(\theta_1{}^*)}{\beta(\theta_2{}^*)} = \frac{\theta_1{}^*}{\theta_2{}^*},\tag{12.33}$$

where θ^* is the perfect gas temperature defined in Eq. 12.31; that is, $\beta(\theta^*)$ is directly proportional to θ^* (Charles's law).

The temperature scale was formerly defined in terms of two fixed points, the temperatures of melting ice and boiling water (each under a pressure of 1 atm). The *Celsius* (*centigrade*) scale was obtained by dividing the interval into 100 degrees, the melting-ice and boiling-water points being designated as 0°C and 100°C, respectively. For a constant-volume gas thermometer, then, one could define (in the perfect gas limit) the Celsius temperature

$$t^* = \lim_{p_0 \to 0} \frac{100(p - p_0)}{p_{100} - p_0}, \quad (V = \text{constant}).\tag{12.34}$$

At that time the absolute temperature θ^* was defined by the equation

$$\theta^* = t^* + \theta_0{}^*\tag{12.35}$$

where $\theta_0{}^*$, the ice-point temperature, had to be chosen such that Eq. 12.33 would be satisfied; the experimental value was about 273.15°, corresponding to 273.16° for the triple point $\theta_3{}^*$. To eliminate the experimental uncertainty, an international agreement in 1954 made $\theta_3{}^* = 273.16$ K by *definition*, as indicated above. The melting point of ice, $\theta_0{}^*$, is then still 273.15 K, but the boiling point of water changes from 373.15 to 373.146 K. Since the Celsius degree is now defined as equal to the kelvin, Eq. 12.35 still holds as a definition of the Celsius temperature t^*, but Eq. 12.34 is no longer exactly valid.[7]

We shall show in Chapter 16 that it is possible to establish a thermodynamic temperature scale independent of the properties of any real or hypothetical thermometric substance. However, it will also be shown that this thermodynamic temperature T is in fact equal to the perfect gas temperature θ^*. To simplify our notation, we shall anticipate this result by regularly using T to designate the perfect gas temperature.

We may as well also say something here about the units of pressure. The SI unit (adopted in 1971) is the *pascal* (Pa), or newton per square meter; closely related are the dyne per square centimeter ($=0.1$ Pa) and the *bar* ($\equiv 10^5$ Pa). However, most scientific measurements are still expressed in terms of the *atmosphere* (atm), which has the defined value

$$1 \text{ atm} \equiv 1.01325 \times 10^5 \text{ Pa},$$

chosen to fall within the range of actual atmospheric pressures near sea level. We shall see in subsequent chapters that thermodynamic quantities are commonly defined in relation to a standard state at 1 atm. For lower pressures one commonly uses the *torr*, defined as $\frac{1}{760}$ atm; this is for practical purposes the same as the conventional "millimeter of mercury" (mm Hg).[8]

---

[7] The 1967 General Conference on Weights and Measures adopted the name "kelvin," symbol "K," for the unit of thermodynamic temperature, with "degree Celsius", symbol "°C", as an alternative unit for a temperature *interval*. Note the omission of the word "degree" and the degree symbol from the kelvin scale.

[8] To be precise, the "millimeter of mercury" is defined as the pressure exerted by a 1-mm column of a fluid with density 13.5951 g/cm^3 (equal to that of mercury at 0°C) at a place where the acceleration of gravity is 9.80665 m/s^2. For the record we note that the torr is named after Torricelli, who was a prominent Renaissance scientist.

We can now combine Eqs. 12.32 and 12.33 to obtain the familiar perfect gas equation of state,[9]

$$pV = nRT, \qquad (12.36)$$

where n is the number of moles of gas in the volume V. The proportionality constant R is known as the (*universal*) *gas constant*, and can be evaluated by measuring the limit

$$R = \lim_{p \to 0} \frac{pV}{nT}, \qquad (12.37)$$

which is the same for all real gases. It has the value

$$R = 0.0820568 \text{ liter atm/mol K}$$
$$\doteq 8.31441 \text{ J/mol K}.$$

Division by Avogadro's number (Section 1.6) gives *Boltzmann's constant*,

$$k_B \equiv \frac{R}{N_A} = 1.38066 \times 10^{-23} \text{ J/K}, \qquad (12.38)$$

which is simply the gas constant per molecule. The perfect gas equation of state can thus also be written as

$$pV = Nk_B T, \qquad (12.39)$$

where N is the number of molecules in the volume V. These equations can be used to give us a thermodynamic definition of a perfect gas—a hypothetical gas that at nonzero pressures obeys the equation of state, Eq. 12.36, that real gases obey only in the limit $p \to 0$. (However, we shall see in the next chapter that this does not suffice for a complete definition of the perfect gas.)

A final note: Given the value of R, one can determine the molar mass ("molecular weight") M of any gas by measuring T, p, and the mass density ρ. From data taken at one temperature and a series of pressures, one makes the extrapolation

$$M = RT \lim_{p \to 0} \frac{\rho}{p}. \qquad (12.40)$$

Although Eq. 12.40 is still used to determine molecular weights in some cases, determinations using a mass spectrometer (Section 1.5) are much more accurate and at least as convenient.

12.6 Comparison of the Microscopic and Macroscopic Approaches

The preceding sections have illustrated some of the differences in approach between the microscopic and macroscopic theories. Let us now try to analyze these differences. The questions we wish to answer include: What are the essential features of each approach? Which arguments are general and which specific to a particular model? And how are our two definitions of the "perfect gas" related? We begin with the last of these questions.

Let us consider just what we mean when we speak of a perfect gas. In the microscopic theory a perfect gas is defined entirely in terms of the properties of the constituent molecules, in general as a gas in which there are no significant interactions between molecules. Our treatment in

---

[9] This is also and perhaps more commonly known as the *ideal gas law*. One can make a formal distinction between a *perfect* gas, which has the microscopic properties described in Section 12.1, and an *ideal* gas, one which satisfies Eq. 12.36. But to the extent that both have the same limiting properties, we can identify the two and call both perfect gases.

Sections 12.1 and 12.2 was even more restrictive than this, being limited to monatomic molecules that collide elastically with the container walls. It is thus not clear to what extent our results are valid for polyatomic molecules, or for molecules that do interact with one another. This illustrates a major limitation of statistical mechanics, that many of its results are applicable only for a particular model of molecular properties. The perfect gas model can in fact be extended to polyatomic molecules with a more sophisticated treatment, but any such extension must be worked through explicitly.

Contrast this with the definition of the perfect gas used in thermodynamics,[10] an abstraction based on the measured properties of real gases in the zero-pressure limit. No assumptions about the properties—or even the existence—of molecules are required; in particular, it does not matter whether the gas molecules are monatomic or polyatomic. Although neither "perfect gas" has the same properties as any real gas, the thermodynamic definition is at least tied directly to experiment.

How are the two definitions related? Consider the equations of state obtained in the two theories. The final result of our kinetic theory analysis was a relationship, Eq. 12.17, among the pressure, volume, and internal energy—the latter meaning in this case the total kinetic energy of the molecules in a monatomic gas. Although the pressure and volume can be measured directly, we have no direct way of determining the molecular kinetic energy. In contrast, the macroscopic perfect gas equation of state, Eq. 12.36, involves only directly measurable variables. Now we take the step joining the two theories: We *assume* that both refer to the same gas. Both equations of state involve the pressure, which is presumably the same physical quantity in each case. If both describe the same gas, then we can equate the two results for the pressure:

$$p = \frac{2}{3}\frac{U}{V} = \frac{nRT}{V}.$$

(12.41)

We thus obtain the important result

$$U = \frac{3}{2}nRT = \frac{3}{2}Nk_BT;$$

(12.42)

that is, the molecular kinetic energy per mole is proportional to the absolute temperature. As we have derived this result it applies only to monatomic gases, but we shall later show that it is generally true, even for gases with intermolecular forces.

What is important about Eq. 12.42 is that it constitutes our first bridge between a microscopically defined quantity (the molecular kinetic energy) and a thermodynamic variable (the temperature). But how do we justify the assumption that led us to this bridge, the assumption that both "perfect gases" are the same? The simplest answer is that they were designed to be the same. In effect the microscopic perfect gas is a model devised to mimic the properties of the thermodynamic perfect gas, an attempt to answer the question, "What kind of molecular model would give macroscopic behavior corresponding to the perfect gas equation of state?" We have shown that the model in question does this, *if* we identify the molecular kinetic energy with the temperature. Here this identity may seem arbitrary, but we shall see that the same identification

---

[10] As mentioned earlier, we have not yet given the full definition. In thermodynamics a perfect gas is one that (1) obeys Eq. 12.36 as its equation of state, and (2) has an internal energy that depends on temperature only. The meaning of the second part of the definition must wait until the next chapter.

(or its equivalent) can be made for many other types of systems, in each case giving correspondence between a properly designed microscopic model and the observed behavior of a macroscopic system. As always, our assumptions are justified by their agreement with experiment.

The approach outlined here can be extended to other systems. We have already mentioned that for a given mass of pure fluid any two of the variables p, V, T constitute a complete set of independent variables. (Remember that this is an experimental fact.) In short, all fluids, no matter how complex their microscopic structure, have an equation of state of the form $f(p, V, T) = 0$; the only example we have yet examined is Eq. 12.36 for the perfect gas case. On the other hand, any parallel equation of state developed from a molecular model must involve p, V, and the molecular energy in some form; in general, one may have to consider the intramolecular (vibrational, rotational, electronic) and intermolecular energies as well as the kinetic energy. Since the two approaches deal with different sets of variables, it is always necessary to establish a bridge of some kind if we wish to apply both kinds of analysis to the same real system. Equation 12.36 is an elementary example of such a bridge, applicable only to a single kind of system; in later chapters we shall develop bridges with more general validity.

Note that we have not actually done any thermodynamics in this chapter, except for introducing the zeroth law and the temperature concept. What we have talked about instead is various ways of obtaining an equation of state: by actual measurement in a real system, by extrapolation from measurement to a hypothetical system (such as the perfect gas), or from a microscopic model via a bridge between two sets of variables. Thermodynamics deals rather with the relationships between equations of state (however determined), other measurable quantities, and functions such as the internal energy and entropy which we have yet to define. These relationships are completely general. For example, the rule that $(\partial U/\partial V)_T = T(\partial p/\partial T)_V - p$ is true for all systems, no matter what their equations of state, but the equation of state of any specific system can be substituted in this rule to obtain $(\partial U/\partial V)_T$ for that system. The existence of such thermodynamic relationships makes it possible to derive a great deal of information about any system from a relatively small amount of data or from a relatively simple molecular model and bridge.

Thus far it would appear that the advantages of generality lie all with thermodynamics. Although this is largely true, the very generality of thermodynamics often precludes it from making specific predictions about particular systems. For example, experiment shows that the equations of state of all gases approach the perfect gas law in the limit $p \to 0$, and thermodynamics can be used to calculate the behavior of any other property in the same limit. But thermodynamics alone cannot tell us at what pressure measurable deviations from the perfect gas law will occur; for this we must rely on experiment. In contrast, a simple modification of the molecular theory to take into account the interactions between molecules enables us to predict how great the deviations from the perfect gas law will be at any pressure—for a particular model of the intermolecular forces. The choice of model is, of course, still based on experiment, but the experiments used in this case may be much farther removed from the quantity we seek (cf. Section 10.2); and in principle, at least, one can calculate directly the intermolecular forces by quantum mechanics. To sum up, then: Thermodynamics establishes general relationships whose validity transcends the nature of particular molecular models, whereas molecular theory is capable of model-based predictions outside the scope of thermodynamics.

As we said in the introduction to Part Two, statistical mechanics and thermodynamics are complementary tools that can be used to study the properties of matter. It is always important to bear in mind the advantages and disadvantages of the two theories. Nevertheless, it should already be clear that an analysis using all the available tools is likely to provide us with greater understanding and more incisive descriptions than an analysis limited to only one approach.

Further Reading

Kennard, E. H., *Kinetic Theory of Gases* (McGraw-Hill, New York, 1938), chap. 1.

Kestin, J., *A Course in Thermodynamics*, Vol. 1 (Blaisdell, Waltham, Mass., 1965), chaps. 2 and 3.

McCullough, J. P., and Scott, D. W. (Eds.), *Experimental Thermodynamics*, Vol. 1 (Plenum Press, New York, 1968), chap. 2.

Swindells, J. F. (Ed.), *Temperature*, NBS Special Publication 300, Vol. 2 (1968).

Zemansky, M. W., *Heat and Thermodynamics*, 4th ed. (McGraw-Hill, New York, 1957), chap. 1.

Problems

1. Explain how momentum and energy are conserved even when a gas molecule sticks to a wall after impact.

2. What happens to the kinetic energy of a body when it undergoes an inelastic collision with another body?

3. A delicate galvanometer mirror, if placed in a gas, may be observed to undergo chaotic oscillation. What is the cause of this oscillation?

4. If the number of molecules with speeds between v and $v + dv$ is given by

$$\text{constant} \times 4\pi v^2 \, dv,$$

how many molecules have kinetic energies between ϵ and $\epsilon + d\epsilon$?

5. The rate of effusion of a gas through a very small hole from a container to an evacuated space is proportional to the average speed of the molecules of the gas. Explain why this is so.

6. Using the results of Problem 5, devise a method for determining the molar weight of a perfect gas.

7. Suppose that all the atoms in a monatomic gas move with the same speed v, but that the directions of motion are randomly distributed. Compute the average number of atoms striking an element of wall area dA per second, and the average momentum imparted per second to dA in terms of N, V, v, and dA. (Hint: Average over all directions of incidence.)

8. Suppose that in a gas the number of atoms with speed between v and $v + dv$ is

$$2N\left(1 - \frac{v}{v_m}\right)\frac{dv}{v_m}$$

for $v < v_m$, where v_m is a constant. As in Problem 7, the directions of motion are randomly distributed. Calculate the average number of molecules striking dA per second and the average momentum imparted to dA per second, in terms of N, V, v_m, and dA.

9. Show that for the gases described in Problems 7 and 8,

$$p = \frac{2}{3}\frac{U}{V}.$$

10. Let $n(v)$ be the average number of molecules per unit volume with speed less than v. Then $n(v)$ increases monotonically as v increases, starting from $n(0)=0$ and approaching $n(\infty)=n$ as $v \to \infty$.

(a) Write an expression for the average number of molecules per unit volume with speed less than v that have velocities directed in the solid angle $d\Omega = 2\pi \sin\theta \, d\theta$.

(b) Write an expression of the average number of molecules per unit volume that have velocities directed between θ and $\theta + d\theta$ and have speed in the range v to $v+dv$.

(c) From the result of part (b), relate the function $n(v)$ to $f(v)$ introduced in Eq. 12.9.

11. If the average number of molecules per unit volume with velocity between $v_x + dv_x$, $v_y + dv_y$, $v_z + dv_z$ is

$$e^{-\frac{1}{2}cv^2} \, dv_x \, dv_y \, dv_z,$$

calculate $n(v)$ defined in the preceding problem. Assume that n, the average number density, is known.

12. Suppose that a box is divided by a thin rigid wall. On one side of the container is gas characterized by number density n_1, mass per molecule m_1, and coefficient c_1 in the exponential of the distribution of Problem 11. On the other side of the container is gas characterized by, respectively, n_2, m_2, and c_2. What condition must be satisfied if the forces exerted on the wall by the gases on the two sides are to balance?

13. The densities of gaseous trimethylamine $[(CH_3)_3N]$ and ammonia (NH_3) as a function of pressure at 0°C are given in the following table:

(CH₃)₃N		NH₃	
p (atm)	ρ (g/liter)	p (atm)	ρ (g/liter)
0.20	0.5336	0.333	0.25461
0.40	1.0790	0.667	0.38293
0.60	1.6363	0.500	0.51182
0.80	2.2054	1.000	0.77169

Determine the molar weights of trimethylamine and ammonia. If we take, on the chemical scale, the atomic weights of H and C to be 1.008 and 12.010, compute the atomic weight of N from both data sets.

14. From the following data, determine the zero of the perfect gas temperature scale (i.e., the temperature at which a hypothetical perfect gas would have zero volume).

Thermal Expansion of N_2 at Constant Volume

p (torr) at 0°C	$\dfrac{p_{100} - p_0}{p_0}$ Relative Increase in Pressure from 0°C to 100°C
333.6	0.36653
449.8	0.36667
599.9	0.36687
750.4	0.36707

p (torr) at 0°C	$\dfrac{p_{100}-p_0}{p_0}$ Relative Increase in Pressure from 0°C to 100°C
998.9	0.36741
5,000	0.3734
10,000	0.3809
20,000	0.3956
30,000	0.4101
40,000	0.4247
50,000	0.4396

15. From the following data, show that the coefficient of thermal expansion,

$$\alpha \equiv \frac{1}{V}\left(\frac{\partial V}{\partial T}\right)_p,$$

is the same for all three gases in the limit as the pressure tends to zero.

He		H$_2$		N$_2$	
p (torr)	$\alpha \times 10^6\,(K^{-1})$	p (torr)	$\alpha \times 10^6\,(K^{-1})$	p (torr)	$\alpha \times 10^6\,(K^{-1})$
504.8	3658.9	508.2	3660.2	511.4	3667.9
520.5	3660.3	1095.3	3659.0	1105.3	3674.2
760.1	3659.1				
1102.9	3658.2				
1116.5	3658.1				

16. Use the following data to show that the perfect gas law is valid only in the limit as $p \to 0$. In this table, V^+ is 1 unit when the temperature is 0°C and $p = 1$ atm.

Behavior of N$_2$ at 0°C		Behavior of C$_2$H$_2$ at 20°C	
p (atm)	V^+	p (atm)	V^+
0.33333	3.00087	1	1
0.66667	1.50016	31.6	0.02892
1.00000	1.00000[a]	45.8	0.01705
19.0215	0.052190	84.2	0.00474
23.7629	0.041712	110.5	0.00411
28.4968	0.034735	176.0	0.00365
37.9526	0.026014	282.2	0.00333
50.000	0.019692	398.7	0.00313
52.2160	0.018856		
100.	0.009846		
200.	0.005181		
400.	0.0031392		
600.	0.0025380		
800.	0.0022449		
1000.	0.0020641		

[a] Reference volume, defined to be 1 unit at 0°C and 1 atm.

17. The accepted value of the molar volume of a perfect gas at 273.15 K and 1 atm is 22.4138 liters. Calculate the volume occupied by 1.0000 mol of each of the following gases at 273.15 K and 1 atm from the densities in grams per liter.

Gas	Density (g/liter)
He	0.1785
Ne	0.9002
Ar	1.7824
Kr	3.708
Xe	5.851
Rn	9.73
O_2	1.4290
O_3	2.144
F_2	1.695
Cl_2	3.214

18. A constant-volume gas thermometer is used to measure a series of reference temperatures. The limiting values of the ratios of pressure at T to pressure at the triple point of water are found to be 0.33017 at the boiling point of oxygen, 0.85770 at the melting point of mercury, and 1.79788 at the boiling point of naphthalene. Calculate the perfect gas temperatures corresponding to these reference points.

19. A Pt-resistance thermometer has a resistance of 9.81 ohms at 0°C, 13.65 ohms at 100°C, and 21.00 ohms at 300°C. Is the thermometer linear over this range? If it is assumed to be linear between 0°C and 100°C, will a temperature of 50°C deduced from a resistance reading of 11.73 ohms be higher or lower than the true temperature?

20. Let X be a general thermometric property of a suitable substance. Following the arguments used in the text, show how a temperature scale may be set up using the property X. What conditions must the property X satisfy?

21. Suppose that for some gas

$$(pV)_0 = A_0 + B_0 p_0,$$
$$(pV)_{100} = A_{100} + B_{100} p_{100},$$
$$(pV)_\theta = A_\theta + B_\theta p_\theta,$$

where the numbers $B_0 p_0$, $B_{100} p_{100}$, $B_\theta p_\theta$ are all small relative to A_0, A_{100}, A_θ, respectively. The numbers A_0, and so on, refer to the limiting perfect gas values of the respective pV products. If we define a thermometric scale in terms of the given pV terms, there will be a difference between temperatures on the defined scale and on the perfect gas scale. If

$$\Delta\theta = \theta^* - \theta_{gas}$$

is that difference in temperature, work out the correction to the temperature (i.e., the difference between the real gas and perfect gas temperatures) when the real gas is used in a constant-volume thermometer.

22. Assume that the electromotive force produced by a given thermocouple, with one junction at 0°C and the other at θ, can be represented by

$$\mathscr{E} = a_1\theta + a_2\theta^2.$$

Taking $\mathscr{E}$ to be the thermometric property, define a Celsius scale in terms of the thermal emfs. Obtain a formula for the correction $\Delta\theta = \theta_{lim} - \theta_{\mathscr{E}}$, where $\theta_{\mathscr{E}}$

is the value of the temperature calculated from the definition. If $a_1 = 0.20$ mV/°C and $a_2 = 5.0 \times 10^{-14}$ mV/(°C)^2, work out the magnitude of the correction at $\theta = 50$°C.

23. It is sometimes stated that there are no collisions between the molecules of a perfect gas. Imagine a volume of a perfect gas in which there is an initial nonequilibrium distribution of molecular velocities. If the molecules can collide elastically with the walls, but not with each other, is there any mechanism by which an equilibrium state can be reached?

24. The molecules of a perfect gas have negligible volume. In a so-called hard-sphere gas, the molecules have nonzero volume and interact with each other through instantaneous (impulsive) collisions. Such a gas has the equation of state

$$p(V - nb) = nRT$$

where the parameter b is proportional to the molecular volume. (a) Find p/RT correct to second order in the molar density. (b) At low density, does the hard-sphere gas approach perfect gas behavior?

The First Law
of Thermodynamics

In this chapter we discuss the first law of thermodynamics, historically the first of the great empirical principles that form the basis of thermodynamics. In simple terms, the first law is the law of conservation of energy; however, it cannot be simply identified with the corresponding mechanical principle. As we did in Chapter 12 for temperature, we must carefully define just what "energy," "work," and "heat" mean in thermodynamics.

Following the pattern of Chapter 12, we carry out a parallel development of the theory from the microscopic point of view, where energy can be given its ordinary (quantum) mechanical meaning. We apply both the thermodynamic and the microscopic models to the perfect gas, still the only system simple enough for us to analyze in detail.

In this chapter we also develop several concepts that are important throughout thermodynamics (rather than being related specifically to the first law). These include the difference between "intensive" and "extensive" variables, the limiting processes called "quasi-static" and "reversible," and the key notion of "constraints." Most of these ideas have already been used implicitly, but for complete understanding we must analyze them further. Although it is true that the foundations of thermodynamics are extremely abstract, we shall do our best to clarify them.

13.1
Microscopic and Macroscopic Energy in a Perfect Gas

As remarked above, the first law of thermodynamics is usually interpreted as a statement of the conservation of energy. Now, the concept of energy conservation is well defined in both classical and quantum mechanics. For a single particle, kinetic energy is $\frac{1}{2}mv^2$, potential energy is defined as in Eq. 2.46, and their sum is constant throughout the particle's motion. For a system of interacting particles, the total energy of all the particles is conserved. But this approach can be used only as long as we are dealing, at least implicitly, with particles. In thermodynamics, it will be recalled, the concepts used are defined only in terms of macroscopic operations, without reference to the microscopic structure of the substance involved. What, then, does "energy" mean in thermodynamics, where it must be some function of the thermodynamic variables discussed in Section 12.3? What similarities and differences are there between the mechanical and thermodynamic concepts with the same name?

As implied by these questions, there does exist an essential difference between the two concepts. The nature of this difference can be seen from the following considerations: In the mechanical description of a system of particles, it is presumed that the positions and velocities (and thus the energies) of individual particles can be controlled by suitable manipulation of the forces available to the experimenter. The work done in displacing a given particle is then exactly determined by the controlled displacements of that particle and all the other particles present. The uncertainty principle limits the accuracy with which this process can be carried out, but the quantum mechanical and classical descriptions are basically the same: One follows the motion of the individual particles as closely as nature allows. Contrast this with the thermodynamic description of a system, in which only macroscopic variables such as the volume are controlled. If work is done by displacing the system's boundaries, the positions and velocities of the individual molecules change in an uncontrolled manner. Only the *average* changes in various molecular quantities are determined by the changes in the macroscopic variable. It is these unavoidable and uncontrolled molecular motions that cause thermodynamic energy to differ from mechanical energy, thermodynamic work from mechanical work, and so on.

Nevertheless, connections between the thermodynamic and mechanical quantities can be found, in the sense that thermodynamic variables correspond to mechanical variables averaged over large numbers of particles. We have already seen one example of this. The temperature T is a characteristic variable of a thermodynamic system, defined in a completely nonmechanical way. Yet according to Eq. 12.42, in a perfect monatomic gas T is directly proportional to the total molecular kinetic energy U, defined as in Eq. 12.10 by summing over individual particle energies. We shall obtain many other correspondences of this sort—including, of course, that between thermodynamic and mechanical energy.

Before going on to the thermodynamic development, let us consider a simple molecular model in an attempt to obtain some intuitive understanding of the interplay between macroscopic and microscopic changes in a system. The model we choose is again that of the perfect monatomic gas, but this time we describe it quantum mechanically. Although this system is simpler than any real substance, the conclusions we shall draw are generally valid, as will be demonstrated more comprehensively in Chapter 16.

Consider a set of N noninteracting particles in a cubical box of volume V. We showed in Chapter 3 that the possible energies of particles in such a box are quantized, with the energy spectrum for any of the independent particles given by Eq. 3.119 as

$$\epsilon_{n_1 n_2 n_3} = \frac{\pi^2 \hbar^2}{2mV^{2/3}} (n_1^2 + n_2^2 + n_3^2), \qquad (n_1, n_2, n_3 = 1, 2, \ldots), \quad (13.1)$$

where n_1, n_2, n_3 are the quantum numbers corresponding to motion along the x, y, z axes of the cube. If the box is of macroscopic size, the spacings between the energy levels are very small. The form of the energy spectrum is the same for each of the particles, because of our assumption that they do not interact. Therefore the total energy of the N particles in the box is simply the number of particles in each quantum state multiplied by the energy of that state, summed over all states,

$$E = \sum_{n_1, n_2, n_3} N_{n_1 n_2 n_3} \epsilon_{n_1 n_2 n_3}, \qquad (13.2)$$

where $N_{n_1 n_2 n_3}$ is the number of particles in the level with energy $\epsilon_{n_1 n_2 n_3}$.

Now suppose that the box containing the particles is changed in size, but in such a way as to remain a cube. If the volume change dV is small relative to V, each of the energy levels of Eq. 13.1 changes by the amount

$$d\epsilon_{n_1 n_2 n_3} = -\frac{\pi^2 \hbar^2}{3mV^{5/3}}(n_1{}^2 + n_2{}^2 + n_3{}^2)\, dV = -\frac{2\epsilon_{n_1 n_2 n_3}}{3V}\, dV. \quad (13.3)$$

Thus if the volume is decreased (dV negative), the energies of all the allowed levels are increased. Given that the microscopic energy spectrum changes when a macroscopic coordinate is altered, what happens to the energy of the system of particles? The answer depends on how the change is made.

In general, a perturbing force such as the motion of the walls of our box will induce transitions between the energy levels of a system. But suppose that the perturbation is carried out very slowly. By the adiabatic principle of quantum mechanics (Section 6.5), we know that a sufficiently slow perturbation cannot induce transitions. More specifically, one can show that transitions between states m and n are unlikely if the rate of change of the Hamiltonian with time is much less than $|E_m - E_n|/\tau_{mn}$, where the uncertainty principle shows that the characteristic period τ_{mn} is of the order of $\hbar/|E_m - E_n|$. In principle, then, one can change the volume of the box slowly enough that no transitions take place; such a change is what we call an *adiabatic perturbation*. Then, whatever the initial distribution of particles over the energy levels, there must still be the same number of particles $N_{n_1 n_2 n_3}$ in each level after the change as there were before. Given that the $N_{n_1 n_2 n_3}$ are constant, combining Eqs. 13.2 and 13.3 tells us that the change in the total energy of the system of particles is

$$dE = \sum_{n_1, n_2, n_3} N_{n_1 n_2 n_3}\, d\epsilon_{n_1 n_2 n_3} = -\frac{2}{3V}\left(\sum_{n_1, n_2, n_3} N_{n_1 n_2 n_3}\epsilon_{n_1 n_2 n_3}\right) dV$$

$$= -\frac{2E}{3V}\, dV \quad (13.4)$$

when the volume is changed very slowly. We shall see that this adiabatic perturbation in the quantum mechanical sense corresponds to what is called a *reversible adiabatic process* in thermodynamics.

How is the energy change of Eq. 13.4 exhibited in our model system? The only energy the particles in the box have is kinetic energy. If the energy levels $\epsilon_{n_1 n_2 n_3}$ increase while the occupation numbers $N_{n_1 n_2 n_3}$ remain unchanged, then clearly the average kinetic energy per particle must increase, as must the total kinetic energy of the system of N particles. But according to Eq. 12.42, this in turn implies an increase in the temperature of the gas. We have thus deduced from purely molecular considerations that in the (reversible) adiabatic compression of a perfect gas the temperature must increase.

But our model is more restrictive than we need. Equation 13.4 applies strictly only to a gas of point-mass particles in a box with perfectly smooth walls. In any real gas the collisions of molecules with the walls and with each other will not be perfectly elastic, and transitions between energy levels will inevitably take place, even if the walls are not moving at all. Yet in either our model or a pure real gas, the molecules are indistinguishable from one another. All one can hope to know about the gas is how many molecules there are in the various energy levels, not which molecules occupy a particular level. And if transitions between levels occur in such a way that there is no net change in the occupation of the individual levels, then the effect on macroscopic properties is the same as if there were no transitions at all. To obtain Eq. 13.4 and similar results, we need only assume that the $N_{n_1 n_2 n_3}$ remain unchanged with time.

Yet clearly this constancy of the energy-level populations can only describe the average behavior of the gas molecules over time. A snapshot of this distribution at some one instant of time will differ from one at some other instant, even if the differences must average out in the long run. Such differences between the instantaneous molecular distribution (of energy, velocity, etc.) and the average distribution are referred to as *fluctuations*. Fluctuations of one sort or another necessarily occur in all collections of molecules. When the fluctuations are small—that is, when the deviations from the average populations of the energy levels are small compared to these average populations—the distribution of molecules is for most purposes adequately described by the average values of the occupation numbers $N_{n_1 n_2 n_3}$. This is the same as saying that a thermo-dynamic description of the system is valid, since the macroscopic variables of thermodynamics correspond to the average properties of the molecular system. But if the fluctuations are not small, then the ordinary thermo-dynamic description is not adequate; it must be either supplemented by the use of additional concepts or completely replaced. Further discussion of the importance and magnitude of fluctuations is deferred to Chapter 15.

To obtain results corresponding to those of thermodynamics, then, we ordinarily need to know only the average values of the $N_{n_1 n_2 n_3}$. Equation 13.4 holds for any process in which these average values remain unchanged and fluctuations can be neglected. We conclude from this equation that, when a box containing N independent particles is compressed adiabatically in the quantum mechanical sense, the total energy of the system of particles increases; when the box is expanded in the same way, the total energy decreases. What are the corresponding energy changes in the surroundings? By the law of conservation of energy, the total energy of the box, the N particles, and the rest of the universe must remain constant. Thus energy in some form must be transferred to or from the particles. One obvious way in which this transfer can take place is in the form of *work*. In order to compress the box, it is in general necessary to exert external forces on it. The product of the force per unit area (the applied pressure p_{app}), the total area of the box, and the (uniform) inward displacement is just the work done on the box by the surroundings; this is the ordinary mechanical definition of work as force times distance. In either mechanics or thermodynamics, work is a form of energy. Thus, changing the volume of the box by dV (compression: $dV < 0$; expansion: $dV > 0$) requires that an amount of energy $-p_{app}\, dV$ be transferred from the surroundings to the box in the form of work. Is this the same as the energy change of the particles inside the box? That depends, of course, on the nature of the box.

The box implied by Eq. 13.1 is of a rather special kind. It has perfectly smooth walls, which affect the particles within it only as boundary surfaces, in that the wave function must vanish at the walls. When the volume of the box is changed, the particles "see" only a change in boundary conditions. In fact, the energies of the particles in such a box can be changed only by moving the walls, that is, only by performing work on the box. Another way to say this is that energy can cross the walls only in the form of work. We shall see that such walls correspond exactly to the adiabatic walls of thermodynamics, which we defined in Section 12.4. If our box has such adiabatic walls, then clearly the increase in energy of the system (box plus N particles) must be exactly equal to the work performed on the box by the surroundings.

How does such a box differ from real boxes? Primarily in that the real boxes are made up of molecules, and these molecules can vibrate. Then it is possible to transfer energy across the walls without moving them. If we impart additional vibrational energy to the molecules on the outside wall, this energy will travel from molecule to molecule until it reaches the inside wall, from which it can be transferred to the gas molecules colliding with that wall. This process, which sounds so complicated, is nothing more than what we call a flow of *heat* across the wall. "Heat," we shall see, is simply a term for energy that crosses the boundary of a system in a form other than that of work. A wall that allows such a process to take place corresponds to the diathermal wall of Section 12.4. The problem we raised above is more complicated in a box with diathermal walls. Clearly the total energy change of the system (box and particles) must be equal and opposite to that of the surroundings, but how much of the energy is transferred as work and how much as heat? It is questions like this that thermodynamics was devised to answer.

We see in all this another of the differences between mechanics and thermodynamics. In mechanics, if we wish to measure the work performed on a system, we measure the properties of the system itself—basically, that is, we follow the trajectories of particles. But in thermodynamics we measure work in terms of what flows across the boundaries of a system, that is, in terms of the properties of the surroundings. The two kinds of "work" are thus not conceptually the same, and we must introduce a new definition of thermodynamic work; Section 13.3 deals with this definition, which in turn leads directly to the thermodynamic definition of energy.

13.2 Description of Thermodynamic States

To discuss the thermodynamic definitions of energy and work, however, we must first make more precise a number of concepts, beginning with that of thermodynamic state. Our definition of thermodynamic state in Section 12.3 is equivalent to saying that two states are identical if each and every member of a specified set of macroscopic variables (pressure, density, etc.) has the same values in the two states. We have not yet stated just how the relevant macroscopic variables are to be chosen. The reason is that no general rules can be given. One must establish by experimentation, for each class of systems, which and how many macroscopic quantities must be fixed, that is, what constraints must be applied, for given properties of the system to be reproduced. Note that each constraint corresponds to fixing one of the macroscopic variables, which implies that the variables chosen must be independently controllable by the experimenter. Only if the latter condition is met can one say that there is an independent set of variables. To sum up: A thermodynamic state is defined uniquely by specifying the values of a complete set of independent variables, each such set of values corresponding to a distinct state of the system. The number and kinds of independent variables needed to describe a given system must be determined by experiment.

There is a convenient and useful way of visualizing thermodynamic states geometrically. In mathematics, it is usual to represent a function $f(x)$ by a graph, that is, by the locus of points satisfying the equation $y = f(x)$ in the plane defined by two perpendicular axes x and y. One can use a similar graphical representation for the states of a fluid. In the absence of external fields, a complete set of independent variables for a fixed mass of a pure fluid is constituted by the pressure p and the density

511

ρ (or the specific volume[1] v). Since these determine the temperature by the equation of state, any two of the variables p, v, T form such a set. In terms of the set v, T, then, any thermodynamic state of the given mass of fluid can be represented as a point on the surface defined by the equation

$$p = f_1(v, T) \qquad (13.5a)$$

in the three-dimensional space defined by the axes p, v, T; such a surface is shown in Fig. 13.1. Of course, in this space the same surface can be equally described by the equations

$$v = f_2(p, T) \qquad (13.5b)$$

or

$$T = f_3(p, v). \qquad (13.5c)$$

Each of the three equations 13.5 arises from solving the equation of state for one of the variables p, v, T, the other two then being the independent variables. Because of the equation of state, fixing any two of the variables p, v, T is sufficient to define a state of the system, represented by a point in the three-dimensional p, v, T space.[2]

We can also construct representations of the equation of state in the (p, v), (v, T), or (p, T) plane. For example, for each possible value of p, Eq. 13.5a defines a different curve in the (v, T) plane; a few such curves are illustrated in Fig. 13.2a. Similarly, families of curves can be obtained for each of the other two variables, as shown in Figs. 13.2b and 13.2c. These families of curves are equivalent to the intersections of the various constant-p, constant-v, or constant-T planes with the surface defined by the equation of state in the full (p, v, T) space.

Every point on the surface defined by Eqs. 13.5 corresponds to a possible equilibrium thermodynamic state of the system, whereas all points in the (p, v, T) space not lying on this surface represent nonexistent states—that is, values of the variables that do not satisfy the equation of state. A change of state corresponds to a shift from one point on the surface to another. Since a change of state is defined simply by the end points (the initial and final thermodynamic states), the same change of state can be carried out by an infinite number of different "paths." What we shall later call a *reversible process* is a change of state that can actually be represented by a path lying entirely in the surface, that is, which proceeds through a continuous sequence of equilibrium states; in Section 13.5 we shall examine the properties of such an idealized process.

What about *non*equilibrium states of the system? Strictly speaking, a nonequilibrium state is not represented by a point in the (p, v, T) space at all. The reason is that a system not in equilibrium requires additional variables for a full description of its state. For example, the system may not be homogeneous; then $p, v,$ and T will vary from point to point,[3] and no set of values of these variables can be assigned to the system as a whole. Other possibilities include a system that is spatially homogeneous

[1] That is, the volume per unit mass, which is the reciprocal of the mass density. The term "specific" is generally applied to extensive quantities divided by mass. We shall use exclusively the mole as unit mass, except in Chapter 20, so v will always be the molar volume.

[2] In general, if n independent variables must be specified to determine the thermodynamic state of a system, the corresponding geometric representation requires an $(n+1)$-dimensional space.

[3] To speak of values of p, v, T "at a point" implies the existence of *local* equilibrium, that is, the uniformity of the molecular averages to which these quantities correspond over a sufficiently large region (and for a sufficiently long time) for a measurement to be performed.

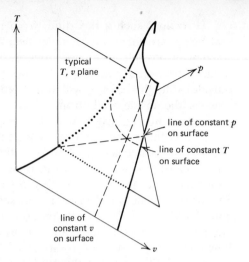

FIGURE 13.1

Part of a surface representing the equation of state in (p, v, T) space. The particular surface shown is for the perfect gas, with the equation of state $pv = RT$. The lines of constant $p, T,$ and v on the surface are the loci of the intersections of the surface with typical (T, v), (p, v), and (T, p) planes, respectively. A typical (T, v) plane is shown; the others are omitted for clarity.

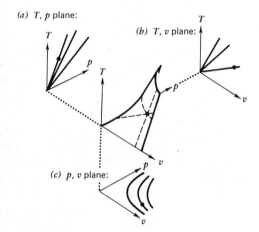

FIGURE 13.2

Intersections of the perfect gas surface of Fig. 13.1 with planes on which $v, p,$ or T is constant. (a) (T, p) plane—each line in this plane is the locus of the intersection of the p, v, T surface and a plane on which v is constant. (b) (T, v) plane—each line in this plane is the locus of the intersection of the p, v, T surface and a plane on which p is constant. (c) (p, v) plane—each curve (hyperbola) in this plane is the locus of the intersection of the p, v, T surface and a plane on which T is constant.

but that has a nonequilibrium velocity distribution; a system in which a chemical reaction is taking place; or a steady-state system with time-independent properties at any given point, but material flowing in and out. However, any system in which a real change of state is occurring at a measurable rate is necessarily a nonequilibrium system. For example, if a gas is expanded by moving a piston, it takes some time for the resulting pressure change to propagate through the entire gas, and during this time the gas is necessarily inhomogeneous. How, then, can we speak of the "path" of a change of states at all? Fortunately, in many cases a system undergoing a change of state is almost in an equilibrium state, especially when the change of state is slow, and can for some purposes be approximated as being in that equilibrium state.

13.3
The Concept of Work in Thermodynamics

We are now ready to make the thermodynamic concept of "work" more precise. The meaning of the term is similar to that used in mechanics; however, since some of the thermodynamic variables we use do not have direct mechanical analogs, we must discuss several new features of the work process.

Let us recall the mechanical definition of work, introduced in Eqs. 2.23 and 2.24: If an object moves through an infinitesimal displacement $d\mathbf{s}$ under the influence of a force $\mathbf{F}$, the work done on the object is

$$dw = \mathbf{F} \cdot d\mathbf{s} = F_s \, ds, \tag{13.6}$$

where F_s is the component of $\mathbf{F}$ in the direction of $d\mathbf{s}$. If the object moves through a finite distance, from point 1 to point 2, the total work done upon it is given by the line integral

$$w_{12} = \int_1^2 \mathbf{F} \cdot d\mathbf{s} = \int_1^2 F_s \, ds. \tag{13.7}$$

Two features of this definition are significant. First, the work depends on the path taken and how the force varies along it; different paths between the same two points entail different amounts of work. Second, work is defined only with respect to motion; the object does not "contain" one amount of work at the beginning of a process, another at its end. Work is a *process variable*, a quantity that passes between systems. We shall see that both these features of mechanical work are also characteristic of thermodynamic work.

The simplest (and most commonly considered) type of thermodynamic work is that involved in the expansion or compression of a fluid. Consider a volume of fluid in a container with movable or deformable walls. The pressure the fluid exerts on the walls is what we call p, whereas the pressure applied by the surroundings to the outside of the container is p_{app}. Note that p_{app} need not be a hydrostatic pressure, but can refer to any force exerted on the container, divided by the area to which this force is applied; one example could be a weight placed on a movable piston. In the absence of any constraint fixing the volume (such as rigid walls, or pegs holding a piston in place), at equilibrium we must have $p = p_{app}$. Suppose now that, from such an initial state of equilibrium, we raise the value of p_{app}. If the walls of the container are free to move, the volume of the fluid must decrease. Consider a small element of the container's surface, of area dA_i, which moves inward an infinitesimal distance dx_i. The work done on the system by p_{app} is simply the product of the applied force and the distance moved or, for the ith area element,

$$dw_i = F_i \, dx_i = (p_{app})_i \, dA_i \, dx_i = -(p_{app})_i \, dV_i, \tag{13.8}$$

where dV_i $(\equiv -dA_i\, dx_i)$ is the change in the volume of the system as a result of the motion of the ith area element; the symbol $đ$ in $đw_i$ will be explained below. Summation over the entire surface of the container gives the total work done on the system for an infinitesimal volume change dV. If we simplify by assuming p_{app} uniform over the surface, this is

$$đw = -p_{app}\, dV. \tag{13.9}$$

The total work done on the system for a measurable volume change is then just

$$w = -\int_{V_1}^{V_2} p_{app}\, dV, \tag{13.10}$$

where V_1 and V_2 are, respectively, the initial and final volumes. Note that w is positive when the volume of the system decreases ($dV < 0$); this is because we have chosen, as a convention, to define w as the work done *on* the system *by* the surroundings. (Later we shall discuss the reasons for this convention.)

We again emphasize that p_{app} is the effective pressure exerted by the surroundings; if work is to be done at a measurable rate, p_{app} cannot be the same as the pressure p within the system. In any case, either p_{app} or p can be varied independently of the volume during the process: p_{app} by adding or removing weights, p by changing the temperature, etc. There is an infinity of such possible variations in p_{app}—that is, of possible paths in the (p_{app}, V) plane—and each such path gives a different value for the integral of Eq. 13.10. This is true even when the paths run between the same initial and final states of the system. The work done on the system is thus incompletely specified if the path by which the process occurs is not specified. In mathematical language we say that $đw$ is an *inexact differential* (cf. Appendix II at the end of the book), in that it is not the differential of some state function $w(p_{app}, V)$; we indicate this by the bar through the d.

When we write a differential in the form $đX$, we imply that the integral $\int_1^2 đX$ between a given pair of end points depends on the path taken. Thus there exists no function X of which $đX$ is the differential, and which could be assigned a unique value in the initial or the final state. Like an object in mechanics, a thermodynamic system in a given state does not "contain" a certain amount of work. To put it another way, work cannot be defined as a function of those variables that determine the thermodynamic state of the system. A variable that can be so expressed is called a *state variable* (or *state function*); examples are p, v, T themselves and (as we shall see) the energy of the system. The change in a state variable between given initial and final states does depend only on those states ($\Delta Y = Y_2 - Y_1$) and is independent of the path followed between them. Remember that in this context "state" refers only to equilibrium states.

Thus far we have considered only the work involved in the expansion or contraction of a fluid (often called "pV work"). As we indicated in Eq. 13.8, this type of work can be interpreted mechanically in a quite straightforward force-times-distance sense. But there are many other possible forms that work can take in thermodynamics, and the relation to mechanical work is often much less clear-cut. To give a single example, the electrical work required to charge a capacitor is found to be

$$w = \int_{Q_1}^{Q_2} \mathscr{E}\, dQ \tag{13.11}$$

where Q is the charge on the capacitor and $\mathscr{E}$ is the applied potential difference.[4] To include this in the same category as mechanical work and pV work, we need a more general definition of work. We find this in the statement that *the performance of work is equivalent to the lifting of a mass in a gravitational field.*

This statement requires some clarification. Consider the system in Fig. 13.3, in which a fluid expands by lifting a mass atop a frictionless piston. The applied pressure on the piston is the gravitational force mg divided by the piston's area A. As the fluid expands from volume V_1 to volume V_2, the work done on the surroundings is

$$-w = \int_1^2 p_{\text{app}}\, dV = \int_1^2 \left(\frac{mg}{A}\right)(A\, dh) = mg \int_1^2 dh = mg\, \Delta h, \quad (13.12)$$

exactly the work done in lifting the mass m a distance Δh. This is, of course, an idealized example, in which the expansion work is applied directly to the lifting of the mass. But imagine next a case of expansion work in which the movable boundary of the system does not move in such a convenient vertical direction. One can still in principal attach to the system walls mechanisms (rods, pulleys, cams, etc.) that will transmit this motion to a vertically moving mass m. If the mechanisms are all ideal (frictionless and weightless), the expansion work can again be completely converted to the work of lifting the mass m. With other kinds of work the process is more complicated. If we discharge the capacitor of Eq. 13.11, for example, we must imagine the resulting current to be operating an ideal electric motor that completely converts the electrical energy to mechanical energy, in turn driving mechanisms of the type already described. For any kind of work one can devise such an idealized process. These processes cannot be carried out perfectly in the real world (where, for example, friction always exists), but for our definition it suffices that they can be described in principle. We can then say that a system performs work on its surroundings if the only effect in the surroundings could be the lifting of a mass. The same reasoning applies when the surroundings perform work on the system, except that the ideal mechanisms must be imagined to be within the system.

The tacit assumption in the above argument is that work is a form of energy, and that various forms of energy can be converted into one another. This is an application of the general physical principle of conservation of energy, which we shall later in this chapter express in thermodynamic terms. The point to note here is that work must have the same dimensions as energy. This is illustrated by the specific examples we have already considered: (force × distance), (pressure × volume), (electric potential × charge) all have the dimensions of energy. There are many other forms of work that may need to be considered in specific chemical problems, and we shall later describe some of these. For now we need only bear in mind that a given system can often perform work (or have work performed on it) in several different ways, and that each case must be examined for its salient features to be determined.

We have stressed the similarity between mechanical and thermodynamic work; just what is the difference? Part of the difference is that thermodynamics can treat a much wider range of variables, but that is not

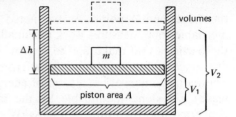

FIGURE 13.3
Equivalence of work with lifting a mass in a gravitational field. The system is a fluid in a cylindrical container with rigid walls, except for a movable piston atop the container. The piston itself is weightless and frictionless, but is held in place by a mass m. The fluid expands in volume from V_1 to V_2 by upward motion of the piston, thereby lifting the mass m through a vertical distance Δh. (See text for computation of work.)

---

[4] The complementary case of a point charge moving through an electric field has been discussed in Section 1.3; for a charge Q in a field $\mathbf{E}$, the work done on the charge in a displacement $d\mathbf{s}$ is

$$đw = \mathbf{F} \cdot d\mathbf{s} = Q\mathbf{E} \cdot d\mathbf{s} = -Q\, d\mathscr{E}$$

(since electric potential is defined by $d\mathscr{E} = -\mathbf{E} \cdot d\mathbf{s}$, or $\mathbf{E} = -\nabla\mathscr{E}$).

the fundamental point. As we noted at the end of Section 13.1, in mechanics one measures work essentially by observing the trajectories of the particles that make up the system. In thermodynamics, however, we are concerned primarily with what crosses the boundaries of the system. Thermodynamic work is a form of energy that we observe as it enters or leaves a system, identifiable by the fact that it can (in principle) be converted into the lifting of a mass. We do not care what happens to this particular energy before or after it crosses the boundary, only with the crossing process itself. As we shall see, the fundamental question is how much of the total energy flow across the boundary consists of work.

13.4 Intensive and Extensive Variables

Continuing our development of concepts needed for the thermodynamic description of a system, let us now take a closer look at the variables used in thermodynamics. The thermodynamic variables we have used can be classified into two types. Some, such as the temperature or the pressure, are independent of the mass of the system; these are called *intensive variables*. Others, such as the volume or the internal energy, are proportional to the mass of the system when the intensive variables are held constant; these are called *extensive variables*. That is, if we combine two identical systems into one by removing a barrier, the temperature and pressure will remain unchanged, but the final volume or energy will be twice that of either original system. Another way to define the difference is that extensive variables are inherently properties of the system as a whole, whereas intensive variables can be measured at a point within the system. In simple though imprecise terms, the bigger a system is, the more one has of the extensive variables; if a system is divided into subsystems, the extensive properties of the whole system are the sums of those of the subsystems. Intensive and extensive variables play different roles in the description of a system, and we shall see that this difference is important in defining the key concept of the reversible process.

Each intensive variable X has what is called a *conjugate* extensive variable Y, the two being related by a particular work process. Suppose that X has the same value inside and outside a system boundary (e.g., $p = p_{app}$); the conjugate extensive variable Y is then a quantity such that an infinitesimal change dY does an amount of work

$$đw = X\,dY. \qquad (13.13)$$

Using this relationship, we can see from Eqs. 13.9 and 13.11 that volume is conjugate to pressure, and electric charge is conjugate to potential. There are many other such pairs,[5] some of which are listed in Table 13.1: Note that the product of each such pair of variables must have the dimensions of energy.

Conjugate **extensive** and intensive variables correspond to the *generalized coordinates* and *generalized forces* of mechanics, respectively. If the potential energy of a mechanical system is expressed as a function of several independent variables q_i (the generalized coordinates), then the generalized force Q_i conjugate to a particular q_i is defined as the negative partial derivative of the potential energy with respect to q_i, with all the other coordinates held fixed. Similarly, we shall see that the intensive variable X conjugate to a given extensive variable Y is defined by $X \equiv -(\partial U/\partial Y)$, where U is the internal energy and all other independent extensive variables are held fixed. The differential work in any mechanical process can be expressed as a sum of terms of the form $Q_i\,dq_i$ (where dq_i

---

[5] The extensive variable conjugate to temperature is the entropy, which we shall not define until Chapter 16.

TABLE 13.1
CONJUGATE INTENSIVE AND EXTENSIVE VARIABLES, AND WORK

System	Intensive Variable	Extensive Variable	Work Done on System[a]
Fluid (or other system that can expand and contract)	Pressure (p)	Volume (V)	$-\int p\,dV$
Surface film	Surface tension (σ)	Area (A)	$\int \sigma\,dA$
Wire	Tension (F)	Length (l)	$\int F\,dl$
Capacitor	Potential ($\mathscr{E}$)	Charge (Q)	$\int \mathscr{E}\,dQ$
Electrochemical cell	emf ($\mathscr{E}$)	Charge (Q)	$\int \mathscr{E}\,dQ$
Paramagnetic solid	Magnetic field strength ($\mathbf{H}$)	Magnetic moment ($\mathbf{m}$)[b]	$\mu_0 \int \mathbf{H} \cdot d\mathbf{m}$

[a] This column gives the work done on the system only when the intensive variable has the same value inside and outside the boundary (i.e., in a reversible process).
[b] This is the same quantity that we have called μ_m for a single atom or molecule (Section 9.6).

is a *generalized displacement*), corresponding to Eq. 13.13. In both cases, the product of the two members of a conjugate pair (XY or $Q_i q_i$) must have the dimensions of energy.[6]

In general, the boundaries of a system (or portions thereof) may or may not allow the performance of various kinds of work. For example, expansion work can be performed at a movable boundary, but not at a rigid one; electrical work can be performed at a conducting boundary, but not at an insulating one; and so forth. The performance of a given type of work is always associated with a change in the corresponding extensive variable (cf. Table 13.1) in the system or the surroundings or both: volume changes in expansion work, flows of charge in electrical work, and so on. Whenever a boundary allows the performance of a particular kind of work, including change in the corresponding extensive variables, we can say that the same boundary "transmits" the conjugate intensive variable. By this we mean that at equilibrium the intensive variable in question has the same value on both sides of the boundary—in the system and its surroundings.[7]

Suppose, for example, that part of a system's boundary is displaceable, allowing the performance of expansion work. When such a boundary moves outward, volume changes occur in both the system and the surroundings. The boundary must therefore transmit pressure, the variable conjugate to volume, so that the equilibrium state has the same pressure on both sides of the boundary. And we know that this is indeed true for any movable boundary in mechanical equilibrium, when the forces on each area element must be in balance.

On the other hand, if a given boundary does not allow the performance of a particular kind of work, then in whatever equilibrium state is attained, the corresponding intensive variable need not have the same value in system and surroundings. To continue our example, if part of a system's boundary is rigid (not allowing expansion work), then at equilibrium the pressures inside and outside the boundary need not be the same.

In order for a system initially at equilibrium to undergo a change of state, its boundaries must be capable of transmitting some intensive

[6] Compare our discussion in Section 3.7 of the *generalized momentum* p_i conjugate to a given generalized coordinate q_i; the product of p_i and q_i, however, must have the dimensions of *action* (energy×time).
[7] This is not a new result, but merely a restatement of our definition of conjugate variables.

variable. That is, any such change of state must involve the passage of either work or heat across the system's boundaries. We have already discussed the various work processes; if none of these is possible, the system can interact with its surroundings only by the flow of heat, that is, by transmission of the intensive variable temperature through a diathermal wall. If neither work nor heat flow is possible, the system is what we have defined as isolated.

13.5
Quasi-static and Reversible Processes

We now return to a consideration of the path by which a change of state takes place. We have earlier pointed out that the change in any state function (volume, internal energy, etc.) depends only on the initial and final states, not on the path between them. Why, then, are we interested in the detailed nature of this path? The answer is that the work and, as we shall see, the heat associated with a given process does depend on the path. The original stimulus for thermodynamics was the study of heat engines, in which the amount of work obtained is clearly all-important. Even when one merely wishes to accomplish a given change of state, one must know *how* best to do it—that is, what routes between the initial and final states are available, and what amounts of work and heat are associated with each route.

So we wish to be able to describe the path followed by a system between its initial and final states. By "describing the path" we mean a specification of the sequence of states through which the system passes, represented by some curve in the kind of space defined in Section 13.2. But this immediately presents a problem. It will be recalled that only points in the surface defined by the equation of state represent states of the system, so that such a path must be entirely in this surface. But every point in the surface represents an *equilibrium* state, defined as one that does not change measurably with time. This means that any process occurring at a measurable rate cannot be described by a path in the equilibrium surface. Since any real change of state must occur at a nonzero rate, how do we describe it?

One obvious way to deal with this problem is by a limiting process. Imagine that in repeated experiments a given change of state, say, the expansion of a fluid, is carried out more and more slowly. In the limit we have a process occurring at an infinitesimal rate of change; such a process is called *quasi-static*. As with more rapid processes, a given change of state can still be carried out quasi-statically by infinitely many routes: The particular limiting behavior one obtains depends on the conditions, or constraints, placed on the system. We shall concern ourselves primarily with a particular class of quasi-static processes known as *reversible processes*.

The essential characteristic of a reversible process is that the system remains at all times infinitesimally close to equilibrium. The slower the process is carried out, the closer the system can be to an equilibrium state at any given time. The limit must thus be an infinitely slow process in which the system passes through a continuous sequence of equilibrium states. Such a process obviously does have a describable path, but how does this help us in the case of real processes occurring at measurable rates? The answer is that a very large class of real processes have characteristics (such as the values of work and heat involved) so close to those for the corresponding reversible processes that the latter can be assumed for purposes of calculation. The implication of this assumption, as we shall see later, is that the deviation from equilibrium brought about by the nonzero rate of change tends to zero more rapidly than the rate of change itself, so that during the process the system is always very close to equilibrium. For any given process, however, the validity of this assumption must ultimately be based on experiment.[8]

All this has been very abstract, so let us now give a concrete example. Consider the expansion of a perfect gas from volume V_1 to volume V_2. How can this process be carried out with the gas close to equilibrium at all times? Assume that the expansion is performed by moving a frictionless piston, as in Fig. 13.3, and that initially the external pressure p_{app} equals the internal pressure p. Then decrease p_{app} by a small amount Δp, and the gas will begin to expand. Maintain the relationship $p_{app} = p - \Delta p$ throughout the expansion (with both p and p_{app} steadily decreasing), and when the desired final volume is reached, set p_{app} again equal to p to stop the expansion. All this is clear enough. Now repeat the process with a smaller value of Δp; since the driving force (equal to Δp times the piston area) is less, the expansion will take place more slowly. In the limit $\Delta p \to 0$, we have an infinitesimal rate of expansion with $p_{app} = p$ at all times. Since the expansion is so slow, the pressure and temperature remain uniform throughout the gas at all times, and the gas passes through a continuous sequence of equilibrium states. In short, the limit we have described is a reversible expansion of the gas.

To calculate the work performed in such a process, we must specify at least one additional condition, since the state of the gas is described by two independent variables, and we already have the condition $p_{app} = p$. The simplest case is that of a constant-temperature expansion, which can be performed by giving the system diathermal walls and immersing it in a thermostat. Then, from Eqs. 13.10 and 12.36, we obtain

$$w = -\int_{V_1}^{V_2} p_{app}\, dV = -\int_{V_1}^{V_2} p\, dV = -\int_{V_1}^{V_2} \frac{nRT}{V}\, dV = -nRT \ln \frac{V_2}{V_1}$$

(13.14)

for the work performed on the system in the reversible, isothermal expansion of a perfect gas. If we represent a fluid's pressure as a function of volume by a curve in the (p, V) plane, then the magnitude of the work performed in a reversible expansion is simply the area under this curve, as shown in Fig. 13.4. For the isothermal expansion of a perfect gas, the curve AB is part of one of the hyperbolas of Fig. 13.2c. Similar graphical representations are easily derived for the other forms of work in Table 13.1. However, Fig. 13.4 is no longer valid if $p_{app} \neq p$, which we shall see means that the process is not reversible.

Note that in the above-described reversible process, (1) expansion work is performed and (2) the intensive variable associated with expansion work, the pressure, has the same value on both sides of the system boundary. This illustrates a general property of reversible processes: The intensive variables corresponding to the types of work being performed (cf. Table 13.1) must be continuous across the boundary at which that work is performed. It is easy to see why this must be so: We pointed out in the last section that at equilibrium an intensive variable has the same value on both sides of a boundary that transmits that variable. But since in a reversible process the system is effectively always in equilibrium, this continuity of intensive variables across the boundary must then also hold at all times. The last column of Table 13.1 therefore gives the work performed on the system in a reversible process of the type specified.

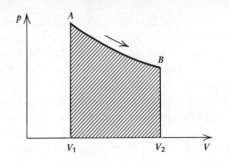

FIGURE 13.4
Work performed in the reversible expansion of a fluid. The heavy line AB gives the value of p as a function of V during the expansion from V_1 to V_2; the work performed on the fluid is then

$$w = -\int_{V_1}^{V_2} p\, dV,$$

equal to the negative of the area under the curve AB.

---

[8] This is true as long as one remains within the bounds of thermodynamics itself. In principle one should be able to determine the limits of validity from a microscopic theory, although this is not always feasible in practice.

We can now give a complete definition of a reversible process: (1) It takes place at an infinitesimal rate, i.e., it is quasi-static; (2) it passes through a continuous sequence of equilibrium states, and can thus be exactly described by a path in, say, (p, v, T) space; (3) at every point in the process, any intensive variable is continuous across a boundary at which the type of work corresponding to that variable is being performed.

Why is such a process called "reversible"? As already indicated, every point along the path is an equilibrium state. But this means that the direction of the process can be reversed by an infinitesimal change in the external constraints. Suppose, for example, that the process is the reversible expansion of a gas, in which p_{app} must equal (or rather, be infinitesimally less than) p at all times; if p_{app} were made infinitesimally *greater* than p, the gas would contract rather than expand. The direction of a reversible process thus cannot be determined from a description of the instantaneous state of the system. This is an idealization, since all real changes of state are *irreversible processes*—processes during which the system passes through nonequilibrium states, with unbalanced "forces" giving the process a unique direction. We shall have more to say about irreversible processes later.

It is clear that a given reversible path can be traversed in either direction, the details of the process being identical except for the sign of change. Thus if the work done in the forward process is $w_f = -\int_{V_1}^{V_2} p\,dV$, the work done in the corresponding reverse process is $w_r = -\int_{V_2}^{V_1} p\,dV$; since the path is the same, p is the same function of V in both cases, and we must have $w_f = -w_r$. We also have $(\Delta X)_f = -(\Delta X)_r$ for any state function X, but this would be true whether the path were the same or not; however, the equality of forward and reverse work is valid only for reversible processes.

One final point: If a reversible process consists of a continuous sequence of equilibrium states and takes place infinitely slowly, then what is the difference between a system actually in equilibrium and a system undergoing a reversible process? The answer is that true equilibrium is a time-independent state of a system that is exchanging neither work nor heat with its surroundings—a system that either is isolated or might as well be.

Let us now look at some processes that are not reversible. Consider again our earlier example of the expansion of a perfect gas, but this time omit the requirement that $p_{app} = p$. Presumably we are able to set p_{app} to any desired value. In general, the driving force, $A(p - p_{app})$, will be appreciable, and the expansion will occur at a nonzero rate—in short, it will be irreversible (cf. Fig. 13.5b). Clearly, Eq. 13.14 no longer gives the work performed, since the substitution $p_{app} = nRT/V$ is not valid unless $p_{app} = p$. Equation 13.10 remains applicable. In computing the work done, however, one must use the actual value of the applied pressure p_{app}. If conditions are such that p_{app} is not actually fixed or measured, calculating the work performed can be a very complicated problem. In particular, the work no longer equals the area under the $p(V)$ curve in Fig. 13.4, but the area under a similar $p_{app}(V)$ curve. It is even possible for *no* work to be performed in an irreversible process, as when a gas expands into a vacuum ($p_{app} = 0$). The nature of irreversible processes requires some careful examination, and we shall return to this problem in Chapter 16.

We have said that reversible processes are only one class of quasi-static processes. This implies that a process can take place at an infinitesimal rate (quasi-statically) and still be irreversible. How can this be? Consider the system in Fig. 13.5c, in which a fluid is impeded from expanding by a series of pegs holding the piston in place. If the lowest of

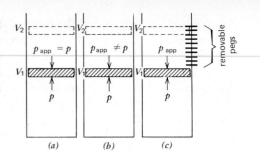

FIGURE 13.5
Reversible, irreversible, and quasi-static expansion of a fluid. (a) Reversible (and thus quasi-static): $p_{app} = p$ at all times, and the expansion proceeds infinitely slowly. (b) Irreversible: $p_{app} \neq p$, and the expansion proceeds at a measurable rate. (c) Quasi-static but irreversible: See text.

these pegs is removed, the fluid will expand irreversibly—but only as far as the next peg; this process can then be repeated over and over. If we imagine more and more pegs installed closer and closer together, the expansion can be carried out as slowly (and apparently continuously) as we like, in the limit quasi-statically. Yet no matter how slowly such a process takes place, it does not become reversible. Rather than passing through a continuous sequence of equilibrium states, the system is in equilibrium only during those moments when the piston is resting against a peg; during the actual expansion steps we presumably have $p_{app} = 0$, and only the pegs hold back the expansion.

A somewhat different type of quasi-static irreversible expansion would be one in which the piston does not move without friction: The work in this case is different from that in a reversible expansion because some of the energy is dissipated as heat. But to demonstrate this we need the first law of thermodynamics, which we are finally ready to introduce.

13.6
The First Law: Internal Energy and Heat

The first law of thermodynamics is an abstraction and generalization of experiment, based on the concepts of work and thermodynamic state. Consider a system enclosed by adiabatic walls, so that its state can be changed only by the performance of work. Since all real materials have some thermal conductivity, no real system can be so enclosed, but real systems exist that closely approximate the ideal one described. Many experiments on such systems lead to the conclusion that the amount of work needed to produce a given change of state is independent of the path. This conclusion is so important that we state it formally:

If the state of an otherwise isolated system is changed from A to B by the performance of work, the amount of work required depends solely on the initial state A and the final state B, and not on the means by which the work is performed, nor on the intermediate stages through which the system passes between the initial and final states.

This is our statement of the first law. The initial and final states referred to must, of course, be equilibrium states.

It is not immediately obvious that this abstract statement is equivalent to the law of conservation of energy. Using it, however, we are at last ready to introduce a thermodynamic definition of energy. In terms of only macroscopic operations, we define the *internal energy U* of a system by the statement:

If an otherwise isolated system is brought from one state to another by performance upon it of an amount of work w_{ad}, the change in the system's internal energy in the process is defined to be an amount ΔU exactly equal to w_{ad}.

(Here w_{ad} stands for "adiabatic work.") Formally, then, for work w_{ad} performed on an otherwise isolated system, we have

$$\Delta U = U_B - U_A = w_{ad}, \tag{13.15}$$

where A and B designate the initial and final (equilibrium) states, respectively. Note that ΔU is positive when work is done *on* the system *by* the surroundings, the condition for which we have set *w* positive by convention.

Two observations must be made here. First, our definition gives only the *change* in internal energy in a given process; it does not establish the zero of energy, which as usual is arbitrary. However, once the internal energy has been fixed arbitrarily at some value U_0 for a given equilibrium state of a system, the value of *U* for every other equilibrium state of that system is uniquely determined. Second, the first law requires that the work needed to produce a given change of state in an otherwise isolated system be independent of the path. Therefore ΔU for the same change of state must also be independent of the path, depending only on the initial and final states. In other words, the internal energy *U* is a state function. Like any other state function, *U* must depend only on the small number of macroscopic variables that define the equation of state. Thus we can write for a fluid

$$U = U_1(p, v) = U_2(p, T) = U_3(v, T), \tag{13.16}$$

where the functions U_1, U_2, U_3 differ in form but, since they represent the same state function, must have the same value in any given equilibrium state of the fluid.

As already mentioned, real systems are never enclosed by perfectly adiabatic barriers. Consider now a system whose containing walls are diathermal. When such a system undergoes a given change of state, the work *w* performed upon it is in general different from the work w_{ad} performed when the same change of state occurs in a system bounded by adiabatic walls. (By "the same change of state," remember, we mean a process between the same initial and final equilibrium states.) Whereas by the first law the adiabatic work w_{ad} is independent of the path, the corresponding diathermal work *w* does depend on the path. We can therefore *define* a quantity *q* by the relationship

$$q \equiv w_{ad} - w; \tag{13.17}$$

we call *q* the *heat* transferred to the system in the diathermal process. Using Eq. 13.15 for the adiabatic process, we can at once write

$$q = \Delta U - w, \tag{13.18a}$$

or, in more common form,

$$\Delta U = q + w. \tag{13.18b}$$

ΔU, the change in a state function, is of course the same in both processes. See Fig. 13.6 for a graphical representation of the relationships among ΔU, *w*, and *q*.

Equations 13.18 uniquely define the heat associated with a given process in terms of the work performed and the change of state produced (which determines ΔU). We have mentioned that work performed on the system by the surroundings is taken to be positive. The corresponding convention for heat, embodied in Eq. 13.17, is that heat transferred *to* the system *from* the surroundings is taken as positive. (More on these sign conventions later.) Heat, like thermodynamic work, is a form of energy that we observe only as it crosses the boundaries of a system. In fact, Eqs. 13.18 define *q* as the amount of (internal) energy that enters a system in a given process in forms other than the performance of work. These equations can thus be taken as a statement of the first law as the law of conservation of energy, in that energy is conserved if both heat and work are taken into account. This was essentially the original formulation of the first law, but in our version the concept of heat is secondary to that of internal energy.

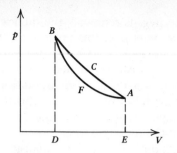

FIGURE 13.6
Graphical representation of internal energy, work, and heat. Consider a given change of state, in this case the reversible compression of a fluid from state A (p_1, V_1) to state B (p_2, V_2). Assume that *ACB* is the path followed when this change of state is carried out *adiabatically*. The adiabatic work w_{ad} associated with this change of state is then equal to the area *ACBDEA*, and by Eq. 13.15 this must equal the internal energy change:

$$\Delta U = U_B - U_A = w_{ad} = \text{area } ACBDEA.$$

Now consider any other path *AFB* along which the same change of state can be carried out diathermally. The work performed on the system in the diathermal process is then equal to the area *AFBDEA*, the internal energy change is the same as in the adiabatic process, and the heat *q* absorbed by the system must equal the area between the two paths:

$$q = w_{ad} - w = \text{area } ACBDEA$$
$$- \text{area } AFBDEA$$
$$= \text{area } ACBFA.$$

In this case *q* and *w* are both positive.

Although the internal energy is a state function, this is clearly true of neither work nor heat. We have already pointed out that the work associated with a given change of state depends on the path; by Eq. 13.18a this must also be true of the heat associated with the same change of state. Thus dU is an exact differential, whereas both work and heat have inexact differentials. In the notation introduced in Section 13.3, we can write

$$ đU = đq + dw \tag{13.19} $$

for an infinitesimal change of state. We can also write for a *cyclical* process (one that returns to its starting point)

$$ \oint dU = 0, \tag{13.20} $$

where the symbol $\oint$ designates integration over a cyclical path; in Chapter 16 we shall study the relationship between work and heat in such processes. Finally, for a given *reversible* path we showed in the previous section that the work done in corresponding forward and reverse processes is related by $w_f = -w_r$; since we must also have $(\Delta U)_f = -(\Delta U)_r$, we immediately obtain $q_f = -q_r$, again for reversible processes only.

For a system undergoing only reversible processes, we can write Eq. 13.19 in the generalized form

$$ dU = đq - \sum_i X_i \, dY_i, \tag{13.21} $$

where X_i, Y_i are the conjugate pairs of intensive and extensive variables involved in the various possible work processes (Table 13.1). We can now see why we can define the intensive variable X_i conjugate to a given extensive variable Y_i by

$$ X_i \equiv -\left(\frac{\partial U}{\partial Y_i}\right)_{\text{rev, ad, } Y_j = \text{const.}}, \tag{13.22} $$

where the subscripts identify a reversible adiabatic process in which all other independent extensive variables Y_j are held fixed.[9]

Remember, however, that the first law is not restricted to reversible processes. The adiabatic work defining ΔU is the same for any process, reversible or irreversible, that joins the same initial and final states—provided only that those states are equilibrium states. This is fortunate for the chemist, since a chemical reaction ordinarily takes place irreversibly. Yet, as we shall see in Chapter 14, the first law can indeed be applied to such reactions.

13.7 Some Historical Notes

We have mentioned that the first law was originally formulated in terms of the conservation of energy. One of the cornerstones of modern thermodynamics was thus the demonstration that work could be converted into heat, with all forms of work giving equivalent amounts of heat. It is of interest to sketch some of the background of this principle.

It was Count Rumford who first showed, in 1798, that mechanical work could be continuously converted to heat (by friction, in the boring of cannon). In 1799 Humphry Davy performed a similar experiment: When two pieces of ice were rubbed together, the ice melted.[10] Later

---

[9] We shall see in Chapter 16 that it is not necessary to make a separate assumption of adiabaticity: in a reversible process we have $đq = T \, dS$, where S is the extensive variable entropy.

experiments made it increasingly clear that heat was in some way a form of energy, where "energy" means roughly the ability to do work. The conservation of energy was already a well-established principle for mechanical systems in the absence of friction (heat). Such systems therefore were called "conservative." It was natural to extend this principle to include heat, but it took some time for the necessary experimental evidence to be obtained. In 1842 J. R. von Mayer first formulated the general principle of the equivalence of different kinds of energy and the conservation of total energy. However, it was primarily the work of J. P. Joule in the next few years that provided the quantitative basis for the law of conservation of energy.

Joule used various ways of producing heat from work, the best known being the rotation of a paddle wheel in a liquid, with the paddle wheel driven by a falling weight. He also measured the heat produced when an electric current passes through a resistance, when bodies are rubbed together, in the expansion and contraction of air, and in other processes. In general, the work done was measured in mechanical or electrical terms, whereas the heat produced was measured by the rise in temperature of a known mass of substance (see Chapter 14). Joule's results established the existence of a consistent proportionality between the mechanical work w dissipated in a system and the heat q produced as a result:

$$w = Jq \tag{13.23}$$

where the quantity J is called the *mechanical equivalent of heat*.[11]

In the older literature J is a conversion factor between work in mechanical units (ergs or joules) and heat in *calories*, one mean 15° calorie ($cal_{15°}$) being the heat required to raise the temperature of 1 g of water from 14.5 to 15.5°C at atmospheric pressure. In terms of this unit Mayer had estimated a value of J equivalent to 3.6 $J/cal_{15°}$, whereas Joule obtained (1849) a value of 4.15 $J/cal_{15°}$; modern measurements give

$$1 \, cal_{15°} = 4.1855 \, J.$$

The modern convention, however, is to define the calorie directly in terms of work, rather than as a function of the properties of water. The definition currently used by chemists, known as the *thermochemical calorie* (cal_{th} or simply cal), is

$$1 \, cal_{th} \equiv 4.1840 \, J,$$

so that $1 \, cal_{15°} = 1.00036 \, cal_{th}$. (Several other slightly different calories have also been used, so one should be careful when using published data.) The existence of separate energy units for heat and work is superfluous, and the calorie is being phased out of use; however, many of the thermochemical data in the literature are still expressed in terms of calories.

The significance of Eq. 13.23 can be expressed in several ways. One of the most incisive is to regard a thermodynamic system as a reservoir of

---

[10] Recent historical research has revealed that, although Davy's conclusion was valid, his experimental technique was faulty, and the observed melting was caused by heat leaks from the surroundings.

[11] For electrical work Joule found that

$$I^2Rt = Jq,$$

where I is current, R is resistance, and J has the same value as for mechanical work. The quantity I^2Rt is still known as the "Joule heat" produced by an electric current.

energy. Then the convertibility of work and heat implies that any energy passing inward across the system's boundaries increases the total amount of energy present, no matter how it enters the system: by performance of mechanical work, transfer of heat, or any other method. This view is equivalent to saying that heat is just energy in transit to or from the system. It is clear that this interpretation is closely associated with the definition of heat given by Eq. 13.17. Furthermore, since heat is defined only for a particular type of energy-transfer process, there is no meaning to the phrase "heat in the system"; this is, of course, the same conclusion we reached earlier for work.

We should say something about the history of the sign conventions for heat and work. In this text, it will be recalled, we use the conventions that w is positive for work done *on* a system, and q is positive for heat absorbed *by* a system. Many textbooks use the opposite sign convention for work, considering work done *by* a system as positive; given this other convention, the first law assumes the form $\Delta U = q - w$. Historically, thermodynamics evolved in large part in response to the stimulus imparted by the introduction of the steam engine and the need to understand the performance of engines. Since engines are constructed for the purpose of performing a useful task, it was natural for work done by an engine to be reckoned positive. This was generalized to the assignment of a positive sign to work done by any system on its surroundings. Today, however, scientists are generally more concerned with the thermodynamic properties of a system itself, particularly its internal energy. The sign of U is fixed by the notion that to do work a system must "expend energy." Thus U is defined by Eq. 13.15 in such a way as to increase when adiabatic work is done on the system and decrease when the system does adiabatic work on the surroundings, and it now becomes natural to reckon positive that work which *increases* a system's internal energy, that is, work done on the system. Although this convention has received international recommendation, the use of the older convention is still widespread, especially among engineers; one should be careful to note which is used in any thermodynamics text one consults.

The history of the sign convention for heat is more straightforward. The engines with which early thermodynamics dealt were driven by the addition of heat to the system, usually by building a fire under the boiler; thus it was natural to consider heat added to a system as positive. The shift in emphasis from engines to system properties required no change in this convention, since addition of heat to a system also increases its internal energy. In short, both w and q are now defined as positive when they tend to increase the internal energy of the system under consideration.

13.8
Microscopic Interpretation of Internal Energy and Heat

Let us now return to the microscopic level of description. Our thermodynamic definition of a system's internal energy (Section 13.6) is based on adiabatic work processes. "Adiabatic" here is used in the thermodynamic sense, implying that energy crosses the system boundary only in the form of work. How does this correspond to the "adiabatic perturbation" of Section 13.1? The process described there, expansion of a perfect gas, must be thermodynamically adiabatic, since the container walls act only as elastic reflectors for the gas molecules. But the adiabatic perturbation is also so slow that no transitions take place between energy levels; the gas is thus always in an equilibrium state, and the process is reversible in the thermodynamic sense. We conclude that the adiabatic perturbation of quantum mechanics corresponds to the reversible adiabatic process of thermodynamics. The discrepancy between the two senses of "adiabatic" is unfortunate, but seldom leads to confusion; henceforth we shall ordinarily use the thermodynamic sense only.

Thus we have a microscopic equivalent of a reversible adiabatic work process, as described for a perfect gas model in Section 13.1. By simply reformulating our definition, we can interpret the change in the system's internal energy as the total change in mechanical energy of the system's constituent particles. For the perfect gas model we saw that a reversible adiabatic compression leads to an increase in the average kinetic energy per particle, but with the distribution of particles over the energy levels remaining unaltered. Using the same model, we can now consider the microscopic interpretation of the heat transferred in nonadiabatic processes.

In general, when a system interacts with its surroundings so that both work and heat are exchanged, both the microscopic energy spectrum and the distribution of particles over the energy levels change. Suppose, for example, that our model perfect gas expands with the temperature fixed. The energy spectrum of the perfect gas, Eq. 13.1, depends only on the dimensions of the box, not on how those dimensions are established; when the box size increases, the energy of each level decreases. We have identified the temperature with the average kinetic energy; for this quantity to remain constant in an expansion, some particles must be promoted to higher energy levels. We showed in Section 13.1 that the average kinetic energy decreases in a reversible adiabatic expansion. For a reversible expansion to the same volume with the average kinetic energy fixed, clearly some energy must be added from the surroundings. It is this added energy, which must be transferred as kinetic energy of the atoms in the box walls, that corresponds to thermodynamic heat.

We have therefore reached these important conclusions: In a reversible adiabatic work process, the distribution of molecules over a system's energy levels remains unchanged; in any other reversible process, the heat transferred across the system boundary is related to the change in the distribution of molecules over energy levels.[12] In an irreversible process, there is in general a change in the molecular distribution whether heat is transferred or not, since the work performed differs from that in the corresponding reversible process. As with the conclusions of Section 13.1, we shall demonstrate later that these statements are valid generally and not merely for the perfect gas model. To quantify the argument, we must first study the second law of thermodynamics and introduce some macroscopic measure of the distribution of molecules. This will be done in Chapters 15 and 16 with the introduction of the concept of entropy.

We can draw one additional macroscopic conclusion from our microscopic perfect gas model. Suppose again that the volume of the box is changed with the temperature held constant. But for a perfect monatomic gas, the only energy attributable to the molecules in the box is their kinetic energy. Since the number of molecules is constant, we conclude that the total energy of the gas does not depend on the box's volume in a constant-temperature process. This result for a quantum mechanical perfect gas is in fact the same as our conclusion in Section 12.6 for a classical perfect gas. By identifying our microscopic model with a macroscopic gas obeying the ideal gas law (an identification that is now confirmed), we obtained Eq. 12.42, stating that

$$U = \tfrac{3}{2}nRT. \tag{13.24}$$

---

[12] Of course, the work done in the two processes is also different. The relationship between work done, heat transferred, and changes in the distribution of molecules over a system's energy levels is discussed in Chapter 15.

In Eq. 12.42, however, U stood only for the kinetic energy of the molecules; now we can identify U with the thermodynamic internal energy.[13] Given this identification, we can say that the internal energy of a perfect gas depends only on temperature, and is independent of density (volume) or pressure. If one wishes to define a perfect gas without any microscopic assumptions, this must be made part of the definition; we shall consider this point further in Chapter 14.

We can now complete a derivation we began in Section 13.5. By Eq. 13.14, the work performed on a perfect gas in a reversible isothermal expansion is $w = -nRT \ln(V_2/V_1)$. But by the above argument, we must have $\Delta U = 0$ for any isothermal process involving a perfect gas. Substituting in Eqs. 13.18, we immediately obtain

$$q = -w = nRT \ln \frac{V_2}{V_1} \qquad (13.25)$$

for the heat absorbed in the reversible isothermal expansion of a perfect gas. The explicit calculation of w and q for other kinds of processes is more complicated; in Chapter 14 we shall see how to treat the general perfect gas case.

One other interesting point: Eq. 13.4 gives the volume dependence of a perfect gas's total energy E in what we can now call a reversible adiabatic expansion. Again identifying E with the thermodynamic internal energy, we obtain

$$dU = -\frac{2U}{3V} dV = -\tfrac{2}{3}(\tfrac{3}{2}nRT) \frac{dV}{V} = -nRT \frac{dV}{V} = -p\,dV, \qquad (13.26)$$

where we have substituted Eq. 13.24 and the perfect gas law. The result $dU = -p\,dV$ is, of course, true for the reversible adiabatic expansion of any fluid (since in any adiabatic process $dU = đw$), but here we have derived it microscopically for the perfect gas model.

13.9 Constraints, Work, and Equilibrium

In Section 12.3 we introduced the notion of constraints, the boundary conditions defining a thermodynamic system. We have made little use of this concept, except in our definition of thermal equilibrium: A new equilibrium is achieved when an adiabatic barrier is replaced by a diathermal one, that is, when a constraint is removed. It is generally true that any change in the constraints imposed on a system changes the equilibrium state of the system. We shall now see that the kind of work done by a system is also defined by the nature of the constraints imposed.

Consider a gas enclosed in a rigid cylinder with adiabatic walls, with a sliding piston dividing the gas into two volumes, V_1 and V_2 (see Fig. 13.7). If the piston is an adiabatic barrier and is fixed in place, then the two subsystems are isolated, and the pressures p_1 and p_2 and temperatures T_1 and T_2 can be completely different; there are four independent variables, two in each subsystem. If the adiabatic barrier is replaced by a fixed diathermal barrier, the two temperatures will equalize but the pressures can still be different; now there are only three independent variables, since $T_1 = T_2$. Finally, if the diathermal piston is allowed to move freely, it will move to a position such that $p_1 = p_2$, while still keeping $T_1 = T_2$; the final system has two independent variables. If in the final process the piston were coupled to an external machine instead of moving freely, external work would be done; assuming that this work

[13] Remember that the zero of energy is always arbitrary. Here, as in Chapter 12, we have implicitly taken $U = 0$ to correspond to zero kinetic energy of the gas molecules.

process is still adiabatic, the system's internal energy and thus its temperature must decrease. We would still have $T_1 = T_2$, but at a lower value.

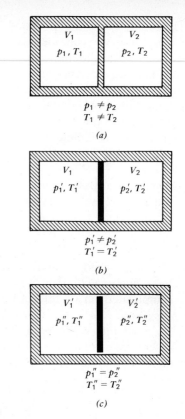

$p_1 \neq p_2$
$T_1 \neq T_2$

(a)

$p_1' \neq p_2'$
$T_1' = T_2'$

(b)

$p_1'' = p_2''$
$T_1'' = T_2''$

(c)

FIGURE 13.7
Interactions of two gaseous subsystems in a rigid adiabatic container. (a) Fixed adiabatic barrier. (b) Fixed diathermal barrier. (c) Movable diathermal piston.

What conclusions can we draw about constraints from these and other observations? It is found that:

1. For each constraint imposed on a system, an additional independent variable is required to describe the system. (This can in fact be used as the definition of a constraint.)

2. The kind of equilibrium achieved by a system is defined by the constraints. When a system is initially in equilibrium subject to a given set of constraints, either addition or removal of a constraint will ordinarily generate a new equilibrium state, described by a different set of independent variables.

3. The kind of work performed by a system is determined by the constraints under which it acts. Stated more precisely, whenever a given constraint is removed, the system becomes capable of spontaneously performing a particular kind of work; to restore the system to its original state including the constraint, work must be performed on it. Thus, in the above example, pV work *can* be performed when the constraint of a fixed barrier is removed, and *must* be performed to restore the barrier to its original position. Similarly, electrical work can be performed when a capacitor or electrochemical cell is discharged (i.e., when an electrically insulating barrier is removed).

Another way to express point 2 is that a system is at equilibrium with respect to change in certain variables, with the constraints operationally defining which such variations are allowable. For example, a one-dimensional harmonic oscillator has its equilibrium at a point where the potential energy is a minimum with respect to the displacement x; the system is constrained not to allow displacements in other directions. In Fig. 13.7c the system reaches equilibrium with respect to variation of (say) $p_1 - p_2$ and $T_1 - T_2$; but what criterion of equilibrium corresponds to the minimum potential energy of the mechanical system? In Chapter 19, where we expand upon this viewpoint, we shall see that there are many such criteria—in fact, one for each combination of constraints on a system.

For another example of the interplay between equilibrium states and constraints, consider a room-temperature mixture of H_2 and O_2. The system is effectively in metastable equilibrium, since the reaction between the two gases is so slow as to be negligible. We can thus say that there is a constraint against chemical reaction. In this situation one can vary the amount of either gas independently of the other; thus for fixed volume there are three independent variables, which can be chosen from among p, T, $n(H_2)$, and $n(O_2)$. Suppose now that the constraint against chemical reaction is removed, either by adding a suitable catalyst or by raising the temperature. The H_2 and O_2 will react, and a new equilibrium will be attained. Corresponding to the removal of one constraint, there is one less independent variable, since $n(H_2)$ and $n(O_2)$ can no longer be varied independently (they are connected through the equilibrium constant).

In this chapter we have developed the first law of thermodynamics and various associated topics, all in rather general and abstract terms. Chapter 14 will be devoted mainly to applying these concepts to practical calculations. The first law enables us to calculate the change in internal energy (and other state functions) in any real or hypothetical process; it does not tell us anything about whether a given process will occur or not. The latter question is extremely important to chemists, (e.g., given a mixture of reactants and products, which way will the reaction go?), and we shall have to come back to it later.

Further Reading

Caldin, E. F., *Introduction to Chemical Thermodynamics* (Oxford University Press, London, 1958), chap. 2.

Zemansky, M. W., *Heat and Thermodynamics*, 4th ed. (McGraw-Hill, New York, 1957), chaps. 3 and 4.

Problems

1. Consider the system pictured below:

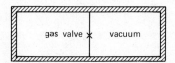

   Gas is confined to a subvolume V_1 in an insulated rigid container. The container has an adjoining subvolume V_2, initially evacuated, which can be connected to V_1 by opening a valve. Suppose that the valve is opened and the gas flows out of V_1, finally filling the entire volume $V_1 + V_2$. Calculate the work done by the gas in this expansion, and the change in the internal energy of the gas.

2. A rock of mass M_R, density ρ_R, is dropped from a height h above the bottom of an insulated beaker containing a viscous fluid. Let the density of the fluid be ρ_F and its height in the beaker d. Assume that no fluid splashes out of the beaker when the rock falls in. Calculate the heat generated as a result of dropping the rock into the beaker.

3. An electric current passes through a resistor that is immersed in running water. Let the resistor be the system under investigation.
   (a) Is there a flow of heat into the resistor?
   (b) Is there a flow of heat into the water?
   (c) Is work done on the resistor?
   (d) Assuming the resistor to be unchanging, apply the first law to this process.

4. A current of 1 A flows across a 2 V potential difference for 1 min. If the energy carried by this current were completely converted to work, calculate how high a 100-g mass could be lifted in the earth's gravitational field.

5. Is the product of two intensive variables an intensive variable or an extensive variable? How about the product of an extensive and an intensive variable?

6. The energy, mass, and volume of a system are extensive. How about the molar energy, molar mass, and molar volume? How about the energy and mass densities?

7. In Fig. 13.5c, how much work is done by the piston as the gas expands? Assume that the pegs are perfectly rigid.

8. In each of the following cases, an ideal gas expands against a constant external pressure. For each case, decide if the expansion is allowed by the first law, and if so, if the expansion can be carried out reversibly:
   (a) Isothermal (constant temperature) and adiabatic
   (b) Isothermal but not adiabatic
   (c) Neither isothermal nor adiabatic

9. Describe how the apparatus shown in Problem 1 above might be used to determine if a gas behaves ideally.

10. A current of 1 A is passed through a resistance of 1 Ω. How long must the current be maintained to raise the temperature of 250 ml of water from 25°C to 100°C?

11. If the force on a particle depends only on the particle's position (conservative force), show that the work done on the particle in the course of a displacement is independent of path. Mechanical work in a conservative force field is analogous to what kind of thermodynamic work?

12. Evaluate the work integral

$$\int_{0,0}^{1,1} p \, dV$$

along the following paths in the (p, V) plane:
   (a) Straight line connecting $0,0$ and $1,1$
   (b) Parabola connecting $0,0$ and $1,1$ $(p = V^2)$
   (c) Rectangular path $(0,0 \rightarrow 0,1 \rightarrow 1,1)$
   (d) Rectangular path $(0,0 \rightarrow 1,0 \rightarrow 1,1)$
   Next, evaluate the integral

$$\int_{0,0}^{1,1} V \, dp$$

along each of the paths (a) through (d); show that the sum of the two integrals is independent of the path of evaluation.

13. Suppose that the isothermal compressibility of a solid, defined by

$$\kappa_T \equiv -\frac{1}{V}\left(\frac{\partial V}{\partial p}\right)_T,$$

is a constant independent of V and T. Write down an equation of state for the solid valid for not too high pressure in terms of κ_T and the volume V_0 ($V = V_0$ when $T = T_0$ and $p = 1$ atm). Calculate the work done in increasing the pressure quasi-statically and isothermally on a solid from p_1 to p_2. Apply this formula to calculate the work done when 10 g of copper is subjected to a pressure increase from 1 atm to 1000 atm at 0°C. For copper we have $\rho = 8.93$ g/cm^3 and $1/\kappa_T = 1.31 \times 10^{12}$ dyn/cm^2 at 0°C.

14. A thin-walled metal bomb of volume V_B contains n mol of a perfect gas at high pressure. Connected to the bomb is a capillary tube and stopcock. When the stopcock is opened slightly, the gas leaks slowly into a cylinder containing a nonleaking frictionless piston. The gas pushes against the piston, and the pressure in the cylinder is maintained at the constant value p_0 by moving the piston.
   (a) Show that, after as much gas as possible has leaked out, an amount of work

$$p_0(V_B - nv_0)$$

has been done. Here, v_0 is the molar volume of the perfect gas at pressure p_0 and temperature T_0.
   (b) How much work would be done if the gas leaked directly into the atmosphere?

15. During a quasi-static adiabatic expansion of a perfect gas the pressure and volume are related by

$$pV^\gamma = \text{constant},$$

where γ is another constant. Show that the work done in expanding from a state (p_i, V_i) to the state (p_f, V_f) is

$$w = \frac{p_f V_f - p_i V_i}{\gamma - 1}.$$

If $p_i = 10$ atm, $p_f = 2$ atm, $V_i = 1$ liter, $V_f = 3.16$ liters, and $\gamma = 1.4$, how much work is done?

16. Calculate the coefficient of thermal expansion,

$$\alpha \equiv \frac{1}{V}\left(\frac{\partial V}{\partial T}\right)_p$$

and the isothermal compressibility

$$\kappa_T \equiv -\frac{1}{V}\left(\frac{\partial V}{\partial p}\right)_T,$$

for a perfect gas. Compute α and κ_T numerically in units of K^{-1} and $(\text{torr})^{-1}$ at $T = 273$ K, $p = 1$ atm.

17. Calculate the work done when:
(a) One mole of water freezes, at 0°C, $p = 1$ atm to form 1 mol of ice. The density of ice is 0.917 g/cm^3 at 0°C, and that of water is 1.000 g/cm^3 at 0°C.
(b) One mole of water is vaporized at 100°C against an applied pressure of 1 atm. At 100°C the vapor pressure of water is 1 atm. Assume that the vapor is a perfect gas. How much error is made if the volume of the liquid is neglected?

18. If the length of a wire in which there is a tension F is changed from L to $L + dL$, what is the infinitesimal amount of work done?

19. The tension in a wire (length L, cross-sectional area A) is increased quasi-statically and isothermally from F_i to F_f. If the isothermal Young's modulus $Y \equiv (L/A)(\partial F/\partial L)_T$ can be regarded as constant during this process, show that the work done is

$$w = \frac{L}{2AY}(F_f^2 - F_i^2).$$

Assume that the stretching of the wire is so small that one can write $F = k(L - L_0)$, with k a constant.

20. Given that the length L of a thin wire under the action of a stretching force F can be expressed as

$$L = a + bF,$$

what experiments would be necessary to obtain an equation of state?

21. Discuss, in general terms, what operations need to be carried out in order to specify the complete set of macroscopic coordinates needed to specify the thermodynamic state of the system. Consider such variables as size, surface area, gravitational field strength, and any others you think relevant.

22. Suppose that the equation of state of a gas is

$$\left(p + \frac{n^2 a}{V^2}\right)(V - nb) = nRT$$

(the van der Waals equation), where a and b are constants. Show that if the gas is isothermally and reversibly compressed from volume V_1 to volume V_2, the work done is

$$w = -nRT \ln\left(\frac{V_2 - nb}{V_1 - nb}\right) - n^2 a\left(\frac{1}{V_2} - \frac{1}{V_1}\right).$$

23. The isothermal compression of a perfect gas from volume V_1 to volume V_2 requires work equal to

$$w = -nRT \ln\left(\frac{V_2}{V_1}\right).$$

Hence, the work computed in Problem 14 differs from the work of compression of a perfect gas by

$$\Delta w = -nRT \ln\left(\frac{1 - nb/V_2}{1 - nb/V_1}\right) - n^2 a\left(\frac{1}{V_2} - \frac{1}{V_1}\right).$$

Given that the constant a is a measure of intermolecular attraction, and that b is a measure of the volume occupied by the molecules, how do you interpret the above equation for Δw?

24. Suppose that the equation of state of a gas is

$$pv = RT\left(1 + \frac{B}{v}\right),$$

where B can vary with temperature, but not with pressure. Repeat the calculation of Problem 14 and the analysis of Problem 15. How would B be related to the a and b of the van der Waals equation?

The following problems are designed to illustrate, by use of a simple model, the (classical mechanical) kinetic theory interpretations of adiabatic and isothermal work.

25. Imagine that a perfect gas is contained in a cylinder that is closed by a frictionless piston. Suppose that all the molecules have the same speed v and

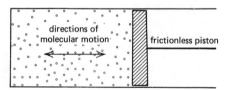

can move only in the direction perpendicular to the piston face. Then the molecular velocity **v** has only two components, namely, $\pm v$. As a result of collisions between the gas molecules and the piston, energy and momentum are exchanged between them. If m is the mass of a gas molecule, and M and v_p are the mass and speed of the piston, show that the change in kinetic energy (KE) of a gas molecule as a result of collision with the piston is

$$\Delta KE = \frac{4mM}{(m+M)^2}\left[\frac{M}{2}v_p^2 - \frac{m}{2}v^2 - \tfrac{1}{2}(M-m)vv_p\right].$$

26. Suppose that the frictionless piston in the system described in the preceding problem is acted on by a force **F**, which has the consequence that the piston moves with the *constant speed* v_p. Furthermore, suppose that energy can be transferred from the piston to the gas molecules, but not to any other part of the system or to the surroundings (adiabatic compression or expansion). Show that if $M \gg m$ and $v \gg v_p$, the rate of gain of kinetic energy of the gas is

$$\frac{d}{dt}(KE) = -2Z_c mv_p v,$$

where Z_c is the number of collisions per second of all molecules with the piston. Interpret this result.

27. Show from the result of Problem 26 that the gain of kinetic energy by the gas in some infinitesimal time interval is just the work

$$đw = -p\,dV.$$

28. In Problem 25 we modeled an adiabatic expansion or compression. To model an isothermal expansion or compression, imagine that a frictionless piston is between two perfect gases, and subjected to collisions from both (see diag-

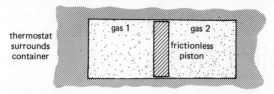

ram). As a result of collisions between the piston and the molecules of gases 1 and 2, the piston jiggles back and forth. When it is balanced between the two gases its average velocity, $\langle v_p \rangle$, is zero, but its mean square velocity $\langle v_p^2 \rangle$, is not zero. The jiggling of the piston is the mechanism by which energy is transferred between gases 1 and 2.

The isothermal compression is imagined to occur as follows. Suppose that there is a very small imbalance of forces on the piston (say, gas 2 pushes on gas 1), so small that the resultant piston velocity is very, very small, indeed sensibly zero. Suppose further that the surroundings of the container are a thermostat, so that the temperatures of gases 1 and 2, and the piston, are fixed throughout the compression. Note that this condition implies that the kinetic energy of the piston cannot change during compression. Show by direct calculation that the kinetic energy of gas 1 is not increased as a result of compression. Use the same simple model for molecular motion as in the preceding problems.

Thermochemistry and Its Applications

We now pause in our development of thermodynamics to see what use can be made of what we have already learned—mainly, the first law. In short, we shall come down from the heights of abstraction and deal for a while with applications. In this chapter we are concerned with the energy and temperature changes that accompany physical and chemical processes of various kinds. The measurement and interpretation of such changes constitutes the science of *thermochemistry*.

To obtain meaningful interpretations of thermochemical measurements, the conditions under which the various processes occur must be carefully defined. We anticipate, for example, that a given chemical reaction will release different amounts of heat when it proceeds under different constraints, since the heat and work associated with any process depend on the path. The first part of the chapter is therefore devoted largely to definitions of the many specific kinds of energy change, and of the standard states relative to which they are defined.

The remainder of the chapter deals with the molecular interpretation of thermochemical data. This is an extremely rich field, in which we finally have the opportunity to combine microscopic and thermodynamic knowledge in a useful way. Although still more can be learned with the help of the second law of thermodynamics, we shall see that the first law alone provides a wealth of fascinating information about such chemically significant topics as bond energies, the shape of molecules, and the stability of oxidation states.

14.1 Heat Capacity and Enthalpy

We begin by introducing some new thermodynamic functions, including those that are particularly applicable to describing energy transfer, the heat capacities. We have had occasion to mention heat capacity before (as early as Section 2.9), but now we are ready for a precise definition. Let us approach this by way of an idealized thermochemical experiment.

Consider a vessel with adiabatic walls, completely filled with fluid and containing a thermometer and an electric heater. Suppose that the walls are rigid, so that the volume of the vessel is fixed. The system is thus isolated except for the heater. At equilibrium in the initial state the temperature of the fluid is T_1. If some current is passed through the electric heater, a new equilibrium will be established at a higher temperature T_2; we define $\Delta T \equiv T_2 - T_1$. The electrical energy supplied is dissipated as a quantity of heat q. The limiting ratio of q to ΔT as both quantities tend to zero,

$$\lim_{\Delta T \to 0} \left(\frac{q}{\Delta T} \right)_V \equiv C_V, \qquad (14.1)$$

is called the *heat capacity at constant volume.*[1] But by the first law of thermodynamics, Eq. 13.18, in the process described we have

$$\Delta U = q_V, \tag{14.2}$$

since no work can be performed by the fluid (i.e., $w_V = 0$). We therefore find that C_V is the rate of change of internal energy with temperature when the system is maintained at constant volume:

$$C_V = \left(\frac{\partial U}{\partial T}\right)_V. \tag{14.3}$$

The constraint of constant volume must always be displayed explicitly.

But C_V is not the only kind of heat capacity. Suppose that we perform the same experiment as before, except that one wall of the vessel is a freely moving (but still adiabatic) piston. The volume of the fluid will no longer be the same after addition of heat. However, we can instead maintain the external pressure p_{app} fixed; if the heat is added so slowly as to approximate a reversible process, the fluid pressure p can also be assumed fixed. The limiting ratio for such a constant-pressure process,

$$\lim_{\Delta T \to 0} \left(\frac{q}{\Delta T}\right)_p \equiv C_p, \tag{14.4}$$

is called the *heat capacity at constant pressure.* (In what follows, pressure–volume work processes should be assumed reversible unless otherwise stated; thus we can consistently take $p = p_{app}$.)

Although one can define many other heat capacities,[2] C_V and C_p are the only ones with which we now need concern ourselves. Both C_V and C_p as we have defined them are extensive quantities, proportional to the mass of fluid present. It is customary to refer them to a specific amount of material, usually the mole, in which case they are called *molar heat capacities.* (The term "molar specific heats" is also used, but it is best to restrict "specific heat" to the meaning of heat capacity per gram.) They can be given in such units as joules per degree (°C or K) per mole. The molar heat capacities are state functions characteristic of the particular substance, and we shall see in later chapters how they can be related to the microscopic energy spectrum. For notational consistency we shall use capital letters for extensive functions, for example U for the total internal energy of an arbitrary mass of substance, and lower case letters for the corresponding molar functions, for example u for the internal energy per mole. However, because the usage is so common, we also use capital letters for the changes in energy, enthalpy, and so forth per mole of product in a chemical reaction.

When the fluid in the constant-pressure experiment is allowed to expand, work is done by the system on the surroundings ($w_p < 0$). By the first law, adding a given quantity of heat must cause a smaller change in the internal energy of the system when the system does work than when no work is performed. In the constant-volume process, none of the energy added to the system as heat is transferred back to the surroundings as work; it all goes into the translational and internal energy of the molecules. By Eq. 13.18, then, for the same positive value of q we must have $(\Delta U)_p < (\Delta U)_V$. Since internal energy increases with temperature,

---

[1] As in the partial derivative notation, we use a subscript (here V) to denote a variable that is held constant in the process described.

[2] For example, in a system whose properties depend on the electric field $\mathbf{E}$, one might need to use $C_{\mathbf{E}} \equiv \lim_{\Delta T \to 0} (q/\Delta T)_{\mathbf{E}}$.

this suggests that for the same q we should have $(\Delta T)_p < (\Delta T)_V$; when this is true (as it usually[3] is), comparison of Eqs. 14.1–14.4 tells us that $C_p > C_V$.

For a constant-volume process we can rewrite Eq. 14.3 in the differential form

$$(dU)_V = C_V \, dT. \tag{14.5}$$

Does there exist a comparably simple relationship for a constant-pressure process? Using the first law in differential form, Eq. 13.19, we have

$$(dU)_p = C_p \, dT - p \, dV \tag{14.6}$$

for a reversible constant-pressure process in a system that can perform only pV work. The heat added at constant pressure, $C_p \, dT$, thus equals the quantity $(dU + p \, dV)_p$. This suggests that we should introduce a new thermodynamic variable,

$$H \equiv U + pV, \tag{14.7}$$

since we can then write

$$(dH)_p = (dU + p \, dV)_p = C_p \, dT, \tag{14.8}$$

analogous to Eq. 14.5; the analog of Eq. 14.3 is thus

$$C_p = \left(\frac{\partial H}{\partial T}\right)_p. \tag{14.9}$$

The function H is called the *enthalpy* of the system. Since U and pV are both unique functions of the thermodynamic state of the system, H must also be a state function; it has the dimensions of energy, and its zero has the same arbitrariness as that of U.

The introduction of the function H may seem arbitrary or even whimsical. This is not so, however. When the system is a simple fluid that can perform only pV work, the natural variables defining the internal energy, those in terms of which the function has the simplest form, are T and V. But many processes are more conveniently carried out at constant pressure than at constant volume,[4] and are thus described more compactly in terms of enthalpy changes than in terms of energy changes. For example, the heat transferred in a reversible constant-pressure process is just $(\Delta H)_p = q_p$.[5] The passage from Eq. 14.6 to Eq. 14.8 is best regarded as a change of independent variable from V to p. If the same change of variable is carried out for a constant-volume process, in which $(\Delta U)_V = q_V$, we obtain

$$(dH)_V = (dU)_V + [d(pV)]_V = C_V \, dT + V \, dp. \tag{14.10}$$

In Section 14.3 we shall see how equations such as these can be used to obtain ΔH for arbitrary changes of state.

The apparatus actually used in experimental thermochemistry is not very different from the idealized models already described. One uses an insulated vessel, called a *calorimeter*, in which heat is delivered to a sample under controlled conditions, usually constant pressure or constant volume. Although our discussion so far has been in terms of a fluid sample, similar measurements can of course be made on solids. The science of

[3] An exception occurs if the fluid *contracts*, rather than expanding, when heat is added at constant pressure ($q_p > 0$, $w_p > 0$); then we have $(\Delta U)_p > (\Delta U)_V$ and thus $C_p < C_V$. This anomaly occurs in liquid water between 0 and 4°C.

[4] An obvious example is any process open to the atmosphere.

[5] In general, the enthalpy change for a constant-pressure process equals the heat that flows across the boundary as long as the pressure–volume work is reversible.

calorimetry has advanced to the point where very small energy changes and very small amounts of a substance can be studied. We shall not go into the fascinating experimental details,[6] but merely mention some general principles that must be taken into account.

The insulation of a calorimeter may be provided by vacuum jacketing or by the use of materials with a very small thermal conductivity, such as a polyurethane foam. Whatever the arrangement, part of the vessel itself will be within the insulation, and one must allow for its heat capacity when calculating that of the sample. The heat capacity of the calorimeter can usually be determined experimentally by measurements in the absence of a sample.

Constant-pressure calorimeters for fluids are inconvenient to work with, and so are seldom used in the laboratory. However, there is no problem if the sample is a nonvolatile solid or liquid, so that the constant-pressure source can be provided by the atmosphere or some other inert medium rather than by a sliding piston. When using a constant-volume calorimeter, one must make sure that the rigid walls are strong enough to withstand the pressure increase usually generated when the temperature of the contents is raised.

The immediate results of calorimetric measurements are values of the heat capacities. Given a sufficient amount of such data, one can obtain C_p and C_V as functions of temperature and pressure (or density) over the whole range of interest. The various equations above can then be integrated to determine ΔU or ΔH for any process; we shall give examples of such calculations in the following sections. Note, however, that to obtain $(\Delta U)_p$ or $(\Delta H)_V$ one must determine $p\,\Delta V$ or $V\,\Delta p$, respectively, for the process in question; this requires either actual measurement or a knowledge of the fluid's equation of state.

Table 14.1 gives data on the heat capacities of various substances. For the same reason that H is more useful than U, most such tables give values of C_p rather than C_V.

For the same substance, C_p and C_V are of course related to one another. One can easily find the relation between the two for the perfect gas. Given Eq. 12.42, $U = \frac{3}{2}nRT$, we can see that $(\partial U/\partial T)_p = (\partial U/\partial T)_V$; we thus have for the perfect gas

$$
\begin{aligned}
C_p - C_V &= \left(\frac{\partial H}{\partial T}\right)_p - \left(\frac{\partial U}{\partial T}\right)_V \\
&= \left[\frac{\partial (U + pV)}{\partial T}\right]_p - \left(\frac{\partial U}{\partial T}\right)_p \\
&= \left[\frac{\partial (pV)}{\partial T}\right]_p = \left[\frac{\partial (nRT)}{\partial T}\right]_p = nR,
\end{aligned}
\tag{14.11}
$$

where we have substituted $pV = nRT$. That is, the molar heat capacities of a perfect gas should differ by the rather sizable quantity R, or about 8.3 J/K mol. The difference is much less in a solid or liquid, where the molar volume is only about 10^{-3} as great as in the gas; as a result, pV is only a small fraction of nRT, so that H and U, and thus C_p and C_V, are nearly the same. Later, in Chapter 17, we shall derive an equation giving the relationship between C_p and C_V for the general case.

Some interesting regularities can be obtained from the data in Table 14.1. For example, near room temperature we have $c_p \approx \frac{5}{2}R$ (20.79 J/K mol) for monatomic gases and $c_p \approx \frac{7}{2}R$ (29.10 J/K mol) for

[6] For a good expositon, see J. P. McCullough and D. W. Scott (Eds.), *Experimental Thermodynamics*, Vol. I (Plenum Press, New York, 1968).

TABLE 14.1
HEAT CAPACITIES OF SOME SUBSTANCES AT 25°C AND 1 ATM

Substance	c_p (J/K mol)	Substance	c_p (J/K mol)
Gases:		*Solids:*	
He	20.79	C (graphite)	8.53
Ar	20.79	C (diamond)	6.12
Xe	20.79	S (rhombic)	22.60
H_2	28.83	S (monoclinic)	23.64
N_2	29.12	I_2	54.44
O_2	29.36	Li	23.64
Cl_2	33.84	K	29.51
HCl	29.12	Cs	31.4
CO	29.15	Ca	26.28
H_2S	34.22	Ba	26.36
NH_3	35.52	Ti	25.00
CH_4	35.79	Hf	25.52
CO_2	37.13	Cr	23.35
SO_2	39.87	W	24.08
PCl_3	72.05	Fe	25.23
C_2H_6	52.70	Ni	26.05
$n\text{-}C_4H_{10}$	98.78	Pt	26.57
		Cu	24.47
Liquids:		Au	25.38
Hg	27.98	Al	24.34
Br_2	75.71	Pb	26.82
H_2O	75.15	U	27.45
CS_2	75.65	NaCl	49.69
N_2H_4	98.93	NaOH	59.45
CCl_4	131.7	$NaNO_3$	93.05
H_2SO_4	138.9	LiF	41.90
C_6H_6	136.1	CsI	51.87
$n\text{-}C_6H_{14}$	195.0	AgCl	50.78
C_2H_5OH	111.4	$BaSO_4$	101.8
$(C_2H_5)_2O$	171.1	$K_4Fe(CN)_6$	335.9
CH_3COOH	123.4	CBr_4	128.0
		SiO_2 (quartz)	44.43
		B_2O_3	62.97
		V_2O_5	127.3
		$C_{10}H_8$ (naphthalene)	165.7
		$n\text{-}C_{33}H_{68}$	912

diatomic gases; in Chapter 21 we shall see how these values can be obtained from microscopic theory. In general, the value of c_p tends to increase with the number of atoms in a compound or, more precisely,

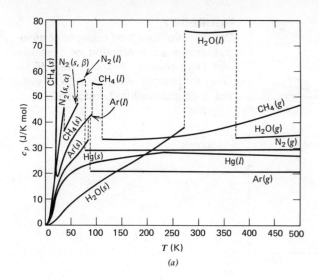

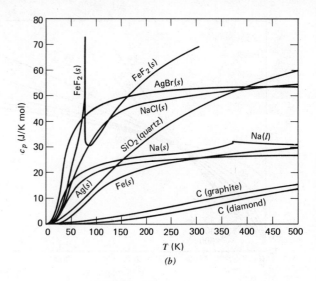

FIGURE 14.1

Temperature dependence of the heat capacity for various substances. (*a*) Room-temperature gases and liquids. (*b*) Room-temperature solids.

with the number of vibrational modes in the molecule. In most solid metals we have to a good approximation $c_p \approx 3R$ (25 J/K mol). This is the Law of Dulong and Petit, which we introduced in Section 2.9; it can also be derived theoretically, as we shall see in Chapter 22. Similarly, for monatomic-ion salts such as the alkali halides, we find $c_p \approx 6R$ (50 J/K mol) or $3R$ per ion.

The temperature dependence of the heat capacity is shown for a variety of substances in Fig. 14.1. Note that for all substances c_p vanishes at absolute zero, rising in most cases to a "plateau" (often the Dulong–Petit value) at higher temperatures. Discontinuities occur at the melting and boiling points, with the liquid usually having a higher heat capacity than the other phases. Some solid–solid phase transitions are also shown: an ordinary discontinuity in N_2, but sharp peaks in CH_4 and FeF_2; the latter are called "lambda points," from the shape of the $c_p(T)$ curve, and represent second-order phase transitions (see Chapter 24). Within a given phase, the temperature dependence is usually well described by an equation of the form $c_p = a + bT + cT^{-2}$, where b and c are typically of the order of 10^{-3} J/K^2 mol and 10^5 J K/mol, respectively. The pressure dependence of the heat capacity is slight in the condensed phases,[7] but can be appreciable for the gas, as shown in Fig. 14.2.

One of the most important applications of the first law of thermodynamics is to the study of the energy changes which accompany chemical reactions. Suppose that we consider a general chemical reaction

$$a\mathrm{A} + b\mathrm{B} + \cdots = l\mathrm{L} + m\mathrm{M} + \cdots. \tag{14.12}$$

For the present purposes, this is more conveniently written as an algebraic equation with all the terms on one side,

$$l\mathrm{L} + m\mathrm{M} + \cdots - a\mathrm{A} - b\mathrm{B} - \cdots = \sum_{i=1}^{r} \nu_i \mathrm{X}_i = 0, \tag{14.13}$$

where the symbols X_i stand for the r species $\mathrm{A}, \mathrm{B}, \ldots$, and the ν_i are the stoichiometric coefficients $a, b, \ldots$. By convention we always subtract

14.2
Energy and Enthalpy Changes in Chemical Reactions

---

[7] In liquids, c_p typically decreases by about 10% in 2000–3000 atm, then increases more slowly; in metals there is an almost negligible decrease, of the order of 10^{-8}/atm.

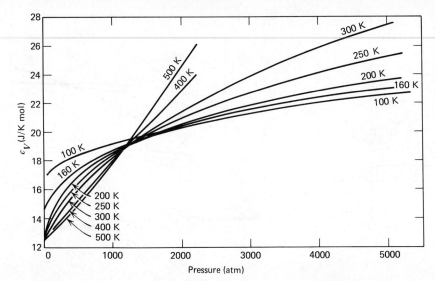

FIGURE 14.2
Constant-volume heat capacity of gaseous argon as a function of pressure and temperature. From F. Din (Ed.), *Thermodynamic Functions of Gases*, Vol. 2 (Butterworths, London, 1962).

reactants from products, so the ν_i must be positive for the products and negative for the reactants. To make this more specific, consider the reaction

$$\tfrac{1}{2}H_2 + \tfrac{1}{2}Cl_2 = HCl,$$

for which we have $\nu_{HCL} = 1$, $\nu_{H_2} = -\tfrac{1}{2}$, $\nu_{Cl_2} = -\tfrac{1}{2}$ in the notation of Eq. 14.13.

How do we go from an equation in terms of chemical formulas to one involving the thermodynamic properties of substances? Consider the process by which we balance a chemical equation. What we actually do is apply the principle that the atoms of each element are conserved in chemical processes. But mass is also conserved, if we neglect relativistic effects.[8] We can therefore write, for the change of mass in a reaction,

$$\Delta M = \sum_{i=1}^{r} \nu_i M_i = 0, \tag{14.14}$$

where M_i is the molar mass (molecular weight) of substance i.

Now consider the internal energy and the enthalpy. They are state functions, and thus defined for any thermodynamic state of a system. Furthermore, they are extensive functions, so that we can define an energy or enthalpy per mole. We can therefore write equations similar to Eq. 14.14 for the total changes in internal energy and enthalpy in a reaction. These changes will not in general be zero, unless the system is isolated. We have

$$\Delta U = \sum_{i=1}^{r} \nu_i u_i \tag{14.15a}$$

and

$$\Delta H = \sum_{i=1}^{r} \nu_i h_i. \tag{14.15b}$$

---

[8] Einstein's fundamental relation of equivalence between mass and energy, $E = mc^2$ (where c is the speed of light), implies that there must be a mass difference between reactants and products if there is an energy change in a reaction. For typical chemical reaction energies (say, 40 kJ/mol), however, the mass change is less than 10^{-9} g/mol, and can be safely neglected.

If u_i and h_i are the internal energy and enthalpy per mole of component i, respectively, then the ΔU and ΔH defined by these equations refer to 1 mol of the stoichiometric reaction. For example, if the reaction is $\frac{1}{2}H_2 + \frac{1}{2}Cl_2 = HCl$, then ΔU is the energy change when 0.5 mol each of H_2 and Cl_2 react to form 1 mol of HCl.

But what are the states for which u_i and h_i are defined? To specify ΔU and ΔH for a reaction, we must clearly know not only the stoichiometric equation but the initial states of the reactants and the final states of the products. By "state," as usual, we mean such things as the temperature and pressure; u_i and h_i in Eqs. 14.15 refer to the initial state for the reactants and to the final state for the products. It is simplest to treat a reaction that occurs entirely at the same T and p, but obviously we cannot expect all real-life reactions to be this simple. A reasonable approximation, however, is a reaction occurring in a thermostat subject to atmospheric pressure.

An additional complication is the fact that changes in energy and enthalpy occur even when substances are mixed without reaction. Since in practice chemical processes are usually directed to the production of pure substances, it is customary to take the initial state as consisting of pure reactants isolated from one another, and the final state as isolated pure products. One can then imagine the overall reaction as a three-step process:

1. Take the pure reactants, in their initial equilibrium states, and physically mix them under reaction conditions.

2. Allow the reaction to proceed.

3. Isolate the individual products from the reaction mixture and bring them to their final equilibrium states.

This is not how a reaction ordinarily takes place, of course; reaction begins while step 1 is still under way, and it is often useful to start step 3 before the reaction is over. But for state functions such as U and H, the total change must be the same for this idealized path as in reality, as long as the initial and final states are the same. If an experimenter wishes to determine the energy change in step 2 alone, then, he must be careful to account for the changes corresponding to steps 1 and 3. Step 2 is of particular interest because it is likely to correspond to the model of any microscopic theory of reaction.

Once we have defined exactly what process we are considering, what is it that Eqs. 14.15 tell us? The first of these equations is in fact a statement of the conservation of energy: The total internal energy of the reactants plus the energy absorbed by the system (ΔU) must equal the total internal energy of the products. That is, the difference between the total energies of the reactants and the products gives just the amount of energy that must be absorbed or released for the reaction to occur under the given conditions. If we have a table of the internal energy per mole as a function of T and p for each of the species involved, we can obtain ΔU directly by subtraction. The same is true of the enthalpy change, which is more commonly used, and expressions like Eqs. 14.15 can also be written for other extensive state functions.

The algebraic properties of chemical equations have led us to Eqs. 14.14 and 14.15. Since a set of chemical equations can be thought of as a set of mass balances, it follows that different equations can be combined by addition and subtraction to obtain any desired mass balance. It does not matter whether or not the final equation represents a real chemical reaction, that is, one observed in the laboratory. For an example of this process, consider the several reactions

(1) $$C + O_2 = CO_2,$$

(2) $$H_2 + \tfrac{1}{2}O_2 = H_2O,$$

(3) $$C_2H_6 + \tfrac{7}{2}O_2 = 2CO_2 + 3H_2O;$$

if we algebraically add $2(\text{Eq. 1}) + 3(\text{Eq. 2}) - \text{Eq. 3}$, multiplying each equation by the indicated numerical factor, we obtain the composite reaction

(4) $$2C + 3H_2 = C_2H_6.$$

But what good is such manipulation? The answer is that it is equally valid for extensive quantities other than mass. Until now we have thought of a substance's chemical formula as representing the substance itself, or in stoichiometric contexts a mole of the substance; but it can equally well represent the internal energy, enthalpy, or any other extensive property of a mole of the substance. This makes possible a tremendous reduction in empiricism, since ΔU, ΔH, and other such quantities can be obtained for reactions without actually performing them.

Let us examine this. The enthalpy balance for Eq. 1 above, namely Eq. 14.15b, can be written as

$$h_C + h_{O_2} = h_{CO_2} - \Delta H,$$

or, in the more usual shorthand,

$$C + O_2 = CO_2 - \Delta H;$$

similarly for the other equations, and with either ΔH or ΔU. Since Eqs. 14.15 are of the same form as the mass balance, Eq. 14.14, the internal energy and enthalpy changes for Eq. 4 can be computed by adding those for Eqs. 1–3 in the same way as we added the equations themselves:

$$\Delta U(\text{Eq. 4}) = 2\,\Delta U(\text{Eq. 1}) + 3\,\Delta U(\text{Eq. 2}) - \Delta U(\text{Eq. 3}),$$
$$\Delta H(\text{Eq. 4}) = 2\,\Delta H(\text{Eq. 1}) + 3\,\Delta H(\text{Eq. 2}) - \Delta H(\text{Eq. 3}).$$

These equations are examples of *Hess's law*, which states that the internal energy and enthalpy changes of such composite reactions are additive, provided that all the reactions are carried out under the same conditions (i.e., at the same temperature and pressure). Like Eqs. 14.15 themselves, Hess's law is a direct consequence of the first law. Observe now what we have shown: If ΔH (or ΔU) has been measured for reactions 1, 2, and 3, the above equations enable us to calculate ΔH (or ΔU) for reaction 4 without ever actually observing that reaction. We thus begin to see the practical value of all the formal structure and laws of thermodynamics, in that they enable us to extend our knowledge beyond what has been measured directly. To do this efficiently, however, it is not enough to organize our knowledge in terms of specific chemical reactions. As we shall see in Section 14.4, a far greater reduction in empiricism is obtained when thermodynamic measurements are analyzed in terms of individual substances.

Hess's law is a consequence of the fact that U and H are state functions, and therefore that ΔU and ΔH are independent of path. It does not matter whether the reaction sequence $2(\text{Eq. 1}) + 3(\text{Eq. 2}) - \text{Eq. 3}$ is an actual pathway or even a physically possible pathway for reaction 4. As long as the net reactants and products of the sequence are, respectively, identical in chemical composition and physical state to the reactants and products of the overall reaction, Hess's law is applicable.

Here is a numerical example of the application of Hess's law. Suppose that for the following reactions we know ΔH (often called the *heat of reaction*, because at constant pressure it is the same as q), all at 25°C and 1 atm:

(5) $\quad PbO(s) + S(s) + \tfrac{3}{2}O_2(g) = PbSO_4(s), \qquad \Delta H = -39.56 \text{ kJ/mol};$

(6) $\quad PbO(s) + H_2SO_4 \cdot 5H_2O(l) = PbSO_4(s) + 6H_2O(l),$

$$\Delta H = -5.57 \text{ kJ/mol};$$

(7) $\quad SO_3(g) + 6H_2O(l) = H_2SO_4 \cdot 5H_2O(l), \quad \Delta H = -9.82 \text{ kJ/mol};$

where the symbols in parentheses following the chemical formulas indicate that the substances are in the solid (s), liquid (l), or gaseous (g) state. What is the heat of reaction for

(8) $$\qquad\qquad S(s) + \tfrac{3}{2}O_2(g) = SO_3(g)$$

at 25°C and 1 atm? In this case, since the product SO_3 is formed directly from the elements S and O_2, the enthalpy change is known as the *enthalpy of formation* (*heat of formation*), ΔH_f, of $SO_3(g)$. We can obtain Eq. 8 by simply subtracting Eqs. 6 and 7 from Eq. 5; the heat of formation of $SO_3(g)$ at 25°C and 1 atm is thus

$$\Delta H_f(SO_3, g) \equiv \Delta H(\text{Eq. 8}) = \Delta H(\text{Eq. 5}) - \Delta H(\text{Eq. 6}) - \Delta H(\text{Eq. 7})$$

$$= [-39.56 - (-5.57) - (-9.82)] \text{ kJ/mol}$$

$$= -24.17 \text{ kJ/mol}.$$

Note how easy the analysis becomes when the chemical formulas are regarded as algebraic symbols, each representing a mole of the corresponding substance; ordinary algebraic manipulations can then be used to simplify complicated sets of equations.

Because of the sign convention for reactions embodied in Eqs. 14.15, a negative value of ΔU means that energy is liberated in a reaction. This energy usually appears mainly as heat; specifically, when only expansion work is possible, the heat liberated is $-\Delta U$ at constant volume and $-\Delta H$ at constant pressure. In the above example, heat is liberated in forming $SO_3(g)$ from the elements at 1 atm. But in general work may also be performed, and under some circumstances this can be the dominant effect. Indeed, both internal combustion engines and electrochemical cells are designed to maximize the amount of work obtained from a chemical reaction.

In the preceding examples we have considered only reactions in which all the products and reactants are at the same temperature and pressure. Of course, this will not always be true, especially if large amounts of energy are liberated. In practice it is often impossible to keep T and p under control during a reaction (and sometimes a change in T or p is exactly what one wants to produce). What one can do to keep the calculation tractable is to assume that the reaction itself occurs at some constant T and p, and to consider the temperature and pressure changes as separate steps in the overall process. This will still give the right values of ΔU and ΔH, as long as one does not change the initial and final states. But to do this one must know how to obtain the internal energy and enthalpy changes accompanying physical processes such as the expansion of a fluid, or a phase change from liquid to gas. The next section is devoted to this subject.

Let us consider first the changes in internal energy and enthalpy when a fluid, originally at equilibrium in the state p_1, V_1, T_1, is brought to the new equilibrium state p_2, V_2, T_2. Since both U and H are state functions, all paths connecting the initial and final states must give the same values of ΔU and ΔH. For convenience we can then imagine the change of state to be carried out in two steps, as shown in Fig. 14.3a. First, the fluid is heated (or cooled) at constant volume V_1 until the temperature T_2 is reached. Second, with the temperature T_2 held fixed in a thermostat, the fluid is expanded (or contracted) until the volume reaches V_2; since only two of the variables p, V, T are independent, this must also bring the pressure to p_2. In short, we split the overall change into an *isochoric* (constant-volume) process and an *isothermal* (constant-temperature) process. We are thus considering V and T as our independent variables, and we shall see that this choice is particularly convenient for calculating ΔU.

If the internal energy U is a function of the independent variables T and V, then by Eq. II.4 in Appendix II at the back of the book we can write any infinitesimal change in internal energy as

$$dU = \left(\frac{\partial U}{\partial T}\right)_V dT + \left(\frac{\partial U}{\partial V}\right)_T dV, \tag{14.16}$$

where dT and dV are the corresponding changes in T and V. To obtain ΔU for the entire process, we must integrate this equation over the path in Fig. 14.3a. In the first step the volume is constant, so the entire contribution to ΔU comes from the first term on the right-hand side of Eq. 14.16; in the second step the temperature is constant, so the entire contribution to ΔU comes from the second term on the right-hand side. Thus, for any change of state in a fluid, we can write

$$\Delta U \equiv U(T_2, V_2) - U(T_1, V_1)$$
$$= \int_{T_1, V_1}^{T_2, V_1} \left(\frac{\partial U}{\partial T}\right)_V dT + \int_{T_2, V_1}^{T_2, V_2} \left(\frac{\partial U}{\partial V}\right)_T dV. \tag{14.17}$$

Now, by Eq. 14.3 the first term is simply $\int_{T_1, V_1}^{T_2, V_1} C_V\, dT$. To evaluate the second term, we must know the value of the derivative $(\partial U/\partial V)_T$ for the fluid; for the present we use without proof the relationship

$$\left(\frac{\partial U}{\partial V}\right)_T = T\left(\frac{\partial p}{\partial T}\right)_V - p, \tag{14.18}$$

which we shall derive later (see Eqs. 17.3–17.17). Substituting in Eq. 14.17, we obtain for the total internal energy change,

$$\Delta U = \int_{T_1, V_1}^{T_2, V_1} C_V(T, V)\, dT + \int_{T_2, V_1}^{T_2, V_2} \left\{ T\left[\frac{\partial p(T, V)}{\partial T}\right]_V - p(T, V) \right\} dV. \tag{14.19}$$

We do not derive Eq. 14.18 at this time only because the second law of thermodynamics is needed for its proof. But why should we make such a substitution at all? The answer is that $(\partial U/\partial V)_T$ cannot be measured directly in the laboratory, whereas p, V, and T can. These variables (and the heat capacities) are the primary measurements from which U and other thermodynamic functions are computed; that is why we call the combined values of p, V, T the equation of state. In other words, we have put our equation for ΔU in terms of directly measurable quantities only.

14.3
Thermochemistry of Physical Processes

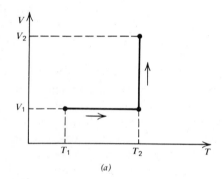

(a)

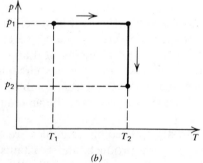

(b)

FIGURE 14.3

Two-step analysis of the change of state of a fluid: (*a*) constant-volume process followed by constant-temperature process, for calculating ΔU; (*b*) constant-pressure process followed by constant-temperature process, for calculating ΔH. In both cases the change is from the initial state p_1, V_1, T_1 to the final state p_2, V_2, T_2.

We shall shortly need another substitution of the same type as Eq. 14.18,

$$\left(\frac{\partial H}{\partial p}\right)_T = V - \left(\frac{\partial V}{\partial T}\right)_p, \tag{14.20}$$

which will be derived as Eq. 17.50.

To obtain the enthalpy change for a general change of state in a fluid, it is most convenient to consider a different two-step process, that shown in Fig. 14.3b. In this case we combine an *isobaric* (constant-pressure) heating step with an isothermal pressure change, using p and T as independent variables. In analogy to Eq. 14.16, we have

$$dH = \left(\frac{\partial H}{\partial T}\right)_p dT + \left(\frac{\partial H}{\partial p}\right)_T dp. \tag{14.21}$$

Each term again vanishes along one step of our path, and integration gives for the total enthalpy change

$$\Delta H \equiv H(T_2, p_2) - H(T_1, p_1)$$

$$= \int_{T_1, p_1}^{T_2, p_1} \left(\frac{\partial H}{\partial T}\right)_p dT + \int_{T_2, p_1}^{T_2, p_2} \left(\frac{\partial H}{\partial p}\right)_T dp$$

$$= \int_{T_1, p_1}^{T_2, p_1} C_p(T, p) \, dT + \int_{T_2, p_1}^{T_2, p_2} \left\{ V(T, p) - T\left[\frac{\partial V(T, p)}{\partial T}\right]_p \right\} dp, \tag{14.22}$$

in which we have substituted Eqs. 14.9 and 14.20.

Inspection of our equations for ΔU and ΔH reveals an important point. The changes in internal energy and enthalpy (and thus U and H themselves) can be thought of as having two components, one a function of temperature only (at some given V or p) and the other describing the change with V or p at a given temperature. Now consider the case of the perfect gas. Substituting $pV = nRT$ into Eqs. 14.18 and 14.20, we see that $(\partial U/\partial V)_T$ and $(\partial H/\partial p)_T$ both become zero; that is, U and H for the perfect gas are both functions of temperature only.[9] The two components just mentioned can thus be considered as a perfect gas term and a correction term for real fluids; this concept will be useful when we consider the properties of real gases (Chapter 21).

We are now ready to apply our results to calculation of the internal energy and enthalpy changes in chemical reactions as functions of temperature and pressure. Here is an example of the kind of question one must know how to answer: Suppose that we know the heat of reaction for

$$SO_2(g) + \tfrac{1}{2}O_2(g) = SO_3(g)$$

at 291 K and 1 atm; what is the heat of reaction at 873 K and 1 atm? Now, the "heat of reaction" is just the difference between the enthalpies of products and reactants. Once ΔH is known at one T and p, it can be easily obtained at any other T and p by applying Eq. 14.22 to each of the substances involved. Substituting in Eq. 14.15b, we find the general

---

[9] As we shall see in Chapter 16, this is not a *consequence* of Eqs. 14.18 and 14.20, but one of the *assumptions* necessary to derive them. We have already reached this conclusion for U from microscopic considerations, in Section 13.5; the result for H follows immediately, because for the perfect gas $H = U + nRT$.

expression for the enthalpy of reaction as a function of T and p:

$$\Delta H_{\text{reac}}(T_2, p_2) = \sum_i \nu_i h_i(T_2, p_2)$$

$$= \Delta H_{\text{reac}}(T_1, p_1)$$

$$+ \sum_i \nu_i \left\{ \int_{T_1,p_1}^{T_2,p_1} c_{p,i}\, dT + \int_{T_2,p_1}^{T_2,p_2} \left[v_i - T\left(\frac{\partial v_i}{\partial T}\right)_p \right] dp \right\},$$

$$(14.23)$$

where $c_{p,i}$ and v_i are, respectively, the molar heat capacity and molar volume of the ith substance in the pure state.[10] Similarly, from Eqs. 14.15a and 14.19 we obtain for the internal energy change as a function of T and p

$$\Delta U_{\text{reac}}(T_2, p_2) = \sum_i \nu_i u_i(T_2, p_2)$$

$$= \Delta U_{\text{reac}}(T_1, p_1)$$

$$+ \sum_i \nu_i \left\{ \int_{T_1,v_{1i}}^{T_2,v_{1i}} c_{V,i}\, dT + \int_{T_2,v_{1i}}^{T_2,v_{2i}} \left[T\left(\frac{\partial p_i}{\partial T}\right)_V - p_i \right] dv \right\},$$

$$(14.24)$$

where v_{1i} and v_{2i} are the initial and final molar volumes corresponding to the states T_1, p_1 and T_2, p_2 for the ith pure substance, in which $p_i = p(T, v_i)$. Thus to compute the heat of reaction at T_2, p_2 we need to know in general (1) the heat of reaction at some T_1, p_1, (2) the heat capacities of all compounds involved, as functions of T and p, and (3) the equations of state of all substances involved. For perfect gases, of course, the pressure integrals vanish, because $v_i = T(\partial v_i/\partial T)_p$. In reactions involving real fluids, however, the pressure term may make an important contribution to the heat of reaction.

Let us now return to the specific example given above and actually carry out the calculations of Eq. 14.23. Since p_1 and p_2 are the same, the pressure integrals drop out (in the vicinity of 1 atm the perfect gas approximation would be valid anyway). The other data that we need are

$$\Delta H\,(291\text{ K, 1 atm}) = -5351\text{ J/mol},$$
$$c_p(SO_2) = 2.72 + 4.10 \times 10^{-4}T - 4.89 \times 10^4 T^{-2},$$
$$c_p(O_2) = 1.80 + 1.93 \times 10^{-4}T - 2.1 \times 10^4 T^{-2},$$
$$c_p(SO_3) = 3.02 + 1.53 \times 10^{-3}T,$$

(T in K, c_p's in J/mol K). Substitution and integration over T then gives

$$\Delta H\,(873\text{ K, 1 atm}) = \Delta H\,(291\text{ K, 1 atm})$$

$$+ \int_{291}^{873} [c_p(SO_3) - c_p(SO_2) - \tfrac{1}{2}c_p(O_2)]\, dT$$

$$= -5351$$

$$+ \int_{291}^{873} [-0.60 + 1.02 \times 10^{-3}T + 5.94 \times 10^4 T^{-2}]\, dT$$

$$= -5219\text{ J/mol}.$$

[10] These are in general not the same as the values of c_p and v assigned to each substance in the reaction mixture (partial molar quantities), because the physical process of mixing also involves energy and enthalpy changes—steps 1 and 3 on p. 541. But we need not consider these effects at this time (cf. Section 14.5), because we have defined ΔH_{reac} as the enthalpy change from pure reactants to pure products.

The equations we have derived in this section are generally valid for fluids or for any system whose state is completely defined by two of the variables p, V, T. The paths we have chosen to connect p_1, V_1, T_1 with p_2, V_2, T_2 were selected for the simplicity of the resulting equations. Other paths are of course possible, and we shall consider some of these later. For all our paths, however, we have assumed that the integrals are continuous—that is, that C_p, U, and so on vary continuously with p, V, and T for all substances. This is clearly not true if a phase change such as freezing or vaporization occurs in the course of the process. We must therefore consider the thermochemistry of phase changes.

14.4
Introduction to Phase Changes

We begin with some definitions. A *phase* is a part of a system, physically distinct, macroscopically homogeneous, and of fixed or variable composition; it is mechanically separable from the rest of the system. Another way to say this is that a phase is a region within which all the intensive variables vary continuously, whereas at least some of them have discontinuities at the borders between phases. For example, ice and water in equilibrium form a two-phase system: The ice and water are mechanically separable (say by sieving), each is internally uniform, and there is a boundary between them at which the density changes discontinuously. It is customary to consider all the ice as a single phase, no matter how many pieces may be in contact with the water. By a *phase change* we mean simply a process that brings a system from a state characterized as phase A to another state characterized as phase B. The transition from water to ice (or vice versa) is an example of a phase change.

Although phase changes in systems with several components can be quite complicated, we are here concerned primarily with one-component systems. For any pure substance, a (p, T) diagram shows the various phases (solid, liquid, gas) as distinct regions separated by sharp boundaries; an example is given in Fig. 14.4. A process that involves a phase change corresponds to a path crossing one of these boundaries. The phase change itself occurs entirely at the point where the path crosses the boundary, and thus at constant temperature and pressure. (Later we shall see why this is so.) In all ordinary phase changes (freezing, vaporization, etc.) there is a difference in density between the two phases involved; whenever this is true, it is found experimentally that a quantity of heat is absorbed or released in the process. This amount of heat is called the *latent heat* of the phase change. Because the pressure is constant, the latent heat is also equal to ΔH for the process, whereas ΔU, of course, equals $\Delta H - p\,\Delta V$.

To illustrate these concepts, suppose that our system is pure water and our path is the isobar $p = 1$ atm. We start with a piece of ice at some temperature below 0°C and gradually add heat. The ice increases in temperature and expands slightly until 0°C is reached. At this point the ice transforms continuously to liquid water, at constant temperature and pressure, with 19.05 J absorbed for each gram of ice melted. Water is unusual in that it expands on freezing, the liquid being 0.083 g/cm³ denser than the solid at 0°C. After all the ice is transformed to liquid, adding more heat will raise the temperature again; the liquid contracts slightly up to 4°C and then again expands. These phenomena are illustrated in Fig. 14.5.

Latent heats of fusion (melting) and vaporization are listed in Tables 14.2 and 14.3, respectively, along with the corresponding (1-atm) melting and boiling points. Note that the heat of vaporization is normally much

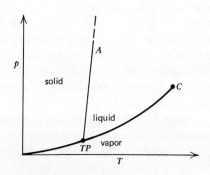

FIGURE 14.4
A typical phase diagram for a simple substance. The lines represent the locus of pressure and temperature for which two phases can coexist, for example liquid and vapor along the line *TP-C*. The points *TP* and *C* are, respectively, the triple point and the critical point for the substance (see Chapter 24).

TABLE 14.2
LATENT HEATS OF FUSION OF SOME SUBSTANCES

Substance	T_{fus} (K)	Δh_{fus} (kJ/mol)	$(\Delta h_{fus}/T_{fus})$ (J/K mol)
Ne	24.55	0.335	13.6
Ar	83.85	1.176	14.0
H_2	13.95	0.117	8.39
O_2	54.34	0.4448	8.19
Cl_2	172	6.406	37.2
HF	189.8	3.928	20.70
HCl	159.0	1.992	12.53
CO	68.1	0.836	12.3
H_2O	273.15	6.007	22.0
NH_3	195.42	5.65	28.9
CH_4	90.6	0.938	10.4
CCl_4	250.3	2.5	10.0
C_6H_6	278.7	9.837	35.30
C_2H_5OH	158.5	5.02	31.7
$(C_2H_5)_2O$	157.0	6.90	44.0
Li	453.7	3.01	6.64
Ca	1123	8.66	7.71
Ba	983	7.66	7.79
Ag	1234.43	11.27	9.13
NaCl	1073	28.8	26.8
LiF	1121	26.4	23.6

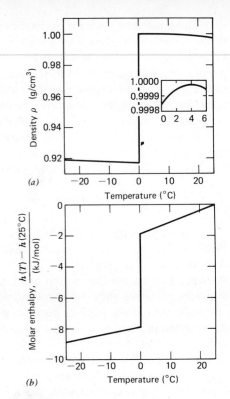

(a)

(b)

FIGURE 14.5
Solid–liquid phase transition for water at 1 atm: (*a*) density; (*b*) molar enthalpy (relative to 25°C). The inset in (*a*) shows the anomalous density behavior between 0 and 4°C. Note in (*b*) that the heat capacity [$c_p = (\partial h/\partial T)_p$] of liquid water is approximately twice that of ice.

larger than the heat of fusion. As can be seen from the table, the ratio of the molar heat of fusion to the melting point, $\Delta h_{fus}/T_{fus}$, is of the order of R (8.3 J/K mol) for simple substances, including metals, but tends to increase with molecular size. For vaporization, the value of $\Delta h_{vap}/T_{vap}$ is close to 85 J/K mol for a wide variety of substances; this generalization is known as *Trouton's rule*.[11] (We shall see in Chapter 17 that $\Delta H/T$ for a phase change is equal to the *entropy* of transition.) Although the data in these tables are all for 1 atm, it should be kept in mind that the transition temperatures and to a lesser extent the latent heats vary with pressure. These and other details of phase equilibria will be discussed at length in Chapter 24.

What concerns us here, however, is the thermochemistry of phase transitions. In general, whenever the path a system follows from state p_1, T_1 to state p_2, T_2 crosses a phase boundary, the latent heat of the transition must be included in ΔH for the process. Thus, consider the above example of the 1-atm ice → water transition: The molar enthalpy change, which would be simply $\int_{T_1}^{T_2} c_p \, dT$ in the absence of a phase transition, must in fact be calculated as

$$h(l, T_2) - h(s, T_1) = \int_{T_1}^{0°C} c_p(s) \, dT + \Delta h_{fus} + \int_{0°C}^{T_2} c_p(l) \, dT.$$

[11] A much more accurate rule is due to Hildebrand. It states that $\Delta h_{vap}/T_{vap}$ is the same for all liquids when the vaporization is carried out at temperatures for which the vapor densities are the same. The value of $\Delta h_{vap}/T_{vap}$ is 93 J/K mol when the vapor density is 2.02×10^{-2} mol/l.

TABLE 14.3
LATENT HEATS OF VAPORIZATION OF SOME SUBSTANCES

Substance	T_{vap} (K)	Δh_{vap} (kJ/mol)	$(\Delta h_{vap}/T_{vap})$ (J/K mol)
Ne	27.07	1.760	65.02
Ar	87.28	6.519	74.69
H_2	20.39	0.904	44.34
O_2	90.15	6.819	75.64
Cl_2	235.11	20.41	86.81
HF	292.61	7.489	25.59
HCl	188.13	16.15	85.84
CO	81.65	6.04	73.97
H_2O	373.15	40.66	109.0
NH_3	239.74	23.35	97.40
CH_4	111.5	8.180	73.40
CCl_4	349.84	30.0	85.75
C_6H_6	353.25	30.76	87.08
C_2H_5OH	351.55	38.74	110.2
$(C_2H_5)_2O$	307.8	26.60	86.43
Li	1590	148.1	93.14
Ca	1760.3	150.0	85.22
Ba	1910	150.9	79.00
Ag	2485	254.0	102.2
NaCl	1734	170	98.0
LiF	1954	213.3	109.2

Clearly, in any other problem involving enthalpy changes, such as the variation of heats of reaction with temperature, one must similarly take account of possible phase changes; in general, one simply adds to Eq. 14.23 a term $\nu_i(\Delta h_{trans})_i$ for each transition along the path of integration.

14.5 Standard States

It should be clear by now that the change in internal energy or enthalpy for a given process measures only the relative properties of the initial and final states. Not only do absolute values of energy and enthalpy not need to be specified, but it is meaningless even to speak of "absolute" values, because the zero of energy is arbitrary. It is convenient, however, to have definite reference points to which the energy and enthalpy of substances can be related, because the values so defined can then be tabulated for substances rather than reactions. (This is much more efficient for storing information, because there are of the order of N^2 possible reactions between N substances.) The choice of such reference states is arbitrary. Once such choices have been made, however, they must be used consistently throughout a given set of calculations.

As we shall see, there are many kinds of reference states, usually called *standard states*. For pure substances, however, it is conventional to choose standard states in the following way:

1. The standard state of a chemical substance at a given temperature is taken to be a state of the pure compound or element at that temperature and a pressure of 1 atm.[12] If no temperature is specified, 25°C is to be assumed.

2. Each element in its most stable form at 25°C (298.15 K) and 1 atm is assigned an enthalpy of zero.[13]

One can define a standard state of a given substance in any of its possible physical forms (phases), even one that is not ordinarily stable at the given temperature and pressure: water as liquid or vapor, carbon as graphite or diamond, and so on. If no phase is specified, "standard state" refers to the most stable form under the given conditions. Thermodynamic quantities referring to a standard state are designated by the superscript ^{0}. Thus the *standard heat of reaction*, the value of ΔH for a reaction in which all the reactants and products are in standard states at the same temperature is designated as ΔH^0 or ΔH_T^0 (where T is the temperature).

It was mentioned in Section 14.2 that ΔH for the formation of a compound from its elements is called the compound's heat (enthalpy) of formation. We can now rigorously define the *standard heat of formation* (ΔH_f^0) as ΔH^0 for the reaction in which the compound is formed from its elements in their most stable forms. The values of ΔH_f^0 have been tabulated for a very large number of compounds, usually at 25°C; a sampling is included in Table 14.4. Given our specification of the standard enthalpy of elements, it is clear that for any species at 25°C the standard enthalpy H^0 has the same value as $\Delta H_f^0(25°C)$; to obtain ΔH_f^0 or H^0 at other temperatures one must apply Eq. 14.23. Since the same elements appear on both sides of any chemical reaction equation, it is obvious that the standard enthalpy change in any reaction can be obtained directly from the standard heats of formation of all the species involved:

$$\Delta H_{\text{reac}}^0 = \sum_{i=1}^{r} \nu_i h_i^0 = \sum_{i=1}^{r} \nu_i (h_i^0 - h_{i,\text{elements}}^0)$$

$$= \sum_{i=1}^{r} \nu_i \,\Delta H_f^0(i). \tag{14.25}$$

A table of ΔH_f^0 thus gives us ΔH_{reac}^0 for all reactions involving only the tabulated substances at the same temperature.

Compounds for which ΔH_f^0 is negative are usually but not always stable with respect to their elements, since energy is released on their formation. A compound with positive ΔH_f^0 can be formed from the elements only when energy is added to the reactive system, and such compounds tend to be unstable. These statements have been hedged because absolute stability is not determined solely by the enthalpy, as we shall see in Chapter 19. Nevertheless, they are correct for nearly all cases in which the magnitude of ΔH_f^0 is large.

Heats of formation are thus very useful, but how are they obtained? For some few compounds one can carry out the formation reaction from the elements directly in a calorimeter. In the many cases where this is not easily feasible, one can obtain ΔH_f^0 by applying Hess's law. In particular,

---

[12] For gases the standard state actually used is that of an equivalent perfect gas at 1 atm (cf. Chapter 21), but in most cases this makes very little difference in thermodynamic quantities.

[13] One could make this assumption for the internal energy or some other energy function, but thermochemical tables and calculations are usually given in terms of the enthalpy. Note that this convention is not suitable in the discussion of nuclear reactions, where transmutation of the elements can occur.

TABLE 14.4
STANDARD HEATS OF FORMATION AT 25°C

Substance	ΔH_f^0 (kJ/mol)	Substance	ΔH_f^0 (kJ/mol)
$AlCl_3$	−695.4	NO	90.37
NH_3	−46.19	NO_2	33.85
$BaCl_2$	−860.1	N_2O_4	9.67
$BaSO_4$	−1465	O_3	142
$BeCl_2$	−511.7	PCl_3	−306.4
$BCl_3(l)$	−418.4	PCl_5	−398.9
HBr	−36.2	KBr	−392.2
CaF_2	−1215	KCNS	−203.4
C(diamond)	1.9	KOH	−425.85
CO_2	−393.5	NaCl	−411.0
HCl	−92.3	SiO_2(quartz)	−859.4
CuCl	−136	AgCl	−127.0
$CuCl_2$	−206	NaBr	−359.9
$H_2O(l)$	−285.9	$NaHCO_3$	−947.7
$H_2O(g)$	−241.8	Na_3PO_4	−1925
$FeCl_3$	−405	$Sr(NO_3)_2$	−975.9
$FeSO_4$	−922.6	S(monoclinic)	0.30
$PbSO_4$	−918.4	SF_6	−1100
LiF	−612.1	H_2S	−20.15
$MgCl_2$	−641.8	SO_2	−296.9
Hg_2Cl_2	−264.9	UF_6	−2110
$HgCl_2$	−230	$ZnCl_2$	−415.9

it is usually easy to measure a compound's heat of combustion, after which one can use the heats of formation of the combustion products to calculate that of the original compound. This is especially convenient for organic compounds, whose principal combustion products are CO_2 and H_2O. Although the method of calculation is identical with that described in Section 14.2, the technique is sufficiently important to justify another worked example.

Suppose, then, that we wish to determine the standard heat of formation of gaseous propane (C_3H_8) at 25°C, the formation reaction being

$$3C(graphite) + 4H_2(g) = C_3H_8(g).$$

The combustion of propane is in turn described by the reaction

(1) $$C_3H_8(g) + 5O_2(g) = 3CO_2(g) + 4H_2O(g),$$

for which the measured value of $\Delta H^0(25°C)$ is −2044.0 kJ/mol. We can also write the formation reactions

(2) $$C(graphite) + O_2(g) = CO_2(g),$$
$$\Delta H^0 = \Delta H_f^0(CO_2, g) = -393.5 \text{ kJ/mol},$$

and

(3) $$H_2(g) + \tfrac{1}{2}O_2(g) = H_2O(g), \qquad \Delta H^0 = \Delta H_f^0(H_2O, g) = -241.8 \text{ kJ/mol},$$

with the 25°C heats of formation obtained from Table 14.4. Now, the formation reaction for $C_3H_8(g)$ can be constructed by taking the sum 3(Eq. 2)$+4$(Eq. 3)$-$(Eq. 1). By Hess's law, the standard heat of formation of $C_3H_8(g)$ at 25°C must then be

$$\Delta H_f^0(C_3H_8, g) = 3\,\Delta H_f^0(CO_2, g) + 4\,\Delta H_f^0(H_2O, g) - \Delta H_{comb}^0(C_3H_8, g)$$
$$= 3(-393.5) + 4(-241.8) - (-2044.0)$$
$$= -103.7 \text{ kJ/mol}.$$

The use of standard states for pure reactants and products is thus quite straightforward. However, many reactions take place entirely in solution, and one often wants to know the heat of reaction without making additional measurements of the heats of mixing and separation—steps 1 and 3 of the analysis on p. 541. This can be done if one develops for species in solution standard states that do not require referring back to the pure components, together with rules for handling the dependence of enthalpy on concentration. The latter point is important, since (as previously mentioned) there are energy changes even when a solution is formed in the absence of reaction; furthermore, these changes depend on the solvent,[14] so that any definition of standard states must include specification of the solvent.

A variety of reference states are in use for solutions, the choice depending on whether the species in question is solvent or solute, whether or not a solute is ionized, and what types of calculations one wishes to carry out. However, there are two principal conventions. The standard state of a solvent is customarily taken to be the pure solvent at the same temperature and pressure. The treatment of solutes is more complicated. The usual standard state is defined to have the limiting properties of infinitely dilute solutions. The rigorous definition of this state involves concepts we have not yet introduced, and is thus deferred to Chapter 25.[15] What we can consider at this time is the nature of the enthalpy changes that can occur in solution, a subject to which we devote the next section.

We are concerned here primarily with the enthalpy changes that occur when substances are mixed to form a solution. These can be referred to by the general term *heats of solution*,[16] but several different quantities are in use to describe these phenomena. In this section we shall define these quantities, say something about their molecular interpretation, and discuss some of the problems involved in calculating heats of reaction in solution.

By "heat of solution" one most often means the *integral heat of solution*, the total change in the system's enthalpy when a given amount of solute is dissolved in a specified amount of solvent at some particular temperature and pressure. For example, dissolving 1 mol (36.5 g) of gaseous HCl in 10 mol (180 g) of liquid water at 25°C and 1 atm releases

14.6
Thermochemistry of Solutions

---

[14] The usual convention is to denote as *solvent* that component of a solution present in largest concentration. Aqueous solutions are in such common use that an unspecified solvent can usually be assumed to be water.

[15] See also Section 21.2 for the treatment of gaseous mixtures, and Chapter 26 for the modifications necessary for electrolytic solutions.

[16] When two pure liquids are mixed to form a solution, the enthalpy change is commonly called the *heat of mixing*. Although this is formally identical with the heat of solution, the two terms are usually distinguished; often the units used are also different.

69.3 kJ of heat; this process can be represented by the equation

$$HCl(g) + 10H_2O(l) = HCl(5.55\ m),$$

$$\Delta H_{sol'n}(HCl,\ 5.55\ m) = -69.3\ kJ/mol\ HCl,$$

where $\Delta H_{sol'n}$ designates the integral heat of solution and $5.55\ m$ is the *molal* concentration (moles per kilogram of solvent) of the HCl. (Since there are 55.5 mol of H_2O to the kilogram, the molality of an aqueous solution is simply $55.5n_2/n_1$, where n_2/n_1 is the solute/solvent mole ratio.) It is useful to write the change of state in a solution process explicitly, as we have done in the above equation; the equation is balanced with respect to mass since "$HCl(5.55\ m)$" designates a solution containing 1 mol of HCl and 10 mol of H_2O.[17] In this section all quantities such as $\Delta H_{sol'n}$ refer to 25°C and a pressure of 1 atmosphere, unless otherwise explicitly stated.

If the solvent is not directly involved in a chemical reaction with the solute, it is convenient to set the enthalpy of the pure liquid solvent (at the temperature and pressure of the solution) equal to zero. In this way we avoid having to manipulate very large and unwieldy numbers for the heats of formation of solvents, which would in any case cancel out when we calculated heats of reaction in solution. Given this convention, we can easily define a heat of formation of any solution for which we know the integral heat of solution. Thus in the above example we have, using Table 14.4,

$$\Delta H_f\ (HCl,\ 5.55\ m) = \Delta H_f^0(HCl,\ g) + \Delta H_{sol'n}(HCl,\ 5.55\ m)$$

$$= (-92.1) + (-69.3) = -161.4\ kJ/mol\ HCl.$$

Later in this section we shall see what to do when the solvent is involved in the reaction.

When additional solvent is added to an already existing solution, resulting in a solution of lower concentration, the total change in the system's enthalpy is called the *integral heat of dilution*. For example, adding another 10 mol of water to our 1 mol of $5.55\ m$ HCl would release an additional 2.5 kJ of heat:

$$HCl(5.55\ m) + 10H_2O(l) = HCl(2.78\ m),$$

$$\Delta H_{dil}(HCl,\ 5.55\ m \rightarrow 2.78\ m) = -2.5\ kJ/mol\ HCl.$$

The heat of dilution is simply the difference between two integral heats of solution, so that

$$\Delta H_{sol'n}(HCl,\ 2.78\ m) = \Delta H_{sol'n}(HCl,\ 5.55\ m) + \Delta H_{dil}(HCl,\ 5.55 \rightarrow 2.78\ m)$$

$$= (-69.3) + (-2.5) = -71.8\ kJ/mol\ HCl.$$

The heat of solution for aqueous HCl is plotted as a function of concentration in Fig. 14.6, from which it can be seen that the heat of dilution falls off rapidly with increasing dilution. The integral heat of solution is thus fairly independent of concentration in dilute solutions,[18] and in fact reaches a limit for infinite dilution, which we shall discuss shortly.

One other quantity sometimes used is the *differential heat of solution*, $\Delta \bar{h}$. This is the enthalpy change when a small quantity of solute is added to so large a quantity of solution that the concentration can be assumed

---

[17] Strictly, one should write "$HCl(aq,\ 5.55\ m)$," where *aq* identifies an aqueous solution, but our notation is complicated enough already, and water as solvent can usually be assumed.

[18] For comparison, ordinary "concentrated HCl" is about $16\ m$, or 3.4 mol H_2O/mol HCl.

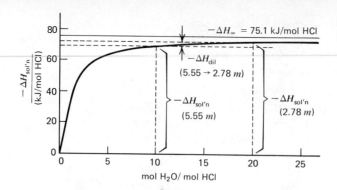

FIGURE 14.6
Heat of solution of HCl in water.

unchanged by the addition, that is,

$$\Delta \bar{h}_i \equiv \left(\frac{\partial H}{\partial n_i}\right)_{T,p,n_j(j \neq i)} = \left(\frac{\partial \Delta H_{sol'n}}{\partial n_i}\right)_{T,p,n_j(j \neq i)}, \qquad (14.26)$$

where n_i is the number of moles of solute i present, and j refers to all other components of the solution.[19] In binary solutions there is a simple relationship between the differential and integral heats of solution: At constant T and p, the integral heat of solution is a function of the amounts of the two components, so one can write

$$d(\Delta H_{sol'n}) = \left(\frac{\partial \Delta H_{sol'n}}{\partial n_1}\right)_{T,p,n_2} dn_1 + \left(\frac{\partial \Delta H_{sol'n}}{\partial n_2}\right)_{T,p,n_1} dn_2$$
$$= \Delta \bar{h}_1 \, dn_1 + \Delta \bar{h}_2 \, dn_2. \qquad (14.27)$$

If we integrate this equation at constant composition, i.e., keeping the ratio n_1/n_2 constant, the $\Delta \bar{h}_i$'s will also remain constant, and we obtain

$$\Delta H_{sol'n} = \Delta \bar{h}_1 n_1 + \Delta \bar{h}_2 n_2. \qquad (14.28)$$

If we designate the solvent and solute as 1 and 2, respectively, then $\Delta \bar{h}_1$, the solvent's differential heat of solution, is simply the slope of a curve like that in Fig. 14.6. For our previous example one can calculate

$$\Delta \bar{h}_2 (\text{HCl}, 5.55 \ m) = -64.0 \ \text{kJ/mol HCl},$$
$$\Delta \bar{h}_2 (\text{HCl}, 2.78 \ m) = -69.4 \ \text{kJ/mol HCl}.$$

In the limit of infinite dilution $\Delta \bar{h}_1$ vanishes and $\Delta \bar{h}_2$ becomes equal to the integral heat of solution per mole of solute.

Why is it that enthalpy changes occur when a solution is formed or diluted? This effect is a result of differences in molecular interactions. Consider for simplicity a liquid–liquid solution. If the energies of molecular interaction were the same in the two pure liquids A and B, then in the A–B mixture a molecule would find all local environments to have the same energy, independent of composition. In such a case there would be no energy change upon mixing. In reality, however, different molecules do interact in different ways (cf. Chapter 10). The heat of solution is therefore a rough measure of the difference in interaction energies between A–A pairs, B–B pairs, and A–B pairs. In a solution where the interactions between solute molecules differ from those between solute and solvent molecules, the addition of more solvent (dilution) changes the relative amounts of the two interactions and thus alters the total interaction energy.

---

[19] The $\Delta \bar{h}_i$ defined here is the same as $\bar{h}_i$, the partial molar enthalpy of component i (see Chapter 19).

Let us examine the solution process more closely. Clearly the average distance between solute molecules in a solution is greater than in the pure solid or liquid solute. In a dilute solution the solute molecules are so far apart that each is surrounded by a nearest-neighbor shell of solvent molecules. In a nonelectrolyte solution the nearest-neighbor interactions are dominant in determining the energy. Let the average (attractive) interaction energy between nearest neighbors be ϵ_{ij},[20] and the number of nearest neighbors be z; then the (integral) solution energy per solute molecule in dilute solution will be

$$\Delta\epsilon_{sol'n} = z[\epsilon_{12} - \tfrac{1}{2}(\epsilon_{11} + \epsilon_{22})]. \qquad (14.29)$$

This is a very crude model, since in general z may be different in pure solute, pure solvent, and solution. The sign of $\Delta\epsilon_{sol'n}$ depends on values of ϵ_{12}, ϵ_{11}, and ϵ_{22}. If $\Delta\epsilon_{sol'n}$ (and thus $\Delta U_{sol'n}$) is negative, as is often the case, energy (heat) will be released upon solution; if $\Delta\epsilon_{sol'n}$ is positive, energy must be added to form the solution. What we have said of the internal energy changes will largely be applicable to the enthalpy changes also, since the change in pV is usually quite small.[21] The simple model outlined here can be greatly refined, as we shall see in Chapter 25. The situation is somewhat more complicated in electrolyte solutions, where there are at least two kinds of solute species interacting by long-range Coulomb forces.

When a solute is dissolved in a sufficiently large amount of solvent, adding more solvent will produce no detectable change in the enthalpy or other thermodynamic variables. A solution of this type is said to be *infinitely dilute* in the solute. Of course, different solutions may appear to become infinitely dilute at different concentrations, depending on their particular properties; a solution may also appear to be infinitely dilute with respect to some properties but not with respect to others. To return again to our earlier example, dissolving 1 mol of gaseous HCl in an effectively infinite amount of water, at 25°C and 1 atm, releases a total of 75.1 kJ of heat, as shown in Fig. 14.6; we can represent this by

$$HCl(g) + \infty H_2O(l) = HCl(aq), \qquad \Delta H_\infty(HCl) = -75.1 \text{ kJ/mol HCl},$$

where ΔH_∞ is the enthalpy of infinite dilution (the dilute-solution limit of $\Delta H_{sol'n}$) and aq used alone denotes the state of infinite dilution. Clearly, the enthalpy change when 5.55 m HCl is infinitely diluted is just

$$\Delta H_{dil}(HCl, 5.55\ m \to 0\ m) = \Delta H_\infty(HCl) - \Delta H_{sol'n}(HCl, 5.55\ m)$$
$$= -75.1 - (-69.3) = -5.8 \text{ kJ/mol HCl}.$$

Thus far we have discussed only the physical process of solution; let us now consider heats of reaction in solution. We shall illustrate some of the principles involved with ionic reactions in aqueous solution, but the same principles can be applied to reactions in any solvent. We examine first the reactions of ions in (effectively) infinitely dilute solutions. More specifically, we shall indicate how thermochemical studies of such solutions can be used in constructing an ionic model of their structure.

When one mixes aqueous solutions of such salts as $LiCl$, $CsNO_3$, and $CaCl_2$, no precipitate forms, and at infinite dilution the heat of mixing goes to zero. On the other hand, mixing solutions of $AgNO_3$ and $NaCl$ yields a precipitate, and a nonzero heat of mixing, even when the initial solutions are "infinitely dilute" with respect to their heats of solution.

[20] ϵ_{11}, ϵ_{22} refer to the interaction of like molecules and ϵ_{12} refers to the solvent–solute interaction.

[21] A solution can form spontaneously even when $\Delta H_{sol'n}$ is positive, because of the entropy of mixing; see Chapter 25.

(We assume, of course, that the solutions are not so dilute as to prevent formation of the precipitate; this is a fairly safe assumption here, because the solubility of AgCl or $BaSO_4$ is only about 10^{-5} m.) Furthermore, it is found that mixing infinitely dilute solutions of $AgNO_3$ and *any* metal chloride releases the *same* amount of heat, 65.5 kJ/mol of AgCl(s) formed. Similarly, mixing infinitely dilute solutions of $Ba(NO_3)_2$ and any metal sulfate releases 19.2 kJ/mol of $BaSO_4$(s) formed. Finally, mixing infinitely dilute solutions of an Arrhenius base (NaOH, KOH) and an Arrhenius acid (HCl, HNO_3) gives the same heat of mixing in all combinations, releasing 55.9 kJ/mol of H_2O formed.

These and many other facts are consistent with the Arrhenius model, in which the species present in dilute aqueous solution are the ions resulting from dissociation of the parent salt, acid, or base. When two infinitely dilute solutions are mixed, provided that the ions undergo no reaction, the environment of each ion remains essentially unchanged (solvent molecules only) and there is no heat of mixing. But suppose that there is a reaction of ions to form a precipitate, water, or some other undissociated compound. In this case ions are removed partially from the solution, and there must be an enthalpy change. In an infinitely dilute aqueous solution the ions are on the average so far apart that each interacts only with the surrounding water molecules.[22] It is therefore to be expected that the enthalpy changes associated with precipitation, acid–base neutralization, and so on, will depend only on the ions removed from solution, and not at all on their original partner ions. We can thus write the reactions described above as

$$Ag^+(aq) + Cl^-(aq) = AgCl(s), \qquad \Delta H^{\ominus}_{reac} = -65.5 \text{ kJ/mol};$$

$$Ba^{2+}(aq) + SO_4^{2-}(aq) = BaSO_4(s), \qquad \Delta H^{\ominus}_{reac} = -19.2 \text{ kJ/mol};$$

$$H^+(aq) + OH^-(aq) = H_2O(l), \qquad \Delta H^{\ominus}_{reac} = -55.9 \text{ kJ/mol}.$$

The superscript $\ominus$ is used to to denote the state of infinite dilution.

We have seen the advantages of analyzing heats of reaction into contributions from individual reactants and products. Can one apply the same method in ionic solutions, splitting the heat effects into separate contributions from the individual ions present? In other words, can one obtain a set of standard ionic heats of formation, analogous to Table 14.4 for neutral compounds? Consider an infinitely dilute aqueous solution of NaCl, which in our model is a mixed solution of Na^+ and Cl^-. We should therefore have

$$\Delta H_f^{\ominus}(\text{NaCl}, aq) = \Delta H_f^{\ominus}(Na^+, aq) + \Delta H_f^{\ominus}(Cl^-, aq).$$

Given our convention that the solvent has zero enthalpy, thermochemical measurements yield

$$\Delta H_f^{\ominus}(\text{NaCl}, aq) = \Delta H_f^0(\text{NaCl}, s) + \Delta H_{\infty}(\text{NaCl})$$
$$= -411.0 + 3.9 = -407.1 \text{ kJ/mol}.$$

But how is this quantity to be divided into contributions from Na^+ and Cl^-? The same problem exists in any other ionic solution: By the condition of electroneutrality, in any macroscopic solution there must be equal amounts of positive and negative charge. We can therefore never isolate and study ions of one species only. The best we can do is to

---

[22] Obviously, ions must come together if any reaction is to take place, but in a solution that is not literally infinitely dilute this can be brought about by random motion. The same principle accounts for reactions in dilute gases. However, these are kinetic effects (see Part Three), and the thermodynamic properties depend only on the average distances between particles.

measure the total enthalpy of a set of ions of opposite signs whose charges add up to zero.

It is nevertheless possible to construct a useful set of standard ionic enthalpies. To do so, however, we must make an additional arbitrary definition. If a value is arbitrarily assigned to the enthalpy of any single ion, a consistent set of values can be obtained for all other ions. The convention actually adopted is the following:

> The enthalpy of formation of the H^+ ion in infinitely dilute aqueous solution is defined to be zero at all temperatures and pressures.

Let us see how this convention is used to construct a table of standard ionic enthalpies. If $\Delta H_f^{\ominus}(H^+, aq)$ is zero by definition, then $\Delta H_f^{\ominus}(Cl^-, aq)$ must be the same as $\Delta H_f^{\ominus}(HCl, aq)$; from the data already given this is

$$\Delta H_f^{\ominus}(Cl^-, aq) = \Delta H_f^{\ominus}(HCl, aq) = \Delta H_f^0(HCl, g) + \Delta H_\infty(HCl)$$
$$= -92.3 + (-75.1) = -167.4 \text{ kJ/mol}.$$

Combining this in turn with the data for NaCl, we have

$$\Delta H_f^{\ominus}(Na^+, aq) = \Delta H_f^{\ominus}(NaCl, aq) - \Delta H_f^{\ominus}(Cl^-, aq)$$
$$= -407.1 - (-167.4) = -239.7 \text{ kJ/mol}.$$

And so forth for other ions. In Table 14.5 we display the standard enthalpies of formation for a number of ions, deduced in the manner just described. Note that by combining such ionic values one can obtain the heat of formation of any infinitely dilute electrolyte solution. For this or any other electrically neutral collection of ions, the effects of the arbitrary choice of zero cancel out.

Heats of reaction in ionic solutions can be calculated in the usual way, by taking the difference between the heats of formation of products and reactants. As already noted, a special problem occurs when the solvent takes part in the reaction. Consider the neutralization reaction between infinitely dilute hydrogen and hydroxyl ions,

$$H^+(aq) + OH^-(aq) = H_2O(l).$$

The water formed in this reaction cannot be treated as having zero heat of formation (our usual convention for the solvent), since it does not appear on both sides of the equation. We must therefore use the standard heat of formation of liquid water from the elements to calculate

$$\Delta H_{reac}^{\ominus} = \Delta H_f^0(H_2O, l) - \Delta H_f^{\ominus}(H^+, aq) - \Delta H_f^{\ominus}(OH^-, aq)$$
$$= -285.84 - 0 - (-229.94) = -55.90 \text{ kJ/mol},$$

which agrees with the experimental value given earlier.

Reactions (or other processes) at ordinary ionic concentrations are more complicated to describe than at infinite dilution. The complication arises because the interactions (and thus the heats of formation) of the ions vary with their concentrations, which are of course changed by reaction. Indeed, the properties of a given ion are affected by the concentrations of other ions as well. The general theory of electrolyte solutions, a quite complex subject, will be surveyed in Chapter 26.

Regardless of these difficulties, one can of course measure the enthalpy change for any given process, as long as all the relevant concentrations are specified. In interpreting such data, one finds again a

TABLE 14.5
STANDARD ENTHALPIES OF FORMATION OF IONS IN INFI-
NITELY DILUTE AQUEOUS SOLUTION AT 25°C AND 1 ATM

Ion	ΔH_f^θ (kJ/mol)	Ion	ΔH_f^θ (kJ/mol)
H^+	0 (convention)	OH^-	−229.94
Li^+	−278.4	F^-	−329.1
Na^+	−239.7	Cl^-	−167.46
Cs^+	−248	HS^-	−17.7
Cu^+	51.9	I^-	−55.94
K^+	−251.21	SO_3^{2-}	−624.3
Ag^+	105.9	SO_4^{2-}	−907.51
NH_4^+	−132.8		
Ba^{2+}	−538.06		
Ca^{2+}	−542.96		
Cu^{2+}	64.39		
Fe^{2+}	−87.9		
Pb^{2+}	1.6		
Mg^{2+}	−461.96		
Mn^{2+}	−219		
Ni^{2+}	−64.0		
Zn^{2+}	−152.4		
Sr^{2+}	−545.51		
Al^{3+}	5470.33		
Cr^{3+}	5480		
Fe^{3+}	−47.7		

special complication for reactions involving the solvent. For the acid–base neutralization reaction in infinitely dilute solution, the addition of 1 mol of water per mole of reaction does not alter the enthalpies of the nonreacting ions; the formation of water as a product is the sole source of the enthalpy change. But at ordinary concentrations one must also consider a dilution effect. If we mix stoichiometrically equivalent amounts of two solutions, one containing x mol of water per mole of HCl and the other containing y mol of water per mole of NaOH, then after neutralization the solution will contain $x + y + 1$ mol of water per mole of NaCl. The total enthalpy change measured thus includes both the heat of reaction and the heat of dilution of Na^+ and Cl^- ions from their hypothetical initial concentration (only $x + y$ mol of water present) to the final concentration after reaction ($x + y + 1$ mol of water present). Formally, one can separate these steps, imagining the overall process to occur in such a way that the mole of water produced by the chemical reaction is removed from the solution, after which an extra mole of "solvent water" is added. This hypothetical sequence of processes is equivalent to the actual reaction, and has the advantage of helping to keep our conventions consistent. As already pointed out, we must say that $\Delta H_f^0 = -285.84$ kJ/mol for the chemically formed water, even though by convention we set ΔH_f^0 equal to zero for the nonreacting solvent water.

In the remainder of this chapter we deal with a variety of phenomena on the molecular level about which information can be derived from thermochemical measurements. The discussion here must be regarded as preliminary, since the second law of thermodynamics is needed for a detailed analysis of these phenomena. We thus give here only those conclusions that can be drawn from the facts already introduced. To begin with, we examine in this section some of the simple physical processes and phase changes discussed in Sections 14.3 and 14.4.

Consider first the vaporization of a liquid. By virtue of the differences in density and thus in average distance between molecules, the molecular interactions are large in a liquid and small in a gas. "Large" and "small" here can be defined with respect to the average kinetic energy in the dilute gas, $\frac{3}{2}RT$ per mole or $\frac{3}{2}k_BT$ per molecule. Thus the heat of vaporization should be a very rough measure of the average intermolecular potential energy in the liquid. By the empirical Hildebrand's rule, we in fact have $\Delta h_{vap} \approx 11RT_{vap}$ per mole, of which RT_{vap} ($\approx pv_{gas}$) can be attributed to the expansion work and the remainder to the potential energy contribution; the latter is thus about $10k_BT_{vap}$ per molecule. If we compare intermolecular potential parameters (see Table 21.13) with boiling points (Table 14.3), we find that ϵ, the depth of the potential well, is typically in the range 1–$2k_BT_{vap}$ per pair of molecules. The discrepancy between these two figures is due to the fact that each molecule in a liquid has many near neighbors.

We noted in Section 14.4 that the heat of fusion is normally much less than the heat of vaporization. We can now see that this is to be expected, since the density difference between a solid and a liquid is relatively small. In terms of the argument just outlined, neither the expansion work nor the potential energy contribution to Δh_{fus} is large. We shall see in Chapter 24 that the entropy of fusion ($\Delta h_{fus}/T_{fus}$) is more significant in interpreting microscopic phenomena.

The change in volume of a substance under external pressure is described in terms of the *isothermal compressibility*,

$$\kappa_T \equiv -\frac{1}{V}\left(\frac{\partial V}{\partial p}\right)_T; \tag{14.30}$$

typical values of κ_T are $10^{-4}\,\text{atm}^{-1}$ for a liquid, $10^{-6}\,\text{atm}^{-1}$ for a solid, and $1/p$ for a perfect gas. For all mechanically stable substances the volume decreases with increasing pressure, so that κ_T is always positive.[23] The work performed on a substance in compressing it isothermally and reversibly is thus

$$w = -\int_{V_1}^{V_2} p\,dV = \int_{p_1}^{p_2} \kappa_T pV\,dp \tag{14.31}$$

which must also be positive. One might therefore expect $(\partial U/\partial p)_T$ to be positive; yet measurements show $(\partial U/\partial p)_T < 0$ for nearly all fluids (water below 4°C is an exception) up to pressures of the order of 10^4 atm. This is not unreasonable, since the quantity of work involved is small[24] and we have no estimate of the heat given off. The molecular interpretation is simple: In a fluid the average distance between near-neighbor molecules is greater than the potential energy minimum, so compression reduces the

---

[23] If it were negative, the substance would spontaneously contract to a state of lower energy.
[24] If $\kappa_T = 10^{-4}\,\text{atm}^{-1}$ and $v = 50\,\text{cm}^3/\text{mol}$ (reasonable values for a liquid), the work required for compression to 100 atm is about 150 J/mol; this is small compared to RT, which at 25°C equals 2478 J/mol.

total energy; it takes a considerable pressure to get past the minimum. In solids, on the other hand, the average distance is near the potential energy minimum, where $dV(R)/dR = 0$; thus we expect $(\partial U/\partial p)_T$ to be close to zero, and κ_T to be small. But all these arguments are over-simplified, since the main effect of very high pressure is *within* molecules: One can show from the virial theorem that the electronic kinetic energy changes much more than the total energy.

Considerable care must be used in developing molecular arguments of the type employed here. For example, it must require work to compress even a perfect gas; substituting in Eqs. 14.30 and 14.31, we obtain the usual $w = nRT \ln(p_2/p_1)$. Yet we have defined a perfect gas as one for which $(\partial U/\partial p)_T = 0$; thus the work of compression must be exactly cancelled by the heat given off. In a real fluid the presence of molecular interactions makes the internal energy a function of pressure. In the perfect gas there are no molecular interactions, and the microscopic effect of compression is simply to shift the gas's spectrum of energy levels, as we have already described in Section 13.1.

14.8
Bond Energies

The interpretation of thermochemical data in terms of molecular structures is an important aspect of chemistry. By the methods already described one can obtain fairly accurate values of the total energy of a given species, but there are many problems in interpreting these energies in terms of molecular parameters. One of these problems, to which we have often alluded in Part One, is that the energies involved in chemical bonding are usually small differences of large quantities. The state of the art of quantum mechanical calculation is not sufficiently advanced for accurate prediction of the properties of molecules with more than a few electrons. By "properties" we mean such things as electronic structures, energy levels, bond lengths and angles, vibrational frequencies, etc.; by "accurate" we mean reproducing the experimental values of these properties. Thus it is still necessary to find methods of bridging the gap between quantum mechanical concepts and macroscopic measurements. Among the most useful bridges are the chemical concepts of bonding and group identity.

The basis of this approach is quite simple: In many molecules the electrons can be described with considerable accuracy as localized in bonds of specific types. Furthermore, a given type of bond has charac-teristic properties (length, energy, etc.) which are much the same whenever it is found—that is, which are "transferable" from molecule to molecule. For example, the C—C and C—H bonds have essentially the same properties in all saturated aliphatic hydrocarbons (CH_4, C_2H_6, C_3H_8, ...); we can thus assume that all the bonds in these molecules are approximately "local" in character. When the bonding is well localized, it is often possible to recognize whole groups of atoms that have the same properties in different compounds: the methyl (—CH_3) group, the —OH group in alcohols, and so on.

Under favorable circumstances, then, a molecule may be viewed as an assemblage of localized bonds or groups, each with an identity sufficiently well defined for its properties to persist in different molecules. The theoretical justification for this model has been sketched in Section 9.2, in terms of localized bond orbitals; here we are concerned with its consequences. The model is not completely accurate: In some cases (NO_3^-, benzene, etc.) the bonding electrons are essentially delocalized over the entire molecule, whereas in other molecules the interactions between bonds or groups may be significant, especially the so-called steric effects. Some of these exceptions have also been discussed in Chapter 9, and in the next section we shall look at them from a thermochemical point of view. But the assumption of localized bonds and groups with transferable properties suffices to explain a wide range of chemical phenomena.

To formulate in detail such an explanation, one must assign a consistent set of properties to the various types of chemical bonds. The property with which we are most concerned in this chapter is of course the bond energy. Let us see how one can establish a standard set of bond energies or enthalpies.

From measured heats of reaction, as already described, one can calculate the total energy or enthalpy for each of the various reactant and product species. But the total energy of, say, liquid water is the sum of intramolecular (bonding) energies and intermolecular energies. We can eliminate the latter by considering reactions in the dilute gaseous phase. Consider, for example, the dissociation of a diatomic molecule:

$$N_2(g) = 2N(g), \qquad \Delta H^0_{298} = 941.69 \text{ kJ/mol}.$$

Since the standard enthalpy of $N_2(g)$ is defined to be zero, the enthalpy of formation of atomic nitrogen is just half the enthalpy of dissociation, or 470.84 kJ/mol. More to the point, the enthalpy of dissociation of N_2 is another name for the enthalpy of the N—N bond in N_2. Thus here we have a direct measurement of a bond enthalpy, and similar measurements can be carried out for other gaseous diatomic molecules. (The "measurements" are usually spectroscopic: cf. Section 7.1.) But how can we extend this method to molecules with more than one bond?

To begin with, we must define our terms. By the *bond energy* we mean the energy required to separate an isolated (i.e., gaseous) molecule into two fragments—atoms or polyatomic groups—at infinite separation from each other, with the molecule and the fragments all in their ground states. In thermodynamic terms, the ground-state restriction means the reaction must be carried out at 0 K, so that the bond energy is ΔU_0 for the dissociation reaction. At all temperatures above absolute zero, however, some fraction of the molecules and fragments will be in excited states. The thermodynamic energy of each species at a given temperature is the average energy of the molecules of that species at a given temperature. Thus the measured value of ΔU_T at any $T > 0$ K contains a temperature-dependent component due to excited states; to obtain the bond energy as we have defined it, one must extrapolate the data to absolute zero.

An additional complication is that one usually measures ΔH rather than ΔU. The *standard enthalpy of a bond* is the measured enthalpy required to break the bond in the gaseous state, usually at 25°C (298.15 K) and 1 atm. This differs from what we have called the bond energy[25] in two respects: the difference between enthalpy and internal energy, and the temperature-dependent contribution just described. If we assume that the reactants and products are all perfect gases, then for any reaction we have

$$\Delta H - \Delta U = \sum (pV)_{\text{products}} - \sum (pV)_{\text{reactants}}$$

$$= (\Delta n)RT; \qquad\qquad (14.32)$$

---

[25] What we refer to as standard enthalpies of bonds are often called bond energies in the literature, and the latter term may be used loosely to cover both concepts.

Δn, the change in the total number of moles, is of course unity for a bond-dissociation reaction. We thus have

$$\Delta H^0_{298} - \Delta U^0_{298} = (1)(8.31441 \text{ J/mol K})(298.15 \text{ K})$$
$$= 2.4789 \text{ kJ/mol}$$

for a perfect gas dissociation at 25°C; this is a relatively small fraction of the total dissociation energy (ΔU or ΔH), which is typically several hundred kilojoules per mole. The other component, $\Delta U_T - \Delta U_0$, can be calculated by Eq. 14.24 if one knows the heat capacities of the species involved (cf. Section 21.5); at 25°C, where there is little vibrational excitation, it should be no more than 10 kJ/mol. Thus the absolute-zero bond energy and the 25°C bond enthalpy usually differ by no more than a few percent.

For diatomic molecules one can define the bond energy or dissociation energy unambiguously and rigorously, and obtain it directly from spectroscopic data. For polyatomic molecules we now also have a rigorous definition of the bond energy, but actually evaluating it is less easy. The motions of the nuclei in a polyatomic molecule are strongly coupled, and one cannot readily extract from the vibrational spectrum a limiting energy corresponding to the breaking of a single bond. Why not then proceed straightforwardly by breaking the molecule in two at the bond in question and measuring the energy required? One way to do this is by bombarding molecules with high-energy electrons until bonds are broken, and then analyzing the masses and energies of the fragments by mass spectrometry or other techniques. Yet even this method is often not feasible, especially for large molecules: The molecules may not exist in the gaseous phase, competing reactions may occur, the bond of interest may not be the one that breaks, the fragments may not be stable, and so on. In practice one must often assign values to bond energies by some averaging method; we shall now look at some methods of this type.

For simple molecules of the form AX_n (cf. Section 9.1) in which there exist only A—X bonds, one can give a rigorous definition of the *average* bond energy, which is simply $1/n$ of the energy required for the reaction

$$AX_n(g) = A(g) + nX(g).$$

To see the significance of this quantity, consider the CH_4 molecule, which is simple enough for the reaction

$$CH_4(g) = CH_3(g) + H(g), \qquad \Delta H^0_{298} = 430 \text{ kJ/mol}$$

to be studied directly. By the definition above, 430 kJ/mol is precisely what we mean by the standard enthalpy of the C—H bond in methane. But one can also carry out the subsequent reactions

$$CH_3(g) = CH_2(g) + H(g), \qquad \Delta H^0_{298} = 473 \text{ kJ/mol};$$
$$CH_2(g) = CH(g) + H(g), \qquad \Delta H^0_{298} = 422 \text{ kJ/mol};$$
$$CH(g) = C(g) + H(g), \qquad \Delta H^0_{298} = 339 \text{ kJ/mol}.$$

The reason these energies differ from one another is that after each reaction the remaining polyatomic fragment undergoes electronic and nuclear rearrangements that make its bonds somewhat different from what they were in the original molecule. For calculations of the structures of these fragments, the individual bond dissociation energies are required. Yet we know that in the CH_4 molecule all four C—H bonds are equivalent. Thus we are interested in the average energy per bond, which is just one-fourth the sum of the previous four quantities, or 416 kJ/mol.

We hypothesize that this is the energy associated with a "typical" C—H bond; is this hypothesis valid for other molecules?

Consider the ethane (C_2H_6) molecule, which has six identical C—H bonds and one central C—C bond. Can we obtain a value for the C—C bond energy without actually measuring it? The enthalpy of complete dissociation (atomization) for C_2H_6 is 2829 kJ/mol; this can be obtained from the ΔH_f^0 values for $C_2H_6(g)$, $C(g)$, and $H(g)$. If the average energy of the C—H bonds were the same as in CH_4, 416 kJ/mol, we could estimate the energy of the C—C bond to be $2829 - 6(416) = 333$ kJ/mol. The dissociation energy of this bond has in fact been measured and found to be 352 kJ/mol, so the C—H bonds in C_2H_6 actually have an average energy of 413 kJ/mol. This is a remarkably good match, and indicates that the properties of the C—H bond are *almost* transferable from CH_4 to C_2H_6. Similar calculations show that the same is true for other saturated hydrocarbons.

By combining data on a large number of molecules, one can obtain average values for the energies of many common types of bonds. Some of these average values are listed in Table 14.6, in terms of both 25°C standard bond enthalpies and absolute-zero bond energies. Although the average energies cannot take the place of measurements on individual molecules, they can be very useful for making predictions in the absence of such measurements. For example, suppose that we wish to compute the 25°C standard heat of formation of gaseous *n*-pentane, which has the structure

$$CH_3CH_2CH_2CH_2CH_3,$$

with four C—C bonds and 12 C—H bonds. We can thus estimate the standard heat of atomization,

$$C_5H_{12}(g) = 5C(g) + 12H(g),$$

at 25°C to be

$$\Delta H_{\text{atom'n}}^0(C_5H_{12}, g) = 4\,\Delta H_{298}^0(C-C) + 12\,\Delta H_{298}^0(C-H)$$
$$= 4(342) + 12(416)$$
$$= 6360 \text{ kJ/mol},$$

with bond enthalpies from Table 14.6. The heat of formation of $C_5H_{12}(g)$ is of course the enthalpy change in the reaction

$$5C(\text{graphite}) + 6H_2(g) = C_5H_{12}(g).$$

Applying Hess's law, one can easily show that

$$\Delta H_f^0(C_5H_{12}, g) = 5\,\Delta H_f^0(C, g) + 12\,\Delta H_f^0(H, g) - \Delta H_{\text{atom'n}}^0(C_5H_{12}, g)$$
$$= 5\,\Delta H_{\text{subl}}^0(C, \text{graphite}) + 6\,\Delta H_{298}^0(H-H)$$
$$\quad - \Delta H_{\text{atom'n}}^0(C_5H_{12}, g)$$
$$= 5(716.7) + 6(435.9) - 6360$$
$$= -161 \text{ kJ/mol},$$

where we have used the measured heat of sublimation of carbon and the heat of atomization just calculated. The value of $\Delta H_f^0(C_5H_{12}, g)$ obtained by more direct thermochemical means (from the heat of combustion) is −146.4 kJ/mol. Our estimate from average bond energies is thus only moderately accurate, but is close enough to be useful if the actual value had not been measured. Note that in calculations of this type the heats of formation of gaseous atoms are crucial. Where the standard state is a gas

TABLE 14.6
SOME AVERAGE BOND ENERGIES AND ENTHALPIES

Bond	ΔU_0 (kJ/mol)	ΔH^0_{298} (kJ/mol)
H—H	432.0	435.9
C—C	337	342
C=C	607	613
C≡C	828	845
N—N	155	85
N≡N (in N_2)	941.7	945.4
O—O	142	139
O=O (in O_2)	493.6	498.3
F—F (in F_2)	154.8	158.0
Cl—Cl (in Cl_2)	239.7	243.3
Br—Br (in Br_2)	190.2	192.9
I—I (in I_2)	149.0	151.2
C—H	411	416
N—H	386	354
O—H	458	463
F—H (in HF)	565	568.2
Cl—H (in HCl)	428.0	432.0
Br—H (in HBr)	362.3	366.1
I—H (in HI)	294.6	298.3
Si—H	318	326
S—H	364	339
C—O		343
C=O		707
C—N		293
C≡N		879
C—Cl	326	328

of diatomic molecules (H_2, O_2, Cl_2, etc.), we have simply

$$\Delta H_f^0(X, g) = \tfrac{1}{2}D_0(X_2, g) = \Delta H^0(X\text{—}X).$$

A key quantity for organic chemistry is $\Delta H_f^0(C, g)$, the heat of sublimation of graphite; measurements of this quantity were uncertain for a long time, but the value 716.7 kJ/mol now seems established. The heat of sublimation of carbon is *defined* as the enthalpy change of the process

$$C(\text{graphite}) \rightarrow C(g),$$

where the gaseous carbon is composed of single atoms. In fact, over a very wide range of temperature, carbon vapor consists of C, C_2, C_3, C_5 and still higher polymers. The existence of these polymers and the lack of knowledge of their bond energies made the determination of the heat of sublimation of carbon a very difficult problem for many years.

We can see that average bond energies must be used with caution. The model of localized bonds with transferable properties is only an approximation, after all, and more valid for some molecules than others. The properties of a particular bond always depend to some extent on its environment in the molecule. Some data illustrating this point are given

TABLE 14.7
EFFECTS OF MOLECULAR ENVIRONMENTS ON SOME BOND ENERGIES

O—H bonds	ΔH^0_{298} (kJ/mol)	C—H bonds	ΔH^0_{298} (kJ/mol)	C—C bonds	ΔH^0_{298} (kJ/mol)
HO—H	498.7	CH_3—H	431.8	CH_3—CH_3	368
HOO—H	374.5	CCl_3—H	377	CF_3—CF_3	406
CH_3O—H	439	CF_3—H	444	CH_3CH_2—CH_3	356
CH_3CH_2O—H	435	$N{\equiv}C$—H	540	$(CH_3)_3C$—CH_3	335
CH_3CO—H $\;\;$ (=O)	469	$HOCH_2$—H	385	CH_3C—CH_3 (=O)	335
		CH_3CH_2—H	410	$N{\equiv}C$—$C{\equiv}N$	603
		$(CH_3)_3C$—H	385	$HC{\equiv}C$—CH_3	490
		HC—H (=O)	364	⬡—CH_3	427
		CH_3C—H (=O)	360	⬡—CH_2—CH_3	301
		⬡—H	469		
		⬡CH_2—H	356		

in Table 14.7; it can be seen that some of the variations are quite large. (See also Table 9.2.) The effects here are of various kinds: change in electronegativity (Cl_3C— versus H_3C—), partial double-bond character (HCN, C_2N_2, butadiene), resonance stabilization of fragments (notably ⬡—CH_2—), steric effects (the $(CH_3)_3C$— compounds), and so on; many of these phenomena have been discussed in Part One. When only the nearby atoms have a significant effect, one can still obtain average values for the energy of a particular bond in a given environment, or alternatively for the contribution of a particular group of atoms to the molecular energy. For example, the correlation of Benson and Buss states that —CH_3 and —CH_2— groups in a hydrocarbon contribute −42.2 and −20.7 kJ/mol, respectively, to ΔH_f^0 at 25°C; this gives for n-pentane

$$\Delta H_f^0(C_5H_{12}, g) = 2\,\Delta H_f^0(-CH_3) + 3\,\Delta H_f^0(-CH_2-)$$
$$= 2(-42.2) + 3(-20.7)$$
$$= -146.5 \text{ kJ/mol,}$$

which is a vast improvement over our previous calculation. There are many empirical schemes of this sort, all involving lengthy tables of group energies for the various types of molecular environments. Such calculations are naturally more accurate than those using Table 14.6, although no method is completely satisfactory in complicated molecules.

Within limits, then, we can accept as useful approximations the additivity of group or bond energies and their transferability from compound to compound. It is thus possible to estimate the energy change in a chemical reaction in terms of the groups separated and joined, or the bonds broken and formed. In this way estimates can be made for processes on which no experimental data are available. The technique is particularly useful for studying reaction mechanisms, which often involve unstable intermediates of unknown energy, or for predicting the properties of compounds one wishes to synthesize. Although at best one can obtain only approximate values by such calculations, they are often the only methods possible; how useful they are depends on the accuracy required to solve the problem at hand.

14.9
Some Energy Effects in
Molecular Structures

The preceding section was based on the assumption that a molecule is an assemblage of localized bonds or groups with additive and transferable properties. The variations in Table 14.7 are among the evidence showing that this assumption is not completely valid. Yet this failure is in itself useful, in that the deviations from bond additivity can provide information about molecular structure. We have given some examples of this in Chapter 9; here we shall review these and discuss other cases in which quantum mechanical calculations can be combined with thermochemical data.

To begin with, we can now see what is meant by the "resonance energy" in such compounds as benzene (Section 9.2). Suppose that we consider the benzene molecule to have one of the Kekulé structures,

with three C—C and three C=C bonds. If we use the average bond energies in Table 14.6 in the way we did for *n*-pentane, we calculate a standard heat of formation of +247 kJ/mol for $C_6H_6(g)$; yet the experimental value of ΔH_f^0 is +82.93 kJ/mol. The discrepancy of 164 kJ/mol (1.70 eV/molecule) is defined as the resonance energy, which must be accounted for by any theory of the molecule's structure. As we explained in Section 9.2, the resonance energy is essentially the energy of delocalization of the π electrons. Similar effects are found in many other compounds, especially those whose nominal structure contains alternating (*conjugated*) multiple bonds. These include the whole range of aromatic ring compounds (cf. Fig. 9.6), graphite, and some catenated compounds: 1,3-butadiene (CH_2=CH—CH=CH_2) has a value of ΔH_f^0 53.6 kJ/mol lower than the nonconjugated 1,2-isomer (CH_2=C=CH—CH_3). Delocalization also occurs with nonconjugated multiple bonds, making the adjacent single bonds stronger than usual (such as the C—H bond in HCN, cf. Table 14.7).

Another area that can be investigated thermochemically is the effect of various kinds of strain within molecules, which we have discussed in Section 9.4. One example we mentioned was the cyclopropane molecule,

in which the C—C—C angles are constrained to be 60° rather than the normal (tetrahedral) 109°28′. Recall (Chapter 9) that the electron distribution in this molecule could be interpreted to imply that the C—C bonds are best imagined as bent, rather than straight. As a result, its ΔH_f^0 is 120 kJ/mol higher than Table 14.7 would predict; this discrepancy is called the *strain energy*. Another type of strain occurs when atoms not directly bonded to one another are close enough together to interact by van der Waals-type forces, as in branched-chain hydrocarbons, where large side chains can get in each other's way. In all such cases the strain forces bonds to be stretched or bent away from their equilibrium positions, with a resulting increase in energy. Calculations using spectroscopic data to obtain the energies of stretching or twisting bonds have given good agreement with experimental strain energies in many cases.

In Section 9.4 we also discussed the problem of hindered rotation in such molecules as ethane, as illustrated in Fig. 9.12. In ethane the three potential energy minima are identical, so there are three equally likely and indistinguishable conformations for the equilibrium form of this molecule. In a less symmetric molecule, such as 1,2-dichloroethane or cyclohexane (Fig. 9.13), one minimum (the ground state) is slightly lower than the others, but at ordinary temperatures an appreciable fraction of molecules are in the higher-energy conformations. The measured heat of formation must then give a weighted average of the energies of the conformational isomers (*conformers*). This effect can be allowed for by the methods of statistical mechanics, as we shall show in Chapter 21.

In the remainder of this section we consider one of the most fundamental applications of thermochemical data to molecular structure, the estimation of electronic correlation energy. The correlation energy was defined in Part One as the difference between the best possible self-consistent-field calculation of the energy (i.e., the Hartree–Fock energy) and the actual molecular energy. It is not possible to obtain an exact solution of the Schrödinger equation for any system with more than one electron, but the development of high-speed computers has led to fairly accurate Hartree–Fock energies for many simple molecules; it is not so easy to calculate the correlation energy *a priori*.

Why should we be concerned about the correlation energy, which is a very small fraction of the total molecular electronic energy? This total energy (defined relative to zero when all the electrons and nuclei are infinitely far from one another) is not really of much interest to us. As chemists we are interested primarily in the energy changes in reactions, which are usually much smaller quantities. If we try to calculate these using our "fairly accurate" theory, we run into the familiar difficulty in determining small differences of large numbers. Let us give an example: The best available theoretical values for the ground-state energies of the N atom and N_2 molecule are, respectively, -1480.25 eV and -2965.69 eV, giving a calculated value of 5.20 eV for the energy of the reaction

$$N_2(g) = 2N(g);$$

the (spectroscopically) measured energies of N and N_2 are -1485.62 eV and -2981.03 eV, with an experimental dissociation energy of 9.79 eV (9446 kJ/mol). The difference between the two sets of values is due to the neglect of correlation energy, the change in which thus accounts for half the dissociation energy. The change in correlation energy is not usually this large in ordinary reactions (the magnitude here being due to the unpaired electrons in atomic N), but some way of predicting its value is clearly desirable.

The correlation energy in a molecule has two components, due to interactions between electrons in the same atom (intraatomic) and in different atoms (interatomic), the first of which should be larger. It will be recalled that in electronic structure calculations a molecular orbital (wave function) is commonly represented as a weighted sum of atomic orbitals. In the separated atoms the correlation energy, obtained from the discrepancy between theory and experiment, can to a good approximation be partitioned into components from each pair of orbitals. It is thus plausible to argue that the intraatomic part of the molecular correlation energy is the sum of these atomic orbital components, with each atomic orbital having the same weighting as in the wave function. Calculations of this kind are now possible for any molecule made up of elements in the first row of the periodic table. This leaves only the interatomic correlation energy to be determined.

The interatomic correlation energy is of essentially the same nature as the van der Waals attraction between separate molecules, which we described in Chapter 10. Such interactions are chemically significant even at intermolecular distances, and thus certainly cannot be neglected within a molecule; an intermediate example is the strain energy mentioned earlier in this section. The change in interatomic correlation energy in a reaction is presumably the residual discrepancy with experiment after one has accounted for all the other energy changes by the methods sketched above. In Fig. 14.7 are shown the results of such calculations for the complete hydrogenation[26] of a number of compounds. Each of these compounds has one central bond between two first-row atoms, the only other bonds being to H atoms; the bulk of the interatomic correlation energy should be that between the two central atoms, and should decrease with increasing distance between them.

The data confirm this hypothesis, the energy falling off approximately as the inverse fourth power of the distance compared with the R^{-6} dependence of the intermolecular van der Waals attraction. On the assumption that the curve of Fig. 14.7 gives the correlation energy between first-row atoms as a function of the distance between them, one can use bond-length data to calculate the change in correlation energy in other reactions. Combining this with all the other calculated values, one obtains the "theoretical" heats of reaction of Table 14.8, which differ from the experimental values by an average of only 9.4 kJ/mol.

At least for simple molecules, then, one can now calculate heats of reaction in almost perfect agreement with experimental values, using thermochemical data to estimate the interatomic correlation energy and

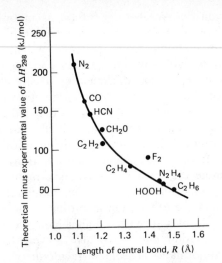

FIGURE 14.7
Discrepancy between theoretical and experimental values for heat of complete hydrogenation of first-row compounds, presumably due to the change in interatomic correlation energy. (The "central bond" in each case is that between two atoms from the set C, N, O, F.) From L. C. Snyder and H. Basch, *J. Am. Chem. Soc.* **91**, 2189 (1969).

TABLE 14.8
THEORETICAL HEATS OF REACTION[a]

Reaction	ΔH^0_{298}, theory (kJ/mol)	ΔH^0_{298}, experiment (kJ/mol)
$H_2 + C_2H_6 \rightarrow 2CH_4$	−65.7	−64.9
$H_2 + H_2O_2 \rightarrow 2H_2O$	−346.0	−348.1
$2H_2 + CH_2O \rightarrow CH_4 + H_2O$	−214.2	−200.8
$3H + HCN \rightarrow CH_4 + NH_3$	−251.5	−251.0
$3H_2 + C_2H_5N \rightarrow 2CH_4 + NH_3$	−311.3	−321.7
$3H_2 + C_2H_4O \rightarrow 2CH_4 + H_2O$	−328.0	−340.2
$4H_2 + C_3H_4 \rightarrow 3CH_4$	−487.4	−502.9
$4H_2 + CO_2 \rightarrow CH_4 + 2H_2O$	−144.3	−164.8

[a] The interatomic correlation energy has been estimated from bond-length data and Fig. 14.7. This energy is combined with other calculated values to obtain the theoretical ΔH. Experimental measurements of ΔH^0 at 298 K are also shown.

[26] That is, the reaction to form only simple hydrides, in these cases CH_4, NH_3, H_2O, and HF. For example, the reaction for formaldehyde is

$$CH_2O + 2H_2 = CH_4 + H_2O.$$

quantum mechanics to calculate all other components of the energy. It would of course be desirable to perform the entire calculation on a purely theoretical basis. Nevertheless, the present results are an impressive achievement, as well as an interesting demonstration of the power of thermochemistry.

14.10
Lattice Energies of Ionic Crystals

In Chapter 11 we described the structure and bonding of crystalline solids, but for the most part left their energies for later consideration. Most of our treatment of the thermodynamics of solids is postponed until Chapter 22, but one topic is particularly appropriate at this time. This is the lattice energy of ionic crystals, which can be determined thermochemically, and which can be applied to obtain such data of chemical interest as electron affinities and the relative stability of oxidation states.

Our model for ionic crystals is that introduced in Sections 11.8 and 11.9: We assume the structural units to be discrete positive and negative ions, interacting primarily by Coulomb (electrostatic) forces and relatively short-ranged repulsive forces. We shall consider in this section only crystals made up of monatomic ions, to avoid the complication contributed by internal degrees of freedom and a multitude of atomic interactions. Our initial concern is the lattice energy or cohesive energy, which thermodynamically equals ΔU for a reaction such as

$$MX(s) = M^+(g) + X^-(g);$$

it is thus the negative of the crystal's energy relative to zero for infinitely separated ions.

To begin with, we assume our crystal to be at a temperature of 0 K,[27] and we neglect the zero-point energy of the lattice vibrations. On these assumptions the ions have no kinetic energy, so that the lattice energy is simply a sum of interionic potential energy terms. Assuming the Born–Mayer pair potential of Eq. 11.38,

$$u(R_{ij}) = \left(\frac{Z_i Z_j e^2}{4\pi\epsilon_0 R_{ij}}\right) + \lambda_{ij} e^{-R_{ij}/\rho}, \qquad (14.33)$$

and making the approximation that all the λ_{ij} have the same value λ_{+-}, we derived the total potential energy of the crystal in the form of Eqs. 11.44 and 11.46. We combine these equations as

$$U_0 = -\frac{NMZ_+Z_-e^2}{4\pi\epsilon_0 R_0} + zN\lambda_{+-}e^{-R_0/\rho}$$

$$= -\frac{NMZ_+Z_-e^2}{4\pi\epsilon_0 R_0}\left(1 - \frac{\rho}{R_0}\right), \qquad (14.34)$$

where N is the number of "molecules" (formula units) in the crystal, M is the Madelung constant, Z_+ and Z_- are the magnitudes of the ionic charges, R_0 is the equilibrium nearest-neighbor distance, z is (approximately) equal to the number of nearest neighbors, and ρ and λ_{+-} are the empirical constants we wish to determine. Designating the energy in Eq. 14.34 as U_0 (the internal energy at 0 K) is simply a convenient choice of energy zero. We could not carry the calculation further in Chapter 11, since we had no way of determining ρ; now, however, we can use thermodynamics to relate ρ to measurable quantities.

[27] See Section 22.2 for the temperature dependence of crystal energies.

To begin with, we again use (still without proof) the relation we introduced as Eq. 14.18. Differentiation with respect to V yields[28]

$$\left(\frac{\partial^2 U}{\partial V^2}\right)_T = T\left[\frac{\partial}{\partial V}\left(\frac{\partial p}{\partial T}\right)_V\right]_T - \left(\frac{\partial p}{\partial V}\right)_T$$

$$= \frac{T}{V}\left[\frac{\partial}{\partial T}V\left(\frac{\partial p}{\partial V}\right)_T\right]_V - \frac{1}{V}\left[V\left(\frac{\partial p}{\partial V}\right)_T\right]$$

$$= T\left[\frac{\partial}{\partial T}\left(-\frac{1}{\kappa_T}\right)\right]_V - \frac{1}{V}\left(-\frac{1}{\kappa_T}\right)$$

$$= \frac{T}{V\kappa_T^2}\left(\frac{\partial \kappa_T}{\partial T}\right)_V + \frac{1}{V\kappa_T}, \tag{14.35}$$

where κ_T is the isothermal compressibility of Eq. 14.30. So far this is a completely general thermodynamic result, valid for any substance at any temperature. For a solid at $0\,K$, however, the first term vanishes and we have simply

$$\left(\frac{\partial^2 U}{\partial V^2}\right)_{T=0} = \frac{1}{V\kappa_0}, \tag{14.36}$$

where κ_0 is $\kappa_T(T=0)$.

Now we obtain from our crystal-structure model a result matching Eq. 14.36. The volume of the crystal must be

$$V = KNR_0^3, \tag{14.37}$$

where K is a constant depending on the lattice (in the rock salt lattice, for example, the volume is $2R_0^3$ per ion pair, so we have $K = 2$). We have, in general,

$$\left(\frac{\partial U}{\partial V}\right)_T = \left(\frac{\partial U}{\partial R_0}\right)_T\left(\frac{\partial R_0}{\partial V}\right)_T, \tag{14.38}$$

$$\left(\frac{\partial^2 U}{\partial V^2}\right)_T = \left(\frac{\partial U}{\partial R_0}\right)_T\left(\frac{\partial^2 R_0}{\partial V^2}\right)_T + \left(\frac{\partial R_0}{\partial V}\right)_T\left[\frac{\partial}{\partial V}\left(\frac{\partial U}{\partial R_0}\right)_T\right]_T$$

$$= \left(\frac{\partial U}{\partial R_0}\right)_T\left(\frac{\partial^2 R_0}{\partial V^2}\right)_T + \left(\frac{\partial^2 U}{\partial R_0^2}\right)_T\left[\left(\frac{\partial R_0}{\partial V}\right)_T\right]^2. \tag{14.39}$$

Now, since R_0 is defined as an equilibrium value, it corresponds to a minimum of the potential energy, and in our model at $0\,K$ also a minimum of the internal energy. Thus $(\partial U/\partial R_0)_{T=0}$ vanishes, and we have

$$\left(\frac{\partial^2 U}{\partial V^2}\right)_{T=0} = \left(\frac{\partial^2 U}{\partial R_0^2}\right)_{T=0}\left[\left(\frac{\partial R_0}{\partial V}\right)_T\right]^2$$

$$= \frac{1}{(3KNR_0^2)^2}\left(\frac{\partial^2 U}{\partial R_0^2}\right)_{T=0}. \tag{14.40}$$

Setting this result equal to Eq. 14.36 gives

$$\left(\frac{\partial^2 U}{\partial R_0^2}\right)_{T=0} = \frac{(3KNR_0^2)^2}{(KNR_0^3)\kappa_0} = \frac{9KNR_0}{\kappa_0}. \tag{14.41}$$

We can now evaluate the parameters ρ and λ_{+-}, the assumed common value of λ_{ij}. Differentiating Eq. 14.34 twice and substituting

---

[28] We use Eq. II.3 in Appendix II at the back of the book to equate

$$\left[\frac{\partial}{\partial V}\left(\frac{\partial p}{\partial T}\right)_V\right]_T = \left[\frac{\partial}{\partial T}\left(\frac{\partial p}{\partial V}\right)_T\right]_V.$$

Eq. 11.45 to clear the exponential, we obtain

$$\left(\frac{\partial^2 U}{\partial R_0{}^2}\right)_{T=0} = -\frac{NMZ_+Z_-e^2}{2\pi\epsilon_0 R_0{}^3} + \frac{zN\lambda_{+-}}{\rho^2}e^{-R_0/\rho}$$

$$= -\frac{NMZ_+Z_-e^2}{2\pi\epsilon_0 R_0{}^3}\left(1-\frac{R_0}{2\rho}\right). \qquad (14.42)$$

Setting this result equal to Eq. 14.41 gives

$$-\frac{NMZ_+Z_-e^2}{2\pi\epsilon_0 R_0{}^3}\left(1-\frac{R_0}{2\rho}\right) = \frac{9KNR_0}{\kappa_0}, \qquad (14.43)$$

which rearranges to

$$\frac{R_0}{\rho} = \frac{36\pi\epsilon_0 KR_0{}^4}{MZ_+Z_-e^2\kappa_0} + 2, \qquad (14.44)$$

with all the quantities on the right-hand side known. Once the value of ρ is known, one can also obtain λ_{+-} by rearranging Eq. 11.46 to read

$$\lambda_{+-} = \frac{MZ_+Z_-e^2\rho}{4\pi\epsilon_0 zR_0{}^2}e^{-R_0/\rho}. \qquad (14.45)$$

Inserting in Eq. 14.44 the experimental value of κ_0 and the appropriate lattice parameters, one can evaluate ρ/R_0, and thus U_0 by Eq. 14.34.

Let us try a sample calculation of the lattice energy of NaCl at 0 K. We have $M = 1.7476$, $K = 2$, $z = 6$ for the rock salt lattice, and experimental values of $R_0 = 2.820$ Å, $\kappa_T(25°C) = 4.02 \times 10^{-6}$ atm^{-1} for NaCl. If we assume the same value for κ_0, substitution in Eq. 14.44 gives

$$\frac{R_0}{\rho} = \frac{(36\pi)(8.854\times10^{-12}\text{ C}^2/\text{N m}^2)(2)(2.820\times10^{-10}\text{ m})^4}{(1.7476)(1^2)(1.6022\times10^{-19}\text{ C})^2(4.02\times10^{-6}\text{ atm}^{-1})}$$

$$\times\left(\frac{1.01325\times10^5\text{ N/m}^2}{\text{atm}}\right) + 2$$

$$= 7.11 + 2 = 9.11,$$

or $\rho = 0.310$ Å. Eq. 14.34 then gives for the energy of the crystal

$$u_0 = -\frac{(6.022\times10^{23}/\text{mol})(1.7476)(1^2)(1.6022\times10^{-19}\text{ C})^2}{(4\pi)(8.854\times10^{-12}\text{ C}^2/\text{N m}^2)(2.820\times10^{-10}\text{ m})}\left(1-\frac{1}{9.11}\right)$$

$$= -766.5\text{ kJ/mol},$$

of which the Coulomb and repulsive terms are -861 and $+94.5$ kJ/mol, respectively. This is to be compared with an experimental value of -784.9 kJ/mol; the lattice energy as we have defined it is the negative of this quantity. The corresponding value of λ_{+-} from Eq. 14.45 is 2.37×10^{-16} J (1.48 keV). With these values of ρ and λ_{+-} we can calculate the pair potential, Eq. 14.33, as a function of distance; the results are plotted in Fig. 14.8.

It should be recalled that this is only an approximate calculation, since we deliberately omitted several corrections to obtain equations in a simple form. These corrections include the zero-point energy of the lattice, the interionic van der Waals attraction, and the repulsive force between nonnearest neighbors; also, the value of κ_T must be corrected from 25°C to 0 K. The inclusion of these effects is straightforward but tedious.[29] When one takes account of them all, the energy of NaCl is

[29] The zero-point and van der Waals energies for NaCl are $+5.9$ and -24 kJ/mol, respectively. The corrected value of the repulsive energy is 104 kJ/mol (9.5 kJ/mol from nonnearest neighbors).

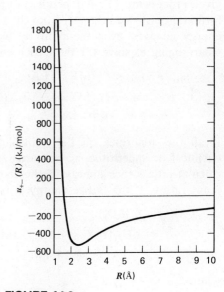

FIGURE 14.8
The Born–Mayer pair potential for the ionic crystal NaCl.

found to be $-773\,\text{kJ/mol}$. This is an improvement over the simple calculation, but even the latter gave quite good agreement with experiment. At either level of approximation, the results are not really full theoretical evaluations, since experimental data for each crystal must be inserted into the theory. Nevertheless, the model is useful to the extent that the results obtained for various properties are consistent. The foremost of these properties is the lattice energy. We shall now describe how its experimental value is determined.

Consider a crystal of an alkali halide, with formula MX. It can be synthesized from the elements by two different pathways, as indicated schematically in the following diagram, known as a *Born–Haber cycle:*

$$
\begin{array}{ccc}
\text{M}(s) & + & \tfrac{1}{2}\text{X}_2(\text{std. state}) \\
\Big\downarrow \Delta U_{\text{subl}}(\text{M}) & & \Big\downarrow \Delta U_f(\text{X, g}) \quad\searrow^{\Delta U_f(\text{MX, }s)} \\
\text{M}(g) & + & \text{X}(g) \qquad\qquad \text{MX}(s) \\
\Big\downarrow I(\text{M}) & & \Big\downarrow -A(\text{X}) \quad\nearrow \\
\text{M}^+(g) & + & \text{X}^-(g) \qquad -U_{\text{lat}}(\text{MX})
\end{array}
$$

Next to each arrow is written the value of ΔU for the process in question: ΔU_f is the internal energy of formation, and U_{lat} the lattice energy, of the MX crystal; the other quantities are the sublimation energy ΔU_{subl}, the ionization potential I, and the electron affinity A. Note that ΔU_f is the energy change for formation from the elements in their standard states, whereas U_{lat} is $-\Delta U$ for formation from isolated ions. One usually measures ΔH rather than ΔU, but one can be converted to the other $(H = U + pV)$ by assuming perfect gas behavior and neglecting pV for solids, so that $H_{\text{gas}} = U_{\text{gas}} + nRT$, $H_{\text{solid}} = U_{\text{solid}}$; we then have, per mole of alkali halide (if X_2 is gaseous), $\Delta u_{\text{subl}} = \Delta h_{\text{subl}} - RT$, $\Delta U_f(\text{X, g}) = \Delta H_f(\text{X, g}) - \tfrac{1}{2}RT$, and $\Delta U_f(\text{MX, }s) = \Delta H_f(\text{MX, }s) + \tfrac{1}{2}RT$, to quite high accuracy. The measurements of these various quantities are of course made at temperatures well above 0 K, but one can convert the results to 0 K by applying Eq. 14.19 or 14.22, using the techniques of Chapters 21 and 22 to calculate the heat capacities of gases and solids, respectively.

The point of all this is that, by the first law, the energy change for a given process must be independent of the path. In this case, the energy of formation of $MX(s)$ directly from the elements must equal the sum of the energy changes over the path by way of separated atoms and ions. Rearranging to solve for the lattice energy, we obtain

$$
\begin{aligned}
u_{\text{lat}}(\text{MX}) &= -\Delta U_f(\text{MX, }s) + \Delta u_{\text{subl}}(\text{M}) + \Delta U_f(\text{X, g}) + I(\text{M}) - A(\text{X}) \\
&= -\Delta H_f(\text{MX, }s) + \Delta h_{\text{subl}}(\text{M}) + \Delta H_f(\text{X, g}) \\
&\quad - 2RT + I(\text{M}) - A(\text{X}).
\end{aligned} \tag{14.46}
$$

If all the quantities on the right-hand side are known from thermochemical or spectroscopic measurements, this equation can be used to calculate the lattice energy. If the thermochemical data are obtained at temperature T, the lattice energy at 0 K is

$$
\begin{aligned}
u_{\text{lat},0}(\text{MX}) &= -u_0(\text{MX}) \\
&= -\Delta H_{f,0}(\text{MX, }s) + \Delta h_{\text{subl},0}(\text{M}) + \Delta H_{f,0}(\text{X, g}) + I(\text{M}) - A(\text{X}) \\
&= u_{\text{lat},T}(\text{MX}) + 2RT + \delta(T),
\end{aligned} \tag{14.47}
$$

where

$$
\delta(T) \equiv \int_0^T \left[c_p(\text{MX, }s) - c_p(\text{M, }g) - c_p(\text{X, }g) \right] dT
$$

$$= \int_0^T c_p(MX, s)\, dT - 5RT, \tag{14.48}$$

since c_p for monatomic gases is $\frac{5}{2}R$ (Chapter 21).[30]

In Table 14.9 we compare the values of u_0 obtained by the above method with those calculated by summing the interionic forces. It can be seen that the agreement with the complete theory is quite good, while even the simplified calculations are adequate for qualitative interpretation. The thermochemical calculation can of course be rearranged to determine any of the quantities in Eq. 14.46 from a knowledge of all the others (including the theoretical lattice energy). This was in fact the method first used to determine the electron affinities of the halogens; it is only in recent years that direct measurements of electron affinities have become sufficiently accurate for the process to be reversed. The data of Table 14.9 give average values of 336.4, 357.3, 336.8, and 307.9 kJ/mol for the A's of F, Cl, Br, and I; all are somewhat higher than the experimental values, showing the existence of a systematic error in even the complete ionic model. This is to be expected, since the real ions in a crystal are not purely spherical charge distributions interacting by simple Coulomb forces, but must be distorted in some way by their neighbors. In addition, the repulsive interaction must be more complicated than the simple Born–Mayer potential. Nevertheless, the ionic model is good enough for interpreting most behavior of chemical interest, and one can anticipate fairly well when it is likely to be in error.

To finish this section, let us now look at one example of how the ionic model can be applied to describe chemical behavior: the relative stabilities of solid metal halides. It is found that for high cation oxidation states the most stable halide is usually the fluoride, that is, that the reaction

$$MX_{n+1}(s) \rightarrow MX_n(s) + \tfrac{1}{2}X_2(\text{std. state})$$

is least likely for $X = F$ (followed by Cl, Br, I in the usual order). To cite some examples, CuI_2 is unstable with respect to CuI at room temperature, whereas all other cupric halides are stable under the same conditions; MnF_4 is stable, but no other tetrahalide of Mn is. To see why this is so, let us begin by setting up a Born–Haber cycle, this time listing the enthalpy change with each step:

$$MX_{n+1}(s) \xrightarrow{\;\;\Delta H_{\text{reac}}\;\;} MX_n(s) \quad + \quad \tfrac{1}{2}X_2(\text{std. state})$$

$$\Big\downarrow u_{\text{lat}}(MX_{n+1}) + (n+2)RT \qquad \Big\downarrow u_{\text{lat}}(MX_n) + (n+1)RT \qquad \Big\downarrow \Delta H_f(X, g)$$

$$M^{(n+1)+}(g) + (n+1)X^-(g) \xleftarrow[I(M^{n+})-A(X)]{} M^{n+}(g) + nX^-(g) + X(g)$$

We immediately obtain

$$\Delta H_{\text{reac}} = u_{\text{lat}}(MX_{n+1}) - u_{\text{lat}}(MX_n) - \Delta H_f(X, g)$$
$$- I(M^{n+}) + A(X) + RT, \tag{14.49}$$

which should become more negative as X goes from F to I. The data for X atoms, which can be obtained from Table 14.9, are in the right direction:

X	F	Cl	Br	I
$-\Delta H_f(X, g) + A(X)$ (kJ/mol)	253.6	227.6	212.5	188.7

[30] We assume that neutral atoms and their ions have the same heat capacities (at temperatures too low for excited electronic states to be important), so that I and A are independent of temperature.

TABLE 14.9
MEASURED AND CALCULATED LATTICE ENERGIES OF ALKALI HALIDES AT 0 K (ALL ENERGIES IN KILOJOULES PER MOLE)[a]

M	$\Delta h_{subl}(M)$ (298 K)	$I(M)$	Data for Elements X	$\Delta H_f(X, g)$ (298 K)	$A(X)$
Li	160.7	520.5	F	79.1	332.6
Na	108.4	495.4	Cl	120.9	348.5
K	90.0	418.4	Br	112.1	324.7
Rb	81.6	402.9	I	106.7	295.4
Cs	78.2	375.3			

MX	$-\Delta H_f(MX, s)$ (298 K)	$\int_0^{298} c_p (MX, s)\, dT$	$-u_0$ (meas)[b]	$-u_0$ (calc) (simple theory)[c]	$-u_0$ (calc) (full theory)[d]
LiF	609.6	6.3	1030.9	1060.6	1032.6
NaF	570.3	7.9	915.9	920.1	915.0
KF	562.7	9.6	814.6	807.5	813.4
RbF	551.5	10.5	780.3	764.8	777.8
CsF	565.3	11.7	764.4	730.9	747.7
LiCl	401.7	7.5	850.2	821.3	845.2
NaCl	410.9	10.5	784.9	761.5	777.8
KCl	436.0	11.7	715.9	693.3	708.8
RbCl	432.6	12.6	689.5	665.7	686.2
CsCl[e]	447.3	13.8	674.5	542.2	652.3
LiBr	350.2	8.4	814.6	771.5	797.9
NaBr	359.8	11.7	750.2	722.6	739.3
KBr	392.0	12.6	687.8	662.3	679.5
RbBr	391.2	13.4	664.0	635.5	659.0
CsBr[e]	408.8	14.6	651.9	600.0	632.2
LiI	271.1	10.0	761.1	707.5	739.7
NaI	287.9	13.0	703.3	666.5	692.0
KI	327.6	13.4	648.1	620.1	640.2
RbI	330.5	13.8	627.6	598.7	622.2
CsI[e]	351.0	14.6	618.0	565.7	601.2

[a] Based on D. Cubicciotti, *J. Chem. Phys.* **31**, 1646 (1959); **33**, 1579 (1960); **34**, 2189 (1961).
[b] $-u_0 = -\Delta H_{f,T}(MX, s) + \Delta h_{subl,T}(M) + \Delta H_{f,T}(X, g) + I(M) - A(X) + \int_0^T c_p(MX, s)\, dT - 5RT$; at 298 K, $5RT = 12.55$ kJ/mol.
[c] From Eqs. 14.34 and 14.44, using values of κ_T at 298 K.
[d] Including zero-point, van der Waals, and complete repulsive energies, and corrected for temperature dependence of κ_T.
[e] These crystals have the CsCl lattice, with eight nearest neighbors; all others tabulated have the NaCl lattice.

However, we must also estimate the effect of varying X on the lattice energy difference between MX_{n+1} and MX_n.

We already know from Table 14.9 that the simple model of Eq. 14.34 gives qualitatively accurate results for u_0, and we can simplify this even further without serious loss of accuracy. For one thing, the value of

ρ does not vary greatly from one crystal to another: The average of the values obtained by Eq. 14.44 for the alkali halides is 0.328 Å, and only one of 20 values differs from this average by more than 0.050 Å. Since this range corresponds to a variation of only $\pm1.5\%$ in u_0, we can assume that the average value is generally valid. A further approximation comes from Table 11.4, which shows that M/ν (where M is the Madelung constant and ν is the number of ions per "molecule") is roughly constant for a wide range of coordination numbers. To a very rough approximation, then, setting $M = 0.83\nu$ and $\rho = 0.33$ Å, we have

$$\frac{U_0}{N} \approx \frac{0.83\nu Z_+ Z_- e^2}{4\pi\epsilon_0 R_0}\left(1 - \frac{0.33 \text{ Å}}{R_0(\text{Å})}\right)$$

$$\approx \left(\frac{1150 \text{ kJ}}{\text{mol}}\right)\frac{\nu Z_+ Z_-}{R_0}\left(1 - \frac{0.33 \text{ Å}}{R_0(\text{Å})}\right). \tag{14.50}$$

The value of Eq. 14.50 is that we can apply it without knowing the lattice type, since in general MX_{n+1} and MX_n may have different lattices. The repulsive energy is only a small fraction of the total lattice energy, so the principal effect of such a change of lattice should be through the Coulomb energy, which will be appreciably affected by a change in ν. Although it is very approximate, Eq. 14.50 should therefore serve to indicate a trend. Specifically, the lattice energy terms of Eq. 14.49 become roughly

$$u_{\text{lat}}(MX_{n+1}) - u_{\text{lat}}(MX_n)$$

$$\approx [u_0(MX_n)]_{\text{Coulomb}} - [u_0(MX_{n+1})]_{\text{Coulomb}}$$

$$\approx \frac{1150(n+1)n(-1)}{r(M^{n+}) + r(X^-)}\left[1 - \frac{0.33}{r(M^{n+}) + r(X^-)}\right]$$

$$- \frac{1150(n+2)(n+1)(-1)}{r(M^{(n+1)+}) + r(X^-)}\left[1 - \frac{0.33}{r(M^{(n+1)+}) + r(X^-)}\right], \tag{14.51}$$

where we introduce ionic radii such that $R_0 = r_+ + r_-$. Next we differentiate with respect to $r(X^-)$, holding all other quantities constant:

$$\frac{\partial}{\partial r(X^-)}[u_{\text{lat}}(MX_{n+1}) - u_{\text{lat}}(MX_n)]$$

$$\approx \frac{1150n(n+1)}{[r(M^{n+}) + r(X^-)]^2}\left[1 - \frac{0.66}{r(M^{n+1}) + r(X^-)}\right]$$

$$- \frac{1150(n+2)(n+1)}{[r(M^{(n+1)+}) + r(X^-)]^2}\left[1 - \frac{0.66}{r(M^{(n+1)+}) + r(X^-)}\right]. \tag{14.52}$$

The above expression would be negative even if $r(M^{n+}) = r(M^{(n+1)+})$, and the fact that in general $r(M^{n+}) > r(M^{(n+1)+})$ makes it even more negative; for $r_+ + r_- \approx 3$ Å, the derivative is roughly

$$\frac{1150}{9}\left(1 - \frac{0.66}{3}\right)[n(n+1) - (n+1)(n+2)] \approx -200(n+1)\frac{\text{kJ/mol}}{\text{Å}}.$$

Since $r(X^-)$ increases as we go from F^- (1.3 Å) to I^- (2.2 Å), we have established that ΔH_{reac} of Eq. 14.49 becomes more negative with this change. This confirms the experimentally observed order of stability.

Many other relative stabilities of compounds can be similarly explained by crude calculations like those we have carried out here. On a slightly more sophisticated level, one can calculate the relative stabilities of the different lattices a given salt might occupy. This also yields results generally in agreement with experiment, i.e., predicting that the actual lattice is most stable. Much more could be said about the applications of lattice energy calculations, but it is about time for us to terminate this long chapter.

Further Reading

Lewis, G. N., and Randall, M., *Thermodynamics*, 2nd ed., revised by K. S. Pitzer and L. Brewer (McGraw-Hill, New York, 1961), chaps. 5 and 6, and pp. 373–392.

There are many specialized data sources. Some commonly used ones are the following:

Din, F. (Ed.), *Thermodynamic Functions of Gases*, Vols. 1, 2, 3 (Butterworths, London, 1962).

Benson, S. W., *Thermochemical Kinetics*, 2nd ed. (Wiley, New York, 1976), chap. 2 and appendices 1–23.

JANAF Thermochemical Tables, 2nd ed., NSRDS-NBS 37 (U.S. Gov't. Printing Office, Washington, D.C., 1971).

Selected Values of Chemical Thermodynamic Properties, NBS Circular 500 (U.S. Govt. Printing Office, Washington, D.C., 1961).

Stull, D. R., Westrum, E. F., and Sinke, G. S., *The Chemical Thermodynamics of Organic Compounds* (Wiley, New York, 1969).

Zwolinski, B. J., and Wilhoit, R. C., *Vapor Pressures and Heats of Vaporization of Hydrocarbons and Related Compounds* (Thermodynamics Research Center Data Distribution Office, College Station, Texas, 1971).

Very detailed tables of thermodynamic properties of many individual substances are available in the form of circulars from the National Bureau of Standards (NBS) or the Commission on Thermodynamics of the International Union of Pure and Applied Chemistry (IUPAC). One example is the following.

Angus, S. and de Reuch, K. M. (Eds.), *International Thermodynamic Tables of the Fluid State* ^4He (Pergamon Press, Elmsford, N.Y., 1977).

Problems

1. When 250 g of water at 30°C is poured into a copper calorimeter of mass 3000 g at 0°C, a final temperature of 11.90°C is observed. Taking the specific heat of water to be 4.2 J/g K, calculate the average specific heat of copper.

2. Suppose that the heat capacity of a calorimeter is 913.8 J/K at 23°C. Further, suppose that a piece of metal of mass 35 g which had previously been heated to 100°C is immersed in the calorimeter. When equilibrium is achieved the temperature of the calorimeter is observed to have changed from 22.45 to 23.50°C. What is the mean heat capacity of the unknown metal in units of J/g K?

3. Calculate the amount of heat required to increase the temperature of 10^{-3} mol of Ag from 0°C to 900°C at constant pressure, if

$$c_p(\text{Ag}) = 21.3 + 8.54 \times 10^{-3} T + 1.51 \times 10^5 T^{-2} \text{ J/K mol.}$$

4. Calculate the heat evolved when 100 g of Au is cooled at constant pressure from 1000 to 0°C, if

$$c_p(\text{Au}) = 23.7 + 5.19 \times 10^{-3} T \text{ J/K mol}.$$

5. The heat capacity of gaseous CO_2 is

$$c_p = 29.3 + 3.0 \times 10^{-2} T - 7.78 \times 10^{-6} T^2 \text{ J/K mol}$$

in the limit of zero pressure. Calculate the amount of heat needed to raise the temperature of 200 g of gaseous CO_2 from 27 to 227°C at constant pressure. Calculate the amount of heat required to change the temperature by the same amount at constant volume. Assume that CO_2 is a perfect gas.

6. Assume that N_2 behaves as a perfect gas with an average specific heat (at constant pressure) of 1.02 J/g K. Calculate the changes in enthalpy and internal energy when 1 mol of N_2 is heated from 0 to 110°C at 2 atm.

7. Calculate the heat absorbed when 3.1 mol of a perfect gas with $c_p =$ 20.9 J/K mol is isothermally expanded from 1 atm to a volume of 100 liters at 15°C.

8. At a constant temperature of 100°C a perfect gas is transferred reversibly from a pressure of 170 torr and a volume of 4 liters to a pressure of 1 atm. Calculate the work of compression and the amount of heat evolved.

9. Suppose that 1.5 mol of a perfect gas is at a pressure of 5 atm at 0°C. After expansion at constant pressure the gas has a volume of 15 liters. Calculate the work of expansion of the gas and the amount of heat absorbed.

10. One mole of a perfect gas, initially with a volume of 20 liters at 200°C, expands adiabatically against a constant pressure of 1 atm until the pressure of the gas drops to 1 atm. Calculate the work done during this process. Note that this is an irreversible expansion.

11. A combustion experiment is performed by burning a specimen in oxygen in a calorimeter bomb of constant volume. As a result of the reaction the temperature of the water bath surrounding the bomb rises. If the mixture of specimen and oxygen is regarded as the system:
    (a) Has heat been transferred?
    (b) Has work been done?
    (c) What is the sign of ΔU?

12. One mole of a gas obeys the equation of state

$$p(v - b) = RT,$$

where b is a constant. The internal energy of this gas is

$$u = CT,$$

where C is a constant. Calculate the heat capacities c_p and c_V.

13. Derive the equations

(a)
$$đq = C_V \, dT + \left[\left(\frac{\partial U}{\partial V} \right)_T + p \right] dV,$$

(b)
$$C_p = C_V + \left[\left(\frac{\partial U}{\partial V} \right)_T + p \right] V\alpha,$$

(c)
$$đq = C_V \, dT + \frac{C_p - C_V}{V\alpha} \, dV,$$

where

$$\alpha = \frac{1}{V} \left(\frac{\partial V}{\partial T} \right)_p.$$

14. For water vapor between $\theta = 0°C$ and $\theta = 650°C$, the following formula is approximately valid:

$$c_p = 36.1 + 0.008\theta + 3.0 \times 10^{-8}\theta^2 \text{ J/K mol.}$$

If $c_p - c_V = R$, determine the increase in internal energy of water when heated from 50 to 650°C.

15. Calculate the changes in internal energy and enthalpy by any two different paths for the following systems and changes of state:

(a) A gas for which the equation of state is

$$p = \frac{RT}{v}\left(1 + \frac{B}{v} + \frac{C}{v^2}\right),$$

where B and C depend on the temperature but not on the volume. The change of state is from (p_1, v_1, T_1) to (p_2, v_2, T_2). Assume c_V is independent of temperature.

(b) A solid for which the equation of state is

$$v = v_0 - Ap,$$

and for which the specific heat at constant volume has the form

$$c_V = C.$$

In the above, A, C, and v_0 are to be considered constants independent of the temperature and volume. As in (a), the change of state is from p_1, v_1, T_1 to p_2, v_2, T_2.

(c) Gaseous CH_4, for which the following data have been obtained:

v (cm^3/mol)	640	640	640
θ (°C)	0	100	200
p (atm)	32.297	46.474	60.486

v (cm^3/mol)	320	320	320
θ (°C)	0	100	200
p (atm)	60.13	91.24	121.81

The change of state is from 30 atm, 0°C to 500 atm, 200°C. Hint: Fit the data to the analytic form $pv = RT(1 + B/v)$ and use 100°C temperature intervals to find the temperature dependence of B. Assume c_V is independent of temperature.

16. Calculate the enthalpy change at 298 K and 1 atm for the reaction

$$NaNO_3(s) + H_2SO_4(l) \rightarrow NaHSO_4(s) + HNO_3(g)$$

from the following heats of formation (at 298 K):

$$NaNO_3(s), \qquad \Delta H_f^0 = -465.26 \text{ kJ/mol}$$
$$H_2SO_4(l), \qquad \Delta H_f^0 = -807.09 \text{ kJ/mol}$$
$$NaHSO_4(s), \qquad \Delta H_f^0 = -1118.8 \text{ kJ/mol}$$
$$HNO_3(g), \qquad \Delta H_f^0 = -143.93 \text{ kJ/mol}$$

17. Given the following heats of reaction:

$$C_2H_4(g) + 3O_2(g) \rightarrow 2CO_2(g) + 2H_2O(l), \qquad \Delta H_{298}^0 = -1411 \text{ kJ/mol}$$
$$H_2(g) + \tfrac{1}{2}O_2(g) \rightarrow H_2O(l), \qquad \Delta H_{298}^0 = -286 \text{ kJ/mol}$$
$$C_2H_6(g) + \tfrac{7}{2}O_2(g) \rightarrow 2CO_2(g) + 3H_2O(l), \qquad \Delta H_{298}^0 = -1560 \text{ kJ/mol}$$

calculate the standard heat of the reaction

$$C_2H_4(g) + H_2(g) \rightarrow C_2H_6(g)$$

at 298 K.

18. Given

$$C_3H_6(g) + H_2(g) \rightarrow C_3H_8(g), \qquad \Delta H^0_{298} = -1246 \text{ kJ/mol}$$

$$C_3H_8(g) + 5O_2(g) \rightarrow 3CO_2(g) + 4H_2O(l), \qquad \Delta H^0_{298} = -2220 \text{ kJ/mol}$$

$$C(s) + O_2(g) \rightarrow CO_2(g), \qquad \Delta H^0_{298} = -393 \text{ kJ/mol}$$

$$H_2(g) + \tfrac{1}{2}O_2(g) \rightarrow H_2O(l), \qquad \Delta H^0_{298} = -286 \text{ kJ/mol}$$

calculate the standard heats of reaction for

$$C_3H_6(g) + \tfrac{9}{2}O_2(g) \rightarrow 3CO_2(g) + 3H_2O(l)$$

and

$$3C(s) + 3H_2(g) \rightarrow C_3H_6(g)$$

at 298 K.

19. (a) Given

$$Sn(s) + O_2(g) \rightarrow SnO_2(s), \qquad \Delta H^0_{298} = -580.82 \text{ kJ/mol}$$

$$SnO(s) + \tfrac{1}{2}O_2(g) \rightarrow SnO_2(s), \qquad \Delta H^0_{298} = -294.85 \text{ kJ/mol}$$

calculate the standard heat of formation of $SnO(s)$.

(b) Given

$$3UO_2(s) + O_2(g) \rightarrow U_3O_8(s), \qquad \Delta H^0_{298} = -318.0 \text{ kJ/mol}$$

$$3U(s) + 4O_2(g) \rightarrow U_3O_8(s), \qquad \Delta H^0_{298} = -3571 \text{ kJ/mol}$$

calculate the standard heat of formation of $UO_2(s)$.

20. (a) On the basis of the energy evolved per gram, which of the following offers the best possibility as a rocket fuel? Assume that gaseous O_2 is the oxidant.
   (a) CH_4
   (b) C_2H_6
   (c) C_3H_8
   (d) B_2H_6
   (e) NH_3
   (f) N_2H_4

(b) Repeat the calculation of part (a), but with gaseous F_2 as the oxidant. Is there any change in your conclusions? Why?

21. Calculate the heat of reaction for

$$2MgO(s) + Si(s) \rightarrow SiO_2(s) + Mg(g)$$

at 1000 K. At 298 K the heats of formation of $MgO(s)$ and $SiO_2(s)$ are -601.83 kJ/mol and -859.4 kJ/mol, respectively. The heat of vaporization of Mg is 132 kJ/mol at 1393 K. The heat capacities of the substances involved in the reaction are (in J/K mol)

$$MgO(s), \qquad c_p = 45.44 + 5.008 \times 10^{-3}T - 8.732 \times 10^5 T^{-2}$$

$$Si(s), \qquad c_p = 24.0 + 2.582 \times 10^{-3}T - 4.226 \times 10^5 T^{-2}$$

$$SiO_2(s), \qquad c_p = 45.48 + 36.45 \times 10^{-3}T - 10.09 \times 10^5 T^{-2}$$

$$Mg(g), \qquad c_p = 20.79$$

$$Mg(s), \qquad c_p = 24.39$$

22. Find ΔH for the following processes:
   (a) $KCl(s) \rightarrow KCl(aq)$
   (b) Precipitation of AgCl from aqueous solution
   (c) Precipitation of PbS from aqueous solution
   (d) $HSO_4^-(aq) \rightarrow H^+(aq) + SO_4^{2-}(aq)$

23. Given the standard heats of formation at 25°C,

$$Na_2SO_4(aq), \qquad \Delta H_f^{\ominus} = -1387 \text{ kJ/mol}$$

$$Na_2SO_4 \cdot 10H_2O(s), \qquad \Delta H_f^0 = -4324 \text{ kJ/mol}$$

$$H_2O(l), \qquad \Delta H_f^0 = -286 \text{ kJ/mol}$$

calculate the heat absorbed in the process

$$Na_2SO_4 \cdot 10H_2O(s) \rightarrow Na_2SO_4(aq).$$

24. One mole of gaseous SO_2 is dissolved in an infinitely dilute solution containing 2 mol of NaOH [i.e., $2NaOH(aq)$] at 25°C and 1 atm. Calculate the enthalpy change for this process from

$$
\begin{array}{lll}
OH^-(aq), & \Delta H_f^{\ominus} = -230\text{ kJ/mol} \\
SO_3^{2-}(aq), & \Delta H_f^{\ominus} = -624\text{ kJ/mol} \\
SO_2(g), & \Delta H_f^0 = -297\text{ kJ/mol} \\
H_2O(l), & \Delta H_f^0 = -286\text{ kJ/mol}
\end{array}
$$

25. When 1 mole of S atoms in rhombic sulfur is dissolved in an infinitely large volume of chloroform at 18°C, 2678 J of heat is evolved. When 1 mole of S atoms as monoclinic sulfur is dissolved in an infinitely large volume of chloroform at 18°C, 2343 J of heat is evolved. What information can be derived from these data?

26. At 7°C the heat of solution of solid acetic acid in an infinite volume of water is 8910 J/mol, whereas the heat of solution of liquid acetic acid under the same conditions is −1670 J/mol. Calculate the heat of fusion of acetic acid at 7°C.

27. Calculate the maximum temperature and pressure that could be produced by the explosion within a constant-volume bomb of 1 mol of H_2, $\frac{1}{2}$ mol of O_2, and 1 mol of N_2 at 100 torr total pressure and 25°C. Assume that the water vapor formed does not dissociate.

28. One mole of ethyl alcohol is boiled off at 78.1°C and 1 atm with the absorption of 858 J/g. What would you expect the heat of vaporization to be for evaporation into a perfect vacuum?

29. The latent heat of vaporization of water is 40,660 J/mol at 100°C. What part of this enthalpy change is spent on the work of expanding the water vapor?

30. A constant-volume insulated container, initially at 25°C, contains 0.1 mol of $CO(g)$ and 0.05 mol of $O_2(g)$. The gases explode, and 0.1 mol of $CO_2(g)$ is produced. The heat capacity of the container and its contents is 10^3 J/K. Assume that the gases are perfect, and that the standard heats of formation at 25°C of $CO(g)$ and $CO_2(g)$ are −110.5 kJ/mol and −393.5 kJ/mol, respectively. Find ΔU, ΔT, and ΔH for the system.

The Concept of Entropy: Relationship to the Energy-Level Spectrum of a System

In Chapters 12, 13, and 14 we introduced two nonmechanical concepts that differentiate thermodynamics from mechanics, namely, temperature and heat, and we showed how two other concepts from mechanics, work and energy, can be extended to apply to the description of processes in which only gross parameters, such as pressure and volume, are controlled. We also showed how the law of conservation of energy could be used to define heat and work in terms of energy changes, leading to the formulation of the first law of thermodynamics. Finally, we found a very fruitful application of the first law in the subject of thermochemistry.

Now, although the first law of thermodynamics restricts the class of possible processes a system can undergo to the subset that conserves energy, it does not further distinguish among these processes. It does not tell us which energy-conserving changes will, and which will not, actually occur. The following examples illustrate situations where the first law alone is inadequate for the purpose of predicting the behavior of the system.

1. Imagine a box divided into two compartments, one filled with one gas, the other with a different gas. Suppose, for simplicity, that each is a perfect gas, and that the pressure and temperature are the same in the two compartments. If a valve is opened between the two compartments, it is *always* observed that the gases mix. Conversely, if the two compartments are initially filled with the same gas mixture, it is *never* observed that demixing occurs. Yet, given that the gases are perfect, and the temperature is the same in the two compartments, mixing and demixing are processes that do not change the energy of the gases. Why does the mixing process always occur and the demixing process never occur?

2. Imagine a bomb with rigid adiabatic walls filled with two reactants, say, H_2 and O_2. The first law of thermodynamics tells us that the energy of the system is conserved, since no work can be done (rigid walls) and no heat can be transferred (adiabatic walls). Yet we know that an infinitesimal perturbation, such as an internal spark, initiates the exothermic reaction

$$H_2 + \tfrac{1}{2}O_2 = H_2O.$$

How can we predict whether or not this reaction, and others, will occur under specified conditions, that is, whether the chemical equilibrium lies on the product or reactant side?

3. The zeroth law of thermodynamics is based on an essential characteristic of macroscopic systems, namely, that two systems in contact through a diathermal wall, and otherwise isolated, always eventually reach a state of thermal equilibrium. Thus, an isolated copper bar, initially in a state with uniform temperature, never spontaneously undergoes a change to a state with different temperatures at the two ends; and conversely, if prepared with a difference in temperatures at the ends, the bar always approaches the state with uniform temperature. Yet, given that the bar is isolated and can do no work on the surroundings, both processes conserve energy.

These commonplace illustrations show that we need to develop some criteria for the equilibrium that exists under given constraints. Those criteria cannot be derived from the first law of thermodynamics alone.

The new concept needed for the development of criteria for equilibrium in various situations is embodied in the definition of a thermodynamic function, called *entropy*, and in the formulation of the *second law of thermodynamics*, which defines the relationship between possible states of equilibrium and the entropy of the system. We shall find that the entropy plays the role of a potential function for a macroscopic system in the sense that derivatives of the entropy with respect to parameters of constraint define the generalized forces acting on the system, and the extremal properties of the entropy determine whether or not an equilibrium is stable. In this chapter, and the next, we study the second law of thermodynamics from two points of view. First, in this chapter we develop further the notion, introduced in Chapter 13, that there is a relationship between the redistribution of molecules over the energy levels of the system and the heat transferred to or from the system in a reversible thermodynamic process. Our study will require consideration of how much information about the microscopic states of a system of many molecules is contained in the specification of only a few macroscopic parameters, for example, the total energy, mass, and volume. We shall see that a macroscopic system with nonzero energy has an energy spectrum with enormously large degeneracies. That is, specification of, say, the energy, only very mildly constrains the possible distribution of molecules over the energy-level spectrum, there being a huge number of distributions with the same energy. The essential idea in the statistical mechanical description of equilibrium states is to find the most probable distribution of molecules over the energy levels, given values of a few macroscopic parameters, and then to relate other macroscopic properties of the system to averages of microscopic quantities over this most probable distribution. We shall show that in the usual cases the most probable distribution of molecules over the energy levels is very sharply peaked. Furthermore, we shall show that macroscopic properties are sufficiently insensitive to many of the details of the energy-level spectrum of the system that complete knowledge of that spectrum is not needed to establish the general features of the macroscopic behavior.

After constructing the microscopic interpretation of entropy, in Chapter 16 we shall study the classical formulation of the second law of thermodynamics, and of the introduction of the entropy. The link between the two formulations, one microscopic and one macroscopic, provides an illuminating interpretation of the nature of thermodynamic equilibrium.

The development of the second law of thermodynamics and the relationship between the microscopic and macroscopic concepts of entropy are two of the greatest intellectual triumphs of science. A careful study of the contents of this chapter and the next will be rewarding, in that the material presented will lay the foundations for detailed analyses of the chemical and physical properties of matter.

Thus far we have accepted as valid the identification of a macroscopic property with the average of a mechanical quantity related to the molecular motion in a system. For example, in the discussion of the kinetic theory of the perfect gas, the average force per unit area exerted on a plane surface by the gas molecules or, equivalently, twice the average momentum transported across a plane per unit area per unit time, was identified with the macroscopic pressure. Three questions immediately arise when we examine this identification. First, why do we use an averaging process? Second, which of several possible averaging processes is the correct one? Third, do different averaging processes lead to the same description of the system?

The answer to the first question has been discussed in several places, but is important enough to be repeated here. A thermodynamic description of a system containing many molecules is characterized by the use of a small number of macroscopic variables, whereas a complete microscopic mechanical description requires a vast number of variables. If a thermodynamic description is to be consistent with a microscopic description, some grouping together, or averaging, or systematic ignoring of microscopic variables must be an inherent part of the connection between the two theories. The thermodynamic description of a system is inevitably coarser than the microscopic description.

The answers to the second and third questions require deeper analysis of how measurements are made and how systems to be studied are prepared. In studying the kinetic theory of the perfect gas we assumed that the system, a container of volume V with N gas molecules, was isolated and in a state of equilibrium. Then the pressure exerted by the gas molecules was calculated by (1) treating all molecules as independent particles, (2) deducing the momentum transported by one molecule across a plane per unit time per unit area, and (3) summing up contributions to the total momentum transport per unit time per unit area from all molecules moving in the appropriate direction. To execute step 3, use was made of the fact that in a gas at equilibrium the velocity distribution function, $f(\mathbf{v})$, is independent of time.

The procedure just described is a calculation of the statistical average of the momentum transport per unit time per unit area. The use of a time-independent velocity distribution function for this purpose implies that averaging the momentum transport for a typical molecular velocity over the distribution of velocities is equivalent to averaging over time the total momentum transported per unit time by all molecules crossing the plane. It is also equivalent to following the trajectory of a *single* molecule for a very long time and averaging the momentum it transports across the plane in successive crossings. All of these procedures, of course, refer to momentum transport in one direction across the plane.

Let us now examine what these equivalences mean. In general, we must expect that the successive crossings of a plane by different molecules are not evenly spaced in time, and that the instantaneous rate of momentum transport varies in time. But in a state of equilibrium the average rate of transport of momentum across a plane by the molecules must be time-independent. Therefore the time average of the rate of transport of momentum, namely, the sum of all momentum transfers in some long interval $\mathcal{T}$ divided by $\mathcal{T}$, must be independent of $\mathcal{T}$ provided that $\mathcal{T}$ is large enough. When $\mathcal{T}$ is small, the differences in the intervals between crossings are comparable to $\mathcal{T}$, so that the rate of momentum transport fluctuates on the time scale of $\mathcal{T}$. On the other hand, when $\mathcal{T}$ is large, differences between the intervals between crossings are small relative to

$\mathcal{T}$, the total momentum transported across the plane becomes proportional to $\mathcal{T}$, hence the time average rate of momentum transport becomes independent of $\mathcal{T}$.

Suppose, now, that we imagine following the trajectory of one molecule in the gas. This molecule occasionally collides with other molecules and with the walls as it travels back and forth across the volume V. If we follow the trajectory long enough, we expect that, by virtue of the infrequent collisions mentioned, the molecule we are following will approach arbitrarily close to every point inside the volume V, and have at some time or other a velocity arbitrarily close to any selected velocity consistent with conservation of energy. This expectation is a special case of the *quasi-ergodic hypothesis*, of which we shall say more later. When this hypothesis is valid, and it is generally expected to be valid, we see that if we wait long enough, the trajectory of the molecule we are following will intersect any reference plane with all possible angles and velocities. The distribution of these angles and velocities will be the same as for the collection of molecules with distribution of velocities $f(\mathbf{v})$ transferring momentum across the same plane at one instant. Then, over a sufficiently long time, the distribution of velocities along any one trajectory must approach the time-independent form $f(\mathbf{v})$, thereby rendering averages over a single trajectory equivalent to averages over $f(\mathbf{v})$.

In the case of the dilute gas, collisions between molecules are infrequent and it is possible, in practice, to follow a sufficient length of trajectory to permit the time average of a mechanical quantity to be computed. For example, for the purpose of calculating momentum transport across a reference plane in the gas, collisions between the gas molecules can be neglected. Then the distance a molecule travels in time t simply grows linearly with t. Thus, summing up all contributions to the momentum transport, dividing by t, and letting t become indefinitely large introduces no change into the calculation made in Chapter 12.

If, on the other hand, collisions are very frequent, as in a dense gas or a liquid, a molecule is rarely if ever out of the force field of some other molecule. In these cases the trajectory of a molecule becomes very complicated and, for all practical purposes, impossible to compute.[1] Since we cannot compute directly the time average of a molecular property without knowledge of the molecular trajectories, a different procedure must be used to relate averages of molecular properties with macroscopic properties.

Our study of the perfect gas suggests the procedure needed. We have noted that in the equilibrium state, for a perfect gas, an average of a mechanical quantity over a trajectory is equivalent to an average of the same quantity with respect to the time-independent velocity distribution function. The proper generalization of this equivalence, applicable to any system, was suggested by Gibbs, and independently by Einstein. The idea is to replace the isolated mechanical system and the required time averaging with a scheme of calculation to which they are equivalent. We have already noted many times that the macroscopic description of a system requires only a few gross variables, whereas the microscopic description requires many variables. This implies that, if our knowledge of a system is represented by a few macroscopic parameters, there must exist

---

[1] The trajectories of individual particles have been followed in computer simulations of molecular motion in systems with $\sim 10^3$ particles. Such calculations are lengthy and the results (trajectories) not amenable to simple analytical representation. At present, the interpretation of the results of computer simulations of molecular dynamics rests on concepts introduced in the statistical theory of matter.

many different sets of microscopic variables consistent with what is known. Gibbs proposed that, for purposes of calculation, the isolated system under study be replaced by a collection of systems, each having the same macroscopic properties, with the total collection encompassing all possible distributions of the molecular variables consistent with the given macroscopic properties. For example, if all we knew about a perfect gas were the total energy of the N molecules in a box of volume V, Gibbs's proposal would call for imagining a collection of boxes, each with N molecules, each of volume V, and within each of which the distributions of molecular positions and velocities could be anything, provided only that in each box the total energy of the N molecules equaled the originally specified value. Any one of the boxes which represents one of the possible microscopic states is called a *replica* system, and the entire collection of boxes an *ensemble*.

Given that there can be many different sets of molecular velocities and positions consistent with our knowledge of a few macroscopic parameters, we can introduce another kind of average, different from a time average. Suppose that we compute the value of some mechanical quantity, say, the kinetic energy per molecule. If we sum up the kinetic energies of the molecules in all of the replicas, and divide by the total number of molecules in the ensemble, we obtain the ensemble average of the kinetic energy per molecule. A similar definition of ensemble average holds for other quantities.[2] Note that the time average would have been obtained by following the trajectories in one isolated replica system for an infinite period of time. It is now postulated that the average of any variable over the distributions of positions and velocities in the ensemble of replica systems (the ensemble average) is equal to the time average of the same variable in any one of the isolated replica systems. If this postulate is valid, time averages, which are difficult to compute, can be replaced by averages over an ensemble, which are easy to compute.

The postulate stated, relating time and ensemble averages, is a direct consequence of what is known as the quasi-ergodic hypothesis. To understand the importance of the quasi-ergodic hypothesis, and for many other reasons as well, we introduce a geometric visualization of the molecular dynamics; for convenience we use the language of classical mechanics. Imagine a many-dimensional coordinate system, the axes of which represent all the components of momentum and position of all N molecules in a volume V. If the molecules have no internal structure, and can therefore be treated as point particles, there will be $6N$ such axes. These axes represent the $3N$ position coordinates, $x_1, y_1, z_1, \ldots, x_N, y_N, z_N$, and the $3N$ momentum components $p_{x1}, p_{y1}, p_{z1}, \ldots, p_{xN}, p_{yN}, p_{zN}$. This space is called the *phase space* of the N-molecule system. For any given value of the energy the possible values of the momenta and coordinates define a surface in the phase space. For example, for a perfect gas the energy depends only on the particle momenta, $E = (m/2) \times (p_{x1}^2 + \cdots + p_{zN}^2)$, which is the equation for the surface of a $3N$-dimensional sphere. When the molecules interact, the energy depends on their instantaneous positions, and the form of the surface is more complicated.

Now, any fully specified set of values for all the coordinates and momenta corresponding to energy E can be represented as a point on the energy surface in the phase space of the system. Since the energy of an isolated system cannot change, the trajectory the system follows can be

---

[2] In general, the ensemble average of a quantity is the sum over all replicas of its value in each replica, divided by the total number of replicas.

represented as a locus of points on the energy surface in phase space. Indeed, it is convenient to think of the evolution in time of an isolated system, after preparation in some initial state, as following a path on the energy surface.

The difference between time averaging and ensemble averaging can be visualized as follows. For any single isolated system, with given E, V, N, the evolution in time is represented as a single path on the energy surface in the system phase space. If we assert that the macroscopic properties of the system are related to the time averages of the corresponding microscopic properties, we are, in effect, following the path representing the mechanical state of the system on the energy surface and then averaging over the sampling of coordinates and momenta along that path. The ensemble average is quite different. Remember that the instantaneous state of a single isolated system is represented by one point on the energy surface, and its evolution by one path. Remember also that the replica systems that make up an ensemble all have the same values of E, V, N, hence the system phase spaces and energy surfaces of all replicas are the same. We can then choose one system phase space, and plot, on one energy surface, representative points for each of the replicas in the ensemble. Each of these points corresponds to a system with the same values of E, V, N, but with different specific instantaneous values of the molecular momenta and coordinates. Clearly, there are so many mechanical states consistent with given E, V, N that it is reasonable to expect the representative points to densely cover the energy surface. The identification of a macroscopic property of the system with the ensemble average of the corresponding microscopic property means adding up the contributions characteristic of each of the representative points on the energy surface and dividing by their number. Unless the path of a single system on the energy surface covers it in a way similar to that achieved by the definition of the ensemble, time averaging and ensemble averaging appear to be different. Despite this apparent difference there are conditions under which the two averages are equivalent. We discuss these conditions in the following paragraphs.

We shall adopt the ensemble average as the fundamental concept of our description of equilibrium. That is, we adopt as a *definition* that *a macroscopic property of a system is to be identified with the ensemble average of the corresponding microscopic dynamical property*. We shall not seek further justification for this identification; we take it as an irreducible concept of the theoretical description. The ultimate justification for our procedure will be based on the comparison of theory and observation. As we shall see, the molecular theory based on ensemble averages is a very successful description of the equilibrium state.

Ergodic theory, at least as applied to statistical mechanics, is devoted to the study of the equivalence of time and ensemble averages for the properties of a system. The ergodic hypothesis, originally advanced by Boltzmann, is now known to be incorrect. Boltzmann postulated that, if only one waited long enough, the path representing the evolution of a system would pass through every point on the energy surface. That this cannot occur in general is a consequence of two facts: First, the equations of mechanics have unique solutions, which implies that the path representing the evolution of a single system cannot cross itself. Second, given the uniqueness of the path, an essentially one-dimensional trajectory can never "fill" an energy surface of dimensionality greater than 1.

The quasi-ergodic hypothesis is a weaker statement about the trajectory on the energy surface. It is based on the *quasi-ergodic theorem* proved by G. Birkhoff in 1931. Briefly, Birkhoff proved that if one cannot draw a boundary on the energy surface such that the trajectory of the representative point lies entirely on one side of the boundary, which implies that the energy surface cannot be divided into sections that have the property that the trajectory of the representative point is confined to lie in only one section, then the trajectory will pass *arbitrarily close to* every point on the energy surface.[3]

The difficulty with applying this theorem, which is valid only in classical mechanics, is that it has not generally been possible to demonstrate whether or not the energy surface corresponding to some given system can be divided as specified. In the absence of such a demonstration we cannot be certain of the equivalence of time and ensemble averages. When the equivalence is postulated to exist, we use the term quasi-ergodic hypothesis.

There are still other problems with attempts to base a description of equilibrium on ergodic theory. The examples cited earlier, namely, mixing of gases, equalization of temperature, and chemical reaction, are typical of macroscopic irreversible processes. Moreover, the idea that the character of an equilibrium depends on constraints, and changes when the constraints are changed, contains the notion of an approach to equilibrium. Clearly, the microscopic theory should at least have the characteristic that, whatever the mechanical state initially, the state to which the system evolves should be time-independent. In this sense the condition of quasi-ergodicity seems necessary but not sufficient. That is, a system may be both quasi-ergodic and quasi-periodic.[4] An example of this behavior is provided by a system of two harmonic oscillators with incommensurable frequencies (ratio of frequencies an irrational number). For this case one can establish that the energy surface is uniformly covered by the trajectory of the representative point, yet any initial state recurs with arbitrary precision. Other more general examples of the same behavior exist.

In the example just given, if two trajectories start from points close together on the energy surface, they remain close together everywhere; that is, the trajectories move apart only smoothly and slowly, the distance between them growing proportional to t. The property needed to avoid reversibility must generate more turbulent trajectories than does quasi-ergodicity. This more general property is called *mixing*. When a system is said to be mixing, two trajectories that start from nearby points on the energy surface diverge from one another at an exponential rate. In these cases the trajectory looks chaotic, and recurrence of the initial state does not occur. Mixing implies quasi-ergodicity, but the reverse is not true. As might be expected, proving that a given system has the mixing property cannot usually be carried out, so its invocation must be considered a hypothesis in most applications.

The are two recent important developments that bear on the foundations of statistical mechanics. First, Sinai has proven that for a system consisting of N (≥ 2) hard spheres in a box, the only conserved mechanical quantity is the energy, and that the system is mixing and quasi-ergodic. Second, some very general analytical studies and some detailed numerical analyses of the solutions of the equations of motion of modeled systems—for example, coupled oscillators—lead to the conclusion that it is generally to be expected that the trajectory of a representative point becomes chaotic, and uniformly fills the energy surface, if the total energy of the system is large enough. The numerical studies suggest that as the number of particles in the system increases, the threshold for this behavior of the trajectory becomes a smaller fraction of the total energy of the system. It is conjectured that in the limit of a system of "thermodynamic size," $N \approx 10^{23}$, the threshold energy becomes insignificantly small and therefore we may anticipate that for all the systems of interest to us it is satisfactory to describe the equilibrium properties in terms of ensemble averages.

---

[3] This is equivalent to treating the energy as the only constant of the motion.

[4] By quasi-periodic we mean that any initial state recurs, although not necessarily with perfect regularity. A distinction is important here. In any closed mechanical system the initial state will recur to arbitrary precision, given enough time (Poincaré recurrence time). In a quasi-periodic system the initial state recurs *before* the energy surface is uniformly covered, so there is not an irreversible approach to equilibrium.

There are three important aspects of the use of ensemble averages in place of time averages not yet mentioned. First, the general dynamical properties of a system, such as the mixing property, have been established only within the framework of classical mechanics. At least to date, quantum ergodic theory is much less well developed. In contrast, the use of ensemble averages to describe equilibrium properties is equally valid in classical and quantum mechanics. Second, the simplest definition of an equilibrium state, namely, that its properties are invariant in time, and the identification of macroscopic properties with the averages of microscopic dynamical properties over an infinite time period, both neglect the existence of fluctuations about the equilibrium state.[5] Yet fluctuations do occur, and have observable consequences. Indeed, macroscopic equilibrium is a concept that cannot be understood fully if fluctuations are not considered. We shall see that the use of ensemble averages to represent equilibrium properties does permit analysis of the properties of fluctuations. Third, the kind of equilibrium that exists—that is, the properties of the equilibrium state—depends on the kinds of constraints imposed on the system. It is easy to take this into account when computing ensemble averages, whereas the procedure based on general dynamics is restricted to describing equilibrium in an isolated system with fixed volume and number of particles.

15.2
Ensembles and Probability Distributions

In the preceding section we commented on the relationship between time averaging and ensemble averaging in the kinetic theory of matter. To understand fully the nature of the averaging procedures used, and their implications, it is necessary to examine more thoroughly the notions of an ensemble and a probability distribution. The manner in which these concepts enter our interpretation of the behavior of matter will become clear in succeeding sections of this chapter.

Probability theory is a useful tool for the description of events in which there is a multitude of possible outcomes. We are all familiar with coin-tossing experiments in which each toss can be either "heads" or "tails." Although it is not possible to state with certainty whether any given toss will be "heads" or "tails," it is possible to state with good accuracy that one-half of a large number of tosses will be "heads" if the coin is not loaded. In fact, although having equal numbers of "heads" and "tails" is the most likely outcome of a large number of tosses of an unloaded coin, other possibilities occur with nonzero frequency. We expect that the larger the number of tosses, the more accurate will be the expectation of equal numbers of "heads" and "tails."

There is a familiar feature of elementary probability theory embodied in the expectations stated above, namely, the definition of probability of a specified result as the ratio of the number of events with the required outcome to the total number of events, the so-called relative frequency of the required outcome. In the coin-tossing experiment the probability that any one toss of the coin will result in "heads" is simply the ratio of the number of possible "heads" to the totality of possible outcomes of the toss, namely, "heads" plus "tails," or $\frac{1}{2}$. The probability that out of any n tosses, n_1 will be heads is $[n!/n_1!(n-n_1)!](1/2)^n$, since there are $n!/n_1!(n-n_1)!$ ways in which n tosses of the coin can give n_1 "heads" irrespective of order, and there are 2^n possible sequences of "heads" and "tails" resulting from n tosses of the coin. The formula quoted illustrates how, given that each toss of the coin is regarded as independent of all others, the probability of an event requiring n tosses is constructed by multiplying together two factors. One is the elementary

[5] This statement refers to an infinite time average of the properties of one replica system.

probabilities per toss for a given sequence $[(1/2)^{n_1}(1/2)^{n-n_1} = (1/2)^n$ for any specified sequence of n_1 "heads" and $n - n_1$ "tails," the same for each such sequence]. The other is a combinatorial factor counting the multiplicity of ways a sequence can yield the required event. The formula also illustrates how deviations from the most probable outcome decrease as the length of the sequence of tosses increases. For example, the probability that 2 of 4 tosses will be "heads" is $\frac{6}{16}$ and that 1 of 4 tosses will be "heads" is $\frac{4}{16}$. Clearly, out of 4 tosses even the most probable event, namely, 2 "heads," is not overwhelmingly more probable than other events. In contrast, the probability that 200 out of 400 tosses will be "heads" is about 5×10^{22} times the probability that 100 out of 400 tosses will be "heads."

To show how similar simple probabilistic ideas can be applied to the description of molecular behavior, consider the following question. Given that the molecules of a gas are in motion, the number in any fixed element of volume must vary from moment to moment. Should we expect there to be considerable variation in the gas density when we examine different volume elements, centered at different points in the container? Consider a perfect gas of N molecules in a volume V. Now focus attention on some volume element ω inside the container. Since ω/V is the fraction of total volume occupied by ω, and the gas is uniformly distributed, the probability of finding any selected molecule in ω is just ω/V, and that of finding all N molecules in ω is just $(\omega/V)^N$. If ω is very small, say, with linear dimensions comparable to the average distance between molecules, we must expect the number of molecules in ω to usually be zero or one, and occasionally two, but in any event fluctuating markedly as molecules move in and out of ω along their trajectories. If ω is large, say $V/10$, we expect the number of molecules in ω to be $(\omega/V)N$, and subject only to small fluctuations, since in this case the number of molecules crossing the boundaries of ω at any instant is small compared to the number in ω. Both limiting cases are described by the probability distribution $P(N_1)$, which gives the probability of finding N_1 molecules in ω and $N - N_1$ in $V - \omega$. This is just $[N!/N_1!(N-N_1)!] \times (\omega/V)^{N_1}(1-\omega/V)^{N-N_1}$, since there are $N!/N_1!(N-N_1)!$ ways of selecting N_1 molecules out of N to place in ω. It is easily shown, given $P(N_1)$ and the methods described later in this section, that the relative deviation in the number of molecules in ω decreases as ω increases, specifically as $(N\omega/V)^{-1/2}$. Therefore, for a perfect gas in equilibrium, the relative deviation in the number of molecules in a small macroscopic volume, say, 1 mm^3, is of the order 10^{-8} at ordinary temperature and pressure. Despite the incessant molecular motion, the macroscopic density scarcely varies at all from point to point in the gas.

As is clear from the example given, the form of probability theory we shall use is based on generalizations of the elementary notions used to describe the coin tossing experiment. In particular, two points merit careful consideration:

1. The relationship between the definition of probability distribution and the class of events that is to be described, that is, the ensemble in which the probability is defined, and

2. How probability theory is related to the fundamental classical or quantum mechanical description of a system (we shall use the quantum mechanical description).

The notions of probability distribution and ensemble are closely related. In the simplest possible terms, as already used in Section 15.1, an ensemble is merely a collection. The nature of the collection, the set of possible states for the objects composing the collection, and other relevant details must be specified before it is possible to use the ensemble to define the probability of finding some one of the objects in a particular condition. As in the elementary theory, the probability of finding a particular characteristic is defined simply as the fraction of all objects in the ensemble having that characteristic. Clearly, this probability depends very much on the nature of the ensemble. For example, the question "What is the probability of finding a molecule in a sample of perfect gas moving with speed v?", is ambiguous until the macroscopic state of the gas, hence also of the ensemble representing the microscopic states of the gas, is specified. It is evident that it makes a significant difference whether the gas has a temperature T_1 or T_2, whether it is at rest or moving, and so forth. However, once the ensemble is defined, the probability defined in that ensemble is uniquely specified.

As already mentioned, the general name assigned to one of the objects that compose an ensemble is a replica. Implicit in this choice of name is the idea that the objects composing the ensemble are to be considered identical within the constraints imposed by the specifications defining the ensemble. Thus, specifying only that an ensemble consists of all pennies does not require any statement about whether the individual pennies lie "head" up or "tail" up. Similarly, specifying only that an ensemble consists of containers of gas, having rigid adiabatic walls, and each having the same density of molecules (temperature unspecified), does not require any statement about whether or not the pressure of the gas is the same in each container.

The fact that there is an intimate connection between the external constraints defining the state of a system and the nature of the corresponding ensemble of replicas is of fundamental importance to our analysis. We shall see how this relationship can be used to establish a correspondence between the microscopic and macroscopic analyses of the behavior of matter.

Finally, it is generally assumed that an ensemble of replicas is so large that the hypothetical limit $\mathcal{N} \to \infty$ may be taken, where $\mathcal{N}$ is the number of replicas in the ensemble. The use of this abstraction will enable us to consider the probability distribution to be, in many instances, a continuous function. For convenience, in those cases where the probability distribution is a function of a continuous variable, we introduce the notion of *probability density*. The reason for introducing this concept is the following. For a variable that takes on only discrete values, it is meaningful to ask for the probability that some one value is attained out of the set of all possible values. On the other hand, for a variable that takes on a continuous range of values, say, between a and b, the set of all possible values is infinite in size, and the only meaningful question is "What is the probability of finding a value between x_1 and x_2, given that the range of x is such that $a \le x \le b$?" This probability must now be expressed as an integral over a "probability density," $\mathcal{P}(x)$:

$$P(x_1 \le x \le x_2) = \int_{x_1}^{x_2} \mathcal{P}(x)\, dx. \qquad (15.1a)$$

If $x_2 = x_1 + dx$,

$$P(x_1 \le x \le x_1 + dx) = \mathcal{P}(x)\, dx. \qquad (15.1b)$$

Note that $\mathcal{P}(x)$ must be multiplied by dx to yield an actual probability and that, in the general case, although $P(x)$ is dimensionless, $\mathcal{P}(x)$ is not (it has the dimensions of $1/x$). It is possible, under certain conditions, to reduce a description in terms of a continuous probability distribution to one in terms of a discrete probability distribution, where the number of possible values of the variable of interest becomes countable. This is done by dividing the accessible range of the variable, say, $a \le x \le b$, into arbitrarily small equal intervals of fixed size, Δ. Each such interval can be labeled by some index i, and the value of x for this entire interval is denoted as x_i. The probability of finding the variable in this range is $P(x_i)$, where clearly

$$P(x_i) = \mathscr{P}(x_i)\Delta. \tag{15.2}$$

The resolution of a continuous probability distribution into a discrete probability distribution is valid only if the variation of $\mathscr{P}(x)$ over the interval Δ is sufficiently small, for all intervals, that such variation as does exist can be neglected.[6]

Let x be a variable that can assume only the discrete values $x_1, \ldots, x_N$, with probabilities $P(x_1), \ldots, P(x_N)$. The probabilities $P(x_i)$ may be used to define the *average value* of x (also called the *mean value* or the *expectation value*) by

$$\langle x \rangle \equiv \frac{x_1 P(x_1) + x_2 P(x_2) + \cdots + x_N P(x_N)}{P(x_1) + P(x_2) + \cdots + P(x_N)} = \frac{\sum_i x_i P(x_i)}{\sum_i P(x_i)}. \tag{15.3}$$

Equation 15.3 is just the definition of an arithmetic average, in which each value of the variable is weighted by the $P(x_i)$ with which that value appears (ratio of frequency of occurrence of x_i to total frequency of all possible values of x). More generally, if $f(x)$ is a function of the variable x, the mean value of the function is

$$\langle f(x) \rangle \equiv \frac{\sum_i f(x_i) P(x_i)}{\sum_i P(x_i)}. \tag{15.4}$$

The average values $\langle x^n \rangle$ are called the *moments* of $P(x)$. For the case that x is a continuous variable and $f(x)$ is a continuous function, an obvious extension of Eqs. 15.3 and 15.4 leads to

$$\langle x \rangle \equiv \frac{\int x \mathscr{P}(x)\, dx}{\int \mathscr{P}(x)\, dx}, \tag{15.5}$$

$$\langle f(x) \rangle \equiv \frac{\int f(x) \mathscr{P}(x)\, dx}{\int \mathscr{P}(x)\, dx}, \tag{15.6}$$

using Eq. 15.1. We can also find the simple relations

$$\langle f(x) + g(x) \rangle = \langle f(x) \rangle + \langle g(x) \rangle, \tag{15.7}$$

$$\langle cf(x) \rangle = c\langle f(x) \rangle, \tag{15.8}$$

where c in Eq. 15.8 is a constant. Note that in Eqs. 15.5 and 15.6 the range of integration is the same as the range of x.

---

[6] The reduction of a continuous probability distribution to a discrete one can be thought of as the use of a histogram, or bar graph, to approximate a smooth curve.

Just as it was convenient to normalize the wave function of a system, it is also convenient to normalize a probability distribution. Let us require that a variable x must have some value between the end points of its defining interval. Then, by our definition of probability as the fraction of events with the required outcome, the total probability that x has a value in the stated range, $\sum_i P(x_i)$ or $\int \mathscr{P}(x)\,dx$, must be equal to unity. It is frequently convenient, however, to deal with relative probabilities of some kind. For such applications $\int \mathscr{P}(x)\,dx$ can have any convenient constant value, since this constant cancels out of ratios such as appear in Eqs. 15.3–15.6. In general, then, $\sum_i P(x_i)$ or $\int \mathscr{P}(x)\,dx$ is equal to some constant, known as the normalization constant. When the probability is normalized to unity, the denominators of Eqs. 15.3, 15.4, 15.5, and 15.6 are replaced by unity.

A measure of the spread of the probability density distribution $\mathscr{P}(x)$ is given by the difference $\langle x^2 \rangle - \langle x \rangle^2$. Except for the trivial case that $\mathscr{P}(x)$ is unity for some value of x (or range dx about x) and zero for all other values,[7] it is necessarily true that

$$\langle x^2 \rangle > \langle x \rangle^2. \tag{15.9}$$

Equation 15.9 can be understood as follows. In Fig. 15.1a is depicted a typical probability density distribution. The important feature of this curve is that it has some width; that is, the distribution of possible values of x is not confined to a single value. The average value of x^2, which is proportional to $\int x^2 \mathscr{P}(x)\,dx$, weights various values of x differently than does the average of x, which is proportional to $\int x \mathscr{P}(x)\,dx$. In the special case that $\mathscr{P}(x)$ is symmetric about $x = 0$ (see Fig. 15.1b), the average value of x is zero. However, since it is always true that $x^2 \geq 0$, clearly $\langle x^2 \rangle$ cannot be zero. Equation 15.9 summarizes these observations for a general probability distribution.[8]

Equation 15.9 has another important implication. Because a probability distribution describes a range of possible values of the variable, there must be some nonzero chance of finding values of the variable that differ from the average value of the variable. The difference between some particular value of a variable and its average value is called a *fluctuation*. A given fluctuation is more or less probable insofar as the deviation from the average is small or large compared with the spread of $\mathscr{P}(x)$. We have already referred to the existence of fluctuations when a

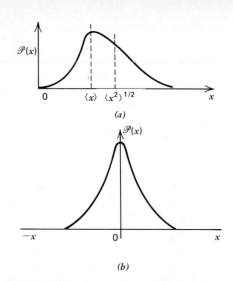

FIGURE 15.1
(a) A typical probability density distribution with $\langle x \rangle \neq 0$. (b) A typical symmetric probability density distribution with $\langle x \rangle = 0$.

---

[7] This exceptional case is not of interest, since there is no distribution of possible values of x.

[8] To demonstrate that Eq. 15.9 is correct, we define

$$\Delta x \equiv x - \langle x \rangle,$$

and calculate the value of $\langle (\Delta x)^2 \rangle$ from

$$\langle (\Delta x)^2 \rangle = \int (x - \langle x \rangle)^2 \mathscr{P}(x)\,dx.$$

Now, we have

$$\langle (x - \langle x \rangle)^2 \rangle = \langle x^2 - 2x\langle x \rangle + \langle x \rangle^2 \rangle$$
$$= \langle x^2 \rangle - 2\langle x \rangle\langle x \rangle + \langle x \rangle^2$$
$$= \langle x^2 \rangle - \langle x \rangle^2.$$

But $(\Delta x)^2$ is always positive or zero, so the right-hand side of the last line above must also be always positive or zero; that is,

$$\langle x^2 \rangle \geq \langle x \rangle^2.$$

Clearly, only when $\langle (\Delta x)^2 \rangle$ vanishes does the equality sign hold; since this can occur only if every value of x is equal to the mean value, Eq. 15.9 is established.

system with many particles is described by only macroscopic variables (see Section 13.1). We shall see later that fluctuations in the values of macroscopic variables are an essential property of matter in bulk, and that the nature of the possible fluctuations around the equilibrium state is related to the properties of that state.

There is one other property of probability distributions that we shall use extensively in later chapters. Suppose that the variable x is distributed as $\mathscr{P}_1(x)$ and the variable y as $\mathscr{P}_2(y)$, where $\mathscr{P}_1(x)$ and $\mathscr{P}_2(y)$ are probability density distributions. If the probability of finding x in the range x to $x+dx$ is independent of the value of y, and vice versa, then the probability of finding x in the range x to $x+dx$ and y in the range y to $y+dy$ simultaneously is the *joint probability distribution*

$$\mathscr{P}(x, y) \, dx \, dy = \mathscr{P}_1(x)\mathscr{P}_2(y) \, dx \, dy. \qquad (15.10)$$

In general, the probability of occurrence of a set of values of independent variables is simply the product of the probabilities of occurrence of the values of each of the variables. When a set of variables is not independent, Eq. 15.10 or its n-variable extension is not valid. The set of variables is then said to be *correlated*. The extent of correlation of the variables can be formulated in a variety of ways. Two convenient formulations of the probability of finding x in the range x to $x+dx$ and y in the range y to $y+dy$ when x and y are not independent are given by

$$\mathscr{P}(x, y) \, dx \, dy = \mathscr{P}_1(x)\mathscr{P}_2(y)g(x, y) \, dx \, dy, \qquad (15.11a)$$

$$\mathscr{P}(x, y) \, dx \, dy = \mathscr{P}_1(x)\mathscr{P}_2(y)[1 + h(x, y)] \, dx \, dy. \qquad (15.11b)$$

Both $g(x, y)$ and $h(x, y)$ are called *correlation functions*. Each is used in the molecular theory; the choice is one of convenience. Note that when x and y are independent, we have $g(x, y) = 1$ but $h(x, y) = 0$.

Most of the ideas described in the preceding paragraphs have been used, in an intuitive fashion, in Chapter 12. There the distribution of molecules with respect to speed was introduced and used to describe the properties of the perfect gas of noninteracting molecules. Note, however, that the speed distribution function $f(v)$, which is a probability density, was normalized to the number density n, and not to unity. As an example of the relationship displayed in Eq. 15.9, we observe that at equilibrium the average velocity of a molecule in the gas is zero (because all directions are equally probable), whereas the root-mean-square velocity, $\langle v^2 \rangle^{1/2}$, is not zero. In fact, the mean square of the velocity defines the average kinetic energy per molecule, which in turn is related directly to the temperature of the gas.

As a final comment, we note that knowledge of the full probability distribution implies complete information about the distribution of values of the variables of interest. Although it is often possible to characterize a system by the mean values, mean square values, and so on, of some small set of variables, it is not possible to determine uniquely the full probability distribution from a knowledge of only a few of its moments.

15.3
Some Properties of a System with Many Degrees of Freedom: Elements of the Statistical Theory of Matter at Equilibrium

Given the concepts of ensemble and probability distribution, we are in possession of the tools necessary for the description of the most important of the equilibrium statistical molecular properties of matter. Consider some sample of matter, containing a large number of molecules, N, in the volume V. From the general theory of quantum mechanics, discussed in earlier chapters of this text, it is clear that there is one feature common to all kinds of matter. This feature is, simply, the existence of an energy spectrum. For our present purposes it does not matter how closely the energy levels of the system are spaced, what the spectral distribution of energy levels is like, or even whether all possible energy levels are discrete. What is important, however, is that in any macroscopic system there exists an enormous number of quantum states for given constraints, say, E, V, N. The degeneracy of the system is exceedingly large. Of course, there are also many different distributions of the molecules over the energy levels of the system consistent with the given E, V, N.

In principle, the exact mechanical behavior of the system is determined once the energies and wave functions for all the molecules are given. However, since we do not know how to solve the equations of motion for N interacting molecules, this information is not available for a macroscopic system. Indeed, it is not even of particular interest to seek a full solution of the N-molecule mechanical problem, at least insofar as the description of macroscopic equilibrium is concerned. This is so because the $\Omega(E, V, N)$ states with energy between E and $E + dE$ will correspond to a range of values of, for example, molecular coordinates and velocities. Although each solution of the N-molecule Schrödinger equation will correspond to one, or a small number, of the $\Omega(E, V, N)$ states, others arise from different initial conditions on the many microscopic coordinates that are macroscopically indistinguishable. To obtain the wave functions for all the $\Omega(E, V, N)$ levels, we must obtain many solutions of the Schrödinger equation, sampling a considerable range of initial conditions on the microscopic coordinates. This is, in fact, unnecessary for our purposes, because the properties of macroscopic equilibrium states are determined by $\Omega(E, V, N)$, and not by the detailed behavior of the wave functions for the levels comprising $\Omega(E, V, N)$.[9]

The fundamental postulate of equilibrium statistical mechanics can be stated as follows:

All possible quantum states of an isolated system consistent with a given set of macroscopic parameters of constraint are to be considered as equally probable.

This postulate of equal *a priori* probability cannot be obtained from more fundamental arguments. Although eminently reasonable, and consistent with the laws of mechanics, the postulate stated must be recognized as a fundamental assumption in the development of the statistical theory—an assumption that can only be indirectly verified *a posteriori* by the success of theoretical calculations in interpreting and reproducing the results of observation.

To take advantage of the fundamental postulate of equal *a priori* probabilities, we consider not one system at a time, but rather an ensemble of systems. The ensemble is constructed by collecting a very large number of replica systems, each of which is defined by the same set of macroscopic parameters, in the present case the set E, V, N. As noted in Section 15.2, those properties of the system not defined by the macroscopic constraints are free to vary, provided only that the macroscopic constraints are never violated. In this way we may suppose that all possible quantum states consistent with the definition of the ensemble are represented among the replica systems. Note that the relative frequency

---

[9] Although $\Omega(E, V, N)$ depends on dE, for the sake of simplicity this dependence will not be displayed explicitly; we show later that for the purposes of obtaining thermodynamic functions the dependence on dE is not important.

definition of probability in an ensemble is in agreement with the assumption of equal *a priori* probability for all quantum states consistent with given E, V, N, for the chance that a random choice from the ensemble will lead to the selection of any one of the replicas is the same, namely, $\mathcal{N}^{-1}$, for each of the $\mathcal{N}$ replicas, and each replica is in one of the quantum states consistent with the given E, V, N.

We must now ascertain which constraints are most suitable for the definition of an ensemble. For the case of interest to us, namely, equilibrium, the macroscopic properties of an isolated system are independent of time. This observation implies that in an isolated system at equilibrium the distribution of molecules over the energy levels is also independent of time, hence a function of only the constants of the motion (see Chapter 3). The most general constants of the motion are the total energy, total linear momentum, and total angular momentum of the system. It is just because each replica system is considered to be isolated that energy, linear momentum, and angular momentum are conserved (no energy may be transferred across the boundaries of an isolated system, and no forces act on an isolated system). Given the constraints that define the ensemble, it is possible to define an appropriate probability distribution. Often only one of the constants of motion is used, and because of its central importance it is usually the total energy of the system that is chosen to define the ensemble. In this case, even though other constants of motion exist, they are unspecified and, just as in the case of the microscopic distribution, are only required to have values consistent with the assigned total energy.

Consider, now, an ensemble of replica systems each with the same volume V and number of molecules N. We complete the specification of the ensemble by requiring that the energy be between E and $E + dE$. Suppose that there are $\Gamma(E, V, N)$ quantum states with energy less than E for this system. The number of quantum states with energy between E and $E + dE$ is defined by the difference

$$\Omega(E, V, N) = \Gamma(E + dE, V, N) - \Gamma(E, V, N). \tag{15.12}$$

The function[10] $\Omega(E, V, N)$ contains information about the degeneracy of the energy-level spectrum of the system, and the dependence of that spectrum on the values of parameters such as V. In general, the parametric dependence of $\Omega(E, V, N)$ on V reflects the influence of boundary conditions on the energy-level spectrum of the isolated system.

This last point is so important that it is necessary to expand on such a terse statement. The Schrödinger equation is a second-order partial differential equation and, depending on the form of the potential, the magnitude of the energy, the shape of the container, and so on, has many possible solutions. The selection of the set of solutions appropriate to a given problem is accomplished by applying the correct boundary conditions. (See Chapter 3.) It is because the boundary conditions depend on the nature of the enclosure (the boundary) that the wave functions and allowed energy levels are functionally dependent on the volume of the system. Consider, for example, a free particle enclosed in a box with infinitely repulsive walls. The solutions to the Schrödinger equation for a free particle are of the forms $\sin kx$ and $\cos kx$, and form a set continuous in the parameter k. Thus before the application of the boundary conditions there is a continuum of possible solutions. However, when it is recognized that the wave function must be zero at an infinitely repulsive

---

[10] For convenience we shall usually suppress the N and V dependence of $\Omega(E, V, N)$, writing only $\Omega(E)$.

boundary, the values that k may take on are limited, as are the energies. In the absence of bounding walls, there is an infinite number of allowed states in the range from E to $E+dE$, but when the particle is confined to a volume V the number of available states in this range becomes finite and *dependent on the value of* V. The change from an infinite number to a finite number of allowed solutions corresponds to a change in the spectrum of energy levels from a continuum to a discrete set.

How does the fact that a macroscopic system contains a very large number of molecules lead to simplifications in the analysis? Suppose that the energy scale is divided into equal intervals of magnitude dE. For a large system even an interval dE that is small by macroscopic standards contains a very large number of energy levels. The number of states between E and $E+dE$, namely, $\Omega(E)$, depends on the magnitude of dE, but there is a wide range over which dE may vary and for which the linear relation

$$\Omega(E) = \left(\frac{\partial \Gamma}{\partial E}\right) dE = \omega(E)\, dE \qquad (15.13)$$

is valid. The function $\omega(E)$ measures the number of states per unit energy interval, and is a characteristic property of the system; $\omega(E)$ is known as the *density of states* of the system. Because of our assertion that $\omega(E)$ is characteristic of a system, it is of obvious interest to examine, even roughly, how sensitive $\Omega(E)$ and $\omega(E)$ are to the energy E of a system of macroscopic size. (See Chapter 11).

The purpose of the following argument is to demonstrate that $\Omega(E)$ is insensitive to the precise nature of the energy spectrum of the system, provided that the system is large enough. In this limit we shall see that $\Omega(E)$ has the characteristics of a macroscopic parameter even though it is defined in terms of the microscopic energy spectrum. Further, because $\Omega(E)$ has the characteristics of a macroscopic parameter, it plays the role of connecting the statistical molecular description with the thermodynamic description of a system.

Let the system with energy E be described by ν quantum numbers, and let the energy per degree of freedom (one for each quantum number) be approximated by

$$\epsilon \approx \frac{E}{\nu}. \qquad (15.14)$$

For example, a perfect gas composed of N monatomic molecules has $3N$ degrees of freedom, three for each molecule. Then, using Eq. 12.42, we find that the approximation 15.14 is equivalent to writing, for a perfect gas,

$$\epsilon_{\text{perfect gas}} \approx \tfrac{1}{2} k_B T. \qquad (15.15)$$

Equation 15.14 has much deeper significance than is apparent from the way it has been introduced. We shall later see that one of the characteristics of a system of many particles described by classical mechanics is that the total energy is divided among the degrees of freedom in such a way that the average energy per degree of freedom is just $\tfrac{1}{2} k_B T$. In a system described by quantum mechanics, the same equipartition of energy occurs when $k_B T$ is large relative to the energy-level spacings of the system, but not when $k_B T$ is smaller than or equal to the spacings between the energy levels. For the present we regard Eq. 15.14 as an approximation, and use it only to obtain an order-of-magnitude estimate of $\Omega(E)$ for an arbitrary system with ν degrees of freedom.

Consider again the total number of quantum states with energy less than E. We consider the energy to be separable into contributions from the ν degrees of freedom. Of course, the total energy of the system may be distributed over the ν degrees of freedom in many different ways, and each of these possible distributions contributes a state to $\Gamma(E)$. Let E_i be the energy in the ith degree of freedom and $\Gamma_i(E_i)$ the number of states of the ith degree of freedom with energy less than E_i. Then $\prod_{i=1}^{\nu} \Gamma_i(E_i)$ is the total number of states with energy less than E for a particular subdivision of the energy. The total number of states of the system with energy less than E when all possible subdivisions of E are accounted for is

$$\Gamma(E) = \sum_{\{E_i\}} \prod_{i=1}^{\nu} \Gamma_i(E_i); \qquad E = \sum_i E_i. \qquad (15.16a)$$

Note that the summation over $\{E_i\}$ is over all possible ways of subdividing the total energy subject to the constraint of conservation of energy. For the purpose of making a crude estimate of $\Gamma(E)$ we approximate it by requiring that each of the E_i be just the average energy ϵ. This corresponds to choosing only one term in the sum over all possible subdivisions of the energy, the one which has a uniform distribution of energy over the degrees of freedom. Then we obtain

$$\Gamma(E) \approx \prod_{i=1}^{\nu} \Gamma_i(\epsilon) \qquad (15.16b)$$

$$\approx [\Gamma_1(\epsilon)]^{\nu}. \qquad (15.16c)$$

Equation 15.16b is only a crude estimate of $\Gamma(E)$, but it is sufficiently accurate for our purposes. The right-hand side of Eq. 15.16c represents still another order-of-magnitude estimate, this time replacing the product of the numbers of states for all degrees of freedom by the νth power of the number of states corresponding to one degree of freedom. The assumption inherent in this estimate is that the ν degrees of freedom are sufficiently alike that some reasonable average behavior can be ascribed to them, at least for the purposes of this argument. We now note that expansion of the right-hand side of Eq. 15.12 in a Taylor's series, followed by substitution of Eq. 15.16c, leads to

$$\Omega(E) = \frac{\partial \Gamma}{\partial E} \, dE \qquad (15.17)$$

$$\approx \frac{\partial}{\partial E} \{[\Gamma_1(\epsilon)]^{\nu}\} \, dE \qquad (15.18)$$

$$= \nu \Gamma_1^{\nu-1} \frac{\partial \Gamma_1}{\partial(\nu\epsilon)} \, dE = \Gamma_1^{\nu-1} \frac{\partial \Gamma_1}{\partial \epsilon} \, dE, \qquad (15.19)$$

where we have used $E = \nu\epsilon$, from Eq. 15.14. Now, $\Gamma_1(\epsilon)$ must increase as ϵ increases. Therefore, when E increases, $\Gamma_1(\epsilon)$ also increases; and because ν is so very large for a macroscopic system, $\Omega(E)$ is a very rapidly increasing function of E. The magnitude of the rate of increase can be appreciated if we rewrite Eq. 15.19 in the form

$$\ln \Omega(E) = (\nu - 1) \ln \Gamma_1(\epsilon) + \ln\left(\frac{\partial \Gamma_1}{\partial \epsilon} \, dE\right). \qquad (15.20)$$

As already mentioned, the energy range dE is very much larger than the separation between the energy levels of the system. It is not important just how much larger, because the term $\ln[(\partial \Gamma_1/\partial \epsilon) \, dE]$ in Eq. 15.20 is negligibly small relative to the term $(\nu - 1) \ln \Gamma_1(\epsilon)$. To show this, let us suppose that dE were as large as ϵ. The order of magnitude of the derivative $(\partial \Gamma_1/\partial \epsilon)$ is Γ_1/ϵ, an estimate that comes from assuming that $\Gamma_1(\epsilon)$ is proportional to ϵ. In fact, studies of simple models, such as the harmonic oscillator or the particle in a box, show that $\Gamma_1(\epsilon)$ increases as some small power of ϵ. For the cases mentioned,

$$[\Gamma_1(\epsilon)]_{\text{harmonic oscillator}} \propto \epsilon,$$

$$[\Gamma_1(\epsilon)]_{\text{particle in box}} \propto \epsilon^{1/2}.$$

In general, then, $\ln[(\Gamma_1/\epsilon)\epsilon]$ will be a factor ν smaller than $(\nu - 1) \ln \Gamma_1(\epsilon)$, and ν is of the order of 10^{24}. Even if our estimates of $\partial \Gamma_1/\partial \epsilon$ and dE were both off by factors of ν, $\ln[(\partial \Gamma_1/\partial \epsilon)\, dE]$ would still be negligibly small relative to $\nu \ln \Gamma_1$, because the logarithm of some small power of ν is a small multiple of $\ln \nu$, which for $\nu = 10^{24}$ means a small multiple of 55. Thus, even multiplying $\Gamma_1(\epsilon)$ by the very large factor ν, or some power of ν, introduces terms in the logarithm that are negligibly small relative to ν itself. An estimate of the magnitude of the remaining term in Eq. 15.20, namely, $(\nu - 1) \ln \Gamma_1(\epsilon)$, can be made as follows: If Δ is the average spacing between levels, then $\Gamma_1(\epsilon)$ must be of the order of magnitude of ϵ/Δ, and $\epsilon/\Delta \gg 1$ when ϵ is far from the ground-state energy. Then $\nu \ln \Gamma_1(\epsilon)$ is of the order of magnitude of ν. We conclude that the relation

$$\ln \Omega(E) \approx \nu \ln \Gamma_1(\epsilon) \tag{15.21}$$

implies that $\ln \Omega(E)$ is of the order of magnitude of ν or larger, provided that the energy E is well above the energy of the ground state of the system. Note that ν is so large that the result cited is insensitive to the rate at which $\Gamma_1(\epsilon)$ increases with increasing ϵ. That is, it is immaterial whether $\Gamma_1(\epsilon)$ increases as ϵ^2, ϵ^3, $\epsilon^{1/2}$, or any other "small" power of ϵ, since our argument is unaffected if there is a coefficient of order 2, 3, $\frac{1}{2}$, and so on, in the relation 15.21.

15.4 The Influence of Constraints on the Density of States

We are now able to examine the role of constraints in defining the density of states, $\omega(E)$, or the number of states, $\Omega(E)$, descriptive of an isolated system with energy between E and $E + dE$. It is convenient to begin with a simple example. Consider, as a model of a perfect gas, a cubical box of volume $V = a^3$ containing N independent point particles. It was shown in Chapter 3 that the wave function for a single particle in a cubical box is

$$\psi_j = A \sin \frac{n_{jx}\pi x}{a} \sin \frac{n_{jy}\pi y}{a} \sin \frac{n_{jz}\pi z}{a}; \quad \begin{array}{l} 0 < x < a \\ 0 < y < a \\ 0 < z < a \end{array} \tag{15.22}$$

where $j = 1, \ldots, N$ labels the particle, A is a normalization constant, and the particle quantum numbers n_{jx}, n_{jy}, n_{jz} satisfy

$$n_{jx}^2 + n_{jy}^2 + n_{jz}^2 = \frac{2m\epsilon_j}{\pi^2 \hbar^2} a^2, \tag{15.23}$$

with ϵ_j the energy of particle j. The values of n_{jx}, n_{jy}, n_{jz} are restricted to the positive integers since the substitutions $n_{jx} \to -n_{jx}$, $n_{jy} \to -n_{jy}$, $n_{jz} \to -n_{jz}$ do not lead to new linearly independent wave functions.[11] The total energy of the N particles in the box is

$$E = \sum_{j=1}^{N} \epsilon_j, \tag{15.24}$$

---

[11] Note that $\sin(-x) = -\sin x$.

which we rewrite in the form

$$\sum_{j=1}^{N} (n_{jx}{}^2 + n_{jy}{}^2 + n_{jz}{}^2) = \frac{2mE}{\pi^2 \hbar^2} a^2. \tag{15.25}$$

To calculate the total number of possible states with energy less than E, we proceed as follows. Imagine a $3N$-dimensional space defined by the orthogonal axes n_{jl} $(j = 1, \ldots, N; l = x, y, z)$. Since every n_{jl} must be a positive integer, each allowed state is represented by a point in the positive orthant[12] $n_{jl} \geq 0$, all j and l. Furthermore, since Eq. 15.25 defines a hyperspherical surface with radius $R_s = (2mEa^2/\pi^2\hbar^2)^{1/2}$, the total number of states with energy less than E is represented by the points with integer coordinates lying between the positive axes n_{jl} (all j and l) and the hyperspherical energy surface. When the volume of the box is large, the spacing between adjacent energy levels is very small; then the number of points with integer coordinates is very well approximated by the volume of the region they occupy. The volume of a $3N$-dimensional sphere can be shown to be

$$\frac{\pi^{3N/2}}{(3N/2)!} R_s{}^{3N} = \frac{\pi^{3N/2}}{(3N/2)!} \left(\frac{2mEa^2}{\pi^2\hbar^2}\right)^{3N/2},$$

and the positive orthant is $(\tfrac{1}{2})^{3N}$ of the total volume. Therefore

$$\Gamma(E) = \frac{1}{(3N/2)!} \left(\frac{mEa^2}{2\pi\hbar^2}\right)^{3N/2} \tag{15.26}$$

Finally, using Eq. 15.13, the number of states between E and $E + dE$ is

$$\Omega(E) = \left[\frac{\partial \Gamma(E)}{\partial E}\right] dE$$

$$= \frac{E^{3N/2-1}}{(3N/2 - 1)!} \left(\frac{ma^2}{2\pi\hbar^2}\right)^{3N/2} dE. \tag{15.27}$$

Given Eq. 15.13, the density of states $\omega(E)$ is just the coefficient of dE in Eq. 15.27. Note that $\Omega(E)$ is proportional to V^N, and that V enters, once for each of the N particles, by virtue of the boundary conditions on the wave function.[13] These boundary conditions define the nature of the energy spectrum.

For the example given, only V influences the boundary conditions on the Schrödinger equation, and thereby the spectrum of energy levels. In more general cases there are several macroscopic variables that influence the energy-level spectrum. Suppose that the macroscopic state of a system is defined by specifying the values of a set of variables, say,

$$y_1 = \alpha_1, y_2 = \alpha_2, \ldots, y_n = \alpha_n.$$

Given these values for the macroscopic variables, the set of possible solutions to the Schrödinger equation is restricted in analytical form, and there is a corresponding dependence of the energy-level spectrum on $y_1 = \alpha_1, y_2 = \alpha_2, \ldots, y_n = \alpha_n$. We define

$$\Omega \equiv \Omega(E, y_1, \ldots, y_n) \tag{15.28}$$

to be the number of states accessible to the system in the range from E to $E + dE$ when each macroscopic parameter of constraint y_i is in the range from y_i to $y_i + dy_i$.

---

[12] The positive orthant is the generalization of the positive octant of a three-dimensional space. The positive octant is defined by $x > 0$, $y > 0$, $z > 0$.

[13] $\psi_j(x, y, z)$ vanishes at $x = 0$ and a, $y = 0$ and a, $z = 0$ and a, for every $j = 1, \ldots, N$.

What general remarks can be made about the dependence of $\Omega(E, y_1, \ldots, y_n)$ on the parameters of constraint $y_1, \ldots, y_n$? First note that when a constraint is removed,

$$\Omega(E, y_1, \ldots, y_{n-1}) > \Omega(E, y_1, \ldots, y_n), \qquad (15.29)$$

since removal of a boundary condition always allows a larger range of possible solutions to the Schrödinger equation.

The inequality 15.29 is descriptive of the difference between initial and final states of a system when the number of constraints is reduced. It is far more common, however, for initial and final states of a system to differ by virtue of having different values for the parameters of constraint, the number remaining the same. How does $\Omega(E, y)$ change in such cases? A simple example is, again, illustrative of the behavior to be expected. Consider N particles in a box. For this system the parameter of constraint is the specified value of the volume of the box. The number of states between E and $E + dE$, displayed in Eq. 15.27, is proportional to V^N. Therefore, increasing V increases $\Omega(E, V, N)$ and decreasing V decreases $\Omega(E, V, N)$. In general, $\Omega(E, y)$ can either increase or decrease, depending on how the parameters of constraint $y_1, \ldots, y_n$ are changed between initial and final states. (See Chapter 3, problem 9.)

To systematize the description of changes in $\Omega(E, y)$ induced by changes in the values of the parameters of constraint, we adopt the definition that *relaxation* of the parameters of constraint always leads to $\Omega_f(E, y) > \Omega_i(E, y)$.

The combination of these observations and the definition lead to the following: Suppose that an isolated system, at equilibrium with a set of constraints, has Ω_i equally accessible states. Suppose, as a result of an external change involving no work done on the system, some constraints are removed or relaxed. A new system is thereby created for which, at equilibrium, there exist Ω_f equally accessible states, with Ω_f and Ω_i related by

$$\Omega_f \geqslant \Omega_i. \qquad (15.30)$$

The preceding remarks bring us to a study of the central point in the statistical molecular theory. Consider an ensemble of systems. We start from a state that is in equilibrium with respect to all constraints acting, and in which all the Ω_i accessible states are equally probable. Now let one constraint be removed or relaxed. At first, none of the replicas of the ensemble will be in states excluded by the original constraints. But in the final equilibrium state, because of the change in the parameters of constraint, there are Ω_f states, and our fundamental postulate requires that these be occupied with equal probability. Thus the state of the system immediately following removal or relaxation of a constraint, one in which only Ω_i/Ω_f of the Ω_f accessible states are occupied, cannot be an equilibrium state of the system. The specific way in which Ω increases is determined by the detailed molecular dynamics associated with the way in which the constraint is changed. Although we cannot give general rules describing how the final equilibrium distribution is achieved, it is possible to assert that either removing or relaxing some of the constraints defining the state of an isolated system causes the system to change until a new equilibrium state, consistent with the remaining constraints, is achieved.

Let us now consider the problem inverse to that posed in the last paragraph. Suppose that, in an ensemble of systems, the final equilibrium has been reached following the relaxation of some constraints.[14] Depending on the nature of the constraints, they may or may not be easy to restore, but in general the constraints and the system cannot both be restored to the initial condition without work being done on the system. Imposition of a constraint without work being done on a system leaves $\Omega(E)$ essentially unchanged.[15] To illustrate this statement we again take as an example the particle-in-a-box model of a perfect gas. Suppose that the initial state is one in which, as shown in the schematic diagram of Fig. 15.2a, perfect gas occupies one-half of the volume of an isolated system because of a rigid barrier that separates the system into two equal volumes. The other half of the system is evacuated. Imagine, further, that the rigid barrier can be removed at will by expenditure of a vanishingly small amount of work. Of course this can be true only in a limiting sense, but it is practical in the laboratory to remove and add barriers in a fashion that approaches the theoretical limit of no expenditure of work. We have already seen that for the perfect gas Ω increases when V increases, so that when the barrier is removed and the gas expands to fill the volume $V = 2V_1$,

$$\Omega_f > \Omega_i \qquad \text{(process 15.2a)}.$$

Consider, now, what happens when the barrier is reinserted after process 15.2a is complete. A sketch of the situation is shown in Fig. 15.2b. Clearly, the final state achieved in process 15.2b is different from the initial state of process 15.2a, since the insertion of the barrier merely divides the box into two parts but leaves gas in each part. How is Ω_f related to Ω_i for process 15.2b? From Eq. 15.27 we have that

$$\Omega(E, V, N) = CV^N \left(\frac{E}{N}\right)^{3N/2}, \qquad (15.31)$$

where C is a constant independent of E and V. To obtain Eq. 15.31 we have used the Stirling approximation to $N!$, namely, $N! \approx N^N \exp(-N)$. In the initial state of process 15.2b, each particle can be anywhere in the volume V, whereas in the final state $N/2$ particles each are in the left- and right-hand volumes $V/2$. The energy per particle, E/N, is the same in the initial and final states. Since $N/2$ particles each can be placed in the left- and right-hand volumes in $N!/[(N/2)!(N/2)!]$ ways, each of which leads to an equivalent state, the use of Eq. 15.31 to describe the initial and final states of process 15.2b leads to

$$\Omega_i(E, V, N) = CV^N \left(\frac{E}{N}\right)^{3N/2}, \qquad (15.32)$$

$$\Omega_f(E, V, N) = C\frac{N!}{(N/2)!(N/2)!} \left(\frac{V}{2}\right)^{N/2} \left(\frac{E/2}{N/2}\right)^{3N/4} \left(\frac{V}{2}\right)^{N/2} \left(\frac{E/2}{N/2}\right)^{3N/4}$$

$$= CV^N \left(\frac{E}{N}\right)^{3N/2}, \qquad (15.33)$$

where we again have used the Stirling approximation to $N!$. We conclude that

$$\Omega_f = \Omega_i \qquad \text{(process 15.2b)}.$$

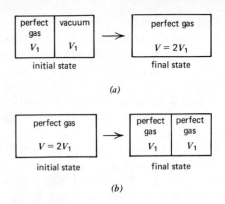

FIGURE 15.2
Schematic illustration demonstrating that restoration of a previously relaxed constraint without execution of work does not regenerate the initial state of the system.

---

[14] From here on we use the term relaxation of constraints to include the possibility of reduction of the number of constraints (removal).

[15] The word "essentially" means to terms of macroscopic significance.

Thus, as asserted earlier, imposition of a constraint without work being done on a system leaves $\Omega(E)$ essentially unchanged.

From the macroscopic point of view, to restore the initial state of process 15.2a, the gas must be swept out of one of the compartments. One way in which this could be achieved is by compressing the gas with a sliding wall, thereby creating a full and an empty compartment, each of volume V_1. But in order to achieve this compression, work must be done on the system, and if the system were thermally isolated the final temperature of the gas would be different from the initial temperature.

Other schemes to recreate the initial state of process 15.2a may be imagined, but it is found after examination of all possible methods of restoring the constraint that if $\Omega_f > \Omega_i$, a simple restoration of the geometry of the system without doing work or allowing energy to cross the boundaries cannot re-create the initial state of the system. Once the molecules are distributed over the Ω_f accessible states, a simple reimposition of boundary conditions cannot lead to a spontaneous rearrangement wherein some of the Ω_f states are vacated and the molecules occupy only the restricted subset of states Ω_i. It is also clear that relaxation of further constraints on the system cannot lead to a restoration of the initial condition, since in each such relaxation even more states become accessible to the system.

We now categorize some of the ways in which a system may change following relaxation of a constraint, so as to lay the groundwork for comparing the statistical molecular and thermodynamic descriptions of reversible and irreversible processes. Consider some process occurring in an isolated system, carrying the system to a final state. If, for an isolated system, the final state is such that imposition or relaxation of constraints without the requirement of external work cannot re-create the initial state, the process undergone by which the final state was achieved is called an *irreversible process*. On the other hand, if the imposition or relaxation of constraints without the requirement of external work on the isolated system can re-create the initial state, the process undergone is said to have been a *reversible process*. In summary, for processes at constant energy and constant total volume:

1. If some of the constraints defining an isolated system are relaxed, $\Omega_f \geqslant \Omega_i$.

2. If $\Omega_f = \Omega_i$, the systems of the ensemble are distributed over the same accessible states before and after relaxation of the constraint. The system remains in equilibrium at all stages of the process, and the transition $i \rightarrow f$ is reversible.

3. If $\Omega_f > \Omega_i$, the distribution of initial systems is over a smaller set of states than is the distribution of final systems. In the transition $i \rightarrow f$ equilibrium does not prevail through all stages of the process, and the process is irreversible.

Statements 2 and 3 merit further comment. In Chapter 13 we defined a reversible process as one in which the intensive variables are continuous across the boundary of the system. This definition refers to the case that the system is in contact with some reservoirs. What is meant by a reversible process in an isolated system, with E, V, and N specified? Consider, as an example, the internally divided system sketched in Fig. 15.3. Subsystems 1 and 2 are gases separated by a movable piston on which weights rest. A pulley arrangement provides a means of removing a weight and depositing it inside the system on a ledge that is at the same height. Therefore, removing a weight from the piston and transferring it

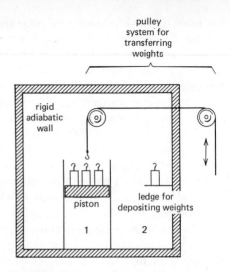

FIGURE 15.3
Schematic illustration of how a reversible process in an isolated system can be made to occur.

to the ledge requires no net work. Accordingly, the pulleys and string are merely a device for changing a parameter of constraint inside an otherwise isolated system. The external walls are rigid and adiabatic. Provided that the weights transferred inside the system are infinitesimally small, the pressure will be continuous across the piston along the path of the expansion of subsystem 1 and the corresponding compression of subsystem 2. Therefore this path describes a reversible process. Note that in this example subsystem 2 plays the role of surroundings for subsystem 1.

In general, the thermodynamic definition of a reversible process requires that every point on the path be infinitesimally close to an equilibrium state of the system, and that everywhere along the path the relevant intensive variables be continuous across the boundaries of the system. These conditions imply that the extensive variables of the system must be infinitesimally close in value in two states that are infinitesimally close, that is, everywhere along the path of a reversible process. Then, returning to statement 2 and using the language of the statistical molecular theory, we say that neighboring points along the path of a reversible process must have $\Omega_f = \Omega_i$, since it is the extensive variables that are usually the parameters of constraint and that determine the boundary conditions defining Ω.

15.5
The Entropy: A Potential Function for the Equilibrium State

In this section we study some of the properties of the number of accessible states Ω for a composite system over the parts of which the total energy may be distributed. It will be shown that the most probable distribution of energy over the parts of the system leads to a condition that is equivalent to the equality of temperatures between systems in thermal equilibrium. On the basis of this result a new function, the *entropy*, will be introduced and shown to be related to the work and internal energy as defined by the first law of thermodynamics. Thus defined, the entropy will provide a bridge enabling us to relate the macroscopic and microscopic descriptions of a system.

To carry through the program described, it is necessary to specify carefully the nature of the ensemble describing a composite system. Consider two macroscopic systems, A and B. Let E_A and E_B be the energies of the two systems.[16] Further, let $\Omega_A(E_A)$ and $\Omega_B(E_B)$ be, respectively, the numbers of states between E_A and $E_A + dE_A$ and between E_B and $E_B + dE_B$. N_A and N_B, the numbers of A and B molecules, are fixed. The ensemble to be considered is specified by the requirement that the sum of the energies of the two systems A and B, $E_T \equiv E_A + E_B$, is a constant. In the ensemble of pairs of systems, different replica pairs have a different distribution of the energy E_T between the systems A and B. In contrast, the volumes V_A and V_B, and any other necessary parameters of constraint, are separately specified for systems A and B, and hence for each replica pair. For simplicity we refer to the ensemble as being defined by the value of E_T.

In this ensemble of composite systems, if system A of a replica pair has energy E_A, then system B must have energy $E_T - E_A$. Now, when system A has energy E_A, it can be in any one of $\Omega_A(E_A)$ states, while simultaneously system B can be in any one of $\Omega_B(E_T - E_A)$ states. Since every possible state of A can be combined with every possible state of B, the total number of distinct states accessible to the combined system

---

[16] This statement is short for: System A has energy between E_A and $E_A + dE_A$, etc.

$A + B$, corresponding to energy E_T and subdivision $E_A = E_T - E_B$, is simply

$$\Omega_T(E_T, E_A) = \Omega_A(E_A)\Omega_B(E_T - E_A). \qquad (15.34)$$

In the ensemble of pairs that we are examining, E_T is fixed. Therefore, $\Omega_T(E_T, E_A)$ varies because E_A may vary, and $\Omega_T(E_T, E_A)$ may be thought of as a function of the one variable E_A. This is a valid simplification provided we remember that it is only true in the ensemble of pairs having the same total energy E_T.

We now proceed by finding the contributions to $\Omega_T(E_A) = \Omega_T(E_T, E_A)$ from different subdivisions of the energy between A and B. The probability of finding $A + B$ in a configuration with the energy of A between E_A and $E_A + dE_A$ is just the ratio

$$\frac{\Omega_T(E_A)}{\sum_{E_A} \Omega_T(E_A)},$$

where $\Omega_T(E_A)$ is the total number of states accessible to the combined system for a given E_A, and the sum is taken over all possible values of E_A. For convenience, let

$$\sum_{E_A} \Omega_T(E_A) \equiv C^{-1}. \qquad (15.35)$$

Using Eqs. 15.34 and 15.35, we can represent the probability of finding the combined system $A + B$ in a state with A having energy between E_A and $E_A + dE_A$ as

$$P(E_A) = C\Omega_A(E_A)\Omega_B(E_T - E_A). \qquad (15.36)$$

As E_A increases, $\Omega_A(E_A)$ increases but $\Omega_B(E_T - E_A)$ decreases. Thus, the product displayed in Eq. 15.36 has a maximum for some value of the energy, say E_A^*. Because $\Omega_A(E_A)$ and $\Omega_B(E_T - E_A)$ are very rapidly varying functions of the energy, this maximum is extremely sharp.[17] To locate the position of the maximum, the most probable state of the combined systems, we regard Ω_A and Ω_B as continuous functions and solve

$$\left(\frac{\partial P(E_A)}{\partial E_A}\right)_{V,N} = 0. \qquad (15.37)$$

Carrying out this operation, with $P(E_A)$ given by Eq. 15.36, leads to

$$C\Omega_B(E_T - E_A)\left(\frac{\partial \Omega_A}{\partial E_A}\right)_{V,N} + C\Omega_A(E_A)\left(\frac{\partial \Omega_B}{\partial E_A}\right)_{V,N} = 0 \qquad (15.38)$$

or, after division by $P(E_A)$ and use of $E_A = E_T - E_B$,

$$\left(\frac{\partial \ln \Omega_A}{\partial E_A}\right)_{V,N} = \left(\frac{\partial \ln \Omega_B}{\partial E_B}\right)_{V,N}. \qquad (15.39)$$

[17] We can illustrate how a sharp maximum develops in the function $P(E_A)$ by examining the simpler function $[4x(1-x)]^n$ in the range $0 \leqslant x \leqslant 1$. This function is the product of two factors, $(2x)^n$ and $(2-2x)^n$, which respectively increase and decrease over this range, thus corresponding to our Ω_A and Ω_B. Clearly, as n increases (corresponding to increasing the number of available states in our model), the increase and decrease of the two factors become steeper and steeper. Whatever the value of n, however, the product of these factors has a maximum value of unity at $x = \frac{1}{2}$. The peak of this function becomes steadily narrower as n increases, sharpening to an infinitely narrow spike in the limit $n \to \infty$. This behavior is illustrated in the accompanying graph.

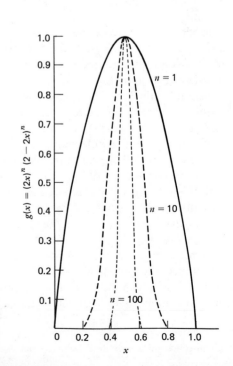

If we define

$$\beta \equiv \left(\frac{\partial \ln \Omega}{\partial E}\right)_{V,N}, \tag{15.40}$$

we see that in the most probable state of two systems over which a given amount of energy is distributed,

$$\beta_A = \beta_B. \tag{15.41}$$

This result is reminiscent of the equality of temperature between two systems in thermal equilibrium. Indeed, we shall see later that we can make the identification $\beta = (k_B T)^{-1}$, where k_B is the Boltzmann constant. The function defined by

$$S \equiv k_B \ln \Omega \tag{15.42}$$

is called the *entropy*. The condition of maximum probability in terms of the entropy, obtained by introducing S in Eq. 15.36, is then

$$(S_A + S_B)_{E_T,V,N} = \text{maximum} \tag{15.43}$$

for the system we have discussed.

We now identify the most probable state of the ensemble of replica pairs with the macroscopic state of thermal equilibrium. As will be shown below, this most probable state is overwhelmingly more likely to occur than are states having even slightly different distributions of energy between the pairs. Thus, for the given macroscopic constraints of constant total volume, total energy, and total mass, a pair of replicas that instantaneously has an energy distribution different from the most probable distribution, as a result of a temporary perturbation, is far more likely to move toward the most probable state than away from it. Another way of saying this is to note that Eq. 15.43 defines a stable equilibrium.

The motivation leading to Eq. 15.42 should now be clear, for we have argued that when a parameter of constraint is relaxed and a system undergoes a change to a new equilibrium state, $\Omega_f > \Omega_i$. Therefore the entropy function can be used to describe the direction in which a system will change upon relaxation of a parameter of constraint, that is, which energy-conserving changes will, and which will not, occur.

Equation 15.43 is of such fundamental importance that we wish to make the following supplementary observations:

1. The derivation given shows that if the ensemble of replica pairs is defined by the requirement that the sum of the energies of two systems is constant, then the most probable distribution of energy in a pair has the feature that $\beta_A = \beta_B$. It is important to note that in this ensemble the systems A and B are not in contact, and there is no flow of energy between them. Thus the energies E_A and E_B of any pair of replica systems are, in our ensemble, constants of the motion and time-independent. Nevertheless, the constraint that $E_A = E_T - E_B$ is sufficient to ensure that in the most probable distribution of energy between A and B the condition $\beta_A = \beta_B$ is satisfied. We emphasize this point because it shows that the most probable distribution of energy is a feature of the equilibrium state that is completely independent of the nature of the approach to equilibrium. Our arguments give no clue whatsoever as to the rate or the mechanism of approach to equilibrium. It is logical to assume, just because the state in which $\beta_A = \beta_B$ represents the overwhelmingly most probable state of a system for which $E_T = E_A + E_B$, that if A and B were in contact, and if energy did then flow between A and B, then the average direction of energy transfer must be such as to establish the state in which $\beta_A = \beta_B$.

2. Although the number of states in the energy range from E to $E + dE$ depends on the size of dE, the entropy function is insensitive to the size of dE, and β is completely independent of dE. To show these two features of the entropy function we rewrite Eq. 15.42, making use of Eq. 15.13, in the form

$$S \equiv k_B \ln \Omega = k_B \ln(\omega \, dE). \tag{15.44}$$

Clearly, since dE is considered to be a fixed energy interval, we must have

$$\beta \equiv \left(\frac{\partial \ln \Omega}{\partial E}\right)_{V,N} = \left(\frac{\partial \ln \omega}{\partial E}\right)_{V,N}, \tag{15.45}$$

so that β is independent of dE. Also, if ΔE is some different subdivision of the energy scale than that defined by dE, an entropy S_Δ can be defined by

$$S_\Delta \equiv k_B \ln \Omega_\Delta(E), \tag{15.46}$$

where $\Omega_\Delta(E)$ is the number of the accessible states in the energy range from E to $E + \Delta E$. The difference between the entropies defined by Eqs. 15.42 and 15.46 is just

$$S - S_\Delta = k_B \ln\left(\frac{dE}{\Delta E}\right), \tag{15.47}$$

and can be shown to be negligibly small by the following argument: Near the ground-state energy of a system there are few states available and $\Gamma_1(\epsilon)$, the number of quantum states of energy less than ϵ corresponding to one degree of freedom, is unity or some small integer. Provided the energy of the system is not too close to that of the ground state, $\Gamma_1(\epsilon)$ greatly exceeds unity. Then, not too close to the ground state, Eq. 15.21 implies that the entropy of the system is of the order of magnitude of νk_B, in the sense that $\ln \Gamma_1$ is some positive number, larger than unity, and very small compared to ν. On the other hand, if ΔE were larger or smaller than dE by a factor of ν, the term $\ln(dE/\Delta E)$ only ranges between $-\ln \nu$ and $+\ln \nu$; since $\ln \nu \ll \nu$ when ν is a large number, the difference $S - S_\Delta$ is negligibly small relative to S. We thus have

$$S = S_\Delta \tag{15.48}$$

with negligible error. The fact that Eq. 15.21 is only a crude estimate does not influence the strength of the argument given.

3. We can approximate the shape of the probability distribution $P(E_A)$ in the vicinity of the maximum as follows: For small displacements from the most probable distribution, we can write the Taylor series expansions of $\ln \Omega_A(E_A)$ and $\ln \Omega_B(E_B)$:

$$\ln \Omega_A(E_A) = \ln \Omega_A(E_A{}^*) + \left(\frac{\partial \ln \Omega_A}{\partial E_A}\right)_{V,N} (E_A - E_A{}^*)$$

$$+ \frac{1}{2}\left(\frac{\partial^2 \ln \Omega_A}{\partial E_A{}^2}\right)_{V,N} (E_A - E_A{}^*)^2 + \cdots, \tag{15.49}$$

$$\ln \Omega_B(E_B) = \ln \Omega_B(E_B{}^*) + \left(\frac{\partial \ln \Omega_B}{\partial E_B}\right)_{V,N} (E_B - E_B{}^*)$$

$$+ \frac{1}{2}\left(\frac{\partial^2 \ln \Omega_B}{\partial E_B{}^2}\right)_{V,N} (E_B - E_B{}^*)^2 + \cdots, \tag{15.50}$$

where $E_A{}^*$ and $E_B{}^*$ are the values of E_A and E_B for which the probability distribution has its maximum value. It will be recalled that

$$E_T = E_A + E_B = E_A{}^* + E_B{}^*,$$

so that

$$E_B{}^* = E_T - E_A{}^*,$$
$$E_A - E_A{}^* = E_B{}^* - E_B,$$

because of the conservation of energy. Adding Eqs. 15.49 and 15.50 yields (see Eq. 15.36)

$$\ln P(E_A) = \ln[C\Omega_A(E_A)\Omega_B(E_B)]$$
$$= \ln[C\Omega_A(E_A{}^*)\Omega_B(E_B{}^*)] + (\beta_A - \beta_B)(E_A - E_A{}^*)$$
$$+ \frac{1}{2}\left[\left(\frac{\partial^2 \ln \Omega_A}{\partial E_A{}^2}\right)_{V,N} + \left(\frac{\partial^2 \ln \Omega_B}{\partial E_B{}^2}\right)_{V,N}\right](E_A - E_A{}^*)^2.$$

$$(15.51)$$

At the maximum of the probability distribution we have $\beta_A = \beta_B$ by Eq. 15.41, and the linear term in Eq. 15.51 vanishes. Equation 15.51 then shows that $P(E_A)$ is the product of two terms (aside from the constant C),

$$F_A(E_A) = \Omega_A(E_A{}^*) \exp\left[\frac{1}{2}\left(\frac{\partial^2 \ln \Omega_A}{\partial E_A{}^2}\right)_{V,N}(E_A - E_A{}^*)^2\right],$$

$$F_B(E_B) = \Omega_B(E_B{}^*) \exp\left[\frac{1}{2}\left(\frac{\partial^2 \ln \Omega_B}{\partial E_B{}^2}\right)_{V,N}(E_B - E_B{}^*)^2\right], \quad (15.52)$$

where the substitution $E_A - E_A{}^* = E_B{}^* - E_B$ has been made because the systems are coupled by the constraint $E_T = E_A + E_B$. Now $(E_A - E_A{}^*)^2$ is always positive, and $\Omega_A(E_A)$ is an increasing function of E_A. Also, $(E_B - E_B{}^*)^2$ is always positive and $\Omega_B(E_B)$ is an increasing function of E_B and thus a decreasing function of E_A. For the product function $P(E_A)$ to have a maximum at $E_A{}^*$, $E_B{}^*$, the extrema of $F_A(E_A)$ and $F_B(E_B)$ at these points must also be maxima rather than minima. It is thus necessary that the coefficients of $(E_A - E_A{}^*)^2$ and $(E_B - E_B{}^*)^2$ be negative, that is, that the second derivatives $(\partial^2 \ln \Omega_A/\partial E_A{}^2)$ and $(\partial^2 \ln \Omega_B/\partial E_B{}^2)$ be negative. To suggest why this is the case, we recall that $\Omega(E)$ is proportional to $[\Gamma_1(\epsilon)]^\nu$, from Eq. 15.21, in connection with which we have argued that $\Gamma_1(\epsilon)$ increases as some small power of ϵ. Thus, at least crudely, $\Omega(E)$ is proportional to $\epsilon_1{}^\nu = (E/\nu)^\nu$, where it is to be understood that the precise exponent may differ from ν by a factor that is of the order of magnitude of unity. Clearly, this error in our estimate of $\Omega(E)$ cannot influence the sign of the second derivative $[\partial^2 \ln \Omega/\partial E^2]$. Using the cited proportionality, we obtain

$$\left(\frac{\partial \ln \Omega}{\partial E}\right)_{V,N} \propto \frac{\nu}{E}, \quad (15.53)$$

$$\left(\frac{\partial^2 \ln \Omega}{\partial E^2}\right)_{V,N} \propto -\frac{\nu}{E^2}. \quad (15.54)$$

Since ν is positive and E^2 is positive, $-(\nu/E^2)$ must be negative, hence the functions displayed in Eq. 15.52 have maxima.

The first derivative of $\ln \Omega$ with respect to E is also of interest. Combining 15.53 and 15.40, we find

$$\frac{E}{\nu} \propto \frac{1}{\beta}, \quad (15.55)$$

which is a form of the principle of equipartition of energy. Equation

15.55 implies that the average energy per degree of freedom is constant and proportional to β^{-1}, or, with the identification to be made later, $\beta^{-1} = k_B T$, to the temperature. The estimate given for $\Omega(E)$ suffices to determine the orders of magnitude and the signs of derivatives and of functions, but is not sufficiently precise to be a complete justification for the equipartition principle. We shall return to that subject later in this text.

4. We now can show that the probability distribution for the energy is extremely sharply peaked about the energies $E_A{}^*$ and $E_B{}^*$. Indeed, the distribution is so sharply peaked that for many purposes the distribution may be replaced by the value of $P(E_A)$ at the maximum. The combination of Eqs. 15.36 and 15.52, followed by the use of the estimate in 15.54, shows that the arguments of the exponential function in Eq. 15.52 are proportional to

$$-\nu \frac{(E-E^*)^2}{E^{*2}}.$$

Then the energy for which the probability has fallen to $1/e$ of the probability of the maximum is given by

$$-\nu \frac{(E-E^*)^2}{E^{*2}} = -1$$

or

$$\frac{E-E^*}{E^*} = \frac{1}{\sqrt{\nu}}. \tag{15.56}$$

For $\nu \approx 10^{24}$, Eq. 15.56 yields $(E-E^*)/E^* \approx 10^{-12}$, and we conclude that in a system of macroscopic size the probability of finding an energy deviating by any substantial amount from the most probable energy is vanishingly small.

5. The probability distribution is so sharply peaked that the entropy can also be defined in terms of $\Gamma(E)$, the total number of states with energy less than E, the result being negligibly different from Eq. 15.42. That is, because of the very sharp maximum in the energy distribution, practically all the accessible states lie within the range $\Delta E^* = E^*/\sqrt{\nu}$ of E^*. Thus Eq. 15.13 becomes

$$\Gamma(E) \approx \omega(E^*)\,\Delta E^*, \tag{15.57}$$

and thereby

$$\ln \Gamma(E) = \ln \Omega(E) + \ln\left(\frac{\Delta E^*}{dE}\right), \tag{15.58}$$

so that

$$S_\Gamma = k_B \ln \Gamma(E) \tag{15.59}$$

differs negligibly from the entropy defined in Eq. 15.42 by application of the same arguments used to establish the equivalence displayed in Eq. 15.48.

There are several properties of the entropy function that can be used to provide a still closer link between the microscopic and macroscopic descriptions of matter. Consider two isolated systems, A and B. At equilibrium in their isolated conditions their energies and entropies are $E_A{}^i$, $E_B{}^i$ and $S_A{}^i$, $S_B{}^i$, respectively. Imagine that a third system, R, is brought into contact with system A. The system R could be, for example, a container of gas. When the contact between R and A is through a rigid diathermal wall, expansion of the system R will extract energy from A while the volume of A remains fixed. Conversely, compression of R will lead to addition of energy to A, at constant volume of A. We shall use system R to transfer energy between A and B without allowing A and B to be in contact.

We have already learned that the most probable state of an ensemble of replicas of A and B has the properties that

$$\beta_A{}^f = \beta_B{}^f \tag{15.60}$$

and

$$E_A{}^{f*} + E_B{}^{f*} = E_A{}^i + E_B{}^i, \tag{15.61}$$

where, as before, the asterisk refers to the most probable state. We now examine further the implications of the identification of the most probable distribution of energy in the ensemble of replica systems A and B with the equilibrium state that is achieved when macroscopic systems A and B are brought into contact through a diathermal wall. Since contact through a diathermal wall implies energy transfer, we must examine how the properties of A, B, and R are related.

Let the system R, in diathermal contact with A, be expanded[18] so as to remove energy

$$-q_A = E_A{}^{f*} - E_A{}^i \tag{15.62}$$

from A. Now let R be isolated from A and placed in diathermal contact with system B, and let R be compressed until energy

$$q_B = E_B{}^{f*} - E_B{}^i \tag{15.63}$$

is added to B. As usual, q_A and q_B are defined as positive for energy added to the system in question. At the end of this sequence of processes, because of the conservation of energy (see Eq. 15.61), we note that

$$q_A + q_B = 0, \tag{15.64}$$

and that the system R has undergone no net change in state. Of course, because of the energy changes in systems A and B the entropies of A and B have been changed. From the fundamental relation between the entropy and the number of accessible states we conclude that

$$S(E_A{}^{f*}) + S(E_B{}^{f*}) \geq S(E_A{}^i) + S(E_B{}^i). \tag{15.65}$$

Equation 15.64 is interesting. Because the volumes of systems A and B have been maintained constant during the diathermal interactions with R, no work is done on or by either A or B. Then, in thermodynamic terminology, the energy changes in A and B are due entirely to heat transfers. It is this observation that motivated the choice of symbol q in Eqs. 15.62 and 15.63. The energy conservation condition, as stated in Eq. 15.64, then says that the heat given off by system A is absorbed by system B.

Further information of interest can be obtained by supposing that system A is very, very much smaller than system B, so that

$$E_A{}^* \ll E_B{}^*. \tag{15.66}$$

Of course, under these conditions system B has many, many more degrees of freedom than does system A. For system B the rate of change of β_B with the energy E_B, namely, $(\partial \beta_B / \partial E_B)_{V,N}$, is of the order of

---

[18] The process can also be run in the opposite direction.

magnitude[19] of $(-\beta_B/E_B{}^*)$. On the other hand, the maximum amount of energy that can be removed from system A is $E_A{}^*$. If system A and system B are brought to a condition of thermal equilibrium, say, by means of the third system R, then because $q_A = -q_B$, it is found that

$$\left|\left(\frac{\partial \beta_B}{\partial E_B}\right)_{V,N} q_B\right| \approx \beta_B \frac{E_A{}^*}{E_B{}^*} \ll \beta_B. \tag{15.67}$$

The importance of Eq. 15.67 is the following: It implies that the characteristic parameter of system B, namely, β_B, remains virtually unchanged no matter how much heat B absorbs from A. When condition 15.67 is satisfied for every energy transfer, system B is considered a reservoir for system A. Thus, if B is a reservoir and condition 15.67 is valid, β_B can be taken as constant throughout any sequence of energy transfers from A to B. Also, and perhaps most important, Eq. 15.67 allows us to establish a direct connection between the amount of energy transferred and the entropy change of the reservoir. After the reservoir has absorbed heat q_B from system A, with $\beta_A = \beta_B$, we can write for the change in the number of states accessible to the reservoir (system B)

$$\ln \Omega_B(E_B + q_B) - \ln \Omega_B(E_B) = \left(\frac{\partial \ln \Omega_B}{\partial E_B}\right)_{V,N} q_B + \cdots$$

$$= \beta_B q_B. \tag{15.68}$$

But the left-hand side of Eq. 15.68 is just the entropy change of the reservoir, so that we reach an important conclusion: If a heat reservoir absorbs heat q_B at constant β_B, then the resultant change in the reservoir's entropy is

$$\Delta S_B = k_B(\beta_B q_B) = \frac{q_B}{T_B}, \tag{15.69}$$

with the final form again using the congruence $\beta_B = (k_B T_B)^{-1}$. We have now established a connection between the macroscopic variable heat and the entropy function that has been defined in terms of the microscopic energy spectrum of the system. A similar relation holds for any system that, at constant β, absorbs an infinitesimal amount of heat $đq$ from some other system. Provided that $đq \ll E$, where E is the energy of the system under study, we have[20]

$$dS = k_B \beta \, đq = \frac{đq}{T}. \tag{15.70}$$

What are the properties of the entropy function when more general interactions between systems are permitted? To answer this question we must first determine the dependence of the variation of the density of

---

[19] In the crude approximation previously used, correct only as to order of magnitude, combining Eqs. 15.53 and 15.54 with Eq. 15.40 yields

$$\beta \propto \frac{\nu}{E},$$

$$\left(\frac{\partial \beta}{\partial E}\right)_{V,N} \propto -\frac{\nu}{E^2},$$

hence

$$\left(\frac{\partial \beta}{\partial E}\right)_{V,N} \approx -\frac{\beta}{E}.$$

[20] The preceding argument shows that Eqs. 15.69 and 15.70 are valid only for reversible heat transfers at the constant temperatures T_B and T, respectively.

states on the variation of the external parameters of constraint defining the system.

Consider, first, the case that the parameter of constraint is the volume V, and let $E_i(V)$ be the energy of the ith state in the range E to $E + dE$. When V is changed by dV, $E_i(V)$ changes by $(\partial E_i(V)/\partial V)\, dV$. It is to be expected that the energies of different states change by different amounts. Now, the number of states between E and $E + dE$ corresponding to some value of V *and having the property that*

$$- (Y + dY) < \left(\frac{\partial E_i}{\partial V} \right)_{\Omega, N} < -Y \qquad (15.71)$$

is denoted[21] $\Omega_Y(E, V, N)$. Since each of the levels corresponds to some range from Y to $Y + dY$, it is necessary that

$$\Omega(E, V, N) = \sum_Y \Omega_Y(E, V, N), \qquad (15.72)$$

with the summation over all values of Y. Note that the derivative in Eq. 15.71 is defined with Ω and N held constant. If there are N_i particles with energy E_i, holding Ω fixed implies holding all the N_i fixed, and by virtue of the definition of the entropy (Eq. 15.42), also holding the entropy constant.

Given the definition of $\Omega_Y(E, V, N)$, we can calculate how many states are altered when $V \to V + dV$ such that their energies change from being less than E to being more than E. Note that those states for which $(\partial E_i/\partial V)_{\Omega, N} = -Y$ undergo an energy change of $-Y\, dV$. Then the total number of states whose energy is changed from a value less than E to a value more than E is simply equal to the product of the number of states per unit energy corresponding to the given values of V and Y, namely, $\Omega_Y(E, V, N)/dE$, and the energy change, $-Y\, dV$, summed over all possible values of Y, that is, to

$$- \sum_Y \frac{\Omega_Y(E, V, N)}{dE} Y\, dV.$$

Consider now the total number of states between E and $E + dE$ irrespective of the value of Y. When $V \to V + dV$, the number of states in this energy range changes by

$$\Omega(E, V + dV, N) - \Omega(E, V, N) = \left(\frac{\partial \Omega(E, V, N)}{\partial V} \right)_{E, N} dV$$

$$= \sum_Y \frac{\Omega_Y(E + dE, V, N)}{dE} Y\, dV$$

$$- \sum_Y \frac{\Omega_Y(E, V, N)}{dE} Y\, dV$$

$$= \sum_Y \left(\frac{\partial \Omega_Y(E, V, N)}{\partial E} \right)_{V, N} Y\, dV.$$

$$(15.73)$$

The second equality of Eq. 15.73 is simply the difference between the net number of states that leave the energy range between E and $E + dE$ by having their energy changed from a value less than $E + dE$ to a value more than $E + dE$, and the net number of states that enter this energy range by having their energy changed from a value less than E to a value

---

[21] We choose to insert the minus sign in the definition of Y for later convenience.

more than E. But the mean value of Y over all states, each state weighted with equal probability, is

$$\langle Y \rangle = \frac{1}{\Omega(E, V, N)} \sum_Y Y \Omega_Y(E, V, N), \qquad (15.74)$$

so that Eq. 15.73 may be written in the form

$$\left(\frac{\partial \Omega}{\partial V}\right)_{E,N} = \frac{\partial}{\partial E}\left(\langle Y \rangle \Omega\right)_{V,N} = \left(\frac{\partial \Omega}{\partial E}\right)_{\langle Y \rangle, V, N} \langle Y \rangle + \left(\frac{\partial \langle Y \rangle}{\partial E}\right)_{\Omega, V, N} \Omega, \quad (15.75)$$

because

$$\langle Y \rangle \Omega(E, V, N) = \sum_Y Y \Omega_Y(E, V, N). \qquad (15.76)$$

If Eq. 15.75 is divided by Ω, we find that

$$\left(\frac{\partial \ln \Omega}{\partial V}\right)_{E,N} = \left(\frac{\partial \ln \Omega}{\partial E}\right)_{\langle Y \rangle, V, N} \langle Y \rangle + \left(\frac{\partial \langle Y \rangle}{\partial E}\right)_{\Omega, V, N}. \qquad (15.77)$$

For a macroscopic system the term $(\partial \langle Y \rangle / \partial E)$ is negligible relative to $(\partial \ln \Omega / \partial E)\langle Y \rangle$, since $(\partial \langle Y \rangle / \partial E)$ is of the order of magnitude of $(\langle Y \rangle / E)$ whereas $(\partial \ln \Omega / \partial E)\langle Y \rangle$ is of the order of magnitude of $(\nu / E)\langle Y \rangle$, which differ by the factor[22] $\nu \approx 10^{24}$. In fact, the larger the system, the more negligible does the second term on the right-hand side of Eq. 15.78 become. We therefore can write

$$\left(\frac{\partial \ln \Omega}{\partial V}\right)_{E,N} = \left(\frac{\partial \ln \Omega}{\partial E}\right)_{\langle Y \rangle, V, N} \langle Y \rangle, \qquad (15.78)$$

or, by Eq. 15.45,

$$\left(\frac{\partial \ln \Omega}{\partial V}\right)_{E,N} = \beta \langle Y \rangle. \qquad (15.79)$$

Since by our definition of Y we have

$$\langle Y \rangle = -\left\langle \left(\frac{\partial E(V)}{\partial V}\right)_{\Omega, N} \right\rangle, \qquad (15.80)$$

it can be seen that $\langle Y \rangle$ plays the role of a generalized force complementary to the volume V. In this case the generalized force is the pressure, as we shall shortly demonstrate.

Consider, as an example, a gas of N noninteracting particles in a box of volume V. We have already shown that (see Eqs. 15.27 and 15.31)

$$\Omega(E) \propto V^N E^{(3N/2)-1}, \qquad (15.81)$$

with the proportionality constant independent of E and V. Then

$$\left(\frac{\partial \ln \Omega}{\partial V}\right)_{E,N} = \frac{N}{V} \qquad (15.82)$$

because the proportionality constant cancels out of the expression for the derivative. We now must evaluate $\langle(\partial E/\partial V)_{\Omega,N}\rangle$. We have already noted that for constant Ω the number of particles per level is a constant. Then for our system of independent particles

$$E = \sum_i N_i E_i, \qquad (15.83)$$

---

[22] To the same, crude, approximation we have used before (cf. discussion after Eq. 15.20), we assume $\langle Y \rangle$ is proportional to E and use Eq. 15.53 for $(\partial \ln \Omega/\partial E)_{V,N}$.

from which we have

$$\left\langle \left(\frac{\partial E}{\partial V}\right)_{\Omega,N}\right\rangle = \left\langle \sum_i N_i \left(\frac{\partial E_i}{\partial V}\right)_{N_i}\right\rangle = -\frac{2}{3}\frac{\langle E\rangle}{V},\qquad(15.84)$$

where the extreme right-hand side of Eq. 15.84 follows from use of Eq. 15.23.[23] By combination of Eqs. 15.82, 15.84, 15.79, and 15.80, we find that the average energy is

$$\langle E\rangle = \frac{3N}{2\beta}.\qquad(15.85)$$

But for a perfect gas, from Eq. 12.42, and identification of the ensemble average energy with the average energy of Chapter 12,

$$\langle E\rangle = \tfrac{3}{2}Nk_BT,\qquad(15.86)$$

so that we can make the congruence

$$\beta = \frac{1}{k_BT},\qquad(15.87)$$

as already used in Eq. 15.69. Furthermore, from the relation

$$pV = \tfrac{2}{3}\langle E\rangle\qquad(15.88)$$

derived in Chapter 12, we find after use of Eq. 15.84 that

$$p = -\left\langle \left(\frac{\partial E}{\partial V}\right)_{\Omega,N}\right\rangle,\qquad(15.89)$$

as asserted above.[24]

The preceding argument can be applied to calculate the change in $\Omega(E, y_1, \ldots, y_n)$ arising from changes in any (or several) parameters of constraint y_i. The algebraic manipulations become more tedious, but the ideas used and the procedure followed are the same. Clearly, when there are n parameters of constraint, $\Omega(E, y_1, \ldots, y_n)$ defines n generalized forces by relations of the form

$$\left(\frac{\partial \ln \Omega}{\partial y_\alpha}\right)_{E,N,y_{i\neq\alpha}} = \beta\langle Y_\alpha\rangle.\qquad(15.90)$$

By a generalized force is meant simply that if the change of parameter of constraint dy is thought of as a "displacement," then $\langle Y\rangle\,dy$ has the dimensions of work. Furthermore, as discussed in Chapter 3, the general definition of force in mechanics is as the negative derivative of the energy with respect to a coordinate.

The notion of generalized force is quite useful. In mechanics the energy is a potential function that defines both the force acting on the system and whether or not the equilibrium state is stable. We have shown in this section that the rate of change of entropy with the parameter of constraint also defines the generalized force acting on the system. By

---

[23] In Eq. 15.23, ϵ_j refers to the energy of particle j, whereas in Eq. 15.83, E_i refers to the energy of level i. Of course, $\sum_{j=1}^N \epsilon_j = \sum_{i=1}^\infty N_iE_i$.

[24] It is easy to verify that $(\partial\langle Y\rangle/\partial E)_{\Omega,V,N}$ is very small compared to $(\partial \ln \Omega/\partial E)_{\langle Y\rangle,V,N}\langle Y\rangle$ as stated after Eq. 15.77. Using Eq. 15.88, and setting $\langle Y\rangle = p$, we see that

$$\left(\frac{\partial p}{\partial E}\right)_{\Omega,V,N} = \frac{2}{3V},$$

which is, indeed, negligible relative to

$$\left(\frac{\partial \ln \Omega}{\partial V}\right)_{E,N} = \frac{N}{V}.\qquad(15.82)$$

analogy with mechanics, we expect the entropy to determine whether or not the equilibrium state is stable. We shall show, in Chapter 19, that this expectation is correct.

We now examine the nature of the changes in entropy corresponding to an interaction between two systems, A and B, in which both work and energy transfers are allowed. We shall use arguments similar to those already outlined. Let the two systems interact through a third system R; A and B are otherwise isolated. Further, let A and B exchange both energy and work through the use of R. This can be accomplished by allowing V_A and V_B to vary (keeping $V_A + V_B$ constant) so that A does work on R and then R does work on B, or vice versa. Clearly, for other kinds of work processes equivalent conditions can be defined. As before, since $E_T = E_A + E_B$ is fixed, the energy E_B is determined if E_A is known. Further, the parameters of constraint $V_A, y_1{}^A, \ldots, y_{n-1}^A$ defining system A are, at equilibrium, some function of the parameters of constraint V_B, $y_1{}^B, \ldots, y_{n-1}^B$ defining system B. For example, $V_T = V_A + V_B$ and, since V_T is a constant, a knowledge of V_A fixes V_B. Analogous relations hold for $y_1{}^A, \ldots, y_{n-1}^A$ and $y_1{}^B, \ldots, y_{n-1}^B$. In the combined system the total number of states is a function of the energy E and the parameters V, $y_1, \ldots, y_{n-1}$. As before, $\Omega_T(E, V, y_1, \ldots, y_{n-1})$ has a very sharp maximum for some values $E = E^*$, $V = V^*$, and $y_\alpha = y_\alpha{}^*$. Also as before, the mean values of E, V, and $y_1, \ldots, y_{n-1}$ differ negligibly from the most probable values.

Consider now a process by which the system A, through interaction with B via the intermediary R, is changed from an equilibrium state characterized by $E, V, y_1, \ldots, y_{n-1}$, to another equilibrium state characterized by $E + dE$, $V + dV$, $y_1 + dy_1, \ldots, y_{n-1} + dy_{n-1}$. What is the change in the entropy? We can write, using Eqs. 15.45 and 15.79,

$$d \ln \Omega = \left(\frac{\partial \ln \Omega}{\partial E}\right)_{N,V,y} dE + \sum_{\alpha=1}^{n-1} \left(\frac{\partial \ln \Omega}{\partial y_\alpha}\right)_{E,V,N} dy_\alpha$$

$$+ \left(\frac{\partial \ln \Omega}{\partial V}\right)_{E,N,y} dV \qquad (15.91)$$

$$= \beta\left(dE + \sum_{\alpha=1}^{n-1} \langle Y_\alpha \rangle \, dy_\alpha + p \, dV\right). \qquad (15.92)$$

But, by our definition of a generalized force, the product $\langle Y_\alpha \rangle \, dy_\alpha$ is the work done in changing the parameter of constraint y_α by dy_α, just as $p \, dV$ is the work done in changing V to $V + dV$. Using the same convention as in Chapter 13, we reckon work done on the system as positive, and work done by the system as negative. With this convention we obtain[25]

$$đw = -\sum_{\alpha=1}^{n-1} \langle Y_\alpha \rangle \, dy_\alpha - p \, dV. \qquad (15.93)$$

Combining Eqs. 15.93 and 15.92, we find that

$$d \ln \Omega = \beta[dE - đw] \equiv \beta \, đq. \qquad (15.94)$$

In Eq. 15.94 the relationship

$$đq = dE - đw$$

has been introduced. By this introduction of $đq$ in terms of dE and $đw$, we use the first law of thermodynamics to provide a bridge between the macroscopic definition of heat and the number of states. Indeed, sub-

---

[25] It was to obtain the correct sign in Eq. 15.93 that the negative sign was introduced in the definition of Y.

stituting Eq. 15.70 into Eq. 15.94, we are led to

$$đq = \frac{dS}{k_B\beta} = dE - đw, \qquad (15.95)$$

a relationship between the entropy and macroscopic functions.

Before using the entropy function to study the macroscopic properties of matter, we turn to another way of introducing the entropy concept, again in terms of the constraints defining the state of a macroscopic system, but now employing only thermodynamic considerations.

Further Reading

Andrews, F. C., *Equilibrium Statistical Mechanics*, 2nd ed. (Wiley, New York, 1975), chaps. 2–6.

Reif, F., *Statistical and Thermal Physics* (McGraw-Hill, New York, 1965), chaps. 2 and 3.

Problems

1. A coin is tossed five times. What is the probability that:
   (a) Head is up all five times?
   (b) Head is up three times?
   (c) The sequence of head and tail is H T H T H?

2. Two dice are thrown. What is the probability of obtaining a total of 7? What is the probability of obtaining 8? What is the probability of obtaining 7 or 8?

3. Two cards are drawn from a deck of well-shuffled cards. What is the probability that both cards drawn are aces?

4. Two cards are drawn from a well-shuffled deck of cards, the first card being returned before the second is taken. What is the probability that both the extracted cards belong to one suit?

5. Consider a box of volume V containing N molecules. What is the probability of finding all the molecules in a portion of the container having volume $\frac{1}{2}V$?
   If $N = 10^{23}$, what is the numerical value of this probability? Assume that the molecules in the box are free of all interactions, that is, that the equation of state is that of a perfect gas.

6. A volume V contains N_A molecules of type A and N_B molecules of type B. A valve is opened and M molecules flow out. What is the probability that among the M molecules there are m_A of species A and m_B of species B?

7. Suppose that a volume V is mentally subdivided into M subvolumes. Let there be N molecules in V. What is the probability that some one subvolume will contain N' molecules?

8. Suppose that a volume V is subdivided into two compartments. One of these contains N_A' A molecules and N_B' B molecules; the other contains N_A'' A molecules and N_B'' B molecules. A valve between the compartments is opened and closed and one molecule transferred from the side marked (') to the side marked ("). After a long time the valve is again opened and closed and one molecule transferred from side (") to side ('). What is the probability that the molecule transferred in both steps is an A molecule?

9. Consider a perfect gas flowing down a tube with constant average flow velocity **u**. Describe in simple terms how the distribution of molecular

velocities differs in this case from that of a stationary perfect gas. If the temperatures of a flowing and a stationary gas are the same, show in a simple diagram how the respective velocity distributions are related.

10. Consider an ensemble of pairs of systems, A and B, coupled together by a frictionless sliding wall that separates the volume V into V_A and V_B. Each

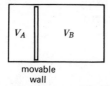

movable
wall

replica in the ensemble has different values of V_A and V_B, but $V_A + V_B = V = $ constant for all replicas. The temperatures in V_A and V_B are the same. Let V_A and V_B each contain 1 mol of a perfect gas.
(a) What is the average value of the pressure difference $p_A - p_B$?
(b) What is the average value of $p_A - p_B$ if the ensemble consisted of the same replicas, but with the wall dividing V_A from V_B fixed in position in each replica?
(c) What is the average value of $p_A - p_B$ if the replicas are so constructed that $V_A < V_B$ in all cases?
(d) Does $\langle p_A - p_B \rangle = \langle (p_A - p_B)^2 \rangle^{1/2}$ in cases (a), (b), and (c)?

11. One simple model of a solid treats the solid as if it were N independent harmonic oscillators of the same frequency ν. Consider such a set of N oscillators. The energy of the N oscillators is

$$E = \tfrac{1}{2}Nh\nu + Mh\nu,$$

The first term is the zero-point energy of the system, and the second term is the energy of M quanta of vibration.
(a) Calculate the number of states that have energy E.
(b) The energy of the system is increased by just $h\nu$, and thereby the number of states with energy E is changed. How large is this change?
(c) In this system, does the number of states with energy E increase indefinitely as E increases? Explain your answer.

12. Suppose that N independent particles can each be in only one of the two energy levels $\pm\Delta$.
(a) How many states of the system have energy

$$E = M\Delta,$$

where M is a number between N and $-N$?
(b) How is the number of states with energy E changed when we replace M by $M+1$?
(c) Discuss how the number of states with energy E varies with M as M changes from very small values to the value N.
(d) How does your answer to part (c) differ from your answer to part (c) of Problem 11? Why should there be a difference in behavior of these two systems?

13. Calculate the entropy corresponding to the system of Problem 11. Express the total energy E in terms of the temperature T.

14. Carry out the calculations of Problem 13, but now for the system of Problem 12.

15. In Problems 13 and 14 the results gave relations involving the energy of the system and the temperature T. From these calculate the specific heats at constant volume corresponding to the two systems.

16. Make a plot of $E/Nh\nu$ versus $k_B T/h\nu$ for the system of Problem 11 and of $E/N\Delta$ versus $k_B T/\Delta$ for the system of Problem 12. Also make plots of C_V

versus $k_B T/h\nu$ and $k_B T/\Delta$ for these two systems. Why are the curves of different shape? How do these shapes reflect the qualitative features of the energy spectra of the two systems?

17. Show that, given the general properties of $\Omega(E, V, N)$, the temperature of a system at equilibrium must be positive.

18. Consider a perfect monatomic gas, consisting of N particles with total energy E, in thermal equilibrium in a box of volume $V = a^3$. From Equation 15.27, show that the probability that a given gas molecule will have energy ϵ is proportional to

$$e^{-\epsilon/k_B T}.$$

Hint: Use the fact that $E/N = \frac{3}{2}k_B T$. The expansion $\ln(1+x) = x$, valid when $|x| \ll 1$, will be found useful.

19. Evaluate $\Omega(E, V, N)$ for a gas of N "hard-sphere" molecules of radius r_0. The molecules are impenetrable and do not attract each other.

20. At 0 K, a perfect crystal of CO has molecules that are all lined up but that can either point "up" or "down." Calculate Ω and S at 0 K for 1 mol of a perfect crystal of CO under the assumption that half the molecules point up and the other half point down. Use Stirling's approximation to obtain a numerical value for S.

The Second Law
of Thermodynamics:
The Macroscopic
Concept
of Entropy

In Chapter 15 we showed that the entropy,

$$S \equiv k_B \ln \Omega(E, V, N), \qquad (16.1)$$

defined with respect to the number of quantum states of a system between the energies E and $E + dE$, is a function that describes a macroscopic property of the system. We should expect, then, that the entropy of a system can be defined solely in terms of its macroscopic behavior, and without any reference to the system's energy-level spectrum or other microscopic properties. In fact, the entropy function was first introduced into thermodynamic theory in this way, and the definition of Eq. 16.1 was proposed later to provide a microscopic interpretation. This chapter is devoted to the analysis of those aspects of the behavior of macroscopic systems that lead to the second law of thermodynamics, and from the second law to the concept of entropy.

The second law of thermodynamics is a statement about the behavior of real macroscopic systems; it must be considered one of the fundamental principles on which the macroscopic description of matter is based. As is true of all of the laws of thermodynamics, the second law is an abstraction from experiment. It is a formulation of an aspect of macroscopic behavior that cannot be derived from, or proved by, appeal to other considerations. Using kinetic theory, we can interpret the meaning and origin of the second law, but that is not a proof.

There are a number of equivalent statements of the second law of thermodynamics. Although we shall not be concerned with proofs of equivalence, it is of interest to set down a few of the superficially different formulations that can be given. The statements most directly representative of experience are the following:

1. *Clausius statement:* It is impossible to devise a continuously cycling engine that produces no effect other than the transfer of heat from a colder to a hotter body.

2. *Kelvin statement:* It is impossible to devise a continuously cycling engine that produces no effect other than the extraction

16.1
The Second Law of Thermodynamics

of heat from a reservoir at one temperature and the performance of an equal amount of mechanical work.

3. *Caratheodory statement:* In the neighborhood of every equilibrium state of a closed system there are states that cannot be reached from the first state along any adiabatic path.

Note that, unlike the first law of thermodynamics, statements 1, 2, and 3 claim the impossibility of something happening. This "statement of impossibility" is a fundamental feature of the second law.

In both the Clausius and Kelvin statements of the second law, an essential role is assigned to cyclic processes. Originally, cyclic processes were singled out for attention because of interest in the use of engines to transform heat into work. By design, an engine is required to repeat exactly some process over and over so that the amount of heat that can be transformed into work is unlimited. More generally, a cyclic process returns a system to its initial state, hence leaves the internal energy of the system unchanged. Implicitly, then, the second law imposes conditions that distinguish between those energy-conserving processes that can occur spontaneously and those that cannot.

It is important to observe that it is relatively simple to invent procedures that transfer heat from a colder to a hotter body, but not in a cyclic process. Consider, for example, a gas contained in a volume that is in thermal contact with a body at some temperature, say, T_1. Let the gas be expanded isothermally to occupy a larger volume, thereby extracting heat from the body with which it is in contact. Now let the container of gas be isolated and then compressed. As a result of the compression, the temperature of the isolated gas is increased, say, to T_c. By sufficient compression the temperature of the isolated gas can be raised above T_1. Finally, let the container of gas be brought into thermal contact with a new body, at a temperature T_2 such that $T_c > T_2 > T_1$. Because $T_c > T_2$, heat will be transferred from the gas to the second body. Thus, in the sequence of steps envisaged, we have transferred heat from a colder body to a hotter body. However, at the end of the sequence of steps the state of the gas is different from its initial state. The process is, clearly, not cyclic, so the heat transfer described is not prohibited by the second law. In general, it is found that no cyclic process can be devised that violates statements 1 or 2.

The Caratheodory statement of the second law, 3, is less obviously an abstraction from experiment than are statements 1 and 2. To illustrate how statement 3 is derived, we consider again how the internal energy of a system is determined. Careful thought reveals that there is a subtle problem connected with our definition of internal energy in terms of adiabatic work processes. It will be recalled that if a system is enclosed by an adiabatic envelope, the state of the system can be altered only by moving part of the envelope, and not by any other external action. Is it possible to reach *all* states of the system by adiabatic means starting from some arbitrary reference state? In other words, are there thermodynamic states of the system that are inaccessible by use of the mechanical processes that define U? The answer to this question comes from experiment. It is concluded that:

1. From any given state of a system, there are other states that cannot be reached by *particular* adiabatic paths. For example, if a paddle wheel is turned in an adiabatically enclosed fluid, the work dissipated in frictional heat can only increase the temperature of the fluid. A state of lower temperature cannot be reached by this particular adiabatic path.

2. It is always possible to connect two thermodynamic states, one of which is inaccessible to the other by an adiabatic work path, by either:

  (a) reversing the direction of some possible adiabatic work path, that is, considering an expansion process in place of a compression process, or

  (b) connecting the two states via separate adiabatic work paths to an identical third state.

Conclusion 1 is clearly related to the Caratheodory statement of the second law. Conclusion 2, in both parts, provides assurance that the internal energy of every possible state of a system can be determined, relative to some reference state, by means of only adiabatic work processes, despite the difficulties posed by the inaccessibility referred to in conclusion 1. It is not implied that the adiabatic work processes used need be of only one kind.

For example, suppose that a gas initially in a state denoted (p_1, V_1, T_1) is expanded adiabatically until the state denoted (p_2, V_2, T_2) is reached. It is characteristic of such an adiabatic expansion that $V_2 > V_1, T_2 < T_1, p_2 < p_1$. Thus, starting from (p_1, V_1, T_1) and using only an adiabatic expansion, one cannot reach a state with $T_2 > T_1$. States with $T_2 > T_1$ can be reached if, in addition to adiabatic expansion work, adiabatic work is done on the gas while the volume is held fixed. Such work could be carried out by rotating a paddle wheel or by rubbing together two plates inside the container of gas, in each case through frictionless seals. The work dissipated in such processes results in an increase in the temperature of the gas. By doing just the right amount of constant-volume adiabatic work, the desired temperature T_2 can be reached. So long as the compound work path is adiabatic, it does not matter how many different ways the work is done, since the relationship between adiabatic work and internal energy change is independent of the nature of the work.

Even with the procedure described, or generalizations of it, there still remain some states inaccessible by an adiabatic path starting from (p_1, V_1, T_1). For example, if $T_3 < T_1, V_3 > V_1, p_3 < p_1$, but T_3 is smaller than the temperature achievable in an adiabatic expansion from the initial state to the state with volume V_3, state 3 cannot be reached from state 1 by an adiabatic path because there is no constant-volume adiabatic work process that can lower the temperature. On the other hand, the state (p_3, V_3, T_3) can be connected to (p_1, V_1, T_1) by reversing the roles of initial and final state. In this example adiabatic compression of the gas from volume V_3 to volume V_1 results in a temperature θ_1 that is less than T_1. Then using, say, a rotating paddle wheel, one can dissipate adiabatic work while keeping the volume fixed until the temperature of the gas rises to T_1. Clearly, if the internal energy of the state (p_1, V_1, T_1) is known, it is still possible to determine that of the state (p_3, V_3, T_3). Thus, as noted in conclusion 2(b), the internal energy of every state of a system can be determined relative to that of some reference state by using only adiabatic work processes, even though it is not possible to transform a given state of a system into every other state along some adiabatic path.

In the example given above, the observation that it is impossible to convert state 1 to state 3 along any adiabatic path does not mean that state 3 cannot be reached from state 1 at all. We could, starting from state 1, reach state 3 by combining an adiabatic expansion with a constant-volume heat transfer, e.g., in this case, removal of heat at the fixed volume V_3. However, the second step of this path is not adiabatic. A study of many systems and possible transformations leads to the generalization of conclusion 1, referring to particular adiabatic processes, into the Caratheodory statement of the second law, referring to general adiabatic processes.

To this point our discussion of the several formulations of the second law makes no reference to the entropy of a system. We now show that the Clausius and Kelvin statements imply both the existence of an entropy function and the Caratheodory statement of the second law.

16.2
The Existence of an Entropy Function for Reversible Processes[1]

We consider, first, how the Clausius and Kelvin statements of the second law can be translated into a mathematical formulation that defines the entropy and its fundamental properties. In this section we analyze reversible processes; the complementary analysis of irreversible processes is deferred to the next section.

Consider a one-component, one-phase closed system. Let the states of this system be displayed on the T, V diagram shown in Fig. 16.1. Suppose that the state of the system is transformed from 1 to 2 by a reversible adiabatic process, and from 2 to 2' by a reversible isothermal process. We seek to close the path in the T, V plane by some process that transforms the state of the system from 2' to 1. *Can we find a reversible adiabatic path connecting 2' to 1?* We proceed by assuming that such a path exists and deducing a contradiction that, *a posteriori*, requires rejection of the assumption.

For any closed cycle in the T, V plane,

$$\Delta U_{cycle} = 0, \tag{16.2}$$

whereupon

$$w = -q. \tag{16.3}$$

In the particular cycle cited, the path $1 \rightarrow 2$ is a reversible adiabatic along which

$$q_{1 \rightarrow 2} = 0, \tag{16.4}$$

the path $2 \rightarrow 2'$ is a reversible isotherm along which the system expands so that

$$q_{2 \rightarrow 2'} > 0, \tag{16.5}$$

and the path $2' \rightarrow 1$ is assumed to be a reversible adiabatic for which

$$q_{2' \rightarrow 1} = 0. \tag{16.6}$$

Combining Eq. 16.3 with Eqs. 16.4, 16.5, and 16.6, we conclude that in the cycle described

$$w < 0. \tag{16.7}$$

Combination of Eqs. 16.3, 16.7, and 16.5 shows that the cycle described converts a quantity of heat drawn from a reservoir at one temperature into an equal amount of work,[2] which is a violation of the Kelvin statement of the second law. We conclude that the path connecting states 2' and 1 cannot be a reversible adiabatic.

The argument just given is valid for every isothermal path between 2 and 2'. Therefore it must be the case that only one reversible adiabatic path passes through each point of the T, V space. Put another way, reversible adiabatics in the T, V space cannot cross.

The relationship between the behavior of reversible adiabatics just described and the Caratheodory statement of the second law follows from a similar argument. Consider again a one-component, one-phase system. All states of the system that can be reached from state 1 by isothermal

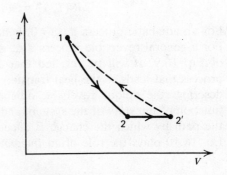

FIGURE 16.1

Existence of an entropy function for reversible processes: If $1 \rightarrow 2$ by a reversible adiabatic process, and $2 \rightarrow 2'$ by a reversible isothermal process, the path $2' \rightarrow 1$ cannot be a reversible adiabatic path (see text).

---

[1] This argument follows an analysis given by J. G. Kirkwood and I. Oppenheim.
[2] Recall that work done by the system is reckoned negative.

transformation lie on the isotherm labeled T_1 in Fig. 16.2; those that can be similarly reached starting from state 2 lie on the isotherm labeled T_2. Through (T_1, V_1), representing state 1 of the system, there can pass one and only one reversible adiabatic; it is shown intersecting the isotherm T_2 at (T_2, V_2). Other reversible adiabatics, each unique, pass through (T_1, V_1'), $(T_1, V_1''), \ldots$. By virtue of the uniqueness of the adiabatics, no adiabatic path through 1 can intersect isotherm T_2 at $2', 2'', \ldots$. Given that there exist unique adiabatics between $1'$ and $2'$, $1''$ and $2'', \ldots$, the points $2, 2', 2'', \ldots$ cannot coincide because reversible adiabatics cannot cross. Although all points on the isotherm T_2 lying to the left of point 2 can be reached via reversible adiabatic paths from points on isotherm T_1 lying to the left of point 1, all points in the T, V plane lying to the left of the reversible adiabatic through point 1 are inaccessible via adiabatic paths from point 1.

If the states of the system depend on more variables than just T and V, the preceding argument still holds for any plane section parallel to the T axis. We conclude that there is a region of space surrounding point 1 which contains states that cannot be reached from 1 along an adiabatic path. This statement will be recognized as equivalent to the Caratheodory formulation of the second law.

We now exploit the uniqueness of the reversible adiabatic through a selected point in the T, V plane in the following fashion. Because of this uniqueness every reversible adiabatic curve can be represented by a function $T = T(V)$ or $f(T, V) = $ constant. Let $S(T, V)$ be such a function, the entropy, which is constant everywhere along the reversible adiabatic through (T, V):

$S(T, V) = $ constant along the reversible adiabatic path through (T, V).

$$(16.8)$$

Clearly, then,

$dS(T, V) = 0$ along the reversible adiabatic path through (T, V).

$$(16.9)$$

The prescription given for the entropy $S(T, V)$ does not specify its detailed form, but its existence is ensured by the uniqueness of the reversible adiabatic. Moreover, because reversible adiabatics cannot cross, they must be monotonic functions of T. To proceed further, we write

$$dS(T, V) = \Theta(T, V)\, đq_{\text{rev}}. \qquad (16.10)$$

For an adiabatic process $đq_{\text{rev}} = 0$, hence Eq. 16.10 reduces to Eq. 16.9. For a general reversible process $đq_{\text{rev}} \neq 0$, and Eq. 16.10 is an extension of Eq. 16.9. It will be recalled that $đq_{\text{rev}}$ depends on the path of the process that leads to the heat transfer. On the other hand, since $S(T, V)$ describes the unique reversible adiabatic through (T, V), it must be a function of the state of the system, and $dS(T, V)$ must be independent of the path by which the change is effected. Then the function $\Theta(T, V)$ in Eq. 16.10 plays the role of an *integrating factor*.[3]

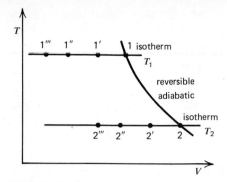

FIGURE 16.2
Diagram illustrating the behavior of reversible adiabatic paths in the T, V plane and the Caratheodory statement of the second law of thermodynamics.

---

[3] An integrating factor is defined as follows: Let $x_1, x_2, \ldots$ be the independent variables in the linear differential form $X_1\, dx_1 + X_2\, dx_2 + \cdots$. The function $\Theta(x_1, x_2, \ldots)$ is an integrating factor for this linear differential form if the product

$$\Theta(x_1, x_2, \ldots)X_1\, dx_1 + \Theta(x_1, x_2, \ldots)X_2\, dx_2 + \cdots = df(x_1, x_2, \ldots)$$

is the exact differential of some function $f(x_1, x_2, \ldots)$.

To determine $\Theta(T, V)$ we examine the behavior of the system shown in Fig. 16.3. A rigid thermally isolated container is divided internally into volumes V_1 and V_2 by a rigid diathermal barrier. At equilibrium the temperatures of the two compartments are the same. Suppose that an infinitesimal amount of heat $đq_{rev}$ is transferred reversibly from one compartment to the other. Then, because the total system is isolated,

$$đq_{1(rev)} = -đq_{2(rev)}, \qquad (16.11)$$

$$dS = 0 = dS_1 + dS_2, \qquad (16.12)$$

with Eq. 16.12 following from Eq. 16.9. By substitution of Eqs. 16.10 and 16.11 in Eq. 16.12, we find

$$[\Theta_1(T, V_1) - \Theta_2(T, V_2)] \, đq_{1(rev)} = 0. \qquad (16.13)$$

Since the argument given holds for arbitrary subdivision into V_1 and V_2, and $đq_{1(rev)} \neq 0$, we conclude that

$$\Theta_1(T, V_1) = \Theta_2(T, V_2) = \Theta(T). \qquad (16.14)$$

In words, the integrating factor Θ depends only on the temperature.

The preceding analysis is a special case of the class of reasoning used to establish the character of the equilibrium state (cf. Chapter 19). Here we have tested the equilibrium state by asking how the system responds when an amount of heat $đq_{rev}$ is transferred between subvolumes while maintaining the equality of temperature characteristic of the original equilibrium. Later we shall reverse the argument; that is, we shall regard Eq. 16.12 as the fundamental relation and deduce the conditions relating intensive variables to one another in the original equilibrium state.

We can take the evaluation of Θ one step further. Our analysis of the behavior of the system shown in Fig. 16.3 did not in any way depend on the nature of the substances in compartments 1 and 2. Accordingly, $\Theta(T)$ must be a universal function of the temperature whose value does not depend on the particular properties of any substance. It is legitimate, then, to choose some very simple substance, evaluate dS for a reversible heat transfer $đq_{rev}$, and so find $\Theta(T)$. The result will be generally valid because $\Theta(T)$ does not depend on the nature of the substance. To carry out this program we choose the ideal gas, for which

$$pV = nRT, \qquad (16.15)$$

$$\left(\frac{\partial U}{\partial V} \right)_T = 0. \qquad (16.16)$$

The first law expression for a change in internal energy,

$$dU = đq_{rev} - p \, dV = C_V \, dT, \qquad (16.17)$$

can be rewritten to read (for the ideal gas)

$$đq_{rev} = C_V \, dT + \frac{nRT}{V} \, dV. \qquad (16.18)$$

Using Eqs. 16.10 and 16.14 with Eq. 16.18 leads to

$$dS = \Theta(T) \, đq_{rev} = \Theta(T) C_V(T) \, dT + \Theta(T) \frac{nRT}{V} \, dV. \qquad (16.19)$$

But $S(T, V)$ is a function of state, and dS is an exact differential, hence

$$\left[\frac{\partial}{\partial V} \Theta(T) C_V(T) \right]_T = \left\{ \frac{\partial}{\partial T} \left[\Theta(T) \frac{nRT}{V} \right] \right\}_V. \qquad (16.20)$$

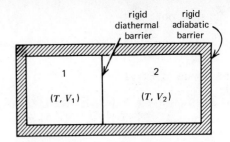

FIGURE 16.3

Schematic diagram illustrating the relationship between $\Theta(T, V)$ and reversible heat transfers (see text).

We now note that

$$\left[\frac{\partial}{\partial V}\Theta(T)C_V(T)\right]_T = \Theta(T)\left(\frac{\partial C_V}{\partial V}\right)_T = 0 \tag{16.21}$$

by virtue of Eq. 16.16. Therefore

$$\left\{\frac{\partial}{\partial T}\left[\Theta(T)\frac{nRT}{V}\right]\right\}_V = \frac{nR}{V}\frac{\partial}{\partial T}[T\Theta(T)]_V = 0 \tag{16.22}$$

by virtue of Eqs. 16.20 and 16.21. We conclude that

$$T\Theta(T) = \text{constant.} \tag{16.23}$$

Any function $\Theta(T)$ that satisfies Eq. 16.23 is an integrating factor.[4] We choose the simplest form, namely,

$$\Theta(T) = \frac{1}{T}. \tag{16.24}$$

This is a good point to pause and examine the implications of our study of $\Theta(T, V)$. One of the most important consequences of the second law is the introduction of a thermodynamic scale of temperature independent of the nature of any particular substance or class of substances. Until now, we have used the ideal gas temperature scale, but the fact that $\Theta(T)$, the integrating factor in Eq. 16.10, is independent of substance implies the existence of an absolute temperature scale. The demonstration that $T\Theta(T)$ is a constant, where T is measured on the ideal gas scale, identifies the absolute temperature with the ideal gas temperature to within a scale factor that, for convenience, we take to be unity. An examination of the arguments used shows that a key role is played by the fact that there is a unique reversible adiabatic through the point (T, V) representing the state of a system. It follows from the existence of this unique reversible adiabatic that $S(T, V)$ is a function of the state of the system and not of how that state was reached. Therefore we have estab-

---

[4] Equation 16.23 can also be deduced from Eq. 16.19 by considering the change of state $(T_1, V_1) \rightarrow (T_2, V_2)$ shown below.

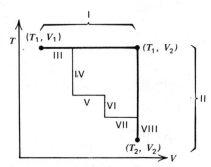

Since S is a state function, the integral of Eq. 16.19 between (T_1, V_1) and (T_2, V_2) is independent of the path. Therefore

$$\int_I \Theta(T)\frac{nRT}{V}dV + \int_{II}\Theta(T)C_V(T)\,dT$$

$$= \int_{III+V+VII}\Theta(T)\frac{nRT}{V}dV + \int_{IV+VI+VIII}\Theta(T)C_V(T)\,dT = \Delta S.$$

Now note that the integrands of the temperature integrals above are independent of the volume. The integral over path II must therefore be equal to the integral over IV+VI+VIII. Then, to maintain the same value of ΔS, it must be true that the volume integral over I equals the volume integral over III+V+VII. In other words, the integrands of the volume integrals must be independent of T. Equation 16.23 follows immediately.

lished that, in an arbitrary reversible process,

$$dS = \frac{\dslash q_{\text{rev}}}{T},\qquad (16.25a)$$

$$S_2 - S_1 = \int_1^2 \frac{\dslash q_{\text{rev}}}{T},\qquad (16.25b)$$

where T is the absolute temperature. For an arbitrary cyclic reversible process,

$$\Delta S_{\text{cycle}} = 0 = \oint \frac{\dslash q_{\text{rev}}}{T}.\qquad (16.26)$$

In general, to compute the entropy difference between two states of the system the integral 16.25b must be taken along a reversible path. It will be recalled that changes in energy are defined by measuring the work done in adiabatic processes. Similarly, changes in entropy are defined, in principle, by measuring the heats absorbed in reversible processes. Even though a reversible process is only the limiting behavior expected as the rate of a real process tends to zero, and therefore is difficult to achieve experimentally, the heat absorbed in a reversible process is readily computed if enough is known about the properties of the substance involved. Examples of calculations of this type will be given in Chapter 17.

We now take up the study of an important consequence of the second law. We shall show that for any change which can in fact take place in an isolated system the inequality

$$\Delta S > 0 \qquad (16.27)$$

must hold. Since it is always observed that a macroscopic system prepared in an arbitrary state approaches a state of equilibrium, and it is never observed that a macroscopic system spontaneously reverts to some original nonequilibrium state even if energy is conserved, expression 16.27 is a statement about the irreversibility of naturally occurring processes.

Before discussing the thermodynamic description of irreversible processes, it is necessary to examine briefly how irreversibility is accounted for by the kinetic molecular theory of matter. In fact, the observation of a ubiquitous trend toward an equilibrium state pertains only to systems containing a very large number of particles. This limiting condition is what we imply when we use the word macroscopic. If the system under examination contains only a few particles—for example, an atom—then time-symmetric (reversible) behavior is observed. In principle, since we believe that the equations of quantum mechanics describe all systems,[5] one might expect all systems to show time-reversible behavior. This is both a correct and a subtle observation. A complete description of a macroscopic system, including all particle coordinates, positional and velocity correlations and so on, would indeed have the mechanical property of time reversibility. But the description used for a macroscopic system is very different, especially in the use of only a few macroscopic nonmechanical properties, for example, the entropy, or the heat transferred. When the description of a system is contracted from the complete detail characteristic of mechanics to only those few properties characteristic of thermodynamics, it can be shown that asymmetric changes toward equilibrium are to be expected.

16.3
Irreversible Processes: The Second-Law Interpretation

[5] We omit discussion of relativistic effects and the creation and annihilation of particles.

One can easily grasp the basic idea in the description of the evolution of a mechanical system of many degrees of freedom from an arbitrary state to the equilibrium state. Suppose that an isolated mechanical system starts out with some distribution of velocities and positions of the particles that deviates from the distribution characteristic of equilibrium. By virtue of the motions of the molecules, many collisions occur. The trajectory of any one molecule includes collisions with other molecules, occasional recollisions with molecules previously encountered, and so forth. After only a few collisions per molecule the mechanical motion of all the molecules has become intertwined in a complicated way; for example, molecule 10 would not have hit molecule 7 at time t if 10 had not previously hit 6, which had not previously hit 2, and 7 had not previously hit 9, ... (see Fig. 16.4). Mechanical reversibility requires that the initial distributions of velocities and positions be re-created at some later time. One way of achieving this, but certainly not the only way, is for the system, after some instant, to retraverse all collisions in the reverse order to that in which they occurred. It is extremely unlikely that such a reversal will arise spontaneously in a system with 10^{24} particles. Nevertheless, a general theorem of mechanics, the Poincaré recurrence theorem, states that an isolated mechanical system will achieve a state with the individual positions and velocities of the particles arbitrarily close to what they were in the initial state, if only enough time is allowed to pass. It is readily shown that the time scale on which this recurrence appears, even for systems that are very small (for example, a chain of 10 oscillators), is much, much longer than the estimated lifetime of the universe. The theory shows that long before any recurrence occurs the system decays from the initial state, whatever that may be, to a state of equilibrium that is independent of the initial state, and dependent only on the temperature, volume, and other macroscopic state variables. In this intermediate time period (which also persists much much longer than the age of the universe), the thermodynamic equilibrium state is the end of the evolution because virtually all possible states of the system are close to the equilibrium state. That this last statement is true was shown in Chapter 15.

A more detailed description of the relationship between the nature of the observations and reversibility will be found in Appendix 16A.

All of the preceding has referred to the properties of an isolated mechanical system. In the real world no system is isolated; even bodies in interstellar space are bombarded by photons, cosmic rays, neutrinos, and other particles. The Poincaré recurrence theorem does not hold for a system that is not isolated. Clearly, if the lack of isolation occurs via random interactions with the surroundings, the delicate correlations in collision chains can be destroyed. If the perturbation comes from outside the system, which is what a violation of isolation means, then mechanical reversibility within the system cannot be attained. In this case, also, an equilibrium state is achieved in due time.

The quantification of the preceding arguments is still a matter of much study. The theory is by no means complete, and algorithms for calculating the rate of approach to equilibrium are not yet well developed.

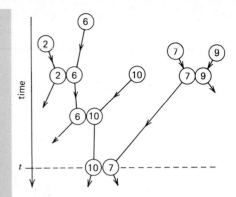

FIGURE 16.4

Schematic representation of a hypothetical correlated sequence of collisions. For convenience all collisions are represented as occurring in one plane, although in a real gas this is not the case. The directions shown are suggestive—the actual angles and distances along each leg of the trajectory are obtained from simultaneous satisfaction of the conditions of conservation of energy, linear momentum, and angular momentum in each binary collision.

The macroscopic description of irreversible processes is based on relation 16.27 and a few phenomenological laws such as Fourier's law of heat conduction and Fick's law of diffusion. The latter laws, which we shall discuss in Chapter 20, describe the temporal evolution toward equilibrium of the properties of a system not at equilibrium. In contrast, the inequality 16.27, which we discuss now, describes the overall consequences of an irreversible process in terms of the differences between some initial state and the final equilibrium state.

Since inequality 16.27 refers to a change that can take place in an isolated system, we must prove that $\Delta S > 0$ for an irreversible adiabatic process. Consider again Fig. 16.1. Recall that $1 \rightarrow 2$ is a reversible adiabatic path and $2 \rightarrow 2'$ is a reversible isothermal path. We showed that $2' \rightarrow 1$ could not be a reversible adiabatic path. We now investigate whether or not $2' \to 1$ can be an irreversible adiabatic path. There are two possibilities:

1. The path $2 \to 2'$ is a reversible isothermal expansion (the case shown in Fig. 16.1). If the path $2' \to 1$ is irreversible and adiabatic, heat is transferred only along the path between 2 and $2'$. We have, as before, $\Delta U = 0$, $q_{2 \to 2'} = -w$, $q_{2 \to 2'} > 0$, and $w < 0$, implying extraction of heat from a reservoir at one temperature and performance of an equal amount of work, which violates the Kelvin statement of the second law. Therefore there can be no adiabatic path, reversible or irreversible, connecting $2'$ and 1 in this case.

2. The path $2 \to 2'$ is a reversible isothermal contraction (see Fig. 16.5). In this case $q_{2 \to 2'} < 0$ and an irreversible adiabatic connecting $2'$ and 1 does exist. One possible irreversible adiabatic path from $2'$ to 1 is shown in Fig. 16.5. Imagine that $2' \to 2''$ is a reversible adiabatic compression that is complete when the volume of the system is equal to that in state 1, V_1. Then adiabatic work at constant volume, for example, rotating a paddle wheel inside the system, is used to traverse $2'' \to 1$, the amount of work done being just enough to achieve the final temperature desired, namely, T_1. In the case that $V_{2'} < V_1$, the path $2' \to 2''$ can be an adiabatic expansion to V_1 followed by constant-volume adiabatic work sufficient to raise the system's temperature to T_1. Consequently, in the case that $2 \to 2'$ is a compression, there does exist an irreversible adiabatic path connecting $2'$ to 1.

Since the entropy depends only on the state of the system, the difference

$$\Delta S = S_1 - S_{2'} \tag{16.28}$$

is defined even when there is no reversible adiabatic path between the two states. We now rewrite Eq. 16.28 to read

$$\Delta S = (S_1 - S_2) + (S_2 - S_{2'}). \tag{16.29}$$

But

$$S_1 - S_2 = 0 \tag{16.30}$$

because the path $1 \to 2$ is a reversible adiabatic. Then, since

$$S_2 - S_{2'} = \frac{q_{2' \to 2}}{T} > 0, \tag{16.31}$$

we have

$$\Delta S > 0 \tag{16.32}$$

for the irreversible process connecting states $2'$ and 1.

The inequality 16.27 is sufficiently important to warrant another, and more general, demonstration. This time we represent the states of the system on a T, S diagram (see Fig. 16.6). A reversible isotherm is a horizontal line, a reversible adiabatic is a vertical line, and the isotherms and adiabatics meet at right angles. we have already established that the reversible adiabatic through a point representing the state of the system is unique, and that two reversible adiabatics cannot cross. We also know that there is not any adiabatic path connecting state 1 to any state lying to the left of the reversible adiabatic through 1; that is, there is no adiabatic path connecting state 1 to any state with smaller entropy. In order for a

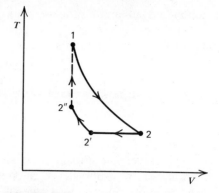

FIGURE 16.5
Schematic diagram illustrating that $\Delta S > 0$ for an irreversible adiabatic process. The path $1 \to 2$ is a reversible adiabatic, $2 \to 2'$ a reversible isothermal compression, $2' \to 2''$ a reversible adiabatic compression, and $2'' \to 1$ involves constant-volume adiabatic work done on the system (see text).

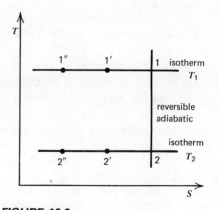

FIGURE 16.6
Diagram demonstrating that $\Delta S > 0$ for an irreversible process (see text).

spontaneous adiabatic process connecting two states to occur, the representative point of the final state on the T, S diagram must lie to the right of that for the initial state, or

$$S_{\text{final}} - S_{\text{initial}} \equiv \Delta S > 0. \tag{16.33}$$

Finally, we now demonstrate that

$$\Delta S > \int \frac{đq}{T_{\text{surroundings}}} \tag{16.34}$$

for any irreversible process. Consider an isolated compound system consisting of two subsystems. Suppose that some spontaneous process occurs in subsystem A, by virtue of which it undergoes an irreversible change of state. The corresponding change in system B, which we regard as the surroundings of system A, will be assumed to be reversible.[6] An example of such a compound system is shown in Fig. 16.7. Now, since the compound system undergoes a spontaneous process, we have

$$\Delta S_{AB} = \Delta S_A + \Delta S_B > 0, \tag{16.35}$$

which implies that

$$\Delta S_A > -\Delta S_B. \tag{16.36}$$

But because the process in part B is reversible, we can write

$$\Delta S_B = \int \frac{đq_B}{T_B}, \tag{16.37}$$

which leads directly to

$$\Delta S_A > -\int \frac{đq_B}{T_B}. \tag{16.38}$$

For the compound isolated system,

$$đq = 0 = đq_A + đq_B, \tag{16.39}$$

so that inequality 16.38 can be rewritten

$$\Delta S_A > \int \frac{đq_A}{T_B}, \tag{16.40}$$

and we have proved inequality 16.34.

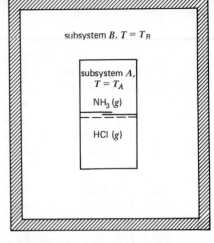

FIGURE 16.7

Schematic diagram of a compound isolated system that can undergo a spontaneous process in one of its component parts. The outer envelope of B is a rigid adiabatic barrier, whereas A is contained by a rigid diathermal envelope. A is also divided internally by a permeable membrane. A slide covers the permeable membrane so that the two sides of A can be loaded. The slide can be opened without expenditure of work. In the initial state, $NH_3(g)$ is put into one side of A and $HCl(g)$ into the other. Once the slide is opened, the membrane permits passage of both gases, so in the final state the NH_3 and HCl have reacted to form NH_4Cl. Some of the heat released in this reaction is transferred through the diathermal envelope to B.

In the figure: subsystem B, $T = T_B$; subsystem A, $T = T_A$; NH_3 (g); HCl (g).

We have shown how the physical statements of the second law given by Clausius and Kelvin lead to the introduction of a new function of state, the entropy, and to a discrimination between reversible and irreversible processes. It is now worthwhile to reverse the argument and use the mathematical formulation of the second law to illuminate the consequences of the physical statements. We shall do this by examining the conversion of heat to work for a system carried through the cyclic processes described in the Clausius and Kelvin statements.

16.4
The Clausius and Kelvin Statements Revisited

---

[6] The assumption that changes in system B are reversible imposes some restrictions. It is usual to require that B be very much larger than A. It is more important to require that B be sufficiently "well stirred" that no local gradients of temperature, pressure, or composition can develop. The stirring, which can be thought of as mechanical but need not be, homogenizes system B very rapidly on the time scale for changes in system A, hence so far as A is concerned system B is always in a state of internal equilibrium.

Consider, first, the Kelvin statement of the second law. The cyclic process referred to draws heat from a reservoir at one temperature, say, T_0. For any cyclic process $\Delta S = 0$, so application of inequality 16.40 leads immediately to

$$0 \geqslant \oint \frac{dq}{T_0}. \tag{16.41}$$

In this cyclic process we also must have $\Delta U = 0$, hence $q = -w$. When the system does work on the surroundings, $w < 0$, requiring that $q > 0$. Since T_0 is positive, inequality 16.41 requires that $q < 0$. We conclude that it is impossible to operate an engine in a cycle so as to produce no effect other than the extraction of heat from a reservoir at one temperature and the performance of an equal amount of work.

Consider, now, the Clausius statement of the second law. We have established that for any process, reversible or irreversible,

$$dS \geqslant \frac{dq}{T}, \tag{16.42}$$

where T is the absolute temperature of the surroundings. The equality sign holds in relation 16.42 only for the case of a reversible process, in which case the temperatures of the system and surroundings are the same. Clearly, for the same infinitesimal entropy change,

$$\frac{dq_{\text{rev}}}{T} \geqslant \frac{dq}{T} \tag{16.43}$$

which, when combined with the first law,

$$dU = dq + dw, \tag{16.44}$$

leads to the conclusion

$$dw_{\text{rev}} \leqslant dw. \tag{16.45}$$

In words, the maximum work extractable from any process is the reversible work.[7] For this reason we proceed by examining the simplest possible reversible cycle for conversion of heat into work. In the design of this cycle it is, of course, necessary to take account of the restrictions imposed by the second law. The Kelvin statement of the second law prohibits continuous conversion of heat into work if the heat is drawn from a source at a single temperature; hence the simplest cycle must use at least two heat sources, say, one at temperature T_1 and the other at T_2 with $T_2 > T_1$. The system undergoing the cycle will then execute two reversible isothermal processes, one at T_1 and the other at T_2. To complete the cycle the two isotherms must be connected by paths along which heat is not transferred, that is, by reversible adiabatics. The combination of isothermal and adiabatic processes just described is known as a (reversible) *Carnot cycle*.

A realization of a Carnot cycle is shown in Fig. 16.8; the working substance need not be specified at this state of the analysis, although the diagram implies that it is a fluid. We start in state A, with the system at equilibrium in contact with a thermostat at temperature T_2. The steps in the Carnot cycle displayed are as follows:

1. $A \to B$: The system undergoes a reversible isothermal expansion during which an amount of heat q_2 is absorbed and an amount of work w_2 is performed on the system, all at temperature T_2.

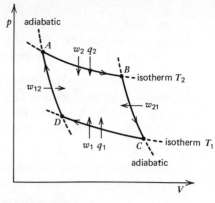

FIGURE 16.8
Representation of a Carnot cycle for a simple fluid as working substance.

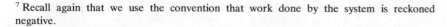

[7] Recall again that we use the convention that work done by the system is reckoned negative.

2. $B \to C$: The system undergoes a reversible adiabatic expansion, the final temperature being $T_1 < T_2$. In carrying out this process the system must be isolated from any thermostat, so that no heat is absorbed or released, but work w_{21} is done on the system.

3. $C \to D$: The system is brought into thermal contact with a reservoir at temperature T_1 and a reversible isothermal compression is carried out. In this process an amount of heat q_1 is absorbed and an amount of work w_1 is done on the system.

4. $D \to A$: The system is removed from contact with the thermostat and, by adiabatic compression, restored to state A. In this last reversible process, work w_{12} is performed on the system.

Note that we have not made reference to the actual properties of any particular substance. In the simplest possible case the substance would be a fluid with equation of state defined by the independent variables p and V. Each of the reversible transformations defined is then a line in the p, V space. In more general cases the number of independent variables required to define the state of the system is greater than two, and then the isothermal and adiabatic processes described define multidimensional surfaces in the appropriate space.

What are the characteristics of the Carnot cycle we have described? The first important deduction is that the ratio q_2/q_1 is independent of the properties of the substance undergoing the cycle, and therefore is the same for all substances undergoing a Carnot cycle between the temperatures T_1 and T_2. To demonstrate this we note that, by the first law, the change in internal energy for the entire cycle must be zero,

$$\Delta U = q_2 + q_1 + w_2 + w_{21} + w_1 + w_{12} = 0, \tag{16.46}$$

so that

$$w \equiv w_2 + w_{21} + w_1 + w_{12} = -q_2 - q_1. \tag{16.47}$$

But since the cyclic process is reversible and heat is transferred only along the isothermal legs, the entropy changes along those legs are

$$\Delta S_2 = \frac{q_2}{T_2}, \tag{16.48}$$

$$\Delta S_1 = \frac{q_1}{T_1}. \tag{16.49}$$

For one complete cycle

$$\Delta S = 0 = \frac{q_2}{T_2} + \frac{q_1}{T_1}, \tag{16.50}$$

or

$$-\frac{q_2}{q_1} = \frac{T_2}{T_1}. \tag{16.51}$$

Thus the ratio q_2/q_1 is independent of the nature of the working substance.

The connection of the preceding with the Clausius statement of the second law is made as follows. Since $T_2 > T_1$, Eq. 16.50 implies that

$$|q_2| > |q_1|. \tag{16.52}$$

But, from Eq. 16.47, if heat is absorbed at the lower temperature T_1, then $q_1 > 0$, $-q_2 - q_1 > 0$, hence $w > 0$, and work must be done by the surroundings on the system. In contrast, if heat is absorbed at the higher temperature T_2, then $q_2 > 0$, $-q_2 - q_1 < 0$, hence $w < 0$, and work is done by the system on the surroundings. We conclude that no cycle can be devised that produces no effect other than the transfer of heat from a colder to a hotter body.

16.5
The Second Law as an Inequality

It is now convenient to again examine that part of the second law which takes the form of an inequality. In Chapter 15 we argued that the entropy of an isolated system is a maximum subject to the constraints that define its state. To understand better the nature of the inequality 16.27, we must consider how one can carry out a process inverse to one that occurs in an isolated system. We define an *inverse process* as one that, by virtue of heat and work transfers to the surroundings, restores a state of the system from which the initial state could have been reached via a natural process in an isolated system. Since any process that does occur in an isolated system must involve no change in the internal energy, we also require that an inverse process involve no change in the internal energy. Note that, in order to carry out the process inverse to one that occurs in an isolated system, we must break the isolation of that system. Then, by doing work and transferring heat in amounts such that the internal energy of the system is maintained constant, we can restore the system to the state it had while isolated and before alteration of the parameters of constraint.

For example, consider the system represented schematically in Fig. 16.9. An isolated box, divided into two equal subvolumes V_1, has perfect gas in one and vacuum in the other. When the diaphragm separating the subvolumes is removed without expenditure of work, the gas spontaneously fills the entire volume $2V_1$. The natural process following removal of the diaphragm is just the irreversible expansion from a volume V_1 to a volume $2V_1$. The inverse process requires compression of the gas from the volume $2V_1$ to the volume V_1, and restoration of the partition. The internal energy of the gas when in V_1 must be the same as when it fills the

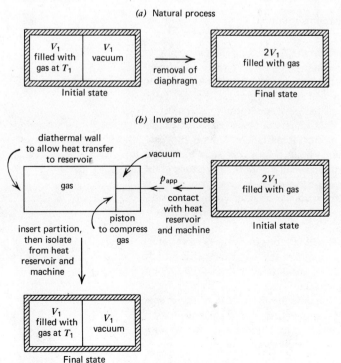

(a) Natural process

(b) Inverse process

FIGURE 16.9
Diagram illustrating the nature of an inverse process.

volume $2V_1$. Since the internal energy of a perfect gas depends only on the temperature, to accomplish the inverse process we couple the system to some machine and reservoir that will compress the perfect gas isothermally. The reversible work required to change the volume from $2V_1$ to V_1 under these conditions is

$$w = -\int_{2V_1}^{V_1} p_{app}\, dV = -\int_{2V_1}^{V_1} \frac{nRT}{V}\, dV = nRT \ln 2. \qquad (16.53)$$

But the internal energy of the gas is to be constant throughout this inverse process, hence the first law requires that the system transfer heat to the surroundings in the amount $nRT \ln 2$. Then, using Eq. 16.25b for an isothermal process, for which T may be removed from the integrand, we find

$$\Delta S_{inv} = \frac{q}{T} = -nR \ln 2 < 0. \qquad (16.54)$$

Thus, in the inverse process described, the entropy of the system is decreased.

It is found to be generally true that, if an isolated system is at equilibrium with respect to some set of constraints, any process inverse to one that could have been involved in reaching that equilibrium state is characterized by

$$\Delta S_{inv} < 0. \qquad (16.55)$$

In terms of the mathematical criteria for a maximum of a function, we state that:

If an isolated system is at equilibrium, then for any infinitesimal variation which is the inverse of a process that could have occurred in an isolated system and during which the original parameters of constraint are maintained, it is necessary that

$$\delta S = 0 \qquad \text{(original constraints retained)}. \qquad (16.56)$$

This statement is to be interpreted as follows. In the original equilibrium state of the isolated system, the entropy was a maximum with respect to the pertinent parameters of constraint. Imagine changing the parameters of constraint so as to create a new equilibrium state, one from which the original state could be obtained in a natural process. The entropy of the new equilibrium state must be smaller than the entropy of the original equilibrium state, since the latter can be obtained from the former in an isolated system (definition of natural process). Suppose that the parameters of constraint are changed in a way such that every state along the path used is displaced only infinitesimally from the preceding state. Evidently the entropy must decrease, because the new state has lower entropy than the original state. But because the entropy in the original equilibrium state was a maximum with respect to its parameters of constraint, the initial changes in the variables of the system start with zero slope, that is, are connected by the differential equation 16.56. When δS is expressed in terms of changes in variables such as T or V, the coefficients of the differentials in this equation refer only to the initial state and not to the final state. The variation that carries the system from one equilibrium state to another equilibrium state derived from it by an inverse process is called a *virtual variation*. Since the precise nature of the change in parameters of constraint is never invoked (remember that the coefficients in 16.56 refer to the initial state), the changes imposed in defining a virtual variation may be very general, provided only that a true equilibrium state can be defined for the final set of constraints.

We now examine, briefly, some relationships between the microscopic and macroscopic theories of the entropy.

First, we observe that the thermodynamic temperature is closely related to, and may be considered to measure a fundamental property of, the density of states of the system. Using Eq. 15.42 as the microscopic definition of the entropy, and making the identification

$$\beta = \frac{1}{k_B T},$$ (16.57)

where T is the thermodynamic (absolute) temperature, we see that the definition

$$\beta \equiv \frac{\partial \ln \Omega}{\partial E}$$ (16.58)

implies that T is positive. For we have noted that Ω is an increasing function of the energy, so that $(\partial \ln \Omega / \partial E)$ must be positive, and hence $T > 0$.[8] Thus, the macroscopic assignment of the sign of the temperature scale such that the dissipation of energy in an isolated system leads to an increase in temperature is fundamentally related to the fact that Ω increases as E increases. If the opposite macroscopic convention had been adopted, the relation between β and $(\partial \ln \Omega / \partial E)$ would require a minus sign to preserve agreement between the two descriptions. Either convention is internally consistent. What is important is to notice that the direction of change of temperature due to dissipation of energy is connected with the direction in which the number of states of the system increases as a function of the energy.

A second point of interest is that the condition that entropy must increase in a natural process is equivalent to the requirement that heat flow from the system with the greater absolute temperature to the system

---

16.6
Some Relationships Between the Microscopic and Macroscopic Theories

---

[8] The statement is true when all degrees of freedom of the system are accounted for. In some situations it is convenient to focus attention on only one or a few degrees of freedom, and to neglect all others. For example, one might want to ascribe thermodynamic properties to an assembly of atoms, neglecting all states of the system except those associated with two levels, say, the ground level and one excited level of each atom. Alternatively, one might want to ascribe thermodynamic properties to an assembly of atoms, neglecting all states of the system except those associated with the nuclear spin. Suppose that the occupation of the set of states singled out for attention is restricted by the Pauli exclusion principle. Furthermore, suppose that the set of states of interest is effectively decoupled from all other states of the system so that they comprise (nearly) an isolated system. The kind of system described differs from an ordinary one in that there is an upper limit to its possible energy, which occurs when all of the atoms (or spins) are in the highest quantum state. When all of the atoms are in their ground state, or in the highest quantum state, the number of quantum states of the whole system with energy between E and $E + dE$ is very small; e.g., if the ground state is nondegenerate, there is only one state of the system with each atom in its ground state. On the other hand, when half the atoms are excited there are many quantum states with energy between E and $E + dE$ differing only in the assignment of which atoms are excited. Thus $\Omega(E)$ increases as E increases from the ground-state energy, decreases as E approaches the upper bound to the energy of the system (all nuclear spins aligned, all atoms excited, etc.), and therefore has a maximum somewhere in between. In these cases it is possible to define negative "isolated-degrees-of-freedom" temperatures. Such behavior occurs in lasers and in nuclear magnetic resonance experiments.

with the lesser absolute temperature. Using the condition[9]

$$\frac{\partial \ln \Omega_A(E_A{}^i)}{\partial E}(E_A{}^f - E_A{}^i) + \frac{\partial \ln \Omega_B(E_B{}^i)}{\partial E}(E_B{}^f - E_B{}^i) \geqslant 0, \quad (16.59)$$

derived from Eq. 15.42 and the second law, we find that

$$(\beta_A{}^i - \beta_B{}^i)q \geqslant 0, \qquad (16.60)$$

so that, if $q > 0$,

$$\beta_A{}^i \geqslant \beta_B{}^i \qquad (16.61)$$

or

$$T_A{}^i \leqslant T_B{}^i, \qquad (16.62)$$

since $T > 0$. We conclude, as stated, that positive heat is always absorbed by the system with lower temperature, and this consequence of the properties of entropy is also related to the fact that Ω increases as E increases.

Finally, we note that the ensemble probability distribution is very sharply peaked near the mean value of the energy. Indeed, it has been shown in connection with Eq. 15.56 that the fractional width of the distribution in energy is proportional to the inverse square root of the number of degrees of freedom of the system. This fractional width is, therefore, so small for a macroscopic system that one nearly always observes the mean value. Although fluctuations (i.e., states of nonmaximal probability) are a characteristic of the state of equilibrium, a completely macroscopic approach that neglects the existence of states of the system other than the most probable state is satisfactory for most problems. There are some properties of matter, for example, the scattering of light by liquids and gases, that are determined by the fluctuations, and to describe these phenomena states other than the most probable state must be included in the description of the processes involved. The nonzero width of the ensemble probability distribution can also be used to prove

---

[9] If two systems, initially with temperatures $T_A{}^i$ and $T_B{}^i$, are brought into thermal contact through a diathermal wall, there is a transfer of heat, and at thermal equilibrium $T_A{}^f = T_B{}^f$. The second law requires that the entropy change for this process be positive,

$$\Delta S_A + \Delta S_B \geqslant 0.$$

Suppose that $T_A{}^i$ and $T_B{}^i$ are only slightly different, and use Eqs. 15.62 and 15.63 to write, for the heats transferred,

$$q_A = E_A{}^f - E_A{}^i; \qquad q_B = E_B{}^f - E_B{}^i.$$

Then, using the fundamental definition 15.42 and the Taylor series expansions

$$\ln \Omega_A(E_A{}^f) = \ln \Omega_A(E_A{}^i) + \frac{\partial \ln \Omega_A(E_A{}^i)}{\partial E}(E_A{}^f - E_A{}^i) + \cdots,$$

$$\ln \Omega_B(E_B{}^f) = \ln \Omega_B(E_B{}^i) + \frac{\partial \ln \Omega_B(E_B{}^i)}{\partial E}(E_B{}^f - E_B{}^i) + \cdots,$$

we obtain the expressions

$$\Delta S_A = k_B \ln\left[\frac{\Omega_A(E_A{}^f)}{\Omega_A(E_A{}^i)}\right],$$

$$\Delta S_B = k_B \ln\left[\frac{\Omega_B(E_B{}^f)}{\Omega_B(E_B{}^i)}\right]$$

for the entropy changes; when added, these give Eq. 16.59 of the text.

that $(\partial T/\partial E) > 0$,[10] which is a stronger condition than $T > 0$, and is an example of a thermodynamic stability condition; that is, no substance can exist for which $(\partial T/\partial E)$ is ever negative or zero.

Because of the importance of entropy in the description of the properties of matter, it is advantageous to gain familiarity with the nature of the entropy changes in various kinds of processes, the magnitudes of the entropy changes, and so on. It is these matters that we shall consider in Chapter 17.

---

[10] Since $\beta \equiv \partial \ln \Omega/\partial E$, and $\beta = 1/k_B T$, we have

$$\frac{\partial \beta}{\partial E} = -\frac{1}{k_B T^2}\frac{\partial T}{\partial E} = \frac{\partial^2 \ln \Omega}{\partial E^2}.$$

But we have shown in Eq. 15.54 that $\partial^2 \ln \Omega/\partial E^2$ is always negative. Then, because T^2 is always positive, it must be true that $(\partial T/\partial E) > 0$.

Appendix 16A

Poincaré Recurrence Times and Irreversibility

In 1896 Poincaré proved the theorem: "In a system of material particles under the influence of forces that depend only on the spatial coordinates, a given initial state (i.e., a representative point in phase space) must, in general, recur, not exactly, but to any desired degree of accuracy, infinitely often, provided that the system always remains in the same finite part of the phase space." The time between two consecutive repetitions is called a *Poincaré recurrence time*. Clearly, the length of a recurrence time depends on how precisely the recurrence condition is specified. A simple but crude estimate can be obtained in the following way. Consider a dilute gas in which each molecule undergoes about 10^{10} collisions per second (Chapter 28). If the number density is 10^{24} molecules per cubic meter, the number of collisions per second in a sample of $1\,\text{cm}^3$ volume is $Z_T = 10^{28}$. If we assume that each collision induces a transition between two states of the system, and note that the original state need not recur until the system has passed through all other states, then the average recurrence time, Θ, is given by

$$\Theta \approx \frac{\Omega}{Z_T}, \qquad (16A.1)$$

where Ω is the total number of states with energy between E and $E + dE$. Ω is easily estimated from the entropy: $\Omega = \exp(S/k_B) \approx 10^{10^{18}}$. Thus,

$$\Theta \approx \frac{10^{10^{18}}}{10^{28}}$$

$$\approx 10^{10^{18}}\,\text{s}. \qquad (16A.2)$$

This is to be compared with the "age" of the universe, which is of order $10^{17}\,\text{s}$. We see immediately that the recurrence time is greater than this by about 10^{18} orders of magnitude.

Boltzmann made a much more detailed calculation by estimating the volume of phase space available to the system, and dividing it into small cells whose dimensions are of the order of magnitude to which a recurrence is specified. In the system discussed in the previous paragraph, the average separation of neighboring molecules is $10^{-8}\,\text{m}$. Boltzmann supposed that the initial state is one in which all the molecules move with velocities of $500\,\text{m/s}$, and that the initial state is reproduced if the same molecules are in the same positions to within $10^{-9}\,\text{m}$ (10% of the separation) with the same velocities to within $1\,\text{m/s}$. Since the gas is dilute, recurrence of the initial state can be discussed by considering the recurrence of velocity and configurational distributions separately. The volume of the velocity space may be calculated in the following way: The subspaces for each molecular velocity are considered in succession and integrations are carried out over each subspace subject to the condition that the total kinetic energy is fixed. Thus, the first molecule may have a speed v_1 anywhere in the range from zero to $a = 500 \times 10^9\,\text{m/s}$,[11] the second anywhere in the range from zero to $(a^2 - v_1^2)^{1/2}$, and so on. Since the length of a side of a cell in the velocity space is $1\,\text{m/s}$, it may be

[11] This value of a exceeds the velocity of light, but we ignore all relativistic effects.

treated as a dimensionless number, and the calculated volume of the velocity space will be given as the number of cells Γ_v. We have, from the preceding,

$$\Gamma_v = (4\pi)^{n-1} \int_0^a dv_1 v_1^2 \int_0^{(a^2-v_1^2)^{1/2}} dv_2 v_2^2 \cdots \int_0^{(a^2-v_1^2-v_2^2-\cdots-v_{n-2}^2)^{1/2}} dv_{n-1} v_{n-1}^2$$

$$= \left\{ \frac{\pi^{(3n-3)/2}}{2 \cdot 3 \cdot 4 \cdot \cdots \cdot [3(n-1)/2]} \right\} a^{3(n-1)}; \qquad n \text{ odd},$$

$$\doteq \left\{ \frac{2(2\pi)^{(3n-4)/2}}{3 \cdot 5 \cdot 7 \cdot \cdots \cdot [3(n-1)/2]} \right\} a^{3(n-1)}; \qquad n \text{ even}. \tag{16A.3}$$

The number of ways of arranging n labeled points in m cells is $[m!/(m-n)!]$, so the number of configurational cells, Ω_{conf}, is

$$\Omega_{\text{conf}} = \frac{m!}{(m-n)!}. \tag{16A.4}$$

If the system passes through all possible states before returning to its initial state, the recurrence time is

$$\Theta = \frac{\Gamma_v \Omega_{\text{conf}}}{Z_T}. \tag{16A.5}$$

The value of Θ is easily estimated for the very large numbers in the example. One finds that[12]

$$\Gamma_v \approx 10^{9n}, \qquad \Omega_{\text{conf}} \approx m^n,$$

so that with $a = 5 \times 10^{11}$ m/s, $n = 10^{18}$, and $m = 10^{21}$ (for 1 cm³),

$$\Theta \approx 10^{10^{19}} \text{ s}, \tag{16A.6}$$

which, although enormously different from 16A.2, nevertheless bears a similar relation to the age of the universe.

The thermodynamic meaning of irreversibility is different from that employed in the study of Poincaré recurrences, because we observe a system only in a gross way. The thermodynamic state of a system corresponds to a large number of trajectories in phase space, as mentioned in Section 16.3. During a Poincaré cycle the trajectory of the point in phase space will pass more or less close to every point in the space, depending on how precisely we define the recurrence, so that a given thermodynamic state will in general recur many times, depending on what fraction of the phase space is occupied by the points corresponding to that state. In general, then, even large fluctuations from a thermodynamic norm will recur, and it is with the frequency of these events that a discussion of irreversibility is concerned.

The problem of fluctuations and their recurrence was analyzed by Smoluchowski[13] in terms of what he called the *probability aftereffect*: Given the state α of a system at some instant, what can we say about α at a time τ later? Consider, for definiteness, that the state of the system is specified by the instantaneous number of particles, ν, in a volume ω. Obviously, when τ is much shorter than the period of the fluctuations of ν, we shall expect to find ν almost unchanged, whereas for very large τ we shall expect the value of ν to be independent of the initial value. For intermediate times we must take into account the speed of the fluctuations explicitly. An example considered by Smoluchowski was the number of colloid particles in a small volume defined within a much larger volume. This number was observed at successive instants at intervals of time τ. In a typical sequence of counts the average number was ≈ 2.0, and

[12]
$$\frac{m!}{(m-n)!} \approx \frac{m^m e^{-m}}{(m-n)^{m-n} e^{-(m-n)}} \approx m^n \qquad \text{for } m \gg n,$$

$$\left\{ \frac{\pi^{(3n-3/2)}}{2 \cdot 3 \cdot 4 \cdot \cdots \cdot [(3n-1)/2]} \right\} a^{3n-3} \approx \frac{\pi^{3n/2} a^{3n}}{(3n/2)!} \approx \left[\left(\frac{2e}{n} \right)^{1/2} a \right]^{3n} \approx (10^3)^{3n}.$$

[13] M. V. Smoluchowski, *Physik. Z.* **17**, 557, 585 (1916).

frequently as many as 5 particles were found. Such sequences, containing frequent large fluctuations from the mean, show none of the expected behavior of an irreversible process, and yet we shall see that they are compatible with the macroscopic irreversible law of diffusion (see Chapter 20).

Consider a system of colloid particles in equilibrium.[14] If the concentration is not too high, the probability that a particle will enter or leave the volume is not influenced by the number in the volume or near it outside. The probability distribution for the number ν will then be a Poisson distribution

$$P(\nu) = \frac{e^{-\gamma} \gamma^{\nu}}{\nu!}, \qquad (16A.7)$$

with

$$\langle \nu \rangle = \gamma \qquad (16A.8)$$

and

$$\langle \nu^2 \rangle = \gamma^2 + \gamma. \qquad (16A.9)$$

This distribution can be regarded as the limit of a binomial distribution as the total number of particles N in the whole system, and its volume V, tend to infinity while (N/V) remains constant: The probability that any particular particle is to be found in the observed volume ω is (ω/V) if all points in V are equiprobable. Thus, the probability that any ν are found in V is

$$P^{(N)}(\nu) = \frac{N!}{\nu!(N-\nu)!} \left(\frac{\omega}{V}\right)^{\nu} \left(1 - \frac{\omega}{V}\right)^{N-\nu}. \qquad (16A.10)$$

We now let $N, V \to \infty$ while $N/V = n$, the average density. Then Eq. 16A.10 becomes

$$P^{(N)}(\nu) = \lim_{N \to \infty} \frac{N!}{\nu!(N-\nu)!} \left(\frac{\gamma}{N}\right)^{\nu} \left(1 - \frac{\gamma}{N}\right)^{N-\nu},$$

where $\gamma = n\omega$. Noting that

$$\frac{N!}{\nu!(N-\nu)!} \to N^{\nu}, \qquad \left(1 - \frac{\gamma}{N}\right)^{N-\nu} \to e^{-\gamma}$$

for large N, we see that

$$\lim_{N \to \infty} P^{(N)}(\nu) = P(\nu). \qquad (16A.11)$$

To discuss the speed of fluctuations, we must calculate the probability that m particles will be observed within ω a time τ after ν were found [called the *transition probability* and denoted $K(\nu|m)$]. We also define the probability aftereffect factor p that a particle initially in ω will have left it during τ. In the present example, individual particles are assumed to obey the macroscopic law of diffusion. The probability that a particle, initially at $\mathbf{r}_1$, will have diffused to $\mathbf{r}_2$ in a time τ is (see Chapter 20)

$$(2\pi D\tau)^{-3/2} \exp\left[-\frac{(\mathbf{r}_1 - \mathbf{r}_2)^2}{4D\tau}\right],$$

where D is the diffusion coefficient. The value of p is found by integrating this probability over values of $\mathbf{r}_1$ inside ω and values of $\mathbf{r}_2$ outside ω. This definition of p again requires both the assumptions made in deriving $P(\nu)$.

Now, let $A_i^{(\nu)}$ denote the probability that if ν particles are initially observed, some i will have left ω after a time τ. $A_i^{(\nu)}$ is the binomial distribution

$$A_i^{(\nu)} = \frac{\nu!}{i!(\nu-i)!} p^i (1-p)^{\nu-i}. \qquad (16A.12)$$

---

[14] We discuss colloid particles as a surrogate for molecules in a dilute gas because experiments have been performed on colloidal suspensions. Think of the colloidal system as a dilute gas of massive particles.

Also, let E_j denote the probability that j particles will enter ω during τ. By assumption, E_j cannot depend on ν, and since the *a priori* probabilities of entry and exit must be the same, we have

$$E_j = \langle A_j^{(\nu)} \rangle$$

$$= \sum_{\nu=j}^{\infty} A_j^{(\nu)} P(\nu)$$

$$= \frac{(\gamma p)^j e^{-\gamma p}}{j!}, \qquad (16A.13)$$

which is a Poisson distribution with mean γp. Consequently, $K(\nu|\nu+k)$ is given by

$$K(\nu|\nu+k) = \sum_{i=0}^{\nu} A_i^{(\nu)} E_{i+k}, \qquad (16A.14a)$$

and also

$$K(\nu|\nu-k) = \sum_{i=k}^{\nu} A_i^{(\nu)} E_{i-k}. \qquad (16A.14b)$$

In Eqs. 16A.14, i cannot be less than k because E_j is not defined for $j < 0$. We see that the transition probability is in general given by

$$K(\nu|m) = \sum_{x+y=m} P_1^{(\nu)}(x) P_2(y), \qquad (16A.15)$$

where the summation is over all values of x and y consistent with $x + y = m$. This form of distribution is called the sum or convolution of the two component distributions $P_1^{(\nu)}$ and P_2. It is easy to show that the mean and mean-square deviations of m for given ν are the sums of the corresponding moments of $P_1^{(\nu)}$ and P_2:

$$\langle x \rangle = \nu(1-p); \qquad \langle (\Delta x)^2 \rangle = \nu p(1-p);$$
$$\langle y \rangle = \gamma p; \qquad \langle (\Delta y)^2 \rangle = \gamma p. \qquad (16A.16)$$

Let a subscript 1 on $\langle \cdots \rangle$ denote an average conditional on the given value of ν. Then

$$\langle (m-\nu) \rangle_1 = \langle m \rangle_1 - \gamma = (\gamma - \nu)p \qquad (16A.17)$$

and

$$\langle (m-\nu)^2 \rangle_1 = [(\gamma - \nu)^2 - \gamma]p^2 + (\gamma + \nu)p. \qquad (16A.18)$$

Equation 16A.17 is important in that it shows that the average tendency is to move toward the mean. This is one of the observed characteristics of irreversible processes, and may be called the *regression law*.

Upon averaging these results over all values of ν, we find

$$\langle \langle (m-\nu) \rangle_1 \rangle = 0, \quad \text{since} \quad \langle \nu \rangle = \gamma, \qquad (16A.19)$$

and

$$\langle \langle (m-\nu)^2 \rangle_1 \rangle = 2\gamma p. \qquad (16A.20)$$

Equation 16A.19 is, of course, to be expected; it simply states the fact that the mean values of two consecutive observations should be the same. This result follows from the initial assumption that the colloid system is in equilibrium.

Both Eqs. 16A.19 and 16A.20 provide a means of testing the theory of the probability aftereffect. Elegant and thorough experiments performed by Svedberg and by Westgren[15] test Smoluchowski's theory. We

[15] T. Svedberg, *Z. Physik. Chem.* **77,** 147 (1911); A. Westgren, *Arkiv Matematik, Astronomi och Fysik* **11,** Nos. 8, 14 (1916); **13,** No. 14 (1918).

shall discuss briefly the results of Westgren's experiments, which were obtained with an experimental arrangement designed specifically with the theory in mind. Westgren set up his experiment in such a way that a volume of fluid with linear dimensions of a few micrometers, and containing colloidal gold particles of radius approximately 5 μm, was viewed in an ultramicroscope. The illumination was steady and the number of particles in view was counted at successive instants measured by the ticking of a metronome. Westgren performed experiments with different concentrations of particles, different sizes of volume ω, and different time intervals. Each experiment consisted of about 1500 observations. Below we give a sample of 114 consecutive observations for which the time interval $\tau = 0.81$ s:

0231320011213231123512142343231110102143232422353413253233

36022354132442342303211111221001232100001121232343120210111.

The average value of ν for this sample is $\gamma = 2.018$. The theoretical numbers of occurrences of different values of ν for a Poisson distribution of this average are shown in Table 16A.1. The agreement is seen to be good. The experimental arrangements under which the sample of observations quoted was obtained correspond to a value of the probability aftereffect factor p of 0.613 (from integrating the diffusion expression described above). From Eq. 16A.20 and the sample, the observed value is 0.582. From the theory of errors we expect a fractional variation in possible values of $\sim (114)^{-1/2} \approx 9\%$, and the observed value is well within this limit.

Despite the fact that regression is very much in evidence, theory shows, and observation confirms, that time symmetry is maintained. Let $H(\nu, \nu + k)$ denote the probability that the pair of values $(\nu, \nu + k)$ is observed consecutively. Then it can be shown that

$$H(\nu, \nu + k) = P(\nu)K(\nu/\nu + k)$$

$$= P(\nu + k)K(\nu + k/\nu)$$

$$= H(\nu + k, \nu), \qquad (16A.21)$$

which is achieved by a rearrangement of factors between the full expressions for $P(\nu)$ and $K(\nu|\nu + k)$. This result can be extended to show that the probability of observing the pair (ν, m) at instants separated by $s\tau$ is also symmetrical. Thus, we conclude that the sequence of observations is statistically unaltered by time reversal, and this is borne out fairly well by the sample of observations displayed, considering its small size, as shown in Table 16A.2. Nevertheless, it is still true that if $\nu > \gamma$, the probability of the pair (ν, m) is larger for $m < \nu$ than for $m > \nu$.

We have seen how a state different from the average tends to regress toward the average, and now come to the question of how soon this state may recur.

Let T_ν be the average lifetime of a state ν, and let Θ_ν be the average recurrence time of that state; T_ν and Θ_ν are related in a simple way. If, in any long time $\mathcal{T}$, the system spends intervals $t_1, t_2, \ldots, t_N$ (N very large) in the state ν, then

$$P(\nu) = \lim_{N \to \infty} \frac{1}{\mathcal{T}} \sum_{i=1}^{N} t_i, \qquad (16A.22)$$

and

$$T_\nu = \lim_{N \to \infty} \frac{1}{N} \sum_{i=1}^{N} t_i \approx \frac{\mathcal{T}}{N} P(\nu); \qquad \mathcal{T}, N \text{ large.} \qquad (16A.23)$$

The time spent not in state ν is

$$\mathcal{T} - \sum_{i=1}^{N} t_i,$$

TABLE 16A.1
EXPERIMENTAL VERIFICATION OF THE POISSON DISTRIBUTION

ν	Obs.	Calc.
0	15	15
1	29	31
2	30	31
3	25	21
4	10	10
5	4	4
6	1	1

TABLE 16A.2
NUMBER OF OCCURRENCES OF PAIRS (ν, m) FOR INTERVAL τ, IN THE SAMPLE QUOTED (TOP ROWS); CALCULATED VALUE FROM EQ. 16A.21 (BOTTOM ROWS)

ν	m 0	1	2	3	4	5
0	4	5	4	1		
	4	5	3	1		
1	5	9	8	3	2	
	5	10	8	4	2	
2	2	8	3	14	2	1
	3	8	9	6	3	1
3	1	4	11	1	4	3
	1	4	6	5	3	1
4		2	4	3	1	
		2	3	3	2	1
5		1		2	1	
			1	1	1	

and therefore,

$$\frac{1}{\mathcal{T}}\left(\mathcal{T} - \sum_{i=1}^{N} t_i\right) = 1 - P(\nu). \qquad (16A.24)$$

The average recurrence time of state ν is, therefore, from Eq. 16A.24,

$$\Theta_\nu = \frac{1}{N}\left(\mathcal{T} - \sum_{i=1}^{N} t_i\right),$$

since the state ν has recurred N times. Hence,

$$\Theta_\nu = \frac{1 - P(\nu)}{P(\nu)} T_\nu. \qquad (16A.25)$$

We now calculate T_ν following Smoluchowski's work. Since $K(\nu|\nu)$ is the probability that the state ν will be observed on two successive occasions, the probability that the state ν will be observed on $(k-1)$ successive occasions and not on the kth is

$$\phi_\nu(k\tau) = [K(\nu|\nu)]^{k-1}[1 - K(\nu|\nu)]. \qquad (16A.26)$$

We can define the average lifetime in a natural way, as

$$\begin{aligned}
T_\nu &= \sum_{i=1}^{\infty} k\tau\phi_\nu(k\tau) \\
&= \tau\left[\sum_{k=1}^{\infty} k(K(\nu/\nu))^{k-1}\right][1 - K(\nu/\nu)] \\
&= \tau[1 - K(\nu/\nu)]^{-1},
\end{aligned} \qquad (16A.27)$$

where we have used

$$\frac{d}{dK}(1-K)^{-1} = (1-K)^{-2} = \sum_{k=1}^{\infty} kK^{k-1}.$$

The observed average lifetimes and recurrence times for the sample of data quoted are shown in Table 16A.3; agreement between theory and experiment is again satisfactory.

So far we have only discussed the case of intermittent observations. In order to study the relation between the recurrence of mechanical and thermodynamic states, we now extend the theory to the case of continuous observations. We must require that the numbers involved be so large that we may ignore the discrete nature of the process, and treat it as continuous. We consider this extension in the limit $\tau \to 0$. Then we must expect $p(\tau) \to 0$ and $K(\nu/\nu) \to 1$. From the formula for $K(\nu/\nu)$ we therefore select the leading term $(i = 0)$:

$$K(\nu|\nu) = e^{-\gamma p}(1-p)^{\nu} + \mathcal{O}(p^2)$$
$$= 1 - (\nu + \gamma)p. \tag{16A.28}$$

Hence, the mean lifetime becomes

$$T_{\nu} = \frac{\tau}{(\nu + \gamma)p}, \tag{16A.29}$$

and in order that this should tend to a finite limit we must have $p(\tau) \to p_0 \tau$ for small τ. This is easily seen to be the case if one considers the example of the number of gas molecules in a given volume. As $\tau \to 0$ we may ignore the effects of collisions and treat the relevant part of the trajectories of the molecules as linear. Then the number leaving the volume element in a time τ is equal to the number striking the surface in that time and is obviously proportional to τ.

In the limit of large numbers the Poisson distribution tends to a Gaussian distribution near its peak:[16]

$$P(\nu) = (2\pi\gamma)^{-1/2} \exp\left[-\frac{(\nu - \gamma)^2}{2\gamma}\right]; \qquad |\nu - \gamma| \ll \gamma. \tag{16A.30}$$

Equation 16A.25 then becomes, with Eq. 16A.29,

$$\Theta_{\nu} \approx \frac{\omega}{\sigma}\left(\frac{m}{k_B T}\right)^{1/2} \gamma^{-1/2} \exp\left[\frac{(\nu - \gamma)^2}{2\gamma}\right], \tag{16A.31}$$

where ω and σ are the volume and surface area of the cell considered, m is the mass of a gas molecule, and T is the absolute temperature. Values of Θ calculated for a 1% deviation in the density of oxygen in a sphere of radius a at 300 K are shown in Table 16A.4. It is at once clear that for volumes just at the limit of unaided visual perception, even quite small fluctuations have large recurrence times, so that diffusion is for all intents

---

TABLE 16A.3

AVERAGE LIFETIME T_{ν} AND RECURRENCE TIME Θ_{ν} OF THE STATE ν FOR THE SAMPLE QUOTED: OBSERVED VALUES (TOP ROWS); CALCULATED VALUES (BOTTOM ROWS).

ν	T_{ν}/τ	Θ_{ν}/τ
0	1.5	11.67
	1.41	9.20
1	1.45	5.20
	1.49	4.06
2	1.125	3.92
	1.42	3.83
3	1.08	4.91
	1.31	5.88
4	1.33	10.
	1.21	11.97
5	1	14.67
	1.13	25.98

---

[16] This is easily seen by writing

$$\ln P(\nu) = \nu \ln \gamma - \gamma - \ln \nu!$$
$$= \nu \ln \gamma - \gamma - (\nu + \tfrac{1}{2}) \ln \nu + \nu - \tfrac{1}{2} \ln 2\pi + \mathcal{O}(\nu^{-1}),$$

where we have used Stirling's formula for log $\nu!$. If

$$\delta = \nu - \gamma$$

and we restrict ν so that $(\delta/\gamma) \ll 1$, then

$$\ln P(\nu) = -(\gamma + \delta + \tfrac{1}{2}) \ln\left(1 + \frac{\delta}{\gamma}\right) + \delta - \tfrac{1}{2} \ln 2\pi\gamma + \mathcal{O}(\nu^{-1})$$

$$\approx -\frac{\delta^2}{2\gamma} - \tfrac{1}{2} \ln 2\pi\gamma,$$

whence

$$P(\nu) = (2\pi\gamma)^{-1/2} \exp\left[-\frac{(\nu - \gamma)^2}{2\gamma}\right]; \qquad |\nu - \gamma| \ll \gamma.$$

TABLE 16A.4
RECURRENCE TIMES FOR A 1% DEVIATION IN THE DENSITY OF
OXYGEN IN A SPHERE OF RADIUS a AT 300 K ($\gamma/\omega = 3 \times 10^{25}$ m^{-3})

a (μm)	10^{-2}	5×10^{-7}	3×10^{-7}	2.5×10^{-7}	10^{-7}
Θ (s)	$10^{10^{14}}$	10^{68}	10^{6}	1	10^{-11}

and purposes an irreversible process. On the other hand, for volumes just at the limit of microscopic vision (~ 0.25 μm), large fluctuations occur rapidly, and there is no question of irreversibility in the usual sense. Nevertheless, we have seen from a study of such fluctuations that the results are, on average, in accordance with the macroscopic theory of diffusion.

In this appendix we have been concerned with only one gross variable, the number density. However small the number of particles considered, it has been possible to discern both the time symmetry of the sequence of occupation numbers and the regression toward the mean. If the initial state is sufficiently far from the mean, then its recurrence time is extremely long. Alternatively, the recurrence time of a fixed fractional deviation from the mean becomes extremely long as the mean itself increases. The case of the number of gas molecules within a spherical volume discussed above is a striking example of this. It is clear, then, that on the gross level of ordinary observation, when the numbers of particles considered are always large enough that local thermodynamic variables may be defined, a significant fluctuation has so long a recurrence time that for practical purposes it decays irreversibly to equilibrium. Given an ensemble of replica systems, in each of which a prescribed volume element initially contains, say, an excess over the average number of particles, a distribution function for the number of particles instantaneously in the volume can be defined. This distribution function approaches its equilibrium form in a time of the order of that taken by a typical fluctuation to decay. The distribution function remains in its equilibrium form, and does not revert to its initial form if the system is undisturbed. The probability of a fluctuation the size of the initial value, as calculated from the equilibrium distribution function, is then related directly to its recurrence time. We then say that the distribution function has approached equilibrium irreversibly.[17]

[17] This does not mean that the system is not reversible, in the sense that reversal of all velocities at any time t after the initial ensemble is set up at 0 leads again to the initial state at $2t$.

Further Reading

Epstein, P. S., *Textbook of Thermodynamics* (Wiley, New York, 1937), chap. 4.

Kestin, J., *A Course in Thermodynamics*, Vol. 1 (Blaisdell, New York, 1966), chaps. 9 and 10.

Pippard, A. B., *Classical Thermodynamics* (Cambridge University Press, London, 1957), chap. 4.

Reiss, H., *Methods of Thermodynamics* (Blaisdell, New York, 1965), chap. 4.

Problems

1. Consider a liquid (at temperature T) in an insulated container. Let the depth of the liquid be d and its density ρ. Above the liquid, at a height h, is suspended a rock of volume v_R and mass M_R. The rock is suddenly dropped into the liquid through a small hole in the top of the insulating container. The rock comes to rest at the bottom of the container. Assume that the mass of the liquid is very much larger than the mass of the rock. What is the entropy change of the liquid?

2. A hot meteorite falls (velocity 200 km/h) into the Atlantic Ocean. The meteorite was originally at a temperature of 1000°C, weighed 1000 g, and had heat capacity of 0.82 J/K g. If the ocean temperature is 15°C, calculate the change in entropy of the universe as a result of this process.

3. An electric current of 10 A is maintained for 1 s in a resistor of 25 Ω while the temperature of the resistor is kept constant at 27°C.
   (a) What is the entropy change of the resistor?
   (b) What is the entropy change of the resistor plus surroundings?
   The same current is maintained for the same time in the same resistor, but now it is thermally insulated. If the resistor has a mass of 10 g and $C_p =$ 0.836 J/K g,
   (c) What is the entropy change of the resistor?
   (d) What is the entropy change of the resistor plus surroundings?

4. A mass of some conducting substance is placed in an evacuated container. The walls of the container are maintained at the constant temperature 500°C. Show that if the conducting substance is heated from 50 to 150°C by passage of an electric current from an external generator, the process undergone is irreversible.

5. A mass of steam at 100°C is irreversibly condensed by contact with a heat sink at temperature T_0. In this process the latent heat of vaporization is extracted from the steam, and its entropy decreased. Does this process violate the second law of thermodynamics? Why?

6. Take a gas with equation of state

$$p(v - b) = R\theta,$$

with C_V a function of θ only, through a Carnot cycle, and show that

$$T = \theta,$$

where T is the thermodynamic temperature and θ the empirical gas scale temperature.

7. The coefficient of volume expansion of water is negative for $0°C < t < 4°C$. Show that over this temperature interval water is cooled by adiabatic compression. Compare this result with the adiabatic compression of a perfect gas.

8. Consider an adiabatic process and an isothermal process, both represented in pV space. Under what conditions would the adiabatic path coincide with the isothermal path?

9. By considering water to be the working fluid in a Carnot cycle, show that the temperature $t = 4°C$ cannot be reached by cooling (heating) in an adiabatic expansion. Assume the temperature at which the density of water is a maximum is independent of the pressure.

10. Consider the first-law expression for a change in internal energy of a simple fluid:

$$dU = đq - p\,dV.$$

Along a constant-volume path $(đq)_V = C_V \, dT$. Along a path with $đq = 0$ we have $dU = -p \, dV$. Thus we can write

$$\Delta U = \int_{V_1 T_1}^{V_1 T_2} C_V(V_1, T) \, dT - \left(\int p_{app} \, dV \right)_{đq=0},$$

but we also know that

$$\Delta U = \int_{V_1 T_1}^{V_1 T_2} C_V(V_1, T) \, dT + \int_{V_1 T_2}^{V_2 T_2} \left(\frac{\partial U}{\partial V} \right)_{T_2} dV.$$

Explain the differences in these two expressions for ΔU. If the isochoric (constant-volume) heating steps are the same, can the final states reached be the same? How is your answer related to the Caratheodory statement of the second law of thermodynamics?

11. If $p \, dV$ were an exact differential for a given substance, could that substance be used as the working fluid in the cylinder of a heat engine undergoing a cyclic process? Explain your answer.

12. Consider a gas for which $U(T)$ is a monotonic increasing function of T, and is independent of V and p. Let this gas be reversibly isothermally compressed. Show that the final state reached in this compression cannot be reached by a reversible adiabatic path starting at the initial state.

13. (a) A sample of gas containing n mol is heated reversibly from T_1 to T_2 at constant volume. Another sample of n mol of the same gas is heated reversibly from T_1 to T_2 at constant pressure. For which process is ΔS larger? Why?
(b) A sample of n mol of Ar is heated reversibly from T_1 to T_2 at constant volume. A sample of n mol of Br_2 is heated reversibly from T_1 to T_2 at constant volume. For which gas is ΔS larger? Why?
(c) A sample of gas containing n mol is expanded from (p_1, V_1) to (p_2, V_2) in a reversible isothermal process. Another sample of n mol of the same gas is expanded from (p, V_1) to (p, V_2') in a reversible isentropic process. Which is the larger, V_2 or V_2'? Why? Draw a graph illustrating your answer.

14. Consider the change of state of a perfect gas $(p_1, T_1) \rightarrow (p_2, T_2)$. Devise two or more reversible paths between states 1 and 2 and show that the heat absorbed by the gas is different for each path, but that the entropy change is the same.

15. Consider 1 mol of a perfect gas which, at 300 K, occupies a volume $V_1 = 10$ liters. Suppose that this gas undergoes an isothermal reversible expansion to a volume $V_2 = 100$ liters.
(a) Calculate the change in entropy of the gas.
(b) Calculate the change in entropy of the surroundings. Demonstrate, by an explicit calculation, that if the isothermal change in volume $V_1 \rightarrow V_2$ is effected by expansion against zero pressure, the expansion is irreversible.

16. Describe qualitatively reversible paths by means of which changes in entropy in the following processes could be calculated.
(a) Superheated steam at 150°C and 2 atm pressure is brought in contact with ice at 0°C and 1 atm pressure, the quantities being chosen so that the final state of the system is only liquid water at 50°C and 1 atm.
(b) A chunk of iron at 500°C is dropped into liquid water at 100°C and 1 atm, steam being evolved at 100°C and 1 atm.

17. Prove that when heat is added to a single phase of any substance, the temperature must rise.

18. Let a closed system be specified in such a fashion that it can only undergo reversible isothermal changes. Show that q for any such change is independent of path.

19. Prove that an isothermal path cannot intersect a reversible adiabatic path twice.

20. Consider a Carnot cycle with 1 mol of a monatomic ideal gas as the working fluid. Calculate w, q, ΔU, and ΔS for each step of the cycle as well as for the entire cycle. Assume that the cycle operates between temperatures T_1 and T_2 and that the volumes at points A, B, C, and D are V_A, V_B, V_C, and V_D, respectively (see Fig. 16.8).

21. Repeat the above problem with 1 mol of the gas described in Problem 12 of Chapter 14 as the working fluid.

22. Calculate the entropy change of 3 mol of an ideal liquid when it is isothermally expanded from 10 liters to 100 liters. By what factor does Ω change during the expansion?

23. In the Carnot cycle shown in Fig. 16.8, heat q_2 is absorbed in the isothermal expansion $A \rightarrow B$ and heat q_1 is absorbed in the isothermal compression $C \rightarrow D$. Is it possible for q_1 to be zero? How is your answer related to the Kelvin statement of the second law?

24. Consider an isolated system consisting of two subvolumes containing perfect gases separated by a frictionless piston (see diagram). As in Problem 25 of

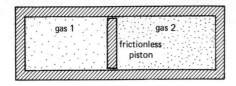

Chapter 13, we imagine that the molecules of gases 1 and 2 have, respectively, the constant speeds v_1 and v_2 and can move only in the direction perpendicular to the piston faces. Because of collisions between the molecules of the two gases and the piston, the piston jiggles back and forth, and thereby transmits energy between the two gases. Suppose that in the initial state the temperatures of gases 1 and 2 are not equal, so that $m_1 v_1^2/2 \neq m_2 v_2^2/2$. Further, assume that any imbalance in forces on the two sides of the piston is so small that the average velocity of the piston, v_p, is sensibly zero, and that any change in kinetic energy of the piston resulting from its motion is vanishingly small. Show that the rate of gain of kinetic energy by gas 1 is

$$\frac{d}{dt}(KE) = \frac{4 Z_{c_1} Z_{c_2} m_1 m_2}{M(Z_{c_1} m_1 + Z_{c_2} m_2)} \left(\tfrac{1}{2} m_2 v_2^2 - \tfrac{1}{2} m_1 v_1^2\right),$$

where Z_{c_1} and Z_{c_2} are the numbers of collisions per second of all the molecules of gases 1 and 2 with the piston, respectively.

25. The result of the preceding problem shows that the direction of energy flow is from the gas with greater kinetic energy per particle to that with lower kinetic energy per particle. This is, of course, the expected macroscopic result: Heat flows from higher to lower temperature. However, the result in Problem 24 was obtained from a microscopic model using a time-reversible equation of motion. Explain how an assumption used in the model leads to the result obtained, and what that assumption means.

Some Applications of the Second Law of Thermodynamics

In this chapter we examine some more consequences of the second law of thermodynamics by computing the entropy changes accompanying a variety of processes. For simplicity we restrict our attention to the case of a one-component closed system, the state of which is defined by specification of any two of the three variables p, V, T.

The combined first and second laws require that the differential change in internal energy be

$$dU = T\,dS - p\,dV. \tag{17.1}$$

Equation 17.1 suggests that interpretation of the energy change in a reversible process is particularly straightforward if the change can be effected by altering the volume with the entropy held fixed, and then altering the entropy with the volume held fixed. Then, for fixed S the slope of U versus V is just the pressure, and for fixed V the slope of U versus S is just the temperature. The simplicity of this interpretation of the rates of change of U with respect to changes of S and V, that is, their identification with variables from the set p, V, T, leads one to classify S and V as the "natural" independent variables for the function U.

Laboratory experiments are usually conducted under conditions that correspond to holding p and T fixed, or sometimes to holding V and T fixed. It is certainly possible to compute the change in U with respect to changes in p and T, or with respect to changes in other pairs of independent variables. However the resultant forms are more complicated than Eq. 17.1 in the sense that the coefficients multiplying the changes in independent variables are combinations of functions related to properties of the system. For example, choosing T and V as independent variables for U leads to

$$dU = \left(\frac{\partial U}{\partial T}\right)_V dT + \left(\frac{\partial U}{\partial V}\right)_T dV \tag{17.2}$$

$$= C_V\,dT + \left[T\left(\frac{\partial S}{\partial V}\right)_T - p\right]dV, \tag{17.3}$$

since

$$\left(\frac{\partial U}{\partial V}\right)_T = T\left(\frac{\partial S}{\partial V}\right)_T - p \tag{17.4}$$

from Eq. 17.1. It will be shown soon (see Eqs. 17.16 and 17.17) that $(\partial S/\partial V)_T = (\partial p/\partial T)_V$, so that the coefficient of dV in Eq. 17.3 can be

17.1
Choice of Independent Variables

expressed in terms of the measurable quantities p, V, T. Even so, the rate of change of U with respect to changes in V is determined by the balance between p and $T(\partial p/\partial T)_V$, which is clearly not as "simple" as when S and V are chosen as the independent variables.

This illustration provides motivation for the introduction of new thermodynamic functions defined so as to simplify the calculation of heat and work exchanges for each particular choice of independent variables. We shall later see that these new thermodynamic functions have another important property, namely, they also play the role of potential functions for the transitions among equilibrium states defined by constraints imposed by different independent variables.

A general method for rewriting a function of one set of independent variables as an equivalent function of other independent variables is the *Legendre transformation*, which we now describe for the case of one variable. Figure 17.1a shows a function $y(x)$ plotted with respect to the variable x. Suppose that, for some reason, we wish to use as independent variable not x but rather the slope of $y(x)$ at x, namely, $y'(x)$. Complete specification of the tangent line $y'(x)$ requires knowledge of its intercept with the y axis, say, $i[y'(x)]$, since both slope and intercept are required to define a straight line uniquely. As shown in Fig. 17.1b, if the slope and intercept are known at every point of $y(x)$, the information available is equivalent to knowledge of $y(x)$. Therefore, the curve $y(x)$ in Fig. 17.1 can also be described by

$$i(y') = y - xy'(x), \tag{17.5}$$

since

$$y'(x) \equiv \frac{y - i(y')}{x}. \tag{17.6}$$

The function $i(y')$ is the Legendre transformation of y; $i(y')$ is equivalent to $y(x)$ but treats y' as the independent variable in place of x. The generalization of Eq. 17.5 to the case that there is more than one independent variable is straightforward, the result being

$$i(y') = y - \sum_{j=1}^{n} \left(\frac{\partial y}{\partial x_j}\right)_{x_i \neq x_j} x_j, \tag{17.7}$$

where the x_j's are the independent variables defining the function $y(x_1, \ldots, x_n)$.

We now apply the Legendre transformation to convert $U(S, V)$ and $H(S, p)$ to equivalent functions in terms of other pairs of independent variables. Consider again choosing T and V as the independent variables. We seek a Legendre transformation of $U(S, V)$ to an equivalent function. Following Eq. 17.5, we find that the required function is

$$A(T, V) = U - \left(\frac{\partial U}{\partial S}\right)_V S$$

$$= U - TS, \tag{17.8}$$

using

$$T = \left(\frac{\partial U}{\partial S}\right)_V \tag{17.9}$$

from Eq. 17.1. The quantity $A(T, V)$ is known as the *Helmholtz free energy*; the reason for the name "free energy" will be made apparent in the next section. Next consider choosing T and p as the independent variables. In this case we make a Legendre transformation of $H(S, p)$.

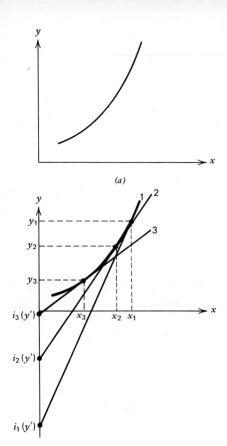

FIGURE 17.1
Diagram illustrating a Legendre transformation (see text).

Recall that $H \equiv U + pV$. Again following Eq. 17.5, we find that the required function is

$$G(T, p) = H - \left(\frac{\partial H}{\partial S}\right)_p S$$
$$= H - TS, \tag{17.10}$$

using

$$T = \left(\frac{\partial H}{\partial S}\right)_p \tag{17.11}$$

from the differential relation for reversible processes

$$dH = dU + p\,dV + V\,dp$$
$$= T\,dS + V\,dp. \tag{17.12}$$

The quantity $G(T, p)$ is known as the *Gibbs free energy*. Both $A(T, V)$ and $G(T, p)$ are, obviously, functions of state.

In summary, for a reversible process described in terms of:

1. Changes in the independent variables S and V,

$$dU = T\,dS - p\,dV, \tag{17.1}$$
$$\left(\frac{\partial T}{\partial V}\right)_S = -\left(\frac{\partial p}{\partial S}\right)_V = \left(\frac{\partial^2 U}{\partial S\,\partial V}\right); \tag{17.13}$$

2. Changes in the independent variables S and p,

$$dH = T\,dS + V\,dp, \tag{17.14}$$
$$\left(\frac{\partial T}{\partial p}\right)_S = \left(\frac{\partial V}{\partial S}\right)_p = \left(\frac{\partial^2 H}{\partial S\,\partial p}\right); \tag{17.15}$$

3. Changes in the independent variables T and V,

$$dA = dU - T\,dS - S\,dT$$
$$= -S\,dT - p\,dV, \tag{17.16}$$
$$\left(\frac{\partial S}{\partial V}\right)_T = \left(\frac{\partial p}{\partial T}\right)_V = -\left(\frac{\partial^2 A}{\partial V\,\partial T}\right); \tag{17.17}$$

4. Changes in the independent variables T and p,

$$dG = dH - T\,dS - S\,dT$$
$$= -S\,dT + V\,dp, \tag{17.18}$$
$$-\left(\frac{\partial S}{\partial p}\right)_T = \left(\frac{\partial V}{\partial T}\right)_p = \left(\frac{\partial^2 G}{\partial p\,\partial T}\right). \tag{17.19}$$

Equations 17.13, 17.15, 17.17, and 17.19 are known as *Maxwell's relations*. They play an important role in practical thermodynamics, since they relate quantities of interest to the measurable variables p, V, T.

17.2
The Available Work Concept

We showed, in Chapter 13, that the work done in a process depends on the path used to connect the initial and final states. There are some special cases for which the work done depends only on the change in state; for example, in an adiabatic work process, $w = \Delta U$. We shall now discuss the limitations imposed by the second law on the conversion of heat into work.

Consider a closed system in contact with a single heat reservoir whose temperature is T_R. Suppose that the system undergoes the change of state $1 \rightarrow 2$. What is the maximum work that can be accomplished by the system? Other systems in the surroundings can be used to assist the change $1 \rightarrow 2$ provided that they are carried around cyclic processes, hence are in the same states before and after $1 \rightarrow 2$.

For any process $1 \rightarrow 2$ in the system, the first law requires that the work done and the internal energy change of the system be related by

$$w = \Delta U - q, \tag{17.20a}$$

or in differential form that

$$đw = dU - đq \tag{17.20b}$$

for each infinitesimal segment of the path $1 \rightarrow 2$. Consider the case that the heat transfer to the system in any infinitesimal segment of the path, $đq$, is reversible. To make this possible, we arrange to interpose a reversible engine between the system and the reservoir, so that the heat transferred to the system comes from the engine, and only indirectly from the reservoir. This engine works between the temperatures T_R and T, where T is the temperature along that part of the path having heat transfer $đq$. Because heat is transferred via an engine, work is done by the engine on the system during the transfer. For the step in which heat $đq$ is transferred to the system at temperature T, the work done by the reversible engine is

$$đw_R = \left(\frac{T - T_R}{T}\right) đq. \tag{17.21}$$

For a reversible transfer of heat the entropy change of the system is $dS = đq/T$, since T is the temperature of the heat source. In general, if the engine supplies heat at T, then for the system

$$dS \geqslant \frac{đq}{T}. \tag{17.22a}$$

Suppose a different engine is used for each infinitesimal segment of the path $1 \rightarrow 2$. Then

$$\Delta S \geqslant \int_1^2 \frac{đq}{T}, \tag{17.22b}$$

where T is the heat source temperature, and differs from T_R by virtue of the engine used to transfer the heat. Using the fact that T_R is a constant, we multiply 17.22b by T_R and rearrange to find

$$T_R \Delta S - T_R \int_1^2 \frac{đq}{T} \geqslant 0, \tag{17.23}$$

which, after combination with 17.20a, assumes the form

$$w \geqslant \Delta U - T_R \Delta S - \int_1^2 \left(\frac{T - T_R}{T}\right) đq. \tag{17.24}$$

Note that the total work done on the system by the reversible engines that altogether deliver heat q to the system is

$$w_R = \int_1^2 \left(\frac{T - T_R}{T}\right) đq, \tag{17.25}$$

so that the total work done by the system and by the engines that serve to transfer heat to the system when it undergoes the change of state $1 \rightarrow 2$ is

$$w + w_R \geqslant \Delta U - T_R \Delta S. \tag{17.26}$$

If all the processes involved are reversible, the equality sign holds in 17.26.

Recall that ΔU and ΔS depend only on the system's change of state, $1 \rightarrow 2$, and that T_R is fixed. Then if all the processes involved are reversible, $w + w_R$ is the maximum possible work that can be done by the system and engines for the system change of state $1 \rightarrow 2$:

$$w_{\max} = \Delta U - T_R \, \Delta S. \tag{17.27}$$

Since the right-hand side of Eq. 17.27 defines a minimum, why is the left hand side labelled $w_{\max}$? The point is that work done by the system is a negative quantity, so Eq. 17.27 defines the most negative amount of work done for the system change of state $1 \rightarrow 2$, and this is just the maximum amount of work done by the system. Moreover, all reversible processes that bring about the same change of state of the system must produce the same maximum work, since $w_{\max}$ depends only on T_R, ΔU, and ΔS. It must be kept in mind that $w_{\max}$ is a composite work, since it includes the work done by the reversible Carnot engines that effect the heat transfer from the reservoir to the system.

Equation 17.27 can be used to relate $w_{\max}$ to changes in the Helmholtz and Gibbs free energies of a system. Indeed, the following cases are of particular interest.

1. Suppose that the volume of the system is held fixed, hence it cannot do any expansion work. The maximum nonexpansion work that can be done, $w_{x,\max}$, is, from Eq. 17.27,

$$w_{x,\max} = \Delta U - T_R \, \Delta S; \qquad V \text{ fixed.} \tag{17.28}$$

2. Suppose that the pressure of the system is held fixed, so the expansion work it does is $-p \, \Delta V$. The maximum nonexpansion work that can be done is, from Eq. 17.27,

$$w_{x,\max} = \Delta U + p \, \Delta V - T_R \, \Delta S$$
$$= \Delta H - T_R \, \Delta S; \qquad p \text{ fixed.} \tag{17.29}$$

3. In cases 1 and 2 no restriction was made concerning the initial and final temperatures of the system. Suppose that the initial and final temperatures of the system are equal to T_R. Then Eq. 17.28 becomes

$$w_{x,\max} = \Delta A; \qquad V \text{ fixed, inital and final temperatures } T_R, \tag{17.30}$$

and Eq. 17.29 becomes

$$w_{x,\max} = \Delta G; \qquad p \text{ fixed, initial and final temperatures } T_R. \tag{17.31}$$

In both Eqs. 17.30 and 17.31 the temperature of the system must be T_R in states 1 and 2, but can vary along the path $1 \rightarrow 2$.

The reason for the nomenclature "free energy" can now be made clear. Although in an adiabatic process the entire energy change in a system is converted to work, under other conditions less than the entire energy change can be converted to work. The amount unavailable for conversion is just $T_R \, \Delta S$ for a change ΔS in the system's entropy. The differences, $\Delta U - T_R \, \Delta S$ or $\Delta H - T_R \, \Delta S$, are then "free energies," in the sense that if all processes used are reversible and the initial and final temperatures of the system are T_R, complete conversion of the free energies to work is possible. Therefore, $w_{x,\max}$ can be thought of as the amount of work available from a reversible change of state in a system coupled to reversible changes in its surroundings. As indicated in the title of this section, it is sometimes called the *available work* or *availability*.

We shall now examine the nature of the temperature dependence of the entropy of a one-component system and then, using this information, we shall calculate the entropy change accompanying a general change of thermodynamic state $1 \to 2$.

It is convenient to begin with the combined first and second laws for an infinitesimal reversible process, namely, Eq. 17.1. From Eq. 17.1 we find

$$\left(\frac{\partial U}{\partial T}\right)_V \equiv C_V = T\left(\frac{\partial S}{\partial T}\right)_V; \tag{17.32}$$

hence C_V/T measures the rate of change of entropy with temperature in a constant-volume process. Suppose, now, that a one-component system undergoes the change of state $1 \to 2$. Suppose, further, that the state of the system is fully defined by any two of the three variables p, V, T. Since the entropy is a function of state, all paths between states 1 and 2 lead to the same entropy change. Consider first a path consisting of an isochoric temperature change followed by an isothermal volume change (see Fig. 17.2a). Along this path we regard the entropy as a function of the independent variables V and T. Thus, a differential change in the entropy can be written as

$$dS = \left(\frac{\partial S}{\partial T}\right)_V dT + \left(\frac{\partial S}{\partial V}\right)_T dV \tag{17.33}$$

or

$$dS = \frac{C_V}{T} dT + \left(\frac{\partial S}{\partial V}\right)_T dV. \tag{17.34}$$

But, as shown in Eq. 17.17,

$$\left(\frac{\partial S}{\partial V}\right)_T = \left(\frac{\partial p}{\partial T}\right)_V, \tag{17.17}$$

so that we can put Eq. 17.34 in the form

$$dS = \frac{C_V}{T} dT + \left(\frac{\partial p}{\partial T}\right)_V dV. \tag{17.35}$$

We then find the entropy difference between states 1 and 2, calculated along the path described, to be

$$\Delta S \equiv S_2 - S_1 = \int_{T_1, V_1}^{T_2, V_1} \frac{C_V}{T} dT + \int_{T_2, V_1}^{T_2, V_2} \left(\frac{\partial p}{\partial T}\right)_V dV. \tag{17.36}$$

The advantage of the form displayed in Eq. 17.36 is that it permits calculation of the entropy change accompanying the process $1 \to 2$ in terms of the equation of state [from which we find $(\partial p/\partial T)_V$] and the heat capacity as a function of T at some one volume, that is, in terms of directly measured quantities. Provided that the change of state is always in a one-phase region of the substance, Eq. 17.36 is a general result applicable to all one-component systems.

The general result embodied in Eq. 17.36 is made more meaningful when we consider the simple case of the perfect gas. In this instance we obtain

$$\left(\frac{\partial p}{\partial T}\right)_V = \left[\frac{\partial}{\partial T}\left(\frac{nRT}{V}\right)\right]_V = \frac{nR}{V}. \tag{17.37}$$

Now assume that the heat capacity per mole at constant volume, c_V, is

17.3
Entropy Changes in Reversible Processes

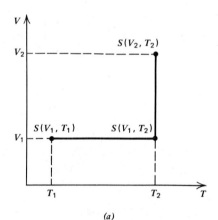

(a)

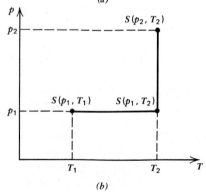

(b)

FIGURE 17.2
Paths along which the entropy change corresponding to the changes in state (a) $(V_1, T_1) \to (V_2, T_2)$ and (b) $(p_1, T_1) \to (p_2, T_2)$ may be computed.

independent of temperature (true for a monatomic gas at temperatures low enough that there is no electronic excitation), and insert Eq. 17.37 into 17.36. The result is, for n mol of perfect gas,

$$\Delta S_{\text{perfect gas}} = nc_V \ln \frac{T_2}{T_1} + nR \ln \frac{V_2}{V_1}. \tag{17.38}$$

It is important to observe how the properties of the equation of state are used to reduce the general equation 17.36 to the specific equation 17.38. Note also how the entropy increases, in general, as the system's temperature and/or volume increases.

Referring back to the discussion of Section 15.4, we easily recognize that in the isothermal expansion of a perfect gas $(\Omega_2/\Omega_1) > 1$, so that the corresponding entropy change should be positive. When the energy of the perfect gas is increased at constant volume we also have $(\Omega_2/\Omega_1) > 1$, and therefore an entropy increase. Equation 17.38 represents the quantitative macroscopic description of the entropy changes accompanying these very processes.

In Chapter 14 we showed that, corresponding to the same change of state $1 \rightarrow 2$, the internal energy change is

$$\Delta U \equiv U_2 - U_1 = \int_{V_1, T_1}^{V_1, T_2} C_V \, dT + \int_{V_1, T_2}^{V_2, T_2} \left(\frac{\partial U}{\partial V} \right)_{T_2} dV \tag{17.39}$$

or, using Eqs. 17.3 and 17.17,

$$\Delta U = \int_{V_1, T_1}^{V_1, T_2} C_V \, dT + \int_{V_1, T_2}^{V_2, T_2} \left[T \left(\frac{\partial p}{\partial T} \right)_V - p \right] dV. \tag{17.40}$$

As might be expected, Eq. 17.40 takes a particularly simple form for the perfect gas. Then we have

$$T \left(\frac{\partial p}{\partial T} \right)_V - p = p - p = 0, \tag{17.41}$$

whereupon, for a perfect gas,

$$\Delta U_{\text{perfect gas}} = \int_{V_1, T_1}^{V_1, T_2} C_V \, dT = C_V (T_2 - T_1). \tag{17.42}$$

The general formula for the energy change, Eq. 17.40, is valid for any one-component, one-phase system characterized by p, V, T. In contrast, Eqs. 17.41 and 17.42 are merely statements of internal consistency. It was stated in Chapter 16 that the specification of the properties of a perfect gas requires, as a supplement to the equation of state, the condition

$$\left(\frac{\partial U}{\partial V} \right)_{T, \text{ perfect gas}} = 0. \tag{17.43}$$

In fact, Eq. 17.43 is a definition necessary for the identification $T_{\text{absolute}} = T_{\text{perfect gas}}$. Equations 17.41 and 17.42 are merely reexpressions of this condition. Stated again, the only kind of energy a perfect gas has is the kinetic energy of the constituent molecules, so that U can only be a function of the temperature and the number of molecules in the sample of gas.

The entropy change represented by Eq. 17.36 displays the result of choosing V and T as the independent variables. Of course, we could have chosen p and T as the independent variables; depending on the experimental arrangement, the choice of one or the other of the pairs (V, T) or (p, T) may lead to greater convenience in interpretation of measurements.

If p and T are chosen as the independent variables describing the state of the system and we consider an isobaric temperature change from T_1 to T_2, followed by an isothermal pressure change from p_1 to p_2, as in Fig. 17.2b, Eq. 17.33 is replaced by

$$dS = \left(\frac{\partial S}{\partial T}\right)_p dT + \left(\frac{\partial S}{\partial p}\right)_T dp$$

$$= \frac{C_p}{T} dT + \left(\frac{\partial S}{\partial p}\right)_T dp. \qquad (17.44)$$

Here the relation

$$C_p = T\left(\frac{\partial S}{\partial T}\right)_p \qquad (17.45)$$

follows from the definition of the enthalpy. The analog of Eq. 17.36 therefore becomes, using Eq. 17.19,

$$\Delta S \equiv S_2 - S_1 = \int_{p_1,T_1}^{p_1,T_2} \frac{C_p}{T} dT - \int_{p_1,T_2}^{p_2,T_2} \left(\frac{\partial V}{\partial T}\right)_p dp. \qquad (17.46)$$

Provided that the change of state is always in a one-phase region of the substance, Eq. 17.46 is also a general result.[1]

Again, the simple case of the perfect gas is of interest. In this case, for n mol of gas, we have

$$\left(\frac{\partial V}{\partial T}\right)_p = \left[\frac{\partial}{\partial T}\left(\frac{nRT}{p}\right)\right]_p = \frac{nR}{p}. \qquad (17.47)$$

Proceeding as before, if we assume C_p to be a constant independent of the temperature (true for monatomic gases under the same conditions

---

[1] As is to be expected from their definitions, the difference between C_p and C_V for a given substance depends on the equation of state of that substance. A general relation for $C_p - C_V$ is easily obtained as follows. By definition,

$$C_p \equiv \left(\frac{\partial H}{\partial T}\right)_p = \left(\frac{\partial U}{\partial T}\right)_p + p\left(\frac{\partial V}{\partial T}\right)_p.$$

To evaluate $(\partial U/\partial T)_p$, we write

$$dU = \left(\frac{\partial U}{\partial T}\right)_V dT + \left(\frac{\partial U}{\partial V}\right)_T dV,$$

from which we find

$$\left(\frac{\partial U}{\partial T}\right)_p = C_V + \left(\frac{\partial U}{\partial V}\right)_T \left(\frac{\partial V}{\partial T}\right)_p,$$

or

$$C_p - C_V = \left[p + \left(\frac{\partial U}{\partial V}\right)_T\right]\left(\frac{\partial V}{\partial T}\right)_p.$$

But from Eqs. 17.1 and 17.17,

$$\left(\frac{\partial U}{\partial V}\right)_T = T\left(\frac{\partial S}{\partial V}\right)_T - p = T\left(\frac{\partial p}{\partial T}\right)_V - p,$$

so that

$$C_p - C_V = T\left(\frac{\partial p}{\partial T}\right)_V = -T\frac{(\partial V/\partial T)_p}{(\partial V/\partial p)_T} = \frac{\alpha^2 VT}{\kappa_T},$$

where we have used the indentity $(\partial p/\partial T)_V(\partial V/\partial p)_T(\partial T/\partial V)_p = -1$ and the definitions $\alpha \equiv (1/V)(\partial V/\partial T)_p$ and $\kappa_T \equiv -(1/V)(\partial V/\partial p)_T$.

that C_V is constant), Eq. 17.47 takes the form

$$\Delta S_{\text{perfect gas}} = nc_p \ln \frac{T_2}{T_1} - nR \ln \frac{p_2}{p_1}. \qquad (17.48)$$

Note how the entropy change increases as the ratio of final to initial pressures decreases.

Finally, to round out our arguments, we consider the calculation of the enthalpy change accompanying the change of state of the system under examination. Starting from Eq. 17.12, the analog of Eq. 17.39 is found to be

$$\Delta H \equiv H_2 - H_1 = \int_{p_1, T_1}^{p_1, T_2} C_p \, dT + \int_{p_1, T_2}^{p_2, T_2} \left(\frac{\partial H}{\partial p}\right)_T dp \qquad (17.49)$$

(see the second part of Fig. 17.3). Of course, we must express the coefficient $(\partial H/\partial p)_T$ in terms of the equation of state of the substance. From Eqs. 17.12 and 17.19,

$$\left(\frac{\partial H}{\partial p}\right)_T = T\left(\frac{\partial S}{\partial p}\right)_T + V$$

$$= -T\left(\frac{\partial V}{\partial T}\right)_p + V, \qquad (17.50)$$

a result we anticipated in Chapter 14. Finally, we obtain for ΔH the relation

$$\Delta H = \int_{p_1, T_1}^{p_1, T_2} C_p \, dT + \int_{p_1, T_2}^{p_2, T_2} \left[V - T\left(\frac{\partial V}{\partial T}\right)_p\right] dp. \qquad (17.51)$$

Equation 17.51 is particularly simple for the perfect gas, because the internal energy and enthalpy differ only by the term pV, which in a perfect gas is just nRT. Since $U_{\text{perfect gas}}$ depends only on T, we conclude that $H_{\text{perfect gas}}$ must likewise depend only on T. Substituting the perfect gas relation

$$T\left(\frac{\partial V}{\partial T}\right)_p = \frac{nRT}{p} = V \qquad (17.52)$$

into Eq. 17.51 and again assuming C_p to be independent of T, we find as the analog of Eq. 17.42 that

$$\Delta H_{\text{perfect gas}} = C_p(T_2 - T_1). \qquad (17.53)$$

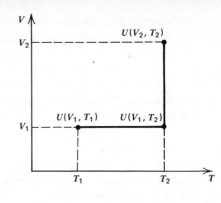

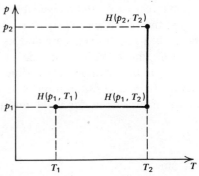

FIGURE 17.3
Paths along which the internal energy and enthalpy changes corresponding to the changes of state, $(V_1, T_1) \rightarrow (V_2, T_2)$ and $(p_1, T_1) \rightarrow (p_2, T_2)$ may be computed.

17.4
Entropy Changes in Irreversible Processes

We return, now, to the discussion of the nature of irreversible processes and the connection between irreversibility and the entropy function. We first expand the discussion of Eq. 16.45. Consider the work done by a fluid in an irreversible process. In general, for a fluid, we can write

$$-w = \int_{\text{path}} p_{\text{app}} \, dV, \qquad (17.54)$$

and the entire difference between w_{rev} and w_{irrev} lies in the path of integration of Eq. 17.54 and the consequent relationship between the applied pressure and the system pressure. Generally, a reversible process represents only the limiting behavior of a system as the rate of the process approaches zero. To compute the work done, in the representation displayed in Eq. 17.54, the pressure to be entered in the integrand is always the external pressure acting on a system. It is only in the limiting case of a reversible process that the difference between the external pressure acting on a system and the pressure exerted on the container by the system is negligible. In that limiting case the external pressure may be replaced by the pressure determined by the equation of state of the system. In contrast, in an irreversible work process the external pressure and the pressure of the system differ by a nonzero amount and work is done at a nonzero rate. For an expansion process this observation implies that $-w_{irrev} < -w_{rev}$ because $p_{app} < p$. Correspondingly, for a compression process the magnitude of the irreversible work required for the given compression exceeds the magnitude of the reversible work required for the same compression, because in this case $p < p_{app}$. Taking into account the sign, we always have $-w_{irrev} < -w_{rev}$. Furthermore, if we consider all possible reversible and irreversible changes of state between the same end points, the first law of thermodynamics implies that, because the internal energy is a function of state and its change thus independent of the path followed, the reversible change of state absorbs a different amount of heat from all possible irreversible changes of state. This can be seen directly from a consideration of the internal energy changes in the two classes of processes, for if the end points are the same we have

$$\Delta U = q_{rev} + w_{rev}$$
$$= q_{irrev} + w_{irrev}, \tag{17.55}$$

and, since

$$-w_{rev} > -w_{irrev} \quad \text{for a compression,} \tag{17.56}$$

$$-w_{rev} > -w_{irrev} \quad \text{for an expansion,} \tag{17.57}$$

subtraction gives

$$q_{rev} > q_{irrev} \quad \text{for a compression,} \tag{17.58}$$

$$q_{rev} > q_{irrev} \quad \text{for an expansion.} \tag{17.59}$$

Although thermodynamic theory clearly distinguishes between reversible and irreversible processes and between their consequences, the source of the irreversibility of a process taking place at a nonzero rate cannot be ascertained from thermodynamics. Rather, to obtain a deeper understanding of the origins of irreversibility in any given process we must appeal to some form of kinetic theory analysis.

The simplest and most obvious example of how conducting a process at a nonzero rate leads to dissipation is provided by a system consisting of a block sliding on a horizontal slab. If there were no friction, lateral displacement of the block on the slab would require no work. However, because of the existence of friction, in order to move the block with constant velocity $\mathbf{v}$ a constant force, proportional to $\mathbf{v}$, must be applied to the block. The work done in moving the block a distance d is transformed into heat. Since the force required to maintain the velocity $\mathbf{v}$ is proportional to $\mathbf{v}$, any motion of the block with nonzero velocity dissipates work as heat, hence is irreversible. For this example it is readily shown that the rate at which heat is generated is proportional to the square of the velocity of the block.[2] Therefore, the rate at which heat is generated

---

[2] The force required to maintain a block of mass M sliding with constant velocity $\mathbf{v}$ is directed parallel to $\mathbf{v}$ and just matches the force arising from friction, which is antiparallel to $\mathbf{v}$, say, $-\zeta\mathbf{v}$. The work done by the constant force $\zeta\mathbf{v}$ does not increase the kinetic energy of the block, which is fixed at $\frac{1}{2}Mv^2$ for constant speed v. The rate at which work is done on the system (considered as the block plus the slab) is simply $\zeta\mathbf{v}$, the magnitude of the external force, times $\mathbf{v}$, the distance traveled per unit of time, or ζv^2. Since the kinetic energy of the block is constant, this must be equal to the rate at which work of displacement is converted to heat.

approaches zero more rapidly than the velocity of the block approaches zero, and the translation of the block on the slab approaches reversibility as its velocity approaches zero.

A similar macroscopic description can be given for slow processes in a liquid system. If the velocity of the fluid is not the same everywhere, there are velocity gradients present. Moreover, there is friction between two infinitesimal adjacent layers of liquid moving with different velocities; work must be done against this friction to maintain any nonzero velocity gradient. It is again found that in the limit of very small fluid velocity the rate of dissipation of work into heat approaches zero more rapidly than does the velocity.

In both the cases described, if the velocity is infinitesimally small, the rate of dissipation of work as heat is an order smaller (e.g., if the velocity is $\mathbf{v}$, the rate at which work is dissipated is proportional to v^2). It is therefore meaningful to say that processes in these systems are irreversible if they proceed at nonzero rates, but approach reversibility as the rates of the processes approach zero.

In the case of thermal processes—for example, the transfer of heat down a nonzero temperature gradient—it is hard to find a macroscopic mechanical interpretation of the irreversibility. Indeed, in cases of this type the macroscopic concept of irreversibility rests on the interpretation of the entropy change in the process, as we now illustrate.

Consider two equal masses of a perfect gas, one maintained at T_{1i} and the other at T_{2i}. Suppose that the containers have rigid walls, so that when the two containers are brought into thermal contact their volumes are maintained constant. Under these conditions no work is done by the gases as they achieve the equilibrium temperature T_f. Since we have specified that the volume is constant and that the masses of gas are equal, the equilibrium temperature is determined by the heat transferred according to the equation

$$C_V(T_{1i} - T_f) = C_V(T_f - T_{2i}), \qquad (17.60)$$

where we have assumed C_V to be independent of T. From Eq. 17.60 we find that

$$T_f = \tfrac{1}{2}(T_{1i} + T_{2i}). \qquad (17.61)$$

But as a result of this process the entropies of the two masses of perfect gas change by the amounts

$$\Delta S_1 = C_V \ln \frac{T_f}{T_{1i}},$$

$$\Delta S_2 = C_V \ln \frac{T_f}{T_{2i}}, \qquad (17.62)$$

so that the total entropy change is

$$\Delta S = \Delta S_1 + \Delta S_2 = C_V \ln \frac{T_f^2}{T_{1i}T_{2i}}. \qquad (17.63)$$

Now, we must have $(T_{2i} - T_{1i})^2 \geq 0$, irrespective of whether T_{1i} or T_{2i} is the greater. By expansion, this becomes

$$(T_{2i} - T_{1i})^2 = (T_{2i})^2 - 2(T_{1i}T_{2i}) + (T_{1i})^2 \geq 0. \qquad (17.64)$$

If we now add $4T_{1i}T_{2i}$ to each side of Eq. 17.64 and collect terms, we find that

$$T_f^2 = \tfrac{1}{4}(T_{1i} + T_{2i})^2 \geq T_{1i}T_{2i}, \qquad (17.65)$$

so that ΔS as defined by Eq. 17.63 is always positive if $T_{1i} \neq T_{2i}$. This is, of course, just the thermodynamic characterization of an irreversible process in an isolated system, namely, $\Delta S > 0$. On the other hand, if $T_{1i} = T_{2i}$, heat may be transferred reversibly, and there is no change in total entropy because the increase in entropy of one mass of gas is exactly balanced by the decrease in entropy of the other mass of gas.

17.5
Entropy Changes in Phase Transitions

There are two other applications of the second law that have great importance, namely, to phase changes and to chemical reactions. We defer discussion of the entropy changes in chemical reactions to the next chapter. In Chapter 14 we mentioned that associated with a phase change, such as liquid→gas, there is a latent heat of transition. Moreover, since the enthalpy change occurs at constant temperature, and since equilibrium prevails between the two phases for any relative proportions of the phases so long as the temperature is maintained at the transition temperature, the entropy change accompanying the phase change is just

$$\Delta S_{\text{transition}} = \frac{\Delta H_{\text{transition}}}{T_{\text{transition}}}. \qquad (17.66)$$

What are the magnitudes of typical entropies of fusion and vaporization, and how can they be interpreted? From the data in Tables 14.3, we can deduce that the molar entropy of vaporization is of the order of 80 J/K mol for many substances. In terms of the statistical molecular theory of matter, this implies that there are many more states accessible to the gas than to the liquid for a given energy range. This is easily understandable, at least qualitatively. Suppose that we approximate the liquid by a perfect gas of the same density as the liquid. Clearly this approximation is very crude, but it will suffice to give some idea of the important role that the volume per molecule plays in determining the entropy of a fluid. Now for many liquids the molar volume is of the order of 50 cm³/mol, whereas at the normal boiling point the molar volume of the vapor is usually of the order of magnitude of $3-5 \times 10^4$ cm³/mol. Using our naive approximation we see that the principal change occurring in the vaporization is the almost 1000-fold increase in the volume per molecule. The isothermal expansion of a perfect gas through such a range of volume leads to a change of entropy per mole of $R \ln(1000) = 6.9R$, or 57 J/K mol. Thus this model suggests that the entropy of vaporization of a liquid has a large contribution arising from the volume change between liquid and gas. Note, however, that there also appears to be a substantial contribution to the entropy of vaporization from the change in the intermolecular force field. That this is so can be verified independently of the crude perfect gas model for the expansion part of the entropy change. Substances that are known to be dimerized in the liquid, such as NO, and substances known to be extensively hydrogen bonded in the liquid, such as H_2O, and that vaporize to give unassociated gases, have large entropies of vaporization. Typical values of the entropies of vaporization of these substances cluster around $13-14R$, or $\sim 108-116$ J/K mol.

It is possible to give a similar simple picture of the enthalpy change on vaporization. At the boiling point the density of the vapor is usually sufficiently low that for the present purposes it can be assumed to be a perfect gas. Then the enthalpy of the vapor is determined only by the kinetic energy of the molecules, whereas the enthalpy of the liquid contains an important contribution from the intermolecular interactions. Since vaporization is an isothermal process, accepting this interpretation implies that the enthalpy of vaporization measures the work required to free a molecule in the liquid from its neighbors, thereby enabling the volume per molecule to increase. In loose terms, the enthalpy of vaporization measures the work necessary to overcome the attractions between molecules so as to be able to expand the fluid. It must be emphasized that this pictorial description is quantitatively crude, and the dense perfect gas is a very poor model of a liquid.

In general the entropies of fusion of substances, which can be obtained from Table 14.2, range over a considerable scale. For very simple solids, such as Ar, Kr, and Xe, the entropies of fusion per mole are close to R, an amount that again may be interpreted as determined mainly by the change in freedom of motion of a molecule, part of which is reflected in the change in volume per molecule. Since the change of total volume per molecule is small (of the order of 10% of the volume) for the solid→liquid transition, what is important is the change in effective volume per molecule. By effective volume we mean the volume available to the center of mass of the molecule. Because of repulsion between molecules this is less than the volume per molecule; it also changes more on fusion of the solid. For some polyatomic molecules the entropy of fusion is very large. For example, for SO_2 the molar entropy of fusion is 37.4 J/K mol. In most cases the large entropy of fusion of a polyatomic molecular solid reflects a change in the freedom of molecular rotation, since in many solids rotation is hindered or even not possible, whereas in most liquids molecular rotation is almost free. The latter deduction is also supported by a comparison of the entropies of vaporization of polyatomic molecular liquids and, say, liquid Ar.

In the preceding remarks we have used the concept of freedom of motion of a molecule, or volume available per molecule, rather loosely. The reason for our choice of analogy will become apparent later in this book. For the present it is sufficient to realize that we have given only an imperfect analogy so as to invest the observed entropy changes with some simple molecular significance, and that a precise description requires a careful analysis of molecular interactions, the real energy spectrum, and other details characteristic of a particular substance.

It is pertinent to mention briefly the computation of entropy changes that accompany chemical reactions. Clearly, the method to be used must be analogous to that introduced in Chapter 14 for computing the energy and enthalpy changes in a reaction. However, the consideration of entropy changes raises interesting questions concerning standard states, and we therefore defer detailed consideration of this topic until the end of the next chapter.

Further Reading

Kestin, J., *A Course in Thermodynamics*, Vol. 1 (Blaisdell, Waltham, Mass., 1965), chaps. 11–13.

Lewis, G. N., and Randall, M., *Thermodynamics*, 2nd ed. revised by K. S. Pitzer and L. Brewer (McGraw-Hill, New York, 1961), chaps. 7–11.

Problems

1. Using the equation of state of the perfect gas, show that Eqs. 17.38 and 17.48 are equivalent. Hint: Start by showing that for a perfect gas with temperature-independent heat capacities, $C_p = C_V + nR$.

2. Show that for a change of state $(p_1, T_1) \rightarrow (p_2, T_2)$

$$\Delta H_{\text{perfect gas}} = \int_{T_1}^{T_2} C_p \, dT.$$

3. Suppose that a perfect gas is taken from an initial state p_1, V_1, T_1 to some other state p_2, V_2, T_2 (with $V_2 > V_1$ and $T_2 > T_1$), by the following reversible processes:
   (a) Adiabatic compression until $T = T_2$ followed by isothermal expansion until $V = V_2$;
   (b) Isobaric cooling until $T = T_1'$ followed by adiabatic compression until $T = T_2$ and then isothermal expansion until $V = V_2$.
   Show that the entropy changes in processes (a) and (b) are equal, and that they are also equal to the entropy change shown in Eqs. 17.48 and 17.38.

4. The constant-volume molar heat capacity of a solid may be approximated by

$$c_V = \frac{12\pi^4}{5} R \left(\frac{T}{\Theta_D} \right)^3.$$

   if $T \ll \Theta_D$. The constant Θ_D is different for each solid. For Al, $\Theta_D = 398$ K. What is the entropy change of a 5-g mass of Al when it is heated at constant volume from 4 K to 176 K?

5. Using the data of Tables 14.2 and 14.3, calculate entropies of fusion and vaporization for a number of compounds. Include some that hydrogen bond in the condensed phase, such as H_2O and HF. Interpret your calculations in terms of simple models.

6. Consider a space in which T and S are taken as the independent variables. Show that an isochoric process is represented in this space by a curve with slope T/C_V, and that an isobaric process is represented by a curve with slope T/C_p.

7. Calculate the entropy increase when 2 mol of NH_3 are heated at constant pressure from 300 to 400 K. Assume that
$$c_p^{NH_3} = 28.0 + 26.3 \times 10^{-3} T \text{ J/K mol}.$$

8. Derive general expressions for the changes in H and S corresponding to the change of state p_1, $T_1 \rightarrow p_2$, T_2 for a gas with equation of state

$$p = \frac{RT}{v - b} - \frac{a}{v^2},$$

   with b and a constants.

9. Calculate the changes in U, H, and S when 1 mol of liquid water at 0°C and 1 atm is converted to steam at 200°C and 3 atm.

   Data: $c_p(\text{liquid}) = 75$ J/K mol

   $c_p(\text{steam}) = 36.9 - 7.9 \times 10^{-3} T + 9.2 \times 10^{-6} T^2$ J/K mol

   $\Delta h_{\text{vap}}(100°C, 1 \text{ atm}) = 40.6$ kJ/mol

   Assume the steam to be a perfect gas and the liquid to have temperature-independent density and c_p.

10. (a) Consider 1 mol of supercooled water at −10°C. Calculate the entropy change of the water and of the surroundings when the supercooled water

freezes at $-10°C$ and 1 atm.

$$\text{Data: } c_p(\text{ice}) = 38 \text{ J/K mol}$$

$$c_p(\text{water}) = 75 \text{ J/K mol}$$

$$\Delta h_{\text{fus}}(0°C) = 6026 \text{ J/mol}$$

Consider $c_p(\text{ice})$ and $c_p(\text{water})$ to be independent of temperature.

(b) Is the freezing of supercooled water at $-10°C$ a reversible or an irreversible process? Justify your answer quantitatively.

11. Show that

(a)
$$\left(\frac{\partial T}{\partial V}\right)_U = \frac{p - T(\partial p/\partial T)_V}{C_V} = -\frac{T^2}{C_V}\left(\frac{\partial(p/T)}{\partial T}\right)_V.$$

(b)
$$\left(\frac{\partial T}{\partial p}\right)_H = -\frac{V - T(\partial V/\partial T)_p}{C_p} = \frac{T^2}{C_p}\left(\frac{\partial(V/T)}{\partial T}\right)_p.$$

12. From the following data, estimate the molar entropy of C_6H_6 imagined as a perfect gas at $25°C$ and 1 atm. At $25°C$ the vapor pressure of C_6H_6 is 95.13 torr and its heat of vaporization is 33850 J/mol. The heat of fusion is 9866 J/mol at the freezing point, which is $5.53°C$. The average value of $c_p(l)$ between 5.53 and $25°C$ is 134.0 J/K mol, and the entropy of the solid at $5.53°C$ is 128.8 J/K mol.

13. Derive the Gibbs–Helmholtz equation,

$$\left(\frac{\partial(G/T)}{\partial T}\right)_p = -\frac{H}{T^2}.$$

14. The heat capacity of solid iodine between $0°C$ and the melting temperature $113.6°C$ is represented by the equation (with t in $°C$)

$$c_p = 54.68 + 13.4 \times 10^{-4}(t-25)^2 \text{ J/K mol.}$$

The molar heat of fusion is 15650 J/mol at the melting point. The entropy of solid iodine is 117 J/K mol at $25°C$. What is the entropy of liquid iodine at the melting point?

15. At thermal equilibrium between two parts of an isolated system, $\beta_1 = \beta_2 = \beta$. Given the expression for β (Eq. 15.45) and the facts that for each part,

$$\Omega \propto U^{3N/2}V^N,$$

$$U = \tfrac{3}{2}Nk_BT \quad \text{(valid for ideal gases),}$$

(a) Show that $\beta = 1/k_BT$.

(b) For fixed V, show that $dU = T\,dS$.

(c) Add the known work term $-p\,dV$ to dU of (b), and show that for an ideal gas

$$\left(\frac{\partial S}{\partial V}\right)_U = \left(\frac{\partial S}{\partial V}\right)_T = \frac{p}{T}.$$

(d) Show that, in general,

$$\left(\frac{\partial S}{\partial V}\right)_T = \frac{p}{T} + \frac{1}{T}\left(\frac{\partial U}{\partial V}\right)_T.$$

(e) Consider an isolated system for which

$$\left(\frac{\partial U}{\partial V}\right)_T = 0.$$

Prove that if the system is divided into two parts by a movable partition, at equilibrium

$$\gamma_1 = \gamma_2,$$

where

$$\gamma \equiv \left(\frac{\partial \ln \Omega}{\partial V}\right)_U$$

for each system. Finally, using all of your results, prove

$$pV = Nk_B T.$$

16. Evaluate $(\partial U/\partial V)_T$ and $(\partial H/\partial p)_T$ for 1 mol of a van der Waals gas:

$$\left(p + \frac{a}{v^2}\right)(v - b) = RT.$$

17. Calculate $C_p - C_V$ for a van der Waals gas. (See Problem 13 in Chapter 14.)

18. The net work done in an irreversible cyclic process is 50 J. Can you say anything about the net heat absorption?

19. Evaluate ΔA and w for the expansion of 1 mol of ideal gas at 300 K from 10 atm pressure to 1 atm:
    (a) Reversibly.
    (b) By free expansion into a vacuum.
    (c) Against constant external pressure of 1 atm.

20. Evaluate $(\partial S/\partial V)_T$ for 1 mol of a van der Waals gas. For a given isothermal expansion, will ΔS be greater for an ideal gas or for a van der Waals gas?

21. Consider a manufacturing process that involves, because of the change from initial to final states, changes in internal energy, volume, and entropy of a system by ΔU, ΔV, ΔS. We need not specify what the system is. Suppose that the input of energy to the system is accomplished by condensation of steam, so that heat q is transferred to the system at temperature T_S. During the manufacturing process the surroundings absorb heat q_0 and are maintained at the constant temperature T_0 and pressure p_0. Show that if the manufacturing process is reversible,

$$q - \frac{T_S}{T_S - T_0}(\Delta U + p_0 \Delta V - T_0 \Delta S) = 0.$$

Show that if the process is irreversible,

$$q - \frac{T_S}{T_S - T_0}(\Delta U + p_0 \Delta V - T_0 \Delta S) > 0,$$

and the amount by which q exceeds the second term represents heat input that is wasted and is unavailable for work.

22. Show that
    (a)
    $$\left(\frac{\partial S}{\partial p}\right)_V \left(\frac{\partial T}{\partial V}\right)_p - \left(\frac{\partial T}{\partial p}\right)_V \left(\frac{\partial S}{\partial V}\right)_p = -1.$$

    (b)
    $$\left(\frac{\partial H}{\partial V}\right)_T = -V^2 \left(\frac{\partial p}{\partial T}\right)_V \left(\frac{\partial (T/V)}{\partial V}\right)_p.$$

23. Calculate the entropy of Pb vapor at 25°C and 1 atm from the following data. Between 900 and 1600 K the vapor pressure of Pb (in torr) is given by

$$\log p = 7.822 - \frac{9854}{T}.$$

The gas contains only Pb atoms. At 25°C the molar entropy of Pb(s) is 64.98 J/K mol. Below 900 K the average value of $c_p(s)$ is 28.0 J/K mol, of $c_p(l)$ is 26.4 J/K mol, and of $c_p(g)$ is 20.8 J/K mol. The melting point of Pb at 1 atm is 600 K, and the entropy of fusion is 7.95 J/K mol.

24. Measurements of the properties of some gaseous substance give

$$\left(\frac{\partial v}{\partial T}\right)_p = \frac{R}{p} + \frac{a}{T^2},$$

$$\left(\frac{\partial v}{\partial p}\right)_T = -Tf(p),$$

where a is a constant and $f(p)$ depends only on the pressure. It is also found that

$$\lim_{p\to 0} c_p = c_p \text{ for monatomic ideal gas} = \tfrac{5}{2}R.$$

Show that:

(a) $f(p) = \dfrac{R}{p^2}$.

(b) $pv = RT - \dfrac{ap}{T}$.

(c) $c_p = \dfrac{2ap}{T^2} + \dfrac{5}{2}R$.

25. The molecular theory of the perfect gas leads to the relations

$$U = an^{5/3}V^{-2/3}e^{2S/3nR},$$

$$A = \frac{3}{2}nRT\ln\frac{2aen^{2/3}}{3RTV^{2/3}},$$

where a is a constant characteristic of the gas, e is the base of natural logarithms, and n is the number of moles. Show that both of these equations yield

$$pV = nRT.$$

26. The molar Gibbs free energy of some gas is

$$g(T, p) = a + bT + cT\ln T + RT\ln p - \left(\frac{d}{T^3}\right)p,$$

where a, b, c, and d are constants. Find the equation of state of this gas.

18

The Third Law
of
Thermodynamics

Thus far we have examined some aspects of the thermodynamic description of a system, how that description is related to its energy-level spectrum, and how changes in thermodynamic functions are measured. With respect to the last point, we have shown that with the equation of state and either C_p or C_V as a function of T at one value of p or V, respectively, we can calculate all changes in the thermodynamic functions of a one-component system. For example, for the change of state p_1, $T_1 \rightarrow p_2$, T_2 we have (see Eqs. 17.51 and 17.46)

$$\Delta H = \int_{p_1, T_1}^{p_1, T_2} C_p \, dT + \int_{p_1, T_2}^{p_2, T_2} \left[V - T \left(\frac{\partial V}{\partial T} \right)_p \right] dp, \qquad (18.1)$$

$$\Delta S = \int_{p_1, T_1}^{p_1, T_2} \frac{C_p}{T} \, dT - \int_{p_1, T_2}^{p_2, T_2} \left(\frac{\partial V}{\partial T} \right)_p dp. \qquad (18.2)$$

Consider, now, an isothermal process in some system, say, at temperature T_2. The Gibbs free-energy change in this process is

$$\Delta G = \Delta H - T_2 \, \Delta S \qquad (18.3)$$

$$= \Delta H + T_2 \left(\frac{\partial \, \Delta G}{\partial T} \right)_p, \qquad (18.4)$$

where Eq. 18.4 follows from Eq. 17.18. Equation 18.4 is known as the *Gibbs–Helmholtz equation*; we shall say more about it, and use it extensively, later.

We shall show, in Chapters 19 and 21, that the equilibrium composition of a reacting mixture is determined by the difference in Gibbs free energies per mole of the products and reactants. Accordingly, it is of interest to be able to calculate the change in Gibbs free energy for the transformation of reactants to products from thermal measurements alone. If that is possible, reaction equilibria can be predicted, and the necessity for sometimes difficult measurements of the composition of reacting mixtures can be avoided. To see what is involved in calculating ΔG from thermal data, we imagine a reaction to be carried out at constant temperature T_2 and constant pressure p_2. Then

$$\Delta G = \Delta H_{T_1} - T_2 \, \Delta S_{T_1} + \int_{T_1}^{T_2} \Delta C_p \, dT - T_2 \int_{T_1}^{T_2} \frac{\Delta C_p}{T} \, dT, \qquad (18.5)$$

where the subscript T_1 refers to evaluation of ΔH and ΔS at that temperature. Note that ΔH_{T_1} and ΔS_{T_1} play the roles of constants of integration; their values must be known in order that ΔG may be computed. It seems natural to choose for the reference temperature $T_1 = 0$, the absolute zero. Then we can write

$$\Delta G = \Delta H_0 - T_2 \, \Delta S_0 + \int_0^T \Delta C_p \, dT - T_2 \int_0^T \frac{\Delta C_p}{T} \, dT. \qquad (18.6)$$

Equation 18.6 will be valid only if ΔH_0, ΔS_0, and the integrands of the two integrals displayed, do not diverge as $T_1 \rightarrow 0$. This condition imposes a restriction on the temperature dependence of the specific heat near the absolute zero. In addition, Eq. 18.6 still contains the limiting changes in enthalpy and entropy at $T_1 = 0$. The possibility that ΔG can be calculated from thermal data alone depends, finally, on the behavior of ΔS_{T_1} and ΔC_p as $T_1 \rightarrow 0$.

We now ask if the changes of entropy and of Gibbs free energy observed to occur show any special pattern as $T \rightarrow 0$. This question was addressed by Nernst in the early years of the twentieth century. He noticed that experimental data for the enthalpy and the Gibbs free-energy changes in reactions appeared to show

$$\lim_{T \rightarrow 0} \Delta G \rightarrow \Delta H. \qquad (18.7)$$

The limited data then available did not extend to very low temperatures, but the trend in the data was obvious. If relation 18.7 is in fact valid, and $\Delta G = \Delta H$ when $T = 0$, it must be true that

$$\lim_{T \rightarrow 0} \Delta S = 0. \qquad (18.8)$$

About the same time, as the technology for obtaining low temperatures improved, measurements of C_p were extended to lower and lower temperatures. The then available data strongly suggested that

$$\lim_{T \rightarrow 0} C_p = 0. \qquad (18.9)$$

Using these observations, Nernst conjectured that the following limiting laws are generally valid for isothermal processes:

$$
\begin{aligned}
&\lim_{T \rightarrow 0} \Delta G = \Delta H, \\[4pt]
&\lim_{T \rightarrow 0} \left(\frac{\partial \Delta G}{\partial T} \right)_p = \left(\frac{\partial \Delta H}{\partial T} \right)_p, \\[4pt]
&\left. \begin{aligned} \lim_{T \rightarrow 0} \Delta S &= 0, \\ \lim_{T \rightarrow 0} \Delta C_p &= 0, \end{aligned} \right\} \quad \begin{aligned} &\text{independent of } p \text{ and any} \\ &\text{other external variable.} \end{aligned}
\end{aligned} \qquad (18.10)
$$

The limiting laws (18.10) represent the original formulation of the *Nernst heat theorem.*

Numerous more recent investigations have substantiated Nernst's conjecture and shown its limitations. From the point of view of practical thermodynamics, the Nernst heat theorem, or the *third law of thermodynamics* as it has come to be known, is of great importance. Its use permits calculation of an absolute entropy and, as we shall see later, equilibrium constants for heterogeneous reactions from thermochemical measurements alone. On the other hand, there remain significant theoretical questions concerning the formulation and use of the third law.

Rather than follow the historical development, we shall open our discussion of the third law by examining the information available from the microscopic spectrum of states of a system. The definition of the entropy in terms of the number of accessible states in the range from E to

$E + dE$ has implications with respect to the properties of the entropy function in the limit as the thermodynamic temperature tends to zero. We shall explore these implications, then set up a scale of absolute values for the entropy and examine the nature of entropy changes in chemical reactions.

As a preliminary step to the discussion of the limiting value of the entropy at $T = 0$, we consider how $\ln \Omega(E)$ varies with E. A plot of $\ln \Omega(E)$ against E has slope $\partial \ln \Omega(E)/\partial E$, which by definition is equal to $1/k_B T$ (Eq. 15.40). Thus for energies close to the ground-state energy, the slope of $\ln \Omega(E)$ versus E is very large, and it decreases continuously as E increases. Furthermore, the curve of $\ln \Omega(E)$ versus E has curvature $\partial^2 \ln \Omega(E)/\partial^2 E$ which, in terms of temperature, is $-(1/k_B T^2)(\partial T/\partial E)$. At the end of Chapter 15 we demonstrated that $\partial T/\partial E$ is positive; therefore the plot of $\ln \Omega(E)$ versus E curves toward the E axis. By combining these two properties we can construct the schematic diagram shown in Fig. 18.1.

Let the ground-state energy of a system be E_0. Usually, corresponding to the energy E_0 there exists only one state, or at most a small number of states, of the system. Thus $\Omega(E)$ approaches unity, or at most some small number, in the limit $E \to E_0$. Suppose that $\Omega(E_0) = 1$. Then the entropy $k_B \ln \Omega(E_0)$ vanishes. Even if $\Omega(E_0) = \eta$, where η is a small number, the entropy $k_B \ln \eta$ is negligibly small relative to the entropy when $E \gg E_0$; the latter entropy is of the order of magnitude of $k_B \nu$, where ν is the number of degrees of freedom of the system (see Eq. 15.21).

We have already indicated that

$$\frac{\partial T}{\partial E} > 0. \tag{18.11}$$

This is just another way of saying that the specific heat of a one-phase system is always positive or, in different words, that adding energy to a system in a one-phase region can only increase its temperature. Conversely, removing energy from a one-phase system can only decrease its temperature. Since the lowest possible energy the system can have is E_0, from Eq. 18.11 and the requirement that T be positive we deduce that

$$\lim_{E \to E_0} T = 0; \tag{18.12}$$

in words, in the limit that the energy of the system approaches the ground-state energy, the temperature of the system approaches zero. This property is usually stated the other way around because it is normal to use the temperature as the independent variable. We therefore cite the complementary relation to Eq. 18.12, namely,

$$\lim_{T \to 0} E = E_0; \tag{18.13}$$

in words, in the limit that the temperature approaches zero, the energy of the system approaches the ground-state energy. Thus, by combining Eq. 18.13 with the assumption that $\Omega(E_0)$ is a small number, we are led to the deduction that

$$\lim_{T \to 0} k_B \ln \eta \approx 0, \tag{18.14}$$

which is to be interpreted to say that in the limit that the temperature

18.1
The Magnitude of the Entropy at $T = 0$

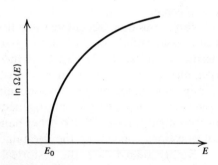

FIGURE 18.1
Schematic plot of the dependence of $\ln \Omega(E)$ on E.

approaches zero, the entropy of the system also effectively approaches zero.

Under what conditions is the assumption that $\Omega(E_0) = \eta$ a valid representation of the states of the system? Clearly, this depends on the nature of the system under examination. If there exists a set of states near to the absolute ground state of the system that span an energy range small compared with the thermal energy $k_B T$, where T is the temperature at which the system is maintained, then the number of accessible states remains very large unless the system is brought to a temperature T_0 such that $k_B T_0$ becomes small compared to the energy spacing of this set of states. Examples of this behavior are numerous. One well-studied case is the set of energy levels corresponding to the possible orientations of N nuclear spins. Now, the nuclear spin–spin interaction is very weak. Yet if the temperature were made low enough, this interaction would become large relative to the thermal energy, and a system of nuclear spins would collapse into a well-defined ground state. However, even at the low temperatures common in modern laboratories, ~ 1 K, the thermal energy available is large relative to the spacings of the spin-state energy levels. To illustrate the consequences of the resultant effective degeneracy of states, consider the limiting case when the interactions between the atoms of the sample are independent of the nuclear spin and there is no external magnetic field; then all allowed orientations of the nuclear spin are of equal energy. If the nuclear spin quantum number is I, there are $2I+1$ states per nucleus, of equal energy; and for a sample of N spins, $\Omega(E_0)$ is just $(2I+1)^N$. The limiting value of the entropy then becomes

$$\lim_{T \to T_0} S = k_B \ln(2I+1)^N = N k_B \ln(2I+1) = S_0. \tag{18.15}$$

Of course the spin–spin interactions are never exactly zero, but for most nonmagnetic solids they are so small that, as indicated above, it is usually a considerable task to achieve a temperature low enough that the thermal energy is less than the energy spacing of the stationary states of nuclear spin. For systems of this type Eq. 18.14 must be replaced by

$$\lim_{T \to T_0} S = S_0, \tag{18.16}$$

where T_0 indicates some very low temperature, but one large enough that the thermal energy exceeds the energy difference between the states, and S_0 is a constant. Since the spins are already distributed over all accessible states at T_0, an increase in the temperature does not affect the states of spin orientation, and S_0 is thus independent of the exact value of T_0.

There is another limiting case of interest: If one set of energy levels is separated from another set by an energy very large relative to the thermal energy, the second set of levels can be ignored for many practical purposes. For example, the spacings of the energy levels of all nuclei are such that at all temperatures of ordinary interest for chemical purposes, say, less than 100,000 K, nuclei remain in their ground states throughout all reactions and changes of state. Insofar as nuclear excitation is concerned, any temperature up to 100,000 K is effectively zero.

We now consider what can be learned about the limiting properties of the entropy solely from thermodynamic arguments. Suppose that a system is constrained to occupy a constant volume. Then the entropy change accompanying a change in temperature $T_1 \to T_2$ is

$$\Delta S \equiv S_2 - S_1 = \int_{T_1}^{T_2} \frac{C_V}{T} \, dT. \tag{18.17}$$

Now consider the limit $T_1 \to 0$. We have

$$\lim_{T_1 \to 0} S_2(T_2) = S_1(0) + \int_0^{T_2} \frac{C_V}{T}\, dT. \qquad (18.18)$$

If the entropy is to be finite at any temperature T, it is necessary that $S_1(0)$ be finite and that $C_V(T)$ approach zero as T approaches zero; C_V must approach zero at least as some positive power of T.[1] In symbolic form, we have

$$\lim_{T \to 0} C_V(T) = 0. \qquad (18.19)$$

Note that the vanishing of the specific heat is a deduction dependent only on the assumption that the entropy function remains finite at $T = 0$, and not on any specific value of the function. If we consider a system constrained to be at constant pressure during the change of state in which $T_1 \to T_2$, reasoning similar to that just outlined leads to the conclusion

$$\lim_{T \to 0} C_p(T) = 0. \qquad (18.20)$$

Thus, specific heats C_V and C_p both vanish in the limit of zero temperature.

Although the results cited thus far follow from the simple argument that the entropy must be finite, many more far-reaching results can be derived from the following postulate, originally due to Nernst:[2]

In any system in internal equilibrium undergoing an isothermal process between two states, the entropy change of the process approaches zero as the temperature of the system approaches zero.

This is one form of the third law of thermodynamics.

The restriction to states of internal equilibrium is important. Frequently, during the approach to $T = 0$, a system develops internal constraints that prevent the achievement of internal equilibrium. These internal constraints are not under the control of the experimenter, and they prevent the achievement of equilibrium with respect to the applied and controlled external constraints. For example, a glass is unstable with respect to crystallization into a solid.[3] The glass is prevented from

---

[1] Unless C_V approaches zero as some positive power of T, the integral term in Eq. 18.18 diverges at the lower limit. For example, were C_V to be of the form

$$C_V = A + BT + CT^2 + \cdots,$$

then substitution into Eq. 18.18 and integration would lead to terms of the form

$$A \ln T|_0^T + BT|_0^T + \tfrac{1}{2}CT^2|_0^T + \cdots.$$

The logarithmic term diverges at $T = 0$; the other terms do not. If C_V is not restricted to depend on integral powers of T, the same argument shows that so long as

$$C_V = DT^m,$$

with $m > 0$, the entropy defined by Eq. 18.18 remains finite at $T = 0$.

[2] As stated, this is not the original form of the Nernst postulate. That form was: *The entropy change in any isothermal process approaches zero as the temperature at which the process occurs approaches zero* (see Eq. 18.10). This form is now known to be inadequate because of the occurrence of long-lived nonequilibrium states, such as glasses.

[3] We choose a glass as an example of a system not at complete equilibrium. Other examples could be cited.

crystallizing, on the usual time scales of interest, by the extremely large viscosity of the medium, which acts to inhibit atomic motion. Usually internal constraints change slowly in time; for example, ordinary silica glass crystallizes (devitrifies) at room temperature on a time scale of many centuries. Since the internal constraints described are not under the control of the experimenter, it cannot be guaranteed that they will remain fixed during some process in which the externally controlled parameters of constraint are varied; the alteration of the internal constraints will, in general, lead to an irreversible process and hence an entropy change. On the other hand, if the internal constraints remain fixed, and if a reversible process can be devised to connect two states of a system with the same internal constraints, then the third law as stated above is valid.

The processes that lead to internal constraints proceed at a variety of rates. It is therefore possible for a system to develop internal constraints with respect to one property, or one degree of freedom, and not another, depending on the method by which the system is prepared. A glass is out of equilibrium with respect to the atomic arrangement that is most favorable, but given the atomic geometry of the glass the motions of the atoms are typical of equilibrium. As the atomic geometry changes very slowly in time, so also do the motions of the atoms. In this particular case the nonequilibrium geometry is created by cooling a molten substance (or mixture) more rapidly than atomic rearrangement to the crystalline form can occur. On the time scales of usual experimental interest, the glass we have just described does not appear to change, and hence its properties can be adequately described thermodynamically provided that the special nature of the uncontrolled internal constraints is recognized. Other systems may change too rapidly for such characterization; each case must be considered separately.

If two glasses are compared it is found, in general, that cooling under different conditions leads to the generation of different internal constraints. If the internal constraints in one glass remain fixed, the entropy change in an isothermal process in that glass will approach zero as the temperature approaches zero. However, the entropy difference between the two glasses, which depends on two sets of uncontrolled internal constraints, will not necessarily approach zero as the temperature approaches zero.

We may summarize the preceding interpretation of the third law by saying that, if a particular set of constraints on a system remains fixed through an isothermal process, then, whether these constraints are controlled or uncontrolled, the entropy change in that process tends to zero as the temperature approaches zero.

What are the consequences of the third law? Consider the derivative $(\partial S/\partial p)_T$, which measures the rate of change of entropy with respect to an isothermal pressure change. From the third law we have

$$\lim_{T \to 0} \left(\frac{\partial S}{\partial p}\right)_T = 0; \tag{18.21}$$

combining this with the expression

$$\left(\frac{\partial S}{\partial p}\right)_T = -\left(\frac{\partial V}{\partial T}\right)_p \tag{18.22}$$

(see Eq. 17.19), we obtain

$$\lim_{T \to 0} \left(\frac{\partial V}{\partial T}\right)_p = 0. \tag{18.23}$$

In other words, the coefficient of thermal expansion,

$$\alpha = \frac{1}{V}\left(\frac{\partial V}{\partial T}\right)_p,$$ (18.24)

must vanish in the limit as the temperature approaches zero.[4]

A similar argument can be given to show that, since by the third law $(\partial S/\partial V)_T$ must be zero in the limit of zero temperature, it is necessary that

$$\lim_{T \to 0}\left(\frac{\partial p}{\partial T}\right)_V = 0.$$ (18.25)

The preceding arguments suggest a simple scheme for defining absolute entropies. The statistical theory leads to the assignment $S = 0$ to a system for which $\Omega(E) = 1$. Suppose that we cannot achieve a temperature low enough to observe temperature variations in the nonzero residual entropy S_0 corresponding to differential excitation of closely spaced energy levels. It is then feasible to adopt a practical scale in which the entropy of a substance is assigned the value zero at the "practical" zero of temperature (i.e., $T = T_0$). It must be remembered, however, that in adopting this practical scale we must restrict the temperature range considered to be such that the smallest thermal energy $(k_B T_0)$ is much greater than the energy spacing of any possible quasi-degenerate states of the system. Since in the consideration of any entropy changes the constant S_0 will cancel out, we can effectively neglect it. In the practical scale to be discussed, S_0 is assigned the value zero.

A different statement of the third law is as follows:

The entropy of any system vanishes in the state for which

$$T \equiv \left(\frac{\partial U}{\partial S}\right)_V = 0.$$

This form, due to Planck, is the usual statement of the third law of thermodynamics, and must be interpreted, together with the practical zero of the temperature scale and the nature of internal constraints and internal equilibrium, in the manner indicated. It is generally believed that if a sufficiently low temperature were achieved and equilibrium established at that temperature, then the Planck statement would be found to be valid. But in the absence of a means to achieve low enough temperatures, and in the absence of knowledge as to whether or not there exist sets of internal states of the system with exceedingly small energy spacing, only the practical scale of absolute entropies can be operationally defined.

---

[4] Note that, accepting the third law as correct, a classical perfect gas cannot exist at $T = 0$, for the coefficient of thermal expansion of a classical perfect gas is

$$\alpha = \frac{1}{T},$$

which obviously does not approach zero as T approaches zero. Similarly, a gas with specific heat independent of temperature near $T = 0$ cannot exist. The quantum properties of matter are of dominant importance near $T = 0$, and the failures just cited arise from the inapplicability of classical mechanics to a system with energy very close to the ground-state energy E_0. Near $T = 0$ or $E = E_0$, a perfect quantum gas (defined by absence of interparticle potential energy) has an equation of state different from the classical equation of state; the new equation of state, and also the specific heat of the gas, satisfy the third law. An example illustrating this behavior is described in Section 22.5.

In the practical scale of absolute entropies we define the *calorimetric entropy* by

$$S(T, p) = \left(\int_{T_0}^{T_f} \frac{C_p}{T} \, dT \right)_{\text{crystal}} + \frac{\Delta H_{\text{fus}}}{T_f} + \left(\int_{T_f}^{T_b} \frac{C_p}{T} \, dT \right)_{\text{liquid}}$$
$$+ \frac{\Delta H_{\text{vap}}}{T_b} + \left(\int_{T_b}^{T} \frac{C_p}{T} \, dT \right)_{\text{vapor}} \tag{18.26}$$

at the temperature T and pressure p. Clearly, if at (T, p) the substance is a solid, the entropies of fusion and vaporization and the entropy changes in the liquid and gaseous phases, all included in the general Eq. 18.26, are to be omitted. In Table 18.1 are shown the practical absolute entropies of a number of substances. Note how large the entropy of a gas is relative to the entropy of a liquid or solid, and also how the entropy increases as the molecule becomes more complicated.

The scheme used to calculate the entropy change in a chemical reaction is identical with that described in Chapter 14 for the calculation of enthalpy changes. Thus, if s_i is the entropy per mole of substance i, the reaction

$$\nu_A A + \nu_B B + \cdots = \nu_M M + \nu_N N + \cdots \tag{18.27}$$

results in an entropy change of

$$\Delta S = \sum_{i=1}^{n} \nu_i s_i. \tag{18.28}$$

Again, as in the calculation of enthalpy changes, it is convenient to choose a reference state for the entropy. We denote the molar entropy of substance i in the reference state by s_i^0. It is usual to choose $T = 298.15 \ K$, $p = 1$ atm as the conditions defining the reference state. Thus, when the reaction

$$\tfrac{1}{2} H_2(g) + \tfrac{1}{2} Cl_2(g) = HCl(g) \tag{18.29}$$

is carried out at 298.15 K and 1 atm, the entropy change is (see Table 18.1)

$$\Delta S = 186.678 - \tfrac{1}{2}(130.587) - \tfrac{1}{2}(222.949)$$
$$= 9.910 \ \text{J/K}. \tag{18.30}$$

The calculation of the entropy change in a reaction under conditions differing from the standard state is carried out in a fashion parallel to the corresponding enthalpy calculation described in Chapter 14. Using the defining relation, Eq. 18.28, we can express the entropy of each substance participating in the reaction in terms of its entropy in the standard state, its specific heat, and its equation of state. Since the procedure followed is similar to that already described in Chapter 14, we shall not provide an explicit illustration here; examples will be found in the problems.

Having worked out procedures for calculating both enthalpy and entropy changes in chemical reactions and other processes, we may ask why the calculations are of interest. We shall see in Chapter 19 that the nature of the equilibrium in a thermodynamic system is defined by the competition between enthalpy (or energy) changes and entropy changes. The precise character of this competition is defined by the constraints on the system. Thus, we shall show that knowledge of the energy and entropy changes for a given thermodynamic system and process enables us to predict the properties of the equilibrium state and its changes as the parameters of constraint vary.

TABLE 18.1
STANDARD ENTROPIES OF SOME SUBSTANCES
($T = 298.15$ K, $p = 1$ atm)

Substance	s^0 (J/K mol)	Substance	s^0 (J/K mol)
Ag(s)	42.702	P(s, white)	44.4
Al(s)	28.321	P(s, red)	63.2
Ba(s)	66.9	Pb(s)	64.81
Be(s)	9.54	S(g)	167.716
Br(g)	174.912	S(s, monoclinic)	32.55
$Br_2(g)$	245.34	Sn(s, white)	44.8
$Br_2(l)$	152.3	Sn(s, gray)	51.5
C(s, diamond)	2.377	Zn(s)	41.21
C(s, graphite)	5.694	AgF(s)	84
Cl(g)	165.088	AgCl(s)	96.11
$Cl_2(g)$	222.949	AgBr(s)	107.11
Cu(s)	33.30	AgI(s)	114.2
F(g)	158.645	$Al_2O_3(s)$	50.986
$F_2(g)$	203.3	$BaCl_2(s)$	125
H(g)	114.612	BeO(s)	14.10
$H_2(g)$	130.587	$CH_4(g)$	202.50
Hg(l)	77.4	$C_2N_2(g)$	242.09
I(g)	180.682	CO(g)	197.907
$I_2(g)$	260.580	$CO_2(g)$	213.639
$I_2(s)$	116.7	$CS_2(g)$	237.82
K(s)	63.856	CuO(s)	42.7
Mg(s)	32.51	$CuSO_4(s)$	113.4
N(g)	153.197	$CuSO_4 \cdot 5H_2O(s)$	305.4
$N_2(g)$	191.489	$D_2O(g)$	198.234
Na(s)	51.417	$D_2O(l)$	75.990
O(g)	144.218	$H_2O(g)$	188.724
$O_2(g)$	205.029	$H_2O(l)$	69.940
$O_3(g)$	237.7	HF(g)	173.51
HCl(g)	186.678	$SnCl_4(l)$	258.6
HBr(g)	198.476	$ZnCl_2(s)$	43.9
HI(g)	206.330	$ZnSO_4(s)$	124.7
$H_2S(g)$	205.64	$CH_4(g)$	186.19
$H_2SO_4(l)$	156.86	$C_2H_6(g)$	229.49
KF(s)	66.57	$C_3H_8(g)$	269.91
KCl(s)	82.68	$n\text{-}C_4H_{10}(g)$	310.12
KBr(s)	96.44	$n\text{-}C_5H_{12}(g)$	348.95
KI(s)	104.35	$n\text{-}C_6H_{14}$	388.40
$KNO_3(s)$	132.93	$C_2H_2(g)$	200.819
$K_2SO_4(s)$	175.7	$C_2H_4(g)$	219.45
MgO(s)	26.8	$C_6H_6(g)$	269.20
$MgSO_4(s)$	95.4	$CH_3OH(l)$	126.8

TABLE 18.1 (CONT.)

Substance	s^0 (J/K mol)	Substance	s^0 (J/K mol)
$NH_3(g)$	195.51	$CH_3OH(g)$	237.7
$NH_4Cl(s)$	94.6	$C_2H_5OH(l)$	160.7
$NO(g)$	210.618	$C_2H_5OH(g)$	282.0
$NO_2(g)$	211.17	$CH_3COOH(l)$	159.8
$N_2O_4(g)$	304.30	$CH_3COOH(g)$	293.3
$PCl_3(g)$	311.67	$CF_4(g)$	262.3
$PCl_5(g)$	352.7	$CCl_4(g)$	309.41
$PH_3(g)$	210.0	$CBr_4(g)$	358.2
$PbS(s)$	91.2	$CHF_3(g)$	223.84
$PbSO_4(s)$	147.3	$CHCl_3(g)$	296.48
$SO_2(g)$	248.53	$CHBr_3(g)$	331.29
$SO_3(g)$	256.23	$CH_3F(g)$	223.01
$SiCl_4(g)$	330.79	$CH_3Cl(g)$	234.18
$SiH_4(g)$	204.14	$CH_3Br(g)$	245.77

We cannot leave the study of the third law without briefly commenting on the *unattainability of absolute zero*. Indeed, to some the statement that the absolute zero cannot be attained in any system using any sequence of processes with a finite number of steps *is* the third law of thermodynamics!

Suppose that we consider some system which is adequately described by two independent variables. Ordinarily—for example, for a fluid—the independent variables chosen would be V and T, or p and V. However, for our present purposes it is convenient to choose as the independent variables T and S. Then any other property of the system is expressed as a function of T and S, for example, $V = V(T, S)$. Consider some arbitrary property of the system, $Z = Z(T, S)$. In Fig. 18.2a are represented two lines along which Z has the constant values Z_1 and Z_2, respectively. The third law requires that in any isothermal process, such as that along the lines P_1P_2 or $P_1'P_2'$, the entropy change must tend to zero as T tends to zero. Therefore the curves of constant Z must come together at the origin

18.2
The Unattainability of Absolute Zero

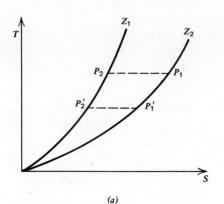

(a)

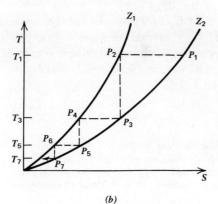

(b)

FIGURE 18.2

(a) Curves of constant Z in the (T, S) plane near $T = 0$. These are determined by the intersection of the surface $Z = Z(T, S)$ with the planes $Z = Z_1$ and $Z = Z_2$. (b) A hypothetical sequence of operations, based on the use of the variation of property Z, to reduce the temperature of a system toward $T = 0$.

where $T = 0$ and $S = 0$. Now suppose that we have available some process, based on the property Z, which can be used to reduce the temperature of the system. For example, the magnetic moments of the ions in a paramagnetic salt can be aligned in a strong magnetic field. Clearly, when the magnetic field is strong enough to align the moments, the thermal energy is small compared to the energy of interaction of an ionic moment with the field. If in the absence of the magnetic field the moments were random in spatial orientation, alignment must result in a reduction of the entropy of the salt. If the magnetization is carried out isothermally, the reduction of entropy must lead to a transfer of heat from the salt to the surroundings. If the magnetic field is then removed adiabatically, randomization of the moments must lead to a decrease in temperature, because no energy can be transferred to the salt. This method of attaining low temperatures, known as *adiabatic demagnetization*, was suggested independently by Debye (1926) and Giauque (1927). The hypothetical process based on Z to which we referred could be adiabatic demagnetization if the system were a paramagnetic salt, or any other process appropriate to the nature of the system; for our present purposes we need only hypothesize that a suitable process exists.

We imagine that it is possible to change Z isothermally from Z_2 to Z_1 along the path P_1P_2 of Fig. 18.2b, and then to change Z from Z_1 to Z_2 along the adiabatic path P_2P_3. If the system is a paramagnetic salt, Z_1 and Z_2 correspond to zero magnetic field and large magnetic field, respectively. As a result of the process described, the temperature of the system has been reduced from T_1 to T_3 (see diagram). But, because the two curves merge at $T = 0$, it is not possible to reach $T = 0$ by the process described in a finite number of steps. As can be easily seen, each successive drop in temperature, corresponding to another cycle of operation, is smaller than the preceding one, and the limit $T = 0$ is approached only as the number of cycles of operation becomes indefinitely large. We therefore conclude that the principle of the unattainability of absolute zero is equivalent to the statement of the third law already given.[5] We leave it as an exercise for the reader to prove that if the third law were not valid, that is, if the isothermal entropy change in a process did not vanish as $T \rightarrow 0$, then absolute zero could be attained in a finite series of steps of some process or processes.

18.3
Experimental Verification of the Third Law

Although we opened this chapter by pointing out that experimental studies led to the postulation of the Nernst heat theorem, our analysis has presented the third law as a postulate that is made plausible by the statistical molecular theory of matter. In this section we return to the empirical side and examine a few experimental data related to verification and demonstration of the limitations of the third law.

Consider first the limiting behavior of C_p as $T \rightarrow 0$. In Fig. 18.3 are shown the values of C_p as a function of T for several substances. In every case it is clear that the natural extrapolation of the observations corresponds to $C_p \rightarrow 0$ as $T \rightarrow 0$. We shall see later, in Chapter 22, that for a nonconducting simple solid such as NaCl

$$\lim_{T \to 0} C_p \propto T^3, \tag{18.31}$$

[5] It is more correct to say that the Nernst heat theorem implies the unattainability of absolute zero, but that the converse implication is not valid; the unattainability principle can be shown to lead to the Nernst heat theorem only under restrictive conditions. There is, then, a residual inequivalence between these two formulations of the behavior of systems as $T \rightarrow 0$.

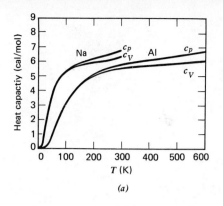

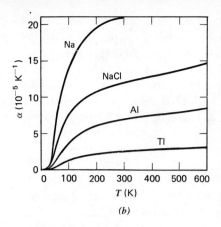

FIGURE 18.3

(a) Typical behavior of c_p and c_V as functions of temperature for two substances. (b) Experimental curves of the thermal expansion coefficient α as a function of T at $p = 0$; α is not necessarily positive, but $\alpha \to 0$ as $T \to 0$. Notable examples of negative thermal expansion are Ge and Si, where α is negative at low T, then becomes positive and continues to increase as T increases. In addition, for anisotropic crystals the thermal expansion may be quite different along the different crystallographic directions, as for example along the c and a axes for Zn and Cd. From D. C. Wallace, *Thermodynamics of Crystals* (Wiley, New York, 1972).

whereas for a simple metal

$$\lim_{T \to 0} C_p \propto \gamma T + \beta T^3. \tag{18.32}$$

Both relations 18.31 and 18.32 satisfy Eq. 18.20 (see also footnote 1 following Eq. 18.18). Even when the low-temperature behavior of the specific heat of the solid is more complex, as shown in Fig. 18.4 for $KCr(SO_4)_2 \cdot 12H_2O$, it is still true that $C_p \to 0$ as $T \to 0$.

It will be shown, in Chapter 21, that the thermodynamic functions of a gas can be calculated from a knowledge of the properties of an isolated molecule. For example, for the case of a diatomic molecule we must know the mass, the moment of inertia, the vibration frequency, and the degeneracy of the electronic ground state. From these we compute the so-called *spectroscopic entropy*. It can be compared with the absolute calorimetric entropy obtained from Eq. 18.26 with $T_0 = 0$. Some typical values are shown in Table 18.2. It is easily seen that for most of the examples cited the residual entropy S_0 (Eq. 18.16) is zero. This is not the case for CO, which provides an example of freezing in of metastable configurations as $T \to 0$. Since CO has a very small dipole moment and is almost symmetric, an individual CO molecule can be distributed in the crystal with two orientations with respect to a fixed reference frame. These two orientations, CO $\cdots$ CO $\cdots$ CO $\cdots$ and CO $\cdots$ OC $\cdots$ OC $\cdots$,[6] have nearly the

TABLE 18.2
COMPARISON OF CALORIMETRIC AND SPECTROSCOPIC ENTROPIES OF SOME DIATOMIC GASES

Substance	Calorimetric Entropy (J/K mol)	Spectroscopic Entropy (J/K mol)	Residual Entropy (J/K mol)
N_2	192.0	191.6	0
O_2	205.4	205.1	0
HCl	186.2	186.8	0
HBr	199.2	198.7	0
HI	207.1	206.7	0
CO	193.3	198.0	$5.76 \equiv R \ln 2$

[6] This notation is not intended to represent the actual molecular distribution, but only to suggest that there is an ordered structure (symbolized by CO $\cdots$ CO $\cdots$ CO $\cdots$) and many disordered structures (symbolized by CO $\cdots$ OC $\cdots$ OC $\cdots$), where the CO can have two orientations at each lattice site.

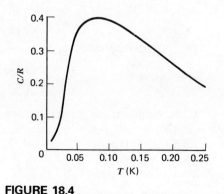

FIGURE 18.4
Specific heat of potassium chrome alum in units of R. Adapted from B. Bleaney, *Proc. Roy. Soc. (Lond.)* **A204**, 216 (1950).

same energy. Presumably, if complete equilibrium is achieved the ordered structure $CO \cdots CO \cdots CO$ is obtained. Since it requires energy to rotate a CO molecule in the force field of its neighbors, a disordered configuration can be frozen in if equilibrium is not established before T drops to where $k_B T$ is less than the energy of reorientation. Assuming that the CO molecules are randomly distributed with respect to orientation, and that all the many disordered distributions have about the same energy, there is an effective ground-state degeneracy of the crystal of 2^N,[7] hence for CO,

$$S_0 = Nk_B \ln 2 = 5.763 \text{ J/K mol.}$$

This value of S_0 is in good agreement with the observed difference between the calorimetric and spectroscopic entropies of CO.

We shall not comment further on the experimental verification of the third law. It is sufficient to say that in addition to the examples cited there are many others that verify its accuracy and utility.

Further Reading

Caldin, E. F., *An Introduction to Chemical Thermodynamics* (Oxford University Press, Oxford, 1958), chap. 8.

Epstein, P. S., *Textbook of Thermodynamics* (Wiley, New York, 1937), chaps. 13–15.

Wilson, A. H., *Thermodynamics and Statistical Mechanics* (Cambridge University Press, Cambridge, 1957), chap. 7.

Problems

1. Show that Eq. 18.25 is valid.

2. Suppose that the equation of state of a solid is

$$pv + G(v) = \Gamma u,$$

where $G(v)$ is a function of volume only, Γ is a constant, and u is the internal energy per mole. Prove that c_V approaches zero as T approaches zero.

3. At $T = 0$, $(\partial S/\partial V)_T = 0$, and also

$$\left[\frac{\partial}{\partial V} \left(\frac{\partial S}{\partial V} \right)_T \right]_{T=0} = 0.$$

Using these facts, prove that

$$\lim_{T \to 0} \left[\frac{\partial (1/\kappa_T)}{\partial T} \right]_V = 0.$$

4. Calculate the standard entropy changes in the following reactions:
   (a) $CH_4(g) + 2H_2S(g) = CS_2(g) + 4H_2(g)$,
   (b) $CO_2(g) + H_2S(g) = OCS(g) + H_2O(g)$,
   (c) $2SO_2(g) + O_2(g) = 2SO_3(g)$,
   (d) $C_2H_4(g) + H_2(g) = C_2H_6(g)$.

5. Calculate the entropy change in the reaction

$$H_2(g) + \tfrac{1}{2}O_2(g) = H_2O(g)$$

[7] Since each molecule has two equivalent orientations.

at 1000 K and 2 atm:

(a) Assuming the equations of state of all substances to be $pv = RT$,

(b) Assuming the equations of state of all substances to be

$$\left(p + \frac{a}{v^2}\right)(v - b) = RT.$$

Data:

H_2	$c_p = 2.87 + 1.17 \times 10^{-3}T - 0.92 \times 10^{-6}T^2$ J/K mol
O_2	$c_p = 36.16 + 0.845 \times 10^{-3}T - 4.310 \times 10^{-5}T^{-2}$ J/K mol
H_2O	$c_p = 30.1 + 11.3 \times 10^{-3}T$ J/K mol
H_2	$a = 0.244 \times 10^{-6}$ cm^6 atm/mol^2
O_2	$a = 1.36 \times 10^{-6}$ cm^6 atm/mol^2
H_2O	$a = 5.46 \times 10^{-6}$ cm^6 atm/mol^2
H_2	$b = 26.6$ cm^3/mol
O_2	$b = 31.8$ cm^3/mol
H_2O	$b = 30.5$ cm^3/mol

6. According to the Debye model of crystals (see Chapter 22), the heat capacity of crystals at low temperatures is given by $C_V = AT^3$, where A is a constant for any crystal sample. Show that $C_p - C_V$ at low temperature is proportional to T^7.

7. Calculate the calorimetric entropy of Na_2SO_4 at 100 K from the following data:

T	c_p (J/K mol)
13.74	0.715
16.25	1.197
20.43	2.619
27.73	6.757
41.11	18.18
52.72	29.42
68.15	43.85
82.96	55.56
95.71	64.14

8. Calculate the entropy of gaseous nitromethane at 298.15 K from the following data:

T	c_p (J/K mol)
15	3.72
20	8.66
30	19.20
40	28.87
50	35.69
60	40.84
70	44.77
80	47.89
90	50.63
100	52.80
120	56.74
140	60.46
160	64.06
180	67.74
200	71.46
220	75.23
240	78.99
260	104.64
280	105.31
300	106.06

The melting point of the solid is 244.7 K with $\Delta h_{fus} = 9703$ J/mol. The vapor pressure at 298.15 K is 36.66 mm Hg, and the heat of vaporization at that temperature is $\Delta h_{vap} = 38{,}271$ J/mol.

9. Which of the following solids do you expect to have nontrivial residual entropies at, say, 1 K? Explain your conclusion.
   (a) O_2
   (b) H_2O
   (c) $AgCl_{0.5}Br_{0.5}$
   (d) $NaCl$

10. Prove that if the isothermal entropy change in a process did not vanish as $T \to 0$, then absolute zero could be attained in a finite series of steps of some process or processes.

11. Three forms of phosphine are known; β-phosphine exists from 0 K to 49.43 K, at which temperature it transforms to α-phosphine, and γ-phosphine exists from 0 K to 30.29 K, at which temperature it also transforms to α-phosphine. Using the data tabulated below [from C. C. Stephenson and W. F. Giauque, *J. Chem. Phys.* **5,** 149 (1937)], devise a test of the third law of thermodynamics based on the computation of the entropy of α-phosphine at 49.43 K by two different routes.

T	c_p (β-phosphine) (J/K mol)	T	c_p (γ-phosphine) (J/K mol)	T	c_p (α-phosphine) (J/K mol)
15.80	4.502	17.71	8.314	31.34	34.31
18.63	6.535	20.58	11.049	32.22	36.53
21.71	8.745	23.65	14.36	33.46	39.25
25.04	11.263	25.55	16.70	34.21	42.51
28.62	13.807	27.62	20.20	34.84	44.14
33.31	17.44	29.05	23.41	35.33	47.49
37.13	20.44	29.29	24.30	35.66	abrupt
41.12	23.21	30.29	transition		decrease
45.10	26.42		$\Delta h = 82.0$ J/mol	36.12	40.21
47.85	29.21			40.87	41.17
49.43	transition			44.17	42.59
	$\Delta h = 777$ J/mol			47.52	43.97
				51.14	45.48

12. The magnetic field on a solid with very low concentration of paramagnetic ions is increased from zero to $\mathbf{H}$. The extra work required to generate the field $\mathbf{H}$ by virtue of the presence of 1 mol of magnetically polarizable material instead of vacuum is (see Table 13.1)

$$\Delta w = \mu_0 \int \mathbf{H} \cdot d\mathbf{M},$$

or, in differential form,

$$\dbar w = \mu_0 \mathbf{H} \cdot d\mathbf{M},$$

where $\mathbf{M}$ is the molar magnetic polarization. Suppose that the volume of the solid is independent of magnetic field, so that no pV work is done as the field is changed. Show that the following are valid:

$$T\,ds = c_{\mathbf{M}}\,dT - \mu_0 T\left(\frac{\partial \mathbf{H}}{\partial T}\right)_{\mathbf{M}} d\mathbf{M},$$

$$T\,ds = c_{\mathbf{H}}\,dT + \mu_0 T\left(\frac{\partial \mathbf{M}}{\partial T}\right)_{\mathbf{H}} d\mathbf{H}.$$

Here μ_0 is the magnetic permeability of free space, $c_{\mathbf{M}}$ is the molar heat capacity at constant magnetization, and $c_{\mathbf{H}}$ is the molar heat capacity at constant field strength.

13. When the field on a paramagnetic solid is held constant and the temperature is varied, it is found that an increase in temperature always leads to a decrease in magnetization **M**. Show, using this fact, that in an isothermal reversible change of magnetic field strength, heat is rejected by the solid when **H** is increased and absorbed when **H** is decreased.

14. Suppose that the magnetic field strength acting on a paramagnetic solid is adiabatically and reversibly changed. If the temperature change produced by this adiabatic process, ΔT, is very small compared to T, and $c_{\mathbf{H}}$ can be regarded as independent of T over that range, show that

$$\Delta T = -\frac{\mu_0 T}{c_{\mathbf{H}}} \int_{\mathbf{H}_1}^{\mathbf{H}_2} \left(\frac{\partial \mathbf{M}}{\partial T}\right)_{\mathbf{H}} d\mathbf{H}.$$

What is the sign of ΔT for an adiabatic decrease of **H**?

15. The magnetization of gadolinium sulfate satisfies Curie's law,

$$\mathbf{M} = \frac{\mathscr{C}_{\mathbf{M}}}{T} \mathbf{H},$$

where $\mathscr{C}_{\mathbf{M}}$, the Curie constant, has the value 780 erg K/mol oersted^2 in cgs emu. In this system of units energy is measured in ergs, H in oersteds, and $\mu_0 = 1$. Given that the molar specific heat of gadolinium sulfate at constant field strength is

$$c_{\mathbf{H}} = (2.66 \times 10^7 + 7.80 H^2) T^{-2} \text{ erg/K mol}$$

for $T < 15$ K, calculate the final temperature achieved if 1 mol is equilibrated at 15 K in a field of 10,000 oersteds and then the field is adiabatically and reversibly reduced to zero.

16. Very low temperatures cannot be easily measured; for example, gas thermometry fails. Suppose that a "magnetic temperature" is defined by

$$T^* \equiv \frac{\mathscr{C}_{\mathbf{M}}}{\chi_{\mathbf{M}}}.$$

$\mathscr{C}_{\mathbf{M}}$ is the Curie constant (see Problem 15), and $\chi_{\mathbf{M}}$, the molar magnetic susceptibility, is defined by

$$\chi_{\mathbf{M}} \equiv \frac{\mathbf{M}}{\mathbf{H}}.$$

Suppose also that a series of demagnetization curves from different fields and starting temperatures is used to calculate the entropy as a function of T^* at zero field and an apparent heat capacity defined by

$$C' = T^* \left(\frac{\partial S}{\partial T^*}\right)_{\mathbf{H}=0}.$$

In a separate series of experiments, by introduction of heat by absorption of γ-rays, a different apparent heat capacity is measured:

$$C'' = \lim_{\Delta T^* \to 0} \left(\frac{q}{\Delta T^*}\right)_{\mathbf{H}=0}.$$

Show that the thermodynamic scale of temperature is related to the magnetic scale by

$$T = T^* \frac{C''}{C'}.$$

The Nature of the Equilibrium State

Thus far we have discussed how an equilibrium state of matter can be described in terms of macroscopic variables and thermodynamic functions, how these thermodynamic functions can be evaluated if the equation of state and heat capacities are known, and what is the nature of the relationship between the entropy and the energy spectrum of the system. We now investigate the nature of the equilibrium state in more detail. Our studies will be of several kinds. First, we examine what the first and second laws imply is characteristic of the equilibrium state, and what relations between thermodynamic variables are imposed by the condition that an equilibrium state be stable. Second, using the general characteristics of the equilibrium state thus obtained, we introduce several new relationships between thermodynamic functions, defined macroscopically, and the energy spectrum of the system. These new relationships act as bridges between the macroscopic and microscopic descriptions of a system; they supplement the fundamental definition $S \equiv k_B \ln \Omega$. Finally, we demonstrate how chemical reaction equilibria are related to the thermodynamic properties of the reactants and products.

To investigate the properties of the equilibrium state of a system we must understand how the thermodynamic functions of the system depend on its mass. We begin with an analysis of the first law including mass transfer from one part of the system to another. Consider a closed system with two parts, I and II. As shown in Fig. 19.1, in part I, with volume V, there are n mol of a pure substance at temperature T and pressure p; in part II there are dn mol of the same substance at the same temperature and pressure, and occupying volume $v \, dn$, where v is the molar volume. Let the boundary between part I and the surroundings be deformable in such a way that as V varies, the pressure of the substance inside never differs more than infinitesimally from the external pressure. Finally, let there be an insulated piston connected to II so that external work can be done on the dn mol of substance in that part of the system. The boundary between parts I and II is permeable to matter; it may be thought of as an imaginary boundary, or it may be real.

We consider two states of the total system. In the initial state of the system the dn mol of substance in part II are outside the envelope defining the volume of part I. The internal energy of the total closed system is then[1]

$$nu + u \, dn,$$

[1] We take advantage of the fact that the state of a one-component system is defined by T and p, so that two samples of the same substance with the same T and p have the same internal energy per mole.

19.1
Properties of the Equilibrium State of a Pure Substance

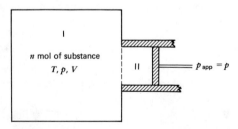

FIGURE 19.1
Schematic representation of a system with two parts. In part I, which has volume V, are n mol of substance at T and p. In part II are dn mol of the same substance, also at T and p. The piston shown permits the volume of part II to be decreased to zero. The dashed line indicates the boundary between parts I and II. Heat can be transferred to part I, but not to part II, and part I can do work on the surroundings.

where u is the molar internal energy of the substance. In the final state of the system the dn mol of substance in part II have been isothermally and reversibly driven at constant pressure p into part I, whose volume thereby changes from V to $V + dV$. In this change of state the surroundings coupled to the piston do work $pv\, dn$ on the system, the expansion of part I does work $+p\, dV$ on the surroundings, and heat $đq$ is transferred to part I from the surroundings. Let the internal energy in the final state be denoted by

$$nu + dU.$$

The first law gives, for the internal energy change of the total system in the process described,

$$dU - u\, dn = đq - p\, dV + pv\, dn. \tag{19.1}$$

Using the definition of the molar enthalpy,

$$h \equiv u + pv, \tag{19.2}$$

we find for the internal energy change of part I,

$$dU = đq + h\, dn - p\, dV, \tag{19.3}$$

which is the form taken by the first law when mass transfer is permitted.

Consider now the form taken by the second law when mass transfer is permitted. In the initial state of the system described above the entropy is

$$ns + s\, dn,$$

where s is the molar entropy. In the final state the entropy is denoted by

$$ns + dS.$$

As a result of the process by which the dn mol of substance are added to part I, there are entropy changes associated with the mass transfer, the work done, and the heat transferred. Since the dn mol of substance were added to part I isothermally and reversibly, the change in entropy of the total system, $dS - s\, dn$, must be

$$dS - s\, dn = \frac{đq}{T} = \frac{dU}{T} - \frac{h}{T}\, dn + \frac{p}{T}\, dV. \tag{19.4}$$

Using the definition of the molar Gibbs free energy,

$$g \equiv h - Ts, \tag{19.5}$$

we find that the generalization of Eq. 17.1 to include mass transfer is

$$dU = T\, dS - p\, dV + g\, dn. \tag{19.6}$$

The molar Gibbs free energy of a pure substance is also called the *chemical potential* of that substance, and is usually designated as μ. A more general definition of the chemical potential is given in Section 19.4.

Having obtained Eq. 19.6, we have the tools necessary to examine some of the most important characteristics of the equilibrium state of a one-component system. We consider, as a concrete example, the equilibrium properties of a one-component isolated system containing two phases. A liquid in contact with solid (e.g., ice and water) or a liquid in contact with gas (e.g., water and water vapor), enclosed in a rigid adiabatic container, are simple particular cases of this situation.

The starting point in our analysis is the second law of thermodynamics. As discussed in Chapter 16, in an isolated system at equilibrium the entropy has the maximum value possible. This fact was expressed, formally, in the two conditions:

1. $\Delta S_{inv} < 0$ for a finite process with the original constraints on the system retained, $\qquad\qquad$ (19.7a)

2. $\delta S \leq 0$ for an infinitesimal process with the original constraints on the system retained. The inequality refers to situations in which virtual displacements in only one direction are allowed. $\quad$ (19.7b)

Note that the equality sign in condition 19.7b implies that the entropy is an extremum with respect to positive and negative virtual displacements, and therefore the change of entropy in any such displacement starts with zero slope. The inequality sign implies that even through the entropy is an extremum, the change in entropy with the allowed unidirectional virtual displacement need not start with zero slope. We now rephrase the conditions 19.7 to make the role of the constraints that determine equilibrium more explicit.

The most general class of isolated systems is that for which only the mass and energy are specified. In such cases it is possible, for example, for the volume to change throughout the natural process that leads to establishment of the equilibrium state. Since the energy of an isolated system cannot change, Gibbs stated the most general criterion for equilibrium of such a system in the form:

For the equilibrium of any isolated system it is necessary and sufficient that, in all possible variations of the state of the system that do not alter its energy, the variation of its entropy shall either vanish or be negative.

Gibbs also stated the criterion for equilibrium in the form:

For the equilibrium of any isolated system it is necessary and sufficient that, in all possible variations of the state of the system that do not alter its entropy, the variation of its energy shall either vanish or be positive.

These statements, when suitably interpreted, include the possibilities of existence of stable, neutral, and metastable equilibria (see Section 19.3). When translated into formal mathematical statements, they read[2]

$$(\Delta S)_U \leq 0, \qquad\qquad (19.8)$$

$$(\Delta U)_S \geq 0. \qquad\qquad (19.9)$$

In keeping with the discussion of inverse processes in Chapter 16, the possible variations considered must not alter the constraints that determined the original equilibrium state. Aside from meeting this requirement, the possible variations include all changes in state produced by removing the system from isolation and operating on it with various processes. The new equilibrium states generated have the property that all will evolve to the original equilibrium state via natural processes when only the constraints that determine that original state are retained.

For some classes of isolated systems auxiliary constraints are specified—for example, that the volume of the system is fixed, or that

---

[2] The equivalence of these statements implies that if an isolated system is at equilibrium, all possible variations in its state must satisfy *both* statements 19.8 and 19.9. It is not possible that some variations in its state satisfy 19.8 but not 19.9, or 19.9 but not 19.8.

there is a given subdivision of the fixed volume into subvolumes. Conditions 19.8 and 19.9 remain valid, the new constraints merely adding extra restrictions on the processes that can be used to generate different equilibrium states.

To understand how statement 19.8 is used, we suppose a system to be described by m variables. We represent the way in which the l constraints acting on the isolated system define its equilibrium state by a set of equations of the form

$$f_1(y_1, \ldots, y_m) = 0,$$
$$f_l(y_1, \ldots, y_m) = 0,$$

(19.10)

where the y_i are the variables of the system. For example, in a one-component, two-phase isolated system the volumes of the two phases, V_1 and V_2, are subject to the constraint $V = V_1 + V_2$. When the isolation of the system is broken and various processes are used to drive it to a new equilibrium state, the values of the y_i may change. However, possible changes in the y_i are restricted by Eqs. 19.10, since in the inverse process the constraints on the isolated system are retained. From Eqs. 19.10 we find that the conditions which changes in the y_i must satisfy are

$$df_1 = 0 = \frac{\partial f_1}{\partial y_1} dy_1 + \frac{\partial f_1}{\partial y_2} dy_2 + \cdots + \frac{\partial f_1}{\partial y_m} dy_m,$$

$$df_2 = 0 = \frac{\partial f_2}{\partial y_1} dy_1 + \frac{\partial f_2}{\partial y_2} dy_2 + \cdots + \frac{\partial f_2}{\partial y_m} dy_m,$$

(19.11)

$$\vdots$$

$$df_l = 0 = \frac{\partial f_l}{\partial y_1} dy_1 + \frac{\partial f_l}{\partial y_2} dy_2 + \cdots + \frac{\partial f_l}{\partial y_m} dy_m.$$

Variations of the state of the system that satisfy Eqs. 19.11 are what we call virtual variations.

We now proceed to the analysis of the one-component, two-phase isolated system. By the definition of an isolated system:

1. The internal energy is constant;

2. The total volume is constant;

3. The total mass of the one component in the system is constant.

Our interpretation of Eq. 19.7*b* requires possible variations of the system, when it is no longer isolated, to satisfy

$$(\delta S)_{U,V,n} \leq 0,$$

(19.12)

where the subscripts denote the constraints that are retained. Let the superscripts (1) and (2) refer to the two phases in the system. Using Eqs. 19.10, we formulate the conditions of constraint in the form

$$U = U^{(1)} + U^{(2)} = \text{constant},$$
$$V = V^{(1)} + V^{(2)} = \text{constant},$$
$$n = n^{(1)} + n^{(2)} = \text{constant}.$$

(19.13)

Then the possible variations of the variables $U^{(1)}$, $U^{(2)}$, $V^{(1)}$, $V^{(2)}$, $n^{(1)}$, and $n^{(2)}$ are connected, as shown in Eqs. 19.11, by the relations

$$dU = 0 = dU^{(1)} + dU^{(2)},$$
$$dV = 0 = dV^{(1)} + dV^{(2)},$$
$$dn = 0 = dn^{(1)} + dn^{(2)}.$$

(19.14)

But, by the definition of an inverse process, the system remains in equilibrium throughout the variation in state imposed by the several processes operating. The combination of the first and second laws including mass transfer between parts of a closed system, Eq. 19.6, becomes in this case

$$T^{(1)} dS^{(1)} \leq dU^{(1)} + p^{(1)} dV^{(1)} - g^{(1)} dn^{(1)}, \qquad (19.15a)$$

$$T^{(2)} dS^{(2)} \leq dU^{(2)} + p^{(2)} dV^{(2)} - g^{(2)} dn^{(2)}. \qquad (19.15b)$$

The total entropy change is the sum of the entropy changes of phases 1 and 2, that is,

$$dS = dS^{(1)} + dS^{(2)}. \qquad (19.16)$$

Equations 19.15a and 19.15b can easily be transformed to provide an explicit form of Eq. 19.16. After division of both sides of Eq. 19.15a by $T^{(1)}$, and of Eq. 19.15b by $T^{(2)}$, substitution into Eq. 19.16 yields

$$dS \leq \frac{dU^{(1)}}{T^{(1)}} + \frac{dU^{(2)}}{T^{(2)}} + \frac{p^{(1)}}{T^{(1)}} dV^{(1)} + \frac{p^{(2)}}{T^{(2)}} dV^{(2)} - \frac{g^{(1)}}{T^{(1)}} dn^{(1)} - \frac{g^{(2)}}{T^{(2)}} dn^{(2)}. \qquad (19.17)$$

Equations 19.14 provide a set of relationships between the variables appearing in Eq. 19.17. Using Eqs. 19.14 to write $dU^{(1)} = -dU^{(2)}$, $dV^{(1)} = -dV^{(2)}$, $dn^{(1)} = -dn^{(2)}$, we reduce Eq. 19.17 to the form

$$dS \leq \left[\frac{1}{T^{(1)}} - \frac{1}{T^{(2)}} \right] dU^{(1)} + \left[\frac{p^{(1)}}{T^{(1)}} - \frac{p^{(2)}}{T^{(2)}} \right] dV^{(1)} - \left[\frac{g^{(1)}}{T^{(1)}} - \frac{g^{(2)}}{T^{(2)}} \right] dn^{(1)}. \qquad (19.18)$$

Consider, first, the case that $dU^{(1)}, \ldots$ can be positive or negative. Then the equality sign describes the variation, and along the particular constrained path defined by Eq. 19.13, we set dS in Eq. 19.18 equal to zero. Now, the conditions of constraint, namely, Eqs. 19.10 or 19.13, are independent. Thus, the variations in Eqs. 19.11 or 19.14 are also independent. Then, if the right-hand side of Eq. 19.18 is to be equal to zero, as required by Eq. 19.12, it is necessary that the coefficients of $dU^{(1)}$, $dV^{(1)}$, and $dn^{(1)}$ separately vanish. We thereby conclude that

$$T^{(1)} = T^{(2)}, \qquad (19.19a)$$

$$p^{(1)} = p^{(2)}, \qquad (19.19b)$$

$$g^{(1)} = g^{(2)}. \qquad (19.19c)$$

In order to obtain Eqs. 19.19b and 19.19c, Eq. 19.19a has been used. The set of conditions 19.19a–19.19c must be characteristic of the original equilibrium state. It is important to observe that the various processes that lead to a change of state of the system need not be specified, and therefore the detailed nature of the new equilibrium state to which the system is displaced also need not be specified. This result is a consequence of the fact that Eqs. 19.11 and 19.14 are differential equations representing how the variables describing a system may change when a specified set of constraints is retained. Note that the retained constraints refer to the original isolated system and that there is one equilibrium condition for each constraint imposed. Provided that the final state is only infinitesimally displaced from the initial state, the conditions 19.19a–19.19c make no demands on the final state.

Consider, now, the case where one or more of $dV^{(1)}, \ldots$ can vary in only one direction. This situation could not arise in the case of equilibrium between two phases, as described above. But in a different case it could arise, for example, if the container holding the system had an internal movable partition that could not pass stops in one direction but could move unopposed in the other direction (giving, say, $dV^{(1)} > 0$ only). The equilibrium condition now requires use of the inequality in Eq. 19.18. Clearly, if $dV^{(1)} > 0$ only, then

$$\frac{p^{(1)}}{T^{(1)}} - \frac{p^{(2)}}{T^{(2)}} < 0$$

in order to satisfy the requirement $dS < 0$. The implication of this condition is that if energy can flow in both directions so $T^{(1)} = T^{(2)}$, then $p^{(1)} < p^{(2)}$. Successive application of this argument leads to the following conditions for the inverse process:

$$\text{if } dU^{(1)} > 0 \text{ only,} \qquad T^{(1)} > T^{(2)}; \qquad (19.19d)$$

$$\text{if } dV^{(1)} > 0 \text{ only,} \qquad \frac{p^{(1)}}{T^{(1)}} < \frac{p^{(2)}}{T^{(2)}}; \qquad (19.19e)$$

$$\text{if } dn^{(1)} > 0 \text{ only,} \qquad \frac{g^{(1)}}{T^{(1)}} > \frac{g^{(2)}}{T^{(2)}}. \qquad (19.19f)$$

In the inverse process to which Eq. 19.18 refers, the system is driven from the original equilibrium state to a new equilibrium state. If the extra constraints that lead to this change of state are relaxed, the system returns to the original state, and an equation just like 19.18 except for reversal of the sign of the inequality describes the entropy change. Clearly, for the case that $dS > 0$ the inequalities in Eqs. 19.19d, 19.19e and 19.19f are reversed. We conclude that in a spontaneous process

$$\text{if } dU^{(1)} > 0 \text{ only,} \qquad T^{(1)} < T^{(2)}; \qquad (19.19g)$$

$$\text{if } dV^{(1)} > 0 \text{ only,} \qquad \frac{p^{(1)}}{T^{(1)}} > \frac{p^{(2)}}{T^{(2)}}; \qquad (19.19h)$$

$$\text{if } dn^{(1)} > 0 \text{ only,} \qquad \frac{g^{(1)}}{T^{(1)}} < \frac{g^{(2)}}{T^{(2)}}. \qquad (19.19i)$$

In words, in the spontaneous approach to an equilibrium state energy flows from higher to lower temperature, volume from lower to higher pressure, and matter from higher to lower molar Gibbs free energy.

We now return to the more common case described by the equality in Eq. 19.18.

It may seem disappointing that two self-evident conditions result from our analysis, namely, the equality of temperatures and pressures in the two phases of the one-component system. However, condition 19.19c, the equality of the chemical potentials of the two phases, is by no means self-evident from earlier considerations. It is true that in the case considered, because the system remains in equilibrium everywhere along the path of the variation, the definitions of thermal and mechanical equilibrium lead directly to Eqs. 19.19a and 19.19b. We have carried through the analysis in this fashion because the technique used is generally valid; in situations where there exist complicated constraints, and where thermal or mechanical equilibrium may be modified by adiabatic walls, or rigid walls, it is not always so easy to write out the equivalents to Eqs. 19.19a–19.19c.[3]

---

[3] For example, consider the equilibrium between a pure liquid substance a and a liquid mixture containing a and other substances $b, c, \ldots$. When the two are separated by a rigid diathermal wall that permits a to pass but not $b, c, d, \ldots$, thermal equilibrium requires that $T^{(1)} = T^{(2)}$. However, determination of the pressures $p^{(1)}$ and $p^{(2)}$ requires use of the methods described, since the pressures are dependent on the permeability of the wall to component a. This phenomenon, known as *osmotic* equilibrium, will be studied in Chapter 25.

The Gibbs criteria, Eqs. 19.8 and 19.9, describe equilibrium in an isolated system. But most of the systems of interest to us, both in nature and in the laboratory, are not isolated. For example, it is common to find systems constrained to have fixed temperature and pressure because of contacts with reservoirs. Clearly, Eqs. 19.8 and 19.9 are not applicable to the equilibria found in these cases. What relations can be used to characterize equilibrium states in systems in contact with their surroundings? This section is devoted to answering that question.

It is convenient to begin by remarking that condition 19.8 can be thought of as a test for the existence of equilibrium in an isolated system. When an isolated system is so tested, that is, when an inverse process is used to drive the system to a new equilibrium state, it is always found that work must be done on the system. Put in formal terms, the inverse process retains the constraint that the system's energy is fixed, so the work necessary to generate the new equilibrium state is

$$w_{inv} = -\int T\,dS. \tag{19.20}$$

Since T is positive, the integral in Eq. 19.20 can be evaluated by use of the mean value theorem. The result is

$$w_{inv} = -\bar{T}\int dS = -\bar{T}\Delta S_{inv}, \tag{19.21}$$

where $\bar{T}$ is some temperature in the range over which T varies during the change of state driven by the inverse process. But the second law of thermodynamics requires that $\Delta S_{inv} < 0$ if the isolated system is at equilibrium. Therefore,

$$w_{inv} > 0. \tag{19.22}$$

Alternatively, an inverse process can be used to test the existence of equilibrium in an isolated system via condition 19.9. In this case the retained constraint requires that the entropy of the system be constant, hence only reversible adiabatic paths can be employed to drive it to a new equilibrium state. In place of Eq. 19.20 we now have

$$w_{inv}^{S} = \int dU, \tag{19.23}$$

so that, using condition 19.9, the work done on the system is

$$w_{inv}^{S} > 0. \tag{19.24}$$

The superscript S is used in Eqs. 19.23 and 19.24 to indicate that in this inverse process the entropy of the system is fixed; a corresponding notation would replace w_{inv} in Eqs. 19.20–19.22 with w_{inv}^{U}.

We have already noted that if the constraints added by an inverse process are removed, an isolated system will spontaneously revert to its original state of equilibrium. This spontaneous change of state could, if properly coupled to the surroundings, be used to do work. If the coupling is such as to retain the constraints on the original isolated system, and the work is done reversibly, the maximum work that can be done by the system in the reversion to the original equilibrium state is $-w_{inv}$.

Consider, now, a system that is not isolated. In this case the nature of the equilibrium state is determined by the kinds of reservoirs with which the system is in contact. To find the criteria that describe this equilibrium state, we examine the maximum work that can be done by the system by

19.2
Alternative Descriptions of the Equilibrium State

virtue of a change in its state. In analogy with our discussion of the behavior of isolated systems, we shall identify this maximum work with the negative of the work done on the system in an appropriately constrained inverse process, $-w_{inv}^C$. The constraints retained in the inverse process are the same as those used to define the equilibrium state of the system studied.

The systems of interest to us are closed, and the influence on them of contact with reservoirs is to fix intensive variables, such as T or p, at values determined by the reservoirs. We represent this as follows. The system of interest, call it I, is imbedded in a very large medium whose temperature and pressure are T_0 and p_0, respectively. This surrounding medium is contained by a rigid adiabatic envelope, and it is assumed to be so large that any change in state in system I does not alter T_0 or p_0 perceptibly. System I plus surroundings, as defined, constitute a composite system. We now calculate the greatest amount of work the composite system can do on its surroundings when system I undergoes the change in state $1 \rightarrow 2$.

It was shown in Chapter 17 that the maximum work that can be done by a system in contact with a single heat reservoir at temperature T_0 is

$$w_{max} = \Delta U - T_0 \Delta S. \qquad (19.25)$$

But in the present case system I is imbedded in a medium that has the fixed pressure p_0. Therefore, if system I expands by ΔV, it must do work $-p_0 \Delta V$ to push back its surroundings, and this work cannot be done on the surroundings of the composite system. We conclude that the maximum work that the composite system can do when system I undergoes the change of state $1 \rightarrow 2$ is

$$w'_{max} = \Delta U + p_0 \Delta V - T_0 \Delta S, \qquad (19.26)$$

which depends only on the change of state $1 \rightarrow 2$ and p_0, T_0 established by contact with the surrounding reservoir.

It is important to notice that when system I is in internal equilibrium and has temperature T_0 and pressure p_0, $w'_{max} = 0$. This follows from the properties of the Gibbs free energy G. We showed in Chapter 17 that for an infinitesimal reversible process in a closed system

$$dG = -S \, dT + V \, dp, \qquad (19.27)$$

so that $dG = 0$ for any reversible process during which p and T are fixed. But $G = U + pV - TS$, so that for a process at fixed p and T,

$$\Delta G = 0 = \Delta U + p \, \Delta V - T \, \Delta S, \qquad (19.28)$$

which, when compared with Eq. 19.26, gives the result $w'_{max} = 0$.

It is proper, then, to interpret w'_{max} as the greatest possible work that system I can do on the environment of the composite system when the change of state $1 \rightarrow 2$ involves removal of a constraint that prevented achievement of internal equilibrium, that is, prevented achievement of equilibrium with respect to *only* the constraints defined by contacts between system I and the surrounding reservoir.

We now turn the argument around. Suppose that our system is at equilibrium while in contact with a reservoir that fixes the pressure and temperature at p_0 and T_0. To test that equilibrium we subject the system to an inverse process during which the constraints $T = T_0$, $p = p_0$ are maintained *and* a reversible work process is used to drive the system to a new state. The new state is in equilibrium with respect to the old constraints and any new ones added by the inverse process. The new state generated is unstable with respect to the original equilibrium state when only the constraints that determine that original state are retained. Hence, if the extra constraints associated with the reversible process that generated the new state are removed, a spontaneous change occurs and the original equilibrium state is regenerated. With the identification

$$w_{inv}^C = -w'_{max} \qquad (19.29)$$

and the definition of a spontaneous process,

$$w'_{max} < 0, \qquad (19.30)$$

we conclude that

$$w_{inv}^C > 0. \qquad (19.31)$$

Combining Eqs. 19.31, 19.29, and 19.26, we have

$$(\Delta U + p_0 \Delta V - T_0 \Delta S)_{inv} > 0. \qquad (19.32)$$

Since Eq. 19.32 must be satisfied for all possible variations of the state of the system that retain the original constraints,

$$(\delta U + p_0 \delta V - T_0 \delta S)_{p_0 T_0} = 0 \qquad (19.33)$$

defines a minimum of the Gibbs free energy of the system.

We have derived Eqs. 19.32 and 19.33 for a particular choice of retained constraints, namely, $p = p_0$, $T = T_0$. Examination of the argument shows that the term $p_0 \Delta V$ represents the amount of work associated with the retained constraints. The generalizations of Eqs. 19.32 and 19.33 to the description of equilibrium determined by an arbitrary set of constraints are

$$\left(\Delta U - w_r - \int T \, dS\right)_{inv} > 0, \qquad (19.34)$$

$$(\delta U - \delta w - T \, \delta S)_r = 0, \qquad (19.35)$$

where the subscript r refers to the retained constraints. The function that is minimized is different for each choice of retained constraints.

Eq. 19.35 defines a different extremum for each set of retained constraints. To determine whether these extrema are maxima or minima we return to the fundamental equation 19.34. We already know that S is a maximum when the energy, volume, and mass are fixed. This follows from Eq. 19.34 since, under these conditions,

$$-\int T \, dS > 0$$

or

$$\Delta S < 0. \qquad (19.36a)$$

Suppose that the system is defined under conditions such that the entropy, volume, and mass are fixed. Then Eq. 19.34 gives

$$\Delta U > 0, \qquad (19.36b)$$

or, in words, the internal energy must be a minimum. Similarly, using Eq. 19.34 and the constraints S, p, n constant, T, V, n constant, T, p, n constant, we deduce that, respectively,

$$\Delta U - w_r > 0, \qquad (19.36c)$$

$$\Delta U - \int T \, dS > 0, \qquad (19.36d)$$

$$\Delta U - w_r - \int T\, dS > 0, \qquad (19.36e)$$

which correspond to the enthalpy, Helmholtz free energy, and Gibbs free energy each being a minimum under the appropriate conditions. All these results are summarized in Table 19.1.

Suppose that the system under study can only do expansion work. The following criteria of equilibrium can be obtained from Eq. 19.35 and the inequality 19.34:

1. The system is defined under conditions of constant energy, volume, and mass:

$$(\delta S)_{U,V,n} \leq 0; \qquad (19.37)$$

2. The system is defined under conditions of constant entropy, volume, and mass:

$$(\delta U)_{S,V,n} \geq 0; \qquad (19.38)$$

3. The system is defined under conditions of constant entropy, pressure, and mass:

$$(\delta U)_{S,p,n} + p(\delta V)_{S,p,n} \geq 0 \qquad (19.39)$$

or

$$(\delta H)_{S,p,n} \geq 0; \qquad (19.40)$$

4. The system is defined under conditions of constant temperature, volume, and mass:

$$(\delta U)_{T,V,n} - T(\delta S)_{T,V,n} \geq 0 \qquad (19.41)$$

or

$$(\delta A)_{T,V,n} \geq 0; \qquad (19.42)$$

5. The system is defined under conditions of constant temperature, pressure, and mass:

$$(\delta U)_{T,p,n} + p(\delta V)_{T,p,n} - T(\delta S)_{T,p,n} \geq 0 \qquad (19.43)$$

or

$$(\delta G)_{T,p,n} \geq 0. \qquad (19.44)$$

We now see that the Helmholtz and Gibbs free energies, A and G, introduced in Chapter 17 to help evaluate derivatives of the entropy function, play a fundamental role in describing the equilibrium state of a system; A is an extremum (a minimum) when the equilibrium is defined

TABLE 19.1
SUMMARY OF CRITERIA FOR STABLE EQUILIBRIUM

Defined Constraints	Function That Is an Extremum	Type of Extremum	For Virtual Variation
U, V, n constant	S	Maximum	$\Delta S_{inv} < 0$
S, V, n constant	U	Minimum	$\Delta U_{inv} > 0$
S, p, n constant	$H \equiv U + pV$	Minimum	$\Delta H_{inv} > 0$
T, V, n constant	$A \equiv U - TS$	Minimum	$\Delta A_{inv} > 0$
T, p, n constant	$G \equiv H - TS$	Minimum	$\Delta G_{inv} > 0$

TABLE 19.2
SUMMARY OF SOME USEFUL RELATIONS FOR A ONE-COMPONENT SYSTEM

Function	Exact Differential Form[a]	Useful Relations	Cross-Differentiation Identities
U	$dU = T\,dS - p\,dV + \mu\,dn$	$C_V = T\left(\dfrac{\partial S}{\partial T}\right)_{V,n}$; $p = -\left(\dfrac{\partial U}{\partial V}\right)_{S,n}$; $\quad T = \left(\dfrac{\partial U}{\partial S}\right)_{V,n}$; $\mu = \left(\dfrac{\partial U}{\partial n}\right)_{S,V}$	$\left(\dfrac{\partial T}{\partial V}\right)_{S,n} = -\left(\dfrac{\partial p}{\partial S}\right)_{V,n}$
H	$dH = T\,dS + V\,dp + \mu\,dn$	$C_p = T\left(\dfrac{\partial S}{\partial T}\right)_{p,n}$; $V = \left(\dfrac{\partial H}{\partial p}\right)_{S,n}$; $\quad T = \left(\dfrac{\partial H}{\partial S}\right)_{p,n}$; $\mu = \left(\dfrac{\partial H}{\partial n}\right)_{S,p}$	$\left(\dfrac{\partial T}{\partial p}\right)_{S,n} = \left(\dfrac{\partial V}{\partial S}\right)_{p,n}$
$A \equiv U - TS$	$dA = -S\,dT - p\,dV + \mu\,dn$	$S = -\left(\dfrac{\partial A}{\partial T}\right)_{V,n}$; $p = -\left(\dfrac{\partial A}{\partial V}\right)_{T,n}$; $\quad \mu = \left(\dfrac{\partial A}{\partial n}\right)_{T,V}$	$\left(\dfrac{\partial S}{\partial V}\right)_{T,n} = \left(\dfrac{\partial p}{\partial T}\right)_{V,n}$
$G \equiv H - TS$	$dG = -S\,dT + V\,dp + \mu\,dn$	$S = -\left(\dfrac{\partial G}{\partial T}\right)_{p,n}$; $V = \left(\dfrac{\partial G}{\partial p}\right)_{T,n}$; $\quad \mu = \left(\dfrac{\partial G}{\partial n}\right)_{T,p}$	$\left(\dfrac{\partial S}{\partial p}\right)_{T,n} = -\left(\dfrac{\partial V}{\partial T}\right)_{p,n}$

[a] In this table we use μ for the Gibbs free energy per mole (chemical potential), previously denoted by g. For a pure component we have $g = G/n = \mu$. As will be seen in the next section, a more general definition, valid in a multicomponent system, is $\mu_i \equiv (\partial G/\partial n_i)_{T,p,n_j}$, where n_j ($j \neq i$) is the number of moles of j in the system.

under the constraints of constant temperature, volume, and mass, and G is an extremum (a minimum) when the equilibrium is defined under the very common constraints of constant pressure, temperature, and mass.

At this point it is also convenient to tabulate some of the more important relations that are useful in manipulating the various thermodynamic functions. This tabulation is given in Table 19.2; the derivations of most of the relations should be obvious.

**19.3
The Stability of the Equilibrium State of a One-Component System**

There still remains the problem of the kind of equilibrium defined by the differential equation 19.35. That is, Eq. 19.35 is a necessary condition for an extremum, but is not in itself sufficient to determine if the equilibrium is stable.

Now, in the discussion leading to Eq. 19.34, the arguments used assumed a case of stable equilibrium. That assumption entered through the use of the inequality 19.7. (The corresponding inequalities 19.36c, 19.36d, and 19.36e also refer to a stable equilibrium.) The importance of 19.35 is that it is a differential equation that can, in principle, be solved. The possible solutions describe the relationships between thermodynamic functions required by the retained constraints in the vicinity of an extremum of the entropy. Such relationships cannot be obtained only from the inequality 19.7. Of the many possible solutions to the differential equation, we select those corresponding to the case of stable equilibrium by using the inequality 19.7 as an extra condition—a generalized boundary condition, as it were.

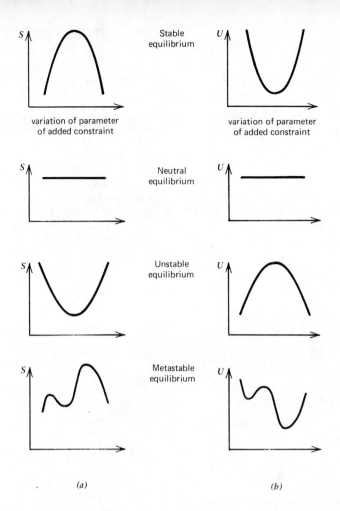

variation of parameter of added constraint

variation of parameter of added constraint

Stable equilibrium

Neutral equilibrium

Unstable equilibrium

Metastable equilibrium

(a)

(b)

FIGURE 19.2

Schematic representations of the variations of S and U with change of a parameter of added constraint for the four classes of equilibrium.

In general, states of equilibrium (i.e., states for which Eq. 19.35 is satisfied) can be divided into four categories:

1. A system is in a state of *stable equilibrium* if, after displacement to a new state and release of the constraint causing the displacement, it always returns to the original state;

2. A system is in a state of *neutral equilibrium* if states that can be reached by virtual variation at constant energy, volume, and mass have the same entropy;

3. A system is in a state of *unstable equilibrium* if, after displacement to a new state and release of the constraint causing the displacement, it never returns to the original state;

4. A system is in a state of *metastable equilibrium* if displacements to nearby states satisfy condition 1, but displacements to distant states satisfy condition 3.

Schematic representations of the change of entropy for these cases are shown in Fig. 19.2a, and for the energy (at constant entropy, volume, and mass) in Fig. 19.2b.

All of the preceding discussion on the properties of the equilibrium state assumed that we were dealing with case 1. Of course, for many purposes—for example, if the variations of interest all lead to states in the neighborhood of the original state—cases 4 and 1 are equivalent. This is true even if there exist variations that demonstrate the metastability of the original state.

This categorization of equilibrium states can be connected with the discussion in Chapter 15 on the reversibility and irreversibility of processes that follow the removal of a constraint. In an equilibrium of type 1 the removal of the constraint causing the virtual variation leads to an irreversible approach to the original state. This is also true in case 4 for states close to the original state. On the other hand, if removal of the constraint results in a reversible process, then the state of equilibrium is neutral. This statement is clearly equivalent to that in Chapter 15, where a reversible process was shown to occur only if $\Omega_i = \Omega_f$. As might be anticipated, unstable equilibria do not actually occur; this will later be shown to be related to the properties of the fluctuations in the system.

If we suppose that a system is in a state of stable equilibrium, satisfying the appropriate one of Eqs. 19.36a–19.36e, the properties the system may have are restricted. Certainly we must expect, for a stable equilibrium, the following common-sense observations to be valid:

1. An increase in the external pressure at constant temperature and mass must lead to a decrease in the volume of the system.

2. The addition of heat (at constant pressure or volume and mass) must lead to an increase in the temperature of the system.

To examine further the consequences of the definition of a stable equilibrium we study the virtual variation depicted in Fig. 19.3. We shall use Eq. 19.36b to describe the virtual variation. We insert a partition at the midpoint of a rigid container of volume $2V$ containing some substance. Initially, each of the new subvolumes has the same volume V, entropy S, and number of moles n, and the system is still in stable equilibrium. We then produce a virtual variation by moving the partition to a point where the two parts of the system have volumes $V+\Delta V$ and $V-\Delta V$, and entropies $S+\Delta S$ and $S-\Delta S$, respectively. Suppose that ΔV and ΔS are, in this particular variation, sufficiently small that the internal energies of the two parts of the system can be accurately represented by

$$U(S+\Delta S, V+\Delta V, n) = U(S, V, n) + \left(\frac{\partial U}{\partial S}\right)_{V,n} (\Delta S) + \left(\frac{\partial U}{\partial V}\right)_{S,n} (\Delta V)$$

$$+ \frac{1}{2}\left[\left(\frac{\partial^2 U}{\partial S^2}\right)_{V,n} (\Delta S)^2 + 2\left(\frac{\partial}{\partial V}\left(\frac{\partial U}{\partial S}\right)_{V,n}\right)_{S,n} (\Delta V\,\Delta S)\right.$$

$$\left. + \left(\frac{\partial^2 U}{\partial V^2}\right)_{S,n} (\Delta V)^2\right] \tag{19.45a}$$

and

$$U(S-\Delta S, V-\Delta V, n) = U(S, V, n) - \left(\frac{\partial U}{\partial S}\right)_{V,n} (\Delta S) - \left(\frac{\partial U}{\partial V}\right)_{S,n} (\Delta V)$$

$$+ \frac{1}{2}\left[\left(\frac{\partial^2 U}{\partial S^2}\right)_{V,n} (\Delta S)^2 + 2\left(\frac{\partial}{\partial V}\left(\frac{\partial U}{\partial S}\right)_{V,n}\right)_{S,n} (\Delta V\,\Delta S)\right.$$

$$\left. + \left(\frac{\partial^2 U}{\partial V^2}\right)_{S,n} (\Delta V)^2\right], \tag{19.45b}$$

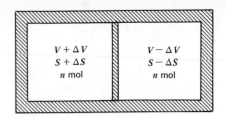

FIGURE 19.3
A system in a state generated by a virtual variation, for the retained constraints $U = $ constant, $V = $ constant, $n = $ constant. The virtual variation is generated by moving the partition from an initial position in which $\Delta V = 0$ and $\Delta S = 0$ to the position shown.

respectively. For a stable equilibrium it is then necessary that

$$\Delta U \equiv U(S+\Delta S, V+\Delta V, n) + U(S-\Delta S, V-\Delta V, n) - 2U(S, V, n) > 0. \tag{19.46}$$

A combination of Eqs. 19.46, 19.45a, and 19.45b shows that the stability condition must be

$$\left(\frac{\partial^2 U}{\partial S^2}\right)_{V,n} (\Delta S)^2 + 2\left(\frac{\partial}{\partial V}\left(\frac{\partial U}{\partial S}\right)_{V,n}\right)_{S,n} (\Delta V \, \Delta S) + \left(\frac{\partial^2 U}{\partial V^2}\right)_{S,n} (\Delta V)^2 > 0. \tag{19.47}$$

The left-hand side of the inequality in 19.47 is a quadratic form in the variables ΔS and ΔV. For that quadratic form to be positive for all possible values of ΔS and ΔV, it is necessary and sufficient that

$$\left(\frac{\partial^2 U}{\partial S^2}\right)_{V,n} > 0, \qquad \left(\frac{\partial^2 U}{\partial S^2}\right)_{V,n}\left(\frac{\partial^2 U}{\partial V^2}\right)_{S,n} - \left(\frac{\partial}{\partial V}\left(\frac{\partial U}{\partial S}\right)_{V,n}\right)_{S,n}^2 > 0. \tag{19.48}$$

Then, using the combined first and second laws and the definitions of C_p, C_V, the isothermal compressibility κ_T, and the coefficient of thermal expansion α, we find that[4]

$$\left(\frac{\partial^2 U}{\partial S^2}\right)_{V,n} = \frac{T}{C_V}, \tag{19.49a}$$

[4] The evaluation of $(\partial^2 U/\partial S^2)_{V,n}$, etc., in terms of measurable quantities is an example of the utility of thermodynamic manipulations. The formulas given are derived as follows:

Eq. 19.49a: $\left(\dfrac{\partial^2 U}{\partial S^2}\right)_{V,n} = \left(\dfrac{\partial}{\partial S}\left(\dfrac{\partial U}{\partial S}\right)_{V,n}\right)_{V,n} = \left(\dfrac{\partial T}{\partial S}\right)_{V,n} = \dfrac{T}{C_V}$.

Eq. 19.49b: $\left(\dfrac{\partial^2 U}{\partial V^2}\right)_{S,n} = \left(\dfrac{\partial}{\partial V}\left(\dfrac{\partial U}{\partial V}\right)_{S,n}\right)_{S,n} = -\left(\dfrac{\partial p}{\partial V}\right)_{S,n}$

$\displaystyle \qquad = \left(\frac{\partial S}{\partial V}\right)_{p,n} \Big/ \left(\frac{\partial S}{\partial p}\right)_{V,n}$ using $\left(\dfrac{\partial p}{\partial V}\right)_{S,n}\left(\dfrac{\partial S}{\partial p}\right)_{V,n}\left(\dfrac{\partial V}{\partial S}\right)_{p,n} = -1$,

$\displaystyle \qquad = -\left(\frac{\partial p}{\partial T}\right)_{S,n} \Big/ \left(\frac{\partial V}{\partial T}\right)_{S,n}$ using the cross-differentiation identities,

$\displaystyle \qquad = -\frac{\left(\dfrac{\partial S}{\partial T}\right)_{p,n} \Big/ \left(\dfrac{\partial S}{\partial p}\right)_{T,n}}{\left(\dfrac{\partial S}{\partial T}\right)_{V,n} \Big/ \left(\dfrac{\partial S}{\partial V}\right)_{T,n}}$ using $\begin{cases}\left(\dfrac{\partial S}{\partial p}\right)_{T,n}\left(\dfrac{\partial p}{\partial T}\right)_{S,n}\left(\dfrac{\partial T}{\partial S}\right)_{p,n} = -1 \\[2ex] \left(\dfrac{\partial S}{\partial V}\right)_{T,n}\left(\dfrac{\partial V}{\partial T}\right)_{S,n}\left(\dfrac{\partial T}{\partial S}\right)_{V,n} = -1\end{cases}$

$\displaystyle \qquad = -\frac{C_p \Big/ -\left(\dfrac{\partial V}{\partial T}\right)_{p,n}}{C_V \Big/ \left(\dfrac{\partial p}{\partial T}\right)_{V,n}}$ using the cross-differentiation identities,

$\displaystyle \qquad = \frac{C_p}{V\kappa_T C_V}$ using $\left(\dfrac{\partial V}{\partial T}\right)_{p,n}\left(\dfrac{\partial T}{\partial p}\right)_{V,n}\left(\dfrac{\partial p}{\partial V}\right)_{T,n} = -1$.

Eq. 19.49c: $\left(\dfrac{\partial}{\partial S}\left(\dfrac{\partial U}{\partial V}\right)_{S,n}\right)_{V,n} = \left(\dfrac{\partial}{\partial V}\left(\dfrac{\partial U}{\partial S}\right)_{V,n}\right)_{S,n} = \left(\dfrac{\partial T}{\partial V}\right)_{S,n}$

$\displaystyle \qquad = -\left(\frac{\partial S}{\partial V}\right)_{T,n} \Big/ \left(\frac{\partial S}{\partial T}\right)_{V,n}$ using $\left(\dfrac{\partial T}{\partial V}\right)_{S,n}\left(\dfrac{\partial V}{\partial S}\right)_{T,n}\left(\dfrac{\partial S}{\partial T}\right)_{V,n} = -1$,

$\displaystyle \qquad = -\left(\frac{\partial p}{\partial T}\right)_{V,n}\frac{T}{C_V}$ using the cross-differentiation identities,

$\displaystyle \qquad = T\left(\frac{\partial V}{\partial T}\right)_{p,n} \Big/ C_V\left(\frac{\partial V}{\partial p}\right)_{T,n}$ using $\left(\dfrac{\partial p}{\partial T}\right)_{V,n}\left(\dfrac{\partial V}{\partial p}\right)_{T,n}\left(\dfrac{\partial T}{\partial V}\right)_{p,n} = -1$,

$\displaystyle \qquad = -\frac{\alpha T}{\kappa_T C_V}$.

$$\left(\frac{\partial^2 U}{\partial V^2}\right)_{S,n} = \frac{C_p}{V\kappa_T C_V}. \qquad (19.49b)$$

$$\left(\frac{\partial}{\partial V}\left(\frac{\partial U}{\partial S}\right)_{V,n}\right)_{S,n} = -\frac{\alpha T}{\kappa_T C_V}. \qquad (19.49c)$$

Thus, the first of conditions 19.48 is equivalent to requiring that

$$\frac{T}{C_V} > 0 \quad \text{or} \quad C_V > 0, \qquad (19.50a)$$

and therefore, since $C_p > C_V$,

$$C_p > 0. \qquad (19.50b)$$

The second part of conditions 19.48 becomes

$$\frac{T}{C_V}\left(\frac{C_p}{V\kappa_T C_V}\right) - \frac{\alpha^2 T^2}{\kappa_T^2 C_V^2} > 0, \qquad (19.50c)$$

implying that

$$\kappa_T > \frac{\alpha^2 TV}{C_p} > 0. \qquad (19.50d)$$

The inequalities 19.50b and 19.50d are equivalent to the "common-sense" requirements for stability mentioned earlier in this section. By choosing different sets of constraints, still other conditions can be derived, for example, that the adiabatic compressibility $\kappa_S \equiv -(1/V)(\partial V/\partial p)_S$ must be positive. The inequalities derived are known collectively as *stability conditions*, since no stable one-phase, one-component system can exist for which they are not valid. This formulation of the properties of a stable equilibrium may be summarized in the following statement, known as *Le Chatelier's principle:*

It is a feature of stable equilibrium that spontaneous processes induced by a deviation from the initial state proceed in such a direction as to restore the system to the initial state.

19.4 The Equilibrium State in a Multicomponent System

Most systems of chemical interest contain more than one component. We must therefore extend our study of the properties of the equilibrium state to examine how, if at all, the presence of several components influences the deductions we have made in earlier sections of this chapter. As in Section 19.1, our first task is to learn how to write the first and second laws when it is possible to transfer some component from one part of a closed system to another.

One of the key concepts in the thermodynamic theory of multicomponent systems is that of the *semipermeable barrier*. A semipermeable barrier is one that permits the passage of one substance but not others. In the idealized thermodynamic concept the selectivity of this barrier is assumed perfect: it offers infinite resistance to the passage of all substances except one. Semipermeable barriers are very common, although real barriers are rarely perfectly selective. For example, hot Pd will permit the passage of gaseous H_2, but is impermeable to other gases; the walls of most living cells are permeable or partially permeable to some but not all organic and inorganic compounds (they are often completely impermeable to the macromolecules synthesized in the cell, while allowing the passage of the chemicals from which the macromolecules are synthesized); liquid films are permeable to gases that dissolve in the liquid, and resist the passage of gases only slightly soluble in the liquid. The thermodynamic concept is, as usual, an abstraction based on the observed behavior of real barriers.

Suppose that the system depicted in Fig. 19.1 is replaced by that in Fig. 19.4. The two systems differ only in that the latter has semipermeable barriers, one for each component, across which the mass of the substances added to the mixture must pass. Across each rigid semipermeable membrane there exists a pressure difference, since there is only one component on one side and r components on the other side. Let the pressures of each of the pure components, each in equilibrium across its own semipermeable membrane with the mixture, be $p_1, p_2, \ldots, p_r$, respectively. We define v_i, u_i, s_i as the molar volume, internal energy, and entropy of the pure component in part II_i. We now proceed, exactly as in Section 19.1, to evaluate the change in internal energy resulting from the transfer of dn_i mol of substance i from part II_i into part I of the system. As before, heat $đq_i$ is transferred to part I from the surroundings, and the expansion of part I does work $(p\,dV)_i$ on the surroundings. We find, corresponding to Eq. 19.1, the internal energy change of the total system

$$dU_i - u_i\,dn_i = đq_i - (p\,dV)_i + p_i v_i\,dn_i. \tag{19.51}$$

Using the definition

$$h_i \equiv u_i + p_i v_i \tag{19.52}$$

we obtain, in analogy with Eq. 19.3, for the internal energy change of part I

$$dU_i = đq_i + h_i\,dn_i - (p\,dV)_i. \tag{19.53}$$

In Eqs. 19.51–19.53 the subscript i refers to that contribution arising from the transfer of dn_i mol of i. Continuing as before, we finally obtain for the total change in internal energy of part I

$$dU = T\,dS - p\,dV + \sum_{i=1}^{r} \mu_i\,dn_i, \tag{19.54}$$

with the definition

$$\mu_i \equiv h_i - Ts_i. \tag{19.55}$$

Equation 19.55 is understood to apply to the substance i in equilibrium with the mixture across the semipermeable barrier that permits passage of i, that is, to substance i in part II_i. Equation 19.54 is the multicomponent-system generalization of Eq. 19.6.

Equations 19.54 and 19.55 are of such fundamental importance that we now give an alternative derivation that emphasizes a different view.

Suppose that the size of a system is increased by a constant factor ξ, for example, by increasing V to ξV, $n_1, \ldots, n_r$ to $\xi n_1, \ldots, \xi n_r$, and S to ξS. Since U is an extensive function, increasing the system size by the factor ξ must also increase U by the same factor. In mathematical form this reads

$$U(\xi S, \xi V, \xi n_1, \ldots, \xi n_r) = \xi U(S, V, n_1, \ldots, n_r). \tag{19.56}$$

An alternative mathematical representation of the properties of an extensive function is derived from examining the rate of change of the function with the "size parameter" ξ. The differentiation of Eq. 19.56 with respect

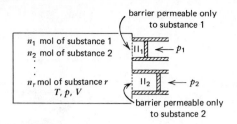

FIGURE 19.4
Schematic representation of an r-component system with $r+1$ parts. In part I, which has volume V, are n_1 mol of substance 1, n_2 mol of substance 2, $\ldots$, n_r moles of substance r, at T and p. In part II_1 are dn_1 mol of substance 1, at T and whatever pressure p_1 represents equilibrium across the semipermeable membrane. In general, this pressure is different from p. In part II_2 are dn_2 mol of substance 2, at T and whatever pressure p_2 represents equilibrium across the semipermeable membrane. There are r semipermeable membranes, each permitting only one of the components to enter part I; only two are shown in the diagram. The semipermeable membranes define the boundary of part I. Heat may be transferred from the surroundings to part I, but not to parts II_1, $II_2, \ldots, II_r$. Also, part I may do work on the surroundings.

to ξ gives the simple result

$$\left[\frac{\partial U}{\partial(\xi S)}\right]_{V,n_j}\frac{\partial(\xi S)}{\partial\xi}+\left[\frac{\partial U}{\partial(\xi V)}\right]_{S,n_j}\frac{\partial(\xi V)}{\partial\xi}+\sum_{j=1}^{r}\left[\frac{\partial U}{\partial(\xi n_j)}\right]_{V,s,n_{i\neq j}}\frac{\partial(\xi n_j)}{\partial\xi}=U,$$

(19.57)

or, after evaluation of $\partial(\xi S)/\partial\xi$, $\partial(\xi V)/\partial\xi$, and $\partial(\xi n_j)/\partial\xi$,

$$\left[\frac{\partial U}{\partial(\xi S)}\right]_{V,n_j}S+\left[\frac{\partial U}{\partial(\xi V)}\right]_{S,n_j}V+\sum_{j=1}^{r}\left[\frac{\partial U}{\partial(\xi n_j)}\right]_{V,S,n_{j\neq i}}n_j=U. \qquad (19.58)$$

But by the definition of an extensive function, Eqs. 19.56, 19.57, and 19.58 must be valid for all values of ξ. For the particular case $\xi=1$, then, we have

$$\left(\frac{\partial U}{\partial S}\right)_{V,n_j}S+\left(\frac{\partial U}{\partial V}\right)_{S,n_j}V+\sum_{j=1}^{r}\left(\frac{\partial U}{\partial n_j}\right)_{V,S,n_{i\neq j}}n_j=U, \qquad (19.59)$$

from which we obtain

$$TS-pV+\sum_{j=1}^{r}\mu_j n_j=U, \qquad (19.60)$$

with the new definition

$$\mu_j\equiv\left(\frac{\partial U}{\partial n_j}\right)_{S,V,n_{i\neq j}}; \qquad (19.61)$$

μ_j is the chemical potential of component j. We shall see that this is the same as the function introduced in Eq. 19.55. In Eq. 19.60 we have assumed that a variation of U with S or with V, when the composition is fixed throughout the variation, must lead to the same results as do the corresponding variations in a one-component system. Therefore we have made the identifications

$$\left(\frac{\partial U}{\partial S}\right)_{V,n_j}=T, \qquad \left(\frac{\partial U}{\partial V}\right)_{S,n_j}=-p.$$

From Eq. 19.60 and the definitions of the enthalpy, Helmholtz free energy, and Gibbs free energy, it follows that

$$H=U+pV=TS+\sum_{j=1}^{r}n_j\mu_j, \qquad (19.62)$$

$$A=U-TS=-pV+\sum_{j=1}^{r}n_j\mu_j, \qquad (19.63)$$

$$G=U-TS+pV=\sum_{j=1}^{r}n_j\mu_j. \qquad (19.64)$$

The chemical potential is particularly important in those common situations in which equilibria are defined at constant temperature and pressure, since under these constraints G is a minimum, and, by Eq. 19.64, G is simply related to the several μ_j.

From Eq. 19.60 it also follows that

$$dU=T\,dS+S\,dT-p\,dV-V\,dp+\sum_{j=1}^{r}\mu_j\,dn_j+\sum_{j=1}^{r}n_j\,d\mu_j. \quad (19.65)$$

But the internal energy is a function of the variables S, V, and all the n_j. Therefore, a differential change in U must have the form

$$dU=\left(\frac{\partial U}{\partial S}\right)_{V,n_j}dS+\left(\frac{\partial U}{\partial V}\right)_{S,n_j}dV+\sum_{j=1}^{r}\left(\frac{\partial U}{\partial n_j}\right)_{S,V,n_{i\neq j}}dn_j, \qquad (19.66)$$

or

$$dU = T\,dS - p\,dV + \sum_{j=1}^{r} \mu_j\,dn_j. \qquad (19.67)$$

Note that Eq. 19.67 is the same as Eq. 19.54. Comparison of Eqs. 19.65 and 19.67 shows that we must have

$$S\,dT - V\,dp + \sum_{j=1}^{r} n_j\,d\mu_j = 0, \qquad (19.68)$$

which is known as the *Gibbs-Duhem equation*. Equation 19.68 plays the role of a restriction on the possible coupled variations of the intensive variables T, p, and μ_j.

Consider now the representation of G given in Eq. 19.64. We note that in a one-component system

$$G = n\mu, \qquad (19.69)$$

or

$$g = h - Ts = \frac{G}{n} = \mu \qquad (19.70)$$

The usual notation for the chemical potential is μ, and we henceforth adopt this symbol. For a mixture, it follows from Eq. 19.67 that (see Table 19.2)

$$dH = T\,dS + V\,dp + \sum_{j=1}^{r} \mu_j\,dn_j, \qquad (19.71)$$

$$dA = -S\,dT - p\,dV + \sum_{j=1}^{r} \mu_j\,dn_j, \qquad (19.72)$$

$$dG = -S\,dT + V\,dp + \sum_{j=1}^{r} \mu_j\,dn_j. \qquad (19.73)$$

But from Eq. 19.64 we have

$$dG = \sum_{j=1}^{r} n_j\,d\mu_j + \sum_{j=1}^{r} \mu_j\,dn_j, \qquad (19.74)$$

and it follows that

$$\sum_{j=1}^{r} n_j\,d\mu_j = \sum_{j=1}^{r} n_j \left[\left(\frac{\partial \mu_j}{\partial T} \right)_{p,n_j} dT + \left(\frac{\partial \mu_j}{\partial p} \right)_{T,n_j} dp \right]. \qquad (19.75)$$

Comparison of Eqs. 19.73, 19.74, and 19.75 shows that

$$-S = \sum_{j=1}^{r} n_j \left(\frac{\partial \mu_j}{\partial T} \right)_{p,n_j} = -\sum_{j=1}^{r} n_j \bar{s}_j, \qquad (19.76)$$

$$V = \sum_{j=1}^{r} n_j \left(\frac{\partial \mu_j}{\partial p} \right)_{T,n_j} = \sum_{j=1}^{r} n_j \bar{v}_j. \qquad (19.77)$$

In a multicomponent system $\bar{s}_j$ and $\bar{v}_j$ are called the *partial molar* entropy and volume. For a one-component system $\bar{s}$ and $\bar{v}$ become the molar entropy and molar volume.

Finally, from Eqs. 19.71–19.73 we have the relations, equivalent to Eq. 19.61,

$$\mu_j = \left(\frac{\partial H}{\partial n_j} \right)_{S,p,n_{i\neq j}} = \left(\frac{\partial A}{\partial n_j} \right)_{T,V,n_{i\neq j}} = \left(\frac{\partial G}{\partial n_j} \right)_{T,p,n_{i\neq j}}; \qquad (19.78)$$

we thus obtain

$$\mu_j = \left[\frac{\partial}{\partial n_j} (H - TS) \right]_{T,p,n_{i\neq j}} = \bar{h}_j - T\bar{s}_j, \qquad (19.79)$$

using the general definition

$$\bar{z}_j \equiv \left(\frac{\partial Z}{\partial n_j}\right)_{T,p,n_{i\neq j}}. \tag{19.80}$$

We are now able to determine what restrictions are placed on the properties of a multicomponent system by the condition that it be in a state of stable equilibrium. Clearly, for any virtual variation in which the composition remains the same at all points in the system, the restrictions derived for a one-component system are valid, for, in this case, the homogeneous mixture behaves like a single substance. When the composition varies, Eq. 19.68 defines a restriction on the possible coupled variations of the intensive variables T, p, μ_i in a fluid system. Thus, at constant temperature and pressure, any change in the composition of a system leads to changes in the chemical potentials such that

$$\left(\sum_{i=1}^{r} n_i \, d\mu_i\right)_{T,p} = 0. \tag{19.81}$$

In a two-component system, addition of dn_1 mol of component 1 requires that

$$n_1 \left(\frac{\partial \mu_1}{\partial n_1}\right)_{T,p,n_2} = -n_2 \left(\frac{\partial \mu_2}{\partial n_1}\right)_{T,p,n_2}. \tag{19.82}$$

Since n_1 and n_2 are positive, the signs of $(\partial \mu_1/\partial n_1)_{T,p,n_2}$ and $(\partial \mu_2/\partial n_1)_{T,p,n_2}$ must be opposite. The extension to an r-component system is straightforward. To determine the sign of $(\partial \mu_1/\partial n_1)_{T,p,n_2}$ we construct a virtual variation for a two-component system defined to be at equilibrium at some T and p. Let the system consist of a deformable container filled with a mixture of $2n_1$ mol of 1 and $2n_2$ mol of 2 (see Fig. 19.5). Imagine that a semipermeable barrier (passing 1 but not 2) is inserted, dividing the container so that there are n_1 mol of 1 and n_2 mol of 2 on each side. Now let the barrier be displaced from the position in which it was inserted. In this displacement component 1 flows freely through the barrier, but however much of component 2 was on each side of the barrier remains there. We assume that during the displacement the temperature and pressure are maintained constant by exchange of heat and work with the surroundings. After the semipermeable barrier is displaced through some distance, an impermeable barrier is slipped in alongside it, and the combined barriers are displaced back to a position dividing the volume equally. In this second displacement the composition of the mixture on each side of the barrier remains fixed. Again we assume that T and p are maintained constant by exchange of heat and work with the surroundings. As a result of the operations described, we have generated a virtual variation in which, although T and p are the same, the composition of the mixture is different on the two sides of the barrier. Because the original state was one of stable equilibrium, in the virtual variation described it is necessary that

$$\Delta G > 0. \tag{19.83}$$

Proceeding as before, we write for the Gibbs free energies on the two sides of the barrier $G(T, p, n_1 + \Delta n_1, n_2)$ and $G(T, p, n_1 - \Delta n_1, n_2)$. Then we must have

$$\Delta G \equiv G(T, p, n_1 + \Delta n_1, n_2) + G(T, p, n_1 - \Delta n_1, n_2) - 2G(T, p, n_1, n_2) > 0. \tag{19.84}$$

Assuming Δn_1 to be small enough that a Taylor series expansion of G

```
┌─────────────────────────────────────┐
│        2n₁ mol of 1 and             │
│        2n₂ mol of 2 at T, p         │
└─────────────────────────────────────┘
```

insertion of semipermeable barrier
(only 1 can pass)

```
┌──────────────────┬──────────────────┐
│  n₁ mol of 1     │  n₁ mol of 1     │
│  n₂ mol of 2     │  n₂ mol of 2     │
│     T, p         │     T, p         │
└──────────────────┴──────────────────┘
```

displacement of semipermeable barrier

```
┌──────────────────┬──────────────────┐
│ n₁−Δn₁ mol of 1  │ n₁+Δn₁ mol of 1  │
│ n₂ mol of 2      │ n₂ mol of 2      │
│     T, p         │     T, p         │
└──────────────────┴──────────────────┘
```
Both sides of the system change volume and exchange heat sufficient to keep T, p constant

insertion of impermeable barrier

```
┌──────────────────┬──────────────────┐
│ n₁−Δn₁ mol of 1  │ n₁+Δn₁ mol of 1  │
│ n₂ mol of 2      │ n₂ mol of 2      │
│     T, p         │     T, p         │
└──────────────────┴──────────────────┘
```

displacement of compound barrier back to original position

```
┌──────────────────┬──────────────────┐
│ n₁−Δn₁ mol of 1  │ n₁+Δn₁ mol of 1  │
│ n₂ mol of 2      │ n₂ mol of 2      │
│     T, p         │     T, p         │
└──────────────────┴──────────────────┘
```
Both sides of the system change volume and exchange heat sufficient to keep T, p constant

may be truncated after the first nonvanishing term, we find that

$$\Delta G = \left(\frac{\partial^2 G}{\partial n_1{}^2}\right)_{T,p,n_2} > 0. \tag{19.85}$$

Using the definition of the chemical potential, we see that Eq. 19.85 is equivalent to

$$\left(\frac{\partial \mu_1}{\partial n_1}\right)_{T,p,n_2} > 0, \tag{19.86}$$

and hence, by Eq. 19.82, to

$$\left(\frac{\partial \mu_2}{\partial n_1}\right)_{T,p,n_2} < 0. \tag{19.87}$$

The stability condition displayed in the inequality 19.85 states that the addition of component i at constant temperature, pressure, and mass of other components must lead to an increase in the chemical potential of component i.

The conditions 19.86 and 19.87 may be used to provide useful information about the chemical potential. Experiments show that $n_2(\partial\mu_2/\partial n_1)_{T,p,n_2}$ is negative, as required, and approaches a constant nonzero value, say, $-B$, for small values of the ratio n_1/n_2. Thus, as $n_1 \to 0$, since $n_2(\partial\mu_2/\partial n_1)_{T,p,n_2}$ becomes constant, $n_1(\partial\mu_1/\partial n_1)_{T,p,n_2}$ must approach the same constant value. But in order that

$$\lim_{n_1 \to 0} n_1\left(\frac{\partial\mu_1}{\partial n_1}\right)_{T,p,n_2}$$

be nonzero, it is necessary that

$$\lim_{n_1 \to 0} \left(\frac{\partial\mu_1}{\partial n_1}\right)_{T,p,n_2} = \infty. \tag{19.88}$$

Let

$$\lim_{n_1 \to 0} n_2\left(\frac{\partial\mu_2}{\partial n_1}\right)_{T,p,n_2} = -B. \tag{19.89}$$

Then for small n_1, Eq. 19.82 gives

$$\left(\frac{\partial\mu_2}{\partial n_1}\right)_{T,p,n_2} = -\frac{B}{n_2} = -\frac{n_1}{n_2}\left(\frac{\partial\mu_1}{\partial n_1}\right)_{T,p,n_2} \tag{19.90}$$

or

$$\left(\frac{\partial\mu_1}{\partial n_1}\right)_{T,p,n_2} = \frac{B}{n_1}. \tag{19.91}$$

The integration of Eq. 19.91 gives for the chemical potential of component 1 the simple formula

$$\mu_1 = B \ln n_1 + f(T, p, n_2), \tag{19.92}$$

where $f(T, p, n_2)$ is an arbitrary function of its arguments. But the chemical potential is an intensive function and cannot depend on the number of moles of 1 or 2, only on some function of their ratio. We note that for the pure component μ_1 can only be a function of T and p. We conclude that Eq. 19.92 should be rewritten as

$$\mu_1 = B \ln \frac{n_1}{n_1 + n_2} + \mu_1{}^0(T, p). \tag{19.93}$$

In the pure fluid $n_1/(n_1 + n_2) \to 1$, and $\mu_1{}^0(T, p)$ is identified as its chemical potential. The constant B may depend on T and p; we shall later show that it has the value RT.

Equation 19.93 plays a fundamental role in chemical thermodynamics, since it gives an explicit representation of the change in Gibbs free energy with composition. We shall see in the next section how this information can be used to describe chemical equilibrium in a reacting mixture.

19.5 Chemical Equilibrium

As a first step in the description of chemical reactions, we must deduce the consequences of the condition stated in Eq. 19.37 when applied to chemical equilibrium.

Consider a closed homogeneous system in which only one chemical reaction occurs. The stoichiometry of the reaction can be expressed in the form

$$\sum_{i=1}^{r} \nu_i X_i = 0, \tag{19.94}$$

where the ν_i are the stoichiometric coefficients and each X_i represents a mole of the ith component. Because of the conservation of mass, the changes in the amounts of the components are all linked through Eq. 19.94, and we can write

$$\frac{dn_1}{\nu_1} = \frac{dn_2}{\nu_2} = \cdots = \frac{dn_r}{\nu_r} = d\lambda, \qquad (19.95)$$

where dn_i is the change in number of moles of component i. As before, the ν_i are positive for products and negative for reactants. The quantity λ is known as the *progress variable* of the reaction. Clearly, if $d\lambda/dt = 0$, no reaction is in progress, and λ (measured relative to some arbitrary zero) gives the equilibrium composition after the reaction is complete. Given Eq. 19.95, Eq. 19.38 can be used to define the nature of the equilibrium state. For a closed homogeneous system in which the masses of the several species are allowed to change, we have from Eq. 19.54, using Eq. 19.95, the condition

$$(dU)_{S,V} = \sum_{i=1}^{r} \mu_i \, dn_i = \sum_{i=1}^{r} \nu_i \mu_i \, d\lambda. \qquad (19.96)$$

The extremum defined by Eq. 19.96 must be a minimum (see Eq. 19.36b), so that for sufficiently large displacements from equilibrium we have

$$\left[\left(\sum_{i=1}^{r} \nu_i \mu_i \right) \Delta\lambda \right]_{S,V} > 0. \qquad (19.97)$$

Clearly $\Delta\lambda \equiv \lambda - \lambda_{\text{equil}}$ can be positive or negative, since there can be a mixture with either more or less product than at equilibrium. Thus

$$\Delta\lambda > 0 \quad \text{implies} \quad \left(\sum_{i=1}^{r} \nu_i \mu_i \right) > 0,$$

$$\Delta\lambda < 0 \quad \text{implies} \quad \left(\sum_{i=1}^{r} \nu_i \mu_i \right) < 0, \qquad (19.98)$$

the two relationships holding on opposite sides of equilibrium; in the equilibrium state, then, we must have

$$\sum_{i=1}^{r} \nu_i \mu_i = 0. \qquad (19.99)$$

Equation 19.99 states a necessary relationship between the chemical potentials of reacting species when equilibrium is attained.

We can now use Eqs. 19.99 and 19.93 to describe the equilibrium composition in a reacting mixture. First note that the generalization of Eq. 19.93 to a mixture with r components is

$$\mu_1 = B \ln \frac{n_1}{n_1 + \cdots + n_r} + \mu_1{}^0(T, p). \qquad (19.100)$$

The quantity

$$x_1 \equiv \frac{n_1}{n_1 + \cdots + n_r} \qquad (19.101)$$

is known as the *mole fraction* of component 1. Using an equation of the form of 19.100 for each component, we can write Eq. 19.99 in the form

$$\sum_{i=1}^{r} \nu_i[B \ln x_i + \mu_i^0(T, p)] = 0, \qquad (19.102)$$

or

$$-\frac{1}{B} \sum_{i=1}^{r} \nu_i \mu_i^0(T, p) = \sum_{i=1}^{r} \ln x_i^{\nu_i}. \qquad (19.103)$$

The left-hand side of this equation depends only on T and p, the right-hand side only on the equilibrium concentrations of each component. The entire equation implies that as the concentration of component i is changed at constant T and p, the concentrations of all other components must change so as to give the same value for $\sum \ln x_i^{\nu_i}$, since the sum must be independent of concentration by its equality with $-(1/B) \sum \nu_i \mu_i^0(T, p)$. As we have defined the chemical potential, each μ_i^0 is the molar Gibbs free energy of the ith pure component. The sum on the left-hand side of Eq. 19.103 is, therefore, just the difference in Gibbs free energy between products and reactants per mole of reaction (see also Chapter 14). Using $B = RT$, as will be shown later, and defining

$$\ln K(T, p) \equiv \sum_{i=1}^{r} \ln x_i^{\nu_i}, \qquad (19.104)$$

we deduce that

$$\sum_{i=1}^{r} \nu_i \mu_i^0 = \Delta G_{\text{reac}} = -RT \ln K(T, p). \qquad (19.105)$$

$K(T, p)$ is known as the *equilibrium constant;* it plays a fundamental role in the description of all chemical equilibria.

As derived, Eq. 19.92 is valid only when n_1/n_2 is small; the assumption that this form for the concentration dependence is correct at all concentrations is not, in general, valid. However, the corrections required do not alter our qualitative conclusions. Similarly, Eq. 19.104 requires corrections when Eqs. 19.92 or 19.100 do, but the qualitative conclusion remains valid that some combination of the concentrations can be found which does not itself vary with concentration, but only with T and p. These topics will be discussed in greater detail in later chapters.

19.6 Thermodynamic Weight: Further Connections Between Thermodynamics and Microscopic Structure

As the last general topic in this chapter we examine the relationship between the various ways of describing an equilibrium state and possible new bridges between thermodynamic variables and the microscopic structure of a system. In Chapter 15 we introduced the fundamental definition

$$S \equiv k_B \ln \Omega. \qquad (19.106)$$

This definition of the entropy in terms of the number of available states can be interpreted in a different way by reversing the roles of S and Ω. We thus rearrange Eq. 19.106 to read

$$\Omega = e^{S/k_B}. \qquad (19.107)$$

In this form we can say that the entropy plays the role of a *thermodynamic weight.* That is, if we treat S as the independent variable, the entropy weights most heavily those microscopic configurations for which $\Omega(E)$ is largest.

The ensemble we considered in defining $\Omega(E)$ is characterized by the statement that each of the isolated replicas has the same values of E, V, and N. This characterization is strikingly similar to that defining a macroscopic state in which S is a maximum at equilibrium. Can the other criteria which define equilibrium states be used to provide formulas analogous to 19.106? This is the problem we now seek to solve.

A hint to the nature of relationships analogous to 19.107 is obtained by transforming the observation just made into the following statement: The entropy is the thermodynamic weight for a microscopic configuration in an ensemble defined by a set of constraints that, when applied to a macroscopic system, lead to a maximum in the entropy. An obvious generalization of this observation is as follows: In an ensemble defined by some set of constraints, the probability of occurrence of a microscopic configuration is proportional to the positive exponential of that thermodynamic function which is a maximum in the corresponding macroscopic system. As noted in Table 19.1, $-A$ is a maximum and A a minimum in a macroscopic system constrained to have constant V, T, and n. Thus, if an ensemble is defined by the requirements that V, T, and N be constant, we guess that the probability of occurrence of a microscopic configuration is proportional to $\exp(-A/k_B T)$. It will be shown below that this guess is correct.

Our analysis has two parts. First, we examine the characteristic features of the most probable state of an ensemble in which the energy and composition of the replica systems differ from replica to replica. Second, we establish the relationships between the probability of occurrence of a microscopic configuration, the energy spectrum of a replica system, and thermodynamic functions.

We consider an ensemble similar to that studied in Section 15.5. Let each replica in the ensemble consist of a pair of systems, A and B. The volumes V_A and V_B, and any other necessary parameters of constraint, are separately specified for systems A and B, and hence for each replica pair. For simplicity we shall assume that V_A and V_B have the common value V. The replica pair is further defined by specifying the total energy $E_T = E_A + E_B$, and the total number of molecules of each species present,

$$N_{T1} = N_{A1} + N_{B1},$$
$$N_{T2} = N_{A2} + N_{B2}, \qquad (19.108)$$
$$\vdots$$
$$N_{Tr} = N_{Ar} + N_{Br}.$$

The replica pair is considered to be an isolated system, and the ensemble of replica pairs is also an isolated system. Furthermore, each of the members of any replica pair is isolated. However, because only the total energy and total numbers of molecules in the pair are specified, different replica pairs have different subdivisions of the energy between A and B, and different numbers of molecules of each species in A and B. Thus, in the ensemble of replica pairs all possible energy distributions and composition distributions are represented. Our problem is to find the most probable distribution of energy and composition.

Let $\Omega_A(E_A, N_{A1}, \ldots, N_{Ar})$ and $\Omega_B(E_B, N_{B1}, \ldots, N_{Br})$ be, respectively, the numbers of states between E_A and $E_A + dE_A$ in a system with numbers of molecules between N_{A1} and $N_{A1} + dN_{A1}, \ldots, N_{Ar}$ and $N_{Ar} + dN_{Ar}$ and between E_B and $E_B + dE_B$ in a system with numbers of molecules between N_{B1} and $N_{B1} + dN_{B1}, \ldots, N_{Br}$ and $N_{Br} + dN_{Br}$. The total number of states accessible to the replica pair with energy E_T, composition $N_{T1}, \ldots, N_{Tr}$, and subdivisions

$$E_A = E_T - E_B,$$

$$N_{A1} = N_{T1} - N_{B1},$$
$$\vdots$$
$$N_{Ar} = N_{Tr} - N_{Br} \tag{19.109}$$

is (see Eq. 15.34)

$$\Omega_T(E_T, E_A, \{N_{Ti}\}, \{N_{Ai}\}) = \Omega_A(E_A, \{N_{Ai}\})\Omega_B(E_T - E_B, \{N_{Ti} - N_{Ai}\}). \tag{19.110}$$

To find the contributions to Ω_T from different distributions of energy and numbers of molecules between A and B we proceed as in Chapter 15. The key observation is that Ω_A increases and Ω_B decreases when E_A and $N_{A1}, \ldots, N_{Ar}$ increase. Thus the product $\Omega_A\Omega_B$ has a maximum for some partition of the energy and composition. Since the composition variables are independent and the energy may be varied at fixed composition, this maximum is located by setting equal to zero, separately, the partial derivatives of Ω_T with respect to E_A and each of the N_{Ai}. It is found, by means similar to the derivation of Eq. 15.41, that the most probable distribution of energy and composition in the ensemble of replica pairs is characterized by the equalities

$$T_A = T_B,$$
$$\mu_{A1} = \mu_{B1}, \tag{19.111}$$
$$\vdots$$
$$\mu_{Ar} = \mu_{Br},$$

where

$$\frac{1}{k_B T_A} \equiv \left(\frac{\partial \ln \Omega_A}{\partial E_A}\right)_{V, N_{Ai}} \qquad (i = 1, \ldots, r), \tag{19.112}$$

$$\frac{\mu_{Ai}}{k_B T_A} \equiv -\left(\frac{\partial \ln \Omega_A}{\partial N_{Ai}}\right)_{V, E_A, N_{Aj}} \qquad (j \neq i = 1, \ldots, r). \tag{19.113}$$

The usual identification of the entropy with $k_B \ln \Omega$, combined with the general thermodynamic formulas of 19.71–19.73, shows that μ_{Ai} is the chemical potential of component i in system A. In other words, the most probable state in an ensemble of replica pairs between which energy and matter may be distributed is characterized by equality between A and B of the temperature and the chemical potentials of all species. We identify the most probable state in the ensemble described with the equilibrium state of the corresponding macroscopic system.

The preceding analysis can be generalized. Let each replica consist of s isolated systems, each with fixed volume V. If only the total energy and total number of molecules of each species in the s systems are specified, different replicas can have different energy and composition distributions. It can then be shown that the most probable state in the ensemble is characterized by equality of temperature and equality of chemical potentials of all species between all s systems in the replica.

To determine the probability of finding a system with a particular energy and composition, we consider an ensemble of $\mathcal{N}$ replica systems, each with volume V. Each of the replica systems is isolated; however, we specify only the total energy and total numbers of molecules in the ensemble. We can now ask what is the most probable state of the ensemble itself, as opposed to a single replica. Imagine that the entire ensemble is itself a replica system in a super-ensemble of ensembles. We then have a situation exactly like that in the previous paragraph, and the corresponding conclusion can be drawn. The most probable state in the super-ensemble, that is, the state in which the largest number of ensembles should be found, is one in which there is equality of temperature and of the chemical potentials of all species between all replicas in a given ensemble.

Let the energy and numbers of molecules of species $1, \ldots, r$ in a given replica be E and $N_1, \ldots, N_r$. We divide the energy spectrum of a replica into levels of width dE, and label the levels with the index k. There are, for each replica, some possible energy levels in the range $E_k \leq E \leq E_k + dE$; the number of such states of course depends on the composition of the replica. We denote by $\Omega(m)$ the number of quantum states available to the entire ensemble when exactly m replicas have the specific composition $N_1, \ldots, N_r$ and an energy in the range $E_k \leq E \leq E_k + dE$.[5] Since all these replicas have the same composition, they also have the same number of states between E_k and $E_k + dE$; we denote this number of states by g_k.

We regard the entire ensemble as an isolated system and compute $\Omega(m)$ for the purpose of calculating the entropy of the ensemble. By the definition of the energy range E_k to $E_k + dE$, each of the g_k states in that range is equally likely to be populated. Therefore, since the first replica of the m with energy between E_k and $E_k + dE$ can be in any of the g_k states, the second in any of the g_k states, and so on, there are $g_k{}^m$ ways in which the m replicas can be distributed over the g_k states. Furthermore, since any of the $\mathcal{N}$ replicas in the ensemble is equally likely to belong to the subset m, the first selected can be any of $\mathcal{N}$, the second any of $\mathcal{N}-1, \ldots,$ the mth any of $\mathcal{N}-m+1$ replicas. However, since it is immaterial in what order the m replicas are selected, we must divide by $m!$, the number of equivalent ways of selecting the same m replicas. Thus, there are

$$\frac{\mathcal{N}(\mathcal{N}-1)(\mathcal{N}-2)\cdots(\mathcal{N}-m+1)}{m!} = \frac{\mathcal{N}!}{m!(\mathcal{N}-m)!} \qquad (19.114)$$

different ways in which m of the $\mathcal{N}$ replicas can be selected. Finally, since each distribution of the m replicas over the g_k states can be combined with each of the possible choices of m replicas out of $\mathcal{N}$, the number of quantum states available to the ensemble with the constraint that exactly m replicas have the specified energy and composition can be written

$$\Omega(m) = \frac{\mathcal{N}!}{m!(\mathcal{N}-m)!} \, g_k{}^m \, \Omega_R(\mathcal{N}-m), \qquad (19.115)$$

where $\Omega_R(\mathcal{N}-m)$ is the number of states available to the remaining $\mathcal{N}-m$ replicas of the ensemble. The quantity $\Omega_R(\mathcal{N}-m)$ must be understood to be evaluated with the ensemble considered as an isolated system, that is, with the total number of molecules of each species and the total energy and volume held constant.

We have already identified the most probable state of the ensemble with the equilibrium state of the corresponding macroscopic system. We have also stated that in the most probable state of the ensemble the systems have equal temperatures and chemical potentials of all species. As in the cases considered in Chapter 15, the distribution of energy and

---

[5] As usual, the ensemble of replicas is assumed to be so large that, for every possible composition, there are enough replicas to give a distribution over all E_k of interest. We treat E and N differently because the possible values of N (the integers) are known independent of E, whereas obtaining E for a given N would require solution of a quantum mechanical problem.

composition is so sharply peaked about the most probable distribution that the entropy of the ensemble can be defined in terms of the properties of the most probable state. The arguments justifying these remarks can be developed systematically using the techniques of Chapter 15.

Of course, since the ensemble is an isolated system, the entropy of the most probable state of the ensemble must be a maximum. The ensemble may be subjected to a virtual variation, as defined previously, by transferring one of the m replicas from the subset in the g_k states and adding it to the remainder of the system. Because m is a large number and the total energy, volume, and composition of the ensemble are fixed, this variation must not alter the entropy of the ensemble. We thus assume that

$$\delta S(m) = S(m-1) - S(m) = 0; \tag{19.116}$$

we can expand $\delta S(m)$, by using Eq. 19.115, into

$$
\begin{aligned}
\delta S(m) &= k_B \ln m! - k_B \ln(m-1)! + k_B \ln(\mathcal{N}-m)! \\
&\quad - k_B \ln(\mathcal{N}-m+1)! - k_B \ln g_k - S_R(\mathcal{N}-m) \\
&\quad + S_R(\mathcal{N}-m+1) \\
&= k_B \ln \frac{m}{(\mathcal{N}-m+1)g_k} + [S_R(\mathcal{N}-m+1) - S_R(\mathcal{N}-m)] \\
&= 0, \tag{19.117}
\end{aligned}
$$

where $S_R \equiv k_B \ln \Omega_R$ is the entropy of the $\mathcal{N}-m$ replicas of the "remainder." But the expression $[S_R(\mathcal{N}-m+1) - S_R(\mathcal{N}-m)]$ is just the increase in entropy of the remainder of the ensemble when one of the m replicas with energy in the range $E_k \le E \le E_k + dE$, volume V, and composition $N_1, \ldots, N_r$ is added to it. We must now evaluate this quantity.

Suppose that $\mathcal{N} \gg m \gg 1$, and consider the limit $\mathcal{N} \to \infty$, in conformity with our initial discussion of the nature of ensembles in Chapter 15. Then we have

$$
\begin{aligned}
\delta S_R &= S_R(\mathcal{N}-m+1) - S_R(\mathcal{N}-m) \\
&= \Delta V_R \left(\frac{\partial S_R}{\partial V}\right)_{E,N} + \Delta E_R \left(\frac{\partial S_R}{\partial E}\right)_{V,N} + \sum_{i=1}^{r} \Delta N_{Ri} \left(\frac{\partial S_R}{\partial N_i}\right)_{E,V,N_{j \ne i}}. \tag{19.118}
\end{aligned}
$$

The use of derivatives is possible because, as $\mathcal{N} \to \infty$, the relative change in S_R arising from addition or subtraction of a single replica becomes infinitesimally small. Also, the changes ΔV_R, ΔE_R, and ΔN_{Ri} are just

$$
\begin{aligned}
\Delta E_R &= E_k, \\
\Delta V_R &= V, \tag{19.119} \\
\Delta N_{Ri} &= N_i,
\end{aligned}
$$

since just one replica is transferred in the virtual variation. A combination of Eqs. 19.118 and 19.119 leads to

$$
\begin{aligned}
\delta S_R &= V \left(\frac{\partial S_R}{\partial V}\right)_{E,N} + E_k \left(\frac{\partial S_R}{\partial E}\right)_{V,N} + \sum_{i=1}^{r} N_i \left(\frac{\partial S_R}{\partial N_i}\right)_{E,V,N_{j \ne i}} \\
&= \frac{pV}{T} + \frac{E_k}{T} - \frac{\sum_{i=1}^{r} N_i \mu_i}{T}, \tag{19.120}
\end{aligned}
$$

where the derivatives have been evaluated on the basis of Eq. 19.54. Note that here μ_i is the chemical potential per molecule.

By our definition of an ensemble, the probability of finding a preselected state of a system is the ratio of the number of replicas in that state to the total number of replica systems in the ensemble; that is, if we select at random one replica system out of the whole ensemble of systems, the probability that it is one of the set m is just $(m/\mathcal{N})$. This ratio is the total probability of finding a system with volume V, energy in the range $E_k \leq E \leq E_k + dE$, and composition $N_1, \ldots, N_r$. With the fundamental hypothesis that each of the g_k states of this energy and composition has equal *a priori* probability, the probability of finding the system in a specified one of the g_k states is just $(m/g_k\mathcal{N})$. We now define the limit

$$P_{k,N}(T, V, \mu_1, \ldots, \mu_r) \equiv \lim_{\substack{\mathcal{N} \to \infty \\ \left(\frac{m}{\mathcal{N}}\right) = \text{constant}}} \left(\frac{m}{\mathcal{N} - m + 1} \right), \qquad (19.121)$$

which becomes, with the use of Eqs. 19.117 and 19.120,

$$P_{k,N}(T, V, \mu_1, \ldots, \mu_r) = g_k \exp\left(-\frac{E_k + pV - \sum_{i=1}^{r} N_i \mu_i}{k_B T} \right). \qquad (19.122)$$

Given that $\mathcal{N} \gg m$, we see that $P_{k,N}$ is the probability $(m/\mathcal{N})$ of finding the system in any one of the g_k states with $E_k \leq E \leq E_k + dE$ and composition $N_1, \ldots, N_r$.

Given the probability 19.122, how can we use it to calculate the thermodynamic properties of a system from the energy spectrum of the system? Because $P_{k,N}$ is the probability of finding a system in the state specified by E_k and $N_1, \ldots, N_r$, and because the system must be in some one of all the possible states, the sum of $P_{k,N}$ over all possible states must be unity. That is, the probability is normalized to unity. Thus we can write

$$\sum_{N_1=0}^{\infty} \sum_{N_2=0}^{\infty} \cdots \sum_{N_r=0}^{\infty} \sum_{k} P_{k,N}(T, V, \mu_1, \ldots, \mu_r) = 1. \qquad (19.123)$$

We again emphasize that the reason the sums in Eq. 19.122 contain the composition variables $N_1, \ldots, N_r$ is that the energy of a replica depends on its composition, and we must sum $P_{k,N}$ over all possible states, even those of very low probability where the composition deviates greatly from the most probable composition. Similarly, the sum over all energy levels includes states of low probability. Equation 19.123 now provides a link between thermodynamic variables and the energy spectrum. For, using Eq. 19.122, we obtain

$$1 = \exp\left(\frac{-pV}{k_B T} \right) \sum_{N_1, \ldots, N_r = 0}^{\infty} \exp\left(\sum_i \frac{N_i \mu_i}{k_B T} \right) \sum_k g_k \exp\left[\frac{-E_k(V, N_1, \ldots, N_r)}{k_B T} \right], \qquad (19.124)$$

or

$$\frac{pV}{k_B T} = \ln \Xi, \qquad (19.125)$$

where

$$\Xi \equiv \sum_{N_1, \ldots, N_r = 0}^{\infty} \exp\left(\sum_i \frac{N_i \mu_i}{k_B T} \right) \sum_k g_k \exp\left[\frac{-E_k(V, N_1, \ldots, N_r)}{k_B T} \right]. \qquad (19.126)$$

The quantity Ξ is called the *grand partition function*, and the ensemble to which it corresponds is called the *grand canonical ensemble*. Despite the formidable appearance of Eq. 19.126, we shall see that it is easy and convenient to use.

It should be noted, in confirmation of the conjecture made at the beginning of this section, that in a macroscopic system defined to have constant T, V, and $\mu_1, \ldots, \mu_r$ the quantity pV is a maximum.

A different ensemble corresponds to a macroscopic system defined to have constant T, V, and $N_1, \ldots, N_r$. In this case each replica in the ensemble has the same composition. We can imagine generating such an ensemble by selecting, from the grand canonical ensemble described above, only those replicas that each have the specified composition $N_1, \ldots, N_r$. As usual, we assume that the number of replicas thus obtained is very large. The subset of replicas defined to have the same volume and composition is now considered to be an isolated system. As such it constitutes an ensemble, called the *canonical ensemble*.

As generated, the canonical ensemble consists of a collection of isolated systems with the same composition but variable energy. Alternatively, the ensemble could have been constructed from scratch by assembling a collection of systems of the same composition and specifying only the total energy of the ensemble. The distribution of energy amongst the isolated replica systems is not specified—hence all possible energies are represented in the ensemble. Then, whichever way the canonical ensemble has been constructed, the most probable state of the ensemble is characterized by equality of the temperatures of all replica systems.

The probability of finding a system in a state of specified energy in a canonical ensemble is obtained from Eq. 19.122 by picking only those terms with the required values of $N_1, \ldots, N_r$, that is, only those terms corresponding to the specified uniform composition of all the replicas. Because a replica system must be in some one of the possible energy states, the normalization condition equivalent to Eq. 19.123 becomes

$$\sum_k P_k^{CE}(T, V, N_1, \ldots, N_r) = 1 \tag{19.127}$$

or

$$\exp\left(\frac{-pV}{k_B T}\right)\exp\left(\sum_i \frac{N_i \mu_i}{k_B T}\right) \sum_k g_k \exp\left(\frac{-E_k}{k_B T}\right) = 1. \tag{19.128}$$

But, using Eqs. 19.63 and 19.64, we find that

$$A = -pV + \sum_{i=1}^r N_i \mu_i. \tag{19.129}$$

Combining Eqs. 19.128 and 19.129, we find that

$$\frac{A}{k_B T} = -\ln Q_{N_1, \ldots, N_r}, \tag{19.130}$$

where

$$Q_{N_1, \ldots, N_r} \equiv \sum_k g_k \exp\left(\frac{-E_k}{k_B T}\right) \tag{19.131}$$

is called the *canonical partition function*.

We again note, in support of our interpretation of thermodynamic weight, that in a macroscopic system constrained to have constant T, V, and $N_1, \ldots, N_r$ the function $-A$ is a maximum.

There is still another ensemble that can be generated from a subset of the systems making up the grand canonical ensemble. Suppose that we choose from the set of all replicas in the grand canonical ensemble that set which both has uniform concentration $N_1, \ldots, N_r$ and has energy in the range $E_k \leqslant E \leqslant E_k + dE$. This is, of course, just the set of m systems originally described in our derivation of Eq. 19.122. The m isolated systems constitute an ensemble known as the *microcanonical ensemble*. Since all m replica systems have the same composition and energy, the normalization condition equivalent to Eq. 19.123 becomes

$$P_k^{MCE}(E, V, N_1, \ldots, N_r) = 1, \tag{19.132}$$

or

$$\exp\left(\frac{-pV}{k_B T}\right)\exp\left(\sum_i \frac{N_i \mu_i}{k_B T}\right) g_k \exp\left(\frac{-E_k}{k_B T}\right) = 1. \tag{19.133}$$

Using Eq. 19.60 we recognize that

$$S = \frac{E_k}{T} + \frac{pV}{T} - \sum_{i=1}^{r} \frac{N_i \mu_i}{T}, \tag{19.134}$$

so that we find, by substitution,

$$S = k_B \ln g_k. \tag{19.135}$$

Of course, in the microcanonical ensemble we have defined, g_k is the same as the function usually denoted as $\Omega(E_k)$.

In all the preceding discussion we have emphasized that the replica systems constituting an ensemble are individually isolated. It is, nevertheless, a property of the most probable state of the ensemble that each replica has the same value of the intensive variable conjugate to an extensive quantity that may be distributed, for example, the temperature when only the total energy of the ensemble is fixed. Thus, these ensembles mimic the corresponding macroscopic situations in which systems in contact can transfer energy and/or mass. However, had energy and/or mass transfer been permitted between replica systems of the ensemble, no replica system would have had mechanical constants of motion, and our description in terms of well-defined energy states would have been inappropriate. Indeed, measurement of an intensive parameter necessarily imposes some uncertainty on the conjugate extensive quantity, since the "meter" must be open to exchange of the extensive quantity with the system whose properties are being measured (e.g., measurement of pressure requires "transfer" of volume, of temperature requires transfer of energy, etc.). Although it is often convenient to think of the ensembles described as if energy and/or mass can be transferred between replicas, the consequences of this interpretation and the properties of the ensemble as a model of a physical situation must always be remembered.

In Chapter 12 we introduced, for the purpose of describing the properties of a perfect gas, a function representing the distribution of molecular speeds. At that point we did not have available the tools necessary to determine the form of the distribution function. We close this chapter by showing how the representation of a perfect gas by a canonical ensemble permits calculation of the speed distribution function.

In using a canonical ensemble to represent a perfect gas, we must decide what the appropriate replica system is. By definition, in a perfect gas collisions between molecules may be neglected. Then each molecule can be considered to be "essentially" isolated; that is, those collisions that occur provide a mechanism for the transfer of energy and maintenance of equilibrium, but they are so infrequent that the energy spectrum of a molecule is unaltered from what it would be in the absence of collisions,

19.7
An Application of the Canonical Ensemble: The Distribution of Molecular Speeds in a Perfect Gas

and the energy of a molecule is constant for periods of time many orders of magnitude larger than the duration of a collision. A canonical ensemble that is a model of the perfect gas can then be constructed by regarding each molecule in the volume V to be an isolated replica system, and the collection of N molecules in the volume V to be an ensemble. Clearly, because the (infrequent) collisions do lead to transfer of energy between molecules, the ensemble representing the gas must allow for unequal distribution of the total energy of the gas amongst the molecules. Thus, the representative ensemble is chosen to be a canonical ensemble. A microcanonical ensemble, in which each replica has the same energy, would be inappropriate, as would a grand canonical ensemble, in which each replica has different composition.

We take as a model of the perfect gas a set of N independent molecules in a cubic box of volume V. The energy of one molecule in the box is

$$\epsilon_{n_1 n_2 n_3} = \frac{\pi^2 \hbar^2}{2mV^{2/3}} (n_1^2 + n_2^2 + n_3^2) \qquad (19.136)$$

or

$$\epsilon_n = \frac{\pi^2 \hbar^2}{2mV^{2/3}} n^2. \qquad (19.137)$$

where $n^2 = n_1^2 + n_2^2 + n_3^2$. In the canonical ensemble we are using, ϵ_n is the energy of one replica system. Now, the probability of finding a replica (molecule) with energy ϵ_n is

$$P_n^{CE}(\epsilon_n) = \frac{g_n \exp(-\epsilon_n/k_B T)}{\sum_n g_n \exp(-\epsilon_n/k_B T)}, \qquad (19.138)$$

which is obtained simply by dividing Eq. 19.122 by the normalization condition 19.127. To evaluate Eq. 19.138 we must first compute g_n, the number of quantum states with energy in the range $\epsilon_n \leq \epsilon \leq \epsilon_n + d\epsilon$. In a macroscopic box the spacings between the energy levels of the spectrum are very small, and at ordinary temperatures n_1, n_2, and n_3 are very large. In what follows we shall find it convenient to take advantage of this fact and treat n_1, n_2, and n_3 as continuous variables.

Let n_1, n_2, and n_3 be three perpendicular axes defining a space. Each state of the molecule is defined by the triplet of numbers n_1, n_2, n_3; hence the energy of each state can be represented as a point in the space described. Equation 19.136, which gives the energy as a function of n_1, n_2, and n_3, describes a spherical surface. Since n_1, n_2, and n_3 can change only by integer values, there is one energy state per unit volume of n_1, n_2, n_3 space. Then the number of states with ϵ in the range $\epsilon_n \leq \epsilon \leq \epsilon_n + d\epsilon$ can be computed from the volume of the spherical shell of radius n and thickness dn.[6] Actually, since n_1, n_2, and n_3 must be positive, only one-eighth the volume of this shell may be counted, namely, that eighth lying in the positive octant of the space. Thus, the volume required is

$$g_n = \tfrac{1}{8} \cdot 4\pi n^2 \, dn. \qquad (19.139)$$

The denominator of Eq. 19.138 can be evaluated by replacing the summation by an integration, treating n as a continuous variable. The result is

$$\sum_n g_n \exp\left(\frac{-\epsilon_n}{k_B T}\right) = \frac{\pi}{2} \int_0^\infty n^2 \exp\left(\frac{-\pi^2 \hbar^2 n^2}{2mV^{2/3} k_B T}\right) dn$$

[6] Note that $d\epsilon$ is proportional to dn.

$$= \left(\frac{mk_BT}{2\pi\hbar^2}\right)^{3/2} V. \tag{19.140}$$

We now convert the variable from n to v, the speed of a molecule. To do so we note that a molecule has only kinetic energy and that Eq. 19.137 can be rewritten to read

$$n^2 = \frac{2mV^{2/3}}{\pi^2\hbar^2}\epsilon = \frac{2mV^{2/3}}{\pi^2\hbar^2}\left(\frac{1}{2}mv^2\right). \tag{19.141}$$

From Eqs. 19.139 and 19.141 we easily find that

$$g_n = \frac{\pi}{2}n^2\,dn = \frac{m^3Vv^2\,dv}{2\pi^2\hbar^3}. \tag{19.142}$$

The substitution of Eqs. 19.142 and 19.140 into Eq. 19.138 gives, for the distribution of molecular speeds,[7]

$$P_v^{CE}(v) = f(v)\,dv = 4\pi v^2\left(\frac{m}{2\pi k_BT}\right)^{3/2}\exp\left(\frac{-mv^2}{2k_BT}\right)dv. \tag{19.143}$$

Equation 19.143, known as the *Maxwell speed distribution*, is one of the most fundamental relations of chemistry and physics. The distribution of molecular speeds plays an important role in determining the rates of chemical reactions and the thermodynamic properties of matter. We shall later see that if the interaction energy between a pair of molecules is independent of the molecular velocity, then Eq. 19.143 is equally valid for condensed phases and gases.

The Maxwell distribution of speeds is so important that we digress to study a few of its features. Note that this distribution predicts there are no molecules with either zero or infinite speed. In between these extremes the distribution function has a maximum (see Fig. 19.6). The most probable speed, obtained by setting $df(v)/dv$ equal to zero, is

$$v^* = \left(\frac{2k_BT}{m}\right)^{1/2}. \tag{19.144}$$

Note how the most probable speed increases as the temperature increases, and also how the breadth of the distribution increases as the temperature increases (see Fig. 19.7).

Since the speed distribution function is not symmetric about the most probable speed, the mean speed $\langle v\rangle$ and the root-mean-square speed $\langle v^2\rangle^{1/2}$ differ from v^*. We have, by the definition of average values,

$$\langle v\rangle \equiv \frac{\int vf(v)\,dv}{\int f(v)\,dv} = \left(\frac{8k_BT}{\pi m}\right)^{1/2}, \tag{19.145}$$

$$\langle v^2\rangle \equiv \frac{\int v^2f(v)\,dv}{\int f(v)\,dv} = \left(\frac{3k_BT}{m}\right). \tag{19.146}$$

Thus

$$\langle v\rangle = \frac{2}{\sqrt{\pi}}v^* = 1.128v^*, \tag{19.147}$$

[7] Note that the "$f(v)$" in Chapter 12 is the number of molecules per unit volume with speeds between v and $v+dv$, while here $f(v)\,dv$ is the fraction of molecules with speeds between v and $v+dv$.

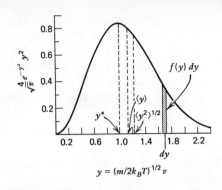

FIGURE 19.6
The Maxwell speed distribution. In the plot we use the reduced variable

$$y = (m/2k_BT)^{1/2}v.$$

In terms of y, the distribution function reads

$$f(y)\,dy = \frac{4}{\sqrt{\pi}}e^{-y^2}y^2\,dy.$$

From F. H. Crawford, *Heat, Thermodynamics and Statistical Physics* (Harcourt Brace Jovanovich, New York, 1963).

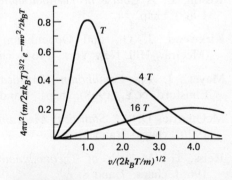

FIGURE 19.7
The Maxwell speed distribution for several temperatures. The unit along the horizontal axis is the most probable speed for the curve at the extreme left. The other two curves are for the temperatures $4T$ and $16T$, respectively. From F. H. Crawford, *Heat, Thermodynamics and Statistical Physics* (Harcourt Brace Jovanovich, New York, 1963).

$$\langle v^2 \rangle^{1/2} = \sqrt{\frac{3}{2}} \, v^* = 1.224 v^*. \qquad (19.148)$$

These two averages are, respectively, 13% and 22% greater than the most probable speed.

Finally, we notice that the average kinetic energy of a molecule is

$$\frac{1}{2} m \langle v^2 \rangle = \frac{3 k_B T}{2}. \qquad (19.149)$$

Recalling that the velocity distribution in a perfect gas must be isotropic, we have

$$\langle v^2 \rangle = \langle v_x^2 + v_y^2 + v_z^2 \rangle = \langle v_x^2 \rangle + \langle v_y^2 \rangle + \langle v_z^2 \rangle \qquad (19.150)$$

with

$$\langle v_x^2 \rangle = \langle v_y^2 \rangle = \langle v_z^2 \rangle. \qquad (19.151)$$

The average kinetic energy per degree of freedom is thus,

$$\frac{m}{2} \langle v_x^2 \rangle = \frac{m}{2} \langle v_y^2 \rangle = \frac{m}{2} \langle v_z^2 \rangle = \frac{k_B T}{2}. \qquad (19.152)$$

This result is one example of the *principle of equipartition of energy*. This principle states that the average energy associated with each degree of freedom in the mechanical motion of the system that is described by a squared term in the mechanical energy is the same, and equal to $k_B T/2$ per molecule. The theorem is valid only for classical mechanics, and only for squared terms in the energy expression. At high temperature, when the thermal energy exceeds all energy spacings in the spectrum, a real quantum mechanical system approaches the behavior of the corresponding classical system, so that the equipartition theorem is a useful general description of many systems of interest to chemists.

Further Reading

Kestin, J., *A Course in Thermodynamics*, Vol. II (Blaisdell, Waltham, 1968), chap. 14.

Kirkwood, J. G., and Oppenheim, I., *Chemical Thermodynamics* (McGraw-Hill, New York, 1961), chap. 6.

Mayer, J. E., *Equilibrium Statistical Mechanics* (Pergamon Press, Elmsford, N.Y., 1968), chaps. 1 and 8.

McQuarrie, D. A., *Statistical Mechanics* (Harper & Row, New York, 1976), chaps. 2 and 3.

Reiss, H., *Methods of Thermodynamics* (Blaisdell, Waltham, Mass., 1965), chaps. 7 and 8.

Problems

1. A perfect gas is contained in a cylinder fitted with a piston. The cylinder is insulated; hence the system is initially isolated, and any changes of state of the gas when isolation is "broken" are adiabatic. Let the pressure acting on the gas be p_0. Consider adding to the system just described new constraints that might change V or T, while retaining the constraint $p_{app} = p_0$. Show *by direct evaluation* that under the original conditions the entropy of the system was a maximum.

2. Consider a solid in equilibrium with its own vapor across an unconstrained interface. Let the vapor pressure be p_0 at temperature T_0. Suppose that a barrier is interposed between the solid and the vapor. The barrier is diathermal and permeable to molecules. If the barrier is rigid, so that the solid can be compressed (or fixed in volume as T increases), is equilibrium between solid and vapor at p_0 and T_0 possible? What differences do you expect between the system with and without the barrier as T changes?

3. Consider a chemical reaction involving perfect gaseous reactants and perfect gaseous products. Let the volume of the products at 1 atm be less than the volume of the reactants at 1 atm and the same temperature. If a closed container with a reaction mixture is compressed at constant temperature, how do you expect the equilibrium concentrations of the components to change?

4. If the reaction described in Problem 3 is constrained to:

   (a) Take place in a fixed volume at constant temperature,

   (b) Take place under constant external pressure and temperature in a vessel of variable volume,

   (c) Take place at fixed volume in a thermally isolated container,

   do you expect the equilibrium concentrations of the components to be the same? How would they change in (b) relative to (a)? In (c) relative to (a)? What are the most convenient thermodynamic descriptions of the equilibria under conditions (a), (b), and (c)?

5. Write down expressions for the average energy of a system represented by (a) a canonical ensemble and (b) a grand canonical ensemble. Show that

$$\langle U \rangle^{CE} = -\left(\frac{\partial \ln Q_N}{\partial(1/k_B T)}\right)_V,$$

$$\langle U \rangle^{GCE} = -\left(\frac{\partial \ln \Xi}{\partial(1/k_B T)}\right)_{\mu/T,V}.$$

6. Show that for a system represented by a grand canonical ensemble the average density, $\langle N \rangle / V$, is given by

$$\frac{\langle N \rangle}{V} = \frac{1}{V}\left(\frac{\partial \ln \Xi}{\partial(\mu/k_B T)}\right)_{T,V}.$$

7. Show that for a system represented by (a) a canonical ensemble and (b) a grand canonical ensemble the pressure is given by

$$p^{CE} = k_B T\left(\frac{\partial \ln Q_N}{\partial V}\right)_T,$$

$$p^{GCE} = k_B T\left(\frac{\partial \ln \Xi}{\partial V}\right)_{T,\mu}.$$

8. Show that in a perfect gas the distribution of kinetic energies has the form

$$f(\epsilon) = \frac{2}{\pi^{1/2}(k_B T)^{3/2}} \epsilon^{1/2} \exp\left(\frac{-\epsilon}{k_B T}\right).$$

9. Calculate the number of molecules in 1 mol of He at 1000 K that have components of velocity in the positive x direction of less than 10^4 cm/s and 10^5 cm/s. Hint: The answer can be expressed in terms of an integral that can be related to

$$\text{erf}(x) \equiv \frac{2}{\sqrt{\pi}} \int_0^x e^{-x^2} \, dx,$$

known as the *error function*. Tables of numerical values of this function are available (see, for example, the *Handbook of Physics and Chemistry*). Cast your answer into a form such that you can use these tables for the numerical work.

10. Calculate the number of molecules in 1 mol of He at 1000 K that have speeds greater than 10^4 cm/s and 10^5 cm/s.

11. Set up the formulas that give the fraction of molecules in a perfect gas that have speeds between $\frac{1}{4}$ and 2 times the most probable speed. Evaluate this fraction for He at 298 K and 1 atm.

12. Evaluate the root-mean-square speed at 300 K for a molecule of UF_6 and a hydrogen atom.

13. Obtain molecular speed distributions, analogous to Eq. 19.143, for a gas whose molecules are constrained to move in two dimensions only, and also for a one-dimensional gas.

14. Equation 19.143 and the above problem deal with distributions of the magnitudes of molecular velocities. The magnitude can, of course, take on only non-negative values: $0 \leqslant v < \infty$. Now consider a gas whose molecules can move only along the x axis. Let v_x be the magnitude and direction of a molecule's velocity, so that v_x has the range $-\infty < v_x < \infty$. Obtain the distribution of v_x, $P(v_x)$, in terms of the one-dimensional speed distribution of the above problem. [Hint: For any v_x, one should have $P(v_x) = P(-v_x)$.]

15. Generalize your expression for $P(v_x)$ in the above problem to obtain an equation for the three-dimensional vector velocity distribution, $P(v_x, v_y, v_z)$. (Hint: v_x, v_y, and v_z should be independent and equivalent to one another.) If you now convert $P(v_x, v_y, v_z)$ to spherical coordinates and integrate over the angular coordinates, you should be able to rederive Eq. 19.143.

16. Check, by direct calculation, the validity of the Euler theorem for the homogeneous function

$$F_n = x^a y^b z^{n-a-b}.$$

The Euler theorem states that, if F is a homogeneous function of its variables of degree n, say, $F(x^n, y^n, \ldots)$, then

$$x \frac{\partial F}{\partial x} + y \frac{\partial F}{\partial y} + \cdots = nF(x, y, \ldots)$$

(see Eqs. 19.56 and 19.57 of the text).

17. The standard Gibbs free energy change at 25°C for the dissociation of 1 mol of iodine vapor is 140.3 kJ. What fraction of iodine vapor at 25°C exists in the atomic state at (a) 1 torr pressure, (b) 10^{-6} torr, (c) 10^{-12} torr? Assume ideality. (760 torr = 1 atm.)

18. Verify Eqs. 19.144–19.146.

19. According to the principle of equipartition, what should C_V be for the following (ideal) gases: Ar, N_2, CO_2, H_2O?

20. The most fundamental condition that must be met for a laser to operate between single-particle energy ϵ_i and single-particle energy ϵ_j (with $\epsilon_j > \epsilon_i$) is that the population be "inverted," that is, the number of molecules with energy ϵ_j must exceed the number of molecules with energy ϵ_i.

    (a) If the population inversion is to be produced by heating to some temperature T, what condition must the degeneracies of the energy levels satisfy?
    (b) Can thermally induced laser action exist between vibrational states of a diatomic molecule?
    (c) Given the fact that vibrational laser action has been produced in the states of product HCl generated in the reaction $H_2 + Cl_2 \rightarrow 2HCl$, can these product states be in thermal equilibrium?

21. The magnetic susceptibility of a linear molecule in a magnetic field is quantized and can have only the values

$$\chi_m = \chi_0 m^2, \qquad m = 1, 0, -1,$$

where χ_0 is a constant. The energy of the molecule in a magnetic field H_0 is

$$W_m = \frac{1}{2} H_0^2 \chi_m.$$

(a) Considering only the magnetic field effect on the molecule, write a formal expression for the canonical partition function in terms of W_m.
(b) Write a formal expression for $\langle \chi \rangle$, the average value of the susceptibility.
(c) Write an expression for $\langle \chi \rangle$ in terms of H_0, k_B, T, and the partition function Q.
(d) Assuming that $W_m \ll k_B T$, obtain an explicit expression for $\langle \chi \rangle$ correct to H_0^2.

22. Show that the Helmholtz free energy is given by

$$A = E^2 \frac{\dfrac{\partial}{\partial E}\left(\dfrac{\ln \Omega}{E}\right)}{\dfrac{\partial \ln \Omega}{\partial E}},$$

where Ω is the microcanonical partition function. Evaluate A for 1 mol of an ideal gas with energy E and volume V.

23. The three levels of a molecule are

$$\epsilon_n = n\gamma,$$

where γ is a constant and $n = -1, 0, 1$. The degeneracies g_n are $g_{-1} = 1$, $g_0 = 2$, and $g_1 = 1$. A system of N such molecules is at equilibrium.

(a) Find the average energy at temperature T.
(b) Evaluate C_V if $\gamma \ll k_B T$.

24. For the reaction

$$C(\text{graphite}) + 2H_2(g) \rightarrow CH_4(g),$$

the enthalpy change at 600°C is -88.052 kJ/mol. The third-law entropies (in J/K mol) at 600°C and 1 atm are 20 for graphite, 163 for H_2, and 237 for CH_4.

(a) Calculate the equilibrium constant at 600°C.
(b) Assume that ΔH for the reaction does not vary significantly and find the equilibrium constant at 750°C. Hint: Use the Gibbs–Helmholtz equation.
(c) What conditions of temperature and pressure favor the yield of CH_4?

25. We showed in Chapter 15 that the distribution of states in a system at equilibrium is extremely sharply peaked, so sharply peaked that for many purposes the distribution can be replaced by its value at the maximum. Show that when there is effectively only one term of importance in the sum defining

$$\Delta \equiv \sum_V \sum_k g_k \exp\left(\frac{-E_k}{k_B T}\right) \exp\left(\frac{-pV}{k_B T}\right),$$

then

$$G = -k_B T \ln \Delta.$$

Δ is called the *isothermal-isobaric partition function*; it is defined in the isothermal-isobaric ensemble in which each replica has the same number of molecules, the total energy and total volume of the ensemble of replicas are fixed, but the energy and volume of an individual replica can differ from those of another replica.

26. Show that

$$P_k^{CE} = \frac{g_k \exp\left(\dfrac{-E_k}{k_B T}\right)}{\sum_k g_k \exp\left(\dfrac{-E_k}{k_B T}\right)},$$

and then that

$$S - -k_B \sum_k P_k^{CE} \ln P_k^{CE}.$$

27. Show that fluctuations of the number of molecules in a volume V at temperature T and with chemical potential μ satisfy

$$\langle N^2 \rangle - \langle N \rangle^2 = \frac{\langle N \rangle^2 k_B T \kappa_T}{V}.$$

Show that the numerical value of the relative fluctuation in number of molecules in V is very small. Hint: Use

$$\langle N \rangle = k_B T \left(\frac{\partial \ln \Xi}{\partial \mu} \right)_{V,T},$$

$$\langle N^2 \rangle = \frac{k_B T}{\Xi} \frac{\partial}{\partial \mu} (\langle N \rangle \Xi).$$

28. Suppose that as a result of some approximate calculation, the following canonical partition function is found for a model of a gas:

$$Q_N = \frac{1}{N!} \left(\frac{2\pi m k_B T}{h^2} \right)^{3N/2} (V - Nb)^N \exp\left(\frac{aN^2}{V k_B T} \right)$$

with a and b constants determined by the model. What is the equation of state of the model gas?

20

An Extension of Thermodynamics to the Description of Nonequilibrium Processes

In the preceding chapters we have developed the foundations of the thermodynamic and statistical molecular descriptions of matter at equilibrium. In doing so we took an uncompromisingly rigid view of how the thermodynamic functions were to be defined—at every stage of the analysis we chose methods that were valid for the case that equilibrium prevailed. A consequence of the approach we have used is that the applicability of the thermodynamic relations derived is restricted to the description of systems at equilibrium. For example, we defined the concept of temperature by use of the zeroth law of thermodynamics, which explicitly requires that a state of thermal equilibrium be established between the thermometer and the system under investigation, including internal equilibrium of each. It would appear, then, that we could not meaningfully talk about the temperature of the reacting gases in a rocket nozzle. Yet it is possible, by a variety of well-defined procedures, to measure the temperature of that reacting, flowing mixture. Moreover, with suitable precautions as to the interpretation, the temperature so measured gives valuable information about the system probed. Numerous other examples of the validity of the measurement of temperatures of systems out of equilibrium can be given. We conclude that it should be possible to extend the definitions of the thermodynamic functions, previously introduced only for matter at equilibrium, to the description of matter not at equilibrium.

There are two key concepts in the extension of thermodynamic arguments to systems not at equilibrium, both related to the notion of separation of time scales. The simpler of the two is the following: If the rate at which a macroscopic inhomogeneity in a system relaxes toward uniformity is small compared to the rate of establishment of equilibrium between the molecular degrees of freedom, local values of thermodynamic functions can be defined. It is then natural to assume that the relations between local values of the thermodynamic functions are the same as those in the state of complete equilibrium (*hypothesis of local equilibrium*). For example, if a pulse of heat is generated at some point in a fluid,[1] a spatially nonuniform distribution of temperature is created. We can speak of such a temperature distribution because the rate at which collisions between molecules establish a canonical distribution of kinetic energy is very large relative to the rate at which collisions transfer energy over a macroscopic distance. The kinetic energy in an element of volume

---

[1] This can be done by passing a pulse of current through a small coil of resistance wire immersed in the fluid.

close to the point of injection is higher than elsewhere, hence so is the temperature. We expect there to be, correspondingly, a distribution of density in the fluid, since an increase in local temperature implies more energetic collisions between molecules, or a larger amplitude of motion, both of which establish a local increase in the volume per molecule. Provided that the time scales for establishment of the local canonical distribution of kinetic energy and the local volume per molecule are both short compared to the time required for relaxation to the uniform state, the local temperature and density will be related by the ordinary macroscopic equation of state.

The second key concept relates to the differences in rates of approach to equilibrium of different degrees of freedom. For example, at room temperature it takes 10^3 to 10^4 collisions to transfer vibrational to translational energy, 10 to 10^2 collisions to transfer rotational to translational energy, whereas almost every collision transfers translational energy between molecules. It is possible, then, by suitable excitation of a system and aging, to have some degrees of freedom at equilibrium and others not. Furthermore, even though it may take $\sim 10^4$ collisions on average to transfer vibrational energy to translational energy, it usually takes only about 10 to 10^2 to establish a canonical distribution of vibrational energy within the vibrational degrees of freedom. When this situation prevails it is possible to assign different temperatures to different degrees of freedom. Unlike the spatial inhomogeneity example described above, the relationships between the defined thermodynamic functions are now, in general, not the same as in the state of equilibrium. Macroscopic concepts can still be used to describe the system, but where their validity requires complete internal equilibrium (e.g., the relationship between compressibility and difference in specific heats), they are inapplicable. Of course, if there is a canonical distribution of energy within the levels of one degree of freedom, macroscopic concepts related only to that degree of freedom can be used, provided that the relaxation time criteria previously mentioned are met (e.g., a relation between vibrational entropy and vibrational temperature is valid if the vibrational energy distribution is canonical). Note that our definition of equilibrium with respect to a degree of freedom requires establishment of a canonical distribution of energy in that degree of freedom. If this is not the case, thermodynamic functions cannot be defined for that degree of freedom.

The most fully developed extension of thermodynamic analysis to the description of nonequilibrium phenomena is in its application to hydrodynamic problems. It is usual to proceed by assuming that the local equilibrium hypothesis is valid. We shall use a variant of the usual procedure and show that, starting with the equations that describe fluid motion, the Fourier law of heat conduction implies that locally defined thermodynamic functions satisfy the same relationships as at equilibrium.

In order to describe a flowing fluid we must use variables whose values are local in space and time. Thus, instead of having state variables that are defined over an entire system, we now regard the pressure, density, and energy to be functions of position within the system, **r**, and time, t. At any particular space-time point these functions are assumed to exist.

In fluid mechanics, when we refer to a point in space we mean a region of space that is small by macroscopic standards. Just as in thermodynamics, however, the molecular nature of matter is ignored. A "point" inside a volume of fluid is therefore sufficiently large that it contains a great many molecules. In other words, matter is considered to be continuous.

20.1
General Form of the Equations of Continuity

The starting point of fluid mechanics is a consideration of conservation laws. The usual conditions of conservation of mass, momentum, and energy must be expressed in a manner that reflects the assumed continuous nature of matter. In this format, the conservation laws are often called *equations of continuity*.

The developments of the equations of continuity for the various conserved quantities in a fluid follow a similar line of reasoning. Let Q represent any conserved quantity. Consider an arbitrary closed fixed volume V inside the fluid. At each point in V the local density of Q, $\rho_Q(\mathbf{r}, t)$, is defined. The total amount of Q contained in V at time t is given by

$$Q(t) = \int_V d\mathbf{r}\rho_Q(\mathbf{r}, t). \tag{20.1}$$

Because Q is a conserved quantity, the only way $Q(t)$ can vary with time is if Q passes across the boundary surface S that defines V. This surface can be divided into infinitesimal sections, with each section labelled by a vector $d\mathbf{s}$ whose magnitude is equal to the area of the section and whose direction is perpendicular to the section and pointing outward from V (see Fig. 20.1). At each point in V there is a vector flux density $\mathbf{J}^Q(\mathbf{r}, t)$ associated with Q. This vector points in the direction of the flow of Q and has the units of Q per unit area per unit time. The total amount of Q *leaving* V per unit time must equal the surface integral of $\mathbf{J}^Q(\mathbf{r}, t)$ over S. We can therefore write

$$\frac{dQ}{dt} = -\int_S d\mathbf{s} \cdot \mathbf{J}^Q(\mathbf{r}, t). \tag{20.2}$$

The minus sign in Eq. 20.2 is required as a result of the convention in assigning the direction of $d\mathbf{s}$.

The surface integral in Eq. 20.2 can be converted to a volume integral if we make use of the Gauss divergence theorem:

$$\int_S d\mathbf{s} \cdot \mathbf{J}^Q(\mathbf{r}, t) = \int_V d\mathbf{r} \, \nabla \cdot \mathbf{J}^Q(\mathbf{r}, t). \tag{20.3}$$

From Eq. 20.1 we also have

$$\frac{dQ}{dt} = \int_V d\mathbf{r} \frac{\partial}{\partial t} \rho_Q(\mathbf{r}, t). \tag{20.4}$$

Because V is an arbitrary volume, the combination of Eqs. 20.2, 20.3, and 20.4 yields

$$\frac{\partial}{\partial t} \rho_Q(\mathbf{r}, t) = -\nabla \cdot \mathbf{J}^Q(\mathbf{r}, t), \tag{20.5}$$

which is the general form of the equations of continuity. In order that Eq. 20.5 be useful, expressions for the flux $\mathbf{J}^Q$ must be introduced, and the particular format each of these expressions takes will vary with the property Q under consideration. In the following sections we shall discuss the detailed forms of Eq. 20.5 for mass, momentum, and energy.

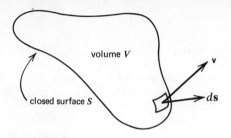

FIGURE 20.1

An elementary area $d\mathbf{s}$ on the surface S enclosing the volume V, and the mass velocity vector **v**. The shape of the volume V is arbitrary.

20.2
Conservation of Mass and the Diffusion Equation

If $Q(t)$ represents the mass contained in an arbitrary volume of a fluid, Q can change only when an actual flow of matter occurs across the boundary surface of the volume. The mass flux $\mathbf{J}^m(\mathbf{r}, t)$ is then given by

$$\mathbf{J}^m(\mathbf{r}, t) = \rho(\mathbf{r}, t)\mathbf{v}(\mathbf{r}, t), \tag{20.6}$$

where ρ is the local mass density and **v** is the local fluid velocity. If we substitute Eq. 20.6 in Eq. 20.5 and use the rule for differentiation of a product,

$$\nabla \cdot (\rho\mathbf{v}) = \rho \, \nabla \cdot \mathbf{v} + \mathbf{v} \cdot \nabla\rho, \tag{20.7}$$

we find

$$\frac{\partial \rho}{\partial t} + \mathbf{v} \cdot \nabla \rho + \rho \nabla \cdot \mathbf{v} = 0 \qquad (20.8)$$

as the local condition expressing conservation of mass.

We can write Eq. 20.8 in a slightly different form if we consider the *total* time derivative of the density. This is

$$\frac{d}{dt} \rho(\mathbf{r}, t) = \frac{\partial \rho}{\partial t} + \sum_i \frac{\partial \rho}{\partial x_i} \frac{dx_i}{dt}$$

$$= \frac{\partial \rho}{\partial t} + \sum_i \frac{\partial \rho}{\partial x_i} v_i$$

$$= \frac{\partial \rho}{\partial t} + \mathbf{v} \cdot \nabla \rho. \qquad (20.9)$$

Here, the index i runs over the three Cartesian coordinates and x_i is the ith component of the position vector $\mathbf{r}$. Thus, Eq. 20.8 is equivalent to

$$\frac{d\rho}{dt} + \rho \nabla \cdot \mathbf{v} = 0. \qquad (20.10)$$

The total derivative $d\rho/dt$ is to be interpreted as the time rate of change of the density of an element of the fluid as it moves in space. By contrast, the partial derivative $\partial\rho/\partial t$ represents the rate of change of the density of the fluid at a fixed point in space.

In the case of incompressible fluids (a very good approximation for most liquids), Eq. 20.10 takes on a particularly simple form. The density is then constant, and the equation of continuity reduces to

$$\nabla \cdot \mathbf{v} = 0. \qquad (20.11)$$

So far, the equation of continuity is still not in a useful form, because it is a differential equation for ρ in terms of an unknown velocity field $\mathbf{v}(\mathbf{r}, t)$. An additional relationship between ρ and $\mathbf{v}$ is needed so that $\mathbf{v}$ can be eliminated from Eq. 20.8. To illustrate this point, suppose that the fluid under consideration is a mixture in which there are not any chemical reactions. Then the same arguments concerning conservation of mass can be applied to the rate of change of ρ_k, the mass per unit volume of component k, as to the total mass density. If $\mathbf{v}_k$ is the velocity of component k, the analog of Eq. 20.8 is

$$\frac{\partial \rho_k}{\partial t} = -\nabla \cdot (\rho_k \mathbf{v}_k). \qquad (20.12)$$

The rate of change of the total density of the fluid is still given by Eq. 20.8 provided that $\mathbf{v} = \sum_k \rho_k \mathbf{v}_k / \rho$ is the velocity of the center of mass. Let

$$\mathbf{J}_k^m = \rho_k (\mathbf{v}_k - \mathbf{v}) \qquad (20.13)$$

be the flux of k relative to that carried along by motion of the center of mass. Substitution of Eq. 20.13 into Eq. 20.12 and use of Eq. 20.9 lead to

$$\frac{d\rho_k}{dt} = -\rho_k \nabla \cdot \mathbf{v} - \nabla \cdot \mathbf{J}_k^m. \qquad (20.14)$$

We now close Eq. 20.14 by introducing the phenomenological relation known as *Fick's law*.

In the absence of convection currents, and if spatial gradients in the density of component k are small, it is reasonable to assume that the flux of k is everywhere proportional to its density gradient. Fick's law expresses this proportionality and introduces a *diffusion coefficient* D_k by the relation

$$\mathbf{J}_k^m(\mathbf{r}, t) = -D_k \nabla \rho_k(\mathbf{r}, t). \qquad (20.15)$$

The minus sign in Eq. 20.15 expresses the observation that flow occurs from a region of high density to a region of low density (D_k is positive). The substitution of Eq. 20.15 in Eq. 20.12 yields, for a fluid at rest ($\mathbf{v} = 0$), a partial differential equation called the *diffusion equation*:

$$\frac{\partial \rho_k}{\partial t} = D_k \nabla^2 \rho_k. \tag{20.16}$$

For given initial conditions and boundary conditions the diffusion equation can be solved for the spatial dependence of the density of k as a function of time. For example, suppose that at $t = 0$ a mass m_k of component k is concentrated at the origin. The solution to Eq. 20.16 for this condition shows that, at time t, component k is distributed in space according to

$$\rho_k(\mathbf{r}, t) = \frac{m_k}{8(\pi D_k t)^{3/2}} \exp\left(\frac{-r^2}{4 D_k t}\right). \tag{20.17}$$

Note that, since the diffusion is isotropic, ρ_k depends only on r, the magnitude of the displacement vector.

Clearly, as time passes the mixture becomes uniform. A schematic representation of how the solution approaches uniform composition is shown in Fig. 20.2.

Equation 20.17 can also be used to calculate the average distance that a "particle" of a diffusing fluid travels in a given time. The mean-square displacement from the origin at time t is found to be

$$\langle r_k^2 \rangle = 6 D_k t. \tag{20.18}$$

The square-root-of-time dependence of the distance traveled is characteristic of diffusive motion, and Eq. 20.18 can be considered a practical definition of the diffusion coefficient.

Of course, Fick's law is not the only way of transforming the equation of mass continuity into a useful form. Under other circumstances there may be mass flux terms due to convection, or gradients in other quantities such as the temperature. The principles involved in the use of other relationships for closure are the same as those described, although details of the analysis will necessarily be different.

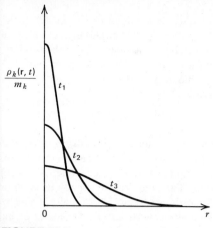

FIGURE 20.2
Schematic representation of the concentration profile as a function of time for the diffusion of a substance originally concentrated at the origin of the coordinates. The times t_1, t_2, and t_3 are ordered such that $t_1 < t_2 < t_3$.

20.3
Conservation of Momentum and the Navier–Stokes Equation

The development of the equation of continuity (Eq. 20.5) for momentum is somewhat more complicated than for mass. First, we must realize that momentum is a vector quantity. The flux conjugate to the scalar mass is a vector; given that momentum is a vector quantity, what is the nature of the momentum flux? To answer this question we select an arbitrary fixed coordinate system and focus on a particular component of the local momentum density. Since the ith component of the momentum of a small region of the fluid with mass dm is

$$dp_i(\mathbf{r}, t) = v_i(\mathbf{r}, t)\, dm(\mathbf{r}, t), \tag{20.19}$$

the momentum density has components of the form

$$\rho_{p_i} = \rho(\mathbf{r}, t) v_i(\mathbf{r}, t). \tag{20.20}$$

The ith component of the momentum density must satisfy an equation of continuity, so we can write

$$\frac{\partial}{\partial t}[\rho(\mathbf{r}, t) v_i(\mathbf{r}, t)] = -\nabla \cdot \mathbf{J}^{p_i}. \tag{20.21}$$

Here, $\mathbf{J}^{p_i}$ is the vector flux for the ith component of the momentum. We can write the divergence of $\mathbf{J}^{p_i}$ in component form:

$$\nabla \cdot \mathbf{J}^{p_i} = \sum_{j=1}^{3} \frac{\partial}{\partial x_j} J_j^{p_i}. \tag{20.22}$$

A typical component of $\mathbf{J}^{p_i}$, say $J_y{}^{p_i}$, is the flow of i component of momentum in the y direction. Then, taking account of all three components of the momentum, we see that nine quantities of the form $J_j{}^{p_i}$ will appear in Eq. 20.22, since the two indices i and j both range from 1 to 3. We now change our notation slightly to emphasize the existence of two indices in the momentum flux. Henceforth we write

$$J_{ji}{}^p = J_j{}^{p_i}. \tag{20.23}$$

The nine quantities $J_{ji}{}^p$ are the components of a new type of quantity, called a tensor, which is seen to be an extension of the vector concept. More formally, the momentum flux is an example of a second-rank tensor, so named because two indices are required to specify its components in a particular coordinate system. An ordinary vector is formally a first-rank tensor; the tensor concept can be generalized to an arbitrary number of indices.

We shall represent the momentum flux tensor by the symbol $\mathbf{J}^p$. The divergence of $\mathbf{J}^p$ is a vector whose ith component is (see Eq. 20.22)

$$(\nabla \cdot \mathbf{J}^p)_i = \sum_{j=1}^{3} \frac{\partial}{\partial x_j} J_{ji}{}^p. \tag{20.24}$$

We can therefore write Eq. 20.21 without reference to components in a particular coordinate system:

$$\frac{\partial}{\partial t} [\rho(\mathbf{r}, t)\mathbf{v}(\mathbf{r}, t)] = -\nabla \cdot \mathbf{J}^p. \tag{20.25}$$

What is the form of the momentum flux tensor? We can start to answer this question by carrying out the differentiation indicated on the left-hand side of Eq. 20.25:

$$\frac{\partial}{\partial t} [\rho(\mathbf{r}, t)\mathbf{v}(\mathbf{r}, t)] = \mathbf{v}\frac{\partial \rho}{\partial t} + \rho\frac{\partial \mathbf{v}}{\partial t}. \tag{20.26}$$

If we substitute Eq. 20.8 for $\partial\rho/\partial t$ in the above equation, we find that

$$\mathbf{v}\frac{\partial \rho}{\partial t} = -\mathbf{v}[\nabla \cdot (\rho\mathbf{v})]. \tag{20.27}$$

The other term on the right-hand side of Eq. 20.26 contains $\partial\mathbf{v}/\partial t$, the acceleration of the fluid at $(\mathbf{r}, t)$. This acceleration has two contributions, one stemming from the pressure within the fluid, and the other resulting from *viscosity*.

For the moment let us suppose that the fluid is nonviscous. (Fluids with no viscosity are sometimes called *ideal fluids*.) If we again consider the fluid volume shown in Fig. 20.1, we can write the force acting on an element $d\mathbf{s}$ of the surface as

$$d\mathbf{f} = -p\,d\mathbf{s}, \tag{20.28}$$

where p is the pressure on the surface element. The force is defined to be positive if it points toward the interior of the volume. The total force acting on the fluid volume is then

$$\mathbf{f} = -\int_S d\mathbf{s}\,p(\mathbf{r}, t). \tag{20.29}$$

The surface integral can be converted to a volume integral in a manner similar to Eq. 20.3:

$$\mathbf{f} = -\int_V d\mathbf{r}\,\nabla p(\mathbf{r}, t). \tag{20.30}$$

We interpret Eq. 20.30 by saying that a force per unit volume equal to $-\nabla p$ acts on any volume element of the fluid. We equate the force per unit volume to the mass per unit volume times the total acceleration of the volume element:

$$\rho \frac{d\mathbf{v}}{dt} = -\nabla p. \qquad (20.31)$$

The total derivative $d\mathbf{v}/dt$ can be expanded in partial derivatives in a manner analogous to Eq. 20.9, and the result is

$$\rho \frac{\partial \mathbf{v}}{\partial t} + \rho \mathbf{v} \cdot \nabla \mathbf{v} = -\nabla p. \qquad (20.32)$$

Note that the gradient $\nabla \mathbf{v}$ is a tensor whose components have the form

$$(\nabla \mathbf{v})_{ij} = \frac{\partial v_j}{\partial x_i}. \qquad (20.33)$$

The dot product $\mathbf{v} \cdot \nabla \mathbf{v}$ is then a vector whose jth component is

$$(\mathbf{v} \cdot \nabla \mathbf{v})_j = \sum_i v_i \frac{\partial v_j}{\partial x_i}. \qquad (20.34)$$

The analogy between Eq. 20.34 and the operations in Eq. 20.9 should be apparent.

Equation 20.32 is known as *Euler's equation*. If we use it and Eq. 20.27 in Eq. 20.26, we find that

$$\frac{\partial}{\partial t}(\rho \mathbf{v}) = -\mathbf{v}[\nabla \cdot (\rho \mathbf{v})] - \rho \mathbf{v} \cdot \nabla \mathbf{v} - \nabla p$$

$$= -\nabla \cdot (\rho \mathbf{v}\mathbf{v}) - \nabla p, \qquad (20.35)$$

which can be checked by calculating the individual components. The components of the tensor $\rho \mathbf{v}\mathbf{v}$ have the form $\rho v_i v_j$. The divergence of this tensor has components

$$(\nabla \cdot \rho \mathbf{v}\mathbf{v})_j = \sum_i \frac{\partial}{\partial x_i}(\rho v_i v_j)$$

$$= v_j \sum_i \frac{\partial}{\partial x_i}(\rho v_i) + \rho \sum_i v_i \frac{\partial v_j}{\partial x_i}, \qquad (20.36)$$

so that

$$\nabla \cdot \rho \mathbf{v}\mathbf{v} = \mathbf{v}(\nabla \cdot \rho \mathbf{v}) + \rho \mathbf{v} \cdot \nabla \mathbf{v} \qquad (20.37)$$

as in Eq. 20.35.

It is convenient to rewrite the pressure gradient in Eq. 20.35 in a somewhat different form. We define the unit tensor $\mathbf{1}$ as one whose diagonal elements are all unity and whose off-diagonal elements are all zero; the components of $\mathbf{1}$ are often denoted by the Kronecker symbol

$$\delta_{ij} = \begin{cases} 1 & \text{if } i = j \\ 0 & \text{if } i \neq j. \end{cases} \qquad (20.38)$$

The jth component of the pressure gradient can then be written

$$(\nabla p)_j = \frac{\partial p}{\partial x_j} = \sum_i \frac{\partial}{\partial x_i}(p\delta_{ij}), \qquad (20.39)$$

or

$$\nabla p = \nabla \cdot (p\mathbf{1}). \qquad (20.40)$$

We can now rewrite Eq. 20.35:

$$\frac{\partial}{\partial t}(\rho \mathbf{v}) = -\nabla \cdot (\rho \mathbf{v}\mathbf{v} + p\mathbf{1}). \qquad (20.41)$$

Comparison with Eq. 20.25 yields (finally!) the momentum flux tensor for a nonviscous fluid:

$$\mathbf{J}^\mathrm{p} = \rho\mathbf{vv} + p\mathbf{1}. \tag{20.42}$$

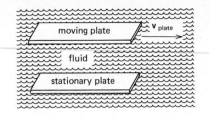

The physical origin of the two terms in $\mathbf{J}^\mathrm{p}$ is easy to see. The first term, $\rho\mathbf{vv}$, represents momentum that is transferred from point to point by an actual flow of fluid. Note that $\rho\mathbf{vv}$ is the local momentum density, $\rho\mathbf{v}$, times the local velocity. This contribution to the momentum flux is in direct analogy to Eq. 20.6 for the mass flux.

In contrast to mass transfer, momentum can be transferred by means other than fluid flow. We express these other contributions to the momentum flux by introducing the *stress tensor*, whose definition is

$$\boldsymbol{\sigma} = -(\mathbf{J}^\mathrm{p} - \rho\mathbf{vv}). \tag{20.43}$$

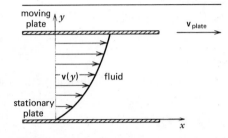

For nonviscous fluids we have already shown that the stress tensor is equal to $-p\mathbf{1}$. This quantity represents momentum that can be transferred from point to point by means of the application of pressure. When viscosity is present, the stress tensor takes on a somewhat more complicated form.

FIGURE 20.3
Schematic representation of the fluid velocity profile between a moving plate and a stationary plate.

What do we mean when we say that a fluid is viscous? Suppose that a region of fluid is contained between two flat parallel plates, as shown in Fig. 20.3. The fluid is supposed to be initially at rest, and one of the plates is then set in uniform motion in its plane while the other plate is held fixed. The moving plate drags a thin layer of fluid along with it, while the layer of fluid adjacent to the stationary plate remains stationary. A velocity gradient has thus been set up in the fluid, and it points perpendicular to the two plates. Now, if the fluid is viscous, momentum will be transferred in the direction opposite to the velocity gradient, so that the stationary plate will feel a force in the direction of the moving plate's motion. This transfer of momentum is behind our intuitive notion of viscosity.

If the velocity gradients in the fluid are small, it is reasonable to assume that the momentum flux due to viscosity is proportional to the local velocity gradient. This assumption is analogous to Eq. 20.15 in our discussion of mass diffusion. Therefore, there must be a term in $\mathbf{J}^\mathrm{p}$ that is proportional to $-\nabla\mathbf{v}$. In Appendix 20A we show that the momentum flux tensor must be symmetric, which means that in any particular coordinate system,

$$J_{ij}^{\,\mathrm{p}} = J_{ji}^{\,\mathrm{p}}. \tag{20.44}$$

Specifically, we demonstrate that every term in $J_{ij}^{\,\mathrm{p}}$ of the form $\partial v_j/\partial x_i$ must be balanced by a term $\partial v_i/\partial x_j$. The velocity gradient must therefore occur in the symmetrized form $(\nabla\mathbf{v})^\mathrm{s}$, where

$$(\nabla\mathbf{v})_{ij}^{\,\mathrm{s}} = \frac{\partial v_j}{\partial x_i} + \frac{\partial v_i}{\partial x_j}. \tag{20.45}$$

In an isotropic fluid the only other possible contribution to the momentum flux tensor components is $(\nabla\cdot\mathbf{v})\delta_{ij}$, which is both symmetric and proportional to elements of the velocity gradient. It is customary to combine these various terms in a certain way, and to define the stress tensor to be

$$\boldsymbol{\sigma} = -p\mathbf{1} + \eta[(\nabla\mathbf{v})^\mathrm{s} - \tfrac{2}{3}(\nabla\cdot\mathbf{v})\mathbf{1}] + \zeta(\nabla\cdot\mathbf{v})\mathbf{1}. \tag{20.46}$$

The stress tensor contains the pressure term and a combination of two viscosity terms; the coefficients η and ζ are called the coefficients of shear and bulk viscosity, respectively. The *shear viscosity* is a measure of the friction experienced in a shear flow, that is, one like that generated in the example of moving parallel plates. The *bulk viscosity* is a measure of the friction experienced in a pure expansion or compression flow. If the liquid is composed of molecules with internal structure, any lag in establishment of internal equilibrium amongst the molecular degrees of freedom contributes to ζ.

Equation 20.35 now takes on the form

$$\frac{\partial}{\partial t}(\rho\mathbf{v}) = -\nabla\cdot(\rho\mathbf{vv} - \boldsymbol{\sigma}). \tag{20.47}$$

If we recall Eq. 20.37, and subtract Eq. 20.27 from Eq. 20.47, we find that

$$\rho \frac{\partial \mathbf{v}}{\partial t} = -\rho \mathbf{v} \cdot \nabla \mathbf{v} + \nabla \cdot \boldsymbol{\sigma} \tag{20.48}$$

or, using Eq. 20.46 in component form,

$$\rho \left(\frac{\partial v_i}{\partial t} + \sum_j v_j \frac{\partial v_i}{\partial x_j} \right) = -\frac{\partial p}{\partial x_i} + \eta \sum_j \left(\frac{\partial}{\partial x_j} \frac{\partial v_i}{\partial x_j} + \frac{\partial}{\partial x_j} \frac{\partial v_j}{\partial x_i} \right) - \tfrac{2}{3} \eta \frac{\partial}{\partial x_i} (\nabla \cdot \mathbf{v}) + \zeta \frac{\partial}{\partial x_i} (\nabla \cdot \mathbf{v}). \tag{20.49}$$

However,

$$\sum_j \frac{\partial}{\partial x_j} \frac{\partial v_j}{\partial x_i} = \frac{\partial}{\partial x_i} (\nabla \cdot \mathbf{v}),$$

and

$$\sum_j \frac{\partial}{\partial x_j} \frac{\partial v_i}{\partial x_j} = (\nabla \cdot \nabla) v_i = \nabla^2 v_i, \tag{20.50}$$

so that

$$\rho \left(\frac{\partial v_i}{\partial t} + \sum_j v_j \frac{\partial v_i}{\partial x_j} \right) = -\frac{\partial p}{\partial x_i} + \eta \nabla^2 v_i + (\zeta + \tfrac{1}{3}\eta) \frac{\partial}{\partial x_i} (\nabla \cdot \mathbf{v}). \tag{20.51}$$

In vector notation, Eq. 20.51 is

$$\rho \left(\frac{\partial \mathbf{v}}{\partial t} + \mathbf{v} \cdot \nabla \mathbf{v} \right) = -\nabla p + \eta \nabla^2 \mathbf{v} + (\zeta + \tfrac{1}{3}\eta) \nabla (\nabla \cdot \mathbf{v}). \tag{20.52}$$

Equation 20.52, called the *Navier–Stokes equation*, is a fundamental hydrodynamic equation of motion; it is a direct consequence of the law of conservation of momentum. The particular form of the stress tensor that appears in the Navier–Stokes equation has been obtained under the assumption that the velocity gradient is small, but is found to be very generally valid. Equation 20.48, where $\boldsymbol{\sigma}$ includes the pressure contribution $-p\mathbf{1}$, is always valid. The Navier–Stokes equation has a great many practical applications to problems of fluid flow in restricted volumes, to effects of viscous drag, and so forth. A consideration of these applications would take us quite far afield. However, the statement of the Navier–Stokes equation will be important in a somewhat different connection in the following pages.

We close this section by noting that for incompressible fluids $\nabla \cdot \mathbf{v} = 0$, and the Navier–Stokes equation becomes

$$\frac{\partial \mathbf{v}}{\partial t} + \mathbf{v} \cdot \nabla \mathbf{v} = -\frac{1}{\rho} \nabla p + \frac{\eta}{\rho} \nabla^2 \mathbf{v}. \tag{20.53}$$

20.4
Conservation of Energy and the Second Law of Thermodynamics

We shall now consider the hydrodynamic statement of the law of conservation of energy and some of the consequences of that law. In a sense, we shall come full circle in this section. By introducing a phenomenological assumption in the energy conservation equation similar to Fick's law, and using the concept of local equilibrium, we shall be able to make contact with the second law of thermodynamics.

Until now, our conservation equations have been derived by considering changes in a quantity Q in a closed fixed volume V (see Eq. 20.1). In this section it will be more convenient to think of a closed volume V' that moves along with the fluid. To emphasize this change we write the total amount of Q contained in V' as Q', where

$$Q'(t) = \int_{V'} d\mathbf{r}\, \rho_Q(\mathbf{r}, t). \tag{20.54}$$

The form of Eq. 20.4 must now be changed slightly. Since V' moves with the fluid, there cannot be any flow of matter across the boundary of V'. Therefore we must subtract from the right-hand side of Eq. 20.4 the contribution to dQ/dt resulting from fluid flow. From Eqs. 20.6 and 20.42 we see that such flow contributes a term to the flux of Q that has the general form $\rho_Q \mathbf{v}$. We can therefore rewrite Eq. 20.4 as

$$\frac{dQ'}{dt} = \int_{V'} d\mathbf{r} \frac{\partial}{\partial t} \rho_Q + \int_{S'} d\mathbf{s} \cdot \rho_Q \mathbf{v}. \qquad (20.55)$$

Note that we have used Eq. 20.2 to remove from Eq. 20.4 that part of dQ/dt which results from flow across the boundary surface.

Suppose that

$$\rho_Q = \rho(\mathbf{r}, t) f_Q(\mathbf{r}, t), \qquad (20.56)$$

where ρ is the mass density and f_Q is some function of space and time. Equation 20.56 has been satisfied by both the properties discussed in the previous two sections. From Eqs. 20.55 and 20.56 we obtain

$$\frac{dQ'}{dt} = \int_{V'} d\mathbf{r} \left[\frac{\partial}{\partial t} (\rho f_Q) + \nabla \cdot (\rho f_Q \mathbf{v}) \right]. \qquad (20.57)$$

If we expand the derivatives in the above equation, and use Eq. 20.8, we find that

$$\frac{dQ'}{dt} = \int_{V'} d\mathbf{r} \left(\rho \frac{\partial f_Q}{\partial t} + \rho \mathbf{v} \cdot \nabla f_Q \right)$$

$$= \int_{V'} d\mathbf{r}\, \rho \frac{df_Q}{dt}, \qquad (20.58)$$

where df_Q/dt is the total derivative discussed in Eq. 20.9.

Equation 20.48 can also be rewritten in terms of a total derivative:

$$\rho \frac{\partial \mathbf{v}}{\partial t} + \rho \mathbf{v} \cdot \nabla \mathbf{v} = \rho \frac{d\mathbf{v}}{dt} = \nabla \cdot \boldsymbol{\sigma}. \qquad (20.59)$$

We now let Q' represent the total kinetic energy, so that

$$f_Q = \tfrac{1}{2} v^2. \qquad (20.60)$$

If we use Eqs. 20.58 and 20.59, we find that

$$\frac{dQ'}{dt} = \int_{V'} d\mathbf{r}\, \rho \mathbf{v} \cdot \frac{d\mathbf{v}}{dt} = \int_{V'} d\mathbf{r}\, \mathbf{v} \cdot \nabla \cdot \boldsymbol{\sigma}. \qquad (20.61)$$

However, from the Gauss divergence theorem,

$$\int_{S'} d\mathbf{s} \cdot \mathbf{v} \cdot \boldsymbol{\sigma} = \int_{V'} d\mathbf{r}\, \nabla \cdot (\mathbf{v} \cdot \boldsymbol{\sigma})$$

$$= \int_{V'} d\mathbf{r}\, \mathbf{v} \cdot \nabla \cdot \boldsymbol{\sigma} + \int_{V'} d\mathbf{r}\, \boldsymbol{\sigma} \cdot \nabla \cdot \mathbf{v}, \qquad (20.62)$$

so that

$$\frac{dQ'}{dt} = \frac{d}{dt} \int_{V'} d\mathbf{r} (\tfrac{1}{2} \rho v^2)$$

$$= \int_{S'} d\mathbf{s} \cdot \mathbf{v} \cdot \boldsymbol{\sigma} - \int_{V'} d\mathbf{r}\, \boldsymbol{\sigma} \cdot \nabla \cdot \mathbf{v}. \qquad (20.63)$$

Recall that $d\mathbf{s} \cdot \boldsymbol{\sigma}$ is a force, and that the scalar product of force with velocity gives power, or rate of energy change. The surface integral in Eq. 20.63 therefore represents the energy change in V' brought about by the work of pressure and viscous forces.

Let $\rho \varepsilon$ be the local density of potential energy in the fluid. Then the law of conservation of total energy (or the first law of thermodynamics) implies that we can write for the time derivative of the total energy in V'

$$\frac{d}{dt}\int_{V'}d\mathbf{r}(\tfrac{1}{2}\rho v^2 + \rho\varepsilon) = \int_{S'}d\mathbf{s}\cdot\mathbf{v}\cdot\boldsymbol{\sigma} - \int_{S'}d\mathbf{s}\cdot\mathbf{q}. \tag{20.64}$$

The first integral on the right of Eq. 20.64 is the work term discussed above. The second integral introduces the *heat flux vector* $\mathbf{q}$. The quantity $-\int_{S'}d\mathbf{s}\cdot\mathbf{q}$ measures the rate at which heat enters V'. The heat flux vector should not be confused with the scalar heat q that has appeared in previous chapters. If V' were fixed in space, Eq. 20.64 would have another term equal to the rate of energy change due to the flow of matter across the boundary surface. We would then write

$$\frac{d}{dt}\int_{V}d\mathbf{r}(\tfrac{1}{2}\rho v^2 + \rho\varepsilon) = \int_{S}d\mathbf{s}\cdot\mathbf{v}\cdot\boldsymbol{\sigma} - \int_{S}d\mathbf{s}\cdot\mathbf{q} - \int_{S}d\mathbf{s}\cdot(\tfrac{1}{2}\rho v^2 + \rho\varepsilon)\mathbf{v}. \tag{20.65}$$

If we subtract Eq. 20.63 from Eq. 20.64, we find that

$$\frac{d}{dt}\int_{V'}d\mathbf{r}\,\rho\varepsilon = \int_{V'}d\mathbf{r}\,\rho\frac{d\varepsilon}{dt} = -\int_{S'}d\mathbf{s}\cdot\mathbf{q} + \int_{V'}d\mathbf{r}\,\boldsymbol{\sigma}\cdot\nabla\cdot\mathbf{v}. \tag{20.66}$$

In the first step above we have used Eq. 20.58. The surface integral of the heat flux can be converted into a volume integral (divergence theorem again), and the fact that V' is arbitrary means that

$$\rho\frac{d\varepsilon}{dt} + \nabla\cdot\mathbf{q} - \boldsymbol{\sigma}\cdot\nabla\cdot\mathbf{v} = 0. \tag{20.67}$$

Let us define the specific volume v to be

$$v \equiv \frac{1}{\rho}. \tag{20.68}$$

Equation 20.10 can then be rewritten as

$$\rho\frac{dv}{dt} = \nabla\cdot\mathbf{v}. \tag{20.69}$$

If we also define $\boldsymbol{\sigma}'$ to be the viscous part of the stress tensor (see Eq. 20.46),

$$\boldsymbol{\sigma}' \equiv \boldsymbol{\sigma} + p\mathbf{1}, \tag{20.70}$$

we see that

$$\boldsymbol{\sigma}\cdot\nabla\cdot\mathbf{v} = -p\,\nabla\cdot\mathbf{v} + \boldsymbol{\sigma}'\cdot\nabla\cdot\mathbf{v}$$

$$= -p\rho\frac{dv}{dt} + \boldsymbol{\sigma}'\cdot\nabla\cdot\mathbf{v}. \tag{20.71}$$

The assumption of local equilibrium implies that ε must be a function of p and v only. We can, therefore, expand the derivative $d\varepsilon/dt$ as follows:

$$\frac{d\varepsilon}{dt} = \frac{\partial\varepsilon}{\partial v}\frac{dv}{dt} + \frac{\partial\varepsilon}{\partial p}\frac{dp}{dt}. \tag{20.72}$$

The substitution of Eqs. 20.72 and 20.71 in Eq. 20.67 yields

$$\rho\left(\frac{\partial\varepsilon}{\partial v} + p\right)\frac{dv}{dt} + \rho\frac{\partial\varepsilon}{\partial p}\frac{dp}{dt} + \nabla\cdot\mathbf{q} - \boldsymbol{\sigma}'\cdot\nabla\cdot\mathbf{v} = 0. \tag{20.73}$$

We can now introduce two thermodynamic functions, $\theta(p, v)$ and $\omega(p, v)$, and require that they satisfy the relations

$$\frac{\partial\varepsilon}{\partial v} + p = \theta\frac{\partial\omega}{\partial v} \tag{20.74a}$$

and

$$\frac{\partial\varepsilon}{\partial p} = \theta\frac{\partial\omega}{\partial p}. \tag{20.74b}$$

Equation 20.73 then becomes

$$\rho\theta\frac{d\omega}{dt}+\nabla\cdot\mathbf{q}=\boldsymbol{\sigma}'\cdot\nabla\cdot\mathbf{v}. \tag{20.75}$$

The functions θ and ω seem arbitrary at this point. However, the theory of differential equations can be used to demonstrate that θ and ω are well determined by Eqs. 20.74 if one additional condition is imposed. That condition is that θ satisfy the relation

$$\mathbf{q}=-\lambda\,\nabla\theta, \tag{20.76}$$

where λ is a positive function of p and v.

Equation 20.76 is reminiscent of another relation between a flux and a gradient. In Fick's law (Eq. 20.15) the assumption was made that the mass flux is proportional to the local density gradient. A similar phenomenological statement can be made about the heat flux. This relation is known as *Fourier's law*, and it asserts that the heat flux is proportional to the temperature gradient. If we identify θ with the temperature, Eq. 20.76 is just a statement of Fourier's law, and the proportionality factor λ is called the *thermal conductivity*.

We can now write

$$\nabla\cdot\frac{\mathbf{q}}{\theta}=\frac{1}{\theta}\nabla\cdot\mathbf{q}-\frac{\mathbf{q}}{\theta^2}\cdot\nabla\theta$$

$$=\frac{1}{\theta}\nabla\cdot\mathbf{q}+\frac{\lambda}{\theta^2}(\nabla\theta)^2. \tag{20.77}$$

The use of Eq. 20.77 in Eq. 20.75 yields

$$\rho\frac{d\omega}{dt}+\nabla\cdot\frac{\mathbf{q}}{\theta}=\frac{\lambda}{\theta^2}(\nabla\theta)^2+\boldsymbol{\sigma}'\cdot\nabla\cdot\mathbf{v}. \tag{20.78}$$

For a moment, suppose that $\nabla\theta=0$ and $\nabla\mathbf{v}=0$, so that the fluid is in uniform motion and the temperature is uniform. The right-hand side of Eq. 20.78 is then zero. If we integrate the left-hand side of the equation over V' and use Eq. 20.58 and the divergence theorem, we find that

$$\frac{d}{dt}\int_{V'}d\mathbf{r}\,\rho\omega+\int_{S'}d\mathbf{s}\cdot\frac{\mathbf{q}}{\theta}=0. \tag{20.79}$$

The surface integral is the rate of heat loss from V' divided by the temperature. Equation 20.79 then is nothing more than the second law of thermodynamics for reversible processes! To make the connection, all we must do is to associate $\int_{V'}d\mathbf{r}\rho\omega$ with the entropy contained in V'. The function ω is thus the local specific entropy. If Eq. 20.79 is then integrated over time, the familiar statement of the second law results.

More generally, the right-hand side of Eq. 20.78 is not equal to zero. However, since λ is positive, we have $(\lambda/\theta^2)(\nabla\theta)^2>0$. Furthermore, if the components of $\boldsymbol{\sigma}'$ given in Eq. 20.46 are used, it will be seen that $\boldsymbol{\sigma}'\cdot\nabla\cdot\mathbf{v}>0$, so we have the general result

$$\frac{d}{dt}\int_{V'}d\mathbf{r}\,\rho\omega+\int_{S'}d\mathbf{s}\cdot\frac{\mathbf{q}}{\theta}\geqslant0. \tag{20.80}$$

in accord with the second law of thermodynamics.

Of course, to derive Eq. 20.80 we have used the phenomenological Fourier's law. The second law of thermodynamics is itself much more general than our development leading to Eq. 20.80. Nevertheless, it has been of considerable interest to demonstrate in this simple example how the equations of hydrodynamics, most often used to describe problems of fluid flow, can be used to make a connection with equilibrium thermodynamics. The key element in the connection is the assumption of local equilibrium.

There are many applications in which the simultaneous solutions to hydro-dynamic and thermodynamic equations are required, for example, supersonic flow of air over an airplane body. We shall not discuss any of these interesting analyses—what is important is that a vast array of physical phenomena can be accurately described using the hypothesis of local equilibrium, and thermodynamic concepts are valid for those cases. The local equilibrium hypothesis fails when it is not possible to separate time scales as described earlier. Each situation must be examined separately and the important dynamic processes assessed to establish whether or not separate time scales exist for the several processes involved in the phenomenon being investigated.

20.5
Negative Temperature

We showed in Chapter 16 that the second law of thermodynamics implied the existence of an absolute temperature scale satisfying the condition $T\Theta(T) = \text{constant}$, with $\Theta(T)$ the integrating factor relating the entropy change to the heat transferred in an infinitesimal reversible process. For our particular choice of absolute temperature scale, T is an everywhere-positive function. This choice of sign has the consequence that a dissipative process in an isolated system leads to an increase in the temperature. We have also shown that the absolute temperature is a measure of the dispersion in energy in a canonical distribution, and that at equilibrium, aside from some minor effects arising from level degeneracy (see Fig. 21.13), the population of energy levels decreases as $\exp(-E/k_B T)$ as the energy increases. We now ask if there are ever circumstances in which it is desirable to extend the domain of the temperature function to allow it to take on negative values.

We note, first, that the introduction of negative absolute temperatures creates an absurdity, for if that negative temperature characterizes a canonical distribution of energy, proportional to $\exp(-E/k_B T)$, the population of the energy levels increases as E increases, and diverges as $E \to \infty$. In all real systems there is no upper bound to the energy scale,[2] although quantum theory does establish the existence of a lower bound to the energy. We conclude that a negative absolute temperature can never exist in an equilibrium state.

The conclusion just reached is valid for the equilibrium states of a system, but it does not preclude the possibility that nonequilibrium states of the system can be characterized by a negative temperature. Consider a system with an energy-level spectrum that can be meaningfully divided into two groups of levels, the interaction between these two groups being very weak. Moreover, let one of the groups consist of a small number of levels, the highest of which has a bounded energy. The fact that the energy of this subgroup has an upper bound is important. One example of a class of systems with the properties described is a solid containing nuclei with nonzero spin. It is often the case that a canonical distribution of energy in the subset of nuclear spin levels can be achieved very much more rapidly than can a canonical distribution of energy over the combined spin and lattice levels. In the case of LiF, for example, at 300 K the former takes $\sim 10^{-5}$ s and the latter $\sim 10^2$–10^3 s.

Suppose that the nuclear spins in some solid are aligned by a magnetic field, and then the magnetic field is reversed so rapidly that the spins cannot follow. The reversed-field situation corresponds to a non-equilibrium state of the system that can persist long enough for many

---

[2] If the constituents of the system have internal degrees of freedom, they can be dissociated into fragments with arbitrarily large kinetic energy. If the constituents are elementary particles, these can have arbitrarily large kinetic energy.

kinds of measurements (~5 min for LiF). Now, at equilibrium in the initial state with, say, magnetic field pointing up, there is a canonical distribution of energy over the levels of the spin system. For a nucleus with spin I, the possible energy levels span the range $-\mu HI$ to μHI, with $\boldsymbol{\mu}$ the nuclear magnetic moment and $\mathbf{H}$ the magnetic field strength. Note that the most positive energy, μHI, is bounded for any given field $\mathbf{H}$. For simplicity, suppose that the nuclear spin is $\frac{1}{2}$, so that the spin energy levels are $-\frac{1}{2}\mu H$ and $\frac{1}{2}\mu H$. At equilibrium the canonical distribution gives, for the ratio of populations of these two levels,

$$\frac{N_+}{N_-} = \exp\left(\frac{\mu H}{k_B T}\right), \qquad (20.81)$$

where T is the *common* temperature of the spin and lattice degrees of freedom and N_+ is the number of nuclei with spin parallel to $\mathbf{H}$. If, as stated, the spins do not follow the reversal of the magnetic field, then the nonequilibrium state that is created has an inverted population ratio, that is, a higher population in the upper energy level than in the lower energy level. When this inverted distribution is canonical, it can be characterized by a negative temperature. Equation 20.81 then becomes, by definition, for the particular situation described,

$$\frac{N_+}{N_-} = \exp\left[\frac{\mu(-H)}{k_B(-T)}\right]. \qquad (20.82)$$

The other degrees of freedom of the system, to which the spins are very weakly coupled, remain at the original temperature, which is positive. As in the case of LiF, because the interaction of the spins with these other degrees of freedom is weak, the spin system can remain in the negative-temperature state described for a considerable period of time, but eventually even the weak interactions are sufficient to drive the system to a new equilibrium state with, as usual, T positive.

Why is the condition that the energy-level spectrum be bounded from above so important? We showed, in Chapter 15, that for all ordinary systems (those with unbounded energy-level schemes), $\Omega(E, V, N)$ and $S \equiv k_B \ln \Omega$ are both everywhere-increasing functions of E. It then follows that the slope $(\partial S/\partial E)_{V,N} = 1/T$ is positive, hence so is T. But if the energy-level spectrum is bounded from above, it is not true that $\Omega(E, V, N)$ is an everywhere-increasing function of E, hence the remainder of the argument fails. Consider, for simplicity, a two-level system with energies ϵ_0 and $2\epsilon_0$. The canonical partition function is $Q_N = (e^{-\beta\epsilon_0} + e^{-2\beta\epsilon_0})^N$, from which we find

$$
\begin{aligned}
A_N = -k_B T \ln Q_N &= -k_B T \ln[e^{-\beta\epsilon_0}(1 + e^{-\beta\epsilon_0})]^N \\
&= N\epsilon_0 - Nk_B T \ln(1 + e^{-\beta\epsilon_0}),
\end{aligned}
\qquad (20.83)
$$

$$S_N = -\left(\frac{\partial A_N}{\partial T}\right)_V = Nk_B \ln(1 + e^{-\beta\epsilon_0}) + Nk_B\beta\epsilon_0 \frac{e^{-\beta\epsilon_0}}{1 + e^{-\beta\epsilon_0}}, \quad (20.84)$$

$$U_N = A_N + TS_N = N\epsilon_0\left(1 + \frac{e^{-\beta\epsilon_0}}{1 + e^{-\beta\epsilon_0}}\right), \qquad (20.85)$$

$$C_V = \left(\frac{\partial U_N}{\partial T}\right)_V = Nk_B(\beta\epsilon_0)^2 \frac{e^{-\beta\epsilon_0}}{(1 + e^{-\beta\epsilon_0})^2}. \qquad (20.86)$$

Note that S_N and U_N are functions of $\tilde{U} \equiv U_N/N\epsilon_0$. In fact,

$$\tilde{S} \equiv \frac{S_N}{Nk_B} = \ln(1 + e^{-\beta\epsilon_0}) + \beta\epsilon_0 \frac{e^{-\beta\epsilon_0}}{1 + e^{-\beta\epsilon_0}}, \qquad (20.87)$$

$$\tilde{U} = 1 + \frac{e^{-\beta\epsilon_0}}{1 + e^{-\beta\epsilon_0}}, \qquad (20.88)$$

and we find, using

$$\tilde{U} + \tilde{U}e^{-\beta\epsilon_0} = 1 + 2e^{-\beta\epsilon_0},$$

that

$$\beta\epsilon_0 = \ln\frac{2-\tilde{U}}{\tilde{U}-1}, \qquad (20.89)$$

$$\tilde{S} = -(\tilde{U}-1)\ln(\tilde{U}-1) - (2-\tilde{U})\ln(2-\tilde{U}), \qquad (20.90)$$

$$\frac{1}{T} \equiv \frac{k_B}{\epsilon_0}\left(\frac{\partial\tilde{S}}{\partial\tilde{U}}\right)_V = \frac{k_B}{\epsilon_0}\ln\frac{2-\tilde{U}}{\tilde{U}-1}. \qquad (20.91)$$

Equation 20.91 shows that this system has $T < 0$ when $\tilde{U} > \frac{3}{2}$; that is, it is in a state of negative temperature (see Fig. 20.4). Thus for energies in the range $(3N\epsilon_0/2) < U \le 2N\epsilon_0$ in this system there is a population inversion relative to that expected from a canonical distribution with positive temperature. The point is that as $T \to 0$ only the lowest level of a system is occupied whether the energy-level spectrum is bounded from above or not. But as $T \to \infty$, if the energy-level spectrum is bounded, all levels become equally populated, whereas if the spectrum is unbounded, this does not happen. Using $1/T$ as the variable rather than T, we see that negative temperatures join on smoothly after $1/T$ passes through zero (see Fig. 20.5). In this sense negative temperatures correspond to "hotter" situations than do positive temperatures.

Another way of ascertaining the importance of an upper bound to the energy-level spectrum for the definition of negative temperature is the following. Given the fact that in a system of energy levels bounded from above the populations of the levels approach equality as $T \to \infty$, the heat capacity associated with that system of levels must approach zero as $T \to \infty$ (see Fig. 20.6). Indeed, the heat capacity of such a system tends to zero sufficiently rapidly that only a finite amount of energy is needed to raise its temperature to infinity. Since the coupling of the subsystem under discussion to the lattice degrees of freedom is very weak, this addition of energy can be effected without much influence on the lattice modes. If still more energy is added to the system, it is forced into the negative-temperature domain. In our first example the energy was forced to become positive by reversing the magnetic field so rapidly that the nuclear spins could not follow, hence they wound up in positive energy states.

The description of more complex systems than the example given is fundamentally the same. The key ideas are that (1) the energy spectrum of some subset of levels of a system is bounded from above, (2) the degrees of freedom corresponding to that subset of levels interact very weakly with all the others in the system, (3) achievement of a canonical distribution of energy within the subset of levels is much more rapid than is equilibration of energy over all the levels of the system, and (4) a means exists for putting the subsystem corresponding to the bounded energy-level spectrum into a very high energy state. The negative-temperature state created in this process is a transient. If there were no contact with the world of positive temperatures it could persist, but because of that contact it is eventually destroyed. It is important to discuss the existence of negative temperatures as an exception to the rule that T is always positive, but it must be emphasized again that such situations never occur with complete systems in equilibrium. It is only for very special transiently isolatable systems that the concept of negative temperature is useful.

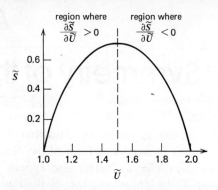

FIGURE 20.4
$\tilde{S}$ as a function of $\tilde{U}$ for the two-level system described in the text.

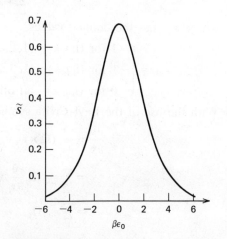

FIGURE 20.5
$\tilde{S}$ as a function of $1/T$ for the two-level system described in the text.

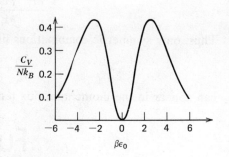

FIGURE 20.6
C_V as a function of $1/T$ for the two-level system described in the text.

Appendix 20A

Symmetry of the Momentum Flux Tensor

In this appendix we show that if the ijth component of the momentum flux tensor has a term of the form $a(\partial v_i/\partial x_j)$, where a is a constant, the term $a(\partial v_j/\partial x_i)$ must also appear. To prove this statement, we note that if the entire volume of fluid under consideration is subjected to uniform rotation, no momentum is transferred within the fluid as a result of viscosity. So, we let $\mathbf{v} = \mathbf{\Omega} \times \mathbf{r}$, where $\mathbf{\Omega}$ is a constant angular velocity, and calculate $\partial v_i/\partial x_j$.

In dealing with vector cross products, it is convenient to introduce the Levi–Civita symbol ε_{ijk}. This quantity requires three indices (it is a third-rank tensor!), each of which ranges from 1 to 3. If $\mathbf{A} = \mathbf{B} \times \mathbf{C}$ is a vector relationship, a little experimentation will convince the reader that

$$A_i = \sum_{j,k} \varepsilon_{ijk} B_j C_k, \qquad (20A.1)$$

where ε_{ijk} has the defined values

$$\varepsilon_{ijk} = \begin{cases} 1 & \text{for } (i, j, k) = (1, 2, 3) \text{ or } (2, 3, 1) \text{ or } (3, 1, 2), \\ -1 & \text{for } (i, j, k) = (1, 3, 2) \text{ or } (2, 1, 3) \text{ or } (3, 2, 1), \\ 0 & \text{for } (i, j, k) = \text{all other combinations.} \end{cases} \qquad (20A.2)$$

With the use of the Levi–Civita symbol, we see that

$$\begin{aligned} \frac{\partial v_i}{\partial x_j} &= \frac{\partial}{\partial x_j} (\mathbf{\Omega} \times \mathbf{r})_i = \frac{\partial}{\partial x_j} \sum_{k,l} \varepsilon_{ikl} \Omega_k x_l \\ &= \sum_{k,l} \varepsilon_{ikl} \Omega_k \frac{\partial}{\partial x_j} x_l = \sum_{k,l} \varepsilon_{ikl} \Omega_k \delta_{jl} \\ &= \sum_k \varepsilon_{ikj} \Omega_k. \end{aligned} \qquad (20A.3)$$

Since the viscous contribution to the stress tensor must be zero, a term like $-\sum_k \varepsilon_{ikj} \Omega_k$ must also be present. Notice now that

$$\frac{\partial v_j}{\partial x_i} = \sum_k \varepsilon_{jki} \Omega_k, \qquad (20A.4)$$

through a sequence of steps analogous to Eq. 20A.4. But by the definition of ε_{ijk},

$$\varepsilon_{ikj} = -\varepsilon_{jki}, \qquad (20A.5)$$

so that

$$\frac{\partial v_i}{\partial x_j} + \frac{\partial v_j}{\partial x_i} = 0. \qquad (20A.6)$$

Thus, only symmetric combinations like

$$\frac{\partial v_i}{\partial x_j} + \frac{\partial v_j}{\partial x_i}$$

can appear in the momentum flux tensor.

Further Reading

Kestin, J., *A Course in Thermodynamics*, Vol. II (Blaisdell, Waltham, Mass., 1968), chap. 24.

Thermodynamics of Irreversible Processes

Prigogine, I., *Introduction to the Thermodynamics of Irreversible Processes* (Charles C. Thomas, Springfield, Ill., 1955).

Fitts, D. D., *Nonequilibrium Thermodynamics* (McGraw-Hill, New York, 1962).

de Groot, S. R., and P. Mazur, *Nonequilibrium Thermodynamics* (North Holland, Amsterdam, 1962).

Nicolis, G., and I. Prigogine, *Self-Organization in Nonequilibrium Systems* (Wiley, New York, 1977).

Abragam, A., *Principles of Nuclear Magnetism* (Oxford University Press, London, 1961), chap. 5.

Loudon, R., *The Quantum Theory of Light* (Oxford University Press, London, 1973), pp. 33–37.

Ramsey, N. F., *Phys. Rev.* **103,** 20 (1956).

Negative Temperature

Problems

1. Verify that Eq. 20.17 is a solution of the diffusion equation.

2. Calculate the integral of $\rho_k(\mathbf{r}, t)$ (Eq. 20.17) over all space. Can you predict the result before doing the integral?

3. Derive Eq. 20.18 from Eq. 20.17.

4. Prove the following vector relations:

   (a) $\nabla \cdot (\nabla S) = \nabla^2 S$,   ($S$ is a scalar);
   (b) $\nabla \cdot (S\mathbf{v}) = S\nabla \cdot \mathbf{v} + \mathbf{v} \cdot \nabla S$.

5. Prove the following tensor relations.

   (a) $\nabla \cdot (\mathbf{v} \cdot \boldsymbol{T}) = \mathbf{v} \cdot (\nabla \cdot \boldsymbol{T}) + \boldsymbol{T} : \nabla \mathbf{v}$.

   The double dot product of two tensors is a scalar whose value is

   $$\boldsymbol{T} : \boldsymbol{X} = \sum_{ij} T_{ij} X_{ji}.$$

   (b) $\boldsymbol{1} : \nabla \mathbf{v} = \nabla \cdot \mathbf{v}$.

6. Use the properties of the Levi–Civita tensor ε_{ijk} to prove the following relations:

   (a) $\nabla \times (\nabla \times \mathbf{v}) = \nabla(\nabla \cdot \mathbf{v}) - \nabla^2 \mathbf{v}$;
   (b) $\nabla \cdot (\mathbf{v} \times \mathbf{u}) = \mathbf{u} \cdot \nabla \times \mathbf{v} - \mathbf{v} \cdot \nabla \times \mathbf{u}$;
   (c) $\mathbf{v} \times (\nabla \times \mathbf{v}) = \frac{1}{2}\nabla v^2 - \mathbf{v} \cdot \nabla \mathbf{v}$.

7. Use Euler's equation, Eq. 20.32, to derive an equation of continuity expressing the conservation of angular momentum for a nonviscous (ideal) fluid.

8. Let s be the entropy per unit mass of an ideal fluid (a fluid with zero viscosity). For this fluid the equation of motion is

$$\frac{\partial \mathbf{v}}{\partial t} + \mathbf{v} \cdot \nabla \mathbf{v} = -\frac{1}{\rho} \nabla p.$$

Show that for adiabatic motion, in which $ds/dt = 0$,

$$\frac{\partial(\rho s)}{\partial t} + \nabla \cdot (\rho s \mathbf{v}) = 0.$$

9. If the motion of the fluid is isentropic, $s = $ constant. Use the combined first and second laws of thermodynamics and the equation of motion to show that, for an ideal fluid,

$$\frac{\partial \mathbf{v}}{\partial t} + \mathbf{v} \cdot \nabla \mathbf{v} = -\nabla h,$$

where h is the enthalpy per unit mass.

10. A fluid can be in mechanical equilibrium without being in thermal equilibrium. A case of this type arises if a small vertical temperature gradient is imposed on the fluid. Show that such a situation can be maintained (be stable) only if

$$\frac{dT}{dz} > -\frac{gT}{c_p v}\left(\frac{\partial v}{\partial T}\right)_p,$$

where g is the acceleration due to gravity. Interpret this result in simple physical terms. Hint: As a result of the temperature gradient, an element of fluid at height z having volume per unit mass $v(p, s)$ tends to rise. Consider this rise to be adiabatic and use the condition of mechanical stability, together with a Taylor series expansion in the entropy difference $(s' - s)$ corresponding to the height difference Δz, to derive the quoted formula.

There are many irreversible processes that can be described by a linear law relating cause and effect. Three examples are Fourier's law of heat conduction,

$$\mathbf{J}^q = -\lambda \nabla T;$$

Fick's law of diffusion,

$$\mathbf{J}_k{}^m = -D_k \nabla \rho_k;$$

and Ohm's law of electric conduction,

$$\mathbf{I} = -\sigma \nabla V = \sigma \mathbf{E}.$$

In each case a flux is generated by the existence of a force. A flux of heat is the response to imposition of a temperature gradient, a flux of matter to a concentration gradient, and a flux of charge to an electric potential gradient in the three examples cited. The forces are the gradients of temperature, concentration, and electric potential, and the coefficients are the thermal conductivity, the diffusion coefficient, and the electrical conductivity, respectively. If two or more linear irreversible processes occur in the same system at the same time, they couple to give rise to new effects. For example, the superposition of a flow of heat and a flow of electric current gives rise to *thermoelectricity*, that is, an electric potential generated by a temperature gradient.

A thermodynamic analysis of linear irreversible processes of the type described is possible if it is assumed that:

(1) At each point in the system the relationships between thermodynamic variables are the same as at equilibrium, but using the local values of temperature, concentration, and so on (hypothesis of local equilibrium).

(2) The entropy depends explicitly only on T, u, and μ, and implicitly on time only insofar as T, u, and μ depend on the time.

(3) The total variation in the entropy of the system as a result of the irreversible process is the sum of the entropy variations at each point in the system.

(4) The flow $\mathbf{J}^i$ caused by the action of forces $\mathbf{X}_k$ is proportional to those

forces. That is,

$$\mathbf{J}^i = \sum_{k=1}^{n} L_{ik}\mathbf{X}_k.$$

The diagonal coefficients L_{kk} are the λ, D, σ, and so on, of the linear laws described above, whereas the off-diagonal coefficients L_{ik} are associated with the coupling of processes, such as the cited thermoelectricity. Onsager showed that if the flows and forces are suitably chosen, then

$$L_{ik} = L_{ki},$$

which is called the *Onsager reciprocal relation*. We shall accept this relation as valid without discussion of how the forces and fluxes must be chosen.

Problems 11–19 are based on use of the ideas sketched above.

11. Consider an isolated system whose state is completely described by the set of n macroscopic variables $a_1, \ldots, a_n$. These variables include the temperature, pressure, and so forth. At equilibrium the values of $a_1, \ldots, a_n$ are denoted $a_1^0, \ldots, a_n^0$. Now suppose that a small deviation from equilibrium is created by changing the values of the a_i's away from a_i^0. Let the force corresponding to the displacement $\alpha_i \equiv a_i - a_i^0$ be defined by

$$X_i \equiv \frac{\partial(\Delta S)}{\partial \alpha_i}$$

and the flux generated by that force be defined by

$$J^i \equiv \frac{d\alpha_i}{dt}.$$

Show that

$$\frac{dS}{dt} = \sum_{i=1}^{n} J^i X_i.$$

12. Show that if $X_1, \ldots, X_m$, where $m < n$, are kept constant, the minimum rate of entropy production corresponding to variable forces $X_{m+1}, \ldots, X_n$ occurs when the system is in a steady state with $J_{m+1} = 0, \ldots, J_n = 0$.

13. Consider an isotropic solid across which a temperature gradient is maintained, say, in the x direction. Show that, if thermal expansion can be neglected, at each point in the interior of the solid the rate of change of entropy per unit volume is

$$\rho\left(\frac{\partial S}{\partial t}\right) + \frac{\partial}{\partial x}\left(\frac{\mathbf{J}^q}{T}\right) = -\frac{\mathbf{J}^q}{T^2}\left(\frac{\partial T}{\partial x}\right),$$

where

$$\mathbf{J}^q = -\lambda\left(\frac{\partial T}{\partial x}\right).$$

14. Use the second law of thermodynamics and the equation for the rate of change of entropy per unit volume derived in Problem 13 to prove that the coefficient of thermal conductivity must be positive ($\lambda > 0$).

Consider the following experimental facts:

(1) Across a junction between two different conductors, each one maintained at a different temperature, there is a difference of electrical potential (*Seebeck effect*).

(2) When an electric current flows through the junction between two conductors, each maintained at the same temperature, at the junction a quantity of heat is absorbed or released that is proportional to the current (*Peltier effect*).

(3) When an electric current flows in a conductor in which there is also a

temperature gradient, in addition to the usual Joule heating (proportional to the resistance and square of the current) there is also heat generated proportional to the temperature gradient and the current (*Thomson effect*).

Neglecting volume changes associated with the variation of temperature, for all of these phenomena the fundamental equation of local equilibrium is

$$T\,dS = dU - (\mu - \mathscr{F}\psi)\,dn,$$

where $\mathscr{F}$ is the charge on 1 mol of electrons, μ is the ordinary density- and temperature-dependent part of the chemical potential, and ψ is the electrostatic potential (see Chapter 26).

15. Use the equations describing conservation of charge and conservation of energy to calculate the rate of change of entropy per unit volume at a point inside the conductor when both a temperature gradient and a potential gradient are present.

16. Let the temperature and potential gradients be along the x axis. The superposition of flows arising from the imposed gradients is represented by

$$\mathbf{J}^q = -\frac{L_{11}}{T}\left(\frac{\partial T}{\partial x}\right) + L_{12}\left[\mathbf{E} + \frac{\partial}{\partial x}\left(\frac{\mu}{\mathscr{F}}\right)\right],$$

$$\mathbf{I} = -\frac{L_{21}}{T}\left(\frac{\partial T}{\partial x}\right) + L_{22}\left[\mathbf{E} + \frac{\partial}{\partial x}\left(\frac{\mu}{\mathscr{F}}\right)\right],$$

with $L_{12} = L_{21}$ by the Onsager reciprocal relation. Solve these equations for $\mathbf{J}^q$ and $\mathbf{E}$, writing the result in the form

$$\mathbf{J}^q = -\lambda\left(\frac{\partial T}{\partial x}\right) - \Pi\mathbf{I},$$

$$\mathbf{E} = \frac{1}{\sigma}\mathbf{I} - \alpha\left(\frac{\partial T}{\partial x}\right) - \frac{\partial}{\partial x}\left(\frac{\mu}{\mathscr{F}}\right).$$

Find expressions for λ, Π, α, and σ in terms of L_{11}, L_{12}, L_{21}, and L_{22}. Note that Π is the Peltier coefficient and α the coefficient of thermoelectric potential.

17. From the form of the equations for $\mathbf{J}^q$ and $\mathbf{E}$ in terms of λ, Π, σ, and α, show that λ is the thermal conductivity and σ the electrical conductivity. Also show that $\Pi = T\alpha$, which relates the Peltier coefficient Π to the coefficient of thermoelectric potential, α.

18. Consider a thermocouple circuit as shown in the accompanying diagram.

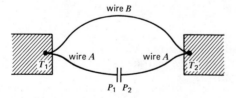

Wires A and B are of different composition, and $T_2 > T_1$. Show that the potential generated across the condenser plates P_1, P_2 is

$$\Delta V = (\alpha_A - \alpha_B)(T_2 - T_1)$$

when α is considered independent of temperature.

19. Consider two wires, A and B, at the same temperature. Let a current I flow through a junction between the wires. Show that the heat released at the junction is $(\Pi_2 - \Pi_1)I$, so that the Peltier heat is proportional to the current passed through the junction.

20. How does the quantum (band) theory of metals explain the Peltier effect, and the relation between the Peltier and Seebeck effects? Hint: The Fermi levels of the electrons of the metals in contact are not the same. See Chapter 11.

21. Suppose that a system consists of N independent identifiable molecules and that the energy levels of each molecule are equally spaced:

$$\epsilon_n = n\epsilon_0, \qquad n = 1, 2, \ldots \to \infty.$$

Show that the entropy of this system is

$$\frac{S}{Nk_B} = \frac{U}{N\epsilon_0} \ln\left(\frac{U}{N\epsilon_0}\right) - \left(\frac{U}{N\epsilon_0} - 1\right) \ln\left(\frac{U}{N\epsilon_0} - 1\right).$$

Demonstrate that S is a monotonic increasing function of U and that the system cannot have a negative temperature.

22. We have introduced the concept of negative temperature from the point of view of the statistical molecular description of matter. This concept can also be introduced by adding a macroscopic condition to the usual laws of thermodynamics. That condition is as follows: A system can attain a state of negative temperature if and only if its energy approaches a finite limit as $T \to \infty$. Discuss the relationship between this condition and the nature of the energy spectrum for a system that can attain a negative temperature.

23. Experimental studies have shown that *in the domain of negative temperatures* it is possible to carry out a cyclic process in which heat is converted to work without altering the thermodynamic states of other systems, and that it is not possible entirely to convert work into heat without altering the thermodynamic states of other systems. Thus, in the negative-temperature domain the relative completeness of conversion of work into heat and heat into work is the reverse of what it is in the positive-temperature domain. Show that this implies that

$$T \, dS \leq đq$$

in the negative temperature domain. Show also that this implies that

$$T \, dS \leq dU - đw$$

and that the entropy is a maximum in the equilibrium state.

24. Consider two bodies with the *negative* temperatures T_1 and T_2, respectively. When these bodies are brought into contact, there is an irreversible flow of heat from 1 to 2. Show that $T_1 > T_2$.

25. Construct a Carnot cycle between the two *negative* temperatures T_1 and T_2, with $T_1 > T_2$. Show that in the domain of negative temperature work must be done to extract heat from a hot body and transfer it to a colder body. Hint: The efficiency of the Carnot cycle is the same function of T_1 and T_2 in the positive- and negative-temperature domains.

26. Given the form of the combined first and second laws for the *negative-*temperature domain,

$$T \, dS \leq dU + p \, dV,$$

show that

$$dA \geq -S \, dT - p \, dV,$$
$$dG \geq -S \, dT + V \, dp.$$

Is the Gibbs free energy a minimum or a maximum in the equilibrium state with *negative* temperature?

27. Show that the restrictions placed on material constants by the criterion that the equilibrium is stable are the same in both positive-temperature and negative-temperature domains. In particular, show that $C_p > 0$, $(\partial V/\partial p)_T < 0$.

21

The Properties of Pure Gases and Gas Mixtures

In the preceding chapters we have developed the tools required for a systematic description of the properties of matter at equilibrium. In this chapter we make use of these tools to analyze the behavior of gases and gas mixtures. Our goal is to answer the following questions:

1. How may observations of the properties of a gas be translated into values of the thermodynamic functions that represent the state of the gas?

2. What is the detailed relationship between the structure of the molecules that compose the gas and the thermodynamic properties of the gas?

3. How do molecular interactions influence the thermodynamic properties of the gas?

4. Which aspects of the molecular structure and the molecular interactions influence chemical equilibrium?

It is amusing to speculate on what the course of development of thermodynamics and of chemistry might have been if the pressure at the surface of the earth were 1000 atm. At a pressure of 1 atm most gases can be described by equations of state that deviate only slightly from the perfect gas equation of state. Indeed, it is this accident of nature that led to early recognition of the existence of the perfect gas equation of state. At a pressure of 1000 atm the volumes of gases are comparable with the volumes of condensed phases, and it is easier to notice the differences in the properties of substances than to recognize their similarities. If 1000 atm were the "normal" earth pressure, it might have taken much longer to discover the unifying simplicity of the perfect gas concept than it actually did. As it is, the perfect gas concept has served four important functions in the development of chemistry and of thermodynamics:

1. It provided the first method of establishing relative molar weights.

2. It provided a realizable temperature scale that could be identified with the thermodynamic temperature scale.

3. It provides a moderately accurate representation of the properties of real gases near 1 atm.

4. It is a statement of a limiting law of behavior for all gases if only the pressure is low enough.

21.1 Thermodynamic Description of a Pure Gas

Functions 1 and 2 are equally well or better executed today without using the properties of gases.[1] On the other hand, function 3 is still useful, and function 4 is still important as an organizing principle when p, V, T measurements on a variety of gases are compared and interpreted.

The properties of real gases may be summarized as follows:

1. At sufficiently low pressure, all gases satisfy the perfect gas equation of state. For simple substances, such as N_2, Ar, or CO, at room temperature, the perfect gas equation of state is accurate to about 1% at pressures less than 15 atm. For more complicated substances, such as H_2O, CH_3OH, or HF, deviations from perfect gas behavior can be observed at lower pressure.

2. At sufficiently low temperature and sufficiently high pressure, all gases condense to a liquid or solid phase. There is a range of temperature and pressure throughout which liquid and gas coexist in equilibrium; there is also a range of temperature and pressure throughout which solid and gas coexist in equilibrium. There is one point at which gas, liquid, and solid coexist at equilibrium (*the triple point*).

3. For each substance there is a temperature, the *critical temperature*, above which it is impossible to condense a gas to a liquid no matter how high the pressure. In a gas–liquid system at equilibrium, the difference in density between the gas and liquid vanishes at the critical temperature. At the point at which the density difference vanishes, the pressure and density are denoted the *critical pressure* and *critical density*, respectively.

4. The properties of many different gaseous chemical substances can be approximately represented by one equation of state if the variables are chosen to be the *reduced temperature*, $\tilde{T} \equiv \dfrac{T}{T_c}$, the *reduced pressure*, $\tilde{p} \equiv \dfrac{p}{p_c}$, and the *reduced volume* or *reduced density*, $\tilde{V} \equiv \dfrac{V}{V_c}$, $\tilde{\rho} \equiv \dfrac{\rho}{\rho_c}$, where T_c, p_c, V_c, and ρ_c are the critical temperature, pressure, volume, and density, respectively. This equation of state is an example of the consequences of the *principle of corresponding states*.

The features of the equation of state of real gases described above are illustrated in Figs. 21.1–21.5.

In this chapter we restrict our attention to the description of the properties of the gas; the nature of the equilibria between solid and gas, liquid and gas, and solid and liquid will be discussed in Chapter 24.

A thermodynamic description of a gas is taken to mean a set of relationships between the derived thermodynamic functions such as U and S, and the measured equation of state and heat capacities. We have already calculated, in Chapters 14 and 17, the energy, enthalpy, and entropy changes accompanying the change of state of a fluid from p_1, V_1,

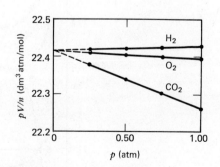

FIGURE 21.1

The pV product for three gases, showing the universal approach to the ideal gas value as the pressure tends to zero. From R. H. Cole and J. S. Coles, *Physical Principles of Chemistry* (Freeman, San Francisco, 1964).

---

[1] Relative molar weights are more conveniently and accurately determined by mass spectrometry, and other temperature scales exist that have a well-defined relationship with the thermodynamic scale—for example, those based on the magnetic properties of a paramagnetic salt.

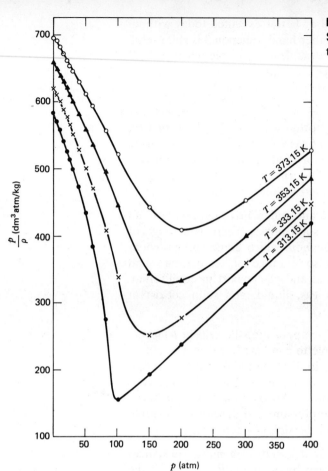

FIGURE 21.2
Some isotherms of CO_2 for temperatures above the critical temperature.

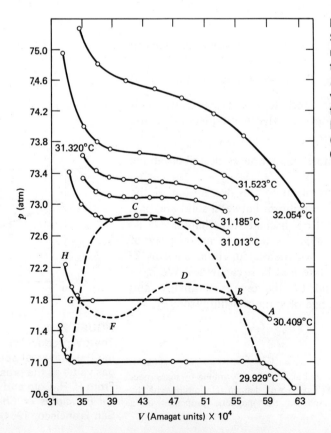

FIGURE 21.3
Some isotherms of CO_2 in the immediate vicinity of the critical temperature. From E. A. Moelwyn-Hughes, *Physical Chemistry* (Cambridge University Press, Cambridge, 1940); data from A. Michels, B. Blaisse, and C. Michels, *Proc. Roy. Soc.* **A160**, 367 (1937). One Amagat unit of volume for CO_2 is 2.2398×10^4 cm³/mol. .

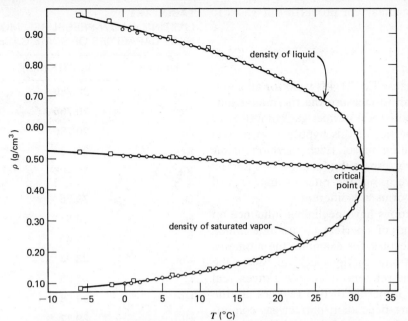

FIGURE 21.4
The coexistence line for CO_2, showing the approach of the gas and liquid densities to a common value at the critical point. From E. A. Moelwyn-Hughes, *Physical Chemistry* (Cambridge University Press, Cambridge, 1940); data from A. Michels, B. Blaisse, and C. Michels, *Proc. Roy. Soc.* **A160**, 367 (1937).

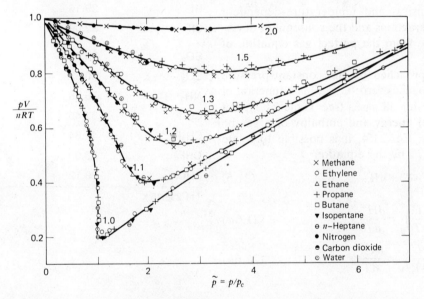

FIGURE 21.5
The compressibility factor pV/nRT for several gases plotted as a function of the reduced pressure p/p_c. The curves are for different values of T/T_c. From G. J. Su, *Ind. Eng. Chem.* **38**, 803 (1946).

T_1 to p_2, T_2, V_2. Here we just quote the previously obtained results:

$$U(T_2, V_2) = U(T_1, V_1) + \int_{T_1, V_1}^{T_2, V_1} C_V(T, V_1)\, dT$$
$$+ \int_{T_2, V_1}^{T_2, V_2} \left[T_2 \left(\frac{\partial p}{\partial T} \right)_V - p \right] dV, \tag{21.1}$$

$$H(T_2, p_2) = H(T_1, p_1) + \int_{T_1, p_1}^{T_2, p_1} C_p(T, p_1)\, dT$$
$$+ \int_{T_2, p_1}^{T_2, p_2} \left[V - T_2 \left(\frac{\partial V}{\partial T} \right)_p \right] dp, \tag{21.2}$$

$$S(T_2, p_2) = S(T_1, p_1) + \int_{T_1, p_1}^{T_2, p_1} \frac{C_p(T, p_1)}{T}\, dT$$
$$- \int_{T_2, p_1}^{T_2, p_2} \left(\frac{\partial V}{\partial T} \right)_p dp. \tag{21.3}$$

We now use statement 1 in the summary of properties of gases to write

$$\lim_{p \to 0} \frac{pV}{nRT} = 1, \tag{21.4}$$

as already indicated in Chapter 12. Because Eq. 21.4 is valid for all gases, a convenient reference state from which to measure the thermodynamic properties of a gas is defined by a hypothetical perfect gas composed of the same molecules. Clearly, the properties of this hypothetical perfect gas can be established only in a limiting sense. Each thermodynamic function[2] of a real gas approaches the corresponding function of a perfect gas when the gas density approaches zero, but the rates of approach of the various functions to their limiting forms are different.

In a perfect gas, molecular interactions have negligible influence on the equation of state. Thus, the properties of a perfect gas are determined by the structure of the component molecules, for example, the moments of inertia, vibrational frequencies, electronic energy spectrum. When the hypothetical perfect gas is chosen as the reference state for a given real gas, it follows that the thermodynamic properties of that real gas are the sum of two contributions: a temperature-dependent perfect gas component arising from the distribution of the molecules over the possible internal molecular states, and a density- and temperature-dependent component arising from molecular interactions and the consequent deviations of the real gas equation of state from the perfect gas equation of state.

Let $C_p^*(T)$ denote the limit approached by the constant-pressure heat capacity as the gas pressure approaches zero; $C_p^*(T)$ is a function of temperature only and is not the same for all gases (see Table 21.1). Let the low-pressure limits of the internal energy and enthalpy of a gas be denoted by $U^*(T)$ and $H^*(T)$. Using Eq. 21.4, it is possible to relate $C_V^*(T)$, defined analogously, to $C_p^*(T)$ by the equation

$$C_p^*(T) = C_V^*(T) + nR, \tag{21.5}$$

since

$$C_p^*(T) \equiv \lim_{p \to 0} \left(\frac{\partial H}{\partial T} \right)_p = \frac{dH^*}{dT}, \tag{21.6a}$$

$$C_V^*(T) \equiv \lim_{p \to 0} \left(\frac{\partial U}{\partial T} \right)_V = \frac{dU^*}{dT}, \tag{21.6b}$$

and

$$H^* = U^* + \lim_{p \to 0} (pV)$$

$$= U^* + nRT. \tag{21.7}$$

It is $C_p^*(T)$ and $C_V^*(T)$ that determine the "perfect gas" contributions to the thermodynamic functions, namely, those that arise from the distribution of molecules over internal molecular states.

We may now transform the general formulas 21.1, 21.2, and 21.3 into a representation in which the properties of the real gas at T_2, p_2, V_2 are referred to the properties of a hypothetical perfect gas at T_2, p_2, V_2. We set $T_1 = T_2$ (eliminating the integrals over T), $p_1 = 0$, and $V_1 = \infty$, so that Eqs. 21.1 and 21.2 take the forms

$$U(T_2, V_2) = U^*(T_2) + \int_{T_2, \infty}^{T_2, V_2} \left[T_2 \left(\frac{\partial p}{\partial T} \right)_V - p \right] dV, \tag{21.8}$$

---

[2] There is an exception: The Joule–Thomson coefficient $\mu_{JT} \equiv (\partial T / \partial p)_H$ does not approach zero as $n/V \to 0$.

TABLE 21.1

LIMITING LOW-PRESSURE MOLAR HEAT CAPACITIES OF SOME GASES

Substance	c_p^* (J/K mol)[a]
He	20.786
Ne	20.786
Ar	20.786
Kr	20.786
Xe	20.786
O_2	29.36
H_2	28.87
F_2	31.46
Cl_2	33.93
I_2	36.86
HF	29.08
HCl	29.12
HI	29.16
ClF	32.09
SO_2	39.79
SO_3	50.63
NH_3	35.66
CH_3F	37.45
CH_2F_2	42.84
CHF_3	53.05
$CHClBr_2$	69.12

[a] at 298.15 K

$$H(T_2, p_2) = H^*(T_2) + \int_{T_2,0}^{T_2,p_2} \left[V - T_2 \left(\frac{\partial V}{\partial T} \right)_p \right] dp. \qquad (21.9)$$

The reduction of Eq. 21.3 to a form referring the entropy to that of a hypothetical perfect gas is a bit tricky. In the general formulas for $U(T, V)$ and $H(T, p)$, namely, Eqs. 21.1 and 21.2, the second integrals on the right-hand sides vanish if the gas obeys the perfect gas equation of state. On the other hand, the second integral on the right-hand side of Eq. 21.3 has the form

$$\int_{T_2,p_1}^{T_2,p_2} \frac{nR}{p} \, dp$$

for the case of a perfect gas. This integral is undefined when $p_1 \to 0$ because the integrand then becomes indefinitely large. To circumvent this difficulty we must eliminate the divergence of the integral term. It is observed that in a real gas

$$\lim_{p \to 0} \left[\left(\frac{\partial V}{\partial T} \right)_p - \frac{nR}{p} \right] = f(T), \qquad (21.10)$$

and that $f(T)$ is bounded. That is, subtraction of the perfect gas value of $(\partial V/\partial T)_p$ from the real gas value eliminates the divergence as $p \to 0$. We exploit this observation by adding and subtracting nR/p in the second integrand of Eq. 21.3, casting the equation into the form

$$S(T_2, p_2) = S(T_1, p_1) + \int_{T_1,p_1}^{T_2,p_1} \frac{C_p(T, p_1)}{T} \, dT$$

$$- \int_{T_2,p_1}^{T_2,p_2} \left[\left(\frac{\partial V}{\partial T} \right)_p - \frac{nR}{p} \right] dp - \int_{T_2,p_1}^{T_2,p_2} \frac{nR}{p} \, dp. \quad (21.11)$$

Let $\mathfrak{p}$ be the unit pressure, that is, the value of the unit in which the pressure is expressed (e.g., 1 atm). Then, letting $p_1 \to 0$, Eq. 21.11 can be rearranged to yield

$$S(T_2, p_2) = S^*(T_2, \mathfrak{p}) + \int_{T_2,0}^{T_2,p_2} \left[\frac{nR}{p} - \left(\frac{\partial V}{\partial T} \right)_p \right] dp - nR \ln \frac{p_2}{\mathfrak{p}}, \qquad (21.12)$$

where $S^*(T_2, \mathfrak{p})$ is the entropy of a perfect gas at $(T_2, \mathfrak{p})$, defined by

$$S^*(T_2, \mathfrak{p}) = \int_{T_{\text{ref}}}^{T_2} \frac{C_p^*(T)}{T} \, dT + S_0^*(T_{\text{ref}}, \mathfrak{p}), \qquad (21.13)$$

and the constant $S_0^*(T_{\text{ref}}, \mathfrak{p})$ depends on the lower limit of integration of Eq. 21.13 (cf. Chapter 18). Note that $p/\mathfrak{p}$ is a dimensionless number with magnitude equal to p in whatever units are chosen: It is important to include $\mathfrak{p}$ explicitly because the logarithm of a dimensional quantity is undefined. Note also that the entropy of a perfect gas depends on the pressure. The integral $\int_{p_1}^{p_2} (nR/p) \, dp$, which diverges as $p_1 \to 0$, is subtracted from both S and S^* to obtain Eq. 21.12.

Equations 21.8, 21.9, and 21.12 for a real gas are more complicated in form than the corresponding equations for a perfect gas. The formulas for the real and perfect gases can be made to look alike as follows. Define the function

$$f \equiv p \exp \left[\frac{1}{RT} \int_0^p \left(v - \frac{RT}{p'} \right) dp' \right], \qquad (21.14)$$

called the *fugacity* of the gas. The fugacity has the same units as pressure, and it plays the same role in a real gas as does pressure in a perfect gas. By definition f contains all of the effects of gas imperfection and has the limiting behavior $f \to p$ as $p \to 0$. With the definition 21.14, the chemical potential of a real gas has a dependence on f like that the chemical potential of a perfect gas has on p. From Eqs. 21.9 and 21.12 we have, for a real gas,

$$\mu(T, p) \equiv \frac{G}{n} = \frac{H - TS}{n}$$

$$= h^*(T_{\text{ref}}) - Ts_0^*(T_{\text{ref}}, \mathfrak{p}) + RT \ln \frac{p}{\mathfrak{p}}$$

$$+ \int_{T_{\text{ref}}}^{T} \left(1 - \frac{T}{T'}\right) C_p^*(T') \, dT' + \int_{0,T}^{p,T} \left(v(p') - \frac{RT}{p'}\right) dp'$$

$$= \mu^*(T, \mathfrak{p}) + RT \ln \frac{p}{\mathfrak{p}} + \int_{0,T}^{p,T} \left(v(p') - \frac{RT}{p'}\right) dp'$$

$$= \mu^*(T, \mathfrak{p}) + RT \ln \frac{f}{\mathfrak{p}}, \tag{21.15}$$

where $\mu^*(T, \mathfrak{p})$ is the chemical potential of a perfect gas at $T, \mathfrak{p}$. For a perfect gas Eq. 21.15 reduces to

$$\mu = \mu^*(T, \mathfrak{p}) + RT \ln \frac{p}{\mathfrak{p}} \qquad \text{(perfect gas)}. \tag{21.16}$$

The entropy, enthalpy, and other thermodynamic functions of the real gas can be obtained directly from Eq. 21.15 by standard manipulations.

Final reduction of the formulas 21.8, 21.9, and 21.12 requires information about the equation of state of the gas. There have been many suggested generalizations of the perfect gas equation of state, all designed to account somehow for the influence of molecular interactions on the behavior of a gas. In general, the interaction between a pair of molecules is conveniently represented as the superposition of a very short-range repulsion and a somewhat longer-range attraction (see Fig. 21.6). Suppose that only interactions between pairs of molecules are important in the density region under study. If the temperature is high enough that the average kinetic energy of a molecule greatly exceeds the molecular attraction, the repulsion between molecules is most important. Then, since the effective volume in which a molecule may move is reduced by the volume excluded by the repulsion of other molecules, we expect the real gas pressure to exceed the perfect gas pressure at the same density and temperature. Conversely, if the temperature is low enough that the average kinetic energy of a molecule is less than the attractive potential, the pressure of the real gas is less than that of a perfect gas at the same density and temperature. A very convenient form of the equation of state, valid for all gases, is the power series

$$\frac{pV}{nRT} = 1 + \frac{nB}{V} + \frac{n^2 C}{V^2} + \cdots, \tag{21.17}$$

known as the *virial equation of state*.[3] The coefficients $B, C, \ldots$ depend on the temperature but not the density, and are known as *virial coefficients*. The behavior of B and C as a function of temperature is shown in Fig. 21.7: Both B and C are negative at low temperature and positive at high temperature. As will be seen in Section 21.7, the virial coefficients are determined by the intermolecular potential. B represents the effects of

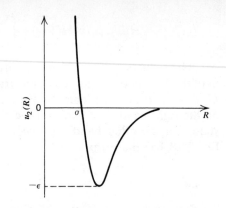

FIGURE 21.6
A representation of the intermolecular potential between a pair of molecules as a function of the distance between them.

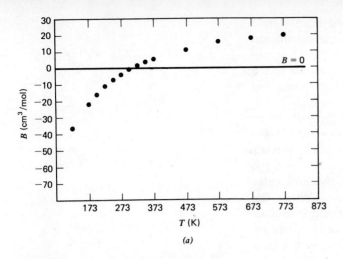

(a)

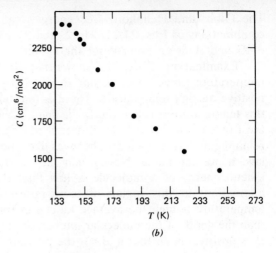

(b)

interactions between pairs of molecules, C interactions among triplets of molecules, and so forth.

For many purposes it is adequate to keep only the first correction term, B, in the virial equation of state. Then we have

$$T\left(\frac{\partial p}{\partial T}\right)_V - p = \frac{nRT}{V} + \frac{n^2 BRT}{V^2} + \frac{n^2 RT^2}{V^2}\frac{dB}{dT} - \frac{nRT}{V} - \frac{n^2 BRT}{V^2}$$

$$= \frac{n^2 RT^2}{V^2}\frac{dB}{dT}, \tag{21.18}$$

$$V - T\left(\frac{\partial V}{\partial T}\right)_p = nB - nT\frac{dB}{dT}, \tag{21.19}$$

$$\left[\left(\frac{\partial V}{\partial T}\right)_p - \frac{nR}{p}\right] = n\frac{dB}{dT}. \tag{21.20}$$

The use of Eqs. 21.18, 21.19, and 21.20 in, respectively, Eqs. 21.8, 21.9, and 21.12 gives for the internal energy

$$U(T_2, V_2) = U^*(T_2) - n\frac{RT_2^2}{V_2}\frac{dB}{dT}, \tag{21.21}$$

for the enthalpy

$$H(T_2, p_2) = H^*(T_2) + n\left(B - T_2\frac{dB}{dT}\right)p_2, \tag{21.22}$$

and for the entropy

$$S(T_2, p_2) = S^*(T_2, \mathfrak{p}) - nR\ln\frac{p_2}{\mathfrak{p}} - n\frac{dB}{dT}p_2. \tag{21.23}$$

FIGURE 21.7
The second and third virial coefficients of Ar.

---

[3] The virial equation of state is sometimes written

$$pV = nRT + nB'p + n^2 C'p^2 + \cdots. \tag{21.17'}$$

This form is less convenient than Eq. 21.17 for comparison with the statistical molecular theory of gases, but is sometimes used in thermodynamic calculations. It may be shown that

$$B = B'$$

but

$$C \neq C'.$$

In calculating the enthalpy we have used the form of the virial equation of state with pressure as the independent variable, namely Eq. 21.17'. Thus the approximations to the equation of state that lead, respectively, to Eqs. 21.21 and 21.22 are almost but not quite the same. The approximations differ by quantities that are small relative to the terms retained.

The Gibbs and Helmholtz free energies can be obtained from simple combinations of Eqs. 21.21, 21.22, and 21.23. We leave the calculation of G and A as an exercise for the reader.

Examination of Fig. 21.7 shows that dB/dT is always positive in the temperature range covered, but B is negative at low temperature and positive at high temperature. There is one temperature at which $B = 0$; this temperature is called the *Boyle point*. At and in the neighborhood of the Boyle point the real gas, if described by an equation of state (21.14) retaining no terms beyond B, behaves like a perfect gas and obeys Boyle's law, hence the name. Now, when the temperature is low the average kinetic energy of a molecule is less than the depth of the molecular interaction; in this region of temperature B is negative. When the temperature is high the average kinetic energy of a molecule is greater than the depth of the molecular interaction; in this region of temperature B is positive. Note that if $B < 0$ the pressure of the real gas is less than that of the corresponding perfect gas, and if $B > 0$ the pressure is greater than that of the corresponding perfect gas. Finally, Fig. 21.7 seems to indicate that dB/dT approaches zero as T increases. Theoretical calculations of B predict that it passes through a maximum at high temperature and then decreases very slowly. For Ar that maximum occurs near 3000 K, and for He near 200 K.

A consequence of molecular interaction is that the distribution of molecules in space cannot be completely random. If the average intermolecular attraction exceeds the average molecular kinetic energy, there is greater than random chance of finding molecules closer than the average separation; if intermolecular repulsion exceeds, on the average, the average molecular kinetic energy, there is less than random chance of finding molecules closer than the average separation. We see from Eq. 21.23 that the entropy of the real gas is less than that of the corresponding perfect gas when $dB/dT > 0$, independent of whether B is positive or negative. This is to be expected, since the nonrandom distribution of molecules in space resulting from molecular interaction decreases the number of configurations of gas molecules with equal energy, hence the number of configurations that can be occupied with equal probability, and thus also the entropy of the gas. Equation 21.21 shows that the internal energy of the gas is also less than that of the corresponding perfect gas. Clearly, if molecular attraction leads to an enhancement of the frequency of less than average molecular separations, there is a negative contribution to the internal energy, as found from Eq. 21.21. At very high temperature, where $dB/dT < 0$, the internal energy exceeds that of the corresponding perfect gas, just as expected if repulsive interactions are dominant. Note that because the repulsive interaction between molecules has a much shorter range than the attractive interaction, $B > 0$ only when the average molecular kinetic energy exceeds the molecular attraction by a factor of two or three. For the average spacing of molecules typical of a gas for which Eq. 21.17 is valid, the dominant interaction between molecules is attractive, not repulsive, and $dB/dT > 0$, so that the internal energy of the real gas is less than that of the corresponding perfect gas. In the hypothetical case that the molecular interaction is everywhere repulsive, $dB/dT < 0$, and the energy of the real gas is greater than that of the corresponding perfect gas, in agreement with intuitive considerations.

A similar discussion of the properties of the enthalpy may be based on Eq. 21.22.

The virial equation of state is not the only form used to relate p, V, and T for a gas. Many empirical equations of state have been proposed, but only the virial equation of state has a simple and unambiguous

TABLE 21.2
EMPIRICAL GAS EQUATIONS OF STATE

Name	General Form	Features	Form of Second Virial Coefficient[a]
Early			
van der Waals (1877)	$p = \dfrac{nRT}{V-nb} - \dfrac{n^2a}{V^2}$	a, b independent of T, V	$B = b - \dfrac{a}{RT}$
Dieterici (1899)	$p = \dfrac{nRT}{V-nb}\, e^{-na/VRT}$	a, b independent of T, V	$B = b - \dfrac{a}{RT}$
Berthelot (1907)	$p = \dfrac{nRT}{V-nb} - \dfrac{n^2a}{TV^2}$	a, b independent of T, V	$B = b - \dfrac{a}{RT^2}$
Later			
Keyes (1914)	$p = \dfrac{nRT}{V-n\delta} - \dfrac{n^2A}{(V+nl)^2}$	$\delta = \beta e^{-n\alpha/V}$; α, β, l, A independent of T, V	$B = \beta - \dfrac{A}{RT}$
Beattie–Bridgeman (1927)	$p = \dfrac{nRT(1-\epsilon)}{V^2}(V+nB) - \dfrac{n^2A}{V^2}$	$B = B_0\left(1 - \dfrac{nb}{V}\right)$ $A = A_0\left(1 - \dfrac{nA}{V}\right)$ $\epsilon = \dfrac{nc}{VT^3}$ c, B_0, b, A_0, a independent of T, V	$B = B_0 - \dfrac{A_0}{RT} - \dfrac{c}{T^3}$
Redlich–Kwong (1949)	$p = \dfrac{nRT}{V-nb} - \dfrac{n^2a}{T^{1/2}V(V+nb)}$	a, b independent of T, V	$B = b - \dfrac{a}{RT^{3/2}}$

[a] The second virial coefficient is obtained, in each case, by expanding the given equation of state in the form of Eq. 21.17.

relationship with the statistical molecular theory of gases. Some suggestions to account for the properties of real gases are listed in Table 21.2. None of the early equations of state gives a good representation of pV/nRT over a wide range of temperature and pressure; however, all reproduce qualitatively the dominant features of pV/nRT when corrections to the perfect gas law are small. All of the equations of state listed give an inaccurate description of the critical point and gas–liquid condensation.

Later equations of state, also displayed in Table 21.2, are more successful in reproducing the volumetric properties of real gases. Note that the Keyes and Beattie–Bridgeman equations of state require four and five empirical constants, respectively. There are other empirical equations of state that require still more constants. At present there are no known analytic formulas capable of reproducing the equation of state of a real gas at all densities and temperatures. Because of its relationship with the molecular theory, the virial equation (21.14) is preferred; from the point of view of experiment, each virial coefficient is a temperature-dependent function to be determined. Thus, the virial equation of state has great mathematical flexibility. Nevertheless, there exists doubt that it can adequately represent the properties of a gas at densities comparable to the liquid density.

21.2
Thermodynamic Description of a Gas Mixture

When considering the properties of a mixture, in addition to the temperature, pressure, and total volume, it is necessary to specify the composition. In general, the composition of a mixture is represented in terms of the masses (in some units) of the several species present. But this simple statement hides an important point. For a one-component fluid any two of the three variables T, p, V determine the third through the equation of state. That is, for a one-component fluid system there are only two independent variables, independent meaning that they can be altered at will and without relationship to each other. Similarly, to specify the composition of a mixture we must use only the set of independent composition variables. For example, suppose that we are given a system containing H_2, I_2, and HI, in the gas phase, at equilibrium. The chemical reaction

$$H_2 + I_2 = 2HI$$

imposes a condition on the way the concentrations of the three molecular species may change relative to one another. Indeed, given the concentrations of H_2 and I_2, and auxiliary thermodynamic information concerning the Gibbs free-energy change in the reaction, the concentration of HI is determined. In this system only the concentrations of H_2 and I_2 (or any two of the set of three concentrations) are independent variables. *In all our considerations we specify the composition of the system in terms of the independently variable concentrations, taking into account the constraints imposed by the conditions of equilibrium.*

Consider a homogeneous gas containing r components. Let this gas be enclosed in one compartment of a container separated into two parts by a semipermeable membrane (see Fig. 21.8). Suppose that compartment I of the container holds pure component i in equilibrium with component i in the gas mixture in compartment II. Suppose also that the semipermeable membrane which allows i to pass is diathermal and therefore permits transfer of energy, but is rigid (nondeformable) and therefore does not allow performance of work. Then, because the system is at equilibrium, we have

$$\mu_i^{\mathrm{I}} = \mu_i^{\mathrm{II}} = \mu_i, \tag{21.24}$$

$$T^{\mathrm{I}} = T^{\mathrm{II}} = T. \tag{21.25}$$

To define the contribution to the total pressure in the gas mixture made by component i, it is necessary to introduce a new extrathermodynamic consideration. If there are $n_1, \ldots, n_r$ mol each of components $1, \ldots, r$ in a mixture, and if

$$x_i \equiv \frac{n_i}{n_1 + n_2 + \cdots + n_r} \tag{21.26}$$

is the mole fraction of component i, then it is observed that

$$\lim_{p \to 0} \frac{p_i}{p} = x_i, \tag{21.27}$$

where p_i is the pressure that pure component i would exert at the same temperature if its chemical potential were the same as that of component i in the mixture. In the perfect gas limit this means that p_i is the contribution to the total pressure from component i, called the *partial pressure* of component i, and is computed just as if no other gas were present. The mathematical representation of this observation is

$$\lim_{p \to 0} p = \frac{RT}{V}(n_1 + n_2 + \cdots + n_r) = p_1 + p_2 + \cdots + p_r. \tag{21.28}$$

Equation 21.28 is a statement of *Dalton's law of partial pressures.*

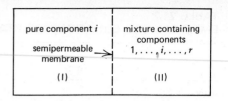

FIGURE 21.8
Schematic diagram of a box with semipermeable membrane separating pure component i from a mixture of gases containing component i. Only i can pass through the membrane.

How do we treat the case of gas mixtures at nonzero pressure? In the pure gas the chemical potential is given by the equation

$$\mu_i^{\text{I}} = \mu_i^*(T, \text{p}) + RT \ln \frac{p^{\text{I}}}{\text{p}} + \int_{0,T}^{p^{\text{I}},T} \left[v_i(p') - \frac{RT}{p'} \right] dp'. \qquad (21.15)$$

As usual, lowercase letters refer to molar quantities; for example, $v_i(p')$ is the molar volume of pure component i at temperature T and pressure p'. Consider again the situation depicted in Fig. 21.8. The equilibrium condition Eq. 21.24 and the definition of the partial pressure lead to $p_i^{\text{I}} = p_i^{\text{II}}$, and in the zero-pressure limit $p_i^{\text{I}} \equiv p^{\text{I}} = x_i^{\text{II}} p^{\text{II}}$. Therefore we can write for the chemical potential of component i in the mixture

$$\mu_i^{\text{II}} = \mu_i^*(T, \text{p}) + RT \ln \frac{x_i^{\text{II}} p^{\text{II}}}{\text{p}} + \int_{0,T}^{p^{\text{II}},T} \left[\bar{v}_i(p') - \frac{RT}{p'} \right] dp', \qquad (21.29)$$

where

$$\bar{v}_i \equiv \left(\frac{\partial \mu_i}{\partial p} \right)_{T,n_i} = \frac{\partial}{\partial p} \left(\frac{\partial G}{\partial n_i} \right)_{T,n_{j \neq i}} = \frac{\partial}{\partial n_i} \left(\frac{\partial G}{\partial p} \right)_{T,n_{j \neq i}} = \left(\frac{\partial V}{\partial n_i} \right)_{T,p,n_{j \neq i}} \qquad (21.30)$$

is the *partial molar volume* of component i.[4]

Just as in the case of the pure gas, evaluation of Eq. 21.29 requires knowledge of the equation of state of the gas. We extend the virial equation 21.17 to read

$$v_m \equiv \frac{V}{\sum\limits_{i=1}^{r} n_i} = \frac{RT}{p} + B(x_1, x_2, \ldots, x_{r-1}, T). \qquad (21.31)$$

Note that v_m is an "average molar volume" for the mixture. Since B is a measure of the interaction between pairs of molecules, we expect it to be proportional to the frequencies of occurrence of possible pairs of all compositions. The probability that a molecule of species i and a molecule of species j will, at random, undergo collision is proportional to the mole fraction of i, x_i, and the mole fraction of j, x_j. Thus we conclude that $B(x_1, \ldots, x_{r-1}, T)$ has a dependence on concentration of the following form:

$$B(x_1, \ldots, x_{r-1}, T) = \sum_{i=1}^{r} \sum_{j=1}^{r} x_i x_j B_{ij}(T). \qquad (21.32)$$

Of course, $B_{ij} = B_{ji}$, since the two molecules in an ij pair are the same as those in a ji pair.

The evaluation of Eq. 21.31 is tedious. For our purposes it suffices to illustrate the evaluation for the specific case of a binary mixture and then to state the general answer. For a binary mixture the equation of state 21.31 becomes

$$\frac{V}{n_1 + n_2} = \frac{RT}{p} + \frac{n_1^2 B_{11} + 2n_1 n_2 B_{12} + n_2^2 B_{22}}{(n_1 + n_2)^2}. \qquad (21.31')$$

[4] The final term on the right-hand side of Eq. 21.29 is just the integral of the $(\partial \mu_i / \partial p)_{T,n_i}$ term in

$$d[\mu_i - \mu_i(\text{perfect gas})] = \left\{ \frac{\partial}{\partial T} [\mu_i - \mu_i(\text{perfect gas})] \right\}_{p,n_i} dT$$

$$+ \left\{ \frac{\partial}{\partial p} [\mu_i - \mu_i(\text{perfect gas})] \right\}_{T,n_i} dp.$$

Here $\mu_i(\text{perfect gas})$ is given by the first two terms on the right side of Eq. 21.29. Note that for a mixture of perfect gases we have

$$\mu_i(\text{mixture}; T, p) = \mu_i(\text{pure } i; T, p) + RT \ln x_i.$$

We multiply both sides of Eq. 21.31' by $(n_1 + n_2)$ and use the definition (see Eq. 21.30)

$$\bar{v}_1 \equiv \left(\frac{\partial V}{\partial n_1}\right)_{T,p,n_2} \tag{21.30'}$$

to obtain

$$\bar{v}_1 = \frac{RT}{p} + \frac{n_1^2 B_{11} + 2n_1 n_2 B_{11} + 2n_2^2 B_{12} - n_2^2 B_{22}}{(n_1 + n_2)^2}. \tag{21.33}$$

We can rewrite Eq. 21.33 in the form

$$\bar{v}_1 = \frac{RT}{p} + x_1^2 (2B_{11} - B_{11}) + x_1 x_2 (2B_{11} - B_{12})$$
$$+ x_2 x_1 (2B_{12} - B_{21}) + x_2^2 (2B_{12} - B_{22}), \tag{21.34}$$

remembering that $B_{12} = B_{21}$. We have rewritten the equation for $\bar{v}_1$ in the form of Eq. 21.34 to provide an illustration of the general result for an r-component mixture, which we now quote:

$$\bar{v}_i = \frac{RT}{p} + \sum_{j=1}^{r} \sum_{k=1}^{r} x_j x_k (2B_{ij} - B_{jk}), \tag{21.34'}$$

The substitution of Eq. 21.34' into Eq. 21.29 leads to

$$\mu_i = \mu_i^*(T) + RT \ln \frac{px_i}{\text{p}} + \sum_{j=1}^{r} \sum_{k=1}^{r} x_j x_k (2B_{ij} - B_{jk})p, \tag{21.35}$$

which is to be compared with

$$\mu_i = \mu_i^*(T) + RT \ln \frac{p}{\text{p}} + B_{ii}p \tag{21.36}$$

for a pure gas.

To use Eq. 21.36, we need equation-of-state data for the mixture as a function of T, p, and composition. One of the classical goals of chemistry has been the prediction of the properties of mixtures from the properties of pure components. But there is no reason to expect a simple quantitative relation to exist between the intermolecular interactions involving, say, species 1 and 2, and those involving species 1 and 1, or 2 and 2. There are, of course, qualitative relations between these interactions, but all studies have revealed their inadequacy for precise predictions of the behavior of mixtures. Nevertheless, because there are so many possible mixtures, it is useful to have even a crude approximation relating the properties of a mixture to those of its pure components. Lewis and Randall studied this problem and proposed the following rule: We define the fugacity of component i in the mixture by

$$f_i \equiv x_i p \exp\left[\int_0^p \left(\bar{v}_i - \frac{RT}{p'}\right)\frac{dp'}{RT}\right]. \tag{21.37}$$

Then, in a real gas mixture, by analogy with Dalton's law of partial pressures for a mixture of perfect gases, it is assumed that

$$f_i(T, p, x_i) = x_i f_i^0(T, p), \tag{21.38}$$

where the superscript (^{0}) refers to the pure component i. A mixture for which this assumption is valid is known as an *ideal solution*.[5] What does

[5] For an ideal solution of perfect gases, f_i approaches p_i, the partial pressure of component i, and

$$p_i(T, p, x_i) = x_i p.$$

Eq. 21.38 imply? In terms of the fugacity, the chemical potential is given by Eqs. 21.15 and 21.29 as

$$\mu_i^0 = \mu_i^*(T, \mathrm{p}) + RT \ln \frac{f_i^0}{\mathrm{p}} \qquad (21.15)$$

for a pure gas and

$$\mu_i = \mu_i^*(T, \mathrm{p}) + RT \ln \frac{f_i}{\mathrm{p}} \qquad (21.39)$$

for a gas mixture. Furthermore, instead of using as a reference state the properties of the perfect gas at unit pressure, we can choose to refer the chemical potential of i in the mixture to the value the chemical potential of pure i has at the same temperature and pressure. This is done simply by subtracting Eq. 21.15 from Eq. 21.39. The result is

$$\mu_i = \mu_i^0 + RT \ln \frac{f_i}{f_i^0}, \qquad (21.40)$$

or for an ideal solution

$$\mu_i = \mu_i^0 + RT \ln x_i, \qquad (21.41)$$

using Eq. 21.38. Thus, in an ideal solution the chemical potential of a component of the mixture, referred to the pure component as reference, has a very simple form, namely, Eq. 21.41. Note that within the approximation described by Eqs. 21.17 and 21.32 for the equations of state, Eqs. 21.15 and 21.17 assume the forms

$$f_i^0 = p \exp\left(\frac{B_{ii}p}{RT}\right) \qquad (21.42)$$

and

$$f_i = x_i p \exp\left[\frac{\sum\limits_{i}^{r} \sum\limits_{k}^{r} x_j x_k (2B_{ij} - B_{jk})p}{RT}\right], \qquad (21.43)$$

and that $f_i^0 \to p_i^0$ and $f_i \to p_i$ as $p \to 0$.

One of the most interesting and fundamental differences between a mixture and a pure gas is the following: The entropy of a mixture is greater than the sum of the entropies of the pure components that constitute the mixture. How large is the difference, and what is its origin?

The generalization of Eq. 21.12 to a mixture is obtained from the relation $\bar{s}_i = -(\partial \mu_i / \partial T)_{p, x_i}$ and Eq. 21.29 for μ_i. We find for the partial molar entropy of component i

$$\bar{s}_i = s_i^*(T, \mathrm{p}) - R \ln \frac{p x_i}{\mathrm{p}} - \int_{0,T}^{p,T} \left[\left(\frac{\partial \bar{v}_i}{\partial T}\right)_{p', n_i} - \frac{R}{p'}\right] dp'. \qquad (21.44)$$

Since the entropy is an extensive function, we can write for the total entropies of the mixture and of the separated pure components

$$S^0 = \sum_{i=1}^{r} n_i s_i^0 \qquad \text{(pure gases)}, \qquad (21.45)$$

$$S^m = \sum_{i=1}^{r} n_i \bar{s}_i \qquad \text{(mixture).}[6] \qquad (21.46)$$

The difference between Eq. 21.46 and Eq. 21.45 is the entropy of mixing.

[6] See Eq. 19.76.

This is

$$\Delta S_{mix} = S^m - S^0 = \sum_{i=1}^{r} n_i(\bar{s}_i - s_i^0)$$

$$= -R \sum_{i=1}^{r} n_i \ln x_i + \sum_{i=1}^{r} n_i \int_{0,T}^{p,T} \left[\left(\frac{\partial v_i}{\partial T} \right)_{p'} - \left(\frac{\partial \bar{v}_i}{\partial T} \right)_{p',n_i} \right] dp'. \tag{21.47}$$

In the perfect gas limit we must have

$$\left(\frac{\partial \bar{v}_i}{\partial T} \right)_{p,n_i} = \left(\frac{\partial v_i}{\partial T} \right)_p, \tag{21.48}$$

so that

$$\Delta S_{mix}(\text{perfect gases}) = -R \sum_{i=1}^{r} n_i \ln x_i. \tag{21.49}$$

We now undertake an examination of the implications of Eq. 21.49. It has often been argued that Eq. 21.49 is difficult to understand because it assigns the same entropy of mixing to all possible mixtures of species (provided that the gases are perfect), but zero entropy of mixing to the case when the "mixture" is made by adding together portions of the same gas. Such a view, which defines what is known as *Gibbs's paradox*, overlooks the important role played by constraints and possible work processes in a thermodynamic system. From the point of view of molecular theory, the mixing of two different perfect gases with the same thermodynamic properties, Ar and Kr, for example, is fundamentally different from mixing two volumes of Ar, because molecules of the same species are indistinguishable, whereas molecules of different species, even though similar, are distinguishable. Even in the extreme case of two isotopes of the same element, the difference in nuclear composition permits a separation procedure to be devised on the basis of the mass difference between the isotopes. How does this property of indistinguishability of identical molecules enter the thermodynamic analysis? Recall that Dalton's law of partial pressures plays an essential role in the description of the gas mixture. According to this law, a gas of one species, when present in a perfect gas mixture, behaves as if there were no other gas present. Consider a container with two compartments, each containing a different perfect gas, as shown in Fig. 21.9. Let the pressure and temperature be the same on both sides of the partition. Then we have

$$n_a = \frac{pV^I}{RT},$$

$$n_b = \frac{pV^{II}}{RT}. \tag{21.50}$$

When the partition is removed, the gases mix and fill the entire volume. From Dalton's law, after mixing we must have

$$p = \frac{RT}{V^I + V^{II}} (n_a + n_b). \tag{21.51}$$

Now, if the gases behave as if they were individually present in the box, that is, if they are indifferent to each other, the process of mixing may be interpreted as an isothermal expansion of each of the gases. As shown in Chapter 17, the entropy change accompanying the isothermal expansion of a perfect gas is

$$\Delta S = nR \ln \frac{V_2}{V_1}, \tag{21.52}$$

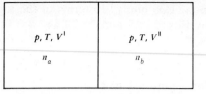

FIGURE 21.9
Schematic representation of a box separated into two compartments I and II, both at the same temperature and pressure, but containing n_a mol of a and n_b mol of b, respectively.

so that in this case we obtain, using Eqs. 21.50 and 21.51,

$$\Delta S_{\text{mix}} = \Delta S_1 + \Delta S_2 = n_a R \ln \frac{V^{\text{I}} + V^{\text{II}}}{V^{\text{I}}} + n_b R \ln \frac{V^{\text{I}} + V^{\text{II}}}{V^{\text{II}}}$$

$$= -n_a R \ln \frac{n_a}{n_a + n_b} - n_b R \ln \frac{n_b}{n_a + n_b}$$

$$= -n_a R \ln x_a - n_b R \ln x_b, \tag{21.53}$$

which is, of course, the same as Eq. 21.49. To understand why Eq. 21.53 represents the effects of molecular distinguishability, let us consider the inverse process of separating a gas mixture. We imagine that there is a suitable semipermeable membrane which allows species a to pass through, but not species b. We insert this membrane at one end of the container holding the gas mixture. The mixture is then compressed using the semipermeable membrane. Let us calculate the work done in this separation process. Since a perfect gas may be compressed to an arbitrarily small volume, we may effect a complete separation of a and b by starting the semipermeable membrane at the volume $V^{\text{I}} + V^{\text{II}}$, and compressing to a volume O^+. (This volume will drop out of the final answer, so we need not worry about taking the logarithm of zero. We use O^+ to denote "essentially" zero volume, in the sense that the expression involves the limit as the volume approaches zero.) For isothermal compression of an ideal gas we have (see Eq. 13.14)

$$w = -nRT \ln \frac{V_2}{V_1}. \tag{21.54}$$

Now in our case the membrane does no work on component a in the compression (a passes through), but this leaves a in a volume $V^{\text{I}} + V^{\text{II}}$ and b in a volume O^+. To restore the initial state we must now compress a and expand b to their original volumes with a nonpermeable piston. The total work done in this sequence of processes is then

$$w = -\left(n_b RT \ln \frac{O^+}{V^{\text{I}} + V^{\text{II}}} + n_a RT \ln \frac{V^{\text{I}}}{V^{\text{I}} + V^{\text{II}}} + n_b RT \ln \frac{V^{\text{II}}}{O^+} \right)$$

$$= -(n_a RT \ln x_a + n_b RT \ln x_b). \tag{21.55}$$

But an isothermal process in a perfect gas is characterized by $\Delta U = 0$, so that the first law of thermodynamics yields

$$T \int dS = \int p \, dV = -w_{\text{sep'n}}, \tag{21.56}$$

and the entropy change accompanying the separation is

$$\Delta S_{\text{sep'n}} = n_a R \ln x_a + n_b R \ln x_b, \tag{21.57}$$

which is just the negative of the entropy of mixing, as required. It is important to note that the inverse process used depends on the constraint that the membrane be semipermeable. But the very concept of a semipermeable membrane requires the existence of some feature distinguishing the molecules of the mixture. Without the existence of such a distinguishing feature, no "separator" can be envisaged. Since the separation process described is just the inverse of the mixing process, we see that the concept of distinguishability, and its embodiment in Dalton's law of partial pressures, leads to a qualitatively new feature in the thermodynamics of mixtures, namely, the entropy of mixing. At the risk of pedanticism, it is worthwhile again to emphasize that the formulation of the entropy of mixing requires information from outside of thermo-dynamics, in this case Dalton's law.

As might be expected, not only the enthalpy but also V, U, G, and A can differ in a mixture from the respective sums of their values in the pure component gases. For ease of discussion we now call the difference between the value some property has in the mixture and the mole-fraction-weighted sum of the corresponding properties of the pure components the *excess value* of that property. For example, the excess volume and enthalpy of a mean mole of mixture are

$$v^E \equiv \frac{V^E}{\sum\limits_{i=1}^{r} n_i} = v_m - \sum_{i=1}^{r} x_i v_i^0 \qquad (21.58a)$$

and

$$h^E \equiv \frac{H^E}{\sum\limits_{i=1}^{r} n_i} = h_m - \sum_{i=1}^{r} x_i h_i^0. \qquad (21.58b)$$

As usual, the quantities with superscript zero refer to the pure compo-nent. Note that the excess entropy of a mixture is just what we have called the entropy of mixing.

There are, in general, two contributions to an excess function. One arises from the difference between intermolecular forces in the mixture and the pure components. The second term on the right-hand side of Eq. 21.47 is of this type. The other contribution arises from the distinguisha-bility of different molecular species, that is, from the perfect gas entropy of mixing. The first term on the right-hand side of Eq. 21.47 is of this type. The excess volume and the excess enthalpy of a perfect gas mixture are zero; the only contributions to a nonzero excess enthalpy and excess volume arise from gas imperfection, hence from intermolecular forces. Using arguments similar to those employed in the calculation of the entropy of mixing, it is found that, in general, for any gas mixture,

$$H^E = \int_0^p \left[V^E - T\left(\frac{\partial V^E}{\partial T}\right)_{p',n_i} \right] dp'. \qquad (21.59)$$

On the other hand, because of the distinguishability of molecules of different chemical species, the excess entropy (i.e., the entropy of mixing) is not zero for a perfect gas mixture, and therefore the excess Gibbs and Helmholtz free energies of a perfect gas mixture are nonzero.

In summary, for a perfect gas mixture we have

$$s^E \equiv \frac{\Delta S_{\text{mix}}}{\sum\limits_{i=1}^{r} n_i} = -R \sum_{i=1}^{r} x_i \ln x_i, \qquad (21.60a)$$

$$h^E = u^E = 0, \qquad (21.60b)$$

$$v^E = 0, \qquad (21.60c)$$

$$g^E \equiv \frac{\Delta G_{\text{mix}}}{\sum\limits_{i=1}^{r} n_i} = RT \sum_{i=1}^{r} x_i \ln x_i, \qquad (21.60d)$$

$$a^E \equiv \frac{\Delta A_{\text{mix}}}{\sum\limits_{i=1}^{r} n_i} = RT \sum_{i=1}^{r} x_i \ln x_i. \qquad (21.60e)$$

The thermodynamic functions of a real gas mixture are obtained by combining Eqs. 21.47, 21.59, and the obvious forms for other mixture properties.

It is interesting to consider the properties of an ideal solution, which we defined in Eq. 21.38. For an ideal solution Eqs. 21.41, 19.76, and 21.47 yield

$$\Delta S_{mix} = \sum_{i=1}^{r} n_i \left[\left(\frac{\partial \mu_i^0}{\partial T} \right)_p - \left(\frac{\partial \mu_i}{\partial T} \right)_{p,n_i} \right]$$

$$= -R \sum_{i=1}^{r} n_i \ln x_i, \qquad (21.60a')$$

the same result obtained in Eq. 21.49 for a mixture of perfect gases. Similarly, it can be shown that the other excess functions for the ideal solution have the same forms as in Eqs. 21.60. An ideal solution is a mixture whose excess functions are the same as those of a perfect gas mixture of the same composition.

21.3 Thermodynamic Description of Gaseous Reactions

With the concepts introduced in Sections 21.1 and 21.2 we can describe the equilibrium composition of a mixture of gases between which chemical reactions occur. It was shown in Chapter 19 that at equilibrium in a mixture of reacting components it is necessary that

$$\sum_{i=1}^{r} \nu_i \mu_i = 0 \qquad (21.61)$$

(see Eq. 19.99). We shall now use Eq. 21.61, together with the expression for the chemical potential in a gas mixture, to evaluate the equilibrium constant in a gaseous reaction (see Eq. 19.81). Substitution of Eq. 21.29 in Eq. 21.61 gives

$$\sum_{i=1}^{r} \nu_i \mu_i^*(T, \mathrm{p}) = -RT \sum_{i=1}^{r} \left[\ln \frac{x_i p}{\mathrm{p}} + \int_{0,T}^{p,T} \left(\bar{v}_i - \frac{RT}{p'} \right) \frac{dp'}{RT} \right] \nu_i. \qquad (21.62)$$

But the left-hand side of this equation is a function of only the temperature. Hence, if the temperature is fixed, the relative composition is also fixed. We write, then, an equilibrium constant $K(T)$, defined by

$$-RT \ln K(T) \equiv \sum_{i=1}^{r} \nu_i \mu_i^*(T, \mathrm{p}), \qquad (21.63)$$

and transform Eq. 21.62 to read

$$K(T) = \prod_{i=1}^{r} \left(\frac{x_i p}{\mathrm{p}} \right)^{\nu_i} \exp \left[\frac{\nu_i}{RT} \int_{0,T}^{p,T} \left(\bar{v}_i - \frac{RT}{p'} \right) dp' \right] \qquad (21.64)$$

$$= \prod_{i=1}^{r} \left(\frac{f_i}{\mathrm{p}} \right)^{\nu_i}, \qquad (21.65)$$

using Eq. 21.37 to introduce the fugacity. In the perfect gas limit we have $f_i \to p x_i$, since the integrand in Eq. 21.37 vanishes. If we then define an equilibrium constant in terms of partial pressures,

$$K_p \equiv \prod_{i=1}^{r} \left(\frac{x_i p}{\mathrm{p}} \right)^{\nu_i}, \qquad (21.66)$$

we have $\lim_{p \to 0} K_p = K(T)$.

The sum $\sum \nu_i \mu_i^*$ is, by definition, the difference in molar free energies of the pure components in their respective hypothetical perfect gas states at pressure $\mathfrak{p}$. That each μ_i^* is given by this prescription is easily verified by substitution of $p = \mathfrak{p}$ into Eq. 21.15. Thus, Eq. 21.63 may also be written in the form

$$-RT \ln K(T) = \Delta G^*(T, \mathfrak{p}), \qquad (21.67)$$

where ΔG^* is the referred-to molar free energy difference between reactants and products. Note that ΔG^* is a function of the temperature but not a function of the pressure. Therefore, the equilibrium constant $K(T)$ depends on T but not on p, whereas K_p is a function of p.

The specific form of the dependence of $K(T)$ on temperature can be determined by examining the derivative $d \ln K / dT$. We find that

$$
\begin{aligned}
-R \frac{d \ln K}{dT} &= \frac{d}{dT}\left(\frac{\Delta G^*}{T}\right) \\
&= -\frac{\Delta G^*}{T^2} + \frac{d \Delta G^*}{dT} \\
&= -\frac{1}{T}\left(\frac{\Delta G^*}{T} + \Delta S^*\right) \\
&= -\frac{\Delta H^*}{T^2}, \qquad (21.68)
\end{aligned}
$$

where we have used the interchange operation

$$\frac{d \Delta G^*}{dT} = \Delta\left(\frac{dG^*}{dT}\right) = -\Delta S^*. \qquad (21.69)$$

Thus, the larger the heat of reaction (for a given T), the larger the rate of change of equilibrium constant with temperature.

Even though it is merely a combination of Eqs. 21.67 and 21.69, the relationship

$$
\begin{aligned}
\Delta S^* &= R \frac{d}{dT}(T \ln K) \\
&= \frac{\Delta H^* - \Delta G^*}{T} \qquad (21.70)
\end{aligned}
$$

is sufficiently important to be worth separate display. Of course, $K(T)$ increases as ΔS^* increases. The extent to which a reaction proceeds is seen to be determined by a balance between the enthalpy required for the chemical transformation and T times the entropy change accompanying the same transformation. If the entropy changes are the same, an exothermic reaction proceeds further to completion than does an endothermic reaction. If the enthalpy changes are the same, a reaction that leads to more product molecules than reactant molecules proceeds further to completion than does a reaction that leads to the same number or fewer product molecules than reactant molecules. In all of these comments we are assuming that "completion" means total conversion of reactants to products. Of course, which chemicals we choose to call reactants and which products is arbitrary; the changes in ΔH^*, ΔS^*, and so on must all be defined consistent with whatever choice is made.

In Eqs. 21.68 to 21.70, ΔH^* and ΔS^* are the differences in enthalpy and entropy between products and reactants in their respective hypothetical perfect gas states. The tabulated thermodynamic properties of gases refer to the hypothetical perfect gas state at 1 atm and 25°C as the standard state of reference. In some cases the equation of state of the gas is not known to sufficient accuracy to distinguish between the actual state of the gas at 1 atm and 25°C and the hypothetical perfect gas state under the same conditions. In fact, for almost all gases there are only very small differences in properties between the actual and hypothetical perfect gas states at 1 atm and 25°C. There is, however, a distinction to be drawn between these states. If the actual state of the gas at 1 atm and 25°C is used as reference state, the equilibrium constant depends very slightly on the pressure because the excess volume of the gas mixture is not exactly zero. Note that the $K(T, p)$ we defined in Section 19.5 was such an equilibrium constant. When the hypothetical perfect gas state is used as reference state, the equilibrium constant is strictly independent of pressure.

Thus, using tables of standard enthalpies, entropies, and Gibbs free energies, the equilibrium composition of a reacting mixture, and the temperature dependence of the composition, may be readily computed. Since data for reactions may be combined to "synthesize" new reactions, the equilibrium properties of reacting mixtures may be predicted from information obtained for other reactions.[7] Of course, because of gas imperfection the equations of state of all components must be known. However, for many applications of the theory of chemical equilibrium, the perfect gas approximation to $K(T)$ is of sufficient accuracy. Indeed, under essentially all conditions, whether some particular species predominates in the reaction mixture can be determined simply from whether or not $\Delta G^* > 0$ or $\Delta G^* < 0$ for the reaction leading to that component. If the former is the case, the component does not predominate in the reaction mixture; if the latter, it does. If the entropy change in the reaction is small enough that $T \Delta S^* \ll \Delta H^*$, similar qualitative information can be obtained from the sign of ΔH^*.

There are, however, important qualitative effects that arise from the interactions between the molecules. For example, if terms up to the second virial coefficient suffice for the equation of state, and if the approximation

$$B_{jk} = \tfrac{1}{2}(B_{jj} + B_{kk}) \tag{21.71}$$

is adopted, it is easily shown that

$$\ln \frac{K_p}{K(T)} = -p \frac{\Delta B}{RT}, \tag{21.72}$$

where

$$\Delta B \equiv \sum_{i=1}^{r} \nu_i B_{ii}(T). \tag{21.73}$$

The approximation of Eq. 21.71 was introduced by Lewis and Randall; it is not an exact representation and cannot be justified from the molecular theory of the second virial coefficient. Nevertheless, Eq. 21.73 is found to be reasonably accurate over a range of gas pressure exceeding the range for which Eq. 21.31 is an accurate equation of state. Equation 21.73 is a useful approximation precisely because it eliminates all reference to the thermodynamic properties of the mixture; for these it substitutes a combination of properties of the pure components. Of course, this equivalence is just what the molecular theory denies is correct if the molecular interactions between jj, kk, and jk pairs are all different.

---

[7] Just as tables of standard enthalpies permit a great economy in the recording of data for the calculation of heats of reaction, tables of standard Gibbs free energies permit a great economy in the recording of data for the calculation of equilibrium constants.

Now suppose that the temperature is such that $B > 0$ for all the reactants and $B < 0$ for all the products, so that $\Delta B < 0$. Then from Eq. 21.72, we see that $K_p/K(T)$ is greater than unity and increases as p increases, implying that the yield of products is greater than would be obtained in a reaction between perfect gases. When might this situation arise? In general, polar molecules, such as NH_3, attract each other strongly and at low temperature have large negative virial coefficients. Thus, when nonpolar reactants, say, H_2 and N_2, react to form a polar product, say, NH_3, we anticipate that $\Delta B < 0$. In fact, the synthesis of NH_3 by this reaction is carried out at high pressure to maximize the yield. The effects of nonideality in this and other cases are by no means small. The study of gas imperfection is one of the important parts of the design of an economically viable manufacturing process, and is part of chemical engineering as well as fundamental chemistry.

Before concluding this section, we must point out a subtle assumption used in the preceding arguments. Equation 19.95 implies that the course of a chemical reaction may be specified at any given point by a value of the progress variable. This further implies that it is possible to stop the chemical reaction at an arbitrary point and measure λ. The reason this supposition is made is that it is necessary, at least in principle, to know the thermodynamic properties of the mixture for all possible compositions. If such knowledge cannot be obtained, at least in principle, the operational interpretation of thermodynamic equilibrium in terms of the changes induced by adding or removing constraints becomes meaningless. There are, indeed, some chemical reactions that can be quenched, and for which the supposition made is a valid extrapolation from experience. On the other hand, an uncatalyzed dimerization reaction such as

$$2NO_2 = N_2O_4$$

cannot be rigorously described by the thermodynamic theory because it is not possible to determine the thermodynamic properties of either pure component as a function of T and p. That is, variation of the pressure or temperature of NO_2 or N_2O_4 is always accompanied by chemical conversion following the stoichiometry shown. Thus it is never possible to determine directly the properties of either pure N_2O_4 or pure NO_2 as functions of T and p. Despite this limitation, it is very useful to apply the methods of thermodynamic analysis to this class of reactions.

21.4
An Example: The Haber Synthesis of NH_3

Much of the thermodynamic theory discussed in the preceding sections can be illustrated by the analysis of the conditions under which the reaction

$$\tfrac{1}{2}N_2(g) + \tfrac{3}{2}H_2(g) = NH_3(g) \tag{21.74}$$

gives a high yield of NH_3. In the Haber process a mixture of H_2 and N_2 under pressure and in the presence of an iron-based catalyst (usually an iron oxide mixed with aluminum oxide and potassium oxide) is caused to combine at 450–600°C. Extensive investigations of the equilibrium composition of mixtures of H_2 and N_2 under a variety of conditions lead to the data in Table 21.3. These data demonstrate that the formation of NH_3 is favored by low temperature and high pressure. Temperatures below 400°C cannot be used in commercial production because the rate of reaction is too small for economic use of the equipment needed. Temperatures above 600°C lead to yields of NH_3 that are too small for economic exploitation.

TABLE 21.3
MOLE PERCENT NH$_3$ IN 3/1 H$_2$/N$_2$ EQUILIBRIUM MIXTURES

T (°C)	p (atm)			
	200	300	400	500
400	38.74	47.85	54.87	60.61
450	27.44	35.93	42.91	48.84
500	18.86	26.00	32.25	37.79
550	12.82	18.40	23.55	28.31
600	8.77	12.93	16.94	20.76

Suppose that the N$_2$, H$_2$, and NH$_3$ form a perfect gas mixture under all the conditions for which data are entered in Table 21.3. Then the equilibrium constant for the formation of NH$_3$ is

$$K_p = \frac{p_{NH_3}/p}{(p_{N_2}/p)^{1/2}(p_{H_2}/p)^{3/2}}$$

$$= \frac{x_{NH_3}}{x_{N_2}^{1/2}x_{H_2}^{3/2}} \cdot \frac{p}{p}. \tag{21.75}$$

If, as supposed, N$_2$, H$_2$, and NH$_3$ do form a perfect gas mixture, then the several mole fractions and p will change in compensatory ways, leaving K_p a function of T alone; if not, the value of K_p computed from experimental values of the mole fractions will vary with p.

Consider the datum that at 500°C and 500 atm there is 37.79 mol% NH$_3$ in the equilibrium reaction mixture. The stoichiometry of the reaction requires that for the formation of y mol of NH$_3$, $\frac{1}{2}y$ mol of N$_2$ and $\frac{3}{2}y$ mol of H$_2$ must be consumed. If we start with a stoichiometric mixture, 0.5 mol of N$_2$ and 1.5 mol of H$_2$, we must have at any point

$$x_{NH_3} = \frac{y}{\frac{1}{2}(1-y)+\frac{3}{2}(1-y)+y} = \frac{y}{2-y}. \tag{21.76}$$

Setting $x_{NH_3} = 0.3779$ and solving for y, we find that under the stated conditions there are 0.5485 mol of NH$_3$ in the equilibrium mixture. Given the value of y, the total numbers of moles of all components are easily computed, and hence also the mole fractions of N$_2$ and H$_2$. In this particular case we find that

$$\text{total number of moles} = \tfrac{1}{2}(1-y)+\tfrac{3}{2}(1-y)+y = 2-y = 1.4515, \tag{21.77a}$$

$$x_{N_2} = \frac{\frac{1}{2}(1-y)}{2-y} = 0.1555, \tag{21.77b}$$

$$x_{H_2} = \frac{\frac{3}{2}(1-y)}{2-y} = 3x_{N_2} = 0.4666. \tag{21.77c}$$

Thus, the equilibrium constant for formation of NH$_3$ at 500°C and 500 atm has the value

$$K_p = 6.014 \times 10^{-3}.$$

Values of the perfect gas equilibrium constant obtained in this way are entered in Table 21.4. Clearly, the computed constant varies with pressure and thus cannot be equal to the equilibrium constant $K(T)$ as we

TABLE 21.4
PERFECT GAS EQUILIBRIUM CONSTANT FOR THE FORMATION OF NH_3

T (°C)	p (atm)			
	200	300	400	500
400	15.897×10^{-3}	18.060×10^{-3}	20.742×10^{-3}	24.065×10^{-3}
450	8.023×10^{-3}	8.985×10^{-3}	10.134×10^{-3}	11.492×10^{-3}
500	4.409×10^{-3}	4.893×10^{-3}	5.408×10^{-3}	6.013×10^{-3}
550	2.598×10^{-3}	2.836×10^{-3}	3.102×10^{-3}	3.392×10^{-3}
600	1.622×10^{-3}	1.751×10^{-3}	1.890×10^{-3}	2.036×10^{-3}

previously assumed. We conclude that the perfect gas approximation is not adequate for the description of the N_2-H_2-NH_3 mixture.

As a first step in our further analysis, we now compute the equilibrium constant $K(T)$ for the formation of NH_3, using thermochemical data. The relevant data are

NH_3: $\Delta H_f^0 = -46.19$ kJ/mol,

$\quad \Delta G_f^0 = -16.64$ kJ/mol,

$$c_p^* = 29.79 + 25.48 \times 10^{-3} T - \frac{1.665 \times 10^5}{T^2} \text{ J/K mol},$$

N_2: $\quad c_p^* = 27.82 + 4.184 \times 10^{-3} T$ J/K mol,

H_2: $\quad c_p^* = 28.66 + 1.17 \times 10^{-3} T - 0.92 \times 10^{-6} T^2$ J/K mol.

Since the temperature dependence of the equilibrium constant is determined by the relation

$$\frac{d}{dT} \ln K(T) = \frac{\Delta H^*}{RT^2}, \tag{21.78}$$

we have, by integration,

$$\ln \frac{K(T_2)}{K(T_1)} = \int_{T_1}^{T_2} \frac{\Delta H^*}{RT^2} \, dT. \tag{21.79}$$

Of course, ΔH^* depends on the temperature. With the definition of standard state we are using, ΔH^* is the heat of reaction not as it actually takes place, but at a sufficiently low pressure for the gases to behave as perfect gases. Since the enthalpy of a perfect gas is independent of pressure, ΔH^* can be identified with the heat of formation in the (hypothetical) perfect gas state at 1 atm. The temperature dependence of ΔH^* is then computed by the method described in Chapter 14. With a reference temperature 298.15 K, for which ΔH^* becomes the standard heat of formation ΔH_f^0, we reduce Eq. 14.23 to

$$\Delta H^*(T_2) = \Delta H_f^0 + \sum_i \nu_i \int_{298.15}^{T_2} c_p^i(T) \, dT.$$

Given the data cited above, for the reaction under consideration,

$$\Delta H^*(T) = \Delta H_f^0 + (29.79 - \tfrac{1}{2} \times 27.82 - \tfrac{3}{2} \times 28.66)(T - 298.15)$$

$$+ \frac{10^{-3}}{2} (25.48 - \tfrac{1}{2} \times 4.184 - \tfrac{3}{2} \times 1.17)[T^2 - (298.15)^2]$$

$$+ 1.665 \times 10^5 \left(\frac{1}{T} - \frac{1}{298.15} \right) - \frac{3}{2} \times \frac{0.92 \times 10^{-6}}{3} [T^3 - (298.15)^3]$$

$$= \Delta H_f^0 + 6576 - 27.11T + 10.82 \times 10^{-3} T^2$$

$$+ \frac{1.665 \times 10^5}{T} - 0.46 \times 10^{-6} T^3. \tag{21.80}$$

Then Eq. 21.79 becomes

$$\ln \frac{K(T_2)}{K(T_1)} = \frac{39614}{R} \left(\frac{1}{T_2} - \frac{1}{298.15} \right) - \frac{27.11}{R} \ln \frac{T_2}{298.15}$$

$$+ \frac{10.82 \times 10^{-3}}{R} (T_2 - 298.15)$$

$$- \frac{23.0 \times 10^{-8}}{R} [T_2^2 - (298.15)^2]$$

$$- \frac{1.99 \times 10^4}{R} \left(\frac{1}{T_2^2} - \frac{1}{(298.15)^2} \right). \tag{21.81}$$

Again for the purpose of illustration we choose $T_2 = 773.15$ K. We find that

$$\ln \frac{K(773.15)}{K(273.15)} = -12.234. \tag{21.82}$$

But from the data given, the equilibrium constant for the formation of NH₃ at 298.15 K is[8]

$$K(298.15) = \exp\left(-\frac{\Delta G_f^0}{RT} \right) = 8.23 \times 10^2, \tag{21.83}$$

so that

$$K(773.15) = 8.23 \times 10^2 \times 10^{-12.234/2.303} = 4.00 \times 10^{-3}. \tag{21.84}$$

Thus, at 773.15 K (500°C), the deviation from perfect gas behavior increases K_{NH_3} by 10% at 200 atm and about 50% at 500 atm (see Table 21.4).

We now turn to calculating the influence of gas imperfection on the value of the equilibrium constant K_p. If the equation of state keeping only the second virial coefficient is adequate, we can apply Eqs. 21.72 and 21.73, using the values of B_{ii} in Table 21.5, to obtain

$$\frac{K_p}{K(T)} = \exp\left(-\frac{p \Delta B}{RT} \right)$$

$$= 1.21 \text{ at } 773 \text{ K and } 200 \text{ atm,}$$

$$= 1.46 \text{ at } 773 \text{ K and } 400 \text{ atm.}$$

The observed values are 1.10 and 1.35, respectively (see the data in Table 21.4).

At higher pressures the limitation of the equation of state to the second virial coefficient term and the use of Eqs. 21.71 and 21.72 result in inaccurate predictions of the equilibrium constants. The limitation imposed by the equation of state can be lifted by (1) using a more accurate equation of state in analytic form, (2) using p, V, T data directly without recourse to an analytic representation (numerical integration), or (3) using the fact that all gases behave almost alike when described by the reduced variables $\tilde{p}$, $\tilde{T}$, and $\hat{V}$. Figure 21.10 shows the ratio of the fugacity to the pressure as a function of the reduced pressure $\tilde{p} = p/p_c$ at several reduced temperatures $\tilde{T} = T/T_c$. These curves may be used to

[8] ΔG_f^0 is defined with respect to the usual standard state, $p = 1$ atm, $T = 298.15$ K.

TABLE 21.5
SECOND VIRIAL COEFFICIENTS OF H₂, N₂, AND NH₃ (cm³/mol)

T (K)	B_{H_2}	B_{N_2}	B_{NH_3}
673	15.6	24.2	−38.2
723	15.7	25.5	−29.3
773	15.7	26.6	−23.4
823	15.7	27.6	−19.8
873	15.7	28.3	−16.7

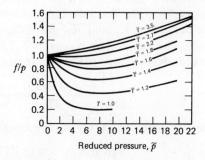

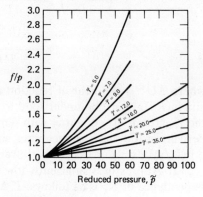

FIGURE 21.10
The ratio f/p of the fugacity to the pressure in reduced units.

estimate accurately the fugacities of the gaseous reactants and products when they are pure components. The only way to improve upon the approximation of Eq. 21.71 is to use equation of state data for the gas mixture. Such data are rarely available, so the approximations cited remain useful.

There remain three topics to be mentioned in this section.

1. Occasionally it is useful to define an equilibrium constant of a different form from that we have used. There are two other forms commonly used: One is

$$K_x \equiv \prod_{i=1} x_i^{\nu_i}, \tag{21.86}$$

so that

$$K_x = K_p \left(\frac{p}{\mathfrak{p}}\right)^{-\Sigma \nu_i} = K_p \left(\frac{p}{\mathfrak{p}}\right)^{-\Delta \nu}, \tag{21.87}$$

where $\Delta \nu$ is the change in the number of moles in the reaction if carried to completion. The other is

$$K_V \equiv \prod_i \left(\frac{n_i}{V}\right)^{\nu_i}. \tag{21.88}$$

In the perfect gas limit we can set

$$p_i = \frac{n_i RT}{V} \tag{21.89}$$

and obtain

$$K_V = K(T) \left(\frac{RT}{\mathfrak{p}}\right)^{-\Delta \nu}. \tag{21.90}$$

Note that K_V has dimensions, unlike either K_x, K_p, or $K(T)$. From the definitions given, it is clear that even for a perfect gas mixture, K_x depends on the pressure unless $\Delta \nu = 0$, $K(T)$ is independent of pressure by definition, K_p and K_V are independent of pressure for perfect gases only. The equilibrium constants K_x, K_p, and $K(T)$ have the same logarithmic derivative with respect to temperature, namely,

$$\frac{d \ln K(T)}{dT} = \left(\frac{\partial \ln K_x}{\partial T}\right)_p = \frac{\Delta H^*}{RT^2}. \tag{21.91}$$

On the other hand, for K_V we obtain, using Eq. 21.68,

$$\left(\frac{\partial \ln K_V}{\partial T}\right)_p = \frac{d \ln K(T)}{dT} - \sum_i \nu_i \frac{d}{dT} \ln\left(\frac{RT}{\mathfrak{p}}\right)$$

$$= \frac{\sum_i \nu_i (h_i^* - RT)}{RT^2} = \frac{\Delta U^*}{RT^2}, \tag{21.92}$$

where ΔU^* is the heat of reaction at constant volume. The terms $\ln K_V$ and $\ln(RT/\mathfrak{p})$ must be understood to have a formal meaning since, strictly, the logarithm of a quantity with dimensions is undefined. However, their derivatives do have a meaning, since one can always set $d \ln y/dx = d \ln (y/y_0)/dx$, where y_0 is an arbitrary constant with the same units as y.

2. At pressures sufficiently low to set $K_x = K(T)(p/\mathfrak{p})^{-\Delta \nu}$, the pressure dependence of K_x is as follows: If $\Delta \nu < 0$, corresponding to a decrease in volume as the forward reaction proceeds, an increase in pressure favors the formation of more product. The use of K_x rather than $K(T)$ yields the clearest indication of the effect of pressure on reaction yield.

Returning to the synthesis of NH_3, we see that for the formation of NH_3 the perfect gas approximation gives

$$K_x = K(T)\frac{p}{\mathrm{p}}; \tag{21.93}$$

that is, the formation of NH_3 is promoted by increasing the pressure. In the case of the dissociation reaction

$$N_2O_4 = 2NO_2, \tag{21.94}$$

we can write, treating the system as a perfect gas mixture,

$$K_x = \frac{x_{NO_2}^2}{x_{N_2O_4}} = K(T)\frac{p}{\mathrm{p}}. \tag{21.95}$$

In this case, with $\Delta \nu > 0$, the forward reaction is promoted by lowering the pressure.

3. It may be advantageous to add an "inert" gas to a reaction mixture, since in that case the yields of various compounds can be altered. If the gases are imperfect, the addition of a gas that does not participate in the reaction can alter the fugacities of all the reacting species. For example, if the added gas has a very strong attraction for one of the reactants, the fugacity of that reactant will be decreased. The effects of an inert gas on the fugacities of the reactants and products are not accounted for by the Lewis and Randall rule. If they are important, analysis of the behavior of the gas mixture must be based on p, V, T measurements of the mixture and cannot be estimated using only pure-component data.

The effect of adding an inert gas is most evident in terms of K_x, since the equilibrium concentration is altered even if the gas mixture is perfect, unless $\Delta \nu = 0$. This happens because the mole fractions of all components are altered by the addition of another species to the mixture, even if that species does not take part in the reaction. For example, in the synthesis of NH_3, if there were initially present a mol of N_2, $3a$ mol of H_2, and b mol of inert gas, and y mol of NH_3 are formed at equilibrium, we find that

$$K_x = \frac{y/(4a+b-y)}{[(a-y/2)/(4a+b-y)]^{1/2}[(3a-3y/2)/(4a+b-y)]^{3/2}}$$

$$= K(T)\frac{p}{\mathrm{p}}, \tag{21.96}$$

since $\Delta \nu = -1$ for this reaction. Given the values of $K(T)$ and p, Eq. 21.96 may be solved for y. In this case, and indeed in general, the algebra of such solutions is complicated and numerical analysis is to be preferred. The calculated mole percent of NH_3 in an equilibrium mixture initially containing 20 mol% inert gas is shown in Table 21.6. Comparison of Tables 21.3 and 21.6 shows how the mole fraction of NH_3 is reduced by the addition of inert gas to the mixture.

Earlier in this book, in the discussion of atomic and molecular structure, we examined how the general principles of quantum mechanics restrict the energy spectrum of a system. The statistical interpretation of the thermodynamic properties of a particular system depends, of course, on the energy spectrum of that system.

A particularly interesting situation exists with respect to the requirements imposed on the symmetry properties of the wave function of a system. It is necessary that the wave function representing a system of

21.5
Statistical Molecular Theory of Gases and Gas Reactions
A. The Partition Function

TABLE 21.6
MOLE PERCENT OF NH$_3$ AT EQUILIBRIUM IN A
3/1 H$_2$/N$_2$ MIXTURE WITH 20% INERT GAS

T (°C)	p (atm)			
	200	300	400	500
400	25.36	31.41	36.14	40.00
450	17.87	23.49	28.14	32.10
500	12.22	16.92	21.05	24.73
550	8.28	11.92	15.31	18.44
600	5.65	8.35	10.97	13.47

many identical particles be either symmetric (i.e., have exactly the same value) or antisymmetric (i.e., be multiplied by −1) after two identical particles are interchanged. Any given system is of one or the other type, and the actual number of accessible states within a particular energy range may vary greatly depending on what symmetry the wave function has (see Appendix 21A).

Consider, now, the description of the motion of the center of mass of a molecule, that is, the translational motion. Let $\Lambda \equiv h/(2\pi m k_B T)^{1/2}$ be the *thermal de Broglie wavelength* associated with a molecule. We expect that if Λ is comparable with the average spacing between molecules, we must take into account the effect of diffraction upon the molecular motion. On the other hand, if $\Lambda \ll (V/N)^{1/3}$, we expect diffraction effects to be negligibly small. Except for He and H$_2$, the masses of all molecular species are sufficiently large that $\Lambda \ll (V/N)^{1/3}$ in the temperature ranges in which they are liquid or gaseous. Thus, for liquid Ar (~85 K), quantum effects do not influence the translational motion of the molecules, whereas for liquid Ne (~20 K), which is a borderline case, they amount to only a few percent of the classical motion contribution. On the other hand, in liquid He (~4 K) the effects of quantum mechanical diffraction dominate the behavior of the substance.

The reader should not derive from these paragraphs the impression that quantization of translational motion is rare in nature. The properties of metals, and of nuclei, bear eloquent testimony to the effects of quantization of the energy levels of an electron gas and of a nuclear gas.

There are other symmetry effects, of similar origin, that restrict the number of states of diatomic and polyatomic molecules (see Appendix 21B). Thus, interchange of two identical nuclei, which leads to changes in the electronic, vibrational, and rotational wave functions of a molecule, must satisfy the restriction of being of even or odd parity. We shall see that in the case of H$_2$ this restriction requires that there be two classes of H$_2$ molecules—one set of molecules that can have only even rotational states, and the other only odd rotational states. Because of the large spacing of the rotational energy levels of H$_2$, this particular restriction has marked thermodynamic consequences.

Consider, now, a one-component perfect gas. We shall analyze the properties of this system using the canonical ensemble to represent the relationship between the energy spectrum and thermodynamic properties. Using Eq. 19.131, we write the canonical partition function for the N-molecule system,

$$Q_N = \sum_i g_i \exp\left(-\frac{E_i}{k_B T}\right), \tag{21.97}$$

where E_i is the energy and g_i the number of states in the ith energy level. In a perfect gas there are no interactions between the molecules, hence it is possible to define the energy of one molecule in meaningful terms. Of course, the total energy of the system, E_i, is just the sum of the energies of the molecules. The energy of any one molecule is denoted by $\epsilon_l(n_x, n_y, n_z, j, v, \ldots)$, where n_x, n_y, n_z are the quantum numbers specifying translational motion, j is the quantum number specifying rotational motion, v is the quantum number specifying molecular vibration, the $\ldots$ implies other quantum numbers, for example, for electronic excitation, and $l = 1, 2, \ldots, N$. For a polyatomic molecule j and v represent several quantum numbers corresponding to the several kinds of rotational and vibrational motion. For simplicity we shall collectively represent the set of all quantum numbers except n_x, n_y, n_z by the symbol s. In a perfect gas the energy of a molecule can be written in the form

$$\epsilon_l(n_x, n_y, n_z, j, v, \ldots) = \epsilon_l^{\text{trans}}(n_x, n_y, n_z) + \epsilon_l^{\text{int}}(s); \qquad (21.98)$$

that is, the translational motion is completely independent of the internal motion of the molecule.

To determine the equation of state from Eq. 21.98, we start by writing the total energy of N molecules as a sum of individual molecule energies:

$$E_i = \sum_{l=1}^{N} [\epsilon_l^{\text{trans}}(n_x, n_y, n_z) + \epsilon_l^{\text{int}}(s)]. \qquad (21.99)$$

Now, a given molecule may be in any one of the possible quantum states provided only that for a given value of E_i the sum of the energies of all the molecules is just E_i. We are, of course, interested in all possible values of E_i (and therefore also in all possible sets of values of the individual energies of the molecules), since the evaluation of Q_N requires that E_i range over all possible values. If molecule l is in an internal state with quantum numbers collectively represented by s, there is a degeneracy[10] $\bar{\omega}_{nsl}$, so that in a gas of independent molecules the overall degeneracy of the state with total energy E_i is

$$g_i \equiv g_i(E_i) = \frac{1}{N!} \sum_{ns} \prod_{l=1}^{N} \bar{\omega}_{nsl}. \qquad (21.100)$$

The factor $(1/N!)$ appears because of the indistinguishability of the molecules—there are $N!$ different ways of assigning N molecules to a specified set of N energies ϵ_l, each of which has the same total energy. The sum $\sum_{ns}$ appears because there are many ways of subdividing internal energy and translational energy that satisfy Eq. 21.99. Each particular subdivision has a degeneracy $(1/N!) \prod_l \bar{\omega}_{nsl}$, but the internal quantum numbers s are, of course, different for each subdivision, so to get the overall degeneracy corresponding to energy E_i we must add up all such contributions. The sum $\sum_{ns}$ is over all sets of values of the internal and translational quantum numbers of the N molecules, but the possible subdivisions of the internal and translational energies of the molecules must satisfy Eq. 21.99. Introducing Eqs. 21.99 and 21.100 into Eq. 21.97, we find that

$$\sum_i g_i(E_i) \exp\left(-\frac{E_i}{k_B T}\right) = \frac{1}{N!} \left\{ \sum_{ns\,1} \bar{\omega}_{ns\,1} \exp\left[-\frac{\epsilon_1^{\text{trans}}(n\,1) + \epsilon_1^{\text{int}}(s\,1)}{k_B T}\right] \right\} \times$$

[10] We use $\bar{\omega}_{nsl}$ for the degeneracy of the nsth state of molecule l so as to distinguish it from the degeneracy of the N-molecule system, g_i. **ns** indicates summation over all molecules.

$$\times\left\{\sum_{ns2}\bar{\omega}_{ns2}\exp\left[-\frac{\epsilon_2^{trans}(n2)+\epsilon_2^{int}(s2)}{k_BT}\right]\right\}$$

$$\times\cdots\times\left\{\sum_{nsN}\bar{\omega}_{nsN}\exp\left[-\frac{\epsilon_N^{trans}(nN)+\epsilon_N^{int}(sN)}{k_BT}\right]\right\},\qquad(21.101)$$

where the numbers $ns1,\ldots,nsN$ stand for all the quantum numbers specifying the states of molecules $1,\ldots,N$. Equation 21.101 includes the fact that, as stated by Eq. 21.100, any allowed state of one molecule can be combined with any allowed states of all other molecules. Of course every molecule in the gas is the same, and therefore the sum over all the states of a molecule must have the same value for each molecule. Since there is one such sum for each of the N molecules, we have

$$\sum_i g_i(E_i)\exp\left(-\frac{E_i}{k_BT}\right)=\frac{1}{N!}\left\{\sum_{ns1}\bar{\omega}_{ns1}\exp\left[-\frac{\epsilon_1^{trans}(n1)+\epsilon_1^{int}(s1)}{k_BT}\right]\right\}^N$$

$$=\frac{Q_1^N}{N!},\qquad(21.102)$$

where

$$Q_1\equiv\sum_{ns1}\bar{\omega}_{ns1}\exp\left[-\frac{\epsilon_1^{trans}(n1)+\epsilon_1^{int}(s1)}{k_BT}\right]\qquad(21.103)$$

is the partition function of a single-molecule system. From here on we drop the subscript 1 in the summand of Eq. 21.103, since Q_1 is the same for each molecule in the perfect gas.

To determine the precise form of the equation of state we must evaluate Q_1. Consider first the translational motion. We know that

$$\epsilon^{trans}=\frac{h^2}{8mV^{2/3}}(n_x^2+n_y^2+n_z^2)\qquad(21.104)$$

for a free particle in a cubical box of volume V. For all cases discussed in this chapter, $h^2/8mV^{2/3}$ is very much smaller than the thermal energy k_BT. This means that the difference in population of two adjacent energy levels is very small, and it is reasonable to replace the sum over energy levels in the definition of Q_1 with an integral over energy levels:

$$\sum_{ns}\bar{\omega}_{ns}\exp\left[-\frac{(\epsilon^{trans}+\epsilon^{int})}{k_BT}\right]=\sum_{n_x,n_y,n_z}\exp\left(-\frac{\epsilon^{trans}}{k_BT}\right)$$

$$\times\sum_s\bar{\omega}_s^{int}\exp\left(-\frac{\epsilon^{int}}{k_BT}\right)$$

$$=Q_1^TQ_1^{RVE},\qquad(21.105)$$

where

$$Q_1^T\equiv\sum_{n_x,n_y,n_z}\exp\left(-\frac{\epsilon^{trans}}{k_BT}\right)$$

$$=\int_0^\infty\int_0^\infty\int_0^\infty dn_x\,dn_y\,dn_z\exp\left[-\frac{h^2(n_x^2+n_y^2+n_z^2)}{8mV^{2/3}k_BT}\right],$$

$$(21.106)$$

and

$$Q_1^{RVE}\equiv\sum_s\bar{\omega}_s^{int}\exp\left(-\frac{\epsilon^{int}}{k_BT}\right),\qquad(21.107)$$

the superscripts T, RVE referring to the translational and rotation-vibration-electronic (internal) components, respectively. The triple integral defining Q_1^T reduces to products of $\int\exp(-ax^2)\,dx$ and is easily found in a table of integrals. Integration gives

$$Q_1{}^T = \left(\frac{2\pi m k_B T}{h^2}\right)^{3/2} V = \Lambda^{-3} V. \qquad (21.108)$$

What are the effects of internal motion? For simplicity we shall consider the case of a diatomic molecule and temporarily neglect possible electronic excitations (see below). In this case, to a good approximation we can write

$$\epsilon^{\text{int}} = (v + \tfrac{1}{2}) h\nu + J(J+1)\frac{h^2}{8\pi^2 I}, \qquad (21.109)$$

as deduced earlier (see Eqs. 7.8 and 7.19). In Eq. 21.109, ν is the frequency of vibration and I is the moment of inertia of the diatomic molecule. The lowest energy state $(v = 0, J = 0)$[11] has energy $h\nu/2$ relative to the point of lowest potential energy for a classical oscillator of frequency ν. This is an example of the consequences of the uncertainty principle, since the molecule cannot be at rest even in its ground state. Let us, as before, take the ground-state energy as the zero of energy. Then we write

$$\epsilon^{\text{int}} - \epsilon_0 = vh\nu + J(J+1)\frac{h^2}{8\pi^2 I}. \qquad (21.110)$$

Before proceeding, it is instructive to explain why we called Eq. 21.109 an approximation (although a very good one). All molecular vibrations are to some extent anharmonic, and all chemical bonds will stretch to some extent as the speed of rotation of the molecule increases. Thus the internal energy of a molecule contains small anharmonicity corrections [proportional to $(v + \tfrac{1}{2})^2$ and higher powers], small centrifugal distortion corrections [proportional to $J^2(J+1)^2$ and higher powers], and a small vibration-rotation interaction correction [whose leading term is proportional to $(v + \tfrac{1}{2}) \times J(J+1)$]. The latter correction arises as follows: If the bond between the two nuclei in a diatomic molecule stretches both because of centrifugal distortion and by virtue of vibrational excitation in an anharmonic potential, then as the speed of rotation is increased, more anharmonic regions of the interatomic potential are reached. It should be noted that for almost all molecules the corrections cited are small, of the order of magnitude of one part per hundred at about 500°C. Having noted their existence, which is important, we shall not need to refer to them any further in this context.

Given Eq. 21.110, we can divide $Q_1{}^{RV}$ into its rotational and vibrational components (recall that we are temporarily neglecting electronic excitation, so $Q_1^{RVE} \to Q_1^{RV}$):

$$Q_1{}^{RV} = \frac{1}{\sigma} \sum_{J=0}^{\infty} (2J+1) \exp\left(-\frac{J(J+1)h^2}{8\pi^2 I k_B T}\right) \sum_{v=0}^{\infty} \exp\left(-\frac{vh\nu}{k_B T}\right) \qquad (21.111)$$

$$\equiv Q_1{}^R Q_1{}^V. \qquad (21.112)$$

The factors $(1/\sigma)$ and $(2J+1)$ in $Q_1{}^R$ will be explained below.[12] We note here only that we have written the degeneracy $\bar{\omega}_s^{\text{int}}$ as the product $\bar{\omega}_s^{\text{int}} = (2J+1)/\sigma$.

---

[11] See Appendix 21B.
[12] See also Appendix 21B.

Since $vh\nu$ is always greater than zero, $\exp(-vh\nu/k_BT)$ is always less than unity, and $Q_1{}^V$ can be simplified as

$$Q_1{}^V \equiv \sum_{v=0}^{\infty} \exp\left(-\frac{vh\nu}{k_BT}\right)$$

$$= \sum_{v=0}^{\infty}\left[\exp\left(-\frac{h\nu}{k_BT}\right)\right]^v = \left[1-\exp\left(-\frac{h\nu}{k_BT}\right)\right]^{-1}. \quad (21.113)$$

To compute $Q_1{}^R$, we note that the degeneracy of a rotational state of quantum number J is $(2J+1)$, which has been entered in Eq. 21.111. But for a homonuclear diatomic molecule, such as I_2, Eq. 21.111 counts every distinguishable configuration twice, since a rotation of a homonuclear diatomic molecule by 180° brings the molecule back to a position indistinguishable from the starting position. This is corrected for by the factor σ, which has the value 1 for heteronuclear diatomic molecules, and 2 for homonuclear diatomic molecules.[13] Because for most molecules (but not H_2) the difference in energy between rotational energy levels is very small relative to the thermal energy k_BT, the argument leading to Eq. 21.106 gives here

$$Q_1{}^R \equiv \frac{1}{\sigma}\sum_{J=0}^{\infty}(2J+1)\exp\left(-\frac{J(J+1)h^2}{8\pi^2Ik_BT}\right)$$

$$= \frac{1}{\sigma}\int_0^{\infty}(2J+1)\exp\left(-\frac{J(J+1)h^2}{8\pi^2Ik_BT}\right)dJ$$

$$= \frac{8\pi^2Ik_BT}{\sigma h^2} \approx 0.0419\frac{IT}{\sigma}. \quad (21.114)$$

This approximation is not valid for the vibrational energy levels, but as seen above, $Q_1{}^V$ can be evaluated in closed form without the use of any approximation. The numerical form on the right-hand side of Eq. 21.114 is valid when I is given in amu Å^2 and T in K. When Eq. 21.114 is compared with the translational partition function, Eq. 21.108, we see that here the molecular mass is replaced by the moment of inertia, the volume is missing, and the exponent is different (1 instead of $\frac{3}{2}$). That I replaces m is a consequence of the mechanical fact that in rotational motion the moment of inertia plays the same role as does the mass in translational motion. The reason the V is missing is that the space in which rotation occurs is fixed by the center of rotation, and has nothing to do with the volume of real space. That is, rotation about the center of mass constrains the atoms to move on a spherical surface; the area of a sphere of unit radius is 4π. Finally, the difference between the exponents arises because translational space has three dimensions, whereas the rotational space of a diatomic molecules has only two dimensions (cf. later discussion of polyatomic molecules).

By combination of Eqs. 21.108, 21.113, and 21.114, and using the relation

$$\frac{A_N}{k_BT} = -\ln Q_N = -N\ln Q_1 - \ln N!, \quad (21.115)$$

(see Eq. 19.130), we find that

$$\frac{A_N}{k_BT} = -N\ln\left\{V\left(\frac{2\pi mk_BT}{h^2}\right)^{3/2}\left(\frac{8\pi^2Ik_BT}{\sigma h^2}\right)\left[1-\exp\left(-\frac{h\nu}{k_BT}\right)\right]^{-1}\right\} + \ln N!. \quad (21.116)$$

[13] See Appendix 21B.

But from the thermodynamic relation

$$p = -\left(\frac{\partial A}{\partial V}\right)_{T,N} \tag{21.117}$$

and Eq. 21.116, we find that

$$p = \frac{Nk_B T}{V}, \tag{21.118}$$

which is just the perfect gas equation of state. The derivation just given shows that, as long as there are no molecular interactions and quantum mechanical diffraction effects may be neglected,[14] the perfect gas equation of state describes equally well gases of monatomic and polyatomic molecules.

The thermodynamic functions of the perfect gas of diatomic molecules are easily obtained from Eq. 21.116. We reduce that equation by using Stirling's approximation,

$$\ln N! = N \ln N - N, \tag{21.119}$$

which is very accurate, and write

$$A_N = -Nk_B T \ln\left\{\frac{V}{N}\left(\frac{2\pi m k_B T}{h^2}\right)^{3/2}\left(\frac{8\pi^2 I k_B T}{\sigma h^2}\right)\left[1 - \exp\left(-\frac{h\nu}{k_B T}\right)\right]^{-1}\right\} - Nk_B T. \tag{21.120}$$

From the thermodynamic relation

$$S = -\left(\frac{\partial A}{\partial T}\right)_{V,N} \tag{21.121}$$

we find that

$$S_N = Nk_B\left\{\frac{7}{2} + \frac{h\nu}{k_B T}\cdot\frac{\exp(-h\nu/k_B T)}{1 - \exp(-h\nu/k_B T)}\right.$$
$$\left. + \ln\left[\frac{V}{N}\left(\frac{2\pi m k_B T}{h^2}\right)^{3/2}\left(\frac{8\pi^2 I k_B T}{\sigma h^2}\right)\right] + \ln\left[\frac{1}{1 - \exp(-h\nu/k_B T)}\right]\right\}, \tag{21.122}$$

whereas from the thermodynamic relations

$$U = A + TS \tag{21.123}$$

and

$$H = A + TS + pV \tag{21.124}$$

we find that

$$U_N = \tfrac{5}{2}Nk_B T + Nh\nu\frac{\exp(-h\nu/k_B T)}{1 - \exp(-h\nu/k_B T)} \tag{21.125}$$

and, with Eq. 21.118,

$$H_N = \tfrac{7}{2}Nk_B T + Nh\nu\frac{\exp(-h\nu/k_B T)}{1 - \exp(-h\nu/k_B T)}. \tag{21.126}$$

Before proceeding to a description of the other properties of a perfect gas in terms of the relationships derived, we consider how the analysis must be extended when electronic excitation of the molecules occurs, or when the molecules are polyatomic.

For most molecules the lowest electronic state is nondegenerate, and the next electronic state has an energy so much higher than the ground-state energy that it is not appreciably populated even at temperatures as high as 1000 K. It is easily verified that, if the first electronic excitation

[14] See Appendix 21A.

requires 1 eV, where $k_B T \approx 0.08$ eV at 1000 K, we find $\exp(-\Delta\epsilon/k_B T) \approx 10^{-6}$. Thus the electronic partition function may be taken to be unity with very small error. There are a few cases in which the first excited electronic state of the molecule is very close in energy to the ground electronic state. For example, the ground state of the F atom is of species $^2P_{3/2}$, and the other member of the doublet state, namely, $^2P_{1/2}$, is only 0.05 eV higher in energy. The next excited state is several electron volts higher in energy, and at all ordinary temperatures does not contribute to the thermodynamic properties of the gas.

As a result of the preceding remarks, we conclude that it is ordinarily sufficiently accurate to approximate the electronic partition function of a molecule by only the first two terms,

$$Q_1{}^E = \bar{\omega}_0{}^E + \bar{\omega}_1{}^E \exp\left(-\frac{\Delta\epsilon^E}{k_B T}\right), \qquad (21.127)$$

where $\bar{\omega}_0{}^E$ and $\bar{\omega}_1{}^E$ are the degeneracies of the ground (0) and first excited (1) electronic states, and $\Delta\epsilon^E$ is the excitation energy of the first electronic state. It should be noted, however, that the separation of the electronic partition function from the rotational and vibrational partition functions is not a good approximation if an excited electronic state is appreciably populated. Different electronic states have, in general, different internuclear separations and different vibration frequencies, so that the rotational and vibrational motion of a molecule is not independent of the electronic state of the molecule. Only when the population of the excited electronic state is small, or the excited electronic state (fortuitously) has the same geometry and vibration spectrum as the ground state, is the separation of the partition functions a good approximation.

Consider, now, a perfect gas of polyatomic molecules. Clearly, the only difference between this and a perfect gas of diatomic molecules arises from the greater number of internal degrees of freedom. That is, a polyatomic molecule has more internal vibrations and, for nonlinear molecules, an extra degree of rotational freedom. For a classically excited rigid rotator, that is, one for which the spacing between rotational energy levels is small relative to $k_B T$, it can be shown that

$$Q_1{}^R = \frac{8\pi^2(2\pi k_B T)^{3/2}}{\sigma h^3}(I_x I_y I_z)^{1/2}, \qquad (21.128)$$

where I_x, I_y, I_z are the principal moments of inertia (see Section 3.12) and σ is the number of equivalent positions of the polyatomic molecule (*symmetry number*). How does Eq. 21.128 differ from Eq. 21.114? Actually, all linear molecules do have a very small moment of inertia about the internuclear axis due to the motion of the electrons (see Section 7.2). For linear molecules $I_x = I_y$, whereas I_z is very small. But the spacing of rotational energy levels is always inversely proportional to the moment of inertia, so it is clear that excitation of rotation about the internuclear axis (corresponding to I_z) is very difficult. Consequently, because the energy required for excitation is so large, only the ground state of rotation about the internuclear axis is of thermodynamic importance.

When all three moments of inertia are different, there is no simple relationship for the rotational energy levels, although they have been worked out in detail for some molecules. For almost all gases the rotational energy spacing is sufficiently small that Eq. 21.128 is valid, so that these complications need not concern us here. It suffices for our purposes to note that they exist and must be taken into account when low-temperature phenomena are discussed.

There is another kind of molecular motion that can be considered a rotation, viz., internal rotation of one part of a molecule relative to some other part. A molecule exhibiting this kind of motion is ethane, C_2H_6. In general, this motion is not free (see Section 9.4). That is, since the energies of the molecule corresponding to non-equivalent positions are somewhat different, the energy of the molecule must change as the angle of internal rotation changes. In passing from one position of minimum potential energy to another, the molecule must surmount some sort of energy barrier because, as can be seen in Fig. 21.11, this is implicit in the notion of energy minimum. When the energy barrier is very small and the energy-level spacing is small relative to $k_B T$, the internal rotation is nearly free, and it is found that

$$Q_1^{IR} = \frac{(8\pi^3 I_{IR} k_B T)^{1/2}}{\sigma_{IR} h},$$

$$(21.129)$$

$$I_{IR} = \frac{I_1 I_2}{I_1 + I_2},$$

where I_1 and I_2 are the moments of inertia about the bond axis of the two counterrotating parts of the molecule. As before, σ_{IR} is the symmetry number, here for the internal rotation. Where there are several bonds around which there can be classical free rotation, there is a contribution to Q_1 of the form of Eq. 21.129 for each of the bonds. On the other hand, if the barrier to rotation is very large, the separate parts of the molecule oscillate relative to one another in a torsional motion. This is very like a vibration, and can be treated as such. For intermediate barrier heights no general results can be stated—direct calculations are needed.

Finally, a polyatomic molecule has many modes of vibrational motion. Provided that the restoring force acting on any atom of the molecule obeys Hooke's law, the total vibrational motion can be separated into a sum of motions along normal coordinates. The number of normal coordinates is equal to the number of vibrational degrees of freedom, and motion along the normal coordinates is harmonic. If ν_i is the frequency of the ith normal mode of motion, the vibrational partition function is just

$$Q_1^V = \prod_{i=1}^{3n-6} \left[1 - \exp\left(-\frac{h\nu_i}{k_B T}\right) \right]^{-1},$$

$$(21.130)$$

because:

1. By resolving the motion of the atoms into a sum of motions along independent normal coordinates, we can express the vibrational energy as a sum of the independent energies of motion along the normal coordinates;

2. For an n-atom molecule, using three coordinates to specify each atom's position and subtracting three coordinates for center-of-mass translational motion and three for rotational motion (nonlinear molecule) leaves $3n-6$ vibrational degrees of freedom (if there is not internal rotation). Thus, the complete partition function for an isolated polyatomic molecule is

$$Q_1 = Q_1^T Q_1^R Q_1^E Q_1^{IR} Q_1^V,$$

$$(21.131)$$

provided that the energies for the separate motions are, to an adequate approximation, independent of one another.

Given the results of the calculations of the last section, it is easy to compute the heat capacity of a perfect gas and to analyze the contributions to it from translational motion, rotational motion, vibrational mo-

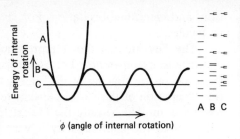

FIGURE 21.11
Potential functions (on the left) and energy levels (on the right) for a restricted rotator B, a harmonic oscillator A, and a free rotator C. When the barrier is large relative to $k_B T$, a parabola can be fitted to the bottom of the potential curve and the internal motion becomes a torsional vibration. When the barrier is small relative to $k_B T$, the internal motion becomes a free rotation. From K. S. Pitzer, *Quantum Chemistry* (Prentice-Hall, Englewood Cliffs, N.J., 1953).

B. The Heat Capacity of a Perfect Gas of Diatomic Molecules

tion, and electronic excitation. For simplicity we consider only diatomic molecules.

The key idea in the interpretation of the heat capacity is that of *characteristic temperature*. For each class of molecular motion we consider the ratio of the spacing of the energy levels to $k_B T$. If the ratio is large, the occupation of the various energy levels corresponding to that class of motion is limited to only the lowest levels, and the corresponding contribution to the heat capacity is very small; if the ratio is small, many levels are occupied and the corresponding contribution to the heat capacity reaches a classical limiting value, the equipartition value. If a characteristic temperature is defined by $\Theta \equiv \Delta \epsilon / k_B$, the former case corresponds to $\Theta / T \gg 1$ and the latter to $\Theta / T \ll 1$.

Using the thermodynamic definition of C_V and Eqs. 21.102 and 21.115,

$$C_V \equiv \left(\frac{\partial U}{\partial T}\right)_{V,N} = T\left(\frac{\partial S}{\partial T}\right)_{V,N} = -T\left(\frac{\partial^2 A}{\partial T^2}\right)_{V,N}$$

$$= Nk_B T\left[\frac{\partial^2}{\partial T^2}(T \ln Q_1)\right]_{V,N}, \qquad (21.132)$$

we can compute the contributions to the heat capacity from translational motion, rotational motion, vibrational motion, and electronic excitation, C_V^T, C_V^R, C_V^V, and C_V^E, respectively. Of course, $C_V = C_V^T + C_V^R + C_V^V + C_V^E$.

1. *Translational heat capacity.* For all gases the spacing of translational energy levels is small relative to $k_B T$, and the heat capacities are

$$C_V^T = \tfrac{3}{2}Nk_B, \qquad C_p^T = \tfrac{5}{2}Nk_B. \qquad (21.133)$$

As can be seen from Table 21.7, there is excellent agreement between theory and experiment for the monatomic gases. Equation 21.133 implies that the contribution to the internal energy and enthalpy of a perfect gas from translational motion is linear in the temperature (Fig. 21.12). We shall later see that such behavior is generally characteristic of the case when the spacing of the energy levels is small relative to the thermal energy $k_B T$, and is not restricted to translational motion.

2. *Rotational heat capacity.* Let

$$\Theta_{\text{rot}} \equiv \frac{h^2}{8\pi^2 I k_B} \qquad (21.134)$$

be called the *rotational temperature*. (The reader should compare Eq. 21.134 with the exponent in Eq. 21.111.) Now, Θ_{rot} is small relative to T for most molecules of interest (see Table 21.8). Then the rotational partition function is accurately given by Eq. 21.114, and the corresponding contribution to the heat capacity is

$$C_V^R = Nk_B. \qquad (21.135)$$

At sufficiently low temperatures ($\Theta_{\text{rot}} \lesssim 1.5$) the sum in the first line of Eq. 21.114 can no longer be approximated by an integral, and Eqs. 21.135 no

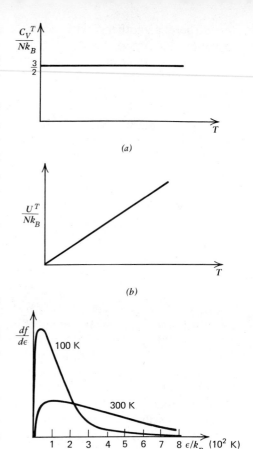

FIGURE 21.12

(a) Translational motion contribution to the heat capacity at constant volume. See Appendix 21A for a description of how C_V is modified near $T = 0$ when the spacing of the translational energy levels becomes comparable to and then smaller than $k_B T$.

(b) Translational motion contribution to the internal energy. The slope of this curve is $\tfrac{3}{2}$.

(c) Molecular translational energy distribution at two temperatures. To illustrate the relationship between the peak position and the energy scale, the latter is expressed as a temperature,

$$\frac{df_{\text{trans}}}{d\epsilon} = \frac{2\pi N}{(\pi k_B T)^{3/2}} \epsilon^{1/2} \exp\left(\frac{-\epsilon}{k_B T}\right).$$

TABLE 21.7
HEAT CAPACITIES OF MONATOMIC GASES AT 298.15 K

Substance	c_V (calc) (J/K mol)	c_V (obs) (J/K mol)	c_p (calc) (J/K mol)	c_p (obs) (J/K mol)
He	12.472	12.519	20.786	20.782
Ar	12.472	12.544	20.786	20.920

longer hold. Q_1^R and C_V^R must then be calculated numerically, the results for diatomic (or linear polyatomic) molecules being shown in Fig. 21.13. Only in the case of H_2 and its isotopic modifications, however, does this behavior occur in the normal gaseous range.

For H_2, D_2, and T_2 the situation is further complicated by the quantum mechanical symmetry requirements applicable to homonuclear molecules.[15] As described in Section 6.7, the wave function for a system of two identical particles must be either symmetric or antisymmetric to exchange of the particles, depending on whether the particles have integral or half-integral spin, respectively. Since the proton has spin $\frac{1}{2}$, the wave function for H_2 must be antisymmetric to exchange of the nuclei. The translational wave function involves only the molecule's center of mass, and at ordinary temperatures only the ground states for electronic and vibrational motion are occupied, and the corresponding wave functions are symmetric. For the total wave function, which is the product of the various contributions, to be antisymmetric, then, either the rotational or the nuclear spin wave function must be antisymmetric.

As described in Section 3.12, the rotational wave function of a symmetric rotor like H_2 is what is called a spherical harmonic. Since we did not examine these in detail, we must state without proof that the spherical harmonics are symmetric for even values of the rotational quantum number J and antisymmetric for odd J with respect to inversion of the coordinates; $x \rightarrow -x$, $y \rightarrow -y$, $z \rightarrow -z$.[16] Even and odd values of J must thus be associated only with antisymmetric and symmetric nuclear spin wave functions, respectively. In the absence of a catalyst such as activated charcoal, there are virtually no transitions between the symmetric and antisymmetric spin states, so that H_2 is effectively a mixture of two stable species. These are called *orthohydrogen* (symmetric spin, J odd) and *parahydrogen* (antisymmetric spin, J even), abbreviated o-H_2 and p-H_2.

The spacing between nuclear spin energy levels is very small relative to $k_B T$ in a gas, so these states may be regarded as degenerate. If a nucleus has $\bar{\omega}$ degenerate spin states, the homonuclear molecule must have $\bar{\omega}^2$, of which it can be shown that $\bar{\omega}(\bar{\omega}+1)/2$ are symmetric and $\bar{\omega}(\bar{\omega}-1)/2$ are antisymmetric. For H_2 we have $\bar{\omega} = 2$ (spin $+\frac{1}{2}$ and $-\frac{1}{2}$), so that $\bar{\omega}_s = 3$, $\bar{\omega}_a = 1$. The rotational partition function is thus

$$Q_1^R = \sum_{J \text{ even}} (2J+1) \exp\left(-\frac{J(J+1)\Theta_{rot}}{T}\right)$$
$$+ 3 \sum_{J \text{ odd}} (2J+1) \exp\left(-\frac{J(J+1)\Theta_{rot}}{T}\right), \quad (21.136)$$

where $\Theta_{rot} \equiv h^2/8\pi^2 I k_B$; this equation applies to an equilibrium mixture (e-H_2) of the para and ortho molecules, in proportions given by the two terms displayed, and yields the curve labeled e-H_2 in Fig. 21.14. As explained above, however, this equilibrium does not exist unless catalyzed, and H_2 prepared at room temperature or higher, where the ortho–para ratio reaches a limiting value of 3, will retain this composition (normal or n-H_2) at lower temperatures. The heat capacity of such a

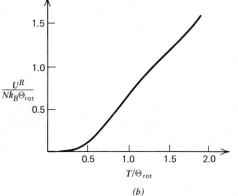

(a)

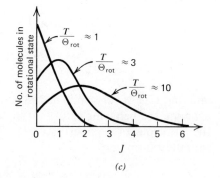

(c)

FIGURE 21.13

(a) Rotational motion contribution to the heat capacity at constant volume.

(b) Rotational motion contribution to the internal energy.

(c) Molecular rotational population distribution,

$$f_{rot} = \frac{N}{Q_1^R}(2J+1) \exp\left(\frac{-J(J+1)\Theta_{rot}}{T}\right).$$

[15] See Appendix 21B.

[16] This can be seen for the examples given in Table 3.1 as follows: Inversion takes θ into $\pi - \theta$ and ϕ into $\pi + \phi$, or $\sin \theta \rightarrow \sin \theta$, $\cos \theta \rightarrow -\cos \theta$, and $\exp(im\phi) \rightarrow (-1)^m \exp(im\phi)$. A function $Y_{J,m}(\theta, \phi)$ has even powers of $\cos \theta$ multiplying even powers of $\exp(i\phi)$ if J is even, and even powers of $\cos \theta$ multiplying odd powers of $\exp(i\phi)$ if J is odd. The symmetric or antisymmetric character of the harmonics $Y_{J,m}(\theta, \phi)$ follows immediately.

mixture is given by

$$C_V^R(n\text{-}H_2) = \tfrac{3}{4}C_V^R(o\text{-}H_2) + \tfrac{1}{4}C_V^R(p\text{-}H_2); \tag{21.137}$$

the ortho and para heat capacities are calculated from the two terms in Eq. 21.136 separately. This result, as shown in Fig. 21.14, does agree with experiment.

Behavior similar to that just described is found for D_2 and T_2. For all other homonuclear diatomic molecules, however, the rotational energy-level spacing is so much smaller than the gas-phase values of $k_B T$ that both sums in the equivalent of Eq. 21.136 can be replaced by integrals. Since the resulting integrals are identical, the differences between ortho and para states vanish under these conditions. The trend toward this high-temperature limit is apparent in Fig. 21.14. Because of the alternation of terms in the sums, the integrals equal one-half the integral in the second line of Eq. 21.114, this being the quantum mechanical explanation of the symmetry factor $\sigma = 2$.

3. *Vibrational heat capacity.* Let

$$\Theta_{\text{vib}} \equiv \frac{h\nu}{k_B} = \frac{hc\tilde{\nu}_e}{k_B} \tag{21.138}$$

be called the *vibrational temperature* of the molecule. For essentially all molecules at ordinary temperatures, $\Theta_{\text{vib}} \gg T$ (see Table 21.8). We then find that

$$C_V^V = Nk_B \left(\frac{h\nu}{k_B T}\right)^2 \frac{\exp(-h\nu/k_B T)}{[1-\exp(-h\nu/k_B T)]^2}$$

$$= Nk_B \left(\frac{\Theta_{\text{vib}}}{T}\right)^2 \frac{\exp(-\Theta_{\text{vib}}/T)}{[1-\exp(-\Theta_{\text{vib}}/T)]^2}, \tag{21.139}$$

and that C_V^V is very small at 298 K for most molecules. It is interesting to examine the behavior of C_V^V at high temperature as another example of the principle of equipartition. As T increases, the heat capacity rises smoothly from zero to

$$\lim_{T\to\infty} C_V^V = \lim_{T\to\infty} Nk_B \left(\frac{\Theta_{\text{vib}}}{T}\right)^2 \frac{\exp(-\Theta_{\text{vib}}/T)}{[1-\exp(-\Theta_{\text{vib}}/T)]^2}$$

$$= Nk_B \left(\frac{\Theta_{\text{vib}}}{T}\right)^2 \frac{1}{(\Theta_{\text{vib}}/T)^2}$$

$$= Nk_B. \tag{21.140}$$

It is interesting that, as shown by a numerical calculation, even when the vibrational heat capacity approaches the value Nk_B per vibration, most of the molecules are still in the $v=0$ and $v=1$ vibrational levels. It is not necessary that the molecules be widely distributed over the available energy levels before the limiting value of C_V^V is achieved. The approach to the limit is shown in Fig. 21.15.

4. *Electronic heat capacity.* Let

$$\Theta_{\text{elec}} \equiv \frac{\epsilon_1 - \epsilon_0}{k_B} = \frac{\Delta\epsilon^{\text{elec}}}{k_B} \tag{21.141}$$

be called the *electronic temperature* of the molecule. As seen from the data in Table 21.8, $\Theta_{\text{elec}} \gg T$ for almost all molecules. For only a few gases is the electronic heat capacity of any consequence, and even for these only two

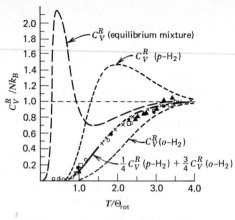

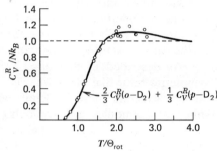

FIGURE 21.14
Rotational contribution to the constant volume molar heat capacity of H_2 and D_2. For H_2 the species with odd values of J is called ortho, but for D_2 it is the species with even values of J that is called ortho. The measured values of C_V show good agreement with the calculated values.

TABLE 21.8
VIBRATIONAL AND ROTATIONAL CONSTANTS OF SOME DIATOMIC MOLECULES

Substance	Electronic Ground State	$\bar{\nu}_e$ (cm^{-1})	r_e (Å)	B_e^a (cm^{-1})	$\dfrac{\epsilon_1-\epsilon_0{}^b}{hc}$ (cm^{-1})	$\Theta_{vib}{}^c$ (K)	$\Theta_{rot}{}^d$ (K)	$\Theta_{elec}{}^e$ (K)
H$_2$	$^1\Sigma_g{}^+$	4395.2	0.7417	60.809	91,690	6320	87.5	129,000
HD	$^1\Sigma_g{}^+$	3817.1	0.7414	45.655	91,711	5500	65.8	129,000
D$_2$	$^1\Sigma_g{}^+$	3118.5	0.7416	30.429	91,698	4490	43.8	129,000
HF	$^1\Sigma^+$	4138.5	0.9171	20.939		5990	31.5	
HCl	$^1\Sigma^+$	2989.7	1.2746	10.591	44,000	4330	15.2	61,800
HBr	$^1\Sigma^+$	2649.7	1.4138	8.473	35,000	3820	12.2	49,200
HI	$^1\Sigma^+$	2309.5	1.6041	6.551	27,500	3340	9.4	38,600
N$_2$	$^1\Sigma_g{}^+$	2359.6	1.094	2.010	69,290	3390	2.89	97,400
CO	$^1\Sigma^+$	2170.2	1.1282	1.9314	65,075	3120	2.78	91,500
NO	$^2\Pi_{1/2}$	1904.0	1.1508	1.7046	43,966	2745	2.45	61,700
O$_2$	$^3\Sigma_g{}^-$	1580.4	1.2074	1.4457	7,918	2278	2.08	11,100
F$_2$	$^1\Sigma_g{}^+$	892.1	1.435		34,500	1290		48,500
Cl$_2$	$^1\Sigma_g{}^+$	564.9	1.988	0.2438	18,311	814	0.35	25,700
Br$_2$	$^1\Sigma_g{}^+$	323.2	2.2836	0.0809	13,814	465	0.12	19,400
I$_2$	$^1\Sigma_g{}^+$	214.6	2.6666	0.0374	11,888	309	0.05	16,700
BeO	$^1\Sigma^+$	1487.3	1.3308	1.6510	9,406	2150	2.38	13,200
C$_2$	$^1\Sigma_g{}^+$	1855	1.2425	1.8195	717	2678	2.62	1006
LiH	$^1\Sigma^+$	1405.6	1.5953	7.5131	26,516	2035	10.8	37,300
SO	$^3\Sigma^-$	1123.7	1.4933	0.7089	39,356	1625	1.02	55,200
ZnH	$^2\Sigma^+$	1607.6	1.5945	6.6794	23,270	2320	9.60	32,500

a $B_e \equiv h/8\pi^2 I_e c$.
b Energy of first electronic excited state relative to the ground state.

c $\Theta_{vib} \equiv \dfrac{hc\bar{\nu}_e}{k_B}$.

d $\Theta_{rot} \equiv \dfrac{h^2}{8\pi^2 I_e k_B}$.

e $\Theta_{elec} \equiv \dfrac{\epsilon_1-\epsilon_0}{k_B}$.

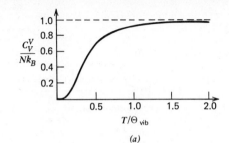

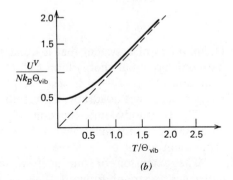

FIGURE 21.15

(a) Vibrational contribution to the constant volume molar heat capacity.

(b) Vibrational contribution to the internal energy.

(c) Molecular vibrational population distribution,

$$f_{vib} = \frac{N}{Q_1{}^V}\exp\left(\frac{-\Theta_{vib}}{T}\right).$$

levels are of importance. From the partition function 21.127, we find

$$C_V^E = Nk_B\frac{\bar{\omega}_0{}^E\bar{\omega}_1{}^E\exp(-\Delta\epsilon^{elec}/k_BT)(\Delta\epsilon^{elec}/k_BT)^2}{[(\bar{\omega}_0{}^E+\bar{\omega}_1{}^E\exp(-\Delta\epsilon^{elec}/k_BT)]^2}. \quad (21.142)$$

If the next excited electronic state lies at very much higher energy, $\epsilon_2 \gg \epsilon_1$, C_V^E ultimately tends to zero as T increases indefinitely (Fig. 21.16). The "few-quantum-states case" is the exception rather than the rule, and even in the example of electronic excitation the limitation to a few states is only apparent—there are other states at high energy, and C_V^E eventually rises again when k_BT becomes comparable to their excitation energies.

Examination of each of the contributions to the heat capacity we have studied shows that when the relevant energy spacing becomes small compared with the thermal energy k_BT, so that the corresponding characteristic temperature is small relative to T, and when there is an effectively infinite set of quantum states available to the system, each degree of freedom of the system contributes just $Nk_B/2$ to the heat capacity (Fig. 21.17). In the classical theory of statistical mechanics this is known as the *equipartition of energy*. It is found that, in general, for every term in the

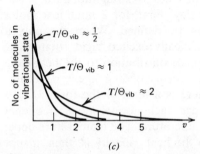

FIGURE 21.16

Electronic contribution to the molar heat capacity of a gas in the case that only one electronic state (labeled 1) is available.

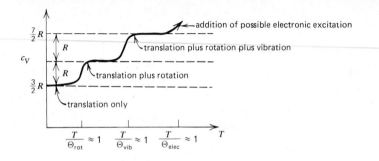

addition of possible electronic excitation

translation plus rotation plus vibration

translation plus rotation

translation only

$\frac{T}{\Theta_{rot}} \approx 1$ $\frac{T}{\Theta_{vib}} \approx 1$ $\frac{T}{\Theta_{elec}} \approx 1$

FIGURE 21.17
Schematic diagram of the variation of c_v with temperature for a perfect gas of diatomic molecules. Note that each increment to c_v becomes important at about its characteristic temperature.

Hamiltonian of a system that is quadratic (e.g., a kinetic energy term, a harmonic potential energy term kx^2, etc.), the limiting contribution to the heat capacity as $T \rightarrow \infty$ is just $Nk_B/2$. When the Hamiltonian contains terms that are not quadratic, the resulting form of C_V is quite different.[17]

The thermodynamic functions of a perfect gas may be easily computed using the formulas shown. Some comparisons between theory and experiment are given in Table 21.9.

The extension of our analysis to the description of polyatomic molecules is straightforward. Two of the resulting changes are worth noting. First, for a nonlinear molecule a temperature analogous to Θ_{rot} can be defined. When the polyatomic molecule can be treated as a classically excited rigid rotator, we find, using 21.128, that the rotational contribution to the heat capacity is $\frac{3}{2}Nk_B$, which is larger than for the corresponding linear molecule. Second, for a polyatomic molecule there is a term of the form 21.139 for each normal mode of vibration. Since each normal-mode frequency is different, so is its characteristic temperature. The temperature dependence of the vibrational contribution to the heat capacity is, then, a superposition of the temperature dependences of the contributions of all the normal vibrations, each with a different value of Θ_{vib}/T.

Given the thermodynamic formulation of the theory of chemical equilibrium in gaseous mixtures, together with the formulas relating the thermodynamic functions of a gas to its molecular properties, it is clear that *a priori* calculations of chemical equilibrium from molecular proper-

21.6
The Statistical Molecular Theory of the Equilibrium Constant

TABLE 21.9
A COMPARISON OF COMPUTED AND CALORIMETRIC ENTROPIES OF SOME GASES

Substance	s^0_{298} (calc) (J/K mol)	s^0_{298} (obs) (J/K mol)
Ne(g)	146.22	146.
Ar(g)	154.74	155.
HCl(g)	186.68	186.
HBr(g)	198.48	199.
Cl₂(g)	222.95	223.1
CO₂(g)	213.64	214.
CH₃Br(g)	243.	242.

[17] See, for example, A. Peterlin, *Am. J. Phys.* **28,** 716 (1960).

ties are feasible in principle. If the gas mixture is imperfect, this program also requires a statistical molecular theory of the deviation from perfect gas behavior, for example, a theory of the second and higher virial coefficients. The theory of gas imperfection is discussed in the next section; for the present we consider equilibrium in a perfect gas mixture.

Starting from Eq. 21.115 and using the thermodynamic definition of the chemical potential per molecule,

$$\mu \equiv \left(\frac{\partial A}{\partial N}\right)_{T,V}, \tag{21.143}$$

we easily find that

$$\mu = -k_B T \ln\left(\frac{Q_1{}^T Q_1{}^R Q_1{}^V Q_1{}^E}{N}\right)$$

$$= -k_B T \ln\left(\frac{Q_1{}^T Q_1^{\text{int}}}{N}\right). \tag{21.144}$$

In all of the preceding discussion of the computation of the partition function of a molecule we have used as the zero of the energy the lowest accessible quantum state of the molecule. Let the energy of that state, relative to the energy of the molecule when dissociated into atoms that are infinitely far apart, be ϵ_0. Clearly, the constant ϵ_0 differs from one species to another. Now, the partition function weight factor $\exp(-\epsilon_j/kT)$ can be evaluated for any choice of the zero of energy. To describe chemical equilibrium between reacting gases we must choose a common zero of energy for all reactant molecules. If we simply write

$$\epsilon_j = (\epsilon_j - \epsilon_0) + \epsilon_0 \tag{21.145}$$

and substitute Eq. 21.145 into the definition of the partition function, we find that

$$Q_1 = \sum_j \bar{\omega}_j \exp\left(-\frac{\epsilon_0}{k_B T}\right) \exp\left(-\frac{\epsilon_j - \epsilon_0}{k_B T}\right)$$

$$= \exp\left(-\frac{\epsilon_0}{k_B T}\right) \sum_j \bar{\omega}_j \exp\left(-\frac{\epsilon_j - \epsilon_0}{k_B T}\right). \tag{21.146}$$

Since ϵ_0 is measured relative to the dissociation limit, the partition functions for all molecules can be referred to a common zero of energy simply by using the lowest accessible quantum state as the internal zero for each class of motion, and then multiplying by $\exp(-\epsilon_0/k_B T)$. For most molecules the only contributions to ϵ_0 are from zero-point vibrational motion and the total bond energy. There are a few exceptions to this generalization. For example, if o-H_2 is involved in a reaction at temperatures comparable to Θ_{rot}, we must note that its lowest rotational state has $J = 1$, whereas the lowest rotational state of most other molecules may be taken as $J = 0$ with the symmetry number σ used to correct the counting of states.

All of the partition functions studied in Section 21.5 were calculated using the lowest quantum state as reference for the energy. Therefore, we need merely multiply the product $Q_1{}^T Q_1{}^R Q_1{}^V Q_1{}^E$ by $\exp(-\epsilon_0/k_B T)$, and return to Eq. 21.144, to find that

$$\mu - \epsilon_0 = -k_B T \ln\left(\frac{Q_1{}^T Q_1^{\text{int}}}{N}\right), \tag{21.147}$$

when calculated for each gas, will refer to a common zero of energy.

We now convert Eq. 21.147 to a form resembling the thermodynamic representation of the chemical potential. Let

$$Q_1^T = \left(\frac{2\pi m k_B T}{h^2}\right)^{3/2} V = q_1^T V,$$

$$q_1^T \equiv \left(\frac{2\pi m k_B T}{h^2}\right)^{3/2} \equiv \Lambda^{-3}, \tag{21.148}$$

and use Eq. 21.118 to replace V/N by $k_B T/p$. Then, multiplying the argument of the logarithm by unity in the form p/p, where p is the unit of pressure, and separating the terms of the argument, we find that

$$\mu = \epsilon_0 - k_B T \ln\left(q_1^T Q_1^{\text{int}} \frac{k_B T}{p}\right) + k_B T \ln \frac{p}{p}$$

$$= \mu^*(T, p) + k_B T \ln \frac{p}{p}, \tag{21.149}$$

where, of course,

$$\mu^*(T, p) = \epsilon_0 - k_B T \ln\left(q_1^T Q_1^{\text{int}} \frac{k_B T}{p}\right). \tag{21.150}$$

The notation used here is the same as in the thermodynamic theory; μ^* is the chemical potential (Gibbs free energy per molecule) in a perfect gas state at unit pressure (1 atm). Using the condition defining chemical equilibrium, Eq. 21.63, we find that the equilibrium constant is

$$K(T) = \prod_{i=1}^{r} \left(q_{1i}^T Q_{1i}^{\text{int}} \frac{k_B T}{p}\right)^{\nu_i} \exp\left(-\frac{\Delta E_0^0}{k_B T}\right), \tag{21.151}$$

where

$$\Delta E_0^0 \equiv \sum_{i=1}^{r} \nu_i \epsilon_{0i} \tag{21.152}$$

is the net change in energy in the conversion of reactants to products, assuming the components to be perfect gases at $T = 0$ K. Of course, when $T = 0$ we also have $\Delta E_0^0 = \Delta H_0^0$, since for a perfect gas H and U differ by just $Nk_B T$. The energy change ΔE_0^0 may easily be related to the dissociation energies of the reacting molecules. We imagine the reaction to occur in two steps:

1. Let all the reactant molecules be dissociated into isolated atoms.

2. Let the atoms be reassembled into the product molecules.

If D_{0i} is the energy required to dissociate into atoms a molecule of the ith species in its lowest quantum state, the sum of the energy changes in steps 1 and 2 is

$$\Delta E_0^0 = -\sum_{i=1}^{r} \nu_i D_{0i}. \tag{21.153}$$

Note again that D_{0i} excludes the zero-point energy of the molecule, and thus does not equal D_{ei}, the dissociation energy measured from the minimum in the potential energy (see Section 7.1).

The form of $\mu^*(T, p)$ in Eq. 21.150 permits easy conversion from one set of units to another. It is written there for the usual choice of standard state, namely, the perfect gas at 1 atm. Then the factor V/N in Eq. 21.147 (the V from Q_1^T, the N already in the denominator) is just $k_B T/p$, as shown in Eq. 21.150. If we choose a standard state with gas concentration of 1 molecule/cm³, the factor V/N becomes 1 cm³; if, instead, we choose a standard state with gas concentration of 1 mol/liter, the factor V/N becomes $(1000\text{ cm}^3/N_A)$. The different standard states are, of course, related to equilibrium constants derived from the standard form $K(T)$. For example, K_V, defined in terms of the concentrations in moles per liter of all the reacting species, is

$$K_V^{\text{perfect gas}} = \exp\left(-\frac{\Delta E_0^{\ 0}}{k_B T}\right) \prod_{i=1}^{r} (q_{1i}{}^T Q_{1i}^{\text{int}}/N_A)^{\nu_i}. \qquad (21.154)$$

Other forms of the equilibrium constant may be easily worked out.

Consider, as an example, the reaction

$$\text{MX(g)} = \text{M}^+(\text{g}) + \text{X}^-(\text{g}). \qquad (21.155)$$

The equilibrium constant for this reaction is, from Eq. 21.151,

$$K(T) = \exp\left(-\frac{\Delta E_0^{\ 0}}{k_B T}\right)$$
$$\times \frac{(2\pi m_{\text{M}^+} k_B T/h^2)^{3/2}(k_B T/\text{p})Q_{\text{M}^+}^{\text{int}} (2\pi m_{\text{X}} k_B T/h^2)^{3/2}(k_B T/\text{p})Q_{\text{X}}^{\text{int}}}{(2\pi m_{\text{MX}} k_B T/h^2)^{3/2}(k_B T/\text{p})Q_{\text{MX}}^{\text{int}}}. \qquad (21.156)$$

Three examples of alkali halides illustrate how the various factors of Eq. 21.156 govern the position of equilibrium. Consider LiF, NaCl, and CsI at 3000 K and at 4000 K. The necessary atomic and molecular constants are collected in Table 21.10. The equilibrium constant we shall compute is that for $\text{MX} \rightleftharpoons \text{M}^+ + \text{X}^-$, so the energy $\Delta E_0^{\ 0}$ of Eq. 21.153 will represent the dissociation of MX to ions. The zero of energy, the reference state for the dissociation energy of MX, is the ground state of the free atoms, so the partition function for the alkali and halide ions must contain Boltzmann factors to reflect the work done to convert M^0 to $\text{M}^+ + \text{e}^-$, and the work released when $\text{X}^0 + \text{e}^-$ goes to X^-. The ions are all in nondegenerate ^1S states, and the electronically excited states of both the ions and the molecules can be neglected because they contribute

TABLE 21.10
DATA TO COMPUTE EQUILIBRIUM CONSTANTS FOR LiF, NaCl, AND CsI

Alkali Atom	Ionization Potential (eV)	
Li	5.390	
Na	5.138	
Cs	3.893	

Halogen Atom	Electron Affinity (eV)	Spin–Orbit Splitting of X^0
F	3.400	404 cm^{-1} equiv. to 581 K
Cl	3.613	881 cm^{-1} equiv. to 1207 K
I	3.081	7603 cm^{-1} equiv. to 10,936 K

Molecule	Moment of Inertia (amu Å^2)	Vibration Frequency (cm^{-1})	Dissociation Energy D_e (eV)
LiF	11.66	910.2	5.96
NaCl	77.33	364.6	4.23
CsI	713.7	119.2	3.55

factors of unity (within the uncertainties of the molecular parameters) to the partition functions. Hence $Q_{M^+}^{int}$ and $Q_{X^-}^{int}$ are simple exponentials. Assume that the interactions between ions in the gas can be neglected at the temperatures of interest.

Let us take the molecules to be rigid rotors and harmonic oscillators, so we can use Eqs. 21.114 and 21.113 for the rotational and vibrational partition functions, respectively. We must note whether the dissociation energy is D_0, the difference in energy of the vibrationless ground states of the molecule and the free atoms at rest, or D_e, the difference in energy between the bottom of the adiabatic potential energy curve and the dissociation limit; D_0 is smaller than D_e by the zero-point energy, $h\nu/2$, of the harmonic oscillator. The values in Table 21.10 are D_e's, not D_0's, so we must multiply Q_1^V (Eq. 21.113) by $\exp(-h\nu/2k_BT)$ or, equivalently, decrease the dissociation energies by $h\nu/2k_BT$. Table 21.11 shows the electronic factors

$$\exp(-\Delta E_0{}^0/k_BT) = \exp\{[D_e(MX) + IP(M) - EA(X)]/k_BT\},$$

the translational factors $[2\pi m_{M^+} m_{X^-} k_B T/(m_{M^+} + m_{X^-})h^2]^{3/2} k_B T/\mathfrak{p}$, which we abbreviate $\bar{Q}_1{}^T$, the rotational factors $Q_1{}^R$, and the vibrational factors $Q_1{}^V$, as well as the equilibrium constants. Note that 1 atm is equivalent to 1.01325×10^5 N/m^2 or 1.01325×10^6 dyn/cm^2, which is the value we take for our reference pressure $\mathfrak{p}$.

In Table 21.11, note first the magnitudes of the four factors. The electronic term is very small and very sensitive to temperature because $\Delta E_0{}^0$ is so large. The translational factors taken together as $\bar{Q}_1{}^T$ give a large and only moderately temperature-sensitive, mass-sensitive number. The rotational partition function, five orders of magnitude smaller, is less sensitive to masses and temperature, and the vibrational partition function is of order unity and the least sensitive term of all. One can look upon the magnitude of $\bar{Q}_1{}^T$ as the quantitative expression of how it is the

TABLE 21.11

TYPICAL EQUILIBRIUM CONSTANTS $K(T) = \exp(-\Delta E_0{}^0/k_BT)$ $\bar{Q}_1{}^T/Q_1{}^R Q_1{}^V$ FOR THREE ALKALI HALIDES

Molecule		3000 K	4000 K
LiF	$\exp(-\Delta E_0{}^0/k_BT)$	4.41×10^{-14}	9.82×10^{-11}
	$\bar{Q}_1{}^T$	1.45×10^8	2.97×10^8
	$Q_1{}^R$	1.47×10^3	1.95×10^3
	$Q_1{}^V$	2.272	3.04
	$K(T)$	1.915×10^{-9}	4.82×10^{-6}
NaCl	$\exp(-\Delta E_0{}^0/k_BT)$	2.15×10^{-10}	5.60×10^{-8}
	$\bar{Q}_1{}^T$	6.58×10^8	1.35×10^9
	$Q_1{}^R$	9.73×10^3	1.297×10^4
	$Q_1{}^V$	5.71	7.62
	$K(T)$	2.54×10^{-6}	7.65×10^{-4}
CsI	$\exp(-\Delta E_0{}^0/k_BT)$	4.35×10^{-8}	3.01×10^{-6}
	$\bar{Q}_1{}^T$	6.605×10^9	1.356×10^{10}
	$Q_1{}^R$	8.97×10^4	1.196×10^5
	$Q_1{}^V$	17.5	23.3
	$K(T)$	1.83×10^{-4}	1.465×10^{-2}

increase of entropy that allows most thermal dissociation processes to occur at temperatures far below D_e/k_B.

In this section we turn to an examination of the influence of intermolecular forces on the properties of a gas. It has already been made clear that the qualitative form of the intermolecular potential is such as to involve repulsion between molecules at small pair separations and attraction between molecules at large pair separations. Corresponding to this interaction between the molecules, the energy spectrum of the system is changed from that characteristic of the perfect gas. In many cases the vibrations and rotations of the molecules are little altered by the presence or absence of neighboring molecules and can, then, still be separated from the translational degrees of freedom. Then that part of the energy of the system associated with internal molecular motion is just the sum of the energies for the separate internal motions. There are also phenomena, however, in which the coupling between the internal motion of the molecule and the intermolecular potential is very important, e.g., rate of loss of vibrational energy from a molecule. One must examine each case carefully to determine if the internal degrees of freedom are indeed independent of the intermolecular potential, but we shall assume here that no such coupling exists.

We consider the case that the spacing between the translational energy levels of the gas is very small relative to $k_B T$. As mentioned previously, this condition is satisfied for all gases. When $k_B T$ is large relative to the energy spacing in a system, and the de Broglie wavelength, $h/(2\pi m k_B T)^{1/2}$, is very small, classical mechanics provides an accurate description of the molecular motion. Given a monatomic gas containing N molecules in the volume V, the classical energy is

$$E_N = \sum_{i=1}^{N} \frac{p_i^2}{2m} + \mathcal{V}_N, \qquad (21.157)$$

where $\mathcal{V}_N$ is the intermolecular potential acting between the N molecules in the gas. Now, $\mathcal{V}_N$ is often thought of as arising from the sum of the interactions of all possible pairs of molecules. If this is correct, the implication is that the potential energy of, say, three molecules is exactly equal to the potential energies of the three pairs of molecules into which the triplet can be broken; that is,

$$u_3(1, 2, 3) = u_2(1, 2) + u_2(1, 3) + u_2(2, 3), \qquad (21.158)$$

where u_i is the interaction energy between i molecules. The best available data indicate that Eq. 21.158 is not completely accurate, and that there is an extra interaction between three molecules over and above that given by Eq. 21.158. Thus, we write

$$u_3(1, 2, 3) = u_2(1, 2) + u_2(1, 3) + u_2(2, 3) + \Delta_3(1, 2, 3). \quad (21.159)$$

Presumably, corresponding to larger groups of molecules there exist other many-molecule contributions to the energy, but almost nothing is known about these at the present time. For our purposes it is not necessary to imagine $\mathcal{V}_N$ as arising from pair interactions, although this picture is so useful that the reader is urged to use it, bearing in mind the limitations mentioned.

The general theory of imperfect gases is complicated even though the physical ideas on which it is based are simple. If we restrict attention to pressures sufficiently small that only the second-virial-coefficient correction to the perfect gas law is important, simple arguments analogous to those employed in Chapter 12 may be used to calculate B. We shall study these simple arguments in this section; a much more general analysis, using the grand partition function, is given in Appendix 21C.

In Chapter 12 we computed the pressure exerted by a gas of N monatomic particles by examining the rate of momentum transferred across an arbitrary plane anywhere inside the gas. We could have computed the pressure from the change of momentum when molecules are reflected from a plane connected to a force-measuring device. The results of the two computations are the same. The reason is, of course, that if molecules that move across the reference plane in the positive direction (say, to the right) transfer positive normal momentum, then molecules that move across the reference plane in the negative direction transfer negative normal momentum. Hence the effect of those molecules moving in the negative direction on the momentum of the gas lying on the positive side of the reference plane is the same as if they had crossed in the positive direction carrying an equal amount of positive momentum. The net effect, then, is to double the rate of transfer of momentum by the molecules moving in the positive direction, just as if a reflecting wall had been present. We thus find that the pressure and mean square velocity are related by $p = \frac{1}{3}nm\langle v^2 \rangle$ (see Eqs. 12.17 and 12.10).[18]

We now assume that the distribution of molecular velocities is unaffected by the intermolecular forces. Provided that $\mathcal{V}_N$ depends only on the molecular positions this is correct, as is verified in Appendix 21C. Then in a dilute gas the rate of transfer of momentum by passage of molecules across a reference plane is the same whether or not there are intermolecular forces. This contribution to the pressure, called the kinetic contribution, is just $p_K = \frac{1}{3}nm\langle v^2 \rangle$. Using the Maxwell velocity distribution to evaluate $\langle v^2 \rangle$, we find, as expected, $p_K = Nk_BT/V$, that is, the perfect gas value.

In addition to the convective flow of momentum just described, the effect of the intermolecular force between a pair of molecules lying on opposite sides of the reference plane is to transfer momentum across the plane, irrespective of whether or not the molecules actually cross the plane. There is, then, a contribution to the pressure from the intermolecular forces. Suppose that a molecule lies on one side of the reference plane. We choose our coordinates so that the origin lies at the center of mass of this molecule. The probability of finding a particular molecule in the volume element dV a distance R from the molecule at the origin would be dV/V if there were no molecular interaction. However, because of the molecular interaction, it is

$$\frac{dV}{V} \exp\left(-\frac{u_2(R)}{k_BT}\right).$$

The probability of finding any molecule (not a selected molecule) in dV a distance R away is

$$\frac{N\,dV}{V} \exp\left(-\frac{u_2(R)}{k_BT}\right),$$

since there are N molecules in the volume V and 1 is negligibly small relative to N.

As shown in Fig. 21.18, a molecule in dV will exert a force on the molecule at the origin with a component $-(du_2/dR)\cos\theta$ normal to the reference plane, where θ is the angle between the line joining the centers of the two molecules and the normal to the reference plane. On the

---

[18] Here n is the number density of molecules.

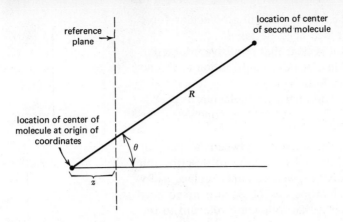

FIGURE 21.18
Schematic representation of the distances and angle referred to in the derivation of Eq. 21.160.

average there are $(N/V)\,dz$ molecules lying at a distance between z and $z+dz$ from a given unit area of the reference plane. The total normal component of the force exerted by the $n\,dz$ molecules between z and $z+dz$ on all molecules lying on the other side of the reference plane is

$$-n^2\,dz\int \frac{du_2}{dR}\cos\theta\,\exp\left(-\frac{u_2}{k_BT}\right)dV,$$

integrated over the whole space beyond the reference plane. Thus, the contribution to the pressure from intermolecular forces between all pairs of molecules is

$$p_{\text{int}}=-2\pi n^2\int_0^\infty dz\int_z^\infty \frac{du_2}{dR}\exp\left(-\frac{u_2}{k_BT}\right)R^2\,dR\int_0^{\cos^{-1} z/R}\cos\theta\sin\theta\,d\theta$$

$$=-2\pi n^2\int_0^\infty R^2\frac{du_2}{dR}\exp\left(-\frac{u_2}{k_BT}\right)dR\int_0^{\pi/2}\cos\theta\sin\theta\,d\theta\int_0^{R\cos\theta}dz,$$

$$\tag{21.160}$$

or

$$p_{\text{int}}=-\frac{2\pi n^2}{3}\int_0^\infty R^3\frac{du_2}{dR}\exp\left(-\frac{u_2}{k_BT}\right)dR. \tag{21.161}$$

By requiring that

$$\lim_{R\to\infty} R^3\frac{du_2}{dR}=0, \tag{21.162}$$

and integrating Eq. 21.161 by parts, we find that

$$p_{\text{int}}=2\pi n^2 k_BT\int_0^\infty R^2\left[1-\exp\left(-\frac{u_2}{k_BT}\right)\right]dR. \tag{21.163}$$

We now note that p_{int} is proportional to n^2 and, therefore, can be related to the second virial coefficient. In fact, since $p_{\text{int}}=p-Nk_BT/V$, when only the second-virial-coefficient term in the equation of state is significant, $p_{\text{int}}=N^2Bk_BT/N_AV^2$ and we find the simple formulas

$$B=2\pi N_A\int_0^\infty [1-\exp(-u_2/k_BT)]R^2\,dR \tag{21.164}$$

and

$$B=-\frac{2\pi N_A}{3k_BT}\int_0^\infty R^3\frac{du_2}{dR}\exp\left(-\frac{u_2}{k_BT}\right)dR, \tag{21.165}$$

which are related by an integration by parts and the use of Eq. 21.162. Equation 21.164 gives the more usual representation of B; Eq. 21.165 is often useful for interpretative purposes, since the intermolecular force appears in the integrand and no difference need be taken.

From the above expressions for B it is clear that repulsive interaction between a pair of molecules will result in a positive contribution to B, and attractive interaction in a negative contribution. Thus, if B is found to be positive, we may deduce that the interaction between molecules is predominantly repulsive, and vice versa. This justifies the conclusion we reached by qualitative arguments in Section 21.1.

Since B is determined by the interaction between a pair of molecules, it is easy to generalize Eq. 21.164 or 21.165 to describe the several second virial coefficients of a mixture; we need only replace u_2 by the appropriate potential. The general properties of B_{ij} are in no way different from those of B_{ii}, so the remarks that follow are relevant to the interpretation of the behavior of both pure gases and gas mixtures.

Theoretical calculations of B, based on Eq. 21.164 or 21.165, have been made for many assumed forms of the pair potential u_2. Some insight into the behavior of B as a function of u_2 can be obtained from the following simple models.

1. *The hard-sphere potential.* Suppose that

$$
\begin{aligned}
u_2(R) &= +\infty \qquad (R \leq \sigma), \\
u_2(R) &= 0 \qquad\;\; (R > \sigma).
\end{aligned} \tag{21.166}
$$

A representation of this potential is presented in Fig. 21.19a. It can only be considered an idealized model because no real repulsive force can be infinitely large. However, the hard-sphere potential is useful in illustrating the influence on the equation of state of the volume occupied by the molecules. We note that

$$
\begin{aligned}
\exp\left(-\frac{u_2}{k_B T}\right) - 1 &= 0 \qquad\;\; (R > \sigma), \\
\exp\left(-\frac{u_2}{k_B T}\right) - 1 &= -1 \qquad (R \leq \sigma),
\end{aligned} \tag{21.167}
$$

for the potential given by Eq. 21.166. Substituting Eq. 21.167 into Eq. 21.164, we find that

$$
\begin{aligned}
B &= 2\pi N_A \int_0^\sigma R^2 \, dR \\
&= \frac{2\pi N_A \sigma^3}{3}.
\end{aligned} \tag{21.168}
$$

Since the volume of one hard-sphere molecule is $\pi\sigma^3/6$, B is just four times the volume of one hard sphere. This is just one-half the volume excluded to the center of mass of one hard sphere of diameter σ by another of diameter σ (see Fig. 21.19b).

2. *Point centers of repulsion.* Suppose that

$$
u_2(R) = \frac{d}{R^\delta} \qquad (\delta > 3). \tag{21.169}
$$

A representation of this potential is presented in Fig. 21.19c. The evaluation of Eq. 21.165 with Eq. 21.169 involves nonelementary functions. It may be shown that

$$
B = \frac{2\pi N_A}{3}\left(\frac{d}{k_B T}\right)^{3/\delta} \Gamma\!\left(\frac{\delta - 3}{\delta}\right), \tag{21.170}
$$

where $\Gamma(x)$ is called the *gamma function.* Tables of $\Gamma(x)$ are available; for

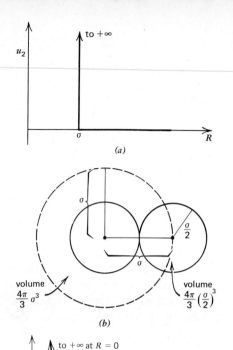

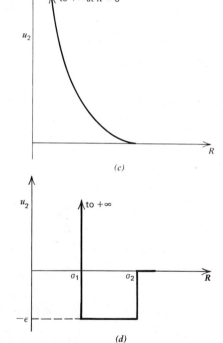

FIGURE 21.19

(a) The potential energy of interaction between two hard-sphere molecules of diameter σ.

(b) The volume excluded to the center of one hard sphere by another of equal diameter σ. Note that the excluded volume is eight times the molecular volume.

(c) The potential energy of interaction between two point centers of repulsion.

(d) The potential energy between two molecules with a square-well interaction.

our purposes all that matters is that $\Gamma[(\delta-3)/\delta]$ is a positive number and is not a function of the temperature.

Equation 21.170 shows that when the intermolecular force is "softer" than that between two hard spheres, the effective excluded volume becomes temperature-dependent. This is the expected result, since as T increases so does the kinetic energy of a molecule, and hence its ability to penetrate closer to the center of mass of the other molecule with which it interacts. Note that for this potential $B > 0$ everywhere and $(dB/dT) < 0$.

3. *The square-well potential.* Some idea of the relative importance of the attractive and repulsive parts of the intermolecular potential can be obtained from examining the potential

$$
\begin{aligned}
u_2(R) &= +\infty && (R \leq \sigma_1), \\
u_2(R) &= -\epsilon && (\sigma_1 < R \leq \sigma_2), \\
u_2(R) &= 0 && (\sigma_2 < R),
\end{aligned}
\tag{21.171}
$$

which is shown in Fig. 21.19d. From Eq. 21.164 we obtain

$$
B = 2\pi N_A \left\{ \int_0^{\sigma_1} \left[1 - \exp\left(-\frac{u_2}{k_B T} \right) \right] R^2 \, dR + \int_{\sigma_1}^{\sigma_2} \left[1 - \exp\left(-\frac{u_2}{k_B T} \right) \right] R^2 \, dR \right.
$$

$$
\left. + \int_{\sigma_2}^{\infty} \left[1 - \exp\left(-\frac{u_2}{k_B T} \right) \right] R^2 \, dR \right\}
$$

$$
= \frac{2\pi N_A \sigma_1^3}{3} + 2\pi N_A \frac{[1 - \exp(\epsilon/k_B T)](\sigma_2^3 - \sigma_1^3)}{3}.
\tag{21.172}
$$

When $k_B T \gg \epsilon$, the second term in Eq. 21.172 tends to zero, and B approaches the positive value $2\pi N_A \sigma_1^3/3$, which is one-half the volume excluded to the center of one sphere by another sphere of diameter σ_1 (see Fig. 21.19b), the same value as for hard spheres of diameter σ_1. On the other hand, when $\epsilon \gg k_B T$, B approaches $-(2\pi N_A/3)(\sigma_2^3 - \sigma_1^3) \times \exp(\epsilon/k_B T)$, which is negative. In a restricted range of temperature, where it is adequate to write $1 + \epsilon/k_B T$ for $\exp(\epsilon/k_B T)$, B simulates in form the second virial coefficient characteristic of a van der Waals gas (see Table 21.1). Note, however, that even for this very simple potential the van der Waals form is not an accurate representation of B over the entire temperature range.

4. *The Lennard–Jones potential.* The potential

$$
u_2 = 4\epsilon \left[\left(\frac{\sigma}{R} \right)^{12} - \left(\frac{\sigma}{R} \right)^6 \right]
\tag{21.173}
$$

has been used extensively in the theoretical interpretation of the properties of gases. The term proportional to R^{-6} represents the London dispersion interaction between closed-shell atoms and/or molecules (see Chapter 10). The term proportional to R^{-12} is an approximation to the strong short-range repulsion between molecules. As shown in Fig. 21.6, which is in fact a Lennard–Jones potential curve, ϵ measures the depth of the potential well and σ is that point for which $u_2 = 0$.

The calculation of B using Eq. 21.173 for the potential can only be carried out numerically. Some results are listed in Table 21.12 as a function of the variable $T^* = k_B T/\epsilon$. Just as for the square-well potential, B is negative at low temperature and positive at high temperature. Furthermore, just as in the case of the point centers of repulsion, the effective excluded volume of Lennard–Jones molecules decreases with T when repulsive forces are dominant (see Table 21.12 for $T^* > 30$).

TABLE 21.12
THE SECOND VIRIAL COEFFICIENT CALCULATED FROM THE LENNARD–JONES POTENTIAL[a]

$T^* \equiv (k_B T / \epsilon);\quad b_0 \equiv (2\pi/3) N_A \sigma^3;\quad B(T) \equiv b_0 B^*(T^*)$

T^*	B^*	T^*	B^*
0.30	−27.8806	2.00	−0.6276
0.35	−18.7549	2.50	−0.3126
0.40	−13.7988	3.00	−0.1152
0.45	−10.7550	3.50	+0.0190
0.50	−8.7202	4.00	+0.1154
0.55	−7.2741	4.50	+0.1876
0.60	−6.1980	5.00	+0.2433
0.65	−5.3682	10.00	+0.4609
0.70	−4.7100	20.00	+0.5254
0.80	−3.7342	30.00	+0.5269
0.90	−3.0471	50.00	+0.5084
1.00	−2.5381	100.00	+0.4641
1.20	−1.8359	200.00	+0.4114
1.40	−1.3758	300.00	+0.3801
1.60	−1.0519	400.00	+0.3584
1.80	−0.8120		

[a] Adapted from J. O. Hirschfelder, C. F. Curtiss, and R. B. Bird, *Molecular Theory of Gases and Liquids* (Wiley, New York, 1954).

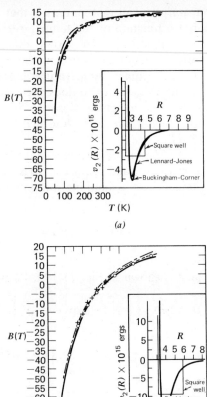

FIGURE 21.20

(a) Second virial coefficients for neon calculated for several molecular models. The potential functions obtained from the experimental $B(T)$ data are also shown. The experimental data are those of L. Holborn and J. Otto, *Z. Physik* **33**, 1 (1925). (b) Second virial coefficients for argon calculated for several molecular models. The potential functions obtained from the experimental $B(T)$ data are also shown. The experimental data are those of L. Holborn and J. Otto, *Z. Physik* **33**, 1 (1925), and A. Michels, H. Wijker, and H. Wijker, *Physica*, **15**, 627 (1949). From J. O. Hirschfelder, C. F. Curtiss and R. B. Bird, Molecular Theory of Gases and Liquids (Wiley, New York, 1954).

There are only a few experimental methods available for the determination of molecular interactions. The most direct of these, namely, the study of the scattering of one molecule by another, is only now coming to be used with any frequency. For many years what little we learned about intermolecular forces came from studies of the second virial coefficients of gases, and from studies of other gas properties. Unfortunately, B is not very sensitive to the form of the intermolecular potential. In Fig. 21.20 are plotted experimental data and three theoretical curves: B as calculated from the Lennard–Jones potential, from the square-well potential, and from

$$u_2(R) = b \exp\left[-\alpha\left(\frac{R}{R_m}\right)\right] - \left(\frac{C}{R^6} + \frac{C'}{R^8}\right)\exp\left[-4\left(\frac{R_m}{R}-1\right)^3\right] \quad (R \leqslant R_m),$$

$$u_2(R) = b \exp\left[-\alpha\left(\frac{R}{R_m}\right)\right] - \left(\frac{C}{R^6} + \frac{C'}{R^8}\right) \quad (R > R_m),$$

where

$$b = [-\epsilon + (1+\beta)CR_m^{-6}]\exp(\alpha),$$

$$C = \epsilon\alpha R_m^6[\alpha(1+\beta) - 6 - 8\beta]^{-1},$$

$$C' = \beta R_m^2 C, \tag{21.174}$$

known as the Buckingham–Corner potential. As can be seen, all fit the data reasonably well, so that deducing the form of $u_2(R)$ from measurements of B requires measurements of very high precision over an extended range of temperature. The following general conclusions can be drawn from experiment and theory.

1. In order to reproduce the observed temperature dependence of B, $u_2(R)$ must have a strongly repulsive region (be positive) at small pair separations, and have an attractive region (be negative) at larger pair separations. This form of the intermolecular potential is that expected from fundamental quantum mechanical arguments (Chapter 10), as well as from physical arguments based on the existence and small compressibility of solids and liquids.

2. Because B is negative at low temperatures and positive at high temperatures, B must be zero at some intermediate temperature. The temperature at which $B = 0$ is called the Boyle temperature, denoted T_B. Near T_B a gas behaves nearly like a perfect gas over a limited range of pressure.

3. The intermolecular potential must have a repulsive portion that is "softer" than a hard core; the precise form of the repulsion cannot be fixed from studies of B alone.

4. Studies of the third virial coefficient establish that the potential energy is not precisely the sum of pair interaction energies, that is, that Δ_3 in Eq. 21.159 is nonzero. This conclusion is supported by a variety of other studies. Aside from the fact that $\Delta_3 \neq 0$, little is known about the functional form. Although higher-order deviations from additivity of potentials must, in principle, also occur, nothing is known of these.

5. The intermolecular potential between unlike pairs of molecules, even when fitted to the same functional form, does not exactly satisfy any simple additivity rules. For example, if 1 and 2 are the two components of a mixture, it is found that

$$\epsilon_{12} \neq (\epsilon_{11}\epsilon_{22})^{1/2},$$
$$\sigma_{12} \neq \tfrac{1}{2}(\sigma_{11} + \sigma_{22}), \tag{21.175}$$

although the deviations are usually small. The combining rules defined by the equal signs in Eq. 21.175 are what would be expected if the repulsive forces were like those between hard spheres (additivity of radii) and if the attractive forces could be approximated by the crudest version of the London dispersion formula using average ionization energies.

6. There is no fundamental reason to believe that a two-parameter potential such as Eq. 21.173 can adequately represent the intermolecular pair potential.

For further details, the reader is referred to the comprehensive monograph, *Molecular Theory of Gases and Liquids*, by J. O. Hirschfelder, C. F. Curtiss, and R. B. Bird (Wiley, New York, 1954).

As the last topic in this section, we analyze the justification for the observation that the properties of all gases are the same when expressed in the reduced variables $\tilde{T}$, $\tilde{p}$, and $\tilde{V}$. Although a much more general argument can be used, we shall demonstrate the *principle of corresponding states*, as this observation is known, through study of the second virial coefficient. Suppose that the intermolecular potential can be written in the form

$$u_2 = \epsilon f\left(\frac{\sigma}{R}\right), \tag{21.176}$$

where ϵ and σ are constants with the dimensions of energy and distance, respectively. Some values of ϵ and σ for common gases and gas mixtures are presented in Table 21.13 and Table 21.14, respectively. All of the

TABLE 21.13
PARAMETERS OF THE LENNARD–JONES POTENTIAL

Substance	σ (Å)	ϵ/k_B (K)
He	2.556	10.22
H_2	2.928	37.00
D_2	2.928	37.00
Ne	2.749	35.60
Ar	3.405	119.8
Kr	3.60	171
Xe	4.100	221
O_2	3.58	117.5
CO	3.763	100.2
N_2	3.698	95.05
CH_4	3.817	148.2

TABLE 21.14
LENNARD–JONES PARAMETERS FOR MIXTURES

Pair	σ_{12} (Å)	ϵ_{12}/k_B (K)
He–H_2	2.74	19.4
He–Ne	2.65	19.1
He–N_2	3.13	31.3
He–O_2	3.07	34.7
He–Ar	2.98	61.2
Ne–H_2	2.84	36.3
Ne–N_2	3.23	58.4
Ne–O_2	3.16	64.7
Ne–Ar	3.08	65.9
N_2–O_2	3.64	106
Ar–H_2	3.16	67.2
Ar–O_2	3.49	120

potentials we have thus far considered, except the Buckingham–Corner potential, can be written in the form of Eq. 21.176. Introduction of Eq. 21.176 into Eq. 21.164 leads to the form

$$B = 2\pi N_A \int_0^\infty \left\{ 1 - \exp\left[-\frac{\epsilon f(\sigma/R)}{k_B T} \right] \right\} R^2 \, dR. \qquad (21.177)$$

We now define three reduced variables by the equations

$$T^* \equiv \frac{k_B T}{\epsilon}, \qquad (21.178a)$$

$$V^* \equiv \frac{V}{N\sigma^3}, \qquad (21.178b)$$

$$p^* \equiv \frac{p\sigma^3}{\epsilon}. \qquad (21.178c)$$

Introducing Eq. 21.178a into Eq. 21.177 leads to

$$B = 2\pi N_A \sigma^3 \int_0^\infty \left\{ 1 - \exp\left[-\frac{\epsilon f(\sigma/R)}{k_B T} \right] \right\} \left(\frac{R}{\sigma} \right)^2 d\left(\frac{R}{\sigma} \right)$$

$$= \frac{2\pi N_A \sigma^3}{3} B^*(T^*). \qquad (21.179)$$

Thus, if the intermolecular potential can be written in the form of Eq. 21.176, the reduced virial coefficient B^* is a function only of T^*. The equation of state

$$p = \frac{N k_B T}{V} \left(1 + \frac{N}{N_A} \frac{B}{V} \right) \qquad (21.180)$$

can then be rewritten in the form

$$p^* = \frac{N k_B T}{V} \cdot \frac{\sigma^3}{\epsilon} \left(1 + \frac{N}{N_A} \frac{B}{V} \right)$$

$$= \frac{T^*}{V^*} \left(1 + \frac{2\pi}{3} \frac{B^*}{V^*} \right). \qquad (21.181)$$

Eq. 21.181 shows that the equations of state of all gases are the same when expressed in the variables T^*, V^*, and p^*. These variables may be related to T_c, V_c, and p_c (see Chapter 24), thereby completing the rationalization of the observed behavior of gases.

The analysis given may be extended to all the virial coefficients, and to other equivalent general forms of the equation of state. Provided Eq. 21.176 is valid, and that the total potential energy is the sum of pair energies, that is, that Eq. 21.158 is valid, it is found that the law of corresponding states is also valid. Deviations from Eq. 21.176 arise when the molecular force field is not spherical, as must be the case, for example, in hexane. The law of corresponding states can be extended to cover this class of potentials by introduction of a new parameter that measures the asymmetry of the molecular field. Deviations from Eq. 21.158 certainly occur, but little is now known about these cases except that they exist. Despite the existence of these classes of deviations, the law of corresponding states is an organizing principle of great value, since it provides a standard of expected behavior. In the evolution of a science it is often the study of deviations from some well-defined model behavior that leads to deeper understanding of the properties of matter. In this chapter we have seen how deviations from the perfect gas law can be interpreted in terms of the properties of intermolecular forces and, to a lesser extent, how deviations from models based on pair interactions lead to information about many-molecules interactions. Despite the efforts of many scientists over a period of more than 100 years, there is much still to be learned from careful study of the properties of gases.

Appendix 21A
Influence of Symmetry of the Wave Function on the Distribution over States: Fermi-Dirac and Bose-Einstein Statistics

We stated in Section 21.5 that the symmetry properties of the wave function of a system impose a restriction on the number of accessible states. In general, the thermodynamic consequences of these symmetry induced restrictions are important only when $k_B T$ is comparable to or smaller than the typical spacing of energy levels. The importance of such quantum mechanical effects is not limited to improvement of the numerical accuracy of prediction of thermodynamic functions. Rather, it is only when the appropriate quantum mechanical description of a system is used that all the general laws of macroscopic behavior are satisfied. For example, under almost all conditions the spacing of translational energy levels in a perfect gas is small compared to $k_B T$. Then Eq. 21.133 is valid. But as $T \rightarrow 0$, so also must $C_V \rightarrow 0$, by the third law of thermodynamics. Clearly, Eq. 21.133 must fail when $k_B T$ becomes comparable to the spacing of translational levels, and quantum mechanical effects must then also change the forms of the equation of state and thermodynamic functions such as U or S.

To illustrate the influence on thermodynamic properties of the symmetry of the system's wave function, we examine a perfect gas of point particles. The microscopic state of the gas is defined by the number of particles with momentum $\mathbf{p}$, n_p, and the associated kinetic energy

$$\epsilon_p = \frac{p^2}{2m}. \tag{21A.1}$$

For the particle-in-a-box model we have used so often,

$$\mathbf{p} = \frac{2\pi\hbar}{L}\mathbf{n}, \tag{21A.2}$$

where $\mathbf{n}$ is a vector with integer components and $V = L^3$. Note that the particles are not identified; only the number with momentum $\mathbf{p}$ is specified.

Suppose that the wave function for the system is antisymmetric under interchange of two particles. This is the case for electrons and other elementary particles with half-integral spin. Then because only two spin states are allowed for each value of $\mathbf{p}$,

$$n_p = 0, 1, \tag{21A.3}$$

which we recognize as the Pauli exclusion principle discussed in Section 5.2. In contrast, if the wave function for the system is symmetric under interchange of two particles, as is the case for ^{4}He and elementary particles with integer spin, it is found that any number of particles can occupy a given state, so that

$$n_p = 0, 1, 2, 3, \ldots. \tag{21A.4}$$

Systems for which Eq. 21A.3 is valid are said to obey *Fermi–Dirac statistics*, and those for which Eq. 21A.4 is valid are said to obey *Bose–Einstein statistics*.

To see how Eqs. 21A.3 and 21A.4 influence the thermodynamic properties of the perfect gas, we shall evaluate the grand partition function (Eq. 19.126) for the two cases. By definition,

$$\Xi = \sum_{N \geq 0} \lambda^N Q_N = \sum_{N \geq 0} \sum_{\substack{\{n_p\} \\ \sum_p n_p = N}} \lambda^{\sum n_p} \exp\left(-\beta \sum n_p \epsilon_p\right), \qquad (21A.5)$$

where $\lambda \equiv \exp(\beta\mu)$ and $\beta \equiv (k_B T)^{-1}$. The second form on the right-hand side of Eq. 21A.5 follows from the facts that for a perfect gas the energy is the sum of individual particle energies, $E = \sum_p n_p \epsilon_p$, and the total number of molecules is the sum over all the numbers with all possible values of momentum, $N = \sum_p n_p$. The notation $\{n_p\}$, $\sum_p n_p = N$ means that we include in the summation every possible set of values for the various occupation numbers n_p subject to the restriction $\sum_p n_p = N$. We now note that when the summation over all possible values of N is carried out, each n_p will range over all possible values. Equation 21A.5 can then be rewritten

$$\Xi = \sum_{\substack{N \geq 0 \\ \sum n_p = N}} \sum_{\{n_p\}} \prod_p [\lambda \exp(-\beta\epsilon_p)]^{n_p}$$

$$= \sum_{n_0 = 0}^{n^{\max}} \sum_{n_1 = 0}^{n^{\max}} \sum_{n_2 = 0}^{n^{\max}} \cdots \prod_p [\lambda \exp(-\beta\epsilon_p)]^{n_p}$$

$$= \prod_p \left\{ \sum_{n_p = 0}^{n^{\max}} [\lambda \exp(-\beta\epsilon_p)]^{n_p} \right\}. \qquad (21.A.6)$$

To obtain an explicit form for the equation of state, we must evaluate Eq. 21A.6. As will be seen, this is very easy for the Bose–Einstein (BE) and Fermi–Dirac (FD) cases.

For the BE case, $n_p = 0, 1, 2, \ldots$, hence

$$\Xi = \prod_p \frac{1}{1 - \lambda \exp(-\beta\epsilon_p)}, \qquad (21A.7)$$

since if we define $x \equiv \lambda \exp(-\beta\epsilon_p)$ it is easily recognized that

$$\sum_n x^n = \frac{1}{1 - x}$$

when $x < 1$. Indeed, $x < 1$ for the BE perfect gas, since $\exp(-\beta\epsilon_p) < 1$ by virtue of the fact that $\epsilon_p > 0$, and $\lambda = \exp(\beta\mu) < 1$ when $\mu < 0$, which is the case.

For the FD case, $n_p = 0, 1$ only, hence

$$\Xi = \prod_p [1 + \lambda \exp(-\beta\epsilon_p)]. \qquad (21A.8)$$

We then find for the equations of state,

$$\frac{pV}{k_B T} = \ln \Xi = -\sum_p \ln[1 - \lambda \exp(-\beta\epsilon_p)] \qquad \text{(BE)}, \qquad (21A.9a)$$

$$\frac{pV}{k_B T} = \ln \Xi = \sum_p \ln[1 + \lambda \exp(-\beta\epsilon_p)] \qquad \text{(FD)}. \qquad (21A.9b)$$

Although explicit and exact, Eqs. 21A.9a and 21A.9b are not convenient, because they give the equation of state as a function of λ, that is, as a function of the chemical potential. To eliminate λ we calculate, first, the average number of particles in the gas. This is, by definition,

$$\langle N \rangle = \frac{\sum\limits_{N \geq 0} N \lambda^N Q_N}{\Xi} = \lambda \frac{\partial}{\partial \lambda} \ln \Xi, \qquad (21A.10)$$

so that

$$\langle N \rangle = \sum_p \frac{\lambda \exp(-\beta \epsilon_p)}{1 - \lambda \exp(-\beta \epsilon_p)} \qquad \text{(BE)}, \qquad (21A.11a)$$

$$\langle N \rangle = \sum_p \frac{\lambda \exp(-\beta \epsilon_p)}{1 + \lambda \exp(-\beta \epsilon_p)} \qquad \text{(FD)}. \qquad (21A.11b)$$

We can also find the average occupation numbers, $\langle n_p \rangle$, from the definition

$$\langle n_p \rangle = \frac{\sum\limits_{N \geq 0} \lambda^{\sum n_p} \sum\limits_{\{n_p\}, \sum n_p = N} n_p \exp\left(-\beta \sum n_p \epsilon_p\right)}{\Xi} \qquad (21A.12)$$

$$= -\frac{1}{\beta} \frac{\partial}{\partial \epsilon_p} \ln \Xi, \qquad (21A.13)$$

so that

$$\langle n_p \rangle = \frac{\lambda \exp(-\beta \epsilon_p)}{1 - \lambda \exp(-\beta \epsilon_p)} \qquad \text{(BE)}, \qquad (21A.14a)$$

$$\langle n_p \rangle = \frac{\lambda \exp(-\beta \epsilon_p)}{1 + \lambda \exp(-\beta \epsilon_p)} \qquad \text{(FD)}, \qquad (21A.14b)$$

and

$$\langle N \rangle = \sum_p \langle n_p \rangle \qquad (21A.15)$$

as expected.

For a macroscopic system the spacing between adjacent translational energy levels is very small, hence the sum over momentum states can be replaced with an integral over momentum. Using Eq. 21A.2, the conversion reads

$$\sum_p \cdots \rightarrow \frac{V}{h^3} \int d\mathbf{p} \cdots$$

and is valid if the summand is finite for all p. This is the case for the perfect Fermi–Dirac gas, so we can write

$$\frac{p}{k_B T} = \frac{4\pi}{h^3} \int_0^\infty dp\, p^2 \ln\left[1 + \lambda \exp\left(-\frac{\beta p^2}{2m}\right)\right], \qquad (21A.16)$$

$$n \equiv \frac{\langle N \rangle}{V} = \frac{4\pi}{h^3} \int_0^\infty dp\, p^2 \frac{\lambda}{\lambda + \exp(\beta p^2/2m)}. \qquad (21A.17)$$

In contrast, for the perfect Bose–Einstein gas the summands diverge as $\lambda \rightarrow 1$ because the single term corresponding to $p = 0$ diverges. To treat this case we split off the term with $p = 0$ and replace the rest of the sum by an integral. The result is

$$\frac{p}{k_B T} = -\frac{4\pi}{h^3} \int_0^\infty dp\, p^2 \ln\left[1 - \lambda \exp\left(-\frac{\beta p^2}{2m}\right)\right] - \frac{1}{V} \ln(1 - \lambda), \quad (21A.18)$$

$$n = \frac{4\pi}{h^3} \int_0^\infty dp\, p^2 \frac{\lambda}{\exp(\beta p^2/2m) - \lambda} + \frac{1}{V} \frac{\lambda}{1 - \lambda}, \qquad (21A.19)$$

where the last terms on the right-hand sides of Eq. 21A.18 and 21A.19 come from the $p = 0$ terms of Eqs. 21A.9a and 21A.14a, respectively.

The term

$$\frac{\langle n_0 \rangle}{V} = \frac{1}{V} \frac{\lambda}{1-\lambda}$$

becomes significant only when a finite fraction of the particles occupy the level with $p = 0$, and is otherwise infinitesimal, as are the terms of the form $\langle n_p \rangle / V$ for all conditions; $\langle n_0 \rangle / V$ is not infinitesimally small when the Bose–Einstein gas condenses.

To calculate $p/k_B T$ as a function of n, the ordinary equation of state, we must eliminate λ between Eqs. 21A.16 and 21A.17 for the FD gas, and between Eqs. 21A.18 and 21A.19 for the BE gas. We shall show how this is done for the FD gas. Consider the behavior of λ as determined from

$$n\Lambda^3 = f_{3/2}(\lambda), \tag{21A.20}$$

$$f_{3/2}(\lambda) \equiv \frac{4}{\sqrt{\pi}} \int_0^\infty dx \frac{\lambda x^2}{e^{x^2} + \lambda}. \tag{21A.21}$$

Equation 21A.20 is obtained from Eq. 21A.17, using the new variable $x^2 \equiv p^2/2mk_B T$ and the definitions of the de Broglie wavelength $\Lambda = (h^2/2\pi mk_B T)^{1/2}$ and of $f_{3/2}(\lambda)$ given by Eq. 21A.21. We now make the expansion

$$\frac{\lambda x^2}{e^{x^2} + \lambda} = \frac{\lambda x^2}{e^{x^2}(1 + \lambda e^{-x^2})}$$

$$= \lambda x^2 e^{-x^2}(1 - \lambda e^{-x^2} + \lambda^2 e^{-2x^2} - \cdots),$$

valid when $\lambda e^{-x^2} < 1$. Then, using the integral

$$\int_0^\infty dx\, x^2 e^{-ax^2} = \frac{1}{4}\left(\frac{\pi}{a^3}\right)^{1/2},$$

we find

$$f_{3/2}(\lambda) = \lambda - \frac{\lambda^2}{2^{3/2}} + \frac{\lambda^3}{3^{3/2}} - \cdots$$

$$= \sum_{l=1}^\infty \frac{(-1)^{l+1}}{l^{3/2}} \lambda^l. \tag{21A.22}$$

When $n\Lambda^3 \ll 1$, corresponding to the average interparticle separation $\mathcal{O}(n^{-3})$ being much larger than Λ,

$$n\Lambda^3 \approx \lambda - \frac{\lambda^2}{2^{3/2}},$$

which may be solved to give

$$\lambda \approx n\Lambda^3 + \frac{1}{2^{3/2}}(n\Lambda^3)^2$$

and the equation of state

$$\frac{p}{k_B T} = \frac{4\pi}{h^3} \int_0^\infty dp\, p^2 \ln\left[1 + \lambda \exp\left(-\frac{\beta p^2}{2m}\right)\right]$$

$$= \frac{4\pi}{h^3} \int_0^\infty dp\, p^2 \left[\lambda \exp\left(-\frac{\beta p^2}{2m}\right) - \tfrac{1}{2}\lambda^2 \exp\left(-\frac{\beta p^2}{m}\right) + \cdots\right]$$

$$= n + \frac{1}{2^{5/2}} n^2 \Lambda^3. \tag{21A.23}$$

Thus, the pressure of a perfect FD gas exceeds that of a perfect gas of particles obeying classical mechanics. This is true for all densities, even though we have shown it only in the low-density limit. The origin of the excess pressure is an effective repulsion between particles generated by the symmetry-imposed requirement that $n_p = 0, 1$ only.

Proceeding in a similar fashion, we find for the perfect BE gas that

$$\frac{p}{k_B T} = n - \frac{1}{2^{5/2}} n^2 \Lambda^3, \tag{21A.24}$$

which is smaller than the pressure of a corresponding perfect gas of particles obeying classical mechanics. In this case the symmetry requirements on the system's wave function give rise to an effective attraction. This effective attraction eventually leads to condensation of the BE gas; no such condensation can occur for the perfect FD gas.

Both the FD and BE gases approach classical behavior as $n\Lambda^3 \to 0$. We find from Eqs. 21A.14a and 21A.14b that

$$\lim_{n\Lambda^3 \to 0} \langle n_p \rangle = n\Lambda^3 \exp(-\beta\epsilon_p), \tag{21A.25}$$

which defines what is sometimes called the *Boltzmann gas*. Then, since $n\Lambda^3 \ll 1$ implies that $\lambda \ll 1$,

$$\frac{p}{k_B T} = \frac{4\pi n \Lambda^3}{h^3} \int_0^\infty dp\, p^2 \exp\left(-\frac{\beta p^2}{2m}\right)$$
$$= n, \tag{21A.26}$$

which is the equation of state of the perfect classical gas. We can also evaluate the grand partition function for the perfect classical gas. Using the semiclassical canonical partition function (see Appendix 21C),

$$Q_N = \frac{Q_1{}^N}{N!}, \tag{21A.27}$$

we write

$$\Xi = \sum_{N \geq 0} \lambda^N Q_N = \sum_{N \geq 0} \lambda^N \frac{Q_1{}^N}{N!} = e^{\lambda Q_1}. \tag{21A.28}$$

Since

$$Q_1 = V\Lambda^{-3} \tag{21A.29}$$

and

$$\frac{pV}{k_B T} = \lambda Q_1, \tag{21A.30}$$

we find

$$\mu = k_B T \ln \frac{p\Lambda^3}{k_B T} \tag{21A.31}$$

and

$$V = \left(\frac{\partial G}{\partial p}\right)_{T,N} = \frac{N k_B T}{p} \tag{21A.32}$$

for the perfect classical gas chemical potential and equation of state.

In this appendix we have briefly described what are called *weakly degenerate* FD and BE gases. To show consistency of thermodynamic behavior with the third law we must study strongly degenerate quantum gases, for which $n\Lambda^3 \approx 1$. The FD case, exemplified by electrons in a metal, will be briefly considered in Chapter 22; the BE case, although very important (e.g., for the understanding of superconductivity and superfluidity), will not be considered further in this text.

Appendix 21B
Symmetry Properties of the Molecular Wave Function: Influence of Nuclear Spin on the Rotational Partition Function

We discussed, in Appendix 21A, how the symmetry of the wave function for a perfect gas of point particles influences the equation of state. When the perfect gas is composed of molecules that have internal structure, new symmetry properties of the wave function must be considered; these are associated with the effects of permutation of the nuclei within a molecule. Because of their spin, the nuclei of a molecule must behave as bosons (obeying BE statistics) or fermions (obeying FD statistics), as the case may be. The point is that the overall symmetry of the molecular wave function must include the properties of the nuclear wave function when the nuclei are permuted within the molecule. As will be seen, nuclear spin effects, acting through the corresponding wave function symmetry, have an important influence on the system's properties.

Consider, first, a heteronuclear diatomic molecule with nuclei a and b. By virtue of the nuclear spin I, there are several nuclear spin states, namely, $\bar{\omega}^{(n)} = 2I + 1$. Then nucleus a has spin eigenfunctions $\psi_1(a), \ldots, \psi_{\bar{\omega}_a^{(n)}}(a)$ and nucleus b has spin eigenfunctions $\psi_1(b), \ldots, \psi_{\bar{\omega}_b^{(n)}}(b)$. We have, therefore $\bar{\omega}_a^{(n)} \bar{\omega}_b^{(n)}$ combined nuclear eigenfunctions of the type $\psi_r(a)\psi_s(b)$ and, correspondingly, a degeneracy $\bar{\omega}_a^{(n)} \bar{\omega}_b^{(n)}$ in the partition function.

Consider, next, a homonuclear diatomic molecule. Of course each nucleus has the same eigenfunctions and degeneracy $\bar{\omega}^{(n)}$. For the molecule there are

$\frac{1}{2}\bar{\omega}^{(n)}(\bar{\omega}^{(n)} - 1)$ antisymmetric nuclear wave functions of the type

$$\psi_r(a)\psi_s(b) - \psi_s(a)\psi_r(b),$$

$\frac{1}{2}\bar{\omega}^{(n)}(\bar{\omega}^{(n)} - 1)$ symmetric nuclear wave functions of the type

$$\psi_r(a)\psi_s(b) + \psi_s(a)\psi_r(b),$$

$\bar{\omega}^{(n)}$ symmetric nuclear wave functions of the type

$$\psi_r(a)\psi_r(b).$$

Thus there are $\frac{1}{2}\bar{\omega}^{(n)}(\bar{\omega}^{(n)} - 1)$ antisymmetric nuclear wave functions and $\frac{1}{2}\bar{\omega}^{(n)}(\bar{\omega}^{(n)} + 1)$ symmetric nuclear wave functions, for a total of $(\bar{\omega}^{(n)})^2$, the same as the number of nuclear eigenfunctions of a heteronuclear diatomic molecule. We now must combine the nuclear wave functions with the rest of the molecular wave function and make the total wave function antisymmetric with respect to exchange of protons and neutrons (both of which obey FD statistics).

What are the symmetry properties of the wave functions corresponding to the non-nuclear molecular degrees of freedom? These are as follows:

1. The eigenfunction descriptive of translation of the center of mass does not contain the coordinates of individual nuclei relative to the center of mass. Therefore the translational eigenfunction is totally symmetric in the nuclear coordinates.

2. The rotational eigenfunctions with even quantum numbers, $J = 0, 2, 4, \ldots$, are symmetric in the nuclear coordinates, and those with odd quantum numbers, $J = 1, 3, 5, \ldots$, are antisymmetric in the nuclear coordinates.

3. The vibrational eigenfunctions are symmetric in the nuclear coordinates (they depend on the displacement from the equilibrium separation).

Since the total molecular eigenfunction must be antisymmetric with respect to proton and neutron permutation, it must be antisymmetric under permutation of the nuclei when they have odd mass numbers, and must be symmetric under permutation of the nuclei when they have even mass numbers.

Consider the case when there is no vibrational excitation of the molecule ($v = 0$). Then, because the vibrational eigenfunction is symmetric in the nuclear coordinates, the following combinations must be considered:

1. Nuclei with odd mass numbers—symmetric rotational wave functions with antisymmetric nuclear wave functions, or antisymmetric rotational wave functions with symmetric nuclear wave functions. Therefore, the terms with even J, including $J = 0$, in the rotational partition function have degeneracy $\frac{1}{2}\bar{\omega}^{(n)} \times (\bar{\omega}^{(n)} - 1)$ from the nuclear spin states, and those with odd J have degeneracy $\frac{1}{2}\bar{\omega}^{(n)}(\bar{\omega}^{(n)} + 1)$.

2. Nuclei with even mass numbers—symmetric rotational wave functions with symmetric nuclear wave functions, or antisymmetric rotational wave functions with antisymmetric nuclear wave functions. Therefore, the terms with even J, including $J = 0$, in the rotational partition function have degeneracy $\frac{1}{2}\bar{\omega}^{(n)} \times (\bar{\omega}^{(n)} + 1)$ from the nuclear spin states, and those with odd J have degeneracy $\frac{1}{2}\bar{\omega}^{(n)}(\bar{\omega}^{(n)} - 1)$.

We conclude that the partition function for rigid-body rotation of a diatomic molecule in the vibrational ground state, including the effects of nuclear spin, takes the form

$$\bar{\omega}_a^{(n)}\bar{\omega}_b^{(n)} \sum_{J=0}^{\infty} (2J+1) \exp\left[-\frac{J(J+1)\Theta_{\rm rot}}{T}\right]$$

$$\text{(heteronuclear molecule)}, \quad (21B.1)$$

or

$$\frac{1}{2}\bar{\omega}^{(n)}(\bar{\omega}^{(n)} - 1) \sum_{J=0,2,\ldots}^{\infty} (2J+1) \exp\left[-\frac{J(J+1)\Theta_{\rm rot}}{T}\right]$$

$$+ \frac{1}{2}\bar{\omega}^{(n)}(\bar{\omega}^{(n)} + 1) \sum_{J=1,3,\ldots}^{\infty} (2J+1) \exp\left[-\frac{J(J+1)\Theta_{\rm rot}}{T}\right]$$

(homonuclear diatomic molecule, odd nuclear mass numbers), (21B.2)

or

$$\frac{1}{2}\bar{\omega}^{(n)}(\bar{\omega}^{(n)} + 1) \sum_{J=0,2,\ldots}^{\infty} (2J+1) \exp\left[-\frac{J(J+1)\Theta_{\rm rot}}{T}\right]$$

$$+ \frac{1}{2}\bar{\omega}^{(n)}(\bar{\omega}^{(n)} - 1) \sum_{J=1,3,\ldots}^{\infty} (2J+1) \exp\left[-\frac{J(J+1)\Theta_{\rm rot}}{T}\right]$$

(homonuclear diatomic molecule, even nuclear mass numbers). (21B.3)

Consider now the case when $\Theta_{rot} \ll T$. Then, to great accuracy, the sum over even rotational quantum numbers is equal to the sum over odd rotational quantum numbers, because the important terms in the sum involve large values of J. Therefore,

$$\sum_{J=0,2,...}^{\infty} (2J+1) \exp\left[-\frac{J(J+1)\Theta_{rot}}{T}\right] = \sum_{J=1,3,...}^{\infty} (2J+1) \exp\left[-\frac{J(J+1)\Theta_{rot}}{T}\right]$$
$$= \frac{1}{2} \sum_{J=0}^{\infty} (2J+1) \exp\left[-\frac{J(J+1)\Theta_{rot}}{T}\right].$$

(21B.4)

The factor $\frac{1}{2}$ that arises naturally in this limit is just the reciprocal of the symmetry number of the homonuclear diatomic molecule. From the standpoint of classical mechanics, the appearance of the factor $\frac{1}{2}$ means that configurations generated by rotating the homonuclear diatomic by 180° are to be considered identical, so that the number of distinguishable states is only half the classical total. That is, the classical value of the rotational partition function must be divided by the symmetry number, which is 1 for a heteronuclear diatomic and 2 for a homonuclear diatomic molecule.

The case of polyatomic molecules is similar. Just as for a diatomic molecule, the symmetry of the molecule imposes requirements on the symmetry of the molecular wave function when nuclei are permuted. Consideration must then be given to the construction of appropriate eigenfunctions of nuclear spin and molecular rotation. The effect is, as before, to restrict the rotational quantum states available to a molecule. In the high-temperature limit, which almost always accurately describes a polyatomic molecule, we again use a symmetry number σ to correct the classical partition function. The symmetry number is the number of indistinguishable positions into which the molecule may be carried by rigid body rotation. For example, $\sigma = 12$ for CH_4, since the molecule can be rotated into three indistinguishable positions about any CH axis, and there are four such axes in the molecule. For benzene $\sigma = 12$, because there are six indistinguishable positions obtained by rotation about an axis perpendicular to the ring, and two indistinguishable positions obtained by 180° rotation about an axis through the 1,4 carbon atoms and lying in the plane of the ring. Clearly, for polyatomic molecules with high symmetry σ can be quite large, hence considerably influences the rotational contribution to the Helmholtz free energy.

When $\Theta_{rot} > T$, no simplification of the combined rotation–nuclear spin partition function is possible, and the full forms of Eq. 21B.2 or 21B.3 must be used to calculate the thermodynamic properties of the diatomic gas. The best-known case where this is necessary is H_2, as described in Section 21.5B.

Appendix 21C

The Semiclassical Partition Function; The Equation of State of an Imperfect Gas

In setting up the classical equivalent of the partition functions already introduced, we are led to the following question: How precisely can we define the energy level in a continuous spectrum of states? According to the uncertainty principle, a particle constrained to one-dimensional motion cannot have its momentum and position simultaneously specified to better than $\Delta p \, \Delta x \approx h$. Thus, if p and x are chosen as coordinate axes, the point representing a one-dimensional system cannot be located more precisely than within a box with sides Δp and Δx and area h. If there are N molecules moving in a three-dimensional space, we need $3N$ position coordinates and $3N$ momentum coordinates to specify the system. Imagine a space of $6N$ dimensions, by which is meant that the coordinates of one point require the specification of $6N$ numbers. When the $6N$ coordinates are chosen to be the $3N$ position coordinates and the $3N$ momenta, the space is called the *phase space* of the system (cf. Chapter 2). Clearly, in a classical mechanical description, the mechanical state of our system at any instant of time is given by *one* point in the $6N$-dimensional phase space. But if the product of each coordinate and the corresponding momentum is uncertain to the amount $\Delta p_i \, \Delta x_i \approx h$, then from the point of view of quantum mechanics, the finest allowable subdivision of phase space is a cell of hypervolume h^{3N}. According to quantum theory, we cannot distinguish any variation in any function over a hypervolume less than h^{3N}. Then, from Eq. 19.131,

$$Q_N = \sum_j g_j \exp\left(-\frac{E_j}{k_B T}\right), \qquad (21C.1)$$

we see that each term in the summation corresponds to

$$\exp\left(-\frac{E_j}{k_B T}\right) = \frac{1}{N! h^{3N}} \underset{E \leqslant E_j \leqslant E + dE}{\int \cdots \int} \exp\left[-\frac{1}{k_B T}\left(\sum \frac{p_i^2}{2m} + \mathscr{V}_N\right)\right] \times$$

$$dx_1 \, dy_1 \, dz_1 \cdots dp_{x_N} \, dp_{y_N} \, dp_{z_N}. \quad (21C.2)$$

The factor $N! \, h^{3N}$ appears because the shell $E \leqslant E_j \leqslant E + dE$ in phase space has volume $N! \, h^{3N}$. The term h^{3N} arises because each product $dx \, dp_x$ ranges over h, and there are $3N$ such products (all subject to the uncertainty relation). The term $N!$ arises because there are $N!$ permutations of the molecules in the cell among the momenta and positions for which the energy is unchanged. The weighting factor g_j, originally chosen to represent the number of quantum mechanically distinct states with energies in the interval $E \leqslant E_j \leqslant E + dE$, is taken to be unity, because dE

is now the smallest distinguishable energy difference. Thus, we have

$$Q_N = \frac{1}{N! \, h^{3N}} \int \cdots \int \exp\left[-\frac{1}{k_B T}\left(\sum \frac{p_i^2}{2m} + \mathcal{V}_N\right)\right] dx_1 \cdots dp_{z_N}, \quad (21C.3)$$

where the integral is over the entire range of variation of p and x (the entire phase space), for the semiclassical form of the canonical partition function.[19] The range of the space variables x_i is limited by the volume of the system. Although the momenta cannot be infinite if the energy is finite, $\exp(-p^2/2mk_B T)$ tends to zero so rapidly as $p^2 \to \infty$ that no error is incurred by allowing the range of momentum to be infinite.

If, as is usually the case, $\mathcal{V}_N$ depends only on the position variables, we obtain

$$Q_N = \frac{1}{N! \, h^{3N}} \left[\int_{-\infty}^{\infty} \exp\left(-\frac{p^2}{2mk_B T}\right) dp\right]^{3N} \int \cdots \int \exp\left(-\frac{\mathcal{V}_N}{k_B T}\right) dx_1 \cdots dz_N$$

$$= \left(\frac{2\pi mk_B T}{h^2}\right)^{3N/2} \frac{1}{N!} \int \cdots \int \exp\left(-\frac{\mathcal{V}_N}{k_B T}\right) dx_1 \cdots dz_N$$

$$= \left(\frac{Q_1^T}{V}\right)^N \frac{Z_N}{N!}, \quad (21C.4)$$

where

$$Z_N \equiv \int \cdots \int \exp\left(-\frac{\mathcal{V}_N}{k_B T}\right) dx_1 \cdots dz_N \quad (21C.5)$$

and Q_1^T is given by Eq. 21.108; Z_N is known as the *configurational partition function* for N molecules.

We may now readily obtain the general virial equation of state from the grand partition function. From Eq. 19.125 we find that

$$\frac{pV}{k_B T} = \ln \Xi, \quad (21C.6)$$

$$\frac{pV}{k_B T} = \ln\left[1 + \exp\left(\frac{\mu}{k_B T}\right) Q_1 + \exp\left(\frac{2\mu}{k_B T}\right) Q_2 + \cdots\right], \quad (21C.7)$$

and by expanding the logarithm,

$$\frac{pV}{k_B T} = \lambda Q_1 + \lambda^2 (Q_2 - \tfrac{1}{2} Q_1^2) + \cdots, \quad (21C.8)$$

where $\lambda \equiv \exp(\mu/k_B T)$ and Q_i is the canonical partition function for i molecules in the volume V. We also have, from the definition of average value,

$$\langle N \rangle \equiv \frac{\sum_N N \lambda^N Q_N}{\sum_N \lambda^N Q_N}$$

$$= \frac{\sum_N \lambda \left(\frac{\partial}{\partial \lambda} \lambda^N Q_N\right)_{T,V}}{\sum_N \lambda^N Q_N}$$

---

[19] The form 21C.3 is called the semiclassical partition function because the energy is described by classical mechanics but the factor $N! \, h^{3N}$ corrects the counting of states for the effects of molecular indistinguishability and the uncertainty principle.

$$= \left(\frac{\partial \ln \Xi}{\partial \ln \lambda}\right)_{T,V}$$

$$= \lambda Q_1 + \lambda^2 (2Q_2 - Q_1^2) + \cdots. \qquad (21C.9)$$

Introducing Eq. 21C.4, we find that Eqs. 21C.8 and 21C.9 become

$$\frac{pV}{k_B T} = (\lambda Q_1{}^T) Z_1 + (\lambda Q_1{}^T)^2 \frac{Z_2 - Z_1{}^2}{2} + \cdots \qquad (21C.10)$$

and

$$\langle N \rangle = (\lambda Q_1{}^T) Z_1 + (\lambda Q_1{}^T)^2 (Z_2 - Z_1{}^2) + \cdots, \qquad (21C.11)$$

which are to be regarded as two simultaneous equations from which $\lambda Q_1{}^T$ is to be eliminated. The general problem of solving Eq. 21C.11 for $\lambda Q_1{}^T$ in terms of $\langle N \rangle$ is algebraically complicated, but if we keep only the terms displayed (up to λ^2), the quadratic equation is easily solved, giving $\lambda Q_1{}^T$ as an explicit function of $\langle N \rangle$. Thus, subtraction of Eq. 21C.11 from Eq. 21C.10, followed by elimination of $\lambda Q_1{}^T$ in terms of $\langle N \rangle$, leads to

$$\frac{pV}{k_B T} = \langle N \rangle - \frac{(Z_2 - Z_1{}^2)\langle N \rangle^2}{2Z_1{}^2} + \cdots. \qquad (21C.12)$$

But, by definition, we must have

$$Z_1 = \int_V dx\, dy\, dz = V, \qquad (21C.13)$$

since Z_1 refers to one molecule in a volume V. Note that, using Eq. 21C.13, Eq. 21C.12 is already in the form of the virial equation of state:

$$\frac{pV}{\langle N \rangle k_B T} = 1 + \frac{(Z_1{}^2 - Z_2)}{2V} \frac{\langle N \rangle}{V} + \cdots. \qquad (21C.14)$$

Indeed, if we compare coefficients of the terms in Eqs. 21C.14 and 21.17, we see that

$$B = \frac{N_A}{2V}(Z_1{}^2 - Z_2)$$

$$= \frac{N_A}{2V}\left[\left(\int dx_1\, dy_1\, dz_1\right)^2 - \int \exp\left(-\frac{u_2}{k_B T}\right) dx_1\, dy_1\, dz_1\, dx_2\, dy_2\, dz_2\right]$$

$$= \frac{N_A}{2V}\int\int\left[1 - \exp\left(-\frac{u_2}{k_B T}\right)\right] dx_1 \cdots dz_2. \qquad (21C.15)$$

To show that B is independent of the volume, we take advantage of the fact that $u_2(1, 2)$ depends only on the distance between the monatomic molecules. Then, introducing the new coordinates

$$X = \tfrac{1}{2}(x_1 + x_2), \qquad Y = \tfrac{1}{2}(y_1 + y_2), \qquad W = \tfrac{1}{2}(z_1 + z_2),$$

$$x = x_2 - x_1, \qquad y = y_2 - y_1, \qquad z = z_2 - z_1, \qquad (21C.16)$$

$$\mathbf{R} = \mathbf{i}x + \mathbf{j}y + \mathbf{k}z,$$

we find that

$$B = 2\pi N_A \int_0^\infty \left[1 - \exp\left(-\frac{u_2}{k_B T}\right)\right] R^2\, dR, \qquad (21C.17)$$

since keeping R fixed implies that u_2 is fixed, and the integrand is constant for all possible values of X, Y, W. The integration of the position of the center of mass over V has just the value V because the integrand has the same value everywhere and the total domain of integration is V.

Further Reading

Davidson, N., *Statistical Mechanics* (McGraw-Hill, New York, 1962), chaps. 8–11, 13, and 15.

Hirschfelder, J. O., C. F. Curtiss, and R. B. Bird, *Molecular Theory of Gases and Liquids* (Wiley, New York, 1954), chaps. 3, 4, and 6.

Mason, E. A., and T. H. Spurling, *The Virial Equation of State* (Pergamon Press, Elmsford, N.Y., 1969).

Mayer, J. E., and M. G. Mayer, *Statistical Mechanics*, 2nd ed. (Wiley, New York, 1977), chaps. 7, 8, and 11.

Rushbrooke, G. S., *Introduction to Statistical Mechanics* (Oxford University Press, 1949), chaps. 6–12, 15, and 16.

Problems

1. Plot $B(T)$ versus T for a van der Waals gas. Provide an interpretation for the behavior shown on your graph.

2. For real gases, there is a temperature at which $B(T)$ passes through a maximum. Why is this behavior not shown by a van der Waals gas? Would you expect a maximum value of $B(T)$ for a gas whose molecules interact through the potential shown in Fig. 21.6?

3. Show that at low density the Dieterici equation of state reduces to the van der Waals equation.

4. Write out expressions for the Gibbs and Helmholtz free energies of a real gas correct to terms including the second virial coefficient.

5. Show that if $(\partial U/\partial V)_T = 0$ and $(\partial H/\partial p)_T = 0$, the equation of state must have the form $pV = aT$, where a is a constant independent of p, V, and T.

6. Consider the following experiment (see diagram). Gas contained in a cylinder is initially present only on one side of a porous plug that divides the cylinder into two volumes. The entire system is thermally isolated from its surroundings. The gas is pushed through the porous plug by the pistons in such a way that both p_1 and p_2 are maintained constant but not equal to one another. The flow is so slow that the gas emerging from the plug is equilibrated at p_2, so p_2 can be taken as uniform throughout the expansion on side 2. Show that for this throttling process

$$h_1 = h_2.$$

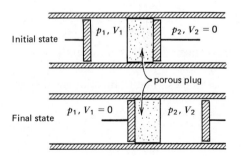

7. When the experiment described in Problem 6 is carried out with a real gas as a continuous-flow process that is so slow there is no appreciable kinetic energy associated with the gas flow, the steady-state temperatures on sides 1 and 2 are different. The temperature change is proportional to $\mu_{JT} \equiv (\partial T/\partial p)_H$, called the Joule–Thomson coefficient. Show that

$$\mu_{JT} = \frac{1}{c_p}\left[T\left(\frac{\partial v}{\partial T}\right)_p - v\right] = \frac{v}{c_p}[\alpha T - 1] = -\frac{1}{c_p}\left(\frac{\partial h}{\partial p}\right)_T.$$

8. Evaluate μ_{JT} for a real gas correct to terms including the second virial coefficient.

9. At 480°C and 723 torr total pressure, 76% of the HCl in a mixture, which originally contained 49% HCl and 51% O_2 by volume, reacts with oxygen to form chlorine and water. Calculate the equilibrium constant of the reaction with atmospheres as units.

10. Calculate the enthalpy and entropy changes for the reaction

$$I_2 = 2I$$

   at 1000 K from the following data:

$T(K)$	1274	1173	1073	973	872
$K(T)$	1.678×10^{-1}	4.803×10^{-2}	1.088×10^{-2}	1.801×10^{-3}	1.81×10^{-4}

11. Calculate the thermal de Broglie wavelength for argon at 85 K, neon at 20 K, and helium at 4 K.

12. Consider the ground state and lowest excited electronic state of a molecule. Suppose that both states are nondegenerate and the characteristic electronic temperature of the molecule is 50,000 K. Find the probability that a molecule is in the excited state at 300 K.

13. Why does the electronic heat capacity depend on the difference of electronic energies, rather than their actual values?

14. Calculate a classical (high-temperature) vibrational partition function by approximating Eq. 21.113 by an integral, in a manner analogous to Eq. 21.106 or Eq. 21.114. Show that Eq. 21.140 results directly from the classical partition function.

15. The entropy of CH_3OH at 298.15 K and 1 atm is calculated to be 244.26 J/K mol from spectroscopic data, and found to be 236.94 J/K mol from calorimetric data. Suggest an interpretation of the residual entropy (7.32 J/K mol) in terms of possible properties of the CH_3OH molecule.

16. The entropy of H_2 at 298.15 K and 1 atm is calculated to be 130.67 J/K mol from spectroscopic data, and found to be 124.43 J/K mol from calorimetric data. Offer an explanation for the residual entropy of 6.28 J/K mol in terms of the properties of o-H_2 and p-H_2. Hint: o-H_2 rotates in the crystal at the lowest temperature used in the calorimetric measurements.

17. Consider the gas-phase isomerization reaction

$$\begin{array}{ccc} \underset{\text{F}}{\overset{\text{H}}{>}}C=C\underset{\text{F}}{\overset{\text{H}}{<}} & = & \underset{\text{H}}{\overset{\text{F}}{>}}C=C\underset{\text{F}}{\overset{\text{H}}{<}} \\ \text{A} & & \text{B} \\ \text{(cis)} & & \text{(trans)} \end{array}$$

   In the presence of a catalyst, A and B are in equilibrium. The energy levels for A and B are

$$\begin{array}{ll}
\underline{\quad\quad} & \\
\underline{\quad\quad}\ \epsilon_i(A) & \underline{\quad\quad}\ \epsilon_i(B) \\
\underline{\quad\quad} & \\
\underline{\quad\quad} & \underline{\quad\quad}\ \epsilon_1(B) \\
\underline{\quad\quad}\ \epsilon_1(A) & \underline{\quad\quad}\ \epsilon_0(B) \\
\underline{\quad\quad}\ \epsilon_0(A)\text{- - - - - -} & \Delta\epsilon_0\updownarrow
\end{array}$$

   In the absence of the catalyst, the partition function for pure A is

$$Q_A = \sum_i \exp\left[-\frac{\epsilon_i(A)}{k_B T}\right]$$

and similarly, for pure B,

$$Q_B = \sum_j \exp\left[-\frac{\epsilon_j(B)}{k_B T}\right].$$

Now let the catalyst be present. Consider $\epsilon_0(A)$ as the energy zero of all $C_2H_2F_2$ molecules. Let N_0 be the number of $C_2H_2F_2$ molecules in the state with energy $\epsilon_0(A)$ and let $N_j(A)$ be the number of molecules in the state with energy $\epsilon_j(A)$.

(a) Write an expression for $N_j(A)$ in terms of N_0 and $\epsilon_j(A)$. Use this to find the total number of *cis* molecules, N_A.

(b) Express $N_j(B)$ in terms of N_0 and the total number of *trans* molecules, N_B.

(c) Find the equilibrium constant for the reaction.

18. Show that in a canonical ensemble for an ideal gas

$$\langle (E - \langle E \rangle)^2 \rangle = k_B T^2 C_V = \frac{3N}{2}(k_B T)^2,$$

where the brackets $\langle \cdots \rangle$ indicate an ensemble average.

19. Suppose that N molecules of an ideal gas are in volume V. Prove that the probability of there being n particles in a subvolume v is

$$P(n) = (\lambda Q_1)^n \frac{e^{-\lambda Q_1}}{n!}.$$

Equation 21A.28 will prove helpful. The distribution $P(n)$ is called the *Poisson distribution*. Find the average value of n.

20. The prescription of accounting for molecular indistinguishability by dividing by $N!$ (see Eq. 21.100) is strictly a semiclassical approximation and is valid only under limiting conditions. To see why, consider a system of two particles where each particle can be in either of two nondegenerate energy levels, ϵ_1 and ϵ_2.

(a) If the particles are distinguishable, prove that the canonical partition function Q_2 for the system is

$$Q_2 = Q_1^2,$$

where Q_1 is the single-particle partition function

$$Q_1 = \exp\left(-\frac{\epsilon_1}{k_B T}\right) + \exp\left(-\frac{\epsilon_2}{k_B T}\right).$$

(b) Suppose that the particles obey Fermi–Dirac statistics. Remember that there can be only one possible state of the system, and evaluate Q_2^{FD}.

(c) Suppose that the particles obey Bose–Einstein statistics. Take careful account of the indistinguishability of the particles and evaluate Q_2^{BE}.

(d) Show that the application of the 1/2! correction factor for indistinguishability to Q_1^2 gives neither Q_2^{FD} nor Q_2^{BE} but rather their average:

$$\frac{1}{2!} Q_1^2 = \frac{1}{2}(Q_2^{BE} + Q_2^{FD}).$$

(e) Suppose that it were exceedingly unlikely for two particles to have the same energy. Show that under this condition

$$Q_2^{BE} = Q_2^{FD} = \frac{1}{2!} Q_1^2.$$

(f) For most systems (but not the one in this problem!) at ordinary temperatures and densities, the number of available molecular energies is vastly greater than the number of particles. Explain why $Q_1^N/N!$ is then a good approximation for the partition function of N indistinguishable particles. Chapter 4 of *Statistical Mechanics* by D. A. McQuarrie (Harper & Row, New York, 1976) contains an excellent discussion of this point.

21. Show that for D_2 gas (nuclear spin = 1), the rotational contribution to the heat capacity is

$$C_V^R = \tfrac{1}{3}C_V^R(p) + \tfrac{2}{3}C_V^R(o),$$

where p (para) refers to odd rotation states and o (ortho) refers to even rotation states.

22. Show that when Q_1 has the form $f(T)V$, the ideal gas equation of state follows.

23. In Section 21.7 it is stated that the probability of finding two molecules in a moderately dense gas separated by a distance R is proportional to $\exp[-u_2(R)/k_BT]$. Sketch this function for various forms of $u_2(R)$ (hard-sphere, square-well, Lennard–Jones) and interpret the features of the graphs.

24. Calculate the second virial coefficient for the potential shown in the accompanying diagram.

25. A barometric formula allows one to calculate the pressure of a gas in a gravitational field. For an ideal gas, derive the barometric formula

$$p(z) = p(0) \exp\left(-\frac{mgz}{k_BT}\right),$$

where z is the height above sea level, m is the mass of a gas molecule, and g is the acceleration of gravity.

26. In Chapter 19, the Maxwell speed distribution was derived for an ideal gas (Eq. 19.143). Now suppose that the molecules interact through a potential energy function that is independent of velocity. Write the canonical partition function (Eq. 21C.3) and show that the Maxwell distribution applies equally well to a nonideal system.

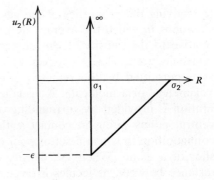

22

Thermodynamic Properties of Solids

In studying the properties of the gas phase, we have thus far dealt with situations in which the average density of molecules, N/V, is very small relative to the density of closest packing. In a state of closest packing, molecules are as close to one another as is permitted by their size, that is, as is permitted by the repulsive forces that provide a measure of the "diameter" of a molecule. A visualization of the closest-packing configuration is provided by surrounding a penny with six other pennies, the central penny being in contact with each of its neighbors, or by the configuration in which a set of pool balls is arranged before the opening shot of a game. When the density of molecules is small, the average distance between molecules is large relative to the molecular diameter, and it is possible to think of the motion of a molecule as being essentially undisturbed by the presence of other molecules except for occasional binary collisions (even these being ignored in the perfect gas). Such a configuration is similar to that encountered when the set of pool balls is distributed at random over the pool table. Now it is clear that, since molecules have both nonzero size and range of interaction, as the density is increased and correspondingly the average distance between molecules decreased, simultaneous interactions between larger and larger groups of molecules become important. It is also easy to see that when the density becomes so large that the average volume per molecule, $V/N = 1/n$, is only slightly larger than the volume of a molecule, then every molecule must be in continuous interaction with a large number of neighboring molecules. Densities for which this situation occurs are characteristic of the condensed (liquid and solid) phases of matter.

There is another characteristic difference between a gas and a liquid or solid. The observation that in a dilute gas the motion of a molecule is only occasionally disturbed is equivalent to saying that in the gas the kinetic energy corresponding to a molecule's translational motion is large relative to its potential energy. In the liquid and solid states the translational energy is of the same order as or less than the height of the localized potential energy maxima that separate the positions at which the molecules are located. As a result of this alteration in the relative importance of translational kinetic energy and potential energy, completely new considerations must be invoked in the study of liquids and solids, the most important of which is the concept of structure. As we use the term, the concept of structure includes any spatial arrangement of atoms and molecules that differs from a random distribution; the term is used in the same fashion as in our earlier discussions of molecular structure and crystal structure (see Chapters 9 and 11). Note that we include in the concept of structure the partially regular spatial arrangements such as are characteristic of the liquid phase. We have already discussed, in Chapter 11, some aspects of crystal structure and their relationship to types of chemical bonding. In this chapter we study the thermodynamic properties of solids.

22.2
Some Thermal Properties of Crystalline Solids

In Chapter 11 we focused attention on those properties of the solid state concerned with the nature of the crystalline order and the nature of the bonding. Corresponding to the several broad categories of bonding type there are characteristic ranges of electrical conductivity, thermal conductivity, strength, and so forth. In an elementary text such as this one, it is not possible to paint even a crude picture of the many properties of the various classes of crystalline solids and how these are related to the electronic and molecular structure of the solid. We shall therefore examine only a few subjects, hoping that they are sufficient to whet the student's appetite to learn more.

Perhaps the oldest aspect of the theory of the solid state is the study of its equation of state and thermodynamic properties. Let us consider first the case of an electrically insulating solid, such as crystalline Ar, NaCl, or diamond. Althought the nature of the bonding differs greatly among these crystals, all share the feature that at ordinary temperatures the concentration of free electrons in the solid is negligibly small. Given that the lattice point is the position of minimum potential energy, it is to be expected that the atoms or ions, which have kinetic energy, will oscillate back and forth about their lattice sites. The determination of the modes of motion of a body containing N particles is obviously a very difficult chore, but this analysis may be made, at least in principle, by using the same techniques as described in our discussion of the internal motions of a molecule. Thus, suppose that we regard a perfect crystalline solid as one huge molecule. If there are N atoms in the crystal, there must be $3N-6$ internal degrees of freedom, and six degrees of translational and rotational freedom of the crystal as a rigid body. Clearly, when $N \approx 10^{24}$, the translation and rotation of the entire crystal may be disregarded, and we need consider only the internal degrees of freedom. If the components of the solid are basically monatomic, as are Ar, NaCl, and diamond (the several structural units are single particles with only electronic degrees of freedom), then all internal degrees of freedom must correspond to vibrations. If the solid contains polyatomic species, as do crystalline CH_4, N_2 and others, then some of the internal degrees of freedom will correspond to rotation (or torsion) about a lattice site. If the molecular species is very complex, say, $CH_3CH_2CH_2CH_2CH_3$, then there are complicated internal modes that have some rotational and some vibrational character. The moral of these observations is simple—each solid must be examined separately to see if there are any special features that must be considered in the theoretical description.

In the simplest case, when all the internal degrees of freedom may be considered to be vibrations, it is possible to imagine a normal-coordinate analysis carried cut. The internal motion of the crystal is then completely described by the normal frequencies and the normal modes. Let ν_i be the frequency of the ith normal mode. Because, by definition, each normal mode of vibration is independent of all others, the canonical partition function of the crystal is of the form, similar to Eq. 21.113,

$$Q_N^{\text{crystal}} = \prod_{i=1}^{3N-6} \left[1 - \exp\left(-\frac{h\nu_i}{k_B T}\right) \right]^{-1}. \qquad (22.1)$$

Because the crystal contains N atoms, we have labeled the partition function with a subscript N. However, Q_N^{crystal} really corresponds to the single-molecule partition function Q_1 of Chapter 21 because all the atoms are in one crystal.

Now, since $N \approx 10^{24}$, and the spectrum of frequencies is dense, we may consider them to be a continuous distribution in the variable ν. Let $\mathscr{P}(\nu)\,d\nu$ be the fraction of frequencies lying in the range between ν and $\nu + d\nu$. We shall later associate one degree of freedom with each frequency, so that $\mathscr{P}(\nu)$ is just proportional to the density of vibrational states of the crystal. Clearly we have

$$\int_0^\infty \mathscr{P}(\nu)\,d\nu = 1, \tag{22.2}$$

and the partition function can be written in the form

$$\ln Q_N^{\text{crystal}} = -\sum_{i=1}^{3N-6} \ln\left[1 - \exp\left(-\frac{h\nu_i}{k_B T}\right)\right]$$

$$= -3N \int_0^\infty \mathscr{P}(\nu) \ln\left[1 - \exp\left(-\frac{h\nu}{k_B T}\right)\right] d\nu. \tag{22.3}$$

For convenience we now choose to measure the energy of each normal mode of vibration relative to the energy of its lowest quantum state. As discussed in Section 21.6, to effect this change of energy zero we multiply through by $\exp(-\epsilon_{0i}/k_B T)$ for each normal mode. In the present case $\epsilon_{0i} = \frac{1}{2}h\nu_i$. Proceeding as usual, we then can obtain from Q_N^{crystal} the thermodynamic properties of the crystal. In the case of the perfect crystal we have

$$U - E_0 = k_B T^2 \left(\frac{\partial \ln Q_N^{\text{crystal}}}{\partial T}\right)_V$$

$$= -3N k_B T^2 \int_0^\infty \mathscr{P}(\nu)\left\{\frac{\partial}{\partial T}\ln\left[1 - \exp\left(-\frac{h\nu}{k_B T}\right)\right]\right\} d\nu$$

$$= -3N k_B T^2 \int_0^\infty \mathscr{P}(\nu)\left[\frac{-(h\nu/k_B T^2)\exp(-h\nu/k_B T)}{1 - \exp(-h\nu/k_B T)}\right] d\nu$$

$$= 3Nh \int_0^\infty \frac{\nu \mathscr{P}(\nu)}{\exp(h\nu/k_B T) - 1}\,d\nu, \tag{22.4}$$

where E_0 is the energy of the crystal at absolute zero, the sum of all oscillator zero-point energies.

From Eq. 22.4 for $U - E_0$, we find, for heat capacity at constant volume,

$$C_V \equiv \left(\frac{\partial U}{\partial T}\right)_{V,N} = 3Nh \int_0^\infty \nu \mathscr{P}(\nu)\left\{\frac{\partial}{\partial T}\left[\frac{1}{\exp(h\nu/k_B T) - 1}\right]\right\} d\nu$$

$$= 3N k_B \int_0^\infty \left(\frac{h\nu}{k_B T}\right)^2 \mathscr{P}(\nu) \frac{\exp(h\nu/k_B T)}{[\exp(h\nu/k_B T) - 1]^2}\,d\nu. \tag{22.5}$$

The other thermodynamic functions and the equation of state can be obtained in the usual manner (cf. Section 21.5).

Equation 22.5 is typical of the representations of the thermodynamic properties of a crystal in terms of a molecular model. All of these representations involve integration over $\mathscr{P}(\nu)\,d\nu$. Clearly, to reduce the general equations to forms suitable for specific calculation, one must determine the frequency spectrum $\mathscr{P}(\nu)$. It is at this point that the description of the thermal properties of a solid becomes difficult. In general, the computation of $\mathscr{P}(\nu)$ is long and tedious, and before discussing its general features it is worthwhile to examine the consequences of assuming particular simple forms for this distribution function.

The theory of the heat capacity of a crystalline solid played an important role in the evolution of the quantum theory. If a solid is modeled as a collection of harmonic oscillators, then the equipartition theorem of classical theory requires that each vibration contribute k_B to the heat capacity. Thus the classical prediction is that the heat capacity of a solid should be equal to $3Nk_B$ and independent of temperature.[1] At about the turn of the century, measurements of the heat capacities of many solids showed that the classical theory was incorrect; the heat capacity decreased as the temperature decreased and appeared to approach zero as the temperature approached absolute zero. These observations could not be reconciled with the classical theory of crystals. In 1907 Einstein suggested that the motions of the atoms in a crystal are quantized. Then the typical characteristic vibrational temperature of a solid could be larger than room temperature (e.g., diamond), about equal to room temperature (e.g., Al), or smaller than room temperature (e.g., Pb). When the temperature is much higher than the solid's characteristic temperature, the heat capacity becomes $3Nk_B$ (see Section 21.5). When the temperature decreases below the solid's characteristic temperature and ultimately approaches zero, the heat capacity of the solid decreases below $3Nk_B$ and ultimately approaches zero. The Einstein model thereby explains the outstanding qualitative features of the heat capacity-versus-temperature curve of a solid.

The Einstein model of a solid is very simple. It is based on the assumption that all the vibrational frequencies of the solid have the common value ν_E. Then $\mathscr{P}(\nu)\,d\nu$ has the value unity when $\nu = \nu_E$, and zero for all other values of ν. Returning to Eq. 22.5 and inserting the form of $\mathscr{P}(\nu)\,d\nu$ just described, the integrand is nonzero only at the frequency ν_E. Thus for the Einstein model we find that

$$C_V^{\text{Einstein}} = 3Nk_B\left(\frac{h\nu_E}{k_BT}\right)^2 \frac{\exp(h\nu_E/k_BT)}{[\exp(h\nu_E/k_BT)-1]^2}$$

$$= 3Nk_B\left(\frac{\Theta_E}{T}\right)^2 \frac{\exp(\Theta_E/T)}{[\exp(\Theta_E/T)-1]^2}. \qquad (22.6)$$

A plot of C_V^{Einstein} versus T/Θ_E is shown in Fig. 22.1. The quantity Θ_E is of course defined by

$$\Theta_E \equiv \frac{h\nu_E}{k_B}, \qquad (22.7)$$

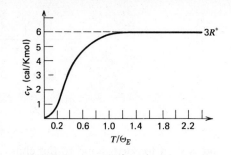

FIGURE 22.1
The heat capacity corresponding to the Einstein model of a solid $[\mathscr{P}(\nu)\,d\nu = 1$ for $\nu = \nu_E,\ \mathscr{P}(\nu) = 0$ otherwise].

[1] This is known as the *law of Dulong and Petit*. It was inferred from experiment about 1819 and only much later discovered to be a general prediction of the classical theory. The law of Dulong and Petit has been used to obtain approximate atomic weights. The idea is the following: If the molar heat capacity of any elemental solid is $3N_Ak_B = 3R = 25.1$ J/K mol, then a measurement of the specific heat of 1 g of unknown elemental substance can be scaled to give the molar heat capacity. For example, the heat capacity of 1 g of Bi is 0.1229 J/K at room temperature. By dividing this into 25.1 we obtain 204 as the rough value of the atomic weight of Bi. The currently accepted value of the atomic weight of Bi is 209.

A general rule, related to the law of Dulong and Petit, is due to Kopp. It states that the molar heat capacity of a solid is the sum of its atomic heat capacities, with the value about 25.9 for all atoms except light ones, for which the contributions are H 10.5; Be 12.5; B 10.5; C 8.4; N 12.5; O 16.7; F 20.9; all in J/K mol. This generalization by Kopp recognizes in an empirical fashion the consequences of quantum mechanics and the fact that for light atoms the characteristic vibrational temperatures are often higher than room temperature. As an application, we note that the heat capacity of 1 g of $CaCO_3$ is 0.869 J/K at room temperature. Using Kopp's rule, we predict the molar heat capacity to be 84.4 J/K mol; the observed value (at room temperature) is 86.9 J/K mol.

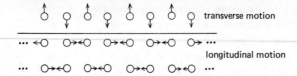

transverse motion

longitudinal motion

FIGURE 22.2
Schematic representation of the small-est-wavelength atomic vibration in a one-dimensional crystal.

and is seen to correspond to the characteristic vibrational temperature of a molecule introduced in Section 21.5.

Notwithstanding its general success, the Einstein model has failings that are attributable to the oversimplified form chosen for $\mathscr{P}(\nu)$. For example, as $T \to 0$, the heat capacity approaches zero exponentially. To show this we examine Eq. 22.6, note that $\exp(\Theta_E/T) \to \infty$ as $T \to 0$, so that the term unity in the denominator becomes negligibly small relative to $\exp(\Theta_E/T)$, and obtain after division of common factors the simple form

$$\lim_{T \to 0} C_V^{\text{Einstein}} = 3Nk_B \left(\frac{\Theta_E}{T}\right)^2 \exp\left(-\frac{\Theta_E}{T}\right). \tag{22.8}$$

Experiments show that, although the heat capacity approaches zero as T approaches zero, it does so more slowly than Eq. 22.8 predicts. Thus the theory requires improvement.

The assumption that $\mathscr{P}(\nu)\,d\nu$ is unity for $\nu = \nu_E$ and zero for all other values treats the atomic oscillators as if they were independent of one another. Because of the strong interactions between the atoms this is, of course, not true. It is convenient, now, to examine some of the general properties of $\mathscr{P}(\nu)$, without actually calculating this function. In the process, we shall see how simple ideas can be used to improve upon the Einstein model of the crystal.

1. In a crystalline medium with atomic structure, and hence a regular lattice spacing, the smallest-wavelength vibration that can occur describes the antiphase motion of two adjacent molecules. By antiphase motion of adjacent molecules we mean that they move in opposite directions either parallel or perpendicular to the line between them, as shown schematically in Fig. 22.2. Why is this the shortest-wavelength motion allowed? Normal modes are defined by the patterns of phased displacements of groups of atoms. Since atoms cannot be closer than one lattice spacing, there can be no simple phase-coherent[2] motion of the atoms with a wavelength less than their separation. Note how a hypothetical continuous medium differs from a real, molecularly coarse medium in this respect. We thus conclude that, for a crystal,

$$\mathscr{P}(\nu) = 0 \qquad \text{for} \qquad \nu > \nu_{\text{max}}, \tag{22.9}$$

where ν_{max} is the frequency corresponding to the shortest-wavelength normal mode of the crystal.

2. Just as for a molecule, the total number of degrees of freedom of a crystal is determined by the number of atoms in the crystal. Neglecting the six degrees of freedom corresponding to translation and rotation of the entire crystal, if the particles at the lattice sites are monatomic, the total number of degrees of freedom must be $3N$.

---

[2] By coherent we mean exactly in phase for all time. This includes the notion of exactly out of phase for all time, since "out of phase" just specifies the direction of motion.

3. Since we have

$$\exp\left(\frac{h\nu}{k_B T}\right) = 1 + \frac{h\nu}{k_B T} + \cdots, \tag{22.10}$$

in the high-temperature limit Eq. 22.5 becomes

$$\lim_{T \to \infty} C_V = 3Nk_B \int_0^\infty \left(\frac{h\nu}{k_B T}\right)^2 \mathscr{P}(\nu) \left[\frac{1}{(h\nu/k_B T)^2}\right] d\nu$$

$$= 3Nk_B, \tag{22.11}$$

where we have used Eq. 22.2. Thus, in the high-temperature limit the contribution to the specific heat from each of the $3N$ modes of vibration is just k_B. The result cited is, again, an example of the equipartition of energy in the limit $T \to \infty$. (See Section 21.5.)

4. Consider now the limit in which the wavelength of the normal vibration is very large relative to the spacing between the particles, such as is shown schematically for a one-dimensional lattice in Fig. 22.3. It is clear that if the wavelength is large enough, so that many atoms are displaced in the normal vibration, the atomic structure of the crystal is of little relevance. In this limit the medium behaves as if there were no atomic structure, that is, as if it were continuous. The vibrations of a continuous medium are well understood. Each possible characteristic vibration of an elastic[3] medium corresponds to a sound wave of a single frequency. That this is so is most easily visualized for the continuous one-dimensional case, which is the violin string. Each possible characteristic vibration of the violin string leads to a sound wave (because the string "drives" the surrounding air) of a single frequency. The superposition of vibrations created when the bow is drawn across the string is the one-dimensional analog of the superposition of vibrations created in an elastic solid when it is struck. Just as an arbitrary displacement of the violin string can be represented as the sum of the displacements corresponding to different characteristic vibrations, so can a displacement in a solid be represented as a sum of normal-mode displacements. It is this possibility of representing a distribution of compressions and expansions in terms of the normal modes of vibration corresponding to sound waves that permits us, in the long-wavelength limit, to associate the distribution of atomic amplitudes with the corresponding continuous density distribution, and thereby with the frequency spectrum of sound waves. Now, it can be shown that for a continuous medium[4]

$$\mathscr{P}(\nu) \propto \nu^2. \tag{22.12}$$

Since $\mathscr{P}(\nu)$ is proportional to the density of vibrational states in the crystal, Eq. 22.12 states that for $\nu \ll \nu_{max}$ the density of vibrational states is proportional to ν^2. Debye suggested (1912) that the approximation

---

[3] By elastic we mean that the force and displacement are related, as they are for a perfect spring, by Hooke's law.

[4] The argument is very similar to that used to derive Eq. 15.25. The numbers of longitudinal and transverse modes with frequencies less than ν in a cube of elastic continuum with edge length L are just the numbers of positive integers that satisfy, respectively,

$$\frac{c_l^2}{L^2}(n_1^2 + n_2^2 + n_3^2) \leq \nu^2$$

$$\frac{2c_t^2}{L^2}(n_1^2 + n_2^2 + n_3^2) \leq \nu^2$$

where c_l and c_t are the velocities of longitudinal and transverse waves. The remainder of the argument follows ...at given in Chapter 15.

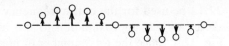

FIGURE 22.3
Schematic representation of a long-wavelength normal mode in a one-dimensional crystal.

22.12 be used for the entire frequency distribution, that is, up to ν_{max}. A sketch of this approximation to the frequency spectrum is displayed in Fig. 22.4. Since the total number of normal modes of vibration is $3N$, ν_{max} is fixed by a normalization condition. We now examine the limiting behavior of the heat capacity as $T \to 0$. At low temperature we have $h\nu_{max}/k_B T \gg 1$, and only the lowest-frequency vibrations of the solid are thermally excited. It is in this temperature range that setting $\mathscr{P}(\nu) \propto \nu^2$ is a good approximation. If $x \equiv (h\nu/k_B T)$, then Eq. 22.5 is easily transformed to read

$$C_V = 3Nk_B \int_0^{x_{max}} x^2 \mathscr{P}(x) \frac{e^x}{(e^x - 1)^2} \left(\frac{k_B T}{h}\right) dx, \qquad (22.13)$$

so that, if we substitute $\mathscr{P}(x) \propto x^2 T^2$,

$$C_V \propto T^3 \int_0^{x_{max}} \frac{x^4 e^x}{(e^x - 1)^2} \, dx. \qquad (22.14)$$

But if $x_{max} \gg 1$, the integrand tends to zero for $x \ll x_{max}$, and therefore the upper limit of integration may be made infinity without introducing appreciable error. The integral is then just a number, independent of temperature, and we can write[5]

$$\lim_{T \to 0} C_V \propto T^3. \qquad (22.15)$$

The specific value of the proportionality constant depends on the properties of the solid, and need not concern us. What is important is the functional form of the predicted temperature dependence, which is verified for many solids when the temperature is low enough. Of course, Eq. 22.15 depends on the assumption embodied in Eq. 22.12, and that is a valid low-frequency approximation only for three-dimensional systems. Crystals of polymer molecules, or crystals containing strong linear or strong planar interactions, are, in a certain sense, "weakly three-dimensional." For these cases, Eq. 22.12 cannot be used and, correspondingly, a different low-temperature limit is found for C_V.

5. In general, except in the very-long-wavelength region, the distribution of vibrations of a crystal differs considerably from that of the Debye model. A qualitative understanding of the nature of the vibrational spectrum of a crystal can be built up as follows.

Consider the one-dimensional lattice shown in Fig. 22.2. As indicated, there are two kinds of atomic motion: along the chain, say, the z axis (*longitudinal modes*), and perpendicular to the chain, say, along the x and y axes (*transverse modes*). Each class of atomic motion gives rise to a *branch* of the vibrational spectrum. Clearly, in the case illustrated in Fig. 22.2, the transverse vibrations are degenerate (x and y are equivalent). The characteristic vibrations of the linear lattice are described by their frequencies $\nu_1, \nu_2, \ldots$ and by the pattern of displacements of the atoms that correspond to a wave on the lattice in which the displacements of adjacent atoms differ by only a phase factor. Therefore, the pattern of displacements associated with a vibration of frequency ν_i can be described by the wave vector $k_i = 2\pi/\lambda_i$ for that vibration. The dependence of vibrational frequency on k is known as the *dispersion relation*. The vibrations of a crystal have dispersion; that is, they have frequencies that are different for different wavelengths, for the same reason that the symmetric and antisymmetric stretches in the H_2O molecule have different frequencies, namely, interaction between the atoms in different oscil-

[5] The thermodynamic consequences of Eq. 22.15 were discussed in Chapter 18.

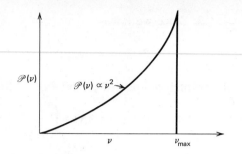

FIGURE 22.4
Schematic representation of a Debye spectrum.

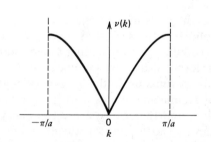

FIGURE 22.5
Dispersion relation, $\nu(k)$ versus k, for a linear lattice with nearest-neighbor spacing a and only nearest-neighbor interaction. The branch shown corresponds to the longitudinal mode. For this lattice, with one atom per unit cell, this is the longitudinal acoustic mode—note that $\nu(k) \to 0$ as $k \to 0$. For a mode of frequency ν, the time dependence of the vibrational displacement of a lattice point is proportional to $\cos(ka - 2\pi\nu t)$. Note that $\nu(k)$ must be periodic with period $2\pi/a$, since the substitution $k \to k + 2\pi/a$ introduces no change in the displacement pattern. It is therefore customary to restrict k to the range $-\pi/a \leqslant k \leqslant \pi/a$.

lators.[6] In a molecule of n atoms, the $3n-6$ vibrations are usually well separated in frequency. In contrast, as already pointed out when $\mathscr{P}(\nu)$ was introduced, in a crystal of N atoms the allowed wavelengths are so close together that we can take k to be a continuous variable. The dispersion relation for longitudinal vibrations of a linear lattice with one atom per unit cell, lattice spacing a, is shown in Fig. 22.5.

Consider next a linear lattice with two atoms per unit cell, as shown in Fig. 22.6. In general, the restoring force between the two atoms in one unit cell will be different from that between adjacent atoms in different unit cells, hence the vibrational spectrum is more complicated than that for a linear lattice with only one atom per unit cell. It is easy to distinguish two classes of atomic motions. In one class the unit cells, that is, the pairs of A and B atoms, are displaced with respect to the lattice points. In the other class the atoms vibrate with respect to one another within the unit cells. These internal vibrations can occur, and have a definite phase with respect to the internal vibrations in other unit cells, even when the centers of mass of the unit cells are undisplaced. The lattice with only one atom per unit cell does not have this degree of freedom. Modes involving the displacement of the center of mass of the unit cell are called *acoustic*, whereas those involving internal motions of the unit cell are called *optic* (see Fig. 22.7). The two classes of modes are most easily distinguished in the long-wavelength limit.

It is easy to see that there is a fundamental difference between acoustic and optic modes when their wavelengths become indefinitely large, at $k=0$. For the case of the acoustic mode, in this limit the centers of mass of all the unit cells move in phase, and there is no longer a restoring force. Consequently, for the acoustic mode we expect that $\nu(k) \to 0$ as $k \to 0$. In contrast, for the optic mode, even if all internal displacements of all unit cells are in phase, there is a nonvanishing restoring force internal to the unit cell. Consequently, for the optic mode we expect that $\nu(k) \to \nu_0 \neq 0$ as $k \to 0$. In fact, ν_0 is usually the highest-frequency vibration of the system. The longitudinal acoustic and optic mode dispersion relations for a linear lattice are shown in Fig. 22.8.

In general, for a three-dimensional lattice, if there is only one atom per unit cell there are three branches in the vibrational spectrum, one corresponding to longitudinal modes and two to transverse modes. These three are acoustic modes and are really purely longitudinal or transverse only for waves propagated along symmetry directions in the crystal structure. When there is more than one atom per unit cell, intracell vibrations give rise to optic branches in the vibrational distribution. If there are s atoms per unit cell, there are $3s-3$ optic branches in the vibrational spectrum. It is generally the case that for the acoustic modes $\nu(k) \to 0$ as $k \to 0$, whereas for the optic modes $\nu(k) \to \nu_0$ as $k \to 0$. Figure 22.9 shows a set of calculated frequency dispersion curves for crystalline Ar. In this solid there is one atom per unit cell, hence only three acoustic branches to the spectrum. Figure 22.10 shows a set of frequency dispersion curves for crystalline Ge. In this case there are two atoms per unit cell, hence three acoustic and three optic branches in the vibrational spectrum.

FIGURE 22.6
Schematic representation of a linear lattice with two atoms, A and B, per unit cell.

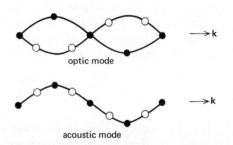

FIGURE 22.7
Schematic representation of the transverse optic and transverse acoustic modes of a diatomic linear lattice. The atomic displacements are illustrated for a common value of the wave vector (hence also the wavelength).

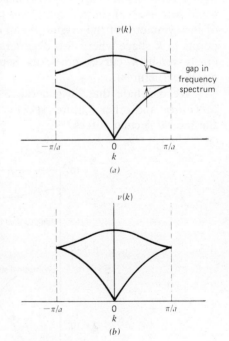

FIGURE 22.8
Dispersion relations for a linear lattice with two atoms per unit cell. Case (a): $m_A \neq m_B$, $a = 2b$. Case (b): $m_A = m_B$, $a = 2b$. Nearest-neighbor interactions only in these cases.

---

[6] If the O atom of H_2O did not move, and the two OH stretching motions were independent, they would have the same frequency. But the O atom does move when a H atom moves, so the two OH motions are not independent. Also, on stretching the OH there is a contribution to the change in potential energy from their interaction. The net result is the replacement of two independent OH motions with symmetric and antisymmetric combined motions of different frequencies.

Finally, it is of interest that there are some points on the frequency dispersion curves where $(\partial \nu / \partial k) = 0$. For example, one such point is the maximum in the longitudinal acoustic branch shown in Fig. 22.9. At the frequencies for which the gradient in the dispersion relation vanishes, there are discontinuities in the frequency derivative of the density of vibrational states, namely, in $\partial \mathcal{P}(\nu) / \partial \nu$. These are called *Van Hove singularities*, and there is at least one in the density of vibrational states of a crystal. They show up as kinks in $\mathcal{P}(\nu)$, as seen in Fig. 22.11.

The reason these occur is as follows. Imagine two surfaces of constant ν in k space (see Section 11.11), corresponding to ν and $\nu + d\nu$. These two surfaces enclose a volume dV_k. The volume dV_k can be obtained by multiplying an element of area on one surface, $d\sigma$, by the distance between surfaces, Δk, and integrating the contributions over the entire constant frequency surface. Then we have

$$\mathcal{P}(\nu)\, d\nu = \frac{1}{8\pi^3}\, dV_k = \frac{1}{8\pi^3} \int \Delta k\, d\sigma,$$

where the factor $1/8\pi^3$ is the number of allowed values of k per unit volume (see Eq. 15.25). We now write

$$d\nu = \left| \left(\frac{\partial \nu}{\partial k} \right) \right| \Delta k + \text{higher-order terms},$$

which, on substitution in the equation for $\mathcal{P}(\nu)$, gives

$$\mathcal{P}(\nu) = \frac{1}{8\pi^3} \int \frac{d\sigma}{|(\partial \nu / \partial k)|}.$$

Then, if $(\partial \nu / \partial k)$ is zero, or if it discontinuously changes sign, the next-higher term in the Taylor series must be kept and we expect a cusp of some sort in $\mathcal{P}(\nu)$. In general, $(\partial \nu / \partial k) = 0$ for special values of k because of the symmetry of the crystal; it can change sign discontinuously only at points in k space where ν is degenerate (degeneracy occurs when different ν-versus-k curves cross, as happens for some three-dimensional crystal structures).

We conclude that, in general, $\mathcal{P}(\nu)$ has maxima, discontinuities in $d\mathcal{P}(\nu)/d\nu$, and other features specific to the type of lattice and range of the intermolecular forces.

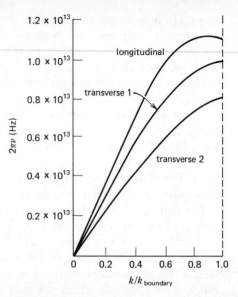

FIGURE 22.9

Typical dispersion relation in a face-centered cubic crystal in the [210] direction (data are for solid argon). From J. Reissland, *The Physics of Phonons* (Wiley, New York, 1973).

FIGURE 22.10

Dispersion relations for the [100] and [111] directions in Ge. From W. Cochran, *Proc. Roy. Soc. A* **253**, 260 (1959).

FIGURE 22.11

Vibrational frequency distribution for crystalline Ar. From D. N. Batchelder, M. F. Collins, B. C. G. Haywood, and R. G. Sidney, *J. Phys. C. Solid State Phys.* **3**, 249 (1970).

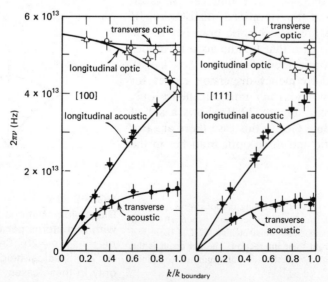

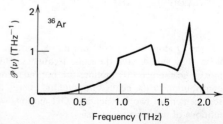

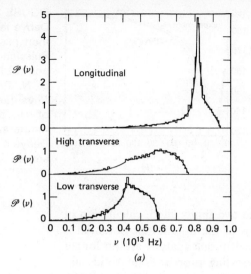

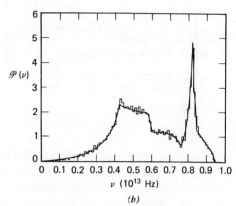

FIGURE 22.12

(a) Vibrational frequency spectra for the three branches of Al. The histograms are obtained from computed eigenfrequencies for 2791 wave vectors. The smooth curves are obtained from the histograms together with the inclusion of the singularities caused by the various critical points.

(b) Total vibrational frequency spectrum for Al at 300 K. The histogram is obtained from 8373 calculated eigenfrequencies. The smooth curve is obtained from the histogram after inclusion of the singularities arising from the critical points of the various branches. From C. B. Walker, *Phys. Rev.* **103**, 547 (1956).

A comparison of the Debye frequency spectrum (Fig. 22.4) with either of the frequency spectra shown in Fig. 22.11 or Fig. 22.12 shows that it does not reproduce at all well the high-frequency end of the true spectrum. Nevertheless, C_V is not very sensitive to the frequency spectrum, and the Debye model provides a good approximation to the specific heat of a monatomic solid. In Fig. 22.13b we see that the specific heats of a variety of solids may be reasonably well described in terms of a characteristic Debye temperature, $\Theta_D \equiv h\nu_{max}/k_B$. Very precise measurements show that Eq. 22.14 does not fit the heat capacity data perfectly; the deviations can be expressed as an empirical variation of Θ_D with temperature which arises from the deviations of the Debye spectrum from the true vibrational spectrum of the solid (Fig. 22.14), and can be accounted for by use of the correct $\mathscr{P}(\nu)$.

It should now be clear that, by using the considerations outlined, one can interpret the thermodynamic properties and equation of state of a crystalline solid in terms of the potential between its constituent atoms or molecules. Of course, the motion of the atoms in a solid can be represented by the harmonic approximation only when the amplitudes of vibration are small. Near the melting point the average amplitude of atomic displacement is large, anharmonic components of the force are important, and all our considerations must be supplemented. Similarly, when the solid is severely compressed or extended, beyond the domain in which Hooke's law is valid, inelastic components of the response to

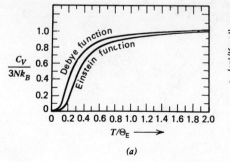

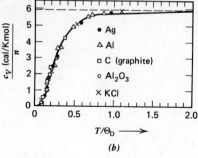

FIGURE 22.13

(a) A comparison of the Einstein and Debye heat capacity curves. The frequency associated with the Einstein curve is equal to the maximum Debye frequency.

(b) Comparison of the Debye heat capacity curve and the observed heat capacities of a number of simple substances. From F. Seitz, *Modern Theory of Solids* (McGraw-Hill, New York, 1940).

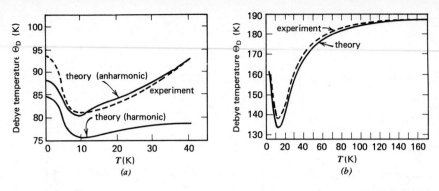

FIGURE 22.14
Debye temperature Θ_D versus T for (a) inert gas solids [J. W. Leech and J. A. Reissland, *Phys. Lett.* **14**, 305 (1965)] and (b) NaI [R. A. Cowley, W. Cochran, B. N. Brockhouse, and D. B. Woods, *Phys. Rev.* **131**, 1036 (1963)]. The value of Θ_D plotted is calculated at each temperature by requiring that the Debye model exactly reproduce the observed value of C_V.

external force must be considered. In the two limits cited, calculation of the equation of state of the solid in terms of the interatomic potential is very difficult, since the simplifications used in our analysis are no longer adequate. Since many of the most interesting properties of solids are those observed at high temperature or pressure, it is worthwhile to examine some aspects of the influence of anharmonicity on crystal properties

22.3
The Contribution of Anharmonicity to the Properties of a Crystal

The potential energy surface of a crystal can be represented by the harmonic approximation only for small-amplitude motion, and this approximation cannot be used to interpret many important crystal properties, such as thermal expansion. Although the general theory of anharmonic crystals is too complex to describe in this text, an idea of the nature of anharmonic contributions to the crystal properties can be obtained from two simple examples.

Consider a one-dimensional anharmonic oscillator that has the Hamiltonian

$$H = \frac{p^2}{2m} + \tfrac{1}{2}m\omega^2 x^2 + a_3 x^3 + a_4 x^4. \tag{22.16}$$

For illustrative purposes we assume that $a_3 x^3$ and $a_4 x^4$ are small enough relative to the other terms and to $k_B T$ that first-order perturbation theory can be used to evaluate their influence on the oscillator's partition function. For our purposes it is sufficient to evaluate the classical partition function. Just as in quantum mechanical perturbation theory, the first-order correction to the classical partition function is obtained by averaging the perturbation energy over the distribution characteristic of the unperturbed system. We must, then, evaluate

$$Q = \frac{1}{h} \int_{-\infty}^{\infty} dp \exp\left(-\frac{p^2}{2mk_B T}\right) \int_{-\infty}^{\infty} dx \exp\left(-\frac{m\omega^2 x^2}{2k_B T}\right)$$

$$\times \left(1 - \frac{a_3 x^3}{k_B T} - \frac{a_4 x^4}{k_B T} + \frac{a_3^2 x^6}{2(k_B T)^2} + \cdots\right)$$

$$= \frac{2\pi k_B T}{h\omega}\left(1 - \frac{3a_4 k_B T}{m^2\omega^4} + \frac{15}{2}\frac{a_3^2 k_B T}{m^3\omega^6} + \cdots\right). \tag{22.17}$$

As usual, $u = k_B T^2(\partial \ln Q/\partial T)$, from which we find

$$u = k_B T - \left[3\frac{a_4}{(m\omega^2)^2}k_B^2 - \frac{15}{2}\frac{a_3^2}{(m\omega^2)^3}k_B^2\right]T^2, \tag{22.18}$$

and

$$c_V = k_B - \left[6 \frac{a_4}{(m\omega^2)^2} k_B^2 - \frac{15}{2} \frac{a_3^2}{(m\omega^2)^3} k_B^2 \right] T. \tag{22.19}$$

Of course, the classical harmonic contribution to c_V is just k_B, so that anharmonicity of the type shown leads to an extra contribution to the heat capacity linear in T. The quantum mechanical prediction gives the same temperature dependence.

The results of a detailed calculation of anharmonic contributions to the heat capacities of crystalline Ne, Ar, Kr, and Xe are shown in Table 22.1. Clearly, the prediction that the anharmonicity adds a term to C_V that grows linearly with T is also valid in the general case. When this anharmonic contribution to C_V is accounted for, the observed and predicted variations of Θ_D with temperature are in good agreement (see Fig. 22.14).

A much simpler, yet qualitatively useful, way of accounting for some of the effects of anharmonicity on the properties of a crystal can be based on the following notion. Although thermal expansion of the crystal leads to changes in atomic separations, the displacement of an atom from any position of equilibrium is adequately approximated by a harmonic potential surface. That is, the crystal is harmonic at all temperatures, but the frequencies change with T through a dependence on the volume. To see how this idea is quantified, we return to Eqs. 22.3 and 22.4 and note that if $\mathscr{P}(\nu)$ is the Debye distribution, then the Helmholtz free energy A_{Debye} and the internal energy U_{Debye} are functions of the variable Θ_D/T. We now write

$$p_{Debye} = -\left(\frac{\partial A_{Debye}}{\partial V} \right)_T$$

$$= -\left(\frac{\partial A_{Debye}}{\partial \Theta_D} \right)_T \left(\frac{\partial \Theta_D}{\partial V} \right)_T. \tag{22.20}$$

Noting that U_{Debye} is of the form $\Theta_D f(\Theta_D/T)$, and using the change of variables

$$-T\left[\frac{\partial}{\partial T} f\left(\frac{\Theta_D}{T} \right) \right]_{\Theta_D} = \Theta_D \left[\frac{\partial}{\partial \Theta_D} f\left(\frac{\Theta_D}{T} \right) \right]_T, \tag{22.21}$$

we find from the thermodynamic relation

$$U = -T^2 \left[\frac{\partial}{\partial T} \left(\frac{A}{T} \right) \right]_V \tag{22.22}$$

that

$$U_{Debye} = \Theta_D \left(\frac{\partial A_{Debye}}{\partial \Theta_D} \right)_T, \tag{22.23}$$

since holding V constant is equivalent to holding Θ_D constant. Then, using Eq. 22.20, we have (simplifying the subscript to D)

$$p_D = -\frac{U_D}{\Theta_D} \left(\frac{\partial \Theta_D}{\partial V} \right)_T \tag{22.24}$$

$$= \frac{U_D}{V} \gamma_G, \tag{22.25}$$

where

$$\gamma_G \equiv -\left(\frac{\partial \ln \Theta_D}{\partial \ln V} \right)_T \tag{22.26}$$

TABLE 22.1

ANHARMONIC CONTRIBUTIONS TO THE HEAT CAPACITY, SHOWING THE APPROACH TO THE LINEAR LAW AT HIGH TEMPERATURES[a]

$T(K)$	$\dfrac{c_V^0}{3k_B}$	$\dfrac{\Delta c_{V(anh)}}{3k_B}$	$-\dfrac{10^4\,\Delta c_{V(anh)}}{3k_B T}\ (K^{-1})$
(a) Neon			
4	0.020	−0.015	35.5
8	0.165	−0.058	72.5
12	0.368	−0.123	102.5
16	0.536	−0.156	97.5
20	0.655	−0.175	87.5
24	0.738	−0.193	80.4
(b) Argon			
10	0.158	−0.014	14.0
20	0.523	−0.040	20.0
30	0.727	−0.050	16.7
40	0.830	−0.059	14.8
50	0.885	−0.072	14.4
60	0.918	−0.085	14.2
70	0.938	−0.098	14.0
80	0.952	−0.111	13.9
(c) Krypton			
20	0.647	−0.021	10.53
40	0.886	−0.038	9.45
60	0.946	−0.055	9.09
80	0.969	−0.072	9.05
100	0.979	−0.090	9.04
120	0.986	−0.108	9.03
(d) Xenon			
20	0.699	−0.014	7.02
40	0.907	−0.025	6.32
60	0.957	−0.037	6.16
80	0.975	−0.049	6.14
100	0.984	−0.061	6.13
120	0.988	−0.074	6.13
140	0.991	−0.086	6.13

[a] c_V^0 is the harmonic part of the heat capacity. The calculations all refer to constant volume corresponding to that at $T = 0$ K. Reproduced, with permission, from J. W. Leech and J. A. Reissland, *J. Phys. Chem. Solid State Phys.* **3**, 997 (1970), copyright The Institute of Physics.

is called the Grüneisen constant. Clearly, the Grüneisen constant is a measure of the effects of anharmonicity inasmuch as it represents the rate of change of crystal vibrational frequencies with the volume. Now, differentiating Eq. 22.25 with respect to T, holding V fixed,

$$\left(\frac{\partial p_D}{\partial T}\right)_V = \gamma_G \frac{C_V}{V}, \qquad (22.27)$$

and using the identity

$$\left(\frac{\partial p}{\partial T}\right)_V = -\left(\frac{\partial p}{\partial V}\right)_T \left(\frac{\partial V}{\partial T}\right)_p, \qquad (22.28)$$

gives

$$\gamma_G = \frac{\alpha V}{\kappa_T C_V}, \qquad (22.29)$$

where α and κ_T are, as usual, the coefficients of thermal expansion and isothermal compressibility, respectively. Equation 22.29 provides a means of calculating γ_G and thereby generating an equation of state for the solid (i.e., Eq. 22.25). According to Eq. 22.29, γ_G is temperature-independent to the extent that α, κ_T, and C_V are.

It is generally found that the Grüneisen equation of state gives a reasonable but not exact representation of the properties of a crystal; some values of γ_G are listed in Table 22.2. Given the crudity of the physical ideas, it could not be expected to be better.

The considerations of the preceding two sections suffice to provide an interpretation of the thermodynamic properties of solids composed of monatomic molecules, but are inadequate to interpret the rich spectrum of properties exhibited by solids composed of polyatomic molecules. In this section we mention a few of the new features introduced into the description when the molecules interact with directional forces and are polyatomic.

Consider a crystal composed of polyatomic molecules. In addition to the spectrum of vibrations describing the motions of the centers of mass of the molecules, there are vibrations corresponding to the internal degrees of freedom of the molecules. Given nonzero interactions between

22.4 Some Properties of Complex Solids and of Imperfect Solids

TABLE 22.2
EXAMPLES OF THE GRÜNEISEN PARAMETER FOR CUBIC CRYSTALS[a]

Crystal	Temperature Range (K)	γ_G Range
Ne	0–25	3.1–3.5
Ar	0–50	2.8–3.1
Kr	0–50	2.75–3.05
Xe	0–50	2.75–3.0
Cu	0–600	1.7–1.95
Ag	0–300	2.7–2.8
Au	0–300	2.3±0.05
Al	0–400	2.3–2.7
Na	0–200	0.9–1.1
K	0–150	1.05–1.2

[a] Reproduced from J. A. Reissland, *The Physics of Phonons* (Wiley, New York, 1973).

the molecules, it is not strictly correct to separate the vibration spectrum of the crystal into intramolecular vibrations and intermolecular vibrations, since all are coupled together and, in principle, of mixed character. In practice, however, the forces within a molecule are so much larger than the forces between molecules that it is an excellent approximation to regard the intramolecular vibrations as being distinct from the intermolecular vibrations, and to assume that the former are only slightly influenced by the latter in the solid. In the simplest cases, then, the vibrational spectrum of a crystal of polyatomic molecules consists of the discrete set of molecular vibrations superimposed on the continuous spectrum arising from center-of-mass motion. Of course, it is also necessary to account for the rotation of the polyatomic molecules. When the rotation is unhindered, the contribution to the energy and other thermodynamic functions is the same as that corresponding to rotation in the dilute gas. When the rotation is hindered, it may be necessary to include in the vibrational spectrum a set of torsional oscillations. Clearly, all the considerations discussed in the study of the internal motion of polyatomic molecules in Chapter 21 are relevant. It must also be remembered that, because of the intermolecular forces in the crystal, the properties of the molecule may be somewhat altered from what they were in the dilute gas. For example, experiments show that the internal vibration frequencies of molecules in the crystalline phase often differ by $\sim 10\ \mathrm{cm}^{-1}$ from the corresponding values in the gas phase, even when there is not specific bonding between molecules. If such bonding exists, as when hydrogen bonding occurs, the shifts in frequencies of bonded groups can be much larger ($\sim 100\ \mathrm{cm}^{-1}$). Of course, if there is any form of covalent bonding between molecules in the crystal (e.g., in I_2), large alterations in the intramolecular vibrational frequencies are found.

It is easy to construct a useful approximation to the thermodynamic properties of a crystal of polyatomic molecules when hydrogen bonding and covalency can be neglected. Then the spectrum of vibrations may be regarded as the superposition of several Einstein spectra, one each for the several intramolecular vibration frequencies, and a Debye spectrum for the centers of mass. If there exist low-frequency molecular torsional modes, these also may be represented by Einstein terms in the spectrum (see Fig. 22.15). Consequently, a first approximation to the heat capacity of a complex solid is a superposition of Einstein heat capacity terms (one for each internal vibration and each torsion) and a Debye heat capacity term. All other thermodynamic functions may be constructed in the same way.

Thus far in our description of the thermal properties of crystals it has been assumed that the solid state has a perfectly ordered lattice in which the environment of any selected molecule is the same as that of any other molecule in the lattice. This is an idealization; naturally occurring crystalline matter has several different kinds of inherent imperfections.[7] These imperfections make a significant contribution to a number of the properties of a crystal—for example, the strength, coefficient of self-diffusion, magnitude of ionic conductivity (in alkali halides), and so forth. The study of the properties of imperfections is an important part of the science of metallurgy, and much of the recent progress in the design of new alloys has arisen from increased understanding of imperfections and how to

[7] That defects must be present follows from the observation that they represent atomic configurations whose energies differ from that of an ideal perfect crystal by only *finite* amounts, and they are not prohibited from occurring by the constraints that define the equilibrium state.

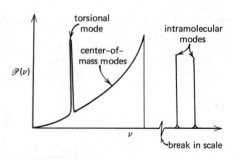

FIGURE 22.15
Schematic representation of a simple approximation to the frequency spectrum of a polyatomic solid. The torsional mode often has a frequency comparable with those of center-of-mass modes, whereas intramolecular modes usually have much higher frequencies. Note the break in the ν axis.

control them. In this book we shall give only a brief qualitative description of several typical kinds of imperfections.

The first and most obvious deviation from perfect lattice symmetry is the existence of crystal surfaces. Molecules at the surface of a crystal obviously cannot have the same environment as a molecule deep inside the lattice. At the level of thermal properties, the differences in interaction energies and vibrational motions between a molecule in the surface and one in the bulk leads to a *surface free energy*. The magnitude of the surface free energy plays a role in determining the shape of a crystal—a perfect crystal at equilibrium has a shape that minimizes the surface free energy. At the level of electronic structure, the existence of a surface leads to energy levels associated with the surface and differing from those characteristic of the bulk crystal. Such energy levels are important in phenomena as diverse as catalysis and electrical conduction in composite semiconductors. Of course, the ratio of numbers of surface to bulk atoms, $N^{2/3}/N \approx N^{-1/3} \approx 10^{-8}$ in 1 mol, implies that the specific properties of the surface of the crystal need be taken into account only when the phenomena of importance are limited to the surface zone (as in catalysis).

Two common types of crystal imperfections are vacant lattice sites and molecules in interstitial positions. Vacancies occur at random throughout a crystal. In general, the concentration of vacancies is extremely small even close to the melting point of the crystal. We can see why this is so as follows. A vacancy can be imagined as generated by removing a molecule from the bulk of the crystal and placing it on the surface. If each molecule in the crystal interacts with Z neighbors, the energy of sublimation of the crystal is $NZ\epsilon/2$, where ϵ is the strength of the interaction between two molecules at lattice sites. The divisor 2 arises from the need to avoid counting each interaction twice. Now, a molecule on the surface has, approximately, half as many neighbors with which to interact as does one in the bulk. Thus the change in energy of interaction on moving one molecule from the bulk to the surface is $Z\epsilon/2$. To make 1 mol of vacancies, therefore, requires energy about equal to the energy of sublimation. To estimate the entropy of formation of a vacancy, we consider vacancies to be a chemical species and compute the entropy of mixing of atoms and vacancies. This is of the form $-N_{atom}k_B \ln x_{atom} - N_{vacancy}k_B \ln x_{vacancy}$. Furthermore, because of changes in the local motions of molecules surrounding a vacancy, there is a vibrational contribution to the entropy of formation. Ordinarily the latter contribution is not large compared to the entropy-of-mixing term. The equilibrium concentration of vacancies is determined by the balance between the energy and entropy changes just described. For all solids the energy of sublimation is so large that $x_{vacancy}$ is very small.

Although the concentration of vacancies is small, it has an important influence on the properties of the crystal. When a molecule moves into a vacancy, it leaves a new vacancy behind (see Fig. 22.16). In this way the existence of vacancies permits diffusion of molecules and, in ionic crystals, conduction of an electric current by the ions. Some special electronic

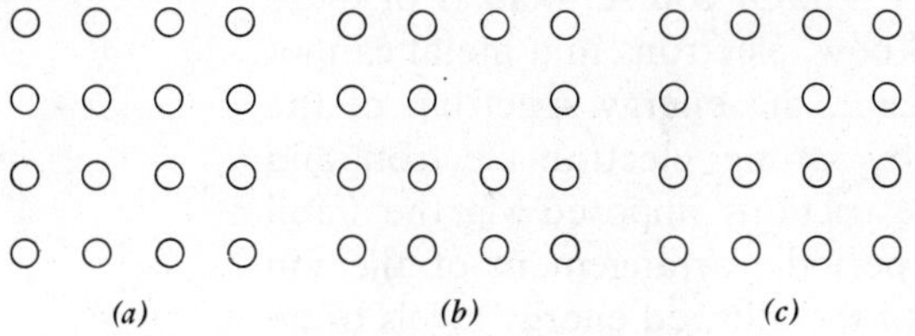

(a) $\qquad\qquad$ (b) $\qquad\qquad$ (c)

FIGURE 22.16
Effect of vacancies on atomic displacement in crystal. (a) Perfect lattice. (b) Lattice with one vacancy. (c) Same lattice as in (b), but with vacancy moved one column.

states can be associated with vacancies; for example, an electron can be trapped in a negative-ion vacancy in an alkali halide, thereby generating new electronic states in the crystal.

It is possible to estimate the concentration of interstitial molecules by the method outlined above. Because of the strong repulsive forces between molecules when close together, formation of an interstitial is usually much harder than is formation of a vacancy. Clearly, interstitials are easier to form if the molecule is small than if it is large. In some crystals this effect is quite pronounced. For example, AgI is a typical salt and behaves as such below 146°C. In that low-temperature region the conductivity of the salt is small, as is the rate of diffusion of Ag^+ ions. Between 146 and 552°C the crystal has the I^- ions arranged in a body-centered cubic lattice, but the very small Ag^+ ions move so freely through the interstices of this lattice that the electrical conductivity of the solid is very high. On going from 143°C (where the Ag^+ ions are on a lattice) to 146°C, the conductivity increases 5000-fold, but on going from 146 to 540°C it increases only twofold. A more thorough discussion of transport in solids will be found in Section 29.5.

There also occur, mainly in the inorganic oxides but also in other compounds, deviations from perfect stoichiometry. During the preparation of a compound there is a nonzero probability that perfect stoichiometry is not maintained. To maintain overall electrical neutrality in an ionic solid, the "missing" ions must go either on the surface of the crystal or into interstitial positions. Electrical neutrality can also be maintained by substitution of a different ion for the one "missing." A well-studied case of this type is ZnO. It is found, for example, that the conductivity of ZnO is sensitive to the pressure of a surrounding O_2 atmosphere, since at equilibrium that pressure regulates the concentration of O^{2-} defects in the ZnO lattice.

There are also many kinds of crystal imperfections that result when crystal planes are mislocated relative to one another. These are classed as *dislocations*; they influence the plasticity and strength of materials. A schematic diagram of an edge dislocation is shown in Fig. 22.17.

For further information on the fascinating chemistry of imperfect crystals, the student is referred to *Imperfections in Crystals* by H. G. van Bueren (North Holland, Amsterdam, 1960).

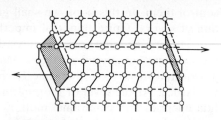

FIGURE 22.17
Schematic representation of an edge dislocation. The shaded area is the dislocation plane. The arrows indicate the direction of slippage that produces the dislocation.

22.5 Electronic Heat Capacity of Metals

As a class of materials, metals are distinguished from nonmetals by having a large electrical conductivity. Shortly after the discovery of the electron, it was realized that the conductivity of a metal implied the existence, in its interior, of mobile ("free") electrons. That idea, slightly modified, remains an inherent feature of the modern quantum mechanical description of a metal, aspects of which were discussed in Section 11.11.

In the classical theory of metals the free electrons are treated as a perfect gas. An immediate implication of this supposition is that the N free electrons of a monovalent metal contribute $3Nk_B$ to its heat capacity. Therefore, the heat capacity of a monovalent metal should approach $6Nk_B$ at high temperature, and not vanish at $T = 0$. Of course, neither of these predictions is correct since, as we now know, electrons in a metal cannot be described as a classical gas. In general, the energy spectrum of the electrons in a metal is complicated by strong electron–electron and electron–ion interactions and by the restrictions imposed via the Pauli exclusion principle. For example, the periodic arrangement of the ion cores on a lattice leads to a separation of the allowed energy levels of an electron into bands. Although the detailed structure of these energy bands (the dispersion relation between electron energy and electron momentum) influences the electron contribution to the heat capacity, the effects are usually not large, and we shall not discuss them. For the present it is sufficiently accurate to assume that shielding of the electron–electron and electron–ion interactions is efficient enough that the electrons can be treated as a perfect Fermi–Dirac gas occupying the molar volume of the metal. However, because the mass of the electron is very small, the deviations of the properties of the gas from classical behavior are very large. As we shall show below, the electronic contribution to the heat capacity is proportional to T. Since the lattice motion contribution to the heat capacity is proportional to T^3 at low temperatures (Eq. 22.15), there is a low-temperature regime where the electronic contribution is dominant. Otherwise, for most of the temperature range where the solid metal is stable, the electronic contribution to the heat capacity is very small. The analysis we present was first made in 1928 by Sommerfeld, who applied the theory of the Fermi–Dirac gas to the earlier free-electron theory of metals worked out by Drude and Lorentz.

Consider, first, the distribution of electrons in a perfect Fermi–Dirac gas at $T = 0$. Clearly, all the electrons occupy the lowest possible energy levels. But for a Fermi–Dirac gas the Pauli exclusion principle prohibits there being more than two electrons per energy level, one with z component of spin angular momentum $+\frac{1}{2}\hbar$, the other with $-\frac{1}{2}\hbar$. The lowest energy of the gas is obtained by filling each energy level with two electrons, hence the energy at $T = 0$ is very different from zero.

Let ϵ_F be the energy of the electrons in the highest quantum state filled at $T = 0$. Since all states of lower energy are filled, and all states with higher energy are empty (see Fig. 22.18), we can obtain the value of ϵ_F as follows. For a perfect Fermi–Dirac gas of electrons, from Eq. 21A.17, we have[8]

$$\langle N \rangle = 2 \cdot \frac{4\pi V}{h^3} \int_0^\infty dp\, p^2 \frac{\exp(\mu/k_B T)}{\exp(\mu/k_B T) + \exp(-p^2/2mk_B T)}. \quad (22.30)$$

But for $T = 0$, $\exp(-p^2/2mk_B T) = 0$, and the upper limit of integration becomes $p_F = (2m\epsilon_F)^{1/2}$. Therefore,

$$\langle N \rangle = \frac{8\pi V}{3h^3}(2m\epsilon_F)^{3/2}, \quad (22.31)$$

or

$$\epsilon_F = \frac{h^2}{8m}\left(\frac{3n}{\pi}\right)^{2/3}, \quad (22.32)$$

with $n \equiv \langle N \rangle / V$. The quantity ϵ_F is known as the *Fermi energy* of the system. The total energy of the $\langle N \rangle$ electrons at $T = 0$ is

$$\begin{aligned} U_0 &= 2 \cdot \frac{4\pi V}{h^3}\int_0^{p_F} dp\, p^2\left(\frac{p^2}{2m}\right) \\ &= \tfrac{3}{2}\langle N \rangle \epsilon_F \int_0^{\epsilon_F} d\left(\frac{\epsilon}{\epsilon_F}\right)\left(\frac{\epsilon}{\epsilon_F}\right)^{3/2} \\ &= \tfrac{3}{5}\langle N \rangle \epsilon_F. \end{aligned} \quad (22.33)$$

Note that the average energy per electron at $T = 0$ is $\frac{3}{5}\epsilon_F$, which is a very large value since for typical metals $\epsilon_F \approx 5$–$10\,\text{eV}$. As might be expected, the contribution of the electrons to the pressure is large because U_0 is

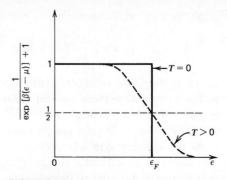

FIGURE 22.18
The distribution of occupied levels in a perfect Fermi–Dirac gas.

[8] The extra factor of 2 in Eq. 22.30 accounts for the two spin states of an electron.

large. In fact, at $T=0$ we have

$$p_0 = -\left(\frac{\partial U_0}{\partial V}\right)_T = -\tfrac{3}{5}\langle N\rangle\frac{\partial\epsilon_F}{\partial V} = \tfrac{2}{5}n\epsilon_F = \frac{2}{3}\frac{U_0}{V}, \tag{22.34}$$

which is the same relation between pressure and internal energy density as for a classical perfect gas (Eq. 12.41), but in this case neither p nor U is zero at $T=0$. Typically, $p_0 \approx 10^5$–10^6 atm for metals. Clearly, in a stable metal the electron gas pressure must be balanced by forces arising from the electron–ion attraction.

In order to calculate the heat capacity of the electron gas, we must first evaluate the energy from

$$U = 2 \cdot \frac{4\pi V}{h^3}\int_0^\infty dp\, p^2\left(\frac{p^2}{2m}\right)\frac{\exp(\mu/k_BT)}{\exp(\mu/k_BT)+\exp(-p^2/2mk_BT)}$$

$$= \frac{3\langle N\rangle}{2\epsilon_F^{3/2}}\int_0^\infty d\epsilon\,\frac{\epsilon^{3/2}}{\exp[(\epsilon-\mu)/k_BT]+1}. \tag{22.35}$$

Since $\epsilon_F \gg k_BT$, the distribution of electrons differs very little from what it is for $T=0$. Consequently, U and all other thermodynamic properties of the dense perfect Fermi–Dirac gas can be determined by expanding the integrand about its value for $T=0$ and integrating term by term. The integral in Eq. 22.35 is of the form

$$I = \int_0^\infty d\epsilon\,\frac{f(\epsilon)}{\exp[(\epsilon-\mu)/k_BT]+1}. \tag{22.36}$$

Using the expansion procedure mentioned, it is shown in Appendix 22A that

$$I = \int_0^\mu f(\epsilon)\,d\epsilon + \frac{(\pi k_BT)^2}{6}\left(\frac{df}{d\epsilon}\right)_{\epsilon=\mu} + \cdots. \tag{22.37}$$

We calculate the chemical potential of the gas first. Rewriting Eq. 22.30 in the form

$$\langle N\rangle = \frac{3\langle N\rangle}{2\epsilon_F^{3/2}}\int_0^\infty d\epsilon\,\frac{\epsilon^{1/2}}{\exp[(\epsilon-\mu)/k_BT]+1} \tag{22.38}$$

enables us to apply Eq. 22.36 with $f(\epsilon) = \epsilon^{1/2}$. Then we find

$$1 = \left(\frac{\mu}{\epsilon_F}\right)^{3/2}\left[1 + \frac{\pi^2}{8}\left(\frac{k_BT}{\mu}\right)^2 + \cdots\right], \tag{22.39}$$

or

$$\mu = \epsilon_F\left[1 - \frac{\pi^2}{12}\left(\frac{k_BT}{\mu}\right)^2 + \cdots\right] \tag{22.40}$$

$$= \epsilon_F\left[1 - \frac{\pi^2}{12}\left(\frac{k_BT}{\epsilon_F}\right)^2 + \cdots\right], \tag{22.41}$$

with Eq. 22.41 following from Eq. 22.40 by iterative substitution of the expansion for μ in the term $(k_BT/\mu)^2$.

To evaluate Eq. 22.35 for the energy we set $f(\epsilon) = \epsilon^{3/2}$. Applying Eq. 22.37 again, we find that

$$U = \frac{3}{2}\frac{\langle N\rangle}{\epsilon_F^{3/2}}\left[\frac{2\mu^{5/2}}{5} + \frac{(\pi k_BT)^2}{6}\left(\frac{3\mu^{1/2}}{2}\right) + \cdots\right]$$

$$= \frac{3\langle N\rangle}{5}\left(\frac{\mu}{\epsilon_F}\right)^{3/2}\mu\left[1 + \frac{5\pi^2}{8}\left(\frac{k_BT}{\mu}\right)^2 + \cdots\right] \tag{22.42}$$

$$= \frac{3\langle N \rangle}{5} \epsilon_F \left[1 + \frac{5\pi^2}{12} \left(\frac{k_B T}{\epsilon_F} \right)^2 + \cdots \right], \tag{22.43}$$

the last line (22.43) following from use of Eq. 22.41 to eliminate μ in favor of ϵ_F. From Eq. 22.43 we immediately obtain the electronic heat capacity,

$$C_V^{\text{el}} = \frac{\pi^2}{2} \langle N \rangle k_B \left(\frac{k_B T}{\epsilon_F} \right), \tag{22.44}$$

which is linear in T as stated earlier. For typical metals, Eq. 22.44 is of the order of 1% of $3Nk_B$ at 300 K, neatly accounting for the qualitative aspects of the observed behavior.

The physical basis for the result displayed in Eq. 22.44 is easily understood. Because of the restrictions imposed by the Pauli exclusion principle, electrons with energy much less than ϵ_F cannot be excited into neighboring states because these are already occupied. The excitation of an electron requires that there be an empty state of proper energy available. Consequently, only those electrons lying within about $k_B T$ of ϵ_F can be excited. The fraction of these is of the order of $(k_B T/\epsilon_F)$, so the increase in energy corresponding to thermal excitation is of the order of $N(k_B T/\epsilon_F)(k_B T)$, and the contribution to the heat capacity is proportional to $Nk_B(k_B T/\epsilon_F)$, which is of the form displayed in Eq. 22.44.

It is also clear that Eq. 22.44 satisfies the third law of thermodynamics, whereas the classical value for the electronic heat capacity does not.

Appendix 22A

Evaluation of Fermi-Dirac Integrals

This appendix is concerned with evaluation of integrals of the type

$$I = \int_0^\infty d\epsilon \frac{f(\epsilon)}{\exp[(\epsilon - \mu)/k_B T] + 1}. \tag{22A.1}$$

As indicated in the text of Chapter 22, when $T = 0$ the function $\{\exp[(\epsilon - \mu)/k_B T] + 1\}^{-1}$ is a step function, having the value unity for $\epsilon \leq \epsilon_F$ and zero for $\epsilon > \epsilon_F$. Moreover, since $\epsilon_F/k_B \approx 5 \times 10^4$ K, the distribution of electrons will be very like that at $T = 0$ for all T of interest. Evaluation of Eq. 22A.1 can then be based on an expansion of the integrand about its value for $T = 0$.

We proceed as follows. Integration of Eq. 22A.1 by parts gives

$$I = \int_0^\infty d\epsilon \frac{\beta \exp[\beta(\epsilon - \mu)]}{\{\exp[\beta(\epsilon - \mu)] + 1\}^2} \int_0^\epsilon d\epsilon' f(\epsilon'); \tag{22A.2}$$

the other terms vanish at the limits of integration. Let $y \equiv \beta(\epsilon - \mu)$ and $F(\epsilon) \equiv \int_0^\epsilon d\epsilon' f(\epsilon')$. Then, expanding $F(y)$ as a Taylor series about $y = 0$, which corresponds to $T = 0$ where $\mu = \epsilon_F$, we find that

$$F(y) = F(0) + \sum_{n=1}^\infty \frac{y^n}{n!} \left(\frac{\partial^n F}{\partial y^n} \right)_{y=0}$$

$$= \int_0^\mu d\epsilon \, f(\epsilon) + \sum_{n=1}^\infty \frac{(y k_B T)^n}{n!} \left(\frac{\partial^{n-1} f(\epsilon)}{\partial \epsilon^{n-1}} \right)_{\epsilon = \mu}. \tag{22A.3}$$

Substitution of Eq. 22A.3 into Eq. 22A.2 then gives

$$I = F(0) \int_0^\infty d\epsilon \frac{\beta \exp[\beta(\epsilon - \mu)]}{\{\exp[\beta(\epsilon - \mu)] + 1\}^2}$$

$$+ \sum_{n=1}^\infty \frac{(k_B T)^n}{n!} \left(\frac{\partial^{n-1} f(\epsilon)}{\partial \epsilon^{n-1}} \right)_{\epsilon = \mu} \int_{y = -\beta\mu}^\infty d\epsilon \, y^n \frac{\beta \exp[\beta(\epsilon - \mu)]}{\{\exp[\beta(\epsilon - \mu)] + 1\}^2}. \tag{22A.4}$$

Consider the first term on the right-hand side of Eq. 22A.4. We have

$$\int_0^\infty d\epsilon \frac{\beta \exp[\beta(\epsilon - \mu)]}{\{\exp[\beta(\epsilon - \mu)] + 1\}^2} = \left\{ \frac{1}{\exp[\beta(\epsilon - \mu)] + 1} \right\}_0^\infty \approx 1, \tag{22A.5}$$

since $\exp(-\beta\mu) \ll 1$. Next consider the second term on the right-hand side of Eq. 22A.4. We rewrite the term

$$\frac{\beta \exp[\beta(\epsilon - \mu)]}{\{\exp[\beta(\epsilon - \mu)] + 1\}^2} = \frac{\beta}{(e^y + 1)(e^{-y} + 1)} \tag{22A.6}$$

to show that it is symmetric in y, and that it approaches zero exponentially as $y \to +\infty$ and $y \to -\infty$. Then the lower limit of the integral can be extended to $-\infty$, since the error made is exponentially small. Since Eq. 22A.6 is symmetric, and the extended range of integration ($\pm\infty$) is symmetric, all those contributions to the second term on the right-hand side of Eq. 22A.4 for which n is odd vanish. Finally, by making the expansion

$$\frac{\beta e^y}{(e^y + 1)^2} = \frac{\beta e^{-y}}{(1 + e^{-y})^2} = -\beta \sum_{m=1}^\infty (-)^m m e^{-my}, \tag{22A.7}$$

and integrating term by term, we find that

$$\int_{-\infty}^{\infty} dy \frac{\beta y^n}{(e^y+1)(e^{-y}+1)} = -2\beta \sum_{m=1}^{\infty} (-)^m m \int_0^{\infty} y^n e^{-my} \, dy$$

$$= -2\beta n! \sum_{m=1}^{\infty} \frac{(-)^m}{m^n}. \qquad (22A.8)$$

The sums in Eq. 22A.8 are known; they have the values

$$n=2: \qquad -\sum_{m=1}^{\infty} \frac{(-)^m}{m^2} = \frac{\pi^2}{12};$$

$$n=4: \qquad -\sum_{m=1}^{\infty} \frac{(-)^m}{m^4} = \frac{7\pi^4}{720}. \qquad (22A.9)$$

$$\vdots$$

Combining Eqs. 22A.4, 22A.5, 22A.8, and 22A.9 we finally obtain

$$I = \int_0^{\mu} d\epsilon \, f(\epsilon) + \frac{(\pi k_B T)^2}{6} \left(\frac{\partial f}{\partial \epsilon}\right)_{\epsilon=\mu} + \frac{7}{360} (\pi k_B T)^4 \left(\frac{\partial^3 f}{\partial \epsilon^3}\right)_{\epsilon=\mu} + \cdots, \qquad (22A.10)$$

which is the form quoted in the text.

Further Reading

Reissland, J. A., *The Physics of Phonons* (Wiley, New York, 1973).

Smith, R. A., *Wave Mechanics of Crystalline Solids* (Wiley, New York, 1961).

More advanced texts:

Wallace, D. C., *Thermodynamics of Crystals* (Wiley, New York, 1972).

Maradudin, A. A., E. W. Montroll, G. H. Weiss, and I. P. Ipatova, *Theory of Lattice Dynamics in the Harmonic Approximation* (Academic Press, New York, 1971).

Chaquard, P., *The Anharmonic Crystal* (W. A. Benjamin, Reading, Mass., 1967).

Born, M., and K. Huang, *Dynamical Theory of Crystal Lattices* (Oxford University Press, London, 1954).

Stoneham, A. M., *Theory of Defects in Solids* (Oxford University Press, London, 1975).

van Bueren, H. G., *Imperfections in Crystals* (North Holland, Amsterdam, 1960).

Problems

1. Suppose that the interaction between two atoms has the form

$$u_2(R) = -\frac{C_1}{R^n} + \frac{C_2}{R^m}.$$

Show that a necessary condition for $u_2(R)$ to have a minimum is that the repulsive force be of shorter range than the attractive force (i.e., $m > n$).

2. Consider a one-dimensional chain of alternating monovalent cations and anions. Show that the Madelung constant (see Chapter 14) for this linear lattice is $2 \ln 2$.

3. Suppose that the molar internal energy at T can be written in the form

$$u(v, T) = e_0(v) + e_{zp}(v) + e_{vib}(v, T),$$

where $e_0(v)$ is the static lattice energy, $e_{zp}(v)$ is the zero-point energy, and $e_{vib}(v, T)$ is the energy associated with thermally excited vibrations. For a close-packed cubic crystal in which the interaction is of the Lennard–Jones type, we have

$$\frac{e_0(v)}{4N_A\epsilon} = 6.006\left(\frac{v_0}{v}\right)^4 - 14.454\left(\frac{v_0}{v}\right)^2,$$

where $v_0 = N_A\sigma^3$, and

$$\frac{e_{zp}(v)}{4N_A\epsilon} = 2.291\Lambda^*\left(\frac{v_0}{v}\right)^{7/3}\left[4.283 - 2\left(\frac{v}{v_0}\right)^2\right],$$

where $\Lambda^* \equiv h/\sigma(4m\epsilon)^{1/2}$. Calculate the pressure as a function of volume at $T = 0$ K. From the formula obtained, evaluate v/v_0 as a function of p/p_0 for H_2, for which $\Lambda^* = 1.73$ Å, $v_0 = 15.12$ cm^3/mol, and $p_0 \equiv 4\epsilon/\sigma^3 = 203$ atm. Compare your calculations with the measurements of J. W. Stewart and C. A. Swenson, *Phys. Rev.* **94,** 1069 (1954); **97,** 578 (1955).

4. In ionic crystals $e_{zp}(v)$ is usually very small compared with $e_0(v)$ (both defined as in Problem 3). At high pressures the repulsive part of the non-Coulomb interaction and the Coulomb attraction dominate the ion–ion potential. Show that if the repulsive part of the non-Coulomb interaction in an ionic crystal is of the form R^{-n}, then at high pressure

$$p \propto v^{-(n+3)/3}.$$

5. For a one-dimensional chain of particles with mass m, lattice spacing a, and displacement force constant f, show that the dispersion relation is

$$\omega = 2\pi\nu = 2\sqrt{\frac{f}{m}}\sin\left(\frac{ka}{2}\right).$$

6. For a one-dimensional chain with lattice spacing a, the number of standing waves in the wave vector interval dk is $L\,dk/\pi$, where L is the length of the chain and $0 \leq k \leq \pi/a$. Show that the energy of the chain is

$$U = \frac{2L}{\pi a}\int_0^{\omega_{max}}\frac{\hbar\omega\,d\omega}{[\exp(\hbar\omega/k_BT) - 1](\omega_{max}^2 - \omega^2)^{1/2}},$$

where $\omega_{max}^2 = 4f/m$. Draw the curve of C_V versus T for this lattice for $\hbar\omega_{max}/k_B = 200$ K.

7. The continuum approximation to the linear chain represents the number of vibrational modes in the range $d\omega$ as $L\,d\omega/\pi c_s$, where c_s is the velocity of "sound wave" propagation on the continuous linear chain. Use the Debye method to calculate C_V for this model and compute C_V versus T for the case $\Theta_D = 200$ K, where $\Theta_D \equiv \hbar\omega'_{max}/k_B$. Compare the two curves of C_V versus T from the calculation of Problem 6 and this one. Note that ω'_{max} here is not the same as ω_{max} in Problem 6.

8. Consider a crystal of N atoms where each atom has one vibrational degree of freedom. Assume that the vibrational energy of each atom can be written as

$$\epsilon = x^2 h\nu,$$

where x can take on all continuous positive values. As in the Einstein treatment, each atom has only one vibrational frequency ν.
(a) Calculate the partition function for this crystal.
(b) Compute the average vibrational energy.
(c) Compute the heat capacity of the crystal.

9. Consider a one-dimensional diatomic lattice such as shown in Fig. 22.6. For this problem let $2b = a$, and let the two masses be M and m. Show that the vibrational frequencies satisfy the dispersion relation

$$\omega^2 = f\left(\frac{1}{m} + \frac{1}{M}\right) \pm f\left[\left(\frac{1}{m} + \frac{1}{M}\right)^2 - \frac{4\sin^2(kb)}{Mm}\right]^{1/2},$$

where f is the Hooke's law force constant for displacement of either atom. Show that the plus sign generates the optical branch and the minus sign the acoustical branch of the spectrum.

10. Discuss the heat capacity of a two-dimensional square lattice with nearest-neighbor separation a using the Debye approximation. The information needed with respect to the dispersion of frequencies can be obtained from the two-dimensional analog of the analysis of footnote 4 on page 809.

11. Show that at low temperature the heat capacity of a one-dimensional chain is proportional to T, and that of a square-planar lattice is proportional to T^2.

12. Show that for an Einstein crystal, when $T \gg \Theta_E$,

$$S \approx 3Nk_B\left[1 + \ln\left(\frac{k_BT}{\hbar\omega_E}\right)\right].$$

13. Show that for a crystalline solid containing N particles

$$\int_0^\infty [3Nk_B - C_V(T)]\,dT = U(0),$$

where $U(0)$ is the internal energy at $T = 0\text{ K}$.

14. The Debye model is an essentially exact treatment of the low-frequency modes of a solid but it does not properly treat the high-frequency modes. Why should this model be a good approximation at low temperature?

15. Prove that as long as there is a maximum vibrational frequency in a crystal, the high-temperature limit of the heat capacity is $3Nk_B$.

16. At 60 K the heat capacity of copper is 8.70 J/K mol. Predict the heat capacity at 4 K using the Einstein and Debye low-temperature limits. Which result agrees better with the experimental value of 5.78×10^{-3} J/K mol?

17. Consider a (hypothetical) perfect lattice containing N atoms. Suppose that we create N_v vacancies, where $N_v \ll N$. Let the energy of creation of a vacancy be w_v, and the change in vibrational contribution to the entropy per vacancy be ΔS_v. Show that

$$\frac{N_v}{N} = \exp\left(\frac{\Delta S_v}{k_B}\right)\exp\left(-\frac{w_v}{k_BT}\right).$$

18. A very simple model can be used to estimate the magnitude of ΔS_v in Problem 17. Suppose that the crystal is described in the Einstein approximation. Suppose further that the atoms adjacent to a vacancy have frequency of oscillation ν', all others frequency of oscillation ν. Show that

$$\Delta S_v = 3Zk_B \ln\left(\frac{\nu}{\nu'}\right),$$

where Z is the number of nearest neighbors to a vacancy.

19. Frenkel defects are formed when atoms leave lattice sites and occupy interstitial positions in the lattice. A Frenkel defect, therefore, consists of an interstitial atom and a vacancy. Show that the number of Frenkel defects in equilibrium at T in a crystal is

$$N_{Fr} = (NN_i)^{1/2}\exp\left(\frac{\Delta S_v}{2k_B}\right)\exp\left(-\frac{w_{Fr}}{2k_BT}\right),$$

where N is the number of atoms, N_i is the number of possible interstitial positions, ΔS_v is the change in vibrational entropy per Frenkel defect, and w_{Fr} is the energy of formation of a Frenkel defect. Note that in chemical language the formation of a Frenkel defect is equivalent to the reaction

$$\text{occupied lattice site} + \text{vacant interstitial site} = \text{vacancy} + \text{interstitial}.$$

20. Estimate the number of vacancies per atom in a state of equilibrium in a crystal at $T = 100$ K, 500 K, and 1000 K for the cases that $w_v = 80$ kJ/mol and $w_v = 150$ kJ/mol.

21. Suppose that the valence electrons of solid Na form a perfect degenerate Fermi gas. Calculate the average speed of an electron in this gas. What would the corresponding "classical temperature" be? What is the electronic contribution to the heat capacity of Na at 300 K?

22. Suppose that the energy of a free electron "at rest" inside a metal is zero. In order to escape from the metal, the electron must have an energy of motion perpendicular to the surface of at least ϵ_s. Then if z is the coordinate perpendicular to the surface, the electron must have momentum $p_z > p_{z_0}$ to escape, where $p_{z_0}^2/2m = \epsilon_s$. Show that the emission current density is

$$I = \frac{4\pi e m k_B^2}{h^3} T^2 \exp\left(-\frac{\phi}{k_B T}\right),$$

which is known as the Dushman–Richardson equation. Hint: $\phi \gg k_B T$ for all metals below the melting point, so only the tail of the Fermi distribution contributes to I.

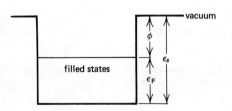

23. At a very high pressure, say, $p > 10^6$ atm, the volumetric properties of a metal are determined by the compressibility of the degenerate electron gas. Indeed, in this pressure range the distinction between solid, liquid, and gas becomes meaningless. Show that for a perfect degenerate electron gas

$$p = \frac{1}{5}\left(\frac{3}{8\pi}\right)^{2/3} \frac{h^2}{m} n^{5/3}.$$

For all elements except hydrogen the ratio of the atomic weight to atomic number is close to 2. (The nucleus has approximately equal numbers of protons and neutrons.) Calculate the pressure corresponding to a mass density of 100 g/cm³ using an atomic weight/atomic number ratio of 2.

24. A solid contains N distinguishable and noninteracting "magnetic sites." At each site is a magnetic moment $\boldsymbol{\mu}$ which can either point along an applied magnetic field $\mathbf{H}$ or opposite to the field. If N_1 of the moments point along the field, the total energy of the solid is

$$E = -N_1 \boldsymbol{\mu} \cdot \mathbf{H} + (N - N_1)\boldsymbol{\mu} \cdot \mathbf{H}.$$

(Each moment pointing along the field contributes energy $-\boldsymbol{\mu} \cdot \mathbf{H}$, whereas each moment pointing opposite to the field contributes energy $+\boldsymbol{\mu} \cdot \mathbf{H}$.) Suppose that the solid is isolated and has energy E. Compute the entropy S as a function of E, $\mathbf{H}$, $\boldsymbol{\mu}$, and N.

25. Suppose that the solid of Problem 24 is maintained at a constant temperature T. Compute its entropy.

26. Suppose that a perfect free-electron gas has n electrons per unit volume. In the presence of a magnetic field $\mathbf{H}$, let there be n_p with spin magnetic moment parallel to $\mathbf{H}$ and n_a with spin magnetic moment antiparallel to $\mathbf{H}$. The magnetization (magnetic moment per unit volume) is

$$\mathbf{M} = (n_p - n_a)\boldsymbol{\mu}_B$$

where $\boldsymbol{\mu}_B$ is the Bohr magneton (see Chapter 5). Show that if the electron distribution were classical and $\boldsymbol{\mu}_B \cdot \mathbf{H} \ll k_B T$, then

$$\mathbf{M} = \frac{n\mu_B^2}{k_B T}\mathbf{H} \equiv \chi_p^{cl}\mathbf{H}.$$

27. The result of Problem 26 disagrees with experimental data for simple metals because the electron distribution *must* be described by Fermi–Dirac statistics. As a result of applying the field **H**, the distribution of electrons over the energy levels is changed as shown in the accompanying diagram. All levels below ϵ_F are filled, as usual. However, electrons with spin magnetic moment parallel to **H** have their energies lowered by $\mu_B \cdot \mathbf{H}$, those antiparallel have it raised by $\mu_B \cdot \mathbf{H}$, where $\mu_B \cdot \mathbf{H} \ll \epsilon_F$. The energy ϵ_F is common to both the spin states of the electrons. Show that now

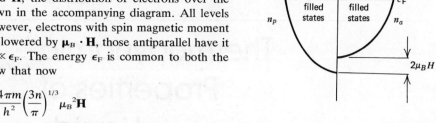

$$\mathbf{M} = \frac{4\pi m}{h^2}\left(\frac{3n}{\pi}\right)^{1/3} \mu_B{}^2 \mathbf{H}$$

$$\equiv \chi_p \mathbf{H},$$

where $\chi_p \equiv 3n\mu_B{}^2 / 2k_B T_F$. Compare this result with $\chi_p{}^{cl}$ of Problem 26.

23

Thermodynamic Properties of Liquids

Our universe contains matter subjected to an enormous range of conditions—from the near-zero temperature and pressure of space to the extremely high temperatures and pressures in the interiors of stars. Liquids exist over only a small part of this range. That domain is, however, of extraordinary importance for technology and for life processes. In this chapter we examine how the thermodynamic properties of a liquid can be understood in terms of the forces between its constituent molecules. The characteristic flow properties of a liquid are associated with nonequilibrium phenomena. These are discussed in Part Three of this text.

23.1 Bulk Properties of Liquids

How do the equation of state and the thermal properties of liquids differ from those of gases and solids?

Consider, first, the pressure exerted by a liquid at temperature T. It was shown in Chapter 19 that two parts of a one-component system in contact through a deformable boundary that permits transfer of energy and matter must have the same pressure, temperature, and chemical potential. The boundary separating a liquid and its vapor is of this type, hence the pressure of the vapor must be equal to that of the liquid. We denote the pressure along the line of equilibrium for coexisting liquid and vapor phases by p_σ.

We shall consider details of the nature of the liquid–vapor equilibrium in Chapter 24, but two points, related to the dependence of p_σ on the state of the liquid, deserve mention here. First, in a liquid–vapor system containing only one component, p_σ is a function of only the temperature (see Chapter 24). If, however, a uniform pressure is applied to the liquid by, say, an inert gas that is insoluble in the liquid, the vapor phase consists of two components (the inert gas and the vapor from the liquid), and p_σ is a function of both the temperature and the applied pressure. Indeed, since the chemical potentials of the vapor and liquid are equal,[1]

$$\mu^L = \mu^*(T) + RT \ln \frac{p_\sigma}{p}. \tag{23.1}$$

Moreover, since the pressure derivative of the chemical potential of the liquid is given by

$$\left(\frac{\partial \mu^L}{\partial p}\right)_T = v^L, \tag{23.2}$$

---

[1] Equation 23.1 is written for the case that the vapor is a perfect gas. When gas imperfection is important, the fugacity replaces the pressure in Eqs. 23.1 and 23.3.

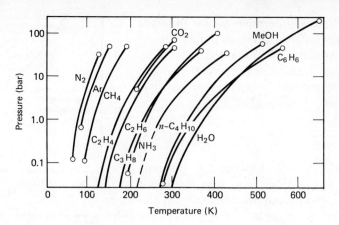

FIGURE 23.1
The logarithms of the vapor pressures of 12 liquids as functions of the absolute temperature. The circles mark the triple and critical points (see Chapter 24). From J. S. Rowlinson, *Liquids and Liquid Mixtures.* 2nd ed. (Butterworths, London, 1969).

we find that the rate of change of vapor pressure with applied pressure is given by

$$RT\left(\frac{\partial \ln p_\sigma}{\partial p}\right)_T = v^L, \qquad (23.3)$$

which is often called the *Poynting equation*. The molar volume of a liquid is of the order of 100 cm^3, so at room temperature $(\partial \ln p_\sigma/\partial p)_T$ is small, of the order of $4 \times 10^{-3} \text{ atm}^{-1}$. That is, the vapor pressure changes on the order of 0.5% per atmosphere of added pressure. Second, the temperature dependence of p_σ is determined by the differences in entropies and molar volumes of the vapor and liquid. Consider an alteration of the common temperature and pressure along the liquid–vapor coexistence line. Since $\mu^L = \mu^G$ in both the initial and final states, it is necessary that their changes, $d\mu^L$ and $d\mu^G$, also be equal:

$$\left.\begin{array}{r} d\mu^G = d\mu^L, \\ -s^G\, dT + v^G\, dp_\sigma = -s^L\, dT + v^L\, dp_\sigma, \end{array}\right\} \begin{array}{l} \text{along the liquid–vapor} \\ \text{coexistence line.} \end{array} \quad (23.4)$$

Equation 23.4 is immediately rearranged to

$$\left(\frac{\partial p}{\partial T}\right)_\sigma = \frac{s^G - s^L}{v^G - v^L}. \qquad (23.5)$$

Since $s^G > s^L$ and $v^G \gg v^L$, we expect $p_\sigma(T)$ to be a strongly increasing function of the temperature. That this is so is shown in Fig. 23.1. Note that along the liquid–vapor coexistence line the volume of the liquid varies.

The formulas used to compute thermodynamic functions are usually expressed in terms of constant-pressure or constant-volume properties of the fluid. Consider, then, the *thermal pressure coefficient*

$$\gamma_V \equiv \left(\frac{\partial p}{\partial T}\right)_V, \qquad (23.6)$$

which measures the rate of change of pressure with temperature at fixed liquid volume. The functional behavior of γ_V determines the volume dependence of the internal energy through (see Eq. 17.40)

$$\left(\frac{\partial u}{\partial v}\right)_T = T\left(\frac{\partial p}{\partial T}\right)_V - p. \qquad (23.7)$$

A schematic representation of the dependence of γ_V on temperature is shown in Fig. 23.2a. It is observed that γ_V changes smoothly with temperature along the liquid–vapor coexistence line (see Chapter 24). On the right-hand side of Fig. 23.2a are lines representing γ_V for different values of v; these lines intersect the coexistence curve at points corresponding to the respective values of the volumes. Note that the slope of γ_V versus T is always small. γ_V can be either negative or positive, depending on whether the coefficient of thermal expansion is negative or positive, since

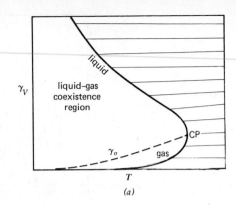
(a)

$$\gamma_V \equiv \left(\frac{\partial p}{\partial T}\right)_V = -\frac{(\partial v/\partial T)_p}{(\partial v/\partial p)_T} = \frac{\alpha}{\kappa_T}. \tag{23.8}$$

Thermodynamic stability requires that $\kappa_T > 0$ (see Chapter 19), but only that γ_V and α have the same sign. Finally, also note that the vapor pressure curve, defined by $\gamma_\sigma \equiv (\partial p/\partial T)_\sigma$, crosses the two-phase region and cuts the coexistence curve at the critical point, at which $\gamma_\sigma = \gamma_V$. Some values of γ_V along the liquid–vapor coexistence line of argon are shown in Fig. 23.2b, and values of $(\partial u/\partial v)_T$ as a function of molar volume for several liquids are shown in Fig. 23.2c.

Direct measurements of the p, v, T properties of liquids lead to the following generalizations:

1. The isothermal compressibility κ_T decreases with increasing pressure, the rate of decrease becoming smaller as the pressure increases.

2. The isothermal compressibility increases as the temperature increases, the effect being largest at low pressure and diminishing as the pressure increases.

3. The coefficient of thermal expansion, α, increases as the temperature increases.

A tabulation of data for liquid argon along the liquid–vapor coexistence line is given in Table 23.1.

The values of γ_V discussed and Eq. 23.7 can be used to reach an interesting conclusion. Since T is positive, and α is usually positive, we expect γ_V to be positive. Then at low pressure u must decrease as v decreases. If the pressure is increased a point is reached where the term $[T(\partial p/\partial T)_V - p]$ in Eq. 23.7 becomes zero, and then for larger pressures this term becomes negative. At the pressure where this change takes place, the net internal forces change from attractive to repulsive. In the low-pressure region energy actually flows out of a liquid during an isothermal compression.

There is no generally valid, simple analytic equation of state for a liquid. Just as for the case of imperfect gases, empirical equations of state are used to represent observations in a compact form. The two most widely used are the *Tait equation*,

$$\frac{v_0 - v}{v_0 p} = \frac{A}{B + p}, \tag{23.9}$$

and the *Huddleston equation*,

$$\ln\left(\frac{pv^{2/3}}{v_0^{1/3} - v^{1/3}}\right) = A + B(v_0^{1/3} - v^{1/3}), \tag{23.10}$$

where v_0 is the molar volume at "zero" pressure and A and B are positive constants (different in the two equations). For the high-density region above the critical temperature, a virial expansion (see Chapter 21) with as many as six, seven, or even more terms has sometimes been used to represent pv/RT. As we saw in Chapter 21, the terms in the virial expansion can be related to the interaction of pairs, triplets, . . . of

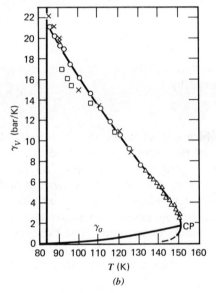
(b)

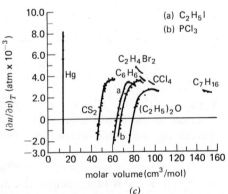
(c)

FIGURE 23.2
(a) The variation of the thermal pressure coefficient with temperature for the saturated gas and liquid, and for the homogeneous fluid at temperatures above saturation. (b) Thermal pressure coefficient of argon. From J. S. Rowlinson, *Liquids and Liquid Mixtures*, 2nd ed. (Butterworths, London, 1969). (c) $(\partial u/\partial v)_T$ as a function of molar volume. From J. H. Hildebrand and R. L. Scott, *Regular Solutions*, (Prentice-Hall, Englewood Cliffs, N.J., 1962).

TABLE 23.1
SOME PROPERTIES OF LIQUID ARGON[a,b]

T (K)	p_σ (bar)	v (cm³/mol)	α_σ ($\times 10^3$/K)	α_p	κ_S ($\times 10^{-4}$/bar)	κ_T	γ_σ (bar/K)	γ_V	c_σ	c_p (J/K mol)	c_V
83.81[t]	0.68905	28.19	4.37	4.39	0.948	2.033	0.080	21.6	41.8	41.9	19.5
85	0.789	28.34	4.40	4.42	0.970	2.089	0.087	21.2	41.9	42.0	19.5
87.29[b]	1.013	28.66	4.47	4.49	1.022	2.217	0.107	20.3	42.1	42.2	19.4
90	1.336	29.00	4.55	4.58	1.077	2.37	0.133	19.3	42.2	42.4	19.3
95	2.149	29.67	4.75	4.80	1.211	2.74	0.193	17.5	42.4	42.7	19.0
100	3.25	30.38	5.00	5.08	1.363	3.16	0.256	16.1	43.1	43.5	18.7
105	4.73	31.15	5.29	5.41	1.556	3.71	0.333	14.6	44	45	19
110	6.67	32.03	5.7	5.9	1.80	4.47	0.435	13.2	45	46	18
120	12.17	34.25	7.3	7.8	2.49	7.40	0.670	10.5	49	51	17
130	20.28	37.50	10	11	3.99	14.1	0.965	7.9	55	60	17
140	31.67	42.42	16	21	7.59	38	1.32	5.6	70	86	18
150.86[c]	48.98	74.6	∞	∞	∞?	∞	1.8	1.8	∞	∞	∞?

[a] From J. S. Rowlinson, *Liquids and Liquid Mixtures*, 2nd ed. (Butterworths, London, 1969).
[b] $t \equiv$ triple point $b \equiv$ boiling point $c \equiv$ critical point

$$\alpha_\sigma \equiv \frac{1}{V}\left(\frac{\partial V}{\partial T}\right)_\sigma,$$

$$\alpha_p \equiv \frac{1}{V}\left(\frac{\partial V}{\partial T}\right)_p,$$

$$\kappa_T \equiv -\frac{1}{V}\left(\frac{\partial V}{\partial p}\right)_T,$$

$$\kappa_S \equiv -\frac{1}{V}\left(\frac{\partial V}{\partial p}\right)_S.$$

molecules, and thereby each term is given an interpretation in terms of the properties of the molecules of the system. No such interpretation is known, even formally, for the two empirical equations 23.9 and 23.10.

Finally, we note that c_p for a liquid increases as the temperature increases, but c_V is much less sensitive to temperature (see Table 23.1 and Fig. 23.3).

We showed, in Chapter 17, that the changes in entropy, internal energy, and other thermodynamic functions of a fluid can be calculated from knowledge of the heat capacity as a function of temperature and the equation of state. For a real gas, a convenient reference from which to measure changes is provided by the properties of an ideal gas at the same temperature and pressure. Since interpretation of the properties of the liquid state depends on understanding the role of intermolecular forces, it is convenient again to choose the perfect gas to define the reference state from which changes in thermodynamic properties are measured. The residual value of any function f, denoted f^E, is defined by

$$f^E(T, v) = \int_\infty^v \left[\left(\frac{\partial f}{\partial v}\right)_T - \left(\frac{\partial f}{\partial v}\right)_T^{\text{perfect gas}}\right] dv. \tag{23.11}$$

We can imagine the integral in Eq. 23.11 to be evaluated in two steps. Let the initial state be one with the system at effectively zero pressure,

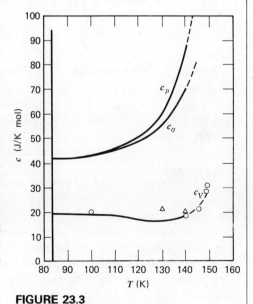

FIGURE 23.3
The heat capacities of liquid argon. From J. S. Rowlinson, *Liquids and Liquid Mixtures*, 2nd ed. (Butterworths, London, 1969).

TABLE 23.2
RESIDUAL PROPERTIES OF LIQUID ARGON[a,b]

T (K)	Δh_{vap}	$h^E + RT$ (J/mol)	u^E	c_p^E	c_V^E
				(J/K mol)	
83.81[t]	6610	−5940	−5940	29.4	7.0
85	6570	−5910	−5910	29.5	7.0
87.29[b]	6517	−5840	−5840	29.7	6.9
90	6440	−5760	−5760	29.9	6.8
95	6290	−5590	−5590	30.2	6.5
100	6110	−5420	−5430	31.2	6.2
105	5900	−5230	−5240	32	6
110	5660	−5030	−5050	34	5
120	5100	−4630	−4670	39	5
130	4350	−4170	−4250	48	5
140	3350	−3610	−3750	74	6
150.86[c]	0	−2030	−2400	∞	∞?

[a] From J. S. Rowlinson, *Liquids and Liquid Mixtures*, 2nd ed. (Butterworths, London, 1969).
[b] $t \equiv$ triple point $b \equiv$ boiling point $c \equiv$ critical point

hence effectively infinite volume, and imagine an isothermal compression from $v = \infty$ to v^G, where v^G is the volume of the vapor at pressure p_σ and temperature T along the coexistence line for liquid–vapor equilibrium. Next, isothermally condense the gas to form a liquid with volume v^L, and add the consequent change in f less any change in $f^{\text{perfect gas}}$. For example, the residual enthalpy of a liquid is

$$h^E(T, v) = \int_\infty^v \left[\left(\frac{\partial h}{\partial v}\right)_T - \left(\frac{\partial h}{\partial v}\right)_T^{\text{perfect gas}} \right] dv$$

$$= \int_\infty^v \left[T\left(\frac{\partial p}{\partial T}\right)_V + v\left(\frac{\partial p}{\partial v}\right)_T \right] dv + \Delta h_{vap}. \qquad (23.12)$$

A plot of the typical variation of h^E with temperature is shown in Fig. 23.4, and the residual enthalpy and energy of argon are listed in Table 23.2. Note how the heat of vaporization decreases continuously and becomes zero at the critical temperature, above which liquid and gas cannot be distinguished.

An important use for the computed residual thermodynamic properties of a liquid is to assess the existence of correlations in the behavior of widely different types of liquids, for removal of the perfect gas contributions leaves only the influence of intermolecular forces on the thermodynamic functions, and it is differences in these that define the behavior of different liquids. Comparisons of residual properties of liquids are thereby free of differences resulting from peculiarities in molecular structure, internal molecular motion, and the like, provided these are independent of density. The most important tool in the comparison of different liquids is the principle of corresponding states, an example of which was discussed in Chapter 21. This principle is consistent with the assumptions that (1) the internal degrees of freedom of the molecules are independent of the density, (2) the translational motion of the molecules can be described by classical mechanics, (3) the total potential energy of

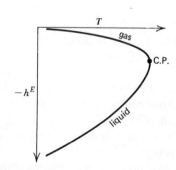

FIGURE 23.4
Sketch of the variation with temperature of the residual enthalpy of gas and liquid along the liquid–vapor coexistence curve. The difference between the two curves is the latent heat of evaporation. From J. S. Rowlinson, *Liquids and Liquid Mixtures*, 2nd ed. (Butterworths, London, 1969).

the liquid can be expressed as the product of an energy parameter, ϵ, and a function of distance between molecular centers i and j scaled by the molecular diameter, R_{ij}/σ. When the principle of corresponding states is valid, the thermodynamic functions of the liquid depend only on the reduced temperature $T^* \equiv k_B T/\epsilon$, reduced pressure $p^* \equiv p\sigma^3/\epsilon$, and reduced density $n^* \equiv n\sigma^3$. The principle is valid when the molecular interaction is spherically symmetric, as in simple liquids like argon. In liquids with more complicated molecules, the potential energy must depend on the orientation of the molecules and other factors, so the simple form of the principle of corresponding states is inadequate. Several schemes have been proposed for generalizing the corresponding-states concept. These are essentially empirical in nature, and all have the feature of adding extra parameters with which the observed p, v, T data can be correlated. The difficulty with such approaches is not inaccuracy— they tend to work rather well as interpolation schemes. Rather, their principal defect is the lack of defined criteria to determine whether or not a given liquid lies outside the domain of applicability of the correlation.

To see what is involved in extending the concept of corresponding states to complicated molecules, consider possible relations between the equations of state of straight-chain hydrocarbons. These constitute an important class of industrial chemicals, and their equations of state have been thoroughly studied. We suppose that the linear hydrocarbons can be thought of as CH_2 groups linked by valence bonds that somewhat alter their properties. In particular, relative to a liquid of unlinked CH_4 molecules, valence bonding in a hydrocarbon chain reduces the number of nearest-neighbor groups with which a given CH_2 can interact at a specified temperature and density [from ~10 in liquid CH_4 to some smaller number in CH_3—$(CH_2)_{n-2}$—CH_3]. Valence bonding also reduces the effective number of degrees of freedom relative to liquid CH_4 with the same number density of C atoms, since the C—C bond has very high stretching frequency relative to the center-of-mass motions in liquid CH_4 and this motion is frozen out at ordinary temperatures (see Section 21.5, where similar phenomena are discussed in terms of the relative sizes of T and the reduced temperatures Θ_{vib} and Θ_{elec}). If these notions are valid, it should be possible to reduce the equations of state of all linear hydrocarbons to a universal form. To test this idea, define the reduced variables

$$\tilde{\tilde{v}} = \frac{v}{v_0(n)},$$

$$\tilde{\tilde{p}} = \frac{pv_0(n)}{u_0(n)}, \qquad (23.13)$$

$$\tilde{\tilde{T}} = \frac{Ts_0(n)}{u_0(n)},$$

where $v_0(n)$, $u_0(n)$, and $s_0(n)$ are the molar volume, intermolecular contribution to the molar internal energy, and intermolecular contribution to the molar entropy of an n-carbon linear hydrocarbon, respectively. If the reduced volume, pressure, and temperature are properly defined, the reduced internal energy, entropy, and other thermodynamic properties should be linear functions of n. This is found to be the case for $n=1$ to $n=20$. It is also necessary that relative functions such as

$$V_{rel} \equiv \frac{\tilde{\tilde{v}}(\tilde{\tilde{p}})}{\tilde{\tilde{v}}(\tilde{\tilde{p}}=0)} \qquad (23.14)$$

be independent of chain length, since all dependence on n is incorporated in the definition of $\tilde{\tilde{v}}$. Figure 23.5 shows that this expectation is fulfilled.

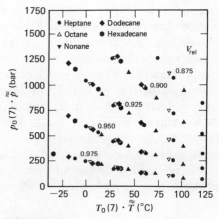

FIGURE 23.5
Reduced pressure versus reduced temperature curves at constant relative volume for n-heptane, n-octane, n-nonane, n-dodecane, and n-hexadecane. Units have been chosen relative to heptane. From J. Hijmans, *Physica* **27**, 433 (1961).

What do we mean by the structure of a liquid? As briefly hinted at the beginning of Chapter 22, we shall use the term structure to describe the arrangement of molecules in space. In the case of a simple liquid, such as Ar, only the relative positions of the centers of the atoms need be specified; in the case of a liquid of polyatomic molecules, for example, H_2O, both the relative positions of the molecular centers of mass and the relative molecular orientations must be specified.

Consider first a simple liquid of monatomic molecules. A representative form for the interaction between a pair of molecules is shown in Fig. 23.6. In liquids of this class, as in most liquids, the average volume per molecule is only about 20–30% larger than the volume of a molecule.[2] Also, every molecule is surrounded by from 8 to 12 nearest neighbors, but the very strong repulsion between any pair of molecules prevents the distance between their centers from becoming much less than about one molecular diameter.[3] Put in other words, the probability of finding a pair of molecules separated by much less than σ is vanishingly small. This geometric restriction implies that in a liquid there must be a shell of molecules at a distance slightly greater than σ from any given molecule, because in the high-density condensed phase the average spacing between molecules is only slightly greater than σ. Continuing the argument, the existence of such a shell of molecules virtually prevents there being any other molecules with centers between one and two molecular diameters from the central molecule, and so forth. The shells of molecules so defined are spread over a range of molecular separations, and the rate of exchange of molecules in any one shell with the surrounding fluid is very large. Consequently, beyond a distance of about three molecular diameters from the central molecule the range over which a shell is spread becomes so large, and the corresponding increase in the local density over the average density so small, that all vestigial structure has vanished.

The kind of structure described above may be represented as follows. In general, in a fluid (gas or liquid), a molecule can be found with equal probability in equal-sized volume elements anywhere in the volume. In fact, if the volume element has magnitude ΔV and the total volume is V, the probability of finding a molecule in ΔV is just $\Delta V / V$. *In the absence of intermolecular forces, each molecule moves throughout the entire volume independent of the presence of other molecules.* Hence the probability of finding *one particular* molecule in the volume ΔV at the point $\mathbf{R}_1$, and *another particular* molecule in the volume ΔV at the point $\mathbf{R}_2$, is simply the product $(\Delta V / V)^2$, and is independent of $\mathbf{R}_1$ and $\mathbf{R}_2$. Since there are N molecules in the volume V, the probability of simultaneously finding *any* molecule in ΔV at $\mathbf{R}_1$ and *any other* molecule in ΔV at $\mathbf{R}_2$ is $N(N-1)(\Delta V / V)^2$, and is independent of $\mathbf{R}_1$ and $\mathbf{R}_2$. Then the joint probability density of finding any molecule at $\mathbf{R}_1$ and any other molecule at $\mathbf{R}_2$ is, for noninteracting molecules, $N(N-1)/V^2 = n^2$, and the conditional probability density of finding a molecule a distance R from the center of a given molecule is n. When there are no intermolecular forces,

23.2
The Structure of Liquids

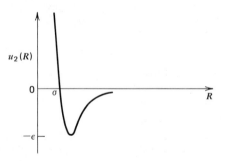

FIGURE 23.6
A typical molecular pair interaction as a function of internuclear separation.

[2] We take, for present purposes, the volume of a molecule to be defined by $\pi\sigma^3/6$, where σ is the distance at which the interaction between a pair of molecules is zero (see Fig. 23.6). When the separation of a pair of molecules is less than σ, the interaction between them is very strongly repulsive. Indeed, the molecules behave almost like billiard balls with diameter σ.

[3] Although this distance is not fixed, because the distance to which one molecule can penetrate another's force field depends on the kinetic energy available, it is sufficiently well defined to be useful in constructing our pictorial representation.

we say that the system has no structure, since the probability of finding a pair of molecules with given separation is independent of that separation.

Consider, now, a liquid in which molecular interactions exist. Let the probability per unit volume of finding a molecule a distance R from the center of a given molecule be denoted as $ng_2(R)$. Then, for a random distribution, such as is characteristic of noninteracting molecules, we would have $g_2(R) = 1$ everywhere (see preceding paragraph). But for a liquid with molecular interactions, $g_2(R)$ has the form displayed in Fig. 23.7. It is particularly important to notice the overwhelming role played by the repulsive interaction between a pair of molecules in determining the structure of the liquid. The function $g_2(R)$, known as the *pair-correlation function*, may be determined from studies of the intensity of scattered x rays and thermal neutrons as a function of scattering angle.[4] (See Chapter 11 for the use of pair-correlation functions in the description of solids.)

The determination of the structure of a liquid from x-ray and/or neutron diffraction experiments is based on the same fundamental principle as is the determination of the structure of a crystal. The important point is this: If the wavelength of the x rays or neutrons is about the same as the average spacing between atoms, there is interference between the secondary waves originating at each of the atoms in the incident beam of radiation. Because the order in the liquid is only of short range, the diffraction pattern from a liquid does not have the sharp reflections characteristic of the Bragg spectrum of a solid. In the case of a solid, the reflections occur within very narrow ranges about a small set of angles precisely because phase coherence between the secondary waves originating at each of the atoms is ensured by the long-range order characteristic of the crystal structure. In the liquid the diffraction pattern is diffuse, and there is some scattering at all angles. Indeed, it can be shown[5] that the intensity of x rays or neutrons scattered without change in energy is, as a function of the scattering angle θ,

$$I(\theta) \propto \tfrac{1}{2}(1 + \cos^2 \theta)I_0\left\{1 + 4\pi n \int_0^\infty [g_2(R) - 1]\frac{\sin sR}{sR} R^2 \, dR\right\} \quad (23.15)$$

where

$$s \equiv \frac{4\pi}{\lambda} \sin \frac{\theta}{2} \quad (23.16)$$

and I_0 is the intensity of the incident radiation of wavelength λ. As written, Eq. 23.15 refers to scattering from monatomic particles; the effects of interference within one atom, because of its internal electronic structure, are incorporated in the proportionality factor. The extension of Eq. 23.15 to describe scattering from a liquid of polyatomic molecules requires two kinds of generalizations. First, account must be taken of the internal structure of the polyatomic molecule. The effects of this structure on the scattering are important, since intramolecular atom–atom separations, especially the intramolecular non-nearest neighbor distances, are comparable in magnitude with intermolecular separations. Second, there are new interference terms corresponding to the addition of the scattered waves from two centers of different atomic number. In principle, the inversion of the scattering data for a polyatomic fluid requires as many independent scattering experiments as there are kinds of pairs of atoms, for example, in H_2O three experiments to get the H–H, H–O, and O–O

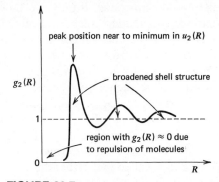

FIGURE 23.7
A typical pair-correlation function for a simple liquid, for example, Ar.

distribution functions. Rarely are these data available. Instead, approximations are made,[6] such as neglecting the scattering from H atoms, neglecting cross correlations of the X–Y type, and so on. Consequently, inferences concerning the structure of a polyatomic liquid must be examined carefully to filter out errors introduced when observations are reduced using such approximations.

Finally, Eq. 23.15 is valid only for that component of the scattered radiation which has the same wavelength (and thus the same energy) as the incident radiation—this is the case of elastic scattering. It is also possible to study the intensity of the radiation scattered with different wavelength, called inelastic scattering. This is more conveniently done with neutrons than with x rays, because the change in energy of the scattered radiation relative to the energy of the incident radiation is more easily measured for neutrons. The elastic scattering gives information about the spatial distribution of atoms in the liquid, whereas the inelastic scattering gives information about the dynamics of particle motion. For the present we confine attention to the elastic scattering and the pair-correlation function deduced therefrom.

Systematic studies of the pair-correlation function, $g_2(R)$, as a function of the thermodynamic state of the fluid, show (see Figs. 23.8 and 23.9):

1. The structural features of the liquid are predominantly determined by the liquid density, and vary little with temperature.

2. The structure of a simple fluid varies smoothly and continuously in going from a high-density (liquid) state to a low-density (gas) state.

3. As the density increases, the ordering in the liquid extends farther and farther from the central molecule.

4. If a coordination number is defined by $Z \equiv 4\pi n \int_0^{R_m} g_2(R) R^2\, dR$, where R_m is the first minimum in the integrand,[7] then it is found that Z varies markedly with the thermodynamic state of the fluid.[8] The value of the coordination number is primarily determined by the density of the fluid, as implied by point 1 above. At any given density the fluctuation in the value of Z arising from exchanges of molecular position is of the same order of magnitude as Z itself. That is, the short-range order in the liquid is not static. Molecules readily enter or leave specific shells, and there is a continuous distribution of molecular separations. In this sense the order in the liquid differs fundamentally from that of the solid, where the atoms are almost immobile.

5. In confirmation of a point made in 4 above, numerical experiments[9] show that the first peak in the pair-correlation function decays on a time scale of the order of magnitude of 10^{-12} s (see Fig. 23.10).

To understand Fig. 23.10, picture some instantaneous arrangement of atoms of the liquid, and call the time at which that arrangement exists $t = 0$. Select a typical atom as the origin of a coordinate system. At $t = 0$ the average density of atoms located a distance R from the one at the origin is just $n g_2(R)$. Now consider some later time. Because the atoms can move, we expect the spatial arrangement to change. In general, every

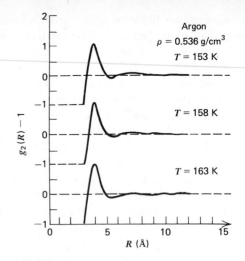

FIGURE 23.8
Pair-correlation functions of liquid Ar at three different temperatures, but with the same density. From C. J. Pings, in H. N. V. Temperley, J. S. Rowlinson, and S. Rushbrooke (Eds.), *Physics of Simple Fluids* (North Holland, Amsterdam, 1968).

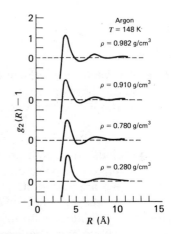

FIGURE 23.9
Pair-correlation functions of liquid Ar at four different densities, but with the same temperature. From C. J. Pings, in H. N. V. Temperley, J. S. Rowlinson, and S. Rushbrooke (Eds.), *Physics of Simple Fluids* (North Holland, Amsterdam, 1968).

---

[6] Different approximations are used in the interpretation of x-ray and neutron diffraction data.

[7] If there is no minimum in the integrand, as happens at high temperature and low density, the concept of coordination number loses all value.

[8] Other definitions of coordination number may be used. However, the qualitative aspects of the deductions cited are not altered by other choices of definition than the one given.

[9] The numerical experiments, carried out by Rahman (1965), consist of using a large computer to solve Newton's equations for 864 molecules interacting with a Lennard–Jones potential corresponding to Ar–Ar. The density chosen was like that in liquid Ar at 85 K.

atom will tend to move away from where it was at $t = 0$. Clearly, close to $t = 0$ the spatial arrangement must be very similar to that at $t = 0$, but as t increases it will resemble that arrangement less and less. Figure 23.10 displays, for several times, the conditional probability, given an atom is at the origin at $t = 0$, that another atom will be a distance R from the origin. For t small, say, 0.5×10^{-12} s, the geometric constraints that lead to the nearest-neighbor peak of $g_2(R)$ dominate; but as t increases, both the probability of finding the original particle at the origin and the probability of finding a different particle at R change. In the long-time limit, an atom can be found with equal probability in all volume elements of equal size, so the peak at $R = \sigma$ decreases and the low average density near $R = 0$ increases, until a uniform density is established.

The preceding comments are valid for simple liquids, in which the constituent particles have spherically symmetric force fields. It is to be expected that a liquid of polyatomic molecules, in which the constituent particles have complicated nonspherical force fields, will have a structure somewhat different from that of a simple liquid. Consider, as an extreme case, liquid water. The water molecule has a dipole moment and higher multipole moments, reflecting the deviation from spherical symmetry of its electronic charge distribution. Of course, the interaction between two polar molecules depends on the angle between their dipole moments, quadrupole moments, and so on. More important, if the molecular charge distribution is not spherical, then neither will be the short-range repulsive forces between molecules. Hence the local packing can deviate considerably from that shown in Fig. 23.9, which is fundamentally like that in a fluid of hard spheres. Finally, and for H_2O most important, water molecules form hydrogen bonds with one another. The hydrogen bond is the dominant interaction between water molecules; it is strongly directional and has a depth very much larger than $k_B T$.

Although the electronic charge distribution of a water molecule is not spherical, the deviation is too small to show up in the x-ray diffraction pattern from the liquid. Therefore, the scattering of x rays from water depends almost completely on only the distribution of molecular centers, that is, on the positions of the O atoms. Figure 23.11 displays O–O pair-correlation functions for liquid H_2O at various temperatures and constant density. These data show that:

1. There is perceptible structure in the local environment of a water molecule to about 8 Å distance from the molecule at the origin.

2. The nearest-neighbor separation is only weakly temperature-dependent at constant density, changing from 2.82 Å at 4°C to 2.94 Å at 200°C.

3. The O–O pair-correlation function is sensibly the same in H_2O and D_2O.

4. The average coordination number in liquid water is sensibly constant at about 4 from 4°C to 200°C (constant density of 1 g/cm^3).

Clearly, water differs appreciably from liquid argon, in which the number of nearest neighbors is about 10, and which is "loosely close-packed." Observations 1 through 4 imply that water is more "open" than is liquid argon, and also that there is tetrahedral coordination in the local structure surrounding a given water molecule. Tetrahedral coordination implies the existence of a correlation between the orientations of water molecules. How far does this orientational correlation extend? This question can be answered by examining neutron scattering from D_2O.

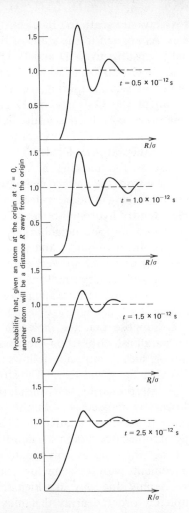

FIGURE 23.10
The probability that, given an atom is at the origin of coordinates at $t = 0$, a different atom will be a distance R away from the origin, as a function of time. The distance R is measured in units of the size parameter σ of the Lennard–Jones potential. From A. Rahman, *Phys. Rev.* **136A,** 405 (1964).

x Rays are scattered by electrons, whereas neutrons are scattered by nuclei. As applied to the study of liquid D_2O, neutron scattering is about equally sensitive to O–D and D–D pair correlations, and rather insensitive to O–O pair correlations. Using the O–O pair-correlation function obtained by x-ray scattering, we can learn, by difference, something about the sum of the O–D and D–D pair correlations, and therefore about orientational correlations in the liquid. The data displayed in Fig. 23.12 show that in liquid D_2O orientational correlations vanish beyond the first shell of four water molecules.

Just as the packing in liquid argon is a blurred and imperfect derivative from the pattern characteristic of closest packing of spheres, so the local packing in liquid water is a blurred and imperfect derivative from some extended hydrogen-bonded network in which each water molecule is coordinated to four neighboring molecules by hydrogen bonds. Even though both are liquids, and macroscopically isotropic, the local structures of Ar and H_2O are very different, reflecting the very different kinds of interaction between molecules in the two systems.

It is worthwhile to examine a few more examples of the relationship between molecular interactions and the structure of a liquid.

1. Suppose that the molecule is very asymmetric, such as a long-chain paraffin hydrocarbon—for example, dodecane. The force field surrounding such a molecule follows the contours of the carbon skeleton, and is also "long and thin." The structure of liquid paraffin hydrocarbons can be interpreted by first considering a collection of randomly oriented hydrocarbon molecules.[10] In this random configuration the density of packing is very low relative to what it would be in an ordered parallel array of molecules. When the density of molecules is low, the total energy of attraction is necessarily small; there are many possible disordered arrangements with the same low energy of attraction. On the other hand, there is only one way in which the molecules can be arranged in the completely ordered array that maximizes the attractive interactions between the molecules. The structure of liquid paraffins is determined by the competition between the tendency toward disorder induced by collisions between molecules and other thermal motions, and the order corresponding to the lowest-energy arrangement, that is, by the competition between energy and entropy terms in the free energy. Thus, the tendency for the molecules to become aligned is stronger the longer the molecular chain, but the ordering is never as complete as in the crystalline phase. As in simple liquids, there is a distribution of possible near-neighbor distances within the constraint of partial parallel alignment enforced by the balance of repulsive and attractive interactions between the molecules.

2. The paraffin hydrocarbons typify a class of substances with asymmetric interactions. What happens when, in addition to these, there are strong local directional interactions, such as are induced by hydrogen bonding? In liquids of the long-chain alcohols the combination of molecular asymmetry and directional hydrogen bonding leads to an accentuation of the features discussed above. It is found that the structure of liquid linear alcohols is in accord with a picture in which the alcohol molecules are linked in chains by hydrogen bonds between hydroxyl groups, while the aliphatic groups remain approximately parallel as in the corresponding hydrocarbons.

[10] A visualization of the random configuration is a disorderly heap of matchsticks. In the disorderly heap there are many "holes," and the density of packing is low. The configuration with closest packing (highest density) is with all the matches parallel and touching as in the matchbox.

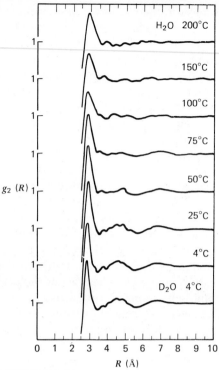

FIGURE 23.11

Pair-correlation functions for liquid H_2O at various temperatures and for liquid D_2O at 4°C, as determined by Narten *et al.* (1967). Note that the baseline of each curve is one unit above that for the curve below. From A. H. Narten, ORNL-4578 UC4, Oak Ridge National Laboratory, 1970.

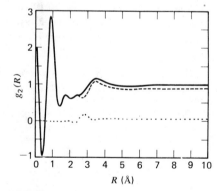

FIGURE 23.12

Atom pair-correlation functions for liquid D_2O. The solid curve is the weighted sum of contributions from O–O, O–D, and D–D correlations, the dotted curve is the O–O pair-correlation function, and the dashed line is the appropriately weighted difference pair-correlation function. In the difference pair-correlation functions shown, the contributions from O–D and D–D pairs have nearly equal weight. From A. H. Narten, *J. Chem. Phys.* **56**, 5681 (1972).

3. In liquid sulfur there is evidence that there exist a wide variety of chain and ring molecules, and that the distribution of atoms among chains and rings depends on the temperature. The structure of liquid sulfur is determined by a complex balance between chemical bonding forces and intermolecular forces of the type found in simple liquids.

4. In all the types of liquids thus far described, the range of the interaction is about equal to the molecular diameter. But both molten salts and liquid metals contain charged particles, and the range of the Coulomb interaction between charged particles is very large relative to the particle diameter. Moreover, the electrostatic interaction is so large relative to the thermal energy that it is difficult to separate positive and negative charges. The consequences of these characteristics of the potential are somewhat different in molten salts and in liquid metals.

In molten salts each positive ion is surrounded by negative ions, and vice versa (see Fig. 23.13). The primary effect of the electrostatic interaction between ions is to lower the total energy of the liquid. Given the overall charge alternation of that structure, short-range forces like those in argon then influence the ionic motion and play an important role in determining details of the local ionic environment.

In a liquid metal the two charged species are ions and electrons. Just as in a molecule, the electrons move so much more rapidly than do the ions that the latter move in an effective potential generated by the distribution of electrons that minimizes the energy for each ion–ion separation R. Consequently, the effective ion–ion potential is not the Coulomb potential, but rather some relatively short-range screened potential. Moreover, because the electron wavelength and ion–ion separation are comparable, diffraction of the electrons influences the form of the ion–ion potential—it has oscillations (see Fig. 23.14a). However, despite the differences between the detailed shapes of the ion–ion interaction and, say, the Ar–Ar interaction, the fact that both are short-range is enough to ensure that the local structure in a liquid metal is very much like that in a simple liquid (see Fig. 23.14b).

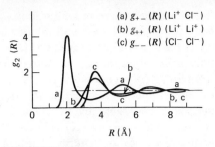

FIGURE 23.13
Pair-correlation functions from a computer simulation of liquid LiCl, $T = 833$ K, $v = 28.3$ cm^3/mol. From L. Woodcock, *Chem. Phys. Lett.* **10**, 257 (1971).

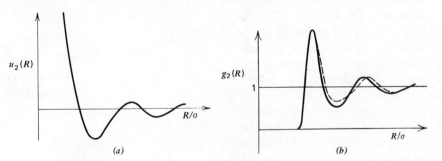

(a)   *(b)*

FIGURE 23.14
(*a*) General shape of the effective pair potential between ion cores in a liquid metal. (*b*) Pair-correlation functions for liquid Ar (– – –) and liquid Cs (———), scaled so that the positions of the first maxima coincide. From A. Paskin, *Adv. Phys.* **16**, 223 (1971).

How can the thermodynamic properties of a liquid be related to its structure? Consider the calculation of the internal energy, and suppose that the total potential energy of molecular interaction can be accurately represented as a sum of pair interactions (see Section 21.7). As usual, we suppose that the pair interactions are independent of the molecular velocities. Given these conditions, the contribution to the internal energy from the kinetic energy of the molecules is $\frac{3}{2}Nk_BT$, just the same as for a perfect gas of monatomic particles. However, in the case of the liquid (or

**23.3
Relationships Between the Structure and the Thermodynamic Properties of a Simple Liquid**

the dense gas) there is an additional contribution to the internal energy from the potential energy of interaction. Since two molecules a distance R apart have a potential energy $u_2(R)$, the average potential energy of the fluid is simply obtained by adding together the potential energy arising from all $\frac{1}{2}N(N-1)$ possible distinct pairs of molecules.[11] The structure of the liquid enters the description when we specify the number of pairs of molecules separated by the distance R. Now, on the average, a given molecule has $4\pi R^2 n g_2(R)\,dR$ neighbors in the range R to $R+dR$. Thus, its average potential energy of interaction with those molecules between R and $R+dR$ is $u_2(R) \cdot 4\pi R^2 n g_2(R)\,dR$, and with all molecules is $\int_0^\infty u_2(R) \cdot 4\pi R^2 n g_2(R)\,dR$. Since there are N molecules in the liquid, to obtain the total energy we multiply the integral expression by N and then divide by 2 to avoid counting each pair interaction twice (once when the molecule is selected as the "given" molecule and once when it is a neighbor of some other "given" molecule). Thus, the internal energy of the liquid is

$$U = \tfrac{3}{2}Nk_BT + 2\pi Nn \int_0^\infty u_2(R)g_2(R)R^2\,dR. \qquad (23.17)$$

Of course, from U as determined by Eq. 23.17 we can obtain the heat capacity and other thermodynamic functions, but such calculations require knowledge of $(\partial g_2(R)/\partial T)_V$, and so on.

Consider now the calculation of the equation of state. The problem to be solved is that of sorting out the effects of collisions with a wall (as in a dilute gas) from the effects of interactions between molecules. The fundamental ideas needed were introduced in Section 21.7 in the calculation of the second virial coefficient. However, before we can use those ideas to calculate the equation of state of a liquid, we must examine what physical picture underlies them.

In Section 21.7 we asserted that the probability density for finding another molecule a distance R from a molecule at the origin of the coordinates is $n\exp(-u_2/k_BT)$. This assertion is correct only when the density of molecules is so low that interactions between all other molecules and the two mentioned can be neglected. Then $n\exp(-u_2/k_BT)$ is seen to be a special case of the probability density for finding a given configuration (separation of a pair of molecules by R) in a canonical ensemble.[12] Now, in a dilute gas the average separation between the molecules is much greater than the range of the potential u_2, and in almost all instances two molecules can be moved toward one another along the line connecting their centers without coming close to any of the other molecules in the gas. Thus at very low density $u_2(R)$ is also just the work required to bring the two molecules to a separation R from a distance large enough that their interaction is vanishingly small. Consider now a liquid. In this case there are many molecules within the range of the pair potential $u_2(R)$, and we cannot isolate any one pair of molecules from interaction with the other molecules. A necessary implication of this observation is that the work required to bring two molecules to a separation R from a distance at which the direct interaction u_2 is vanishingly small, denoted as $w_2(R)$, is different from $u_2(R)$. The point is that in a dense system moving two molecules must result in some rearrangement in the positions of all the other molecules. Thus, when two

---

[11] It does not matter whether the pair AB is constructed by first choosing A or first choosing B. Hence the $N(N-1)$ possible choices of pairs include each pair, for example, the pair AB, twice; we then divide by 2 to obtain the number of distinct pairs.

[12] See Section 19.6.

molecules are brought to a separation R from some initial large separation, there are uncontrolled changes in the rest of the system. Just as in our discussion of the first law of thermodynamics (see Chapter 13), this characteristic behavior of the system of molecules makes w_2 different from simple mechanical work. In general, w_2 is both temperature- and density-dependent; it is known as the *potential of average force*, since the rate of change of w_2 with displacement is the average force acting between the two molecules; for example, $-\partial w_2/\partial x$ is the x component of the average force between the molecule at the origin and the molecule a distance R away.

The generalization of the arguments of Section 21.7 to a dense fluid consists in asserting that the probability density for finding another molecule a distance R from a molecule at the origin of the coordinates is

$$n \exp\left(-\frac{w_2(R)}{k_BT}\right). \qquad (23.18)$$

But we have also defined $ng_2(R)$ to be the probability density of finding another molecule a distance R from a molecule at the origin of the coordinates. Comparison of these two definitions establishes the identity

$$g_2(R) \equiv \exp\left(-\frac{w_2(R)}{k_BT}\right). \qquad (23.19)$$

A few words about w_2 are appropriate. First, when the density is very low, w_2 must approach the pair-interaction potential u_2, since in the low-density limit two molecules may be moved without influencing any of the other molecules in the system. Second, in a dense system w_2 becomes large and positive for small pair separation [the region where $g_2(R)$ is essentially zero], is negative at a distance corresponding to the first peak in $g_2(R)$, and then oscillates between positive and negative values while decaying nearly to zero after three molecular diameters (see Fig. 23.15). Note that the range of w_2 is longer than that of u_2. This difference arises because w_2 incorporates the effects of rearrangement in the medium, hence of geometric restraints. To illustrate how geometric restraints influence w_2, imagine moving two hard-sphere molecules (see Fig. 23.16)

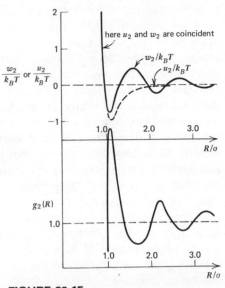

FIGURE 23.15
The potential of mean force, w_2, and the pair potential, u_2, for a dense fluid. The pair-correlation function g_2 of the fluid to which w_2 and u_2 refer is plotted in the lower part of the figure.

FIGURE 23.16
Schematic representation of the origin of w_2 for hard spheres. Note that a position and a direction of motion are associated with each sphere.

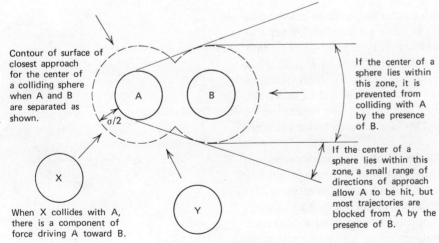

Contour of surface of closest approach for the center of a colliding sphere when A and B are separated as shown.

$\sigma/2$

If the center of a sphere lies within this zone, it is prevented from colliding with A by the presence of B.

If the center of a sphere lies within this zone, a small range of directions of approach allow A to be hit, but most trajectories are blocked from A by the presence of B.

When X collides with A, there is a component of force driving A toward B.

Y can collide with A only in a restricted zone of directions. Simultaneous collision of Y with A and B is very infrequent; hence, in the average, collisions of X with A have more influence than those of Y with A, and there is a net force pushing A toward B.

in a dense fluid of hard-sphere molecules. When the two are far apart, collisions with the other molecules are uniformly distributed in all directions, so the net force acting on each vanishes, as does the net force between them. But if the two hard spheres are less than two diameters apart (see Fig. 23.16), other molecules can collide only with the "outside" region, and none can get between them. Thus each molecule is subjected to a net force, that molecule on the left in Fig. 23.16 to a force acting to the right, and the reverse for the other molecule. The net force between the molecules is the source of an effective attraction. We see that even though two isolated hard spheres never attract one another, two hard spheres in a dense fluid of hard spheres do attract one another when $R < 2\sigma$. More generally, we find that in a real fluid w_2 takes on positive and negative values reflecting the balance between the effects of the real potential energy u_2 and the geometric restraints imposed by "impenetrability" of the molecules.

The calculation of the equation of state of a liquid proceeds in the same way as the calculation in Chapter 21. Whenever the potential energy of interaction of the molecules is independent of the molecular velocities, the contribution to the pressure from collisions with the wall, the *kinetic pressure*, is

$$p_K = \frac{Nk_BT}{V}. \tag{23.20}$$

That this is so is easily seen from the arguments preceding Eq. 21.160 or, more formally, from the separation of the partition function into translational and configurational contributions, as in Eq. 21C.4. The calculation of the molecular interaction contribution to the pressure starts with the substitution of $n^2 \exp(-w_2/k_BT)$ for $n^2 \exp(-u_2/k_BT)$ in Eq. 21.160, and then proceeds through all the same steps. Of course, the direct force exerted by the molecule a distance R from the one at the origin is still $-(du_2/dR)\cos\theta$. Integrating as before, adding p_K from Eq. 23.20 above, and using Eq. 23.19 to replace $\exp(-w_2/k_BT)$ with $g_2(R)$, we find that the equation of state of the liquid is

$$p = nk_BT - \frac{2\pi}{3}n^2 \int_0^\infty R^3 \frac{du_2}{dR} g_2(R)\, dR. \tag{23.21}$$

From Eqs. 23.17 and 23.21 we can obtain the other thermodynamic functions, at least in principle.[13]

If we consider a realistic intermolecular potential, such as is shown in Fig. 23.6, we see that du_2/dR is negative at small intermolecular separations and positive at large separations, where it very rapidly ($\propto R^{-7}$) falls again to zero. The dominant contribution to the integral in Eq. 23.21 is thus from the region near the molecule at the origin. At high densities there are enough neighbors in the region where $du_2/dR < 0$ to make the integral negative and hence p higher than the perfect gas value, but at moderate densities the contrary is true. This can be seen from the data presented in Fig. 21.5.

The equations we have obtained thus far involve the pair-correlation function $g_2(R)$. In order to construct an interpretation of liquid properties based entirely on theoretical considerations, it is necessary to relate $g_2(R)$ to the intermolecular potential (as a function of temperature and density, of course). There are various theoretical approaches directed to this end, differing in the use of various approximations, most of which cannot be easily visualized. We shall make no attempt to describe these theories,

[13] Note that the integral in Eq. 23.21 is not simply proportional to the second virial coefficient B, since $g_2(R)$ itself incorporates density dependence. At low density, however, $w_2(R) \rightarrow u_2(R)$ and $g_2(R) \rightarrow \exp(-u_2(R)/k_BT)$, and we are led back to Eq. 21.165.

but shall discuss some of the physical ideas involved and illustrate some of the results obtained.

We have seen that a principal difference between a dilute gas and a liquid can be stated as follows. In a dilute gas a typical molecule is usually outside the force fields of all other molecules and only occasionally in the force field of one other molecule (binary collision), whereas in a liquid a typical molecule is usually within the force fields of, say, 10 nearest-neighbor molecules and is never completely free of the influence of other molecules. One consequence of the multiplicity of simultaneous interactions in the liquid is that the work required to bring a molecule to a distance R from another molecule at the origin of the coordinates is the potential of average force $w_2(R)$, and *not* the pair potential $u_2(R)$. We can think of $w_2(R)$ as the potential of the force acting on two molecules, say, 1 and 2, averaged over the possible configurations of the remaining molecules of the liquid. In symbols, the average force acting on molecule 1 at $\mathbf{R}_1$, when molecule 2 is known to be at $\mathbf{R}_2$, is[14] $-\nabla_1 w_2(R_{12})$, where $R_{12} \equiv |\mathbf{R}_1 - \mathbf{R}_2|$ is the distance between the two molecules. Note that by using Eq. 23.19 we can rewrite this average force as $k_B T \nabla_1 \ln g_2(R_{12})$. The average force can be thought of as the resultant of that force due to molecule 2 at $\mathbf{R}_2$ alone and the average force due to all the other particles. But the probability of finding a molecule in the volume element $d\mathbf{R}_3$ at $\mathbf{R}_3$, when it is known that other molecules are located at $\mathbf{R}_1$ and $\mathbf{R}_2$, is, using the argument of Section 23.2,

$$\frac{n^3 g_3(\mathbf{R}_1, \mathbf{R}_2, \mathbf{R}_3)}{n^2 g_2(R_{12})} \, d\mathbf{R}_3.$$

The denominator of this expression is just the probability density for finding two molecules separated a distance R_{12}, including the effect of intermolecular forces. The numerator is, by definition, the probability density for finding three molecules at positions $\mathbf{R}_1$, $\mathbf{R}_2$, and $\mathbf{R}_3$ multiplied by the volume element $d\mathbf{R}_3$. The *triplet correlation function* $g_3(\mathbf{R}_1, \mathbf{R}_2, \mathbf{R}_3)$ accounts for the effect of intermolecular forces in the medium just as does $g_2(R_{12})$. We can now write for the average force on the molecule at $\mathbf{R}_1$ due to all other molecules except the molecule at $\mathbf{R}_2$ the expression

$$\int d\mathbf{R}_3 \left[-\nabla_1 u_2(R_{13}) \frac{n g_3(\mathbf{R}_1, \mathbf{R}_2, \mathbf{R}_3)}{g_2(R_{12})} \right].$$

Adding the contributions, we have

$$\nabla_1 w_2(R_{12}) = \nabla_1 u_2(R_{12}) + n \int d\mathbf{R}_3 \, \nabla_1 u_2(R_{13}) \frac{g_3(\mathbf{R}_1, \mathbf{R}_2, \mathbf{R}_3)}{g_2(R_{12})}, \qquad (23.22)$$

and using Eq. 23.19 we can rewrite Eq. 23.22 to be an equation for $g_2(R_{12})$:

$$-k_B T \nabla_1 \ln g_2(R_{12}) = \nabla_1 u_2(R_{12}) + n \int d\mathbf{R}_3 \, \nabla_1 u_2(R_{13}) \frac{g_3(\mathbf{R}_1, \mathbf{R}_2, \mathbf{R}_3)}{g_2(R_{12})}. \quad (23.23)$$

If we could solve Eq. 23.23, we would find the pair-correlation function in terms of the pair potential and the temperature and density, with the thermodynamic variables entering via the averaging process. However, to solve Eq. 23.23 we must know the function $g_3(\mathbf{R}_1, \mathbf{R}_2, \mathbf{R}_3)$, and we do not! Several of the theories of simple liquids are based on Eq. 23.23 and a supplementary assumption concerning the form of $g_3(\mathbf{R}_1, \mathbf{R}_2, \mathbf{R}_3)$.

In Eq. 23.23 the terms $\nabla_1 u_2(R_{12})$ and $\nabla_1 u_2(R_{13})$ are "direct" forces, whereas the effect of the averaging in the second term on the right-hand side is to account for "indirect" effects that influence the distribution of molecules. A different way of representing the structure of a liquid also relies on the idea of direct and

---

[14] ∇_1 is shorthand notation for differentiation with respect to the position of molecule 1.

indirect effects. The idea is that $[g_2(R) - 1]$ measures the total correlation between pairs of particles. To see the motivation for this statement, notice that $[g_2(R) - 1]$ tends to zero as $R \to \infty$, since there is no correlation between the positions of two widely separated molecules. We now *define* $[g_2(R) - 1]$ to be the sum of two contributions. The first contribution, $C(R)$, consists of correlations transmitted directly between the two molecules; we need not specify how this occurs. The second contribution comes from correlations transmitted via a third molecule, and is made up of direct correlation between molecules 1 and 3 multiplied by the total correlation function between molecules 2 and 3 and the probability of finding a third molecule in the volume element $d\mathbf{R}_3$, integrated over all possible positions of molecule 3. Adding these two contributions we are led to the *definition*[15]

$$g_2(R_{12}) - 1 \equiv C(R_{12}) + n \int d\mathbf{R}_3 C(R_{23})[g_2(R_{23}) - 1]. \qquad (23.24)$$

If we could solve Eq. 23.24, we would find the pair-correlation function required for the description of the structure of the liquid. But to do so we must know $C(R_{12})$, and we do not! Several of the theories of simple liquids are based on Eq. 23.24 and a supplementary hypothesis concerning the form of $C(R_{12})$.

Equation 23.23 and Eq. 23.24 are exact, yet they cannot be used without extra information that must be obtained from considerations other than those used to derive these equations. We shall not examine these considerations or the approximate theories of liquid structure to which they lead, since such detailed analysis is out of place in an elementary textbook. We shall, however, quote some of the results of approximate theories to illustrate the accuracy with which the structure and thermodynamic properties of a liquid can be predicted.

Before proceeding further in the discussion of the theory of liquid structure, it is appropriate to mention briefly the technique of computer simulation. Although fundamentally a theoretical procedure, it has much in common with experiment, as will become evident from the following description.

There are two computer simulation techniques of importance for the study of liquids. In the method of *molecular dynamics* the classical equations of motion of N interacting particles are solved by step-by-step numerical integration, thereby determining the trajectories of all particles over a suitable time interval. Typical calculations, involving $100 < N < 1000$ particles, use time steps of 10^{-14} s and give the complete history of the system for $10^{-14} < t < 10^{-11}$ s; occasionally longer intervals are covered. The thermodynamic properties are obtained, using the quasi-ergodic hypothesis, from the time average of appropriate molecular properties (see Section 15.1). In the *Monte Carlo* method one generates a sequence of particle configurations in such a way that the probability of a particular configuration, of potential energy $\mathscr{V}_N$, appearing in the sequence is proportional to $\exp(-\mathscr{V}_N/k_B T)$. Then the unweighted average of a molecular property over the configurations of the sequence gives the canonical average of that function (see Section 19.6). Typical calculations involve sequences of 10^5–10^6 configurations. The method of molecular dynamics provides information about the structure and properties of the liquid at equilibrium, and about transport properties and the approach to equilibrium. It is particularly valuable for the study of dynamical properties of single molecules, or groups of molecules, in strong interaction with others. The Monte Carlo method is restricted to the study of equilibrium properties of the system. When properly executed, both methods yield the true properties of the system unbiased by errors arising from approximations.

The molecular dynamics and Monte Carlo methods are similar to experiments in the sense that they generate data which must be analyzed using theoretical concepts introduced by other methods. Unlike real experiments, a computer simulation can test the effects of adding and subtracting particular forces to the intermolecular force, and can provide tests of theoretical ideas by

---

[15] Equation 23.24 defines the function $C(R_{12})$ in terms of $g_2(R_{12})$; $C(R_{12})$ is called the *direct correlation function*.

computing the exact behavior of the model used in developing those ideas. It is a powerful addition to the tools available for the study of matter.

Consider what can be learned about $g_2(R)$ from the study of a simple model. We have already noted that the experimental data indicate that $g_2(R)$ is relatively insensitive to the temperature if the density of the liquid is kept fixed. We have also deduced the general dependence of $g_2(R)$ on R from arguments based on the consideration of only the repulsive forces acting between molecules. Suppose that we treat the molecules as if they were billiard balls; that is, suppose that the hard-sphere potential is used to represent the repulsive interaction between molecules, with the effects of the attractive interaction ignored. Clearly this potential cannot lead to a stable liquid because there is no attraction between the molecules. Nevertheless, for fixed density we can examine the extent to which our previous arguments are verified. In particular, we wish to know if $g_2(R)$ for a liquid of hard spheres behaves as proposed and resembles the distribution of molecules in a real liquid.

The earliest answer to this question was given by Morrell and Hildebrand, who constructed a model of a liquid by placing several hundred gelatin spheres, including a few colored ones, in a vessel containing a boiled gelatin solution of the same density and refractive index. The vessel was shaken and, from photographs of the contents, the distribution of distances between a pair of colored spheres determined. The observed distribution is very much like that observed in a real liquid. Using the Monte Carlo or molecular dynamics computer simulation methods described earlier, a more sophisticated version of the same experiment can be conducted. Typical results are shown in Fig. 23.17a. Alternatively, one of the approximate theories we mentioned can be used; typical results of such a calculation are also shown in Figure 23.17a. Aside from the discontinuous fall of $g_2(R)$ to zero for $(R/\sigma) < 1$, the pair-correlation function of a rigid-sphere fluid does show the characteristic features of the distribution of matter in a real liquid. The discontinuous change in $g_2(R)$ at $(R/\sigma) = 1$ is a consequence of the discontinuous change in potential at $(R/\sigma) = 1$ that is characteristic of the hard-sphere interaction. For a more realistic, "softer," continuous repulsive potential, the pair-correlation function is substantially the same as for hard spheres everywhere, but does not have the unphysical sudden change in value at $(R/\sigma) = 1$. This is shown in Fig. 23.17b. These results for fluids in which there are only repulsive forces, together with the observation that $g_2(R)$ for a rigid-sphere fluid depends only on the fluid's density and not on its temperature, confirm our analysis of the overwhelming importance of repulsive forces in determining the structure of a liquid.

What about the properties of a real liquid? There are, fundamentally, two uncertainties in the theoretical analysis. First, it is now known that the interactions in a liquid cannot be satisfactorily represented as a sum of interactions between pairs of molecules (see Eq. 21.159). Although the contribution to the potential energy by three-body terms is not large, they are required for quantitative prediction of the properties of the liquid. Higher-order, many-body interactions appear to be unimportant for liquid argon, and probably also for other simple liquids. Given an accurate expression for the potential energy, the method of computer simulation provides excellent predictions of the thermodynamic properties of a simple liquid (see Tables 23.3 and 23.4). Indeed, viewed as an experiment, the method of computer simulation tests the accuracy of a potential energy expression by calculating the properties of the liquid without approximation, these calculated properties to be compared with those of the real liquid.

Useful as computer simulations are, theoretical concepts and understanding are usually derived from analytical, albeit approximate, theories of the liquid state. We must acknowledge that the use of approximations in the solution of Eq. 23.23 or 23.24 introduces error in the predicted properties of the liquid—this is the second of the uncertainties in theoretical analysis. Modern theories of the structure of a liquid are actually rather good, although not exact. The error in prediction of the pair-correlation function for a typical liquid density can be ascertained by examination of Fig. 23.17a and Fig. 23.18. The latter figure

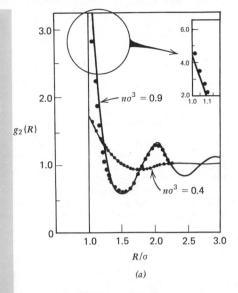

$g_2(R)$

$n\sigma^3 = 0.9$

$n\sigma^3 = 0.4$

R/σ

(a)

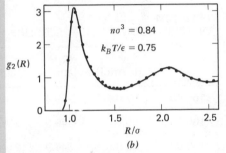

$n\sigma^3 = 0.84$

$k_B T/\epsilon = 0.75$

$g_2(R)$

R/σ

(b)

FIGURE 23.17

(a) Pair-correlation functions for a hard-sphere fluid at two densities, $n\sigma^3 = 0.9$ and $n\sigma^3 = 0.4$. ● Computer simulation results; —— PY theory. The PY theory is obtained by making the approximation

$$C(R) = \left[1 - \exp\left(\frac{u_2(R)}{k_B T} \right) \right] g_2(R)$$

and then solving Eq. 23.24 for $g_2(R)$. From R. O. Watts and I. J. McGee, *Liquid State Chemical Physics* (Wiley, New York, 1976). (b) Pair-correlation function for a fluid with purely repulsive pair interactions of the form $u_2 = 4\epsilon[(\sigma/R)^{1/2} - (\sigma/R)^6] + \epsilon$ for $R \leq R_m \equiv 2^{1/6}\sigma$, $u_2 = 0$ for $R > R_m$. From J. P. Hansen and I. R. McDonald, *Theory of Simple Liquids* (Academic Press, New York, 1976).

TABLE 23.3

CONTRIBUTIONS TO THE INTERNAL ENERGY OF LIQUID ARGON[a]

T (K)	v (cm³/mol)	$u_1(2b)$[b] (cal/mol)	$u_1(3b)$[c] (cal/mol)	$u_1(Q)$[d] (cal/mol)	u_1(theor)[e] (cal/mol)	u_1(exp) (cal/mol)
100.00	27.04	−1525.2	87.1	15.6	−1423	−1432
100.00	29.66	−1393.6	67.9	12.5	−1313	−1324
140.00	30.65	−1284.7	62.8	9.3	−1213	−1209
140.00	41.79	−951.4	39.5	6.4	−906	−922
150.87	70.73	−603.8	26.6	4.6	−573	−591

[a] From J. A. Barker and D. Henderson, *Rev. Mod. Phys.* **48**, 587 (1976).
[b] Pair potential (two-body) contribution.
[c] Triplet potential (three-body) contribution.
[d] Quantum correction (the de Broglie wavelength Λ is, for the temperature listed, large enough to make about a 1% quantum correction to the internal energy necessary).
[e] $u_1(\text{theor}) = u_1(2b) + u_1(3b) + u_1(Q)$.

TABLE 23.4

CONTRIBUTIONS TO THE PRESSURE OF LIQUID ARGON[a]

T (K)	v (cm³/mol)	$p(2b)$[b] (atm)	$p(3b)$[c] (atm)	$p(Q)$[d] (atm)	p(theor)[e] (atm)	p(exp) (atm)
100.00	27.04	+239.9	364.2	42.2	646	652
100.00	29.66	−148.0	238.8	23.3	116	105
140.00	30.65	+348.9	214.3	16.7	580	583
140.00	41.79	−33.7	49.0	2.7	18	37
150.87	70.73	+34.5	13.2	1.2	49	49

[a] From J. A. Barker and D. Henderson, *Rev. Mod. Phys.* **48**, 587 (1976).
[b] Pair potential (two-body) contribution.
[c] Triplet potential (three-body) contribution.
[d] Quantum correction (the de Broglie wavelength Λ is, for the temperatures listed, large enough to make about a 1% quantum correction to the internal energy necessary).
[e] $p(\text{theor}) = p(2b) + p(3b) + p(Q)$.

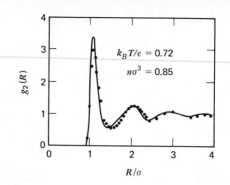

FIGURE 23.18

The pair-correlation function of the Lennard–Jones 6–12 liquid near its triple point. The points are determined by the method of molecular dynamics, the line is the prediction of the PY theory (see caption to Fig. 23.17a). From J. A. Barker and D. Henderson, *Rev. Mod. Phys.* **48**, 587 (1976).

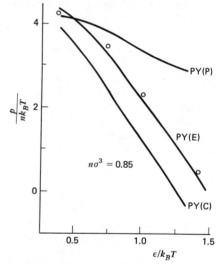

FIGURE 23.19

The equation of state, p/nk_BT, predicted by PY theory (see the caption to Fig. 23.17a).

PY(P): Calculated from Eq. 23.21.

PY(E): Calculated by first integrating the energy equation 23.17 with respect to temperature to obtain the Helmholtz free energy, then differentiating this Helmholtz free energy with respect to volume to obtain the pressure.

PY(C): Calculated from

$$k_BT(\partial n/\partial p)_T$$
$$= 1 + n\int [g_2(R) - 1]\, d\mathbf{R}$$

and integration with respect to the density to obtain the pressure. From J. A. Barker and D. Henderson, *Rev. Mod. Phys.* **48**, 587 (1976).

displays the pair-correlation function for a liquid with pair interaction of the form of Eq. 21.173, the so-called Lennard–Jones 6-12 potential. We see that in the two cases illustrated the overall agreement between theory and computer experiment is very good, but it is equally evident that there are small discrepancies with respect to prediction of the height of the first peak of $g_2(R)$ and with respect to the position and shape of the second peak. The thermodynamic functions computed from Eqs. 23.17 and 23.21 are very sensitive to the precise shapes of $u_2(R)$ and $g_2(R)$. Consequently these small discrepancies lead to appreciable errors in the predicted thermodynamic properties of the liquid, particularly the pressure. In addition, because of the form of the approximation used to solve Eq. 23.23 or 23.24, it is usually found that self-consistency is lost, and the pressures calculated in different ways do not have the same value. An example of this behavior is shown in Fig. 23.19. Clearly, even though the available theory is rather good at predicting $g_2(R)$, it must be improved before fully quantitative predictions of the thermodynamic properties of the liquid can be made. That improvement is the subject of much current research, and remains to be accomplished.

TABLE 23.5
VALUES OF THE REDUCED HELMHOLTZ FREE ENERGY, A_N/Nk_BT, FOR A LENNARD–JONES 6–12 LIQUID[a]

k_BT/ϵ	$n\sigma^3$	Molecular Dynamics	PY(E)[b]	BH	WCA
2.74	0.60	−0.34	−0.33	−0.33	−0.33
	0.70	+0.01	+0.01	+0.01	+0.01
	0.80	0.43	0.43	0.42	0.41
	0.90	0.93	0.95	0.95	0.92
	1.00	1.59	1.61	1.62	1.56
1.35	0.60	−1.77	−1.75	−1.75	−1.74
	0.70	−1.65	−1.62	−1.63	−1.63
	0.80	−1.41	−1.37	−1.41	−1.41
	0.90	−1.02	−0.99	−1.01	−1.02
	0.95	−0.72	−0.72	−0.72	(−0.74)
1.15	0.60	−2.29	−2.28	−2.30	−2.26
	0.70	−2.25	−2.23	−2.26	−2.24
	0.80	−2.06	−2.06	−2.10	−2.09
	0.90	−1.79	−1.74	−1.76	−1.77
0.75	0.60	−4.24		−4.29	−4.18
	0.70	−4.53	−4.50	−4.28	−4.51
	0.80	−4.69	−4.63	−4.74	−4.69
	0.90		−4.55	−4.67	(−4.62)

[a] From J. A. Barker and D. Henderson, *Rev. Mod. Phys.* **48**, 587 (1976).

[b] PY(E) denotes the results obtained from a calculation using the pair-correlation function derived from the theory of J. K. Percus and G. J. Yevick, *Phys. Rev.* **110**, 1 (1958).

The thrust of the preceding argument is that the structure of a liquid is mostly, but not entirely, determined by the repulsion between molecules. The very close resemblance between the pair-correlation functions of a rigid-sphere fluid, a Lennard–Jones 6–12 fluid, and liquid argon suggests the following question. Can we construct an analysis of the properties of a real liquid in which the attraction between molecules acts only as a small perturbation to the distribution induced by the repulsion between molecules? Just as in the application of perturbation theory to problems of quantum mechanical behavior, the success of this approach depends on the suitability of the choice of unperturbed reference system, and on the rate of convergence of the calculated corrections. For example, given what we have learned about the similarity between the pair-correlation functions of a hard-sphere fluid and a real liquid, we might assume that to calculate the influence of attractive interactions on, say, the Helmholtz free energy of the liquid, it is sufficiently accurate to imagine the molecules distributed as in a hard-sphere fluid, and average the attractive potential over the possible molecular separations in a hard-sphere fluid (BH theory).[16] Of course, real molecules do not repel like hard spheres; the repulsive energy of real molecules is something like $\epsilon(\sigma/R)^{12}$, and the distance of closest approach depends on the temperature. Furthermore, there can be large fluctuations in the environment of a molecule in a liquid. Assuming that the attractive contribution can be computed

[16] J. A. Barker and D. Henderson, *J. Chem. Phys.* **47**, 2856, 4714 (1967); *Can. J. Phys.* **45**, 3959 (1967).

TABLE 23.6
VALUES OF p/nk_BT FOR A LENNARD–JONES 6–12 LIQUID[a]

k_BT/ϵ	$n\sigma^3$	Molecular Dynamics	Monte Carlo	PY(E)	BH	WCA
2.74	0.65	2.22		2.23	2.22	2.18
	0.75	3.05		3.11	3.10	3.04
	0.85	4.38		4.42	4.44	4.30
	0.95	6.15		6.31	6.40	6.10
1.35	0.10	0.72		0.72	0.74	0.77
	0.20	0.50		0.51	0.52	0.53
	0.30	0.35		0.36	0.36	0.31
	0.40	0.26		0.29	0.26	0.17
	0.50	0.30		0.33	0.27	0.18
	0.55	0.41		0.43	0.35	0.27
	0.65	0.80		0.85	0.74	0.71
	0.75	1.73		1.72	1.64	1.64
	0.85	3.37		3.24	3.36	3.28
	0.95	6.32		5.65	6.32	(5.90)
1.00	0.65	−0.25		−0.22	−0.36	−0.50
	0.75	+0.58	0.48	+0.57	+0.53	+0.40
	0.85	2.27	2.23	2.14	2.25	2.20
	0.90	3.50		3.33	3.53	(3.55)
0.72	0.85	0.40	0.25	0.33	0.25	0.26
	0.94		1.60	1.59	1.63	(1.83)

[a] From J. A. Barker and D. Henderson, *Rev. Mod. Phys.* **48**, 587 (1976).

using only the hard-sphere distribution of distances is equivalent to replacing the distribution of potential energies of the molecules by an average potential that is the same for every molecule. Both the "softness" of the repulsive interaction and the deviations from average potential behavior can be accounted for approximately. Alternatively, instead of choosing the rigid-sphere fluid as reference and then correcting for the "softness" of the real repulsive interaction, one can divide the molecular interaction into a part containing all the *repulsive forces* and another part containing all the *attractive forces*. The structure of the reference fluid is then defined by the repulsive force component, and the effect of the attractive force component is treated via perturbation theory (WCA theory).[17] The BH form of perturbation theory has the advantage that it uses the same reference system irrespective of the form of the potential, and the properties of that reference system are very well known from computer simulation studies. It has the disadvantage that the perturbation must be carried to second order, or higher, to obtain high accuracy of prediction. The WCA form of perturbation theory has the disadvantage that the reference system is, or can be, different for different forms of the potential. In addition, the properties of the reference system depend on both temperature and density and are less well known than are those of a hard-sphere fluid, whose properties depend only on density. On the other hand, the WCA form of perturbation theory converges very rapidly, and only

[17] J. D. Weeks, D. Chandler and H. C. Andersen, *J. Chem. Phys.* **54**, 5237 (1971); **55**, 5422 (1971).

first-order corrections need be included to obtain very accurate predictions of liquid properties. The two forms of theory are complementary in a useful fashion. Both can be extended to include three-molecule interactions—the term Δ_3 in Eq. 21.159.

To illustrate the use of perturbation theory in the theory of liquids, we show in Fig. 23.20 the region of the peak of the pair-correlation function of a Lennard–Jones 6–12 liquid. Note how the fit to the molecular dynamics calculations improves. A comparison of predicted and (computer) experimental thermodynamic properties of the 6–12 liquid is given in Tables 23.5 and 23.6.

Just as for theories based on Eq. 23.23 or 23.24, application of the BH or WCA perturbation theory to the description of real liquids requires an accurate molecular interaction function. Given that potential energy function, one can expect predictions of the properties of real liquids as satisfactory as we show is possible for the model Lennard–Jones 6–12 liquid.

From the preceding considerations we conclude that there exists a reasonably accurate representation of the properties of a simple liquid in terms of the interactions between the molecules in the liquid. In particular, the statistical description of the ordering is fairly accurate. Unfortunately, the kind of theory sketched above does not describe accurately the critical region of a fluid, nor does it lead to an understanding of the molecular origins of freezing. The generalization of the theory of liquids to achieve an accurate description of phase transitions is one of the most challenging of the unsolved problems of the statistical molecular theory of matter. Noting the difficulties in the theory of simple liquids, it is not surprising that there does not presently exist any comparably accurate analysis of the properties of complicated liquids. Nor is there much information from experiment about the structure of those liquids. Since most chemistry is carried out in solution, and the solvents used are always composed of complicated molecules, the development of a theory of the behavior of complicated liquids is a challenging and important problem for chemists.

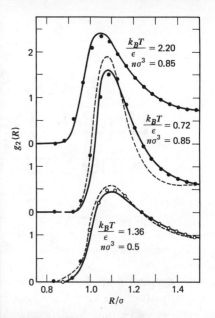

FIGURE 23.20

Pair-correlation functions of the Lennard–Jones 6–12 liquid. The points give the results of molecular-dynamics calculations, the broken curve the prediction of PY theory, and the solid curve the prediction of BH perturbation theory using the rigid-sphere fluid as reference. From J. A. Barker and D. Henderson, *Rev. Mod. Phys.* **48**, 587 (1976).

Appendix 23A

x-Ray Scattering from Liquids: Determination of the Structure of a Liquid

This appendix is concerned with the derivation of Eq. 23.15. Although the analysis will be couched in terms descriptive of x-ray scattering, the same principles can be used to describe electron and neutron scattering.

x Rays are scattered by the electrons in a sample of matter. In the classical picture the very high-frequency x-ray electromagnetic radiation induces an oscillating dipole moment in an atom. This oscillating dipole, in turn, emits radiation at the frequency at which it is driven. The net effect is that some energy is removed from the incident field and scattered (reradiated) in different directions. The quantum mechanical picture is similar, but now the atomic polarization and reradiation is described in terms of the mixture of excited states and ground state induced by the electric field rather than in terms of a classical harmonically driven bound electron.

The geometry of a typical scattering event is shown in Fig. 23A.1.

Consider first, for simplicity, the scattering of x rays by N structureless point particles. We suppose the scattering to be so weak that only single scattering events need be considered. Under these conditions a scattered x-ray photon is not scattered again before it reaches the detector. Furthermore, we suppose that the scattering is elastic. This assumption is an excellent approximation in the case of x-ray photons, since the amount of energy that can be gained from or lost to the translational motion of the atoms is a very small fraction of the x-ray photon energy $(\approx k_B T/h\nu \approx 10^{-2}\,\text{eV}/10^4\,\text{eV} = 10^{-6})$. Then if the initial radiation has wave vector $\mathbf{k}_0 = (2\pi/\lambda)\mathbf{u}_0$, and the scattered radiation has wave vector $\mathbf{k}_s = (2\pi/\lambda)\mathbf{u}_s$, where $\mathbf{u}_0$ and $\mathbf{u}_s$ are unit vectors specifying the directions of the incident and scattered waves, we have

$$|\mathbf{k}_0| = |\mathbf{k}_s| = \frac{2\pi}{\lambda}. \tag{23A.1}$$

Consider two particles, at $\mathbf{R}_i$ and $\mathbf{R}_j$, in the incident field of radiation (see Fig. 23A.1). Each will act as a secondary emitting center. The extent of the interference between the waves originating at these two centers is determined by the difference in path lengths traversed, since this determines the phase difference between the two waves. Let $\mathbf{R}_{ij} = \mathbf{R}_j - \mathbf{R}_i$ be the vector separation of the molecules. As readily seen from Fig. 23A.1, the extra path length referred to is just $\mathbf{R}_{ij} \cdot \mathbf{u}_0 - \mathbf{R}_{ij} \cdot \mathbf{u}_s$, so that the phase difference is

$$\Delta\phi = \frac{2\pi}{\lambda}(\mathbf{R}_{ij} \cdot \mathbf{u}_0 - \mathbf{R}_{ij} \cdot \mathbf{u}_s)$$

$$= -\mathbf{R}_{ij} \cdot (\mathbf{k}_s - \mathbf{k}_0)$$

$$= -\mathbf{R}_{ij} \cdot \mathbf{s}. \tag{23A.2}$$

As shown in Fig. 23A.2, $|\mathbf{s}| = 2|\mathbf{k}_0| \sin(\theta/2)$, where θ is the scattering

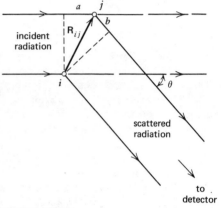

FIGURE 23A.1
Geometry of a typical scattering event. Particle i is at $\mathbf{R}_i$, particle j at $\mathbf{R}_j$, and their vector separation is $\mathbf{R}_{ij} = \mathbf{R}_j - \mathbf{R}_i$. The path-length difference between waves scattered from i and j is just $a + b$. Since $\mathbf{u}_0$ is a unit vector specifying the direction of the incident radiation, $a = \mathbf{R}_{ij} \cdot \mathbf{u}_0$. Since $\mathbf{u}_s$ is a unit vector specifying the direction of the scattered radiation, and the angle between $\mathbf{u}_s$ and $\mathbf{R}_{ij}$ is greater than 90° for all θ greater than 0°, $b = -\mathbf{R}_{ij} \cdot \mathbf{u}_s$. Here θ is the scattering angle.

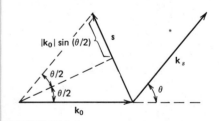

FIGURE 23A.2
Relations between the incident wave vector $\mathbf{k}_0$, the scattered wave vector $\mathbf{k}_s$, and the difference vector $\mathbf{s} = \mathbf{k}_s - \mathbf{k}_0$. Note that

$$s = |\mathbf{s}| = 2|\mathbf{k}_0| \sin\left(\frac{\theta}{2}\right)$$

$$= \frac{4\pi}{\lambda} \sin\left(\frac{\theta}{2}\right).$$

angle. Let ϕ_j be the phase of the wave scattered from the particle at $\mathbf{R}_j$. If there were no interference between particles, the electric field of this scattered wave would fall off as the reciprocal of the distance from particle j. But because of the other particle at $\mathbf{R}_i$, the electric field at some point $\mathbf{R}$ of the scattered wave of amplitude A is

$$\frac{A}{|\mathbf{R}_i - \mathbf{R}_j|} \exp(i\phi_j) = \frac{A}{|\mathbf{R}_i - \mathbf{R}_j|} \exp(i\phi_i) \exp(-\mathbf{s} \cdot \mathbf{R}_{ij}), \qquad (23A.3)$$

using $\phi_j = \phi_i + \Delta\phi$ and $\Delta\phi$ from Eq. 23A.2. To obtain the effects of interference between all the scattering centers, we must sum Eq. 23A.3 over all molecules. When the detector is very far from the scattering region, so that $|\mathbf{R}|$ is large compared to all dimensions of the scattering region, the magnitude of the electric field of the scattered radiation is

$$E_s = \frac{A}{R} \exp(i\phi_i) \exp(i\mathbf{s} \cdot \mathbf{R}_i) \sum_{j=1}^{N} \exp(-i\mathbf{s} \cdot \mathbf{R}_j), \qquad (23A.4)$$

and its intensity is

$$I_s = \frac{E_s E_s^*}{4\pi} = \frac{A^2}{4\pi R^2} \left[N + \sum_{i=j} \exp(-i\mathbf{s} \cdot \mathbf{R}_{ij}) \right]. \qquad (23A.5)$$

To obtain Eq. 23A.4 we have used the condition $|\mathbf{R}| \gg |\mathbf{R}_i|$, hence to an excellent approximation $|\mathbf{R} - \mathbf{R}_i| \to R$ for all $\mathbf{R}_i$. To obtain Eq. 23A.5 we have separated out the term $i = j$ from $[\sum_{j=1}^{N} \exp(-i\mathbf{s} \cdot \mathbf{R}_j)]$ $[\sum_{i=1}^{N} \exp(-i\mathbf{s} \cdot \mathbf{R}_i)]^* = \sum_{i,j} \exp(-i\mathbf{s} \cdot \mathbf{R}_{ij})$.

Clearly, since the intensity of the scattered wave depends on all the values of $\mathbf{R}_{ij}$, it depends on the distribution of pairs of molecules in the scattering region. For the case of interest to us, namely, a uniform liquid, the probability density for finding molecules at $\mathbf{R}_1$ and $\mathbf{R}_2$ is just $n^2 g_2(R_{12})$. Then the sum in Eq. 23A.5 can be replaced by an integral over $d\mathbf{R}_1 \, d\mathbf{R}_2$ weighted by the distribution of pairs of molecules:

$$I_s = \frac{A^2}{4\pi R^2} \Bigg\{ N + n^2 \iint d\mathbf{R}_1 \, d\mathbf{R}_2 \exp(-i\mathbf{s} \cdot \mathbf{R}_{12})$$

$$+ n^2 \iint d\mathbf{R}_1 \, d\mathbf{R}_2 \exp(-i\mathbf{s} \cdot \mathbf{R}_{12})[g_2(R_{12}) - 1] \Bigg\}. \qquad (23A.6)$$

The first term in Eq. 23A.6 is the intensity of scattering from N independent molecules (without interference), the second term is the intensity of scattering when $g_2(R_{12}) = 1$ at the surface of the volume, and the third term is the portion of the scattered intensity that is modulated by interference between the scattering centers. For a macroscopic system the surface scattering term is negligibly small. When the volume element $d\mathbf{R}_1 \, d\mathbf{R}_2$ is rewritten in terms of center-of-mass and relative coordinates, the integration over the center of mass gives a factor of V (see Eqs. 21C.16 and 21C.17), which when multiplied by n gives N. After this manipulation we can rewrite Eq. 23A.6 in the form

$$I_s = I_m \left[1 + 4\pi n \int \frac{\sin sR_{12}}{sR_{12}} [g_2(R_{12}) - 1] R_{12}^2 \, dR_{12} \right], \qquad (23A.7)$$

where I_m is the intensity of scattering from N independent molecules. Aside from a proportionality factor to be discussed below, this is Eq. 23.15.

x Rays are scattered from electrons and we must not forget that atoms have an internal electronic structure. Inclusion of this internal structure of the atom modifies Eq. 23A.7. In addition, the amplitude of x rays scattered by electrons depends on the polarization of the incident beam. For an unpolarized beam the intensity contains a factor $(1 + \cos^2 \theta)$, where θ is the angle between the scattered and incident radiation.

This term appears as a prefactor as shown in Eq. 23.15. Returning to the influence of the structure of the atom, the distribution of electrons about an atom leads to interference effects in the scattered radiation. If $\mathbf{x}_i^{(n)}$ is the position of the nth electron bound to the ith atom, this interference leads to a term of the form

$$F(s) = \left\langle \left| \sum_{i=1}^{Z} \exp[i\mathbf{s} \cdot (\mathbf{x}_i^{(n)} - \mathbf{R}_i)] \right|^2 \right\rangle_{QM},$$

where the subscript QM means a quantum mechanical expectation value (the electron distribution is not influenced by the atomic distribution). This so-called *atomic form factor* can be evaluated using atomic wave functions. It too appears as a prefactor in the equation for the scattered intensity (it was omitted from Eq. 23.15, hence the proportionality sign).

There are serious technical problems associated with inverting Eq. 23.15 to determine $g_2(R)$ from the measured intensity of scattered x rays, but we shall not discuss that matter here.

Further Reading

Barker, J. A., and D. Henderson, *Rev. Mod. Phys.* **40,** 587 (1976).

Croxton, C. A., *Liquid State Physics* (Cambridge University Press, Cambridge, 1974), chaps. 2, 3, and 5.

Hansen, J. P., and I. R. McDonald, *Theory of Simple Liquids* (Academic Press, New York, 1976), chaps. 1–6.

Hirschfelder, J. O., C. F. Curtiss, and R. B. Bird, *Molecular Theory of Gases and Liquids* (Wiley, New York, 1954), chaps. 4 and 6.

McQuarrie, D. A., *Statistical Mechanics* (Harper & Row, New York, 1976), chaps. 13 and 14.

Marcus, Y., *Introduction to Liquid State Chemistry* (Wiley, New York, 1977), chaps. 1–3.

Rice, S. A., and P. Gray, *The Statistical Mechanics of Simple Liquids* (Wiley, New York, 1965), chap. 2.

Rowlinson, J. S., *Liquids and Liquid Mixtures,* 2nd ed. (Butterworths, London, 1969), chaps. 1, 2, 7, and 8.

Watts, R. O., and I. J. McGee, *Liquid State Chemical Physics* (Wiley, New York, 1976), chaps. 3, 5–10.

Problems

1. Suppose that the coefficient of thermal expansion of a liquid in equilibrium with its vapor is measured. Since the vapor pressure increases as the temperature increases, along the liquid–vapor equilibrium line one measures

$$\alpha_\sigma \equiv \frac{1}{V}\left(\frac{\partial V}{\partial T}\right)_\sigma$$

and not

$$\alpha \equiv \frac{1}{V}\left(\frac{\partial V}{\partial T}\right)_p,$$

which refers to a process at constant pressure. Show that

$$\alpha = \alpha_\sigma + \kappa_T\left(\frac{\partial p}{\partial T}\right)_\sigma,$$

where, as above, a subscript σ implies that the process is along the liquid–vapor equilibrium line.

2. Derive the following relationships between the several coefficients descriptive of the properties of a liquid:

   (a) $\alpha = \kappa_T \gamma_V$

   (b) $\gamma_V - \gamma_\sigma = \dfrac{\alpha_\sigma}{\kappa_T}$

   (c) $\alpha_\sigma = \alpha\left(1 - \dfrac{\gamma_\sigma}{\gamma_V}\right)$

   (d) $\left(\dfrac{\partial H}{\partial T}\right)_\sigma = C_\sigma + V\gamma_\sigma$

   (e) $\left(\dfrac{\partial U}{\partial T}\right)_\sigma = C_\sigma - pV\alpha_\sigma$

3. The relative volume of water as a function of pressure at 0°C is given below:

p (atm)	v/v_0
1	1.00000
501	0.97668
1501	0.93924
2001	0.92393
2501	0.91065
3001	0.89869

   Calculate the constants A and B in the Tait equation,

   $$\frac{v_0 - v}{v_0 p} = \frac{A}{B + p}.$$

4. Calculate the isothermal compressibility for a liquid that is described by the Tait equation. Using the values obtained in Problem 3, compute κ_T at 1, 250, 1000, 2000, and 3000 atm and 0°C for water, and compare these with the observed values, 46, 43, 37, 30, and 25×10^{-12} cm^2/dyn, respectively.

5. The specific volume of liquid Hg as a function of pressure at 0°C is tabulated below:

p (atm)	v (cm^3/g)
1	0.073554
1000	0.073270
2000	0.072993
3000	0.072724
4000	0.072463
5000	0.072213
6000	0.071973
7000	0.071744

   Calculate the isothermal compressibility of Hg at 0°C for several pressures in the range of the data given. How do these values compare with those for water? How can you explain the difference found?

6. Suppose that the equation of state of a liquid is of the form

   $$p = nk_B T F_1(n) + F_2(n),$$

where $F_1(n)$ and $F_2(n)$ are functions of density but not of temperature. Show that γ_V is a function only of the molar volume, and that the entropy of the liquid is determined solely by $F_1(n)$.

7. Let v_0 be the molar volume of a liquid at $p = 1$ atm and $T = T_0$. Suppose that the equation of state can be represented by

$$v = v_0[1 + a_0(t) + a_1(t)p + a_2(t)p^2 + \cdots],$$

where $t \equiv T - T_0$ and the various a's are functions of temperature alone.
   (a) Calculate the coefficients of thermal expansion and isothermal compressibility in terms of a_0, a_1, and a_2.
   (b) Show that

$$c_p = c_p{}^0 - v_0 T\left(\frac{d^2 a_0}{dT^2}p + \frac{1}{2}\frac{d^2 a_1}{dT^2}p^2 + \cdots\right),$$

where $c_p{}^0$ is the specific heat at a constant pressure of 1 atm (essentially zero pressure for our purposes).

8. An alternative form for the equation of state of a liquid is

$$p = p_0(T) + b_1(T)\left(\frac{v_0 - v}{v_0}\right) + b_2(T)\left(\frac{v_0 - v}{v_0}\right)^2 + \cdots,$$

where $c_p{}^0$ is the specific heat at a constant pressure of 1 atm (essentially zero to v_0 from v. Show that the coefficients of this and the equation of state in problem 7 are related by

$$p_0 = -\frac{a_0}{a_1} - \frac{a_2 a_0{}^2}{a_1{}^3}$$

$$b_1 = -\frac{1}{a_1}\left(\frac{2 a_2 a_0}{a_1{}^2} + 1\right)$$

$$b_2 = -\frac{a_2}{a_1{}^3}.$$

9. Let $c_V{}^0$ be the specific heat at constant volume v_0.
   (a) Show that

$$c_V = c_V{}^0 - v_0 T\left[\frac{d^2 p_0}{dT^2}\left(\frac{v_0 - v}{v_0}\right) + \frac{1}{2}\frac{d^2 b_1}{dT^2}\left(\frac{v_0 - v}{v_0}\right)^2 + \cdots\right].$$

   (b) Show that

$$c_p{}^0 - c_V{}^0 = \frac{v_0 T}{2 a_1}\frac{d^2}{dT^2}(a_0{}^2).$$

   (c) Compare the result (b) with the exact relation

$$c_p - c_V{}^0 = Tv\frac{\alpha^2}{\kappa_T},$$

using the values of a_0 and a_1 previously obtained, and show that your calculations are internally consistent.

10. Show that the residual entropy of a liquid can be expressed in the form

$$s^E = -\int_v^\infty \left[\left(\frac{\partial p}{\partial T}\right)_v - \frac{R}{v}\right]dv.$$

11. Consider the following model of a one-dimensional liquid. The "molecules" are line segments of length σ; they are constrained to move on a line of length L. In all there are N such molecules on the line, and we imagine that their interaction has the following form:

$$u_2(x) = \infty, \qquad x \leq \sigma,$$
$$u_2(x) = 0, \qquad x > \sigma,$$

At the ends of the line there are impenetrable barriers (hard walls). Show that the equation of state is

$$p(l - \sigma) = k_B T,$$

$$l \equiv \frac{L}{N}.$$

Hint: The molecules cannot pass through one another, so once an ordering is established it must stay the same. There are $N!$ equivalent orderings of the molecules on the line.

12. A simple model of a liquid, called the cell model, is described as follows. Imagine that each monatomic molecule is constrained, because of repulsive forces exerted by neighbors, to move only in the volume v_f. The idea is that the motion is irregular, unlike that in a crystal, but that v_f is of molecular dimensions. Let the canonical partition function be written as

$$Q_N = \left(\frac{2\pi m k_B T}{h^2} \right)^{3N/2} \frac{Z_N}{N!}.$$

The molecules are localized, so we must account for the $N!$ ways of arranging N molecules in N cells.

(a) Show that the equation of state is

$$p = \frac{N k_B T}{v_f} \left(\frac{\partial v_f}{\partial V} \right)_T.$$

(b) Suppose that the volume of the cell in which the molecule moves is $v \equiv V/N$, and the molecule is a hard sphere of diameter σ. Then if the cell is represented as a hollow sphere, the center of the molecule can approach the wall no closer than $\sigma/2$. Show that the equation of state is

$$pv = k_B T \left[1 - \left(\frac{\pi \sigma^3}{6v} \right)^{1/3} \right]^{-1}.$$

Hint:

$$v_f = \frac{4\pi}{3} \left(\frac{d}{2} - \frac{\sigma}{2} \right)^3,$$

where d is the diameter of the cell of volume v.

(c) Take the limit $\sigma \to 0$, but still allowing motion only in the cell of volume v. Compare the Helmholtz free energies and entropies of this limiting model and a perfect gas. What is the nature of the difference?

13. An explicit and simple way of seeing how attractive interactions influence the properties of a liquid is generated from the assumption that the potential energy inside a cell is constant and negative, but rises to infinity at the walls so that a molecule cannot escape from the cell. Suppose that there are 12 nearest neighbors that define a cell, and that $u_2(R)$ is the Lennard–Jones potential,

$$u_2(R) = 4\epsilon \left[\left(\frac{\sigma}{R} \right)^{12} - \left(\frac{\sigma}{R} \right)^6 \right].$$

Counting only nearest-neighbor interactions, show that the equation of state is

$$pv = k_B T \left[1 - \left(\frac{v_0}{v} \right)^{1/3} \right]^{-1} + 48\epsilon \left[2 \left(\frac{v_0}{v} \right)^4 - \left(\frac{v_0}{v} \right)^2 \right],$$

with $v_0 \equiv \sigma^3/\sqrt{2}$.

Hint: Use the Lennard–Jones potential to evaluate the potential energy when all molecules are at the cell centers. By hypothesis, this is the value of the constant potential energy in the cell.

14. A more sophisticated version of the cell model improves the calculation of v_f with the following idea. Imagine some reference lattice with the same

molecules as the liquid, having static lattice energy E_L. Write the potential energy of the liquid in the form

$$\mathcal{V}_N \equiv (\mathcal{V}_N - E_L) + E_L.$$

Then $\mathcal{V}_N - E_L$ can be regarded as the potential energy of irregular displacement from the reference lattice sites. If the total potential is the sum of pair potentials, show that

$$v_f = 4\pi \int_{\text{cell}} e^{-\beta[\epsilon(r)-\epsilon(0)]} r^2 \, dr,$$

where $\epsilon(r) - \epsilon(0)$ is the energy corresponding to displacement r from the cell center.

15. Using the general principles described in the text, and the known properties of a liquid, criticize the ideas used in formulating the cell model of a liquid.

16. Show that the equation of state of a rigid-sphere fluid can be written in the form

$$\frac{p}{nk_BT} = 1 + \frac{2\pi n\sigma^3}{3} g_2(\sigma),$$

where $g_2(\sigma)$ is the value of the pair-correlation function at contact. Hint: For rigid spheres, $u_2(R) = 0$ if $R > \sigma$, and $u_2(R) = \infty$ if $R \leq \sigma$. Then $\exp[-\beta u_2(R)] = 1$ if $R > \sigma$ and 0 if $R \leq \sigma$. Differentiation of $\exp[-\beta u_2(R)]$ with respect to R gives $-\beta(du_2/dR)\exp[-\beta u_2(R)]$, which is unity when $R = \sigma$ and zero everywhere else.

17. The residual chemical potential, internal energy, and entropy of a liquid can be represented by

$$\mu^E = \mu - k_BT \ln\left(\frac{p\Lambda^3}{k_BT}\right),$$

$$u^E = u - \tfrac{3}{2}k_BT,$$

$$s^E = s - \tfrac{5}{2}k_B + k_B \ln\left(\frac{p\Lambda^3}{k_BT}\right).$$

Show that

$$Ts^E = u^E - \mu^E + (pv - k_BT).$$

18. In the case of a rigid-sphere fluid, $u^E = 0$. Show that

$$\frac{\mu^E}{k_BT} = -\ln\left(\frac{pv}{k_BT}\right) + \left(\frac{pv}{k_BT} - 1\right) + \int_v^\infty \left(\frac{p}{k_BT} - \frac{1}{v}\right) dv$$

and

$$\frac{s^E}{k_B} = \ln\left(\frac{pv}{k_BT}\right) - \int_v^\infty \left(\frac{p}{k_BT} - \frac{1}{v}\right) dv.$$

Hint: For a rigid-sphere fluid the Helmholtz free energy per molecule is

$$a^0 - a(v) = -\int_v^{v^0} p \, dv = (\mu^0 - k_BT) - (\mu - pv).$$

19. A very good approximate equation of state for the rigid-sphere fluid is the Carnahan–Starling form,

$$\frac{p}{nk_BT} = \frac{1 + y + y^2 - y^3}{(1-y)^3},$$

with $y = \pi n\sigma^3/6$. Calculate the residual entropy and chemical potential for a rigid-sphere fluid using this equation of state.

20. Suppose that two atoms interact by a square-well potential (see Fig. 21.19d). Consider a gas of such atoms in the low-density region where $u_2(R) \approx w_2(R)$.

(a) Prove that in this low-density limit we have

$$U = \tfrac{3}{2} N k_B T - \frac{N^2 \epsilon}{V} \exp\left(\frac{\epsilon}{k_B T}\right)\left[\frac{2\pi}{3}(\sigma_2{}^3 - \sigma_1{}^3)\right].$$

(b) Sketch $g_2(R)$ versus R for the low-density gas.

(c) How will $g_2(R)$ for the liquid differ from your answer to (b)?

21. Suppose that the potential energy of a liquid is given by

$$\mathscr{V}_N = \sum_{i>j} u_2(R_{ij}) + \sum_{i>j>k} \Delta_3(R_{ij}, R_{ik}, R_{jk})$$

as in Eq. 21.159. Express U in terms of $g_2(R)$ and an analogously defined three-particle correlation function g_3.

22. A general, exact equation of state is the virial equation of state (derived from the virial theorem of mechanics), which can be written

$$pV = N k_B T + \tfrac{1}{3}\left\langle \sum_{i=1}^{N} \mathbf{R}_i \cdot \mathbf{F}_i \right\rangle,$$

where $\mathbf{R}_i$ and $\mathbf{F}_i$ are the (vector) position of particle i and the total force on particle i due to all the other particles, and the angular brackets indicate an ensemble average. For the case of a fluid containing identical spherical particles, and where the intermolecular potential is pair-additive, derive Eq. 23.21 from the virial equation.

23. Find the value of $4\pi\int_0^\infty g_2(R)R^2\,dR$ for a fluid containing N particles in a volume V.

24

Phase Equilibria in One-Component Systems

Among the most dramatic of observable phenomena are the sudden alterations in the properties of a substance when it changes from solid to liquid form, or from liquid to gaseous form. In the cases just cited there are discontinuous changes in the properties of the substance as the transition is traversed. Equally dramatic are the appearance of super-conductivity in many metals at low temperature (less than 20 K) and of superfluidity in liquid helium at about 2.2 K, but in these cases the more obvious properties of the substance undergo continuous transformations as the transition is traversed. Why and how do phase transitions occur, and what changes in properties are to be expected for a given kind of transition? It is these questions that we study in this chapter. Wherever possible we shall examine both the macroscopic thermodynamic description and the corresponding statistical molecular interpretation. We shall see that, although it is possible to formulate a convenient and systematic classification of phase transitions by the methods of thermodynamics, the present understanding of the underlying molecular processes is both meager and imperfect.

24.1 General Survey of Phase Equilibria

We have previously cited, as examples of phase transitions, the evaporation of a liquid and the melting of a solid. Recalling the definition of a phase given in Section 14.4, one can easily see that the characteristic features of the processes mentioned are their discontinuities in some, but not all properties. That is, when a solid melts, for example, ice $\rightarrow$ water, the process occurs at constant temperature and pressure with a marked change in the density, heat capacity, and so on, of the substance. Note that in referring to the solid phase as distinct from the liquid phase, we imply that there exists a well-defined crystal structure in the solid phase. An amorphous solid, such as glass, gradually softens as it is heated and becomes liquid without undergoing a discontinuous change of properties. This behavior can be traced to the fact that the entropies of a crystalline solid and a liquid are different, reflecting largely the difference in ordering between the two phases, whereas a glass has positional disorder very much like that in a liquid. It is then to be expected that there will not be a sharp transition from solid glass to liquid, and that there will be a continuous transformation between these two states. In fact, some of the properties of a substance undergoing the glass $\rightarrow$ liquid transition do change considerably over a small temperature range, but they do not change discontinuously. Thus, glass and liquid are not distinct phases.

It is observed that a phase transition occurs at a fixed temperature for any specified constant pressure, but that the transition temperature changes when the pressure is changed. Also, in the transformation from one phase to another, there is an enthalpy change. Thus, in the absence of external influences, two phases can coexist indefinitely long at the transition temperature corresponding to a specified pressure. However, at the same pressure and temperatures above or below that transition temperature, only one or the other phase is stable. For example, the line representing the pressure at which liquid and vapor coexist in equilibrium, called the vapor pressure curve, divides the (p, T) plane into two regions (see Fig. 24.1), one where the liquid is the stable phase, the other where the gas is the stable phase. The points defining the vapor pressure curve represent states in which an arbitrary quantity of liquid may coexist with an arbitrary quantity of gas. The latter property of a phase transition is easily appreciated when the (V, T) plane is examined (see Fig. 24.2). Suppose that a gas with volume V_a is compressed at constant temperature until condensation begins, which occurs at $V = V_G$. As the system is compressed further, the amount of gas in the system decreases while the amount of liquid increases. When $V = V_L$, the substance is entirely converted to liquid. Consider the point D in Fig. 24.2. Let the fractions of gas and liquid at D be x and $1 - x$. Then the total volume of the system is

$$V = xV_G + (1 - x)V_L, \tag{24.1}$$

so that

$$x = \frac{V - V_L}{V_G - V_L}, \tag{24.2}$$

$$1 - x = \frac{V_G - V}{V_G - V_L}. \tag{24.3}$$

The ratio of the amounts of gas and liquid is, from Eqs. 24.2 and 24.3,

$$\frac{x}{1 - x} = \frac{V - V_L}{V_G - V_L} = \frac{\text{length } BD}{\text{length } AD}, \tag{24.4}$$

the final form referring to the lines in Fig. 24.2. This result is known as the *lever rule*.

The phase diagram in the (p, V) plane is similar to that in the (T, V) plane, and both differ from the phase diagram in the (T, p) plane. Note that each point on the line AB of Fig. 24.2 represents a different thermodynamic state, while in a (T, p) diagram all these states are represented by a single point on the vapor pressure curve. The region of separation into two coexisting phases is only a line in the (T, p) plane, but occupies an area in the (T, V) and (p, V) planes. This difference results from the requirement that coexisting phases have the same temperature and pressure, although they need not have the same density. These remarks are easily verified by examination of the (p, V, T) surface of a substance, as, for example, that in Fig. 24.3.

As the temperature rises, the pressure at which there is coexistence between liquid and vapor also rises. Corresponding to the increase in vapor pressure of the substance, the difference in densities of the liquid and gaseous phases decreases. There exists a temperature and pressure at which the densities of liquid and gas are the same and the two phases can no longer be distinguished from one another. At higher temperatures or pressures, liquid and gas are everywhere indistinguishable; that is, there is only one phase. Thus, the locus of points of coexistence of liquid and gas in the (p, T) plane terminates at a point T_c, p_c, called the *critical point*. (See Section 21.1.)

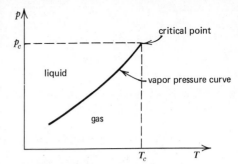

FIGURE 24.1
Schematic diagram of a vapor pressure curve.

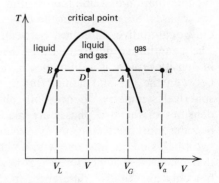

FIGURE 24.2
Schematic diagram of the coexistence region of liquid–gas equilibrium in the (V, T) plane. The line $aADB$ represents the condensation path described in the text.

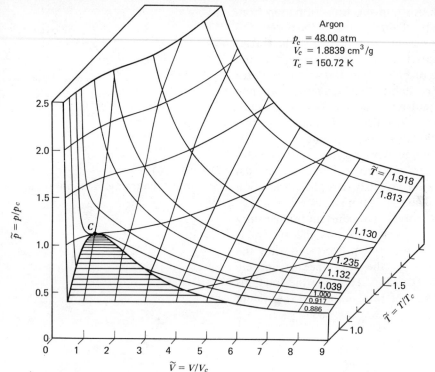

Argon
$p_c = 48.00$ atm
$V_c = 1.8839$ cm^3/g
$T_c = 150.72$ K

$\tilde{p} = p/p_c$

$\tilde{T} = 1.918$
1.813
1.130
1.235
1.132
1.039
1.000
0.917
0.886

1.5
1.0

$\tilde{T} = T/T_c$

$\tilde{V} = V/V_c$

FIGURE 24.3
Part of the (p, V, T) surface of argon, showing lines of constant p, V, and T.

In the (T, V) and (p, V) planes the approach to the critical point is manifested by the approach to equality of the densities of the two phases as the temperature of the system increases. (See Fig. 21.4.) The manner in which the pressure approaches the critical pressure in the (p, V) plane, as shown in Fig. 24.4, and the manner in which the critical volume is approached in the (T, V) plane, are very different from that expected from the equations of state listed in Table 21.2. Indeed, the thermo-dynamic state of the system near the critical point has some very peculiar characteristics. Furthermore, the existence of a critical point in the gas–liquid phase transition strongly suggests that there is no fundamental difference between the liquid and gaseous states. It is in fact possible to pass from the liquid to the gaseous state around the critical point without undergoing any discontinuous change of state. This observation can be made the basis of a classification of the kinds of transitions in which critical points can be expected to occur. Note that a critical point marks the termination of a line representing the locus of points of coexistence of two phases. Suppose, as in the case of a gas and a liquid, that there are only quantitative difference between the properties of the two phases. The quantitative differences arise because, accompanying the difference in densities, there is a difference in the averaged molecular environments between the gas and liquid. But in both phases the local structure has spherical symmetry,[1] hence both phases are isotropic. Thus, when the densities of the two phases are made to approach equality, the phases become more and more alike. When the densities are made equal, the two phases merge into one with no change in local symmetry. It is found that when there is no difference between the symmetries of two phases, but only quantitative differences in the densities and associated proper-ties, the coexistence line between the phases has a critical point.

---

[1] We neglect here the case of liquid crystals, which can exist in several fluid phases, some of which are anisotropic and only one of which is isotropic. The transitions between the several anisotropic phases of a liquid crystal are analogous to solid–solid transitions from one crystal structure to another and not to the gas–liquid transition.

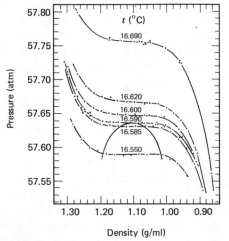

Pressure (atm)

t (°C)

16.690

16.620
16.600
16.590
16.585

16.550

Density (g/ml)

FIGURE 24.4
Experimental pressure against density isotherms in the critical region of xenon. From H. W. Habgood and W. G. Schneider, *Can. J. Chem.* **32**, 98 (1954).

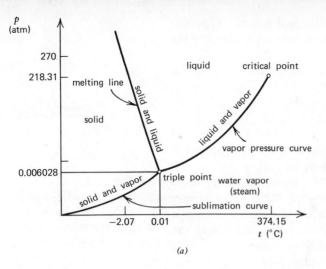

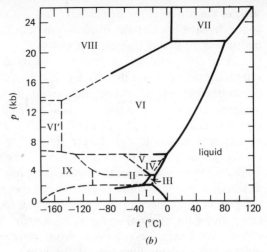

FIGURE 24.5

Phase diagrams for water. (*a*) Low-pressure ice–water vapor–liquid equilibrium. From J. Kestin, *A Course in Thermodynamics* (Blaisdell, Waltham, Mass., 1966). (*b*) High-pressure equilibria among various forms of ice. From B. Kamb, in E. Whalley, S. Jones, and L. Gold (Eds.), *Physics and Chemistry of Ice* (University of Toronto Press, Toronto, 1973).

On the other hand, if the two phases under consideration have different internal symmetries, as is the case for a liquid and a crystal, they must be regarded as qualitatively different. Adjusting the densities of, for instance, a crystal and a liquid to be equal, if it could be done, would still leave a difference in local symmetries between them. In cases of this type it is found that there is no critical point on the coexistence curve representing equilibrium between the two phases. Either the coexistence curve in the (p, T) plane goes to infinity or it terminates by intersecting the coexistence curves of still other phases.

The preceding argument is based on the observation that a particular symmetry property exists or does not exist, and that it belongs to all the molecules of the phase as a whole. A symmetry property can only appear or disappear entirely; it cannot be "partially present." It follows that if we are able to recognize that a symmetry property is present, we can always determine to which of two phases it belongs. Therefore, two phases that have different symmetry properties can, in principle, always be distinguished from one another, and no critical point can terminate their coexistence line.

Although the most widely known phase transitions are melting and vaporization, there exist many other kinds. For example, it is found that at different temperatures and pressures a solid may exist in different crystalline states, each with a specific structure. These different crystalline forms, called crystal modifications, are to be regarded as different phases of the same substance. The different polymorphic crystalline forms usually have different densities; these different modifications can be in equilibrium with one another only along certain lines in the (p, T) plane. In general, when such a coexistence line is crossed, and one crystal form is changed into another, there is absorption (release) of heat. Examples of polymorphism are found in elements (diamond and graphite are forms of C) and compounds (quartz, tridymite, and cristobalite are forms of SiO_2). The phase diagram of water, a portion of which is shown in Fig. 24.5, displays many crystal modifications, most of which are stable only at high pressures. In this diagram, ordinary ice is stable in region I.

The liquid–gas, solid–gas, and liquid–solid transitions, and most crystal–crystal transformations, involve a discontinuous change in density across the coexistence line. Along with the discontinuity in density there are also discontinuities in the molar entropy, internal energy, and specific heat. The class of phase transitions just described is known as *transforma-*

tions of the first kind (or *first order*).[2] It is a transformation of this type that connects each of the pairs of phases in both the ordinary-pressure and the high-pressure phase diagrams of water.

It is also possible, in principle, to have a discontinuous change of symmetry type without there being discontinuities in the volume, entropy, or internal energy of the substance. As an example of how this might happen, consider a tetragonal crystal, and suppose that the lengths of the two unequal edges of the unit cell, a and c, change with the temperature at different rates. If c is larger than a at some temperature, and if $(\partial a/\partial T)_p > (\partial c/\partial T)_p$, then as T increases, the lengths of the sides of the unit cell become more nearly equal. Suppose that there is a temperature at which a and c become equal, so that the unit cell is cubic. On further heating all three sides of the cubic unit cell increase in length at the same rate, therefore remaining equal in length. Thus, when $a = c$ and the crystal symmetry has changed from tetragonal to cubic, we have a different crystal modification of the same substance. The characteristic feature of this kind of transition is that there is *not* a discontinuous change in the state of the crystal, because the configuration of the atoms in the crystal changes continuously as T changes. So long as all three sides of the unit cell are equal in length, the phase remains cubic. However, when the temperature drops, the appearance of an infinitesimal difference between a and c converts the phase to tetragonal symmetry. A transition of the type just described is called *a transition of the second kind* (or *second order*).

The example given, although a possible transformation, is not common. A well-studied class of compounds, the perovskites ABO_3, have the crystal structure shown in Fig. 24.6. Phase transitions between crystal modifications of these compounds involve rotations of the BO_6 octahedra. In $SrTiO_3$ the cubic high-temperature structure undergoes a tetragonal distortion at the transition, corresponding to a rotation of the BO_6 octahedra about a cube axis. In $LaAlO_3$ the cubic high-temperature phase undergoes a trigonal distortion to a rhombohedral structure described by a rotation of the AlO_6 octahedra about a cube diagonal. Still other kinds of distortions occur in other perovskites.

We have emphasized that in a phase transition of the second kind there is a discontinuous change of symmetry type. Nevertheless, the transition is continuous in the sense that the volume, internal energy, and entropy change continuously from one phase to the other.[3] Thus, such a

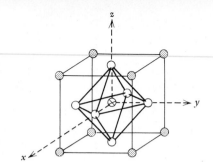

FIGURE 24.6
Schematic diagram of the unit cell of the cubic perovskite structure with the B ion at the center: ⊚ A, ⊗ B, ○ O. The general chemical formula is ABO_3.

---

[2] Transitions of the *n*th kind are also known as *n*th-order transitions. We shall use both names, choosing one or the other for stylistic reasons.

[3] The usual definition of the order of a transition is in terms of the behavior of derivatives of the Gibbs free energy. An *n*th-order transition is one in which discontinuities appear only in the *n*th and higher derivatives of G with respect to T and p. Thus, the ordinary first-order transitions have discontinuities in

$$V = \left(\frac{\partial G}{\partial p}\right)_T, \qquad S = -\left(\frac{\partial G}{\partial T}\right)_p$$

and thus also in H, U, etc. Second-order transitions have discontinuities in

$$\alpha = \frac{1}{V}\left(\frac{\partial V}{\partial T}\right)_p = \frac{1}{V}\left[\frac{\partial}{\partial T}\left(\frac{\partial G}{\partial p}\right)_T\right]_p,$$

$$\kappa_T = -\frac{1}{V}\left(\frac{\partial V}{\partial p}\right)_T = -\frac{1}{V}\left(\frac{\partial^2 G}{\partial p^2}\right)_T,$$

$$C_V = T\left(\frac{\partial S}{\partial T}\right)_V = -T\left[\frac{\partial}{\partial T}\left(\frac{\partial G}{\partial T}\right)_p\right]_V$$

$$C_p = T\left(\frac{\partial S}{\partial T}\right)_p = -T\left(\frac{\partial^2 G}{\partial T^2}\right)_p,$$

and so forth. The order of a particular transition must be determined experimentally.

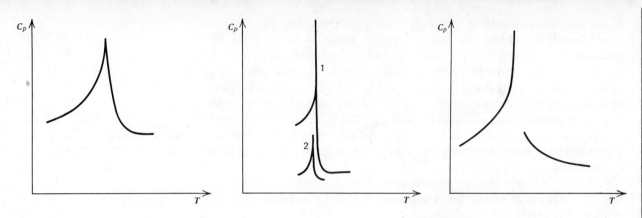

FIGURE 24.7
Some schematic heat capacity versus temperature curves near transition points.

(*a*) Order–disorder transition in a binary alloy.

(*b*) Liquid–gas transition; curves 1 and 2 are approximately to scale for argon and helium, respectively.

(*c*) Order–disorder transition in a magnet; the curve is for nickel chloride hexahydrate.

transition is not accompanied by the absorption or evolution of heat. There are, however, discontinuous changes in the temperature and pressure derivatives of the volume, energy, and entropy when the transition occurs. In the expository model described above, the thermal expansion of the crystal is different in the two phases. In the cubic phase the crystal expands isotropically, whereas in the tetragonal phase the expansion is anisotropic. The symmetry properties of the two crystalline phases enter the problem because the definition of anisotropy requires a nonzero difference between properties along different directions in space. That is, there cannot be a continuous change in the coefficient of thermal expansion across the transition temperature because above the transition temperature $\alpha_{a_1} = \alpha_{a_2} = \alpha_{a_3} = \alpha_c$, whereas below the transition temperature $\alpha_{a_1} = \alpha_{a_2} \neq \alpha_{a_3}$, hence the difference $\alpha_{a_3} - \alpha_c$ must be nonzero at the transition temperature.

The existence of discontinuities in the temperature derivatives of, for example, the internal energy is a principal characteristic of transitions of the second kind. In Fig. 24.7 is displayed a typical variation of the heat capacity with temperature in the vicinity of a transition temperature. It should be noted that the available experimental data suggest that there are very few second-order transitions. Nevertheless, many observed transitions have major features in common with the hypothesized properties of second-order transitions.

There are many other kinds of transitions, but the examples cited suffice for illustrative purposes. Indeed, as will be seen, in the study of the properties of liquid solutions we deal almost exclusively with first-order phase transitions.

We now turn to an examination of how thermodynamic and statistical molecular analyses can be used to provide quantitative descriptions of the possible equilibria between phases.

We have previously shown, in Section 19.1, that at equilibrium between two phases of a one-component system the temperatures, pressures, and chemical potentials of the two phases are equal. In this section we deduce the consequences of this characteristic feature of phase equilibrium.

Consider a transition of the first kind, say, the liquid–gas transition. For each temperature and pressure at which the liquid and gas coexist, we must have

24.2
Thermodynamics of Phase Equilibria in One-Component Systems
A. The Clausius–Clapeyron Equation

$$\mu^L(T, p) = \mu^G(T, p). \tag{24.5}$$

If the equations of state of the two phases are known, Eq. 24.5 may be used directly to obtain the vapor pressure curve by solving for the pressure as a function of the temperature. In general, however, the equations of state of the two coexisting phases are not known. How, then, can Eq. 24.5 be used? From Eq. 24.5 it is possible to derive a differential equation whose solution represents the coexistence curve, in our example the vapor pressure curve. It will be seen that this differential equation is simple in form and permits us to calculate the rate of change of pressure with temperature without complete knowledge of the equations of state of the two phases.

If Eq. 24.5 is valid everywhere along the coexistence curve of the two phases in equilibrium, then it must be true that for a change of T and p along this curve

$$d\mu^L(T, p) = d\mu^G(T, p); \tag{24.6}$$

that is, for the phases to remain in equilibrium a differential change in the chemical potential of one phase must be matched by an equal change in the other phase. But we have in a one-component closed system

$$d\mu = \left(\frac{\partial \mu}{\partial T}\right)_p dT + \left(\frac{\partial \mu}{\partial p}\right)_T dp$$
$$= -s\, dT + v\, dp, \tag{24.7}$$

where, as before, s and v are the entropy and volume per mole. Then

$$-s_L\, dT + v_L\, dp = -s_G\, dT + v_G\, dp, \tag{24.8}$$

or

$$\left(\frac{dp}{dT}\right)_\sigma = \frac{s_L - s_G}{v_L - v_G}. \tag{24.9}$$

Equation 24.9, known as the *Clapeyron equation*, describes the rate of change of pressure with temperature along the coexistence curve (which is what the subscript σ denotes). Since

$$s_G - s_L = \Delta s_{vap} = \frac{\Delta h_{vap}}{T_{vap}}, \tag{24.10}$$

we can rewrite Eq. 24.9 in the form

$$\left(\frac{dp}{dT}\right)_\sigma = \frac{\Delta h_{vap}}{T\, \Delta v_{vap}}. \tag{24.11}$$

A similar argument may be used to describe the coexistence line between the liquid and solid phases. The analog of Eq. 24.11 is

$$\left(\frac{dp}{dT}\right)_{\sigma f} = \frac{\Delta h_{fus}}{T\, \Delta v_{fus}}, \tag{24.12}$$

where the subscript *fus* refers to the fusion process, and σf to the solid–liquid coexistence line.

Equations 24.11 and 24.12 cannot be integrated unless we know the temperature and pressure dependences of the latent heats of vaporization and fusion, and of the volume changes on vaporization and fusion. It is at just this point that information about the equations of state of the two phases is required. Nevertheless, Eqs. 24.11 and 24.12 may be used backwards to determine Δh_{vap} and Δh_{fus} if the corresponding volume changes and rates of change of pressure with temperature along the coexistence lines are known. Measurements of dp/dT along the coexistence lines are commonly made, and density determinations are often not difficult, so that Eqs. 24.11 and 24.12 are, in fact, widely used to determine Δh_{vap} and Δh_{fus}.

We can also use Eqs. 24.11 and 24.12 in qualitative arguments. For example, since $\Delta h_{vap} > 0$ always, and $\Delta v_{vap} > 0$ always, it follows that $(dp/dT)_\sigma > 0$; that is, the boiling temperature always increases when the pressure is increased. Although $\Delta h_{fus} > 0$ always, Δv_{fus} can be either positive or negative, and correspondingly we have $(dp/dT)_{\sigma f} > 0$ when $\Delta v_{fus} > 0$ and $(dp/dT)_{\sigma f} < 0$ when $\Delta v_{fus} < 0$. The most famous example for which $(dp/dT)_{\sigma f} < 0$ is the ice–water transition (cf. Fig. 24.5). It can now be seen that the depression of the freezing point of water with increase of pressure is a consequence of the fact that the density of ice near 0°C is less than the density of water near 0°C. This relationship between the sign of $(dp/dT)_{\sigma f}$ and the sign of Δv_{fus} is not an "explanation" of the phenomenon, but just a description of the interrelationship of macroscopic properties required by the constraint of phase equilibrium.

Suppose that the vaporization of a liquid is carried out at a low enough temperature that the vapor formed is at low pressure. Then, as a first approximation, we can take the equation of state of the vapor to be that of a perfect gas. In this limit it is also true that $v_G \gg v_L$, so that $\Delta v_{vap} \approx v_G$. Equation 24.11 now assumes the approximate form (using $p = RT/v_G$)

$$\left(\frac{dp}{dT}\right)_\sigma = \frac{\Delta h_{vap}}{T v_G}$$

$$= \frac{\Delta h_{vap}}{RT^2} \cdot p, \tag{24.13}$$

which is known as the *Clausius–Clapeyron equation*. Equation 24.13 may be integrated to give

$$\ln\left[\frac{p_2(T_2)}{p_1(T_1)}\right]_\sigma = \int_{T_1}^{T_2} \frac{\Delta h_{vap}}{RT^2} \, dT. \tag{24.14}$$

Finally, if the temperature range $T_1 \to T_2$ is sufficiently small that the variation of Δh_{vap} with temperature may be neglected, we obtain

$$\ln\left[\frac{p_2(T_2)}{p_1(T_1)}\right]_\sigma = -\frac{\Delta h_{vap}}{R}\left(\frac{1}{T_2} - \frac{1}{T_1}\right), \tag{24.15}$$

so that a measurement of the pressure of vapor in equilibrium with liquid at two temperatures will permit calculation of Δh_{vap} for the average temperature. Of course, Δh_{vap} does vary with temperature, and indeed becomes zero at the critical temperature,[4] so that Eq. 24.15 should be regarded as an approximation to Eq. 24.14.

Example. For an illustration of the use of the Clapeyron and Clausius–Clapeyron equations, consider the calculation of the heat of vaporization of water. At the temperatures 0, 2, 99, and 101°C we have the following information:

T (°C)	p (torr)
0	4.579
2	5.294
99	733.24
101	787.57

[4] At the critical temperature the molar volumes of the two phases are equal, hence Eq. 24.13 is invalid because of the neglect of the volume of the liquid phase.

Therefore, using Eq. 24.13 directly, we obtain

$$\Delta h_{vap}(1°C) = \frac{(8.314 \text{ J/K mol})(274.2 \text{ K})^2(0.358 \text{ torr/K})}{(4.936 \text{ torr})}$$

$$= 45.32 \text{ kJ/mol},$$

$$\Delta h_{vap}(100°C) = \frac{8.314 \times (373.2)^2 \times 27.17}{760} = 41.40 \text{ kJ/mol}.$$

A graphical representation of the variation of Δh_{vap} with temperature is displayed in Fig. 24.8.

We may also use Eq. 24.12 to calculate dp/dT. Consider again the water–ice equilibrium. At 0°C we have $v_S = 19.66 \text{ cm}^3/\text{mol}$, $v_L = 18.02 \text{ cm}^3/\text{mol}$, and $\Delta h_{fus} = 6037 \text{ J/mol}$, so that

$$\left(\frac{dp}{dT}\right)_{\sigma f} = \frac{\Delta h_{fus}}{T \Delta v_{fus}} = -\frac{6037 \times 9.870}{273 \times 1.640} \text{ atm/K}$$

$$= -133 \text{ atm/K},$$

where the number 9.870 derives from the observation that $R = 8.314 \text{ J/K mol}$ and $82.06 \text{ cm}^3 \text{ atm/K mol}$, hence $1 \text{ J} \leftrightarrow 9.870 \text{ cm}^3 \text{ atm}$. Thus, in order to lower the freezing point of water by 1°C, it is necessary to impose on the system an external pressure of 133 atm.

Before considering other properties of the liquid–vapor equilibrium, it is pertinent for us to examine the possibility of equilibrium among three phases of a pure substance. Let the three phases be numbered 1, 2, 3. At equilibrium between the phases we must have

$$\mu^{(1)} = \mu^{(2)},$$
$$\mu^{(1)} = \mu^{(3)}, \tag{24.16}$$
$$\mu^{(2)} = \mu^{(3)},$$

where the third line may also be considered to be a consequence of the zeroth law of thermodynamics. Corresponding to the three pairs of equilibria, we can find three Clapeyron equations,

$$\left(\frac{dp_{12}}{dT}\right)_{\sigma_{12}} = \frac{\Delta h_{12}}{T \Delta v_{12}},$$

$$\left(\frac{dp_{13}}{dT}\right)_{\sigma_{13}} = \frac{\Delta h_{13}}{T \Delta v_{13}}, \tag{24.17}$$

$$\left(\frac{dp_{23}}{dT}\right)_{\sigma_{23}} = \frac{\Delta h_{23}}{T \Delta v_{23}},$$

only the first two of which are independent, the third being a consequence of them.

A plot of the pressure of the system versus temperature looks like Fig. 24.9. Each of the three curves in Fig. 24.9 corresponds to one of the three equations of Eq. 24.17. Any point in the (p, T) plane not on one of the lines OA, OB, or OC corresponds to a single phase of the system, and the lines OA, OB, OC to the points of coexistence of pairs of phases. The point O represents a state of the system in which all three phases are in equilibrium; it is called a *triple point*.

The ordinary triple points of most substances lie at pressures that are small relative to atmospheric pressure. When such a substance is heated, at a constant pressure that exceeds the triple-point pressure but is less than the critical pressure, the change of state can be plotted as a horizontal line above the triple point in Fig. 24.9. In this case the crystal

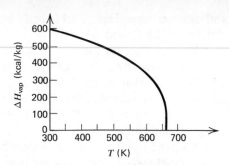

FIGURE 24.8
The heat of vaporization of water as a function of temperature. From J. Kestin, *A Course in Thermodynamics* (Blaisdell, Waltham, Mass., 1966).

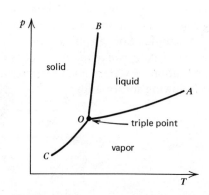

FIGURE 24.9
Schematic diagram showing the triple point and the equilibrium of solid, liquid, and vapor.

TABLE 24.1
TRIPLE POINTS OF PURE H_2O

Phases in Equilibrium	p	T (°C)
Ice I, liquid, vapor	4.579 torr	+0.01
Ice I, liquid, ice III	2,115 kg/cm²	−22.0
Ice I, ice II, ice III	2,170 kg/cm²	−34.7
Ice II, ice III, ice V	3,510 kg/cm²	−24.3
Ice III, liquid, ice V	3,530 kg/cm²	−17.0
Ice V, liquid, ice VI	6,380 kg/cm²	+0.16
Ice VI, liquid, ice VII	22,400 kg/cm²	+81.6

melts to a liquid, and then later vaporizes to a gas. Heating the substance at a pressure less than the triple-point pressure leads to a direct conversion of the crystal to vapor. This change of state can be represented by a horizontal line below the triple point in Fig. 24.9. Triple points of various phases of pure water are shown in Table 24.1.

The phase diagram in Fig. 24.9 shows that each of the three phases is stable in a planar region within which the temperature and the pressure may be varied independently of each other. Any two of the phases can coexist only on a line on the diagram, and along such a line either the temperature or the pressure, but not both, can be considered an independent variable. Finally, all three phases can coexist only at a single point, the triple point, at which the temperature and pressure are both fixed. Thus, if the number of independent variables is called f and the number of coexisting phases is called P, the above observations can be summarized by writing $f = 3 - P$.

The equation $f = 3 - P$ is a simple example of a general and useful relation known as the *Gibbs phase rule*. Consider a system containing P coexisting phases, with C components in each phase. By a *component* we mean a chemical species whose concentration in a phase can be varied independently of the other species' concentrations. The number of components is not necessarily equal to the number of chemical species present. For example, if two species are interconvertible through a chemical reaction, and are in chemical equilibrium, their concentrations are not independent of each other. Two species in chemical equilibrium are therefore considered to form a single component. In order to characterize a system completely, we must specify the temperature and pressure, and the mole fractions of each of the C components in each of the P phases, a total of $CP + 2$ variables.

These $CP + 2$ variables are not all mutually independent. In each phase the mole fractions of the various components must add to unity. This condition means that one variable for each of the P phases is dependent on the other variables. In addition, the existence of phase equilibrium requires that each component satisfy a set of equations like Eq. 24.17. If a component is present in P phases, there will be $P - 1$ independent equalities involving the chemical potential. For example, for a component present in three phases, Eq. 24.17 represents two independent conditions that guarantee phase equilibrium. The third equation in Eq. 24.17 is only a consequence of the zeroth law. For C components there are therefore $C(P - 1)$ relationships among the chemical potentials.

To find the number of independent variables describing the system, we must subtract from $CP + 2$ the number of equations relating these variables. In the previous paragraph, $P + C(P-1)$ such equations were found. The number of independent variables (also called the number of *degrees of freedom*) is therefore $f = C - P + 2$. This is the Gibbs phase rule; it predicts how many variables are required to describe a system, and it also places a limit on the number of phases that can coexist in a system containing C components. For one-component systems, that limit is three.

We have previously discussed in detail the heat capacity of a one-phase system. When two phases must be kept in equilibrium, the addition of heat to the system is described by neither c_V nor c_p, since the constraint of phase equilibrium is not the same as the constraints of constant volume or constant pressure. We now define a third heat capacity as the limiting ratio of the heat absorbed to the temperature rise generated under the constraint that the system consist of two phases maintained in equilibrium. It is clear that the existence of a nonzero latent heat of transition will influence the magnitude of the heat capacity of a two-phase system.

Consider, for example, a two-phase system consisting of liquid and vapor in equilibrium. The first law of thermodynamics,

$$\dbar q = du - \dbar w, \tag{24.18}$$

can be used to define the heat capacity of the two-phase system. With the definition cited, we obtain

$$c_\sigma \equiv \left(\frac{\dbar q}{dT}\right)_{\text{phase equilibrium}} = \left[\left(\frac{du}{dT}\right) - \left(\frac{\dbar w}{dT}\right)\right]_{\text{phase equilibrium}} \tag{24.19}$$

For the case of the liquid–vapor transition, the expansion of the system results in the performance of pressure-volume work only, so that we have (per mole)

$$c_\sigma = \left(\frac{du}{dT} + p\frac{dv}{dT}\right)_{\text{vapor pressure curve}} \tag{24.20}$$

Let $c_{\sigma L}$ and $c_{\sigma G}$ be the heat capacities of saturated liquid and vapor, respectively. Then, using the equalities $p_L = p_G = p_\sigma$ and $T_L = T_G$, both valid because of the constraint of phase equilibrium, we find that

$$
\begin{aligned}
c_{\sigma L} &= \left(\frac{du_L}{dT}\right)_\sigma + p_\sigma\left(\frac{dv_L}{dT}\right)_\sigma, \\
c_{\sigma G} &= \left(\frac{du_G}{dT}\right)_\sigma + p_\sigma\left(\frac{dv_G}{dT}\right)_\sigma.
\end{aligned}
\tag{24.21}
$$

By subtracting one of equations 24.21 from the other, we obtain

$$c_{\sigma G} - c_{\sigma L} = \left[\frac{d}{dT}(u_G - u_L)\right]_\sigma + p_\sigma\left[\frac{d}{dT}(v_G - v_L)\right]_\sigma. \tag{24.22}$$

Now, by definition, the latent heat of vaporization per mole is

$$\Delta h_{\text{vap}} \equiv h_G - h_L \equiv u_G - u_L + p_\sigma(v_G - v_L), \tag{24.23}$$

so that, by combining Eqs. 24.22 and 24.23, we find

$$
\begin{aligned}
\frac{d\,\Delta h_{\text{vap}}}{dT} &= \left[\frac{d}{dT}(u_G - u_L)\right]_\sigma + p_\sigma\left[\frac{d}{dT}(v_G - v_L)\right]_\sigma + (v_G - v_L)_\sigma\frac{dp_\sigma}{dT} \\
&= c_{\sigma G} - c_{\sigma L} + (v_G - v_L)_\sigma\frac{dp_\sigma}{dT}. \tag{24.24}
\end{aligned}
$$

But the last term on the right-hand side of Eq. 24.24 may be evaluated from the Clapeyron equation, because the constraint of phase equilibrium is implicit in Eq. 24.24. Thus, introducing Eq. 24.11 in Eq. 24.24, we obtain

$$c_{\sigma G} - c_{\sigma L} = \frac{d\,\Delta h_{vap}}{dT} - \frac{\Delta h_{vap}}{T}. \tag{24.25}$$

If the temperature is low, $c_{\sigma L}$ is almost equal to c_{pL}. With this approximation entered into Eq. 24.25, we find for the heat capacity of the saturated vapor

$$c_{\sigma G} = c_{pL} + \frac{d\,\Delta h_{vap}}{dT} - \frac{\Delta h_{vap}}{T}. \tag{24.26}$$

Consider, as an example, water vapor in equilibrium with liquid water at 100°C. Using tables of the thermodynamic properties of water, we find $\Delta h_{vap} = 40.67$ kJ/mol, $(d\Delta h_{vap}/dT) = -47.93$ J/K mol, $c_{pL} = 75.86$ J/K mol. Thus in this case $c_{\sigma G} = -80.91$ J/K mol; the heat capacity of saturated water vapor at 100°C is negative. What does this result mean? If saturated water vapor is compressed adiabatically, its temperature increases with pressure more rapidly than along the coexistence curve between liquid and vapor. Conversely, if saturated water vapor is adiabatically expanded, its temperature decreases with decreasing pressure more rapidly than along the coexistence curve, so some condensation to liquid must occur. In our numerical example, if water vapor at 100°C and 760 torr were adiabatically compressed to 787.6 torr (the vapor pressure at 101°C) its temperature would rise to about 103°C; 80.91 J/mol would then have to be withdrawn at constant pressure to bring the vapor back to the coexistence curve at 101°C. To an external observer this appears as a negative heat capacity of the saturated vapor.

It has been pointed out in Section 21.1 that there exists a critical temperature, T_c, above which gas and liquid cannot be distinguished. The conditions describing phase equilibrium, namely, the equality of chemical potentials and hence of pressures and temperatures between the two phases, in principle permit there to be a critical point between any two coexisting phases. Given the equations of state of the two phases and the constraint of phase equilibrium, we have five equations in six variables:

B. The Critical Point

$$\left.\begin{array}{l} p_1 = p_2 \\ T_1 = T_2 \\ \mu_1(p_1, T_1) = \mu_2(p_2, T_2) \end{array}\right\} \quad \text{constraint of phase equilibrium,}$$

$$\left.\begin{array}{l} p_1 = p_1(v_1, T_1) \\ p_2 = p_2(v_2, T_2) \end{array}\right\} \quad \text{equations of state.}$$

Thus, five of the six variables p_1, T_1, v_1, p_2, T_2, v_2 are determined by these five equations, leaving one relation between the remaining two variables. This is, for example, the equation connecting the vapor pressure and temperature along the coexistence line between vapor and liquid. Now, if the values of v_1 and v_2 become equal for some value of the temperature, the two phases become identical if they have the same symmetry. This temperature is defined to be the critical temperature, and the corresponding values of p and v to be the critical pressure (p_c) and critical volume (v_c), respectively.

It may happen that no temperature can be found for which $v_1 = v_2$ given that $p_1 = p_2$ and $T_1 = T_2$. In this case there will be no critical point on the coexistence line between the two phases. It may also happen that the equality of temperature, pressure, and chemical potential between the two phases does not provide a complete description of the phase equilibrium. For example, in the description of crystal–crystal transitions it is necessary to know the symmetries of the crystal lattices for the two phases. Although the volumes of the two crystals might be the same, unless the lattice structures were also the same the phases would not be identical, and the temperature at which $v_1 = v_2$ would not be a critical point. The role of symmetry properties in defining the properties of transitions of the second kind was described in Section 24.1. Our use of the concept of symmetry is related to the concept of distinguishability. Just as in the description of the entropy of mixing (see Section 21.2), it is distinguishability that determines the qualitative features of the phenomenon described.

We now restrict attention to the study of the gas–liquid critical point and some of the properties of the system in the vicinity of that point. We begin by defining conditions that locate the critical point of a fluid. A study of the isotherms of pressure versus volume, as shown in Fig. 24.4, reveals that one isotherm passes through a point in the (p, v) plane at which it is horizontal, and at which it also has a point of inflection (change from positive to negative curvature). These two properties,

$$\left(\frac{\partial p}{\partial v}\right)_{T=T_c} = 0 \qquad \text{(horizontal tangent)},$$

$$\left(\frac{\partial^2 p}{\partial v^2}\right)_{T=T_c} = 0 \qquad \text{(point of inflection)}, \tag{24.27}$$

define the critical point of the fluid.

What are the properties of a fluid at and near the critical point?

1. Since $(\partial v/\partial p)_T < 0$ for any substance at equilibrium, it follows that

$$\lim_{\substack{T \to T_c \\ v \to v_c}} \left[-\frac{1}{v}\left(\frac{\partial v}{\partial p}\right)_T\right] = \infty, \tag{24.28}$$

hence at the critical point the isothermal compressibility becomes positive infinite.

2. Suppose that $(\partial u/\partial v)_T$ is finite. Then from

$$\left(\frac{\partial u}{\partial v}\right)_T = T\left(\frac{\partial p}{\partial T}\right)_V - p,$$

it follows that $(\partial p/\partial T)_V$ is finite. But along the gas–liquid coexistence line we have, by Eqs. 24.9 and 24.11,

$$\left(\frac{dp}{dT}\right)_\sigma = \frac{s_G - s_L}{v_G - v_L},$$

so that we may write

$$\lim_{\substack{T \to T_c \\ v \to v_c}} \left(\frac{dp}{dT}\right)_\sigma = \left(\frac{\partial s}{\partial v}\right)_{T=T_c} \tag{24.29}$$

or, using Eq. 17.17,

$$\left(\frac{dp}{dT}\right)_{\sigma,\text{ critical point}} = \left(\frac{\partial p}{\partial T}\right)_{v=v_c}. \tag{24.30}$$

Since $(\partial p/\partial T)_V$ is always finite, $(dp/dT)_{\sigma, T=T_c}$ is finite. Then, using

$$\left(\frac{\partial v}{\partial T}\right)_p = -\left(\frac{\partial v}{\partial p}\right)_T\left(\frac{\partial p}{\partial T}\right)_V,$$

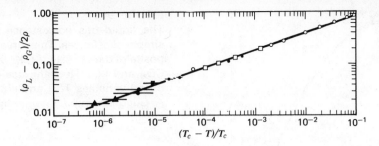

FIGURE 24.10
The dependence of $\rho_L - \rho_G$ on $T - T_c$ near T_c for CO_2. From P. Heller, *Rep. Prog. Phys.* **30** (II), 731 (1967).

one can easily show that

$$\lim_{\substack{T \to T_c \\ v \to v_c}} \frac{1}{v}\left(\frac{\partial v}{\partial T}\right)_p = \infty. \qquad (24.31)$$

In words, the coefficient of thermal expansion at the critical point is positive and infinite.

3. Since the Clapeyron equation can be written in the form

$$\Delta h_{vap} = T(v_G - v_L)\left(\frac{dp}{dT}\right)_\sigma,$$

then at the critical point, when $T = T_c$, $v_G = v_L$, and $(dp/dT)_\sigma$ is finite, we must have

$$\lim_{T \to T_c} \Delta h_{vap} = 0. \qquad (24.32)$$

It can also be shown that $(d\Delta h_{vap}/dT)_{T=T_c} = -\infty$. A glance at Fig. 24.8 shows that the predicted behavior of Δh_{vap} and $d\Delta h_{vap}/dT$ as $T \to T_c$ is in agreement with the observed behavior.

The fact that so many thermodynamic properties of a system become either indefinitely large or zero at the critical point is sufficient evidence that the region of the critical point of a fluid has some very peculiar characteristics. At the present time we do not understand well the relationship between the behavior of a fluid in the critical region and the underlying intermolecular interactions. Some of the observations that require interpretation follow:

1. As the critical point is approached, the liquid and gas densities converge toward one another as[5]

$$(\rho_L - \rho_G) \propto |T_c - T|^\beta; \qquad \beta = 0.355 \pm 0.001 \qquad (24.33)$$

(see Figs. 24.10 and 24.11). Note that to a first approximation $\beta = \frac{1}{3}$, and the coexistence curve near the critical point is cubic. However, this is not true to within the experimental error. Because of the algebraic form of Eq. 24.33, β is called a *critical exponent* (as are δ, α, and γ defined below).

2. The critical point is associated with a striking thermal anomaly. In fact, at a constant density equal to the critical density (see Fig. 24.12),

$$c_V(T) \to \infty \qquad (24.34)$$

as $T \to T_c$ from above or below with $v = v_c$.

3. The rates at which various thermodynamic functions approach the critical point are given by

$$(p - p_c) \propto |\rho - \rho_c|^{\delta-1}(\rho - \rho_c); \qquad \delta = 4.35 \pm 0.1, \qquad (24.35)$$

$$c_V(T) \propto \ln\left|1 - \frac{T}{T_c}\right|$$

[5] For the present we do not distinguish between approaching the critical point from $T > T_c$ or $T < T_c$. The available experimental evidence suggests that the critical exponents are the same for both approaches to the critical point (see Table 24.4).

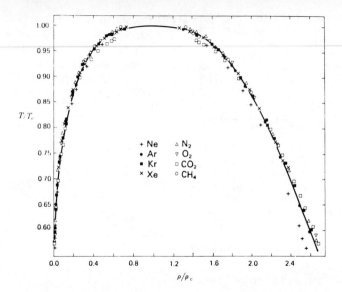

FIGURE 24.11
The liquid–gas coexistence curve for simple substances. The curve is a composite of data for Ne, Ar, Kr, Xe, N₂, O₂, CO₂, and CH₄. Note the use of the reduced variables T/T_c and ρ/ρ_c. From E. A. Guggenheim, *J. Chem. Phys.* **13**, 253 (1945).

or

$$c_V(T) \propto |T - T_c|^{-\alpha}; \qquad \alpha = 0.1 \pm 0.04, \qquad (24.36)$$

$$\frac{1}{\kappa_T} \equiv -V\left(\frac{\partial p}{\partial V}\right)_T \propto |T - T_c|^{\gamma}; \qquad \gamma = 1.2 \pm 0.04. \qquad (24.37)$$

There is no simple equation of state of a fluid that predicts the detailed behavior described in Eqs. 24.33–24.37.

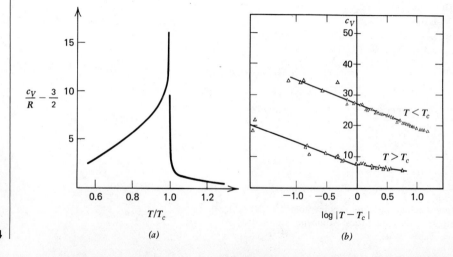

(a) (b)

FIGURE 24.12
(a) Schematic diagram of the constant-volume heat capacity of argon near the critical point.

(b) Variation of the constant-volume heat capacity of argon along the critical isochore. The data show that the variation of c_V near T_c is logarithmic. From P. A. Egelstaff and J. W. Ring [data of M. I. Bagatskil, A. V. Voronel, and B. G. Gusak, *Zh. Eksperim. i Teor. Fiz.* **43**, 728 (1962); *JETP* **16**, 517 (1963)], in H. N. V. Temperley, J. S. Rowlinson, and G. S. Rushbrooke (Eds.), *Physics of Simple Liquids* (North Holland, Amsterdam, 1968).

The chemical potential of a pure substance depends only on the pressure and the temperature, the functional forms being different for the solid, liquid, and vapor phases. When the chemical potential of one phase, say, the liquid, becomes less than that of another, say, the vapor, then the latter is unstable with respect to the former. Viewed this way, the condition of phase equilibrium, $\mu^L = \mu^G$ in the particular case mentioned, specifies the boundary of stability between the two phases, and we can think of the phase transition as the system's response to an instability. In this section we pursue this idea.

It is convenient, for the purpose of visualizing the way a thermodynamic instability develops, to represent the equilibrium properties of a one-component system by a surface in a suitable space of variables. Consider a specified mass, say, 1 mol, of a pure substance. Suppose that we know the energy as a function of the entropy and volume, $u(s, v)$. Then, if we construct a three-dimensional space with orthogonal axes u, s, and v, the function $u(s, v)$ is a surface in this space. Each point on this surface is associated with an equilibrium state of the system, and the slopes of $u(s, v)$ at such a point parallel to the s and v axes are simply

$$\left(\frac{\partial u}{\partial s}\right)_v = T, \tag{24.38}$$

$$\left(\frac{\partial u}{\partial v}\right)_s = -p. \tag{24.39}$$

Therefore any plane parallel to the u, v plane intersects the surface $u(s, v)$ in a curve that has negative slope everywhere (since $p > 0$) and any plane parallel to the u, s plane intersects the surface $u(s, v)$ in a curve that has positive slope everywhere (since $T > 0$). These features of the surface $u(s, v)$ are sketched in Fig. 24.13.

The surface $u(s, v)$ has some interesting properties. Consider some point Γ_0 on this surface, and the tangent plane to the surface that passes through it. The equation of that tangent plane is

$$u = u_0 + T_0(s - s_0) - p_0(v - v_0), \tag{24.40}$$

the subscript zero referring to values at Γ_0. Now consider the plane $v = v_0$. This plane is parallel to the (u, s) plane and passes through Γ_0. Moreover, it intersects the plane tangent at Γ_0 in a straight line the equation of which is

$$u = u_0 + T_0(s - s_0) \tag{24.41}$$

(remember that the intersection also lies in the plane for which $v = v_0$), and this straight line intersects the u axis at a value

$$u = u_0 - T_0 s_0 \equiv a_0 \tag{24.42}$$

(see Fig. 24.13). Similarly, the plane $s = s_0$ passes through Γ_0 and intersects the tangent plane at Γ_0 in the straight line

$$u = u_0 - p_0(v - v_0), \tag{24.43}$$

which intercepts the u axis at the point

$$u = u_0 + p_0 v_0 \equiv h_0 \tag{24.44}$$

(see Fig. 24.13). Finally, the tangent plane at Γ_0 itself intercepts the u axis at

$$u = u_0 - T_0 s_0 + p_0 v_0 \equiv g_0 \tag{24.45}$$

(see Fig. 24.13).

Since a one-component system with fixed mass has definite values of u, s, and v irrespective of the number of phases present, the surface $u(s, v)$ must exist for all the system's equilibrium states. We now examine what the surface looks like when several phases are present (the case of heterogeneous equilibrium).

Consider a one-component system at its triple point. There is some point of the surface $u(s, v)$ representing the properties of the solid at the triple point, call

24.3
Phase Transitions Viewed as Responses to Thermodynamic Instabilities

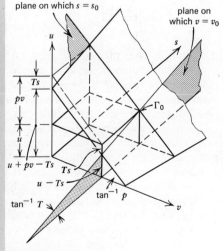

FIGURE 24.13
The tangent plane to the thermodynamic surface. Drawing courtesy of Prof. J. A. Beattie.

it Γ_3^S. Similarly, there are points Γ_3^L and Γ_3^G on $u(s, v)$ representing the states of the liquid and the vapor at the triple point. Now construct the three tangent planes at the points Γ_3^S, Γ_3^L, and Γ_3^G. By Eq. 24.45 the intercepts of these tangent planes with the u axis are g_3^S, g_3^L, and g_3^G, respectively. But at the triple point $g_3^S = g_3^L = g_3^G$, since for a one-component system μ is just the Gibbs free energy per mole. Moreover, since $T_S = T_L = T_G$ and $p_S = p_L = p_G$ at the triple point, the slopes of the three tangent planes parallel to the s and v axes are the same at Γ_3^S, Γ_3^L, and Γ_3^G. We conclude that the three tangent planes coincide or, better put, that Γ_3^S, Γ_3^L, and Γ_3^G have a common tangent plane. Therefore, a mixture of solid, liquid, and vapor in equilibrium at the triple point will have its state represented by a point on the so-called *derivative surface*, which is just the region of the common tangent plane to $\Gamma_3^S, \Gamma_3^L, \Gamma_3^G$ lying within the triangle formed by these three points. The position of the point representing the equilibrium mixture of the three phases is determined by

$$u = x_S u_S + x_L u_L + x_G u_G,$$
$$s = x_S s_S + x_L s_L + x_G s_G, \qquad (24.46)$$
$$v = x_S v_S + x_L v_L + x_G v_G,$$

where x_i is the mole fraction of phase i.

Consider now liquid–vapor equilibrium in a one-component system of fixed mass. A similar argument leads to the surface shown in Fig. 24.14. The derived surface is the set of straight lines tangent simultaneously at Γ_2^L and Γ_2^G representing the coexisting liquid and vapor phases. These two points of tangency on $u(s, v)$ approach one another and eventually coincide at the critical point. Notice that there is a gap in the surface drawn that is spanned by the derived surface. The gap coincides with that region of $u(s, v)$ which would have negative curvature.

To understand how an instability is represented, and the reason for the gap in Fig. 24.14, we examine the nature of internal stability in a homogeneous phase. Suppose that we have a homogeneous pure substance isolated in a volume v; its energy and entropy are u and s. We test the stability of this phase with respect to variations that keep the energy and entropy fixed. This is just the situation treated in Chapter 19 (see Fig. 19.4, Eqs. 19.45–19.50). We saw there that for the system to be stable it is necessary and sufficient that

$$\left(\frac{\partial^2 u}{\partial s^2}\right)_v = \frac{T}{c_V} > 0, \qquad (24.47a)$$

$$\left(\frac{\partial^2 u}{\partial s^2}\right)_v \left(\frac{\partial^2 u}{\partial v^2}\right)_s - \left(\frac{\partial^2 u}{\partial v \, \partial s}\right)_{s,v}^2 = \frac{T}{c_V}\left(\frac{c_p}{v \kappa_T c_V}\right) - \frac{\alpha^2 T^2}{\kappa_T^2 c_V^2} > 0, \qquad (24.47b)$$

the latter implying that $\kappa_T > \alpha^2 T v / c_p > 0$. At least for infinitesimal displacements, a one-phase system is stable so long as Eqs. 24.47a and 24.47b are satisfied. However, the stability of the system against the formation of a new phase has not been tested by the class of displacements considered. This is an important observation, as we shall now show.

Consider a region of the surface $u(s, v)$ representing two-phase equilibrium, say, between liquid and vapor. The intersection of this region of $u(s, v)$ and a plane in which s is constant is shown in Fig. 24.15. The straight line that is the common tangent to the points labeled Γ_2^L and Γ_2^G is the intersection of the derived surface and the same plane; it has slope $p = -(\partial u / \partial v)_s$, as required by its definition. In either of the pure phases the pressure is a monotonic function of the volume since, by Eq. 24.47b, $v \kappa_T = -(\partial v / \partial p)_T > 0$. This condition can be satisfied in a homogeneous phase beyond the temperature and pressure at which a new phase emerges. Hence, *if we imagine that it is possible to prevent formation of a new phase*, the curve of internal energy versus volume may be continued until Eqs. 24.47a and 24.47b become invalid. The continuations are shown by dots in Fig. 24.15; they stop at the points of inflection where $(\partial^2 u / \partial v^2)_s = 0$. There can be no continuation of u versus v between these points of inflection since in that

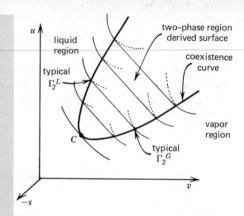

FIGURE 24.14
Schematic representation of the surface $u(s, v)$ in a region of vapor–liquid equilibrium. The derived surface lies below $u(s, v)$; each line shown in the two-phase region is an example of the common tangent to $u(s, v)$ at points Γ_2^L and Γ_2^G. The locus of the points of tangency defines the coexistence curve. C is the critical point. The dotted portions of $u(s, v)$ show regions of metastability.

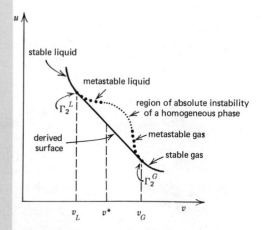

FIGURE 24.15
Schematic representation of the intersection of the two-phase region of $u(s, v)$ and a plane on which $s =$ constant. The dotted lines are continuations of the homogeneous-phase energies; they represent metastable states of the system. If we attempt to prepare a homogeneous system with the volume v^* shown, it breaks up into liquid and gas, this being the region of absolute instability of a homogeneous phase.

domain $(\partial^2 u/\partial v^2)_s < 0$ and a homogeneous system is absolutely unstable. Notice that the continuation of $u(s, v)$ for the liquid corresponds to a pressure less than that along the common tangent, and the continuation of $u(s, v)$ for the vapor corresponds to a pressure greater than that along the common tangent. The continuations shown correspond to metastable states of a homogeneous vapor or liquid. They can sometimes be observed. For example, the formation of the clusters of molecules that convert a gas to a liquid can occur slowly enough that supersaturated vapor[6] can persist long enough to be observed. The supersaturated vapor must be prepared in the absence of the liquid phase, since the homogeneous system is stable only to infinitesimal displacements that do not include the appearance of a new phase. Once the liquid phase appears, the supersaturated vapor cannot persist as it is unstable with respect to that displacement which creates the liquid.

The region of the surface $u(s, v)$ that would have negative curvature is just a gap; this gap is spanned by the derived surface described earlier. Because of the absolute instability of systems whose representative point falls in the gap region of $u(s, v)$, if we were to imagine a hypothetical homogeneous system prepared with volume v^*, $v_L < v^* < v_G$, it would immediately decompose into liquid and vapor, the amounts of each being determined by Eq. 24.1 (see Fig. 24.15). That is, the energy of the system is lowest along the common tangent from the derived surface, and along this tangent $v = x_L v_L + x_G v_G$.

A convenient way of portraying the relationship between the coexistence curves along which phase equilibria exist, and the limit of absolute stability for a system, is to project the surface $u(s, v)$ onto an (s, v) plane, as shown in Fig. 24.16.

As an example of the use of stability criteria in discussing phase transitions, consider the van der Waals equation of state,

$$p = \frac{RT}{v - b} - \frac{a}{v^2}, \tag{24.48}$$

the isotherms of which are shown in Fig. 24.17a. Clearly, the isotherms for $T < T_c$ have a region with $(\partial v/\partial p)_T > 0$. But the conditions for stability of a homogeneous phase discussed in Chapter 19 and summarized in Eqs. 24.47a and 24.47b show that unless $(\partial v/\partial p)_T < 0$, a homogeneous system is unstable. Consequently, the van der Waals gas cannot remain one homogeneous phase at all temperatures. Consider a typical isotherm along which condensation occurs. In Fig. 24.17b we plot the volume as a function of pressure along the isotherm, and the chemical potential along the same isotherm. From the relation

$$d\mu = -s\, dT + v\, dp, \tag{24.49}$$

we see that

$$\mu(p_2) - \mu(p_1) = \int_{p_1}^{p_2} v(p)\, dp \tag{24.50}$$

along the isotherm. In the unstable region between p_B and p_C, $v(p)$ is triple-valued. For every value of p in this region, two of the values of $v(p)$ lie on curves for which $(\partial v/\partial p)_T < 0$, whereas the third does not. The branches where a homogeneous system would be stable include regions where each is metastable with respect to formation of the other phase. The branches of $v(p)$ where the liquid is stable or metastable are DE' and $E'C$, respectively. The branches of $v(p)$ where the vapor is stable or metastable are AE and EB, respectively. The tie line EE' is the common tangent from the derived surface associated with $u(s, v)$ in the liquid–vapor equilibrium region. The prediction of a third value of $v(p)$ is a major failing of the van der Waals equation of state, hence of its underlying model of the molecular interactions.

In the (p, v) plane we now have the following picture of the liquid–gas phase transition: At the pressure p_E the two phases have the same chemical potential. When the pressure is changed isothermally, the locus of equilibrium points lies on

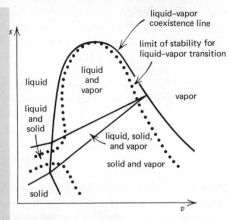

FIGURE 24.16
Projection of the surface $u(s, v)$ onto an (s, v) plane, showing the several coexistence curves for equilibria involving pairs of phases and also the limit-of-stability curve (spinodal curve) for the liquid–gas transition. The area labels show projections of regions of $u(s, v)$ where the specified phases are stable, and of the derived surfaces where mixtures of phases are stable. Drawing courtesy of Prof. J. A. Beattie.

---

[6] Vapor at a pressure in excess of the vapor pressure of the liquid at the same temperature.

either the liquid branch DE' or the gas branch AE. If we start at A in the diagram and make an isothermal compression, the path followed is $AEE'D$, the system being entirely gaseous along AE, entirely liquid along $E'D$, and a mixture of gas and liquid along EE'. (Eqs. 24.2–24.4 give the relative fractions of gas and liquid along EE'.)

Thus far we have discussed the stability conditions with respect to conditions away from the critical point. At the critical point we have $(\partial p/\partial v)_T = 0$ and $(\partial^2 p/\partial v^2)_T = 0$, hence the condition $(\partial v/\partial p)_T < 0$ cannot be satisfied, and the stability criteria deduced in Chapter 19 are insufficient to describe the system. If one higher order in the expansion 19.45 is carried, it may be deduced that when $(\partial p/\partial v)_T = 0$, $(\partial^3 p/\partial v^3)_T < 0$ is a stability requirement.

At the critical point the isothermal compressibility becomes indefinitely large. This implies that an isothermal compression of the fluid requires vanishingly little work to be done, which further implies that vanishingly little work need be done to effect a compression or expansion along an arbitrary thermodynamic path. Imagine that the volume of a system at its critical point is divided into subvolumes that can exchange matter and energy with their surroundings. Because $(\partial p/\partial v)_T = 0$, we anticipate that fluctuations in the densities of the subvolumes are large. However, since the average density of the entire system is fixed by the amount of matter in the total volume, subvolumes with higher than average density must be compensated for by the existence of subvolumes with lower than average density. This observation implies that the fluctuations in the densities of the subvolumes cannot be independent, since the constraint that the average density of the total volume is fixed leads to a coupling of fluctuations. That the fluctuations in density at and near the critical point are very large is easily demonstrated experimentally. Corresponding to the fluctuations in density, there exist fluctuations in the fluid's refractive index that lead to scattering of a light beam passed through the fluid. The larger the fluctuations in refractive index, the more intense the scattering becomes. At the critical point the fluid scatters so much of the light that it appears milky and opaque. This phenomenon is known as *critical opalescence*.

Now, it is also true that $(\partial p/\partial v)_T = 0$ in the two-phase region of the liquid–gas transition. By analogy with the argument given above, should there not be large fluctuations in density in the two-phase region? The answer to this question is, basically, yes. In the two-phase region the density fluctuates from point to point just because the system is not homogeneous, and in moving from point to point it is possible to go into and out of each of the two phases present. Thus, along any path in the volume containing the system, the density does fluctuate between that characteristic of the liquid and that characteristic of the gas. Also, during the isothermal condensation process an increase or decrease in volume changes the relative proportions of gas and liquid, since the equilibrium pressure is fixed. In this case, because the pressure is fixed at p_σ, the work done is just $-p_\sigma \Delta v$, where Δv is the volume change. Of course, at the critical point we have

$$\lim_{p_\sigma \to p_c} p_\sigma \Delta v \to 0.$$

It should now be clear that the properties of a fluid at and near the critical point are very strange. Besides the divergences displayed in Eqs. 24.28, 24.31, and 24.33–24.37, it can be seen that, although the vapor pressure line in the (p, T) plane is the locus of points representing vapor–liquid equilibrium in a transition of the first kind, it terminates in a critical point that has some of the characteristics of a transition of the second kind ($\Delta s = 0, \Delta v = 0$). Also, the Clapeyron equation defines the slope $(dp/dT)_\sigma$ at all temperatures $T < T_c$, but the ratio $(\Delta s/\Delta v)$ becomes indeterminate when $T = T_c$. As usual, to evaluate $(\Delta s/\Delta v)$ in the limit that numerator and denominator both vanish, we use l'Hospital's rule:

$$\lim_{T \to T_c} \frac{\Delta s}{\Delta v} = \lim_{T \to T_c} \frac{(\partial \Delta s/\partial T)_p}{(\partial \Delta v/\partial T)_p}$$

$$= \frac{1}{Tv}\left(\frac{c_{pG} - c_{pL}}{\alpha_G - \alpha_L}\right). \tag{24.51}$$

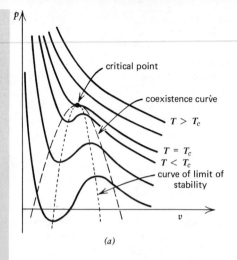

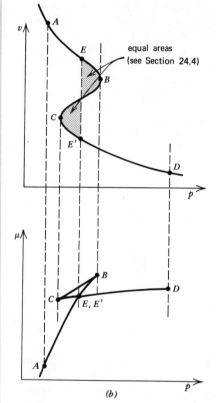

FIGURE 24.17

(a) Schematic representation of isotherms of the van der Waals equation of state.

(b) Representation of the relation between v, p and μ, p plots for a van der Waals gas.

Experiment shows that $(dp/dT)_\sigma$ is finite up to and at the critical point, whereas from Eqs. 24.31 and 24.36 we see that each term in the numerator and denominator of Eq. 24.51 tends to infinity. It is possible that the differences $c_{pG} - c_{pL}$ and $\alpha_G - \alpha_L$ are finite at T_c even though the individual terms are not, but that is not presently known. Proceeding once more to use l'Hospital's rule on the right-hand side of Eq. 24.51, we find another ratio that is indeterminate, this time of the form 0/0. Further application of l'Hospital's rule leads to functions with unknown behavior. Presumably the very strange behavior of the thermodynamic functions at the critical point reflect the extraordinary circumstance that the critical point lies on both the coexistence curve *and* the curve of the limit of stability (spinodal curve), yet is a stable state of the system.

Despite the difficulties just sketched, there are two fruitful applications of thermodynamics to the description of critical phenomena. The first of these provides inequalities that the critical exponents α, β, γ, δ of Eqs. 24.33–24.37 must satisfy, and the second provides an equation of state in the immediate vicinity of the critical point. We shall discuss an example of each type of analysis.

Consider the pV diagram shown in Fig. 24.18. The critical isotherm is $A'CA$, and BCB' is the liquid–vapor coexistence curve. Imagine that the system executes the cycle $CABDC$. Since no work is done along the legs AB and DC, and $\Delta U_{cycle} = 0$, we can write for an arbitrary mass of substance

$$q + w = 0 = T_c(S_A - S_C) - \int_C^A p\, dV + \int_{T_c}^{T_1} C_{VG}\, dT$$

$$+ T_1(S_D - S_B) - \int_B^D p\, dV + \int_{T_1}^{T_c} (C_V)_{V=V_c}\, dT, \quad (24.52)$$

where the two-phase heat capacity $(C_V)_{V=V_c}$ is evaluated along the path DC where $V = V_c$. Using the identity

$$S_A - S_C \equiv S_A - S_B + S_B - S_D + S_D - S_C,$$

and Eq. 24.9 in the form[7]

$$S_B - S_D = (V_G - V_c)\left(\frac{dp}{dT}\right)_{T=T_1},$$

we can rewrite Eq. 24.52 as

$$\int_{T_1}^{T_c} \frac{T_c - T}{T}(C_V)_{V=V_c}\, dT = \int_{T_1}^{T_c} \frac{T_c - T}{T} C_{VG}\, dT + \int_C^A (p_c - p)\, dV$$

$$+ (T_c - T_1)(V_G - V_c)\left[\left(\frac{dp}{dT}\right)_{T=T_1} - \frac{p_c - p(T_1)}{T_c - T_1}\right]. \quad (24.53)$$

By the same reasoning, when the system executes the cycle $A'CDB'A'$,

$$\int_{T_1}^{T_c} \frac{T_c - T}{T}(C_V)_{V=V_c}\, dT = \int_{T_1}^{T_c} \frac{T_c - T}{T} C_{VL}\, dT + \int_{A'}^C (p - p_c)\, dV$$

$$- (T_c - T_1)(V_c - V_L)\left[\left(\frac{dp}{dT}\right)_{T=T_1} - \frac{p_c - p(T_1)}{T_c - T_1}\right]. \quad (24.54)$$

Now divide Eq. 24.53 by $V_G - V_c$ and Eq. 24.54 by $V_c - V_L$ and add. The result

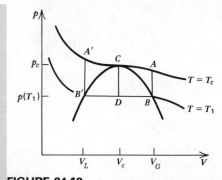

FIGURE 24.18
Schematic representation of the coexistence curve and isotherms near the critical point.

---

[7] Since $S_D = x_L S_{B'} + x_G S_B$ and $V_c = x_L V_L + x_G V_G$, we have

$$\frac{S_B - S_D}{V_G - V_c} = \frac{(S_B - S_{B'})x_L}{(V_G - V_L)x_L} = \left(\frac{dp}{dT}\right)_{T=T_1}.$$

is

$$\left(\frac{1}{V_G - V_c} + \frac{1}{V_c - V_L}\right) \int_{T_1}^{T_c} \frac{T_c - T}{T}(C_V)_{V=V_c}\, dT$$

$$= \frac{1}{V_G - V_c} \int_{T_1}^{T_c} \frac{T_c - T}{T} C_{VG}\, dT + \frac{1}{V_c - V_L} \int_{T_1}^{T_c} \frac{T_c - T}{T} C_{VL}\, dT$$

$$+ \frac{1}{V_G - V_c} \int_{V_c}^{V_G} (p_c - p)_{T_c}\, dV + \frac{1}{V_c - V_L} \int_{V_L}^{V_c} (p - p_c)_{T_c}\, dV. \quad (24.55)$$

The integral on the left-hand side of Eq. 24.55 follows the path DC. Along this path, $x_L = (V_G - V_c)/(V_G - V_L)$ and $x_G = (V_c - V_L)/(V_G - V_L)$ both vary. In contrast, on the right-hand side of Eq. 24.55 the integrals with respect to temperature follow the paths BA and BA', along which V_L and V_G are fixed. We can then rewrite Eq. 24.55 in an illustrative form by explicitly displaying this difference:

$$\int_{T_1}^{T_c} \frac{T_c - T}{T}(C_V)_{V=V_c}\, dT$$

$$= \int_{T_1}^{T_c} \frac{T_c - T}{T}[x_L(T_1)C_{VL} + x_G(T_1)C_{VG}]\, dT$$

$$+ x_G(T_1) \int_{V_c}^{V_G} (p_c - p)_{T_c}\, dV + x_L(T_1) \int_{V_L}^{V_c} (p - p_c)_{T_c}\, dV, \quad (24.56)$$

where $x_G(T_1) = [V_c - V_L(T_1)]/[V_G(T_1) - V_L(T_1)]$ and $x_L(T_1) = [V_G(T_1) - V_c]/[V_G(T_1) - V_L(T_1)]$ are considered fixed values. In the integral on the left-hand side of Eq. 24.56, the amounts of vapor and liquid adjust appropriately as T is varied; but in the integral on the right-hand side of Eq. 24.56, no adjustment is permitted, just as if phase equilibrium were suppressed.

We now note that the first two terms on the right-hand side of Eq. 24.55 are always positive since $T_c > T_1$, $V_G > V_c$, $V_c > V_L$, $C_{VG} > 0$, $C_{VL} > 0$ in the integrals considered. Therefore, removing this positive term from the right-hand side of Eq. 24.55 leaves the inequality

$$\left(\frac{1}{V_G - V_c} + \frac{1}{V_c - V_L}\right) \int_{T_1}^{T_c} \frac{T_c - T}{T}(C_V)_{V=V_c}\, dT$$

$$> \frac{1}{V_G - V_c} \int_{V_c}^{V_G} (p_c - p)_{T_c}\, dV + \frac{1}{V_c - V_L} \int_{V_L}^{V_c} (p - p_c)_{T_c}\, dV. \quad (24.57)$$

If we assume that[8]

$$|V - V_c| = B(T_c - T)^\beta, \quad (24.58)$$

$$p - p_c = -D\,|V - V_c|^{\delta-1}\,(V - V_c), \quad (24.59)$$

$$(C_V)_{V=V_c} = A(T_c - T)^{-\alpha} \quad (24.60)$$

for $T < T_c$ near the critical temperature, we find from Eq. 24.57 that

$$\alpha + \beta(\delta + 1) \geq 2, \quad (24.61)$$

known as *Griffith's inequality*.

Using other thermodynamic arguments it is also possible to show that

$$\alpha + 2\beta + \gamma \geq 2, \quad (24.62)$$

$$\gamma(\delta + 1) \geq (2 - \alpha)(\delta - 1), \quad (24.63)$$

$$\gamma \geq \beta(\delta - 1). \quad (24.64)$$

Using the values of $\alpha, \beta, \gamma, \delta$ from experiment listed in Eqs. 24.33–24.37, we find

$$(24.61): \quad 1.999 \pm 0.09 \sim \alpha + \beta(\delta + 1) \geq 2$$

---

[8] Since we consider only the region very close to the critical point, Eqs. 24.58–24.60 are equivalent to Eqs. 24.33–24.37. For example, $(\rho_L - \rho_G) = v_L^{-1} - v_G^{-1} = (v_G - v_L)(v_G v_L)^{-1}$. Clearly, the variation is all in the difference $v_G - v_L$, as $v_G v_L = v_c^2$ to very high precision close to v_c. Different investigators use Eqs. 24.58–24.60 or 24.33–24.37 as convenient for the analysis in question.

$$(24.62): \quad 2.01 \pm 0.08 \sim \alpha + 2\beta + \gamma \geqslant 2$$

$$(24.63): \quad 6.42 \pm 0.34 \sim \gamma(\delta+1) \geqslant (2-\alpha)(\delta-1) \sim 6.37 \pm 0.32$$

$$(24.64): \quad 1.2 \ \pm 0.04 \sim \gamma \geqslant \beta(\delta-1) \sim 1.19 \pm 0.04$$

so that it appears that the equality sign is valid. This is, indeed, a remarkable result.

Motivated by the observation just described, we examine what properties of the equation of state are sufficient to make Eqs. 24.61–24.64 hold with the equality sign. Let $M(T)$ be the chemical potential along the critical isochore $\rho = \rho_c$. At the critical point the chemical potential satisfies the conditions[9]

$$\left(\frac{\partial \mu}{\partial \rho}\right)_{T=T_c} = 0, \tag{24.65}$$

$$\left(\frac{\partial^2 \mu}{\partial \rho^2}\right)_{T=T_c} = 0. \tag{24.66}$$

Let the difference in value between $\mu(\rho, T)$ and its value on the critical isochore, $M(T)$, be expressed as a Taylor series expansion in ρ. By virtue of Eqs. 24.65 and 24.66, the first two terms in this series vanish. If $T = \tau(\rho)$ is the equation of the vapor–liquid coexistence curve in the (T, ρ) plane, the first nonvanishing term in the series is

$$\mu(\rho, T) - M(T) = (\rho - \rho_c)[T - \tau(\rho)]\Phi, \tag{24.67}$$

because $(\partial^2 \mu / \partial \rho \, \partial T) \neq 0$. The van der Waals equation, and the others described in Chapter 21 (Table 21.1), require that Φ be a constant at the critical point. None of these predicts the correct values of α, β, γ, and δ. That Φ is a constant at the critical point is not, however, a necessary condition on the equation of state. In view of the failure of the classical equations of state in the description of the critical region, B. Widom suggested that the expansion 24.67 still be used to describe that region, but with Φ a function of density and temperature. That function is not known, which is another way of saying that the equation of state in the critical region is not known. But *if Φ is assumed to be a homogeneous function of its variables*, the equation of state can be shown to satisfy Eqs. 24.61–24.64 with the equality signs. This is such a remarkable prediction, and it agrees so well with otherwise puzzling observations, that we give one example of the form of analysis used.

Let

$$z = T - T_c, \tag{24.68}$$

$$y = T_c - \tau(\rho), \tag{24.69}$$

$$\Phi = \Phi(z, y), \tag{24.70}$$

so that Eq. 24.67 becomes

$$\mu(\rho, T) - M(T) = (\rho - \rho_c)(z + y)\Phi(z, y). \tag{24.71}$$

If $\Phi(z, y)$ is a homogeneous function of degree $\gamma - 1$, then

$$\Phi(z, y) = y^{\gamma-1}\Phi\left(\frac{z}{y}, 1\right) \tag{24.72}$$

---

[9] Starting from

$$d\mu = -s \, dT + v \, dp$$
$$= -s \, dT + \rho^{-1} \, dp,$$

we easily find

$$\left(\frac{\partial \mu}{\partial \rho}\right)_T = \frac{1}{\rho}\left(\frac{\partial p}{\partial \rho}\right)_T = -v^3\left(\frac{\partial p}{\partial v}\right)_T,$$

and by definition $(\partial p / \partial v)_{T=T_c} = 0$. A similar argument establishes the validity of Eq. 24.66 at $T = T_c$.

$$= z^{\gamma-1}\Phi\left(1, \frac{y}{z}\right) \qquad \text{if } z > 0 \tag{24.73}$$

$$= (-z)^{\gamma-1}\Phi\left(-1, -\frac{y}{z}\right) \qquad \text{if } z < 0. \tag{24.74}$$

Equation 24.71 determines the chemical potential in the one-phase region, $T > T_c$. From the assumed form

$$T_c - \tau(\rho) = b\,|\rho - \rho_c|^{1/\beta} \tag{24.75}$$

(see Eqs. 24.33 and 24.71), it follows that:

1. On the critical isochore, where $y = 0$ and $z > 0$,

$$\left(\frac{\partial\mu}{\partial\rho}\right)_T = \Phi(1, 0)z^{\gamma}. \tag{24.76}$$

2. On the critical isotherm, where $z = 0$,

$$\left(\frac{\partial\mu}{\partial\rho}\right)_T = \left(1 + \frac{\gamma}{\beta}\right)\Phi(0, 1)y^{\gamma}. \tag{24.77}$$

3. On the coexistence curve, where $z + y = 0$,

$$\left(\frac{\partial\mu}{\partial\rho}\right)_T = \left(\frac{1}{\beta}\right)\Phi(-1, 1)y^{\gamma}$$

$$= \left(\frac{1}{\beta}\right)\Phi(-1, 1)(-z)^{\gamma}. \tag{24.78}$$

Equations 24.76 and 24.78 show that the compressibility becomes infinite proportional to $|T - T_c|^{-\gamma}$ as the critical point is approached along either the critical isochore or the coexistence curve. The degree-of-homogeneity index γ can, therefore, be identified with the critical index γ introduced in Eq. 24.37. The same exponent γ appears in Eqs. 24.76 and 24.78 because of the assumed homogeneity of the function $\Phi(z, y)$.

Using Eq. 24.72 in Eq. 24.71 gives for $\mu - M(T)$ along the critical isotherm [where $z = 0$ and $M(T_c) \equiv \mu(\rho_c, T_c)$],

$$\mu - \mu(\rho_c, T_c) = \Phi(0, 1)(\rho - \rho_c)y^{\gamma}, \tag{24.79}$$

which is just the integrated form of Eq. 24.77. But the coexistence curve itself is of algebraic degree $1/\beta$ (see Eq. 24.75), so that the algebraic degree of the critical isotherm must be[10]

$$\delta = 1 + \frac{\gamma}{\beta}, \tag{24.80}$$

or

$$\gamma = \beta(\delta - 1), \tag{24.81}$$

which is just Eq. 24.64 *with an equality sign*. By similar methods of analysis Widom finds the equality sign to hold in the rest of the expressions previously displayed.

The assumption that $\Phi(z, y)$ is a homogeneous function of its variables provides a powerful tool for thermodynamic analysis. This assumption also implies that $\Phi(z, y)$ is a universal function, that is, the same for all substances. A

[10] The right-hand side of Eq. 24.79 is proportional to $(\rho - \rho_c)^{1+1/\beta}$. Integration of the Gibbs–Duhem equation and use of Eq. 24.59 shows that $\mu - \mu_c \propto (\rho - \rho_c)^{\delta}$, from which Eq. 24.80 follows.

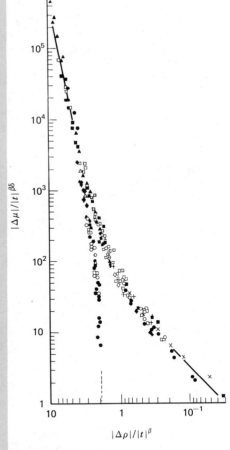

FIGURE 24.19

A scaled plot of $|\Delta\mu|/|t|^{\beta\delta}$ versus $|\Delta\rho|/|t|^{\beta}$ in the critical regions of N_2O, $CCIF_3$, CO_2, SF_6, and Xe. The quantity t is defined as $\Delta T/T_c$. From M. S. Green, M. Vincenti-Missioni, and J. M. H. Levett-Sengers, *Phys. Rev. Lett.* **18**, 1113 (1967).

test of this implication is shown in Fig. 24.19.[11] It is seen that the experimental data available are in agreement with Widom's hypotheses concerning $\Phi(z, y)$.

We have already remarked that the existence of a phase transition signals the instability of one phase relative to some other phase. In this section we examine how the characteristic features of a phase transition can be related to molecular properties. The molecular theory of phase transitions is complex and only partially developed, so we shall be content to examine the key ideas and some qualitative indications of their implications.

There are three paths that can be followed in constructing a molecular theory of phase transitions. The simplest of these uses directly the thermodynamic condition defining the equilibrium state under a given set of constraints, for example, that the Gibbs free energy must be a minimum for fixed temperature and pressure. Suppose that one molecular model is proposed for phase I, say, the crystal phase, and a different model for phase II, say, the liquid phase. It is, in principle, possible to calculate the Gibbs free energies of phases I and II assuming that they are isolated from one another and also that they are stable over the range of temperature and pressure of interest. The techniques employed in these calculations are just those sketched in Chapters 22 and 23.

Suppose that the pressure is assumed constant and the Gibbs free energies of phases I and II are computed as functions of the temperature. A schematic plot of the results is displayed in Fig. 24.20. As shown in Fig. 24.20, the crystal phase (I) has lower chemical potential for $T < T_m$, whereas the liquid phase (II) has lower chemical potential for $T > T_m$. The temperature T_m can then be identified as the melting point predicted by the statistical molecular model considered.

Models of the type just described have frequently been used to describe phase transitions. However, they are deficient because they do not describe the phase transition itself at all. What the model does describe are the sets of real and hypothetical stable and metastable states of the separate phases. A purely thermodynamic criterion is then used to decide which of the phases is the more stable. No information is obtained about the special character of the distribution of molecules near the transition point, about the presence or absence of special kinds of molecular ordering, or about the "origin" of the phase transition.

24.4
The Statistical Molecular Description of Phase Transitions

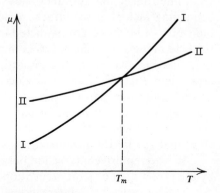

FIGURE 24.20
Schematic representation of the location of a transition point by comparing the chemical potentials for two models, one for each of the phases. When phases I and II are the solid and liquid, respectively, T_m is the melting point.

---

[11] The test is carried out by rewriting Eq. 24.67 as

$$\Delta\mu(T) = \mu(\rho, T) - M(T)$$

$$= \Delta\rho(\Delta T + b\,|\Delta\rho|^{1/\beta})\,|\Delta\rho|^{(\gamma-1)/\beta}\,\Phi\!\left(\frac{\Delta T}{|\Delta\rho|^{1/\beta}}, 1\right)$$

$$= \Delta\rho\,|\Delta\rho|^{\gamma/\beta}\,\Psi\!\left(\frac{\Delta T}{|\Delta\rho|^{1/\beta}}\right),$$

using $z + y = T - \tau(\rho) = \Delta T + b\,|\Delta\rho|^{1/\beta}$ from Eqs. 24.68 and 24.69, and $\Phi(\Delta T, b\,|\Delta\rho|^{1/\beta}) = |\Delta\rho|^{(z-1)/\beta}\,\Phi(\Delta T/b\,|\Delta\rho|^{1/\beta}, 1)$ from Eq. 24.72. In this form $\Delta\mu(T)$ is expressed as a function Ψ of the single variable $\Delta T/|\Delta\rho|^{1/\beta}$. Using $\delta = 1 + \gamma/\beta$ and dividing leads to

$$\frac{|\Delta\mu(T)|}{|\Delta T|^{\beta\delta}} = \left(\frac{|\Delta\rho|^{1/\beta}}{|\Delta T|}\right)\Psi\!\left(\frac{\Delta T}{|\Delta\rho|^{1/\beta}}\right),$$

which predicts that $|\Delta\mu(T)|/|\Delta T|^{\beta\delta}$ is a universal function of $\Delta T/|\Delta\rho|^{1/\beta}$. The function has two branches, for $\Delta T > 0$ and $\Delta T < 0$, which become asymptotically identical as $\Delta T \to 0$.

An alternative approach is based on the postulation of very simple models that retain only the skeleton of the real properties of a system. These models are specifically designed so that an exact mathematical analysis can be carried to completion. It is then hoped that the general features of the phase transition deduced will be independent of the simplifications of the basic model. Actually, the mathematical description of the statistical mechanics of phase transitions is so difficult that there are very few models for which exact solutions have been obtained: moreover, these models tend to be bizarre. Nevertheless, the results obtained are of great importance, because they show how delicate is the balance between competing factors in the critical region, and also because they are often indicative of behavior not predicted by approximate calculations for more realistic models. Consider, for example, the model known as the *lattice gas*. Imagine that there are $\mathcal{N}$ lattice sites in some regular crystalline array, and that N of these are occupied by particles and $\mathcal{N} - N$ are vacant. Only one particle can be at a given lattice site. Finally, if two particles occupy near-neighbor lattice sites, there is an attractive interaction of magnitude w, whereas particles farther apart do not interact. It is clear that the lattice gas model has some of the features of real matter:

1. The repulsive interaction between a pair of molecules is mimicked by the prohibition of joint occupation of one lattice site by two particles.

2. The attractive interaction between a pair of molecules is mimicked by the attraction between near neighbors, with the range of the potential such that there is zero interaction between pairs separated by more than a near-neighbor distance.

3. Changes in density are represented by changes in the number of vacant lattice sites, $\mathcal{N} - N$. When the density of particles is high, $\mathcal{N} - N$ is small relative to $\mathcal{N}$, and the lattice gas mimics the liquid. When the density of particles is low, $\mathcal{N} - N$ is almost as large as $\mathcal{N}$, and the lattice gas mimics the dilute gas. The major difference between the lattice gas and real liquids or gases, and the principal source of simplification in the mathematical analysis, is the replacement of the continuous space of real systems with the discrete space of a lattice. That is, in real space molecules may be any distance apart, whereas in the lattice gas their separation is determined by the lattice spacing and lattice structure. Clearly, although the lattice gas has some skeletal features in common with real fluids, it also has some bizarre characteristics not found in any real system.

From our point of view the important characteristic of the lattice gas is that, if the interaction between near-neighbor particles is attractive, there is a temperature below which the system separates into two phases, one with large density and the other with small density. The model thereby mimics a two-phase liquid–gas system with a critical temperature. If the near-neighbor interaction is repulsive, particles will tend to occupy alternate lattice sites at low temperature, but without splitting into two phases. In this case there can still be a critical temperature.

With great insight and mathematical dexterity, L. Onsager has evaluated exactly the partition function for a *two*-dimensional infinite square-lattice gas. He obtained the following results:

1. For a half-filled lattice, with an attractive interaction between near neighbors, the system splits into two phases below a critical temperature T_c.

2. The internal energy is continuous at the critical point, but has an infinite slope there. The heat capacity near the critical point diverges at a rate proportional to $\ln |(T - T_c)/T_c|$.

3. Let ρ_G be the density of the gas phase and ρ_L the density of the liquid phase along the coexistence line between liquid and gas. Then for the two-dimensional lattice gas model one has

$$(\rho_L - \rho_G) \propto (T - T_c)^{1/8} \tag{24.82}$$

near the critical point. Also,

$$(p - p_c) \propto (\rho_L - \rho_G)^8 \tag{24.83}$$

and $(\partial p/\partial T)_{V = V_c}$ is continuous at the critical point.

For the moment we merely call the reader's attention to the form of the approach to the critical state. A discussion of the nature of the results is deferred until we have examined the third approach to the study of phase transitions.

This approach also makes use of simple models of the substances of interest. In this case, however, the models are designed to incorporate as many of the real-system properties as is possible without overly complicating the necessary mathematical analysis. There is a fine line dividing models that have qualitative features in common with those of real systems and for which the statistical mechanics *cannot*, at present, be done exactly, from models that are so simple that the corresponding statistical mechanics *can* be done exactly. Since much of our qualitative understanding of phase transitions (but not of critical phenomena) comes from studies of models of the former class, we consider a few examples.

We examine first crystal–crystal transitions. It appears to be the case that for most transitions in which the crystal possesses translational ordering in both phases the structural change is associated with an instability in some lattice vibration. To be more specific, the frequency of the mode corresponding to the displacement that generates the structural change decreases substantially as the transition temperature is approached from below. This "soft mode" generates a mechanical instability in the sense that displacement of the original atomic configuration toward the new atomic configuration is resisted less and less as the temperature approaches the transition point. Our description of the transition must include this behavior.

Clearly, what is needed is an expression for the thermodynamic properties of the crystal as a function of the displacement that generates the transition. Landau suggested that we find this as follows. Let $\rho(\mathbf{r})$ represent the atomic distribution in the crystal phase with lower symmetry and $\rho_0(\mathbf{r})$ the atomic distribution in the crystal phase with higher symmetry. Then, writing

$$\rho(\mathbf{r}) = \rho_0(\mathbf{r}) + \Delta\rho(\mathbf{r}), \tag{24.84}$$

we have

$$\Delta\rho(\mathbf{r}) = \sum_i c_i \xi_i, \tag{24.85}$$

where the ξ_i are normal coordinates of the high-symmetry lattice. Let η be a normalized linear combination of the coefficients c_i. Then η describes the amplitude of the mean displacements of the atoms from their high-symmetry positions. We assume that the Gibbs free energy per mole of the crystal can be represented as a Taylor series expansion in powers of η:

$$G = G_0 + \alpha\eta + A\eta^2 + B\eta^3 + C\eta^4 + \cdots, \tag{24.86}$$

where $A, B, C, \ldots$ are functions of p and T. Since for fixed p and T the

Gibbs free energy is a minimum, we impose the conditions

$$\left(\frac{\partial G}{\partial \eta}\right)_{T,p} = 0, \tag{24.87}$$

$$\left(\frac{\partial^2 G}{\partial \eta^2}\right)_{T,p} > 0, \tag{24.88}$$

and immediately find

$$\alpha = 0, \tag{24.89}$$

$$A > 0 \text{ for } T > T_c, \ A < 0 \text{ for } T < T_c, \tag{24.90}$$

so that there is no term linear in η in the expansion of G about G_0. Evaluation of B and C requires another consideration. Suppose that the transition occurs at $T = T_c$. Since η represents the amplitude of the *mean* displacements of the atoms from their high-symmetry lattice positions, we must have $\eta = 0$ in the high-symmetry phase, say, for $T > T_c$. Then, starting at some $T > T_c$, $\eta = 0$ and remains zero until $T = T_c$, and then continuously increases as T decreases in the region $T < T_c$. At the transition point, $G(p, T, \eta)$ must be a minimum with respect to variation of η, which implies that (since $A(T_c) = 0$, Eq 24.90),

$$B = 0, \tag{24.91}$$
$$C > 0.$$

We consider only the case that $B \equiv 0$ by virtue of the symmetry properties of the crystal. Then from

$$\left(\frac{\partial G}{\partial \eta}\right)_{T,p} = 0 = 2A\eta + 4C\eta^3 \tag{24.92}$$

we have

$$\eta^2 = -A/2C \qquad (T < T_c) \tag{24.93}$$

and, of course, $\eta = 0$ if both $A > 0$ and $C > 0$. This is as far as a general macroscopic analysis of the crystal–crystal transition can carry us. To proceed further we must either introduce some hypothesis concerning the pressure and temperature dependences of the coefficients A and C, or calculate their properties from a statistical molecular theory. We consider just the major ideas from the latter.

As we remarked in Chapter 22, a perfectly harmonic lattice has zero thermal expansion or, put another way, on average the atoms execute thermally excited motion centered on the same lattice sites for all T. Thermal expansion, that is, the change of lattice spacing with change in temperature, arises from anharmonicity of the potential surface. We also saw in Chapter 22 that a simple picture of the effect of thermal expansion could be constructed on the basis that at each T the motion of the atoms was harmonic, but that the centers about which oscillation took place changed as T changed. Now, the existence of a crystal–crystal phase transition, which can be characterized as involving a change in positions of atoms, necessarily implies that the potential energy surface is anharmonic with respect to displacements about either set of the crystal lattice sites. Although complete solution of the dynamics of coupled anharmonic oscillators is not possible at present, we can construct a physically plausible picture of the main features to be expected from such an analysis. Suppose that the potential energy is expanded as a power series in the displacement of an atom from a lattice site, but retaining terms that are cubic, quartic, etc., in the displacement. Suppose further that we seek a way of representing the averaged effects of the anharmonic terms in the potential energy, much in the spirit of the analysis of Section 23.3. One such approximate representation is based on the notion that a principal effect of the anharmonicity is to weaken the restoring force returning a displaced atom to its lattice site. To make this clear, consider a one-dimensional example. Suppose that the potential energy of a one-dimensional anharmonic oscillator is $V_2 x^2 - V_4 x^4$. We represent the average effect of the $V_4 x^4$ term by making the approximation $x^4 \to \langle x^2 \rangle x^2$, where the thermally averaged mean-square displacement is evaluated by, say, perturbation theory. In first-order perturbation theory $\langle x^2 \rangle$ is determined by weighting x^2 with the distribution of x for a harmonic oscillator; the result is $\langle x^2 \rangle = k_B T / V_2$. Substituting this value into the potential energy expression for the anharmonic oscillator gives

$$V_2 x^2 - V_4 x^4 \approx (V_2 - V_4 \langle x^2 \rangle) x^2$$

$$\approx \left(V_2 - \frac{k_B T V_4}{V_2} \right) x^2, \qquad (24.94)$$

from which one derives an effective temperature-dependent frequency

$$\omega \propto \left(1 - \frac{k_B T V_4}{V_2^2} \right)^{1/2} V_2^{1/2} \propto \omega_0 (1 - aT)^{1/2}, \qquad (24.95)$$

which decreases as T increases. This soft mode will, eventually, generate the instability that leads to the phase transition.

In a real crystal the atomic dynamics is much more complicated than the one-dimensional model mentioned. Nevertheless, when the anharmonic contributions to the potential energy are partially averaged, the main result is the same, namely, the prediction of the existence of a soft mode. The precise form of the temperature dependence is related to the nature of correlations near the transition point. If the interactions between particles are very long-range, then $\omega \propto (T_c - T)^{1/2}$, but if they are short-range, then $\omega \propto (T_c - T)^{\beta}$ with $\beta \approx \frac{1}{3}$. This difference in behavior is similar to that observed for various theoretical models of the gas–liquid critical point, and we shall see later that it has the same origin.

Does the description just given agree with experiment? We have already mentioned the existence of second-order transitions in the perovskite structures. Figure 24.21a shows the soft mode in $SrTiO_3$, Fig. 24.21b the continuous decrease of the vibrational frequency as T_c is approached, and Fig. 24.21c the corresponding continuous displacement of the oxygen octahedron. In each case the line represents the predicted behavior. Clearly, there is generally good agreement between theory and observation.

Actually, the behavior observed and to be expected is even richer than so far discussed. The theoretical description sketched is based on the idea that anharmonicity leads to the softening of one or more vibrational modes and eventually to an instability that generates a phase transition. In this analysis the atomic motion is progressively influenced by the anharmonicity as T changes. There is theoretical and computer experiment evidence that when the anharmonicity is very large, as when there is a double-well potential for an atom, there are two classes of elementary excitations (see Fig. 24.22). One, already discussed, is quasi-harmonic motion influenced, on average, by anharmonic coupling. The energy required to create one of these excitations is spread throughout the lattice. The other corresponds to a localized large-amplitude displacement, and the energy to create it is relatively localized in the lattice. This

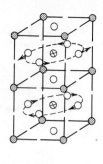

(a)

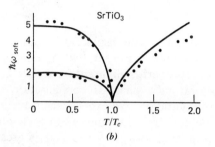

(b)

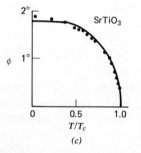

(c)

FIGURE 24.21
(a) Soft-mode motion in $SrTiO_3$ transition.

(b) Soft-mode frequencies as a function of T/T_c both above and below T_c. Note that there are two soft modes in the low-temperature phase.

(c) Angle of rotation, ϕ, of oxygen octahedron in $SrTiO_3$ transition. From J. F. Scott, *Rev. Mod. Phys.* **46**, 83 (1974).

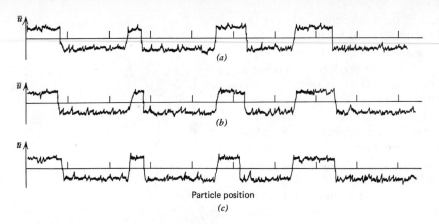

FIGURE 24.22
Particle displacements along a chain of local double-well potentials. Nearest-neighbor coupling only,

Hamiltonian H =

$$\sum_i \frac{p_i^2}{2m} + \sum_i \left[\frac{A}{2} u_i^2 + \frac{B}{4} u_i^4 + \frac{1}{2} (u_{i+1} - u_i)^2 \right].$$

The horizontal axis measures position along the chain (1000 particles long), and the vertical axis the displacement in units of $(|A|/B)^{1/2}$, that is, $\bar{u} = u/u_0$. Curves (a), (b), and (c) are for times $\bar{t} \equiv (|A|/m)^{1/2}$ of 0, 100, and 200 units. The temperature is $\bar{T} \equiv k_B T B/A^2 = 0.117$. The mean energy per particle is less than the barrier height. From T. R. Koehler, A. R. Bishop, J. A. Krumhansl, and J. R. Schrieffer, *Solid State Comm.* **17**, 1515 (1975).

large-amplitude displacement can move through the lattice, and has been taken to be representative of "domain walls" that bound clusters of atoms with properties different from their surroundings, for example, a group of atoms in a phase-II configuration surrounded by atoms in a phase-I configuration. We expect such traveling cluster waves to exist in measurable concentration only when their energy of formation is not very large compared to $k_B T$. When $k_B T$ is greater than the energy of domain wall formation, we expect thermal fluctuations to destroy them. Both of these expectations are supported by the computer simulations shown in Figs. 24.22 and 24.23. The traveling cluster waves described represent an entirely different class of excitations than do the ordinary vibrational excitations—they arise from the strong nonlinearity (strong anharmonicity) of the interaction. Their existence does not modify those averaged structural or dynamical aspects of the crystal–crystal transition discussed, but does explain other phenomena and, presumably, influences the mechanism of the crystal–crystal transition.

The preceding discussion has concerned the dynamics of the crystal–crystal transition and relied on information about the two structures involved, presumably obtained from experimental data. How can we understand, in terms of the molecular interactions, the structures of different crystalline phases of the same substance? The general theory of crystal–crystal structural transformations has yet to be developed, but a good physical picture of the relative influence of energetic and entropic

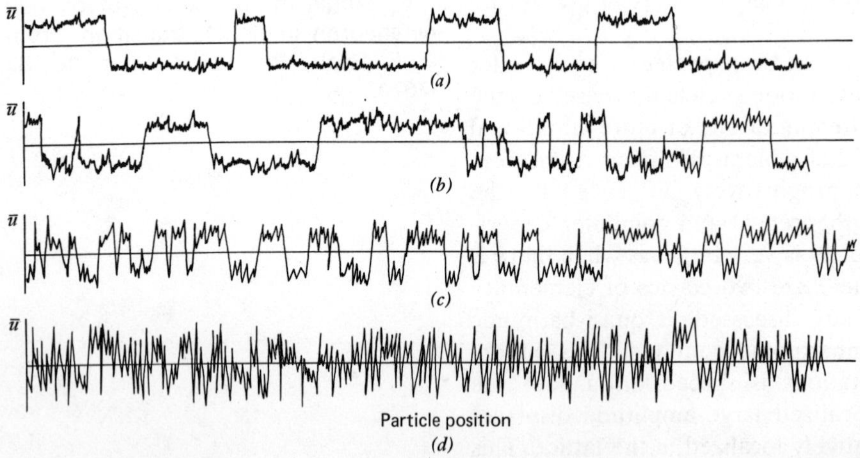

FIGURE 24.23
Particle displacements along a chain of double-well potentials. The model is described in the caption to Fig. 24.22. (a) $\bar{T} = 0.117$. (b) $\bar{T} = 0.19$. (c) $\bar{T} = 0.26$. (d) $\bar{T} = 1.68$. From T. R. Koehler, A. R. Bishop, J. A. Krumhansl, and J. R. Schrieffer, *Solid State Comm.* **17**, 1515 (1975).

effects, and their dependence on the molecular interactions, can be obtained from a simple model. Consider, for definiteness, a crystal of methane. The interaction between CH_4 molecules is angle-dependent, and we anticipate that there can exist different crystal phases with different orientational orderings of the CH_4's. Now, the interaction between two CH_4's depends on the angles θ_1, ϕ_1, ψ_1 and θ_2, ϕ_2, ψ_2 that describe their orientations. In the lowest order of approximation we suppose that each CH_4 feels only the average field generated by all surrounding molecules. If $f_i(\omega_i)\,d\omega_i$ is the probability for finding a molecule with orientation between θ_i, ϕ_i, ψ_i and $\theta_i + d\theta_i, \phi_i + d\phi_i, \psi_i + d\psi_i$ (which we denote as between ω_i and $\omega_i + d\omega_i$, where ω_i stands for the set of angles θ_i, ϕ_i, ψ_i), then the average interaction of molecule i with orientation ω_i and all its neighbors is

$$V_i(\omega_i) = \sum_j \int d\omega_j \, u_{ij}(\omega_i, \omega_j) f_j(\omega_j), \qquad (24.96)$$

where $u_{ij}(\omega_i, \omega_j)$ is the direct pair potential. Note that $V_i(\omega_i)$ depends on the distribution of orientations of the neighbors, $f_j(\omega_j)$. To be consistent we must require that each molecule feel the same average field, and this in turn requires that each distribution function $f_i(\omega_i)$ be the same and be determined by

$$\begin{aligned}
f_i(\omega_i) &= \frac{\exp[-V_i(\omega_i)/k_B T]}{\displaystyle\int d\omega_i \, \exp[-V_i(\omega_i)/k_B T]} \\[2mm]
&= \frac{\exp\left[-(1/k_B T)\displaystyle\int d\omega_j \, u_{ij}(\omega_i, \omega_j) f_j(\omega_j)\right]}{\displaystyle\int d\omega_i \, \exp\left[-(1/k_B T)\displaystyle\int d\omega_j \, u_{ij}(\omega_i, \omega_j) f_j(\omega_j)\right]}. \qquad (24.97)
\end{aligned}$$

That is, $f_i(\omega_i)$ is just the ordinary canonical distribution of orientation for a molecule subject to the effective field $V_i(\omega_i)$. Equation 24.97 is a *nonlinear* equation for $f_i(\omega_i)$.

Thus far we have treated the orientation of a CH_4 molecule by the methods of classical mechanics, and Eq. 24.97 is the proper expression for the distribution function using classical statistical mechanics. In the case of CH_4 one must use quantum mechanics to describe the orientations, since the moment of inertia is small and the temperatures of interest quite low. The formulation of the problem is the same, and the resulting quantum statistical equation for the distribution function also the same except for summation over energy levels in place of integration over angles. For our purposes it is sufficient to use Eq. 24.97, since we seek only a qualitative picture of the possible phase transitions.

We now note again that Eq. 24.97 is a nonlinear equation for $f_i(\omega_i)$. It is a property of nonlinear equations of the type shown that the nature of the solution can change suddenly when the temperature passes a characteristic value. Just this happens for the potential between CH_4 molecules. The method used to solve Eq. 24.97 is complex because of the need to take into account quantization of rotation and the nature of the splittings in the energy levels induced by the field acting on a molecule. In Fig. 24.24 is plotted the heat capacity of CH_4 as a function of temperature. The thermal anomaly at the transition temperature is relatively well described. The high-temperature phase is predicted to be orientationally disordered, and in the low-temperature phase there is partial ordering as shown in Fig. 24.25. These predictions concerning order have been confirmed by neutron diffraction experiments.

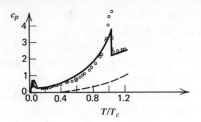

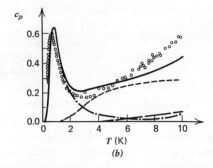

FIGURE 24.24

(a) Heat capacity of CH_4 versus T/T_c. The solid curve is the theoretical prediction derived from the quantum mechanical version of Eq. 24.97 with spin species equilibration.

(b) Heat capacity of CH_4 versus T for $T < 10$ K. The solid curve is the theoretical prediction derived from the quantum mechanical version of Eq. 24.97; the dot-dash and dashed lines are the contributions from molecules in phase-I and phase-II environments, respectively. From T. Yamamoto, Y. Kataoka, and K. Okada, *J. Chem. Phys.* **66**, 2701 (1977).

In fact there are three phases of solid CH_4 that appear in order of decreasing temperature at low pressure, and a fourth that is stable only at high pressure and low temperature. Although the theory described accurately predicts the properties of phase I (the disordered phase) and phase II (the partially ordered phase shown in Fig. 24.25), the predicted structure of phase III is incorrect. The theory described is a rigid lattice model (i.e., vibrational motion is not considered), and it neglects any correlation in the rotational motions of molecules on neighboring lattice sites [only the average over all neighbor orientations is used in defining the effective potential $V_i(\omega_i)$]. These omissions, and lack of complete knowledge of the intermolecular potential, are probably responsible for the erroneous predictions concerning the structure of phase III.

We now examine the liquid–solid transition. This is a first-order transition (ΔV_{fus}, ΔS_{fus}, ΔH_{fus} nonzero) and is less well understood than are second-order transitions. One model of the melting process is based on the analog of the soft-mode instability discussed above. M. Born assumed that when the solid can no longer withstand a force that tends to make it flow, a transition to the liquid state occurs. This condition does not account for the thermodynamic condition of equality of chemical potentials of the liquid and solid phases and leads to an overestimate of the stability of the solid. A different model of the freezing process is based on the assumption that at some temperature the liquid becomes unstable to a density fluctuation that has a wavelength equal to the lattice separation characteristic of the solid. This condition also does not account for the thermodynamic condition of equality of chemical potentials of the liquid and solid phases and leads to an overestimate of the stability of the liquid. Although neither of these "one-sided stability" theories is correct, each has an element that represents an important physical observation. To see this we now consider the results of some computer experiments.

J. G. Kirkwood and E. Monroe predicted[12] that a fluid of hard spheres becomes unstable with respect to a crystal of hard spheres when the density exceeds a critical value. This prediction was verified many years later by computer simulation studies. The fluid → crystal transition for hard spheres is of the first order; it occurs at a reduced density $n\sigma^3 = 0.943$, where σ is the hard-sphere diameter. The reduced density of the solid is 1.041, so there is a 10.3% change of volume when the fluid freezes. Since there is no direct attractive interaction between hard spheres, the fluid → solid transition must be a consequence of the advantages of the geometry of the crystal. It is easy to imagine (but has not yet been proven) that a disordered array of hard spheres will eventually jam together in a way that prevents any motion. This configuration would have very small entropy. Therefore, if a regular array of hard spheres allowed small amplitude motion about the lattice positions, that configuration would have larger entropy than the jammed disordered array, hence be thermodynamically stable relative to it.

To what extent does the crystallization of a fluid of hard spheres mimic crystallization in a real liquid? To answer this question we display in Tables 24.2 and 24.3 the melting parameters for particles interacting with an inverse power repulsion of the form R^{-n}, and with a Lennard–Jones interaction, respectively. We see that as the exponent n decreases, so does the relative change of volume on freezing, whereas the entropy change on freezing is very weakly dependent on the nature of the repulsion between particles. When attractive interactions are added to the

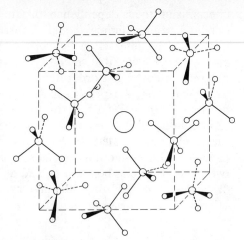

FIGURE 24.25
Phase relations between the orientations of CH_4 molecules in the partially ordered phase II. Dashed lines represent bonds in back of the plane of the C atom, doubled wedge lines bonds in front of the C atom, and solid lines bonds in the plane of the C atom parallel to the plane of this page. From T. Yamamoto, Y. Kataoka, and K. Okada, *J. Chem. Phys.* **66**, 2701 (1977).

---

[12] The basis for this prediction is the mean field equation discussed a little further on in this section.

TABLE 24.2
MELTING CHARACTERISTICS OF INVERSE POWER POTENTIALS[a]

n	$\left[\left(\dfrac{\epsilon}{k_BT}\right)^{3/n}n\sigma^3\right]_L$	$\left[\left(\dfrac{\epsilon}{k_BT}\right)^{3/n}n\sigma^3\right]_S$	$\dfrac{\Delta V}{V_S}$ (%)	$\left(\dfrac{\epsilon}{k_BT}\right)^{3/n}\dfrac{p\sigma^3}{k_BT}$	$\dfrac{\Delta S}{Nk_B}$	$\dfrac{\beta\,\Delta U}{N}$	θ
∞	0.943	1.041	10.3	11.7	1.16	0	0.13
12	1.141	1.193	3.8	22.6	0.90	0.18	0.15
9	1.333	1.373	3.0	31.1	0.84	0.21	0.16
6	2.17	2.21	1.3	86.3	0.75	0.25	0.17
4	4.53	4.57	0.5	602	0.80	0.35	0.18

[a] From J. P. Hansen and I. R. McDonald, *Theory of Simple Liquids* (Academic Press, New York, 1976).

TABLE 24.3
DATA ON THE FLUID–SOLID TRANSITION IN THE LENNARD–JONES (LJ) SYSTEM AND IN ARGON[a]

	$\dfrac{k_BT}{\epsilon}$	$\dfrac{p\sigma^3}{\epsilon}$	$(n\sigma^3)_L$	$(n\sigma^3)_S$	$\dfrac{\Delta V}{V_S}$ (%)	$\dfrac{\Delta S}{Nk_B}$	θ
LJ	0.75	0.67	0.875	0.973	13.1	1.75	0.145
argon	0.75	0.59	0.856	0.967	12.9	1.64	
LJ	1.15	5.68	0.936	1.024	9.3	1.44	0.139
argon	1.15	6.09	0.947	1.028	8.4	1.25	
LJ	1.35	9.00	0.964	1.053	9.1	1.39	0.137
argon	1.35	9.27	0.982	1.056	7.6	1.20	
LJ	2.74	32.2	1.113	1.179	5.9	0.98	0.149
argon	2.74	37.4	1.167	1.230	5.3	1.04	

[a] From J. P. Hansen and I. R. McDonald, *Theory of Simple Liquids* (Academic Press, New York, 1976).

potential, $\Delta V/V$ and $\Delta S/Nk_B$ both decrease as the temperature is raised, as is to be expected from the increase in densities of both fluid and solid phases at the transition as the temperature increases. Note that although the Lennard–Jones potential is not a very accurate representation of the Ar–Ar interaction, the computer-simulated melting parameters are very close to those of Ar.

We now turn to a test of the idea that a mechanical instability of the solid triggers the transition to the liquid phase. An early version of this idea was proposed by F. Lindemann, who speculated that a crystal melts when the mean-square amplitude of vibration of the atoms exceeds some critical fraction of the atomic separation. We display in Table 24.2 the Lindemann ratio θ, equal to the root-mean-square displacement of an atom from its lattice position, $\mathbf{R}_i$, divided by the nearest-neighbor distance a:

$$\theta = \frac{\langle u^2\rangle^{1/2}}{a} = \frac{1}{a}\left(\frac{1}{N}\sum_{i=1}^{N}\langle|\mathbf{r}_i - \mathbf{R}_i|^2\rangle\right)^{1/2}. \tag{24.98}$$

We find that θ is practically independent of the nature of the repulsive force between the atoms, and is scarcely affected by adding an attractive force (see Table 24.3).

Turning to the criterion for instability of a liquid, computer simulations show that the amplitude of the first peak of the function

$$S(\mathbf{k}) = 1 + n \int d\mathbf{R} e^{-i\mathbf{k} \cdot \mathbf{R}} g_2(R) \qquad (24.99)$$

for a variety of model liquids with different interactions between the atoms is practically constant at 2.85 ± 0.1 everywhere along the line of equilibrium between liquid and crystal.

The outstanding characteristic of a second-order phase transition is the appearance at the same T and p of both a mechanical instability (e.g., $\omega_{\text{soft}} \to 0$) and the thermodynamic condition $\mu^{\text{I}} = \mu^{\text{II}}$. It appears to be the case that in a first-order phase transition, such as melting, the condition $\mu^L = \mu^S$ is reached *before* any purely mechanical instability occurs. Why this happens is not at present known, but the fact that it does prevents us from interpreting the information on the behavior of $S(\mathbf{k})$ or θ along the melting line in any fundamental fashion.

The basic thermodynamic condition of equality of chemical potential along the melting line can be incorporated into a mean field theory for the distribution of atoms in space only approximately at present. The idea is like that used to describe the CH_4 transition. We first write the average energy of interaction of a molecule at $\mathbf{R}_1$ with all the others. This is

$$V(\mathbf{R}_1) = \int d\mathbf{R}_2 u_2(\mathbf{R}_1, \mathbf{R}_2) n(\mathbf{R}_2), \qquad (24.100)$$

where $n(\mathbf{R}_2)$ is the number density of molecules a distance $\mathbf{R}_2 - \mathbf{R}_1$ from the one fixed at $\mathbf{R}_1$. In terms of the pair-correlation function defined in Chapter 23, $n(\mathbf{R}_2) = n g_2(\mathbf{R}_1, \mathbf{R}_2)$. Since each molecule must feel the same average potential, we now demand that

$$n(\mathbf{R}_1) = \frac{N \exp[-\beta V(\mathbf{R}_1)]}{\int d\mathbf{R}_1 \exp[-\beta V(\mathbf{R}_1)]}$$

$$= \frac{N \exp\left[-(1/k_B T) \int d\mathbf{R}_2 u_2(\mathbf{R}_1, \mathbf{R}_2) n(\mathbf{R}_2)\right]}{\int d\mathbf{R}_1 \exp\left[-(1/k_B T) \int d\mathbf{R}_2 u_2(\mathbf{R}_1, \mathbf{R}_2) n(\mathbf{R}_2)\right]}, \qquad (24.101)$$

which is the same as Eq. 24.97. In a fluid we know that $n = N/V$ is constant everywhere, hence clearly not a function of $\mathbf{R}_1$. However, in a crystal the probability of finding a molecule in the volume element $d\mathbf{R}_1$ depends on whether or not that element contains a lattice point, hence n is a function of $\mathbf{R}_1$. The basic problem to be solved is finding solutions to Eq. 24.101 that have the triply periodic character of a crystal lattice. This was first done by Kirkwood and Monroe, but they used many approximations that make precise interpretation difficult. The two most important of these are (1) the assumption that at the melting line the pair-correlation functions of the solid and liquid are enough alike to be equated for purposes of reducing Eq. 24.101, and (2) the assumption of a particular lattice geometry for the crystal without verifying that it has lower chemical potential than other lattices. Nevertheless, the physical ideas are sound—because of the nonlinearity of Eq. 24.101, density variations with periodic character are unstable relative to $n = $ constant for high T, but become stable when T reaches T_{fus} and remain stable for $T < T_{\text{fus}}$. The onset of this change in $n(\mathbf{R}_1)$ is sudden because of the feedback inherent to the nonlinear form. That is, a small-amplitude fluctuation that is favored is exponentially amplified and dominates the behavior of the solution to the equation.

Finally, we examine briefly the theory of the first-order vapor–liquid transition. In this case there is an enormous difference between the structures of the two phases, and none of the considerations used in the analysis of crystal–crystal transitions, or even of melting, can be usefully applied. There is not any theoretical indication yet available of the nature of the microscopic motion that leads to the instability that generates the phase transition. From the point of view of evolution of a structural change, the mechanism of condensation in a fluid system appears to involve the following. In any distribution of the molecules in the system (e.g., distribution of occupied and unoccupied lattice sites in the lattice gas, or positions in space in a real gas), there will be present isolated molecules, pairs of molecules close together, groups of three molecules close together, and so forth. These clusters are in mutual equilibrium associating and disassociating, and even fairly large clusters resembling droplets of liquid have some small but nonzero probability of occurring. The energy and entropy of each cluster depend on both the number of molecules composing the cluster and the amount of "surface" the cluster has. The existence of a free energy associated with the surface tends to minimize the amount of surface, hence to contract the cluster. This tendency is opposed by the increase in the free energy of the bulk that accompanies compression. For example, the bulk contribution to the entropy of a cluster will be larger if the molecules have more internal volume in which to move. At low temperature the minimization of the surface free energy dominates the behavior of the system, and it is advantageous to combine clusters to form ever larger groupings, since in this way the surface-to-volume ratio of the system is diminished, and thereby the total free energy of the system minimized. Indeed, if conditions are sufficiently favorable, we can imagine the clusters growing to macroscopic size. A macroscopic droplet represents the liquid, and its presence indicates that condensation has occurred. In contrast, if the temperature is high enough, all the clusters dissociate, because the decrease in bulk free energy from the gain in translational entropy more than counterbalances the increase in "surface free energy." In this limit the system is completely dispersed as single molecules and no condensation can occur. The highest temperature at which a balance in the competing factors leads to some condensation, that is, the highest temperature at which some condensation can occur, is infinitesimally below the critical temperature. The critical temperature T_c is itself the limit above which, and at which, condensation cannot occur. The description of condensation just given can be made consistent with what is now known of the peculiar approach to the critical region, that is, the values of α, β, γ, and δ.

All approximate theories of the equation of state of a fluid that are known predict isotherms in the two-phase region that are similar to those of the van der Waals equation (they have loops), hence contain an inherent error. Nevertheless, if an equation of state is accurate in the one-phase regions where liquid or gas are stable, an accurate prediction of the coexistence curve can be made by the *Maxwell equal areas construction*. Consider again the isotherm displayed in Fig. 24.17b. As discussed earlier, the isotherm should look like $AEE'D$ and not have loops. Imagine that the fluid is carried around the closed cycle $EBCE'E$. For this closed cycle $\Delta u = 0$ and $\Delta s = 0$, and since only pV work is allowed we conclude that $\oint p\,dv = 0$, hence the two areas marked are equal.[13] Approximate theories of the liquid state, such as are discussed in Chapter 23, when supplemented with this equal areas construction, give a tolerably accurate description of the vapor–liquid coexistence curve. This is shown in Fig. 24.26.

The several molecular models of phase transitions described are useful, but each contains an inherent approximation and hence the predictions must be viewed with some reserve. This is certainly the case for the theory of critical behavior, which we have seen is peculiar. What can be said about critical phenomena as a result of studies of models that can be treated exactly?

1. As the critical point is approached, the pair-correlation function becomes long-range. A hint that this must occur was given in Section 24.3 in the description of the coupling of density fluctuations near T_c. It is found that in all systems that exhibit a critical point the range of pair correlations becomes very large as T_c is approached, no matter what the form of the molecular interaction. For example, in this limit the correlation between directions of magnetic moments in a magnet behaves like the correlation between positions of atoms in a fluid. The general form found is

$$g_2(R) - 1 \propto \frac{e^{-R/\xi}}{R^{d-2+\eta}}, \qquad (24.102)$$

where ξ is the scale length of the correlations, d is the number of dimensions (three for real space), and η is a critical exponent like α, β, γ, and δ. The scale length of correlations, ξ, diverges as $T \to T_c$.

2. Near T_c, ξ is very much greater than the range of the pair potential energy $u_2(R)$, hence the properties of the system become independent of $u_2(R)$. For this reason the behavior of magnets, fluids, and liquid mixtures is expected to be the same near T_c.

3. Provided that condition **2** is valid, the true approach to the critical region is very different from that predicted by the van der Waals and other classical equations of state. These classical theories predict that the critical exponents have the following values: $\alpha = 0$ and c_V has a jump discontinuity at T_c, $\beta = \frac{1}{2}$, $\gamma = 1$, $\delta = 3$, and $\eta = 0$. The experimental values of α, β, γ, and δ have already been mentioned (see Eqs. 24.33–24.37 and Table 24.4), and the best available data suggest that η is small and positive, but definitely not zero.

4. When condition **2** is not valid, that is, when the range of the interaction exceeds ξ, the behavior in the critical region is classical with $\beta = \frac{1}{2}$, $\gamma = 1$, $\delta = 3$, and $\eta = 0$. For example, a gas for which $u_2(R)$ becomes indefinitely long-range but weak, so that $\int d\mathbf{R}u_2(R)$ remains finite, satisfies the van der Waals equation of state supplemented by the Maxwell equal areas construction. This statement means that the supplementary condition is part of the equation of state and it determines the shape of an isotherm in the two-phase region. One example of a real system where $u_2(R)$ is very long-range is a solution of hydrogen in a transition metal. The insertion of the H atom causes displacements that strain the metal lattice, and the elastic energy associated with these displacements falls off only as R^{-1}. Two H atoms interact when the displacements they generate overlap, which occurs at very great distances. Consequently, the range of the interaction exceeds ξ for the hydrogen–hydrogen positional correlations. The hydrogen–metal system has a criti-

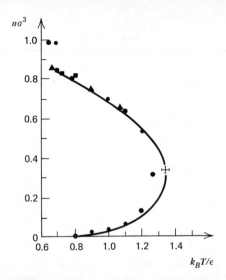

FIGURE 24.26
Densities of coexisting phases of a Lennard–Jones 6–12 fluid, calculated using the PY approximation mentioned in Chapter 23. The points represent the results of computer simulation calculations and experimental data for Ar, and the cross the critical point estimated from the simulations. From R. O. Watts and I. J. McGee, *Liquid State Chemical Physics* (Wiley, New York, 1976).

[13] Note that this argument assumes it is legitimate to use the laws of thermodynamics in the region of absolute instability, i.e., along the hypothetical curve CB, where $(\partial p/\partial v)_T > 0$.

TABLE 24.4
CRITICAL EXPONENTS PREDICTED BY SOME STATISTICAL MODELS

Property	Exponent	Classical Theory Mean Field Theory van der Waals Equation	Two-Dimensional Lattice Gas	Three-Dimensional Lattice Gas[a]	Experiment (Gas–Liquid)
Heat capacity[b]	α	0, finite discontinuity at $T = T_c$	0, logarithmic divergence at $T = T_c$	$1/8$[c]	0.1
Coexistence curve	β	1/2	1/8	$5/16$[d]	0.35[e]
Compressibility[f]	γ	1	7/4	$5/4$[g]	1.21[h]
Critical isotherm	δ	3	15	5[i]	4.45[j]

[a] From long-series expansions of the partition function in powers of $(1/T)$.
[b] $T > T_c$, $V = V_c$. When $T < T_c$ the exponent is labeled α'. Experimental data suggest that $\alpha' = \alpha$.
[c] For $V = V_c$, $T > T_c$. When $V = V_c$, $T < T_c$ the calculations give only the range $0 \to \frac{1}{8}$, and the exact value is considered uncertain.
[d] $0.307 < \beta < 0.317$ from the calculations.
[e] The data lie in the range $0.345 \leqslant \beta \leqslant 0.365$, with most confined to $0.347 \leqslant \beta \leqslant 0357$.
[f] $T > T_c$, $V = V_c$.
[g] For $T > T_c$ the calculations are considered to strongly suggest this value. For $T < T_c$, the corresponding value is considered to be consistent with $\frac{5}{4}$, with some uncertainty.
[h] The data lie in the range $1.17 \leqslant \gamma \leqslant 1.26$.
[i] Considered to be a good value.
[j] For ^4He, $\delta = 4.34$; for Xe, $\delta = 4.53$; for CO_2, $\delta = 4.44$; for H_2O, $\delta = 4.50$.

cal temperature for phase separation. It is found that the approach to the critical region is classical.

5. The Widom hypothesis has a basis in the molecular dynamics. The idea, due to L. Kadanoff and greatly improved by K. Wilson, is that it is possible systematically to reduce the total number of degrees of freedom contained in a volume ξ^3 (or ξ^d for d dimensions) to a smaller number of effective degrees of freedom by iteratively removing fluctuations with wavelengths smaller than ξ. Since $\xi \to \infty$ as $T \to T_c$, the successive transformations that reduce the number of degrees of freedom transform the Hamiltonian of the system into one that, eventually, is invariant to further transformations. The relations between critical exponents we have described earlier are a consequence of these transformations, and the theory permits calculation of their values. A thorough discussion is given by K. G. Wilson and J. B. Kogut, *Physics Reports* **12C,** 2 (1974).

It should be clear from the discussion of this chapter that much remains to be done before a completely satisfactory molecular theory of phase transitions is created.

Further Reading

Brout, R., *Phase Transitions* (W. A. Benjamin, Reading, Mass., 1965).

Epstein, P. S., *Textbook of Thermodynamics* (Wiley, New York, 1937), chaps. 6 and 7.

Kestin, J., *A Course in Thermodynamics*, Vol. II (Blaisdell, Waltham, Mass., 1968), chap. 22.

Rao, C. N. R., and Rao, K. J., *Phase Transitions in Solids* (McGraw-Hill, New York, 1978).

Scott, J. F., *Rev. Mod. Phys.* **46,** 83 (1974).

Stanley, H. E., *Introduction to Phase Transitions and Critical Phenomena* (Oxford University Press, London, 1971).

Stephenson, J., in H. Eyring, D. Henderson, and W. Jost (Eds.), *Physical Chemistry: An Advanced Treatise*, Vol. VIII B, chap. 10.

Advanced Treatments

Brey, J., and Jones, R. B. (Eds.), *Critical Phenomena* (Springer-Verlag, Berlin, 1976).

Domb, C., and Green, M. S. (Eds.), *Phase Transitions and Critical Phenomena* (Academic Press, New York, 7 vols. published from 1972 to 1976).

Pfeuty, P., and Toulouse, G., *Introduction to the Renormalization Group and to Critical Phenomena* (Wiley, New York, 1977).

Problems

1. At the normal boiling point of isopropanol, 82.3°C, its heat of vaporization is 665 J/g. Estimate its vapor pressure at 27°C.

2. Calculate the pressure at which water will boil at 95°C. The heat of vaporization of water is 40.6 kJ/mol.

3. The vapor pressure of supercooled liquid water is 2.149 torr at −10.0°C. The vapor pressure of ice at that temperature is 1.950 torr. The triple point of water is at 0.01°C. Calculate the heat of fusion of ice.

4. Calculate the change in freezing point per atmosphere of benzene at its normal freezing point, 5.5°C. The density of liquid benzene is 0.894 g/cm^3 and that of solid benzene is 1.014 g/cm^3 at 5.5°C. The heat of fusion of benzene is 9791 J/mol. Repeat this calculation for water, whose heat of fusion is 6000 J/mol. The density of liquid water at 0°C is 0.998 g/cm^3 and of ice, 0.9168 g/cm^3. Would a lake of frozen benzene be suitable for "ice" skating?

5. Prove that

$$c_\sigma = c_p - \alpha v \frac{\Delta h}{\Delta v},$$

where c_σ is defined in Eq. 24.19 and

$$\alpha \equiv \frac{1}{v}\left(\frac{\partial v}{\partial T}\right)_P.$$

6. The melting point of iodine (I_2) is 114°C. At this temperature, the vapor pressure of the solid increases 4.34 torr for a 1° rise in the temperature, and the value of the vapor pressure is 88.9 torr. Calculate the heat of sublimation of iodine.

7. The vapor pressure of liquid naphthalene is 10 torr at 85.8°C, and 40 torr at 119.3°C. Calculate the heat of vaporization and the boiling point. Calculate the vapor pressure at the melting point, 80°C. The vapor pressure of the solid is 1 torr at 52.6°C. Assume that the triple point is also 80°C, and calculate the heat of sublimation and the heat of fusion of the solid.

8. The vapor pressure of liquid arsenic depends on temperature as follows:

$$\log \frac{p_L}{\text{atm}} = -\frac{2460 \text{ K}}{T} + 3.81.$$

For solid arsenic, the vapor pressure is given by

$$\log \frac{p_S}{\text{atm}} = -\frac{6947 \text{ K}}{T} + 7.92.$$

Calculate the triple-point pressure and temperature.

9. Trouton's rule (see Section 14.4) states that the entropy of vaporization of most liquids is close to 85 J/K mol. Obtain a formula for the approximate vapor pressure of any liquid as a function of temperature and the boiling point of the liquid.

10. The vapor pressure of liquid iodine at 389.7 K is 100 torr and its boiling point is 456.2 K. For solid iodine, the vapor pressure is 1 torr at 311.9 K and the heat of fusion is 15.6 kJ/mol. Calculate the triple-point temperature and pressure of iodine.

11. At the normal melting point of mercury, 234.28 K, the density of the liquid is 13.69 g/cm^3 and the density of the solid is 14.19 g/cm^3. If the heat of fusion is 9.75 J/g, calculate the melting point of mercury at 20 atm.

12. The heat of vaporization of water is 44.8 kJ/mol and the heat of fusion of ice is 6.0 kJ/mol. If the vapor pressure of ice at 0°C is 4.58 torr, calculate its vapor pressure at −20°C.

13. It is found that the vapor pressure of a particular liquid varies with temperature by the equation

$$\log \frac{p_L}{\text{atm}} = -\frac{4849 \text{ K}}{T} - 14.70 \log \frac{T}{\text{K}} + 50.24.$$

How does the heat of vaporization vary with temperature? What is the heat of vaporization at 20°C?

14. Can a system containing three components in six phases be in equilibrium?

15. The discussion of the Gibbs phase rule in the text assumes that each component of the mixture occurs in each phase. How does the rule change if one or more components is absent from one or more phases?

16. Find the critical temperature, pressure, and volume of 1 mol of a van der Waals gas.

17. Show that

$$\frac{1}{v}\left(\frac{\partial v}{\partial T}\right)_p \quad \text{and} \quad -\frac{1}{v}\left(\frac{\partial v}{\partial p}\right)_T$$

for a van der Waals gas become infinite at the critical point.

18. Suppose that the virial equation of state,

$$\frac{p}{k_B T} = n + \sum_{l=2}^{\infty} B_l n^l,$$

accurately describes a hard-sphere gas. For this case the virial coefficients B_l are independent of T. It is believed that all of the B_l are positive. Assume that this is correct and show that a consequence is that a hard-sphere gas cannot have a critical point. How do you interpret this result?

19. Find the critical temperature, pressure, and volume of a gas for which

$$\frac{p}{k_B T} = n + B_2 n^2 + B_3 n^3,$$

where B_2 and B_3 depend on the temperature but not on the density.

The following comments provide background material for the next eight problems.

The classical theory of the critical region assumes that near the critical point the following expansions are valid:

$$\tilde{A} = \sum_{m=0}^{\infty} \sum_{l=0}^{\infty} (m!l!)^{-1} A_{ml} (\Delta \tilde{n})^m (\Delta \tilde{T})^l,$$

$$\tilde{\mu} = \sum_{m=0}^{\infty} \sum_{l=0}^{\infty} (m!l!)^{-1} \mu_{ml} (\Delta \tilde{n})^m (\Delta \tilde{T})^l,$$

$$\tilde{p} = \sum_{m=0}^{\infty} \sum_{l=0}^{\infty} (m!l!)^{-1} p_{ml} (\Delta \tilde{n})^m (\Delta \tilde{T})^l,$$

where $\tilde{T} \equiv T/T_c$, $\tilde{n} \equiv n/n_c$, and $\tilde{p} \equiv p/p_c$ are the reduced temperature, number density, and pressure, respectively. The reduced Helmholtz free energy is defined by $\tilde{A} \equiv A/Vp_c$ and the reduced chemical potential by $\tilde{\mu} \equiv \mu n_c/p_c$, whereas the expansion variables $\Delta \tilde{n}$ and $\Delta \tilde{T}$ are defined by $\Delta \tilde{n} \equiv (n - n_c)/n_c$ and $\Delta \tilde{T} \equiv (T - T_c)/T$. We shall also need the reduced functions $\tilde{\kappa}_T \equiv \kappa_T p_c/n_c^2$, $\tilde{S} \equiv ST_c/Vp_c$, $\tilde{C}_V \equiv C_V T_c/Vp_c$, and $\Delta \tilde{\mu} \equiv (n_c/p_c)[\mu(n, T) - \mu(n_c, T)]$. Note that because $\tilde{\mu} = (\partial \tilde{A}/\partial \tilde{n})_T$ and $\tilde{p} = \tilde{\mu}\tilde{n} - \tilde{A}$, the expansion coefficients A_{ml}, μ_{ml}, and p_{ml} are related by

$$\mu_{ml} = A_{m+1,l}, \qquad\qquad m \geqslant 0;$$

$$p_{ml} = A_{m+1,l} + (m-1)A_{ml}, \qquad m \geqslant 0;$$

$$p_{ml} = \mu_{ml} + (m-1)\mu_{m-1,l}, \qquad m \geqslant 1.$$

The usually stated conditions for a critical point, namely, $(\partial p/\partial V)_T = 0$ and $(\partial^2 p/\partial V^2)_T = 0$, imply that $A_{20} = 0$, $A_{30} = 0$, $\mu_{10} = 0$, $\mu_{20} = 0$, $p_{10} = 0$, and $p_{20} = 0$. It is, furthermore, usually assumed that $p_{30} = \mu_{30} = A_{30} \neq 0$.

20. Show that the condition of mechanical stability requires that for $T > T_c$

$$p_{30} = \mu_{30} = A_{40} > 0 \quad \text{and} \quad p_{11} > 0.$$

21. Show that the condition of thermal stability requires that $A_{02} < 0$ when $T > T_c$.

22. Show that the classical value of the exponent in Eq. 24.35 is

$$\delta_{\text{classical}} = 3.$$

23. Show that the classical value of the exponent in Eq. 24.36 is, for $\Delta \tilde{T} > 0$,

$$\alpha_{\text{classical}} = 0.$$

24. Show that the classical value of the exponent in Eq. 24.37 is, for $\Delta \tilde{T} > 0$,

$$\gamma_{\text{classical}} = 1.$$

25. Show that the coexistence curve near T_c satisfies

$$\frac{(\Delta \tilde{n}_L + \Delta \tilde{n}_G)^2 - (\Delta \tilde{n}_L)(\Delta \tilde{n}_G)}{\Delta \tilde{T}} = -\frac{6p_{11}}{p_{30}}.$$

26. From the result stated in Problem 25, derive the law of rectilinear diameters,

$$\frac{\Delta \tilde{n}_L + \Delta \tilde{n}_G}{2} = B_c |\Delta \tilde{T}| + \cdots,$$

where

$$B_c = \frac{\mu_{21}}{\mu_{30}} - \frac{3\mu_{11}\mu_{40}}{5\mu_{30}^2}$$

$$= \frac{p_{21}}{p_{30}} + \frac{4p_{11}}{5p_{30}} - \frac{3p_{11}p_{40}}{5p_{30}^2}.$$

27. A different theory of the critical region is generated if it is assumed that the first *four* isothermal derivatives of p with respect to V vanish at $T = T_c$. This theory is still classical in that the power series expansions set out above are

assumed to be valid. In this case we have

$$A_{20} = A_{30} = A_{40} = A_{50} = 0,$$

$$\mu_{10} = \mu_{20} = \mu_{30} = \mu_{40} = 0,$$

$$p_{10} = p_{20} = p_{30} = p_{40} = 0,$$

$$p_{50} = \mu_{50} = A_{60} > 0,$$

$$p_{11} = \mu_{11} > 0.$$

(a) Show that under these assumptions the exponents δ and γ, defined in Eqs. 24.35 and 24.37, have the values 5 and 1, respectively.

(b) Show that under these assumptions the law of rectilinear diameters is no longer valid.

28. Use Eq. 24.50 to derive the Maxwell equal areas construction.

The following problems consider some of the thermodynamic implications of the free-energy function (Eq. 24.86).

29. Given Eq. 24.89 applied to a system that can undergo a second-order transition,

$$G(p, T) = G_0 + A\eta^2 + C\eta^4,$$

suppose that $A = a(T - T_c)$. Show that:

(a) This form for the coefficient A implies that A changes sign at $T = T_c$.

(b) There are two minima of G for $T < T_c$, located at

$$\eta = \pm \left(\frac{|A|}{2C}\right)^{1/2},$$

and one maximum at $\eta = 0$.

(c) The entropy has the form

$$S = S_0 + \frac{a^2}{2C}(T - T_c).$$

(d) The specific heat at constant pressure has the form

$$C_p = T\left(\frac{\partial S_0}{\partial T}\right)_p + T\left(\frac{a^2}{2C}\right).$$

(e) S is continuous and C_p discontinuous at $T = T_c$.

30. Suppose that the free energy of a system is described by

$$G = G_0 + A\eta^2 + B\eta^3 + C\eta^4,$$

with $B > 0$. In the case that $B = 0$, as in Problem 29, the free energy describes a second-order transition. When $B > 0$, as here, the free-energy function describes a different kind of transition.

(a) Draw several curves of G versus η, some for $T > T_c$, one for $T = T_c$, and some for $T < T_c$.

(b) Show that at $T = T_c$ there is one minimum at

$$\eta = -\frac{3B}{4C},$$

at lower temperatures there are two minima at

$$\eta = \frac{-3B \pm (9B^2 - 32AC)^{1/2}}{8C},$$

and at high temperature there is only one minimum at $\eta = 0$.

(c) Show that $(\partial G/\partial T)_p$ is discontinuous at $T = T_c$.

(d) Trace the behavior of the system around a closed cycle starting from $T > T_c$ and going through $T < T_c$. Show that the system exhibits hysteresis.

Hint: Plot the entropy of the system along the cyclic path chosen.

31. Plot the extrema of G found in Problem 29 as a function of $-T$ for the following cases:
    (a) $A > 0$ and $C > 0$.
    (b) $A < 0$ and $C > 0$.
    (c) Show that in case (a) the system is stable when $\eta = 0$, and that in case (b) the system is unstable when $\eta = 0$.

32. Suppose that some substance undergoes a solid–solid transition from a low-temperature form α to a high-temperature form β at $T = T_{\text{trans}}$. Construct a simple molecular description of the transition by assuming that solids α and β can be described by the Einstein model (see Chapter 22); the frequencies and binding energies of α and β are $\nu_\alpha, -\epsilon_\alpha$ and $\nu_\beta, -\epsilon_\beta$, respectively. Show that

$$T_{\text{trans}} = \frac{\epsilon_\beta - \epsilon_\alpha}{3k_B \ln(\nu_\beta/\nu_\alpha)},$$

and that there is only one temperature at which α and β are in equilibrium.

33. Imagine a solid in equilibrium with its vapor, which is an ideal gas of atoms. Imagine, for simplicity, that the solid can be described by the Einstein model (Chapter 22), with binding energy per atom ϵ, frequency ν, and atomic mass m. Find an expression for the equilibrium vapor pressure of the solid.

34. Consider the same solid as in Problem 33, but suppose now that the vapour is an ideal gas mixture of atoms and diatomic molecules. If the binding energy of the diatomic molecule is D, its frequency ν', its moment of inertia I, and its mass $2m$, derive an expression for the vapor pressure of the solid.

25

Solutions of Nonelectrolytes

A great many chemical phenomena are associated with the properties of solutions. For example, the solubility of a compound may vary greatly from solvent to solvent, reaction equilibria may depend on the solvent medium in which the reaction occurs, and so forth. It is these and related properties that we examine in this chapter. As will be seen, much of the information available about the properties of solutions is derived from studies of the nature of the equilibria between a solution and other phases with which it may be in contact, say, a pure solid, vapor, or an electrode.

The study of the thermodynamic properties of solutions has two complementary aspects, reflecting alternatives in the analysis. If we assume that we know the dependences on temperature, pressure and concentration of the chemical potentials of all components in all the phases involved, the purpose of the thermodynamic analysis is to predict the characteristics of the equilibrium state. For example, knowing the forms of the chemical potentials in both liquid and vapor phases of a binary solution permits us to predict the vapor composition for a given composition of the liquid. This kind of information is of importance in the discussion of the practicality of a distillation procedure for the separation of the two components in the liquid. The alternative analysis uses the measured compositions, temperature, and pressure to deduce the chemical potentials of the components. The deviations of the chemical potentials from ideal behavior can then be interpreted in terms of molecular interactions. As in previous chapters, whenever possible we shall examine both the thermodynamic and the statistical molecular interpretation of solution properties.

It has been shown, in Chapter 19, that knowledge of a substance's chemical potential is a necessary prerequisite to developing a quantitative description of the properties characteristic of phase equilibrium. Furthermore, it has been shown that the equilibrium composition of a reacting gas mixture can be predicted once the chemical potentials of all reacting species are known. It should, therefore, be obvious that the first step in studying the thermodynamics of solutions is to determine the functional dependence of the chemical potential of each species present on the composition, temperature, and pressure.

An examination of the procedure employed in Section 21.2 to determine this dependence for a component in a gaseous mixture reveals that knowledge of the equation of state of the gas plays an important role in the analysis. Starting from the general thermodynamic relationship connecting a change in chemical potential with changes in entropy and pressure, we used the fact that the real gas approaches perfect gas behavior in the zero-pressure limit, together with Dalton's law defining the partial pressure of a component in the gas mixture, to provide a representation of the chemical potential of the component in terms of

25.1
The Chemical Potential of a Component in an Ideal Solution

composition and pressure. Once the chemical potentials of all components of the mixture were specified, the properties of the mixture and the relationships of these properties to those of the pure components could be easily analyzed.

The specific procedure used successfully for gaseous mixtures cannot now be used to analyze the behavior of liquid or solid mixtures. The reason is simple: At present there is not available any limiting form of the equation of state of a solid or liquid suitable for use in the same way as the perfect gas limit is used in the study of gases. For this reason we must modify the procedure used previously, so as to avoid requiring specific knowledge of the equation of state of the condensed phase. As already shown in Chapter 19 (see Eqs. 19.88–19.93), the composition dependence of the chemical potential can be deduced without knowledge of the equation of state, but determination of the temperature and pressure dependence of the chemical potential does require such knowledge.

Consider a mixture in which no chemical reactions occur. It is observed that when two liquids are mixed at constant temperature and pressure, there is usually a change in volume and evolution or absorption of heat. There are some mixtures for which these enthalpy and volume changes are very small, occasionally zero within the precision of measurement. Clearly, the more nearly alike the molecules of the several components, the smaller will be the changes in enthalpy and volume on mixing. For example, a mixture of perprotobenzene (C_6H_6) and perdeuterobenzene (C_6D_6) can be formed at constant temperature and pressure with negligibly small volume and enthalpy changes. We define an *ideal solution*, as an abstraction from observations, by the conditions

$$(\Delta H_{\text{mixing}})_{T,p} = (H^E)_{T,p} = 0 \qquad (25.1)$$

and

$$(\Delta V_{\text{mixing}})_{T,p} = (V^E)_{T,p} = 0, \qquad (25.2)$$

as a result of which we have

$$(\Delta U_{\text{mixing}})_{T,p} = (U^E)_{T,p} = 0 \qquad (25.3)$$

and

$$(\Delta G_{\text{mixing}})_{T,p} = -T(\Delta S_{\text{mixing}})_{T,p}. \qquad (25.4)$$

We shall now use these conditions to determine the composition dependence of the chemical potential for a component of an ideal solution. Note that this is the reverse of the procedure we used to study the properties of gaseous mixtures (Section 21.2). In that study we first established the composition dependence of the chemical potentials, and then deduced the limiting properties of a perfect gas mixture. It is the lack of knowledge of the equation of state of a liquid (or solid) that forces us to reverse the procedure in the study of condensed-phase mixtures.

Let the superscript zero (^{0}) refer to pure components. The change in Gibbs free energy on formation of a solution is related to the chemical potentials of the components as follows:

$$\Delta G_{\text{mixing}} \equiv G - G^0 = \sum_{i=1}^{r} n_i(\mu_i - \mu_i^0), \qquad (25.5)$$

where n_i is the number of moles of component i. Equation 25.2 can be rewritten to read

$$V^E \equiv V - V^0 = \left(\frac{\partial \Delta G_{\text{mixing}}}{\partial p}\right)_{T,n} = 0, \qquad (25.6)$$

and Eq. 25.1 can be rewritten to read

$$H^E \equiv H - H^0 = -T^2 \left[\frac{\partial}{\partial T} \left(\frac{\Delta G_{\text{mixing}}}{T} \right) \right]_{p,n} = 0. \qquad (25.7)$$

We now assume that the conditions on the right-hand sides of Eqs. 25.6 and 25.7, which pertain to $\sum_{i=1}^{r} n_i(\mu_i - \mu_i^0)$, are equally valid for each of the $\mu_i - \mu_i^0$, that is, that

$$\left[\frac{\partial(\mu_i - \mu_i^0)}{\partial p} \right]_{T,n} = 0, \qquad (25.8)$$

$$\left[\frac{\partial}{\partial T} \left(\frac{\mu_i - \mu_i^0}{T} \right) \right]_{p,n} = 0. \qquad (25.9)$$

Equations 25.8 and 25.9 may be thought of as differential equations for the variable $(\mu_i - \mu_i^0)$; together they imply that $\mu_i - \mu_i^0$ can only be a function of the form

$$\mu_i - \mu_i^0 = T w_i(n_1, \ldots, n_r), \qquad (25.10)$$

where $w_i(n_1, \ldots, n_r)$ is a function of only the composition variables $n_1, \ldots, n_r$ and μ_i^0 a function only of T and p. Substitution of Eq. 25.10 into Eqs. 25.8 and 25.9 verifies that it is a solution of the differential equations. Our remaining task is to determine the form of the composition function $w_i(n_1, \ldots, n_r)$.

It has been shown in Chapter 19 (Eq. 19.69) that for an ideal solution

$$w_i(n_1, \ldots, n_r) = R \ln x_i, \qquad (25.11)$$

so that Eq. 25.10 becomes[1]

$$\mu_i = \mu_i^0 + RT \ln x_i. \qquad (25.12)$$

The form 25.12 arises when it is asserted that the chemical potentials of all species in solution have the same analytical form. Although Eq. 25.12 is written with the mole fraction x_i as the concentration unit, other choices are possible. For mixtures of small molecules the mole fraction is most often chosen, but for solutions of polymers the volume fraction, ϕ_i, is a far better choice. Finally, the most important aspect of the form 25.12 is its relationship to the properties of the pure component, displayed explicitly in the term $\mu_i^0(T, p)$. The function $\mu_i^0(T, p)$ plays a role in the

---

[1] We can also obtain Eq. 25.12 from the following argument. From Eqs. 25.4, 25.5, and 25.10 we can write

$$(\Delta S_{\text{mixing}})_{T,p} = -\sum_i n_i w_i(n_1, \ldots, n_r)$$

for the entropy of mixing. Note that nothing in our argument thus far has assumed any particular form of the equation of state, or any particular kind of phase. The equation written for the entropy of mixing is completely general. Since w_i is a function of composition only, an ideal solution of given composition must have the same value of ΔS_{mixing} for any T and p, independent of the state of aggregation. At sufficiently high temperature and low pressure any such mixture will approach the perfect gas limit, where we already know from Eq. 21.49 that

$$(\Delta S_{\text{mixing}})_{T,p} = -R \sum_i n_i \ln x_i,$$

x_i being the mole fraction of component i. A perfect gas mixture is ideal by our present definition ($V^E = 0$, $H^E = 0$). The two forms in which we have written ΔS_{mixing} are then identical if

$$w_i(n_1, \ldots, n_r) = R \ln x_i.$$

The chemical potential of a component in an ideal solution is then given by Eq. 25.12.

theory of liquid mixtures analogous to the hypothetical perfect gas chemical potential, $\mu_i^*(T)$, in the theory of gaseous mixtures.

The procedure used to derive Eq. (25.12) emphasizes the lack of a generally valid equation of state for condensed systems and starts with the definition of an ideal mixture, embodied in Eqs. 25.1–25.4. Alternatively, noting that the chemical potential of a component i in an ideal gas mixture is proportional to $RT \ln x_i$, one can by analogy define the chemical potential of i in an ideal liquid or solid mixture by Eq. 25.12 and then deduce as consequences the conditions 25.1–25.4. The difference between these approaches is, obviously, a matter of emphasis. The one we have used takes as fundamental the (putative) observed properties of an ideal liquid or solid mixture, whereas the alternative mentioned takes as fundamental the definition formulated by analogy with the behavior of an ideal gas mixture.

As already noted, when liquids are mixed there is ordinarily some volume change and some heat absorbed or released. Thus, however useful the concept of an ideal solution as an abstraction from and extrapolation of observed behavior, it is necessary to analyze the properties of the chemical potential further if we wish to describe most real solutions.

There are many different formal ways of extending the representation of the chemical potential. For example, if the chemical potential of component i is written in the form

$$\mu_i = \mu_i^0(T, p) + RT \ln x_i + \mu_i^E, \tag{25.13}$$

then μ_i^E, known as the *excess chemical potential* of i, is directly related to the observable heat of mixing and volume change of mixing. (See Eqs. 25.6 and 25.7.) In fact, from the fundamental thermodynamic relations connecting the enthalpy and volume changes with the temperature and pressure derivatives of the corresponding Gibbs free energy change, we find

$$V^E = \left(\frac{\partial \Delta G_{\text{mixing}}}{\partial p}\right)_{T, n_1, \ldots, n_r} = \sum_{i=1}^{r} n_i \left(\frac{\partial \mu_i^E}{\partial p}\right)_{T, n_1, \ldots, n_r}, \tag{25.14}$$

$$H^E = -T^2 \left(\frac{\partial (\Delta G_{\text{mixing}}/T)}{\partial T}\right)_{p, n_1, \ldots, n_r}$$

$$= -T^2 \sum_{i=1}^{r} n_i \left(\frac{\partial (\mu_i^E/T)}{\partial T}\right)_{p, n_1, \ldots, n_r}. \tag{25.15}$$

The first two terms on the right-hand side of Eq. 25.13, which correspond to the "ideal contribution" to the chemical potential of component i, make no contribution to the heat of mixing and volume change on mixing. As might be surmised from the use of the words *excess chemical potential* and the definition of an ideal solution, ΔV_{mixing} and ΔH_{mixing} are sometimes called the *excess volume* and *excess enthalpy* of the solution. It is for this reason that we introduced the notation V^E, H^E in Eqs. 25.2 and 25.1. Clearly, if we have $\Delta V_{\text{mixing}} \neq 0$ and $\Delta H_{\text{mixing}} \neq 0$, the entropy of mixing will in general also deviate from the ideal form. Again using standard thermodynamic relations (see Table 19.2), we obtain

$$\Delta S_{\text{mixing}} = -\left(\frac{\partial \Delta G_{\text{mixing}}}{\partial T}\right)_{p, n_1, \ldots, n_r}$$

$$= -R \sum_{i=1}^{r} n_i \ln x_i - \sum_{i=1}^{r} n_i \left(\frac{\partial \mu_i^E}{\partial T}\right)_{p, n_1, \ldots, n_r}. \tag{25.16}$$

25.2
The Chemical Potential of a Component in a Real Solution

The *excess entropy* of mixing is defined to be the difference $\Delta S_{\text{mixing}} - (-R \sum_i n_i \ln x_i)$. Thus, all the excess thermodynamic functions are referred to the properties of a hypothetical ideal solution with the same composition, temperature, and pressure. This definition of the excess entropy and the excess free energy differs from that introduced in Chapter 21. In Chapter 21 we defined the excess value of X as the X of mixing ($G^E = RT \sum_i n_i \ln x_i$) but in this chapter it is defined as the difference from the ideal solution value. For U, H and V the two definitions are equivalent, while for S and G they differ by the ideal entropy of mixing ($-R \sum_i n_i \ln x_i$) and ideal free energy of mixing ($RT \sum_i n_i \ln x_i$), respectively.

For convenience, instead of using the extensive excess functions $V^E, H^E, \ldots$, we shall often describe a mixture in terms of deviations from its *mean molar properties*. For example, the mean molar volume is

$$v_m \equiv \frac{V}{\sum_i n_i}, \tag{25.17}$$

and the excess volume per mean mole of mixture is

$$v^E \equiv \frac{V}{\sum_i n_i} - \frac{\sum_i n_i v_i^0}{\sum_i n_i} = \frac{V}{\sum_i n_i} - \sum_i x_i v_i^0, \tag{25.18}$$

where v_i^0 is the molar volume of component i at the same temperature and pressure as is the mixture. Corresponding definitions hold for the mean molar excess enthalpy, h^E, entropy, s^E, and other properties.

In the discussion of the properties of imperfect gases we introduced the fugacity of the gas. When written in terms of the fugacity instead of the pressure, the algebraic forms of the chemical potentials of a perfect gas and an imperfect gas are the same. Similarly, the chemical potential of a component of a real solution can be written in forms that are different from Eq. 25.13. One form, much used in the chemical literature, consists of writing the chemical potential of component i as

$$\mu_i = \mu_i^0(T, p) + RT \ln \gamma_i x_i. \tag{25.19}$$

Then

$$\gamma_i x_i \equiv a_i \tag{25.20}$$

is known as the *activity* of component i; γ_i is called the *activity coefficient*. This language probably stems from the notion that the "effective concentration" of a component determines the extent to which it influences, or *acts* in determining, the equilibrium properties of the solution. The product $\gamma_i x_i$ replaces the mole fraction x_i in the analytical form of the chemical potential of i in an ideal solution, and therefore may be thought of as the effective concentration of i. Clearly γ_i must depend on temperature, pressure, and composition. Also, comparison of Eqs. 25.13 and 25.19 shows that

$$\mu_i^E = RT \ln \gamma_i, \tag{25.21}$$

so that ΔH_{mixing}, ΔV_{mixing}, ΔS_{mixing} may all be expressed in terms of derivatives of γ_i.

The two formulations displayed in Eqs. 25.13 and 25.19 do not exhaust the possibilities for formal analysis of the chemical potential,[2] but

[2] For example, the fugacity can be extended to condensed phases, being defined in relation to the activity by $a_i = f_i/f_i^0$, where f_i^0 is the value in the standard state.

we shall not discuss any of the other representations of nonideality in this book. In this chapter we discuss the properties of solutions mostly in terms of the excess functions μ^E, h^E, s^E, and v^E, whereas in Chapter 26 we mostly use the activity and activity coefficient representation. Different choices of representation are made in the two chapters to conform to the respective common usages in studies of nonelectrolyte and electrolyte solutions.

Before we proceed further, it is necessary to make some comments about the term $\mu_i^0(T, p)$ in Eqs. 25.13 and 25.19. This term, usually called the chemical potential of the standard state, defines the reference level with respect to which changes in the chemical potential of component i are measured. There are two common conventions used to choose the standard state of a component in a solution. Since $\mu_i^0(T, p)$ is independent of composition, one obvious choice is the chemical potential of the pure liquid or solid at the same temperature and pressure,

$$\mu_i^0(T, p) = \mu_i^L(T, p) \qquad (25.22a)$$

or

$$\mu_i^0(T, p) = \mu_i^S(T, p). \qquad (25.22b)$$

If the pure phase chosen as the standard state for a given T and p is not stable at that T and p, then an extrapolation of the known properties of the liquid (or solid) is necessary to define the standard state. Although this seems an unnecessarily complicated way to choose a standard state, it is actually quite convenient in practical computation and is widely used. When this standard state is employed we denote it by the superscript $(^0)$. A second convention used to define a standard state distinguishes between the component present in largest amount (solvent, labeled 1) and all other components. The standard state for component 1 is selected to be the pure liquid (or solid),

$$\mu_1^0(T, p) = \mu_1^L(T, p) \qquad \text{(solvent)}, \qquad (25.23a)$$

whereas for all the other components a *hypothetical* standard state is chosen: We imagine that there exists a fictitious state of component i with the properties that pure i would have if its limiting low-concentration properties in solution were to be retained in a pure substance. In mathematical terms this is equivalent to writing

$$\mu_i^0 \equiv \mu_i^\oplus(T, p) = \lim_{x_1 \to 1}(\mu_i - RT \ln x_i) \qquad \text{(solute)}. \qquad (25.23b)$$

Note that the standard state in this convention, which we denote by the superscript $(^\oplus)$, depends on temperature, pressure, *and* the nature of the solvent (component 1). How does this hypothetical standard state differ from the state of the true pure component? Consider a very dilute solution, one in which each solute molecule of species i is surrounded only by solvent molecules, and in which the average separation between molecules of i is so large that they do not interact with one another. Further dilution will not then alter the nature of the interaction of a molecule of i and the solvent. If the interaction between molecules of i and solvent is different from that between molecules of i alone, the motion and properties of an i molecule surrounded by solvent molcules must be different from those of an i molecule surrounded by only other i molecules. A hypothetical pure liquid constructed of particles imagined to interact as do i and solvent will, clearly, be different from pure liquid i. In macroscopic terms we expect that, for example, in a very dilute solution the partial molar volume $\bar{v}_i \equiv (\partial V / \partial n_i)_{T,p,n_{i'}}$ becomes independent of composition, but not equal to the molar volume v_i^0 of pure component i. Extrapolation of the limiting low-concentration solution behavior then leads to a fictitious state with component i occupying the volume $v_i^\oplus$ and not the molar volume v_i^0. Similar discrepancies may exist for all of the thermodynamic functions. The standard state defined by the second convention is most useful in the description of dilute solutions. It is often called, misleadingly, the standard state of infinite dilution. The concentration of component i in this standard state is not infinitesimally small. Rather, we repeat, there is defined a fictitious state of matter that has the same properties per mole as does component i when present in dilute solution.

One of the principal goals of the description of real gases and real gas mixtures was the representation of the chemical potential in terms of the equation of state and specific heat (see Chapter 21). The form of the chemical potential used to describe solid and liquid mixtures, Eq. 25.13 or Eq. 25.19, has not yet been reduced to comparable form. Of course, from the definition of the excess properties of the solution we know that[3]

$$\left[\frac{\partial(\mu_i^E/T)}{\partial p}\right]_{T,n_1,\ldots,n_r} = R\left(\frac{\partial \ln \gamma_i}{\partial p}\right)_{T,n_1,\ldots,n_r} = \frac{\bar{v}_i - v_i^0}{T},$$

$$\left[\frac{\partial(\mu_i^E/T)}{\partial T}\right]_{p,n_1,\ldots,n_r} = R\left(\frac{\partial \ln \gamma_i}{\partial T}\right)_{p,n_1,\ldots,n_r} = -\frac{\bar{h}_i - h_i^0}{T^2}. \tag{25.24}$$

Equations 25.24 can be regarded as differential equations determining the behavior of the excess chemical potential at constant composition. From the Gibbs–Duhem equation (Eq. 19.68), at constant temperature and pressure we have

$$\left(\sum_{i=1}^r n_i \, d\mu_i\right)_{T,p} = 0, \tag{25.25}$$

which yields an equation defining the composition dependence of the excess chemical potential at constant T and p:

$$\left(\sum_{i=1}^r n_i \, d\mu_i^E\right)_{T,p} = RT\left(\sum_{i=1}^r n_i \, d\ln \gamma_i\right)_{T,p} = 0. \tag{25.26}$$

Equation 25.26 follows from Eq. 25.25 directly after substitution of Eqs. 25.13 and 25.21 into Eq. 25.25 and evaluation of the differentials. For a two-component system, Eq. 25.26 can be rewritten in the form

$$n_1(d\ln \gamma_1)_{T,p} + n_2(d\ln \gamma_2)_{T,p} = 0,$$

$$(1-x_2)\left(\frac{\partial \ln \gamma_1}{\partial x_2}\right)_{T,p} + x_2\left(\frac{\partial \ln \gamma_2}{\partial x_2}\right)_{T,p} = 0. \tag{25.27}$$

Therefore, at constant temperature and pressure, knowledge of μ_1^E enables us to compute μ_2^E:

$$\mu_2^E = RT \ln \gamma_2 = RT \int_{x_2}^1 \frac{1-x_2'}{x_2'} \frac{\partial \ln \gamma_1}{\partial x_2'} dx_2' = \int_{x_2}^1 \frac{1-x_2'}{x_2'} \frac{\partial \mu_1^E}{\partial x_2'} dx_2'. \tag{25.28}$$

Simultaneous solution of the three differential equations in Eqs. 25.24 and 25.27 completely determines μ_i^E.

---

[3] As written, Eqs. 25.24 use pure liquid or solid of component i as the standard state. If, instead, the standard state is defined by Eq. 25.23b, v_i^0 and h_i^0 are to be replaced by $v_i^\oplus$ and $h_i^\oplus$.

It is clear that actual evaluation of Eqs. 25.24 and 25.28 requires knowledge of the composition dependence of the equation of state, the solution heat capacity, and the activity coefficient. This requirement is analogous to the need to know the composition dependence of the equation of state of a gaseous mixture in order to evaluate Eq. 21.29. In that case it proved advantageous to use a model equation of state, such as the van der Waals equation, or the virial equation of state for more general results. What analogous relations exist for liquid and solid solutions?

A model for the behavior of solutions that is analogous to the van der Waals model of the gas is defined by the following two statements:

1. The entropy of mixing has the ideal form $-R\sum_{i=1}^{r} n_i \ln x_i$.

2. The excess chemical potentials have the form (binary mixture)

$$\mu_1^E = RT \ln \gamma_1 = wx_2^2, \tag{25.29}$$

$$\mu_2^E = RT \ln \gamma_2 = wx_1^2 = w(1-x_2)^2. \tag{25.30}$$

For simplicity we shall assume that w is an energy that is independent of temperature, pressure, and composition. From these statements defining the model mixture, known as a *regular mixture*, we deduce that the heat of mixing has the form

$$\begin{aligned}
\Delta H_{\text{mixing}} &= -T^2\left[n_1\left(\frac{\partial}{\partial T}\frac{\mu_1^E}{T}\right)_{p,n_2} + n_2\left(\frac{\partial}{\partial T}\frac{\mu_2^E}{T}\right)_{p,n_1}\right] \\
&= n_1 wx_2^2 + n_2 wx_1^2 \\
&= n_1 wx_2(1-x_1) + n_2 wx_1(1-x_2) \\
&= w(n_1+n_2)x_2(1-x_2). \tag{25.31}
\end{aligned}$$

Notice that ΔH_{mixing} is symmetrical in the mole fractions x_1 and x_2.

We shall show later that the regular mixture model predicts that if w/RT is sufficiently large a one-phase mixture separates into two phases with different compositions. Also, just as in the case of condensation of a gas, there exists a critical temperature above which phase separation does not occur.

Instead of using a specific model, such as a regular solution, we can employ a representation of solution properties more closely related to the concept of the virial expansion. It is obtained by noticing that, at constant temperature and pressure, the general solution of Eq. 25.27 is

$$\ln \gamma_1 = \int x_2 \varphi(x_2)\, dx_2, \tag{25.32a}$$

$$\ln \gamma_2 = -\int (1-x_2)\varphi(x_2)\, dx_2, \tag{25.32b}$$

where $\varphi(x_2)$ is some function of x_2. That Eqs. 25.32a and 25.32b are indeed a solution of Eq. 25.27 can be verified by direct substitution. In the spirit of the virial expansion we now write for a binary solution

$$\varphi(x_2) = \sum_j \sum_k a_{jk}(T, p)x_2^j(1-x_2)^k, \tag{25.33}$$

where the coefficients $a_{jk}(T, p)$ are independent of composition. Consider the lowest-order approximation, which retains only the term $a_{00}(T, p)$ in the series expansion. Then we have

$$\ln \gamma_1 = \int x_2 a_{00}(T, p)\, dx_2 = \frac{1}{2} a_{00}x_2^2, \tag{25.34a}$$

$$\ln \gamma_2 = -\int (1-x_2)a_{00}(T, p)\, dx_2 = \frac{1}{2}a_{00}(1-x_2)^2. \qquad (25.34b)$$

Comparison with Eq. 25.30 shows that, to this order, if $a_{00}(T, p)$ is independent of temperature and pressure, retention of only the term a_{00} in the series generates what we have called regular mixture theory. Clearly, just as in the virial equation representation, higher-order terms can be retained to provide a better and better description of the excess chemical potential.

It is natural, when examining the properties of a mixture, to try to assess the contributions made by the several components. If the excess volume, excess enthalpy, ... of the mixture are not zero, then the total volume, enthalpy, ... are not simply the sums of the volumes, enthalpies, ... of the pure components. Let Z be some extensive thermodynamic function of the mixture. The partial molar function $\bar{z}_i$, defined by

25.3
Partial Molar Quantities

$$\bar{z}_i \equiv \left(\frac{\partial Z}{\partial n_i}\right)_{T,p,n_{j\neq i}}, \qquad (25.35)$$

can be used to provide a systematic subdivision of the extensive function Z into a sum of component contributions. Suppose that the amount of each component in the mixture is increased, at constant temperature and pressure, by a factor ξ. Then, because Z is an extensive function,

$$Z(\xi n_1, \xi n_2, \ldots, T, p) = \xi Z(n_1, n_2, \ldots, T, p). \qquad (25.36)$$

Differentiation of $Z(\xi n_1, \xi n_2, \ldots, T, p)$ with respect to ξ, followed by setting $\xi = 1$, leads to[4]

$$Z = \sum_{i=1}^{r} n_i \bar{z}_i. \qquad (25.37)$$

Partial molar quantities are intensive properties of the mixture; they depend on T, p and the relative composition of the mixture. To illustrate how Eq. 25.37 provides the wanted subdivision into component contributions, consider forming a binary solution of n_1 mol of 1 and n_2 mol of 2 in a total volume V. We imagine that, at constant temperature and pressure, we mix δn_1 mol of 1 with δn_2 mol of 2, and repeat this procedure as many times as needed while keeping the ratio $\delta n_1/\delta n_2 = n_1/n_2$ the same for each addition. Then, since the partial molar volumes $\bar{v}_1$ and $\bar{v}_2$ depend only on the relative composition for fixed T and p, adding up all the increments in volume gives

$$V = n_1 \bar{v}_1 + n_2 \bar{v}_2, \qquad (25.38)$$

which is just a special case of Eq. 25.37.

---

[4] We have

$$\left[\frac{\partial Z(\xi n_1, \ldots, T, p)}{\partial \xi}\right]_{T,p,n_i} = Z(n_1, n_2, \ldots, T, p)$$

$$= \sum_{i=1}^{r} \left(\frac{\partial Z}{\partial(\xi n_i)}\right)_{T,p,n_i} \left(\frac{\partial(\xi n_i)}{\partial \xi}\right)_{T,p,n_j}$$

$$= \sum_{i=1}^{r} \left(\frac{\partial Z}{\partial(\xi n_i)}\right)_{T,p,n_i} n_i,$$

which becomes, for $\xi = 1$,

$$Z(n_1, \ldots, n_r, T, p) = \sum_{i=1}^{r} n_i \bar{z}_i.$$

A differential change in the extensive property Z, at constant temperature and pressure, is given by

$$(dZ)_{T,p} = \left(\sum_{i=1}^{r} n_i \, d\bar{z}_i\right)_{T,p} + \left(\sum_{i=1}^{r} \bar{z}_i \, dn_i\right)_{T,p}. \qquad (25.39)$$

Comparison of Eqs. 25.39 and 25.35 leads to a constraint on the possible changes of the several $\bar{z}_i$, namely,

$$\left(\sum_{i=1}^{r} n_i \, d\bar{z}_i\right)_{T,p} = 0. \qquad (25.40)$$

When $\bar{z}_i = \mu_i$, Eq. 25.40 becomes the Gibbs–Duhem relation (see Eqs. 19.68 and 25.25). In general, then, the possible changes in the partial molar properties of a mixture satisfy, individually, Gibbs–Duhem type equations of constraint.

The experimental determination of partial molar properties is based on the following transformation of Eq. 25.35. Let z_m be the mean molar value defined by

$$z_m \equiv \frac{Z}{\sum_i n_i} = \sum_i x_i \bar{z}_i. \qquad (25.41)$$

Consider a binary mixture, and differentiate z_m with respect to the mole fraction of component 1:

$$\left(\frac{\partial z_m}{\partial x_1}\right)_{T,p} = \left(\frac{\partial}{\partial x_1}\left[x_1\bar{z}_1 + (1-x_1)\bar{z}_2\right]\right)_{T,p}$$

$$= \bar{z}_1 - \bar{z}_2 + x_1\left(\frac{\partial \bar{z}_1}{\partial x_1}\right)_{T,p} + x_2\left(\frac{\partial \bar{z}_2}{\partial x_1}\right)_{T,p}$$

$$= \bar{z}_1 - \bar{z}_2, \qquad (25.42)$$

the last line following from an application of Eq. 25.40. Combining Eqs. 25.42 and 25.41 leads to

$$\bar{z}_1 = z_m - x_2\left(\frac{\partial z_m}{\partial x_2}\right)_{T,p},$$

$$\bar{z}_2 = z_m - x_1\left(\frac{\partial z_m}{\partial x_1}\right)_{T,p}. \qquad (25.43)$$

Similar but algebraically more complex expressions relate the partial molar and mean molar properties of multicomponent mixtures.

To use Eq. 25.41 we plot the mean molar property z_m as a function of mole fraction of, say, component 1. Then, as shown in Fig. 25.1, the tangent to the curve at $z_m(x_1')$ has intercepts $\bar{z}_2(x_1')$ at $x_1 = 0$ and $\bar{z}_1(x_1')$ at $x_1 = 1$.

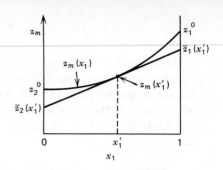

FIGURE 25.1
The mean molar function $z_m \equiv Z/\sum_i n_i$ as a function of the mole fraction of component 1 in a binary mixture. The tangent to $z_m(x_1)$ at $x_1 = x_1'$ has intercepts $\bar{z}_2(x_1')$ at $x_1 = 0$ and $\bar{z}_1(x_1')$ at $x_1 = 1$.

We begin our study of nonelectrolyte mixtures with an examination of some of the characteristic features of liquid–vapor equilibrium. For each component of the mixture we must have

$$\mu_i(\text{liquid}) = \mu_i(\text{vapor}), \qquad (25.44)$$

25.4
Liquid–Vapor Equilibrium

and, therefore, choosing the pure component as reference state,

$$\mu_i{}^0 + RT \ln \gamma_i x_i = \mu_i{}^*(T) + RT \ln \frac{f_i}{p}. \tag{25.45}$$

A simple rearrangement now gives

$$\gamma_i x_i = \frac{f_i}{p} \exp\left[\frac{1}{RT}(\mu_i{}^* - \mu_i{}^0)\right]. \tag{25.46}$$

But when $x_i = 1$ we have $\mu_i = \mu_i{}^0$ and $\gamma_i = 1$, since the pure component has been chosen as the reference state even for solutes, and for $x_i = 1$, $\mu_i = \mu_i{}^0$ no matter what the reference state is. Therefore, putting $x_i = 1$ in Eq. 25.46, we obtain

$$1 = \frac{f_i{}^0}{p} \exp\left[\frac{1}{RT}(\mu_i{}^* - \mu_i{}^0)\right]. \tag{25.47}$$

A combination of Eqs. 25.46 and 25.47 gives the simple result

$$\gamma_i x_i = \frac{f_i}{f_i{}^0}. \tag{25.48}$$

In the particular case that the solution is ideal ($\gamma_i = 1$) and the vapor phase is ideal ($f_i = p_i$), Eq. 25.48 reduces to

$$p_i = x_i p_i{}^0 \qquad \left(\begin{array}{c}\text{solution ideal,}\\ \text{gas phase ideal}\end{array}\right), \tag{25.49}$$

which is known as *Raoult's law*. Equation 25.48 can be used in two ways: The activity coefficient of component i in the condensed phase can be deduced from measurements of the fugacity of component i in the vapor in equilibrium with the solution; on the other hand, if γ_i is known the vapor composition can be predicted.

Consider, first, equilibrium between an ideal liquid (or solid) mixture and an ideal gas mixture. Let y_i be the mole fraction of component i in the vapor phase. Then Dalton's law gives

$$p_i = y_i p \qquad \text{(gas phase)}, \tag{25.50}$$

where p is the total pressure of the gaseous mixture. By hypothesis the liquid mixture considered is ideal, and Eq. 25.49 thus describes the relation between the partial pressure of component i and its concentration in solution. Combining Eqs. 25.49 and 25.50, we have

$$y_i p = x_i p_i{}^0,$$
$$y_i = x_i \frac{p_i{}^0}{p}. \tag{25.51}$$

Thus the vapor and liquid compositions can be very different if the vapor pressures of the several pure components differ considerably. For example, at 80°C the vapor pressures of the monomethyl and monoethyl ethers of ethylene glycol are 154.4 torr and 102.2 torr, respectively. When mixed, these substances form an ideal solution. When the solution mole fractions are $x_1 = x_2 = 0.50$, the vapor mole fractions at 80°C are $y_1 = 0.602$, $y_2 = 0.398$, component 1 being the higher-vapor-pressure monomethyl ether. That is, the vapor is richer in the more volatile component.

Example. The vapor pressures of pure benzene and chlorobenzene are as follows:

$T(°C)$	$p^0_{benzene}$ (torr)	$p^0_{chlorobenzene}$ (torr)
90	1013	208
100	1340	293
110	1744	403
120	2235	542
132	2965	760

These substances form an ideal mixture. Calculate the boiling point of a mixture containing 0.40 mole fraction benzene and 0.60 mole fraction chlorobenzene.

The boiling point is the temperature at which the total vapor pressure is equal to 1 atm or 760 torr. Since the solution is ideal, we have

$$p = x_1 p_1^0 + x_2 p_2^0 \qquad \begin{matrix} (1 = \text{benzene}, \\ 2 = \text{chlorobenzene}) \end{matrix}$$

at all temperatures. We must solve the equation

$$760 = 0.40 p_1^0(T) + 0.60 p_2^0(T).$$

The simplest way to do this is to use the data given and plot $p = 0.40 p_1^0 + 0.60 p_2^0$ versus T, and find that value of T corresponding to $p = 760$ torr. We find $T = 102.6°C$.

It is useful to examine further liquid–vapor equilibrium in an ideal binary mixture. The total vapor pressure

$$p = p_1 + p_2 = x_1 p_1^0 + x_2 p_2^0 = x_1(p_1^0 - p_2^0) + p_2^0 \qquad (25.52)$$

is, at constant temperature, a linear function of x_1 and/or x_2, as shown in Fig. 25.2a. On the other hand, because of the difference in composition between vapor and liquid, p is not a linear function of y_1. In fact, from Eq. 25.51 we find, for the concentration ratios in vapor and liquid,

$$\frac{(y_1/y_2)}{(x_1/x_2)} \equiv \alpha_{12} = \frac{y_1(1 - x_1)}{x_1(1 - y_1)} = \frac{p_1^0}{p_2^0}. \qquad (25.53)$$

FIGURE 25.2

(a) Schematic representation of the isothermal dependence of the total pressure of an ideal binary mixture on the mole fraction of component 1 in the liquid, x_1, and on the mole fraction of component 1 in the vapor, y_1.

(b) Schematic representation of the separation of a mixture into vapor and liquid phases when compressed to $p = p'$ isothermally from a state where p is less than the pressure along the dew-point curve, or when expanded isothermally from a state where p is greater than the pressure along the bubble-point curve.

(c) Schematic diagram of the bubble-point and dew-point curves on a temperature-composition diagram at constant pressure (ideal binary mixture).

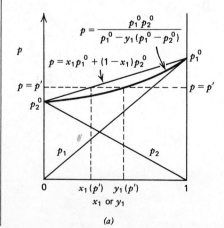

(a)

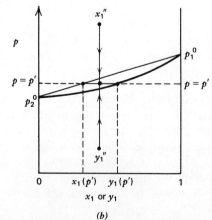

(b)

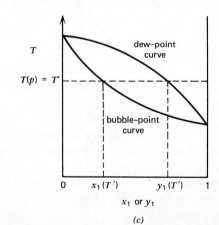

(c)

For an ideal mixture α_{12}, which measures the relative volatility of components 1 and 2, is independent of composition. If we now eliminate x_1 between Eqs. 25.53 and 25.52 we find for the total pressure as a function of y_1

$$p = \frac{p_1^{\,0}}{\alpha_{12} - y_1(\alpha_{12} - 1)} = \frac{p_1^{\,0} p_2^{\,0}}{p_1^{\,0} - y_1(p_1^{\,0} - p_2^{\,0})}, \qquad (25.54)$$

which is a section of a rectangular hyperbola that is concave upward (see Fig. 25.2). Since vapor and liquid have the same pressure, the difference in their compositions at, say, $p = p'$, is obtained from the intersection of the line $p = p'$ with the two curves representing p as a function of x_1 and p as a function of y_1. As shown in Fig. 25.2, these intersections are $x_1(p')$ and $y_1(p')$, respectively. The line $p = x_1 p_1^{\,0} + (1 - x_1) p_2^{\,0}$ is called the *bubble-point curve*, and $p = p_1^{\,0} p_2^{\,0} / [p_1^{\,0} - y_1(p_1^{\,0} - p_2^{\,0})]$ is called the *dew-point curve*. If a mixture is subjected to a pressure greater than that on its bubble-point curve, only the liquid phase will be stable; if it is subjected to a pressure less than that on its dew-point curve, only the vapor phase will be stable. If the pressure of the mixture falls between the dew-point and bubble-point pressures, it will separate into coexisting liquid and vapor phases. For example, if we imagine taking a gas mixture of composition y_1'', $(1 - y_1'')$, at a pressure below the dew-point pressure for that composition, and compressing that mixture isothermally until the pressure is p', then as shown in Fig. 25.2b the mixture separates into a liquid phase with composition $x_1(p')$, $[1 - x_1(p')]$ and a vapor phase with composition $y_1(p')$, $[1 - y_1(p')]$. The ratio of the number of moles of liquid to vapor is $[y_1(p') - y_1'']/[y_1'' - x_1(p')]$.[5] Liquid first appears when the vertical line representing the fixed composition y_1'' intersects the dew-point curve. Had we started with a liquid mixture with composition x_1'', $(1 - x_1'')$, at a pressure greater than its bubble-point pressure, and lowered the pressure to p', vapor would first appear when the vertical line through x_1'' intersects the bubble-point curve.

We now note that, because $p_1^{\,0}$ and $p_2^{\,0}$ are, in general, different functions of T, the boiling point of the mixture is not a linear function of the liquid composition. The composition of the vapor changes with temperature according to

$$y_1(T) = \frac{x_1 \alpha_{12}(T)}{1 + x_1[\alpha_{12}(T) - 1]}, \qquad (25.55)$$

where for an ideal mixture $\alpha_{12}(T) \equiv p_1^{\,0}(T)/p_2^{\,0}(T)$. Figure 25.2c displays the bubble-point and dew-point curves on a temperature-composition diagram. Since the vapor and liquid are at the same temperature, the difference in their compositions when the liquid is boiled at pressure p is obtained from the intersections of $T(p)$ with the bubble-point and dew-point curves, as shown in Fig. 25.2c. Of course, as the composition of the liquid changes, so does the boiling temperature, hence also the enrichment of the vapor in one component.

As an introduction to the study of nonideality in mixtures, we now consider a special case, but one that is very interesting. Specifically, we shall examine the nature of the equilibrium between a gas and a liquid in which the gas is only slightly soluble. When the gas and liquid do not react, for example, $H_2(g)$ and $H_2O(l)$, it is generally observed that the solubility of the gas is proportional to the gas pressure (strictly, to the gas fugacity) and often increases as the temperature is lowered. To make a

[5] See the discussion of the lever rule, Chapter 24, Eq. 24.4.

thermodynamic analysis of this gas–liquid system we start with the observation that in the phase equilibrium between liquid and gas the boundary (the surface of the liquid) is permeable to transfer of the solute and the solvent. Consider a binary solution. Since both components are present in both phases, we know that

$$\mu_1(\text{gas phase}) = \mu_1(\text{liquid phase}),$$
$$\mu_2(\text{gas phase}) = \mu_2(\text{liquid phase}). \tag{25.56}$$

We shall use the convention that the component normally a liquid is labeled 1, and the component normally a gas is labeled 2. Equation 21.39 gives the chemical potentials of the components in the gas phase. In the liquid phase we have, for a small range of concentration near $x_1 = 1$,

$$\mu_1(\text{liquid phase}) = \mu_1{}^0 + RT \ln x_1$$
$$\mu_2(\text{liquid phase}) = \mu_2{}^\ominus + RT \ln x_2. \tag{25.57}$$

Note that in writing Eqs. 25.57 for the liquid phase we have assumed that the solution is ideal. We have chosen as the standard state for the liquid solvent, component 1, the pure liquid at the same temperature and pressure, and for the solute, component 2, the hypothetical state with the same properties per mole as an infinitely dilute solution ($\mu_2{}^\ominus$ is given by Eq. 25.23b). Using Eq. 25.56 and Eq. 21.39, we obtain

$$\mu_2{}^*(T) + RT \ln \frac{f_2}{\mathfrak{p}} = \mu_2{}^\ominus(T, p, \text{solvent}) + RT \ln x_2, \tag{25.58}$$

which can be rearranged to read

$$\frac{f_2}{\mathfrak{p}} = x_2 \exp\left(\frac{\mu_2{}^\ominus - \mu_2{}^*}{RT}\right) = x_2 k_2(T, p, \text{solvent}). \tag{25.59}$$

Equation 25.59 (or more often Eq. 25.60 just below) is known as *Henry's law*, and $k_2(T, p, \text{solvent})$ as the *Henry's law constant*. If the vapor phase is at sufficiently low pressure to be described as an ideal gas mixture, f_i can be replaced by p_i, the partial pressure of component i. In that limit we have

$$p_2 = x_2 \mathfrak{p} k_2(T, p, \text{solvent}) \qquad \text{(gas phase ideal)}. \tag{25.60}$$

Equation 25.60 is in agreement with the observation that the solubility of the gas is proportional to its partial pressure.

Corresponding to Eq. 25.59, we find for the solvent the result

$$\frac{f_1}{\mathfrak{p}} = x_1 \exp\left(\frac{\mu_1{}^0 - \mu_1{}^*}{RT}\right), \tag{25.61}$$

or, when the vapor phase is ideal,

$$\frac{p_1}{\mathfrak{p}} = x_1 \exp\left(\frac{\mu_1{}^0 - \mu_1{}^*}{RT}\right). \tag{25.62}$$

Equations 25.59 and 25.61 differ in an important way despite their superficial similarity. The difference arises from the choices of standard states for solute and solvent, and thereby illustrates an important point. We shall now show that the partial pressure of the solvent is proportional to the pressure a hypothetical pure phase would have. To show this we need merely evaluate the right-hand side of Eq. 25.61 or 25.62. The pure solvent vapor–liquid equilibrium is described by

$$\mu_1{}^*(T) + RT \ln \frac{f_1{}^0}{\mathfrak{p}} = \mu_1{}^0(T, p), \tag{25.63}$$

(see Eq. 21.15) or, on rearrangement,

$$\frac{f_1^0}{\mathfrak{p}} = \exp\left(\frac{\mu_1^0 - \mu_1^*}{RT}\right). \qquad (25.64)$$

But we have chosen the standard state for the solvent to be the pure liquid. Therefore, the exponentials on the right-hand sides of Eqs. 25.64 and 25.61 are equal, and we have

$$f_1 = x_1 f_1^0, \qquad (25.65)$$

or, when the gas phase is ideal,

$$p_1 = x_1 p_1^0, \qquad (25.66)$$

with p_1^0 the vapor pressure of pure component 1. In contrast, $k_2(T, p, \text{solvent})$ is *not* the vapor pressure of pure component 2. We illustrate in Fig. 25.3 how $k_2(T, p, \text{solvent})$ defines the vapor pressure (fugacity) of a hypothetical pure phase which has the same properties per mole as does the dilute solution.

Why does f_2^0 differ greatly from the vapor pressure the pure component would have at the same temperature and pressure when the concentration dependence of the chemical potential is of ideal form? The answer to this question becomes evident once it is realized that in a very dilute solution, such as the one we have described by Eqs. 25.57, every solute molecule is surrounded by solvent molecules only. Moreover, the solute molecules are far enough apart that solute–solute interaction is negligibly small. We expect that the solute molecule in the liquid can move only in an effective volume much smaller than that available in the gas, and therefore that the entropy of solution will be negative. However, the effective volume per molecule in solution need not be, and in general is not, the same as the effective volume per molecule in the pure liquid solute. In fact, if the interaction between two isolated solute molecules differs from that between an isolated solute–solvent molecule pair, there is no reason to expect the properties of the hypothetical state defined by extrapolating the low-concentration properties to be like those of the pure solute. For example, the solute–solvent attractive interaction will ordinarily be intermediate in magnitude between the solvent–solvent and solute–solute attractive interactions. Thus, there should be some excess or deficiency of binding of the solute to the solvent relative to the binding in the pure solute; that is, the enthalpy per solute molecule will be more negative or positive in the solution than in the pure solute. Moreover, in addition to the very large loss of translational entropy when a gas molecule enters the liquid phase, and the likelihood that the effective volume per solute molecule in solution differs from that in the pure solute, differences in the solute–solute and solute–solvent attractive and repulsive interactions can lead to changes in the range of local ordering around a solute molecule, and thereby to further changes in the entropy per molecule. The balance between the enthalpy and entropy differences per molecule is rarely exactly the same in the hypothetical standard state and the true pure component, so that in general we have $k_2(T, p, \text{solvent}) \neq p_2^0/\mathfrak{p}$. As indicated in the argument, the very existence of a difference between solute–solute and solute–solvent interactions also implies that k_2 and the chemical potential of the hypothetical standard state of the solute must depend on the nature of the solvent. Finally, the reason that the concentration dependence of the chemical potential, and of the entropy of mixing, is of the ideal form is that each solute molecule is effectively isolated in an environment of solvent. Changes in solute concentration in the range considered do not alter this environment.

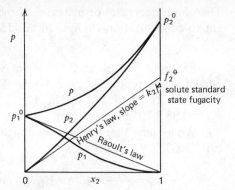

FIGURE 25.3
Schematic display of the Henry's law representation and the extrapolation to the standard state of the solute.

Therefore, decreases or increases of the solute concentration are equivalent to concentration changes in an ideal mixture.

Example. Water at 20°C is in equilibrium with gaseous H_2 at a partial pressure of H_2 of 200 torr. How many grams of H_2 are dissolved in 100 g of H_2O? The Henry's law constant for this system is $k_{H_2} = 6.83 \times 10^4$ for $p = 1$ atm.

We assume that the gaseous H_2 can be treated as ideal so that, from Eq. 25.60,

$$x_{H_2} = \frac{p_{H_2}}{k_{H_2}p} = 1.46 \times 10^{-5} p_{H_2}/1 \text{ atm} = 1.46 \times 10^{-5} \times \frac{200}{760}$$

$$= 3.85 \times 10^{-6}.$$

To compute the mass of dissolved H_2 we convert the mole fractions of H_2 and H_2O to masses. The mole fraction of H_2 in solution is so small relative to the mole fraction of H_2O that we can write, to a good approximation,

$$x_{H_2} = \frac{n_{H_2}}{n_{H_2} + n_{H_2O}} \approx \frac{w_{H_2}/M_{H_2}}{w_{H_2O}/M_{H_2O}},$$

where w_i is the mass in grams and M_i is the molecular weight of component i. We obtain

$$w_{H_2} = x_{H_2} \cdot \frac{w_{H_2O}M_{H_2}}{M_{H_2O}} = 3.85 \times 10^{-6} \times \frac{100 \times 2}{18}$$

$$= 4.28 \times 10^{-5} \text{ g}.$$

Example. The Henry's law constant for solution of N_2 in liquid H_2O varies with temperature as follows:

T (°C)	$k_{N_2} \times 10^{-7}$
0	4.08
5	4.57
10	5.07
15	5.55
20	6.00
25	6.43
30	6.85
35	7.23
40	7.61
45	7.99
50	8.37

What is the heat of solution of N_2 in water at 5°C? At 45°C?

At constant partial pressure of the solute gas, the solubility is defined by

$$\ln x_2 = \ln \frac{p_2}{p} - \ln k_2,$$

yielding for the temperature dependence

$$\left(\frac{\partial \ln x_2}{\partial T}\right)_{p_2} = -\left(\frac{\partial \ln k_2}{\partial T}\right)_{p_2} = -\left[\frac{\partial}{\partial T}\left(\frac{\mu_2^{\ominus} - \mu_2^*}{RT}\right)\right]_{p_2} = \frac{h_2^{\ominus} - h_2^*}{RT^2},$$

using Eqs. 25.59 and 25.7. The quantity $h_2^{\ominus} - h_2^*$ is the heat of solution of N_2 in liquid H_2O, that is, the difference in enthalpy between the hypothetical solution reference state and the perfect gas reference state. In slightly different words, $h_2^* - h_2^{\ominus}$ is the latent heat per mole required to drive a dissolved gas out of dilute solution.

From the data given, approximating derivatives by differences, we obtain

$$\left(\frac{\partial \ln k_2}{\partial T}\right)_{p_2} = \frac{1}{k_2}\left(\frac{\partial k_2}{\partial T}\right)_{p_2} \approx \frac{1}{k_2}\frac{\Delta k_2}{\Delta T} = \frac{0.99 \times 10^7}{4.57 \times 10^7} \times \frac{1}{10}$$

$$= 2.17 \times 10^{-2}\,\text{K}^{-1}\ \text{at } 5°C,$$

and

$$\left(\frac{\partial \ln k_2}{\partial T}\right)_{p_2} \approx 0.95 \times 10^{-2}\,\text{K}^{-1}\ \text{at } 45°C.$$

Then we find

$$h_2^* - h_2^{\ominus} = RT^2\left(\frac{\partial \ln k_2}{\partial T}\right)_{p_2} = 13.97\ \text{kJ/mol at } 5°C,$$

and

$$(h_2^* - h_2^{\ominus}) = 7.99\ \text{kJ/mol at } 45°C.$$

Note that the sign of $h_2^* - h_2^{\ominus}$ is positive, so that the gas solubility decreases as the temperature increases. Note also that there is a nonzero cohesive energy of the N_2 in the liquid, which decreases rapidly as the temperature increases. We easily obtain $\mu_2^* - \mu_2^{\ominus} = -RT \ln k_2$, and with the enthalpies of solution just calculated find for the entropies of solution the values

$$s_2^{\ominus} - s_2^* = -\frac{\mu_2^{\ominus} - \mu_2^* - h_2^{\ominus} + h_2^*}{T}$$

$$= -97.1\ \text{J/K mol at } 5°C,$$

and

$$s_2^{\ominus} - s_2^* = -128\ \text{J/K mol at } 45°C.$$

It is the large loss of entropy when the gas molecule is dissolved that is the principal factor limiting the solubility. This agrees with the qualitative arguments we advanced concerning the role of ordering in the liquid and the loss of translational motion.

We now turn to the study of nonideality in mixtures that are not dilute. In a nonideal mixture of volatile liquids (or solids), the partial pressures may be greater or less than those predicted by Raoult's law (see Fig. 25.4). Which of the two kinds of deviation occurs depends on molecular interactions in the condensed phase. To get a feeling for what happens, assume the entropy of mixing to be ideal. If unlike molecules attract one another less than do the repective pure-component molecular pairs, the mixture has a higher vapor pressure than predicted from Raoult's law; if unlike molecules are more strongly attracted to one another than like molecules, the mixture has a lower vapor pressure than predicted from Raoult's law. Deviations from the ideal form of the entropy of mixing also lead to deviations from Raoult's low, but these are often of less importance than the energetic changes mentioned, except when the unlike molecules are very different in size. Great differences in

FIGURE 25.4
Partial pressure as a function of composition for carbon disulfide–acetone and chloroform–acetone solutions. From R. H. Cole and J. S. Coles, *Physical Principles of Chemistry* (Freeman, San Francisco, 1964).

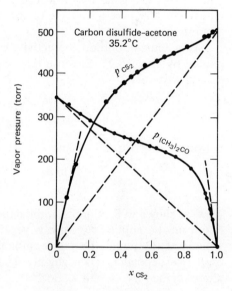

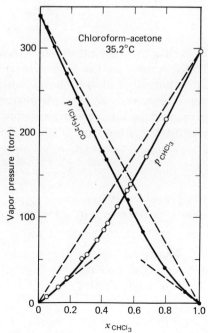

molecular size lead to considerable deviations from the mole-fraction form of the ideal entropy of mixing, and hence from Raoult's law.

Some examples of the concentration dependence of the excess enthalpy and entropy are shown in Fig. 25.5.

There are a number of different ways of determining the activity coefficients (hence the excess chemical potentials by Eq. 25.21) of the components of a mixture. Sticking to the use of liquid–vapor equilibrium for the present, we see that if enough is known about the properties of the vapor, Eq. 25.48 can be employed to determine the several γ_i. The measurements required are the total pressure p and composition y_1, $(1-y_1)$ of the vapor in equilibrium with a liquid of known composition x_1, $(1-x_1)$ at a given temperature, and the equation of state of the vapor. Regarding the latter, it is usually sufficiently accurate to retain only the second virial coefficient (see Chapter 21) because the vapor pressure is not large. Although this prescription for measurements sounds simple, it is actually difficult to carry through with high precision. For this reason much effort has gone into reworking the analysis of activity coefficients into forms that take advantage of those measurements that can be made accurately.

In a binary solution the activity coefficients of species 1 and 2 are related by Eqs. 25.27. Suppose that the vapor is ideal, so that Eq. 25.48 can be written

$$p_1 = \gamma_1 x_1 p_1{}^0,$$
$$p_2 = \gamma_2 x_2 p_2{}^0. \tag{25.67}$$

Substitution of Eq. 25.67 into Eq. 25.48 gives

$$(1-x_2)\left(\frac{\partial \ln p_1}{\partial x_2}\right)_{T,p} + x_2\left(\frac{\partial \ln p_2}{\partial x_2}\right)_{T,p} = 0. \tag{25.68}$$

Now, as shown in Section 23.1, the dependence of the vapor pressure of a liquid on the applied pressure is very small because of the small molar volume of the liquid. Therefore, although p_1 and p_2 in principle depend on p, that dependence is negligibly small and Eq. 25.68 can be accurately approximated by

$$(1-x_2)\left(\frac{\partial \ln p_1}{\partial x_2}\right)_{T} + x_2\left(\frac{\partial \ln p_2}{\partial x_2}\right)_{T} = 0, \tag{25.69}$$

in which the derivatives are at constant temperature only. Equation 25.69, known as the *Duhem–Margules equation*, is valid for arbitrary deviations from ideality in the liquid or solid mixture, provided that the vapor behaves as an ideal gas mixture and that the pressure dependences of p_1 and p_2 are indeed negligibly small. If we now use Eq. 25.52 to eliminate p_2 from Eq. 25.69, we find

$$\left(\frac{\partial p_1}{\partial x_2}\right)_T = \frac{(1-x_1)p_1}{p_1 - x_1 p}\left(\frac{\partial p}{\partial x_2}\right)_T, \tag{25.70}$$

which enables us to calculate p_1, p_2, γ_1, and γ_2 from a knowledge of the total vapor pressure as a function of liquid composition.

Example. Suppose that the deviations from ideality in some mixture are not large, and that the form of the activity coefficient given in Eq. 25.29 is valid. Then

$$p = p_1 + p_2 = \gamma_1 x_1 p_1{}^0 + \gamma_2 x_2 p_2{}^0 = x_1 p_1{}^0 \exp\left(\frac{wx_2{}^2}{RT}\right) + x_2 p_2{}^0 \exp\left(\frac{wx_1{}^2}{RT}\right).$$

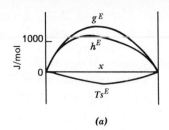

(a)

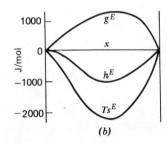

(b)

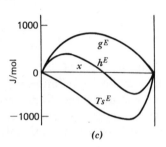

(c)

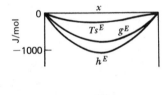

(d)

FIGURE 25.5

(a) The excess functions of the system acetonitrile + carbon tetrachloride at 45°C as a function of the mole fraction of acetonitrile.

(b) The excess functions of water + pyridine at 80°C as a function of the mole fraction of water.

(c) The excess functions of water + dioxane at 25°C as a function of the mole fraction of water.

(d) The excess functions of water + hydrogen peroxide at 25°C as a function of the mole fraction of water. From J. S. Rowlinson, *Liquids and Liquid Mixtures*, 2nd ed. (Butterworths, London, 1969).

We use the assumption that γ_1 and γ_2 are not far from unity to expand the exponentials to order x_2^2 and x_1^2, respectively:

$$p = x_1 p_1^0 \left(1 + \frac{w x_2^2}{RT}\right) + x_2 p_2^0 \left(1 + \frac{w x_1^2}{RT}\right).$$

Solving for w/RT gives

$$\frac{w}{RT} = \frac{p - (x_1 p_1^0 + x_2 p_2^0)}{x_1 x_2 [p_1^0 + (p_2^0 - p_1^0) x_1]}.$$

Since the total vapor pressure over an ideal mixture is just $x_1 p_1^0 + x_2 p_2^0 = p^{\text{ideal}}$, we can also write

$$\ln \gamma_1 = \frac{(p - p^{\text{ideal}}) x_2^2}{x_1 x_2 [p_1^0 + (p_2^0 - p_1^0) x_1]},$$

permitting easy evaluation from experimental measurements of p as a function of x_1.

We have seen how activity coefficients can be calculated from measurements of the composition dependence of the liquid–vapor equilibrium of a mixture. Consider, now, the inverse problem, namely, the calculation of the bubble-point and dew-point curves of a binary mixture from the known composition dependence of the chemical potentials, including the effects of nonideality. We start, as usual, with the condition for phase equilibrium. For equilibrium at temperature T and pressure p we have, from Eq. 25.63,

$$\mu_i^*(T) - \mu_i^0(T, p) + RT \ln \frac{f_i/p}{\gamma_i x_i}$$

$$= \mu_i^* - \mu_i^0(T, p) + RT \ln \frac{p}{p} + RT \ln \frac{f_i}{p} - RT \ln \gamma_i x_i = 0. \qquad (25.71)$$

Suppose that we consider equilibrium at the adjacent temperature $T + dT$ and pressure $p + dp$. Then

$$\mu_i^V(T + dT, p + dp) = \mu_i^L(T + dT, p + dp), \qquad (25.72)$$

which yields on expansion, using Eq. 25.71,

$$-R \, d \ln \frac{f_i/p}{\gamma_i x_i} = \left[\frac{\partial}{\partial T} \frac{\mu_i^*(T) - \mu_i^0(T, p)}{T} \right]_p dT$$

$$+ \left[\frac{\partial}{\partial p} \left(\frac{\mu_i^*(T) - \mu_i^0(T, p)}{T} \right)_T + R \ln \frac{p}{p} \right] dp$$

$$= \frac{h_i^0 - h_i^*}{T^2} dT - \frac{v_i^0(T, p) - RT/p}{T} dp. \qquad (25.73)$$

Equation 25.73 must be integrated to yield predictions concerning the bubble-point and dew-point curves.

Consider first the composition dependence of isothermal liquid–vapor equilibrium. From Eq. 25.73,

$$\ln \left(\frac{\gamma_1^V y_1}{\gamma_1 x_1} \right) = -\frac{1}{RT} \int_{p_1^0}^{p} \Delta v_1^0 \, dp \qquad (25.74a)$$

and

$$\ln \left(\frac{\gamma_2^V y_2}{\gamma_2 x_2} \right) = -\frac{1}{RT} \int_{p_2^0}^{p} \Delta v_2^0 \, dp, \qquad (25.74b)$$

where $\gamma_i^V \equiv f_i/y_i p$ is the activity coefficient of i in the vapor. Since the molar volume of the liquid is very small relative to that of the gas, Δv_1^0 and Δv_2^0 become effectively the molar volumes of 1 and 2 in the gas phase. If we now make the approximation that the vapor phase is an ideal gas mixture,

$$\gamma_i^V \equiv \frac{f_i}{y_i p} = \frac{y_i p}{y_i p} = 1 \qquad (i = 1, 2),$$

and set

$$\Delta v_1^0 = \Delta v_2^0 = \frac{RT}{p},$$

direct integration gives

$$\ln\left(\frac{y_1}{\gamma_1 x_1}\right) = \ln\left(\frac{p_1^0}{p}\right) \qquad (25.75a)$$

and

$$\ln\left(\frac{y_2}{\gamma_2 x_2}\right) = \ln\left(\frac{p_2^0}{p}\right), \qquad (25.75b)$$

which can be solved to give

$$x_2 = \frac{p_1^0 \gamma_1 - p}{p_1^0 \gamma_1 - p_2^0 \gamma_2} \qquad \text{(bubble-point curve)} \qquad (25.76)$$

and

$$y_2 = \frac{p_1^0 p_2^0 \gamma_1 \gamma_2 - p p_2^0 \gamma_2}{p p_1^0 \gamma_1 - p p_2^0 \gamma_2} \qquad \text{(dew-point curve)}. \qquad (25.77)$$

Note that Eq. 25.76 is just a rearrangement of the familiar $p = p_1 + p_2 = \gamma_1 x_1 p_1^0 + \gamma_2 x_2 p_2^0$.

To calculate the composition dependence of the constant pressure liquid–vapor equilibrium we transform Eq. 25.73 to read

$$\ln\left(\frac{\gamma_1^V y_1}{\gamma_1 x_1}\right) = \int_{T_{1b}}^{T} \frac{h_1^* - h_1^0}{RT'^2} \, dT' \qquad (25.78a)$$

and

$$\ln\left(\frac{\gamma_2^V y_2}{\gamma_2 x_2}\right) = \int_{T_{2b}}^{T} \frac{h_2^* - h_2^0}{RT'^2} \, dT'. \qquad (25.78b)$$

Note that $h_1^* - h_1^0$ is just the latent heat of vaporization of component 1, and T_{1b} its boiling point at the given fixed pressure. Consider the approximation that $h_1^* - h_1^0$ and $h_2^* - h_2^0$ are independent of temperature over the range $T_{1b}, T_{2b} \to T$. Then the integrals in Eqs. 25.78 are immediately found to be

$$\int_{T_{1b}}^{T} \frac{h_1^* - h_1^0}{RT'^2} \, dT' = \frac{h_1^* - h_1^0}{R}\left(\frac{1}{T_{1b}} - \frac{1}{T}\right) \equiv \Lambda_{1b} \qquad (25.79a)$$

and

$$\int_{T_{2b}}^{T} \frac{h_2^* - h_2^0}{RT'^2} \, dT' = \frac{h_2^* - h_2^0}{R}\left(\frac{1}{T_{2b}} - \frac{1}{T}\right) \equiv \Lambda_{2b}. \qquad (25.79b)$$

With the additional approximation that the vapor is an ideal gas mixture, $\gamma_1^V y_1 \to y_1$, $\gamma_2^V y_2 \to y_2$, and we find

$$y_2 = \frac{\exp(\Lambda_{1b})\gamma_2 - \gamma_1\gamma_2}{\exp(\Lambda_{1b})\gamma_2 - \exp(-\Lambda_{2b})\gamma_1}, \qquad (25.80)$$

$$x_2 = \frac{\exp(\Lambda_{1b}) - \gamma_1}{\exp(\Lambda_{1b})\exp(\Lambda_{2b})\gamma_2 - \gamma_1}. \qquad (25.81)$$

Example. Again suppose that

$$RT \ln \gamma_1 = wx_2{}^2,$$

$$RT \ln \gamma_2 = wx_1{}^2.$$

In this case Eq. 25.81 can be written

$$x_2 = \frac{\exp(\Lambda_{1b}) - \exp(wx_2{}^2/RT)}{\exp(\Lambda_{1b})\exp(\Lambda_{2b})\exp(wx_1{}^2/RT) - \exp(-wx_2{}^2/RT)},$$

which can be solved for x_2 by iteration from an initial guess. For the mixture acetone(1)–ether(2) at 30°C, $w = 1.88$ kJ/mol, $h_1{}^* - h_1{}^0 = 33.97$ kJ/mol, $T_{1b} = 329.8$ K, $h_2{}^* - h_2{}^0 = 26.19$ kJ/mol, $T_{2b} = 308.0$ K. The calculated and observed values of x_2 are as follows:

T (K)	x_2 (calc.)	x_2 (obs.)
308.0	1	1
313.0	0.45	0.45
317.0	0.28	0.28
321.0	0.16	0.16
325.0	0.007	0.007
329.8	0	0

25.5 Liquid–Solid Equilibrium

Henry's law describes the relationship between the solubility and the partial pressure or fugacity of a gas. It is also possible to use measurements of the solubility of a solid or liquid component in a solid or liquid solution to obtain information about the thermodynamic properties of the solution. Suppose that some pure solid, component 2, is in equilibrium with a liquid solution in which component 2 is a solute. The equilibrium is described by

$$\mu_2{}^S(T, p) = \mu_2{}^\ominus(T, p, \text{solvent}) + RT \ln \gamma_2 x_2. \qquad (25.82)$$

Consider now the solution–solid equilibrium at the same pressure but at a different temperature, T_0. We shall assume that at the temperature T_0 component 2 has a solubility x_{20}. Then we have

$$\mu_2{}^S(T_0, p) = \mu_2{}^\ominus(T_0, p, \text{solvent}) + RT \ln \gamma_{20} x_{20}. \qquad (25.83)$$

If Eq. 25.82 is divided by T and Eq. 25.83 by T_0, and the resultant equations are subtracted, we find that

$$R \ln \gamma_2 x_2 - R \ln \gamma_{20} x_{20} = \frac{\mu_2{}^S(T) - \mu_2{}^\ominus(T)}{T} - \frac{\mu_2{}^S(T_0) - \mu_2{}^\ominus(T_0)}{T_0}. \qquad (25.84)$$

The right-hand side of Eq. 25.84 is easily evaluated. To do so we use the fundamental relation connecting the partial molar enthalpy and the chemical potential (see Table 19.2):

$$\left[\frac{\partial(\mu_2/T)}{\partial T}\right]_p = -\frac{\bar{h}_2}{T^2}. \qquad (25.85)$$

The integral form of Eq. 25.85 is

$$\frac{\mu_2(T)}{T} - \frac{\mu_2(T_0)}{T_0} = -\int_{T_0}^{T} \frac{\bar{h}_2}{T'^2}\, dT'. \tag{25.86}$$

The use of Eq. 25.86 on the right-hand side of Eq. 25.84 finally yields

$$\ln \gamma_2 x_2 - \ln \gamma_{20} x_{20} = \int_{T_0}^{T} \frac{h_2^{\ominus} - h_2^{S}}{RT'^2}\, dT'$$

$$= \int_{T_0}^{T} \frac{L_{2f}^{\ominus}\, dT'}{RT'^2}. \tag{25.87}$$

If the solubility at T_0 is sufficiently small that $\gamma_{20} = 1$, then

$$\ln \gamma_2 x_2 - \ln x_{20} = \int_{T_0}^{T} \frac{L_{2f}^{\ominus}}{RT'^2}\, dT'. \tag{25.88}$$

The quantity $L_{2f}^{\ominus} \equiv h_2^{\ominus} - h_2^{S}$ is the molar heat of fusion of pure solid 2 into the hypothetical state with the same properties per mole as exhibited by the infinitely dilute solution of component 2. Equation 25.87 enables us to compute the ratio of the activity coefficients of component 2 at two temperatures from solubility measurements at those temperatures. If one of the temperatures is sufficiently low that the chemical potential of the solute has the ideal form, then Eq. 25.88 enables us to calculate the absolute value of $\gamma_2(T)$.

Example. The latent heat of fusion of p-dibromobenzene is 13220 J/mol, and the melting point is 86.9°C. The difference in heat capacity between solid and liquid is 24.8 J/K mol. Assuming that mixtures of benzene and p-dibromobenzene are ideal, calculate the solubility of p-dibromobenzene in benzene at 20°C.

When a mixture is ideal, the standard states of the solute species are identical with the states of the pure liquids at the same temperature and pressure. Thus for an ideal solution $L_{2f}^{\ominus}$ as defined in Eq. 25.87 is the molar heat of fusion of the solid to the supercooled liquid at the same temperature and pressure. This quantity must be calculated from the heat of fusion at the melting point and the difference in molar heat capacity between solid and liquid. We have

$$L_{2f}^{\ominus}(T) = \Delta h_{2f}(T_{2f}) + \int_{T_{2f}}^{T} (c_p^{L} - c_p^{S})\, dT'.$$

Returning to Eq. 25.87, we set $\gamma_2 = 1$ for an ideal solution. If we take the temperature T_0 to be the melting temperature T_{2f}, we have $\gamma_{20} x_{20} = 1$; this is easily seen from Eq. 25.83, since $\mu_2^{\ominus} = \mu_2^{L}$ for an ideal solution and at the melting point $\mu_2^{S} = \mu_2^{L}$. Finally, substituting the expansion of $L_{2f}^{\ominus}$ into Eq. 25.87 with $T_0 = T_{2f}$ gives

$$\ln x_2 = \int_{T_{2f}}^{T} \frac{1}{RT'^2} \left[\Delta h_{2f}(T_{2f}) + \int_{T_{2f}}^{T'} \Delta c_p\, dT'' \right] dT'.$$

When Δc_p is independent of temperature, this reduces to

$$\ln x_2 = \frac{(\Delta h_{2f} - T_{2f}\Delta c_p)(T - T_{2f})}{RTT_{2f}} + \frac{\Delta c_p}{R} \ln \frac{T}{T_{2f}}.$$

Using the figures given, we obtain

$$\ln x_2 = -0.936$$

or

$$x_2 = 0.392.$$

As already noted, when the solution is ideal we have $\mu_2^{\Theta} = \mu_2^{L}$, so that the calculated solubility at a given temperature is independent of the solvent. When the solution is not ideal this statement is not valid. The fact that the solubility of a component is the same in all ideal solutions is easily rationalized in terms of molecular interactions, because in an ideal solution solute–solute and solute–solvent interactions are identical.

As shown in the preceding example, when we set T_0 equal to T_{2f}, the melting temperature of pure solid 2, Eq. 25.87 becomes

$$\ln \gamma_2 x_2 = \int_{T_{2f}}^{T} \frac{L_{2f}^{\Theta} \, dT'}{RT'^2}, \tag{25.89}$$

which describes the temperature dependence of the coexistence curve between pure solid 2 and the liquid mixture with which it is in equilibrium.

How does the solubility of a pure solid in contact with a liquid mixture change when the applied pressure is changed at constant temperature? By virtue of the phase equilibrium at the two pressures, Eq. 25.82 is valid at both p and $p + dp$. We have, then, for this isothermal change,

$$\mu_2^{S}(T, p+dp) - \mu_2^{\Theta}(T, p+dp, \text{solvent})$$
$$- \mu_2^{S}(T, p) + \mu_2^{\Theta}(T, p, \text{solvent})$$
$$= RT \ln (\gamma_2 x_2)_{p+dp} - RT \ln (\gamma_2 x_2)_p. \tag{25.90}$$

Expansion of Eq. 25.90 as a Taylor series gives

$$\frac{\partial}{\partial p} [\mu_2^{S}(T, p) - \mu_2^{\Theta}(T, p, \text{solvent})]_T \, dp$$
$$= RT\left[\left(\frac{\partial \ln \gamma_2}{\partial p}\right)_{T,x_2} dp + \left(\frac{\partial \ln \gamma_2 x_2}{\partial x_2}\right)_{T,p} dx_2\right], \tag{24.91}$$

since γ_2 is a function of p and x_2. The left-hand side of Eq. 25.91 is the difference between the molar volume of the solid and the molar volume in the standard state, $v_2^{S} - v_2^{\Theta}$. The term $RT(\partial \ln \gamma_2/\partial p)_{T,x_2}$ on the right-hand side is, from Eq. 25.24, just $\bar{v}_2 - v_2^{\Theta}$. Introducing these terms and rearranging Eq. 25.91 leads to

$$(v_2^{S} - \bar{v}_2)_T \, dp + RT\left(\frac{\partial \ln \gamma_2 x_2}{\partial x_2}\right)_{T,p} dx_2 = 0 \tag{25.92}$$

and

$$\left(\frac{\partial x_2}{\partial p}\right)_T = \frac{\bar{v}_2 - v_2^{S}}{RT\left(\dfrac{\partial \ln \gamma_2 x_2}{\partial x_2}\right)_{T,p}}. \tag{25.93}$$

The denominator must be positive (see Eq. 19.86), so $(\partial x_2/\partial p)_T$ is positive or negative according to the sign of $\bar{v}_2 - v_2^{S}$. For an ideal mixture $\bar{v}_2 = v_2^{S}$, and the solubility is independent of the applied pressure.

Thus far we have examined the case that a pure solid is in equilibrium with a liquid mixture. In general we must expect that both the solid and liquid phases will be mixtures. What does the liquid–solid equilibrium coexistence curve look like in the more general case?

Suppose that both the liquid and solid mixtures exist over the entire composition range. The coexistence line at constant pressure can be calculated exactly as in the case of liquid–vapor equilibrium. All that need be done is to replace $\mu_i{}^*(T)$ and $RT \ln f_i/\mathfrak{p}$ in Eq. 25.71 with the corresponding quantities for a solid mixture, namely, $\mu_i{}^{OS}(T, p)$ and $RT \ln \gamma_i{}^S x_i{}^S$. Equation 25.78 then becomes

$$\ln \frac{\gamma_1{}^S x_1{}^S}{\gamma_1{}^L x_1{}^L} = \int_{T_{1f}}^{T} \frac{h_1{}^{OS} - h_1{}^{OL}}{RT'^2} \, dT' \qquad (25.94a)$$

and

$$\ln \frac{\gamma_2{}^S x_2{}^S}{\gamma_2{}^L x_2{}^L} = \int_{T_{2f}}^{T} \frac{h_2{}^{OS} - h_2{}^{OL}}{RT'^2} \, dT', \qquad (25.94b)$$

and $h_1{}^{OS} - h_1{}^{OL}$ is the latent heat of crystallization of component 1, with a similar definition for component 2. If $h_1{}^{OS} - h_1{}^{OL}$ and $h_2{}^{OS} - h_2{}^{OL}$ are independent of temperature over the range T_{1f}, $T_{2f} \to T$ and we define, analogous to Eq. 25.79, the following variables,

$$\frac{h_1{}^{OS} - h_1{}^{OL}}{R}\left(\frac{1}{T_{1f}} - \frac{1}{T}\right) \equiv \Lambda_{1f}, \qquad (25.95a)$$

$$\frac{h_2{}^{OS} - h_2{}^{OL}}{R}\left(\frac{1}{T_{2f}} - \frac{1}{T}\right) \equiv \Lambda_{2f}, \qquad (25.95b)$$

then

$$x_2{}^L = \frac{\exp(\Lambda_{1f})\gamma_1{}^L\gamma_2{}^S - \gamma_2{}^S\gamma_1{}^L}{\exp(\Lambda_{1f})\gamma_1{}^L\gamma_2{}^S - \exp(-\Lambda_{2f})\,\gamma_2{}^L\gamma_1{}^S} \qquad (25.96)$$

and

$$x_2{}^S = \frac{\exp(\Lambda_{1f})\,\gamma_1{}^L\gamma_2{}^L - \gamma_2{}^L\gamma_1{}^S}{\exp(\Lambda_{1f})\exp(\Lambda_{2f})\,\gamma_1{}^L\gamma_2{}^S - \gamma_2{}^L\gamma_1{}^S}. \qquad (25.97)$$

The line $T(x_2{}^L)$ is called the *liquidus*; it is analogous to the dew-point curve. The line $T(x_2{}^S)$ is called the *solidus*; it is analogous to the bubble-point curve. An example of a T, x_2 diagram is displayed in Fig. 25.6. Note that if both the liquid and solid mixtures are ideal, and all activity coefficients are unity, then Eqs. 25.96 and 25.97 reduce to

$$x_2{}^L = \frac{\exp(\Lambda_{1f}) - 1}{\exp(\Lambda_{1f}) - \exp(-\Lambda_{2f})}, \qquad (25.98)$$

and

$$x_2{}^S = \frac{\exp(\Lambda_{1f}) - 1}{\exp(\Lambda_{1f})\exp(\Lambda_{2f}) - 1}. \qquad (25.99)$$

Usually there are gross deviations from the ideal mixture behavior described by Eqs. 25.98 and 25.99.

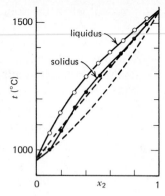

FIGURE 25.6
Phase diagram for liquid–solid equilibrium in the system Ag(1)–Pd(2). The dashed curves are the calculated liquidus and solidus lines for ideal solutions (Eqs. 25.98 and 25.99).

Thus far we have considered the thermodynamic properties of mixtures in which both components are volatile. If one of the two components of a binary mixture is nonvolatile, for example, sugar in water, special simplifications of the analysis are possible. Consider, for example, the equilibrium between the vapor and solution phases when only the solvent has appreciable vapor pressure. Equation 25.44 describes the solvent equilibrium. However, by presumption the vapor phase now contains only the solvent species. Proceeding as in Eqs. 25.45–25.48, we see that the net effect of adding a nonvolatile solute to a volatile solvent is to lower the vapor pressure. Put the other way around, because the vapor pressure is proportional to the mole fraction of the solvent and

25.6
The Colligative Properties of Solutions: Boiling-Point Elevation, Freezing-Point Depression, and Osmotic Pressure

increases with temperature, the temperature at which the solution boils must be higher than the temperature at which the pure solvent boils. This boiling-point elevation can be used to determine the molecular weight of the nonvolatile solute or, if that is known, the activity coefficient of the solvent.

As usual, let component 1 be the solvent and component 2 the nonvolatile solute. As stated, Eq. 25.44 describes the gas–liquid phase equilibrium for the solvent in the solution. When the solvent is pure, gas–liquid equilibrium at the boiling point is described by

$$\mu_1{}^V(T_{1b}, p) = \mu_1{}^L(T_{1b}, p). \tag{25.100}$$

Proceeding as in Eqs. 25.82 et seq., we divide Eq. 25.44 by T, Eq. 25.100 by T_{1b}, and subtract to find

$$\frac{\mu_1{}^V(T, p)}{T} - \frac{\mu_1{}^V(T_{1b}, p)}{T_{1b}} = \frac{\mu_1{}^L(T, p)}{T} - \frac{\mu_1{}^L(T_{1b}, p)}{T_{1b}} + R \ln \gamma_1 x_1. \tag{25.101}$$

Using Eq. 25.86, we can rewrite Eq. 25.101 in the form

$$\ln \gamma_1 x_1 = -\int_{T_{1b}}^{T} \frac{h_1{}^V - h_1{}^L}{RT'^2} \, dT' = -\int_{T_{1b}}^{T} \frac{L_{1v} \, dT'}{RT'^2}, \tag{25.102}$$

where it is understood that the integration is along a path at constant pressure, with $L_{1v} \equiv h_1{}^V - h_1{}^L$ the latent heat of vaporization of pure solvent. Of course, the latent heat of vaporization depends on the temperature because the heat capacities of the vapor and liquid are different. Setting

$$L_{1v} = \Delta h_{1v} + [c_p{}^V - c_p{}^L](T - T_{1b}), \tag{25.103}$$

we find

$$\begin{aligned}
\ln \gamma_1 x_1 &= -\int_{T_{1b}}^{T} \frac{\Delta h_{1v} + \Delta c_p(T' - T_{1b})}{RT'^2} \, dT' \\
&= -\int_{0}^{\theta} \frac{\Delta h_{1v} + \Delta c_p \theta'}{R(T_{1b} + \theta')^2} \, d\theta',
\end{aligned} \tag{25.104}$$

where $\theta \equiv T - T_{1b}$ is the boiling-point difference between the solution and the solvent. Ordinarily the boiling-point change is small compared to the boiling point, so that we have $(\theta/T_{1b}) \ll 1$ and

$$\frac{1}{(T_{1b} + \theta)^2} \approx \frac{1}{T_{1b}{}^2}\left(1 - \frac{2\theta}{T_{1b}} + \cdots\right) \tag{25.105}$$

can be used in the integrand of Eq. 25.104. Finally, by substituting Eq. 25.105 in Eq. 25.104 and carrying out the integration, we obtain

$$\ln \gamma_1 x_1 = -\frac{\Delta h_{1v} \theta}{RT_{1b}{}^2} + \left(\frac{\Delta c_p}{2R} + \frac{\Delta h_{1v}}{RT_{1b}}\right)\left(\frac{\theta}{T_{1b}}\right)^2 + \mathcal{O}(\theta^3) \tag{25.106}$$

(where $\mathcal{O}$ means "order of"). The second term on the right-hand side of Eq. 25.106 is ordinarily small relative to the first term. Note that all the variables on the right-hand side except θ are functions of the solvent only. Thus, all nonvolatile solutes elevate the boiling point of a volatile solvent by the same amount when their mole fractions are the same.

Suppose that the solution we are studying is ideal and dilute so that, with only the term linear in θ retained, Eq. 25.106 reduces to

$$\ln x_1 = -\frac{\Delta h_{1v} \theta}{RT_{1b}{}^2}. \tag{25.107}$$

But in a dilute solution we can write

$$\ln x_1 = \ln(1 - x_2) \approx -x_2, \tag{25.108}$$

so that

$$x_2 = \frac{\Delta h_{1v} \theta}{RT_{1b}^2}. \tag{25.109}$$

We also have, when $x_2 \ll 1$,

$$x_2 = \frac{w_2/M_2}{w_1/M_1 + w_2/M_2} \approx \frac{w_2/M_2}{w_1/M_1}, \tag{25.110}$$

so that the molecular weight of the solute can be determined from

$$M_2 = \frac{w_2}{(w_1/M_1)} \cdot \frac{RT_{1b}^2}{\Delta h_{1v} \theta} = 1000 \frac{w_2}{w_1} \cdot \frac{K_b}{\theta}. \tag{25.111}$$

For convenience, it is usual to tabulate, for various solvents, the quantity $K_b \equiv (M_1/1000)(RT_{1b}^2/\Delta h_{1v})$. This constant, known as the *ebullioscopic constant*, gives the boiling-point increase when 1 mol of nonvolatile solute is dissolved in 1000 g of solvent. This concentration can be described by saying that the *molality* (number of moles in 1000 g of solvent) is unity. For the case of CCl_4, the ebullioscopic constant is 5.03°C kg/mol.

Example. 0.250 g of an unknown nonvolatile substance is mixed with 96.8 g of CCl_4, and the boiling point of the mixture is determined. It is found that the boiling point of the solution is 0.055°C higher than that of pure CCl_4. What is the molecular weight of the solute?

Now, 0.250 g of solute per 96.8 g of solvent is equivalent to 2.59 g of solute per 1000 g of solvent. Thus, if 2.59 g produces a boiling-point elevation of 0.055°C, then 2.69 g must be

$$\frac{0.055}{5.03} = 0.0109 \text{ mol};$$

that is, the molecular weight of the solute is

$$M_2 = 2.59 \times \frac{1}{0.0109} = 236.$$

Note that in terms of the molality the arithmetic described is equivalent to writing

$$\theta = K_b m_2,$$

where $m_2 \equiv 1000 w_2/M_2 w_1$ is the molality of the solute. Values of K_b for several solvents are given in Table 25.1.

The reason that the boiling-point elevation of a solution of a non-volatile solute in a volatile solvent depends directly on the activity of the solvent is that there is no solute in the vapor. Similarly, if we consider the equilibrium between a solution and a crystalline phase of the solvent in which the solute is insoluble, the difference in freezing point between the solution and the pure solvent is related directly to the activity of the solvent. The analysis of this freezing-point depression parallels that for the boiling-point elevation. In fact, if in Eq. 25.100 *et seq.* we every-where replace functions referring to the solvent in the gas phase with functions referring to the solvent in the soiid phase, we obtain

$$\ln \gamma_1 x_1 = -\frac{\Delta h_{1f} \theta}{RT_{1f}} \cdot \frac{1}{T_{1f}} - \left(\frac{\Delta c_p}{2R} + \frac{\Delta h_{1f}}{RT_{1f}}\right)\left(\frac{\theta}{T_{1f}}\right)^2 + \mathcal{O}(\theta^3), \quad (25.112)$$

where the subscript f refers to fusion (or freezing) and here $\theta \equiv T_{1f} - T$ is

TABLE 25.1 EBULLIOSCOPIC CONSTANTS

	T_b (°C)	K_b (°C kg/mol)
Benzene	80.15	2.53
Ethanol	78.26	1.22
Ethyl iodide	72.5	5.05
n-Heptane	98.42	3.43
Naphthalene	218.0	5.65
Water	100.0	0.512

TABLE 25.2 CRYOSCOPIC CONSTANTS

	T_f (°C)	K_f (°C kg/mol)
Benzene	5.5	5.12
Bromoform	7.8	14.4
Diphenylamine	52.9	8.6
Naphthalene	80.2	6.9
Stannic bromide	31	28.0
Water	0	1.86

the freezing-point depression. As before, θ measures the activity of the solvent in the presence of a given amount of solute. When the solute concentration is low and $(\theta/T_{1f}) \ll 1$, for an ideal solution we have

$$x_2 = \frac{\Delta h_{1f}\theta}{RT_{1f}^2}, \qquad (25.113)$$

analogous to Eq. 25.109 for the boiling-point elevation. If the solute concentration is expressed in molality, a *cryoscopic constant*, K_f, can be defined such that

$$\theta = K_f m_2, \qquad (25.114)$$

and freezing-point depression measurements can be used to determine solute molecular weights, and so on. The values of K_f for a few solvents are given in Table 25.2.

Example. The freezing-point depressions of a series of solutions of urea in water have been determined as follows:

m_2	θ(°C)
0.510	0.928
1.850	3.260
3.660	6.190
7.140	11.400

Calculate the activity of the water at the freezing temperature for each of these concentrations.

From[6]

$$\theta = K_f m_2$$

we have

$$K_f \equiv \frac{M_1}{1000} \cdot \frac{RT_{1f}^2}{\Delta h_{1f}}$$

for dilute enough solutions that $x_1 \gg x_2$. Thus, from Eq. 25.112 we obtain

$$\ln \gamma_1 x_1 = -\frac{\Delta h_{1f}\theta}{RT_{1f}^2} = -\frac{\theta}{K_f} \cdot \frac{M_1}{1000}.$$

[6] Careful examination of the tabulated data shows that θ/m_2 is not a constant, hence the quadratic term in Eq. 25.112 cannot be neglected. For purposes of illustration we shall use only the linear approximation.

927

Using the value of K_f given in Table 25.2, we can reduce this to

$$\ln \gamma_1 x_1 = -0.968 \times 10^{-2}\theta.$$

We obtain the results tabulated below using the data given for the urea–water solutions:

m_2	$\theta(°C)$	$\gamma_1 x_1$	x_1	γ_1	θ/m_2
0.510	0.928	0.9908	0.9911	1.000	1.8196
1.850	3.260	0.9689	0.9677	1.001	1.762
3.660	6.190	0.9419	0.9382	1.004	1.691
7.140	11.400	0.8957	0.8861	1.011	1.597

Note how the solvent's activity coefficient increases as the solute concentration increases.

There is one other simple method that can be used to exclude a component from one of two phases in equilibrium. Suppose that two phases are separated by a membrane permeable to solvent but not to solute (see Fig. 25.7). At equilibrium, assuming that the membrane is rigid and therefore the confined system cannot do mechanical work, but that energy can cross the membrane, we have

$$p^{\text{I}} \neq p^{\text{II}} \quad \text{(rigid membrane)},$$
$$T^{\text{I}} = T^{\text{II}} \quad \text{(membrane permeable to energy)}, \quad (25.115)$$
$$\mu_1^{\text{I}} = \mu_1^{\text{II}} \quad \text{(membrane permeable to solvent)}.$$

In general, the chemical potential can be related to the pressure in a system by (see Table 19.2)

$$\left(\frac{\partial \mu_i}{\partial p}\right)_{T,x_i} = \bar{v}_i. \quad (25.116)$$

Integration of Eq. 25.116 between p_0 and $p_0 + \pi$ gives

$$\mu_i(T, p_0 + \pi, x) - \mu_i(T, p_0, x) = \int_{p_0}^{p_0 + \pi} \bar{v}_i\, dp. \quad (25.117)$$

In the case under consideration the solvent chemical potential on side I, where there is no solute, is $\mu_1^L(T, p_0)$, whereas on side II the solvent chemical potential is $\mu_1^L(T, p_0 + \pi) + RT \ln \gamma_1 x_1$. The pressure π is known as the *osmotic pressure*. Applying Eq. 25.117 to the particular case of the solvent we find, using Eq. 25.115,

$$RT \ln \gamma_1 x_1 = \mu_1^L(T, p_0) - \mu_1^L(T, p_0 + \pi)$$
$$= -\int_{p_0}^{p_0 + \pi} \bar{v}_1\, dp. \quad (25.118)$$

Suppose that the partial molar volume of solvent varies smoothly with pressure, so that we can write

$$\bar{v}_1(p) = \bar{v}_1(p_0) + (p - p_0)\left(\frac{\partial \bar{v}_1}{\partial p}\right)_{T,x,p_0} + \cdots. \quad (25.119)$$

The compressibilities of most liquids and the osmotic pressures generated in dilute solutions are so small that under most circumstances $\bar{v}_1(p_0) \gg (p - p_0)(\partial \bar{v}_1/\partial p)_{T,x,p_0}$. In this limit we have

$$RT \ln \gamma_1 x_1 = -\pi \bar{v}_1(p_0). \quad (25.120)$$

Once again consider the case that $x_2 \ll x_1$ and the solution is ideal. Then

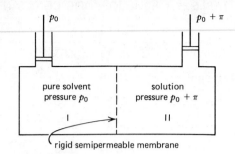

FIGURE 25.7
Schematic representation of an osmotic pressure experiment.

we obtain

$$RT \ln x_1 \approx -RTx_2 = -\pi \bar{v}_1. \tag{25.121}$$

But in a dilute solution we have

$$x_2 \approx \frac{w_2/M_2}{w_1/M_1} = \frac{n_2}{n_1} \tag{25.122}$$

and

$$\frac{w_1}{M_1} \bar{v}_1 = n_1 \bar{v}_1 \approx V, \tag{25.123}$$

where V is the total volume of the solution. Combining Eqs. 25.121, 25.122, and 25.123, we find for an ideal dilute solution

$$\pi V = n_2 RT, \tag{25.124}$$

which closely resembles the perfect gas law. In fact, the resemblance can be made even clearer if we use a different representation of the effects of nonideality in the solution. Suppose that we write

$$-\pi \bar{v}_1(p_0) = RT \ln \gamma_1 x_1 \equiv g_1 RT \ln x_1. \tag{25.125}$$

Equation 25.125 defines the quantity g_1 in terms of γ_1. It is equally valid to use γ_1 or g_1 to describe deviations from ideality, although in osmotic pressure experiments g_1, known as the *osmotic coefficient*, is more useful. Now suppose that g_1 varies with the concentration, $c_2 = n_2/V$, as

$$g_1 = 1 + Bc_2 + Cc_2{}^2 + \cdots. \tag{25.126}$$

Then Eq. 25.125 becomes, in a dilute solution,

$$\pi = RTc_2(1 + Bc_2 + \cdots), \tag{25.127}$$

which closely resembles the virial equation of state. The solution virial coefficients can be calculated from the statistical molecular theory of solutions. It is found that the solution virial coefficients depend on the effective interaction potential (potential of mean force) in the solution in the same way as do the gas virial coefficients on the true intermolecular potential.

Clearly, just as the freezing-point depression and boiling-point elevation can be used to determine solute molecular weight, solvent activity, and so on, so can the osmotic pressure of the solution be used. These three phenomena are known as the *colligative properties* of a solution.

Example. Determine the molecular weight of the enzyme chymotrypsinogen from the datum that 1.553 g dissolved in 100 ml of solution has, at 25°C, an osmotic pressure of 157.0 mm H_2O. We write for the osmotic pressure, assuming ideal solution behavior,

$$\pi = \frac{w_2}{M_2} \cdot \frac{RT}{V}$$

or

$$M_2 = \frac{w_2}{\pi} \frac{RT}{V}.$$

Converting the pressure in mm H_2O to the standard units of (torr) by dividing by 13.596, the density of Hg in g/cm^3, we find

$$M_2 = \frac{1.553 \times 298 \times 0.08206}{\left(\dfrac{157.0}{13.596} \times \dfrac{1}{760}\right) \times 0.100} = 25,000.$$

Osmotic pressure experiments are frequently used to measure the molecular weights of polymers and proteins. Ordinarily solutions of these materials are not ideal, and it is necessary to use Eq. 25.127, plotting π/c_2 against c_2 and extrapolating to $c_2 = 0$ in order to determine the molecular weight.

25.7
Chemical Reactions in Nonelectrolyte Solutions

Given the concentration dependences and the standard states for the chemical potentials of all components of a solution, the treatment of chemical equilibrium is very simple. The general conditions describing chemical equilibrium in a mixture were derived in Section 19.5. There it was shown that at chemical equilibrium we have

$$\sum_{i=1}^{r} \nu_i \mu_i = 0, \qquad (25.128)$$

where the ν_i are the stoichiometric coefficients for the reaction under consideration. Proceeding as in Section 19.5, but using for the chemical potentials the forms appropriate to a condensed phase, we find that

$$\Delta G^0 = -RT \ln K(T, p, \text{solvent}), \qquad (25.129)$$

where

$$\Delta G^0 \equiv \sum_{i=1}^{r} \nu_i \mu_i^0 \qquad (25.130)$$

and

$$K(T, p, \text{solvent}) \equiv \prod_{i=1}^{r} (\gamma_i x_i)^{\nu_i}. \qquad (25.131)$$

When the standard state for each component is chosen as the pure liquid or solid, the equilibrium constant is a function only of the temperature and pressure. However, when the standard state is the hypothetical state with the same properties per unit mass as an infinitely dilute solution, the equilibrium constant also depends on the solvent. Clearly, the temperature and pressure derivatives of $K(T, p, \text{solvent})$ depend on the enthalpy change and the volume change in the reaction, just as in the gaseous reactions discussed in Section 21.3.

For an ideal solution we can write

$$K_{\text{ideal solution}} = \prod_{i=1}^{r} x_i^{\nu_i}. \qquad (25.132)$$

Note that the equilibrium constant in Eq. 25.132 is dimensionless, as is the corresponding equilibrium constant in a gas (Eqs. 21.65 and 21.66).

There is a commonly used, but misleading, convention in which the equilibrium constant appears to have units. This convention arises as follows. Suppose that, instead of mole fraction units, we use some other concentration variable to describe the composition dependence of the chemical potential. When the molality is chosen as the concentration unit, the only standard state that may be used is the one referring to the hypothetical substance with the same properties per unit mass as the solute at infinite dilution. The reason for this is that the molal scale presumes that one component is singled out as the solvent. The mole fraction of a component is

$$x_i = \frac{n_i}{n_1 + \sum_{j=2}^{r} n_j} = \frac{w_i/M_i}{(w_1/M_1) + \sum_{j=2}^{r} (w_j/M_j)}. \qquad (25.133)$$

To obtain molalities we fix $w_1 = 1000$ g; the value of w_j/M_j in the solution is then the molality of j, denoted m_j. That is, m_j is the number of moles of j in 1000 g of solvent. Thus, Eq. 25.133 becomes

$$x_i = \frac{m_i}{(1000/M_1) + \sum\limits_{j=2}^{r} m_j}. \tag{25.134}$$

We now define the activity coefficients on the molality scale from the equation

$$\mu_i = \mu_i^0(T, p) + RT \ln \left\{ \frac{\gamma_i m_i}{(1000/M_1)\left[1 + (M_1/1000) \sum\limits_{j=2}^{r} m_j\right]} \right\}$$

$$= \mu_i^0(T, p) + RT \ln \left\{ \left[\frac{\gamma_i}{1 + (M_1/1000) \sum\limits_{j=2}^{r} m_j}\right]\left(\frac{m_i}{1000/M_1}\right) \right\}$$

$$= \mu_i^0(T, p) - RT \ln \mathfrak{m}_1 + RT \ln \gamma_i' m_i, \tag{25.135}$$

where

$$\gamma_i' \equiv \frac{\gamma_i}{1 + (M_1/1000) \sum\limits_{j=2}^{r} m_j} = \frac{\gamma_i}{1 + \sum\limits_{j=2}^{r} (m_j/\mathfrak{m}_1)} \tag{25.136}$$

is the activity coefficient of component i and

$$\mathfrak{m}_1 \equiv \frac{1000}{M_1}. \tag{25.137}$$

Note that, despite the change in concentration units, we have retained the dimensionless form of the argument of the logarithm, because the molality is a dimensionless number referred to a standard mass of solvent. Clearly, $\mathfrak{m}_1$ is the molality of the solvent, a different number for every different solvent but the same number for all solutions in the same solvent ($\mathfrak{m}_1 = 55.51$ for H_2O). It is usual for the factor $RT \ln \mathfrak{m}_1$ to be absorbed into the chemical potential of the standard state, thereby defining a new standard-state chemical potential. With reference to the molality concentration scale and the new standard state for which $\mu_i^{0m}(T, p, \text{solvent}) = \mu_i^0(T, p) - RT \ln \mathfrak{m}_1$, the equilibrium constant becomes

$$K_m(T, p, \text{solvent}) = \prod_{i=1}^{r} (\gamma_i' m_i)^{\nu_i} \tag{25.138}$$

and is a dimensionless number.

A similar analysis can be used to write $K(T, p, \text{solvent})$ in terms of the molar concentrations, c_i (*molarity* c, number of moles per liter of solution). In this case the new standard-state chemical potential differs from $\mu_i^0(T, p)$ by the added term $RT \ln(v_1^0/1000)$ where v_1^0 is the molar volume of pure solvent, and the 1000 has units cm^3 and refers to the standard 1 liter of solution for which the molarity is defined. Of course, the activity coefficients γ_i'' on the molarity scale differ from γ_i and γ_i', as K and K_m differ from

$$K_c(T, p, \text{solvent}) = \prod_{i=1}^{r} (\gamma_i'' c_i)^{\nu_i}. \tag{25.139}$$

The important point is that, whatever concentration unit is chosen, the argument of the logarithm in the chemical potential must be dimensionless. Then, strictly, $K(T, p, \text{solvent})$ will be dimensionless. However, because different standard states correspond to different choices of concentration units, it is common (as a convenience) to attach to $K(T, p, \text{solvent})$ units indicating the concentration scale in which it is defined.

Brushing aside questions of which concentration units are most convenient, we note that all the features of chemical equilibrium previously described are also relevant to the description of reactions in liquid or solid solution. Thus, an arbitrary reaction can be subdivided into different sequences of reactions, and from suitable tables of data the changes in free energy, enthalpy, and entropy for the reaction of interest can be computed as described in Chapters 14, 19, and 21.

The phenomena associated with phase equilibrium in mixtures are surprisingly numerous and diverse in character. And, of course, the more components there are in the mixture, the greater the range of possible behavior. Consequently, the study of phase equilibrium in mixtures is a very large subject, and we shall be able only to touch on a few points and illustrate a few important principles of interpretation.

Although we have examined liquid–vapor and liquid–solid equilibria for mixtures, in each case the focus of the argument was on exploitation of the features of the phase equilibrium in a fashion that permitted extraction of information about the properties of the components of the mixture. We now consider the stability of a mixture against phase separation. Unlike a pure component, even at constant temperature and pressure a mixture may be unstable with respect to separation into two phases with different compositions. Now, within one phase the Gibbs free energy at constant temperature and pressure satisfies the equation (see Table 19.2)

$$dG = \mu_1 \, dn_1 + \mu_2 \, dn_2 \qquad (T, p, \text{constant}), \qquad (25.140)$$

or, in terms of the mole fractions,

$$dG = (n_1 + n_2)\mu_1 \, dx_1 + (n_1 + n_2)\mu_2 \, d(1 - x_1)$$
$$= (n_1 + n_2)(\mu_1 - \mu_2) \, dx_1$$
$$= (n_1 + n_2)(\mu_2 - \mu_1) \, dx_2. \qquad (25.141)$$

Thus, a plot of G versus x_2 has slope proportional to $\mu_2 - \mu_1$ at every point. Provided that phase separation does not occur, G versus x_2 is a continuous curve (see Fig. 25.8a). If phase separation occurs and phases I and II with different compositions are in equilibrium, we have

$$\mu_1^{\text{I}} = \mu_1^{\text{II}} \qquad (25.142a)$$

and

$$\mu_2^{\text{I}} = \mu_2^{\text{II}}, \qquad (25.142b)$$

or

$$\mu_1^{\text{I}} - \mu_2^{\text{I}} = \mu_1^{\text{II}} - \mu_2^{\text{II}}. \qquad (25.143)$$

In a plot of G versus x_2 for a system of two phases, there will be separate lines for each phase (see Fig. 25.8b). Equation 25.143 implies that the slopes of the two curves are equal when the phases are in equilibrium. Hence the compositions of the two phases can be determined by drawing the common tangent to the two curves. For the example shown in Fig.

25.8
More About Phase Equilibrium in Mixtures

FIGURE 25.8
Schematic representation of the Gibbs free energy as a function of composition for a binary mixture. (a) One-phase region. (b) Two-phase region.

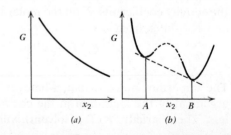

25.8, the two phases are in equilibrium when their compositions are those at points A and B, respectively.

Consider a mixture for which the Gibbs free energy resembles the curve in Fig. 25.8a for temperatures $T > T_c$, and resembles the curve in Fig. 25.8b for temperatures $T < T_c$. The temperature T_c, called the *critical temperature of mixing*, characterizes that one isotherm for which the change from curve a behavior to curve b behavior occurs. Clearly, T_c plays a role in the thermodynamics of mixtures analogous to that of the critical temperature in the thermodynamics of pure fluids.

Everywhere along curve a of Fig. 25.8 we have $(\partial^2 G/\partial x_2^2) > 0$ (G is concave upward). If the separate parts of curve b are joined, then there is a region for which $(\partial^2 G/\partial x_2^2) < 0$ (G is concave downward) connecting two regions in which $(\partial^2 G/\partial x_2^2) > 0$. Therefore, there must be two points between those labeled A and B for which $(\partial^2 G/\partial x_2^2) = 0$. These are points of inflection of the free-energy curve. At T_c the points just defined merge into one point at which

$$\left(\frac{\partial^2 G}{\partial x_2^2}\right)_{T_c, p} = 0 \qquad (25.144a)$$

and

$$\left(\frac{\partial^3 G}{\partial x_2^3}\right)_{T_c, p} = 0. \qquad (25.144b)$$

Using 25.141 in Eqs. 25.144a and 25.144b, we find that at the critical temperature

$$\left(\frac{\partial \mu_1}{\partial x_2}\right)_{T_c, p} = \left(\frac{\partial \mu_2}{\partial x_2}\right)_{T_c, p} \qquad (25.145a)$$

and

$$\left(\frac{\partial^2 \mu_1}{\partial x_2^2}\right)_{T_c, p} = \left(\frac{\partial^2 \mu_2}{\partial x_2^2}\right)_{T_c, p}. \qquad (25.145b)$$

We also have, from the Gibbs–Duhem relation at constant temperature and pressure,

$$(1 - x_2)\left(\frac{\partial \mu_1}{\partial x_2}\right)_{T, p} + x_2\left(\frac{\partial \mu_2}{\partial x_2}\right)_{T, p} = 0. \qquad (25.146)$$

After differentiation with respect to x_2, this yields

$$(1 - x_2)\left(\frac{\partial^2 \mu_1}{\partial x_2^2}\right)_{T, p} + x_2\left(\frac{\partial^2 \mu_2}{\partial x_2^2}\right)_{T, p} - \left(\frac{\partial \mu_1}{\partial x_2}\right)_{T, p} + \left(\frac{\partial \mu_2}{\partial x_2}\right)_{T, p} = 0. \quad (25.147)$$

It follows from Eqs. 25.145–25.147 that at T_c we must have

$$\left(\frac{\partial \mu_1}{\partial x_2}\right)_{T_c, p} = \left(\frac{\partial \mu_2}{\partial x_2}\right)_{T_c, p} = 0 \qquad (25.148a)$$

and

$$\left(\frac{\partial^2 \mu_1}{\partial x_2^2}\right)_{T_c, p} = \left(\frac{\partial^2 \mu_2}{\partial x_2^2}\right)_{T_c, p} = 0. \qquad (25.148b)$$

When represented in terms of activity coefficients, the conditions 25.148 assume the form

$$\left(\frac{\partial \ln \gamma_1}{\partial x_2}\right)_{T_c, p} - \frac{1}{1 - x_2} = 0, \qquad (25.149a)$$

$$\left(\frac{\partial \ln \gamma_2}{\partial x_2}\right)_{T_c, p} + \frac{1}{x_2} = 0, \qquad (25.149b)$$

$$\left(\frac{\partial^2 \ln \gamma_1}{\partial x_2{}^2}\right)_{T_c,p} - \frac{1}{(1-x_2)^2} = 0, \qquad (25.149c)$$

$$\left(\frac{\partial^2 \ln \gamma_2}{\partial x_2{}^2}\right)_{T_c,p} - \frac{1}{x_2{}^2} = 0. \qquad (25.149d)$$

To illustrate the preceding analysis of phase separation in a mixture, we evaluate T_c for the regular mixture model. For a regular mixture we have

$$RT \ln \gamma_1 = wx_2{}^2, \qquad (25.29)$$

so that, using Eqs. 25.149a and 25.149c, we find

$$\frac{w}{RT_c}(2x_2)_c = \frac{1}{(1-x_2)_c} \qquad (25.150a)$$

and

$$\frac{w}{RT_c} \cdot 2 = \frac{1}{(1-x_2)_c{}^2}. \qquad (25.150b)$$

The solution to these simultaneous equations is

$$(x_2)_c = \frac{1}{2}, \qquad \frac{w}{RT_c} = 2. \qquad (25.151)$$

Thus, in a regular mixture, if $w/RT > 2$ (i.e., if $T < T_c$), there is incomplete mixing.

Consider some temperature and pressure at which a binary liquid mixture separates into two liquid phases with compositions $x_1{}^I$, $x_2{}^I$ and $x_1{}^{II}$, $x_2{}^{II}$. How do the compositions of these phases change when the temperature and pressure are altered? This problem is a variant of the calculation of the bubble-point and dew-point curves, or of the calculation of the solidus and liquidus curves (see Eqs. 25.71–25.78 and 25.94–25.97). The key point is that $\mu_1{}^I = \mu_1{}^{II}$ and $\mu_2{}^I = \mu_2{}^{II}$ at both T, p and $T+dT$, $p+dp$, so along the coexistence line $d\mu_1{}^I = d\mu_1{}^{II}$ and $d\mu_2{}^I = d\mu_2{}^{II}$. By expansion we then find

$$\left(\frac{\partial \mu_1}{\partial p}\right)_{T,x}^I dp + \left(\frac{\partial \mu_1}{\partial T}\right)_{p,x}^I dT + \left(\frac{\partial \mu_1}{\partial x_2}\right)_{T,p}^I dx_2$$
$$= \left(\frac{\partial \mu_1}{\partial p}\right)_{T,x}^{II} dp + \left(\frac{\partial \mu_1}{\partial T}\right)_{p,x}^{II} dT + \left(\frac{\partial \mu_1}{\partial x_2}\right)_{T,p}^{II} dx_2, \qquad (25.152)$$

and similarly for component 2. Equation 25.152 is the analog of Eq. 25.73. Evaluation of the derivatives in Eq. 25.152, with use of Eq. 25.43 with $z_m = g_m$, followed by rearrangement leads to

$$(\bar{v}_1{}^I - \bar{v}_1{}^{II}) \, dp - (\bar{s}_1{}^I - \bar{s}_1{}^{II}) \, dT$$
$$- x_2{}^I\left(\frac{\partial^2 g_m}{\partial x_2{}^2}\right)_{T,p}^I dx_2{}^I + x_2{}^{II}\left(\frac{\partial^2 g_m}{\partial x_2{}^2}\right)_{T,p}^{II} dx_2{}^{II} = 0 \qquad (25.153)$$

and

$$(\bar{v}_2{}^I - \bar{v}_2{}^{II}) \, dp - (\bar{s}_2{}^I - \bar{s}_2{}^{II}) \, dT$$
$$+ x_1{}^I\left(\frac{\partial^2 g_m}{\partial x_2{}^2}\right)_{T,p}^I dx_2{}^I - x_1{}^{II}\left(\frac{\partial^2 g_m}{\partial x_2{}^2}\right)_{T,p}^{II} dx_2{}^{II} = 0. \qquad (25.154)$$

Consider the equilibrium states that lie along the constant-pressure coexistence line. In this case we find

$$\left(\frac{\partial x_2}{\partial T}\right)_p^I = -\frac{(\bar{h}_1{}^I - \bar{h}_1{}^{II})x_1{}^{II} + (\bar{h}_2{}^I - \bar{h}_2{}^{II})x_2{}^{II}}{T(\partial^2 g_m/\partial x_2{}^2)_{T,p}^I (x_2{}^I - x_2{}^{II})} \qquad (25.155)$$

and

$$\left(\frac{\partial x_2}{\partial T}\right)_p^{II} = -\frac{(\bar{h}_1^{I} - \bar{h}_1^{II})x_1^{I} + (\bar{h}_2^{I} - \bar{h}_2^{II})x_2^{I}}{T(\partial^2 g_m/\partial x_2^2)_{T,p}^{II}(x_2^{I} - x_2^{II})}. \qquad (25.156)$$

Within the stable phases $(\partial^2 g_m/\partial x_2^2) > 0$, so the sign of $(\partial x_2/\partial T)_p$ is determined by the balance between $x_2^{I} - x_2^{II}$ and the sign of the numerator. The interpretation of the numerator is straightforward; it is, in the case of Eq. 25.155, the molar heat of solution of an infinitesimal amount of phase II in phase I at constant temperature and pressure, and similarly for Eq. 25.156.

If instead of the constant-pressure coexistence line we examine the constant-temperature coexistence line, we find

$$\left(\frac{\partial x_2}{\partial p}\right)_T^{I} = \frac{x_2^{II}(\bar{v}_2^{II} - \bar{v}_2^{I}) + x_1^{II}(\bar{v}_1^{II} - \bar{v}_1^{I})}{(x_2^{II} - x_2^{I})(\partial^2 g_m/\partial x_2^2)_{T,p}^{I}} \qquad (25.157)$$

and

$$\left(\frac{\partial x_2}{\partial p}\right)_T^{II} = \frac{x_2^{I}(\bar{v}_2^{II} - \bar{v}_2^{I}) + x_2^{I}(\bar{v}_1^{II} - \bar{v}_1^{I})}{(x_2^{II} - x_2^{I})(\partial^2 g_m/\partial x_2^2)_{T,p}^{II}}. \qquad (25.158)$$

Just as for $(\partial x_2/\partial T)_p$, the sign of $(\partial x_2/\partial p)_T$ depends on a balance between the composition difference $x_2^{II} - x_2^{I}$ and the sign of the numerator. The numerator of Eq. 25.157 is the volume change on solution of an infinitesimal amount of phase I in phase II, and similarly for Eq. 25.158.

Somewhat different conditions are satisfied along the constant-composition coexistence lines. If phase I has constant composition, so that $dx_1^{I} = 0$, $dx_2^{I} = 0$, we find

$$\left(\frac{\partial p}{\partial T}\right)_{x_2^{I}} = \frac{x_2^{II}(\bar{h}_2^{II} - \bar{h}_2^{I}) + x_1^{II}(\bar{h}_1^{II} - \bar{h}_1^{I})}{T[x_2^{II}(\bar{v}_2^{II} - \bar{v}_2^{I}) + x_1^{II}(\bar{v}_1^{II} - \bar{v}_1^{I})]} \qquad (25.159)$$

and

$$\left(\frac{\partial x_2}{\partial T}\right)_{x_2^{I}}^{II} = \frac{(\bar{v}_1^{II} - \bar{v}_1^{I})(\bar{h}_2^{II} - \bar{h}_2^{I}) - (\bar{v}_2^{II} - \bar{v}_2^{I})(\bar{h}_1^{II} - \bar{h}_1^{I})}{T(\partial^2 g_m/\partial x_2^2)_{T,p}^{II}[x_2^{II}(\bar{v}_2^{II} - \bar{v}_2^{I}) + x_1^{II}(\bar{v}_1^{II} - \bar{v}_1^{I})]}. \qquad (25.160)$$

Equation 25.159 is the analog of the Clausius–Clapeyron equation for this system in that it gives the variation of the total pressure with temperature of a phase of fixed composition in equilibrium with another phase.

The results just obtained lead to two important conclusions:

1. If $x_1^{I} = x_1^{II}$, $x_2^{I} = x_2^{II}$, then from Eqs. 25.155 and 25.156 we find that

$$\left(\frac{\partial T}{\partial x_2}\right)_p^{I} = 0, \qquad \left(\frac{\partial T}{\partial x_2}\right)_p^{II} = 0. \qquad (25.161)$$

Therefore, if at constant pressure the temperature along the coexistence line between two phases of a binary mixture has a maximum, a minimum, or a point of inflection with horizontal tangent, the two phases have the same composition, and *vice versa*.

2. If in a series of isothermal equilibrium states of a two-phase binary system the composition of the phases becomes the same, Eq. 25.157 shows that the pressure of the system must pass through an extremum, and *vice versa*.

These conclusions are known as the *Gibbs–Konovalov theorems*. They are important in the study of azeotropy—at an *azeotropic point* the compositions of a binary liquid and its vapor are the same. More generally, an azeotropic transformation is said to occur if, in a closed system with several phases in equilibrium, the mass of one of the phases increases at the expense of the others without any change in the composition of any of the phases. Figure 25.9 shows T, x_2 diagrams for typical azeotropic systems. Systems with a maximum in the total vapor pressure curve have a positive excess Gibbs free energy, those with a minimum a negative excess Gibbs free energy.

We now consider again the stability of a mixture, but from a more general point of view than expressed at the beginning of this section. In Section 24.3 we introduced the energy surface $u(s, v)$ to aid in the visualization of the behavior of a one-component system in the region of phase separation. Alternatively, we could have used the surface $g(T, p)$ for that purpose. Since the properties of mixtures are usually studied via projections of the surface $g(T, p, x_1, \ldots)$, we start by examining that surface for a one-component system.

A schematic representation of the surface $g(T, p)$ for a one-component system is presented in Fig. 25.10. A point on that surface, say Γ_0, represents the Gibbs free energy of one mole of substance at pressure p_0 and temperature T_0. The plane tangent to Γ_0 has slope $(\partial g/\partial T)_{p=p_0} = -s_0$ along the line of intersection with the plane $p = p_0$, and slope $(\partial g/\partial p)_{T=T_0} = v_0$ along the line of intersection with the plane $T = T_0$, hence has the equation

$$g - g_0 = -s_0(T - T_0) + v_0(p - p_0). \tag{25.162}$$

This tangent plane cuts the g axis at

$$g = g_0 + T_0 s_0 - p_0 v_0 = u_0. \tag{25.163}$$

We conclude that the surfaces $u(s, v)$ and $g(T, p)$ for the same system have a reciprocal relationship. That is, the quantities represented by the tangent plane to $g(T, p)$ are on the surface $u(s, v)$ and *vice versa*. This reciprocal relationship is fundamentally a consequence of the Legendre transformation used in Section 17.1 to define $g(T, p)$.

There are differences in the representations afforded by $u(s, v)$ and $g(T, p)$. Whereas there is only one surface $u(s, v)$, with any gap bridged by a derived surface, it is convenient to think of $g(T, p)$ as having several sheets, one for each homogeneous phase of the substance. Strictly, if we omit consideration of metastable states, $g(T, p)$ is one surface formed from the intersection and breaking off of pieces from three surfaces. That is, the intersection of $g^L(T, p)$ and $g^S(T, p)$ is a curve defining the melting line for the equilibrium between liquid and solid, and $g^S(T, p)$ does not extend beyond that intersection into the domain where one would have $g^S(T, p) > g^L(T, p)$. A corresponding statement can be made about $g^L(T, p)$. If, however, we include one-phase metastable states just as we did in the $u(s, v)$ representation,[7] then the surfaces do penetrate somewhat beyond their several intersections. In Fig. 25.10 the projections of the intersections EM, MA, and MD onto the (T, p) plane are the stable solid–vapor, solid–liquid, and liquid–vapor coexistence lines, respectively, and the projection of M is the triple point. The continuations $M'F'$, $M'B'$, and $M'C'$ are projections of the intersections of a metastable part of one surface with a metastable part of another surface. For example, $M'F'$ is the extension of the sublimation curve into the region where the liquid is the stable phase. Note that the extended curve describes coexistence between two phases each of which is unstable relative to the third phase in the region of the extension. In terms of the geometry of Fig. 25.10, above the space $E'M'B'$ $g^V < g^S < g^L$; above the space $B'M'D'$ $g^V < g^L < g^S$; above the space $D'M'F'$ $g^L < g^V < g^S$; above the space $F'M'A'$ $g^L < g^S < g^V$; above the space $A'M'C'$ $g^S < g^L < g^V$, and above the space $C'M'E'$ $g^S < g^V < g^L$.

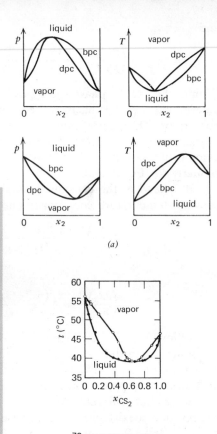

(a)

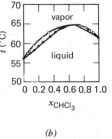

(b)

FIGURE 25.9

(*a*) Minimum- and maximum-boiling-point binary system phase diagrams. The bubble-point curve, labeled bpc, represents the liquid composition; the dew-point curve, labeled dpc, represents the vapor composition.

(*b*) Phase diagrams for the binary systems acetone–carbon disulfide and acetone–chloroform. The partial pressures of components in these two systems are shown, at 35.2°C, in Fig. 25.4. From J. Hildebrand and R. Scott, *Regular Solutions* (Van Nostrand Reinhold, New York, 1950).

[7] The states on the continuation of $u(s, v)$ above the derived surface are metastable.

Given that every point on the surface $g(T, p)$ represents a stable state of the system, this surface is concave downward in its two principal directions, as can be verified by noting that

$$\left(\frac{\partial^2 g}{\partial T^2}\right)_p = -\left(\frac{\partial s}{\partial T}\right)_p = -\frac{c_p}{T} < 0 \qquad (25.164a)$$

and

$$\left(\frac{\partial^2 g}{\partial p^2}\right)_T = \left(\frac{\partial v}{\partial p}\right)_T = -v\kappa_T < 0. \qquad (25.164b)$$

Since $c_p > 0$, $T > 0$, $v > 0$, and $k_T > 0$ for a stable system. Furthermore each point on the surface $g(T, p)$ must be a minimum with respect to virtual variations (Section 19.3); these virtual variations generate different surfaces $g'(T, p)$, $g''(T, p), \ldots$, that are everywhere above $g(T, p)$ and above the tangent plane to $g(T, p)$ at the particular T, p point.

$$G = \sum_{i=1}^{r} n_i \mu_i \qquad (25.165)$$

can be written

$$g_m \equiv \frac{G}{\sum_i n_i} = \sum_{i=1}^{r} x_i \mu_i$$

$$= \left(1 - \sum_{i=2}^{r} x_i\right)\mu_1 + \sum_{i=2}^{r} x_i \mu_i. \qquad (25.166)$$

If $T, p, x_2, \ldots, x_r$ are considered to be orthogonal axes, $g_m(T, p, x_2, \ldots, x_r)$ is a surface in an $(r+2)$-dimensional space. Although it is difficult to visualize this surface because of the large number of dimensions required for the representation, it is a useful construct in the analysis of the restrictions imposed by the conditions that define a stable equilibrium state. As shown in Section 19.4, an r-component system will be stable at given fixed temperature and pressure if $g_m(T, p, x_2, \ldots, x_r)$ is a minimum.

Let $\Gamma_0 \equiv (T_0, p_0, x_2^0, \ldots, x_r^0)$ be a point on the surface $g_m(T, p, x_2, \ldots, x_r)$. Given that this point corresponding to a stable equilibrium state, virtual variations that generate new states in the vicinity of Γ_0 have Gibbs free energies that lie on surfaces $g_m', g_m'', \ldots$, each of which is above $g_m(T_0, p_0, x_2^0, \ldots, x_r^0)$ and also the tangent plane at Γ_0, measured parallel to the g_m-axis. Therefore, for small displacements $\Delta x_2, \ldots, \Delta x_r$ at fixed temperature and pressure.

$$\Delta g_m - (\Delta g_m)_{\text{tangent plane}} > 0 \qquad (\text{constant } T, p), \qquad (25.167)$$

where $(\Delta g_m)_{\text{tangent plane}}$ is given by the extension of Eq. 25.162. If Δg_m is expanded in a Taylor series about Γ_0, then $(\Delta g_m)_{\text{tangent plane}}$ is equal to the collection of terms linear in the Δx_i. Keeping terms quadratic in the Δx_i, we find that Eq. 25.167 is equivalent to

$$\left[\left(\frac{\partial^2 g_m}{\partial x_2^2}\right)_{T,p,n}(\Delta x_2)^2 + \left(\frac{\partial^2 g_m}{\partial x_2 \partial x_3}\right)_{T,p,n}\Delta x_2 \Delta x_3 + \cdots + \left(\frac{\partial^2 g_m}{\partial x_2 \partial x_r}\right)_{T,p,n}\Delta x_2 \Delta x_r\right.$$

$$+ \left(\frac{\partial^2 g_m}{\partial x_3 \partial x_2}\right)_{T,p,n}\Delta x_3 \Delta x_2 + \cdots \qquad\qquad + \left(\frac{\partial^2 g_m}{\partial x_3 \partial x_r}\right)_{T,p,n}\Delta x_3 \Delta x_r$$

$$\vdots$$

$$\left. + \left(\frac{\partial^2 g_m}{\partial x_r \partial x_2}\right)_{T,p,n}\Delta x_r \Delta x_2 + \cdots \qquad\qquad + \left(\frac{\partial^2 g_m}{\partial x_r^2}\right)_{T,p,n}(\Delta x_r)^2\right] > 0. \qquad (25.168)$$

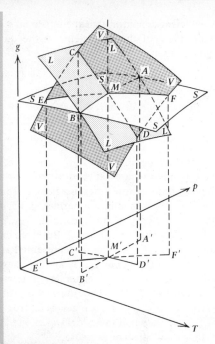

FIGURE 25.10

Free energy-temperature-pressure surface for a substance. The surfaces labelled V, L and S refer to the vapor, liquid and solid, respectively. (Courtesy of Prof. J. A. Beattie.)

A sufficient condition that the left-hand side of Eq. 25.168 be positive is that the determinant of the coefficients in the quadratic form be positive definite. This requires that

$$\Delta_{r-1} \equiv \begin{vmatrix} \left(\dfrac{\partial^2 g_m}{\partial x_2^2}\right)_{T,p,n} & \cdots & \left(\dfrac{\partial^2 g_m}{\partial x_2 \, \partial x_r}\right)_{T,p,n} \\ \vdots & & \\ \left(\dfrac{\partial^2 g_m}{\partial x_r \, \partial x_2}\right)_{T,p,n} & \cdots & \left(\dfrac{\partial^2 g_m}{\partial x_r^2}\right)_{T,p,n} \end{vmatrix} > 0 \qquad (25.169)$$

and that all the minors of Δ_{r-1} be positive.[8] Any system that is in a state for which these conditions are satisfied is stable with respect to so-called adjacent states, namely, those whose free energies are spanned by Δg_m of Eq. 25.167.

We now examine the surface $g_m(T, p, x_2)$ which represents the equilibrium states of a binary mixture. Even in this simple case we need a four-dimensional space to represent $g_m(T, p, x_2)$. But if we examine the free-energy surface for fixed pressure, $g_m(T, p, x_2)$ can be represented as a surface in a three-dimensional space defined by orthogonal g_m, T, and x_2 axes. Projection of the surface $g_m(T, p, x_2)$ onto the (T, x_2) plane generates a temperature-composition diagram such as shown in Fig. 25.11a. The intersections of $g_m(T, p, x_2)$ with the several planes on which $T = T_1$, $T = T_2, \ldots$ generate free energy-composition diagrams for T_1, $T_2, \ldots$, as shown in Figs. 25.11a–25.11j. All of the diagrams 25.11a–25.11j refer to the same constant pressure.

For the case of a binary mixture, $\Delta_{r-1} = \Delta_1$ reduces to a single term. We find that the binary mixture will be stable if

$$\left(\frac{\partial^2 g_m}{\partial x_2^2}\right)_{T,p} > 0 \qquad (25.170)$$

and unstable if

$$\left(\frac{\partial^2 g_m}{\partial x_2^2}\right)_{T,p} < 0. \qquad (25.171)$$

The limit of stability occurs when

$$\left(\frac{\partial^2 g_m}{\partial x_2^2}\right)_{T,p} = 0. \qquad (25.172)$$

Condition 25.170 can be expressed in terms of the chemical potential. Using Eq. 25.165 and the Gibbs–Duhem equation (25.25) in the form

$$x_1 \left(\frac{\partial \mu_1}{\partial x_2}\right)_{T,p} + x_2 \left(\frac{\partial \mu_2}{\partial x_2}\right)_{T,p} = 0, \qquad (25.173)$$

we readily find that condition 25.170 requires

$$\left(\frac{\partial \mu_2}{\partial x_2}\right)_{T,p} > 0, \qquad (25.174)$$

which also implies that

$$\left(\frac{\partial \mu_1}{\partial x_2}\right)_{T,p} < 0 \qquad (25.175)$$

and

$$\left(\frac{\partial \mu_2}{\partial x_1}\right)_{T,p} < 0. \qquad (25.176)$$

Thus, in a stable mixture μ_2 must increase when x_2 is increased and T and p are held constant.

[8] The minors of a determinant are the smaller determinants formed when a row and column are deleted.

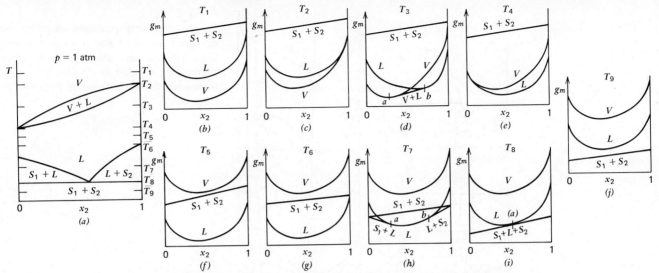

Typical temperature composition diagram for various temperatures

FIGURE 25.11

(a) Temperature-composition diagram for a binary mixture in which neither compounds nor solid solutions are formed, there is no miscibility gap, and there are no minimum- or maximum-boiling mixtures.

(b) T_1 is greater than the boiling points of both components, hence the vapor is the stable state for all compositions.

(c) The temperature is, in this case, equal to the boiling temperature of component 2, hence $g_m^L = g_m^V$ at $x_2 = 1$. Everywhere else the vapor is the most stable state.

(d) T_3 intersects the region where vapor and liquid coexist. For large x_2 the liquid phase is most stable, for small x_2 the vapor phase is most stable. The common tangent is the line along which g_m is least, hence describes the equilibrium situation of coexisting liquid and vapor between domains where either only liquid ($x_2 \geqslant b$) or only vapor ($x_2 \leqslant a$) is stable. The phase coexistence domain is $a < x_2 < b$.

(e) T_4 coincides with the boiling point of component 1, so this case is the analog of that in Fig. 25.11c and $g_m^L = g_m^V$ at $x_2 = 0$.

(f, g) The analogs of Figs. 25.11b and 25.11c with liquid and solid replacing vapor and liquid in the descriptions.

(h) T_7 intersects the coexistence lines for liquid–solid 1 and liquid–solid 2, hence two common tangents need to be drawn in the g_m, x_2 diagram.

(i) T_8 touches the liquid–solid coexistence curve at only one point, where solids 1 and 2 and liquid are in equilibrium. This is called the eutectic temperature.

(j) T_9 lies below the melting points of both solids, hence the two solids are coexisting phases.

All of the preceding also apply to solid solutions; only the free energy-composition diagram needs relabeling for the appropriate phases. (Courtesy of Prof. J. A. Beattie.)

We also note that at constant temperature and pressure the slope of the free energy-composition curve is (see also Eq. 25.42)

$$\left(\frac{\partial g_m}{\partial x_1}\right)_{T,p} = \mu_1 - \mu_2 = -\left(\frac{\partial g_m}{\partial x_2}\right)_{T,p}. \qquad (25.177)$$

When $x_1 \to 0$, $g_m \to \mu_2^{\,0}$ and $(\partial g_m/\partial x_1)_{T,p,n} \to -\infty$, whereas when $x_1 \to 1$, $g_m \to \mu_1^{\,0}$ and $(\partial g_m/\partial x_1)_{T,p,n} \to \infty$. The behavior of the derivatives follows from the dependence of the chemical potential on the logarithm of the mole fraction (see Eq. 25.12).

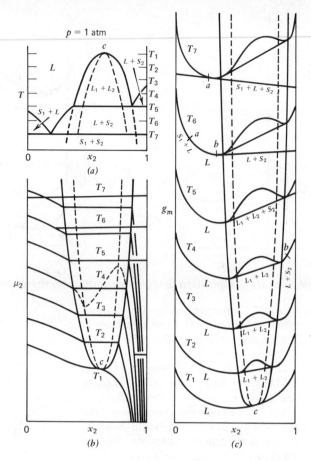

$p = 1$ atm

(a)

(b)

(c)

FIGURE 25.12
Temperature-composition diagram at constant pressure and the corresponding molar free energy-composition diagrams at constant pressure and temperature. (Courtesy of Prof. J. A. Beattie.)

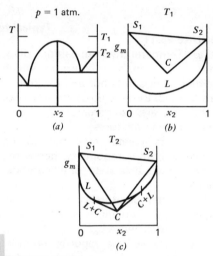

$p = 1$ atm.

(a)

(b)

(c)

FIGURE 25.13
The components form a compound C that melts congruently. (Courtesy of Prof. J. A. Beattie.)

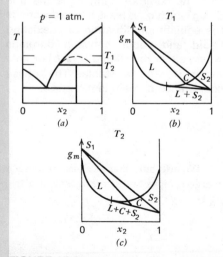

$p = 1$ atm.

(a)

(b)

(c)

FIGURE 25.14
The components form a compound C that melts incongruently. (Courtesy of Prof. J. A. Beattie.)

From the preceding we deduce that the free energy-composition diagram has the following properties:

1. For a homogeneous system g_m is a simple curve extending from $x_2 = 0$ to $x_2 = 1$; it is tangent to the axes $x_2 = 0$ and $x_2 = 1$, the slopes there being such that g_m is concave upward.

2. Corresponding to the existence of three phases of a homogeneous system there will be at least three lines on the g_m, x_2 diagram, one each for the vapor, liquid, and solid.

3. As shown by Eq. 25.143, points on the g_m, x_2 diagram that represent two coexisting phases have a common tangent.

The free energy-composition diagrams for a system are, of course, different at different temperatures. Figures. 25.11a, 25.12a, and 25.13a–25.16a display a variety of temperature-composition diagrams. The simplest of these, Fig. 25.11a, represents a binary system in which neither compounds nor solid solutions are formed, in which there is not any region of liquid–liquid insolubility (miscibility gap), and in which there are not either minimum- or maximum-boiling solutions (azeotropic mixtures). Even in this case the several free energy-composition diagrams for different temperatures, Figs. 25.11b–25.11j, look very different. The examples displayed are described in the caption, so only one will be cited here. Figure 25.11d illustrates a g_m, x_2 diagram when T_3 intersects the region where vapor and liquid coexist. For large x_2, namely, $x_2 \geqslant b$, the liquid phase is most stable; for small x_2, namely, $x_2 \leqslant a$, the vapor phase is most stable. The common tangent to g_m at $x_2 = a$ and $x_2 = b$ is the line along which g_m is least, hence describes the equilibrium between coexisting liquid and vapor when $a < x_2 < b$.

Figure 25.12 illustrates possible free energy-composition diagrams corresponding to a temperature-composition diagram for a binary mixture with a miscibility gap in the liquid region, and Figs. 25.13–25.16 illustrate the effects on the free energy-composition diagram of formation of a compound that melts at constant composition (congruently), formation of a compound that melts with a change of composition (incongruently), formation of a maximum-boiling liquid, and the case that the liquid components are not miscible in all proportions, respectively. In each case two temperatures are selected for display of the corresponding free energy-composition diagrams. A few comments about the diagrams are to be found in the respective figure captions.

We now consider, briefly, some properties of three-component mixtures. For this case $\Delta_{r-1} = \Delta_2$ reduces to (see Eq. 25.169)[9]

$$\left(\frac{\partial^2 g_m}{\partial x_1^2}\right)_{T,p,n} \left(\frac{\partial^2 g_m}{\partial x_2^2}\right)_{T,p,n} - \left[\left(\frac{\partial^2 g_m}{\partial x_1 \, \partial x_2}\right)_{T,p,n}\right]^2 > 0, \qquad (25.178)$$

and the condition on the minors of Δ_2 to

$$\left(\frac{\partial^2 g_m}{\partial x_1^2}\right)_{T,p,n} > 0,$$
$$\left(\frac{\partial^2 g_m}{\partial x_2^2}\right)_{T,p,n} > 0. \qquad (25.179)$$

The ternary mixture is stable when Eqs. 25.178 and 25.179 are satisfied. These conditions can, of course, be rewritten in terms of derivatives of the chemical potentials of the components.

To find the analog of Eq. 25.177, we write

$$g_m = x_1\mu_1 + x_2\mu_2 + (1 - x_1 - x_2)\mu_3 \qquad (25.180)$$

and calculate (at constant temperature and pressure)

$$\left(\frac{\partial g_m}{\partial x_1}\right)_{T,p,x_2} = \mu_1 - \mu_2,$$
$$\left(\frac{\partial g_m}{\partial x_2}\right)_{T,p,x_1} = \mu_2 - \mu_3. \qquad (25.181)$$

The subscripts 1, 2, 3 can be interchanged to find relations for other derivatives of $g_m(T, p, x_1, x_2)$. Then when $x_1 \to 1$, $g_m \to \mu_1^0$ and $(\partial g_m/\partial x_2)_{T,p,x_3} \to -\infty$, $(\partial g_m/\partial x_1)_{T,p,x_2} \to -\infty$; when $x_2 \to 1$, $g_m \to \mu_2^0$ and $(\partial g_m/\partial x_3)_{T,p,x_1} \to -\infty$, $(\partial g_m/\partial x_1)_{T,p,x_3} \to -\infty$; when $x_3 \to 1$, $g_m \to \mu_3^0$ and $(\partial g_m/\partial x_1)_{T,p,x_2} \to -\infty$, $(\partial g_m/\partial x_2)_{T,p,x_1} \to -\infty$. Also, when

$$x_1 \to 0: \qquad g_m = x_2\mu_2 + (1 - x_2)\mu_3,$$
$$\left(\frac{\partial g_m}{\partial x_2}\right)_{T,p,x_3} \to \infty,$$
$$\left(\frac{\partial g_m}{\partial x_3}\right)_{T,p,x_2} \to \infty;$$
$$x_2 \to 0: \qquad g_m = x_1\mu_1 + (1 - x_1)\mu_3,$$
$$\left(\frac{\partial g_m}{\partial x_3}\right)_{T,p,x_1} \to \infty,$$
$$\left(\frac{\partial g_m}{\partial x_1}\right)_{T,p,x_3} \to \infty;$$
$$x_3 \to 0: \qquad g_m = x_1\mu_1 + x_2\mu_2,$$
$$\left(\frac{\partial g_m}{\partial x_1}\right)_{T,p,x_2} \to \infty,$$
$$\left(\frac{\partial g_m}{\partial x_2}\right)_{T,p,x_1} \to \infty. \qquad (25.182)$$

[9] The labeling of the components is arbitrary.

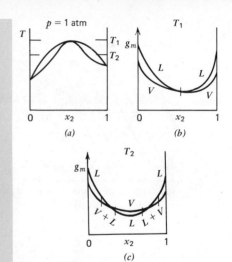

FIGURE 25.15
The components form a maximum-boiling liquid. (Courtesy of Prof. J. A. Beattie.)

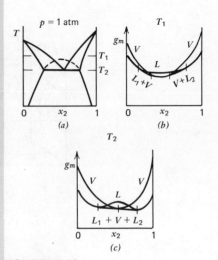

FIGURE 25.16
The liquid components are not miscible in all proportions. (Courtesy of Prof. J. A. Beattie.)

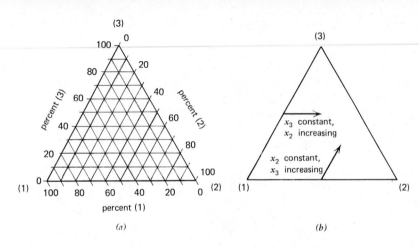

FIGURE 25.17
Triangular coordinates.

The equilibrium states of a ternary system can be represented using a triangular prism. The base of this prism is formed by the coordinate axes x_1, x_2, and x_3 arranged as an equilateral triangle (see Fig. 25.17), and the height above the base is g_m, of course all for fixed temperature and pressure. The various limiting forms of g_m given in Eqs. 25.182 represent the free energy-composition diagrams for the three binary systems into which a ternary system can be subdivided. These limiting curves appear on the appropriate faces of the triangular prism of coordinates just described. In general, the surface $g_m(T, p, x_1, x_2)$ extends over the entire composition triangle, touching the edges $x_1 = 1$, $x_2 = 1$, and $x_3 = 1$ at μ_1^0, μ_2^0, and μ_3^0, respectively. Moreover, $g_m(T, p, x_1, x_2)$ is tangent to the three edges and three sides of the coordinate prism, being concave upward in its two principle curvatures. There will be, in general, three or more sheets of $g_m(T, p, x_1, x_2)$, one each for the solid, liquid, and vapor phases, each covering the composition triangle. As before, points representing coexisting phases have a common tangent plane.

The preceding observations on the properties of $g_m(T, p, x_1, x_2)$ are seen to parallel those for a binary system. Ternary system diagrams, even in very simple cases, present a complicated appearance. Some examples are displayed in Fig. 25.18. The caption describes the features of interest.

FIGURE 25.18

(a) The temperature-composition surface at $p = 1$ atm for the condensed phases of a ternary system with no compound formation, no solid solutions, and no miscibility gap in the liquid phase. Points in the surfaces (1, 4, 7, 6), (2, 5, 7, 4), (3, 6, 7, 5) give the compositions of the liquid phase in equilibrium at various temperatures with the solid components 1, 2, 3, respectively. Points on the curves (4, 7), (5, 7), (6, 7) give the composition of the liquid in equilibrium at various temperatures with the two solids 1 and 2, 2 and 3, 1 and 3, respectively. The point 7 gives the temperature and composition of the liquid in equilibrium with the three solids 1, 2, 3. Points 4, 5, and 6 are binary eutectic points, 7 is a ternary eutectic point. When the curves (4, 7), (5, 7), and (6, 7) are projected onto the composition triangle, an arrow is appended to show the direction of decreasing temperature.

(b) Molar free energy-composition surface corresponding to (a).

(c), (d), (e), (f) Projections of (b) on the composition triangle for different temperatures. The labels on the temperatures correspond to the numbers in part (a). (Courtesy of Prof. J. A. Beattie.)

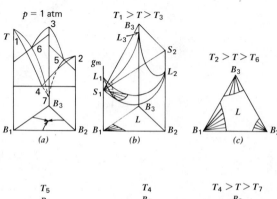

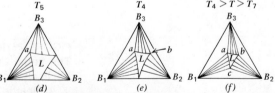

We saw in Chapter 24 that at the critical point of a pure fluid the differences between the vapor and liquid phases vanish. For a pure fluid the critical point lies on both the coexistence curve representing the vapor–liquid equilibrium and the curve of the limit of stability. The one point where both conditions are met occurs where $(\partial p/\partial v)_T = 0$ and $(\partial^2 p/\partial v^2)_T = 0$ simultaneously.

In general, we expect a critical state to occur in a system when the differences between two coexisting phases vanish. Consider the case of equilibrium between two phases, each of which has the same r components. The requirements of thermal and mechanical equilibrium ensure that the temperatures and pressures of the two phases are equal. To specify the concentrations of the r components in each phase (a total of $2r$ concentrations), we must specify r variables.[10] Suppose that T, p, and $r-1$ concentrations are fixed at values characteristic of a critical state, and that the rth is varied. The set of states so generated consists of pairs of coexisting phases whose properties become identical when the rth concentration achieves the value characteristic of the critical state. Put another way, for fixed T, p, $n_1/V, \ldots, n_{r-1}/V$, variation of n_r/V defines a coexistence line for two-phase equilibrium that terminates in the critical state. Suppose now that T, p, and $n_1/V, \ldots, n_{r-1}/V$ are varied infinitesimally, then fixed, and n_r/V is again varied. This procedure generates a new set of pairs of coexisting phases, each of which differs infinitesimally from the corresponding pairs of coexisting phases in the first variation of n_r/V, hence the new two-phase coexistence curve must terminate in a critical state differing infinitesimally from the first critical state. This argument shows that two-phase equilibrium in an r-component system leads to a critical state with $r-1$ degrees of freedom. For a one-component system there is only a critical point ($r-1=0$), but for a binary mixture there is a critical line ($r-1=1$), and so forth. In the case of the binary mixture, in which at a fixed pressure two liquids do not mix at some temperatures but do at other temperatures, the critical solution temperature T_c is a function of the pressure; this function defines the critical line.

Just as for a one-component system, the critical state of a mixture lies on both the coexistence curve for two-phase equilibrium *and* the limit-of-stability curve, yet is itself a stable state of the system. For a binary mixture the limit of stability occurs when $\Delta_1 = 0$, that is, when

$$\left(\frac{\partial^2 g_m}{\partial x_2^2}\right)_{T,p} = \left[\frac{\partial(\mu_2 - \mu_1)}{\partial x_2}\right]_{T,p} = 0. \tag{25.172}$$

Furthermore, again just as for a one-component system, Eq. 25.172 defines the stability of a homogeneous phase against small displacements not including formation of a new phase. Therefore, for portions of the coexistence region away from the critical point, phase separation occurs before Eq. 25.172 is satisfied, and there is a portion of the g_m, x_2 curve that lies inside the coexistence curve and describes metastable states (see Fig. 25.12 and, for the one-component analog, the discussion following Fig. 24.15). There is a gap in the free energy-composition surface where $\Delta_1 < 0$; this gap is spanned by the derived surface generated from the common tangent to the two portions of g_m representing stable phases. The gap disappears and the points of tangency on the two portions of

---

[10] The condition $\mu_i^I = \mu_i^{II}$, $i = 1, \ldots, r$, fixes the ratio of the concentrations in the two phases.

$g_m(T, p, x_2)$ coincide at the critical point $T_c(p)$. Therefore, a critical state exists when *both* Eq. 25.172 and

$$\left(\frac{\partial^3 g_m}{\partial x_2^3}\right)_{T=T_c,p} = \left[\frac{\partial^2(\mu_2 - \mu_1)}{\partial x_2^2}\right]_{T=T_c,p} = 0 \qquad (25.183)$$

hold. The extension of this definition of the critical state to r-component mixtures is algebraically tedious, but involves no new principles.

Because of the analogy between Eqs. 25.172 and 25.183 and the corresponding conditions for a pure fluid, $(\partial p/\partial v)_T = -(\partial^2 a/\partial v^2)_T = 0$ and $(\partial^2 p/\partial v^2)_T = (\partial^3 a/\partial v^3)_T = 0$, respectively, we expect that the behavior of a mixture in the region near its critical state will be very much like that of a pure fluid near its critical point. Indeed, this is the case. Just as for the one-component system, the assumption that $g_m(T, p, x_2)$ can be expressed in a Taylor series expansion about the critical point (that is, about a point on the critical line) and the use of mean field-type equations of state predict incorrect behavior in the critical region. In summary form, it is found that:

1. At constant pressure,

$$x_2^{II} - x_2^{I} \propto |T_c - T|^\beta, \qquad (25.184)$$

   with the observed value $\beta \approx 0.35$; the classical theory predicts that $\beta = 0.50$ (see Fig. 25.19).

2. Along a line of constant temperature $T = T_c$ and constant pressure $p = p_c$,

$$(\mu_1 - \mu_2) - (\mu_1 - \mu_2)_c \propto (x_2 - x_{2c})^\delta, \qquad (25.185)$$

   with the observed value $4 < \delta < 5$; the classical theory predicts that $\delta = 3$.

3. Along a line of constant pressure $p = p_c$ and constant composition $x_2 = x_{2c}$,

$$\left[\frac{\partial(\mu_2 - \mu_1)}{\partial x_2}\right]_{T, p_c, n} \propto (T - T_c)^\gamma, \qquad (25.186)$$

   with the observed value $1.1 < \gamma < 1.4$; the classical theory predicts that $\gamma = 1$.

4. Along a line of constant pressure $p = p_c$ and constant composition $x_2 = x_{2c}$,

$$c_{px_2} \propto |T - T_c|^{-\alpha}, \qquad (25.187)$$

   with the observed value $0 < \alpha < 0.1$; the classical theory predicts that $\alpha = 0$ and that there is a jump discontinuity in c_{px_2}.

The forms displayed in Eqs. 25.184–25.187 do not distinguish between the values of the critical exponents for approaches to the critical state from $T > T_c$ and $T < T_c$. Just as in the case of the pure fluid, the limited information available suggests that the values for the two approaches are the same, and just this is suggested by model calculations and also by the assumption that $g_m(T, p, x_2)$ is a homogeneous function of $(T - T_c)$, $(p - p_c)$, and $(x_2 - x_{2c})$ very near the critical state (see also the discussion in Chapter 24).

Clearly, just as is the case for a pure fluid, the classical predictions of α, β, γ, and δ are incorrect. A comparison of the experimental values of α, β, γ, δ in Eqs. 25.184–25.187 with the entries in Table 24.4, referring there to the critical behavior of a pure fluid, shows that a pure fluid and a

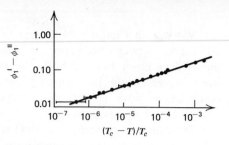

FIGURE 25.19
Coexistence curve for the liquid mixture C_7H_{14}–CCl_4. The critical exponent in

$$\phi_1{}^I - \phi_1{}^{II} \propto |T_c - T|^\beta$$

is $\beta = 0.335 \pm 0.02$. ϕ_1 is the volume fraction of component 1. Either the volume fraction or the mole fraction can be used to define the approach to T_c. Only the coefficient, but not the exponent, is changed when $x \to \phi$ in Eq. 25.184. From P. Heller, *Rept. Prog. Phys.* **30**, 731 (1967).

binary mixture have strikingly similar critical behavior. Since the data for binary mixtures even include results from studies of alloy systems and solutions with water and other polar substances, the uniformity of behavior near a critical state must be regarded as a universal characteristic of condensed matter, one not much dependent on the nature of the molecular interactions provided they are short-range. We showed in Chapter 24 that the assumption that the singular part of the chemical potential near the critical point is a homogeneous function of $T - T_c$ and $T_c - \tau(\rho)$ led to a universal equation of state near the critical point (Fig. 24.19). For a binary mixture the assumption that the singular part of $g_m(T, p, x_2)$ is a homogeneous function of $T - T_c$, $p - p_c$, and $x_2 - x_{2c}$ very near the critical state leads to an analogous universal equation of state. The available tests of this equation show, within the uncertainties of the data, about as good agreement as does the test for pure fluids (Fig. 24.19).

This description of the failure of the classical theory of the critical region should not be allowed to obscure the fact that away from the critical region, where the range of correlations is not very much larger than the range of the molecular interactions, simple equations of state give very useful approximate descriptions of the phase equilibria and describe correctly all the qualitative features of such equilibria. It is only in the critical region that there is a qualitative breakdown of the classical analysis. Since a very large part of the interpretation of phase equilibria is unrelated to critical phenomena, there is ample justification for development of statistical molecular models without the refinements necessary for the analysis of the critical region. Moreover, a very significant fraction of the effort in the study of mixtures is concerned with the role of molecular interactions in the one-phase region of complete miscibility, which further reinforces the need for model statistical molecular theories of mixtures.

The statistical molecular theory of mixtures has three principal goals. These are (1) accurate prediction of the thermodynamic properties of the homogeneous phases of a mixture, (2) accurate prediction of the location of phase boundaries and the compositions of coexisting phases, and (3) interpretation of the properties of a mixture in terms of the properties of its parent pure components. Clearly, the existence of different interactions between pairs of like and unlike molecules makes the statistical molecular theory of mixtures more difficult than the corresponding theory of the pure components. The many theoretical investigations that have been made range from elementary treatments of very crude models to sophisticated generalizations of the theory of liquids. As in Chapter 23, we shall discuss the important ideas in the theoretical analyses but not work out the mathematical details; a few of the ideas will be illustrated by model calculations. We consider first crystalline mixtures, and then liquid mixtures.

We have shown in Chapter 22 that, if the binding energy at 0 K, the vibrational spectrum, and the anharmonicity of vibration of a one-component crystalline solid are known, its free energy and all other thermodynamic properties can be computed. Construction of a similar description of the thermodynamic states of a crystalline mixture requires consideration of the following.

1. The several components are distributed in some way over the total number of lattice sites. Consider, for example, a binary mixture of A and B. If the AA, BB, and AB interactions are all the same, then in the nonvibrating lattice A and B will be randomly distributed. On the other hand, if AB interactions are more attractive than AA or BB interactions,

25.10
The Statistical Molecular Theory of Solutions of Nonelectrolytes

A. Crystalline Mixtures

then in the static lattice there will be preferential clustering and a larger-than-random fraction of AB near neighbors. Thus, depending on the nature of the intermolecular interactions, there will be some contribution to the free energy of the mixed crystal arising from the static distribution of components.

2. In general, irrespective of the nature of the components and their distribution over the lattice sites, the vibrational spectrum of the mixed crystal differs from the vibrational spectra of the pure components. When the AB interaction differs from the AA and BB interactions, it is easy to see why this is so, since in this case it is likely that the curvatures of the pair potentials differ at their respective minima, hence so do the force constants that determine the vibrational spectra. It is also likely that the anharmonicities are different in the mixture and the pure components.

3. Even when the pair potentials AA, AB, and BB are the same,[11] if the masses of A and B differ then the vibrational spectra of pure A, pure B, and the AB mixture differ. The reason for this is also easy to see. The vibrational frequency of a mass on a spring is proportional to $m^{-1/2}$, and the vibrational spectrum of a collection of masses also depends on the magnitudes of the masses, although in a more complicated way. For example, if only one light particle is substituted in a heavy-particle lattice and all the force constants remain the same, the frequency spectrum changes as follows: First, one normal mode and the corresponding normal frequency are removed from the dense set of frequencies characteristic of the pure crystal. Second, there appears a new frequency, larger than ν_{max} (the maximum frequency for the pure crystal), which corresponds to the symmetric ("breathing") motion of the light particle and the surrounding nearest-neighbor particles in antiphase motion. For this motion the effective mass is dominated by the smallest mass, which in the case under discussion is the mass of the substituent atom. Then, because the force constant is the same as in the pure lattice, the corresponding frequency must be larger than that in the pure lattice (which would be ν_{max}). Finally, the frequencies of all the other normal modes are changed very little, in fact by a fraction of order N^{-1}, where N is the number of particles in the crystal. Clearly, when there are many particles with different masses in the crystal, and in addition the force constants differ and the distribution of atoms over the lattice sites is not as regular as is the crystal lattice itself, the distribution of vibrational frequencies in the mixed crystal will differ considerably from that in a pure crystal. It is to be expected that the thermal properties of the mixed crystal will then differ somewhat from those of the pure components.

4. The distribution of A and B atoms over the lattice sites is not independent of the spectrum of vibrations. As shown in Chapter 22, a typical normal mode in a one-component crystal involves coherent motion of many atoms (see Fig. 22.3). In a mixture the corresponding group of many atoms can have a range of compositions, and even for fixed composition many different arrangements of A and B within the group. Each of these arrangements has, in general, a different influence on the vibrational spectrum. At equilibrium that distribution of atoms is favored which, other things being equal, minimizes the vibrational contribution to the free energy. The effect of this dependence of vibrational spectrum on atomic arrangement is to generate an effective interaction between atoms. This effective interaction is temperature-dependent and is in addition to the ordinary interaction between stationary atoms.

5. An AB mixed crystal can have a different density of imperfections than pure A or B crystals.

[11] As, for example, in a mixture of isotopes.

6. Although conditions **1–5** have been phrased in terms of the distribution of A and B over lattice sites, it must not be assumed that the lattice sites of the mixed crystal are the same as those of pure A or B. Put another way, we expect that when the mixed crystal is formed there will be some change in lattice structure. For a dilute solution of A and B this will take the form of displacement of the B atoms near an A atom from the positions they would have occupied in a pure B crystal. In an equimolar AB mixed crystal, all atoms are displaced from their nominal positions in pure A or B crystals. This effect, which depends mostly on the repulsive part of the AA, AB, and BB interactions, leads to changes in the energy of the static lattice and the vibrational spectrum of the lattice.

Many theories of the thermodynamic properties of crystalline mixtures completely neglect points **2**, **3**, **4**, **5**, and **6**. When point **6** is neglected, the complete partition function can be separated into the product of a vibrational partition function and a configurational partition function. It is usually the case that the effective interaction described in point **4** is negligibly small compared to other interactions, except in some special instances,[12] so factorization of the mixed-crystal partition function is usually a good approximation. However, the neglect of points **2** and **6** leads to inaccurate predictions of the properties of the mixture. As we shall see, the structure imposed by short-range repulsive forces is very important in the determination of the excess thermodynamic properties of the mixed crystal. As an illustration of the effects to be expected, we consider the calculation of the entropy of mixing from a very simple model.

Consider a crystal lattice with N_A A atoms and N_B B atoms, and suppose that the AA, AB, and BB interactions, both repulsive and attractive, are identical. Then, neglecting effects of types **3** and **4**, it is clear that the energy and the volume of the mixture are independent of composition, hence there is no heat of mixing and no volume change on mixing. Clearly, the model described is an ideal mixture, the thermodynamics of which is completely described by the entropy of mixing.

Because all compositions have the same energy, we can use the simple Boltzmann definition of the entropy,

$$S = k_B \ln \Omega, \qquad (25.188)$$

to compute the entropy of mixing. If the $N = N_A + N_B$ particles occupy all the sites of the lattice, the number of distinguishable arrangements is just

$$\Omega = \frac{N!}{N_A! N_B!}, \qquad (25.189)$$

because the particles can be permuted among the N sites in $N! \equiv (N_A + N_B)!$ ways, but $N_A!$ permutations of the A's among themselves and $N_B!$ permutations of the B's among themselves lead to indistinguishable distributions of A and B over the lattice. The numbers of distinguishable arrangements of A and B in their pure lattices are, of course, just $N_A!/N_A! = 1$ and $N_B!/N_B! = 1$. Therefore, in this static crystal model the entropy of the mixed crystal is just the entropy of mixing. Using Stirling's

[12] If the vacancies in a crystal are thought of as a component, and a crystal with vacancies as a mixture, then the effective interaction induced by vibrational motion leads to clustering of vacancies at low temperature. This phenomenon occurs, for example, in metals exposed for long periods to radiation from nuclear reactors, such as fuel rod casings.

approximation, $\ln N! \approx N \ln N - N$, Eq. 25.189 is readily reduced to

$$\Delta S_{\text{mixing}} = -N_A k_B \ln \frac{N_A}{N_A + N_B} - N_B k_B \ln \frac{N_B}{N_A + N_B}$$

$$= -N_A k_B \ln x_A - N_B k_B \ln x_B. \qquad (25.190)$$

As expected, the entropy of the model crystal considered is just the ideal entropy of mixing.

In the model just examined, each lattice site could be occupied by only one molecule, and each molecule could occupy only one lattice site, so there are no geometric constraints on the occupation of the sites neighboring a given lattice site just because that site is occupied. What happens to the distribution of molecules when A and B have different volumes and, say, a B molecule occupies more than one lattice site? So as to avoid unnecessary complications, we again consider a model crystal in which there is no difference between the attractive parts of the AA, AB, and BB interactions. The repulsive parts of these interactions must differ by virtue of the difference in molecular volumes. Even with this simplification, counting exactly the number of distinguishable configurations available to the system is very difficult. It is fairly easy, however, to evaluate the entropy of mixing when the concentration of the component with larger molecular volume is very small. The argument goes as follows: Suppose that component B, with molecular volume $v_B{}^0$, occupies many lattice sites per molecule, and component A, with molecular volume $v_A{}^0$, occupies only one lattice site per molecule (see Fig. 25.20). Neglecting edge effects, we find that the number of ways in which the center of mass of a B molecule can be put into an empty lattice of N sites is just

$$\omega_1 = N = \left(\frac{1}{\Delta}\right) V, \qquad (25.191)$$

where $\Delta \equiv V/N$ is the volume of a lattice cell. As shown in Fig. 25.20, the presence of a molecule of B results in the exclusion of A from possible occupation of a number of lattice sites, say z of them. The volume excluded thereby is $z\Delta$. The number of ways of introducing a second molecule of species B, given that the first one is somewhere in the lattice, is

$$\omega_2 = N - z = \left(\frac{1}{\Delta}\right)(V - b), \qquad (25.192)$$

where $b \equiv z\Delta$. Similarly, the third molecule will be excluded from $N - 2z$ sites, and we have

$$\omega_3 = N - 2z = \left(\frac{1}{\Delta}\right)(V - 2b). \qquad (25.193)$$

Equation 25.193 involves the use of an approximation. When the two molecules already present are as close together as possible, they exclude from occupation fewer than $2z$ sites, because the exclusion zones overlap in the region between the molecules. (See Fig. 25.20.) In writing Eq. 25.193 we are assuming that the number of configurations in which two B molecules are adjacent is negligibly small relative to the number of configurations in which they are far apart. If this approximation is made at each step as a B particle is added to the lattice, the number of ways the ith B molecule can be placed in a lattice already occupied by $i - 1$ other B molecules is

$$\omega_i = N - (i-1)z = \left(\frac{1}{\Delta}\right)[v - (i-1)b]. \qquad (25.194)$$

This approximate calculation of the excluded-volume factors is only

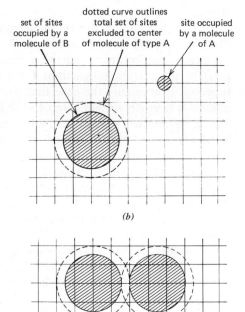

FIGURE 25.20
Schematic representation of excluded-volume effects in a static lattice.

(a) Each molecule occupies only one site and does not exclude occupation of neighboring sites.

(b) A molecule of species B occupies more than one lattice site and blocks the center of an A molecule from neighboring sites that are closer to the center of B than the sum of the radii of an A and a B atom.

(c) The volume excluded by two nearest-neighbor particles is less than the sum of the excluded volumes of each would be if they were far apart.

accurate for very dilute solutions, for which $N_B v_B^0 \ll V$. We shall assume that this restriction is met.

After placing N_B B molecules on the lattice we have left, in our approximation, $N - zN_B$ empty lattice sites. These can be filled in $N_A!$ ways with A molecules. The total number of distinguishable ways of arranging N_A A molecules and N_B B molecules on the lattice is thus

$$\Omega = \frac{N_A!}{N_A!} \frac{1}{N_B!} \prod_{i=1}^{N_B} \omega_i = \frac{1}{N_B!} \left(\frac{1}{\Delta}\right)^{N_B} \prod_{i=1}^{N_B} [V - (i-1)b]. \quad (25.195)$$

As in the case of Eq. 25.189, the factors $1/N_A!$ and $1/N_B!$ account for the indistinguishability of A and B molecules—these may be interchanged among themselves $N_A!$ and $N_B!$ ways, respectively, without altering the state of the system.

In a dilute solution we have $N_B b \ll V$, so that

$$\ln\left\{\prod_{i=1}^{N_B} [V - (i-1)b]\right\} = \ln\left[\prod_{j=0}^{N_B-1} (V - jb)\right]$$

$$= \ln\left[\prod_{j=0}^{N_B-1} V\left(1 - \frac{jb}{V}\right)\right]$$

$$\approx N_B \ln V - \frac{b}{V} \sum_{j=0}^{N_B-1} j. \quad (25.196)$$

Therefore, the entropy of the mixed crystal is just

$$S_{\text{mixture}} = k_B \ln \Omega = N_B k_B \ln \frac{V}{\Delta} - k_B \ln N_B! - \frac{k_B b}{V} \sum_{j=0}^{N_B-1} j. \quad (25.197)$$

The sum in the last term of Eq. 25.197 is equal to the number of terms, N_B, multiplied by the average value of the terms, $(N_B - 1)/2$. Expressing the total volume in terms of the volumes of the components,

$$V = N_A v_A^0 + N_B v_B^0, \quad (25.198)$$

and assuming that $N_B \gg 1$, we convert Eq. 25.197 to the form

$$S_{\text{mixture}} = N_B k_B \ln\left(\frac{N_A v_A^0 + N_B v_B^0}{\Delta}\right) - k_B \ln N_B! - \frac{N_B^2 k_B b}{2(N_A v_A^0 + N_B v_B^0)}. \quad (25.199)$$

Equation 25.199 represents, for the model considered, the entropy of the mixture. We can obtain the entropy of pure A and pure B on the basis of this same model by setting $N_B = 0$ and $N_A = 0$, respectively.[13] We then

---

[13] It is necessary to add a few words to the description of the calculation of the entropy of mixing. For the model under discussion the vibrational contributions to the entropies of the mixture and the pure components have been neglected. We then expect to find $S_A = 0$ and $S_B = 0$, respectively, when N_B is set equal to zero and when N_A is set equal to zero in the expression for the entropy of the mixture. Eq. 25.199 gives the correct value for S_A when $N_B = 0$ because in that case the number of configurations available to the system has been correctly evaluated. However, Eq. 25.199 does not give $S_B = 0$ when $N_A = 0$ because of the approximate treatment of the number of configurations of the system when one molecule blocks the occupancy of z lattice sites. To make the analysis more consistent we require that $S_B = 0$, which implies that b has the value

$$b = zv_B^0\left(1 + \ln \frac{v_B^0}{\Delta}\right).$$

Since $b = z\Delta$, we can substitute $\Delta = b/z$ in the argument of the logarithm and iterate the procedure to successively remove Δ from the expression for b. This leads to the approximate formula

$$b \approx zv_B^0\left(1 + \ln \frac{z}{2}\right),$$

which varies very slowly with z and is consistent with the simple idea that the excluded volume per sphere is larger than the volume of one sphere. For the purposes of this discussion we assume that b is constrained to have the value that leads to $S_B = 0$ when $N_A = 0$.

find for the entropy of mixing

$$\Delta S_{\text{mixing}} = N_B k_B \ln\left(\frac{N_A v_A{}^0 + N_B v_B{}^0}{N_B v_B{}^0}\right) - \frac{N_B{}^2 k_B b}{2(N_A v_A{}^0 + N_B v_B{}^0)} + \frac{N_B k_B b}{2 v_B{}^0}.$$

(25.200)

Equation 25.200 differs in two important ways from Eq. 25.190. First, the entropy of mixing depends on the volume fraction $\phi_B = N_B v_B{}^0/(N_A v_A{}^0 + N_B v_B{}^0)$, and not on the mole fraction. Second, because of the disparity in sizes of A and B molecules, there is a nonideality contribution to ΔS_{mixing}. These points are best illustrated when Eq. 25.200 is written in the form

$$\Delta S_{\text{mixing}} = -N_B k_B \ln \phi_B - \tfrac{1}{2} N_B k_B \phi_B \frac{b}{v_B{}^0} + \tfrac{1}{2} N_B k_B \frac{b}{v_B{}^0}$$

$$= -N_B k_B \ln \phi_B + \frac{N_B k_B b}{2 v_B{}^0} \phi_A$$

(25.201)

and compared with Eq. 25.190 for a solution dilute in component B:

$$\Delta S_{\text{mixing}}^{\text{ideal}} = -N_B k_B \ln x_B - N_A k_B \ln(1 - x_B)$$

$$\approx -N_B k_B \ln x_B + N_A k_B x_B$$

$$= -N_B k_B \ln x_B + N_B k_B x_A.$$

(25.202)

We may now ask which of the features characteristic of these simple models are of general validity. First, note that the absence of a heat of mixing does not imply the absence of intermolecular interactions, but only that the AA, AB, and BB interactions are the same; conversely, even if this is true there can still be a nonzero heat of mixing if $\Delta V_{\text{mixing}} \neq 0$. Second, when the sizes of the two component molecules are very different, the entropy of mixing depends on the volume fractions of the components, and not on the mole fractions. For molecules of the same size, of course, the volume fraction and mole fraction are identical. Third, mixtures in which one component is much larger than the other are intrinsically nonideal because of the volume from which the center of mass of the smaller particle is excluded by the larger particle. In the particular case we have examined, the nonideality increases with the excluded volume. Although our deductions are derived from a crude static lattice model, it is found that in real solutions all of the conclusions cited remain valid.

We have thus far not considered explicitly the influence of molecular interactions on the distributions of molecules over the lattice sites, except for the influence of dissimilar molecular volumes. In a sense, the formulation of the excluded volume used in Eqs. 25.191 *et seq.* is a crude representation of the effect of dissimilar repulsive interactions between AA, AB, and BB molecular pairs. In general, the AA, AB, and BB pair interactions differ in both their repulsive and attractive components, and the distribution of molecules over the lattice sites is a complicated function of the interaction energies. To simplify our description, let us now assume that the molecular volumes of A and B are the same, so that attention can be focused on differences in the attractive interactions. The crudest approach to the evaluation of the lattice partition function is to assume that the distribution of components over the lattice is random despite the differences in interactions. When carried through, this assumption leads to the theory of regular mixtures, the thermodynamics of which we have outlined in Sections 25.2 and 25.8. In this crude model the entropy of mixing is of ideal form because the distribution of components is defined to be random and other possible contributions to the entropy of mixing are ignored. The energy difference $\epsilon_{AB} - \frac{1}{2}\epsilon_{AA} - \frac{1}{2}\epsilon_{BB}$ determines the heat of mixing.

Three different approaches have been used to obtain a more exact description of the molecular distribution for this model. First, there are a variety of approximations designed to account for the major effects of dissimilar molecular interaction without being so mathematically clumsy as to make the evaluation of the partition function very difficult. For example, if we think of the AA, AB, and BB near-neighbor pairs of a binary solution as diatomic molecules in a mixed gas, their relative concentrations can be related through an "equilibrium constant." In the nonvibrating lattice model we are considering, there is no entropy difference between the several "diatomic molecules," so that the "equilibrium constant" $N_{AB}^2/N_{AA}N_{BB}$ is determined by the near-neighbor energy difference $\epsilon_{AB} - \frac{1}{2}\epsilon_{AA} - \frac{1}{2}\epsilon_{BB}$. The distribution of A and B molecules over the lattice is then calculated, in this approximation, subject to the subsidiary constraint that the number of near-neighbor AB pairs satisfy the equilibrium $N_{AB}^2 = KN_{AA}N_{BB}$ at each temperature.[14] This approximation leads to some minor changes in the thermodynamics of the mixture, but in most cases the predicted behavior is close to that of the regular solution with the same interaction energies. In other words, although the entropy of mixing differs from the ideal form, for reasonable values of ϵ_{AB}, ϵ_{AA}, ϵ_{BB} the deviation is not large.

A second approach is based on an analysis similar to the virial expansion used in Chapter 21 to describe a gas. In place of an integration over the translational space in which gas molecules move, for this model there is a summation over the lattice points at which molecules may be placed. The results of the analysis are a set of virial expansions (the concentration replaces the gas density) for the thermodynamic functions, with virial coefficients defined by lattice sums instead of integrations. Just as in the theory of the gas, this calculation is useful for dilute solutions of, say, A in B, for which only terms of lowest order in the mole fraction need be retained. The lattice virial coefficients are difficult to calculate, and this theory is not much used.

A third scheme is based on term-by-term evaluation of the partition function after the exponential term is expanded as a series in powers of $(1/T)$. This procedure is tedious, and requires the evaluation of many terms to get an accurate evaluation of the partition function. If enough terms are calculated, and if the properties of the full partition function can be guessed from the properties of the terms calculated, valuable information about the approach to phase separation can be obtained. In Section 24.4 we described a lattice gas, which can be thought of as a mixture of occupied and unoccupied lattice sites, or of molecules and "holes." The results quoted there are typical of what can be learned from this type of model and this scheme of calculation. The qualitative results confirm the picture given by the quasi-chemical equilibrium approximation to the molecular distribution: When unlike molecules repel more strongly than like molecules, AB near-neighbor pairs occur with less than random frequency, there is heat absorbed on mixing, and the total vapor pressure is greater than that of an ideal mixture. On the other hand, when unlike molecules attract more strongly than like molecules, AB near-neighbor pairs occur with greater than random frequency, heat is evolved on mixing, and the vapor pressure is less than for an ideal mixture. Corresponding to the decreased or increased frequency of AB pairs relative to the random distribution, the entropy of mixing deviates from the ideal form. In this crude model, in which all vibrational motion is

[14] This is known as the quasi-chemical approximation.

neglected, the entropy of mixing is less than the ideal entropy of mixing in both cases. In real mixtures vibrational effects alter this result, and the deviation from the ideal entropy of mixing can be in either direction. Comparison of the several calculations shows that, although differences in attractive interaction do not change the entropy of mixing much, differences in molecular volume have a large effect on the entropy of mixing. This result is similar to that obtained for the effects of attractive (a-term) and repulsive (b-term) interactions on the entropy of a van der Waals gas, and are confirmed by the general virial expansion results. Finally, all theories agree that in the limit $T \to \infty$ the molecular distribution approaches a random distribution. This is a simple consequence of the thermal energy $k_B T$ becoming large relative to $\epsilon_{AB} - \frac{1}{2}\epsilon_{AA} - \frac{1}{2}\epsilon_{BB}$, so that the Boltzmann factor approaches unity and all configurations of A and B on the lattice have equal weight.

All of the preceding discussion was based on the analysis of a model mixture in which vibrational motion was neglected. In any real mixture the difference in vibrational spectrum from the spectra of the pure components can lead to considerable alteration of the thermodynamic functions. The calculation of a mixture's vibrational spectrum is extremely complicated because the disorderly distribution of molecules over the lattice prevents the use of many of the simplifications of the theory of pure-crystal lattices. There are also insufficient experimental data relevant to the spectra of mixed crystals, so that simplified analyses have not yet emerged in response to experimental findings. In general, we expect the thermodynamic functions to be affected most when new low-frequency vibrations appear in the mixed crystal (say, when very heavy atoms are substituted in the crystal), or when AB pairs are so tightly bound that many crystal vibrations appear as new high-frequency "quasi-molecular" vibrations. The total number of vibrational modes is determined by the number of atoms, so that the appearance of new modes must mean the loss of modes elsewhere in the frequency spectrum if the number of atoms is the same. When the mixed crystal has more low-frequency components in its spectrum than do the pure crystals the vibrational entropy of the mixture will be greater than that of the pure crystals, and *vice versa*. Corresponding to these entropy differences will be differences in other thermodynamic functions, for example, the heat capacities. Although the alterations in vibrational spectra described are also accompanied by energy differences between the pure and mixed crystals greater than those expected in the absence of vibration, in most cases the principal influence of the changed vibrational spectrum will be on the entropy of mixing.

To illustrate and test the ideas about the thermodynamic properties of crystalline mixtures just described, we now examine alkali halide solid solutions. These are, of course, ionic crystals. We showed in Section 14.10 how the lattice energy of a pure ionic crystal can be computed from a knowledge of the ion–ion potential energy function; the extension of the calculation to treat mixed crystals is straightforward.

The lattice energy of an ionic crystal is very large compared to its zero-point and thermal vibrational energies, and very large compared to the heat of mixing. Also, the vibrational spectra of the mixture and the pure components are expected to be only slightly different, because of the similarity in the ion–ion potentials. In this case it is likely that the static lattice energy of the mixture is very nearly the same for all distributions of, say, Li$^+$ and Na$^+$ in the mixed crystal LiF–NaF, hence to a very good approximation the distribution of Li$^+$ and Na$^+$ can be taken to be random. Of course the static lattice contribution to the entropy of mixing is then of the ideal solution form, but the full entropy of mixing, which includes a contribution from the small change in the vibrational spectrum, is not the ideal solution entropy of mixing.

Although the Coulomb interaction is the same for every ion pair with the same charges, the other contributions to the ion–ion potential energy are different for each pair of ions. When one ion is substituted for another in an ionic lattice there are three major components of the change in energy:

1. The ion–ion repulsion, of the form $\lambda \exp(-R/\rho)$, with λ and ρ constants, is different for each pair of ions. This, of course, corresponds to the observation that the ionic diameters are all different, hence so must be the possible distances of closest approach. We expect the change in repulsive interaction on substitution to be an important contributor to the force that leads to a change in ionic positions.

2. When ions are displaced from the lattice positions characteristic of the parent pure component, there is a change in the electrostatic field acting on each ion. Because the displacements are small, the change in electrostatic energy per ion can be calculated in terms of the interactions between all of the effective dipole moments created by the displacements of the ions from their lattice sites.

3. The ion polarizabilities are all different. Since the electric field of one ion distorts the charge cloud of a neighboring ion, the interaction energy being proportional to the induced dipole moment, this contribution to the energy will change when the ion polarizabilities and positions change.

If we consider an isolated substitution of one ion for another, so that the environment of the solute ion is entirely defined by the host, the change in energy is obtained by just summing contributions **1**, **2**, and **3** for the one geometry of substitution. If we consider a mixed crystal in the "middle concentration range," an ion of one type can be found in many different environments. For the case that there is random mixing in the static lattice, the concentration of each type of environment is constructed from the known bulk concentrations. For example, let x_A and x_B be the mole fractions of the two cations in the cubic mixed crystal $A_{x_A}B_{x_B}X$, and note that in the cubic crystal the six nearest neighbors of an A or B cation are always X anions. Then the differences between the environments of different A's and B's arise from differences in the makeup of the 12 second-neighbor ions. To take one configuration, the concentration of sites each surrounded by a second-neighbor shell with seven A's and five B's is proportional to $x_A^7 x_B^5$. The energy associated with each such configuration must be evaluated, then weighted with the appropriate concentration, and all terms summed to give the change in energy. The energy must then be minimized with respect to the ion displacements.

Calculations of the type described have been made both for dilute and concentrated alkali halide mixtures. Consider the dilute solutions first. Table 25.3 shows the calculated heats of mixing (or formation) of 1% mixtures. The results illustrate the common-sense idea that the energy of formation of the mixture increases with increasing degree of deformation required. For example, KBr–Na has a smaller heat of mixing than does NaBr–K because it is easier to put a small ion in a large site than *vice versa*. Also, the series of mixtures KCl–Na, KBr–Na, KI–Na shows a decreasing heat of formation, as expected, since the cation site increases in volume steadily as the anion size increases.

Table 25.4 contains a comparison of the observed and calculated heats of mixing, and we show in Fig. 25.21 a comparison of the observed and calculated concentration dependences of the heats of mixing. The

TABLE 25.3a
HEAT OF MIXING FOR 1% CATION DEFECT CONCENTRATION (in J/mol)[a]

LiF–Na	206	NaF–Li	556	LiF–K	2130	KF–Li	1540
LiCl–Na	156	NaCl–Li	120	LiCl–K	1100	KCl–Li	498
LiBr–Na	126.4	NaBr–Li	100	LiBr–K	812	KBr–Li	427
LiI–Na	92.5	NaI–Li	73	LiI–K	590	KI–Li	315
LiF–Rb	3490	RbF–Li	1940	NaF–K	494	KF–Na	305
LiCl–Rb	1720	RbCl–Li	686	NaCl–K	266	KCl–Na	165
LiBr–Rb	1260	RbBr–Li	598	NaBr–K	197	KBr–Na	143
LiI–Rb	1020	RbI–Li	460	NaI–K	143	KI–Na	103
NaF–Rb	1020	RbF–Na	531	KF–Rb	55.6	RbF–K	46.8
NaCl–Rb	536	RbCl–Na	296	KCl–Rb	39.1	RbCl–K	19.2
NaBr–Rb	401	RbBr–Na	265	KBr–Rb	23.0	RbBr–K	27.0
NaI–Rb	324	RbI–Na	202	KI–Rb	23.8	RbI–K	21.6

TABLE 25.3b
HEAT OF MIXING FOR 1% ANION DEFECT CONCENTRATION (in J/mol)[a]

LiF–Cl	933	LiCl–F	858	LiF–Br	1950	LiBr–F	1080
NaF–Cl	946	NaCl–F	485	NaF–Br	1850	NaBr–F	736
KF–Cl	686	KCl–F	351	KF–Br	1230	KBr–F	561
RbF–Cl	540	RbCl–F	289	RbF–Br	946	RbBr–F	481
LiF–I	4070	LiI–F	1070	LiCl–Br	60.6	LiBr–Cl	43.9
NaF–I	3550	NaI–F	1070	NaCl–Br	52.7	NaBr–Cl	44.3
KF–I	2590	KI–F	870	KCl–Br	43.9	KBr–Cl	30.2
RbF–I	2220	RbI–F	774	RbCl–Br	36.6	RbBr–Cl	28.9
LiCl–I	402	LiI–Cl	150	LiBr–I	102	LiI–Br	59.8
NaCl–I	347	NaI–Cl	194	NaBr–I	101	NaI–Br	69.9
KCl–I	284	KI–Cl	158	KBr–I	79	KI–Br	61.1
RbCl–I	264	RBI–Cl	151	RbBr–I	71	RbI–Br	61.1

[a] From D. L. Fancher and G. R. Barsch, *J. Phys. Chem. Solids* **30**, 2503 (1969).

TABLE 25.4
COMPARISON OF OBSERVED AND COMPUTED HEATS OF FORMATION OF 1% MIXTURES OF ALKALI HALIDES (all in J/mol)[a]

x		$Na_xK_{1-x}Cl$	$Na_xK_{1-x}Br$	$K_xRb_{1-x}Cl$	$K_xRb_{1-x}I$
0.99	Exp.	192.5–230.1	192.5	108.8	37.6
	Theory	266.1	197.5	142.7	39.3
0.01	Exp.	163.2–192.5	144.3	92.1	30.5
	Theory	164.8	143.1	102.9	19.2
x		$NaCl_xBr_{1-x}$	$NaBr_xI_{1-x}$	KCl_xBr_{1-x}	KBr_xI_{1-x}
0.99	Exp.	45.2–59.8	92.	35.6	62.7
	Theory	52.7	100.8	43.9	79.0
0.01	Exp.	43.9–52.3	75.3	34.7	69.0
	Theory	44.3	69.8	30.1	61.1

[a] From D. L. Fancher and G. R. Barsch, *J. Phys. Chem. Solids* **30**, 2503 (1969).

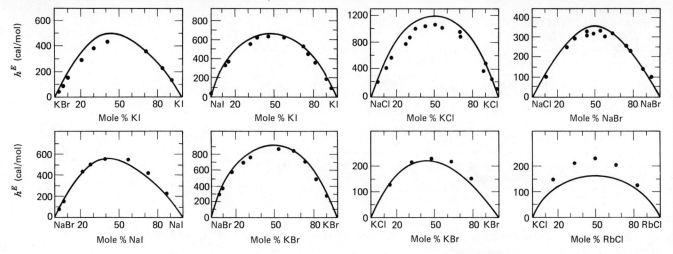

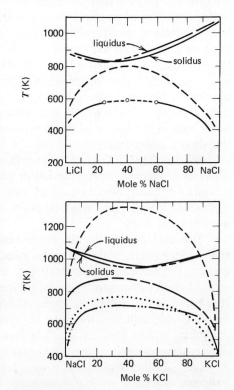

FIGURE 25.21

Comparison of theoretical and observed heats of mixing for alkali halide systems: ——— Theory; ● experiment. From D. L. Fancher and G. R. Barsch, *J. Phys. Chem. Solids* **30**, 2503 (1969).

agreement is seen to be good but not perfect.

Heat capacity studies of 1:1 KCl:KBr mixtures reveal a residual entropy (see Chapter 18) of 5.77 J/K mol, which is the same, within experimental error, as the entropy of a random mixture. This observation confirms the hypothesis that the static distribution of ions is accurately described as random. On the other hand, for $T > 0$ the same experiments show that the entropy of mixing deviates from the ideal value, so there must be a vibrational contribution. For $KCl_{0.5}Br_{0.5}$ at 300 K, calorimetric experiments reveal there is a contribution of 0.46 J/K mol to the entropy of mixing from the change in the vibrational spectrum. This is actually a rather large effect, as can be seen from the change in solubility from that predicted if the entropy of mixing were ideal. We saw, earlier, that at equilibrium between a mixture and pure components the chemical potential of each species must be the same in the mixture and the pure component. The difference in Gibbs free energy between the solid solution and the pure components is, of course, $\Delta G_{mix} = \Delta H_{mix} - T \Delta S_{mix}$. The heat of mixing of the solid solution can be calculated from the ionic lattice model as described. Then separating ΔS_{mix} into an ideal mixture component and a vibrational component, and assuming these to be independent, $\Delta S_{mix} = \Delta S_{mix}^{ideal} + \Delta S_{mix}^{vib}$, so the solubility of one alkali halide in another can be computed if ΔS_{mix}^{vib} can be computed. As shown in Chapter 22, the entropy of a crystal is rather well approximated by the value predicted from a Debye spectrum of vibrations. The Debye spectrum of a mixture differs from that of its parent components because of changes in the elastic constants. When such changes are taken into account, and ΔS_{mix}^{vib} calculated and combined with ΔH_{mix} obtained earlier, one predicts the solubility curves shown in Fig. 25.22. It is clear that the agreement with the observed solubility is better when the vibrational contribution to ΔS_{mix} is taken into account. The estimated magnitudes of the vibrational contribution to ΔS_{mix} are given in Table 25.5.

We conclude that a proper theoretical accounting for the thermodynamic properties of a mixture requires (1) very accurate pair potentials, (2) knowledge of how the structure changes when the mixture is formed, and (3) inclusion in the entropy of mixing of contributions arising from the change in the vibrational spectrum of the solid.

We shall soon see that each of these requirements has a counterpart in the theory of liquid mixtures.

FIGURE 25.22

Solid solubility curves for two alkali halide systems. – – – Theoretical curve calculated using ΔS_{mix}^{ideal} only; ——— Theoretical curve calculated using $\Delta S_{mix}^{ideal} + \Delta S_{mix}^{vib}$; — · · · — · · · Semitheoretical curve calculated from the experimental ΔH_{mix} and $\Delta S_{mix}^{ideal} + \Delta S_{mix}^{vib}$; ∘ ∘ ∘ and · · · experimental data. From D. L. Fancher and G. R. Barsch, *J. Phys. Chem. Solids* **30**, 2517 (1969).

TABLE 25.5

VIBRATIONAL CONTRIBUTION TO THE ENTROPY OF MIXING
(in J/K mol)—1:1 MIXTURES[a]

	LiCl–NaCl	NaCl–KCl	NaBr–KBr	NaCl–NaBr	KCl–KBr	KBr–KI
Δs_{mix}^{ideal}	5.77	5.77	5.77	5.77	5.77	5.77
Δs_{mix}^{vib}(calc.)[b]	1.17	1.42	1.04	1.33	0.83	0.83
Δs_{mix}^{vib}(exp.)	—	—	—	—	0.46	—

[a] From D. L. Fancher and G. R. Barsch, *J. Phys. Chem. Solids* **30**, 2517 (1969).
[b] From application of the Debye model.

Examination of the contents of Chapters 22 and 23 reveals that the concept of structure is employed somewhat differently in the theoretical descriptions of crystals and liquids. In the case of a crystalline solid the regularity of the atomic arrangement permits straightforward calculation of both the static lattice energy and the energy of vibrational motion. To decide whether one or another lattice is more stable, the Gibbs free energy of each is computed and the one with the lowest Gibbs free energy chosen. The point is that the lattice structure is known, the distribution of molecules is therefore also known, and the focus of the theoretical analysis is the influence on the thermodynamic properties of displacements of the atoms from specfied lattice sites. In the case of a liquid the distribution of atoms is not *a priori* known, and one of the functions of the theory is to determine that distribution. The same difference in use of the concept of structure separates the theory of crystalline mixtures from the theory of liquid mixtures. In this section we outline a few of the main ideas used in the theory of liquid mixtures, and discuss one application of the theory as an example.

We start our discussion by considering a very crude model, since it provides a pictorial representation of some important characteristics of a mixture. It must be emphasized that this model is not an accurate representation of real liquid mixtures, yet has features that real liquid mixtures also have.

The essence of the difference between liquids and solids can be traced to the nature of the molecular motions in the two classes of condensed matter. The simplest approximation to a liquid mixture that retains some of the features distinguishing a liquid from a solid is called the *cell model*. It is based on the following ideas: Although molecular displacement in a liquid is not difficult, the rate at which a molecule diffuses is small compared to the rate at which a free molecule would move. At the density characteristic of the liquid phase, each molecule is surrounded by a "cage" of other molecules. If the "cage" molecules confine the central molecule to a small volume element for a long time compared to the time the molecule takes to move across the "cage," we can think of the central molecule as restricted to move only within a cell centered about some point. Of course the "cage" molecules move also, the "cage" decays in time, and the center of the cell defined by the "cage" moves. The cell model of liquids adopts the idea that a "cage" exists, but neglects the decay of the "cage" and the drift of the cell center in time. In other words, every molecule is assumed to be confined to a cell in which its motion is defined by the intermolecular potential of the "caging" molecules. In this model a molecule cannot move throughout the entire liquid, but only in a small volume about some point in space. The

B. Liquid Mixtures

distribution of component molecules, say, A and B, throughout the liquid is now calculated on the basis of two ideas: First, it is imagined that some regular lattice can be passed through the centers of the cells occupied by the molecules. Clearly, this approximation overestimates the long-range order in the liquid (see Chapter 23) and makes the model too solid-like. The distribution of molecules over the cell centers (lattice sites) is calculated as if the lattice were rigid. Second, a molecule can move about its cell in an irregular way. The trajectory in the cell is weighted at each point by a Boltzmann factor involving the instantaneous energy of interaction of the molecule with its surroundings. That is, vibrational motion (as in a crystal) is replaced by a thermal motion weighted according to the instantaneous interaction energy. Each molecule is assumed to move in its cell independently of all others, except insofar as the near neighbors (assumed to be stationary for the calculation of the field in the cell) provide the field in which it moves. Because of the assumed independence of motion, the model is an analogue of the Einstein model of a crystal. Like the Einstein model, it errs by neglecting correlation in the motions of neighboring particles.

The canonical partition function for a binary liquid mixture has the form

$$Q_{N_A N_B} = (N_A! N_B! h^{3(N_A+N_B)})^{-1} \int_V \cdots \int dx_1 \cdots dp_{zN}$$

$$\times \exp\left[-\frac{1}{k_B T}\left(\sum_i \frac{p_i^2}{2m_i} + \mathscr{V}_N \right) \right], \quad (25.203)$$

where $N = N_A + N_B$ molecules are contained in the volume V. Equation 25.203 should be compared with Eq. 21C.3 of Appendix 21C, which is the canonical partition function of a pure fluid. Just as there, Eq. 25.203 can be reduced to

$$Q_{N_A N_B} = \frac{1}{N_A! N_B!}\left(\frac{2\pi m_A k_B T}{h^2} \right)^{3N_A/2}\left(\frac{2\pi m_B k_B T}{h^2} \right)^{3N_B/2}$$

$$\times \int_V \cdots \int dx_1 \cdots dz_N \exp\left(-\frac{\mathscr{V}_N}{k_B T} \right). \quad (25.204)$$

The cell model we have described represents the total potential energy, $\mathscr{V}_N$, in the form

$$\mathscr{V}_N = N_A \epsilon_A(0) + N_B \epsilon_B(0) + \sum_{i=1}^N [\epsilon_i(r) - \epsilon_i(0)]. \quad (25.205)$$

That is, the total potential energy is thought of as the sum of (1) the lattice energy that would exist if each molecule were stationary at the center of its cell, $N_A \epsilon_A(0) + N_B \epsilon_B(0)$, and (2) the energy associated with the deviation of a molecule from its lattice position in the field generated by the near-neighbor molecules at the centers of their cells, $\epsilon_i(r) - \epsilon_i(0)$, summed over all molecules. Using Eq. 25.205 in Eq. 25.204, we find that

$$Q_{N_A N_B} = \left(\frac{2\pi m_A k_B T}{h^2} \right)^{3N_A/2}\left(\frac{2\pi m_B k_B T}{h^2} \right)^{3N_B/2} \cdot \frac{(N_A + N_B)!}{N_A! N_B!}$$

$$\times J_A^{N_A} J_B^{N_B} \exp\left\{ -\frac{1}{k_B T}[N_A \epsilon_A(0) + N_B \epsilon_B(0)] \right\}, \quad (25.206)$$

because there are, neglecting near-neighbor correlations, $(N_A + N_B)!$ ways of distributing $N_A + N_B$ molecules over $N_A + N_B$ sites. The cell partition functions are

$$J_A = 4\pi \int_\Delta \exp\left\{ -\frac{1}{k_B T} [\epsilon_A(r) - \epsilon_A(0)] \right\} r^2 \, dr, \qquad (25.207)$$

$$J_B = 4\pi \int_\Delta \exp\left\{ -\frac{1}{k_B T} [\epsilon_B(r) - \epsilon_B(0)] \right\} r^2 \, dr, \qquad (25.208)$$

where Δ is the cell volume and $N\Delta = V$. Note that the cell partition functions appear to the powers N_A and N_B because each molecule is considered to move independently of the others. From $Q_{N_A N_B}$ and cell model approximations to $Q_{N_A}{}^0$ and $Q_{N_B}{}^0$, the canonical partition functions of pure liquid A and B, one easily finds the excess entropy of mixing to be

$$s^E \equiv \Delta s_{\text{mixing}} - \Delta s_{\text{mixing}}^{\text{ideal}}$$

$$= x_A k_B \left[\ln\left(\frac{J_A}{J_A{}^0}\right) + T \frac{\partial}{\partial T} \ln\left(\frac{J_A}{J_A{}^0}\right) \right]$$

$$+ x_B k_B \left[\ln\left(\frac{J_B}{J_B{}^0}\right) + T \frac{\partial}{\partial T} \ln\left(\frac{J_B}{J_B{}^0}\right) \right], \qquad (25.209)$$

where the superscript zero refers to the pure component. We have already assumed that the assignment of molecules to cells is random, hence s^E should vanish for a static lattice. When the molecular volumes are the same and no displacement from the lattice sites is allowed, we have $J_A = J_A{}^0$, $J_B = J_B{}^0$, and Eq. 25.209 gives $s^E = 0$ as required. The cell partition functions depend on the composition because the field in which a molecule moves is determined by the composition of the surrounding medium. Without making a detailed calculation of the J's, we can still deduce the following interesting information:

1. Suppose that the average composition-dependent potential is such that the molecule in the cell in the solution can move over a larger fraction of the cell volume than in either pure component. Then we have $J_A > J_A{}^0$, $J_B > J_B{}^0$, and there is a positive contribution to the excess entropy of mixing. If the fraction of the cell volume available in solution is less than in the pure components, the contribution to s^E is negative. Note how this entropy change is related to the ratio of volumes available for center-of-mass motion, a result that can be compared with the entropy change when a gas is expanded isothermally.

2. Examination of Eqs. 25.207 and 25.208 shows that the temperature derivative of the cell partition function is related to the average energy of a particle in its cell. Since, in thermodynamic terms, the temperature derivative of the average energy is related to the specific heat, the integral of c_p/T to the entropy, and finally the entropy to the density of states for motion in the cell, differences in average energy per particle should be expected to contribute to an excess entropy of mixing.

In sum, part of the entropy change on mixing comes from the change in "available volume," and part from the way a molecule moves in the "available volume."

The cell model provides a qualitative understanding of some of the principal contributions to the deviations of the properties of a mixture from ideal behavior, but is much too crude to provide a quantitative description. As mentioned earlier, the approximations used to simplify the calculation of the partition function make the liquid too much like a crystal in the sense of assuming too much long-range order, as well as neglecting important correlations in the motion of near-neighbor molecules. As we shall see, the cell model also does not make explicit the role of structure change in the determination of the properties of the mixture, although it does implicitly account crudely for some of the effects of structure change (e.g., by the change in cell volume).

An exact theory of mixtures can be based on the same arguments as used in Chapter 23 to relate the internal energy and pressure of a liquid to the distribution of molecules and the intermolecular potential. In a binary mixture of A and B there are three pair distribution functions (for AA, AB, and BB pairs), six triplet distribution functions (for AAA, ABA and AAB, ABB, BAA, BBA and BAB, and BBB triplets, but only four of these are independent functions), and so forth. This theory provides a correct description of the liquid mixture for all compositions, provided only that the densities of A and B and the interaction potentials are known, from which we must calculate all the pair-correlation functions $g_{ij}(R)$. These pair-correlation functions are then used to calculate the equation of state and internal energy of the mixture from the exact relations

$$\frac{p}{k_B T} = n - \frac{1}{6} \sum_{i=1}^{2} \sum_{j=1}^{2} n_i n_j \int \mathbf{R} \cdot \nabla \left(\frac{u_{ij}}{k_B T} \right) g_{ij}(R) \, d\mathbf{R} \qquad (25.210)$$

and

$$\frac{U}{k_B T V} = \frac{3}{2} n + \frac{1}{2} \sum_{i=1}^{2} \sum_{j=1}^{2} n_i n_j \int \left(\frac{u_{ij}}{k_B T} \right) g_{ij}(R) \, d\mathbf{R}, \qquad (25.211)$$

which are obvious extensions of Eqs. 23.21 and 23.17 to a two-component liquid. In practice the pair-correlation functions $g_{ij}(R)$ can be calculated only after introduction of approximations.

Just as in the case of pure liquids, computer simulation of the properties of mixtures provides invaluable information obtainable in no other fashion. Figure 25.23 shows the three pair-correlation functions that characterize a binary equimolar mixture of molecules that interact with a Lennard–Jones 6–12 potential. The parameters chosen are appropriate to the Ar–Kr mixture at $T = 118\,\text{K}$ and $v_m = 33.3\,\text{cm}^3/\text{mol}$. Note how the peaks are separated even though the molecular diameters are only 7% different. Figure 25.24 shows the effect of changes in the ratio of molecular diameters on the excess Gibbs free energy for equimolar binary mixtures of Lennard–Jones molecules. The parameters chosen correspond to $p = 0$, $T = 117\,\text{K}$, $\epsilon_{22}/\epsilon_{11} = 1.11$, and $\epsilon_{12} = (\epsilon_{11}\epsilon_{22})^{1/2}$. The results of computer simulation studies of mixtures of hard spheres and mixtures of Lennard–Jones molecules lead to the following conclusions:

1. Just as for pure fluids, the repulsive part of the molecular interaction determines the structure of the liquid. The addition of an attractive component to the interaction has little influence on the liquid's structure.

2. Because of conclusion **1**, the packing geometry in a liquid mixture will be very like that in a mixture of hard spheres with the same ratio of diameters. When that ratio differs much from unity, there is considerable deviation from random mixing.

3. The assumption that the AB attractive interaction scales as

$$\epsilon_{AB} = \xi (\epsilon_{AA} \epsilon_{BB})^{1/2}$$

with $\xi = 1$ is not accurate enough to be useful. Although ξ deviates from unity by only a few percent, the excess thermodynamic properties of the mixture are themselves only a few

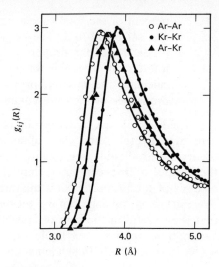

FIGURE 25.23
Pair-correlation functions from a molecular dynamics simulation of an equimolar liquid mixture of argon and krypton ($T = 118\,\text{K}$, $v_m = 33.3\,\text{cm}^3/\text{mol}$). The three main peaks are well separated. From I. R. McDonald and K. Singer, *Chemistry in Britain* **9**, 56 (1973).

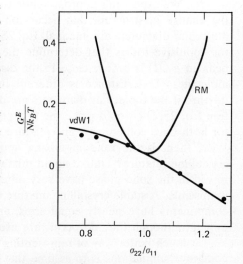

FIGURE 25.24
The effect of changes in the molecular diameter ratio on the excess Gibbs free energy of binary mixtures of Lennard–Jones liquids ($T = 117\,\text{K}$, $p = 0$) for the case when $\epsilon_{22}/\epsilon_{11} = 1.11$. The points give the results of Monte Carlo calculations and the curves show the predictions of various theories: random mixing (RM) and one-fluid (vdW1) versions of the van der Waals model. From I. R. McDonald and K. Singer, *Chemistry in Britain* **9**, 56 (1973).

percent of the total values, so inaccurate predictions result when ξ is set equal to 1. The consequence of this observation is that prediction of the properties of a mixture requires very accurate interaction potentials for all pairs of molecules, and one cannot calculate the mixture properties from knowledge of only the properties of the pure components.

Theories of the pair-correlation functions in a mixture can be generated by the same class of approximations as used to develop a theory of pure liquids (see Chapter 23). A more direct approach is to use the results of the computer simulation studies to construct a simple analysis of the properties of a mixture.

First, what mileage can we get out of the concept of random mixing that is so useful in the case of crystalline mixtures? The random mixing assumption, applied to a liquid mixture, amounts to assuming that all $g_{ij}(R)$ have the same dependence on R. Then, for a binary mixture,

$$g_{11}(R) = g_{22}(R) = g_{12}(R) \equiv g_2{}^x(R) \tag{25.212}$$

and

$$u_2{}^x(R) = \sum_{i,j} x_i x_j u_{ij}(R), \tag{25.213}$$

so that the energy, Eq. 25.211, becomes

$$\frac{U}{k_B T V} = \frac{3}{2} n + \frac{2\pi n^2}{k_B T} \int u_2{}^x(R) g_2{}^x(R) R^2 \, dR. \tag{25.214}$$

In the random mixing approximation $g_2{}^x(R)$ is the pair-correlation function of a hypothetical fluid with the composition-dependent potential $u_2{}^x(R)$. We expect this approximation to be inaccurate because it neglects the change in structure that occurs on mixing. Put another way, if the molecular diameters are unequal, Eq. 25.212 is very inaccurate. Since it is the repulsive forces that determine the structure of the liquid, the first peak in $g_2(R)$ is always close to the distance of closest approach of two molecules. This distance is different in the pure components and the mixture if the molecular diameters are different, whereas the approximation Eq. 25.212 would make the first peak of $g_2(R)$ occur at the same R for both pure components and mixture regardless of the molecular sizes. Why, then, is the random mixing approximation useful to describe a crystalline mixture? Unlike liquid mixtures, if the two components to be mixed in the solid phase have very different molecular sizes, they will be immiscible. A stable crystalline mixture can be formed only when the two components have nearly equal sizes, and then deviations from random mixing, although they do occur, are usually not large.

A much better way of representing the consequences **1**, and **2** of the computer simulations is to assume that

$$g_{11}(R/\sigma_{11}) = g_{12}(R/\sigma_{12}) = g_{22}(R/\sigma_{22}) \equiv g_2{}^x(R/\sigma_x). \tag{25.215}$$

This form for $g_2{}^x(R/\sigma_x)$ preserves the key feature that its first peak occurs near to the distance of closest approach. In other words, once scaled by the distance of closest approach, we assume the form of $g_2(R)$ is the same for all pairs of species, as it is in a hard-sphere fluid. Using Eq. 25.215 in Eq. 25.211, we find that

$$\begin{aligned}
\frac{U}{k_B T V} &= \frac{3}{2} n + \frac{2\pi n^2}{k_B T} \sum_{i,j} x_i x_j \epsilon_{ij} \sigma_{ij}{}^3 \int F(R/\sigma_{ij}) g_{ij}(R/\sigma_{ij}) \left(\frac{R}{\sigma_{ij}}\right)^2 d\left(\frac{R}{\sigma_{ij}}\right) \\
&= \frac{3}{2} n + \frac{2\pi n^2}{k_B T} \epsilon_x \sigma_x{}^3 \int F(\tilde{R}) g_2{}^x(\tilde{R}) \tilde{R}^2 \, d\tilde{R}, \tag{25.216}
\end{aligned}$$

where $u_2(R) = \epsilon F(R/\sigma) = \epsilon F(\tilde{R})$ is the form assumed for all pair interactions and

$$\epsilon_x \sigma_x^3 = \sum_{i,j} x_i x_j \epsilon_{ij} \sigma_{ij}^3. \qquad (25.217)$$

In this approximation $g_2^x(\tilde{R})$ is the pair correlation function for a fluid whose depth and range potential parameters satisfy Eq. 25.217. The theory of mixtures just described is known as a van der Waals one-fluid theory (vdW1). As shown in Fig. 25.24, it agrees rather well with the observed excess free energy as a function of molecular diameter ratio, whereas the random mixing assumption agrees only at the one point $\sigma_{11} = \sigma_{22} = \sigma_{12}$. In Table 25.6 we compare the observed and calculated excess thermodynamic functions of some simple mixtures. The random mixing assumption is again shown to be very poor. The van der Waals one-fluid theory is in good agreement with the computer simulation results, and both of these are in modest agreement with experimental data, with some glaring inaccuracies. It is likely that a principal source of the errors in the predictions is the uncertainty in the pair potentials for unlike molecules, since the approximate theory agrees well with the computer experiments based on the same interactions.

We have also included in Table 25.6 the predictions of a perturbation theory of mixtures just like the perturbation theory of pure liquids described briefly in Chapter 23. The perturbation theory is seen to be about as accurate as the van der Waals one-fluid theory.

TABLE 25.6
COMPARISON OF EXPERIMENTAL AND THEORETICAL EXCESS THERMODYNAMIC FUNCTIONS FOR SOME LIQUID MIXTURES

T (K)	Ar+Kr 116	Ar+N$_2$ 84	Ar+O$_2$ 84	Ar+CO 84	Ar+CH$_4$ 91	Xe+Kr 161	N$_2$+O$_2$ 84	CO+CH$_4$ 91
(a) g^E (J/mol)								
Exp.	84	34	37	57	74	115	39	115
Comp.	45	35	0	27	−10	28	38	76
RM	203	98	—	99	277	—	143	137
VDW1	46	39	—	29	−17	—	43	83
Pert'n	39	32	—	25	−12	—	36	69
(b) h^E (J/mol)								
Exp.	—	51	60	—	103	—	42	105
Comp.	−18	40	0	34	−35	−52	48	13
RM	254	142	—	149	406	—	220	115
VDW1	−30	43	—	35	−55	—	52	26
Pert'n	−28	35	—	30	−30	—	42	24
(c) v^E (cm^3/mol)								
Exp.	−0.52	−0.18	0.14	0.10	0.17	—	−0.31	−0.32
Comp.	−0.60	−0.23	0.00	−0.17	−0.13	−0.62	−0.25	−0.68
RM	0.42	0.11	—	0.23	1.34	—	0.35	−0.32
VDW1	−0.68	−0.25	—	−0.19	−0.23	—	−0.28	−0.71
Pert'n	−0.62	−0.23	—	−0.17	−0.12	—	−0.25	−0.63

In summary, the structure of a liquid mixture must be known accurately before its excess thermodynamic properties can be predicted. That structure is determined by short-range repulsive forces, hence the use of models that accurately represent the equation of state of mixed hard spheres is likely to be fruitful. It is also necessary to know accurately the mixed-pair interaction—this cannot be adequately approximated by the geometric mean of the pure component interactions.

There is one other approach to the statistical molecular theory of solutions that must be mentioned because it is exact (i.e., it does not use models and avoids mathematical approximations) and is very useful for describing the properties of dilute solutions. This theory is constructed so as to exploit the analogy between the pressure of a pure gas and the osmotic pressure of a solution in equilibrium with pure solvent through a semipermeable membrane. The fundamental idea is to consider the effective interactions between pairs, triplets, quadruplets, ... of solute molecules in a solvent. The effective interaction is determined by averaging the total interaction of all molecules, solute and solvent, over the positions of the solvent molecules. The result of this procedure is a representation of the deviations from ideal solution behavior in terms of a set of solution virial coefficients, which are related to the potential of mean force between solute molecules in pure solvent by the same formulas as relate the virial coefficients in a gas to the intermolecular potential (see Chapter 21). To establish the result cited requires a straightforward but tedious analysis, which we shall not discuss.

In order to use the solution virial coefficients to describe the properties of solute molecules, we must know the potentials of mean force. At present little is known about these potentials in condensed phases, and the theory cannot now be used to predict the properties of a solution from intermolecular potentials. This is, however, a practical difficulty similar to that encountered in the theory of pure liquids, and not a difficulty in principle. The solution virial coefficients can be calculated by adopting models for the potentials of mean force. In essence, this is the approach adopted in the theory of dilute electrolyte solutions. To illustrate the utility of models of the solution virial coefficients, consider a solution of long rod-shaped molecules in some solvent composed of small molecules. Example of systems of this type are tobacco mosaic virus in water or a dilute electrolyte solution, and helical polypeptides in organic solvents. As a simple model of the potential of mean force between rodlike molecules, we shall treat the solvent as a continuum and the solute molecule as a "hard" right circular cylinder. Thus, by analogy with the behavior of hard spheres in a gas (see Eq. 21.168), the second solution virial coefficient is determined by the volume excluded to the center of one hard rod by another hard rod. Clearly, this depends on the length of the rod. The calculation of the virial coefficients for even this model is difficult and tedious. In the limit that the length-to-diameter ratio is very large, $(l/d) \gg 1$, the second solution virial coefficient is proportional to $l^2 d$, and higher coefficients to higher powers of $(l^2 d)$. Indeed, the concentration dependence of the osmotic pressure is represented by a series of terms of the form $(cl^2 d)^n$ with various numerical coefficients (where $c \equiv N_{rods}/V$). It is found that the solute–solute interactions are so effective that above a critical concentration the solution becomes unstable with respect to separation into two phases, each a solution, one concentrated and one dilute with respect to solute. The calculated solution virial coefficients lead to the prediction that the phase transition is first-order, so is accompanied by a volume change.

How does calculating the set of solution virial coefficients help us to understand the predicted phase transition? What is the nature of that transition? As the concentration of solute is increased, solute–solute interactions become more and more important. Now, a long rodlike molecule sweeps out a large volume when it rotates about its center of mass. At a relatively low concentration the number of solute molecules multiplied by the volume swept out per molecule equals or exceeds the volume of the solution. When this concentration is exceeded, the ends of the rods tend to get in each other's way and solute–solute interaction results in partial parallelism of the solute molecules, since when the volumes swept by two rods intersect, collision can still be avoided by correctly orienting the molecules. However, as the orientation of the molecules becomes more and more parallel, the entropy per solute molecule decreases. Of course, the energy per solute molecule also decreases, so that the free energy is minimized. The important point is that the loss of entropy per particle eventually becomes so large that the chemical potential of a dilute solution with no hindrance of orientation becomes equal to that of the concentrated solution with partial parallel orientation. At that concentration the solution separates into two phases, one with a small concentration of solute and no restriction on orientation, and one with a large concentration of solute and all solute molecules partially parallel (see the discussion of paraffin hydrocarbon liquids in Chapter 23). This predicted phase separation has been observed in tobacco mosaic virus solutions in water, and in certain polypeptide solutions in organic solvents.

In this section we have surveyed only the simplest features of the statistical molecular theory of solutions, focusing attention on those aspects of molecular interaction connected with the translational motion of the molecules. Of course, in real systems the internal degrees of freedom of the molecules can be different in solution and in the pure component, different solvents may stabilize different molecular geometries of the solute molecule, local directional bonding of the solute to the solvent can occur, and so forth. Many of the subtler aspects of the chemistry of condensed phases are connected with just such phenomena. As of this date the statistical molecular interpretation of chemical effects arising from specific interactions in solution is restricted to the predictions obtained from very simple models. The theory of solutions of polyatomic molecules with directional and nondirectional interactions is so complicated that little progress has been made in deducing generally valid descriptions. What is needed is a set of new concepts into which the complications can be swept and on the basis of which a simple and incisive theoretical description can be based.

Further Reading

Findlay, A., *The Phase Rule and Its Applications*, 9th ed., rev. by A. N. Campbell and N. O. Smith (Dover, New York, 1951).

Hildebrand, J. H., Prausnitz, J. M., and Scott, R. L., *Regular and Related Solutions* (Van Nostrand Reinhold, New York, 1970).

Lewis, G. N., and Randall, M., *Thermodynamics*, 2nd ed., rev. by K. S. Pitzer and L. Brewer (McGraw-Hill, New York, 1961), chaps. 17–21, 26, and 34.

Prigogine, I., *The Molecular Theory of Solutions* (North Holland, Amsterdam, 1957).

Rushbrooke, G. S., *Introduction to Statistical Mechanics* (Oxford University Press, London, 1949), chaps. 13, 14, 17, and 18.

Henderson, D., and Leonard, P. J., in D. J. Henderson (Ed.), *Physical Chemistry; An Advanced Treatise*, Vol. VIIIB, p. 414 (Academic Press, New York, 1971).

McDonald, I. R., *Specialist Periodical Report, Statistical Mechanics* **1**, 134 (1973), The Chemical Society, London.

Münster, A., *Statistical Thermodynamics*, Vol. II (Springer-Verlag, Berlin, 1974), chap. 17.

Palatnik, L. S., and Landau, A. I., *Phase Equilibria in Multicomponent Systems* (Holt, Rinehart and Winston, New York, 1964).

Rowlinson, J. S., *Liquids and Liquid Mixtures*, 2nd ed. (Butterworths, London, 1969).

Advanced Treatments

Problems

1. A mixture of $C_6H_5CH_3$ and C_6H_6 contains 30% by weight of $C_6H_5CH_3$. At 30°C the vapor pressure of pure $C_6H_5CH_3$ is 36.7 torr and that of pure C_6H_6 is 118.2 torr. Assuming ideality, calculate the total pressure and the partial pressure of each component above the solution at 30°C.

2. Consider a system that contains two components in two phases, say, liquid and vapor. From the Gibbs phase rule, the system has two degrees of freedom, so that p and n_2^V may be taken as independent variables. The temperature can therefore be written $T = T(p, n_2^V)$. If the system is azeotropic, however, there is an additional constraint $(x_2^L = x_2^V)$ that results in the loss of one degree of freedom. The temperature of the system then depends only on the pressure. Evaluate dT/dp for the azeotropic mixture.

3. Mixtures of C_6H_5Cl and C_6H_5Br form nearly ideal solutions. At 136.7°C the vapor pressure of pure C_6H_5Cl is 863 torr and that of C_6H_5Br is 453 torr. Find the composition of a mixture whose total vapor pressure is 760 torr. Find the composition of a mixture whose vapor is equimolar in C_6H_5Br and C_6H_5Cl.

4. Assuming that methane forms ideal solutions in C_8H_{18}, calculate the change in chemical potential at 298 K when 3 mol of methane at 1 atm pressure is dissolved in an infinite amount of a CH_4-C_8H_{18} solution, in which the partial pressure of CH_4 is 27 torr.

5. Calculate the change in free energy per gram-atom when monoclinic sulfur is changed to rhombic sulfur at 298 K. Sulfur forms ideal solutions of S_8 molecules in benzene, and the solubilities of monoclinic and rhombic sulfur in 1000 g of benzene are 27.1 g and 21.2 g, respectively.

6. The partial vapor pressure of chloroform in a mixture with acetone at 300 K is 20.0 torr when its mole fraction is 0.12, and 220 torr when its mole fraction is 0.80. How does the chemical potential of the chloroform in the second solution differ from the chemical potential in the first solution? What would your answer be if the solutions were ideal?

7. Consider a binary, nonideal solution defined by the equations (see Eqs. 25.29–25.30)

$$\mu_1 = \mu_1{}^0 + RT \ln x_1 + WT^2(1-x_1)^2,$$
$$\mu_2 = \mu_2{}^0 + RT \ln(1-x_1) + WT^2 x_1{}^2,$$

where x_1 is the mole fraction of component 1 and W is a constant. Calculate Δh_{mixing}, Δs_{mixing}, and Δv_{mixing}.
The reader is cautioned that the form WT^2 is not a realistic description of the temperature dependence of the interactions in any real solution.

8. A mixture contains 1 mol each of liquid benzene and liquid toluene. At 300 K the vapor pressure of pure toluene is 32.06 torr and that of pure benzene is 103.01 torr.
   (a) If the pressure over the mixture is reduced at 300 K, at what pressure does boiling start?
   (b) What is the composition of the first trace of vapor boiled off?
   (c) If the pressure is lowered further, at what pressure does the last trace of liquid disappear?
   (d) What is the composition of the last trace of liquid?

9. For nondilute solutions, show that Eq. 25.107 should be replaced by

$$\ln x_1 = -\frac{\Delta h_{1v}\theta}{RT_{1b}T}$$

if Δh_{1v} is considered constant. An analogous expression is found for the freezing-point depression. If the heat of fusion of water is 6.01 kJ/mol, calculate the freezing point in solutions where the mole fraction of water is 0.8, 0.6, 0.4, 0.2.

10. Two grams of benzoic acid dissolved in 25 g of benzene ($K_f = 4.90°C$ kg/mol) produce a freezing point depression of 1.62°C. Calculate the molecular weight and compare your answer with the molecular weight obtained from the formula C_6H_5COOH.

11. At 0°C the Henry's law constant for argon in water is 2.17×10^4 and at 30°C the constant is 3.97×10^4 for $p = 1$ atm. Find the approximate heat of solution over this temperature range.

12. Find the composition of the vapor that boils off from the benzene–chlorobenzene mixture discussed in the example of page 912.

13. 10.0 g of an unknown nonvolatile solid was dissolved in 100.0 g of H_2O. At 26°C, the vapor pressure of the solution was observed to be 24.75 torr. At that same temperature, pure water has a vapor pressure of 25.20 torr. What is the molecular weight of the solid?

14. At 30°C a $1M$ solution of a sugar has a vapor pressure of 31.207 torr, whereas pure water has a vapor pressure of 31.824 torr. Calculate the osmotic pressure of the solution using Eq. 25.124 and Eq. 25.120. In the latter case, approximate the partial molar volume of the water by the molar volume of pure water.

15. The Henry's law constant for N_2 in H_2O at 25°C is 6.43×10^7, and that for O_2 is 3.30×10^7 for $p = 1$ atm. If air contains 80% N_2 and 20% O_2 by volume, calculate the percentage composition of air dissolved in water.

16. Verify that the chemical potentials for an ideal two-component solution (Eq. 25.12) satisfy the Gibbs–Duhem relation.

17. Suppose that the excess chemical potentials of a binary solution are given by

$$\mu_1{}^E = a_1 x_2 + b_1 x_2{}^2 + c_1 x_2{}^3,$$
$$\mu_2{}^E = a_2 x_1 + b_2 x_1{}^2 + c_2 x_1{}^3.$$

Prove that $a_1 = 0$, $a_2 = 0$, $c_2 = -c_1$, $b_2 = b_1 + \frac{3}{2}c_1$.

18. Consider an ideal solution of two volatile components. The partial pressures of the components in the vapor phase are p_1 and p_2, and the vapor occupies a volume V. If the partial molar volumes of the gases are $\bar{v}_1 = RT/p_1^0$ and $\bar{v}_2 = RT/p_2^0$, show that Raoult's law requires that $V = n_1\bar{v}_1 + n_2\bar{v}_2$, where n_1 and n_2 are the number of moles of the components in the vapor.

19. Calculate the partial molar volumes of methanol and water in a water–methanol solution that is 12.5% methanol by weight, from the following data obtained at 15°C and 1 atm: A 12.0% methanol solution has a density of 0.97942 g/cm³, and a 13.0% methanol solution has a density of 0.97799 g/cm³.

20. The vapor pressure of pure ethanol at 35°C is 102.8 torr and that of pure chloroform is 295.1 torr. In an ethanol–chloroform solution in which the mole fraction of ethanol is 0.20, the total vapor pressure is 304.2 torr, and the mole fraction of ethanol in the vapor is 0.138. Calculate the activity coefficients of the ethanol and the chloroform, and the excess free energy of mixing.

21. When phenol (C_6H_5OH) is dissolved in a particular solvent, it is partially dimerized. If 2.58 g of phenol is dissolved in 100 g of the solvent, the freezing point of the solvent ($K_f = 14.1°C\,kg/mol$) is depressed by 2.37°C. Find how much of the phenol is dimerized in solution.

22. Use the Gibbs phase rule to show that Fig. 25.11a cannot have a eutectic line or area.

23. Is it possible for one component of a binary solution to behave ideally but not the other component?

24. The volume of an aqueous solution of NaCl at 25°C in which m mol of NaCl are dissolved in 1 kg of water is found to be

$$V = 1003 + 16.6m + 1.77m^{3/2} + 0.119m^2,$$

where the volume is given in cubic centimeters. Find the partial molal volumes of NaCl and H_2O when $m = 0.1$.

25. In an acetone–chloroform solution where the mole fraction of acetone is 0.531, the partial molar volume of acetone is 74.2 cm³/mol and that of chloroform is 80.2 cm³/mol. What volume of solution weighs 100 g? The densities of the pure components are 1.351×10^{-2} mol/cm³ for acetone and 1.240×10^{-2} mol/cm³ for chloroform. What is the total volume of the un-mixed components of the 100-g solution?

26. Suppose that you are given a binary solution with components A and B, and the partial molar volume of A is known as a function of $x_A : \bar{v}_A = f(x_A)$. Obtain an equation for $\bar{v}_B$ in terms of x_A.

27. Polybenzyl glutamate (PBG) is a polypeptide that has an internally hydrogen-bonded rigid helical conformation in chloroform, and a nonhydrogen-bonded, flexible, randomly coiled conformation in trichloracetic acid. If trichloracetic acid is added to a solution of PBG in chloroform, the PBG can be made to undergo the transformation helix→random coil. The entropy change in this transformation is found to be negative. But the random coil conformation of PBG has many more configurations accessible for energy between E and $E + dE$ than does the helix conformation, so we expect the entropy change for the transformation helix→random coil to be positive. Provide a molecular model-type explanation for the observation.

28. Consider a mixed liquid solvent containing components 1 and 2, in equilibrium with a gas labeled component 3. Suppose that the solubility of the gas is proportional to the gas fugacity. For the case discussed in Section 25.4, where the solvent has only one component, the change in solvent chemical potential when the concentration of solute is changed can be obtained directly by use of Henry's law. Show that for the case that the solvent is a mixture, knowledge of the Henry's law constant is insufficient to determine the

separate changes of the chemical potentials of components 1 and 2 when the solute concentration is changed.

29. Consider a system of two immiscible liquids, in each of which there is dissolved some third component. An example of this situation is the two-phase system consisting of alcohol in water in contact with alcohol in benzene. The Nernst distribution law asserts that, in a system such as that described, the ratio of the concentrations of solute in the two phases is a constant.
    (a) Derive the Nernst distribution law for the case that the solution in each of the immiscible liquid phases is ideal.
    (b) Derive the generalization of the Nernst distribution law when the solutions in each of the phases are not ideal. Is the ratio of solute concentrations in the two phases necessarily a constant?
    (c) Describe the relationship between the Nernst distribution law as formulated above and Henry's law.

30. In Section 25.6 we discussed the freezing-point depression of a solution when the solid formed is pure (has no solute). Consider, now, the case that an ideal dilute liquid solution freezes to form an ideal dilute solid solution of different concentration. Let the ratio of the concentrations in the solid and liquid phases be K. Show that the rate of change of freezing point with solute concentration is

$$\frac{dT}{dx_2} = (K-1)\frac{RT_{1f}^2}{\Delta h_{1f}}$$

for $\theta \equiv T_{1f} - T$ very small, which corresponds to $x_2 \ll 1$.

31. Suppose that a two-component liquid mixture is in equilibrium with a two-component solid mixture, neither mixture being ideal. Show that

$$RT_{1f}\ln\frac{\gamma_1^L x_1^L}{\gamma_1^S x_1^S} = -\frac{\Delta\bar{h}_{1f}\theta}{T_{1f}} - \left(\frac{\Delta\bar{h}_{1f}}{T_{1f}^2} + \frac{\Delta\bar{c}_p}{2T_{1f}}\right)\theta^2 + \cdots,$$

where, as usual, solvent $\equiv 1$ and solute $\equiv 2$. What is the meaning of $\Delta\bar{h}_{1f}$ and $\Delta\bar{c}_p$ in this situation?

32. Consider a multicomponent mixture in which there are many solutes, denoted $2, 3, \ldots, r$, each at very low concentration. It is found that the activity coefficients of components $2, \ldots, r$ can be represented in the form

$$\ln\gamma_2 = \ln\gamma_2^0 + k_{22}x_2 + k_{23}x_3 + \cdots + k_{2r}x_r,$$

$$\vdots$$

$$\ln\gamma_r = \ln\gamma_r^0 + k_{r2}x_2 + k_{r3}x_3 + \cdots + k_{rr}x_r.$$

The coefficients k_{ij} represent the influence of interactions between the different solute species. Show that

$$k_{ij} = k_{ji}.$$

33. In the case of a binary solution, knowledge of the chemical potential of one component for all concentrations can be used to determine completely the chemical potential of the other component (see Eq. 25.28). Consider now a ternary solution. Show that the knowledge of the chemical potential of one component for all concentrations, both of that component and the other two components, determines the chemical potentials of the other components. *Hint:* Fix the ratio of concentrations of two components and treat the mixture as a pseudo-binary system. Use the Gibbs–Duhem equation to integrate along different pseudo-binary system lines.

34. A slight generalization of the regular mixture model allows the quantity w defined by Eqs. 25.29 and 25.30 to be temperature-dependent. Show that, when w depends on T,

$$S^E = -(n_1 + n_2)x_1 x_2\frac{dw}{dT},$$

$$H^E = (n_1 + n_2)x_1x_2\left(w - T\frac{dw}{dT}\right),$$

$$C_p^E = -(n_1 + n_2)x_1x_2T\frac{d^2w}{dT^2}.$$

For the equimolar benzene–cyclohexane mixture at 293 K, $w = 1270$ J/mol, $dw/dT = -7.0$ J/K mol, $d^2w/dT^2 = 0.046$ J/K^2 mol. Calculate $\ln \gamma_1$, H^E, S^E, and C_p^E, and compare these with the observed values, which are 0.130, 836 J/mol, 1.76 J/K mol, and -3.3 J/K mol, respectively.

35. The simplest useful description of a polymer solution (solvent $\equiv 1$, polymer solute $\equiv 2$) is the Flory–Huggins model, which gives for the Gibbs free energy of mixing

$$\Delta G_{\text{mixing}} = RT(n_1 \ln \phi_1 + n_2 \ln \phi_2) + wn_1\phi_2,$$

with ϕ_1 and ϕ_2 the volume fractions of solvent and solute, and w a temperature-dependent interaction energy.
(a) Show that the heat of mixing is

$$\Delta H_{\text{mixing}} = \left(w - T\frac{dw}{dT}\right)n_1\phi_2.$$

(b) Show that the solvent activity is

$$\ln a_1 = \ln(1 - \phi_2) + \phi_2\left(1 - \frac{v_1^0}{v_2^0}\right) + \phi_2^2\left(\frac{w}{RT}\right),$$

with v_1^0, v_2^0 the molar volumes of solvent and solute.
(c) Calculate the partial molar entropy of the solvent.
(d) For the case $v_2^0 = 1000v_1^0$, $M_2 = 1000M_1$, plot on the same graph $\bar{s}_1 - s_1^0$ as a function of ϕ_2 using the Flory–Huggins model and $\bar{s}_1 - s_1^0$ as a function of x_2 using Raoult's law. *Hint:* Convert x_2 to ϕ_2 assuming additivity of molecular volumes.

36. Imagine a crystal of N_A A atoms and N_B B atoms, where $N_B \ll N_A$. Suppose that the following model can be used: (i) The interaction between molecules is limited to nearest neighbors, of which there are z; (ii) the mass of a B atom is much smaller than the mass of an A atom, so that the vibrational spectrum of the dilute solid solution has one high-frequency mode ν', which lies above all other crystal frequencies; (iii) all other crystal frequencies are essentially unchanged by the substitution of a B atom for an A atom; (iv) the vibrational spectrum of the "A part" of the mixed crystal is adequately described by the Einstein model (Chapter 22).
(a) Calculate the chemical potential, partial molecular enthalpy, and partial molecular entropy of the solute B.
(b) Suppose that the mixed crystal just described is in equilibrium with an ideal gas of B atoms. Calculate the Henry's law constant for the system.

37. In a classical mechanical description the only difference between two isotopes of the same element, say, ^{3}He and ^{4}He, is the difference in their masses; the pair interaction, $u_2(R)$, is the same for both isotopic species. In a quantum mechanical description this is no longer true, because the effective interaction must include the consequences of the lack of commutativity of the kinetic and potential energy operators (Section 3.4). It can be shown that the chemical potential of a monatomic species has the form

$$\mu = \mu_{\text{classical}} + \frac{\hbar^2}{24mk_BT}\langle(\nabla u_2)^2\rangle$$

to order $\hbar^2$. The term $\langle(\nabla u_2)^2\rangle$ is the mean square force on one atom arising from all the other atoms in the system.
(a) Show that in the classical limit the difference in chemical potentials of isotopic species of masses m_1 and m_2 is

$$\mu_1 - \mu_2 = \frac{3}{2}k_BT \ln \frac{m_2}{m_1}.$$

(b) Show that, to order $\hbar^2$, the difference between the vapor pressures of the pure isotopic liquids (or solids) is

$$p_1 - p_2 = p_{\text{classical}} \frac{\hbar^2 \langle (\nabla u_2)^2 \rangle}{24(k_B T)^2} \left(\frac{1}{m_1} - \frac{1}{m_2} \right),$$

where $p_{\text{classical}}$ is the common limiting value of the vapor pressure of each isotopic species.

(c) Assuming that the form for the quantum correction to the chemical potential is valid to arbitrarily low temperature, predict whether or not an isotopic mixture will remain one homogeneous phase in the limit $T \to 0$. Justify your conclusion.

38. Suppose that hydrogen is slightly soluble in some crystal in which it occupies sites that prohibit molecular rotation. Estimate the ratio of solubilities of $o\text{-}H_2$ and $p\text{-}H_2$ in this crystal at 60 K, assuming that conversion of $o\text{-}H_2$ to $p\text{-}H_2$ and *vice versa* does not occur.

39. Consider a liquid mixture of monatomic species B in monatomic host A, with $N_B \ll N_A$. Suppose that this mixture is in equilibrium with a perfect gas of B atoms. Using the cell model to describe the liquid phase, derive a relationship for the Henry's law constant. Compare your result with that of Problem 36(b); what are the differences and similarities?

40. The principle of corresponding states is a powerful tool in the study of mixtures. Consider some pure reference substance, denoted 0, in which the pair potential is

$$u_{00}(R) = \epsilon_{00} F\left(\frac{R}{\sigma_{00}} \right),$$

and some other pure substance, denoted 1, with pair potential

$$u_{11}(R) = \epsilon_{11} F\left(\frac{R}{\sigma_{11}} \right).$$

(a) Show that the potential energy contribution (superscript c for "configurational") to the Helmholtz free energy of substance 1 is given by

$$A_1^c(T, V) = f_{11} A_0^c\left(\frac{T}{f_{11}}, \frac{V}{g_{11}^{\;3}} \right) - 3Nk_B T \ln g_{11},$$

where

$$f_{11} \equiv \frac{\epsilon_{11}}{\epsilon_{00}},$$

$$g_{11} \equiv \frac{\sigma_{11}}{\sigma_{00}}.$$

(b) Show also that the equation of state of substance 1 can be expressed in the form

$$p_1^c(T, V) = \frac{f_{11}}{g_{11}^{\;3}} p_0^c\left(\frac{T}{f_{11}}, \frac{V}{g_{11}^{\;3}} \right).$$

(c) If $f_{11} - 1 \ll 1$ and $g_{11} - 1 \ll 1$, show that

$$A_1^c(T, V) - A_0^c(T, V) = U_0^c(f_{11} - 1) - 3(Nk_B T - p_0^c V)(g_{11} - 1),$$

where U_0^c is the potential energy contribution to the internal energy of the reference substance.

41. Imagine a hypothetical pure substance in which the pair potential is the average, with random mixing, of the pair potentials in a mixture. Explicitly, define

$$\langle u(R_{ij}) \rangle = \sum_\alpha \sum_\beta x_\alpha x_\beta u_{\alpha\beta}(R_{ij}).$$

Then the potential energy of a mixture, averaged over all assignments of molecules to positions with random distribution, is

$$\langle \mathcal{V} \rangle = \sum_{i>j} \sum \sum_{\alpha} \sum_{\beta} x_\alpha x_\beta u_{\alpha\beta}(R_{ij})$$

$$= \sum_{i>j} \sum \langle u(R_{ij}) \rangle.$$

Substitution of $\langle \mathcal{V} \rangle$ into the configurational partition function

$$Z = \frac{1}{\prod N_\alpha!} \int \cdots \int \exp\left(-\frac{\mathcal{V}}{k_B T}\right) d\mathbf{R}_1 \ldots d\mathbf{R}_N \qquad (\textstyle\sum N_\alpha = N),$$

defines the hypothetical pure substance for the particular composition. Call it the equivalent substance, and note that its properties depend on the composition.

(a) Suppose that the molecules in the mixture interact via a Lennard–Jones potential. Show that

$$A_x{}^c(T, V) = f_x A_0{}^c\left(\frac{T}{f_x}, \frac{V}{g_x{}^3}\right) - 3Nk_B T \ln g_x,$$

where the zero refers to a reference substance as in Problem 39 and

$$f_x = \left(\sum_\alpha \sum_\beta x_\alpha x_\beta f_{\alpha\beta} g_{\alpha\beta}{}^6\right)^2 \left(\sum_\alpha \sum_\beta x_\alpha x_\beta f_{\alpha\beta} g_{\alpha\beta}{}^{12}\right)^{-1},$$

$$g_x = \left(\sum_\alpha \sum_\beta x_\alpha x_\beta f_{\alpha\beta} g_{\alpha\beta}{}^6\right)^{-1/6} \left(\sum_\alpha \sum_\beta x_\alpha x_\beta f_{\alpha\beta} g_{\alpha\beta}{}^{12}\right)^{1/6},$$

$$f_{\alpha\beta} = \left(\frac{\epsilon_{\alpha\beta}\sigma_{\alpha\beta}{}^6}{\epsilon_{00}\sigma_{00}{}^6}\right)^2 \left(\frac{\epsilon_{\alpha\beta}\sigma_{\alpha\beta}{}^{12}}{\epsilon_{00}\sigma_{00}{}^{12}}\right)^{-1},$$

$$g_{\alpha\beta} = \left(\frac{\epsilon_{\alpha\beta}\sigma_{\alpha\beta}{}^6}{\epsilon_{00}\sigma_{00}{}^6}\right)^{-1/6} \left(\frac{\epsilon_{\alpha\beta}\sigma_{\alpha\beta}}{\epsilon_{00}\sigma_{00}}\right)^{1/6}.$$

(b) Show that the equation of state is

$$p_x{}^c(T, V) = \frac{f_x}{g_x{}^3} p_0{}^c\left(\frac{T}{f_x}, \frac{V}{g_x{}^3}\right).$$

(c) Show that the excess Gibbs free energy of the mixture is

$$G^E(T, p, x) = f_x G_0{}^c\left(\frac{T}{f_x}, \frac{pg_x{}^3}{f_x}\right) - \sum_\alpha x_\alpha f_{\alpha\alpha} G_0{}^c\left(\frac{T}{f_{\alpha\alpha}}, p\frac{g_{\alpha\alpha}{}^3}{f_{\alpha\alpha}}\right)$$

$$- 3Nk_B T \sum_\alpha x_\alpha \ln\left(\frac{g_x}{g_{\alpha\alpha}}\right).$$

(d) For the particularly simple case that $g_{\alpha\beta} = 1$ for all α, β, show that to first order in $f_x - \sum_\alpha x_\alpha f_{\alpha\alpha}$ the excess thermodynamic functions of the mixture are given by

$$G^E = U_0{}^c\left(f_x - \sum_\alpha x_\alpha f_{\alpha\alpha}\right) = U_0\left[\sum_{\alpha>\beta}\sum x_\alpha x_\beta(2f_{\alpha\beta} - f_{\alpha\alpha} - f_{\beta\beta})\right],$$

$$H^E = (U^c - TC_p{}^c)_0\left(f_x - \sum_\alpha x_\alpha f_{\alpha\alpha}\right),$$

$$S^E = -(C_p{}^c)_0\left(f_x - \sum_\alpha x_\alpha f_{\alpha\alpha}\right),$$

$$V^E = (-VT\alpha^c)_0\left(f_x - \sum_\alpha x_\alpha f_{\alpha\alpha}\right).$$

These formulas describe a *conformal solution*.

42. The characteristic feature of a gravitational field is that the work required to move a mass M from a position where the gravitational potential is ϕ^I to

another position where it is ϕ^{II} is simply

$$w = M(\phi^{II} - \phi^{I}),$$

and is independent of the chemical nature of the matter. This leads us to define the chemical potential of species i in a solution in the gravitational potential ϕ by

$$\mu_i = \mu_i^0 + RT \ln \gamma_i x_i + M_i \phi$$

(see also Chapter 26).

(a) Show that at constant temperature and pressure

$$RT \, d\ln \gamma_i x_i + \left(M_i - \bar{v}_i \frac{M_m}{v_m} \right) d\phi = 0,$$

where M_m and v_m are the mean molar mass and mean molar volume, respectively.

(b) Show that for an ideal binary mixture

$$\frac{x_2^{II}}{x_2^{I}} = \exp\left[-(M_2 - \rho\bar{v}_2) \frac{\phi^{II} - \phi^{I}}{RT} \right],$$

where ρ is the mass density.

26

Equilibrium Properties of Solutions of Electrolytes

In Chapter 25 we studied the equilibrium properties of nonelectrolyte mixtures. These mixtures are operationally distinguishable from those containing electrolytes—the latter conduct electricity whereas the former do not. The most elementary implication of this observation is that some of the constituents of an electrolyte solution are charged. Now, the Coulomb interaction between two charges is of very long range and is very strong; it differs profoundly from the short-range van der Waals-type interaction between uncharged molecules. Of course, the interaction between two charged molecules, or between a charged and a neutral molecule, also has a short-range van der Waals-type component. Nevertheless, the effect of the Coulomb interactions on the properties of the solution is the feature that distinguishes electrolyte from nonelectrolyte solutions.

In this chapter we examine the equilibrium properties of electrolyte solutions. Although most of the analysis will be devoted to dilute solutions, we shall also consider briefly mixtures of molten salts. Since water is the most frequently used solvent for electrolytes, special attention will be paid to the nature of aqueous solutions.

26.1
The Chemical Potential

Just as in the study of nonelectrolyte mixtures, the first step in a thermodynamic analysis of electrolyte solutions is the determination of the form of the chemical potential. We adopt the convention that in solutions of strong electrolytes the solute is completely dissociated into ions. From the point of view of thermodynamic analysis, it is not necessary to adopt this convention. However, the complete-dissociation model, although an oversimplification, is reasonably accurate as a description of the state of ionization of strong electrolytes. When the properties of a real electrolyte solution are compared with the properties of a hypothetical ideal solution containing independent ions, the deviations are found to be relatively small. In contrast, if we chose to describe the strong electrolyte solution in terms of the concentration of neutral molecules, it would be found that deviations from the hypothetical ideal solution behavior were much larger. The thermodynamic analysis can be formulated either way, but the availability of a statistical molecular interpretation and the possibility of minimizing "nonideal" effects lead us to choose the convention that the solute exists in a completely dissociated form in solution.

Despite the fact that we shall use concentrations of ions in the thermodynamic formalism, it is not possible to make a solution of ions of one charge type. All we can do is make an electroneutral solution in which the amount of positive charge equals the amount of negative charge. Thus, the chemical potential of the "electroneutral" electrolyte can be measured, but not that of the positive ion or the negative ion separately. For example, consider a salt—say, NaCl—dissolved in water. We cannot determine

$$\mu_{Na^+} \equiv \left(\frac{\partial G}{\partial n_{Na^+}} \right)_{T,p,n_{Cl^-},n_{H_2O}}, \tag{26.1}$$

because we cannot vary n_{Na^+} and n_{Cl^-} independently. However, if we add to the solution dn mol of NaCl, the change of Gibbs free energy at constant T and p is

$$dG = \left(\frac{\partial G}{\partial n_{NaCl}} \right)_{T,p,n_{H_2O}} dn_{NaCl} \equiv \mu_{NaCl}\, dn_{NaCl}. \tag{26.2}$$

It is often convenient to define individual ion chemical potentials by Eq. 26.1, but it must be kept in mind that only

$$\mu_{NaCl} \equiv \mu_{Na^+} + \mu_{Cl^-} \tag{26.3}$$

is operationally definable and experimentally measurable.

For an electrolyte of the $CaCl_2$ type, we have, similarly,

$$dG = \mu_{CaCl_2}\, dn_{CaCl_2}$$
$$= (\mu_{Ca^{2+}} + 2\mu_{Cl^-})\, dn,$$
$$dn_{CaCl_2} = dn_{Ca^{2+}} = \tfrac{1}{2}\, dn_{Cl^-}. \tag{26.4}$$

In general, for an electrolyte of the type $A_{\nu_+}B_{\nu_-}$, the definition corresponding to Eq. 26.3 is

$$\mu_{A_{\nu_+}B_{\nu_-}} = \nu_+\mu_+ + \nu_-\mu_-, \tag{26.5}$$

because dissolution of 1 mol of $A_{\nu_+}B_{\nu_-}$ creates ν_+ mol of A^{Z_+} and ν_- mol of B^{Z_-}, where Z_+ and Z_- are the charges on the A and B ions. The electroneutrality condition for this electrolyte is

$$\nu_+ Z_+ + \nu_- Z_- = 0. \tag{26.6}$$

We now restrict attention to solutions in which the concentration of electrolyte is small compared with the concentration of solvent. For this reason, it is convenient to adopt the molality concentration scale, and to use as the reference state for the electrolyte the hypothetical state in which the properties per mole are the same as in the infinitely dilute electrolyte solution (see Chapter 25). Recalling that μ_{Na^+} and μ_{Cl^-} are not independently definable, but only the sum $\mu_{Na^+} + \mu_{Cl^-}$, we write for the chemical potential of NaCl

$$\mu_{NaCl} = \mu_{NaCl}^{\ominus} + RT \ln a_{Na^+} a_{Cl^-}$$
$$\equiv \mu_{NaCl}^{\ominus} + RT \ln a_{\pm}^2$$
$$= \mu_{NaCl}^{\ominus} + 2RT \ln \gamma_{\pm} m_{\pm}. \tag{26.7}$$

Note that we have used the convention defined in Eq. 25.23b to represent the chemical potential. The product $a_{Na^+} a_{Cl^-}$ is defined to be the activity of NaCl, and its square root $a_{\pm} = \gamma_{\pm} m_{\pm}$ is the *mean ionic activity*. In terms of activity coefficients and molalities, we have

$$a_{Na^+} = \gamma_{Na^+} m_{Na^+},$$
$$a_{Cl^-} = \gamma_{Cl^-} m_{Cl^-},$$
$$\gamma_{\pm}^2 \equiv \gamma_+ \gamma_-, \tag{26.8}$$
$$m_{\pm}^2 \equiv m_+ m_-,$$
$$m_+ = m_- = m_{NaCl}.$$

Similarly, the chemical potential of $CaCl_2$ is

$$\mu_{CaCl_2} = \mu_{Ca^{2+}} + 2\mu_{Cl^-}$$
$$= \mu_{CaCl_2}^{\ominus} + RT \ln a_{Ca^{2+}} a_{Cl^-}^{2}$$
$$\equiv \mu_{CaCl_2}^{\ominus} + RT \ln a_{\pm}^{3}$$
$$= \mu_{CaCl_2}^{\ominus} + RT \ln \gamma_{\pm}^{3} m_{\pm}^{3}, \qquad (26.9)$$

where, in this case,

$$\gamma_{\pm}^{3} = \gamma_{Ca^{2+}} \gamma_{Cl^-}^{2}$$
$$m_{\pm}^{3} = m_{Ca^{2+}} m_{Cl^-}^{2} \qquad (26.10)$$

It is easily seen that for the general salt the formula is

$$\mu_{salt} = \nu_+ \mu_+ + \nu_- \mu_-$$
$$= \mu_{salt}^{\ominus} + RT \ln a_+^{\nu_+} a_-^{\nu_-}$$
$$\equiv \mu_{salt}^{\ominus} + RT \ln a_{\pm}^{(\nu_+ + \nu_-)}, \qquad (26.11)$$

with the corresponding mean activity coefficients and mean ionic molality:

$$\gamma_{\pm}^{(\nu_+ + \nu_-)} = \gamma_+^{\nu_+} \gamma_-^{\nu_-},$$
$$m_{\pm}^{(\nu_+ + \nu_-)} = m_+^{\nu_+} m_-^{\nu_-} = \nu_+^{\nu_+} \nu_-^{\nu_-} m_{salt}^{(\nu_+ + \nu_-)}. \qquad (26.12)$$

The reason we have introduced these extra complications in the activity coefficient and concentration scales is that they are necessary if the properties of $\mu_{salt}^{\ominus}$ are to be obtainable by extrapolation of experimental data. For example, in the case of NaCl-type electrolytes,

$$\lim_{m_{NaCl} \to 0} (\mu_{NaCl} - RT \ln m_{NaCl}) \qquad (26.13)$$

does not exist, whereas

$$\lim_{m_{\pm} \to 0} (\mu_{NaCl} - 2RT \ln m_{\pm}) \qquad (26.14)$$

does exist. Clearly, if $\mu_{NaCl}^{\ominus}$ is to be determined by extrapolation of the properties of an infinitely dilute NaCl solution, a limit such as Eq. 26.13 or 26.14 must be used. Now, why does Eq. 26.14 exist as a limit when Eq. 26.13 diverges? In simple terms, the reason that Eq. 26.14 is finite whereas Eq. 26.13 diverges is that the electrolyte is dissociated into ions. To determine the thermodynamic functions in the hypothetical limit of ideal behavior (infinitely dilute solution), it is necessary to count the number of solute particles correctly. But Eq. 26.14 is nothing more than $(\mu_{NaCl} - RT \ln m_+ - RT \ln m_-)$, which accounts for all solute particles, whereas Eq. 26.13 uses a concentration scale that is the same whether NaCl gives one, two, . . . , ten particles in solution. Thus $(\mu_{NaCl} - RT \ln m_{NaCl})$ does not give the correct concentration dependence for the chemical potential of a solution of independent positive ions of concentration $m_+ = \nu_+ m$ and negative ions of concentration $m_- = \nu_- m$. In an ideal solution each solute particle makes the same contribution to the chemical potential. Hence, by omitting some of the particles from the particle count, Eq. 26.13 in effect neglects a term of the form, say, $RT \ln m_+$, and this term diverges as $m \to 0$.

All the colligative properties of solutions discussed in Chapter 25 can be used to study the properties of solutions of strong electrolytes. The major new property of electrolyte solutions, the fact that they conduct an electric current, can be used to derive electrical work from a chemical reaction. To describe the operation of electrochemical cells, we must extend somewhat our previous considerations.

26.2
Cells, Chemical Reactions, and Activity Coefficients

Thus far we have described how a phase containing ions can be treated as if it were a phase containing only electrically neutral particles. The only significant difference between the analysis of electrolyte and nonelectrolyte solutions is the imposition of the electroneutrality condition

$$\sum_i n_i Z_i = 0. \tag{26.15}$$

What happens if the electroneutrality condition is relaxed?

We have remarked many times that electrostatic interactions are very strong. The charge on the electron is 4.803×10^{-10} esu, or 1.602×10^{-19} C. Thus, 1 mol of charge is just $6.022 \times 10^{23} \times 1.602 \times 10^{-19} = 0.9648 \times 10^5$ C. This quantity of charge is known as the *Faraday*, and denoted $\mathscr{F}$. Suppose that a sphere consisting of 1 mol of matter deviates from electroneutrality by one part in 10^{10}. The excess charge Q in the sphere will collect on the surface, and in vacuum will generate a potential of magnitude $Q/4\pi\epsilon_0 R$ at a distance R from the center of the sphere. In our case, an amount of charge equal to $10^{-10}\mathscr{F}$ generates a potential of 0.87×10^6 V at a point 10 cm from the center of the sphere. This example shows that minuscule deviations from electroneutrality generate very large electric potentials and fields. In particular, a departure from electroneutrality that cannot be detected chemically leads to significant electrical effects: The mass balance and charge balance implied by Eq. 26.15 can be met to one part in 10^{15} and sufficient charge imbalance still exist to give potentials in the volt range.

We conclude that it is meaningful to use as shorthand notation the statement that two phases can have the same chemical composition but different electrical potentials. It is true that the origin of the potential differences is charge imbalance, but the corresponding composition differences are not important chemically except insofar as they generate electrical potentials.

Given the above definition, we see that the chemical potentials of all solutes in a solution must depend on the electrical potential. The difference in chemical potential of component i between two phases of identical chemical composition is proportional to the charge on the ions of component i, and is independent of all other properties of the ion. For two phases, say I and II, we write

$$\mu_i^{II} - \mu_i^{I} = Z_i \mathscr{F}(\psi^{II} - \psi^{I}), \tag{26.16}$$

where the ψ's are the electrical potentials of the phases of identical chemical composition. The quantity $\mathscr{E} \equiv \psi^{II} - \psi^{I}$ is the electrical potential difference between the two phases.

Following our previous analysis of phase equilibrium, one can clearly see that if the boundary between phases I and II is permeable to charge, at thermodynamic equilibrium $\psi^{I} = \psi^{II}$. On the other hand, if the boundary is *not* permeable to charge, it is *not* necessary that the electrical potentials of the two phases be equal. This observation about phase equilibrium suggests that, by suitable selection of the boundaries between phases, measurements of the electrical potential difference between phases can provide information about the equilibrium state. Finally, we note that only potential differences such as $\psi^{II} - \psi^{I}$ are measurable. Therefore, separation of the chemical potential into a "chemical" contribution and an "electrical" contribution, however useful for thinking, must be considered to be arbitrary. There is, in fact, one particular decomposition of the chemical potential in solution that simplifies the treatment of equilibria in electrochemical cells so much that it is extensively employed. We write formally

$$\mu_i = \mu_i^{\ominus}(T, p) + RT \ln \gamma_i m_i + Z_i \mathscr{F}\psi. \tag{26.17}$$

We shall use the form of Eq. 26.17, but again remind the reader that it is the chemical potential that is fundamental, that ψ cannot be measured, and that the right-hand side of Eq. 26.17, although eminently reasonable in form, is arbitrary in its subdivision into chemical and electrical terms.

A galvanic cell is a device, such as that displayed in Fig. 26.1, in which a chemical reaction can be made to result directly in the performance of electrical work. We seek to relate the voltage difference across the cell, $\psi^{VI} - \psi^{I}$, measured between two wires of the *same metal* attached to the two electrodes, to the properties of the chemical reaction in the cell. First, we note that the boundary between two metals is permeable to electrons, so that, when two different metals are placed in contact electrons flow until the chemical potential of the electron is the same in both metals. Although this results in a net excess or deficiency of electrons in each piece of metal, this concentration change is negligibly small. Similarly, when a Zn electrode is dipped into a solution containing ions of Zn^{2+}, the boundary between electrode and solution must be considered permeable to Zn^{2+}. Thus, Zn^{2+} ions dissolve from the electrode and enter the solution, or deposit onto the electrode from the solution, until the chemical potential of Zn^{2+} is the same in the two phases. Again, charge transfer is sufficiently small to cause negligible change in the electron/ion ratio of the Zn electrode. In the Zn metal we have

$$2\mu_{e^-}^{II} + \mu_{Zn^{2+}}^{II} = \mu_{Zn}^{II}, \tag{26.18}$$

using the strong-electrolyte convention. That is, we regard the interior of the Zn metal as dissociated into Zn^{2+} ions and electrons. For the electrode–solution equilibrium, we have

$$\mu_{Zn^{2+}}^{II} = \mu_{Zn^{2+}}^{III}. \tag{26.19}$$

Combining Eqs. 26.18 and 26.19, we find that

$$\mu_{e^-}^{II} = \tfrac{1}{2}(\mu_{Zn}^{II} - \mu_{Zn^{2+}}^{III}). \tag{26.20}$$

This describes the "half-cell" of Zn plus $ZnCl_2$. Consider now the full cell of Fig. 26.1. At the Ag electrode the following equations are valid:

$$\mu_{Cl^-}^{IV} = \mu_{Cl^-}^{III}, \tag{26.21}$$

$$\mu_{Ag^+}^{V} = \mu_{Ag^+}^{IV}, \tag{26.22}$$

$$\mu_{Ag^+}^{IV} + \mu_{Cl^-}^{IV} = \mu_{AgCl}^{IV}, \tag{26.23}$$

$$\mu_{Ag^+}^{V} + \mu_{e^-}^{V} = \mu_{Ag}^{V}. \tag{26.24}$$

These can be combined to yield

$$\mu_{e^-}^{V} = \mu_{Ag}^{V} - \mu_{AgCl}^{IV} + \mu_{Cl^-}^{III}. \tag{26.25}$$

Finally, combining Eqs. 26.20 and 26.25, we obtain

$$\begin{aligned} \mu_{e^-}^{V} - \mu_{e^-}^{II} &= \mu_{Ag}^{V} - \mu_{AgCl}^{IV} + \mu_{Cl^-}^{III} - \tfrac{1}{2}\mu_{Zn}^{II} + \tfrac{1}{2}\mu_{Zn^{2+}}^{III} \\ &= \mu_{Ag}^{V} + \tfrac{1}{2}\mu_{ZnCl_2}^{III} - \tfrac{1}{2}\mu_{Zn}^{II} - \mu_{AgCl}^{IV}. \end{aligned} \tag{26.26}$$

Clearly, the last line of Eq. 26.26 shows that the difference in electron chemical potential between phases II and V is determined by the Gibbs free energy change in the reaction

$$\tfrac{1}{2}Zn(II) + AgCl(IV) = \tfrac{1}{2}Zn^{2+}(III) + Cl^-(III) + Ag(V). \tag{26.27}$$

Now, each of the phase boundaries in this cell is only semipermeable: I–II to electrons but not ions, II–III to Zn^{2+} but not Cl^-, III–IV to Cl^- but not Zn^{2+} or Ag^+, IV–V to Ag^+ but not Cl^-, and V–VI to electrons but not ions. None of the potential differences I–II, . . . , V–VI is measurable,

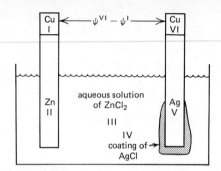

FIGURE 26.1
Schematic representation of a typical galvanic cell without liquid junctions.

because the phases have different chemical compositions. However, the potential $\psi^{VI} - \psi^I$ is between two Cu wires and hence measurable.

Let us derive the chemical information contained in Eq. 26.27 by working backwards. Consider what happens if we connect the two Cu wires through some device that can do work, such as a motor. Suppose that a current is made to flow through the cell and motor. At the Zn–solution boundary there occurs the reaction

$$\text{Zn(II)} \rightarrow \text{Zn}^{2+}(\text{III}) + 2e^-(\text{II}), \tag{26.28}$$

and the resulting electrons move from the Zn in phase II to the Cu in phase I, through the motor to the Cu in phase VI, and then to the Ag of phase V. The two extra electrons then react with AgCl to form Ag,

$$2e^-(\text{V}) + 2\text{AgCl(IV)} \rightarrow 2\text{Ag(V)} + 2\text{Cl}^-(\text{III}), \tag{26.29}$$

so that the net change is just

$$\text{Zn(II)} + 2\text{AgCl(IV)} \rightarrow \text{Zn}^{2+}(\text{III}) + 2\text{Cl}^-(\text{III}) + 2\text{Ag(V)}, \tag{26.30}$$

which is the same as Eq. 26.27 written for the transfer of two electrons.

Finally, let us use the representation Eq. 26.16 of the chemical potential to derive a relationship between the cell reaction and the electrical potential. Noting that the chemical potentials of the electrons in phases I and VI satisfy the equations

$$\mu_{e^-}^{I} = \mu_{e^-}^{II}, \tag{26.31}$$

$$\mu_{e^-}^{VI} = \mu_{e^-}^{V}, \tag{26.32}$$

and using Eq. 26.16, we find that

$$\psi^{VI} - \psi^{I} = -\frac{1}{\mathscr{F}}(\mu_{e^-}^{V} - \mu_{e^-}^{II}), \tag{26.33}$$

since for electrons $Z = -1$. With Eqs. 26.19, 26.21, and 26.22, this becomes

$$\mathscr{E} = \psi^{VI} - \psi^{I} - \frac{1}{\mathscr{F}}(\mu_{e^-}^{V} + \mu_{Ag^+}^{V}) + \frac{1}{\mathscr{F}}(\mu_{Ag^+}^{IV} + \mu_{Cl^-}^{IV})$$

$$- \frac{1}{\mathscr{F}}(\mu_{Cl^-}^{III} + \tfrac{1}{2}\mu_{Zn^{2+}}^{III}) + \frac{1}{\mathscr{F}}(\tfrac{1}{2}\mu_{Zn^{2+}}^{II} + \mu_{e^-}^{II}). \tag{26.34}$$

Only the third term on the right-hand side of Eq. 26.34 refers to the solution phase. All the other terms can be lumped into a constant, say, $\mathscr{E}'$. Then we have

$$\mathscr{E} = \mathscr{E}' - \frac{1}{2\mathscr{F}}(2\mu_{Cl^-}^{III} + \mu_{Zn^{2+}}^{III})$$

$$= \mathscr{E}' - \frac{1}{2\mathscr{F}}\mu_{ZnCl_2}^{\ominus} - \frac{RT}{2\mathscr{F}}\ln(a_{Zn^{2+}}a_{Cl^-}^{-2}). \tag{26.35}$$

The term $\mathscr{E}'$ is nothing but

$$\mathscr{E}' = -\frac{1}{\mathscr{F}}(\mu_{Ag}^{V} - \mu_{AgCl}^{IV} = \tfrac{1}{2}\mu_{Zn}^{II}), \tag{26.36}$$

so that

$$\mathscr{E}^{\ominus} \equiv \mathscr{E}' - \frac{\mu_{ZnCl_2}^{\ominus}}{2\mathscr{F}} = -\frac{1}{2\mathscr{F}}(2\mu_{Ag}^{0} - 2\mu_{AgCl}^{0} - \mu_{Zn}^{0} + \mu_{ZnCl_2}^{\ominus}), \tag{26.37}$$

where we have used the fact that the standard-state chemical potential of the Zn is just that of the solid, and so on. Note that, in the cell considered, $ZnCl_2$ is in solution and all other chemical components are in the solid state. The quantity $\mathscr{E}^\ominus$ defined in Eq. 26.37 is clearly $-\Delta G^\ominus/2\mathscr{F}$ for the reaction 26.30. Thus, the electrical potential of the cell is

$$\mathscr{E} = \mathscr{E}^\ominus - \frac{RT}{2\mathscr{F}} \ln(a_{Zn^{2+}} a_{Cl^-}{}^2). \tag{26.38}$$

Equation 26.38 is known as the *Nernst equation*.

In general, if the reaction

$$aA + bB + \cdots = xX + yY + \cdots \tag{26.39}$$

can be made to occur in a cell that has no interfaces where two solutions of different composition are in contact, the cell potential is

$$\mathscr{E} = \mathscr{E}^\ominus - \frac{RT}{z\mathscr{F}} \ln \frac{a_X{}^x a_Y{}^y \cdots}{a_A{}^a a_B{}^b \cdots}, \tag{26.40}$$

where z is the number of electrons transferred in the chemical reaction. Following the procedure used to define the reference state for a component in solution, it is conventional to set equal to unity the activity of a pure solid or a pure liquid. Thus, if pure solids or liquids are either reactants or products in the cell reaction, their activities do not appear in the second term on the right-hand side of Eq. 26.40.

Having deduced the relationships among cell potential, activity, and chemical equilibrium, we can use suitable cells to determine free energies of reaction, activity coefficients, and so on. We shall give a few examples here, leaving others for the problems. In all of the following we shall use the convention that the symbol | denotes a solid/solution interface. Thus, the cell we have discussed would be represented by

$$\text{Cu, Zn} | ZnCl_2(aq) | \text{AgCl, Ag, Cu.}$$

The usual convention is that the cell potential is positive if the left-hand electrode (as written) *supplies electrons* to the outside circuit when the cell acts as a source of energy. With this convention, when the cell potential is positive the left-hand electrode is negative.

Consider the cell

$$\text{Cu, Pt, } H_2(g) | HX(aq) | \text{AgX, Ag, Cu,}$$

where X represents any of the halogens. The Pt electrode is usually coated with colloidal Pt so as to catalyze the reaction

$$\tfrac{1}{2}H_2(g) = H^+ + e^-.$$

This is the reaction at the left-hand electrode, which serves as a source of electrons for the outside circuit. At the right-hand electrode the reaction is

$$e^- + AgX = Ag + X^-,$$

which serves as a sink for electrons. The overall chemical reaction in the cell is

$$\tfrac{1}{2}H_2(g) + AgX = Ag + H^+ + X^-.$$

Thus, the cell potential (electromotive force) is, from Eq. 26.40,

$$\mathscr{E} = \mathscr{E}^\ominus - \frac{RT}{\mathscr{F}} \ln \frac{a_\pm{}^2}{a_{H_2}{}^{1/2}}$$

$$= \mathscr{E}^\ominus + \frac{RT}{2\mathscr{F}} \ln a_{H_2} - \frac{2RT}{\mathscr{F}} (\ln \gamma_\pm + \ln m_\pm). \tag{26.41}$$

A. Determination of Solute Activities

In this reaction H_2 is gaseous. Examination of the mathematical form of the chemical potential of a real gas, Eq. 21.39, shows that we may identify $f_i/\mathfrak{p}$ with the activity. In general, when gases are involved in the cell reaction, $a_i = f_i/\mathfrak{p}$.

The constant $\mathscr{E}^\ominus$ can be deduced from measurements of $\mathscr{E}$ versus m and a_{H_2} by the extrapolation

$$\mathscr{E}^\ominus = \lim_{m_\pm \to 0}\left(\mathscr{E} - \frac{RT}{2\mathscr{F}}\ln a_{H_2} + \frac{2RT}{\mathscr{F}}\ln m_\pm\right). \qquad (26.42)$$

Of course, the activity of gaseous H_2 is determined separately. Once $\mathscr{E}^\ominus$ and a_{H_2} are known, measurements of $\mathscr{E}$ for given concentrations provide a direct means of evaluating the solute activity coefficients. Measurements of this type have been made for many electrolyte solutions. The method is limited only by the availability of suitable reversible electrodes.

Instead of using a chemical reaction, we can generate a potential from the free energy of dilution, if matters are arranged properly. Consider the cell

Ag, AgCl $|$KCl$(aq, m_1)|$ K, Hg$(l,$ amalgam$)$ $|$KCl$(aq, m_2)|$ AgCl, Ag.

In this cell the two solution phases have different concentrations. They are separated from one another by a solution of K in Hg that will permit transfer of K^+ but not of Cl^-. At the left-hand electrode the reaction is

$$Ag(s) + Cl^-(aq, m_1) + K^+(aq, m_1) = AgCl(s) + K(\text{in Hg}).$$

At the right-hand electrode we have

$$AgCl(s) + K(\text{in Hg}) = Ag(s) + Cl^-(aq, m_2) + K^+(aq, m_2),$$

which is the same reaction in reverse, but creates ions at a different concentration. The net "reaction" is[1]

$$K^+(aq, m_1) + Cl^-(aq, m_1) = K^+(aq, m_2) + Cl^-(aq, m_2).$$

Since there is no chemical change, we have $\mathscr{E}^\ominus = 0$, and Eq. 26.40 gives

$$\mathscr{E} = -\frac{2RT}{\mathscr{F}}\ln\frac{a_\pm(m_2)}{a_\pm(m_1)}. \qquad (26.43)$$

If the concentration m_1 is held fixed while m_2 is varied, we can calculate $a_\pm(m_1)$ from

$$\ln a_\pm(m_1) = \lim_{m_2 \to 0}\left(\frac{\mathscr{F}\mathscr{E}}{2RT} + \ln m_2\right). \qquad (26.44)$$

A cell of the type just considered is called a *concentration cell*. Another example of a concentration cell is shown in Fig. 26.2.

Consider the reaction

$$H_2O = H^+ + OH^-,$$

for which the equilibrium constant is

$$K_w \equiv \frac{a_{H^+}a_{OH^-}}{a_{H_2O}}. \qquad (26.45)$$

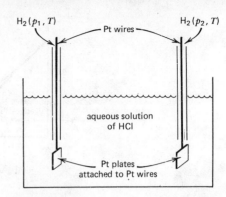

FIGURE 26.2
Schematic representation of a concentration cell. The hydrogen electrodes operate under different pressures. If oxidation occurs at the left electrode the reaction is

$$H_2(p_1) \to H_2(p_2)$$

and the cell potential is

$$\mathscr{E} = -\frac{RT}{2\mathscr{F}}\ln\frac{p_2}{p_1}.$$

B. Determination of an Equilibrium Constant

---

[1] Alternatively, we can write the following set of reactions:
Left-hand electrode: $Ag(s) + Cl^-(aq, m_1) \to AgCl + e^-$; Right-hand electrode: $AgCl(s) + e^- \to Ag(s) + Cl^-(aq, m_2)$; K/Hg interface: $K^+(aq, m_1) + K(\text{in Hg}) \to K(\text{in Hg}) + K^+(aq, m_2)$; Net reaction: $K^+(aq, m_1) + Cl^-(aq, m_1) \to K^+(aq, m_2) + Cl^-(aq, m_2)$.

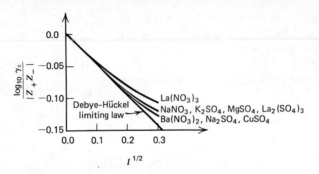

We wish to construct a cell with which K_w can be measured. In the cell

$$\text{Pt, } H_2(g) \,|\, \text{NaOH}(m_1), \text{NaCl}(m_2)| \, \text{AgCl, Ag, Pt}$$

the cell reaction is

$$\tfrac{1}{2}H_2(g) + \text{AgCl} = \text{Ag} + H^+ + Cl^-,$$

although the H^+ concentration is controlled by the supplementary equilibrium 26.45. In this cell NaOH is deliberately added so as to vary m_{OH^-}. Suppose that conditions are arranged so that $a_{H_2} = 1$. Then Eq. 26.40 gives

$$\mathscr{E} = \mathscr{E}^{\ominus} - \frac{RT}{\mathscr{F}} \ln a_{H^+} a_{Cl^-} \tag{26.46}$$

or

$$\mathscr{E} = \mathscr{E}^{\ominus} - \frac{RT}{\mathscr{F}} \ln K_w \cdot \frac{a_{H_2O} a_{Cl^-}}{a_{OH^-}}, \tag{26.47}$$

where we have used Eq. 26.45 to eliminate a_{H^+} in Eq. 26.46. Thus, K_w can be evaluated from

$$\ln K_w = \lim_{\substack{m_{OH^-} \to 0 \\ m_{Cl^-} \to 0}} \left[\frac{\mathscr{F}}{RT}(\mathscr{E}^{\ominus} - \mathscr{E}) - \ln \frac{m_{Cl^-}}{m_{OH^-}} \right]. \tag{26.48}$$

By making measurements with $(m_{Cl^-}/m_{OH^-}) = 1$, the last term in Eq. 26.48 can be dropped. It is found that $K_w = 1.008 \times 10^{-14}$ at 25°C and 1 atm.

We close this section with a brief description of what has been learned about the thermodynamic properties of electrolyte solutions. All of the techniques described, as well as several others, have been used to study electrolyte solutions. Figures 26.3, 26.4, and 26.5 show the variation with concentration of the activity coefficients of a few electrolytes. We note that:

1. In dilute solution the activity coefficient decreases with increasing concentration.

2. In very dilute solution $-\ln \gamma_\pm$ is proportional to the square root of the ionic concentration (see Fig. 26.3). More specifically, $-\ln \gamma_\pm$ is proportional to the square root of the *ionic strength* I, defined by

$$I \equiv \frac{1}{2} \sum_i Z_i^2 m_i, \tag{26.49}$$

which weights the contribution of each ionic species by the square of its charge.

3. Although the excess free energy of the electrolyte solution is proportional to the square root of the ionic strength in very dilute

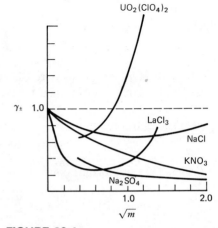

FIGURE 26.4
Schematic representation of the variation of the mean activity coefficients of several electrolytes at high concentration.

solutions, the concentration range over which that proportionality is valid is very small (see Fig. 26.4), and among several electrolytes is smallest for the highest-valence types.

4. For some electrolytes the activity coefficient plotted against concentration shows a minimum, and at high solute concentration the value of $\gamma_\pm$ may be very large. For example, in uranyl perchlorate we have $\gamma_\pm = 1510$ at $5.5\ m$. An extreme example in the opposite direction is CdI_2, for which $\gamma_\pm = 0.0168$ at $2.5\ m$.

5. The preceding observations suggest that non-Coulombic interactions are also important in electrolyte solutions, and that they become dominant when the solute concentration is large. Thus, very low activity coefficients can often be interpreted in terms of the formation of complex ions, new species in which non-Coulombic binding plays an important role. On the other hand, very large activity coefficients can often be interpreted in terms of short-range ion–ion repulsion, much as in a fluid of rigid spheres. Intermediate cases, that is, moderately low activity coefficients, are sometimes interpreted as arising from ion–pair formation. By ion–pair formation is meant the existence of an equilibrium of the type

$$A^+ + B^- = (A^+ \cdots B^-)$$

and sometimes higher multiple ions,

$$(B^- \cdots A^+ \cdots B^-),\ (A^+ \cdots B^- \cdots A^+),$$

and so on. In a few instances there is evidence for such ion pairing, but often the concept is used as a cloak for our ignorance of the true interactions. An example of short-range non-Coulombic contributions to the excess free energy is displayed in Fig. 26.5. Were there no non-Coulombic interactions, all the curves in Fig. 26.5 would fall on top of one another.

6. Electrolytes with polyvalent cations usually have much larger activity coefficients than electrolytes of analogous valency type containing a polyvalent anion. Lanthanum chloride and potassium ferricyanide are a contrasting pair of this type. If, as discussed later in this chapter, polyvalent cations form complexes with water and thereby are "large hard spheres," whereas polyvalent anions do not form such complexes, the observation can be rationalized, for the repulsion between hard-sphere ions increases the activity, the more so the larger the radius of the hard sphere.

7. Numerous empirical formulas have been used to describe the concentration dependence of the activity coefficient. All have in common:

(a) An approach to the limiting form

$$\lim_{m \to 0} \ln \gamma_\pm = -AI^{1/2} = -Bm^{1/2}. \tag{26.50}$$

We shall see, in Section 26.5, that this limiting law is the form predicted by the statistical molecular theory of Debye and Hückel.

(b) The addition of some representation of specific ion interactions to the limiting-law expression. Non-Coulombic short-range interactions lead to changes in the shielded Coulomb contribution to $\ln \gamma_\pm$, and also to "nonelectrolyte"-type contributions that depend on m, m^2, and so on, but not on $m^{1/2}$.

(c) At higher concentrations there are Coulombic contributions to the excess free energy that depend on $m^{3/2}$, $\ln m$, and other nonintegral powers of the solute concentration.

8. Temperature and pressure derivatives of $\ln \gamma_\pm$ also confirm the observations and deductions listed above.

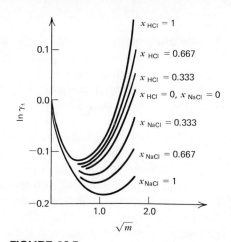

FIGURE 26.5
Schematic representation of the mean activity coefficients of HCl and NaCl in mixed electrolyte solutions. Along a vertical line m is fixed, but the proportions of NaCl and HCl vary as shown.

We discussed, in Chapter 23, the structure of liquid water as revealed by x-ray and neutron diffraction studies. The data available are consistent with a picture in which each water molecule is surrounded by four others, to which it is hydrogen-bonded, with approximately tetrahedral coordination. There is, of course, dispersion in the distribution of molecular separations, and also in the distribution of relative molecular orientations. Indeed, translational ordering extends only about two molecular diameters from a given molecule, and orientational ordering does not extend beyond the first shell of neighbors.

Figure 26.6 displays two instantaneous configurations from a computer simulation of the structure of water. It is necessary to exercise some caution in interpreting figures such as these because they are derived from a classical mechanical model and there is much evidence that the translational and librational motions of water molecules cannot be quantitatively described without use of quantum mechanical methods. Nevertheless, the classical mechanical model has most of the structural features that are inferred from the x-ray and neutron diffraction experiments. It is easy to see the dispersion in molecular separations and relative orientations. Given that the range of separations over which hydrogen bonding is effective in water is about 2.7–3.0 Å, it is also possible to trace a continuous network of hydrogen bonds among the molecules portrayed in the figure. Because of the inherent positional and orientational disorder we shall refer to this kind of connectivity as a continuous random network. Clearly, the hydrogen bonds that make up the network are not all equally strong—in addition to the obvious dependence on the $O \cdots O$ separation, the further the $OH \cdots O$ angle deviates from 180°, the weaker the hydrogen bond. It is probably only a semantic distinction to assert that there are broken and unbroken hydrogen bonds, since there is a continuous range of variation in the hydrogen-bond energy with $O \cdots O$ separation and $OH \cdots O$ angle.

Figure 26.6 reveals that there are very many environments generated by local differences in hydrogen bonding. The observed range of variation can be taken as indicative of the extent to which a continuous hydrogen-bonded network can be deformed and still retain connectivity.

26.3
Comments on the Structure of Water

FIGURE 26.6
Typical molecular configurations from a Monte Carlo simulation of liquid water. The numbers inside the O-atom circles are values of the z coordinate (out of the plane of the page) in angstroms. From G. N. Sarkisov, V. G. Dashevsky, and V. G. Malenkov, *Mol. Phys.* **27**, 1249 (1974).

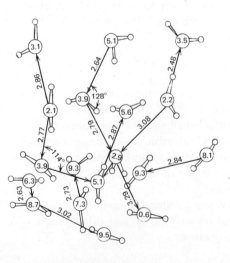

(a)

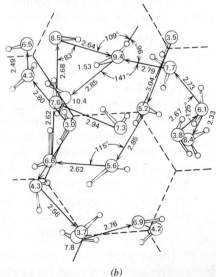

(b)

This characteristic of the network of hydrogen bonding must in some way influence the interaction between water and solute molecules, hence also the thermodynamic properties of aqueous solutions. It is this subject that we now examine.

26.4
The Influence of Solutes on the Structure of Water

When a substance is dissolved in water, the arrangement of molecules near the solute is different from that far away. Of course, the alteration of local structure induced by the solute depends on the nature and strength of its interaction with water. We have shown that in a simple liquid, whether pure or mixed, the repulsive interaction between molecules plays the dominant role in determining their spatial distribution (see Chapters 23 and 25). Although the repulsive force between water molecules does not play so direct a role in determining the structure, that structure is influenced locally by the size and shape of a solute molecule. Moreover, if the solute is charged, the combining of a strong electric field acting on the water molecules with the difference in sizes and shapes of the solute and water molecules can produce a striking difference between the local and distant arrangements in the liquid. There is also a more subtle change in the liquid water structure that extends some distance from the ion. In extreme cases, for example, a trivalent transition metal ion, the solute ion and its near-neighbor shell of water molecules can be considered to be a single chemical species.

To examine how solutes influence the structure of water, we first consider the properties of aqueous solutions of gases. The thermodynamic data for such systems are usually derived from a study of the temperature dependence of the Henry's law constant (see Section 25.4). Suppose that an infinitesimal quantity of substance is transferred from the gas to the solution phase. Accompanying this transfer are changes in enthalpy, entropy, and volume, which we denote $\Delta \bar{h}_2, \Delta \bar{s}_2$, and $\Delta \bar{v}_2$, respectively. For the dilute solutions we consider the solute–solute interaction is negligibly small and the derived thermodynamic functions are properly interpreted solely in terms of the effects of solute–solvent interaction. The natural reference state is then the hypothetical pure substance which has the same properties as the solute in an infinitely dilute solution.

We list in Table 26.1 values of the enthalpy, entropy, and volume changes that accompany the transfer of 1 mol of solute from the hypothetical pure state defined above to a gas at 1 atm. These are denoted $\Delta \bar{h}_2{}^\ominus, \Delta \bar{s}_2{}^\ominus$, and $\Delta \bar{v}_2{}^\ominus$, respectively. From the data presented we deduce the following:

1. The increase in entropy associated with transfer of the solute from solution to the gas phase is greater than the entropy of vaporization of the pure liquid solute at its normal boiling point.

2. The enthalpy change associated with transfer of the solute from solution to the gas phase is greater than the enthalpy of vaporization of the pure liquid solute at its normal boiling point.

3. The enthalpy and entropy of transfer from solution to gas phase are strongly temperature-dependent.

4. The volume of the hypothetical pure-component reference state is about equal to the molar volume of the real pure liquid solute at its normal boiling point. Furthermore, the volume of the reference state is much less sensitive to variation of the temperature than are $\Delta \bar{h}_2{}^\ominus$ and $\Delta \bar{s}_2{}^\ominus$.

TABLE 26.1
THERMODYNAMIC PROPERTIES OF SOLUTIONS OF SIMPLE GASES IN WATER

Solute	Properties of Solute in Reference State									Properties of Solute as Pure Liquid		
	$\Delta \bar{h}_2^{\ominus}$ (J/mol)			$\Delta \bar{s}_2^{\ominus}$ (J/K mol)			$\Delta \bar{v}_2^{\ominus}$ (cm³/mol)			$\Delta h_{2(\text{vap})}$ (J/mol)	$\Delta s_{2(\text{vap})}$ (J/K mol)	v_2^0 (cm³/mol)
	25°C	50°C	80°C	25°C	50°C	80°C	25°C	50°C	80°C			
He	3,510	2,790	2,300	111	109	107	(15.5)[a]					32
Ne	7,870	5,360	2,430	120	111	103				1,740	64.0	17
Ar	11,420	7,700	2,970	126	114	98				6,530	74.9	29
Kr	14,850	9,830	3,100	135	116	97				9,030	75.3	34
Xe	18,790	11,210	3,010	141	117	92				12,640	76.6	43
Rn	21,130	12,790	3,010	144	118	88				16,400	77.8	50
H_2	5,360	1,780[b]	711	109	99[b]	91	24	26	24			28
N_2	8,950	8,330[b]	753	125	123[b]	100	44	40	38	5,560	72.0	35
CO	16,360	8,240[b]	42	125	119[b]	95	37	36	32	6,040	74.1	35
O_2	12,510	8,660[b]		131	126[b]		31	31	32	6,820	75.7	28
CH_4	13,310			133			36	37	38	8,180	72.8	39

[a] Probably too small. An estimate based on extrapolation of the error in other (similar) measurements would indicate a value of 20 cm³/mol.
[b] At 40°C.

Interpretation of the listed inferences is, with present knowledge, necessarily speculative. The following arguments are intended to illustrate the reasoning used to convert macroscopic observations into structural concepts that can be incorporated into a molecular model of aqueous solutions.

Because the entropy of vaporization of the pure liquid solute is less than the entropy of transfer from the hypothetical standard state of the solute to the gas, and since in both cases the gas is taken to be in the same thermodynamic state, we must conclude that the entropy of the aqueous solution is less than the entropy of pure water. But in the hypothetical standard state of the solute there are no solute–solute interactions, so the difference in entropies must be attributed to the influence of the solute on the behavior of the solvent water. Then the decrease in entropy implies that the number of configurations of equal energy available to the water molecules must be smaller in solution than in the pure liquid. Some chemists have used this observation as the basis for the conclusion that introduction of a solute molecule of the type under discussion creates a well-defined structure in the solvent proximate to the solute molecule. Indeed, there does exist a class of compounds known as *clathrates*, examples of which are simple crystalline hydrates of Xe, Cl_2, $CHCl_3$, and other similar small molecules. The structures of many clathrate compounds are known. Each can be characterized as a complex cage of water molecules enclosing a cavity in which the guest resides. They are usually less dense than ice, and are unstable with respect to ice if the guest molecule is removed. The clathrate compound is rendered stable by virtue of nonspecific, nonsaturable dispersion interactions between the guest and the surrounding water molecules. It is tempting to think that the structure of water near a given solute molecule resembles that in the corresponding clathrate compound.

Computer simulation studies of the properties of an inert gas atom in water have been carried out. In these studies the model inert gas atom and the model rigid water molecule are assigned the same diameter, and their interaction is taken to be of the Lennard-Jones form (Eq. 21.173) with a well depth intermediate between those expected for Ne-H_2O and Ar-H_2O. The results of the simulations show that the solute is surrounded by a water cage whose structure differs from that of pure water. The nearest-neighbor hydration shell resembles a highly strained net of six hexagonal rings of hydrogen-bonded waters. The orientational structure of this shell persists to some extent, but only weakly, in the surrounding water molecules of the second hydration shell. The differences between the water structure around the solute and that in pure water are everywhere small, the hydrogen bonding between a water molecule and its four nearest neighbors is preserved, and the hydration structure blends continuously into that of the surrounding liquid. Even though some of the preferred water orientations in the hydration shell resemble those in a clathrate, there are profound differences between the two structures. The most important of these differences, which is also the one responsible for the continuous blending of the hydration shell into the surrounding water, is that the hydration-shell water is not immobilized as it is in a crystalline clathrate. The computer simulations indicate that there is only a small difference between the mobilities of water molecules in the hydration shell and in pure water. In summary, the computer simulations show that introduction of an inert gas atom into water alters the structure of the liquid immediately around the solute, but that the local environment of the solute is neither rigid nor precisely ordered as in a crystalline arrangement of water molecules.

Even the limited local structure modifications found in the computer simulations described can be expected to lower the entropy per water molecule, as is observed in real solutions. Moreover, because of the continuous blending of the hydration structure into the liquid structure it is reasonable to expect $\Delta \bar{v}_2^{\theta}$ to be much less sensitive to temperature than $\Delta \bar{h}_2^{\theta}$ and $\Delta \bar{s}_2^{\theta}$, both of which depend on the thermal excitation of quasi-vibrational motions and on the extent of the difference between the local and the bulk liquid structures. The data of Table 26.1 support this view.

Given the extreme sensitivity of the structure of a liquid to repulsive interactions between the molecules, we must expect some changes from the picture presented by the computer simulations when the solute and water molecule have different diameters. There are no computer simulations of this class of solutions available, but the data of Table 26.1 provide some clue which can be used for the construction of crude structural models.

Proceeding with our illustrative arguments, we can understand the data in Table 26.1 qualitatively within the framework of the continuous-random-network, distorted-hydrogen-bond model of water. Insertion of a He atom leads to a reference state with $\Delta \bar{v}_2^{\theta} < v_2^0$ because of the existence of a distribution of $O \cdots O$ separations in water. Geometric constraints imposed by the hydrogen bonding often leave voids too small for other water molecules or solutes as large or larger, but of sufficient size to accommodate a He atom.[2] These voids are not "vacuum"; there is, in each void, an intermolecular potential field generated by the surrounding water molecules. The field is not uniform. All we are asserting is that a He atom interacts so weakly with the molecules surrounding a void that occupancy of the void is possible, whereas the interaction of a larger atom

---

[2] Helium diffuses through glass for the same reason. The He atom is sufficiently small to be able to migrate between atoms that would strongly repel a larger atom.

is sufficiently repulsive to make occupation of the void unlikely. In both cases our conception includes rearrangement of the water. We expect the equilibrium configuration of an occupied void to be different from that of an unoccupied void, and the change in configuration to reflect the effect of He–H_2O repulsion. The values of $\Delta \bar{h}_2^{\ominus}$ observed for He as solute greatly exceed the expected heat of vaporization of a He atom from liquid Ne (which is isoelectronic with H_2O). Thus, the combination of the dispersion attraction of He for H_2O with the relaxation of the water about a He atom is energetically favored. Presumably, since $|\Delta \bar{s}_2^{\ominus}| > |\Delta s_{2(vap)}^0|$, the water around a He atom is more ordered than in the bulk liquid. This picture of the structure of a He-atom-in-water solution requires that $\Delta \bar{h}_2^{\ominus}$ and $\Delta \bar{s}_2^{\ominus}$ be insensitive to temperature because preexisting voids are the initial configurations from which relaxation starts. This conclusion is supported by the data in Table 26.1. The values of $\Delta \bar{h}_2^{\ominus}$ and $\Delta \bar{s}_2^{\ominus}$ for He change much less with increasing temperature than do those for other gaseous solutes.

Insertion of solutes larger than He into the water structure requires much more extensive alteration of that structure, primarily because of the lack of suitable voids for placement of a solute molecule. In the simplest picture the process of inserting a solute molecule can be thought of as proceeding in two steps. First, a cavity of suitable size must be created in the liquid. In general, it requires work to create such a cavity. Typical model calculations for water at 298 K indicate that creation of a cavity to hold an Ar atom requires about 2.9 kJ/mol and leads to an entropy change of −52 J/K mol. Second, the solute molecule is inserted in the cavity and the liquid allowed to relax in the new potential field. Calculation of the changes in thermodynamic functions due to this relaxation is difficult, and the results rather dependent on the model chosen. The first step described (cavity formation) is an approximate way of accounting for the effect on liquid structure of repulsion between solute and solvent molecules. We have already indicated that such repulsive interactions play a role in He/H_2O solutions, and have shown how they determine the local structure in simple liquid mixtures, so we expect them to be very important, possibly dominant, in the cases now under discussion. If this view is taken seriously, and only enthalpy changes considered in the second step (solvent relaxation around the inserted solute molecule), reasonable agreement between calculation and experiment is obtained. For example, for Ar-H_2O at 298 K one finds:

$$\Delta \bar{h}_2^{\ominus}(\text{calc.}) = 10.2 \text{ kJ/mol}; \qquad \Delta \bar{h}_2^{\ominus}(\text{exp.}) = 11.4 \text{ kJ/mol}$$

$$\Delta \bar{s}_2^{\ominus}(\text{calc.}) = 120 \text{ J/K mol}; \qquad \Delta \bar{s}_2^{\ominus}(\text{exp.}) = 126 \text{ J/K mol.}$$

The model described does not provide any information about the nature of changes in hydrogen bonding in water, or other details—it is semimacroscopic in character. There is not, at present, a satisfactory microscopic theory of structure changes in aqueous solutions of uncharged species.

Ionic solutes interact very strongly with water, and we therefore expect these to generate a greater relative alteration of the local water structure than do neutral solutes. If electrostatic effects in the interaction are of dominant importance, we expect there to be a regular variation of the thermodynamic properties of individual ions with ionic radius. If nonelectrostatic effects in the interaction are of importance, we expect there to be differences between the properties of ions with different electronic structures but the same charge and size.

Following the method used in our discussion of solutions of simple gases, we first examine the available experimental data. As noted in

TABLE 26.2
GIBBS FREE ENERGIES, ENTHALPIES, AND ENTROPIES OF FORMATION OF SOME IONS IN WATER AT 25°C

Ion	$\Delta g_f^{\ominus}$(kJ/mol)	$\Delta h_f^{\ominus}$(kJ/mol)	$\Delta s_f^{\ominus}$(J/K mol)
H^+	0(convention)	0(convention)	0(convention)
OH^-	−157.3	−230.0	−10.5
F^-	−276.5	−329.1	−9.6
Cl^-	−131.2	−167.4	+55.2
Br^-	−102.8	−120.9	+80.7
I^-	−51.7	−55.9	+109.4
HS^-	+12.6	−17.2	+62.8
S^{2-}	+8.6	+32.6	−17.
SO_4^{2-}	−742.0	−907.5	+17.2
HSO_4^-	−752.9	−885.8	+127.7
NO_3^-	−110.6	−206.6	+125.
NH_4^+	−79.5	−132.8	+112.8
Li^+	−293.8	−278.4	+14.
Na^+	−261.9	−239.7	+60.2
K^+	−282.3	−251.2	+103.
Rb^+	−283.0	−248.5	+120.
Cs^+	−296.0	−261.9	+133.
Tl^+	−32.5	+5.8	+127.
Ag^+	+77.1	+105.9	+73.93
Sn^{2+}	−24.3	−10.0	+21.
Pb^{2+}	−24.3	+1.6	+21.
Zn^{2+}	−147.2	−152.4	−106.5
Cd^{2+}	−77.7	−72.4	−61.1
Hg^{2+}	+164.8	174.0	−23.
Fe^{2+}	−84.9	−87.9	−113.
Fe^{3+}	−1060.	−47.7	−293.
Mg^{2+}	−456.0	−462.0	−118.
Ca^{2+}	−553.0	−543.0	−55.2
Sr^{2+}	−557.3	−545.5	−39.
Ba^{2+}	−560.7	−538.4	+13.

Section 14.6, it is not possible to determine the properties of individual ions, only those of neutral salts. Following the convention introduced in Section 14.6, Table 26.2 displays thermodynamic functions for individual ions computed under the *ad hoc* assumption that the Gibbs free energy of formation, the enthalpy of formation, and the entropy of formation of the hydrogen ion at 25°C and 1 atm in water are *all* taken to be zero:

$$\Delta g_f^{\ominus}(H^+, aq) \equiv 0, \tag{26.51}$$

$$\Delta h_f^{\ominus}(H^+, aq) \equiv 0, \tag{26.52}$$

$$\Delta s_f^{\ominus}(H^+, aq) \equiv 0. \tag{26.53}$$

TABLE 26.3
SOME ENTHALPIES AND ENTROPIES OF HYDRATION AT 25°C

	Δh^0_{cryst} [a] (kJ/mol)	$\Delta h^\ominus_{hydration}$ [b] (kJ/mol)	$\Delta s^\ominus_{hydration}$ [b] (J/mol K)
LiF	−1032	−1005	−276
NaCl	−779	−760	−185
KF	−813	−819	−207
SrCl$_2$	−2110	−2161	−400
AlF$_3$	−6140	−6161	−930
AlCl$_3$	−5452	−5774	−757
AgCl	−916	−851	−191

[a] For the reaction

$$A^+(g) + B^-(g) = AB(crystal).$$

[b] For the reaction

$$A^+(g) + B^-(g) = A^+(aq) + B^-(aq).$$

Table 26.2 contains data also included in Table 14.5; these are repeated here for convenience. By combination of these quantities, as illustrated in Sections 14.6 and 18.1, we can deduce the thermodynamics of reactions. Note that for the purpose of computing changes that occur in a chemical reaction the arbitrary choice of standard state (Eqs. 26.51–26.53) is irrelevant.

A quantitative measure of the absolute magnitude of the ion–solvent interaction is provided by the heat of the reaction

$$A^+(g) + B^-(g) = A^+(aq) + B^-(aq). \tag{26.54}$$

When the final state in the reaction is the pure crystal instead of the ionic solution, the enthalpy change is just the heat of formation of the crystal from the ions. This is, of course, also just the enthalpy required to separate the crystal into a set of infinitely separated ions. Using tables of energies of ionization, heats of sublimation, and so on, one can easily compute the values given in Table 26.3.

In an ionic crystal the energy of stabilization comes primarily from the ion–ion Coulomb interactions. These interactions are sensitive to the size of the ions, since it is the ionic size that determines the minimum separation of the ions. In turn, the ionic size is determined by the strong repulsions generated by the overlap between closed-shell charge distributions. An examination of the data in Table 26.3 shows that Δh^0_{cryst} and $\Delta h^\ominus_{hyd}$ are nearly equal in magnitude, and that they change in a similar fashion when the ionic size is changed. This suggests that it might be useful to decompose $\Delta h^\ominus_{hyd}$ and $\Delta h_f^\ominus$ into contributions from separate ions. We have already made such a decomposition for $\Delta h_f^\ominus$ using the convention cited in Eqs. 26.51–26.53. These are the values in Table 26.2. What we now seek are "absolute" ionic enthalpies, that is, the assignment of "realistic values" for the enthalpies of hydration of ions. From the point of view of thermodynamics, all choices of convention are equally valid, and there is no set of observations that can be used, without theoretical modeling, to obtain single-ion properties. However, from the point of view of a molecular theory, it is clearly desirable to choose a convention that, for a simple model, gives sensible results. In the earliest attempts to construct tables of ionic properties, the enthalpy of hydration of KF was equally divided into an enthalpy of hydration of K$^+$ and an enthalpy of hydration of F$^-$. This choice was based on the observation that the ionic radii of K$^+$ and F$^-$ are almost the same, hence the heat of reaction of each with water should be the same. Of course, given this new convention, the tables of thermodynamic data based on Eqs. 26.51–26.53 can easily be converted, since once one value has been defined all others can be worked out by hypothetical reaction sequences.

Other schemes besides the one based on equality of enthalpies of hydration of K$^+$ and F$^-$ have been proposed. We select for discussion just one of these schemes.

A set of values that represent a close approach to absolute single-ion free energies of hydration can be obtained from the concept of *absolute half cell potential*. The idea is as follows. Consider two half-cells, the left one consisting of an electrode M$_1$ immersed in a solution of M$_1{}^+$ and X$^-$ ions, and the right one consisting of an electrode M$_2$ immersed in a solution of M$_2{}^+$ and X$^-$ ions. Electrical contact between these two half-cells is via a liquid junction, that is, a boundary region in which the concentrations of M$_1{}^+$, M$_2{}^+$, and X$^-$ vary continuously from those characteristic of the left-hand half-cell to those characteristic of the right-hand half-cell. In general, cells that have a liquid junction have a potential that depends on the nature of that junction. Conditions can be found under which the potential difference attributed to the liquid junction is very small. We shall assume that in the case under discussion it is zero, recognizing that this assumption is one of the inherent approximations in the scheme of determining absolute half-cell potentials. Now, in the left-hand cell the reaction is M$_1(s) \rightarrow$ M$_1{}^+(sol'n) +$ e$^-$(M$_1$), and in the right-hand cell it is M$_2{}^+(sol'n) +$ e$^-$(M$_2$) $\rightarrow$ M$_2(s)$. Let the Gibbs free energy change for the left-hand cell reaction be separated into the following steps:

(i)   M$_1(s) \rightarrow$ M$_1(g)$; $\quad\quad\quad\quad\quad\quad\quad\quad$ $\Delta g = \Delta g^0_{1(\text{vap})}$

(ii) M$_1(g) \rightarrow$ M$_1{}^+(g) +$ e$^-$(M$_1$); $\quad\quad\quad\quad$ $\Delta g = \mathscr{F}(I_1 - \phi_1)$

(iii) M$_1{}^+(g) \rightarrow$ M$_1{}^+(sol'n$ adjacent to M$_1$); $\quad$ $\Delta g = \Delta g^\ominus_{1(\text{hyd})}$

(iv) M$_1{}^+(sol'n$ adjacent to M$_1$) $\rightarrow$ M$_1{}^+(bulk\ sol'n$); $\quad$ $\Delta g = \mathscr{F} V_{\text{M}_1 s}$

where $\Delta g^0_{1(\text{vap})}$ is the Gibbs free energy of vaporization of neutral M$_1$, I_1 and ϕ_1 the ionization potential of the metal and the work function[3] of the metal in contact with the solution, respectively, $\Delta g^\ominus_{1(\text{hyd})}$ the Gibbs free energy of hydration of M$_1{}^+$, and $V_{\text{M}_1 s}$ the electrostatic potential difference between M$_1$ and the solution. Similarly, for the right-hand half-cell:

(i)   M$_2{}^+(bulk\ sol'n) \rightarrow$ M$_2{}^+(sol'n$ adjacent to M$_2$); $\quad$ $\Delta g = -\mathscr{F} V_{\text{M}_2 s}$

(ii) M$_2{}^+(sol'n$ adjacent to M$_2$) $\rightarrow$ M$_2{}^+(g)$; $\quad$ $\Delta g = -\Delta g^\ominus_{2(\text{hyd})}$

(iii) M$_2{}^+(g) +$ e$^-$(M$_2$) $\rightarrow$ M$_2(g)$; $\quad\quad\quad$ $\Delta g = -\mathscr{F}(I_2 - \phi_2)$

(iv) M$_2(g) \rightarrow$ M$_2(s)$; $\quad\quad\quad\quad\quad\quad\quad$ $\Delta g = -\Delta g^0_{2(\text{vap})}$

The total changes in Gibbs free energy for the left and right half-cells are obtained by summation of all contributions. For the overall cell reaction M$_1 +$ M$_2{}^+ \rightarrow$ M$_1{}^+ +$ M$_2$, we have, at equilibrium,

$$(\Delta g^0_{1(\text{vap})} - \Delta g^0_{2(\text{vap})}) + (\Delta g^\ominus_{1(\text{hyd})} - \Delta g^\ominus_{2(\text{hyd})}) + \mathscr{F}(I_1 - I_2)$$

$$= -[\mathscr{F}(V_{\mu_1 s} - \phi_1) - \mathscr{F}(V_{\mu_2 s} - \phi_2)]. \quad (26.55)$$

The left-hand side of Eq. 26.55 is just the Gibbs free energy change for the cell reaction carried out nonelectrically, so the right-hand side is $-\mathscr{F}\mathscr{E}^\ominus$. Note that because M$_1$ and M$_2$ are different metals, the cell

---

[3] The work function is the free-energy change on transferring an electron from the inside of the metal to the outside; it depends on the nature of the medium in contact with the metal.

potential is the difference of the electrostatic potentials $V_{M_1s} - V_{M_2s}$ minus the work function difference $\phi_1 - \phi_2$. We now define a half-cell potential as

$$\mathscr{E}_{1/2} \equiv V_{Ms} - \phi = -\frac{(\Delta g^0_{vap} + \Delta g^\ominus_{hyd} + \mathscr{F}I)}{\mathscr{F}}, \qquad (26.56a)$$

which is the difference between the mean electrostatic energy of an electron in solution (but not hydrated) and the chemical potential of the electrons in the metal in contact with the solution.

How can one measure $\mathscr{E}_{1/2}$? Suppose that another electrode, say R, is connected by a wire to M_1. The electrode R is in contact with air, and does not come into contact with the solution. Since M_1 and R are connected with a wire, the chemical potential of the electrons in M_1 must be the same as that of the electrons in R. Now consider the reaction $M_1 \rightarrow M_1^+(bulk\ sol'n) + e^-(M_1)$ to consist of the following steps:

(i) $M_1(s) \rightarrow M_1(g,\ at\ sol'n–air\ interface)$
(ii) $M_1(g) \rightarrow M_1^+ + e^-(both\ in\ gas\ at\ the\ sol'n–gas\ interface)$
(iii) $e^-(gas\ at\ sol'n–gas\ interface) \rightarrow e^-(M_1)$
(iv) $M_1^+(gas\ at\ sol'n–gas\ interface) \rightarrow M_1^+(bulk\ sol'n)$

At equilibrium the sum of the Gibbs free energy changes for these processes is

$$\Delta g^0_{1(vap)} + \Delta g^\ominus_{1(hyd)} + \mathscr{F}I_1 - (-\mathscr{F}\mathscr{E}_{1/2}) = 0, \qquad (26.56b)$$

since the transfer of an electron from a region of electrostatic potential corresponding to that just outside the solution's free surface to the metal M_1 is given by $-(-\mathscr{F}\mathscr{E}_{1/2})$. Note that $\Delta g^\ominus_{1(hyd)}$ corresponds to the transfer of an ion to the interior of the solution from a gas phase with the same electrostatic potential as the mean electrostatic potential in the bulk solution.

Since R and M_1 are connected by a wire and the chemical potentials of the electrons in them are equal, we have

$$V_{M_1s} - \phi_1 = V_{Rs} - \phi_R = \mathscr{E}_{1/2}, \qquad (26.56c)$$

with ϕ_R the work function of R in contact with air. Then, if ϕ_R is known and the electrostatic potential is the same in the bulk and surface, we need only measure the potential difference between the solution and an electrode R outside it (say, just over the solution surface) in order to obtain $\mathscr{E}_{1/2}$ for M_1 immersed in a solution of M_1^+ and X^- ions. In fact, that potential difference can be readily measured. However, we must expect that orientation of molecules at the solution surface will lead to a potential difference between the bulk and surface of the solution. This potential difference depends on temperature, electrolyte concentration, and other variables. Its sign and magnitude can be reasonably estimated from theoretical and experimental studies, hence the measured potential difference between R and the solution surface corrected, and $\mathscr{E}_{1/2}$ thereby determined. The use of this correction is a second approximation in this method of determining absolute half-cell potentials.

R. Gomer has used the method described and found that $\mathscr{E}_{1/2}$ for the reaction $H_2 \rightarrow 2H^+ + 2e^-$ is -4.73 ± 0.05 V. Hence absolute half-cell potentials can be obtained from tabulations of standard half-cell potentials by adding -4.73 V. With these values of $\mathscr{E}_{1/2}$, absolute values of the Gibbs free energy of hydration can be obtained from Eq. 26.56b; some values are shown in Table 26.4. The values obtained in this way are seen to be rather different from those based on Eq. 26.51.

Reviews of the uses to which single ion thermodynamic data can be put are presented in the books by Y. Marcus and N. S. Hush listed at the end of this chapter. Our present interest in these data is limited to examination of how they illuminate the nature of ion hydration and its variation from ion to ion. For this purpose we have included in Table 26.4 thermodynamic data for ions obtained by methods other than the one just discussed.

In Figs. 26.7 and 26.8 are plotted the variation of the "absolute" enthalpy and entropy of hydration of individual ions as a function of the reciprocal of the ionic radius. Note that there is near proportionality of $\Delta h^{\ominus}_{\text{hydration}}$ and $\Delta s^{\ominus}_{\text{hydration}}$ to r^{-1}_{ion}, but there are also significant deviations from linearity. The data for transition metal ions fall so far off the curves for the corresponding closed-shell simple ions that for these ions the ion–water interaction must contain a significant contribution from other than electrostatic interactions. A similar conclusion can be drawn for the case of simple ions: The data displayed show a clear dependence of $\Delta h^{\ominus}_{\text{hydration}}$ and $\Delta s^{\ominus}_{\text{hydration}}$ on the ionic charge, but the differences in properties between oppositely charged ions of like size show that more than just electrostatic interaction between an ion and water is involved in determining the thermodynamic properties of the ion. Thus the simple electrostatic model of hydration does not adequately account for the dependence of the thermodynamic properties of ions on the diameter, electronic structure, or sign of the charge. Incidentally, the fact that the thermodynamic functions for an ion depend on the sign of its charge points up another difficulty in defining a satisfactory convention for the decomposition of observable changes into separate ionic contributions.

The roles of the sign of the electrostatic interaction and of non-Coulombic interactions within water are particularly obvious when the entropies of hydration of the ions are compared with the entropy change associated with the solution of an inert gas of the same radius. To make this comparison we must use the same standard states. The data in Table 26.1 use as standard state the hypothetical pure substance retaining the same properties as the solute in dilute solution. The data in Table 26.4 use as standard state the hypothetical $1\,m$ solution retaining the same properties as the solute in dilute solution. Thus the entropies of these standard states differ by $R\ln(18/1000)$, or 33 J/K mol. We must subtract 33 J/K mol from the entries in Table 26.1 to compare them with those in Table 26.4. With reference to a hypothetical perfect $1\,m$ solution, the entropy change associated with solution of 1 mol of inert gas is about -92 J/K mol. The data in Table 26.4 show that for small ions the entropy of hydration is more negative, and for large ions less negative, than would correspond to the solution of the same number of uncharged inert gas atoms. For example, we have $\Delta s^{\ominus}_{\text{hyd}}(\text{LiF}) = -275$ J/K mol, which is more negative than $2\times(-92) = -184$ J/K mol, and $\Delta s^{\ominus}_{\text{hyd}}(\text{RbI}) = -101$ J/K mol, which is more positive than -184 J/K mol. There are at least two effects that compete in influencing the enthalpy and entropy of hydration. First, the ion–dipole and ion–quadrupole interactions become stronger with decreasing ion–water molecule separation. It is then to be expected that the orientation of the water dipoles in the field of a small ion will be more nearly complete than in the field of a large ion. Second, insertion of the ion into the hydrogen-bond network of water alters that network. Using the data of Tables 26.1 and 26.4, we see that the first effect tends to produce larger negative entropies of hydration for smaller ions, and the second effect to produce larger negative entropies of hydration for larger ions.

More detailed interpretation of the thermodynamics of hydration and other ionic properties requires consideration of the nature of the ion–solvent interaction. At present, we know very little about the structure of

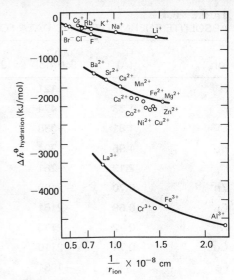

FIGURE 26.7
The computed heat of hydration of ions as a function of ionic size.

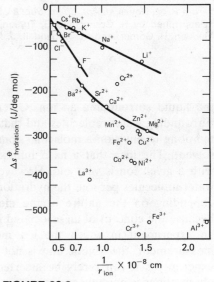

FIGURE 26.8
The computed entropy of hydration of ions as a function of ionic size.

TABLE 26.4
ABSOLUTE THERMODYNAMIC DATA FOR AQUEOUS IONS (298 K)

Ion	r_{ion} (Å)	$\Delta h_{hyd}^{\ominus}$ (kJ/mol)[a]	$\Delta s_{hyd}^{\ominus}$ (J/mol K)[a]	$\Delta g_{hyd}^{\ominus}$ (kJ/mol)[a]	$\Delta g_{hyd}^{\ominus}$ (kJ/mol)[b]
H^+		−1121	−123	−1084	−1059
F^-	1.33	−474	−143	−431	−463
Cl^-	1.81	−338	−85	−313	−341
Br^-	1.96	−304	−68	−284	−318
I^-	2.19	−261	−47	−247	
Zn^{2+}	0.70	−2105	−295	−2017	
Cu^{2+}	0.69	−2161	302	−2071	2030
Ni^{2+}	0.72	−2174	−369	−2056	
Co^{2+}	0.74	−2110	−366	−2004	
Fe^{2+}	0.76	−1981	−321	−1885	−1846
Fe^{3+}	0.64	−4343	−506	−4192	
Mn^{2+}	0.80	−1905	−286	−1820	
Cr^{2+}	0.84	−1910	−183	−1856	
Cr^{3+}	0.69	−4368	−531	−4209	
Al^{3+}	0.45	−4750	−507	−4599	−4533
La^{3+}	1.15	−3374	−404	−3253	
Mg^{2+}	0.65	−1982	−294	−1895	−1850
Ca^{2+}	0.94	−1653	−238	−1582	−1514
Sr^{2+}	1.10	−1482	−222	−1416	−1376
Ba^{2+}	1.29	−1364	−187	−1308	−1259
Li^+	0.68	−545	−133	−505	−494
Na^+	0.98	−436	−100	−406	−379
K^+	1.33	−351	−66	−332	−306
Rb^+	1.48	−326	−54	−310	
Cs^+	1.67	−293	−51	−278	

[a] From R. M. Noyes, *J. Am. Chem. Soc.* **84**, 513 (1962); **86**, 971 (1964).
[b] Determined by R. Gomer and G. Tryson, *J. Chem. Phys.* **66**, 4413 (1977); W. G. Madden, R. Gomer, and M. J. Mandell, *J. Phys. Chem.* **81**, 2652 (1977).

the liquid surrounding an ion. Given that the ion tends to orient the surrounding water molecules and that the interaction energy is large, it is tempting to envision a molecular complex made up of the ion plus some solvent. The idea that a fixed number of water molecules is associated with a given ion is very old. However, measurements of the number of water molecules per ion, the hydration number, give very different results depending on the nature of the measurement. This is not surprising, because the concept of an associated ion–water complex is ambiguous. In an experiment in which the ion moves, for example, a conductance measurement, the kinetic unit is not properly defined, whereas in other experiments it can rarely be asserted that no other explanation of the observations is possible and that ionic hydration must be present. Despite these uncertainties, there is good evidence for ionic hydration in some instances. In general, the hydration numbers of univalent ions are small, particularly those of anions. For these ions the concept of a solvent–ion complex is not useful. In the case of multivalent ions (such as Cr^{3+}), studies of the rate of exchange of isotopically labeled water show that the exchange of a water molecule adjacent to an ion for a water molecule from the outer solvent is very slow (half-life of many hours). Also, the spectra of some multivalent rare earth ions, as well as their heats of hydration, suggest local field effects that alter the electronic structure of the ion (see later and also the discussion in Chapter 9). Thus, multivalent transition metal ions probably are intimately associated with some solvent molecules, and in these cases it is useful to consider the hydrated ion as a specific complex with well-defined structure.

Pertinent information about ion–water interactions can be obtained from numerical simulation experiments; a few of these have been reported. E. Clementi and his co-workers studied polymeric clusters of the type $Li^+(H_2O)_n$, $F^-(H_2O)_n$, ... for n up to 10. In these quantum mechanical calculations the arrangement of the waters around an ion was varied until the minimum energy configuration was found. For $n < 5$ the minimum energy configuration consists of a symmetric arrangement of waters adjacent to the ion, but for $n \geqslant 5$ some of the water molecules form a second solvation layer (see Fig. 26.9). The hydration numbers for ion–water clusters are found to be about 4 for Li^+, between 5 and 6 for Na^+, between 4 and 6 for F^-, and between 6 and 7 for Cl^-. On average, the oxygen atoms of the water molecules are closer to the cations than are the hydrogen atoms, the reverse holding for the anions (see Fig. 26.10).

Some calculations by R. Watts and co-workers provide an even more detailed picture of the structure induced by ion–water interactions. Figures 26.11 and 26.12 show, respectively, the distribution of oxygen atoms and hydrogen atoms around Li^+ and F^- ions separated by 4 Å. Clearly, both the Li^+ and F^- ions have a strongly bound nearest-neighbor shell of oxygens, with some evidence for a diffuse second-neighbor shell of oxygens. The center of the $Li^+ \cdots O$ distance distribution is about 2 Å, that of the $F^- \cdots O$ distribution about 2.7 Å, but each has considerable dispersion. The hydrogen atom distributions show a complementary pattern—there is a strongly bound nearest-neighbor shell around each ion and a second-neighbor shell about the F^- ion. The H atoms are, on average, closer to the F^- (~ 1.7 Å) than to the Li^+ (~ 3.0 Å). Obviously, the differences in atom locations shown in Figs. 26.11 and 26.12 reflect the influence of the charge distribution of the water molecule on the

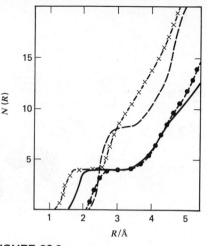

FIGURE 26.9

Number of solvent atoms within a distance R of Li^+ and F^- ions: ——Li^+—O; ————Li^+—H; —●—●— F^-—O; —x—x— F^-—H. From R. O. Watts, J. Chem. Phys. **61**, 2550 (1974).

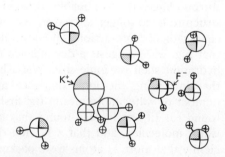

FIGURE 26.10

Perspective drawing showing the relative orientation of water molecules in the nearest-neighbor shell around KF. From R. O. Watts and I. J. McGee, *Liquid State Chemical Physics* (Wiley, New York, 1976).

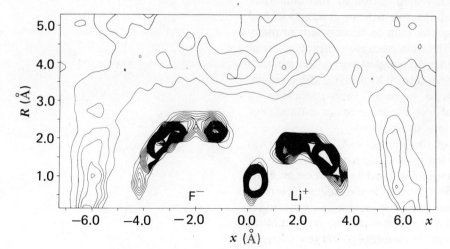

FIGURE 26.11

Distribution of oxygen atoms around the ion pair LiF for an ionic separation of 4.0 Å at 298 K. From R. O. Watts, J. Chem. Phys. **61**, 2550 (1974).

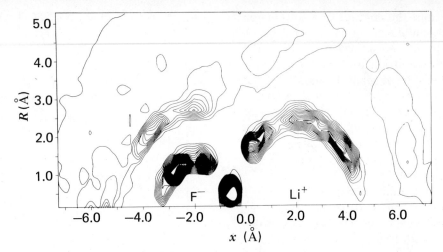

FIGURE 26.12
Distribution of hydrogen atoms around the ion pair LiF for an ionic separation of 4.0 Å at 298 K. From R. O. Watts, *J. Chem. Phys.* **61**, 2550 (1974).

directionality of the ion–water interaction. Any given H_2O will orient so that the O atom is toward the cation and at least one of the H atoms is toward the anion. In fact, each of the four water molecules around the anion has only one H pointed toward it, and the second-neighbor distribution of H's arises from the other H on each water molecule. These second-neighbor H atoms are oriented so that hydrogen bonding to the surrounding water is possible. Beyond two coordination layers the water structure is no longer altered by the presence of the ion pair, although a charge–dipole interaction still contributes to the energy. Finally, we note that Fig. 26.9 suggests a dependence of the number of molecules in the hydration shell on ionic size. Watts has shown, by similar calculations, that increasing the ionic diameter at fixed charge does increase the number of water molecules in the first coordination shell relative to that found for Li^+ or F^-. He found that a Cl^- is associated with about five water molecules, and that a K^+ ion disturbed the water structure sufficiently that more H atoms were packed around the co-ion F^-. Moreover, when two ions are close together, as at high solute concentration, their respective solvation shells are different than when the ions are far apart. Put another way, the hydration structures of the ions in an electrolyte solution are not always independent of one another.

These very recent studies open new possibilities for the understanding of the interaction between ions and water, particularly because of the wealth of structural information revealed. Prior to the numerical simulations, several simple theories of the ion–solvent interaction had been proposed. Although these can now be seen to be deficient in many respects, they are still worthy of consideration because of the fashion in which different important ionic properties are singled out as contributory to the thermodynamic properties of the solution. Moreover, these simple models permit categorization of ions in terms of such properties as diameter and charge, and give order-of-magnitude or better estimates of the thermodynamic properties of the ions in solution.

A very simple molecular model of the ion–solvent interaction was proposed by Born. The ion is regarded as a charged sphere, and the solvent as a dielectric continuum. Although this model cannot be an exact representation of the solution, and the effects of molecular structure in the solvent are not accounted for, the model is useful in providing a measure of the influence of the very strong electrostatic ion–solvent interaction. The potential at a distance R from the center of a charged sphere in vacuum is just $q/4\pi\epsilon_0 R$. The work required to bring up an element of charge dq from infinity to a distance R is then $q\,dq/4\pi\epsilon_0 R$. Thus, the total work required to charge up a sphere of radius a from zero to Ze is just

$$\int_0^{Ze} \frac{q\,dq}{4\pi\epsilon_0 a} = \frac{Z^2 e^2}{8\pi\epsilon_0 a}, \tag{26.57}$$

since the potential at the surface of the sphere is $q/4\pi\epsilon_0 a$. Now let the sphere be placed in a continuous dielectric. We shall assume that there is no change in the structure of the dielectric because of insertion of the ion, and that the dielectric constant of the medium, D, is constant everywhere. Then, since in the dielectric the potential at a distance R from a sphere with charge q is $q/4\pi\epsilon_0 DR$, the work of charging up the ion at constant temperature and pressure is

$$\int_0^{Ze} \frac{q\,dq}{4\pi\epsilon_0 Da} = \frac{Z^2 e^2}{8\pi\epsilon_0 Da}. \tag{26.58}$$

The dielectric constant may depend on temperature and pressure, which was the reason for our stating the constancy of T and p. The difference between Eqs. 26.57 and 26.58 is the difference between the amounts of work for creating an ion in vacuum and in solution, at least according to this model. This difference is, then, the Gibbs free energy of hydration. Thus, we can write

$$\Delta g_{\text{hyd}}^{\ominus} = -\frac{N_A Z^2 e^2}{8\pi\epsilon_0 a}\left(1 - \frac{1}{D}\right), \tag{26.59}$$

where N_A is Avogadro's number. Equation 26.59 predicts an inverse dependence of the free energy of hydration on ionic radius, in qualitative agreement with observation. However, quantitative predictions based on Eq. 26.59 are not very good. For example, the free energies of hydration of Na^+, Ca^{2+}, Al^{3+}, and F^- in H_2O are predicted to be -699, -2920, -12340, and $-515\,$kJ/mol, whereas the observed values are -406, -1582, -4599, and $-431\,$kJ/mol, respectively. Although the sign and the order of magnitude are correct, discrepancies by a factor of 2 or 3 between prediction and observation are common.

It is not hard to point out a number of gross oversimplifications in the Born model: the use of a structureless dielectric continuum in place of the true structure of the solvent, the neglect of all bonding effects except the electrostatic ion–solvent continuum interaction, the assumption that the dielectric properties of the solvent are independent of distance from the ion, and so on. It is hard, however, to improve the model. It has been suggested that the ionic radii be adjusted empirically (by adding 0.75 Å to the radii of cations and 0.1 Å to the radii of anions) on the basis of the rationalization that the effective radius should be some combination of ionic radius and solvated layer. However, this adjustment then neglects the very strong ion–layer interaction augmenting the radius of the complex, and so cannot be considered satisfactory. Reversing the procedure, one can use empirical values of $\Delta g_{\text{hyd}}^{\ominus}$ to derive an effective dielectric constant. The effective dielectric constants so derived are less than the bulk dielectric constant, and vary with ion size in the expected manner, i.e., they increase toward the bulk value as the ion size increases. These effective dielectric constants are independent of charge for closed-shell cations, but for open-shell cations and anions new effects appear. There is indirect evidence of covalent bonding in the case of the open-shell cations, and probably the local-structure-disrupting effects of anions are too complex to be summed up into a simple effective dielectric constant.

There are a number of more detailed models of the local structure of the water surrounding an ion that have been used to calculate ionic enthalpies and entropies. In these it is assumed that the ion and the first-neighbor shell of water molecules form a chemical species with well-defined geometry.[4] Both tetrahedral and octahedral complexes have been suggested. Although the several models emphasize different aspects of the local ion–water structure, all agree that charge–dipole and charge–quadrupole interactions are important. Calculations suggest that the dipole moment of a water molecule bound to a monovalent ion is larger by a factor of 2 than the dipole moment of the free water molecule. The agreement between the calculated and experimental thermodynamic properties of a hydrated ion obtained with the more complicated models is better than that given by the simple Born model, but it is still not completely satisfactory.

We previously noted that the properties of the transition metal ions differ considerably from those of simple closed-shell ions of the same size, that is, ions that have the ground-state term $^1S_0$. In all the models considered so far, it has been assumed that the field of the water molecules does not alter the electronic state of the ion. Somewhat better stated, it has been assumed that no degeneracies of the ionic electronic states are altered by the field of the water molecules, and that the lowest excited states of the ion are so high in energy that they do not influence the thermodynamic properties of the solution. But the transition metal ions have d^n electron configurations, and the degeneracy of the d-orbital states of the free ion can be lifted by a field of suitable symmetry. In fact, the lattice energies of the transition metal halides do deviate from those expected on the basis of the simple Born-Mayer ionic model (see Chapters 9 and 14).

Spectroscopic studies of transition metal ions in solution are consistent with the existence of hexaaquo octahedral complexes of the ion and water. In these cases there appears to be ample evidence that on solution a reaction occurs and a well-defined compound is formed. The ligand-field stabilization energies in an octahedral field are tabulated in Table 26.5. Typical values of Δ_{oct} are 85–200 kJ/mol (see Table 26.6).

To analyze the enthalpy of hydration of transition metal ions we assume that, except for the existence of the ligand-field splitting, there would be a smooth variation of $\Delta h^{\ominus}_{hyd}$ with atomic number as the ionic size varies. As shown in Fig. 26.13, when the extra stabilization arising from the lowering of the energies of some d orbitals is subtracted, the heats of hydration of the divalent transition metal ions vary smoothly across the series. The properties of the trivalent ions still deviate from the expected behavior. Probably some covalent bonding is important in these cases. This simply means that the elementary electrostatic interpretation of the properties of the hexaaquo ion complex must be replaced by a more refined model.

To sum up, all of the interpretations of the alteration of the structure of water caused by inserting a simple closed-shell ion agree that the factors leading to a large negative enthalpy of hydration also lead to a large negative entropy of hydration. These factors are, primarily, small

TABLE 26.5
LIGAND-FIELD STABILIZATION ENERGIES IN AN OCTAHEDRAL-FIELD

Configuration	Stabilization[a]
d^0	0
d^1	$\frac{2}{5}\Delta_{oct}$
d^2	$\frac{4}{5}\Delta_{oct}$
d^3	$\frac{6}{5}\Delta_{oct}$
d^4	$\frac{3}{5}\Delta_{oct}$
d^5	0
d^6	$\frac{2}{5}\Delta_{oct}$
d^7	$\frac{4}{5}\Delta_{oct}$
d^8	$\frac{6}{5}\Delta_{oct}$
d^9	$\frac{3}{5}\Delta_{oct}$
d^{10}	0

[a] Δ_{oct} is the total splitting in an octahedral field of the degenerate d orbitals into orbitals with symmetries e_g and t_{2g}.

[4] The ions H_3O^+ and OH^- require special consideration because of the possibility of hydrogen bonding to other water molecules. M. Eigen has proposed a model in which the principal species present in these solutions are $H_9O_4^+$ and $H_7O_4^-$. In each case there is assumed to be approximate tetrahedral coordination about the O atom, as in ice, except that only three hydrogen bonds are formed. In this model the species H_3O^+ hydrogen-bonds only through its own OH bonds, whereas the species OH^- hydrogen-bonds only through the hydrogens of the other water molecules.

TABLE 26.6
VALUES OF OCTAHEDRAL-FIELD STABILIZA-
TION ENERGIES OF HEXAAQUO COMPLEXES
OF TRANSITION METAL IONS

Configuration	Ion	Δ_{oct} (kJ/mol)[a]
d^1	Ti^{3+}	245
d^2	V^{3+}	215
d^3	$\begin{cases} V^{2+} \\ Cr^{3+} \end{cases}$	$\begin{cases} 146 \\ 211 \end{cases}$
d^4	$\begin{cases} Cr^{2+} \\ Mn^{3+} \end{cases}$	$\begin{cases} 167 \\ 251 \end{cases}$
d^5	$\begin{cases} Mn^{2+} \\ Fe^{3+} \end{cases}$	$\begin{cases} 90 \\ 167 \end{cases}$
d^6	$\begin{cases} Fe^{2+} \\ Co^{3+} \end{cases}$	$\begin{cases} 120 \\ - \end{cases}$
d^7	Co^{2+}	120
d^8	Ni^{2+}	103
d^9	Cu^{2+}	160
d^{10}	Zn^{2+}	0

[a] Δ_{oct} is the total splitting, in an octahedral field, of the degenerate free-ion d orbitals into two groups of orbitals with symmetries t_{2g} and e_g.

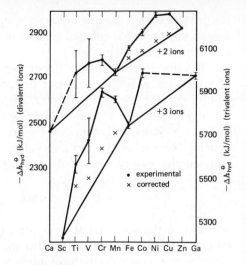

FIGURE 26.13
The heats of hydration of transition metal ions, showing the influence of ligand-field stabilization on the bulk thermodynamic properties. From D. S. McClure, in T. M. Dunn, D. S. McClure, and R. G. Pearson (Eds.), *Aspects of Crystal Field Theory* (Harper & Row, New York, 1965).

ionic radius and high ionic charge. Deductions from all the theories also suggest that ions cause an alteration of the structure of the water, over and above that describable as dipole orientation or restriction of rotational motion. In solutions of small ions the orienting effects of high field outweigh those associated with altering the hydrogen-bond network of water. In solutions of large ions the solvent orientation is diminished relative to that in a solution of small ions because the field is weaker; however, the effects on the hydrogen-bond network are increased because of the greater size of the ion. None of the theories available adequately describes nonelectrostatic interactions, except in very special cases. In one such case, the properties of transition metal ions in solution are qualitatively described by a model in which the ion reacts with water to form a hexaaquo complex. Clearly, this reaction leads to exaggeration of the alteration of the structure of water arising from electrostatic and nonbonding forces.

26.5
The Statistical Molecular Theory of Electrolyte Solutions

We have already argued that the existence of a Coulomb potential between the ions of an electrolyte solution grossly differentiates that solution from a nonelectrolyte solution. The counterpart of this macroscopic difference occurs in the molecular theory; the long range of the ion–ion interaction prevents use of the kind of analysis developed for nonelectrolyte solutions. In fact, because the ions in a solution are mobile and the Coulomb interaction at small separations is large relative to the thermal energy $k_B T$, positive ions tend to be surrounded by negative ions, and *vice versa*. When the ions are arranged as described, the effective interaction between them is altered and becomes short-range. To represent the properties of electrolyte solutions we shall use a primitive model, first introduced by P. Debye and E. Hückel in 1923. This model is a mixture of microscopic and macroscopic concepts, based on a picture of ions as impenetrable charged spheres in a solvent treated as a continuous and structureless dielectric.

Because the interactions depend on the relative distribution of pairs of molecules, it is clear that our first step must be to examine the pair distribution function (see Chapter 23). Consider an electrolyte solution occupying volume V and containing $N_1, N_2, \ldots, N_r$ ions of species $1, 2, \ldots, r$. Suppose that an ion of species i is at the origin of a coordinate system. The probability per unit volume of finding an i ion at the origin and, simultaneously, an ion of species j at the point $\mathbf{R}_j$ is

$$\frac{N_i}{V} \frac{N_j}{V} \exp\left(-\frac{w_{ij}}{k_B T}\right), \qquad (26.60)$$

where w_{ij} is the potential of mean force (see Section 23.3) acting between ions i and j. Recall that w_{ij} is also the work required to bring i and j from infinite separation to the required configuration, in the presence of all the other ions and solvent. Because of the nature of interionic forces, w_{ij} must have some short-range repulsive part and some longer-range modified Coulombic part. The analysis assumes that:

1. The short-range repulsive part of w_{ij} can be treated as if the ions were impenetrable spheres of radius a.

2. The long-range modified Coulombic part of w_{ij} can be represented as $eZ_j\Phi_i(\mathbf{R}_j)$, where $\Phi_i(\mathbf{R}_j)$ is the average electric potential at $\mathbf{R}_j$ arising from the ion i at the origin and all the other ions in the system.

3. The electrostatic potential Φ_i satisfies the Poisson equation when $|\mathbf{R}| \geq a$,

$$\nabla^2 \Phi_i = -\frac{1}{\epsilon_0 D} q_i(\mathbf{R}) \qquad (|\mathbf{R}| \geq a), \qquad (26.61)$$

and the Laplace equation when $|\mathbf{R}| < a$,

$$\nabla^2 \Phi_i = 0 \qquad (|\mathbf{R}| < a). \qquad (26.62)$$

In Eq. 26.61, D is the dielectric constant of the medium and $q_i(\mathbf{R})$ is the charge density at $\mathbf{R}$, given that ion i is at the origin of coordinates.

4. The charge density results from the distribution of the charges in the medium. We do not yet know this distribution, but whatever it is it must be consistent with Eq. 26.60. We assume that, when $|\mathbf{R}| > a$,

$$q_i(\mathbf{R}) = \sum_\alpha eZ_\alpha \frac{N_\alpha}{V} \exp\left[-\frac{eZ_\alpha \Phi_i}{k_B T}\right] \qquad (|\mathbf{R}| \geq a). \quad (26.63)$$

Note that this description of the charge distribution neglects all short-range ion–ion interactions, since it is obtained from Eq. 26.60 by using just the long-range component of w_{ij}.

Combining Eqs. 26.61 and 26.63, we obtain the *Poisson–Boltzmann equation*,

$$\nabla^2 \Phi_i = -\frac{1}{\epsilon_0 D} \sum_\alpha eZ_\alpha \frac{N_\alpha}{V} \exp\left[-\frac{eZ_\alpha \Phi_i}{k_B T}\right] \qquad (|\mathbf{R}| \geq a). \qquad (26.64)$$

The solution of Eq. 26.64, in principle, permits calculation of the ion–ion distribution function and then the thermodynamic properties of the electrolyte solution. However, Eq. 26.64 cannot be solved analytically, so numerical methods must be used.

There is one case, fortunately of some chemical interest, for which Eq. 26.64 can be reduced to a simpler form and solved. Moreover, that simplified equation is consistent in its treatment of w_{ij}, whereas Eq. 26.64 is not.[5] Suppose that the ionic solution is sufficiently dilute that the average interionic spacing corresponds to $\Phi_i/k_BT \ll 1$. In this limit, expanding the exponential in Eq. 26.64, we find that

$$\nabla^2\Phi_i(R) = -\frac{e}{\epsilon_0 D}\sum_\alpha \frac{N_\alpha Z_\alpha}{V}$$
$$+\left[\frac{e^2}{\epsilon_0 Dk_BT}\sum_\alpha Z_\alpha{}^2\frac{N_\alpha}{V}\right]\Phi_i(R) \qquad (|\mathbf{R}|\geqslant a). \quad (26.65)$$

The condition that the solution be electrically neutral requires that

$$\sum_\alpha eZ_\alpha\frac{N_\alpha}{V}=0. \qquad (26.66)$$

Thus, using the definition

$$\kappa^2 \equiv \frac{e^2}{\epsilon_0 Dk_BT}\sum_\alpha Z_\alpha{}^2\frac{N_\alpha}{V}, \qquad (26.67)$$

we can reduce Eq. 26.65 to

$$\nabla^2\Phi_i(R)=\kappa^2\Phi_i(R) \qquad (|\mathbf{R}|\geqslant a). \quad (26.68)$$

In our model the ions are spheres, and the only distance dependence is through the ionic separation R. In this case, Eq. 26.68 becomes

$$\frac{1}{R}\frac{d^2(R\Phi_i)}{dR^2}=\kappa^2\Phi_i \qquad (|\mathbf{R}|\geqslant a). \quad (26.69)$$

To solve Eq. 26.69 we rewrite it in the form

$$\frac{d^2}{dR^2}(R\Phi_i)=\kappa^2(R\Phi_i) \qquad (|\mathbf{R}|\geqslant a). \quad (26.70)$$

Direct substitution shows that

$$R\Phi_i = C_1 e^{-\kappa R}+C_2 e^{\kappa R} \qquad (26.71)$$

is a solution. But the potential must approach zero as the distance from the central ion gets indefinitely large,

$$\lim_{R\to\infty}\Phi_i=0, \qquad (26.72)$$

so that $C_2=0$ and we find that

$$\Phi_i=\frac{C_1 e^{-\kappa R}}{R} \qquad (|\mathbf{R}|\geqslant a) \qquad (26.73)$$

is the proper solution.

To complete the solution we must evaluate the constant C_1. This can be done by solving Eq. 26.62 and requiring that the electrostatic potential and the electrostatic field be continuous at the surface of the ion. We use a different method here. Our method is based on the conservation of charge. If a positive ion is at the origin of coordinates, in any spherical shell at distance R from it there will, on average, be more negative ions than positive ions; the spherical shell will therefore have net negative

---

[5] See later comments.

charge. The totality of all shells surrounding the cation at the origin must carry a negative charge just equal to the positive charge of the ion. This condition of overall electroneutrality may be written in the form

$$\int_a^\infty 4\pi R^2 q_i(R)\, dR = -Z_i e. \tag{26.74}$$

The charge density is, of course, related to the electrostatic potential by Eq. 26.63. However, we are solving Eq. 26.61 in the domain in which the electrostatic interaction is small compared with $k_B T$ (see Eq. 26.68). Comparison of Eqs. 26.61, 26.68, and 26.73 shows that

$$q_i(R) = -\kappa^2 \epsilon_0 D \Phi_i$$

$$= -C_1 \kappa^2 \epsilon_0 D \frac{e^{-\kappa R}}{R}. \tag{26.75}$$

Using Eq. 26.75 in Eq. 26.74 leads to

$$4\pi C_1 \kappa^2 \epsilon_0 D \int_a^\infty R e^{-\kappa R}\, dR = Z_i e \tag{26.76}$$

or

$$C_1 = \frac{Z_i e}{4\pi\epsilon_0 D} \cdot \frac{\exp(\kappa a)}{1 + \kappa a}, \tag{26.77}$$

so that the potential Φ_i is given by

$$\Phi_i = \frac{Z_i e}{4\pi\epsilon_0 D} \cdot \frac{\exp(\kappa a)}{1 + \kappa a} \cdot \frac{\exp(-\kappa R)}{R}. \tag{26.78}$$

One of the fundamental principles of electrostatics is that of linear superposition of potentials. This principle states that the total electric potential at some point may be treated as a sum of the potentials arising from all sources in the medium. In our case, Eq. 26.78 represents the total electrostatic potential at a distance R from an ion with charge $Z_i e$. We now treat this total electrostatic potential as the sum of the direct potential of the ion at the origin,

$$\Phi_i'' = \frac{Z_i e}{4\pi\epsilon_0 D R}, \tag{26.79}$$

and the potential due to all the other ions in the solution,

$$\Phi_i' = \frac{Z_i e}{4\pi\epsilon_0 D R} \left[\frac{\exp(\kappa a)\exp(-\kappa R)}{1 + \kappa a} - 1 \right]. \tag{26.80}$$

Equation 26.80 is obtained by using the superposition principle

$$\Phi_i = \Phi_i' + \Phi_i'' \tag{26.81}$$

and subtracting Eq. 26.79 from Eq. 26.78. In our model Eq. 26.80 is valid for all R down to $R = a$. No ions can penetrate the region $R < a$, and for all $R < a$ the potential due to the spherical distribution of ions outside the central ion is constant and equal to its value at $R = a$. We write, then,

$$\Phi_i'(a) \equiv \Phi_i^{\text{atm}} = -\frac{Z_i e}{4\pi\epsilon_0 D} \frac{\kappa}{1 + \kappa a}. \tag{26.82}$$

Thus, the electrostatic energy of the central ion itself is reduced by the product of its charge $Z_i e$ and the potential 26.82 because of the interaction with all other ions in the solution. Corresponding to the existence of ionic electrostatic interaction energy, the chemical potential of the ions is different from that of molecules in an ideal solution. We now calculate the activity coefficients of the ions under the assumption that the solution would be ideal if there were no electrostatic interactions, that is, that the short-range ion–solvent and ion–ion forces do not lead to nonideality.

From the discussion of Section 19.4 we can infer that μ_i is the Gibbs free energy change when 1 mol (or 1 molecule, depending on units) of solute is added to the solution at constant temperature and pressure. In the standard representation of the chemical potential of a solute, $\mu^\ominus + RT \ln m_i$ is the part of μ_i that does not involve solute–solute interaction and $RT \ln \gamma_i$ is the part of the chemical potential arising from solute–solute interaction. In other words, $RT \ln \gamma_i$ is the work (at constant T and p) done against intermolecular forces when 1 mol of solute is added to the solution.

Suppose that it were possible to discharge an ion continuously. We assume that the uncharged ion, which we call a ghost ion, does not interact with the other ions. If the ghost ion is recharged, the work done in charging up the ion, allowing the other ions to adjust to the new equilibrium at every point along the charging process, is just $RT \ln \gamma_i$. This follows because, under our assumptions, addition of a ghost ion to a solution does not contribute to the interaction energy. But the final state obtained when either a fully charged ion is added to the solution, or a ghost ion is added and then charged, is the same. Hence the Gibbs free energy changes for the two processes must be the same.

Suppose that it were possible, as just suggested, to discharge an ion. If the fractional charge on ion i is ξ_i, then the atmosphere contribution to the potential is

$$\Phi_i^{\text{atm}}(\xi_i) = -\frac{\xi_i Z_i e \kappa}{4\pi\epsilon_0 D(1+\kappa a)}. \tag{26.83}$$

To bring up an amount of charge $Z_i e \, d\xi_i$ to the ion i, the work done is just the product of the charge and the ion–atmosphere potential, $\Phi_i^{\text{atm}} Z_i e \, d\xi_i$. Therefore we obtain

$$
\begin{aligned}
k_B T \ln \gamma_i &= \int_0^1 \Phi_i^{\text{atm}}(\xi_i) Z_i e \, d\xi_i \\
&= -\int_0^1 \frac{\xi_i Z_i^2 e^2 \kappa}{4\pi\epsilon_0 D(1+\kappa a)} \, d\xi_i \\
&= -\frac{Z_i^2 e^2 \kappa}{8\pi\epsilon_0 D(1+\kappa a)}.
\end{aligned} \tag{26.84}
$$

Here γ_i is, by definition, the activity coefficient of the ion i. The single-ion activity coefficients must be combined as described earlier in this chapter before comparison with experiment is possible. The result is, using Eq. 26.12 and $|\nu_+ Z_+| = |\nu_- Z_-|$,

$$
\begin{aligned}
\ln \gamma_\pm &= -\frac{e^2 \kappa}{8\pi\epsilon_0 D k_B T} \left(\frac{\nu_+ Z_+^2 + \nu_- Z_-^2}{\nu_+ + \nu_-} \right) \\
&= -|Z_+ Z_-| \frac{e^2 \kappa}{8\pi\epsilon_0 D k_B T(1+\kappa a)}.
\end{aligned} \tag{26.85}
$$

Equation 26.84 is the principal result of the Debye-Hückel theory of electrolyte solutions. From it we deduce that:

1. Since κ is proportional to the square root of the concentration (see Eq. 26.67), we recover the empirically discovered rules concerning ionic strength.

2. In the limit of very small concentration, we obtain

$$\lim_{\kappa a \to 0} k_B T \ln \gamma_i = -\frac{Z_i^2 e^2 \kappa}{8\pi\epsilon_0 D},\tag{26.86}$$

which is known as the *limiting-law value*. It is found that the limiting-law form for the activity coefficient (and other derived quantities) is exact for the model considered. On the other hand, Eq. 26.84 and similar extensions are not completely correct because of the approximations made to w_{ij}. Only in the limit of infinite dilution are the cited assumptions valid.

3. We can verify the qualitative picture of the screening given earlier by the following arguments. First, expansion of the exponential of Eq. 26.60 shows that the distribution of pairs of ions is of the form

$$\frac{N_i N_j}{V^2}\left(1 - \frac{Z_i Z_j e^2}{4\pi\epsilon_0 D k_B T} \cdot \frac{e^{-\kappa R}}{R}\right),\tag{26.87}$$

so that the correlation between two ions extends over a distance of the order of $(1/\kappa)$, the *Debye screening length*. When $\kappa R \gg 1$, the ions are independent, since then their interaction is screened out. Moreover, as anticipated, the screening arises because the ions distribute themselves in such a way that a cloud of charge is interposed between any two ions. To see how the screening charge arises, we calculate the total charge $Q_i(B)$ outside the ion i but inside a sphere of radius B. This is, in the limiting-law case,

$$\begin{aligned}
Q_i(B) &= -\int_0^B d\mathbf{R} \sum_\alpha \frac{N_\alpha}{V} e^3 Z_i Z_\alpha^2 \frac{e^{-\kappa R}}{4\pi\epsilon_0 D R}\\
&= -eZ_i\kappa^2 \int_0^B dR \cdot R e^{-\kappa R}\\
&= -eZ_i[1 - e^{-\kappa B}(1 + \kappa B)].
\end{aligned}\tag{26.88}$$

Clearly, as $\kappa B \to \infty$, we find that $Q_i(B) \to -Z_i e$, just equal and opposite to the charge of ion i. For any given value of B, an ion "outside" B sees a net charge less than $Z_i e$, a result that shows how interposition of other ions reduces the effective charge from $Z_i e$ to some smaller value.

A comparison of some predicted and observed thermodynamic properties of dilute electrolyte solutions is displayed in Fig. 26.14. Note the good agreement near zero concentration.

We have remarked that Eq. 26.64 is not self-consistent. The inconsistency arises from two approximations in the derivation. The first of these is easy to see. As shown in Chapter 23, the pair-correlation function for simple spherical molecules has oscillatory structure for a range of separations of the order of a few molecular diameters. This structure in the pair-correlation function is suppressed by the assumption that the potential of mean force is $eZ_j\Phi_i(R_j)$ for all $|\mathbf{R}| \geqslant a$, hence the influence of short-range interactions on the ionic distribution is omitted except for the region $|\mathbf{R}| < a$. The second approximation is more subtle. The mean electrostatic potential depends on the distribution of pairs of ions. Just as in the theory of simple liquids, the determination of the pair-correlation function requires knowledge of the distribution of triplets of ions, and so on (see Eq. 23.23). The assumption that the pair potential of mean force $w_2 = w_{ij}$ is $eZ_j\Phi_i(R_j)$ is equivalent to the assumption that the potential of

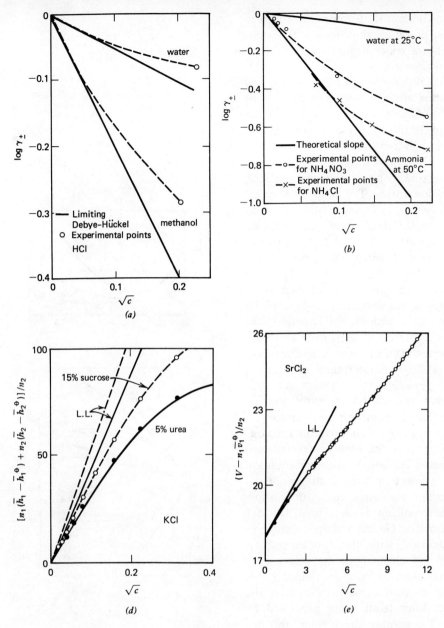
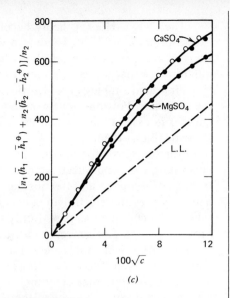

FIGURE 26.14

(a) Activity coefficients of hydrochloric acid in water (upper curves) and methanol (lower curves) at 25°C.

(b) Activity coefficients of ammonium chloride and nitrate in ammonia at −50°C.

(c) Relative apparent molal enthalpies of calcium sulfate and magnesium sulfate in water at 25°C. L.L. denotes the Debye–Hückel limiting law behavior.

(d) Relative apparent molal enthalpy of KCl in mixed solvents at 25°C; the straight lines represent the theoretical limiting law

(e) Apparent molal volume of strontium chloride in water at 25°C; note the departure from the theoretical limiting law at moderate dilutions.

In all these figures c is the molality of the solution.

(a) and (b) from R. A. Robinson and R. H. Stokes, *Electrolyte Solutions* (Butterworths, London, 1955); (c), (d), and (e) from H. Harned and B. B. Owen, *Physical Chemistry of Electrolytic Solutions* (Van Nostrand Reinhold, New York, 1958).

mean force between a triplet of molecules is the sum of the three pair potentials of mean force between them: $w_3(1, 2, 3) = w_2(1, 2) + w_2(1, 3) + w_2(2, 3)$. This approximation neglects interference effects in the distribution of three molecules relative to the superposed distributions of pairs of molecules. It is exact in the limit that the density of molecules becomes vanishingly small, but otherwise in error.

There are a variety of ways of showing that the Debye–Hückel approximation becomes exact in the limit of zero electrolyte concentration. Although the detailed analysis is complicated, it is possible to see that this is likely the case by noting that both of the approximations used to derive Eq. 26.64 become exact in the limit of zero concentration, the first because the pair correlations induced by short-range interactions vanish, and the second because interference between the configurations of widely separated pairs of molecules vanishes.

The Debye–Hückel theory becomes inaccurate for concentrations in excess of about $10^{-3} M$. There is a vast literature that deals with extensions of the theory of electrolytes to more concentrated solutions. The different approaches can be sorted into four categories.

First, and simplest, are theories that retain the essential structure of the linearized Poisson–Boltzmann equation and use models of one sort or another to extend the range of its validity. The best known of these is the Bjerrum–Fuoss theory of ion association. In this theory it is assumed that some of the ions form neutral pairs and thereby do not participate in generating the mean electrostatic potential in the solution. The concentrations of ion pairs and free ions are related by an equilibrium constant, hence vary as the electrolyte concentration varies. The simplest form of this theory considers as paired only those ions in contact; other versions broaden the ion-pair concept to include oppositely charged ions separated by only one solvent molecule, or all oppositely charged ions that have mutual attraction larger than $2k_BT$. By adjustment of ionic radii, which are the parameters that determine the ion-pair equilibrium constant, the activity coefficients and other thermodynamic properties of electrolyte solutions can be fitted over a much larger concentration range than by Debye–Hückel theory. There are some observations, for example, absorption spectra, that are reasonably interpreted as indicating that ion-pairing occurs for high-valency electrolytes even at room temperature in water, for lower-valency electrolytes at high temperature in aqueous solution, and for many electrolytes in low-dielectric-constant solvents. Nevertheless, the theory of ion pairing has *ad hoc* features that in practice make it a parametric fit more than a predictive methodology.

Second, the Poisson–Boltzmann equation can be improved by including corrections that account for the omissions or the approximations used in its derivation. Also, the full uncorrected Poisson–Boltzmann equation can be integrated numerically, thereby avoiding the linearization approximation that leads to Eq. 26.65. Concerning the latter, calculations show that if $2a = 4\,\text{Å}$, up to $0.1\,M$ (1) ion pairing is negligible for 1:1 electrolytes but important for 2:2 electrolytes; (2) the activity coefficients of 1:1 electrolytes are practically identical with those predicted by Debye–Hückel theory; (3) for 1:2 and 2:1 electrolytes the values of γ predicted differ from the Debye–Hückel values by no more than 2%; (4) for 2:2 electrolytes the values of γ agree with those predicted by the ion-pairing theory to within 2%, but differ from those predicted by Debye–Hückel theory by up to 20%. Concerning the former, the improved Poisson–Boltzmann equation leads to predictions of the ionic activity coefficient that are superior to those of Debye-Hückel theory and the original Poisson–Boltzmann equation 26.64, as shown in Table 26.7. Note that in this table the nonideality contribution is represented by the osmotic coefficient (see Chapter 25, Eq. 25.93).

Third, the electrolyte solution can be described using the generalization of the virial expansion mentioned in Chapter 25. This theory, originally due to J. E. Mayer, is mathematically complex because of the need to take special measures to account for the influence of the long-range Coulomb potential between ions. Mayer's theory permits accurate calculation of the activity coefficient or osmotic coefficient up to about $0.1\,M$. An optimized virial expansion theory, due to H. C. Andersen and D. Chandler, extends the range of accurate prediction considerably (see Table 26.7).

Fourth, the methods used in the development of the theory of liquids (see Chapter 23) can be extended to models of electrolyte solutions. The so-called method of integral equations has been largely developed by H.

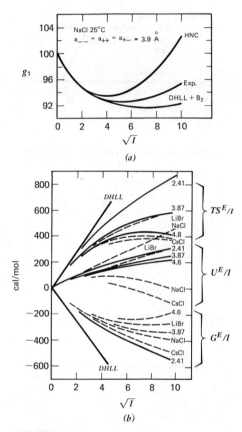

FIGURE 26.15

(*a*) The osmotic coefficient g_1 for the primitive model compared with experimental data for NaCl in H_2O at 25°C. The curve marked HNC is computed by an integral equation approach, that marked DHLL+B_2 is the Debye–Hückel limiting law plus the first cluster correction (analogous to the second virial coefficient of a gas).

(*b*) The thermodynamic functions G^E/I, U^E/I, and TS^E/I for the primitive model with equal anion and cation sizes, compared with experimental data for aqueous LiBr, NaCl, and CsCl. The numbers entered on the right hand side of the figure are the ion size parameters. From J. C. Rasaiah, *J. Chem. Phys.* **52**, 704 (1970).

TABLE 26.7
THE OSMOTIC COEFFICIENT g_1 GIVEN BY THE VARIOUS THEORIES DISCUSSED FOR THE 1:1 PRIMITIVE ELECTROLYTE MODEL (HARD-CORE IONS IN A CONTINUOUS DIELECTRIC FLUID) COMPARED WITH NUMERICAL SIMULATIONS FOR THE SAME MODEL

M^a	DH[b]	PB[c]	MPB[d]	PY(p)[e]	PY(C)[f]	AC[g]	Simulation[h]
0.00911	0.968	0.970	0.970	0.9703	0.9705	0.9707	0.9701 ± 0.0008
0.10376	0.933	0.946	0.945	0.9452	0.9461	0.9452	0.9445 ± 0.0012
0.42502	0.949	0.981	0.980	0.9765	0.9844	0.9787	0.9774 ± 0.0046
1.0001	1.03	1.08	1.09	1.079	1.108	1.091	1.094 ± 0.005
1.9676	1.20	1.26	1.34	1.311	1.386	1.342	1.346 ± 0.009

[a] Moles per liter.
[b] Debye–Hückel theory.
[c] Full Poisson–Boltzmann equation (26.64).
[d] Modified Poisson–Boltzmann equation with correction for short-range correlations and deviations from additivity of pair potential of mean force.
[e] See caption to Fig. 23.18.
[f] See caption to Fig. 23.18.
[g] Andersen–Chandler optimized virial expansion theory.
[h] Numerical simulation results of D. N. Card and J. P. Valleau, *J. Chem. Phys.* **52**, 6232 (1970); **56**, 248 (1972).

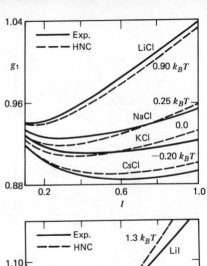

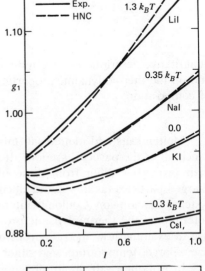

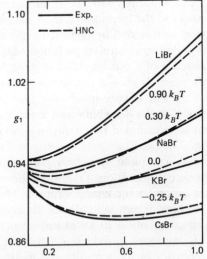

FIGURE 26.16
Theoretical osmotic coefficients for a model of the electrolyte solution in which there is a square-well interaction between the ions—otherwise the model is the same as the primitive model. The well depths used to fit the data are shown next to each curve. From J. C. Rasaiah, *J. Chem. Phys.* **52**, 704 (1970).

Friedman and his co-workers. The problems discussed in Chapter 23 with respect to finding an accurate and tractable description of the distribution of triplets of molecules, or of the direct correlation function, occur again in the theory of electrolyte solutions, and the methods of solution used are fundamentally the same. As shown in Table 26.7, the predicted osmotic coefficients are in much better agreement with numerical simulations of an electrolyte solution than are those from the Debye–Hückel, Poisson–Boltzmann, or modified Poisson–Boltzmann theories; the optimized virial expansion theory is about as good.

Despite the success of these extended theories in describing the primitive model of an electrolyte solution (hard-core ions in a continuous dielectric fluid), they do less well in predicting the properties of real electrolyte solutions (see Fig. 26.15). Given the agreement between theory and the primitive model, we conclude that a more refined model of the electrolyte solution is needed. There is at least one glaring omission in the primitive model—the van der Waals-type interactions between the ion and the solvent have been neglected. Of course, so have the individual ion–dipole interactions, but these are at least crudely accounted for by the introduction of the dielectric constant of the fluid. In fact, when any particle is inserted in a dielectric, whether charged or not, the electric field in the medium is altered because the dielectric is no longer homogeneous. Moreover, we have seen that in the immediate neighborhood of an ion the solvent structure is altered from that characteristic of the bulk. These phenomena contribute to the effective interaction (the potential of mean force) between the ions. Some extensions of the integral equation theory to include short-range attractions have been studied. As shown in Fig. 26.16, inclusion of a square-well attraction does improve the agreement between theory and experiment for simple electrolytes. A more complicated multiparameter potential is needed for salts such as the tetraalkylammonium halides. That a consistent set of these parameters can be found is shown in Fig. 26.17.

Clearly, the statistical molecular theory of electrolyte solutions needs considerable development before it can be trusted to give a realistic and

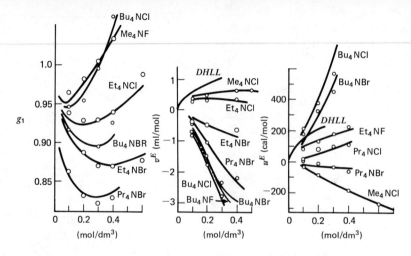

FIGURE 26.17

(*a*) Comparison of model osmotic coefficients with experiment in typical cases. The curves are experimental values of g_1, whereas the model values are shown as discrete data.

(*b*) Comparison of model excess volumes with experiment in typical cases. The curves are experimental values of v^E, whereas the model values are shown as discrete data.

(*c*) Comparison of model excess energies with experiment in typical cases. The curves are experimental values of u^E, whereas the model values are shown as discrete data. From P. S. Ramanathan, C. V. Krishnan, and H. L. Friedman, *J. Solution Chem.* **1**, 237 (1972).

quantitative description of the influence of ions on the structure of water and on the thermodynamic properties of the solution for the entire range of composition.

26.6 Molten Salts and Molten Salt Mixtures

Molten salts and molten salt mixtures are representative of fluids in which all of the particles are ions. It is to be expected, and is found, that their properties differ from those of a dilute mixture of ions and uncharged particles (as in a dilute electrolyte solution). Yet in each class of systems the ion–ion Coulomb interaction is the dominant influence in determining the thermodynamic properties. For this reason it is instructive to examine how the properties of a molten salt depend on the ionic size, charge, temperature, and other variables.

In a molten salt, as in all liquids, there is positional disorder with respect to the locations of the ions. Now, in an ionic crystal every positive ion is surrounded by nearest-neighbor negative ions and *vice versa*. And in a dilute electrolyte solution, Debye–Hückel theory establishes the existence of a diffuse cloud of negative ions about each positive ion, and *vice versa*. It should not be surprising, then, that the local structure of a molten salt is very much like that of a simple liquid, such as Ar, with respect to the distribution of ionic centers of mass, but that every positive ion is surrounded by nearest-neighbor negative ions, and *vice versa* (see Fig. 23.12).

A complete statistical molecular theory of molten salts involves subtle considerations related to the way the long-range Coulomb potential generates the charge ordering observed in the liquid state. That class of theory is not suitable for description in this text. Instead, we shall restrict attention to an extension of the theory of corresponding states. This approach provides a satisfying unification and overview of many properties of molten salts and molten salt mixtures.

Consider, as a simple case, a pure binary molten salt, such as NaCl. Recall that a positive ion has only negative-ion nearest neighbors, and *vice versa*. It is to be expected, then, that two like-charged ions are always far enough apart that any short-range repulsion between them can be neglected, and that the dominant part of their interaction is the Coulomb repulsion. In contrast, the short-range repulsion between unlike ions that are nearest neighbors plays an important role in defining the structure of the liquid. As in the case of simple liquids, it is expected that a qualitatively correct model of that short-range repulsion is the hard-core interaction (see Chapter 23). We now assume that the following simple model potential will adequately describe the principal properties of a molten salt:

$$u_2^{++}(R) = u_2^{--}(R) = \frac{(Ze)^2}{R} \qquad (R > 0),$$

$$u_2^{+-}(R) = \infty \qquad\qquad (R \leq \sigma), \qquad (26.89)$$

$$u_2^{+-}(R) = \frac{Z_+ Z_- e^2}{R} \qquad (R > \sigma).$$

In Eqs. 26.89, σ is the sum of unlike ion radii. The model potential 26.89 treats the ions as unpolarizable rigid spheres without any form of multipole interaction or dispersion interaction. We must expect the neglect of these interactions and the use of a hard-core repulsion to lead to some error in the predicted thermodynamic properties. However, drawing on the results of our study of ionic crystals, we expect that error to be small enough that qualitative properties of the molten salt are accounted for.

To introduce a principle of corresponding states, we define the reduced distance

$$R^* \equiv \frac{R}{\sigma} \qquad\qquad (26.90)$$

and rewrite Eqs. 26.89 in the form

$$u_2^{++}(R^*) = u_2^{--}(R^*) = \frac{(Ze)^2}{R^*\sigma} = \frac{(Ze)^2}{\sigma} u_2^*(R^*) \qquad (R^* > 0),$$

$$u_2^{+-}(R^*) = \infty \qquad\qquad (R^* \leq 1), \quad (26.91)$$

$$u_2^{+-}(R^*) = -\frac{(Ze)^2}{\sigma} u_2^*(R^*) \qquad (R^* > 1).$$

The potential energy of interaction of all pairs of ions can then be written as

$$\mathcal{V}_{2N} = \sum_{i>j} u_{ij}^{++} + \sum_{i>j} u_{ij}^{--} + \sum_{i>j} u_{ij}^{+-}$$

$$= \frac{(Ze)^2}{\sigma} \mathcal{V}_{2N}^*(\mathbf{R}_1^* \ldots, \mathbf{R}_{2N}^*), \qquad (26.92)$$

where we have assumed that there are a total of $2N$ ions in the system, with $N_- = N_+ = N$. The canonical partition function for the ionic melt is just

$$Q_{2N} = \frac{1}{N!}\left(\frac{2\pi m_+ k_B T}{h^2}\right)^{3N/2} \frac{1}{N!}\left(\frac{2\pi m_- k_B T}{h^2}\right)^{3N/2}$$

$$\times \int_V \exp\left(-\frac{\mathcal{V}_{2N}}{k_B T}\right) d\mathbf{R}_1 \cdots d\mathbf{R}_{2N}, \quad (26.93)$$

where V is the volume of the melt. Introduction of the reduced distance of Eq. 26.90 then gives

$$Q_{2N} = \frac{\sigma^{6N}}{(N!\Lambda_+^{3N/2})(N!\Lambda_-^{3N/2})} \int_{V/\sigma^3} \exp\left(-\frac{\mathcal{V}_{2N}^*}{k_B T}\right) d\mathbf{R}_1^* \cdots d\mathbf{R}_{2N}^*$$

$$(26.94)$$

$$= \frac{\sigma^{6N}}{(N!\Lambda_+^{3N/2})(N!\Lambda_-^{3N/2})} \mathscr{I}(T^*, V^*, N), \qquad (26.95)$$

where $N!\Lambda_+^{3N/2}$ and $N!\Lambda_-^{3N/2}$ are the kinetic energy contributions to Q_{2N} for positive and negative ions and $\mathscr{I}$ depends only on the number of ions in the reduced volume V^* and the reduced temperature T^*, defined by

$$V^* \equiv \frac{V}{\sigma^3}, \tag{26.96}$$

$$T^* \equiv \frac{\sigma k_B T}{(Ze)^2}. \tag{26.97}$$

From the fundamental relation between the Helmholtz free energy and the canonical partition function,

$$A_{2N} = -k_B T \ln Q_{2N},$$

we readily find for the pressure

$$\begin{aligned}
p &= k_B T \left[\frac{\partial \ln \mathscr{I}(T^*, V^*, N)}{\partial V} \right]_{T,N} \\
&= \left(\frac{Z^2 e^2}{\sigma^4} \right) T^* \left(\frac{\partial \ln \mathscr{I}}{\partial V^*} \right)_{T^*,N} \\
&= \left(\frac{Z^2 e^2}{\sigma^4} \right) p^*(T^*, V^*, N), \tag{26.98}
\end{aligned}$$

using the definition

$$p^*(T^*, V^*, N) \equiv T^* \left(\frac{\partial \ln \mathscr{I}}{\partial V^*} \right)_{T^*,N}. \tag{26.99}$$

The reduced pressure p^* is simply calculated from the system pressure p by inverting Eq. 26.98, that is,

$$p^* \equiv \left(\frac{\sigma^4}{Z^2 e^2} \right) p. \tag{26.100}$$

The reduced equation of state

$$p^* = p^*(T^*, V^*) \tag{26.101}$$

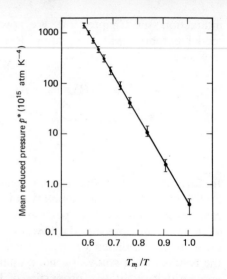

FIGURE 26.18
Mean reduced vapor pressure curve for the halides of Na, K, Rb, and Cs. Average deviation from the mean is shown by the vertical lines. T_m is the melting temperature of an alkali halide. From M. Blander, *Adv. Chem. Phys.* **11**, 83 (1967).

should be a universal function for all molten salts with pair interactions of the form 26.89. A test of the validity of this approach is shown in Fig. 26.18, which is a plot of the reduced vapor pressure as a function of reduced temperature for the fluorides, chlorides, bromides, and iodides of Na, K, Rb, and Cs. Although at a given value of the absolute temperature T the vapor pressures of these salts are very different, all of the values conform to one reduced equation of state. The ideas sketched above can be, and have been, extended to treat unsymmetrical electrolytes, such as CaF_2. The reduced equation of state found is a universal function within a particular valence class, but is not necessarily the same for different valence classes. For example, using the methods described, it is possible to compare the properties of two different unidivalent salts, but not those of a unidivalent salt with those of a unitrivalent salt.

Many different tests of the applicability of the theorem of corresponding states to molten salts have been made, all rather successful. We shall, therefore, use this approach to describe the excess thermodynamic properties of molten salt mixtures.

The extension of the corresponding-states principle to the description of molten salt mixtures, although straightforward, involves some new considerations. First, since there are more than two cations and anions in the mixture, a single distance of anion–cation closest approach cannot provide the necessary parameterization for a reduced equation of state.

Instead, several distances of closest approach, one for each unlike anion–cation pair, are required. Second, it is necessary to make some assumption concerning the distribution of different cations amongst the positions occupied by all cations, and similarly for the anions. A plausible zeroth-order approximation is that the cations mix randomly among themselves, and the anions mix randomly among themselves. A test of this assumption can be made by the method of computer simulation; it will be discussed later. Third, the excess energy, entropy, and volume of molten salt mixtures are very small relative to the internal energy, entropy, and volume of either the pure molten salts or the molten mixture. It is not obvious, then, that the neglect of dispersion interactions between the ions is acceptably accurate. A better model, still within the framework of the corresponding-states approach, must include these interactions.

To apply these ideas to the description of a molten salt mixture we write for the anion–cation interaction

$$u_2^{+-}(R) = u_{+-}^{\text{Coulomb}} + \xi_{+-} u_{+-}^{\text{non-Coulomb}} \qquad (R > \sigma),$$
$$u_2^{+-}(R) = \infty \qquad\qquad\qquad\qquad\quad (R \leqslant \sigma), \qquad (26.102)$$

and for the anion–anion and cation–cation interactions

$$u_2^{--}(R) = u_{--}^{\text{Coulomb}} + \xi_{--} u_{--}^{\text{non-Coulomb}},$$
$$u_2^{++}(R) = u_{++}^{\text{Coulomb}} + \xi_{++} u_{++}^{\text{non-Coulomb}}, \qquad (26.103)$$

where ξ is a measure of the strength of the non-Coulomb potential. When we use the form $\xi u^{\text{non-Coulomb}}$ we imply that $u^{\text{non-Coulomb}}$ is a function of R only and all the coefficients are contained in ξ. For example, if $\xi u^{\text{non-Coulomb}}$ were to be a Lennard-Jones attraction, we would have $u^{\text{non-Coulomb}} = -R^{-6}$ and $\xi = 4\epsilon\sigma^6$ (see Eq. 21.173).

Following the spirit of the analysis of a pure molten salt, Eqs. 26.102 and 26.103 should be reduced by introduction of the anion–cation contact separation as a reference length. Instead of using several such contact separations, one for each unlike anion–cation pair in the mixture, we suppose that differences in ionic size are sufficiently small that the properties of a melt can be expanded in a Taylor series about the properties of a reference melt with anion–cation contact separation σ_0, the expansion parameter being $(\sigma_{+-} - \sigma_0)$. In the case of a mixture, there must be terms in this expansion for all possible ion combinations. The non-Coulomb part of the interaction is handled similarly. Let ξ_0 be a single value of ξ for the reference melt. Then the effect of differences among ξ_{++}, ξ_{+-}, and ξ_{--} is described by a Taylor series expansion about the properties of the reference melt, the expansion parameter being $(\xi_{ij} - \xi_0)$, where i, j run over all ion species. Writing down the Taylor series expanion of the partition function is now very simple, and evaluation of the many derivatives that appear is straightforward but tedious. For a molten salt mixture with a common anion and the same charge on both cations (call the cations 1 and 2), and for which all non-Coulomb interactions are the same in the pure melts and the mixture ($\xi_{++} = \xi_{+-} = \xi_{--}$), the excess Helmholtz free energy of mixing is

$$A^E = N_1 N_2 \left(\frac{k_B T}{2}\right)\left(\frac{\sigma_1 - \sigma_2}{\sigma_0}\right)^2 \times \text{function dependent on the two, three,}$$
$$\text{and four particle distribution functions}$$
$$\text{of the melt.} \qquad (26.104)$$

It is not necessary, for our purposes, to specify the nature of the several distribution functions that appear in Eq. 26.104. The crux of the matter is that a difference in ionic size generates local structural changes that lead to a nonzero excess free energy of mixing. If the non-Coulombic terms in the interaction are not all the same, there is an additional contribution to the excess Helmholtz free energy of mixing of the form

$$-N_1 N_2\left[\left(\frac{\xi_{12}-\xi_0}{\xi_0}\right)a_{12}-\frac{1}{2}\left(\frac{\xi_{11}-\xi_0}{\xi_0}\right)a_{11}-\frac{1}{2}\left(\frac{\xi_{22}-\xi_0}{\xi_0}\right)a_{22}\right], \quad (26.105)$$

where the functions a_{ij} are the averages of the R-dependent part of the non-Coulomb interaction, $u_{ij}^{non\text{-}Coulomb}$, over the distribution of ions. The form 26.105 implies that in the first approximation the non-Coulombic part of the excess free energy of mixing arises from the difference in averaged non-Coulombic energies of ions with the configurations M_1XM_2, M_1XM_1, and M_2XM_2, where X is the anion and M_1, M_2 the two cations.

In molten salt mixtures the excess entropy of mixing is sufficiently small that a^E is very nearly the same as h^E. Figure 26.19 shows experimental data for mixtures of alkali nitrates plotted versus $(\sigma_1-\sigma_2)^2/(\sigma_1{}^2\sigma_2{}^2)$. For the values of σ_1 and σ_2 characteristic of these ions, the product $\sigma_1\sigma_2$ varies very little compared to $\sigma_1-\sigma_2$, so the functional form predicted in Eq. 26.104 is verified over a wide range by the observations. In fact, the fit of the observations to theory is even better when the non-Coulombic contribution 26.105 is included. That contribution is small for the alkali nitrates, but much larger in mixtures with highly polarizable ions, for example, Ag^+ or Tl^+. In such systems there can also occur partial covalent bonding, which simply means that a description of the system as simple polarizable ions is not completely accurate.

It is interesting to compare the simple ideas of the corresponding-states representation of molten salt properties with the results of computer simulations. Consider, first, pure salts. It was shown in Chapter 14 that the lattice energy of alkali halide crystals is well accounted for if the ion–ion interaction is of the Born–Mayer form (Eq. 14.33). This potential is semiempirical in nature; it is an extreme test to require that it account for material properties at temperatures far removed from that where it was fitted. Nevertheless, since the Born–Mayer potential or slight var-

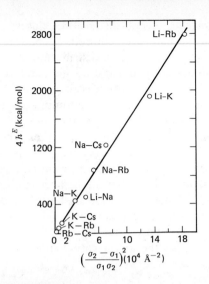

FIGURE 26.19
Plot of $4h^E$ for the alkali nitrates versus the quantity $[(\sigma_2-\sigma_1)/\sigma_1\sigma_2]^2$. The cation pair to which each point corresponds is shown next to each point. From M. Blander, *Adv. Chem. Phys.* **11**, 83 (1967).

TABLE 26.8
THERMODYNAMIC PROPERTIES OF ALKALI HALIDE CRYSTALS AT 298 K AND ZERO PRESSURE[a,b]

Salt	R_0 (Å)		u (kJ/mol)		κ_T (10^{-12} cm²/dyn)		α (10^{-4}/K)		c_p (J/K mol)	
	exp.	MC	exp.	MC	exp.	MC	exp.	MC	exp.	MC
NaF	2.317	2.322	897.0	910.3	2.15	2.88	0.98	1.37	46.8	51.6
KF	2.674	2.681	794.1	798.9	3.28	3.99	1.00	1.20	50.0	49.0
RbF	2.815	2.840	759.0	773.2	3.81	4.41	0.95	1.17	50.5	50.0
LiCl	2.570	2.573	832.2	837.0	3.36	3.67	1.22	1.41	50.3	52.5
NaCl	2.820	2.842	764.0	762.6	4.17	4.36	1.10	1.20	50.8	52.2
KCl	3.147	3.151	693.7	693.1	5.73	6.42	1.01	1.43	51.5	53.5
RbCl	3.291	3.297	666.5	673.3	6.40	6.32	0.99	1.07	51.2	49.9
NaBr	2.989	2.997	726.3	730.5	5.02	5.51	1.19	1.14	52.3	50.1
KBr	3.298	3.314	663.2	663.8	6.75	7.34	1.10	1.30	52.5	50.1
RbBr	3.445	3.444	638.5	651.7	7.69	7.69	1.04	1.01	51.7	46.6

[a] exp. = experimental results; MC = results of Monte Carlo computations.
[b] From D. J. Adams and I. R. McDonald, *J. Phys. C (Solid State Phys.)* **7**, 2761 (1974).

iants of it are the best available ion–ion potentials, we must make just that test. Tables 26.8 and 26.9 show calculated and observed properties of alkali halide crystals, at 298 K and near the melting points of the several substances. The computer simulations are based on a version of the Born–Mayer potential like Eq. 14.33 that neglects the charge-induced dipole part of the interaction but does include the dispersion interactions. Other simulations including the ion-induced dipole term show that it contributes about 0.2% of the lattice energy at 298 K. These calculations also show that inclusion of the charge-induced dipole interaction considerably improves the values of the predicted lattice vibration frequencies. Examination of the entries in Tables 26.8 and 26.9 shows that agreement between computer simulation and experiment is generally better at 298 K than near 1000 K. Note that the volume of the crystal is systematically overestimated. Also, the rather good agreement between computed and observed lattice energies masks the fact that the remaining error is confined to the non-Coulombic contribution. This is so because, for a given lattice, the Coulombic contribution to the lattice energy can be computed exactly. Since it is very much larger than the non-Coulombic contribution, the relative error in the latter is larger than meets the eye on examining the table entries. All in all, the Born–Mayer potential provides a fair description of the thermodynamic properties of the solid, but one that worsens as T increases.

Table 26.9 also has entries for liquid alkali halides. It is seen that the predicted properties are less accurate than was the case for the crystals. The estimated contribution of the charge-induced dipole interaction is 0.5–0.8% of the internal energy of the liquid. Despite the small contribution of the charge-induced dipole interaction to the energy, it does influence the distribution of ions quite a bit, as shown in Fig. 26.20.

We now turn to the case of molten salt mixtures. Given the documented inaccuracy of the Born–Mayer potential, we cannot expect a completely quantitative description to emerge from a computer simulation. Nevertheless, that potential is adequate to test the qualitative notions used in formulating the corresponding-states analysis of molten

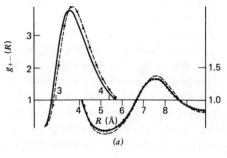

(a)

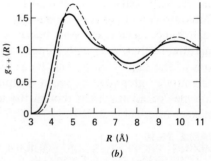

(b)

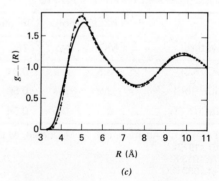

(c)

FIGURE 26.20
The pair-correlation functions for the three pairs of ions in a model of liquid KI. (a) $g_{+-}(R)$: —— rigid-ion model; —— polarizable-ion model. (b) $g_{++}(R)$: —— rigid-ion model; —— polarizable-ion model. (c) $g_{--}(R)$: —— rigid-ion model; —— polarizable-ion model. From G. Jacucci, I. R. McDonald, and A. Rahman, *Phys. Rev.* **A13**, 1581 (1976).

TABLE 26.9
THERMODYNAMIC PROPERTIES OF ALKALI HALIDES AT ZERO PRESSURE AND TEMPERATURE NEAR THE EXPERIMENTAL MELTING POINTS[a,b]

| Salt | T (K) | v (cm³/mol) | | | | −u (kJ/mol) | | | |
| | | crystal | | liquid | | crystal | | liquid | |
		exp.	MC	exp.	MC	exp.	MC	exp.	MC
LiF	1120	11.1		14.3		990.2		963.4	
KF	1130	26.0	27.91	30.4	36.09	768.8	775.3	740.6	758.3
CsF	976	31.8		41.6		701.6		679.8	
LiCl	883	22.4	22.78	28.2	30.35	814.9	819.0	795.0	801.1
NaCl	1073	30.0	31.37	37.6	39.72	739.0	744.2	711.0	717.3
KCl	1045	41.6	42.52	48.8	53.10	669.7	676.3	643.1	649.5
RbCl	995	47.2	47.91	53.9	58.66	645.7	658.4	627.3	629.9
KBr	1007	48.0	49.12	56.0	60.26	639.9	648.3	614.4	622.0
RbBr	953	53.6	54.68	60.9	65.85	618.6	629.8	603.1	603.9

[a] exp. = experimental results; MC = results of Monte Carlo computations.
[b] From D. J. Adams and I. R. McDonald, *J. Phys. C (Solid State Phys.)* **7**, 2761 (1974).

salt mixtures. Consider, for example, the hypothesis that the cations mix randomly amongst themselves in a mixture with a common anion. If this is correct, the pair-correlation functions $g_{Na^+K^+}$, $g_{Na^+Na^+}$, and $g_{K^+K^+}$ (see Chapter 23) should be the same for all compositions of the mixture. As shown in Fig. 26.21, this is not the case; the size difference between Na^+ and K^+ necessarily generates some deviation from random mixing of the Na^+ and K^+ among themselves. The same effect of ionic size can be seen in the distribution of angles between triplets of ions. Let $P_\theta(\theta, R)2\pi R^2 \sin\theta \, dR \, d\theta$ be the probability of finding a third ion, say k, in the volume element $2\pi R^2 \sin\theta \, dR \, d\theta$ around the pair of ions ij when their separation is in the range R to $R + dR$ and the angle between $\mathbf{R}_{ij}$ and $\mathbf{R}_{ik}$ is between θ and $\theta + d\theta$. Figure 26.22 shows, for a simulation of an equimolar KCl/NaCl melt, the $Na^+K^+Na^+$ and $K^+Na^+K^+$ angular distributions. If mixing among the K^+ and Na^+ ions were random, these distributions should peak at the same angle, which they obviously do not. We must conclude, then, that the cations in a molten salt mixture with common anion do not mix randomly among themselves, although the deviation from random mixing is not large.

What can be learned from the computer simulations about the energetics of molten salt mixtures? Because of the inadequacy of the ion–ion potential, only qualitative conclusions can be drawn from the calculated results. These do show that dispersion-energy contributions to the excess enthalpy of mixing are important, and that nearest- and next-nearest-neighbor interactions contribute about equally to h^E because of the delicate balance between the changes in the Coulomb and non-Coulomb components of the energy. The contributions to the entropy by each ionic species are very nearly the same in the pure melt and the mixture, so that, as already hinted, deviations from random mixing among ions of the same charge must be a small effect. Finally, there is order-of-magnitude agreement between computed and observed excess heats and volumes of mixing even though the ion–ion potential energy function is inexact (see Table 26.10).

TABLE 26.10
MIXING PROPERTIES OF EQUIMOLAR ALKALI CHLORIDE MELTS AT $T = 1073$ K, $p = 0$[a]

System	h^E (kJ/mol)		v^E (cm³/mol)	
	MC	exp.	MC	exp.
LiCl/KCl	−3.5	−4.4[b]	0.31	0.30
LiCl/RbCl	−3.9	−5.3[b]	0.50	0.36
NaCl/KCl	0	−0.5[c]	0.54	0.25

[a] From D. J. Adams and I. R. McDonald, *Mol. Phys.* **34**, 287 (1977).
[b] At 1013 K.
[c] At 1083 K.

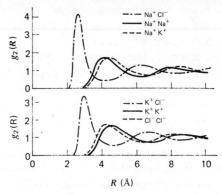

FIGURE 26.21
Pair-correlation functions for NaCl-KCl(*l*) at $T = 1083$ K and zero pressure. Composition of the mixture: $x_{NaCl} = x_{KCl} = 0.5$. From B. Larsen, T. Førland, and K. Singer, *Mol. Phys.* **26**, 1521 (1973).

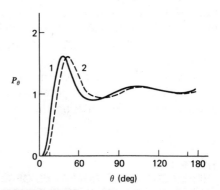

FIGURE 26.22
Selected angular distribution functions for NaCl–KCl(*l*) at $T = 1083$ K and zero pressure. Composition of the mixture: $x_{NaCl} = x_{KCl} = 0.5$. 1: $Na^+K^+Na^+$ distribution, K^+ as central ion; 2: $K^+Na^+K^+$ distribution, Na^+ as central ion. From B. Larsen, T. Førland, and K. Singer, *Mol. Phys.* **26**, 1521 (1973).

Further Reading

Gordon, J. E., *The Organic Chemistry of Electrolyte Solutions* (Wiley, New York, 1975).

Gurney, R. W., *Ionic Processes in Solution* (Dover, New York, 1953).

Hush, N. S., Ed., *Reactions of Molecules at Electrodes* (Wiley, New York, 1971), chap. 2.

Lewis, G. N., and Randall, M., *Thermodynamics*, 2nd ed., rev. by K. S. Pitzer and L. Brewer (McGraw-Hill, New York, 1961), chaps. 22–26 and 34.

McQuarrie, D. A., *Statistical Mechanics* (Harper & Row, New York, 1976), chap. 15.

Marcus, Y., *Introduction to Liquid State Chemistry* (Wiley, New York, 1977), chaps. 6 and 7.

Robinson, R. A., and Stokes, R. H., *Electrolyte Solutions*, 2nd ed. (Butterworths, London, 1959).

Watts, R. O., and McGee, I. J., *Liquid State Chemical Physics* (Wiley, New York, 1976), chaps. 9 and 10.

More Advanced Treatments

Ben-Naim, A., *Water and Aqueous Solutions* (Plenum, New York, 1974).

Friedman, H. L., *Ionic Solution Theory* (Wiley, New York, 1962).

Harned, H. S., and Owen, B. B., *The Physical Chemistry of Electrolytic Solutions*, 3rd ed. (Van Nostrand Reinhold, New York, 1958).

Outhwaite, C. W., *Specialist Periodical Report, Statistical Mechanics* **2,** 188 (1975), The Chemical Society, London.

Resibois, P., *Electrolyte Theory* (Harper & Row, New York, 1968).

Problems

1. Describe how the concentration cell shown in Fig. 26.2 might be used to measure the fugacity of hydrogen at various pressures.

2. Below are given the molalities of a saturated TlCl solution at two temperatures, and the corresponding activity coefficients. Calculate the heat of solution of TlCl.

T (K)	m_{TlCl}	$\gamma_\pm$
298.1	0.0161	0.81
312.8	0.0252	0.77

3. The dissociation pressure of oxygen over silver oxide, Ag_2O, at 298 K is 5×10^{-4} atm. Calculate $\mathscr{E}$ for the cell

$$Ag \,|\, Ag_2O(sat.) \,|\, Pt, O_2(0.2\ atm).$$

4. Calculate the solubility of silver iodide, AgI, in pure water at 298 K. The following cell generates 0.807 V:

$$Ag \,|\, AgI(sat.), KI(0.1\ m) \,|\, AgNO_3(0.1\ m) \,|\, Ag.$$

Regard the solutions as ideal.

5. A cell

$$Cu(amalgam, activity\ a') \,|\, CuSO_4(aq.) \,|\, Cu(amalgam, activity\ a'')$$

generates 0.124 V at 298 K. In the more dilute amalgam, which may be regarded as an ideal solution, the concentration of copper is $1.664 \times 10^{-5}\%$ by weight, and in the other amalgam the copper concentration is 0.04472% by weight. Calculate the activity coefficient of Cu in the more concentrated amalgam.

6. Calculate the cell potential of the cell

$$Cl_2(g, 1 \text{ atm}), Pt \,|\, HCl(10 \, m) \,|\, Pt, O_2(g, 1 \text{ atm})$$

at 298 K. The equilibrium constant of the gas phase reaction

$$4HCl + O_2 \rightarrow 2H_2O + 2Cl_2$$

is $K_p = 10^{13} \text{ atm}^{-1}$, and the partial vapor pressures over the HCl solution in the cell are $p_{H_2O} = 9.4$ torr and $p_{HCl} = 4.2$ torr.

7. The cell potential of a particular cell has the temperature dependence

$$\mathscr{E}(\text{volts}) = 0.96466 + 1.74 \times 10^{-4}(t - 25) + 3.8 \times 10^{-7}(t - 25)^2,$$

where t is the Celsius temperature. Find the heat of reaction at 25°C.

8. How does the potential of a cell depend on the external pressure?

9. A solution of KI has an ionic strength of 0.32. Calculate its concentration. What concentrations of K_2SO_4 and $MgSO_4$ have the same ionic strength?

10. Using the Debye-Hückel limiting law, Eq. 26.86, calculate the ionic activity coefficients γ_+ and γ_- and the mean activity coefficient $\gamma_\pm$ for $10^{-4} M$ and $10^{-3} M$ solutions of NaCl and $CaCl_2$.

11. Obtain a limiting expression for the solubility of a salt (say, AgCl) in the presence of a soluble electrolyte, in terms of the sparingly soluble salt's solubility-product constant and the ionic strength.

12. The value of K_{sp} for AgCl at 25°C is 1.71×10^{-10}. Calculate the solubility of AgCl in pure water and in $10^{-3} M$ $ZnSO_4$.

13. For a reaction such as Eq. 26.39 which can occur in a cell at constant temperature and pressure, obtain a relation between the cell potential $\mathscr{E}$ and the Gibbs free energy change for the reaction.

14. Consider a membrane that is permeable to K^+ but not to Cl^- or H_2O. The membrane separates two solutions of KCl, one of which is $0.1 \, m$ and the other is $0.01 \, m$. A voltage is found to arise spontaneously across the membrane.
(a) What is the source of the voltage?
(b) Calculate the magnitude of the voltage. (*Hint:* Consider Eq. 26.17 and disregard activity coefficients.)
(c) The charge separation Q necessary to produce a potential difference across an electric capacitor is given by $Q = CV$, where C is called the capacitance. For a biological membrane whose area is 1 cm², the value of C is of the order 1 μF. (If C is given in F (farads), then V is in volts and Q in coulombs.) Calculate the number of ions that must pass through a 1-cm² membrane to produce the voltage found in part (b).
(d) Suppose that the two KCl solutions have volumes of 1 cm³ each. How does the number of ions found in part (c) compare to the total number of ions present? Is the charge imbalance implied by the potential difference detectable by chemical means?

15. The following values of $\gamma_\pm$ have been found for NaCl at 25°C:

m	0.001	0.002	0.005	0.10	0.20
$\gamma_\pm$	0.9649	0.9519	0.9275	0.9024	0.8712

Show that the above data are consistent with the limiting law, Eq. 26.86.

16. For a particular reaction occurring in an electrochemical cell, the cell poten-

tial was found to vary with temperature according to

$$\mathscr{E} = 0.134 - 2.94 \times 10^{-4}t - 4.20 \times 10^{-6}t^2,$$

where $\mathscr{E}$ is in volts and t is the Celsius temperature. Find ΔH, ΔS, and ΔG for the reaction at 25°C.

17. The potential of the cell

$$\text{Pt, } H_2(1 \text{ atm}) \,|\, HCl(0.01 \text{ m}) \,|\, Cl_2(p \text{ atm}), \text{ Pt}$$

was measured at 25°C for various values of p, the pressure of chlorine. The results, with $\mathscr{E}$ in volts, were

p	1	50	100
$\mathscr{E}$	1.5962	1.6419	1.6451

What is the fugacity of Cl_2 at each pressure? What assumption must you make about the hydrogen fugacity to do this problem? How might you test this assumption?

18. What is the potential of the cell

$$\text{Pt, } H_2(1 \text{ atm}) \,|\, HCl(0.1 \text{ m}) \,|\, HCl(0.2 \text{ m}) \,|\, H_2(10 \text{ atm}), \text{ Pt?}$$

19. What is the mean ionic activity of K_2SO_4 in terms of the molality of the salt?

20. Calculate the ionic strength (Eq. 26.49) of the following solutions: 0.1 m KCl, 0.1 m K_2SO_4, 0.1 m $MgSO_4$.

21. The potential of the following cell was measured at various HCl molalities at 20°C:

$$\text{Pt, } H_2(\text{unit activity}) \,|\, HCl(\text{sol'}n) \,|\, AgCl \,|\, Ag.$$

The voltages obtained were

m	0.005314	0.008715	0.013407	0.021028	4.0875
$\mathscr{E}$	0.49395	0.46987	0.44899	0.42726	0.12307

Find $\mathscr{E}^\theta$ for this cell and $\gamma_\pm$ for HCl at $m = 4.0875$.

22. Show that $\mathscr{E}^\theta$ for the cell described in Eq. 26.42 can be obtained more accurately by the extrapolation

$$\mathscr{E}^\theta = \lim_{m_\pm \to 0} \left(\mathscr{E} - \frac{RT}{2\mathscr{F}} \ln a_{H_2} + \frac{2RT}{\mathscr{F}} \ln m_\pm - \text{constant } m_\pm^{1/2} \right).$$

23. Use the data of Problem 21 to calculate K_{sp} for AgCl.

24. Suppose that two electrolyte solutions, I and II, are separated by a rigid semipermeable membrane that allows passage of some, but not all, ions. At equilibrium we expect both a pressure difference and an electric potential difference between I and II. Using the method introduced to discuss osmotic pressure, show that

$$\ln \frac{(m_i \gamma_i)^{\text{II}}}{(m_i \gamma_i)^{\text{I}}} + \frac{Z_i \mathscr{F}}{RT} (\psi^{\text{II}} - \psi^{\text{I}}) = \frac{M_1}{1000} \frac{\bar{v}_i}{\bar{v}_1} \left(g_1{}^{\text{I}} \sum_i m_i{}^{\text{I}} - g_1{}^{\text{II}} \sum_i m_i{}^{\text{II}} \right).$$

25. Show that, at constant temperature and pressure, the osmotic coefficient and ionic activity coefficients are related by

$$-(\nu_+ + \nu_-)m \, dg_1 + (1 - g_1)(\nu_+ + \nu_-) \, dm + \nu_+ m \, d \ln \gamma_+ + \nu_- m \, d \ln \gamma_- = 0,$$

where m is the molality of the salt $M_{\nu_+}X_{\nu_-}$. Then show that

$$-\ln \gamma_\pm = (1 - g_1) + \int_0^m (1 - g_1) \, d \ln m.$$

Hint: Use the Gibbs–Duhem equation.

26. Given that the osmotic coefficient for an electrolyte solution has the form

$$1 - g_1 = a_1 m^{1/2} + a_2 m + a_3 m^{3/2},$$

what is the mean activity coefficient of the solute?

27. Suppose that a 1:1 ionic salt is accurately described by the Born–Mayer model, and that the Born model for ionic hydration accurately describes the changes that occur when the salt dissolves. Show that the expected Gibbs free energy of solution of a series of 1:1 ionic salts with common anion has a maximum with respect to cation radius. What are the consequences of this result for ionic solutions? *Hint:* Treat the ionic solution as ideal.

28. Consider a dilute gas of positive and negative ions (a plasma). Show that the second virial coefficient for such a gas does not exist (diverges). The second virial coefficient of a gas is defined by (see Chapter 21)

$$B(T) = -2\pi \int_0^\infty \left[\exp\left(-\frac{u_2(R)}{k_B T} \right) - 1 \right] R^2 \, dR.$$

29. The McMillan–Mayer theory of solutions leads to a virial expansion for the osmotic pressure. The virial coefficients found have the same functional form as for a gas, but the direct intermolecular potential is replaced by the potential of mean force. Show that when the Debye–Hückel approximation to the potential of mean force is used in

$$B(T) = -2\pi \int_0^\infty \left[\exp\left(-\frac{w_2(R)}{k_B T} \right) - 1 \right] R^2 \, dR,$$

the Debye–Hückel limiting law is recovered.

30. The charge density around the ith ion is, according to Debye–Hückel theory,

$$q_i(R) = -\frac{Z_i e \kappa^2}{4\pi R} \cdot \frac{e^{-\kappa(R-a)}}{(1+\kappa a)}.$$

Find the most probable value of R, and the mean value of R, for this distribution of charge. What physical insight into the importance of κ does this calculation give?

31. The pair-correlation function in Debye–Hückel theory is, for very low concentration where $\kappa a \to 0$,

$$g_{ij}(R) = 1 - \frac{Z_i Z_j e^2}{4\pi \epsilon_0 D k_B T} \cdot \frac{e^{-\kappa R}}{R}.$$

Use this result in the general equation

$$\frac{U}{V} = \frac{3}{2} n k_B T + 2\pi \sum_{i=1}^2 n_i n_j \int_a^\infty g_{ij}(R) u_{ij}(R) R^2 \, dR$$

to derive the limiting-law properties of the ionic solution. Note that $u_{ij} \to \infty$ for $R \le a$, and u_{ij} is the direct Coulomb interaction $Z_i Z_j e^2 / 4\pi \epsilon_0 DR$ for $R > a$.

32. Show that the osmotic coefficient, g_1, for a 1:1 electrolyte solution is, according to Debye–Hückel theory,

$$1 - g_1 = \frac{\sum n_i Z_i^2 e^2 \kappa}{24\pi \epsilon_0 D k_B T \sum n_i}.$$

33. Show that the total electrical work required to charge simultaneously all of the ions in an electrolyte solution is

$$w = -\sum \frac{N_i Z_i^2 e^2 \kappa}{12\pi \epsilon_0 D} = -\frac{V k_B T}{12\pi} \kappa^3.$$

Use the Gibbs–Helmholtz equation to calculate, from w, the integral heat of dilution of the electrolyte solution. On what solution parameters does the integral heat of dilution depend? Is it altered if the dielectric constant D depends on temperature?

34. The Bjerrum–Fuoss theory of ion association was originally introduced to extend the domain of validity of the analysis of electrolyte solutions. The idea is as follows: The number of ions j a distance R from one at the origin is $n_j g_{ij} \, d\mathbf{R}_j$ or $4\pi R^2 n_j \exp(-w_{ij}/k_B T) \, dR_j$. For ions that are very close together, Bjerrum and Fuoss make the approximation $w_{ij} = Z_i Z_j e^2/4\pi\epsilon_0 DR$. The number of ions j surrounding i goes through a minimum at $q_0 \equiv |Z_i Z_j|/8\pi\epsilon_0 D k_B T$. It is now asserted that all ions separated by less than q_0 are "associated," and all others free. The associated ions are treated as ideal neutral solute particles, whereas the free ions interact as in Debye–Hückel theory.

(a) Let $1 - \alpha$ be the degree of association of ions. Show that for a 1:1 electrolyte

$$1 - \alpha = \frac{4\pi N}{V} \left(\frac{e^2}{4\pi\epsilon_0 D k_B T} \right)^3 Q(b),$$

where

$$Q(b) = \int_2^b e^y y^{-4} \, dy,$$

$$b \equiv \frac{e^2}{4\pi\epsilon_0 D a k_B T},$$

$$y \equiv \frac{e^2}{4\pi\epsilon_0 D R k_B T}.$$

(b) Take the Debye–Hückel ionic size parameter a to be equal to q_0, and treat the relation between free and associated ions as an equilibrium. Show that

$$\frac{1-\alpha}{\alpha^2} = \gamma_\pm^2 \frac{4\pi N}{V} \left(\frac{e^2}{4\pi\epsilon_0 D k_B T} \right)^3 Q(b).$$

(c) Given the following values for $1 - \alpha$ with $a = 1.76$ Å, calculate $\gamma_\pm$ for the electrolyte solution and compare these values with those deduced from Debye–Hückel theory.

m	0.001	0.005	0.01	0.05	1.10	0.50
$1-\alpha$	0.001	0.007	0.012	0.046	0.072	0.204

The theory of association has been very considerably extended beyond the simple model considered in this problem.

35. The arguments used in this chapter suggest that assuming

$$g_{ij}(R) = g_{ij}^{HS}(R) \exp\left(-\frac{w_{ij}^{DH}(R)}{k_B T} \right)$$

is probably a good approximation for the primitive model of an electrolyte solution. Here $g_{ij}^{HS}(R)$ is the pair-correlation function for hard spheres of diameter $2a$, and

$$w_{ij}^{DH}(R) = \frac{Z_i Z_j e^2 e^{-\kappa(R-a)}}{4\pi\epsilon_0 DR(1+\kappa a)}.$$

Assume that all ions have the same size, and calculate, for a 1:1 electrolyte, the excess energy and free energy of the solution. For the concentrations considered (say, $0.1\,m$ or less), it is adequate to use the approximation

$$g_{ij}^{HS}(R) = 0 \qquad \text{for } R \leqslant 2a,$$

$$= 1 + \frac{4\pi}{3} (8na^3)\left[1 - \frac{3}{8}\left(\frac{R}{a}\right) + \frac{1}{128}\left(\frac{R}{a}\right)^3 \right]$$

$$\text{for } 2a < R \leqslant 4a,$$

$$= 1 \qquad \text{for } R \geqslant 4a.$$

PART THREE

Physical and Chemical Kinetics

In the first part of this book we studied the structure of atoms and molecules. We investigated the electronic structure of such systems and found it necessary to use quantum mechanics. Solutions of the Schrödinger equation gave us the maximum allowable amount of information about the mechanical system, usually consisting of a small number of particles. For the most part we were interested in stationary solutions of the Schrödinger equation, which can yield information about energies, force constants, bond strengths, molecular geometry, and other time-independent properties. However, we were also concerned with time-dependent behavior of mechanical systems, and here we sought solutions for transition probabilities of a given system from one stationary state to another.

In the second part, on statistical thermodynamics, we studied aggregates of atoms and molecules for the purpose of finding the equilibrium properties of macroscopic systems. In order to do so we needed some information on the mechanical behavior of the system, such as the energy spectrum of the harmonic oscillator, for instance, or in general the density of energy states. For some of the discussion we found classical mechanics sufficient, in which case we could replace summations over energy states by integrations over phase space. Although we found it necessary to define reversible and irreversible processes, we carefully calculated thermodynamic changes along reversible paths only.

In this third part of the book, we turn to the evolution in time of processes in aggregates of molecules. This subject is called *kinetics*.

Let us discuss a few examples of such processes. Consider water in a container open to the atmosphere at room temperature. We know that in time the water will evaporate, and we know how to calculate the changes in thermodynamic state functions for the evaporation. Such changes depend only on the initial and final states of the process, and not on the particular path or the time required to complete the process. But often we also want to know such things as how quickly the evaporation occurs, that is, what the *rate* of the process is. On what variables does this rate depend? It is easy to guess that the rate of evaporation depends on the temperature of the water; does it depend on the humidity of the air in the room?

Suppose that a gas of diatomic molecules, for example nitrogen, is heated very rapidly to a high temperature. The heating can be accomplished so rapidly, say by the passage of a shock wave through the gas, that the average kinetic energy of the molecules is increased within 10^{-6} s; within that interval, however, the average vibrational energy of the molecules will not change. How long will it take for complete equilibrium to be achieved by the transfer of energy from translational to vibrational (or other) degrees of freedom? On what variables does this process depend? Does it depend on temperature, pressure, structure of the molecules, the surface of the container? Is it sensitive to impurities present in the gas?

Suppose H_2 and O_2 gases are present in a vessel at 20°C and a given pressure. We can calculate the Gibbs free energy change for conversion of these gases to water vapor at the same temperature and pressure. Since the Gibbs free energy of the water vapor is lower than that of the reactants, the process is "spontaneous" and "irreversible." But one would grow impatient standing in front of that vessel waiting for the reaction to occur, because essentially nothing will happen. Make an electric spark in the mixture, however, or put into it some platinum powder, and the reaction will occur quickly, even explosively. In any case, what we often want to know is the rate of the reaction, how it depends on temperature, on the structure of the molecules, on the presence of foreign substances. In still more detail, how many molecules have to collide, and in what particular way, in order to bring about the reaction? We ask, "What is the mechanism of the reaction?" The net stoichiometric equation

$$2H_2(g) + O_2(g) = 2H_2O(g)$$

may or may not tell us anything about the mechanism. Do two hydrogen molecules collide with one oxygen molecule simultaneously, and two water molecules leave the collision zone? If so, the stoichiometric equation tells us all the chemical species involved in the reaction process. Does an oxygen molecule have to dissociate first to atoms, and these then collide with H_2? If so, then we have a two-step mechanism,

$$O_2 = 2O,$$

$$O + H_2 = H_2O,$$

but the stoichiometric equation tells us nothing about the possible presence of oxygen atoms. What are the forces between the various species? How do these forces determine the collision dynamics? What is the rate of each reaction step in the mechanism, and how is this *average* rate related to the nature of the individual collisions?

The student will see, from just the few examples cited, how complicated time-dependent processes can be. The study of such processes involves a number of important topics which we shall discuss: first, the simple geometric aspects of molecular motion (kinematics), forces between molecules, collisions of molecules, and other subjects properly classified under mechanics; second, the average rate of change of properties such as the concentration, temperature, pressure, entropy (all representative of many-molecule systems), and other subjects belonging to a time-dependent statistical thermodynamics.

27

Molecular Motion and Collisions

The three processes cited in the introduction, and others like them, can be described in terms of the change with time of some macroscopic observable quantity: the amount of liquid evaporating per unit time, the change of translational temperature per unit time, and the change of concentration of chemical reactant per unit time. In each case the change in the macroscopic quantity is descriptive of the averaged molecular behavior. If there were no molecular motion and no molecular collisions, no macroscopic change could occur. If the molecules of a vapor do not collide with the surface of a liquid, there can be no condensation. Or if a molecule in a liquid does not acquire, through collision with other molecules, enough kinetic energy to escape from the surface, there can be no evaporation. Transfer of energy from one degree of freedom to another, for example, from translation to vibration, can occur only by collisions. Chemical reactions occur upon collisions between molecules of reactants, collisions that usually must be relatively violent if the reactant molecules are to approach each other closely enough for rearrangements of chemical bonds to occur.

It is clear that the detailed dynamics of molecular collisions must determine the rate of a process. Therefore, unlike the study of the equilibrium properties of matter, where we needed only the stationary energy spectrum of a system, the study of time-dependent properties requires a detailed study of molecular dynamics.

We consider the simplest case first: a molecule represented as a point moving in a given direction with a given speed. A collision of one molecule with another is defined as an event during which the velocity of each particle is changed. If we consider the motion of the molecules to be described by classical mechanics, then Newton's laws imply that an interaction must occur between the particles; that is, a force must act between them for any change of velocity to occur. A change of velocity implies an acceleration $d\mathbf{v}/dt$, which, according to Newton's second law of motion, is proportional to the force acting on the particle.

27.1 Kinematics

Consider now two point particles, one of mass m_1, the other m_2, with corresponding velocities $\mathbf{v}_1$, $\mathbf{v}_2$. These velocities are measured relative to some fixed frame of reference, and so are called *laboratory* velocities. What can be said about a collision between these two particles without knowing the particular forces between them? The statements that follow, taken to be postulates, form the content of *kinematics*. First, we assume conservation of mass; the total mass $m_1 + m_2$ is the same before and after collision,

$$m_1 + m_2 = m_1' + m_2' \qquad (27.1)$$

(primes are used to designate quantities after a collision), and also we must have

$$m_1 = m_1', \qquad m_2 = m_2' \qquad (27.2)$$

if no reaction takes place. This is a trivial observation once the conservation of mass is accepted. (Actually mass and energy are jointly conserved, with the changes related by Einstein's equation $\Delta m = \Delta E/c^2$. But the mass change Δm is negligible in ordinary chemical reactions, becoming significant only in high-energy nuclear reactions.) Second, we assume the conservation of linear momentum, implying that

$$m_1\mathbf{v}_1 + m_2\mathbf{v}_2 = m_1'\mathbf{v}_1' + m_2'\mathbf{v}_2'. \qquad (27.3)$$

By using the conservation of mass, we can rewrite Eq. 27.3 in the form

$$m_1\mathbf{v}_1 + m_2\mathbf{v}_2 = m_1\mathbf{v}_1' + m_2\mathbf{v}_2'. \qquad (27.4)$$

Of course, a vector equation such as Eq. 27.4 is shorthand for three equations:

$$m_1 v_{1x} + m_2 v_{2x} = m_1 v_{1x}' + m_2 v_{2x}',$$
$$m_1 v_{1y} + m_2 v_{2y} = m_1 v_{1y}' + m_2 v_{2y}', \qquad (27.5)$$
$$m_1 v_{1z} + m_2 v_{2z} = m_1 v_{1z}' + m_2 v_{2z}'.$$

Third, we assume the conservation of energy. Before the collision, the two particles are sufficiently far apart that there is no interaction between them, and the two-particle system has only kinetic energy. The same is true after the collision has occurred and the particles have separated. Thus we have

$$\tfrac{1}{2}m_1 v_1^2 + \tfrac{1}{2}m_2 v_2^2 = \tfrac{1}{2}m_1 v_1'^2 + \tfrac{1}{2}m_2 v_2'^2. \qquad (27.6)$$

Note that energies are scalar quantities. A fourth set of conditions can be obtained from the principle of conservation of angular momentum, but we are not yet ready to consider this.

We can transform the description of the motion of two particles into that of one particle in a field of force. Let the position coordinate of particle 1, relative to some arbitrary origin, be $\mathbf{r}_1$, and the force exerted (by particle 2) on particle 1 be $\mathbf{F}_1$, both of these quantities being vectors; we similarly define $\mathbf{r}_2$ and $\mathbf{F}_2$. Newton's third law requires that

$$\mathbf{F}_1 = -\mathbf{F}_2, \qquad (27.7)$$

and by Newton's second law we can write the equations of motion

$$\mathbf{F}_1 = m_1 \frac{d^2\mathbf{r}_1}{dt^2}, \qquad \mathbf{F}_2 = m_2 \frac{d^2\mathbf{r}_2}{dt^2}, \qquad (27.8)$$

$d^2\mathbf{r}_1/dt^2$ and $d^2\mathbf{r}_2/dt^2$ being the accelerations. By combining Eqs. 27.7 and 27.8, we obtain

$$\frac{d^2\mathbf{r}_2}{dt^2} - \frac{d^2\mathbf{r}_1}{dt^2} = \frac{d}{dt}\left(\frac{d\mathbf{r}_2}{dt} - \frac{d\mathbf{r}_1}{dt}\right) = \left(\frac{1}{m_2} + \frac{1}{m_1}\right)\mathbf{F}_2. \qquad (27.9)$$

At this point it is convenient to define the relative position vector

$$\mathbf{r} \equiv \mathbf{r}_2 - \mathbf{r}_1, \qquad (27.10)$$

as shown in Fig. 27.1, and its time derivative, the relative velocity

$$\mathbf{v} \equiv \mathbf{v}_2 - \mathbf{v}_1. \qquad (27.11)$$

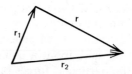

FIGURE 27.1
Definition of relative position vector **r**.

If we now drop the subscript on $\mathbf{F}_2$, and understand $\mathbf{F}$ to be the force between particles 1 and 2, we obtain the simple result

$$\mu\frac{d\mathbf{v}}{dt}=\mathbf{F},\qquad(27.12)$$

where

$$\mu\equiv\frac{m_1 m_2}{m_1+m_2}\qquad(27.13)$$

is the reduced mass. Now, what does Eq. 27.12 tell us? It is the equation of motion of a single particle with mass μ in the field of force $\mathbf{F}$. In fact, we can go a bit further along this line. Recall that linear momentum is conserved (cf. Eq. 27.3). Let us consider a point of mass $M\equiv m_1+m_2$ moving with a velocity $\mathbf{V}$ such that its momentum equals the total momentum of our two particles:

$$m_1\mathbf{v}_1+m_2\mathbf{v}_2=(m_1+m_2)\mathbf{V}.\qquad(27.14)$$

The motion of this point corresponds to that of the center of mass of the two-particle system. The center-of-mass motion is usually uninteresting. It is the *relative* motion of the two particles that counts in a collision. Having started with velocities $\mathbf{v}_1$ and $\mathbf{v}_2$, and having defined the velocities $\mathbf{v}$ and $\mathbf{V}$, we can solve Eqs. 27.11 and 27.14 for one set of variables in terms of the other set, obtaining

$$\mathbf{v}_1=\mathbf{V}-\frac{m_2}{m_1+m_2}\mathbf{v},$$
$$\mathbf{v}_2=\mathbf{V}+\frac{m_1}{m_1+m_2}\mathbf{v}.\qquad(27.15)$$

These vector equations have an easy pictorial interpretation. Consider the simple case that $\mathbf{v}_1$ and $\mathbf{v}_2$ are perpendicular to each other. Then Fig. 27.2 shows the vector relations embodied in Eqs. 27.15.

Next let us substitute Eqs. 27.15 into the expression for the total energy,

$$\tfrac{1}{2}m_1 v_1{}^2+\tfrac{1}{2}m_2 v_2{}^2,$$

where v_1 and v_2 are given by equations of the form

$$v_1{}^2=v_{1x}{}^2+v_{1y}{}^2+v_{1z}{}^2.\qquad(27.16)$$

We find that

$$\tfrac{1}{2}m_1 v_1{}^2+\tfrac{1}{2}m_2 v_2{}^2=\tfrac{1}{2}(m_1+m_2)V^2+\tfrac{1}{2}\mu v^2.\qquad(27.17)$$

The kinetic energy of the two-particle system is just the sum of the kinetic energy arising from the center-of-mass motion and the kinetic energy arising from the relative motion. For our purposes, we can often neglect the center-of-mass motion and concentrate on the relative motion, because the center-of-mass velocity is not changed in a collision between the two particles. The velocity $\mathbf{V}$ can change only if outside forces are brought to bear on the two-particle system. Thus the relative kinetic energy is itself conserved,

$$\tfrac{1}{2}\mu v^2=\tfrac{1}{2}\mu v'^2,\qquad(27.18)$$

so that $v=v'$.

What *is* changed in a collision? In a collision between two point particles, only their velocities can be changed, because they have no internal structure in which to store energy. The magnitude of the relative

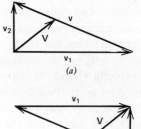

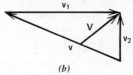

FIGURE 27.2
Relation of laboratory velocities $\mathbf{v}_1$, $\mathbf{v}_2$ to center-of-mass velocity $\mathbf{V}$ and relative velocity $\mathbf{v}$; parts (a) and (b) are equivalent representations.

velocity is invariant, by Eq. 27.18. We can draw, as in Fig. 27.3, the velocities of particles 1 and 2 before and after a collision. Remember that **V** remains the same, as does the magnitude of **v**. In general, the direction and magnitude of both $\mathbf{v}_1$ and $\mathbf{v}_2$ are changed by the collision. Thus, in a collision between structureless point particles, momentum and kinetic energy can be transferred from one particle to another. Any collision in which *only* momentum and kinetic energy are transferred is called "elastic." In the elastic collision shown in Fig. 27.3, particle 1 changes its direction of motion through an angle α_1 and particle 2 through an angle α_2. If we could observe a single particle (as in principle can be done in classical mechanics), we would see particle 1 before the collision traveling in the direction of $\mathbf{v}_1$ and after the collision in the direction of $\mathbf{v}_1'$. We say that particle 1 is *scattered*[1] through the laboratory angle α_1. The relative velocity vector is turned by the collision through an angle χ, the relative scattering angle.

To illustrate the advantage of using relative coordinates, let us plot a collision as shown in Fig. 27.4. (Compare Fig. 2C.1, which illustrates the limiting case of one particle much heavier than the other.) We again show the relative velocities **v** and **v′** and the relative scattering angle χ. Now, however, we have placed these vectors in alternative positions corresponding to the actual collision trajectory. Initially, before the collision, a fictitious particle with mass μ and velocity **v** moves in the direction indicated toward a center of force located at O. This particle then may be pulled toward O by attractive forces or pushed away by repulsive forces, but the net effect of the collision is the deflection of the velocity vector **v** into **v′** through an angle χ. If there were no forces of interaction, the fictitious particle would pass point O (or the real particles each other) in a straight line. The distance of closest approach would then be some value b, which we call the *impact parameter*; similarly, we can define b' by extrapolating the final trajectory backwards.

Now, what role does the angular momentum play in the collision process? We have mentioned that angular momentum is conserved in a collision. The initial relative angular momentum around point O has a magnitude $\mu v b$ (cf. Appendix 2C). Since both μ and v are the same before and after the collision, this must also be true for the impact parameter b, if angular momentum is to be conserved. We thus have $b = b'$.

It is clear that the impact parameter is an important quantity in the specification of a collision. If b is large relative to the range of the force between the molecules, then the molecules hardly interact and are but little deflected from their original paths. If the impact parameter is small, the interaction and deflection may be large.

In summary, if we have two point particles of masses m_1 and m_2, what other variables are at our disposal for studying their collisions? The velocities $\mathbf{v}_1$ and $\mathbf{v}_2$, or, equivalently, the center-of-mass velocity **V** and relative velocity **v**; and the impact parameter b. If we are not interested in how fast the two particles move through space but only in the collision itself, there remain only two variables: v and b, or $E = \frac{1}{2}\mu v^2$ and b.

Finally, we note that for elastic scattering there are only three quantities whose sums for the two particles remain invariant before and after the collision. These quantities are the mass, Eq. 27.1, the linear momentum, Eq. 27.3, and the kinetic energy, Eq. 27.6; they are called *summational invariants*.

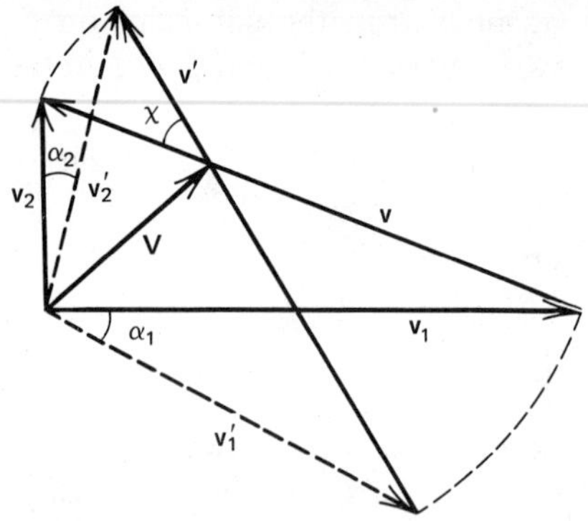

FIGURE 27.3
Velocities before and after an elastic collision.

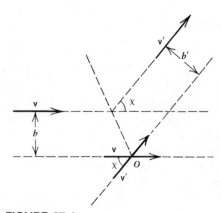

FIGURE 27.4
Relative velocity vector before (**v**) and after (**v′**) an elastic collision. The relative scattering angle is χ, and the impact parameter is b.

[1] *Scattering* is the general term used to describe the results of collisions.

We have gone about as far as we can with the pure kinematics of the two-particle system; to proceed further, we must give the particles more properties, so that we can discuss forces between molecules and the effect of these forces on collisions. This is the subject of *dynamics*.

We shall use the concepts of forces and potentials (of interaction); let us now review their meaning. Suppose that two particles repel each other, that is, they interact with a force of repulsion. Let that force be zero when the two particles are infinitely far apart, and vary with distance as they are brought together. For this example we take the initial relative kinetic energy to be zero. It is clear that work must be done to bring the two particles together from infinity to some distance R. Thus the energy of the two-particle system at a separation R is greater than the energy at infinite separation, by an amount exactly equal to the work done on the system. The zero of potential energy is conveniently chosen at infinite separation of the pair, so that

$$V(R) = E(R) - E(\infty), \qquad (27.19)$$

where $E(R)$ is the energy of the system at separation R. This equation of course applies equally well if the interaction is attractive. The force required to bring the particles together is in the opposite direction to the force of interaction between them. Hence the work done on the system (cf. Eq. 2.55) is

$$W = -\int_{\infty}^{R} F(R)\, dR = V(R), \qquad (27.20)$$

equal to the potential energy. The relation between force and potential energy is thus

$$F(R) = -\frac{dV(R)}{dR}. \qquad (27.21)$$

Note that we have assumed the force and potential energy to depend only on the distance R between the particles, with the force vector directed along the line between them; such a force is called a *central force*.

In principle, as we outlined in Section 10.1, one can calculate the potential energy of two molecules by means of quantum mechanics. Consider two helium atoms. At infinite separation one can calculate the electronic energy of each atom. Even this calculation cannot be expressed in terms of known mathematical functions, but can be carried to completion numerically. If this calculation is repeated for a finite separation between the atoms, account must be taken of the Coulomb interaction of each electron, not with one nucleus, but rather with two nuclei. The system under consideration consists of two nuclei at a fixed separation and four electrons around them. Second, there is a nucleus–nucleus Coulomb repulsion, and electron–electron Coulomb repulsions between all possible electron pairs. Thus the interaction is quite complex, and it is not surprising that an accurate calculation of the energy at a given internuclear distance is an elaborate procedure requiring high-speed computers. Calculations have been made, to various approximations, for simple systems such as He–He. For more complicated cases theoretical predictions, although still meager, are becoming increasingly available. Information about intermolecular forces can be obtained from experimental studies of scattering; and many macroscopic properties, which can be expressed in terms of averages over molecular interactions and collisions, can be used to obtain information about the intermolecular potential.

If you know little about a subject, you make guesses, construct models, try them out, and see how they work. Moreover, in your guesses you will usually be forced to make drastic approximations, at least at first. Refinements come later on.

We do in fact know something about molecular interactions, to which we devoted Chapter 10. (The remainder of this section is basically a brief review of material already covered there and elsewhere.) All substances can be condensed from the gaseous to the liquid and solid states, which implies that molecules must attract each other in order to form more or less permanent aggregates. Also, all substances in the liquid and solid state resist compression. Thus we may postulate that at small intermolecular distances molecules repel each other. In pictorial terms, if one neon atom is forced to be very close to another neon atom, the net force between the two atoms is one of repulsion. Yet neon can be condensed to a liquid, so that at the average distance between neon atoms in the liquid there must be some attraction between the atoms. When two neon atoms are infinitely far apart, their interaction energy is zero. The potential energy, then, must behave more or less like the actual Ne–Ne curve shown in Fig. 27.5. The depth of the potential well, ϵ, is not very large in the case of neon atoms. It takes only about 0.03 eV to separate two neon atoms from the distance R_m (at which the energy is a minimum) to infinity. Work must be done in order to bring about this separation, but not much work. One might say that the "bond" holding two neon atoms together is weak. In general, forces of attraction between chemically inert substances (van der Waals forces) are weak.

Let us compare the neon–neon interaction potential with that of two hydrogen atoms, also shown in Fig. 27.5. We take the spin of the electron on one H atom to be opposite to that of the other. In this case the potential well has a depth of 4.8 eV, more than three orders of magnitude larger than for Ne–Ne. Thus H_2 is indeed a very stable molecule. Chemical valence forces are generally stronger than van der Waals forces.

Since it is so difficult to calculate intermolecular forces, we attempt to represent them by guesses. We must guess both the analytic form of the potential and the numerical value of any parameters that appear in the potential. (We shall see later how one can estimate potential parameters by comparing the predictions made for an assumed potential with the measured properties.) In Sections 10.2 and 21.7 we described some of the most commonly used empirical potentials, that is, guesses as to the form of the potential. Perhaps the simplest guess, and the one most widely used so far, consists simply of supposing that molecules behave like billiard balls or hard elastic spheres. The only parameter in this potential is the diameter of the sphere, d, as shown in Fig. 27.6. Clearly, the hard-sphere potential represents only the repulsive forces between molecules and neglects completely the attractive forces. A more realistic guess is the Lennard–Jones potential, of which the most common form is

$$V(R) = 4\epsilon\left[\left(\frac{\sigma}{R}\right)^{12} - \left(\frac{\sigma}{R}\right)^{6}\right]. \tag{27.22}$$

In Eq. 27.22 there are two parameters, ϵ and σ, where σ is the value of R for which $V(R) = 0$ (see Fig. 27.6), and the potential gives a reasonable representation of both attractive and repulsive forces. The choice of the index 6 in Eq. 27.22 has some theoretical justification, but the repulsive index 12 is based in part on mathematical convenience.

Thus far we have discussed the potential energy of two molecules at a fixed distance. We now once more consider a collision. Let the initial relative kinetic energy be $E = \frac{1}{2}\mu v^2$, which is the total energy of the two

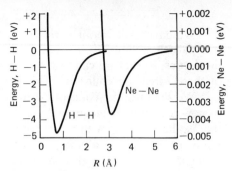

FIGURE 27.5
Potential energy curves for typical physical (Ne–Ne) and chemical (H–H) interactions. Note the difference in scale for the two curves.

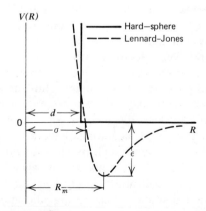

FIGURE 27.6
Comparison of hard-sphere potential and Lennard–Jones potential.

particles if each is considered to be structureless. As the particles come together and begin to interact, they will gain or lose potential energy, depending on the force between them. We assume that there is no outside work done on the system, so that the total energy must be conserved. The kinetic energy then must change in accordance with the equation

$$E = \text{kinetic energy} + \text{potential energy}. \qquad (27.23)$$

If, for instance, the particles repel each other at a given distance, the kinetic energy will be less than at infinite separation.

For the simple case of the hard-sphere potential we can easily obtain the relations between the scattering angle, the initial relative kinetic energy, and the impact parameter. We consider the scattering of one hard sphere from another, each of diameter d and mass m. We represent the collision in relative coordinates in Fig. 27.7; this is similar to Fig. 27.4, but for the hard-sphere case we can immediately draw the entire trajectory. At impact the centers of the hard spheres must be a distance d apart (Fig. 27.8), so that the relative motion is equivalent to that of a mass point bouncing off a stationary hard sphere of *radius d*, centered at O. In Fig. 27.7 we define the angles β_1 and β_2 by extending the line OA beyond the impact point A, and the angle α (OAB) by drawing AB perpendicular to $\mathbf{v}$.

What relations can we obtain among the various collision parameters? First we note that if the impact parameter b exceeds the distance d, no collision can occur. Second, until the very instant of impact there are no forces between the two hard spheres, and hence they travel in straight lines; after the (instantaneous) collision, the trajectory is again linear. Third, one need not be a veteran billiard player to recognize that the angles β_1 and β_2 must be equal. The "reflection" of one sphere from the other is specular. (That's just how you aim in a "straight" billiard shot.) Hence we have simply

$$\beta_1 = \frac{\pi - \chi}{2}, \qquad (27.24)$$

and, since

$$\pi = \alpha + \frac{\pi}{2} + \beta_1, \qquad (27.25)$$

we obtain

$$\alpha = \frac{\chi}{2} \qquad (27.26)$$

and, finally, for $b \le d$,

$$b = d \cos \frac{\chi}{2}. \qquad (27.27)$$

If the impact parameter equals the hard-sphere diameter, then the scattering angle is zero, and there is no deflection. But if the impact parameter is zero, there is a head-on collision, and $\chi = \pi$, as expected. It is interesting that no term in Eq. 27.27 depends on the initial relative kinetic energy. This simplicity does not hold for more complicated interaction potentials.

What happens to the kinetic and potential energies on impact? To answer this question we draw Fig. 27.9, showing the relative velocity vectors immediately before and after impact. We can decompose the

27.3
Collision Dynamics

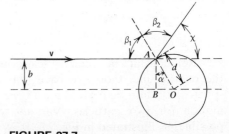

FIGURE 27.7
Scattering of hard spheres of diameter d, in relative coordinates.

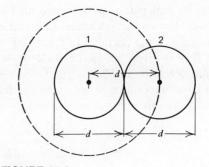

FIGURE 27.8
Impact between two hard spheres. The dashed line (circle of radius d) indicates the minimum distance to which the center of sphere 2 can approach the center of sphere 1.

relative velocity at any time into two components, one in the direction OA, the other in the perpendicular direction AC—that is, respectively parallel and perpendicular to the line of centers at impact. These components are so chosen that the effect of the collision is to reverse the sign of v_{OA} but to leave v_{AC} unchanged. At the very instant of impact, we can say that v_{OA} has a value of zero,[2] because its contribution to the kinetic energy is momentarily coverted entirely into potential energy. Hence the kinetic energy at the instant of impact is $\frac{1}{2}\mu v_{AC}^2$ or

$$\frac{1}{2}\mu v^2 \cos^2 \frac{\chi}{2}.$$

But the total energy must still equal the initial relative kinetic energy, $\frac{1}{2}\mu v^2$, so that the potential energy at impact (at the distance d) is

$$V(d) = \frac{1}{2}\mu v^2 \left(1 - \cos^2 \frac{\chi}{2}\right) = \frac{1}{2}\mu v^2 \sin^2 \frac{\chi}{2}; \qquad (27.28)$$

as we indicated above, this equals the value of $\frac{1}{2}\mu v_{OA}^2$ away from impact. For the general case of "soft" potentials, the quantity corresponding to $V(d)$ is the potential energy at the point of closest approach, or the energy along the line of centers. We note that if $\chi = 0$, then there is no collision and $V(d) = 0$; if $\chi = \pi$, there is a head-on collision, $V(d) = \frac{1}{2}\mu v^2 = E$, and the entire initial kinetic energy goes into potential energy on impact. Finally, we can rewrite Eq. 27.28, making use of Eq. 27.27, as

$$V(d) = E\left(1 - \frac{b^2}{d^2}\right). \qquad (27.29)$$

Relationships between the scattering angle, the impact parameter, and the energy, like those we have obtained for the special case of hard spheres, can also be derived for a general potential from the equations of motion for two particles. Various types of collision trajectories are possible, as illustrated in Fig. 27.10. We need a method of describing the shape of any particular trajectory, and for this purpose we identify any point P on the trajectory by the polar coordinates R, θ shown in Fig. 27.11. The distance of closest approach, OA in Fig. 27.11, is designated as R_0; the corresponding value of θ must of course equal $(\pi - \chi)/2$ (cf. Eq. 27.24).

To summarize, in our relative-coordinate representation a fictitious particle of reduced mass μ and initial relative velocity $\mathbf{v}$ impinges on a scattering center at O with impact parameter b, and is scattered at an angle χ. If the relative motion is initially in the plane of the page, it must stay in that plane, as there are no forces to push the particle out of the plane. Hence two coordinates, in our case R and θ, suffice to describe the position of the particle. In this polar coordinate system the kinetic energy at R, θ is given by[3] $\frac{1}{2}\mu\dot{R}^2 + \frac{1}{2}\mu R^2\dot{\theta}^2$, where $\dot{R}$, $\dot{\theta}$ are abbreviations for the time derivatives dR/dt, $d\theta/dt$. Since we have assumed the potential energy to be a function of R only, and the total energy E must always equal the initial kinetic energy, we can write

$$E = \frac{1}{2}\mu v^2$$
$$= \frac{1}{2}\mu\dot{R}^2 + \frac{1}{2}\mu R^2\dot{\theta}^2 + V(R). \qquad (27.30)$$

Similarly, the angular momentum at the point R, θ is given by $\mu R^2\dot{\theta}$ (Eq.

---

[2] Strictly speaking, v_{OA} changes discontinuously at impact. However, this purely mathematical objection is readily overcome by imagining the impact to last an extremely short time, rather than being instantaneous.

[3] See the derivation of Eq. 3.125.

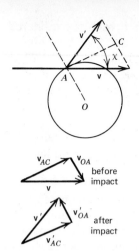

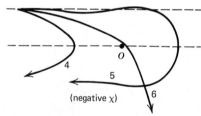

FIGURE 27.9
Relative velocity relationships at impact in an elastic hard-sphere collision.

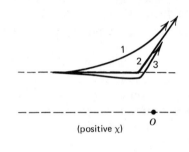

FIGURE 27.10
Typical collision trajectories for various types of interaction: 1, purely repulsive; 2, hard-sphere; 3, 4, 5, attractive at large distances, repulsive at small; 6, purely attractive.

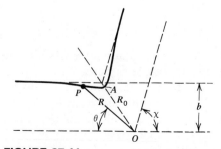

FIGURE 27.11
Coordinates used to describe relative collision trajectory.

3.128), which must also equal the initial value,

$$p_\theta = \mu v b = \mu R^2 \dot{\theta}, \qquad (27.31)$$

that is,

$$\frac{d\theta}{dt} = \frac{vb}{R^2}. \qquad (27.32)$$

We seek to eliminate the variable t; this can be accomplished by solving Eq. 27.30 for dR/dt. First we rewrite that equation, using Eq. 27.32, as

$$\frac{1}{2}\mu v^2 = \frac{1}{2}\mu \dot{R}^2 + \frac{1}{2}\mu v^2 \frac{b^2}{R^2} + V(R), \qquad (27.33)$$

and then solve to obtain

$$\frac{dR}{dt} = \sqrt{v^2 - v^2 \frac{b^2}{R^2} - \frac{V(R)}{\frac{1}{2}\mu}}. \qquad (27.34)$$

Now, division of Eq. 27.32 by Eq. 27.34 leads to

$$\frac{d\theta}{dR} = \frac{b}{R^2 \sqrt{1 - \dfrac{b^2}{R^2} - \dfrac{V(R)}{\frac{1}{2}\mu v^2}}}, \qquad (27.35)$$

the equation describing the trajectory.

Suppose we had decided to derive Eq. 27.35 using not the initial values of E and b, but instead the final values of relative kinetic energy and impact parameter. From the conservation of energy and angular momentum we have already shown that these quantities are the same before and after a collision. Hence, if we were to try to derive Eq. 27.35 by beginning the collision with the relative velocity $-\mathbf{v}'$ and impact parameter b', that is, our previous collision run in reverse, we would obtain an equation indistinguishable from Eq. 27.35. Therefore we conclude (as we have previously assumed intuitively) that the collision trajectory must be symmetric about the line OA (Fig. 27.11), which intercepts the trajectory at the distance of closest approach. At that extremum point, where $R = R_0$, we must have $dR/dt = 0$, so that Eq. 27.34 can be solved to give

$$1 - \frac{b^2}{R_0^2} = \frac{V(R_0)}{E}. \qquad (27.36)$$

It is interesting to compare Eq. 27.36 with Eq. 27.29 for the hard-sphere case, where $R_0 = d$.

The scattering angle χ can be obtained by integrating Eq. 27.35. What are the limits of integration? The value of R changes from infinity to R_0 and then out to infinity again, since we wish to begin and end the collision at infinite separation of the two particles. For these limits on R, the variable θ can be seen from Fig. 27.11 to vary from zero to $\pi - \chi$. Because of the trajectory's symmetry, we can integrate from R_0 to infinity and multiply the integral by 2,[4]

---

[4] More rigorously, the negative and positive square roots in Eq. 27.35 describe the incoming and outgoing halves of the trajectory, respectively. Thus if we write the equation as $d\theta = \pm g(R)\, dR$, integration gives

$$\int_0^{\pi-\chi} d\theta = \int_\infty^{R_0} -g(R)\, dR + \int_{R_0}^\infty + g(R)\, dR = 2\int_{R_0}^\infty g(R)\, dR,$$

the positive root being understood in Eq. 27.37.

$$\chi = \pi - 2b \int_{R_0}^{\infty} \frac{dR}{R^2 \sqrt{1 - \dfrac{b^2}{R^2} - \dfrac{V(R)}{E}}}. \qquad (27.37)$$

This is the desired result, R_0 being the value of R that satisfies Eq. 27.36. We see that, in general, the scattering angle χ depends parametrically on both E (or v) and b. It is a great pity that with some exceptions, such as for the hard-sphere potential, the Coulomb potential, and a very few others, the integral in Eq. 27.37 is too complicated to solve analytically.

For the hard-sphere potential, we easily recover the results previously derived. The distance of closest approach of two hard spheres that do collide can only be d, the diameter of the hard sphere. For $R > d$, the potential is everywhere zero, so that we have simply

$$\chi = \pi - 2b \int_{d}^{\infty} \frac{dR}{R^2 \sqrt{1 - \dfrac{b^2}{R^2}}}. \qquad (27.38)$$

Integration then gives (consult a table of integrals)

$$\chi = \pi - 2b \left[\frac{1}{b} \left(\frac{\pi}{2} - \cos^{-1} \frac{b}{d} \right) \right] = 2 \cos^{-1} \frac{b}{d}, \qquad (27.39)$$

or

$$b = d \cos \frac{\chi}{2} \qquad (27.40)$$

as in Eq. 27.26.[5]

27.4
Types of Collisions

So far we have discussed only one kind of collision, an elastic collision between two structureless particles (where by "structureless" we mean particles having no internal energy states). It was shown in Section 27.3 that two particles can exchange both kinetic energy and momentum in an elastic collision. This important feature will be used later in discussing the molecular basis of the attainment of equilibrium in certain irreversible processes, such as heat conduction in a monatomic gas. There are, however, a number of other kinds of collisions that we wish to define.

Let us first consider a diatomic molecule, such as N_2. The molecule can exist in a variety of internal states, defined by a set of rotational, vibrational, electronic, and nuclear quantum numbers. In general, for each degree of freedom of the molecule we have a quantum number, these numbers in combination specifying a particular eigenstate. Let the symbol i denote the entire set of quantum numbers necessary to describe the internal state of N_2, and let the molecule have a velocity v_1. If N_2 in the state (v_1, i) collides with another molecule, say, Ar with velocity v_2, then an elastic collision may occur, in which the internal state is unaltered; we can write a stoichiometric equation,[6]

$$N_2(v_1, i) + Ar(v_2) \rightarrow N_2(v_1', i) + Ar(v_2'), \qquad (27.41)$$

which represents this situation. The relative kinetic energy before and after the collision is unaltered.

---

[5] If $b > d$, no collision occurs, and R_0 is simply b. We then obtain $b = b \cos (\chi/2)$, and $\chi = 0$ as expected.

[6] In this and subsequent equations we assume (for simplicity) no change in the internal state of Ar, which of course in the general case may undergo an electronic transition in the collision.

There is another possible type of collision, in which the internal quantum state of N_2 does change in the encounter with Ar; we then say that an *inelastic collision* has occurred. For instance, N_2 initially in the ground vibrational state may be excited to its first vibrational state. We may have collisions in which the energy is increased (excitation) or decreased (deexcitation) in the internal degrees of freedom of the molecule. In either case the final relative kinetic energy must differ from its initial value in order that the total energy be conserved. Again we can write a stoichiometric equation,

$$N_2(\mathbf{v}_1, i) + Ar(\mathbf{v}_2) \rightarrow N_2(\mathbf{v}_1', j) + Ar(\mathbf{v}_2'), \qquad (27.42)$$

for the inelastic collision in which the internal quantum numbers of N_2 are altered from the set i to the set j. If N_2 has internal energy ε_i in state i, and ε_j in state j, then the conservation of energy requires that

$$\varepsilon_i + \tfrac{1}{2}\mu v^2 = \varepsilon_j + \tfrac{1}{2}\mu v'^2, \qquad (27.43)$$

where v and v' are the initial and final relative speeds.[7] All possible kinds of inelastic collisions are known to occur, each, of course, with its own characteristic probability. Energy exchange in collisions may occur between the degrees of freedom, for example, translational–rotational, translational–vibrational, translational–electronic, or vibrational–rotational. We may also generalize our example by considering a collision of two molecules in which the internal states of both are changed, or the collision of a molecule with a photon, in which there may occur absorption of the photon and excitation of the molecule.

Finally, we consider *reactive collisions*, those in which the chemical identity of one or more of the colliding species is altered. The dissociation of I_2 by collision with Ar is one such example:

$$I_2(\mathbf{v}_1, i) + Ar(\mathbf{v}_2) \rightarrow I(\mathbf{v}_3, j) + I(\mathbf{v}_4, l) + Ar(\mathbf{v}_2'). \qquad (27.44)$$

The letter i denotes the initial internal state of the I_2 molecule, and the letters j and l the electronic states of the iodine atoms produced in the dissociation. The rearrangement collision of H_2 with I_2 is another example:

$$H_2(\mathbf{v}_1, i) + I_2(\mathbf{v}_2, j) \rightarrow HI(\mathbf{v}_3, l) + HI(\mathbf{v}_4, m). \qquad (27.45)$$

In all cases there must be conservation of energy and of linear and angular momentum. The distribution of angular momentum presents a difficult problem, because there is the possibility of exchange of the angular momentum of the relative motion (orbital angular momentum) with angular momentum of the internal motions (rotational angular momentum), all vector quantities. The formalism required for a proper treatment of the conservation of angular momentum is known, but too complicated to be worth introducing here. The effects of the conservation of energy are always simple to discern. For the reaction (27.45) we can write either

$$\varepsilon_{H_2,i} + \varepsilon_{I_2,j} + \tfrac{1}{2}m_{H_2}v_1^2 + \tfrac{1}{2}m_{I_2}v_2^2 = \varepsilon_{HI,l} + \varepsilon_{HI,m} + \tfrac{1}{2}m_{HI}v_3^2 + \tfrac{1}{2}m_{HI}v_4^2$$
$$(27.46)$$

or, after separating out the center-of-mass motion,

$$\varepsilon_{H_2,i} + \varepsilon_{I_2,j} + \tfrac{1}{2}\mu v^2 = \varepsilon_{HI,l} + \varepsilon_{HI,m} + \tfrac{1}{2}\mu' v'^2. \qquad (27.47)$$

---

[7] Suppose that during a collision N_2 undergoes a transition between two degenerate quantum states, that is, a transition from one internal state to another of the same energy. Then kinetic energy is again conserved. However, unless the degeneracy is removed, for instance by placing the molecule in an external field, there is no way of determining whether such a transition has taken place.

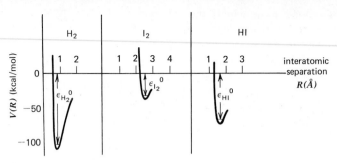

FIGURE 27.12
Potential energy curves for the molecules H_2, I_2, and HI.

Note that in this case the reduced mass of the products, μ', differs from that of the reactants (see Eq. 27.13).

Suppose that in the last example H_2 and I_2 are initially in their ground states, and that both HI molecules are produced in their ground states. What relation exists among the ground-state internal energies (which we designate as ε^0)? If, as usual, we take the zero of energy for each molecule as that configuration in which the constituent atoms are in their ground states at infinite separation, then $\varepsilon_{H_2}{}^0$ is simply the negative of the dissociation energy of H_2 from the ground state. In Fig. 27.12 we show the potential curves for the individual isolated molecules H_2, I_2, and HI. The internal energy change from reactants in their ground states to products in their ground states we shall call $\Delta E_0{}^0$, which must be

$$\Delta E_0{}^0 = 2\varepsilon_{HI}{}^0 - (\varepsilon_{H_2}{}^0 + \varepsilon_{I_2}{}^0). \tag{27.48}$$

It is also the thermodynamic energy change for the reaction at 0 K.

Other examples of reactive collisions are ionization reactions with particles, such as

$$Ar + Ar \rightarrow Ar^+ + e^- + Ar, \tag{27.49}$$

or with photons, such as

$$Ar + h\nu \rightarrow Ar^+ + e^-, \tag{27.50}$$

and reactions between neutral molecules and ions, or between ions.

It might be thought that we could specify collisions in still greater detail than we have done. For instance, consider the rotational motion of I_2 and H_2. The rotational energy of each isolated molecule can be determined, and is in fact described by the sets of quantum numbers i and j of Eq. 27.45. If the motion were classical, then we could also, at least in principle, measure the orientation of each molecule in space as a function of time. Quantum mechanics tells us, however, that it is impossible to measure the motion of real molecules in such detail.

It is very difficult to isolate detectable numbers of molecules in specified quantum states, react them, and isolate products in specified quantum states. It is much easier to mix 1 mol of $H_2(g)$ with 1 mol of $I_2(g)$ and measure the rate of disappearance of $I_2(g)$—which can be accomplished easily, because $I_2(g)$ is purple and $H_2(g)$ and HI(g) are colorless. In that case, in the macroscopic experiment, we examine the reaction

$$H_2(g) + I_2(g) = 2HI(g),$$

which is an average over all possible types of reaction 27.45, the different types being distinguished by different internal quantum numbers of reactants and products, and by different relative kinetic energies. We shall return to this point below.

27.5
Scattering Cross Sections

We have seen that collisions between molecules depend on the forces between them. In this section we define the experimental quantities called collision cross sections and indicate how they are determined; later we shall show how they are related to the intermolecular forces for the simple case of the hard-sphere interaction. For simplicity, we begin with collisions between single atoms.

Consider the following experiment, illustrated in Fig. 27.13: We form a beam of atoms, say of argon, by making a small puncture in a container of gas surrounded by a vacuum. If the orifice is large, the resultant flow is similar to a stream of water from a hose, a hydrodynamic flow. But if the orifice is small enough, the argon atoms will emerge singly, independently of one another, leaving the orifice in all directions with all possible speeds. Let us consider the latter type of flow. To obtain a beam of atoms, it is necessary to collimate the flow by the placement of slits, just as one would collimate a source of light in order to obtain a light beam. Then the argon atoms pass through a velocity selector, which may consist of a set of slotted disks mounted on a shaft. When the shaft turns at a given speed, only atoms with a given velocity will pass through the slots. (The atom must move at a speed enabling it to pass from one slot to another in the time between the two slots' rotation onto the line of the beam.) Atoms with smaller or larger speeds may pass through a few disks, but will then hit a solid portion of some disk. In this way there is produced a velocity-selected atomic beam, that is, within experimental error, a collision-free, unidirectional flow of atoms all moving with the same speed, say v_1. We can specify the density of this beam, $n_1(v_1)$, the number of molecules per unit volume with speed v_1; and the flux density, $v_1 n_1$, the number of such molecules that cross unit area per unit time.

We also produce a second velocity-selected beam, say of xenon atoms, with density $n_2(v_2)$ and flux density $v_2 n_2$. Finally, the apparatus is arranged so that the two beams collide in a volume element τ, the so-called scattering volume, and a detector of some kind is used to determine the angular distribution of the scattered atoms. (Although either species could be studied, we shall confine our discussion to the detection of scattered argon.) In a typical experiment, each beam may be about 1 cm^2 in cross section, so that for beams at right angles τ is 1 cm^3. The densities are of the order of 10^9 atoms/cm^3, and the flux densities

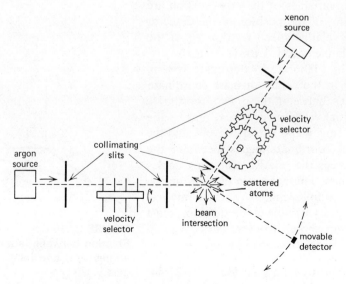

FIGURE 27.13
Experimental arrangement for a molecular-beam scattering experiment. All parts are in an evacuated enclosure.

10^{13} atoms/cm^2 s. Some of the argon atoms will collide with xenon atoms and be scattered, we assume elastically, out of the argon beam in all possible directions. Other argon atoms will never come close enough to a xenon atom for detectable interaction, and will leave the collision volume undeflected. Experimental conditions are usually so chosen that one can neglect the probability of an argon atom's colliding successively with two xenon atoms within the volume τ; this requires a low density of xenon.

The quantity measured in this experiment is the flux of scattered argon atoms arriving at the detector, which is situated in space a given distance L from the scattering volume at laboratory angles α, β, as shown in Fig. 27.14. The detector area A subtends at the origin O a small element of solid angle,

$$d\Omega' = \frac{A}{L^2} \qquad (27.51)$$

(cf. Appendix 2B). In the spherical coordinate system specified by the angles α, β, this solid-angle element is

$$d\Omega' = \sin \alpha \, d\alpha \, d\beta. \qquad (27.52)$$

The *scattering cross section* is now defined by the experiment. The number of argon atoms arriving at the detector (which could be a mass spectrometer, for instance) per unit solid angle per second, designated as $dN_{Ar}(\alpha, \beta)$, is determined by the number of collisions that occur per unit time. The number of collisions is proportional to the densities of argon and xenon, to the relative speed with which an argon atom approaches a xenon atom, and to the size of the beam-intersection volume τ. All these quantities are experimentally adjustable. In addition, the number of argon atoms scattered and their angular distribution depend on the force between argon and xenon. If the range of the force is large, more collisions will occur than if the range is small. That, of course, is just the information we want, and is contained in the cross section σ defined by the equation

$$dN_{Ar}(\alpha, \beta) = \sigma(v, \alpha, \beta)vn_1(v_1)n_2(v_2)\tau \, d\Omega'. \qquad (27.53)$$

The units of σ are cm^2/atom, and the cross section can be thought of as the effective area that the argon and xenon atoms present to each other, such that, if they approach within that area, argon will be scattered into the specified angles. Note also that the variables of the cross section are the operational variables of the experiment: the position of the detector at the laboratory angles α, β and the relative speed $v = |\mathbf{v}_2 - \mathbf{v}_1|$. If the beams intersect at right angles, the relative speed is $v = (v_1^2 + v_2^2)^{\frac{1}{2}}$.

We measure scattering processes in a laboratory coordinate system, but usually analyze them theoretically in the center-of-mass coordinate system. Let us look at some relations between these two coordinate systems. In Fig. 27.15 we show the initial velocity vectors, $\mathbf{v}_1$ and $\mathbf{v}_2$, the relative velocity before and after scattering, $\mathbf{v}$ and $\mathbf{v}'$ respectively, and the position of the detector. The laboratory coordinate angles specifying the position of the detector are α, β, whereas the relative scattering angles are denoted by χ and ϕ, where ϕ is the azimuthal angle of $\mathbf{v}'$ around $\mathbf{v}$. The important point about all this is the following: Any scalar physical quantity, in this case the number of particles arriving at the detector, must be independent of the coordinate system chosen to describe it. We can thus write the equivalent of Eq. 27.53 in relative coordinates,

$$dN_{Ar}(\chi, \phi) = \sigma(v, \chi)vn_1(v_1)n_2(v_2)\tau \sin \chi \, d\chi \, d\phi, \qquad (27.54)$$

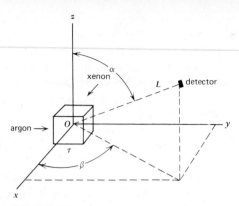

FIGURE 27.14
Laboratory coordinates for scattering from crossed molecular beams. It is assumed that $L \gg \tau^{1/3}$, so that the angles α, β do not depend on the actual point of collision within the volume τ.

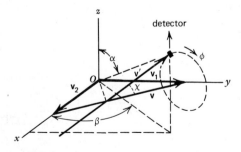

FIGURE 27.15
Relation between laboratory scattering angles α, β and relative scattering angles χ, ϕ.

and set $dN_{Ar}(\alpha, \beta)$ equal to $dN_{Ar}(\chi, \phi)$, yielding the relation

$$\sigma(v, \alpha, \beta) \sin \alpha \, d\alpha \, d\beta = \sigma(v, \chi) \sin \chi \, d\chi \, d\phi \qquad (27.55)$$

between the cross sections in the laboratory and center-of-mass coordinates. In the last two equations we have omitted the azimuthal angle ϕ as one of the variables of the relative cross section. Examination of Fig. 27.16 shows that, for scattering from a central force field with origin at O, there cannot exist any preferred value of ϕ in the absence of external forces. Hence all values of ϕ are equally probable, and the scattering cross section is independent of ϕ.

It is easy to imagine (though not always easy to carry out) more complicated experiments than the elastic scattering of one atom from another. Suppose that we are able to form a beam of nitrogen molecules, all in the same known vibrational, rotational, and electronic state, collide them with a beam of argon atoms, and detect N_2 molecules with only the vibrational quantum number (here V) changed from the initial state. (As before, the internal state of Ar is assumed unchanged.) We then have a scattering experiment described by the stoichiometric equation

$$N_2(\mathbf{v}_1, V) + Ar(\mathbf{v}_2) \rightarrow N_2(\mathbf{v}_1', V') + Ar(\mathbf{v}_2'), \qquad (27.56)$$

which is only a special case of Eq. 27.42. The cross section for this scattering process is defined by an equation very similar to Eq. 27.53 or Eq. 27.54,

$$dN_{N_2}(V', \alpha, \beta) = \sigma(V|V'; v, \alpha, \beta)vn_{N_2}(v_1, V)n_{Ar}(v_2)\tau \, d\Omega', \qquad (27.57)$$

where the notation $\sigma(V|V'; v, \alpha, \beta)$ can be read as "the cross section for conversion of V to V' at collision parameters v, α, β." The variables of the cross section now include, in addition to v, α, β, the initial and final vibrational states of N_2.

Similarly, we can devise, at least in principle, a scattering experiment for a chemical reaction. Let us consider a somewhat simpler case than reaction 27.44, for example,

$$H(\mathbf{v}_1, 0) + I_2(\mathbf{v}_2, j) \rightarrow HI(\mathbf{v}_3, l) + I(\mathbf{v}_4, 0), \qquad (27.58)$$

the reaction of a H atom in its ground state with I_2 in its (rotational, vibrational, and electronic) quantum state j, yielding HI in quantum state l and an I atom in its ground state (0, j, or l represents all nonvibrational quantum numbers). The cross section for this reaction is defined by the equation

$$dN_{HI}(l, \alpha, \beta) = \sigma_R(0, j \mid l, 0; v, \alpha, \beta)vn_H(v_1)n_{I_2}(v_2, j)\tau \, d\Omega', \qquad (27.59)$$

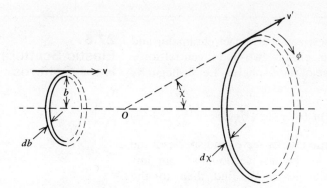

FIGURE 27.16
Relation of impact parameter b and relative speed v. The vectors $\mathbf{v}$ and $\mathbf{v}'$ define a plane with azimuthal angle ϕ, all values of ϕ being physically equivalent for a central force field.

where $dN_{HI}(l, \alpha, \beta)$ is the number of HI molecules in quantum state l arriving in solid angle $d\Omega'$ per unit time at the detector situated at laboratory angles α, β. Note that the final velocity of either product of the reaction is not an independent variable. Conservation of energy fixes the length of the final relative velocity vector, and the angles α, β determine its orientation. Hence neither the final velocity nor the final relative speed is listed as a variable of the cross section.

In each case we have discussed so far, it is possible to make simpler but still useful measurements in which less than maximal information is obtained. Suppose that in the elastic scattering of argon by xenon we measure not the flux of scattered argon atoms at given laboratory angles, but only the total amount of argon scattered. This can be determined either by measuring the total flux at all angles and then summing or integrating, or by measuring the decrease in the flux of the argon beam on passage through the xenon beam. The measurement thus yields the integral

$$\int_{\alpha,\beta} dN(\alpha, \beta) \equiv N, \qquad (27.60)$$

the total number of molecules scattered per unit time in the volume τ; according to Eq. 27.53, this must equal

$$N = \left[\int\int_{\Omega'} \sigma(v, \alpha, \beta) d\Omega' \right] vn_1(v_1)n_2(v_2)\tau. \qquad (27.61)$$

The integral in this equation is a function of v only, and is called the *total scattering cross section*, $\sigma(v)$, whereas $\sigma(v, \alpha, \beta)$ is called the *differential scattering cross section*; they are connected by the relation

$$\sigma(v) = \int_{\alpha=0}^{\pi} \int_{\beta=0}^{2\pi} \sigma(v, \alpha, \beta) \sin \alpha \, d\alpha \, d\beta. \qquad (27.62)$$

Again, for the reaction of Eq. 27.58, suppose that we measure at given laboratory angles not a specified quantum state of HI, but all possible quantum states. The cross section determined in such an experiment is simply that of Eq. 27.59 summed over all values of l,

$$\sum_l \sigma_R(0, j \mid l, 0; v, \alpha, \beta).$$

One can, of course, go on to sum over all possible initial states j of I_2, and then to consider states of H and I other than the ground states. A variety of cross sections can similarly be defined for any scattering process, depending on the information one has available and wishes to apply. The quantities measured can also be further varied: for example, one could record the number of particles arriving at the detector with a given laboratory speed.

We shall now derive the relation between the impact parameter and the scattering cross section for classical, elastic, two-body, central-force scattering, and then specialize to the case of hard spheres. Let us begin by writing again Eq. 27.54,

27.6
Elastic Scattering of Hard Spheres

$$dN_1(\chi, \phi) = \sigma(v, \chi)vn_1(v_1)n_2(v_2)\tau \sin \chi \, d\chi \, d\phi; \qquad (27.63)$$

here species 1 is the one whose distribution we observe, and species 2 the "scatterer." Following the discussion in the last section, we can immediately integrate Eq. 27.63 over all angles ϕ. We find, then, for the total number of particles scattered into angles between χ and $\chi + d\chi$,

$$dN_1(\chi) = \int_{\phi=0}^{2\pi} dN_1(\chi, \phi) = 2\pi\sigma(v, \chi)vn_1(v_1)n_2(v_2)\tau \sin \chi \, d\chi.$$

(27.64)

Now suppose that we have unit density of scatterers $n_2(v_2)$ and unit scattering volume τ; that is, we consider the number of particles scattered per scatterer. Of the incoming particles, what fraction are scattered between χ and $\chi + d\chi$? For a given relative speed, the scattering angle is determined by the impact parameter, as indicated by Eq. 27.37 and in Fig. 27.16, where db is the range of b from which particles are scattered into $d\chi$. The number of such particles per unit time is the product of the flux density $vn_1(v_1)$ and the area element $2\pi b \, db$. Setting this product equal to $dN_1(\chi)$ (for a single scatterer) gives us

$$2\pi b \, db = 2\pi\sigma(v, \chi) \sin \chi \, d\chi,$$

(27.65)

an equation directly relating the impact parameter to the differential cross section. Equation 27.65 can be solved for $\sigma(v, \chi)$, to yield

$$\sigma(v, \chi) = \frac{b}{\sin \chi} \left| \frac{db}{d\chi} \right|;$$

(27.66)

we have taken the absolute magnitude of the derivative, which may be negative, because scattering can only lead to a positive flux at the detector.

For the case of hard spheres of diameter d, we showed in Eq. 27.27 that

$$b = d \cos \frac{\chi}{2}.$$

(27.67)

For this case b is a single-valued function of χ, so that

$$\frac{db}{d\chi} = -\frac{d}{2} \sin \frac{\chi}{2}.$$

(27.68)

The differential cross section thus must be

$$\sigma(v, \chi) = \frac{d^2}{2} \cos \frac{\chi}{2} \sin \frac{\chi}{2} \frac{1}{\sin \chi} = \frac{d^2}{4}.$$

(27.69)

The differential cross section for the elastic scattering of hard spheres thus depends neither on v nor on χ! The total cross section is obtained by integrating the differential cross section over all angles:

$$\sigma(v) = 2\pi \int_0^\pi \sigma(v, \chi) \sin \chi \, d\chi$$

$$= \frac{\pi d^2}{2} \int_0^\pi \sin \chi \, d\chi = \pi d^2.$$

(27.70)

This equation displays in transparent form the simple geometric interpretation of the cross section. If one hard sphere approaches another within the distance d, then scattering into some angle will occur. Hence the effective area one hard sphere presents to another for scattering into *any* angle is πd^2.

Real molecules, of course, do not interact in accordance with such a simple model. In Fig. 27.17 is shown the measured differential elastic cross section for K–Kr, and for comparison the same quantity for hard

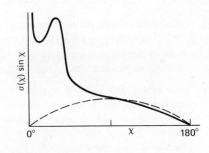

FIGURE 27.17
Elastic scattering cross section for K–Kr (solid line). The dotted line represents the same quantity for the model of hard spheres, fitted to the solid line (arbitrarily) at 90°.

spheres, arbitrarily fitted to the measurements at 90°. The large-angle scattering, determined primarily by the repulsive forces, is fairly well represented by the hard-sphere model. The small-angle scattering, however, is governed mostly by the attractive forces, so that the hard-sphere model would be expected to fail. The oscillations in the measured cross section are due to quantum-mechanical interference effects in the scattering process, a fascinating topic beyond the scope of the present discussion.

Further Reading

Fluendy, M. A. D., and Lawley, K. P., *Chemical Applications of Molecular Beam Scattering* (Chapman and Hall, London, 1973).

Lawley, K. P. (Ed.), *Molecular Scattering: Physical and Chemical Applications* (Advances in Chemical Physics, Vol. 30) (Wiley, New York, 1975).

Levine, R. D., and Bernstein, R. B., *Molecular Reaction Dynamics* (Oxford University Press, 1974).

Ross, J. (Ed.), *Molecular Beams* (Advances in Chemical Physics, Vol. 10) (Wiley, New York, 1966).

Schlier, C. (Ed.), *Molecular Beams and Reaction Kinetics* (Proc. Intern. School of Phys. "Enrico Fermi," Course 44) (Academic Press, New York, 1970).

Scattering Theory (Physical and Chemical)

Bederson, B., and Fite, W. L. (Eds.), *Atomic Interactions*, Vol. 7 of L. Marton (Ed.), *Methods of Experimental Physics* (Academic Press, New York, 1968), chaps. 3 and 4.

Experimental Methods

See references for Chapter 10.

Intermolecular Potentials

Problems

1. Two argon atoms have laboratory velocities with (x, y, z) components of $(2 \times 10^3 \, \text{m/s}, 0, 0)$ and $(0, 10^3 \, \text{m/s}, 0)$. What are the magnitude and direction of the center-of-mass momentum? What are the magnitude and direction of the relative momentum?

2. Two argon atoms have a relative velocity of $10^3 \, \text{m/s}$ and approach one another with an impact parameter of $2 \, \text{Å}$. What is the orbital angular momentum of this system? Assume that the atoms are hard spheres with diameter $2.86 \, \text{Å}$; what are the magnitude and direction of linear momentum transfer during the collision?

3. Use the table below to calculate the equilibrium internuclear separation for the rare-gas dimers. Compare these separations to the average monomer-monomer internuclear distance in gases at standard temperature and pressure. Compare these separations to the internuclear separations in liquids and solids.

Atom	σ (Å)	ϵ (J)	ρ_{solid} (g/cm³)	ρ_{liquid} (g/cm³)
He	2.58	1.41×10^{-22}	0.205	0.125
Ne	2.79	4.93×10^{-22}	1.51	1.20
Ar	3.42	1.71×10^{-21}	1.77	1.37

4. Compare the binding energies of the rare-gas dimers and that of H_2 (which is 4.476 eV) to $k_B T$ at 20 K, 300 K, and 10,000 K. What does this comparison indicate about the relative stability of the molecules at the given temperatures?

5. Define the duration of a collision on a LJ surface as the time during which $V(R) > 0.1\epsilon$. For high energies one can approximate the distance of closest approach of two species by $R_0 = b$, and the time dependence of a trajectory by $R^2 = v^2 t^2 + b^2$. Using this information, calculate the duration of a N_2–N_2 collision at $v = 1 \times 10^3$ m/s as a function of b. For "round" N_2 molecules $\epsilon/k_B = 47.6$ K and $\sigma = 3.85$ Å. Compare these times to the rotational and vibrational frequencies of N_2.

6. Calculate the force, $F(R)$, between two He atoms interacting with a LJ potential. What is the maximum force in the range $\sigma \leqslant R < \infty$? Compare this force (a) to the force of gravity on the He atom, (b) to the force of attraction of a singly charged ion at $R = \sigma$. (*Hint*: The potential of an ion–atom interaction is

$$V(R) = -\frac{e^2 \alpha}{32\pi^2 \epsilon_0^2 R^4},$$

where α is the polarizability of the atom; for He $\alpha = 1.37 a_0^3$.)

7. The distance of closest approach between two particles, R_0, is the solution of the equation

$$E_T = V(R_0) + \frac{E_T b^2}{R_0^2},$$

where E_T is the relative translational energy. How does R_0 behave (a) for very small b; (b) for very large b? What is R_0 for an argon–argon collision with relative translational energy of 1.7×10^{-21} J and an impact parameter of 6 Å? (*Hint*: Solve the above equation graphically.)

8. A simple intermolecular potential, intermediate between a hard-sphere potential and a realistic potential with van der Waals attraction, is the square-well potential shown in the accompanying diagram. Derive an expression for

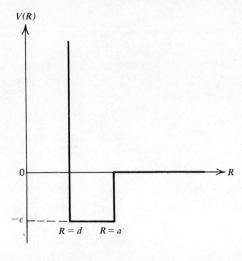

the deflection function $\chi(b, E)$ for this potential. Comment on the behavior of $\chi(b, E)$ as E becomes large.

9. Other potentials for which the deflection function $\chi(b, E)$ has an analytic form are (a) $V(r) = A/r^2$; and (b) $V(r) = A/r$. Derive $\chi(b, E)$ for these potentials.

10. Consider the collision

$$N_2(\mathbf{v}_1, i) + Ar(\mathbf{v}_2) \rightarrow N_2(\mathbf{v}_1', j) + Ar(\mathbf{v}_2'),$$

where i and j refer to the oscillator quantum numbers. Assume that N_2 is a harmonic oscillator with frequency $7.08 \times 10^{13}\,\mathrm{s}^{-1}$.
(a) Based on conservation of energy alone, what is the maximum value of j allowed if $i = 0$, $|\mathbf{v}_1| = 1.4 \times 10^3\,\mathrm{m/s}$, and $|\mathbf{v}_2| = 2 \times 10^3\,\mathrm{m/s}$?
(b) What is the maximum value of j if $i = 2$ and the velocities are the same as in part (a)?

11. Consider a velocity selector consisting of two slotted disks, 5 cm apart; both rotate at the same speed of 1000 Hz, but with the second lagging by $\pi/5$ radians (see the accompanying diagram). If the width of the slots is 0.5 cm, what are (a) the median, (b) the minimum, and (c) the maximum velocities of particles that pass through the selector? Comment on what would happen if each disk had two or more slots; that is, what are the advantages and disadvantages of using multislotted disks?

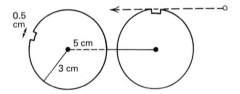

28

The Kinetic Theory
of Gases

In the last chapter we discussed the kinematics and dynamics of molecular collisions by considering molecules to be mechanical systems, in the simplest case point particles with forces between them. The kinetic theory of gases is based on the hypotheses that molecules exist, that their motions and collisions can be analyzed according to the laws of either classical or quantum mechanics, and that the macroscopic properties of gases can be related to certain averages of molecular motions and intermolecular forces.

Historically, the development of the kinetic theory of gases began much earlier than that of statistical mechanics. This may have resulted partly from the fact that in a dilute gas one only needs to consider a molecule's collisions either with the wall of the vessel or with a single other molecule (binary collisions). We can solve the two-particle problem in mechanics, but we do not have similarly explicit solutions to the equations of motion of three or more particles colliding with one another. From an understanding of the nature of collisions there follows, as we shall see, an interpretation of both equilibrium properties (such as the equation of state) and nonequilibrium properties (such as the viscosity and thermal conductivity). For the statistical mechanics of equilibrium properties we do not need specific solutions of the equations of motion. We do need, as we saw in Part Two, certain averages over the energy spectrum of the system. Although the calculation of such averages is difficult, at least we have a generally valid equation for, say, the Helmholtz free energy in terms of the partition function (and thus the molecular energy levels). We must now develop similarly general expressions for the nonequilibrium properties of a gas.

Thus the kinetic theory of gases, although of separate origin, can now be considered a part of statistical mechanics, the part concerned with the equilibrium and nonequilibrium properties of dilute gases. The analysis of nonequilibrium phenomena in solids, liquids, and dense gases is much more difficult, and is the subject of much present-day research.

28.1
Distribution Functions

We have said that we shall picture a gas as consisting of molecules, which move and collide with one another according to the laws of mechanics. As noted in our earlier discussion of molecular motion, we neither hope nor aspire to solve the equations of motion and write down their solutions by listing at each second (or microsecond) the velocities and positions of 6×10^{23} molecules. A high-speed computer can do this job for about 1000 particles with simple forces between them, such as hard-sphere or Lennard–Jones interactions. But we really could not possibly use or even examine all this information. The macroscopic state of the system that we wish to investigate is defined by different variables, those of thermodynamics. We may want to know the pressure and temperature of the gas, each perhaps as a function of time, or the rate at which the gas conducts heat between two walls at different temperatures. We usually seek answers to questions about a small number of variables for nonequilibrium conditions, just as we did in equilibrium thermodynamics. As described in Part Two, we proceed from a large number of variables (such as the velocities and positions of all the molecules at each instant of time) to a small number of newly defined macroscopic variables, by a process of integration or averaging.

We have previously (Section 12.2) defined the velocity distribution function for molecules in a gas, $f(\mathbf{v})$, where $f(\mathbf{v}) \, d\mathbf{v}$ is the number of molecules per unit volume with velocities in the range from $\mathbf{v}$ to $\mathbf{v}+d\mathbf{v}$ (i.e., v_x to v_x+dv_x, etc.); in the three-dimensional velocity space the volume element $d\mathbf{v}$ is

$$d\mathbf{v} = dv_x \, dv_y \, dv_z \qquad (28.1)$$

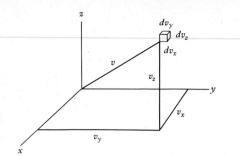

FIGURE 28.1
Cartesian coordinates in velocity space.

in Cartesian coordinates, as shown in Fig. 28.1. Our original assumptions were that the state of the gas was spatially homogeneous and unchanging with time. If we remove these restrictions and include the necessary additional variables, we must replace $f(\mathbf{v})$ by $f(\mathbf{r}, \mathbf{v}, t)$, with the same definition as before. Clearly, $f(\mathbf{r}, \mathbf{v}, t) \, d\mathbf{r} \, d\mathbf{v}$ is the total number of molecules with positions in the real-space volume element $dx \, dy \, dz$ and velocities in $dv_x \, dv_y \, dv_z$, so that $f(\mathbf{r}, \mathbf{v}, t)$ is a density, the number of molecules per unit "volume" in the six-dimensional space defined by $d\mathbf{r} \, d\mathbf{v}$.

Although the generality of $f(\mathbf{r}, \mathbf{v}, t)$ is useful, our primary concern here will be with the velocity distribution alone. We can therefore integrate over the position (configuration) coordinates to obtain

$$f(\mathbf{v}, t) \equiv \iiint\limits_{V} f(\mathbf{r}, \mathbf{v}, t) \, dx \, dy \, dz, \qquad (28.2)$$

the density of molecules in velocity space, where the integration is carried out over the entire volume V of the gas.[1]

If we further integrate $f(\mathbf{v}, t)$ over all possible values of $\mathbf{v}$, then we must obtain the total number of molecules,

$$N = \int_{v_x=-\infty}^{+\infty} \int_{v_y=-\infty}^{+\infty} \int_{v_z=-\infty}^{+\infty} f(\mathbf{v}, t) \, dv_x \, dv_y \, dv_z. \qquad (28.3)$$

It is often convenient to work with a distribution function normalized to unity. Suppose that we divide both sides of Eq. 28.3 by N. Since the number of molecules is a constant, N can be taken inside the integral. Thus, if we define the distribution function density as

$$f(\mathbf{v}, t) \equiv \frac{1}{N} f(\mathbf{v}, t), \qquad (28.4)$$

Eq. 28.3 becomes

$$1 = \int_{v_x=-\infty}^{+\infty} \int_{v_y=-\infty}^{+\infty} \int_{v_z=-\infty}^{+\infty} f(\mathbf{v}, t) \, dv_x \, dv_y \, dv_z. \qquad (28.5)$$

The function $f(\mathbf{v}, t)$ is therefore a *probability density*, since its integral over all possibilities, in this case all velocities, is unity. (Cf. Section 15.2.) The function $f(\mathbf{v}, t) \, d\mathbf{v}$ is the probability of finding in the gas a molecule with velocity between $\mathbf{v}$ and $\mathbf{v}+d\mathbf{v}$. If one could somehow pick, at

---

[1] Note that $f(\mathbf{v}, t)$ is not the same as the $f(\mathbf{v})$ of Chapter 12, which was merely shorthand for $f(\mathbf{r}, \mathbf{v}, t)$ in the case of a spatially uniform gas. In the latter case we have $f(\mathbf{v}, t) = V f(\mathbf{r}, \mathbf{v}, t)$.

random, 1000 molecules out of the gas at time t, then on the average $1000 f(\mathbf{v}, t)\, d\mathbf{v}$ molecules would have a velocity in this range.

Suppose that we know the probability density $f(\mathbf{v}, t)$ and wish to calculate the probability of finding in the gas a molecule with a given *speed*. If we are not interested in the direction of the vector $\mathbf{v}$ but only in its magnitude v, it is advantageous to transform from Cartesian to spherical coordinates. These new coordinates are the length v and the two spherical angles θ and ϕ, Fig. 28.2a. The relations between the Cartesian and spherical coordinates are

$$v_x = v \sin \theta \cos \phi,$$
$$v_y = v \sin \theta \sin \phi, \qquad (28.6)$$
$$v_z = v \cos \theta,$$

and between the volume elements (Fig. 28.2b)

$$d\mathbf{v} = dv_x\, dv_y\, dv_z = v^2 \sin \theta\, d\theta\, d\phi\, dv. \qquad (28.7)$$

(See also Appendix 2B.) Hence the probability $f(\mathbf{v}, t)\, d\mathbf{v}$ becomes

$$f(\mathbf{v}, t)\, d\mathbf{v} = f(v, \theta, \phi, t) v^2 \sin \theta\, d\theta\, d\phi\, dv. \qquad (28.8)$$

Integration of the velocity distribution function over all possible directions, that is, all possible values of θ and ϕ, yields the probability of finding a particle with speed between v and $v + dv$. We denote the latter quantity by $f(v, t)\, dv$,

$$f(v, t)\, dv \equiv \int_{\theta=0}^{\pi} \int_{\phi=0}^{2\pi} f(\mathbf{v}, t)\, d\mathbf{v} = v^2\, dv \int_{\theta=0}^{\pi} \int_{\phi=0}^{2\pi} f(v, \theta, \phi, t) \sin \theta\, d\theta\, d\phi; \qquad (28.9)$$

this is the same as the $f(v)\, dv$ that we introduced in Section 19.7. If, as we assumed in Chapter 12, the gas is isotropic, then $f(\mathbf{v}, t)$ must be independent of the angular variables and can be taken outside the integral above. We then have

$$f(v, t) = v^2 f(\mathbf{v}, t) \int_{\theta=0}^{\pi} \int_{\phi=0}^{2\pi} \sin \theta\, d\theta\, d\phi = 4\pi v^2 f(\mathbf{v}, t), \qquad (28.10)$$

a result that we anticipated in Eq. 12.9.

We can, of course, also define a probability density corresponding to the more general distribution function $f(\mathbf{r}, \mathbf{v}, t)$, by the equation

$$f(\mathbf{r}, \mathbf{v}, t) \equiv \frac{1}{N} f(\mathbf{r}, \mathbf{v}, t). \qquad (28.11)$$

As before, we have a normalization condition,

$$1 = \int_{v_x=-\infty}^{+\infty} \int_{v_y=-\infty}^{+\infty} \int_{v_z=-\infty}^{+\infty} \iiint_V f(\mathbf{r}, \mathbf{v}, t)\, dv_x\, dv_y\, dv_z\, dx\, dy\, dz, \qquad (28.12)$$

and $f(\mathbf{r}, \mathbf{v}, t)$ is the probability density in the six-dimensional position-velocity space. By analogy with Eq. 28.2, integration over the position coordinates yields

$$f(\mathbf{v}, t) = \iiint_V f(\mathbf{r}, \mathbf{v}, t)\, dx\, dy\, dz. \qquad (28.13)$$

Alternatively, if we integrate $f(\mathbf{r}, \mathbf{v}, t)$ over all velocities, we obtain the

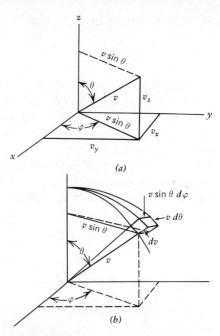

FIGURE 28.2
(a) Spherical coordinates in velocity space. (b) The differential volume element in spherical coordinates.

probability density in configuration space,

$$\digamma(\mathbf{r}, t) \equiv \int_{v_x=-\infty}^{+\infty} \int_{v_y=-\infty}^{+\infty} \int_{v_z=-\infty}^{+\infty} \digamma(\mathbf{r}, \mathbf{v}, t)\, dv_x\, dv_y\, dv_z. \qquad (28.14)$$

The function $\digamma(\mathbf{r}, t)\, d\mathbf{r}$ is the probability of finding a particle between $\mathbf{r}$ and $\mathbf{r}+d\mathbf{r}$; thus, $f(\mathbf{r}, t) \equiv N\digamma(\mathbf{r}, t)$ is the local number density at $\mathbf{r}$.

For a final example of a distribution function, let us formulate one for the internal degrees of freedom. Say that we have a system of N molecular oscillators, each with g quantum states labeled by the integers $0, 1, 2, \ldots, g-1$. The function $f(j, t)$ is defined to be the number of oscillators in quantum state j at time t.[2] The sum of $f(j, t)$ over all j equals the total number of oscillators,

$$N = \sum_{j=0}^{g-1} f(j, t). \qquad (28.15)$$

Now the function $N^{-1}f(j, t)$ is the mole fraction of oscillators in quantum state j at time t, $x(j, t)$, and again we must have the normalization

$$1 = \sum_j x(j, t). \qquad (28.16)$$

In the kinetic theory of matter we picture gas molecules as moving and colliding with other molecules or with the container wall. Imagine a volume element well removed from the walls. At equilibrium, the spatial distribution of the molecules is uniform, and the velocity distribution is Maxwellian (see Section 19.7, especially Eq. 19.143). Suppose that we were to take the molecules in this small volume and somehow give them a nonequilibrium velocity distribution, while maintaining the uniform spatial distribution. How will the nonequilibrium distribution of velocities be converted in time to the equilibrium distribution? What is required to effect this change? In order to change a distribution, the numbers of molecules with given velocities must change. In the assumed absence of collisions with the walls of the vessel, the molecules can change their velocities only by colliding with one another. Since we are interested in the rate at which the equilibrium velocity distribution is attained (we speak of the *relaxation time* of the process), we need to know the frequency with which such collisions take place.

The hypothetical example of a nonequlibrium distribution we have just discussed is difficult to achieve in a physical system. However, it is possible to "experiment" with a computer. In 1955 Alder and Wainwright programmed a computer to solve the following problem: Take 100 hard spheres of a given diameter d, and distribute them uniformly in a volume V. Now start off all the spheres with the same speed but moving in random directions. The machine calculates and follows the motion of each particle. When any particle approaches another within a distance d, a collision occurs. The computer has in storage ('knows') the velocity and position of each sphere at each instant. Hence it can calculate the impact parameter before the collision, the scattering angle (cf. Section 27.6), and the magnitude and direction of each sphere's velocity after the collision. Thus, laboriously, the machine can print out a table and even a picture of the distribution functions for both velocity and speed at any time. It can also calculate the average time between collisions for a single particle. One of Alder and Wainwright's results is shown in Fig. 28.3, where the

28.2 Collision Frequency in a Dilute Gas

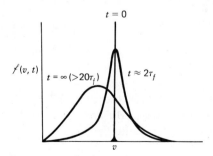

FIGURE 28.3
Calculated distribution of speeds as a function of time; τ_f is the average time between collisions. Schematic, after B. Alder and T. Wainwright. (Not to scale.)

---

[2] Note that we could have written $f(j, t)\, \Delta j$, in closer analogy to the velocity distribution. Since j can have only integral values, Δj is unity!

distribution of speeds is drawn at several time intervals, measured in units of the average time between collisions, τ_f. The assumed initial distribution is a sharp spike at one speed. After just a few multiples of τ_f, that is, after a few collisions per molecule, the distribution has altered radically and already closely approaches an asymptotic limit. This limit is called the equilibrium distribution. We say that the initial nonequilibrium distribution of speeds has relaxed to its equilibrium form in a few *mean free times* (average times between collisions). Note carefully that it is only by the mechanism of collisions that the "gas" attains equilibrium.

In this and the next few sections we wish to discuss a part of the theory associated with such an experiment. We begin with a derivation of the collision frequency in a dilute gas. For this purpose we consider a spatially uniform mixture of two gases, each composed of structureless particles, say argon (1) and xenon (2). Let the number densities of the two gases be n_1 and n_2, each constant with time, and let the corresponding probability distributions of velocities be $f_1(\mathbf{v}_1, t)$ and $f_2(\mathbf{v}_2, t)$, each possibly a function of time. Thus the number density of argon atoms with velocities in the range from $\mathbf{v}_1$ to $\mathbf{v}_1 + d\mathbf{v}_1$ is $n_1 f_1(\mathbf{v}_1, t)\, d\mathbf{v}_1$. In Section 27.5 we discussed collisions between molecules in molecular beams. The beams as such are only experimental devices to achieve a collection of molecules all with essentially the same velocity, and our results there can be easily generalized. Thus, according to Eq. 27.55, the number of collisions per unit time in a volume τ between argon atoms with velocities from $\mathbf{v}_1$ to $\mathbf{v}_1 + d\mathbf{v}_1$, whose density is $n_1 f_1(\mathbf{v}_1, t)\, d\mathbf{v}_1$, and xenon atoms with velocities from $\mathbf{v}_2$ to $\mathbf{v}_2 + d\mathbf{v}_2$, whose density is $n_2 f_2(\mathbf{v}_2, t)\, d\mathbf{v}_2$, is

$$\tau Z_{12}(\mathbf{v}_1, \mathbf{v}_2) = 2\pi v \int_{\chi=0}^{\pi} \sigma(v, \chi) n_1 n_2 f_1(\mathbf{v}_1, t) f_2(\mathbf{v}_2, t)\tau\, d\mathbf{v}_1\, d\mathbf{v}_2 \sin \chi\, d\chi.$$

(28.18)

The relative speed v equals $|\mathbf{v}_2 - \mathbf{v}_1|$, and is thus determined in the beam experiment by the beam intersection angle for given speeds v_1, v_2. Once we have established the collision rate between $\mathrm{Ar}(n_1, \mathbf{v}_1)$ and $\mathrm{Xe}(n_2, \mathbf{v}_2)$, we can obtain the total collision rate between argon and xenon atoms in the gas by summing Eq. 28.18 over all possible values of $\mathbf{v}_1$ and $\mathbf{v}_2$.

First, however, we must call attention to one crucial assumption made here, concerning the free and independent motion of the atoms prior to collision. How can we be sure, for example, that the velocities of the colliding atoms are not somehow correlated with one another? Such a correlation might cause the values of $f(\mathbf{v}, t)$ for colliding atoms to differ from the values in the gas as a whole. We can only be reasonably sure by limiting our discussion to gases where, *on the average*, most of the molecules are independent of each other most of the time. We know that this must be true in the ideal-gas limit. Note, however, that we have to make a *statistical* assumption regarding the independence of the atoms, on the average, or most of the time. In dense gases and liquids this assumption of independence is invalid, and a different formulation must be used.

The average number of collisions per unit volume per unit time between argon atoms with velocities from $\mathbf{v}_1$ to $\mathbf{v}_1 + d\mathbf{v}_1$ and xenon atoms of all possible velocities is simply obtained from Eq. 28.18 by dividing by τ and integrating over all possible values of $\mathbf{v}_2$:

$$Z_{12}(\mathbf{v}_1) = 2\pi n_1 n_2\, d\mathbf{v}_1 \int_{\mathbf{v}_2} \int_{\chi=0}^{\pi} v\sigma(v, \chi) f_1(\mathbf{v}_1, t) f_2(\mathbf{v}_2, t)\, d\mathbf{v}_2 \sin \chi\, d\chi.$$

(28.19)

Finally, the total number of collisions per unit volume per unit time between argon atoms of all velocities and xenon atoms of all velocities is the integral of Eq. 28.19 over all possible values of $\mathbf{v}_1$. This number we shall call Z_{12}, the *collision frequency:*

$$Z_{12} \equiv 2\pi n_1 n_2 \int_{\mathbf{v}_1} \int_{\mathbf{v}_2} \int_{\chi=0}^{\pi} v\sigma(v, \chi) f_1(\mathbf{v}_1, t) f_2(\mathbf{v}_2, t) \, d\mathbf{v}_2 \, d\mathbf{v}_1 \sin\chi \, d\chi.$$

$$(28.20)$$

28.3
The Evolution of Velocity Distributions in Time

Let us continue with the example of the previous section. We assume that in a gas, either pure or a mixture, there exists at some initial time a nonequilibrium velocity distribution. Further, as time goes on the velocity distribution changes and approaches the equilibrium form. We recognize that molecular collisions are the mechanism by which energy and momentum are exchanged between the molecules, so that the collision frequency will determine the rate of change of the velocity distribution. We shall now derive an equation for this rate of change.

The number of molecules per unit volume with velocities in the range from $\mathbf{v}_1$ to $\mathbf{v}_1 + d\mathbf{v}_1$ is $n f(\mathbf{v}_1, t) \, d\mathbf{v}_1$, the product of the number density, the velocity probability density, and the volume element in velocity space. If we consider a pure gas uniformly distributed in a volume V, then the density n is a constant. The expression

$$n\left[\frac{df(\mathbf{v}_1, t)}{dt}\right]_{\mathbf{v}_1} d\mathbf{v}_1$$

describes the increase per unit time in the number of molecules per unit volume with velocities in the range from $\mathbf{v}_1$ to $\mathbf{v}_1 + d\mathbf{v}_1$. The subscript $\mathbf{v}_1$ on the brackets is to remind the reader that we are looking for a rate of change in a given range $d\mathbf{v}_1$ at a fixed (constant) $\mathbf{v}_1$. Any such change in the number of molecules in the range from $\mathbf{v}_1$ to $\mathbf{v}_1 + d\mathbf{v}_1$ must be a balance of two processes: A molecule entering a collision with velocity $\mathbf{v}_1$ will have a different velocity afterwards; conversely, a molecule with a different initial velocity can have velocity $\mathbf{v}_1$ after a collision. In the last section we obtained an expression, Eq. 28.19, for the total number of collisions per unit volume per unit time between atoms of a given velocity $\mathbf{v}_1$ and those with all possible values of $\mathbf{v}_2$; we rewrite that result here for a pure gas, omitting the subscripts on the densities and distribution functions,

$$Z_{12}(\mathbf{v}_1) = 2\pi n^2 \, d\mathbf{v}_1 \int_{\chi} \int_{\mathbf{v}_2} v\sigma(v, \chi) f(\mathbf{v}_1, t) f(\mathbf{v}_2, t) \, d\mathbf{v}_2 \sin\chi \, d\chi. \quad (28.21)$$

What we must now do is find the corresponding expression for the number of collisions with *final* velocity $\mathbf{v}_1$.

Let us again consider the dynamics of a collision. This time we suppose that the collision occurs with given velocities $\mathbf{v}_1, \mathbf{v}_2$ and a given scattering angle χ. Such a choice also implies a given initial impact parameter b, since for given $\mathbf{v}_1, \mathbf{v}_2$ the relative velocity $\mathbf{v}$ is fixed, and in classical mechanics the scattering angle is completely determined by v and b. As before, we designate the final velocities of the particles after the collision by $\mathbf{v}'_1, \mathbf{v}'_2$. Remember that in an elastic collision the final relative speed v' must equal v, and the final impact parameter b' must equal b (see Section 27.1). Next, look at Fig. 27.3, which illustrates the velocity relationships in an elastic collision, and imagine a collision in which $\mathbf{v}'_1$ and $\mathbf{v}'_2$ are the initial velocities. What must be the final velocities if the scattering angle χ is to be the same as in the process $(\mathbf{v}_1, \mathbf{v}_2 \rightarrow \mathbf{v}'_1, \mathbf{v}'_2)$? Clearly, they can only be $\mathbf{v}_1, \mathbf{v}_2$. Notice that in Fig. 27.3 there is no way of distinguishing between initial and final velocities, the choice of which set of velocities is designated as initial or final being up to the observer.

We can write, therefore, for the number of collisions between atoms of velocities $\mathbf{v}_1'$ and $\mathbf{v}_2'$ resulting in scattering through a relative scattering angle from χ to $\chi + d\chi$ per unit volume per unit time,

$$Z_{12}(\mathbf{v}_1', \mathbf{v}_2', \chi) = 2\pi n^2 v' \sigma(v', \chi) f(\mathbf{v}_1', t) f(\mathbf{v}_2', t) \, d\mathbf{v}_1' \, d\mathbf{v}_2' \sin \chi \, d\chi, \quad (28.22)$$

according to Eq. 27.54. For such (elastic) collisions we know that v' equals v. Furthermore, we can obtain[3] a relationship between the differential volume elements in velocity space,

$$d\mathbf{v}_1' \, d\mathbf{v}_2' = d\mathbf{V}' \, d\mathbf{v}', \quad (28.23)$$

and similarly for the unprimed velocities,

$$d\mathbf{v}_1 \, d\mathbf{v}_2 = d\mathbf{V} \, d\mathbf{v}. \quad (28.24)$$

Since the center-of-mass vector is unchanged by the collision, $d\mathbf{V}$ is the same in Eqs. 28.23 and 28.24. One might also expect that $d\mathbf{v} = d\mathbf{v}'$, and we shall now show this explicitly.

In order to prove the equivalence of the volume elements $d\mathbf{v}$ and $d\mathbf{v}'$, let us consider Fig. 28.4. This resembles Fig. 27.3, but we have added a line called the *apse line*, which lies in the plane determined by $\mathbf{v}$, $\mathbf{v}'$ and bisects the relative scattering angle χ. The relative velocity vector, which always lies in the same plane, can be decomposed into two components, one parallel and one perpendicular to the apse line. As we previously found for the hard-sphere case (cf. discussion in connection with Fig. 27.9), the effect of the collision is to leave the former component unchanged and to reverse the sign of the latter. Since the magnitudes of

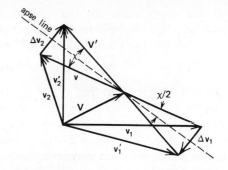

FIGURE 28.4
Velocity relationships for an elastic collision in a pure gas. (Note that in a pure gas **V** bisects **v**, as is apparent from Eqs. 27.15.)

---

[3] We can derive these relations by considering the Cartesian velocity components separately. The x components of $\mathbf{v}$ and $\mathbf{V}$ are given by

$$v_x = |v_{2x} - v_{1x}|$$

and

$$(m_1 + m_2)V_x = m_1 v_{1x} + m_2 v_{2x},$$

respectively (cf. Eqs. 27.11 and 27.14). We thus obtain from Eqs. 27.15

$$v_{1x} = V_x - \frac{m_2}{m_1 + m_2} v_x,$$

$$v_{2x} = V_x + \frac{m_1}{m_1 + m_2} v_x.$$

By Eq. II.17, the differential elements are connected by the relation

$$dv_{1x} \, dv_{2x} = J(v_{1x}, v_{2x}/v_x, V_x) \, dv_x \, dV_x,$$

where the Jacobian J is given by

$$J(v_{1x}, v_{2x}/v_x, V_x) = \left(\frac{\partial v_{1x}}{\partial V_x}\right)_{v_x} \left(\frac{\partial v_{2x}}{\partial v_x}\right)_{V_x} - \left(\frac{\partial v_{1x}}{\partial v_x}\right)_{V_x} \left(\frac{\partial v_{2x}}{\partial V_x}\right)_{v_x}$$

$$= (1)\left(\frac{m_1}{m_1 + m_2}\right) - \left(\frac{-m_2}{m_1 + m_2}\right)(1) = 1.$$

The Jacobian is thus unity, and we have

$$dv_{1x} \, dv_{2x} = dV_x \, dv_x;$$

identical arguments hold for the y and z coordinates. Since for each of the vectors the volume element is multiplicative, for example,

$$d\mathbf{v}_1 = dv_{1x} \, dv_{1y} \, dv_{1z},$$

we immediately obtain Eq. 28.24. Obviously, the procedure leading to Eq. 28.23 is identical.

all the components are thus the same before and after collision, we conclude that the volume elements must also be equal, $d\mathbf{v} = d\mathbf{v}'$.

Inserting this result in Eqs. 28.23 and 28.24, we deduce the equality

$$d\mathbf{v}'_1 \, d\mathbf{v}'_2 = d\mathbf{v}_1 \, d\mathbf{v}_2, \qquad (28.25)$$

which holds provided that $\mathbf{v}_1$ and $\mathbf{v}_2$ are related to $\mathbf{v}'_1$ and $\mathbf{v}'_2$ as shown in Fig. 28.4, that is, related by the equation of motion for a given b or χ. To show this relation more clearly, we shall replace $\mathbf{v}'_1$ by the expression $\mathbf{v}_1 - \Delta\mathbf{v}_1$, Fig. 28.4, where $\Delta\mathbf{v}_1(v, \chi)$ is completely determined by v and χ, and is in the direction perpendicular to the apse line. Similarly, we write $\mathbf{v}'_2 = \mathbf{v}_2 + \Delta\mathbf{v}_2(v, \chi)$.[4]

Now we can introduce in Eq. 28.22 the transformation of differential volume elements, giving

$$Z_{12}(\mathbf{v}_1 - \Delta\mathbf{v}_1, \mathbf{v}_2 - \Delta\mathbf{v}_2, \chi) = 2\pi n^2 v\sigma(v, \chi) f[\mathbf{v}_1 - \Delta\mathbf{v}_1(v, \chi), t]$$
$$\times f[\mathbf{v}_2 + \Delta\mathbf{v}_2(v, \chi), t] \, d\mathbf{v}_1 \, d\mathbf{v}_2 \sin \chi \, d\chi; \quad (28.26)$$

by integrating this expression over all possible values of $\mathbf{v}_2$ and χ, we obtain the total number of collisions per unit volume per unit time in which one atom emerges from the collision with a velocity between $\mathbf{v}_1$ and $\mathbf{v}_1 + d\mathbf{v}_1$,

$$Z_{12}(\mathbf{v}_1 - \Delta\mathbf{v}_1) = 2\pi n^2 \, d\mathbf{v}_1 \int_\chi \int_{\mathbf{v}_2} v\sigma(v, \chi) f[\mathbf{v}_1 - \Delta\mathbf{v}_1(v, \chi), t]$$
$$\times f[\mathbf{v}_2 + \Delta\mathbf{v}_2(v, \chi), t] \, d\mathbf{v}_2 \sin \chi \, d\chi. \quad (28.27)$$

The difference between the increase (per unit volume per unit time) in the number of molecules with velocity from $\mathbf{v}_1$ to $\mathbf{v}_1 + d\mathbf{v}_1$, Eq. 28.27, and the corresponding decrease, Eq. 28.21, must equal the net increase $n[d f(\mathbf{v}_1, t)/dt]_{\mathbf{v}_1} \, d\mathbf{v}_1$. Cancelling the common factors $n \, d\mathbf{v}_1$, we obtain the *Boltzmann equation*,

$$\left[\frac{d f(\mathbf{v}_1, t)}{dt}\right]_{\mathbf{v}_1} = 2\pi n \int_\chi \int_{\mathbf{v}_2} v\sigma(v, \chi)\{f[\mathbf{v}_1 - \Delta\mathbf{v}_1(v, \chi), t] f[\mathbf{v}_2 + \Delta\mathbf{v}_2(v, \chi), t]$$
$$- f(\mathbf{v}_1, t) f(\mathbf{v}_2, t)\} \, d\mathbf{v}_2 \sin \chi \, d\chi, \quad (28.28)$$

which describes the evolution in time of the velocity distribution function in a spatially homogeneous pure gas. It is a very complicated type of equation, in which the unknown, the velocity distribution $f(\mathbf{v}_1, t)$, appears both as a differential and under an integral sign. (Such an equation is called an *integro-differential equation*.) Its general solutions are unknown.

However, we shall be interested here only in a very special solution, the limit for infinite time. In that limit we can assume that the velocity distribution function has approached its equilibrium form for virtually all initial nonequilibrium distributions. And for almost all such initial distributions the assumption of the independence of the particles prior to collision, which was essential to our formulation of the collision frequency, may be expected to hold.[5] Note also that, in the absence of

---

[4] We shall retain this more general notation, but note that in a collision between identical molecules we have $\Delta\mathbf{v}_1 = -\Delta\mathbf{v}_2$, as is apparent from Fig. 28.4.

[5] An example of a system where these assumptions do not hold is a box containing a few particles, all moving with the same velocity normal to two opposite, perfectly smooth walls. They bounce back and forth between the two walls, but can never collide with one another. The equilibrium distribution will never be attained, even though the distribution function does not change with time! The motions and velocities of the molecules are not independent: If you know the velocity of one, then you know the velocity of *all* the particles. There is thus a perfect correlation between the velocities of all the molecules.

external fields or hydrodynamic flow, the equilibrium distribution function cannot have a preferred direction, hence must be isotropic; $f(\mathbf{v}_1, \infty)$ thus depends only on the magnitude of $\mathbf{v}$. In the limit $t \to \infty$ we expect, therefore, that

$$\left[\frac{df(\mathbf{v}_1, t)}{dt}\right]_{\mathbf{v}_1} = 0. \tag{28.29}$$

A sufficient (though not necessary) condition for Eq. 28.29, that is, for the right-hand side of Eq. 28.28 to vanish, is that

$$f[\mathbf{v}_1 - \Delta\mathbf{v}_1(v, \chi)]f[\mathbf{v}_2 + \Delta\mathbf{v}_2(v, \chi)] - f(\mathbf{v}_1)f(\mathbf{v}_2) = 0 \tag{28.30}$$

for all values of $\mathbf{v}_1$, $\mathbf{v}_2$, and χ. What does this equation tell us? It says that at equilibrium there are as many collisions per unit time of the type

$$\mathbf{v}_1, \mathbf{v}_2 \to \mathbf{v}_1 - \Delta\mathbf{v}_1(v, \chi), \mathbf{v}_2 + \Delta\mathbf{v}_2(v, \chi)$$

as there are inverse collisions of the type

$$\mathbf{v}_1 - \Delta\mathbf{v}_1(v, \chi), \mathbf{v}_2 + \Delta\mathbf{v}_2(v, \chi) \to \mathbf{v}_1, \mathbf{v}_2.$$

It says that the equilibrium distribution of velocities is maintained by a "balance of active tendencies," as J. Willard Gibbs put it; that is, by a *detailed balance* between each possible collision and its inverse. Although we shall not prove so here, the necessity for this balance to exist can be obtained from the laws of mechanics.

There is still more information in Eq. 28.30. If we rearrange that equation and take the logarithm of both sides, we have

$$\ln f[\mathbf{v}_1 - \Delta\mathbf{v}_1(v, \chi)] + \ln f[\mathbf{v}_2 + \Delta\mathbf{v}_2(v, \chi)] = \ln f(\mathbf{v}_1) + \ln f(\mathbf{v}_2). \tag{28.31}$$

Thus the logarithm of the velocity distribution function is a summation invariant of the collision! But at the end of Section 27.1 we pointed out that there are only three summation invariants for an elastic collision: the mass, the linear momentum, and the kinetic energy. Hence the logarithm of $f(\mathbf{v}_1)$ must be some linear combination of the summation invariants. Any such combination will satisfy Eq. 28.31; but since we wish $f(\mathbf{v}_1)$ to depend only on the magnitude v_1, we omit the momentum (which contains $\mathbf{v}_1$ vectorially) and include the velocity only in the form of the kinetic energy. We thus assume the form

$$\ln f(\mathbf{v}_1) = \alpha + \beta(\tfrac{1}{2}mv_1^2), \tag{28.32}$$

where α and β are constants independent of v_1; this result is clearly consistent with Eq. 28.31. As in Section 15.5, we shall make the identification $\beta = 1/k_B T$, which can again be justified by deriving the perfect gas law. The resulting distribution function can be written

$$f(\mathbf{v}_1) = Ae^{-mv_1^2/2k_B T}; \tag{28.33}$$

the normalization factor A will be evaluated (and the substitution $\beta = 1/k_B T$ justified) in the next section. This result is called the *Maxwell–Boltzmann distribution* of velocities for a gas at rest, and can be seen to correspond to Eq. 19.143, which we derived from statistical mechanics. If we were also to include a term proportional to the linear momentum in Eq. 28.32, we would obtain the (nonisotropic) equilibrium velocity distribution for a gas moving through space; we shall also come back to this point in the next section.

In summary, in this section we have derived an equation for the variation with time of the velocity distribution in a dilute, spatially homogeneous gas. The velocity distribution function will attain its equilibrium (invariant) form if the integrand in the Boltzmann equation is everywhere zero. This condition can be satisfied only if the distribution function depends on the collisional invariants in a certain way. For a gas at rest only the Maxwell–Boltzmann distribution meets the requirements.

The Maxwell–Boltzmann distribution is one of the most important relationships in all of physical science. Historically, it predated the more general Boltzmann distribution of energy-level populations, one of the cornerstones of applied statistical mechanics, of which it is of course a special case. We shall find many applications of the Maxwell–Boltzmann distribution to the nonequilibrium properties of gases, but in the present section we only consider some of its simpler consequences.

First we must evaluate the normalization constant of Eq. 28.33. By our definition of the probability density f, we must have[6]

$$\int_{\mathbf{v}} f(\mathbf{v})\, d\mathbf{v} = 1, \qquad (28.34)$$

where the integral is taken over all possible values of the velocity $\mathbf{v}$. This equation is satisfied by

$$f(\mathbf{v}) = \left(\frac{m}{2\pi k_B T}\right)^{3/2} e^{-mv^2/2k_B T}, \qquad (28.35)$$

as we shall now show. Substituting this expression in the integral of Eq. 28.34 and expanding in Cartesian coordinates, we have

$$\int_{\mathbf{v}} f(\mathbf{v})\, d\mathbf{v} = \left(\frac{m}{2\pi k_B T}\right)^{3/2} \int_{v_x} \int_{v_y} \int_{v_z} e^{-m(v_x^2+v_y^2+v_z^2)/2k_B T}\, dv_x\, dv_y\, dv_z. \qquad (28.36)$$

We see that the triple integration can be separated into three independent integrations,

$$\int_{\mathbf{v}} f(\mathbf{v})\, d\mathbf{v} = \left(\frac{m}{2\pi k_B T}\right)^{3/2} \left[\int_{v_x=-\infty}^{+\infty} e^{-mv_x^2/2k_B T}\, dv_x\right]\left[\int_{v_y=-\infty}^{+\infty} e^{-mv_y^2/2k_B T}\, dv_y\right]$$
$$\times \left[\int_{v_z=-\infty}^{+\infty} e^{-mv_z^2/2k_B T}\, dv_z\right], \qquad (28.37)$$

so that we have the product of three identical factors of the form

$$\left(\frac{m}{2\pi k_B T}\right)^{1/2} \int_{u=-\infty}^{+\infty} e^{-mu^2/2k_B T}\, du = 1. \qquad (28.38)$$

Equation 28.34 is thus satisfied, and Eq. 28.35 represents the normalized probability density.

Because of the separability of the integral 28.36, we can write

$$f(\mathbf{v})\, d\mathbf{v} = [f(v_x)\, dv_x][f(v_y)\, dv_y][f(v_z)\, dv_z], \qquad (28.39)$$

where $f(v_x)$, $f(v_y)$, $f(v_z)$ are the probability densities for the velocity components v_x, v_y, v_z. That is, the fraction of molecules with x component of velocity between v_x and $v_x + dv_x$ is given by

$$f(v_x)\, dv_x = \left(\frac{m}{2\pi k_B T}\right)^{1/2} e^{-mv_x^2/2k_B T}\, dv_x, \qquad (28.40)$$

and the distributions for v_y and v_z are of the same form. The significance

28.4
The Maxwell–Boltzmann Distribution

---

[6] To simplify the notation, in this section we use the unsubscripted $\mathbf{v}$ for the velocity of an individual molecule, rather than the relative velocity.

of this separability is that the distribution of velocity components in one direction is independent of the components in the other directions. The one-dimensional probability distribution given by Eq. 28.40 is plotted in Fig. 28.5. It is symmetric about the ordinate for zero velocity, at which the maximum value is proportional to $T^{-1/2}$. The "spread" of the distribution, which can be described by, say, its width at half the maximum height, is a function of both the temperature and the molecular mass. For a given gas, as the temperature increases the spread of the distribution increases, but the area under the curve remains unity (normalized distribution). The Maxwell–Boltzmann distribution is an example of what is called (in the field of statistics) a *Gaussian* or *normal distribution*, proportional to $e^{-a\delta^2}$, where δ is a measure of the deviation from some mean and a is a positive constant. Such distributions give a good description of the behavior of many random processes.

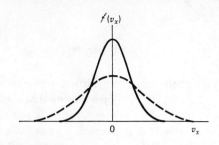

FIGURE 28.5
Probability density for velocity component in one dimension. Solid curve: $T = 300$ K; broken curve: $T = 1000$ K.

Next let us evaluate the average velocity in a gas at rest. The average velocity is defined as

$$\langle \mathbf{v} \rangle \equiv \int_{\mathbf{v}} \mathbf{v} f(\mathbf{v})\, d\mathbf{v}, \tag{28.41}$$

and its x component as

$$\langle \mathbf{v}_x \rangle = \int_{v_x=-\infty}^{+\infty} \int_{v_y=-\infty}^{+\infty} \int_{v_z=-\infty}^{+\infty} v_x f(\mathbf{v})\, dv_x\, dv_y\, dv_z. \tag{28.42}$$

If we substitute for $f(\mathbf{v})$ the Maxwell–Boltzmann distribution, as expanded in Eq. 28.36, we can carry out the integrations over v_y and v_z in the same way as in Eq. 28.38. This leaves

$$\langle v_x \rangle = \left(\frac{m}{2\pi k_B T}\right)^{1/2} \int_{v_x=-\infty}^{+\infty} v_x e^{-mv_x^2/2k_B T}\, dv_x. \tag{28.43}$$

In the integrand v_x is an odd function (negative for $v_x < 0$, positive for $v_x > 0$), and $\exp(-mv_x^2/2k_B T)$ is an even function (positive for all v_x, symmetric around $v_x = 0$). The complete integrand is thus an odd function, so that the integral from $v_x = -\infty$ to $v_x = 0$ and that from $v_x = 0$ to $v_x = +\infty$ are of the same magnitude but opposite signs, and the total integral vanishes:

$$\langle v_x \rangle = 0. \tag{28.44}$$

The same argument holds for $\langle v_y \rangle$ and $\langle v_z \rangle$, and since all three components of $\langle \mathbf{v} \rangle$ are zero, then $\langle \mathbf{v} \rangle$ itself must have zero magnitude in a gas at rest.

For a gas in macroscopic motion, say in the x direction with a given speed v_{x0}, the average velocity in that direction is not zero but v_{x0}. We still have a Maxwell–Boltzmann velocity distribution, but now symmetric around not $v_x = 0$ but $v_x = v_{x0}$. The form of the distribution is then

$$f(v_x) = \left(\frac{m}{2\pi k_B T}\right)^{1/2} e^{-m(v_x-v_{x0})^2/2k_B T}, \tag{28.45}$$

which, however, is still consistent with the equilibrium solution of the Boltzmann equation, Eq. 28.31, in this case with a momentum term added to Eq. 28.32.

From the probability density of velocities, Eq. 28.35, we can readily calculate the probability density of speeds (cf. Eq. 28.9 and the accompanying discussion). We have

$$f(v) = v^2 \int_{\theta=0}^{\pi} \int_{\phi=0}^{2\pi} \left(\frac{m}{2\pi k_B T}\right)^{3/2} e^{-mv^2/2k_B T} \sin\theta\, d\theta\, d\phi; \tag{28.46}$$

our isotropic $f(\mathbf{v})$ can be taken outside the integral, and we immediately obtain

$$f(v) = 4\pi v^2 f(\mathbf{v}) = 4\pi\left(\frac{m}{2\pi k_B T}\right)^{3/2} v^2 e^{-mv^2/2k_B T}. \qquad (28.47)$$

This probability distribution is shown in Fig. 28.6, again for two temperatures, and the reader will note that the distribution is not symmetric about any speed. As the temperature is increased, the fraction of molecules with large speeds is increased. The area under each distribution is unity. (See also Fig. 19.7.)

The average *speed* of the molecules in the gas is not zero, but is

$$\langle v\rangle = \int_{v=0}^{\infty} v\, f(v)\, dv = 4\pi\left(\frac{m}{2\pi k_B T}\right)^{3/2}\int_{v=0}^{\infty} v^3 e^{-mv^2/2k_B T}\, dv$$
$$= \left(\frac{8k_B T}{\pi m}\right)^{1/2}. \qquad (28.48)$$

the same result we obtained as Eq. 19.145. As the temperature increases, the average speed increases, but only slowly, proportional to $T^{1/2}$.

For any distribution we can, of course, define and calculate not only the average of the independent variable, say speed, but also the average of the square, the cube, and so on, of that variable. Each such average is called a "moment" of the distribution (Section 15.2), the average speed being the first moment of the probability distribution of speeds. Suppose that we wish to calculate the average kinetic energy of the molecules in the gas; this is related to the second moment of the distribution, and is evaluated as

$$\langle \tfrac{1}{2}mv^2\rangle = 4\pi\left(\frac{m}{2\pi k_B T}\right)^{3/2}\int_{v=0}^{\infty}(\tfrac{1}{2}mv^2)v^2 e^{-mv^2/2k_B T}\, dv \qquad (28.49)$$
$$= \tfrac{3}{2}k_B T$$

(cf. Eq. 19.146). Since the energy of an ideal gas is the average kinetic energy per molecule multiplied by the number of molecules, the energy per mole is

$$u = \tfrac{3}{2}N_A k_B T = \tfrac{3}{2}RT. \qquad (28.50)$$

We can combine this result with that of Eq. 12.17,

$$p = \frac{2}{3}\frac{U}{V} \qquad (28.51)$$

(which was derived assuming only that the gas is homogeneous and isotropic), and obtain the perfect gas law,

$$pV = Nk_B T. \qquad (28.52)$$

It is this conclusion that confirms the correctness of our replacing β by $1/k_B T$ in Eq. 28.33—or, to put it another way, which proves that the $k_B T$ introduced there means the same thing as in earlier chapters.

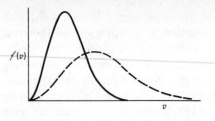

FIGURE 28.6
Equilibrium probability distribution of speeds in a gas at rest. Solid curve: $T = 300$ K; broken curve: $T = 1000$ K.

28.5 Collision Frequency for Hard-Sphere Molecules

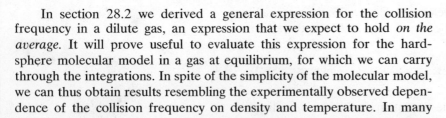

In section 28.2 we derived a general expression for the collision frequency in a dilute gas, an expression that we expect to hold *on the average*. It will prove useful to evaluate this expression for the hard-sphere molecular model in a gas at equilibrium, for which we can carry through the integrations. In spite of the simplicity of the molecular model, we can thus obtain results resembling the experimentally observed dependence of the collision frequency on density and temperature. In many applications the numerical predictions of this model are of the right order of magnitude, and sometimes even better.

As in the discussion preceding Eq. 28.20, we shall consider a gas mixture of argon and xenon, with the respective number densities n_1 and n_2. The total number of collisions per unit volume per unit time between argon and xenon atoms is

$$Z_{12} \equiv 2\pi n_1 n_2 \int_{\mathbf{v}_1} \int_{\mathbf{v}_2} \int_{\chi=0}^{\pi} v\sigma(v, \chi) f_1(\mathbf{v}_1, t) f_2(\mathbf{v}_2, t) \, d\mathbf{v}_1 \, d\mathbf{v}_2 \sin \chi \, d\chi.$$

For the hard-sphere model the differential elastic scattering cross section $\sigma(v, \chi)$ is independent of both angle and relative speed, and is simply $d^2/4$, as we found in Eq. 27.69. Furthermore, in a gas at equilibrium we know the velocity probability distribution, given by Eq. 28.35. Neither the cross section nor the distribution functions depend on the relative scattering angle χ, so that this integration can be performed at once:

$$Z_{12} = \pi d^2 n_1 n_2 \left(\frac{m_1}{2\pi k_B T}\right)^{3/2} \left(\frac{m_2}{2\pi k_B T}\right)^{3/2} \int_{\mathbf{v}_1} \int_{\mathbf{v}_2} v e^{-(m_1 v_1^2 + m_2 v_2^2)/2k_B T} \, d\mathbf{v}_1 \, d\mathbf{v}_2. \tag{28.53}$$

Care must be used in integrating over the velocities $\mathbf{v}_1$ and $\mathbf{v}_2$, since the relative speed v depends on both, $v = |\mathbf{v}_1 - \mathbf{v}_2|$. It is simplest to convert to the relative and center-of-mass velocities. First we convert the constant factors,

$$\left(\frac{m_1}{2\pi k_B T}\right)^{3/2} \left(\frac{m_2}{2\pi k_B T}\right)^{3/2} = \left[\frac{m_1 m_2}{(m_1 + m_2)2\pi k_B T}\right]^{3/2} \left[\frac{(m_1 + m_2)}{2\pi k_B T}\right]^{3/2}$$

$$= \left(\frac{\mu_{12}}{2\pi k_B T}\right)^{3/2} \left(\frac{m_1 + m_2}{2\pi k_B T}\right)^{3/2}, \tag{28.54}$$

introducing the reduced mass, here called μ_{12}. We also need the equivalence of kinetic energies given by Eq. 27.17,

$$\tfrac{1}{2} m_1 v_1^2 + \tfrac{1}{2} m_2 v_2^2 = \tfrac{1}{2}(m_1 + m_2)V^2 + \tfrac{1}{2}\mu_{12} v^2, \tag{28.55}$$

and the equivalence of the volume elements,

$$d\mathbf{v}_1 \, d\mathbf{v}_2 = d\mathbf{V} \, d\mathbf{v} \tag{28.56}$$

(see Eq. 28.24). With the introduction of the last three equations, the expression for the collision frequency becomes

$$Z_{12} = \pi d^2 n_1 n_2 \int_{\mathbf{v}} \left[\int_{\mathbf{V}} \left(\frac{m_1 + m_2}{2\pi k_B T}\right)^{3/2} e^{-(m_1 + m_2)V^2/2k_B T} \, d\mathbf{V} \right]$$

$$\times \left(\frac{\mu_{12}}{2\pi k_B T}\right)^{3/2} v e^{-\mu_{12} v^2/2k_B T} \, d\mathbf{v}. \tag{28.57}$$

Note that, if we have Maxwell–Boltzmann distributions for both $\mathbf{v}_1$ and $\mathbf{v}_2$, then we can convert the product of those distributions into a product of Maxwell–Boltzmann distributions for the center-of-mass velocities and the relative velocities.

The integration over $\mathbf{V}$, being of the same form as Eq. 28.36, yields unity. The integration over $\mathbf{v}$ yields the average relative speed (not velocity, since v in Eq. 28.53 is a *speed*), here denoted by

$$\langle v_{12} \rangle = \left(\frac{8k_B T}{\mu_{12}}\right)^{1/2}. \tag{28.58}$$

The only difference between the average speed and the average relative speed is in the mass factors of Eqs. 28.48 and 28.58. The collision frequency between argon and xenon atoms per unit volume per unit time

thus becomes

$$Z_{12} = \pi d^2 n_1 n_2 \langle v_{12} \rangle. \tag{28.59}$$

The number of collisions per unit time of a single argon atom with xenon atoms is just

$$\frac{Z_{12}}{n_1} = \pi d^2 n_2 \langle v_{12} \rangle. \tag{28.60}$$

If we want to know the collision frequency in a pure gas, we must amend Eq. 28.59. First, for identical atoms we have $m_1 = m_2 \equiv m$, so that $\mu_{12} = m/2$; Eq. 28.58 thus becomes

$$\langle v_{12} \rangle = \left[\frac{8k_B T}{\pi (m/2)} \right]^{1/2} = \sqrt{2} \langle v \rangle, \tag{28.61}$$

where $\langle v \rangle$ is the pure-gas average speed of Eq. 28.48. Second, if we begin with the number of collisions per unit time for a single atom, Eq. 28.60, then for a mixture we need only multiply by the number of atoms per unit volume to obtain the collision frequency (Eq. 28.59), but for like atoms we must also divide by a factor of 2, since we have counted every argon–argon collision twice: once for each of the atoms involved. Hence the collision frequency in a pure gas is

$$Z = \frac{\pi d^2}{\sqrt{2}} n^2 \langle v \rangle. \tag{28.62}$$

The collision frequency in a dilute gas thus increases with increasing density and increasing temperature, as intuitive arguments suggest.

For the hard-sphere model there is a simple alternative way of obtaining the collision frequency, with a bit of hand-waving. Picture one atom moving through the gas, with relative speed $\langle v_{12} \rangle$. The atom sweeps out a cylinder of cross-sectional area πd^2 (see Fig. 28.7), and hence a volume of $\pi d^2 \langle v_{12} \rangle$ per second, such that any other atom in this volume will collide with the first atom. If the density is n, then in the volume $\pi d^2 \langle v_{12} \rangle$ there are $\pi d^2 n \langle v_{12} \rangle$ atoms and hence that is the number of collisions per atom per unit time.

It is is useful sometimes to know the average distance a molecule travels between collisions, which is called the *mean free path*. If in time t one atom in a pure gas makes $\pi d^2 n \langle v_{12} \rangle t$ collisions and travels a distance $\langle v \rangle t$, where $\langle v \rangle$ is the average speed, then the mean free path l is

$$l \equiv \frac{\langle v \rangle t}{\pi d^2 n \langle v_{12} \rangle t} = \frac{1}{\sqrt{2} \pi d^2 n}. \tag{28.63}$$

The mean free path thus decreases with increasing density and is independent of temperature. The concept of a mean free path is often useful for an order-of-magnitude calculation; it is not, however, directly measureable, as is the cross section (πd^2 for hard spheres).

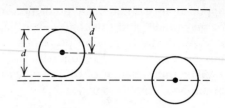

FIGURE 28.7
Illustrating the calculation of collision frequency for the molecular model of hard spheres.

So far we have discussed collisions between molecules, and the establishment of the equilibrium velocity distribution by collisions in a spatially homogeneous gas. Next, we turn to the subject of *molecular fluxes*, net flows of physical quantities in a given direction. Examples of such quantities are the number of particles, the momentum, and the energy; flows of molecules produce fluxes of these quantities. The *flux* will be defined as the amount of a given quantity that flows across some surface per unit time; the *flux density* is the flow per unit time per unit

28.6
Molecular Fluxes of Density, Momentum Density, and Energy Density

area. We know from experience that a gas between two walls at different temperatures will conduct heat from the hotter to the colder wall. We can formulate a statistical expression for this flow of heat, called *thermal conduction*, which corresponds on the molecular level to a flux of energy. Similarly, a net flux of particles is called *diffusion*, and a net flux of momentum produces *viscous flow*.[7] (The various fluxes are also defined in somewhat different form in Chapter 20, where the treatment is macroscopic, and the emphasis on hydrodynamic applications.)

To derive an expression for the molecular flux, we consider a dilute gas and imagine somewhere in the gas a hypothetical plane surface (see Fig. 28.8; this derivation differs somewhat from the similar one in Section 12.2). The orientation of the plane is given by a perpendicular unit vector **k** pointing in what we shall call the positive direction. Molecules will cross the plane in both directions, and we wish to calculate the net flow of molecules across the plane per unit area per unit time. Let us consider molecules with velocity **v**, such that the angle between the vectors **v** and **k** is θ. In time dt a molecule with speed v moves a distance $v\,dt$. We thus construct a parallelepiped of length $v\,dt$, base area dA, and inclination θ with respect to the negative prolongation of **k**. The volume of this parallelepiped can be seen from the figure to be $v\cos\theta\,dA\,dt$. If we ignore collisions between molecules within the parallelepiped, we can say that all molecules in this volume with velocity **v** will cross the surface at dA within time dt. How many molecules with velocity **v** (i.e., from **v** to $\mathbf{v}+d\mathbf{v}$) are there per unit volume? We have already designated this number as $f(\mathbf{r}, \mathbf{v}, t)\,d\mathbf{v}$ (see Section 28.1), the notation indicating that it may vary with position, velocity, or time. Hence

$$v\cos\theta\, f(\mathbf{r}, \mathbf{v}, t)\, d\mathbf{v} \qquad (28.64)$$

is the number of molecules with velocities from **v** to $\mathbf{v}+d\mathbf{v}$ that cross the plane per unit area per unit time.

If we integrate the expression 28.64 over all velocities, we obtain a number flux density Γ_n, the number of molecules crossing unit area of the plane per unit time in the positive direction:

$$\Gamma_n(\mathbf{r}, t) \equiv \int_{\mathbf{v}} v\cos\theta\, f(\mathbf{r}, \mathbf{v}, t)\, d\mathbf{v}. \qquad (28.65)$$

Note that the factor $\cos\theta$, which may be positive or negative depending on the direction of **v** with respect to **k**, keeps track of the *net* flow across the plane in the positive direction (the direction of **k**). In Eq. 28.65 the term $v\cos\theta$ can also be written $\mathbf{v}\cdot\mathbf{k}$, the dot or scalar product of the two vectors (cf. Section 2.7). Hence we can define a flux density vector $\mathbf{\Gamma}_n$,

$$\mathbf{\Gamma}_n \equiv \int_{\mathbf{v}} \mathbf{v}\, f(\mathbf{r}, \mathbf{v}, t)\, d\mathbf{v}, \qquad (28.66)$$

such that the flux density in the **k** direction, a scalar, is given by

$$\Gamma_n = \mathbf{k}\cdot\mathbf{\Gamma}_n. \qquad (28.67)$$

Thus $\mathbf{\Gamma}_n$ is the flux density, in whatever direction it happens to be, and $\Gamma_n = \mathbf{k}\cdot\mathbf{\Gamma}_n$ is the component of that flux density in the **k** direction.

The flow of mass per unit area per unit time, Γ_m (equivalent to the $\mathbf{J}^m$ of Section 20.2), is simply

$$\Gamma_m = m\Gamma_n. \qquad (28.68)$$

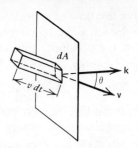

FIGURE 28.8
Molecular flux across a plane surface. The orientation of the plane is given by the unit vector **k**.

---

[7] To speak more precisely, viscous flow is a flow of particles interacting in such a way that momentum flows at right angles to the mass flow, i.e., down a velocity gradient.

To derive the flow of energy across our plane, we return to Eq. 28.64. If that expression is the number of molecules with velocities from $\mathbf{v}$ to $\mathbf{v}+d\mathbf{v}$ that cross the plane per unit area per unit time, then we can multiply by the kinetic energy per molecule to obtain

$$(\tfrac{1}{2}mv^2)v\cos\theta\, f(\mathbf{r},\mathbf{v},t)\,d\mathbf{v}, \qquad (28.69)$$

the corresponding kinetic energy flux density. Hence the net energy carried by molecules crossing unit area of the plane in unit time in the positive direction is

$$\Gamma_E \equiv \tfrac{1}{2}m\int_{\mathbf{v}} v^3\cos\theta\, f(\mathbf{r},\mathbf{v},t)\,d\mathbf{v}. \qquad (28.70)$$

A flux density vector $\mathbf{\Gamma}_E$ (corresponding to the $\mathbf{q}$ of Section 20.4) can again be simply related to Γ_E by the equation

$$\Gamma_E = \mathbf{k}\cdot\mathbf{\Gamma}_E, \qquad (28.71)$$

where

$$\mathbf{\Gamma}_E = \tfrac{1}{2}m\int_{\mathbf{v}} v^2\mathbf{v}\, f(\mathbf{r},\mathbf{v},t)\,d\mathbf{v}. \qquad (28.72)$$

Mass and kinetic energy are scalar quantities, and their fluxes are vectors. However, the momentum is itself a vector, and its flux is a still more complicated quantity, called a tensor. By analogy with the flux density vectors for mass and energy, we write the flux density for momentum (equivalent to the $\mathbf{J}^p$ of Section 20.3) as

$$\mathbf{\Gamma}_{mv} \equiv m\int_{\mathbf{v}} \mathbf{v}\mathbf{v}\, f(\mathbf{r},\mathbf{v},t)\,d\mathbf{v}. \qquad (28.73)$$

This somewhat strange integral contains two vectors multiplying each other. (Remember that $d\mathbf{v}$, the volume element in velocity space, is not a vector.) This momentum flux density is an example of a second-rank tensor, a quantity that has nine components in comparison with the three of a vector (a first-rank tensor). The components of $\mathbf{\Gamma}_{mv}$ are labeled by *two* Cartesian coordinates, one for each vector $\mathbf{v}$. Thus the xy component is

$$(\mathbf{\Gamma}_{mv})_{xy} = m\int_{\mathbf{v}} v_x v_y\, f(\mathbf{r},\mathbf{v},t)\,d\mathbf{v}. \qquad (28.74)$$

If the two indices labeling a component are different, we speak of an off-diagonal component; if they are alike, of a diagonal component. This terminology has its origin in the practice of arranging the components in an array.[8]

Each component of the momentum flux tensor has a simple physical interpretation. Suppose that we define a Cartesian coordinate system with x axis parallel to the vector $\mathbf{k}$ (thus $v_x = v\cos\theta$). Now what is $(\mathbf{\Gamma}_{mv})_{zz}$? It is a molecule's momentum component in the z direction, mv_z, times the number of molecules with velocity from $\mathbf{v}$ to $\mathbf{v}+d\mathbf{v}$ crossing unit area of the plane in unit time, $v_z f(\mathbf{r},\mathbf{v},t)\,d\mathbf{v}$, integrated over all possible velocities:

$$(\mathbf{\Gamma}_{mv})_{zz} = m\int_{\mathbf{v}} v_z v_z\, f(\mathbf{r},\mathbf{v},t)\,d\mathbf{v}; \qquad (28.75)$$

it is thus the total amount of momentum in the z direction transferred

8

$$\mathbf{\Gamma}_{mv}:\quad \begin{matrix} (\Gamma_{mv})_{xx} & (\Gamma_{mv})_{xy} & (\Gamma_{mv})_{xz} \\ (\Gamma_{mv})_{yx} & (\Gamma_{mv})_{yy} & (\Gamma_{mv})_{yz} \\ (\Gamma_{mv})_{zx} & (\Gamma_{mv})_{zy} & (\Gamma_{mv})_{zz}. \end{matrix}$$

It is clear from Eq. 28.74 that the tensor is symmetric, that is, that $(\Gamma_{mv})_{xy} = (\Gamma_{mv})_{yx}$, etc.

across the plane per unit area per unit time. The component $(\boldsymbol{\Gamma}_{mv})_{zz}$ of the flux density constitutes a stress normal to the surface. Since it is a rate of transfer of momentum per unit area, it has the dimensions of pressure,

$$\frac{1}{\text{area}} \frac{\Delta(mv)}{\Delta t} = \frac{\text{force}}{\text{area}} = \text{pressure};$$

for this reason $\boldsymbol{\Gamma}_{mv}$ is often called the *pressure tensor* (or *stress tensor*), and in a gas at rest the diagonal components are in fact equal to the hydrostatic pressure.

What about $(\boldsymbol{\Gamma}_{mv})_{xz}$? This can be similarly analyzed as the total amount of momentum in the x direction transferred across the plane per unit area per unit time,

$$(\boldsymbol{\Gamma}_{mv})_{xz} = m \int_v v_x v_z f(\mathbf{r}, \mathbf{v}, t) \, d\mathbf{v}. \tag{28.76}$$

The component $(\boldsymbol{\Gamma}_{mv})_{xz}$ constitutes a shear stress on the surface. Picture a gas layer flowing parallel to the plane with a smaller velocity in the x direction on one side of the plane than on the other. Because of molecular motion in the z direction (across the plane), there will be a net transfer of momentum from the faster-moving to the slower-moving layer, tending to equalize their velocities. (Cf. Fig. 20.3 and accompanying discussion.)

We can easily calculate the various flux densities in a gas at equilibrium. In that case the number of molecules per unit volume with velocities from $\mathbf{v}$ to $\mathbf{v} + d\mathbf{v}$ is

$$f(\mathbf{r}, \mathbf{v}, t) \, d\mathbf{v} = n\left(\frac{m}{2\pi k_B T}\right)^{3/2} e^{-mv^2/2k_B T} \, d\mathbf{v}, \tag{28.77}$$

where the distribution is not a function of time or position since the gas is spatially uniform. The number density is n and we have used the equilibrium probability density for the velocity, Eq. 28.35. If we substitute Eq. 28.77 into the expression for the flux density of molecules, Eq. 28.66, we find $\boldsymbol{\Gamma}_n$ to be

$$\boldsymbol{\Gamma}_n = n\left(\frac{m}{2\pi k_B T}\right)^{3/2} \int_v \mathbf{v} e^{-mv^2/2k_B T} \, d\mathbf{v}. \tag{28.78}$$

But since the average velocity $\langle \mathbf{v} \rangle$ in a gas at equilibrium is zero, as we showed in Section 28.4, then $\boldsymbol{\Gamma}_n$ must be zero. There is thus no net flow of molecules in any direction in a gas at equilibrium, which is no surprise. The same is of course true for $\boldsymbol{\Gamma}_m$, the mass flux density. We can convince ourselves that the energy flux density must also be zero,

$$\boldsymbol{\Gamma}_E = \tfrac{1}{2} mn\left(\frac{m}{2\pi k_B T}\right)^{3/2} \int_v v^2 \mathbf{v} e^{-mv^2/2k_B T} \, d\mathbf{v}, \tag{28.79}$$

since $v^2 e^{-mv^2/2k_B T}$ is an even function, but each component of $\mathbf{v}$ is an odd function.

Similarly, the off-diagonal components of the momentum flux density tensor, which must describe the coupling of motion in two directions, for example,

$$(\boldsymbol{\Gamma}_{mv})_{xy} = mn\left(\frac{m}{2\pi k_B T}\right)^{3/2} \int_v v_x v_z e^{-mv^2/2k_B T} \, dv_x \, dv_y \, dv_z, \tag{28.80}$$

involve integrals odd in two velocity components, and thus vanish. However, the diagonal components, for example,

$$(\boldsymbol{\Gamma}_{mv})_{xx} = mn\left(\frac{m}{2\pi k_B T}\right)^{3/2} \int_v v_x{}^2 e^{-mv^2/2k_B T} \, dv_x \, dv_y \, dv_z, \tag{28.81}$$

are not zero. The three diagonal components are equal to each other at equilibrium,

$$(\Gamma_{mv})_{xx} = (\Gamma)_{mv})_{yy} = (\Gamma_{mv})_{zz}; \qquad (28.82)$$

this is just what should be expected, since each equals the pressure, which at equilibrium is the same in all directions. Thus we can write for the pressure

$$p = \tfrac{1}{3}[(\Gamma_{mv})_{xx} + (\Gamma_{mv})_{yy} + (\Gamma_{mv})_{zz}], \qquad (28.83)$$

or, using the equivalent of Eq. 28.80 for each of the diagonal components,

$$p = \tfrac{1}{3}mn\left(\frac{m}{2\pi k_B T}\right)^{3/2} \int_{\mathbf{v}} (v_x{}^2 + v_y{}^2 + v_z{}^2)e^{-mv^2/2k_B T}\, dv_x\, dv_y\, dv_z. \qquad (28.84)$$

It is possible to integrate this equation directly, most easily by transforming to spherical coordinates. The sum $v_x{}^2 + v_y{}^2 + v_z{}^2$ equals v^2; the volume element $dv_x\, dv_y\, dv_z$ becomes $v^2\, dv \sin\theta\, d\theta\, d\phi$; and the integration over angles yields 4π. Hence we have

$$p = \tfrac{1}{3}mn\left(\frac{m}{2\pi k_B T}\right)^{3/2} 4\pi \int_{v=0}^{\infty} v^4 e^{-mv^2/2k_B T}\, dv. \qquad (28.85)$$

This result corresponds to that of Eq. 12.16, which we obtained by a different route but with the same assumptions about the state of the gas. Now, however, we can carry out the integration directly, yielding the perfect gas law,[9]

$$p = nk_B T. \qquad (28.86)$$

(Remember that n here is the number density of molecules, N/V, not the number of moles.)

What assumptions are crucial to the derivation leading us to the equation of state of a perfect gas, Eq. 28.86? There are two such assumptions that deserve analysis. The first was made just before Eq. 28.64, where we agreed to ignore collisions between molecules about to cross our reference plane and other molecules present in the gas. If such collisions occur, some will increase and others decrease the number of molecules with any given velocity $\mathbf{v}$; thus, in Fig. 28.8, the number crossing the plane at dA with velocity $\mathbf{v}$ in time dt may differ from the number with velocity $\mathbf{v}$ initially present in our parallelepiped, causing Eq. 28.64 to be in error. If the gas is in equilibrium, the opposing effects should exactly cancel each other, and the assumption is rigorously valid.

The second assumption is a little more subtle. We evaluated the flux densities of particles, energy, and momentum by calculating the net amount of these quantities crossing a given surface per unit time as a result of the actual crossing of molecules (molecular transport). Let us agree that this is the only way that particles can be transferred across the surface. But energy and momentum can also be transferred in collisions. Suppose that we have two molecules colliding as shown in Fig. 28.9. Neither molecule crosses the surface, so the molecular flux density is unaffected by such collisions. Nevertheless, in such a collision, energy and momentum can be transported across the surface by means of the intermolecular forces. This mechanism, called *collisional transfer*, depends on the simultaneous presence of two molecules near the surface S within a region determined by the range of the forces. If the average distance

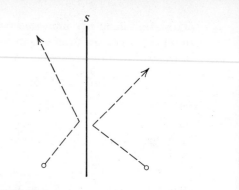

FIGURE 28.9
Collisional transfer across a surface S.

[9] See the discussion at the end of Section 28.4.

between molecules is large compared to the range of the forces between them, a condition attainable at sufficiently low density, then the probability of the occurrence of collisions contributing to collisional transfer is low. In that case we have a perfect gas, the equation of state for which we obtained by neglecting collisional transfer in the formulation of the density vectors. As the density of the gas is increased, collisional transfer becomes more important. In a liquid, where the density is so high that each molecule is in continuous interaction with neighboring molecules, the dominant mechanism is collisional transfer.

28.7 Effusion

We have evaluated the flux density vectors for the transport of molecules and molecular quantities. In this and the next section we shall illustrate the use of these quantities with a number of important applications.

Take a vessel, fill it with gas at low pressure, surround it with a vacuum, and punch a small hole in it, as shown in Fig. 28.10. Molecules will escape from the gas vessel into the vacuum chamber. We assume that the pump is good enough to maintain a vacuum, by which we mean a pressure so much lower than the pressure in the gas vessel that we can neglect the flow of molecules from the vacuum chamber back to the gas vessel. We wish to calculate the rate of flow of gas from the vessel. The process of course corresponds to the molecular beam sources we described at the beginning of Section 27.5; as we did there, we wish to consider the gas molecules as emerging independently of one another. That is, we make the density so low that the mean free path, Eq. 28.63, is large compared to the diameter of the orifice. In that case we can neglect collisions at the orifice, and then we know how to calculate the molecular flux.

The plane of Fig. 28.8 is now that of the orifice (i.e., the continuation of the surrounding wall). According to Eq. 28.64, the number of molecules with velocities in the range from $\mathbf{v}$ to $\mathbf{v}+d\mathbf{v}$ that cross this surface per unit area per unit time is

$$\Gamma_n(\mathbf{r}, \mathbf{v}, t) = v \cos \theta\, f(\mathbf{r}, \mathbf{v}, t)\, d\mathbf{v}. \tag{28.87}$$

As before, θ is the angle between the velocity vector $\mathbf{v}$ and the (outward) normal to the surface. Next we assume that the trickle of molecules from the orifice is so small that the gas in the vessel not only remains spatially uniform but also maintains its equilibrium velocity distribution, Eq. 28.77. These conditions define *effusive flow*, or *effusion*. The vector $\mathbf{v}$ gives the direction of motion both inside and outside the vessel. It is convenient to introduce spherical coordinates for the differential volume element $d\mathbf{v}$,

$$d\mathbf{v} = v^2\, dv\, d\Omega = v^2\, dv \sin \theta\, d\theta\, d\phi, \tag{28.88}$$

where the polar angle θ and azimuthal angle ϕ are as indicated in Fig. 28.10.

Let us evaluate first the total flux density through the orifice, comprising molecules with all possible speeds and with directions of emergence corresponding to the ranges $0 \leqslant \phi \leqslant 2\pi$, $0 \leqslant \theta < \pi/2$. Note that the range of θ is only half the usual range, since we are counting only molecules going in *one* direction through the orifice. It is precisely this step that yields a nonzero flux in spite of our assuming an equilibrium distribution inside the vessel. (Compare again the present situation with that pictured in Fig. 28.8, where molecules cross the plane in *both* directions and at equilibrium the pressure is the same on both sides.) The

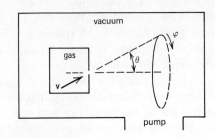

FIGURE 28.10
Simple experimental apparatus for the study of effusion.

total flux density of molecules is then, by Eqs. 28.65 and 28.77,

$$\Gamma_n = n\left(\frac{m}{2\pi k_B T}\right)^{3/2} \int_{v=0}^{\infty} \int_{\theta=0}^{\pi/2} \int_{\phi=0}^{2\pi} v \cos\theta \, e^{-mv^2/2k_B T} v^2 \sin\theta \, d\theta \, d\phi \, dv.$$

(28.89)

The angular integrals are easily evaluated (with the change of variable $\sin\theta = x$),

$$\int_{\theta=0}^{\pi/2} \int_{\phi=0}^{2\pi} \cos\theta \sin\theta \, d\theta \, d\phi = 2\pi \int_0^1 x \, dx = \pi,$$

(28.90)

and the integral over v is the same as that in Eq. 28.48 for the average speed. The total flux density due to effusive flow per unit area of the orifice is thus simply

$$\Gamma_n = \tfrac{1}{4} n \langle v \rangle.$$

(28.91)

The density in the vessel is related to the pressure by the perfect-gas equation of state, $p = nk_B T$, so that Γ_n can be expressed as

$$\Gamma_n = \frac{p}{(2\pi m k_B T)^{1/2}}$$

(28.92)

with the help of Eq. 28.48.

Suppose that we want to measure not the total flux density but only a part of it, say, that part which emerges in a given direction and hits a detector of a size subtending a solid angle element $d\Omega$ at the orifice. The number of molecules striking the detector per unit area of the orifice is

$$n\left(\frac{m}{2\pi k_B T}\right)^{3/2} \int_{v=0}^{\infty} v \cos\theta \, e^{-mv^2/2k_B T} v^2 \, dv \, d\Omega = n\langle v \rangle \cos\theta \frac{d\Omega}{4\pi}.$$

(28.93)

If we also place a velocity selector (Fig. 27.12) in front of the detector, then we can determine the fraction of molecules with speed from v to $v + dv$; the number of such molecules entering the detector per unit time per unit area of the orifice is

$$\Gamma_n(v, \Omega) \, dv \, d\Omega = n\left(\frac{m}{2\pi k_B T}\right)^{3/2} v \cos\theta \, e^{-mv^2/2k_B T} v^2 \, dv \, d\Omega.$$

(28.94)

Equation 28.94 is fundamental to direct experimental tests of the Maxwell–Boltzmann velocity distribution.

Such experiments have in fact been performed in a number of laboratories. A sketch of a typical apparatus is shown in Fig. 28.11. Suppose that we put potassium metal in the gas vessel and heat it until the vapor pressure of liquid K is about 0.1 torr. The mean free path of K atoms is then of the order of 0.1 mm, so that the diameter of the orifice should be smaller, say 0.01 mm, for effusive flow. The K atoms emerge from the orifice with all speeds and in all directions. (How do we know that only K atoms emerge? Could it be that K_2 or higher polymers are also present? Such species can exist, and one must test for their presence, say by analyzing the flux with a mass spectrometer; can you think of another way? For potassium, it is found that essentially only atoms exist in the gas at this pressure and temperature.) The K flux is collimated, velocity-selected, and then measured by a detector. A good detector for this purpose is a hot platinum wire: The K atoms impinging on the wire come off as K^+ ions, the positive ions are drawn to a nearby negative plate, and the resulting current measures the K flux. In Fig. 28.12 we

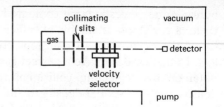

FIGURE 28.11
Schematic drawing of apparatus for experimental test of Maxwell–Boltzmann velocity distribution.

show a comparison of a measured velocity distribution with one calculated from the theoretical expression Eq. 28.94 (taking account of necessary apparatus corrections). The agreement between the two is very good.

The process of effusion has been used extensively to determine the vapor pressures of solids and liquids as functions of temperature, and to study the equilibrium composition of vapors. Let us go back again to the example of the effusion of potassium. Suppose that we measure the total flux of K emerging from an orifice of known size, and the temperature of the vessel containing the potassium. If the temperature is above the melting point but below the boiling point of K, then there will be present in the vessel both liquid and vapor. From this measurement we know the vapor pressure of liquid K at that temperature, according to Eq. 28.92, provided that we know the mass of the effusing species. If all the K emerges from the orifice as atoms, as is very nearly the case, then the mass m is known, of course. If we repeat the measurement at a series of temperatures, we can obtain the variation of the vapor pressure of potassium with temperature, which by the Clausius–Clapeyron equation, Eq. 24.13, gives us the heat of vaporization. Similar measurements of the vapor pressure of a solid yield the heat of sublimation.

As mentioned already, however, in such measurements one must be sure of the mass of the effusing species. Suppose that we performed an effusion experiment on graphite at a temperature below its melting point. If we were to interpret the results on the assumption that only C atoms effuse from the vessel, we would be making a serious error, especially serious since the heat of sublimation of graphite is such an important quantity in thermochemical calculations (cf. Section 14.8). This mistake was actually made for many years until a mass-spectrometric analysis was carried out on the effusing vapor. A variety of species are actually present in the vapor: C_1, C_2, C_3, C_4, C_5, ... (especially those with odd numbers of C atoms). These species are in equilibrium with each other in the vessel, with the equilibrium mole fractions varying with temperature. Thus the interpretation of the experiment is much more difficult than for a monatomic vapor, and can be made only if the mole fractions are known.

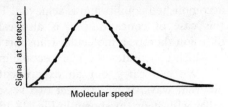

FIGURE 28.12
Experimental test of the Maxwell–Boltzmann velocity distribution, not to scale. Solid line: predicted by the Maxwell–Boltzmann distribution. Points: experimental measurements [schematic, but cf. R. C. Miller and P. Kusch, *Phys. Rev.* **99**, 1314 (1955)].

28.8
Transport Properties of Gases

The transfer from one part of a system to another of molecules, momentum, energy, or some other physical quantity such as charge, polarization, and so on, is called a *transport process*. The study of these processes is of importance for learning by what mechanisms physical systems in nonequilibrium states attain equilibrium, how these mechanisms change with the state of aggregation, and how the mechanisms depend on the structure and arrangement of molecules and the forces between them. We begin with some examples of transport processes in gases, then present a brief macroscopic description, and finally turn to a simple molecular description based on the kinetic theory.

Many examples of transport processes are well known to the reader already. Suppose we have two containers, each filled with a pure gas, say argon and neon. If the two containers are connected with each other, a natural, irreversible process will take place between them, the mixing of the two gases. The detectable process, consisting of the diffusion of each gas through the other, ceases when the composition of the gas mixture is uniform. We say that a concentration gradient, that is, a variation of concentration with position, gives rise to diffusion.[10] The process of diffusion tends to remove the concentration gradient, and when that is

[10] This statement is adequate for gases at low density and dilute solutions. Strictly, it is a chemical potential gradient that gives rise to diffusion.

accomplished equilibrium is achieved. The removal of spatial gradients, in this case of concentration, is also called *hydrodynamic relaxation*. In diffusion there is a molecular transport of particles (or mass), the average of which constitutes the macroscopic mass flow.

We can carry out an experiment on diffusion in two somewhat different ways. If we connect two vessels, one filled with argon and the other with neon, as shown in Fig. 28.13a (i.e., by completely removing a large partition), then the concentration at each point changes with time. The process occurs under transient or non-steady conditions. However, if we connect the two large vessels by a small capillary, as shown in Fig. 28.13b, then the flow through the capillary due to diffusion is small and the concentration in each of the two vessels will vary only slowly with time.

In the limit that the vessels acting as mass reservoirs are very large, we can study the process of diffusion at fixed concentration in each vessel. As long as the concentrations in the two vessels are different, diffusion will proceed. In the capillary itself there will still be concentration gradients, but now the concentration of each species is to a good approximation constant in time everywhere. If one looks at a particular point in the capillary, it will be found that all macroscopic properties (such as the thermodynamic functions) are virtually invariant with time. However, the system consisting of the capillary is not in equilibrium, because it is not a closed system: mass crosses the boundaries between the capillary and the vessels. Since the concentration at each point is nearly time-independent, we say rather that the system is in a *steady state*.

Other transport processes are easily envisioned. The presence of a temperature gradient in a system gives rise to a molecular transport of energy, the average of which constitutes a flow of heat. Again, the process is irreversible and proceeds until the gradient is removed. The presence of a velocity gradient gives rise to a molecular transport of momentum, the effect of which constitutes viscous drag. In each case we can design steady-state or non-steady-state experiments. The three transport processes (diffusion, viscosity, and thermal conduction) that we have described are not the only ones possible, even though we shall fix our attention primarily on these. For instance, if charged particles are present in a gas, then we can have electric conduction in the presence of a potential gradient produced by an electric field. Transport processes in dense gases, liquids, and solids are discussed in Chapter 29.

Coupling between two transport processes inevitably occurs. Consider a uniform mixture of two gases in a vessel. Now impose and maintain a temperature gradient across the vessel. In time a concentration gradient will be established, which will be maintained as long as the temperature gradient is present. This phenomenon, called *thermal diffusion*, is a coupling between the two transport processes, diffusion and thermal conduction. The reciprocal coupled process is also known. Suppose that we carry out a diffusion experiment, of the type pictured in Fig. 28.13a, under adiabatic conditions; that is, we let the vessel containing the two gases be thermally insulated. At the beginning of the experiment there is a large concentration gradient in the middle of the vessel, and the presence of this concentration gradient gives rise to a temperature gradient. Many other examples of coupled transport processes have been observed. (See Problems 11 to 19 of Chapter 20.)

So far we have discussed the molecular transport of energy caused by a temperature gradient. However, we can visualize another mechanism for energy transfer, namely, by inelastic collisions in which energy is transferred between an internal molecular motion and either another such

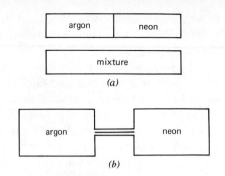

FIGURE 28.13
Diffusion experiments: (a) non-steady-state; (b) steady-state.

motion or translational motion. Suppose that we introduce a set of oscillators, all in the same quantum state, into a large excess of inert gas. The oscillators are uniformly distributed in the vessel so that no spatial gradients exist. However, the oscillators will collide with the inert-gas molecules, with both elastic and inelastic collisions occurring. As a result of the inelastic collisions, the populations of the various oscillator quantum states will change with time until a Boltzmann equilibrium distribution is established, one in which the number of oscillators in quantum state j is proportional to $g_j e^{-\varepsilon_j/k_B T}$, where ε_j is the energy of the jth oscillator level, with degeneracy g_j, and T is the temperature of the inert gas. If the energy of the initial quantum state is larger than the average energy of the oscillators in their equilibrium distribution at the temperature T, then energy will be transferred from the oscillators to the inert gas. It is the difference between the actual and equilibrium concentrations of the oscillators in the individual quantum states that gives rise to the irreversible process, in this case called *vibrational relaxation*, and to the energy transfer. Again, many similar processes are known, including energy exchange between translation and rotation, translation and electronic motion, vibration and electronic motion, and so on.

A chemical reaction can also be viewed as a transport process, the transport of atoms from the configuration of reactants to that of products. At constant temperature and pressure, chemical equilibrium is attained when the Gibbs free energy change of the reaction, $\sum_i \nu_i \mu_i$, is zero (cf. Eq. 19.99). A Gibbs free energy difference between products and reactants gives rise to the irreversible process of the reaction.[11] Again, the molecular mechanism by which chemical equilibrium is achieved simply consists of collisions in which a reaction occurs.

In experiments on many different systems it has been observed experimentally that the flux of particles, momentum, or energy due to the existence of a gradient (of concentration, velocity, or temperature, respectively) is simply proportional to the gradient, provided that the gradient is not too large. The proportionality coefficients are called, respectively, the coefficients of *diffusion*, D, *viscosity*, η, and *thermal conductivity*, λ. (To simplify the discussion we shall consider only *self-diffusion*, the diffusion of two identical species. This process cannot be observed directly, but can only be approximated—for instance, by the diffusion of $^{14}CO_2$ in $^{12}CO_2$.) These three empirical, macroscopic laws are summarized in Table 28.1, the last column of which constitutes the definitions of D, η, and λ. (For a

TABLE 28.1

SUMMARY OF RELATIONS FOR TRANSPORT PROCESSES (GRADIENTS TAKEN TO BE IN THE z DIRECTION)

Molecular Flux Density	Transport Process	Gradient in z Direction	Macroscopic Law
Particles, Γ_n	Diffusion	Density, $\dfrac{dn}{dz}$	$(\Gamma_n)_z = -D\dfrac{dn}{dz}$
Momentum, Γ_{mv}	Viscosity	Average velocity, $\dfrac{d\langle v_y\rangle}{dz}$	$(\Gamma_{mv})_{yz} = -\eta\dfrac{d\langle v_y\rangle}{dz}$
Energy, Γ_E	Thermal conductivity	Temperature, $\dfrac{dT}{dz}$	$(\Gamma_E)_z = -\lambda\dfrac{dT}{dz}$

[11] By the phrase "irreversible process" we mean that the reaction proceeds by itself, spontaneously. Any reaction can always be reversed by doing work on it, just as a uniform gas mixture can be separated into its components or heat transported against a temperature gradient (as in an air conditioner) by doing work on these systems. (See Section 17.4.)

more detailed formulation of these laws, see the discussions leading to Eqs. 20.15, 20.46, and 20.76, respectively.) The coefficients η and λ, and the product nD, are experimentally found to depend on temperature, to be nearly independent of density in a dilute gas, but to vary appreciably at higher densities.

Let us look in more detail at a gaseous system in which there occurs a transport process, say diffusion. Again we take the experimental arrangement pictured in Fig. 28.13a. Initially we have equal volumes of gaseous neon and argon, at the same pressure and temperature, in separate containers. At a given time ($t = 0$) the partition separating the two volumes is removed and diffusion begins. A plot of the mole fraction of argon as a function of position is shown in Fig. 28.14 for both the initial situation and some later time prior to the attainment of equilibrium. As long as the concentration gradient exists, diffusion takes place, producing a net flow, or flux, of argon from the left to the right side of the vessel. Of course, there is simultaneously an opposite flux of neon. We have seen in Section 28.6, in deriving the expressions for flux vectors, that if the velocity distribution in a gas is the equilibrium Maxwell–Boltzmann distribution, the flux vectors must vanish. Hence, if a transport process is going on, the velocity distribution cannot have its equilibrium form. We say that the existence of a spatial gradient, such as the concentration gradient in our example here, perturbs the equilibrium Maxwell–Boltzmann velocity distribution. Strictly, only the deviation of the velocity distribution from its equilibrium form contributes to the flux. How large is the perturbation, the deviation from the equilibrium distribution? It can't be very large, because, as we pointed out in discussing the computer experiment on the velocity distribution (Section 28.2), the velocity (or momentum) relaxation time is very short, typically of the order of a few mean times between collisions, say about 10^{-9} s. Thus, if we could extract a small sample of gas from the vessel in which the diffusion process takes place, a sample of so small a volume that the macroscopic spatial gradient in the sample is itself negligibly small, then the perturbed velocity distribution in this sample would approach its equilibrium form in about 10^{-9} s. The approach of the entire mixture to spatial homogeneity usually takes much longer, of the order of seconds or minutes, depending on the density of the mixture and the size of the vessel. This characteristic time is called the *hydrodynamic relaxation time*. The slow process of hydrodynamic relaxation continually keeps the velocity distribution perturbed, but only slightly so because the velocity relaxation is very much faster. Because of the large difference between the velocity and hydrodynamic relaxation times, we expect that in each small volume element of the vessel we have nearly an equilibrium velocity distribution, what we called a *local equilibrium distribution* in the introduction to Chapter 20. The word "local" reminds us that the variables (density or concentration, temperature, average velocity) that occur in the velocity distribution vary with position, since there must be gradients in these quantities for there to be a transport process. The gradients are usually so small that the variations in these quantities over distances of the order of a few mean free paths are negligible. Thus, in volume elements only a few mean free paths wide we must have a local near-equilibrium velocity distribution.[12] The gradients in concentration and other variables are too small to cause much perturbation, and the velocity relaxation is too rapid to maintain a highly perturbed distribution.

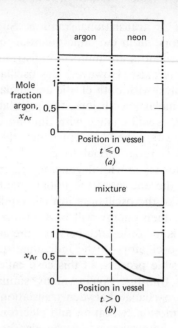

FIGURE 28.14
Concentration profile in diffusion experiment: (*a*) initially; (*b*) at a later time, but prior to attainment of equilibrium.

---

[12] Indeed, if this were not the case, there would be no temperature as we have defined it in this chapter, cf. Eq. 28.33. However, one can give a more general definition in terms of the local mean-square speed, as in Eq. 28.49.

From what we have just said, it is clear that we must obtain information on the perturbation of the velocity distribution. To do this theoretically, we need to obtain an equation like the Boltzmann equation for the velocity distribution, solve it, and then calculate flux vectors with the solution. However, the Boltzmann equation we derived in Section 28.3 is not sufficient, because it describes the velocity distribution only in the spatially homogeneous case. How to amend that equation for systems in which gradients exist, how to solve it in various approximations and use the solution—all these things are known but beyond the scope of this presentation.

Nevertheless, we can obtain some indication of the correct results by making more drastic approximations. Let us recall the expressions for the flux density across a plane, such as Eqs. 28.65 and 28.72, which can be summed up in one equation,

$$\Gamma_\psi = \int_v v \cos \theta \, \psi f(\mathbf{r}, \mathbf{v}, t) \, d\mathbf{v}, \qquad (28.95)$$

where ψ is 1, $m\mathbf{v}$, and $\frac{1}{2}mv^2$ for the flux densities of particles, momentum, and energy, respectively. Now picture a one-dimensional transport process in the z direction, as shown in Fig. 28.15. We want to calculate the net flux of density across the plane $z = 0$. This consists of the difference between the values in the positive and negative directions, that is,

$$\Gamma_\psi = \Gamma_\psi^{+z} - \Gamma_\psi^{-z}. \qquad (28.96)$$

We have for Γ_ψ^{+z}, for instance, introducing a factor n/n for later convenience,

$$\Gamma_\psi^{+z} = \int_v \int_{\theta, \phi} (n\psi \cos \theta) \frac{v}{n} f(\mathbf{r}, \mathbf{v}, t) \, d\mathbf{v}; \qquad (28.97)$$

as usual, θ is the angle between the vector $\mathbf{v}$ and the positive z axis, with a range of integration $0 \le \theta \le \pi/2$ for Γ_ψ^{+z} and $\pi/2 \le \theta \le \pi$ for Γ_ψ^{-z} (see Section 28.7).

Now to the approximations: First we "break the average" in Eq. 28.97; that is, we suppose that the average value of the quantity $n\psi \cos \theta$ can be immediately taken outside the integral. This is equivalent to assuming either that every molecule has the same value of $n\psi \cos \theta$, or that the mean of this quantity is also the most probable value. Second, in the remainder of the integral we forget about perturbation of the distribution function, using instead the Maxwell–Boltzmann local-equilibrium form. Since this does not depend explicitly on either time or position, we can use Eqs. 28.11 and 28.13 to make the substitution

$$\frac{1}{n} f(\mathbf{r}, \mathbf{v}, t) \to f(\mathbf{v}). \qquad (28.98)$$

Combining these approximations, we can write for the flux in the $+z$ direction

$$\Gamma_\psi^{+z} \approx \langle \psi n \cos \theta \rangle \int_v v f(\mathbf{v}) \, d\mathbf{v}, \qquad (28.99)$$

or, by Eq. 28.41,

$$\Gamma_\psi^{+z} \approx \langle \psi n \cos \theta \rangle \langle v \rangle, \qquad (28.100)$$

where the average $\langle \psi n \cos \theta \rangle$ still needs to be evaluated. Suppose now that we consider the molecules as moving only in the directions of the coordinate axes; that is, one-sixth of the molecules move in the $+z$ direction, one-sixth in the $-z$ direction. Thus we can take $\cos \theta$ to be $+1$ in Γ_ψ^{+z} and -1 in Γ_ψ^{-z}. If the molecules of the system are hard spheres,

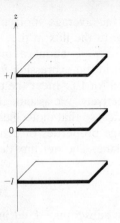

FIGURE 28.15
Calculation of flux across the plane $z = 0$ for a general transport process.

then on the average molecules that collide near the plane $z = 0$ and contribute to the flux in the $+z$ direction will have collided last near the plane at $z = -l$, about a mean free path away. Hence the average $\langle \psi n \cos \theta \rangle$ in the expression for Γ_ψ^{+z} should be evaluated near the plane $z = -l$ and for Γ_ψ^{-z} near the plane $z = +l$, with $\cos \theta$ set equal to $+1$ and -1, respectively. We assume that a condition of local equilibrium exists at these planes, so that molecules leaving the plane $z = -l$, for instance, have the average properties of density, momentum, and energy characteristic of that plane. The net flux density across the plane $z = 0$ then becomes

$$\Gamma_\psi \approx \frac{\langle v \rangle}{6} (\langle \psi n \rangle_{-l} - \langle \psi n \rangle_{+l}), \tag{28.101}$$

where the subscripts $-l$, $+l$ indicate the positions at which $\langle \psi n \rangle$ is to be calculated. Next we expand each function on the right-hand side of Eq. 28.101 in a Taylor series,

$$\langle \psi n \rangle_{-l} = \langle \psi n \rangle_0 - l \left(\frac{d \langle \psi n \rangle}{dz} \right)_{z=0}, \tag{28.102}$$

$$\langle \psi n \rangle_{+l} = \langle \psi n \rangle_0 + l \left(\frac{d \langle \psi n \rangle}{dz} \right)_{z=0}, \tag{28.103}$$

retaining first-order terms only, and thus obtain the flux density at the plane $z = 0$:

$$\Gamma_\psi \approx -\frac{\langle v \rangle}{3} l \left(\frac{d \langle \psi n \rangle}{dz} \right)_{z=0}. \tag{28.104}$$

The first-order terms in the expansion are sufficient if the gradients are such that changes in $\langle \psi n \rangle$ over a mean free path are small.

For the case of diffusion we have $\psi = 1$, so that the calculated flux density is

$$\Gamma_n = -\frac{\langle v \rangle}{3} l \frac{dn}{dz}. \tag{28.105}$$

By comparison of this derived, statistical expression for the flux with the empirical, macroscopic law listed in Table 28.1, we obtain a statistical expression for the diffusion coefficient

$$D = \tfrac{1}{3} \langle v \rangle l. \tag{28.106}$$

For viscous flow we have $\psi = m\mathbf{v}$, and the flux density in the z direction of the y component of momentum is

$$(\boldsymbol{\Gamma}_{mv})_{yz} = -\tfrac{1}{3} nm \langle v \rangle l \frac{d \langle v_y \rangle}{dz}. \tag{28.107}$$

The statistical expression for the viscosity coefficient is thus

$$\eta = \tfrac{1}{3} nm \langle v \rangle l. \tag{28.108}$$

Finally, for thermal conduction we have $\psi = \tfrac{1}{2} mv^2$ and $\langle \psi n \rangle = nC_V T$, since for a perfect gas by Eq. 28.49 $\langle \tfrac{1}{2} mv^2 \rangle = \tfrac{3}{2} k_B T = C_V T$, where C_V is the constant-volume heat capacity per molecule. Hence the energy flux density is

$$\Gamma_E = -\tfrac{1}{3} \langle v \rangle l n C_V \frac{dT}{dz}, \tag{28.109}$$

yielding for the coefficient of thermal conductivity

$$\lambda = \tfrac{1}{3} n C_V \langle v \rangle l. \tag{28.110}$$

Let us summarize what went into the above derivation: We assumed small spatial gradients, so that in the expansion 28.102 only first-order terms needed to be retained; this simplification makes the calculated flux proportional to the gradient. In evaluating Eq. 28.97 we broke the average and used a local-equilibrium approximation for the distribution function. A molecular model of hard-sphere interactions was in effect assumed by the introduction of the mean free path. We restricted the calculation to molecular transport (no collisional transfer) in perfect gases. Finally, we greatly oversimplified the angular distribution of molecular velocities. In Table 28.2 we list the results of this simple derivation, with the $\langle v \rangle$ and l of the above equations evaluated by Eqs. 28.48 and 28.63 (the latter making the hard-sphere assumption explicit). Also given in the same table are the results obtained for hard-sphere molecules from a much more rigorous derivation, based on the calculation of perturbed velocity distribution functions from solutions of a Boltzmann equation. (Such solutions have also been obtained for more complicated intermolecular potentials, but the results cannot be written down in simple algebraic form.) For hard spheres, then, the differences between the simple and the more rigorous theory are not large, amounting only to numerical factors of order unity. The essential temperature and density dependence of the transport coefficients is the same in either theory.

The molecular model of hard spheres predicts a simple $T^{1/2}$ dependence for all three transport coefficients at constant density. The coefficients of viscosity and thermal conductivity are predicted to be independent of density, and the diffusion coefficient inversely proportional to density (and thus proportional to $T^{3/2}$ at constant *pressure*, since $n = p/k_B T$). These important results were obtained by Maxwell in 1860, in a derivation essentially the same as presented here. This was a brilliant analysis, which he confirmed by experiments showing that the viscosity was indeed independent of density in the region considered by the theory.[13] If a tremendous theoretical advance is made by very clever approximations, then frequently additional progress is slow. The first attempts at improvement usually involve the removal of some of the assumptions. In the kinetic theory of gases, however, it was necessary to return to the fundamental postulate regarding the velocity distribution and its perturbation by a transport process. Thus, perhaps it is not surprising that the more rigorous theory, which considers the perturbation properly, was not developed until more than 50 years after Maxwell.

TABLE 28.2
TRANSPORT COEFFICIENTS FOR HARD SPHERES

Transport Coefficient	Simple Theory	Rigorous Theory
Self-diffusion, D	$\dfrac{2}{3}\left(\dfrac{k_B T}{\pi m}\right)^{1/2}\dfrac{1}{\pi d^2 n}$	$\dfrac{3\pi}{8}\left(\dfrac{k_B T}{\pi m}\right)^{1/2}\dfrac{1}{\pi d^2 n}$
Viscosity, η	$\dfrac{2}{3}\left(\dfrac{m k_B T}{\pi}\right)^{1/2}\dfrac{1}{\pi d^2}$	$\dfrac{5\pi}{16}\left(\dfrac{m k_B T}{\pi}\right)^{1/2}\dfrac{1}{\pi d^2}$
Thermal conduction, λ	$\dfrac{2}{3}C_V\left(\dfrac{k_B T}{\pi m}\right)^{1/2}\dfrac{1}{\pi d^2}$	$\dfrac{25\pi}{32}C_V\left(\dfrac{k_B T}{\pi m}\right)^{1/2}\dfrac{1}{\pi d^2}$

[13] If the density is too high, then both triple collisions and collisional transfer have to be taken into account. If the density is so low that the mean free path becomes of the order of the dimensions of the vessel in which the experiment is made, then transfer of energy, for instance, depends not on molecule–molecule collisions but on molecule–surface collisions. That is a new problem.

How good is the theory? In Fig. 28.16 we show some experimental measurements of viscosity (the most accurately known transport coefficient) as a function of temperature. All the measurements were made at low densities, corresponding to pressures of about 1 atm. Since we are plotting $\eta/T^{1/2}$, hard-sphere molecules should give a horizontal line (see Table 28.2). Substituting the experimental value for any one temperature (as shown in the figure for $T = 273$ K) into our theoretical expression yields an estimate of the hard-sphere diameter d. Remember that doing this amounts to forcing the experiment to fit the hard-sphere model. One can find published tables of "hard-sphere diameters" for various molecules, but these in no way prove that molecules are really hard spheres. They are merely tables of numbers obtained by *assuming* a molecular model. The moral holds not only for one, but for all molecular models. A comparison of the measured temperature dependence with that predicted by the hard-sphere model permits an estimate of the deviation of the real intermolecular potential from that for hard spheres.

In fact, as can be seen from the figure, the hard-sphere model is not entirely accurate in describing the exact temperature dependence of the viscosity. As we pointed out in Section 27.2, a Lennard–Jones type of potential, corresponding to repulsion at small distances and attraction farther away, describes actual molecular interactions better than a hard-sphere potential. At high temperatures, where the average kinetic energy, of the order of $k_B T$, is large compared to the measure of the attractive part of the potential, say the depth of the potential well ϵ, the role of the attractive part becomes minor. Hence the behavior predicted by the hard-sphere model ($\eta \propto T^{1/2}$) can be expected to be reasonable at temperatures such that $k_B T \gg \epsilon$, and the data in Fig. 28.16 in fact tend to level off at higher temperatures. At lower temperatures, however, where we have $k_B T \approx \epsilon$, the attractive part of the potential cannot be neglected and the hard-sphere model is poor. It is also possible, as we have mentioned, to solve the rigorous theory for more complicated potentials, and we have included in Fig. 28.16 such a solution for the Lennard–Jones potential. Once again, the actual curve drawn is obtained by forcing the experimental data to fit a theoretical expression, in this case much more complex than that in Table 28.2. The resulting curve, although not a perfect fit, is far more satisfactory than the hard-sphere prediction, indicating that the trouble with the latter lies much more in the potential model than in the transport theory itself. It is from such fits as this that the constants in Tables 10.1 and 21.10 can be obtained. The variation of the transport coefficients with density will be discussed in Section 29.1.

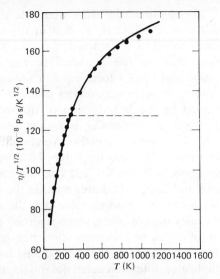

FIGURE 28.16
Temperature dependence of 1-atm viscosity for argon. Solid line: Lennard–Jones potential behavior. Dashed line: hard-sphere behavior (fitted to η for 273 K). The circles are experimental points.

28.9
Energy Exchange Processes

In the preceding sections we have limited our attention to collisions between structureless particles, and thus to elastic collisions. What issues and problems arise when inelastic collisions occur, and what information becomes available from their study? We have given earlier some illustrations of the factors that must be taken into account, including the cross section defined in Eq. 27.57 and the conservation of energy (Eq. 27.43). Obviously, the theoretical treatment of such collisions is much more complicated than for elastic collisions, and we shall make no attempt to develop it here. In this section we merely give a qualitative description of some of the interesting processes that occur and what can be learned about them from experiment.

Consider an isolated molecule with internal degrees of freedom and a given amount of energy. Subtract from that energy the translational (kinetic) energy, and the remainder is internal energy. Frequently we separate the internal energy into its components of rotational, vibrational, and electronic energy (we neglect here small contributions from interactions involving nuclear magnetic moments); but we recognize that this kind of separation is an approximation that does not always hold. For instance, when the interaction of rotational and vibrational degrees of freedom is large, one can only speak of individual rovibrational energy levels, rather than a manifold of rotational states within a vibrational level. Similarly, it may not be possible to speak of distinct electronic states: for instance, the excitation of NO_2 with 488-nm radiation leads to strong coupling of an excited electronic state and a highly excited vibrational level of the ground electronic state, a condition best described by a wave function that is a linear combination of the two states. The Born–Oppenheimer approximation of separating electronic and nuclear degrees of freedom is inapplicable in this case. Each case must be examined for the appropriateness of the separation of degrees of freedom. Although interactions between various modes of motion do occur, we can nevertheless often use the idealization of partitioning the internal energy into separate contributions from each of the degrees of freedom.

Excitation of a molecule, with consequent increase of internal energy, may occur either by absorption of a photon (see Section 4.5) or by collision with another molecule. Energy transfer to a molecule by collision may take place in the gas phase (binary collision), in the liquid (binary or multibody collision), on a surface, or in a solid. Similarly, deexcitation of a molecule may occur either by emission of a photon or by collision; which of these events is most likely depends on the lifetime of the excited state and the collision frequency. We need to remember that the radiative lifetimes of electronically excited states are typically from 10^{-8} to 10^{-6} s, except that metastable states (those that can radiate only by forbidden transitions) may last as long as 1 s; of vibrationally excited states, 10^{-3} to 10^{-1} s; and of rotationally excited states, 10^{-1} s and longer. In a gas at atmospheric pressure and room temperature, the time between collisions is about 10^{-8} s. As a rough measure, the efficiency of energy transfer by collision varies inversely with the amount of energy transferred. Thus deexcitation of rotational and vibrational states usually occurs by collision, unless the density of the gas is extremely low; deexcitation from electronic states may occur by radiation or by collision.

All possible kinds of energy transfer occur by collision with varying probabilities (cross sections): from translation of one molecule to that of another (the process by which momentum relaxation occurs); from translation of one molecule to rotation, vibration, or electronic excitation of another; from rotation or vibration in one molecule to the same degree of freedom in another; and so on. Energy can readily be transferred to produce a transition between two levels ΔE apart if ΔE is small compared to the thermal energy $k_B T$; otherwise the transition will take place only in some small fraction of all collisions. Given the typical energy-level spacings, the efficiency of energy transfer thus usually follows the sequence

translational > rotational > vibrational > electronic.

These simple guidelines are useful, but sometimes fail because of other features of the transfer process, such as Franck–Condon restrictions (in the excitation of a molecule by a photon) or the requirement of conservation of angular momentum.

There are three additional features of excitation and inelastic collisions that require explanation. First, there is the possibility of *resonant energy transfer*. Suppose that nitrogen in the $v = 1$ level of the ground electronic state collides with a N_2 molecule in the ground ($v = 0$) vibrational state. A quantum of vibrational energy can then be transferred directly from one molecule to the other, with the energy given up by the first molecule exactly equaling that required by the second (which is what we mean by "resonant"). As one might expect, the probability of such transfer is high. This enhancement also occurs in resonant transfer from one degree of freedom to another (say, from electronic motion to vibration). In this and most other cases, however, the resonance is not likely to be exact. For example, the process

$$N_2(v = 2) + N_2(v = 0) \rightarrow N_2(v = 1) + N_2(v = 1)$$

is not quite resonant, since N_2 for low vibrational energy is a slightly anharmonic oscillator; but it requires only 0.0034 eV (equivalent to 28 cm^{-1}) transferred from translational energy to satisfy conservation of energy.

Second, consider the excitation of a molecule, say SF_6, by light under collision-free conditions. The density of vibrational energy levels of the molecule in its ground electronic state is shown in Fig. 28.17. For low excitation energy, the density of vibrational states is low, and one can view the molecule as a collection of independent normal modes. For example, if by Raman absorption at 776 cm^{-1} SF_6 is excited to the $v = 1$ level of the symmetrical stretch mode, then there is no other state of nearly the same energy, and the molecule remains in that well-defined state until it loses energy by emission of a photon. But if the SF_6 molecule absorbs about five times as much energy (say 4000 cm^{-1}, equivalent to 48 kJ/mol), it is excited to a region where the density of energy levels is very high, about 200/cm^{-1}. In this region the line widths may be large enough for these levels to overlap, and for a large number of states to be excited simultaneously. The net effect of the excitation is then a simultaneous deposition of energy in many vibrational degrees of freedom. In either case, nearly every collision of an excited SF_6 molecule with one in the ground state will produce a near-resonant transfer of vibrational energy.

Third, we need to distinguish between two kinds of relaxation processes in excited molecules. Consider a gas of molecules in which initially there exists vibrational excitation. In most cases the excitation will be *incoherent*, that is, there will be no relationships among the vibrational *phases* of the excited molecules; incoherent excitation can be produced by either collisions or incoherent (black-body) radiation. Let the temperature of the gas be such that collisions will produce a net transfer of energy from vibration to translation (that is, the initial vibrational energy exceeds that in the final equilibrium distribution). The process of equilibration follows a simple *relaxation equation*, in which the concentration c_n of the nth vibrational state changes at a rate proportional to the deviation from equilibrium:

$$-\frac{d[c_n(t) - c_n(\infty)]}{dt} = \frac{c_n(t) - c_n(\infty)}{\tau}, \qquad c_n(t) - c_n(\infty) = [c_n(0) - c_n(\infty)]e^{-t/\tau},$$

$$(28.111)$$

where $c_n(\infty)$ is the equilibrium value. The proportionality coefficient τ is called the *relaxation time* for vibration; it is an averaged function of the cross sections that contribute to the vibrational energy transfer. Relaxa-

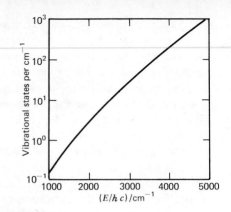

FIGURE 28.17
Density of vibrational states for SF_6 (energy E measured relative to ground state).

tion of the type just described, in incoherently excited molecules, occurs only by inelastic collisions. Suppose, however, that one vibrationally excites a number of molecules in such a way (the method will be described shortly) that the phase of vibration for all the molecules is initially the same. Both elastic and inelastic collisions can destroy such phase coherence, and the former are generally much more probable.

Next we consider a few typical experiments on energy transfer and their results.

A molecular-beam apparatus is in principle well suited for the determination of inelastic (energy-exchange) cross sections, although in practice the measurements are quite difficult. Consider an experiment in which a beam of "reactant" O_2^+ ions is prepared in such a way that all atoms have nearly the same translational energy, in the range 10–20 eV. For charged particles this is easier to do than for neutrals (how would you do it?). The O_2^+ beam is made to collide with a much slower cross beam of Ar atoms, and the scattering of O_2^+ ions is measured as a function of angle and energy in the usual manner. If an exchange from translational to internal energy occurs, then the translational energy of the scattered O_2^+ will be less than its initial value. Some typical measurements are shown in Fig. 28.18, which gives the energy-loss spectrum (i.e., probability distribution for various amounts of translational energy loss) at several angles. As described in the figure caption, the energy losses corresponding to particular vibrational excitations are indicated by arrows; the peaks can thus be identified with specific excitations. The total cross section for each such transition can be obtained by summing the scattering intensities over all angles. Table 28.3 gives the orders of magnitude of typical cross sections for various kinds of energy-transfer processes.

Kinetic spectroscopy is an effective method for studying energy transfer, and we describe a representative experiment. The irradiation of I_2 vapor with the 514.5-nm line of an Ar^+ ion laser produces a fluorescence spectrum that can be measured easily and accurately. For a given intensity of irradiation and a given gas density, a known number of quanta are absorbed per unit time. Each 514.5-nm quantum absorbed excites an I_2 molecule in the vibrationless ground state to the 43rd excited vibrational

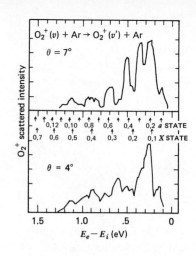

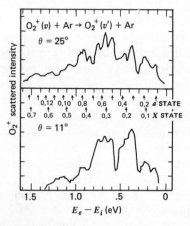

FIGURE 28.18
Measurements of conversion of translational to internal energy in O_2^+–Ar scattering. The intensity of scattered O_2^+ is plotted as a function of translational energy loss (from the initial value of 13.25 eV) at several angles. The energy losses corresponding to specific vibrational excitations are indicated by arrows: "0,1" refers to the transition $(v = 0) \rightarrow (v = 1)$, and so on. (The two sets of arrows refer to the $X^2\Pi_g$ and $a^4\Pi_u$ electronic states of O_2^+, both of which may be present in the beam.) From P. C. Cosby and J. F. Moran, *J. Chem. Phys.* **52**, 6157 (1970).

TABLE 28.3
TYPICAL PARAMETERS FOR ENERGY-EXCHANGE PROCESSES IN THE GAS PHASE

Type of Energy Exchange	Energy Change (kcal/mol)	Total Cross Section (cm²)	Average Number of Collisions	Relaxation Times (s)
Translation-translation	fraction of $k_B T$	10^{-15}–10^{-13}	1	10^{-8}
Translation-rotation	0.1	10^{-15}	5–10	10^{-7}
Translation-vibration	1–5	10^{-17}–10^{-16}	10^3–10^6	10^{-5}–10^{-2}
Translation-electronic	10–50	10^{-19}–10^{-16}	10^6—	$>10^{-2}$

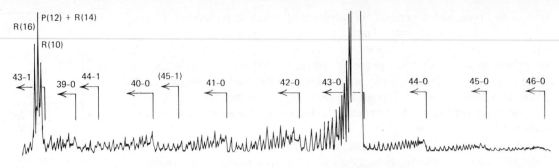

FIGURE 28.19
Part of the fluorescence spectrum of I_2 vapor irradiated with the 514.5-nm line of an Ar$^+$ laser. The labels 43–0 and so on identify the peaks corresponding to various transitions: 43–0 is a drop from the 43rd excited vibrational level to the ground state. From R. B. Kurzel, J. I. Steinfeld, D. A. Hatzenbuhler, and G. E. Leroi, *J. Chem. Phys.* **55**, 4822 (1971).

level of the first excited electronic state. A fraction of the molecules so excited undergo fluorescence, emitting a 514.5-nm quantum and returning to the ground state. Some of the laser light is not absorbed, but is scattered elastically; the large peak labeled 43–0 in Fig. 28.19 is due to a combination of resonance fluorescence and elastic scattering of the 514.5-nm radiation. The next largest peak (labeled 43–1) is fluorescence from the 43rd excited vibrational level to the first excited level. The remaining fluorescence spectrum in this region is due to vibrational energy transfer on collision. A molecule in the 43rd excited vibrational level may collide with any other molecule, and as a result of the collision may end up in the 44th or 42nd or any other vibrational level. The I_2 molecule then may fluoresce from any vibrational level thus reached; the structures labeled 44–0, 42–0, . . . are due to fluorescence from the 44th, 42nd, . . . excited vibrational levels to the ground state. The small peaks within each structure show the participation of various rotational states within the vibrational manifolds.

The spectrum shown in Fig. 28.19 is obtained after irradiating $I_2(g)$ for some time, until stationary conditions are reached. By this we mean that the concentration of each vibrational and rotational state is constant in time. Of course, we do not have equilibrium: The laser produces a continuous pumping from the ground state to the 43rd excited vibrational state, and there is continuous relaxation (energy redistribution and losses) due to collisions and fluorescence. Measurement of these steady-state concentrations, with the use of Boltzmann-type kinetic equations for the energy-transfer and radiation losses, leads to information about inelastic energy-transfer cross sections. A few experiments have been made with short pulses of light to excite a given vibrational state, followed by measurements of fluorescence (and hence concentration) of vibrational states as a function of time. These studies are more difficult, but lead more directly to information on energy-transfer cross sections and probabilities.

An intense pulse of laser light may induce multilevel absorption. The first few energy levels for some of the vibrational modes of CH_3F are shown in Fig. 28.20. (Here ν_3 is essentially a C—F stretch, ν_6 an H—C—F bend, $\nu_{2,5}$ an H—C—H bend, and $\nu_{1,4}$ a C—H stretch.) A CO_2 laser emits 9.6-μm photons. These photons excite not only the first, but, by subsequent excitations, also the second and third vibrational levels of the ν_3 mode. Vibration–vibration transfer due to collisions within the ν_3 manifold is measured to occur on a time scale of microseconds. Near-resonant energy transfers from the third level of ν_3 to the second level of $\nu_{2,5}$, and from the first and second levels of ν_3 to the corresponding levels of ν_6, all take place on the same time scale. Hence within microseconds after irradiation a stationary state of vibrational

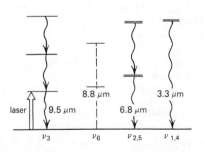

FIGURE 28.20
Some of the lower vibrational energy levels of CH_3F. (See discussion in text.) From R. E. McNair, S. F. Fulghum, G. W. Flynn, M. S. Feld, and B. J. Feldman, *Chem. Phys. Lett.* **48**, 241 (1977).

excitation is achieved. The relaxation from vibration to translation takes much longer, of the order of milliseconds.

In order to measure phase relaxation of vibrational excitation, it is necessary to set up coherent oscillations. This can be accomplished by the following ingenious method: Let 18,910-cm^{-1} (green) laser light impinge on liquid nitrogen. This radiation is produced by a frequency-doubling device from 9455-cm^{-1} (red) laser light, and the two frequencies are in phase. The green radiation causes some N_2 molecules to undergo Raman scattering (Section 7.3), leaving them in the first excited vibrational level (Fig. 28.21). The intensity of the green radiation is sufficient to produce *stimulated* Raman scattering in phase with the green radiation. The stimulated Raman scattering is much more intense than the spontaneous Raman scattering, so most of the $N_2(v = 1)$ molecules oscillate in phase. These coherent vibrations are now probed by the red radiation after an interval of the order of picoseconds. The red radiation produces *coherent anti-Stokes Raman scattering* (CARS) as shown in the figure. The intensity of the CARS signal depends on the density of the coherently oscillating $N_2(v = 1)$ molecules. We expect that collisions will destroy the coherence of the oscillations, so the CARS intensity is expected to decrease as one increases the delay time between the stimulating (green) pulse and the probing (red) pulse. This decay is indeed observed, as is seen in Fig. 28.22. The phase relaxation time is about 75 ps. In contrast, the relaxation time for transfer from vibrational to translational energy in liquid N_2 is enormously longer, namely, 1.5s!

Let us summarize a few observations on inelastic energy transfer. In doing so it is convenient to speak of the number of collisions required to achieve a given energy transfer. The relaxation time introduced in Eq. 28.111 is a measure of the characteristic time required; hence the number of collisions is simply τ/τ_f, where τ_f is the average time between collisions, usually defined for an (effective) hard-sphere model. Translational relaxation (translation–translation energy exchange) is attained in a few collisions per molecule. The interchange of rotational and translational energies is almost as fast, requiring 5 or 10 collisions on the average, since the rotational energy levels are closely spaced. The process is even more rapid for polar molecules, which can interact at a greater distance. Direct interchange of rotational energy also seems to take place very quickly. These generalizations do not apply to H_2 and D_2, for which the rotational levels are much farther apart (cf. Section 21.5.B) and a few hundred collisions may be required for the molecules to reach equilibrium.

Vibration–translation energy exchange is relatively slow near room temperature, requiring thousands to millions of collisions in many instances. (But at 2000 K, say, tens or hundreds of collisions *may* suffice.) Vibration–vibration exchange, either within a normal mode (in different molecules) or under near-resonant conditions among normal modes, is fast and may need only a few collisions. Thus the vibrational energy levels can reach equilibrium among themselves much more rapidly than with translational energy. When this occurs one can define a vibrational temperature T_{vib}, the parameter characterizing a Boltzmann distribution proportional to $e^{-E_{\text{vib}}/k_B T_{\text{vib}}}$. Such a temperature can be introduced for any system of energy levels that attain internal equilibrium more rapidly than external equilibrium; thus there can also be rotational and electronic temperatures. (Cf. the discussion of nuclear-spin temperatures in Section 20.5.)

Electronic energy levels are generally very widely spaced, so the probability of electronic–translational energy exchange is small. For resonant conditions, however, electronic–vibrational exchange can occur with high probability.

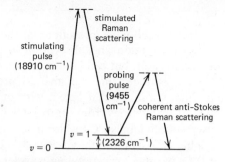

FIGURE 28.21
Schematic illustration of the process leading to CARS scattering (see text).

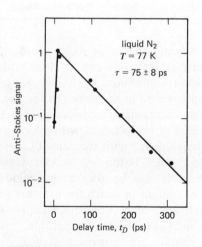

FIGURE 28.22
Intensity of CARS signal described in the text as a function of delay time between the stimulating and probing pulses. From A. Laubereau, *Chem. Phys. Lett.* **27**, 600 (1974).

28.10
Sound Propagation and Absorption

In this section we present an elementary formulation of the theory of sound propagation and absorption in fluids. The subject is not only of interest in itself, but can be used to study inelastic and reactive processes. "Sound" is the name we give to pressure or compression waves, in particular those we can detect with our ears. Audible sound covers the range of frequencies from 20 to 20,000 Hz (with some variation in range between individuals). Consider a sound wave with frequency $\nu = 500$ Hz (roughly B above middle C): The speed of sound in air is about 330 m/s (at 0°C), so the corresponding wavelength is

$$\lambda = \frac{c}{\nu} = \frac{300 \text{ m/s}}{500 \text{ s}^{-1}} = 0.66 \text{ m.}$$

For our 500-Hz wave the time interval between passage of a crest and a trough at a given point is just 1 μs. We can therefore assume that the pressure variation occurs rapidly enough for heat conduction to be negligible, that is, that the compression and expansion in the sound wave are nearly adiabatic processes.

A sound wave is a longitudinal wave, one in which the particles of the medium move back and forth in the direction of propagation (rather than transverse to it, as in water waves). In this section we shall consider only *plane* waves, one-dimensional waves in which the pressure and other properties of the fluid vary only in the direction of propagation. In Fig. 28.23a we show the propagation through a gas of a single pressure or density pulse (as opposed to a periodic wave), represented by the traveling region of greater density. Such a pressure or density perturbation must be propagated by collisions between molecules; thus it is not surprising that the speed of sound (c) is comparable to the average molecular speed $\langle v \rangle$, as shown by the following data[14] for several gases at 0°C:

	H_2	He	N_2	Air	O_2	Cl_2
c (m/s)	1269.5	970	337	331.45	317.2	205.3
$\langle v \rangle$ (m/s)	1693.8	1202	454.3		425.1	285.6

Assume that a periodic sound wave is transmitted through a gas initially at rest; this can be done with a tuning fork or other periodic mechanical vibration. In the absence of the wave, the average molecular velocity is zero in all regions of the gas. The effect of the wave is to cause the gas molecules to oscillate about their equilibrium positions in the direction of propagation, with a local average velocity $u(x, t)$; this should not be confused with the velocity c at which the wave itself propagates. Since the gas is moving backward and forward in alternating regions, there must also be alternating regions of compression and rarefaction, that is, regions in which the pressure (or density) is above and below the equilibrium value. Adiabatic compression is accompanied by a temperature increase, so the temperature must also vary periodically. These relationships are shown schematically in Fig. 28.23b.

Energy must, of course, be imparted to the gas to set up this oscillation. The kinetic energy per unit volume is $\frac{1}{2}\rho u^2$, where ρ is the local density and u is the local average velocity (both are functions of x

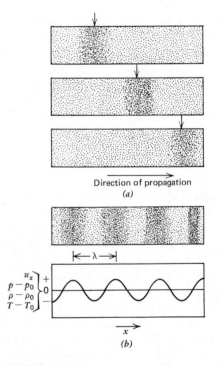

Direction of propagation
(a)

(b)

FIGURE 28.23

(a) Propagation of a pressure or density pulse (indicated by arrows) through a gas; the dots represent molecules. (b) Schematic representation of a periodic sound wave, in terms of molecules (density variation exaggerated) and of local variables: average velocity in the direction of propagation (x direction), excess pressure, excess density, excess temperature (p_0, ρ_0, T_0 are the equilibrium values); one can show (though we shall not) that all these variables are in phase.

[14] The values of c are experimental; those for $\langle v \rangle$ are calculated by Eq. 28.48.

and *t*); the corresponding potential energy is equal to the *pV* work of (nearly) adiabatic compression. What happens to this energy? In a perfectly elastic medium it would all travel onward with the wave, which would thus never decrease in intensity.[15] But we know that sound does not travel infinitely far, so we must consider dissipative processes, in particular viscosity and thermal conduction. The compression (heating) and expansion (cooling) cycle is not quite adiabatic. On each cycle some of the local velocity gradient is lost due to viscous flow, and some of the local temperature gradient is lost due to thermal conduction. Each such loss reduces the energy, and thus the amplitude, of the wave, which eventually is completely dissipated. The energy lost is converted into random motion of the gas molecules, that is, into an overall rise in temperature. We refer to the process described as the *absorption* of sound by the medium.

We see that the properties of sound waves must have some relation to the nature of molecular motion: The velocity of propagation is related to the average molecular speed, and the process of absorption involves the transport properties of viscosity and thermal conductivity. For these reasons, and for certain later applications of the results, we shall give here an elementary discussion of the macroscopic theory of sound propagation and absorption. We begin with the informative exercise of setting up the hydrodynamic equations of motion in a one-dimensional system.

Hydrodynamics,[16] like thermodynamics, is a theory of macroscopic variables. There are normally a limited number of such variables of interest, which (also as in thermodynamics) have to be found empirically. Those with which we shall be dealing include the mass velocity (what we have been calling the local average velocity), the pressure, the energy density, and other thermodynamic functions, all of which vary in time and space. We seek the equations describing this variation.

It is clear that we have come a long way from equilibrium thermodynamics, and even from kinetic theory. The principal difference from thermodynamics is that we are dealing with time-dependent processes; as in thermodynamics, however, we are concerned with a level of description far less detailed than in kinetic theory. In the latter we derived the Boltzmann equation, Eq. 28.28, which describes the time dependence of the velocity distribution function $f(\mathbf{v}, t)$ for homogeneous systems. Now we are considering an inhomogeneous system, so we would have to use a more general distribution function $f(\mathbf{r}, \mathbf{v}, t)$ depending on spatial coordinates as well as velocity and time; the function $f(\mathbf{r}, \mathbf{v}, t)$ obeys a more general form of the Boltzmann equation. Integration of the distribution function over all velocities yields the local density,

$$\rho(\mathbf{r}, t) = Nm \int f(\mathbf{r}, \mathbf{v}, t) \, d\mathbf{v}; \qquad (28.112)$$

a similar integration of the Boltzmann equation over velocities should lead to the macroscopic hydrodynamic equations—and so it does. We shall not follow this route, however. Instead we shall proceed in a heuristic and entirely macroscopic way.

[15] The *intensity* of a sound wave is the rate at which energy is transmitted by the wave per unit area of wave front; if the wave has a velocity *c* and an energy per unit volume of *W*, the intensity is simply *cW*. ("Loudness" as perceived by the ear is approximately proportional to the logarithm of the sound intensity. This is why loudness is measured in the logarithmic decibel scale, in which a difference of 10 decibels between two sounds corresponds to an intensity ratio of 10.)

[16] The equations of hydrodynamics have been presented in Chapter 20. With some purposeful repetition, we reestablish these equations here with simpler physical arguments and apply them to sound propagation and absorption.

We obtain the equations of hydrodynamics by combining the conservation laws of mechanics (conservation of mass, momentum, energy) with the laws of thermodynamics. But how valid is the application of thermodynamics to time-dependent processes? We must make the standard assumption of nonequilibrium thermodynamics: that each volume element which is small compared to the dimensions of the fluid, but large enough to contain very many molecules, is in "local equilibrium." By this phrase we mean a condition in which a local temperature, density, mass velocity, and so on can be usefully defined for the volume element. Of course, these quantities vary from point to point within the fluid, but in the neighborhood of any point there is a region containing many molecules over which they are effectively constant. (Even 20,000-Hz sound has a wavelength of about 1.6 cm in air, so the gradients due to the wave should be negligible over a distance of, say, 1 μm; and a 1-μm cube at STP contains 2.7×10^7 molecules.) It is the variation over longer ranges that gives rise to irreversible processes: flow, heat conduction, diffusion, etc. We may be able to apply the laws of thermodynamics to local conditions if local equilibrium is attained very rapidly, while the large-scale dissipative processes occur much more slowly. This assumes that the slower processes do not significantly perturb the local equilibrium, the same assumption we made in our earlier discussion of the transport processes.

Having reassured ourselves as to our premises, let us now derive the hydrodynamic equations. We restrict ourselves to a one-dimensional system, in which flows, pressure or density gradients, and dissipative processes occur only in the x direction. Thus all local variables are functions only of x and time: mass velocity $u(x, t)$, density $\rho(x, t)$, and so on. Consider a volume element with cross-sectional area A (perpendicular to the x axis) and thickness dx, located at the point x. The density within this volume element will vary with time at the rate $[\partial\rho(x, t)/\partial t]_x$, with x held constant because the position is fixed. Since mass is conserved, this change in density can occur only by a net mass flow into or out of the volume element. The rate of mass flow across unit area is simply ρu, where u is defined as positive in the positive x direction. Thus the net mass flow into the volume element $A\,dx$ is the sum of the inward flows across the surfaces at x and $x + dx$, which is

$$A[\rho(x, t)u(x, t) - \rho(x + dx, t)u(x + dx, t)].$$

To obtain the rate of density increase, we divide by the volume $A\,dx$; going to the limit $dx \to 0$ in the usual manner gives us the derivative:

$$\left[\frac{\partial\rho(x, t)}{\partial t}\right]_x = \lim_{dx \to 0} \frac{[\rho(x, t)u(x, t) - \rho(x + dx, t)u(x + dx, t)]}{dx}$$
$$= -\left\{\frac{\partial[\rho(x, t)u(x, t)]}{\partial x}\right\}_t. \quad (28.113)$$

Rearranging and omitting the functional notation, we obtain

$$\left(\frac{\partial\rho}{\partial t}\right)_x + \left[\frac{\partial(\rho u)}{\partial x}\right]_t = 0, \quad (28.114)$$

which is the one-dimensional version of the *continuity equation* (or mass-conservation equation). It is the mathematical expression of the simple statement that, because mass is conserved, the density at a given location can be changed only by flow of mass. The general form of this equation was obtained as Eq. 20.12.

Next we consider the conservation of linear momentum (which in this model is in the x direction only). The mass of gas within the volume element $A\,dx$ will be accelerated if there is a difference in the forces acting on the two surfaces of the volume element. Since pressure is force per unit area, the net force difference in the positive x direction between the surfaces at x and $x + dx$ is

$$A[p(x, t) - p(x + dx, t)].$$

The acceleration produced by this net force is given by Newton's second law ($\mathbf{F} = m\mathbf{a}$), which expresses the conservation of momentum. The mass in the volume $A\,dx$ is $\rho A\,dx$, and its acceleration is du/dt, the derivative of the mass velocity. We thus have

$$F_x(x, t) = \lim_{dx \to 0} A[p(x, t) - p(x + dx, t)] = -A\,dx\left[\frac{\partial p(x, t)}{\partial x}\right]_t$$

$$\tag{28.115}$$

$$= \rho A\,dx\,\frac{du}{dt}.$$

Dividing through by $A\,dx$ and rearranging, we have

$$\left(\frac{\partial p}{\partial x}\right)_t + \rho\,\frac{du}{dt} = 0, \tag{28.116}$$

the one-dimensional *equation of motion* (momentum-conservation equation); it corresponds to the general Eq. 20.31. Note that in this equation we have written du/dt rather than $(\partial u/\partial t)_x$; this is because the acceleration described is that of the moving mass of gas, rather than the gradient of the mass velocity at a fixed point. We can relate the two representations by expanding du/dt in terms of partial derivatives,

$$\frac{du}{dt} = \left(\frac{\partial u}{\partial t}\right)_x + \left(\frac{\partial u}{\partial t}\right)_x\frac{\partial x}{\partial t} = \left(\frac{\partial u}{\partial t}\right)_x + u\left(\frac{\partial u}{\partial x}\right)_t, \tag{28.117}$$

so that the equation of motion can be written as

$$\left(\frac{\partial p}{\partial x}\right)_t + \rho\left(\frac{\partial u}{\partial t}\right)_x + \rho u\left(\frac{\partial u}{\partial x}\right)_t = 0; \tag{28.118}$$

the quantity $u(\partial u/\partial x)_t$ will be small, and we shall later neglect this term. Note that by using Newton's second law we assume that all the net force produces a change in mass velocity, with none lost in overcoming viscosity; we are still dealing with a system without dissipation.

Finally we introduce the conservation of energy. It is possible to obtain a hydrodynamic energy equation analogous to Eqs. 28.114 and 28.116 for mass and momentum, but we shall not need this. Instead we introduce conservation of energy in thermodynamic terms, using the first law of thermodynamics in the form

$$dU = đq + đw \tag{28.119}$$

(we continue to use the convention that q is positive for heat absorbed by the system, w positive for work done on the system). We assume that the compression of the gas is adiabatic and reversible (again ignoring dissipative processes), and thus have $đq = 0$, $đw = -p\,dV$. If we make the further restriction that the gas is ideal, we can write

$$dU = C_V\,dT = -p\,dV, \tag{28.120}$$

where C_V is the constant-volume heat capacity. Since $\rho \equiv m/V$, we have

$$dV = d\left(\frac{m}{\rho}\right) = -\left(\frac{m}{\rho^2}\right)d\rho \tag{28.121}$$

and thus

$$C_V \, dT = \frac{mp}{\rho^2} \, d\rho; \qquad (28.122)$$

we can replace m by the molecular weight M, if we stipulate that C_V is the *molar* heat capacity.

Thus far our derivation has been rather general, applying to any type of nonviscous, one-dimensional fluid flow. Now we shall consider the specific case of a periodic wave. Consider an ideal gas initially at rest, with pressure p_0, density ρ_0, temperature T_0. Upon this gas we impose a pressure disturbance in the form of a plane traveling wave, that is, a periodic pressure variation of the type shown in Fig. 28.23b. We describe the pressure variation in terms of the excess pressure $\mathscr{P}$, which obeys the standard wave equation

$$\mathscr{P}(x, t) \equiv p(x, t) - p_0 = \mathscr{P}_0 e^{i(\omega t - kx)}, \qquad (28.123)$$

where $\mathscr{P}_0$ is the amplitude of the pressure variation, $\omega (= 2\pi\nu)$ is the angular frequency of the wave, and $k (= 2\pi/\lambda)$ is the wave vector. We write similar equations for the excess density,

$$\mathscr{R}(x, t) \equiv \rho(x, t) - \rho_0 = \mathscr{R}_0 e^{i(\omega t - kx)}, \qquad (28.124)$$

the excess temperature,

$$\mathscr{T}(x, t) = T(x, t) - T_0 = \mathscr{T}_0 e^{i(\omega t - kx)}, \qquad (28.125)$$

and the mass velocity (which is zero in the gas at rest),

$$u(x, t) = u_0 e^{i(\omega t - kx)}. \qquad (28.126)$$

Now we substitute these expressions into the conservation equations. In the continuity equation, Eq. 28.126, we have

$$\left(\frac{\partial \rho}{\partial t} \right)_x = \left(\frac{\partial \mathscr{R}}{\partial t} \right)_x = i\omega \mathscr{R} \qquad (28.127)$$

and

$$\left[\frac{\partial(\rho u)}{\partial x} \right]_t = \rho \left(\frac{\partial u}{\partial x} \right)_t + u \left(\frac{\partial \rho}{\partial x} \right)_t = (\rho_0 + \mathscr{R}) \left(\frac{\partial u}{\partial x} \right)_t + u \left(\frac{\partial \mathscr{R}}{\partial x} \right)_t. \qquad (28.128)$$

We assume that the perturbations due to the sound wave are small, so that we need consider only first-order deviations from the conditions in the gas at rest. Thus we can neglect the terms containing $\mathscr{R}(\partial u/\partial x)_t$ and $u(\partial \mathscr{R}/\partial x)_t$ in Eq. 28.128, which reduces to

$$\left[\frac{\partial(\rho u)}{\partial x} \right]_t = \rho_0 \left(\frac{\partial u}{\partial x} \right)_t = -ik\rho_0 u; \qquad (28.129)$$

combining this with Eq. 28.127 gives the equation of continuity in the forms

$$\left(\frac{\partial \mathscr{R}}{\partial t} \right)_x + \rho_0 \left(\frac{\partial u}{\partial x} \right)_t = 0 \quad \text{or} \quad i\omega\mathscr{R} - ik\rho_0 u = 0. \qquad (28.130)$$

Applying the same method to the equation of motion, Eq. 28.118, we obtain

$$\left(\frac{\partial \mathscr{P}}{\partial x} \right)_t + \rho_0 \left(\frac{\partial u}{\partial t} \right)_x = 0 \quad \text{or} \quad i\omega\rho_0 u - ik\mathscr{P} = 0; \qquad (28.131)$$

the $\rho u (\partial u/\partial x)_t$ term is a second-order perturbation to begin with, and so drops out. To substitute in the energy equation, Eq. 28.122, we must assume that dT and $d\rho$ can be replaced by our difference variables $\mathscr{T}$ and $\mathscr{R}$, respectively; this is legitimate as long as $\mathscr{T}$ and $\mathscr{R}$ are small perturbations from equilibrium. We thus obtain

$$C_V \mathcal{T} = \frac{Mp_0}{\rho_0{}^2} \mathcal{R}. \tag{28.132}$$

Since we have four variables ($\mathcal{P}$, $\mathcal{R}$, $\mathcal{T}$, and u), we need a fourth equation connecting them to obtain a solution. For this purpose we can use the equation of state,

$$p = p(\rho, T), \qquad dp = \left(\frac{\partial p}{\partial \rho}\right)_T d\rho + \left(\frac{\partial p}{\partial T}\right)_\rho dT; \tag{28.133}$$

again substituting the difference variables (with the small-perturbation assumption) gives

$$\mathcal{P} - \left(\frac{\partial p}{\partial \rho}\right)_T \mathcal{R} - \left(\frac{\partial p}{\partial T}\right)_\rho \mathcal{T} = 0 \tag{28.134}$$

(the two derivatives are state functions that we can treat as constants).

We have thus obtained four simultaneous linear equations in the variables $\mathcal{P}$, $\mathcal{R}$, $\mathcal{T}$, and u, which we can arrange in the form:

$$\begin{aligned}
iku \qquad &- \frac{i\omega}{\rho_0} \mathcal{R} &&= 0, \\
i\omega u - \frac{ik}{\rho_0} \mathcal{P} \qquad &&&= 0, \\
&- \frac{Mp_0}{\rho_0{}^2} \mathcal{R} \; + \; && C_V \mathcal{T} = 0, \\
\mathcal{P} \qquad &- \left(\frac{\partial p}{\partial \rho}\right)_T \mathcal{R} - && \left(\frac{\partial p}{\partial T}\right)_\rho \mathcal{T} = 0.
\end{aligned} \tag{28.135}$$

The simultaneous equations can be solved by the standard methods, either trial-and-error substitution or evaluating the determinant of the coefficients. The latter is particularly convenient because of the large number of zeros in the determinant; expanding by cofactors in the usual way, we obtain

$$\begin{vmatrix}
ik & 0 & -\dfrac{i\omega}{\rho_0} & 0 \\[2mm]
i\omega & -\dfrac{ik}{\rho_0} & 0 & 0 \\[2mm]
0 & 0 & -\dfrac{Mp_0}{\rho_0{}^2} & C_V \\[2mm]
0 & 1 & -\left(\dfrac{\partial p}{\partial \rho}\right)_T & -\left(\dfrac{\partial p}{\partial T}\right)_\rho
\end{vmatrix}$$

$$= ik\left\{ -\frac{ik}{\rho_0}\left[\frac{Mp_0}{\rho_0{}^2}\left(\frac{\partial p}{\partial T}\right)_\rho + C_V\left(\frac{\partial p}{\partial \rho}\right)_T\right]\right\} + i\omega\left(-\frac{i\omega C_V}{\rho_0}\right) = 0. \tag{28.136}$$

We have a particular reason for expressing the result in this form. For any traveling wave of the form $Ae^{i(\omega t - kx)}$, the velocity with which the wave propagates is simply

$$c = \frac{\omega}{k}, \tag{28.137}$$

which in this case is the speed of sound. We can thus solve for c by rearranging Eq. 28.136:

$$c^2 = \frac{\omega^2}{k^2} = \left(\frac{\partial p}{\partial \rho}\right)_T + \frac{Mp_0}{C_V\rho_0{}^2}\left(\frac{\partial p}{\partial T}\right)_\rho. \tag{28.138}$$

We assumed in deriving this equation that the gas was ideal, specifically in introducing Eq. 28.122. Thus we can now evaluate the derivatives in Eq. 28.138 by substituting the ideal-gas equation of state,

$$p = \frac{\rho RT}{M}. \tag{28.139}$$

We obtain

$$c^2 = \frac{p_0}{\rho_0}\left(1 + \frac{R}{C_V}\right) = \frac{p_0}{\rho_0}\gamma \qquad \left(\gamma \equiv \frac{C_p}{C_V}\right), \tag{28.140}$$

using the thermodynamic equality $C_p - C_V = R$ (for an ideal gas). Thus we have achieved our first goal, the evaluation of the speed of sound. The reason for the relationship between the speed of sound and the average molecular speed is now obvious: By Eq. 28.140 we have $c = (\gamma RT/M)^{1/2}$ (recall that γ is $\frac{5}{3}$ for monatomic, $\frac{7}{5}$ for diatomic ideal gases), whereas by Eq. 28.48 $\langle v \rangle = (8RT/\pi M)^{1/2}$. This result predicts that c is a function of temperature only, but experiment shows that c does vary with pressure (as the other transport properties do: see Section 29.1). If one removes the restriction to ideal gases, one can derive the expression

$$c^2 = \left(\frac{\partial p}{\partial \rho}\right)_S = \frac{1}{\rho \kappa_S} \qquad \left[\kappa_S \equiv -\frac{1}{V}\left(\frac{\partial V}{\partial p}\right)_S\right] \tag{28.141}$$

(where κ_S is the adiabatic compressibility), which is the generalization of Eq. 28.138. This equation in fact gives the speed of sound in any medium, as long as the assumption of no dissipation holds.

In the remainder of this section we shall see what changes in the theory are necessary to account for dissipative processes; we consider only viscosity explicitly. Since conservation of mass cannot be violated, the continuity equation, Eq. 28.114, remains unaffected. The equation of state is also unchanged, and we shall continue to use the ideal-gas Eq. 28.122.[17] However, the equation of motion does need to be amended, since a certain amount of the momentum flux is required to overcome the viscosity of the gas. The net force accelerating the gas in our volume element is not simply the gradient in the static pressure p, as we adequately assumed in writing Eq. 28.115, but the gradient in $(\boldsymbol{\Gamma}_{mv})_{xx}$, the term in the pressure tensor (Section 28.6) that gives the flux of the x component of momentum in the x direction. When one has flow in the x direction only (as we continue to assume), this term is given by[18]

---

[17] Strictly speaking, one cannot have dissipative processes in an ideal gas, since viscosity and other such processes require interactions between molecules. But the assumption is legitimate, since at and below atmospheric pressure the correction due to van der Waals forces is quite small.

[18] To derive Eq. 28.142 one needs the full three-dimensional theory of fluid flow. Briefly, the assumptions made are that the stresses in a flowing fluid are linear in the velocity gradients, and that there is no stress in uniform rotational flow. Given these assumptions, symmetry requires that the elements of the pressure tensor have the form

$$(\boldsymbol{\Gamma}_{mv})_{ij} = p\delta_{ij} - \lambda\left(\sum_k \frac{\partial v_k}{\partial x_k}\right)\delta_{ij} - \mu\left(\frac{\partial v_i}{\partial x_j} + \frac{\partial v_j}{\partial x_i}\right),$$

where δ_{ij} is a Kronecker delta, the x_i and v_i are Cartesian components of position and velocity, and λ, μ are two constants characteristic of the fluid. For $i \neq j$ we have only the term in μ, which is thus the viscosity coefficient introduced in Table 28.1. But for $i = j$ we obtain

$$(\boldsymbol{\Gamma}_{mv})_{xx} = p - \lambda\left(\frac{\partial u}{\partial x} + \frac{\partial v}{\partial y} + \frac{\partial w}{\partial z}\right) - 2\mu\frac{\partial u}{\partial x}$$

(where u, v, w are equivalent to v_x, v_y, v_z), with similar equations for the yy and zz elements. If Eq. 28.83 is to hold, we must have

$$3\lambda + 2\mu = 0,$$

and setting $v = w = 0$ then gives Eq. 28.142. (Actually, the quantity $3\lambda + 2\mu$ is the so-called *bulk viscosity*, which governs the rate at which translational and internal energy can exchange in compression or expansion; but this quantity is negligibly small in a gas at ordinary pressures. See also the discussion in Section 20.3.)

$$(\boldsymbol{\Gamma}_{mv})_{xx} = p - \tfrac{4}{3}\eta\left(\frac{\partial u}{\partial x}\right)_t, \qquad (28.142)$$

where η is the viscosity. For the equation of motion we must thus replace Eq. 28.116 by

$$\left(\frac{\partial p}{\partial x}\right)_t - \tfrac{4}{3}\eta\left(\frac{\partial^2 u}{\partial x^2}\right)_t + \rho\frac{du}{dt} = 0, \qquad (28.143)$$

in which du/dt can be expanded as before.

Substituting in the equation of motion the plane-wave expressions for p, ρ, u as before, we now obtain

$$\left(\frac{\partial \mathscr{P}}{\partial x}\right)_t - \tfrac{4}{3}\eta\left(\frac{\partial^2 u}{\partial x^2}\right)_t + \rho_0\left(\frac{\partial u}{\partial t}\right)_x = 0 \quad \text{or} \quad (i\omega\rho_0 - \tfrac{4}{3}\eta k^2)u - ik\mathscr{P} = 0 \qquad (28.144)$$

instead of Eq. 28.130. The other three simultaneous equations in Eqs. 28.135 remain unchanged. Carrying through the solution as before, we find the dispersion relation

$$c^2 = \frac{\omega^2}{k^2} = c_0{}^2 + \frac{4}{3}\frac{i\eta\omega}{\rho_0}, \qquad (28.145)$$

where c_0 is the no-dispersion value given by Eq. 28.138; this result can also be written as

$$\frac{k^2}{\omega^2} = \left(c_0{}^2 + \frac{4}{3}\frac{i\eta\omega}{\rho_0}\right)^{-1}. \qquad (28.146)$$

Thus if we choose the angular frequency ω to be real, the wave vector k must be complex. If we then set

$$k \equiv k_r - i\alpha, \qquad (28.147)$$

where k_r is the real part of k, and substitute into Eq. 28.146, equating the real and imaginary parts on each side of the resulting equation yields

$$k_r{}^2 - \alpha^2 = \frac{\omega^2 c_0{}^2}{c_0{}^4 + \frac{16}{9}\frac{\eta^2\omega^2}{\rho_0{}^2}}, \qquad 2k_r\alpha = \frac{\frac{4}{3}\frac{\eta\omega^3}{\rho_0}}{c_0{}^4 + \frac{16}{9}\frac{\eta^2\omega^2}{\rho_0{}^2}}. \qquad (28.148)$$

When the imaginary part of the wave vector is nonzero, there must be an attenuation of the sound wave with distance, because of the dissipative process of viscosity. For example, if we substitute Eq. 28.147 into the wave equation for the mass velocity, Eq. 28.126, we obtain

$$u(x, t) = u_0 e^{-\alpha x} e^{i(\omega t - k_r x)}. \qquad (28.149)$$

That is, as the distance of propagation increases, the amplitude of the wave (which is $u_0 e^{-\alpha x}$) steadily decreases. The same thing happens to the amplitudes of variations in pressure, density, and temperature. We refer to α as the *attenuation constant*. Normally the attenuation due to viscosity is small in gases ($\alpha^2 \ll k_r{}^2$), and to a reasonable approximation we can write

$$k_r \approx \frac{\omega}{c_0}; \qquad (28.150)$$

that is, the attenuation does not affect the dispersion relation to lowest order. We then have, from Eqs. 28.148,

$$\alpha = \frac{\frac{2}{3}\frac{\eta}{\rho_0}c_0\omega^2}{c_0^4 + \frac{16}{9}\frac{\eta^2}{\rho_0^2}\omega^2} \approx \frac{8}{3}\frac{\pi^2\eta\nu^2}{\rho_0c_0^3} \qquad (28.151)$$

(since in a gas the second term in the denominator is negligibly small relative to c_0^4). For room-temperature air α/ν^2 is about 10^{-11} s^2/m, so for the 500-Hz wave discussed at the beginning of this section we have $\alpha \approx 2 \times 10^{-6}$/m (or 0.2%/km), quite small attenuation indeed. Note that measuring attenuation as a function of sound frequency in a dissipative medium provides a method of determining the viscosity coefficient, by solving Eq. 28.151 for η.

The effect of dissipative processes is most directly expressed in terms of the energy lost by the sound wave. The rate at which energy is lost per unit volume due to viscosity in a plane wave is simply the viscous term in the equation of motion times the mass velocity, or

$$\frac{d}{dt}\left(\frac{U}{V}\right) = -\frac{4}{3}\eta\left(\frac{\partial^2 u}{\partial x^2}\right)_t u = \tfrac{4}{3}\eta k^2 u^2. \qquad (28.152)$$

As we indicated earlier, this energy is converted to heat. Why, then, did we not include this term as a correction to Eq. 28.142 when calculating the effect of attenuation? The answer is that this power loss is second-order in u and could thus be neglected, whereas the viscous force itself is only first-order in u and had to be included in the equation of motion.

We have treated only the case of dissipation due to viscosity, but the analysis of the thermal-conductivity effect is not dissimilar. One obtains an expression analogous to Eq. 28.151 for the attenuation coefficient,

$$\alpha_\lambda \approx \frac{2\pi^2\nu^2}{c_0^3}\left(\frac{\gamma-1}{\gamma}\right)\frac{\lambda'}{\rho_0 C_V}, \qquad (28.153)$$

where λ' is the thermal conductivity (the prime to distinguish it from wavelength) and $\gamma = C_p/C_V$. The attenuation coefficients for viscosity and thermal conductivity are of the same order of magnitude.

Besides determining the transport coefficients, measurements of sound absorption can also be used to obtain information about inelastic energy transfer. Suppose that we pass a sound wave through gaseous N_2. The wave is propagated by mass flow, which on the microscopic level occurs by molecular motion and collisions. But in these collisions some of the translational energy will be transferred to the internal degrees of freedom (vibration and rotation) of the N_2 molecules. This process is basically irreversible: Even if the energy is reconverted to translation on subsequent collisions, the resulting motion will be in random directions, since the time between collisions is much greater than the vibrational or rotational periods. Thus the effect of the inelastic energy transfer is twofold: to decrease the amplitude of the sound wave, and to increase the temperature. If the N_2 molecules are in a large excess of Ar molecules, which have no internal degrees of freedom, then the argon can act as a constant-temperature bath and conduct the heat produced away from the N_2 molecules.

Now suppose that the inelastic energy transfer is adequately described by a relaxation equation (cf. Eq. 28.111)

$$\frac{d(E_i)}{dt} = -\frac{E_i - E_{i0}}{\tau_i}, \qquad (28.154)$$

where E_i is the energy in the ith quantum state (of the gas as a whole) and E_{i0} is the energy in that state at equilibrium, that is, in the absence of any perturbation. This says that after a disturbance the system returns to equilibrium by a first-order or exponential decay process ($E_i - E_{i0} = Ae^{-t/\tau_i}$). The quantity τ_i is the relaxation time (or *excitation time*) for the ith state, and is what we wish to determine. Let us accept Eq. 28.154 for the time being and see if it is consistent with experiment.

In a periodic sound wave, the equilibrium energy levels at a given point x themselves become periodic functions of time. (Recall our discussion in Section 13.1 of the changing of energy levels in an adiabatic compression.) We can thus write E_{i0} as $a \sin \omega t$ (taking the mean value as our zero of energy), where a is a constant and ω is the angular frequency of the sound wave. Equation 28.154 then becomes

$$\tau_i \frac{dE_i}{dt} + E_i = a \sin \omega t. \tag{28.155}$$

The solution to this linear differential equation, as can be verified by substitution, is

$$E_i = a \sin(\omega t - \phi_i), \tag{28.156}$$

where ϕ_i is called a *phase shift* (or *lag*); an alternative way of writing this solution is

$$E_i = a \cos \phi_i \sin \omega t - a \sin \phi_i \cos \omega t. \tag{28.157}$$

In either case we have

$$\cos \phi_i = \frac{1}{1 + \tau_i^2 \omega^2}, \qquad \sin \phi_i = \frac{\omega \tau_i}{1 + \tau_i^2 \omega^2} \qquad \tan \phi_i = \omega \tau_i. \tag{28.158}$$

Obviously, the amplitude of the variation in E_i is less than that for E_{i0}. At the peaks of the sound wave we have $\cos \omega t = 0$, so that E_i is less than E_{i0} by the factor $1 + \tau_i^2 \omega^2$. The amount of energy deposited in the ith quantum state is less than the equilibrium value (at the temperature corresponding to the crest of the wave) because there is insufficient time for translation and the internal degrees of freedom to come to equilibrium with each other. For very low sound frequencies, however, we have $\omega \tau_i \ll 1$ and $E_i \approx E_{i0}$; that is, the time interval between crests is long enough for equilibrium to be attained. For very high frequencies ($\omega \tau_i \gg 1$), E_i does not vary at all, since there is no time to transfer any energy.

Now let us look at the second term in Eq. 28.157, which is out of phase with the sound wave. For $\omega \tau_i \ll 1$ we have $\sin \phi_i \approx \omega \tau_i$, whereas for $\omega \tau_i \gg 1$ we have $\sin \phi_i \approx (\omega \tau_i)^{-1}$; thus $\sin \phi_i$ vanishes at both extremes. The second term has its maximum amplitude when $\omega \tau_i \approx 1$, that is, when the time scales of the sound wave and the dissipative energy-transfer process are comparable and resonance can occur. This term is in fact related to the net absorption of energy from the sound wave by inelastic energy transfer. One can see this by comparing Eqs. 28.151 and 28.158: The variation of α/ω in viscous absorption is similar to that for $\sin \phi_i$. The relaxation time plays a role in inelastic energy transfer analogous to that of the viscosity η in viscous dissipation.

It is possible to generate sound—that is, pressure waves—with frequencies as high as 10^9 Hz (the frequencies above the audible range are referred to as *ultrasonic*). By the argument just given, relaxation times as low as 10^{-9} s can be usefully investigated by sound-absorption techniques. We shall return to this point in our study of chemical kinetics.

Further Reading

Kauzmann, W., *Kinetic Theory of Gases* (*Thermal Properties of Matter*, Vol. I) (Benjamin, Reading, Mass., 1966).

Kennard, E. H., *Kinetic Theory of Gases* (McGraw-Hill, New York, 1938).

Present, R. D., *Kinetic Theory of Gases* (McGraw-Hill, New York, 1958).

Reif, F., *Fundamentals of Statistical and Thermal Physics* (McGraw-Hill, New York, 1965), chaps. 12–14.

Chapman, S., and Cowling, T. G., *The Mathematical Theory of Non-Uniform Gases*, 3rd ed. (Cambridge University Press, Cambridge, 1970).

Hirschfelder, J. O., Curtiss, C. F., and Bird, R. B., *Molecular Theory of Gases and Liquids*, 2nd Ed. (Wiley, New York, 1964).

Callear, A. B., and Lambert, J. D., "The Transfer of Energy Between Chemical Species," in C. H. Bamford and C. F. H. Tipper (Eds.), *Comprehensive Chemical Kinetics*, Vol. 3 (Elsevier, New York, 1969), pp. 182–273.

Cottrell, T. L., and McCoubrey, J. C., *Molecular Energy Transfer in Gases* (Butterworths, London, 1961).

Herzfeld, K., and Litovitz, T. A., *Absorption and Dispersion of Ultrasonic Waves* (Academic Press, New York, 1959).

Texts on Kinetic Theory

Advanced Treatises

Energy Exchange

Sound Propagation and Absorption

Problems

1. A distribution function of some variable x can be uniquely described in terms of its moments M_n, defined by

$$M_n \equiv \langle x^n \rangle = \int_{-\infty}^{\infty} dx\, x^n f(x).$$

For the Maxwell–Boltzmann distribution, Eq. 28.40, evaluate
   (a) M_1
   (b) M_2
   (c) M_{2n+1}
   (d) M_{2n}

2. By direct insertion of the Maxwell–Boltzmann distribution into the Boltzmann equation 28.28, show that the equilibrium solution is indeed a solution to Eq. 28.28. Assume that the interactions between particles are spherically symmetric.

3. For a system subject to no external forces, the time dependence of the distribution function is given by

$$\frac{\partial f_1}{\partial t} = \int (f'_1 f'_2 - f_1 f_2) v \sigma(v, \chi)\, d\mathbf{v}_2 \sin \chi\, d\chi,$$

where

$$f_1 \equiv f(\mathbf{v}_1, t), \qquad f_2 \equiv f(\mathbf{v}_2, t),$$
$$f'_1 \equiv f(\mathbf{v}_1 + \Delta\mathbf{v}_1, t), \qquad f'_2 \equiv f(\mathbf{v}_2 + \Delta\mathbf{v}_2, t).$$

Consider the quantity $H \equiv \sum_i \int f_i \ln f_i \, d\mathbf{v}_i$, and write the expression for dH/dt. Note that H depends only on time and that $v\sigma$ depends only on the relative velocity. Consequently, H will not change if $\mathbf{v}_1$ and $\mathbf{v}_2$ are interchanged. Write the expression for dH/dt with these velocities interchanged. Next, in each of these expressions, interchange the primed and unprimed variables and use the fact that

$$v\sigma \, d\mathbf{v}_1 \, d\mathbf{v}_2 \sin \chi \, d\chi = v'\sigma' \, d\mathbf{v}_1' \, d\mathbf{v}_2' \sin \chi' \, d\chi'.$$

This symmetrization has yielded four equivalent expressions for dH/dt. Add these four together and divide by four to prove the Boltzmann H *theorem*, namely, that

$$\frac{dH}{dt} \leq 0.$$

How does the H theorem show that the Maxwell–Boltzmann distribution is a necessary (as well as sufficient) solution to the equilibrium Boltzmann equation? This theorem has played an enormously important role in kinetic theory, since it attempts to establish the validity of the second law of thermodynamics from kinetic theory. Explain.

4. What is the average velocity of each of the following:
   (a) H_2 at 300 K and 1200 K
   (b) Xe at 300 K and 1200 K
   The average kinetic energy of an atom or molecule is $\frac{3}{2}k_B T$. Is the average kinetic energy equal to $\frac{1}{2}m\langle v \rangle^2$? Explain.

5. Calculate the velocity necessary for H_2, O_2, and CO_2 to escape the pull of the earth's gravity. What fraction of the molecules at 300 K have sufficient velocity to escape the gravitational field of the earth? Why is there very little H_2 in the earth's atmosphere?

6. Calculate the collision frequency and mean free path of argon atoms ($d = 2.86$ Å)
   (a) At standard temperature and pressure
   (b) At 1000 K and 1 atm
   How many collisions take place per atom per second? How many collisions take place per second in 1 cm³? Molecular-dynamics calculations on a system of 100 hard spheres indicate that the momentum distribution of the system relaxes to its equilibrium form in about 5 collision times per atom. How long is that?

7. Calculate Z for N_2 at 1 atm and 300 K by assuming that the molecule is spherical and that $d = 3.85$ Å. What is the average time between collisions? During this average time, how many oscillations does a N_2 molecule make? (Assume that N_2 is a harmonic oscillator with $\bar{\nu}_e = 2360$ cm^{-1}.) How many rotations does the molecule make during this time? (Assume that the rotational temperature is 300 K, and that the equilibrium bond length is 1.094 Å.)

8. Assume that the binary-collision approximation still holds in the liquid phase, and calculate Z and the mean free path for Ar with $\rho = 1.37$ g/cm³ and $T = 90$ K. Compare the mean free path to the average internuclear distance in the liquid.

9. What is the average kinetic energy of an effusing beam at temperature T? (*Hint:* Use the variable change $\cos \theta = x$.) Is the average kinetic energy greater than, equal to, or less than that of the molecules inside the chamber? Why?

10. If an effusing beam consists of monomers and dimers, then the velocity distribution of the beam will be the sum of the velocity distributions of each species. If the fraction of dimers is a, derive an expression for the observed velocity distribution. Calculate the maximum of this distribution. (Assume that $a \ll 1$.) Calculate the fraction of dimers, given the following data:

$$M = 168 \text{ g/mol (for monomers)},$$
$$T = 800 \text{ K},$$
$$v = 340 \text{ m/s (at maximum of distribution)}.$$

11. An effusion chamber has a 1-mm pinhole and is loaded with gas to a pressure of 0.01 torr. Calculate the effusive flux out of the chamber if it is filled with (a) H_2 at 300 K, (b) H_2 at 800 K, (c) Ar at 300 K, (d) Ar at 800 K. Consider the case of Ar at 800 K and calculate the density of particles in the beam. Compare this to the density of particles inside the chamber. Calculate the average number of collisions per molecule for the case of two of these Ar beams intersecting at right angles.

12. Read the following articles, which describe some of the experimental techniques used to study energy transfer. In each case, outline the major components of the apparatus and discuss the types of information obtained in the experiment.

   (a) Lasers: "Applications of Lasers to Chemical Research" by Stephen R. Leone, *Journal of Chemical Education* **53,** 12 (1976).

   (b) Ion beams: "Reactions of Accelerated Carbon Ions and Atoms" by Richard M. Lemmon, *Accounts of Chemical Research* **6,** 65 (1973).

   (c) Picosecond spectroscopy: "Ultrafast Phenomena in Liquids and Solids" by R. R. Alfano and S. L. Shapiro, *Scientific American* **228,** no. 6, 43 (June 1973).

13. Stimulated Raman fluorescence data give a rate constant for deactivation by HCl of $H_2(v=1)$ to ground-state H_2 of $1510 \, s^{-1} torr^{-1}$ at 296 K. Calculate the average number of collisions that it takes to achieve deactivation. (*Hint:* For collisions of different "hard spheres," replace m by the reduced mass and d by the distance between centers.) Take d as 2.9 Å for H_2 and 3.3 Å for HCl.

14. Sound-absorption measurements yield rate constants for the rotational transition of $D_2(j=2)$ to $D_2(j=0)$ at 90.5 K. The experimental rate constants are described by

$$k = (8 \times 10^7 + 1.15 \times 10^8 x) \, atm^{-1} \, s^{-1},$$

   where x is the mole fraction of Ar in the system. Calculate the average number of collisions with Ar necessary to achieve relaxation of the $j=2$ level of D_2. Calculate the self-relaxation collision number. Calculate these same quantities for the transition from $j=0$ to $j=2$.

15. The speed of sound in solids is given by $c = (B/\rho)^{1/2}$, where B, the bulk modulus, is the reciprocal of the compressibility. For argon at 77 K we have $B = 1.6 \times 10^{10} \, dyn/cm^2$ and $\rho = 1.77 \, g/cm^3$. For iron at 298 K we have $B = 1.683 \times 10^{12} \, dyn/cm^2$ and $\rho = 7.86 \, g/cm^3$. Compute the speed of sound in these solids and compare to the values obtained assuming that the ideal-gas equation of state holds. Does the ideal-gas expression give a larger or smaller result, and why?

16. How far does a 1000-Hz sound wave travel in an oxygen atmosphere at standard temperature and pressure before its intensity, given by the square of the amplitude, decreases by 50%? Do the same calculation for liquid O_2 at $T = -183°C$ by assuming that the ideal-gas relationships still hold. Take the hard-sphere radius to be 3.4 Å and the density at $-183°C$ to be 1.149 g/cm^3.

17. The energy density of a wave is related to the amplitude by the expression

$$\frac{U}{V} = \tfrac{1}{4}\rho(|u|_{max})^2,$$

   where ρ is the density of the medium. Calculate the rate of energy loss of a 1000-Hz sound wave with an initial amplitude of 240 m/s in an oxygen atmosphere at standard temperature and pressure.

18. Two possible explanations for the way a bat uses sonar to determine the distance of an object are that the bat measures the intensity of the echo from the object or that it measures the time delay of the echo. Estimate the accuracy needed in each case to locate to within ±1 cm an object about 10 m away. Use a sonar frequency of 40 kHz. What is the smallest object the bat can detect at this frequency?

29

The Kinetic Theory of Dense Phases

So far our discussion of kinetic theory and transport properties has been restricted to dilute gases. These are remarkably simple systems for theoretical purposes: Collisions between molecules are infrequent and only involve two molecules at a time, so that in principle one can solve the equations of motion exactly. Successive binary collisions are statistically independent; that is, they behave as if there were no correlations among them, at any time. As might be expected, the situation in denser media is much more complicated. In the next few sections we consider the transport properties of dense gases, liquids, and solids (including a brief treatment of ionized media). We begin with some qualitative features, and then turn to a statistical analysis that includes elements of the theory of Brownian motion and leads to relationships between transport coefficients and correlation functions.

29.1 Transport Properties in Dense Fluids

Consider what happens in a gas when we increase its density. The closer the molecules are to one another, the shorter is the mean free path and the greater the rate of collisions. Although a part of this is taken into account in the dilute-gas theory (the mean free path is inversely proportional to density, the collision frequency per molecule proportional to density), the assumptions of that theory start to break down when the gas becomes dense enough. In a dense gas, several molecules are likely to be interacting simultaneously, so that triple and higher-order collisions must be included in the description. The mechanism of collisional transfer (discussed at the end of Section 28.6) comes into play, and at sufficiently high densities becomes the dominant mode of transport of energy and momentum (but not mass). Finally, the assumption that successive collisions are uncorrelated (discussed after Eq. 28.18) becomes invalid: at high densities a molecule may be reduced to bouncing around in a small space among its neighbors, and the sequence of collisions is then not random.

We have just spoken of increasing the density of a gas, but the dense gas is, of course, continuous with the liquid above the critical point. (It is only recently, however, that measurements of transport properties have begun to cover the entire range.) In any ordinary liquid, all the effects just described are fully present. A molecule is essentially in continuous interaction with its nearest neighbors, and the notion of a mean free path becomes inapplicable.

As in the dilute gas, the viscosity is the transport property that has been most extensively studied. Figure 29.1 illustrates the variation of the viscosity of a dense fluid with temperature, pressure, and density. Since the principle of corresponding states applies quite well to transport properties, the behavior of all simple fluids is qualitatively similar. Note that at constant pressure the viscosity increases with temperature in the

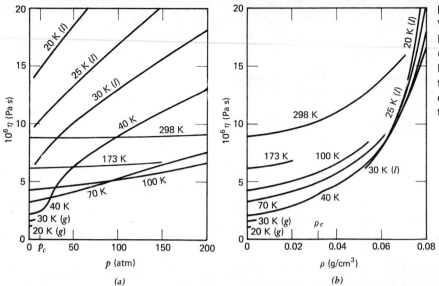

FIGURE 29.1

Viscosity of dense gaseous and liquid hydrogen as a function of (a) pressure, (b) density. (The data at 100 K and below are actually for p-H_2.) Note that the curves for gas and liquid are discontinuous below the critical temperature (33 K).

gas, but decreases with temperature in the liquid. The viscosity of a liquid (at low pressure) usually has a temperature dependence fairly well represented by an equation of the form

$$\eta = A e^{\Delta E_{vis}/RT}, \tag{29.1}$$

where ΔE_{vis} plays the role of an activation energy (cf. Section 30.5), often of the order of one-third of the heat of vaporization. This is illustrated in Fig. 29.2, which gives the viscosities of several liquids as a function of temperature.

The thermal conductivity λ varies with temperature and density in much the same way as does the viscosity,[1] except near the critical point, where λ may become very large (cf. Fig. 29.3). Both coefficients rise more and more sharply as the density increases, the rate of increase being greatest at low temperatures. In a normal liquid η and λ are both typically 10 to 100 times greater than in the dilute gas. The larger values of the transport coefficients are due primarily to collisional transfer, which is an extremely efficient method of transporting energy and momentum. This is easily seen in the hard-sphere model, where a collision results in an instantaneous transfer over the distance d between the centers of the two molecules. (For the same reason, the speed of sound in liquids is larger than the average speed of molecular motion.)

Diffusion, by contrast, must take place by transport of the molecules themselves, and thus cannot involve collisional transfer at all. This process therefore becomes more and more difficult as the density increases, since the molecules get in one another's way. In a dilute gas the diffusion coefficient D is inversely proportional to density (cf. Table 28.2); at higher densities the product $D\rho$ increases, behaving somewhat similarly to η and λ.

A number of empirical correlations, theoretical approaches, and models have been used to gain an understanding of transport in dense media. If one thinks of the methods used to study equilibrium properties, such as the equation of state, then the technique of a virial (density) expansion seems promising. For a transport coefficient such as the viscosity,

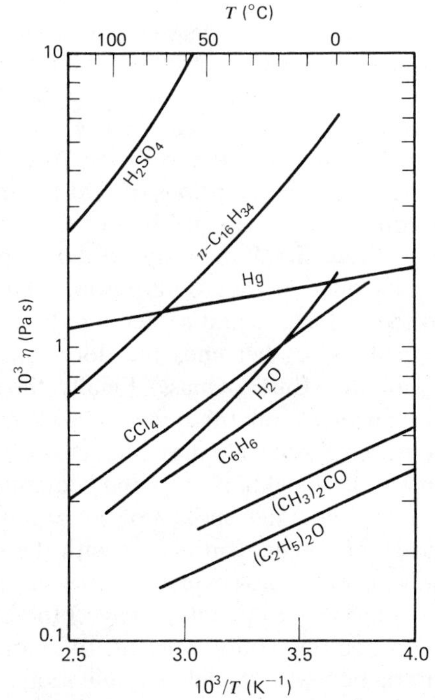

---

[1] Liquid water is, as usual, anomalous: Its thermal conductivity *increases* with increasing temperature.

FIGURE 29.2

Viscosities of some common liquids at 1 atm: A curve obeying Eq. 29.1 is a straight line with slope $\Delta E_{vis}/R$. (In the same units, the 25°C viscosity of H_2SO_4 is 23.5; of SAE 10 oil, about 70; of glycerol, 945.)

the dilute-gas limit (η_0) depends on binary collisions; the coefficient a of the first density-dependent term in a virial expansion,

$$\eta(T, \rho) = \eta_0(T) + a\rho + b\rho^2 + \ldots, \qquad (29.2)$$

depends on triple collisions; and so on. For many cases such expressions are useful for correlating experimental data. However, there is not yet a satisfactory theory to calculate the "virial coefficients" $a, b, \ldots$ in the same way as can be done for their equation-of-state counterparts. Some theoretical analysis suggests that the power-series expansion itself is invalid, in that there should be a term proportional to $\rho^2 \ln \rho$; this question has not yet been settled experimentally.

The main problem in the analysis of transport in dense systems is the difficulty of solving the equations of motion for three or more particles. Such a solution is needed to determine the appropriate averages that constitute the coefficients $a, b, \ldots$. It might be possible to get around this difficulty by asking for less! If we view a dense fluid we find, first in our imagination and then in confirming molecular-dynamics calculations, that a molecule interacts very strongly with a number of neighboring molecules most of the time. These interactions are mostly in the range of the attraction due to van der Waals forces; repulsive (hard-core) interactions occur mostly between two molecules at a time. The motion of a molecule may be confined to a "cage" of neighboring molecules for a short interval of about 10 collisions; after that, on the average, either the cage breaks up due to the motion of the neighboring molecules, or the central molecule escapes. Over much longer periods of time, the motion of the molecule looks like a random walk, so a statistical approach is possible. In that time regime we do not require details of the dynamics of a molecule's interaction with its neighbors, but rather treat the interactions in a statistical way. This topic is taken up in the next section.

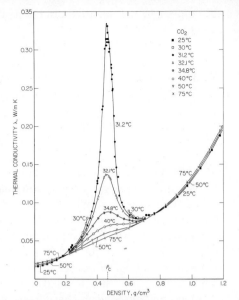

FIGURE 29.3
Thermal conductivity of CO_2 near the critical point: The critical temperature is 31.04°C. From J. V. Sengers, *Ber. Bunsenges. physik. Chem.* **76,** 234 (1972).

29.2
Some Basic Aspects of Brownian Motion

In our discussion of the qualitative behavior of motion and transport in dense systems, we remarked on the random, zigzag motion of molecules. The motion of a macroscopic, visible particle (a smoke particle in air or a colloidal particle in a liquid) reflects this random motion of molecules. If one observes such a macroscopic particle over a period of time, one finds that it follows a chaotic trajectory like that shown in Fig. 29.4; this behavior is called *Brownian motion*. The particle's motion is determined by collisions with the molecules of the surrounding medium, and these collisions are uncorrelated (i.e., random) over intervals long enough to be observed. A theoretical description of this phenomenon was first given by Albert Einstein in 1905. We shall summarize here some elements of this description, which has applications to a wide variety of phenomena, including the motion of colloidal particles, many-particle systems, chemical reactions, and hydrodynamics.

Let us investigate a simple one-dimensional model of random motion. (All the significant aspects of the three-dimensional problem are present in the one-dimensional case.) Consider a particle initially ($t = 0$) situated at the position $x = 0$, Fig. 29.5. Suppose now that at regular intervals τ the particle is displaced by one length unit l either in the positive x direction, with probability p, or in the negative x direction, with probability q. If p and q are equal, that is, $\frac{1}{2}$ each, then on the average half of the displacements are in each direction. Let n_1 be the actual (not average) number of displacements in the positive direction, n_2

the number in the negative direction; then the total number of displacements is

$$N = n_1 + n_2, \tag{29.3}$$

and the *net* number of displacements in the positive direction is

$$m = n_1 - n_2. \tag{29.4}$$

From these equations we have

$$m = n_1 - (N - n_1) = 2n_1 - N; \tag{29.5}$$

if N is an odd (even) number, then m is also odd (even).

We now make the assumption that, as we construct a *sequence* of displacements, the probabilities p and q at each step are independent of the previous displacements. That is, the displacements are *statistically independent*. If we view smoke or dust particles in a light beam, their motion indeed seems chaotic on a time scale corresponding, for instance, to taking a photograph every 0.1 s. The assumption of random displacements seems poor when applied to a single molecule surrounded by 8 or 10 others and trapped in that "cage" for perhaps 10^{-11} s: On that time scale successive displacements are surely not statistically independent. However, when we inquire about the displacement of a macroscopic particle over a period 10^{10} as long, then it is clear that the correlation of events from one location to the next probably has disappeared and statistical independence is a reasonable assumption.

With that assumption the probability of n_1 positive steps is p^{n_1}, the probability of n_2 negative steps is q^{n_2}, and the probability of a *given* sequence of n_1 positive and n_2 negative steps is $p^{n_1}q^{n_2}$. However, many sequences of N $(= n_1 + n_2)$ steps can lead to the net displacement $(n_1 - n_2)l$. The number of such sequences is the same as the number of ways of dividing N distinguishable objects into two piles, containing n_1 and n_2 objects respectively, independent of the arrangement of the n_1 objects in one pile and n_2 objects in the other pile; as we showed in Eq. 25.189, this number is

$$\frac{N!}{n_1! n_2!}.$$

Therefore the probability of a net displacement $(n_1 - n_2)l$ after n_1 positive

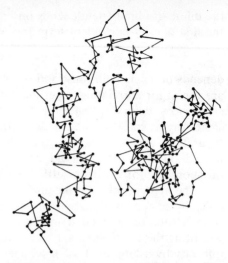

FIGURE 29.4
Observations of Brownian motion: Each dot represents a position of the same particle, at intervals of 30 s. (The true trajectory between each pair of dots is, of course, not a straight line, but a smaller-scale jagged curve.) From J. Perrin, *Atoms* (2nd English ed., rev., Constable, London, 1923), p. 116.

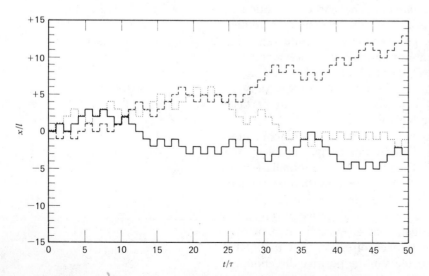

FIGURE 29.5
Illustration of one-dimensional random motion. The particle starts at the position $x = 0$ at time $t = 0$, and randomly jumps a distance l in either direction at time intervals τ. In this figure the jumps were generated from a random number table, with equal probability of jumping in either direction ($p = q = \frac{1}{2}$); three runs are shown.

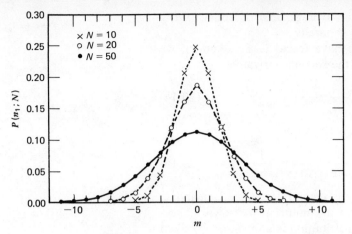

and n_2 ($= N - n_1$) negative displacements, independent of the sequence of steps, is

$$P(n_1; N) = \frac{n!}{n_1!(N - n_1)!} p^{n_1} q^{N - n_1}. \tag{29.6}$$

Note that this is one term in the binomial expansion of $(p + q)^N$.

A plot of the probability distribution of displacements is shown in Fig. 29.6. Even for the small values of N shown here, a line connecting the points closely approximates the shape of a Gaussian distribution. Therefore we suspect that $P(n_1; N)$ tends toward a Gaussian distribution in the limit $N \to \infty$. To show this, we take the logarithm of the probability,

$$\ln P(n_1; N) = \ln N! - \ln n_1! - \ln(N - n_1)! + n_1 \ln p + (N - n_1) \ln q, \tag{29.7}$$

and consider n_1 to be a continuous, rather than discrete, variable for sufficiently large N and n_1. The value of n_1 that maximizes P, and hence $\ln P$, is obtained by setting the derivative equal to zero:

$$\left[\frac{\partial \ln P(n_1; N)}{\partial n_1} \right]_N = 0. \tag{29.8}$$

In this differentiation we use Stirling's approximation,

$$\ln n_1! \approx n_1 \ln n_1 - n_1, \tag{29.9}$$

so that

$$\frac{\partial \ln n_1!}{\partial n_1} \approx \ln n_1 + 1 - 1 = \ln n_1, \tag{29.10}$$

and similarly for the derivative of $\ln(N - n_1)!$. Hence we have

$$\left[\frac{\partial \ln P(n_1; N)}{\partial n_1} \right]_N \approx -\ln n_1 + \ln(N - n_1) + \ln p - \ln q. \tag{29.11}$$

We denote the value of n_1 at the maximum of P by n_1^*; to find this quantity, we set the right side of Eq. 29.11 equal to zero and obtain

$$(N - n_1^*)p = n_1^* q, \tag{29.12}$$

or, since $q = 1 - p$,

$$n_1^* = Np. \tag{29.13}$$

If $p = q = \frac{1}{2}$, then $n_1^* = \frac{1}{2}N$, and the net displacement ml at the maximum of the probability distribution is zero as expected. Since the distribution is then symmetric around its maximum, the average of n_1 is equal to n_1^*. We shall see that this equality also holds generally.

To be sure that the extremum we have found is a maximum, we differentiate Eq. 29.11 again to obtain the second derivative

$$\left[\frac{\partial^2 \ln P(n_1; N)}{\partial n_1^2}\right]_N = -\frac{1}{n_1} + \frac{1}{N - n_1}. \qquad (29.14)$$

At the point $n_1^* = Np$, this derivative has the value

$$\left[\frac{\partial^2 \ln P(n_1; N)}{\partial n_1^2}\right]_{n_1 = n_1^*} = -\frac{1}{Np} + \frac{1}{Nq} = -\frac{1}{Npq}; \qquad (29.15)$$

since N, p, q are all positive quantities, the second derivative is negative. The distribution $P(n_1; N)$ therefore has a maximum at $n_1^* = Np$.

We anticipate that the probability distribution is also sharply peaked around its maximum value. If this is the case, then a Taylor-series expansion around the maximum n_1^* should be useful; we thus write

$$\ln P(n_1; N) = \ln P(n_1^*; N) + (n_1 - n_1^*)\left(\frac{\partial \ln P}{\partial n_1}\right)_{n_1 = n_1^*}$$
$$+ \frac{(n_1 - n_1^*)^2}{2}\left(\frac{\partial^2 \ln P}{\partial n_1^2}\right)_{n_1 = n_1^*} + \dots . \qquad (29.16)$$

At the maximum, $(\partial \ln P/\partial n_1)_{n_1 = n_1^*}$ is zero; hence, up to and including second-order terms, we have

$$P(n_1; N) = P(n_1^*; N)e^{-(n_1 - n_1^*)^2/2Npq}, \qquad (29.17)$$

where Eq. 29.15 has been used to evaluate the second derivative. Thus the probability distribution is clearly of the Gaussian form.

We demand that the probability distribution $P(n_1; N)$ be normalized, that is, that

$$\int_0^N P(n_1; N)\, dn_1 \approx \int_{-\infty}^{\infty} P(n_1; N)\, d(n_1 - n_1^*) = 1. \qquad (29.18)$$

We have here again taken n_1 to be continuous, changed the variable of integration to $n_1 - n_1^*$, and extended the integration over all values of $n_1 - n_1^*$, since the integrand is expected to be near zero everywhere except in the vicinity of n_1^*. If we carry out the integration, we have

$$P(n_1^*; N)\int_{-\infty}^{\infty} e^{-(n_1 - n_1^*)^2/2Npq}\, d(n_1 - n_1^*) = 1; \qquad (29.19)$$

since the integral here is of standard form ($\int_{-\infty}^{\infty} e^{-a^2x^2}\, dx = \pi^{1/2}/a$), we find the normalization constant to be

$$P(n_1^*; N) = (2\pi Npq)^{-1/2}. \qquad (29.20)$$

We can now evaluate the average value of n_1, defined by

$$\langle n_1 \rangle \equiv \int_0^N n_1 P(n_1; N)\, dn_1. \qquad (29.21)$$

Making the same approximations as in Eq. 29.18, we have

$$\langle n_1 \rangle - n_1^* = (2\pi Npq)^{-1/2}\int_{-\infty}^{\infty} (n_1 - n_1^*)\, e^{-(n_1 - n_1^*)^2/2Npq}\, d(n_1 - n_1^*) = 0; \qquad (29.22)$$

the integral vanishes since the integrand is the product of an odd and an even function of the variable $n_1 - n_1^*$. To the extent that Eq. 29.17 is valid, then, the average $\langle n_1 \rangle$ and the most probable value $n_1{}^*$ are identical for any values of p, q. The first moment of the Gaussian distribution, the average deviation from the maximum, is thus zero; the second moment or mean squared deviation, now defined as

$$\langle (\Delta n_1)^2 \rangle \equiv \int_{-\infty}^{\infty} (n_1 - n_1{}^*)^2 P(n_1; N) \, d(n_1 - n_1{}^*) \qquad (29.23)$$

(where $\Delta n_1 \equiv n_1 - n_1{}^*$ and the brackets indicate averaging), has the value

$$\langle (\Delta n_1)^2 \rangle = (2\pi Npq)^{-1/2} \int_{-\infty}^{\infty} (n_1 - n_1{}^*)^2 \, e^{-(n_1 - n_1{}^*)^2/2Npq} \, d(n_1 - n_1{}^*) = Npq. \qquad (29.24)$$

Note that the mean squared deviation increases linearly with the number of steps in the random walk; its square root, the root-mean-square (rms) deviation, therefore varies as $N^{1/2}$. If we rewrite the Gaussian distribution in the form

$$P(n_1; N) = (2\pi \langle (\Delta n_1)^2 \rangle)^{-1/2} \, e^{-(n_1 - n_1{}^*)^2/2\langle (\Delta n_1)^2 \rangle}$$
$$= (2\pi Npq)^{-1/2} \, e^{-(n_1 - n_1{}^*)^2/2Npq}, \qquad (29.25)$$

we see that the distribution narrows as N increases. Note that $\langle (\Delta n_1)^2 \rangle$ is the analog of a temperature in the formally similar one-dimensional Maxwell–Boltzmann distribution, Eq. 28.40.

To connect the concept of a random walk with the diffusion process, let us consider the following problem: Imagine a one-dimensional medium (a line, as in Fig. 29.5), and calculate the rate at which particles diffuse along it. At $t = 0$ let all the particles be at the same point x_0, with none elsewhere; in other words, the initial "concentration" (number of particles per unit length) is a δ function,

$$c(x, t) = c_0 \delta(x - x_0). \qquad (29.26)$$

Beginning at $t = 0$, let the process of diffusion occur according to *Fick's second law*,[2]

$$\frac{\partial c(x, t)}{\partial t} = D \frac{\partial^2 c(x, t)}{\partial x^2}, \qquad (29.27)$$

with diffusion coefficient D. One solution of the differential equation 29.27, as can be verified by differentiation, is

$$c(x, t) = c_0 (4\pi Dt)^{-1/2} \, e^{-(x - x_0)^2/4Dt}. \qquad (29.28)$$

This particular solution also satisfies the initial condition 29.26. That is, as

---

[2] This can be easily derived from Fick's first law, which is the diffusion law in Table 28.1 (and Eq. 20.15). In one dimension, the flux of particles through a point x is

$$(\Gamma_n)_x = -D \frac{\partial c(x, t)}{\partial x},$$

whereas the flux at $x + \delta x$ is

$$(\Gamma_n)_{x+\delta x} = (\Gamma_n)_x + \delta x \frac{\partial (\Gamma_n)_x}{\partial x} = (\Gamma_n)_x - D \frac{\partial^2 c(x, t)}{\partial x^2} \, \delta x.$$

But the rate at which the "concentration" between x and $x + \delta x$ changes is simply the difference of these two fluxes (one in and the other out) divided by the length δx; thus for $\delta x \to 0$ we have

$$\frac{\partial c(x, t)}{\partial t} = \lim_{\delta x \to 0} \frac{(\Gamma_n)_x - (\Gamma_n)_{x+\delta x}}{\delta x} = D \frac{\partial^2 c(x, t)}{\partial x^2}.$$

(In three dimensions this becomes the diffusion equation 20.18.)

$t \rightarrow 0$, the function $c(x, t)$ becomes more and more sharply peaked, but the area under the curve remains constant:

$$\int_{-\infty}^{\infty} c(x, t) \, dx = c_0 \qquad \text{(for all } t\text{)}, \qquad (29.29)$$

so that

$$\lim_{t \to 0} c(x, t) = c_0 \delta(x - x_0)$$

by the definition of the δ function. At any time the distribution of the diffusing particles is Gaussian in the one-dimensional space, and the Gaussian becomes broader (diffuses) as time passes. The average position of the particles, $\langle x \rangle \equiv \int_{-\infty}^{\infty} x c(x, t) \, dx$, is therefore x_0 at all times, and the mean squared displacement from this position is

$$\langle (x - x_0)^2 \rangle = \frac{\displaystyle\int_{-\infty}^{\infty} (x - x_0)^2 c(x, t) \, dx}{\displaystyle\int_{-\infty}^{\infty} c(x, t) \, dx}$$

$$= \int_{-\infty}^{\infty} (x - x_0)^2 \frac{1}{(4xDt)^{1/2}} e^{-(x-x_0)^2/4Dt} \, d(x - x_0) = 2Dt. \qquad (29.30)$$

This is the diffusion law first derived by Einstein. (See Eqs. 20.17 and 20.18 for the three-dimensional equivalents of Eqs. 29.28 and 29.30, respectively.)

It is clear that Eqs. 29.24 and 29.30 are of the same form; they also describe the same process. Diffusion of particles through a medium occurs by motions in random directions, uncorrelated with one another after periods of time corresponding to many molecular collisions. In the random-walk problem, we start with unit probability at the origin of the one-dimensional coordinate system and ask for the probability distribution of positions in space after N steps. To correspond with our intuition about diffusion in an isotropic medium, we choose the *a priori* probabilities p and q to be equal. Thus, *on the average*, we have $\langle n_1 \rangle = \langle n_2 \rangle = N/2$, and hence $\langle m \rangle = \langle n_1 \rangle - \langle n_2 \rangle = 0$. The average displacement from the origin is zero. For any one sequence of steps, however, the net displacement from the origin may differ from zero, with a probability distribution given by $P(n_1; N)$, Eq. 29.25. We can alter that equation by substituting

$$n - n_1{}^* = n_1 - \frac{N}{2} = \frac{m}{2}, \qquad (29.31)$$

so that Eq. 29.24 becomes

$$\left\langle \left(\frac{m}{2}\right)^2 \right\rangle = Npq \quad \text{or} \quad \langle m^2 \rangle = 4Npq. \qquad (29.32)$$

But for constant step size l, the value of m (the net number of steps in the positive direction) is simply $\Delta x/l$, where Δx is the net distance traversed. Thus in the isotropic case $(p = q = \frac{1}{2})$, the mean squared displacement for the random-walk problem is

$$\langle (\Delta x)^2 \rangle = Nl^2 = \frac{l^2 t}{\tau}, \qquad (29.33)$$

where τ is the interval between displacements. By making the identification $D = l^2/\tau$, we find the solutions of the diffusion and random-walk problems to be identical.

Equation 29.30 relates the mean squared displacement of a particle in a given time t to the diffusion coefficient D. This equality (or rather its three-dimensional equivalent) can be used directly to determine the diffusion coefficients of visible colloid or aerosol particles in their respective media from observations of Brownian motion. Since we have not had to specify any restriction about the mass of the diffusing particle, the equation is generally valid, provided only that the time t exceeds the time over which correlations of molecular velocity can persist. This is the same as saying that the equation holds if the diffusion occurs by a random-walk process. We shall have more to say about correlations (and their relation to transport coefficients) in the next section; see also the discussion of probability aftereffects in Appendix 16A.

The concept of a random-walk process has been applied to a variety of physical and chemical processes besides diffusion. As an example, consider the growth of a polymer by addition of monomer segments. Suppose that we are interested in the configuration of the polymer, which consists of a given number of monomer units. If we view each monomer segment simply as a vector, then the configuration of the polymer is obtained by addition of the vectors representing monomer segments. The sequence of these vectors will resemble a Brownian-motion trajectory like that in Fig. 29.4. The vector addition process, if not constrained by the fixed bond angles between the monomer segments, is a random-walk process in three dimensions, provided that we neglect the probability of two segments of the chain occupying the same space (excluded-volume effect). The rms separation between the ends of the chain, which provides a measure of the "size" of the polymer molecule, can thus be estimated with simple Brownian-motion methods. However, a realistic calculation requires that one correct for the constraints just mentioned (bond angles and excluded volume), and then the analysis becomes much harder.

Another example suitable for analysis by a random-walk process is the dissociation of a molecule. Consider a diatomic molecule, say Cl_2, colliding with Ar atoms. By these collisions the Cl_2 molecule may be excited from the ground vibrational state to one of the higher vibrational states. Deexcitation by collision may, of course, also take place. Collisions thus provide the mechanism for a random walk over the vibrational and rotational energy levels of Cl_2. On reaching the dissociation energy, the Cl_2 molecule breaks apart. The rate of dissociation may then be calculated from the average time for the first passage of the random walk across a designated energy level.

29.3
Stochastic Approach to Transport

The kinetic theory of transport processes in a dilute gas, as presented in Section 28.8, is totally dependent on the dynamics of a two-body mechanical system, in particular such details as the scattering angle and the impact parameter. We did not even attempt to follow the complicated route by which the transport coefficients (defined in Table 28.1) can be calculated. The full calculation requires expressing the gradients in a fluid in terms of deviations from the Maxwell-Boltzmann velocity distribution. These deviations must satisfy the general form of the Boltzmann equation, the spatially homogeneous form of which we gave as Eq. 28.28. And this equation in turn depends on scattering cross sections, that is, on collision dynamics. We short-circuited this intricate procedure in Section 28.8 by resorting to a simple argument in terms of mean free paths, and thus of the hard-sphere model. But neither the simple nor the complex procedure can be extended to dense fluids. The mean free path cannot be meaningfully defined when the molecules are in continuous interaction, and the Boltzmann equation likewise fails because its derivation assumes binary collisions: There are no analytic solutions for n-particle scattering problems with $n > 2$. The prospects for developing a theory look bleak, but in the very complexity of the problem lies the origin of a simpler approach.

Consider a macroscopic, Brownian particle immersed in a fluid. It undergoes zigzag random motion due to collisions with a number of molecules of the fluid at one time. If there are interactions with many particles, perhaps a statistical (stochastic) rather than an exact mechanical analysis is possible. We now briefly outline such a stochastic approach in order to present some of its basic ideas.

Consider a Brownian particle moving with a large initial velocity through a fluid; the particle's velocity decreases with time, very much like that of a marble shot into molasses. The decrease in velocity is due to molecular "friction," the net effect of many random collisions of the particle with the ambient fluid. The frictional deceleration, in many systems, can be taken to be at least proportional to the velocity; this is an approximation, useful and probably sufficient. Yet the frictional deceleration cannot be absolutely smooth: since the force on the particle is due to random collisions with the molecules of the fluid, it fluctuates and only on the average is directly opposed to the particle's velocity. Indeed, such a random, fluctuating force would be acting on the particle even if it were held in place (with zero velocity). We therefore write the equation of motion of the Brownian particle as

$$m \frac{dv}{dt} = -\zeta v + f(t), \qquad (29.34)$$

where m is the particle's mass; v is its velocity; ζ is the so-called friction coefficient, representing the mean frictional force per unit velocity; and $f(t)$ is a fluctuating, statistical force that represents the *random* component of the particle's interactions with the fluid. This is known as the *Langevin equation* (after Paul Langevin, a French physicist who worked on the problem of Brownian motion at the turn of the century). Since the average of the interactions with the medium leads by definition to the frictional term $-\zeta v$, the average of the *deviations* from this average force must vanish:

$$\langle f(t) \rangle = 0. \qquad (29.35)$$

A plot of such a fluctuating force might resemble Fig. 29.7.

If we take an ensemble average of Eq. 29.34, that is, an average over the motions of many independent Brownian particles, we have

$$m \frac{d\langle v \rangle}{dt} = -\zeta \langle v \rangle; \qquad (29.36)$$

from this equation we see that a large initial velocity will, on the average, decay to zero with a relaxation time m/ζ. The initial kinetic energy of the Brownian particle is thereby converted to random thermal motion of the fluid (heat). This conversion, called *dissipation*, occurs because of collisions of the particle with the molecules of the medium. We surmise that the friction coefficient must be related to the fluctuating force, and we shall return to this point shortly.

Let us next use the Langevin equation to derive the mean squared displacement of a Brownian particle. From our discussion of the random-walk problem, we know that the mean squared displacement at time t is equal to $2Dt$, where D is the diffusion coefficient. The combination of this result with the Langevin equation yields an important relationship between the diffusion coefficient and the friction coefficient ζ. We begin with

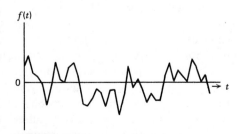

FIGURE 29.7
Schematic representation of a randomly fluctuating force (in one dimension). Over a sufficiently long time, $f(t)$ should average to zero.

Eq. 29.34, multiplying both sides by the coordinate variable x to obtain

$$mx\frac{d\dot{x}}{dt} = -\zeta x\dot{x} + xf(t), \tag{29.37}$$

where $\dot{x} \equiv dx/dt = v$. The first term can be written as

$$mx\frac{d\dot{x}}{dt} = m\left[\frac{d}{dt}(x\dot{x}) - \dot{x}^2\right], \tag{29.38}$$

making the ensemble average of Eq. 29.37

$$m\left\langle\frac{d}{dt}(x\dot{x})\right\rangle = -\zeta\langle x\dot{x}\rangle + m\langle\dot{x}^2\rangle. \tag{29.39}$$

Note that we have omitted the fluctuating term $\langle xf(t)\rangle$. This term can be written as $\langle x\rangle\langle f(t)\rangle$, since we assume the forces on the particle to be independent of its position in the fluid; and since $\langle f(t)\rangle = 0$, the product also vanishes. The term $m\langle\dot{x}^2\rangle$ should equal $k_B T$ for a particle with one degree of freedom (by equipartition of energy), and we can thus rewrite Eq. 29.39 in the form

$$m\frac{d}{dt}\langle x\dot{x}\rangle = -\zeta\langle x\dot{x}\rangle + k_B T. \tag{29.40}$$

(Note that throughout this argument we have assumed the equivalence of ensemble averages and time averages: see Section 15.1 on ergodic theory.) The solution of Eq. 29.40 is simply

$$\langle x\dot{x}\rangle = Ae^{-\zeta t/m} + \frac{k_B T}{\zeta}, \tag{29.41}$$

as can be easily checked by differentiation. To determine the integration constant A one needs an initial condition, for which we choose (without restriction of generality) $x = 0$ at $t = 0$; we thus obtain

$$A = -\frac{k_B T}{\zeta}. \tag{29.42}$$

Rewriting $\langle x\dot{x}\rangle$ as

$$\frac{1}{2}\frac{d}{dt}\langle x^2\rangle,$$

we have

$$\frac{1}{2}\frac{d}{dt}\langle x^2\rangle = \frac{k_B T}{\zeta}(1 - e^{-\zeta t/m}); \tag{29.43}$$

on integrating this equation, we finally arrive at the desired expression for the mean squared displacement of a Brownian particle,

$$\langle x^2\rangle = \frac{2k_B T}{\zeta}\left[t - \frac{m}{\zeta}(1 - e^{-\zeta t/m})\right]. \tag{29.44}$$

Let us take two limiting cases that provide insight into the meaning of this equation. For periods much shorter than the relaxation time, that is, for $t \ll m/\zeta$, we can expand the exponential through the term in t^2,

$$\langle x^2\rangle = \frac{k_B T}{m}t^2. \tag{29.45}$$

Thus the root-mean-square displacement at time t is $(k_B T/m)^{1/2}t$, where the factor $(k_B T/m)^{1/2}$ is the rms speed. We recover the expected fact that for very short times the Brownian particle undergoes free motion. The time scale for "very short" is determined by the condition $t \ll m/\zeta$, that is, by the friction coefficient.

At the other extreme, for periods long compared to the relaxation time $(t \gg m/\zeta)$, Eq. 29.44 simply reduces to

$$\langle x^2 \rangle = \frac{2k_BT}{\zeta} t. \tag{29.46}$$

Over such a long period the Brownian particle suffers myriads of collisions, so many that the correlation of interactions disappears and the displacement is a sequence of uncorrelated events, a random walk. We can thus equate Eq. 29.46 to our result for the mean squared displacement in a random walk, Eq. 29.30, to obtain the important relationship

$$\langle x^2 \rangle = \frac{2k_BT}{\zeta} t = 2Dt, \tag{29.47}$$

or

$$D = \frac{k_BT}{\zeta}, \tag{29.48}$$

connecting the friction coefficient ζ and the diffusion coefficient D. We have derived this result in one dimension, but it is valid in three dimensions as well. Both of these coefficients, in turn, can be related to the viscosity of the fluid medium. As was shown originally by Stokes, the frictional force on a large particle moving through a continuum fluid with velocity $\mathbf{v}$ is

$$\mathbf{F} = -\zeta \mathbf{v} = -6\pi\eta a\mathbf{v}, \tag{29.49}$$

where a is the radius of the particle (assumed spherical) and η is the viscosity. This result, *Stokes's law*, may be applicable under less stringent conditions, and is used frequently for particles of molecular size. Substituting the value of ζ thus obtained into Eq. 29.48 yields the Stokes-Einstein law of diffusion,

$$D = \frac{k_BT}{6\pi\eta a}. \tag{29.50}$$

It is time now to give some numerical estimates. Consider a Brownian particle, say, a colloidal particle of radius 100 nm, in water. The viscosity of water is almost exactly 10^{-3} Pa s $(= 10^{-3}$ kg/m s$)$, so by Stokes's law the friction coefficient is

$$\zeta = 6\pi(10^{-3} \text{ kg/m s})(10^{-7} \text{ m}) \approx 1.9 \times 10^{-9} \text{ kg/s}.$$

Introducing the value of Boltzmann's constant, we obtain for the diffusion coefficient of the particle

$$D = \frac{k_BT}{\zeta} = \frac{1.38 \times 10^{-23} \text{ J/K} \times 300 \text{ K}}{1.9 \times 10^{-9} \text{ kg/s}} = 2.2 \times 10^{-12} \text{ m}^2/\text{s}.$$

If the density of the particle is close to that of water, its mass is about 4×10^{-18} kg; the relaxation time is then

$$\frac{m}{\zeta} = \frac{4 \times 10^{-18} \text{ kg}}{1.9 \times 10^{-9} \text{ kg/s}} \approx 2 \times 10^{-9} \text{ s}.$$

Thus it is only for $t \ll 10^{-9}$ s that the velocity of the Brownian particle is essentially constant. For $t \gg 10^{-9}$ s, on the other hand, the particle's motion is described by a random-walk or diffusion process, a process random in position *and* velocity.

Note that according to Eq. 29.36 the average velocity of a Brownian particle at first approaches zero exponentially with a time constant m/ζ; that is, for a particle of the size described here a given initial velocity is reduced by a factor of e every 2×10^{-9} s, until v is so small (for $t \gg m/\zeta$) that the motion is random. In the latter limit the velocity should have its thermal value, as given, say, by Eq. 28.49; for the particle in our example this yields

$$\langle v^2 \rangle^{1/2} = \left(\frac{k_B T}{m}\right)^{1/2} = \left(\frac{1.38 \times 10^{-23} \text{ J/K} \times 300 \text{ K}}{4 \times 10^{-18} \text{ kg}}\right)^{1/2} \approx 30 \text{ m/s}$$

for the rms speed at 300 K. However, this is merely the average speed *between* collisions; by Eq. 29.47 the *net* distance traversed in a second of random motion should be only about

$$\langle x^2 \rangle^{1/2} = (2Dt)^{1/2} = (2 \times 2.2 \times 10^{-12} \text{ m}^2/\text{s} \times 1 \text{ s})^{1/2} \approx 2 \times 10^{-6} \text{ m,}$$

or 10 times the particle's diameter.

We can use Langevin's equation to obtain yet another important relationship. Suppose that we have a charged Brownian particle in a fluid. If the charge on the particle is q, then in an electric field $\mathbf{E}$ the force on the particle is $q\mathbf{E}$. Thus the Langevin equation must be amended to read (still in one dimension)

$$m\frac{dv}{dt} = -\zeta v + f(t) + qE. \qquad (29.51)$$

We again take an ensemble average and consider the stationary-state condition defined by zero acceleration, $dv/dt = 0$; Eq. 29.50 then becomes

$$qE = \zeta \langle v \rangle. \qquad (29.52)$$

The *mobility* u of a charged particle is the ratio between its velocity and the electric field; the ensemble average of mobility is thus

$$\langle u \rangle = \frac{\langle v \rangle}{E}. \qquad (29.53)$$

From these two equations and Eq. 29.48 we obtain the important Einstein relation between mobility and diffusion coefficient,

$$\frac{\langle u \rangle}{D} = \frac{q}{k_B T}. \qquad (29.54)$$

29.4
Autocorrelation Functions and Transport Coefficients

In our discussion of transport processes, we have several times noted the existence of multiple time scales for events in a given system. There is a molecular time scale concerned with momentum relaxation (say, in a dilute gas) or the *local* equilibration of any quantity such as energy density. We expect that in a dense system, where a Brownian particle interacts with many molecules at once, the force on the particle must vary rapidly. But one can expect that over a period of 10^{-13}–10^{-12} s, during which the particle can collide with many molecules,[3] this force will essentially average out to zero; thus the random force $f(t)$ should be uncorrelated over any time longer than 10^{-12} s. Note that this is an interval during which an initial velocity imparted to the particle hardly changes at all. According to our previous calculation, the relaxation time for the motion of a Brownian particle is of the order of 10^{-9} s; the random force can thus be assumed uncorrelated on the time scale of the particle's motion, and Eq. 29.35 applies. On a still longer time scale ($t \gg 10^{-9}$ s), the particle's velocity itself becomes uncorrelated with its

[3] As we shall see later, molecular-dynamics calculations confirm the assumption that time correlations of molecular velocity decay in about 10^{-12} s.

initial value, and the particle executes random motion (diffusion). In this section we discuss the interesting fact that transport coefficients can be expressed in terms of *correlation functions* of molecular dynamical quantities. We shall define these functions and show in detail that the friction coefficient for a Brownian particle is the autocorrelation function of the fluctuating force on that particle, whereas the diffusion coefficient is the autocorrelation function of its velocity.

Let us begin with the Langevin equation, Eq. 29.34, and seek a so-called autocorrelation expression for the friction coefficient. Formal integration of this equation yields

$$v(t) = e^{-\zeta t/m} v(0) + \frac{1}{m} \int_0^t e^{-\zeta(t-t')/m} f(t') \, dt'. \tag{29.55}$$

In the limit $t \to \infty$ the first term on the right-hand side vanishes; in that limit we expect the Brownian particle to come to equilibrium with the medium, so that $\frac{1}{2}m\langle v^2 \rangle \to \frac{1}{2}k_B T$ for one-dimensional motion. Let us first change integration variables, defining $\rho \equiv t - t'$, so that

$$\int_0^t e^{-\zeta(t-t')/m} f(t') \, dt' = \int_0^t e^{-\zeta\rho/m} f(t-\rho) \, d\rho. \tag{29.56}$$

To evaluate $\langle v^2 \rangle$ at equilibrium ($t \to \infty$), we then take the square of the nonvanishing term in Eq. 29.55—using a second dummy variable σ for the second integration—and take the ensemble average:

$$\langle v^2 \rangle = \frac{k_B T}{m} = \lim_{t \to \infty} \frac{1}{m^2} \int_0^t d\rho \int_0^t d\sigma \, e^{-\zeta\rho/m} e^{-\zeta\sigma/m} \langle f(t-\rho)f(t-\sigma) \rangle. \tag{29.57}$$

If $f(t)$ is truly random, then the ensemble average $\langle f(t-\rho)f(t-\sigma) \rangle$ can depend only on the time difference $|\rho - \sigma|$ and not on t ($\to \infty$) itself; we thus write

$$g(|\rho - \sigma|) \equiv \langle f(t-\rho)f(t-\sigma) \rangle. \tag{29.58}$$

From arguments given before, we expect this average to become uncorrelated for intervals longer than 10^{-13}–10^{-12} s, where by "uncorrelated" we mean that

$$\langle f(t-\rho)f(t-\sigma) \rangle = 0. \tag{29.59}$$

Any such function of the form $\langle X(t_1) X(t_2) \rangle$ is called the *autocorrelation function* of the variable X over the time interval $|t_2 - t_1|$. To reduce Eq. 29.57 we need a bit of mathematical fiddling. Let us change from the variable σ to θ, where

$$\theta \equiv \sigma - \rho, \tag{29.60}$$

so that the double integral becomes

$$\frac{k_B T}{m} = \frac{1}{m^2} \int_0^\infty d\rho \int_{-\rho}^\infty d\theta \, e^{-2\zeta\rho/m} e^{-\zeta\theta/m} g(|\theta|). \tag{29.61}$$

We wish next to change the order of integration, but to do so we need to note carefully the range of integration, shown in Fig. 29.8. Thus to integrate over ρ before σ we rewrite the integral as

$$\frac{k_B T}{m} = \frac{1}{m^2} \left[\int_{-\infty}^0 d\theta \int_{-\theta}^\infty d\rho \, e^{-2\zeta\rho/m} e^{-\zeta\theta/m} g(|\theta|) \right.$$
$$\left. + \int_0^\infty d\theta \int_0^\infty d\rho \, e^{-2\zeta\rho/m} e^{-\zeta\theta/m} g(|\theta|) \right], \tag{29.62}$$

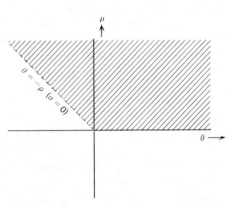

FIGURE 29.8
The range of integration of Eq. 29.61 is indicated by the shaded area; it is the same as the region for which both ρ and σ are positive.

with separate terms for the regions with $\theta < 0$ and $\theta > 0$. The integration over ρ yields

$$\frac{k_B T}{m} = \frac{1}{m^2} \left[\int_{-\infty}^{0} d\theta \, \frac{m}{2\zeta} e^{-\zeta|\theta|/m} g(|\theta|) + \int_{0}^{\infty} d\theta \, \frac{m}{2\zeta} e^{-\zeta\theta/m} g(|\theta|) \right], \quad (29.63)$$

or

$$\zeta = \frac{1}{2k_B T} \int_{-\infty}^{\infty} d\theta \, e^{-\zeta|\theta|/m} g(|\theta|). \quad (29.64)$$

Since θ is a time variable depending only on the difference between ρ and σ, we can arbitrarily set $\rho = 0$, $\sigma = \theta$ to obtain

$$g(|\theta|) = \langle f(0) f(\theta) \rangle, \quad (29.65)$$

and thereby

$$\zeta = \frac{1}{2k_B T} \int_{-\infty}^{\infty} d\theta \, e^{-\zeta|\theta|/m} \langle f(0) f(\theta) \rangle. \quad (29.66)$$

Since we expect the correlation function of the fluctuating force to have decayed to zero in an interval of the order of 10^{-12} s, whereas the relaxation time m/ζ is of the order of 10^{-9} s, we can set the factor $e^{-\zeta|\theta|/m}$ to unity over the entire range for which the integrand is significantly different from zero. Thus we formally obtain for the friction coefficient

$$\zeta = \frac{1}{2k_B T} \int_{-\infty}^{\infty} d\theta \, \langle f(0) f(\theta) \rangle. \quad (29.67)$$

We see that the friction coefficient ζ is a time integral of the autocorrelation function of the fluctuating force $f(t)$. The friction coefficient, which describes a kind of transport of kinetic energy to heat, is related to the correlation of fluctuations of a dynamical quantity, a force, in an *equilibrium* ensemble. This equivalence is an example of the so-called *fluctuation-dissipation theorem*. We shall return to the significance of using an equilibrium ensemble after we present the corresponding derivation for the diffusion coefficient.

To a first approximation, the autocorrelation function of the fluctuating force is essentially a δ function, $\delta(\theta - 0)$, since the correlation time is so short. More realistically, one expects the autocorrelation function to approach zero in a time interval of $10^{-13} - 10^{-12}$ s. Molecular-dynamics calculations (cf. Section 23.4) indicate that this is true for the force on a molecule interacting with other molecules of the same mass; although one cannot carry out such calculations explicitly for a Brownian particle (which interacts with thousands of molecules at once), the same time scale is probably appropriate.

Next we proceed to obtain an expression for the diffusion coefficient in terms of an autocorrelation function. We return to Eq. 29.30 for the diffusion coefficient, writing the random displacement of a Brownian particle as

$$x(t) = \int_{0}^{t} v(t') \, dt', \quad (29.68)$$

so that

$$\langle [\Delta x(t)]^2 \rangle = \int_{0}^{t} dt' \int_{0}^{t} dt'' \langle v(t') v(t'') \rangle. \quad (29.69)$$

where the brackets $\langle \; \rangle$ again denote an average over an equilibrium ensemble. If the particle's motion is completely random (any initial velocity having been lost), the correlation function $\langle v(t')v(t'')\rangle$ can depend only on the time interval $\sigma \equiv t'' - t'$ and not on the absolute time itself. Hence we write the *velocity autocorrelation function* as

$$g_v(\sigma) \equiv \langle v(t')v(t'+\sigma)\rangle = \langle v(0)v(\sigma)\rangle. \qquad (29.70)$$

Note that, since we assume $g_v(\sigma)$ to be unaffected by shifts of the time scale, we have

$$\langle v(t')v(t'+\sigma)\rangle = \langle v(t'-\sigma)v(t')\rangle \qquad (29.71)$$

or

$$g_v(\sigma) = g_v(-\sigma). \qquad (29.72)$$

Changing the variables in Eq. 29.69 from t', t'' to t', σ, we obtain

$$\langle [\Delta x(t)]^2 \rangle = \int_0^t dt' \int_{-t'}^{t-t'} d\sigma \, g_v(\sigma). \qquad (29.73)$$

We wish to integrate first over t'; to do so we again note the range of integration, as plotted in Fig. 29.9, and rewrite the integral with separate terms for the regions with $\sigma > 0$ and $\sigma < 0$:

$$\langle [\Delta x(t)]^2 \rangle = \int_0^t d\sigma \int_0^{t-\sigma} dt' \, g_v(\sigma) + \int_{-t}^0 d\sigma \int_{-\sigma}^t dt' \, g_v(\sigma)$$

$$= \int_0^t d\sigma \, g_v(\sigma)(t-\sigma) + \int_{-t}^0 d\sigma \, g_v(\sigma)(t+\sigma). \qquad (29.74)$$

If we change the variable in the second integral from σ to $-\sigma$ and apply Eq. 29.72, this becomes

$$\langle [\Delta x(t)]^2 \rangle = 2t \int_0^t d\sigma \left(1 - \frac{\sigma}{t}\right) g_v(\sigma). \qquad (29.75)$$

Substituting this result into Eq. 29.30, we find that the Brownian-particle diffusion coefficient in the limit $t \to \infty$ is

$$D = \int_0^\infty d\sigma \langle v(0)v(\sigma)\rangle. \qquad (29.76)$$

We can approximate the velocity autocorrelation function by using the Langevin equation, Eq. 29.34. Neglecting the fluctuating force (as we have indicated to be appropriate on the time scale of Brownian motion), we have

$$m\frac{dv}{dt} = -\zeta v, \qquad v(t) = v(0)e^{-\zeta t/m}, \qquad (29.77)$$

and therefore

$$\langle v(0)v(\sigma)\rangle = \langle [v(0)]^2 \rangle e^{-\zeta\sigma/m} = \frac{k_B T}{m} e^{-\zeta\sigma/m}. \qquad (29.78)$$

Note that in this approximation the velocity autocorrelation function decays exponentially on the time scale m/ζ: after a few times m/ζ the velocity autocorrelation has disappeared. For $t \gg m/\zeta$ we then recover the expression for the diffusion coefficient,

$$D = \frac{k_B T}{m} \int_0^\infty d\sigma \, e^{-\zeta\sigma/m} = \frac{k_B T}{\zeta}, \qquad (29.79)$$

in agreement with Eq. 29.48.

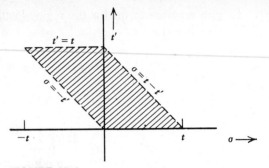

FIGURE 29.9
The range of integration in Eq. 29.73 is indicated by the shaded area.

Thus far our discussion has been entirely in terms of the motion of Brownian particles, for which the autocorrelation of the velocity decays on a time scale much longer than that of the random force; this separation of time scales makes many simplifying assumptions possible. But we are more interested in the transport properties of a homogeneous fluid, which involve the interaction of a molecule with other molecules of the same size. The same approach can be used, but the simplifying assumptions are no longer valid. For example, calculations of the friction coefficient ζ in liquid argon indicate that m/ζ for an Ar molecule is of the order of 10^{-13} s, comparable to the fluctuations in the random force. On this scale, however, one can use computers to make direct calculations of molecular motion in systems of up to a few hundred molecules. Figure 29.10 illustrates the result of such a calculation for the velocity autocorrelation function in liquid argon. Note that $\langle v(t)v(0) \rangle$ initially resembles the rapidly decreasing function predicted by Eq. 29.78, but that the function then becomes negative (this essentially corresponds to velocity reversal after "bouncing" off a nearest-neighbor molecule) and trails off to zero.

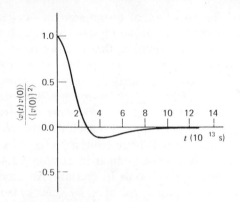

FIGURE 29.10
The velocity autocorrelation function for liquid argon near its triple point, as obtained from molecular-dynamics calculations by D. Levesque, L. Verlet, and J. Kürkijarvi, *Phys. Rev.* **A7**, 1690 (1973), assuming a Lennard–Jones intermolecular potential.

The theory we have outlined can be extended to give all the transport coefficients in terms of autocorrelation functions, as Eq. 29.76 does for the diffusion coefficient. We cannot go into further derivations here, but we give the result for the viscosity as an example:

$$\eta = \frac{1}{Vk_BT} \int_0^\infty d\sigma \langle J^{xy}(0)J^{xy}(\sigma) \rangle, \qquad (29.80)$$

where

$$J^{xy} \equiv \sum_i \left(\frac{p_{ix}p_{iy}}{m_i} \right) + R_{ix}F_{iy} \qquad (29.81)$$

($\mathbf{R}_i$ is the position of the ith molecule, $\mathbf{F}_i$ the force exerted on it by all the other molecules, and the sum goes over all the molecules in the fluid).

One final remark on the structure of the correlation-function expressions for transport coefficients, specifically the use of equilibrium ensemble averages: Let us recall how a transport coefficient is defined in the regime of small gradients (of mass, momentum, etc.); for diffusion, for instance, we define D by

$$\text{flux of mass} = -D \text{ (gradient of concentration)}$$

(cf. Table 28.1). We assume in this expression that the transport *coefficient* does not itself depend on the gradient. Hence we can formulate the transport coefficient under the condition of zero concentration gradient, that is, equilibrium. It suffices, therefore, to take the *equilibrium* ensemble average of the appropriate dynamical quantity, in this case $v(0)v(\sigma)$.

29.5
Transport in Solids

We continue this chapter on transport in condensed phases with a brief discussion of transport in solids. We focus primarily on the process of diffusion, a topic that enables us to present a few basic points, problems, and qualitative approaches.

In Chapters 11 and 22 we analyzed the structure and some equilibrium properties of perfect solids, that is, substances that are periodic, faultless arrays of recurring structures. We dealt with metals, ionic crystals, and molecular crystals. In most of our previous discussion, every lattice site had to be occupied by precisely its prescribed atom or ion; this was an essential assumption in our treatment of the determination of structure by x-ray crystallography. If, however, every lattice point in a crystal is occupied, that is, if we have a perfect crystal, then transport of mass or charge through the crystal cannot occur (except by quantum-mechanical exchange, a very slow process in crystals). Although the idea of a perfect crystal served us well for some purposes, it is too idealized to be used in analyzing transport processes. We must look for deviations from a perfect array, called *defects*. (Note that this terminology implies a reluctant departure from a preconceived standard of perfection.)

As we mentioned in Section 22.4, a number of different types of defects may occur in crystals. An atom or ion may leave a lattice site when it attains the necessary energy from thermal fluctuations, which are always present at any temperature. The atom or ion so freed moves through the interstices of the crystal. The empty lattice site, called a *vacancy*, can also move through the crystal—for example, by transfer of an atom or ion from a neighboring lattice site to that occupied by the vacancy, so that the atom or ion changes places with the vacancy. Figure 29.11a shows the initial position and subsequent motion of such a vacancy, and Fig. 29.11b the motion of an interstitial atom. Two other types of defect motion may occur: (1) an *interstitialcy* mechanism, in which an atom or ion acquires enough energy to leave a lattice site and then knocks a neighboring atom or ion out of *its* site, and so on, Fig. 29.11c; (2) the possible but rarely occurring mechanism of site exchange among the atoms or ions on two or more neighboring sites, Fig. 29.11d.

The energy needed to create an interstitial atom and a vacancy may be supplied locally at a lattice site by a fluctuation in thermal energy. The magnitude of energy fluctuations increases with temperature, and hence we expect the concentrations of vacancies to increase with temperature also. This conclusion is confirmed by x-ray diffraction experiments, some results of which are shown in Fig. 29.12.

To any given number of vacancies in the crystal there corresponds a particular contribution to the entropy, due to the number of arrangements (configurations) of the vacancies among the lattice sites. If the vacancy mole fraction is small, then to a reasonable approximation we may neglect interactions between vacancies; in short, we are dealing with a dilute solution. The entropy of mixing of this "solution" of vacancies and occupied sites is, in the ideal-solution limit, given by Eq. 25.190. since mixing is a spontaneous process, the entropy change upon mixing is positive. Hence for the change in state

$$\text{crystal (no vacancies)} \rightarrow \text{crystal } (N_v \text{ vacancies}),$$

the energy change is $N_v E_v$, where E_v is the energy necessary for formation of one vacancy-interstitial atom pair (typically about 1 eV in a metal). If we can neglect the difference between the enthalpy and energy changes for this process, we obtain the Gibbs free energy change,

$$\Delta G = \Delta H - T\,\Delta S \approx \Delta U - T\,\Delta S = N_v E_v - k_B T \ln \frac{N!}{(N-N_v)!\,N_v!},$$
(29.82)

where N is the total number of sites and ΔS is obtained from Eq. 25.189. At a given temperature, the equilibrium number of vacancies must be that which gives a minimum in ΔG,

$$\left(\frac{\partial \Delta G}{\partial N_v}\right)_T = 0;$$
(29.83)

using Stirling's approximation for the factorials, we differentiate Eq.

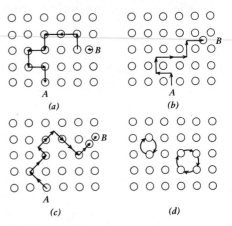

FIGURE 29.11

The atomic rearrangements caused by various migration mechanisms. (a) A vacancy enters the region at A and leaves a trail of displaced atoms along its path to B. (b) Interstitial migration causes larger displacements of single atoms. (c) The interstitialcy mechanism leaves atoms displaced to sites neighboring an adjacent interstice. (d) Simple exchange and ring mechanisms. From C. P. Flynn, *Point Defects and Diffusion* (Oxford University Press, London, 1972), p. 4.

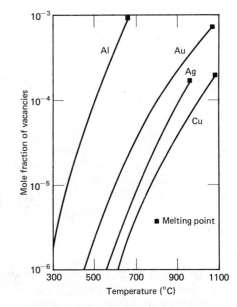

FIGURE 29.12

Vacancy concentrations in several metals as a function of temperature (data of R. O. Simmons and R. W. Balluffi). From A. G. Guy, *Essentials of Materials Science* (McGraw-Hill, New York, 1976), p. 127.

29.82 to obtain

$$\ln\left(\frac{N_v}{N - N_v}\right) = \frac{E_v}{k_B T}. \qquad (29.84)$$

For $N_v \ll N$ (dilute "solution") we thus have

$$\frac{N_v}{N} = e^{-E_v/k_B T} \qquad (29.85)$$

for the equilibrium concentration of vacancies at the temperature T.

The term *point defect* refers to a site imperfection such as an interstitial atom or a vacancy. Crystal imperfections that involve many sites may also occur. *Dislocations* are lines along which discontinuities in the sequence of crystal planes occur; two types are illustrated in Fig. 29.13. Imperfections at the surface of a crystal may create "ledges" and "kinks," such as are shown in Fig. 29.14. This figure also shows the formation of a vacancy by motion of a kink, a process that requires less energy than formation of a vacancy in the bulk of the crystal. In the reverse process, the surface may serve as a sink (destroyer) for vacancies.

The presence of defects provides a mechanism of mobility for atoms, molecules, and ions within a crystal. It is through these mechanisms that transport of mass and charge, and hence diffusion and electrical conduction, can occur. Other transport processes are of course known, and we shall mention a few later.

Measurements of diffusion in a solid can be made in a variety of ways. One effective method is based on the use of radioactive tracers. Suppose that two bars of the same metal are joined to each other by a thin layer of a radioactive isotope of the same metal (Fig. 29.15a). The assembly is then heated to a chosen temperature, and diffusion of the radioactive species takes place. After a given time interval the process is halted by cooling; assays of the radioactive concentrations are then made by slicing thin sections of the metal cylinder at various distances from the original juncture and measuring their radioactivity. A plot of such a concentration distribution is shown in Fig. 29.15b. If the diffusion of the radioactive tracer occurs by a molecular mechanism that can be described as a random walk, then we can apply the theory of Section 29.2. Since the problem is one of one-dimensional diffusion from an initial δ-function concentration, as in our model, the concentration at time t should be given by Eq. 29.28. The measurements plotted in Fig. 29.15b are indeed an excellent fit to this Gaussian curve, and thus can be used to determine the diffusion coefficient D at the chosen temperature; for the 920°C measurements shown, one obtains $D = 2 \times 10^{-13}$ m²/s. This is a typical order of magnitude for a metallic diffusion coefficient, and should be compared with the corresponding values in gases (about 10^{-4} m²/s) and liquids (about 10^{-9} m²/s).

In an ionic crystal the diffusion coefficients of the positive and negative species are frequently very different. In NaCl diffusion of both ions occurs via formation and motion of vacancies and interstitial ions. The measured diffusion coefficient of Na^+ (as obtained by tracer techniques) is about 10^{-13} m²/s, whereas that of Cl^- is 10^{-14} m²/s. The positive ion is commonly the smaller, and thus has a larger diffusion coefficient. In AgCl crystals, the difference between ionic diffusion coefficients is even larger: $D(Ag^+) \approx 10^{-10}$ m²/s, $D(Cl^-) \approx 10^{-13}$ m²/s. The silver ion is thought to diffuse by an interstitialcy mechanism.

The diffusion coefficient of a species in a solid increases exponentially with temperature, as shown in Fig. 29.16. This is easily understandable: The rate of diffusion increases approximately linearly with increasing

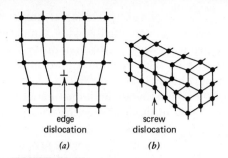

FIGURE 29.13
Two types of crystal dislocations. (a) *Edge dislocation* (shown end-on), a line along which a crystal plane terminates. (b) *Screw dislocation* (shown in cutaway perspective), a line along which a plane bifurcates.

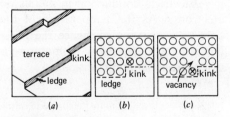

FIGURE 29.14
Creation of a vacancy at a kink in the terrace-ledge-kink surface structure of a metal. (a) Terrace-ledge-kink model of a surface. (b) Initial condition of the metal. (c) Condition after formation of the vacancy. From A. G. Guy, *Essentials of Materials Science* (McGraw-Hill, New York, 1976), p. 128.

number of vacancy–interstitial atom pairs, the concentration of which increases exponentially with temperature. The exponential variation stems from the "activation process" of forming a vacancy–interstitial atom pair. This activation process consists of providing sufficient energy (from thermal fluctuations) to remove an atom from its site, simultaneously accommodate an interstitial atom in the neighboring region, and set the atom in motion after the vacancy is formed.[4]

For comparison, note that at constant density the diffusion coefficient of a gas varies as $T^{1/2}$, roughly proportional to the mean molecular speed, with no evidence that any activation process is necessary. For diffusion in a liquid, the transport coefficient at constant density is essentially independent of temperature.

Let us end this section with a few words about other transport processes in solids. The static coefficient of viscosity does not exist as such for a crystal, since the phenomenological law of Table 28.1 does not hold. A crystal subjected to a moderate shearing stress will simply bend elastically; a higher stress will cause it to either break or undergo plastic deformation, but even in the latter case the flow cannot be described in terms of a viscosity alone. Deformation and flow in dense media is a very complicated subject by itself, constituting the science of *rheology*.

Thermal excitations in solids are dissipated by interactions among the many vibrational degrees of freedom of the crystal. We have discussed the spectrum of these vibrations in Chapter 22. The vibrations are quantized, a quantum of vibrational energy being known as a phonon. If the vibrational degrees of freedom are uncoupled (as in the hypothetical Einstein-model solid, in which all are represented by harmonic oscillators with a common frequency), then when one degree of freedom is excited, no dissipation of the excitation energy to the other degrees of freedom is possible. Hence excitation of a vibrational degree of freedom at one end of an ideal crystal propagates without loss to the other end; in this limit the thermal conductivity is infinite. For a real solid, however, anharmonic terms are always present, coupling exists among the degrees of freedom, excitation of one degree of freedom is followed by flow of energy to the others, and the thermal conductivity is finite. It is through these processes that thermal conduction takes place.

The situation in metals is complicated, and in fact dominated, by the presence of free electrons which contribute to the process of heat conduction. To a first approximation, the electrons may be considered as an electron gas (see Sections 11.11 and 22.5); the electrons are not free, but collide with phonons, which again leads to a finite thermal conductivity.

Electrical conduction in metals occurs predominantly by motion of electrons.[5] As the temperature of the metal is increased, the phonon density also increases; hence the resistivity, the inverse of conductivity, increases with increasing temperature because of the increased frequency of electron–phonon collisions.

---

[4] This process is similar to that observed in certain (unimolecular) chemical reactions such as isomerization or decomposition, in which a molecule may have sufficient energy to react, but still requires some time for that energy to be localized in the appropriate bonds to be rearranged or ruptured.

[5] In terms of Table 28.1, the macroscopic law defining the conductivity σ is

$$(\Gamma_q)_z = -\sigma \frac{dV}{dz},$$

where Γ_q is the flux of charge (i.e., the current density) and V is the electrical potential. With the assumption that σ is independent of the current, this equation is a form of Ohm's law.

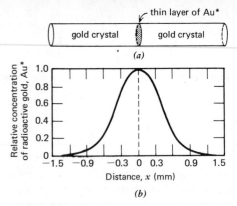

FIGURE 29.15

Illustration of the principle of tracer diffusion and of the planar-source method for determining the self-diffusion coefficient of gold. (a) Initial diffusion couple with planar source of radioactive gold, Au*. (b) Distribution of Au* after diffusion for 100 h at 920°C (data of H. M. Gilder and D. Lazarus). From A. G. Guy, *Principles of Materials Science* (McGraw-Hill, New York, 1976), p. 133.

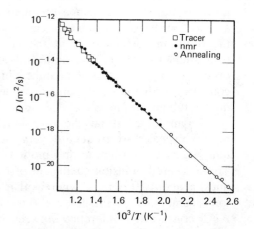

FIGURE 29.16

Some measurements of self-diffusion in Al. From C. P. Flynn, *Point Defects and Diffusion* (Oxford University Press, London, 1972), p. 784.

When two electrodes (platinum strips, say) are immersed in a solution containing ions, imposition of an electric potential across the electrodes leads to passage of electric current. The current is carried by both positive and negative ions in the solution, but not necessarily in equal portions. Negative ions are attracted to the anode, positive ions to the cathode, at each of which electrode reactions take place (see Chapter 26). In order to minimize the effects of the electrode reactions on the measurement of ionic conductivity, the polarity of the potential is alternated (say 1000 Hz) to obtain an AC conductance. The quantity normally measured is the resistance of the electrolyte solution, R, from which the *electrical conductivity* σ (SI unit $\Omega^{-1}\,m^{-1}$) is calculated as

$$\sigma = \frac{l}{RA}, \qquad (29.86)$$

where, for an idealized geometry, l is the distance between the electrodes and A is the area of each electrode exposed to the other. (Pure water free of ionized solutes has $\sigma \approx 4 \times 10^{-6}\,\Omega^{-1}\,m^{-1}$; 1 M HCl at 25°C has $\sigma = 33\,\Omega^{-1}\,m^{-1}$.)

It is reasonable to suppose that in infinitely dilute solutions the migration of each ion, and hence its contribution to electrical conduction, is independent of all the other ions. If this is so, we may write the electric current density $\mathbf{j}$ (SI unit A/m^2) as

$$\mathbf{j} = n_+\mathbf{v}_+Z_+e + n_-\mathbf{v}_-Z_-e, \qquad (29.87)$$

where n_+, n_- are the number densities of positive and negative ions, $\mathbf{v}_+$, $\mathbf{v}_-$ the corresponding average velocities of the ions, and Z_+e, Z_-e the respective electric charges. Any macroscopic portion of the electrolyte solution must be electrically neutral, which implies that

$$n_+Z_+e = n_-|Z_-|\,e = c^*\mathscr{F}, \qquad (29.88)$$

where c^* is the "equivalent" concentration (moles of positive or negative charge per unit volume) and $\mathscr{F}$ is the Faraday constant (96,487 C/mol, the charge per mole of electrons). If we introduce the ionic mobilities u_+, u_-, defined in an imposed electric field $\mathbf{E}$ by the equations

$$\mathbf{v}_+ = u_+\mathbf{E}, \qquad \mathbf{v}_- = -u_-\mathbf{E}, \qquad (29.89)$$

then the current density can be written as

$$\mathbf{j} = c^*\mathscr{F}(u_+ + u_-)\mathbf{E}. \qquad (29.90)$$

We now introduce Ohm's law in the form

$$j = \frac{I}{A} = \frac{\Delta V}{RA}, \qquad (29.91)$$

where I is the current between the electrodes and ΔV is the potential difference between the electrodes. Combining the last two equations with Eq. 29.86 (remember that the electric field is the gradient of the applied potential, $E_x = \partial V/\partial x$, so $\Delta V = lE$) leads to the expression for the electrical conductivity,

$$\sigma = c^*\mathscr{F}(u_+ + u_-). \qquad (29.92)$$

The *molar conductance* Λ is defined as the conductivity divided by the molar concentration,

$$\Lambda = \frac{\sigma}{c}; \qquad (29.93)$$

we add the superscript zero ($\Lambda°$) to designate the value of the conductance at infinite dilution. When the concentration is expressed in terms of "equivalents" (moles of positive or negative charge: e.g., for Cu^{2+} or SO_4^{2-}, 1 mole = 2 equivalents), one may refer to the "equivalent conductance" $\Lambda^* = \sigma/c^*$.

Measurements of electrical conductance in dilute solutions of strong (fully ionized) electrolytes show the following dependence on electrolyte concentration:

$$\Lambda = \Lambda° - a\sqrt{c}, \qquad (29.94)$$

where a is an empirical constant characteristic of the solute. (In a mixed solution the total conductance shows a similar dependence on the ionic strength.) For weak electrolytes different behavior is observed: The values of Λ^0 are comparable to those for strong electrolytes, but with increasing concentration Λ rapidly falls to a very low value (for 1 M acetic acid, $\Lambda/\Lambda^0 = 0.0038$). This can be accounted for, however, by changes in degree of dissociation of the weak electrolyte with concentration. Equation 29.94 is still valid if one takes the variable c to stand for the concentration of *ions* (not the total concentration of ionized plus un-ionized molecules). Figure 29.17 shows the variation of conductance with concentration for several electrolytes.

Let us turn to a qualitative discussion of the molar conductance in very dilute electrolyte solutions. At infinite dilution the ions are independent of one another, and in this limit the molar conductance Λ is as expected independent of concentration. In dilute solution, at equilibrium, we know from the Debye-Hückel theory (Section 26.5) that any ion is surrounded by a spherically symmetric ion atmosphere of opposite charge. The same central ion, say a positive ion, is itself a member of the ion atmospheres of the surrounding negative ions. The imposition of an electric field exerts a force on each ion. Picture now the motion of a central positive ion and its surrounding negative-ion atmosphere in the electric field. As the positive ion moves in one direction because of the field, it exerts a force tending to drag along the negative-ion atmosphere. Because of the frictional force of the solvent on all ions, there is a time lag in the response of the ion atmosphere; thus the atmosphere undergoes a distortion, Fig. 29.18. This distortion in turn exerts a force on the positive ion opposite to that of the imposed field, thus reducing that ion's mobility and hence the molar conductance. This contribution to the reduction in molar conductance is called the *relaxation effect*; the second major contribution to that reduction is the *electrophoretic effect*. As an ion moves under the imposed field, it tends to set the surrounding solvent into motion. This hydrodynamic motion is opposite to that of the surrounding, oppositely charged ions, and thus reduces their mobility.

If the ions of the electrolyte are infinitely far apart, then their interaction vanishes. At dilute but nonzero electrolyte concentrations, a measure of the deviation from ideality (non-interaction) is given by the activity coefficient $\gamma_\pm$. At low concentration (Eq. 26.50) $\ln \gamma_\pm$ is found to be proportional to the square root of ionic strength, or for a 1–1 electrolyte to the square root of the electrolyte concentration; both the relaxation effect and the electrophoretic effect depend on deviations from the behavior at infinite dilution, and therefore may by expected to have this concentration dependence.

The mathematical analyses of these effects are complex, and beyond the scope of this presentation; we simply cite the result, which for a

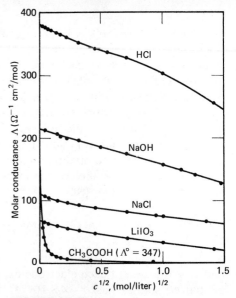

FIGURE 29.17
Molar conductance of some electrolytes in aqueous solution at 18°C. Note that for strong electrolytes Λ is approximately linear in $c^{1/2}$.

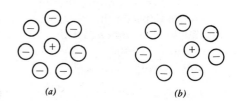

FIGURE 29.18
Schematic illustration of the asymmetry effect in ionic atmospheres: (*a*) no applied electric field; (*b*) with field.

uni-univalent electrolyte is of the form

$$\Lambda = \Lambda^0 - \left[\frac{A}{(DT)^{1/2}\eta} + \frac{B\Lambda^0}{(DT)^{3/2}} \right] \sqrt{c}, \qquad (29.95)$$

where D is the dielectric constant of the solvent, T is the temperature, η is the viscosity, and A, B are constants; this confirms the empirical Eq. 29.94. The assumptions made in deriving Eq. 29.95 are valid only in dilute solution; at higher concentrations the value of Λ levels off (in time to keep it from becoming negative). Measurements of the electrical conductivity cannot determine the relative contribution of the relaxation and electrophoretic effects. However, the two effects have different AC frequency dependences, and a separation of the two effects can be obtained on the basis of measurements of those dependences.

Further Reading

Chandrasekhar, S., "Stochastic Problems in Physics and Astronomy," *Rev. Mod. Phys.* **15,** 1 (1943).

Brownian Motion

Reif, F., *Fundamentals of Statistical and Thermal Physics* (McGraw-Hill, New York, 1965), chaps. 1, 15.

Berne, B. J., "Time-Dependent Properties of Condensed Media," in H. Eyring, D. Henderson, and W. Jost (Eds.), *Physical Chemistry: An Advanced Treatise*, Vol. VIIIB, *Liquid State* (Academic Press, New York, 1971), pp. 540–716.

Autocorrelation Functions and Transport in Liquids

Croxton, C. A., *Introduction to Liquid State Physics* (Wiley, New York, 1975).

Egelstaff, P. A., *An Introduction to the Liquid State* (Academic Press, New York, 1967).

(See also the references for Chapter 23.)

Guy, A. G., *Essentials of Materials Science* (McGraw-Hill, New York, 1976).

Transport in Solids

Jost, W., *Diffusion in Solids, Liquids, Gases*, 2nd ed. (Academic Press, New York, 1960).

Flynn, C. P., *Point Defects and Diffusion* (Oxford University Press, 1972).

Harned, H. S., and Owen, B. B., *The Physical Chemistry of Electrolytic Solutions*, 3rd ed., (Van Nostrand Reinhold, New York, 1958).

Electrical Conductivity in Solution

Robinson, E. A., and Stokes, R. H., *Electrolyte Solutions*, 2nd ed. (Academic Press, New York, 1959).

Problems

1. Two drunks start out from a common point (the origin) and independently take a one-dimensional random walk with $p = q = \frac{1}{2}$. What is the probability that they will meet again when each has taken N steps?

2. A random walker starts out 10 paces to the left of a fence. Normally $p = q = \frac{1}{2}$ except at the fence, where $q = 1$. What is the probability that the walker will be back at his starting point after N paces?

3. Calculate the transport coefficients D, η, and λ for argon (ideal gas with $d = 2.86$ Å) at standard temperature and pressure. Assuming that the binary-collision approximation still holds in the liquid phase, calculate D, η, and λ for liquid argon at $T = 90 \, K$, $\rho = 1.37 \, g/cm^3$.

4. Calculate the diffusion coefficient D for a dust particle suspended in air. Take $d(\text{air}) = 3.6$ Å, $d(\text{dust}) = 1 \, \mu m$, and the density of the dust particle as $2 \, g/cm^3$. What is the root-mean-square distance the dust particle will travel in $10 \, s$?

5. Calculate the diffusion coefficient for a hemoglobin molecule (molecular weight 63,000, $d = 50$ Å) in aqueous solution at 20°C. Use the formulas derived in the stochastic approach to transport processes, and compare the result to that obtained if the binary-collision approximation for gases is assumed to hold in the liquid phase.

6. Describe the difference in mechanism between the transfer of momentum (and energy) in the gas phase and in the liquid phase. Account for the fact that, at constant pressure, η increases with increasing temperature in the gas phase and decreases in the liquid phase.

7. Read the article "Viscoelastic Properties of Macromolecules in Dilute solution," by John D. Ferry, in *Accounts of Chemical Research* **6,** 60 (1973). Outline the simple model used to account for the behavior of polymers and discuss the unique features characteristic of polymer viscosity.

8. Consider a fluid forced through a pipe of radius R and length l by a pressure differential $p_2 - p_1$ between the ends of the pipe (see accompanying diagram).

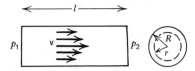

What is the frictional force in terms of $\partial v/\partial r$ due to viscous drag on a cylinder of fluid of radius r and length l? In order to have steady flow through the pipe, this force must balance the force driving the fluid. Equate these forces to obtain the expression for $\partial v/\partial r$. Integrate this equation to find $v(r)$. The integration constant is fixed by requiring the velocity at the boundary of the pipe, $v(R)$, to be zero. Calculate the volume of fluid flowing through the pipe, which is given by

$$\frac{dV}{dt} = \int_0^R dr \, 2\pi r v(r).$$

9. A reasonable model potential for a particle diffusing in a solid is shown in the accompanying diagram, where the minima refer to interstitial sites. By approximating the potential for a site by a harmonic-oscillator well (as depicted by the dashed curve):
   (a) Calculate the probability (at temperature T) that the oscillator has sufficient energy to overcome the barrier.
   (b) Calculate (in terms of frequency of the harmonic well) the rate of diffusion from one site to another.
   (c) The above potential assumes that diffusion is occurring in a static lattice. How will the diffusion rate change as the lattice energy increases?

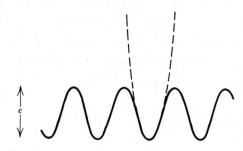

30

Chemical Kinetics

The kinetics of chemical reactions—the study of *how fast* reactions take place—is a topic of central importance to the chemist. A knowledge of the phenomenological rates of reactions is of course important for its own sake, in industrial chemistry, in understanding life processes, and in other applied fields. Thus we must discuss how reaction rates are defined and measured, and what relationships exist between rates and equilibria. But beyond this, we wish to know *how* reactions take place, what are the microscopic processes corresponding to a change of chemical species. Since nearly all reactions occur by molecular collisions of one kind or another, we must study the nature of those collisions.

We begin with a survey of the major concepts of the field (rate, order, molecularity, etc.) and some of the simpler types of reactions. Then we proceed to develop the microscopic theory of bimolecular reactions, those that occur by simple two-molecule collisions. We discuss the interactions between two reactive molecules and the types of collisions they can undergo, then examine the dynamics of a reactive collision—first in terms of a simple hard-sphere model, then (after introducing some thermodynamic ideas) by the widely used activated-complex theory.

In the remainder of the chapter we consider various specific types of reactions. Most reactions take place by more than a single microscopic step, and it is important to study the various combinations of processes that can occur, what is usually called the "mechanism" of the reaction. There are many different types of reaction mechanisms, each with its own points of interest. We must also examine the effects of such variables as the medium (gases, solutions, solid surfaces, etc.), the temperature, and the presence of catalysts. Chemical kinetics is as wide a field as chemistry itself, and in this chapter we can only give the barest outline of it.

30.1 General Concepts of Kinetics

Although much of this chapter will deal with microscopic processes, we must begin by defining some macroscopic concepts. The first of these, of course, is the *rate* of a reaction. Consider a reaction such as

$$a\mathrm{A} + b\mathrm{B} = c\mathrm{D} + d\mathrm{D},$$

where a, b, c, d are the stoichiometric coefficients, or, more generally,

$$\sum_i \nu_i \mathrm{X}_i = 0 \qquad (30.1)$$

(in the notation of Section 14.2: the ν_i are positive for products, negative for reactants). It has long been customary to refer to the time derivative of a given substance's concentration, $d[\mathrm{X}_i]/dt$, as the "rate of the reaction." However, this has the disadvantage of being in general different for different substances in a reaction; for example, in the reaction

$$\mathrm{N}_2 + 3\mathrm{H}_2 = 2\mathrm{NH}_3$$

we have

$$\frac{d[NH_3]}{dt} = -2\frac{d[N_2]}{dt} = -\frac{2}{3}\frac{d[H_2]}{dt}.$$

The International Union of Pure and Applied Chemistry has therefore recommended that the term "rate of reaction" be restricted to the quantity[1]

$$\dot{\xi} \equiv \frac{1}{\nu_i}\frac{dn_i}{dt}, \tag{30.2}$$

where n_i is the amount (usually in moles) of substance X_i present in the system; $\dot{\xi}$ has the advantage of being independent of both the choice of X_i and the volume of the system. Nevertheless, most of the theoretical expressions we shall introduce involve the time derivative of concentration, and we shall continue the practice of referring to this loosely as the "rate."

However one defines the reaction rate, it ordinarily has to be obtained experimentally. Measurement of the rate involves determining the concentration of one or more chemical species as a function of time. Such determinations can be made by either chemical or physical methods. Chemical analysis for the concentration of a species has the advantages of specificity and relatively high accuracy; it has the disadvantages of being slow and disturbing the system. A reaction cannot in general be brought to a stop while one makes an analysis;[2] thus any method of measuring concentration must be capable of being carried out in a time within which the concentration will not change significantly. Physical methods of measurement are more rapid than chemical ones, and usually do not perturb the reactive system, but are seldom specific for a given substance. For instance, measuring the index of refraction requires a calibration of the concentration of each species present against index of refraction; impurities or competing reactions may also complicate such a measurement. For a more detailed discussion of various methods of measuring reaction rates, see Section 30.10.

The variables that affect reaction rates are for the most part the same as those important in equilibrium thermodynamics: temperature, pressure, and concentrations (or activities) of the substances present. We shall see that reaction rates are often strongly dependent on the temperature, which must thus be carefully controlled if one is to make accurate measurements. The medium is also important, since some reactions occur by quite different mechanisms in, say, the gaseous phase and aqueous solution.

A reaction that occurs entirely in a single phase is said to be *homogeneous*; one that occurs partly or entirely at an interface between phases is *heterogeneous*. The significance of the latter is that many reactions have rates strongly influenced by the presence of solid surfaces: some crucial step in the reaction mechanism must occur on the surface,

---

[1] Here $\dot{\xi}$ is the time derivative of the *extent of reaction* ξ (equivalent to the progress variable λ of Section 19.5), defined by

$$n_i = (n_i)_0 + \nu_i\xi$$

relative to some arbitrary zero point.

[2] In some special cases, reactions can be stopped almost instantaneously, provided that the rate is a sensitive function of some externally adjustable parameter. Many reactions are so strongly temperature-dependent that they can be "quenched" by lowering the temperature; and others can be effectively stopped by removing a catalyst or even turning off a light.

and the rate may depend on the speed with which reactants can diffuse to the surface. To complicate matters further, some reactions proceed by competing homogeneous and heterogeneous mechanisms. One way to distinguish the two is by varying the surface-to-volume ratio of the vessel, either by altering its shape or, more drastically, by filling it with something like small chips of the same material. The heterogeneous rate increases with the surface area, while the homogeneous rate remains constant. Another way is by varying the temperature: The homogeneous rate, as we shall see, varies exponentially with temperature, whereas a heterogeneous reaction limited by diffusion should have a rate roughly proportional to $T^{1/2}$ (cf. Table 28.2 for hard-sphere molecules).[3]

Both heterogeneous and homogeneous reaction rates can be influenced by *catalysis*. A *catalyst* has been defined as a substance that changes the rate of a chemical reaction without affecting the final equilibrium. In *surface catalysis* the catalyst is the solid surface itself, which acts as a substrate upon which the reactants can interact more readily. In *homogeneous catalysis* the catalyst somehow takes part directly in the sequence of elementary reactions, although its overall concentration does not change in the course of the reaction. We shall have much more to say about catalysts in Chapter 31.

Of all the variables that affect reaction rates, the greatest emphasis is usually laid upon the concentrations of the substances taking part. When temperature, pressure, and so on are held constant, it is usually found experimentally that the reaction rate is a simple function of these concentrations. In many cases the rate is directly proportional to the concentrations raised to powers, that is,

$$-\frac{d[A]}{dt} = k[A]^\alpha[B]^\beta[C]^\gamma \cdots, \tag{30.3}$$

where $\alpha, \beta, \gamma, \ldots$ are constants (commonly integers) and k is a proportionality constant known as the *rate coefficient* (or *rate constant*). Equation 30.3 is an example of an *experimental rate expression*. The *order* of a reaction with respect to a given species is the exponent of that species's concentration in the experimental rate expression; thus the reaction in Eq. 30.3 is said to be of order α in A, β in B, and so on. The overall order of the reaction is the sum of the orders with respect to each of the species appearing in the rate expression, here $\alpha + \beta + \gamma + \cdots$. The species that appear in an experimental rate expression need not be reactants in the strict sense: A reaction rate may depend on the concentration of any substance present, including products or even substances not appearing in the stoichiometric equation for the reaction (either transient intermediates or catalysts). Negative orders are possible (i.e., rate expressions with concentrations appearing in the denominator), and some rate expressions are much more complicated than the simple product 30.3, often having two or more terms on the right-hand side. Let us now examine some of the simpler types of reactions as classified by reaction order.

In a first-order reaction the reaction rate is directly proportional to the concentration of a single species A and not dependent on the concentration of any other species. Thus we can write

A. First-Order Reactions

$$-\frac{d[A]}{dt} = k[A], \tag{30.4}$$

[3] The distinction between homogeneous and heterogeneous reactions is not absolute, breaking down in such intermediate cases as a reaction that proceeds at certain sites in a large biological molecule. (The best examples of this are enzymes, which are biological catalysts.)

where k is the rate coefficient. A well-known example is the decomposition of gaseous nitrogen pentoxide,

$$2N_2O_5 \rightarrow 4NO_2 + O_2, \qquad -\frac{d[N_2O_5]}{dt} = k[N_2O_5];$$

most first-order reactions are decompositions or isomerizations.

Given the rate expression 30.4, how does the concentration [A] vary with time? Integration immediately gives

$$[A] = [A]_0 e^{-kt}, \tag{30.5}$$

where $[A]_0$ is the concentration of A at $t = 0$, that is, at the beginning of the experiment. This behavior is illustrated in Fig. 30.1a, in which the curved line represents the function defined by Eq. 30.5. There are various methods of treating kinetic data to obtain rate coefficients, and for first-order reactions this is most easily done by taking the logarithm of the concentration. Since Eq. 30.5 is equivalent to

$$\ln[A] = \ln[A]_0 - kt, \tag{30.6}$$

a plot of log[A] against time should give a straight line with slope $-k/2.303$; this is illustrated in Fig. 30.1b.

It is clear, if one rearranges Eq. 30.4 to

$$-\frac{d[A]}{[A]} = k \, dt, \tag{30.7}$$

that the fractional change in [A] for a given time interval is the same at all times. For first-order reactions it is therefore convenient to define the time interval necessary for any concentration [A] to reach half its initial value as the *half-life* of the reaction, $t_{1/2}$. The half-life can be simply related to the rate coefficient: If we substitute $t = t_{1/2}$, $[A]/[A]_0 = \frac{1}{2}$ in Eq. 30.6, we at once obtain

$$t_{1/2} = \frac{\ln 2}{k}. \tag{30.8}$$

The half-life here has the same meaning as in radioactive decay processes (which are inherently first-order, since the decay rate of a given nuclide depends only on the amount of that nuclide present in the sample). It can be seen from any of Eqs. 30.4–30.8 that the first-order rate coefficient has the dimensions $(\text{time})^{-1}$, and is thus independent of the units used to measure concentration.

There are two principal types of second-order reactions. In the simplest form, the reaction rate is proportional to the square of the concentration of a single species, that is,

$$-\frac{d[A]}{dt} = k[A]^2. \tag{30.9}$$

An example is the aqueous-solution reaction

$$2ClO^- \rightarrow 2Cl^- + O_2(g), \qquad -\frac{d[ClO^-]}{dt} = k[ClO^-]^2.$$

To obtain the variation of concentration with time, we rearrange Eq. 30.9 to

$$-\frac{d[A]}{[A]^2} = k \, dt, \tag{30.10}$$

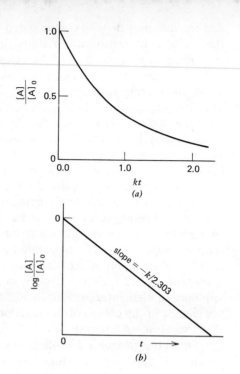

FIGURE 30.1
Concentration of A versus time for a first-order rate equation, according to Eqs. 30.5 and 30.6. (a) gives the exponential decay curve, (b) gives the logarithmic plot used to obtain the rate coefficient k.

B. Second-Order Reactions

integration of which gives

$$\frac{1}{[A]} - \frac{1}{[A]_0} = kt, \tag{30.11}$$

where $[A]_0$ is again the concentration at $t = 0$. Thus a plot of $[A]^{-1}$ against t gives a straight line with slope k, as shown in Fig. 30.2.

The other simple type of second-order rate expression is of the form

$$\text{rate} = k[A][B], \tag{30.12}$$

that is, first-order in each of two species. To integrate this equation we must know the stoichiometry of the reaction, which we suppose to be[4]

$$a A + b B \rightarrow \text{products.} \tag{30.13}$$

An example is the reaction between persulfate and iodide ions,

$$S_2O_8^{2-} + 2I^- \rightarrow 2SO_4^{2-} + I_2, \qquad \frac{d[S_2O_8^{2-}]}{dt} = -k[S_2O_8^{2-}][I^-].$$

If we take the rate expression to be

$$-\frac{1}{a}\frac{d[A]}{dt} = -\frac{1}{b}\frac{d[B]}{dt} = k[A][B], \tag{30.14}$$

which is in fact a definition of k, and define a variable x by[5]

$$[A] = [A]_0 - ax, \qquad [B] = [B]_0 - bx, \tag{30.15}$$

the rate expression becomes

$$\frac{dx}{dt} = -\frac{1}{a}\frac{d[A]}{dt} = k([A]_0 - ax)([B]_0 - bx). \tag{30.16}$$

We shall not go into the details of the integration of Eq. 30.16, which can be readily found in standard tables of integrals. With the lower limit $x = 0$, $t = 0$, the integration yields

$$\frac{1}{b[A]_0 - a[B]_0} \ln\left\{\frac{[B]_0([A]_0 - ax)}{[A]_0([B]_0 - bx)}\right\} = kt \tag{30.17}$$

or

$$\frac{[A]}{[B]} = \frac{[A]_0}{[B]_0} e^{(b[A]_0 - a[B]_0)kt}, \tag{30.18}$$

and a plot of log $([A]/[B])$ against t gives a straight line whose slope is proportional to k, as shown in Fig. 30.3.

We could give other examples of rate expressions and their integration, but our point has been made: Given a knowledge of the empirical rate expression (and in particular its order), determining the rate coefficient from the data is a straightforward procedure. Finding the rate expression in the first place is, of course, a matter of trial and error: One plots concentration against time in various ways until a straight-line or other simple result is obtained. This is not always a trivial process, since some reactions have quite complicated rate expressions; for example, the

---

[4] Sometimes a rate expression such as Eq. 30.12 is obtained for a reaction of the form

$$A \rightarrow B + \cdots.$$

A reaction whose rate is proportional to the concentration of a product is said to be *autocatalytic*.

[5] Thus x is the extent of reaction per unit volume, ξ/V.

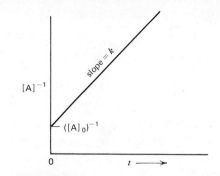

FIGURE 30.2
Straight-line plot giving k for a second-order reaction of the type

$$2A \xrightarrow{k} \text{products,}$$

according to Eq. 30.11.

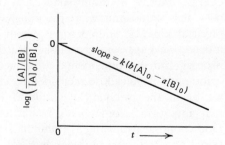

FIGURE 30.3
Straight-line plot giving k for a second-order reaction of the type

$$a A + b B \rightarrow \text{products,}$$

according to Eq. 30.18.

simple-appearing reaction

$$H_2 + Br_2 = 2HBr$$

is found experimentally to have the strange rate expression

$$\frac{d[HBr]}{dt} = \frac{k[H_2][Br_2]^{1/2}}{1 + k'[HBr]/[Br_2]} \tag{30.19}$$

(in Section 31.3 we shall see how this sort of thing can come about).

The example just given illustrates an important point: There is in general *no relation* between the stoichiometric coefficients of a reaction and the reaction order with respect to the various substances. Furthermore, neither stoichiometry nor order tells us how the reaction actually takes place. One might think that a second-order reaction occurs by a simple encounter between two reactant molecules, and sometimes this is indeed the case; but in general there is no such relationship. To see why this is so, we must examine the dynamics of reactions on the molecular level.

On the molecular level, a chemical reaction is a process in which one or more chemical species—molecules, atoms, ions, free radicals, or electrons—are transformed into one or more different chemical species. (In what follows we shall use "molecule" to include all the types of species mentioned.) The overall reaction may involve a single microscopic process or a series of such processes; in either case, the individual processes are called *elementary reactions*. What happens in an elementary reaction? Sometimes a single molecule can react by itself (after gaining energy by absorption of radiation or by inelastic collision), but most elementary reactions involve molecular collisions of some kind.

We use the term *molecularity* to designate the number of molecules taking part in an elementary reaction: A *unimolecular* elementary reaction is one in which a single molecule undergoes decay or rearrangement, a *bimolecular* elementary reaction is one in which two molecules collide and react, and so forth. It was originally supposed that molecularity corresponded to the empirical order of a reaction; but as we have already pointed out, things are more complicated than this. Molecularity, order, and stoichiometry are in general different from one another, and the task of the kineticist is to discover the relations between them.

But the kineticist also wishes to know how the individual elementary reactions take place. He is interested in the details of the molecular collisions leading to reaction—the dynamics of the collision; the forces between molecules; the scattering induced by these forces; the reaction cross section (or reaction probability) and its dependence on the relative energy, internal quantum states, and orientation of reactants and products. This sounds much like the things we have considered in our study of physical kinetics, and to some extent we can use the same formalism; indeed, we have said a chemical reaction can be viewed as a transport process. But the fundamental difference is that in a chemical reaction the colliding molecules lose their identity. By the "dynamics" of the reaction we mean the mechanical details of how the process occurs: Do the reactants form a complex (an association with a relatively long lifetime)? Does bond rearrangement occur in a time of the order of typical rotational or vibrational periods? Is there need for localization of energy in certain degrees of freedom? More practically, given a knowledge of the molecular structure of the reactants, can one predict the kinds of reaction products to be expected, the reaction probability (and thus the rate coefficient), the energy necessary for reaction, and other quantities usually obtained from experiment? These are only some of the points of interest in the dynamics of elementary reactions, which can be studied both theoretically (either *a priori* or by correlating macroscopic data) and experimentally (particularly by molecular-beam and spectroscopic techniques).

All this is complicated by the fact that most chemical reactions do not occur in a single elementary step, but require a series of steps for completion. For all but the simplest molecules, the bond rearrangements required to give stable products are too complicated to be brought about by a single collision. Thus one must determine a complete sequence of elementary reactions, which is referred to as the *mechanism* of the overall reaction. Later we shall have much to say about the types of mechanisms and how they are identified; here let us just give some examples without proof.

Consider the reaction of potassium atoms with bromine molecules,

$$K + Br_2 = KBr + Br.$$

The reaction has been carried out with crossed molecular beams, and is thus known to occur in a single step, that is, a single bimolecular collision. In fact, we shall see later that the probability of reaction on collision is quite large. This is an example of a reaction in which the mechanism does correspond to the stoichiometric equation.

The reaction between gaseous H_2 and I_2,

$$H_2 + I_2 = 2HI,$$

is second-order and was long thought to involve a single-step bimolecular collision. But in recent years evidence has been found suggesting that the reaction proceeds by the mechanism

$$I_2 \rightleftharpoons 2I,$$

$$H_2 + 2I \rightarrow 2HI,$$

supplemented at high temperatures by the mechanism

$$I_2 \rightleftharpoons 2I,$$

$$H_2 + I \rightleftharpoons HI + H,$$

$$I_2 + H \rightleftharpoons HI + I.$$

This illustrates that a single reaction can often occur by more than one mechanism proceeding simultaneously. Another interesting point is that the first mechanism predicts the same second-order rate equation as for a single bimolecular collision: in general, there is no unique mechanism corresponding to a given empirical rate law.

In still another example, the formation and destruction of ozone in the upper atmosphere,

$$3O_2 = 2O_3,$$

is an extremely complex process involving at least the reactions

$$O_2 + h\nu \rightarrow 2O,$$

$$O + O_2 + M \rightarrow O_3 + M,$$

$$O_3 + h\nu \rightarrow O_2 + O,$$

$$O_3 + O \rightarrow 2O_2$$

(where M is any molecule taking part in a collision); other species that may be present can also play an important role.

These examples may suggest that the elucidation of reaction mechanisms is a difficult task, and indeed most mechanisms are guessed (from a knowledge of which elementary reactions are likely in a given system) rather than deduced. Of course, such guesses should then be tested by further experiments; as we have noted, often more than one mechanism is consistent with a given set of data. Plausible mechanisms have been proposed for many, but by no means all, reactions on which kinetic studies have been made.

In the next few sections we shall concentrate on the theory of bimolecular elementary reactions, which are by far the most common. Even these are of many types; although we cannot discuss them in detail, we shall close this section with a listing of some of these types (using examples actually observed). One way to classify reactions is by the nature of the reactants; for example, we may have *neutral–neutral* reactions,

$$H + O_2 \rightarrow OH + O;$$

ion–neutral reactions,

$$Ar^+ + H_2 \rightarrow ArH^+ + H;$$

or *neutralization* reactions,

$$N^+ + O^- \rightarrow N + O$$

(reactions between species with the same charge are also possible but energetically unlikely). Within or across these categories, we can also classify by the type of effect produced. Perhaps the most common type are *transfer* (or *metathetical*) reactions, in which an atom or group of atoms breaks one bond and forms a new one. The $H + O_2$ and $Ar^+ + H_2$ reactions above involve transfer of O atoms and H atoms respectively, whereas

$$H_2^+ + Ar \rightarrow ArH^+ + H$$

is an example of *proton transfer*. In other cases more complex rearrangements occur, as in

$$CH_3^+ + CH_4 \rightarrow C_2H_5^+ + H_2,$$

a *condensation* reaction, or in

$$C_2H_4 + O \rightarrow \begin{array}{c} H_2C\text{———}CH_2 \\ \diagdown \diagup \\ O \end{array},$$

which can be called *atom insertion*. Sometimes the species transferred is only an electron, and we have *charge-exchange* (or *electron-transfer*) reactions such as

$$O^+ + N_2 \rightarrow O + N_2^+;$$

charge exchange with neutral reactants leads to *ion-pair formation*,

$$O + O \rightarrow O^+ + O^-.$$

The latter is an extreme case of *collisional ionization*, which more usually leads to the formation of a free electron,

$$Ar + Ar \rightarrow Ar^+ + Ar + e^-.$$

Here the ionization energy comes from the kinetic energy of the reactants, whereas in *associative ionization* it comes from the formation of a new bond,

$$CH + O \rightarrow HCO^+ + e^-;$$

finally, in *Penning ionization* the energy is furnished by an excited state of a reactant,

$$He^* + O_2 \rightarrow He + O_2^+ + e^-$$

(where * designates an excited state, in this case $1s2s$). The ion–molecule equivalent of associative ionization is called *associative detachment*.

$$O_2^- + O \rightarrow O_3 + e^-.$$

The destruction of a bond, as in the simple *dissociation* reaction

$$Br_2 + Ar \rightarrow Br + Br + Ar$$

(ionic dissociation is also possible), merely requires the supplying of sufficient kinetic energy. Whenever a new stable bond is formed, however, the excess energy must be removed from the molecule in some way. (When the product molecule has many degrees of freedom, energy can be transferred from the reaction site into a variety of vibrational modes; this temporarily stabilizes the energized product until some later deexcitation step can remove the energy.) One way of removing energy is by the release of an electron; another is the emission of radiation from an excited state, as in the *recombination* reaction

$$Br + Br \rightarrow Br_2^* \rightarrow Br_2 + h\nu$$

(whether such a process is considered to be one step depends on the lifetime of the excited state). The emission of radiation by a product of chemical reaction is called *chemiluminescence*; another example, in which the emission is not required for stability, is

$$F + H_2 \rightarrow H + HF^* \rightarrow H + HF + h\nu.$$

Excited states can also be produced by *collisional excitation*,

$$H + Ar \rightarrow H + Ar^*,$$

or by *excitation transfer*,

$$N_2^* + NO \rightarrow N_2 + NO^*.$$

Although we still have not exhausted the types of bimolecular reactions, this listing should be enough to give an idea of the possibilities.[6] The examples given here have all been of gas-phase reactions, but many of the same types (proton-transfer, charge-exchange, etc.) can also occur in solution.

30.2 Interactions Between Reactive Molecules

We have previously discussed the forces between nonreacting molecules in some detail. Accurate theoretical calculations of inter-molecular forces are long and expensive, so that few such calculations have been made. But extensive information is becoming available from elastic-scattering measurements, in addition to what has been learned from measuring such properties as the temperature dependence of the virial coefficients and transport coefficients. Nevertheless, it is still difficult to get accurate results for the intermolecular potential, which as we have seen depends only weakly on measurable quantities, and is probably much more complicated than any simple form with only a few parameters.

The problem of determining the interactions between reacting molecules is even more difficult, so we know even less about these

[6] Note that much of the terminology used here is also applicable to binary *nuclear* reactions (in which, however, the energies involved are much higher).

interactions. There are both theoretical and experimental reasons for these difficulties, but most important is the large number of species involved. An elastic collision ordinarily involves only two particles (molecules, atoms, ions), which retain their identity throughout; to a good approximation they can be treated as spherically symmetric, with the force between them dependent on a single intermolecular distance. But the occurrence of a reaction means the disappearance of one or more chemical species, and the appearance of one or more new ones. This means that the internal structure of reactants and products (and whatever state exists in the interim) must be taken into account, and thus that a greater number of variables are required for a complete description.[7] For example, in a reaction of the common type

$$AB + C \rightarrow A + BC$$

we need at least three variables to describe the relative positions of A, B, and C, and perhaps more to describe their orientations if they are themselves polyatomic.

The methods used to study reactive interactions are much the same as for nonreactive interactions. The *a priori* calculation of the interaction potential is an extremely difficult task, being essentially the same as the calculation of molecular energy for given nuclear coordinates—in a "molecule" containing *all* the reacting species; such calculations have been made for only a few cases as yet, one of which we shall consider. On the experimental side, obtaining molecular information from macroscopic reaction-rate data is the analog of deducing molecular parameters from transport properties. A more useful approach is again the direct study of reactive collisions by scattering experiments, the results of which we shall discuss later.

With this pessimistic prologue to inspire us with caution, let us now examine what *is* known about the interactions between reactive molecules. As an example, we shall consider the reaction

$$F + H_2 \rightarrow HF + H.$$

The interaction between a F atom and two H atoms will certainly depend on the distances R_{H-F} (the shorter of the two H–F distances) and R_{H-H} (see Fig. 30.4), and on the angle α, as well as on the internal states of F and H_2, or HF and H. For simplicity of discussion we assume all species to be in their ground electronic states. If the Born–Oppenheimer approximation holds, the energy of the system then depends only on the relative positions of the nuclei; and to make the problem more tractable we neglect the energy's dependence on the angle α. (It is easy to make approximations when one doesn't know much.) The latter assumption allows us to make our calculations for the simple case that all three atoms are in a straight line ($\alpha = \pi$). And it is for just this configuration that a very good *a priori* calculation has been made.[8] We shall examine the results of this calculation.

Since the linear F–H–H system has only two independent variables (the distances R_{F-H} and R_{H-H}), its energy can be represented by a

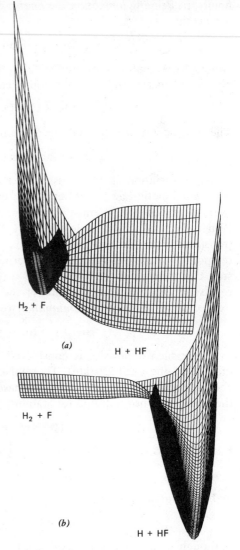

FIGURE 30.4
Two views of the potential energy surface for the linear F—H—H system: (*a*) from the $H_2 + F$ channel; (*b*) from the HF + H channel. The horizontal axes represent R_{H-F} and R_{H-H}, and the vertical axis represents energy. Each small "square" on the surface represents a square region in space $0.05a_0$ on a side (where a_0 is the Bohr radius, 0.5292 Å). From C. F. Bender, S. V. O'Neil, P. K. Pearson, and H. F. Schaefer, III, *Science* **176**, 1412 (1972).

---

[7] There may appear to be an exception to this when two particles combine into one,

$$A + B \rightarrow AB,$$

but such reactions have complications of their own. As already noted, just as energy must be supplied to a stable molecule to split it into separate particles, so to form a stable composite from the two particles energy must be given off—by collision with a third particle (making the reaction really $A + B + M \rightarrow AB + M$) or with a surface, or by emission of a photon.
[8] C. F. Bender, S. V. O'Neil, P. K. Pearson, and H. F. Schaefer, III, *Science* **176**, 1412 (1972).

three-dimensional surface. The potential energy surface calculated by Bender et al. is illustrated here by two perspective views (Fig. 30.4) and the equivalent "contour map" (Fig. 30.5). Let us examine first the limiting regions. When all three atoms are far apart, their energy is independent of their location; this is represented by the flat plateau at the upper right of the contour map. When two atoms are close together and one far away, we have a diatomic molecule and a separate free atom. If $R_{H-H} \gg R_{H-F}$, for example, a cross section of the surface at constant R_{H-H} essentially[9] reproduces the potential energy curve for the HF molecule; this is represented by the line $A-A'$ in Fig. 30.5. Similarly, for $R_{H-F} \gg R_{H-H}$ a cross section along the line $B-B'$ essentially reproduces the H_2 potential energy curve.

Thus the F–H–H potential energy surface has two "valleys" or *channels*, corresponding respectively to the limiting cases HF+H and H_2+F. What happens in the region where these valleys intersect, where the reaction must take place? There must be a point corresponding to a mountain pass, whose height is a maximum if one is traveling along a road from one valley to the other, but a minimum if one travels in the perpendicular direction along the mountain crests. Such a point is called a *saddle point*, since the curvature of the surface corresponds to that of a saddle; the one for this surface is indicated by "$\times$" in Fig. 30.5. In this case the location of the saddle point is clearly asymmetric, and the deeper of the two valleys must "turn a corner" to reach the pass; however, it is possible to have a symmetric surface (as in the exchange reaction $H_A H_B + H_C \rightarrow H_A + H_B H_C$), with the saddle point on the 45° line.

Although the entire shape of the potential energy surface is of interest, it is often useful to concentrate on particular features. Of special importance is the minimum-energy path or *reaction coordinate*, which is simply a line drawn along the deepest parts of the two valleys (corresponding to where one might build a road in the geographical analogy). The reaction coordinate is illustrated by the heavy dashed line in Fig. 30.5; note that it necessarily passes through the saddle point $\times$. The energy along the reaction coordinate is shown in Fig. 30.6a; this is simply the surface's cross section along the dashed line. To a first approximation, the reaction coordinate is the path the system must follow in the course of a reaction (in either direction). As can be seen in the figure, reaction thus involves crossing a potential energy barrier. The *barrier height*[10] is the energy difference between the saddle point (or transition state) and the initial state of the system, that is, between the top of the pass and the bottom of the valley in which the reaction begins. The barrier height is in general different for reactions in opposite directions: In our example the height is 36.1 kcal/mol (151 kJ/mol) for HF+H$\rightarrow$H$_2$+F, but only 1.7 kcal/mol (7 kJ/mol) for H_2+F$\rightarrow$HF+H. The energy change for the reaction, ΔE_{reac}, is the difference between these two quantities, or ± 34.7 kcal/mol (± 144 kJ/mol); the sign, of course, depends on the direction of the reaction. If one is not interested in the exact shape of the barrier, the barrier heights and energy change can be represented schematically as in Fig. 30.6b.

It is not always practical (or necessary) to compute an exact potential energy surface *a priori*, and a variety of semiempirical methods have therefore been developed. One of the best of these is usually called the *LEPS* (for London-Eyring-Polyani-Sato) *method*. This assumes that the

---

[9] We say "essentially" because the potential energy corresponds exactly to that of HF only in the limit $R_{H-H} \rightarrow \infty$.

[10] The barrier height should not be confused with the *activation energy*, an experimental quantity to be discussed later, although the two are often similar.

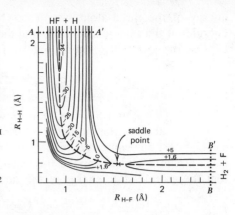

FIGURE 30.5

"Contour map" of the linear F—H—H potential energy surface, as calculated by Bender et al. The zero of energy is taken to be the limiting energy for H_2+ F, and the energy contours are labeled in kilocalories per mole (1 kcal/mol = 4.184 kJ/mol = 0.043364 eV). The cross sections indicated by dotted lines are discussed in the text; the heavy dashed line represents the reaction coordinate.

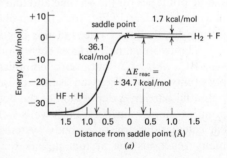

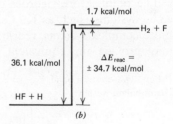

FIGURE 30.6

(a) Potential energy along the reaction coordinate for the reaction

$$HF+H \rightleftharpoons H_2+F.$$

(b) Schematic representation of the potential barrier.

energy of the three-atom system can be calculated from the two-atom energies at the same distances. Using the valence bond approach, one can express the lowest two energy levels of a two-atom system in the form

$$E_{AB} = \frac{J_{AB} \pm K_{AB}}{1 \pm S_{AB}^2},$$

(30.20)

where J_{AB}, K_{AB}, and S_{AB} are quantities respectively called the Coulomb, exchange, and overlap integrals;[11] using spectroscopic data or *a priori* calculations, one can evaluate these quantities as functions of the distance R_{AB}. Given J and K for the three pairs, the LEPS approximation says that the ground-state energy of the three-atom system is

$$E_{ABC} = \frac{J_{AB} + J_{BC} + J_{CA} - \{\frac{1}{2}[(K_{AB} - K_{BC})^2 + (K_{BC} - K_{CA})^2 + (K_{CA} - K_{AB})^2]\}^{1/2}}{1 + \kappa}$$

(30.21)

where κ is a complicated but small overlap integral that one evaluates empirically (usually from activation energies). Using this method, one can obtain reasonably accurate potential energy surfaces.

A significant point exhibited by Eqs. 30.20 and 30.21 is that the three-body energy is not the sum of the pair energies. To put it another way, the interatomic forces (gradients of the energy) are *nonadditive*: cf. Eq. 21.159. (The forces between individual electrons and nuclei must be additive, of course, but this does not carry over to the atomic level.) The extent of this nonadditivity varies with the system: The force between two Ar atoms is largely independent of the location of a third, but in the H–Br–K system the force between H and Br can depend very much on where the K is. It is precisely this effect that makes bond rearrangements possible.

In general, the potential energy surface for a three-body reaction has two channels that meet at a saddle point. Except for a reaction such as $H_A H_B + H_C \rightarrow H_A + H_B H_C$, the surface is usually asymmetric; that is, the saddle point lies off the 45° line and one of the channels must bend to reach it, as we have seen in Fig. 30.5. In that figure the "shorter" (unbent) channel is also the shallower one, but this is not always the case.

The two-channel surfaces correspond to simple substitution reactions $(A + BC = AB + C)$, but more complicated surfaces are possible. One possibility often discussed is the presence of a "well" on top of the potential barrier, a depression at the position corresponding to the saddle point in simpler surfaces. Such wells were predicted by early semiempirical calculations for three-body systems, reinforced by evidence that in many reactions the system lingers for a relatively long time near the saddle point. More rigorous calculations show that wells are present in some systems (such as $K + CsBr$) but not in others (such as $H + H_2$). In more complex systems, redistribution of energy over internal coordinates may provide the equivalent of a well (more like a multidimensional

---

[11] For a two-electron bond, if $\varphi_A(i)$ is the orbital for electron i on atom A, etc., and H is the Hamiltonian, we have

$$J_{AB} \equiv \int \int \varphi_A^*(1)\varphi_B^*(2) H \varphi_A(1)\varphi_B(2) \, d\tau_1 \, d\tau_2,$$

$$K_{AB} \equiv \int \int \varphi_A^*(1)\varphi_B^*(2) H \varphi_A(2)\varphi_B(1) \, d\tau_1 \, d\tau_2,$$

$$S_{AB}^2 \equiv \int \int \varphi_A^*(1)\varphi_A^*(2)\varphi_B(1)\varphi_B(2) \, d\tau_1 \, d\tau_2,$$

in accord with the definitions in Section 8.4.

bottle); but the point to be made here is that a system with more than two significant coordinates cannot be adequately described by a simple surface.

Thus far we have considered only *adiabatic* reactions, those corresponding to motion on a single potential energy surface. But different electronic states of the system correspond to different surfaces, and transitions between these are possible. Such phenomena are commonly referred to as *curve-crossing*, a term strictly applicable only to systems with one independent variable. Consider a diatomic system two of whose electronic states have, in first approximation, the $E(R)$ curves shown in Fig. 30.7a, in which the two curves cross. If the two states have the same symmetry, however, the noncrossing rule applies (Section 7.8): The actual electronic states can be represented by linear combinations of the two assumed approximate states, and the $E(R)$ curves for these states avoid crossing, as shown in Fig. 30.7b. However, the noncrossing curves often do approach each other closely, and at these points transitions between them are likely if R is changing so rapidly that the Born–Oppenheimer approximation ceases to hold. For multidimensional potential surfaces the situation is more complicated (for one thing, the noncrossing rule does not in general hold[12]), but the same basic principle applies: Some pairs of surfaces approach each other closely (or even intersect) in restricted regions, and in such regions electronic transitions are likely.

One reaction involving nonadiabatic transitions is the quenching of sodium fluorescence, in which excited Na atoms (which emit 5890-Å light in the $3p \to 3s$ transition) are deexcited by collision:

$$Na(3^2P) + M \to Na(3^2S) + M + 2.09\,eV.$$

The reaction appears to proceed via an ionic intermediate state which crosses both excited and ground states, as shown in Fig. 30.8a. (If M has two or more atoms, the situation is complicated by the multiplicity of vibrational levels associated with each electronic level. This produces many more crossing points and allows a variety of routes between initial and final states. Quenching by diatomic molecules is much more efficient than that by single atoms for this reason; electronic-vibrational energy interchange is generally easier than electronic-translational.) Calculations on the probability of the transitions give reaction rates in reasonable agreement with experiment. Another example is the reaction

$$H^+ + H_2 \to H_2^+ + H.$$

Figure 30.8b shows the limiting curves for H_2 (with H^+ far away) and H_2^+ (with H far away), which cross at $R \approx 2.5a_0$; these are cross sections through limiting regions of the H_3^+ energy surface(s). When all three atoms are fairly close together, one finds a "seam" (avoided crossing) between the two lowest $^1\Sigma^+$ surfaces along the line $R_{AB} \approx 2.5a_0$ (for $R_{BC} \gtrsim 6a_0$). Calculations of the "surface-hopping" probability give a steep maximum along this seam, and again give good agreement with experiment.

For most of this section we have restricted ourselves to the consideration of linear configurations. Some studies of the variation of the potential energy surface with angle have been made for the $H + H_2$

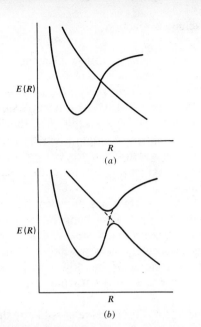

FIGURE 30.7
Interaction between two potential curves of the same symmetry. (*a*) Crossing of approximate curves. (*b*) Avoidance of crossing of exact curves.

[12] The noncrossing rule is based on the principle that two states of the same symmetry cannot be made degenerate by varying a single parameter (the intermolecular distance). When there is more than one independent parameter, however, one can show that such a crossing is possible; but the two surfaces then intersect in a double cone, meeting at a single point and "repelling" each other elsewhere. Of course, crossings of states with *different* symmetry are unimpeded in either case.

exchange reaction,

$$H_A + H_B H_C \rightarrow H_A H_B + H_C.$$

The surface is, of course, symmetric for all values of the angle $\theta = \angle ABC$, with two channels connected by a saddle point at $R_{AB} = R_{BC}$. The height of the barrier on the minimum-energy path, however, varies from about 0.5 eV for $\theta = 180°$ (the linear configuration) to 1.2 eV for $\theta = 90°$ and 2.8 eV for $\theta = 60°$; in this case, therefore, reaction is most likely to occur for near-linear configurations.[13] In general, however, calculations of this sort are too difficult for present-day quantum chemistry, and one must rely on chemical intuition to predict the probable configurations for reaction.

The interactions between isolated atoms and molecules are complicated enough; what happens in condensed phases, where each molecule interacts with many others? Consider, for example, what happens when a gas molecule strikes a solid surface. There are several possible cases. If the gas molecule has low kinetic energy and interacts weakly (van der Waals interaction) with the surface atoms, the only likely "reaction" is *physical adsorption*; an example is the adsorption of helium on tungsten. The adsorption is made possible by the transfer of the adsorbed molecule's kinetic energy of approach into many vibrational modes of the solid. With low-energy molecules that interact strongly with the surface, on the other hand, one can obtain *chemisorption*, in which the adsorbed molecule is actually chemically bonded to one or a few surface atoms; the dissociative adsorption of hydrogen on tungsten is of this type. Molecules with high kinetic energy are not likely to be adsorbed at all, but simply to rebound, seeing only the repulsive portion of the interaction potential; to a first approximation, one can treat such rebounding as an elastic collision with a single surface atom. The details of the interaction in all these cases depend on the specific substances involved. In many instances there seems to be little or no potential-energy barrier to adsorption, the interaction being strongly attractive; the only major difficulty is then the dissipation of the adsorbed molecule's energy.

Reactions can take place between molecules adsorbed on a solid surface. Although little is known about the interactions involved, the weaker interactions at least seem to be similar to those between the same molecules in the gas phase. The significant difference is that the adsorbed molecules are much less free to move; if they are held in favorable orientations relative to one another, this can have the effect of catalyzing reaction between them.

And many reactions take place in liquid solution, where a wide variety of interactions is possible. Some solutes interact very strongly with the solvent, others hardly at all; an obvious example of the former is given by ionic solutes in a polar solvent such as water. Where the solute–solvent interaction is weak, reactions between solute molecules are little different from the same reactions in gases: in dilute solution only two solute molecules are likely to collide at once, and if the barrier height for reaction is fairly high ($\gg k_B T$), the solvent interactions ($\approx k_B T$) should make little difference in the potential energy surface. The case of strong solute–solvent interaction is much more difficult to treat, since one must recalculate the potential energy surface. Finally, of course, we have the case in which the solvent is *itself* a reactant, and one cannot avoid considering many-body interactions. Similar problems make it difficult to analyze reactions in solids.

[13] We have discussed the same problem earlier (Section 8.1) in terms of calculating the molecular energy of H_3. Cf. Fig. 8.2c (energy as a function of θ for fixed $R_{AB} + R_{BC}$) and Fig. 8.3 (the linear potential energy surface).

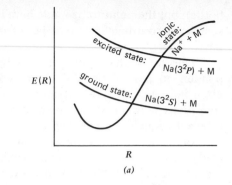

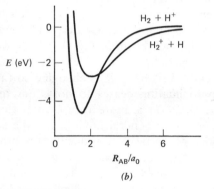

FIGURE 30.8
Some examples of curve-crossing reactions. (*a*) Schematic diagram of the sodium quenching process, in which the system proceeds from the excited state to the ground state via the intermediate ionic state. (For clarity the crossing points are shown unmodified.) (*b*) Limiting curves for $H_2^+ + H$ and $H_2 + H^+$, each with the third atom far distant; in the complete H_3^+ system, the crossing point at $2.5a_0$ becomes a "seam" between two potential energy surfaces (see text).

Having described the interactions between reactive molecules in some detail, we turn now to the dynamics of collisions between such molecules. In Section 30.1 we defined "dynamics" as the details of how the collision process leading to reaction occurs, and mentioned some of the questions this raises; now we must try to answer these questions. For example, we are interested in how the initial and final states of the reactant molecules—including the internal state (vibrational, rotational, electronic) as well as the kinetic energy—affect the reactivity. Another significant question is how long the colliding molecules remain in each other's vicinity, relative to such characteristic times as vibrational or rotational periods; it turns out that one can classify reaction mechanisms into two distinct types, depending on whether the duration of the encounter is long or short. What is the distribution of energy, both during the collision and in the products? Which degrees of freedom take part in the reaction, and which do not? These and other questions are of interest.

In principle the answers to such questions are to be found in quantum mechanics, say by solving the (time-dependent) Schrödinger equation for the entire reacting system. We have seen that much can be learned in this way about the structures of stable molecules, and it is possible to extend the method to collisions for relatively simple cases. Yet calculations of this type are not only difficult, but often unnecessary. To an excellent approximation one can treat collisions classically, in terms of trajectories on a potential energy surface. That is, given a surface that describes the interaction between reactive molecules for all possible relative positions, we wish to examine the trajectories on this surface that correspond to the course of actual collisions. We shall as usual consider primarily binary collisions between gas molecules.

We have earlier defined the three types of molecular collisions: *elastic*, in which only the translational motion of the molecules is altered by the collision; *inelastic*, in which the internal state (vibrational, rotational, electronic) of one or both molecules is changed but the molecules themselves remain the same; and *reactive*, in which the chemical species present are not the same before and after collision. Depending on the initial conditions, any of these may occur in an encounter between a given pair of molecules. We shall illustrate this with the potential surface for the F–H–H system described in the previous section.

Consider the collision of a H atom with an HF molecule in its ground vibrational state. (Since we shall again limit ourselves to linear encounters, we necessarily neglect rotational motion.) On the potential energy surface of Fig. 30.5, this system is described initially by a point moving inward along the left channel. If the initially available (kinetic or internal) energy is less than the minimum necessary for reaction, then only elastic or inelastic collisions can occur. As we have seen, the barrier height from the H + HF channel is 36.1 kcal/mol (1.57 eV); thus any collision with an initial energy less than this value should not lead to reaction (if we neglect the possibility of tunneling).

In Fig. 30.9a the heavy solid line shows a possible trajectory for an elastic collision, with the HF molecule having the same vibrational level in the initial and final states. The broken line in Fig. 30.9a shows a possible inelastic-collision trajectory for the same initial kinetic energy. We assume that a quantity of translational energy is converted to vibrational energy of the HF molecule, which thus recedes from the collision with an increased amplitude of vibration. Both these trajectories are for an energy insufficient for reaction. Next consider a collision with initial energy greater than 36.1 kcal/mol; reaction may now occur, as shown by

30.3
Collisions Between Reactive Molecules

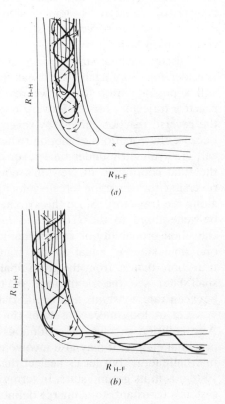

FIGURE 30.9
Some possible trajectories (schematic) in a collision between HF and H. (a) Insufficient energy for reaction: elastic (——) and inelastic (———) collisions. (b) Sufficient energy for reaction: reactive (——) and elastic nonreactive (———) collisions.

the solid-line trajectory in Fig. 30.9*b*. However, not every collision with sufficient energy for reaction in fact leads to reaction. Elastic and inelastic collisions are still possible, the former illustrated by the broken line in Fig. 30.9*b*. (Note that whether reaction occurs largely depends on the phase of the HF molecule's vibration when it approaches the barrier.) On some potential surfaces it is even possible for a trajectory to cross the barrier and rebound into the original channel. These examples should make it clear that more than energy is needed for a successful reaction; one of the kineticist's tasks is to determine the reaction probability as a function of energy.

We can see that, when two (or more) reactive molecules collide, many final states are possible. In the example just discussed, the collision of H and HF, the final state may be (1) H and HF in their initial states, (2) H and HF with the latter in a different vibrational state, or (3) H_2 and F with the H_2 in any of several vibrational states—and we have not touched on the possibility of rotational or electronic transitions. Where reaction occurs, more than one product channel may be available; one example is the collisional dissociation

$$Ar + CsBr \rightarrow \begin{cases} Ar + Cs + Br \\ Ar + Cs^+ + Br^- \end{cases},$$

in which the ionic reaction is dominant. (Given sufficient kinetic energy, the $H + HF$ reaction can also produce $H + H + F$.) Thus a complete analysis of a reactive system must include the distribution of products over *all* the accessible final states, subject to the constraints of the conservation laws.

Discussing final states brings us to another important point. If we consider Fig. 30.9 again, it is clear that any of the trajectories shown is still a possible trajectory if we reverse the arrows. In particular, any reactive trajectory for the forward reaction can be converted into one for the reverse reaction merely by reversing its direction. This must be so, because the equations of motion (whether classical or quantum-mechanical) are symmetric under time reversal. Classically, changing the sign of the time coordinate in Newton's equations of motion is equivalent to reversing the direction of all velocities, and thus to reversing motion along the trajectory. Since this symmetry holds for *all* trajectories, it can be generalized to the very useful *law of microscopic reversibility*: The transition probability of any mechanical system from one initial state to one final state is equal to the transition probability for the reverse transition, that is, from the given final state to the given initial state. We shall later see the significance of this principle for the relationship between rate constants and equilibrium constants.

Let us look more closely at the energy requirements for reaction. Any collision involves a continuous interchange between kinetic and potential energy; it may also involve exchange between translational and internal energy, but let us neglect this for the moment. Assume that the system is in its ground state. In terms of the potential energy surface, the system's (constant) total energy defines a plane intersecting the surface at a fixed "height" (distance along the energy axis); the potential energy is the height of the surface at a given point; and the kinetic energy is the height of the total-energy plane above the surface at that point. Regions of the surface higher than the total-energy plane are classically inaccessible, corresponding as they do to negative kinetic energy. In the accessible regions, the kinetic energy—and thus the speed of the point representing the system—decreases as the potential energy increases. In particular, the higher the barrier to reaction, the more slowly systems can cross that barrier; or in macroscopic terms, the higher the barrier, the smaller the rate constant. Things are not quite that simple, however.

We must in fact distinguish several energy quantities that are often confused. The meaning of the *barrier height* should be clear by now: It is simply the energy difference between the saddle point of the potential surface (the maximum potential energy on the reaction coordinate) and the ground state of the separated reactants, as shown in Fig. 30.6. If we designate the barrier heights for the forward and reverse reactions as $\Delta E_f{}^*$ and $\Delta E_r{}^*$ respectively, we obtain the simple relationship

$$\Delta E_f{}^* - \Delta E_r{}^* = [E(\text{saddle point}) - E^0(\text{reac})] - [E(\text{saddle point}) - E^0(\text{prod})]$$

$$= E^0(\text{prod}) - E^0(\text{reac}) = \Delta E^0_{\text{reac}}, \qquad (30.22)$$

where E^0 designates the ground-state energy, and ΔE_{reac} is the energy of reaction; this should also be obvious from Fig. 30.6. The *threshold energy*, an experimental quantity, is the minimum initial relative kinetic energy actually necessary for reaction. The threshold energy is sometimes found to be less than the barrier height ΔE^*, which can be attributed to tunneling through the barrier. Frequently, however, the threshold energy exceeds ΔE^*; this illustrates the point that merely satisfying the energy requirements is not sufficient for reaction to occur, as we have already seen from Fig. 30.9. Finally, the *activation energy* E_a is an empirical parameter derived from the temperature dependence of the rate constant, which for many reactions is approximately proportional to e^{-E_a/k_BT} (where k_B is the Boltzmann constant and T is the absolute temperature); we shall discuss this in Sections 30.5 ff. Some writers apply the term "activation energy" loosely to all of these quantities[14], largely because certain simple theories predict that they *should* be the same; but this is an oversimplification.

Let us now remove the limitation that the system be in its ground state. We then have not only the continuous interchange between kinetic and potential energy, but the possibility of transitions between internal energy states. The total energy must remain constant, but its "height" above the ground-state potential energy surface in general has both translational (kinetic energy) and internal (electronic, vibrational, rotational) components. Usually one assumes a potential energy surface for each electronic state; as long as there are no electronic transitions, the translational and vibrational motion can be described by a trajectory on one such surface, as we have shown in Fig. 30.9. The inelastic collision of Fig. 30.9b is an example of an interchange between translational and vibrational energy (represented, respectively, by motion along and across the HF channel). It is clear that the possibility of internal transitions greatly complicates the analysis of reaction dynamics.

Thus far we have largely been defining terms. Now let us look at some of the details of chemical dynamics—specifically, at the measurable characteristics of reactive collisions and the conclusions that can be drawn from these measurements. Among the most suitable measurements for this purpose are scattering experiments. With suitable instrumentation one can determine the angular distribution of the product molecules, their kinetic energies, and even the distribution over internal states. What can such measurements tell us?

Consider first the angular distribution of scattered products. Figure 30.10 is an example of such a measurement, for the reaction

$$N_2{}^+ + D_2 \rightarrow N_2D^+ + D.$$

[14] And to the quantity in the transition-state theory (Section 30.6) which we shall call the *energy of activation*, $\Delta E^\ddagger$.

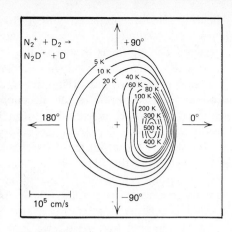

FIGURE 30.10

Intensity contour map for N_2D^+ formed in $N_2{}^+ + D_2$ collisions at 3.12 eV initial relative kinetic energy. The magnitude of the product velocity is indicated by distance from the origin ($+$). The asymmetry of the map (relative to the $\pm 90°$ line) indicates a direct-interaction mechanism. From B. H. Mahan, *Acc. Chem. Res.* **3**, 393 (1970).

The N_2^+ ions are prepared with a given kinetic energy (3.12 eV in the figure) and scattered from a cross beam of D_2 molecules at thermal energies. The figure is drawn in the center-of-mass coordinate system: the N_2^+ ions are initially moving in the direction designated as $0°$; the reactants collide at the origin (marked by a cross), and the resulting products are scattered in all directions. In this case the intensity of scattered N_2D^+ was measured as a function of both angle and speed; the results were converted from laboratory to center-of-mass coordinates, and plotted as a contour map. The scattering angle is represented by the polar angle, and the final relative speed by distance from the origin, so that each point represents the magnitude and direction of a given product velocity; the contours give the relative probability of such velocities, each contour line being a locus of constant N_2D^+ intensity.

In examining Fig. 30.10, we look first at the angular distribution, which is very asymmetric. Although some products are found at all scattering angles, the intensity of N_2D^+ clearly has a maximum in the forward (0°) direction, the direction in which the initial N_2^+ beam was moving. This immediately suggests that in attaching a D atom the N_2^+ ion essentially retains its initial velocity—that is, that the collision is basically a grazing one, in which the heavy N_2^+ ion is only slightly deflected by the light D atom. Such a collision must be of short duration, with the bond rearrangement occurring very rapidly with respect to the nuclear motion. To see why this is so, suppose that the alternative occurred: that the reactants remained together for a relatively long time. This would be dynamically possible only if the fragments (initially N_2^+ and D_2) orbited each other for one or more revolutions—to put it another way, only if the aggregate $N_2D_2^+$ existed for a time longer than its own rotational period (say, 10^{-12} s). But the result of such orbiting would be the loss of information on *how* the reactants came together: When the aggregate eventually broke up, the products would be emitted at essentially random angles. The products would be distributed, if not isotropically, at least symmetrically in the forward and backward directions (symmetric about the ±90° line in a center-of-mass plot). But as we have seen in Fig. 30.10, what is actually observed is a strong asymmetry. We conclude that the effective lifetime of the $N_2D_2^+$ aggregate—which is the same as saying the duration of the collision—is in fact short compared to its rotational period. In other words, the reactants do not orbit around each other, but merely interact in passing.

A reaction such as that in Fig. 30.10 is said to take place by a *direct-interaction mechanism*. We can define this as one in which the duration of the collision (the time the reactants spend near each other in some region of the potential energy surface) is short, compared to typical rotational or vibrational periods in either the separate reactants or the interacting aggregate of reactants. If the duration of the collision is long by the same standards, we speak of a *complex-formation mechanism* and call the long-lived aggregate a *complex*. Although there are many reactions intermediate between these extremes, a number are known that fit well into one category or the other. And as our discussion has already suggested, the angular distribution of reaction products gives the best indication of where a given reaction fits on this scale.

We should now know what to expect in a reaction that occurs by complex formation. An example of such a reaction is

$$O_2^+ + D_2 \rightarrow DO_2^+ + D,$$

measurements for which are shown in Fig. 30.11. In this case we find that

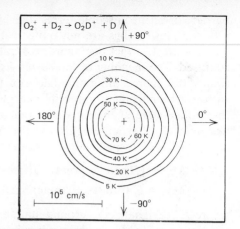

FIGURE 30.11
Intensity contour map for DO_2^+ formed in $O_2^+ + D_2$ collisions at 2.76 eV initial relative kinetic energy. The intensity is nearly isotropic, indicating the formation of a long-lived $D_2O_2^+$ complex. From B. H. Mahan, *Acc. Chem. Res.* **3**, 393 (1970).

the intensity of scattered DO_2^+ is virtually symmetric between the forward and backward directions, and in fact nearly isotropic. By the same reasoning as already outlined, we conclude that the $D_2O_2^+$ aggregate is a complex, lasting for a long time relative to its rotational period. Complex formation is most likely to occur in a reaction whose potential energy surface has a minimum in the strong-interaction region. The lifetime of the complex is particularly enhanced if the surface has a structure that inhibits departure from this region. As we noted in the previous section, however, the interactions in such a region are likely to be too complicated to describe in terms of a trajectory on a simple surface: for example, the rotations we have just been describing cannot be represented at all on a surface like Fig. 30.5, which assumes a linear configuration of the reactants.

There is much additional information that can be obtained from measurements such as we have been describing. One of the most obvious applications involves the energy balance. In each experiment the initial relative kinetic energy is known, and the final relative kinetic energy corresponding to any point on the contour map can easily be calculated. For example, the intensity peak in Fig. 30.10 corresponds to a kinetic energy 1.3 eV less than that of the initial beam. Since the reaction between N_2^+ and D_2 is exothermic by about 1 eV, and since energy must be conserved, we conclude that (in the reaction's most probable outcome) about 2.3 eV has been converted into internal energy of the products. (Unfortunately, one cannot determine from these measurements just *which* form of excitation this energy appears as.) In Fig. 30.11, where the intensity peak is at the origin, the entire initial kinetic energy *and* exothermicity must be converted into internal energy. With sufficiently high initial energy, the intensity maps change from peaks to craters, as in Fig. 30.12: Low final kinetic energies of N_2D^+ become impossible, because the corresponding internal energy would be higher than the dissociation limit.

It is clearly desirable to learn more about the distribution of energy in reaction products; fortunately, there are other types of measurements that yield such detailed information. As already noted, the total energy available to the products consists of the initial relative kinetic energy of the reactants plus the exothermicity of the reaction. This energy may appear in the products as either translational, rotational, vibrational, or electronic energy. The distribution of the internal energy can best be determined by taking spectra at high speed, so as to observe the state of the products before relaxation can occur; the relative populations of the various vibrational and rotational states are obtained from the intensities of the spectral lines. The way of carrying out such measurements is to work with a beam of one reactant directed into a low-density gas of the other reactant. Low densities are necessary to minimize collisional quenching so that the emission spectra of excited states produced in the reaction may be observed.

When electronic excitation can be neglected, the energy distribution is commonly displayed in a "triangular plot" like those of Fig. 30.13. The ordinate is the product vibrational energy, the abscissa the product rotational energy; since the sum of vibrational, rotational, and translational energy is constant (equal to the total available energy), the distance of a given point from the diagonal line

$$E_{vib} + E_{rot} = E_{total}$$

yields the translational energy. The relative amounts of various states are again indicated by contours. For example, Fig. 30.13*a* is the triangular

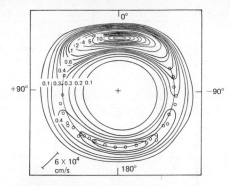

FIGURE 30.12
Intensity contour map for N_2D^+ formed in $N_2^+ + D_2$ collisions at 8.1 eV initial relative kinetic energy. As in Fig. 30.10, the map is asymmetric about the $\pm 90°$ line; however, the high initial energy produces a "crater" about the origin. The small circles are experimental intensity maxima. From B. H. Mahan, *Acc. Chem. Res.* **1**, 217 (1968).

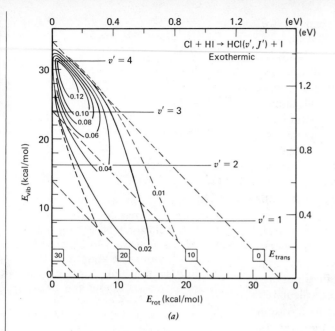

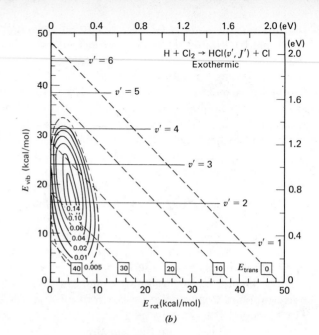

FIGURE 30.13
Triangular plots for the reactions (a) Cl+HI, (b) H+Cl₂. In each case the ordinate and abscissa, respectively, give the vibrational and rotational energy of the products; the broken diagonal lines give the translational energy on the same scale. The contours show the relative intensity of the various states observed. (The horizontal solid lines indicate the amounts of energy corresponding to the various product vibrational levels.) From D. H. Maylotte, J. C. Polanyi, and K. B. Woodall, *J. Chem. Phys.* **57**, 1547 (1972); K. G. Anlauf, D. S. Horne, R. G. Macdonald, J. C. Polanyi, and K. B. Woodall, *ibid.*, 1561.

plot for the reaction

$$Cl + HI \rightarrow HCl + I,$$

showing the distribution of internal energy in HCl; the total available energy is 34 kcal/mol. The peak of intensity shows that nearly all this energy is likely to appear as vibration, primarily in the levels $v = 3$ and $v = 4$. In contrast, Fig. 30.13b for the reaction

$$H + Cl_2 \rightarrow HCl + Cl$$

(available energy 48.4 kcal/mol) has a peak corresponding to slightly more translational than vibrational energy; in neither case is the amount of rotational energy large. Similar analyses have been made for many reactions.

In the reactions of Fig. 30.13 and some other cases, the products are formed primarily in excited states. For example, in the reaction

$$F + D_2 \rightarrow FD + D$$

the highest proportion of the product FD is found in the vibrational state $v = 3$, with decreasing amounts in the states $v = 2$, $v = 1$, $v = 0$. In other words, immediately after reaction (and before any further inelastic collisions) the FD has a highly nonequilibrium, i.e., non-Maxwell–Boltzmann, distribution. Especially at low pressures (where few collisions occur), this state will tend to decay by emission of radiation; we have already noted the use of this emission to measure the relative populations of excited states. This process, the spontaneous emission of radiation from excited states produced by reaction, is known as *chemiluminescence*, as seen in most flames.

A different type of emission from excited states is utilized in the *chemical laser*. A laser (from "*l*ight *a*mplification by *s*timulated *e*mission of *r*adiation") is a device in which an intense beam of coherent monochromatic radiation[15] is produced by transitions to a lower-energy state.

---

[15] "Coherent" means here that the radiation is in phase over the entire width of the beam (which thus does not spread appreciably); "monochromatic" means that the radiation is essentially of a single frequency.

If two states are separated by an energy $h\nu$ and the upper state is sufficiently populated, then radiation of frequency ν can induce transitions from the upper to the lower state; this releases still more radiation of frequency ν, and with suitable optics (reflecting the emitted light back and forth through the sample) one can trigger a massive transition to the lower state accompanied by intense emission at frequency ν. But to produce this effect, one must first have the higher population in the upper state, a so-called population inversion. In physical lasers the population inversions are produced by irradiation, electric discharge, and so on; in a chemical laser the same effect is produced by chemical reaction, as just described. For example, it should be possible to produce laser action from the $F + D_2$ reaction mixture with radiation corresponding to such transitions as $(v = 3) \rightarrow (v = 2)$, or $(v = 2) \rightarrow (v = 1)$; in fact, about a dozen vibrational-rotational transitions have been observed in each of these bands. Laser action has been produced in most of the well known diatomic and triatomic gases. At first it was necessary to initiate the reactions in chemical lasers with external energy (as in physical lasers), but in recent years lasers driven completely by chemical energy have been developed.

Lasers can also be used to study the excited states produced by reactions. In the older chemiluminescence technique we described above, one is limited to observing the radiation spontaneously emitted from these excited states, some of which will decay by other means. Faster and more accurate results can now be obtained by the method of *laser-induced fluorescence*: One shines a tunable laser (one in which the beam frequency can be varied) upon the reaction products, varying the frequency across an electronic absorption band; each vibrational-rotational state in the sample is excited in turn, and the intensity of the subsequent fluorescence is measured. If the spectrum is known, each of the resulting fluorescence lines can be correlated with the state excited to produce it, and the relative populations of these states can be calculated.

Let us consider the $F + D_2$ reaction again, and suppose that we wish to carry out the reverse reaction,

$$FD + D \rightarrow F + D_2.$$

Assume that the same total energy is available as in the forward reaction we have described. If we want the $FD + D$ reaction to proceed with the highest probability, where should we put this available energy? If we put it all into relative translation (kinetic energy), we would be tracing the reverse of the reaction

(1) $$F + D_2 \rightarrow FD(v = 0) + D.$$

Alternatively, if we put as much of the available energy as possible into vibration, we would be tracing the reverse of, say,

(2) $$F + D_2 \rightarrow FD(v = 3) + D.$$

At this point the principle of microscopic reversibility comes into play: We know from experiment that the *forward* reaction (2) is much more probable than (1). But by microscopic reversibility a given reaction between specified states has the same probability as its reverse. Therefore the *reverse* of reaction (2) must also be much more probable than the reverse of (1), and the reverse reaction will go most rapidly when the FD molecules are vibrationally excited. This effect is known as *vibrational enhancement*, and can be remarkably large in some cases. In one extreme example, the reaction

$$HCl + Br \rightarrow HBr + Cl,$$

the rate constants for HCl in the states $v = 0$, $v = 1$, $v = 2$ have been found to be, respectively, 1×10^{-2}, 2×10^4, and 1×10^9 liter/mol s.

How can we understand a requirement like this, that the energy be localized in one specific degree of freedom to make reaction likely, in terms of potential energy surfaces? We have already hinted at the answer in our discussion of Fig. 30.9b, which we shall now consider again (the surface for FD + D is, of course, similar to that for HF + H). The reactive trajectory in that figure is in fact one with a large vibrational amplitude, and can easily make the turn into the output channel if the vibration has the correct phase (i.e., if the H and F atoms in the reactant HF molecule are moving apart just as the HF collides with the initially free H atom). In contrast, a system with only translational energy would be likely to come down the center of the input channel and rebound (probably inelastically), rather than making the turn into the output channel. And conversely, a reactive trajectory in the opposite direction (H$_2$ + F) can hardly avoid producing a vibrationally excited HF molecule.

In the examples just described, the products are formed in non-equilibrium distributions as the result of a direct-interaction mechanism. To what extent does this occur in reactions that proceed by complex formation? A number of theories of such mechanisms assume that the complexes are *statistical*—that is, that equilibration within a long-lived complex is sufficiently rapid to produce a Maxwell–Boltzmann distribution over the energy levels of the complex. Studies of this question have been made for such reactions as

$$F + H_2C{=}CH_2 \rightarrow H_2C{=}CHF + H,$$

by both molecular-beam and chemiluminescence techniques. The reaction clearly involves a long-lived complex, since the intensity contour map for C$_2$H$_3$F is symmetric about 90° (cf. Fig. 30.11). But the energy distribution of product molecules implies that only 5 of C$_2$H$_3$F's 12 vibrational modes are activated, that is, that the complex is nonstatistical. The reason apparently lies in the geometry of the complex: C$_2$H$_4$ and C$_2$H$_3$F both being planar, there are only certain angles at which F can approach and H leave. In contrast, more nearly statistical behavior is observed in the reaction

$$F + H_2C{=}CHCH_3 \rightarrow H_2C{=}CHF + CH_3,$$

in which the complex has a higher density of energy levels. The complete interpretation of these classes of reactions is difficult, but it is clear that a nonequilibrium distribution can be produced even from a long-lived complex.

Still another way to produce nonequilibrium distributions is in photofragmentation reactions, in which a molecule is brought to an activated state (or complex) by absorption of a photon and then splits into fragments. Since only photons of particular wavelengths will be absorbed, the activated state must initially have a nonstatistical energy distribution; whether or not this is retained in the reaction products depends on how rapidly the energy is redistributed compared to the lifetime of the complex. This behavior makes it interesting to carry out "laser-induced chemistry": using a laser, the radiation from which is of a specific wavelength only, one can selectively excite a particular energy level of a reactant molecule and thereby follow the ensuing reaction with much greater specificity than with conventional methods of excitation.

Earlier in this section we discussed the symmetry between dissociation and recombination reactions. It is clear that in any dissociation reaction the energy for bond breaking has to come from somewhere:

translational energy from a collision, internal excitation (perhaps resulting from a previous collision), or irradiation. If we count electrons and photons as "third bodies" (admittedly not the usual terminology), then all dissociation reactions must involve third-body collisions at some stage. And by the principle of microscopic reversibility, the recombination reactions inverse to these must also be third-body processes in this wider sense. For example, the iodine recombination reaction $(2I \rightarrow I_2)$ must involve a third body to take the energy away; but what are the details of this process? It is highly improbable (at least in a gas) for three molecules to reach the same point simultaneously; thus we seek a mechanism in which two molecules can join together and wait for the third to arrive. The only obvious way in which this can occur is for the first two atoms to form a complex orbiting each other. What happens after the third body does show up? It is most likely that such a collision will be glancing and remove only a little of the required energy; the result is the formation of I_2 in a highly excited state. The fate of such an excited product depends on the next few collisions: vibrational-translational or rotational-translational interactions may remove additional energy and thus stabilize the molecule; or enough energy may be added to dissociate it again. One finds that dissociation and recombination reactions are both best treated in terms of "ladder-climbing" models: a series of collisions in each of which only a small amount of energy is transferred, moving the system up or down the "ladder" of internal states.

The last two sections dealt with some aspects of experimental studies of reaction dynamics. We have seen that a great deal can be learned from experiment about the details of molecular interactions and collisions. As always, however, one would like to be able to obtain the same results *a priori*. In principle this would involve solving the complete time-dependent Schrödinger equation for the reacting system. Although one can go a long way in this direction for simple systems (such as $H + H_2$), in practice one must usually rely on simplified models. In this section we shall begin with the simplest model of all, that of hard-sphere collisions.

We discussed the collision dynamics of hard spheres in Section 27.3 (for elastic collisions), but we need to go a bit farther now. Suppose that we represent each reactant molecule, say the HF and H of Section 30.2, by a hard sphere. Upon impact, which for hard spheres is instantaneous, we assume that there is an immediate transformation into two new hard spheres representing H_2 and F molecules. Both before and after the impact there are no interactions between the molecules in this model. In view of all we have said about the dynamics of real molecular collisions, it is clear that the model is drastically oversimplified; let us nevertheless see what can be deduced from it.

Consider a general reaction

$$A + B \rightarrow C + D.$$

If the diameters of the hard spheres representing the molecules are d_A, d_B, and so on, then no collision between A and B can occur unless the impact parameter b is less than the average diameter

$$d = \tfrac{1}{2}(d_A + d_B) \tag{30.23}$$

30.4
Hard-Sphere Collision Theory: Reactive Cross Sections

(cf. Fig. 27.8). At the point of impact we have the relationship

$$b = d \cos \frac{\chi}{2}, \qquad (30.24)$$

where χ is the scattering angle in relative coordinates (Fig. 27.7); this corresponds to Eq. 27.27 for the case $d_A = d_B$. We also have the equation

$$V(d_{AB}) = E\left(1 - \frac{b^2}{d^2}\right), \qquad (30.25)$$

corresponding to Eq. 27.29, which connects the potential energy at the point of impact with the initial relative kinetic energy E.

The potential energy surface is, of course, greatly simplified in this model. We assume that the potential energy is constant both before and after impact, with values corresponding to the ground-state energy of reactants and products, respectively; at the point of impact we insert an impenetrable barrier of finite height and zero width. The reaction coordinate thus has the form shown in Fig. 30.14, similar to the schematic representation of Fig. 30.6b. The barrier serves to put some measure of realism into the model. We now assume that reaction is possible only when the potential energy at the point of impact equals or exceeds a threshold energy denoted by ΔE_f^* (for the forward reaction). The value of ΔE_f^* corresponds to the maximum energy along the reaction coordinate on a more realistic potential energy surface, measured relative to the ground state of the separated reactants. All the quantities shown in Fig. 30.14 thus correspond to those defined in connection with Fig. 30.6.

We assume, then, that reaction can occur only when $V(d) \geq \Delta E_f^*$. But what is the *probability* of reaction for collisions that satisfy this criterion? As we saw in discussing Fig. 30.9, even collisions with sufficient energy do not always lead to reaction, and our model must reflect this fact. Clearly, the simplest possible assumption is that the probability of reaction is constant for $V(d) > \Delta E_f^*$. This need not be true, since the probability may vary with the amount by which the available energy exceeds the threshold (and experiment shows that this does happen). For the moment, however, let us adopt a simple model with a constant reaction probability p; that is, we assume that of all collisions with sufficient energy, only the fraction p lead to reaction.

Given these assumptions, we can calculate the cross section for reactive collisions of hard spheres. The maximum impact parameter that can lead to reaction, which we call b^*, is the value of b for which $V(d) = \Delta E_f^*$; solving Eq. 30.25 for b, we obtain

$$b^* = d\left(1 - \frac{\Delta E_f^*}{E}\right)^{1/2}. \qquad (30.26)$$

That is, all impact parameters in the range $b^* < b \leq d$ can contribute only to elastic scattering, whereas values in the range $b \leq b^*$ can lead to reaction with a probability $P(b, E)$; we assume $P(b, E)$ to have the constant value p. As we showed in Eq. 27.69, the differential cross section for elastic scattering of nonreactive hard spheres is simply $\sigma(v, \chi) = d^2/4$. Under our present assumptions, the differential *reactive* cross section is simply this value multiplied by the probability of reaction:

$$\sigma_R(v, \chi) = \frac{pd^2}{4} \qquad (b \leq b^*)$$

$$= 0 \qquad (b > b^*). \qquad (30.27)$$

We are more interested in the *total* reactive cross section, which is the integral of the differential reactive cross section over all impact parame-

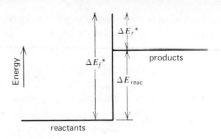

FIGURE 30.14
Reaction coordinate for the hard-sphere model; the barrier is of zero thickness. Here ΔE_{reac} is the energy change in the (forward) reaction, whereas ΔE_f^* and ΔE_r^* are the barrier heights for the forward and reverse reactions, respectively.

ters; by analogy with Eq. 27.70, we have[16]

$$\sigma_R(E) = 2\pi \int_0^\pi \sigma_R(v, \chi) \sin \chi \, d\chi$$

$$= 2\pi \int_0^{b^*} \frac{\sigma_R(v, \chi)}{\sigma(v, \chi)} b \, db = 2\pi \int_0^{b^*} pb \, db \qquad (30.28)$$

$$= 2\pi p\left(\frac{b^{*2}}{2}\right) = \pi p d^2 \left(1 - \frac{\Delta E_f^*}{E}\right).$$

Note that, since $E \geq V(d)$ by Eq. 30.25 and we must have $V(d) \geq \Delta E_f^*$ for reaction, both $\sigma_R(v, \chi)$ and $\sigma_R(E)$ must vanish whenever $E < \Delta E_f^*$. We shall later use Eq. 30.28 to derive the temperature dependence of reaction rates. Before we can do this, however, we must obtain the general relationship between reaction rates and cross sections; the next section is devoted to this problem.

Although we are thus not yet ready to apply the hard-sphere model to calculating reaction rates, we can give a comparison with experiment at this stage. It is possible to evaluate cross sections directly from molecular-beam measurements, and the results of such measurements for the T–H₂ reaction are shown in Fig. 30.15. Two output channels are observed, the abstraction reaction

$$\text{T} + \text{H}_2 \rightarrow \text{HT} + \text{T}$$

and the dissociation reaction

$$\text{T} + \text{H}_2 \rightarrow \text{T} + \text{H} + \text{H}.$$

It can be seen from the figure that only the dissociation reaction has a cross section like that predicted by Eq. 30.28 (zero below a threshold value, rising to a constant value at high E); the abstraction reaction has a quite different energy dependence. We shall see that this comparison is indicative: Some reactions are well described by the hard-sphere model, but many more are not.

Nevertheless, the use of the hard-sphere model has been very extensive. A caution is therefore in order with respect to the model's basic limitation, which can be seen clearly *a priori*: that it represents the inherently more-than-two-particle process of a chemical reaction by a two-body collision. Yet this same simplicity is the model's great virtue, for treatments attempting to take full account of the interactions and trajectories of three or more particles are generally quite complex. We shall thus have to see to what extent the accuracy of the results vindicates the simplicity of their calculation.

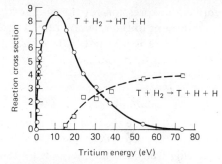

FIGURE 30.15
Total reaction cross sections for the T–H₂ reaction. (The cross sections are given in "atomic units," that is, in terms of a_0^2, where a_0 is the Bohr radius, 0.529 Å.) From J. Dubrin, *Ann. Rev. Phys. Chem.* **24**, 97 (1973).

30.5
Hard-Sphere Collision Theory: The Rate Coefficient

We are now ready to develop a theoretical expression for the rate of a chemical reaction, that is, an expression giving the rate constant k in terms of microscopic quantities. The method we introduce first is parallel to that used for physical transport processes in Chapter 28; thus it uses kinetic theory to obtain the frequency of collisions between reactive molecules, and is valid only for dilute gases. As we have seen, not all such collisions actually lead to reaction, so we must use some model to estimate the fraction of reactive collisions; in this section we use the simple hard-sphere model just developed. From these principles we can

[16] By Eq. 27.66, we can make the substitution $\sigma(v, \chi) \sin \chi \, d\chi = b \, db$.

derive an expression for the rate at which reactive collisions occur, which is equivalent to the reaction rate. This expression should indicate how the reaction rate is affected by temperature and other variables.

We consider the simplest possible case, a bimolecular elementary reaction

$$A + B \rightarrow C + D,$$

in a gas so dilute that we need consider only binary collisions. We further assume that each species has available only one internal quantum state, so that only translational energy is present. (Note that we are speaking of a single *elementary* reaction; in a multistep mechanism, one would have to obtain a separate rate expression for each step.) In Section 28.2 we derived an expression for the frequency of elastic collisions in such a dilute gas; we shall now see how this can be modified to apply to reactive collisions.

As before, we designate the number density of species A by n_A, and the probability density of A velocities by $f_A(\mathbf{v}_A)$. Although we have not included t explicitly in the notation, both these quantities may be functions of time; this is surely the case for n_A, since the A molecules disappear by reaction. But as for $f_A(\mathbf{v}_A)$, we shall see later that the Maxwell–Boltzmann distribution is essentially unaffected by the occurrence of reaction in most cases; we can therefore assume that $f_A(\mathbf{v}_A)$ has the Maxwell–Boltzmann form, Eq. 28.35, and is invariant with time. Basically this amounts to assuming that the chemical reaction rate is much smaller than the rate at which the velocity distribution reaches equilibrium.

As usual, we shall obtain the collision rate in terms of a scattering experiment like Fig. 27.13. That is, we suppose that beams of the reactants A and B collide within a volume τ, and that the flux of the product C can be measured as a function of the scattering angles (α, β in laboratory coordinates or χ, ϕ in center-of-mass coordinates) and the initial velocities $\mathbf{v}_A, \mathbf{v}_B$. The relationship between these quantities defines a reactive cross section σ_R, like that introduced in Eq. 27.59. In our present model the reactants and products each have only one internal state, so the reactive cross section can depend only on the initial relative speed $v \equiv |\mathbf{v}_A - \mathbf{v}_B|$ (or the energy $E = \frac{1}{2}\mu v^2$) and on the scattering angles. We can thus *define* a reactive cross section $\sigma_R(v, \alpha, \beta)$ by the equation

$$dN_C(\alpha, \beta, \mathbf{v}_A, \mathbf{v}_B) = \sigma_R(v, \alpha, \beta)v n_A n_B f_A(\mathbf{v}_A) f_B(\mathbf{v}_B)\, d\mathbf{v}_A\, d\mathbf{v}_B\, \tau\, d\Omega', \tag{30.29}$$

where $dN_C(\alpha, \beta, \mathbf{v}_A, \mathbf{v}_B)$ is the number of C molecules arriving per unit time at a detector of solid angle $d\Omega'$ (at laboratory angles α, β) due to collisions within the volume τ of A and B molecules in the velocity ranges $\mathbf{v}_A$ to $\mathbf{v}_A + d\mathbf{v}_A$, $\mathbf{v}_B$ to $\mathbf{v}_B + d\mathbf{v}_B$. This differs from Eq. 27.59 in the omission of internal quantum states and in the generalization from velocity-selected beams to a range of possible velocities.

In a macroscopic rate experiment we measure the *total* rate of appearance or disappearance of a chemical species, not the amount due to collisions at particular angles or energies. Since we wish to obtain a result comparable to a macroscopic rate equation, we must integrate Eq. 30.29 over all laboratory angles ($d\Omega' = \sin \alpha\, d\alpha\, d\beta$) and all values of $\mathbf{v}_A$, $\mathbf{v}_B$. We thus obtain the total flux of product molecules due to collisions in the volume τ,

$$dN_C = \int_{\alpha, \beta} \int_{\mathbf{v}_A} \int_{\mathbf{v}_B} dN_C(\alpha, \beta, \mathbf{v}_A, \mathbf{v}_B), \tag{30.30}$$

with the dimensions of molecules per unit time. Dividing this expression by τ should give us the number of reactive collisions producing C per unit volume per unit time; but this quantity is the same as the rate of change of concentration (number density) of C due to reaction in the forward direction,

$$\left(\frac{dn_C}{dt}\right)_f = n_A n_B \int_{\alpha,\beta} \int_{\mathbf{v}_A} \int_{\mathbf{v}_B} \sigma_R(v, \alpha, \beta) v f_A(\mathbf{v}_A) f_B(\mathbf{v}_B)\, d\mathbf{v}_A\, d\mathbf{v}_B\, d\Omega'. \tag{30.31}$$

Note the similarity of this equation to Eq. 28.20, which gives the collision frequency for elastic scattering. We can simplify by introducing the total reactive cross section, here

$$\sigma_R(v) = \int_{\alpha,\beta} \sigma_R(v, \alpha, \beta)\, d\Omega', \tag{30.32}$$

to obtain the theoretical rate equation

$$\left(\frac{dn_C}{dt}\right)_f = n_A n_B \int_{\mathbf{v}_A} \int_{\mathbf{v}_B} \sigma_R(v) v f_A(\mathbf{v}_A) f_B(\mathbf{v}_B)\, d\mathbf{v}_A\, d\mathbf{v}_B. \tag{30.33}$$

The resemblance of Eq. 30.33 to a second-order rate expression like Eq. 30.12 is obvious, and can be emphasized by rewriting it in the form

$$\left(\frac{dn_C}{dt}\right)_f = \kappa_f n_A n_B, \tag{30.34}$$

where the theoretical rate coefficient κ_f is defined by

$$\kappa_f \equiv \int_{\mathbf{v}_A} \int_{\mathbf{v}_B} \sigma_R(v) v f_A(\mathbf{v}_A) f_B(\mathbf{v}_B)\, d\mathbf{v}_A\, d\mathbf{v}_B. \tag{30.35}$$

These equations give the fundamental form for the rate of a bimolecular elementary reaction, and κ_f should be proportional to the experimental rate constant for the corresponding reaction step. Although our derivation applies only to dilute gas reactions, it turns out that the same form of rate expression empirically describes many kinds of reactions in gases, liquid solutions, and even solids.

Thus we have obtained, as in so many previous cases, an expression connecting a macroscopically measurable quantity (the rate constant) with the statistical average of a molecular quantity (the reactive cross section). Equation 30.35 is thus analogous with our expressions for the transport coefficients in Section 28.8, and for equilibrium properties in Part Two. As always, to use such relationships we must assume some model for the molecular quantities. In this case we evaluate the rate coefficient for the simple model of reactive hard spheres developed in the previous section.

The model assumes, it will be recalled, that reaction occurs with a probability p when the initial relative kinetic energy E exceeds the threshold value ΔE_f^*, and not at all for $E < \Delta E_f^*$. These assumptions yield the total reactive cross section of Eq. 30.28, in which $\sigma_R(E)$ is equivalent to the $\sigma_R(v)$ of Eq. 30.35 (with $E = \frac{1}{2}\mu v^2$). For the probability densities $f_A(\mathbf{v}_A)$, $f_B(\mathbf{v}_B)$ we use the Maxwell–Boltzmann equilibrium velocity distribution of Eq. 28.35, assuming that both beams have the same temperature T. Do not confuse "equilibrium velocity distribution" with *chemical* equilibrium: One is a description of the relative numbers of molecules of a given species with different velocities, the other of the relative total concentrations of molecules of different species. As already noted, we assume that the chemical reaction does not significantly perturb the velocity distribution, which will be true whenever the rate of reaction is small compared with the rate of attaining velocity equilibrium. This is ordinarily the case, since all collisions can lead to exchange of translational energy, but only a small fraction of energetic collisions can lead to reaction. Few exceptions to this limiting assumption are known. When we generalize our analysis to molecules with internal degrees of freedom, we must also consider the effects of vibrational and rotational relaxation. If we substitute these expressions into Eq. 30.35, we obtain

$$\kappa_f = \pi p d^2 \left(\frac{m_A}{2\pi k_B T}\right)^{3/2} \left(\frac{m_B}{2\pi k_B T}\right)^{3/2}$$
$$\times \int_{\mathbf{v}_A} \int_{\mathbf{v}_B} v\left(1 - \frac{\Delta E_f^*}{E}\right) e^{-(m_A v_A^2 + m_B v_B^2)/2k_B T} d\mathbf{v}_A \, d\mathbf{v}_B, \quad (30.36)$$

where the integration is restricted to regions where $E > \Delta E_f^*$, the reactive cross section being zero elsewhere. The integral here closely resembles that in Eq. 28.53; applying the same transformations as were used to convert the latter to Eq. 28.57, we have

$$\kappa_f = \pi p d^2 \left(\frac{\mu}{2\pi k_B T}\right)^{3/2} \int_{\mathbf{V}} \left(\frac{m_A + m_B}{2\pi k_B T}\right)^{3/2} e^{-(m_A + m_B)V^2/2k_B T} \, d\mathbf{V}$$
$$\times \int_{\mathbf{v}} v\left(1 - \frac{\Delta E_f^*}{E}\right) e^{-\mu_{AB} v^2/2k_B T} d\mathbf{v}. \quad (30.37)$$

The integration over the center-of-mass velocity $\mathbf{V}$ yields unity, since we are using normalized probability distributions: cf. Eqs. 28.34 and 28.35. To integrate over the relative velocity $\mathbf{v}$, we first express the volume element $d\mathbf{v}$ in spherical coordinates:

$$\int_{\mathbf{v}} v\left(1 - \frac{\Delta E_f^*}{E}\right) e^{-\mu_{AB} v^2/2k_B T} d\mathbf{v}$$
$$= \int_{\theta=0}^{\pi} \int_{\phi=0}^{2\pi} \int_{v=(2\Delta E_f^*/\mu_{AB})^{1/2}}^{\infty} v^3 \left(1 - \frac{\Delta E_f^*}{E}\right) e^{-\mu_{AB} v^2/2k_B T} \sin\theta \, d\theta \, d\phi \, dv$$
$$= 4\pi \int_{v=(2\Delta E_f^*/\mu_{AB})^{1/2}}^{\infty} v^3 \left(1 - \frac{\Delta E_f^*}{E}\right) e^{-\mu_{AB} v^2/2k_B T} \, dv. \quad (30.38)$$

The lower limit on v corresponds to $E = \Delta E_f^*$, since for lower values of E no reaction can occur; we have integrated at once over the angles θ and ϕ. Next we change the variables from the initial relative speed v to the initial relative kinetic energy E,

$$E = \tfrac{1}{2}\mu_{AB} v^2, \qquad dE = \mu_{AB} v \, dv, \qquad (30.39)$$

so that Eq. 30.37 becomes

$$\kappa_f = \pi p d^2 \left(\frac{\mu_{AB}}{2\pi k_B T}\right)^{3/2} 4\pi \left(\frac{2}{\mu_{AB}^2}\right) \int_{\Delta E_f^*}^{\infty} (E - \Delta E_f^*) e^{-E/k_B T} \, dE.$$
$$(30.40)$$

Integration of this equation is straightforward and finally yields the result[17]

$$\kappa_f = \pi p d^2 \left(\frac{8k_B T}{\pi \mu_{AB}}\right)^{1/2} e^{-\Delta E_f^*/k_B T}. \qquad (30.41)$$

[17] This result applies when A and B are *different* species. If A and B are the same, we must divide by 2, since each collision is counted twice: Cf. the discussion leading to Eq. 28.62.

This equation can be given a simple interpretation. From Eqs. 28.58 and 28.59, the factor $\pi d^2 (8k_B T/\pi \mu_{AB})^{1/2}$ is the number of collisions per unit time per unit volume between hard-sphere molecules of A and B, for unit concentrations of A and B; this is simply the elastic cross section (πd^2) times the average relative speed. Only those collisions for which $V(d) \geqslant \Delta E_f^*$ are capable of leading to reaction, and our derivation is equivalent to a proof that $\exp(-\Delta E_f^*/k_B T)$ is the fraction of all collisions meeting this requirement.[18] Finally, we have assumed that only the fraction p of those collisions with sufficient energy do lead to reaction. The total number of reactive collisions per unit time per unit volume—which is the rate of reaction—is obtained by multiplying κ_f by $n_A n_B$, as in Eq. 30.34.

The probability factor p is sometimes called a *steric factor*, since it can be thought of as the fraction of collisions with sufficient energy that also have the proper orientation for reaction. We have taken p to be a constant; if we remove this restriction, p must go back inside the integral in Eqs. 30.36 ff. But the idea of proper orientation goes somewhat beyond the model of hard spheres, and to evaluate this effect we must use a more realistic model. In a simple theory it is often useful to have a "fudge factor," which is not intellectually harmful as long as one recognizes it as such.

Equation 30.41 predicts that the rate coefficient should vary with temperature, and we shall find that the prediction is well confirmed by experiment. Although this result was derived for two colliding molecular beams with the same temperature T, the assumption that only binary collisions occur makes it generally applicable to a dilute gas mixture at temperature T. The exponential factor $\exp(-\Delta E_f^*/k_B T)$ is dominant when $\Delta E_f^* \gg k_B T$, as is commonly the case. Such an exponential factor appears in virtually all theories; in general the *preexponential factor*, here proportional to $T^{1/2}$, is relatively unimportant in determining the rate coefficient's temperature dependence. When the temperature is increased, the average relative speed $\langle v \rangle = (8k_B T/\pi \mu_{AB})^{1/2}$ also increases, and hence so does the collision frequency. But if the threshold energy ΔE_f^* is appreciably larger than $k_B T$, then a small increase in temperature increases the fraction of collisions with enough energy for reaction, $\exp(-\Delta E_f^*/k_B T)$, much more rapidly than the total number of collisions. This can be easily seen from the following table:

$k_B T/\Delta E_f^*$	0	0.01	0.05	0.1	0.2	0.3	0.4	0.5
$e^{-\Delta E_f^*/k_B T}$	0	3.7×10^{-44}	1.9×10^{-22}	4.5×10^{-5}	0.0067	0.0357	0.0821	0.1353

However, most reactive collisions have energies only slightly higher than the threshold value. Examine the integrand of Eq. 30.40, which is the product of the two factors $1 - (\Delta E_f^*/E)$ and $Ee^{-E/k_B T}$; the first is from the total reactive cross section, Eq. 30.28, and the second weights the contribution of collisions with energy E to the average relative speed.[19] Although the cross section increases steadily with E, the factor $Ee^{-E/k_B T}$ decreases much more rapidly at large E. The product of the two factors,

---

[18] The fraction of all collisions for which $E > \Delta E_f^*$ is somewhat larger; but from Eq. 4.30 we see that such collisions must still be nonreactive if

$$b > d\left(1 - \frac{\Delta E_f^*}{E}\right)^{1/2},$$

i.e., for small scattering angles.

[19] Cf. Eq. 28.48: Except for constant factors, $v^3 \, dv$ is equivalent to $E \, dE$.

which gives the contribution to κ_f of collisions with energy E, thus reaches a peak at an energy not far above ΔE_f^* and declines rapidly thereafter. These relationships are illustrated in Fig. 30.16.

How well does the theory work? Very well in some cases, especially those in which there is no threshold energy. For example, in recombination reactions of polyatomic[20] free radicals, such as

$$CH_3 + CH_3 \rightarrow C_2H_6,$$

Eq. 30.41 with $\Delta E_f^* = 0$ and $p = \frac{1}{4}$ usually gives the observed rate constant within a factor of 2; this tells us that such radicals react on virtually every collision, so that a pure kinetic-theory approach is workable. (We assume $p = \frac{1}{4}$ because only one collision in four leads to the bonded singlet; the other three yield nonbonded triplets, as with $H + H$.) If the hard-sphere model is replaced by the ion–induced dipole potential, Eq. 10.14, and the derivation is otherwise carried through in the same way,[21] the resulting equation also gives excellent results for ion–molecule reactions with no threshold energy. But for reactions with appreciable threshold energies, the agreement is not nearly as good.

The temperature dependence is predicted better than the preexponential factor; the reason, as already noted, is that an exponential temperature dependence appears even in more advanced theories, whereas the preexponential factor has a comparatively weak temperature dependence. Thus it was observed empirically at an early date that to a good approximation most reaction rates vary with temperature according to the *Arrhenius equation*,

$$k = Ae^{-E_a/RT}, \tag{30.42}$$

where A is a constant often called the *frequency factor*, and E_a is known as the *activation energy*. Unlike the barrier heights and threshold energies we have discussed thus far, E_a is a purely empirical quantity derived from macroscopic measurements. A somewhat better fit can often be obtained with an equation of the form

$$k \propto T^n e^{-E_a/RT}, \tag{30.43}$$

where n is a small number (positive or negative) and E_a is again an empirical activation energy. We can see that our hard-sphere result, Eq. 30.41, corresponds to this equation with $n = \frac{1}{2}$; we shall see in the next section that the activated-complex theory gives $n = 1$. But when $E_a \gg RT$, as is usually the case, the exponential factor dominates the temperature dependence, which thus can tell us little about the value of n unless very precise measurements are made over a wide temperature range. For most reactions, in fact, the rate constant is known only well enough for an estimate of the activation energy.

To illustrate this insensitivity of the rate constant to the preexponential factor, let us look at some experimental data. The reaction in question

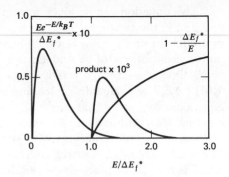

FIGURE 30.16
Factors contributing to the hard-sphere rate constant, Eq. 30.40, as functions of energy. We plot the reactive-cross-section factor $1 - (\Delta E_f^*/E)$, the relative-speed frequency factor $E\exp(-E/k_B T)$, and their product, all for $\Delta E_f^*/k_B T = 5$. In general, $E\exp(-E/k_B T)$ has its peak at $E = k_B T$, and the product has its peak at $E/\Delta E_f^* = 1 + (k_B T/\Delta E_f^*)$.

---

[20] This theory does not apply to atomic recombinations (e.g., $H + H \rightarrow H_2$), for which a third body is always necessary to remove the excess energy.

[21] The maximum impact parameter for reaction, b^*, is assumed to be the value of b that produces an orbiting trajectory. One can show that in this case the total reactive cross section is

$$\sigma_R(v) = 2\pi\left[\frac{\alpha Q^2}{(4\pi\epsilon_0)^2 \mu_{AB} v^2}\right]^{1/2} = \frac{Q}{2\epsilon_0 v}\left(\frac{\alpha}{\mu_{AB}}\right)^{1/2}$$

for reaction between an ion of charge Q and a molecule of polarizability α. Substitution in Eq. 30.35 gives

$$\kappa_f = \frac{Q}{2\epsilon_0}\left(\frac{\alpha}{\mu_{AB}}\right)^{1/2}.$$

is

$$H_2 + I_2 = 2HI,$$

the forward rate of which is here assumed to be

$$-\frac{d[I_2]}{dt} = k_f[H_2][I_2].$$

If Eq. 30.42 holds, a plot of $\ln k_f$ versus T^{-1} should give a straight line with slope $-E_a/R$. Such a plot is shown in Fig. 30.17a; the data do appear to fit a straight line, with a slope of -1.99×10^4 K, so that

$$E_a = -R \times \text{slope} = (8.314 \text{ J/mol K})(1.99 \times 10^4 \text{ K}) = 165 \text{ kJ/mol}.$$

But if Eq. 30.43 holds with $n = \frac{1}{2}$, then a plot of $\ln(k_f/T^{1/2})$ versus T^{-1} should give a straight line; and in this case, as shown in Fig. 30.17b, we obtain $E_a = 163$ kJ/mol, with the fit not significantly better than before. That is, we clearly have a good value of E_a, but we cannot distinguish between $n = 0$ and $n = \frac{1}{2}$ within the accuracy of these data.[22]

But although the preexponential factor has little effect on the temperature dependence of the rate constant, it is obviously significant in determining the absolute value of k. If we assume that the experimental activation energy E_a can be set equal to the threshold energy ΔE_f^* in Eq. 30.41, we can compare the calculated preexponential factor with the empirical frequency factor A. For the H_2–I_2 reaction just discussed, calculation (with $d_{AB} = 3.2$ Å, $p = 1$) gives a preexponential factor of 8.1×10^{-13} liter/s $= 4.9 \times 10^{11}$ liters/mol s at 600 K; the corresponding experimental value of A (from the lines in Fig. 30.17) is 1×10^{11} liters/mol s. (The actual *rates* at 600 K are 2.6×10^{-3} liter/mol s calculated, 5.4×10^{-4} liter/mol s experimental.) This result is typical: The calculated frequency factor and rate are appreciably larger than the experimental values. It is at this point that one invokes the probability factor p, which in this case would have to be about 0.2; but as already noted, there is no way to predict the value of p in the hard-sphere model. Furthermore, the values of p obtained from other comparisons of this type extend over a range of several orders of magnitude; values of 10^{-2} or 10^{-3} are typical. It is thus clear that a more accurate theory is needed, and in the next section we develop such a theory.

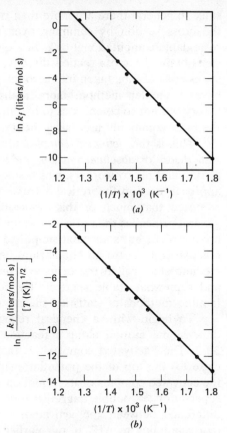

FIGURE 30.17

Temperature dependence of the rate constant for the reaction $H_2 + I_2 \rightarrow 2HI$. (a) Plot of $\ln k_f$ versus T^{-1}, linear if Eq. 30.42 is correct. (b) Plot of $\ln(k_f/T^{1/2})$ versus T^{-1}, linear if Eq. 30.43 is correct with $n = \frac{1}{2}$. In each case the slope as plotted should be equal to $-E_a/R$; the lines drawn in (a) and (b) give E_a values of 165 kJ/mol and 163 kJ/mol, respectively.

30.6
Activated-Complex Theory

A satisfactory theory of chemical kinetics should be one capable of predicting rate constants from the properties of molecules. No completely satisfactory theory in this sense yet exists. For gaseous reactions the kinetic-theory result, Eq. 30.35, is formally valid but leaves us with the problem of calculating the reactive cross section $\sigma_R(v)$; this becomes significantly difficult for anything more realistic than the simple hard-sphere model. For a complete classical calculation one would have to construct an accurate potential energy surface (or rather hypersurface,

[22] The data plotted in Fig. 30.17 are from the classic work of Bodenstein in the 1890s. As we mentioned in Section 30.1, more recent work suggests that $H_2 + I_2$ is not a simple bimolecular reaction, but has a competing chain-reaction mechanism at higher temperatures; the effect is that the activation energy increases with temperature, i.e., that the straight lines of Fig. 30.17 are really curved. However, our conclusion remains valid: that the preexponential factor has little effect on the temperature dependence.

since in general there are more than two significant variables) and obtain the cross section by summing over all possible trajectories; this is a formidable computational task. In a quantum-mechanical calculation one can obtain the cross section directly, but transitions between quantum states must also be taken into account. For ordinary use one would like to have a simpler method, more realistic than the hard-sphere kinetic theory but not so complex as to be intractable. In this section we describe the most commonly used such theory.

This is the *activated-complex theory* (also known as the *transition-state theory* or *absolute-reaction-rate theory*), developed mainly by Henry Eyring and his co-workers, which since the 1930s has been the principal approach used in theoretical discussion of chemical kinetics. (It must be admitted that much of this discussion has been more in the area of general principles than of concrete application.) To describe the theory in a convenient but somewhat simplified way, it reduces the dynamics of a reactive collision to an equilibrium between the reactants and an intermediate state called the *activated complex*, conceived as a well-defined and somewhat stable species; this equilibrium is then treated by the familiar methods of statistical mechanics.

To begin with, a chemical reaction is assumed to occur by one-dimensional motion along a reaction coordinate defined as in Section 30.2. The "activated complex" is then the state of the system at (and close to) the top of the potential energy barrier. Quantities referring to the activated complex or its formation are designated in this theory by the superscript ‡ ("double dagger"); thus, for example, we write the energy difference between the activated complex and the reactants as $\Delta E^{\ddagger}$ (corresponding to ΔE^{*} in our earlier notation). A bimolecular reaction can be written as

$$A + B \rightarrow (AB)^{\ddagger} \rightarrow products,$$

where $(AB)^{\ddagger}$ is the activated complex; although the theory can be formulated more generally, we shall consider only the bimolecular case.

The fundamental assumption is that an equilibrium is established between the reactants and the activated complex (and among the degrees of freedom within the complex). The equilibrium constant in terms of concentrations, defined in Eq. 21.90, is then

$$K_v^{\ddagger} = \frac{[(AB)^{\ddagger}]}{[A][B]}, \tag{30.44}$$

so that the concentration of activated complexes is

$$[(AB)^{\ddagger}] = K_v^{\ddagger}[A][B]. \tag{30.45}$$

The reaction coordinate of the activated complex has not been included in the degrees of freedom of $AB^{\ddagger}$.

This concentration is ordinarily negligibly small relative to the concentrations of reactants and products. The rate of the forward reaction must then be the rate at which complexes break up to yield products, which we can express as $[(AB)^{\ddagger}]/\tau_f$, where τ_f is the average time required for a complex to break up in this way. There are two complications here: First, the reverse reaction is in general occurring simultaneously, with the same activated complex; but the two processes are statistically independent (uncorrelated), and can thus be considered separately. Second, and more serious, not all systems that reach the top of the barrier in the forward direction leave it in the same direction: Some go back the way they came, with the activated complex separated into the original reactants.

We therefore introduce a *transmission coefficient* κ, formally defined as the fraction of those systems forming an activated complex that pass directly through it to form products. The value of κ depends on the shape of the potential energy surface; direct trajectory calculations for some elementary reactions have given values of κ only a few percent less than unity. For most reactions, however, κ is as yet nothing more than another "fudge factor" to cover the discrepancy between theory and experiment. But this discrepancy is not as great as one might expect: Many reactions do have empirical values of κ effectively equal to unity, although lower values are known. (One obvious class of exceptions are two-atom recombination reactions, which effectively have $\kappa = 0$ in the absence of a third body. Low values of κ are also found in reactions that involve changes of electronic state.)

Assuming for the moment that $\kappa = 1$, we return to the question of how rapidly activated complexes break up. The actual mechanism of breakup must involve the breaking of a bond in the complex, that is, one vibrational degree of freedom's becoming a translation. Since an activated complex is a very unstable species, the bond in question must be extremely weak. It is therefore plausible to argue that the complex will break up the first time the bond is stretched, that is, within a single vibrational period after its formation. This gives us a breakup rate of approximately $\nu[(AB)^{\ddagger}]$, where ν is the frequency of the vibration in question. This vibration is assumed to be completely excited, so that the partition function for this degree of freedom is

$$\frac{k_B T}{h\nu} \tag{30.46}$$

Combining our assumptions, we have for the breakup rate

$$-\left(\frac{d[A]}{dt}\right)_f = \nu[(AB)^{\ddagger}] = \frac{k_B T}{h} K_V^{\ddagger}[A][B], \tag{30.47}$$

where the value of $[(AB)^{\ddagger}]$ has been introduced from Eq. 30.45; the forward rate coefficient is therefore

$$k_f = \frac{k_B T}{h} K_V^{\ddagger}. \tag{30.48}$$

(Given equivalent assumptions, the same result with appropriate forms of $K_V^{\ddagger}$ is found for all elementary processes.) Finally, we allow for the fact that not all complexes break up in this way by reintroducing the transmission coefficient,

$$k_f = \kappa \frac{k_B T}{h} K_V^{\ddagger}. \tag{30.49}$$

This is the formal result of activated-complex theory for the rate constant.

The obvious question now is how to determine $K_V^{\ddagger}$. This is not difficult in principle *if* one knows the structure of the activated complex. One simply uses the standard techniques of statistical mechanics to calculate the partition function and energy of each species, and thus the equilibrium constant. For a perfect-gas reaction we have, from Eq. 21.89,

$$K_V = \left(\frac{RT}{\mathrm{p}}\right)^{-\Delta\nu} K(T), \tag{30.50}$$

where p is the standard pressure (usually 1 atm), $\Delta\nu$ is the difference in stoichiometric coefficients between products and reactants, and $K(T)$ is the equilibrium constant in terms of fugacities (the low-pressure limit of K_p, the constant in terms of partial pressures), as given by Eq. 21.65:

$$K(T) \equiv \prod_i \left(\frac{f_i}{\mathrm{p}}\right)^{\nu_i} = \lim_{\mathrm{p}\to 0} \prod_i \left(\frac{p_i}{\mathrm{p}}\right)^{\nu_i}; \tag{30.51}$$

remember that the stoichiometric coefficient ν_i is positive for products, negative for reactants. And from statistical mechanics we can derive

$$K(T) = \prod_i \left(q_{1i} \frac{k_B T}{p} \right)^{\nu_i} e^{\Delta E_0{}^0/RT}, \tag{30.52}$$

where q_{1i} is the single-molecule partition function per unit volume for species i $(= V^{-1} \sum_j g_{ij} e^{-\epsilon_{ij}/k_B T})$ and $\Delta E_0{}^0$ is the ground-state energy difference between reactants and products (assumed to be perfect gases at absolute zero). The "reaction" we are concerned with here is simply the formation of the bimolecular-reaction activated complex,

$$A + B \rightarrow (AB)^{\ddagger},$$

for which $\Delta \nu = -1$. Substituting in Eqs. 30.50 and 30.52, we obtain

$$K_V = N_A \frac{q_1[(AB)^{\ddagger}]}{q_1(A)q_1(B)} e^{-(\Delta E_0{}^0)^{\ddagger}/RT}, \tag{30.53}$$

where N_A is Avogadro's number.

The problem now is how to evaluate the factors in Eq. 30.53. The partition functions of the reactants are easily evaluated by the standard techniques (see Section 21.5), but that of the activated complex is more difficult. As usual, one factors q_1 into translational, rotational, and vibrational components, and makes the assumption that the energy is statistically distributed among these degrees of freedom. The translational part is straightforward; so is the rotational part, if one knows the structure of the complex (from chemical intuition and/or the potential energy surface). But estimation of the vibrational partition function is particularly difficult for the complex: it has one less degree of freedom than usual, since in writing Eq. 30.47 we assumed one vibration to be effectively a translation along the reaction coordinate;[23] yet some complexes have more than one weak bond (depending on the shape of the potential energy surface), so that several vibrational frequencies may be hard to guess. In any case, all the partition functions go into the preexponential part of the rate coefficient, which, as noted earlier, is relatively unimportant. But what can one say about the exponential factor $e^{-(\Delta E_0{}^0)^{\ddagger}/RT}$?

The energy $(\Delta E_0{}^0)^{\ddagger}$ is the difference between the ground-state energies of the complex $(AB)^{\ddagger}$ and the reactants; if $(AB)^{\ddagger}$ is the state at the top of the potential energy barrier, then $(\Delta E_0{}^0)^{\ddagger}$ is effectively the barrier height. One must thus try to calculate the potential energy surface, or at least its cross section along the reaction coordinate. Complete calculation of the surface a priori is difficult, and also unnecessary for this purpose, given the other limitations of the activated-complex theory. One can use a semiempirical approach such as the LEPS method outlined in Section 30.2. A still simpler technique is the purely empirical *BEBO* (*bond energy–bond order*) *method* developed by H. S. Johnston et al. The BEBO method is based on the relation proposed by Pauling between bond length R and bond order n,

$$R = R_s - 0.26 \ln n, \tag{30.54}$$

where R_s is the single-bond length; a similar relationship is assumed for bond energy,

$$E = E_s n^p, \tag{30.55}$$

where E_s is the single-bond energy and p is an empirical constant (≈ 1)

---

[23] To put it another way, the partition function of a fully excited vibration is $k_B T/h\nu$, which we have by definition factored out of the $K_V^{\ddagger}$ in Eq. 30.47.

for each type of bond. In reactions where one bond is broken and another bond formed, it is assumed that the *total* bond order remains constant: By Eqs. 30.54 and 30.55, this gives the reaction coordinate (one bond length as a function of the other) and the energy along it. For the $H_A H_B + H_C$ reaction, the BEBO method predicts a barrier height of 41 kJ/mol, compared with an experimental activation energy of 36 kJ/mol; for many other hydrogen-transfer reactions the method gives agreement within 10 kJ/mol. (It can also be used to estimate vibrational constants, and thus to help evaluate the preexponential factor.)

By a variety of methods, then, one can predict with reasonable accuracy an energy corresponding to the activated-complex theory's $(\Delta E_0{}^0)^{\ddagger}$. What the theory itself can predict is the preexponential factor A, but with much less accuracy. Generally, however, the prediction is correct within an order of magnitude, and this is significantly better than the results of the simple kinetic theory (which often gives values of A several orders of magnitude too high). Thus we see that even this simple model can give approximately correct results for the absolute magnitude of rate constants, even though it completely neglects the dynamics of the reactive collision and considers only a few critical parameters of the potential energy surface. We found earlier that the kinetic-theory model, crude as it is, does fairly well at predicting the temperature dependence of rate constants; what does the activated-complex theory have to say on this topic? Before answering this question, we shall introduce some of the relationships between thermodynamics and kinetics.

30.7 Activated-Complex Theory: Thermodynamic Interpretation

We begin by considering the relationship between reaction rate constants and equilibrium constants. This will allow us to apply some of the tools of thermodynamics to kinetics, and in particular to describe the temperature dependence of reaction rates in terms of quasi-thermodynamic quantities.

We consider again the general dilute-gas reaction

$$A + B = C + D.$$

In Section 30.5 we made a kinetic-theory analysis of the reaction in the forward direction, due to collisions between A and B molecules. But we know that C and D molecules will also collide, sometimes with enough energy to bring about the reverse reaction producing A and B. By analogy with Eq. 30.34, the rate of formation of A by such reverse collisions must be

$$\left(\frac{dn_A}{dt}\right)_r = \kappa_r n_C n_D, \tag{30.56}$$

where the reverse rate coefficient κ_r is defined as in Eq. 30.35, with C and D replacing A and B; of course, the forward and reverse reactive cross sections $\sigma_R(v)$ will in general be different. Since the forward and reverse reactions occur simultaneously, the *net* reaction rate should be given by

$$-\frac{dn_A}{dt} = -\left(\frac{dn_A}{dt}\right)_f - \left(\frac{dn_A}{dt}\right)_r = \kappa_f n_A n_B - \kappa_r n_C n_D. \tag{30.57}$$

Remember that the n_i are number densities (molecules per unit volume). The corresponding equation in terms of molar concentrations is

$$-\frac{d[A]}{dt} = k_f[A][B] - k_r[C][D], \tag{30.58}$$

where k_f, k_r are the macroscopic rate coefficients; if [A], [B], ... are molar concentrations, we should have $k_f/\kappa_f = k_r/\kappa_r = N_A$ (Avogadro's number).

Something should be said here about the nature of these equations. We have assumed here that the overall reaction involves a single reversible elementary reaction that is bimolecular in both directions. (But one can, of course, generalize to more complicated mechanisms.) Under these conditions the macroscopic reaction should be observed to be second-order in both directions; k_f and k_r are the empirical rate constants based on this assumption (later we shall see how such constants can be extracted from the data), and Eq. 30.58 is then a description of experimental fact. Equation 30.57, on the other hand, we derived from our kinetic-theory model, which implies a particular theoretical form for the coefficients κ_f, κ_r. But if Eq. 30.58 holds empirically, then an equation of the *form* 30.57 must also hold, since the number densities and concentrations are proportional to each other. It is only the *values* of κ_f and κ_r that need not agree between theory and experiment—and as we have seen earlier, they often disagree significantly.

Suppose that we mix the gases A and B in a vessel. Reaction will occur, and for some time the concentrations of A and B will decrease, while those of C and D increase. The forward reaction rate will thus decrease and the reverse reaction rate increase, until a point is approached at which the two rates are equal. Then the disappearance of A due to the forward reaction will exactly balance the production of A by the reverse reaction, and the net rate of change of concentration of A (and all other species) will be zero. This is our standard definition of the state of *chemical equilibrium*. Since at equilibrium $d[A]/dt = 0$, the right-hand side of Eq. 30.58 gives us

$$k_f[A]_{eq}[B]_{eq} - k_r[C]_{eq}[D]_{eq} = 0, \qquad (30.59)$$

where the subscript "eq" designates equilibrium; note that the individual terms (the forward and reverse reaction rates) do not vanish at equilibrium, only their difference. We can rearrange this equation into the form

$$\frac{k_f}{k_r} = \frac{[C]_{eq}[D]_{eq}}{[A]_{eq}[B]_{eq}} = K_V, \qquad (30.60)$$

where K_V is the standard equilibrium constant in terms of concentrations. For the microscopic rate coefficients we should similarly have

$$\frac{\kappa_f}{\kappa_r} = \frac{n_{C,eq} n_{D,eq}}{n_{A,eq} n_{B,eq}} = K_V, \qquad (30.61)$$

provided that the theory used to calculate κ_f, κ_r is correct (or at least gives both wrong by the same constant factor). Equation 30.60 or 30.61 is the fundamental relationship connecting kinetics with equilibrium thermodynamics.

We see clearly that chemical equilibrium, like all equilibria, is maintained by a "balance of active tendencies"; in this case the active tendencies are the reactions in the forward and reverse directions. We have considered a case involving only a single elementary reaction, which at equilibrium must have the same rates in both directions. In a more complex system, one can show that overall equilibrium can exist only when *all* elementary reaction steps have the same rates in both directions. This is the principle of *detailed balance*, which is a consequence of microscopic reversibility. (Whether one or many reactions are involved, remember that chemical equilibrium is essentially a macroscopic concept.

On the microscopic level fluctuations must occur, and only the *average* reaction rate can be zero.)

Note that we have assumed k_f, k_r (and κ_f, κ_r) to be constants, with the same values whether the system is at equilibrium or not. For the empirical rate constants this must be based on observation: If k_f, k_r *seem* to vary with concentration, then probably the reaction mechanism is not as simple as one has assumed. And for κ_f, κ_r in the kinetic-theory model, as we noted earlier, this is equivalent to assuming the reaction sufficiently slow that the distributions of velocity and internal energy are close to their equilibrium values. Experiment ordinarily does confirm the constancy of k_f, k_r: That is, Eq. 30.60 is found to be valid even when k_f, k_r are measured far from equilibrium. Another limitation of this model is that it applies only to dilute gases.

Despite these limitations, the basic principles of Eqs. 30.60 and 30.61 are of general applicability. In any reaction whatever, at equilibrium (by definition) the forward rate must equal the reverse rate. If each of these rates can be expressed as some product of concentrations multiplied by a rate coefficient, then the ratio of forward and reverse coefficients must equal some unique ratio of equilibrium concentrations, which we designate as the equilibrium constant. But note that this argument applies only to individual *elementary* reactions. For a more complex mechanism the ratio of concentrations given by the stoichiometry has in general no simple relationship to the rate constants (for the same reason that order and molecularity are in general different).

For those cases where Eq. 30.59 does hold, we can proceed to draw thermodynamic conclusions. We know from ordinary thermodynamics that the standard Gibbs free energy change of a reaction is related to the equilibrium constant by the equation

$$K = e^{-\Delta G^0/RT}, \tag{30.62}$$

where K is the equilibrium constant in terms of activities. In the dilute-gas limit K is the same as K_p (the equilibrium constant in terms of pressure), and for our reaction $A + B = C + D$ (with $\Delta \nu = 0$) also the same as K_V. We can thus write Eq. 30.60 in the form

$$K = \frac{k_f}{k_r} = e^{-(\Delta G_f^{0*} - \Delta G_r^{0*})/RT}, \tag{30.63}$$

where ΔG_f^{0*} and ΔG_r^{0*} are given by

$$k_f = A^* e^{-\Delta G_f^{0*}/RT} \quad \text{and} \quad k_r = A^* e^{-\Delta G_r^{0*}/RT}, \tag{30.64}$$

with A^* a constant, so that

$$\Delta G^0 = \Delta G_f^{0*} - \Delta G_r^{0*}. \tag{30.65}$$

But this is only a formal manipulation unless we can assign some meaning to ΔG_f^{0*} and ΔG_r^{0*}. Unless we can define the value of A^*, only their difference is physically significant, since we can add the same quantity to both without affecting the above equations. To make these quantities meaningful, we must consider the temperature dependence of the rate constants.

Remember that rate constants are often found empirically to vary with temperature according to the Arrhenius equation, Eq. 30.42. If, as usual, the measurements of $k(T)$ are carried out at constant pressure, the activation energy E_a is actually an enthalpy change of some sort, which we could as well call ΔH^*. Now, since at a given temperature $\Delta G = \Delta H - T\,\Delta S$, we can rewrite Eqs. 30.64 in the form

$$k_f = A^* e^{\Delta S_f^{0*}/R} e^{-\Delta H_f^{0*}/RT} \tag{30.66}$$

(and so for k_r), with ΔG_f^{0*}, ΔS_f^{0*}, and ΔH_f^{0*} all referring to the same (as yet undefined) process. This is not inconsistent with the Arrhenius equation if one recalls that entropy changes quite slowly with temperature, so that $e^{\Delta S^{0*}/R}$ should be nearly a constant. Thus $A^* e^{\Delta S^{0*}/R}$ corresponds approximately to the preexponential part of the rate constant, and ΔH^{0*} to the activation energy.

The obvious interpretation of the process to which ΔG_f^{0*} and so on refer is that leading from the initial state of the reactants to the top of the potential energy barrier—in other words, the formation of the activated complex. Thus we can now finally revert to the question at the end of the last section: What has the activated-complex theory to say about the temperature dependence of reaction rates? This is most fruitfully answered in terms of the thermodynamic language we have introduced.

The convenient assumption we made in deriving the activated-complex theory was that an equilibrium exists between the reactants and the activated complex. We can therefore write

$$\Delta G^{0\ddagger} = \Delta H^{0\ddagger} - T\,\Delta S^{0\ddagger} = -RT \ln K^{\ddagger}, \qquad (30.67)$$

with the double-dagger quantities referring to the reaction $A + B \rightarrow (AB)^{\ddagger}$. Ordinarily the equilibrium constant in such an equation is that in terms of activities (or pressures), in accordance with Eq. 30.62; but in the activated-complex theory it is customary to use instead the $K_V^{\ddagger}$ defined in Eq. 30.44. For Eq. 30.67 to be valid in terms of $K_V^{\ddagger}$, unit concentration must be taken as the standard state for all components.[24] With this definition in mind, we can rewrite Eq. 30.48 as

$$k_f = \frac{k_B T}{h} e^{-\Delta G^{0\ddagger}/RT} = \frac{k_B T}{h} e^{\Delta S^{0\ddagger}/R} e^{-\Delta H^{0\ddagger}/RT}, \qquad (30.68)$$

with the same form as Eq. 30.66. One usually refers to $\Delta G^{0\ddagger}$, $\Delta H^{0\ddagger}$, and $\Delta S^{0\ddagger}$, respectively, as the *free energy, enthalpy,* and *entropy of activation.* Since $\Delta \nu^{\ddagger} = 1 - n$, where n is the molecularity of the reaction, we have for perfect gases

$$\Delta H^{0\ddagger} = \Delta E^{0\ddagger} + \Delta(pV)^{0\ddagger} = \Delta E^{0\ddagger} - (n-1)RT. \qquad (30.69)$$

$\Delta E^{0\ddagger}$ is the *energy of activation,* not to be confused with the empirical *activation energy* E_a. Assuming $\Delta E^{0\ddagger}$ and $\Delta S^{0\ddagger}$ to be independent of temperature (often a fairly good approximation), we can differentiate the rate coefficient to obtain

$$\frac{d \ln k_f}{dT} = \frac{1}{T} - \frac{d}{dT}\left(\frac{\Delta H^{0\ddagger}}{RT}\right) = \frac{1}{T} + \frac{\Delta H^{0\ddagger}}{RT^2} + \frac{n-1}{T} = \frac{\Delta H^{0\ddagger} + nRT}{RT^2}.$$

$$(30.70)$$

Comparing this with the derivative of the Arrhenius equation, Eq. 30.42, we see that the empirical activation energy should be

$$E_a = \Delta H^{0\ddagger} + nRT = \Delta E^{0\ddagger} + RT \qquad (30.71)$$

for perfect gases.[25]

Equation 30.71 enables us to obtain $\Delta H^{0\ddagger}$ from experimental data (provided that the data fit the Arrhenius equation), and substitution into

---

[24] More precisely, the $K^{\ddagger}$ in Eq. 30.67 should then be $\Pi_i(c_i/c_0)$, where c_0 is unit concentration.

[25] For condensed phases we can neglect $\Delta(pV)^{0\ddagger}$ in Eq. 30.69, so that $\Delta H^{0\ddagger} = \Delta E^{0\ddagger}$; we then obtain

$$E_a = \Delta H^{0\ddagger} + RT.$$

TABLE 30.1
ENTHALPIES AND ENTROPIES OF ACTIVATION FOR SOME GASEOUS REACTIONS (CALCULATED FOR 300°C, STANDARD STATE OF 1 MOL/CM³)

Reaction	$\Delta H^{0\ddagger}$ (kJ/mol)	$\Delta S^{0\ddagger}$ (J/mol K)
Bimolecular:		
$H_2 + I_2 \rightarrow 2HI$	158	+10
$2CH_2{=}CH_2 \rightarrow CH_2{=}CHCH_2CH_3$	148	−50
$NO + O_3 \rightarrow NO_2 + O_2$	1	−35
$CH_3 + CH_3 \rightarrow C_2H_6$	0	+75
Unimolecular:		
$trans\text{-}(ClCH{=}CHCl) \rightarrow cis\text{-}$	171	−5
$\begin{array}{c} H_2C{-}CH_2 \\ \backslash \quad / \\ C \\ H_2 \end{array} \rightarrow CH_2{=}CHCH_3$	267	+45
$cis\text{-}\left(\begin{array}{c} CH_3C{-}COOH \\ \| \\ HC{-}COOH \end{array}\right) \rightarrow trans\text{-}$	106	−135

Eq. 30.68 then yields a value for $\Delta S^{0\ddagger}$ (and thus for $\Delta G^{0\ddagger}$ and $K_V^{\ddagger}$). These quantities can be used to make deductions about the nature of the activated complex. For example, $\Delta H^{0\ddagger}$ and $\Delta E^{0\ddagger}$ are closely related to the barrier height, and we have already discussed some of the ways of obtaining them from bond energies (such as the BEBO method). The entropy of activation gives some measure of the number of available configurations of the complex, relative to the reactants. One speaks of "tight" and "loose" complexes, a more loosely bound one obviously having a higher entropy. Since the formation of a complex from two molecules necessarily involves a loss of translational and rotational degrees of freedom, $\Delta S^{0\ddagger}$ is generally negative for bimolecular reactions. The factor $e^{\Delta S^{0\ddagger}/R}$ is largely analogous to the steric factor p of the hard-sphere collision theory; we can thus interpret the small value of p in most reactions as an entropy effect. Some typical values of $\Delta H^{0\ddagger}$ and $\Delta S^{0\ddagger}$ are listed in Table 30.1.

In general, as we noted in Section 30.5, the temperature dependence of rate constants does not exactly follow the Arrhenius equation. The activated-complex theory predicts a non-Arrhenius temperature dependence, since a combination of Eqs. 30.49 and 30.53 yields for a bimolecular reaction

$$k_f = \kappa \frac{k_B T}{h} N_A \frac{q_1[(AB)^{\ddagger}]}{q_1(A)q_1(B)} e^{-(\Delta E_0^{0})^{\ddagger}/RT}. \qquad (30.72)$$

Besides the factor $k_B T/h$, the partition functions q_{1i} have their own temperature dependence. Each q_{1i} is a product of translational, rotational, and vibrational factors: Each translational or rotational degree of freedom contributes a factor proportional to $T^{1/2}$, whereas the vibrational partition function is $(1 - e^{-h\nu/k_B T})^{-1}$ per degree of freedom. Thus the preexponential part of the rate constant should indeed be proportional to T^m (where m is a small integer or half-integer), as indicated by the empirical Eq. 30.43, whenever the vibrational partition function can be approximated by either unity ($h\nu \gg k_B T$) or $k_B T/h\nu$ ($\gg 1$); the actual value of m depends on the specific molecules involved. (Unfortunately, the empirical value of m is not always that predicted by the theory.) If we thus have a predicted rate constant of the form

$$k = AT^m e^{-(\Delta E_0^0)^\ddagger/RT}, \tag{30.73}$$

differentiation gives

$$\frac{d \ln k}{dT} = \frac{(\Delta E_0^0)^\ddagger}{RT^2} + \frac{m}{T} = \frac{E_a}{RT^2}, \tag{30.74}$$

where E_a is by definition the activation energy obtained from an Arrhenius-equation fit. We thus have

$$(\Delta E_0^0)^\ddagger = E_a - mRT; \tag{30.75}$$

comparison of this with Eq. 30.71 shows that the quantum-mechanical $(\Delta E_0^0)^\ddagger$ and the thermodynamic $\Delta E^{0\ddagger}$ are not in general the same, but one can easily convert from one to the other (or to $\Delta H^{0\ddagger}$).

It should be pointed out that there is much looseness in the terminology of this field. One is apt to find "activation energy" used indiscriminately for the empirical quantity $E_a (\equiv RT^2 \, d \ln k/dT)$, for the energy $(\Delta E_0^0)^\ddagger$ obtained from either an empirical equation of the form 30.73 or quantum-mechanical calculation, or for the "thermodynamic" quantities $\Delta E^{0\ddagger}$ and $\Delta H^{0\ddagger}$. Since all of these are in general different, one should be careful to note which definition (and which standard state for $\Delta E^{0\ddagger}$, etc.) a particular author is using. (However, the differences are not very great, amounting to a few times RT, which at 300 K is 2.5 kJ/mol; activation energies are typically of the order of 100 kJ/mol.) As we mentioned earlier, one should avoid confusing any of these quantities with the barrier height (the height of the peak in the potential energy curve along the reaction coordinate) or with the threshold energy (the difference between the barrier height and the initial energy levels of the reactants): The threshold energy is the minimum energy that must be supplied for reaction to occur from given initial states, whereas the various macroscopic activation energies must represent thermal averages of threshold energies.

We have not said anything yet about the pressure dependence of the rate constant, but this is quite easy to develop. If we assume that Eq. 30.68 applies, we can immediately write

$$\left(\frac{\partial k}{\partial p}\right)_T = -\frac{k}{RT}\left(\frac{\partial \Delta G^{0\ddagger}}{\partial p}\right)_T = \frac{k}{RT}\Delta V^{0\ddagger}, \tag{30.76}$$

since $V = (\partial G/\partial p)_T$. Here $\Delta V^{0\ddagger}$ is the volume change corresponding to $\Delta G^{0\ddagger}$, that is, presumably the volume difference between the activated complex and the reactants. The values of $\Delta S^{0\ddagger}$—obtained similarly from $(\partial k/\partial T)_p$, since $S = -(\partial G/\partial T)_p$—and $\Delta V^{0\ddagger}$ can usefully be compared to interpret the structure of the activated complex. In bimolecular reactions between neutral species, the activation process normally involves the formation of a new bond and thus a decrease in total molecular volume; typical values of $\Delta V^{0\ddagger}$ for such reactions are around -10 cm^3/mol (in solution). Unimolecular reactions, on the other hand, necessarily proceed by bond breaking, and thus usually have positive $\Delta V^{0\ddagger}$. Volumes of activation and entropies of activation tend to be correlated, with both positive or both negative; this corresponds to our earlier observation that $e^{\Delta S^{0\ddagger}/R}$ is a steric factor.

Thus we see that a good deal can be deduced about the properties of the activated complex by interpreting kinetic data in a "thermodynamic" way. However, all these interpretations are somewhat arbitrary, since the activated complex itself cannot really be so well defined a species as the theory supposes. The problem is that the dynamics of the reactive collision are essentially ignored. The assumption of equilibrium requires that the energy of the complex be distributed statistically among its various degrees of freedom, which may be true for some but not all reactions. In addition, real collisions do not always occur neatly along the optimum reaction coordinate; in an off-center collision the system's angular momentum must be taken into account, and this the theory completely ignores. It is also incapable of saying anything about the distribution of products when the complex can break up in more than one way (or yield products in various excited states).

In view of all these objections, it is remarkable that the activated-complex theory has been used as much as it has. The theory does give a good empirical representation of the kinetic data for a large number of reactions, and is useful in giving a simple interpretation in terms of molecular pictures. Thus the results of the theory used are apparently more valid than the assumptions used to derive them (as we also found for the even simpler kinetic theory). Indeed, similar expressions for the rate coefficient have been obtained from more advanced theories (which do not need to assume the actual existence of the activated complex), so the accuracy may be regarded as fortuitous. One has to use the activated-complex theory with caution, and in general it is better for interpreting or correlating experimental reaction rates (e.g., explaining the trends over a series of similar reactions) than for deriving them *a priori.*

30.8 Theory of Reaction Kinetics in Solution

So far our discussion of the theory of reaction rates has been essentially limited to gaseous reactions—in fact, to perfect gases. Most reactions of interest to the chemist, however, occur in liquid (especially aqueous) solutions. Obviously the situation is then more complicated. There is a very large body of theoretical and experimental information on reactions in solution, but in this section we can consider only a few major points. In particular, we discuss the similarities and differences between gas and solution kinetics, then outline some special features of the kinetics of ionic reactions.

Readers should begin by recalling what they know about the equilibrium properties of liquid solutions (Chapters 23, 25, and 26), and about the transport properties of liquids (Chapter 29). The key feature of the microscopic structure of a liquid is that each molecule is in continuous interaction with its neighbors. As a result of this, one cannot in general consider a chemical reaction as a sequence of independent two-body collisions. Since even the simpler physical transport processes[26] cannot be treated accurately in liquids, it is not surprising that there is no satisfactory general theory of solution kinetics. One would expect such a theory to be extremely complicated.

Remarkably enough, things are *not* this bad. In many cases the theories we have already described work almost as well in solution as in the gas phase. "Almost as well," of course, can mean an error of one or two orders of magnitude in the rate constant. Unfortunately, few reactions have been studied in both phases (partly because so many reactions

[26] The physical processes are simpler because lower energies are involved—for example, it takes less energy on the average to move one molecule past another (diffusion) than to transfer atoms between them (reaction). Yet both can be formulated in similar terms, involving potential barriers and activation energies: Compare Eq. 29.1 with Eq. 30.42.

in solution are ionic), so few direct comparisons can be made; but research in the area is very active and the situation is improving. In some cases, the rate coefficients and activation energies are nearly the same in gas and solution; in others, the rates differ by only an order of magnitude or so; in still others, the differences are much greater. For those reactions known only in solution, the rates predicted from the gas-phase theories are often accurate within an order of magnitude. Why should theories based on single-collision models be so accurate in a medium where intuition suggests that these models might well be inappropriate?

There are plausible answers to this question. For one thing, the energies involved in reactions are usually appreciably greater than the energy of interaction with neighboring solvent molecules. (Activation energies, which roughly correspond to barrier heights, are typically about 50–100 kJ/mol, whereas interaction energies between nonreacting molecules might be 1–2 kJ/mol.) One might thus expect the influence of neighbors to be a relatively small perturbation, making it reasonable to consider for reactions two-molecule collisions only. This hypothesis gains further support from the empirical fact that reaction mechanisms in solution, as well as those in gases, can usually be analyzed into simple bimolecular steps.

More to the point is a consideration of the mechanics of collision. One elementary model of liquid structure considers each molecule to be contained within a cell or "cage" of its neighbors. In a reactive collision, however, we must effectively have *two* molecules within such a cage—close enough to each other to interact strongly, relative to their interactions with the surrounding molecules. (This model assumes that the "walls" of the cage are made of nonreactive solvent molecules, and thus should be applicable only in dilute solution.) When a reactive pair of molecules get into such a situation, they cannot easily get away from each other: they will tend to bounce back and forth within the cage, perhaps colliding with each other many times before escaping. The process can be treated in terms of kinetic theory, and a crude calculation (not included here) predicts that the total frequency of collisions between reactive solute molecules should be very roughly—give or take an order of magnitude—the same as in a gas at the same concentration. (Note that a gas at STP has a concentration of 0.045 mol/liter, which is comparable to a dilute solution.)

One can also look at reactions in solution in terms of the activated-complex theory. Consider Eq. 30.68 for the rate constant. The energy of activation, $\Delta H^{0\ddagger}$, does not differ significantly between the gas and solution phases whenever our assumption holds, that interactions between reactant and solvent molecules are relatively weak. But the entropy of activation, $\Delta S^{0\ddagger}$, is another matter. Applying Hess's law (Section 14.2), one can write

$$(\Delta S^{0\ddagger})_{\text{solution}} - (\Delta S^{0\ddagger})_{\text{gas}} = \sum_i \nu_i (\Delta S^0_{\text{soln}})_i, \qquad (30.77)$$

where the sum goes over product (the activated complex) and reactants, and ΔS^0_{soln} is the entropy of solution. (A similar equation holds for $\Delta H^{0\ddagger}$, but by our assumption the ΔH^0_{soln}'s nearly cancel out.) And a crude model of liquids gives the molar entropy of solution as $R \ln(V_f/V)$, where V_f is the "free volume," effectively the total volume minus that of the molecules (as in the van der Waals equation of state). If the ratio V/V_f is the same for all species, we obtain

$$(\Delta S^{0\ddagger})_{\text{solution}} - (\Delta S^{0\ddagger})_{\text{gas}} = (n-1)R \ln (V/V_f), \qquad (30.78)$$

where n is the molecularity of the reaction. For a bimolecular reaction the preexponential factor $\exp(\Delta S^{0\ddagger}/R)$, and thus (roughly) the reaction rate, should be V/V_f times higher in solution than in the gas phase. The value of V/V_f is typically about 100, and a number of bimolecular reactions do seem to have rates one to two orders of magnitude higher in solution. For unimolecular reactions, on the other hand, Eq. 30.78 predicts no change, and unimolecular reactions are found to have very nearly the same gas and solution rates.

Not all reactions show the similarity between gas and solution kinetics that we have been talking about, for a variety of reasons. For example, if the activation energy is low enough, two molecules are likely to react as soon as they are in the same "cage." In such "fast reactions," the rate of reaction is controlled not by the reaction process itself, but by the rate at which reactive molecules can get into the same cage—that is, the walls of the cage present a higher potential energy barrier than the formation of the activated complex. Such reactions have a limiting rate coefficient proportional to the rate of encounters between the reactants, and thus to the sum of the reactants' diffusion coefficients; they are said to be *diffusion-controlled*. We shall return to this topic later in this section.

Among other factors causing differences between gas and solution kinetics, the limited volume of a cage can make steric factors important. That is, it may be very difficult in solution for two molecules to achieve the proper orientation for reaction; this physical fact would assert itself as a strongly negative contribution to $\Delta S^{0\ddagger}$ (outweighing the positive free-volume effect discussed above) and a lowered value of the rate constant. In many other cases, what appears to be a single reaction may have a completely different nature in solution from that in the gas phase. An obvious example of this is any reaction involving the H^+ ion, which in most solvents is completely solvated (forming H_3O^+, etc., in water) and does not exist as a bare proton; similar considerations apply for any reactant that interacts strongly with the solvent.

Obviously, most of what we have just said does not apply when the solvent itself is one of the reactants. In particular, the model of an inert cage of solvent molecules is invalid in such a case, and the reactant molecules themselves must be in continuous interaction. For reactions of this type our original remarks about the complicated nature of solution kinetics come back into full effect. The solvent does affect the reaction rate even when it is not one of the reactants, although, as we have noted, these effects are usually small; we can see by looking at Eq. 30.77 that such effects must exist, since ΔH^0_{soln} and ΔS^0_{soln} do depend on the nature of the solvent. Specifically, the rates of reactions between ions and/or polar molecules must vary with the solvent, since the electrostatic energy between two such species is inversely proportional to the dielectric constant of the intervening medium (see Section 26.5).

Indeed, most of the complicating effects we have mentioned are especially significant in ionic reactions. Reactions between oppositely charged ions of course tend to be faster, and those between like-charged ions slower, than the simple kinetic theory would predict. Indeed, such reactions as $H_3O^+ + OH^-$ or $Ag^+ + Cl^-$ have virtually zero activation energies, and their rates may thus be diffusion-limited; however, another possibility is that charge transfer through the solvent molecules limits the rate. In general, one must expect a complete treatment of kinetics in electrolytic solutions to involve special concepts, like those one must introduce in studying the equilibrium properties (Chapter 26). We shall consider only some special aspects of the theory that have been extensively developed.

The first of these is the so-called *primary salt effect*, derived by Brønsted and Bjerrum on the basis of activated-complex theory. Consider a general bimolecular reaction between ions,

$$A^{Z_A} + B^{Z_B} \rightarrow [(AB)^{\ddagger}]^{Z_A + Z_B} \rightarrow \text{products},$$

where $Z_A e$ and $Z_B e$ are the charges on ions A and B, and $(AB)^{\ddagger}$ is the activated complex. Since ionic solutions often deviate strongly from ideality, we must formulate the theory in terms of activities, so that the $K_V^{\ddagger}$ of Eq. 30.44 is replaced by

$$K^{\ddagger} = \frac{a_{(AB)^{\ddagger}}}{a_A a_B} = \frac{[(AB)^{\ddagger}]}{[A][B]} \frac{\gamma^{\ddagger}}{\gamma_A \gamma_B}, \tag{30.79}$$

where the γ's are activity coefficients. The breakup rate of the activated complex is still defined in terms of its concentration, and the assumptions we used to derive Eq. 30.47 now give

$$-\left(\frac{d[A]}{dt}\right)_f = \frac{k_B T}{h}[(AB)^{\ddagger}] = \frac{k_B T}{h} K^{\ddagger} \frac{\gamma_A \gamma_B}{\gamma^{\ddagger}} [A][B] \tag{30.80}$$

or

$$k_f = \frac{k_B T}{h} K^{\ddagger} \frac{\gamma_A \gamma_B}{\gamma^{\ddagger}}. \tag{30.81}$$

This differs from Eq. 30.49 only in the activity coefficients, which can be derived from the equilibrium theory of electrolytic solutions. In dilute solution we can use the Debye-Hückel limiting law, Eq. 26.50, which in water at 25°C is

$$\log \gamma_i = -0.509 Z_i^2 I^{1/2} \qquad \left(I \equiv \tfrac{1}{2}\sum_i Z_i^2 m_i\right), \tag{30.82}$$

I being the ionic strength defined in Eq. 26.49 (m_i = molality). Substituting in Eq. 30.81, we have

$$\begin{aligned}
\log k_f &= \log\left(\frac{k_B T}{h} K^{\ddagger}\right) + \log\left(\frac{\gamma_A \gamma_B}{\gamma^{\ddagger}}\right) \\
&= \log k_0 + 0.509 I^{1/2}[(Z_A + Z_B)^2 - Z_A^2 - Z_B^2] \\
&= \log k_0 + 1.018 Z_A Z_B I^{1/2}, \tag{30.83}
\end{aligned}$$

where $k_0 \equiv (k_B T/h) K^{\ddagger}$ is the rate coefficient at infinite dilution.

According to Eq. 30.83, the logarithm of the rate coefficient should be linear in $I^{1/2}$ at concentrations where the limiting law is valid. This is confirmed by experiment for a wide variety of reactions (including those between ions and neutral molecules, where $Z_A Z_B = 0$ and there is no salt effect). The effect is to increase the rate when Z_A and Z_B have the same sign and to decrease it when they have opposite signs; the screening of reactants by other ions counteracts the simple electrostatic effects mentioned above. The density of the ionic atmosphere, of course, increases with the total ionic concentration. Since I is summed over *all* the ions in the solution, kinetic measurements can be taken over a wide range of I by adding a large quantity of a salt that takes no part in the reaction: This produces a high ionic strength that changes little as the reaction proceeds, and has little effect on other factors of the reaction.

Another phenomenon, known as the *secondary salt effect*, appears mainly in reactions catalyzed by acids or bases. Suppose, for example, that an aqueous solution contains a weak acid HA, with the ionization equilibrium

$$HA + H_2O = H_3O^+ + HA^-.$$

The equilibrium constant in terms of activities (with $a_{H_2O} = 1$) is

$$K_a = \frac{a_{H_3O^+} a_{A^-}}{a_{HA}} = \frac{[H_3O^+][A^-]}{[HA]} \frac{\gamma_{H_3O^+} \gamma_{A^-}}{\gamma_{HA}}. \qquad (30.84)$$

If we substitute Eq. 30.82 for the activity coefficients and solve for $[H_3O^+]$, we obtain

$$\log[H_3O^+] = \log\left(\frac{K_a[HA]}{[A^-]}\right) - \log\left(\frac{\gamma_{H_3O^+} \gamma_{A^-}}{\gamma_{HA}}\right)$$

$$= \log\left(\frac{K_a[HA]}{[A^-]}\right) + 1.018 I^{1/2}, \qquad (30.85)$$

showing that the concentration of acid increases with the total ionic strength. If the H_3O^+ ion takes part in or catalyzes any reaction in the solution, this effect increases the rate (but not the rate coefficient) of that reaction. Similar effects are found involving the ionization of a weak base, or indeed in any other reaction whose equilibrium can be shifted by a change in ionic strength.

Next we shall develop the theory of diffusion-controlled reactions, which we mentioned earlier in this section. Whenever the activation energy for a reaction is negligibly small, the limiting factor in the reaction rate must be the rate at which the reactants encounter each other. In a gas this rate can be calculated by the kinetic-theory argument of Section 30.5, but in a solution the rate is governed by diffusion. We shall make an estimate of this rate by a crude argument that mixes macroscopic and microscopic concepts. (One can do better, but the theory requires much formal machinery, and the results do not give easy physical insight.)

The reactants in diffusion-controlled reactions may be either neutral or ionic, but we shall begin with the motion of an ion in an electric field. If ions of species i have a concentration c_i and mass velocity $\mathbf{v}_i$ at a given point, then

$$f_i = v_i c_i \qquad (30.86)$$

is the molar flux density (number of moles crossing unit area per unit time) across a surface normal to the mass velocity. The mobility u_i of ion i is defined as the proportionality factor between its velocity and the electric field $\mathbf{E}$,

$$\mathbf{v}_i = u_i \mathbf{E}, \qquad (30.87)$$

so we can write the flux density as

$$f_i = \frac{Z_i}{|Z_i|} u_i c_i E, \qquad (30.88)$$

where $Z_i e$ is the charge[27] on ion i. (The unit-magnitude factor $Z_i/|Z_i|$ is introduced to fix the sign. A positive ion has a positive flux in the direction of $\mathbf{E}$.) We have assumed here that the flux is due entirely to the electric field, and parallel to it; if in addition there is a concentration gradient, then we must add a diffusion-law term (Table 28.1) to obtain the total flux density

$$f_i = -\left(D_i \frac{dc_i}{dx} - \frac{Z_i}{|Z_i|} u_i c_i E\right), \qquad (30.89)$$

where D_i is the diffusion coefficient for ion i relative to the solvent. We

[27] Note that the contribution of ion i to the electric current density $\mathbf{j}$ ($=$ charge per unit time per unit area) has the magnitude $Z_i e N_A f_i$ ($N_A =$ Avogadro's number).

are assuming that the concentration gradient and the electric field have the same direction (the x direction). At equilibrium the flux density must be everywhere zero, so that

$$\frac{dc_i}{c_i} = -\frac{Z_i}{|Z_i|}\frac{u_i}{D_i}\,d\phi, \tag{30.90}$$

where the electric potential ϕ is defined by $E_x = -\partial\phi/\partial x$. Integration of Eq. 30.90 yields

$$\ln\frac{c_i}{c_i^0} = -\frac{Z_i}{|Z_i|}\frac{u_i}{D_i}\phi, \tag{30.91}$$

where the concentration c_i^0 corresponds to the (arbitrary) zero of electric potential. From statistical mechanical considerations we assume that the concentration must obey a Boltzmann law,

$$c_i(x) = c_i^0 e^{-V_i(x)/k_BT}, \tag{30.92}$$

where $V_i(x)$ is the potential energy of ion i. But the potential energy in an electric field is simply given by $V = q\phi$, which here becomes

$$V_i(x) = Z_i e\phi(x). \tag{30.93}$$

We can thus combine Eqs. 30.91 and 30.92 to obtain the important relation between an ion's mobility and its diffusion coefficient,

$$\frac{u_i}{D_i} = \frac{|Z_i e|}{k_BT}, \tag{30.94}$$

which we derived by a different route as Eq. 29.54.

Suppose now that we consider two ions which undergo a diffusion-controlled reaction:

$$A^{Z_A} + B^{Z_B} \rightarrow \text{products}.$$

This is where we begin to mix up microscopic and macroscopic points of view. Presumably the solution is macroscopically uniform, but we suppose that there are local concentration gradients produced by the ions' electric fields—that is, that there is a gradient in c_B about each A ion and vice versa. We shall assume that these concentration gradients are described by the above equations; this may not be valid, since the macroscopic concept of diffusion may not hold on the scale of a few atomic diameters, but we shall see how it works. (One makes a similar approximation in the equilibrium theory of ionic solutions, by using the dielectric constant in the expression for the potential energy of two ions.) Thus we assume that the flux density of B ions across a sphere of radius r around each A ion is given by

$$f_B(r) = -(D_A + D_B)\left[\frac{dc_B(r)}{dr} + \frac{c_B(r)}{k_BT}\frac{dV(r)}{dr}\right], \tag{30.95}$$

in which we have combined Eqs. 30.89, 30.93, and 30.94. Since ions of both species are diffusing through the solvent simultaneously, the rate at which they approach each other is proportional to the sum of the two diffusion coefficients, $D_A + D_B$.

The concentration gradient does not extend all the way to $r = 0$. We stipulate that the reaction occurs whenever a B ion approaches within a critical distance r^* of an A ion. Thus all B ions reaching the radius r^* are lost, and the total flux (flux density times area) of B across the sphere of radius r^* must equal the reaction rate:

$$-\frac{d[B]}{dt} = k_f[A][B] = -N_A[A][4\pi r^{*2}f_B(r^*)]; \tag{30.96}$$

the expression in square brackets gives the total molar flux of B ions to each A ion, and the number of A ions per unit volume is $N_A[A]$. The $[A]$ and $[B]$ in this equation are the bulk concentrations, with respect to which the rate coefficient k_f is defined. At steady state the flux of B across a sphere of radius r ($r \geqslant r^*$) will be independent of r, so we can replace r^* by r in Eq. 30.96. Substitution of Eq. 30.95 gives

$$k_f[B] = 4\pi r^2 N_A(D_A + D_B)\left[\frac{dc_B(r)}{dr} + \frac{c_B(r)}{k_B T}\frac{dV(r)}{dr}\right], \qquad (30.97)$$

in which the right-hand side must be independent of r, in spite of its apparent functional dependence on r.

To integrate Eq. 30.97 for $c_B(r)$, we note that the function

$$B(r) \equiv c_B(r)e^{V(r)/k_B T} \qquad (30.98)$$

can be differentiated to yield

$$\frac{dB(r)}{dr} = \left[\frac{dc_B(r)}{dr} + \frac{c_B(r)}{kT}\frac{dV(r)}{dr}\right]e^{V(r)/k_B T}$$

$$= \frac{k_f[B]}{4\pi N_A(D_A + D_B)}\frac{1}{r^2}e^{V(r)/k_B T}. \qquad (30.99)$$

Now we can integrate Eq. 30.99 from the critical distance r^*, at which reaction occurs, to infinity:

$$B(\infty) - B(r^*) = \frac{k_f[B]}{4\pi N_A(D_A + D_B)}\int_{r^*}^{\infty}\frac{e^{V(r)/k_B T}}{r^2}\,dr. \qquad (30.100)$$

But if we set $V(\infty)$, the potential energy between two ions infinitely far apart, equal to zero, then $B(\infty)$ is simply the bulk concentration $[B] \equiv c_B(\infty)$. If we now define

$$\lambda^{-1} \equiv \int_{r^*}^{\infty}\frac{e^{V(r)/k_B T}}{r^2}\,dr \qquad (30.101)$$

(where λ has the dimensions of a length), we find that the concentration of B at the critical distance r^* is

$$c_B(r^*) = [B]e^{-V(r^*)/k_B T}\left[1 - \frac{k_f}{4\pi N_A(D_A + D_B)\lambda}\right]. \qquad (30.102)$$

We now introduce a quantity k_f^*, the local rate coefficient at the critical distance r^*, defined by

$$-\frac{d[B]}{dt} = k_f[A][B] = k_f^*[A]c_B(r^*), \qquad (30.103)$$

so that $c_B(r^*) = (k_f/k_f^*)[B]$. Substituting this result in Eq. 30.102 and solving for k_f, we obtain

$$k_f = \frac{4N_A(D_A + D_B)\lambda}{1 + \dfrac{4N_A(D_A + D_B)\lambda}{k_f^* e^{-V(r^*)/k_B T}}}. \qquad (30.104)$$

The coefficient k_f^* is the rate constant that would be obtained from, say, an activated-complex analysis of the A–B encounter. But, if the activation energy is indeed negligibly small, then k_f^* must be very large. In the limit $k_f^* \to \infty$, the macroscopic rate coefficient k_f approaches the so-called diffusion-controlled limit, that is, the maximum value

$$k_D = 4\pi N_A(D_A + D_B)\lambda. \qquad (30.105)$$

To evaluate λ, we must use the other quasi-macroscopic approximation we mentioned earlier: We assume that the potential energy between two ions in solution is

$$V_{AB}(r) = \frac{Z_A Z_B e^2}{4\pi\epsilon_0 \mathscr{D} r}, \tag{30.106}$$

where we have used $\mathscr{D}$ for the dielectric constant of the solvent (D being already in use). Substituting this expression into Eq. 30.101 and integrating, we eventually obtain

$$\lambda = \frac{-Z_A Z_B e^2}{4\pi\epsilon_0 \mathscr{D} k_B T[1 - \exp(Z_A Z_B e^2 / 4\pi\epsilon_0 \mathscr{D} k_B T r^*)]}. \tag{30.107}$$

Note that the quantity $e^2/4\pi\epsilon_0 \mathscr{D} k_B T$ itself has the dimensions of a length; for water at 25°C, it is about 7 Å. The length λ is positive for any combination of ions: the exponential in the denominator is small when the ions are oppositely charged, large when they have the same charge (since ordinarily r^* is less than $e^2/4\pi\epsilon_0 \mathscr{D} k_B T$). For the neutralization reaction

$$H^+(aq) + OH^-(aq) \rightarrow H_2O(l),$$

the rate coefficient measured by relaxation techniques is 1.4×10^{11} liters/mol s; the value calculated from Eqs. 30.105 and 30.107, with an assumed value of 5 Å for r^*, is 1.2×10^{11} liters/mol s. This near agreement is interpreted as evidence that the reaction is in fact diffusion-controlled. Similar agreement is found for a number of other fast reactions between oppositely charged ions, all with rates of similar magnitude (some examples: $H^+ + CH_3COO^-$, $k = 4.5 \times 10^{10}$ liters/mol s; $OH^- + NH_4^+$, $k = 3.4 \times 10^{10}$ liters/mol s). Agreement is not as good when the ions have the same charge, in which case the exact value of r^* becomes significant in Eq. 30.107.

The results of the above analysis can also be applied to fast reactions between neutral species. In this case we can assume that $V_{AB}(r) = 0$ for all $r > r^*$, so that Eq. 30.101 yields $\lambda = r^*$. The diffusion-controlled maximum rate is then

$$k_D = 4\pi N_A (D_A + D_B) r^*, \tag{30.108}$$

which also gives good agreement with experiment when r^* is calculated from van der Waals radii.

The formalism developed here can also be used to advantage for studying reaction rates considerably smaller than the diffusion-controlled ones. Although one would not have proceeded in this way by choice, it is interesting to discover that the result we derived for fast reactions, Eq. 30.104, also gives reasonable results for slow reactions. That is, in the limit that the local rate coefficient k_f^* is small, so that

$$k_f^* e^{-V(r^*)/k_B T} \ll k_D, \tag{30.109}$$

Eq. 30.104 reduces to

$$k_f = k_f^* e^{-V(r^*)/k_B T} \tag{30.110}$$

for the macroscopic rate coefficient. For reactions between ions, substitution of Eq. 30.106 for $V(r^*)$ gives a reasonably good prediction of the dependence of reaction rate on dielectric constant (over a range of composition in water–alcohol solutions, for example).[28] Similarly, for

[28] A similar dependence on $\mathscr{D}$ is found for the primary salt effect. If one uses the full Debye-Hückel expression for γ_i rather than Eq. 30.82, one finds that $\log \gamma_i$ and thus $\log(k_f/k_0)$ are proportional to $\mathscr{D}^{-3/2}$.

reactions between ions and polar molecules one can use the ion–dipole potential

$$V(r) = -\frac{Z_A e \mu_B \cos\theta}{4\pi\epsilon_0 \mathscr{D}r^2};\qquad (30.111)$$

however, this does not predict the dielectric-constant dependence as accurately as does Eq. 30.106 for ionic reactions. For reactions between neutral species this solvent effect is negligible, as we implicitly assumed in writing Eq. 30.108.

The material in the next section also applies almost exclusively to reactions in solution, but has been split off because it deals with empirical correlations rather than microscopic theory.

30.9 Linear Free Energy Relationships

We have stated that one of the objects of chemical kinetics is the prediction of reaction rates and mechanisms from a knowledge of the structure and properties of the isolated reactants and products. We have gotten some feeling for how difficult this is, especially for reactions involving complex molecules. But even when prediction is impossible, we are not helpless: As in other areas of science where *a priori* calculations have not yet succeeded, one can find empirical correlations. This section describes a particular type of correlation that describes trends in rate coefficients and equilibrium constants over sets of similar reactions.

Consider the hydrolysis of substituted ethyl benzoates,

in which the substituent X may be halogen, NO_2, and so on, in either meta or para position on the benzene ring. Rate coefficients and equilibrium constants have been measured for dozens of reactions of this type. Furthermore, similar measurements have been made for many different sets of reactions of the substituted ethyl benzoates or other substituted aromatic compounds. All told, data are available for hundreds of reactions of the general type

$$X(\text{aromatic}) + Y \rightarrow \text{products}.$$

The correlation making this mass of data meaningful was developed by L. P. Hammett. The *Hammett relations* are usually expressed in the form

$$\log k = \log k_0 + \sigma\rho,$$
$$\log K = \log K_0 + \sigma\rho,\qquad (30.112)$$

where k is the rate coefficient for a reaction of a substituted aromatic compound; k_0 is the rate coefficient for the same reaction of the unsubstituted compound; K and K_0 are the corresponding equilibrium constants; and σ, ρ are empirical parameters. The parameter ρ is constant for a given set of reactions (i.e., a group of reactions differing only in the substituent X) in the same solvent, but varies when either the reaction type or the solvent is changed. And the parameter σ varies only with the substituent group X and its position. Hammett and later workers found that values of σ and ρ could be obtained that satisfy Eqs. 30.112 for the great majority of reactions studied.

To determine the constants σ and ρ, one begins by setting ρ arbitrarily equal to unity for the set of reactions

that is, the ionization of the substituted benzoic acids in aqueous solution. The value of σ for a given substituent X and position could then be obtained from the equation

$$\sigma = \log\left(\frac{K_i}{K_{i0}}\right), \tag{30.113}$$

where K_i and K_{i0} are, respectively, the ionization constants of the substituted and unsubstituted benzoic acids. Having obtained this set of σ's, which are assumed to depend only on the substituent, one can insert them back in Eqs. 30.112 to determine ρ empirically for the same set of reactions in other solvents, or for different sets of reactions in various solvents. A listing of some of the values of σ and ρ obtained in this way is given in Table 30.2.

Thus it is clear that just a few values of σ and ρ can summarize a large body of equilibrium and rate measurements, and can help to predict rate coefficients and equilibrium constants for reactions not yet studied. But impressive as this correlation is, its range is relatively limited. Although the Hammett rules work reasonably well for meta- and para-substituted aromatic compounds, they frequently fail for other aromatic and aliphatic compounds. (Other correlation techniques do exist for different classes of reactions, but we shall not consider them here; we shall come back to the subject in our discussion of reaction mechanisms.)

The term "linear free energy relationship" in the title of this section derives from the following considerations. In the thermodynamic formulation of the activated-complex theory, the rate coefficient is written in the form of Eq. 30.68, in terms of the free energy of activation $\Delta G^{0\ddagger}$ and related quantities. Now if Eq. 30.112 for the rate coefficient holds for a given set of reactions with a particular value of ρ, we have, from Eq. 30.68,

$$\Delta G^{0\ddagger} = (\Delta G^{0\ddagger})_0 - 2.303 RT\sigma\rho. \tag{30.114}$$

A similar equation will hold for another set of reactions with a different value of ρ, say ρ'. If we divide each of these equations by the respective value of ρ and subtract one from the other, we obtain

$$\frac{\Delta G^{0\ddagger}}{\rho} - \frac{(\Delta G^{0\ddagger})'}{\rho'} = \frac{(\Delta G^{0\ddagger})_0}{\rho} - \frac{(\Delta G^{0\ddagger})'_0}{\rho'} = 0. \tag{30.115}$$

(We assume here, of course, that $\Delta G^{0\ddagger}$ and $(\Delta G^{0\ddagger})'$ refer to reactions of the same substituted compound, so that σ is the same for both and the right-hand side cancels out.) We can therefore write a linear equation between the free energies of activation of one set of reactions and those of a corresponding set,

$$\Delta G^{0\ddagger} - \frac{\rho}{\rho'}(\Delta G^{0\ddagger})' = 0, \tag{30.116}$$

where the coefficient ρ/ρ' has the same value for all reactions in the set.

TABLE 30.2

SOME VALUES OF THE HAMMETT CONSTANTS AT 25°C

Reaction constant, ρ:
(Ar $\equiv$ substituted phenyl group)

Reaction	ρ
$ArCOOH \xrightarrow{H_2O} ArCOO^- + H^+$	1.00
$ArCOOH \xrightarrow{EtOH} ArCOO^- + H^+$	1.96
$ArCH_2COOH \xrightarrow{H_2O} ArCH_2COO^- + H^+$	0.56
$ArCOOH + CH_3OH \xrightarrow{H_2O,\ H^+}$ $ArCOOCH_3 + H_2O$	-0.58
$ArOH \xrightarrow{H_2O} ArO^- + H^+$	2.26
$ArNH_3 \xrightarrow{H_2O} ArNH_2 + H^+$	2.94
$Ar_2CHCl + EtOH \xrightarrow{EtOH} Ar_2CHOEt + HCl$	-4.03

Substituent constant, σ:
(σ_m, in meta position; σ_p, in para position)

Substituent	σ_m	σ_p
H	0.00	0.00
CH_3	-0.07	-0.13
F	0.34	0.06
Cl	0.37	0.24
NH_2	-0.04	-0.17
NO_2	0.71	0.78
COOH	0.35	0.40
CN	0.61	0.67

One can write an equation similar to Eq. 30.114 for the total free energy of reaction,

$$\Delta G^0 = (\Delta G^0)_0 - 2.303 RT \sigma \rho \qquad (30.117)$$

(since $\Delta G^0 = -RT \ln K$), and therefore obtain

$$\Delta G^{0\ddagger} = \Delta G^0 + [(\Delta G^{0\ddagger})_0 - (\Delta G^0)_0]. \qquad (30.118)$$

Since the term in brackets is a constant, this is a restatement of the Hammett relations as an assumption about the relationship between the free energy of activation and the free energy of reaction.

It is interesting to inquire about the explanation of such empirical correlations as Hammett's, where by "explanation" we mean a theoretical basis, profound or modest. No satisfactory explanation has in fact been given. Some of the discussions in a developing field such as this make fascinating reading. They show the speculations, the guesses, the frequently awkward groping for clues, the working hypotheses, the simple models, all of which are the tools by which science usually progresses. A science in development has an appearance far different from that of a well-formulated construct such as thermodynamics.

We have discussed at some length the general concepts of the theoretical approach to chemical kinetics. Before we go on to the interpretation of individual reaction mechanisms, it is time for us to obtain a more balanced picture of the science of kinetics by studying the experimental methods used to measure reaction rates. We summarized these methods very briefly at the start of Section 30.1, but a fuller treatment is now warranted. In doing so we have two purposes: first, to give some indication of the great variety of techniques employed; second, to illustrate the vast range of time scales that can be studied.

The study of reaction kinetics is to a large extent based on measurements of the rate of disappearance of one or more reactants and the rate of appearance of one or more products. One may or may not know the stoichiometry of the overall reaction (which is the sum of all the steps in the reaction mechanism). When the stoichiometry is fully known, then one usually also knows the equilibrium constant and therefore (as we shall see) the ratio of forward and reverse rate coefficients. But sometimes one must deal with a reaction in which not all the products are known. Suppose, for example, that you are interested in the rate of ozonization of a hydrocarbon such as octane—a problem of some importance in the pollution of urban atmospheres—and that not all the products have yet been identified; measurements of the rate of disappearance of both reactants would provide important, but not complete, information on the reaction rate and mechanism.

Thus measurements of the variation of concentrations of chemical species with time are central to kinetics. When possible, such measurements are usually made at constant temperature, and if possible with all other variables that may affect the rate coefficient held constant. Since rate coefficients are strongly temperature-dependent (cf. Section 30.5), a high degree of control may be required. For instance, to determine a rate coefficient to a precision of 5% at 25°C, the temperature must be held constant to about 0.4°C for an activation energy of about 100 kJ/mol.

As we said in Section 30.1, experimental techniques for measuring reaction rates can be divided into two categories, physical and chemical methods. By *chemical methods* we mean the standard gravimetric or volumetric techniques of quantitative analysis, which can be made specific for one or more reactants or products. Chemical analysis has the advantage of specificity and relatively high accuracy: A very good quantitative analysis can yield an accuracy of about 0.1% in the concentration of a given species. But to determine a rate coefficient we must have a concentration that changes in time; this leads to a loss in accuracy, because one is interested in the *differences* between concentrations. Moreover, whatever method is used for measuring concentrations, one must be able to complete a measurement in a sufficiently short time that the concentrations do not change significantly by reaction. Chemical analysis has the disadvantages of being slow and possibly disturbing the reacting system (when one must withdraw a physical sample). These problems can sometimes be overcome by taking a sample from the reacting mixture and at once arresting or slowing the reaction rate in the sample—which can be done by removing a catalyst, adding an inhibitor, or lowering the temperature—to a point where the analysis can be completed before the concentration changes significantly.

Physical methods of measuring concentrations usually do not disturb the reaction rate, and can be much more rapid than chemical methods. Thus they are preferred when they can provide the information desired. Often, however, a physical method may not be specific. As an example of

30.10
Experimental Methods in Kinetics

this we mentioned earlier the measurement of the index of refraction, which can be significantly changed if the system contains an unknown impurity with a high refractivity. But this problem can sometimes be overcome by using different physical methods on the same reaction. The great advantage of physical methods is the range of time scales to which they can be applied. Chemical methods are limited to reactions with characteristic times of a few milliseconds or longer. (By the characteristic reaction time we mean an order-of-magnitude value of the time interval available for measurements on the reaction, say, the time necessary for the reaction to go 90% to completion.) But physical measurements are now possible for reactions with time scales as short as 10^{-12} s![29] If one wishes to have any hope of detecting short-lived intermediates such as free radicals, then only physical methods are applicable.

Let us now review some of the physical methods suitable for kinetic measurements. The simplest involve the measurement of macroscopic properties such as *pressure* and *volume*. Consider a gaseous reaction in a constant-volume vessel when the number of moles of products produced differs from the number of moles of reactants consumed, as in

$$2NO_2 \rightarrow 2NO + O_2.$$

As the reaction proceeds, the pressure increases at a rate virtually proportional to the reaction rate: if the gas mixture can be considered ideal, we have

$$\frac{d[O_2]}{dt} = \frac{1}{2}\frac{d[NO]}{dt} = -\frac{1}{2}\frac{d[NO_2]}{dt} = \frac{1}{V}\frac{dn_{total}}{dt} = \frac{1}{RT}\frac{dp}{dt}. \quad (30.119)$$

Pressure measurements (e.g., with strain gauges) take of the order of a millisecond, so the method, although simple, is limited to reactions with longer time scales.

At constant pressure, most easily obtained in liquid solutions open to the atmosphere, one can measure the change in the volume of a reacting system. The total volume is the sum of the partial volumes of the species present, $V = \sum_i n_i \bar{v}_i$; both the numbers of moles and the partial molar volumes $\bar{v}_i$ may change as the reaction proceeds. Such volume changes are most likely to be measurable when the reactants and products differ significantly in density; but since the change is small (especially in dilute solution), it is magnified by locating the liquid–air interface in a fine capillary, in an instrument called a *dilatometer*. (The temperature must be very carefully controlled, to eliminate volume changes due to thermal expansion.) One example of a reaction studied by dilatometry is the hydrolysis of acetic anhydride,

$$(CH_3CO)_2O + H_2O \rightarrow 2CH_3COOH.$$

The method is especially useful for polymerization reactions. Since the movement of a liquid in a fine capillary is slow, dilatometry is limited to reactions with time scales longer than about a second.

The variation of *transport properties* with the progress of a reaction has also been used to measure reaction rates. For example, the electrical conductivity of a solution will vary significantly in such reactions as the hydrolysis of an alkyl halide,

$$C_2H_5Cl + 2H_2O \rightarrow C_2H_5OH + H_3O^+ + Cl^-,$$

or any ordinary acid–base neutralization. In polymerization reactions the

[29] Either physical or chemical methods may conveniently be used at the other end of the time-scale spectrum. If you want to consider radioactive decay as a chemical reaction, half-lives as long as 10^{15} years have been measured (the age of the earth is only 4.6×10^9 years).

viscosity of the solution increases along with the average size and density of the polymers formed, and this change in viscosity (measured in samples withdrawn from the system) can be used to determine the reaction rate.

The *mass spectrometer* (two types of which were described in Section 1.5) is a sensitive detector of atoms and molecules, and has been applied to rate measurements. The method is most useful for gas-phase reactions: One allows a capillary leak from the reaction vessel so that a small flow of the reacting mixture enters the mass spectrometer, which provides a continuous mass analysis. A complete mass scan takes at best about a millisecond; but if only a single mass peak characteristic of a given species is measured, the time can be cut to about a microsecond.

There are two important techniques based on the electromagnetic properties (other than charge) of individual molecules: *electron spin resonance* and *nuclear magnetic resonance*, both of which we discussed in Part One (cf. Sections 5.1 and 9.8). Electron spin resonance spectroscopy, it will be recalled, is applicable to species with unpaired electron spins; it is thus a powerful method for studying reactions that involve free radicals as reactants, products, or even intermediates in the reaction mechanism. Nuclear magnetic resonance spectroscopy is similarly applicable to species containing atoms with nonzero nuclear spins (especially ^1H). The nuclear spin energy levels depend on the electron density around the nucleus, as exhibited in the so-called *chemical shift*, the dependence on chemical site of the proportionality factor between magnetic field and resonance frequency. For example, the nmr spectrum due to a proton in a $—CH_3$ group differs from that due to a proton in an alcohol's $—OH$ group. Thus nmr is a sensitive tool for structural analysis, and can in principle be used to measure concentrations.

But how does one obtain information about kinetics from nmr and esr spectra? One commonly used method is to calibrate the intensity of the spectral lines against concentration; a more subtle approach involves the spectral line broadening due to interactions. The natural line width of a spectral line (Section 7.3) is given by the uncertainty principle as $\Delta E \approx \hbar/\Delta t$, where Δt is roughly the lifetime of the less stable state. The natural line width is determined by the rate at which energy flows from the excited molecule by emission of radiation. Spectral lines can be broadened by interactions that change the emission rate, such as those with electric fields. They can also be broadened by processes that shift the energy levels of the excited species; *pressure broadening* by collisions in gases is an example. In nmr spectra the principal source of broadening is the interaction between the spins of neighboring nuclei; this spin–spin relaxation is a quite slow process, with relaxation times of the order of seconds in aqueous solution (corresponding to line widths of the order of 1 Hz). The shape of the spectral lines will thus be strongly affected if reaction occurs on the same time scale as the spin relaxation.

These principles are especially applicable to rapid exchange reactions. Consider the reaction

$$HA + B^- \rightleftharpoons A^- + HB.$$

Suppose that in the absence of reaction the nmr spectrum of the protons in HA and HB is as shown in Fig. 30.18a: separate peaks for each species, separated by a chemical shift Δv. If the exchange rate is very small relative to Δv, the nmr spectrum will not be much affected by the reaction. But when the exchange rate becomes comparable to Δv—that is, when the average lifetime of a proton in the HA or HB state is of the order of $(\Delta v)^{-1}$—the uncertainty principle blurs the distinction between the two states, and the two peaks coalesce into a single broad line, as shown in Fig. 30.18b. In the limit of very fast exchange, with the reaction

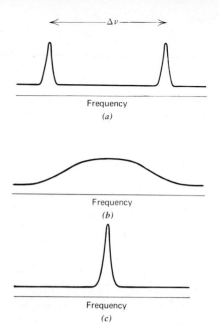

$\longleftarrow \Delta v \longrightarrow$

Frequency
(a)

Frequency
(b)

Frequency
(c)

FIGURE 30.18
Nuclear magnetic resonance spectra of exchange reactions: (*a*) no reaction, two peaks separated by chemical shift Δv; (*b*) exchange rate comparable to Δv, two peaks coalesce into one broad line; (*c*) exchange rate much greater than Δv, single sharp line.

rate much greater than $\Delta\nu$, the single line becomes sharp again, as in Fig. 30.18c; one can say that in this case the proton shifts back and forth many times within a single cycle of the electromagnetic field, and is thus observed in the average of the HA and HB environments. The kinetic significance is this: The resonance equations for such systems have been solved for the complete range of values of the reaction rate, and one can thus determine the rate coefficient by measuring the shape of the spectrum (especially the line widths). Chemical shifts for protons are of the order of 100 Hz, but one can distinguish line shapes corresponding to much larger exchange rates; it is easy to obtain qualitative rate measurements on a time scale of milliseconds. A typical reaction studied by nmr techniques is

$$NH_4^+ + NH_3 \rightleftharpoons NH_3 + NH_4^+,$$

for which $k_f = k_r = 1.1 \times 10^9$ liters/mol s. Similar techniques are used in esr measurements on reactions in which a suitable electron exchange (not altering the total spin) occurs, for example,

$$C_{10}H_8^- + C_{10}H_8 \rightleftharpoons C_{10}H_8 + C_{10}H_8^-$$

in solvents such as 1,2-dimethoxyethane: interaction between the electron and proton spins produces a hyperfine spectrum whose lines are broadened by the exchange.

Conventional *spectroscopic techniques* are useful and popular for kinetic measurements. The basis is quite simple: One merely measures the intensity of some spectral line characteristic of a given species as a function of time, and calibrates that intensity against concentration. The infrared, visible, and ultraviolet spectral ranges have all been used for this purpose, as have both absorption and emission techniques. The limiting factor is the time necessary for measuring the intensity of a spectral line. This time was of the order of 1 s for ordinary spectrometers, but was reduced to microseconds with the introduction of fast-recording infrared spectrometers. The last 10 years have seen extensive development of picosecond (10^{-12} s) spectroscopy, using pulses of laser light on that time scale: reactions can be initiated by one such intense pulse and followed in time by succeeding weaker pulses. One system studied in this way is that of excess electrons in water: Free electrons generated by photoionization of $Fe(CN)_6^{3-}$ or I^- ions are hydrated (to H_2O^-, etc.) in about 4 ps. Another example is the recombination of iodine atoms: I_2 molecules (in such solvents as CCl_4) are excited by the initial laser pulse, subsequently dissociated to ground-state I atoms, and reformed from these atoms. The recombination process is found to have two stages, a rapid one ($\approx 10^{-10}$ s) in which I atoms trapped within a "cage" rejoin their original partners, and a slower one ($\gtrsim 10^{-8}$ s) for those atoms that escape from the "cage."

Other optical methods have also been applied to kinetics, including *optical rotation* (Section 9.6) and *light scattering*.

For a long time a major limitation in measuring the rates of fast reactions was the simple fact that it takes about a second to mix the reactants in a flask or beaker. One way around this is the application of a stimulus to an already homogeneous system (as in the laser experiments just described), but this cannot be done in all systems. Another increasingly common solution is the use of *flow systems*, rather than static reaction systems. In its simplest form a flow system consists of three tubes in a Y-arrangement, as shown in Fig. 30.19. Reactants A and B flow through the correspondingly labeled tubes, mix at the junction O, and react as they flow down the outlet tube. Measurements of concentration at given distances x from O correspond to specific time intervals from the commencement of reaction. Since the volume at O is very small, mixing

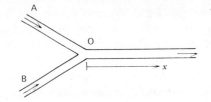

FIGURE 30.19
Schematic diagram of a flow system.

times can be as short as 0.1 ms; for a typical flow rate of 10 m/s, 1 cm in the outlet tube corresponds to an interval of 1 ms. This technique, with various improvements (such as the "stopped-flow method," in which the reactants are mixed rapidly and then observed at rest), is useful for studying rates on the millisecond scale.

Two other methods, *flash photolysis* and *relaxation techniques*, considerably extended the available time scale for static measurements before the development of picosecond spectroscopy. Flash photolysis covers the microsecond and nanosecond ranges for reactions in both gases and solutions; relaxation techniques are used over the same range primarily for solutions. Flash photolysis is relatively simple: A premixed gaseous system is irradiated briefly ($\approx 1\ \mu s$) with an intense pulse of light, reactive species are formed (either excited states or free radicals formed by photodissociation), and the subsequent reaction is followed by spectroscopic methods; the picosecond techniques already described are merely an extreme application of the same principles. The method can be used to study physical processes (e.g., the return of excited molecules to the ground state, or energy transfer in collisions) as well as chemical reactions in the strict sense. Among the most thoroughly studied examples of the latter are recombination reactions, such as

$$I + I \rightarrow I_2$$

(which actually proceeds by one or more third-body mechanisms), in which the I atoms are produced by the photolysis reaction

$$I_2 + h\nu \rightarrow 2I.$$

Other methods fundamentally similar to flash photolysis include *pulse radiolysis*, in which the initial excitation is produced by ionizing radiation (most commonly a beam of electrons) rather than light, and the use of *shock waves* (produced, for example, by rupturing a diaphragm across which a large pressure difference exists).

Relaxation techniques avoid the time limitation of mixing reactants in a different way, by starting with a system already in chemical equilibrium. Consider the general reaction

$$A + B \underset{k_r}{\overset{k_f}{\rightleftharpoons}} C + D,$$

with equilibrium concentrations a, b, c, d. Suppose that the system at equilibrium is perturbed, on a time scale either shorter than or comparable to the time required for the reaction to return to equilibrium.

Let us begin for simplicity with the case of a single-pulse perturbation—for example, a temperature jump induced by discharging a capacitor through an ionic solution. The ionic species attempt to move in response to the varying electrical field, and the electrical resistance of the solution produces heating. If ΔH for the reaction studied is not zero, then an increase in temperature alters the equilibrium constant. The system is thus no longer in equilibrium at the new temperature, and reaction must occur to bring the system to a new equilibrium. This reaction can be followed by spectroscopic or other rapid physical measurements.

Let a', b', c', d' be the equilibrium concentrations to be attained at the new temperature; we can then define a reaction variable x by

$$x \equiv a' - [A] = b' - [B] = [C] - c' = [D] - d', \tag{30.120}$$

where [A], [B], [C], [D] are the concentrations at a given time. The reaction rate is thus

$$\frac{dx}{dt} = -\frac{d[A]}{dt}, \text{ etc.} = k_f[A][B] - k_r[C][D]$$

$$= k_f(a' - x)(b' - x) - k_r(c' + x)(d' + x). \tag{30.121}$$

If the temperature jump has a relatively small effect on the equilibrium constant, then the change in equilibrium concentrations is small and x is always a small quantity compared to a', b', c', d'. We can thus linearize Eq. 30.121 by multiplying out the parentheses and neglecting terms in x^2, which yields

$$\frac{dx}{dt} = k_f a'b' - k_r c'd' - x[k_f(a'+b') + k_r(c'+d')]. \qquad (30.122)$$

Since the equilibrium constant at the new temperature is (cf. Section 30.7)

$$K(T') = \frac{k_f(T')}{k_r(T')} = \frac{c'd'}{a'b'}, \qquad (30.123)$$

the first two terms in Eq. 30.122 cancel. We are thus left (in the linear approximation) with a relaxation-type equation,

$$\frac{dx}{dt} = -\frac{x}{\tau}, \qquad (30.124)$$

where the relaxation time τ is defined by

$$\tau^{-1} \equiv k_f(a'+b') + k_r(c'+d'). \qquad (30.125)$$

One can measure τ by following the concentration of one of the four species (usually by spectroscopic methods) and substituting in Eq. 30.124. If one also knows the equilibrium concentrations a', b', c', d', which are usually quite close to those at the initial temperature, then Eqs. 30.123 and 30.125 can be combined to solve for the rate coefficients k_f and k_r.

Temperature perturbations are not, of course, the only kind that can be applied. For example, a reacting system can be subjected to a pressure disturbance by means of a sound wave, which is a periodic perturbation rather than a single pulse. If a volume change is associated with the reaction under study, then ΔG and thus the equilibrium constant vary with pressure; thus a pressure perturbation on a system at equilibrium will again produce a nonequilibrium state. We assume the same general reaction as in our previous example, but in this case (for a periodic perturbation) we consider the difference from the *initial* concentrations,

$$y \equiv a - [A] = b - [B] = [C] - c = [D] - d. \qquad (30.126)$$

The rate equation is thus

$$\frac{dy}{dt} = k_f(a-y)(b-y) - k_r(c+y)(d+y), \qquad (30.127)$$

which in the linearized approximation becomes

$$\frac{dy}{dt} = k_f ab - k_r cd - y[k_f(a+b) + k_r(c+d)]. \qquad (30.128)$$

All this is formally as before, but now we introduce the periodic perturbation, which we shall define as a sound wave with frequency ω. The equilibrium concentrations corresponding to the pressure at time t will vary with the same frequency, so that we have

$$y_{eq}(t) = \mathscr{A} \sin \omega t, \qquad (30.129)$$

where the amplitude $\mathscr{A}$ depends on the amplitude of the sound wave. In general, the actual concentrations will not correspond to this equilibrium value, but will lag behind. However, the two curves will cross from time to time, and the crossings must be at the extrema of y (since y is increasing when $y < y_{eq}$, decreasing when $y > y_{eq}$). Thus we have $dy/dt = 0$ when $y = y_{eq}$; subtracting this result from Eq. 30.128 yields the relaxation equation

$$\frac{dy}{dt} = -\frac{y - y_{eq}}{\tau},$$ (30.130)

with the relaxation time given by

$$\tau^{-1} \equiv k_f(a+b) + k_r(c+d).$$ (30.131)

Rearrangement and combination with Eq. 30.129 gives the linear differential equation

$$y_{eq} = \tau \frac{dy}{dt} + y = \mathcal{A} \sin \omega t,$$ (30.132)

which has the solution

$$y(t) = \frac{\mathcal{A}}{1 + \omega^2 \tau^2} \sin \omega t - \frac{\mathcal{A}\omega t}{1 + \omega^2 \tau^2} \cos \omega t.$$ (30.133)

Equation 30.133 is of a form we have seen before. It is directly analogous to Eq. 28.157, which we derived for the energy variations due to a periodic sound wave in a gas in which inelastic energy transfer can occur. (Obviously, chemical reaction is an extreme case of inelastic energy transfer.) Carrying over the analysis made previously, we find that the first term on the right-hand side is in phase with the sound wave and thus with the variation of y_{eq}, but has an amplitude reduced from that of y_{eq} by the factor $1 + \omega^2 \tau^2$; the second, out-of-phase term corresponds to attenuation of the sound wave: it takes energy to displace the reaction from equilibrium, and the phase lag prevents this energy from being completely returned to the wave. The experiment is so designed that $\omega \tau \approx 1$, the condition under which the second term has its maximum and the energy transfer is the greatest. The variation with time of y_{eq} and y is shown in Fig. 30.20; by analogy with Eq. 28.156, the phase lag between y_{eq} and y is

$$\phi = \cos^{-1}\left(\frac{1}{1 + \omega^2 \tau^2}\right) = \sin^{-1}\left(\frac{\omega \tau}{1 + \omega^2 \tau^2}\right).$$ (30.134)

Since ω is known, the phase lag gives a direct measurement of the relaxation time τ and hence of the forward and reverse rate coefficients, provided that the equilibrium constant is known. In principle one could measure the phase lag directly by following pressure and concentration as functions of time; however, it is simpler to measure the sound absorption coefficient, which as we have seen is closely related to the out-of-phase term in Eq. 30.133.

The relaxation method is very powerful, and has been applied to the study of many reactions. One of the earliest important reactions studied by the temperature-jump method was

$$H^+ + OH^- \rightarrow H_2O,$$

which was found to have $\tau = 35 \ \mu s$ and $k_f = 1.5 \times 10^{11}$ liters/mol s; the limiting factor is the time over which a temperature jump can be produced, which can be as short as a microsecond for a jump of several degrees. Similar results can be obtained from a sudden jump in electric field (the *Wien effect*). By sound-absorption techniques, as we noted at the end of Section 28.11, it is possible to measure relaxation times as

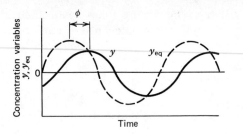

FIGURE 30.20
Variation of concentration with time in a reaction system perturbed by a periodic sound wave: – – –, equilibrium concentration y_{eq} (in phase with pressure variation); ——, actual concentration y. The two sinusoidal curves differ in amplitude by the factor $1 + \omega^2 \tau^2$, and in phase by the lag ϕ defined in Eq. 30.134.

short as 10^{-9} s. An example of a reaction studied by this method is

for which $k_f = 2.8 \times 10^8$ liters/mol s and $k_r = 1.4 \times 10^9$ liters/mol s. Studies have also been made with alternating electromagnetic fields, which are especially useful for ionization equilibria.

Isotopes, both stable and radioactive, can be advantageously used as tracers in studying reaction rates and mechanisms. In a reaction some of whose products are effectively removed from the system, for example, when one product is a precipitate, the amount of radioactivity in the system can be a direct measure of concentration (provided that the radioactive atoms wind up in the precipitate). More commonly, however, isotopic substitution is used to answer mechanistic questions. The most obvious use is in the study of exchange reactions: for example, the rate at which hydrogen atoms exchange between two compounds can be studied by substituting deuterium in one reactant, yielding a reaction of the type

$$HX + DY \rightleftharpoons HY + DX.$$

An example of the more detailed questions that can be answered involves the combustion of butane to CO_2. To test the hypothesis that the product CO_2 was formed by way of intermediate CO, radioactive ^{14}CO was added to the system; since no significant amount of $^{14}CO_2$ was formed, the proposed mechanism had to be rejected. Carbon-14 is, of course, especially useful in studying the rates of biochemical reactions. (Note that isotopic-substitution measurements are complicated by the fact that isotopes are not perfectly interchangeable: There are isotopic effects on reaction rates that must be taken into account.)

Thus we have seen that there is a wide variety of methods available for measuring reaction rates. But remember that no single rate measurement can completely explain the kinetics of a reaction. Just to obtain the order of a reaction with respect to a given species, one must vary the concentration of that species and observe the consequent change in reaction rate (although for simple mechanisms this information can be obtained in a single experiment). Furthermore, the rate coefficient may depend on a number of variables, such as temperature, pressure, and ionic strength. To have a complete understanding of the kinetics, one must measure the rate as a function of all such variables—a substantial task that is not often accomplished. Finally, the reaction rate is not the only type of information useful in guessing reaction mechanisms; we shall have more to say on this subject in Section 30.12.

30.11 Analysis of Data for Complex Reactions

In Section 30.1 we introduced the concepts of experimental rate expression, order, and molecularity, and we illustrated how rate coefficients can be obtained from experimental data for simple first- and second-order reactions. We shall now return to this topic by considering some more complicated possibilities, in preparation for our study of reaction mechanisms.

A. Third-Order Reactions

A third-order reaction is, of course, one obeying the rate law

$$-\frac{d[A]}{dt} = k[A][B][C], \tag{30.135}$$

where A, B, C may be the same or different. Once one knows the stoichiometry, obtaining the rate coefficient from experimental data is straightforward and need not be discussed in detail. Of more interest is the question of how third-order reactions can exist. If such a reaction proceeds in a single elementary reaction step, then it is termolecular; but since the probability of three molecules colliding simultaneously is much smaller than for bimolecular collision, termolecular reactions are found to be rare. More commonly, a third-order rate expression results from a sequence of bimolecular reaction steps, as we shall illustrate in Section 30.15.

Thus far most of our analysis has been carried out as if the rate of a chemical reaction depended on only one rate coefficient. Although such cases do (effectively) exist, generally things are more complicated. The remainder of this section is devoted to a formal analysis of the more common types of complications, beginning with the case of *reversible reactions*.

In a sense all reactions are reversible. If a general (forward) reaction

$$A+B \xrightarrow{k_f} C+D$$

occurs, then the reverse reaction

$$C+D \xrightarrow{k_r} A+B$$

must also be possible. Whether we begin by mixing A and B or by mixing C and D, eventually a state of chemical equilibrium will be established, in which the rates of the forward and reverse reactions are equal and the composition of the system does not change further with time. Many reactions do essentially "go to completion"; that is, the equilibrium constant is so large that at equilibrium the concentrations of products far exceed those of reactants. For such cases the reaction rate is hardly influenced by the existence of the reverse reaction. But for reactions with equilibrium constants of the order of unity (within a factor of 10), the net reaction rate depends significantly on both the forward and the reverse rate coefficients.

Consider the reaction

$$H_2+I_2 \underset{k_r}{\overset{k_f}{\rightleftharpoons}} 2HI,$$

which at moderate temperatures (say, 300–500°C) has significant rates in both directions. The forward and reverse reactions are both observed to be second-order, as if each proceeded by a single bimolecular step, although the actual mechanism is probably more complicated. The reaction rate is thus given by

$$-\frac{d[I_2]}{dt} = k_f[H_2][I_2] - k_r[HI]^2. \qquad (30.136)$$

Since this is one equation with two unknown constants, we need an additional relation between them before they can be evaluated; this is provided by the equilibrium constant. At equilibrium we must have $d[I_2]/dt = 0$, and Eq. 30.136 then reduces to

$$\frac{k_f}{k_r} = \left(\frac{[HI]^2}{[H_2][I_2]}\right)_{eq} = K_V, \qquad (30.137)$$

where K_V is the equilibrium constant in terms of concentrations; this is the

B. Reversible Reactions

result we have already obtained in general terms as Eq. 30.60. One can thus evaluate both k_f and k_r by measuring concentration as a function of time and fitting the data to Eqs. 30.136 and 30.137.

Calculations such as that just described are simpler if one formally solves the equations beforehand. We shall carry out such a derivation here for the simple case of opposing first-order reactions:

$$A \underset{k_r}{\overset{k_f}{\rightleftharpoons}} B.$$

Suppose that we begin with only A present, so that $[B] = [A]_0 - [A]$, where $[A]_0$ is the value of $[A]$ at $t = 0$. The rate equation is then

$$-\frac{d[A]}{dt} = k_f[A] - k_r([A]_0 - [A]), \tag{30.138}$$

and integration over time from 0 to t yields

$$\ln\left(\frac{k_f[A]_0}{(k_f + k_r)[A] - k_r[A]_0}\right) = (k_f + k_r)t. \tag{30.139}$$

At equilibrium we must have $d[A]/dt = 0$, so, by Eq. 30.138,

$$k_f[A]_{eq} - k_r([A]_0 - [A]_{eq}) = 0, \qquad [A]_{eq} = \frac{k_r[A]_0}{k_f + k_r}. \tag{30.140}$$

Substitution into Eq. 30.139 then gives the simple result

$$\ln\left(\frac{[A]_0 - [A]_{eq}}{[A] - [A]_{eq}}\right) = (k_f + k_r)t. \tag{30.141}$$

This is essentially the same form as Eq. 30.6, so $k_f + k_r$ can be obtained from a plot of $\ln([A] - [A]_{eq})$ against time; adding the value of the equilibrium constant $(= k_f/k_r)$ enables one to evaluate k_f and k_r separately. Similar analyses can be made for more complex reversible reactions.

A given system of reactants can often yield more than one set of products; this situation is variously described as one of *simultaneous*, *parallel*, or *competitive reactions*. For example, the photolysis of formaldehyde has been found to occur by two processes:

$$\text{HCHO} \overset{h\nu}{\longrightarrow} \Big\langle {\text{H}_2 + \text{CO} \atop \text{H} + \text{CHO}} \,,$$

C. Simultaneous Reactions

with the ratio between the two dependent on the wavelength of the incident light. (We shall discuss similar examples in the section on photochemical reactions.) In general there need be no relation between the rate constants, or even the orders, of the simultaneous reactions, and the ratio of the reaction rates may vary with temperature, pressure, and so on. One usually has to study the reactions separately, which of course involves measuring the concentrations of the different products. Matters are somewhat simpler for the case in which the reactions are of the same order, since then the ratio of products is invariant with time. For example, in the reaction system

$$A \overset{k_1}{\longrightarrow} B,$$

$$A \overset{k_2}{\longrightarrow} C$$

(with no other reactions involving A, B, C taking place), the ratio of B and C produced is simply the ratio of the rate coefficients; if only A is initially present, then we have

$$\frac{[B]}{[C]} = \frac{k_1}{k_2}.$$ (30.142)

Whatever reactions take place, of course, the rate of change of a given species is always the sum of the rates of all the processes involving it; in the example just mentioned, we have

$$-\frac{d[A]}{dt} = k_1[A] + k_2[A].$$ (30.143)

One must usually know (or guess) what all the significant reactions are, and make some measurements distinguishing them, before one can evaluate the individual terms in an equation like 30.143.

One can also apply the terms "simultaneous reactions" and so on to the common situation in which a given substance reacts simultaneously with more than one other reactant. This can get even more complicated, but the same principles apply.

Most reactions proceed not by a single elementary reaction step, but by a sequence of such steps (called a mechanism). We shall look at such sequences in detail in the next and succeeding sections; here we shall only point out some consequences of the existence of *consecutive* (or *series*) *reactions*.

Suppose that species A reacts unimolecularly to yield B, which in turn reacts unimolecularly to yield C:

$$A \xrightarrow{k_1} B \xrightarrow{k_2} C.$$

(Few chemical systems are this simple, but sequences like this are common in radioactive decay series.) If both reaction steps are first-order, we can write the rate equations

$$-\frac{d[A]}{dt} = k_1[A],$$

$$\frac{d[B]}{dt} = k_1[A] - k_2[B],$$ (30.144)

$$\frac{d[C]}{dt} = k_2[B].$$

Note that we have applied the principle just described: The rate at which [B] changes is the algebraic sum of all the rates involving species B. It is not difficult to integrate Eqs. 30.144, but we do not need to integrate them to see qualitatively how the concentrations vary with time. Say that we start with only A present, and end with only C present (i.e., that each reaction "goes to completion"). Then the concentration of A decreases steadily from $[A]_0$ to zero, while that of C increases steadily from zero to $[C]_\infty = [A]_0$. What about the intermediate B? Since no B is present at either the beginning or the end, [B] must rise to a maximum and then fall off again. In fact integration gives the results:

$$[A] = [A]_0 e^{-k_1 t},$$

$$[B] = \frac{k_1[A]_0}{k_2 - k_1} e^{-k_1 t} - e^{-k_2 t},$$ (30.145)

$$[C] = [A]_0 - [A] - [B].$$

The exact variation of [A], [B], [C] with time depends on the ratio of the rate coefficients k_1 and k_2. In Fig. 30.21 we plot the concentrations of

D. Consecutive Reactions

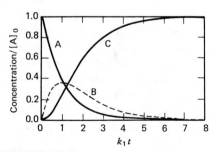

FIGURE 30.21
Variation of concentration with time for the consecutive reactions

$$A \xrightarrow{k_1} B \xrightarrow{k_2} C,$$

with $k_1 = k_2$.

the three species for the case $k_1 = k_2$. Note that the concentration of B stays relatively small if the rate coefficients are about equal. This is, of course, what you might expect, since before very long (after the point when [B] = [A]) B is being destroyed faster than it is being formed. The higher the ratio of k_2 to k_1, the more rapidly B will be depleted and the smaller will be its maximum concentration. If [B] is always small, then $d[B]/dt$ must also be small in absolute terms; we shall make use of this observation in the next section.

If k_1 is larger than k_2, on the other hand, then for a long time the concentration of B will be built up faster than it is depleted, and the maximum value of [B] will be relatively high. The change from A to B occurs quickly, the change from B to C slowly. The rate of formation of C is then determined almost exclusively by the slow second reaction, as we can see from Fig. 30.22 for the case $k_1 = 20k_2$: for $k_1 t \gtrsim 5$ the curves for [B] and [C] are essentially identical to those for a simple first-order process beginning with [B] = [A]$_0$ (which we would obtain in the limit $k_1/k_2 \to \infty$, with all the A instantaneously transformed to B). Whenever one step in a sequence of elementary reaction steps is considerably slower than the others, it dominates the kinetics in this way and is called the *rate-determining step*.[30] This property will also be called upon in the next section.

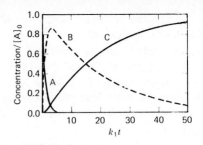

FIGURE 30.22
Same as Fig. 30.21, but for the case $k_1 = 20k_2$.

30.12
Mechanisms of Chemical Reactions

As we have already indicated, most chemical reactions proceed by a series of elementary reaction steps, rather than by a single step. The sequence of stoichiometric equations defining these steps is called the *mechanism* of the reaction. In the last section we began to look at some of the formal ways of analyzing a reaction whose mechanism is known; but how does one find out the mechanism in the first place? In principle it should be possible to obtain either empirical rules or a full theory for predicting reaction mechanisms, given a knowledge of the molecular structures of the reactants, products, and solvent and the interactions among all these species. Although much progress has been made toward this goal, the science of chemical kinetics is still a long way from attaining it. At the present time one is usually reduced to guessing the mechanism of a reaction; we shall see, however, that such guesses need not be blind.

Once a guess has been made about a reaction mechanism, one must then see if the guess is consistent with what is known about the reaction: the stoichiometry, the experimental rate expression, its temperature dependence, and any other measurements that have been made (e.g., the detection of free radicals, isotopic tracer studies, or stereochemical evidence). Alas, it is frequently not too difficult to assign a reaction two different mechanisms, both consistent with the available measurements. One can then make more measurements, of course; but this is not the only recourse. Although what is now known about mechanisms consists largely of empirical classifications (usually the first approach to interpreting experimental results), this is often enough for one to recognize the similarities between reactions of related types. With enough experience of this sort, one can guess a plausible mechanism with some assurance.

As you may have noticed, we still haven't given any indication of *how* these guesses are made. Unfortunately, there are no universal rules

---

[30] For the case with $k_2 \gg k_1$ discussed in the previous paragraph, the reaction A → B is the rate-determining step.

one can follow; if there were, it wouldn't be guessing. But there do exist quite a few indications that one can use as clues. Although the only way to learn all of these is by actual experience with a wide variety of reactions, some of the most important ones can be derived quite easily. Let us therefore look at some standard types of mechanisms, and see what sorts of empirical rate expressions they should give—for the nature of the empirical rate expression is usually the most important clue to the mechanism.

We saw in the last section how rate equations can be derived from a mechanism, but the resulting equations tend to be unwieldy (even when the reaction steps are all of the form A→B). In many cases, however, one can make simplifying assumptions; usually one makes either of the following two assumptions:

1. There is one elementary reaction step in the mechanism with a rate much smaller than all preceding or following steps. The rate of the reaction is then essentially controlled by the rate of the slow step, which is therefore called the *rate-determining step.*

2. The sequential elementary reaction steps have comparable rates, but the rates of change of concentration for intermediate species (which do not appear in the stoichiometric equation for the overall reaction) are small. This is called the *stationary-state hypothesis.*

(Note that assumptions 1 and 2 correspond to the two limiting cases mentioned in our discussion of consecutive reactions.) The two assumptions are not incompatible: if an early step in the mechanism is rate-determining, subsequent steps necessarily reach a stationary state. But usually it suffices to make one or the other. Let us look at some examples.

A reaction with a rate-determining step is

$$I^- + OCl^- \rightarrow Cl^- + OI^-,$$

the rate of which is found empirically to be

$$-\frac{d[I^-]}{dt} = k \frac{[I^-][OCl^-]}{[OH^-]}. \tag{30.146}$$

Since $[OH^-]$ appears in the rate law but not the stoichiometry, it cannot be a simple bimolecular transfer. Suppose that we assume the mechanism

$$OCl^- + H_2O \underset{}{\overset{K_1}{\rightleftharpoons}} HOCl + OH^- \quad \text{(rapid equilibrium)}$$

$$HOCl + I^- \overset{k_2}{\longrightarrow} HOI + Cl^- \quad \text{(slow)}$$

$$HOI + OH^- \overset{K_3}{\rightleftharpoons} OI^- + H_2O \quad \text{(rapid equilibrium)},$$

in which K_1, K_3 are equilibrium constants and k_2 is a rate constant. The slow second reaction is the rate-determining step; it cannot reach equilibrium, since the HOI formed is immediately consumed by the faster third reaction. But the first and third reactions are in "rapid equilibrium," a term that requires some explanation. It means simply that the forward and reverse rates of the first and third reaction steps are so large that quasi-equilibrium among the species in these steps is established in a time (determined by k_2) during which the concentration of I^- hardly changes. On a longer time scale the concentration of I^- does slowly decrease, but the quasi-equilibria in the first and third steps are continuously maintained. (The type of approximation used here, based on the existence of multiple time scales in the system, will be found quite useful in subsequent sections.) The I^- ion appears only in the second step, so we have

$$-\frac{d[I^-]}{dt} = k_2[HOCl][I^-]. \qquad (30.147)$$

Since the first step is essentially at equilibrium (the withdrawal of HOCl by the slow second step is only a small perturbation), we can write

$$K_1 = \frac{[HOCl][OH^-]}{[OCl^-]} \quad \text{or} \quad [HOCl] = \frac{K_1[OCl^-]}{[OH^-]}. \qquad (30.148)$$

Substitution of this value for [HOCl] gives

$$-\frac{d[I^-]}{dt} = k_2 \frac{K_1[OCl^-]}{[OH^-]}[I^-], \qquad (30.149)$$

which corresponds to the empirical rate law if we set $k = k_2 K_1$. We thus conclude that the assumed mechanism is probably correct. Note that the third reaction step has no effect on the empirical rate law: This is generally true for any step that follows a rate-determining step. Furthermore, we see that the appearance of a concentration ($[OH^-]$ here) in the denominator of the rate expression can result from a rapid equilibrium prior to a rate-determining step. (This is one of the "indications" to which we referred above.)

Next let us consider an application of the stationary-state hypothesis, with a mechanism similar to that just described. Suppose that the reaction

$$RCl + OH^- \rightarrow ROH + Cl^-$$

(where R may be, for example, the t-butyl group) has the mechanism

$$RCl \underset{k_1'}{\overset{k_1}{\rightleftharpoons}} R^+ + Cl^-$$

$$R^+ + OH^- \xrightarrow{k_2} ROH.$$

If the second step were rate-determining, this would resemble the previous case, and one would predict the rate equation $d[ROH]/dt = k_1 k_2[RCl]/k_1'[Cl^-]$; nothing this simple is observed, however. Thus let us suppose instead that the rates of the two steps are such that the first step cannot be assumed in equilibrium; this will be true if the second step has a rate at least comparable to that of the first. Formally, we make the assumption that the concentration of the intermediate R^+ is always very small, so that $d[R^+]/dt \approx 0$. This does not mean that $[R^+]$ is constant; in fact, R^+ will presumably change with time much like the intermediate B in Fig. 30.21. But if $[R^+]$ is low enough, its derivative will be negligible on an absolute scale, that is, relative to the overall reaction rate. Given this assumption, we can write

$$\frac{d[R^+]}{dt} = k_1[RCl] - k_1'[R^+][Cl^-] - k_2[R^+][OH^-] = 0 \qquad (30.150)$$

and solve for the stationary-state value of $[R^+]$,

$$[R^+] = \frac{k_1[RCl]}{k_1'[Cl^-] + k_2[OH^-]}. \qquad (30.151)$$

If $[R^+]$ is always small, then ROH must be formed virtually at the same rate as RCl is consumed; the predicted rate equation is thus

$$-\frac{d[RCl]}{dt} = \frac{d[ROH]}{dt} = k_2[R^+][OH^-] = \frac{k_1 k_2[RCl][OH^-]}{k_1'[Cl^-] + k_2[OH^-]}. \qquad (30.152)$$

This complex rate law is indeed confirmed by experiment. Its most striking feature is that the concentration of the product Cl$^-$ appears in the denominator of the rate expression; this shows the importance of measuring a reaction rate as a function of both reactant and product concentrations if one wishes to determine the mechanism. The technique developed here can be easily extended to reactions with more than one stationary-state intermediate, as we shall see in the section on chain reactions.

It is possible for a chemical species to appear in a rate equation without taking part in the overall reaction, in the phenomenon we call catalysis. For example, consider the electron-transfer reaction

$$V^{III} + Fe^{III} \rightarrow V^{IV} + Fe^{II}.$$

(The superscript numerals designate oxidation states; the actual reactant species are complexes, but the ligands do not concern us here.) In the presence of copper ions, this reaction has the empirical rate equation

$$\frac{d[V^{IV}]}{dt} = k[V^{III}][Cu^{II}]. \tag{30.153}$$

We say that the Cu^{II} acts as a catalyst. We earlier defined a catalyst as a species that affects the reaction rate but not the equilibrium. A more general definition would be: any species that appears in the rate equation to an order larger than its stoichiometric coefficient—which here is zero. How does this occur? Suppose that the reaction has the mechanism

$$V^{III} + Cu^{II} \xrightarrow{k_1} V^{IV} + Cu^{I}$$

$$Cu^{I} + Fe^{III} \xrightarrow{k_2} Cu^{II} + Fe^{II}.$$

If we apply the stationary-state assumption to Cu^{I}, we have

$$\frac{d[Cu^{I}]}{dt} = k_1[V^{III}][Cu^{II}] - k_2[Cu^{I}][Fe^{III}], \qquad [Cu^{I}] = \frac{k_1[V^{III}][Cu^{II}]}{k_2[Fe^{III}]}, \tag{30.154}$$

for a rate equation of

$$-\frac{d[V^{III}]}{dt} = \frac{d[Fe^{II}]}{dt} = k_2[Cu^{I}][Fe^{III}] = k_1[V^{III}][Cu^{II}], \tag{30.155}$$

in accord with experiment.

Note that in both the last two mechanisms the stationary-state rate equation is the same as would be obtained by simply assuming the first step to be rate-determining. This equivalence is generally true for any mechanism made up only of consecutive reaction steps, since to assume the intermediate concentration always small is the same as saying that $k_1 \ll k_2$.

We have mentioned that some reactions have rate equations containing fractional powers of concentrations; let us see how this can occur. For example, the hydrolysis of esters,

$$RCOOR' + H_2O \rightarrow RCOOH + R'OH,$$

in the presence of acetic acid (HAc) is found to obey the rate law

$$\frac{d[RCOOH]}{dt} = k[RCOOR'][HAc]^{1/2}. \tag{30.156}$$

Since HAc does not appear in the stoichiometry, it must act catalytically;

the true catalyst, however, is the hydrogen ion.[31] Suppose that the mechanism is as follows:

$$RCOOR' + H^+ \underset{k_{-1}}{\overset{k_1}{\rightleftharpoons}} \left(\begin{matrix} RCOOR' \\ | \\ H \end{matrix}\right)^+ \qquad \text{(fast)}$$

$$\left(\begin{matrix} RCOOR' \\ | \\ H \end{matrix}\right)^+ + H_2O \xrightarrow{k_2} \text{complex} \qquad \text{(rate-determining)}$$

$$\text{complex} \longrightarrow RCOOH + R'OH + H^+$$

(the breakup of the complex probably occurs in two steps, but since these are after the rate-determining step they do not affect the overall rate). If we designate RCOOR' and the protonated intermediate by S and SH$^+$, respectively, the bimolecular rate-determining step yields a rate law of

$$\frac{d[RCOOH]}{dt} = k_2[H_2O][SH^+]. \qquad (30.157)$$

If the initial step is effectively in equilibrium, we can find the value of [SH$^+$] from its equilibrium constant,

$$[SH^+] = \frac{k_1}{k_{-1}}[S][H^+]. \qquad (30.158)$$

Furthermore, if the only source of protons is the weak acid HAc, the value of [H$^+$] is governed by its ionization constant

$$K_a = \frac{[H^+][Ac^-]}{[HAc]}; \qquad (30.159)$$

for small degree of ionization we have [H$^+$] = [Ac$^-$] and

$$[H^+]^2 = K_a[HAc]. \qquad (30.160)$$

Combining all these equations, we have the rate equation

$$\frac{d[RCOOH]}{dt} = k_2 \frac{k_1}{k_{-1}} K_a^{1/2}[S][H_2O][HAc]^{1/2}; \qquad (30.161)$$

this is in agreement with the experimental Eq. 30.156 if the solvent concentration [H$_2$O] can be taken as constant. As in this case, the appearance of fractional powers of concentrations in a rate expression usually indicates a mechanism in which there is molecular dissociation.

At this point let us pause and sum up. The guessing of a mechanism is an example of the formulation of a scientific hypothesis. Let us suppose that a reaction (or better still a set of homologous reactions) has been studied experimentally: the stoichiometry is known, at least in part; the reaction rate has been measured as a function of the concentrations of the known participating species; the temperature dependence of the rate is known over a given interval. No doubt, as the measurements proceeded, working hypotheses were formulated, tested, and then rejected, accepted, or altered to fit newly obtained data. To design an experiment usually requires some preconceived notion of the outcome; the outcome then influences and alters the original notion. Thus eventually the time comes to guess a mechanism and make predictions that can be tested by experiment.

---

[31] Actually, the acid species is H$_3$O$^+$ (or something more complicated) rather than H$^+$, but we shall neglect these details here.

We have seen the ways in which a rate equation can be derived from an assumed mechanism. But do not forget that it is usually the experimental rate equation that comes first, with the mechanism being designed to lead to it. The order of the reaction with respect to any species can be known only if the reaction rate has been measured as a function of that species' concentration. In general there is no relation between the stoichiometry of a reaction and its mechanism, so one must be careful to consider the possibility that species not appearing in the stoichiometric equation do affect the rate. One example of this is the dependence of the rates of ionic reactions on the solvent (cf. Section 30.8); another, which we shall discuss in Section 30.14, is the effect of inert-gas concentration on the rates of unimolecular reactions. In each case one must first have a suspicion that the solvent or the inert gas makes a difference in the mechanism, and then carry out measurements to test that suspicion.

For mechanisms in which a rate-determining step is postulated, the atoms of the reactants in that step must also appear in the predicted rate expression. Note that we say "the atoms," not the species: The species in the rate-determining step may be involved in rapidly established equilibria with other species whose concentrations are more easily measured; but atoms are, of course, conserved in any such equilibria, and the concentrations are proportional to one another. See, for example, the reaction between I^- and OCl^- discussed earlier, in which the rate-determining step involves HOCl but the rate equation 30.146 instead contains $[OCl^-]$, which is related to [HOCl] by the equilibrium constant 30.148.

We have mentioned a number of clues to guessing a mechanism, but none are completely clear-cut. For example, the appearance of a concentration in the denominator of the rate law, as with $[OH^-]$ in Eq. 30.146, *may* indicate that the species is produced in a rapid equilibrium preceeding a rate-determining step. On the other hand, the complex denominator in Eq. 30.152 indicates a stationary-state intermediate that can react in two different ways. There is no substitute for experience and "chemical intuition" in guessing what reaction steps are likely in such systems. The appearance of a species that takes no part in the stoichiometric equation, such as Cu^{II} in Eq. 30.153, is a definite indication of catalysis; but one still has to determine *how* the catalyst participates in the mechanism, and how it is regenerated. Finally, the presence of fractional powers of concentration, as in Eq. 30.156, suggests a dissociation step (in this case, however, the mechanism could be determined more directly by measuring the rate as a function of $[H^+]$ rather than [HAc]). We have two more such clues to describe.

When the overall order in an experimental rate expression exceeds three, then it is likely that quickly established equilibrium reaction steps precede the rate-determining step. Consider the reaction

$$BrO_3^- + 5Br^- + 6H^+ \rightarrow 3Br_2 + 3H_2O,$$

for which the observed rate equation is

$$\frac{d[Br_2]}{dt} = k[BrO_3^-][Br^-][H^+]^2. \tag{30.162}$$

It is very unlikely that four ions come together to react simultaneously. A plausible mechanism is

$$BrO_3^- + H^+ \rightleftharpoons HBrO_3$$

$$HBrO_3 + H^+ \rightleftharpoons H_2BrO_3^+$$

$$H_2BrO_3^+ + Br^- \xrightarrow{k'} \quad \text{(rate-determining step).}$$

The predicted rate equation is then simply

$$\frac{d[Br_2]}{dt} = k'[H_2BrO_3^+][Br^-]; \qquad (30.163)$$

using the two equilibrium steps preceding the rate-determining step to solve for $[H_2BrO_3^+]$, we can readily rearrange this to the observed rate equation. Note that in this, as in all other rate studies, the participation of solvent is unknown unless measured.

If a stoichiometric coefficient for any reactant (including the solvent) exceeds the reaction order with respect to that reactant, then there are one or more intermediates following the rate-determining step. For example, in the reaction of aniline with peracetic acid to form nitrosobenzene,

$$C_6H_5NH_2 + 2CH_3CO_3H \rightarrow C_6H_5NO + 2CH_3CO_2H + H_2O,$$

the observed rate equation is

$$\frac{d[C_6H_5NO]}{dt} = k[C_6H_5NH_2][CH_3CO_3H]. \qquad (30.164)$$

The postulated mechanism consists of an initial rate-determining step to produce the intermediate N-phenylhydroxylamine,

$$C_6H_5NH_2 + CH_3CO_3H \rightarrow C_6H_5NHOH + CH_3CO_2H,$$

followed by a fast step,

$$C_6H_5NHOH + CH_3CO_3H \rightarrow C_6H_5NO + CH_3CO_2H + H_2O.$$

The second step can be confirmed independently: The intermediate N-phenylhydroxylamine can be prepared as a stable species, and its reaction with peracetic acid is indeed fast compared to the first step.

To complicate matters, a reaction may proceed by different mechanisms under different conditions. For example, if the temperature dependence of the rate coefficient is such as to indicate an abrupt change in activation energy (i.e., if the Arrhenius plot gives different slopes above and below a certain temperature), then it is likely that the mechanism of the reaction has changed. The most common examples include reactions that proceed explosively above a given temperature, and reactions that change from homogeneous to heterogeneous mechanisms at a given temperature. Some reactions also have different mechanisms depending on whether the activation energy is produced thermally or photochemically. We shall discuss all these examples in later sections.

30.13 Bimolecular Reactions

We have defined a bimolecular reaction step as one involving the collision of two molecules (atoms, ions, etc.). A complex reaction is sometimes referred to as bimolecular if the rate-determining step in the mechanism is bimolecular. All the theoretical treatments of reaction rates presented here have assumed bimolecularity (although they can with some difficulty be extended to other cases). This emphasis is reasonable, since a bimolecular collision is by far the most likely type of molecular encounter, at least in gases. In the present section we shall review some general principles about bimolecular reactions and discuss a few special cases in detail.

As we illustrated with our list at the end of Section 30.1, there are many types of bimolecular reaction steps. They can be classified according to the kinds of species involved (molecules, ions, atoms, free radicals), the type of process occurring (transfer, addition, dissociation, etc.), or a combination of the two. We shall mention only a few of these types here.

The most rapid reactions tend to be those involving free radicals (including atoms), which can readily form new bonds. This occurs most easily in transfer reactions, in which one bond is broken and another formed, for example,

$$CH_3 + Br_2 \rightarrow CH_3Br + Br.$$

Processes of this type ordinarily occur as intermediate steps in complex mechanisms (after some other step has generated the free radicals), but one can study them directly by such techniques as molecular-beam collisions; we have described such studies in Section 30.3. For exothermic reactions of free radicals the activation energy is usually very low or even zero; that is, one has only to bring the reactants together for reaction to occur. In general, a low activation energy corresponds to a large reaction cross section, as indicated by Eq. 30.28. Many bimolecular transfer reactions of free radicals occur in various parts of the atmosphere; some examples are

$$H + NO_2 \rightarrow OH + NO,$$
$$N + NO \rightarrow N_2 + O,$$
$$N + OH \rightarrow NO + H,$$
$$OH + CH_4 \rightarrow H_2O + CH_3.$$

The last of these is important in the chemistry of pollution: Hydrocarbons attacked by OH radicals yield alkyl radicals, which are highly reactive and attack other species present

Addition reactions of free radicals and atoms are less likely, since there is the problem of stabilizing the addition product. Thus one might expect atoms to recombine bimolecularly, for example,

$$H + Cl \rightarrow HCl,$$

but this is almost impossible as a single-step process. As we have noted on several occasions, in such a reaction the energy corresponding to the bond formation must be removed from the product molecule; otherwise the two atoms will simply come together to the distance of closest approach, reverse their radial motion, and recoil as rapidly as they approached. (The pair of atoms could in principle radiate energy during their encounter; but collisions typically last only $10^{-13} - 10^{-12}$ s, while 10^{-9} s or more is required for even the fastest available radiation processes.) But if, while in close proximity, the pair of atoms can transfer enough energy to a third particle, a surface, or an emitted photon, then a stable molecule can be formed. With more complicated radicals, as in

$$CH_3 + CH_3 \rightarrow C_2H_6,$$

the product can be temporarily stabilized by distribution of the energy over several degrees of freedom. Although energy removal still must eventually occur for a stable product to exist, the rates of polyatomic-radical recombinations are generally much greater than those involving single atoms.

A collision between stable molecules can result in dissociation, excitation, or bond rearrangement. An example of collisional dissociation is

$$Ar + CsBr \rightarrow \begin{cases} Ar + Cs + Br \\ Ar + Cs^+ + Br^- \end{cases},$$

which, as indicated, may occur by either ionic or neutral channels. Depending on the conditions (especially the concentration of inert molecules), such a dissociation reaction may yield a "unimolecular" rate equation, as we shall see in the next section. Collisional ionization is a special case of collisional dissociation, as in

$$Ar + Ar \rightarrow Ar + Ar^+ + e^-;$$

it is necessarily strongly endothermic, since electron-binding energies usually are greater than bond energies.

Collisions leading to excitation have been discussed in Section 28.10. In general, excitation need not lead to chemical reaction, but sometimes it does: The excited molecule may dissociate, or the excitation may furnish the activation energy for a subsequent reactive collision. (The reactive excited states produced by collision behave much the same as those resulting from photoexcitation, which we shall treat in Section 31.2.) Excitation and bond rearrangement are often combined, with the products of reaction formed in excited states; we have surveyed the energy distribution in reaction products in Section 30.3. As we noted there, the excited states produced by reaction may decay by emission of radiation (chemiluminescence); an example is

$$NO + O_3 \rightarrow NO_2^* + O_2$$
$$NO_2^* \rightarrow NO_2 + h\nu,$$

a transfer reaction followed by emission,

Four-center exchange reactions, such as

$$H_2 + I_2 \rightarrow 2HI,$$

generally proceed only with large activation energies,[32] if at all bimolecularly. (Note that *two* bonds must be broken, rather than one in a simple transfer reaction.) In both the H_2–I_2 reaction and its H_2–D_2 analog, there is evidence that vibrational excitation can lead to reaction, whereas translational excitation does not. This brings up the interesting question, on which we touched in Section 30.3, of the need for "specificity" of activation energy in certain degrees of freedom. The issue is probably important only for reactant molecules with few degrees of freedom. If a molecule has many degrees of freedom, it is unlikely that a specific one will remain excited long enough for bond rearrangement to occur, since intramolecular energy transfer occurs on the time scale of a vibrational period; exceptions to this rule may occur, and further experiments are needed to make certain.

An interesting example of a bimolecular reaction with a number of product channels is that between Ba atoms and Cl_2, which can proceed by neutral transfer, ionic transfer, or addition followed by chemiluminescent emission:

$$Ba + Cl_2 \rightarrow \begin{cases} BaCl + Cl \\ BaCl^+ + Cl^- \\ BaCl_2^* \rightarrow BaCl_2 + h\nu. \end{cases}$$

The first channel is found to be the most probable.

A special class of bimolecular reactions are the so-called *hot-atom reactions*, in which one reactant is an atom with very large kinetic energy. Such atoms may be produced by nuclear reactions, for example,

$$^3He + n \rightarrow p + {}^3T,$$

[32] For example, the activation energy for the H_2–I_2 reaction is 163 kJ/mol. The exchange reaction

$$CO + NO_2 \rightarrow CO_2 + NO$$

has an activation energy of 116 kJ/mol, although its ΔH_{reac} is -226 kJ/mol.

in which the tritium nucleus is ejected with kinetic energy of the order of 1 MeV ($\approx 10^8$ kJ/mol). As the fast nucleus enters matter, it suffers many collisions, and its velocity is steadily decreased; it also probably picks up an electron to become a neutral atom. Reaction is unlikely while the kinetic energy is very high, say above 20 eV, since the atom is moving too quickly for bond rearrangement to occur.[33] Unlike the prediction of the hard-sphere model (Fig. 30.16), the evidence for real reactions is that the reactive cross section rises to a peak and then decreases with increasing energy. But reaction does occur when the energy of the tritium atom is low enough. One example is

$$T + CH_4 \rightarrow CH_3T + H,$$

which proceeds by a "knock-out process": the T atom hits one of the methane H atoms, transfers its kinetic energy to the H, and remains bonded in the latter's place; the process occurs so rapidly that the configuration of the rest of the molecule is unaltered. Another possible type of reaction is a "pick-up process," such as

$$T + RH \rightarrow R + HT.$$

We have been discussing only the reactions of neutral species, but there are many examples of bimolecular *ion–molecule reactions*. Some of them are of fundamental importance in the kinetics of combustion, the upper atmosphere, and plasmas (all of which are systems with substantial concentrations of charged particles). For example, an important reaction in flames is

$$CHO^+ + H_2O \rightarrow CO + H_3O^+,$$

whereas a reaction leading to loss of NO in the upper atmosphere (20–100 km altitude) is

$$NO + O_2^+ \rightarrow NO^+ + O_2.$$

(See Section 30.3 for some other examples.) Ion–molecule reactions usually take place quite readily: Most exothermic reactions have little or no activation energy, whereas for most endothermic reactions the activation energy is just the endothermicity.

Up to this point we have been considering the reactions of isolated molecules in the gas phase. We turn now to examples of bimolecular reactions in solution, where, as we have seen (Section 30.8), conditions may be significantly different. For example, such a simple association reaction as

$$Ag^+ + Cl^- \rightarrow AgCl$$

would be impossible in the gas phase, because of the usual problem of removing the reaction energy. Yet this reaction occurs very rapidly in solution, since the ions are solvated and the energy can be easily dissipated to the solvent. In the remainder of this section we shall discuss some classes of bimolecular reactions that are especially characteristic of solution chemistry.

One very important class of reactions is usually denoted as S_N2, for "*s*ubstitution, *n*ucleophilic, *bi*molecular." These are displacement reactions in which the substituent is nucleophilic (i.e., an electron donor), such as

$$F^- + CH_3Cl \rightarrow Cl^- + CH_3F;$$

---

[33] However, the molecules hit *by* the hot atom (or other high-energy particle) are not unaffected by these collisions, being likely to undergo ionization or excitation. It is the ions thus produced (and the condensations or evaporations they nucleate) that make a particle's track visible in a cloud chamber or bubble chamber.

the empirical rate equation is first-order in each of the reactants, second-order overall. This seems simple enough, and the reaction is indeed a single-step process; what is of interest is the geometry of that process. For the reaction just described, it is thought that the substituent approaches on the "back side" of the molecule, with inversion of configuration occurring upon reaction (*Walden inversion*):

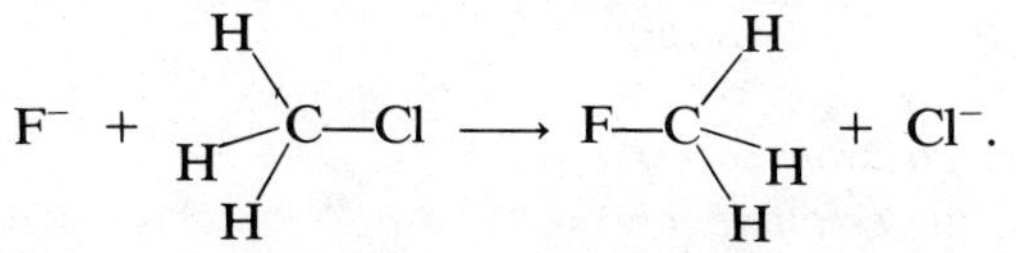

This model is confirmed by optical-activity measurements when the larger reactant has an asymmetric carbon atom (one with four different substituents). The bond rearrangement in S_N2 reactions is usually thought to occur more or less synchronously, but there is some evidence for complex formation: The same reactions have in some cases been found to occur much faster in the gas phase than in solution, where the rate is strongly solvent-dependent. If the reaction indeed occurs by the postulated back-side attack on carbon, then the rate should be sensitive to steric hindrance. This is in fact the case: Replacing the H atoms in CH_3Cl or similar compounds by bulky alkyl groups drastically reduces the rate at which S_N2 reactions occur.

A typical example of an inorganic S_N2 reaction in that of square-planar four-coordinated Pt^{2+} with various ions, such as

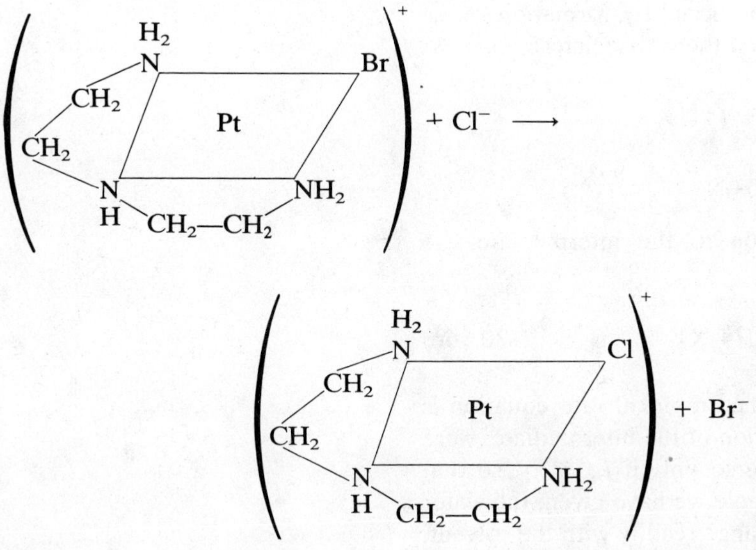

(the ligand is diethylenetriamine). In this case the kinetics are less simple: If we designate the reactant complex as ABr^+, the empirical rate equation is

$$-\frac{d[ABr^+]}{dt} = (k_1 + k_2[Cl^-])[ABr^+]. \qquad (30.165)$$

The term proportional to $[Cl^-]$ arises from that part of the reaction which proceeds bimolecularly. It is believed that the Cl^- ion approaches the complex to form a trigonal-bipyramidal intermediate, which then dissociates to the products. For a given complex the rate depends strongly on the other reactant, being nearly 400 times as much for I^- as for Cl^-.

The k_1 term in Eq. 30.165 represents a parallel mechanism involving the solvent:

$$ABr^+ + H_2O \xrightarrow{k_{1'}} A(H_2O)^{2+} + Br^-$$

$$A(H_2O)^{2+} + Cl^- \longrightarrow ACl^+ + H_2O,$$

with the first step rate-determining. This is still a bimolecular mechanism, but the kinetics are pseudo-first-order, with k_1 in Eq. 30.165 equivalent to $k_1'[H_2O]$.

Another important class of bimolecular reactions are *electron-transfer reactions*. This broad category is usually divided into *inner-sphere* and *outer-sphere* electron-transfer reactions; inner-sphere reactions in turn may occur with or without *ligand transfer*, outer-sphere reactions only without ligand transfer. We shall give an example of each of the three types thus defined.

Inner-sphere electron-transfer reactions with ligand transfer may generally be written as

$$M + M'\!-\!X \longrightarrow M\!-\!X + M',$$

where the oxidation states of the metals M and M′ change in going from reactants to products (this change of oxidation states is what is meant by "electron transfer"). An example is

$$[Cr(H_2O)_6]^{2+} + [Co(NH_3)_5(H_2{}^{18}O)]^{3+} + 5H^+ + 6H_2O \rightarrow$$
$$[Cr(H_2O)_5(H_2{}^{18}O)]^{3+} + [Co(H_2O)_6]^{2+} + 5NH_4{}^+$$

(the ^{18}O is used to demonstrate that a ligand is in fact transferred from Co to Cr). It is postulated that these reactions occur by formation of an intermediate, the *inner-sphere complex*, which then dissociates:

$$M + N\!-\!X \underset{k_{-1}}{\overset{k_1}{\rightleftharpoons}} M \cdots X \cdots N$$

$$M \cdots X \cdots N \xrightarrow{k_2} M\!-\!X + N.$$

Applying the stationary-state approximation to the intermediate, we derive the rate equation

$$-\frac{d[M]}{dt} = \frac{k_1 k_2}{k_{-1} + k_2}[M][N\!-\!X], \tag{30.166}$$

which agrees with experiment. (Note that the empirical rate equation is the same as if the activation step, the formation of the intermediate, were rate-determining; but this would actually be so only if $k_{-1} \ll k_2$, so that the empirical k would equal k_1.) In the example we have given, the water molecules in the $[Cr(H_2O)_6]^{2+}$ complex exchange readily with the solvent, but this is not true of the more tightly bound ligands in $[Cr(H_2O)_6]^{3+}$ or the Co^{3+} complex. Thus the presence of $H_2{}^{18}O$ in the Cr^{3+} complex is evidence of direct transfer of H_2O from the cobalt to the chromium; this direct transfer, accompanied by the transfer of an electron from Cr to Co, requires close proximity of the central ions, and is thus called an inner-sphere process.

We can most conveniently treat the remaining two classes of electron-transfer reactions in parallel. Consider the reaction types

(1) $\quad [Cr(H_2O)_6]^{2+} + [Co(NH_3)_5X]^{2+} \rightarrow [Cr(H_2O)_6]^{3+} + [Co(NH_3)_5X]^+$

and

(2) $\quad [Cr(bipy)_3]^{2+} + [Co(NH_3)_5X]^{2+} \rightarrow [Cr(bipy)_3]^{3+} + [Co(NH_3)_5X]^+$

(where X is a halogen or similar ion with charge -1, and "bipy" = bipyridine). These reactions appear similar, and obviously in neither case is there ligand transfer. Although the two complexes must of course approach each other for the electron transfer to take place, the bulky bipyridine ligands in reaction (2) make it impossible for the central ions to come as close together as in the inner-sphere case. Reaction (2) is thus necessarily outer-sphere, but what about reaction (1)? The criteria actually used to classify reactions as inner-sphere or outer-sphere do not rely on guesswork, but on analysis of kinetic measurements; in this case variation of the ligand X provides the key.

The rates of the electron-transfer reactions (1) and (2) have been measured with $X = N_3^-$ (azide) and $X = NCS^-$ (thiocyanate). For reaction (1) the ratio of rate coefficients $k_{N_3^-}/k_{NCS^-}$ is about 10^4, whereas for reaction (2) the corresponding ratio is only 4. We conclude that reaction (1) is highly sensitive to the nature of the ligand X, whereas reaction (2) is not. This is interpreted to mean that reaction (1) occurs by inner-sphere, reaction (2) by outer-sphere, electron transfer. In the inner-sphere case, the azide ion can form a symmetric bridge directly linking the chromium and cobalt ions,

$$(H_2O)_6Cr^{2+}\!\!-\!\!N\!\!=\!\!N\!\!=\!\!N\!\!-\!\!Co^{2+}(NH_3)_5,$$

which provides comparatively strong bonding and thus facilitates electron transfer. The NCS^- ion, however, forms an asymmetric and thus much weaker bridge, so that the rate of electron transfer is much lower. But if electron transfer occurs by an outer-sphere mechanism, then the potential bonding characteristics of the ligand X are unimportant (since it does not bind directly to both central ions) and the ratio of rates is much less than with the inner-sphere mechanism.

A bimolecular reaction mechanism frequently leads to a second-order rate equation, but does not necessarily have to do so. Often there are secondary processes that complicate the kinetics and give what appears to be a different order. Examples include the pseudo-fractional order resulting from dissociation, as in Eq. 30.161, and the pseudo-first order due to reaction with the solvent, as in Eq. 30.165. A third example is the combination of two bimolecular steps to give a third-order reaction, which we shall analyze in Section 30.15.

30.14
Unimolecular Reactions

In a unimolecular reaction step, a single reagent molecule is transformed to one or more products; if this is the rate-determining step of the overall reaction, the kinetics will be first-order. Unimolecular processes must be either isomerizations or decompositions (depending on how many products are formed). Among the best-known examples are cis-trans isomerization such as

which is, of course, unimolecular in both directions. An example of unimolecular decomposition is

$$C_2H_6 \rightarrow 2CH_3,$$

which takes place in ethane (either pure or mixed with an inert gas) at high temperatures.

The formulation of a workable mechanism for unimolecular reactions presented difficulties for many years, until in 1922 F. A. Lindemann devised a simple but useful model. What was the problem? Unimolecular reaction rates increase with temperature much as do those of bimolecular reactions, so it seems probable that some activation process is part of the mechanism. (Cf. Eq. 30.31 and the discussion following.) The puzzle was: What can the activation process be if only one molecule is involved? As so often happens in science, the answer was difficult because the question was poor. Only one molecule is involved in the rate-determining step, perhaps, but that does not preclude earlier bimolecular steps. Let us now look at the Lindemann model in its simplest form; later (Section 31.1) we shall see how it can be substantially improved.

Lindemann proposed the following mechanism: Consider the gas-phase dissociation of molecule A into fragments B and C,

$$A \rightarrow B + C.$$

At any given temperature, suppose that only a small fraction of the A molecules have sufficient energy in their internal degrees of freedom to react. How do they get that energy? By inelastic collisions, either with other A molecules or with inert molecules M present in the gas; let us assume for generality that "M" includes both possibilities. In an inelastic collision relative kinetic energy can be transferred to the internal degrees of freedom, as we have seen in Section 28.10. Let us designate a molecule of A with sufficient energy to dissociate as A^*, and write the *excitation* process as

$$(1) \qquad\qquad A + M \xrightarrow{k_1} A^* + M.$$

Note that we are not very specific here. In particular, we do not indicate the internal quantum state of A, but are content to divide those states into two classes: The molecules in states of one class (A) will not react, whereas those in the other class (A^*) *may* react. The alternative, of course, is for the excited molecule to undergo another inelastic collision, transferring energy back from the internal degrees of freedom to relative kinetic energy of A and M. We write this *deexcitation* step as

$$(2) \qquad\qquad A^* + M \xrightarrow{k_2} A + M.$$

Some of the excited molecules, however, will react before they can be deexcited; thus we also have the *reaction* step

$$(3) \qquad\qquad A^* \xrightarrow{k_3} B + C.$$

The three equations (1)–(3) constitute the mechanism proposed by Lindemann for unimolecular reactions.

What are the relative probabilities of deexcitation and reaction? The deexcitation rate can be estimated from bimolecular collision theory, but the rate of dissociation is more difficult to guess. The shortest time characteristic of molecular dissociation is a vibrational period, typically about 10^{-13} s. But the dissociation time of A^* (which is essentially k_3^{-1}) may be considerably longer, especially if the molecule has many internal degrees of freedom. Say that it takes an amount of energy α to dissociate A into the fragments B and C. To effect the dissociation, this energy may have to be in specific internal degrees of freedom, for example, in the vibrational mode corresponding to stretching of a particular bond. However, reaction usually will not occur if the same amount of energy is merely distributed over all the internal degrees of freedom; for the energy to be effective, it has to be localized. Thus a process of *intramolecular energy transfer* must occur. This can happen if the internal degrees of freedom are coupled to one another, in much the same way as translational energy can be transformed into internal energy in a collision; but the characteristic time for getting the energy into the right degree(s) of freedom may be much longer than a typical vibrational period.

One can think about the same problem instructively in a somewhat different way. Suppose that we consider the molecule, with n internal degrees of freedom, as a system to be treated by statistical mechanics. That is, we treat the individual degrees of freedom themselves as elements in a statistical ensemble. If there are enough degrees of freedom, say about 20, this might be a reasonable representation. Now we give the system an amount of energy β. Statistical mechanics predicts that *on the average* this energy will be distributed evenly over all the degrees of freedom (equipartition of energy). But as we know from Part Two, *fluctuations* from the average can take place. For dissociation to occur, we have to wait for a fluctuation in which an amount of energy α appears in the specific degree(s) of freedom required. The characteristic time for such a fluctuation determines the rate coefficient k_3. We shall see in Section 31.1 that such calculations as described here have actually been carried out.

For now, however, let us take coefficients k_1, k_2, k_3 as given and see what kind of rate law the Lindemann mechanism predicts. The rate of dissociation of A^*, which equals the rate of formation of B, is simply

$$\frac{d[B]}{dt} = k_3[A^*], \qquad (30.167)$$

and the total rate of change of $[A^*]$ is

$$\frac{d[A^*]}{dt} = k_1[A][M] - k_2[A^*][M] - k_3[A^*]. \qquad (30.168)$$

If we apply the stationary-state hypothesis to the excited species A^*, setting $d[A^*]/dt = 0$ and solving for $[A^*]$, we obtain

$$[A^*] = \frac{k_1[A][M]}{k_2[M] + k_3}. \qquad (30.169)$$

Combining Eqs. 30.167 and 30.169, we have the rate equation

$$\frac{d[B]}{dt} = \frac{k_1 k_3[A][M]}{k_2[M] + k_3} = -\frac{d[A]}{dt}. \qquad (30.170)$$

Equation 30.170 does not look first-order, but let us examine it more closely. Suppose that we vary the concentration of M, which corresponds to the total gas pressure. If $[M]$ is made sufficiently large that $k_2[M] \gg k_3$ (which can always be achieved), then the equation reduces to the high-density limit

$$-\frac{d[A]}{dt} = \frac{k_3 k_1}{k_2}[A] \qquad \text{(high density)}. \qquad (30.171)$$

This equation clearly is first-order. As a mere limiting case, this result would be of minor interest; what makes it significant is that for many reactions the "high density" necessary for Eq. 30.171 to hold is achieved at less than atmospheric pressure! If this were not the case, the reactions would never have been characterized as "unimolecular" in the first place.

At the other extreme, if the pressure is so low that $k_2[M] \ll k_3$ (which can also always be achieved), then Eq. 30.170 reduces to

$$-\frac{d[A]}{dt} = k_1[A][M] \qquad \text{(low density)}, \qquad (30.172)$$

which is *second*-order.

Thus we find that the supposedly unimolecular reactions actually have a second-order rate equation at low density, changing to first-order at high (but not too high) density. Experiment bears this out, as can be seen in Fig. 30.23, where k_{uni} is the *effective unimolecular rate coefficient* defined by

$$-\frac{d[A]}{dt} \equiv k_{uni}[A] \qquad (30.173)$$

at each pressure. Equation 30.170 predicts that this coefficient should have the form

$$k_{uni} = \frac{k_1 k_3 / k_2}{1 + k_3/(k_2[M])} = \frac{k_\infty}{1 + k_3/(k_2[M])}, \qquad (30.174)$$

where k_∞ ($\equiv k_3 k_1/k_2$) is the high-pressure limit; as we shall see, however, there are problems with this equation.

What are the physical reasons for this change from second-order to first-order behavior with increasing density? At low density the time between collisions is so long that most excited A* molecules have time to dissociate before being deexcited. The rate-determining step is then the bimolecular excitation of A, and the coefficients k_2, k_3 (referring to steps following the rate-determining step) do not appear in the rate equation 30.172. At high density, however, the time between collisions is short, and most excited A molecules are deexcited by collisions before they can dissociate. In that case the dissociation process depletes [A*] in only a minor way, and to a first approximation [A*] is given by an equilibrium condition. Note that in the high-density limit ($k_2[M] \gg k_3$) Eq. 30.169 becomes

$$[A^*] = \frac{k_1}{k_2}[A] \qquad \text{(high density)}, \qquad (30.175)$$

precisely the equilibrium condition obtained for steps (1) and (2) in the absence of dissociation. The rate-determining step is then the dissociation of A*; since this is a first-order process, the overall reaction is first-order at high density.

Let us now return to the effective unimolecular rate constant predicted by Eq. 30.174. As we saw in Fig. 30.23, the predicted fall-off of the increase of k_{uni} with pressure is in fact observed. However, the prediction of the pressure at which this occurs is quite poor. For instance, for the cis-trans isomerization of 2-butene ($CH_3CH{=}CHCH_3$) at 469°C, experiment yields the high-density rate coefficient $k_\infty = 1.9 \times 10^{-5}\ s^{-1}$; if we write

$$k_\infty = A_\infty e^{-E_\infty/RT}, \qquad (30.176)$$

the temperature dependence gives $E_\infty = 263\ kJ/mol$. For a rough calculation, let us assume that k_1, the rate coefficient for excitation, is given by the hard-sphere model. Using Eq. 30.41 with $d = 5$ Å, and simply setting $\Delta E_f^* = E_\infty$, we obtain

$$\frac{k_1}{RT} = 2.46 \times 10^{-12}\ torr^{-1}\ s^{-1}$$

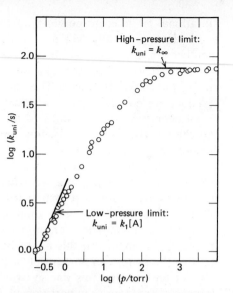

FIGURE 30.23
Effective unimolecular rate coefficient for the reaction

$$CH_3NC \rightarrow CH_3CN$$

at 199.4°C. From R. E. Weston, Jr., and H. A. Schwarz, *Chemical Kinetics* (Prentice-Hall, Englewood Cliffs, N.J., 1972), p. 120. The low-pressure and high-pressure limiting curves are of the form predicted by Eq. 30.174.

for binary collisions of 2-butene molecules at 469°C. (We calculate k_1/RT rather than k_1 itself, since the data are in terms of pressure rather than density.) Now, the value of [M] at which k_{uni} has half its high-pressure value is given by Eq. 30.174 as

$$\frac{k_\infty}{2} = \frac{k_\infty}{1 + k_3/(k_2[M]_{1/2})} \quad \text{or} \quad [M]_{1/2} = \frac{k_3}{k_2}. \qquad (30.177)$$

Since $k_\infty \equiv k_3 k_1/k_2$, the corresponding pressure is

$$p_{1/2} = RT[M]_{1/2} = RT\frac{k_\infty}{k_1} \qquad (30.178)$$

(assuming perfect-gas behavior). Substituting the above numerical values of k_∞ and k_1/RT, we obtain $p_{1/2} \approx 8 \times 10^6$ torr. Unfortunately for our theory, the experimental value of $p_{1/2}$ is only 0.04 torr. This is quite an impressive discrepancy, and clearly indicates that the theory is inadequate for quantitative calculations.

It is also instructive to estimate the value of k_3, the rate coefficient for dissociation of A* molecules. We rewrite Eq. 30.41 in the form

$$k_i = Z_i e^{-E_i^*/RT}, \qquad (30.179)$$

where Z_i is the collision frequency for step i. If we assume that the deexcitation step has no activation energy ($E_2^* = 0$), we have

$$k_3 = k_\infty \frac{k_2}{k_1} = A_\infty e^{-E_\infty/RT} \frac{Z_2}{Z_1 e^{-E_1^*/RT}}; \qquad (30.180)$$

if $Z_1 = Z_2$ (as should be true for an isomerization) and $E_1^* = E_\infty$ (as assumed above), this reduces to $k_3 = A_\infty$. For the 2-butene reaction the date already given correspond to $A_\infty = 6 \times 10^{13}\,s^{-1}$, which is of the order of a vibrational frequency. The implication is that the simple Lindemann model does not allow for a significant time lag between excitation and reaction. We have now established sufficiently well that the Lindemann model is not adequate for a complete understanding of unimolecular reactions, and in Section 31.1 we shall discuss some improvements in the theory.

Before leaving our preliminary discussion of unimolecular reactions, we briefly consider an important class of reactions observed primarily in solution. Analogous to the S_N2 reactions described in the last section, there are also nucleophilic reactions that proceed by a unimolecular mechanism, designated as S_N1 (substitution, nucleophilic, unimolecular). For example, the hydrolysis of t-butyl chloride,

$$(CH_3)_3CCl + H_2O \rightarrow (CH_3)_3COH + HCl,$$

has the observed rate expression

$$-\frac{d[(CH_3)_3CCl]}{dt} = k[(CH_3)_3CCl]. \qquad (30.181)$$

The rate-determining step is assumed to be the unimolecular carbonium-ion formation,

$$(CH_3)_3CCl \rightarrow (CH_3)_3C^+ + Cl^-$$

(after an activating step, of course), followed by the rapid transfer reaction

$$(CH_3)_3C^+ + H_2O \rightarrow (CH_3)_3COH + H^+.$$

(See the discussion leading to Eq. 30.152 for the corresponding hydrolysis in basic solution.)

An analogous inorganic S_N1 reaction is

$$[Co(CN)_5(H_2O)]^{2-} + X^- \rightarrow [Co(CN)_5X]^{3-} + H_2O,$$

where X^- may be I^-, Br^-, N_3^-, SCN^-, or the like. This reaction is believed to proceed by the mechanism

$$[Co(CN)_5(H_2O)]^{2-} \underset{k_{-1}}{\overset{k_1}{\rightleftharpoons}} [Co(CN)_5]^{2-} + H_2O$$

$$[Co(CN)_5]^{2-} + X^- \overset{k_2}{\longrightarrow} [Co(CN)_5X]^{3-}.$$

Applying the stationary-state approximation for the species $[Co(CN)_5]^{2-}$, we obtain the rate expression

$$-\frac{d[\{Co(CN)_5(H_2O)\}^{2-}]}{dt} = \frac{k_2 k_1 [X^-][\{Co(CN)_5(H_2O)\}^{2-}]}{k_2[X^-] + k_{-1}},$$
(30.182)

which agrees with experiment. In the limit $k_2[X^-] \gg k_{-1}$, the unimolecular dissociation of $[Co(CN)_5(H_2O)]^{2-}$ becomes rate-determining, and the rate equation is then first-order.

30.15
Termolecular Reactions

If a third-order reaction proceeds in a single elementary step, it is termolecular. Termolecular reactions are rare, since simultaneous collisions of three molecules occur far less often than binary collisions. As we have pointed out, however, atomic recombination reactions are necessarily termolecular, since a third body is required to remove energy and yield a stable product. At the end of Section 30.3 we discussed the dynamics of the reaction

$$I + I + M \rightarrow I_2 + M,$$

where M may be either another I atom or a different molecule. The reaction may begin with an encounter either between the two I atoms or between I and M; in either case the first two reactants form an orbiting complex, and the third must arrive almost immediately if reaction is to occur. The I_2 molecule is formed in a highly excited state, and will dissociate again unless it is stabilized by removal of more energy in inelastic collisions. On the average, M will carry away an energy only in the thermal range, say a couple of times $k_B T$; to yield I_2 in the ground state, M would have to remove the entire bond energy of 149 kJ/mol.

The frequency of termolecular collisions can be affected by the forces among the reactant molecules. For example, the collision cross section for the above reaction increases substantially with increasing attractive force between I and M. As the temperature of the system increases, the average kinetic energy of the colliding molecules is also increased, and the relative importance of the attractive forces is decreased. The point of closest approach then falls higher on the repulsive part of the potential energy curve, which is relatively independent of the nature of the molecule. Hence we expect the rate of recombination to decrease with increasing temperature, and this is in fact observed.

There is another interesting point with regard to atomic recombination reactions (and their inverse dissociation reactions). The fate of the highly excited I_2 molecule, either dissociation or stabilization, depends on its subsequent collisions. If the time required for the excited molecule to form and dissociate again is comparable to the time scale for inelastic energy transfer, then a nonequilibrium distribution of internal states in the I_2 molecule will result. This in turn affects the reaction rate (see the discussion of vibrational enhancement in Section 30.3).

It is to some extent arbitrary whether one considers a third-order reaction as proceeding in a single step (termolecular) or in a rapid sequence of bimolecular steps, that is,

$$A + B \rightarrow (AB)^*$$
$$(AB)^* + M \rightarrow AB + M,$$

where $(AB)^*$ is an orbiting complex. Since such complexes are quite short-lived, the third reactant must in any case arrive very rapidly; it is more convenient to consider the process as a single step when the initial encounter may be between any two of the three. When A and B are polyatomic, however, the spread of energy over the internal degrees of freedom may stabilize the complex long enough for a two-step interpretation to be useful.[34] In some cases the role of the third body may be simply to excite one of the actual reactants. Thus the reaction

$$CH_4 + D_2 \rightarrow CH_3D + HD,$$

which in an excess of argon has the rate equation

$$\frac{d[CH_3D]}{dt} = k[CH_4][D_2][Ar], \qquad (30.183)$$

has been postulated to occur with initial excitation of D_2 to an excited vibrational state,

$$D_2 + Ar \rightarrow D_2^* + Ar,$$

followed by a bimolecular exchange step.

In the examples given above, the "third body" acts merely as a catalyst and does not enter the stoichiometric equation. The only well-known third-order reactions of which this is not true are those of nitric oxide,

$$2NO + X_2 \rightarrow 2NOX$$

(where X is either O or a halogen), which obey the rate equation

$$\frac{d[NOX]}{dt} = k[NO]^2[X_2]. \qquad (30.184)$$

For the $NO + O_2$ reaction the activation energy is zero, presumably because both molecules have unpaired electrons. Activated-complex calculations for these reactions, assuming a termolecular collision, give reasonable agreement with experiment.

Termolecular reactions are more likely in solution, where the duration of a collision can be long (with the molecules bouncing about inside a "cage") and many solvent molecules are at hand to carry off excess energy.

Further Reading

Amdur, I., and Hammes, G. G., *Chemical Kinetics: Principles and Selected Topics* (McGraw-Hill, New York, 1966).

Benson, S. W., *The Foundations of Chemical Kinetics* (McGraw-Hill, New York, 1960).

General Texts on Chemical Kinetics

[34] Similarly, polyatomic M molecules enhance the likelihood of reaction, since they can carry off more energy.

Frost, A. A., and Pearson, R. G., *Kinetics and Mechanism*, 2nd ed. (Wiley, New York, 1961).

Gardiner, W. C., Jr., *Rates and Mechanisms of Chemical Reactions* (Benjamin, Reading, Mass., 1969).

Laidler, K. J., *Chemical Kinetics*, 2nd ed. (McGraw-Hill, New York, 1965).

Weston, R. E., Jr., and Schwarz, H. A., *Chemical Kinetics* (Prentice-Hall, Englewood Cliffs, N.J., 1972).

Bamford, C. H., and Tipper, C. F. H. (Eds.), *Comprehensive Chemical Kinetics* (Elsevier, New York, 1969–): especially Vol. 1, *The Practice of Kinetics*; Vol. 2, *The Theory of Kinetics*; Vol. 3, *Formation and Decay of Excited Species*; Vol. 18, *Selected Elementary Reactions*. **Advanced Treatises**

Eyring, H., Henderson, D., and Jost, W. (Eds.), *Physical Chemistry: An Advanced Treatise:* Vols. VIA, VIB, *Kinetics of Gas Reactions;* Vol. VII, *Reactions in Condensed Phases* (Academic Press, New York, 1974–1975).

See references for Chapter 27. **Reactive Collisions**

Kondrat'ev, V. N., *Chemical Kinetics of Gas Reactions* (Addison-Wesley, Reading, Mass., 1964). **Gas Kinetics**

Moelwyn-Hughes, E. A., *The Chemical Statics and Kinetics of Solutions* (Academic Press, New York, 1971). **Solution Kinetics**

Wells, P. R., *Linear Free Energy Relationships* (Academic Press, New York, 1968).

Lewis, E. S. (Ed., Part I), and Hammes, G. G. (Ed., Part II), *Investigation of Rates and Mechanisms of Reactions*, 3rd ed. (Wiley, New York, 1974), Vol. VI of Arnold Weissberger (Ed.), *Techniques of Chemistry*. **Experimental Methods**

Basolo, F., and Pearson, R. G., *Mechanisms of Inorganic Reactions*, 2nd ed. (Wiley, New York, 1967). **Reaction Mechanisms**

Benson, D., *Mechanisms of Inorganic Reactions in Solution* (McGraw-Hill, London, 1968).

Breslow, R., *Organic Reaction Mechanisms*, 2nd ed. (Benjamin, Reading, Mass., 1969).

Ingold, C. K., *Structure and Mechanism in Organic Chemistry*, 2nd ed. (Cornell University Press, Ithaca, N.Y., 1969).

Lowry, T. H., and Richardson, K. S., *Mechanism and Theory in Organic Chemistry* (Harper & Row, New York, 1976).

Problems

1. The second-order rate constant for the reaction

$$2NO_2 \rightarrow 2NO + O_2$$

is 6.3×10^2 cm³/mol s at 600 K. If the initial pressure of NO_2 is 760 torr, how long will it take for 20% of the NO_2 to decompose?

2. For the decomposition of di-t-butyl peroxide,

$$(CH_3)_3COOC(CH_3)_3 \rightarrow 2(CH_3)_2CO + C_2H_6,$$

the following table gives the data for total pressure of the system versus time [*J. Am. Chem. Soc.* **70**, 88 (1948)]:

t (min)	p (torr)
0	173.5
3	193.4
6	211.3
9	228.6
12	244.4
15	259.2
18	273.9
21	286.8

What is the order of this reaction? What is the rate constant for this reaction? What is the half-life?

3. For the reaction

$$NO + H_2 \rightarrow \tfrac{1}{2}N_2 + H_2O,$$

one has the following data [*J. Chem. Soc.* **129**, 730 (1926)]:

$p_{initial}$ (torr)	177	144	122
$t_{1/2}$ (s)	81	140	176

where $p_{initial}$ is the initial partial pressure of NO and H_2 each. What are the order and rate constant for this reaction?

4. If all reactants in a given chemical reaction have initial concentration a and the reaction is of order n, show that the half-life $t_{1/2}$ is given by

$$t_{1/2} = \frac{2^{n-1} - 1}{(n-1)a^{n-1}k} \qquad (\text{for } n \neq 1).$$

5. For the catalytic decomposition of SbH_3 there are the following data:

t (min)	11.8	28.6	52.2	85.7
Percent reaction	50	75	87.5	93.75

What is the order of this reaction? Speculate about the fact that the order is nonintegral.

6. Write a computer program (or use a straightedge and protractor) to calculate the classical reaction probability for the potential shown in the accompanying diagram. The reaction probability is defined as the number of trajectories that reach the product region divided by the total number of (randomly chosen)

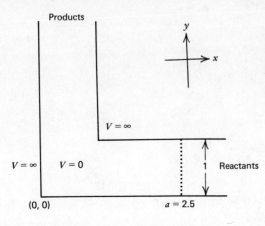

trajectories run. Use the initial values $|v_y| = 1$ and $v_x = -1$. To obtain a fair estimate of a group of random trajectories, use for the initial position of the classical particle 10 equally spaced points with x, y coordinates (a, y), $0 < y < 1$. Any value of $a > 1$, for example, $a = 2.5$, may be used. Also allow the initial value of v_y to be $\pm|v_y|$ in order to have a random selection over the direction of the initial velocity.

7. Use the data given in Figs. 30.10 and 30.11 and plot the velocity distribution of products in each case. (For the $N_2^+ + D_2$ velocity distribution use the distribution in the forward direction. Why is this a good estimate of the total velocity distribution?) For each system estimate the most probable relative velocity of products and the most probable internal energy of products. In Fig. 30.10 the experimental initial relative energy is 3.12 eV and the exothermicity is 24.93 eV. In Fig. 30.11 these values are 2.76 eV and 24.8 eV, respectively.

8. Use the data in Figs. 30.13 and plot the following:
   (a) The rotational energy distributions for $v' = 1, 2, 3, 4$.
   (b) The vibrational-level populations for $E_{trans} = 5$ kcal/mol in Fig. 30.13a and $E_{trans} = 30$ kcal/mol in Fig. 30.13b.
   Estimate in each case $\langle E_{trans} \rangle$, $\langle E_{vib} \rangle$, $\langle E_{rot} \rangle$. Assume that each degree of freedom can be associated with a temperature. Calculate the translational, vibrational, and rotational temperatures corresponding to these average energies. Use $\bar{\nu}_e(HCl) = 2990$ cm^{-1}, $R_{HCl} = 1.275$ Å.

9. Calculate the hard-sphere reactive cross section for the reaction

$$F + CH_3D \rightarrow CH_3 + FD$$

at $T = 300$ K, 500 K, 1000 K. Take hard-sphere $d_F = 1.4$Å, $d_{CH_3D} = 3.8$Å, and $E_a = 5$ kJ/mol. Estimate the steric factor p and justify your estimate.

10. For a reaction of hard spheres, take $d = 2$ Å, $\Delta E_f^* = 50$ kJ/mol, and $p = \frac{1}{2}$. Plot the *elastic* total cross section for $E = 25, 50, 100$, and 250 kJ/mol. In each case compare to the elastic cross section that would be observed if no reaction took place (i.e., $p = 0$). Point out how the parameters of the reaction (d, ΔE_f^*, and p) are related to the features of the elastic differential cross section.

11. Suppose for a reaction $A + BC \rightarrow AB + C$ that the reactive cross section is (approximately) given by

$$\sigma_R = \begin{cases} 100 \text{ Å}^2 & \text{if } v > 5 \times 10^3 \text{ m/s} \\ 0 & \text{if } v < 5 \times 10^3 \text{ m/s.} \end{cases}$$

What is the rate at 300 K if the reduced mass of the reactants is 1 amu? How long will it take to form a mole of C?

12. From the following data [*J. Chem. Phys.* **45**, 3871 (1966)] for the reaction $T + H_2 \rightarrow T + H + H$, calculate the hard-sphere collision-theory parameters ΔE_f^* and pd^2.

Tritium Energy (eV)	$\sigma/\pi a_0^2$
15	0.2
20	0.9
31	2.3
38	2.8
54	3.8
74	4.0

Take the hard-sphere diameter of H_2 to be $5.6a_0$ and that of H to be $2a_0$; estimate p for this reaction.

13. A treatment of kinetics alternative to the deterministic approach described in the text is the stochastic approach. Read the article "Stochastic Approach to Reaction and Physico-Chemical Kinetics" by E. A. Boucher, *Journal of Chemical Education* **51,** 580 (1974). Outline the differences between the two methods and between their applications.

14. A certain reaction is 20% complete in 15 min at 40°C and in 3 min at 60°C. Estimate the activation energy for this reaction.

15. Use the following data [*J. Chem. Phys.* **38,** 1518 (1963)] for the reaction

$$Ar + O_2 \xrightarrow{k} Ar + O + O$$

to calculate the activation energy.

T (K)	5,000	10,000	15,000	18,000
$k \left(\dfrac{cm^3}{mol\ s} \right)$	5.49×10^9	9.86×10^{11}	5.09×10^{12}	8.60×10^{12}

H. Johnston and J. Birks, in *Accounts of Chemical Research* **5,** 327 (1972), give a good discussion of why this activation energy is less than the bond dissociation energy of 495 kJ/mol.

16. Derive the expressions for the reactive cross section and rate coefficient of ion–molecule reactions governed by an ion-induced dipole attraction

$$V(R) = -\frac{\alpha}{2R^4}.$$

(*Hint:* $\sigma = \pi b_{max}^2$, where b_{max} is the largest value of b for which

$$\dot{R}^2 = \frac{2}{\mu} \left[E_T - V(R) - \frac{E_T b^2}{R^2} \right]$$

is positive at the maximum of the effective barrier. For all $b < b_{max}$ the system reaches the reaction zone.)

17. For the reaction

$$H_2 + I_2 \underset{k_{-1}}{\overset{k_1}{\rightleftharpoons}} 2HI,$$

$k_1 = 10^{14} e^{-(165\ kJ/mol)/RT}$ cm³/mol s and $k_{-1} = 6 \times 10^{13} e^{-(185\ kJ/mol)/RT}$ cm³/mol s. What is the equilibrium constant K for a mixture of H_2, I_2, and HI at $T = 300$ K and 2000 K?

18. Plot the energy along the reaction coordinate for the following reactions using the BEBO method:

(a) $H + H_2 \rightarrow H_2 + H$

(b) $F + H_2 \rightarrow FH + H$

Use the values $R_s = 1.4a_0$, $p = 1.041$, and $E_s = 435$ kJ/mol for H_2; $R_s = 1.76a_0$, $p = 1.136$, and $E_s = 565$ kJ/mol for FH. In each case, what are the values of n, R_{AB}, R_{BC}, and E at the maximum point along the reaction coordinate?

19. Use the BEBO activation energy to calculate the activated-complex-theory rate of reaction for $H + H_2 \rightarrow H_2 + H$ at $T = 300$ K, 1000 K, and 5000 K. Use vibrational wavenumbers of 4500 cm^{-1} for H_2 and 5000 cm^{-1} for the activated complex.

20. For the potential surface described in Problem 6, assume that the activated complex corresponds to the configuration $R_{AB} = R_{BC}$. Draw typical trajectories that reach the activated-complex region, yet do not react. If κ is a parameter defining what fraction of the trajectories that reach the

activated complex react, then will κ increase, decrease, or stay the same as the translational energy of the particle increases? Explain.

21. Predict the value of m (Eq. 30.73) in both the high- and low-temperature regions for the following reactions:

    (a) $H + Cl_2 \rightarrow HCl + Cl$ via a linear intermediate

    (b) $H + Cl_2 \rightarrow HCl + Cl$ via a nonlinear intermediate

    (c) $F + CH_4 \rightarrow H + CH_3F$ via an S_N2 mechanism

    (d) $cis\text{-}C_2H_2BrCl \rightarrow trans\text{-}C_2H_2BrCl$

22. How does increasing the ionic strength affect the rates of the following reactions?

    (a) $2[Co(NH_3)_3Br]^{2+} + Hg^{2+} + 2H_2O \rightarrow 2[Co(NH_3)_5H_2O]^{3+} + HgBr_2$

    (b) $OH^- + CH_3Br \rightarrow CH_3OH + Br^-$

    (c) $[Co(NH_3)_5Br]^{2+} + NO_2^- \rightarrow [Co(NH_3)_5NO_2]^{2+} + Br^-$

    (d) $H_2O_2 + 2H^+ + 2Br^- \rightarrow 2H_2O + Br_2$

In each case, what ionic strength will cause a change in rate of 25%?

23. The decomposition of diazoacetic ester,

$$CHN_2COOC_2H_5 + H_2O \rightarrow CH_2OHCOOC_2H_5 + N_2,$$

which is catalyzed by acid (the reaction is first-order with respect to $[H^+]$) has a rate coefficient of $0.0134\ \text{min}^{-1}$ in $0.001\ M$ nitric acid. The following table [*J. Phys. Chem.* **28,** 579 (1924)] gives the rate of decomposition in $0.05\ M$ acetic acid and various concentrations of potassium nitrate:

$[KNO_3]$ (mol/liter)	0	0.01	0.05	0.100
$k\ (\text{min}^{-1})$		0.0127	0.0135	0.0142	0.0146

What are the concentrations of H_3O^+ and CH_3COO^- in each case? ($pK_a = 4.8$ for acetic acid.)

24. What is the temperature dependence of the macroscopic rate coefficient for diffusion-controlled neutralization reactions? What is the temperature dependence of the rate coefficient for reactions in solution that are far from the diffusion-controlled limit? For these latter reactions, what is the difference in activation energies between the rate coefficient k_f and the local rate coefficient k_f^* (see Eq. 29.103), for reactions in water at 25°C and $r^* = 3$ Å? Compare this difference to typical activation energies in the gas phase.

25. Discuss the differences expected between rates of reaction in polar solvents and in nonpolar solvents. Compare magnitudes and signs of the various thermodynamic variables. There is a good treatment of this subject in *Journal of Chemical Education* **51,** 231 (1974), "The Influence of Solvents on Chemical Reactivity" by Michael R. J. Dack.

26. A simple model to depict the interdependence of $\log k$ and $\log K$ is shown in the accompanying figure. One set of curves depicts the energy along the reaction coordinate for $XH + Y \rightarrow X + HY$ and the other set for $XH + Y' \rightarrow X + HY'$. Assume that each curve is a harmonic well and that the two product curves are merely shifted vertically. Show that for small changes

$$\Delta E^{\ddagger} = B\ \Delta E,$$

where B is a constant. What are the limits of the magnitude of B?

27. If the ammonium proton-exchange reaction,

$$NH_3 + NH_4^+ + OH^- \rightleftharpoons NH_3 + NH_3 + H_2O \rightleftharpoons NH_4^+ + NH_3 + OH^-,$$

is catalyzed by bases, how will the nmr spectrum change as the pH increases? How could isotopic substitution be used to test the above mechanism?

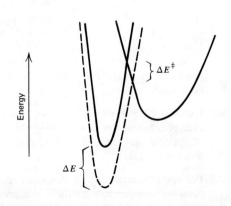

28. Consider a system for which the forward and reverse rate coefficients are known, for example,

$$A \xrightarrow{k_f} B + C$$

$$B + C \xrightarrow{k_r} A.$$

Take the initial concentration of A as $[A]_0$ and the initial concentrations of B and C as zero. Write the equation for $d[A]/dt$ for the total system. Describe qualitatively the time dependence of $[A]$.

29. Consider the following set of competitive, consecutive reactions:

$$NaOH + C_2H_4(CO_2C_2H_5)_2 \xrightarrow{k_1} C_2H_4(CO_2C_2H_5)CO_2Na + C_2H_5OH$$

$$NaOH + C_2H_4(CO_2C_2H_5)CO_2Na \xrightarrow{k_2} C_2H_4(CO_2Na)_2 + C_2H_5OH.$$

Evaluate the rate expressions under the limiting conditions $k_1 \gg k_2$, $k_1 \ll k_2$, and $k_1 = 2k_2$.

30. What is the rate law for the following mechanism?

$$[Cr(H_2O)_6]^{3+} \underset{k_{-1}}{\overset{k_1}{\rightleftharpoons}} [Cr(H_2O)_5]^{3+} + H_2O$$

$$[Cr(H_2O)_5]^{3+} + NCS^- \xrightarrow{k_2} [Cr(H_2O)_5NCS]^{2+}.$$

What are the apparent orders of the reaction for $[NCS^-]$ very small and for $[NCS^-]$ very large?

31. It is possible to treat a termolecular reaction as a sequence of bimolecular steps, for example,

$$A + A \rightarrow A_2$$

$$A_2 + B \rightarrow C,$$

in which case it is important to know the steady-state concentration of dimers, which is related to the monomer concentration by $K = [A_2]/[A]^2$. For a Lennard–Jones potential, it can be shown that the reduced equilibrium constant, $K^* \equiv 3K/2\pi\sigma^3$, is approximately related to the reduced temperature, $T^* \equiv k_BT/\epsilon$, as $K^* = 5(T^*)^{-3/2}$. Calculate the concentration of $(O_2)_2$ dimers at $T = 300$ K and at pressures of 1 atm and 10 atm. Take $\epsilon/k_B = 113$ K and $\sigma = 3.433$ Å. Why does K^* decrease with increasing T^*?

31

Some Advanced Topics in Chemical Kinetics

In Section 30.14 we explained unimolecular reactions in terms of the simple Lindemann model, involving a bimolecular excitation (and deexcitation) process followed by the unimolecular decomposition step. This model was a significant achievement, providing a qualitatively correct picture of the change in mechanism from unimolecular at high pressure to bimolecular at low pressure, and the related fall-off with pressure of the effective first-order rate coefficient. But the simplicity of the Lindemann model was bought at the price of a number of restrictive approximations: (1) The process by which molecule A is excited yields A^* with some average excitation energy, but no account is taken of how this energy is distributed over the various degrees of freedom of A. (2) Every collision is assumed to lead to deexcitation, independent of the amount of energy in A^*. This approximation is reasonable for molecules with many degrees of freedom (triatomic and larger) that have been excited to an energy close to the dissociation energy; but it does not necessarily hold for low excitation, such as suffices for some isomerization reactions. (3) Once the molecule is given sufficient energy to react, the rate coefficient k_3 is assumed independent of the amount by which the energy exceeds the minimum value. This is probably a poor approximation; a molecule with twice the energy necessary to react is likely to have a higher reaction probability than one with barely enough energy, especially when the distribution of energy within the molecule is taken into account.

Since the work of Lindemann, substantial improvements in the theory of unimolecular reactions have been made,[1] based on the removal of some of the approximations just listed.

Let us begin by describing the Lindemann reaction mechanism in greater detail. Consider the scheme

$$A + M \xrightarrow{\delta k_1(E)} A^*(E, \delta E) + M,$$

$$A^*(E, \delta E) + M \xrightarrow{k_2} A + M, \qquad (31.1)$$

$$A^*(E, \delta E) \xrightarrow{k_3(E)} \text{products}.$$

We now assign to the first step a differential rate coefficient $\delta k_1(E)$ describing the rate at which A is excited to the energy range from E to $E + \delta E$. The deexcitation rate (k_2) is again taken to be independent of the

31.1
More About Unimolecular Reactions

---

[1] The names of C. N. Hinshelwood, O. K. Rice, H. C. Ramsperger, L. S. Kassel, and R. A. Marcus are associated with the more important developments.

amount of energy in A*; deexcitation is assumed to occur with no activation energy. In the third step we recognize the possibility—in fact the likelihood—that the rate coefficient k_3 does depend on the energy of A*. We now proceed exactly as before: We set up the kinetic equations for $d[A*]/dt$ and the rate of product formation, and make a stationary-state assumption for $[A*]$. In complete analogy with Eq. 30.174, we find the effective unimolecular rate coefficient for the energy range from E to $E + \delta E$ to be

$$\delta k(E) = \frac{k_3(E)\{\delta k_1(E)/k_2\}}{1 + \{k_3(E)/k_2[M])\}};$$ (31.2)

hence the effective total unimolecular rate coefficient at a given temperature is the integral of Eq. 31.2 over all energies above E_0, the threshold energy for reaction,

$$k_{uni} = \int_{E=E_0}^{\infty} \frac{k_3(E)\{\delta k_1(E)/k_2\}}{1 + \{k_3(E)/(k_2[M])\}}.$$ (31.3)

Thus far we have in effect assumed that the A molecule can have any (continuous) amount of energy; that is, we have taken the internal degrees of freedom to be classical. It is preferable, and indeed necessary, to give these degrees of freedom a quantum-mechanical description. This can be done with a relatively crude model, representing the molecule as a system of oscillators all of frequency ν: the only change made is to replace the continuous variable E by $ih\nu$, where i is an integer giving the total number of quanta in the molecule. Equation 31.3 then becomes

$$k_{uni} = \sum_{i=m}^{\infty} \frac{k_3(i)\{\delta k_1(i)/k_2\}}{1 + \{k_3(i)/(k_2[M])\}},$$ (31.4)

where $mh\nu = E_0$. This remains a purely formal result; to make it useful we must investigate how $\delta k_1(i)$ varies with energy.

We recognize at once that the quantity $\delta k_1(i)/k_2$ is the equilibrium constant for the first two reactions in Eqs. 31.1, that is, the excitation and deexcitation of the excited state $A*(i)$. Let us assume that the molecule consists of s equivalent harmonic oscillators of frequency ν. (Of course, this can be only an approximation: By the very fact that A* decomposes, some bond or bonds must be stretched to the point that the harmonic approximation fails. Moreover, obviously not all the oscillators in a molecule have the same frequency; one can formulate the model in terms of a set of frequencies, but this is tedious and makes no major difference.) We thus have i quanta distributed over s oscillators, and a total energy $(i-m)h\nu$ in the states with $i \geq m$ which we define as A*. If energy is distributed statistically among the molecules, the fraction of molecules in the ith state (i.e., containing i quanta of internal energy) is given by

$$\frac{N_i}{N} = \frac{g_i e^{-ih\nu/k_BT}}{\sum_{j=0}^{\infty} g_j e^{-jh\nu/k_BT}},$$ (31.5)

where g_i is the degeneracy of the ith state; the sum in the denominator is simply the partition function for a set of independent harmonic oscillators all of frequency ν. (Note that in this equation the oscillators are the molecules themselves, rather than the internal oscillators of our model.) The degeneracy g_i is the number of ways in which i identical quanta can be distributed over the s internal oscillators (Bose–Einstein statistics), which is

$$g_i = \frac{(i+s-1)!}{i!(s-1)!}$$ (31.6)

The partition function in Eq. 31.5 is thus easily shown to be the sum of a binomial series,

$$\sum_{j=0}^{\infty} \frac{(j+s-1)!}{j!(s-1)!} e^{-jh\nu/k_BT} = (1-e^{-h\nu/k_BT})^{-s}. \tag{31.7}$$

If the total fraction of A molecules in excited states is small, we can make the approximation

$$\frac{\delta k_1(i)}{k_2} = \frac{[A^*(i)]}{[A]} = \frac{N_i}{\sum\limits_{i<m} N_i} \approx \frac{N_i}{N} = \frac{(i+s-1)!}{i!(s-1)!} e^{-ih\nu/k_BT}(1-e^{-h\nu/k_BT})^s, \tag{31.8}$$

which gives us the ratio appearing in the numerator of Eq. 31.3.

Next we shall evaluate the decomposition coefficient $k_3(i)$. Assume that the total energy $ih\nu$ is distributed statistically within each A molecule. For reaction to occur, the critical energy $mh\nu$ must be concentrated in a single oscillator (i.e., bond). The assumption we make is that $k_3(i)$ is proportional to the probability that m quanta are localized in that oscillator, while the remaining $i-m$ quanta are distributed randomly over the oscillators. We thus have

$$k_3(i) = \alpha \frac{(i-m+s-1)!/(i-m)!(s-1)!}{(i+s-1)!/i!(s-1)!}, \tag{31.9}$$

where α is the proportionality constant; the numerator is the number of ways in which $i-m$ quanta can be distributed over s oscillators, and the denominator is the number of ways in which the total i quanta could be distributed over the same s oscillators. Substituting this result and Eq. 31.8 into Eq. 31.4, we find that the effective unimolecular rate coefficient is

$$k_{uni} = \alpha k_2[M] \sum_{i=m}^{\infty} \frac{\{(i-m+s-1)!/(i-m)!(s-1)!\}e^{-ih\nu/k_BT}(1-e^{-h\nu/k_BT})^s}{k_2[M]+\alpha(i!)(i-m+s-1)!/(i-m)!(i+s-1)!}. \tag{31.10}$$

Let us discuss Eq. 31.10 piecemeal. First we evaluate the fraction of molecules in the ith excited state, Eq. 31.8, in the classical limit. If $i \gg s$, so that the number of quanta in each oscillator is large, we have

$$\frac{(i+s-1)!}{i!(s-1)!} = \frac{(i+1)(i+2)\cdots(i+s-1)}{(s-1)!} \approx \frac{i^{s-1}}{(s-1)!}. \tag{31.11}$$

The classical limit also implies that $h\nu/k_BT \ll 1$, so that

$$1-e^{-h\nu/k_BT} \approx \frac{h\nu}{k_BT}. \tag{31.12}$$

We thus have in the classical limit

$$\frac{\delta k_1(i)}{k_2} \approx \frac{i^{s-1}}{(s-1)!}\left(\frac{h\nu}{k_BT}\right)^s e^{-ih\nu/k_BT}. \tag{31.13}$$

If we wish to go back to our original continuous-energy assumption, we simply set $E = ih\nu$; the fraction of molecules in the range from E to $E+\delta E$ is then

$$\frac{\delta k_1(E)}{k_2} \approx \frac{(E/k_BT)^{s-1}}{k_BT(s-1)!} e^{-E/k_BT} \delta E. \tag{31.14}$$

Note the way in which this expression depends on the value of s, the number of oscillators in the molecule: For $s = 1$ (a diatomic molecule) the population of molecules with energy E decreases exponentially with E; but for large s the energy dependence is less extreme, as a result of the many ways in which the energy can be distributed over the oscillators. A typical value for E_0, the threshold energy corresponding to $mh\nu$, is $20k_BT$ or more.

To get some insight into the meaning of these formal expressions, let us make a simple calculation. We return to the Lindemann model for the effective unimolecular rate coefficient, expressing the result in the form

$$k_{uni} = \frac{k_\infty}{1 + k_3/(k_2[M])}, \tag{31.15}$$

where

$$k_\infty \equiv \frac{k_1 k_3}{k_2} \quad \text{or} \quad k_3 = \frac{k_2 k_\infty}{k_1}. \tag{31.16}$$

Suppose we now assume that k_1/k_2 can be obtained by integrating Eq. 31.14 over the entire energy range from the threshold energy E_0 to infinity:

$$\frac{k_1}{k_2} = \frac{1}{(s-1)!} \int_{E_0}^{\infty} \frac{1}{k_BT} \left(\frac{E}{k_BT}\right)^{s-1} e^{-E/k_BT} \, dE$$

$$= e^{-E_0/k_BT} \left\{ \frac{(E_0/k_BT)^{s-1}}{(s-1)!} + \frac{(E_0/k_BT)^{s-2}}{(s-2)!} + \cdots \right\}$$

$$\approx \frac{1}{(s-1)!} \left(\frac{E_0}{k_BT}\right)^{s-1} e^{-E_0/k_BT}. \tag{31.17}$$

(We have assumed that $E_0/k_BT \gg 1$ and thus neglected lower powers.) Assume next that the coefficient k_∞ has an Arrhenius temperature dependence,

$$k_\infty = A_\infty e^{-E_\infty/k_BT}, \tag{31.18}$$

with the typical experimental values $A_\infty = 10^{13}\,\text{s}^{-1}$ and $E_\infty = 40k_BT$. If we can take k_3 to be temperature-independent, we have

$$E_\infty = k_BT^2 \frac{d\ln k_\infty}{dT} \approx k_BT^2 \frac{d\ln(k_1/k_2)}{dT} \approx E_0 - (s-1)k_BT. \tag{31.19}$$

Given this string of approximations, we have sufficient information to calculate k_3 as a function of s, the number of participating degrees of freedom: We obtain E_0 by Eq. 31.19, k_1/k_2 by Eq. 31.17, and finally k_3 by Eq. 31.16, with the values:

s	1	5	10
$k_3(\text{s}^{-1})$	10^{13}	3.5×10^9	1.8×10^7

It will be recalled that k_3 is the rate coefficient for reaction of the excited molecules A^*; if we interpret k_3^{-1} as a reaction (relaxation) time, we see that the lifetime of A^* increases sharply with increasing s. As the number of degrees of freedom increases, it takes longer for a given fraction of the energy to collect in the particular degree(s) of freedom where reaction can occur. Therefore, as s increases, the effective rate coefficient k_{uni} approaches its high-pressure-limit value k_∞ at lower pressures. This appears in Fig. 31.1 as a shift to the left of the curves marked HL (for Hinshelwood and Lindemann, by whose names the model developed here is known); the reaction illustrated is the decomposition of azomethane,

$$CH_3N{=}NCH_3 \rightarrow 2CH_3 + N_2.$$

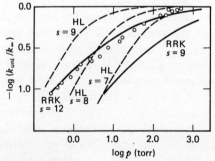

FIGURE 31.1

Comparison of theory and experiment for the unimolecular decomposition of azomethane. The curves shown are for the Hinshelwood–Lindemann (HL) and Rice–Ramsperger–Kassel (RRK) theories, whereas the open circles give Ramsperger's experimental data. From P. J. Robinson and K. A. Holbrook, *Unimolecular Reactions* (Wiley, London, 1972), p. 57.

Now let us return to our analysis of Eq. 31.10. We have evaluated k_1/k_2 in the classical limit, on the assumption that k_3 is constant; the classical limit for k_3 also has a simple interpretation. If we again make the assumption that $i \gg s$, we can rewrite Eq. 31.9 as

$$k_3(i) \approx \alpha \frac{(i-m)^{s-1}/(s-1)!}{i^{s-1}/(s-1)!} = \alpha \left(\frac{i-m}{i}\right)^{s-1} \tag{31.20}$$

by exactly the same reasoning as that leading to Eq. 31.11. With $E = ih\nu$ and $E_0 = mh\nu$, this is equivalent to

$$k_3(E) = \alpha \left(\frac{E-E_0}{E}\right)^{s-1} \tag{31.21}$$

in the continuous-energy model. Remember that in this equation $[(E-E_0)/E]^{s-1}$ is the probability that one of the s oscillators has an energy greater than E_0 if the total energy E is distributed statistically over all the oscillators. Thus the coefficient k_3 is a function of energy (a fact we mentioned previously but neglected); it increases with increasing E, but at a given E decreases with increasing s. The more energy the excited state A^* has, the more likely that the amount E_0 will be in the degree of freedom essential for reaction. But for a given energy the probability of finding E_0 in a given degree of freedom decreases with the number of degrees of freedom available.

Although the Hinshelwood–Lindemann model is the only one we shall consider in detail in this section, we must say something about later developments. The *Rice–Ramsperger–Kassel (RRK) theory* involves the full evaluation of Eq. 31.10 or its classical-limit equivalent. Figure 31.1 also includes the classical RRK results for the decomposition of azomethane. The quantum-mechanical RRK theory gives better results than the classical version, which in turn is better than the Hinshelwood–Lindemann approach (as can be seen from the experimental data in the figure). The best value of the parameter s for a given molecule is obtained by fitting to the data. The quantum-mechanical version of the RRK theory usually (but not always) yields a value of s close to the total number of degrees of freedom in the excited molecule; the classical version gives a value of s only half as large.

A further improvement in the theory (designated as *RRKM*, where the added initial stands for *Marcus*) involves consideration of more than just the statistical distribution of energy in A^* when one calculates k_3. Once the molecule has enough energy for reaction, it is still necessary for it to reach some critical configuration before reaction can occur; activated-complex theory is used to calculate the probability of this, and gives a way of evaluating the probability constant α. But we cannot go into the ramifications of this approach.

None of the theories of unimolecular reactions we have described addresses the possibility of *non*statistical energy distribution in the original A molecule or of nonstatistical reaction dynamics. Except for the dissociation of diatomic molecules, no theory of such an encompassing nature is yet available. Much recent work is thus directed toward finding the extent to which nonstatistical behavior actually exists.

A key factor affecting the energy distribution is the mode of excitation of A. Thus far we have concentrated on excitation by collisions, that is, *thermal* excitation, which it seems should produce a statistical energy distribution. But other methods can be used. For instance, *chemical activation* occurs in reactions such as

$$CH_2 + CH_3CH=CH_2 \rightarrow C_4H_8 \text{ (various isomers)}$$

or

$$CH_2 + CH_2 \!-\! CH_2 \rightarrow C_4H_8 \text{ (various isomers).}$$
$$\underset{\displaystyle CH_2}{\diagdown \diagup}$$

The relative amounts of the various products formed are the same for both reactions; one thus concludes that both proceed via the same excited species, with the same (most likely statistical) energy distribution prior to the final reaction step. Another method of excitation is by nonthermal collisions; for example, collision of an alkane with electronically excited Hg transfers energy to the alkane, which then dissociates. Still another method is excitation by photons, which we shall discuss in some detail later; suffice it to say here that the energy absorbed is randomly distributed in molecules with more than a few degrees of freedom. But note that, for both chemical and photochemical excitation, there is also evidence for nonstatistical energy distributions in many instances.

One potential method of obtaining information about unimolecular reactions with nonthermal activation is the use of molecular beams. (The reactions studied in this way are, of course, bimolecular, but can be the activation steps of "unimolecular" mechanisms; as we have seen, this bimolecular step is rate-determining at low pressures.) For instance, molecular-beam studies have been made of the reactions between F atoms and olefins, which involve complexes with lifetimes much longer than rotational periods ($\tau_{rot} \approx 10^{-12}$ s). Various subsequent reaction paths are available, such as splitting off a H atom or a CH_3 radical. The translational energy distribution of these fragments is found to be nonstatistical, contrary to the predictions of (say) the RRKM theory; this implies that not all degrees of freedom participate in the fragmentation of the complex. On the other hand, when I atoms react with chlorohydrocarbons, the Cl atoms ejected from the reaction complexes show a statistical energy distribution. The chemiluminescence spectrum of C_2H_3F produced by the reaction $F + C_2H_4$ is different from that of thermally excited C_2H_3F, again implying a nonstatistical energy distribution; however, similar studies on the reaction of O atoms with cyclooctene indicate a statistical distribution. Trajectory calculations on the isomerization of CH_3CN indicate nonstatistical reaction dynamics.

Thus we see that the evidence is mixed. The reaction dynamics of molecules with many degrees of freedom is far from being fully understood, and is an area of active research. The apparently random collection of facts in the last few paragraphs has been included to give the student some feeling for the nature of a research frontier.

31.2 Kinetics of Photochemically Induced Reactions

The absorption of photons by atoms or molecules provides the source of energy for many reactions. We have mentioned a number of aspects of photochemistry before; in this section we give a more unified account of the field, with primary emphasis on the kinetics.

The key step in any photochemically induced reaction is the excitation of an atom or molecule by a photon. The excitation may be in electronic, vibrational, or rotational degrees of freedom; the absorption spectrum of the species is an integrated measure of the probabilities of all such processes. One example of a reaction triggered by radiation is

$$NO_2 \xrightarrow{\;h\nu\;} NO_2{}^* \rightarrow NO + O.$$

When the initial reactant is an isolated NO_2 molecule in its ground state, there is sufficient energy for dissociation only when the exciting radiation has a wavelength below 398 nm; this means that the excitation is an electronic one. The lifetime of the excited NO_2 molecule ranges from a few microseconds up to about 100 μs, depending on the particular state attained in the excitation. Not all the excited molecules dissociate, of course: Some return to the ground state by *resonance fluorescence* (emitting photons of the same wavelength as the exciting radiation), whereas others drop to various lower states by ordinary *fluorescence* processes (yielding a distribution of photons with wavelengths different from that of the exciting radiation). Whatever the outcome, any remaining excitation energy is distributed by collisions until equilibrium is reached.

Most molecules[2] have a singlet electronic ground state, that is, a state with zero net electron spin. In general, the selection rule $\Delta S = 0$ applies, so that allowed excitation from a singlet ground state can go only to another singlet electronic state. This excitation is represented by process A in Fig. 31.2, which also illustrates several of the possible subsequent photoemission processes: case B is that of resonance fluorescence, in which the final vibrational-rotational state of the molecule is the same as that before excitation; case C yields fluorescence of longer wavelength from the same excited state; and case D (also of longer wavelength than B) fluorescence from a lower vibrational-rotational level of the excited state. Case D requires an intermediate process of *energy transfer* from the molecule; in the gas phase this can occur only by collision with another molecule or a surface, but in solutions or solids energy transfer to the lowest level of the excited electronic state takes place very rapidly ($\approx 10^{-12}$ s).

If the molecule has an excited triplet state of lower energy than the lowest excited singlet state, then a *radiationless transition* can take place from the excited singlet to the triplet state. (This can occur only if spin–orbit coupling is present, so that the rule $\Delta S = 0$ does not hold rigorously.) The two levels involved in such a transition will not have exactly the same energy (E, Fig. 31.2); the energy difference can be transferred to other degrees of freedom in the molecule (vibrations, say), to other molecules, or to phonons (lattice vibrations) in a crystal. Since spin–orbit coupling is at best small, the singlet–triplet transition is "forbidden" and takes place relatively slowly. This is also true of the transition from the triplet state back to the singlet ground state (F, Fig. 31.2), which occurs from milliseconds to seconds after the original excitation; the radiation thus emitted is called *phosphorescence*.

An excited molecule may lose part or all of its excitation energy without emission of radiation, in a collision with another molecule or a surface; this process is called *quenching*. Electronic excitation may be transferred as such from one molecule to another, or may be converted to vibrational excitation. An example of the latter is the reaction

$$Hg(^3P_1) + CO(v = 0) \rightarrow Hg(^3P_0) + CO(v = 1);$$

the energy difference between the two Hg states is 0.218 eV, whereas the first vibrational quantum of CO is 0.161 eV; the remaining 0.047 eV goes into translational energy. Also, an excited molecule may form a complex, either by dimerization with a ground-state molecule of the same species (an *excimer*) or by combination with a molecule of a different species (an *exciplex*), and such a complex can emit radiation of a wavelength different from either fluorescence or phosphorescence.

After this review of nonchemical deexcitation processes, we finally come to photochemistry in the strict sense. By a photochemical reaction

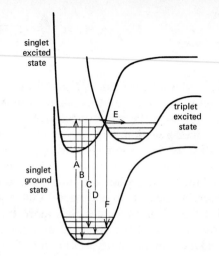

FIGURE 31.2

Schematic diagram of photoexcitation and photoemission processes: A, excitation; B, resonance fluorescence; C, longer-wavelength fluorescence; D, fluorescence after energy transfer; E, radiationless transition to triplet state; F, phosphorescence.

[2] But not, as it happens, our example of NO_2, which has an unpaired electron and thus a doublet ground state (designated $^2A_1$).

we mean a chemical reaction whose activation energy is provided by absorption of a photon (in either the reactant itself or some precursor). Consider again the dissociation of NO_2. The bimolecular reaction

$$2NO_2 \rightarrow 2NO + O_2$$

has an activation energy of 1.17 eV, and at room temperature its rate is negligibly small. But upon irradiation with light in the range 450–600 nm, a significant amount of dissociation (by various processes) takes place within minutes, whereas below 398 nm the exciting photon provides sufficient energy for immediate dissociation into $NO + O$.

Dissociation of a molecule upon photoexcitation may occur in either of two ways: direct dissociation of the excited state, or *predissociation*. The first of these is shown in Fig. 31.3a, in which we picture schematically two electronic states. In case A the excitation is to a bound state, yielding a discrete absorption spectrum but no dissociation; in case B a higher excitation energy brings the molecule to a point above the dissociation limit of the excited state, yielding a continuous absorption spectrum and dissociation. In Fig. 31.3b we illustrate predissociation: a photon absorbed by a molecule in electronic state 1 excites it to state 2; a transition can then take place (usually within a few picoseconds) to the repulsive state 3, with subsequent dissociation. Note that in predissociation the absorption spectrum is discrete (since excitation is to a definite bound state), but broadened according to the uncertainty principle ($\Delta E \approx \hbar/\tau$, where τ is the lifetime of state 2).

We can now discuss some of the typical competitive kinetic events that can follow upon excitation. Consider the sequence of elementary steps for a mixture of NO_2 and Xe irradiated with light less energetic than needed for direct dissociation:

(1) $$NO_2 + h\nu \xrightarrow{k_1} NO_2^*,$$

(2) $$NO_2^* \xrightarrow{k_F} NO_2 + h\nu',$$

(3) $$NO_2^* + NO_2 \xrightarrow{k_{Q1}} 2NO_2,$$

(4) $$NO_2^* + Xe \xrightarrow{k_{Q2}} NO_2 + Xe,$$

(5) $$NO_2^* + NO_2 \xrightarrow{k_2} 2NO + O_2.$$

Step (1) is the absorption of a single photon by a NO_2 molecule, producing a transition to an excited electronic state labeled NO_2^*. (Absorption of two or more photons is possible, but likely only at high light intensity; we shall not consider it just now.) Step (2) encompasses the various possible fluorescence processes, whereas steps (3) and (4) are quenching by NO_2 itself or a third body, respectively. Finally, step (5) is the bimolecular chemical reaction. A complete kinetic description must include all these processes, each of which has its own rate constant. For a given light intensity I_0, the rate equation for excited NO_2 is

$$\frac{d[NO_2^*]}{dt}$$

$$= k_1 I_0 [NO_2] - k_F [NO_2^*] - (k_{Q1} + k_2)[NO_2^*][NO_2] - k_{Q2}[NO_2^*][Xe].$$

(31.22)

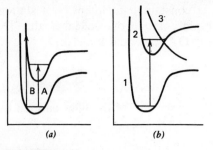

FIGURE 31.3
Mechanisms of photodissociation. (a) Direct dissociation, by excitation above the dissociation limit of the excited state (B), compared with excitation to a bound excited level (A). (b) Predissociation, with excitation from ground state 1 to excited state 2 followed by transition to unstable state 3.

At constant light intensity, a stationary-state concentration of NO_2^* is rapidly attained; we label this $[NO_2^*]_s$ and evaluate it by setting $d[NO_2^*]/dt$ equal to zero:

$$[NO_2^*]_s = \frac{k_1 I_0 [NO_2]}{k_F + (k_{Q1} + k_2)[NO_2] + k_{Q2}[Xe]}. \qquad (31.23)$$

The intensity[3] of the light emitted by fluorescence in the stationary state is

$$I_F = k_F [NO_2^*]_s, \qquad (31.24)$$

and the *fluorescence yield* (or efficiency) Φ_F is the ratio of the number of photons emitted to the number of photons absorbed:

$$\Phi_F = \frac{I_F}{k_1 I_0 [NO_2]} = \frac{k_F}{k_F + (k_{Q1} + k_2)[NO_2] + k_{Q2}[Xe]}. \qquad (31.25)$$

The *quantum yield* Φ for product formation is similarly defined as the ratio of the number of product molecules of a given species formed to the number of photons absorbed; for O_2 it is

$$\Phi(O_2) = \frac{d[O_2]/dt}{k_1 I_0 [NO_2]} = \frac{k_2 [NO_2^*]_s [NO_2]}{k_1 I_0 [NO_2]} = \frac{k_2 [NO_2]}{k_F + (k_{Q1} + k_2)[NO_2] + k_{Q2}[Xe]} \qquad (31.26)$$

in the stationary state. By measuring the fluorescence yield and quantum yield as functions of the reactant concentrations and light intensity, one can eventually evaluate all the rate coefficients; the values obtained at 488 nm are

$$k_1 = 1.9 \text{ s}^{-1} \text{ W}^{-1},$$
$$k_F = 2 \times 10^4 \text{ s}^{-1},$$
$$k_{Q1} = 4.4 \times 10^{10} \text{ liters/mol s},$$
$$k_{Q2} = 1.6 \times 10^9 \text{ liters/mol s},$$
$$k_2 = 2.2 \times 10^6 \text{ liters/mol s}.$$

If we define a quasi-first-order rate coefficient covering all the processes by which $[NO_2^*]$ decreases (i.e., rate of decrease $\equiv k'[NO_2^*]$), then the reciprocal of that coefficient is the effective lifetime of the excited state,

$$\tau = \frac{1}{k_F + (k_{Q1} + k_2)[NO_2] + k_{Q2}[Xe]}. \qquad (31.27)$$

Let us now say something more about how the various rate coefficients are determined. Fluorescence decay coefficients such as k_F are obtained by irradiating a sample (in this case, of NO_2) with a pulse of light of nanosecond duration (e.g., from a laser), then measuring the intensity of light emitted by fluorescence as a function of time after the pulse. Data from such a measurement are plotted in Fig. 31.4. If reaction (2) in our mechanism, the fluorescence decay, were the only event taking place, and if (as our notation implies) only a single excited state were involved, then the fluorescence measurements would give a simple exponential curve (first-order decay); but as can be seen from Fig. 31.4, this is not the case. Although the interpretation is tentative, the experimental curve appears to be the sum of *two* exponential decay processes with different relaxation times; this probably results from different types of

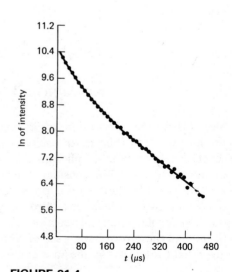

FIGURE 31.4
Decay of 593.67-nm fluorescence in a sample of NO_2 irradiated by a laser source. From V. M. Donnelly and F. Kaufman, *J. Chem. Phys.* **66**, 4100 (1977).

---

[3] The light intensity (I_0 or I_F) can be defined in whatever units are convenient, as long as the product $k_1 I_0$ has the dimensions of time^{-1}.

coupling between the excited electronic state and high vibrational levels of the ground state.

Quenching coefficients, as we have indicated, can be obtained by measuring the intensity of fluorescence as a function of the partial pressure of the quenching species. For example, suppose that we have $[NO_2] \ll [Xe]$, so that Eq. 31.23 can be approximated by

$$[NO_2^*]_s = \frac{k_1 I_0 [NO_2]}{k_F + k_{Q2}[Xe]}. \qquad (31.28)$$

Then Eq. 31.25 becomes

$$(\Phi_F)^{-1} = 1 + \frac{k_{Q2}}{k_F}[Xe], \qquad (31.29)$$

and a plot of the inverse of the fluorescence yield versus $[Xe]$, called a *Stern–Volmer plot*, yields a straight line from whose slope one can obtain k_{Q2}. Alternatively, one can obtain quenching coefficients from quantum-yield measurements for products. We can simplify Eq. 31. 26 by neglecting k_F in the denominator, so that

$$[\Phi(O_2)]^{-1} = 1 + \frac{k_{Q1}}{k_2} + \frac{k_{Q2}}{k_2} \frac{[Xe]}{[NO_2]}; \qquad (31.30)$$

Fig. 31.5 shows Stern–Volmer plots of this equation for the two quenching gases Xe and CO_2. The ratio of slope to intercept yields k_{Q2}/k_{Q1}; thus if the self-quenching coefficient k_{Q1} is known, the quenching coefficient for other gases can be determined. Finally, once both quenching coefficients in Eq. 31.30 are known, then the reaction rate coefficient k_2 can readily be obtained from the quantum-yield measurements.

Having learned all we can from the NO_2 decomposition, let us now turn to some other types of photochemical reactions. *Photoionization*, the photochemical production of ions, is a common process in the ionosphere. One important reaction is the photoionization of NO by Lyman-alpha radiation from the sun ($\lambda = 121.6$ nm, in the far ultraviolet),

$$NO + h\nu \rightarrow NO^+ + e^-.$$

Free radicals can also be generated by photochemical means, for example, the production of methyl radicals from acetone,

$$CH_3COCH_3 + h\nu(280 \text{ nm}) \rightarrow 2CH_3 + CO,$$

or methylene radicals from ketene,

$$CH_2CO + h\nu \rightarrow CH_2 + CO.$$

The experimental technique of flash photolysis (Section 30.10) commonly involves the generation of free radicals as reactants. Also, since free radicals can act as intermediates in polymer formation, the process of polymerization can be initiated photochemically.

The products of a photochemical reaction need not be the same as those of the thermal reaction between the same reactants. For instance, if one heats a mixture of

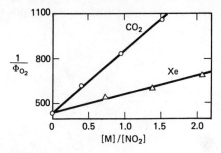

FIGURE 31.5
Stern–Volmer plots for quenching of NO_2 fluorescence by Xe and CO_2; the straight lines are fitted to Eq. 31.30. From C. L. Creel and J. Ross, *J. Chem. Phys.* **64**, 3560 (1976).

the pyridine does not react, but the first two compounds yield

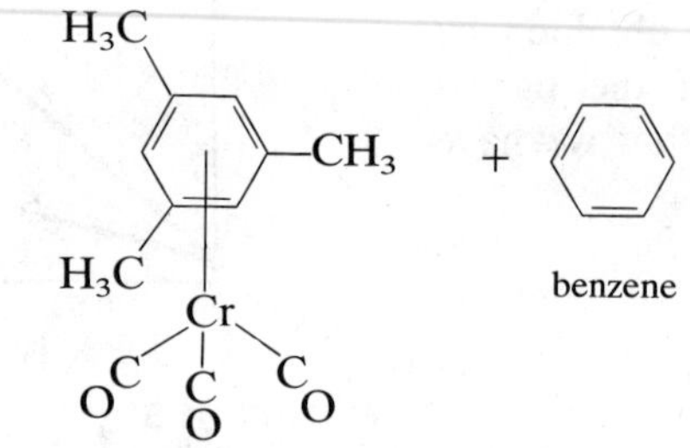

tricarbonyl(η^6-mesitylene)-
chromium

benzene

This thermal reaction proceeds by a complicated bimolecular mechanism. If, however, the same mixture is irradiated with ultraviolet light, then the products are

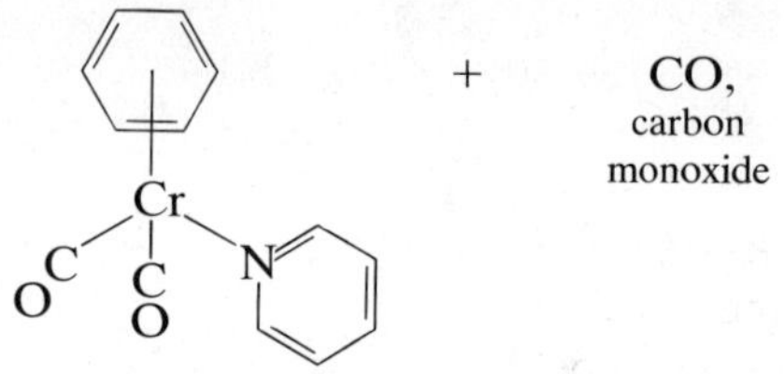

(η^6-benzene)dicarbonyl-
(pyridine)chromium

CO,
carbon
monoxide

with the mesitylene not reacting; the photochemical mechanism is also bimolecular.

Not only can corresponding photochemical and thermal reactions have different products, but the products of a photochemical reaction may also vary with the wavelength (and thus energy) of the incident radiation, as we have seen from the case of NO_2. Another example involves the six-coordinated tungsten complex $W(CO)_5NH_3$: Upon irradiation with 436-nm light, only NH_3 is split off, but at 366 nm both NH_3 and CO are liberated.

In Section 30.3 we discussed the reaction

$$F + D_2 \rightarrow FD + D,$$

and noted that the product FD is preferentially produced in the highest vibrational state energetically available ($v = 3$). A product $FD(v = 3)$ molecule will subsequently emit a photon corresponding to a transition to a lower vibrational state (usually $v = 2$, but sometimes $v = 1$ or even the ground state), unless the vibrational excitation is first quenched by collision. The emission process is an example of *chemiluminescence*, the emission of radiation from an excited state produced by reaction. The overall process (reaction followed by emission) is the reverse of a photoinduced reaction mechanism (excitation followed by reaction). Since by microscopic reversibility the FD + D reaction is most efficient from the $FD(v = 3)$ state (vibrational enhancement), it follows that one should use light of an appropriate frequency to excite the FD molecules to that state. This is an example of *specific excitation*: In this case vibrational excitation is a more efficient source of activation energy than is the translational energy furnished by simply heating the reaction system.

The existence of the specific excitation effect is clear in FD, where there is only one vibrational degree of freedom to excite; but specificity is more problematical in complex molecules. A question under much debate is the extent to which specific excitation occurs in multiphoton excitation by high-intensity lasers. That is, when reaction is induced by the absorption of many photons in a short time, is the energy thus supplied distributed statistically over all the degrees of freedom or concentrated in a few of them? (The reaction rate will presumably be higher in the latter case.) For the reaction

$$SF_6 \xrightarrow{h\nu} SF_5 + F,$$

about 20 infrared photons must be absorbed for the dissociation to occur; the available evidence thus far supports the interpretation that the energy of these photons is distributed statistically. But in the lower energy range ($\lesssim 0.7\,D_e$) only a few of the many modes dominate the dynamics, and the distribution is not statistical; this influences some details of the kinetics.

To conclude this section, let us briefly mention some of the ways in which photochemical reactions dominate our lives. The most significant of these is *photosynthesis*, the process by which plants utilize the energy of sunlight to convert water and carbon dioxide into oxygen and carbohydrates. The overall reaction can be written as

$$n\,CO_2 + n\,H_2O \xrightarrow{h\nu} n\,O_2 + (CH_2O)_n,$$

but the mechanism is extremely complicated; the initial step is the absorption of light, primarily by the chlorophyll molecule, which then transfers the energy to other molecules. This process has produced plant material for billions of years, and a small fraction of the products have been preserved as what we call fossil fuels. Thus photosynthesis is the primary source of not only our food but most of the energy we use.

Having used up a large part of the fossil fuel reserves, we now seek additional ways of converting the sun's energy to our use. One such method uses light to induce an electrode reaction on an n-type semiconductor, as shown in Fig. 31.6. The reaction electrolyzes water to generate H_2 and O_2, and the H_2 can be collected and used for fuel. (The mechanism of the electrode reactions is not known.) Many other attempts to derive photochemical energy from sunlight are under development.

Another class of photochemical reactions of much concern for public policy in recent years involves the layer of ozone in the stratosphere, which shields the earth from most of the sun's ultraviolet radiation. The ozone layer is maintained by the reaction cycle

$$O_2 \xrightarrow{h\nu} 2O,$$

$$O + O_2 + M \longrightarrow O_3 + M,$$

$$O + O_3 \longrightarrow 2O_2,$$

$$O_3 \xrightarrow{h\nu} O_2 + O,$$

with a steady-state concentration of O_3 high enough to absorb most of the ultraviolet light. However, this balance is in danger of being upset by contaminants. For example, the chlorinated hydrocarbons (e.g., CH_2Cl_2) used in aerosol spray cans are sufficiently inert to migrate to the upper atmosphere, where photolysis can release ozone-scavenging free radicals; also, the oxides of nitrogen (especially NO) produced in high-altitude jet

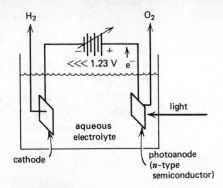

FIGURE 31.6
Example of a photochemical cell used to electrolyze water. From M. S. Wrighton, A. B. Ellis, and S. W. Kaiser, in J. B. Goodenough and M. S. Whittingham (Eds.), *Solid State Chemistry of Energy Conversion and Storage* (Advances in Chemistry Series 163, 1977), p. 71.

flights can remove ozone by such processes as

$$NO + O_3 \rightarrow NO_2 + O_2.$$

Our final example of a photochemical reaction significant for life is that by which you are reading this text. The process of vision involves the absorption of light by a pigment in the eye's receptor cells, which then convert the absorbed energy to an electrical signal transmitted to the brain. The details of the mechanism, which must include the encoding of information on the incident light's intensity and color, are a fascinating topic for current photochemical research.

31.3 Chain Reactions

Chain reactions are those that proceed by a mechanism consisting of a small number of elementary reaction steps some of which are repeated many times. For an example, consider the reaction of methane and fluorine, which is thought to have the mechanism:

$$CH_4 + F_2 \rightarrow CH_3 + HF + F \qquad \text{(initiation)},$$

$$\left. \begin{array}{l} CH_3 + F_2 \rightarrow CH_3F + F \\ CH_4 + F \rightarrow CH_3 + HF \end{array} \right\} \quad \text{(propagation)},$$

$$CH_3 + F + M \rightarrow CH_3F + M \qquad \text{(termination)}.$$

The first, or *initiation*, step in a chain reaction is one in which one or more very reactive species (in this case the free radicals CH_3 and F) are generated. It is followed by *propagation* steps, in which the number of very reactive species either remains constant or increases. In the example here, each of the two propagation steps consumes one free radical and produces another; the propagation steps are thus repeated as long as the free radicals and the original reactants remain. Finally, there is a *termination* step that removes free radicals from the reaction; in our example, M may denote either a molecule in the gas phase or the surface of the reaction vessel.

The CH_4–F_2 reaction is an example of a *stationary* chain reaction, by which is meant a reaction that proceeds at a constant rate (for a given temperature) determined by the propagation steps. We expect a constant rate in this case, since the propagation steps leave the number of free radicals unaltered. Contrast this behavior with that of a nonstationary or *branching* chain reaction, in which the propagation steps increase the number of free radicals. An example is the explosive reaction of gaseous hydrogen and oxygen. The mechanism is complex and will not be discussed in full here; there are several possible initiation steps, variously leading to the formation of H, O, and OH radicals. The OH radicals participate in the stationary propagation step

$$OH + H_2 \rightarrow H_2O + H.$$

Of more interest to us, however, are the branching propagation steps

$$H + O_2 \rightarrow OH + O,$$

$$O + H_2 \rightarrow OH + H,$$

in each of which two free radicals are produced for every one consumed. The effect is to increase the reaction rate rapidly, since the concentration of free radicals increases. In principle, the rate may increase without limit as long as there is a supply of reactants, this being the condition we describe as an explosion.

It must be pointed out here that branching-chain mechanisms are not the only causes of explosions: A *thermal* explosion is possible with any exothermic reaction. If such a reaction is proceeding in an unthermostat-ted vessel, then the temperature will rise because of the heat evolved by the reaction. As the temperature increases, the reaction rate and thus the rate of heat evolution also increase, forming a positive feedback loop. If the heat is not carried off rapidly enough, the rate may become so large that explosion occurs. The thermal effect may, of course, also be present in a branching-chain reaction, where it contributes to the acceleration of the reaction rate.[4]

One speaks of *chain carriers*, which are conserved in the propagation steps. In the examples given thus far, the chain carriers are free radicals; such cases are also called *material chains*. There are also *energy chains*, in which the chain carriers are energized (and thus highly reactive) molecules of normally stable species. An example is the reaction of hydrogen and chlorine; this reaction proceeds very slowly in the absence of light, but upon irradiation with blue light ($\leqslant 400\,nm$) the rate is greatly increased by the following sequence of reactions:

$$Cl_2 + h\nu \rightarrow Cl_2{}^* \qquad \text{(initiation)},$$

$$\left.\begin{array}{l} Cl_2{}^* + H_2 \rightarrow HCl + HCl^* \\ HCl^* + Cl_2 \rightarrow HCl + Cl_2{}^* \end{array}\right\} \quad \text{(propagation)},$$

$$\left.\begin{array}{l} Cl_2{}^* \rightarrow Cl_2 + h\nu \\ Cl_2{}^* + M \rightarrow Cl_2 + M \end{array}\right\} \quad \text{(termination)},$$

The initiation step is the photochemical formation of an electronically (or possibly vibrationally) excited Cl_2 molecule. The chain propagation then requires two steps: the actual reaction with H_2, and an energy transfer that regenerates $Cl_2{}^*$. In the termination steps, the chain carrier $Cl_2{}^*$ may lose energy by either radiation or collision with a molecule or a wall.

Let us now analyze the kinetics of a material chain reaction. For this purpose we choose the thermal decomposition of acetone, $(CH_3)_2CO$, a case that has been studied extensively. The observed rate expression is deceptively simple:

$$-\frac{d[A]}{dt} = k_f[A] \tag{31.31}$$

(here and in what follows, A is used as an abbreviation for acetone). But even the stoichiometry is complicated, since many products are obtained, principally CH_4, CH_2CO (ketene), and $CH_3COC_2H_5$. There must be free-radical intermediates, since scavengers of free radicals (compounds such as NO or propylene that react very rapidly with free radicals) reduce the reaction rate. It is clear that we are dealing with a complex situation.

We shall discuss only some elements of the reaction mechanism, which account for a significant part, if not all, of the observations. We begin by writing down a proposed mechanism (which is, of course, a guess based on many experimental studies) and developing the rate expression corresponding to it. The proposed mechanism is as follows, including the estimated activation energies:

Initiation: $\qquad CH_3COCH_3 \xrightarrow{k_1} CH_3 + CH_3CO \quad (E_1 = 350\,\text{kJ/mol})$

Propagation: $\quad CH_3 + CH_3COCH_3 \xrightarrow{k_2} CH_4 + CH_3COCH_2$
$$(E_2 = 65\,\text{kJ/mol})$$

[4] Since the H_2–O_2 reaction is both branching-chain and exothermic, how is it that a hydrogen–oxygen mixture is not explosive under all conditions? The answer lies in the termination steps of the mechanism: If free radicals can be removed from the system fast enough, then their multiplication by branching will not occur. Later in this section we shall discuss the processes by which this takes place.

Propagation:
$$CH_3COCH_2 \xrightarrow{k_3} CH_3 + CH_2CO \quad (E_3 = 200 \text{ kJ/mol})$$

Termination:
$$CH_3 + CH_3COCH_2 \xrightarrow{k_4} CH_3COC_2H_5 \quad (E_4 = 20 \text{ kJ/mol}).$$

The radical CH_3CO formed in the initiation step decomposes to CH_3 and CO and need not be considered further. Given this mechanism, the rate of disappearance of acetone is

$$-\frac{d[A]}{dt} = k_1[A] + k_2[CH_3][A]; \quad (31.32)$$

although the first term is written in unimolecular form, we recognize that this thermal decomposition must actually follow excitation by bimolecular collisons. Next we set up rate equations for the change of the free-radical concentrations:

$$\frac{d[CH_3]}{dt} = 2k_1[A] - k_2[CH_3][A] + k_3[CH_3COCH_2] - k_4[CH_3][CH_3COCH_2], \quad (31.33)$$

$$\frac{d[CH_3COCH_2]}{dt} = k_2[CH_3][A] - k_3[CH_3COCH_2] - k_4[CH_3][CH_3COCH_2]. \quad (31.34)$$

The factor of 2 in the term $2k_1[A]$ includes both the CH_3 radicals formed directly in the initiation step and those produced by rapid decomposition of CH_3CO. Assuming that we can apply the stationary-state hypothesis to the free-radical concentrations, we write

$$\frac{d[CH_3]}{dt} = \frac{d[CH_3COCH_2]}{dt} = 0; \quad (31.35)$$

solution of the second equation for $[CH_3COCH_2]$ then yields

$$[CH_3COCH_2] = \frac{k_2[CH_3][A]}{k_3 + k_4[CH_3]}. \quad (31.36)$$

Adding Eqs. 31.33 and 31.34, we have in the stationary-state approximation

$$2k_1[A] - 2k_4[CH_3][CH_3COCH_2] = 0; \quad (31.37)$$

substituting the value of $[CH_3COCH_2]$ from Eq. 31.36 and rearranging gives us

$$k_3 + k_4[CH_3] = \frac{k_2 k_4}{k_1}[CH_3]^2. \quad (31.38)$$

If we make the reasonable assumption that $k_3 \gg k_4[CH_3]$, that is, that the rate of termination is much smaller than the rate of propagation, we find that

$$[CH_3] = \left(\frac{k_1 k_3}{k_2 k_4}\right)^{1/2}. \quad (31.39)$$

Substituting this result back in Eq. 31.32, we find that the rate of disappearance of acetone is

$$-\frac{d[A]}{dt} = \left\{ k_1 + \left(\frac{k_1 k_2 k_3}{k_4}\right)^{1/2} \right\}[A], \quad (31.40)$$

where the expression in braces corresponds to the empirical k_f of Eq.

31.30. This result can be further simplified by omission of the k_1 term, since we expect k_1 to be small compared to the other rate coefficients. In either case, we have deduced the observed rate expression from the proposed reaction mechanism, with only what we believe to be justifiable handwaving.

If we thus take the effective rate coefficient to be

$$k_f = \left(\frac{k_1 k_2 k_3}{k_4}\right)^{1/2}, \tag{31.41}$$

then the corresponding effective overall activation energy, defined by

$$\frac{d \ln k_f}{dT} = \frac{E_a}{RT^2}, \tag{31.42}$$

is related simply to the activation energies of the individual reaction steps:

$$E_a = \tfrac{1}{2}(E_1 + E_2 + E_3 - E_4) = 298 \text{ kJ/mol}, \tag{31.43}$$

where we have used the estimated values of the E_i listed earlier. We see here an important property of chain reactions: The effective activation energy of the overall reaction can be less than the largest activation energy of an individual elementary reaction step. The reason in this case is basically that the initiation step, which as usual has the highest activation energy, produces two chain carriers.

A rough measure of the *chain length* is the number of times a propagation step is repeated before a termination step occurs, which is equivalent to the rate of the propagation step divided by that of the termination step. For the acetone-decomposition mechanism, we can take the rate of propagation to be that of the first propagation step, which is simply $k_2[CH_3][A]$. The rate of the termination step is similarly given by $k_4[CH_3][CH_3COCH_2]$. Combining these two values and substituting Eqs. 31.36 and 31.39, we have for the chain length

$$\nu = \frac{k_2[CH_3][A]}{k_4[CH_3][CH_3COCH_2]} = \frac{k_2[CH_3][A]}{k_4[CH_3]} \frac{(k_3 + k_4[CH_3])}{k_2[CH_3][A]} \tag{31.44}$$

$$= \frac{k_3 + k_4[CH_3]}{k_4[CH_3]} \approx \frac{k_3}{k_4[CH_3]} = \frac{k_3}{k_4}\left(\frac{k_2 k_4}{k_1 k_3}\right)^{1/2} = \left(\frac{k_2 k_3}{k_1 k_4}\right)^{1/2}.$$

An order-of-magnitude estimate of ν can be made by assuming that the preexponential factors of the somewhat similar rate coefficients k_1 and k_3 cancel, and likewise for the also similar pair k_2, k_4. Given these assumptions, the chain length is

$$\nu = \left(\frac{k_2 k_3}{k_1 k_4}\right)^{1/2} \approx e^{-(E_2 + E_3 - E_1 - E_4)/2RT}, \tag{31.45}$$

since $k_i = A_i e^{-E_i/RT}$; for the activation energies given earlier, we obtain chain lengths of about 11,500 at 700 K, 700 at 1000 K. That is, in this temperature range something like 10^3–10^4 propagation steps occur for every termination step.

Let us now see how the predicted rate equation varies if one lowers the pressure sufficiently to make the initiation step effectively bimolecular:

$$2CH_3COCH_3 \xrightarrow{k_1'} CH_3 + CH_3CO + CH_3COCH_3;$$

note that k_1' is not the same as our earlier k_1, which is the high-pressure k_{uni} for this reaction step (cf. Section 30.14). With the rest of the mechanism unchanged, we replace Eq. 31.32 for the rate of disappearance of acetone with

$$-\frac{d[A]}{dt} = k_1'[A]^2 + k_2[CH_3][A], \qquad (31.46)$$

and Eq. 31.33 with

$$\frac{d[CH_3]}{dt} =$$

$$2k_1'[A]^2 - k_2[CH_3][A] + k_3[CH_3COCH_2] - k_4[CH_3][CH_3COCH_2]. \quad (31.47)$$

Carrying through the stationary-state derivation as before, we now obtain

$$[CH_3] = \left(\frac{k_1'k_3}{k_2k_4}\right)^{1/2}[A]^{1/2} \qquad (31.48)$$

instead of Eq. 31.39. Thus the rate of disappearance of acetone (with the small initiation term $k_1'[A]^2$ neglected) should be

$$-\frac{d[A]}{dt} = \left(\frac{k_1'k_2k_3}{k_4}\right)^{1/2}[A]^{3/2}, \qquad (31.49)$$

and the reaction is in fact observed to be of 3/2 order in acetone at low pressure (≤ 100 torr). The appearance of fractional powers in an observed rate expression always indicates the dissociation of a reactant, and is thus suggestive of free-radical formation and a material chain-reaction mechanism.

The kinetics of branching chain reactions has some additional features of interest. It is simplest to derive these in terms of a general reaction mechanism, let us say for an energy chain with an excited species A^* as chain carrier:

$$A + M \xrightarrow{k_1} A^* + M$$

$$A^* + M \xrightarrow{k_2} A + M$$

$$A^* + B \xrightarrow{k_3} (\alpha + 1)A^* + C + \cdots$$

$$A^* \xrightarrow{k_4} A.$$

The first two steps are simply the excitation and deexcitation of A by collision with M (which may be another A molecule), just as in the Lindemann mechanism for unimolecular reactions. The third reaction is the branching step, in which $\alpha + 1$ chain carriers A^* ($\alpha > 1$) are produced for each one lost. (Cf. the H_2–O_2 reaction discussed earlier, in which each of the two branching steps had $\alpha = 1$, that is, two chain carriers produced for each one lost.[5]) The final step is the deexcitation of A^* on the wall of the reaction vessel (we cannot here use M, which is reserved for gas molecules that collide with A). The rate of change of the concentration of excited molecules is

$$\frac{d[A^*]}{dt} = k_1[A][M] - k_2[A^*][M] + \alpha k_3[A^*][B] - k_4[A^*]. \quad (31.50)$$

[5] Another branching-chain reaction of great significance is that which occurs in nuclear fission: A slow neutron splits a uranium nucleus, resulting in the emission of more than one neutron—and a very large amount of energy. A nuclear reactor is so designed as to keep the value of $\alpha + 1$ barely greater than unity; in a nuclear bomb α is allowed to attain significant values.

Applying the stationary-state hypothesis, we find that

$$[A^*] = \frac{k_1[A][M]}{k_2[M] + k_4 - \alpha k_3[B]}. \tag{31.51}$$

The rate of disappearance of the reactant B is thus

$$-\frac{d[B]}{dt} = k_3[A^*][B] = \frac{k_1 k_3[A][M][B]}{k_2[M] + k_4 - \alpha k_3[B]}. \tag{31.52}$$

Since we have $\alpha > 0$, it is possible for the denominator of this expression to become zero; this corresponds to an infinite reaction rate, that is, explosion. This behavior may occur for two entirely different reasons. Suppose first that the total pressure is low enough for deexcitation of A^* at a wall to be much more likely than deexcitation by collision ($k_2[M] \ll k_4$); this means the rate of diffusion of A^* to the walls must exceed the reaction rate. If at this low pressure the concentration of B is such that $k_4 > \alpha k_3[B]$, the reaction rate is finite; but as the pressure is increased the diffusion rate (k_4) decreases proportionally, and the reaction rate may approach infinity. The second possibility occurs at pressures so high that deexcitation by collision is the much more likely alternative ($k_2[M] \gg k_4$); in this case the reaction rate is finite when $k_2[M] > \alpha k_3[B]$, but may approach infinity when one decreases the pressure (and thus [M]).

For a reaction involving free radicals, we may write the following general branching-chain mechanism:

$$A + M \xrightarrow{k_1} 2R + M$$

$$2R + M \xrightarrow{k_2} A + M$$

$$R + B \xrightarrow{k_3} (\alpha + 1)R + C + \cdots$$

$$R \xrightarrow{k_4} \text{removal on surface.}$$

It can easily be seen that this is exactly parallel to the mechanism just described for an energy chain. In the termination step we again assume that the rate-determining process is the diffusion of chain carriers (R) to the wall, rather than the bimolecular recombination of radicals, which presumably occurs after their adsorption on the surface.

We can see that the two effects described above allow for the possibility of two *explosion limits*, boundaries between the regions of the p, T plane in which explosion will or will not occur. At low pressures the reaction rate is finite because of the high diffusion rate, at high pressures because of the high collision rate, both mechanisms removing chain carriers as fast as they are formed; between these limits is the region in which explosion can occur. The explosive pressure range widens with increasing temperature; at sufficiently low temperatures the upper and lower limits overlap, and there is no explosive region (this explains why a H_2–O_2 mixture does not explode at room temperature). At high temperatures, on the other hand, the thermal-explosion effect becomes dominant, producing explosive behavior at all but the lowest pressures. This behavior is illustrated in Fig. 31.7.

To close this section, we shall briefly mention two of the many important types of reactions that proceed by free-radical chain mechanisms. The first of these is the pyrolysis (thermal decomposition) of hydrocarbons. With butane, for example, the initiation step is

$$C_4H_{10} \to \begin{cases} 2C_2H_5 \\ CH_3 + C_3H_7. \end{cases}$$

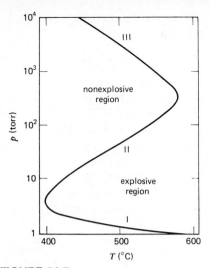

FIGURE 31.7
Explosion limits for a stoichiometric H_2–O_2 mixture (approximate only: the exact curve depends strongly on the nature of the vessel). The curves I and II are the low-pressure and high-pressure explosive limits discussed the text. The third explosion limit (III) is generally caused by the dominance of the thermal-explosion mechanism.

Each radical thus formed can attack a butane molecule in a propagation step such as

$$C_2H_5 + C_4H_{10} \rightarrow C_2H_6 + C_4H_9;$$

the C_4H_9 radical formed may be either primary or secondary, decomposing to form stable hydrocarbons and smaller radicals (primary: $C_2H_4 + C_2H_5$; secondary: $C_3H_6 + CH_3$). Obviously, there are a great variety of such propagation steps. The chains are terminated either on the surface or by free-radical recombination. Most chain-reaction mechanisms of interest are like this one: simple in principle, but with a large number of alternate paths possible.

The second class of reactions we shall mention is free-radical polymerization. For instance, styrene,

$$\text{C}_6\text{H}_5\text{—CH}=\text{CH}_2,$$

polymerizes rapidly when the process is initiated by a compound such as benzoyl peroxide, which decomposes thermally to yield two free radicals:

$$\text{C}_6\text{H}_5\text{—C(=O)—O—O—C(=O)—C}_6\text{H}_5 \longrightarrow 2\,\text{C}_6\text{H}_5\text{—C(=O)—O}\cdot$$

One of the benzoxy radicals adds to styrene,

$$\text{C}_6\text{H}_5\text{—C(=O)—O} + \text{C}_6\text{H}_5\text{—CH}=\text{CH}_2 \longrightarrow \text{C}_6\text{H}_5\text{—C(=O)—OCH}_2\text{CH—C}_6\text{H}_5,$$

and so forth. The chain is terminated by free-radical recombination.

31.4 Oscillatory Reactions

As you read these words, the electrical signal controlling your heartbeat is oscillating, similar to that shown in Fig. 31.8. This electrical signal is generated by chemical reactions that therefore also are likely to be oscillatory. Note that the repetitive pattern is far from being a simple sine wave, as one might expect for a harmonic oscillator; the oscillations appear to be highly *nonlinear* (remember that Newton's equation for the harmonic oscillator is linear in the spatial coordinate). In this section we shall discuss the mechanisms that make this type of behavior possible.

The characteristic feature of oscillatory reactions is nonlinearity. One can, of course, say that all but first-order reactions are nonlinear, in that concentrations (or products of concentrations) appear in the rate equation with exponents other than unity. But most such reactions nevertheless seem to behave as though they were linear, in that we usually observe them near enough to equilibrium for the kinetics to be nearly linear. This is indeed the basic assumption of the relaxation techniques we described in Section 30.10: that in the vicinity of equilibrium one can neglect the nonlinear terms in the rate expression, which then reduces to an equation

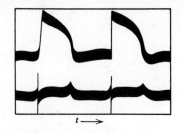

of the form

$$\frac{dx}{dt} = -\frac{x}{\tau},$$

(31.53)

where x is a measure of the deviation from equilibrium—typically, $x = [A] - [A]_{eq}$ for some reactant A—and the relaxation time τ is real and positive. For any system obeying this equation, the relaxation to equilibrium is exponential ($x \propto e^{-t/\tau}$) and monotonic in the time (no oscillations). Whatever the reaction mechanism, linear behavior of this sort must occur when the system is sufficiently close to equilibrium. Hence oscillatory behavior in reactive systems can occur only away from the linear region, a condition we simply call "far from equilibrium."

Clearly it is not possible to maintain an isolated system indefinitely in a far-from-equilibrium state; in due time such a system will come to equilibrium. To maintain nonequilibrium of any kind, one must subject a system to a continuing external force, which may be energy flow through the boundary of a closed system and/or mass flow through the boundary of an open system. For example, the oscillatory behavior of your heartbeat is made possible by the fact that you yourself are a thermodynamically open system, with intakes of food and oxygen and elimination of wastes. Without an external source of energy (or free energy), any such system must go toward a static equilibrium, sometimes called the dead state.

Although many reactions display nonlinear behavior when far from equilibrium, most of them do not exhibit oscillations. Nonlinearity is necessary for oscillatory behavior, but not every kind of nonlinearity suffices to provide it. As a first example, let us consider reactions with the interesting property of multiple stationary states.

In treating such topics as the Lindemann mechanism for unimolecular reactions or the mechanism of chain reactions, we have made the stationary-state assumption: We obtain an expression for the rate of change of some intermediate X (A^* in the Lindemann mechanism), set $d[X]/dt = 0$, and solve for $[X]$. Thus far we have assumed that the equation $d[X]/dt = 0$ has only a single solution, that is, that at a given time there is only one stationary-state concentration of X. Consider, however, the hypothetical reaction mechanism

$$A + X \underset{k_2}{\overset{k_1}{\rightleftharpoons}} 2X$$

$$X + E \underset{k_4}{\overset{k_3}{\rightleftharpoons}} C$$

$$C \underset{k_6}{\overset{k_5}{\rightleftharpoons}} E + B;$$

this is artificial but very much like the gas-phase process

$$Na + e^- \rightarrow Na^+ + 2e^-$$

$$I + e^- + M \rightarrow I^- + M$$

$$I^- + Na^+ \rightarrow I + Na^* (\rightarrow Na + h\nu),$$

which does have multiple steady states and displays periodic behavior. The first step is one of autocatalysis, and the next two correspond to Michaelis–Menten enzyme catalysis converting X to B (cf. Section 31.7). The overall reaction is simply the conversion of A to B. Let us set the concentrations of A and B constant at values such that the system is far from equilibrium; reaction must then take place continuously. (This implies an open system.) Also let $[E] + [C]$, the total enzyme concentration,

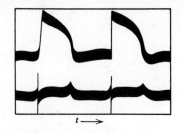

FIGURE 31.8
Electrical potential controlling the heartbeat of a frog: the upper curve is the potential of a single ventricular fiber, the lower curve the averaged potential measured by an electrocardiograph. Based on Samson Wright, *Applied Physiology*, 11th ed. (Oxford, London, 1965), p. 96.

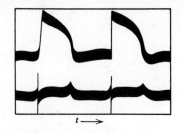

$t \longrightarrow$

equal a constant ε. The assumed reaction mechanism gives us the rate expressions

$$\frac{d[X]}{dt} = k_1[A][X] - k_2[X]^2 - k_3[X][E] + k_4[C], \qquad (31.54)$$

$$\frac{d[E]}{dt} = -k_3[X][E] + (k_4 + k_5)[C] - k_6[E][B]. \qquad (31.55)$$

Making the stationary-state approximation $d[E]/dt = 0$, and letting $[E] + [C] = \varepsilon$, we obtain, from Eq. 31.55,

$$-k_3[X][E] - k_6[E][B] + (k_4 + k_5)(\varepsilon - [E]) = 0 \qquad (31.56)$$

or

$$[E] = \frac{(k_4 + k_5)\varepsilon}{k_4 + k_5 + k_3[X] + k_6[B]}. \qquad (31.57)$$

Substituting the values of $[E]$ and $[C]$ $(= \varepsilon - [E])$ thus obtained into Eq. 31.54, and setting $d[X]/dt = 0$, we obtain a cubic equation of the form

$$[X]^3 + \rho[X]^2 + \mu[X] - \lambda = 0; \qquad (31.58)$$

the parameters ρ, μ, λ (which we shall not evaluate explicitly) are functions of the rate coefficients k_i and the fixed concentrations $[A]$, $[B]$, ε.

A cubic equation such as Eq. 31.58 may have one or three real roots. In this case the roots, which we designate as X^*, are the values of $[X]$ for which a stationary state is achieved. Their number depends on the values of the parameters ρ, μ, λ; although the single-real-root case is more common, it is not difficult to find regions in which three real (i.e., physically meaningful) roots exist. For simplicity let us take the case in which $\rho = 0$; then we have only two parameters to vary, and one can determine X^* as a function of λ for various fixed values of μ. The result of such a calculation[6] is shown schematically in Fig. 31.9. For $\rho = 0$, $\mu \geqslant 0$ the system has only one stationary state for all values of λ; but for $\rho = 0$, $\mu < 0$ there is a range of λ for which three stationary states exist. Note the formal similarity of Fig. 31.9 to the pressure–volume graph of the van der Waals equation for a pure substance near its critical point; it will be useful to keep this similarity in mind in the following discussion.

The existence of multiple stationary states implies the possibility of a system's switching from one to another. We must therefore ask: how stable are such multiple stationary states? Equation 31.53 expresses the observed fact that small perturbations away from equilibrium always decay back to the equilibrium state, which is by definition stable. But these are far-from-equilibrium states. Consider now a small linear perturbation away from one of the stationary states, so that the concentration is $X^* + \delta[X]$. If we write a rate expression corresponding to Eq. 31.53 in the general form.

$$\frac{d[X]}{dt} = F([X]), \qquad (31.59)$$

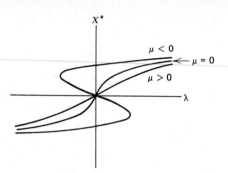

FIGURE 31.9
Real roots of Eq. 31.58 as a function of the parameter λ, for $\rho = 0$ and various fixed values of μ. (In this case the roots with $X^* < 0$ are not physically meaningful, but in the general case $\rho \neq 0$ one can obtain regions with three positive roots.)

[6] The simplest way to solve the equation

$$[X]^3 + \mu[X] - \lambda = 0$$

for given values of μ, λ is to plot the functions $y_1 \equiv [X]^3 + \mu[X]$ and $y_2 \equiv \lambda$ on the same graph; the roots X^* are the values of $[X]$ for which $y_1 = y_2$, i.e., where the two curves cross.

then the leading term in a Taylor-series expansion of the rate equation, with $\delta[X] \equiv [X] - X^*$, is

$$\frac{d\delta[X]}{dt} = \left\{\frac{dF([X])}{d[X]}\right\}_{[X]=X^*} \delta[X], \qquad (31.60)$$

since $F(X^*) = 0$, that is, $d[X]/dt = 0$ at a stationary state. For the model we have been considering, $F([X]) = 0$ is a cubic equation[7] in $[X]$ which for some values of the parameters has three real roots X^*. Since $F([X])$ is a continuous function, the slope $\{dF([X])/d[X]\}_{[X]=X^*}$ must be positive for some of these three roots, negative for the others. If the slope is negative, then by Eq. 31.60 a small perturbation $\delta[X]$ decays back toward the stationary state X^*, which means that X^* is stable. But if the slope is positive, the stationary state is unstable: The slightest perturbation (such as may be caused by fluctuations) grows with time, and the system eventually makes a transition to one of the stable stationary states.

In Fig. 31.9 it is the S-shaped curve with $\mu < 0$ that illustrates this behavior; Fig. 31.10 shows such a curve in greater detail. When one works through the algebra, it turns out that the stability condition,

$$\left\{\frac{dF([X])}{d[X]}\right\}_{[X]=X^*} < 0 \qquad (31.61)$$

for stable stationary states, corresponds to $(\partial X^*/\partial\lambda)_\mu > 0$ (for $\rho = 0$). Thus the central branch of the S-shaped curve, between the extrema B and C, is unstable. Small perturbations of the stationary state will decay anywhere on the stable branches $A–B$ and $C–D$, but will grow on the unstable branch $B–C$. Another way of putting this is to say that Eq. 31.53 applies, with the relaxation time τ positive on the stable branches, negative on the unstable branch; thus at the points B and C, which are called *marginal stability points*, the relaxation time is zero.

Suppose now that we have a system in the stable stationary state A, Fig. 31.10. If we increase the parameter λ, the system should follow the branch AB. For values of λ beyond B, the only stable stationary states are those along the branch ED (where E has the same value of λ as B); thus increasing λ beyond this point forces the system to make a transition from B to E. In fact we are never likely to get as far as B on the original branch: the curve $ABCD$ gives the stationary states calculated from the macroscopic rate equations, but in a real system these equations describe only the average behavior; fluctuations are always present. The closer one gets to the marginal stability points, where $(\partial X^*/\partial\lambda)_\mu$ becomes infinite, the larger and longer-lasting the fluctuations become, just as occurs near the critical point in a liquid–gas system. Before the marginal stability point is reached, one of these fluctuations is likely to be large enough to trigger a transition to the other stable branch. If it were possible to maintain stationary-state conditions at all times, however, increasing λ would cause the system to follow the path $ABED$. But if we reverse the process and decrease λ in the same manner, the system would follow the path $DCFA$, remaining on the upper branch until the marginal stability point C. This phenomenon, a system's following different paths depending on whether a parameter is increasing or decreasing,

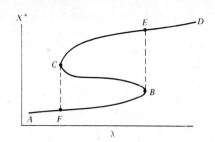

FIGURE 31.10
More detailed look at a curve like that in Fig. 31.9 for $\mu < 0$. (See discussion in text.)

---

[7] Note that the left side of Eq. 31.58 is not itself $F([X])$, i.e, $d[X]/dt$, which by Eqs. 31.54 and 31.57 has the form

$$F([X]) = \frac{[X]^3 + \rho[X]^2 + \mu[X] - \lambda}{\alpha + \beta[X]},$$

where $\rho, \mu, \lambda, \alpha, \beta$ are constants. However, the two functions behave similarly enough for illustrative purposes, and $F([X])$ is unwieldly.

is called *chemical hysteresis*. Such behavior has an electrical analog (i.e., a current–voltage curve resembling Fig. 31.10) in systems containing a negative resistance element or a flip-flop switch. Magnetic hysteresis occurs in a ferromagnet on varying the direction strength of an applied magnetic field.

Quite a large number of reactions exhibiting multiple stationary states and chemical hysteresis have been investigated (especially by chemical engineers). One fairly simple example of such a system is the oxidation of thiosulfate by peroxide,

$$2S_2O_3^{2-} + 4H_2O_2 \rightarrow S_3O_6^{2-} + SO_4^{2-} + 4H_2O.$$

Figure 31.11 illustrates some experimental results for this reaction, showing the steady-state temperature as a function of coolant flow rate (which corresponds to our earlier parameter λ).

We have seen how multiple stationary states can exist, and how a system can change from one set of such states to another when the external parameters are altered. But to obtain oscillatory behavior like that of the heartbeat, we need a mechanism that will switch the system back and forth without external intervention. One much-studied example of an oscillatory chemical reaction is the bromination of malonic acid in the presence of an oxidation–reduction couple (such as Ce^{III}, Ce^{IV}),

$$5CH_2(COOH)_2 + 3BrO_3^- + 3H^+ \rightarrow$$

$$3BrCH(COOH)_2 + 2HCOOH + 4CO_2 + 5H_2O,$$

known as the *Belousov–Zhabotinskii reaction*. Figure 31.12 shows the oscillatory behavior observed in this reaction. The oscillations persist in a well-stirred beaker for nearly an hour, as long as the concentration of reactants is such as to keep the system far from equilibrium; in an open system, fed with reactants and depleted of products, they can persist indefinitely. But the mechanism of this reaction is very complicated; to get some insight into the nature of oscillatory reactions, let us consider a simpler example.

Consider an enzyme E that can undergo two stages of ionization:

$$E \rightleftharpoons E' + H^+,$$

$$E' \rightleftharpoons E'' + H^+.$$

Suppose that form E is catalytically active, but the ionized forms E', E'' are inactive. One enzyme with such properties is papain, which catalyzes the hydrolysis of an ester (the substrate S):

$$S + E \rightarrow E + P^- + H^+.$$

Let the enzyme be restricted to a particular volume, and let the reactant and product concentrations be maintained constant at far-from-equilibrium values outside this reaction volume. Suppose now that initially the enzyme is mostly ionized, with a low concentration of the active form E. The substrate S enters the reaction volume by diffusion, since its concentration outside the reaction volume is higher than within it. At first the reaction rate is small, since [E] is small. But as the products are formed (more rapidly than the H^+ can diffuse out of the reaction volume), $[H^+]$ increases and shifts the rapidly attained enzyme equilibrium to the left, increasing the concentration of the active form. This product-enhanced catalysis in turn increases the rate of the hydrolysis, until a point is reached where S is consumed by reaction more rapidly than it can diffuse into the reaction volume. When this happens, the reaction volume runs out of S, the reaction rate (proportional to [E][S]) falls, the H^+ and P^- already formed diffuse out of the reaction volume, the enzyme

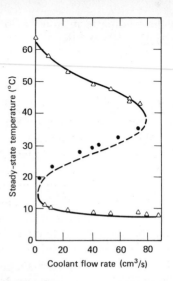

FIGURE 31.11

Stable (△) and unstable (•) states for the thiosulfate–peroxide reaction; the solid and dashed lines are the corresponding theoretical curves. Based on R. Schmitz, *Chemical Reaction Engineering Reviews* (ed. Hugh M. Hulburt), p. 156 (American Chemical Society, 1975).

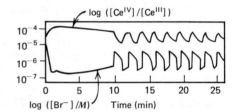

FIGURE 31.12

Oscillatory behavior in the Belousov–Zhabotinskii reaction. From R. J. Field, E. Körös, and R. M. Noyes, *J. Am. Chem. Soc.*, **94**, 8649 (1972).

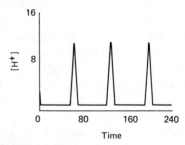

FIGURE 31.13

Relaxation oscillations in the enzyme reaction described in the text, calculated from the rate equations. From H.-S. Hahn, A. Nitzan, P. Ortoleva, and J. Ross, *Proc. Nat. Acad. Sci. (USA)* **71**, 4067 (1974).

equilibrium shifts to the right again, lowering [E], and we are back to our starting point. When the rate equations are set up with appropriate values of the rate constants and fixed concentrations, one can obtain a numerical solution like that in Fig. 31.13. The similarity of this variation to that of the heartbeat (Fig. 31.8) is striking: in both cases a rapid variation is followed by a slow relaxation, then a rapid return to nearly the original point and a very slow variation to complete the cycle. This behavior is called a *relaxation oscillation*. (No inference should be made from this resemblance about the mechanism controlling the heartbeat, which is most likely far more complex, though not known at present.)

The relaxation oscillations in the system just described have two principal causes: (1) the positive feedback loop of product-enhanced catalysis, through the effect of H^+ on the enzyme equilibrium; (2) the limitation of the reaction rate by diffusion into the reaction volume, followed by flushing out of products to complete a cycle.

There are a number of interesting examples of oscillatory reactions, particularly in biochemical systems. For instance, Fig. 31.14 illustrates oscillatory behavior in glycolysis (the conversion of glucose to alcohol); the mechanism is complicated, but known to contain several feedback loops. In Fig. 31.15 we show several types of oscillations observed in mitochondria.

To conclude this section, we note that reactions can exhibit periodic behavior in space as well as time. We normally expect diffusion to work toward the eradication of any concentration gradients existing in a system. However, a nonlinear reaction mechanism, under far-from-equilibrium conditions, can interact with diffusion processes to establish concentration gradients in an originally uniform system. For example, if the Belousov–Zhabotinskii reaction is started in an unstirred vessel, there may appear a macroscopic spatial pattern, either stationary or time-dependent, as shown in Fig. 31.16. The light and dark regions correspond to the extremes of Br^- concentration shown in Fig. 31.12. It has been suggested that this remarkable phenomenon of reaction–diffusion interaction may be of importance in understanding biomorphogenesis, the creation of structure in biological systems.

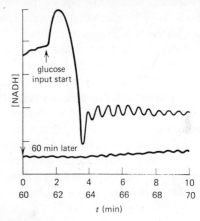

FIGURE 31.14

Oscillations in glycolysis in yeast cells, when glucose is injected at a constant rate from the indicated starting point. (The concentration of the enzyme NADH is plotted.) From A. Boiteux and B. Hess, in *Physical Chemistry of Oscillatory Phenomena* (Faraday Symposium of the Chemical Society No. 9, 1974), p. 202.

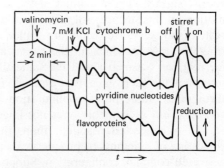

FIGURE 31.15

Some oscillatory phenomena observed in mitochondria. From A. Boiteux and B. Hess (cf. Fig. 31.14).

FIGURE 31.16

Spatial wave patterns in the Belousov–Zhabotinskii reaction. From A. T. Winfree, *Sci. Am.* **230** (no. 6), 82 (June 1974).

The concept of symmetry plays an important role in various areas of physical chemistry: in the analysis of electronic structure of atoms and molecules, and hence in electronic spectra; in vibrational and rotational spectra as well; in crystal structure and the electronic theory of solids. In

31.5
Symmetry Rules for Chemical Reactions

this section we discuss briefly the ways in which symmetry affects chemical kinetics, in particular by providing "propensity" rules and selection rules for certain classes of reaction.

Consider first the coupling of electron spin to other types of motion, such as the translational degrees of freedom of the electrons. This coupling is frequently small and can be neglected, in which case the electron spin is said to be conserved: that is, the spin angular momentum is taken to be constant. In general, of course, only the *total* angular momentum of the system, that is, the vector sum of all the electronic and nuclear angular momentum components, is conserved. But for many systems (those with weak spin–orbit interactions, where the magnetic moment of the electrons interacts weakly with other magnetic moments), there is a propensity for electron spin conservation. Consider, for instance, the reaction

$$N^+(^3P) + O_2(^3\Sigma_g^-) \rightarrow NO^+(^3\Sigma^+) + O(^3P),$$

in which the sum of spin multiplicities (indicated by the left superscripts on the term symbols) and thus of spin angular momenta is the same for both reactants and products. Exceptions to this *Wigner–Witmer rule* are rare but do occur, in particular for heavy atoms. Another such rule applies to the approximate conservation of orbital angular momentum (and also holds in the cited reaction of N^+ with O_2).

Of greater interest, however, are the approximate correlation rules governing electronic orbitals. Let us think of the species taking part in a reaction as a collection of electrons and nuclei, and make the Born–Oppenheimer assumption that the electronic and nuclear motions are separable. By calculating the electronic energy for various values of the nuclear coordinates, one can, as usual, obtain the (adiabatic) potential energy surface. The dynamics of the reaction then describe the system's motion on this surface. All this is routine by now. However, we shall now restrict ourselves to what are called *concerted* reactions; this means roughly those in which all the chemical bond rearrangements necessary for reaction occur at about the same time, without the formation of any intermediate such as a free radical. (A concerted reaction does pass through a saddle-point configuration that can be called an activated complex, but presumably does not linger there.)

In a concerted reaction, then, the system is clearly definable as consisting of reactants or products except in the immediate vicinity of the activated-complex configuration. Suppose now that we consider the electronic wave function of the *reactants* at or near the activated-complex configuration—that is, in first approximation, we consider the reactants as isolated and noninteracting in that configuration—and make a simple molecular orbital analysis. Next we do the same thing for the *products* of the reaction, considered as noninteracting in the same nuclear configuration. The two models differ, of course, in the bonding interactions assumed.

This approach assumes that a reaction will proceed readily if the overlap of the molecular orbitals of the reactants is large and this large overlap is maintained continuously in the transition from reactants to products. If there is such continuity of overlap, the reaction is said to be allowed; if not, the reaction is said to be forbidden. If the overlap is not maintained, then the activated-complex configuration presumably involves fewer strong bonds and higher energy than the reactants or products, so that the potential barrier and activation energy are high. In effect, the electrons of the reactants cannot easily rearrange to the distribution described by the MOs of the products. Such a rearrangement may take place if some excited electronic state (of reactants or products or both) is involved, bringing in MOs that do overlap; in such a case the activation energy of the reaction is essentially that required to promote electrons to the appropriate excited orbital at some intermediate configuration. Note that some excited electronic states of reactants can be reached by photoexcitation, so that it is possible for forbidden reactions to be photochemically allowed. And remember that even those reactions which are allowed by this rule may still have substantial activation energies, though not so high as to prevent a reasonable reaction rate.

All this is very well, but we have passed over the key question: How does one know if significant overlap exists? The beauty of this method is that one can determine this without a complete calculation of the MOs, simply by making qualitative orbital constructions and applying symmetry considerations. It is best to illustrate this with an example, and for this purpose we choose the well-studied *electrocyclic reactions*, the reactions of substituted butadienes to form cyclobutenes. One of the simplest of these reactions is

$$\underset{H_2C}{}^{CH-CH}\underset{CH_2}{} \longrightarrow \square \, ,$$

but it is easier to follow the argument for the reactant 2,4-hexadiene,

$$\underset{H_3C}{}^{CH-CH}\underset{H}{}\underset{H}{}\underset{CH_3}{}$$

From this reactant we may expect either of two products, the *cis*- and *trans*-3,4-dimethylcyclobutenes

$$\text{(I)} \qquad \text{and} \qquad \text{(II)} \, ;$$

can we use an overlap principle to predict which of these reactions will occur? We can, and in this case molecular symmetry makes the prediction especially easy.

Remember first that in the reactant the CH$_3$— and H— groups at each end are in the plane of the molecule (sp^2 bonding, bond angles about 120°). In the product molecules, however, these same groups take part in a strained sp^3 hybridization, and must be either above or below the plane of the ring. Product I would be obtained from the reactant by what is called a *disrotatory* motion (rotating in opposite directions) of the two methyl groups, as indicated by arrows in Fig. 31.17a, whereas product II would be obtained by a *conrotatory* motion (rotating in the same direction), Fig. 31.17b. Note that the original molecule has two major symmetry properties besides reflection in its own plane: reflection through a plane perpendicular to the molecular plane at the dashed line in the figure (this symmetry operation is designated by the symbol σ_v) and rotation through 180° about the same dashed line (designated by the symbol C_2).[8] The two reaction mechanisms are thus distinguished by the fact that the disrotatory motion preserves the symmetry operation σ_v, whereas the conrotatory motion preserves the symmetry operation C_2.

Next we construct molecular orbitals for the reactant. It is a conjugated molecule (one with alternating double and single bonds), so we

[8] The symbols σ_v and C_2 are among a large number of symmetry-operation symbols that we have not had occasion to use; they are introduced here merely as shorthand.

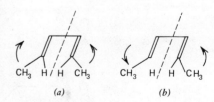

FIGURE 31.17
Conversion of 2,4-hexadiene to 3,4-dimethylcyclobutene with (*a*) disrotatory, (*b*) conrotatory motion of the methyl groups, as indicated by the arrows; the dashed line indicates the location of the original molecule's plane of symmetry.

follow a method similar to that used for benzene in Section 9.2. That is, we define a π orbital extending over the four central carbon atoms,

$$\psi = c_1\varphi_1 + c_2\varphi_2 + c_3\varphi_3 + c_4\varphi_4, \tag{31.62}$$

where each of the φ_i is the $2p_z$ orbital of one of the carbon atoms. We then use the standard approximations of the Hückel method,

$$S_{jk} = \int \varphi_j{}^*\varphi_k \, d\tau = \delta_{jk} \tag{31.63}$$

and

$$H_{jk} = \int \varphi_j{}^*\mathbf{H}\varphi_k \, d\tau = \begin{cases} \alpha & \text{(if } j = k) \\ \beta & \text{(if } j \text{ is bonded to } k) \\ 0 & \text{(otherwise)}, \end{cases} \tag{31.64}$$

where α and β are arbitrary parameters. When these values are substituted into the secular equation,

$$\det(H_{jk} - \epsilon S_{jk}) = 0, \tag{31.65}$$

one can solve for the four eigenvalues ϵ_i and their corresponding eigenfunctions. We shall merely list the solutions here, together with the results of applying the symmetry operations σ_v and C_2 to each eigenfunction:

$$\epsilon_1 = \alpha + 1.618\beta \qquad (\alpha, \beta < 0);$$

$$\psi_1 = 0.37\varphi_1 + 0.6\varphi_2 + 0.6\varphi_3 + 0.37\varphi_4; \qquad \sigma_v\psi_1 = \psi_1; \qquad C_2\psi_1 = -\psi_1;$$

$$\epsilon_2 = \alpha + 0.618\beta;$$

$$\psi_2 = 0.6\varphi_1 + 0.37\varphi_2 - 0.37\varphi_3 - 0.6\varphi_4; \qquad \sigma_v\psi_2 = -\psi_2; \qquad C_2\psi_2 = \psi_2;$$

$$\epsilon_3 = \alpha - 0.618\beta;$$

$$\psi_3 = 0.6\varphi_1 - 0.37\varphi_2 - 0.37\varphi_3 + 0.6\varphi_4; \qquad \sigma_v\psi_3 = \psi_3; \qquad C_2\psi_3 = -\psi_3;$$

$$\epsilon_4 = \alpha - 1.618\beta;$$

$$\psi_4 = 0.37\varphi_1 - 0.6\varphi_2 + 0.6\varphi_3 - 0.37\varphi_4; \qquad \sigma_v\psi_4 = -\psi_4; \qquad C_2\psi_4 = \psi_4.$$

Figure 31.18 gives a rough representation of these eigenfunctions and their components: The two lobes on each carbon atom represent its $2p_z$

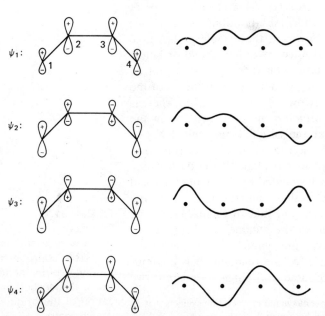

FIGURE 31.18
Schematic representations of the π-orbital eigenfunctions of 2,4-hexadiene (or any other substituted butadiene) according to the HMO calculation described in the text. On the left are shown the four $2p_z$-orbital components in proportion to their contributions, while on the right is shown the approximate value of ψ_i along the carbon chain.

orbital, with the positive lobe up or down according to the sign, and the size of the lobes in proportion to the coefficient. In applying the symmetry operations, σ_v is equivalent to exchanging φ_1 with φ_4, φ_2 with φ_3; C_2 performs the same exchanges with reversal of all the signs (φ_1 becomes $-\varphi_4$, etc.). The wave function as a whole is either symmetric (unchanged) or antisymmetric (only the sign changed) under each of these operations; these symmetry properties are represented by S and A, respectively, in the following table:

	σ_v	C_2
ψ_1	S	A
ψ_2	A	S
ψ_3	S	A
ψ_4	A	S

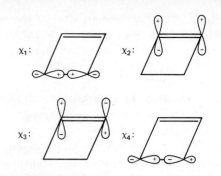

FIGURE 31.19
Schematic representation of the four molecular orbitals formed from carbon $2p_z$ orbitals in a cyclobutene. The orbitals χ_1, χ_2, χ_3, χ_4 are in order of increasing energy.

Since α and β are both negative, the eigenvalues are ϵ_1, ϵ_2, ϵ_3, ϵ_4 in order of increasing energy. There are four π electrons to be distributed, so the ground state must have two electrons each in the orbitals ψ_1 and ψ_2. (We did not really need to solve the secular equation to determine the order of the eigenvalues; all we had to remember was the simple fact that the number of nodes of an orbital increases with increasing energy.)

All this has only given us the MOs of the reactant molecule; now we must repeat the process for the products. Of the four $2p_z$ electrons that occupy the π orbitals of the reactant, two remain in the π orbital of the ring's double bond, but the other two are in the newly formed σ bond that closes the ring. If we again set up four molecular orbitals ($\chi_j = \sum_i c_i' \varphi_i$) using the $2p_z$ orbitals, the eigenfunctions obtained are schematically as shown in Fig. 31.19. Their symmetry properties are as follows:

	σ_v	C_2
χ_1	S	S
χ_2	S	A
χ_3	A	S
χ_4	A	A

In the ground state of either product there are two electrons each in the orbitals χ_1 and χ_2. (Note, by the way, that all our calculations for both reactant and product involve only the central four carbon atoms, and are thus applicable to *any* electrocyclic reaction.)

Now we are finally in a position to follow the correlation of the molecular orbitals from reactant to product. Remember that the two possible products are distinguished by the symmetry elements preserved in their formation. We consider first the conrotatory ring closure, which preserves the symmetry operation C_2. If we set up a correlation diagram for the MOs of reactant and product, those that have the same symmetry (S or A) under the operation C_2 must be correlated, as shown in Fig. 31.20*a*. Although both sets of orbitals are ordered according to energy, no quantitative energy scale is implied. Note that two of the reactant electrons are in the orbital ψ_1, which has the same symmetry with respect to C_2 as the product orbital χ_2; thus the orbitals ψ_1 and χ_2 overlap (+ lobe to + lobe). Similarly, the orbitals ψ_2 and χ_1 overlap. Thus the reaction

$$(\psi_1)^2 (\psi_2)^2 \rightarrow (\chi_1)^2 (\chi_2)^2,$$

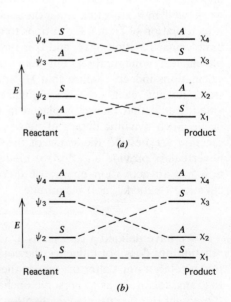

FIGURE 31.20
Correlation diagram for the formation of cyclobutenes from substituted butadienes (including only π orbitals of the reactant). (*a*) Conrotatory ring closure, C_2 symmetry preserved. (*b*) Disrotatory ring closure, σ_v symmetry preserved.

that is, from the ground state of the reactant to the ground state of the product, can take place with conservation of orbital symmetry. The energy required for this reaction must be relatively low, so we predict that conrotatory ring closure (leading to the *trans-* product) should be thermally allowed.

Consider, in contrast, the disrotatory ring closure, in which the symmetry operation σ_v is preserved. The corresponding correlation diagram is shown in Fig. 31.20b. In this case there is overlap between ψ_1 and χ_1, but not between ψ_2 and χ_2; thus the four electrons in ψ_1 and ψ_2 cannot make the ground-state transition to χ_1 and χ_2 with conservation of orbital symmetry, and the disrotatory ring closure (leading to the *cis-* product) is not thermally allowed. Note, however, that the reaction can proceed if one electron is brought to the orbital ψ_3 by photoexcitation: ψ_3 does overlap with χ_2, and the excitation makes available enough energy to bring the other electron in ψ_2 to χ_3:

$$(\psi_1)^2(\psi_2)^2 \xrightarrow{h\nu} (\psi_1)^2(\psi_2)(\psi_3) \rightarrow (\chi_1)^2(\chi_2)(\chi_3) \xrightarrow{-E} (\chi_1)^2(\chi_2)^2.$$

Our theory predicts that in the thermal reaction only conrotatory ring closure should occur, and thus only the *trans*-cyclobutene should be formed. Experiment confirms this prediction triumphantly: Only the *trans-* isomer has ever been obtained in thermal electrocyclic reactions.

The maximum-overlap rule, sometimes expressed as the "rule of conservation of orbital symmetry," is extremely useful and powerful. Like the symmetry or selection rules in optical spectroscopy, this rule is an approximation: In our discussion we neglected any interaction between electron spin and other degrees of freedom, or between electronic and nuclear motion; we assumed that a single nuclear configuration, that of the transition state, is all that need be considered; and we neglected any effects of vibrational or rotational motion. Yet the rule frequently works well, indicating that either the approximations are reasonable or there is cancellation of errors. Exceptions to the rules do exist; for example, the bimolecular reaction of H_2 and I_2 is symmetry-forbidden for any reasonable nuclear configuration, but takes place anyway. Classical trajectory calculations indeed confirm that H_2 and I_2 near their vibrational ground states do not react, even when sufficient relative kinetic energy is provided; but if that energy is placed in the vibration of I_2, reaction does occur. Such dynamic details of nuclear motion and its consequences for reaction are beyond the scope of the orbital-symmetry rules. The rules nevertheless provide an effective method for predicting whether a reaction has an activation energy of thermal ($\sim$50–100 kJ/mol) or photochemical ($\gtrsim$200 kJ/mol) magnitude.[9]

We have defined a catalyst as a substance that changes the rate of a reaction but does not influence the equilibrium of the reaction. It follows that a catalyst must alter the rate in the forward and reverse directions by the same factor, so as to preserve unchanged the ratio of rate coefficients, the equilibrium constant. We therefore have

31.6 Introduction to Catalysis

$$K = \frac{k_f}{k_r} = \frac{k_f'}{k_r'}, \tag{31.66}$$

---

[9] For more about symmetry rules and their applications, see H. E. Simmons and J. F. Bunnett (Eds.), *Orbital Symmetry Papers* (American Chemical Society, Washington, D.C., 1974); R. B. Woodward and R. Hoffmann. *The Conservation of Orbital Symmetry* (Verlag Chemie/Academic Press, New York, 1970).

where the unprimed and primed rate coefficients are those in the absence and the presence of a catalyst, respectively. Thus the addition of a catalyst does not affect microscopic reversibility.

The question to be addressed is, "How does a catalyst work?" This is usually answered in a simple way: A catalyst works by providing an alternative reaction mechanism that changes the rate, usually by altering the activation energy. For instance, the activation energy for the thermal decomposition

$$(CH_3)_3COH \rightarrow (CH_3)_2C{=}CH_2 + H_2O$$

is 274 kJ/mol, whereas the presence of HBr as a catalyst reduces the activation energy to 127 kJ/mol. Although the activation energy is not identical with the barrier height along the reaction coordinate, we know that the two quantities are closely related. A catalyst is thus a substance that, by interacting with one or more of the reactants, enlarges the number of coordinates describing the potential surface, and thereby permits a path from reactants to products with a different barrier height than that of the uncatalyzed reaction (a lower barrier height for positive, rate-increasing catalysis). For the example just cited, a possible, but not confirmed, mechanism is the direct addition of HBr to $(CH_3)_3COH$ to form the intermediate

with a subsequent splitting off of H_2O and HBr.

It is possible for a catalyst to operate, not by altering the barrier height, but merely by increasing the fraction of collisions leading to reaction. In that case one can say, in the language of the activated-complex theory, that the catalyst lowers the entropy of activation. (However, it is often difficult to disentangle the energy and entropy effects.) One clear-cut example is the metal-ion–catalyzed carboxylation of nitroparaffins,

$$RCH_2NO_2 + CO_2 \xrightarrow{Mg^{2+}} RCHCOOH,$$
$$\underset{NO_2}{}$$

in which the Mg^{2+} catalyst apparently works by forming the chelate

What we have said thus far is extremely general. The mechanism of catalysis depends strongly on the type of reaction involved, as well as on the nature of the catalyst. In the remainder of this chapter we shall discuss several common types of catalysis in detail, with primary emphasis on how they affect the reaction kinetics. We consider first *homogeneous* catalysis, in which the catalyst is in the same phase as the reactants and takes part in the reaction mechanism directly; and then *heterogeneous* catalysis, which takes place on a surface. Our first example, however, is a borderline case.

An extraordinarily important class of catalytic reactions consists of the reactions involving *enzymes*. An enzyme is a biological catalyst: Enzymes are protein molecules, and thus typically have molecular weights of the order of 10^4–10^6 g/mol; enzymes catalyze a huge number of reactions of biological interest. The size of the enzyme molecule, about 10–100 nm in diameter, blurs the distinction between homogeneous and heterogeneous reactions. An enzyme immobilized in a membrane, and serving as a catalyst for substrates in a fluid phase bounded by the membrane, is clearly in the heterogeneous category. An enzyme in solution is usually considered a homogeneous catalyst. There are two broad categories of enzymes: *hydrolytic* enzymes, which catalyze processes such as ionic reactions, proton transfer, and so on, and act in a role similar to acid–base catalysts (see the next section); and *oxidation–reduction* enzymes, for electron-transfer reactions. Carbonic anhydrase, for instance, catalyzes the biologically important reaction

$$CO_2 + H_2O \rightleftharpoons H^+ + HCO_3^-;$$

tyrosinase oxidizes catechols to *o*-quinones.

The basic mechanism of enzyme reactions, named after Michaelis and Menten, is simple: An enzyme E and a *substrate* S (the compound to be altered chemically) react to form a complex ES,

$$E + S \underset{k_{-1}}{\overset{k_1}{\rightleftharpoons}} ES,$$

with equilibrium rapidly established. The complex then dissociates,

$$ES \underset{k_{-2}}{\overset{k_2}{\rightleftharpoons}} E + P,$$

forming one or more products P and regenerating the enzyme. The reaction usually goes to completion, so the rate coefficient k_{-2} is effectively zero.

The rate of change of concentration of the complex is

$$\frac{d[ES]}{dt} = k_1[E][S] - k_{-1}[ES] - k_2[ES]. \tag{31.67}$$

If we make the stationary-state approximation, $d[ES]/dt = 0$, then the concentration of the enzyme–substrate complex is

$$[ES] = \frac{k_1[E][S]}{k_{-1} + k_2}. \tag{31.68}$$

Since the total enzyme concentration (both free and complex-bound) is fixed, say $[E]_0$, we have

$$[E]_0 = [E] + [ES] = [E]\left(1 + \frac{k_1[S]}{k_{-1} + k_2}\right). \tag{31.69}$$

We solve this equation for $[E]$,

$$[E] = \frac{[E]_0(k_{-1} + k_2)}{k_{-1} + k_2 + k_1[S]}, \tag{31.70}$$

and substitute in Eq. 31.68,

$$[ES] = \frac{k_1[E]_0[S]}{k_{-1} + k_2 + k_1[S]}. \tag{31.71}$$

This enables us to obtain the rate in terms of the measurable quantities $[E]_0$ and $[S]$. Since $[ES]$ is small, the rate of disappearance of the substrate can be assumed equal to the rate of the second reaction step,

$$-\frac{d[S]}{dt} = k_2[ES] = \frac{k_1k_2[E]_0[S]}{k_{-1}+k_2+k_1[S]}; \qquad (31.72)$$

this is called the *Michaelis-Menten equation*. The similarity to the simple model of unimolecular reactions is clear.

A plot of the reaction rate as a function of the substrate concentration is interesting (Fig. 31.21). At low concentration, such that $k_{-1}+k_2 \gg k_1[S]$, the reaction rate is second-order; which step is rate-determining depends on the relative magnitudes of k_{-1} and k_2. If $k_{-1} \ll k_2$, then the initial rate is

$$-\frac{d[S]}{dt} = k_1[E]_0[S], \qquad (31.73)$$

that is, the rate-determining step is the rate of complex formation. But if $k_{-1} \gg k_2$, then, since

$$\frac{k_1}{k_{-1}}[E]_0[S] = [ES] \qquad (31.74)$$

by our assumption of rapid equilibrium, the rate-determining step is the unimolecular decomposition of ES. At high substrate concentration, such that $k_1[S] \gg k_{-1}+k_2$, the rate is zero-order in the substrate. The limiting step is again the decomposition of the complex, but in this approximation the rate depends only on the total enzyme concentration,

$$-\frac{d[S]}{dt} = k_2[E]_0. \qquad (31.75)$$

This, incidentally, is the maximum rate; one speaks of *saturation* kinetics when this rate is approached. A convenient way to test the Michaelis-Menten mechanism is to rewrite Eq. 31.72 in the form

$$\frac{1}{-d[S]/dt} = \frac{k_{-1}+k_2}{k_1k_2[E]_0[S]} + \frac{1}{k_2[E]_0}; \qquad (31.76)$$

a plot of the reciprocal of the reaction rate against the reciprocal of the substrate concentration should thus give a straight line, with intercept $(k_2[E]_0)^{-1}$ and slope $(k_{-1}+k_2)/k_1k_2[E]_0$. Since $[E]_0$ is known, both k_2 and k_{-1} can thus be determined from these measurements. Figure 31.22 shows an example of such a *Lineweaver–Burk plot*.

Enzyme catalysis has two remarkable features: *specificity* and *efficiency*. Although some enzymes, such as tyrosinase, can catalyze a number of different reactions, these usually involve structurally similar substrates. Other enzymes are completely specific catalysts; for instance, urease is specific for the hydrolysis of urea, and does not act as a catalyst even for the seemingly quite similar hydrolysis of methylurea. By efficiency we mean here simply the increase in reaction rate produced per unit concentration of the catalyst. Many enzymes are such powerful catalysts that they increase the rate by many orders of magnitude even at quite low concentrations.

What is the origin of these remarkable characteristics? The enzyme is a protein with a well-defined (and increasingly well-established) primary and secondary structure. The primary structure is determined by the sequence of constituent amino acids, the secondary structure by their geometric arrangement in space, an arrangement much influenced by hydrogen bonding and other types of interactions (such as dipolar attrac-

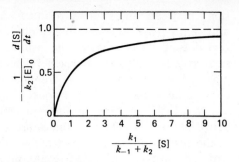

FIGURE 31.21
Reaction rate as a function of substrate concentration as predicted by the Michaelis–Menten equation, Eq. 31.72. The dashed line represents the maximum (saturation) rate, Eq. 31.75.

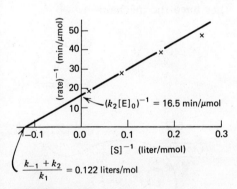

FIGURE 31.22
Example of a Lineweaver–Burk plot, Eq. 31.76. The data of Wratten and Cleland for the alcohol dehydrogenase from yeast are shown. The reaction is

$$C_2H_5OH + NAD^+ \xrightarrow{enzyme}$$

$$CH_3CHO + NADH + H^+,$$

with $S = C_2H_5OH$.

tion) between amino acid residues. Most of the enzyme's structure apparently plays no role in the catalytic activity. The relatively small part that does, called the *active site*, usually consists of a nonproteinlike part such as a heme group (the general term is a "prosthetic" group) or at most a few amino acids, and frequently a metal atom. This part of the enzyme takes the shape of a pocket, into which the substrate fits like a key in a lock. Substituting different amino acids in regions far from the active site hardly influences the catalytic property; but such a substitution at the active site, or in a part of the amino acid chain that can interact with the substrate on the active site, may change the rate of the catalyzed reaction by many orders of magnitude. Some fundamental questions are still being investigated in current research on enzyme catalysis: the mechanism of the lowering of activation energies, the energy barriers to the substrate's arrival at the active site and the products' departure from that site, and the structural features that lead to specificity. One enzyme whose structure has been determined is carboxypeptidase A, which catalyzes the hydrolysis of the amide linkage to a peptide chain's terminal amino acid:

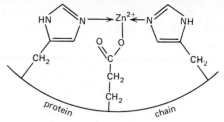

FIGURE 31.23
Structure of the active site in carboxypeptidase A. The Zn^{2+} ion has approximately tetrahedral coordination, with the fourth coordination site open.

$$
\underset{H}{\overset{R_2}{\underset{|}{N}}}-\underset{H}{\overset{|}{C}}-\underset{O}{\overset{\|}{C}}-\underset{H}{\overset{R_1}{\underset{|}{N}}}-\underset{H}{\overset{|}{C}}-\underset{O}{\overset{\|}{C}}-OH+H_2O \rightarrow -\underset{H}{\overset{R_2}{\underset{|}{N}}}-\underset{H}{\overset{|}{C}}-\underset{O}{\overset{\|}{C}}-OH+H_2N-\underset{H}{\overset{R_1}{\underset{|}{C}}}-\underset{O}{\overset{\|}{C}}-OH.
$$

In Fig. 31.23 we show the structure of carboxypeptidase A's active site. Figure 31.24 gives a two-dimensional representation of the "lock-and-key" arrangement of the enzyme and the polypeptide substrate, together with the likely mechanism of the hydrolysis reaction; note that not only the active site itself, but several nearby amino acids in the enzyme chain take part in the reaction.

Many enzyme-catalyzed reactions proceed by more complicated mechanisms than the simple Michaelis-Menten model. Consider a reaction catalyzed by an enzyme E that follows a Michaelis-Menten mechanism. However, suppose there is an *inhibitor* species I that can combine with the enzyme to produce a catalytically inert complex EI. The inhibitor species acts as a control device that prevents the catalyzed reaction from proceeding too quickly; this is called *competitive* enzyme inhibition. We thus have the mechanism

$$ E+S \underset{k_{-1}}{\overset{k_1}{\rightleftharpoons}} ES \overset{k_2}{\longrightarrow} P+E $$

$$ E+I \underset{k_{-3}}{\overset{k_3}{\rightleftharpoons}} EI; $$

FIGURE 31.24
Schematic representation of the action of carboxypeptidase A in cleaving a peptide linkage. In (a) we show how the polypeptide is positioned on the enzyme: Arginine and tyrosine form hydrogen bonds with the terminal amino acid's carboxylate and amino groups, while a nearby "pocket" in the enzyme chain accommodates any bulky side groups (e.g., R_1); the oxygen of the peptide linkage coordinates with the Zn^{2+} ion at the active site; and glutamate forms an incipient polar bond with the carbon of the peptide linkage. In (b) all these partial bonds are shown, the peptide C–N bond has been broken, and the elements of water (circled) are being added. In (c) the products separate from the enzyme and move away.

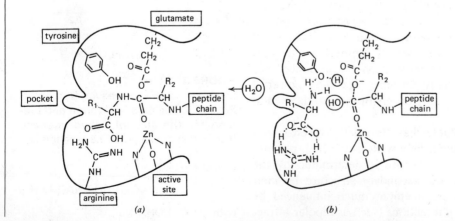

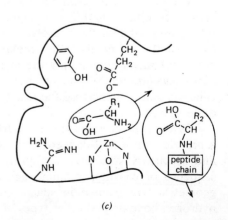

we assume that the last equilibrium is established rapidly, and define

$$K_I \equiv \frac{k_3}{k_{-3}} = \frac{[EI]}{[E][I]}.$$ (31.77)

Since the total enzyme concentration is now

$$[E]_0 = [E] + [ES] + [EI],$$ (31.78)

the reaction rate is easily shown to be

$$-\frac{d[S]}{dt} = \frac{k_1 k_2 [E]_0 [S]}{(k_{-1} + k_2)(1 + K_I[I]) + k_1[S]}.$$ (31.79)

A plot of the reciprocal rate versus the reciprocal substrate concentration is still linear, but the slope now depends on the inhibitor concentration.

An alternative form of inhibition, called *noncompetitive*, is shown in the mechanism

$$E + S \rightleftharpoons ES \rightarrow E + P$$

$$ES + I \rightleftharpoons ESI \text{ (catalytically inert)}.$$

31.8 Acid–Base Catalysis

Many reactions are catalyzed by acids and bases. "Acids" and "bases" here can be taken to include any of the common definitions: substances that yield H^+ and OH^- ions in aqueous solution (Arrhenius); proton donors and acceptors (Brønsted); or electron acceptors and donors (Lewis).

As an example, consider the bromination of acetone in aqueous solution. The reaction is thought to occur with an initial enolization process,

$$CH_3-\overset{\overset{\displaystyle O}{\|}}{C}-CH_3 \rightarrow CH_3-\overset{\overset{\displaystyle OH}{|}}{C}=CH_2,$$

followed by bromination of the enol,

$$CH_3-\overset{\overset{\displaystyle OH}{|}}{C}=CH_2 + Br_2 \rightarrow CH_3-\overset{\overset{\displaystyle O}{\|}}{C}-CH_2Br + HBr.$$

When the reaction is carried out in neutral solution, the rate is observed to be

$$-\frac{d[CH_3COCH_3]}{dt} = k_0[CH_3COCH_3],$$ (31.80)

provided that the bromine concentration is sufficiently large to make the enolization step rate-determining. Note that the solvent concentration does not appear explicitly. The addition of either acid or base accelerates the rate of enolization, and hence the overall reaction. In the enolization process, one proton must be removed from the methyl group and another added to the carbonyl oxygen. In neutral solution this can be accomplished with water alone, but obviously the process occurs more easily with stronger acids and bases:

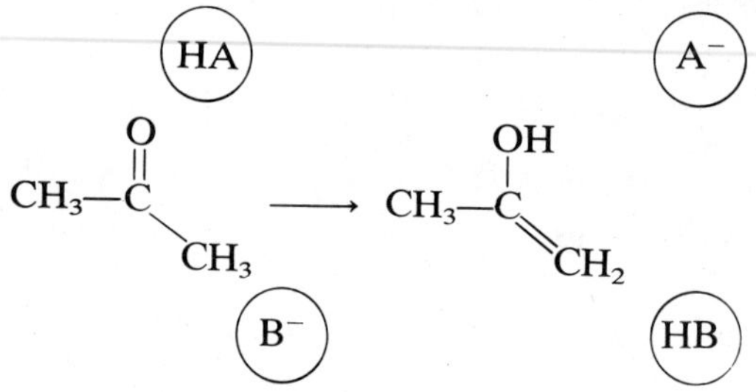

In the presence of a weak acid HA (say, acetic acid), the two species HA and H_3O^+ are both present, and each makes its own contribution to the rate,

$$-\left(\frac{d[CH_3COCH_3]}{dt}\right)_{acid} = k_1[CH_3COCH_3][H_3O^+] + k_2[CH_3COCH_3][HA].$$
(31.81)

Similarly, in basic solution one may have the two species OH^- and B^- (say, acetate ion) present, giving for the total rate contribution

$$-\left(\frac{d[CH_3COCH_3]}{dt}\right)_{base} = k_3[CH_3COCH_3][OH^-] + k_4[CH_3COCH_3][B^-].$$
(31.82)

When both acid and base are present (say, in a buffer solution of acetic acid and sodium acetate), we must combine all these effects to obtain the complete rate expression

$$\begin{aligned}
\frac{-d[CH_3COCH_3]}{dt} = &\ k_0[CH_3COCH_3] \\
&+ k_1[CH_3COCH_3][H_3O^+] + k_2[CH_3COCH_3][HA] \\
&+ k_3[CH_3COCH_3][OH^-] + k_4[CH_3COCH_3][B^-] \\
&+ k_5[CH_3COCH_3][HA][B^-].
\end{aligned}$$
(31.83)

The last (k_5) term accounts for simultaneous attacks of the acid on the oxygen and the base on a methyl group; simultaneous attacks of other species in equilibrium with HA and B respectively are not distinguishable, and are therefore included in this term.

The mechanisms and kinetics of acid–base reactions can take various forms. In most cases there occurs first a proton transfer to one of the reactant molecules (S), say,

$$S + BH \underset{k_{-1}}{\overset{k_1}{\rightleftharpoons}} SH^+ + B^-$$

for the example of an acid-catalyzed reaction. This is then followed by one of two possibilities, proton transfer to the solvent (*protolytic* mechanism), say,

$$SH^+ + H_2O \xrightarrow{k_2} P + H_3O^+,$$

or proton transfer to the base B^- (*prototropic* mechanism), say,

$$SH^+ + B^- \rightarrow P + BH.$$

In the first case the reaction rate is taken to be

$$-\frac{d[S]}{dt} = k_2[SH^+],$$
(31.84)

with the protolytic step rate-determining. Since the variation of [SH$^+$] with time is

$$\frac{d[SH^+]}{dt} = k_1[S][BH] - k_{-1}[SH^+][B^-] - k_2[SH^+], \qquad (31.85)$$

application of the stationary-state approximation to [SH$^+$] gives the rate expression

$$-\frac{d[S]}{dt} = \frac{k_1 k_2[S][BH]}{k_{-1}[B^-] + k_2}. \qquad (31.86)$$

We are assuming that the concentrations [SH$^+$] and [B$^-$] are small compared to [S] and [BH], respectively. In this limit, note the similarity to the Michaelis–Menten rate expression for enzyme catalysis.

If $k_2 \ll k_{-1}[B^-]$, then Eq. 31.86 becomes

$$-\frac{d[S]}{dt} = \frac{k_1 k_2[S][BH]}{k_{-1}[B^-]}; \qquad (31.87)$$

this can be rewritten as

$$-\frac{d[S]}{dt} = \frac{k_1 k_2}{k_{-1}K}[S][H_3O^+], \qquad (31.88)$$

where K is the equilibrium constant for the acid–base reaction

$$BH + H_2O \rightleftharpoons B^+ + H_3O^+.$$

Thus when the acid–base reaction is essentially in equilibrium, the derived rate expression for the catalyzed reaction becomes first-order in H_3O^+. Since the species H_3O^+ appears in the derived rate expression, we say that the catalysis is (H_3O^+)-*specific*. In the reverse limit, when $k_2 \gg k_{-1}[B^-]$, Eq. 31.86 reduces to

$$-\frac{d[S]}{dt} = k_1[S][BH]. \qquad (31.89)$$

The catalysis is now brought about by the species BH, which appears in the derived rate expression to first order; the catalysis is said to be *general*, since it occurs for any acid BH.

For the second possibility, the prototropic mechanism, the rate equation is

$$-\frac{d[S]}{dt} = k_2[SH^+][B^-], \qquad (31.90)$$

with the second step still rate-determining. The rate equation for SH$^+$ is now

$$\frac{d[SH^+]}{dt} = k_1[S][BH] - k_{-1}[SH^+][B^-] - k_2[SH^+][B^-],$$

and with the stationary-state approximation we have

$$-\frac{d[S]}{dt} = \frac{k_2 k_1[S][BH]}{k_{-1} + k_2}. \qquad (31.91)$$

Since the concentration [BH] appears in the derived rate expression, we see that general acid catalysis occurs.

The rates of protolytic reactions extend over quite a wide range (for $H_3O^+ + H_2O$, one finds $k = 1 \times 10^{10}$ liters/mol s; $HF + H_2O$, $k = 7.1 \times 10^7$ liters/mol s; $H_2S + H_2O$, $k = 4.3 \times 10^3$ liters/mol s; $H_3BO_3 + H_2O$, $k = 10$ liters/mol s). Prototropic reactions between oppositely charged ions, however, can have typical diffusion-controlled rates of the orders 10^{10}–10^{11} liters/mol s (Section 30.8).

The development for base catalysis is parallel to that given for acid catalysis. It is also possible that a reaction may be both acid- and base-catalyzed; the rate for the overall reaction of the substrate S is then, for specific catalysis,

$$-\frac{d[S]}{dt} = k_0[S] + k_a[S][H_3O^+] + k_b[S][OH^-]. \tag{31.92}$$

Here k_0 is the rate coefficient for the uncatalyzed reaction, k_a corresponds to the coefficient in Eq. 31.88, and k_b is its counterpart for base catalysis. At low pH the second term dominates, at high pH the third term is most important. If one makes rate measurements at various values of pH, then the maximum rate can be determined by differentiating with respect to $[H_3O^+]$:

$$\frac{d}{d[H_3O^+]}\left(-\frac{d[S]}{dt}\right) = k_a[S] - k_b\frac{K_w}{[H_3O^+]^2}[S], \tag{31.93}$$

where $K_w(\equiv[H_3O^+][OH^-])$ is the ion product of water. The rate is a maximum at the value of $[H_3O^+]$ for which the derivative vanishes,

$$[H_3O^+] = \left(\frac{k_b K_w}{k_a}\right)^{1/2}. \tag{31.94}$$

Note that such measurements can also be used to determine K_w.

There exists an empirical correlation for acid- and base-catalyzed reactions, which in fact is a form of linear free energy relationship (Section 30.9). For an acid-catalyzed reaction, for instance, Brønsted proposed the equation

$$\log\frac{k_a}{p} = \alpha + \beta\log\left(\frac{K_a q}{p}\right), \tag{31.95}$$

which relates the rate coefficient k_a to the acid's ionization constant K_a. The quantities α and β are empirical parameters; p is the number of equivalent protons on the acid; and q is the number of equivalent proton-acceptor sites on the conjugate base. (For example, p and q are respectively 2 for oxalic acid, $(COOH)_2$, and oxalate ion, but are unity for monobasic acids and bases.)

Now consider the Hammett correlation for the rate coefficient, Eq. 30.112, in the form

$$\log k_a = \log k_{a0} + \sigma\rho, \tag{31.96}$$

for the set of reactions

$$S \xrightarrow{BH} P,$$

catalyzed by a series of substituted acids BH. For the same series of acids we can also consider the ionization reaction

$$BH \rightleftharpoons B^- + H^+,$$

and write the corresponding equation for the equilibrium constants,

$$\log K_a = \log K_{a0} + \sigma\rho'. \tag{31.97}$$

Note that the two reactions cited here are different, and thus have different reaction parameters, ρ and ρ'; however, the parameter σ depends only on the substituents, and hence is the same in Eqs. 31.96 and 31.97. These two equations can now easily be rearranged into the form

$$k_a = k_{a0}(K_{a0})^{-\rho/\rho'}(K_a)^{\rho/\rho'} = A(K_a)^c, \tag{31.98}$$

which is equivalent to Eq. 31.95; in any of its forms, this equation is known as the *Brønsted correlation*.

31.9
Metal–Ion Complex, and Other Types of Homogeneous Catalysis

Many different types of reactions are catalyzed under homogeneous conditions by metal ions and their complexes, in particular by ions of the transition metals. This subject is of great scientific interest, and has tremendous importance for industrial processes. As we know from Chapter 9, metal ions in solution are often complexed with groups called ligands, which may include the solvent itself (for instance, in $[Ni(H_2O)_6]^{2+}$). When a metal-ion catalyst acts on a substrate (reactant), the latter usually itself becomes a ligand, in whole or in part, in the course of the mechanism.

Transition metal complexes can be grouped for our purposes into two major categories. The first are called *organometallic complexes*, and contain metals in low oxidation states (Fe^0, Co^I), with the bonding to the ligands primarily covalent and delocalized over the complex. If that bonding involves fewer than 18 electrons, then as a rule the catalytic activity involves addition of substrate to the complex to bring the number of valence-shell electrons up to 18. The second category are called *Werner complexes*, which follow classical coordination chemistry: The metal is in a higher oxidation state (Fe^{II}, Fe^{III}), and the bonding is localized. In these complexes the metal acts as a Lewis acid (it is electropositive), and the complex catalyzes oxidation–reduction (electron-transfer) reactions. Enzymes with metal atoms in the active site are usually Werner complexes (one exception is vitamin B_{12}, which is an organometallic complex).

As a rule, metal-ion complexes displaying catalytic activity either have labile (freely movable) ligands that can be readily displaced by a substrate, or permit the substrate to add directly to the metal ion. Ligands in active catalysts include unsaturated hydrocarbons, CO, H^-, O_2, and similar reactive substances. It follows that homogeneous catalysts are available for reactions with these as substrates. Ligands can influence the lability of other ligands in the same complex, particularly in the *trans*-position. The introduction of hydrocarbon radicals, CO, or H^-, for instance, makes the ligand in the *trans*- position more labile and therefore catalytically more active.

Now let us discuss some examples, first of organometallic catalysts. The hydrogenation of olefins is catalyzed by the square-planar rhodium complex $[RhCl(PPh_3)_3]^+$ ($Ph \equiv$ phenyl), which has 16 valence-shell electrons. The mechanism is thought to be

In the first step of the mechanism, addition of H_2 leads to a complex with 18 valence electrons. The delocalized bonding in this complex affects the binding of the olefin through its π bonds.

Another example of organometallic catalysis occurs in some symmetry-forbidden reactions (Section 31.5). The isomerization of cubane, with a rhodium catalyst like that just described, is thought to proceed by the mechanism

in which each step is symmetry-allowed. In the absence of a catalyst, however, the reaction is symmetry-forbidden.

Next we examine some examples of catalysis by Werner complexes. Consider the reaction

$$H_2(g) + Tl^{3+} \rightarrow 2H^+ + Tl^+;$$

without catalysis the mechanism is thought to be

$$H_2(g) \rightarrow H^+ + H^-$$

$$H^- + Tl^{3+} \rightarrow H^+ + Tl^+.$$

The first step has an endothermicity of about 150 kJ/mol, and is thus probably the rate-determining step. With catalysis by Cu^{2+} complexes, however, the rate-determining step is

$$Cu^{2+} + H_2(g) \rightarrow CuH^+ + H^+,$$

which has the substantially lower endothermicity of about 110 kJ/mol; the catalyst stabilizes the active hydride ion, and this lowers the activation energy and increases the reaction rate.

The nucleophilic hydrolysis of amino acid esters,

is effectively catalyzed by metal complexes according to the proposed mechanism

The metal ion can act as a catalyst because it is positively charged, which increases the density of positive charge on a given ligand. This effect makes the ligand more susceptible to nucleophilic attack, so catalytic activity in hydrolytic reactions is to be expected.

The existence of complexes with different oxidation states of a given metal ion is important for the catalysis of oxidation–reduction reactions. One example is the reaction

$$V^{III} + Fe^{III} \rightarrow V^{IV} + Fe^{II},$$

which is catalyzed by Cu^{2+} complexes. We have already given the mechanism of this reaction as an example in Eqs. 30.153 ff. Another example is the reaction

$$Tl^{I} + 2Ce^{IV} \rightarrow Tl^{III} + 2Ce^{III},$$

which is catalyzed by Ag^{+} ions according to the mechanism

$$Ag^{I} + Ce^{IV} \underset{k_2}{\overset{k_1}{\rightleftharpoons}} Ag^{II} + Ce^{III} \qquad \text{(fast)}$$

$$Tl^{I} + Ag^{II} \xrightarrow{k_3} Tl^{II} + Ag^{I} \qquad \text{(slow)}$$

$$Tl^{II} + Ce^{IV} \longrightarrow Tl^{III} + Ce^{III} \qquad \text{(fast)}.$$

If one makes the stationary-state assumption for $[Ag^{II}]$, one can derive the observed rate expression

$$\frac{d[Tl^{III}]}{dt} = \frac{k_1 k_3 [Tl^{I}][Ce^{IV}]}{k_2[Ce^{III}] + k_3[Tl^{I}]}. \qquad (31.99)$$

It may be possible for a given metal complex to act as either an organometallic or a Werner catalyst. Consider the reaction

for which the following mechanism has been suggested:

Although $[PdCl_4]^{2-}$ is ordinarily a Werner catalyst, the ligand substitution in the first step of this mechanism indicates a bond delocalization characteristic of organometallic catalysts.

Finally, we briefly mention two areas of catalysis not yet discussed:[10] *selectivity* and *stereospecificity*. A catalyst can act to accelerate a given reaction along one of a number of competitive reaction paths, and thus lead selectively to one of many possible products. Stereospecificity is another example of selectivity; for instance, Rh^I complexes with chiral (optically active) phosphine ligands have been used as catalysts in the synthesis of optically active amino acids, yielding up to 90% of the product with the desired asymmetry.

Before closing our discussion of homogeneous catalysis, let us mention some examples that do not fit in well anywhere else. Nearly every substance can act as a catalyst in some reaction. Thus we conclude with two examples, chosen essentially at random (one involving free radicals as catalysts, the other involving ions), to show the diversity of possibilities.

The protective layer of ozone in the stratosphere shields all life on earth from a great deal of ultraviolet radiation, which the ozone absorbs. Both chlorine atoms and nitric oxide molecules are catalysts for the decomposition of ozone:

$$Cl + O_3 \rightarrow ClO + O_2$$
$$ClO + O_3 \rightarrow Cl + 2O_2,$$

and

$$NO + O_3 \rightarrow NO_2 + O_2$$
$$NO_2 + O_3 \rightarrow NO_3 + O_2$$
$$NO_3 + h\nu \rightarrow NO + O_2.$$

Both sets of reactions pose a threat to the ozone layer, as the catalytic species are introduced into the stratosphere by humans: Cl from the chlorinated hydrocarbons used in spray cans, transported to the upper atmosphere and photolyzed to give Cl atoms; NO from stratospheric jet flights.

Many ions in solution may act as catalysts. The formation of benzoin from benzaldehyde is catalyzed by cyanide ions according to the proposed mechanism

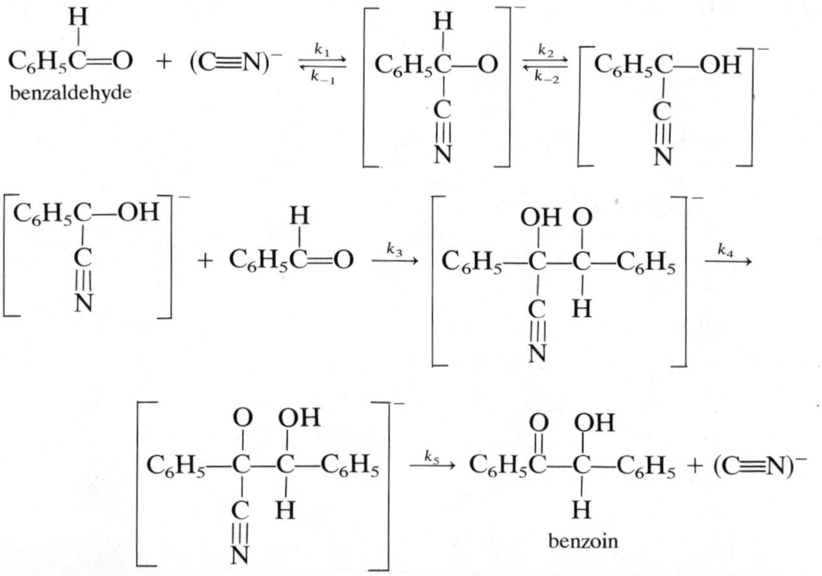

---

[10] In Section 31.7 we mentioned the "specificity" of enzyme catalysts, i.e., their effectiveness in only a single reaction. The properties discussed here are even more "specific," referring to particular reaction paths within a given system.

31.10
Heterogeneous Reactions: Adsorption of Gas on a Surface

Up to this point our discussion of chemical kinetics has been essentially restricted to homogeneous reactions, that is, reactions that occur in a single phase. Now we turn to heterogeneous reactions, those that occur at the interface of two phases. Of all the possible combinations of two phases, gas–solid and liquid (solution)–solid interfaces occur most often, and we shall concentrate on these cases.[11] Since most heterogeneous reactions are in fact examples of catalysis, our study not only includes but concentrates on the topic of heterogeneous catalysis. Heterogeneous reactions present a challenge to several branches of science—chemical kinetics, surface and solid-state physics, and surface chemistry; this is a good example of a problem requiring interdisciplinary study and research. Aside from their fundamental importance to science, heterogeneous reactions play a significant role in many industrial chemical processes such as the cracking of oil, polymerizations, hydrogenations, conversion of nitrogen to ammonia, and so on. The study of heterogeneous catalysis is still as much an art as a science, especially with regard to choosing an effective catalyst for a given reaction.

In the remaining sections of this chapter we present a discussion of chemical reactions and catalysis at interfaces, concluding with a brief consideration of electrochemical kinetics, that is, the rates of reactions at electrodes.

Chemical reactions at the interface of two phases must involve a number of distinct processes. Picture a solid immersed in a solution: the reactants in the solution must diffuse to the interface, be adsorbed there, and participate in a given reaction mechanism on the solid surface (in which reaction the solid itself may be either a catalyst or a reactant); the products on the surface must then desorb and diffuse into the solution. Any one of these steps may be rate-determining. The rate of diffusion to and from the surface can be accelerated by stirring. The rate and extent of adsorption can be increased by increasing the surface area of the solid; this can be accomplished by preparing the solid in small particles, sometimes supported on the surface of another solid. The rate of desorption can be increased by removing the product from the region near the surface. Frequently the rate-determining step is the actual reaction on the surface. Since this rate must depend on the concentration of chemical species adsorbed on the surface, we begin with a simple analysis of the adsorption of gases on surfaces.

We need to distinguish between *physical adsorption* and *chemical adsorption* (*chemisorption*). In physical adsorption the energy released as an atom or molecule becomes bound to the surface is of the order of a few times $k_B T$, that is, the binding is relatively weak; in chemisorption the energy released is much larger, of the order of chemical bond energies. The simplest model imaginable consists of a surface on which gas molecules are adsorbed at individual sites. All sites are the same; the adsorption at each site is independent of that at any other site. If the surface is initially bare of adsorbed species, then adsorption begins when the surface is exposed to gas at a given temperature and pressure. The rate of adsorption is proportional to the flux of molecules to the surface (which in turn is proportional to the pressure: cf. Eq. 28.92 for molecular effusion), and to the fraction of unoccupied sites (on the assumption that

---

[11] The other cases are important, but much less is known about them; for instance, interesting reactions occur in and on the surface of tiny water droplets (aerosol particles) produced by wind and wave action above the surface of oceans.

only one gas molecule can be adsorbed at a given site). The rate of desorption is proportional to the fraction θ of surface covered by adsorbed molecules. At equilibrium the rates of adsorption and desorption are equal,

$$\theta = bp(1-\theta), \tag{31.100}$$

where p is the pressure and b contains all proportionality constants. Solving this equation for θ yields

$$\theta = \frac{bp}{1+bp}, \tag{31.101}$$

which is called the *Langmuir adsorption isotherm* (after Irving Langmuir, who presented this analysis in 1916). A plot of Eq. 31.101 is shown in Fig. 31.25. At low pressure the fraction of sites covered by adsorbed molecules is proportional to pressure, but as the fraction approaches unity it becomes nearly independent of pressure.

We can obtain some further insight by considering the equilibrium between the gaseous molecules and those adsorbed on surface sites. We formally write the "reaction"

$$M + S \rightleftharpoons MS$$

and the equilibrium constant

$$K = \frac{[MS]}{[M][S]}, \tag{31.102}$$

where $[M]$ is the concentration of molecules in the gas phase, $[S]$ is the concentration of unoccupied sites, and $[MS]$ is the concentration of occupied sites (the latter two "concentrations" are expressed as the number of sites per unit area of surface). For an ideal gas we have $[M] = p/k_B T$; for $[S]$ we write $(1-\theta)N$, where N is the total number of sites per unit area; and for $[MS]$ we therefore have θN. On substitution of these quantities into Eq. 31.102, we obtain

$$K = \frac{\theta k_B T}{(1-\theta)p}; \tag{31.103}$$

this is equivalent to the Langmuir adsorption isotherm, Eq. 31.101, if we make the identification

$$b = \frac{K}{k_B T}. \tag{31.104}$$

The statistical-mechanical expression for the equilibrium constant (cf. Section 21.6) is

$$K = \frac{q_{MS}}{q_M q_S} e^{-E_{ad}/RT}, \tag{31.105}$$

where the q's are the partition functions of the indicated species, and E_{ad} is the difference in ground-state energy between the molecules in the gaseous and adsorbed states, that is, the energy of adsorption. E_{ad} is negative, since energy is released upon adsorption. Thus, combining Eqs. 31.104 and 31.105 provides us with a molecular interpretation of the constant b in terms of the adsorption energy and the molecular quantities that determine the partition function: vibrational and rotational energies, temperature, masses.

The simple Langmuir model of adsorption can be improved in various ways. For example, all sites on the surface may not be equivalent:

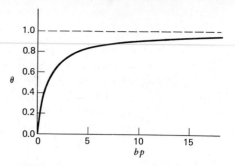

FIGURE 31.25
The Langmuir adsorption isotherm: fraction θ of occupied sites as a function of pressure p.

there may be two or more types of sites with different adsorption energies. The probability of occupation of a given site may depend on the state (occupied or not) of neighboring sites. And finally, the number of molecules adsorbed on a given site may exceed unity (more than monolayer coverage), and if so the adsorption energy per molecule may vary with the occupation number. By the time a few molecules are stacked upon one another at a given site, we expect the adsorption energy to approach the heat of vaporization. This is in fact observed, and lends credence to the picture of a multilayer of physically adsorbed species resembling a liquid.

Although these extensions are interesting, let us now return to the simple Langmuir model and use it to derive rate expressions for various mechanisms and conditions in a heterogeneous reaction. Suppose that we have reactants and products in the gas phase, and that reactants adsorbed on a solid surface react to form products that then desorb. The reaction on the surface is assumed to be the rate-determining step.

If a single substance M reacts on the surface (isomerization or dissociation), then the rate is simply proportional to the fraction of the surface covered (the surface concentration of adsorbed species):

$$\text{rate} = k\theta = \frac{kbp}{1+bp}. \tag{31.106}$$

At low values of bp, that is, either low partial pressure of the reactant in the gas phase or weak adsorption, the rate is first-order in the reactant pressure. At large values of bp, that is, high pressure or strong adsorption, the rate is zero-order in the reactant pressure.

Next suppose that there is more than one adsorbable species in the gas phase, say M and P. Let the fractions of surface covered by M and P be θ_M and θ_P, respectively. The rates of adsorption and desorption of M at equilibrium are equal, and in analogy to Eq. 31.100 we have

$$\theta_M = b_M p_M (1 - \theta_M - \theta_P); \tag{31.107}$$

similarly, for the species P we have

$$\theta_P = b_P p_P (1 - \theta_M - \theta_P). \tag{31.108}$$

If we solve Eq. 31.108 for θ_P and substitute that result in Eq. 31.107, we can rearrange to obtain

$$\theta_M = \frac{b_M p_M}{1 + b_M p_M + b_P p_P}; \tag{31.109}$$

a similar expression holds for θ_P. Comparing Eqs. 31.101 and 31.109, we see that the adsorption of P reduces the fraction of surface covered by M, which is obvious but has interesting consequences.

If P does not participate in the reaction, then it merely acts as an inhibitor (or poisoner) of the surface reaction. The rate of the reaction M $\rightarrow$ products is then

$$\text{rate} = \frac{k b_M p_M}{1 + b_M p_M + b_P p_P}; \tag{31.110}$$

for sufficiently large values of $b_P p_P$, this rate is inversely proportional to the partial pressure of P. Suppose, however, that P is a product of the reaction; in this case we see that adsorption of the product decreases the reaction rate. In fact, if the term $b_P p_P$ sufficiently exceeds the others in the denominator of Eq. 31.110, then the rate is simply $k b_M p_M / b_P p_P$. This is the type of rate expression observed for the decomposition reaction

$$2NH_3 \rightarrow N_2 + 3H_2$$

on a platinum surface:

$$\text{rate} = \frac{k[NH_3]}{[H_2]}; \tag{31.111}$$

the effect of adsorbed H_2 is much greater than that of adsorbed NH_3 or N_2. Similarly, for the decomposition of N_2O the observed rate expression has the form

$$\text{rate} = \frac{k[N_2O]}{1 + c[N_2O] + c'[O_2]} \tag{31.112}$$

in analogy to Eq. 31.110.

So far we have cited examples of a single molecule reacting unimolecularly on a surface. Now let us consider bimolecular reactions in which two species, say A and B, need to be adsorbed on a surface to react with each other. The fractions of surface covered by A and B are both of the form 31.109; the reaction rate is thus simply

$$\text{rate} = k\theta_A\theta_B = \frac{kb_Ab_Bp_Ap_B}{(1 + b_Ap_A + b_Bp_B)^2}. \tag{31.113}$$

Various limiting cases may occur; for weak adsorption (or low pressure) the rate is expectedly just $kb_Ab_Bp_Ap_B$.

For a bimolecular reaction in which both A and B are adsorbed, but the rate-determining step is the collision of a gaseous B molecule with an adsorbed A molecule, the rate law is

$$\text{rate} = k\theta_Ap_B = \frac{kb_Ap_Ap_B}{1 + b_Ap_A + b_Bp_B}. \tag{31.114}$$

The reaction of ethylene and hydrogen,

$$C_2H_4 + H_2 \rightarrow C_2H_6,$$

which in the gas phase is symmetry-forbidden, takes place on various metal surfaces. The mechanism is complex, but on a copper surface, under certain conditions, the experimental rate expression is

$$\text{rate} = \frac{k[H_2][C_2H_4]}{(1 + c[C_2H_4])^2}; \tag{31.115}$$

this is a limiting form of Eq. 31.113 for the case of C_2H_4 being adsorbed more strongly than H_2.

As we have seen before, fractional exponents in experimental rate expressions usually indicate dissociation of a reactant. Let a reactant A_2 be adsorbed on a surface as A atoms. The rate of the adsorption process, which can be formally expressed as

$$A_2 + 2S \rightarrow 2AS,$$

is proportional to $p_{A_2}(1-\theta)^2$, and the rate of desorption to θ^2. At equilibrium we thus have

$$\theta^2 = b_{A_2}p_{A_2}(1-\theta)^2, \tag{31.116}$$

or

$$\theta = \frac{b_{A_2}^{1/2}p_{A_2}^{1/2}}{1 + b_{A_2}^{1/2}p_{A_2}^{1/2}}. \tag{31.117}$$

All the previous examples of rate laws can occur in many different combinations. For instance, the isotope exchange reaction

$$NH_3 + D_2 \rightarrow NH_2D + HD$$

on an iron surface has the rate expression

$$\text{rate} = \frac{k[D_2]^{1/2}[NH_3]}{(1 + c[NH_3])^2}. \tag{31.118}$$

The interpretation is straightforward: D_2 is adsorbed on the surface and dissociates into D atoms; a D atom then reacts with an adsorbed NH_3 to form H and NH_2D (which may further react to form NHD_2 and ND_3). Ammonia is much more strongly adsorbed on the surface than D_2 or H_2, and thus $[NH_3]$ appears alone in the denominator.

Finally, we must mention at least one case in which the solid acts as a reactant rather than as a catalyst: chlorine atoms impinging on a clean nickel surface react to give the product NiCl, with the rate proportional to the concentration of Cl atoms.

31.11
Heterogeneous Catalysis

Having discussed adsorption and simple rate expressions for heterogeneous reactions, we can now consider the important topic of heterogeneous catalysis in greater detail. Heterogeneous catalysis, like its homogeneous counterpart, changes the rate of a reaction by providing an alternative reaction mechanism. Normally we are interested in increasing a reaction rate; this is usually accomplished by finding an alternative reaction path with a lower overall activation energy. For example, the decomposition of N_2O in the gas phase without a catalyst requires an activation energy of about 250 kJ/mol, whereas on catalysts such as platinum or gold the overall activation energy is only about 125 kJ/mol. Even more striking, the conversion of p-H_2 to o-H_2 in the gas phase without a catalyst has an activation energy of 255 kJ/mol, but on palladium this is reduced to only 17 kJ/mol! There are numerous issues and problems to be studied with regard to heterogeneous catalysis; we merely attempt a brief consideration of a few important points.

The substances most often studied and used as heterogeneous catalysts are metals (usually transition metals), alloys, and semiconducting oxides and sulfides. The effectiveness of a catalyst can be measured by the amount of product formed per unit time per unit surface area of the catalyst (since we expect the effect of a catalyst to be at least proportional to its surface area). An indication of the surface-to-volume ratio in a dispersed catalytic solid is given in Fig. 31.26. A catalyst is said to be *structure-insensitive* if the reaction rate is proportional to the catalyst's surface area, that is, to the number of atoms exposed to the reactants (in the gas phase or solution). For a *structure-sensitive* catalyst the reaction rate increases with surface area more rapidly than a simple proportionality. The model proposed for such behavior is this: Suppose that not every surface site on a solid can act as a catalyst, but only special sites, such as point defects or lattice dislocations (see Section 29.5) on the crystal surface. If this is the case, then we might expect an increase in such active sites on very small catalytic particles, say with diameters of the order of 5 nm. For such small particles the fraction of atoms on the surface increases sharply, as we have seen in Fig. 31.26. Examples of the two mechanisms just described are provided by the competitive reactions of

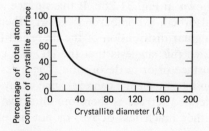

FIGURE 31.26
Surface-to-volume ratio for platinum crystallites as a function of crystallite size; the quantity plotted is the fraction of atoms considered to be on the surface. From P. Stonehart and P. H. Zucks, *Electrochim. Acta* **17**, 2333 (1972).

neopentane on platinum,

$$CH_3-\underset{\underset{CH_3}{|}}{\overset{\overset{CH_3}{|}}{C}}-CH_3 \quad \xrightarrow{H_2} \quad \begin{array}{l} \underset{CH_3}{\overset{CH_3}{|}}CHCH_2CH_3 \\[2em] \underset{CH_3}{\overset{CH_3}{|}}CHCH_3 + CH_4 \end{array}$$

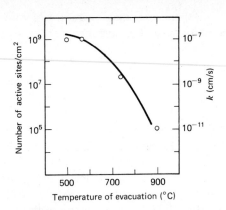

FIGURE 31.27
Correlation of catalytic activity with density of active sites (holes trapped on the surface) for the H_2-D_2 exchange reaction on MgO at 78 K. The density of holes (number per unit area) and the rate constant are both plotted as a function of the temperature to which the catalyst is heated beforehand. From M. Boudart, A. Delbouille, E. G. Derouane, U. Indovina, and A. B. Walters, *J. Am. Chem. Soc.* **94**, 6622 (1972).

The isomerization to isopentane is structure-insensitive, whereas the cracking reaction to yield isobutane and methane is reported to be structure-sensitive.

It is difficult to characterize active sites for heterogeneous catalysis, and many ingenious experiments have been devised to obtain either correlative or direct evidence. Metal-oxide and metal-sulfide catalysts have been analysed with more success than have metal and-alloy catalysts. Consider the exchange reaction

$$H_2 + D_2 \rightarrow 2HD.$$

In the gas phase this reaction is symmetry-forbidden and very slow; however, the reaction is catalyzed by MgO(s). MgO is a semiconductor, and there exists in the solid (depending on its method of preparation) a very small concentration of free electrons and an equal number of positive sites (holes). When a sample of MgO(s) is heated in vacuum, the number of such electron–hole pairs decreases, due to an increased recombination rate. The density of holes trapped on the surface after heating and quenching to 78 K (the temperature of the H_2-D_2 exchange experiment) can be followed by esr measurements, since a hole corresponds to an unpaired electron spin. In Fig. 31.27 we show the correlation between the surface density of trapped holes and the rate of the exchange reaction.

The same exchange reaction has been studied by molecular-beam scattering from Pt surfaces. The (111) surface of Pt may be pictured as in Fig. 31.28a. When the surface is exposed to gaseous D_2, some of the D_2 molecules are adsorbed in the form of dissociated D atoms. When this surface is then bombarded with a H_2 beam, both H_2 and desorbed D_2 can be detected in the scattering products but no HD is produced, Fig. 31.28b. However, if the Pt surface is machined at an angle to the (111) layer, then on the microscopic scale stepped terraces are produced, as shown in Fig. 31.28c. If this surface, on which again deuterium has been adsorbed, is bombarded with H_2, then HD *is* observed, Fig. 31.28d. The angular distribution of the desorbed HD is cosinelike, similar to effusive flow; this suggests that the HD has come to thermal equilibrium on the surface prior to desorption. These results suggest that D atoms adsorbed on a terrace edge can exchange easily with H_2 molecules. The edge of the stepped terrace is therefore an active site; the sites on the flat plane are either much less active, or not all effective in catalyzing the exchange reaction.

One crucial goal in the study of heterogeneous catalysis is the development of a correlation, a model, a theory describing the effectiveness of a catalyst in terms of its electronic structure and those of the reactants and products. The task is quite difficult: Can the electronic properties of a bulk metal be indicative of its catalytic activity? We would like very much to be able to answer this. In Fig. 31.29 are shown plots of catalytic activity and "% *d* character" for some families of transition

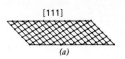

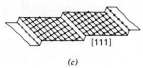

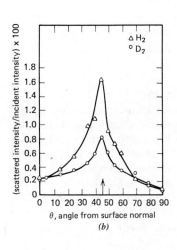

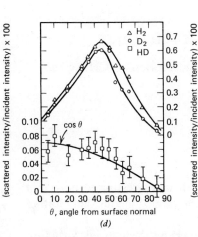

FIGURE 31.28
Scattering results for the H_2–D_2 exchange reaction when D_2 adsorbed on Pt is bombarded with beams of H_2. (*a*) The (111) surface (schematic). (*b*) Scattering intensity from a pure (111) surface. (*c*) Surface machined at a small angle to the (111) layer. (*d*) Scattering intensity from the surface in (*c*). Note that in (*d*) the HD scattering fairly well fits a $\cos\theta$ dependence. From S. L. Bernasek and G. A. Somorjai, *Surface Sci.* **48**, 204 (1975).

metals. The quantity "% d character" refers to the metallic bond within the solid metal, and is calculated by Pauling's valence bond theory (cf. Chapter 11). Although there is some correlation within each transition series, there is none between different series: Ni and Pt have about the same amount of d character, but differ by a factor of 6 in catalytic activity. The problem is, of course, even more complicated: the bulk properties of the metal are not necessarily the same as its surface properties; furthermore, the surface properties may be substantially altered by strong adsorption (chemisorption, if the energy of adsorption is typically that of a chemical bond). For alloys and semiconductors all these problems are still more complex. Usually the best one can do is to obtain an empirical correlation with some measurable property; some examples are shown in Figs. 31.30 and 31.31.

The sort of challenge with which both correlation and theory are faced is illustrated by the following example. Isopropanol undergoes dehydrogenation on a surface of ZnO, which is an *n*-type semiconductor:

$$(CH_3)_2CHOH \xrightarrow{ZnO(s)} CH_3-\overset{\displaystyle O}{\overset{\|}{C}}-CH_3 + H_2(g).$$

The mechanism probably involves adsorption of $(CH_3)_2CHOH$ with loss of an H atom, schematically

$$\begin{array}{c} CH_3 \\ | \\ CH_3-C-OH \\ | \\ -Zn-O-Zn-O-Zn- , \end{array}$$

the H atom itself being adsorbed on a neighboring site. Subsequently another H atom is removed to yield

$$\begin{array}{c} CH_3 \\ | \\ CH_3-C-O \\ | \\ -Zn-O-Zn-O-Zn- . \end{array}$$

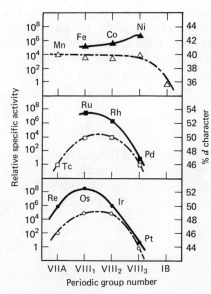

FIGURE 31.29
Catalytic activity (solid points) and "percentage of d character" after Pauling (open points) for some transition metal series; the data are for the reaction

$$H_2 + C_2H_6 \rightarrow 2CH_4.$$

From J. H. Sinfelt, *Catal. Rev.* **9**, 147 (1974).

In time two H atoms recombine to form H_2, which desorbs to yield $H_2(g)$; the $(CH_3)_2CO$ fragment also desorbs to give the other product, acetone. Suppose, however, that $(CH_3)_2CHOH$ were adsorbed instead on the catalyst $Al_2O_3(s)$; this surface is an ionic insulator, and acts as a Lewis acid in the reaction

$$(CH_3)_2CHOH \xrightarrow{Al_2O_3(s)} CH_3CH{=}CH_2 + H_2O.$$

The likely mechanism in this case is first the adsorption of $(CH_3)_2CHOH$,

$$\begin{array}{c} H \\ | \\ CH_3{-}C{-}CH_3 \\ | \\ OH \\ \overset{..}{} \\ {-}O{-}Al{-}O{-} \end{array} \quad,$$

followed by dehydration to yield the olefin. It is interesting to note that this example is a heterogeneous analog of the homogeneous dehydration of an alcohol by HBr (Section 31.6).

There is probably a close connection between the electronic structure of the surface complex formed between an adsorbed molecule and the surface atoms of a transition-metal catalyst, and the structure of the analogous homogeneous catalytic complex. The search for this connection has brought much theory and experimentation to the study of heterogeneous catalysis. Some typical examples of the topics under study are the role of d orbitals in organometallic compounds of transition metals; calculations of electronic structure; and the conditions needed for circumvention of symmetry restrictions. Orbitals of the d type have a relatively narrow energy range, are well localized in space, and have good overlap with π electrons (as in the common ligands CO and C_2H_4). These characteristics serve to give good ligand bonding in homogeneous catalysis, good chemisorption in heterogeneous catalysis. Orbitals of the same symmetry, when bonded to ligands such as C_2H_4, lead to conservation of orbital symmetry in a reaction such as

$$\begin{array}{ccccc} CH_2 & & CH_2 & & CH_2{-}CH_2 \\ \| & + & \| & \longrightarrow & | \qquad | \\ CH_2 & & CH_2 & & CH_2{-}CH_2 \end{array} \quad;$$

in the absence of these d orbitals, the reaction is symmetry-forbidden. An interesting experimental example of the homogeneous–heterogeneous relationship is the following: The infrared spectrum of the transition metal complex $Rh_2(CO)_4Cl_2$, with structure

$$\begin{array}{c} OC\diagdown \qquad \diagup Cl \diagdown \qquad \diagup CO \\ \qquad Rh \qquad \qquad Rh \\ OC\diagup \qquad \diagdown Cl \diagup \qquad \diagdown CO \end{array} \quad,$$

is shown in Fig. 31.32a for the carbonyl-stretch region of the spectrum. Two somewhat broader, but nonetheless similar, peaks are observed in the infrared spectrum of CO adsorbed on a Rh catalyst, Fig. 31.32b. This allows the inference that the structures are similar, that is, that two CO molecules are chemisorbed by a single Rh atom on the surface.

As in homogeneous catalysis, so in the heterogeneous case, it is possible to affect the stereochemistry of the reaction product(s). For instance, some transition metals are heterogeneous catalysts for the polymerization of propylene, in a stereospecific manner such that every

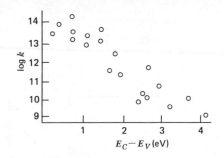

FIGURE 31.30
Dependence of the rate constant k for dehydrogenation of isopropanol at 200°C on the width of the bandgap (cf. Chapter 11) of III-V, IV-IV, II-VI, and I-VII semiconductors. From J. E. Madey, J. T. Yates, Jr., D. R. Sandstrom, and R. J. H. Voorhoeve, in N. B. Hannay (Ed.), *Treatise on Solid State Chemistry*, Vol. 6B (Plenum, New York, 1976), p. 1.

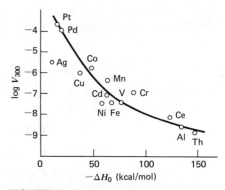

FIGURE 31.31
Rate of oxidation of propylene at 300°C (V_{300}) in an excess of oxygen as a function of the enthalpy of formation per oxygen atom (ΔH_0) of the oxide catalyst. From Y. Morooka and A. Ozaki, *J. Catal.* **5**, 116 (1966).

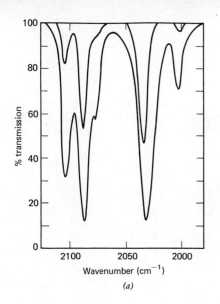

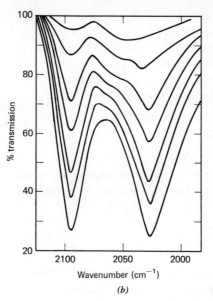

FIGURE 31.32
The carbonyl-stretch region of the spectrum for (a) the complex $Rh_2(CO)_4Cl_2$ in two solvents, (b) CO chemisorbed on a Rh catalyst at various pressures. From A. C. Yang and C. W. Garland, *J. Phys. Chem.* **61**, 1504 (1957).

second carbon atom in the chain is asymmetric,

$$-\overset{H}{\underset{CH_3}{C^*}}-\overset{H}{\underset{H}{C}}-\overset{H}{\underset{CH_3}{C^*}}-\overset{H}{\underset{H}{C}}-\overset{H}{\underset{CH_3}{C^*}}-$$

(where the asymmetric carbons are marked by asterisks), yet the configuration of all the asymmetric carbons is the same!

31.12 Kinetics of Electrode Reactions

The study of chemical reactions proceeding at electrodes is called *electrochemistry*. In Chapter 26 we discussed the equilibrium properties of electrolytic solutions; now we shall briefly consider their kinetics. There are many reactions that can be observed under ordinary conditions as well as in electrochemical cells. For instance, the reaction

$$H_2(g) + Cl_2(g) \rightarrow HCl(aq),$$

in which the dissolved HCl is of course fully ionized, can be initiated photochemically in the gas phase: gaseous HCl is formed, and can then be dissolved in water. Both rate and equilibrium properties can be measured. This reaction, however, can also be studied in an electrochemical cell. Two electrodes are immersed in a solution of HCl at a given concentration; at one electrode there occurs the reaction

$$H_2(g) \rightarrow 2H^+[\text{in } HCl(aq)] + 2e^-,$$

and at the other electrode the reaction

$$Cl_2(g) \rightarrow 2Cl^-[\text{in } HCl(aq)] - 2e^-.$$

At equilibrium the standard Gibbs free energy change for the overall reaction is, as usual,

$$\Delta G^0 = -RT \ln K \tag{31.119}$$

(cf. Eq. 19.81); the standard electrochemical potential of the cell is

$$\mathscr{E}^{\ominus} = \frac{-\Delta G^0}{z\mathscr{F}}, \tag{31.120}$$

where z is the number of electrons transferred in the stoichiometric equation as written (two in this case) and $\mathscr{F}$ is the Faraday constant. At chemical equilibrium the rates of the reactions in the forward and reverse directions are equal; the Gibbs free energy change of the reaction is then zero, and the electrochemical potential is also zero. For arbitrary (non-equilibrium) concentrations of reactants and products, the Gibbs free energy change of the reaction becomes

$$\Delta G = \Delta G^0 - RT \ln Q, \tag{31.121}$$

where Q is the ratio of the actual concentrations (or rather activities) of products and reactants, in the form that appears in the equilibrium constant. The electrochemical cell thus has in general the potential (Nernst equation)

$$\mathscr{E} = \mathscr{E}^{\ominus} + \frac{RT}{z\mathscr{F}} \ln Q; \tag{31.122}$$

if this potential is applied to the cell externally, it produces equilibrium at the given concentrations. But if the applied potential $\mathscr{E}'$ is changed from the specific value $\mathscr{E}$, then an irreversible reaction takes place. If the applied potential is lowered, the net reaction is such that the cell acts as a battery supplying current; if the applied potential is raised, the net reaction corresponds to the battery's "being charged." As of course is expected, a net reaction can occur in a cell in a finite time only under nonequilibrium conditions, that is, for a nonzero value of $\mathscr{E} - \mathscr{E}'$. All that we have said thus far is merely repeated from Chapter 26; now let us move on.

We note that an electrode reaction has similarities to the heterogeneous reactions of the last two sections. At least two phases are involved: an electrode (usually a solid), immersed in a liquid solution that contains reactants and products. Some of the points we made for general heterogeneous reactions apply here as well: Reactants need to diffuse to the electrode, may need to be adsorbed, exchange electrons, desorb, and diffuse away. In addition, oxidation or reduction of an ion may occur. The diffusion processes are more complicated than in many other heterogeneous reactions, since an electric double layer of charges, of opposite signs, surrounds each electrode.[12]

As a brief illustration of some of the concepts and arguments used in electrochemical kinetics, let us first consider an electrode reaction in which a diffusion process is the rate-determining step. Think of a concentration cell with two hydrogen electrodes; at one electrode there occurs the half-cell reaction

$$H_2(g, p) \rightarrow 2H^+(aq) + 2e^-,$$

and at the other electrode the reaction

$$2H^+(aq) + 2e^- \rightarrow H_2(g, p'),$$

where p and p' are the respective partial pressures of H_2. The Gibbs free energy change, assuming the gases to be ideal, is then

$$\Delta G = RT \ln\left(\frac{p'}{p}\right) \tag{31.123}$$

(since $K = 1$ and thus $\Delta G^0 = 0$), and the reaction can be balanced by

---

[12] At the cathode there is a slight excess of electrons; hence there is a layer of positive ions adjacent to the cathode, surrounded in turn by a more diffuse layer of negative ions.

applying an external potential $\mathscr{E}'$ equal to

$$\mathscr{E} = \frac{RT}{2\mathscr{F}} \ln\left(\frac{p'}{p}\right). \qquad (31.124)$$

Now change the applied potential $\mathscr{E}'$ so that a net cell reaction occurs and a net current flows through the cell. Suppose that the direction of current is such as to produce $H_2(g)$ at the pressure p'. At the hydrogen electrode H^+ ions are converted to $H_2(g)$, and because of the nonzero rate (which depends on the potential difference $\mathscr{E}' - \mathscr{E}$) there is a depletion of H^+ ions near that electrode. If the concentration of H^+ ions in the bulk of the solution is c_0, while the concentration near the electrode is c, then the number of ions arriving at the electrode per unit time, dn/dt, can be expressed as the flux of ions per unit area produced by the concentration gradient (Table 28.1) times the electrode's area A:

$$\frac{dn}{dt} = AD\frac{c_0 - c}{\delta}, \qquad (31.125)$$

where D is the diffusion coefficient and we have approximated the gradient as a concentration difference divided by some typical distance δ over which that difference exists. The current in the cell and its outside connections due to this flux of ions is

$$i = |Z|e\mathscr{F}\frac{dn}{dt}, \qquad (31.126)$$

where $|Z|e$ is the charge transferred in the electrode reaction per ion diffusing to the electrode.

The difference in H^+ concentration at the electrode and in the bulk necessarily corresponds to a potential difference η, called the *overpotential*; this is assumed to be given by the Nernst equation as

$$\eta = \frac{RT}{z\mathscr{F}} \ln\frac{c\gamma}{c_0\gamma_0}, \qquad (31.127)$$

where $z = 2$ and γ is an activity coefficient. We use the word "assumed" here, since we are applying an equation derived for equilibrium conditions to a cell in which an irreversible process is occurring. We can now combine Eqs. 31.125–31.127 to obtain

$$\eta = \frac{RT}{z\mathscr{F}} \ln\left[\frac{\gamma}{\gamma_0}\left(1 - \frac{i\delta}{Ac_0D|Z|e\mathscr{F}}\right)\right]. \qquad (31.128)$$

This gives a relation between the net current i in the cell and the overpotential η *in a diffusion-controlled reaction*. Since we established earlier that the potential applied to a cell must differ from the cell's electrochemical potential in order for an irreversible process (such as a net flow of current) to take place, we identify the overpotential with this difference between applied and cell potentials.

Next let us turn to the case in which a chemical reaction at an electrode is the rate-determining step. We again take the cell reactions to be those of a concentration cell,

$$M \rightarrow M^+(c_1) + e^- \qquad \text{(oxidation at anode)}$$
$$M^+(c_2) + e^- \rightarrow M \qquad \text{(reduction at cathode)},$$

with the two ionic concentrations c_1 and c_2. Let there be a net current flowing through the cell, in a direction such that there is a net deposit of M at the cathode. Therefore we expect the local concentration of M^+ to diminish from the bulk concentration c_2 to some lower value, say c_2'.

Hence there is a potential difference between the bulk of the solution and the surface of the cathode, which we assume can be written as

$$\frac{RT}{z\mathscr{F}}\ln\frac{c_2'}{c_2}.$$

The sum of this potential difference and a similar one at the anode (where c_1', the local concentration of M^+, will be larger than the concentration c_1 in the bulk of the solution around the anode) is the overpotential of the reaction. It is customary to split up the overpotential into two parts, $\alpha\eta$ and $(1-\alpha)\eta$, with the unsubstantiated assumption that α is independent of either cell current or overpotential. Therefore let $\alpha\eta$ be equal to the cathode overpotential,

$$\alpha\eta = \frac{RT}{z\mathscr{F}}\ln\frac{c_2'}{c_2}. \tag{31.129}$$

The current through the cell is written as

$$i = z\mathscr{F}kc_2', \tag{31.130}$$

where the rate of reaction is assumed to be kc_2', and z is the number of electrons transferred. On substitution of Eq. 31.129, we find that

$$i = z\mathscr{F}kc_2 e^{\alpha z\mathscr{F}\eta/RT}. \tag{31.131}$$

At equilibrium η is zero, and the current $z\mathscr{F}kc_2$ in one direction is balanced by an equal and opposite current, calculated in a similar way, at the other electrode (anode). Equation 31.131 can be rearranged to give

$$\ln i = a + b\eta, \tag{31.132}$$

which *is* a good empirical representation of the kinetics of many electrode reactions; it is called the *Tafel equation*. The "derivation" we have given is clearly of the hand-waving type, but has some value for thinking about irreversible processes in electrochemical cells.

Further Reading

Forst, W., *Theory of Unimolecular Reactions* (Academic Press, New York, 1973). — Unimolecular Reactions

Robinson, P. J., and Holbrook, K. A., *Unimolecular Reactions* (Wiley, New York, 1972).

Calvert, J. G., and Pitts, J. N., Jr., *Photochemistry* (Wiley, New York, 1966). — Photochemistry

Wayne, R. P., *Photochemistry* (Butterworths, London, 1970).

Dainton, F. S., *Chain Reactions: An Introduction*, 2nd ed. (Methuen, London, 1966). — Chain Reactions

Chance, B., Pye, E. K., Ghosh, A. K., and Hess, B. (Eds.), *Biological and Biochemical Oscillators* (Academic Press, New York, 1973). — Oscillatory Reactions

Winfree, A. T., "Rotating Chemical Reactions," *Sci. Am.* **230** (no. 6), 82 (June 1974).

Woodward, R. B., and Hoffmann, R., *The Conservation of Orbital Symmetry* (Academic Press, New York, 1970). — Symmetry Rules

Bell, R. P., *Acid-Base Catalysis* (Oxford University Press, London, 1941); *The Proton in Chemistry* (Cornell University Press, Ithaca, N.Y., 1959).

Bender, M. L., *Mechanisms of Homogeneous Catalysis from Protons to Proteins* (Wiley, New York, 1971).

Bender, M. L., and Brubacher, L., *Catalysis and Enzyme Action* (McGraw-Hill, New York, 1973).

Jencks, W. P., *Catalysis in Chemistry and Enzymology* (McGraw-Hill, New York, 1969).

Bond, G. C., *Catalysis by Metals* (Academic Press, New York, 1962).

Thomas, J. M., and Thomas, W. J., *Introduction to the Principles of Heterogeneous Catalysis* (Academic Press, New York, 1967).

Thomson, S. J., and Webb, G., *Heterogeneous Catalysis* (Oliver & Boyd, Edinburgh, 1968).

Conway, B. E., *Theory and Principles of Electrode Processes* (Ronald Press, New York, 1965).

Damaskin, B. B., *The Principles of Current Methods for the Study of Electrochemical Reactions* (McGraw-Hill, New York, 1967).

(More up-to-date reviews of these and other topics in chemical kinetics can be found in such periodical volumes as *Advances in Catalysis*, *Advances in Chemical Physics*, *Annual Review of Physical Chemistry*, *Organic Reaction Mechanisms*, etc.)

Homogeneous Catalysis

Heterogeneous Catalysis

Electrode Reactions

Problems

1. Discuss qualitatively how isotopic substitution affects unimolecular reactions. Consider changes in average vibrational level spacing and in zero-point energies. How will a heavier isotopic atom in the bond that breaks affect the quantities α and k_3 of Eq. 31.9?

2. The usual expression for k_{uni} assumes that the internal degrees of freedom relax to a statistical distribution infinitely rapidly on the time scale of dissociation. If this assumption does not hold, one can still write $k_{uni} = (A_0 + A_1 + \cdots)^{-1}$, where A_0^{-1} is the usual unimolecular rate coefficient (Eq. 31.3). The relative importance of the correction terms, which are monitored by evaluating A_1/A_0, decreases as s increases (at fixed E) and increases as E increases (at fixed s). Give a qualitative account of these trends.

3. Write the Hückel Hamiltonian matrices for reactants and products for the ring-opening reaction of *trans*-dimethylcyclopropyl ion,

or other isomer

determine the molecular orbitals and eigenvalues in each case. How does each molecular orbital transform under the symmetry operations C_2 and σ_v?

Determine whether this thermal ring-opening reaction proceeds via disrotatory or conrotatory motion.

4. Determine the rate equation for the radical polymerization of styrene ($S \equiv CH_2 = CHPh$), which is initiated by the thermal decomposition of azoisobutyronitrile, $R-N=N-R$ [where R is $(CH_3)_2CCN$]. The mechanism is

$$R-N=N-R \xrightarrow{k_i} 2R + N_2$$

$$R + S \xrightarrow{k_p} RS$$

$$RS + S \xrightarrow{k_p} RS_2, \text{ etc.}$$

$$RS_n + RS_m \xrightarrow{k_t} \text{recombination.}$$

Take k_p to be independent of the size of the polymer chain. For the typical values $k_i - 1.5 \times 10^{-4} \, s^{-1}$, $k_p = 10^3 \, M^{-1} \, s^{-1}$ and $k_t = 10^7 \, M^{-1} \, s^{-1}$, calculate the average chain length for $10^{-3} \, M$ $R-N=N-R$ and $1 \, M$ styrene monomer. How does the average chain length change as the reaction proceeds and the monomer is depleted?

5. Consider the following chain-reaction mechanism for the ortho–para hydrogen conversion:

initiation $\qquad M + H_2(o \text{ or } p) \xrightarrow{k_i} 2H + M$

propagation $\qquad H + H_2(p) \xrightarrow{k_p} H_2(o) + H$

termination $\qquad M + 2H \xrightarrow{k_t} H_2(o \text{ or } p) + M.$

Determine the rate of formation of o-H_2. Determine the apparent activation energy for the reaction. Point out why the apparent activation energy may be less than the activation energy for initiation.

6. For the Lotka–Volterra scheme with all forward rate coefficients equal and all reverse rate coefficients zero,

$$A + X \xrightarrow{k} 2X$$

$$X + Y \xrightarrow{k} 2Y$$

$$Y \xrightarrow{k} E,$$

determine the kinetic equations for [X] and [Y] and find the steady-state solution. Is this solution stable, unstable, or marginally stable? What will happen if this system is slightly perturbed from the stationary state? Obtain the equation for $d[Y]/d[X]$, and plot typical "trajectories" of $[X](t)$ and $[Y](t)$ in the [X], [Y] plane.

7. For the hypothetical reaction mechanism discussed on page 1223, set all the rate coefficients equal to 1 and obtain the expression for the stationary values of [X]. Determine the stationary-state concentrations and the stability of each state under the conditions $[A] = 8.5$, $[B] = 0.2$, and $\varepsilon = 30$.

8. The relative vibrational distribution of product molecules in an electronic-to-vibrational quenching process such as

$$Hg(^3P_1) + CO(v = 0) \rightarrow Hg(^3P_0) + CO(v = 1)$$

can be obtained from the square of the Franck-Condon overlap,

$$\left| \int_{-\infty}^{\infty} dr \, \psi_{i0}(r) \psi_{fv}(r) \right|^2,$$

of vibrational wave functions. Take the CO of the reactant excimer to be a harmonic oscillator with $r_{eq} = 1.32$ Å and wavenumber 2170 cm^{-1}; take the product CO to be a harmonic oscillator with $r_{eq} = 1.13$ Å and wavenumber 2170 cm^{-1}. Calculate the relative populations of product vibrational states 0 to 9. Explain qualitatively the form of the distribution obtained.

9. Read the article, "The Absorption of Light in Photosynthesis" by Govindjee and Rajni Govindjee, *Scientific American* **231** (no. 6), 68 (December 1974), and discuss how various aspects of photochemical kinetics come into play in photosynthesis.

10. Two articles treating the connection between photography and photochemistry are "The Chemistry of Color Photography" by Wayne C. Guida and Douglas J. Raber, *Journal of Chemical Education* **52,** 622 (1975) and "The Dynamic Interplay Between Photochemistry and Photography" by Samuel A. Forman, *Journal of Chemical Education* **52,** 629 (1975). Read these articles and discuss how photochemical kinetics and diffusion processes come into play in photography.

11. An article entitled "Reduction of Stratospheric Ozone by Nitrogen Oxide Catalysts from Supersonic Transport Exhaust" by Harold Johnston, *Science* **173,** 517 (1971) discusses catalytic reactions and photochemically induced reactions in the atmosphere. After reading this paper, derive Eq. 2 in the article, which is the rate equation for $[O_3]+[O]$. Also discuss how the author obtains Eq. 7, the zero-order approximations for the steady-state concentrations of O_3, O, NO_2, and NO.

12. For the Michaelis-Menten enzyme reaction, what is the maximum reaction rate for a given $[E_0]$ and initial $[S]_0$? What is $[S]$ when the rate is a maximum? What is $[S]$ when the rate equals one-half of its maximum value?

13. Consider an acid–base-catalyzed reaction that proceeds according to the following scheme:

$$
\begin{array}{ccc}
\text{HE} + \text{S} & \xrightarrow{k_1} & \text{HES}^{\ddagger} \longrightarrow \text{products} \\
K_E \updownarrow & & \updownarrow K_{ES}^{\ddagger} \\
\text{E} + \text{S} & \xrightarrow{k_2} & \text{ES}^{\ddagger} \longrightarrow \text{products}.
\end{array}
$$

Derive the expression for the observed rate constant in terms of k_1, k_2, K_E, $K_{ES}^{\ddagger}$, and $[H^+]$. Plot $\log k_{obs}$ versus pH for the parameters $k_1 = 10^7$ $M^{-1} s^{-1}$, $k_2 = 10^3 M^{-1} s^{-1}$, $pK_E = 5$, and $pK_{ES}^{\ddagger} = 9$.

14. A metal ion may act as a catalyst for several reasons. Among these are: (1) it increases reactivity of an atom by withdrawing electrons; (2) it can act as a bridging group between nucleophile and substrate; and (3) it can change the pK and thus the reactivity of the nucleophile. Suggest possible experiments to elucidate the mechanism and discover the role of a metal ion in a given catalytic reaction.

15. Consider a situation where a catalytic reaction on a metal surface takes place at a point defect and one contribution to the rate of reaction comes from the rate at which the reactants on the surface diffuse to the active site. Qualitatively discuss the differences between diffusion on a metal surface and diffusion in the gas or liquid phase. Also discuss how one might go about calculating the rate of migration on a surface.

16. The equilibrium constant K for adsorption of gas molecules on a surface can, of course, be written as $K = k_a/k_d$, where k_a is the rate of adsorption and k_d is the rate of desorption. Derive an expression for k_a by calculating the rate at which ideal gas particles strike the metal surface. Then use K and k_a to derive an expression for k_d. Finally, draw the energy along a one-dimensional reaction coordinate perpendicular to the surface for this model.

17. Read the article, "How Is Muscle Turned On and Off?" by Graham Hoyle, *Scientific American* **222** (no. 4), 84 (April 1970), and discuss how the various

concepts related to the kinetics of electrode reactions come into play in this biochemical process.

18. Discuss qualitatively the behavior of overpotentials that you would expect under the following conditions:
    (a) The polarizing current is interrupted.
    (b) The solution is stirred.
    (c) Different electrodes are used.
    (d) A sequence of ions with increasing charge is used (i.e., M^+, M^{2+}, etc.).
    (*Hint:* Consider changes due to the dependence of the diffusion coefficient on the parameters of the system and to salt effects on the reaction rate.)

Appendix I

Systems of Units

The measurement of physical and chemical properties requires the definition of units. Frequently the chosen units in a given investigation are first the commonly used ones, such as inches, pounds, and so on. The complexity and multiplicity of common units for weights and measures is analogous to that of any other aspect of culture, such as language. The need for easy comparison of measurements requires a common "language," an agreed-upon set of units. Just as an agreed-upon language evolves in time, so do agreed-upon units. The currently recommended units are called the *International System of Units* (abbreviated *SI*, for *Système International d'Unités*). This system is the result of much discussion and compromise. For a fuller description of the process of development of units, see M. L. McGlashan, *Physicochemical Quantities and Units*, 2nd ed. (Royal Institute of Chemistry, London, 1971), and *Ann. Rev. Phys. Chem.* **24,** 51 (1973); F. D. Rossini, *Fundamental Measures and Constants for Science and Technology* (CRC Press, Cleveland, 1974);

The International System of Units (SI)

Base Units

Quantity	Unit	Symbol	Definition (abridged)
Length	meter	m	1 650 763.73 wavelengths in vacuum of the radiation corresponding to the transition $2p_{10}$–$5d_5$ of the ^{86}Kr atom.
Mass	kilogram	kg	The mass of the international prototype of the kilogram (a piece of platinum-iridium kept in Sèvres, France).
Time	second	s	9 192 631 770 periods of the radiation corresponding to the transition between the hyperfine levels of the ^{133}Cs ground state.
Electric current	ampere	A	The current that produces a force of 2×10^{-7} newton per meter of length between two infinitely long, negligibly thick parallel conductors 1 meter apart in a vacuum.
Thermodynamic temperature	kelvin	K	The fraction 1/273.16 of the thermodynamic temperature of the triple point of water.
Amount of substance	mole	mol	The amount of substance of a system containing as many specified elementary entities (atoms, molecules, ions, etc.) as there are atoms in 12 grams of ^{12}C.
Luminous intensity	candela	cd	The perpendicular luminous intensity of a surface of 1/60 cm^2 of a black body at the freezing point of platinum (ca. 2042 K).

The International System of Units (SI), National Bureau of Standards (U.S.) Special Publication 330 (Washington, D. C., 1977); M. A. Paul, J. Chem. Soc. **11,** 3 (1971). The following tables list the essentials of the SI units, and give their relationship to some other commonly used units (including the older cgs and mks systems).

Prefixes

Fraction	Prefix	Symbol	Multiple	Prefix	Symbol
10^{-1}	deci	d	10	deca	da
10^{-2}	centi	c	10^2	hecto	h
10^{-3}	milli	m	10^3	kilo	k
10^{-6}	micro	μ	10^6	mega	M
10^{-9}	nano	n	10^9	giga	G
10^{-12}	pico	p	10^{12}	tera	T
10^{-15}	femto	f			
10^{-18}	atto	a			

Derived Units

Quantity	Unit	Symbol	Definition
Frequency	hertz	Hz	s^{-1}
Energy	joule	J	$kg\ m^2/s^2$
Force	newton	N	$kg\ m/s^2 = J/m$
Power	watt	W	$kg\ m^2/s^3 = J/s$
Pressure	pascal	Pa	$kg/m\ s^2 = N/m^2 = J/m^3$
Electric charge	coulomb	C	$A\ s$
Electric potential difference	volt	V	$kg\ m^2/s^3\ A = J/A\ s = J/C$
Electric resistance	ohm	Ω	$kg\ m^2/s^3\ A^2 = V/A$
Electric conductance	siemens	S	$s^3\ A^2/kg\ m^2 = \Omega^{-1}$
Electric capacitance	farad	F	$A^2\ s^4/kg\ m^2 = A\ s/V = C/V$
Magnetic flux	weber	Wb	$kg\ m^2/s^2\ A = V\ s$
Inductance	henry	H	$kg\ m^2/s^2\ A^2 = V\ s/A = Wb/A$
Magnetic flux density (magnetic induction)	tesla	T	$kg/s^2\ A = V\ s/m^2$

The SI is an extension of the *meter-kilogram-second (mks) system*. A parallel system of units formerly predominant in scientific work is the *centimeter-gram-second (cgs) system*, in which the indicated three units (cm, g, s) are used as base units. The cgs mechanical units with special names include:

Quantity	Unit	Symbol	Definition	Relation to SI Unit
Energy	erg	erg	$g\,cm^2/s^2$	$10^{-7}\,J$
Force	dyne	dyn	$g\,cm/s^2$	$10^{-5}\,N$
Viscosity	poise	P	$g/cm\,s$	$10^{-1}\,Pa\,s$
Kinematic viscosity	stokes	St	cm^2/s	$10^{-4}\,m^2/s$

The following named units are decimal multiples of SI or cgs units:

Quantity	Unit	Symbol	SI Definition	cgs Definition
Length	angstrom	Å	$10^{-10}\,m$	$10^{-8}\,cm$
Length	micron	μ	$10^{-6}\,m$	$10^{-4}\,cm$
Volume	liter	l	$1\,dm^3\,(=10^{-3}\,m^3)$	$10^3\,cm^3$
Mass	tonne	t	$10^3\,kg$	$10^6\,g$
Pressure	bar	bar	$10^5\,Pa$	$10^6\,dyn/cm^2$
Concentration	(molarity)	M	mol/dm^3	$10^{-3}\,mol/cm^3\,(=mol/liter)$

The degree Celsius (°C) is equal in magnitude to the kelvin, and may be used in either system for temperature *differences* or for the *Celsius temperature*, defined as $T-273.15\,K$, where T is the thermodynamic temperature (cf. Section 12.5).

Electric and magnetic units present a problem, since several inconsistent systems have been used, with even the basic equations (and thus the unit dimensions) differing between them. The SI is a so-called four-quantity system, with the ampere (on which other electromagnetic units are based) defined independently of the mechanical units; a four-quantity cgs system is also possible, for example, with the franklin (Fr), equal to the esu of charge, as the fourth unit. However, the usual practice was to use a three-quantity cgs system, defining electric and magnetic units in terms of mechanical units by suppressing the constants in either Coulomb's law (*electrostatic system*) or Ampere's law (*electromagnetic system*); the *Gaussian system* uses electrostatic units for electric quantities and electromagnetic units for magnetic quantities. Some of the units in these systems include:

Quantity	Unit	Definition	Relation to SI Unit
Electrostatic units:			
Electric charge	statcoulomb ("esu of charge")	$g^{1/2} cm^{3/2}/s$	3.33564×10^{-10} C
Electric current	statampere	statcoulomb/s	3.33564×10^{-10} A
Electric potential	statvolt	erg/statcoulomb	299.793 V
Electric resistance	statohm	statvolt/statampere	8.98758×10^{11} Ω
Dipole moment	debye (D)	10^{-18} statcoulomb cm	3.33564×10^{-30} C m
Electromagnetic units:			
Electric current	abampere ("emu of current")	$g^{1/2} cm^{1/2}/s$	2.99793×10^{9} A
Magnetic field strength	oersted (Oe)	abampere/cm	79.5775 A/m
Magnetic flux density	gauss (G)	abampere/cm	10^{-4} T
Magnetic flux	maxwell (Mx)	abampere cm	10^{-8} Wb

The following units belong to neither the SI nor any variation of the cgs system, but have been in sufficiently common use to require definition (a few of the major English-system units are included):

Quantity	Unit	Symbol	Value	Other Equivalents
Length	inch	in.	2.54 cm	
Length	bohr	(a_0)	5.291771×10^{-11} m	0.5292 Å $(\equiv \epsilon_0 h^2 / \pi m_e e^2)$
Mass	pound	lb	0.45359237 kg	
Mass	atomic mass unit	amu	1.660531×10^{-27} kg	$(\equiv 1\,g/N_A)$
Time	minute	min	60 s	
Time	hour	h	3600 s	60 min
Energy	calorie[a]	cal	4.184 J	
Energy	electron volt	eV	1.602192×10^{-19} J	$(\equiv e \times 1\,V)$
Energy	hartree	$(R_\infty hc)$	4.35981×10^{-18} J	27.2116 eV $(\equiv m_e e^4 / 8\epsilon_0^2 h^2)$
Energy/hc (wavenumber)	——	cm^{-1}	1.98648×10^{-23} J/hc	1.23985×10^{-4} eV/hc
Pressure	torr	Torr	133.3224 Pa	(1/760) atm ≈ 1 mm Hg
Pressure	atmosphere	atm	1.01325×10^{5} Pa	1.01325 bar = 760 Torr
Celsius temperature	degree Fahrenheit	°F	(5/9)°C	
Thermodynamic temperature	degree Rankine	°R	(5/9) K	$(T/°R = \theta/°F + 459.688)$

[a] This is the defined thermochemical calorie (cf. Section 13.7); note that several other "calories," with slightly different values, are also in use.

Appendix II

Partial Derivatives

Suppose that a given function $u(x, y)$ depends on the values of two independent variables, x and y, and that we are interested in just how u varies with x and y. Such a problem can be conveniently analyzed by separating the effects of the variables, that is, varying only one at a time. By a generalization of the process used to define the derivative in elementary calculus, we can define the *partial derivative* of u with respect to x at constant y:

$$\left(\frac{\partial u}{\partial x}\right)_y = \lim_{\Delta x \to 0} \frac{u(x + \Delta x, y) - u(x, y)}{\Delta x}. \tag{II.1}$$

The symbol ∂ (rather than d) is used for partial differentiation; a subscript (here y) is used to designate a variable held constant. Similarly, we have

$$\left(\frac{\partial u}{\partial y}\right)_x = \lim_{\Delta y \to 0} \frac{u(x, y + \Delta y) - u(x, y)}{\Delta y}. \tag{II.2}$$

The derivatives of various functions have the same form as in ordinary differentiation; for example, if $u(x, y) = x^2 y$, the partial derivatives are

$$\left(\frac{\partial u}{\partial x}\right)_y = y \frac{d}{dx}(x^2) = 2xy \quad \text{and} \quad \left(\frac{\partial u}{\partial y}\right)_x = x^2 \frac{d}{dy}(y) = x^2.$$

The extension to functions of three or more variables is obvious: the function $v(x, y, z)$ has the partial derivatives $(\partial v/\partial x)_{y,z}$, $(\partial v/\partial y)_{x,z}$, $(\partial v/\partial z)_{x,y}$. Second and higher derivatives can be taken in the usual way; one can easily show that

$$\left[\frac{\partial}{\partial x}\left(\frac{\partial u}{\partial y}\right)_x\right]_y = \left[\frac{\partial}{\partial y}\left(\frac{\partial u}{\partial x}\right)_y\right]_x = \frac{\partial^2 u}{\partial x\, \partial y}, \tag{II.3}$$

that is, that the order of differentiation is immaterial, as long as u and its derivatives are continuous.

As with ordinary differentiation, we can obtain a simple geometric interpretation of the partial derivative. A function of only one variable can be represented by a simple curve in a plane; this curve, so long as it is smooth, has at any point a single well-defined slope, equivalent to the derivative of the function. To plot all the possible values of a function like $u(x, y)$ we need a three-dimensional surface, as shown in Fig. II.1. Such a surface has an infinite number of slopes at a given point, corresponding to the infinite number of directions in which the slope can be taken. As can be seen from the figure, the derivatives defined in Eqs. II.1 and II.2 correspond to the slopes parallel to the x and y axes. More generally, for a function of n variables, the partial derivatives are equivalent to slopes on an $(n + 1)$-dimensional surface.

The partial derivatives we have defined correspond to slopes parallel to the principal axes representing the independent variables. However, it is possible and sometimes important to take slopes in other directions. For example, one might want to know how $u(x, y)$ varies when we impose

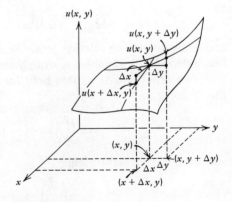

FIGURE II.1

Geometric interpretation of partial differentiation. The figure shows a portion of the surface $u(x, y)$. The partial derivatives $(\partial u/\partial x)_y$ and $(\partial u/\partial y)_x$ for given values of x and y are the slopes of this surface parallel to the x and y axes, respectively, at the point $u(x, y)$.

A5

the condition that $x = \alpha y^2$ at all times; this will be given by the slope along the curve $x = \alpha y^2$. The derivative of u in the direction corresponding to the steepest slope is called the *gradient* of u; it equals the magnitude of a vector (in the xy plane) whose x and y components are $(\partial u/\partial x)_y$ and $(\partial u/\partial y)_x$, respectively.[1]

How does $u(x, y)$ change when x and y vary simultaneously? For infinitesimal changes in x and y, the resulting infinitesimal change in u is clearly

$$du = \left(\frac{\partial u}{\partial x}\right)_y dx + \left(\frac{\partial u}{\partial y}\right)_x dy; \tag{II.4}$$

this expression is called the *total differential* of u. Suppose we vary x and y together in such a way that u remains constant; then $du = 0$ and we have

$$\left(\frac{\partial u}{\partial x}\right)_y dx + \left(\frac{\partial u}{\partial y}\right)_x dy = 0 \qquad (u = \text{const.}). \tag{II.5}$$

The ratio of dx to dy at constant u is simply $(\partial x/\partial y)_u$, so we can rearrange Eq. II.5 in the forms

$$\left(\frac{\partial x}{\partial y}\right)_u = -\frac{(\partial u/\partial y)_x}{(\partial u/\partial x)_y} \quad \text{or} \quad \left(\frac{\partial u}{\partial x}\right)_y\left(\frac{\partial x}{\partial y}\right)_u\left(\frac{\partial y}{\partial u}\right)_x = -1. \tag{II.6}$$

(The second form is easy to remember because the variables appear in cyclic order.) Note that u is as legitimate an independent variable as x or y: If we have an equation $f(u, x, y) = 0$ connecting the values of u, x, and y, we can solve for any one in terms of the other two.

Suppose that there is a set of related variables x, y, z, u, ... of which only two—*any* two—are independent. (An example would be the pressure, temperature, density, energy, ... of a fluid.) We can then, for example, express the variable u as a function of x and y, of x and z, of y and z, and so forth. By differentiating Eq. II.4 we can obtain

$$\left(\frac{\partial u}{\partial x}\right)_z = \left(\frac{\partial u}{\partial x}\right)_y + \left(\frac{\partial u}{\partial y}\right)_x\left(\frac{\partial y}{\partial x}\right)_z$$

(since $\partial x/\partial x = 1$, no matter what is held constant),

$$\left(\frac{\partial u}{\partial z}\right)_v = \left(\frac{\partial u}{\partial x}\right)_y\left(\frac{\partial x}{\partial z}\right)_v + \left(\frac{\partial u}{\partial y}\right)_x\left(\frac{\partial y}{\partial z}\right)_v, \tag{II.8}$$

and many other such relationships. Generalization of all this to more than two independent variables is straightforward.

Two other equations that hold for any combination of variables are the "chain rule,"

$$\left(\frac{\partial u}{\partial y}\right)_z = \left(\frac{\partial u}{\partial x}\right)_z\left(\frac{\partial x}{\partial y}\right)_z, \tag{II.9}$$

and the inverse relationship,

$$\left(\frac{\partial x}{\partial y}\right)_z = \frac{1}{(\partial y/\partial x)_z}; \tag{II.10}$$

we used the latter in deriving Eq. II.6.

---

[1] In three dimensions the gradient corresponds to the vector **grad** u or ∇u defined in footnote 9 on page 94.

Consider a general differential expression

$$df = M(x, y)\, dx + N(x, y)\, dy, \qquad (\text{II}.11)$$

where M and N are some functions of x and y. If this expression is actually the differential of some function f, it is called an *exact differential*. According to Eqs. II.3 and II.4, a differential of the form II.1 is exact if

$$\left(\frac{\partial M}{\partial y}\right)_x = \left(\frac{\partial N}{\partial x}\right)_y. \qquad (\text{II}.12)$$

For the two-variable case, an inexact differential can always be converted to an exact differential by multiplying by some *integrating factor*. For example, $2y\, dx + x\, dy$ is inexact; multiplication by x converts it to $2xy\, dx + x^2\, dy$, which is the exact differential of $x^2 y$.

We often have occasion to transform from one set of variables to another—to introduce new coordinates in mechanics, or new independent variables in thermodynamics. Given the equations relating the new to the old variables, it is a straightforward process to transform an algebraic expression; for an example, see the derivation of Eq. 3.125. Things are more difficult when one has to differentiate or integrate with respect to the variables in question.

Suppose that we want to transform from the variables $x, y, z, \ldots$ to the variables $u, v, w, \ldots$. Derivatives with respect to one of the old variables can be transformed by an equation of the form

$$\left(\frac{\partial}{\partial x}\right)_{y,z,\ldots} = \left(\frac{\partial u}{\partial x}\right)_{y,z,\ldots} \left(\frac{\partial}{\partial u}\right)_{v,w,\ldots} + \left(\frac{\partial v}{\partial x}\right)_{y,z,\ldots} \left(\frac{\partial}{\partial v}\right)_{u,w,\ldots} + \ldots, \quad (\text{II}.13)$$

which is simply the generalization of Eq. II.8 in operator form (cf. Section 3.4). For an example, consider the transformation from Cartesian coordinates x, y to circular polar coordinates r, ϕ. We must evaluate the operators

$$\left(\frac{\partial}{\partial x}\right)_y = \left(\frac{\partial r}{\partial x}\right)_y \left(\frac{\partial}{\partial r}\right)_\phi + \left(\frac{\partial \phi}{\partial x}\right)_y \left(\frac{\partial}{\partial \phi}\right)_r,$$
$$\left(\frac{\partial}{\partial y}\right)_x = \left(\frac{\partial r}{\partial y}\right)_x \left(\frac{\partial}{\partial r}\right)_\phi + \left(\frac{\partial \phi}{\partial y}\right)_x \left(\frac{\partial}{\partial \phi}\right)_r. \qquad (\text{II}.14)$$

From the transformation equations given with Fig. 3.9, we obtain

$$\left(\frac{\partial r}{\partial x}\right)_y = \frac{x}{(x^2+y^2)^{1/2}} = \frac{x}{r} = \cos\phi, \qquad \left(\frac{\partial r}{\partial y}\right)_x = \frac{y}{(x^2+y^2)^{1/2}} = \frac{y}{r} = \sin\phi,$$

$$\left(\frac{\partial \phi}{\partial x}\right)_y = \frac{-y}{x^2+y^2} = -\frac{y}{r^2} = -\frac{\sin\phi}{r}, \qquad \left(\frac{\partial \phi}{\partial y}\right)_x = \frac{x}{x^2+y^2} = \frac{x}{r^2} = \frac{\cos\phi}{r}.$$

$$(\text{II}.15)$$

Substitution in Eqs. II.14 then yields

$$\left(\frac{\partial}{\partial x}\right)_y = \cos\phi \left(\frac{\partial}{\partial r}\right)_\phi - \frac{\sin\phi}{r}\left(\frac{\partial}{\partial \phi}\right)_r,$$
$$\left(\frac{\partial}{\partial y}\right)_x = \sin\phi \left(\frac{\partial}{\partial r}\right)_\phi + \frac{\cos\phi}{r}\left(\frac{\partial}{\partial \phi}\right)_r; \qquad (\text{II}.16)$$

these expressions are used in Appendix 3B to obtain the Hamiltonian operator in circular coordinates.

When transforming integrals, one needs to know the "volume element" in the new variables. That is, for any quantity f that depends on the variables, one must know the factor J for which

$$\int_V f(x, y, \ldots) \, dx \, dy \ldots = \int_V f(u, v, \ldots) J \, du \, dv \ldots \qquad \text{(II.17)}$$

when the two integrals are taken over corresponding ranges of coordinates. The expression $J \, du \, dv \ldots$ has the same dimensions as $dx \, dy \ldots$, which for Cartesian coordinates is a volume element. We state without proof the rule giving the value of J:

$$J\left(\frac{x, y, \ldots}{u, v, \ldots}\right) = \begin{vmatrix} \left(\dfrac{\partial x}{\partial u}\right)_{v,w,\ldots} & \left(\dfrac{\partial x}{\partial v}\right)_{u,w,\ldots} & \cdots \\[2ex] \left(\dfrac{\partial y}{\partial u}\right)_{v,w,\ldots} & \left(\dfrac{\partial y}{\partial v}\right)_{u,w,\ldots} & \cdots \\[2ex] \multicolumn{3}{c}{\cdots\cdots\cdots\cdots\cdots} \end{vmatrix}; \qquad \text{(II.18)}$$

the determinant J is called the *Jacobian* of the transformation. For two-variable systems the Jacobian has the simple form

$$J(x, y/u, v) = \left(\frac{\partial x}{\partial u}\right)_v \left(\frac{\partial y}{\partial v}\right)_u - \left(\frac{\partial x}{\partial v}\right)_u \left(\frac{\partial y}{\partial u}\right)_v. \qquad \text{(II.19)}$$

For the transformation from Cartesian to circular coordinates, we have (again using Fig. 3.9)

$$\begin{aligned} J(x, y/r, \phi) &= \left(\frac{\partial x}{\partial r}\right)_\phi \left(\frac{\partial y}{\partial \phi}\right)_r - \left(\frac{\partial x}{\partial \phi}\right)_r \left(\frac{\partial y}{\partial r}\right)_\phi \\ &= (\cos\phi)(r\cos\phi) - (-r\sin\phi)(\sin\phi) \qquad \text{(3A.20)} \\ &= r(\cos^2\phi + \sin^2\phi) = r, \end{aligned}$$

and the area element corresponding to $dx\,dy$ is $r\,dr\,d\theta$. One can also deduce this result directly from a diagram, as shown in Fig. II.2. On the other hand, the volume element of Fig. 2B.2 could have been obtained by calculating

$$J(x, y, z/r, \theta, \phi) = r^2 \sin\theta \qquad \text{(II.21)}$$

from the transformation relations.

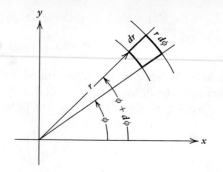

FIGURE II.2
Area element in circular polar coordinates:

$$dA = (dr)(r\,d\phi) = r\,dr\,d\phi.$$

Appendix III

Glossary of Symbols

Symbol	Meaning	Where used
A	Helmholtz free energy	general
	action	Part I; Eq. 2.25
	area	general
	frequency factor in rate coefficient	Part III; Eq. 30.42
A	operator for a classical property A	Part I; Sec. 3.4
$A(X)$	electron affinity of X	Eq. 14.46
A_∞	high-pressure frequency factor for unimolecular reactions	Part III; Eq. 30.176
$\mathscr{A}$	amplitude of perturbation	Part III; Eq. 30.129
a	molar Helmholtz free energy	Part II; general
	ionic radius	Eq. 26.69
	radius of Brownian particle	Eq. 29.49
$\mathbf{a}$	acceleration	general
a_i	activity of component i	Parts II and III; general
a_0	Bohr radius	general
$a_\pm$	mean ionic activity	Eq. 26.8
$\mathbf{a, b, c}$	primitive vectors of a crystal	p. 430
a, b, c	primitive distances in a crystal	p. 430
a_x, a_y, a_z	x-, y-, z-components of acceleration	general
B	second virial coefficient of a gas	Part II; Eq. 21.17
	second virial coefficient in the concentration expansion of the osmotic pressure of a solution	Part II; Eq. 25.127
	magnitude of the magnetic field strength $\mathbf{B}$	general
$\mathbf{B}$	magnetic field strength	general
B_e	rotational constant for a rigid rotor	Eq. 7.21
B_v	rotational constant for a molecule in the vibrational state designated by quantum number v	Eq. 7.21
b	impact parameter	general
	excluded volume term for the van der Waals equation of state	Table 21.2
	excluded volume in a lattice model of a mixture	Eq. 25.193
$C_p; C_v$	heat capacity at constant pressure; heat capacity at constant volume	general
C_σ	heat capacity along the liquid-vapor coexistence line	Eq. 24.19
C	the Coulomb; SI unit of charge	Table 1.2
C_2	symmetry operator for 180° rotation	Section 31.5

Symbol	Meaning	Where used
$C(R)$	direct correlation function of a liquid	Eq. 23.24
C	third virial coefficient of a gas	Eq. 21.17
	third virial coefficient in the concentration expansion of the osmotic pressure of a solution	Eq. 25.127
c	speed of light	general
	speed of sound	Section 28.10
c^*	"equivalent" concentration	Eq. 29.88
c_i	concentration of component i	general
c_{ij}	coefficient of the jth basis orbital in the expression for the ith molecular orbital	Part I; Eq. 8.17
c_p ; c_v	molar heat capacity at constant pressure; molar heat capacity at constant volume	general
c_v^x	contribution of x (for example, vibration) to the constant volume molar heat capacity	Eq. 22.44
D	dielectric constant ($\mathscr{D}$ in Section 30.8)	Eq. 26.58
	Debye unit	Eq. 4.29 ff
	(self-) diffusion coefficient	Table 28.1
D_k	(mutual) diffusion coefficient of component k	Eq. 20.15; App. 16A
D_0	dissociation energy measured from the zero point energy level	general
D_{0i}	energy required to dissociate a molecule of species i into atoms, measured from the zero point level	general
D_e	dissociation energy measured from the minimum in the potential well (well depth)	general
d	spacing between lattice planes	Fig. 11.7
	hard-sphere diameter	Part III; Fig. 27.7
$d(\mathbf{r}_2, \theta_{12} \| \mathbf{r}_1)$	conditional probability density for finding electron 2 at $\mathbf{r}_2, \theta_{12}$ when electron 1 is at $\mathbf{r}_1$	Part I; Eq. 5.17
E	energy	general
E	operator for energy	Part I; Eq. 3.32
$\mathbf{E}$	electric field strength	general; Eq. 1.8
$E(R)$	eigenvalue of the Schrödinger equation based on the Hamiltonian $H_{elec}(r, R)$ for fixed nuclei	Part I; Eq. 6.3
E_{rot}	rotational energy; an eigenvalue of the Schrödinger equation for rotational energy of a molecule	Eq. 7.19
E_{vib}	vibrational energy; an eigenvalue of the Schrödinger equation for vibrational energy of a molecule	Eq. 7.6
E_a	activation energy	Part III; Eq. 30.42
E_v	energy for formation of a vacancy	Eq. 29.82
E_∞	high-pressure activation energy for a unimolecular reaction	Part III; Eq. 30.176
ΔE_{reac}	energy change in a reaction	general
ΔE_0^0	net change in energy in conversion of reactants to products at $T = 0$	Eq. 21.152
$\Delta E_f^*, \Delta E_r^*$	barrier height in forward and reverse directions of reaction	Part III; Fig. 30.6
$\Delta E^\ddagger$	energy of activation (energy difference between activated complex and reactants)	Part III; Section 30.6
e	elementary electronic charge	general; Table 1.1

Symbol	Meaning	Where used
$\mathscr{E}$	electromotive force	Eq. 26.38
	electric potential	Table 13.1
$\mathscr{E}^{\theta}$	standard emf	Eq. 26.37
$\mathscr{E}_{1/2}$	half-cell potential	Part II; Eq. 25.566
F	structure factor	Eq. 11.28
	magnitude of the force $\mathbf{F}$	general
	tension	Part II; Table 13.1
$\mathbf{F}$	force	general
$F_{bonding}$	magnitude of the force bonding two nuclei	Part I; Eq. 6.8
$\mathscr{F}$	The Faraday constant	general
f	number of degrees of freedom of the macroscopic variables in a multicomponent multiphase system	Part II; Chapter 24
	flux density of a beam	Eq. 2.13; Eq. 30.86
$f(\mathbf{r}, \mathbf{v}, t)$	velocity and space distribution function; number of molecules per unit "volume" in phase space at time t	Part III; Section 28.1
$f(\mathbf{r}, t)$	density of molecules in configuration space	Part III; Eq. 28.14
$f(\mathbf{v}, t)$	density of molecules in velocity space	Part III; Eq. 28.2
$f(\mathbf{r}, \mathbf{v}, t), \mathscr{f}(\mathbf{r}, t)$ $\mathscr{f}(\mathbf{v}, t)$	probability densities corresponding to $f(\mathbf{r}, \mathbf{v}, t)$, $f(\mathbf{r}, t)$ and $f(\mathbf{v}, t)$, $\mathscr{f}(\mathbf{r}, \mathbf{v}, t) \equiv 1/N\, f(\mathbf{r}, \mathbf{v}, t)$, etc. Note that $\mathscr{f}(\mathbf{v}, t)$ corresponds to the $f(\mathbf{v})$ of Section 19.7	Part III; Eqs. 28.4 ff
$f(\mathbf{v})$	equilibrium velocity distribution function	Part II; Eq. 12.8
$f(v)$	equilibrium speed distribution function	Part II; Eq. 12.10
$f(\omega)$	distribution of orientations of a molecule	Eq. 24.96
$f(t)$	randomly fluctuating force on a Brownian particle	Eq. 29.34
f_i^0	fugacity of pure component i	Eq. 21.14
f_i	fugacity of component i in a mixture	Eq. 21.37
f_s	atomic scattering factor	Part I; Eq. 11.26
$f(\epsilon)$	one-particle energy distribution function	general; Eq. 11.56
$f(j, t)$	number of oscillators in quantum state j at time t	Part III; Eq. 28.15
G	Gibbs free energy	general
$G(k)$	density of momentum states	Part I; Eq. 11.52
g	molar Gibbs free energy	Part II; general
	acceleration due to gravity	general; Eq. 1.22
g_m	mean molar Gibbs free energy of a mixture	Chapter 25
g_k	degeneracy of the level with energy E_k; the number of states with energy in the range $E_k \le E \le E_k + dE$	Eq. 19.115
g_1	osmotic coefficient	Part II; Eq. 25.125
g_s	Landé magnetic g factor	Part I; Eq. 5.7
g_{eff}	effective magnetic g-factor	Part I; Eq. 9.15
$g(\epsilon)$	density of one-electron energy states	Eq. 11.54
$g_2(\mathbf{R}_1, \mathbf{R}_2)$	pair correlation function of a liquid	Eq. 11.2; Eq. 23.15
$g_3(\mathbf{R}_1, \mathbf{R}_2, \mathbf{R}_3)$	triplet correlation function of a liquid	Eq. 23.23
$g(t)$	autocorrelation function	Part III; Eq. 29.58
$g_v(t)$	velocity autocorrelation function	Part III; Eq. 29.70

Symbol	Meaning	Where used
H	enthalpy	general
H	Hamiltonian operator	general; Eq. 3.402
H_{AB}	integral of the Hamiltonian over the orbital functions $\varphi_A{}^*$ and φ_B	Part I; Eq. 6.1
H_{HF}	Hartree-Fock Hamiltonian operator	Part I; Eq. 8.16
H_{nuc}	Hamiltonian operator for motion of the nuclei of a molecule in the effective potential of the electrons $E(R)$	Eq. 6.1
H_{elec}	Hamiltonian operator for electronic energy and nuclear potential energy	Eq. 6.1
$H_n(z)$	nth Hermite polynomial	Eq. 4.9
H	magnetic field strength	Table 13.1
h	Planck's constant ($\hbar \equiv h/2\pi$)	general
h	molar enthalpy	general
h, k, l	Miller indices	Part I; Eq. 11.3
h_1	1s hydrogenic orbital on atom 1	Part I; p. 360
h_m	mean molar enthalpy of a mixture	Chapter 25
Δh_{fus}	molar heat of fusion	general
Δh_{vap}	molar heat of vaporization	general
Δh_{subl}	molar heat of sublimation	general
$\bar{h}_i$	partial molar enthalpy of component i	general
$\Delta \bar{h}_i \equiv \bar{h}_i$	partial molar enthalpy of component i	only in Section 14.6
ΔH	heat of formation	general
$\Delta H°$	standard heat of formation per mole of product	general
$\Delta H^0_{atom\,'n}$	enthalpy change on atomization of one mole of a compound under standard state conditions	Section 14.8
$\Delta H_{sol\,'n}$	integral heat of solution	Section 14.6
ΔH_∞	enthalpy change when one mole of solute is added to an effectively infinite amount of solvent	Section 14.6
ΔH_{dil}	integral heat of dilution	Section 14.6
I	intensity of a beam	Eq. 2.8
	electric current	general
	ionic strength	Eq. 26.49
	moment of inertia of diatomic molecule	Eq. 3.129
	nuclear spin quantum number	Eq. 18.15
I_x, I_y, I_z	principal moments of inertia of a polyatomic	Eq. 3.173
I_{IR}	moment of inertia for internal rotation	Eq. 21.129
I_0	intensity of incident light	Eq. 31.22
$I(M)$	ionization potential of M	Eq. 14.46
i, j, k	unit vectors in the x-, y-, z-directions	Eq. 3.54
i	net current in an electrolytic cell	Eq. 31.126
$\mathscr{I}(x, t)$	intensity of a wave at the point x at time t	Eq. 3.47
J	total angular momentum	general, Eq. 5.18
J	rotational quantum number	general. Eq. 7.19
	cell partition function	Part II; Eq. 25.207
J	Joules	general

Symbol	Meaning	Where used
J_{AB}	Coulomb integral over atoms A and B	Eq. 6.40
$J(x, y, .../u, v, ..)$	Jacobian of the transformation from variables $x, y, ...$ to variables $u, v, ...$	Eq. 11.18
$\mathbf{J}^Q(\mathbf{r}, t)$	vector flux of Q at $\mathbf{r}, t$	Part II; Eq. 20.2
$\mathbf{J}^m(\mathbf{r}, t)$	mass flux at $\mathbf{r}, t$	Part II; Eq. 20.6
$\mathbf{J}_k^m(\mathbf{r}, t)$	mass flux of component k, relative to the center of mass, at $\mathbf{r}, t$	Part II; Eq. 20.13
$\mathbf{J}^p$	momentum flux	Part II; Eq. 20.21
J_{ij}^p	component ij of the momentum flux tensor	Part II; Eq. 20.23
j	rotational quantum number (usually J in Parts I and II)	Part III; general
$\mathbf{j}$	electric current density	Part II; Eq. 29.87
$K(T)$	(equilibrium constant for a gas phase reaction independent of pressure)	Eq. 21.64
K_p	equilibrium constant in terms of component pressures	Eq. 21.75
K_v	equilibrium constant in terms of component concentrations	Eq. 21.88
K_x	equilibrium constant in terms of component mole fractions	Eq. 21.86
$K(T, p, \text{solvent})$	equilibrium constant for a reaction in a condensed phase	Eq. 25.131
K_a	equilibrium constant in terms of component activities	Eq. 30.84
K_w	ion product of water ($\equiv [H_3O^+][OH^-]$)	Eq. 31.93
$K_v^{\ddagger}$	value of K_v for formation of an activated complex ($K^{\ddagger}$, Eq. 30.79, is the corresponding equilibrium constant in terms of activities)	Part III; Eq. 30.44
K_{AB}	exchange integral over atoms A and B (sometimes K)	Eq. 6.40
K_b	boiling point constant	Part II; Eq. 25.111
K_f	freezing point constant	Part II; Eq. 25.114
k	force constant of a harmonic oscillator	Eq. 2.32
	wave number ($\equiv 2\pi/\lambda$)	Eq. 3.3
	rate coefficient of a chemical reaction	Part III; general
$\mathbf{k}$	wave vector	general
	unit normal to a surface	general
k_B	Boltzmann's constant	general
$k_2(T, p, \text{solvent})$	Henry's law constant	Part II; Eq. 25.60
k_{α}	force constant for the αth normal mode of vibration of a molecule	Part I; Eq. 8.38
k_a, k_b	rate coefficients for acid and base catalysis	Part II; Eq. 31.92
k_D	diffusion-controlled limit of rate coefficient	Part III; Eq. 30.105
k_f, k_r	rate coefficients for forward and reverse reactions	Part III; general
k_{uni}	effective unimolecular rate coefficient	Part III; Eq. 30.173
k_{∞}	high-pressure limit of k_{uni}	Part III; Eq. 30.174
$k_1, k_2, ...$	rate coefficients for steps $1, 2, ...$ in a reaction mechanism (k_{-1} is the coefficient for the step inverse to step 1)	Part III; general

Symbol	Meaning	Where used
$\mathbf{L}$	orbital angular momentum	Eq. 3.151
L	magnitude of angular momentum $\mathbf{L}$	Eq. 2.31
L^2	operator for the square of the angular momentum $\mathbf{L}$	Eq. 3.176
L_z	operator for the z-component of the angular momentum $\mathbf{L}$	Eq. 3.176
$\mathbf{L}_i$	orbital angular momentum of the ith electron	Eq. 5.20
l	length	general
	mean free path	Part III; Eq. 28.63
	orbital angular momentum quantum number	Eq. 3.162
M	Madelung constant	Eq. 11.41
	total mass (in a system of two or more particles)	general
$M(T)$	chemical potential along the critical isochore	Part II; Eq. 24.67
M_i	molecular weight of component i	Eq. 14.14
$\mathbf{M}$	magnetic moment	Table 13.1
M_L	quantum number for the z-component of the total orbital angular momentum $\mathbf{L}$	Eq. 5.22
M_S	quantum number for the z-component of the total spin angular momentum $\mathbf{S}$	Eq. 5.22
m	mass	general
m_e, m_n, m_p	mass of the electron, neutron, proton	Table 1.1
m_{eff}	effective mass	Eq. 11.95
m_i	molality of component i	general
$m_\pm$	mean ionic molality	Eq. 26.8
$\mathfrak{m}_1$	molality of pure solvent	Part II; Eq. 25.137
m	quantum number of angular momentum for a particle in a circular box, or for the z-component of angular momentum of an electron	Eq. 3.144; Eq. 3.161
m_s	spin quantum number; the z-component of the spin angular momentum of an electron	Eq. 5.2
m_{li}, m_{si}	quantum number for the z-components of the orbital and spin angular momentum of electron i	Eq. 5.22
N	number of molecules in a system	general
N_A	Avogadro's number	general
$\mathcal{N}$	number of replicas in an ensemble	Part II; Chapter 15
$N(E_n)$	number of states with energy E_n	Part I; Eq. 4.17
N_v	number of vacancies in a lattice	Eq. 29.82
n	number density	general
	number of moles	general
	integer, a general quantum number molecularity of a reaction	Eq. 2.53 Part III; Eq. 30.69
n_i	number of moles of component i	general
	principal quantum number for the ith one-particle quantum state	Part I; Eq. 2.10
n_i^*	effective principal quantum number for the ith one electron quantum state	Part I; Eq. 2.12
$n(v)$	number density of molecules with speed v	Part III; Section 27.5
$P(x)$	probability distribution of x	general

Symbol	Meaning	Where used
$\mathscr{P}(x)$	probability density of x	general
$\mathscr{P}$	excess pressure in a perturbation	Part III; Eq. 28.123
P_{12}	operator indicating permutation of particles 1 and 2	Eq. 6.53
p	pressure	general
	magnitude of momentum $\mathbf{p}$	general
	probability aftereffect factor	Part II; Eq. 16.A.12
	molecular orbital bond order	Part I; Eq. 9.4
	reaction probability factor	Part III; Eq. 30.27
$\mathfrak{p}$	unit of pressure	Part II; Eq. 21.12
$\mathbf{p}$	momentum	general
p	momentum operator	general, Eq. 3.36
$\mathbf{p}_\theta, \mathbf{p}_\phi$	angular momentum	general; Eq. 3.128
p_ϕ	angular momentum operator	general; Eq. 3.138
p_{app}	applied pressure	general
p_σ	vapor pressure of a liquid along the liquid-vapor coexistence line	Eq. 23.1
p^*	reduced pressure (with respect to pair potential parameters)	Eq. 21.178
$\bar{p}$	reduced pressure (with respect to the critical pressure)	Chapter 21
p, q	probabilities of positive and negative displacements in Brownian-motion model	Part III; Section 29.2
	numbers of protons on acid and proton-acceptor sites on conjugate base	Part III; Eq. 31.95
Q	electric charge	general
	ratio of activities in an electrochemical-cell reaction	Part III; Eq. 31.121
Q_α	αth normal coordinate	Eq. 8.38
Q_N	canonical partition function for N particles in a volume V at temperature T	general; Eq. 19.31
q	heat absorbed by a system	general
	spatial coordinate	Part I; Eq. 3.98
	electric charge	Eq. 1.5
$\mathbf{q}$	quadrupole moment	Part I; Eq. 10.6
	heat flux vector	Part II; Eq. 20.64
$q_i(\mathbf{R})$	charge density around ion i	Part II; Eq. 26.74
q_1	single-molecule partition function per unit volume	Parts II and III; Eq. 21.148, Eq. 30.52
R	gas constant	general
	interatomic or intermolecular distance	general
	electrical resistance	Eq. 29.86
R_e	internuclear distance for which $E(R)$ is a minimum	general
R_H	Rydberg constant for hydrogen	Eq. 2.7
R_∞	Rydberg constant for a hydrogenic atom with infinite nuclear mass	Eq. 2.69
$R_{nl}(r)$	solution of the radial wave equation for the hydrogen atom	Eq. 4.18

Symbol	Meaning	Where used
R^*	reduced distance (with respect to pair potential parameters)	Eq. 26.90
$\mathscr{R}$	excess density in a perturbation	Part III; Eq. 28.124
r	radial distance; radial polar coordinate	general
$\mathbf{r}$	position vector	
r^*	critical distance for a diffusion-controlled reaction	Part III; Eq. 30.96
r_{kl}	distance between electrons k and l	Eq. 8.14
r_{Kk}	distance between nucleus K and electron k	Eq. 8.14
r_{max}	distance at which the electronic probability density has its maximum	Eq. 4.23
S	entropy	general
	screening constant	Part I; Eq. 2.6
$\mathbf{S}$	spin angular momentum	general; Eq. 5.1
$\mathbf{S}_i$	spin angular momentum of the ith electron	general; Eq. 5.20
S_{AB}	overlap integral over atoms A and B	general; Eq. 6.30
s	molar entropy	general
	number of internal oscillators in a molecule	Part III; Section 31.1
s_m	mean molar entropy of a mixture	Chapter 25
$\bar{s}_i$	partial molar entropy of component i	
$\mathbf{s}$	scattering vector	Eq. 23.16
T	absolute temperature	general
T	kinetic energy operator	general
T_e	Kinetic energy operator of a molecule when $R = R_e$	Part I; p. 212
T_{vap}, T_{fus}	temperature of vaporization, fusion	general
T^*	reduced temperature (with respect to pair potential parameters)	Eq. 21.178
$\tilde{T}$	reduced temperature (with respect to the critical temperature)	Chapter 21
$\mathscr{T}$	excess temperature in a perturbation	Part III; Eq. 28.125
t	time	general
	centigrade (Celsius) temperature	general
t^*	limiting zero-pressure temperature on the centigrade scale	Eq. 12.34
$t_{1/2}$	half-life for a first-order reaction	Part III; Eq. 30.8
U	internal energy	general
	potential energy of an ionic crystal	Part I; Eq. 11.39
u	molar internal energy	general
	magnitude of the mass velocity (local average velocity)	Part III; Section 28.10
	mobility of a charged particle	Part III; Eq. 29.53
$\mathbf{u}$	velocity	App. 2C
u_m	mean molar internal energy of a mixture	general
$u_2(R)$	pair interaction potential	Part II; general
$u_3(\mathbf{R}_1, \mathbf{R}_2, \mathbf{R}_3)$	potential energy of three molecules	Part II; Eq. 21.159

Symbol	Meaning	Where used
V	volume	general
	electric potential difference	general
$V(r), V(R)$	potential energy as a function of r, R	general
V	potential energy operator	general, Eq. 3.40b
V^*	reduced volume (with respect to pair potential parameters)	Eq. 21.178
$\tilde{V}$	reduced volume (with respect to the critical volume)	Chapter 21
$\mathbf{V}$	center-of-mass velocity	Eq. 27.14
$\mathbf{v}(\mathbf{r}, t)$	local fluid velocity at $\mathbf{r}$, t	Eq. 20.6
$\mathscr{V}_N$	total potential energy of N molecules	Eq. 21.157
$V_{ind}(R)$	induction energy of interaction between charge distributions	Eq. 10.16
V_f	free volume in solution	Part III; Eq. 30.78
v	molar volume	general
	quantum number for vibrational energy	general
	volume per unit mass	Chapter 20
	speed (magnitude of velocity $\mathbf{v}$)	general
$\bar{v}_i$	partial molar volume of component i	Eq. 25.24
v_m	mean molar volume of a mixture	Eq. 25.17
v_m^E	excess volume per mean mole of mixture	Eq. 25.18
v_x, v_y, v_z	x-, y-, z-components of velocity $\mathbf{v}$	general
$v_{eff}(k)$	effective velocity	Part I; Eq. 11.94
w	work done on a system (sometimes W)	general
	parameter in regular solution model	Part II; Eq. 25.29
w_{ad}	adiabatic work	general
w_2	pair potential of average force in a liquid	Eq. 23.18
X	intensive generalized force	Eq. 13.13
x, y, z	Cartesian coordinates	general
x_i	mole fraction of component i	general
Y	extensive generalized displacement	Eq. 13.13
$Y_{lm}(\theta, \phi)$	spherical harmonic	Eq. 3.166
y_i	mole fraction of component i in the vapor over a liquid (two-phase, two-component system)	Eq. 25.20
Z	atomic number	general
Z_i	number of elementary charges on ion i	general
Z_N	configurational partition function for N molecules in a volume V at temperature T	Eq. 21.C.5
Z_{12}	collision frequency between species 1 and 2	Eq. 28.18
z	number of electrons transferred in an electro-chemical reaction	Chapters 26, 31
α	coefficient of thermal expansion	general
	polarizability	general
	diagonal element of the energy matrix of the Hückel Hamiltonian	Part I; Eq. 9.6
	critical exponent for the temperature dependence of the heat capacity near the critical temperature	Part II; Eq. 24.36

Symbol	Meaning	Where used
α	attenuation constant for sound waves	Part III; Eq. 28.147
	proportionality constant in Hinshelwood-Lindemann mechanism	Part III; Eq. 31.9
	net number of chain carriers produced in a branching step	Part III; Eq. 31.50
$\alpha(1), \beta(1)$	spin functions for electron 1, with $m_s = \frac{1}{2}$ and $-\frac{1}{2}$, respectively	general; Eq. 6.47
β	$1/k_B T$	general
	nearest-neighbor off-diagonal element of the energy matrix of the Hückel Hamiltonian	Part I; Eq. 9.7
	critical exponent for the temperature dependence of the density near the critical temperature along the coexistence line	Part II; Eq. 24.33
γ	ratio of heat capacities, c_p/c_v	general
	critical exponent for the temperature dependence of the compressibility near the critical temperature along the isochore $\rho = \rho_c$, $T > T_c$	Part II; Eq. 24.37
$\gamma_i, \gamma_i', \gamma_i''$	activity coefficient of component i, mole fraction scale, molality scale, molarity scale, respectively	general; Eq. 25.19 Eq. 25.136; Eq. 25.139
$\gamma_\pm$	mean ionic activity coefficient	general; Eq. 26.8
γ_v	$(\partial p/\partial T)_v$	Eq. 23.6
γ_G	Grüneisen constant	Eq. 22.26
$\Gamma(E, V, N)$	total number of quantum states with energy less than E for given N, V	Eq. 15.12
$\Gamma(x)$	Gamma function of x	Eq. 21.170
$\Gamma_n, \Gamma_m, \Gamma_E$	flux densities for numbers of molecules, mass, and energy crossing a plane	Part III: Eqs. 28.65 ff
$\boldsymbol{\Gamma}_n, \boldsymbol{\Gamma}_m, \boldsymbol{\Gamma}_E$	the flux density vectors corresponding to Γ_n, Γ_m and Γ_E ($\Gamma_n = \mathbf{k} \cdot \boldsymbol{\Gamma}_n$, etc.)	Eqs. 28.66 ff
$\boldsymbol{\Gamma}_{mv}$	flux density for momentum: pressure tensor (or stress tensor)	Part III; Eq. 28.73
δ	quantum defect	Part I; Eq. 2.12
	critical exponent for the density dependence of the pressure near the critical density along the critical isotherm	Part II; Eq. 24.35
	range of concentration gradient near an electrode	Part III; Eq. 31.125
δ_{ij}	Kronecker delta	general
$\delta(x)$	delta function	general
Δ	volume of a lattice cell	Eq. 25.191
Δ_{r-1}	determinant of the derivatives $(\partial^2 g_m/\partial x_i \partial x_j)$, $i, j = 2, \ldots, r$	Eq. 25.169
$\Delta_3(1, 2, 3)$	excess of the energy of three molecules over the sum of the pair interaction energies	Eq. 21.159
ϵ	energy parameter (well depth) in the intramolecular potential	general; Eq. 10.23
ε	energy per molecule	general
ϵ_0	permittivity of free space	general; Eq. 1.5
ϵ_F	Fermi energy	Eq. 11.58

Symbol	Meaning	Where used
$\epsilon_{n_1 n_2 n_3}$	energy of a molecule in a state with quantum numbers n_1, n_2, n_3	general
ε_{ijk}	Levi-Civita symbol	Eq. 20.A.1
ζ	spin-orbit coupling constant	Eq. 5.24
	bulk viscosity	Eq. 20.46
	friction coefficient	Eq. 29.34
η	shear viscosity	general; Eq. 1.21
	ground state degeneracy	Part II; Eq. 18.14
	critical exponent for the distance dependence of the pair correlation function near the critical temperature	Part II; Eq. 24.102
	overpotential	Part III; Eq. 31.127
θ	gas scale temperature	Eq. 12.29
	empirical temperature	Chapter 12
	Lindemann ratio	Eq. 24.98
	angular polar coordinate	general
	fraction of surface sites covered by adsorbed molecules	Part III; Eq. 31.100
θ^*	gas scale temperature in the limit of zero gas pressure	Eq. 12.31
$\Theta(\theta)$	factor of the wave function, for a particle in a spherically symmetric potential which contains the θ dependence of that function	Eq. 3.155
$\Theta_{rot}, \Theta_{vib}, \Theta_{elec}$	characteristic temperature for rotational, vibrational and electronic excitation, respectively	Eq. 21.134, Eq. 21.138, Eq. 21.141
Θ_E, Θ_D	characteristic temperature of the Einstein and Debye models of a solid, respectively	Chapter 22
κ	Debye screening length	Part II; Eq. 26.67
	transmission coefficient in activated-complex theory	Part II; Eq. 30.49
κ_T	isothermal compressibility	general
κ_S	adiabatic compressibility	general
κ_f, κ_r	theoretical rate coefficients for forward and reverse reactions	Part III; Eq. 30.34
λ	wavelength	general
	quantum number for orbital angular momentum of an electron along the axis of a linear molecule	general; Eq 6.43
	absolute activity	Part II; Eq. 21.A.5
	thermal conductivity	general
	progress variable	Eq. 19.95
	length parameter in theory of diffusion-controlled reactions	Part III; 30.101
λ_{ij}	parameter in the Born-Mayer repulsion between ions i and j	Eq. 14.33
Λ	quantum number for total orbital angular momentum along the axis of a linear molecule	general
	molar conductance	Part III; Eq. 29.93
μ	reduced mass	general; Eq. 2.74
μ, μ_e	electric dipole moment	Eq. 4.29

Symbol	Meaning	Where used
μ_0	magnetic permittivity of free space	general; Chapter 18
$\boldsymbol{\mu}_l$	orbital magnetic moment	Eq. 5.4
μ_N	nuclear magneton	Eq. 5.9
μ_B	Bohr magneton	Eq. 5.6
$\boldsymbol{\mu}_m$	magnetic dipole moment	Eq. 5.3
$\boldsymbol{\mu}_s$	spin magnetic dipole moment	Eq. 5.7
$\boldsymbol{\mu}_{n'n}$	transition dipole moment between states characterized by quantum numbers n' and n	Eq. 4.33
μ_i	chemical potential of component i	general
$\mu_i^{\,0}$	chemical potential of the reference state of component i (the reference state is the pure liquid or solid)	general; Eq. 25.5
$\mu_i^{\,\theta}$	chemical potential of the reference state of component i (the reference state is a hypothetical substance having the same properties as does the solute at infinite dilution)	general; Eq. 25.23
$\mu_i^{\,E}$	excess chemical potential of component i	Eq. 25.13
μ_{JT}	Joule-Thomson coefficient	Prob. 21.7
ν	frequency in cycles per second (Hz)	general
	number of degrees of freedom of a system	Part II; Chapter 15
$\tilde{\nu}$	wavenumber	general
$\tilde{\nu}_e$	first coefficient in the power series expansion of E_{vib} as a function of the quantum number v (approximately equal to the fundamental vibration frequency in cm^{-1})	Eq. 7.10
$\tilde{\nu}_e x_e$	second coefficient in the power series expansion of E_{vib} as a function of the quantum number v	Eq. 7.10
ν_i	stoichiometric coefficient of the ith species in a chemical reaction	general
ν_+, ν_-	numbers of positive and negative ions in a salt	Eq. 26.6
ξ	distance scaled to dimensionless units	Part I; Eq. 3.65
	extent of reaction ($\dot{\xi}$, rate of reaction)	Part III; Eq. 30.2
	scale factor in the geometric mean combining law for mixed species pair interaction	Part II; Eq. 25.211
	correlation length in the critical region of a fluid	Part II; Eq. 24.102
ξ_i	ith normal coordinate	Part II; Eq. 24.84
Ξ	grand partition function	general; Eq. 19.126
π_g, π_u	designations of orbitals of a linear molecule with quantum numbers $\lambda = 1$ and with even and odd parity, respectively	general; Fig. 6.8
π	osmotic pressure	general; Eq. 25.121
Π	designation of an electronic state of a linear molecule for which $\Lambda = 1$	general; p. 287
ρ	mass density	general
	Hammett reaction parameter	Part III; Eq. 30.112
	scale length in Born-Mayer repulsion between ions	Eq. 14.34
$\tilde{\rho}$	reduced density (with respect to the critical density)	Chapter 21
$\rho(\mathbf{r}, t)$	local mass density	Part III; Eq. 28.112
$\rho_Q(\mathbf{r}, t)$	local density of Q at point $\mathbf{r}$ and time t	Eq. 20.1

Symbol	Meaning	Where used
$\rho_k(\mathbf{r}, t)$	mass density of component k at $\mathbf{r}$, t	Eq. 20.12
ρ_ε	local density of potential energy in a fluid	Eq. 20.64
σ	total scattering cross section	general; Eq. 2.9
	size parameter in the pair interaction between molecules	Eq. 10.23
	electrical conductivity	Eq. 29.86
	time interval	Part III; Eq. 29.70
	Hammett substitution parameter	Part III; Eq. 30.112
	symmetry number	Eq. 21.111
	surface tension	Table 13.1
$\boldsymbol{\sigma}$	stress tensor	Part II; Eq. 20.43
σ_g, σ_u	designations of orbitals of a linear molecule with quantum number $\lambda = 0$ and with even and odd parity, respectively	general, Fig. 6.8
σ_v	symmetry operator for reflection through a plane	Section 31.5
$\sigma(v)$	total scattering cross section	Part III; Eq. 27.62
$\sigma(v, \alpha, \beta)$	differential scattering cross section in laboratory coordinates	Eq. 27.53
$\sigma(v, \chi)$	differential scattering cross section, in relative (center-of-mass) coordinates	Eq. 23.54
$\sigma_R(v)$	cross section for reactive scattering	Eq. 30.29 ff
Σ	designation for an electronic state of a linear molecule for which $\Lambda = 0$	general; p. 232
τ	period	general; Eq. 2.38
	relaxation time	general; Eq. 28.111
	effective lifetime of photoexcited state	Part III; Eq. 31.27
	scattering volume	Section 27.5
τ_f	mean free time	Fig. 28.3
ϕ	angle; azimuthal polar coordinate	general
	work function for electron emission	Part II; Eq. 26.55
	phase shift	Part III; Eq. 28.156
	electric potential	Eq. 30.90
ϕ_A	volume fraction of component A	Eq. 25.20
φ_A, φ_B	atomic orbitals centered on nuclei A, B	Eq. 6.18
Φ	quantum yield for a photochemical reaction	Eq. 31.25
Φ_i	electrostatic potential acting on ion i	Eq. 26.61
Φ_A, Φ_B	wave functions for molecules A, B	Eq. 11.35
$\Phi(\phi)$	factor of the wave function for a particle in a spherically symmetric potential which contains the dependence of that function on the angle ϕ	Eq. 3.155
χ	magnetic susceptibility	Fn. 3, p. 167
	relative scattering angle	Fig. 27.3
χ_i	ith product orbital	Section 31.5
ψ	wave function	general
	electrostatic potential	Part II; general
	general transport property	Part III

Symbol	Meaning	Where used
$\psi(x)$	spatial part of the wave amplitude along the x-direction	Eq. 3.59
$\psi(1,2)$	molecular orbital wave function for electrons 1 and 2	Eq. 6.46
$\psi_{elec}(r, R)$	solution of the electronic wave equation for fixed nuclei (separated by R)	Eq. 6.3
$\psi_{nucl}(R)$	factor of the molecular wave function describing rotation and vibration	Eq. 6.2
ψ_{MO}	molecular orbital wave function	Section 6.6
ψ_{VB}	valence bond wave function	Section 6.8
Ψ	wave amplitude	Eq. 3.14
Ψ^*	complex conjugate of Ψ	general
ω	angular frequency ($\equiv 2\pi\nu$)	general
	small element of volume	Part II; Chapter 15
$\boldsymbol{\omega}$	angular velocity	general
	set of angles defining the orientation of a molecule	Part II; Eq. 24.96
$\omega(E)$	density of states at energy E	general
$\bar{\omega}_{sl}$	degeneracy of the sth state of molecule 1	Eq. 21.100
Ω	solid angle	general; Eq. 2.15
$\Omega(E, V, N)$	number of quantum states in the energy range between E and $E+dE$ for given V, N	general; Eq. 15.12

Commonly Used Superscripts and Subscripts

0	Generally used to denote a standard state, or a change in which both initial and final states are standard states.	Part II
0	Denotes pure component	
⊖	Denotes the hypothetical reference state defined to be a pure substance which has the same properties as does the solute at infinite dilution.	
*	Denotes the perfect gas reference state, or the limiting value of the gas property when the pressure approaches zero.	
E	Denotes, for a mixture, the excess value of a thermodynamic function over that expected for an ideal mixture, and, for a pure component, the residual value of a thermodynamic function over that expected for a perfect gas at the same temperature and density (the latter is used only in Chapter 23).	
$\bar{z}_i$	Denotes partial molar "zee" of component i.	
i	component i	
$V, p, T, \ldots$	constant volume, pressure, temperature, ...	
σ	along the coexistence curve	
0	ground state	
m	mean molar value for a mixture	
$\pm$	mean ionic quantity	
reac, fus vap, subl inv	value of the thermodynamic function for the process indicated	
'	value (of velocity, impact parameter, etc.)	Part III; Eq. 27.1

Symbol	Meaning	Where used
′ ″	designates quantum numbers in the upper and lower states of a transition	Eq. 7.28
*	complex conjugate	general
	excited state	Section 5.3
	unusual configuration	Section 5.3
	value at the maximum of probability distribution	Eq. 15.49
	value corresponding to top of potential energy barrier	Eq. 30.22
‡	value for activated complex	Part III; Eq. 30.44

Index